D0745746

McGraw-Hill

Dictionary of
Electrical and Computer Engineering

McGraw-Hill

New York Chicago San Francisco Lisbon London Madrid
Mexico City Milan New Delhi San Juan Seoul Singapore
Sydney Toronto

The McGraw·Hill Companies

1 2 3 4 5 6 7 8 9 0   DOC/DOC   0 9 8 7 6 5 4

ISBN 0-07-144210-3

 This book is printed on recycled, acid-free paper containing a minimum of 50% recycled, de-inked fiber.

This book was set in Helvetica Bold and Novarese Book by TechBooks, Fairfax, Virginia. It was printed and bound by RR Donnelley, The Lakeside Press.

McGraw-Hill books are available at special quantity discounts to use as premiums and sales promotions, or for use in corporate training programs. For more information, please write to the Director of Special Sales, Professional Publishing, McGraw-Hill, Two Penn Plaza, New York, NY 10121-2298. Or contact your local bookstore.

Library of Congress Cataloging-in-Publication Data

McGraw-Hill dictionary of electrical and computer engineering.
      p.   cm.
   ISBN 0-07-144210-3
   1. Computer engineering—Dictionaries.   2. Electric engineering—Dictionaries.

TK7885.A2M37   2004
004'.03—dc22   2004049888

Contents

Preface

The *McGraw-Hill Dictionary of Electrical and Computer Engineering* provides a compendium of more than 18,000 terms that are central to these fields as well as related fields. In addition to computer science, electronics, electricity, and electrical engineering, coverage includes terminology in control systems, engineering acoustics, systems engineering, and communications.

The definitions are drawn from the *McGraw-Hill Dictionary of Scientific and Technical Terms*, Sixth Edition (2003). Each one is classified according to the field with which it is primarily associated. The pronunciation of each term is provided along with synonyms, acronyms, and abbreviations where appropriate. A guide to the use of the *Dictionary* is included, explaining the alphabetical organization of terms, the format of the book, cross referencing, and how synonyms, variant spellings, abbreviations, and similar information are handled. A pronunciation key is also provided to assist the reader. An extensive appendix provides conversion tables for commonly used scientific and technical units as well as charts, a "family tree" of programming languages, and listings of useful mathematical, engineering, and scientific data, laws, and equations.

It is the editors' hope that this dictionary will serve the needs of scientists, engineers, specialists in information technology, students, teachers, librarians, and writers for high-quality information, and that it will contribute to scientific literacy and communication.

Mark D. Licker
Publisher

Staff

Mark D. Licker, Publisher—Science

Elizabeth Geller, Managing Editor
Jonathan Weil, Senior Staff Editor
David Blumel, Staff Editor
Alyssa Rappaport, Staff Editor
Charles Wagner, Digital Content Manager
Renee Taylor, Editorial Assistant

Roger Kasunic, Vice President—Editing, Design, and Production

Joe Faulk, Editing Manager
Frank Kotowski, Jr., Senior Editing Supervisor

Ron Lane, Art Director

Thomas G. Kowalczyk, Production Manager
Pamela A. Pelton, Senior Production Supervisor

Henry F. Beechhold, Pronunciation Editor
Professor Emeritus of English
Former Chairman, Linguistics Program
The College of New Jersey
Trenton, New Jersey

How to Use the Dictionary

ALPHABETIZATION. The terms in the McGraw-Hill *Dictionary of Electrical and Computer Engineering* are alphabetized on a letter-by-letter basis; word spacing, hyphen, comma, and solidus in a term are ignored in the sequencing. For example, an ordering of terms would be:

absolute-value computer   **airborne radar**
absolute vector   **air capacitor**
accuracy control system
ac/dc receiver

FORMAT. The basic format for a defining entry provides the term in boldface, the field in small capitals, and the single definition in lightface:

term [FIELD] Definition.

A field may be followed by multiple definitions, each introduced by a boldface number:

term [FIELD] **1.** Definition. **2.** Definition. **3.** Definition.

A term may have difinitions in two or more fields:

term [COMMUN] Definition. [COMPUT SCI] Definition.

A simple cross-reference entry appears as:

term *See* another term.

A cross reference may also appear in combination with definitions:

term [COMMUN] Definition. [COMPUT SCI] *See* another term.

CROSS REFERENCING. A cross-reference entry directs the user to the defining entry. For example, the user looking up "chroma band-pass amplifier" finds:

chroma band-pass amplifier *See* burst amplifier.

The user then turns to the "B" terms for the definition. Cross references are also made from variant spellings, acronyms, abbreviations, and symbols.

ACK *See* acknowledge character.
A-O-I gate *See* AND-OR-INVERT gate.
bps *See* bit per second.
chip *See* microchip.

ALSO KNOWN AS ..., etc. A definition may conclude with a mention of a synonym of the term, a variant spelling, an abbreviation for the term, or other such information, introduced by "Also known as ...," "Also spelled ...," "Abbreviated ...," "Symbolized ...," "Derived from ...." When a term has more than one definition, the positioning of any of these phrases conveys the extent of applicability. For example:

term [COMPUT SCI] **1.** Definition. Also known as synonym. **2.** Definition. Symbolized T.

In the above arrangement, "Also known as ..." applies only to the first definition; "Symbolized ..." applies only to the second definition.

term [COMMUN] **1.** Definition. **2.** Definition. [COMPUT SCI] Definition. Also known as synonym.

In the above arrangement, "Also known as ..." applies only to the second field.

term [COMMUN] Also known as synonym. **1.** Definition. **2.** Definition. [COMPUT SCI] Definition.

In the above arrangement, "Also known as ..." applies only to both definitions in the first field.

term Also known as synonym. [COMMUN] **1.** Definition. **2.** Definition. [COMPUT SCI] Definition.

In the above arrangement, "Also known as ..." applies to all definitions in both fields.

Fields and Their Scope

[COMMUN] **communications**—The science and technology by which information is collected from an originating source; converted into a form suitable for transmission; transmitted over a pathway such as a satellite channel, underwater acoustic channel, telephone cable, or fiber-optic link; and reconverted into a form suitable for interpretation by a receiver.

[COMPUT SCI] **computer science**—The study of computing, including computer hardware, software, programming, networking, database systems, information technology, interactive systems, and security.

[CONT SYS] **control systems**—The study of those systems in which one or more outputs are forced to change in a desired manner as time progresses.

[ELEC] **electricity**—The science of physical phenomena involving electric charges and their effects when at rest and when in motion.

[ELECTROMAG] **electromagnetism**—The branch of physics dealing with the observations and laws relating electricity to magnetism, and with magnetism produced by an electric current.

[ELECTR] **electronics**—The technological area involving the manipulation of voltages and electric currents through the use of various devices for the purpose of performing some useful action with the currents and voltages; this field is generally divided into analog electronics, in which the signals to be manipulated take the form of continuous currents or voltages, and digital electronics, in which signals are represented by a finite set of states.

[ENG] **engineering**—The science by which the properties of matter and the sources of power in nature are made useful to humans in structures, machines, and products.

[ENG ACOUS] **engineering acoustics**—The field of acoustics that deals with the production, detection, and control of sound by electrical devices, including the study, design, and construction of such things as microphones, loudspeakers, sound recorders and reproducers, and public address sytems.

[GEOPHYS] **geophysics**—The branch of geology in which the principles and practices of physics are used to study the earth and its environment, that is, earth, air, and (by extension) space.

[MATER] **materials**—A multidisciplinary field concerned with the properties and uses of materials in terms of composition, structure, and processing.

[MATH] **mathematics**—The deductive study of shape, quantity, and dependence; the two main areas are applied mathematics and pure mathematics, the former arising from the study of physical phenomena, the latter involving the intrinsic study of mathematical structures.

[NAV] **navigation**—The science or art of directing the movement of a craft, such as a ship, small marine craft, underwater vehicle, land vehicle, aircraft, missile, or spacecraft, from one place to another with the assistance of onboard equipment, objects, or devices, or of systems external to the craft.

[OPTICS] **optics**—The study of phenomena associated with the generation, transmission, and detection of electromagnetic radiation in the spectral range extending from the long-wave edge of the x-ray region to the short-wave edge of the radio region; and the science of light.

[PHYS] **physics**—The science concerned with those aspects of nature which can be understood in terms of elementary principles and laws.

[SOLID STATE] **solid-state physics**—The branch of physics centering on the physical properties of solid materials; it is usually concerned with the properties of crystalline materials only, but it is sometimes extended to include the properties of glasses or polymers.

[STAT] **statistics**—The science dealing with the collection, analysis, interpretation, and presentation of masses of numerical data.

[SYS ENG] **systems engineering**—The branch of engineering dealing with the design of a complex interconnection of many elements (a system) to maximize an agreed-upon measure of system performance.

Pronunciation Key

Vowels

a	as in b**a**t, th**a**t
ā	as in b**ai**t, cr**a**te
ä	as in b**o**ther, f**a**ther
e	as in b**e**t, n**e**t
ē	as in b**ee**t, tr**ea**t
i	as in b**i**t, sk**i**t
ī	as in b**i**te, l**igh**t
ō	as in b**oa**t, n**o**te
ȯ	as in b**ough**t, t**au**t
u̇	as in b**oo**k, p**u**ll
ü	as in b**oo**t, p**oo**l
ə	as in b**u**t, sof**a**
au̇	as in cr**ow**d, p**ow**er
ȯi	as in b**oi**l, sp**oi**l
yə	as in form**u**la, spectac**u**lar
yü	as in f**ue**l, m**u**le

Semivowels/Semiconsonants

w	as in **w**ind, t**w**in
y	as in **y**et, on**i**on

Stress (Accent)

ˈ precedes syllable with primary stress

ˌ precedes syllable with secondary stress

ˌ̣ precedes syllable with variable or indeterminate primary/ secondary stress

Consonants

b	as in **b**i**b**, dri**bb**le
ch	as in **ch**arge, stre**tch**
d	as in **d**og, ba**d**
f	as in **f**ix, sa**f**e
g	as in **g**ood, si**g**nal
h	as in **h**and, be**h**ind
j	as in **j**oint, di**g**it
k	as in **c**ast, bri**ck**
k̲	as in Ba**ch** (used rarely)
l	as in **l**oud, be**ll**
m	as in **m**ild, su**mm**er
n	as in **n**ew, de**n**t
n̲	indicates nasalization of preceding vowel
ŋ	as in ri**ng**, si**ng**le
p	as in **p**ier, sli**p**
r	as in **r**ed, sca**r**
s	as in **s**ign, po**s**t
sh	as in **s**ugar, **sh**oe
t	as in **t**imid, ca**t**
th	as in **th**in, brea**th**
t̲h̲	as in **th**en, brea**th**e
v	as in **v**eil, wea**v**e
z	as in **z**oo, crui**s**e
zh	as in bei**g**e, trea**s**ure

Syllabication

· Indicates syllable boundary when following syllable is unstressed

xi

A

a *See* ampere.

A *See* ampere.

aΩ *See* abohm.

(aΩ)⁻¹ *See* abmho.

A+ *See* A positive.

aA *See* abampere.

aA/cm² *See* abampere per square centimeter.

A AND NOT B gate *See* AND NOT gate. { 'ā an nöt 'bē ‚gāt }

abampere [ELEC] The unit of electric current in the electromagnetic centimeter-gram-second system; 1 abampere equals 10 amperes in the absolute meter-kilogram-second-ampere system. Abbreviated aA. Also known as Bi; biot. { ab'am·pēr }

abampere per square centimeter [ELEC] The unit of current density in the electromagnetic centimeter-gram-second system. Abbreviated aA/cm². { ab'am·pēr pər 'skwer 'sen·tə‚mēd·ər }

A battery [ELECTR] The battery that supplies power for filaments or heaters of electron tubes in battery-operated equipment. { 'ā ‚bat·ə·rē }

abbreviated dialing [COMMUN] A feature which requires less than the usual number of dialing operations to connect two or more subscribers. { ə'brē·vē·ād·əd 'dī·liŋ }

ABC *See* automatic brightness control.

abcoulomb [ELEC] The unit of electric charge in the electromagnetic centimeter-gram-second system, equal to 10 coulombs. Abbreviated aC. { ab'kü·lōm }

abcoulomb centimeter [ELEC] In the electromagnetic centimeter-gram-second system of units, the unit of electric dipole moment. Abbreviated aCcm. { ab'kü·lōm 'sen·tə‚mēd·ər }

abcoulomb per cubic centimeter [ELEC] The electromagnetic centimeter-gram-second unit of volume density of charge. Abbreviated aC/cm³. { ab'kü·lōm pər 'kyü·bik 'sen·tə‚mēd·ər }

abcoulomb per square centimeter [ELEC] The electromagnetic centimeter-gram-second unit of surface density of charge, electric polarization, and displacement. Abbreviated aC/cm². { ab'kü·lōm pər skwer 'sen·tə‚mēd·ər }

abeam *See* on the beam. { a'bēm }

abend [COMPUT SCI] An unplanned program termination that occurs when a computer is directed to execute an instruction or to process information that it cannot recognize. Also known as blow up; bomb; crash. { 'ab·end }

abfarad [ELEC] A unit of capacitance in the electromagnetic centimeter-gram-second system equal to 10⁹ farads. Abbreviated aF. { ab'far·ad }

abhenry [ELEC] A unit of inductance in the electromagnetic centimeter-gram-second system of units which is equal to 10⁻⁹ henry. Abbreviated aH. { ab'hen·rē }

able [COMPUT SCI] A name for the hexadecimal digit whose decimal equivalent is 10. { 'ā·bəl }

abmho [ELEC] A unit of conductance in the electromagnetic centimeter-gram-second system of units equal to 10⁹ mhos. Abbreviated (aΩ)⁻¹. Also known as absiemens (aS). { 'ab‚mō }

Abney level *See* clinometer. { 'ab·nē 'lev·əl }

abnormal glow discharge [ELECTR] A discharge of electricity in a gas tube at currents somewhat higher than those of an ordinary glow discharge, at which point the glow covers the entire cathode and the voltage drop decreases with increasing current. { ab'nȯr·məl ‚glō 'dis·chärj }

abnormal propagation [COMMUN] Phenomena of unstable or changing atmospheric or ionospheric conditions acting upon transmitted radio waves, preventing such waves from following their normal path, thereby causing difficulties and disruptions of communications. { ab'nȯr·məl ‚präp·ə'gā·shən }

abnormal statement [COMPUT SCI] An element of a FORTRAN V (UNIVAC) program which specifies that certain function subroutines must be called every time they are referred to. { ab'nȯr·məl 'stāt·mənt }

abohm [ELEC] The unit of electrical resistance in the centimeter-gram-second system; 1 abohm equals 10⁻⁹ ohm in the meter-kilogram-second system. Abbreviated aΩ. { a'bōm }

abohm centimeter [ELEC] The centimeter-gram-second unit of resistivity. Abbreviated aΩcm. { a'bōm 'sen·tə‚mē·dər }

abort [COMPUT SCI] To terminate a procedure, such as the running of a computer program or the printing of a document, while it is still in progress. { ə'bȯrt }

abort branch [CONT SYS] A branching instruction in the program controlling a robot that causes a test to be performed on whether the tool-center point is properly positioned, and to reposition it if it drifts out of the acceptable range. { ə'bȯrt ‚branch }

AB power pack [ELEC] **1.** Assembly in a single unit of the A battery and B battery for a battery-operated vacuum-tube circuit. **2.** Unit that supplies the necessary A and B direct-current voltages from an alternating-current source of power. { ā¦bē 'paú·ər ‚pak }

abrupt junction [ELECTR] A *pn* junction in which the concentration of impurities changes suddenly from acceptors to donors. { ə'brəpt 'jəŋk·shən }

abs [COMPUT SCI] A special function occurring in ALGOL, which yields the absolute value, or modulus, of its argument.

absiemens *See* abmho. { ab'sē·mənz }

absolute address [COMPUT SCI] The numerical identification of each storage location which is wired permanently into a computer by the manufacturer. { 'ab·sə‚lüt ə'dres }

absolute addressing [COMPUT SCI] The identification of storage locations in a computer program by their physical addresses. { 'ab·sə ‚lüt ə'dres·iŋ }

absolute category rating mean opinion score [COMMUN] Methodology for subjectively testing audio quality where participants are presented with sound samples, one at a time, and are asked to grade them on a 5-point scale. For the NRSC FM IBOC tests, the MOS scale used was 5 = excellent, 4 = good, 3 = fair, 2 = poor, 1 = bad. Abbreviated ACR-MOS. { ¦ab·sə‚lut kad·ə‚gór·ē räd·iŋ mēn 'ə·'pin·yən ‚skōr }

absolute cell reference [COMPUT SCI] A cell reference used in a formula in a spreadsheet program that does not change when the formula is copied or moved. { ¦ab·sə‚lüt 'sel ‚ref·rəns }

absolute code [COMPUT SCI] A code used when the addresses in a program are to be written in machine language exactly as they will appear when the instructions are executed by the control circuits. { 'ab·sə‚lüt 'kōd }

absolute efficiency [ENG ACOUS] The ratio of the power output of an electroacoustic transducer, under specified conditions, to the power output of an ideal electroacoustic transducer. { 'ab·sə ‚lüt ə'fish·ən·sē }

absolute electrometer [ELEC] A very precise type of attracted disk electrometer in which the attraction between two disks is balanced against the force of gravity. { 'ab·sə‚lüt ə‚lek'träm·əd·ər }

absolute gain of an antenna [ELECTROMAG] Gain in a given direction when the reference antenna is an isotropic antenna isolated in space. Also known as isotropic gain of an antenna. { 'ab·sə ‚lüt ‚gān əv ən an'ten·ə }

absolute index of refraction *See* index of refraction. { 'ab·sə‚lüt 'in‚deks əv ri'frak·shən }

absolute instruction [COMPUT SCI] A computer instruction in its final form, in which it can be executed. { 'ab·sə‚lüt in'strək·shən }

absolute programming [COMPUT SCI] Programming with the use of absolute code. { 'ab·sə ‚lüt 'prō·gram·iŋ }

absolute refractive constant *See* index of refraction. { 'ab·sə‚lüt ri'frak·tiv 'kän·stənt }

absolute-value computer [COMPUT SCI] A computer that processes the values of the variables rather than their increments. { 'ab·sə‚lüt 'val·yü kəm'pyüd·ər }

absolute vector [COMPUT SCI] In computer graphics, a vector whose end points are given in absolute coordinates. { 'ab·sə‚lüt 'vek·tər }

absorbed charge [ELEC] Charge on a capacitor which arises only gradually when the potential difference across the capacitor is maintained, due to gradual orientation of permanent dipolar molecules. { əb'sórbd 'chärj }

absorber [ELECTR] A material or device that takes up and dissipates radiated energy; may be used to shield an object from the energy, prevent reflection of the energy, determine the nature of the radiation, or selectively transmit one or more components of the radiation. { əb'sór·bər }

absorber control *See* absorption control. { əb'sór·bər kən'trōl }

absorption [ELEC] The property of a dielectric in a capacitor which causes a small charging current to flow after the plates have been brought up to the final potential, and a small discharging current to flow after the plates have been short-circuited, allowed to stand for a few minutes, and short-circuited again. Also known as dielectric soak. [ELECTROMAG] Taking up of energy from radiation by the medium through which the radiation is passing. { əb'sórp·shən }

absorption circuit [ELECTR] A series-resonant circuit used to absorb power at an unwanted signal frequency by providing a low impedance to ground at this frequency. { əb'sórp·shən 'sər·kət }

absorption control *See* absorption modulation. { əb'sórp·shən kən'trōl }

absorption current [ELEC] The component of a dielectric current that is proportional to the rate of accumulation of electric charges within the dielectric. { əb'sórp·shən 'kər·ənt }

absorption fading [COMMUN] Slow type of fading, primarily caused by variations in the absorption rate along the radio path. { əb'sórp·shən 'fād·iŋ }

absorption loss [COMMUN] That part of the transmission loss due to the dissipation or conversion of either sound energy or electromagnetic energy into other forms of energy, either within the medium or attendant upon a reflection. { əb'sórp·shən ‚lòs }

absorption meter [ENG] An instrument designed to measure the amount of light transmitted through a transparent substance, using a photocell or other light detector. { əb'sórp·shən 'mēd·ər }

absorption modulation [ELECTR] A system of amplitude modulation in which a variable-impedance device is inserted in or coupled to the output circuit of the transmitter. Also known as absorption control; loss modulation. { əb'sórp·shən mäd·yü'lā·shən }

absorption wavemeter [ELECTR] A frequency- or wavelength-measuring instrument consisting of a calibrated tunable circuit and a resonance indicator. { əb'sórp·shən 'wāv,mēd·ər }

abstract automata theory [COMPUT SCI] The mathematical theory which characterizes automata by three sets: input signals, internal states, and output signals; and two functions: input functions and output functions. { 'abz·trakt ȯ'tam·ə·tə 'thē·ə·rē }

abstract data type [COMPUT SCI] A mathematical model which may be used to capture the essentials of a problem domain in order to translate it into a computer program; examples include queues, lists, stacks, trees, graphs, and sets. Abbreviated ADT. { 'abz·trakt 'dad·ə ,tīp }

abvolt [ELEC] The unit of electromotive force in the electromagnetic centimeter-gram-second system; 1 abvolt equals 10^{-8} volt in the absolute meter-kilogram-second system. Abbreviated aV. { 'ab₁vōlt }

abvolt per centimeter [ELEC] In the electromagnetic centimeter-gram-second system of units, the unit of electric field strength. Abbreviated aV/cm. { 'ab₁vōlt pər 'sen·tə,mēd·ər }

abwatt [ELEC] The unit of electrical power in the centimeter-gram-second system; 1 abwatt equals 1 watt in the absolute meter-kilogram-second system. { 'ab,wät }

ac See alternating current.

aC See abcoulomb.

ACAS See airborne collision avoidance system.

accelerated graphics port [COMPUT SCI] A personal computer graphics bus that transfers data at a greater rate than a PCI bus. { ak,sel·ə,rād·əd 'graf·iks ,pȯrt }

accelerated test [ELEC] A test of the serviceability of an electric cable in use for some time by applying twice the voltage normally carried. { ak'sel·ər,ā·dəd 'test }

accelerating electrode [ELECTR] An electrode used in cathode-ray tubes and other electron tubes to increase the velocity of the electrons that contribute the space current or form a beam. { ak'sel·ər,ād·iŋ i'lek,trōd }

accelerating potential [ELECTR] The energy potential in electron-beam equipment that imparts additional speed and energy to the electrons. { ak'sel·ər,ād·iŋ pə'ten·shəl }

accelerating relay [ELEC] Any relay that is used to assist in starting a motor or increasing its speed. { ak'sel·ə,rād·iŋ 'rē,lā }

acceleration-error constant [CONT SYS] The ratio of the acceleration of a controlled variable of a servomechanism to the actuating error when the actuating error is constant. { ak,sel·ə'rā·shən 'er·ər 'kän·stənt }

acceleration switch [ELEC] A switch that opens or closes in the presence of acceleration that 0 exceeds a certain value. { ak,sel·ə'rā·shən ,swich }

acceleration time [COMPUT SCI] The time required for a magnetic tape transport or any other mechanical device to attain its operating speed. { ak,sel·ə'rā·shən ,tīm }

acceleration tolerance [ENG] The degree to which personnel or equipment withstands acceleration. { ak,sel·ə'rā·shən 'täl·ər·əns }

acceleration voltage [ELECTR] The voltage between a cathode and accelerating electrode of an electron tube. { ak,sel·ə'rā·shən 'vōl·təj }

accentuation [ELECTR] The enhancement of signal amplitudes in selected frequency bands with respect to other signals. { ak,sen·chə'wā·shən }

accentuator [ELECTR] A circuit that provides for the first part of a process for increasing the strength of certain audio frequencies with respect to others, to help these frequencies override noise or to reduce distortion. Also known as accentuator circuit. { ak'sen·chə,wād·ər }

accentuator circuit See accentuator. { ak'sen·chə,wād·ər 'sər·kət }

accept [COMPUT SCI] A data transmission statement which is used in FORTRAN when the computer is in conversational mode, and which enables the programmer to input, through the teletypewriter, data the programmer wishes stored in memory. { ak'sept }

acceptor [SOLID STATE] An impurity element that increase the number of holes in a semiconductor crystal such as germanium or silicon; aluminum, gallium, and indium are examples. Also known as acceptor impurity; acceptor material. { ak'sep·tər }

acceptor circuit [ELECTR] A series-resonant circuit that has a low impedance at the frequency to which it is tuned and a higher impedance at all other frequencies. { ak'sep·tər 'sər·kət }

acceptor impurity See acceptor. { ak'sep·tər im 'pyür·ə·dē }

acceptor material See acceptor. { ak'sep·tər mə 'tir·ē·əl }

access [COMPUT SCI] The reading of data from storage or the writing of data into storage. { 'ak ,ses }

access arm [COMPUT SCI] The mechanical device which positions the read/write head on a magnetic storage unit. { 'ak,ses ,ärm }

access code [COMMUN] **1.** Numeric identification for internetwork or facility switching. **2.** The preliminary digits that a user must dial to be connected through an automatic PBX to the serving switching center. [COMPUT SCI] A sequence of characters which a user must enter into a terminal in order to use a computer system. { 'ak,ses ,kōd }

access control [COMPUT SCI] A restriction on the operations that a user of a computer system may perform on files and other resources of the system. { 'ak,ses kən,trōl }

access-control list [COMPUT SCI] A column of an access matrix, containing the access rights of various users of a computer system to a given file or other resource of the system. { 'ak,ses kən,trōl ,list }

access-control mechanism See reference monitor. { |ak,ses kən'trōl |me·kə·ni·zəm }

access-control register [COMPUT SCI] A storage device which controls the word-by-word transmission over a given channel. { 'ak,ses kən'trōl ,rej·ə·stər }

3

access-control words |COMPUT SCI| Permanently wired instructions channeling transmitted words into reserved locations. {ak,ses kən'trōl ,wərdz }

access gap See memory gap. { 'ak,ses ,gap }

access line |COMMUN| Four-wire circuit between a subscriber or a local PBX to the serving switching center. { 'ak,ses ,līn }

access management |COMPUT SCI| The use of techniques to allow various components of a computer's operating system to be used only by authorized personnel. { 'ak,ses ,man·ij·mənt }

access matrix |COMPUT SCI| A method of representing discretionary authorization information, with rows representing subjects or users of the system, columns corresponding to objects or resources of the system, and cells (intersections of rows and columns) composed of allowable operations that a subject may apply to an object. { 'ak,ses ,mā·triks }

access mechanism |COMPUT SCI| The mechanism of positioning reading or writing heads onto the required tracks of a magnetic disk. { 'ak,ses 'mek·ə,niz·əm }

access method |COMMUN| The procedures required to obtain access to a communications network. |COMPUT SCI| A set of programming routines which links programs and the data that these programs transfer into and out of memory. { 'ak,ses ,meth·əd }

access mode |COMPUT SCI| A programming clause in COBOL which is required when using a random-access device so that a specific record may be read out of or written into a mass storage bin. { 'ak,ses ,mōd }

access privileges |COMPUT SCI| The extent to which a user of a computer in a network is allowed to use and read, write to, and execute files in other computers in the network. { 'ak ,ses ,priv·ə·ləj·əs }

access protocol |COMMUN| A set of rules observed by all nodes in a local-area network so that one node can get the attention of another and its data packet can be transferred, and so that no two data packets can be simultaneously transmitted over the same medium. { 'ak,ses ,prōd·ə,kȯl }

access provider See service provider. { 'ak,ses prə,vīd·ər }

access time |COMPUT SCI| The time period required for reading out of or writing into the computer memory. { 'ak·ses ,tīm }

access type |COMPUT SCI| One of the allowable operations that a given user of a computer system governed by access controls may perform on a file or other resource of the system, such as own, read, write, or execute. { 'ak·ses ,tīp }

aCcm See abcoulomb centimeter.

aC/cm² See abcoulomb per square centimeter.

aC/cm³ See abcoulomb per cubic centimeter.

accommodation |CONT SYS| Any alteration in a robot's motion in response to the robot's environment; it may be active or passive. { ə,käm·ə'dā·shən }

accommodation time |ELECTR| The time from the production of the first electron to the production of a steady electric discharge in a gas. { ə,käm·ə'dā·shən ,tīm }

accordion cable |ELEC| A flat, multiconductor cable prefolded into a zigzag shape and used to make connections to movable equipment such as a chassis mounted on pullout slides. { ə'kȯrd·ē·ən 'kā·bəl }

accounting package |COMPUT SCI| A set of special routines that allow collection of information about the usage level of various components of a computer system by each production program. { ə'kaůnt·iŋ 'pak·ij }

accumulator |COMPUT SCI| A specific register, in the arithmetic unit of a computer, in which the result of an arithmetic or logical operation is formed; here numbers are added or subtracted, and certain operations such as sensing, shifting, and complementing are performed. Also known as accumulator register; counter. |ELEC| See storage battery. { ə'kyü·myə,lād·ər }

accumulator battery See storage battery. { ə'kyü·myə,lād·ər 'bad·ə·rē }

accumulator jump instruction |COMPUT SCI| An instruction which programs a computer to ignore the previously established program sequence depending on the status of the accumulator. Also known as accumulator transfer instruction. { ə'kyü·myə,lād·ər ,jəmp in'strək·shən }

accumulator register See accumulator. { ə'kyü·myə,lād·ər 'rej·ə·stər }

accumulator shift instruction |COMPUT SCI| A computer instruction which causes the word in a register to be displaced a specified number of bit positions to the left or right. { ə'kyü·myə,lād·ər 'shift in'strək·shən }

accumulator transfer instruction See accumulator jump instruction. { ə'kyü·myə,lād·ər 'trans·fər in'strək·shən }

accuracy control system |COMPUT SCI| Any method which attempts error detection and control, such as random sampling and squaring. { 'ak·yə·rə·sē kən'trōl ,sis·təm }

ac/dc motor See universal motor. { ,ā·sē,dē·sē 'mōd·ər }

ac/dc receiver |ELECTR| A radio receiver designed to operate from either an alternating- or direct-current power line. Also known as universal receiver. { ,ā·sē,dē·sē ri'sēv·ər }

ACK See acknowledge character.

acknowledge character |COMPUT SCI| A signal that a receiving station transmits in order to indicate that a block of information has been received and that its validity has been checked. Also known as acknowledgement. Abbreviated ACK. { ak'näl·ij 'kar·ək·tər }

acknowledgement See acknowledge character. { ak'näl·ij·mənt }

aΩcm See abohm centimeter.

acorn tube |ELECTR| An ultra-high-frequency electron tube resembling an acorn in shape and size. { 'ā,kȯrn ,tüb }

acoustic amplifier |ELECTR| A device that amplifies mechanical vibrations directly at audio and ultrasonic frequencies. Also known as acoustoelectric amplifier. { ə'küs·tik 'am·plə,fī·ər }

acoustic array |ENG ACOUS| A sound-transmitting or sound-receiving system whose elements are arranged to give desired directional characteristics. { ə'küs·tik ə'rā }

acoustic bridge |ELECTR| A device, based on the principle of the electrical Wheatstone bridge, used for analysis of deafness. { ə'küs·tik 'brij }

acoustic center |ENG ACOUS| The center of the spherical sound waves radiating outward from an acoustic transducer. { ə'küs·tik 'sen·tər }

acoustic clarifier |ENG ACOUS| System of cones loosely attached to the baffle of a loudspeaker and designed to vibrate and absorb energy during sudden loud sounds to suppress these sounds. { ə'küs·tik 'klar·ə,fī·ər }

acoustic convolver See convolver. { ə'küs·tik kən'välv·ər }

acoustic coupler |ENG ACOUS| A device used between the modem of a computer terminal and a standard telephone line to permit transmission of digital data in either direction without making direct connections. { ə'küs·tik 'kəp·lər }

acoustic delay |ENG ACOUS| A delay which is deliberately introduced in sound reproduction by having the sound travel a certain distance along a pipe before conversion into electric signals. { ə'küs·tik di'lā }

acoustic delay line |ELECTR| A device in which acoustic signals are propagated in a medium to make use of the sonic propagation time to obtain a time delay for the signals. Also known as sonic delay line. { ə'küs·tik di'lā ,līn }

acoustic detector |ELECTR| The stage in a receiver at which demodulation of a modulated radio wave into its audio component takes place. { ə'küs·tik di'tek·tər }

acoustic feedback |ENG ACOUS| The reverberation of sound waves from a loudspeaker to a preceding part of an audio system, such as to the microphone, in such a manner as to reinforce, and distort, the original input. Also known as acoustic regeneration. { ə'küs·tik 'fēd ,bak }

acoustic filter See filter. { ə'küs·tik 'fil·tər }

acoustic generator |ENG ACOUS| A transducer which converts electrical, mechanical, or other forms of energy into sound. { ə'küs·tik 'jen·ə ,rād·ər }

acoustic hologram |ENG| The phase interference pattern, formed by acoustic beams, that is used in acoustical holography; when light is made to interact with this pattern, it forms an image of an object placed in one of the beams. { ə'küs·tik 'häl·ə,gram }

acoustic horn See horn. { ə'küs·tik 'hórn }

acoustic jamming |ENG ACOUS| The deliberate radiation or reradiation of mechanical or electroacoustic signals with the objectives of obliterating or obscuring signals which the enemy is attempting to receive and of deterring enemy weapons systems. { ə'küs·tik 'jam·iŋ }

acoustic labyrinth |ENG ACOUS| Special baffle arrangement used with a loudspeaker to prevent cavity resonance and to reinforce bass response. { ə'küs·tik 'lab·ə,rinth }

acoustic line |ENG ACOUS| The acoustic equivalent of an electrical transmission line, involving baffles, labyrinths, or resonators placed at the rear of a loudspeaker and arranged to help reproduce the very low audio frequencies. { ə'küs·tik 'līn }

acoustic radiator |ENG ACOUS| A vibrating surface that produces sound waves, such as a loudspeaker cone or a headphone diaphragm. { ə'küs·tik 'rād·ē,ād·ər }

acoustic radiometer |ENG| An instrument for measuring sound intensity by determining the unidirectional steady-state pressure caused by the reflection or absorption of a sound wave at a boundary. { ə'küs·tik ,rād·ē'ä·məd·ər }

acoustic ratio |ENG ACOUS| The ratio of the intensity of sound radiated directly from a source to the intensity of sound reverberating from the walls of an enclosure, at a given point in the enclosure. { ə'küs·tik 'rā·shō }

acoustic receiver |ELECTR| The complete equipment required for receiving modulated radio waves and converting them into sound. { ə'küs·tik rə'sēv·ər }

acoustic reflex enclosure |ENG ACOUS| A loudspeaker cabinet designed with a port to allow a low-frequency contribution from the rear of the speaker cone to be radiated forward. { ə'küs·tik 'rē,fleks in,klō·zhər }

acoustic regeneration See acoustic feedback. { ə'küs·tik rē,jen·ə'rā·shən }

acoustic seal |ENG ACOUS| A joint between two parts to provide acoustical coupling with low losses of energy, such as between an earphone and the human ear. { ə'küs·tik 'sēl }

acoustic spectrometer |ENG ACOUS| An instrument that measures the intensities of the various frequency components of a complex sound wave. Also known as audio spectrometer. { ə'küs·tik spek'träm·əd·ər }

acoustic transducer |ENG ACOUS| A device that converts acoustic energy to electrical or mechanical energy, such as a microphone or phonograph pickup. { ə'küs·tik tranz'dü·sər }

acoustic transformer |ENG ACOUS| A device, such as a horn or megaphone, for increasing the efficiency of sound radiation. { ə'küs·tik tranz 'fòr·mər }

acoustic-wave amplifier |ELECTR| An amplifier in which the charge carriers in a semiconductor are coupled to an acoustic wave that is propagated in a piezoelectric material, to produce amplification. { ə'küs·tik wāv 'am·plə,fī·ər }

acoustoelectric amplifier See acoustic amplifier. { ə¦küs·tō·ə¦lek·trik 'am·plə,fī·ər }

acoustoelectric effect |ELECTR| **1.** The development of a direct-current voltage in a semiconductor or metal by an acoustic wave traveling parallel to the surface of the material. Also known as electroacoustic effect. **2.** The amplification of a sound wave propagating in a piezoelectric semiconductor subject to a steady electric field that is strong enough that the resulting electron drift velocity exceeds the speed of sound. { ə¦küs·tō·ə'lek·trik i¸fekt }

acoustoelectronics |ENG ACOUS| The branch of electronics that involves use of acoustic waves at microwave frequencies (above 500 megahertz), traveling on or in piezoelectric or other solid substrates. Also known as pretersonics. { ə¦küs·tō·ə¸lek¦trän·iks }

acoustooptical cell |ELEC| An electric-to-optical transducer in which an acoustic or ultrasonic electric input signal modulates or otherwise acts on a beam of light. { ə¦küs·tō¦äp·tə·kəl 'sel }

acoustooptic interaction |OPTICS| A way to influence the propagation characteristics of an optical wave by applying a low-frequency acoustical field to the medium through which the wave passes. { ə¦küs·tō¦äp·tik ¸in·tə'rak·shən }

acoustooptic modulator |OPTICS| A device utilizing acoustooptic interaction ultrasonically to vary the amplitude or the phase of a light beam. Also known as Bragg cell. { ə¦küs·tō¦äp·tik 'mäd·yə¸lād·ər }

acoustooptics |OPTICS| The science that deals with interactions between acoustic waves and light. { ə¦küs·tō¦äp·tiks }

acquire |ELECTR| **1.** Of acquisition radars, the process of detecting the presence and location of a target in sufficient detail to permit identification. **2.** Of tracking radars, the process of positioning a radar beam so that a target is in that beam to permit the effective employment of weapons. Also known as target acquisition. { ə'kwīr }

acquisition |ELECTR| Also known as target acquisition. **1.** Of acquisition radars, the process of detecting and locating a target so as to permit reliable tracking and possible identification of it or other determinations about it. **2.** Of precision tracking radars, the detecting and tracking of a target designated to it by another radar or other initial data source to support continued intended action. |ENG| The process of pointing an antenna or a telescope so that it is properly oriented to allow gathering of tracking and telemetry data from a satellite or space probe. { ¸ak·wə'zish·ən }

acquisition and tracking radar |ENG| A radar set capable of locking onto a received signal and tracking the object emitting the signal; the radar may be airborne or on the ground. { ¸ak·wə'zish·ən ən 'trak·iŋ ¸rā¸där }

acquisition tone |COMPUT SCI| An audible tone that verifies entry into a computer. { ¸ak·wə'zish·ən ¸tōn }

ACR-MOS *See* absolute category rating mean opinion score.

ACSR *See* aluminum cable steel-reinforced.

actinodielectric |ELEC| Of a substance, exhibiting an increase in electrical conductivity when electromagnetic radiation is incident upon it. { ¸ak·tə·nō¸dī·ə'lek·trik }

actinoelectricity |ELEC| The electromotive force produced in a substance by electromag-

netic radiation incident upon it. { ¸ak·tə·nō·i ¸lek'tris·ə·dē }

action entries |COMPUT SCI| The lower right-hand portion of a decision table, indicating which of the various possible actions result from each of the various possible conditions. { 'ak·shən ¸en·trēz }

action period |ELECTR| The period of time during which data in a Williams tube storage device can be read or new data can be written into this storage. { 'ak·shən ¸pir·ē·əd }

action portion |COMPUT SCI| The lower portion of a decision table, comprising the action stub and action entries. { 'ak·shən ¸pȯr·shən }

action stub |COMPUT SCI| The lower left-hand portion of a decision table, consisting of a single column listing the various possible actions (transformations to be done on data and materials). { 'ak·shən ¸stəb }

activate |ELEC| To make a cell or battery operative by addition of a liquid. |ELECTR| To treat the filament, cathode, or target of a vacuum tube to increase electron emission. { 'ak·tə¸vāt }

activated cathode |ELECTR| A thermionic cathode consisting of a tungsten filament to which thorium has been added, and then brought to the surface, by a process such as heating in the absence of an electric field in order to increase thermionic emission. { 'ak·tə¸vād·əd 'kath¸ōd }

activation |ELEC| The process of adding liquid to a manufactured cell or battery to make it operative. |ELECTR| The process of treating the cathode or target of an electron tube to increase its emission. Also known as sensitization. { ¸ak·tə'vā·shən }

activation record |COMPUT SCI| A variable part of a program module, such as data and control information, that may vary with different instances of execution. { ¸ak·tə'vā·shən 'rek·ərd }

active accommodation |CONT SYS| The alteration of preprogrammed robotic motions by the integrated effects of sensors, controllers, and the robotic motion itself. { 'ak·tiv ə¸käm·ə'dā·shən }

active area |ELECTR| The area of a metallic rectifier that acts as the rectifying junction and conducts current in the forward direction. { 'ak·tiv 'er·ē·ə }

active array |ELECTROMAG| A radar antenna composed of many radiating elements, each of which contains an amplifier, generally solid state in nature, for the final amplification of the signal transmitted; when the elements are also phased controlled for electronic beam steering, the term active phased array is used. { ¦ak·tiv ə'rā }

active balance |COMMUN| Summation of all return currents, in telephone repeater operation, at a terminal network balanced against the impedance of the local circuit or drop. { 'ak·tiv 'bal·əns }

active cell |COMPUT SCI| The cell that continues the value being used or modified in a spreadsheet program, and that is highlighted by the cell pointer. Also known as current cell.   { ˌak·tiv 'sel }

active communications satellite |ENG| Satellite which receives, regenerates, and retransmits signals between stations.   { 'ak·tiv kə ˌmyü·nə'kā·shənz 'sad·ə,līt }

active component |ELEC| In the phasor representation of quantities in an alternating-current circuit, the component of current, voltage, or apparent power which contributes power, namely, the active current, active voltage, or active power. Also known as power component. |ELECTR| See active element.   { 'ak·tiv kəm'pō·nənt }

active computer |COMPUT SCI| When two or more computers are installed, the one that is on-line and processing data.   { 'ak·tiv kəm'pyüd·ər }

active current |ELEC| The component of an electric current in a branch of an alternating-current circuit that is in phase with the voltage. Also known as watt current.   { 'ak·tiv 'kə·rənt }

active detection system |ENG| A guidance system which emits energy as a means of detection; for example, sonar and radar.   { 'ak·tiv di'tek·shən ˌsis·təm }

active device |ELECTR| A component, such as an electron tube or transistor, that is capable of amplifying the current or voltage in a circuit.   { 'ak·tiv di'vīs }

active electric network |ELEC| Electric network containing one or more sources of energy.   { 'ak·tiv ə'lek·trik 'net,wərk }

active electronic countermeasures |ELECTR| The major subdivision of electronic countermeasures that concerns electronic jamming and electronic deceptions.   { 'ak·tiv ə,lek'trän·ik 'kaünt·ər,mezh·ərz }

active element |ELECTR| Any generator of voltage or current in an impedance network. Also known as active component.   { 'ak·tiv 'el·ə·mənt }

active file |COMPUT SCI| A collection of records that is currently being used or is available for use.   { 'ak·tiv 'fīl }

active filter |ELECTR| A filter that uses an amplifier with conventional passive filter elements to provide a desired fixed or tunable pass or rejection characteristic.   { 'ak·tiv 'fil·tər }

active jamming See jamming.   { 'ak·tiv 'jam·iŋ }

active leg |ELECTR| An electrical element within a transducer which changes its electrical characteristics as a function of the application of a stimulus.   { 'ak·tiv 'leg }

active logic |ELECTR| Logic that incorporates active components which provide such functions as level restoration, pulse shaping, pulse inversion, and power gain.   { 'ak·tiv 'läj·ik }

active master file |COMPUT SCI| A relatively active computer master file, as determined by usage data.   { 'ak·tiv 'mas·tər 'fīl }

active master item |COMPUT SCI| A relatively active item in a computer master file, as determined by usage data.   { 'ak·tiv 'mas·tər 'ī·təm }

active material |ELEC| **1.** A fluorescent material used in screens for cathode-ray tubes. **2.** An energy-storing material, such as lead oxide, used in the plates of a storage battery. **3.** A material, such as the iron of a core or the copper of a winding, that is involved in energy conversion in a circuit. **4.** In a battery, the chemically reactive material in either of the electrodes that participates in the charge and discharge reactions. |ELECTR| The material of the cathode of an electron tube that emits electrons when heated.   { 'ak·tiv mə'tir·ē·əl }

active-matrix liquid-crystal display |ELEC| A liquid-crystal display that has an active element, such as a transistor or diode, on every picture element. Abbreviated AMLCD.   { ˌak·tiv ˌmā·triks ˌlik·wid 'kris·təl di,splā }

active power |ELEC| The product of the voltage across a branch of an alternating-current circuit and the component of the electric current that is in phase with the voltage.   { 'ak·tiv 'paů·ər }

active-RC filter |ELEC| An active filter whose frequency-sensitive mechanism is the charging of a capacitor (C) through a resistor (R), giving a characteristic frequency at which the impedances of the resistor and the capacitor are equal.   { ˌak·tiv ˌär¦sē 'fil·tər }

active region |ELECTR| The region in which amplifying, rectifying, light emitting, or other dynamic action occurs in a semiconductor device.   { 'ak·tiv 'rē·jən }

active-RLC filter |ELEC| An integrated-circuit filter that uses both inductors (L), made as spirals of metallization on the top layer, and amplifiers, connected to simulate negative resistors (R), that enhance the performance of the inductors as well as capacitors (C).   { ˌak·tiv ˌär¦el'sē fil·tər }

active satellite |ENG| A satellite which transmits a signal.   { 'ak·tiv 'sad·ə,līt }

active sonar |ENG| A system consisting of one or more transducers to send and receive sound, equipment for the generation and detection of the electrical impulses to and from the transducer, and a display or recorder system for the observation of the received signals.   { 'ak·tiv 'sō,när }

active system |ENG| In radio and radar, a system that requires transmitting equipment, such as a beacon or transponder.   { 'ak·tiv 'sis·təm }

active termination |COMPUT SCI| A means of ending a chain of peripheral devices connected to a small computer system interface (SCSI) port, suitable for longer chains, where it can reduce electrical interference. { ¦ak·tiv ‚tər·mə'nā·shən }

active transducer |ELECTR| A transducer whose output is dependent upon sources of power, apart from that supplied by any of the actuating signals, which power is controlled by one or more of these signals. { 'ak·tiv tranz'düs·ər }

active voltage |ELEC| In an alternating-current circuit, the component of voltage which is in phase with the current. { 'ak·tiv 'vōl·tij }

active window |COMPUT SCI| In a windowing environment, the window in which the user is currently working and which receives keyboard input. { ¦ak·tiv 'win‚dō }

activity |COMPUT SCI| The use or modification of information contained in a file. { ‚ak'tiv·əd·ē }

activity level |COMPUT SCI| **1.** The value assumed by a structural variable during the solution of a programming problem. **2.** A measure of the number of times that use or modification is made of the information contained in a file. { ‚ak'tiv·əd·ē 'lev·əl }

activity ratio |COMPUT SCI| The ratio between used or modified records and the total number of records in a file. { ‚ak'tiv·əd·ē ‚rā·shō }

activity sequence method |COMPUT SCI| A method of organizing records in a file so that the records most frequently used are located where they can be found most quickly. { ak'tiv·əd·ē 'sē·kwəns ‚meth·əd }

actual argument |COMPUT SCI| The variable which replaces a dummy argument when a procedure or macroinstruction is called up. { 'ak·chə·wəl 'är·gyə·mənt }

actual decimal point |COMPUT SCI| The period appearing on a printed report as opposed to the virtual point defined only by the data structure within the computer. { 'ak·chə·wəl 'des·məl 'pȯint }

actual instruction See effective instruction. { 'ak·chə·wəl in'strək·shən }

actual key |COMPUT SCI| A data item in COBOL computer language which can be used as an address. { 'ak·chə·wəl 'kē }

actuating system |CONT SYS| An electric, hydraulic, or other system that supplies and transmits energy for the operation of other mechanisms or systems. { 'ak·chə‚wād·iŋ ‚sis·təm }

actuator |CONT SYS| A mechanism to activate process control equipment by use of pneumatic, hydraulic, or electronic signals. |ENG ACOUS| An auxiliary external electrode used to apply a known electrostatic force to the diaphragm of a microphone for calibration purposes. Also known as electrostatic actuator. { 'ak·chə‚wād·ər }

acyclic feeding |COMPUT SCI| A method employed by alphanumeric readers in which the trailing edge or some other document characteristic is used to activate the feeding of the succeeding document. { ā'sik·lik 'fēd·iŋ }

acyclic machine See homopolar generator. { ā'sik·lik mə'shēn }

Ada |COMPUT SCI| A computer language that was chosen by the United States Department of Defense to support the development of embedded systems, and uses the language Pascal as a base to meet the reliablity and efficiency requirements imposed by these systems. { 'ā·də }

adapter |COMPUT SCI| A device which converts bits of information received serially into parallel bit form for use in the inquiry buffer unit. |ENG| A device used to make electrical or mechanical connections between items not originally intended for use together. { ə'dap·tər }

adapter transformer |ELEC| A transformer designed to supply a single electric lamp; its primary terminals are designed to fit into an ordinary lampholder, its secondary terminals into a lampholder of a low-voltage lamp. { ə'dap·tər tranz‚fȯr·mər }

adaptive antenna |ELECTROMAG| An antenna that adjusts its pattern automatically to be the inverse to any nonuniform distribution in angle of offending interference sources, tending to "whiten" or make appear uniform the noise in angle and minimizing the effects of strong jamming. { ə'dap·tiv an'ten·ə }

adaptive branch |CONT SYS| A branch instruction in the computer program controlling a robot that may lead the robot to execute a series of instructions, depending on external conditions. { ə'dap·tiv 'branch }

adaptive communications |COMMUN| A communications system capable of automatic change to meet changing inputs or changing characteristics of the device or process being controlled. Also known as self-adjusting communications; self-optimizing communications. { ə'dap·tiv kə‚myü·nə'kā·shənz }

adaptive control |CONT SYS| A control method in which one or more parameters are sensed and used to vary the feedback control signals in order to satisfy the performance criteria. { ə'dap·tiv kən'trōl }

adaptive differential pulse-code modulation |COMMUN| A method of compressing speech and music signals in which the transmitted signals represent differences between input signals and predicted signals, and these predicted signals are synthesized by predictors with response functions representative of the short- and long-term correlation inherent in the signal. Abbreviated ADPCM. { ə¦dap·tiv ‚dif·ə¦ren·chəl 'pəls ‚cōd ‚mäj·ə‚lā·shən }

adaptive equalization |COMMUN| A signal-processing technique designed to compensate for impairments in received signals over a communications channel resulting from imperfect transmission characteristics. { ə'dap·tiv ‚ē·kwə·lə‚zā·shən }

adaptive filter |ELECTR| An electric filter whose frequency response varies with time, as a function of the input signal. { ə‚dap·tiv 'fil·tər }

adaptive robot |CONT SYS| A robot that can alter its responses according to changes in the environment. { ə'dap·tiv 'rō‚bät }

adaptive signal processing [COMMUN] The design of adaptive systems for signal-processing applications. { ə¦dap·tiv ¦sig·nəl 'prä·sə·siŋ }

adaptive structure [ENG] A structure whose geometric and inherent structural characteristics can be changed beneficially in response to external stimulation by either remote commands or automatic means. { ə,dap·tiv 'strək·chər }

adaptive system [SYS ENG] A system that can change itself in response to changes in its environment in such a way that its performance improves through a continuing interaction with its surroundings. { ə'dap·tiv 'sis·təm }

adaptive system theory [COMPUT SCI] The branch of automata theory dealing with adaptive, or self-organizing, systems. { ə'dap·tiv 'sis·təm ,the·ə·rē }

adaptor [COMPUT SCI] A printed circuit board that is plugged into an expansion slot in a computer to communicate with an external peripheral device. { ə'dap·tər }

Adcock antenna [ELECTROMAG] A pair of vertical antennas separated by a distance of one-half wavelength or less and connected in phase opposition to produce a radiation pattern having the shape of a figure eight. { 'ad·käk ,an'ten·ə }

Adcock direction finder [NAV] A radio direction finder utilizing one or more pairs of Adcock antennas. { 'ad·käk də'rek·shən ,fīn·dər }

ADCON See address constant. { 'ad,kän }

adconductor cathode [ELECTR] A cathode in which adsorbed alkali metal atoms provide electron emission in a glow or arc discharge. { ¦ad·kən¦dək·tər 'kath,ōd }

add See add operation. { ad }

adder [COMPUT SCI] A computer device that can form the sum of two or more numbers or quantities. [ELECTR] A circuit in which two or more signals are combined to give an output-signal amplitude that is proportional to the sum of the input-signal amplitudes. Also known as adder circuit. { 'ad·ər }

adder circuit See adder. { 'ad·ər ,sər·kət }

add-in [COMPUT SCI] An electronic component that can be placed on a printed circuit board already installed in a computer to enhance the computer's capability. { 'ad ,in }

adding circuit [ELECTR] A circuit that performs the mathematical operation of addition. { 'ad·iŋ 'sər·kət }

adding machine [COMPUT SCI] A device which performs the arithmetical operation of addition and subtraction. { 'ad·iŋ mə,shēn }

add-in program [COMPUT SCI] A computer program that enhances the capabilities of a particular application. { 'ad,in ,prō·grəm }

addition item [COMPUT SCI] An item which is to be filed in its proper place in a computer. { ə'di·shən 'īd·əm }

addition record [COMPUT SCI] A new record inserted into an updated master file. { ə'di·shən ,rek·ərd }

addition table [COMPUT SCI] The part of memory that holds the table of numbers used in addition

in a computer employing table look-up techniques to carry out this operation. { ə'di·shən ,tā·bəl }

additive synthesis [ENG ACOUS] A method of synthesizing complex tones by adding together an appropriate number of simple sine waves at harmonically related frequencies. { ¦ad·ə·div 'sin·thə·səs }

additive white Gaussian noise [COMMUN] Noise that contains equal energy per frequency across the spectrum of the noise employed. Also known as white noise. Abbreviated AWGN. { 'ad·əd·iv wīt ¦gaù·sē·ən 'nóiz }

add-on [COMPUT SCI] A peripheral device, such as a printer or disk drive, that is added to a basic computer. { 'ad,ón }

add-on memory [COMPUT SCI] Computer storage that is added to the original main storage to enhance the computer's processing capability. { 'ad,ón 'mem·rē }

add operation [COMPUT SCI] An operation in computer processing in which the sum of two or more numbers is placed in a storage location previously occupied by one of the original numbers. Also known as add. { 'ad ,äp·ə,rā·shən }

address [COMPUT SCI] The number or name that uniquely identifies a register, memory location, or storage device in a computer. { 'ad·res }

addressable [COMPUT SCI] Capable of being located by a computer through an addressing technique. { ə'dres·ə·bəl }

addressable cursor [COMPUT SCI] A cursor that can be moved by software or keyboard controls to any point on the screen. { ə'dres·ə·bəl 'kər·sər }

address book [COMPUT SCI] A feature in an e-mail program for storing e-mail addresses. { 'ad·rəs ,bùk }

address bus [COMPUT SCI] An internal computer communications channel that carries addresses from the central processing unit to components under the unit's control. { 'ad·res ,bəs }

address computation [COMPUT SCI] The modification by a computer of an address within an instruction, or of an instruction based on results obtained so far. Also known as address modification. { 'ad·res ,käm·pyə'tā·shən }

address constant [COMPUT SCI] A value, or its expression, used in the calculation of storage addresses from relative addresses for computers. Abbreviated ADCON. Also known as base address; presumptive address; reference address. { 'ad·res ,kän·stənt }

address conversion [COMPUT SCI] The use of an assembly program to translate symbolic or relative computer addresses. { 'ad·res kən,vər·zhən }

address counter [COMPUT SCI] A counter which increments an initial memory address as a block of data is being transferred into the memory locations indicated by the counter. { 'ad·res ,kaùnt·ər }

address field [COMPUT SCI] The portion of a computer program instruction which specifies where a particular piece of information is located in the computer memory. { 'ad·res ,fēld }

address format |COMPUT SCI| A description of the number of addresses included in a computer instruction. { 'ad·res ‚fȯr·mat }

address-free program |COMPUT SCI| A computer program in which all addresses are represented as displacements from the expected contents of a base register. { 'ad·res ¦frē 'prō·grəm }

address generation |COMPUT SCI| An addressing technique which facilitates addressing large storages and implementing dynamic program relocation; the effective main storage address is obtained by adding together the contents of the base register of the index register and of the displacement field. { 'ad·res ‚jen·ə'rā·shən }

addressing |COMPUT SCI| **1.** The methods of locating and gaining access to information in a computer's storage. **2.** The methods of selecting a particular peripheral device from several that are available at a given time. { ə'dres·iŋ }

addressing mode |COMPUT SCI| The specific technique by means of which a memory reference instruction will be spelled out if the computer word is too small to contain the memory address. { ə'dres·iŋ ‚mōd }

addressing system |COMPUT SCI| A labeling technique used to identify storage locations within a computer system. { ə'dres·iŋ ‚sis·təm }

address interleaving |COMPUT SCI| The assignment of consecutive addresses to physically separate modules of a computer memory, making possible the very-high-speed access of a sequence of contiguously addressed words, since all modules operate nearly simultaneously. { 'ad·res ‚in·tər'lēv·iŋ }

addressless instruction format See zero-address instruction format. { ə'dres·ləs ‚in'strək·shən 'fȯr·mat }

address modification See address computation. { 'ad·res ‚mäd·ə·fə'kā·shən }

address part |COMPUT SCI| That part of a computer instruction which contains the address of the operand, or of the result, or of the next instruction. { 'ad·res ‚pärt }

address register |COMPUT SCI| A register wherein the address part of an instruction is stored by a computer. { 'ad·res ‚rej·ə·stər }

address resolution |COMPUT SCI| **1.** The process of obtaining the actual machine address needed to perform an operation. **2.** The process by which the address used to identify a workstation on a local-area network is translated to an address that can be handled on the Internet. { 'ad·res ‚rez·ə‚lü·shən }

address sort routine |COMPUT SCI| A debugging routine which scans all instructions of the program being checked for a given address. { 'ad·res 'sȯrt ‚rü'tēn }

address space |COMPUT SCI| The number of storage locations available to a computer program. { 'ad·rəs ‚spās }

address track |COMPUT SCI| A path on a magnetic tape, drum, or disk on which are recorded addresses used in the retrieval of data stored on other tracks. { 'ad·res ‚trak }

address translation |COMPUT SCI| The assignment of actual locations in a computer memory to virtual addresses in a computer program. { 'ad·res tranz'lā·shən }

add-subtract time |COMPUT SCI| The time required to perform an addition or subtraction, exclusive of the time required to obtain the quantities from storage and put the sum or difference back into storage. { 'ad səb'trakt ‚tīm }

add time |COMPUT SCI| The time required by a computer to perform an addition, not including the time needed to obtain the addends from storage and put the sum back into storage. { 'ad ‚tīm }

add-to-memory technique |COMPUT SCI| In direct-memory-access systems, a technique which adds a data word to a memory location; permits linear operations such as data averaging on process data. { ¦ad tə ¦mem·rē 'tek·nēk }

adequacy |ELEC| The existence of sufficient facilities within an electric power system to satisfy the customer load requirement under static system conditions. { 'ad·ə·kwə·sē }

ADF See automatic direction finder.

ad hoc inquiry |COMPUT SCI| A single request for a piece of information, such as a report. { 'ad ¦häk in'kwī·rē }

A-display |ELECTR| A radar display in cartesian coordinates; the targets appear as vertical deflection lines; their Y coordinates are proportional to signal intensity; their X coordinates are proportional to distance to targets. Also known as A-indicator; A-scan; A-scope. { 'ādi‚splā }

adjacency |COMPUT SCI| A condition in character recognition in which two consecutive graphic characters are separated by less than a specified distance. { ə'jās·ən·sē }

adjacent-channel interference |COMMUN| Interference that is caused by a transmitter operating in an adjacent channel. Also known as A-scan; A-scope. { ə'jās·ənt 'chan·əl in·tər'fir·əns }

adjacent-channel selectivity |ELECTR| The ability of a radio receiver to respond to the desired signal and to reject signals in adjacent frequency channels. { ə'jās·ənt 'chan·əl sə‚lek'tiv·əd·ē }

adjustable resistor |ELEC| A resistor having one or more sliding contacts whose position may be changed. { ə'jəs·tə·bəl ri'zis·tər }

adjustable transformer See variable transformer. { ə'jəs·tə·bəl tranz'fȯr·mər }

adjusted decibel |ELECTR| A unit used to show the relationship between the interfering effect of a noise frequency, or band of noise frequencies, and a reference noise power level of −85 dBm. Abbreviated dBa. Also known as decibel adjusted. { ə'jəs·təd 'des·ə‚bel }

admittance |ELEC| A measure of how readily alternating current will flow in a circuit; the reciprocal of impedance, it is expressed in siemens. { əd'mit·əns }

admittance matrix [ELEC] A matrix Y whose elements are the mutual admittances between the various meshes of an electrical network, it satisfies the matrix equation I = YV, where I and V are column vectors whose elements are the currents and voltages in the meshes. { əd'mit·əns 'mā·triks }

ADP See automatic data processing.

ADPCM See adaptive differential pulse-code modulation.

ADR studio [ENG ACOUS] A sound-recording studio used in motion-picture and television production to allow an actor who did not intelligibly record his or her speech during the original filming or video recording to do so by watching himself or herself on the screen and repeating the original speech with lip synchronism; it is equipped with facilities for recreating the acoustical liveness and background sound of the environment of the original dialog. Derived from -automatic dialog replacement studio. Also known as postsynchronizing studio. { 'ā¦dē'är ¦stüd·ē·ō }

ADSEL See Mode S.

ADSL See asymmetric digital subscriber line; asynchronous digital subscriber loop. { a·dē·es 'el or 'ad·səl }

ADT See abstract data type.

advanced battery [ELEC] A large battery storage system designed to harness solar or wind energy or to store excess electricity during low-demand periods for use during higher-demand periods. { əd'vanst 'bad·ə·rē }

Advanced Research Projects Agency Network [COMPUT SCI] The computer network developed by the U.S. Department of Defense in 1969 from which the Internet originated. Abbreviated ARPANET. { əd,vanst ri'sərch ,prä,jeks ,ā·jən·sē ,net,wərk }

advanced signal-processing system [COMPUT SCI] A portable data-processing system for military use; its complete configuration may consist of the analyzer unit, a postprocessing unit (for data-processing and control tasks), and an advanced signal-processing display unit. Also known as Proteus. { əd'vanst 'sig·nəl 'präs·əs·iŋ ,sis·təm }

Advanced Television Technology Center [COMMUN] A private, nonprofit corporation organized by members of the television broadcasting and consumer products industries to test and recommend technologies for the delivery and reception of new U.S. digital services. Abbreviated ATTC. { əd'vanst 'tel·ə,vizh·ən tek'näl·ə·jē sen·tər }

aerial See antenna. { 'e·rē·əl }

aerogenerator [ELEC] A generator that is driven by the wind, designed to utilize wind power on a commercial scale. { ,e·rō'jen·ə,rād·ər }

aeronautical mobile satellite service [COMMUN] A mobile satellite service in which the mobile earth stations are located on board aircraft. Abbreviated AMSS. { ,er·ə¦nód·ə·kəl ,mō·bəl 'sad·əl,īt ,sər·vəs }

aeronautical mobile service [COMMUN] A mobile service between aircraft stations and land stations, or between aircraft stations, in which survival craft stations may also participate. { ,er·ə ¦nòd·ə·kəl ¦mō·bəl 'sər·vəs }

aerophare See radio beacon. { 'e·rə'fer }

aerospace electronics [ELECTR] The field of electronics as applied to aircraft and spacecraft. { ¦e·rō¦spās i,lek'trän·iks }

af See abfarad.

AFC See automatic frequency control.

affinity [COMPUT SCI] A specific relationship between data processing elements that requires one to be used with the other, where a choice might otherwise exist. { ə'fin·əd·ē }

a format [COMPUT SCI] A nonexecutable statement in FORTRAN which permits alphanumeric characters to be transmitted in a manner similar to numeric data. { 'ā 'fȯr,mat }

AGC See automatic gain control.

age coating [ELEC] The black deposit that is formed on the inner surface of an electric lamp by material evaporated from the filament. { āj 'kōd·iŋ }

agenda [COMPUT SCI] **1.** The sequence of control statements required to carry out the solution of a computer problem. **2.** A collection of programs used for manipulating a matrix in the solution of a problem in linear programming. { ə'jen·də }

aggregate data type See scalar data type. { 'ag·rə·gət 'da·də tīp }

aggregate function [COMPUT SCI] A command in a database management program that performs an arithmetic operation on the values in a specified column or field in all the records in the database, such as computing their sum or average or counting the number of records that satisfy particular criteria. { ¦ag·rə·gət 'fəŋk·shən }

aggressive device [COMPUT SCI] A unit of a computer that can initiate a request for communication with another device. { ə'gres·iv di'vīs }

aging [ELEC] Allowing a permanent magnet, capacitor, meter, or other device to remain in storage for a period of time, sometimes with a voltage applied, until the characteristics of the device become essentially constant. [ENG] **1.** The changing of the characteristics of a device due to its use. **2.** Operation of a product before shipment to stabilize characteristics or detect early failures. { 'āj·iŋ }

AGP See accelerated graphics port.

agricultural robot [CONT SYS] A robot used to pick and harvest farm products and fruits. { ¦ag·rə ¦kəl·chə·rəl 'rō,bät }

aH See abhenry.

Ah See ampere-hour.

aided tracking [ENG] A system of radar-tracking a target signal in bearing, elevation, or range, or any combination of these variables, in which the rate of motion of the tracking equipment is machine-controlled in collaboration with an operator so as to minimize tracking error. { 'ād·əd 'trak·iŋ }

aided-tracking mechanism [ENG] A device consisting of a motor and variable-speed drive

11

which provides a means of setting a desired tracking rate into a director or other fire-control instrument, so that the process of tracking is carried out automatically at the set rate until it is changed manually. { 'ād·əd 'trak·iŋ ‚mek·ə ‚niz·əm }

aided-tracking ratio |ENG| The ratio between the constant velocity of the aided-tracking mechanism and the velocity of the moving target. { 'ād·əd 'trak·iŋ ‚rā·shō }

A/in.² *See* ampere per square inch.

A-indicator *See* A-display. { 'ā ‚in·də‚kād·ər }

air battery |ELEC| A connected group of two or more air cells; also, a single air cell. { 'er 'bad· ə·rē }

airblast circuit breaker |ELEC| An electric switch which, on opening, utilizes a high-pressure gas blast (air or sulfur hexafluoride) to break the arc. { 'er‚blast 'sər·kət 'brāk·ər }

airborne collision avoidance system |NAV| A navigation system for preventing collisions between aircraft that relies primarily on equipment carried on the aircraft itself, but which may make use of equipment already employed in the ground-based air-traffic control system. Abbreviated ACAS. { 'er‚bórn kə'lizh·ən ə'vóid·əns ‚sis·təm }

airborne collision warning system |ENG| A system such as a radar set or radio receiver carried by an aircraft to warn of the danger of possible collision. { 'er‚bórn kə'lizh·ən 'wórn·iŋ ‚sis·təm }

airborne detector |ENG| A device, transported by an aircraft, whose function is to locate or identify an air or surface object. { 'er‚bórn di 'tek· tər }

airborne electronic survey control |ENG| The airborne portion of very accurate positioning systems used in controlling surveys from aircraft. { 'er‚bórn i‚lek'trän·ik 'sər·vā kən'trōl }

airborne intercept radar |ENG| Airborne radar used to track and "lock on" to another aircraft to be intercepted or followed. { 'er‚bórn 'in·tər ‚sept ‚rā‚där }

airborne profile recorder |ENG| An electronic instrument that emits a pulsed-type radar signal from an aircraft to measure vertical distances between the aircraft and the earth's surface. Abbreviated APR. Also known as terrain profile recorder (TPR). { 'er‚bórn 'prō‚fīl ri‚kórd·ər }

airborne radar |ENG| Radar equipment carried by aircraft to assist in navigation by pilotage, to determine drift, and to locate weather disturbances; a very important use is locating other aircraft either for avoidance or attack. { 'er‚bórn 'rā‚där }

airborne self-protection jammer |ELECTR| An electronic system carried by an aircraft to prevent detection by enemy radar by emitting signals that deceive the radar, causing confusion and uncertainty. { 'er‚bórn ¦self·prə'tek·shən ‚jam·ər }

air-break switch *See* air switch. { 'er¦brāk ‚swich }

air capacitor |ELEC| A capacitor having only air as the dielectric material between its plates. Also known as air condenser. { 'er kə'pas·əd·ər }

air cell |ELECTR| A cell in which depolarization at the positive electrode is accomplished electrically by reduction of the oxygen in the air. { 'er‚sel }

air check |ENG ACOUS| A recording made of a live radio broadcast for filing purposes at the broadcasting facility. { 'er ‚chek }

air condenser *See* air capacitor. { 'er ‚kən'dens ·ər }

air-control center |COMMUN| An area set aside in a submarine for the control of aircraft; it is the equivalent of a combat information center on an aircraft or a ship. { 'er kən'trōl ‚sent·ər }

air-cooled condenser *See* air condenser. { 'er ‚küld kən'dens·ər }

air-core coil |ELECTR| An inductor without a magnetic core. { 'er ‚kòr ‚kòil }

aircraft antenna |ELECTR| An airborne device used to detect or radiate electromagnetic waves. { 'er‚kraft an'ten·ə }

aircraft detection |ENG| The sensing and discovery of the presence of aircraft; major techniques include radar, acoustical, and optical methods. { 'er‚kraft di'tek·shən }

air-depolarized battery |ELEC| A primary battery which is kept depolarized by atmospheric oxygen rather than chemical compounds. Also known as metal-air battery. { 'er dē'pōl·ə‚rīzd 'bad·ə·rē }

air gap |ELECTR| **1.** A gap or an equivalent filler of nonmagnetic material across the core of a choke, transformer, or other magnetic device. **2.** A spark gap consisting of two electrodes separated by air. **3.** The space between the stator and rotor in a motor or generator. { 'er ‚gap }

air-ground communication |COMMUN| Two-way communication between aircraft and stations on the ground. { ¦er ¦graùnd kə‚myü·nə'kā·shən }

air-insulated substation |ELEC| An electric power substation that has the busbars and equipment terminations generally open to air and utilizes insulation properties of ambient air for insulation to ground. { 'er 'in·sə'lād·əd 'səb‚stā·shən }

air mileage indicator |ENG| An instrument on an airplane which continuously indicates mileage through the air. { ¦er ‚mī·lij 'in·də'kād·ər }

air mileage unit |ENG| A device which derives continuously and automatically the air distance flown, and feeds this information into other units, such as an air mileage indicator. { 'er ‚mī·lij ‚yü·nət }

air navigation |NAV| The process of directing and monitoring the progress of an aircraft between selected geographic points or with respect to some predetermined plan. Also known as avigation. { ¦er ‚nav·ə'gā·shən }

airport surface detection equipment |ENG| Radar and other equipment specifically designed to assist in the control of aircraft and the many other vehicles that must use taxiways and other surface routes in the airport area. Also known as surface movement radar. { 'er‚pòrt 'sər·fəs di'tek·shən i‚kwip·mənt }

airport surveillance radar |ENG| Radar designed for air surveillance and to assist in air traffic management in the area of airports; designated as ASR in the United States nomenclature;

usually composed of both primary and secondary radars. { 'er₁pȯrt sər'vā·ləns ₁rā₁där }

air-route surveillance radar |ENG| Radar designed for air surveillance along established air routes to assist, through netted data operation, in air traffic management. Often in rather remote locations, such radars are designed for minimum on-site operator and maintenance attention. Abbreviated ARSR. { ¦er ₁rüt sər¦vā·ləns ₁rā₁där }

air-spaced coax |ELECTROMAG| Coaxial cable in which air is basically the dielectric material; the conductor may be centered by means of a spirally wound synthetic filament, beads, or braided filaments. { 'er ₁spāst 'kō₁aks }

airspeed indicator |ENG| A device that computes and displays the speed of an aircraft relative to the air mass in which the aircraft is flying. { ¦er ₁spēd ₁in·də₁kād·ər }

air surveillance |ENG| Systematic observation of the airspace by visual, electronic, or other means, primarily for identifying all aircraft in that airspace, and determining their movements. { 'er sər'vā·ləns }

air surveillance radar |ENG| Radar of moderate range providing position of aircraft by azimuth and range data without elevation data; used for air-traffic control. { ¦er sər¦vā·ləns ¦ra₁där }

air survey See aerial survey. { ¦er ¦sər₁vā }

air switch |ELEC| A switch in which the breaking of the electric circuit takes place in air. Also known as air-break switch. { 'er ₁swich }

air terminal |ELEC| A structure, such as a tower, that serves as a lightning arrester. { 'er ₁tərm·ən·əl }

air-traffic control radar beacon system |NAV| A system adopted by the Federal Aviation Agency for use in controlling air traffic over the United States; the aircraft carry identification transponders designed to transmit an airplane identity code, altitude, and additional message when interrogated by an air-traffic controller's equipment. Abbreviated ATCRBS. { 'er ¦traf·ik kən'trōl ¦rā₁där 'bē·kən ₁sis·təm }

air-variable capacitor |ELEC| A device with one rotating and one fixed set of metal plates positioned in meshed fashion and separated by air; capacitance is varied by rotating one set of plates to vary the overlap with the fixed plates. { 'er ¦ver·ē·ə·bəl kə'pas·əd·ər }

airwave |ELECTR| A radio wave used in radio and television broadcasting. { 'er₁wāv }

alarm signal |ELECTR| The international radio-telegraph alarm signal transmitted to actuate automatic devices that sound an alarm indicating that a distress message is about to be broadcast. { ə'lärm ₁sig·nəl }

alarm system |ENG| A system which operates a warning device after the occurrence of a dangerous or undesirable condition. { ə'lärm ₁sis·təm }

ALC See automatic level control.

alert box |COMPUT SCI| A dialog box that warns of an existing condition or the consequences of a command that has been given, or explains why a command cannot be executed. { ə'lərt ₁bäks }

alerting signal |COMMUN| Specific signal that is applied to subscriber access lines to indicate an incoming call. { ə'lərt·iŋ 'sig·nəl }

Alexanderson antenna |ELECTROMAG| An antenna, used at low or very low frequencies, consisting of several base-loaded vertical radiators connected together at the top and fed at the bottom of one radiator. { ₁al·ig'zan·dər·sən an₁ten·ə }

Alford loop |ELECTROMAG| An antenna utilizing multielements which usually are contained in the same horizontal plane and adjusted so that the antenna has approximately equal and in-phase currents uniformly distributed along each of its peripheral elements and produces a substantially circular radiation pattern in the plane of polarization; it is known for its purity of polarization. { 'ȯl·fərd ₁lüp }

algebraic computation system See symbolic system. { al·jə₁brā·ik ₁käm·pyə'tā·shən ₁sis·təm }

algebraic manipulation language |COMPUT SCI| A programming language used in the solution of analytic problems by symbolic computation. { ₁al·jə¦brā·ik mə·ni·pyə'lā·shən ₁laŋ·gwij }

Algol |COMPUT SCI| An algorithmic and procedure-oriented computer language used principally in the programming of scientific problems. { 'al₁gȯl }

algorithmic error |COMPUT SCI| An error in computer processing resulting from imprecision in the method used to carry out mathematical computations, usually associated with either rounding or truncation of numbers. { ¦al·gə¦rith·mik 'er·ər }

algorithmic language |COMPUT SCI| A language in which a procedure or scheme of calculations can be expressed accurately. { ¦al·gə¦rith·mik 'laŋ·gwij }

algorithm translation |COMPUT SCI| A step-by-step computerized method of translating one programming language into another programming language. { 'al·gə₁rith·əm tranz'lā·shən }

alias |COMPUT SCI| **1.** An alternative entry point in a computer subroutine at which its execution may begin, if so instructed by another routine. **2.** An alternative name for a file or device. { 'ā·lē·əs }

aliasing |COMPUT SCI| In computer graphics, the jagged appearance of diagonal lines on printouts and on video monitors. { 'āl·yəs·iŋ }

alignment |ELECTR| The process of adjusting components of a system for proper interrelationship, including the adjustment of tuned circuits for proper frequency response and the time synchronization of the components of a system. { ə'līn·mənt }

alignment wire See ground wire. { ə'līn·mənt ₁wīr }

alive See energized. { ə'līv }

alkaline cell |ELEC| A primary cell that uses an alkaline electrolyte, usually potassium hydroxide, and delivers about 1.5 volts at much higher current rates than the common carbon-zinc cell. Also known as alkaline-manganese cell. { 'al·kə₁līn ₁sel }

alkaline-manganese cell See alkaline cell. { |al ·kə,līn |maŋ·gə,nēs ,sel }

alkaline storage battery [ELEC] A storage battery in which the electrolyte consists of an alkaline solution, usually potassium hydroxide. { 'al·kə ,līn 'stòr·ij ,bad·ə·rē }

all-channel tuning [COMMUN] The ability of a television set to receive ultra-high-frequency as well as very-high-frequency channels. { 'òl ,chan·əl 'tün·iŋ }

all-diffused monolithic integrated circuit [ELECTR] Microcircuit consisting of a silicon substrate into which all of the circuit parts (both active and passive elements) are fabricated by diffusion and related processes. { |òl də|fyüzd ,män·ə'lith·ik 'in·tə,grād·əd 'sər·kət }

all-digital AM IBOC [COMMUN] The final mode of the AM IBOC system approved by the Federal Communications Commission for use in the United States that increases data capacity by increasing signal power and adjusting the bandwidth of the digital sidebands to minimize adjacent channel interference; uses four frequency partitions and no analog carrier. In this mode, the digital audio data rate can change from 40 to 60 kbits/s, and the corresponding ancillary data rate will remain at 0.4 kbits/s. { |òl dij·əd·əl 'ā ,em 'ī,bäk }

all-digital FM IBOC [COMMUN] The third of three modes in the FM IBOC system approved by the Federal Communications Commission for use in the United States that increases data capacity by adding additional digital carriers; uses four frequency partitions and no analog carrier. In this mode, the digital audio data rate can range from 64 to 96 kbits/s, and the corresponding ancillary data rate can range from 213 kbits/s for 64-kbits/s audio to 181 kbits/s for 96-kbits/s audio. { |òl dij·əd·əl 'ef,em 'ī,bäk }

alligator clip [ELEC] A long, narrow spring clip with meshing jaws; used with test leads to make temporary connections quickly. Also known as crocodile clip. { 'al·ə,gād·ər ,klip }

allocate [COMPUT SCI] To place a portion of a computer memory or a peripheral unit under control of a computer program, through the action of an operator, program instruction, or executive program. { 'a·lō,kāt }

allotter [COMMUN] A telephone term referring to a distributor, which allots an idle line-finder in preparation for an additional call. { ə'läd·ər }

alloy junction [ELECTR] A junction produced by alloying one or more impurity metals to a semiconductor to form a p or n region, depending on the impurity used. Also known as fused junction. { 'a,lòi ,jəŋk·shən }

alloy-junction diode [ELECTR] A junction diode made by placing a pill of doped alloying material on a semiconductor material and heating until the molten alloy melts a portion of the semiconductor, resulting in a pn junction when the dissolved semiconductor recrystallizes. Also known as fused-junction diode. { 'a,lòi |jəŋk·shən 'dī ,ōd }

alloy-junction transistor [ELECTR] A junction transistor made by placing pellets of a p-type impurity such as indium above and below an n-type wafer of germanium, then heating until the impurity alloys with the germanium to give a pnp transistor. Also known as fused-junction transistor. { 'a,lòi |jəŋk·shən tranz'is·tər }

all-pass network [ELECTR] A network designed to introduce a phase shift in a signal without introducing an appreciable reduction in energy of the signal at any frequency. { |òl |pas 'net ,wərk }

all-translational system [CONT SYS] A simple robotic system in which there is no rotation of the robot or its components during any movements of the robot's body. { |òl ,tranz'lā·shən·əl 'sis·təm }

all-wave receiver [ELECTR] A radio receiver capable of being tuned from about 535 kilohertz to at least 20 megahertz; some go above 100 megahertz and thus cover the FM band also. { |òl |wāv ri'sē·vər }

aloha [COMMUN] A radio-channel random-access technique that depends on positive acknowledgement of correct receipt for error control. { ə'lō·ə }

alpha [ELECTR] The ratio between the change in collector current and the change in emitter current of a transistor. { 'al·fə }

alphabetic character [COMPUT SCI] A letter or other symbol used to form data, other than a digit. { |al·fə|bed·ik 'kar·ik·tər }

alphabetic coding [COMPUT SCI] **1.** Abbreviation of words for computer input. **2.** A system of coding with a number system of base 26, the letters of the alphabet being used instead of the cardinal numbers. { |al·fə|bed·ik 'kōd·iŋ }

alphabetic string See character string. { |al·fə |bed·ik 'striŋ }

alpha cutoff frequency [ELECTR] The frequency at the high end of a transistor's range at which current amplification drops 3 decibels below its low-frequency value. { 'al·fə 'kəd,óf ,frē·kwən·sē }

alphageometric technique See alphamosaic technique. { |al·fə,jē·ə'me·trik ,tek,nēk }

alphameric characters See alphanumeric characters. { |al·fə|mer·ik 'kar·ik·tərz }

alphameric typebar [COMPUT SCI] A metal bar containing the alphabet, the ten numerical characters, and the ampersand, in use in electromechanical accounting machines. { |al·fə|mer·ik 'tīp,bär }

alphamosaic technique [COMPUT SCI] In computer graphics, a technique for displaying very-low-resolution images by constructing them from a set of elementary graphics characters. Also known as alphageometric technique. { |al·fə·mō'zā·ik ,tek,nēk }

alphanumeric characters [COMPUT SCI] All characters used by a computer, including letters, numerals, punctuation marks, and such signs as $, @, and #. Also known as alphameric characters. { ¦al·fə·nü¦mer·ik 'kar·ik·tərz }

alphanumeric display device [ELECTR] A device which visibly represents alphanumeric output information from some signal source. { ¦al·fə·nü ¦mer·ik dis'plā di,vīs }

alphanumeric instruction [COMPUT SCI] The name given to instructions which can be read equally well with alphabetic or numeric kinds of fields of data. { ¦al·fə·nü¦mer·ik in'strək·shən }

alphanumeric pager [COMMUN] A receiver in a radio paging system that contains a device which can display text or numeric messages. { ¦al·fə·nü¦mer·ik 'pā·jər }

alphanumeric reader [ELECTR] A device capable of reading alphabetic, numeric, and special characters and punctuation marks. { ¦al·fə·nü ¦mer·ik 'rēd·ər }

alpha test [COMPUT SCI] A test of software carried out at the user's location and using actual data. { 'al·fə ,test }

alpha test site [COMPUT SCI] A place where a complete computer system is tested with actual data and transactions. { 'al·fə ,test ,sīt }

alternate-channel interference [COMMUN] Interference that is caused in one communications channel by a transmitter operating in the next channel beyond an adjacent channel. Also known as second-channel interference. { ¦ól·tər·nət ¦chan·əl in·tər'fir·əns }

alternate index See secondary index. { 'ól·tər· nət 'in,deks }

alternate key [COMPUT SCI] A key on a computer keyboard that does not itself generate a character but changes the nature of the character generated by another key when depressed simultaneously with it; similar to the control and shift keys. Abbreviated ALT key. { 'ól·tər·nət ,kē }

alternate routing [COMMUN] The operation of a switching center when all circuits are found busy in a programmed route to the destination, and the call is offered to another programmed route. { 'ól·tər·nət 'rüt·iŋ }

alternate track [COMPUT SCI] The disk track used if, after a disk volume is initialized, a defective track is sensed by the system. { 'ól·tər·nət 'trak }

alternating current [ELEC] Electric current that reverses direction periodically, usually many times per second. Abbreviated ac. { ¦ól·tər,nād· iŋ ¦kər·ənt }

alternating-current circuit theory [ELEC] The mathematical description of conditions in an electric circuit driven by an alternating current or sources. { ¦ól·tər,nād·iŋ ¦kər·ənt 'sər·kət ,thē·ə·rē }

alternating-current coupling [ELECTR] A coupling which passes alternating-current signals but blocks direct-current signals. { ¦ól·tər ,nād·iŋ ¦kər·ənt 'kəp·liŋ }

alternating-current/direct-current [ELECTR] Pertaining to electronic equipment capable of operation from either an alternating-current or

direct-current primary power source. { ¦ól·tər ,nād·iŋ ¦kər·ənt di¦rekt ¦kər·ənt }

alternating-current dump [ELECTR] The removal of all alternating-current power from a computer intentionally, accidentally, or conditionally. { ¦ól·tər,nād·iŋ ¦kər·ənt 'dəmp }

alternating-current erase [ELECTR] The use of an alternating current to energize a tape recorder erase head in order to remove previously recorded signals from a tape. { ¦ól·tər,nād·iŋ ¦kər·ənt ə'rās }

alternating-current erasing head [ELECTR] In magnetic recording, an erasing head which uses alternating current to produce the magnetic field necessary for erasing. { ¦ól·tər,nād·iŋ ¦kər·ənt ə'rās·iŋ ,hed }

alternating-current generator [ELEC] A machine, usually rotary, which converts mechanical power into alternating-current electric power. { ¦ól·tər ,nād·iŋ ¦kər·ənt 'jen·ə,rād·ər }

alternating-current magnetic biasing [ELECTR] Biasing with alternating current, usually well above the signal frequency range, in magnetic tape recording. { ¦ól·tər,nād·iŋ ¦kər·ənt mag'ned·ik 'bī·əs·iŋ }

alternating-current motor [ELEC] A machine that converts alternating-current electrical energy into mechanical energy by utilizing forces exerted by magnetic fields produced by the current flow through conductors. { ¦ól·tər,nād·iŋ ¦kər·ənt 'mōd·ər }

alternating-current network [ELEC] An electrical network that has elements with both resistance and reactance. { ¦ól·tər,nād·iŋ ¦kər·ənt 'net ,wərk }

alternating-current power supply [ELEC] A power supply that provides one or more alternating-current output voltages, such as an ac generator, dynamotor, inverter, or transformer. { ¦ól·tər ,nād·iŋ ¦kər·ənt 'paú·ər sə,plī }

alternating-current resistance See high-frequency resistance. { ¦ól·tər,nād·iŋ ¦kər·ənt ri'zis·təns }

alternating-current transmission [ELECTR] In television, that form of transmission in which a fixed setting of the controls makes any instantaneous value of signal correspond to the same value of brightness for only a short time. { ¦ól·tər,nād·iŋ ¦kər·ənt tranz'mish·ən }

alternating voltage [ELEC] Periodic voltage, the average value of which over a period is zero. { 'ól·tər,nād·iŋ 'vōl·tij }

alternator [ELEC] A mechanical, electrical, or electromechanical device which supplies alternating current. { 'ól·tər,nād·ər }

altitude delay [ELECTR] Synchronization delay introduced between the time of transmission of the radar pulse and the start of the trace on the indicator to eliminate the altitude/height hole on the plan position indicator-type display. { 'al·tə ,tüd di'lā }

altitude hole [ELECTR] The blank area in the center of a plan position indicator-type radarscope display caused by the time interval between transmission of a pulse and the receipt of the first ground return. { 'al·tə,tüd ,hōl }

altitude signal |ELECTR| The radio signals returned to an airborne electronics device by the ground or sea surface directly beneath the aircraft. { 'al·tə,tüd ,sig·nəl }

ALT key *See* alternate key. { 'ȯlt 'kē }

ALU *See* arithmetical unit.

aluminum arrester *See* aluminum-cell arrester. { ə'lüm·ə·nəm ə'res·tər }

aluminum cable steel-reinforced |ELEC| A type of power transmission line made of an aluminum conductor provided with a core of steel. Abbreviated ACSR. { ə'lüm·ə·nəm 'kā·bəl 'stēl ,rē·in'fȯrst }

aluminum-cell arrester |ELEC| A lightning arrester consisting of a number of electrolytic cells in series formed from aluminum trays containing electrolyte. Also known as aluminum arrester; electrolytic arrester. { ə'lüm·ə·nəm ,sel ə'res·tər }

aluminum conductor |ELEC| Any of several aluminum alloys employed for conducting electric current; because its weight is one-half that of copper for the same conductance, it is used in high-voltage transmission lines. { ə'lüm·ə·nəm kən'dək·tər }

A/m² *See* ampere per square meter.

AM *See* amplitude modulation.

amateur bands |COMMUN| Bands of frequencies assigned to licensed radio amateurs. { 'a·mə·chər ,banz }

amateur radio |ELECTR| A radio used for two-way radio communications by private individuals in leisure-time activity. Also known as ham radio. { 'a·mə·chər 'rād·ē,ō }

ambiguity |ELECTR| The condition in which a synchro system or servosystem seeks more than one null position. |ELECTROMAG| In radar, the consequence of using a periodic waveform in estimating a target's range and, in coherent radar, its radial velocity by Doppler sensing; deliberate change of periodicity is used to help resolve these ambiguities. { ,am·bə'gyü·əd·ē }

ambiguity error |COMPUT SCI| An error in reading a number represented in a digital display that can occur when this representation is changing; for example, the number 699 changing to 700 might be read as 799 because of imprecise synchronization in the changing of digits. { ,am·bə'gyü·əd·ē ,er·ər }

ambiguous name |COMPUT SCI| A name of a file or other item which is only partially specified; it is useful in conducting a search of all the items to which it might apply. { am'big·yə·wəs 'nām }

AMC *See* automatic modulation control.

Amdahl's law |COMPUT SCI| A law stating that the speed-up that can be achieved by distributing a computer program over p processors cannot exceed $1/\{f + [1 - f]/p]\}$, where f is the fraction of the work of the program that must be done in serial mode. { 'am,dälz ,lȯ }

amendment record *See* change record. { ə'mend·mənt ,rek·ərd }

American Standard Code for Information Interchange |COMMUN| Coded character set to be used for the general interchange of information among information-processing systems, communications systems, and associated equipment; the standard code, comprising characters 0 through 127, includes control codes, upper- and lower-case letters, numerals, punctuation marks, and commonly used symbols; an additional set is known as extended ASCII. Abbreviated ASCII. { ə'mer·ə·kən 'stan·dərd 'kōd fər in·fər'mā·shən 'in·tər,chānj }

AM field signature |ELECTR| The characteristic pattern of an alternating magnetic field, as displayed by detection and classification equipment. { |ā¦em 'fēld ,sig·nə·chər }

A min *See* ampere-minute.

AML *See* automatic modulation limiting.

AMLCD *See* active-matrix liquid-crystal display.

ammeter |ENG| An instrument for measuring the magnitude of electric current flow. Also known as electric current meter. { 'a,mēd·ər }

amorphous memory array |COMPUT SCI| An array of memory switches made of amorphous material. { ə'mȯr·fəs 'mem·rē ə,rā }

amortisseur winding *See* damper winding. { a¦mȯrd·ə¦sər 'wīnd·iŋ }

amp *See* amperage; ampere. { amp }

ampacity |ELEC| Current-carrying capacity in amperes; used as a rating for power cables. { am'pas·əd·ē }

amperage |ELEC| The amount of electric current in amperes. Abbreviated amp. { 'am·prij }

ampere |ELEC| The unit of electric current in the rationalized meter-kilogram-second system of units; defined in terms of the force of attraction between two parallel current-carrying conductors. Abbreviated A; amp. { 'am,pir }

Ampère balance *See* current balance. { 'äm,per ,bal·əns }

ampere-hour |ELEC| A unit for the quantity of electricity, obtained by integrating current flow in amperes over the time in hours for its flow; used as a measure of battery capacity. Abbreviated Ah; amp-hr. { 'am,pir ¦aü·ər }

ampere-hour capacity |ELEC| The charge, measured in ampere-hours, that can be delivered by a storage battery up to the limit to which the battery may be safely discharged. { 'am,pir ¦aü·ər kə'pas·əd·ē }

ampere-hour meter |ENG| A device that measures the total electric charge that passes a given point during a given period of time. { 'am,pir ¦aü·ər ,mēd·ər }

ampere-minute |ELEC| A unit of electrical charge, equal to the charge transported in 1 minute by a current of 1 ampere, or to 60 coulombs. Abbreviated A min. { ¦am,pir ¦min·ət }

ampere per square inch |ELEC| A unit of current density, equal to the uniform current density of a current of 1 ampere flowing through an area of 1 square inch. Abbreviated A/in². { 'am,pir pər ,skwer 'inch }

ampere per square meter |ELEC| The SI unit of current density. Abbreviated A/m². { 'am,pir pər ,skwer 'mēd·ər }

amp-hr *See* ampere-hour.

amplidyne |ELEC| A rotating magnetic amplifier having special windings and brush connections so that small changes in power input to the field coils produce large changes in power output. { 'am·plə,dīn }

amplification factor |ELECTR| In a vacuum tube, the ratio of the incremental change in plate voltage to a given small change in grid voltage, under the conditions that the plate current and all other electrode voltages are held constant. { ,am·plə·fə'kā·shən ,fak·tər }

amplification noise |ELECTR| Noise generated in the vacuum tubes, transistors, or integrated circuits of an amplifier. { ,am·plə·fə'kā·shən ,nóiz }

amplified back bias |ELECTR| Degenerative voltage developed across a fast time-constant circuit within a stage of an amplifier and fed back into a preceding stage. { 'am·plə,fīd 'bak ,bī·əs }

amplifier |ENG| A device capable of increasing the magnitude or power level of a physical quantity, such as an electric current or a hydraulic mechanical force, that is varying with time, without distorting the wave shape of the quantity. { 'am·plə,fī·ər }

amplifier-type meter |ENG| An electric meter whose characteristics have been enhanced by the use of preamplification for the signal input eventually used to actuate the meter. { 'am·plə ,fī·ər ,tīp 'mēd·ər }

amplify |ENG ACOUS| To strengthen a signal by increasing its amplitude or by raising its level. { 'am·plə,fī }

amplifying delay line |ELECTR| Delay line used in pulse-compression systems to amplify delayed signals in the super-high-frequency region. { 'am·plə,fī·iŋ di'lā ,līn }

amplitron |ELECTR| Crossed-field continuous cathode reentrant beam backward-wave amplifier for microwave frequencies. { 'am·plə ,trän }

amplitude discriminator See pulse-height discriminator. { 'am·plə,tüd dis'krim·ə,nād·ər }

amplitude distortion See frequency distortion. { 'am·plə,tüd di'stòr·shən }

amplitude fading |COMMUN| Fading in which the amplitudes of frequency components of a modulated carrier wave are uniformly attenuated. { 'am·plə,tüd ,fād·iŋ }

amplitude-frequency distortion See frequency distortion. { 'am·plə,tüd 'frē·kwən·sē di'stòr·shən }

amplitude-frequency response See frequency response. { 'am·plə,tüd 'frē·kwən·sē ri'späns }

amplitude gate |ELECTR| A circuit which transmits only those portions of an input signal which lie between two amplitude boundary level values. Also known as slicer; slicer amplifier. { 'am·plə ,tüd ,gāt }

amplitude limiter See limiter. { 'am·plə,tüd ,lim· əd·ər }

amplitude-limiting circuit See limiter. { 'am·plə ,tüd ¦lim·əd·iŋ ,sər·kət }

amplitude-modulated indicator |ENG| A general class of radar indicators, in which the sweep of the electron beam is deflected vertically or horizontally from a base line to indicate the existence of an echo from a target. Also known as deflection-modulated indicator; intensity-modulated indicator. { 'am·plə,tüd ¦mäj·ə,lād· əd ¦in·də,kād·ər }

amplitude modulation |ELECTR| Abbreviated AM. **1.** Modulation in which the aplitude of a wave is the characteristic varied in accordance with the intelligence to be transmitted. **2.** In telemetry, those systems of modulation in which each component frequency f of the transmitted intelligence produces a pair of sideband frequencies at carrier frequency plus f and carrier minus f. { 'am·plə ,tüd ,maj·ə'lā·shən }

amplitude-modulation noise |COMMUN| Noise produced by undesirable amplitude variations of a radio-frequency signal. { 'am·plə,tüd ,maj· ə'lā·shən ,nóiz }

amplitude-modulation radio |COMMUN| Also known as AM radio. **1.** The system of radio communication employing amplitude modulation of a radio-frequency carrier to convey the intelligence. **2.** A receiver used in such a system. { 'am·plə,tüd ,maj·ə'lā·shən 'rād·ē,ō }

amplitude modulator |PHYS| Any device which imposes amplitude modulation upon a carrier wave in accordance with a desired program. { 'am·plə,tüd ,maj·ə,lād·ər }

amplitude noise |ELECTROMAG| Effect on radar accuracy of the fluctuations in the amplitude of the signal returned by the target; these fluctuations are caused by any change in aspect if the target is not a point source. { 'am·plə,tüd ,nóiz }

amplitude resonance |PHYS| The frequency at which a given sinusoidal excitation produces the maximum amplitude of oscillation in a resonant system. { 'am·plə,tüd 'rez·ə·nəns }

amplitude response |ELECTR| The maximum output amplitude obtainable at various points over the frequency range of an instrument operating under rated conditions. { 'am·plə,tüd ri'späns }

amplitude selector See pulse-height selector. { 'am·plə,tüd si,lek·tər }

amplitude separator |ELECTR| A circuit used to isolate the portion of a waveform with amplitudes above or below a given value or between two given values. { 'am·plə,tüd 'sep·ə ,rād·ər }

amplitude shift keying |COMMUN| A method of transmitting binary coded messages in which a sinusoidal carrier is pulsed so that one of the binary states is represented by the presence of the carrier while the other is represented by its absence. Abbreviated ASK. { 'am·plə,tüd ¦shift ,kē·iŋ }

amplitude suppression ratio |ELECTR| Ratio, in frequency modulation, of the undesired output to the desired output of a frequency-modulated receiver when the applied signal has simultaneous amplitude and frequency modulation. { 'am·plə,tüd sə'presh·ən ,rā·shō }

17

amplitude-versus-frequency distortion [ELECTR] The distortion caused by the nonuniform attenuation or gain of the system, with respect to frequency under specified terminal conditions. { 'am·plə,tüd ¦vər·səs ¦frē·kwən·sē di'stȯr·shən }

AM radio See amplitude-modulation radio. { ¦ā ¦em 'rād·ē,ō }

AM signature [COMMUN] A graphic representation of the significant identifying characteristics of an amplitude-modulated signal. { ¦ā¦em 'sig·nə·chər }

AMSS See aeronautical mobile satellite service.

analog [ELECTR] **1.** A physical variable which remains similar to another variable insofar as the proportional relationships are the same over some specified range; for example, a temperature may be represented by a voltage which is its analog. **2.** Pertaining to devices, data, circuits, or systems that operate with variables which are represented by continuously measured voltages or other quantities. { 'an·əl,äg }

analog adder [ELECTR] A device with one output voltage which is a weighted sum of two input voltages. { 'an·əl,äg 'ad·ər }

analog channel [ELECTR] A channel on which the information transmitted can have any value between the channel limits, such as a voice channel. { 'an·əl,äg 'chan·əl }

analog communications [COMMUN] System of telecommunications employing a nominally continuous electric signal that varies in frequency, amplitude, or other characteristic, in some direct correlation to nonelectrical information (sound, light, and so on) impressed on a transducer. { 'an·əl,äg kə,myü·nə'kā·shənz }

analog comparator [ELECTR] **1.** A comparator that checks digital values to determine whether they are within predetermined upper and lower limits. **2.** A comparator that produces high and low digital output signals when the sum of two analog voltages is positive and negative, respectively. { 'an·əl,äg kəm'par·əd·ər }

analog computer [COMPUT SCI] A computer in which quantities are represented by physical variables; problem parameters are translated into equivalent mechanical or electrical circuits as an analog for the physical phenomenon being investigated. { 'an·əl,äg kəm'pyüd·ər }

analog data [COMPUT SCI] Data represented in a continuous form, as contrasted with digital data having discrete values. { 'an·əl,äg 'dad·ə }

analog-digital computer See hybrid computer. { 'an·əl,äg 'dij·ə·təl kəm,pyüd·ər }

analog indicator [ELECTR] A device in which the result of a measurement is indicated by a pointer deflection or other visual quantity. { 'an·əl,äg 'in·də,kād·ər }

analog monitor [ELECTR] A display unit that accepts only analog signals, which must be converted from digital signals by the computer's video display board. { 'an·əl·äg ,män·əd·ər }

analog multiplexer [ELECTR] A multiplexer that provides switching of analog input signals to allow use of a common analog-to-digital converter. { 'an·əl,äg 'məl·tə,plek·sər }

analog multiplier [ELECTR] A device that accepts two or more inputs in analog form and then produces an output proportional to the product of the input quantities. { 'an·əl,äg 'məl·tə ,plī·ər }

analog network [ELECTR] A circuit designed so that circuit variables such as voltages are proportional to the values of variables in a system under study. { 'an·əl,äg 'net,wərk }

analog output [CONT SYS] Transducer output in which the amplitude is continuously proportional to a function of the stimulus. { 'an·əl,äg 'aüt,pút }

analog recording [ELECTR] Any method of recording in which some characteristic of the recording signal, such as amplitude or frequency, is continuously varied in a manner analogous to the time variations of the original signal. { 'an·əl,äg ri'kȯrd·iŋ }

analog signal [ELECTR] A nominally continuous electrical signal that varies in amplitude or frequency in response to changes in sound, light, heat, position, or pressure. { 'an·əl,äg 'sig·nəl }

analog simulation [COMPUT SCI] The representation of physical systems and phenomena by variables such as translation, rotation, resistance, and voltage. { 'an·əl,äg ,sim·yə'lā·shən }

analog switch [ELECTR] **1.** A device that either transmits an analog signal without distortion or completely blocks it. **2.** Any solid-state device, with or without a driver, capable of bilaterally switching voltages or current. { 'an·əl,äg ,swich }

analog-to-digital converter [ELECTR] A device which translates continuous analog signals into proportional discrete digital signals. { ¦an·əl,äg tə ¦dij·ət·əl kən'vərd·ər }

analog-to-frequency converter [ELECTR] A converter in which an analog input in some form other than frequency is converted to a proportional change in frequency. { ¦an·əl,äg tə ¦frē·kwən·sē kən'vərd·ər }

analog voltage [ELECTR] A voltage that varies in a continuous fashion in accordance with the magnitude of a measured variable. { 'an·əl,äg 'vōl·tij }

analysis by synthesis [COMMUN] A method of determining the parameters of a speech coder in which the consequence of choosing a particular value of a coder parameter is evaluated by locally decoding the signal and comparing it to the original input signal. { ə¦nal·ə·sis ,bī 'sin·thə·səs }

analytical engine [COMPUT SCI] An early-19th-century form of mechanically operated digital computer. { ,an·əl'id·ə·kəl 'en·jən }

analytical function generator [ELECTR] An analog computer device in which the dependence of an output variable on one or more input variables is given by a function that also appears in a physical law. Also known as natural function generator; natural law function generator. { ,an·əl'id·ə·kəl 'faŋk·shən ,jen·ə,rād·ər }

analytic hierarchy [MATH] A systematic procedure for representing the elements of any problem which breaks down the problem into its smaller constituents and then calls for only simple pairwise comparison judgments to develop priorities at each level.   { ˌan·əl'id·ik 'hī·ər ˌär·kē }

analyzer [COMPUT SCI] **1.** A routine for the checking of a program. **2.** One of several types of computers used to solve differential equations. [ENG] A multifunction test meter, measuring volts, ohms, and amperes. Also known as set analyzer.   { 'an·ə,līz·ər }

anchor [COMPUT SCI] A tag that indicates either the source or destination of a hyperlink; for example, HTML anchors are used to create links within a document or to another document.   { 'aŋ·kər }

anchored graphic [COMPUT SCI] A picture or graph that remains at a fixed position on a page of a document rather than being attached to the text.   { ¦aŋ·kərd 'graf·ik }

anchor frame [COMMUN] In MPEG-2, a video frame that is used for prediction. I-frames and P-frames are generally used as anchor frames, but B-frames are never anchor frames.   { 'aŋ·kər ˌfrām }

AND circuit *See* AND gate.   { 'and ˌsər·kət }

Anderson bridge [ELECTR] A six-branch modification of the Maxwell-Wien bridge, used to measure self-inductance in terms of capacitance and resistance; bridge balance is independent of frequency.   { 'an·dər·sən ˌbrij }

AND gate [ELECTR] A circuit which has two or more input-signal ports and which delivers an output only if and when every input signal port is simultaneously energized. Also known as AND circuit; passive AND gate.   { 'and ˌgāt }

AND/NOR gate [ELECTR] A single logic element whose operation is equivalent to that of two AND gates with outputs feeding into a NOR gate.   { ¦and ¦nȯr ˌgāt }

AND NOT gate [ELECTR] A coincidence circuit that performs the logic operation AND NOT, under which a result is true only if statement A is true and statement B is not. Also known as A AND NOT B gate.   { ¦and ¦nät ˌgāt }

AND-OR circuit [ELECTR] Gating circuit that produces a prescribed output condition when several possible combined input signals are applied; exhibits the characteristics of the AND gate and the OR gate.   { ¦and ¦ȯr ˌsər·kət }

AND-OR-INVERT gate [ELECTR] A logic circuit with four inputs, a_1, a_2, b_1, and b_2, whose output is 0 only if either a_1 and a_2 or b_1 and b_2 are 1. Abbreviated A-O-I gate.   { ¦and ¦ȯr in'vərt ˌgāt }

angel echo [ENG] A radar echo from a region where there are no visible targets; may be caused by insects, birds, or refractive index variations in the atmosphere.   { 'ān·jəl ˌek·ō }

angle diversity [COMMUN] Diversity reception in which beyond-the-horizon tropospheric scatter signals are received at slightly different angles, equivalent to paths through different scatter volumes in the troposphere.   { 'aŋ·gəl də'vər·səd·ē }

angle jamming [ELECTR] Electronic countermeasures used to introduce large errors in angle-measuring radars; methods involve producing a false echo with pulse-to-pulse modulation that is inverse to that otherwise produced by a radar using conical scanning, or the generation of multiple interfering signals that may confuse monopulse radars.   { 'aŋ·gəl ˌjam·iŋ }

angle marker *See* azimuth marker.   { 'aŋ·gəl ˌmärk·ər }

angle modulation [ELECTR] The variation in the angle of a sine-wave carrier; particular forms are phase modulation and frequency modulation. Also known as sinusoidal angular modulation.   { 'aŋ·gəl mäj·ə'lā·shən }

angle of deflection [ELECTR] The angle through which the electron beam in a cathode-ray tube is diverted from a straight path.   { 'aŋ·gəl əv di'flek·shən }

angle of departure *See* angle of radiation.   { 'aŋ·gəl əv di'pär·chər }

angle of divergence [ELECTR] The angular spread of an electron beam in an oscilloscope.   { 'aŋ·gəl əv də'vərj·əns }

angle tracking noise [ELECTR] Deviation of the tracking axis or other angle estimate from the true angle of a radar target; it results from target reflective behavior and propagation path characteristics (such as fluctuation, glint, and scintillation) and also from the radar's own receiver, mechanical or computational noise.   { 'aŋ·gəl ¦trak·iŋ ˌnȯiz }

angular error of closure *See* error of closure.   { 'an·gyə·lər 'er·ər əv 'klōzh·ər }

angular resolver *See* resolver.   { 'aŋ·gyə·lər ri'zälv·ər }

ANL *See* automatic noise limiter.

annotation [COMPUT SCI] Any comment or note included in a program or flow chart in order to clarify some point at issue.   { ˌan·ə'tā·shən }

annual service availability index [ELEC] The ratio of customer-hours of service supplied by an electrical utility during one year to the customer-hours requested, expressed as a percentage.   { ¦an·yə·wəl ¦sər·vəs ə,vāl·ə'bil·əd·ē ˌin ˌdeks }

annular conductor [ELEC] A number of wires stranded in three reversed concentric layers around a saturated hemp core.   { 'an·yə·lər kən'dək·tər }

annular transistor [ELECTR] Mesa transistor in which the semiconductor regions are arranged in concentric circles about the emitter.   { 'an·yə·lər tran'zis·tər }

annunciator [ENG] A signaling apparatus which operates electromagnetically and serves to indicate visually, or visually and audibly, whether a current is flowing, has flowed, or has changed direction of flow in one or more circuits.   { ə'nən·sē·ād·ər }

anode [ELEC] The terminal at which current enters a primary cell or storage battery; it is positive with respect to the device, and negative with respect to the external circuit. [ELECTR] **1.** The collector of electrons in an electron tube.

Also known as plate; positive electrode. **2.** In a semiconductor diode, the terminal toward which forward current flows from the external circuit. { 'a,nōd }

anode balancing coil |ELEC| A set of mutually coupled windings used to maintain approximately equal currents in anodes operating in parallel from the same transformer terminal. { 'a,nōd ¦bal·əns·iŋ ,kȯil }

anode characteristic |ELECTR| Relationship of anode current to anode voltage in a vacuum tube. { 'a,nōd ,kar·ik·tə'ris·tik }

anode circuit |ELECTR| Complete external electrical circuit connected between the anode and the cathode of an electron tube. Also known as plate circuit. { 'a,nōd ,sər·kət }

anode-circuit detector |ELECTR| Detector functioning by virtue of a nonlinearity in its anode-circuit characteristic. Also known as plate-circuit detector. { 'a,nōd ¦sər·kət di,tek·tər }

anode current |ELECTR| The electron current flowing through an electron tube from the cathode to the anode. Also known as plate current. { 'a,nōd ,kər·ənt }

anode dark space |ELECTR| A thin, dark region next to the anode glow in a glow-discharge tube. { 'a,nōd 'därk ,spās }

anode detector |ELECTR| A detector in which rectification of radio-frequency signals takes place in the anode circuit of an electron tube. Also known as plate detector. { 'a,nōd di,tek·tər }

anode dissipation |ELECTR| Power dissipated as heat in the anode of an electron tube because of bombardment by electrons and ions. { 'a,nōd dis·ə'pā·shən }

anode drop *See* anode fall. { 'a,nōd ,dräp }

anode efficiency |ELECTR| The ratio of the ac load circuit power to the dc anode power input for an electron tube. Also known as plate efficiency. { 'a,nōd i,fish·ən·sē }

anode fall |ELECTR| **1.** A very thin space-charge region in front of an anode surface, characterized by a steep potential gradient through the region. **2.** The voltage across this region. Also known as anode drop. { 'a,nōd ,fȯl }

anode glow |ELECTR| A thin, luminous layer on the surface of the anode in a glow-discharge tube. { 'a,nōd ,glō }

anode impedance |ELECTR| Total impedance between anode and cathode exclusive of the electron stream. Also known as plate impedance; plate-load impedance. { 'a,nōd im ,pēd·əns }

anode input power |ELECTR| Direct-current power delivered to the plate (anode) of a vacuum tube by the source of supply. Also known as plate input power. { 'a,nōd 'in,pu̇t ,pau̇·ər }

anode modulation |ELECTR| Modulation produced by introducing the modulating signal into the anode circuit of any tube in which the carrier is present. Also known as plate modulation. { 'a,nōd ,mäj·ə'lā·shən }

anode neutralization |ELECTR| Method of neutralizing an amplifier in which the necessary 180° phase shift is obtained by an inverting network in the plate circuit. Also known as plate neutralization. { 'a,nōd ,nü·trə·lə'zā·shən }

anode pulse modulation |ELECTR| Modulation produced in an amplifier or oscillator by application of externally generated pulses to the plate circuit. Also known as plate-pulse modulation. { 'a,nōd 'pəls ,mäj·ə'lā·shən }

anode rays |ELECTR| Positive ions coming from the anode of an electron tube; generally due to impurities in the metal of the anode. { 'a,nōd ,rāz }

anode resistance |ELECTR| The resistance value obtained when a small change in the anode voltage of an electron tube is divided by the resulting small change in anode current. Also known as plate resistance. { 'a,nōd ri,zis·təns }

anode saturation |ELECTR| The condition in which the anode current of an electron tube cannot be further increased by increasing the anode voltage; the electrons are then being drawn to the anode at the same rate as they are emitted from the cathode. Also known as current saturation; plate saturation; saturation; voltage saturation. { 'a,nōd ,sach·ə'rā·shən }

anode sheath |ELECTR| The electron boundary which exists in a gas-discharge tube between the plasma and the anode when the current demanded by the anode circuit exceeds the random electron current at the anode surface. { 'a,nōd ,shēth }

anodized dielectric film |ELEC| An insulating film produced on a conducting surface by anodizing; used for producing thin-film capacitors, trimming resistor values, and passivation in the manufacture of integrated circuits. { 'an·ə,dīzd dī·ə¦lek·trik 'film }

anomalous Funkel effect |ELECTR| Current fluctuations in an electron tube resulting from positive ions entering the space-charge region in front of the cathode. { ə¦näm·ə·ləs 'fəŋ·kəl i ,fekt }

anomalous skin effect |ELEC| The skin effect at very low temperatures and high frequencies at which the thickness of the conducting skin layer is less than the electron mean free path, so that the classical theory of electrical conductivity breaks down. { ə¦näm·ə·ləs 'skin i,fekt }

anomaly detection |COMPUT SCI| The technology that seeks to identify an attack on a computer system by looking for behavior that is out of the norm. { ə'näm·ə·lē di,tek·shən }

anonymous FTP |COMPUT SCI| A public FTP (file transfer protocol) site at which users can log in and download documents by entering "anonymous" as their user ID, and their e-mail address as password.   { ə‚nän·ə·məs ¦ef¦tē'pē }

anotron |ELECTR| A cold-cathode glow-discharge diode having a copper anode and a large cathode of sodium or other material.   { 'an·ə‚trän }

A-N radio range |NAV| A type of radio beacon station whose signals provide definite track guidance for aircraft by establishing four radial lines of position which can be identified by a continuous-tone signal made up of keyed pulses of equal amplitude representing the Morse code letters A and N.   { ¦ā ¦en 'rād·ē‚ō ‚rānj }

answer back |COMPUT SCI| The ability of a device such as a computer or terminal to automatically identify itself when it is contacted by another communicating device.   { 'an·sər ¦bak }

answering cord |ELEC| Cord nearest the face of the switchboard which is used for answering subscribers' calls and incoming trunks.   { 'an·sər·iŋ ‚kȯrd }

answering jack |ELEC| Jack on which a station calls in and is answered by an operator.   { 'an·sər·iŋ ‚jak }

answer lamp |ELEC| Telephone switchboard lamp that lights when an answer cord is plugged into a line jack; the lamp goes out when the call is completed.   { 'an·sər ‚lamp }

answer-only modem |COMMUN| A modem that can answer but not initiate a call.   { ¦an·sər ‚ōn·lē 'mō‚dem }

antenna |ELECTROMAG| A device used for radiating or receiving radio waves. Also known as aerial; radio antenna.   { an'ten·ə }

antenna amplifier |ELECTROMAG| One or more stages of wide-band electronic amplification placed within or physically close to a receiving antenna to improve signal-to-noise ratio and mutually isolate various devices receiving their feed from the antenna.   { an'ten·ə 'am·plə‚fī·ər }

antenna circuit |ELECTR| A complete electric circuit which includes an antenna.   { an'ten·ə ‚sər·kət }

antenna coil |ELECTROMAG| Coil through which antenna current flows.   { an'ten·ə ‚kȯil }

antenna counterpoise   See counterpoise.{ an'ten·ə 'kaúnt·ər‚pȯiz }

antenna coupler |ELECTROMAG| A radio-frequency transformer, tuned line, or other device used to transfer energy efficiently from a transmitter to a transmission line or from a transmission line to a receiver.   { an'ten·ə ‚kəp·lər }

antenna crosstalk |ELECTROMAG| The ratio or the logarithm of the ratio of the undesired power received by one antenna from another to the power transmitted by the other.   { an'ten·ə 'krȯs‚tȯk }

antenna directive gain |ELECTROMAG| The ratio of the spatial power density on transmit, or sensitivity on receive, experienced at a distant point for using an idealized (lossless) directive antenna, as in radar, to that density of sensitivity experienced had an imaginary

isotropic antenna been used.   { ¦an'ten·ə 'də'rek·tiv ‚gān }

antenna directivity diagram |ELECTROMAG| Curve representing, in polar or cartesian coordinates, a quantity proportional to the gain of an antenna in the various directions in a particular plane or cone.   { an'ten·ə di·rek'tiv·əd·ē 'dī·ə‚gram }

antenna effect |ELECTROMAG| A distortion of the directional properties of a loop antenna caused by an input to the direction-finding receiver which is generated between the loop and ground, in contrast to that which is generated between the two terminals of the loop. Also known as electrostatic error; vertical component effect.   { an'ten·ə i'fekt }

antenna efficiency |ELECTROMAG| The ratio of the amount of power radiated into space by an antenna to the total energy received by the antenna.   { an'ten·ə i‚fish·ən·sē }

antenna field |ELECTROMAG| A group of antennas placed in a geometric configuration.   { an'ten·ə ‚fēld }

antenna gain |ELECTROMAG| A measure of the effectiveness of a directional antenna as compared to a standard nondirectional antenna. Also known as gain.   { an'ten·ə ‚gān }

antenna loading |ELECTR| **1.** The amount of inductance or capacitance in series with an antenna, which determines the antenna's electrical length.   **2.** The practice of loading an antenna in order to increase its electrical length.   { an'ten·ə ‚lōd·iŋ }

antenna matching |ELECTROMAG| Process of adjusting impedances so that the impedance of an antenna equals the characteristic impedance of its transmission line.   { an'ten·ə ‚mach·iŋ }

antenna pair |ELECTROMAG| Two antennas located on a base line of accurately surveyed length, sometimes arranged so that the array may be rotated around an axis at the center of the base line; used to produce directional patterns and in direction finding.   { an'ten·ə ‚per }

antenna pattern   See radiation pattern.   { an'ten·ə ‚pad·ərn }

antenna polarization |ELECTROMAG| The orientation of the electric field lines in the electromagnetic field radiated or received by the antenna.   { an'ten·ə ‚pō·lə·rə'zā·shən }

antenna power |ELECTROMAG| Radio-frequency power delivered to an antenna.   { an'ten·ə ‚paú·ər }

antenna power gain |ELECTROMAG| The ratio of the spatial power density on transmit, or sensitivity on receive, experienced at a distant point for using an actual directive antenna, as in radar, to that density or sensitivity experienced had an imaginary isotropic antenna been used. Power gain will, then, be slightly less than directive gain, differing by the insertion loss of the actual antenna, and is the gain actually measured in constructed antennas and used in most calculations about radar performance.   { an'ten·ə 'paú·ər ‚gān }

antenna resistance |ELECTROMAG| The power supplied to an entire antenna divided by the square of the effective antenna current measured at the point where power is supplied to the antenna. { an'ten·ə ri,zis·təns }

antenna scanner |ELECTROMAG| A microwave feed horn which moves in such a way as to illuminate sequentially different reflecting elements of an antenna array and thus produce the desired field pattern. { an'ten·ə ,skan·ər }

antenna tilt error |ENG| Angular difference between the tilt angle of a radar antenna shown on a mechanical indicator, and the electrical center of the radar beam. { an'ten·ə 'tilt ,er·ər }

antialiasing technique |COMPUT SCI| In computer graphics, a technique for smoothing the jagged appearance of diagonal lines on printouts and on video monitors. { ,an·tē'āl·ē·əs·iŋ ,tek,nēk }

anticapacitance switch |ELECTR| A switch designed to have low capacitance between its terminals when open. { ,an·tē·kə'pas·ə·təns ,swich }

anticathode |ELECTR| The anode or target of an x-ray tube, on which the stream of electrons from the cathode is focused and from which x-rays are emitted. { ¦an·tē'kath,ōd }

anticipatory staging |COMPUT SCI| Moving blocks of data from one storage device to another prior to the actual request for them by the program. { 'an'tis·ə·pə,tȯr·ē 'stāj·iŋ }

anticlutter gain control |ELECTR| Device which automatically and smoothly increases the gain of a radar receiver from a low level to the maximum, within a specified period after each transmitter pulse, so that short-range echoes producing clutter are amplified less than long-range echoes. { ,an·tē'kləd·ər 'gān kən,trōl }

anticoincidence circuit |ELECTR| Circuit that produces a specified output pulse when one (frequently predesignated) of two inputs receives a pulse and the other receives no pulse within an assigned time interval. { ,an·tē,kō'in·sə·dəns ,sər·kət }

anticollision radar |ENG| A radar set designed to give warning of possible collisions during movements of ships or aircraft. { ,an·tē·kə'li·zhən ,rā,där }

antifading antenna |ELECTR| An antenna designed to confine radiation mainly to small angles of elevation to minimize the fading of radiation directed at larger angles of elevation. { ¦an·tē¦fād·iŋ an'ten·ə }

antiglare shield |COMPUT SCI| A sheet of nonreflective material placed over the screen of an electronic display to reduce the amount of light reflected from the screen. { 'an·tē,gler 'shēld }

anti-g suit See g suit. { ¦an·tē¦jē ,süt }

antihunt circuit |ELECTR| A stabilizing circuit used in a closed-loop feedback system to prevent self-oscillations. { 'an·tē,hənt ,sər·kət }

anti-intrusion technology |COMPUT SCI| One of the different ways in which an attack on a computer system can be detected and countered, including prevention, deterrence, detection, deflection, and diminution. { ,an·tē,in'trü·zhən ,tek¦nāl·ə·jē }

antijamming |ELECTR| Any system or technique used to counteract the jamming of communications or of radar operation; part of electronic protection. { ,an·tē'jam·iŋ }

antimagnetic |ENG| Constructed so as to avoid the influence of magnetic fields, usually by the use of nonmagnetic materials and by magnetic shielding. { ,an·tē,mag'ned·ik }

antinoise microphone |ENG ACOUS| Microphone with characteristics which discriminate against acoustic noise. { ¦an·tē¦nȯiz 'mī·krə ,fōn }

antireflection coating |ENG| The application of a thin film of dielectric material to a surface to reduce its reflection and to increase its transmission of light or other electromagnetic radiation. { ,an·tē·ri'flek·shən ,kōd·iŋ }

antiresonance See parallel resonance. { ,an· tē'res·ən·əns }

antiresonant circuit See parallel resonant circuit. { ,an·tē'rez·ən·ənt 'sər·kət }

anti-sidetone circuit |ELEC| Telephone circuit which prevents sound, introduced in the local transmitter, from being reproduced in the local receiver. { ,an·tē¦sīd,tōn ,sər·kət }

antistatic mat |COMPUT SCI| A floor mat placed in front of a device such as a tape drive that is sensitive to discharges of static electricity to safeguard against loss of data from such discharges during human handling of the device. { ¦an·tē¦stad·ik 'mat }

anti-transmit-receive tube |ELECTR| A switching tube that prevents the received echo signal from being dissipated in the transmitter. { ¦an·tē·tranz¦mit ri¦sēv ,tüb }

antivirus software |COMPUT SCI| Software that is designed to protect against computer viruses. { ¦an·tē,vī·rəs 'sȯf,wer }

A-O-I gate See AND-OR-INVERT gate. { ,ā,ō'ī ,gāt }

APC See automatic phase control.

aperiodic antenna |ELECTROMAG| Antenna designed to have constant impedance over a wide range of frequencies because of the suppression of reflections within the antenna system; includes terminated wave and rhombic antennas. { ¦a,pir·ē¦äd·ik an'ten·ə }

aperiodic waves |ELEC| The transient current wave in a series circuit with resistance R, inductance L, and capacitance C when $R^2C = 4L$. { ¦a,pir·ē¦äd·ik 'wāvz }

aperture |ELECTR| An opening through which electrons, light, radio waves, or other radiation can pass. { 'ap·ə,chər }

aperture antenna |ELECTROMAG| Antenna in which the beam width is determined by the dimensions of a horn, lens, or reflector. { 'ap·ə ,chər an'ten·ə }

aperture grill picture tube |ELECTR| An in-line gun-type picture tube in which the shadow mask is perforated by long, vertical stripes and the screen is coated with vertical phosphor stripes. { 'ap·ə,chər ,gril 'pik·chər ,tüb }

aperture mask See shadow mask. { 'ap·ə,chər ,mask }

aperture plate |ELECTR| A small part of a piece of perforated ferromagnetic material that forms a magnetic cell. { 'ap·ə,chər ,plāt }

API *See* application program interface.

APL |COMPUT SCI| An interactive computer language whose operators accept and produce arrays with homogeneous elements of type number or character.

apodization |ELECTR| A technique for modifying the response of a surface acoustic wave filter by varying the overlap between adjacent electrodes of the interdigital transducer. { ,a·pə·də'zā·shən }

A positive |ELEC| Symbolized A+. **1.** Positive terminal of an A battery or positive polarity of other sources of filament voltage. **2.** Denoting the terminal to which the positive side of the filament voltage source should be connected. { ¦ā 'päz·əd·iv }

A power supply *See* A supply. { 'ā 'pau̇·ər sə,plī }

apparent power |ELEC| The product of the root-mean-square voltage and the root-mean-square current delivered in an alternating-current circuit, no account being taken of the phase difference between voltage and current. { ə'pa·rənt 'pau̇·ər }

apparent source *See* effective center. { ə'pa·rənt 'sȯrs }

Applegate diagram |ELECTR| A graph of the electron paths in a two-cavity klystron tube, showing how electron bunching occurs. { 'ap·əl,gāt 'dī·ə ,gram }

applet |COMPUT SCI| A small program, typically written in Java. { 'ap·lət }

appliance |ENG| A piece of equipment that draws electric or other energy and produces a desired work-saving or other result, such as an electric heater, a radio, or an electronic range. { ə'plī·əns }

appliance panel |ENG| In electric systems, a metal housing containing two or more devices (such as fuses) for protection against excessive current in circuits which supply portable electric appliances. { ə'plī·əns ,pan·əl }

application |COMPUT SCI| A computer program that performs a specific task, for example, a word processor, a Web browser, or a spread sheet. { ,ap·lə'kā·shən }

application development language |COMPUT SCI| A very-high-level programming language that generates coding in a conventional programming language or provides the user of a database management system with a programming language that is easier to implement than conventional programming languages. { ,ap·lə'kā·shən di'vel·əp·mənt ,laŋ·gwij }

application development system |COMPUT SCI| An integrated group of software products used to assist in the efficient development of computer programs and systems. { ,ap·lə'kā·shən di'vel·əp·mənt ,sis·təm }

application generator |COMPUT SCI| A commercially prepared software package used to create applications programs or parts of such programs. { ,ap·lə'kā·shən ,jen·ə,rād·ər }

application package |COMPUT SCI| A combination of required hardware, including remote inputs and outputs, plus programming of the computer memory to produce the specified results. { ,ap·lə'kā·shən ,pak·ij }

application processor |COMPUT SCI| A computer that processes data. { ,ap·lə¦kā·shən 'prä ,ses·ər }

application program |COMPUT SCI| A program written to solve a specific problem, produce a specific report, or update a specific file. { ,ap·lə'kā·shən ,prō·grəm }

application program interface |COMPUT SCI| A language that enables communication between computer programs, in particular between application programs and control programs. Abbreviated API. { ,ap·lə¦kā·shən ¦prō·grəm 'in·tər,fās }

application server |COMPUT SCI| A computer that executes commands requested by a Web server to fetch data from databases. Also known as app server. { ,ap·lə'kā·shən ,ser·vər }

application-specific integrated circuit |ELECTR| An integrated circuit that is designed for a particular application by integrating standard cells from a library, making possible short design times and rapid production cycles. Abbreviated ASIC. { ,ap·lə,kā·shən spi¦sif·ik ,int·i,grād·əd 'sər·kət }

application study |COMPUT SCI| The detailed process of determining a system or set of procedures for using a computer for definite functions of operations, and establishing specifications to be used as a base for the selection of equipment suitable to the specific needs. { ap·lə'kā·shən ,stəd·ē }

application system |COMPUT SCI| A group of related applications programs designed to perform a specific function. { ,ap·lə'kā·shən ,sis·təm }

application window |COMPUT SCI| In a graphical user interface, the chief window of an application program, with a title bar, a menu bar, and a work area. { ,ap·lə'kā·shən ,win,dō }

applicative language |COMPUT SCI| A programming language in which functions are repeatedly applied to the results of other functions and, in its pure form, there are no statements, only expressions without side effects. { 'ap·lə,kād·iv 'laŋ·gwij }

applied epistemology |COMPUT SCI| The use of machines or other models to simulate processes such as perception, recognition, learning, and selective recall, or the application of principles assumed to hold for human categorization, perception, storage, search, and so on, to the design of machines, machine programs, scanning, storage, and retrieval systems. { ə'plīd i¦pis·tə ¦mäl·ə·jē }

appliqué circuit |ELEC| Special circuit which is provided to modify existing equipment to allow for special usage; for example, some carrier telephone equipment designed for ringdown manual operation can be modified through the use of an appliqué circuit to allow for use between points having dial equipment. { ¦ap·lə ¦kā ¦sər·kət }

approach vector |CONT SYS| A vector that describes the orientation of a robot gripper and points in the direction from which the gripper approaches a workpiece. { ə'prōch ‚vek·tər }

app server See application server. { 'ap ‚sər·vər }

APT See Automatic Programming Tool.

APT system See automatic picture-transmission system. { ‚ā‚pē'tē ‚sis·təm }

aquadag |ELECTR| Graphite coating on the inside of certain cathode-ray tubes for collecting secondary electrons emitted by the face of the tube. { 'ak·wə‚dag }

arbiter |COMPUT SCI| A computer unit that determines the priority sequence in which two or more processor inputs are connected to a single functional unit such as a multiplier or memory. { 'är·bəd·ər }

arbitrary function generator See general-purpose function generator. { 'är·bə‚trer·ē 'fəŋk·shən ‚jen·ə‚rād·ər }

arbitration |COMPUT SCI| The set of rules in a computer's operating system for allocating the resources of the computer, such as its peripheral devices or memory, to more than one program or user. { ‚ar·bə'trā·shən }

arc See electric arc. { ärk }

arcback |ELECTR| The flow of a principal electron stream in the reverse direction in a mercury-vapor rectifier tube because of formation of a cathode spot on an anode; this results in failure of the rectifying action. Also known as backfire. { 'ärk ‚bak }

arc chute |ELEC| A collection of insulating barriers in a circuit breaker for confining the arc and preventing it from causing damage. { 'ärk ‚shüt }

arc converter |ELECTR| A form of oscillator using an electric arc as the generator of alternating or pulsating current. { 'ärk kən‚vər·dər }

arc discharge |ELEC| A direct-current electrical current between electrodes in a gas or vapor, having high current density and relatively low voltage drop. { 'ärk 'dis‚chärj }

Archie |COMPUT SCI| A system of file servers that searches for specific files that are publicly available in File Transfer Protocol archives on the Internet. { 'är·chē }

archival storage |COMPUT SCI| Storage of infrequently used or backup information that cannot be readily or immediately accessed by a computer system. { 'är‚kīv·əl 'stòr·ij }

archiving |COMPUT SCI| The storage of files in auxiliary storage media for very long periods, in the event it is necessary to regenerate the file due to subsequent errors introduced. { 'är‚kīv·iŋ }

arcing contacts |ELEC| Special contacts on which the arc is drawn after the main contacts of a switch or circuit breaker have opened. { 'ärk·iŋ ‚kän‚taks }

arcing ring |ELEC| A metal ring attached to an insulator to protect it from damage by a power arc. { 'ärk·iŋ ‚riŋ }

arcing time |ELEC| **1.** Interval between the parting, in a switch or circuit breaker, of the arcing contacts and the extension of the arc. **2.** Time elapsing, in a fuse, from the severance of the fuse

link to the final interruption of the circuit under a specified condition. { 'ärk·iŋ ‚tīm }

arc lamp |ELEC| An electric lamp in which the light is produced by an arc made when current flows through ionized gas between two electrodes. Also known as electric-arc lamp. { 'ärk ‚lamp }

arc-over |ELEC| An unwanted arc resulting from the opening of a switch or the breakdown of insulation. { 'ärk ‚ō·vər }

arc resistance |ELEC| **1.** A measure of the durability of an insulating or dielectric material against the formation of conductive paths along the surface by arc discharges. **2.** The ratio of the voltage that gives rise to an arc discharge to the current in the arc. { 'ärk ri‚zis·təns }

arc-suppression coil |ELEC| A grounding reactor, used in alternating-current power transmission systems, which is designed to limit the current flowing to ground at the location of a fault almost to zero by setting up a reactive current to ground that balances the capacitive current to ground flowing from the lines. Also known as Petersen coil. { ‚ärk sə'presh·ən ‚kòil }

arc-through |ELECTR| Of a gas tube, a loss of control resulting in the flow of a principal electron stream in the normal direction during a scheduled nonconducting period. { 'ärk ‚thrü }

area |COMPUT SCI| A section of a computer memory assigned by a computer program or by the hardware to hold data of a particular type. { 'er·ē·ə }

area code |COMMUN| A three-digit prefix used in dialing long-distance telephone calls in the United States and Canada. { 'er·ē·ə ‚kōd }

area effect |ELECTR| In general, the condition of the dielectric strength of a liquid or vacuum separating two electrodes being higher for electrodes of smaller area. { 'er·ē·ə i'fekt }

areal density |COMPUT SCI| The amount of data that can be stored on a unit area of the surface of a hard disk, floppy disk, or other storage device. { ‚er·ē·əl 'den·səd·ē }

area search |COMPUT SCI| A computer search that examines only those records which satisfy some broad criteria. { 'er·ē·ə ‚sərch }

A register See arithmetic register. { 'ā ‚rej·ə·stər }

argument |COMPUT SCI| A value applied to a procedure, subroutine, or macroinstruction which is required in order to evaluate any of these. { 'är·gyə·mənt }

argument separator |COMPUT SCI| A comma or other punctuation mark that separates successive arguments in a command or statement in a computer program. { 'är·gyü·mənt ‚sep·ə ‚rād·ər }

arithmetic address |COMPUT SCI| An address in a computer program that results from performing an arithmetic operation on another address. { ‚a·rith'med·ik ə'dres }

arithmetical element See arithmetical unit. { ‚a·rith'med·ə·kəl 'el·ə·mənt }

arithmetical instruction |COMPUT SCI| An instruction in a computer program that directs the computer to perform an arithmetical operation (addition, subtraction, multiplication, or division) upon specified items of data. { ¦a·rith ¦med·ə·kəl ˌin'strək·shən }

arithmetical operation |COMPUT SCI| A digital computer operation in which numerical quantities are added, subtracted, multiplied, divided, or compared. { ¦a·rith¦med·ə·kəl ˌäp·ə'rā·shən }

arithmetical unit |COMPUT SCI| The section of the computer which carries out all arithmetic and logic operations. Also known as arithmetical element; arithmetic-logic unit (ALU); arithmetic section; logic-arithmetic unit; logic section. { ¦a·rith¦med·ə·kəl 'yü·nət }

arithmetic check |COMPUT SCI| The verification of an arithmetical operation or series of operations by another such process; for example, the multiplication of 73 by 21 to check the result of multiplying 21 by 73. { ə'rith·mə,tik ,chek }

arithmetic circuitry |COMPUT SCI| The section of the computer circuitry which carries out the arithmetic operations. { ¦a·rith¦med·ik 'sər·kə·trē }

arithmetic coding |COMMUN| A method of data compression in which a long character string is represented by a single number whose value is obtained by repeatedly partitioning the range of possible values in proportion to the probabilities of the characters. { ¦a·rith¦med·ik 'cōd·iŋ }

arithmetic-logic unit See arithmetical unit. { ə¦rith·mə,tik 'läj·ik ,yü·nət }

arithmetic processor See numeric processor extension. { ə'rith·mə,tik ,präs,es·ər }

arithmetic register |COMPUT SCI| A specific memory location reserved for intermediate results of arithmetic operations. Also known as A register. { ¦a·rith¦med·ik 'rej·ə·stər }

arithmetic scan |COMPUT SCI| The procedure for examining arithmetic expressions and determining the order of execution of operators, in the process of compilation into machine-executable code of a program written in a higher-level language. { ¦a·rith¦med·ik ,skan }

arithmetic section See arithmetical unit. { ¦a·rith ¦med·ik ,sek·shən }

arithmetic shift |COMPUT SCI| A shift of the digits of a number, expressed in a positional notation system, in the register without changing the sign of the number. { ¦a·rith¦med·ik 'shift }

arithmetic symmetry |ELECTR| Property of a band-pass or band-rejection filter whose graph of amplitude versus frequency is symmetrical around a center frequency; that is, the left-hand side of the response is a mirror image of the right-hand side. { ¦a·rith¦med·ik 'sim·ə·trē }

arm |CONT SYS| A robot component consisting of an interconnected set of links and powered joints that move and support the wrist socket and end effector. |ELEC| See branch. |ENG ACOUS| See tone arm. { ärm }

armature contact See movable contact. { 'är·mə ,chər 'kän,takt }

armature resistance |ELEC| The ohmic resistance in the main current-carrying windings of an electric generator or motor. { 'är·mə,chər ri'zis·təns }

armor |ELEC| Metal sheath enclosing a cable, primarily for mechanical protection. { 'är·mər }

armored cable |ELEC| An electrical cable provided with a sheath of metal primarily for mechanical protection. { 'är·mərd 'kā·bəl }

arm solution |CONT SYS| The computation performed by a robot controller to calculate the joint positions required to achieve desired tool positions. { ärm sə,lü·shən }

Armstrong oscillator |ELECTR| Inductive feedback oscillator that consists of a tuned-grid circuit and an untuned-tickler coil in the plate circuit; control of feedback is accomplished by varying the coupling between the tickler and the grid circuit. { 'ärm,strȯŋ 'äs·ə,lād·ər }

ARPA See automated radar plotting aid. { 'är·pə }

ARQ See automatic repeat request.

array |COMPUT SCI| A collection of data items with each identified by a subscript or key and arranged in such a way that a computer can examine the collection and retrieve data from these items associated with a particular subscript or key. |ELECTR| A group of components such as antennas, reflectors, or directors arranged to provide a desired variation of radiation transmission or reception with direction. { ə'rā }

array element |COMPUT SCI| A single data item in an array. { ə'rā, el·ə·mənt }

array processor |COMPUT SCI| A multiprocessor composed of a set of identical central processing units acting synchronously under the control of a common unit. { ə'rā 'präs,es·ər }

array radar |ENG| A radar incorporating a multiplicity of phased antenna elements. { ə'rā 'rā ,där }

array sonar |ENG| A sonar system incorporating a phased array of radiating and receiving transducers. { ə'rā 'sō,när }

arrester See lightning arrester. { ə'res·tər }

ARSR See air-route surveillance radar.

articulation |COMMUN| The percentage of speech units understood correctly by a listener in a communications system; it generally applies to unrelated words, as in code messages, in distinction to intelligibility. |CONT SYS| The manner and actions of joining components of a robot when connecting parts or links that allow motion. { är ,tik·yə'lā·shən }

articulation equivalent |COMMUN| Of a complete telephone connection, a measure of the articulation of speech reproduced over it, expressed numerically in terms of the trunk loss of a working reference system when the latter is adjusted to give equal articulation. { är,tik·yə'lā·shən i'kwiv·ə·lənt }

artifact |COMMUN| Any component of a signal that is extraneous to the variable represented by the signal. { 'ärd·ə,fakt }

artificial antenna See dummy antenna. { ¦ärd·ə ¦fish·əl an'ten·ə }

artificial atom |ELECTR| A structure, typically 50–100 nanometers in diameter, that is fabricated in a semiconductor crystal and holds a small number of electrons which are trapped in a bowllike potential well. { ,ärd·ə,fish·əl 'ad·əm }

artificial crystal See superlattice. { ¦ärd·ə¦fish·əl 'krist·əl }

artificial delay line See delay line. { ¦ärd·ə¦fish·əl di'lā ‚līn }

artificial ear [ENG ACOUS] A device designed to duplicate the frequency response, acoustic impedance, threshold sensitivity, and relative perception of loudness, consisting of a special microphone enclosed in a box with properties similar to those of the human ear. { ¦ärd·ə ¦fish·əl 'ir }

artificial ground [ELEC] A common correction for a radio-frequency electrical or electronic circuit that is not directly connected to the earth. { ¦ärd·ə¦fish·əl 'graünd }

artificial intelligence [COMPUT SCI] The property of a machine capable of reason by which it can learn functions normally associated with human intelligence. { ¦ärd·ə¦fish·əl in'tel·ə·jəns }

artificial ionization [COMMUN] Introduction of an artificial reflecting or scattering layer into the atmosphere to permit beyond-the-horizon communications. { ¦ärd·ə¦fish·əl ‚ī·ə·nə'zā·shən }

artificial language [COMPUT SCI] A computer language that is specifically designed to facilitate communication in a particular field, but is not yet natural to that field; opposite of a natural language, which evolves through long usage. { ¦ärd·ə¦fish·əl 'laŋ·gwij }

artificial line [ELEC] Circuit made up of lumped constants, which is used to simulate various characteristics of a transmission line. { ¦ärd·ə ¦fish·əl 'līn }

artificial line duct [ELEC] Balancing network simulating the impedance of the real line and distant terminal apparatus, which is employed in a duplex circuit to make the receiving device unresponsive to outgoing signal currents. { ¦ärd·ə ¦fish·əl 'līn ‚dəkt }

artificial load [ELEC] Dissipative but essentially nonradiating device having the impedance characteristics of an antenna, transmission line, or other practical utilization circuit. { ¦ärd·ə¦fish·əl 'lōd }

artificially layered structure See superlattice. { ¦ärd·ə¦fish·əl·ē ¦lā·ərd 'strək·chər }

artificial radio aurora [COMMUN] Modification of the ionosphere by high-power high-frequency radio transmitters to improve scatter and auroral long-distance communication. Also known as radio aurora. { ¦ärd·ə¦fish·əl 'rād·ē‚ō ə'rȯr·ə }

artificial reality See virtual reality. { ‚ärd·ə'fish·əl rē'al·əd·ē }

artificial voice [ENG ACOUS] **1.** Small loudspeaker mounted in a shaped baffle which is proportioned to simulate the acoustical constants of the human head; used for calibrating and testing close-talking microphones. **2.** Synthetic speech produced by a multiple tone generator; used to produce a voice reply in some real-time computer applications. { ¦ärd·ə¦fish·əl 'vȯis }

aS See abmho.

A-scan See A-display. { 'ā ‚skan }

ascending sort [COMPUT SCI] The arrangement of records or other data into a sequence running from the lowest to the highest in a specified field. { ə'send·iŋ 'sȯrt }

ASCII See American Standard Code for Information Interchange. { 'as‚kē }

ASCII file [COMPUT SCI] A data or text file that contains only codes that constitute the 128-character ASCII set. { ¦as‚kē ¦fīl }

ASCII protocol [COMMUN] A protocol for the simplest mode of transmitting ASCII data, with little or no error checking. { ¦as‚kē ¦prōd·ə‚kȯl }

ASCII sort order [COMPUT SCI] A sort order determined by the numbering of characters in the American Standard Code for Information Interchange. { ¦as‚kē 'sȯrt ‚ȯrd·ər }

A-scope See A-display. { 'ā skōp }

asdic [ELECTR] British term for sonar and underwater listening devices. Derived from Anti-Submarine Detection Investigation Committee. { 'az‚dik }

ASIC See application-specific integrated circuit. { 'ā‚sik or ¦ā¦es¦ī'sē }

ASK See amplitude shift keying.

aspect ratio [COMPUT SCI] In computer graphics, the ratio between the width and height of an image. [ENG] The ratio of frame width to frame height in television; in the United States and Britain it is 4:3 for standard television and 16:9 for high-definition television. { 'a‚spekt ‚rā·shō }

assembler [COMPUT SCI] A program designed to convert symbolic instruction into a form suitable for execution on a computer. Also known as assembly program; assembly routine. { ə'sem·blər }

assembler directive [COMPUT SCI] A statement in an assembly-language program that gives instructions to the assembler and does not generate machine language. { ə'sem·blər di‚rek·tiv }

assembler language See assembly language. { ə'sem·blər ‚laŋ·gwij }

assembler program [COMPUT SCI] A program that is written in assembly language. { ə'sem·blər ‚prō·grəm }

assembly [COMPUT SCI] The automatic translation into machine language of a computer program written in symbolic language. { ə'sem·blē }

assembly language [COMPUT SCI] A symbolic, nonbinary format for instructions (human-readable version of machine language) that allows mnemonic names to be used for instructions and data; for example, the instruction to add the number 39321 to the contents of register D1 in the central processing unit might be written as ADD#39321, D1 in assembly language, as opposed to a string of 0's and 1's in machine language. { ə'sem·blē ‚laŋ·gwij }

assembly list [COMPUT SCI] A printed list which is the by-product of an assembly procedure; it lists in logical instruction sequence all details of a routine, showing the coded and symbolic notation next to the actual notations established by the assembly procedure; this listing is highly useful in the debugging of a routine. { ə'sem·blē ,list }

assembly program *See* assembler. { ə'sem·blē 'prō·grəm }

assembly robot [COMPUT SCI] A robot that positions, mates, fits, and assembles components or parts and adjusts the finished product to function as intended. { ə'sem·blē ,rō,bät }

assembly routine *See* assembler. { ə'sem·blē rü'tēn }

assembly system [COMPUT SCI] An automatic programming software system with a programming language and machine-language programs that aid the programmer by performing different functions such as checkout and updating. { ə'sem·blē ,sis·təm }

assembly unit [COMPUT SCI] **1.** A device which performs the function of associating and joining several parts or piecing together a program. **2.** A portion of a program which is capable of being assembled into a larger program. { ə'sem·blē ,yü·nət }

assign [COMPUT SCI] A control statement in FORTRAN which assigns a computed value *i* to a variable *k*, the latter representing the number of the statement to which control is then transferred. { ə'sīn }

assignment problem [COMPUT SCI] A special case of the transportation problem in a linear program, in which the number of sources (assignees) equals the number of designations (assignments) and each supply and each demand equals 1. { ə'sīn·mənt 'präb·ləm }

assignment statement [COMPUT SCI] A statement in a computer program that assigns a value to a variable. { ə'sīn·mənt ,stāt·mənt }

assisted panel [COMPUT SCI] In an interactive system, a screen that explains a question the computer has asked, the available options, the expected format, and so forth. { ə'sis·təd 'pan·əl }

associated document [COMPUT SCI] A file that is linked to the application program in which it was created, so that the application can be started by choosing such a file. { ə,sō·sē,ād·əd 'däk·yə·mənt }

association trail [COMPUT SCI] A linkage between two or more documents or items of information, discerned during the process of their examination and recorded with the aid of an information retrieval system. { ə,sō·sē'ā·shən ,trāl }

associative dimensioning system [COMPUT SCI] A system for making automatic changes in the dimensions of workpieces manufactured by machine tools. { ə'sō·sē,ād·iv di'men·shən·iŋ 'sis·təm }

associative key [COMPUT SCI] In a computer system with an associative memory, a field used to reference items through comparing the value of the field with corresponding fields in each memory cell and retrieving the contents of matching cells. { ə'sō·sē,ād·iv 'kē }

associative memory [COMPUT SCI] A data-storage device in which a location is identified by its informational content rather than by names, addresses, or relative positions, and from which the data may be retrieved. Also known as associative storage. { ə'sō·sē,ād·iv 'mem·rē }

associative processor [COMPUT SCI] A digital computer that consists of a content-addressable memory and means for searching rapidly changing random digital data stored within, at speeds up to 1000 times faster than conventional digital computers. { ə'sō·sē,ād·iv 'präs,es·ər }

associative storage *See* associative memory. { ə'sō·sē,ād·iv 'stòr·ij }

associator [COMPUT SCI] A device for bringing like entities into conjunction or juxtaposition. { ə'sō·sē,ād·ər }

assumed decimal point [COMPUT SCI] For a decimal number stored in a computer or appearing on a printout, a position in the number at which place values change from positive to negative powers of 10, but to which no location is assigned or at which no printed character appears, as opposed to an actual decimal point. Also known as virtual decimal point. { ə'sümd 'des·məl ,pòint }

astable circuit [ELECTR] A circuit that alternates automatically and continuously between two unstable states at a frequency dependent on circuit constants; for example, a blocking oscillator. { 'ā'stā·bəl 'sər·kət }

astable multivibrator [ELECTR] A multivibrator in which each active device alternately conducts and is cut off for intervals of time determined by circuit constants, without use of external triggers. Also known as free-running multivibrator. { ā'stā·bəl ,məlt·i'vī,brād·ər }

astatic wattmeter [ENG] An electrodynamic wattmeter designed to be insensitive to uniform external magnetic fields. { ā'stad·ik 'wät,mēd·ər }

A station [NAV] In loran, the designation applied to one transmitting station of a pair, the signal of which always occurs less than half a repetition period after the preceding signal and more than half a repetition period before the succeeding signal of the other station, designated a B station. { 'ā ¦stā·shən }

astigmatism [ELECTR] In an electron-beam tube, a focus defect in which electrons in different axial planes come to focus at different points. { ə'stig·mə,tiz·əm }

Aston dark space [ELECTR] A dark region in a glow-discharge tube which extends for a few millimeters from the cathode up to the cathode glow. { 'as·tən ¦därk ,spās }

astrionics [ELECTR] The science of adapting electronics to aerospace flight. { ,as·trē'än·iks }

A supply [ELECTR] Battery, transformer filament winding, or other voltage source that supplies power for heating filaments of vacuum tubes. Also known as A power supply. { 'ā sə,plī }

asymmetrical cell [ELECTR] A cell, such as a photoelectric cell, in which the impedance to the flow of current in one direction is greater than in the other direction. { ¦ā·sə¦me·tri·kəl 'sel }

asymmetrical conductivity

asymmetrical conductivity [ELEC] A variation in the conductivity of a conductor over its cross section that is not symmetric about the conductor's central axis. { ¦ā·sə¦me·tri·kəl ¸kän ¸dək'tiv·əd·ē }

asymmetrical deflection [ELECTR] A type of electrostatic deflection in which one deflector plate is maintained at a fixed potential and the deflecting voltage is supplied to the other plate. { ¦ā·sə¦me·tri·kəl di'flek·shən }

asymmetrical modem [COMMUN] A modem that simultaneously transmits and receives data, but at different speeds. { ¸ā·si¦me·trə·kəl 'mō¸dem }

asymmetrical-sideband transmission See vestigial-sideband transmission. { ¦ā·sə¦me·tri·kəl 'sīd¸band ¸tranz'mish·ən }

asymmetric digital subscriber line [COMMUN] A broadband communication technology designed for use on conventional telephone lines, which reserves more bandwidth for receiving data than for sending data. Abbreviated ADSL. { ¦ā·sə'me·trik ¦dij·ə·dəl ¸səb'skrī·bər ¸līn }

asynchronous [COMPUT SCI] Operating at a speed determined by the circuit functions rather than by timing signals. { ā'siŋ·krə·nəs }

asynchronous communications [COMMUN] The transmission and recognition of a single character at a time. { ā'siŋ·krə·nəs kə¸myü·nə'kā·shənz }

asynchronous communications adaptor [COMPUT SCI] A device connected to a computer to allow it to carry out asynchronous communications over a telephone line. { ā'siŋ·krə·nəs kə¸myü·nə'kā·shənz ə¸dap·tər }

asynchronous computer [COMPUT SCI] A computer in which the performance of any operation starts as a result of a signal that the previous operation has been completed, rather than on a signal from a master clock. { ā'siŋ·krə·nəs kəm'pyüd·ər }

asynchronous control [CONT SYS] A method of control in which the time allotted for performing an operation depends on the time actually required for the operation, rather than on a predetermined fraction of a fixed machine cycle. { ā'siŋ·krə·nəs kən'trōl }

asynchronous data [COMPUT SCI] Information which is sampled at irregular intervals with respect to another operation. { ā'siŋ·krə·nəs 'dad·ə }

asynchronous device [CONT SYS] A device in which the speed of operation is not related to any frequency in the system to which it is connected. { ā'siŋ·krə·nəs di'vīs }

asynchronous digital subscriber loop See asymmetric digital subscriber line. { ā'siŋ·krə·nəs 'dij·əd·əl səb'skrī·bər ¸lüp }

asynchronous input/output [COMPUT SCI] The ability to receive input data while simultaneously outputting data. { ā'siŋ·krə·nəs 'in¸pút 'aút¸pút }

asynchronous inputs [ELECTR] The terminals in a flip-flop circuit which affect the output state of the flip-flop independently of the clock. { ā'siŋ·krə·nəs 'in¸púts }

asynchronous logic [ELECTR] A logic network in which the speed of operation depends only on the signal propagation through the network. { ā'siŋ·krə·nəs 'läj·ik }

asynchronous machine [ELEC] An ac machine whose speed is not proportional to the frequency of the power line. { ā'siŋ·krə·nəs mə'shēn }

asynchronous operation [ELECTR] An operation that is started by a completion signal from a previous operation, proceeds at the maximum speed of the circuits until finished, and then generates its own completion signal. { ā'siŋ·krə·nəs ¸äp·ə'rā·shən }

asynchronous tie [ELEC] An installation at which power is transmitted between two alternating-current power systems, operating at the same nominal frequency but with different frequency controls, by a direct-current link. { ā'siŋ·krə·nəs 'tī }

asynchronous time-division multiplexing [COMMUN] A data-transmission technique in which several users utilize a single channel by means of a system which assigns time slots only to active channels. { ā'siŋ·krə·nəs 'tīm də'vi·zhən 'məlt·i ¸pleks·iŋ }

asynchronous transfer mode [COMMUN] A high-speed packet-switching technology based on cell-oriented switching and multiplexing that uses 53-byte packets to transfer different types of information, such as voice, video, and data, over the same communications network at different speeds. Abbreviated ATM. { ¸a¦siŋ·krə·nəs 'tranz·fər ¸mōd }

asynchronous transmission [COMMUN] Data transmission in which each character contains its own start and stop pulses and there is no control over the time between characters. { ā'siŋ·krə·nəs ¸tranz'mish·ən }

asynchronous working [COMPUT SCI] The mode of operation of a computer in which an operation is performed only at the end of the preceding operation. { ā'siŋ·krə·nəs 'wərk·iŋ }

asyndetic [COMPUT SCI] **1.** Omitting conjunctions or connectives. **2.** Pertaining to a catalog without cross references. { ¦as·ən¦ded·ik }

ATCRBS See air-traffic control radar beacon system.

ATDM See asynchronous time-division multiplexing.

ATM See asynchronous transfer mode; automatic teller machine.

atmospheric attenuation [GEOPHYS] The loss of radar or radio signals sent through earth's (or other) atmosphere due to the thermal agitation of various gas molecules as the electromagnetic wave passes through; oxygen and water vapor are the two most sensitive gases in the microwave region, with severity generally, but very linearly, increasing with frequency. { ¦at·mə¦sfir·ik ə¸ten·yə'wā·shən }

atmospheric noise [ELECTR] Noise heard during radio reception due to atmospheric interference. { ¦at·mə¦sfir·ik 'nóiz }

atmospheric radio wave [ELECTROMAG] Radio wave that is propagated by reflection in the

28

atmosphere; may include either the ionospheric wave or the tropospheric wave, or both. { ¦at·mə ¦sfir·ik 'rād·ē·ō ,wāv }

atom |COMPUT SCI| A primitive data element in a data structure. { 'ad·əm }

atomic fallout See fallout. { ə'täm·ik 'fȯl,aút }

atomic operation |COMPUT SCI| An operation that cannot be broken up into smaller parts that could be performed by different processors. { ə'täm·ik ,äp·ə'rā·shən }

A trace |ELECTR| The first trace of an oscilloscope, such as the upper trace of a loran indicator. { 'ā ,trās }

ATR tube See anti-transmit-receive tube. { 'ā'tē ,är ,tüb }

attached processing |COMPUT SCI| A method of data processing in which several relatively inexpensive computers dedicated to specific tasks are connected together to provide a greater processing capability. { ə'tacht 'präs,es·iŋ }

attached processor |COMPUT SCI| A computer that is electronically connected to and operates under the control of another computer. { ə'tacht 'präs,es·ər }

attaching gas |ELECTR| A gas in which electron attachment takes place. { ə'tach·iŋ ,gas }

attachment |COMPUT SCI| An additional file sent with an e-mail message. { ə'tach·mənt }

attachment coefficient |ELECTR| The probability that an electron drifting through a gas under the influence of a uniform electric field will undergo electron attachment in a unit distance of drift. { ə'tach·mənt ,kō·ə,fish·ənt }

attachment plug |ELEC| A device having an attached flexible cord containing conductors, and capable of being inserted in a receptacle so as to form an electrical connection between the conductors in the cord and conductors permanently connected to the receptacle. { ə'tach·mənt ,pləg }

attachment unit interface |COMMUN| A 15-pin connector on an Ethernet card for connecting a network cable. Abbreviated AUI. { ə¦tach·mənt ,yü·nət 'in·tər,fās }

attack director |COMPUT SCI| An electromechanical analog computer which is designed for surface antisubmarine use and which computes continuous solution of several lines of submarine attack; it is part of several antisubmarine fire control systems. { ə'tak di'rek·tər }

ATTC See Advanced Television Technology Center.

attendant's switchboard |COMMUN| Switchboard of one or more positions in a central-office location which permits the central-office operator to receive, transmit, or cut in on a call to or from one of the lines which the office services. { ə'ten·dəns 'swich,bȯrd }

attended time |COMPUT SCI| The time in which a computer is either switched on and capable of normal operation (including time during which it is temporarily idle but still watched over by computer personnel) or out of service for maintenance work. { ə'tend·əd ¦tīm }

attenuate |ENG ACOUS| To weaken a signal by reducing its level. { ə'ten·yə,wāt }

attenuation |ELEC| The exponential decrease with distance in the amplitude of an electrical signal traveling along a very long uniform transmission line, due to conductor and dielectric losses. { ə,ten·yə'wā·shən }

attenuation constant |PHYS| A rating for a line or medium through which a plane wave is being transmitted, equal to the relative rate of decrease of an amplitude of a field component, voltage, or current in the direction of propagation, in nepers per unit length. { ə,ten·yə'wā·shən ,kän·stənt }

attenuation distortion |COMMUN| 1. In a circuit or system, departure from uniform amplification or attenuation over the frequency range required for transmission. 2. The effect of such departure on a transmitted signal. { ə,ten·yə'wā·shən dis ,tȯr·shən }

attenuation equalizer |ELECTR| Corrective network which is designed to make the absolute value of the transfer impedance, with respect to two chosen pairs of terminals, substantially constant for all frequencies within a desired range. Also known as attenuation factor. { ə,ten·yə 'wā·shən 'ē·kwə,līz·ər }

attenuation network |ELECTR| Arrangement of circuit elements, usually impedance elements, inserted in circuitry to introduce a known loss or to reduce the impedance level without reflections. { ə,ten·yə'wā·shən 'net,wərk }

attenuator |ELECTR| An adjustable or fixed transducer for reducing the amplitude of a wave without introducing appreciable distortion. { ə'ten·yə,wād·ər }

attracted-disk electrometer |ELEC| A type of electrometer in which the attraction between two oppositely charged disks is measured. { ə¦trak· təd ¦disk i,lek'träm·əd·ər }

attraction gripper |CONT SYS| A robot component that uses adhesion, suction, or magnetic forces to grasp a workpiece. { ə'trak·shən ,grip·ər }

attribute |COMPUT SCI| 1. A data item containing information about a variable. 2. A characteristic of computer-generated characters, such as underline, boldface, or reverse image. { 'a·trə ,byüt }

audible feedback |COMPUT SCI| A feature of a computer keyboard that generates sound each time a key is depressed sufficiently to generate a character on the screen. { ,ȯd·ə·bəl 'fēd ,bak }

audio adapter See sound board. { ,ȯd·ē·ō ə'dap·tər }

audio amplifier See audio-frequency amplifier. { 'ȯd·ē·ō 'am·plə,fī·ər }

audio-frequency amplifier |ELECTR| An electronic circuit for amplification of signals within, and in some cases above, the audible range of frequencies in equipment used to record and reproduce sound. Also known as audio amplifier. { 'ȯd·ē·ō ¦frē·kwən·sē ¦am·plə,fī·ər }

audio-frequency meter |ENG| One of a number of types of frequency meters usable in the audio range; for example, a resonant-reed frequency meter. { 'ȯd·ē·ō ¦frē·kwən·sē ,mēd·ər }

29

audio-frequency oscillator [ELECTR] An oscillator circuit using an electron tube, transistor, or other nonrotating device to produce an audio-frequency alternating current. Also known as audio oscillator. { 'ȯd·ē·ō ¦frē·kwən·sē 'äs·ə ‚lād·ər }

audio-frequency peak limiter [ELEC] A circuit used in an audio-frequency system to cut off signal peaks that exceed a predetermined value. Also known as audio peak limiter. { 'ȯd·ē·ō ¦frē·kwən·sē 'pēk ‚lim·əd·ər }

audio-frequency shift modulation [COMMUN] System of facsimile transmission over radio, in which the frequency shift required is applied through a change in audio signal, rather than shifting the radio transmitter frequency; the radio signal is modulated by the shifting audio signal, usually at 1500 to 2300 hertz. { 'ȯd·ē·ō ¦frē·kwən·sē ‚shift mäj·ə'lā·shən }

audio-frequency transformer [ELEC] An iron-core transformer that is used for coupling audio-frequency circuits. Also known as audio transformer. { 'ȯd·ē·ō ¦frē·kwən·sē tranz'fȯr·mər }

audio oscillator See audio-frequency oscillator. { 'ȯd·ē·ō 'äs·ə·lād·ər }

audio patch bay [ENG ACOUS] Specific patch panels provided to terminate all audio circuits and equipment used in a channel and technical control facility; this equipment can also be found in transmitting and receiving stations. { 'ȯd·ē·ō ¦pach ‚bā }

audio peak limiter See audio-frequency peak limiter. { 'ȯd·ē·ō 'pēk ‚lim·ə·dər }

audio response [COMMUN] A form of computer output in which prerecorded spoken syllables, words, or messages are selected and put together by a computer as the appropriate verbal response to a keyboarded inquiry on a time-shared on-line information system. { 'ȯd·ē·ō ri'späns }

audio response unit [COMMUN] A system that provides voice response to an inquiry; the inquiry is typically made using the dual-tone multifrequency (DTMF) dial on a telephone set. { 'ȯd·ē·ō ri'späns ‚yü·nət }

audio spectrometer See acoustic spectrometer. { 'ȯd·ē·ō spek'träm·əd·ər }

audio system See sound-reproducing system. { 'ȯd·ē·ō ‚sis·təm }

audio taper [ENG ACOUS] A special type of potentiometer used in a volume-control apparatus to compensate for the nonlinearity of human hearing and give the impression of a linear increase in audibility as volume is raised. Also known as linear taper. { 'ȯd·ē·ō ‚tā·pər }

audio transformer See audio-frequency transformer. { 'ȯd·ē·ō tranz'fȯr·mər }

audiovisual [COMMUN] Pertaining to methods of education and training that make use of both hearing and sight. { ¦ȯd·ē·ō¦vizh·ə·wəl }

audiphone [ENG ACOUS] A device that enables persons with certain types of deafness to hear, consisting of a plate or diaphragm that is placed against the teeth and transmits sound vibrations to the inner ear. { 'ȯd·ə‚fōn }

audit [COMPUT SCI] The operations developed to corroborate the evidence as regards authenticity and validity of the data that are introduced into the data-processing problem or system. { 'ȯd·ət }

audit total [COMPUT SCI] A count or sum of a known quantity, calculated in order to verify data. { 'ȯd·ət ‚tōd·əl }

audit trail [COMPUT SCI] A system that provides a means for tracing items of data from processing step to step, particularly from a machine-produced report or other machine output back to the original source data. { 'ȯd·ət ‚trāl }

augmented operation code [COMPUT SCI] An operation code which is further defined by information from another portion of an instruction. { 'ȯg‚men·təd äp·ə'rā·shən ‚kōd }

AUI See attachment unit interface.

auralization See virtual acoustics. { ‚ȯr·əl·ə'zā·shən }

aural radio range [ELECTR] A radio-range station providing lines of position by virtue of aural identification or comparison of signals at the output of a receiver. { 'ȯr·əl 'rād·ē‚ō ‚rānj }

aural transmitter [COMMUN] Radio equipment used for transmitting aural (sound) signals from a television broadcast station. { 'ȯr·əl ‚tranz'mid·ər }

aurora See corona discharge. { ə'rȯr·ə }

aurora gating [ELECTR] Operator-controlled gating to eliminate undesirable radar returns from aurora. { ə'rȯr·ə ¦gād·iŋ }

auroral propagation [COMMUN] The propagation of radio waves that are reflected from the aurora in the presence of unusual solar activity. { ə'rȯr·əl ‚präp·ə'gā·shən }

authentication [COMMUN] Security measure designed to protect a communications system against fraudulent transmissions and establish the authenticity of a message. { ə‚thent·ə'kā·shən }

authenticator [COMMUN] Letter, numeral, or groups of letters or numerals attesting to the authenticity of a message or transmission. { ə'thent·ə‚kād·ər }

authoring language [COMPUT SCI] A programming language designed to be convenient for authors of computer-based learning materials. { 'ȯ·thər·iŋ 'laŋ·gwij }

authorization code [COMPUT SCI] A password or identifying number that is used to gain access to a computer system. { ‚ȯth·ə·rə'zā·shən ‚kōd }

authorized carrier frequency [COMMUN] A specific carrier frequency authorized for use, from which the actual carrier frequency is permitted to deviate, solely because of frequency instability, by an amount not to exceed the frequency tolerance. { 'ȯ·thə‚rīzd 'kar·ē·ər ‚frē·kwən·sē }

authorized library [COMPUT SCI] A group of authorized programs. { 'ȯ·thə‚rīzd 'li‚brer·ē }

authorized program [COMPUT SCI] A computer program that can alter the fundamental operation or status of a computer system. { 'ȯ·thə ‚rizd 'prō·grəm }

auto-abstract |COMPUT SCI| **1.** To select key words from a document, commonly by an automatic or machine method, for the purpose of forming an abstract of the document. **2.** The material abstracted from a document by machine methods. { ¦ȯd·ō 'ab,strakt }

autoadaptivity |CONT SYS| The ability of an advanced robot to sense the environment, accept commands, and analyze and execute operations. { ¦ȯd·ō·,ə,dap'tiv·əd·ē }

autoalarm See automatic alarm receiver. { 'ȯd·ō-ə,lärm }

auto answer |COMMUN| The feature of a modem that receives the telephone ring for an incoming call and accepts the call to establish a connection. { ¦ȯd·ō 'an·sər }

auto bypass |COMPUT SCI| The ability of a computer network to bypass a terminal or other device if it fails, allowing other devices connected to the network to continue operation. { ¦ȯd·ō 'bī ,pas }

autocall |COMPUT SCI| The automatic placing of a telephone call by a computer or a computer-controlled modem. Also known as automatic call origination. { 'ȯd·ō,kȯl }

autocode |COMPUT SCI| The process of using a computer to convert automatically a symbolic code into a machine code. Also known as automatic code. { 'ȯd·ō,kōd }

autocoder |COMPUT SCI| A person or machine producing or using autocode as a part or the whole of a task. { 'ȯd·ō,kōd·ər }

autocorrelation |ELECTR| A technique used to detect cyclic activity in a complex signal. { ¦ȯd·ō ,kär·ə,lād·ər }

autocorrelator |ELECTR| A correlator in which the input signal is delayed and multiplied by the undelayed signal, the product of which is then smoothed in a low-pass filter to give an approximate computation of the autocorrelation function; used to detect a nonperiodic signal or a weak periodic signal hidden in noise. { ,ȯd·ō'kär·ə,lād·ər }

autodecrement addressing |COMPUT SCI| An addressing mode of computers in which the register is first decremented and then used as a pointer. { ,ȯd·ō'dek·rə·mənt ə'dres·iŋ }

auto dial |COMMUN| The feature of a modem that automatically opens a telephone line and dials the telephone of a receiving computer to establish a connection. { ¦ȯd·ō 'dīl }

autodyne circuit |ELECTR| A circuit in which the same tube elements serve as oscillator and detector simultaneously. { 'ȯd·ō,dīn ,sər·kət }

autodyne reception |COMMUN| System of heterodyne reception through the use of a device which is both an oscillator and a detector. { 'ȯd·ō,dīn ri'sep·shən }

autoincrement addressing |COMPUT SCI| An addressing mode of minicomputers in which the operand address is gotten from the specified register which is then incremented. { ¦ȯd·ō'in,krə·mənt ə'dres·iŋ }

autoindexing See automatic indexing. { ¦ȯd·ō'in,deks·iŋ }

automata theory |MATH| A theory concerned with models used to simulate objects and processes such as computers, digital circuits, nervous systems, cellular growth and reproduction. { ȯ'täm·əd·ə 'thē·ə·rē }

automated decision making |COMPUT SCI| The use of computers to carry out tasks requiring the generation or selection of options. { ¦ȯd·ə ,mād·əd di'sizh·ən ,māk·iŋ }

automated guided vehicle system |CONT SYS| A computer-controlled system that uses pallets and other interface equipment to transport workpieces to numerically controlled machine tools and other equipment in a flexible manufacturing system, moving in a predetermined pattern to ensure automatic, accurate, and rapid work-machine contact. { 'ȯd·ə,mād·əd ¦gīd·əd 've·ə·kəl ,sis·təm }

automated identification system |COMPUT SCI| In a data processing system, the use of a technology such as bar coding, image recognition, or voice recognition instead of keyboarding for data entry. { ,ȯd·ə¦mād·əd ī,den·tə·fə'ka·shən ,sis·təm }

automated radar plotting aid |NAV| A marine computer-based anticollision system that automatically processes time coordinates of radar echo signals into space coordinates in digital form, determines consecutive coordinates and motion parameters of targets, calculates the predicted closest point of approach and time to closest point of approach and presents them in graphic or alphanumeric form on the radar display, and switches on alarms if there is a danger of collision. { 'ȯd·ə,mād·əd ¦rä,där 'pläd·iŋ ,ād }

automated tape library |COMPUT SCI| A computer storage system consisting of several thousand magnetic tapes and equipment under computer control which automatically brings the tapes from storage, mounts them on tape drives, dismounts the tapes when the job is completed, and returns them to storage. { 'ȯd·ə,mād·əd 'tāp ¦lī,brer·ē }

automatic |ENG| Having a self-acting mechanism that performs a required act at a predetermined time or in response to certain conditions. { ¦ȯd·ə¦mad·ik }

automatic abstracting |COMPUT SCI| Techniques whereby, on the basis of statistical properties, a subset of the sentences in a document is selected as representative of the general content of that document. { ¦ȯd·ə¦mad·ik 'ab,strakt·iŋ }

automatic acceleration See dynamic resolution. { ¦ȯd·ə¦mad·ik ik,sel·ə'rā·shən }

automatic alarm receiver |ELECTR| A complete receiving, selecting, and warning device capable of being actuated automatically by intercepted radio-frequency waves forming the international automatic alarm signal. Also known as autoalarm. { ¦ȯd·ə¦mad·ik ə'lärm ri,sē·vər }

automatic-alarm-signal keying device |COMMUN| A device capable of automatically keying

31

the radiotelegraph transmitter on board a vessel to transmit the international automatic-alarm signal, or to respond to receipt of an internationally agreed-upon distress signal and wake up the radio operator on ships not having a 24-hour radio watch. { ¦öd·ə¦mad·ik ə'lärm ¸sig·nəl 'kē·iŋ di¸vīs }

automatic back bias [ELECTR] Radar technique which consists of one or more automatic gain control loops to prevent overloading of a receiver by large signals, whether jamming or actual radar echoes. { ¦öd·ə¦mad·ik 'bak ¸bī·əs }

automatic background control See automatic brightness control. { ¦öd·ə¦mad·ik 'bak¸graúnd kən¸tröl }

automatic bass compensation [ELECTR] A circuit related to the volume control in some radio receivers and audio amplifiers to make bass notes sound properly balanced, in the audio spectrum, at low volume-control settings. { ¦öd·ə¦mad·ik 'bās käm·pən'sā·shən }

automatic bias [ELECTR] A method of obtaining the correct bias for a vacuum tube or transistor through use of a resistor, usually in the cathode or emitter circuit. { ¦öd·ə¦mad·ik 'bī·əs }

automatic brightness control [ELECTR] A circuit used in an analog television receiver to keep the average brightness of the reproduced image essentially constant. Abbreviated ABC. Also known as automatic background control. { ¦öd·ə¦mad·ik 'brīt·nəs kən¸tröl }

automatic calibration [ENG] A process in which an electronic device automatically performs the recalibration of a measuring range of a weighing instrument, for example an electronic balance. { ¦öd·ə¦mad·ik ¸kal·ə'brā·shən }

automatic calling unit [COMPUT SCI] A device that enables a business machine or computer to automatically dial calls over a communications network. { ¦öd·ə¦mad·ik 'kól·iŋ ¸yü·nət }

automatic call origination See autocall. { ¦öd·ə ¦mad·ik 'kól ə¸rij·ə'nā·shən }

automatic carriage [COMPUT SCI] Any mechanism designed to feed continuous paper or plastic forms through a printing or writing device, often using sprockets to engage holes in the paper. { ¦öd·ə¦mad·ik 'kar·ij }

automatic C bias See self-bias. { ¦öd·ə¦mad·ik 'sē ¸bī·əs }

automatic character recognition [COMPUT SCI] The technology of using special machine systems to identify human-readable symbols, most often alphanumeric, and then to utilize this data. { ¦öd·ə¦mad·ik 'kar·ik·tər ¸rek·ig'nish·ən }

automatic check [COMPUT SCI] An error-detecting procedure performed by a computer as an integral part of the normal operation of a device, with no human attention required unless an error is actually detected. { ¦öd·ə¦mad·ik 'chek }

automatic check-out system [CONT SYS] A system utilizing test equipment capable of automatically and simultaneously providing actions and information which will ultimately result in the efficient operation of tested equipment while keeping time to a minimum. { ¦öd·ə¦mad·ik 'chek¸aút ¸sis·təm }

automatic chroma control See automatic color control. { ¦öd·ə¦mad·ik 'kröm·ə kən¸tröl }

automatic chrominance control See automatic color control. { ¦öd·ə¦mad·ik 'kröm·ə·nəns kən ¸tröl }

automatic code See autocode. { ¦öd·ə¦mad·ik 'köd }

automatic coding [COMPUT SCI] Any technique in which a computer is used to help bridge the gap between some intellectual and manual form of describing the steps to be followed in solving a given problem, and some final coding of the same problem for a given computer. { ¦öd·ə¦mad·ik 'köd·iŋ }

automatic color control [ELECTR] A circuit used in an analog color television receiver to keep color intensity levels essentially constant despite variations in the strength of the received color signal; control is usually achieved by varying the gain of the chrominance band-pass amplifier. Also known as automatic chroma control; automatic chrominance control. { ¦öd·ə¦mad·ik 'kəl·ər kən¸tröl }

automatic computer [COMPUT SCI] A computer which can carry out a special set of operations without human intervention. { ¦öd·ə¦mad·ik kəm'pyüd·ər }

automatic connection [ELECTR] Ability of electronic switching equipment to make a connection between users without human intervention. { ¦öd·ə¦mad·ik kə'nek·shən }

automatic contrast control [ELECTR] A circuit that varies the gain of the radio-frequency and video intermediate-frequency amplifiers in such a way that the contrast of the television picture is maintained at a constant average level. { ¦öd·ə ¦mad·ik 'kän¸trast kən¸tröl }

automatic control [CONT SYS] Control in which regulating and switching operations are performed automatically in response to predetermined conditions. Also known as automatic regulation. { ¦öd·ə¦mad·ik kən¸tröl }

automatic-control block diagram [CONT SYS] A diagrammatic representation of the mathematical relationships defining the flow of information and energy through the automatic control system, in which the components of the control system are represented as functional blocks in series and parallel arrangements according to their position in the actual control system. { ¦öd·ə¦mad·ik kən'tröl 'bläk ¸dī·ə ¸gram }

automatic-control error coefficient [CONT SYS] Three numerical quantities that are used as a measure of the steady-state errors of an automatic control system when the system is subjected to constant, ramp, or parabolic inputs. { ¦öd·ə¦mad·ik kən'tröl 'er·ər ¸kō·ə'fish·ənt }

automatic-control frequency response [CONT SYS] The steady-state output of an automatic control system for sinusoidal inputs of varying frequency. { ¦öd·ə¦mad·ik 'frē·kwən·sē ri ¸späns }

automatic controller [CONT SYS] An instrument that continuously measures the value of a variable quantity or condition and then automatically acts on the controlled equipment to correct any deviation from a desired preset value. Also known as automatic regulator; controller. { ¦òd·ə¦mad·ik kən¦trōl·ər }

automatic-control servo valve [CONT SYS] A mechanically or electrically actuated servo valve controlling the direction and volume of fluid flow in a hydraulic automatic control system. { ¦òd·ə ¦mad·ik kən'trōl 'sər·vō ,valv }

automatic-control stability [CONT SYS] The property of an automatic control system whose performance is such that the amplitude of transient oscillations decreases with time and the system reaches a steady state. { ¦òd·ə¦mad·ik kən'trōl stə,bil·ə·dē }

automatic control system [CONT SYS] A control system having one or more automatic controllers connected in closed loops with one or more processes. Also known as regulating system. { ¦òd·ə¦mad·ik kən'trōl ,sis·təm }

automatic cutout [ELEC] A device, usually operated by centrifugal force or by an electromagnet, that automatically shorts part of a circuit at a particular time. { ¦òd·ə¦mad·ik 'kəd,aút }

automatic data processing [ENG] The machine performance, with little or no human assistance, of any of a variety of tasks involving informational data; examples include automatic and responsive reading, computation, writing, speaking, directing artillery, and the running of an entire factory. Abbreviated ADP. { ¦òd·ə¦mad·ik ¦dad·ə 'präs,əs·iŋ }

automatic degausser [ELECTR] An arrangement of degaussing coils mounted around a color television picture tube, combined with a circuit that energizes these coils only while the set is warming up; demagnetizes any parts of the receiver that have been affected by the magnetic field of the earth or of any nearby devices. { ¦òd·ə¦mad·ik dē'gaús·ər }

automatic detection [ELECTR] A computer-based process in radar wherin the receiver's output video is examined, compared to appropriate thresholds and contacts (detections) reported; augments or replaces the similar role played by the human operator viewing an analog display of the video in more elementary radar. { ¦òd·ə ¦mad·ik di'tek·shən }

automatic dialer [ELECTR] A device in which a telephone number up to some maximum number of digits can be stored in a memory and then activated, directly into the line, by the caller's pressing a button. { ¦òd·ə¦mad·ik 'dīl·ər }

automatic dictionary [COMPUT SCI] Any table within a computer memory which establishes a one-to-one correspondence between two sets of characters. { ¦òd·ə¦mad·ik 'dik·shə,ner·ē }

automatic direction finder [ELECTR] A direction finder that without manual manipulation indicates the direction of arrival of a radio signal. Abbreviated ADF. Also known as radio compass. { ¦òd·ə¦mad·ik di'rek·shən ,fīnd·ər }

automatic error correction [COMMUN] A technique, usually requiring the use of special codes or automatic retransmission, which detects and corrects errors occurring in transmission; the degree of correction depends upon coding and equipment configuration. { ¦òd·ə¦mad·ik 'er·ər kə'rek·shən }

automatic exchange [ELECTR] A telephone, teletypewriter, or data-transmission exchange in which communication between subscribers is effected, without the intervention of an operator, by devices set in operation by the originating subscriber's instrument (for example, the dial n a telephone). Also known as automatic switching system; machine switching system. { ¦òd·ə ¦mad·ik iks'chanj }

automatic fine-tuning control [ELECTR] A circuit used in a color television receiver to maintain the correct oscillator frequency in the tuner for best reception by compensating for drift and incorrect tuning. { ¦òd·ə¦mad·ik ,fīn 'tün·iŋ kən,trōl }

automatic frequency control [ELECTR] Abbreviated AFC. **1.** A circuit used to maintain the frequency of an oscillator within specified limits, as in a transmitter. **2.** A circuit used to keep a superheterodyne receiver tuned accurately to a given frequency by controlling its local oscillator, as in an FM receiver. **3.** A circuit used in radar superheterodyne receivers to vary the local oscillator frequency so as to compensate for changes in the frequency of the received echo signal. **4.** A circuit used in television receivers to make the frequency of a sweep oscillator correspond to the frequency of the synchronizing pulses in the received signal. { ¦òd·ə¦mad·ik 'frē·kwən·sē kən,trōl }

automatic gain control [ELECTR] A control circuit that automatically changes the gain (amplification) of a receiver or other piece of equipment so that the desired output signal remains essentially constant despite variations in input signal strength. Abbreviated AGC. { ¦òd·ə¦mad·ik 'gān kən,trōl }

automatic grid bias See self-bias. { ¦òd·ə¦mad·ik 'grid ,bī·əs }

automatic head parking [COMPUT SCI] A feature that moves the read/write head of a hard disk over the landing zone whenever power is shut off to ensure against a head crash. { ¦òd·ə¦mad·ik 'hed ,pärk·iŋ }

automatic indexing [COMPUT SCI] Selection of key words from a document by computer for use as index entries. Also known as autoindexing. [CONT SYS] The procedure for determining the orientation and position of a workpiece with respect to an automatically controlled machine, such as a robot manipulator, that is to perform an operation on it. { ¦òd·ə¦mad·ik 'in ,deks·iŋ }

automatic intercept [COMMUN] Telephone service that automatically records messages a caller may leave when the called party is away from his telephone. This may be an answering machine or a function provided by an automatic exchange. { ¦òd·ə¦mad·ik 'in·tər,sept }

automatic interrupt |COMPUT SCI| Interruption of a computer program brought about by a hardware device or executive program acting as a result of some event which has occurred independently of the interrupted program. { ¦öd·ə¦mad·ik 'in·tə ‚rəpt }

automatic level compensation |COMMUN| System which automatically compensates for amplitude variations in a circuit. { ¦öd·ə¦mad·ik 'lev·əl ‚käm·pen'sā·shən }

automatic level control |ELECTR| A circuit that keeps the output of a radio transmitter, tape recorder, or other device essentially constant, even in the presence of large changes in the input amplitude. Abbreviated ALC. { ¦öd·ə¦mad·ik 'lev·əl kən‚trōl }

automatic light control |ELECTR| Automatic adjustment of illumination reaching a film, television camera, or other imaging device as a function of scene brightness. { ¦öd·ə¦mad·ik 'līt kən‚trōl }

automatic mathematical translator |COMPUT SCI| An automatic-programming computer capable of receiving a mathematical equation from a remote input and returning an immediate solution. { ¦öd·ə¦mad·ik ‚math·ə'mad·ə·kəl 'tranz ‚lād·ər }

automatic message accounting |COMMUN| System whereby toll calls are automatically recorded and timed. { ¦öd·ə¦mad·ik 'mes·ij ə‚kaúnt·iŋ }

automatic message-switching center |COMMUN| A center in which messages are automatically routed according to information in them. { ¦öd·ə¦mad·ik 'mes·ij ‚swich·iŋ ‚sen·tər }

automatic modulation control |ELECTR| A transmitter circuit that reduces the gain for excessively strong audio input signals without affecting the strength of normal signals, thereby permitting higher average modulation without overmodulation. Abbreviated AMC. { ¦öd·ə¦mad·ik ‚mäj·ə'lā·shən kən‚trōl }

automatic modulation limiting |COMMUN| A circuit that prevents overmodulation in some citizen-band radio transmitters by reducing the gain of one or more audio amplifier stages when the voice signal becomes stronger. Abbreviated AML. { ¦öd·ə¦mad·ik mäj·ə'lā·shən ‚lim·əd·iŋ }

automatic noise limiter |ELECTR| A circuit that clips impulse and static noise peaks, and sets the level of limiting or clipping according to the strength of the incoming signal, so that the desired signal is not affected. Abbreviated ANL. { ¦öd·ə¦mad·ik 'nöiz ‚lim·əd·ər }

automatic peak limiter See limiter. { ¦öd·ə¦mad·ik 'pēk ‚lim·əd·ər }

automatic phase control |ELECTR| **1.** A circuit used in color television receivers to reinsert a 3.58-megahertz carrier signal with exactly the correct phase and frequency by synchronizing it with the transmitted color-burst signal. **2.** An automatic frequency-control circuit in which the difference between two frequency sources is fed to a phase detector that produces the required control signal. Abbreviated APC. { ¦öd·ə¦mad·ik 'fāz kən‚trōl }

automatic picture control |ELECTR| A multiple-contact switch used in some color television receivers to disconnect one or more of the regular controls and make connections to corresponding preset controls. { ¦öd·ə¦mad·ik 'pik·chər kən ‚trōl }

automatic picture-transmission system |ELECTR| A system in which a meteorological satellite continuously scans and transmits a view of a transverse swath directly beneath it; transmissions can be recorded by simple ground equipment to reconstruct an image of the cloud patterns within a thousand kilometers of the ground station. Abbreviated APT system. { ¦öd·ə¦mad·ik 'pik·chər tranz'mish·ən ‚sis·təm }

automatic programming |COMPUT SCI| The preparation of machine-language instructions by use of a computer. { ¦öd·ə¦mad·ik 'prō‚gram·iŋ }

Automatic Programming Tool |COMPUT SCI| A computer language used to program numerically controlled machine tools. Abbreviated APT. { ¦öd·ə¦mad·ik 'prō‚gram·iŋ ‚tül }

automatic regulation See automatic control. { ¦öd·ə¦mad·ik ‚reg·yə'lā·shən }

automatic regulator See automatic controller. { ¦öd·ə¦mad·ik 'reg·yə‚lād·ər }

automatic relay |COMMUN| Means of selective switching which causes automatic equipment to record and retransmit communications. { ¦öd·ə ¦mad·ik 'rē‚lā }

automatic repeat request |COMPUT SCI| A request from a receiving device to retransmit the most recent block of data. Abbreviated ARQ. { ¦öd·ə¦mad·ik ri'pēt ri‚kwest }

automatic routine |COMPUT SCI| A routine that is executed independently of manual operations, but only if certain conditions occur within a program or record, or during some other process. { ¦öd·ə¦mad·ik rü'tēn }

automatic scanning receiver |ELECTR| A receiver which can automatically and continuously sweep across a preselected frequency, either to stop when a signal is found or to plot signal occupancy within the frequency spectrum being swept. { ¦öd·ə¦mad·ik 'skan·iŋ ri‚sē·vər }

automatic sensitivity control |ELECTR| Circuit used for automatically maintaining receiver sensitivity at a predetermined level; it is similar to automatic gain control, but it affects the receiver constantly rather than during the brief interval selected by the range gate. { ¦öd·ə ¦mad·ik sen·sə'tiv·əd·ē kən‚trōl }

automatic sequences |COMPUT SCI| The characteristic of a computer that can perform successive operations without human intervention. { ¦öd·ə ¦mad·ik 'sē·kwon·səs }

automatic short-circuiter |ELEC| Device designed to automatically short-circuit the commutator bars in some forms of single-phase commutator motors. { ¦öd·ə¦mad·ik ‚shórt 'sər·kəd·ər }

automatic shutdown |COMPUT SCI| A procedure whereby a network or computer system stops work in an orderly fashion with as little data loss and other damage as possible when the system's software determines that it has encountered unacceptable conditions.   { ¦òd·ə¦mad·ik 'shət ‚daùn }

automatic speed sensing |COMPUT SCI| The capability of a modem to automatically determine the maximum rate of data transfer over a connection.   { ¦òd·ə¦mad·ik 'spēd ‚sen·siŋ }

automatic stop |COMPUT SCI| An automatic halting of a computer processing operation as the result of an error detected by built-in checking devices.   { ¦òd·ə¦mad·ik 'stäp }

automatic switchboard |COMMUN| Telephone switchboard in which the connections are made by using remotely controlled switches.   { ¦òd·ə ¦mad·ik 'swich‚bòrd }

automatic switching system   See automatic exchange.   { ¦òd·ə¦mad·ik 'swich·iŋ ‚sis·təm }

automatic teller machine |COMPUT SCI| A banking terminal that is activated by inserting a magnetic card containing the user's account number, and that accepts deposits, dispenses cash, provides information about current balances, and may perform other services such as making payments and transfers and providing account statements. Abbreviated ATM.   { ¦òd·ə¦mad·ik 'tel·ər mə ‚shēn }

automatic threshold variation |ELECTR| Constant false-alarm rate scheme that is an open-loop of automatic gain control in which the decision threshold is varied continuously in proportion to the incoming intermediate frequency and video noise level.   { ¦òd·ə¦mad·ik 'thresh‚hòld ‚ver·ē'ā·shən }

automatic time switch |ENG| Combination of a switch with an electric or spring-wound clock, arranged to turn an apparatus on and off at predetermined times.   { ¦òd·ə¦mad·ik ‚tīm ‚swich }

automatic tint control |ELECTR| A circuit used in color television receivers to maintain correct flesh tones by correcting phase errors before the chroma signal is demodulated.   { ¦òd·ə¦mad·ik 'tint kən‚trōl }

automatic tracking |ELECTR| A computer-based process in radar wherein successive contacts (detections) are associated and tracks of targets are estimated and updated with further observations. |NAV| **1.** Tracking in which a servomechanism autpmatically follows some characteristic of the signal; specifically, a process by which tracking or data-acquisition systems are enabled to keep their antennas continuously directed at a moving target without manual operation. **2.** An instrument which displays the actual course made good through the use of navigation derived from several sources.   { ¦òd·ə¦mad·ik 'trak·iŋ }

automatic track shift |ENG ACOUS| A system used with multiple-track magnetic tape recorders to index the tape head, after one track is played, to the correct position for the start of the next track.   { ¦òd·ə¦mad·ik 'trak ‚shift }

automatic transfer equipment |ELEC| Equipment which automatically transfers a load so that a source of power may be selected from one of several incoming lines.   { ¦òd·ə¦mad·ik 'tranz‚fər i‚kwip·mənt }

automatic tuning system |CONT SYS| An electrical, mechanical, or electromechanical system that tunes a radio receiver or transmitter automatically to a predetermined frequency when a button or lever is pressed, a knob turned, or a telephone-type dial operated.   { ¦òd·ə¦mad·ik 'tün·iŋ ‚sis·təm }

automatic video noise leveling |ELECTR| Constant false-alarm rate scheme in which the video noise level at the output of the receiver is sampled at the end of each range sweep and the receiver gain is readjusted accordingly to maintain a constant video noise level at the output.   { ¦òd·ə¦mad·ik ¦vid·ē·ò 'nòiz ‚lev·əl·iŋ }

automatic voltage regulator   See voltage regulator.   { ¦òd·ə¦mad·ik 'vol·tij ‚reg·yə‚lād·ər }

automatic volume compressor   See volume compressor.   { ¦òd·ə¦mad·ik 'väl·yəm kəm‚pres·ər }

automatic volume control |ELECTR| An automatic gain control that keeps the output volume of a radio receiver essentially constant despite variations in input-signal strength during fading or when tuning from station to station. Abbreviated AVC.   { ¦òd·ə¦mad·ik 'väl·yəm kən‚trōl }

automatic volume expander   See volume expander.   { ¦òd·ə¦mad·ik 'väl·yəm ik‚spand·ər }

automation |ENG| **1.** The use of technology to ease human labor or extend the mental or physical capabilities of humans. **2.** The mechanisms, machines, and systems that save or eliminate labor, or imitate actions typically associated with human beings.   { ‚òd·ə'mā·shən }

automaton |COMPUT SCI| A robot which functions without step-by-step guidance by a human operator.   { ò'täm·ə‚tän }

automechanism |CONT SYS| A machine or other device that operates automatically or under control of a servomechanism.   { ¦òd·ō'mek·ə ‚niz·əm }

automonitor |COMPUT SCI| A computer program used in debugging which instructs a computer to make a record of its own operations.   { ¦òd·ō ¦män·əd·ər }

automotive alternator |ELEC| An ac generator used in an automotive vehicle to provide current for the vehicle's electrical systems.   { ¦òd·ə ¦mōd·iv 'òl·tə‚nād·ər }

automotive voltage regulator |ELEC| A device in the automotive electrical system to prevent generator or alternator overvoltage.   { ¦òd·ə'mōd·iv 'vōl·tij ‚reg·yə‚lād·ər }

autonomous channel operation |COMPUT SCI| The rapid transfer of data between computer peripherals and the main store in which an entire block of data is transferred, word by word; the cycles of storage time for the word transfer are stolen from those available to the central processing unit.   { ò'tän·ə·məs ¦chan·əl ‚äp·ə'rā·shən }

autonomous robot |ENG| A robot that not only can maintain its own stability as it moves, but also can plan its movements.   { ò¦tän·ə·məs 'rō ‚bät }

autonomous vehicle |ENG| A vehicle that is able to plan its path and to execute its plan without human intervention. { ȯ¦tän·ə·məs 'vē· ə·kəl }

autopatch |ELECTR| A device for connecting radio transceivers to telephone lines by remote control, generally through the use of repeaters. { 'ȯd·ō,pach }

autoplotter |COMPUT SCI| A machine which automatically draws a graph from input data. { 'ȯd·ō ,pläd·ər }

autopolarity |ELECTR| Automatic interchanging of connections to a digital meter when polarity is wrong; a minus sign appears ahead of the value on the digital display if the reading is negative. { ,ȯd·ō·pə'lär·əd·ē }

autostability |CONT SYS| The ability of a device (such as a servomechanism) to hold a steady position, either by virtue of its shape and proportions, or by control by a servomechanism. { ¦ȯd·ō·stə'bil·əd·ē }

autostarter |ELEC| **1.** Automatic starting and switchover generating system consisting of a standby generator coupled to the station load through an automatic power transfer control unit. **2.** See autotransformer starter. { 'ȯd·ō ,stärd·ər }

autostart routine |COMPUT SCI| A set of instructions that is permanently stored in a computer memory and activated when the computer is turned on, to perform diagnostic tests and then load the operating system. { 'ȯd·ō,stärt rü,tēn }

autotest program |COMPUT SCI| A computer program within the operating system that aids in testing and debugging programs. { 'ȯd·ȯ,test 'prō·grəm }

autotrace |COMPUT SCI| A routine that locates outlines of raster graphics images and transforms them into vector graphics, usually at higher resolution. { 'ȯd·ō,trās }

autotransformer |ELEC| A power transformer having one continuous winding that is tapped; part of the winding serves as the primary and all of it serves as the secondary, or vice versa; small autotransformers are used to start motors. { ¦ȯd·ō·tranz¦fȯr·mər }

autotransformer starter |ELEC| Motor starter having an autotransformer to furnish a reduced voltage for starting; includes the necessary switching mechanism. Also known as autostarter. { ¦ȯd·ō·tranz¦fȯr·mər ,stärd·ər }

auxiliary channel |COMMUN| A secondary path for low-speed communication that uses the same circuit as a higher-speed stream of data. { ȯg'zil·yə·rē 'chan·əl }

auxiliary contacts |ELEC| Contacts, in a switching device, in addition to the main circuit contacts, which function with the movement of the latter. { ȯg'zil·yə·rē 'kän,taks }

auxiliary equipment See off-line equipment. { ȯg'zil·yə·rē ə'kwip·mənt }

auxiliary instruction buffer |COMPUT SCI| A section of storage in the instruction unit, 16 bytes in length, used to hold prefetched instructions. { ȯg'zil·yə·rē in'strək·shən ,bəf·ər }

auxiliary memory |COMPUT SCI| **1.** A high-speed memory that is in a large main frame or

supercomputer, is not directly addressable by the central processing unit, and is connected to the main memory by a high-speed data channel. **2.** See auxiliary storage. { ȯg'zil·yə·rē 'mem·rē }

auxiliary operation |COMPUT SCI| An operation performed by equipment not under continuous control of the central processing unit of a computer. { ȯg'zil·yə·rē ,äp·ə'rā·shən }

auxiliary processor |COMPUT SCI| Any equipment which performs an auxiliary operation in a computer. { ȯg'zil·yə·rē 'präs,es·ər }

auxiliary relay |ELEC| Relay that operates in response to the opening or closing of its operating circuit to assist another relay or device in performing a function. { ȯg'zil·yə·rē 'rē,lā }

auxiliary routine |COMPUT SCI| A routine designed to assist in the operation of the computer and in debugging other routines. { ȯg'zil·yə·rē rü'tēn }

auxiliary storage |COMPUT SCI| Storage device in addition to the main storage of a computer; for example, magnetic tape, magnetic or optical disk, or magnetic drum. Also known as auxiliary memory. { ȯg'zil·yə·rē 'stȯr·ij }

auxiliary switch |ELEC| A switch actuated by the main device (such as a circuit breaker) for signaling, interlocking, or other purposes. { ȯg'zil·yə·rē 'swich }

aV See abvolt.

availability |COMPUT SCI| Of data, data channels, and input-output devices in computers, the condition of being ready for use and not immediately committed to other tasks. { ə,vāl·ə'bil·ə·dē }

available line |ELECTR| Portion of the length of the scanning line which can be used specifically for picture signals in a facsimile system. { ə'vāl·ə·bəl 'līn }

available power |ELECTR| The power which a linear source of energy is capable of delivering into its conjugate impedance. { ə'vāl·ə·bəl 'pau̇·ər }

available-power gain |ELECTR| Ratio, in an electronic transducer, of the available power from the output terminals of the transducer, under specified input termination conditions, to the available power from the driving generator. { ə'vāl·ə·bəl 'pau̇·ər ,gān }

available space list |COMPUT SCI| A pool of inactive memory cells, available for use in a list-processing system, to which cells containing items deleted from data lists are added, and from which cells needed for newly inserted data items are removed. { ə'vāl·ə·bəl 'spās ,list }

available time See up time. { ə'vāl·ə·bəl 'tīm }

avalanche |ELECTR| **1.** The cumulative process in which an electron or other charged particle accelerated by a strong electric field collides with and ionizes gas molecules, thereby releasing new electrons which in turn have more collisions, so that the discharge is thus self-maintained. Also known as avalanche effect; cascade; cumulative ionization; electron avalanche; Townsend avalanche; Townsend ionization. **2.** Cumulative multiplication of carriers in a semiconductor as a result of avalanche breakdown. Also known as avalanche effect. { 'av·ə,lanch }

avalanche breakdown |ELECTR| Nondestructive breakdown in a semiconductor diode when the electric field across the barrier region is strong enough so that current carriers collide with valence electrons to produce ionization and cumulative multiplication of carriers. { 'av·ə ‚lanch 'brāk‚daùn }

avalanche diode |ELECTR| A semiconductor breakdown diode, usually made of silicon, in which avalanche breakdown occurs across the entire *pn* junction and voltage drop is then essentially constant and independent of current; the two most important types are IMPATT and TRAPATT diodes. { 'av·ə‚lanch 'dī‚ōd }

avalanche effect *See* avalanche. { 'av·ə‚lanch i ‚fekt }

avalanche impedance |ELECTR| The complex ratio of the reverse voltage of a device that undergoes avalanche breakdown to the reverse current. { 'av·ə‚lanch im'pēd·əns }

avalanche-induced migration |ELECTR| A technique of forming interconnections in a field-programmable logic array by applying appropriate voltages for shorting selected base-emitter junctions. { 'av·ə‚lanch in¦düsd ‚mī'grā·shən }

avalanche noise |ELECTR| **1.** A junction phenomenon in a semiconductor in which carriers in a high-voltage gradient develop sufficient energy to dislodge additional carriers through physical impact; this agitation creates ragged current flows which are indicated by noise. **2.** The noise produced when a junction diode is operated at the onset of avalanche breakdown. { 'av·ə‚lanch ‚nȯiz }

avalanche oscillator |ELECTR| An oscillator that uses an avalanche diode as a negative resistance to achieve one-step conversion from direct-current to microwave outputs in the gigahertz range. { 'av·ə‚lanch ¦äs·ə‚lād·ər }

avalanche photodiode |ELECTR| A photodiode operated in the avalanche breakdown region to achieve internal photocurrent multiplication, thereby providing rapid light-controlled switching operation. { 'av·ə‚lanch ‚fōd·ō'dī‚ōd }

avalanche transistor |ELECTR| A transistor that utilizes avalanche breakdown to produce chain generation of charge-carrying hole-electron pairs. { 'av·ə‚lanch tran'zis·tər }

avalanche voltage |ELECTR| The reverse voltage required to cause avalanche breakdown in a *pn* semiconductor junction. { 'av·ə‚lanch ‚vōl·tij }

avatar |COMPUT SCI| A virtual representation of a person or a person's interactions with others in a virtual environment, conveying a sense of someone's presence (known as telepresence) by providing the location (position and orientation) and identity; examples include the graphical human figure model, the talking head, and the real-time reproduction of a three-dimesional human image. { 'av·ə‚tär }

AVC *See* automatic volume control.

aV/cm *See* abvolt per centimeter.

average acoustic output |ENG ACOUS| Vibratory energy output of a transducer measured by a radiation pressure balance; expressed in terms of watts per unit area of the transducer face. { 'av·rij ə'kü·stik 'aùt‚pùt }

average-calculating operation |COMPUT SCI| A common or typical calculating operation longer than an addition and shorter than a multiplication; often taken as the mean of nine additions and one multiplication. { 'av·rij ¦kal·kyə‚lād·iŋ ‚äp·ə‚rā·shən }

average-edge line |COMPUT SCI| The imaginary line which traces or smooths the shape of any written or printed character to be recognized by a computer through optical, magnetic, or other means. { 'av·rij ¦ej ‚līn }

average effectiveness level *See* effectiveness level. { 'av·rij i'fek·tiv·nəs ‚lev·əl }

average information content |COMMUN| The average of the information content per symbol emitted from a source. { 'av·rij ‚in·fər'mā·shən ‚kän·tent }

average noise figure |ELECTR| Ratio in a transducer of total output noise power to the portion thereof attributable to thermal noise in the input termination, the total noise being summed over frequencies from zero to infinity, and the noise temperature of the input termination being standard (290 K). { 'av·rij 'nȯiz ‚fig·yər }

average power output |ELECTR| Radio-frequency power, in an audio-modulation transmitter, delivered to the transmitter output terminals, averaged over a modulation cycle. { 'av·rij 'paù·ər 'aùt‚pùt }

averaging |CONT SYS| The reduction of noise received by a robot sensor by screening it over a period of time. { 'av· rij·iŋ }

avigation *See* air navigation. { ‚a·və'gā·shən }

avionics |ENG| The design and production of airborne electrical and electronic devices; term is derived from aviation electronics. { ‚ā·vē'än·iks }

AWGN *See* additive white Gaussian noise.

axial lead |ELEC| A wire lead extending from the end along the axis of a resistor, capacitor, or other component. { 'ak·sē·əl 'lēd }

axial ratio |ELECTR| The ratio of the major axis to the minor axis of the polarization ellipse of a waveguide. Also known as ellipticity. { 'ak·sē·əl 'rā·shō }

Ayrton-Jones balance |ELEC| A type of balance with which force between current-carrying conductors is measured; uses single-layer solenoids as the fixed and movable coils. { ¦er·tən ¦jōnz 'bal·əns }

Ayrton-Perry winding |ELEC| Winding of two wires in parallel but opposite directions to give better cancellation of magnetic fields than is obtained with a single winding. { ¦er·tən ¦per·ē ‚wind·iŋ }

Ayrton shunt |ELEC| A shunt used to increase the range of a galvanometer without changing the damping. Also known as universal shunt. { 'er·tən ‚shənt }

azel display |ELECTR| Modified type of plan position indicator presentation showing two separate radar displays on one cathode-ray screen; one display presents bearing information and the other shows elevation. { 'az·el dis‚plä }

azimuth |ELECTR| Horizontal direction on the earth's surface, as represented by a radar plan position indicator. { 'az·ə·məth }

azimuth alignment |ENG ACOUS| The condition whereby the center lines of the playback- and recording-head gaps are exactly perpendicular to the magnetic tape and parallel to each other. { 'az·ə·məth ə'līn·mənt }

azimuth blanking |ELECTR| Blanking (disabling) either the radar receiver or transmitter or both in selected azimuth regions, to reduce interference or lessen radiation hazards. { 'az·ə·məth ,blaŋk·iŋ }

azimuth error |ENG| An error in the indicated azimuth of a target detected by radar. { 'az·ə·məth ,er·ər }

azimuth gain reduction |ELECTR| Technique which allows control of the radar receiver system throughout any two azimuth sectors. { 'az·ə·məth 'gān ri,dək·shən }

azimuth gating |ENG| The practice of selectively brightening and enhancing the gain-desired sectors of a radar plan position indicator display, usually by applying a step waveform to the automatic gain control circuit, or similar data separation by sectors in more automated systems. { 'az·ə·məth ,gād·iŋ }

azimuth indicator |ENG| An approach-radar scope which displays azimuth information. { 'az·ə·məth ,in·də,kād·ər }

azimuth marker |ELECTR| On a radar plan position indicator, a bright rotatable radial line used for bearing determination. Also known as angle marker; bearing marker. { 'az·ə·məth ,mär·kər }

azimuth resolution |ELECTROMAG| Angle or distance by which two targets must be separated in azimuth to be distinguished by a radar set, when the targets are at the same range. { 'az·ə·məth ,rez·ə'lü·shən }

azimuth-stabilized plan position indicator |ENG| A north-upward plan position indicator (PPI), a radarscope, which is stabilized by a gyrocompass so that either true or magnetic north is always at the top of the scope regardless of vehicle orientation. { 'az·ə·məth ¦sta·bə,līzd 'plan pə'zish·ən 'in·də,kād·ər }

azimuth versus amplitude |ELECTR| Electronic protection technique using a plan position indicator to display strobes due to jamming sources, particularly useful in making passive fixes when two or more radar sites operate together. { 'az·ə·məth ,vər·səs 'am·plə,tüd }

Azusa |ENG| A continuous-wave, high-accuracy, phase-comparison, single-station tracking system operating at C-band and giving two direction cosines and slant range which can be used to determine space position and velocity of a vehicle (usually a rocket or a missile). { ə'züs·ə }

B

babble [COMMUN] **1.** Aggregate crosstalk from a large number of channels. **2.** Unwanted disturbing sounds in a carrier or other multiple-channel system which result from the aggregate crosstalk or mutual interference from other channels. { 'bab·əl }

babs See blind approach beacon system. { babz }

baby spot [ELEC] A small spotlight, usually equipped with a hood, used (as in the theater) to concentrate light on an area or an object a small distance from the spotlight. { ¦bā·bē 'spät }

back bias [ELECTR] **1.** Degenerative or regenerative voltage which is fed back to circuits before its originating point; usually applied to a control anode of a tube or other device. **2.** Voltage applied to a grid of a tube (or tubes) or electrode of another device to reduce a condition which has been upset by some external cause. { 'bak ¸bī·əs }

backbone [COMPUT SCI] The portion of a communication network that handles the largest volume of traffic, usually employing a high-speed, high-capacity medium designed to transmit data over long distances. { 'bak¸bōn }

back contact [ELEC] Normally closed stationary contact on a relay that is opened when the relay is energized. { 'bak ¦kän¸takt }

back diode [ELECTR] A special type of tunnel diode operated at low levels of reverse bias at which the device has negative resistance. { 'bak ¸dī¸ōd }

back echo [ELECTROMAG] An echo signal produced on a radar screen by one of the minor back lobes of a search radar beam. { 'bak ¸ek·ō }

back-echo reflection [ELECTR] A radar echo produced by radiation reflected to the target by a large, fixed obstruction; that is, the ray path is from the antenna to obstruction to target and back similarly, giving a false indication of target position; an indirect-path echo. { 'bak ¸ek·ō ri'flek·shən }

back-emission electron radiography [ELECTR] A technique used in microradiography to visualize, among other things, the presence of material of different atomic numbers in the surface of the specimen being observed; the polished side of the specimen is facing and in close contact with the emulsion side of a fine-grain photographic plate; a light-tight cover holds the specimen and plate in place to be subjected to hardened x-rays. { 'bak i'mish·ən i'lek¸trän ¸rād·ē·'äg·rə·fē }

back-end system [COMPUT SCI] A computer that operates on data which have been previously processed by another computer system. { 'bak ¦end ¸sis·təm }

backfire See arcback. { 'bak¸fīr }

backfire antenna [ELECTROMAG] An antenna which exhibits significant gain in a direction 180° from its principal lobe. { 'bak¸fīr an'ten·ə }

backflow preventer See vacuum breaker. { 'bak ¸flō pri'ven·tər }

background [COMMUN] **1.** Picture white of the facsimile copy being scanned when the picture is black and white only. **2.** Undesired printing in the recorded facsimile copy of the picture being transmitted, resulting in shading of the background area. **3.** Noise heard during radio reception caused by atmospheric interference or the operation of the receiver at such high gain that inherent circuit noises become noticeable. { 'bak¸graúnd }

background discrimination [ENG] The ability of a measuring instrument, circuit, or other device to distinguish signal from background noise. { 'bak¸graúnd dis¸krim·ə'nā·shən }

background ink [COMPUT SCI] In optical character recognition, a highly reflective ink used to print the parts of a document that are to be ignored by the scanner. { 'bak¸graúnd ¸iŋk }

background noise [ENG] The undesired signals that are always present in an electronic or other system, independent of whether or not the desired signal is present. { 'bak¸graúnd ¸nóiz }

background processing [COMPUT SCI] **1.** The execution of lower-priority programs when higher-priority programs are not being handled by a data-processing system. **2.** Computer processing that is not interactive or visible on the display screen. { 'bak¸graúnd 'prä·ses·iŋ }

background program [COMPUT SCI] A computer program that has low priority in a multiprogramming system. { 'bak¸graúnd 'prō·grəm }

background reflectance [COMPUT SCI] The reflectance, relative to a standard, of the surface on which a printed or handwritten character has been inscribed in optical character recognition. { 'bak¸graúnd ri'flek·təns }

background returns [ENG] **1.** Signals on a radar screen from objects which are of no interest. **2.** *See* clutter. { 'bak,graůnd ri'tərnz }

backhaul [COMMUN] Point-to-point satellite transmission of video from a remote site to a network distribution center in real time. { 'bak,hól }

backing [ELECTR] Flexible material, usually cellulose acetate or polyester, used on magnetic tape as the carrier for the oxide coating. { 'bak·iŋ }

backing storage [COMPUT SCI] A computer storage device whose capacity is larger, but whose access time is slower, than that of the computer's main storage or immediate access storage; usually slower than main storage. Also known as bulk storage. { 'bak·iŋ ,stór·ij }

backlash [ELECTR] A small reverse current in a rectifier tube caused by the motion of positive ions produced in the gas by the impact of thermoelectrons. { 'bak,lash }

backlit display [ELECTR] An electronic display that incorporates a light source in back of a liquid-crystal or other electronic display to increase readability, especially in daylight. { |bak ,lit di'splā }

back lobe [ELECTROMAG] The three-dimensional portion of the radiation pattern of a directional antenna that is directed away from the intended direction. { 'bak ,lōb }

backout [COMPUT SCI] To remove a change that was previously made in a computer program. { 'bak,aůt }

backplane [ELECTR] A wiring board, usually constructed as a printed circuit, used in computers to provide the required connections between logic, memory, input/output modules, and other printed circuit boards which plug into it at right angles. { 'bak,plān }

backplate lamp holder [ENG] A lamp holder, integrally mounted on a plate, which is designed for screwing to a flat surface. { 'bak ,plāt 'lamp ,hōl·dər }

back porch [ELECTR] The period of time in a television circuit immediately following a synchronizing pulse during which the signal is held at the instantaneous amplitude corresponding to a black area in the received picture. { 'bak |pórch }

back radiation *See* backscattering. { 'bak ,rād·ē'ā·shən }

back resistance [ELECTR] The resistance between the contacts opposing the inverse current of a metallic rectifier. { 'bak ri'sis·təns }

backscatter gage [ENG] A radar instrument used to measure the radiation scattered at 180° to the direction of the incident wave. { 'bak |skad·ər ,gaj }

backscattering [COMMUN] Propagation of extraneous signals by F- or E-region reflection in addition to the desired ionospheric scatter mode; the undesired signal enters the antenna through the back lobes. [ELECTROMAG] **1.** Radar echoes from a target. **2.** Undesired radiation of energy to the rear by a directional antenna. **3.** Also known as back radiation; backward scattering. { 'bak|skad·ə·riŋ }

back solution [CONT SYS] The calculation of the tool-coordinated positions that correspond to specified robotic joint positions. { 'bak sə ,lü·shən }

backspace [COMPUT SCI] To move a recording medium one unit in the reverse or background direction. { 'bak,spās }

back-surface field [ELECTR] A p^+ layer that is added to a silicon solar cell to reduce electronhole recombination at the cell's back surface and thereby increase the cell's efficiency. { 'bak ,sər·fəs ,fēld }

backtalk [COMPUT SCI] Passage of information from a standby computer to the active computer. { 'bak,tók }

backtracking [COMPUT SCI] A method of solving problems automatically by a systematic search of the possible solutions; the invalid solutions are eliminated and are not retried. { 'bak ,trak·iŋ }

backup [COMPUT SCI] **1.** Logical or physical facilities to aid the process of restarting a computer system and recovering the information in it following a failure. **2.** The provision of such facilities. { 'bak,əp }

backup arrangement *See* cascade. { 'bak,əp ,ə'rānj·mənt }

backup relay [ELEC] A relay designed to protect a power system in case a primary relay fails to operate as desired. { 'bak,əp 'rē·lā }

backup system [SYS ENG] A system, normally redundant but kept available to replace a system which may fail in operation. { 'bak,əp ,sis·təm }

Backus-Naur form [COMPUT SCI] A metalanguage that specifies which sequences of symbols constitute a syntactically valid program language. Abbreviated BNF. { |bäk·əs |naůr ,fórm }

backward-acting regulator [ELECTR] Transmission regulator in which the adjustment made by the regulator affects the quantity which caused the adjustment. { 'bak·wərd 'ak·tiŋ 'reg·yə,lād·ər }

backward chaining [COMPUT SCI] In artificial intelligence, a method of reasoning which starts with the problem to be solved and repeatedly breaks this goal into subgoals that are more readily solvable with the relevant data and the system's rules of inference. { |bak·wərd 'chān·iŋ }

backward compatibility *See* downward compatibility. { |bak·wərd kəm,pad·ə'bil·əd·ē }

backward diode [ELECTR] A semiconductor diode similar to a tunnel diode except that it has no forward tunnel current; used as a low-voltage rectifier. { 'bak·wərd 'dī,ōd }

backward error analysis [COMPUT SCI] A form of error analysis which seeks to replace all errors made in the course of solving a problem by an equivalent perturbation of the original problem. { 'bak·wərd 'er·ər ə,nal·ə·səs }

backward read [COMPUT SCI] The transfer of data from a magnetic tape to computer storage when the tape is running in reverse. { 'bak·wərd 'rēd }

backward scattering *See* backscattering. { 'bak·wərd |skad·ə·riŋ }

backward search |COMPUT SCI| A search of a document or database that starts at the cursor's location and moves backwards toward the beginning of the document or database.   { ¦bak·wərd 'sərch }

backward wave |ELECTROMAG| An electromagnetic wave traveling opposite to the direction of motion of some other physical quantity in an electronic device such as a traveling-wave tube or mismatched transmission line.   { 'bak·wərd ¸wāv }

backward-wave magnetron |ELECTR| A magnetron in which the electron beam travels in a direction opposite to the flow of the radio-frequency energy.   { 'bak·wərd ¸wāv 'mag·nə ¸trän }

backward-wave oscillator |ELECTR| An electronic device which amplifies microwave signals simultaneously over a wide band of frequencies and in which the traveling wave produced is reflected backward so as to sustain the wave oscillations. Abbreviated BWO. Also known as carcinotron.   { 'bak·wərd ¸wāv 'äs·ə ¸lād·ər }

backward-wave tube |ELECTR| A type of microwave traveling-wave electron tube in which electromagnetic energy on a slow-wave circuit flows opposite in direction to the travel of electrons in a beam.   { 'bak·wərd ¸wāv ¸tüb }

bad branch |COMPUT SCI| An error in which execution of a computer program jumps to an incorrect instruction, usually as a result of errors in the program.   { ¦bad 'branch }

bad page break |COMPUT SCI| A soft page break at an inappropriate location in a document, such as one that splits a table or leaves a single line of text at the top or bottom of a page.   { ¦bad 'pāj ¸brāk }

bad sector |COMPUT SCI| An area of disk storage that does not record data reliably and therefore is not used.   { ¸bad 'sek·tər }

bad track |COMPUT SCI| A disk track that contains a bad sector.   { ¸bad 'trak }

bad track table |COMPUT SCI| A listing of the bad sectors on a disk, which is packaged with or attached to a disk.   { ¦bad 'trak ¸tā·bəl }

baffle |ELEC| Device for deflecting oil or gas in a circuit breaker. |ELECTR| An auxiliary member in a gas tube used, for example, to control the flow of mercury particles or deionize the mercury following conduction. |ENG| A plate that regulates the flow of a fluid, as in a steam-boiler flue or a gasoline muffler. |ENG ACOUS| A cabinet or partition used with a loudspeaker to reduce interaction between sound waves produced simultaneously by the two surfaces of the diaphragm.   { 'baf·əl }

balance |ELEC| The state of an electrical network when it is adjusted so that voltage in one branch induces or causes no current in another branch. |ENG| An instrument for measuring mass or weight.   { 'bal·əns }

balance coil |ELEC| An iron-core solenoid with adjustable taps near the center; used to convert a two-wire circuit to a three-wire circuit, the taps furnishing a neutral terminal for the latter.   { 'bal·əns ¸kȯil }

balance control |ELECTR| A control used in a stereo sound system to vary the volume of one loudspeaker system relative to the other while maintaining their combined volume essentially constant.   { 'bal·əns kən'trōl }

balanced amplifier |ELECTR| An electronic amplifier in which there are two identical signal branches connected so as to operate with the inputs in phase opposition and with the output connections in phase, each balanced to ground.   { 'bal·ənst 'am·plə¸fī·ər }

balanced armature unit |ENG ACOUS| Driving unit used in magnetic loudspeakers, consisting of an iron armature pivoted between the poles of a permanent magnet and surrounded by coils carrying the audio-frequency current; variations in audio-frequency current cause corresponding changes in armature magnetism and corresponding movements of the armature with respect to the poles of the permanent magnet.   { 'bal·ənst 'ärm·ə·chər ¸yü·nət }

balanced bridge |ELEC| Wheatstone bridge circuit which, when in a quiescent state, has an output voltage of zero.   { 'bal·ənst 'brij }

balanced circuit |ELEC| **1.** A circuit whose two sides are electrically alike and symmetrical with respect to a common reference point, usually ground. **2.** An electric circuit that has been adjusted to neutralize the mutual induction of an adjacent circuit.   { 'bal·ənst 'sər·kət }

balanced converter See balun.   { 'bal·ənst kən 'vərd·ər }

balanced currents |ELEC| Currents flowing in the two conductors of a balanced line which, at every point along the line, are equal in magnitude and opposite in direction. Also known as push-pull currents.   { 'bal·ənst 'kər·əns }

balanced detector |ELECTR| A detector used in frequency-modulation receivers; in one form the audio output is the rectified difference between voltages produced across two resonant circuits, one being tuned slightly above the carrier frequency and one slightly below.   { 'bal·ənst di'tek·tər }

balanced input |ELECTR| A symmetrical input circuit having equal impedance from both input terminals to reference.   { 'bal·ənst ¦in¸pút }

balanced line |ELEC| A transmission line consisting of two conductors capable of being operated so that the voltages of the two conductors at any transverse plane are equal in magnitude and opposite in polarity with respect to ground.   { 'bal·ənst ¸līn }

balanced load

balanced load |ELEC| A load that presents the same impedance, with respect to ground, at both ends or terminals. { 'bal·ənst 'lōd }

balanced merge |COMPUT SCI| A merge or sort operation in which the data involved are divided equally between the available storage devices. { 'bal·ənst 'mərj }

balanced method |ENG| Method of measurement in which the reading is taken at zero; it may be a visual or audible reading, and in the latter case the null is the no-sound setting. { 'bal·ənst ¦meth·əd }

balanced modulator |ELECTR| A modulator in which the carrier and modulating signal are introduced in such a way that the output contains the two sidebands without the carrier. { 'bal·ənst 'maj·ə,lād·ər }

balanced network |ELEC| Hybrid network in which the impedances of the opposite branches are equal. { 'bal·ənst ¦net,wərk }

balanced oscillator |ELECTR| Any oscillator in which, at the oscillator frequency, the impedance centers of the tank circuits are at ground potential, and the voltages between either end and their centers are equal in magnitude and opposite in phase. { 'bal·ənst 'äs·ə,lād·ər }

balanced output |ELECTR| A three-conductor output (as from an amplifier) in which the signal voltage alternates above and below a third, neutral wire. { 'bal·ənst ¦aùt,pùt }

balanced ring modulator |ELECTR| A modulator that uses tubes or diodes to suppress the carrier signal while providing double-sideband output. { 'bal·ənst ¦riŋ ,mäj·ə,lād·ər }

balanced set |ELECTR| Two or more components, such as tubes or transistors, connected in parallel or push-pull configuration, that have been chosen on the basis of identical, or nearly identical, gain and load characteristics. { 'bal·ənst ,set }

balanced transmission line |ELEC| Transmission line having equal conductor resistances per unit length and equal impedances from each conductor to earth and to other electrical circuits. { 'bal·ənst tranz'mish·ən ,līn }

balanced-tree |COMPUT SCI| A system of indexes that keeps track of stored data, and in which data keys are stored in a hierarchy that is continually modified in order to minimize access times. Abbreviated B-tree. { 'bal·ənst 'trē }

balanced voltages |ELEC| Voltages that are equal in magnitude and opposite in polarity with respect to ground. Also known as push-pull voltages. { 'bal·ənst ,vōl·tij·əz }

balanced wire circuit |ELEC| Circuit wherein the two sides are electrically alike and symmetrical with respect to ground and other conductors. { 'bal·ənst ¦wīr ,sər·kət }

balance error |COMPUT SCI| An error voltage that arises at the output of analog adders in an analog computer and is directly proportional to the drift error. { 'bal·əns ,er·ər }

balance method See null method. { 'bal·əns ,meth·əd }

balancer |ELEC| A mechanism for equalizing the loads on the outer lines of a three-wire system for electric power distribution, consisting of two similar shunt or compound machines which are coupled together with the armatures connected in series across the outer lines. { 'bal·ən·sər }

balancer set |ELEC| Two coupled direct-current generators or motors that are used to equalize the voltage on each side of a three-wire system. { 'bal·ən·sər ,set }

balance-to-unbalance transformer |ELEC| Device for matching a pair of lines, balanced with respect to earth, to a pair of lines not balanced with respect to earth. { 'bal·əns tü ¦ən,bal·əns tranz'fer·mər }

balancing |COMPUT SCI| The distribution of workload among computing resources to optimize performance. { 'bal·əns·iŋ }

balancing capacitor |ELECTR| A variable capacitor used to improve the accuracy of a radio direction finder. Also known as compensating capacitor. { 'bal·əns·iŋ kə'pas·əd·ər }

balancing unit |ELEC| **1.** Antenna-matching device used to permit efficient coupling of a transmitter or receiver having an unbalanced output circuit to an antenna having a balanced transmission line. **2.** Device for converting balanced to unbalanced transmission lines, and vice versa, by placing suitable discontinuities at the junction between the lines instead of using lumped components. { 'bal·əns·iŋ,yü·nət }

ballast |ELEC| A circuit element that serves to limit an electric current or to provide a starting voltage, as in certain types of lamps, such as in fluorescent ceiling fixtures. { 'bal·əst }

ballast factor |ELEC| The ratio of the luminous output of a lamp when operated on a ballast to its luminous output when operated under standardized rating conditions. { 'bal·əst ,fak·tər }

ballast lamp |ELEC| A light-producing electrical resistance device which maintains nearly constant current by increasing in resistance as the current increases. { 'bal·əst ,lamp }

ballast reactor |ELEC| A coil wound on an iron core and connected in series with a fluorescent lamp to compensate for the negative-resistance characteristics of the lamp by providing an increased voltage drop as the current through the lamp is increased. { 'bal·əst rē'ak·tər }

ballast resistor |ELEC| A resistor that increases in resistance as current through it increases, and decreases in resistance as current decreases. Also known as barretter (British usage). { 'bal·əst ri'sis·tər }

ballast tube |ELEC| A ballast resistor mounted in an evacuated glass or metal envelope, like that of a vacuum tube, to reduce radiation of heat from the resistance element and thereby improve the voltage-regulating action. { 'bal·əst ,tüb }

ball bonding |ENG| The making of electrical connections in which a flame is used to cut a wire, the molten end of which solidifies as a ball, which is pressed against the bonding pad on an integrated circuit. { 'bȯl ,bänd·iŋ }

ballistic galvanometer [ELEC] A galvanometer having a long period of swing so that the deflection may measure the electric charge in a current pulse or the time integral of a voltage pulse. { bə'lis·tik ,gal·və'näm·əd·ər }

ballistic magnetometer [ENG] A magnetometer designed to employ the transient voltage induced in a coil when either the magnetized sample or coil are moved relative to each other. { bə'lis·tik ,mag·nə'täm·əd·ər }

ballistic tracking See dynamic resolution. { bə ,lis·tik 'trak·iŋ }

ballistic transport [ELECTR] The passage of electrons through a semiconductor whose length is less than the mean free path of electrons in the semiconductor, so that most of the electrons pass through the semiconductor without scattering. { bə'lis·tik 'tranz,pórt }

ballistic vehicle [ENG] A nonlifting vehicle; a vehicle that follows a ballistic trajectory. { bə'lis·tik 'vē·ə·kəl }

balun [ELEC] A device used for matching an unbalanced coaxial transmission line or system to a balanced two-wire line or system. Also known as balanced converter; bazooka; line-balance converter. { 'ba,lən }

banana jack [ELEC] A jack that fits a banana plug; generally designed for panel mounting. { bə'nan·ə ,jak }

banana plug [ELEC] A plug having a spring-metal tip shaped like a banana and used on test leads or as terminals for plug-in components. { bə'nan·ə,pləg }

band [COMMUN] A range of electromagnetic-wave frequencies between definite limits, such as that assigned to a particular type of radio service. [COMPUT SCI] A set of circular or cyclic recording tracks on a storage device such as a magnetic drum, disk, or tape loop. { band }

bandage [ELEC] Rubber ribbon about 4 inches (10 centimeters) wide for temporarily protecting a telephone or coaxial splice from moisture. { 'ban·dij }

band-elimination filter See band-stop filter. { ¦band i,lim·ə'nā·shən 'fil·tər }

band-pass [ELECTR] A range, in hertz or kilohertz, expressing the difference between the limiting frequencies at which a desired fraction (usually half power) of the maximum output is obtained. { 'band ,pas }

band-pass amplifier [ELECTR] An amplifier designed to pass a definite band of frequencies with essentially uniform response. { 'band ,pas ¦am·plə,fī·ər }

band-pass filter [ELECTR] An electric filter which transmits more or less uniformly in a certain band, outside of which the frequency components are attenuated. { 'band ,pas ,fil·tər }

band-pass response [ELECTR] Response characteristics in which a definite band of frequencies is transmitted uniformly. Also known as flat-top response. { 'band ,pas ri'späns }

band-pass system [ENG ACOUS] A loudspeaker system, often used for subwoofers, in which the speaker is mounted inside an enclosure on a shelf that divides the enclosure into two parts, and one or both parts are coupled to the outside by a vent; the frequency response of the system is that of a fourth-order band-pass filter (one vent) or an asymmetrical sixth-order band-pass filter (two vents). { 'band,pas ,sis·təm }

band printer [COMPUT SCI] A line printer that uses a band of type characters as its printing mechanism. { 'band ¦print·ər }

band-rejection filter See band-stop filter. { 'band ri'jek·shən ,fil·tər }

band selector [ELECTR] A switch that selects any of the bands in which a receiver, signal generator, or transmitter is designed to operate and usually has two or more sections to make the required changes in all tuning circuits simultaneously. Also known as band switch. { 'band sə'lek· tər }

band spreading [COMMUN] Method of double-sideband transmission in which the frequency band of the modulating wave is shifted upward in frequency so that the sidebands produced by modulation are separated in frequency from the carrier by an amount at least equal to the bandwidth of the original modulating wave, and second-order distortion products may be filtered from the demodulator output. { 'band ,spred·iŋ }

band-spread tuning control [ELECTR] A tuning control provided on some shortwave receivers to spread the stations in a single band of frequencies over an entire tuning dial. { 'band ,spred 'tün·iŋ kən'trōl }

band-stop filter [ELECTR] An electric filter which transmits more or less uniformly at all frequencies of interest except for a band within which frequency components are largely attenuated. Also known as band-elimination filter; band-rejection filter. { 'band ,stäp ,fil·tər }

band switch See band selector. { 'band ,swich }

bandwidth [COMMUN] **1.** The difference between the frequency limits of a band containing the useful frequency components of a signal. **2.** A measure of the amount of data that can travel a communications path in a given time, usually expressed as thousands of bits per second (kbps) or millions of bits per second (Mbps). { 'band ,width }

bang-bang circuit [ELECTR] An operational amplifier with double feedback limiters that drive a high-speed relay (1–2 milliseconds) in an analog computer; involved in signal-controlled programming. { ¦baŋ ¦baŋ ,sər·kət }

bang-bang control [COMPUT SCI] Control of programming in an analog computer through a bang-bang circuit. [CONT SYS] A type of automatic control system in which the applied control signals assume either their maximum or minimum values. { ¦baŋ ¦baŋ kən'trōl }

bang-bang-off control See bang-zero-bang control. { ¦baŋ ¦baŋ 'óf kən,trōl }

bang-bang robot [CONT SYS] A simple robot that can make only two types of motions. { ¦baŋ ¦baŋ 'rō,bät }

43

bang-zero-bang control |CONT SYS| A type of control in which the control values are at their maximum, zero, or minimum. Also known as bang-bang-off control. { ¦baŋ ‚zir·ō 'baŋ kən ‚trōl }

bank |ELEC| **1.** A number of similar electrical devices, such as resistors, connected together for use as a single device. **2.** An assemblage of fixed contacts over which one or more wipers or brushes move in order to establish electrical connections in automatic switching. { baŋk }

bank-and-wiper switch |ELEC| Switch in which electromagnetic ratchets or other mechanisms are used, first, to move the wipers to a desired group of terminals, and second, to move the wipers over the terminals of the group to the desired bank contacts. { ¦baŋk ən 'wī·pər ‚swich }

banked winding |ELECTR| A radio-frequency coil winding which proceeds from one end of the coil to the other without return by having, side by side, many flat spirals formed by winding single turns one over the other, thereby reducing the distributed capacitance of the coil. { 'baŋkt 'wīnd·iŋ }

bank select |COMPUT SCI| To activate and deactivate blocks of memory or other internal system components using electronic control signals. Also known as bank switch. { 'baŋk si‚lekt }

bank selected memory |COMPUT SCI| Auxiliary blocks of memory in a microcomputer that can be switched in to replace some or all of the internal memory by software-controlled switches located outside the microprocessor. { 'baŋk si¦lek·təd 'mem·rē }

bank switch See bank select. { 'baŋk ‚swich }

bantam tube |ELECTR| Vacuum tube having a standard octal base, but a considerably smaller glass tube than a standard glass tube. { 'ban·təm ¦tüb }

bar code |COMPUT SCI| The representation of alphanumeric characters by series of adjacent stripes of various widths, for example, the universal product code. { 'bär ‚kōd }

bar-code reader See bar-code scanner. { 'bär ‚kōd 'rēd·ər }

bar-code scanner |COMPUT SCI| An optical scanning device that reads texts which have been converted into a special bar code. Also known as bar-code reader. { 'bär ‚kōd 'skan·ər }

bare board |ELECTR| A printed circuit board with conductors but no electronic components. { ¦ber 'bórd }

bare disk |ELECTR| A floppy-disk drive without electronic control circuits. { ¦ber 'disk }

bar generator |ELECTR| Generator of pulses or repeating waveforms that are equally separated in time; these pulses are synchronized by the synchronizing pulses of a television system, so that they can produce a stationary bar pattern on a television screen. { 'bär ¦jen·ə‚rād·er }

BARITT diode See barrier injection transit-time diode. { 'bar·ət ¦dī‚ōd }

barium fuel cell |ELEC| A fuel cell in which barium is used with either oxygen or chlorine to convert chemical energy into electrical energy. { 'bar·ē·əm 'fyül ‚sel }

Barkhausen criterion |ELECTR| A criterion used to determine the stability of an oscillator circuit which states that, if the circuit is seen as a loop consisting of an amplifier with gain A and a linear circuit whose gain $\beta(j\omega)$ depends on frequency ω, then the loop will oscillate with a perfect sine wave at some frequency ω_0 if at that frequency $A\beta(j\omega_0) = 1$ exactly, that is, if the magnitude of $A\beta(j\omega_0)$ is exactly 1 and its phase is $0°$ or $360°$. { 'bärk‚haůz·ən krī‚tir·ē·ən }

Barkhausen interference |COMMUN| Interference caused by Barkhausen oscillations. { 'bärk‚haůz·ən in·tər'fir·əns }

Barkhausen-Kurz oscillator |ELECTR| An oscillator of the retarding-field type in which the frequency of oscillation depends solely on the transit time of electrons oscillating about a highly positive grid before reaching the less positive anode. Also known as Barkhausen oscillator; positive-grid oscillator. { 'bärk‚haůz·ən ¦kərts 'äs·ə‚lād·ər }

Barkhausen oscillation |ELECTR| Undesired oscillation in the horizontal output tube of a television receiver, causing one or more ragged dark vertical lines on the left side of the picture. { 'bärk‚haůz·ən ‚äs·ə'lā·shən }

Barkhausen oscillator See Barkhausen-Kurz oscillator. { 'bärk‚haůz·ən 'äs·ə‚lād·ər }

barometric fuse |ENG| A fuse that functions as a result of change in the pressure exerted by the surrounding air. { bar·ə'met·rik 'fyüz }

bar pattern |ELECTR| Pattern of repeating lines or bars on a television screen. { 'bär ‚pad·ərn }

bar printer |COMPUT SCI| An impact printer in which the character heads are mounted on type bars. { 'bär ‚print·ər }

barrage jamming |COMMUN| The simultaneous jamming of a number of radio frequencies or even multiple radar bands of frequencies. { bə'räzh ‚jam·iŋ }

barrel printer |COMPUT SCI| A computer printer in which the entire set of characters is placed around a rapidly rotating cylinder at each print position; computer-controlled print hammers opposite each print position strike the paper and press it against an inked ribbon between the paper and the cylinder when the appropriate character reaches a position opposite the print hammer. { 'bar·əl ‚prin·tər }

barretter |ELEC| Bolometer that consists of a fine wire or metal film having a positive temperature coefficient of resistivity, so that resistance increases with temperature; used for making power measurements in microwave devices. See ballast resistor. { bə'red·ər }

barrier capacitance |ELECTR| The capacitance that exists between the p-type and n-type semiconductor materials in a semiconductor pn junction that is reverse-biased so that it does not conduct. Also known as depletion-layer capacitance; junction capacitance. { 'bar·ē·ər kə‚pas·əd·əns }

barrier-grid storage tube   *See* radechon.
{ 'bar·ē·ər ‚grid 'stòr·ij ‚tüb }
barrier injection transit-time diode |ELECTR| A microwave diode in which the carriers that traverse the drift region are generated by minority carrier injection from a forward-biased junction instead of being extracted from the plasma of an avalanche region. Abbreviated BARITT diode.
{ 'bar·ē·ər in'jek·shən 'trans·ət ‚tīm 'dī‚ōd }
barrier layer   *See* depletion layer.   { 'bar·ē·ər ‚lā·ər }
barrier-layer cell   *See* photovoltaic cell.   { 'bar·ē·ər ‚lā·ər ‚sel }
barrier-layer photocell   *See* photovoltaic cell.
{ 'bar·ē·ər ‚lā·ər 'fōd·ō‚sel }
barrier-layer rectification   *See* depletion-layer rectification.   { 'bar·ē·ər ‚lā·ər ‚rek·tə·fə'kā·shən }
barrier strip |ELECTR| A device for connecting two cables without using plugs in which bare wires from one cable are connected to lugs of screws on one side of the strip and wires from the other cable are attached at corresponding points on the opposite side.   { 'bar·ē·ər ‚strip }
barrier voltage |ELECTR| The voltage necessary to cause electrical conduction in a junction of two dissimilar materials, such as *pn* junction diode.
{ 'bar·ē·ər ‚vōl·tij }
bar winding |ELEC| An armature winding made up of a series of metallic bars connected at their ends.   { 'bär ‚wīnd·iŋ }
base |COMPUT SCI| *See* root. |ELECTR| **1.** The region that lies between an emitter and a collector of a transistor and into which minority carriers are injected.   **2.** The part of an electron tube that has the pins, leads, or other terminals to which external connections are made either directly or through a socket.   **3.** The plastic, ceramic, or other insulating board that supports a printed wiring pattern.   { bās }
base address   *See* address constant.   { bās ə'dres }
baseband |COMMUN| The band of frequencies occupied by all transmitted signals used to modulate the radio wave.   { 'bās‚band }
baseband frequency response |COMMUN| Frequency response characteristics of the frequency band occupied by all of the signals used to modulate a transmitted carrier.   { 'bās‚band 'frē·kwən·sē ri'späns }
baseband system |COMMUN| A communications system in which information is transmitted over a single unmodulated band of frequencies.   { 'bās ‚band ‚sis·təm }
base bias |ELECTR| The direct voltage that is applied to the majority-carrier contact (base) of a transistor.   { 'bās ‚bī·əs }
base-displacement |COMPUT SCI| In machine-language programming, a technique in which addresses are specified relative to a base address where the beginning of the program is stored.
{ 'bās dis‚plās·mənt }
base electrode |ELECTR| An ohmic or majority carrier contact to the base region of a transistor.
{ 'bās i'lek‚trōd }

base font |COMPUT SCI| The font used in a document if none other is specified.   { 'bās ‚fänt }
base insulator |ELEC| Heavy-duty insulator used to support the weight of an antenna mast and insulate the mast from the ground or some other surface.   { 'bās 'in·sə‚lād·ər }
base language |COMPUT SCI| The component of an extensible language which provides a complete but minimal set of primitive facilities, such as elementary data types, and simple operations and control constructs.   { 'bās 'laŋ·gwij }
base line |ELECTR| The line traced on amplitude-modulated indicators which corresponds to the power level of the weakest echo detected by the radar; it is retraced with every pulse transmitted by the radar but appears as a nearly continuous display on the scope. Abbreviated BL.   { 'bās ‚līn }
baseline |ENG| The geographic line between transmitter and receiver locations in bistatic radar, or between pairs of radars or radio receivers in a network, used in calculations relative to the data. Abbreviated BL.   { 'bās ‚līn }
base-line break |ELECTR| Technique in radar which uses the characteristic break in the base line on an A-scope display due to a pulse signal of significant strength in noise jamming.   { 'bās ‚līn ‚brāk }
base-line check   *See* ground check.   { 'bās ‚līn ‚chek }
baseload |ELEC| Minimum load of a power generator over a given period of time.   { 'bās‚lōd }
base-loaded antenna |ELECTROMAG| Vertical antenna having an impedance in series at the base for loading the antenna to secure a desired electrical length.   { 'bās ‚lōd·əd an'ten·ə }
base modulation |ELECTR| Amplitude modulation produced by applying the modulating voltage to the base of a transistor amplifier.   { 'bās ‚mäj·ə'lā·shən }
base pin   *See* pin.   { 'bās ‚pin }
base rate area |COMMUN| Area within which service is given without mileage charges.   { 'bās ‚rāt ¦er·ē·ə }
base register   *See* index register.   { 'bās ‚rej·ə·stər }
base-spreading resistance |ELECTR| Resistance which is found in the base of any transistor and acts in series with it, generally a few ohms in value.   { 'bās ¦spred·iŋ ri'zis·təns }
base station |COMMUN| **1.** A land station, in the land mobile service, carrying on a service with land mobile stations (a base station may secondarily communicate with other base stations incident to communications with land mobile stations).   **2.** A station in a land mobile system which remains in a fixed location and communicates with the mobile stations.   { 'bās ‚stā·shən }
base system |COMPUT SCI| A computer system containing only program modules that carry out basic functions.   { 'bās ‚sis·təm }
BASIC |COMPUT SCI| A procedure-level computer language designed to be easily learned and used by nonprofessionals, and well suited for

an interactive, conversational mode of operation. Derived from Beginners All-purpose Symbolic Instruction Code. { 'bā·sik }

basic batch |COMPUT SCI| The least complex level of computer processing, in which application systems are normally made up of small programs that are run through the computer one at a time and that can process transactions only from sequential files. { 'bā·sik 'bach }

basic disk operating system |COMPUT SCI| The part of a computer's operating system that handles the transfer of data between programs and disk units and the control of files. Abbreviated BDOS. { 'bā·sik ¦disk ¸äp·ə'rād·iŋ 'sis·təm }

basic input/output system |COMPUT SCI| The part of a computer's operating system that handles communications between a program and external devices such as printers and electronic displays. Abbreviated BIOS. { 'bā·sik 'in¸put 'aút¸pút ¸sis·təm }

basic instruction |COMPUT SCI| An instruction in a computer program which is systematically changed by the program to obtain the instructions which are actually carried out. Also known as presumptive instruction; unmodified instruction. { 'bā·sik in'strək·shən }

basic linkage |COMPUT SCI| Computer coding that provides a standard means of connecting a given routine or program with other routines and that can be used repeatedly according to the same rules. { 'bā·sik 'liŋ·kij }

basic processing unit |COMMUN| Principal controller and data processor within the communications system. { 'bā·sik 'präs¸es·iŋ ¸yü·nət }

basic Q *See* nonloaded Q. { 'bā·sik 'kyü }

basic software |COMPUT SCI| Software requirements that are taken into account in the design of the data-processing hardware and usually are provided by the original equipment manufacturer. { 'bā·sik 'sóft¸wer }

basic telecommunications access method |COMPUT SCI| A method of controlling data transmission between a computer's main storage and its terminals and of providing applications programs with the capability of communicating with printers, terminals, and other devices. Abbreviated BTAM. { 'bā·sik ¸tel·ə·kə¸myü·nə¦kā·shənz 'ak¸ses ¸meth·əd }

basic variables |COMPUT SCI| The *m* variables in a basic feasible solution for a linear programming model. { 'bā·sik 'ver·ē·ə·bəlz }

basket coil *See* basket winding. { 'bas·kət ¸kóil }

basket winding |ELECTR| A crisscross coil winding in which successive turns are far apart except at points of crossing, giving low distributed capacitance. Also known as basket coil. { 'bas·kət ¸wīnd·iŋ }

bass boost |ELECTR| A circuit that emphasizes the lower audio frequencies, generally by attenuating higher audio frequencies. { ¦bās ¦büst }

bass compensation |ELECTR| A circuit that emphasizes the low-frequency response of an audio amplifier at low volume levels to offset the lower sensitivity of the human ear to weak low frequencies. { 'bās ¸käm·pən'sā·shən }

bass control |ELECTR| A manual tone control that attenuates higher audio frequencies in an audio amplifier and thereby emphasizes bass frequencies. { 'bās kən'trōl }

bass reflex baffle |ENG ACOUS| A loudspeaker baffle having an opening of such size that bass frequencies from the rear of the loudspeaker emerge to reinforce those radiated directly forward. { ¦bas 'rē¸fleks ¸baf·əl }

bass response |ELECTR| A measure of the output of an electronic device or system as a function of an input of low audio frequencies. { 'bās ri¸späns }

bass trap |ENG ACOUS| Any device used in a sound-recording studio to absorb sound at frequencies less than about 100 hertz. { 'bās ¸trap }

bassy |ENG ACOUS| Pertaining to sound reproduction that overemphasizes low-frequency notes. { 'bās·ē }

batch |COMPUT SCI| A set of items, records, or documents to be processed as a single unit. { bach }

batch-and-forward system |COMPUT SCI| A data-processing system in which data are collected for a time and then transmitted as a unit to a computer. { 'bach ən 'fór·wərd ¸sis·təm }

batching |COMPUT SCI| Grouping records for the purpose of processing them in a computer. { 'bach·iŋ }

batch job |COMPUT SCI| One of a group of jobs that are executed together by batch-processing techniques. { 'bach ¸jäb }

batch-oriented applications |COMMUN| Applications of data communications that involve the transfer of thousands or even millions of bytes of data and are usually point-to-point and computer-to-computer. { 'bach ¸ór·ē ¦ent·əd ¸ap·lə¦kā·shənz }

batch processing |COMPUT SCI| A technique that uses a single program loading to process many individual jobs, tasks, or requests for service. { 'bach ¸präs·es·iŋ }

batch stream |COMPUT SCI| A group of batch processing programs that are scheduled to run on a computer. { 'bach ¸strēm }

batch system |COMPUT SCI| A computer system that uses batch processing. { 'bach ¸sis·təm }

batch total |COMPUT SCI| The total for a specified constituent quantity in a batch; used to verify the accuracy of operations on the batch. { 'bach ¦tōd·əl }

bat-handle switch |ELEC| A toggle switch having an actuating lever shaped like a baseball bat. { 'bat ¸hand·əl ¸swich }

bathtub capacitor |ELEC| A capacitor enclosed in a metal housing having broadly rounded corners like those on a bathtub. { 'bath¸təb kə'pas·əd·ər }

battery |ELEC| A direct-current voltage source made up of one or more units that convert chemical, thermal, nuclear, or solar energy into electrical energy. { 'bad·ə·rē }

battery charger |ELEC| A rectifier unit used to change alternating to direct power for charging a storage battery. Also known as charger. { 'bad·ə·rē ,chär·jər }

battery clip [ELEC] A terminal of a connecting wire having spring jaws that can be quickly snapped on a terminal of a device, such as a battery, to which a temporary wire connection is desired. { 'bad·ə·rē ,klip }

battery eliminator [ELECTR] A device which supplies electron tubes with voltage from electric power supply mains. { 'bad·ə·rē ə'lim·ə,nād·ər }

battery, overvoltage, ringing, supervision, coding, hybrid and test access See BORSCHT. { 'bad·ə·rē ¦ō·vər¦vōl·tij 'riŋ·iŋ ,sü·pər'vizh·ən 'kōd·iŋ 'hī·brid ən 'test ,ak,ses }

battery separator [ELEC] An insulating plate inserted between the positive and negative plates of a battery to prevent them from touching. { 'bad·ə·rē ,sep·ə,rād·ər }

baud [COMMUN] A unit of telegraph signaling speed equal to the number of code elements (pulses and spaces) per second or twice the number of pulses per second. { bȯd }

Baudot code [COMMUN] A teleprinter code that uses a combination of five or six marking and spacing intervals of equal duration for each character; no longer in extensive use since it has been replace by ASCII code. { bȯ'dō ,kōd }

bay [COMPUT SCI] See drive bay. [ELECTROMAG] One segment of an antenna array. { bā }

Bayard-Alpert ionization gage [ELECTR] A type of ionization vacuum gage using a tube with an electrode structure designed to minimize x-ray-induced electron emission from the ion collector. { ¦bā·ərd ¦al,pərt ī·ən·ə'zā·shən ,gāj }

bayonet base [ELEC] A tube base or lamp base having two projecting pins on opposite sides of a smooth cylindrical surface to engage in corresponding slots in a bayonet socket and hold the base firmly in the socket. { ¦bā·ə'net ,bās }

bayonet Neil-Concelman connector See BNC connector. { ,bā·ə'net 'nēl 'käns·əl·mən kə,nek·tər }

bazooka See balun. { bə'zü·kə }

B battery [ELECTR] The battery that furnishes required direct-current voltages to the plate and screen-grid electrodes of the electron tubes in a battery-operated circuit. { 'bē ,bad·ə·rē }

BBD See bucket brigade device.

B box See index register. { 'bē ,bäks }

BBS See bulletin board system.

BCAS See beacon collision avoidance system.

BCD system See binary coded decimal system. { ¦bē¦sē'dē ,sis·təm }

B-display [ELECTR] The presentation of radar output data in rectangular coordinates in which range and azimuth are plotted on the coordinate axes. Also known as B-indicator; B-scan; B-scope; range-bearing display. { 'bē dis'plā }

BDOS See basic disk operating system. { 'bē,dȯs }

beacon [ELECTR] A radio transmitter and antenna used to indicate its location or that of the vehicle carrying it; a beacon that responds to an interrogation, as in secondary radar, is more properly called a transponder. { 'bē·kən }

beacon collision avoidance system [NAV] An airborne collision avoidance system that makes use of the air-traffic control radio beacon system (ATCRBS) transponders. Abbreviated BCAS. { 'bē·kən kə'lizh·ən ə'vȯid·əns ,sis·təm }

beacon delay [ELECTR] The amount of transponding delay within a beacon, that is, the time between the arrival of a signal and the response of the beacon. { 'bē·kən di'lā }

beacon presentation [ELECTR] The radar display resulting from receipt of signals from a beacon. { 'bē·kən ,prē·zən'tā·shən }

beacon skipping [ELECTR] A condition where transponder return pulses from a beacon are missing at the interrogating radar. { 'bē·kən ,skip·iŋ }

beacon stealing [ELECTR] Loss of beacon tracking by one radar due to stronger signals from other beacons, transponders, or interfering radars. { 'bē·kən ,stēl·iŋ }

beacon tracking [ENG] The tracking of a moving object by means of signals emitted from a transmitter or transponder within or attached to the object. { 'bē·kən ,trak·iŋ }

beacon-tracking radar [NAV] Radar equipment used in air-traffic control facilities for beacon tracking. { 'bē·kən ,trak·iŋ ¦rä,där }

bead [COMPUT SCI] A small subroutine. [ELECTROMAG] A glass, ceramic, or plastic insulator through which passes the inner conductor of a coaxial transmission line and by means of which the inner conductor is supported in a position coaxial with the outer conductor. { bēd }

beaded transmission line [ELECTROMAG] Line using beads to support the inner conductor in coaxial transmission lines. { 'bēd·əd tranz'mish·ən ,līn }

bead thermistor [ELEC] A thermistor made by applying the semiconducting material to two wire leads as a viscous droplet, which cements the leads upon firing. { 'bēd thər'mis·tər }

beam angle See beam width. { 'bēm ,aŋ·gəl }

beam antenna [ELECTROMAG] An antenna that concentrates its radiation into a narrow beam in a definite direction. { 'bēm an'ten·ə }

beam approach beacon system See blind approach beacon system. { 'bēm ə'prōch 'bē·kən ,sis·təm }

beam blank See blank. { 'bēm blaŋk }

beam box See wall box. { 'bēm ,bäks }

beam coupling [ELECTR] The production of an alternating current in a circuit connected between two electrodes that are close to, or in the path of, a density-modulated electron beam. { 'bēm ,kəp·liŋ }

beam current [ELECTR] The electric current determined by the number and velocity of electrons in an electron beam. { 'bēm ,kər·ənt }

beam-deflection tube [ELECTR] An electron-beam tube in which the current to an output electrode is controlled by transversely moving the electron beam. { 'bēm di'flek·shən ,tüb }

beam efficiency [ELECTROMAG] The fraction of the total radiated energy from an antenna contained in a single beam. { 'bēm i,fish·ən·sē }

beam-forming electrode [ELECTR] Electron-beam focusing elements in power tetrodes and cathode-ray tubes. { 'bēm ,fȯrm·iŋ i'lek,trōd }

beamguide [ELECTROMAG] A set of elements arranged and spaced so as to form and conduct a beam of electromagnetic radiation. { 'bēm ,gīd }

beam holding [ELECTR] Use of a diffused beam of electrons to regenerate the charges stored on the screen of a cathode-ray storage tube. { 'bēm ,hōl·diŋ }

beam-indexing tube [ELECTR] A single-beam color television picture tube in which the color phosphor strips are arranged in groups of red, green, and blue. { 'bēm 'in,dek·siŋ ,tüb }

beam lead [ELECTR] A flat thick-film lead, sometimes of gold, deposited on a semiconductor chip chemically or by evaporation, as a connecting lead for a semiconductor device or integrated circuit. { 'bēm ,lēd }

beam lobe switching [ELECTR] Method of determining the direction of a remote object by comparison of the signals corresponding to two or more successive beam angles, differing slightly from the direction of the object. { 'bēm ,lōb ,swich·iŋ }

beam magnet See convergence magnet. { 'bēm ,mag·nət }

beam parametric amplifier [ELECTR] Parametric amplifier that uses a modulated electron beam to provide a variable reactance. { 'bēm ,par·ə'me·trik 'am·plə,fī·ər }

beam pattern See directivity pattern. { 'bēm ,pad·ərn }

beam power tube [ELECTR] A vacuum tube, most often an amplifier, used in radar and other microwave transmitters in which the electrons travel from the cathode in a well-focused beam, to interact with the electromagnetic signal being amplified. { 'bēm ¦paù·ər ,tüb }

beam recording [ELECTR] A method of using an electron beam to write data generated by a computer directly on microfilm. { 'bēm ri'kȯrd·iŋ }

beam splitting [ELECTR] Process for increasing angle accuracy in locating targets by radar by noting the azimuths at which one radar scan first discloses a target and at which the echoes cease, revealing the azimuth center, or by similarly intended algorithms in more automated systems. { 'bēm ,splid·iŋ }

beam spread [ENG] The angle of divergence from the central axis of an electromagnetic or acoustic beam as it travels through a material. { 'bēm ,spred }

beam steering [ELECTR] Changing the direction of the major lobe of a radiation pattern, usually by switching antenna elements. { 'bēm ,stir·iŋ }

beam switching [ELECTR] Method of obtaining more accurately the bearing or elevation of an object by comparing the signals received when the beam is in directions differing slightly in bearing or elevation; when these signals are equal, the object lies midway between the beam axes. Also known as lobe switching. { 'bēm ,swich·iŋ }

beam-switching tube [ELECTR] An electron tube which has a series of electrodes arranged around a central cathode and in which an electron beam is switched from one electrode to another. Also known as cyclophon. { 'bēm ,swich·iŋ ,tüb }

beam tetrode See beam power tube. { 'bēm 'te ,trōd }

beam width [ELECTROMAG] The angle, measured in a horizontal plane, between the directions at which the intensity of an electromagnetic beam, such as a radar or radio beam, is one-half its maximum value. Also known as beam angle. { 'bēm ,width }

bearing cursor [ENG] Of a radar set, the radial line inscribed on a transparent disk which can be rotated manually about an axis coincident with the center of the plan position indicator; used for bearing determination. Also known as mechanical bearing cursor. { 'ber·iŋ ,kər·sər }

bearing loss [ELEC] Loss of power in a machine caused by friction between the shaft and the bearing. { 'ber·iŋ ,lȯs }

bearing marker See azimuth marker. { 'ber·iŋ ,märk·ər }

bearing resolution [ELECTR] Minimum angular separation in a horizontal plane between two targets at the same range that will allow an operator to obtain data on either target. { 'ber·iŋ ,rez·ə ,lü·shən }

beat frequency [ELECTR] The frequency of a signal equal to the difference in frequencies of two signals which produce the signal when they are combined in a nonlinear circuit. { 'bēt ,frē·kwən,sē }

beat-frequency oscillator [ELECTR] An oscillator in which a desired signal frequency, such as an audio frequency, is obtained as the beat frequency produced by combining two different signal frequencies, such as two different radio frequencies. Abbreviated BFO. Also known as heterodyne oscillator. { 'bēt ,frē·kwən·sē 'äs·ə ,lād·ər }

beating-in [ELECTR] Interconnecting two transmitter oscillators and adjusting one until no beat frequency is heard in a connected receiver; the oscillators are then at the same frequency. { 'bēd·iŋ ¦in }

beat note [ELECTR] The beat frequency whose signal is produced by two signals having waves that are sinusoidal. { 'bēt ,nōt }

beat reception See heterodyne reception. { 'bēt ri'sep·shən }

beat-time programming [COMPUT SCI] A type of programming which requires that data be made available to the computer during some ongoing process prior to a particular point in time. { ¦bēt 'tīm 'prō,gram·iŋ }

beat tone [ENG ACOUS] Musical tone due to beats, produced by the heterodyning of two high-frequency wave trains. { 'bēt ,tōn }

beavertail |ELECTROMAG| Fan-shaped radar beam, wide in the horizontal plane and narrow in the vertical plane, which is swept up and down for height finding. { 'be·vər,tāl }

Beck effect |ELEC| An increase in the light intensity of an arc lamp whose carbon anode has been treated with rare-earth salts when a certain current is exceeded. { bek i'fekt }

Becquerel effect |ELEC| The phenomenon of a current flowing between two unequally illuminated electrodes of a certain type when they are immersed in an electrolyte. { ¦bek·ə¦rel *or* be'krel i'fekt }

bedspring array *See* billboard array. { 'bed,spriŋ ə'rā }

beetle *See* rammer. { 'bēd·əl }

BEGIN |COMPUT SCI| An enclosing statement of ALGOL used to indicate the beginning of a block; any variable in a block enclosed by BEGIN and END is normally local to this block. { bi'gin }

beginning-of-information marker |COMPUT SCI| A section of magnetic tape covered with reflective material that indicates the beginning of the area on which information is to be recorded. { bi'gin·iŋ əv ,in·fər'mā·shən ,mär·kər }

B eliminator |ELECTR| Power pack that changes the alternating-current powerline voltage to the direct-current source required by plant circuits of vacuum tubes or semiconductor devices. { 'bē i'lim·ə,nād·ər }

bell character |COMPUT SCI| A control character that activates a bell, alarm, or other audio device to get someone's attention. { 'bel ,kar·ik·tər }

bells and whistles |COMPUT SCI| Special hardware features that are likely to attract attention but may not be important or even practical. { 'belz ən 'wis·əlz }

bell transformer |ELEC| An iron-core, step-down transformer with a voltage step-down ratio of approximately 6 to 1 or 12 to 1, used in low-current power supplies and frequently in circuits for doorbells, alarm bells, and buzzers. { 'bel tranz,fór·mər }

bell wire |ELEC| A copper wire, usually solid rather than stranded, and soft-drawn rather than hard-drawn, used in low-current, low-voltage applications. { 'bel ,wīr }

belt printer |COMPUT SCI| A type of impact printer similar to a chain printer in which the characters are carried on a moving belt rather than a chain. { 'belt ,print·ər }

benchmark problem |COMPUT SCI| A problem to be run on computers to evaluate their performances relative to one another. { 'bench,märk ,präb·ləm }

benchmark test |COMPUT SCI| A test of computer software or hardware that is generally run on a number of products to compare their performance. { 'bench,märk ,test }

bender element |ELECTR| A combination of two thin strips of different piezoelectric materials bonded together so that when a voltage is applied, one strip increases in length and the other becomes shorter, causing the combination to bend. { 'ben·dər 'el·ə·mənt }

bent-pipe system |COMMUN| A transponder on board a communications satellite that performs no signal processing other than heterodyning (frequency-changing) the uplink frequency bands to those of the downlinks. { ,bent 'pīp ,sis·təm }

bergy-bit *See* growler. { 'bərg·ē ,bit }

beta |ELECTR| The current gain of a transistor that is connected as a grounded-emitter amplifier, expressed as the ratio of change in collector current to resulting change in base current, the collector voltage being constant. { 'bād·ə }

beta circuit |ELEC| The part of an amplifier circuit that is responsible for the feedback. { 'bād·ə ,sər·kət }

beta-cutoff frequency |ELECTR| The frequency at which the current amplification of an amplifier transistor drops to 3 decibels below its value at 1 kilohertz. { 'bād·ə 'kəd,óf ,frē·kwən·sē }

beta rule *See* reduction rule. { 'bād·ə ,rül }

beta software |COMPUT SCI| An application or program that is in development and undergoing testing. Also known as beta version; betaware. { ,bād·ə 'sóf,wer }

beta test |COMPUT SCI| The first test of a computer system outside the laboratory, in its actual working environment. { 'bād·ə ,test }

beta test site |COMPUT SCI| An organization or company that tests a software or hardware product under actual working conditions and reports the results to the vendor. { 'bād·ə ¦test ,sīt }

beta version *See* beta software. { 'bād·ə ,vər·zhən }

betaware *See* beta software. { 'bād·ə,wer }

Beverage antenna *See* wave antenna. { 'bev·rij an'ten·ə }

beyond-the-horizon communication *See* scatter propagation. { bə'yänd thə hə'rīz·ən kə ,myü·nə'kā·shən }

Bézier curve |COMPUT SCI| A curve in a drawing program that is defined mathematically, and whose shape can be altered by dragging either of its two interior determining points with a mouse. { ¦bāz·yā 'kərv }

BFL *See* buffered FET logic.

BFO *See* beat-frequency oscillator.

B-frames *See* bidirectional pictures. { 'bē ,frāmz }

B-H meter |ENG| A device used to measure the intrinsic hysteresis loop of a sample of magnetic material. { ¦bē¦āch ,mēd·ər }

Bi *See* abampere.

bias |ELEC| **1.** A direct-current voltage used on signaling or telegraph relays or electromagnets to secure desired time spacing of transitions from marking to spacing. **2.** The restraint of a relay armature by spring tension to secure a desired time spacing of transitions from marking to spacing. **3.** The effect on teleprinter signals produced by the electrical characteristics of the line and equipment. **4.** The force applied to a relay to hold it in a given position. |ELECTR| **1.** A direct-current voltage applied to a transistor control electrode to establish the desired operating point. **2.** *See* grid bias. { 'bī·əs }

bias cell |ELECTR| A small dry cell used singly or in series to provide the required negative bias for

bias current

the grid circuit of an electron tube. Also known as grid-bias cell. { 'bī·əs ,sel }

bias current |ELECTR| **1.** An alternating electric current above about 40,000 hertz added to the audio current being recorded on magnetic tape to reduce distortion. **2.** An electric current flowing through the base-emitter junction of a transistor and adjusted to set the operating point of the transistor. { 'bī·əs ,kər·ənt }

bias distortion |ELECTR| Distortion resulting from the operation on a nonlinear portion of the characteristic curve of a vacuum tube or other device, due to improper biasing. { 'bī·əs dis 'tȯr·shən }

biased automatic gain control See delayed automatic gain control. { 'bī·əst ȯd·ə'mad·ik 'gān kən,trōl }

bias meter |COMMUN| A meter used in teletypewriter work for measuring signal bias directly in percent; a positive reading indicates a marking signal bias; a negative reading, a spacing signal bias. { 'bī·əs ,mēd·ər }

bias oscillator |ELECTR| An oscillator used in a magnetic recorder to generate the alternating-current signal that is added to the audio current being recorded on magnetic tape to reduce distortion. { 'bī·əs ,äs·ə,lād·ər }

bias register |COMPUT SCI| A computer device that stores a number that is added to the memory address each time the computer memory is referenced by the program, thus offsetting the program addresses by a fixed amount. { 'bī·əs ,rej·ə·stər }

bias resistor |ELECTR| A resistor used in the cathode or grid circuit of an electron tube to provide a voltage drop that serves as the bias. { 'bī·əs ri'sis·tər }

bias voltage |ELECTR| A voltage applied or developed between two electrodes as a bias. { 'bī·əs ,vōl·tij }

bias winding |ELEC| A control winding that carries a steady direct current which serves to establish desired operating conditions in a magnetic amplifier or other magnetic device. { 'bī·əs ,wīn·diŋ }

BiCMOS technology |ELECTR| An integrated circuit technology that combines bipolar transistors and CMOS devices on the same chip. { ¦bī'sē ,mȯs tek,näl·ə·jē }

biconditional gate See equivalence gate. { ,bī·kən'dish·ən·əl 'gāt }

biconical antenna |ELECTROMAG| An antenna consisting of two metal cones having a common axis with their vertices coinciding or adjacent and with coaxial-cable or waveguide feed to the vertices. { bī'kän·ə·kəl an'ten·ə }

bidirectional |ENG| Being directionally responsive to inputs in opposite directions. { ,bī·də'rek·shən·əl }

bidirectional antenna |ELECTROMAG| An antenna that radiates or receives most of its energy in only two directions. { ,bī·də'rek·shən·əl an'ten·ə }

bidirectional clamping circuit |ELECTR| A clamping circuit that functions at the prescribed time irrespective of the polarity of the signal source at the time the pulses used to actuate the clamping action are applied. { ,bī·də'rek·shən·əl 'klam·piŋ ,sər·kət }

bidirectional clipping circuit |ELECTR| An electronic circuit that prevents transmission of the portion of an electrical signal that exceeds a prescribed maximum or minimum voltage value. { ,bī·də'rek·shən·əl 'klip·iŋ ,sər·kət }

bidirectional counter See forward-backward counter. { ,bī·də'rek·shən·əl 'kaun·tər }

bidirectional data bus |COMPUT SCI| A channel over which data can be transmitted in either direction within a computer system. { ,bī·də'rek·shən·əl 'dad·ə ,bəs }

bidirectional microphone |ENG ACOUS| A microphone that responds equally well to sounds reaching it from the front and rear, corresponding to sound incidences of 0 and 180°. { ,bī·də'rek·shən·əl 'mī·krə,fōn }

bidirectional parallel port |COMPUT SCI| A parallel port that can transfer data in both directions, and at speeds much greater than a standard parallel port. { ,bī·də,rek·shən·əl ,par·ə,lel 'pȯrt }

bidirectional pictures |COMMUN| In MPEG-2, pictures that use both future and past pictures as a reference. This technique is termed bidirectional prediction; bidirectional pictures provide the most compression and do not propagate coding errors as they are never used as a reference. Also known as B-frames; B-pictures. { ,bī·də'rek·shən·əl 'pik·chərz }

bidirectional printer |COMPUT SCI| A printer in which printing can be done in both a left-to-right and a right-to-left direction. { ,bī·də'rek·shən·əl 'print·ər }

bidirectional pulse-amplitude modulation See double-polarity pulse-amplitude modulation. { ,bī·də'rek·shən·əl ¦pəls ¦am·plə,tüd ,mäj·ə'lā·shən }

bidirectional transducer |ELECTR| A transducer capable of measuring in both positive and negative directions from a reference position. Also known as bilateral transducer. { ,bī·də'rek·shən·əl tranz'dü·sər }

bidirectional transistor |ELECTR| A transistor that provides switching action in either direction of signal flow through a circuit; widely used in telephone switching circuits. { ,bī·də'rek·shən·əl tran'zis·tər }

bidirectional triode thyristor |ELECTR| A gate-controlled semiconductor switch designed for alternating-current power control. { ,bī·də'rek·shən·əl 'trī,ōd thī'ris·tər }

bifilar electrometer |ENG| An electrostatic voltmeter in which two conducting quartz fibers, stretched by a small weight or spring, are separated by their attraction in opposite directions toward two plate electrodes carrying the voltage to be measured. { bī'fi·lər i·lek'träm·əd·ər }

bifilar resistor |ELEC| A resistor wound with a wire doubled back on itself to reduce the inductance. { bī'fi·lər ri'zis·tər }

bifilar transformer |ELEC| A transformer in which wires for the two windings are wound side by side to give extremely tight coupling. { bī'fi·lər tranz'fór·mər }

bifilar winding |ELEC| A winding consisting of two insulated wires, side by side, with currents traveling through them in opposite directions. { bī'fi·lər 'wīn·diŋ }

bifurcated contact |ELEC| A contact having a forked shape such that it can slide over and interlock with an identical mating contact. { 'bī·fər ‚kād·əd 'kän‚takt }

bigit See bit. { 'bij·ət }

big LEO system |COMMUN| A system of relatively large satellites in low earth orbit (LEO) to provide global mobile handheld telephony and other services. { ‚big 'lē·ō ‚sis·təm }

big M method |COMPUT SCI| A technique for solving linear programming problems in which artificial variables are assigned cost coefficients which are a very large number M, say, M = 10^{35}. { ‚big 'em ‚meth·əd }

bilateral |ELECTR| Having a voltage current characteristic curve that is symmetrical with respect to the origin. { bī'lad·ə·rəl }

bilateral amplifier |ELECTR| An amplifier capable of receiving as well as transmitting signals; used primarily in transceivers. { bī'lad·ə·rəl 'am·plə ‚fī·ər }

bilateral antenna |ELECTROMAG| An antenna having maximum response in exactly opposite directions, 180° apart, such as a loop. { bī'lad·ə·rəl an'ten·ə }

bilateral circuit |ELEC| Circuit wherein equipment at opposite ends is managed, operated, and maintained by different services. { bī'lad·ə·rəl 'sər·kət }

bilateral network |ELEC| A network or circuit in which the magnitude of the current remains the same when the voltage polarity is reversed. { bī'lad·ə·rəl 'net‚wərk }

bilateral transducer See bidirectional transducer. { bī'lad·ə·rəl tranz'dü·sər }

billboard array |ELECTROMAG| A broadside antenna array consisting of stacked dipoles spaced one-fourth to three-fourths wavelength apart in front of a large sheet-metal reflector. Also known as bedspring array; mattress array. { 'bil‚bórd ə'rā }

bimag core See bistable magnetic core. { 'bī‚mag ‚kòr }

bimorph cell |ELECTR| Two piezoelectric plates cemented together in such a way that an applied voltage causes one to expand and the other to contract so that the cell bends in proportion to the applied voltage; conversely, applied pressure generates double the voltage of a single cell; used in phonograph pickups and microphones. { 'bī ‚mórf ‚sel }

bin |COMPUT SCI| A magnetic-tape memory in which a number of tapes are stored in a single housing. { bin }

binary |COMPUT SCI| Possessing a property for which there exists two choices or conditions, one choice excluding the other. { 'bīn·ə·rē }

binary arithmetic operation |COMPUT SCI| An arithmetical operation in which the operands are in the form of binary numbers. Also known as binary operation. { 'bīn·ə·rē ‚ar·ith'med·ik äp·ə'rā·shən }

binary cell |COMPUT SCI| An elementary unit of computer storage that can have one or the other of two stable states and can thus store one bit of information. { 'bīn·ə·rē ‚sel }

binary chain |COMPUT SCI| A series of binary circuit elements so arranged that each can change the state of the one following it. { 'bīn·ə·rē ‚chān }

binary chop See binary search. { 'bīn·ə·rē 'chäp }

binary code |COMPUT SCI| A code in which each allowable position has one of two possible states, commonly 0 and 1; the binary number system is one of many binary codes. { 'bīn·ə·rē ‚kōd }

binary coded character |COMPUT SCI| One element of a notation system representing alphanumeric characters such as decimal digits, alphabetic letters, and punctuation marks by a predetermined configuration of consecutive binary digits. { 'bīn·ə·rē ‚kōd·əd 'kar·ik·tər }

binary coded decimal system |COMPUT SCI| A system of number representation in which each digit of a decimal number is represented by a binary number. Abbreviated BCD system. { 'bīn·ə·rē ‚kōd·əd 'des·məl ‚sis·təm }

binary coded decimal-to-decimal converter |COMPUT SCI| A computer circuit which selects one of ten outputs corresponding to a four-bit binary coded decimal input, placing it in the 0 state and the other nine outputs in the 1 state. { 'bīn·ə·rē ‚kōd·əd 'des·məl tə 'des·məl kən'vərd·ər }

binary coded octal system |COMPUT SCI| Octal numbering system in which each octal digit is represented by a three-place binary number. { 'bīn·ə·rē ‚kōd·əd 'äk·təl ‚sis·təm }

binary component |ELECTR| An electronic component that can be in either of two conditions at any given time. Also known as binary device. { 'bīn·ə·rē kəm'pō·nənt }

binary conversion |COMPUT SCI| Converting a number written in binary notation to a number system with another base, such as decimal, octal, or hexadecimal. { 'bīn·ə·rē kən'vər·zhən }

binary counter See binary scaler. { 'bīn·ə·rē 'kaùnt·ər }

binary decision |COMPUT SCI| A decision between only two alternatives. { 'bīn·ə·rē di'sizh·ən }

binary device See binary component. { 'bīn·ə·rē di'vīs }

binary digit See bit. { 'bīn·ə·rē 'dij·ət }

binary dump |COMPUT SCI| The operation of copying the contents of a computer memory in binary form onto an external storage device. { 'bīn·ə·rē ‚dəmp }

binary encoder |ELECTR| An encoder that changes angular, linear, or other forms of input data into binary coded output characters. { 'bīn·ə·rē en'kōd·ər }

binary field |COMPUT SCI| A field that contains data in the form of binary numbers. { 'bīn·ə·rē 'fēld }

51

binary file |COMPUT SCI| A computer program in machine language that can be directly executed by the computer. { 'bīn·ə·rē 'fīl }

binary incremental representation |COMPUT SCI| A type of incremental representation in which the value of change in a variable is represented by one binary digit which is set equal to 1 if there is an increase in the variable and to 0 if there is a decrease. { 'bīn·ə·rē ‚iŋ·krə'men·təl ‚rep·ri‚zen'tā·shən }

binary large object |COMPUT SCI| In a database management system, a file-storage system used most often for multimedia files (large files). Abbreviated BLOB. { ¦bīn·ə·rē ¦lärj 'äb‚jekt }

binary loader |COMPUT SCI| A computer program which transfers to main memory an exact image of the binary pattern of a program held in a storage or input device. { 'bīn·ə·rē ¦lōd·ər }

binary logic |ELECTR| An assembly of digital logic elements which operate with two distinct states. { 'bīn·ə·rē 'läj·ik }

binary operation See binary arithmetic operation. { 'bīn·ə·rē äp·ə'rā·shən }

binary phase-shift keying |COMMUN| Keying of binary data or Morse code dots and dashes by ±90° phase deviation of the carrier. Abbreviated BPSK. { 'bīn·ə·rē 'fāz ‚shift 'kē·iŋ }

binary point |COMPUT SCI| The character, or the location of an implied symbol, that separates the integral part of a numerical expression from its fractional part in binary notation. { 'bīn·ə·rē 'pȯint }

binary scaler |ELECTR| A scaler that produces one output pulse for every two input pulses. Also known as binary counter; scale-of-two circuit. { 'bīn·ə·rē ¦skā·lər }

binary search |COMPUT SCI| A dichotomizing search in which the set of items to be searched is divided at each step into two equal, or nearly equal, parts. Also known as binary chop. { 'bīn·ə·rē 'sərch }

binary signal |ELECTR| A voltage or current which carries information by varying between two possible values, corresponding to 0 and 1 in the binary system. { 'bīn·ə·rē 'sig·nəl }

binary system |ENG| Any system containing two principal components. { 'bīn·ə·rē 'sis·təm }

binary word |COMPUT SCI| A group of bits which occupies one storage address and is treated by the computer as a unit. { 'bīn·ə·rē ¦wərd }

B-indicator See B-display. { ¦bē ¦in·də‚kād·ər }

binding post |ELEC| A manually turned screw terminal used for making electrical connections. { 'bīn·diŋ ‚pōst }

binding time |COMPUT SCI| **1.** The instant when a symbolic expression in a computer program is reduced to a form which is directly interpretable by the hardware. **2.** The instant when a variable is assigned its data type, such as integer or string. { 'bīn·diŋ ‚tīm }

binistor |ELECTR| A silicon *npn* tetrode that serves as a bistable negative-resistance device. { ‚bī'nis·tər }

binode |ELECTR| An electron tube with two anodes and one cathode used as a full-wave rectifier. Also known as double diode. { 'bī‚nōd }

binomial array antenna |ELECTROMAG| Directional antenna array for reducing minor lobes and providing maximum response in two opposite directions. { bī'nō·mē·əl ə'rā an'ten·ə }

biochemical fuel cell |ELEC| An electrochemical power generator in which the fuel source is bioorganic matter; air is the oxidant at the cathode, and microorganisms catalyze the oxidation of the bioorganic matter at the anode. { ¦bī·ō'kem·ə·kəl 'fyül ‚sel }

biochip |ELECTR| An experimental type of integrated circuit whose basic components are organic molecules. { 'bī·ō‚chip }

bioinformatics |COMPUT SCI| The use of computers to study biological systems. { ‚bī·ō‚in·fər'mad·iks }

bioinstrumentation |ENG| The use of instruments attached to animals and humans to record biological parameters such as breathing rate, pulse rate, body temperature, or oxygen in the blood. { ¦bī·ō‚in·strə·mən'tā·shən }

biomedical engineering |ENG| The application of engineering technology to the solution of medical problems; examples are the development of prostheses such as artificial valves for the heart, various types of sensors for the blind, and automated artificial limbs. { ‚bī·ō'med·ə·kəl ‚en·jə'nir·iŋ }

biometric device |COMPUT SCI| A device that identifies persons seeking access to a computing system by determining their physical characteristics through fingerprints, voice recognition, retina patterns, pictures, weight, or other means. { ‚bī·ō¦me·trik di¦vīs }

bionics |ENG| The study of systems, particularly electronic systems, which function after the manner of living systems. { bī'än·iks }

BIOS See basic input/output system.

biot See abampere. { 'bī·ät }

biotechnical robot |CONT SYS| A robot that requires the presence of a human operator in order to function. { ¦bī·ō¦tek·nə·kəl 'rō‚bät }

biotelemetry |ENG| The use of telemetry techniques, especially radio waves, to study behavior and physiology of living things. { ¦bī·ō·tə'lem·ə·trē }

bipolar amplifier |ELECTR| An amplifier capable of supplying a pair of output signals corresponding to the positive or negative polarity of the input signal. { bī'pō·lər 'am·plə‚fī·ər }

bipolar circuit |ELECTR| A logic circuit in which zeros and ones are treated in a symmetric or bipolar manner, rather than by the presence or absence of a signal; for example, a balanced arrangement in a square-loop-ferrite magnetic circuit. { bī'pō·lər 'sər·kət }

bipolar electrode |ELEC| Electrode, without metallic connection with the current supply, one face of which acts as anode surface and the opposite face as a cathode surface when an electric current is passed through a cell. { bī'pō·lər i'lek‚trōd }

bipolar format |COMPUT SCI| A method of representing binary data in which 0 bits have zero voltage and each 1 bit has a polarity opposite that of the preceding 1 bit. { bī'pō·lər 'fór,mat }

bipolar integrated circuit |ELECTR| An integrated circuit in which the principal element is the bipolar junction transistor. { bī'pō·lər 'in·tə ‚grād·əd 'sər·kət }

bipolar junction transistor |ELECTR| A bipolar transistor that is composed entirely of one type of semiconductor, silicon. Abbreviated BJT. Also known as silicon homojunction. { ¦bī,pōl·ər ‚jəŋk·shən tran'zis·tər }

bipolar magnetic driving unit |ENG ACOUS| Headphone or loudspeaker unit having two magnetic poles acting directly on a flexible iron diaphragm. { bī'pō·lər mag'ned·ik 'driv·iŋ ‚yü·nət }

bipolar memory |COMPUT SCI| A computer memory employing integrated-circuit bipolar junction transistors as bistable memory cells. { bī'pō·lər 'mem·rē }

bipolar power supply |ELEC| A high-precision, regulated, direct-current power supply that can be set to provide any desired voltage between positive and negative design limits, with a smooth transition from one polarity to the other. { bī'pō·lər 'paú·ər sə'plī }

bipolar signal |COMMUN| A signal in which different logical states are represented by electrical voltages of opposite polarity. { bī'pō·lər 'sig·nəl }

bipolar spin device See magnetic switch. { ¦bī ‚pō·lər 'spin di,vīs }

bipolar spin switch See magnetic switch. { ¦bī ‚pō·lər 'spin ‚swich }

bipolar transistor |ELECTR| A transistor that uses both positive and negative charge carriers. { bī'pō·lər tranz'is·tər }

bipolar video See coherent video. { bī'pō·lər 'vid·ē·ō }

bipotential electrostatic lens |ELECTR| An electron lens in which image and object space are field-free, but at different potentials; examples are the lenses formed between apertures of cylinders at different potentials. Also known as immersion electrostatic lens. { ¦bī·pə'ten·chəl i ‚lek·trə'stad·ik 'lenz }

biquartic filter |ELECTR| An active filter that uses operational amplifiers in combination with resistors and capacitors to provide infinite values of Q and simple adjustments for band-pass and center frequency. { ¦bī¦kwórd·ik 'fil·tər }

birefringence |OPTICS| 1. Splitting of a light beam into two components, which travel at different velocities, by a material. 2. For a light beam that has been split into two components by a material, the difference in the indices of refraction of the components within the material. Also known as double refraction. { ‚bī·ri'frin·jəns }

biscuit See preform. { 'bis·kət }

bistable circuit |ELECTR| A circuit with two stable states such that the transition between the states cannot be accomplished by self-triggering. { ¦bī ¦stā·bəl ‚sar·kət }

bistable magnetic core |ELECTR| A magnetic core that can be in either of two possible states of magnetization. Also known as bimag core. { ¦bī ¦stā·bəl mag'ned·ik 'kòr }

bistable multivibrator |ELECTR| A multivibrator in which either of the two active devices may remain conducting, with the other nonconducting, until the application of an external pulse. Also known as Eccles-Jordan circuit; Eccles-Jordan multivibrator; flip-flop circuit; trigger circuit. { ¦bī¦stā·bəl məl·ti'vī,brād·ər }

bistable optical device |OPTICS| A device which can be in either of two stable states of optical transmission for a single value of the input light intensity. { ¦bī¦stā·bəl 'äp·tə·kəl di'vīs }

bistable unit |ENG| A physical element that can be made to assume either of two stable states; a binary cell is an example. { ¦bī¦stā·bəl 'yü·nət }

bistatic radar |ENG| Radar in which the transmitter and receiver are not located in the same place; the line between their positions is called the baseline. { 'bī,stad·ik 'rā,där }

bisynchronous transmission |COMMUN| A set of procedures for handling synchronous transmission of data and, in particular, for handling a block of data, called a message format, that is transmitted in a single operation. { bī'siŋ·krə·nəs tranz'mish·ən }

bit |COMPUT SCI| 1. A unit of information content equal to one binary decision or the designation of one of two possible and equally likely values or states of anything used to store or convey information. 2. A dimensionless unit of storage capacity specifying that the capacity of a storage device is expressed by the logarithm to the base 2 of the number of possible states of the device. { bit }

bit block transfer |COMPUT SCI| In computer graphics, a hardware function that moves a rectangular block of bits from the main memory to the display memory at high speed. Abbreviated bitblt. { ¦bit ‚bläk 'tranz·fər }

bitblt See bit block transfer.

bit buffer unit |COMMUN| A unit that terminates bit-serial communications lines coming from and going to technical control. { ¦bit 'bəf·ər,yü·nət }

bit cone See roller cone bit. { 'bit ‚kōn }

bit count appendage |COMPUT SCI| One of the two-byte elements replacing the parity bit stripped off each byte transferred from main storage to disk volume (the other element is the cyclic check); these two elements are appended to the block during the write operation; on a subsequent read operation these elements are calculated and compared to the appended elements for accuracy. { 'bit ,kaúnt ə'pen·dij }

bit density |COMPUT SCI| Number of bits which can be placed, per unit length, area, or volume, on a storage medium; for example, bits per inch of magnetic tape. Also known as record density. { 'bit 'den·səd·ē }

bit depth |COMPUT SCI| In a digital file, the number of colors for an image; calculated as 2 to the power of the bit depth; for example, a bit depth of

8 supports up to 256 colors, and a bit depth of 24 supports up to 16 million colors. { 'bit ˌdepth }

bit flipping *See* bit manipulation. { 'bit ˌflip·iŋ }

bit location |COMPUT SCI| Storage position on a record capable of storing one bit. { 'bit lō'kā·shən }

bit manipulation |COMPUT SCI| Changing bits from one state to the other, usually to influence the operation of a computer program. Also known as bit flipping. { 'bit mə,nip·yə'lā,shən }

bit-mapped font |COMPUT SCI| A font that is specified by a complete set of dot patterns for each character and symbol. { ¦bit ,mapt 'fänt }

bit-mapped graphics *See* raster graphics. { ¦bit ,mapt 'graf·iks }

bit mapping |COMPUT SCI| The assignment of each location in a computer's storage to a physical location on an electronic display. { 'bit 'map·iŋ }

bit-oriented protocol |COMMUN| A communications protocol in which individual bits within a byte are used as control codes. { ¦bit ,ȯr·ē ,ent·əd 'prōd·ə,kȯl }

bit pattern |COMPUT SCI| A combination of binary digits arranged in a sequence. { 'bit ,pad·ərn }

bit per second |COMMUN| A unit specifying the instantaneous speed at which a device or channel transmits data. Abbreviated bps. { 'bit pər 'sek·ənd }

bit position |COMPUT SCI| The position of a binary digit in a word, generally numbered from the least significant bit. { 'bit pa'zish·ən }

bit rate |COMMUN| Quantity, per unit time, of binary digits (or pulses representing them) which will pass a given point on a communications line or channel in a continuous stream. { 'bit ,rāt }

bit serial |COMMUN| Sequential transmission of character-forming bits. { ¦bit 'sir·ē·əl }

bit-sliced microprocessor |COMPUT SCI| A microprocessor in which the major logic of the central processor is partitioned into a set of large-scale-integration circuits, as opposed to being placed on a single chip. { 'bit ,slīst ,mī·krō'präs·əs·ər }

bit stream |COMPUT SCI| **1.** A consecutive line of bits transmitted over a circuit in a transmission method in which character separation is accomplished by the terminal equipment. **2.** A binary signal without regard to grouping by character. { 'bit ,strēm }

bit-stream generator |COMMUN| An algorithmic procedure for producing an unending sequence of binary digits to implement a stream. { 'bit ,strēm 'jen·ə,rād·ər }

bit string |COMPUT SCI| A set of consecutive binary digits representing data in coded form, in which the significance of each bit is determined by its position in the sequence and its relation to the other bits. { 'bit ,striŋ }

bit stuffing |COMMUN| The insertion of extra bits in a transmitted message in order to fill a frame to a fixed size or to break up a pattern of bits that could be mistaken for control codes. { 'bit ,stəf·iŋ }

bit synchronization |COMMUN| Element of a message header used to synchronize all of the bits and characters that follow. { 'bit ,siŋ·krə·nə'zā·shən }

bit test |COMPUT SCI| A check by a computer program to determine the status of a particular bit. { 'bit ,test }

bit zone |COMPUT SCI| **1.** One of the two left-most bits in a commonly used system in which six bits are used for each character; related to overpunch. **2.** Any bit in a group of bit positions that are used to indicate a specific class of items; for example, numbers, letters, special signs, and commands. { 'bit ,zōn }

BJT *See* bipolar junction transistor.

BL *See* base line.

black *See* black signal. { blak }

black-and-white television *See* monochrome television. { ¦blak ən ¦wīt 'tel·ə,vizh·ən }

black box |ENG| Any component, usually electronic and having known input and output, that can be readily inserted into or removed from a specific place in a larger system without knowledge of the component's detailed internal structure. { 'blak ,bäks }

blacker-than-black level |COMMUN| In television, a level of greater instantaneous amplitude than the black level, used for synchronization and control signals. { 'blak·ər ¦than 'blak ,lev·əl }

black hole *See* stale link. { ¦blak 'hōl }

black level |ELECTR| The level of the television picture signal corresponding to the maximum limit of black peaks. { 'blak ,lev·əl }

blackout *See* radio blackout. { 'blak,aut }

black peak |COMMUN| A peak excursion of the television picture signal in the black direction. { 'blak ,pēk }

black scope |ELECTR| Cathode-ray tube operating at the threshold of luminescence when no video signals are being applied. { ¦blak 'skōp }

black signal |COMMUN| Signal at any point in a facsimile system produced by the scanning of a maximum density area of the subject copy. Also known as black; picture black. { 'blak ,sig·nəl }

black-surface field |ELECTR| A layer of p^+ material which is applied to the back surface of a solar cell to reduce hole-electron recombinations there and thereby increase the cell's efficiency. { 'blak ,sər·fəs ,fēld }

black transmission |COMMUN| The amplitude-modulated transmission of facsimile signals in which the maximum signal amplitude corresponds to the greatest copy density or darkest shade. { 'blak tranz'mish·ən }

blade |ELEC| A flat moving conductor in a switch. { blād }

blank |ELECTR| To cut off the electron beam of a television picture tube or camera tube during the process of retrace by applying a rectangular pulse voltage to the grid or cathode during each retrace interval. Also known as beam blank. { blaŋk }

blank cell |COMPUT SCI| A cell of a spreadsheet that contains no text or numeric values, and for which no formatting is specified other than the global formats of the spreadsheet. { 'blaŋk ,sel }

blank character |COMPUT SCI| A character, either printed or appearing as a blank, used to denote a blank space among printed characters. Also known as space character. { 'blaŋk 'kar·ik·tər }

blanketing |COMMUN| Interference due to a nearby transmitter whose signals are so strong that they override other signals over a wide band of frequencies. { 'blaŋ·kəd·iŋ }

blank form See blank medium. { 'blaŋk ‚fòrm }

blanking |ELECTR| The act, useful in adapting a radar to its environment, of disabling selected apparatus at specified times or of deleting certain data from further treatment. { 'blaŋk·iŋ }

blanking circuit |ELECTR| A circuit preventing the transmission of brightness variations during the horizontal and vertical retrace intervals in television scanning. { 'blaŋk·iŋ ‚sər·kət }

blanking level |ELECTR| The level that separates picture information from synchronizing information in a composite television picture signal; coincides with the level of the base of the synchronizing pulses. Also known as pedestal; pedestal level. { 'blaŋk·iŋ ‚lev·əl }

blanking pulse |ELECTR| A control pulse used to switch off a part of a television or radar set electronically for a predetermined length of time. { 'blaŋk·iŋ ‚pəls }

blanking signal |ELECTR| The signal rendering the return trace invisible on the picture tube of a television receiver. { 'blaŋk·iŋ ‚sig·nəl }

blanking time |ELECTR| The length of time that the electron beam of a cathode-ray tube is shut off. { 'blaŋk·iŋ ‚tīm }

blank medium |COMPUT SCI| An empty position on the medium concerned, such as a column without holes on a punch tape, used to indicate a blank character. Also known as blank form. { ¦blaŋk 'mēd·ē·əm }

blank tape |COMPUT SCI| A portion of a paper tape having sprocket holes only, to indicate a blank character. { ¦blaŋk 'tāp }

blank tape halting problem |COMPUT SCI| The problem of finding an algorithm that, for any Turing machine, decides whether the machine eventually stops if it started on an empty tape; it has been proved that no such algorithm exists. { ¦blaŋk 'tāp 'hòl·tiŋ ‚präb·ləm }

blast |COMPUT SCI| To release internal or external memory areas from the control of a computer program in the course of dynamic storage allocation, making these areas available for reallocation to other programs. { blast }

blast freezer |ENG| An upright freezer in which very cold air circulated by blowers is used for rapid freezing of food. { 'blast ‚frē·zər }

bleed |COMPUT SCI| In optical character recognition, the flow of ink in printed characters beyond the limits specified for their recognition by a character reader. { blēd }

bleeder |ELECTR| A high resistance connected across the dc output of a high-voltage power supply which serves to discharge the filter capacitors after the power supply has been turned off, and to provide a stabilizing load. { 'blēd·ər }

bleeder current |ELEC| Current drawn continuously from a voltage source to lessen the effect of load changes or to provide a voltage drop across a resistor. { 'blēd·ər ‚kər·ənt }

bleeder resistor |ELEC| A resistor connected across a power pack or other voltage source to improve voltage regulation by drawing a fixed current value continuously; also used to dissipate the charge remaining in filter capacitors when equipment is turned off. { 'blēd·ər ri'zis·tər }

blended data |ENG| Q point that is the combination of scan data and track data to form a vector. { ¦blen·dəd 'dad·ə }

blend to analog |COMMUN| The point at which the block error rate of an AM/FM IBOC receiver falls below some predefined threshold and the digital audio is faded out while simultaneously the analog audio is faded in, preventing the received audio from simply muting when the digital signal is lost. The receiver audio will also blend to digital upon reacquisition of the digital signal. { ¦blend tə 'an·əl‚äg }

blend to mono |COMMUN| The process of progressively attenuating the left-right component of a stereo decoded signal as the received radio frequency signal decreases, with the net result of lowering the audible noise. { ¦blend tə 'män·ō }

BLER See block error rate.

blind approach beacon system |NAV| A pulse-type, ground-based navigation beacon used for runway approach at airports, which sends out signals that produce range and runway position information on the L-scan cathode-ray indicator of an aircraft making an instrument approach. Also known as beam approach beacon system (British usage). Abbreviated babs. { ¦blīnd ə'prōch 'bē·kən ‚sis·təm }

blind controller system |CONT SYS| A process control arrangement that separates the in-plant measuring points (for example, pressure, temperature, and flow rate) and control points (for example, a valve actuator) from the recorder or indicator at the central control panel. { ¦blīnd kən'trōl·ər ‚sis·təm }

blind drilling |ENG| Drilling in which the drilling fluid is not returned to the surface. { 'blīnd 'dril·iŋ }

blind flange |ENG| A flange used to close the end of a pipe. { ¦blīnd 'flanj }

blind hole |ENG| A hole which does not pass completely through a workpiece. |ENG| A type of borehole that does not have the drilling mud or other circulating medium carry the cuttings to the surface. { ¦blīnd 'hōl }

blinding |ENG| **1.** A thin layer of lean concrete, fine gravel, or sand that is applied to a surface to smooth over voids in order to provide a cleaner, drier, or more durable finish. **2.** A layer of small rock chips applied over the surface of a freshly tarred road. **3.** See blanking. { 'blīn·diŋ }

blind joint |ENG| A joint which is not visible from any angle. { ¦blīnd 'joint }

blind spot

blind spot [ENG] An area on a filter screen where no filtering occurs. Also known as dead area. { 'blīnd ,spät }

blind zone [COMMUN] Area from which echoes cannot be received; generally, an area shielded from the transmitter by some natural obstruction and therefore from which there can be no return. { 'blīnd ,zōn }

B line *See* index register. { 'bē ,līn }

blinking [COMMUN] Method of providing information in pulse systems by modifying the signal at its source so that signal presentation on the display scope alternately appears and disappears; in loran, this indicates that a station is malfunctioning. [ELECTR] Electronic-attack technique employed by two aircraft separated by a short distance and not resolved in azimuth so as to appear as one target to a tracking radar; the two aircraft alternately spot-jam, causing the radar system to oscillate from one place to another, greatly degrading the fire-control accuracy. [NAV] Regular shifting right and left or alternate appearance and disappearance of a loran signal to indicate that the signals of a pair of stations are out of synchronization. { 'bliŋ·kiŋ }

blip [ELECTR] The display of a received pulse on the screen of a cathode-ray tube. Also known as pip. { blip }

blip-scan ratio [ELECTR] The ratio of the number of times a target is detected (a contact generated, or a display clearly evident) to the number of times of opportunities to do so provided by the radar routine; provides a rough estimate of the probability of detection occurring during the detection process. { 'blip ,skan 'rā·shō }

bloatware *See* fatware. { 'blōt,wer }

BLOB *See* binary large object. { bläb *or* 'bē¦el ¦ō'bē }

block [COMMUN] An 8-by-8 array of pel values or discrete cosine transform coefficients representing luminance or chrominance information. [COMPUT SCI] A group of information units (such as records, words, characters, or digits) that are transported or considered as a single unit by virtue of their being stored in successive storage locations; for example, a group of logical records constituting a physical record. { bläk }

block body [COMPUT SCI] A list of statements that follows the block head in a computer program with block structure. { 'bläk ,bäd·ē }

block chaining *See* chained block encryption. { 'bläk ,chān·iŋ }

block check character [COMMUN] A character that is added to a block of data to check its accuracy, and consists of parity bits each of which is set by observing a specified set of bits in the block. { 'bläk ¦chek ,kar·ik·tər }

block cipher [COMMUN] A cipher that transforms a string of input bits of fixed length into a string of output bits of fixed length. { 'bläk ,sī·fər }

block code [COMMUN] An error-correcting code generated by an encoder that produces a fixed-length code word with each incoming fixed-length message block. { 'bläk ,kōd }

block data [COMPUT SCI] A statement in FOR-TRAN which declares that the program following is a data specification subprogram. { 'bläk ,dad·ə }

block diagram [ENG] A diagram in which the essential units of any system are drawn in the form of rectangles or blocks and their relation to each other is indicated by appropriate connecting lines. { 'bläk ,dī·ə,gram }

blocked F-format data set *See* FB data set. { 'bläkt ¦ef'fòr,mat 'dad·ə ,set }

blocked impedance [ELEC] The impedance at the input of a transducer when the impedance of the output system is made infinite, as by blocking or clamping the mechanical system. { 'bläkt im'pēd·əns }

blocked impurity band detector [ELECTR] A detector of long-wavelength infrared radiation consisting of a heavily doped extrinsic photoconductor on which an undoped intrinsic layer is grown epitaxially to prevent dark current from flowing in the impurity band. { 'bläkt im'pyür·əd·ē ¦band di,tek·tər }

blocked process [COMPUT SCI] A program that is running on a computer but is temporarily prevented from making progress because it requires some resource (such as a printer or user input) that is not immediately available. { ¦bläkt 'prä,ses }

blocked resistance [ENG ACOUS] Resistance of an audio-frequency transducer when its moving elements are blocked so they cannot move; represents the resistance due only to electrical losses. { 'bläkt ri'zis·təns }

block encryption [COMMUN] The use of a block cipher, usually employing the data encryption standard (DES), in which each 64-bit block of data is enciphered or deciphered separately, and every bit in a given output block depends on every bit in its respective input block and on every bit in the key, but on no other bits. Also known as electronic codebook mode (ECB). { 'bläk en'krip·shən }

block error rate [COMMUN] A ratio of the number of data blocks received with at least one uncorrectable bit to the total number of blocks received. Abbreviated BLER. { 'bläk 'er·ər ,rāt }

blockette [COMPUT SCI] A subdivision of a group of consecutive machine words transferred as a unit, particularly with reference to input and output. { 'blä'ket }

block head [COMPUT SCI] A list of declarations at the beginning of a computer program with block structure. { 'bläk ¦hed }

block identifier [COMPUT SCI] A means of identifying an area of storage in FORTRAN so that this area may be shared by a program and its subprograms. { 'bläk ī'den·tə,fī·ər }

block ignore character [COMPUT SCI] A character associated with a block which indicates the presence of errors in the block. { 'bläk ig'nòr ,kar·ik·tər }

blocking [COMPUT SCI] Combining two or more computer records into one block. [ELECTR] **1.** Applying a high negative bias to the grid of

56

an electron tube to reduce its anode current to zero. **2.** Overloading a receiver by an unwanted signal so that the automatic gain control reduces the response to a desired signal. **3.** Distortion occurring in a resistance-capacitance-coupled electron tube amplifier stage when grid current flows in the following tube. { 'bläk·iŋ }

blocking capacitor *See* coupling capacitor. { 'bläk·iŋ kə'pas·əd·ər }

blocking factor [COMPUT SCI] The largest possible number of records of a given size that can be contained within a single block. { 'bläk·iŋ ,fak·tər }

blocking layer *See* depletion layer. { 'bläk·iŋ ,lā·ər }

blocking oscillator [ELECTR] A relaxation oscillator that generates a short-time-duration pulse by using a single transistor or electron tube and associated circuitry. Also known as squegger; squegging oscillator. { 'bläk·iŋ 'äs·ə,lād·ər }

blocking oscillator driver [ELECTR] Circuit which develops a square pulse used to drive the modulator tubes, and which usually contains a line-controlled blocking oscillator that shapes the pulse into the square wave. { 'bläk·iŋ 'äs·ə ,lād·ər 'drī·vər }

block input [COMPUT SCI] **1.** A block of computer words considered as a unit and intended or destined to be transferred from an internal storage medium to an external destination. **2.** *See* output area. { 'bläk 'in,pút }

block length [COMPUT SCI] The total number of records, words, or characters contained in one block. { 'bläk ,leŋkth }

block loading [COMPUT SCI] A program loading technique in which the control sections of a program or program segment are loaded into contiguous positions in main memory. { 'bläk ,lōd·iŋ }

block mark [COMPUT SCI] A special character that indicates the end of a block. { 'bläk ,märk }

block move *See* cut and paste. { ¦bläk 'müv }

block multiplexor channel [COMPUT SCI] A transmission channel in a computer system that can simultaneously transmit blocks of data from several high-speed input/output devices by interleaving the data. { 'bläk ¦məlt·i,plek·sər ,chan·əl }

block operation [COMPUT SCI] An editing or formatting procedure that is carried out on a selected block of text in a word-processing document. { 'bläk ,äp·ə¦rā·shən }

block parity [COMMUN] An error-checking technique involving the comparison of a transmitted block check character with one calculated by the receiving device. { 'bläk 'par·əd·ē }

block protection [COMPUT SCI] An instruction in a word-processing or page-layout program that prevents a soft page break from being inserted in a specified block of text, ensuring against a bad page break. { 'bläk prə,tek·shən }

block protector [ELEC] Rectangular piece of carbon, bakelite with a metal insert, or porcelain with a carbon insert which, in combination with

each other, make one element of a protector; they form a gap which will break down and provide a path to ground for excessive voltages. { 'bläk prə'tek·tər }

block signal system [CONT SYS] An automatic railroad traffic control system in which the track is sectionalized into electrical circuits to detect the presence of trains, engines, or cars. { 'bläk 'sig·nəl ,sis·təm }

block standby [COMPUT SCI] Locations always set aside in storage for communication with buffers in order to make more efficient use of such buffers. { 'bläk ,stand,bī }

block structure [COMPUT SCI] In computer programming, a conceptual tool used to group sequences of statements into single compound statements and to allow the programmer explicit control over the scope of the program variables. { 'bläk ,strək·chər }

block transfer [COMPUT SCI] The movement of data in blocks instead of by individual records. { 'bläk ¦trans·fər }

blooming [ELECTR] **1.** Defocusing of television picture areas where excessive brightness results in enlargement of spot size and halation of the fluorescent screen. **2.** An increase in radar display spot size due to a particularly strong signal exciting the phosphorus material. **3.** The wide spatial dispersion of chaff after being dispensed in small bundles. { 'blüm·iŋ }

blow [COMPUT SCI] To write data or code into a programmable read-only memory chip by melting the fuse links corresponding to bits that are to be zero. { blō }

blow-lifting gripper [CONT SYS] A robot component that uses compressed air to lift objects. { 'blō ¦lift·iŋ ,grip·ər }

blown-fuse indicator [ELEC] A neon warning light connected across a fuse so that it lights when the fuse is blown. { ¦blōn ¦fyüz 'in·də ,kād·ər }

blowout [ELEC] The melting of an electric fuse because of excessive current. { 'blō,aút }

blow up *See* abend. { 'blō ,əp }

blue glow [ELECTR] A glow normally seen in electron tubes containing mercury vapor, due to ionization of the mercury molecules. { 'blü ,glō }

Bluetooth [COMMUN] A technical specification for the wireless connection over short distances of digital devices, such as cellular telephones, portable computers, and computer peripheral equipment, utilizing the unlicensed 2.4-GHz radio frequency spectrum. { 'blü,tüth }

BNC connector [ELEC] A small device for connecting coaxial cables, used frequently in low-power, radio-frequency and test applications. Abbreviation for bayonet Neil-Concelman connector. { ,bē,en'sē kə,nek·tər }

BNF *See* Backus-Naur form.

Board of Trade unit *See* kilowatt-hour. { ¦bórd əv 'trād ,yü·nət }

bobbing [ELECTR] Fluctuation of the strength of a radar echoand its display, due to alternate constructive and destructive interference of the received signal as in a multipath propagation situation.   { 'bäb·iŋ }

bobtail curtain antenna   [ELECTROMAG] A bidirectional, vertically polarized, phased-array antenna that has two horizontal sections, each 0.5 electrical wavelength long, that connect three vertical sections, each 0.25 electrical wavelength long.   { 'bäb,tāl 'kərt·ən an,ten·ə }

Bode diagram   [ELECTR] A diagram in which the phase shift or the gain of an amplifier, a servomechanism, or other device is plotted against frequency to show frequency response; logarithmic scales are customarily used for gain and frequency.   { 'bōd ,dī·ə,gram }

body capacitance   [ELEC] Capacitance existing between the human hand or body and a circuit.   { 'bäd·ē kə'pas·ə·təns }

body rotation   [CONT SYS] An axis of motion of a pick-and-place robot.   { 'bäd·e rō,tā·shən }

Boersch effect   [ELECTR] The deviation of the energy distribution of electrons emitted from a cathode from a Maxwellian distribution, due to broadening of the distribution by a space-charge region in front of the cathode.   { 'bersh i,fekt }

boiler plate   [COMPUT SCI] A commonly used expression or phrase that is stored in memory and can be copied into a word-processing document as needed.   { 'bȯil·ər ,plāt }

bolograph   [ENG] Any graphical record made by a bolometer; in particular, a graph formed by directing a pencil of light reflected from the galvanometer of the bolometer at a moving photographic film.   { 'bōl·ə,graf }

bolometer   [ENG] An instrument that measures the energy of electromagnetic radiation in certain wavelength regions by utilizing the change in resistance of a thin conductor caused by the heating effect of the radiation. Also known as thermal detector.   { bə'läm·əd·ər }

bomb   See abend.   { bäm }

bombardment   [ELECTR] The use of induction heating to heat electrodes of electron tubes to drive out gases during evacuation.   { bäm'bärd·mənt }

bond   [ELEC] The connection made by bonding electrically.   { bänd }

bonded NR diode   [ELECTR] An n+ junction semiconductor device in which the negative resistance arises from a combination of avalanche breakdown and conductivity modulation which is due to the current flow through the junction.   { ¦bän·dəd ,en¦är 'dī,ōd }

bonded strain gage   [ENG] A strain gage in which the resistance element is a fine wire, usually in zigzag form, embedded in an insulating backing material, such as impregnated paper or plastic, which is cemented to the pressure-sensing element.   { ¦bän·dəd 'strān ,gāj }

bonded transducer   [ENG] A transducer which employs a bonded strain gage for sensing pressure.   { ¦bän·dəd tranz'dü·sər }

bonding   [ELEC] The use of low-resistance material to connect electrically a chassis, metal shield cans, cable shielding braid, and other supposedly equipotential points to eliminate undesirable electrical interaction resulting from high-impedance paths between them.   [ENG] **1.** The fastening together of two components of a device by means of adhesives, as in anchoring the copper foil of printed wiring to an insulating baseboard.   **2.** See cladding.   { 'bän·diŋ }

bonding pad   [ELECTR] A metallized area on the surface of a semiconductor device, to which connections can be made.   { 'bän·diŋ ,pad }

bonding wire   [ELEC] Wire used to connect metal objects so they have the same potential (usually ground potential).   { 'bän·diŋ ,wīr }

bond strength   [ENG] The amount of adhesion between bonded surfaces measured in terms of the stress required to separate a layer of material from the base to which it is bonded.   { 'bänd ,streŋkth }

Böning effect   [ELEC] The displacement of associated ions that have been bound to capturing ions in fine channels in a dielectric medium when an electric field is applied.   { 'bən·iŋ i,fekt }

Book A   See DVD-read-only.   { ¦bùk 'ā }

Book B   See DVD-video.   { ¦bùk 'bē }

book capacitor   [ELEC] A trimmer capacitor consisting of two plates which are hinged at one end; capacitance is varied by changing the angle between them.   { 'bùk kə'pas·əd·ər }

Book D   See DVD-write once.   { ¦bùk 'dē }

Book E   See DVD-rewritable.   { ¦bùk 'ē }

bookkeeping operation   [COMPUT SCI] A computer operation which does not directly contribute to the result, that is, arithmetical, logical, and transfer operations used in modifying the address section of other instructions in counting cycles and in rearranging data. Also known as red-tape operation.   { 'bùk,kēp·iŋ äp·ə'rā·shən }

bookmark   [COMPUT SCI] **1.** Any method of halting the processing of a transaction and holding it, as far as it has been completed, until processing resumes.   **2.** A code that is inserted at a particular place in a document or that is associated with a particular document so that the user can easily return to the specified insertion point or document.   **3.** A Web page location (URL) which is saved by a user for quick reference.   { 'bùk ,märk }

Boolean   [COMPUT SCI] A scalar declaration in ALGOL defining variables similar to FORTRAN's logical variables.   { 'bü·lē·ən }

Boolean algebra   [MATH] An algebraic system with two binary operations and one unary operation important in representing a two-valued logic.   { 'bü·lē·ən 'al·jə·brə }

Boolean calculus   [MATH] Boolean algebra modified to include the element of time.   { 'bü·lē·ən 'kal·kyə·ləs }

Boolean data type   See logical data type.   { 'bü·lē·ən 'dad·ə ,tīp }

Boolean determinant   [MATH] A function defined on Boolean matrices which depends on the elements of the matrix in a manner analogous to the manner in which an ordinary determinant depends on the elements of an ordinary matrix, with the operation of multiplication replaced by intersection and the operation of addition replaced by union. { ¦bül·ē·ən di'tər·mə·nənt }

Boolean function [MATH] A function $f(x,y,\ldots,z)$ assembled by the application of the operations AND, OR, NOT on the variables $x, y,\ldots, z$ and elements whose common domain is a Boolean algebra. { 'bü·lē·ən 'fəŋk·shən }

Boolean matrix [MATH] A rectangular array of elements each of which is a member of a Boolean algebra. { ¦bül·ē·ən 'mā,triks }

Boolean operation table [MATH] A table which indicates, for a particular operation on a Boolean algebra, the values that result for all possible combination of values of the operands; used particularly with Boolean algebras of two elements which may be interpreted as "true" and "false." { ¦bül·ē·ən ,äp·ə'rā·shən ,tā·bəl }

Boolean operator [MATH] A logic operator that is one of the operators AND, OR, or NOT, or can be expressed as a combination of these three operators. { ¦bül·ē·ən 'äp·ə,rād·ər }

Boolean ring [MATH] A commutative ring with the property that for every element a of the ring, $a \times a$ and $a + a = 0$; it can be shown to be equivalent to a Boolean algebra. { ¦bül·ē·ən 'riŋ }

Boolean search [COMPUT SCI] A search for selected information, that is, information satisfying conditions that can be expressed by AND, OR, and NOT functions. { 'bü·lē·ən 'sərch }

boost [ELECTR] To augment in relative intensity, as to boost the bass response in an audio system. { büst }

boost charge [ELEC] Partial charge of a storage battery, usually at a high current rate for a short period. { 'büst ,chärj }

booster [ELEC] A small generator inserted in series or parallel with a larger generator to maintain normal voltage output under heavy loads. [ELECTR] **1.** A separate radio-frequency amplifier connected between an antenna and a television receiver to amplify weak signals. **2.** A radio-frequency amplifier that amplifies and rebroadcasts a received television or communication radio carrier frequency for reception by the general public. { 'büs·tər }

booster battery [ELECTR] A battery which increases the sensitivity of a crystal detector by maintaining a certain voltage across it and thereby adjusting conditions to increase the response to a given input. { 'büs·tər ,bad·ə·rē }

booster voltage [ELECTR] The additional voltage supplied by the damper circuit to the horizontal output, horizontal oscillator, and vertical output circuits of a television receiver to give greater sawtooth sweep output. { 'büs·tər ,vōl·tij }

boot [COMPUT SCI] To load the operating system into a computer after it has been swi-

tched on; usually applied to small computers. { büt }

boot button See bootstrap button. { 'büt ,bət·ən }

boot record [COMPUT SCI] A special area of a floppy diskette or hard drive which is used by the computer during system startup. { 'büt ,rek·ərd }

bootstrap [COMPUT SCI] The procedures for making a computer or a program function through its own actions. [ENG] A technique or device designed to bring itself into a desired state by means of its own action. { 'büt,strap }

bootstrap button [COMPUT SCI] The first button pressed when a computer is turned on, causing the operating system to be loaded into memory. Also known as boot button; initial program load button; IPL button. { 'büt,strap ,bət·ən }

bootstrap circuit [ELECTR] A single-stage amplifier in which the output load is connected between the negative end of the anode supply and the cathode, while signal voltage is applied between grid and cathode; a change in grid voltage changes the input signal voltage with respect to ground by an amount equal to the output signal voltage. { 'büt,strap ,sər·kət }

bootstrap driver [ELECTR] Electronic circuit used to produce a square pulse to drive the modulator tube; the duration of the square pulse is determined by a pulse-forming line. { 'büt ,strap ,drīv·ər }

bootstrap instructor technique [COMPUT SCI] A technique permitting a system to bring itself into an operational state by means of its own action. Also known as bootstrap technique. { 'büt ,strap in'strək·tər tek'nēk }

bootstrap integrator [ELECTR] A bootstrap sawtooth generator in which an integrating amplifier is used in the circuit. Also known as Miller generator. { 'büt,strap 'in·tə,grād·ər }

bootstrap loader [COMPUT SCI] A very short program loading routine, used for loading other loaders in a computer; often implemented in a read-only memory. { 'büt,strap 'lōd·ər }

bootstrap memory [COMPUT SCI] A device that provides for the automatic input of new programs without erasing the basic instructions in the computer. { 'büt,strap 'mem·rē }

bootstrapping [ELECTR] A technique for lifting a generator circuit above ground by a voltage value derived from its own output signal. { 'büt ,strap·iŋ }

bootstrap program See loading program. { 'büt ,strap ,prō·grəm }

bootstrap sawtooth generator [ELECTR] A circuit capable of generating a highly linear positive sawtooth waveform through the use of bootstrapping. { ¦büt,strap ¦sȯ,tüth 'jen·ə,rād·ər }

bootstrap technique See bootstrap instructor technique. { 'büt,strap tek'nēk }

boot virus [COMPUT SCI] A virus that infects the boot records on floppy diskettes and hard drives and is designed to self-replicate from one disk to another. { 'büt ,vī·rəs }

boresighting [ENG] Initial alignment of a directional microwave or radar antenna system by using an optical procedure or a fixed target at a known location. { 'bȯr,sīd·iŋ }

BORSCHT [COMMUN] An interface circuit between ordinary telephone lines carrying analog voice signals and digital time-division multiplex facilities, which digitizes voice signals, assigns them time slots, and then multiplexes them. Acronym for battery, overvoltage, ringing, supervision, coding, hybrid and test access. { bȯrsht }

bottleneck analysis [COMPUT SCI] A detailed study of the manner in which elements of a computer system are related to find out where bottlenecks arise, so that the system's performance can be improved. { 'bäd·əl,nek ə ,nal·ə·səs }

bottle thermometer [ENG] A thermoelectric thermometer used for measuring air temperature; the name is derived from the fact that the reference thermocouple is placed in an insulated bottle. { 'bäd·əl thər'mäm·əd·ər }

bottom [COMPUT SCI] The termination of a file. { 'bäd·əm }

bottom-up analysis [COMPUT SCI] A reductive method of syntactic analysis which attempts to reduce a string to a root symbol. { ¦bäd·əm·əp ¦ə'nal·ə·səs }

bounced message [COMPUT SCI] An electronic mail message that is returned to sender because attempts to deliver it have been unsuccessful. { ,baúnst 'mes·ij }

boundary [ELECTR] An interface between *p*- and *n*-type semiconductor materials, at which donor and acceptor concentrations are equal. { 'baún·drē }

boundary-layer photocell See photovoltaic cell. { 'baún·drē ,lā·ər 'fō·dō,sel }

bound charge [ELEC] Electric charge which is confined to atoms or molecules, in contrast to free charge, such as metallic conduction electrons, which is not. Also known as polarization charge. { ¦baúnd 'chärj }

bounds register [COMPUT SCI] A device which stores the upper and lower bounds on addresses in the memory of a given computer program in a time-sharing system. { 'baúnz ,rej·ə·stər }

Bourne shell [COMPUT SCI] The original Unix shell. { 'búrn ,shel }

bowtie antenna [ELECTROMAG] An antenna that consists of two triangular pieces of stiff wire or two triangular flat metal plates, arranged in the configuration of a bowtie, with the feed point at the gap between the apexes of the triangles. { 'bō,tī an,ten·ə }

boxcar [COMMUN] One of a series of long signal-wave pulses that are separated by very short intervals of time. { 'bäks,kär }

boxcar circuit [ELECTR] A circuit used in radar for sampling voltage waveforms and storing the latest value sampled; the term is derived from the flat, steplike segments of the output voltage waveform. { 'bäks,kär ,sər·kət }

B-pictures See bidirectional pictures. { 'bē 'pik·chərz }

B power supply See B supply. { 'bē 'paú·ər ,sə·plī }

bps See bit per second.

BPSK See binary phase-shift keying.

brachiating motion [CONT SYS] A type of robotic motion that employs legs or other equipment to help the manipulator move in its working environment. { ¦brā·kē'ād·iŋ 'mō·shən }

brachiating robot [CONT SYS] A robot that is capable of moving over the surface of an object. { ¦brā·kē'ād·iŋ 'rō,bät }

Bragg cell See acoustooptic modulator. { 'brag ,sel }

braided wire [ELEC] A tube of fine wires woven around a conductor or cable for shielding purposes or used alone in flattened form as a grounding strap. { 'brād·əd ,wīr }

branch [COMPUT SCI] **1.** Any one of a number of instruction sequences in a program to which computer control is passed, depending upon the status of one or more variables. **2.** See jump. [ELEC] A portion of a network consisting of one or more two-terminal elements in series. Also known as arm. { branch }

branch circuit [ELEC] A portion of a wiring system in the interior of a structure that extends from a final overload protective device to a plug receptacle or a load such as a lighting fixture, motor or heater. { ¦branch ¦sər·kət }

branch-circuit distribution center [ELEC] Distribution center at which branch circuits are supplied. { ¦branch ¦sər·kət dis·trə'byü·shən ,sen·tər }

branch cutout [ELEC] The holder for a fuse that protects a branch circuit in an interior wiring system. { 'branch 'kəd,aút }

branch gain See branch transmittance. { 'branch ,gān }

branching [COMPUT SCI] The selection, under control of a computer program, of one of two or more branches. { 'branch·iŋ }

branch instruction [COMPUT SCI] An instruction that makes the computer choose between alternative subprograms, depending on the conditions determined by the computer during the execution of the program. { 'branch in'strək·shən }

branch joint [ELEC] Joint used for connecting a branch conductor or cable, where the latter continues beyond the branch. { 'branch ,jóint }

branch point [COMPUT SCI] A point in a computer program at which there is a branch instruction. [ELEC] A terminal in an electrical network that is common to more than two elements or parts of elements of the network. Also known as junction point; node. { 'branch ,póint }

branch prediction [COMPUT SCI] A method whereby a processor guesses the outcome of a branch instruction so that it can prepare in advance to carry out the instructions that follow the predicted outcome. { 'branch prə,dik·shən }

branch transmittance [CONT SYS] The amplification of current or voltage in a branch of an electrical network; used in the representation of such a network by a signal-flow graph. Also known as branch gain. { ¦branch trans'mit·əns }

Branley-Lenard effect |ELECTR| The strong ionization of air and other gases by ultraviolet radiation with wavelengths in the range 120–150 nanometers. { 'bran·lē 'len·ərd i‚fekt }

Braun tube *See* cathode-ray tube. { 'braún ‚tüb }

breadboard |ELECTR| A printed circuit board designed so that the user can mount and wire whatever circuitry is desired. { 'bred‚bórd }

breadboarding |ELECTR| Assembling an electronic circuit in the most convenient manner, without regard for final locations of components, to prove the feasibility of the circuit and to facilitate changes when necessary. { 'bred‚bórd·iŋ }

breadboard model |ENG| Uncased assembly of an instrument or other piece of equipment, such as a radio set, having its parts laid out on a flat surface and connected together to permit a check or demonstration of its operation. { 'bred‚bórd ‚mäd·əl }

break |COMPUT SCI| **1.** To interrupt processing by a computer, usually by depressing a key. **2.** A place in a file of records where one or more of the values in the records change. { brāk }

break-before-make contact |ELEC| One of a pair of contacts that interrupt one circuit before establishing another. { ¦brāk bə‚fór ¦māk 'kän ‚takt }

break contact |ELEC| The contact of a switching device which opens a circuit upon the operation of the device. { 'brāk ‚kän‚takt }

breakdown |ELEC| A large, usually abrupt rise in electric current in the presence of a small increase in voltage; can occur in a confined gas between two electrodes, a gas tube, the atmosphere (as lightning), an electrical insulator, and a reverse-biased semiconductor diode. Also known as electrical breakdown. { 'brāk‚daún }

breakdown diode |ELEC| A semiconductor diode in which the reverse-voltage breakdown mechanism is based either on the Zener effect or the avalanche effect. { 'brāk‚daún¦dī‚ōd }

breakdown impedance |ELECTR| Of a semiconductor, the small-signal impedance at a specified direct current in the breakdown region. { 'brāk ‚daún im'pēd·əns }

breakdown potential *See* breakdown voltage. { 'brāk‚daún pə'ten·shəl }

breakdown region |ELECTR| Of a semiconductor diode, the entire region of the volt-ampere characteristic beyond the initiation of breakdown for increasing magnitude of bias. { 'brāk ‚daún ‚rē·jən }

breakdown torque |ELEC| The maximum torque that a motor can develop at its rated applied voltage and frequency without an abrupt drop in speed. { 'brāk‚daún ‚tórk }

breakdown voltage |ELEC| **1.** The voltage measured at a specified current in the electrical breakdown region of a semiconductor diode. Also known as Zener voltage. **2.** The voltage at which an electrical breakdown occurs in a dielectric. **3.** The voltage at which an electrical breakdown occurs in a gas. Also known as breakdown potential; sparking potential; sparking voltage. { 'brāk‚daún ‚vól·tij }

breaker-and-a-half |ELEC| A substation switching arrangement that involves two buses between which three breaker bays are installed. { ¦brā·kər ən ə 'haf }

breaker-and-a-third |ELEC| A substation switching arrangement having four breakers and three connections per bay. { ¦brā·kər ən ə 'thərd }

breaker points |ELEC| Low-voltage contacts used to interrupt the current in the primary circuit of a gasoline engine's ignition system. { 'brā·kər ‚póints }

break frequency |CONT SYS| The frequency at which a graph of the logarithm of the amplitude of the frequency response versus the logarithm of the frequency has an abrupt change in slope. Also known as corner frequency; knee frequency. { 'brāk ‚frē·kwən·sē }

break-in device |ELECTR| A device in a radiotelegraph communication system allowing an operator to receive signals in intervals between his own transmission signals. { 'brāk‚in di'vīs }

break-in operation |COMMUN| A method of radio communication in which it is possible for the receiving operator to interrupt or break into the transmission. { 'brā¦kin ‚äp·ə‚rā·shən }

break key |COMPUT SCI| A key on a computer keyboard whose depression causes processing to be interrupted. { 'brāk ‚kē }

breakout |ELEC| A joint at which one or more conductors are brought out from a multiconductor cable. { 'brā‚kaút }

breakout box |ELECTR| A device connected to a multiconductor cable that provides terminal connections to test the signals in a transmission. { 'brāk‚aút ‚bäks }

breakoutput |COMPUT SCI| An ALGOL procedure which causes all bytes in a device buffer to be sent to the device rather than wait until the buffer is full. { ¦brā¦kaút‚pút }

breakover |ELECTR| In a silicon controlled rectifier or related device, a transition into forward conduction caused by the application of an excessively high anode voltage. { 'brā ‚kō·vər }

breakover voltage |ELECTR| The positive anode voltage at which a silicon controlled rectifier switches into the conductive state with gate circuit open. { 'brā‚kō·vər ‚vól·tij }

break period |COMMUN| Of a rotary dial telephone, the time interval during which the circuit contacts are open. { 'brāk ‚pir·ē·əd }

breakpoint |COMPUT SCI| A point in a program where an instruction, instruction digit, or other condition enables a programmer to interrupt the run by external intervention or by a monitor routine. { 'brāk‚póint }

breakpoint switch |COMPUT SCI| A manually operated switch which controls conditional operation at breakpoints, used primarily in debugging. { 'brāk‚póint ‚swich }

breakpoint symbol |COMPUT SCI| A symbol which may be optionally included in an instruction, as an indication, tag, or flag, to designate it as a breakpoint. { 'brāk‚póint ‚sim·bəl }

breakthrough

breakthrough |COMPUT SCI| An interruption in the intended character stroke in optical character recognition. { 'brāk,thrü }

B register See index register.

bridge |COMMUN| A device that joins two networks of the same type. |ELEC| **1.** An electrical instrument having four or more branches, by means of which one or more of the electrical constants of an unknown component may be measured. **2.** An electrical shunt path. { brij }

bridge circuit |ELEC| An electrical network consisting basically of four impedances connected in series to form a rectangle, with one pair of diagonally opposite corners connected to an input device and the other pair to an output device. { 'brij ,sər·kət }

bridged tap |ELEC| Portion of a cable pair connected to a circuit which is not a part of the useful path. { ¦brijd 'tap }

bridged-T network |ELEC| A T network with a fourth branch connected between an input and an output terminal and across two branches of the network. { ¦brijd ¦tē 'net,wərk }

bridge hybrid See hybrid junction. { 'brij 'hī·brəd }

bridge limiter |ELECTR| A device employed in analog computers to keep the value of a variable within specified limits. { 'brij ¦lim·əd·ər }

bridge magnetic amplifier |ELECTR| A magnetic amplifier in which each of the gate windings is connected in series with an arm of a bridge rectifier; the rectifiers provide self-saturation and direct-current output. { ¦brij mag'ned·ik 'am·plə ,fī·ər }

bridge oscillator |ELECTR| An oscillator using a balanced bridge circuit as the feedback network. { 'brij äs·ə'lād·ər }

bridge rectifier |ELECTR| A full-wave rectifier with four elements connected as a bridge circuit with direct voltage obtained from one pair of opposite junctions when alternating voltage is applied to the other pair. { 'brij ,rek·tə,fī·ər }

bridgeware |COMPUT SCI| Software or hardware that translates programs or converts data from one format to another. { 'brij,wer }

bridging |ELEC| **1.** Connecting one electric circuit in parallel with another. **2.** The action of a selector switch whose movable contact is wide enough to touch two adjacent contacts so that the circuit is not broken during contact transfer. |MATH| The operation of carrying in addition or multiplication. { 'brij·iŋ }

bridging amplifier |ELECTR| Amplifier with an input impedance sufficiently high so that its input may be bridged across a circuit without substantially affecting the signal level of the circuit across which it is bridged. { 'brij·iŋ ,am·plə,fī·ər }

bridging connection |ELECTR| Parallel connection by means of which some of the signal energy in a circuit may be withdrawn frequently, with imperceptible effect on the normal operation of the circuit. { 'brij·iŋ kə,nek·shən }

bridging contacts |ELEC| A contact form in which the moving contact touches two

stationary contacts simultaneously during transfer. { 'brij·iŋ ,kän,taks }

bridging loss |ELECTR| Loss resulting from bridging an impedance across a transmission system; quantitatively, the ratio of the signal power delivered to that part of the system following the bridging point, and measured before the bridging, to the signal power delivered to the same part after the bridging. { 'brij·iŋ ,lòs }

brightness control |ELECTR| A control that varies the luminance of the fluorescent screen of a cathode-ray tube, for a given input signal, by changing the grid bias of the tube and hence the beam current. Also known as brilliance control; intensity control. { 'brīt·nəs kən'trōl }

brilliance |ELECTR| **1.** The degree of brightness and clarity of the display of a cathode-ray tube. **2.** The degree to which the higher audio frequencies of an input sound are reproduced by a sound system. { 'bril·yəns }

brilliance control See brightness control. { 'bril·yəns kən'trōl }

broaching bit See reaming bit. { 'brōch·iŋ bit }

broadband |COMMUN| A band with a wide range of frequencies. { 'bród,band }

broadband amplifier |ELECTR| An amplifier having essentially flat response over a wide range of frequencies. { 'bród,band ¦am·plə,fī·ər }

broadband antenna |ELECTROMAG| An antenna that functions satisfactorily over a wide range of frequencies, such as for all 12 very-high-frequency television channels. { 'bród,band an'ten·ə }

broadband channel |COMMUN| A data transmission channel that can handle frequencies higher than the normal voice-grade line limit of 3 to 4 kilohertz; it can carry many voice or data channels simultaneously or can be used for high-speed single-channel data transmission. { 'bród,band ¦chan·əl }

broadband klystron |ELECTR| Klystron having three or more resonant cavities that are externally loaded and stagger-tuned to broaden the bandwidth. { 'bród,band ¦klī,strän }

broadband path |COMMUN| A path having a bandwidth of 20 kilohertz or greater. { 'bród ,band ,path }

broadcast |COMMUN| A television, radio, or data transmission intended for public reception. { 'bród,kast }

broadcast band |COMMUN| The band of frequencies extending from 535 to 1605 kilohertz, corresponding to assigned radio carrier frequencies that increase in multiples of 10 kHz between 540 and 1600 kHz for the United States. Also known as standard broadcast band. { 'bród,kast ,band }

broadcast message |COMMUN| A message that is sent to all users of a computer network when they log on to the network. { ¦bród,kast ¦mes·ij }

broadcast station |COMMUN| A television or radio station used for transmitting programs to the general public. Also known as station. { 'bród ,kast ,stā·shən }

62

broadcast transmitter |ELECTR| A transmitter designed for use in a commercial amplitude-modulation, frequency-modulation, or television broadcast channel. { 'bród,kast tranz'mid·ər }

broadside array |ELECTROMAG| An antenna array whose direction of maximum radiation is perpendicular to the line or plane of the array. { 'bród ‚sīd ə'rā }

broad tuning |ELECTR| Poor selectivity in a radio receiver, causing reception of two or more stations at a single setting of the tuning dial. { ¦bród ¦tün·iŋ }

Brooks variable inductometer |ELEC| An inductometer providing a nearly linear scale and consisting of two movable coils, side by side in a plane, sandwiched between two pairs of fixed coils. { ¦brúks 'ver·ē·ə·bəl ‚in‚dək'täm·əd·ər }

brownout |ELEC| **1.** A restriction of electrical power usage during a power shortage, especially for advertising and display purposes. **2.** An extinguishing of some of the lights in a city as a defensive measure against enemy bombardment. { 'braún‚aút }

browse mode |COMPUT SCI| A mode of operation in which data in a document or database are conveniently displayed for rapid, on-screen review. { 'braúz ‚mōd }

browser |COMPUT SCI| An interactive program (client) that requests, retrieves, and displays pages from the World Wide Web. { 'braúz·ər }

brush |ELEC| A conductive metal or carbon block used to make sliding electrical contact with a moving part. { brəsh }

brush discharge |ELEC| A luminous electric discharge that starts from a conductor when its potential exceeds a certain value but remains too low for the formation of an actual spark. { ¦brəsh ¦dis‚chärj }

brush encoder |ELECTR| An encoder in which brushes that make contact with conductive segments on a rotating or linearly moving surface convert positional information to digitally encoded data. { 'brəsh en'kōd·ər }

brush holder |ELEC| A structure in which a brush can slide in a direction perpendicular to the moving surface of a motor, generator, or other device. { 'brəsh ‚hōl·dər }

brush lag |ELEC| The distance that the brushes on a motor are displaced in a direction opposite to the motor's rotation in order to overcome the effect of armature reaction. { 'brəsh ‚lag }

brush lead |ELEC| The distance that the brushes on a generator are displaced in the direction of the motor's rotation in order to overcome the effect of armature reaction. { 'brəsh ‚lēd }

brush rocker |ELEC| A yoke to which the brush holders in an electrical machine are attached, and which can be moved to adjust the positions of the brushes. Also known as brush rocker ring. { 'brəsh ‚rä·kər }

brush rocker ring See brush rocker. { 'brəsh ‚rä·kər ‚riŋ }

brush-shifting motor |ENG| A category of alternating-current motor in which the brush

contacts shift to modify operating speed and power factor. { 'brəsh ‚shif·tiŋ ‚mōd·ər }

brute force attack |COMPUT SCI| An attempt to gain unauthorized access to a computing system by generating and trying all possible passwords. { ¦brüt ¦fórs ə'tak }

brute-force filter |ELEC| Type of powerpack filter depending on large values of capacitance and inductance to smooth out pulsations rather than on resonant effects of tuned filters. { ¦brüt ‚fórs 'fil·tər }

brute-force technique |COMPUT SCI| Any method that relies chiefly on the advanced processing capabilities of a large computer to accomplish a task. { ¦brüt ‚fórs tek'nēk }

brute supply |ELEC| A type of power supply that is completely unregulated, employing no circuitry to maintain output voltage constant with changing input line or load variations. { ¦brüt sə'plī }

B-scan See B-display. { 'bē ‚skan }

B-scope See B-display. { 'bē ‚skōp }

b-spline |COMPUT SCI| A curve that is generated by a computer-graphics program, guided by a mathematical formula which ensures that it will be continuous with other such curves; it is mathematically more complex but easier to blend than a Bézier curve. { 'bē‚splīn }

B station |NAV| In loran, the designation applied to one transmitting station of a pair, the signal of which always occurs more than half a repetition period after the succeeding signal and less than half a repetition period before the preceding signal from the other station of the pair, designated an A station. { 'bē ‚stā·shən }

B store See index register. { 'bē ‚stór }

B supply |ELECTR| Anode high voltage and screen-grid power source in vacuum tube circuits. Also known as B power supply. { 'bē sə'plī }

B trace |ELECTR| In loran the second trace of an oscilloscope which corresponds to the signal from the B station. { 'bē ‚trās }

B-tree See balanced-tree. { 'bē ‚trē }

B+-tree |COMPUT SCI| A version of the balanced-tree that maintains a hierarchy of indexes while linking the data sequentially. { ¦bē ¦pləs ‚trē }

bubble |COMPUT SCI| A circle that represents data in a data flow diagram. { 'bəb·əl }

bubble chart See data flow diagram. { 'bəb·əl ‚chärt }

bubble memory |COMPUT SCI| A computer memory in which the presence or absence of a magnetic bubble in a localized region of a thin magnetic film designates a 1 or 0; storage capacity can be well over 1 megabit per cubic inch. Also known as magnetic bubble memory. { 'bəb·əl ¦mem·rē }

bubble sort |COMPUT SCI| A procedure for sorting a set of items that begins by sequencing the first and second items, then the second and third, and so on, until the end of the set is reached, and then repeats this process until all items are correctly sequenced. { 'bəb·əl ˌsȯrt }

Buchholz protective device [ELEC] A protective relay which is attached to an oil-filled tank containing a transformer and which is activated either by gas produced by faults or by oil surges produced by explosive faults in the transformer. Also known as gas bubble protective device. { 'būkˌhōls prə'tek·tiv di'vīs }

bucket [COMPUT SCI] A name usually reserved for a storage cell in which data may be accumulated. { 'bək·ət }

bucket brigade device [ELECTR] A semiconductor device in which majority carriers store charges that represent information, and minority carriers transfer charges from point to point in sequence. Abbreviated BBD. { 'bək·ət briˈgād di'vīs }

bucking transformer [ELEC] A transformer whose voltage opposes that of a second transformer. { 'bək·iŋ tranz'fȯr·mər }

bucking voltage [ELEC] A voltage having a polarity opposite to that of another voltage against which it acts. { 'bək·iŋ ˌvōl·tij }

buffer [ELEC] An electric circuit or component that prevents undesirable electrical interaction between two circuits or components. [ELECTR] **1.** An isolating circuit in an electronic computer used to prevent the action of a driven circuit from affecting the corresponding driving circuit. **2.** See buffer amplifier. { 'bəf·ər }

buffer amplifier [ELECTR] An amplifier used after an oscillator or other critical stage to isolate it from the effects of load impedance variations in subsequent stages. Also known as buffer; buffer stage. { ˈbəf·ər 'am·plə̄ˌfī·ər }

buffer capacitor [ELECTR] A capacitor connected across the secondary of a vibrator transformer or between the anode and cathode of a cold-cathode rectifier tube to suppress voltage surges that might otherwise damage other parts in the circuit. { 'bəf·ər kə'pas·əd·ər }

buffered computer [COMPUT SCI] A computer having a temporary storage device to compensate for differences in transmission speeds. { 'bəf·ərd kəm'pyüd·ər }

buffered device [COMPUT SCI] A piece of peripheral equipment, such as a printer, that is equipped with a buffer storage so that it can accept information more rapidly than it can process it. { 'bəf·ərd di'vīs }

buffered FET logic [ELECTR] A logic gate configuration used with gallium-arsenide field-effect transistors operating in the depletion mode, in which the level shifting required to make the input and output voltage levels compatible is achieved with Schottky barrier diodes. Abbreviated BFL. { 'bəf·ərd ˈefˈēˈtē 'läj·ik }

buffered I/O channel [COMPUT SCI] A storage device located between input/output (I/O) channels and main storage control to free the channels for use by other operations. { 'bəf·ərd ˈīˌō ˌchan·əl }

buffered terminal [COMPUT SCI] A computer terminal which contains storage equipment so that

the rate at which it sends or receives data over its line does not need to agree exactly with the rate at which the data are entered or printed. { 'bəf·ərd 'tər·mən·əl }

buffer element [ELEC] A low-impedance inverting driver circuit. { 'bəf·ər ˌel·ə·mənt }

buffer pooling [COMPUT SCI] A technique for receiving data in an input/output control system in which a number of buffers are available to the system; when a record is produced, a buffer is taken from the pool, used to hold the data, and returned to the pool after data transmission. { 'bəf·ər ˌpül·iŋ }

buffer stage See buffer amplifier. { 'bəf·ər ˌstāj }

buffer storage [COMPUT SCI] A synchronizing element used between two different forms of storage in a computer; computation continues while transfers take place between buffer storage and the secondary or internal storage. Also known as buffer. { 'bəf·ər ˌstȯr·ij }

buffer zone [COMPUT SCI] An area of main memory set aside for temporary storage. { 'bəf·ər ˌzōn }

bug [COMPUT SCI] A defect in a program code or in designing a routine or a computer. [ELECTR] **1.** A semiautomatic code-sending telegraph key in which movement of a lever to one side produces a series of correctly spaced dots and movement to the other side produces a single dash. **2.** An electronic listening device, generally concealed, used for commercial or military espionage. [ENG] A defect or imperfection present in a piece of equipment. { bəg }

build [ELECTR] To increase in received signal strength. { bild }

building-out circuit [ELEC] Short section of transmission line, or a network which is shunted across a transmission line, for the purpose of impedance matching. { ˈbil·diŋ ˌaút 'sər·kət }

building-out network [ELEC] Network designed to be connected to a based network so that the combination will simulate the sending-end impedance, neglecting dissipation, of a line having a termination other than that for which the basic network was designed. { ˈbil·diŋ ˌaút ˌnet·wərk }

building-out section [ELEC] Short section of transmission line, either open or short-circuited at the far end, shunted across another transmission line for use on an impedance-matching transformer. { ˈbil·diŋ ˌaút ˌsek·shən }

built-in antenna [ELECTROMAG] An antenna that is located inside the cabinet of a radio or television receiver. { 'biltˌin an'ten·ə }

built-in check [COMPUT SCI] A hardware device which controls the accuracy of data either moved or stored within the computer system. { 'biltˌin 'chek }

built-in function [COMPUT SCI] A function that is available through a simple reference and specification of arguments in a given higher-level programming language. Also known as built-in procedure; intrinsic procedure; standard function. { 'biltˌin 'fəŋk·shən }

built-in pointing device |COMPUT SCI| A trackball or pointing stick that is built into the case of a portable computer and used to move an on-screen pointer. { ¦bilt¸in 'pȯint·iŋ di¸vīs }

built-in procedure See built-in function. { 'bilt ¸in prə'sēj·ər }

bulb See envelope. { bəlb }

bulk-acoustic-wave delay line |ELECTR| A delay line in which the delay is determined by the distance traveled by a bulk acoustic wave between input and output transducers mounted on a piezoelectric block. { ¦bəlk ə'kü·stik ¦wāv di'lā ¸līn }

bulk diode |ELECTR| A semiconductor microwave diode that uses the bulk effect, such as Gunn diodes and diodes operating in limited space-charge-accumulation modes. { ¦bəlk 'dī ¸ōd }

bulk effect |ELECTR| An effect that occurs within the entire bulk of a semiconductor material rather than in a localized region or junction. { 'bəlk i'fekt }

bulk-effect device |ELECTR| A semiconductor device that depends on a bulk effect, as in Gunn and avalanche devices. { 'bəlk i'fekt di'vīs }

bulk memory |COMPUT SCI| A high-capacity memory used in connection with a computer for bulk storage of large quantities of data. { ¦bəlk 'mem·rē }

bulk photoconductor |ELECTR| A photoconductor having high power-handling capability and other unique properties that depend on the semiconductor and doping materials used. { ¦bəlk ¦fō·dō·kən¦dək·tər }

bulk resistor |ELECTR| An integrated-circuit resistor in which the *n*-type epitaxial layer of a semiconducting substrate is used as a noncritical high-value resistor; the spacing between the attached terminals and the sheet resistivity of the material together determine the resistance value. { 'bəlk ri'zis·tər }

bulk storage See backing storage. { ¦bəlk 'stȯr·ij }

bulletin board |COMPUT SCI| A collection of information that is stored in a computer system and can be accessed either by a specified group of people or the general public, usually by dialing a number on the public telephone system. { 'bül·ət·ən ¸bȯrd }

bulletin board system |COMPUT SCI| A computer system that enables its users, usually members of a particular interest group, to leave messages and to share information and software. Abbreviated BBS. { 'bül·ət·ən ¸bȯrd ¸sis·təm }

bump contact |ELECTR| A large-area contact used for alloying directly to the substrate of a transistor for mounting or interconnecting purposes. { 'bəmp ¸kän¸takt }

bunched pair |ELEC| Group of pairs tied together or otherwise associated for identification. { ¦bəncht 'per }

buncher See buncher resonator. { 'bən·chər }

buncher resonator |ELECTR| The first or input cavity resonator in a velocity-modulated tube, next to the cathode; here the faster electrons catch up with the slower ones to produce bunches of electrons. Also known as buncher; input resonator. { 'bən·chər ¸rez·ən¸ād·ər }

bunching |ELECTR| The flow of electrons from cathode to anode of a velocity-modulated tube as a succession of electron groups rather than as a continuous stream. { 'bən·chiŋ }

bunching voltage |ELECTR| Radio-frequency voltage between the grids of the buncher resonator in a velocity-modulated tube such as a klystron; generally, the term implies the peak value of this oscillating voltage. { 'bən·chiŋ ¸vōl·tij }

bundled program |COMPUT SCI| A computer program written, maintained, and updated by the computer manufacturer, and included in the price of the hardware. { ¦bənd·əld 'prō·grəm }

bundling |COMMUN| The provision of a combination of services, such as cable television and telephone service, over a single communications system. |COMPUT SCI| The provision of hardware and software as a single product or the combination of different software packages for sale as a single unit. { 'bən·dliŋ }

burden |ELEC| The amount of power drawn from the circuit connecting the secondary terminals of an instrument transformer, usually expressed in volt-amperes. { 'bərd·ən }

burglar alarm |ENG| An alarm in which interruption of electric current to a relay, caused, for example, by the breaking of a metallic tape placed at an entrance to a building, deenergizes the relay and causes the relay contacts to operate the alarm indicator. Also known as intrusion alarm. { 'bər·glər ə¦lärm }

buried set-point method |CONT SYS| A procedure for guiding a robot manipulator along a template, in which low-gain servomechanisms apply a force along the edge of the template, while the manipulator's tool is parallel to, and buried below, the template surface. { 'ber·ēd 'set ¸pȯint ¸meth·əd }

burn-in |ELECTR| Operation of electronic components before they are applied in order to stabilize their characteristics and reveal defects. { 'bərn ¸in }

burnout |ELEC| Failure of a device due to excessive heat produced by excessive current. { 'bərn ¸aȯt }

burnthrough |ELECTR| **1.** An electronic-protection effort by a radar to overcome the obscuration effect of jamming signals by using the highest energy transmission and longest possible dwell in the direction of the jamming or other direction of specific interest being affected. **2.** See jammer finder. { 'bərn¸thrü }

burst |COMMUN| **1.** A sudden increase in the strength of a signal being received from beyond line-of-sight range. **2.** A group of bits of characters that are transmitted together as a unit. **3.** A group of errors that occur together in a communication and alter its content. **4.** See color burst. |COMPUT SCI| **1.** To separate a continuous roll of paper into stacks of individual sheets by means of a burster. **2.** The transfer of a collection of records in a storage device, leaving an interval in which data for other requirements can be obtained from or entered into the device. **3.** A sequence of signals regarded as a unit in data transmission. { bərst }

burst amplifier [COMMUN] An amplifier stage in an analog color television receiver that is keyed into conduction and amplification by a horizontal pulse at the instant of each arrival of the color burst. Also known as chroma band-pass amplifier. { 'bərst ,am·plə,fī·ər }

burster [COMPUT SCI] An off-line device in a computer system used to separate the continuous roll of paper produced as output from a printer into individual sheets, generally along perforations in the roll. { 'bər·stər }

burst mode [COMPUT SCI] A method of transferring data between a peripheral unit and a control processing unit in a computer system in which the peripheral unit sends the central processor a signal to receive data until the peripheral unit signals that the transfer is completed. { 'bərst ,mōd }

burst pedestal [COMMUN] Rectangular pulselike analog television signal which may be part of the color burst; the amplitude of the color burst pedestal is measured from the alternating-current axis of the sine-wave portion to the horizontal pedestal. { 'bərst ,ped·ə·stəl }

burst separator [ELECTR] The circuit in a color television receiver that separates the color burst from the composite video signal. { 'bərst sep·ə'rād·ər }

bus [COMPUT SCI] The circuitry and wiring connecting the various components of a computer through which data are transmitted; for example, in a personal computer the system bus interconnects the CPU, memory, and input/output devices. [ELEC] A set of two or more electric conductors that serve as common connections between load circuits and each of the polarities (in direct-current systems) or phases (in alternating-current systems) of the source of electric power. [ELECTR] One or more conductors in a computer along which information is transmitted from any of several sources to any of several destinations. { bəs }

bus architecture [COMPUT SCI] A structure for handling data transmission in a computer system or network, in which components are all linked to a common bus. { 'bəs 'är·kə,tek·chər }

busbar [ELEC] A heavy, rigid metallic conductor, usually uninsulated, used to carry a large current or to make a common connection between several circuits. Also known as bus. { 'bəs,bär }

bus cable [ELECTR] An electrical conductor that can be attached to a bus to extend it outside the computer housing or join it to another bus within the same computer. { 'bəs ,kā·bəl }

bus cycle [COMPUT SCI] A single transaction between the main memory and the CPU. { 'bəs ,sī·kəl }

bus duct [ELEC] An enclosed metal unit containing copper or aluminum busbars for distribution of large amounts of power between components of the distribution system. { 'bəs,dəkt }

bus extender [ELECTR] A printed circuit board that can be joined to a bus to increase its capacity. { 'bəs ik,sten·dər }

bushing See sleeve. { 'bùsh·iŋ }

bus mouse [COMPUT SCI] A mouse that is plugged into a printed circuit board inserted into the computer's bus. { 'bəs ,maús }

bus network [COMMUN] A communications network whose components are joined together by a single cable. { 'bəs 'net,wərk }

bus reactor [ELEC] An air-core inductor connected between two buses or two sections of the same bus in order to limit the effects of voltage transients on either bus. { 'bəs re'ak·tər }

busway [ELEC] A prefabricated assembly of standard lengths of busbars rigidly supported by solid insulation and enclosed in a sheet-metal housing. { 'bəs,wā }

busy test [COMMUN] A test, in telephony, made to find out whether certain facilities which may be desired, such as a subscriber line or trunk, are available for use. { 'biz·ē ,test }

busy tone [COMMUN] Interrupted low tone returned to the subscriber as an indication that the party's line is busy. { 'biz·ē ,tōn }

Butler oscillator [ELEC] Oscillator in which a piezoelectric crystal is connected between the cathode of two tubes, one functioning as a cathode follower, and the other as a grounded-grid amplifier. { 'bət·lər 'äs·ə,lād·ər }

butt contact [ELEC] A hemispherically shaped contact designed to mate against a similarly shaped contact. { 'bət ,kän,takt }

butterfly capacitor [ELEC] A variable capacitor having stator and rotor plates shaped like butterfly wings, with the stator plates having an outer ring to provide an inductance so that both capacitance and inductance may be varied, thereby giving a wide tuning range. { 'bəd·ər,flī kə'pas·əd·ər }

butterfly network [COMPUT SCI] A scheme that connects the units of a multiprocessing system and needs n stages to connect 2^n processors; at each stage a switch is thrown, depending on a particular bit in the addresses of the processors being connected. { ¦bəd·ər,flī ¦net,wərk }

Butterworth filter [ELECTR] An electric filter whose pass band (graph of transmission versus frequency) has a maximally flat shape. { 'bəd·ər ,wərth 'fil·tər }

butt joint [ELEC] A connection formed by placing the ends of two conductors together and joining them by welding, brazing, or soldering. { 'bət ,jóint }

button [COMPUT SCI] A small circle or rectangle on a graphical user interface, such that moving the pointer to it and clicking the mouse initiates some action. [ELECTR] **1.** A small, round piece of metal alloyed to the base wafer of an alloy-junction transistor. Also known as dot. **2.** The container that holds the carbon granules of a carbon microphone. Also known as carbon button. { 'bət·ən }

buttonhook contact [ELEC] A curved, hooklike contact often used on feed-through terminals of headers to facilitate soldering or unsoldering of leads. { 'bət·ən‚húk 'kän‚takt }

buzz [CONT SYS] *See* dither. [ELECTR] The condition of a combinatorial circuit with feedback that has undergone a transition, caused by the inputs, from an unstable state to a new state that is also unstable. { bəz }

BWO *See* backward-wave oscillator.

BX cable [ELEC] Insulated wires in flexible metal tubing used for bringing electric power to electronic equipment. { ¦bē¦eks ¦kā·bəl }

bypass [COMMUN] The use of alternative systems, such as satellite and microwave, to transmit data and voice signals, avoiding use of the communication lines of the local telephone company. [ELEC] A shunt path around some element or elements of a circuit. { 'bī ‚pas }

bypass capacitor [ELEC] A capacitor connected to provide a low-impedance path for radio-frequency or audio-frequency currents around a circuit element. Also known as bypass condenser. { 'bī‚pas kə'pas·əd·ər }

bypass condenser *See* bypass capacitor. { 'bī ‚pas kən'den·sər }

bypass filter [ELECTR] Filter which provides a low-attenuation path around some other equipment, such as a carrier frequency filter used to bypass a physical telephone repeater station. { 'bī‚pas ‚fil·tər }

byte [COMPUT SCI] A sequence of adjacent binary digits operated upon as a unit in a computer and usually shorter than a word. { bīt }

byte addressable computer [COMPUT SCI] A computer in which each byte of memory can be addressed independently of the others. { ¦bīt ə¦dres·ə·bəl kəm'pyüd·ər }

byte-aligned [COMMUN] A bit in a coded bit stream is byte-aligned if its position is a multiple of 8 bits from the first bit in the stream. { 'bīt ə'līnd }

bytecode [COMPUT SCI] Compiled Java programs that can be transferred across a network and executed by the Java virtual machine. { 'bīt‚kōd }

byte mode [COMPUT SCI] A method of transferring data between a peripheral unit and a central processor in which one byte is transferred at a time. { bīt ‚mōd }

byte multiplexor channel [COMPUT SCI] A transmission channel in a computer system that can transmit data simultaneously from several devices and only one byte at a time. { bīt 'məlt·i ‚plek·sər ‚chan·əl }

byte-oriented protocol [COMPUT SCI] A communications protocol in which full bytes are used as control codes. Also known as character-oriented protocol. { ¦bīt ‚ór·ē‚ent·əd 'prōd·ə‚kól }

C

C |COMPUT SCI| A programming language designed to implement the Unix operating system. |ELEC| *See* capacitance; capacitor; coulomb.

C++ |COMPUT SCI| An object-oriented language that was created as an extension to the C language. { 'sē,pləs,pləs }

cable |ELEC| Strands of insulated electrical conductors laid together, usually around a central core, and surrounded by a heavy insulation. { 'kā·bəl }

cable-and-trunk schematic |ELEC| A drawing which shows, in block form, the interconnection between all major electric circuits in an office. { ¦kā·bəl ən 'traŋk skə'mad·ik }

cable armor |ELEC| One or more layers of extra-strength material, such as steel wire or tape, to reinforce the usual lead wall in cable construction. { 'kā·bəl ,är·mər }

cable bridge |ELEC| A rubber tube that encloses cables running over a floor or other surface. { 'kā·bəl ,brij }

cable code *See* Morse cable code. { 'kā·bəl ,kōd }

cable complement |ELEC| Group of wire pairs in a cable having some common distinguishing characteristic. { 'kā·bəl ,käm·plə·mənt }

cable delay |COMPUT SCI| The time required for one bit of data to go through a cable, about 1.5 nanoseconds per foot of cable. { 'kā·bəl di'lā }

cable fill |ELEC| Ratio of the number of wire pairs in use to the total number of pairs in a cable. { 'kā·bəl ,fil }

cable matcher *See* gender changer. { 'kā·bəl ,mach·ər }

cable messenger |ELEC| Stranded group of wires supported above the ground at intervals by poles or other structures and employed to furnish, within these intervals, frequent points of support for conductors or cables. { 'kā·bəl ,mes·ən·jər }

cable modem |ELEC| A device that converts the signals used in a computer to signals that can be transmitted over cable television networks, and vice versa. { ¦kā·bəl ¦mō,dem }

cable noise |ELECTR| Electrical noise that is picked up by the conductors in a cable. { 'kā·bəl ,nóiz }

cable run |ELEC| Path occupied by a cable on cable racks or other support from one termination to another. { 'kā·bəl ,rən }

cable running list |ELEC| Drawing showing the code of cable, terminations, circuit names, and numbering of cables appearing in an office. { 'kā·bəl ¦rən·iŋ ,list }

cable shield |ELEC| A metallic layer applied over insulation covering a cable, composed of woven or braided wires, foil wrap, or metal tube, which acts to prevent electromagnetic or electrostatic interference from affecting conductors within. { 'kā·bəl ,shēld }

cable television |COMMUN| A television program distribution system in which signals from all local stations and usually a number of distant stations and program services are picked up by one or more high-gain antennas amplified on individual channels, then fed directly to individual receivers of subscribers by overhead or underground coaxial cable. Also known as community antenna television (CATV). { 'kā·bəl 'tel·ə,vizh·ən }

cabletext |COMMUN| Any videotex service that uses coaxial cable. { 'kā·bəl,tekst }

cable trough |ELEC| An enclosed channel, usually beneath a floor, that provides a path for cables. { 'kā·bəl ,tróf }

cable vault |ELEC| Vault in which the outside plant cables are spliced to the tipping cables. { 'kā·bəl ,vólt }

cache |COMPUT SCI| A small, fast storage buffer integrated in the central processing unit of some large computers. { kash }

CAD *See* computer-aided design. { kad }

CADD *See* computer-aided design and drafting. { kad }

caddy |COMPUT SCI| In certain types of disk drives, a plastic tray in which a CD-ROM disk is placed before loading. { 'kad·ē }

cadmium cell |ELEC| A standard cell used as a voltage reference; at 20°C its voltage is 1.0186 volts. { 'kad·mē·əm ,sel }

cadmium lamp |ELEC| A lamp containing cadmium vapor; wavelength (6438.4696 international angstroms, or 643.84696 nanometers) of light emitted is a standard of length. { 'kad·mē·əm ,lamp }

cadmium-nickel storage cell *See* nickel-cadmium battery. { ¦kad·mē·əm ¦nik·əl 'stór·ij ,sel }

cadmium selenide cell |ELECTR| A photoconductive cell that uses cadmium selenide as the semiconductor material and has a fast response time and high sensitivity to longer wavelengths of light.   { 'kad·mē·əm 'sel·ə,nīd ,sel }

cadmium silver oxide cell   [ELEC] An alkaline-electrolyte cell that may be used without recharging in primary batteries or that may be recharged for secondary-battery use.   { 'kad·mē·əm 'sil·vər 'äk,sīd ,sel }

cadmium sulfide cell   [ELECTR] A photoconductive cell in which a small wafer of cadmium sulfide provides an extremely high dark-light resistance ratio.   { 'kad·mē·əm 'səl,fīd ,sel }

cadmium telluride detector   [ELECTR] A photoconductive cell capable of operating continuously at ambient temperatures up to 750°F (400°C); used in solar cells and infrared, nuclear-radiation, and gamma-ray detectors.   { 'kad·mē·əm 'tel·yə,rīd di'tek·tər }

cadmium yellow   See cadmium sulfide.   { 'kad·mē·əm 'yel·ō }

cage antenna   [ELECTROMAG] Broad-band dipole antenna in which each pole consists of a cage of wires whose overall shape resembles that of a cylinder or a cone.   { 'kāj an'ten·ə }

CAI   See computer-assisted instruction.

CAL   [COMPUT SCI] A higher-level language, developed especially for time-sharing purposes, in which a user at a remote console typewriter is directly connected to the computer and can work out problems on-line with considerable help from the computer. Derived from Conversational Algebraic Language.   { kal }

calculated address   See generated address.   { 'kal·kyə,lād·əd 'ad,res }

calculating machine   See calculator.   { 'kal·kyə ,lād·iŋ mə'shēn }

calculator   [COMPUT SCI] A device that performs logic and arithmetic digital operations based on numerical data which are entered by pressing numerical and control keys. Also known as calculating machine.   { 'kal·kyə,lād·ər }

calculus of enlargement   See calculus of finite differences.   { 'kal·kyə·ləs əv in'lärj·mənt }

calculus of finite differences   [MATH] A method of interpolation that makes use of formal relations between difference operators which are, in turn, defined in terms of the values of a function on a set of equally spaced points. Also known as calculus of enlargement.   { 'kal·kyə·ləs əv 'fī,nīt 'dif·rən·səs }

calibration curve   [ENG] A plot of calibration data, giving the correct value for each indicated reading of a meter or control dial.   { 'kal·ə,brā·shən ,kərv }

calibration markers   [ENG] On a radar display, electronically generated marks which provide numerical values for the navigational parameters such as bearing, distance, height, or time.   { 'kal·ə,brā·shən ,mär·kərz }

call   [COMPUT SCI] **1.** To transfer control to a specified closed subroutine. **2.** A statement in a computer program that references a closed subroutine or program.   { kȯl }

call announcer   [ELECTR] Device for receiving pulses from an automatic telephone office and audibly reproducing the corresponding number

in words, so that it may be heard by a manual operator.   { 'kȯl ə'naún·sər }

call by location   [COMPUT SCI] A method of transferring arguments from a calling program to a subprogram in which the referencing program provides to the subprogram the memory location at which the value of the argument can be found, rather than the value itself. Also known as call by reference.   { 'kȯl bī ,lō'kā·shən }

call by name   [COMPUT SCI] A method of transferring arguments from a calling program to a subprogram in which the actual expression is passed to the subprogram.   { 'kȯl bī 'nām }

call by reference   See call by location.   { 'kȯl bī 'ref·rəns }

call by value   [COMPUT SCI] A method of transferring arguments from a calling program to a subprogram in which the subprogram is provided with the values of the argument and on path leads back to the referencing program.   { 'kȯl bī 'val·yü }

call circuit   [ELEC] Communications circuit between switching points used by traffic forces for transmitting switching instructions.   { 'kȯl ,sər·kət }

called routine   [COMPUT SCI] A subroutine that is accessed by a call or branch instruction in a computer program.   { 'kȯld rü,tēn }

call forwarding   [COMMUN] A telephone service that automatically transfers incoming calls to a designated number.   { 'kȯl 'fȯr·wərd·iŋ }

call in   [COMPUT SCI] To transfer control of a digital computer, temporarily, from a main routine to a subroutine that is inserted in the sequence of calculating operations, to fulfill an ancillary purpose.   { 'kȯl ,in }

call indicator   [ELECTR] Device for receiving pulses from an automatic switching system and displaying the corresponding called number before an operator at a manual switchboard.   { 'kȯl 'in·də,kād·ər }

calling device   [ELECTR] Apparatus which generates signals, either dual-tone multifrequency (DTMF) or the pulse required for establishing connections in an automatic telephone switching system.   { 'kȯl·iŋ di'vīs }

calling program   [COMPUT SCI] A computer program that initiates a call to another program.   { 'kȯl·iŋ ,prō·gram }

calling routine   [COMPUT SCI] A subroutine that initiates a call to another subroutine.   { 'kȯl·iŋ rü,tēn }

calling sequence   [COMPUT SCI] A specific set of instructions to set up and call a given subroutine, make available the data required by it, and tell the computer where to return after the subroutine is executed.   { 'kȯl·iŋ ,sē·kwəns }

call letters   [COMMUN] Identifying letters, sometimes including numerals, assigned to radio and television stations by the Federal Communications Commission and other regulatory authorities throughout the world. Also known as call sign.   { 'kȯl ,led·ərz }

call number   [COMPUT SCI] In computer operations, a set of characters identifying a subroutine, and containing information concerning parameters to be inserted in the subroutine, or information to be used in generating the subroutine, or information related to the operands. { 'kȯl ‚nəm·bər }

call setup time |COMMUN| The period of time between the lifting of a handset to make a telephone call and the start of voice or data transmission. { 'kȯl 'sed,əp ‚tīm }

call sign See call letters. { 'kȯl ‚sīn }

call up |COMPUT SCI| To retrieve data from computer memory, especially for display and user interaction. { 'kȯl ‚əp }

Calzecchi-Onesti effect |ELEC| A change in the conductivity of a loosely aggregated metallic powder caused by an applied electric field. { ‚kält'se·kē ‚ȯ'nes·tē i'fekt }

CAM See computer-aided manufacturing. { ¦sē ¦ā'em or kam }

camcorder |ELECTR| A one-piece hand-held television camera with built-in videocassette recorder, microphone, and battery pack, utilizing a charge-coupled device array as its light-sensitive element. { 'kam,córd·ər }

camera See television camera. { 'kam·rə }

camera cable |ELEC| Cable or group of wires that carries the picture from the television camera to the control room. { 'kam·rə ‚kā·bəl }

camera chain |COMMUN| A television camera, associated amplifiers, a monitor, and the cable needed to bring the camera output signal to the control room. { 'kam·rə 'chān }

camera tube |ELECTR| An electron-beam tube used in a television camera to convert an optical image into a corresponding charge-density electric image and to scan the resulting electric image in a predetermined sequence to provide an equivalent electric signal. Also known as pickup tube; television camera tube. { 'kam·rə ‚tüb }

Campbell bridge |ELEC| **1.** A bridge designed for comparison of mutual inductances. **2.** A circuit for measuring frequencies by adjusting a mutual inductance, until the current across a detector is zero. { 'kam·əl ‚brij }

camp-on system |COMMUN| A circuit control feature whereby a user attempting to establish a telephone call and encountering a busy station will hold the connection for a preset time, to the exclusion of other callers, in case the original conversation should terminate. { 'kamp ¦ȯn ‚sis·təm }

canceler |ELECTR| A circuit used in providing moving-target indication in radar, in which small sets of successive pulses are compared such that invariant returns, presumed indicative of stationary objects, are cancelled and ignored; a primitive form of Doppler processing. Usually cited as a "two-pulse" or "threee-pulse canceler," for example. { 'kan·səl·ər }

cancellation circuit |ELECTR| A circuit used in providing moving-target indication on a plan position indicator scope; cancels constant-amplitude fixed-target pulses by subtraction of successive pulse trains. { kan·sə'lā·shən ‚sər·kət }

canned cycle |COMPUT SCI| Any set of operations, either software or hardware, that is activated by a single command. { 'kand 'sī·kəl }

canned program |COMPUT SCI| A program which has been written to solve a particular problem, is available to users of a computer system, and is usually fixed in form and capable of little or no modification. { ¦kand 'prō·grəm }

canonical form |CONT SYS| A specific type of dynamical system representation in which the associated matrices possess specific row-column structures. { kə'nän·ə·kəl ‚fȯrm }

canonical schema |COMPUT SCI| A model that represents the structure and interrelationships of data within a database. { kə'nän·ə·kəl 'skē·mə }

capability |COMPUT SCI| A permission that is given to a user of a computing system in advance to access a particular object in the system in a particular way, and that the user can later present to a reference monitor as a prevalidated ticket to gain access. { ‚kāp·ə'bil·ə·dē }

capability list |COMPUT SCI| A row of an access matrix that contains the access rights of a given user to various files and other resources of a computer system. { ‚kā·pə'bil·əd·ē ‚list }

capacitance |ELEC| The ratio of the charge on one of the conductors of a capacitor (there being an equal and opposite charge on the other conductor) to the potential difference between the conductors. Symbolized C. Formerly known as capacity. { kə'pas·ə·təns }

capacitance altimeter |ENG| An absolute altimeter which determines height of an aircraft aboveground by measuring the variations in capacitance between two conductors on the aircraft when the ground is near enough to act as a third conductor. { kə'pas·ə·təns al'tim·əd·ər }

capacitance box |ELEC| An assembly of capacitors and switches which permits adjustment of the capacitance existing at the terminals in nominally uniform steps, from a minimum value near zero to the maximum which exists when all the capacitors are connected in parallel. { kə'pas·ə·təns ‚bäks }

capacitance bridge |ELEC| A bridge for comparing two capacitances, such as a Schering bridge. { kə'pas·ə·təns ‚brij }

capacitance hat |ELECTROMAG| A network of wires that is placed at the top of an antenna either to increase its bandwidth or to lower its resonant frequency. { kə'pas·əd·əns ‚hat }

capacitance level indicator |ENG| A level indicator in which the material being monitored serves as the dielectric of a capacitor formed by a metal tank and an insulated electrode mounted vertically in the tank. { kə'pas·ə·təns ¦lev·əl 'in·də,kād·ər }

capacitance meter |ENG| An instrument used to measure capacitance values of capacitors or of circuits containing capacitance. { kə'pas·ə·təns ‚mēd·ər }

capacitance-operated intrusion detector |ENG| A boundary alarm system in which the approach of an intruder to an antenna wire encircling the protected area a few feet above ground changes the antenna-ground capacitance and sets off the alarm. { kə¦pas·ə·təns¦;aop·ə,räd·əd in'trü·zhən di'tek·tər }

capacitance relay [ELECTR] An electronic relay that responds to a small change in capacitance, such as that created by bringing a hand near a pickup wire or plate. { kə¦pas·ə·təns 'rē,lā }

capacitance standard See standard capacitor. { kə¦pas·ə·təns ,stan·dərd }

capacitive coupling [ELEC] Use of a capacitor to transfer energy from one circuit to another. { kə¦pas·ə·təns ,kəp·liŋ }

capacitive diaphragm [ELECTROMAG] A resonant window used in a waveguide to provide the equivalent of capacitive reactance at the frequency being transmitted. { kə¦pas·əd·iv 'dī·ə,fram }

capacitive-discharge ignition [ELECTR] An automotive ignition system in which energy is stored in a capacitor and discharged across the gap of a spark plug through a step-up pulse transformer and distributor each time a silicon controlled rectifier is triggered. { kə¦pas·əd·iv ¦dis,chärj ig ¦nish·ən }

capacitive-discharge pilot light [ELECTR] An electronic ignition system, operating off an alternating-current power line or battery power supply, that produces a spark for lighting a gas flame. { kə¦pas·əd·iv ¦dis,chärj 'pī·lət ,līt }

capacitive divider [ELEC] Two or more capacitors placed in series across a source, making available a portion of the source voltage across each capacitor; the voltage across each capacitor will be inversely proportional to its capacitance. { kə¦pas·əd·iv di'vīd·ər }

capacitive electrometer [ENG] An instrument for measuring small voltages; the voltage is applied to the plates of a capacitor when they are close together, then the voltage source is removed and the plates are separated, increasing the potential difference between them to a measurable value. Also known as condensing electrometer. { kə¦pas·əd·iv ,i,lek'träm·əd·ər }

capacitive feedback [ELECTR] Process of returning part of the energy in the plate (or output) circuit of a vacuum tube (or other device) to the grid (or input) circuit by means of a capacitance common to both circuits. { kə¦pas·əd·iv 'fēd ,bak }

capacitive loading [ELECTROMAG] **1.** Raising the resonant frequency of an antenna by connecting a fixed capacitor or capacitors in series with it. **2.** Lowering the resonant frequency of an antenna by installing a capacitance hat. { kə¦pas·əd·iv 'lōd·iŋ }

capacitive post [ELECTROMAG] Metal post or screw extending across a waveguide at right angles to the E field, to provide capacitive susceptance in parallel with the waveguide for tuning or matching purposes. { kə¦pas·əd·iv ¦pōst }

capacitive pressure transducer [ENG] A measurement device in which variations in pressure upon a capacitive element proportionately change the element's capacitive rating and thus the strength of the measured electric signal

from the device. { kə¦pas·əd·iv 'presh·ər tranz ,dü·sər }

capacitive reactance [ELECTROMAG] Reactance due to the capacitance of a capacitor or circuit, equal to the inverse of the product of the capacitance and the angular frequency. { kə ¦pas·əd·iv rē'ak·təns }

capacitive tuning [ELECTR] Tuning involving use of a variable capacitor. { kə¦pas·əd·iv 'tün·iŋ }

capacitive window [ELECTROMAG] Conducting diaphragm extending into a waveguide from one or both sidewalls, producing the effect of a capacitive susceptance in parallel with the waveguide. { kə¦pas·əd·iv 'win·dō }

capacitor [ELEC] A device which consists essentially of two conductors (such as parallel metal plates) insulated from each other by a dielectric and which introduces capacitance into a circuit, stores electrical energy, blocks the flow of direct current, and permits the flow of alternating current to a degree dependent on the capacitor's capacitance and the current frequency. Symbolized C. Also known as condenser; electric condenser. { kə¦pas·əd·ər }

capacitor antenna [ELECTROMAG] Antenna consisting of two conductors or systems of conductors, the essential characteristic of which is its capacitance. Also known as condenser antenna. { kə¦pas·əd·ər an'ten·ə }

capacitor bank [ELEC] A number of capacitors connected in series or in parallel. { kə¦pas·əd·ər ,baŋk }

capacitor box [ELECTR] A box-shaped structure in which a capacitor is submerged in a heat-absorbing medium, usually water. Also known as condenser box. { kə¦pas·əd·ər ,bäks }

capacitor color code [ELEC] A method of marking the value on a capacitor by means of dots or bands of colors as specified in the Electronic Industry Association color code. { kə¦pas·əd·ər 'kəl·ər ,kōd }

capacitor-input filter [ELECTR] A power-supply filter in which a shunt capacitor is the first element after the rectifier. { kə¦pas·əd·ər ¦in,püt ,fil·tər }

capacitor loudspeaker See electrostatic loudspeaker. { kə¦pas·əd·ər 'laüd,spēk·ər }

capacitor microphone [ENG ACOUS] A microphone consisting essentially of a flexible metal diaphragm and a rigid metal plate that together form a two-plate air capacitor; sound waves set the diaphragm in vibration, producing capacitance variations that are converted into audio-frequency signals by a suitable amplifier circuit. Also known as condenser microphone; electrostatic microphone. { kə¦pas·əd·ər 'mī· krə,fōn }

capacitor motor [ELEC] A single-phase induction motor having a main winding connected directly to a source of alternating-current power and an auxiliary winding connected in series with a capacitor to the source of ac power. See capacitor-start motor. { kə¦pas·əd·ər ,mōd·ər }

capacitor-resistor unit See rescap. { kə¦pas·əd·ər ri'zis·tər ,yü·nət }

capacitor-start motor |ELEC| A capacitor motor in which the capacitor is in the circuit only during the starting period; the capacitor and its auxiliary winding are disconnected automatically by a centrifugal switch or other device when the motor reaches a predetermined speed. Also known as capacitor motor. { kə'pas·əd·ər ¦stärt ‚mōd·ər }

capacitor start-run motor See permanent-split capacitor motor. { kə'pas·əd·ər ¦stärt ¦rən ‚mōd·ər }

capacity See capacitance; storage capacity. { kə'pas·əd·ē }

capacity cell |ELEC| **1.** Capacitance-type device used to measure the dielectric constants of gases, liquids, or solids. **2.** Capacitance-type device used to monitor certain composition changes in flowing streams. { kə'pas·əd·ē ‚sel }

capacity-rate product |COMMUN| The product of the capacity of a data-storage device in gigabytes and the data rate in megabits per second. { kə'pas·ə·dē ‚rāt ‚präd·əkt }

capristor See rescap. { ka'pris·tər }

capstan |ENG| A shaft which pulls magnetic tape through a machine at constant speed. { 'kap·stən }

capture effect |ELECTR| The effect wherein a strong frequency-modulation signal in an FM receiver completely suppresses a weaker signal on the same or nearly the same frequency. { 'kap·chər i'fekt }

capture ratio |COMMUN| A measure of the ability of a frequency-modulation tuner to reject the weaker of two stations that are on the same frequency; the lower the ratio of desired to undesired signals, the better the performance of the tuner. { 'kap·chər ‚rā·shō }

CAR See computer-assisted retrieval. { kär }

carbon arc |ELEC| An electric arc between two electrodes, at least one of which is made of carbon; used in welding and high-intensity lamps, such as in searchlights and photography lamps. { ¦kär·bən ¦ärk }

carbon-arc lamp |ELEC| An arc lamp in which an electric current flows between two electrodes of pure carbon, with incandescence at one or both electrodes and some light from the luminescence of the arc. { ¦kär·bən ¦ärk 'lamp }

carbon brush |ELEC| A rod made of carbon that bears against a commutator, collector ring, or slip ring to provide passage for the electric current from a dynamo through an outside circuit or for an external current through a motor. { ¦kär·bən ¦brəsh }

carbon button See button. { ¦kär·bən ¦bət·ən }

carbon-film hygrometer element |ELEC| An electrical hygrometer element constructed of a plastic strip coated with a film of carbon black dispersed in a hygroscopic binder; variations in atmospheric moisture content vary the volume of the binder and thus change the resistance of the carbon coating. { 'kär·bən ‚film hī'gräm·əd·ər ‚el·ə·mənt }

carbon-film resistor |ELEC| A resistor made by depositing a thin carbon film on a ceramic form. { 'kär·bən ‚film ri'zis·tər }

carbon lamp |ELEC| An arc lamp with carbon electrodes. { 'kär·bən ‚lamp }

carbon microphone |ENG ACOUS| A microphone in which a flexible diaphragm moves in response to sound waves and applies a varying pressure to a container filled with carbon granules, causing the resistance of the microphone to vary correspondingly. { ¦kär·bən 'mī·krə‚fōn }

carbon pile |ELEC| A variable resistor consisting of a stack of carbon disks mounted between a fixed metal plate and a movable one that serve as the terminals of the resistor; the resistance value is reduced by applying pressure to the movable plate. { 'kär·bən ‚pīl }

carbon-pile pressure transducer |ENG| A measurement device in which variations in pressure upon a conductive carbon core proportionately change the core's electrical resistance, and thus the strength of the measured electric signal from the device. { 'kär·bən ‚pīl 'presh·ər tranz ‚dü·sər }

carbon resistor |ELECTR| A resistor consisting of carbon particles mixed with a binder, molded into a cylindrical shape, and baked; terminal leads are attached to opposite ends. Also known as composition resistor. { 'kär·bən ri'zis·tər }

carbon transducer |ENG| A transducer consisting of carbon granules in contact with a fixed electrode and a movable electrode, so that motion of the movable electrode varies the resistance of the granules. { 'kär·bən tranz'dü·sər }

carcinotron See backward-wave oscillator. { 'kärs·ən·ə‚trän }

card |COMPUT SCI| See punch card. |ELECTR| A printed circuit board or other arrangement of miniaturized components that can be plugged into a computer or peripheral device. { kärd }

card cage |ELECTR| A rack built into a computer to hold printed circuit boards and allow them to be installed or removed easily. { 'kärd ‚kāj }

card dialer |COMMUN| A telephone in which a number can be dialed automatically and almost instantly by inserting a coded card for that number in a slot on the dialer; now obsolete, having been replaced by automatic dialers using electronic memory. { 'kärd ‚dī·lər }

card-edge connector |ELEC| A connector that mates with printed-wiring leads running to the edge of a printed circuit board on one or both sides. Also known as edgeboard connector. { 'kärd ‚ej kə'nek·tər }

card holder |ELECTR| A U-shaped slot designed to hold the edge of a printed circuit board securely in a card cage. { 'kärd ‚hōl·dər }

cardinal point effect |ELECTR| The increased intensity of a line or group of returns on the radarscope occurring when the radar beam is perpendicular to the rectangular surface of a line or group of similarly aligned features in the ground pattern. { 'kärd·nəl ‚point i'fekt }

cardioid microphone |ENG ACOUS| A microphone having a heart-shaped, or cardioid, response pattern, so it has nearly uniform response for a range of about 180° in one direction and

cardioid pattern

minimum response in the opposite direction. { 'kärd·ē,óid 'mī·krə,fōn }

cardioid pattern [ENG] Heart-shaped pattern obtained as the response or radiation characteristic of certain directional antennas, or as the response characteristic of certain types of microphones. { 'kärd·ē,óid ,pad·ərn }

card key access [ENG] A physical security system in which doors are unlocked by placing a badge that contains magnetically coded information in proximity to a reading device; some systems also require the typing of this information on a keyboard. { 'kärd ,kē 'ak,ses }

card slot [ELECTR] A groove where a printed circuit board fits into a card cage or backplane. { 'kärd ,slät }

carriage return [COMPUT SCI] The operation that causes the next character to be printed at the extreme left margin, and usually advances to the next line at the same time. { 'kar·ij ri'tərn }

carrier [COMMUN] **1.** The radio wave produced by a transmitter when there is no modulating signal, or any other wave, recurring series of pulses, or direct current capable of being modulated. Also known as carrier wave; signal carrier. **2.** A wave generated locally at a receiver that, when combined with the sidebands of a suppressed-carrier transmission in a suitable detector, produces the modulating wave. **3.** *See* carrier system. [SOLID STATE] *See* charge carrier. { 'kar·ē·ər }

carrier amplifier [ELECTR] A direct-current amplifier in which the dc input signal is filtered by a low-pass filter, then used to modulate a carrier so it can be amplified conventionally as an alternating-current signal; the amplified dc output is obtained by rectifying and filtering the rectified carrier signal. { 'kar·ē·ər ,am·plə,fī·ər }

carrier amplitude regulation [COMMUN] Change in amplitude of the carrier wave in an amplitude-modulated transmitter when modulation is applied under conditions of symmetrical modulation. { 'kar·ē·ər 'am·plə,tüd reg·yə'lā·shən }

carrier beat [COMMUN] An undesirable heterodyne of facsimile signals, each synchronous with a different stable reference oscillator, causing a pattern in received copy. { 'kar·ē·ər ,bēt }

carrier channel [COMMUN] The equipment and lines that make up a complete carrier-current circuit between two or more points. { 'kar·ē·ər ,chan·əl }

carrier chrominance signal *See* chrominance signal. { 'kar·ē·ər 'krō·mə·nəns ,sig·nəl }

carrier current [COMMUN] A higher-frequency alternating current superimposed on ordinary telephone, telegraph, and power-line frequencies for communication and control purposes. { 'kar·ē·ər ,kər·ənt }

carrier detect [COMPUT SCI] A signal sent by a modem to a computer or a terminal to indicate that it is receiving a character. { 'kar·ē·ər di ,tekt }

carrier frequency [COMMUN] The frequency generated by an unmodulated radio, radar, carrier

communication, or other transmitter, or the average frequency of the emitted wave when modulated by a symmetrical signal. Also known as center frequency; resting frequency. { 'kar·ē·ər ,frē·kwən·sē }

carrier leak [COMMUN] Carrier remaining after carrier suppression in a suppressed-carrier transmission system. { 'kar·ē·ər ,lēk }

carrier level [COMMUN] The strength or level of an unmodulated carrier signal at a particular point in a radio system, expressed in decibels in relation to some reference level. { 'kar·ē·ər ,lev·əl }

carrier line [ELEC] Any transmission line used for multiple-channel carrier communication. { 'kar·ē·ər ,līn }

carrier loading [ELECTROMAG] The addition of lumped inductances to the cable section of a transmission line specifically designed for carrier transmission; it serves to minimize impedance mismatch between cable and open wire and to reduce the cable attenuation. { 'kar·ē·ər ,lōd·iŋ }

carrier noise [COMMUN] Noise produced by undesired variation of a radio-frequency signal in the absence of any intended modulation. Also known as residual modulation. { 'kar·ē·ər ,nóiz }

carrier power output rating [COMMUN] Power available at the output terminals of a transmitter when the output terminals are connected to the normal-load circuit or to a circuit equivalent thereto. { 'kar·ē·ər ¦pau·ər ,aut,pút ,rād·iŋ }

carrier repeater [ELECTR] Equipment designed to raise carrier signal levels to such a value that they may traverse a succeeding line section at such amplitude as to preserve an adequate signal-to-noise ratio; while the heart of a repeater is the amplifier, necessary adjuncts are filters, equalizers, level controls, and so on, depending upon the operating methods. { 'kar·ē·ər ri'pēd·ər }

carrier sense multiple access with collision detection *See* CSMA/CD. { 'kar·ē·ər ¦sens 'məl·tə·pəl 'ak,ses with kə'lizh·ən di,tek·shən }

carrier shift [COMMUN] **1.** Transmission of information by radio through shifting the carrier frequency in one direction for a mark signal and in the opposite direction for a spacing signal. **2.** Condition resulting from imperfect modulation whereby the positive and negative excursions of the envelope pattern are unequal, thus effecting a change in the power associated with the carrier. { 'kar·ē·ər ,shift }

carrier signaling [COMMUN] Method by which busy signals, ringing, or dial signaling relays are operated by the transmission of a carrier-frequency tone. { 'kar·ē·ər ,sig·nəl·iŋ }

carrier suppression [COMMUN] **1.** Suppression of the carrier frequency after conventional modulation at the transmitter, with reinsertion of the carrier at the receiving end before demodulation. **2.** Suppression of the carrier when there is no modulation signal to be transmitted; used on ships to reduce interference between transmitters. { 'kar·ē·ər sə'presh·ən }

carrier swing |COMMUN| The total deviation of a frequency-modulated or phase-modulated wave from the lowest instantaneous frequency to the highest instantaneous frequency. { 'kar·ē·ər ˌswiŋ }

carrier system |COMMUN| A system permitting a number of simultaneous, independent communications over the same circuit. Also known as carrier. { 'kar·ē·ər ˌsis·təm }

carrier telegraphy |COMMUN| Telegraphy in which a single-frequency carrier wave is modulated by the transmitting apparatus for transmission over wire lines. { 'kar·ē·ər tə'leg·rə·fē }

carrier telephony |COMMUN| Telephony in which a single-frequency carrier wave is modulated by a voice-frequency signal for transmission over wire lines. { 'kar·ē·ər tə'lef·ə·nē }

carrier terminal |ELECTR| Apparatus at one end of a carrier transmission system, whereby the processes of modulation, demodulation, filtering, amplification, and associated functions are effected. { 'kar·ē·ər ˌtərm·ən·əl }

carrier-to-noise ratio |COMMUN| The ratio of the magnitude of the carrier to that of the noise after specified band limiting and before any nonlinear process such as amplitude limiting and detection. { ¦kar·ē·ər tə ¦nȯiz ˌrā·shō }

carrier transfer filters |ELECTR| Filters arranged as a carrier-frequency crossover or bridge between two transmission circuits. { 'kar·ē·ər ¦tranz·fər ˌfil·tərz }

carrier transmission |COMMUN| Transmission in which the transmitted electric wave is a wave resulting from the modulation of a single-frequency wave by a modulating wave. { 'kar·ē·ər tranz'mish·ən }

carrier wave See carrier. { 'kar·ē·ər ˌwāv }

carry |MATH| An arithmetic operation that occurs in the course of addition when the sum of the digits in a given position equals or exceeds the base of the number system; a multiple m of the base is subtracted from this sum so that the remainder is less than the base, and the number m is then added to the next-higher-order digit. { 'kar·ē }

carry-complete signal |COMPUT SCI| A signal generated by a digital parallel adder, indicating that all carries from an adding operation have been generated and propagated, and that the addition operation is completed. { ¦kar·ē kəm ¦plēt ˌsig·nəl }

carry flag |COMPUT SCI| A flip-flop circuit which indicates overflow in arithmetic operations. { 'kar·ē ˌflag }

carrying capacity |ELEC| The maximum amount of current or power that can be safely handled by a wire or other component. { 'kar·ē·iŋ kə'pas·əd·ē }

carry lookahead |COMPUT SCI| A circuit which allows low-order carries to ripple through all the way to the highest-order bit to output a completed sum. { 'kar·ē 'lúk·ə,hed }

carry-save adder |COMPUT SCI| A device for the rapid addition of three operands; consists of a sequence of full adders, in which one of the operands is entered in the carry inputs, and the carry outputs, instead of feeding the carry inputs of the following full adders, form a second output word which is then added to the ordinary output in a two-operand adder to form the final sum. { ¦kar·ē ¦sāv 'ad·ər }

carry signal |COMPUT SCI| A signal produced in a computer when the sum of two digits in the same column equals or exceeds the base of the number system in use or when the difference between two digits is less than zero. { 'kar·ē ˌsig·nəl }

carry time |COMPUT SCI| The time needed to transfer all carry digits to the next higher column. { 'kar·ē ˌtīm }

Cartesian-coordinate robot |CONT SYS| A robot having orthogonal, sliding joints and supported by a nonrotary base as the axis. { kär'tē·zhən kō ¦örd·ən·ət 'rō,bät }

cartridge |COMPUT SCI| A self-contained module that contains disks, magnetic tape, or integrated circuits for storing data. { 'kär·trij }

cartridge disk |COMPUT SCI| A type of disk storage device consisting of a single disk encased in a compact container which can be inserted in and removed from the disk drive unit; used extensively with computer systems. { 'kär·trij ˌdisk }

cartridge font |COMPUT SCI| A font for a computer printer that is stored on a read-only memory chip within a cartridge (a module that is inserted in a slot in the printer). { 'kär·trij ˌfänt }

cartridge fuse |ELEC| A type of electric fuse in which the fusible element is connected between metal ferrules at either end of an insulating tube. { 'kär·trij ˌfyüz }

cartridge lamp |ELEC| A pilot or dial lamp that has a tubular glass envelope with metal-ferrule terminals at each end. { 'kär·trij ˌlamp }

cartridge tape drive |COMPUT SCI| A tape drive which will automatically thread the tape on the takeup reels without human assistance. Formerly known as hypertape drive. { 'kär·trij ˌtāp ˌdrīv }

cascade |COMPUT SCI| A series of actions that take place in the course of data processing, each triggered by the previous action in the series. |ELEC| An electric-power circuit arrangement in which circuit breakers of reduced interrupting ratings are used in the branches, the circuit breakers being assisted in their protection function by other circuit breakers which operate almost instantaneously. Also known as backup arrangement. |ELECTR| See avalanche. { ka'skād }

cascade amplifier |ELECTR| A vacuum-tube amplifier containing two or more stages arranged in the conventional series manner. Also known as multistage amplifier. { ka'skād ˌam·plə,fī·ər }

cascade-amplifier klystron |ELECTR| A klystron having three resonant cavities to provide increased power amplification and output; the extra resonator, located between the input and output resonators, is excited by the bunched beam emerging from the first resonator gap and produces further bunching of the beam. { ka'skād ˌam·plə,fī·ər 'klī,strän }

cascade compensation |CONT SYS| Compensation in which the compensator is placed in series

75

cascade connection

with the forward transfer function. Also known as series compensation; tandem compensation. { ka'skād käm·pən'sā·shən }

cascade connection [ELECTR] A series connection of amplifier stages, networks, or tuning circuits in which the output of one feeds the input of the next. Also known as tandem connection. { ka'skād kə'nek·shən }

cascade control [CONT SYS] An automatic control system in which various control units are linked in sequence, each control unit regulating the operation of the next control unit in line. { ka'skād kən,trōl }

cascade converter [ELEC] A rotary converter that is powered from the secondary of an induction motor that is connected to the same shaft. { ka'skād kən,vərd·ər }

cascaded [ENG] Of a series of elements or devices, arranged so that the output of one feeds directly into the input of another, as a series of dynodes or a series of airfoils. { ka'skād·əd }

cascaded carry [COMPUT SCI] A carry process in which the addition of two numerals results in a sum numeral and a carry numeral that are in turn added together, this process being repeated until no new carries are generated. { ka'skād·əd 'kar·ē }

cascaded feedback canceler [ELECTR] Sophisticated moving-target-indicator canceler which provides clutter and chaff rejection. Also known as velocity shaped canceler. { ka'skād·əd 'fēd ,bak ,kan·slər }

cascade image tube [ELECTR] An image tube having a number of sections stacked together, the output image of one section serving as the input for the next section; used for light detection at very low levels. { ka'skād 'im·ij ,tüb }

cascade junction [ELECTR] Two pn semiconductor junctions in tandem such that the condition of the first governs that of the second. { ka'skād 'jəŋk·shən }

cascade limiter [ELECTR] A limiter circuit that uses two vacuum tubes in series to give improved limiter operation for both weak and strong signals in a frequency-modulation receiver. Also known as double limiter. { ka 'skād 'lim·əd·ər }

cascade mixing [ELEC] A mechanism for ion-beam mixing of a film and a substrate in which the recoil of an atom from a collision with an incident ion initiates a series of secondary collisions among the film and substrate atoms, leading to transfer of atoms from the substrate into the film as well as from the film into the substrate. { ka'skād ,mik·siŋ }

cascade networks [ELEC] Two networks in tandem such that the output of the first feeds the input of the second. { ka'skād 'net,wərks }

cascade noise [ELECTR] The noise in a communications receiver after an input signal has been subjected to two tandem stages of amplification. { ka'skād 'nȯiz }

cascade transformer [ELEC] A source of high voltage that is made up of a collection of step-up transformers; secondary windings are in series, and primary windings, except the first, are supplied from a pair of taps on the secondary winding of the preceding transformer. { ka'skād tranz'fȯr·mər }

cascading [ELEC] An effect in which a failure of an electrical power system causes this system to draw excessive amounts of power from power systems which are interconnected with it, causing them to fail, and these systems cause adjacent systems to fail in a similar manner, and so forth. { ka'skād·iŋ }

cascading menu [COMPUT SCI] A menu that appears next to a pull-down menu as the result of selecting a choice on the latter. { ka,skād·iŋ 'men·yü }

cascading windows [COMPUT SCI] Two or more windows displayed so that they overlap but their title bars are still visible. { ka,skād·iŋ 'win,dōz }

cascode amplifier [ELECTR] An amplifier consisting of a grounded-emitter input stage that drives a grounded-base output stage; advantages include high gain and low noise; widely used in television tuners. { 'ka,skōd 'am·plə,fī·ər }

case [COMPUT SCI] **1.** In computers, a set of data to be used by a particular program. **2.** The metal box that houses a computer's circuit boards, disk drives, and power supply. Also known as system unit. { kās }

CASE See computer-aided software engineering. { kās }

case-sensitive language [COMPUT SCI] A programming language in which upper-case letters are distinguished from lower-case letters. { ¦kās ,sens·ə·tiv 'laŋ·gwij }

case structure [COMPUT SCI] A group of program statements in which a condition is tested and, according to the results of the test, one of at least three specific groups of program statements is executed, after which the program returns to the original location. { 'kās ,strək·chər }

Cassegrain antenna [ELECTROMAG] A microwave antenna in which the feed radiator is mounted at or near the surface of the main reflector and aimed at a mirror at the focus; energy from the feed first illuminates the mirror, then spreads outward to illuminate the main reflector. { kas·gran an'ten·ə }

cassette [ENG ACOUS] A small, compact container that holds a magnetic tape and can be readily inserted into a matching tape recorder for recording or playback; the tape passes from one hub within the container to the other hub. { kə'set }

cassette cartridge system [COMPUT SCI] An input system often used in computers; its low cost and ease in mounting often offset its slow access time. { kə'set ,kär·trij ,sis·təm }

cassette memory [COMPUT SCI] A removable magnetic tape cassette that stores computer programs and data. { kə'set 'mem·rē }

catalog [COMPUT SCI] **1.** All the indexes to data sets or files in a system. **2.** The index to all other indexes; the master index. **3.** To add an entry to an index or to build an entire new index. **4.** A list of items in a data storage device, usually

76

arranged so that a particular kind of information can be located easily.   { 'kad·əl,äg }

catalog-order device [ELECTR] A logic circuit element that is readily obtainable from a manufacturer, and can be combined with other such elements to provide a wide variety of logic circuits.   { 'kad·əl,äg ¦ȯr·dər di'vīs }

catastrophic error [COMPUT SCI] A situation in which so many errors are detected in a computer program that its compilation or execution is automatically terminated.   { ,kad·ə¦sträf·ik 'er·ər }

catastrophic failure [ENG] **1.** A sudden failure without warning, as opposed to degradation failure. **2.** A failure whose occurrence can prevent the satisfactory performance of an entire assembly or system.   { ,kad·ə'sträf·ik 'fāl·yər }

catcher [ELECTR] Electrode in a velocity-modulated vacuum tube on which the spaced electron groups induce a signal; the output of the tube is taken from this element.   { 'kach·ər }

catching diode [ELECTR] Diode connected to act as a short circuit when its anode becomes positive; the diode then prevents the voltage of a circuit terminal from rising above the diode cathode voltage.   { 'kach·iŋ ,dī,ōd }

categorization [COMPUT SCI] Process of separating multiple addressed messages to form individual messages for singular addresses.   { ,kad·ə·gə·rə'zā·shən }

catena [COMPUT SCI] A series of data items that appears in a chained list.   { kə'tē·nə }

catenate [COMPUT SCI] To arrange a collection of items in a chained list or catena.   { 'kat·ən,āt }

cathode [ELEC] The terminal at which current leaves a primary cell or storage battery; it is negative with respect to the device, and positive with respect to the external circuit. [ELECTR] **1.** The primary source of electrons in an electron tube; in directly heated tubes the filament is the cathode, and in indirectly heated tubes a coated metal cathode surrounds a heater. Designated K. Also known as negative electrode. **2.** The terminal of a semiconductor diode that is negative with respect to the other terminal when the diode is biased in the forward direction.   { 'kath,ōd }

cathode bias [ELECTR] Bias obtained by placing a resistor in the common cathode return circuit, between cathode and ground; flow of electrode currents through this resistor produces a voltage drop that serves to make the control grid negative with respect to the cathode.   { 'kath,ōd ,bī·əs }

cathode-coupled amplifier [ELECTR] A cascade amplifier in which the coupling between two stages is provided by a common cathode resistor.   { ¦kath,ōd ¦kəp·əld 'am·plə,fī·ər }

cathode coupling [ELECTR] Use of an input or output element in the cathode circuit for coupling energy to another stage.   { 'kath,ōd ,kəp·liŋ }

cathode crater [ELECTR] A depression formed in the surface of a cathode by sputtering.   { 'kath,ōd ,krād·ər }

cathode dark space [ELECTR] The relatively non-luminous region between the cathode glow and the negative flow in a glow-discharge cold-cathode tube. Also known as Crookes dark space; Hittorf dark space.   { 'kath,ōd 'därk ,spās }

cathode disintegration [ELECTR] The destruction of the active area of a cathode by positive-ion bombardment.   { 'kath,ōd dis,int·ə'grā·shən }

cathode drop [ELECTR] The voltage between the arc stream and the cathode of a glow-discharge tube. Also known as cathode fall.   { 'kath,ōd ,dräp }

cathode emission [ELECTR] A process whereby electrons are emitted from the cathode structure.   { 'kath,ōd i'mish·ən }

cathode fall *See* cathode drop.   { 'kath,ōd ,fȯl }

cathode follower [ELECTR] A vacuum-tube circuit in which the input signal is applied between the control grid and ground, and the load is connected between the cathode and ground. Also known as grounded-anode amplifier; grounded-plate amplifier.   { 'kath,ōd ,fäl·ə·wər }

cathode glow [ELECTR] The luminous glow that covers all or part of the cathode in a glow-discharge cold-cathode tube.   { 'kath,ōd ,glō }

cathode interface capacitance [ELECTR] A capacitance which, when connected in parallel with an appropriate resistance, forms an impedance approximately equal to the cathode interface impedance. Also known as layer capacitance.   { ¦kath,ōd ¦in·tər,fās kə'pas·əd·əns }

cathode interface impedance [ELECTR] The impedance between the cathode base and coating on an electron tube, due to a high-resistivity layer or a poor mechanical bond. Also known as layer impedance.   { ¦kath,ōd ¦in·tər ,fās im'pēd·əns }

cathode keying [ELECTR] Transmitter keying by means of a key in the cathode lead of the keyed vacuum-tube stage, opening the direct-current circuits for the grid and anode simultaneously.   { 'kath,ōd ,kē·iŋ }

cathode layers [ELECTR] One or more faint layers next to, and on the anode side of, the Aston dark space in a glow-discharge tube.   { 'kath,ōd ,lā·ərz }

cathode modulation [ELECTR] Amplitude modulation accomplished by applying the modulating voltage to the cathode circuit of an electron tube in which the carrier is present.   { 'kath,ōd ,mäj·ə'lā·shən }

cathode ray [ELECTR] A stream of electrons, such as that emitted by a heated filament in a tube, or that emitted by the cathode of a gas-discharge tube when the cathode is bombarded by positive ions.   { 'kath,ōd ¦rā }

cathode-ray oscillograph [ELECTR] A cathode-ray oscilloscope in which a photographic or other permanent record is produced by the electron beam of the cathode-ray tube.   { 'kath,ōd ¦rā ä'sil·ə,graf }

cathode-ray oscilloscope [ELECTR] A test instrument that uses a cathode-ray tube to make visible on a fluorescent screen the instantaneous values and waveforms of electrical quantities that are rapidly varying as a function of time or another quantity. Abbreviated CRO. Also known as oscilloscope; scope. { 'kath‚ōd ¦rā ä'sil‚ə ‚skōp }

cathode-ray storage tube |ELECTR| A storage tube in which the information is written by means of a cathode-ray beam. { 'kath‚ōd ¦rā 'stȯr‚ij ‚tüb }

cathode-ray tube |ELECTR| An electron tube in which a beam of electrons can be focused to a small area and varied in position and intensity on a surface. Abbreviated CRT. Originally known as Braun tube; also known as electron-ray tube. { 'kath‚ōd ¦rā ‚tüb }

cathode-ray tuning indicator |ELECTR| A small cathode-ray tube having a fluorescent pattern whose size varies with the voltage applied to the grid; used in radio receivers to indicate accuracy of tuning and as a modulation indicator in some tape recorders. Also known as electric eye; electron-ray indicator; magic eye; tuning eye. { 'kath‚ōd ¦rā 'tün·iŋ in·də'kād·ər }

cathode-ray voltmeter |ELEC| An instrument consisting of a cathode-ray tube of known sensitivity, whose deflection can be used to measure voltages. { 'kath‚ōd ¦rā 'vōlt‚mēd·ər }

cathode resistor |ELECTR| A resistor used in the cathode circuit of a vacuum tube, having a resistance value such that the voltage drop across it due to tube current provides the correct negative grid bias for the tube. { 'kath‚ōd ri'zis·tər }

cathode spot |ELECTR| The small cathode area from which an arc appears to originate in a discharge tube. { 'kath‚ōd ‚spät }

cathode sputtering *See* sputtering. { 'kath‚ōd 'spəd·ə·riŋ }

cathodoluminescence |ELECTR| Luminescence produced when high-velocity electrons bombard a metal in vacuum, thus vaporizing small amounts of the metal in an excited state, which amounts emit radiation characteristic of the metal. Also known as electroluminescence. { ¦kath·ə‚dō‚lüm·ə'nes·əns }

cathodophosphorescence |ELECTR| Phosphorescence produced when high-velocity electrons bombard a metal in a vacuum. { ¦kath·ə‚dō ‚fas·fə'res·əns }

CATT *See* controlled avalanche transit-time triode. { kat }

CATV *See* cable television.

catwhisker |ELECTR| A sharply pointed, flexible wire used to make contact with the surface of a semiconductor crystal at a point that provides rectification. { 'kat‚wis·kər }

Cauer filter *See* elliptic-integral filter. { 'kau̇·ər ‚fil·tər }

Cauer form |ELEC| A continued fraction expansion of the impedance used in the network synthesis for a driving point function resulting in a ladder network. { 'kau̇·ər ‚fȯrm }

causal system |CONT SYS| A system whose response to an input does not depend on values of the input at later times. Also known as nonanticipatory system; physical system. { 'kȯ·zəl ‚sis·təm }

cautious control |CONT SYS| A control law for a stochastic adaptive control system which hedges

and uses lower gain when the estimates are uncertain. { 'kȯ·shəs kən'trōl }

cavity *See* cavity resonator. { 'kav·əd·ē }

cavity coupling |ELECTROMAG| The extraction of electromagnetic energy from a resonant cavity, either waveguide or coaxial, using loops, probes, or apertures. { 'kav·əd·ē ‚kəp·liŋ }

cavity filter |ELECTROMAG| A microwave filter that uses quarter-wavelength-coupled cavities inserted in waveguides or coaxial lines to provide band-pass or other response characteristics at frequencies in the gigahertz range. { 'kav·əd·ē ‚fil·tər }

cavity frequency meter |ENG| A device that employs a cavity resonator to measure microwave frequencies. { 'kav·əd·ē 'frē·kwən·sē ‚mēd·ər }

cavity impedance |ELECTR| The impedance of the cavity of a microwave tube which appears across the gap between the cathode and the anode. { 'kav·əd·ē im'pēd·əns }

cavity magnetron |ELECTR| A magnetron having a number of resonant cavities forming the anode; used as a microwave oscillator. { 'kav·əd·ē 'mag·nə‚trän }

cavity oscillator |ELECTR| An ultra-high-frequency oscillator whose frequency is controlled by a cavity resonator. { 'kav·əd·ē 'äs·ə‚lād·ər }

cavity resonance |ELECTROMAG| The resonant oscillation of the electromagnetic field in a cavity. |ENG ACOUS| The natural resonant vibration of a loudspeaker baffle; if in the audio range, it is evident as unpleasant emphasis of sounds at that frequency. { 'kav·əd·ē 'rez·ən·əns }

cavity resonator |ELECTROMAG| A space totally enclosed by a metallic conductor and excited in such a way that it becomes a source of electromagnetic oscillations. Also known as cavity; microwave cavity; microwave resonance cavity; resonant cavity; resonant chamber; resonant element; rhumbatron; tuned cavity; waveguide resonator. { 'kav·əd·ē 'rez·ən‚ād·ər }

cavity tuning |ELECTROMAG| Use of an adjustable cavity resonator as a tuned circuit in an oscillator or amplifier, with tuning usually achieved by moving a metal plunger in or out of the cavity to change the volume, and hence the resonant frequency of the cavity. { 'kav·əd·ē ‚tün·iŋ }

cavity-type diode amplifier *See* diode amplifier. { 'kav·əd·ē ‚tīp 'dī‚ōd ‚am·plə‚fī·ər }

CAW *See* channel address word.

C band |COMMUN| A band of radio frequencies extending from 4 to 8 gigahertz. { 'sē ‚band }

C-band fixed satellite service |COMMUN| Satellite communication at frequencies in and near the C band, with the uplink frequency in a band from 5.85 to 7.075 gigahertz and the downlink frequency in bands from 3.4 to 4.2 gigahertz and 4.5 to 4.8 gigahertz. { 'sē ‚band ¦fikst ¦sad·ə‚līt ‚sər·vəs }

C-band waveguide |ELECTROMAG| A rectangular waveguide, with dimensions 3.48 by 1.58 centimeters, which is used to excite only the dominant mode (TE_{01}) for wavelengths in the range 3.7–5.1 centimeters. { 'sē ,band 'wāv ,gīd }

C battery |ELEC| The battery that supplies the steady bias voltage required by the control-grid electrodes of electron tubes in battery-operated equipment. Also known as grid battery. { 'sē ,bad·ə·rē }

CBC See cipher block chaining.

C bias See grid bias. { 'sē ,bī·əs }

CBX See computerized branch exchange.

CCD See charge-coupled device.

CCIS See common-channel interoffice signaling.

CCIT 2 code |COMMUN| A printing-telegraph code in which each character is represented by five binary digits. Also known as international telegraph alphabet; International Telegraphic Consultative Committee code 2. { ,sē,sē,ī,tē 'tü ,kōd }

CCTV See closed-circuit television.

CCU See communications control unit.

CCW See channel command word.

CD See compact disk.

CD-4 sound See compatible discrete four-channel sound. { ¦sē¦de 'fōr ,saůnd }

C-display |ELECTR| A radar display format in which targets appear as spots with azimuth angle as the horizontal axis, and elevation angle as the vertical. Also known as C-indicator; C-scan; C-scope. { 'sē di'splā }

CDM See code-division multiplex.

CDMA See code-division multiple access.

CD-R |COMMUN| A compact-disk format that allows users to record audio or other digital data in such a way that the recording is permanent (nonerasable) and may be read indefinitely. Derived from compact-disk recordable. Also known as compact-disk write-once (CD-WO). { ¦sē¦dē¦'är }

CD-ROM See compact-disk read-only memory. { ¦sē¦dē 'räm }

CD-RW |COMMUN| A compact-disk format that allows audio or other digital data to be written, read, erased, and rewritten. Derived from compact-disk rewritable. Also known as compact-disk erasable.

CDTV See conventional definition television.

CD-WO See CD-R.

cell |COMPUT SCI| 1. An elementary unit of data storage. 2. In a spreadsheet, the intersection of a row and a column. |ELEC| A single unit of a battery. { sel }

cell address |COMPUT SCI| A combination of a letter and a number that specifies the column and row in which a cell is located on a spreadsheet. { 'sel ə,dres }

cellar See push-down storage. { 'sel·ər }

cell pointer |COMPUT SCI| A rectangular highlight that indicates the active cell in a spreadsheet program. { 'sel ,pȯint·ər }

cell protection |COMPUT SCI| A format applied to a cell or range of cells in a spreadsheet, or to the

entire spreadsheet, that prevents the contents of the cells in question from being altered. { 'sel prə,tek·shən }

cell reference |COMPUT SCI| The address of a cell that contains a value that is needed to solve a formula in a spreadsheet program. { 'sel ,ref·rəns }

cell-type tube |ELECTR| Gas-filled radio-frequency switching tube which operates in an external resonant circuit; a tuning mechanism may be incorporated in either the external resonant circuit or the tube. { 'sel ,tīp ,tüb }

cellular automaton |COMPUT SCI| A theoretical model of a parallel computer which is subject to various restrictions to make practicable the formal investigation of its computing powers. |MATH| A mathematical construction consisting of a system of entities, called cells, whose temporal evolution is governed by a collection of rules, so that its behavior over time may appear highly complex or chaotic. { 'sel·yə·lər ȯ'täm·ə·tən }

cellular chain |COMPUT SCI| A chain which is not allowed to cross a cell boundary. { 'sel·yə·lər 'chān }

cellular horn See multicellular horn. { 'sel·yə·lər 'hȯrn }

cellular mobile radio |COMMUN| A system that serves portable and mobile radio receivers in which the service area is subdivided into multiple cells or zones, and unique radio channel frequencies are assigned to each cell. { 'sel·yə·lər 'mō·bəl 'rād·ē·ō }

cellular multilist |COMPUT SCI| A type of multilist organization composed of cellular chains. { 'sel·yə·lər 'məl·ti,list }

cellular splitting |COMPUT SCI| A method of adding records to a file in which the records are grouped into cells and each cell is divided into two when it becomes full. { 'sel·yə·lər 'splid·iŋ }

CELP coder See code-excited linear predictive coder. { ¦sē¦ē¦el'pē ,kōd·ər or 'selp ,kōd·ər }

center-coupled loop |ELECTR| Coupling loop in the center of one of the resonant cavities of a multicavity magnetron. { 'sen·tər ,kup·əld 'lüp }

center frequency See carrier frequency. { 'sen·tər 'frē·kwən·sē }

centering control |ELECTR| One of the two controls used for positioning the image on the screen of a cathode-ray tube; either the horizontal centering control or the vertical centering control. { 'sen·tə·riŋ kən'trōl }

center line See stroke center line. { 'sen·tər ,līn }

center loading |ELECTROMAG| Alteration of the resonant frequency of a transmitting antenna by inserting an inductance or capacitance about halfway between the feed point and the end of the antenna. { 'sen·tər 'lōd·iŋ }

center tap |ELEC| A terminal at the electrical midpoint of a resistor, coil, or other device. Abbreviated CT. { 'sen·tər ,tap }

centimetric waves |COMMUN| Microwaves having wavelengths between 1 and 10 centimeters, corresponding to frequencies between 3 and 30 gigahertz. { ¦sent·ə¦me·trik 'wāvz }

central-battery system |COMMUN| A telephone or telegraph system which obtains all the energy for signaling (and for speaking, in the case of the telephone) from a single battery of secondary cells located at the main exchange. { 'sen·trəl 'bad·ə·rē ,sis·təm }

central control |SYS ENG| Control exercised over an extensive and complicated system from a single center. { 'sen·trəl kən'trōl }

centralized configuration See star network. { 'sen·trə,līzd kən,fig·yə'rā·shən }

centralized database |COMPUT SCI| A database at a single physical location, usually employed in conjunction with centralized data processing. { 'sen·trə,līzd 'dad·ə ,bās }

centralized data processing |COMPUT SCI| The processing of all the data concerned with a given activity at one place, usually with fixed equipment within one building. { 'sen·trə,līzd 'dad·ə 'präs,əs·iŋ }

central office |COMMUN| A switching unit, installed in a telephone system serving the general public, having the necessary equipment and operating arrangements for terminating and interconnecting lines and trunks. Also known as telephone central office. { 'sen·trəl 'ȯ·fəs }

central office line See subscriber line. { ¦sen·trəl ¦ȯ·fəs ,līn }

central processing unit |COMPUT SCI| The part of a computer containing the circuits required to interpret and execute the instructions. Abbreviated CPU. { 'sen·trəl 'präs,əs·iŋ ,yü·nət }

central-processing-unit time |COMPUT SCI| The time actually required to process a set of instructions in the logic unit of a computer. { 'sen·trəl 'präs,es·iŋ ,yü·nət ,tīm }

central terminal |COMPUT SCI| A communication device which queues tellers' requests for processing and which channels answers to the consoles originating the transactions. { 'sen·trəl 'tər·mən·əl }

centrifugal cutout |ELEC| A switch that is opened by centrifugal force and is usually closed by a spring when the centrifugal force is reduced. { sen'trif·ə·gəl 'kəd,aút }

centroid |NAV| In radar, the estimate of a contact's position as a single point, whereas the echoes may have occupied adjacent beam positions and-or range cells on successive pulses; the result of a centroiding algorithm in a radar contact generator. { 'sen,tróid }

centroid of asymptotes |CONT SYS| The intersection of asymptotes in a root-locus diagram. { 'sen,tróid əv 'as·əm,tōd·ēz }

cepstrum vocoder |ENG ACOUS| A digital device for reproducing speech in which samples of the cepstrum of speech, together with pitch information, are transmitted to the receiver, and are then converted into an impulse response that is convolved with an impulse train generated from the pitch information. { 'sep·trəm 'vō ¦kōd·ər }

ceramic amplifier |ELECTR| An amplifier that utilizes the piezoelectric properties of semiconductors such as silicon. { sə'ram·ik 'am·plə,fī·ər }

ceramic-based microcircuit |ELECTR| A microminiature circuit printed on a ceramic substrate. { sə'ram·ik,bāst 'mī·krō,sər·kət }

ceramic capacitor |ELEC| A capacitor whose dielectric is a ceramic material such as steatite or barium titanate, the composition of which can be varied to give a wide range of temperature coefficients. { sə'ram·ik kə'pas·əd·ər }

ceramic cartridge |ENG ACOUS| A device containing a piezoelectric ceramic element, used in phonograph pickups and microphones. { sə'ram·ik 'kär·trij }

ceramic earphones See crystal headphones. { sə'ram·ik 'ir,fōnz }

ceramic filter |ELECTR| A type of mechanical filter that uses a series of resonant ceramic disks to obtain a band-pass response. { sə'ram·ik 'fil·tər }

ceramic microphone |ENG ACOUS| A microphone using a ceramic cartridge. { sə'ram·ik 'mī·krə,fōn }

ceramic pickup |ENG ACOUS| A phonograph pickup using a ceramic cartridge. { sə'ram·ik 'pik·əp }

ceramic transducer See electrostriction transducer. { sə'ram·ik tranz'dü·sər }

ceramic tube |ELECTR| An electron tube having a ceramic envelope capable of withstanding operating temperatures over 500°C, as required during reentry of guided missiles. { sə'ram·ik 'tüb }

ceraunograph |ENG| An instrument that detects radio waves generated by lightning discharges and records their occurrence. { sə'rȯn·ə,graf }

Cerenkov rebatron radiator |ELECTR| Device in which a tightly bunched, velocity-modulated electron beam is passed through a hole in a dielectric; the reaction between the higher velocity of the electrons passing through the hole and the slower velocity of the electromagnetic energy passing through the dielectric results in radiation at some frequency higher than the frequency of modulation of the electron beam. { chə'reŋ·kəf ¦rē·bə,trän ¦rād·ē,ād·ər }

cermet resistor |ELEC| A metal-glaze resistor, consisting of a mixture of finely powdered precious metals and insulating materials fired onto a ceramic substrate. { 'sər,met ri'zis·tər }

certainty equivalence control |CONT SYS| An optimal control law for a stochastic adaptive control system which is obtained by solving the control problem in the case of known parameters and substituting the known parameters with their estimates. { 'sərt·ən·tē i'kwiv·ə·ləns kən'trōl }

certificate |COMMUN| A data record containing an identification, a digital signature from a third party who is believed to be trustworthy, attesting to the authenticity of the identity, and an encryption key which provides a basis for two unknown entities to establish a shared encryption. { sər'tif·i·kət }

cesium-antimonide photocathode |ELECTR| A photocathode obtained by exposing a thin layer of antimony to cesium vapor at elevated temperatures; has a maximum sensitivity in the

blue and ultraviolet regions of the spectrum.
{ 'sē·zē·əm 'an·tə·mə,nīd ,fōd·ō'kath,ōd }

cesium-beam sputter source |ELECTR| A source of negative ions in which a beam of positive cesium ions, accelerated through a potential difference on 20–30 kilovolts, sputters the cesium-coated inner surface of a hollow cone fabricated from or containing the element whose negative ion is required, and an appreciable fraction of the negative ions leaving the surface are extracted from the rear hole of the sputter cone.
{ 'sē·zē·əm ,bēm 'spəd·ər ,sȯrs }

cesium-beam tube See cesium electron tube.
{ 'sē·zē·əm ,bēm ,tüb }

cesium electron tube |ELECTR| An electronic device used as an atomic clock, producing electromagnetic energy that is accurate and stable in frequency. Also known as cesium beam tube.
{ 'sē·zē·əm i'lek,trän ,tüb }

cesium hollow cathode |ELECTR| A cathode in which cesium is heated at the bottom of a cylinder serving as the cathode of an electron tube, to give current densities that can be as high as 800 amperes per square centimeter.
{ 'sē·zē·əm ¦häl·ō 'ka,thōd }

cesium magnetometer |ENG| A magnetometer that uses a cesium atomic-beam resonator as a frequency standard in a circuit that detects very small variations in magnetic fields. { 'sē·zē·əm ,mag·nə'täm·əd·ər }

cesium phototube |ELECTR| A phototube having a cesium-coated cathode; maximum sensitivity in the infrared portion of the spectrum.
{ 'sē·zē·əm 'fōd·ō,tüb }

cesium thermionic converter |ELECTR| A thermionic diode in which cesium vapor is stored between the plates to neutralize space charge and to lower the work function of the emitter. { 'sē·zē·əm thər·mē'än·ik kən'vərd·ər }

cesium-vapor lamp |ELECTR| A lamp in which light is produced by the passage of current between two electrodes in ionized cesium vapor.
{ 'sē·zē·əm ¦vā·pər ,lamp }

cesium-vapor Penning source |ELECTR| A conventional Penning source modified for negative-ion generation through the introduction or a third, sputter cathode, made from or containing the element of interest, which is the source of negative ions, and through the introduction of cesium vapor into the arc chamber. { 'sē·zē·əm ¦vā·pər 'pen·iŋ ,sȯrs }

cesium-vapor rectifier |ELECTR| A gas tube in which cesium vapor serves as the conducting gas and a condensed monatomic layer of cesium serves as the cathode coating. { 'sē·zē·əm ¦vā·pər 'rek·tə,fī·ər }

CFIA See component-failure-impact analysis.

CGI See common gateway interface.

CGI script |COMPUT SCI| A program, written in a language such as Perl, that is used for creating interactive Web pages; for example, it allows a Web server to process a request from a user, communicate with a database, and reply to the user by creating a Web page. { ¦sē¦jē'ī ,skript }

CGM See computer graphics metafile.

chad |COMPUT SCI| The piece of material removed when forming a hole or notch in a punched tape or punched card. Also known as chip.
{ chad }

chaff |ELECTROMAG| Reflective particulate matter, such as tiny strips of coated films or of metallic foil, that can be dispensed by aircraft in the airspace covered by an enemy radar, so as to create such an echo density that echoes of interest to that radar are obscured or the radar is distracted by the chaff return. { chaf }

chain |COMMUN| A network of radio, television, radar, navigation, or other similar stations connected by telephone lines, coaxial cables, or radio relay links so all can operate as a group for broadcast purposes, communication purposes, or determination of position. |COMPUT SCI| **1.** A series of data or other items linked together in some way. **2.** A sequence of binary digits used to construct a code. |ELECTR| A series of amplifiers in a transmitter, achieving a higher overall gain than any one amplifier could reasonably achieve. { chān }

chain code |COMPUT SCI| A binary code consisting of a cyclic sequence of some or all of the possible binary words at a given length such that each word is derived from the previous one by moving the binary digits one position to the left, dropping the leading bit, and inserting a new bit at the end, in such a way that no word recurs before the cycle is complete. { 'chān ,kōd }

chain command |COMPUT SCI| Any input/output command in a sequence of input/output commands such as WRITE, READ, SENSE. { 'chān kə'mand }

chain data flag |COMPUT SCI| A value of 1 given to a specific bit of a channel command word, commonly used with scatter read or scatter write operations. { 'chān 'dad·ə ,flag }

chained block encryption |COMMUN| The use of a block cipher in which the bits of a given output block depend not only on the bits in the corresponding input block and in the key, but also on any or all prior data bits, either inputted to or produced during the enciphering or deciphering process. Also known as block chaining. { ¦chānd 'bläk in'krip·shən }

chained list |COMPUT SCI| A collection of data items arranged in a sequence so that each item contains an address giving the location of the next item in a computer storage device. Also known as linked list. { ¦chānd 'list }

chained records |COMPUT SCI| A file of records arranged according to the chaining method. { ¦chānd 'rek·ərdz }

chaining |COMPUT SCI| A method of storing records which are not necessarily contiguous, in which the records are arranged in a sequence and each record contains means to identify its successor. { 'chān·iŋ }

chaining search |COMPUT SCI| A method of searching for a data item in a chained list in which an initial key is used to obtain the location of either the item sought or another item in the list, and the search then progresses through the chain until the required item is obtained or the chain is completed. { 'chān·iŋ ‚sərch }

chain pointer |COMPUT SCI| The part of a data item in a chained list that gives the address of the next data item. { 'chān 'pȯint·ər }

chain printer |COMPUT SCI| A high-speed printer in which the type slugs are carried by the links of a revolving chain. { 'chān ‚print·ər }

chain printing |COMPUT SCI| The printing of a group of linked files by placing commands at the end of each file that direct the program to continue printing the next one. { ¦chān 'print·iŋ }

chain radar beacon |COMMUN| A beacon with a fast recovery time to permit simultaneous interrogation and tracking of the beacon by a number of radars. { ¦chān 'rā‚där ‚bē·kən }

chain radar system |ENG| A number of radar stations located at various sites on a missile range to enable complete radar coverage during a missile flight; the stations are linked by data and communication lines for target acquisition, target positioning, or data-recording purposes. { ¦chān 'rā‚där ‚sis·təm }

challenge |COMMUN| To cause an interrogator to transmit a signal which puts a transponder into operation. { 'chal·ənj }

challenger See interrogator. { 'chal·ən·jər }

challenge-response |COMPUT SCI| A method of identifying and authenticating persons seeking access to a computing system; each user is issued a device resembling a pocket calculator and is given a different problem to solve (the challenge), to which the calculator provides part of the answer, each time the person seeks authentication. { 'chal·ənj ri'späns }

challenging signal See interrogation. { 'chal·ən·jiŋ ‚sig·nəl }

chance-constrained programming |COMPUT SCI| Type of nonlinear programming wherein the deterministic constraints are replaced by their probabilistic counterparts. { ¦chans kən'strānd 'prō‚gram·iŋ }

changed memory routine |COMPUT SCI| A selective memory dump routine in which only those words that have been changed in the course of running a program are printed. { ¦chānjd 'mem‚rē rü‚tēn }

change dump |COMPUT SCI| A type of dump in which only those locations in a computer memory whose contents have changed since some previous event are copied. { 'chānj ‚dəmp }

change file |COMPUT SCI| A transaction file that is used to update a master file. { 'chānj ‚fīl }

change of control |COMPUT SCI| **1.** A break in a series of records at which processing of the records may be interrupted and some predetermined action taken. **2.** See jump. { 'chānj əv kən'trōl }

changeover switch |ELEC| A means of moving a circuit from one set of connections to another. { 'chān‚jō·vər ‚swich }

change record |COMPUT SCI| A record that is used to alter information in a corresponding

master record. Also known as amendment record; transaction record. { 'chānj ‚rek·ərd }

change tape |COMPUT SCI| A paper tape or magnetic tape carrying information that is to be used to update filed information; the latter is often on a master tape. Also known as transaction tape. { 'chānj ‚tāp }

channel |COMMUN| **1.** A band of radio frequencies allocated for a particular purpose; a standard broadcasting channel is 10 kilohertz wide, an FM channel is 200 kHz wide, and a television channel 6 megahertz wide. **2.** A path through which electrical transmission of information takes place. |COMPUT SCI| A path along which digital or other information may flow in a computer. |ELECTR| **1.** A path for a signal, as an audio amplifier may have several input channels. **2.** The main current path between the source and drain electrodes in a field-effect transistor or other semiconductor device. { 'chan·əl }

channel adapter |COMPUT SCI| Equipment that allows devices operating at different rates of speed to be connected and data to be transferred at the slower data rate. { 'chan·əl ə‚dap·tər }

channel address word |COMPUT SCI| A four-byte code containing the protection key and the main storage address of the first channel command word at the start of an input/output operation. Abbreviated CAW. { 'chan·əl 'ad‚res ‚wərd }

channel-attached device |COMPUT SCI| Equipment that is directly connected to a computer by a channel. { 'chan·əl ə¦tacht di‚vīs }

channel bank |ELECTR| Part of a carrier-multiplex terminal that performs the first step of modulation of the transmitting voice frequencies into a higher-frequency band, and the final step in the demodulation of the received higher-frequency band into the received voice frequencies. { 'chan·əl ‚baŋk }

channel capacity |COMMUN| The maximum number of bits or other information elements that can be handled in a particular channel per unit time. { 'chan·əl kə'pas·əd·ē }

channel command |COMPUT SCI| The step, equivalent to a program instruction, required to tell an input/output channel what operation is to be performed, and where the data are or should be located. { 'chan·əl kə'mand }

channel command word |COMPUT SCI| A code specifying an operation, one or more flags, a count, and a storage location. Abbreviated CCW. { 'chan·əl kə'mand ‚wərd }

channel configuration |COMPUT SCI| The types, number, and logical relationships of devices connected to a given computer channel. { 'chan·əl kən‚fig·yə‚rā·shən }

channel control command |COMPUT SCI| An order to a control unit to perform a non-data input/output operation. { 'chan·əl kən'trōl kə'mand }

channel design |COMPUT SCI| The type of channel, characterized by the tasks it can perform, available to a computer. { 'chan·əl di'zīn }

channel director |COMPUT SCI| A unit in some very large computers that controls the functioning of several channels. { 'chan·əl ,rek·tər }

channel effect |ELECTR| A leakage current flowing over a surface path between the collector and emitter in some types of transistors. { 'chan·əl i'fekt }

channel electron multiplier |ELECTR| A single-particle detector which consists of a hollow glass or ceramic tube with a semiconducting inner surface; it responds to one or more primary particle impact events at its entrance by producing, in a cascade multiplication process, a charge pulse of typically 104–108 electrons. { ¦chan·əl i¦lek,trän 'məl·tə,plī·ər }

channel-end condition |COMPUT SCI| A signal indicating that the use of an input/output channel is no longer required. { 'chan·əl ,end kən'dish·ən }

channel FET microphone |ENG ACOUS| A microphone in which a membrane is used as the gate to a field-effect transistor (FET) located just below it, and motion of the membrane modulates the current between the source and drain of the transistor. { ¦chan·əl ¦fet 'mī·krə,fōn or ¦ef¦ē¦tē }

channeling |COMMUN| A type of multiplex transmission in which the separation between communication channels is accomplished through the use of carriers or subcarriers. { 'chan·əl·iŋ }

channelization |COMMUN| The division of a single wide-band (high-capacity) communications channel into many relatively narrow-band (lower-capacity) channels. { ,chan·əl·ə'zā·shən }

channelizing |COMMUN| The process of subdividing a wide-band transmission facility so as to handle a number of different circuits requiring comparatively narrow bandwidths. { 'chan·əl ,īz·iŋ }

channel mask |COMPUT SCI| A portion of a program status word indicating which channels may interrupt the task by their completion signals. { 'chan·əl ,mask }

channel miles |COMMUN| The summation, in miles, of the electrical path of individual channels between two points; these points may be connected by wire or radio, or a combination of both. { 'chan·əl ,mīlz }

channel plate multiplier *See* microchannel plate. { 'chan·əl ¦plāt 'məl·tə,plī·ər }

channel program |COMPUT SCI| The set of steps, called channel commands, by means of which an input/output channel is controlled. { 'chan·əl ,prō·grəm }

channel read-backward command |COMPUT SCI| A command to transfer data from tape device to main storage while the tape is moving backward. { 'chan·əl 'rēd ¦bak·wərd kə,mand }

channel read command |COMPUT SCI| A command to transfer data from an input/output device to main storage. { ¦chan·əl 'rēd kə'mand }

channel reliability |COMMUN| The percent of time a channel was available for use in a specific direction during a specified period of time. { 'chan·əl ri,lī·ə'bil·əd·ē }

channel selector |ELEC| A control used to tune in the desired channel in a radio or television receiver. { 'chan·əl si'lek·tər }

channel sense command |COMPUT SCI| A command commonly used to denote an unusual condition existing in an input/output device and requesting more information. { 'chan·əl 'sens kə'mand }

channel shifter |ELECTR| Radiotelephone carrier circuit that shifts one or two voice-frequency channels from normal channels to higher voice-frequency channels to reduce cross talk between channels; the channels are shifted back by a similar circuit at the receiving end. { 'chan·əl ,shif·tər }

channel skip |COMPUT SCI| A control character that causes a printer to skip down to a specified line on a page or to the top of the next page. { 'chan·əl ,skip }

channel spacing |COMMUN| The difference in frequency between successive radio or television channels. { 'chan·əl ,spās·iŋ }

channel status table |COMPUT SCI| A table that is set up by an executive program to show the status of the various channels that connect the central processing unit with peripheral units, enabling the program to control input/output operations. { ¦chan·əl 'stad·əs ,tā·bəl }

channel status word |COMPUT SCI| A storage register containing the status information of the input/output operation which caused an interrupt. Abbreviated CSW. { ¦chan·əl 'stad·əs ,wərd }

channel synchronizer |ELECTR| An electronic device providing the proper interface between the central processing unit and the peripheral devices. { 'chan·əl 'siŋ·krə,nīz·ər }

channel-to-channel adapter |COMPUT SCI| A device which provides two computer systems with interchannel communications. { ¦chan·əl tə ¦chan·əl ə'dap·tər }

channel write command |COMPUT SCI| A command which transfers data from main storage to an input/output device. { ¦chan·əl 'wrīt kə'mand }

character |COMPUT SCI| **1.** An elementary mark used to represent data, usually in the form of a graphic spatial arrangement of connected or adjacent strokes, such as a letter or a digit. **2.** A small collection of adjacent bits used to represent a piece of data, addressed and handled as a unit, often corresponding to a digit or letter. { 'kar·ik·tər }

character-addressable computer |COMPUT SCI| A computer that processes data as single characters, and is therefore able to handle words of varying length. { 'kar·ik·tər ə¦dres·ə·bəl kəm'pyüd·ər }

character adjustment |COMPUT SCI| An address modification affecting a specific number of characters of the address part of the instruction. { 'kar·ik·tər ə'jəs·mənt }

character boundary |COMPUT SCI| In character recognition, a real or imaginary rectangle which serves as the delimiter between consecutive characters or successive lines on a source document. { 'kar·ik·tər ,baún·drē }

character cell |COMPUT SCI| A matrix of dots that is used to form a single character on a printer or display screen. { 'kar·ik·tər ,sel }

character code |COMMUN| A bit pattern assigned to a particular character in a coded character set. { 'kar·ik·tər ,kōd }

character data type |COMPUT SCI| A scalar data type which provides an internal representation of printable characters. { 'kar·ik·tər 'dad·ə ,tīp }

character density |COMPUT SCI| The number of characters recorded per unit of length or area. Also known as record density. { 'kar·ik·tər ,den·səd·ē }

character display terminal |COMPUT SCI| A console that can display only alphanumeric characters, and cannot show arbitrary lines or curves. { 'kar·ik·tər di'splä ,tərm·ə·nəl }

character emitter |COMPUT SCI| In character recognition, an electromechanical device which conveys a specimen character in the form of a time pulse or group of pulses. { 'kar·ik·tər i'mid·ər }

character fill |COMPUT SCI| To fill one or more locations in a computer storage device by repeated insertion of some particular character, usually blanks or zeros. { 'kar·ik·tər ,fil }

character generator |COMPUT SCI| A hard-wired subroutine which will display alphanumeric characters on a screen. { 'kar·ik·tər ,jen·ə,rād·ər }

character graphics |COMPUT SCI| A collection of special symbols that can be strung together like letters of the alphabet to generate graphics. { 'kar·ik·tər ,graf·iks }

characteristic |ELECTR| A graph showing how the voltage or current between two terminals of an electronic device varies with the voltage or current between two other terminals. { ,kar·ik·tə'ris·tik }

characteristic frequency |COMMUN| Frequency which can be easily identified and measured in a given emission. { ,kar·ik·tə'ris·tik 'frē·kwən·sē }

characteristic impedance |COMMUN| The impedance that, when connected to the output terminals of a transmission line of any length, makes the line appear to be infinitely long, for there are then no standing waves on the line, and the ratio of voltage to current is the same for each point on the line. Also known as surge impedance. { ,kar·ik·tə'ris·tik im'pēd·əns }

characteristic overflow |COMPUT SCI| An error condition encountered when the characteristic of a floating point number exceeds the limit imposed by the hardware manufacturer. { ,kar·ik·tə'ris·tik 'ō·vər,flō }

characteristic underflow |COMPUT SCI| An error condition encountered when the characteristic of a floating point number is smaller than the smallest limit imposed by the hardware manufacturer. { ,kar·ik·tə'ris·tik 'ən·dər,flō }

character mode |COMPUT SCI| A mode of computer operation in which only text is displayed. { 'kar·ik·tər ,mōd }

character-oriented computer |COMPUT SCI| A computer in which the locations of individual characters, rather than words, can be addressed. { 'kar·ik·tər ¦ór·ē,en·təd kəm,pyüd·ər }

character-oriented protocol See byte-oriented protocol. { 'kar·ik·tər ,ór·ē,ent·əd 'prōd·ə,kòl }

character outline |COMPUT SCI| The graphic pattern formed by the stroke edges of a printed or handwritten character in character recognition. { 'kar·ik·tər 'aút,līn }

character reader |COMPUT SCI| In character recognition, any device capable of locating, identifying, and translating into machine code the handwritten or printed data appearing on a source document. { 'kar·ik·tər ,rēd·ər }

character recognition |COMPUT SCI| The technology of using a machine to sense and encode into a machine language the characters which are originally written or printed by human beings. { 'kar·ik·tər ,rek·ig'nish·ən }

character set |COMMUN| A set of unique representations called characters, for example, the 26 letters of the English alphabet, the Boolean 0 and 1, the set of signals in Morse code, and the 128 characters of the USASCII. { 'kar·ik·tər ,set }

character skew |COMPUT SCI| In character recognition, an improper appearance of a character to be recognized, in which it appears in a tilted condition with respect to a real or imaginary horizontal base line. { 'kar·ik·tər ,skyü }

character string |COMPUT SCI| A sequence of characters in a computer memory or other storage device. Also known as alphabetic string. { 'kar·ik·tər 'striŋ }

character string constant |COMPUT SCI| An arbitrary combination of letters, digits, and other symbols which, in the processing of nonnumeric data involving character strings, performs a function analogous to that of a numeric constant in the processing of numeric data. { 'kar·ik·tər ,striŋ ,kän·stənt }

character stroke See stroke. { 'kar·ik·tər ,strōk }

character style |COMPUT SCI| In character recognition, a distinctive construction that is common to all members of a particular character set. { 'kar·ik·tər ,stīl }

character terminal |COMPUT SCI| A screen that can display only text. { 'kar·ik·tər ,tər·mə·nəl }

character-writing tube |ELECTR| A cathode-ray tube that forms alphanumeric and symbolic characters on its screen for viewing or recording purposes. { 'kar·ik·tər ,rīd·iŋ ,tüb }

charge |ELEC| **1.** A basic property of elementary particles of matter; the charge of an object may be a positive or negative number or zero; only integral multiples of the proton charge occur, and the charge of a body is the algebraic sum of the charges of its constituents; the value of the charge may be inferred from the Coulomb force between charged objects. Also known as electric charge; quantity of electricity. **2.** To convert electrical energy to chemical energy in a secondary battery. **3.** To feed electrical energy to a capacitor or other device that can store it. |ENG| The material or part to be heated by induction or dielectric heating. { 'chärj }

charge carrier |SOLID STATE| A mobile conduction electron or mobile hole in a semiconductor. Also known as carrier. { 'chärj ‚kar·ē·ər }

charge collector |ELEC| The structure within a battery electrode that provides a path for the electric current to or from the active material. Also known as current collector. { 'chärj kə ‚lek·tər }

charge conservation *See* conservation of charge. { 'chärj ‚kän·sər'vā·shən }

charge-coupled device |ELECTR| A semiconductor device wherein minority charge is stored in a spatially defined depletion region (potential well) at the surface of a semiconductor and is moved about the surface by transferring this charge to similar adjacent wells. Abbreviated CCD. { 'chärj ¦kəp·əld di'vīs }

charge-coupled image sensor |ELECTR| A device in which charges are introduced when light from a scene is focused on the surface of the device; image points are accessed sequentially to produce a television-type output signal. Also known as solid-state image sensor. { 'chärj ¦kəp·əld 'im·ij ‚sen·sər }

charge-coupled memory |COMPUT SCI| A computer memory that uses a large number of charge-coupled devices for data storage and retrieval. { 'chärj ¦kəp·əld 'mem·rē }

charge coupling |COMPUT SCI| Transfer of all electric charges within a semiconductor storage element to a similar, nearby element by means of voltage manipulations. { 'chärj ‚kəp·liŋ }

charge density |ELEC| The charge per unit area on a surface or per unit volume in space. { 'chärj ‚den·səd·ē }

charge-exchange source |ELECTR| A source of negative ions, generally negative helium ions, in which positive ions generated in a duoplasmatron are directed through a donor canal, usually containing lithium vapor, where they pick up sequentially two electrons to form negative ions. { 'chärj iks‚chānj ‚sòrs }

charge-injection device |ELECTR| A charge-transfer device used as an image sensor in which the image points are accessed by reference to their horizontal and vertical coordinates. Abbreviated CID. { 'chärj in‚jek·shən di'vīs }

charge-mass ratio |ELEC| The ratio of the electric charge of a particle to its mass. { ‚chärj ‚mas 'rā·shō }

charge quantization |ELEC| The principle that the electric charge of an object must equal an integral multiple of a universal basic charge. { 'chärj ‚kwan·tə'zā·shən }

charger *See* battery charger. { 'chär·jər }

charger-eliminator |ELEC| A battery charger with a low-noise, low-impedance output which can either charge a storage battery or supply a dc load directly, without a storage battery in parallel. { 'chär·jər ə'lim·ə‚nād·ər }

charge-storage transistor |ELECTR| A transistor in which the collector-base junction will charge when forward bias is applied with the base at

a high level and the collector at a low level. { 'chärj ‚stòr·ij tranz'is·tər }

charge-storage tube |ELECTR| A storage tube in which information is retained on a surface in the form of electric charges. { 'chärj ‚stòr·ij ‚tüb }

charge-storage varactor |ELECTR| A varactor that uses semiconductor techniques to achieve power outputs above 50 watts at ultra-high and microwave frequencies. { 'chärj ‚stòr·ij və'rak·tər }

charge-transfer device |ELECTR| A semiconductor device that depends upon movements of stored charges between predetermined locations, as in charge-coupled and charge-injection devices. { 'chärj ‚tranz·fər di'vīs }

charging current |ELEC| The current that flows into a capacitor when a voltage is first applied. { 'chär·jiŋ ‚kər·ənt }

chassis |ENG| **1.** A frame on which the body of an automobile or airplane is mounted. **2.** A frame for mounting the working parts of a radio or other electronic device. { 'chas·ē }

chassis ground |ELEC| A connection made to the metal chassis on which the components of a circuit are mounted, to serve as a common return path to the power source. { 'chas·ē ‚graünd }

chat mode |COMPUT SCI| A communications option that allows two or more computers to conduct a conversation by typing in turn. { 'chat ‚mōd }

chat room |COMPUT SCI| A Web site or server space on the Internet where live keyboard conversations (usually organized around a specific topic) with other people occur. { 'chat ‚rüm }

chatter |ELEC| Prolonged undesirable opening and closing of electric contacts, as on a relay. Also known as contact chatter. |ENG ACOUS| Vibration of a disk-recorder cutting stylus in a direction other than that in which it is driven. { 'chad·ər }

chattering |CONT SYS| A mode of operation of a relay-type control system in which the relay switches back and forth infinitely fast. { 'chad·ə· riŋ }

Chebyshev filter |ELECTR| A filter in which the transmission frequency curve has an equal-ripple shape, with very small peaks and valleys. { 'cheb·ə·shəf ‚fil·tər }

Chebyshev filter |ELECTR| A filter in which the transmission frequency curve has an equalripple shape, with very small peaks and valleys. { 'cheb·ə·shəf ‚fil·tər }

check |COMPUT SCI| A test which is necessary to detect a mistake in computer programming or a computer malfunction. { chek }

check bit |COMPUT SCI| A binary check digit. { 'chek ‚bit }

check box |COMPUT SCI| In a graphical user interface, a small box on which an x or check mark appears when the option indicated next to the box is turned on, and disappears when the option is turned off. { 'chek ‚bäks }

check character |COMPUT SCI| A redundant character used to perform a check. { 'chek ‚kar· ik·tər }

check digit |COMPUT SCI| A redundant digit used to perform a check. { 'chek ,dij·ət }

check indicator |COMPUT SCI| A console device, usually a light, informing the operator that an error has occurred. { 'chek ,in·də,kād·ər }

check indicator instruction |COMPUT SCI| A computer instruction which directs that a signal device is turned on to call the operator's attention to the fact that there is some discrepancy in the instruction now in use. { 'chek ,in·də,kād·ər in'strək·shən }

checking program |COMPUT SCI| A computer program which detects and determines the nature of errors in other programs, particularly those that involve incorrect coding or punching of wrong characters. Also known as checking routine. { 'chek·iŋ ,prō·grəm }

checking routine See checking program. { 'chek· iŋ rü'tēn }

check number |COMPUT SCI| A number denoting a specific type of hardware malfunction. { 'chek ,nəm·bər }

checkout |COMPUT SCI| A collection of routines that are built into a compiler to test and debug programs. { 'chek,aút }

checkout compiler |COMPUT SCI| A special compiler designed specifically to test and debug programs by using checkout routines. { 'chek ,aút kəm,pī·lər }

checkpoint |COMPUT SCI| That place in a routine at which the entire state of the computer (memory, registers, and so on) is written out on auxiliary storage from which it may be read back into the computer if the program is to be restarted later. |NAV| Geographical location on land or water above which the position of an aircraft in flight may be determined by observation or by electronic means. { 'chek,póint }

checkpoint/restart |COMPUT SCI| The procedures for resuming a processing run after it has been halted either accidentally or deliberately. { 'chek,póint 'rē,stärt }

check problem See check routine. { 'chek ,präb· ləm }

check protect symbol |COMPUT SCI| A character, usually an asterisk, that is printed in place of leading zeros in a number, such as a dollar amount on a check. { 'chek prə'tekt ,sim·bəl }

check register |COMPUT SCI| A register in which transferred data are temporarily stored so that they may be compared with a second transfer of the same data, to verify the accuracy of the transfer. { 'chek ,rej·ə·stər }

check routine |COMPUT SCI| A routine or problem designed primarily to indicate whether a fault exists in a computer, without giving detailed information on the location of the fault. Also known as check problem; test program; test routine. { 'chek rü'tēn }

check row |COMPUT SCI| A row (or one of two or more rows) on a paper tape which contains the cumulated sum of existing rows, column by column, resulting in either 1 or 0 by column, thus verifying that all rows have been properly read. { 'chek ,rō }

check sum |COMPUT SCI| A sum of digits or numbers used in a summation check. { 'chek ,səm }

check symbol |COMPUT SCI| One or more digits generated by performing an arithmetic check or summation check on a data item which are then attached to the item and copied along with it through various stages of processing, allowing the check to be repeated to verify the accuracy of the copying processes. { 'chek ,sim·bəl }

check word |COMPUT SCI| A computer word, containing data from a block of records, that is joined to the block and serves as a check symbol during transfers of the block between different locations. { 'chek ,wərd }

cheese antenna |ELECTROMAG| An antenna having a parabolic reflector between two metal plates, dimensioned to permit propagation of more than one mode in the desired direction of polarization. { 'chēz an'ten·ə }

chemical film dielectric |ELEC| An extremely thin layer of material on one or both electrodes of an electrolytic capacitor, which conducts electricity in only one direction and thereby constitutes the insulating element of the capacitor. { 'kem·i·kəl ,film ,dī·ə'lek·trik }

chemically sensitive field-effect transistor |ELECTR| A field-effect transistor in which the ordinary gate electrode is replaced by a chemically sensitive membrane so that the gain of the transistor depends on the concentration of chemical substances. { 'kem·ik·lē |sen·səd·iv 'fēld i|fekt tran,zis·tər }

child |COMPUT SCI| **1.** An element that follows a given element in a data structure. **2.** In object-oriented programming, a subclass. { chīld }

Child-Langmuir equation See Child's law.

Child-Langmuir-Schottky equation See Child's law. { |chīld |laŋ·myür 'shät,kē i'kwā·zhən }

child process |COMPUT SCI| One of the subsidiary processes that branches out from the root task in the fork-join model of programming on parallel machines. { 'chīld ,präs·es }

Child's law |ELECTR| A law stating that the current in a thermionic diode varies directly with the three-halves power of anode voltage and inversely with the square of the distance between the electrodes, provided the operating conditions are such that the current is limited only by the space charge. Also known as Child-Langmuir equation; Child-Langmuir-Schottky equation; Langmuir-Child equation. { 'chīldz ,ló }

chimney |ELECTR| A pipelike enclosure that is placed over a heat sink to improve natural upward convection of heat and thereby increase the dissipating ability of the sink. { 'chim,nē }

chip |COMPUT SCI| See chad. |ELECTR| **1.** The shaped and processed semiconductor die that is mounted on a substrate to form a transistor, diode, or other semiconductor device. **2.** An integrated microcircuit performing a significant number of functions and constituting a subsystem. Also known as microchip. { chip }

chip capacitor |ELECTR| A single-layer or multilayer monolithic capacitor constructed in chip form, with metallized terminations to facilitate direct bonding on hybrid integrated circuits. { 'chip kə'pas·əd·ər }

chip card See smart card. { 'chip ,kärd }

chip circuit See large-scale integrated circuit. { 'chip ,sər·kət }

chip resistor |ELECTR| A thick-film resistor constructed in chip form, with metallized terminations to facilitate direct bonding on hybrid integrated circuits. { 'chip ri'zis·tər }

chipset |COMPUT SCI| A number of integrated circuits, packaged as one unit, which perform one or more related functions. { 'chip,set }

Chireix antenna |ELECTROMAG| A phased array composed of two or more coplanar square loops, connected in series. Also known as Chireix-Mesny antenna. { ki'rāks an,ten·ə }

Chireix-Mesny antenna See Chireix antenna. { ki'rāks ,mez,nē an,ten·ə }

chirp |COMMUN| **1.** An undesirable variation in the frequency of a continuous-wave carrier when it is keyed. **2.** The sound heard in a code receiver when the transmitted carrier frequency is increased linearly for the duration of a pulse code. { chərp }

chirp modulation |COMMUN| A modulation of the carrier frequency from a lover to a higher frequency, or vice versa, often linearly, used in radar pulse compression. { ¦chərp mäj·ə'lā·shən }

chirp radar |ENG| Radar in which a swept-frequency signal is transmitted, received from a target, then compressed in time to give a narrow pulse called the chirp signal. { 'chərp ,rā,där }

chisel bond |ENG| A thermocompression bond in which a contact wire is attached to a contact pad on a semiconductor chip by applying pressure with a chisel-shaped tool. { 'chiz·əl ,bänd }

choke |ELEC| An inductance used in a circuit to present a high impedance to frequencies above a specified frequency range without appreciably limiting the flow of direct current. Also known as choke coil. |ELECTROMAG| A groove or other discontinuity in a waveguide surface so shaped and dimensioned as to impede the passage of guided waves within a limited frequency range. { chōk }

choke coil See choke. { 'chōk ,kȯil }

choke coupling |ELECTROMAG| Coupling between two parts of a waveguide system that are not in direct mechanical contact with each other. { 'chōk ,kəp·liŋ }

choke filter See choke input filter. { 'chōk ,fil·tər }

choke flange |ELECTROMAG| A waveguide flange having in its mating surface a slot (choke) so shaped and dimensioned as to restrict leakage of microwave energy within a limited frequency range. { 'chōk ,flanj }

choke input filter |ELEC| A power-supply filter in which the first filter element is a series choke. Also known as choke filter. { ¦chōk 'in,pu̇t ,fil·tər }

choke joint |ELECTROMAG| A connection between two waveguides that uses two mating choke flanges to provide effective electrical continuity without metallic continuity at the inner walls of the waveguide. { 'chōk ,jȯint }

choke piston |ELECTROMAG| A piston in which there is no metallic contact with the walls of the waveguide at the edges of the reflecting surface; the short circuit for high-frequency currents is achieved by a choke system. Also known as noncontacting piston; noncontacting plunger. { 'chōk ,pis·tən }

chopper amplifier |ELECTR| A carrier amplifier in which the direct-current input is filtered by a low-pass filter, then converted into a square-wave alternating-current signal by either one or two choppers. { 'chäp·ər 'am·plə,fī·ər }

chopper-stabilized amplifier |ELECTR| A direct-current amplifier in which a direct-coupled amplifier is in parallel with a chopper amplifier. { ¦chäp·ər ¦stā·bə,līzd 'am·plə,fī·ər }

chopper transistor |ELECTR| A bipolar or field-effect transistor operated as a repetitive "on/off" switch to produce square-wave modulation of an input signal. { 'chäp·ər tran'zis·tər }

chopping |ELECTR| The removal, by electronic means, of one or both extremities of a wave at a predetermined level. { 'chäp·iŋ }

chroma band-pass amplifier See burst amplifier. { 'krō·mə 'band ,pas 'am·plə,fī·ər }

chroma control |ELECTR| The control that adjusts the amplitude of the carrier chrominance signal fed to the chrominance demodulators in an analog color television receiver, so as to change the saturation or vividness of the hues in the color picture. Also known as color control; color-saturation control. { 'krō·mə kən'trōl }

chroma oscillator |ELECTR| A crystal oscillator used in analog color television receivers to generate a 3.579545-megahertz signal for comparison with the incoming 3.579545-megahertz chrominance subcarrier signal being transmitted. Also known as chrominance-subcarrier oscillator; color oscillator; color-subcarrier oscillator. { 'krō·mə 'äs·ə,lād·ər }

chromatic aberration |ELECTR| An electron-gun defect causing enlargement and blurring of the spot on the screen of a cathode-ray tube, because electrons leave the cathode with different initial velocities and are deflected differently by the electron lenses and deflection coils. { krō'mad·ik ab·ə'rā·shən }

chromatron |ELECTR| A single-gun color picture tube having color phosphors deposited on the screen in strips instead of dots. Also known as Lawrence tube. { 'krō·mə'trän }

chrominance carrier See chrominance subcarrier. { 'krō·mə·nəns ,kar·ē·ər }

chrominance-carrier reference |COMMUN| A continuous signal having the same frequency as the chrominance subcarrier in a color television system and having fixed phase with respect to the color burst; this signal is the reference with which the phase of a chrominance signal is compared for the purpose of modulation or demodulation. Also known as chrominance-subcarrier reference; color-carrier reference; color-subcarrier reference. { 'krō·mə·nəns ¦kar·ē·ər ‚ref·rəns }

chrominance channel |COMMUN| Any path that is intended to carry the chrominance signal in an analog color television system. { 'krō·mə·nəns ‚chan·əl }

chrominance demodulator |ELECTR| A demodulator used in an analog color television receiver for deriving the I and Q components of the chrominance signal from the chrominance signal and the chrominance-subcarrier frequency. Also known as chrominance-subcarrier demodulator. { 'krō·mə·nəns dē'mäj·ə‚lād·ər }

chrominance frequency |COMMUN| The frequency of the chrominance subcarrier, equal to 3.579545 megahertz. { 'krō·mə·nəns ‚frē·kwən·sē }

chrominance gain control |ELECTR| Variable resistors in red, green, and blue matrix channels that individually adjust primary signal levels in an color television receiver. { 'krō·mə·nəns 'gān kən'trōl }

chrominance modulator |ELECTR| A modulator used in an analog color television transmitter to generate the chrominance signal from the video-frequency chrominance components and the chrominance subcarrier. Also known as chrominance-subcarrier modulator. { 'krō·mə·nəns 'mäj·ə‚lād·ər }

chrominance signal |COMMUN| One of the two components, called the I signal and Q signal, that add together to produce the total chrominance signal in an analog color television system. Also known as carrier chrominance signal. { 'krō·mə·nəns ‚sig·nəl }

chrominance subcarrier |COMMUN| The 3.579545-megahertz carrier whose modulation sidebands are added to the monochrome signal to convey color information in an analog color television system. Also known as chrominance carrier; color carrier; color subcarrier; subcarrier. { 'krō·mə·nəns səb'kar·ē·ər }

chrominance-subcarrier demodulator See chrominance demodulator. { 'krō·mə·nəns səb'kar·ē·ər dē'mäj·ə‚lād·ər }

chrominance-subcarrier modulator See chrominance modulator. { 'krō·mə·nəns səb'kar·ē·ər 'mäj·ə‚lād·ər }

chrominance-subcarrier oscillator See chroma oscillator. { 'krō·mə·nəns səb'kar·ē·ər 'äs·ə‚lād·ər }

chrominance-subcarrier reference See chrominance-carrier reference. { 'krō·mə·nəns səb 'kar·ē·ər 'ref·rəns }

chrominance video signal |ELECTR| Voltage output from the red, green, or blue section of a color television camera or receiver matrix. { 'krō·mə·nəns 'vid·ē·ō ‚sig·nəl }

chromium dioxide tape |ELECTR| A magnetic recording tape developed primarily to improve quality and brilliance of reproduction when used in cassettes operated at 1⅞ inches per second (4.76 centimeters per second); requires special recorders that provide high bias. { 'krō·mē·əm dī'äk‚sīd 'tāp }

chromium-gold metallizing |ELECTR| A metal film used on a silicon or silicon oxide surface in semiconductor devices because it is not susceptible to purple plague deterioration; a layer of chromium is applied first for adherence to silicon, then a layer of chromium-gold mixture, and finally a layer of gold to which bonding contacts can be applied. { ¦krō·mē·əm ¦gōld 'med·əl·ī·z·iŋ }

chronistor |ELECTR| A subminiature elapsed-time indicator that uses electroplating principles to totalize operating time of equipment up to several thousand hours. { krə'nis·tər }

chronometric encoder |ELECTR| An encoder that uses an electronic counter to time or count electrical events and deliver in digital form a number equivalent to the input magnitude. { 'krän·ə‚me·trik en'kōd·ər }

chronopher |ELECTR| Instrument for emitting standard time signal impulses from a standard clock or timing device. { 'krän·ə·fər }

chronotron |ELECTR| A device that measures millimicrosecond time intervals between pulses on a transmission line to determine the time between the events which initiated the pulses. { 'krän·ə ‚trän }

chute blades |COMPUT SCI| Thin metal bands which form channels to the various pockets of a sorter. { 'shüt ‚blādz }

C³I See command, control, communications, and intelligence. { 'sē 'thrē'ī }

CID See charge-injection device.

CIM See computer input from microfilm; computer-integrated manufacturing.

cinching |COMPUT SCI| Creases produced in magnetic tape when the supply reel is wound at low tension and suddenly stopped during playback. { 'sin·chiŋ }

C-indicator See C-display. { 'sē ‚in·də‚kād·ər }

cipher |COMMUN| A transposition or substitution code for transmitting secret messages. { 'sī·fər }

cipher block chaining |COMMUN| A technique for block chaining in which each block of ciphertext is produced by adding, through the EXCLUSIVE OR operation, the previous block of ciphertext to the current block of plaintext. Abbreviated CBC. { 'sī·fər ‚bläk ‚chān·iŋ }

cipher feedback |COMMUN| An implementation of ciphertext autokey cipher in which the leftmost n bits of the data encryption standard (DES) output are added by the EXCLUSIVE OR operation to N bits of plaintext to produce N bits of ciphertext (where N is the number of bits enciphered at one time), and these N bits of ciphertext are fed back into the algorithm by first shifting the current DES input N bits to the left, and then appending the N bits of ciphertext to the right-hand side of the shifted input to produce a new DES input used for the next iteration of the algorithm. { ¦sī·fər ¦fēd¦bak }

cipher machine |COMMUN| Mechanical or electrical apparatus for enciphering and deciphering. { 'sī·fər mə'shēn }

ciphertext |COMMUN| A message which has been transformed by a cipher so that it can be read only by those privy to the secrets of the cipher. { 'sī·fər¦tekst }

ciphertext autokey cipher |COMMUN| A stream cipher in which the cryptographic bit stream generated at a given time is determined by the ciphertext generated at earlier times. { 'sī·fər ¦tekst 'ȯd·ō¦kē ¦si·fər }

ciphony |COMMUN| A technique by which security is accomplished by converting speech into a series of on-off pulses and mixing these with the pulses supplied by a key generator; to recover the original speech, the identical key must be subtracted and the resultant on-off pulses reconverted into the original speech pattern; unauthorized listeners are unable to reconstruct the plain text unless they have an identical key generator and the daily key setting. { 'sī·fə·nē }

ciphony equipment |ELECTR| Any equipment attached to a radio transmitter, radio receiver, or telephone for scrambling or unscrambling voice messages. { 'sī·fə·nē i¦kwip·mənt }

circle diagram |ELEC| A diagram which gives a graphical solution of equations for a transmission line, giving the input impedance of the line as a function of load impedance and electrical length of the line. { ¦sər·kəl ¦dī·ə¦gram }

circle-dot mode |ELECTR| Mode of cathode-ray storage of binary digits in which one kind of digit is represented by a small circle of excitation of the screen, and the other kind by a similar circle with a concentric dot. { ¦sər·kəl ¦dät ¦mōd }

circuit |ELEC| See electric circuit. |ELECTROMAG| A complete wire, radio, or carrier communications channel. { 'sər·kət }

circuit analyzer See volt-ohm-milliammeter. { 'sər·kət ¦an·ə¦līz·ər }

circuit board See printed circuit board. { 'sər·kət ¦bȯrd }

circuit breaker |ELEC| An electromagnetic device that opens a circuit automatically when the current exceeds a predetermined value. { 'sər·kət ¦brāk·ər }

circuit capacity |COMMUN| Number of communications channels which can be handled by a given circuit at the same time. { 'sər·kət kə'pas·əd·ē }

circuit conditioning |ELECTR| Test, analysis, engineering, and installation actions to upgrade a communications circuit to meet an operational requirement; includes the reduction of noise, the equalization of phase and level stability and frequency response, and the correction of impedance discontinuities, but does not in-clude normal maintenance and repair activities. { 'sər·kət kən'dish·ə·niŋ }

circuit design |ELEC| The art of specifying the components and interconnections of an electrical network. { 'sər·kət də'zīn }

circuit diagram |ELEC| A drawing, using standardized symbols, of the arrangement and interconnections of the conductors and components of an electrical or electronic device or installation. Also known as schematic circuit diagram; wiring diagram. { 'sər·kət ¦dī·ə¦gram }

circuit efficiency |ELECTR| Of an electron tube, the power delivered to a load at the output terminals of the output circuit at a desired frequency divided by the power delivered by the electron stream to the output circuit at that frequency. { 'sər·kət i'fish·ən·sē }

circuit element See component. { 'sər·kət ¦el·ə·mənt }

circuit grade |COMMUN| A circuit rating defining the ability to carry information; grades include telegraph, voice, and broad-band. { 'sər·kət ¦grād }

circuit interrupter |ELEC| A device in a circuit breaker to remove energy from an arc in order to extinguish it. { 'sər·kət ¦in·tə¦rəp·tər }

circuit loading |ELEC| Power drawn from a circuit by an electric measuring instrument, which may alter appreciably the quantity being measured. { 'sər·kət ¦lȯd·iŋ }

circuit noise |COMMUN| In telephone practice, the noise which is brought to the receiver electrically from a telephone system, excluding noise picked up acoustically by telephone transmitters. { 'sər·kət ¦nȯiz }

circuit noise level |COMMUN| Ratio of the circuit noise at that point to some arbitrary amount of circuit noise chosen as a reference; usually expressed in decibels above reference noise, signifying the reading of a circuit noise meter, or in adjusted decibels, signifying circuit noise meter reading adjusted to represent interfering effect under specified conditions. { 'sər·kət ¦nȯiz ¦lev·əl }

circuit protection |ELECTR| Provision for automatically preventing excess or dangerous temperatures in a conductor and limiting the amount of energy liberated when an electrical failure occurs. { 'sər·kət prə'tek·shən }

circuit reliability |COMMUN| The percent of time a circuit was available to the user during a specified period of time. { 'sər·kət ri¦lī·ə'bil·əd·ē }

circuitron |ELECTR| Combination of active and passive components mounted in a single envelope like that used for tubes, to serve as one or more complete operating stages. { 'sər·kyə¦trän }

circuitry |ELEC| The complete combination of circuits used in an electrical or electronic system or piece of equipment. { 'sər·kə·trē }

circuit shift See cyclic shift. { 'sər·kət ¦shift }

circuit switching |COMMUN| **1.** The method of providing communication service through a switching facility, either from local users or from other switching facilities. **2.** A method of transmitting messages through a communications network in which a path from the sender to the receiver of fixed bandwidth or speed is set up for the entire duration of a communication or call. { 'sər·kət ˌswich·iŋ }

circuit testing |ELEC| The testing of electric circuits to determine and locate an open circuit, or a short circuit or leakage. { 'sər·kət ˌtes·tiŋ }

circuit theory |ELEC| The mathematical analysis of conditions and relationships in an electric circuit. Also known as electric circuit theory. { 'sər·kət ˌthē·ə·rē }

circular antenna |ELECTROMAG| A folded dipole that is bent into a circle, so the transmission line and the abutting folded ends are at opposite ends of a diameter. { 'sər·kyə·lər an'ten·ə }

circular arc See arc. { 'sər·kyə·lər 'ärk }

circular buffering |COMPUT SCI| A technique for receiving data in an input-output control system which uses a single buffer that appears to be organized in a circle, with data wrapping around it. { 'sər·kyə·lər 'bəf·ə·riŋ }

circular current |ELEC| An electric current moving in a circular path. { 'sər·kyə·lər 'kər·ənt }

circular file |COMPUT SCI| An organized collection of records, generally with a high turnover, in which new records are inserted by replacing the oldest records. { 'sər·kyə·lər 'fīl }

circular horn |ELECTROMAG| A circular-waveguide section that flares outward into the shape of a horn, to serve as a feed for a microwave reflector or lens. { 'sər·kyə·lər 'hórn }

circular polarized loop vee |ELECTROMAG| Airborne communications antenna with an omnidirectional radiation pattern to provide optimum near-horizon communications coverage. { 'sər·kyə·lər 'pō·lə,rīzd 'lüp ,vē }

circular polling |COMMUN| A form of polling in which each terminal is interrogated exactly once in every pass, regardless of its level of activity. { 'sər·kyə·lər 'pōl·iŋ }

circular reference |COMPUT SCI| A situation created by a programming error in which two or more entities each refer to the other so that the execution of the program is carried on endlessly with no resolution. { 'sər·kyə·lər 'ref·rəns }

circular scanning |ENG| Radar scanning in which the direction of maximum radiation describes a right circular cone. { 'sər·kyə·lər 'skan·iŋ }

circular shift See cyclic shift. { 'sər·kyə·lər 'shift }

circular sweep generation |ELECTR| The use of electronic circuits to provide voltage or current which causes an electron beam in a device such as a cathode-ray tube to move in a circular deflection path at constant speed. { 'sər·kyə·lər 'swēp ,jen·ə,rā·shən }

circular wait See mutual deadlock. { 'sər·kyə·lər 'wāt }

circular waveguide |ELECTROMAG| A waveguide whose cross-sectional area is circular. { 'sər·kyə·lər 'wāv,gīd }

circulating memory |ELECTR| A digital computer device that uses a delay line to store information in the form of a pattern of pulses in a train;

the output pulses are detected electrically, amplified, reshaped, and reinserted in the delay line at the beginning. Also known as delay-line memory; delay-line storage; circulating storage. { 'sər·kyə,lād·iŋ 'mem·rē }

circulating register |COMPUT SCI| A shift register in which data move out of one end and reenter the other end, as in a closed loop. { 'sər·kyə ,lād·iŋ 'rej·ə·stər }

circulating storage See circulating memory. { 'sər·kyə,lād·iŋ 'stór·ij }

circulator |ELECTROMAG| A waveguide component having a number of terminals so arranged that energy entering one terminal is transmitted to the next adjacent terminal in a particular direction. Also known as microwave circulator. { ,sər·kyə·'lād·ər }

CISC See complex instruction set computer. { sisk }

citizens' band |COMMUN| A frequency band allocated for citizens' radio service (462.550–467.425, 72–76, or 26.965–27.405 megahertz). { 'sit·ə·zənz ,band }

citizens' radio service |COMMUN| A radio communication service intended for private or personal radio communication, including radio signaling and control of objects by radio. { 'sit·ə·zənz 'rād·ē·ō ,sər·vəs }

cladding |COMMUN| A plastic or glass sheath that is fused to and surrounds the core of an optical fiber. |ENG| Process of covering one material with another and bonding them together under high pressure and temperature. Also known as bonding. { 'klad·iŋ }

clamp See clamping circuit. { klamp }

clamper See direct-current restorer. { 'klamp·ər }

clamping |ELECTR| The introduction of a reference level that has some desired relation to a pulsed waveform, as at the negative or positive peaks. Also known as direct-current reinsertion; direct-current restoration. { 'klamp·iŋ }

clamping circuit |ELECTR| A circuit that reestablishes the direct-current level of a waveform; used in the dc-restorer stage of an analog television receiver to restore the dc component to the video signal after its loss in capacitance-coupled alternating-current amplifiers, to reestablish the average light value of the reproduced image. Also known as clamp. { 'klamp·iŋ ,sər·kət }

clamping diode |ELECTR| A diode used to clamp a voltage at some point in a circuit. { 'klamp·iŋ ,dī,ōd }

clamping gripper |CONT SYS| A robot element that uses two-link movements, parallel-jaw movements, and combination movements to grasp and handle objects. { 'klamp·iŋ 'grip·ər }

clamp-on |COMMUN| A method of holding a call for a line that is in use and of signaling when it becomes free. { 'klamp ,ón }

clamp-on ammeter See snap-on ammeter. { 'klamp,ón 'a,mēd·ər }

clapper |ELEC| A hinged or pivoted relay armature. { 'klap·ər }

Clapp oscillator |ELECTR| A series-tuned Colpitts oscillator, having low drift. { ¦klap ¦äs·ə ,lād·ər }

90

Clark cell [ELEC] An early form of standard cell, having 1.433 volts at 15°C, now largely replaced by the Weston standard cell as a voltage standard. { 'klärk ,sel }

class [COMPUT SCI] In object-oriented programming, a description of the structure and operations of an object. A new class is defined by stating how it differs from an existing class. The new (more specific) class is said to inherit from the original (general) class and is referred to as a subclass of the original class. The original class is referred to as the superclass of the new class. { klas }

class A amplifier [ELECTR] 1. An amplifier in which the grid bias and alternating grid voltages are such that anode current in a specific tube flows at all times. 2. A transistor amplifier in which each transistor is in its active region for the entire signal cycle. { ,klas 'ā 'am·plə,fī·ər }

class AB amplifier [ELECTR] 1. An amplifier in which the grid bias and alternating grid voltages are such that anode current in a specific tube flows for appreciably more than half but less than the entire electric cycle. 2. A transistor amplifier whose operation is class A for small signals and class B for large signals. { ,klas ¦ā¦bē 'am·plə ,fī·ər }

class A modulator [ELECTR] A class A amplifier used to supply the necessary signal power to modulate a carrier. { ,klas 'ā 'mäj·ə,lād·ər }

class A push-pull sound track [ENG ACOUS] Two single photographic sound tracks side by side, the transmission of one being 180° out of phase with the transmission of the other; both positive and negative halves of the sound wave are linearly recorded on each of the two tracks. { ,klas 'ā ¦push¦pul 'saùn ,trak }

class B amplifier [ELECTR] 1. An amplifier in which the grid bias is approximately equal to the cutoff value, so that anode current is approximately zero when no exciting grid voltage is applied, and flows for approximately half of each cycle when an alternating grid voltage is applied. 2. A transistor amplifier in which each transistor is in its active region for approximately half the signal cycle. { ,klas 'bē 'am·plə ,fī·ər }

class B auxiliary power [ELEC] Standby power plant to cover extended outages (days) of primary power. { ,klas 'bē òg'zil·yə·rē 'paù·ər }

class B modulator [ELECTR] A class B amplifier used to supply the necessary signal power to modulate a carrier; usually connected in push-pull. { ,klas 'bē 'mäj·ə,lād·ər }

class B push-pull sound track [ENG ACOUS] Two photographic sound tracks side by side, one of which carries the positive half of the signal only, and the other the negative half; during the inoperative half-cycle, each track transmits little or no light. { ,klas 'bē ¦push¦pul 'saùn ,trak }

class C amplifier [ELECTR] 1. An amplifier in which the bias on the control element is appreciably greater than the cutoff valve, so that the output current in each device is zero when no alternating control signal is applied, and flows for appreciably less than half of each cycle when an alternating control signal is applied. 2. A transistor amplifier in which each transistor is in its active region for significantly less than half the signal cycle. { ,klas 'sē 'am·plə,fī·ər }

class C auxiliary power [ELEC] Quick start (10–60 seconds) power unit to cover short-term outages (hours) of primary power. { ,klas 'sē òg'zil·yə·rē 'paù·ər }

class D amplifier [ELECTR] A power amplifier that employs a pair of transistors that are connected in push-pull and driven to act as a switch, and a series-tuned output filter, which allows only the fundamental-frequency component of the resultant square wave to reach the load. { ,klas 'dē 'am·plə,fī·ər }

class D auxiliary power [ELEC] Uninterruptible (no-break) power unit using stored energy to provide continuous power within specified voltage and frequency tolerances. { ,klas 'dē òg'zil·yə·rē 'paù·ər }

class E amplifier [ELECTR] A power amplifier that employs a single transistor driven to act as a switch, and an output filter selected to bring the drain voltage to zero at the instant the transistor is switched on. { ,klas 'ē 'am·plə,fī·ər }

class F amplifier [ELECTR] A power amplifier that employs a single transistor and a multiple-resonance output circuit. { ,klas 'ef 'am·plə ,fī·ər }

class NP problems [COMPUT SCI] Problems that cannot necessarily be solved in polynomial time on a sequential computer but can be solved in polynomial time on a nondeterministic computer which, roughly speaking, guesses in turn each of 2N possible values of some N-bit quantity. { 'klas ¦en¦pē ,präb·ləmz }

class P problems [COMPUT SCI] Problems that can be solved in polynomial time on a conventional sequential computer. { 'klas 'pē ,präb·ləmz }

class S modulator [ELECTR] A modulator that is based on pulse-width modulation with a switching frequency several times the highest output frequency, and in which the pulse-width modulated signal is boosted to the desired power level by switching amplifiers, after which the desired audio output is obtained by a low-pass filter. { ,klas 'es 'mäj·ə,lād·ər }

clause [COMPUT SCI] A part of a statement in the COBOL language which may describe the structure of an elementary item, give initial values to items in independent and group work areas, or redefine data previously defined by another clause. { klòz }

Clausius-Mosotti equation [ELEC] An expression for the polarizability γ of an individual molecule in a medium which has the relative dielectric constant ϵ and has N molecules per unit volume: $\gamma = (3/4\pi\,N)\,[(\epsilon - 1)/(\epsilon + 2)]$ (Gaussian units). { ¦klòz·ē·əs mə'zäd·ē i'kwā·zhən }

clean and certify [COMPUT SCI] To prepare a magnetic tape for a computer system by running it through a machine that cleans it, writes a data test pattern on it, and checks it for errors. { 'klēn ən 'sərd·ə,fī }

clean compile |COMPUT SCI| Conversion of a computer program from source to object language with no detection of significant errors by the compiler; logic errors not identified by the compiler may exist. { 'klēn kəm'pīl }

clean track |ENG ACOUS| A sound track having no leakage from other tracks. { ¦klēn ¦trak }

cleanup |ELECTR| Gradual disappearance of gases from an electron tube during operation, due to absorption by getter material or the tube structure. { 'klē,nəp }

clear |COMPUT SCI| **1.** To restore a storage device, memory device, or binary stage to a prescribed state, usually that denoting zero. Also known as reset. **2.** A function key on calculators, to delete an entire problem or just the last keyboard entry. { klir }

clear area |COMPUT SCI| In optical character recognition, any area designated to be kept free of printing or any other extraneous markings. { 'klir ,er·ē·ə }

clear band |COMPUT SCI| In character recognition, a continuous horizontal strip of blank paper which must be obtained between consecutive code lines on a source document. { 'klir ,band }

clear channel |COMMUN| A standard broadcast channel in which the dominant station or stations render service over wide areas; stations are cleared of objectionable interference within their primary service areas and over all or a substantial portion of their secondary service areas. { ¦klir 'chan·əl }

clear text |COMMUN| Text or language which conveys an intelligible meaning in the language in which it is written with no hidden meaning. { 'klir ,tekst }

clear-voice override |COMMUN| The ability of a speech scrambler to receive a clear message even when the scrambler is set for scrambler operation. { ¦klir ¦vȯis 'ō·vǝ,rīd }

click |COMMUN| A short-duration electric disturbance, such as that sometimes produced by a code-sending key or a switch. |COMPUT SCI| To select an object when the pointer is touching it by pressing and quickly releasing a button on a mouse. |ENG ACOUS| A perforation in a sound track which produces a clicking sound when passed over the projector sound head. { klik }

click filter |ELECTR| A capacitor connected across a switch, relay, or key to lengthen the decay time from the closed to the open condition when the device is opened or closed. { 'klik ,fil·tǝr }

click track |ENG ACOUS| A sound track containing a series of clicks, which may be spaced regularly (uniform click track) or irregularly (variable click track). { 'klik ,trak }

client |COMPUT SCI| A hardware or software entity that requests shared services from a server. { 'klī·ǝnt }

client-based application |COMPUT SCI| An application that runs on a work station or personal computer in a network and is not available to others in the network. { 'klī·ǝnt ,bäst ,ap·lǝ ¦kā·shǝn }

client-server system |COMPUT SCI| A computing system composed of two logical parts: a server, which provides information or services, and a client, which requests them. On a network, for example, users can access server resources from their personal computers using client software. { ¦klī·ǝnt 'sǝr·vǝr ,sis·tǝm }

clip art |COMPUT SCI| A collection of graphic images that are stored on a computer disk for use in desktop publishing, word processing, and presentation graphics programs. { 'klip ,ärt }

clipboard |COMPUT SCI| An area in memory or a file where cut or copied material is held temporarily before being inserted elsewhere in the same document or in another document. { 'klip,bȯrd }

clip lead |ELEC| A short piece of flexible wire with an alligator clip or similar temporary connector at one or both ends. { 'klip ,lēd }

clipper See limiter. { 'klip·ǝr }

Clipper Chip |COMPUT SCI| A chip proposed by the United States government to be used in all devices that might use encryption, such as computers and communications devices, for which the government would have at least some access or control over the decryption key for purposes of surveillance. { 'klip·ǝr ,chip }

clipper diode |ELECTR| A bidirectional breakdown diode that clips signal voltage peaks of either polarity when they exceed a predetermined amplitude. { 'klip·ǝr ,dī,ōd }

clipper-limiter |ELECTR| A device whose output is a function of the instantaneous input amplitude for a range of values lying between two predetermined limits but is approximately constant, at another level, for input values above the range. { ¦klip·ǝr ¦lim·ǝd·ǝr }

clipping |COMMUN| The perceptible mutilation of signals or speech syllables during transmission, often due to limiting. |COMPUT SCI| See scissoring. |ELECTR| See limiting. { 'klip·iŋ }

clipping circuit See limiter. { 'klip·iŋ ,sǝr·kǝt }

clipping level |ELECTR| The level at which a clipping circuit is adjusted; for example, the magnitude of the clipped wave shape. { 'klip·iŋ ,lev·ǝl }

CLIST |COMPUT SCI| A file containing a series of commands that are processed in the order given when the file is entered. Acronym for command list. { 'sē,list }

clobber |COMPUT SCI| To write new data and thereby erase good data in a file, or to otherwise destroy data. { 'kläb·ǝr }

clock |ELECTR| A source of accurately timed pulses, used for synchronization in a digital computer or as a time base in a transmission system. { kläk }

clock control system |CONT SYS| A system in which a timing device is used to generate the control function. Also known as time-controlled system. { 'kläk kǝn'trōl ,sis·tǝm }

clock-doubled |COMPUT SCI| Describing a microprocessor that operates at twice the clock speed of the bus or motherboard to which it is attached. { 'kläk ¦dǝb·ǝld }

clocked flip-flop [ELECTR] A flip-flop circuit that is set and reset at specific times by adding clock pulses to the input so that the circuit is triggered only if both trigger and clock pulses are present simultaneously. { 'kläkt 'flip ,fläp }

clocked logic [ELECTR] A logic circuit in which the switching action is controlled by repetitive pulses from a clock. { 'kläkt ¦läj·ik }

clock frequency [ELECTR] The master frequency of the periodic pulses that schedule the operation of a digital computer. Also known as clock rate; clock speed. { 'kläk ,frē·kwən·sē }

clock motor *See* timing motor. { 'kläk ,mōd·ər }

clock oscillator [ELECTR] An oscillator that controls an electronic clock. { 'kläk 'äs·ə ,lād·ər }

clock pulses [COMPUT SCI] Electronic pulses which are emitted periodically, usually by a crystal device, to synchronize the operation of circuits in a computer. Also known as clock signals. { 'kläk ,pəl·səz }

clock rate *See* clock frequency. { 'kläk ,rāt }

clock signals *See* clock pulses. { 'kläk ,sig·nəlz }

clock speed *See* clock frequency. { 'kläk ,spēd }

clock time *See* internal cycle time. { 'kläk ,tīm }

clock track [COMPUT SCI] A track on a magnetic recording medium that generates clock pulses for the synchronization of read and write operations. { 'kläk ,trak }

clock-tripled [COMPUT SCI] Describing a microprocessor that operates at three times the clock speed of the bus or motherboard to which it is attached. { 'kläk ¦trip·əld }

clone [COMPUT SCI] A hardware or software product that closely resembles another product created by a different manufacturer or developer, in operation, appearance, or both. { klōn }

close [COMPUT SCI] To make a file unavailable to a computer program which previously had access to it. { klōs }

close coupling [ELEC] **1.** The coupling obtained when the primary and secondary windings of a radio-frequency or intermediate-frequency transformer are close together. **2.** A degree of coupling that is greater than critical coupling. Also known as tight coupling. { ¦klōs 'kəp·liŋ }

closed architecture [COMPUT SCI] A computer architecture whose detailed, technical specifications are published only to those authorized by the manufacturer. { ¦klōzd 'ärk·ə,tek·chər }

closed-box system [ELECTR] A loudspeaker system in which the woofer is mounted in a sealed box. { ,klōzd 'bäks ,sis·təm }

closed-bus system [COMPUT SCI] A computer that lacks receptacles for expansion boards and is difficult to upgrade. { ¦klōd 'bəs ,sis·təm }

closed-caption television [COMMUN] A method of captioning or subtitling television programs by coding captions as a vertical-interval data signal in an analog television system or in the transport of a digital television system that is decoded at the receiver and superimposed on the normal television picture. { ¦klōzd ¦kap·shən 'tel·ə,vizh·ən }

closed circuit [COMMUN] Program source that is not broadcast for general consumption but is fed to remote monitoring units. { ¦klōzd 'sər·kət }

closed-circuit communications system [COMMUN] A communications systems which is entirely self-contained, and does not exchange intelligence with other facilities and systems. { ¦klōzd ¦sər·kət kə,myü·nə'kā·shənz ,sis·təm }

closed-circuit signaling [COMMUN] Signaling in which current flows in the idle condition, and a signal is initiated by increasing or decreasing the current. { ¦klōzd ¦sər·kət 'sig·nə·liŋ }

closed-circuit telegraph system [COMMUN] Telegraph system in which, when no station is transmitting, the circuit is closed and current flows through the circuit. { ¦klōzd ¦sər·kət 'tel·ə ,graf ,sis·təm }

closed-circuit television [COMMUN] Any application of television that does not involve broadcasting for public viewing; the programs can be seen only on specified receivers connected to the television camera by circuits, which include microwave relays and coaxial cables. Abbreviated CCTV. { ¦klōzd ¦sər·kət 'tel·ə,vizh·ən }

closed-coil armature [ELEC] The configuration of an armature in which the connection of all the coils forms a closed circuit. { 'klōzd ¦kȯil 'är·mə·chər }

closed-cycle fuel cell [ELEC] A fuel cell in which the reactants are regenerated by an auxiliary process, such as electrolysis. { ¦klōzd ¦sī·kəl 'fyül ,sel }

closed file [COMPUT SCI] A file that cannot be accessed for reading or writing. { ¦klōzd 'fil }

closed loop [COMPUT SCI] A loop whose execution continues indefinitely in the absence of any external intervention. [CONT SYS] A family of automatic control units linked together with a process to form an endless chain; the effects of control action are constantly measured so that if the controlled quantity departs from the norm, the control units act to bring it back. { ¦klōzd 'lüp }

closed-loop control system *See* feedback control system. { ¦klōzd ¦lüp kən'trōl ,sis·təm }

closed-loop telemetry system [ENG] **1.** A telemetry system which is also used as the display portion of a remote-control system. **2.** A system used to check out test vehicle or telemetry performance without radiation of radio-frequency energy. { ¦klōzd ¦lüp tə'lem·ə·trē ,sis·təm }

closed-loop voltage gain [ELECTR] The voltage gain of an amplifier with feedback. { ¦klōzd ¦lüp 'vōl·tij ,gān }

closed shop [COMPUT SCI] A data-processing center so organized that only professional programmers and operators have access to the center to meet the needs of users. { ¦klōzd 'shäp }

closed subroutine [COMPUT SCI] A subroutine that can be stored outside the main routine and can be connected to it by linkages at one or more locations. { ¦klōzd 'səb·rü,tēn }

closefile |COMPUT SCI| A procedure call in time sharing which enables an ALGOL program to close a file no longer required. { 'klōz,fīl }

close-out file |COMPUT SCI| A file created at the end of a processing cycle, usually encompassing a specified period of time. { 'klōz ,aút ,fīl }

close routine |COMPUT SCI| A computer program that changes the state of a file from open to closed. { 'klōz rü'tēn }

close-talking microphone |ENG ACOUS| A microphone designed for use close to the mouth, so noise from more distant points is suppressed. Also known as noise-canceling microphone. { 'klōs ,tók·iŋ 'mī·krə,fōn }

cloud pulse |ELECTR| The output resulting from space charge effects produced by turning the electron beam on or off in a charge-storage tube. { 'klaúd ,pəls }

cloverleaf antenna |ELECTROMAG| Antenna having radiating units shaped like a four-leaf clover. { 'klō·vər,lēf an 'ten·ə }

cluster |COMPUT SCI| **1.** In a clustered file, one of the classes into which records with similar sets of content identifiers are grouped. **2.** A grouping of hardware devices in a distributed processing system. **3.** A group of disk sectors that is treated as a single entity by the operating system. { 'kləs·tər }

cluster controller |COMPUT SCI| A control unit to which several peripheral devices are assigned. { 'kləs·tər kən,trōl·ər }

clustered file |COMPUT SCI| A collection of records organized so that items which exhibit similar sets of content identifiers are automatically grouped into common classes. { 'kləs·tərd 'fīl }

clustering algorithm |COMPUT SCI| A computer program that attempts to detect and locate the presence of groups of vectors, in a high-dimensional multivariate space, that share some property of similarity. { ¦kləs·tə·riŋ ¦al·gə ,rith·əm }

clutter |ELECTROMAG| Unwanted echoes on a radar screen, such as those caused by the ground, sea, rain, stationary objects, chaff, enemy jamming transmissions, and grass. Also known as background returns; radar clutter. { 'kləd·ər }

clutter gating |ELECTR| A technique which provides switching between moving-target-indicator and normal videos; this results in normal video being displayed in regions with no clutter and moving-target-indicator video being switched in only for the clutter areas. { 'kləd·ər ,gad·iŋ }

clutter suppression |ELECTR| Technique of reducing, by various means integral to the radar system, the effects of echoes from scatterers such as rain and surface features among the received signals. { 'kləd·ər sə,presh·ən }

CMI See computer-managed instruction.

CML See current-mode logic.

CMOS device |ELECTR| A device formed by the combination of a PMOS (p-type-channel metal oxide semiconductor device) with an NMOS (n-type-channel metal oxide semiconductor device). Derived from complementary metal oxide semiconductor device. { 'se,mòs di'vīs }

CMRR See common-mode rejection ratio.

CNC See computer numerical control.

C network |ELECTR| Network composed of three impedance branches in series, the free ends being connected to one pair of terminals, and the junction points being connected to another pair of terminals. { 'sē ,net,wərk }

coast |ENG| A memory feature on a radar which, when activated, causes the range and angle systems to continue to move in the same direction and at the same speed as that required to track an original target. { kōst }

coastal refraction |ELECTROMAG| An apparent change in the direction of travel of a radio wave when it crosses a shoreline obliquely. Also known as land effect. { 'kōs·təl ri'frak·shən }

coated cathode |ELECTR| A cathode that has been coated with compounds to increase electron emission. { 'kōd·əd 'kāth,ōd }

coated filament |ELECTR| A vacuum-tube filament coated with metal oxides to provide increased electron emission. { 'kōd·əd 'fil·ə·mənt }

coax See coaxial cable. { 'kō,aks }

coaxial antenna |ELECTROMAG| An antenna consisting of a quarter-wave extension of the inner conductor of a coaxial line and a radiating sleeve that is in effect formed by folding back the outer conductor of the coaxial line for a length of approximately a quarter wavelength. { kō'ak·sē·əl an'ten·ə }

coaxial attenuator |ELECTROMAG| An attenuator that has a coaxial construction and terminations suitable for use with coaxial cable. { kō'ak·sē·əl ə'ten·yə,wād·ər }

coaxial bolometer |ELECTR| A bolometer in which the desired square-law detection characteristic is provided by a fine Wollaston wire element that has been thoroughly cleaned before being axially located and soldered in position in its cylinder. { kō'ak·sē·əl bə'läm·əd·ər }

coaxial cable |ELECTROMAG| A transmission line in which one conductor is centered inside and insulated from an outer metal tube that serves as the second conductor. Also known as coax; coaxial line; coaxial transmission line; concentric cable; concentric line; concentric transmission line. { kō'ak·sē·əl 'kā·bəl }

coaxial capacitor See cylindrical capacitor. { kō'ak·sē·əl kə'pas·əd·ər }

coaxial cavity |ELECTROMAG| A cylindrical resonating cavity having a central conductor in contact with its pistons or other reflecting devices. { kō'ak·sē·əl 'kav·əd·ē }

coaxial cavity magnetron |ELECTR| A magnetron which achieves mode separation, high efficiency, stability, and ease of mechanical tuning by coupling a coaxial high Q cavity to a normal set of quarter-wavelength vane cavities. { kō'ak·sē·əl ,kav·əd·ē 'mag·nə,trän }

coaxial connector |ELECTROMAG| An electric connector between a coaxial cable and an equipment circuit, so constructed as to maintain the conductor configuration, through the separable connection, and the characteristic impedance of the coaxial cable. { kō'ak·sē·əl kə'nek·tər }

coaxial-cylinder magnetron |ELECTR| A magnetron in which the cathode and anode consist of coaxial cylinders. { kō'ak·sē·əl ,sil·ən·dər 'mag·nə,trän }

coaxial diode |ELECTR| A diode having the same outer diameter and terminations as a coaxial cable, or otherwise designed to be inserted in a coaxial cable. { kō'ak·sē·əl 'dī,ōd }

coaxial filter |ELECTROMAG| A section of coaxial line having reentrant elements that provide the inductance and capacitance of a filter section. { kō'ak·sē·əl 'fil·tər }

coaxial hybrid |ELECTROMAG| A hybrid junction of coaxial transmission lines. { kō'ak·sē·əl 'hī ,brəd }

coaxial isolator |ELECTROMAG| An isolator used in a coaxial cable to provide a higher loss for energy flow in one direction than in the opposite direction; all types use a permanent magnetic field in combination with ferrite and dielectric materials. { kō'ak·sē·əl 'ī·sə,lād·ər }

coaxial line See coaxial cable. { kō'ak·sē·əl 'līn }

coaxial-line resonator |ELECTROMAG| A resonator consisting of a length of coaxial line short-circuited at one or both ends. { kō'ak·sē·əl ,līn 'rez·ən,ād·ər }

coaxial speaker |ENG ACOUS| A loudspeaker system comprising two, or less commonly three, speaker units mounted on substantially the same axis in an integrated mechanical assembly, with an acoustic-radiation-controlling structure. { kō'ak·sē·əl 'spēk·ər }

coaxial stub |ELECTROMAG| A length of nondissipative cylindrical waveguide or coaxial cable branched from the side of a waveguide to produce some desired change in its characteristics. { kō'ak·sē·əl 'stəb }

coaxial switch |ELEC| A switch that changes connections between coaxial cables going to antennas, transmitters, receivers, or other high-frequency devices without introducing impedance mismatch. { kō'ak·sē·əl 'swich }

coaxial transistor |ELECTR| A point-contact transistor in which the emitter and collector are point electrodes making pressure contact at the centers of opposite sides of a thin disk of semiconductor material serving as base. { kō'ak·sē·əl tran'zis·tər }

coaxial transmission line See coaxial cable. { kō'ak·sē·əl tranz'mish·ən ,līn }

coaxial wavemeter |ENG| A device for measuring frequencies above about 100 megahertz, consisting of a rigid metal cylinder that has an inner conductor along its central axis, and a sliding disk that shorts the inner conductor and the cylinder. { kō'ak·sē·əl 'wāv,mēd·ər }

COBOL |COMPUT SCI| A business data-processing language that can be given to a computer as a series of English statements describing a complete business operation. Derived from common business-oriented language. { 'kō ,ból }

cochannel cells |COMMUN| Two cells in a cellular mobile radio system that use the same frequency. { 'kō,chan·əl 'selz }

cochannel interference |COMMUN| Interference caused on one communication channel by a transmitter operating in the same channel. { 'kō,chan·əl ,in·tər'fir·əns }

cochannel interference reduction factor |COMMUN| The ratio of the minimum separation between two cochannel cells without interference to the radius of a cell. { 'kō,chan·əl ,in·tər,fir·əns ri'dək·shən ,fak·tər }

codan |ELECTR| A device that silences a receiver except when a modulated carrier signal is being received. { 'kō,dan }

Coddington shape factor See shape factor. { 'käd·iŋ·tən 'shāp ,fak·tər }

code |COMMUN| **1.** A system of symbols and rules for expressing information, such as the Morse code, **2.** Electronic Industries Association color code, and the binary and other machine languages used in digital computers. { kōd }

code book |COMMUN| A book containing a large number of plaintext words, phrases, and sentences and their codetext equivalents. { 'kōd ,búk }

codec |ELECTR| A device that converts analog signals to digital form for transmission and converts signals traveling in the opposite direction from digital to analog form. Derived from coder-decoder. { 'kō,dek }

code-check |COMPUT SCI| To remove mistakes from a coded routine or program. { 'kōd ,chek }

code checking time |COMPUT SCI| Time spent checking out a problem on the computer, making sure that the problem is set up correctly and that the code is correct. { 'kōd ,chek·iŋ ,tīm }

code converter |COMPUT SCI| A converter that changes coded information to a different code system. { 'kōd kən'vərd·ər }

coded character set |COMPUT SCI| A set of characters together with the code assigned to each character for computer use. { 'kōd·əd 'kar·ik·tər ,set }

coded decimal See decimal-coded digit. { 'kōd·əd 'des·məl }

coded interrogator |COMMUN| An interrogator whose output signal forms the code required to trigger a specific radio or radar beacon; part of an address-selective system. { 'kōd·əd in'ter·ə ,gād·ər }

code-division multiple access |COMMUN| The transmission of messages from a large number of transmitters over a single channel by assigning each transmitter a pseudorandom noise code (typically more than 2000 symbols long for each bit of information) so that the codes are mathematically independent of each other. Abbreviated CDMA. { 'kōd dəịvizh·ən 'məl·tə·pəl 'ak,ses }

code-division multiplex |COMMUN| Multiplex in which two or more communication links occupy the entire transmission channel simultaneously, with code signal structures designed so a given receiver responds only to its own signals and treats the other signals as noise. Abbreviated CDM. { 'kōd də'vizh·ən 'məlt·ə,pleks }

coded passive reflector antenna [ELECTROMAG] An object intended to reflect Hertzian waves and having variable reflecting properties according to a predetermined code for the purpose of producing an indication on a radar receiver. { 'kōd·əd 'pas·iv ri'flek·tər an,ten·ə }

coded program [COMPUT SCI] A program expressed in the required code for a computer. { 'kōd·əd 'prō·grəm }

coded stop [COMPUT SCI] A stop instruction built into a computer routine. { 'kōd·əd 'stäp }

code element [COMMUN] One of the separate elements or events constituting a coded message, such as the presence or absence of a pulse, dot, dash, or space. { 'kōd ,el·ə·mənt }

code error [COMPUT SCI] A surplus or lack of a bit or bits in a machine instruction. { 'kōd ,er·ər }

code-excited linear predictive coder [COMMUN] A speech coder that uses both short-term and long-term predictors, vector quantization techniques, and an analysis-by-synthesis approach to search for the best combination of coder parameters. Abbreviated CELP coder. { ¦kōd i¦sīd·əd ¦lin·ē·ər prə¦dik·tiv 'kōd·ər }

code extension [COMPUT SCI] A method of increasing the number of characters that can be represented by a code by combining characters into groups. { 'kōd ik,sten·chən }

code group [COMMUN] A combination of letters or numerals or both, assigned to represent one or more words of plain text in a coded message. { 'kōd ,grüp }

code line [COMPUT SCI] In character recognition, the area reserved for the inscription of the printed or handwritten characters to be recognized. { 'kōd ,līn }

code practice oscillator [ELECTR] An oscillator used with a key and either headphones or a loudspeaker to practice sending and receiving Morse code. { ¦kōd ¦prak·təs 'äs·ə,lād·ər }

coder [COMMUN] A device that generates a code by producing pulses having varying lengths or spacings, as required for radio beacons and interrogators. Also known as moder; pulse coder; pulse-duration coder. [COMPUT SCI] A person who translates a sequence of computer instructions into codes acceptable to the machine. { 'kōd·ər }

coder-decoder See codec. { ¦kōd·ər dē¦kōd·ər }

code reader [COMPUT SCI] A scanning device used for automated identification of a two-dimensional pattern, one part after the other, and generation of either analog or digital signals that correspond to the pattern. Also known as code scanner. { 'kōd ,rēd·ər }

code ringing [COMMUN] In telephone switching, party-line ringing wherein the number or duration of rings indicates which station is being called. { 'kōd ,riŋ·iŋ }

code scanner See code reader. { 'kōd ,skan·ər }

code sensitivity [COMPUT SCI] Property of hardware or software that can handle only data presented in a particular code. { 'kōd ,sen·sə ,tiv·əd·ē }

code signal [COMMUN] A sequence of discrete conditions or events corresponding to a coded message. { 'kōd ,sig·nəl }

codetext [COMMUN] A message which has been transformed by a code into a form which can be read only by those privy to the secrets of the code. { 'kōd,tekst }

code translation [COMMUN] Conversion of a directory code or number into a predetermined code for controlling the selection of an outgoing trunk or line. { 'kōd tranz,lā·shən }

code transparency [COMPUT SCI] Property of hardware or software that can handle data regardless of what form it is in. { 'kōd tranz ,par·ən·sē }

coding [COMPUT SCI] **1.** The process of converting a program design into an accurate, detailed representation of that program in some suitable language. **2.** A list, in computer code, of the successive operations required to carry out a given routine or solve a given problem. { 'kōd·iŋ }

coding disk [COMMUN] Disk with small projections for operating contacts to give a certain predetermined code to a transmission. { 'kōd·iŋ ,disk }

coding form See coding sheet. { 'kōd·iŋ ,fòrm }

coding line See instruction word. { 'kōd·iŋ ,līn }

coding sheet [COMPUT SCI] A sheet of paper printed with a form on which one can conveniently write a coded program. Also known as coding form. { 'kōd·iŋ ,shēt }

codistor [ELECTR] A multijunction semiconductor device which provides noise rejection and voltage regulation functions. { kō'dis·tər }

coefficient of capacitance [ELEC] One of the coefficients which appears in the linear equations giving the charges on a set of conductors in terms of the potentials of the conductors; a coefficient is equal to the ratio of the charge on a given conductor to the potential of the same conductor when the potentials of all the other conductors are 0. { ¦kō·ə'fish·ənt əv kə'pas·ə·təns }

coefficient of induction [ELEC] One of the coefficients which appears in the linear equations giving the charges on a set of conductors in terms of the potentials of the conductors; a coefficient is equal to the ratio of the charge on a given conductor to the potential on another conductor, when the potentials of all the other conductors equal 0. { ¦kō·ə'fish·ənt əv in'dək·shən }

coefficient of potential [ELEC] One of the coefficients which appears in the linear equations giving the potentials of a set of conductors in terms of the charges on the conductors. { ¦kō·ə'fish·ənt əv pə'ten·chəl }

coercion [COMPUT SCI] A method employed by many programming languages to automatically convert one type of data to another. { kō'ər·shən }

cog [ELEC] A fluctuation in the torque delivered by a motor when it runs at low speed, due to electromechanical effects. Also known as torque ripple. { käg }

COGO [COMPUT SCI] A higher-level computer language oriented toward civil engineering, enabling one to write a program in a technical vocabulary familiar to engineers and feed it to the computer; several versions have been implemented. Derived from coordinated geometry. { 'kō͵gō }

cohered video [ELECTR] The video detector output signal in a coherent moving-target indicator radar system. { kō'hird 'vid·ē·ō }

coherent [ELECTR] Referring to radar signals and signal processing and related equipment wherein attention is given to both the amplitude and the phase of the signal; many valuable processes in radar operation are coherent in nature. { kō'hir·ənt }

coherent carrier system [NAV] Transponder system in which the interrogating carrier is retransmitted at a definite multiple frequency for comparison. { kō'hir·ənt 'kar·ē·ər ͵sis·təm }

coherent detector [ELECTR] A detector used in coherent radar giving an output-signal amplitude that depends on the phase of the echo signal (rather than only its amplitude) relative to the phase of that which was transmitted, as required for sensing the radial velocity of targets. Also known as phase detector. { kō'hir·ənt di'tek·tər }

coherent echo [ELECTR] A radar echo whose phase and amplitude at a given range remain relatively constant. { kō'hir·ənt 'ek·ō }

coherent integration [ELECTR] A radar signal processing technique in which the phase relationships among successive pulses being echoed from a target are interpreted, usually to estimate or to separate signals based on the apparent Doppler shift of the signals. { kō'hir·ənt ͵int·ə'grā·shən }

coherent interrupted waves [COMMUN] Interrupted continuous waves occurring in wave trains in which the phase of the waves is maintained through successive wave trains. { kō'hir·ənt in·tə'rəp·təd 'wāvz }

coherent light communications [COMMUN] Communications using the optical band as a transmission medium by modulating a laser in amplitude or pulse frequency. { kō'hir·ənt 'līt kə͵myü·nə'kā·shənz }

coherent moving-target indicator [ENG] A radar system in which the Doppler frequency of the target echo is compared to a local reference frequency generated by a coherent oscillator. { kō'hir·ənt ͵müv·iŋ ͵tär·gət ͵in·də͵kād·ər }

coherent noise [ENG] Noise that affects all tracks across a magnetic tape equally and simultaneously. { kō'hir·ənt 'nȯiz }

coherent oscillator [ELECTR] An oscillator locked in phase to the transmitted signal as used in coherent radar to provide a reference by which changes in the phase of successively received pulses may be recognized. Abbreviated coho. { kō'hir·ənt 'äs·ə͵lād·ər }

coherent processing interval [ELECTR] That period of time over which radar return signals are coherently integrated, permitting a resolution in Doppler shift being sensed as great as the reciprocal of the interval. { kō'hir·ənt 'präs·əs·iŋ 'in·tər·vəl }

coherent-pulse radar [ELECTR] A radar in which the radio-frequency oscillations of recurrent pulses bear a constant phase relation to those of a continuous oscillation. { kō'hir·ənt ͵pəls 'rā͵där }

coherent pulses [ELECTR] Characterizing pulses in which the phase of the radio-frequency waves is maintained through successive pulses. { kō'hir·ənt 'pəl·səz }

coherent radar [ELECTR] A radar capable of comparing the phase of received signals with the phase of the transmitted signal, generally with the object of sensing pulse-to-pulse phase changes, indicative of radial motion, and hence the Doppler shift, of the target. { kō'hir·ənt 'rā ͵där }

coherent reference [ELECTR] A reference signal, usually of stable frequency, to which other signals are phase-locked to establish coherence throughout a system. { kō'hir·ənt 'ref·rəns }

coherent side-lobe canceler [ELECTR] A radar feature in which interfering signals in the side lobes of the radar antenna are cancelled by adaptively adjusting the phase and amplitude of signals received in a number of auxiliary antennas and subtracting those from the signal in the main antenna. { kō'hir·ənt 'sid ͵lōb 'kan·səl·ər }

coherent signal [ELECTR] In coherent radar, a signal having a known phase, often constant, as that produced by the coherent oscillator to be mixed in the coherent detector with the echo signal to detect pulse-to-pulse phase changes indicative of target radial motion. { kō'hir·ənt 'sig·nəl }

coherent system [NAV] A navigation system in which the signal output is obtained by demodulating the received signal after mixing with a local signal having a fixed phase relation to that of the transmitted signal, to permit use of the information carrier by the phase of the received signal. { kō'hir·ənt 'sis·təm }

coherent transponder [ELECTR] A transponder in which a fixed relation between frequency and phase of input and output signals is maintained. { kō'hir·ənt tranz'pänd·ər }

coherent video [ELECTR] The video signal produced in a coherent radar by combining in a coherent detector a radar echo signal with the output of the continuous wave coherent oscillator. Also called bipolar video. { kō'hir·ənt 'vid·ē·ō }

coherer [ELEC] A cell containing a granular conductor between two electrodes; the cell becomes highly conducting when it is subjected to an electric field, and conduction can then be stopped only by jarring the granules. { kō'hir·ər }

coho *See* coherent oscillator. { 'kō͵hō }

coil [CONT SYS] Any discrete and logical result that can be transmitted as output by a programmable controller. [ELECTROMAG] A number of turns of wire used to introduce inductance into an electric circuit, to produce magnetic flux, or to react mechanically to a changing magnetic flux; in high-frequency circuits a coil may be only a fraction of a turn. Also known as electric coil; inductance coil; inductor. { kȯil }

coil antenna [ELECTROMAG] An antenna that consists of one or more complete turns of wire. { 'kȯil an'ten·ə }

coil loading [COMMUN] Loading in which inductors, commonly called loading coils, are inserted in a line at intervals. { 'kȯil ‚lōd·iŋ }

coil neutralization See inductive neutralization. { 'kȯil nü·trə·lə'zā·shən }

coil serving See serving. { 'kȯil ‚sərv·iŋ }

coincidence amplifier [ELECTR] An electronic circuit that amplifies only that portion of a signal present when an enabling or controlling signal is simultaneously applied. { kō'in·sə·dəns ‚am·plə‚fī·ər }

coincidence circuit [ELECTR] A circuit that produces a specified output pulse only when a specified number or combination of two or more input terminals receives pulses within an assigned time interval. Also known as coincidence counter; coincidence gate. { kō'in·sə·dəns ‚sər·kət }

coincidence counter See coincidence circuit. { kō'in·sə·dəns ‚kaůnt·ər }

coincidence gate See coincidence circuit. { kō'in·sə·dəns ‚gāt }

coincident-current selection [ELECTR] The selection of a particular magnetic cell, for reading or writing in computer storage, by simultaneously applying two or more currents. { kō‚in·sə·dənt 'kər·ənt si'lek·shən }

cold [ELEC] Pertaining to electrical circuits that are disconnected from voltage supplies and at ground potential; opposed to hot, pertaining to carrying an electrical charge. { kōld }

cold boot [COMPUT SCI] To turn the power on and boot a computer. { ¦kōld 'büt }

cold cathode [ELECTR] A cathode whose operation does not depend on its temperature being above the ambient temperature. { 'kōld 'kath‚ōd }

cold-cathode counter tube [ELECTR] A counter tube having one anode and three sets of 10 cathodes; two sets of cathodes serve as guides that direct the flow discharge to each of the 10 output cathodes in correct sequence in response to driving pulses. { 'kōld 'kath‚ōd 'kaůnt·ər ‚tüb }

cold-cathode discharge See glow discharge. { 'kōld 'kath‚ōd 'dis‚chärj }

cold-cathode ionization gage See Philips ionization gage. { 'kōld 'kath‚ōd ‚ī·ən·ə'zā·shən ‚gāj }

cold-cathode rectifier [ELECTR] A cold-cathode gas tube in which the electrodes differ greatly in size so electron flow is much greater in one direction than in the other. Also known as gas-filled rectifier. { 'kōld 'kath‚ōd 'rek·tə‚fī·ər }

cold-cathode tube [ELECTR] An electron tube containing a cold cathode, such as a cold-

cathode rectifier, mercury-pool rectifier, neon tube, phototube, or voltage regulator. { 'kōld 'kath‚ōd ‚tüb }

cold emission See field emission. { 'kōld i'mish·ən }

cold junction [ELECTR] The reference junction of thermocouple wires leading to the measuring instrument; normally at room temperature. { 'kōld 'jəŋk·shən }

cold link [COMPUT SCI] A linking of information in two documents in which updating the link requires recopying the information from the source document to the target document. { 'kōld 'liŋk }

cold start [COMPUT SCI] To start running a computer program from the very beginning, without being able to continue the processing that was occurring previously when the system was interrupted. { 'kōld 'stärt }

Cole-Cole plot [ELEC] For a substance displaying orientation polarization, a graph of the imaginary part versus the real part of the complex relative permittivity that is a circular arc, with its center below the abscissa. { 'kōl 'kōl ‚plät }

Cole-Davidson plot [ELEC] For a substance displaying orientation polarization, a graph of the real part versus the imaginary part of the complex relative permittivity that is a skewed arc which approximates a straight line at the high-frequency end and a circular arc at the low-frequency end. { 'kōl 'dā·vəd·sən ‚plät }

collate [COMPUT SCI] To combine two or more similarly ordered sets of values into one set that may or may not have the same order as the original sets. { 'kä‚lāt }

collating sequence [COMPUT SCI] The ordering of a set of items such that sets in that assigned order can be collated. { 'kä‚lād·iŋ ‚sē·kwəns }

collector [ELECTR] **1.** A semiconductive region through which a primary flow of charge carriers leaves the base of a transistor; the electrode or terminal connected to this region is also called the collector. **2.** An electrode that collects electrons or ions which have completed their functions within an electron tube; a collector receives electrons after they have done useful work, whereas an anode receives electrons whose useful work is to be done outside the tube. Also known as electron collector. { kə'lek·tər }

collector capacitance [ELECTR] The depletion-layer capacitance associated with the collector junction of a transistor. { kə'lek·tər kə'pas·əd·əns }

collector current [ELECTR] The direct current that passes through the collector of a transistor. { kə'lek·tər ‚kər·ənt }

collector cutoff [ELECTR] The reverse saturation current of the collector-base junction. { kə'lek·tər ‚kəd‚óf }

collector junction [ELECTR] A semiconductor junction located between the base and collector electrodes of a transistor. { kə'lek·tər ‚jəŋk·shən }

collector modulation [ELECTR] Amplitude modulation in which the modulator varies the collector voltage of a transistor. { kə'lek·tər ,mäj·ə'lā·shən }

collector plate [ELEC] One of several metal inserts that are sometimes embedded in the lining of an electrolyte cell to make the resistance between the cell lining and the current leads as small as possible. { kə'lek·tər ,plāt }

collector resistance [ELECTR] The back resistance of the collector-base diode of a transistor. { kə'lek·tər ri'zis·təns }

collector ring See slip ring. { kə'lek·tər ,riŋ }

collector voltage [ELECTR] The direct-current voltage, obtained from a power supply, that is applied between the base and collector of a transistor. { kə'lek·tər ,vōl·tij }

colliding-beam source [ELECTR] A device for generating beams of polarized negative hydrogen or deuterium ions, in which polarized negative hydrogen or deuterium atoms are converted to negative ions through charge exchange during collisions with cesium atoms. { kə'līd·iŋ ,bēm ,sòrs }

collimation error [ENG] 1. Angular error in magnitude and direction between two nominally parallel lines of sight. 2. Specifically, the angle by which the line of sight of a radar differs from what it should be. { ,käl·ə'mā·shən ,er·ər }

collimation tower [ENG] Tower on which a visual and a radio target are mounted to check the electrical axis of an antenna. { ,käl·ə'mā·shən ,taù·ər }

collinear array See linear array. { kə'lin·ē·ər ə'rā }

collinear heterodyning [ELECTR] An optical processing system in which the correlation function is developed from an ultrasonic light modulator; the output signal is derived from a reference beam in such a way that the two beams are collinear until they enter the detection aperture; variations in optical path length then modulate the phase of both signal and reference beams simultaneously, and phase differences cancel out in the heterodyning process. { kə'lin·ē·ər 'hed·ə·rə,dīn·iŋ }

collision-avoidance radar [ENG] Radar equipment utilized in a collision-avoidance system. { kə'lizh·ən ə'vòid·əns ,rā,där }

collision-avoidance system [ENG] Electronic devices and equipment used by a pilot to perform the functions of conflict detection and avoidance. { kə'lizh·ən ə'vòid·əns ,sis·təm }

collision detection [COMPUT SCI] A procedure in which a computer network senses a situation where two computer devices attempt to access the network at the same time and blocks the messages, requiring each device to resubmit its message at a randomly selected time. { kə'lizh·ən di,tek·shən }

color aberration See chromatic aberration. { 'kəl·ər ab·ə'rā·shən }

color balance [ELECTR] Adjustment of the circuits feeding the three electron guns of a television color picture tube to compensate for differences in light-emitting efficiencies of the three color phosphors on the screen of the tube. { 'kəl·ər ,bal·əns }

color-bar generator [ELECTR] A signal generator that delivers to the input of a video system the signal needed to produce a color-bar test pattern on a device or system. { 'kəl·ər ,bär 'jen·ə ,rād·ər }

color-bar test pattern [COMMUN] A test pattern of different colors of vertical bars, used to check the performance of a video system. { 'kəl·ər ,bär 'test ,pad·ərn }

color breakup [COMMUN] A transient or dynamic distortion of the color in an analog color television picture that can originate in videotape equipment, a television camera, or a receiver. { 'kəl·ər ,brāk,əp }

color burst [ELECTR] The portion of an analog composite color television signal consisting of a few cycles of a sine wave of chrominance subcarrier frequency. Also known as burst; reference burst. { 'kəl·ər ,bərst }

color carrier See chrominance subcarrier. { 'kəl·ər ,kar·ē·ər }

color-carrier reference See chrominance-carrier reference. { 'kəl·ər ,kar·ē·ər ,ref·rəns }

color code [ELEC] A system of colors used to indicate the electrical value of a component or to identify terminals and leads. { 'kəl·ər ,kōd }

color coder See matrix. { 'kəl·ər ,kōd·ər }

color contamination [ELECTR] An error in the color rendition of an analog color television picture that results from incomplete separation of the paths that carry different color components of a picture. { 'kəl·ər kən,tam·ə'nā·shən }

color control See chroma control. { 'kəl·ər kən 'trōl }

color decoder See matrix. { 'kəl·ər dē'kōd·ər }

color-difference signal [ELECTR] A signal that is added to the monochrome signal in an analog color television receiver to obtain a signal representative of one of the three tristimulus values needed by the color picture tube. { 'kəl·ər ¦dif·rəns ,sig·nəl }

color encoder See matrix. { 'kəl·ər en'kōd·ər }

color facsimile [COMMUN] A facsimile system for transmission of color photographs, in which three separate facsimile transmissions are made from the original color print, using color-separation filters in the optical system of the facsimile transmitter. { 'kəl·ər ,fak'sim·ə·lē }

color fringing [ELECTR] Spurious chromaticity at boundaries of objects in a television picture. { 'kəl·ər 'frinj·iŋ }

color killer circuit [ELECTR] The circuit in an analog color television receiver that biases chrominance amplifier tubes to cutoff during reception of monochrome programs. Also known as killer stage. { 'kəl·ər ,kil·ər ,sər·kət }

color kinescope See color picture tube. { 'kəl·ər 'kin·ə·skōp }

color oscillator See chroma oscillator. { 'kəl·ər ,äs·ə,lād·ər }

color phase [COMMUN] The difference in phase between components (I or Q) of a chrominance signal and the chrominance-carrier reference in an analog color television receiver. { 'kəl·ər ,fāz }

color-phase alternation |COMMUN| The periodic changing of the color phase of one or more components of the chrominance subcarrier between two sets of assigned values after every field in an analog color television system. Abbreviated CPA. { 'kəl·ər ,fāz ȯl·tər'nā·shən }

color-phase detector |ELECTR| The analog color television receiver circuit that compares the frequency and phase of the incoming burst signal with those of the locally generated 3.579545-megahertz chroma oscillator and delivers a correction voltage to ensure that the color portions of the picture will be in exact register with the black-and-white portions on the screen. { 'kəl·ər ,fāz di'tek·tər }

color picture signal |COMMUN| The electric signal that represents complete color picture information, excluding all synchronizing signals. { 'kəl·ər ,pik·chər ,sig·nəl }

color picture tube |ELECTR| A cathode-ray tube having three different colors of phosphors, so that when these are appropriately scanned and excited, a color picture is obtained. Also known as color kinescope; color television picture tube; tricolor picture tube. { 'kəl·ər ,pik·chər ,tüb }

color purity |ELECTR| Absence of undesired colors in the spot produced on the screen by each beam of a color picture tube. { 'kəl·ər ,pyür·əd·ē }

color-saturation control See chroma control. { 'kəl·ər sach·ə'rā·shən kən'trōl }

color signal |COMMUN| Any signal that controls the chromaticity values of a color picture in a video system. { 'kəl·ər ,sig·nəl }

color subcarrier See chrominance subcarrier. { 'kəl·ər səb'kar·ē·ər }

color-subcarrier oscillator See chroma oscillator. { 'kəl·ər səb'kar·ē·ər 'ä·sə,lād·ər }

color-subcarrier reference See chrominance-carrier reference. { 'kəl·ər səb'kar·ē·ər 'ref·rəns }

color sync signal |COMMUN| A signal that is transmitted with each line of an analog color television broadcast to ensure that the color relationships in the transmitted signal are established and maintained in the receiver. { 'kəl·ər 'siŋk ,sig·nəl }

color television |COMMUN| A television system that reproduces an image approximately in its original colors. { |kəl·ər |tel·ə,vizh·ən }

color television picture tube See color picture tube. { |kəl·ər |tel·ə,vizh·ən 'pik·chər ,tüb }

color transmission |COMMUN| In television, the transmission of a signal waveform that represents both the brightness values and the chromaticity values in the picture. { 'kəl·ər tranz'mish·ən }

Colpitts oscillator |ELECTR| An oscillator in which a parallel-tuned tank circuit has two voltage-dividing capacitors in series, with their common connection going to the cathode in the electron-tube version and the emitter circuit in the transistor version. { 'kōl,its ,äs·ə,lād·ər }

column |COMPUT SCI| A vertical arrangement of characters or other expressions, usually referring to a specific print position on a printer. { 'käl·əm }

column order |COMPUT SCI| The storage of a matrix $a(m,n)$ as $a(1,1)$, $a(2,1),....,a(m,1)$, $a(1,2),....$ { 'kä l·əm ,ȯr·dər }

column printer |COMPUT SCI| A small line printer used with some calculators to provide hardcopy printout of input and output data; typically consists of 20 columns of numerals and a limited number of alphabetic or other identifying characters. { 'käl·əm ,print·ər }

COM See computer output on microfilm.

coma |ELECTR| A cathode-ray tube image defect that makes the spot on the screen appear comet-shaped when away from the center of the screen. { 'kō·mə }

coma lobe |ELECTROMAG| Side lobe that occurs in the radiation pattern of a microwave antenna when the reflector alone is tilted back and forth to sweep the beam through space because the feed is no longer always at the center of the reflector; used to eliminate the need for a rotary joint in the feed waveguide. { 'kō·mə ,lōb }

comb antenna |ELECTROMAG| A broad-band antenna for vertically polarized signals, in which half of a fishbone antenna is erected vertically and fed against ground by a coaxial line. { 'kōm an,ten·ə }

comb filter |ELECTR| A wave filter whose frequency spectrum consists of a number of equispaced elements resembling the teeth of a comb. { 'kōm ,fil·tər }

combinational circuit |ELECTR| A switching circuit whose outputs are determined only by the concurrent inputs. { ,käm·bə'nā·shən·əl 'sər·kət }

combination cable |ELEC| A cable having conductors grouped in both quads and pairs. { ,käm·bə'nā·shən |kā·bəl }

combination distributing frame |ELEC| Frame which combines the functions of a main distributing frame and an intermediate distributing frame. { ,käm·bə'nā·shən dis'trib·yəd·iŋ ,frām }

combined head See read/write head. { kəm'bīnd 'hed }

combiner circuit |ELECTR| The circuit that combines the luminance and chrominance signals with the synchronizing signals in a color television camera chain. { kəm'bīn·ər ,sər·kət }

combining network |COMPUT SCI| A switching system for accessing memory modules in a multiprocessor, in which each switch remembers the memory addresses it has used, and can then satisfy several requests with a single memory access. { kəm'bīn·iŋ 'net,wərk }

comfort control |ENG| Control of temperature, humidity, flow, and composition of air by using heating and air-conditioning systems, ventilators, or other systems to increase the comfort of people in an enclosure. { 'kəm·fərt kən'trōl }

COMIT |COMPUT SCI| A user-oriented, general-purpose, symbol-manipulation programming language for computers. { 'kō,mit }

command |COMPUT SCI| A signal that initiates a predetermined type of computer operation that is defined by an instruction. |CONT SYS| An independent signal in a feedback control system, from which the dependent signals are controlled in a predetermined manner. { kə'mand }

command button [COMPUT SCI] A small rectangle on a graphical user interface with a command, such as open, close, OK, or print, that is immediately activated upon selection of the button. { kə'mand ,bət·ən }

command code *See* operation code. { kə'mand ,kōd }

command control program [COMPUT SCI] The interface between a time-sharing computer and its users by means of which they can create, edit, save, delete, and execute their programs. { kə'mand kən,trōl ,prō·grəm }

command-driven program [COMPUT SCI] A computer program that accepts command words and statements typed in by the user. { kə¦mand ,driv·ən 'prō·grəm }

command interpreter [COMPUT SCI] A program that processes commands and other input and output from an active terminal in a time-sharing system. { kə'mand ,in'tər·prə·tər }

command language [COMPUT SCI] The language of an operating system, through which the users of a data-processing system describe the requirements of their tasks to that system. Also known as job control language. { kə'mand ,laŋ·gwij }

command level [COMPUT SCI] The ability to control a computer's operating system through the use of commands, normally available only to computer operators. { kə'mand ,lev·əl }

command line [COMPUT SCI] On a display screen, the space following a prompt (such as $) where a text instruction to a computer or device is typed. { kə'mand ,līn }

command list *See* CLIST. { kə'mand ,list }

command mode [COMPUT SCI] The status of a terminal in a time-sharing environment enabling the programmer to use the command control program. { kə'mand ,mōd }

command processor [COMPUT SCI] A computer program that converts a limited number of user commands into the machine commands that direct the operating system. Also known as command shell. { kə'mand 'prä,ses·ər }

command pulses [ELECTR] The electrical representations of bit values of 1 or 0 which control input/output devices. { kə'mand ,pəl·səs }

command set [COMMUN] A radio set used to receive or give commands, as between one aircraft and another or between an aircraft and the ground. { kə'mand ,set }

command shell *See* command processor. { kə'mand ,shel }

comment [COMPUT SCI] An expression identifying or explaining one or more steps in a routine, which has no effect on execution of the routine. { 'käm,ent }

comment code [COMPUT SCI] One or more characters identifying a comment. { 'käm,ent ,kōd }

comment out [COMPUT SCI] To render a statement in a computer program inactive by making it a comment. { 'kä,ment 'aut }

common area [COMPUT SCI] An area of storage which two or more routines share. { ¦käm·ən ¦er·ē·ə }

common-base connection *See* grounded-base connection. { ¦käm·ən 'bās kə'nek·shən }

common-base feedback oscillator [ELECTR] A bipolar transistor amplifier with a common-base connection and a positive feedback network between the collector (output) and the emitter (input). { ¦käm·ən 'bās 'fēd,bak ,äs·ə,lād·ər }

common battery [COMMUN] System of current supply where all direct current energy for a unit of a telephone system is supplied by one source in a central office or exchange. { ¦käm·ən ¦bäd·ə·rē }

common branch [ELEC] A branch of an electrical network which is common to two or more meshes. Also known as mutual branch. { 'käm·ən 'branch }

common business-oriented language *See* COBOL. { ¦käm·ən ¦biz·nəs ¦òr·ē,ent·əd ,laŋ·gwij }

common carriage *See* transmission access. { 'käm·ən 'kar·ij }

common-channel interoffice signaling [COMMUN] A method of signaling in a telecommunications switching system in which a network of separate data communication paths separate from the communications transmission is used for transmitting all signaling information between offices. Abbreviated CCIS. { ¦käm·ən ¦chan·əl ,in·tər,ò·fəs 'sig·nəl·iŋ }

common-collector connection *See* grounded-collector connection. { ¦käm·ən kə'lek·tər kə 'nek·shən }

common control unit [COMPUT SCI] Control unit that is shared by more than one machine. { ¦käm·ən kən'trōl ,yü·nət }

common declaration statement [COMPUT SCI] A nonexecutable statement in FORTRAN which allows specified arrays or variables to be stored in an area available to other programs. { ¦käm·ən ,dek·lə'rā·shən ,stät·mənt }

common-drain amplifier [ELECTR] An amplifier using a field-effect transistor so that the input signal is injected between gate and drain, while the output is taken between the source and drain. Also known as source-follower amplifier. { ¦käm·ən 'drān 'am·plə,fī·ər }

common-emitter connection *See* grounded-emitter connection. { ¦käm·ən i'mid·ər kə'nek·shən }

common-gate amplifier [ELECTR] An amplifier using a field-effect transistor in which the gate is common to both the input circuit and the output circuit. { ¦käm·ən 'gāt 'am·plə,fī·ər }

common gateway interface [COMPUT SCI] A protocol that allows the secure data transfer to and from a server and a network user by means of a program which resides on the server and handles the transaction. For example, if an intranet user sent a request with a Web browser for database information, a CGI program would execute on the server, retrieve the information from the database, format it in HTML, and send it back to the user. Abbreviated CGI. { ,käm·ən ,gāt,wā 'in·tər,fās }

common language [COMPUT SCI] A machine-readable language that is common to a group of computers and associated equipment. { ¦käm·ən ¦laŋ·gwij }

common mode [ELECTR] Having signals that are identical in amplitude and phase at both inputs, as in a differential operational amplifier. { ¦käm·ən ˌmōd }

common-mode error [ELECTR] The error voltage that exists at the output terminals of an operational amplifier due to the common-mode voltage at the input. { ¦käm·ən ˌmōd 'er·ər }

common-mode gain [ELECTR] The ratio of the output voltage of a differential amplifier to the common-mode input voltage. { ¦käm·ən ˌmōd 'gān }

common-mode input capacitance [ELECTR] The equivalent capacitance of both inverting and noninverting inputs of an operational amplifier with respect to ground. { ¦käm·ən ˌmōd 'in¸pút kə'pas·əd·əns }

common-mode input impedance [ELECTR] The open-loop input impedance of both inverting and noninverting inputs of an operational amplifier with respect to ground. { ¦käm·ən ˌmōd 'in ¸pút im'ped·əns }

common-mode input resistance [ELECTR] The equivalent resistance of both inverting and noninverting inputs of an operational amplifier with respect to ground or reference. { ¦käm·ən ˌmōd 'in¸pút ri'zis·təns }

common-mode rejection [ELECTR] The ability of an amplifier to cancel a common-mode signal while responding to an out-of-phase signal. Also known as in-phase rejection. { ¦käm·ən ˌmōd ri'jek·shən }

common-mode rejection ratio [ELECTR] The ratio of the gain of an amplifier for difference signals between the input terminals, to the gain for the average or common-mode signal component. Abbreviated CMRR. { 'käm·ən ˌmōd ri'jek·shən 'rā·shō }

common-mode signal [ELECTR] A signal applied equally to both ungrounded inputs of a balanced amplifier stage or other differential device. Also known as in-phase signal. { ¦käm·ən ˌmōd 'sig·nal }

common-mode voltage [ELECTR] A voltage that appears in common at both input terminals of a device with respect to the output reference (usually ground). { ¦käm·ən ˌmōd 'vōl·tij }

common object request broker [COMPUT SCI] A system that provides interoperability among objects in a heterogeneous, distributed, object-oriented environment in a way that is transparent to the programmer; its design is based on the OMG object model. Abbreviated CORBA. { ¦käm·ən ¦äb·jekt ri'kwest ¸brō·kər }

common return [ELECTR] A return conductor that serves two or more circuits. { ¦käm·ən ri'tərn }

common-source amplifier [ELECTR] An amplifier stage using a field-effect transistor in which the input signal is applied between gate and source and the output signal is taken between drain and source. { ¦käm·ən ¸sórs 'am·plə¸fī·ər }

common storage [COMPUT SCI] A section of memory in certain computers reserved for temporary storage of program outputs to be used as input for other programs. { ¦käm·ən 'stór·ij }

common-user channel [COMMUN] Any of the communications channels which are available to all authorized agencies for transmission of command, administrative, and logistic traffic. { ¦käm·ən ¸yü·zər ¸chan·əl }

common-user circuit [ELEC] A circuit designated to furnish a communications service to a number of users. { ¦käm·ən ¸yü·zər ¸sər·kət }

communicating word processor [COMPUT SCI] A word processor that can be linked to other word processors to exchange information. { kə'myü·nə¸kād·iŋ 'wórd ¸prä¸ses·ər }

communication [COMMUN] The transmission of intelligence between two or more points over wires or by radio; the terms telecommunication and communication are often used interchangeably, but telecommunication is usually the preferred term when long distances are involved. { kə¸myü·nə'kā·shən }

communication band [COMMUN] The band of frequencies effectively occupied by a radio transmitter for the type of transmission and the speed of signaling used. { kə¸myü·nə'kā·shən ¸band }

communication bus [COMMUN] A device that transfers control, timing, and data signals between switching processor subsystems; designed to provide physical and electrical isolation, to provide for simple addition of units on an in-service basis, and to provide pluggable connection for efficient factory testing, installation, and maintenance. { kə¸myü·nə'kā·shən ¸bəs }

communication cable [COMMUN] A metallic wire or fiber-optic material used in the telephone industry to connect customers to their local switching centers and to interconnect local and long-distance switching centers. { kə¸myü·nə'kā·shən ¸kā·bəl }

communication channel [COMMUN] The wire or radio channel that serves to convey intelligence between two or more terminals. { kə¸myü·nə'kā·shən ¸chan·əl }

communication countermeasure [COMMUN] Any electronic countermeasure against communications, such as jamming. { kə¸myü·nə'kā·shən 'kaúnt·ər¸mezh·ər }

communication engineering [COMMUN] The design, construction, and operation of all types of equipment used for radio, wire, or other types of communication. { kə¸myü·nə'kā·shən en·jə'nir·iŋ }

communication link See data link. { kə¸myü·nə'kā·shən ¸liŋk }

communication protocol [COMPUT SCI] Procedures that enable devices within a computer network to exchange information. Also known as protocol. { kə¸myü·nə'kā·shən 'prōd·ə¸kól }

communication receiver [ELECTR] A receiver designed especially for reception of voice or code messages transmitted by radio communication systems. { kə¸myü·nə'kā·shən ri'sē·vər }

communications [ENG] The science and technology by which information is collected from an originating source, transformed into electric currents or fields, transmitted over electrical networks or space to another point, and reconverted into a form suitable for interpretation by a receiver. { kə,myü·nə'kā·shənz }

communications control unit |COMMUN| A device that handles data transmission between components of a communications network, and performs related functions such as multiplexing, message switching, and code conversion. Abbreviated CCU. { kə,myü·nə'kā·shənz kən'trōl ,yü·nət }

communications intelligence |COMMUN| Technical and intelligence information derived from communications by other than the intended recipients. { kə,myü·nə'kā·shənz in'tel·ə·jəns }

communications language |COMMUN| A language structure complete with conventions, syntax, and character set, used primarily for conveying knowledge of processes between two participants. { kə,myü·nə'kā·shənz ,laŋ·gwij }

communications network |COMMUN| Organization of stations capable of intercommunications but not necessarily on the same channel. { kə ,myü·nə'kā·shənz ,net,wərk }

communications package |COMPUT SCI| A software product that specifies communications protocols for data transmission within a computer network or between a computer and its peripheral equipment. { kə,myü·nə'kā·shənz ,pak·ij }

communication speed |COMMUN| The rate at which information is transmitted over a communications channel, adjusted for redundancies. { kə,myü·nə'kā·shən ,spēd }

communications program |COMPUT SCI| A computer program that transmits data to and receives data from local and remote terminals and other computers. { kə,myü·nə'kā·shənz ,prō·grəm }

communications relay station |COMMUN| Facility for rapidly passing message traffic from one tributary to another by automatic, semiautomatic, or manual means, or by electrically connecting circuits (circuit switching) between two tributaries for direct transmission. { kə ,myü·nə'kā·shənz 'rē,la ,stā·shən }

communications satellite |ENG| An orbiting, artificial earth satellite that relays radio, television, and other signals between ground terminal stations thousands of miles apart. Also known as radio relay satellite; relay satellite. { kə ,myü·nə'kā·shənz 'sad·ə,līt }

communications traffic |COMMUN| All transmitted and received messages. { kə,myü·nə'kā·shənz ,traf·ik }

communication system |COMMUN| A telephone, radio, television, data transmission, or other system in which information-bearing signals originated at one place are reproduced at a distant point. { kə,myü·nə'kā·shən ,sis·təm }

communications zone indicator |ELECTR| Device to indicate whether or not long-distance high-frequency broadcasts are successfully reaching their destinations. { kə,myü·nə'kā·shənz ,zōn 'in·də,kād·ər }

communication theory |COMMUN| The mathematical theory of the communication of information from one point to another. { kə ,myü·nə'kā·shən ,thē·ə·rē }

community antenna television See cable television. { kə'myü·nə·dē an'ten·ə 'tel·ə,vizh·ən }

community dial office |COMMUN| Small dial office with no employees located in the building serving an exchange area. { kə'myü·nə·dē 'dīl ,óf·əs }

commutating capacitor |ELECTR| A capacitor used in gas-tube rectifier circuits to prevent the anode from going highly negative immediately after extinction. { 'käm·yə,tād·iŋ kə'pas·əd·ər }

commutating reactance |ELECTR| An inductive reactance placed in the cathode lead of a three-phase mercury-arc rectifier to ensure that tube current holds over during transfer of conduction from one anode to the next. { 'käm·yə,tād·iŋ rē'ak·təns }

commutating reactor |ELEC| A reactor found primarily in silicon controlled rectifier (SCR) converters where it is connected in series with a commutation capacitor to form a highly efficient resonant circuit used to cause a current oscillation which turns off (commutates) the conducting SCR. { 'käm·yə,tād·iŋ rē'ak·tər }

commutation |COMMUN| The sampling of various quantities in a repetitive manner for transmission over a single channel in telemetering. { ,käm·yə'tā·shən }

commutator head |ELEC| The butt end of a commutator. { 'käm·yə,tād·ər ,hed }

commutator motor |ELEC| An electric motor having a commutator. { 'käm·yə,tād·ər ,mōd·ər }

commutator pulse |COMPUT SCI| One of a series of pulses indicating the beginning or end of a signal representing a single binary digit in a computer word. Also known as position pulse; P pulse. { 'käm·yə,tād·ər ,pəls }

commutator switch |ELEC| A switch that performs a set of switching operations in repeated sequential order, such as is required for telemetering many quantities. Also known as sampling switch; scanning switch. { 'käm·yə,tād·ər ,swich }

compact disk |COMMUN| A nonmagnetic (optical) disk, usually 4¾ inches (12 centimeters) in diameter, used for audio or video recording or for data storage; information is recorded using a laser beam to burn microscopic pits into the surface and is accessed by means of a lower-power laser to sense the presence or absence of pits. Abbreviated CD. { 'käm,pak 'disk }

compact-disk erasable See CD-RW. { ¦käm,pak ,disk i'rās·ə·bəl }

compact-disk read-only memory |COMPUT SCI| A compact disk used for the permanent storage of up to approximately 500 megabytes of data. Abbreviated CD-ROM. { 'käm,pakt ¦disk ¦rēd ¦ōn·lē 'mem·rē }

compact-disk recordable See CD-R. { ¦käm,pak ,disk ri'kórd·ə·bəl }

compact-disk rewritable See CD-RW. { ¦käm‚pak ‚disk ‚rē¦rīd·ə·bəl }

compact-disk write-once See CD-R. { ¦käm‚pak ‚disk ¦rīt 'wəns }

compacting garbage collection |COMPUT SCI| The physical rearrangement of data cells so that those cells whose contents are no longer useful (garbage) are compressed into a contiguous array. { ‚käm'pak·tiŋ 'gär·bij kə'lek·shən }

compaction |COMPUT SCI| A technique for reducing the space required for data storage without losing any information content. Also known as squishing. { kəm'pak·shən }

companded single-sideband system |COMMUN| A long-haul microwave telecommunications system that employs repeaters and single-sideband amplitude modulation and achieves subjective noise improvement by companding to reduce circuit noise between syllables and during pauses in speech. Abbreviated CSSB system. { kəm'pan·dəd ¦siŋ·gəl ¦sīd‚band ‚sis·təm }

companding |ELECTR| A process in which compression is followed by expansion; often used for noise reduction in equipment, in which case compression is applied before noise exposure and expansion after exposure. { kəm 'pand·iŋ }

compandor |ELECTR| A system for improving the signal-to-noise ratio by compressing the volume range of the signal at a transmitter or recorder by means of a compressor and restoring the normal range at the receiving or reproducing apparatus with an expander. { kəm'pand·ər }

comparator |COMPUT SCI| A device that compares two transcriptions of the same information to verify the accuracy of transcription, storage, arithmetical operation, or some other process in a computer, and delivers an output signal of some form to indicate whether or not the two sources are equal or in agreement. |CONT SYS| A device which detects the value of the quantity to be controlled by a feedback control system and compares it continuously with the desired value of that quantity. { kəm'par·əd·ər }

comparator circuit |ELECTR| An electronic circuit that produces an output voltage or current whenever two input levels simultaneously satisfy predetermined amplitude requirements; may be linear (continuous) or digital (discrete). { kəm 'par·əd·ər ‚sər·kət }

comparator probe |COMPUT SCI| A component of a hardware monitor that is used to sense the number of bits that appear in parallel, as in an address register. { kəm'par·əd·ər ‚prōb }

comparing unit |ELECTR| An electromechanical device which compares two groups of timed pulses and signals to establish either identity or nonidentity. { kəm'per·iŋ ‚yü·nət }

comparison |COMPUT SCI| A computer operation in which two numbers are compared as to identity, relative magnitude, or sign. { kəm'par· ə·sən }

comparison bridge |ELECTR| A bridge circuit in which any change in the output voltage with respect to a reference voltage creates a cor-responding error signal, which, by means of negative feedback, is used to correct the output voltage and thereby restore bridge balance. { kəm'par·ə·sən ‚brij }

comparison indicators |COMPUT SCI| Registers, one of which is activated during the comparison of two quantities to indicate whether the first quantity is lower than, equal to, or greater than the second quantity. { kəm'par·ə·sən ‚in·də ‚kād·ərz }

compatibility |COMPUT SCI| The ability of one device to accept data handled by another device without conversion of the data or modification of the code. |SYS ENG| The ability of a new system to serve users of an old system. { kəm ‚pad·ə'bil·ə·dē }

compatibility mode |COMPUT SCI| A feature of a computer or operating system that enables it to run programs written for another system. { kəm ‚pad·ə'bil·əd·ē ‚mōd }

compatible color television system |COMMUN| A color television system that permits substantially normal monochrome reception of the transmitted color picture signal on a typical unaltered monochrome receiver. { kəm¦pad·ə·bəl 'kəl·ər 'tel·ə‚vizh·ən ‚sis·təm }

compatible discrete four-channel sound |ENG ACOUS| A sound system in which a separate channel is maintained from each of the four sets of microphones at the recording studio or other input location to the four sets of loudspeakers that serve as the output of the system. Abbreviated CD-4 sound. { kəm'pad·ə·bəl dis'krēt ¦fōr ¦chan·əl 'saund }

compatible monolithic integrated circuit |ELECTR| Device in which passive components are deposited by thin-film techniques on top of a basic silicon-substrate circuit containing the active components and some passive parts. { kəm'pad·ə·bəl ‚män·ə'lith·ik 'in·tə‚grād·əd 'sər· kət }

compatible single-sideband system |COMMUN| A single-sideband system that can be received by an ordinary amplitude-modulation radio receiver without distortion. { kəm'pad·ə·bəl ‚siŋ·gəl'sīd ‚band ‚sis·təm }

compensated amplifier |ELECTR| A broadband amplifier in which the frequency range is extended by choice of circuit constants. { 'käm·pən‚sād·əd 'am·plə‚fī·ər }

compensated-loop direction finder |ELECTR| A direction finder employing a loop antenna and a second antenna system to compensate for polarization error. { 'käm·pən‚sād·əd ‚lüp də'rek·shən ‚find·ər }

compensated semiconductor |ELECTR| Semiconductor in which one type of impurity or imperfection (for example, donor) partially cancels the electrical effects on the other type of impurity or imperfection (for example, acceptor). { 'käm·pən‚sād·əd 'sem·i·kən'dək·tər }

compensated volume control See loudness control. { 'käm·pən‚sād·əd 'väl·yəm kən'trōl }

compensating capacitor See balancing capacitor. { 'käm·pən‚sād·iŋ kə'pas·əd·ər }

compensating leads |ENG| A pair of wires, similar to the working leads of a resistance thermometer or thermocouple, which are run alongside the working leads and are connected in such a way that they balance the effects of temperature changes in the working leads. { 'käm·pən,sād·iŋ 'lēdz }

compensating network |CONT SYS| A network used in a low-energy-level method for suppression of excessive oscillations in a control system. { 'käm·pən,sād·iŋ 'net,wərk }

compensation |CONT SYS| Introduction of additional equipment into a control system in order to reshape its root locus so as to improve system performance. Also known as stabilization. |ELECTR| The modification of the amplitude-frequency response of an amplifier to broaden the bandwidth or to make the response more nearly uniform over the existing bandwidth. Also known as frequency compensation. { ,käm·pən'sā·shən }

compensation signals |ENG| In telemetry, signals recorded on a tape, along with the data and in the same track as the data, used during the playback of data to correct electrically the effects of tape-speed errors. { ,käm·pən'sā·shən ,sig·nəlz }

compensator |CONT SYS| A device introduced into a feedback control system to improve performance and achieve stability. Also known as filter. |ELECTR| A component that offsets an error or other undesired effect. { 'käm·pən,sād·ər }

compile |COMPUT SCI| To prepare a machine-language program automatically from a program written in a higher programming language, usually generating more than one machine instruction for each symbolic statement. { kəm'pīl }

compile-and-go |COMPUT SCI| A continuous sequence of steps that combine compilation, loading, and execution of a computer program. { kəm'pīl ən 'gō }

compiler |COMPUT SCI| A program to translate a higher programming language into machine language. Also known as compiling routine. { kəm'pīl·ər }

compiler-level language |COMPUT SCI| A higherlevel language normally supplied by the computer manufacturer. { kəm'pīl·ər ,lev·əl ,laŋ·gwij }

compiler listing |COMPUT SCI| A report that is produced by a compiler and contains an annotated printout of the source program together with other useful information. { kəm'pīl·ər ,list·iŋ }

compiler system |COMPUT SCI| The set consisting of a higher-level language, such as FORTRAN, and its compiler which translates the program written in that language into machine-readable instructions. { kəm'pīl·ər ,sis·təm }

compiler toggle |COMPUT SCI| A piece of information transmitted to a compiler to activate some special feature or otherwise control the way in which the compiler operates. { kəm'pī·lər ,täg·əl }

compiling routine See compiler. { kəm'pil·iŋ rü ,tēn }

complementary |ELECTR| Having pnp and npn or p- and n-channel semiconductor elements on or within the same integrated-circuit substrate or working together in the same functional amplifier state. { ,käm·plə'men·trē }

complementary constant-current logic |ELECTR| A type of large-scale integration used in digital integrated circuits and characterized by high density and very fast switching times. Abbreviated CCCL; C³L. { ,käm·plə¦men·trē ¦kän·stənt ¦kə·rənt 'läj·ik }

complementary logic switch |ELECTR| A complementary transistor pair which has a common input and interconnections such that one transistor is on when the other is off, and vice versa. { ,käm·plə'men·trē 'läj·ik ,swich }

complementary metal oxide semiconductor device See CMOS device. { ,käm·plə¦men·trē ¦med·əl ¦äk,sīd 'sem·i·kən,dək·tər di'vīs }

complementary symmetry |ELECTR| A circuit using both pnp and npn transistors in a symmetrical arrangement that permits push-pull operation without an input transformer or other form of phase inverter. { ,käm·plə'men·trē 'sim·ə·trē }

complementary transistors |ELECTR| Two transistors of opposite conductivity (pnp and npn) in the same functional unit. { ,käm·plə'men·trē tran'zis·tərs }

complement number system |COMPUT SCI| System of number handling in which the complement of the actual number is operated upon; used in some computers to facilitate arithmetic operations. { 'käm·plə·mənt 'nəm·bər ,sis·təm }

complete carry |COMPUT SCI| In parallel addition, an arrangement in which the carries that result from the addition of carry digits are allowed to propagate from place to place. { kəm'plēt 'kar·ē }

complete operation |COMPUT SCI| An operation which includes obtaining all operands from storage, performing the operation, returning resulting operands to storage, and obtaining the next instruction. { kəm'plēt äp·ə'rā·shən }

complete routine |COMPUT SCI| A routine, generally supplied by a computer manufacturer, which does not have to be modified by the user before being applied. { kəm'plēt rü'tēn }

complex data type |COMPUT SCI| A scalar data type which contains two real fields representing the real and imaginary components of a complex number. { 'käm,pleks 'dad·ə ,tīp }

complex declaration statement |COMPUT SCI| A nonexecutable statement in FORTRAN used to specify that the type of identifier appearing in the program is of the form $a + bi$, where i is the square root of -1. { 'käm,pleks ,dek·lə'rā·shən ,stāt·mənt }

complex frequency |ENG| A complex number used to characterize exponential and damped sinusoidal motion in the same way that an ordinary frequency characterizes simple harmonic motion; designated by the constant s corresponding to a motion whose amplitude is given by Ae^{st}, where A is a constant and t is time. { 'käm ,pleks 'frē·kwən·sē }

complex impedance

complex impedance *See* electrical impedance. { 'käm,pleks im'pēd·əns }

complex instruction set computer |COMPUT SCI| A computer in which relatively high-level or complex hardware incorporating microcode is used to implement a relatively large number of instructions. Abbreviated CISC. { ¦käm,pleks in'strək·shən ,set kəm,pyüd·ər }

complexity |COMPUT SCI| The number of elementary operations used by a program or algorithm to accomplish a given task. { kəm'plek·səd·ē }

complex permittivity |ELEC| A property of a dielectric, equal to $\epsilon_0(C/C_0)$, where C is the complex capacitance of a capacitor in which the dielectric is the insulating material when the capacitor is connected to a sinusoidal voltage source, and C_0 is the vacuum capacitance of the capacitor. { 'käm,pleks ,pər·mə'tiv·əd·ē }

complex reflector |ENG| A structure or group of structures having many radar-reflecting surfaces facing in different directions. { 'käm,pleks ri'flek·tər }

complex relative attenuation |ELECTR| The ratio of the peak output voltage, in complex notation, of an electric filter to the output voltage at the frequency being considered. { 'käm,pleks ¦rel·əd·iv ə,ten·yə'wā·shən }

complex target |ENG| A radar target composed of a number of reflecting surfaces that, in the aggregate, are smaller in all dimensions than the resolution capabilities of the radar. { 'käm ,pleks 'tär·gət }

compliant substrate |ELECTR| A semiconductor substrate into which an artificially formed interface is introduced near the surface which makes the substrate more readily deformable and allows it to support a defect-free semiconductor film of essentially any lattice constant, with dislocations forming in the substrate instead of in the film. Also known as sacrificial compliant substrate. { kəm¦plī·ənt 'səb,strāt }

component |ELEC| Any electric device, such as a coil, resistor, capacitor, generator, line, or electron tube, having distinct electrical characteristics and having terminals at which it may be connected to other components to form a circuit. Also known as circuit element; element. { kəm'pō·nənt }

component-failure-impact analysis |SYS ENG| A study that attempts to predict the consequences of failures of the major components of a system. Abbreviated CFIA. { kəm'pō·nənt ¦fāl·yər 'im ,pakt ə,nal·ə·səs }

component name *See* metavariable. { kəm'pō·nənt ,nām }

component symbol |ELEC| A graphical design used to represent a component in a circuit diagram. { kəm'pō·nənt ,sim·bəl }

composite |ENG ACOUS| A re-recording consisting of at least two elements. { kəm'päz·ət }

composite balance |ELEC| An electric balance made by modifying the Kelvin balance to measure amperage, voltage, or wattage. { kəm'päz·ət 'bal·əns }

composite cable |ELEC| Cable in which conductors of different gages or types are combined under one sheath. { kəm'päz·ət 'kā·bəl }

composite circuit |ELECTR| A circuit used simultaneously for voice communication and telegraphy, with frequency-discriminating networks serving to separate the two types of signals. { kəm'päz·ət 'sər·kət }

composite color signal |COMMUN| The analog color television picture signal plus all blanking and synchronizing signals. Also known as composite picture signal. { kəm'päz·ət 'kəl·ər ,sig·nəl }

composite color sync |COMMUN| The signal comprising all the synchronization signals necessary for proper operation of an analog color television receiver. { kəm'päz·ət 'kəl·ər ,siŋk }

composite filter |ELECTR| A filter constructed by linking filters of different kinds in series. { kəm'päz·ət 'fil·tər }

composite picture signal *See* composite color signal. { kəm'päz·ət 'pik·chər ,sig·nəl }

composite pulse |ELECTR| A pulse composed of a series of overlapping pulses received from the same source over several paths in a pulse navigation system. { kəm'päz·ət 'pəls }

composite set |ELECTR| Assembly of apparatus designed to provide one end of a composite circuit. { kəm'päz·ət 'set }

composite video signal |COMMUN| The video-only portion of the analog color television signal used in the United States, in which red, green, and blue signals are encoded. { kəm'päz·ət 'vid·ē·ō ,sig·nəl }

composite wave filter |ELECTR| A combination of two or more low-pass, high-pass, band-pass, or band-elimination filters. { kəm'päz·ət 'wāv ,fil·tər }

composition resistor *See* carbon resistor. { ,käm·pə'zish·ən ri'zis·tər }

compound cryosar |ELECTR| A cryosar consisting of two normal cryosars with different electrical characteristics in series. { 'käm,paúnd 'krī·ō ,sär }

compound document |COMPUT SCI| A document that contains two or more different data structures, such as text, graphics, and sound. { ,käm ,paúnd 'däk·yə·mənt }

compound field winding |ELEC| A winding composed of shunt and series coils that act either together or against each other. { 'käm,paúnd 'fēld ,wind·iŋ }

compound generator |ELEC| A direct-current generator which has both a series field winding and a shunt field winding, both on the main poles with the shunt field winding on the outside. { 'käm,paúnd 'jen·ə'rād·ər }

compound magnet |ELEC| A permanent magnet that is constructed from a number of thin magnets having the same shape. { 'käm,paúnd 'mag·nət }

compound modulation *See* multiple modulation. { 'käm,paúnd ,mäj·ə'lā·shən }

compound motor |ELEC| A direct-current motor with two separate field windings, one connected in parallel with the armature circuit, the other connected in series with the armature circuit. { 'käm,paund 'mōd·ər }

compound statement |COMPUT SCI| A single program instruction that contains two or more instructions which could stand alone. { 'käm ,paund 'stāt·mənt }

compound winding |ELEC| A winding that is a combination of series and shunt winding. { 'käm,paund 'wīnd·iŋ }

compressed-air loudspeaker [ENG ACOUS| A loudspeaker having an electrically actuated valve that modulates a stream of compressed air. { kəm¦prest ¦er 'laud,spēk·ər }

compressed file See packed file. { kəm,prest 'fīl }

compression |COMPUT SCI| See data compression. |ELECTR| **1.** Reduction of the effective gain of a device at one level of signal with respect to the gain at a lower level of signal, so that weak signal components will not be lost in background and strong signals will not overload the system. **2.** See compression ratio. { kəm'presh·ən }

compression cable See pressure cable. { kəm'presh·ən ,kā·bəl }

compression ratio |ELECTR| The ratio of the gain of a device at a low power level to the gain at some higher level, usually expressed in decibels. Also known as compression. { kəm'presh·ən ,rā·shō }

compressive intercept receiver |ELECTR| An electromagnetic surveillance receiver that instantaneously analyzes and sorts all signals within a broad radio-frequency spectrum by using pulse compression techniques which perform a complete analysis up to 10,000 times faster than a superheterodyne receiver or spectrum analyzer. { kəm'pres·iv 'in·tər,sept ri'sē·vər }

compressor |COMPUT SCI| A routine or program that reduces the number of binary digits needed to represent data or information. |ELECTR| The part of a compandor that is used to compress the intensity range of signals at the transmitting or recording end of a circuit. { kəm'pres·ər }

compromise network |ELEC| **1.** Network employed in conjunction with a hybrid coil to balance a subscriber's loop; adjusted for an average loop length or an average subscriber's set, or both, to secure compromise (not precision) isolation between the two directional paths of the hybrid. **2.** Hybrid balancing network which is designed to balance the average of the impedances that may be connected to the switchboard side of a hybrid arrangement of a repeater. { 'käm·prə,mīz 'net,wərk }

compromising emanations [COMMUN | Unintentional data-related or intelligence-bearing signals which, if intercepted and analyzed by any technique, could disclose the classified information transmitted, received, handled, or otherwise processed by equipments. { 'käm·prə,miz·iŋ ,em·ə'nā·shənz }

computational numerical control See computer numerical control. { ,käm·pyə'tā·shən·əl nü'mer·ə·kəl kən'trōl }

compute-bound program See CPU-bound program. { kəm'pyüt ¦baund 'prō·grəm }

computed go to |COMPUT SCI| A control procedure in FORTRAN which allows the transfer of control to the *i*th label of a set of *n* labels used as statement numbers in the program. { kəm'pyüd·əd 'gō ,tü }

computed path control |CONT SYS| A control system designed to follow a path calculated to be the optimal one to achieve a desired result. { kəm'pyüd·əd ¦path kən'trōl }

compute mode |COMPUT SCI| The operation of an analog computer in which input signals are used by the computing units to calculate a solution, in contrast to hold mode and reset mode. { kəm'pyüt ,mōd }

computer |COMPUT SCI| A device that receives, processes, and presents data; the two types are analog and digital. Also known as computing machine. { kəm'pyüd·ər }

computer-aided design |CONT SYS| The use of computers in converting the initial idea for a product into a detailed engineering design. Computer models and graphics replace the sketches and engineering drawings traditionally used to visualize products and communicate design information. Abbreviated CAD. { kəm'pyüd·ər ,ād·əd də'zīn }

computer-aided design and drafting |COMPUT SCI| The carrying out of computer-aided design with a system that has additional features for the drafting function, such as dimensioning and text entry. Abbreviated CADD. { kəm'pyüd·ər ,ād·əd di'zīn ən 'draft·iŋ }

computer-aided engineering |ENG| The use of computer-based tools to assist in solution of engineering problems. { kəm'pyüd·ər ,ād·əd ,en·jə'nir·iŋ }

computer-aided instruction See computer-assisted instruction. { kəm'pyüd·ər ,ād·əd in'strək·shən }

computer-aided management of instruction See computer-managed instruction. { kəm'pyüd·ər ,ād·əd 'man·ij·mənt əv in'strək·shən }

computer-aided manufacturing |CONT SYS| The use of computers in converting engineering designs into finished products. Computers assist managers, manufacturing engineers, and production workers by automating many production tasks, such as developing process plans, ordering and tracking materials, and monitoring production schedules, as well as controlling the machines, industrial robots, test equipment, and systems that move and store materials in the factory. Abbreviated CAM. { kəm'pyüd·ər ,ād·əd ,man·ə'fak·chə·riŋ }

computer-aided software engineering |COMPUT SCI| The use of software packages to assist in all phases of the development of an information system, including analysis, design, and programming. Abbreviated CASE. { kəm'pyüd·ər ,ād·əd ,sóft,wer en·jə'nir·iŋ }

computer algebra system See symbolic system. { kəm¦pyüd·ər 'al·jə·brə ,sis·təm }

computer analyst |COMPUT SCI| A person who defines a problem, determines exactly what is required in the solution, and defines the

107

outlines of the machine solution; generally, an expert in automatic data processing applications. { kəm'pyüd·ər 'an·ə‚list }

computer animation |COMPUT SCI| The use of a computer to present, either continuously or in rapid succession, pictures on a cathode-ray tube or other device, graphically representing a time developing system at successive times. { kəm'pyüd·ər an·ə'mā·shən }

computer architecture |COMPUT SCI| The art and science of assembling logical elements to form a computing device. { kəm'pyüd·ər 'är·kə‚tek·chər }

computer-assisted instruction |COMPUT SCI| The use of computers to present drills, practice exercises, and tutorial sequences to the student, and sometimes to engage the student in a dialog about the substance of the instruction. Abbreviated CAI. Also known as computer-aided instruction; computer-assisted learning. { kəm'pyüd·ər ə'sis·təd in 'strək·shən }

computer-assisted learning See computer-assisted instruction. { kəm'pyüd·ər ə'sis·təd 'lərn·iŋ }

computer-assisted retrieval |COMPUT SCI| The use of a computer to locate documents or records stored outside of the computer, on paper or microfilm. Abbreviated CAR. { kəm'pyüd·ər ə'sis·təd ri'trē·vəl }

computer center See electronic data-processing center. { kəm'pyüd·ər ‚sen·tər }

computer code |COMPUT SCI| The code representing the operations built into the hardware of a particular computer. { kəm'pyüd·ər ‚kōd }

computer conferencing See computer networking. { kəm'pyüd·ər 'kän·frəns·iŋ }

computer control |CONT SYS| Process control in which the process variables are fed into a computer and the output of the computer is used to control the process. { kəm'pyüd·ər kən'trōl }

computer control counter |COMPUT SCI| Counter which stores the next required address; any counter which furnishes information to the control unit. { kəm'pyüd·ər kən'trōl ‚kaúnt·ər }

computer-controlled system |CONT SYS| A feedback control system in which a computer operates on both the input signal and the feedback signal to effect control. { kəm'pyüd·ər kən'trōld ‚sis·təm }

computer control register See program register. { kəm'pyüd·ər kən'trōl rej·ə·stər }

computer efficiency |COMPUT SCI| **1.** The ratio of actual operating time to scheduled operating time of a computer. **2.** In time-sharing, the ratio of user time to the sum of user time plus system time. { kəm'pyüd·ər i'fish·ən·sē }

computer graphics |COMPUT SCI| The process of pictorial communication between humans and computers, in which the computer input and output have the form of charts, drawings, or appropriate pictorial representation; such devices as cathode-ray tubes, mechanical plotting boards, curve tracers, coordinate digitizers, and light pens are employed. { kəm'pyüd·ər 'graf·iks }

computer graphics interface |COMPUT SCI| A standard format for writing graphics drivers. Abbreviated CGI. { kəm¦pyüd·ər ¦graf·iks 'in·tər‚fās }

computer graphics metafile |COMPUT SCI| A standard device-independent graphics format that is used to transfer graphics images between computer programs and storage devices. Abbreviated CGM. { kəm¦pyüd·ər ¦graf·iks 'med·ə‚fīl }

computer input from microfilm |COMPUT SCI| The technique of reading images on microfilm and transforming them into a form which is understandable to a computer. Abbreviated CIM. { kəm'pyüd·ər 'in‚pút frəm 'mī·krə‚film }

computer-integrated manufacturing |ENG| A computer-automated system in which individual engineering, production, marketing, and support functions of a manufacturing enterprise are organized; functional areas such as design, analysis, planning, purchasing, cost accounting, inventory control, and distribution are linked through the computer with factory floor functions such as materials handling and management, providing direct control and monitoring of all process operations. Abbreviated CIM. { kəm'pyüd·ər ¦int·ə‚grād·əd ‚man·ə'fak·chər·iŋ }

computerized branch exchange |COMMUN| A computer-controlled telephone switching system that supports such services as conference calling, least-cost routing, direct inward dialing, and automatic reringing of a busy line. Abbreviated CBX. { kəm'pyüd·ə‚rīzd 'branch iks'chānj }

computer-limited |COMPUT SCI| Pertaining to a situation in which the time required for computation exceeds the time required to read inputs and write outputs. { kəm'pyüd·ər ‚lim·əd·əd }

computer literacy |COMPUT SCI| Knowledge and understanding of computers and computer systems and how to apply them to the solution of problems. { kəm'pyüd·ər 'lit·rə·sē }

computer-managed instruction |COMPUT SCI| The use of computer assistance in testing, diagnosing, prescribing, grading, and record keeping. Abbreviated CMI. Also known as computer-aided management of instruction. { kəm'pyüd·ər ¦man·ijd in'strək·shən }

computer memory See memory. { kəm'pyüd·ər 'mem·rē }

computer modeling |COMPUT SCI| The use of a computer to develop a mathematical model of a complex system or process and to provide conditions for testing it. { kəm'pyüd·ər 'mäd·əl·iŋ }

computer network |COMPUT SCI| A system of two or more computers that are interconnected by communication channels. Also known as network. { kəm'pyüd·ər 'net‚wərk }

computer networking |COMMUN| The use of a network of computers and computer terminals by individuals at various locations to interact with each other by entering data into the computer system. Also known as computer conferencing. { kəm'pyüd·ər 'net‚wərk·iŋ }

computer numerical control |CONT SYS| A control system in which numerical values corresponding to desired tool or control positions are generated by a computer. Abbreviated CNC. Also known as computational numerical control; soft-wired numerical control; stored-program numerical control. { kəm'pyüd·ər nü'mer·i·kəl kən'trōl }

computer operation |COMPUT SCI| The electronic action that is required in a computer to give a desired computation. { kəm'pyüd·ər äp·ə'rā·shən }

computer-oriented language |COMPUT SCI| A low-level programming language developed for use on a particular computer or line of computers produced by a specific manufacturer. Also known as machine-oriented language. { kəm'pyüd·ər ¦ór·ē¸ent·əd 'laŋ·gwij }

computer output on microfilm |COMPUT SCI| The generation of microfilm which displays information developed by a computer. Abbreviated COM. { kəm'pyüd·ər 'aút¸pút ón 'mī·krə¸film }

computer part programming |CONT SYS| The use of computers to program numerical control systems. { kəm'pyüd·ər 'pärt 'prō¸gram·iŋ }

computer performance evaluation |COMPUT SCI| The measurement and evaluation of the performance of a computer system, aimed at ensuring that a minimum amount of effort, expense, and waste is incurred in the production of data-processing services, and encompassing such tools as canned programs, source program optimizers, software monitors, hardware monitors, simulation, and bench-mark problems. Abbreviated CPE. { kəm'pyüd·ər pər'fór·məns i¸val·yə'wā·shən }

computer programming See programming. { kəm'pyüd·ər 'prō¸gram·iŋ }

computer science |COMPUT SCI| The study of computers and computing, including computer hardware, software, programming, networking, database systems, information technology, interactive systems, and security. { kəm'pyüd·ər 'sī·əns }

computer security |COMPUT SCI| Measures taken to protect computers and their contents from unauthorized use. { kəm'pyüd·ər sə'kyùr·əd·ē }

computer storage device See storage device. { kəm'pyüd·ər 'stòr·ij di'vīs }

computer system |COMPUT SCI| **1.** A set of related but unconnected components (hardware) of a computer or data-processing system. **2.** A set of hardware parts that are related and connected, and thus form a computer. { kəm'pyüd·ər ¸sis·təm }

computer systems architecture |COMPUT SCI| The discipline that defines the conceptual structure and functional behavior of a computer system, determining the overall organization, the attributes of the component parts, and how these parts are combined. { kəm'pyüd·ər ¦sis·təmz 'ar·kə¸tek·chər }

computer theory |COMPUT SCI| A discipline covering the study of circuitry, logic, micro-programming, compilers, programming languages, file structures, and system architectures. { kəm'pyüd·ər ¸thē·ə·rē }

computer utility |COMPUT SCI| A computer that provides service on a time-sharing basis, generally over telephone lines, to subscribers who have appropriate terminals. { kəm'pyüd·ər yü'til·əd·ē }

computer vision |COMPUT SCI| The use of digital computer techniques to extract, characterize, and interpret information in visual images of a three-dimensional world. Also known as machine vision. { kəm'pyüd·ər 'vizh·ən }

computer word See word. { kəm'pyüd·ər ¸wərd }

computing machine See computer. { kəm 'pyüd·iŋ mə'shēn }

computing power |COMPUT SCI| The number of operations that a computer can carry out in 1 second. { kəm'pyüd·iŋ ¸paú·ər }

computing unit |COMPUT SCI| The section of a computer that carries out arithmetic, logical, and decision-making operations. { kəm'pyüd·iŋ ¸yü·nət }

concatenate |COMPUT SCI| To unite in a sequence, link together, or link to a chain. { kən'kat·ən¸āt }

concatenation |COMPUT SCI| **1.** An operation in which a number of conceptually related components are linked together to form a larger, organizationally similar entity. **2.** In string processing, the synthesis of longer character strings from shorter ones. |ELEC| A method of speed control of induction motors in which the rotors of two wound-rotor motors are mechanically coupled together and the stator of one motor is supplied with power from the rotor slip rings of the first motor. |ENG ACOUS| The linking together of phonemes to produce meaningful sounds. { kən¸kat·ən'ā·shən }

concentrator |ELECTR| Buffer switch (analog or digital) which reduces the number of trunks required. { 'kän·sən¸trād·ər }

concentric cable See coaxial cable. { kən'sen·trik 'kā·bəl }

concentric line See coaxial cable. { kən'sen·trik 'līn }

concentric slip ring |ELEC| A large slip-ring assembly consisting of concentrically arranged insulators and conducting materials. { kən'sen·trik 'slip ¸riŋ }

concentric transmission line See coaxial cable. { kən'sen·trik tranz'mish·ən ¸līn }

concentric windings |ELEC| Transformer windings in which the low-voltage winding is in the form of a cylinder next to the core, and the high-voltage winding, also cylindrical, surrounds the low-voltage winding. { kən'sen·trik 'wīnd·iŋz }

conceptual modeling |COMPUT SCI| Writing a program by means of which a given result will be obtained, although the result is incapable of proof. Also known as heuristic programming. { kən'sep·chə·wəl 'mäd·liŋ }

conceptual schema |COMPUT SCI| The logical structure of an entire data base. { kən'sep·chə·wəl 'skē·mə }

concurrency |COMPUT SCI| Referring to two or more tasks of a computer system which are in progress simultaneously. { kən'kər·ən·sē }

concurrent input/output |COMPUT SCI| The simultaneous reading from and writing on different media by a computer. { kən'kər·ənt |in,pùt |aùt ,pùt }

concurrent operations control |COMPUT SCI| The supervisory capability required by a computer to handle more than one program at a time. { kən'kər·ənt äp·ə'rā·shənz kən'trōl }

concurrent processing |COMPUT SCI| The conceptually simultaneous execution of more than one sequential program on a computer or network of computers. { kən'kər·ənt 'präs,əs·iŋ }

concurrent real-time processing |COMPUT SCI| The capability of a computer to process simultaneously several programs, each of which requires responses within a time span related to its particular time frame. { kən'kər·ənt 'rēl ,tīm ,präs,əs·iŋ }

condensation |ELEC| An increase of electric charge on a capacitor conductor. { ,kän·dən 'sā·shən }

condenser See capacitor. { kən'den·sər }

condenser antenna See capacitor antenna. { kən'den·sər an'ten·ə }

condenser box See capacitor box. { kən'den·sər ,bäks }

condenser bushing |ELEC| An insulation made up of alternate layers of insulating material and metal foil placed between the conductor and outer casing in terminals of transformers and other high-voltage equipment such as switchgears. { kən'den·sər ,bùsh·iŋ }

condenser microphone See capacitor microphone. { kən'den·sər 'mī·krə,fōn }

condenser transducer See electrostatic transducer. { kən'den·sər ,tranz'dü·sər }

condensing electrometer See capacitive electrometer. { kən|dens·iŋ ə,lek'träm·əd·ər }

conditional |COMPUT SCI| Subject to the result of a comparison made during computation in a computer, or subject to human intervention. { kən'dish·ən·əl }

conditional assembly |COMPUT SCI| A feature of some assemblers which suppresses certain sections of code if stated program conditions are not met at assembly time. { kən'dish·ən·əl ə'sem·blē }

conditional branch See conditional jump. { kən'dish·ən·əl 'branch }

conditional breakpoint |COMPUT SCI| A conditional jump that, if a specified switch is set, will cause a computer to stop; the routine may then be continued as coded or a jump may be forced. { kən'dish·ən·əl 'brāk,pòint }

conditional expression |COMPUT SCI| A COBOL language expression which is either true or false, depending upon the status of the variables within the expression. { kən'dish·ən·əl ik'spresh·ən }

conditional jump |COMPUT SCI| A computer instruction that will cause the proper one of two or more addresses to be used in obtaining the next instruction, depending on some property of a numerical expression that may be the result of some previous instruction. Also known as conditional branch; conditional transfer; decision instruction; discrimination; IF statement. { kən'dish·ən·əl 'jəmp }

conditionally stable circuit |ELECTR| A circuit which is stable for certain values of input signal and gain, and unstable for other values. { kən'dish·ən·əl·ē |stā·bəl ,sər·kət }

conditional replenishment |COMMUN| A form of differential pulse-code modulation in which the only information transmitted consists of addresses specifying the locations of picture samples in the moving area, and information by which the intensities of moving area picture samples can be reconstructed at the receiver. { kən'dish·ən·əl ri'plen·ish·mənt }

conditional statement |COMPUT SCI| A statement in a computer program that is executed only when a certain condition is satisfied. { kən'dish·ən·əl 'stāt·mənt }

conditional transfer See conditional jump. { kən'dish·ən·əl 'tranz·fər }

condition code |COMPUT SCI| Portion of a program status word indicating the outcome of the most recently executed arithmetic or boolean operation. { kən'dish·ən ,kōd }

conditioned line |COMPUT SCI| A communications channel, usually a telephone line, that has been adapted for data transmission. { kən'dish·ənd 'līn }

conditioned stop instruction |COMPUT SCI| A computer instruction which causes the execution of a program to stop if some given condition exists, such as the specific setting of a switch on a computer console. { kən'dish·ənd 'stäp in'strek·shən }

condition entries |COMPUT SCI| The upper-right-hand portion of a decision table, indicating, for each of the conditions, whether the condition satisfies various criteria listed in the condition stub, or the values of various parameters listed in the condition stub. { kən'dish·ən ,en,trēz }

conditioning |ELECTR| Equipment modifications or adjustments necessary to match transmission levels and impedances or to provide equalization between facilities. { kən'dish·ən·iŋ }

condition portion |COMPUT SCI| The upper portion of a decision table, comprising the condition stub and condition entires. { kən'dish·ən ,pòr·shən }

condition stub |COMPUT SCI| The upper-left-hand portion of a decision table, consisting of a single column listing various criteria or parameters which are used to specify the conditions. { kən'dish·ən ,stəb }

conductance |ELEC| The real part of the admittance of a circuit; when the impedance contains no reactance, as in a direct-current circuit, it is the reciprocal of resistance, and is thus a measure of the ability of the circuit to conduct electricity. Also known as electrical conductance. Designated G. { kən'dək·təns }

conductance-variation method [ELEC] A technique for measuring low admittances; measurements in a parallel-resonance circuit with the terminals open-circuited, with the unknown admittance connected, and then with the unknown admittance replaced by a known conductance standard are made; from them the unknown can be calculated. { kən'dək·təns ver·ē'ā·shən ,meth·əd }

conducted interference [COMMUN] Interfering signals arriving by direct coupling such as on communications and power lines. { kən'dək·təd ,in·tər'fir·əns }

conduction [ELEC] The passage of electric charge, which can occur by a variety of processes, such as the passage of electrons or ionized atoms. Also known as electrical conduction. { kən'dək·shən }

conduction cooling [ELECTR] Cooling of electronic components by carrying heat from the device through a thermally conducting material to a large piece of metal with cooling fins. { kən'dək·shən ,kül·iŋ }

conductive coupling [ELEC] Electric connection of two electric circuits by their sharing the same resistor. { kən'dək·tiv 'kəp·liŋ }

conductive gasket [ELEC] A flexible metallic gasket used to reduce radio-frequency leakage at joints in shielding. { kən'dək·tiv 'gas·kət }

conductive interference [ELECTR] Interference to electronic equipment that orginates in power lines supplying the equipment, and is conducted to the equipment and coupled through the power supply transformer. { kən'dək·tiv ,in·tər'fir·əns }

conductivity [ELEC] The ratio of the electric current density to the electric field in a material. Also known as electrical conductivity; specific conductance. { ,kän,dək'tiv·əd·ē }

conductivity bridge [ELEC] A modified Kelvin bridge for measuring very low resistances. { ,kän,dək'tiv·əd·ē ,brij }

conductivity cell [ELEC] A glass vessel with two electrodes at a definite distance apart and filled with a solution whose conductivity is to be measured. { ,kän,dək'tiv·əd·ē ,sel }

conductivity ellipsoid [ELEC] For an anisotropic material, an ellipsoid whose axes are the eigenvectors of the conductivity tensor. { ,kän,dək ¦tiv·əd·ē i'lip,sòid }

conductivity modulation [ELECTR] Of a semiconductor, the variation of the conductivity of a semiconductor through variation of the charge carrier density. { ,kän,dək'tiv·əd·ē ,mäj·ə'lā·shən }

conductivity modulation transistor [ELECTR] Transistor in which the active properties are derived from minority carrier modulation of the bulk resistivity of the semiconductor. { ,kän ,dək 'tiv·əd·ē ,mäj·ə'lā·shən tran'zis·tər }

conductivity tensor [ELEC] A tensor which, when multiplied by the electric field vector according to the rules of matrix multiplication, gives the current density vector. { ,kän,dək'tiv·əd·ē ,ten·sər }

conductor [ELEC] A wire, cable, or other body or medium that is suitable for carrying electric current. Also known as electric conductor. { kən'dək·tər }

conductor skin effect See skin effect. { kən 'dək·tər ,skin i'fekt }

conduit [ELEC] Solid or flexible metal or other tubing through which insulated electric wires are run. { 'kän·də·wət }

cone [ENG ACOUS] The cone-shaped paper or fiber diaphragm of a loudspeaker. { kōn }

cone antenna See conical antenna. { 'kōn an'ten·ə }

cone loudspeaker [ENG ACOUS] A loudspeaker employing a magnetic driving unit that is mechanically coupled to a paper or fiber cone. Also known as cone speaker. { 'kōn 'laúd ,spēk·ər }

cone speaker See cone loudspeaker. { 'kōn ,spēk·ər }

conference communications [COMMUN] Communications facilities whereby direct speech conversation may be conducted between three or more locations simultaneously. { 'kän·frəns kə,myü·nə'kā·shənz }

configuration [COMPUT SCI] For a computer system, the relationship of hardware elements to each other, and the manner in which they are electronically connected. [SYS ENG] A group of machines interconnected and programmed to operate as a system. { kən,fig·yə'rā·shən }

confirmation message [COMPUT SCI] A message that appears on a computer screen asking the user to confirm an action that could have destructive effects, such as loss of data. { ,kän·fər'mā·shən ,mes·ij }

conformable optical mask [ELECTR] An optical mask made on a flexible glass substrate so that it can be pulled down under vacuum into intimate contact with the substrate for accurate circuit fabrication. { kən'fòr·mə·bəl ¦äp·tə·kəl 'mask }

conformal array [ELECTR] An array-type antenna in which the radiating elements are mounted on a surface shaped for other purposes, such as aerodynamics, or on a surface more convenient of beneficial than a plane. Circular or cylindrical arrays provide an antenna-pattern consistency particularly valuable in TACAN, IFF, and secondary radar applications. { kən'fòr·məl ə'rā }

confusion jamming [ELECTR] An electronic countermeasure technique in which the signal from an enemy tracking radar is amplified and retransmitted with distortion to create a false echo that affects accuracy of target range, azimuth, and velocity data. { kən'fyü·zhən ,jam·iŋ }

confusion matrix [COMPUT SCI] In pattern recognition, a matrix used to represent errors in assigning classes to observed patterns in which the ijth element represents the number of samples from class i which were classified as class j. { kən'fyü·zhən ,mā·triks }

congruential generator [COMPUT SCI] A method of generating a sequence of random numbers x_0, x_1, x_2, . . . , in which each member is generated from the previous one by the formula $x_{i+1} \equiv ax_i + b$ modulus m, where a, b, and m are constants. { ¦kän,grü¦en·chəl 'jen·ə,rād·ər }

conical antenna

conical antenna [ELECTROMAG] A wide-band antenna in which the driven element is conical in shape. Also known as cone antenna. { 'kän·ə·kəl an'ten·ə }

conical beam [ELECTR] The radar beam produced by conical scanning methods. { 'kän·ə·kəl 'bēm }

conical-horn antenna [ELECTROMAG] A horn antenna having a circular cross section and straight sides. { 'kän·ə·kəl ,hȯrn an'ten·ə }

conical monopole antenna [ELECTROMAG] A variation of a biconical antenna in which the lower cone is replaced by a ground plane and the upper cone is usually bent inward at the top. { 'kän·ə·kəl 'män·ə,pōl an'ten·ə }

conical scanning [ELECTR] Scanning in radar in which the direction of maximum radiation generates a cone, the vertex angle of which is of the order of the beam width; may be either rotating or nutating, according to whether the direction of polarization rotates or remains unchanged. Done to effect accurate angle measurement in precision tracking radars. { 'kän·ə·kəl 'skan·iŋ }

conjugate branches [ELEC] Any two branches of an electrical network such that a change in the electromotive force in either does not result in a change in current in the other. Also known as conjugate conductors. { 'kän·jə·gət 'bran·chəz }

conjugate bridge [ELECTR] A bridge in which the detector circuit and the supply circuits are interchanged, as compared with a normal bridge of the given type. { 'kän·jə·gət 'brij }

conjugate conductors See conjugate branches. { 'kän·jə·gət kən'dək·tərz }

conjugate impedances [ELEC] Impedances having resistance components that are equal, and reactance components that are equal in magnitude but opposite in sign. { 'kän·jə·gət im'pēd·ən·səz }

conjunctive search [COMPUT SCI] A search to identify items having all of a certain set of characteristics. { kən'jəŋk·tiv 'sərch }

connected load [ELEC] The sum of the continuous power ratings of all load-consuming apparatus connected to an electric power distribution system or any part thereof. { kə'nek·təd 'lōd }

connect function [COMPUT SCI] A signal sent over a data line to a selected peripheral device to connect it with the central processing unit. { kə'nekt ,fəŋk·shən }

connecting circuit [ELECTR] A functional switching circuit which directly couples other functional circuit units to each other to exchange information as dictated by the momentary needs of the switching system. { kə'nekt·iŋ ,sər·kət }

connectionless transmission [COMMUN] Data transmission by packets that include addresses of the source and destination, so that a direct connection between these nodes is unnecessary. { kə,nek·shən·ləs tranz'mish·ən }

connection-oriented transmission [COMMUN] Data transmission in which a physical path between the source and destination must be established and maintained for the duration of the transmission. { kə¦nek·shən ,ȯr·ē,ent·əd tranz'mish·ən }

connector [COMPUT SCI] In database management, a pointer or link between two data structures. [ELECTR] A switch, or relay group system in old electromechanical central offices, which found the telephone line being called as a result of digits being dialed; it also caused interrupted ringing voltage to be placed on the called line or returned a busy tone to the calling party if the line were busy. [ENG] **1.** A detachable device for connecting electrical conductors. **2.** A symbol on a flowchart indicating that the flow jumps to a different location on the chart. { kə'nek·tər }

connector block [ELECTR] A device for connecting two cables without using plugs, similar to a barrier strip but larger, in which wires from one cable are attached to lugs of screws on one side, and wires from the other cable are fastened to corresponding points on the opposite side. { kə'nek·tər ,bläk }

connect time [COMPUT SCI] The time that a user at a terminal is signed on to a computer. { kə'nekt ,tīm }

conode See tie line. { 'kō,nōd }

consequence finding program [COMPUT SCI] A computer program that attempts to deduce mathematical consequences from a set of axioms and to select those consequences that will be significant. { 'kän·sə·kwəns ¦fīnd·iŋ ,prō·grəm }

conservation of charge [ELEC] A law which states that the total charge of an isolated system is constant; no violation of this law has been discovered. Also known as charge conservation. { ,kän·sər'vā·shən əv 'chärj }

consistency routine [COMPUT SCI] A debugging routine which is used to determine whether the program being checked gives consistent results at specified check points; for example, consistent between runs or with values calculated by other means. { kən'sis·tən·sē rü'tēn }

console [COMPUT SCI] **1.** The section of a computer that is used to control the machine manually, correct errors, manually revise the contents of storage, and provide communication in other ways between the operator or service engineer and the central processing unit. Also known as master console. **2.** A display terminal together with its keyboard. [ENG] **1.** A main control desk for electronic equipment, as at a radar station, radio or television station, or airport control tower. Also known as control desk. **2.** A large cabinet for a radio or television receiver, standing on the floor rather than on a table. **3.** A grouping of controls, indicators, and similar items contained in a specially designed model cabinet for floor mounting; constitutes an operator's permanent working position. { 'kän,sōl }

console display [COMPUT SCI] The visible representation of information, whether in words, numbers, or drawings, on a console screen connected to a computer. { 'kän,sōl di'splā }

console file adapter [COMPUT SCI] A special input/output device which allows the operator to

112

load reloadable control storage from the system console. { 'kän¦sōl 'fīl ə'dap·tər }

console receiver |ELECTR| A television or radio receiver in a console. { 'kän¦sōl ri'sēv·ər }

console switch |COMPUT SCI| A switch on a computer console whose setting can be sensed by a computer, so that an instruction in the program can direct the computer to use this setting to determine which of various alternative courses of action should be followed. { 'kän¦sōl ¦swich }

constancy See persistence. { 'kän·stən·sē }

constant-amplitude recording |ENG ACOUS| A sound-recording method in which all frequencies having the same intensity are recorded at the same amplitude. { ¦kän·stənt 'am·plə¦tüd ri ¦kòrd·iŋ }

constant area |COMPUT SCI| A part of storage used for constants. { 'kän·stənt ¦er·ē·ə }

constant bit rate |COMMUN| A mode of operation in a digital system where the bit rate is constant from start to finish of the compressed bit stream. { 'kän·stənt 'bit ¦rāt }

constant-conductance network See constant-resistance network. { ¦kän·stənt kən'dək·təns ¦net¸wərk }

constant-current characteristic |ELECTR| The relation between the voltages of two electrodes in an electron tube when the current to one of them is maintained constant and all other electrode voltages are constant. { ¦kän·stənt 'kər·ənt ¸kar·ik·tə'ris·tik }

constant-current dc potentiometer |ELEC| A potentiometer in which the unknown electromotive force is balanced by a constant current times the resistance of a calibrated resistor or slide-wire. Also known as Poggendorff's first method. { ¦kän·stənt 'kər·ənt ¦dē¦sē pə¸ten·chē'am·əd·ər }

constant-current filter |ELECTR| A filter network intended to be connected to a source whose internal impedance is so high it can be assumed as infinite. { ¦kän·stənt 'kər·ənt ¦fil·tər }

constant-current generator |ELECTR| A vacuum-tube circuit, generally containing a pentode, in which the alternating-current anode resistance is so high that anode current remains essentially constant despite variations in load resistance. { ¦kän·stənt 'kər·ənt 'jen·ə¸rād·ər }

constant-current modulation |COMMUN| System of amplitude modulation in which output circuits of the signal amplifier and the carrier-wave generator or amplifier are connected via a common coil to a constant-current source. Also known as Heising modulation. { ¦kän·stənt 'kər·ənt ¸mäj·ə'lā·shən }

constant-current source |ELECTR| A circuit which produces a specified current, independent of the load resistance or applied voltage. { ¦kän·stənt 'kər·ənt ¸sòrs }

constant-current supply |ELEC| The power supply for repeatered submarine telephone cables; the voltage is varied automatically to maintain a constant current through the use of variable-voltage rectifiers and constant-current regulators at each shore station. { ¦kän·stənt 'kər·ənt sə'plī }

constant-current transformer |ELEC| A transformer that automatically maintains a constant current in its secondary circuit under varying loads, when supplied from a constant-voltage source. { ¦kän·stənt 'kər·ənt tranz'fór·mər }

constant-distance sphere |ENG ACOUS| The relative response of a sonar projector to variations in acoustic intensity, or intensity per unit band, over the surface of a sphere concentric with its center. { ¦kän·stənt 'dis·təns ¸sfir }

constant-false-alarm rate |ELECTR| Radar system devices used to prevent receiver saturation and overload so as to present clean video information to the display, and to present a constant noise level to an automatic detector. { ¦kän·stənt ¸fòls ə'lärm ¸rāt }

constant-false-alarm-rate detection |ELECTR| Radar detection in which the sensitivity threshold is adjusted to adapt to a changing and uncertain background of clutter or interference. { 'kän·stənt fòls ə'lärm rāt di¸tek·shən }

constant instruction |COMPUT SCI| A nonexecutable instruction. { 'kän·stənt in¦strək·shən }

constant-k filter |ELECTR| A filter in which the product of the series and shunt impedances is a constant that is independent of frequency. { ¦kän·stənt ¦kā 'fil·tər }

constant-k network |ELECTR| A ladder network in which the product of the series and shunt impedances is independent of frequency within the operating frequency range. { ¦kän·stənt ¦kā 'net¸wərk }

constant-luminance transmission |COMMUN| Type of transmission in which the transmission primaries are a luminance primary and two chrominance primaries. { ¦kän·stənt ¦lü·mə·nəns tranz'mish·ən }

constant radio code |COMMUN| Code in which all characters are represented by combinations having a fixed ratio of ones to zeros. { 'kän·stənt 'rād·ē·ō ¸kōd }

constant-resistance dc potentiometer |ELEC| A potentiometer in which the ratio of an unknown and a known potential are set equal to the ratio of two known constant resistances. Also known as Poggendorff's second method. { ¦kän·stənt ri'zis·təns ¦dē¦sē pə¸ten·chē'äm·əd·ər }

constant-resistance network |ELECTR| A network having at least one driving-point impedance that is a positive constant. Also known as constant-conductance network. { ¦kän·stənt ri'zis·təns 'net¸wərk }

constant-velocity recording |ENG ACOUS| A sound-recording method in which, for input signals of a given amplitude, the resulting recorded amplitude is inversely proportional to the frequency; the velocity of the cutting stylus is then constant for all input frequencies having that given amplitude. { ¦kän·stənt və'läs·əd·ē ri ¸kòrd·iŋ }

constant-voltage generator |ELEC| An axle generator that is equipped with a regulator which keeps voltage constant. { ¦kän·stənt 'vōl·tij 'jen·ə¸rād·ər }

constant-voltage transformer |ELEC| A power transformer which will supply a constant voltage to an unvarying load, even with changes in the primary voltage. { ¦kän·stənt 'vōl·tij tranz'fȯr·mər }

constraint matrix |COMPUT SCI| The set of equations and inequalities defining the set of admissible solutions in linear programming. { kən'strānt ,mā·triks }

constraint programming language |COMPUT SCI| A programming language in which constraints (relationships that must hold among a number of variables) are directly usable as programming constructs. { kən¦strānt 'prō,gram·iŋ ,laŋ·gwij }

construction operator |COMPUT SCI| The part of a data structure which is used to construct composite objects from atoms. { kən'strək·shən 'äp·ə,rād·ər }

contact |ELEC| See electric contact. |ENG| A report of a target of interest in a radar's data processing; a detection. Also known as plot. { 'kän,takt }

contact arc |ELEC| A spark that occurs immediately after the breaking of an electric contact carrying a current. { 'kän,takt ,ärk }

contact block |ELEC| A block of conducting material such as carbon, used in a relay. { 'kän,takt ,bläk }

contact bounce |ELEC| The uncontrolled making and breaking of contact one or more times, but not continuously, when relay contacts are moved to the closed position. { 'kän,takt ,bauns }

contact chatter See chatter. { 'kän,takt ,chad·ər }

contact clip |ELEC| The clip which the blade of a knife switch is clamped to in the closed condition. { 'kän,takt ,klip }

contact drop |ELEC| The voltage drop across the terminals of an electric contact. { 'kän,takt ,dräp }

contact electricity |ELEC| An electric charge at the surface of contact of two different materials. { 'kän,takt i,lek'tris·əd·ē }

contact electromotive force See contact potential difference. { 'kän,takt i¦lek·trə'mōd·iv 'fȯrs }

contact follow |ELEC| The distance two contacts travel together after just touching. Also known as contact overtravel. { 'kän,takt ,fäl·ō }

contact force |ELEC| The force exerted by the moving contact of a switch or relay on a stationary contact. { 'kän,takt ,fȯrs }

contact head |COMPUT SCI| A read/write head that remains in contact with the recording surface of a hard disk, rather than hovering above it. { 'kän,takt ,hed }

contact-making meter See instrument-type relay. { 'kän,takt ,māk·iŋ ,mēd·ər }

contact-mask read-only memory See last-mask read-only memory. { 'kän,takt ,mask 'rēd ,ōn·lē 'mem·rē }

contact microphone |ENG ACOUS| A microphone designed to pick up mechanical vibrations directly and convert them into corresponding electric currents or voltages. { 'kän,takt 'mī·krə,fōn }

contact modulation |ELEC| The use of a fast-acting relay, whose contacts make and break

at a certain threshold current, to generate square waves from a sine-wave, rectified sine-wave or direct-current source. { 'kän,takt ,mäj·ə'lā·shən }

contactor |ELEC| A heavy-duty relay used to control electric power circuits. Also known as electric contactor. { 'kän,tak·tər }

contactor control system |CONT SYS| A feedback control system in which the control signal is a discontinuous function of the sensed error and may therefore assume one of a limited number of discrete values. { 'kän,tak·tər kən'trōl ,sis·təm }

contact overtravel See contact follow. { 'kän,takt 'ō·vər ,trav·əl }

contact piston |ELECTROMAG| A waveguide piston that makes contact with the walls of the waveguide. Also known as contact plunger. { 'kän,takt ,pis·tən }

contact plunger See contact piston. { 'kän,takt ,plən·jər }

contact point |ELEC| In the ignition system of an internal combustion engine, any of the stationary and movable electrically conducting metal points that open and close to complete or break an electric circuit. { 'kän,takt ,pȯint }

contact potential See contact potential difference. { 'kän,takt pə'ten·chəl }

contact potential difference |ELEC| The potential difference that exists across the space between two electrically connected materials. Also known as contact electromotive force; contact potential; Volta effect. { 'kän,takt pə'ten·chəl 'dif·rəns }

contact pressure |ELEC| The amount of pressure holding a set of contacts together. { 'kän,takt ,presh·ər }

contact protection |ELEC| Any method for suppressing the surge which results when an inductive circuit is suddenly interrupted; the break would otherwise produce arcing at the contacts, leading to their deterioration. { 'kän ,takt prə'tek·shən }

contact rectifier See metallic rectifier. { 'kän,takt 'rek·tə ,fī·ər }

contact resistance |ELEC| The resistance in ohms between the contacts of a relay, switch, or other device when the contacts are touching each other. { 'kän,takt ri'zis·təns }

contact sparking |ELEC| The formation of a spark or arc at the contact points when a circuit is opened while it is carrying a current. { 'kän,takt ,spärk·iŋ }

contamination |COMPUT SCI| Placement of data at incorrect locations in storage, where it generally overlays valid information or a program code and produces bizarre results. { kən,tam·ə'nā·shən }

content analysis |COMPUT SCI| A method of automatically assigning words that identify the content of information items or search requests in an information retrieval system. { 'kän,tent ə'nal·ə·səs }

content indicator |COMPUT SCI| Display unit that indicates the content in a computer, and the program or mode being used. { 'kän,tent ,in·də ,kād·ər }

contention |COMMUN| A method of operating a multiterminal communication channel in which any station may transmit if the channel is free; if the channel is in use, the queue of contention requests may be maintained in predetermined sequence. |COMPUT SCI| **1.** The condition arising when two or more units attempt to transmit over a time-division-multiplex channel at the same time. **2.** Competition for the same computer resources by two or more devices or programs, such as an attempt by several programs to use the same disk drive simultaneously, or by several users in a multiaccess system to use the system's resources. { kən'ten·chən }

contention resolver |COMPUT SCI| A device that enables a central processing unit, memory, or channel whose attention is being requested over several pathways to give its attention to one pathway and ignore all others. { kən'ten·chən ri'zäl·vər }

contents |COMPUT SCI| The information stored at any address or in any register of a computer. { 'kän,tens }

context-driven line editor |COMPUT SCI| A line editor in which the user need not know or keep track of line numbers but can call up text by line content; the computer will then search for the indicated pattern. { 'kän,tekst ,driv·ən 'līn ,ed·əd·ər }

context-free grammar |COMPUT SCI| A grammar in which any occurrence of a metavariable may be replaced by one of its alternatives. { 'kän ,tekst ,frē 'gram·ər }

context-sensitive grammar |COMPUT SCI| A grammar in which the rules are applicable only when a metavariable occurs in a specified context. { 'kän,tekst ,sen·səd·iv 'gram·ər }

context-sensitive help |COMPUT SCI| A help screen that provides specific information about the current status or mode of a computer program or instructions for dealing with a particular error condition that has just occurred. { 'kän,tekst ,sen·səd·iv 'help }

context switch |COMPUT SCI| The action of a central processing unit that suspends work on one process to work on another. { 'kän,text ,swich }

context switching See task switching. { 'kän,text ,swich·iŋ }

contextual analysis |COMPUT SCI| A phase of natural language processing, following semantic analysis, whose purpose is to elaborate the semantic representation of what has been made explicit in the utterance with what is implicit from context. { kən'teks·chə·wəl ə'nal·ə·səs }

contextual search |COMPUT SCI| A search for documents or records based upon the data they contain, rather than their file names or key fields. { kən'teks·chə·wəl 'sərch }

contiguous data |COMPUT SCI| Data that are stored in a collection of adjacent locations in a computer memory device. { kən'tig·yə·wəs 'dad·ə }

continental code |COMMUN| The code commonly used for manual telegraph communication, con-sisting of short (dot) and long (dash) symbols, but not the various-length spaces used in the original Morse code. Also known as international Morse code. { ¦känt·ən¦ent·əl 'kōd }

contingency interrupt |COMPUT SCI| A processing interruption due to an operator's action or due to an abnormal result from the system or from a program. { kən'tin·jən·sē 'in·tə,rəpt }

continue statement |COMPUT SCI| A nonexecutable statement in FORTRAN used principally as a target for transfers, particularly as the last statement in the range of a do statement. { kən'tin·yü ,stāt·mənt }

continuity |ELEC| Continuous effective contact of all components of an electric circuit to give it high conductance by providing low resistance. { ,känt·ən'ü·əd·ē }

continuity test |ELEC| An electrical test used to determine the presence and location of a broken connection. { ,känt·ən'ü·əd·ē ,test }

continuous carrier |COMMUN| A carrier signal that is transmitted at all times during maintenance of a communications link, whether or not data are being transmitted. { kən¦tin·yə·wəs 'kar·ē·ər }

continuous clamp See voltage-amplitude-controlled clamp. { kən¦tin·yə·wəs 'klamp }

continuous comparator See linear comparator. { kən¦tin·yə·wəs kəm'par·əd·ər }

continuous control |CONT SYS| Automatic control in which the controlled quantity is measured continuously and corrections are a continuous function of the deviation. { kən¦tin·yə·wəs kən'trōl }

continuous-duty rating |ELEC| The rating that defines the load which can be carried for an indefinite time without exceeding a specified temperature rise. { kən¦tin·yə·wəs ,düd·ē 'rād·iŋ }

continuous film scanner |ELECTR| A television film scanner in which the motion picture film moves continuously while being scanned by a flying-spot device. { kən¦tin·yə·wəs 'film ,skan·ər }

continuous forms |COMPUT SCI| **1.** In character recognition, any batch of source information that exists in reel form, such as tally rolls or cash-register receipts. **2.** Preprinted forms that repeat on each page, with the bottom of one page joined to the top of the next by a perforated attachment, so that they can be fed through a printer. { kən¦tin·yə·wəs 'fórmz }

continuous loading |ELEC| Loading in which the added inductance is distributed uniformly along a line by wrapping magnetic material around each conductor. { kən¦tin·yə·wəs 'lōd·iŋ }

continuously adjustable transformer See variable transformer. { kən¦tin·yə·wəs·lē ə'jəs·tə·bəl tranz'fór·mər }

continuous stationery |COMPUT SCI| A continuous ribbon of paper consisting of several hundred or more sheets separated by perforations and folded to form a pack, used to feed a computer printer and generally having sprocket holes along the margin for this purpose. { kən¦tin·yə·wəs 'stā·shə,ner·ē }

continuous stationery reader |COMPUT SCI| A type of character reader which processes only continuous forms of predefined dimensions. { kən¦tin·yə·wəs 'stā·shə,ner·ē 'rēd·ər }

continuous system |CONT SYS| A system whose inputs and outputs are capable of changing at any instant of time. Also known as continuous-time signal system. { kən¦tin·yə·wəs 'sis·təm }

continuous-time signal system See continuous system. { kən¦tin·yə·wəs ¦tīm 'sig·nəl ,sis·təm }

continuous-tone squelch |ELECTR| Squelch in which a continuous subaudible tone, generally below 200 hertz, is transmitted by frequency-modulation equipment along with a desired voice signal. { kən¦tin·yə·wəs ¦tōn 'skwelch }

continuous variable |COMPUT SCI| A variable that can take on any of a range of values. { kən ¦tin·yə·wəs 'ver·ē·ə·bəl }

continuous wave |ELECTROMAG| A radio or radar wave whose successive sinusoidal oscillations are identical under steady-state conditions. Abbreviated CW. Also known as type A wave. { kən ¦tin·yə·wəs 'wāv }

continuous-wave Doppler radar See continuous-wave radar. { kən¦tin·yə·wəs ¦wāv 'däp·lər ,rā ,där }

continuous-wave jammer |ELECTR| An electronic jammer that emits a single frequency continuously, giving the appearance of a picket or rail fence on an elementary radar display. Also known as rail-fence jammer. { kən¦tin·yə·wəs ¦wāv ¦'jam·ər }

continuous-wave modulation |COMMUN| Modulation of a continuous wave by modification of its amplitude, frequency, or phase, in contrast to pulse modulation. { kən¦tin·yə·wəs ¦wāv ,mäj·ə'lā·shən }

continuous-wave radar |ENG| A radar system in which a transmitter sends out a continuous flow of radio energy; the target reradiates a small fraction of this energy to a separate receiving antenna. Also known as continuous-wave Doppler radar. { kən¦tin·yə·wəs ¦wāv 'rā ,där }

continuous-wave tracking system |ELECTR| Tracking system which operates by keeping a continuous radio beam on a target and determining its behavior from changes in the antenna necessary to keep the beam on the target. { kən¦tin·yə·wəs ¦wāv 'trak·iŋ ,sis·təm }

contour analysis |COMPUT SCI| In optical character recognition, a reading technique that employs a roving spot of light which searches out the character's outline by bouncing around its outer edges. { 'kän,tùr ə'nal·ə·səs }

contouring control |COMPUT SCI| The guidance by a computer of a machine tool along a programmed path by interpolating many intermediate points between selected points. { 'kän ,tùr·iŋ kən'trōl }

contour model |COMPUT SCI| A model for describing the run-time execution of programs written in block-structured languages, consisting of a program component, the data component, and the control component. { 'kän,tùr ,mäd·əl }

contourograph |ELECTR| Device using a cathode-ray oscilloscope to produce imagery that has a three-dimensional appearance. { ,kän'tùr·ə ,graf }

contracted code sonde See code-sending radiosonde. { kən'trak·təd ¦kōd ,sänd }

contrast |COMMUN| The degree of difference in tone between the lightest and darkest areas in a video or facsimile picture. |COMPUT SCI| In optical character recognition, the difference in color, reflectance, or shading between two areas of a surface, for example, a character and its background. { 'kän,trast }

contrast control |ELECTR| A manual control that adjusts the range of brightness between highlights and shadows on the reproduced image of a display device. { 'kän,trast kən'trōl }

contrast ratio |ELECTR| The ratio of the maximum to the minimum luminance values in a video image. { 'kän,trast ,rā·shō }

control |COMPUT SCI| **1.** The section of a digital computer that carries out instructions in proper sequence, interprets each coded instruction, and applies the proper signals to the arithmetic unit and other parts in accordance with this interpretation. **2.** A mathematical check used with some computer operations. |CONT SYS| A means or device to direct and regulate a process or sequence of events. |ELECTR| An input element of a cryotron. { kən'trōl }

control accuracy |CONT SYS| The degree of correspondence between the ultimately controlled variable and the ideal value in a feedback control system. { kən'trōl ,ak·yə·rə·sē }

control and read-only memory |COMPUT SCI| A read-only memory that also provides storage, sequencing, execution, and translation logic for various microinstructions. Abbreviated CROM. { kən'trōl ən ¦rēd ,ōn·lē 'mem·rē }

control bit |COMPUT SCI| A bit which marks either the beginning or the end of a character transmitted in asynchronous communication. { kən'trōl ,bit }

control block |COMPUT SCI| A storage area containing (in condensed, formalized form) the information required for the control of a task, function, operation, or quantity of information. { kən'trōl ,bläk }

control board |ELEC| A panel at which one can make circuit changes, as in lighting a theater. |ENG| A panel in which meters and other indicating instruments display the condition of a system, and dials, switches, and other devices are used to modify circuits to control the system. Also known as control panel; panel board. { kən'trōl ,bòrd }

control break |COMPUT SCI| **1.** A key change which takes place in a control data field, especially in the execution of a report program. **2.** A suspension of computer operation that is accomplished by simultaneously depressing the control key and the break key. { kən'trōl ,brāk }

control character |COMPUT SCI| A character whose occurrence in a particular context initiates, modifies, or stops a control operation in a computer or associated equipment. { kən'trōl ,kar·ik·tər }

control characteristic [ELECTR] **1.** The relation, usually shown by a graph, between critical grid voltage and anode voltage of a gas tube. **2.** The relation between control ampere-turns and output current of a magnetic amplifier. { kən'trōl ,kar·ik·tə'ris·tik }

control circuit [COMPUT SCI] One of the circuits that responds to the instructions in the program for a digital computer. [ELEC] A circuit that controls some function of a machine, device, or piece of equipment. [ELECTR] The circuit that feeds the control winding of a magnetic amplifier. { kən'trōl ,sər·kət }

control code [COMPUT SCI] A special code that is entered by a user to carry out a particular function, such as the moving or deleting of text in a word-processing program. { kən'trōl ,kōd }

control computer [COMPUT SCI] A computer which uses inputs from sensor devices and outputs connected to control mechanisms to control physical processes. { kən'trōl kəm'pyüd·ər }

control counter [COMPUT SCI] A counter providing data used to control the execution of a computer program. { kən'trōl ,kaùn·tər }

control data [COMPUT SCI] Data used for identifying, selecting, executing, or modifying another set of data, a routine, a record, or the like. { kən'trōl ,dad·ə }

control desk See console. { kən'trōl ,desk }

control diagram See flow chart. { kən'trōl ,dī·ə ,gram }

control electrode [ELECTR] An electrode used to initiate or vary the current between two or more electrodes in an electron tube. { kən'trōl i'lek ,trōd }

control element [CONT SYS] The portion of a feedback control system that acts on the process or machine being controlled. { kən'trōl ,el·ə· mənt }

control flow graph [COMPUT SCI] A graph describing the logic structure of a software module, in which the nodes represent computational statements or expressions, the edges represent transfer of control between nodes, and each possible execution path of the module has a corresponding path from the entry to the exit node of the graph. { kən¦trōl 'flō ,graf }

control grid [ELECTR] A grid, ordinarily placed between the cathode and an anode, that serves to control the anode current of an electron tube. { kən'trōl ,grid }

control-grid bias [ELECTR] Average direct-current voltage between the control grid and cathode of a vacuum tube. { kən'trōl ¦grid ,bī·əs }

control-grid plate transconductance [ELECTR] Ratio of the amplification factor of a vacuum tube to its plate resistance, combining the effects of both into one term. { kən'trōl ¦grid ¦plāt ,tranz·kən'dək·təns }

control handle See handle. { kən'trōl ,hand·əl }

control head gap [COMPUT SCI] The distance maintained between the read/write head of a

disk drive and the disk surface. { kən'trōl ¦hed ,gap }

control hierarchy See hierarchical control. { kən'trōl 'hī·ər,är·kē }

control inductor See control winding. { kən'trōl in'dək·tər }

control instructions [COMPUT SCI] Those instructions in a computer program which ensure proper sequencing of instructions so that a programmed task can be performed correctly. { kən'trōl in'strək·shənz }

control key [COMPUT SCI] A special key on a computer keyboard which, when depressed together with another key, generates a different signal than would be produced by the second key alone. { kən'trōl ,kē }

controllability [CONT SYS] Property of a system for which, given any initial state and any desired state, there exists a time interval and an input signal which brings the system from the initial state to the desired state during the time interval. { kən,trōl·ə'bil·əd·ē }

control lead [COMPUT SCI] A character or sequence of characters indicating that the information following is a control code and not data. { kən'trōl ,lēd }

controlled avalanche device [ELECTR] A semiconductor device that has rigidly specified maximum and minimum avalanche voltage characteristics and is able to operate and absorb momentary power surges in this avalanche region indefinitely without damage. { kən¦trōld 'av·ə ,lanch di'vīs }

controlled avalanche rectifier [ELECTR] A silicon rectifier in which carefully controlled, nondestructive internal avalanche breakdown across the entire junction area protects the junction surface, thereby eliminating local heating that would impair or destroy the reverse blocking ability of the rectifier. { kən¦trōld 'av·ə,lanch 'rek·tə,fī·ər }

controlled avalanche transit-time triode [ELECTR] A solid-state microwave device that uses a combination of IMPATT diode and npn bipolar transistor technologies; avalanche and drift zones are located between the base and collector regions. Abbreviated CATT. { kən¦trōld 'av·ə ,lanch ¦tranz·ət ,tīm 'trī,ōd }

controlled carrier modulation [COMMUN] System of modulation wherein the carrier is amplitude-modulated by the signal frequencies and, in addition, the carrier is amplitude-modulated according to the envelope of the signal so that the modulation factor remains constant regardless of the amplitude of the signal. Also known as floating carrier modulation; variable carrier modulation. { kən¦trōld 'kar·ē·ər ,mäj·ə'lā·shən }

controlled mercury-arc rectifier [ELECTR] A mercury-arc rectifier in which one or more electrodes control the start of the discharge in each cycle and thereby control output current. { kən¦trōld ¦mər·kyə·rē ¦ärk 'rek·tə,fī·ər }

controlled parameter [ENG] In the formulation of an optimization problem, one of the parameters

controlled rectifier

whose values determine the value of the criterion parameter. { kən¦trōld pə'ram·əd·ər }

controlled rectifier [ELECTR] A rectifier that has provisions for regulating output current, such as with thyratrons, ignitrons, or silicon controlled rectifiers. { kən¦trōld 'rek·tə,fī·ər }

controlled variable [CONT SYS] In process automatic-control work, that quantity or condition of a controlled system that is directly measured or controlled. { kən¦trōld 'ver·ē·ə·bəl }

controller *See* automatic controller. { kən'trōl·ər }

controller-structure interaction [CONT SYS] Feedback of an active control algorithm in the process of model reduction; this occurs through observation spillover and control spillover. { kən'trōl·ər ,strək·chər in·tər'ak·shən }

control limits [ELECTR] In radar evaluation, upper and lower control limits are established at those performance figures within which it is expected that 95% of quality-control samples will fall when the radar is performing normally. { kən'trōl ,lim·əts }

control logic [COMPUT SCI] The sequence of steps required to perform a specific function. { kən'trōl ,läj·ik }

control mark *See* tape mark. { kən'trōl ,märk }

control-message display [COMPUT SCI] A device, such as a console typewriter, on which control information, such as information on the progress of a running computer program, is displayed in ordinary language. { kən'trōl ,mes·ij di'splā }

control module [COMPUT SCI] The set of registers and circuitry required to carry out a specific function. { kən'trōl ,mä·jül }

control operation [COMPUT SCI] Any action that affects data processing but is not directly included, such as managing input/output operations or determining job sequence. { kən'trōl ,äp·ə,rā·shən }

control panel [COMPUT SCI] An array of jacks or sockets in which wires (or other elements) may be plugged to control the action of an electromechanical device in a data-processing system such as a printer. Also known as plugboard; wiring board. [ELEC] *See* control board; panel board. { kən'trōl ,pan·əl }

control point [COMPUT SCI] **1.** The numerical value of the controlled variable (speed, temperature, and so on) which, under any fixed set of operating conditions, an automatic controller operates to maintain. **2.** One of the hardware locations at which the output of the instruction decoder of the processor activates the input to and output from specific registers as well as operational resources of the system. { kən'trōl ,póint }

control program [COMPUT SCI] A program which carries on input/output operations, loading of programs, detection of errors, communication with the operator, and so forth. { kən'trōl ,prō·grəm }

control record [COMPUT SCI] A special record added to the end of a file to provide information about the file and the records in it. { kən'trōl ,rek·ərd }

control register [COMPUT SCI] Any one of the registers in a computer used to control the execution of a computer program. { kən'trōl ,rej·ə·stər }

control room [COMMUN] A room from which engineers and production people control and direct a video or audio program or a recording session. { kən'trōl ,rüm }

control section [COMPUT SCI] **1.** The smallest integral subsection of a program, that is, the smallest unit of code that can be separately relocated during loading. **2.** The part of a central processing unit that controls other sections of the unit. { kən'trōl ,sek·shən }

control sequence [COMPUT SCI] The order in which a set of executions are carried to perform a specific function. { kən'trōl ,sē·kwəns }

control signal [COMPUT SCI] A set of pulses used to identify the channels to be followed by transferred data. [CONT SYS] The signal applied to the device that makes corrective changes in a controlled process or machine. { kən'trōl ,sig·nəl }

control spillover [CONT SYS] The excitation by an active control system of modes of motion that have been omitted from the control algorithm in the process of model reduction. { kən'trōl 'spil ,ō·vər }

control state [COMPUT SCI] The operating mode of a system which permits it to override its normal sequence of operations. { kən'trōl ,stāt }

control statement [COMPUT SCI] A statement in a computer program that controls program execution, such as a GOTO statement, conditional jump, or a loop. { kən'trōl ,stāt·mənt }

control supervisor [COMPUT SCI] The computer software which controls the processing of the system. { kən'trōl ¦sü·pər,vī·zər }

control switching point [COMMUN] A telephone office which is an important switching center in the routing of long-distance calls in the direct distance dialing system. Abbreviated CSP. { kən'trōl 'swich·iŋ ,póint }

control symbol [COMPUT SCI] A symbol which, coded into the machine memory, controls certain steps in the mechanical translation process; since control symbols are not contextual symbols, they appear neither in the input nor in the output. { kən'trōl ,sim·bəl }

control synchro *See* control transformer. { kən'trōl ,siŋ·krō }

control system [ENG] A system in which one or more outputs are forced to change in a desired manner as time progresses. { kən'trōl ,sis·təm }

control-system feedback [CONT SYS] A signal obtained by comparing the output of a control system with the input, which is used to diminish the difference between them. { kən'trōl ,sis·təm 'fēd,bak }

control systems equipment [COMPUT SCI] Computers which are an integral part of a total facility or larger complex of equipment and have the primary purpose of controlling, monitoring, analyzing, or measuring a process or other equipment. { kən'trōl ,sis·təmz i'kwip·mənt }

control total |COMPUT SCI| The sum of the numbers in a specified record field of a batch of records, determined repetitiously during computer processing so that any discrepancy from the control indicates an error. { kən'trōl ,tōd·əl }

control track |ENG ACOUS| A supplementary sound track, usually containing tone signals that control the reproduction of the sound track, such as by changing feed levels to loudspeakers in a theater to achieve stereophonic effects. { kən'trōl ,trak }

control transformer |ELEC| A synchro in which the electrical output of the rotor is dependent on both the shaft position and the electric input to the stator. Also known as control synchro. { kən'trōl tranz'fȯr·mər }

control unit |COMPUT SCI| An electronic device containing data buffers and logical circuitry, situated between the computer channel and the input/output device, and controlling data transfers and such operations as tape rewind. { kən'trōl ,yü·nət }

control unit terminal emulation |COMPUT SCI| A technique that enables a personal computer to imitate a terminal of a main frame. Abbreviated CUT emulation. { kən'trōl ,yü·nət ¦tər·mə·nəl ,em·yə'lā·shən }

control variable |CONT SYS| One of the input variables of a control system, such as motor torque or the opening of a valve, which can be varied directly by the operator to maximize some measure of performance of the system. { kən'trōl ,ver·ē·ə·bəl }

control winding |ELECTR| A winding used on a magnetic amplifier or saturable reactor to apply control magnetomotive forces to the core. Also known as control inductor. { kən'trōl ,wīnd·iŋ }

control word |COMPUT SCI| A computer word specifying a certain action to be taken. { kən'trōl ,wərd }

convection current |ELECTR| The time rate at which the electric charges of an electron stream are transported through a given surface. { kən'vek·shən ,kər·ənt }

convective current See convection current. { kən'vek·div ,kər·ənt }

convective discharge |ELECTR| The movement of a visible or invisible stream of charged particles away from a body that has been charged to a sufficiently high voltage. Also known as electric wind; static breeze. { kən'vek·div 'dis,chärj }

convenience receptacle See outlet. { kən'vēn·yəns ri'sep·tə·kəl }

conventional algorithm |COMMUN| A cryptographic algorithm in which the enciphering and deciphering keys are easily derivable from each other, or are identical, and both must be kept secret. { kən'ven·chən·əl 'al·gə,rith·əm }

conventional current |ELEC| The concept of current as the transfer of positive charge, so that its direction of flow is opposite to that of electrons which are negatively charged. { kən'ven·chən·əl 'kər·ənt }

conventional definition television |COMMUN| The analog NTSC (National Television Standards Committee) television system. Abbreviated CDTV. { kən'ven·chən·əl 'def·ə,nish·ən 'tel·ə,vizh·ən }

conventional programming |COMPUT SCI| The use of standard programming languages, as opposed to application development languages, financial planning languages, query languages, and report programs. { kən'ven·chən·əl 'prō ,gram·iŋ }

convergence |ELECTR| A condition in which the electron beams of a multibeam cathode-ray tube intersect at a specified point, such as at an opening in the shadow mask of a three-gun color television picture tube; both static convergence and dynamic convergence are required. { kən'vər·jəns }

convergence circuit |ELECTROMAG| An auxiliary deflection system in a color television receiver which maintains convergence, having separate convergence coils for electromagnetic controls of the positions of the three beams in a convergence yoke around the neck of the kinescope. { kən'vər·jəns ,sər·kət }

convergence coil |ELECTR| One of the coils used to obtain convergence of electron beams in a three-gun color television picture tube. { kən'vər·jəns ,kȯil }

convergence control |ELECTR| A control used in a color display device to adjust certain parameters of the three-gun color picture tube to achieve convergence. { kən'vər·jəns kən'trōl }

convergence electrode |ELECTR| An electrode whose electric field converges two or more electron beams. { kən'vər·jəns i'lek,trōd }

convergence magnet |ELECTR| A magnet assembly whose magnetic field converges two or more electron beams; used in three-gun color picture tubes. Also known as beam magnet. { kən'vər·jəns ,mag·nət }

Conversational Algebraic Language See CAL. { kän·vər¦sā·shən·əl al·jə¦brā·ik 'laŋ·gwij }

conversational compiler |COMPUT SCI| A compiler which immediately checks the validity of each source language statement entered to the computer and informs the user if the next statement can be entered or if a mistake must be corrected. Also known as interpreter. { kän·vər'sā·shən·əl kəm'pīl·ər }

conversational mode |COMMUN| A computer operating mode that permits queries and responses between the computer and human operators at keyboard terminals. { kən·vər'sā·shən·əl ,mōd }

conversational processing |COMPUT SCI| The operating mode of a computer system which enables a user to have each statement he keys into the system processed immediately. { kän·vər'sā·shən·əl 'präs·əs·iŋ }

conversational time-sharing |COMPUT SCI| The simultaneous utilization of a computer system by multiple users, each user being equipped with a remote terminal with which he communicates with the computer in conversational mode. { kän·vər'sā·shən·əl 'tīm ,sher·iŋ }

conversion See data conversion. { kən'vər·zhən }

conversion gain |ELECTR| **1.** Ratio of the intermediate-frequency output voltage to the input signal voltage of the first detector of a superheterodyne receiver. **2.** Ratio of the available intermediate-frequency power output of a converter or mixer to the available radio-frequency power input. { kən'vər·zhən ˌgān }

conversion program |COMPUT SCI| A set of instructions which allows a program written for one system to be run on a different system. { kən'vər·zhən ˌprō·grəm }

conversion rate |COMPUT SCI| The number of complete conversions an analog-to-digital converter can perform per unit time, usually specified in cycles (or conversions) per second. { kən'vər·zhən ˌrāt }

conversion routine |COMPUT SCI| A flexible, self-contained, and generalized program used for data conversion, which only requires specifications about very few facts in order to be used by a programmer. { kən'vər·zhən rü'tēn }

conversion time |COMPUT SCI| The time required to read in data from one code into another code. { kən'vər·zhən ˌtīm }

convert |COMPUT SCI| To transform the representation of data. { kən'vərt }

converter |COMPUT SCI| A computer unit that changes numerical information from one form to another, as from decimal to binary or vice versa, from fixed-point to floating-point representation, from magnetic tape to disk storage, or from digital to analog signals and vice versa. Also known as data converter. |ELECTR| **1.** The section of a superheterodyne radio receiver that converts the desired incoming radio-frequency signal to an intermediate-frequency value; the converter section includes the oscillator and the mixer-first detector. Also known as heterodyne conversion transducer; oscillator-mixer-first-detector. **2.** An auxiliary unit used with a television or radio receiver to permit reception of channels or frequencies for which the receiver was not originally designed. **3.** In facsimile, a device that changes the type of modulation delivered by the scanner. **4.** Unit of a radar system in which the mixer of a superheterodyne receiver and usually two stages of intermediate-frequency amplification are located; performs a preamplifying operation. { kən'vərd·ər }

converter substation |ELEC| An electric power substation whose main function is the conversion of power from ac to dc, and vice versa. { kən'vərd·ər 'səb,stā·shən }

converter tube |ELECTR| An electron tube that combines the mixer and local-oscillator functions of a heterodyne conversion transducer. { kən'vərd·ər ˌtüb }

convolutional code |COMMUN| An error-correcting code that processses incoming bits serially rather than in large blocks. { ˌkän·və·lü·shən·əl 'kōd }

convolver |ELECTR| A surface acoustic-wave device in which signal processing is performed by a nonlinear interaction between two waves traveling in opposite directions. Also known as acoustic convolver. { kən'väl·vər }

cookbook |COMPUT SCI| A document that describes how to install and use a software product or carry out other complex tasks in step-by-step fashion. { 'kůk,bůk }

cookie |COMPUT SCI| A data file written to a hard drive by some Web sites, contains information the site can use to track such things as passwords, login, registration or identification, user preferences, online shopping cart information, and lists of pages visited. { 'kůk·ē }

cooled infrared detector |ELECTR| An infrared detector that must be operated at cryogenic temperatures, such as at the temperature of liquid nitrogen, to obtain the desired infrared sensitivity. { 'küld ˌin·frə'red di'tek·tər }

cooperative multitasking |COMPUT SCI| A method of running more than one program on a computer at a time in which the program currently in control of the processor retains control until it yields the control to another program voluntarily, which it can do only at certain points in the program. Also known as nonpreemptive multitasking. { kō,äp·rəd·iv 'məl·tə,task·iŋ }

coordinate addressing |COMPUT SCI| The use of cartesian coordinates to specify a location, such as the position of a character in an electronic display. { kō'örd·ən·ət 'ad,res·iŋ }

coordinate data receiver |ELECTR| A receiver specifically designed to accept the signal of a coordinate data transmitter and reconvert this signal into a form suitable for input to associated equipment such as a plotting board, computer, or radar set. { kō'örd·ən·ət 'dad·ə ri ˌsē·vər }

coordinate data transmitter |ELECTR| A transmitter that accepts two or more coordinates, such as those representing a target position, and converts them into a form suitable for transmission. { kō'örd·ən·ət 'dad·ə tranz,mid·ər }

coordinated-axis control |CONT SYS| Robotic control in which the robot axes reach their end points simultaneously, thus giving the robot's motion a smooth appearance. { kō'örd·ən ˌād·əd ¦ak·səs kən,trōl }

coordinated geometry See COGO. { kō'örd·ən ˌād·əd jē'äm·ə·trē }

coordinated transpositions |ELEC| Transpositions which are installed in either electric supply or communications circuits or in both, for the purpose of reducing inductive coupling, and which are located effectively with respect to the discontinuities in both the electric supply and communications circuits. { kō'örd·ən ˌād·əd tranz·pə'zish·ənz }

coordinate indexing |COMPUT SCI| An indexing scheme in which equal-rank descriptors are used to describe a document, for information retrieval by a computer or other means. { kō'örd·ən·ət 'in,deks·iŋ }

coordinate storage See matrix storage. { kō'örd· ən·ət 'stör·ij }

coordination |ELEC| Design of series-connected circuit breakers whereby breakers with lower current ratings trip before those with higher ratings. { kō,ȯrd·ən'ā·shən }

coplanar electrodes |ELECTR| Electrodes mounted in the same plane. { kō'plān·ər i'lek,trōdz }

copper cable |ELEC| A mechanically assembled group of copper wires, used in place of a single, large wire for increased flexibility. { 'käp·ər 'kā·bəl }

copper loss |ELEC| Power loss in a winding due to current flow through the resistance of the copper conductors. Also known as I²R loss. { 'käp·ər ,lȯs }

copper oxide photovoltaic cell |ELECTR| A photovoltaic cell in which light acting on the surface of contact between layers of copper and cuprous oxide causes a voltage to be produced. { 'käp·ər 'äk,sīd ,fōd·ō·vōl'tā·ik 'sel }

copper oxide rectifier |ELECTR| A metallic rectifier in which the rectifying barrier is the junction between metallic copper and cuprous oxide. { 'käp·ər 'äk,sīd 'rek·tə,fī·ər }

copper pair See twisted pair. { 'käp·ər ,per }

copper sulfide rectifier |ELECTR| A semiconductor rectifier in which the rectifying barrier is the junction between magnesium and copper sulfide. { 'käp·ər 'səl,fīd 'rek·tə,fī·ər }

coprocessor |COMPUT SCI| A processing unit that works together with a primary central processing unit to speed a computer's execution of time-consuming operations. { kō'prä,ses·ər }

copy |COMMUN| To transcribe Morse code signals into written form. |COMPUT SCI| A string procedure in Algol by means of which a new byte string can be generated from an existing byte string. { 'käp·ē }

copying program |COMPUT SCI| A system program which copies a data or program file from one peripheral device onto another. { 'käp·ē·iŋ ,prō·grəm }

copy protection See software protection. { 'käp·ē prə,tek·shən }

CORBA See common object request broker. { 'kȯr·bə }

corbinotron |ENG| The combination of a corbino disk, made of high-mobility semiconductor material, and a coil arranged to produce a magnetic field perpendicular to the disk. { kȯr'bē·nə,trän }

cord |ELEC| A small, very flexible insulated cable. { kȯrd }

cord circuit |ELEC| Connecting circuit terminating in a plug at one or both ends and used at switchboard positions in establishing telephone connections. { 'kȯrd ,sər·kət }

cordless telephone |COMMUN| A telephone whose headset and base are equipped with small antennas and are linked by low-power radio instead of a wire. { 'kȯrd·ləs 'tel·ə,fōn }

cordwood module |ELECTR| High-density circuit module in which discrete components are mounted between and perpendicular to two small, parallel printed circuit boards to which

their terminals are attached. { 'kȯrd,wu̇d ,mä·jül }

core See magnetic core. { kȯr }

core array |ELECTR| A rectangular grid arrangement of magnetic cores. { 'kȯr ə'rā }

core bank |ELECTR| A stack of core arrays and associated electronics, the stack containing a specific number of core arrays. { 'kȯr ,baŋk }

core-dump |COMPUT SCI| To copy the contents of all or part of core storage, usually into an external storage device. { 'kȯr ,dəmp }

core hitch |ELEC| Attachment to a cable core to permit pulling it into a duct without damaging the sheath. { 'kȯr ,hich }

core image |COMPUT SCI| 1. A computer program whose storage addresses have been assigned so that it can be loaded directly into main storage for processing. 2. A visual representation of a computer's main storage. { 'kȯr ,im·ij }

core-image library |COMPUT SCI| A collection of computer programs residing on mass-storage device in ready-to-run form. { 'kȯr 'im·ij ,lī,brer·ē }

coreless-type induction heater |ENG| A device in which a charge is heated directly by induction, with no magnetic core material linking the charge. Also known as coreless-type induction furnace. { 'kȯr·ləs ,tīp in'dək·shən ,hēd·ər }

core logic |ELECTR| Logic performed in ferrite cores that serve as inputs to diode and transistor circuits. { 'kȯr ,läj·ik }

core memory See magnetic core storage. { 'kȯr ,mem·rē }

core memory resident |COMPUT SCI| A control program which is in the main memory of a computer at all times to supervise the processing of the computer. { 'kȯr ,mem·rē ,rez·ə·dənt }

core rope storage |COMPUT SCI| Direct-access storage consisting of a large number of doughnut-shaped ferrite cores arranged on a common axis, with sense, inhibit, and set wires threaded through or around individual cores in a predetermined manner to provide fixed storage of digital data; each core rope stores one or more complete words, rather than just a single bit. { 'kȯr ,rōp ,stȯr·ij }

coresident |COMPUT SCI| A computer program or program module that is stored in a computer memory along with other programs. { kō'rez·ə·dənt }

core stack |ELECTR| A number of core arrays, next to one another and treated as a unit. { 'kȯr ,stak }

core storage |COMPUT SCI| The main memory of a computer. { 'kȯr ,stȯr·ij }

corner effect |ELECTR| The departure of the frequency-response curve of a band-pass filter from a perfect rectangular shape, so that the corners of the rectangle are rounded. { 'kȯr·nər i'fekt }

corner frequency See break frequency. { 'kȯr·nər ,frē·kwən·sē }

corner reflector |ELECTROMAG| An antenna consisting of two conducting surfaces intersecting

121

corona

at an angle that is usually 90°, with a dipole or other antenna located on the bisector of the angle.   { 'kȯr·nər ri'flek·tər }

corona See corona discharge.   { kə'rō·nə }

corona current [ELEC] The current of electricity equivalent to the rate of charge transferred to the air from an object experiencing corona discharge.   { kə'rō·nə ¦kər·ənt }

corona discharge [ELEC] A discharge of electricity appearing as a bluish-purple glow on the surface of and adjacent to a conductor when the voltage gradient exceeds a certain critical value; due to ionization of the surrounding air by the high voltage. Also known as aurora; corona; electric corona.   { kə'rō·nə 'dis,chärj }

corona failure [ELEC] High-voltage failure initiated by corona discharge at areas of high-voltage stress such as metal inserts or terminals.   { kə'rō·nə ¦fāl·yər }

corona resistance [ELEC] Ability of a conductor to resist destruction when a high-voltage electrostatic field ionizes within insulation voids.   { kə'rō·nə ri'zis·təns }

corona shield [ELEC] A shield placed about a point of high potential to redistribute electrostatic lines of force.   { kə'rō·nə ,shēld }

corona stabilization [ELEC] The increase in the breakdown voltage of a gas separating two electrodes, where the electric field is very high at one pointed electrode and low at the other, due to the reduction of electric field around the pointed electrode by corona discharge.   { kə'rō·nə ,stā·bə·lə'zā·shən }

corona start voltage [ELEC] The voltage difference at which corona discharge is initiated in a given system.   { kə'rō·nə 'stärt ,vōl·tij }

corona tube [ELEC] A gas-discharge voltage-reference tube employing a corona discharge.   { kə'rō·nə ,tüb }

corona voltmeter [ELEC] A voltmeter in which the crest value of a voltage is indicated by the inception of corona at a known electrode spacing.   { kə'rō·nə 'vōlt,mēd·ər }

coroutine [COMPUT SCI] A program module for which the lifetime of a particular activation record is independent of the time when control enters or leaves the module, and in which the activation record maintains a local instruction counter so that, whenever control enters the module, execution begins at the point where it stopped when control last left that particular instance of execution.   { 'kō·rü,tēn }

correction time [CONT SYS] The time required for the controlled variable to reach and stay within a predetermined band about the control point following any change of the independent variable or operating condition in a control system. Also known as settling time.   { kə'rek·shən ,tīm }

corrective action [CONT SYS] The act of varying the manipulated process variable by the controlling means in order to modify overall process operating conditions.   { kə'rek·tiv 'ak·shən }

corrective maintenance [COMPUT SCI] The maintenance performed as required, on an unscheduled basis, by the contractor following equipment failure. Also known as remedial maintenance.   [ENG] A procedure of repairing components or equipment as necessary either by on-site repair or by replacing individual elements in order to keep the system in proper operating condition.   { kə'rek·tiv mānt·ən·əns }

corrective network [ELEC] An electric network inserted in a circuit to improve its transmission properties, impedance properties, or both. Also known as shaping circuit; shaping network.   { kə'rek·tiv 'net,wərk }

correed relay [ELEC] Hermetically sealed reed capsule surrounded by a coil winding, used as a switching device with telephone equipment.   { 'kō,rēd 'rē,lā }

correlated orientation tracking and range See cotar.   { 'kär·ə,lād·əd ,ȯr·ē·ən'tā·shən 'trak·iŋ ən 'rānj }

correlation detection [ENG] A method of detection of aircraft or space vehicles in which a signal is compared, point to point, with an internally generated reference. Also known as cross-correlation detection.   { ,kär·ə'lā·shən di'tek·shən }

correlation direction finder [ENG] Satellite station separated from a radar to receive jamming signals; by correlating the signals received from several such stations, range and azimuth of many jammers may be obtained.   { ,kär·ə'lā·shən də'rek·shən ,fīnd·ər }

correlation distance [COMMUN] In tropospheric scatter propagation, the minimum spatial separation between antennas which will give rise to independent fading of the received signals.   { ,kär·ə'lā·shən ,dis·təns }

correlation tracking and triangulation See cotat.   { ,kär·ə'lā·shən 'trak·iŋ ən trī,aŋ·gyə'lā·shən }

correlation tracking system [ENG] A trajectory-measuring system utilizing correlation techniques where signals derived from the same source are correlated to derive the phase difference between the signals.   { ,kär·ə'lā·shən 'trak·iŋ ,sis·təm }

correlation-type receiver See correlator.   { ,kär·ə'lā·shən ,tīp ri'sē·vər }

correlator [ELECTR] A device that detects weak signals in noise by performing an electronic operation approximating the computation of a correlation function. Also known as correlation-type receiver.   { 'kär·ə,lād·ər }

correspondence See relation.   { ,kär·ə'spän·dəns }

correspondence printer See letter-quality printer.   { ,kär·ə'spän·dəns ,print·ər }

corrugated conical-horn antenna [ELECTROMAG] A horn antenna that has a circular cross section and a series of equally spaced ridges protruding from otherwise straight sides.   { ¦kär·ə,gād·əd ,kän·ə·kəl ,hȯrn an'ten·ə }

corrupt [COMPUT SCI] To destroy or alter information so that it is no longer reliable.   { kə'rəpt }

cosecant antenna [ELECTROMAG] An antenna that gives a beam whose amplitude varies as the cosecant of the angle of depression below the horizontal; used in navigation radar.   { kō'sē ,kant an'ten·ə }

cosecant-squared antenna |ELECTROMAG| An antenna that has a cosecant-squared pattern. { kō'sē,kant ¦skwerd an'ten·ə }

cosecant-squared pattern |ELECTROMAG| A ground radar-antenna radiation pattern that sends less power to nearby objects than to those farther away in the same sector; the field intensity varies as the square of the cosecant of the elevation angle. { kō'sē,kant ¦skwerd 'pad·ərn }

cosine winding |ELECTR| A winding used in the deflection yoke of a cathode-ray tube to prevent changes in focus as the beam is deflected over the entire area of the screen. { 'kō,sīn ,wīnd·iŋ }

cosmic noise |COMMUN| Radio static caused by a phenomenon outside the earth's atmosphere, such as sunspots. { 'käz·mik 'nóiz }

cost function |SYS ENG| In decision theory, a loss function which does not depend upon the decision rule. { 'kòst ,fəŋk·shən }

cotar |ENG| A passive system used for tracking a vehicle in space by determining the line of direction between a remote ground-based receiving antenna and a telemetering transmitter in the missile, using phase-comparison techniques. Derived from correlated orientation tracking and range. { 'kō,tär }

cotat |ENG| A trajectory-measuring system using several antenna base lines, each separated by large distances, to measure direction cosines to an object; then the object's space position is computed by triangulation. Derived from correlation tracking and triangulation. { 'kō,tat }

Cotton balance |ENG| A device which employs a current-carrying conductor of special shape to determine the strength of a magnetic field. { 'kät·ən 'bal·əns }

coul See coulomb.

coulomb |ELEC| A unit of electric charge, defined as the amount of electric charge that crosses a surface in 1 second when a steady current of 1 absolute ampere is flowing across the surface; this is the absolute coulomb and has been the legal standard of quantity of electricity since 1950; the previous standard was the international coulomb, equal to 0.999835 absolute coulomb. Abbreviated coul. Symbolized C. { 'kü,läm }

Coulomb attraction |ELEC| The electrostatic force of attraction exerted by one charged particle on another charged particle of opposite sign. Also known as electrostatic attraction. { 'kü ,läm ə'trak·shən }

Coulomb field |ELEC| The electric field created by a stationary charged particle. { 'kü,läm ,fēld }

Coulomb force |ELEC| The electrostatic force of attraction or repulsion exerted by one charged particle on another, in accordance with Coulomb's law. { 'kü,läm ,fórs }

Coulomb interactions |ELEC| Interactions of charged particles associated with the Coulomb forces they exert on one another. Also known

as electrostatic interactions. { 'kü,läm in·tər 'ak·shənz }

coulombmeter |ENG| An instrument that measures quantity of electricity in coulombs by integrating a stored charge in a circuit which has very high input impedance. { 'kü,läm,mēd·ər }

Coulomb potential |ELEC| A scalar point function equal to the work per unit charge done against the Coulomb force in transferring a particle bearing an infinitesimal positive charge from infinity to a point in the field of a specific charge distribution. { kü'läm pə'ten·chəl }

Coulomb repulsion |ELEC| The electrostatic force of repulsion exerted by one charged particle on another charged particle of the same sign. Also known as electrostatic repulsion. { kü'läm ri'pəl·shən }

Coulomb's law |ELEC| The law that the attraction or repulsion between two electric charges acts along the line between them, is proportional to the product of their magnitudes, and is inversely proportional to the square of the distance between them. Also known as law of electrostatic attraction. { 'kü'lämz ,lò }

Coulomb's theorem |ELEC| The proposition that the intensity of an electric field near the surface of a conductor is equal to the surface charge density on the nearby conductor surface divided by the absolute permittivity of the surrounding medium. { 'kü,lämz ,thir·əm }

count cycle |COMPUT SCI| An increase or decrease of the cycle index by unity or by an arbitrary integer. { 'kaùnt ,sī·kəl }

countdown |COMMUN| The ratio of the number of interrogation pulses not answered by a transponder to the total number received. { 'kaùnt ,daùn }

counter |COMPUT SCI| 1. A register or storage location used to represent the number of occurrences of an event. 2. See accumulator; scaler. { 'kaùnt·ər }

counter circuit See counting circuit. { 'kaùnt·ər ,sər·kət }

counter coupling |COMPUT SCI| The technique of combining two or more counters into one counter of larger capacity in electromechanical devices by means of control panel wiring. { 'kaùnt·ər ,kəp·liŋ }

counter decade See decade scaler. { 'kaùnt·ər ,dek,ād }

counterelectromotive cell |ELEC| Cell of practically no ampere-hour capacity, used to oppose the line voltage. { ¦kaùnt·ər·i,lek·trō'mōd·iv 'sel }

counter-free machine |COMPUT SCI| A sequential machine that cannot count modulo any integer greater than 1. { 'kaùnt·ər ,frē mə'shēn }

counter/frequency meter [ENG] An instrument that contains a frequency standard and can be used to measure the number of events or the number of cycles of a periodic quantity that occurs in a specified time, or the time between two events. { 'kaunt·ər 'frē·kwən·sē ‚mēd·ər }

countermeasures set [ELECTR] A complete electronic set specifically designed to provide facilities for intercepting and analyzing electromagnetic energy propagated by transmitter and to provide a source of radio-frequency signals which deprive the enemy of effective use of his electronic equipment. { 'kaunt·ər‚mezh·ərz ‚set }

counterpoise [ELEC] A system of wires or other conductors that is elevated above and insulated from the ground to form a lower system of conductors for an antenna. Also known as antenna counterpoise. { 'kaunt·ər‚poiz }

counter tube [ELECTR] An electron tube having one signal-input electrode and 10 or more output electrodes, with each input pulse serving to transfer conduction sequentially to the next output electrode; beam-switching tubes and cold-cathode counter tubes are examples. { 'kaunt·ər ‚tüb }

counter voltage [ELEC] The reverse voltage that appears across an inductor when current through the inductor is shut off. { 'kaunt·ər ‚vōl·tij }

counting circuit [ELECTR] A circuit that counts pulses by frequency-dividing techniques, by charging a capacitor in such a way as to produce a voltage proportional to the pulse count, or by other means. Also known as counter circuit. { 'kaunt·iŋ ‚sər·kət }

counting-down circuit See frequency divider. { 'kaunt·iŋ ‚daun ‚sər·kət }

counting rate-voltage characteristic See plateau characteristic. { 'kaunt·iŋ ‚rāt 'vōl·tij ‚kar·ik·tə'ris·tik }

couple [ELEC] To connect two circuits so signals are transferred from one to the other. [ELECTR] Two metals placed in contact, as in a thermocouple. { 'kəp·əl }

coupled antenna [ELECTROMAG] An antenna electromagnetically coupled to another. { 'kəp·əld an'ten·ə }

coupled circuits [ELEC] Two or more electric circuits so arranged that energy can transfer electrically or magnetically from one to another. { 'kəp·əld 'sər·kəts }

coupled systems [COMPUT SCI] Computer systems that share equipment and can exchange information. { 'kəp·əld 'sis·təmz }

coupled transistors [ELECTR] Transistors connected in series by transformers or resistance-capacitance networks, in much the same manner as electron tubes. { 'kəp·əld tran'zis·tərz }

coupler [ELEC] A component used to transfer energy from one circuit to another. [ELECTROMAG] **1.** A passage which joins two cavities or waveguides, allowing them to exchange energy. **2.** A passage which joins the ends of two waveguides, whose cross section changes continuously from that of one to that of the other. { 'kəp·lər }

coupling [ELEC] **1.** A mutual relation between two circuits that permits energy transfer from one to another, through a wire, resistor, transformer, capacitor, or other device. **2.** A hardware device used to make a temporary connection between two wires. { 'kəp·liŋ }

coupling aperture [ELECTROMAG] An aperture in the wall of a waveguide or cavity resonator, designed to transfer energy to or from an external circuit. Also known as coupling hole; coupling slot. { 'kəp·liŋ ‚ap·ə·chər }

coupling capacitor [ELECTR] A capacitor used to block the flow of direct current while allowing alternating or signal current to pass; widely used for joining two circuits or stages. Also known as blocking capacitor; stopping capacitor. { 'kəp·liŋ kə'pas·əd·ər }

coupling coefficient [ELECTR] The ratio of the maximum change in energy of an electron traversing an interaction space to the product of the peak alternating gap voltage and the electronic charge. { 'kəp·liŋ ‚kō·i'fish·ənt }

coupling hole See coupling aperture. { 'kəp·liŋ ‚hōl }

coupling loop [ELECTROMAG] A conducting loop projecting into a waveguide or cavity resonator, designed to transfer energy to or from an external circuit. { 'kəp·liŋ ‚lüp }

coupling probe [ELECTROMAG] A probe projecting into a waveguide or cavity resonator, designed to transfer energy to or from an external circuit. { 'kəp·liŋ ‚prōb }

coupling slot See coupling aperture. { 'kəp·liŋ ‚slät }

course programmer [CONT SYS] An item which initiates and processes signals in a manner to establish a vehicle in which it is installed along one or more projected courses. { 'kórs 'prō ‚gram·ər }

courseware [COMPUT SCI] Computer programs designed to be used in computer-aided instruction or computer-managed instruction. { 'kórs‚wer }

coverage [ELECTROMAG] A spatial account of the regions of useful sensitivity in a radar's surroundings that can be affected, for example, by multipath propagation or by obscuring terrain. { 'kəv·rij }

COZI [COMMUN] An ionospheric sounding system for determining propagation characteristics of the ionosphere at various angles at any instant; used to determine how well long-distance, high-frequency broadcasts are reaching their intended destinations. Derived from communications zone indicator. { ‚kō‚zī }

CPA See color-phase alternation.

CPE See computer performance evaluation.

CPM See critical path method.

C power supply [ELECTR] A device connected in the circuit between the cathode and grid of a vacuum tube to apply grid bias. { ‚sē 'paur sə ‚plī }

CPU See central processing unit.

CPU-bound program [COMPUT SCI] A computer program that involves a large amount of calculation and internal rearrangement of data, so that the speed of execution depends on the speed of the central processing unit (CPU) and memory. Also known as cycle-bound program; process-bound program. { ˌsēˌpe'yü ¦baùnd ˌprō·grəm }

CPU fan [COMPUT SCI] A fan mounted directly over the integrated-circuit chip containing a computer's central processing unit to prevent overheating. { ¦sē¦pē¦yü 'fan }

crash [COMPUT SCI] **1.** A breakdown, hardware failure, or software problem that renders a computer system inoperative. **2.** See abend. { krash }

crash locator beacon [COMMUN] An automatic radio beacon carried in aircraft to guide searching forces in the event of a crash. { 'krash 'lō,kād·ər ,bē·kən }

crater lamp [ELECTR] A glow-discharge tube used as a point source of light whose brightness is proportional to the signal current sent through the tube; used for photographic recording of facsimile signals. { 'krād·ər ,lamp }

CRC See cyclic redundancy check.

creation operator [COMPUT SCI] The part of a data structure which allows components to be created. { krē'ā·shən ,äp·ə,rād·ər }

credence [ELECTROMAG] In radar, a measure of confidence in a target detection, generally proportional to target return amplitude. { 'krēd·əns }

creep [ELECTR] A slow change in a characteristic with time or usage. { krēp }

creepage [ELEC] The conduction of electricity across the surface of a dielectric. { 'krē·pij }

crest value See peak value. { 'krest ,val·yü }

crest voltmeter [ELEC] A voltmeter reading the peak value of the voltage applied to its terminals. { 'krest 'vōlt,mēd·ər }

crimp contact [ELEC] A contact whose back portion is a hollow cylinder that will accept a wire; after a bared wire is inserted, a swaging tool is applied to crimp the contact metal firmly against the wire. Also known as solderless contact. { 'krimp ,kän,takt }

crippled leap-frog test [COMPUT SCI] A variation of the leap-frog test, modified so the computer tests are repeated from a single set of storage locations rather than a changing set of locations. { ¦krip·əld ¦lēp ,fräg ,test }

crippled mode [COMPUT SCI] The operation of a computer at reduced capacity when certain parts are not working. { 'krip·əld ,mōd }

critical anode voltage [ELECTR] The anode voltage at which breakdown occurs in a gas tube. { 'krid·ə·kəl 'a,nōd ,vōl·tij }

critical area See picture element. { 'krid·ə·kəl 'er·ē·ə }

critical coupling [ELEC] The degree of coupling that provides maximum transfer of signal energy from one radio-frequency resonant circuit to another when both are tuned to the same frequency. Also known as optimum coupling. { 'krid·ə·kəl 'kəp·liŋ }

critical field [ELECTR] The smallest theoretical value of steady magnetic flux density that would prevent an electron emitted from the cathode of a magnetron at zero velocity from reaching the anode. Also known as cutoff field. { 'krid·ə·kəl 'fēld }

critical frequency [ELECTR] See cutoff frequency. [ELECTROMAG] The limiting frequency below which a radio wave will be reflected by an ionospheric layer at vertical incidence at a given time. { 'krid·ə·kəl 'frē·kwən·sē }

critical grid current [ELECTR] Instantaneous value of grid current when the anode current starts to flow in a gas-filled vacuum tube. { 'krid·ə·kəl 'grid ,kər·ənt }

critical grid voltage [ELECTR] The grid voltage at which anode current starts to flow in a gas tube. Also known as firing point. { 'krid·ə·kəl 'grid ,vōl·tij }

critical path method [SYS ENG] A systematic procedure for detailed project planning and control. Abbreviated CPM. { 'krid·ə·kəl 'path ,meth·əd }

critical potential [ELEC] A potential which results in sudden change in magnitude of the current. { 'krid·ə·kəl pə'ten·chəl }

critical voltage [ELECTR] The highest theoretical value of steady anode voltage, at a given steady magnetic flux density, at which electrons emitted from the cathode of a magnetron at zero velocity would fail to reach the anode. Also known as cutoff voltage. { 'krid·ə·kəl 'vōl·tij }

critical wavelength [COMMUN] The free-space wavelength corresponding to the critical frequency. { 'krid·ə·kəl 'wāv,leŋkth }

CR law [ELEC] A law which states that when a constant electromotive force is applied to a circuit consisting of a resistor and capacitor connected in series, the time taken for the potential on the plates of the capacitor to rise to any given fraction of its final value depends only on the product of capacitance and resistance. { ¦sē ¦är ,lò }

CRO See cathode-ray oscilloscope.

crocodile [ELEC] A unit of potential difference or electromotive force, equal to 10^6 volts; used informally at some nuclear physics laboratories. { 'kräk·ə,dīl }

crocodile clip See alligator clip. { 'kräk·ə,dīl ,klip }

CROM See control and read-only memory. { 'sē ,räm }

Crookes dark space See cathode dark space. { ¦krüks 'därk ,spās }

Crookes tube [ELECTR] An early form of low-pressure discharge tube whose cathode was a flat aluminum disk at one end of the tube, and whose anode was a wire at one side of the tube, outside the electron stream; used to study cathode rays. { 'krüks ,tüb }

cross antenna [ELECTROMAG] An array of two or more horizontal antennas connected to a single feed line and arranged in the pattern of a cross. { 'kròs an,ten·ə }

cross assembler

cross assembler [COMPUT SCI] An assembly program that allows a computer program written on one type of computer to be used on another type. { 'krós ə,sem·blər }

crossbar switch [ELEC] A switch having a three-dimensional arrangement of contacts and a magnet system that selects individual contacts according to their coordinates in the matrix. { 'krós,bär ,swich }

crossbar system [COMMUN] Automatic telephone switching system which is generally characterized by the following features: selecting mechanisms are crossbar switches, common circuits select and test the switching paths and control the operation of the selecting mechanisms, and method of operations is one in which the switching information is received and stored by controlling mechanisms that determine the operations necessary in establishing a telephone connection; largely replaced by electronic switching systems using digital switching techniques. { 'krós,bär ,sis·təm }

cross-color [ELECTR] In analog color television, the interference in the receiver chrominance channel caused by cross talk from monochrome signals. { 'krós ,kəl·ər }

cross compiler [COMPUT SCI] A compiler that allows a computer program written on one type of computer to be used on another type. { 'krós kəm,pī·lər }

cross-correlation detection See correlation detection. { 'krós kär·ə'lā·shən di'tek·shən }

cross-correlation function [COMMUN] A function, $\phi_{12}(\tau)$, where τ is a time-delay parameter, equal to the limit, as T approaches infinity, of the reciprocal of 2T times the integral over t from $-T$ to T of $f_1(t)f_2(t-\tau)$, where f_1 and f_2 are functions of time, such as the input and output of a communication system. { 'krós kär·ə'lā·shən ,fəŋk·shən }

cross-correlator [ELECTR] A correlator in which a locally generated reference signal is multiplied by the incoming signal and the result is smoothed in a low-pass filter to give an approximate computation of the cross-correlation function. Also known as synchronous detector. { ¦krós'kär·ə ,lād·ər }

cross-coupling [COMMUN] A measure of the undesired power transferred from one channel to another in a transmission medium. { ¦krós 'kəp·liŋ }

crossed-field amplifier [ELECTR] A forward-wave, beam-type microwave amplifier that uses crossed-field interaction to achieve good phase stability, high efficiency, high gain, and wide bandwidth for most of the microwave spectrum. { 'króst ,fēld 'am·plə,fī·ər }

crossed-field backward-wave oscillator [ELECTR] One of several types of backward-wave oscillators that utilize a crossed field, such as the amplitron and carcinotron. { 'króst ,fēld 'bak,wərd ,wāv 'äs·ə,lād·ər }

crossed-field device [ELECTR] Any instrument which uses the motion of electrons in perpendicular electric and magnetic fields to generate

microwave radiation, either as an amplifier or oscillator. { 'króst ,fēld di'vīs }

crossed-field multiplier phototube [ELECTR] A multiplier phototube in which repeated secondary emission is obtained from a single active electrode by the combined effects of a strong radio-frequency electric field and a perpendicular direct-current magnetic field. { 'króst ,fēld ,məl·tə,plī·ər 'fōd·ō,tüb }

crossed-field tubes [ELECTR] Vacuum tubes often used in radar transmitters, either as oscillators or as amplifiers, in which the electrons leaving the cathode surface travel in a plasma to the anode in paths determined by the crossed electric and magnetic bias fields applied to the tube, so that the density of the plasma can be easily affected by the electromagnetic signal with which the electrons are interacting. { 'króst ,fēld ,tübz }

cross-fade [ENG ACOUS] In dubbing, the overlapping of two sound tracks, wherein the outgoing track fades out while the incoming track fades in. { 'krós ,fād }

cross fire [COMMUN] Interfering current in one telegraph or signaling channel resulting from telegraph or signaling currents in another channel. { 'krós ,fīr }

crossfoot [COMPUT SCI] To add numbers in several different ways in a computer, for checking purposes. { 'krós,fút }

crosshatch generator [ELECTR] A signal generator that generates a crosshatch pattern for adjusting a video display device. { 'krós,hach ,jen·ə,rād·ər }

cross modulation [COMMUN] A type of interference in which the carrier of a desired signal becomes modulated by the program of an undesired signal on a different carrier frequency; the program of the undesired station is then heard in the background of the desired program. { ¦krós ,mäj·ə'lā·shən }

cross-neutralization [ELECTR] Method of neutralization used in push-pull amplifiers, whereby a portion of the plate-cathode alternating-current voltage of each vacuum tube is applied to the grid-cathode circuit of the other vacuum tube through a neutralizing capacitor. { ¦krós ,nü·trə·lə'zā·shən }

cross office switching time [COMMUN] Time required to connect any input through the switching center to any selected output. { 'krós ,óf·əs 'swich·iŋ ,tīm }

crossover [ELEC] A point at which two conductors cross, with appropriate insulation between them to prevent contact. [ELECTR] The plane at which the cross section of a beam of electrons in an electron gun is a minimum. { 'krós,ō·vər }

crossover distortion [ELECTR] Amplitude distortion in a class B transistor power amplifier which occurs at low values of current, when input

126

impedance becomes appreciable compared with driver impedance.   { 'krȯs‚ō·vər dis'tȯr·shən }

crossover frequency [ENG ACOUS] **1.** The frequency at which a dividing network delivers equal power to the upper and lower frequency channels when both are terminated in specified loads. **2.** *See* transition frequency.   { 'krȯs‚ō·vər ‚frē·kwən·sē }

crossover network [ENG ACOUS] A selective network used to divide the audio-frequency output of an amplifier into two or more bands of frequencies. Also known as dividing network; loudspeaker dividing network.   { 'krȯs‚ō·vər ‚net‚wərk }

crossover voltage [ELECTR] In a cathode-ray storage tube, the voltage of a secondary writing surface, with respect to cathode voltage, on which the secondary emission is unity.   { 'krȯs‚ō·vər ‚vōl·tij }

cross-platform computing [COMPUT SCI] The use of very similar user interfaces for versions of programs running on different operating systems and computer architectures.   { ‚krȯs ¦plat‚fȯrm kəm'pyüd·iŋ }

cross-referencing program [COMPUT SCI] A computer program used in debugging that produces indexed lists of both the variable names and the statement numbers of the source program.   { ¦krȯs 'ref·rəns·iŋ ‚prō·grəm }

crosstalk [COMMUN] **1.** The sound heard in a receiver along with a desired program because of cross modulation or other undesired coupling to another communication channel; it is also observed between adjacent pairs in a telephone cable. **2.** Interaction of audio and video signals in an analog television system, causing video modulation of the audio carrier or audio modulation of the video signal at some point. **3.** Interaction of the chrominance and luminance signals in an analog color television receiver. [ELECTR] *See* magnetic printing.   { 'krȯs‚tȯk }

crosstalk coupling [COMMUN] The cross coupling between speech communications channels or their component parts. Also known as crosstalk loss.   { 'krȯs‚tȯk ‚kəp·liŋ }

crosstalk level [COMMUN] Volume of crosstalk energy, measured in decibels, referred to a reference level.   { 'krȯs‚tȯk ‚lev·əl }

crosstalk loss *See* crosstalk coupling.   { 'krȯs‚tȯk ‚lȯs }

crosstalk unit [COMMUN] A measure of the coupling between two circuits; the number of crosstalk units is 1 million times the ratio of the current or voltage at the observing point to the current or voltage at the origin of the disturbing signal, the impedances at these points being equal. Abbreviated cu.   { 'krȯs‚tȯk ‚yü·nət }

crowbar [ELEC] A device or action that in effect places a high overload on the actuating element of a circuit breaker or other protective device, thus triggering it.   { 'krō‚bär }

crowbar voltage protector [ELEC] A separate circuit which monitors the output of a regulated power supply and instantaneously throws a short circuit (or crowbar) across the output terminals

of the power supply whenever a preset voltage limit is exceeded.   { 'krō‚bär 'vōl·tij prə'tek·tər }

crown cell [ELEC] The generic name for alkaline zinc-manganese dioxide dry-cell battery; manganese dioxide-graphite cathode mix is pressed into a steel can onto which a steel cap is spotwelded to contain the amalgamated powdered-zinc anode.   { 'kraün ‚sel }

CRT *See* cathode-ray tube.

cruciform core [ELEC] A transformer core in which all windings are on one center leg, and four additional legs arranged in the form of a cross serve as return paths for magnetic flux.   { 'krü·sə ‚fȯrm ‚kȯr }

cryoelectronics [ELECTR] A branch of electronics concerned with the study and application of superconductivity and other low-temperature phenomena to electronic devices and systems. Also known as cryolectronics.   { ¦krī·ō·i‚lek 'trän·iks }

cryogenic engineering [ENG] A branch of engineering specializing in technical operations at very low temperatures (about 200 to 400°R, or −160 to −50°C).   { ‚krī·ə'jen·ik en·jə'nir·iŋ }

cryogenic film [COMPUT SCI] A storage element using superconducting thin films of lead at liquid-helium temperature.   { ‚krī·ə'jen·ik 'film }

cryogenic transformer [ELECTR] A transformer designed to operate in digital cryogenic circuits, such as a controlled-coupling transformer.   { ‚krī·ə'jen·ik tranz'fȯr·mər }

cryolectronics *See* cryoelectronics.   { ¦krī·ō·i‚lek 'trän·iks }

cryoresistive transmission line [ELEC] An electric power transmission line whose conducting cables are cooled to the temperature of liquid nitrogen, 77 K (-196°C), resulting in a reduction of the resistance of the conductor by a factor of approximately 10, leading to increased transmission capacity.   { ‚krī·ō·ri'zis·tiv tranz'mish·ən ‚līn }

cryosar [ELECTR] A cryogenic, two-terminal, negative-resistance semiconductor device, consisting essentially of two contacts on a germanium wafer operating in liquid helium.   { 'krī·ō‚sär }

cryosistor [ELECTR] A cryogenic semiconductor device in which a reverse-biased *pn* junction is used to control the ionization between two ohmic contacts.   { ¦krī·ə'zis·tər }

cryotron [ELECTR] A switch that operates at very low temperatures at which its components are superconducting; when current is sent through a control element to produce a magnetic field, a gate element changes from a superconductive zero-resistance state to its normal resistive state.   { 'krī·ə‚trän }

cryotronics [ELECTR] The branch of electronics that deals with the design, construction, and use of cryogenic devices.   { ‚krī·ə'trän·iks }

cryptanalysis [COMMUN] Steps and operations performed in converting encrypted messages into plain text without previous knowledge of the key employed.   { ‚krip·tə'nal·ə·səs }

cryptochannel [COMMUN] A complete system of communication that uses electronic

cryptogram

encryption and decryption equipment and has two or more radio or wire terminals. { ¦krip·tō'chan·əl }

cryptogram |COMMUN | Information written in code or cipher. { 'krip·tə₎gram }

cryptographic algorithm |COMMUN| An unchanging set of rules or steps for enciphering and deciphering messages in a cipher system. { ¦krip·tə¦graf·ik 'al·gə₎rith·əm }

cryptographic bitstream |COMMUN| An unending sequence of digits which is combined with ciphertext to produce plaintext or with plaintext to recover ciphertext in a stream cipher system. { ¦krip·tə¦graf·ik 'bit₎strēm }

cryptographic key |COMMUN| A sequence of numbers or characters selected by the user of a cipher system to implement a cryptographic algorithm for enciphering and deciphering messages. Also known as key. { ¦krip·tə¦graf·ik 'kē }

cryptography |COMMUN| The science of preparing messages in a form which cannot be read by those not privy to the secrets of the form. { krip'täg·rə·fē }

cryptology |COMMUN| The science of preparing messages in forms which are intended to be unintelligible to those not privy to the secrets of the form, and of deciphering such messages. { krip'täl·ə·jē }

cryptopart |COMMUN| One of several portions of a cryptotext; each cryptopart bears a different message indicator. { 'krip·tō₎pärt }

cryptotext |COMMUN| In cryptology, a text of visible writing which conveys no intelligible meaning in any language, or which apparently conveys an intelligible meaning that is not the real meaning. { 'krip·tō₎tekst }

crystal |ELECTR| A natural or synthetic piezoelectric or semiconductor material whose atoms are arranged with some degree of geometric regularity. { 'krist·əl }

crystal activity |ELECTR| A measure of the amplitude of vibration of a piezoelectric crystal plate under specified conditions. { 'krist·əl ak 'tiv·əd·ē }

crystal-audio receiver |ELECTR| Similar to the crystal-video receiver, except for the path detection bandwidth which is audio rather than video. { ¦krist·əl ¦ód·ē·ō ri'sē·vər }

crystal blank |ELECTR| The result of the final cutting operation on a piezoelectric or semiconductor crystal. { 'krist·əl ₎blaŋk }

crystal calibrator |ELECTR| A crystal-controlled oscillator used as a reference standard to check frequencies. { ¦krist·əl 'kal·ə₎brād·ər }

crystal cartridge |ENG ACOUS| A piezoelectric unit used with a stylus in a phonograph pickup to convert disk recordings into audio-frequency signals, or used with a diaphragm in a crystal microphone to convert sound waves into af signals. { ¦krist·əl 'kär₎trij }

crystal control |ELECTR| Control of the frequency of an oscillator by means of a quartz crystal unit. { 'krist·əl kən'trōl }

crystal-controlled oscillator |ELECTR| An oscillator whose frequency of operation is controlled by a crystal unit. { ¦krist·əl kən¦trōld 'äs·ə ₎lād·ər }

crystal-controlled transmitter |ELECTR| A transmitter whose carrier frequency is directly controlled by the electromechanical characteristics of a quartz crystal unit. { ¦krist·əl kən¦trōld 'tranz ₎mid·ər }

crystal current |ELECTR| The actual alternating current flowing through a crystal unit. { 'krist·əl ₎kər·ənt }

crystal cutter |ENG ACOUS| A cutter in which the mechanical displacements of the recording stylus are derived from the deformations of a crystal having piezoelectric properties. { 'krist·əl ₎kəd·ər }

crystal detector |ELECTR| **1.** A crystal used to rectify a modulated radio-frequency signal to obtain the audio or video signal directly. **2.** A crystal diode used in a microwave receiver to combine an incoming radio-frequency signal with a local oscillator signal to produce an intermediate-frequency signal. { 'krist·əl di'tek·tər }

crystal diode See semiconductor diode. { ¦krist·əl 'dī₎ōd }

crystal filter |ELECTR| A highly selective tuned circuit employing one or more quartz crystals; sometimes used in intermediate-frequency amplifiers of communication receivers to improve the selectivity. { ¦krist·əl 'fil·tər }

crystal harmonic generator |ELECTR| A type of crystalcontrolled oscillator which produces an output rich in harmonics (overtones or multiples) of its fundamental frequency. { 'krist·əl har ¦män·ik 'jen·ə₎rād·ər }

crystal headphones |ENG ACOUS| Headphones using Rochelle salt or other crystal elements to convert audio-frequency signals into sound waves. Also known as ceramic earphones. { 'krist·əl 'hed₎fōnz }

crystal-lattice filter |ELECTR| A crystal filter that uses two matched pairs of series crystals and a higher-frequency matched pair of shunt or lattice crystals. { ¦krist·əl 'lad·əs ₎fil·tər }

crystal loudspeaker |ENG ACOUS| A loudspeaker in which movements of the diaphragm are produced by a piezoelectric crystal unit that twists or bends under the influence of the applied audio-frequency signal voltage. Also known as piezoelectric loudspeaker. { ¦krist·əl 'laùd₎spēk·ər }

crystal microphone |ENG ACOUS| A microphone in which deformation of a piezoelectric bar by the action of sound waves or mechanical vibrations generates the output voltage between the faces of the bar. Also known as piezoelectric microphone. { ¦krist·əl 'mī·krə₎fōn }

crystal mixer |ELECTR| A mixer that uses the nonlinear characteristic of a crystal diode to mix two frequencies; widely used in radar receivers to convert the received radar signal to a lower intermediate-frequency value by mixing it with a local oscillator signal. { ¦krist·əl 'mik·sər }

crystal operation |ELECTR| Operation using crystal-controlled oscillators. { 'krist·əl 'äp·ə ₎rā·shən }

128

crystal oscillator |ELECTR| An oscillator in which the frequency of the alternating-current output is determined by the mechanical properties of a piezoelectric crystal. Also known as piezoelectric oscillator. { ¦krist·əl 'äs·ə‚lād·ər }

crystal plate |ELECTR| A precisely cut slab of quartz crystal that has been lapped to final dimensions, etched to improve stability and efficiency, and coated with metal on its major surfaces for connecting purposes. Also known as quartz plate. { 'krist·əl ‚plāt }

crystal rectifier See semiconductor diode. { ¦krist·əl 'rek·tə‚fī·ər }

crystal resonator |ELECTR| A precisely cut piezoelectric crystal whose natural frequency of vibration is used to control or stabilize the frequency of an oscillator. Also known as piezoelectric resonator. { ¦krist·əl 'rez·ən‚ād·ər }

crystal set |ELECTR| A radio receiver having a crystal detector stage for demodulation of the received signals, but no amplifier stages. { 'krist·əl ‚set }

crystal-stabilized transmitter |ELECTR| A transmitter employing automatic frequency control, in which the reference frequency is that of a crystal oscillator. { ¦krist·əl ¦stā·bə‚līzd 'tranz‚mid·ər }

crystal transducer |ELECTR| A transducer in which a piezoelectric crystal serves as the sensing element. { 'krist·əl tranz'dü·sər }

crystal unit |ELECTR| A complete assembly of one or more quartz plates in a crystal holder. { ¦krist·əl ¦yü·nət }

crystal video receiver |ELECTR| A broad-tuning radar or other microwave receiver consisting only of a crystal detector and a video or audio amplifier. { ¦krist·əl ¦vid·ē·ō ri'sē·vər }

crystal video rectifier |ELECTR| A crystal rectifier transforming a high-frequency signal directly into a video-frequency signal. { ¦krist·əl ¦vid·ē·ō 'rek·tə‚fī·ər }

C-scan See C-display. { 'sē ‚skan }

C-scope See C-display. { 'sē ‚skōp }

CSMA/CD |COMPUT SCI| A method of controlling multiaccess computer networks in which each station on the network senses traffic and waits for it to clear before sending a message, and two devices that try to send concurrent messages must both step back and try again. Abbreviation for carrier-sense multiple access with collision detection.

CSP See control switching point.

CSSB system See companded single-sideband system. { ¦sē‚es‚es¦bē ‚sis·təm }

CSW See channel status word.

CT See center tap; computerized tomography.

cu See crosstalk unit.

cubical antenna |ELECTROMAG| An antenna array, the elements of which are positioned to form a cube. { 'kyü·bə·kəl an'ten·ə }

cubicle |ENG| An enclosure for high-voltage equipment. { 'kyü·bə·kəl }

Cuccia coupler See electron coupler. { 'kü·chē·ə 'kəp·lər }

cue circuit |ELECTR| A one-way communication circuit used to convey program control information. { 'kyü ‚sər·kət }

cumulative compound generator |ELEC| A compound generator in which the series field is connected to aid the shunt field magnetomotive force. { 'kyü·myə·ləd·iv ‚käm‚paund 'jen·ə ‚rād·ər }

cumulative ionization See avalanche. { 'kyü·myə· ləd·iv ‚ī·ən·ə'zā·shən }

cup electrometer |ENG| An electrometer that has a metal cup attached to its plate so that a charged body touching the inside of the cup gives up its entire charge to the instrument. { 'kəp i‚lek'träm·əd·ər }

Curie balance |ENG| An instrument for determining the susceptibility of weakly magnetic materials, in which the deflection produced by a strong permanent magnet on a suspended tube containing the specimen is measured. { 'kyür·ē ‚bal·əns }

current |ELEC| The net transfer of electric charge per unit time; a specialization of the physics definition. Also known as electric current. { 'kər·ənt }

current amplification |ELECTR| The ratio of output-signal current to input-signal current for an electron tube, transistor, or magnetic amplifier, the multiplier section of a multiplier phototube, or any other amplifying device; often expressed in decibels by multiplying the common logarithm of the ratio by 20. { 'kər·ənt am·plə·fə'kā·shən }

current amplifier |ELECTR| An amplifier capable of delivering considerably more signal current than is fed in. { 'kər·ənt ‚am·plə‚fī·ər }

current antinode |ELEC| A point at which current is a maximum along a transmission line, antenna, or other circuit element having standing waves. Also known as current loop. { 'kər·ənt 'an·tə ‚nōd }

current attenuation |ELECTR| The ratio of input-signal current for a transducer to the current in a specified load impedance connected to the transducer; often expressed in decibels. { 'kər·ənt ə‚ten·yə'wā·shən }

current awareness system |COMPUT SCI| A system for notifying users on a periodic basis of the acquisition, by a central file or library, of information (usually literature) which should be of interest to the user. { 'kər·ənt ə'wer·nəs ‚sis·təm }

current balance |ELEC| An apparatus with which force is measured between current-carrying conductors, with the purpose of assigning the value of the ampere. Also known as ampere balance. { 'kər·ənt ‚bal·əns }

current-carrying capacity |ELEC| The maximum current that can be continuously carried without causing permanent deterioration of electrical or mechanical properties of a device or conductor. { 'kər·ənt ‚kar·ē·iŋ kə'pas·əd·ē }

current cell

current cell *See* active cell. { ˌkər·ənt 'sel }

current collector *See* charge collector. { 'kər·ənt kəˌlek·tər }

current comparator |ELEC| An instrument for determining the ratio of two direct or alternating currents, based on Ampère's laws, in which the two currents are passed through a toroid by two windings of known numbers of turns and the ampere-turn unbalance is measured by a detection winding. { 'kə·rənt kəmˌpar·əd·ər }

current-controlled switch |ELECTR| A semiconductor device in which the controlling bias sets the resistance at either a very high or very low value, corresponding to the "off" and "on" conditions of a switch. { 'kər·ənt kənˌtrōld 'swich }

current density |ELEC| The current per unit cross-sectional area of a conductor; a specialization of the physics definition. Also known as electric current density. { 'kər·ənt ˌden·səd·ē }

current divider |ELEC| A device used to deliver a desired fraction of a total current to a circuit. { 'kər·ənt diˌvīd·ər }

current drain |ELEC| The current taken from a voltage source by a load. Also known as drain. { 'kər·ənt ˌdrān }

current-equalizing reactor |ELEC| A reactor that is used to achieve a desired division of current between several circuits operating in parallel. { 'kər·ənt ˌē·kwəˌlīz·iŋ rē'ak·tər }

current feed |ELECTR| Feed to a point where current is a maximum, as at the center of a half-wave antenna. { 'kər·ənt ˌfēd }

current feedback |ELECTR| Feedback introduced in series with the input circuit of an amplifier. { 'kər·ənt ˌfēdˌbak }

current feedback circuit |ELECTR| A circuit used to eliminate effects of amplifier gain instability in an indirect-acting recording instrument, in which the voltage input (error signal) to an amplifier is the difference between the measured quantity and the voltage drop across a resistor. { 'kər·ənt ˌfēdˌbak ˌsər·kət }

current gain |ELECTR| The fraction of the current flowing into the emitter of a transistor which flows through the base region and out the collector. { 'kər·ənt ˌgān }

current generator |ELECTR| A two-terminal circuit element whose terminal current is independent of the voltage between its terminals. { 'kər·ənt ˌjen·əˌrād·ər }

current hogging |ELECTR| A condition in which the largest fraction of a current passes through one of several parallel logic circuits because it has a lower resistance than the others. { 'kər·ənt ˌhäg·iŋ }

current-instruction register *See* instruction register. { 'kər·ənt in'strak·shən ˌrej·ə·stər }

current intensity |ELEC| The magnitude of an electric current. Also known as current strength. { 'kər·ənt in'ten·səd·ē }

current interrupter |ELEC| Mechanism connected into a current-carrying line to periodically interrupt current flow to allow no-current tests of system components. { 'kər·ənt in·tə'rəp·tər }

current limiter |ELECTR| A device that restricts the flow of current to a certain amount, regardless of applied voltage. Also known as demand limiter. { 'kər·ənt ˌlim·əd·ər }

current-limiting reactor *See* series reactor. { 'kər·ənt ˌlim·əd·iŋ rē'ak·tər }

current-limiting resistor |ELEC| A resistor inserted in an electric circuit to limit the flow of current to some predetermined value; used chiefly to protect tubes and other components during warm-up. { 'kər·ənt ˌlim·əd·iŋ ri'zis·tər }

current location reference |COMPUT SCI| A symbolic expression, such as a star, which indicates the current location reached by the program; a transfer to * + 2 would bring control to the second statement after the current statement. { 'kər·ənt lō'kā·shən ˌref·rəns }

current loop *See* current antinode. { 'kər·ənt ˌlüp }

current margin |COMMUN| Difference between the steady-state currents flowing through a telegraph receiving instrument corresponding respectively to the two positions of the telegraph transmitter. { 'kər·ənt ˌmär·jən }

current measurement |ELEC| The measurement of the flow of electric current. { 'kər·ənt ˌmezh·ər·mənt }

current meter *See* ammeter; velocity-type flowmeter. { 'kər·ənt ˌmēd·ər }

current mirror |ELECTR| An electronic circuit that generates, at a high-impedance output node, an inflowing or outflowing current that is a scaled replica of an input current flowing into or out of a low-impedance input node. { 'kər·ənt ˌmir·ər }

current-mode filter |ELECTR| An integrated-circuit filter in which the signals are represented by current levels rather than voltage levels. { 'kər·əntˌmōd ˌfil·tər }

current-mode logic |ELECTR| Integrated-circuit logic in which transistors are paralleled so as to eliminate current hogging. Abbreviated CML. { 'kər·ənt ˌmōd 'läj·ik }

current node |ELEC| A point at which current is zero along a transmission line, antenna, or other circuit element having standing waves. { 'kər·ənt ˌnōd }

current noise |ELEC| Electrical noise of uncertain origin which is observed in certain resistances when a direct current is present, and which increases with the square of this current. { 'kər·ənt ˌnoiz }

current phasor |ELEC| A line referenced to a point, whose length and angle represent the magnitude and phase of a current. { 'kər·ənt ˌfā·zər }

current regulator |ELECTR| A device that maintains the output current of a voltage source at a predetermined, essentially constant value despite changes in load impedance. { 'kər·ənt ˌreg·yəˌlād·ər }

current relay |ELEC| A relay that operates at a specified current value rather than at a specified voltage value. { 'kər·ənt ˌrē,lā }

current saturation *See* anode saturation. { 'kər·ənt sach·əˈrā·shən }

current source |ELECTR| An electronic circuit that generates a constant direct current into or out of a high-impedance output node. { 'kər·ənt ˌsȯrs }

current strength *See* current intensity. { 'kər·ənt ˌstreŋkth }

current tap *See* multiple lamp holder; plug adapter lamp holder. { 'kər·ənt ˌtap }

current transformer |ELEC| An instrument transformer intended to have its primary winding connected in series with a circuit carrying the current to be measured or controlled; the current is measured across the secondary winding. { 'kər·ənt tranz'fȯr·mər }

current-transformer phase angle |ELEC| Angle between the primary current vector and the secondary current vector reversed; it is conveniently considered as positive when the reversed secondary current vector leads the primary current vector. { 'kər·ənt tranz'fȯr·mər 'fāz ˌaŋ·gəl }

current-voltage dual |ELEC| A circuit which is equivalent to a specified circuit when one replaces quantities with dual quantities; current and voltage impedance and admittance, and meshes and nodes are examples of dual quantities. { ¦kər·ənt ¦vōl·tij ¦dül }

cursor |COMPUT SCI| A movable spot of light that appears on the screen of a visual display terminal and can be positioned horizontally and vertically through keyboard controls to instruct the computer at what point a change is to be made. { 'kər·sər }

cursor arrows |COMPUT SCI| Arrows marked on keys of a computer keyboard that control the movement of the cursor. { 'kər·sər ˌar·ōz }

curtain array |ELECTROMAG| An antenna array consisting of vertical wire elements stretched between two suspension cables. { 'kərt·ən ə'rā }

curtain rhombic antenna |ELECTROMAG| A multiple-wire rhombic antenna having a constant input impedance over a wide frequency range; two or more conductors join at the feed and terminating ends but are spaced apart vertically from 1 to 5 feet (30 to 150 centimeters) at the side poles. { 'kərt·ən 'räm·bik an'ten·ə }

curvature effect |ELECTR| Generally, the condition in which the dielectric strength of a liquid or vacuum separating two electrodes is higher for electrodes of smaller radius of curvature. { 'kər·və·chər i'fekt }

curve follower |COMPUT SCI| A device in which a photoelectric, capacitive or inductive pick-off guided by a servomechanism reads data in the form of a graph, such as a curve drawn on paper with suitable ink. Also known as graph follower. { 'kərv ˌfäl·ə·wər }

curve tracer |ENG| An instrument that can produce a display of one voltage or current as a function of another voltage or current, with a third voltage or current as a parameter. { 'kərv ˌtrā·sər }

custom-designed device |ELECTR| An integrated logic circuit element that is generated

by a series of steps resembling photographic development from highly complicated artwork patterns. { ¦kəs·təm də'zīnd di'vīs }

customer substation |ELEC| A distribution substation located on the premises of a larger customer, such as a shopping center, commercial building, or industrial plant. { 'kəs·tə·mər 'səb ˌstā·shən }

cut and paste |COMPUT SCI| An editing function of a word processing system in which a portion of text is marked with a particular character at the beginning and at the end and is then copied to another location within the text. Also known as block move. { ¦kət ən 'pāst }

cut constraint |SYS ENG| A condition sometimes imposed in an integer programming problem which excludes parts of the feasible solution space without excluding any integer points. { 'kət kən'strānt }

CUT emulation *See* control unit terminal emulation. { 'kət ˌem·yə,lā·shən }

cut form |COMPUT SCI| In optical character recognition, any document form, receipt, or such, of standard dimensions which must be issued a separate read command in order to be recognized. { 'kət ¦fȯrm }

cut-in |CONT SYS| A value of temperature or pressure at which a control circuit closes. |ELEC| An electrical device that allows current to flow through an electric circuit. { 'kət ˌin }

cut-in angle |ELECTR| The phase angle at which a semiconductor diode begins to conduct; it is slightly greater than 0° because the diode requires some forward bias to conduct. { 'kət ˌin ˌaŋ·gəl }

Cutler feed |ELECTROMAG| A resonant cavity that transfers radio-frequency energy from the end of a waveguide to the reflector of a radar spinner assembly. { 'kət·lər ˌfēd }

cut methods |SYS ENG| Methods of solving integer programming problems that employ cut constraints derived from the original problem. { 'kət ˌmeth·ə, }

cutoff |ELECTR| **1.** The minimum value of bias voltage, for a given combination of supply voltages, that just stops output current in an electron tube, transistor, or other active device. **2.** *See* cutoff frequency. { 'kət,ȯf }

cutoff bias |ELECTR| The direct-current bias voltage that must be applied to the grid of an electron tube to stop the flow of anode current. { 'kət,ȯf ˌbī·əs }

cutoff field *See* critical field. { 'kət,ȯf ,fēld }

cutoff frequency |ELECTR| A frequency at which the attenuation of a device begins to increase sharply, such as the limiting frequency below which a traveling wave in a given mode cannot be maintained in a waveguide, or the frequency above which an electron tube loses efficiency rapidly. Also known as critical frequency; cutoff. { 'kət,ȯf ,frē·kwən·sē }

cutoff limiting |ELECTR| Limiting the maximum output voltage of a vacuum tube circuit by driving the grid beyond cutoff. { 'kət,ȯf ,lim·əd·iŋ }

cutoff voltage

cutoff voltage |ELECTR| **1.** The electrode voltage value that reduces the dependent variable of an electron-tube characteristic to a specified low value. **2.** *See* critical voltage. { 'kət,óf ,vól·tij }

cutoff wavelength |ELECTROMAG| **1.** The ratio of the velocity of electromagnetic waves in free space to the cutoff frequency in a uniconductor waveguide. **2.** The wavelength corresponding to the cutoff frequency. { 'kət,óf 'wāv ,leŋkth }

cut-out |CONT SYS| A value of temperature or pressure at which a control circuit opens. { 'kət ,aút }

cutout |ELEC| **1.** Pairs brought out of a cable and terminated at some place other than at the end of the cable. **2.** An electrical device that is used to interrupt the flow of current through any particular apparatus or instrument, either automatically or manually. Also known as electric cutout. { 'kət,aút }

cutout angle |ELECTR| The phase angle at which a semiconductor diode ceases to conduct; it is slightly less than 180° because the diode requires some forward bias to conduct. { 'kət ,aút ,aŋ·gəl }

cutout box |ELEC| A fireproof cabinet or box with one or more hinged doors that contains fuses and switches for various leads in an electrical wiring system. Also known as fuse box. { 'kət ,aút ,bäks }

cut-set |ELEC| A set of branches of a network such that the cutting of all the branches of the set increases the number of separate parts of the network, but the cutting of all the branches except one does not. { 'kət ,set }

cut-sheet printer |COMPUT SCI| A printer designed to print on separate sheets of paper. { 'kət ,shēt ¦print·ər }

cut-signal-branch operation |ELECTR| In systems where radio reception continues without cutting off the carrier, the cut-signal-branch operation technique disables a signal branch in one direction when it is enabled in the other to preclude unwanted signal reflections. { ¦kət ¦sig·nəl ¦branch ,äp·ə,rā·shən }

cutter |ENG ACOUS| An electromagnetic or piezoelectric device that converts an electric input to a mechanical output, used to drive the stylus that cuts a wavy groove in the highly polished wax surface of a recording disk. Also known as cutting head; head; phonograph cutter; recording head. { 'kəd·ər }

cutting head *See* cutter. { 'kəd·iŋ ,hed }

cutting stylus |ENG ACOUS| A recording stylus with a sharpened tip that removes material to produce a groove in the recording medium. { 'kəd·iŋ ,stī·ləs }

CW *See* continuous wave.

cyberspace |COMPUT SCI| The digital realms, including Web sites and virtual worlds. { 'sī·bər ,spās }

cycle-bound program *See* CPU-bound program. { 'sī·kəl ¦baúnd 'prō·grəm }

cycle count |COMPUT SCI| The operation of keeping track of the number of cycles a computer

system goes through during processing time. { 'sī·kəl ,kaúnt }

cycle criterion |COMPUT SCI| Total number of times a cycle in a computer program is to be repeated. { 'sī·kəl krī'tir·ē·ən }

cycle index |COMPUT SCI| **1.** The number of times a cycle has been carried out by a computer. **2.** The difference, or its negative, between the number of executions of a cycle which are desired and the number which have actually been carried out. { 'sī·kəl ,in,deks }

cycle index counter |COMPUT SCI| A device that counts the number of times a given cycle of instructions in a computer program has been carried out. { 'sī·kəl ,in,deks ,kaúnt·ər }

cycle-matching loran *See* low-frequency loran. { 'sī·kəl ,mach·iŋ ,lò'ran }

cycle reset |COMPUT SCI| The resetting of a cycle index to its initial or other specified value. { 'sī·kəl 'rē,set }

cycle skip *See* skip logging. { 'sī·kəl ,skip }

cycle stealing |COMPUT SCI| A technique for memory sharing whereby a memory may serve two autonomous masters, commonly a central processing unit and an input-output channel or device controller, and in effect provide service to each simultaneously. { 'sī·kəl ,stēl·iŋ }

cycle time |COMPUT SCI| The shortest time elapsed between one store (or fetch) and the next store (or fetch) in the same memory unit. Also known as memory cycle. { 'sī·kəl ,tīm }

cycle timer |ELECTR| A timer that opens or closes circuits according to a predetermined schedule. { 'sī·kəl ,tīm·ər }

cycle timing diagram |COMPUT SCI| A diagram showing the activity that occurs in each clock cycle of a computer during the execution of a machine-language instruction. { 'sī·kəl ¦tīm·iŋ ,dī·ə,gram }

cyclic code |COMPUT SCI| A code, such as a binary code, that changes only in one digit when going from one number to the number immediately following, and in that digit by only one unit. { 'sīk·lik 'kōd }

cyclic currents *See* mesh currents. { 'sīk·lik ¦kər·ənts }

cyclic feeding |COMPUT SCI| In character recognition, a system employed by character readers in which each input document is issued to the document transport in a predetermined and constant period of time. { 'sīk·lik 'fēd·iŋ }

cyclic redundancy check |COMPUT SCI| A block check character in which each bit is calculated by adding the first bit of a specified byte to the second bit of the next byte, and so forth, spiraling through the block; used to verify the correctness of data. Abbreviated CRC. { 'sīk·lik ri'dən·dən·sē ,chek }

cyclic shift |COMPUT SCI| A computer shift in which the digits dropped off at one end of a word are returned at the other end of the word. Also known as circuit shift; circular shift; end-around shift; nonarithmetic shift; ring shift. { 'sīk·lik 'shift }

cyclic storage |COMPUT SCI| A computer storage device, such as a magnetic drum, whose storage medium is arranged in such a way that information can be read into or extracted from individual locations at only certain fixed times in a basic cycle. { 'sīk·lik 'stȯr·ij }

cyclic transfer [COMPUT SCI] The automatic transfer of data from some medium to memory or from memory to some medium until all the data are read. { 'sīk·lik 'tranz·fər }

cycling [CONT SYS] A periodic change of the controlled variable from one value to another in an automatic control system. { 'sīk·liŋ }

cycloconverter [ELEC] A device that produces an alternating current of constant or precisely controllable frequency from a variable-frequency alternating-current input, with the output frequency usually one-third or less of the input frequency. { ¦sī·klō·kən'vərd·ər }

cyclomatic complexity [COMPUT SCI] A measure of the complexity of a software module, equal to $e - n + 2$, where e is the number of edges in the control flow graph and n is the number of nodes in this graph (that is, the cyclomatic number of the graph plus one). { ¦sī·klə‚mad·ik kəm'plek·səd·ē }

cyclophon See beam-switching tube. { 'sī·klə ‚fän }

cyclotron-frequency magnetron [ELECTR] A magnetron whose frequency of operation depends on synchronism between the alternating-current electric field and the electrons oscillating in a direction parallel to this field. { 'sī·klə‚trän ¦frē·kwən·sē 'mag·nə‚trän }

cyclotron-resonance maser See gyrotron. { 'sī·klə‚trän 'rez·ən·əns 'mā·zər }

cylinder [COMPUT SCI] **1.** The virtual cylinder represented by the tracks of equal radius of a set of disks on a disk drive. **2.** See seek area. { 'sil·ən·dər }

cylindrical antenna [ELECTROMAG] An antenna in which hollow cylinders serve as radiating elements. { sə'lin·drə·kəl an'ten·ə }

cylindrical array [ELECTR] An antenna, generally using electronic scanning, in which columns of radiating elements are arranged in a circle; used in some secondary radars. { sə'lin·drə·kəl ə'rā }

cylindrical capacitor [ELEC] A capacitor made of two concentric metal cylinders of the same length, with dielectric filling the space between the cylinders. Also known as coaxial capacitor. { sə'lin·drə·kəl kə'pas·əd·ər }

cylindrical-coordinate robot [CONT SYS] A robot in which the degrees of freedom of the manipulator arm are defined chiefly by cylindrical coordinates. { sə'lin·drə·kəl kō¦ȯrd·ən·ət 'rō ‚bät }

cylindrical-film storage [ELECTR] A computer storage in which each storage element consists of a short length of glass tubing having a thin film of nickel-iron alloy on its outer surface. { sə'lin·drə·kəl 'film ‚stȯr·ij }

cylindrical pinch See pinch effect. { sə'lin·drə·kəl 'pinch }

cylindrical winding [ELEC] The current-carrying element of a core-type transformer, consisting of a single coil of one or more layers wound concentrically with the iron core. { sə'lin·drə·kəl 'wīnd·iŋ }

D

DAB _See_ digital audio broadcasting.

DABS _See_ Mode S. { dabz _or_ ,dē,ā,bē¹es }

dac _See_ digital-to-analog converter.

DAC _See_ digital-to-analog converter.

daemon |COMPUT SCI| In Unix, a program that runs in the background, such as a server. { 'dē·mən }

Dahlin's algorithm |CONT SYS| A digital control algorithm in which the requirement of minimum response time used in the deadbeat algorithm is relaxed to reduce ringing in the system response. { 'dä·lənz ,al·gə,rith·əm }

daily keying element |COMMUN| Part of a specific cipher key that changes at predetermined intervals, usually daily. { ¦dā·lē ,kē·iŋ ,el·ə·mənt }

daisy chain |COMPUT SCI| A means of connecting devices (readers, printers, and so on) to a central processor by party-line input/output buses which join these devices by male and female connectors, the last female connector being shorted by a suitable line termination. { 'dāz·ē ,chān }

daisy wheel printer |COMPUT SCI| A serial printer in which the printing element is a plastic hub that has a large number of flexible radial spokes, each spoke having one or more different raised printing characters; the wheel is rotated as it is moved horizontally step by step under computer control, and stops when a desired character is in a desired print position so a hammer can drive that character against an inked ribbon. { 'dāz·ē ,wēl ,print·ər }

damaged pack |COMPUT SCI| A disk drive whose use is impaired by physical damage such as a scratch on the recording surface or by a serious software error that renders control information on the disk unreadable. { 'dam·ijd 'pak }

damper |ELECTR| A diode used in the horizontal deflection circuit of a CRT display device to make the sawtooth deflection current decrease smoothly to zero instead of oscillating at zero; the diode conducts each time the polarity is reversed by a current swing below zero. { 'dam·pər }

damper winding |ELEC| A winding consisting of several conducting bars on the field poles of a synchronous machine, short-circuited by conducting rings or plates at their ends, and used to prevent pulsating variations of the position or magnitude of the magnetic field linking the poles. Also known as amortisseur winding. { 'dam·pər ,wīnd·iŋ }

damping coefficient _See_ resistance. { 'dam·piŋ ,kō·i,fish·ənt }

damping constant _See_ resistance. { 'dam·piŋ ,kän·stənt }

damping resistor |ELEC| **1.** A resistor that is placed across a parallel resonant circuit or in series with a series resonant circuit to decrease the Q factor and thereby eliminate ringing. **2.** A noninductive resistor placed across an analog meter to increase damping. { 'dam·piŋ ri,zis·tər }

dance-hall machine |COMPUT SCI| A multiprocessor in which the memory is spread over several modules, and a switch is used to make connections between memory modules and processors, so that several processors can use the memory simultaneously. { 'dans ,hȯl mə ,shēn }

dangling ELSE |COMPUT SCI| A situation in which it is not clear to which part of a compound conditional statement an ELSE instruction belongs. { ¦daŋ·gliŋ 'els }

daraf |ELEC| The unit of elastance, equal to the reciprocal of 1 farad. { 'da,raf }

dark conduction |ELECTR| Residual conduction in a photosensitive substance that is not illuminated. { ¦därk kən¦dək·shən }

dark current _See_ electrode dark current. { 'därk ,kər·ənt }

dark-current pulse |ELECTR| A phototube dark-current excursion that can be resolved by the system employing the phototube. { 'därk ,kər·ənt ¦pəls }

dark discharge |ELECTR| An invisible electrical discharge in a gas. { ¦därk 'dis,chärj }

dark resistance |ELECTR| The resistance of a selenium cell or other photoelectric device in total darkness. { 'därk ri,zis·təns }

dark space |ELECTR| A region in a glow discharge that produces little or no light. { 'därk ,spās }

dark spot |ELECTR| A spot on a television receiver tube that results from a spurious signal generated in the television camera tube during rescan, generally from the redistribution of secondary electrons over the mosaic in the tube. { 'därk ,spät }

dark-trace tube |ELECTR| A cathode-ray tube with a bright face that does not necessarily luminesce, on which signals are displayed as dark traces or dark blips where the potassium chloride screen is hit by the electron beam. Also known as skiatron. { 'därk ,träs ,tüb }

Darlington amplifier |ELECTR| A current amplifier consisting essentially of two separate transistors and often mounted in a single transistor housing. { 'dar·liŋ·tən ,am·plə,fī·ər }

DARS See direct audio radio service. { ¦dē¦ä¦är'es *or* därz }

d'Arsonval current |ELEC| A current consisting of isolated trains of heavily damped high-frequency oscillations of high voltage and relatively low current, used in diathermy. { 'dars·ən ,völ ,kar·ənt }

d'Arsonval galvanometer |ENG| A galvanometer in which a light coil of wire, suspended from thin copper or gold ribbons, rotates in the field of a permanent magnet when current is carried to it through the ribbons; the position of the coil is indicated by a mirror carried on it, which reflects a light beam onto a fixed scale. Also known as light-beam galvanometer. { 'dars·ən ,völ gal·və'näm·əd·ər }

DASD See direct-access storage device. { 'daz,dē }

DAT See digital audio tape.

data |COMPUT SCI| **1.** General term for numbers, letters, symbols, and analog quantities that serve as input for computer processing. **2.** Any representations of characters or analog quantities to which meaning, if not information, may be assigned. { 'dad·ə, 'dād·ə, *or* 'däd·ə }

data acquisition |COMMUN| The phase of data handling that begins with the sensing of variables and ends with a magnetic recording or other record of raw data; may include a complete radio telemetering link. { 'dad·ə ,ak·wə ,zish·ən }

data acquisition computer |COMPUT SCI| A computer that is used to acquire and analyze data generated by instruments. { 'dad·ə ,ak·wə ,zish·ən kəm'pyüd·ər }

data aggregate |COMPUT SCI| The set of data items within a record. { 'dad·ə ,ag·rə·gət }

data analysis |COMPUT SCI| The evaluation of digital data. { 'dad·ə ə,nal·ə·səs }

data attribute |COMPUT SCI| A characteristic of a block of data, such as the type of representation used or the length in characters. { 'dad·ə ¦a·trə'byüt }

data automation |COMPUT SCI| The use of electronic, electromechanical, or mechanical equipment and associated techniques to automatically record, communicate, and process data and to present the resultant information. { ¦dad·ə öd·ə'mā·shən }

data bank |COMPUT SCI| A complete collection of information such as contained in automated files, a library, or a set of computer disks. { 'dad·ə ,baŋk }

database |COMPUT SCI| A nonredundant collection of interrelated data items that can be shared and used by several different subsystems. { 'dad·ə,bās }

database/data communication |COMPUT SCI| An advanced software product that combines a database management system with data communications procedures. Abbreviated DB/DC. { 'dad·ə,bās 'dad·ə kə,myü·nə'kā·shən }

database machine |COMPUT SCI| A computer that handles the storage and retrieval of data into and out of a database. { 'dad·ə,bās mə,shēn }

database management system |COMPUT SCI| A special data processing system, or part of a data processing system, which aids in the storage, manipulation, reporting, management, and control of data. Abbreviated DBMS. { 'dad·ə,bās 'man·ij·mənt ,sis·təm }

database server |COMPUT SCI| An independently functioning computer in a local-area network that holds and manages the database. { 'dad·ə,bās ,sər·vər }

data break |COMPUT SCI| A facility which permits input/output transfers to occur without disturbing program execution in a computer. { 'dad·ə ,brāk }

data buffering |COMPUT SCI| The temporary collection and storage of data awaiting further processing in physical storage devices, allowing a computer and its peripheral devices to operate at different speeds. { 'dad·ə ,bəf·ə·riŋ }

data bus |ELECTR| An internal channel that carries data between a computer's central processing unit and its random-access memory. { 'dad·ə ,bəs }

data capture |COMPUT SCI| The acquisition of data to be entered into a computer. { 'dad·ə ,kap·chər }

data carrier |COMPUT SCI| A medium on which data can be recorded, and which is usually easily transportable, such as disks or tape. { 'dad·ə ,kar·ē·ər }

data carrier storage |COMPUT SCI| Any type of storage in which the storage medium is outside the computer, such as disks and tape, in contrast to inherent storage. { 'dad·ə ,kar·ē·ər ,stór·ij }

data cartridge |COMPUT SCI| A tape cartridge used for nonvolatile and removable data storage in small digital systems. { 'dad·ə ,kar·trij }

data cell drive |COMPUT SCI| A large-capacity storage device consisting of strips of magnetic tape which can be individually transferred to the read-write head. { 'dad·ə ,sel ,drīv }

data center |COMPUT SCI| An organization established primarily to acquire, analyze, process, store, retrieve, and disseminate one or more types of data. { 'dad·ə ,sen·tər }

data chain |COMPUT SCI| Any combination of two or more data elements, data items, data codes, and data abbreviations in a prescribed sequence to yield meaningful information; for example, "date" consists of data elements year, month, and day. { 'dad·ə ,chān }

data chaining |COMPUT SCI| A technique used in scatter reading or scatter writing in which new storage areas are defined for use as soon as the current data transfer is completed. { 'dad·ə ,chān·iŋ }

data channel |COMPUT SCI| A bidirectional data path between input/output devices and the main memory of a digital computer permitting one or more input/output operations to proceed concurrently with computation. { 'dad·ə ,chan·əl }

data circuit |ELECTR| A telephone facility that allows transmission of digital data pulses with minimum distortion. { 'dad·ə ,sər·kət }

data code |COMPUT SCI| A number, letter, character, symbol, or any combination thereof, used to represent a data item. { 'dad·ə ,kōd }

data collection |COMPUT SCI| The process of sending data to a central point from one or more locations. { 'dad·ə kə,lek·shən }

data communication network |COMPUT SCI| A set of nodes, consisting of computers, terminals, or some type of communication control units in various locations, connected by links consisting of communication channels providing a data path between the nodes. { 'dad·ə kə,myü·nə,kā·shən 'net,wərk }

data communications |COMMUN| The conveying from one location to another of information that originates or is recorded in alphabetic, numeric, or pictorial form, or as a signal that represents a measurement; includes telemetering and facsimile but not voice or television. Also known as data transmission. { 'dad·ə kə,myü·nə'kā·shənz }

data communications processor |COMPUT SCI| A small computer used to control the flow of data between machines and terminals over communications channels. { 'dad·ə kə,myü·nə ¦kā·shənz 'präs,es·ər }

data compression |COMPUT SCI| Reduction in the number of bits used to represent an item of data. Also known as compression. { 'dad·ə kəm ,presh·ən }

data concentrator |ELECTR| A device, such as a microprocessor, that takes data from several different teletypewriter or other slow-speed lines and feeds them to a single higher-speed line. { 'dad·ə kän·sən,trād·ər }

data conversion |COMPUT SCI| The changing of the representation of data from one form to another, as from binary to decimal, or from one physical recording medium to another (as from tape to disk), or from one file format to another, or from one programming language to another. Also known as conversion. { 'dad·ə kən,vər·zhən }

data conversion line |COMPUT SCI| The channel, electronic or manual, through which data elements are transferred between data banks. { 'dad·ə kən,vər·zhən ,līn }

data converter *See* converter. { 'dad·ə kən ,vərd·ər }

data definition |COMPUT SCI| The statements in a computer program that specify the physical attributes of the data to be processed, such as location and quantity of data. { 'dad·ə ,def·ə'nish·ən }

data dependence graph |COMPUT SCI| A chart that represents a program in a data flow language, in which each node is a function and each arc carries a value. { 'dad·ə di,pen·dəns ,graf }

data description language |COMPUT SCI| A programming language used to specify the arrangement of data items within a database. { 'dad·ə di¦skrip·shən ,laŋ·gwij }

data descriptor |COMPUT SCI| A pointer indicating the memory location of a data item. { 'dad·ə di'skrip·tər }

data dictionary |COMPUT SCI| A catalog which contains the names and structures of all data types. { 'dad·ə ,dik·shə,ner·ē }

data display |COMPUT SCI| Visual presentation of processed data by specially designed electronic or electromechanical devices, such as video monitors, through interconnection (either on- or off-line) with digital computers or component equipments. { 'dad·ə di,splā }

data distribution |COMPUT SCI| Data transmission to one or more locations from a central point. { 'dad·ə ,dis·trə,byü·shən }

data division |COMPUT SCI| The section of a program (written in the COBOL language) which describes each data item used for input, output, and storage. { 'dad·ə di,vizh·ən }

data-driven execution |COMPUT SCI| A mode of carrying out a program in a data flow system, in which an instruction is carried out whenever all its input values are present. { 'dad·ə ,driv·ən ,ek·sə'kyü·shən }

data element |COMPUT SCI| A set of data items pertaining to information of one kind, such as months of a year. |COMMUN| An item of data as represented before encoding and after decoding. { 'dad·ə ,el·ə·mənt }

data encryption standard |COMMUN| A cryptographic algorithm of validated strength which is in the public domain and is accepted as a standard. Abbreviated DES. { 'dad·ə en,krip·shən 'stan·dərd }

data entry |COMPUT SCI| The procedures for placing data in a computer system. { 'dad·ə ,en·trē }

data entry program |COMPUT SCI| An application program that receives data from a keyboard or other input device and stores it in a computer system. Also known as input program. { 'dad·ə ¦en·trē ,prō·grəm }

data entry terminal |COMPUT SCI| A portable keyboard and small numeric display designed for interactive communication with a computer. { 'dad·ə ¦en·trē ,tər·mən·əl }

data error |COMPUT SCI| A deviation from correctness in data, usually an error, which occurred prior to processing the data. { 'dad·ə ,er·ər }

data exchange system |COMPUT SCI| A combination of hardware and software designed to accept data from various sources, sort the data according to its destination and priority, carry out any necessary code conversions, and transmit the data to its destination. { 'dad·ə iks¦chānj ,sis·təm }

data expansion |COMPUT SCI| The reproduction in its original form of information that has undergone data compression. { 'dad·ə ik,span·chən }

data field |COMPUT SCI| An area in the main memory of the computer in which a data record is contained. { 'dad·ə ,fēld }

data flow |COMMUN| The route followed by a data message from its origination to its destination, including all the nodes through which it travels. |COMPUT SCI| The transfer of data from an external storage device, through the processing unit and memory, and out to an external storage device.   { 'dad·ə ¦flō }

data flow analysis |COMPUT SCI| The development of models for the movement of information within an organization, indicating the sources and destinations of information and where and how information is transmitted, processed, and stored.   { 'dad·ə ¦flō ə¦nal·ə·səs }

data flow diagram |COMPUT SCI| A chart that traces the movement of data in a computer system and shows how the data is to be processed, using circles to represent data. Also known as bubble chart; system flowchart.   { 'dad·ə ¦flō ¸dī·ə¸gram }

data flow language |COMPUT SCI| A programming language used in a data flow system.   { 'dad·ə ¦flō ¸laŋ·gwij }

data flow system |COMPUT SCI| An alternative to conventional programming languages and architectures which is able to achieve a high degree of parallel computation, in which values rather than value containers are dealt with, and in which all processing is achieved by applying functions to values to produce new values.   { 'dad·ə ¦flō ¸sis·təm }

data flow technique |COMPUT SCI| A method of computer system design in which diagrams and charts that show how data is to be handled by the system are used to prepare detailed specifications from which actual programs can be written.   { 'dad·ə ¦flō tek¸nēk }

data formatting |COMPUT SCI| Structuring the presentation of data as numerical or alphabetic and specifying the size and type of each datum.   { 'dad·ə fór'mad·iŋ }

data fusion |ELECTR| The combining of data as from several radars or other sensors with common fields of view, in order to improve the accuracy of the estimations being made about features of interest.   { 'dad·ə ¸fyü·zhən }

data generator |COMPUT SCI| A specialized word generator in which the programming is designed to test a particular class of device, the pulse parameters and timing are adjustable, and selected words may be repeated, reinserted later in the sequence, omitted, and so forth.   { 'dad·ə ¸jen·ə ¸rād·ər }

datagram |COMPUT SCI| A unit of information in the Internet Protocol (IP) containing both data and address information. In TCP/IP networks, datagrams are referred to as packets.   { 'dad·ə ¸gram }

data-handling system |COMPUT SCI| Automatically operated equipment used to interpret data gathered by instrument installations. Also known as data reduction system.   { 'dad·ə ¸hand·liŋ ¸sis·təm }

data independence |COMPUT SCI| Separation of data from processing, either so that changes in the size or format of the data elements require no change in the computer programs processing them or so that these changes can be made automatically by the database management system.   { 'dad·ə in·də'pen·dəns }

data-initiated control |COMPUT SCI| The automatic handling of a program dependent only upon the value of input data fed into the computer.   { 'dad·ə i¸nish·ē¸äd·əd kən'trōl }

data-intense application |COMPUT SCI| A program or computer system that handles large quantities of data and extremely repetitive tasks.   { 'dad·ə in¦tens ¸ap·lə'kā·shən }

data interchange |COMPUT SCI| Switching of data in and out of storage units.   { 'dad·ə 'in·tər ¸chānj }

data item |COMPUT SCI| A single member of a data element. Also known as datum.   { 'dad·ə¸ī·dəm }

data level |COMPUT SCI| The rank of a data element in a source language with respect to other elements in the same record.   { 'dad·ə ¸lev·əl }

data library |COMPUT SCI| A center for the storage of data not in current use by the computer.   { 'dad·ə lī¸brer·ē }

data line |COMMUN| An individual circuit that transmits data within a communications or computer channel.   { 'dad·ə ¸līn }

data line monitor |COMMUN| A test instrument that analyzes the signals transmitted over a communications line and provides a visual display or stores the results for further analysis, or both.   { ¦dad·ə ¸līn 'män·əd·ər }

data link |COMMUN| The physical equipment for automatic transmission and reception of information. Also known as communication link; information link; tie line; tie-link.   { 'dad·ə ¸liŋk }

data logging |COMPUT SCI| Conversion of electrical impulses from process instruments into digital data to be recorded, stored, and periodically tabulated.   { 'dad·ə ¸läg·iŋ }

data management |COMPUT SCI| The collection of functions of a control program that provide access to data sets, enforce data storage conventions, and regulate the use of input/output devices.   { 'dad·ə ¸man·ij·mənt }

data management program |COMPUT SCI| A computer program that keeps track of what is in a computer system and where it is located, and of the various means to store and access the data efficiently.   { 'dad·ə ¸man·ij·mənt ¸prō·grəm }

data manipulation |COMPUT SCI| The standard operations of sorting, merging, input/output, and report generation.   { 'dad·ə mə¸nip·yə¸lā·shən }

data manipulation language |COMPUT SCI| The interface between a data base and an applications program, which is embedded in the language of the applications program and provides the programmer with procedures for accessing data in the data base.   { 'dad·ə mə¸nip·yə¦lā· shən ¸laŋ·gwij }

data mining |COMPUT SCI| **1.** The identification or extraction of relationships and patterns from data using computational algorithms to reduce, model, understand, or analyze data. **2.** The automated process of turning raw data into useful information by which intelligent computer systems sift and sort through data, with little or no help from humans, to look for patterns or to predict trends.   { 'dad·ə ¸mīn·iŋ }

data module [COMPUT SCI] A sealed disk drive unit that includes mechanical and electronic components for handling data stored on the disk. { 'dad·ə ,mäj·yül }

data move instruction [COMPUT SCI] An instruction in a computer program to transfer data between memory locations and registers or between the central processor and peripheral devices. { 'dad·ə ,müv in'strək·shən }

data name [COMPUT SCI] A symbolic name used to represent an item of data in a source program, in place of the address of the data item. { 'dad·ə ,nām }

data organization [COMPUT SCI] Any one of the data management conventions for physical and spatial arrangement of the physical records of a data set. Also known as data set organization. { 'dad·ə ,ȯr·gə·nə,zā·shən }

data origination [COMPUT SCI] The process of putting data in a form that can be read by a machine. { 'dad·ə ə,rij·ə'nā·shən }

data patch panel [COMMUN] A plugboard used to rearrange communications lines and modems by connecting them with double-ended cables, or to attach monitoring devices to analyze circuit signals. { 'dad·ə 'pach ,pan·əl }

data plotter [COMPUT SCI] A device which plots digital information in a continuous fashion. { 'dad·ə ,pläd·ər }

data processing [COMPUT SCI] Any operation or combination of operations on data, including everything that happens to data from the time they are observed or collected to the time they are destroyed. Also known as information processing. { 'dad·ə 'präs,es·iŋ }

data processing center [COMPUT SCI] A computer installation providing data processing service for others, sometimes called customers, on a reimbursable or nonreimbursable basis. { 'dad·ə ¦präs,es·iŋ ,sent·ər }

data processing inventory [COMPUT SCI] An identification of all major data processing areas in an agency for the purpose of selecting and focusing upon those in which the use of automatic data processing (ADP) techniques appears to be potentially advantageous, establishing relative priorities and schedules for embarking on ADP studies, and identifying significant relationships among areas to pinpoint possibilities for the integration of systems. { 'dad·ə ¦präs,es·iŋ ,in·vən ,tȯr·ē }

data processor [COMPUT SCI] **1.** Any device capable of performing operations on data, for instance, a desk calculator, an analog computer, or a digital computer. **2.** Person engaged in processing data. { 'dad·ə 'präs,es·ər }

data protection [COMPUT SCI] The safeguarding of data against unauthorized access or accidental or deliberate loss or damage. { 'dad·ə prə,tek·shən }

data purification [COMPUT SCI] The process of removing as many inaccurate or incorrect items as possible from a mass of data before automatic data processing is begun. { 'dad·ə pyür·ə·fə'kā·shən }

data rate [COMMUN] The number of digital bits per second that are recorded or retrieved from a data storage device during the transfer of a large data block. { 'dad·ə ,rāt }

data record [COMPUT SCI] A collection of data items related in some fashion and usually contiguous in location. { 'dad·ə ,rek·ərd }

data recorder [COMPUT SCI] A keyboard device for entering data onto magnetic tape. { 'dad·ə ri,kȯr·dər }

data reduction [COMPUT SCI] The transformation of raw data into a more useful form. { 'dad·ə ri ,dək·shən }

data reduction system See data-handling system. { ,dad·ə ri,dək·shən ,sis·təm }

data redundancy [COMPUT SCI] The occurrence of values for data elements more than once within a file or database. { 'dad·ə ri,dən·dən·sē }

data register [COMPUT SCI] A register used in microcomputers to temporarily store data being transmitted to or from a peripheral device. { 'dad·ə ,rej·ə·stər }

data representation [COMPUT SCI] **1.** The way that the physical properties of a medium are used to represent data. **2.** The manner in which data is expressed symbolically by binary digits in a computer. { 'dad·ə ,rep·ri·zen'tā·shən }

data retrieval [COMPUT SCI] The searching, selecting, and retrieving of actual data from a personnel file, data bank, or other file. { 'dad·ə ri'trē·vəl }

data rules [COMPUT SCI] Conditions which must be met by data to be processed by a computer program. { 'dad·ə ,rülz }

data scope [ELECTR] An electronic display that shows the content of the information being transmitted over a communications channel. { 'dad·ə ,skōp }

data security [COMPUT SCI] The protection of data against the deliberate or accidental access of unauthorized persons. Also known as file security. { 'dad·ə sə,kyür·əd·ē }

data set [COMPUT SCI] **1.** A named collection of similar and related data records recorded upon some computer-readable medium. **2.** A data file in IBM 360 terminology. { 'dad·ə ,set }

data set coupler [COMPUT SCI] The interface between a parallel computer input/output bus and the serial input/output of a modem. { 'dad·ə ,set ,kəp·lər }

data set label [COMPUT SCI] A data element that describes a data set, and usually includes the name of the data set, its boundaries in physical storage, and certain characteristics of data items within the set. { 'dad·ə ,set ,lā·bəl }

data set migration [COMPUT SCI] The process of moving inactive data sets from on-line storage to back up storage in a time-sharing environment. { 'dad·ə ,set mī,grā·shən }

data set organization See data organization. { 'dad·ə ,set ,ȯr·gə·nə,zā·shən }

data sink [COMPUT SCI] A memory or recording device capable of accepting data signals from a data transmission device and storing data for future use. { 'dad·ə ,siŋk }

data source |COMPUT SCI| A device capable of originating data signals for a data transmission device. { 'dad·ə ,sȯrs }

data stabilization |ELECTR| Stabilization of the display of radar signals with respect to a selected reference, regardless of changes in radar-carrying vehicle attitude, as in azimuth-stabilized plan-position indicator. { 'dad·ə ,stā·bə·lə,zā·shən }

data statement |COMPUT SCI| An instruction in a source program that identifies an item of data in the program and specifies its format. { 'dad·ə ,stāt·mənt }

data station |COMPUT SCI| A remote input/output device which handles a variety of transmissions to and from certain centralized computers. { 'dad·ə ,stā·shən }

data station control |COMPUT SCI| The supervision of a data station by means of a program resident in the central computer. { 'dad·ə ,stā·shən kən,trōl }

data stream |COMMUN| The continuous transmission of data from one location to another. { 'dad·ə ,strēm }

data striping See disk striping. { 'dad·ə ,strīp·iŋ }

data structure |COMPUT SCI| A collection of data components that are constructed in a regular and characteristic way. { 'dad·ə ,strək·chər }

data switch |COMPUT SCI| A manual or automatic device that connects data-processing machines to one another. { 'dad·ə ,swich }

data system |COMPUT SCI| The means, either manual or automatic, of converting data into action or decision information, including the forms, procedures, and processes which together provide an organized and interrelated means of recording, communicating, processing, and presenting information relative to a definable function or activity. { 'dad·ə ,sis·təm }

data system interface |COMPUT SCI| **1.** A common aspect of two or more data systems involving the capability of intersystem communications. **2.** A common boundary between automatic data-processing systems or parts of a single system. { 'dad·ə ¦sis·təm 'in·tər,fās }

data systems integration |COMPUT SCI| Achievement through systems design of an improved or broader capability by functionally or technically relating two or more data systems, or by incorporating a portion of the functional or technical elements of one data system into another. { 'dad·ə ,sis·təmz ,in·tə,grā·shən }

data system specifications |COMPUT SCI| **1.** The delineation of the objectives which a data system is intended to accomplish. **2.** The data processing requirements underlying that accomplishment; includes a description of the data output, the data files and record content, the volume of data, the processing frequencies, training, and such other facts as may be necessary to provide a full description of the system. { 'dad·ə ,sis·təm ,spes·ə·fə,kā·shən }

data table |COMPUT SCI| An on-screen display of the information in a database management system, presented in columnar format, with field names at the top. { 'dad·ə ,tā·bəl }

data tablet See electronic tablet. { 'dad·ə ,tab·lət }

data tracks |COMPUT SCI| Information storage positions on drum storage devices; information is stored on the drum surface in the form of magnetized or nonmagnetized areas. { 'dad·ə ,traks }

data transcription equipment |COMPUT SCI| Those devices or equipment designed to convey data from its original state to a data processing media. { 'dad·ə tranz,krip·shən i,kwip·mənt }

data transfer |COMPUT SCI| The technique used by the hardware manufacturer to transmit data from computer to storage device or from storage device to computer; usually under specialized program control. { 'dad·ə 'tranz·fər }

data transmission See data communications. { 'dad·ə tranz'mish·ən }

data transmission equipment |COMPUT SCI| The communications equipment used in direct support of data processing equipment. { 'dad·ə tranz'mish·ən i,kwip·mənt }

data transmission line |ELEC| A system of electrical conductors, such as a coaxial cable or pair of wires, used to send information from one place to another or one part of a system to another. { 'dad·ə tranz'mish·ən ,līn }

data transmission-utilization measure |COMMUN| The ratio of useful data output to the sum total of data input. { ¦dad·ə tranz'mish·ən yüd·əl·ə'zā·shən ,mezh·ər }

data type |COMPUT SCI| The manner in which a sequence of bits represents data in a computer program. { 'dad·ə ,tīp }

data under voice |COMMUN| A telephone digital data service that allows digital signals to travel on the lower portion of the frequency spectrum of existing microwave radio systems; digital channels initially available handled speeds of 2.4, 4.8, 9.6, and 56 kilobits per second. Abbreviated DUV. { 'dad·ə ,ən·dər 'vȯis }

data unit |COMPUT SCI| A set of digits or characters treated as a whole. { 'dad·ə ,yü·nət }

data validation |COMPUT SCI| The checking of data for correctness, or the determination of compliance with applicable standards, rules, and conventions. { 'dad·ə val·ə'dā·shən }

data warehouse |COMPUT SCI| **1.** A large specialized database, holding perhaps hundreds of terabytes of data. **2.** A database specifically structured for information access and reporting. { ¦dad·ə 'wer,haüs }

data word |COMPUT SCI| A computer word that is part of the data which the computer is manipulating, in contrast with an instruction word. Also known as information word. { 'dad·ə ,wərd }

date time group |COMMUN| The date and time, expressed in digits and zone suffix, at which the message was prepared for transmission (expressed as six digits followed by the zone suffix; first pair of digits denoting the date, second pair the hours, third pair the minutes). { ¦dāt ¦tīm ,grüp }

datum See data item. { 'dad·əm, 'dād·əm, or 'däd·əm }

daughter board |COMPUT SCI| A small printed circuit board that is attached to another printed circuit board. { 'dȯd·ər ˌbȯrd }

Davisson-Calbick formula |ELECTR| A formula which states that the focal length of a simple electrostatic lens consisting of a circular hole in a conducting plate is equal to four times the potential of the plate divided by the difference in the potential gradients on either side of the plate. { ¦da·və·sən 'kal·bik ˌfȯr·myə·lə }

day clock |COMPUT SCI| An internal binary counter, with a resolution usually of a microsecond and a cycle measured in years, providing an accurate measure of elapsed time independent of system activity. { 'dā ˌkläk }

daylight controls |ENG| Special devices which automatically control the electric power to a lamp, causing the light to operate during hours of darkness and to be extinguished during daylight hours. { 'dāˌlīt kən'trōlz }

daylight lamp |ELEC| An incandescent or fluorescent lamp that emits light whose spectral distribution is approximately that of daylight. { 'dāˌlīt ˌlamp }

dBa See adjusted decibel.

DB/DC See database/data communication.

dBf See decibels above 1 femtowatt.

dBk See decibels above 1 kilowatt.

dBm See decibels above 1 milliwatt.

DBMS See database management system.

dBp See decibels above 1 picowatt.

dBrn See decibels above reference noise.

DBRT diode See double-barrier resonant tunneling diode. { ¦dē¦bē¦är¦tē 'dī‚ōd }

DB server |COMPUT SCI| The database portion of a Web server, which serves as a repository of data and content. { ¦dē¦bē ˌsər·vər }

DBS system See direct broadcasting satellite system. { ¦dē¦bē 'es ˌsis·təm }

dBV See decibels above 1 volt.

dBW See decibels above 1 watt.

dBx See decibels above reference coupling.

dc See direct current.

D cable |ELEC| Two-conductor cable, each conductor having the shape of the letter D, with insulation between the conductors and between the conductors and the sheath. { 'dē ˌkā·bəl }

DCFL See direct-coupled FET logic.

DCT See discrete cosine transform.

DCTL See direct-coupled transistor logic.

dc-to-ac converter See inverter. { ¦dē‚sē tü ¦ā‚sē kən'vərd·ər }

dc-to-ac inverter See inverter. { ¦dē‚sē tü ¦ā‚sē in'vərd·ər }

dc-to-dc converter |ELEC| An electronic circuit which converts one direct-current voltage into another, consisting of an inverter followed by a step-up or step-down transformer and rectifier. { ¦dē‚sē tü ¦dē‚sē kən'vərd·ər }

dcwv See direct-current working volts.

DDA See digital differential analyzer.

D-display |ELECTR| A radar display format in which the coordinates are the same as in the C-display, with target spots extended vertically to indicate range. Also known as D-indicator; D-scan; D-scope. { 'dē di‚splā }

DDR See double data rate.

DDS See digital data service.

deaccentuator |ELECTR| A circuit used in a frequency-modulation receiver to offset the preemphasis of higher audio frequencies introduced at the transmitter. { ˌdē·ak'sen·chə‚wād·ər }

dead |ELEC| Free from any electric connection to a source of potential difference from electric charge; not having a potential different from that of earth; the term is used only with reference to current-carrying parts which are sometimes alive or charged. { ded }

dead band |ELEC| The portion of a potentiometer element that is shortened by a tap; when the wiper traverses this area, there is no change in output. |ENG| The range of values of the measured variable to which an instrument will not effectively respond. Also known as dead zone; neutral zone. { 'ded ‚band }

deadbeat algorithm |CONT SYS| A digital control algorithm which attempts to follow set-point changes in minimum time, assuming that the controlled process can be modeled approximately as a first-order plus dead-time system. { 'ded‚bēt 'al·gə‚rith·əm }

dead-center position |ELEC| Position in which a brush would be placed on the commutator of a direct-current motor or generator if the field flux were not distorted by armature reaction. { ¦ded 'sen·tər pə'zish·ən }

dead code |COMPUT SCI| Statements in a computer program that are not executed, usually as the result of modification of a large program. { 'ded 'kōd }

dead earth |ELEC| A connection between a line conductor and earth by means of a path of low resistance. { ¦ded 'ərth }

dead end |ELEC| The portion of a tapped coil through which no current is flowing at a particular switch position. { 'ded ‚end }

dead-end effect |ELEC| Absorption of energy by unused portions of a tapped coil. { 'ded ‚end i'fekt }

dead-end switch |ELEC| A switch used to short-circuit unused portions of a tapped coil to prevent dead-end effects. { 'ded ‚end ‚swich }

dead ground |ELEC| A low-resistance connection between the ground and an electric circuit. { ¦ded 'graünd }

dead halt See drop-dead halt. { ¦ded 'hȯlt }

dead letter box |COMMUN| A file for storing undeliverable messages in a data communications system, particularly a message switching system. { ¦ded 'led·ər ‚bäks }

deadlock |COMPUT SCI| A situation in which a task in a multiprogramming system cannot proceed because it is waiting for an event that will never occur. Also known as deadly embrace; interlock; knot. { 'ded‚läk }

deadman switch |ELEC| An electrical switch that activates some function if it is turned off. { 'ded ‚man ‚swich }

dead short |ELEC| A short-circuit path that has extremely low resistance. { ¦ded'shȯrt }

dead spot |COMMUN| A geographic location in which signals from a radio or television transmitter are received poorly or not at all. { 'ded ‚spät }

dead time |CONT SYS| The time interval between a change in the input signal to a process control system and the response to the signal. |ENG| The time interval, after a response to one signal or event, during which a system is unable to respond to another. Also known as insensitive time. { 'ded ‚tīm }

dead-time compensation |CONT SYS| The modification of a controller to allow for time delays between the input to a control system and the response to the signal. { 'ded ‚tīm käm·pən'sā·shən }

dead zone See dead band. { 'ded ‚zōn }

dead zone unit |COMPUT SCI| An analog computer device that maintains an output signal at a constant value over a certain range of values of the input signal. { 'ded ‚zōn ‚yü·nət }

deallocation |COMPUT SCI| The release of a portion of computer storage or a peripheral unit from control by a computer program when it is no longer needed. { dē‚al·ə'kā·shən }

debatable time |COMPUT SCI| In the keeping of computer usage statistics, time that cannot be attributed with certainty to any one of various categories of computer use. { di'bād·ə·bəl 'tīm }

deblocking |COMPUT SCI| Breaking up a block of records into individual records. { dē'bläk·iŋ }

debug |COMPUT SCI| To test for, locate, and remove mistakes from a program or malfunctions from a computer. |ELECTR| To detect and remove secretly installed listening devices popularly known as bugs. |ENG| To eliminate from a newly designed system the components and circuits that cause early failures. { dē'bəg }

debugging routine |COMPUT SCI| A routine to aid programmers in the debugging of their routiness; some typical routines are storage printout, tape printout, and drum printout routines. { dē'bəg·iŋ rü‚tēn }

debugging statement |COMPUT SCI| Temporary instructions inserted into a program being tested so as to pinpoint problem areas. { dē'bəg·iŋ ‚stāt·mənt }

debug on-line |COMPUT SCI| **1.** To detect and correct errors in a computer program by using only certain parts of the hardware of a computer, while other routines are being processed simultaneouly. **2.** To detect and correct errors in a program from a console distant from a computer in a multiaccess system. { dē'bəg·iŋ ȯn 'līn }

debunching |ELECTR| A tendency for electrons in a beam to spread out both longitudinally and transversely due to mutual repulsion; the effect is a drawback in velocity modulation tubes. { dē'bənch·iŋ }

debye |ELEC| A unit of electric dipole moment, equal to 10^{-18} Franklin centimeter. { də'bī }

Debye theory |ELEC| The classical theory of the orientation polarization of polar molecules in which the molecules have a single relaxation time, and the plot of the imaginary part of the complex relative permittivity against the real part is a semicircle. { də'bī ‚thē·ə·rē }

decade |ELEC| A group or assembly of 10 units; for example, a decade counter counts 10 in one column, and a decade box inserts resistance quantities in multiples of powers of 10. { de'kād }

decade box |ELEC| An assembly of precision resistors, coils, or capacitors whose individual values vary in submultiples and multiples of 10; by appropriately setting a 10-position selector switch for each section, the decade box can be set to any desired value within its range. { de'kād ‚bäks }

decade bridge |ELECTR| Electronic apparatus for measurement of unknown values of resistances or capacitances by comparison with known values (bridge); one secondary section of the oscillator-driven transformer is tapped in decade steps, the other in 10 uniform steps. { de'kād ‚brij }

decade counter See decade scaler. { de'kād ‚kau̇nt·ər }

decade scaler |ELECTR| A scaler that produces one output pulse for every 10 input pulses. Also known as counter decade; decade counter; scale-of-ten circuit. { de'kād ‚skāl·ər }

decelerating electrode |ELECTR| Of an electron-beam tube, an electrode to which a potential is applied to decrease the velocity of the electrons in the beam. { də'sel·ə‚rād·iŋ i'lek‚trōd }

deceleration time |COMPUT SCI| For a storage medium, such as magnetic tape that must be physically moved in order for reading or writing to take place, the minimum time that must elapse between the completion of a reading or writing operation and the moment that motion ceases. Also known as a stop time. { dē‚sel·ə'rā·shən ‚tīm }

decentralized data processing |COMPUT SCI| An arrangement comprising a data-processing center for each division or location of a single organization. { dē'sen·trə‚lizd 'dad·ə 'präs‚es·iŋ }

deception |ELECTR| The deliberate radiation, reradiation, alteration, absorption, or reflection of electromagnetic energy in a manner intended to mislead an enemy in the interpretation of information received by his electronic systems. { di'sep·shən }

decibel adjusted See adjusted decibel. { 'des·ə ‚bel ə'jəs·təd }

decibel loss |COMMUN| Signal attenuation over a transmission path or a conductor expressed in decibels. { 'des·ə‚bel ‚lȯs }

decibel meter |ENG| An instrument calibrated in logarithmic steps and labeled with decibel units and used for measuring power levels in communication circuits. { 'des·ə‚bel ‚mēd·ər }

decibels above 1 femtowatt |ELEC| A power level equal to 10 times the common logarithm of the ratio of the given power in watts to 1 femtowatt

$(10^{-15}$ watt). Abbreviated dBf. { 'des·ə·bəlz ə ¦bəv ¦wən 'fem·tō,wät }

decibels above 1 kilowatt |ELEC| A measure of power equal to 10 times the common logarithm of the ratio of a given power to 1000 watts. Abbreviated dBk. { 'des·ə·bəlz ə¦bəv ¦wən 'kil·ə ,wät }

decibels above 1 milliwatt |ELEC| A measure of power equal to 10 times the common logarithm of the ratio of a given power to 0.001 watt; a negative value, such as −2.7 dBm, means decibels below 1 milliwatt. Abbreviated dBm. { 'des·ə·bəlz ə¦bəv ¦wən 'mil·i,wät }

decibels above 1 picowatt |ELEC| A measure of power equal to 10 times the common logarithm of the ratio of a given power to 1 picowatt. Abbreviated dBp. { 'des·ə·bəlz ə¦bəv ¦wən 'pē·kō ,wät }

decibels above 1 volt |ELEC| A measure of voltage equal to 20 times the common logarithm of the ratio of a given voltage to 1 volt. Abbreviated dBV. { 'des·ə·bəlz ə¦bəv ¦wən 'vōlt }

decibels above 1 watt |ELEC| A measure of power equal to 10 times the common logarithm of the ratio of a given power to 1 watt. Abbreviated dBW. { 'des·ə·bəlz ə¦bəv ¦wən 'wät }

decibels above reference coupling |ELEC| A measure of the coupling between two circuits, expressed in relation to a reference value of coupling that gives a specified reading on a specified noise-measuring set when a test tone of 90 dBa is impressed on one circuit. Abbreviated dBx. { 'des·ə·bəlz ə¦bəv 'ref·rəns ,kəp·liŋ }

decibels above reference noise |ELEC| Units used to show the relationship between the interfering effect of a noise frequency, or band of noise frequencies, and a fixed amount of noise power commonly called reference noise; a 1000-hertz tone having a power level of −90 dBm was selected as the reference noise power; superseded by the adjusted decibel unit. Abbreviated dBrn. { 'des·ə·bəlz ə¦bəv 'ref·rəns ,nóiz }

decimal attenuator |ELECTR| System of attenuators arranged so that a voltage or current can be reduced decimally. { 'des·məl ə'ten·yə,wād·ər }

decimal-binary switch |ELEC| A switch that connects a single input lead to appropriate combinations of four output leads (representing 1, 2, 4, and 8) for each of the decimal-numbered settings of its control knob; thus, for position 7, output leads 1, 2, and 4 would be connected to the input. { ¦des·məl ¦bīn·ə·rē 'swich }

decimal code |COMPUT SCI| A code in which each allowable position has one of 10 possible states; the conventional decimal number system is a decimal code. { ¦des·məl ¦kōd }

decimal-coded digit |COMPUT SCI| One of 10 arbitrarily selected patterns of 1 and 0 used to represent the decimal digits. Also known as coded decimal. { ¦des·məl ¦kōd·əd 'dij·ət }

decimal processor |COMPUT SCI| A digital computer organized to calculate by decimal arithmetic. { ¦des·məl 'präs,es·ər }

decimal-to-binary conversion |COMPUT SCI| The mathematical process of converting a number

written in the scale of 10 into the same number written in the scale of 2. { ¦des·məl tə ¦bin·ə·re kən'vər·zhən }

decision |COMPUT SCI| The computer operation of determining if a certain relationship exists between words in storage or registers, and taking alternative courses of action; this is effected by conditional jumps or equivalent techniques. { di'sizh·ən }

decision box |COMPUT SCI| A flow-chart symbol indicating a decision instruction; usually diamond-shaped. { di'sizh·ən ,bäks }

decision calculus |SYS ENG| A guide to the process of decision-making, often outlined in the following steps: analysis of the decision area to discover applicable elements; location or creation of criteria for evaluation; appraisal of the known information pertinent to the applicable elements and correction for bias; isolation of the unknown factors; weighting of the pertinent elements, known and unknown, as to relative importance; and projection of the relative impacts on the objective, and synthesis into a course of action. { di'sizh·ən 'kal·kyə·ləs }

decision element |ELECTR| A circuit that performs a logical operation such as "and," "or," "not," or "except" on one or more binary digits of input information representing "yes" or "no" and that expresses the result in its output. Also known as decision gate. { di'sizh·ən ,el·ə·mənt }

decision gate |ELECTR| See decision element. |NAV| In an instrument landing, that point along the path at which the pilot must decide to land or to execute a missed-approach procedure. { di'sizh·ən ,gāt }

decision instruction See conditional jump. { di'sizh·ən in'strək·shən }

decision mechanism |COMPUT SCI| In character recognition, that component part of a character reader which accepts the finalized version of the input character and makes an assessment as to its most probable identity. { di'sizh·ən ,mek·ə ,niz·əm }

decision rule |SYS ENG| In decision theory, the mathematical representation of a physical system which operates upon the observed data to produce a decision. { di'sizh·ən ,rül }

decision support |COMPUT SCI| The process of filtering, optimizing, and organizing mined information to support decision making. { di'sizh·ən sə,pòrt }

decision support system |COMPUT SCI| A computer-based system that enables management to interrogate the computer system on an ad hoc basis for various kinds of information on the organization and to predict the effect of potential decisions beforehand. Abbreviated DSS. { di'sizh·ən sə'pòrt ,sis·təm }

decision table |COMPUT SCI| **1.** A table of contingencies to be considered in the definition of a problem, together with the actions to be taken; sometimes used in place of a flow chart for program documentation. **2.** See DETAB. { di'sizh·ən ,tā·bəl }

decision theory [SYS ENG] A broad spectrum of concepts and techniques which have been developed to both describe and rationalize the process of decision making, that is, making a choice among several possible alternatives. { di'sizh·ən ,the·ə·rē }

deck [ENG] A magnetic-tape transport mechanism. { dek }

deck switch See gang switch. { 'dek ,swich }

declaration See declarative statement. { ,dek·lə'rā·shən }

declarative language [COMPUT SCI] A nonprocedural programming language that allows the programmer to state the task to be accomplished without specifying the procedures needed to carry it out. { di,klar·əd·iv 'laŋ·gwij }

declarative macroinstruction [COMPUT SCI] An instruction in an assembly language which directs the compiler to take some action or take note of some condition and which does not generate any instruction in the object program. { di¦klar·əd·iv ¦mak·rō·in¦strək·shən }

declarative markup language [COMPUT SCI] A system of codes for identifying the subdivisions of a text-processing document, without carrying out the actual formatting. { di,klar·əd·iv 'mär·kəp ,laŋ·gwij }

declarative statement [COMPUT SCI] Any program statement describing the data which will be used or identifying the memory locations which will be required. Also known as declaration. { di ¦klar·əd·iv 'stāt·mənt }

decode [COMMUN] 1. To translate coded characters into a more understandable form. 2. See demodulate. { dē'kōd }

decoded stream [COMMUN] The decoded reconstruction of a compressed bit stream. { dē'kōd·əd 'strēm }

decoder [ELECTR] 1. A matrix of logic elements that selects one or more output channels, depending on the combination of input signals present. 2. See decoder circuit; matrix; tree. { dē'kōd·ər }

decoder circuit [ELECTR] A circuit that responds to a particular coded signal while rejecting others. Also known as decoder. { dē'kōd·ər ,sər·kət }

decoding gate [COMPUT SCI] The use of combinatorial logic in circuitry to select a device identified by a binary address code. Also known as recognition gate. { dē'kōd·iŋ ,gāt }

decollator [COMPUT SCI] A device which separates the sheets of continuous stationery that form the output of a computer printer into separate stacks. { dē'kō,lād·ər }

decometer [ELECTR] An adding-type phasemeter which rotates continuously and adds up the total number of degrees of phase shift between two signals, such as those received from two transmitters in the Decca navigation system. { də'käm·əd·ər }

decommutation [ELECTR] The process of recovering a signal from the composite signal previously created by a commutation process. { dē ,käm·yə'tā·shən }

decommutator [ELECTR] The section of a telemetering system that extracts analog data from a time-serial train of samples representing a multiplicity of data sources transmitted over a single radio-frequency link. { dē'käm·yə,tād·ər }

decoupling [ELEC] Preventing transfer or feedback of energy from one circuit to another. { dē'kəp·liŋ }

decoupling filter [ELECTR] One of a number of low-pass filters placed between each of several amplifier stages and a common power supply. { dē'kəp·liŋ ,fil·tər }

decoupling network [ELEC] Any combination of resistors, coils, and capacitors placed in power supply leads or other leads that are common to two or more circuits, to prevent unwanted interstage coupling. { dē'kəp·liŋ ,net,wərk }

decoy transponder [ELECTR] A transponder that returns a strong signal when triggered directly by a radar pulse, to produce large and misleading target signals on enemy radar screens. { 'dē,kòi tran,spän·dər }

decrement [COMPUT SCI] 1. A specific part of an instruction word in some binary computers, thus a set of digits. 2. For a counter, to subtract 1 or some other number from the current value. { 'dek·rə·mənt }

decrement field [COMPUT SCI] That part of an instruction word which is used to modify the contents of a storage location or register. { 'dek·rə·mənt ,fēld }

decrypt [ELECTR] To convert a cryptogram or series of electronic pulses into plain text by electronic means. { dē'kript }

dedicated file server [COMPUT SCI] A computer that operates solely to provide services to other computers in a particular local-area network and to manage the network operating system. Also known as dedicated server. { ,ded·ə,kād·əd 'fīl ,sər·vər }

dedicated line [COMPUT SCI] A permanent communications link that is used solely to transmit information between a computer and a data-processing system. { 'ded·ə,kād·əd 'līn }

dedicated server See dedicated file server. { ,ded·ə,kād·əd ,sər·vər }

dedicated terminal [COMPUT SCI] A computer terminal that is permanently connected to a data-processing system by a communications link that is used only to transmit information between the two. { 'ded·ə,kād·əd 'tərm·ən·əl }

deemphasis [ENG ACOUS] A process for reducing the relative strength of higher audio frequencies before reproduction, to complement and thereby offset the preemphasis that was introduced to help override noise or reduce distortion. Also known as postemphasis; postequalization. { dē'em·fə·səs }

deemphasis network [ENG ACOUS] An RC filter inserted in a system to restore preemphasized signals to their original form. { dē'em·fə·səs ,net,wərk }

deenergize [ELEC] To disconnect from the source of power. { dē'en·ər,jīz }

deerhorn antenna [ELECTROMAG] A dipole antenna whose ends are swept back to reduce wind resistance when mounted on an airplane. { 'dir ,hórn an'ten·ə }

de facto standard [COMPUT SCI] A set of criteria for software, hardware, or communications procedures that is widely accepted because of the dominance of a particular technology over others rather than the action of a recognized standards organization. { dē 'fak·tō 'stan·dərd }

default [COMPUT SCI] A value automatically used or an action automatically carried out unless another is specified. { di'fȯlt }

default printer [COMPUT SCI] The printer that is automatically used by a program unless another printer is specifically designated. { di'fȯlt ,print·ər }

defect conduction [SOLID STATE] Electric conduction in a semiconductor by holes in the valence band. { 'dē,fekt kə'dək·shən }

defective track [COMPUT SCI] Any circular path on the surface of a magnetic disk which is detected by the system as unable to accept one or more bits of data. { di'fek·tiv 'trak }

deferred addressing [COMPUT SCI] A type of indirect addressing in which the address part of an instruction specifies a location containing an address, the latter in turn specifies another location containing an address, and so forth, the number of iterations being controlled by a preset counter. { di'fərd ə'dres·iŋ }

deferred data item [COMPUT SCI] A quantity or attribute that is assigned a value only at the time it is actually processed. { di'fərd 'dad·ə ,īd·əm }

deferred entry [COMPUT SCI] The passing of control of the central processing unit to a subroutine or to an entry point as the result of an asynchronous event. { di'fərd 'en·trē }

deferred mount [COMPUT SCI] Postponement of the placement of a tape on a tape drive until it is actually needed, rather than when the program starts to run. { di'fərd 'maünt }

deferred processing [COMPUT SCI] The making of computer runs which are postponed until nonpeak periods. { di'fərd 'präs,es·iŋ }

definite network [COMPUT SCI] A sequential network in which no feedback loops exist. { ¦def·ə·nət 'net,wərk }

definition [COMMUN] The fidelity with which an imaging system conveys and reproduces an image. [ELECTR] The extent to which the fine-line details of a printed circuit correspond to the master drawing. { ,def·ə'nish·ən }

deflection [COMPUT SCI] Encouraging a potential attacker of a computer system to direct the attack elsewhere. [ELECTR] The displacement of an electron beam from its straight-line path by an electrostatic or electromagnetic field. { di'flek·shən }

deflection circuit [ELECTR] A circuit which controls the deflection of an electron beam in a cathode-ray tube. { di'flek·shən ,sər·kət }

deflection coil [ELECTR] One of the coils in a deflection yoke. { di'flek·shən ,kȯil }

deflection defocusing [ELECTR] Defocusing that becomes greater as deflection is increased in

a cathode-ray tube, because the beam hits the screen at a greater slant and the beam spot becomes more elliptical as it approaches the edges of the screen. { di'flek·shən de,fō·kəs·iŋ }

deflection electrode [ELECTR] An electrode whose potential provides an electric field that deflects an electron beam. Also known as deflection plate. { di'flek·shən i,lek,trōd }

deflection factor [ELECTR] The reciprocal of the deflection sensitivity in a cathode-ray tube. { di'flek·shən ,fak·tər }

deflection-modulated indicator See amplitude-modulated indicator. { di'flek·shən ¦mäj·ə,lād·əd 'in·də,kād·ər }

deflection plate See deflection electrode. { di'flek·shən ,plāt }

deflection polarity [ELECTR] Relationship between the direction of a displacement of the cathode beam and the polarity of the applied signal wave. { di'flek·shən pə'lar·əd·ē }

deflection sensitivity [ELECTR] The displacement of the electron beam at the target or screen of a cathode-ray tube per unit of change in the deflection field; usually expressed in inches per volt applied between deflection electrodes or inches per ampere in a deflection coil. { di'flek·shən sen·sə'tiv·əd·ē }

deflection voltage [ELECTR] The voltage applied between a pair of deflection electrodes to produce an electric field. { di'flek·shən ,vōl·tij }

deflection yoke [ELECTR] An assembly of one or more electromagnets that is placed around the neck of an electron-beam tube to produce a magnetic field for deflection of one or more electron beams. Also known as scanning yoke; yoke. { di'flek·shən ,yōk }

defocus-dash mode [ELECTR] A mode of cathode-ray tube storage of binary digits in which the writing beam is initially defocused so as to excite a small circular area on the screen; for one kind of binary digit it remains defocused, and for the other kind it is suddenly focused to a concentric dot and drawn out into a dash. { dē ¦fō·kəs ¦dash ,mōd }

defocus-focus mode [ELECTR] A variation of the defocus-dash mode in which the focused dot is drawn out into a dash. { dē¦fō·kəs ¦fō·kəs ,mōd }

defragmentation [COMPUT SCI] A procedure in which portions of files on a computer disk are moved until all parts of each file occupy continuous sectors, resulting in a substantial improvement in disk access times. { ,dē ,frag·mən'tā·shən }

defragmenter [COMPUT SCI] A program that analyzes storage locations of files on a computer disk and then carries out defragmentation. { ,dē ,frag'men·tər }

defruit [ELECTR] To remove random asynchronous replies from the video input of a display unit in a secondary (beacon) radar system by such means as comparing the video signals on successive sweeps. { dē'früt }

degas |ELECTR| To drive out and exhaust the gases occluded in the internal parts of an electron tube or other gastight apparatus, generally by heating during evacuation. { dē'gas }

degauss |ELECTR| To remove, erase, or clear information from a magnetic tape, disk, drum, or core. |ELECTROMAG| To neutralize (demagnetize) a magnetic field of, for example, television tube.

degaussing coil |ELECTROMAG| A plastic-encased coil, about 1 foot (0.3 meter) in diameter, that can be plugged into a 120-volt alternating-current wall outlet and moved slowly toward and away from a color television picture tube to demagnetize adjacent parts. { dē'gaús·iŋ ,kóil }

degenerate amplifier |ELECTR| Parametric amplifier with a pump frequency exactly twice the signal frequency, producing an idler frequency equal to that of the signal input; it is considered as a single-frequency device. { di'jen·ə·rət 'am·plə,fī·ər }

degeneration |ELECTR| The loss or gain in an amplifier through unintentional negative feedback. { di,jen·ə'rā·shən }

deglitcher |ELECTR| A nonlinear filter or other special circuit used to limit the duration of switching transients in digital converters. { dē'glich·ər }

degradation |COMPUT SCI| Condition under which a computer operates when some area of memory or some units of peripheral equipment are not available to the user. { ,deg·rə'dā·shən }

degradation failure |ENG| Failure of a device because of a shift in a parameter or characteristic which exceeds some previously specified limit. { ,deg·rə'dā·shən ,fāl·yər }

degree of current rectification |ELECTR| Ratio between the average unidirectional current output and the root mean square value of the alternating current input from which it was derived. { di'grē əv 'kər·ənt ,rek·tə·fə'kā·shən }

degree of voltage rectification |ELECTR| Ratio between the average unidirectional voltage and the root mean square value of the alternating voltage from which it was derived. { di'grē əv 'vōl·tij ,rek·tə·fə'kā·shən }

deion circuit breaker |ELEC| Circuit breaker built so that the arc that forms when the circuit is broken is magnetically blown into a stack of insulated copper plates, giving the effect of a large number of short arcs in series; each arc becomes almost instantly deionized when the current drops to zero in the alternating current cycle, and the arc cannot reform. { dē'ī,än 'sər·kət ,brāk·ər }

deionization |ELECTR| The return of an ionized gas to its neutral state after all sources of ionization have been removed, involving diffusion of ions to the container walls and volume recombination of negative and positive ions. { dē,ī·ən·ə'zā·shən }

deionization potential |ELECTR| The potential at which ionization of the gas in a gas-filled

tube ceases and conduction stops. { dē,ī·ən· ə'zā·shən pə'ten·chəl }

deionization time |ELECTR| The time required for a gas tube to regain its preconduction characteristics after interruption of anode current, so that the grid regains control. Also called recontrol time. { dē,ī·ən·ə'zā·shən ,tīm }

de la Rue and Miller's law |ELECTR| The law that in a field between two parallel plates, the sparking potential of a gas is a function of the product of gas pressure and sparking distance only. { del·ə¦rü ən 'mil·ərz ,lò }

delay |COMMUN| **1.** Time required for a signal to pass through a device or a conducting medium. **2.** Time which elapses between the instant at which any designated point of a transmitted wave passes any two designated points of a transmission circuit; such delay is primarily determined by the constants of the circuit. { di'lā }

delay circuit See time-delay circuit. { di'lā ,sər· kət }

delay counter |COMPUT SCI| A counter which inserts a time delay in a sequence of events. { di'lā ,kaúnt·ər }

delay distortion |ELECTR| Phase distortion in which the rate of change of phase shift with frequency of a circuit or system is not constant over the frequency range required for transmission. Also called envelope delay distortion. { di'lā di'stòr,shən }

delayed automatic gain control |ELECTR| An automatic gain control system that does not operate until the signal exceeds a predetermined magnitude; weaker signals thus receive maximum amplification. Also known as biased automatic gain control; delayed automatic volume control; quiet automatic volume control. { di'lād ,òd·ə¦mad·ik 'gān kən,trōl }

delayed automatic volume control See delayed automatic gain control. { di'lād ,òd·ə¦mad·ik 'väl·yəm kən,trōl }

delayed plan position indicator |ELECTR| A plan position indicator in which initiation of the time base is delayed a fixed time after each transmitted pulse, to give expansion of the range scale for distant targets so that they show more clearly on the screen. { di'lād 'plan pə'zish·ən ,in·də,kād·ər }

delayed sweep |ELECTR| A sweep whose beginning is delayed for a definite time after the pulse that initiates the sweep. { di'lād 'swēp }

delay equalizer |ELECTR| A corrective network used to make the phase delay or envelope delay of a circuit or system substantially constant over a desired frequency range. { di'lā 'ē·kwə,līz·ər }

delay flip-flop See D flip-flop. { di'lā 'flip,fläp }

delay/frequency distortion |COMMUN| That form of distortion which occurs when the delay of a circuit or system is not constant over the frequency range required for transmissions. { di ¦lā ¦frē·kwən·sē di'stòr·shən }

delay line |ELECTR| **1.** A transmission line (as dissipationless as possible), or an electric network approximation of it, which, if terminated in its characteristic impedance, will reproduce at its output a waveform applied to its input terminals with little distortion, but at a time delayed by an amount dependent upon the electrical length of the line. Also known as artificial delay line. **2.** A circuit component, analog or digital, in a radar system by which pulses may be delayed a controllable amount; used typically for pulse comparisons as in canceler circuits. { di'lā, līn }

delay-line memory *See* circulating memory. { di'lā ,līn 'mem·rē }

delay-line storage *See* circulating memory. { di'lā ,līn 'stór·ij }

delay multivibrator [ELECTR] A monostable multivibrator that generates an output pulse a predetermined time after it is triggered by an input pulse. { di'lā ,məl·tə'vī,brād·ər }

delay relay [ELEC] A relay having predetermined delay between energization and closing of contacts or between deenergization and dropout. { di'lā 'rē,lā }

delay time [CONT SYS] The amount of time by which the arrival of a signal is retarded after transmission through physical equipment or systems. [ELECTR] The time taken for collector current to start flowing in a transistor that is being turned on from the cutoff condition. { di'lā ,tīm }

delay unit *See* transport delay unit. { di'lā ,yü·nət }

deleted representation [COMPUT SCI] In paper tape codes, the superposition of a pattern of holes upon another pattern of holes representing a character, to effectively remove or obliterate the latter. { di'lēd·əd ,rep·rə,zen'tā·shən }

deletion operator [COMPUT SCI] The part of a data structure which allows components to be deleted. { di'lē·shən ,äp·ə,rād·ər }

deletion record [COMPUT SCI] A record which removes and replaces an existing record when it is added to a file. { di'lē·shən ,rek·ərd }

delimiter [COMPUT SCI] A character that separates items of data. { də'lim·əd·ər }

Dellinger fadeout [COMMUN] Type of fadeout that occurs during shortwave reception, believed to be caused by rapid shifting of ionosphere layers during solar eruptions. { 'del·ən·jər 'fād ,aút }

delta [ELECTR] The difference between a partial-select output of a magnetic cell in a one state and a partial-select output of the same cell in a zero state. { 'del·tə }

delta connection [ELEC] A combination of three components connected in series to form a triangle like the Greek letter delta. Also known as mesh connection. { 'del·tə kə'nek·shən }

delta current [ELEC] Electricity going through a delta connection. { 'del·tə ,kər·ənt }

delta-gun tube [ELECTR] A color television picture tube in which three electron guns, arranged in a triangle, provide electron beams that fall on phosphor dots on the screen, causing them to emit light in three primary colors; a shadow mask located just behind the screen ensures that each beam excites only dots of one color. { 'del·tə ,gən ,tüb }

delta matching transformer [ELEC] Impedance device used to match the impedance of an open-wire transmission line to an antenna; the two ends of the transmission line are fanned out so that the impedance of the line gradually increases; the ends of the transmission line are attached to the antenna at points of equal impedance, symmetrically located with respect to the center of the antenna. { 'del·tə ,mach·iŋ tranz,fór·mər }

delta modulation [ELECTR] A pulse-modulation technique in which a continuous signal is converted into a binary pulse pattern, for transmission through low-quality channels. { 'del·tə ,mäj·ə'lā·shən }

delta network [ELEC] A set of three branches connected in series to form a mesh. { 'del·tə ¦net,wərk }

delta pulse code modulation [ELECTR] A modulation system that converts audio signals into corresponding trains of digital pulses to give greater freedom from interference during transmission over wire or radio channels. { 'del·tə ¦pəls ,kōd ,mäj,ə'lā·shən }

delta-sigma converter *See* sigma-delta converter. { ¦del·tə ¦sig·mə kən'vərd·ər }

delta-sigma modulator *See* sigma-delta modulator. { ¦del·tə¦sig·mə 'mä·jə,lād·ər }

delta transformer [ELEC] A three-phase electrical transformer in which the ends of the three windings are connected to form a triangle. { 'del·tə tranz'fór·mər }

delta-Y transformation *See* Y-delta transformation. { 'del·tə ,wī ,tranz·fár'mā·shən }

deltic method [ELECTR] A method of sampling incoming radar, sonar, seismic, speech, or other waveforms along with reference signals, compressing the samples in time, and comparing them by autocorrelation. { 'del·tik ,meth·əd }

demagnetizer [ELECTR] A device for removing undesired magnetism, as from the playback head of a tape recorder or from a recorded reel of magnetic tape that is to be erased. { dē'mag·nə ,tī·zər }

demand *See* demand factor. { də'mand }

demand assignment multiple access [COMMUN] The allocation of bandwidth in a communications system among multiple users based on demand, such as by multiplexing. Abbreviated DAMA. { di¦mand ə,sīn·mənt ¦məl·tə·pəl 'ak,ses }

demand-driven execution [COMPUT SCI] A mode of carrying out a program in a data flow system in which no calculation is carried out until its results are demanded as input to another calculation. Also known as lazy evaluation. { də'mand ,driv·ən ,ek·sə'kyü·shən }

demand factor [ELEC] The ratio of the maximum demand of a building for electric power to the total connected load. Also known as demand. { də'mand ,fak·tər }

demand limiter *See* current limiter. { də'mand ,lim·əd·ər }

demand meter [ENG] Any of several types of instruments used to determine a customer's maximum demand for electric power over an appreciable time interval; generally used for billing industrial users. { də'mand ˌmēd·ər }

demand paging |COMPUT SCI| The characteristic of a virtual memory system which retrieves only that part of a user's program which is required during execution. { də'mand ˌpā·jiŋ }

demand processing |COMPUT SCI| The processing of data by a computer system as soon as it is received, so that it is not necessary to store large amounts of raw data. Also known as immediate processing. { də'mand ˌpräs,es·iŋ }

demand rate |ELEC| The maximum amount of electric power that must be kept available to a customer. { də'mand ˌrāt }

demand reading |COMPUT SCI| A method of carrying out input operations in which blocks of data are transmitted to the central processing unit as needed for processing. { də'mand ˌrēd·iŋ }

demand staging |COMPUT SCI| Moving blocks of data from one storage device to another when programs request them. { də'mand ˌstā·jiŋ }

demand writing |COMPUT SCI| A method of carrying out output operations in which blocks of data are transmitted from the central processing unit as they are needed by the user. { də'mand ˌrīd·iŋ }

Dember effect |ELECTR| Creation of a voltage in a conductor or semiconductor by illumination of one surface. Also known as photodiffusion effect. { däm·bā i'fekt }

demodifier |COMPUT SCI| A data element used to restore part of an instruction which has been modified to its original value. { dē'mäd·ə,fī·ər }

demodulate |COMMUN| To recover the modulating wave from a modulated carrier. Also known as decode; detect. { dē'mäj·ə,lāt }

demodulation |COMMUN| The recovery, from a modulated carrier, of a signal having substantially the same characteristics as the original signal. { dē,mäj·ə'lā·shən }

demodulator See detector. { dē'mäj·ə,lad·ər }

demount |COMPUT SCI| To take out a magnetic storage medium from a device that reads or writes on it. { dē'maúnt }

demountable pack |COMPUT SCI| A disk pack that can be taken out and replaced by another. { dē'maúnt·ə·bəl 'pak }

demountable tube |ELECTR| High-power radio tube having a metal envelope with porcelain insulation; can be taken apart for inspection and for renewal of electrodes. { dē'maúnt·ə·bəl 'tüb }

DEMS See Digital Electronic Message Service.

demultiplexer |ELECTR| A device used to separate two or more signals that were previously combined by a compatible multiplexer and transmitted over a single channel. { dē,məl·tə ,plek·sər }

demultiplexing |COMMUN| The separation of two or more channels previously multiplexed. { dē'məl·tə,pleks·iŋ }

demultiplexing circuit |ELECTR| A circuit used to separate the signals that were combined for transmission by multiplex. { dē'məl·tə,plek·siŋ ,sər·kət }

dense binary code |COMPUT SCI| A code in which all possible states of the binary pattern are used. { 'dens ¦bī·nə·rē 'kōd }

dense list |COMPUT SCI| A list in which all the cells contain records of the file. { ¦dens ¦list }

density modulation |ELECTR| Modulation of an electron beam by making the density of the electrons in the beam vary with time. { 'den·səd·ē ,mäj·ə'lā·shən }

density packing |COMPUT SCI| In computers, the number of binary digit magnetic pulses stored on tape or drum per linear inch on a single track by a single head. { 'den·səd·ē ,pak·iŋ }

density step tablet |COMMUN| Facsimile test chart consisting of a series of areas; density of the areas increases from a low value to a maximum value in steps. Also known as step tablet. { 'den·səd·ē 'step ,tab·lət }

dependency |COMPUT SCI| The necessity for a computer to complete work on some job before execution of another can begin. { di'pen·dən·sē }

dependent segment |COMPUT SCI| In a database management system, a block of data that depends on data at a higher level for its full meaning. { di'pen·dənt 'seg·mənt }

deperm See degauss. { dē'pərm }

depletion |ELECTR| Reduction of the charge-carrier density in a semiconductor below the normal value for a given temperature and doping level. { də'plē·shən }

depletion layer |ELECTR| An electric double layer formed at the surface of contact between a metal and a semiconductor having different work functions, because the mobile carrier charge density is insufficient to neutralize the fixed charge density of donors and acceptors. Also known as barrier layer (deprecated); blocking layer (deprecated); space-charge layer. { də'plē·shən ,lā·ər }

depletion-layer capacitance See barrier capacitance. { di'plē·shən ,lā·ər kə'pas·əd·əns }

depletion-layer rectification |ELECTR| Rectification at the junction between dissimilar materials, such as a pn junction or a junction between a metal and a semiconductor. Also known as barrier-layer rectification. { də'plē·shən ,lā·ər ,rek·tə·fə'kā·shən }

depletion-layer transistor |ELECTR| A transistor that relies directly on motion of carriers through depletion layers, such as spacistor. { də'plē·shən ,lā·ər tran'zis·tər }

depletion mode |ELECTR| Operation of a field-effect transistor in which current flows when the gate-source voltage is zero, and is increased or decreased by altering the gate-source voltage. { də'plē·shən ,mōd }

depletion-mode HEMT |ELECTR| A high-electron mobility transistor (HEMT) in which application of negative bias to the gate electrode cuts off the current between source and drain. Abbreviated D-HEMT. { də'plē·shən ,mōd ,āch,ē,em'tē }

depletion region |ELECTR| The portion of the channel in a metal oxide field-effect transistor in which there are no charge carriers. { də'plē·shən ,rē·jən }

depolarization |ELEC| The removal or prevention of polarization in a substance (for example, through the use of a depolarizer in an electric cell) or of polarization arising from the field due to the charges induced on the surface of a dielectric when an external field is applied. { dē ‚pō·lə·rə'zā·shən }

depolarization factor |ELEC| The ratio of the internal electric field induced by the charges on the surface of a dielectric when an external field is applied to the polarization of the dielectric. { dē‚pō·lə·rə'zā·shən ‚fak·tər }

deposit |COMPUT SCI| To preserve the contents of a portion of a computer memory by copying it in a backing storage. { də'päz·ət }

deposited carbon resistor |ELECTR| A resistor in which the resistive element is a carbon film pyrolytically deposited on a ceramic substrate. { də'päz·əd·əd 'kär·bən ri'zis·tər }

derating |ELECTR| The reduction of the rating of a device to improve reliability or to permit operation at high ambient temperatures. { dē 'rād·iŋ }

derivative action |CONT SYS| Control action in which the speed at which a correction is made depends on how fast the system error is increasing. Also known as derivative compensation; rate action. { də'riv·əd·iv ‚ak·shən }

derivative compensation See derivative action. { də'riv·əd·iv ‚käm·pən'sā·shən }

derivative network |CONT SYS| A compensating network whose output is proportional to the sum of the input signal and its derivative. Also known as lead network. { də'riv·əd·iv 'net‚wərk }

derived sound system |ENG ACOUS| A four-channel sound system that is artificially synthesized from conventional two-channel stereo sound by an adapter, to provide feeds to four loudspeakers for approximating quadraphonic sound. { də'rīvd 'saúnd ‚sis·təm }

DES See data encryption standard.

DeSauty's bridge |ELEC| A four-arm bridge used to compare two capacitances; two adjacent arms contain capacitors in series with resistors, while the other two arms contain resistors only. Also known as Wien-DeSauty bridge. { də'sōd·ēz ‚brij }

descending sort |COMPUT SCI| The arranging of data records from high to low sequence (9 to 0, and Z to A). { di'send·iŋ 'sòrt }

describing function |CONT SYS| A function used to represent a nonlinear transfer function by an approximately equivalent linear transfer function; it is the ratio of the phasor representing the fundamental component of the output of the nonlinearity, determined by Fourier analysis, to the phasor representing a sinusoidal input signal. { di'skrīb·iŋ ‚faŋk·shən }

descriptor |COMPUT SCI| A word or phrase used to identify a document in a computer-based information storage and retrieval system. { di'skrip·tər }

desensitization |COMMUN| Reduction in receiver sensitivity due to the presence of a high-level off-channel signal overloading the radio-frequency amplifier or mixer stages, or causing automatic gain control action. { dē‚sen·sə·tə'zā·shən }

deserialize |COMMUN| To convert a data stream from a serial stream of bits to parallel streams of bits. { dē'sir·ē·ə‚līz }

designation |COMPUT SCI| An item of data forming part of a computer record that indicates the type of record and thus determines how it is to be processed. { ‚dez·əg'nā·shən }

design-oriented system |COMPUT SCI| A computer system developed primarily to maximize performance of hardware and software, rather than ease of use. { di'zīn ¦ōr·ē‚ent·əd ‚sis·təm }

desk calculator |COMPUT SCI| A device that is used to perform arithmetic operations and is small enough to be conveniently placed on a desk. { ¦desk 'kal·kya‚lād·ər }

desk check See dry run. { 'desk ‚chek }

desktop |COMPUT SCI| In a graphical user interface, a screen on which frequently used software resources are represented by icons. { 'desk‚täp }

desktop accessory software |COMPUT SCI| A set of computer programs providing functions that simulate the office accessories normally found on a desktop, such as a notepad, appointment calendar, and calculator. Also known as desktop application; desktop organizer. { ¦desk‚täp ik ¦ses·ə·rē 'sòf‚wer }

desktop application See desktop accessory software. { ¦desk‚täp ‚ap·lə'kā·shən }

desktop organizer See desktop accessory software. { ¦desk‚täp 'òr·gə‚nīz·ər }

desktop publishing |COMPUT SCI| The use of a personal computer to produce printed output of high quality that is camera-ready for a printing facility. { ¦desk‚täp 'pəb·lish·iŋ }

despooler |COMPUT SCI| Software that reads computer output information from a buffer and routes it to a printer. { dē'spül·ər }

despun antenna |ELECTROMAG| Satellite directional antenna pointed continuously at earth by electrically or mechanically despinning the antenna at the same rate that the satellite is spinning for stabilization. { dē'spən an'ten·ə }

destination |COMPUT SCI| The location (record, file, document, program, device, or disk) to which information is moved or copied. { ‚des·tə'nā·shən }

destination address |COMPUT SCI| The location to which a jump instruction passes control in a program. { ‚des·tə'nā·shən ə'dres }

destination time |COMPUT SCI| The time involved in a memory access plus the time required for indirect addressing. { ‚des·tə'nā·shən ‚tīm }

destination warning mark See tape mark. { ‚des·tə'nā·shən ‚wòrn·iŋ ‚märk }

destructive breakdown |ELECTR| Breakdown of the barrier between the gate and channel of a field-effect transistor, causing failure of the transistor. { di'strək·tiv 'brāk‚daún }

destructive memory See destructive readout memory. { di¦strək·tiv 'mem·rē }

destructive read |COMPUT SCI| Reading that partially or completely erases the stored information as it is being read. { di'strək·tiv 'rēd }

destructive readout memory |COMPUT SCI| A memory type in which reading the contents of a storage location destroys the contents of that location. Also known as destructive memory. { di'strək·tiv 'rēd‚aút ‚mem·rē }

destructive testing |ENG| **1.** Intentional operation of equipment until it fails, to reveal design weaknesses. **2.** A method of testing a material that degrades the sample under investigation. { di'strək·tiv 'test·iŋ }

DETAB |COMPUT SCI| A programming language based on COBOL in which problems can be specified in the form of decision tables. Acronym for decision table. { 'dē‚tab }

detachable plugboard |COMPUT SCI| A control panel that can be removed from the computer or other system and exchanged for another without altering the positions of the plugs and cords. Also known as removable plugboard. { di'tach·ə·bəl 'pləg‚bȯrd }

detail chart |COMPUT SCI| A flow chart representing every single step of a program. { 'dē‚tāl ‚chärt }

detail file |COMPUT SCI| A file containing current or transient data used to update a master file or processed with the master file to obtain a specific result. Also known as transaction file. { 'dē‚tāl ‚fīl }

detailing See screening. { 'dē‚tāl·iŋ }

detect See demodulate. { di'tekt }

detection |COMMUN| The recovery of information from an electrical or electromagnetic signal. { di'tek·shən }

detectivity |ELECTR| The normalized radiation power required to give a signal from a photoconductor that is equal to the noise. { ‚dē ‚tek'tiv·əd·ē }

detector |ELECTR| The stage in a receiver at which demodulation takes place; in a superheterodyne receiver this is called the second detector. Also known as demodulator; envelope detector. { di'tek·tər }

detector balanced bias |ELECTR| Controlling circuit used in radar systems for anticlutter purposes. { di'tek·tər ¦bal·ənst 'bī·əs }

determinant |CONT SYS| The product of the partial return differences associated with the nodes of a signal-flow graph. { də'tər·mə·nənt }

deterministic algorithm See static algorithm. { də‚tər·mə'nis·tik 'al·gə‚rith·əm }

deterrence |COMPUT SCI| Making an attack on a computer sufficiently difficult to discourage potential attackers. { di'tər·əns }

detune |ELECTR| To change the inductance or capacitance of a tuned circuit so its resonant frequency is different from the incoming signal frequency. { dē'tün }

detuning stub |ELECTROMAG| Quarter-wave stub used to match a coaxial line to a sleeve-stub antenna; the stub detunes the outside of the coaxial feed line while tuning the antenna itself. { dē'tün·iŋ 'stəb }

deuterium discharge tube |ELECTR| A tube similar to a hydrogen discharge lamp, but with deuterium replacing the hydrogen; source of high-intensity ultraviolet radiation for spectroscopic microanalysis. { dü'tir·ē·əm 'dis‚chärj ‚tüb }

developer's toolkit |COMPUT SCI| A collection of program subroutines that are used to help write an application program in a particular programming language or with a particular operating system. { di¦vel·əp·ərz 'tül‚kit }

development system |COMPUT SCI| The computer and software that are used to create a computer program. { di'vel·əp·mənt ‚sis·təm }

development tool |COMPUT SCI| A piece of hardware or software that is used to help design a computer or write a computer program. { di'vel·əp·mənt ‚tül }

deviation |ENG| The difference between the actual value of a controlled variable and the desired value corresponding to the set point. { ‚dēv·ē'ā·shən }

deviation absorption |COMMUN| Distortion in a frequency-modulated receiver due to inadequate bandwidth, inadequate amplitude-modulation rejection, or inadequate discriminator linearity. { ‚dēv·ē'ā·shən əb‚sȯrp·shən }

deviation ratio |COMMUN| Ratio of the maximum frequency deviation to the maximum modulating frequency of a frequency-modulated system under specified conditions. { ‚dēv·ē'ā·shən ‚rā·shō }

device |COMPUT SCI| A general-purpose term used, often indiscriminately, to refer to a computer component or the computer itself. |ELECTR| An electronic element that cannot be divided without destroying its stated function; commonly applied to active elements such as transistors and transducers. { di'vīs }

device address |COMPUT SCI| The binary code which corresponds to a unique device, referred to when selecting this specific device. { di'vīs ə'dres }

device assignment |COMPUT SCI| The use of a logical device number used in conjunction with an input/output instruction, and made to refer to a specific device. { di'vīs ə'sīn·mənt }

device cluster |COMPUT SCI| A collection of peripheral devices (usually terminals) that have a common control unit. { di'vīs ‚kləs·tər }

device control character |COMPUT SCI| A special character used to direct a peripheral or communications device to perform a specific function. { di'vīs kən'trōl ‚kar·ik·tər }

device dependence |COMPUT SCI| Property of a computer program that will operate only with specified hardware. { di 'vīs de‚pen·dəns }

device driver |COMPUT SCI| A subroutine which handles a complete input/output operation. { di'vīs ‚drīv·ər }

device-end condition |COMPUT SCI| The completion of an input/output operation, such as the transfer of a complete data block, recognized by the hardware in the absence of a byte count. { di'vīs ‚end kən'dish·ən }

device end pending |COMPUT SCI| A hardware error in which a peripheral device does not respond when addressed by the central processing unit, usually because the device has become inoperative. { di'vīs 'end ‚pend·iŋ }

device flag |COMPUT SCI| A flip-flop output which indicates the ready status of an input/output device. { di'vīs ,flag }

device independence |COMPUT SCI| Property of a computer program whose successful execution (without recompilation) does not depend on the type of physical unit associated with a given logical unit employed by the program. { di'vīs ,in·də'pen·dəns }

device-independent colors |COMPUT SCI| Colors produced by printers, monitors, and other output devices that have been modified to conform with a standard method of color description. { di¦vīs ,in·də,pen·dənt 'kəl·ərz }

device-name assignment |COMPUT SCI| The designation of a peripheral device by a symbolic name rather than an address. { di'vīs ¦nām ə ,sīn·mənt }

device number |COMPUT SCI| The physical or logical number which refers to a specific input/output device. { di'vīs ,nəm·bər }

device selector |COMPUT SCI| A circuit which gates data-transfer or command pulses to a specific input/output device. { di'vīs si'lek·tər }

D flip-flop |ELECTR| A flip-flop whose output is a function of the input which appeared one pulse earlier. Also known as delay flip-flop. { ¦dē 'flip ,fläp }

D-frame |COMMUN| A frame coded according to an MPEG-1 mode that uses dc (direct-current or zero-frequency) coefficients only. { 'dē ,frām }

DG synchro amplifier |ELECTR| Synchro differential generator driven by servosystem. { ¦dē¦jē ¦siŋ·krō 'am·plə,fī·ər }

D-HEMT See depletion-mode HEMT.

diac See trigger diode. { 'dī,ak }

diactor |ELEC| Direct-acting automatic regulator for control of shunt generator voltage output. { dī'ak·tər }

diagnosis |COMPUT SCI| The process of locating and explaining detectable errors in a computer routine or hardware component. { ,dī·əg'nō·səs }

diagnostic check See diagnostic routine. { ,dī·əg'näs·tik 'chek }

diagnostic message |COMPUT SCI| A statement produced automatically during some computer processing activity, such as program compilation, that provides information on the status of the computer or its software, particularly errors or potential problems. { ¦dī·əg¦näs·tik 'mes·ij }

diagnostic routine |COMPUT SCI| A routine designed to locate a computer malfunction or a mistake in coding. Also known as diagnostic check; diagnostic subroutine; diagnostic test; error detection routine. { ,dī·əg'näs·tik rü'tēn }

diagnostics |ENG| Information on what tests a device has failed and how they were failed; used to aid in troubleshooting. { ,dī·əg'näs·tiks }

diagnostic subroutine See diagnostic routine. { ,dī·əg'näs·tik 'səb·rü,tēn }

diagnostic test See diagnostic routine. { ,dī·əg'näs·tik 'test }

diagnotor |COMPUT SCI| A combination diagnostic and edit routine which questions un-usual situations and notes the implied results. { ,dī·əg'nōd·ər }

diagonal horn antenna |ELECTROMAG| Horn antenna in which all cross sections are square and the electric vector is parallel to one of the diagonals; the radiation pattern in the far field has almost perfect circular symmetry. { dī'ag·ən·əl 'hȯrn an'ten·ə }

diagram |COMPUT SCI| A schematic representation of a sequence of subroutines designed to solve a problem; it is a coarser and less symbolic representation than a flow chart, frequently including descriptions in English words. { 'dī·ə ,gram }

dial |COMMUN| In automatic telephone switching, either a type of calling device that, when wound up and released, generates pulses required for establishing connections or a pushbutton array that, with associated electronics, generates dualtone multifrequency (DTMF) signals. |ENG| A separate scale or other device for indicating the value to which a control is set. { dīl }

dial backup |COMMUN| A dial telephone line that can be used in case a point-to-point line fails, so that data transmission can continue. { 'dīl 'bak ,əp }

dial central office |COMMUN| Telephone or teletypewriter office where necessary automatic equipment is located for connecting two or more users together by wires for communications purposes. { ¦dīl ¦sen·trəl 'ȯf·əs }

dialect |COMPUT SCI| A version of a programming language that differs from other versions in some respects but generally resembles them. { 'dī·ə ,lekt }

dial exchange |COMMUN| A telephone exchange area in which all subscribers originate their calls by dialing. { 'dīl iks,chānj }

dialing key |COMMUN| Method of dialing in which a set of numerical keys is used to originate dial pulses instead of a dial; generally used in connection with voice-frequency dialing. { 'dī·liŋ ,kē }

dial jacks |ELEC| Strip of jacks associated with and bridged to a regular out-going trunk jack circuit to provide a connection between the dial cords and the outgoing trunks. { 'dīl ,jaks }

dial key |ELEC| Key unit of the subscriber's cord circuit used to connect the dial into the line. { 'dīl ,kē }

dial lamp |ELEC| A small lamp used to illuminate a dial. { 'dīl ,lamp }

dial leg |ELEC| Conductor in a circuit brought out for direct-current dial signaling. { 'dīl ,leg }

dial office |COMMUN| Central office operating on dial signals. { 'dīl ,ȯf·əs }

dialog |COMPUT SCI| A form of data processing involving an interaction between a computer system and a terminal operator who uses a keyboard and electronic display to enter data which the computer edits and may respond to. { 'dī·ə,läg }

dialog box |COMPUT SCI| On a computer screen, a small window that is used to emphasize the importance of some action or to request an answer to a question. { 'dī·ə,läg ,bäks }

dial pulse interpreter [ELECTR] A device that converts the signaling pulses of a dial telephone to a form suitable for data entry to a computer. { 'dīl ,pəls in'tər·prəd·ər }

dial pulsing *See* loop pulsing. { 'dīl ,pəls·iŋ }

dial telephone system [COMMUN] A telephone system in which telephone connections between customers are ordinarily established by electronic and mechanical apparatus, controlled by manipulations of dials operated by calling parties. { 'dīl 'tel·ə,fōn ,sist·əm }

dial tone [COMMUN] A tone employed in a dial telephone system to indicate that the equipment is ready for dialing operation. { 'dīl ,tōn }

dial-up [COMMUN] **1.** The service whereby a dial telephone can be used to initiate and effect station-to-station telephone calls. **2.** In computer networks, pertaining to terminals which must dial up to receive service, as contrasted with those hand-wired or permanently connected into the network. { 'dīl ,əp }

dial-up telephone system [COMMUN] The switched telephone network that is regulated by national governments; operated in the United States by various carriers. { ¦dīl ,əp 'tel·ə,fōn ,sis·təm }

diamagnetic [ELECTROMAG] Having a magnetic permeability less than 1; materials with this property are repelled by a magnet and tend to position themselves at right angles to magnetic lines of force. { ¦dīəmag'ned·ik }

diamond antenna *See* rhombic antenna. { 'dī ,mənd an'ten·ə }

diamond circuit [ELECTR] A gate circuit that provides isolation between input and output terminals in its off state, by operating transistors in their cutoff region; in the on state the output voltage follows the input voltage as required for gating both analog and digital signals, while the transistors provide current gain to supply output current on demand. { 'dī·mənd ,sər·kət }

diaphragm [ELECTROMAG] *See* iris. [ENG ACOUS] A thin, flexible sheet that can be moved by sound waves, as in a microphone, or can produce sound waves when moved, as in a loudspeaker. { 'dī·ə ,fram }

diaphragm horn [ENG ACOUS] A horn that produces sound by means of a diaphragm vibrated by compressed air, steam, or electricity. { 'dī·ə ,fram ,hòrn }

diathermy interference [COMMUN] Television interference caused by diathermy equipment; produces a herringbone pattern in a dark horizontal band across the picture. { 'dī·ə,thər·mē ,in·tər'fir·əns }

diathermy machine [ELECTR] A radio-frequency oscillator, sometimes followed by rf amplifier stages, used to generate high-frequency currents that produce heat within some part of the body for therapeutic purposes. { 'dī·ə,thər·mē mə ,shēn }

dibit [COMPUT SCI] A pair of binary digits, used to specify one of four values. { 'dī,bit }

di-cap storage [ELECTR] Device capable of holding data in the form of an array of charged capacitors and using diodes for controlling information flow. { 'dī,kap 'stòr·ij }

DICE *See* digital intercontinental conversion equipment.

dichotomizing search [COMPUT SCI] A procedure for searching an item in a set, in which, at each step, the set is divided into two parts, one part being then discarded if it can be logically shown that the item could not be in that part. { dī'käd·ə,mīz·iŋ ,sərch }

dichotomy [COMPUT SCI] A division into two subordinate classes; for example, all white and all nonwhite, or all zero and all nonzero. { dī'käd·ə·mē }

dicing [ELECTR] Sawing or otherwise machining a semiconductor wafer into small squares, or dice, from which transistors and diodes can be fabricated. { 'dīs·iŋ }

Dicke radiometer [ELECTR] A radiometer-type receiver that detects weak signals in noise by modulating or switching the incoming signal before it is processed by conventional receiver circuits. { 'dik·ə ,rād·ē'äm·əd·ər }

dictionary [COMPUT SCI] A table establishing the correspondence between specific words and their code representations. { 'dik·shə,ner·ē }

dictionary code [COMPUT SCI] An alphabetical arrangement of English words and terms, associated with their code representations. { 'dik·shə ,ner·ē ,kōd }

dictionary encoding [COMPUT SCI] A method of data compression in which each word is replaced by a number which is the position of that word in a dictionary. { 'dik·shə,ner·ē en'kōd·iŋ }

dictionary sort [COMPUT SCI] A sort algorithm that ignores capitalization, punctuation, and spaces, and treats numbers as if they were spelled out alphabetically. { 'dik·shə,ner·ē ,sòrt }

die [ELECTR] The tiny, sawed or otherwise machined piece of semiconductor material used in the construction of a transistor, diode, or other semiconductor device; plural is dice. { dī }

dielectric *See* dieletric material. { ,dī·ə'lek·trik }

dielectric absorption [ELEC] The persistence of electric polarization in certain dielectrics after removal of the electric field. { ,dī·ə'lek·trik əb'sòrp·shən }

dielectric amplifier [ELECTR] An amplifier using a ferroelectric capacitor whose capacitance varies with applied voltage so as to give signal amplification. { ,dī·ə'lek·trik 'am·plə,fī·ər }

dielectric antenna [ELECTROMAG] An antenna in which a dielectric is the major component used to produce a desired radiation pattern. { ,dī·ə'lek·trik an'ten·ə }

dielectric breakdown [ELECTR] Breakdown which occurs in an alkali halide crystal at field strengths on the order of 10^6 volts per centimeter. { ,dī·ə'lek·trik 'brāk,daùn }

dielectric circuit [ELEC] Any electric circuit which has capacitors. { ,dī·ə'lek·trik 'sər·kət }

dielectric constant [ELEC] **1.** For an isotropic medium, the ratio of the capacitance of a capacitor filled with a given dielectric to that of the same capacitor having only a vacuum as dielectric. **2.** More generally, $1 + \gamma\chi$, where γ is 4π in Gaussian and cgs electrostatic units or 1 in rationalized mks units, and χ is the electric susceptibility tensor. Also known as relative dielectric constant; relative permittivity; specific inductive capacity (SIC). { ‚dī·ə'lek·trik 'kän·stənt }

dielectric crystal |ELEC| A crystal which is electrically nonconducting { ‚dī·ə'lek·trik 'krist·əl }

dielectric current |ELEC| The current flowing at any instant through a surface of a dielectric that is located in a changing electric field. { ‚dī·ə'lek·trik 'kər·ənt }

dielectric displacement See electric displacement. { ‚dī·ə'lek·trik di'splās·mənt }

dielectric ellipsoid |ELEC| For an anisotropic medium in which the dielectric constant is a tensor quantity **K**, the locus of points **r** satisfying $\mathbf{r} \cdot \mathbf{K} \cdot \mathbf{r} = 1$. { ‚dī·ə'lek·trik ə'lip‚sòid }

dielectric fatigue |ELECTR| The property of some dielectrics in which resistance to breakdown decreases after a voltage has been applied for a considerable time. { ‚dī·ə'lek·trik fə'tēg }

dielectric field |ELEC| The average total electric field acting upon a molecule or group of molecules inside a dielectric. Also known as internal dielectric field. { ‚dī·ə'lek·trik 'fēld }

dielectric film |ELEC| A film possessing dielectric properties; used as the central layer of a capacitor. { ‚dī·ə'lek·trik 'film }

dielectric flux density See electric displacement. { ‚dī·ə'lek·trik 'fləks ‚den·səd·ē }

dielectric gas |ELEC| A gas having a high dielectric constant, such as sulfur hexafluoride. { ‚dī·ə'lek·trik 'gas }

dielectric heating |ELEC| Heating of a nominally electrical insulating material due to its own electrical (dielectric) losses, when the material is placed in a varying electrostatic field. { ‚dī·ə'lek·trik 'hēd·iŋ }

dielectric hysteresis See ferroelectric hysteresis. { ‚dī·ə'lek·trik hi·stə'rē·səs }

dielectric leakage |ELEC| A very small steady current that flows through a dielectric subject to a steady electric field. { ‚dī·ə'lek·trik 'lēk·ij }

dielectric lens |ELECTROMAG| A lens made of dielectric material so that it refracts radio waves in the same manner that an optical lens refracts light waves; used with microwave antennas. { ‚dī·ə'lek·trik 'lenz }

dielectric-lens antenna |ELECTROMAG| An aperture antenna in which the beam width is determined by the dimensions of a dielectric lens through which the beam passes. { ‚dī·ə¦lek·trik ¦lenz an'ten·ə }

dielectric loss |ELECTROMAG| The electric energy that is converted into heat in a dielectric subjected to a varying electric field. Also known as dielectric absorption. { ‚dī·ə'lek·trik 'lòs }

dielectric loss angle |ELEC| difference between 90° and the dielectric phase angle. { ‚dī·ə'lek·trik 'lòs ‚aŋ·gəl }

dielectric loss factor |ELEC| Product of the dielectric constant of a material and the tangent of its dielectric loss angle. { ‚dī·ə¦lek·trik ¦lòs ‚fak·tər }

dielectric matching plate |ELECTROMAG| In waveguide technique, a dielectric plate used as an impedance transformer for matching purposes. { ‚dī·ə'lek·trik 'mach·iŋ ‚plāt }

dielectric material |MATER| Also known as dielectric. **1.** A material which is an electrical insulator or in which an electric field can be sustained with a minimum dissipation of power. **2.** In a more general sense, any material other than a condensed state of a metal. { ‚dī·ə'lek·trik mə‚tir·ē·əl }

dielectric phase angle |ELEC| Angular difference in phase between the sinusoidal alternating potential difference applied to a dielectric and the component of the resulting alternating current having the same period as the potential difference. { ‚dī·ə'lek·trik 'fāz ‚aŋ·gl }

dielectric polarization See polarization. { ‚dī·ə'lek·trik ‚pō·lə·rə'zā·shən }

dielectric power factor |ELEC| Cosine of the dielectric phase angle (or sine of the dielectric loss angle). { ‚dī·ə'lek·trik 'paùr ‚fak·tər }

dielectric-rod antenna |ELECTROMAG| A surface-wave antenna in which an end-fire radiation pattern is produced by propagation of a surface wave on a tapered dielectric rod. { ‚dī·ə¦lek·trik ¦räd an'ten·ə }

dielectric shielding |ELEC| The reduction of an electric field in some region by interposing a dielectric substance, such as polystyrene, glass, or mica. { ‚dī·ə'lek·trik 'shēld·iŋ }

dielectric strength |ELEC| The maximum electrical potential gradient that a material can withstand without rupture; usually specified in volts per millimeter of thickness. Also known as electric strength. { ‚dī·ə'lek·trik 'streŋkth }

dielectric susceptibility See electric susceptibility. { ‚dī·ə'lek·trik sə‚sep·tə'bil·əd·ē }

dielectric test |ELEC| A test involving application of a voltage higher than the rated value for a specified time, to determine the margin of safety against later failure of insulating materials. { ‚dī·ə'lek·trik 'test }

dielectric waveguide |ELEC| A waveguide consisting of a dielectric cylinder surrounded by air. { ‚dī·ə'lek·trik 'wāv‚gīd }

dielectric wedge |ELECTROMAG| A wedge-shaped piece of dielectric used in a waveguide to match its impedance to that of another waveguide. { ‚dī·ə'lek·trik 'wej }

dielectric wire |ELECTROMAG| A dielectric waveguide used to transmit ultra-high-frequency raio waves short distances between parts of a circuit. { ‚dī·ə'lek·trik 'wīr }

difference amplifier See differential amplifier. { 'dif·rəns ‚am·plə‚fī·ər }

difference channel |ENG ACOUS| An audio channel that handles the difference between the signals in the left and right channels of a stereophonic sound system. { 'dif·rəns ˌchan·əl }

difference detector [ELECTR] A detector circuit in which the output is a function of the difference between the amplitudes of the two input waveforms. { 'dif·rəns di,tek·tər }

difference encoding [COMPUT SCI] A method of data compression that takes advantage of a sequence of data that differs little from one value to the next by encoding each value as the difference from the previous value. { 'dif·rəns in,kōd·iŋ }

difference equation [MATH] An equation expressing a functional relationship of one or more independent variables, one or more functions dependent on these variables, and successive differences of these functions. { 'dif·rəns i'kwā·zhən }

difference in depth modulation [COMMUN] In directive systems employing overlapping lobes with modulated signals, a ratio obtained by subtracting from the percentage of modulation of the larger signal the percentage of modulation of the smaller signal and dividing by 100. { 'dif·rəns 'in ¦depth ,mäj·ə'lā·shən }

difference mapping [COMMUN] A method of coding information in which a sample value is presented as an error term formed by the difference between the sample and the previous sample. { 'dif·rəns ,map·iŋ }

differential [CONT SYS] The difference between levels for turn-on and turn-off operation in a control system. { ,dif·ə'ren·chəl }

differential amplifier [ELECTR] An amplifier whose output is proportional to the difference between the voltages applied to its two inputs. Also called difference amplifier. { ,dif·ə'ren·chəl 'am·plə,fī·ər }

differential analyzer [COMPUT SCI] A mechanical or electromechanical device designed primarily to solve differential equations. { ,dif·ə'ren·chəl 'an·ə,līz·ər }

differential backup [COMPUT SCI] Backup of only files that have been changed or added since the last backup. { ,dif·ə,ren·chəl 'bak,əp }

differential capacitance [ELECTR] The derivative with respect to voltage of a charge characteristic, such as an alternating charge characteristic or a mean charge characteristic, at a given point on the characteristic. { ,dif·ə'ren·chəl kə'pas·əd·əns }

differential capacitor [ELEC] A two-section variable capacitor having one rotor and two stators so arranged that as capacitance is reduced in one section it is increased in the other. { ,dif·ə'ren·chəl kə'pas·əd·ər }

differential comparator [ELECTR] A comparator having at least two high-gain differential-amplifier stages, followed by level-shifting and buffering stages, as required for converting a differential input to single-ended output for digital logic applications. { ,dif·ə'ren·chəl kəm'par·əd·ər }

differential compound motor [ELEC] A direct-current motor whose speed may be made nearly

constant or may be adjusted to increase with increasing load. { ,dif·ə'ren·chəl 'käm,paůnd ,mōd·ər }

differential delay [COMMUN] The difference between the maximum and minimum frequency delays occurring across a band. { ,dif·ə'ren·chəl di'lā }

differential discriminator [ELECTR] A discriminator that passes only pulses whose amplitudes are between two predetermined values, neither of which is zero. { ,dif·ə'ren·chəl di'skrim·ə ,nād·ər }

differential duplex system [ELECTR] System in which the sent currents divide through two mutually inductive sections of a receiving apparatus, connected respectively to the line and to a balancing artificial line in opposite directions, so that there is substantially no net effect on the receiving apparatus; the received currents pass mainly through one section, or through the two sections in the same direction, and operate the apparatus. { ,dif·ə'ren·chəl 'dü,pleks ,sis·təm }

differential electromagnet [ELEC] An electromagnet having part of its winding opposed to the other part, so that the force exerted by the magnet can be adjusted. { ,dif·ə'ren·chəl i ,lek·trō'mag·nət }

differential encoding [COMMUN] A method of compressing television signals by transmitting only differences between pixels in neighboring lines and successive frames. { ,dif·ə,ren·chəl in'kōd·iŋ }

differential frequency circuit [ELEC] A circuit that provides a continuous output frequency equal to the absolute difference between two continuous input frequencies. { ,dif·ə'ren·chəl ¦frē·kwən·sē ¦sər·kət }

differential frequency meter [ENG] A circuit that converts the absolute frequency difference between two input signals to a linearly proportional direct-current output voltage that can be used to drive a meter, recorder, oscilloscope, or other device. { ,dif·ə'ren·chəl 'frē·kwən·sē ,mēd·ər }

differential gain control [ELECTR] Device for altering the gain of a radio receiver according to expected change of signal level, to reduce the amplitude differential between the signals at the output of the receiver. Also known as gain sensitivity control. { ,dif·ə'ren·chəl ,gān kən,trōl }

differential galvanometer [ELEC] A galvanometer having a magnetic needle which is free to rotate in the magnetic field produced by currents flowing in opposite directions through two separate identical coils, so that there is no deflection when the currents are equal. { ,dif·ə'ren·chəl ,gal·və'näm·əd·ər }

differential game [CONT SYS] A two-sided optimal control problem. { ,dif·ə'ren·chəl 'gām }

differential gap controller [CONT SYS] A two-position (on-off) controller that actuates when the manipulated variable reaches the high or low value of its range (differential gap). { ,dif·ə'ren·chəl 'gap kən,trōl·ər }

differential generator [ELEC] A generator whose shunt and series windings are opposed to each other, to limit the maximum current. { ‚dif·ə'ren·chəl 'jen·ə‚rād·ər }

differential input [ELECTR] Amplifier input circuit that rejects voltages that are the same at both input terminals and amplifies the voltage difference between the two input terminals. { ‚dif·ə'ren·chəl 'in‚pút }

differential-input capacitance [ELECTR] The capacitance between the inverting and noninverting input terminals of a differential amplifier. { ‚dif·ə¦ren·chəl ¦in‚pút kə'pas·əd·əns }

differential-input impedance [ELECTR] The impedance between the inverting and noninverting input terminals of a differential amplifier. { ‚dif·ə¦ren·chəl ¦in‚pút im'ped·əns }

differential-input measurement [ELECTR] A measurement in which the two inputs to a differential amplifier are connected to two points in a circuit under test and the amplifier displays the difference voltage between the points. { ‚dif·ə¦ren·chəl ¦in‚pút 'mezh·ər·mənt }

differential-input resistance [ELECTR] The resistance between the inverting and noninverting input terminals of a differential amplifier. { ‚dif·ə¦ren·chəl ¦in‚pút ri'zis·təns }

differential-input voltage [ELECTR] The maximum voltage that can be applied across the input terminals of a differential amplifier without causing damage to the amplifier. { ‚dif·ə¦ren·chəl ¦in‚pút 'vōl‚tij }

differential instrument [ENG] Galvanometer or other measuring instrument having two circuits or coils, usually identical, through which currents flow in opposite directions; the difference or differential effect of these currents actuates the indicating pointer. { ‚dif·ə'ren·chəl 'in·strə·mənt }

differential keying [ELECTR] Method for obtaining chirp-free break-in keying of continuous wave transmitters by using circuitry that arranges to have the oscillator turn on fast before the keyed amplifier stage can pass any signal, and turn off fast after the keyed amplifier stage has cut off. { ‚dif·ə'ren·chəl 'kē·iŋ }

differentially coherent phase-shift keying See differential phase-shift keying. { ‚dif·ə'ren·chə·lē kō'hir·ənt 'fāz ‚shift ‚kē·iŋ }

differential microphone See double-button microphone. { ‚dif·ə'ren·chəl 'mī·krə‚fōn }

differential-mode gain [ELECTR] The ratio of the output voltage of a differential amplifier to the differential-mode input voltage. { ‚dif·ə¦ren·chəl ¦mōd ‚gān }

differential-mode input [ELECTR] The voltage difference between the two inputs of a differential amplifier. { ‚dif·ə¦ren·chəl ¦mōd ‚in‚pút }

differential-mode signal [ELECTR] A signal that is applied between the two ungrounded terminals of a balanced three-terminal system. { ‚dif·ə¦ren·chəl ¦mōd ‚sig·nəl }

differential modulation [COMMUN] Modulation in which the choice of the significant condition for any signal element is dependent on the choice for the previous signal element. { ‚dif·ə'ren·chəl ‚mäj·ə'lā·shən }

differential motor [ELEC] A direct-current motor whose shunt and series field windings oppose each other to produce a constant speed. { ‚dif·ə'ren·chəl 'mōd·ər }

differential operational amplifier [ELECTR] An amplifier that has two input terminals, used with additional circuit elements to perform mathematical functions on the difference in voltage between the two input signals. { ‚dif·ə'ren·chəl äp·ə'rā·shən·əl 'am·plə‚fī·ər }

differential output voltage [ELECTR] The difference between the values of two ac voltages, 180° out of phase, present at the output terminals of an amplifier when a differential input voltage is applied to the input terminals of the amplifier. { ‚dif·ə'ren·chəl 'aút‚pút ‚vōl·tij }

differential phase [ELECTR] Difference in output phase of a small high-frequency sine-wave signal at two stated levels of a low-frequency signal on which it is superimposed in a video transmission system. { ‚dif·ə'ren·chəl 'fāz }

differential phase-shift keying [COMMUN] Form of phase-shift keying in which the reference phase for a given keying interval is the phase of the signal during the preceding keying interval. Also known as differentially coherent phase-shift keying. { ‚dif·ə'ren·chəl 'fāz ‚shift ‚kē·iŋ }

differential-pressure pickup [ELEC] An instrument that measures the difference in pressure between two pressure sources and translates this difference into a change in inductance, resistance, voltage, or some other electrical quality. { ‚dif·ə¦ren·chəl ¦presh·ər ‚pik‚əp }

differential pulse-code modulation [COMMUN] A type of pulse-code modulation in which an analog signal is sampled and the difference between its actual value and its predicted value, based on a previous sample or samples, is quantized; for example, in television transmission, only the differences between the continuous picture elements on the scanning lines are transmitted, enabling the bandwidth of the signal to be reduced. Abbreviated DPCM. { dif·ə'ren·chəl 'pəls ‚kōd ‚mäj·ə'lā·shən }

differential relay [ELEC] A two-winding relay that operates when the difference between the currents in the two windings reaches a predetermined value. { ‚dif·ə'ren·chəl 'rē‚lā }

differential selsyn [ELEC] Selsyn in which both rotor and stator have similar windings are spread 120° apart; position of the rotor corresponds to the algebraic sum of the fields produced by the stator and rotor. { ‚dif·ə'ren·chəl 'sel·sən }

differential signal [ELECTR] In a circuit, a signal that is the voltage difference between two nodes, neither of which is at ground potential. Also known as floating signal. { ‚dif·ə'ren·chəl 'sig·nəl }

differential stage [ELECTR] A symmetrical amplifier stage with two inputs balanced against each other so that with no input signal or equal input signals, no output signal exists, while a signal to either input, or an input signal unbalance, produces an output signal proportional to the difference. { ‚dif·ə'ren·chəl 'stāj }

differential synchro *See* synchro differential receiver; synchro differential transmitter. { ‚dif·ə'ren·chəl 'siŋ·krō }

differential transducer [ELEC] A transducer that simultaneously senses two separate sources and provides an output proportional to the difference between them. { ‚dif·ə'ren·chəl tranz'dü·sər }

differential transformer [ELEC] A transformer used to join two or more sources of signals to a common transmission line. { ‚dif·ə'ren·chəl tranz'fór·mər }

differential-transformer transducer [ELEC] A transducer in which movement of the iron core of a transformer varies the output voltage across two series-opposing secondary windings. { ‚dif·ə¦ren·chəl tranz¦fór·mər tranz'dü·sər }

differential voltage gain [ELECTR] Ratio of the change in output signal voltage at either terminal, or in a differential device, to the change in signal voltage applied to either input terminal, all voltages being measured to common reference. { ‚dif·ə¦ren·chəl 'vōl·tij ‚gān }

differential voltmeter [ELEC] A voltmeter that measures only the difference between a known voltage and an unknown voltage. { ‚dif·ə'ren·chəl 'vōlt‚mēd·ər }

differential winding [ELEC] A winding whose magnetic field opposes that of a nearby winding. { ‚dif·ə'ren·chəl 'wīnd·iŋ }

differential wound field [ELEC] Type of motor or generator field having both series and shunt coils that are connected to oppose each other. { ‚dif·ə¦ren·chəl ¦waúnd 'fēld }

differentiating circuit [ELEC] A circuit whose output voltage is proportional to the rate of change of the input voltage. Also known as differentiating network. { ‚dif·ə¦ren·chē‚ād·iŋ ¦sər·kət }

differentiating network *See* differentiating circuit. { ‚dif·ə¦ren·chē‚ād·iŋ 'net‚wərk }

differentiator [ELECTR] A device whose output function is proportional to the derivative, or rate of change, of the input function with respect to one or more variables. { ‚dif·ə'ren·chē‚ād·ər }

diffractional pulse-height discriminator *See* pulse-height selector. { di'frak·shən·əl 'pəls ‚hīt di'skrim·ə‚nād·ər }

diffused-alloy transistor [ELECTR] A transistor in which the semiconductor wafer is subjected to gaseous diffusion to produce a nonuniform base region, after which alloy junctions are formed in the same manner as for an alloy-junction transistor; it may also have an intrinsic region, to give a *pnip* unit. Also known as drift transistor. { də¦fyüzd 'al‚ói tran'zis·tər }

diffused-base transistor [ELECTR] A transistor in which a nonuniform base region is produced by gaseous diffusion; the collector-base junction is also formed by gaseous diffusion, while the emitter-base junction is a conventional alloy junction. { də¦fyüzd ¦bās tran'zis·tər }

diffused emitter-collector transistor [ELECTR] A transistor in which both the emitter and collector are produced by diffusion. { də¦fyüzd i'mid·ər kə'lek·tər tran'zis·tər }

diffused junction [ELECTR] A semiconductor junction that has been formed by the diffusion of an impurity within a semiconductor crystal. { də'fyüzd 'jəŋk·shən }

diffused-junction rectifier [ELECTR] A semiconductor diode in which the *pn* junction is produced by diffusion. { də¦fyüzd ¦jəŋk·shən 'rek·tə‚fī·ər }

diffused-junction transistor [ELECTR] A transistor in which the emitter and collector electrodes have been formed by diffusion by an impurity metal into the semiconductor wafer without heating. { də¦fyüzd ¦jəŋk·shən tran'zis·tər }

diffused mesa transistor [ELECTR] A diffused-junction transistor in which an *n*-type impurity is diffused into one side of a *p*-type wafer; a second *pn* junction, required for the emitter, is produced by alloying or diffusing a *p*-type impurity into the newly formed *n*-type surface; after contacts have been applied, undesired diffused areas are etched away to create a flat-topped peak called a mesa. { də¦fyüzd ¦mā·sə tran'zis·tər }

diffused resistor [ELECTR] An integrated-circuit resistor produced by a diffusion process in a semiconductor substrate. { də'fyüzd ri'zis·tər }

diffusion [ELECTR] A method of producing a junction by difusing an impurity metal into a semiconductor at a high temperature. { də'fyü·zhən }

diffusion capacitance [ELECTR] The rate of change of stored minority-carrier charge with the voltage across a semiconductor junction. { də'fyü·zhən kə'pas·əd·əns }

diffusion theory [ELEC] The theory that in semiconductors, where there is a variation of carrier concentration, a motion of the carriers is produced by diffusion in addition to the drift determined by the mobility and the electric field. { də'fyü·zhən ‚thē·ə·rē }

diffusion transistor [ELECTR] A transistor in which current flow is a result of diffusion of carriers, donors, or acceptors, as in a junction transistor. { də¦fyü·zhən tran‚zis·tər }

digicom [COMMUN] A wire communication system that transmits speech signals in the form of corresponding trains of pulses and transmits digital information directly from computers, radar, tape readers, teleprinters, and telemetering equipment. { 'dij·ə‚käm }

digicon [ELECTR] An image tube in which the image produced by electrons from the photocathode is focused directly on a silicon diode array and each incoming photoelectron produces an electrical pulse that is amplified and recorded. { 'dij·ə‚kän }

digit [COMPUT SCI] In a decimal digital computer, the space reserved for storage of one digit of information. { 'dij·ət }

digit absorbing selector [ELECTR] Dial switch arranged to set up and then fall back on the first one of two digits dialed; it then operates on the next digit dialed. { 'dij·ət əb‚sórb·iŋ si'lek·tər }

digital [COMPUT SCI] Pertaining to data in the form of digits. { 'dij·əd·əl }

digital audio broadcasting |COMMUN| The radio broadcasting of audio signals encoded in digital form. Abbreviated DAB. { ¦dij·əd·əl ¦ȯd·ē·ō 'brȯd,kast·iŋ }

digital audio tape |COMPUT SCI| A magnetic tape on which sound is recorded and played back in digital form. Abbreviated DAT. { ¦dij·əd·əl 'ȯd·ē·ō ,tāp }

digital camera |ELECTR| A television camera that breaks up a picture into a fixed number of pixels and converts the light intensity (or the intensities of each of the primary colors) in each pixel to one of a finite set of numbers. { 'dij·əd·əl 'kam·rə }

digital channel |COMMUN| A transmission path that carries only digital signals. { 'dij·əd·əl 'chan·əl }

digital circuit |ELECTR| A circuit designed to respond at input voltages at one of a finite number of levels and, similarly, to produce output voltages at one of a finite number of levels. { 'dij·əd·əl 'sər·kət }

digital circuit multiplication equipment |COMMUN| Equipment that uses digital compression techniques to increase the capacity of digital satellite and cable links carrying voice, facsimile, and voice-frequency modem traffic. { ,dij·əd·əl ,sər·kət ,məl·tə·plə'kā·shən i,kwip·mənt }

digital communications |COMMUN| System of telecommunications employing a nominally discontinuous signal that changes in frequency, amplitude, time, or polarity. { 'dij·əd·əl kə,myü·nə'kā·shənz }

digital comparator |ELECTR| A comparator circuit operating on input signals at discrete levels. Also known as discrete comparator. { 'dij·əd·əl kəm'par·əd·ər }

digital computer |COMPUT SCI| A computer operating on discrete data by performing arithmetic and logic processes on these data. { 'dij·əd·əl kəm'pyüd·ər }

digital control |CONT SYS| The use of digital or discrete technology to maintain conditions in operating systems as close as possible to desired values despite changes in the operating environment. { 'dij·əd·əl kən'trōl }

digital converter |ELECTR| A device that converts voltages to digital form; examples include analog-to-digital converters, pulse-code modulators, encoders, and quantizing encoders. { 'dij·əd·əl kən'vərd·ər }

digital counter |ELECTR| A discrete-state device (one with only a finite number of output conditions) that responds by advancing to its next output condition. { 'dij·əd·əl 'kaúnt·ər }

digital data |COMPUT SCI| Data that are electromagnetically stored in the form of discrete digits. { 'dij·əd·əl 'dad·ə }

digital data modulation system |COMMUN| A digital communications system in which the information source consists of a finite number of discrete messages which are coded into a sequence of waveforms or symbols, each one selected from a specified and finite set. { 'dij·əd·əl 'dad·ə ,mäj·ə'lā·shən ,sis·təm }

digital data recorder |COMPUT SCI| Electronic device that converts continuous electrical analog signals into number (digital) values and records these values onto a data log via a high-speed typewriter. { 'dij·əd·əl ¦dad·ə ri,kȯrd·ər }

digital data service |COMMUN| A telephone communication system developed specifically for digital data, using existing local digital lines combined with data-under-voice microwave transmission facilities. Abbreviated DDS. { 'dij·əd·əl 'dad·ə ,sər·vəs }

digital delayer |ENG ACOUS| A device for introducing delay in the audio signal in a sound-reproducing system, which converts the audio signal to digital format and stores it in a digital shift register before converting it back to analog form. { 'dij·əd·əl di'lā·ər }

digital delay generator |ELECTR| A high-precision adjustable time-delay generator in which delays may be selected in increments such as 1, 10, or 100 nanoseconds by means of panel switches and sometimes by remote programming. { 'dij·əd·əl di¦lā jen·ə'rād·ər }

digital differential analyzer |COMPUT SCI| A differential analyzer which uses numbers to represent analog quantities. Abbreviated DDA. { 'dij·əd·əl ,dif·ə,ren·chəl 'an·ə,līz·ər }

digital display |COMPUT SCI| A display in which the result is indicated in directly readable numerals. { 'dij·əd·əl di'splā }

Digital Electronic Message Service |COMMUN| A communication system whose purpose is to provide efficient means for two-way high-speed data communications, transfer of graphic images (fascimile), and teleconferencing between cities and within a city environment. Abbreviated DEMS. { 'dij·əd·əl i,lek'trän·ik 'mes·ij ,sər·vəs }

digital filter |ELECTR| An electrical filter that responds to an input which has been quantified, usually as pulses. { 'dij·əd·əl 'fil·tər }

digital format |COMPUT SCI| Use of discrete integral numbers in a given base to represent all the quantities that occur in a problem or calculation. { 'dij·əd·əl 'fȯr·mat }

digital frequency meter |ELECTR| A frequency meter in which the value of the frequency being measured is indicated on a digital display. { 'dij·əd·əl 'frē·kwən·sē ,mēd·ər }

digital incremental plotter |COMPUT SCI| A device for converting digital signals in the output of a computer into graphical form, in which the digital signals control the motion of a plotting pen and of a drum that carries the paper on which the graph is drawn. { 'dij·əd·əl ,iŋ·krə,ment·əl 'pläd·ər }

digital integrator |COMPUT SCI| A device for computing definite integrals in which increments in the input variables and output variable are represented by digital signals. { 'dij·əd·əl 'in·tə ,grād·ər }

digital intercontinental conversion equipment |ELECTR| Equipment which uses pulse-code modulation to convert a 525-line, 60-frame-per-second television signal used in the United

digital loop carrier

States into a 625-line, 50-frame-per-second phase-alternation line signal used in Europe; the 525-line signal is sampled and quantized into a pulse-code modulation signal which is stored in shift registers from which the phase-alternation line signal is read out. Abbreviated DICE. { 'dij·əd·əl ,in·tər,känt·ən'ent·əl kən'vər·zhən i,kwip·mənt }

digital loop carrier [COMMUN] A technology for providing 24 or more telephone circuits on many fewer pairs of wires, in which analog input signals are first sampled and digitized, and the binary digital signals from each user is then time-multiplexed into a single bit stream. { 'dij·əd·əl 'lüp ,kar·ē·ər }

digital message entry system [ELECTR] A system that encodes formatted messages in digital form; it enters the encoded digital information into a voice communications transceiver by frequency shift techniques. { ¦dij·əd·əl ¦mes·ij 'en·trē ,sis·təm }

digital microwave radio [COMMUN] Transmission of voice and data signals in digital form on microwave links, as in the 2-gigahertz common-carrier bands; pulse-code modulation is used. { ¦dij·əd·əl ¦mī·krō,wāv 'rād·ē·ō }

digital modulation [COMMUN] A method of placing digital traffic on a microwave system without use of modems, by transmitting the information in the form of discrete phase or frequency states determined by the digital signal. { 'dij·əd·əl ,mäj·ə'lā·shən }

digital monitor [ELECTR] A display unit that accepts digital signals and converts them to analog signals internally in order to illuminate the screen. { 'dij·əd·əl 'män·əd·ər }

Digital Multiplexed Interface [COMPUT SCI] A cost-effective, high-speed interconnection between terminals and host computers in a private branch exchange environment. { 'dij·əd·əl 'məl·tə,plekst 'in·tər,fās }

digital multiplier [ELECTR] A multiplier that accepts two numbers in digital form and gives their product in the same digital form, usually by making repeated additions; the multiplying process is simpler if the numbers are in binary form wherein digits are represented by a 0 or 1. { 'dij·əd·əl 'məl·tə,plī·ər }

digital object identifier [COMPUT SCI] A system for identifying and exchanging intellectual properties (including, for example, physical objects as well as digital files) in the digital environment. { ¦dij·əd·əl ¦äb,jekt ī'den·tə,fī·ər }

digital output [ELECTR] An output signal consisting of a sequence of discrete quantities coded in an appropriate manner for driving a printer or digital display. { 'dij·əd·əl 'aút,pút }

digital phase shifter [ELECTR] Device which provides a signal phase shift by the application of a control pulse; a reversal or phase shift requires a control pulse of opposite polarity. { 'dij·əd·əl 'fāz ,shif·tər }

digital plotter [ELECTR] A recorder that produces permanent hard copy in the form of a graph from digital input data. { 'dij·əd·əl 'pläd·ər }

digital printer [COMPUT SCI] A printer that provides a permanent readable record of binary-coded decimal or other coded data in a digital form that may include some or all alphanumeric characters and special symbols along with numerals. Also known as digital recorder. { 'dij·əd·əl 'print·ər }

digital private automatic branch exchange [COMMUN] A central communications switching system for a local-area network, which employs existing telephone wires in a building for the connection of telephones and computer terminals and systems. { 'dij·əd·əl ¦prīv·ət ¦öd·ə¦mad·ik 'branch iks,chānj }

digital radio [COMMUN] The microwave transmission of digital signals through space or the atmosphere. { ¦dij·əl·əl 'rād·ē·ō }

digital recorder See digital printer. { 'dij·əd·əl ri'kòrd·ər }

digital recording [ELECTR] Magnetic recording in which the information is first coded in a digital form, generally with a binary code that uses two discrete values of residual flux. { 'dij·əd·əl ri'kòrd·iŋ }

digital representation [COMPUT SCI] The use of discrete impulses or quantities arranged in coded patterns to represent variables or other data in the form of numbers or characters. { 'dij·əd·əl ,rep·rə,zen'tā·shən }

digital resolution [COMPUT SCI] The ability of a digital computer to approach a truly correct answer, generally established by the number of places expressed, and the value of the least significant digit in a digitally coded representation. { 'dij·əd·əl ,rez·ə'lü·shən }

digital set-top box [COMMUN] A device that is attached to a television receiver and can collect, store, and output digitally compressed television signals. { ,dij·əd·əl 'set,täp ,bäks }

digital signal analyzer [ELECTR] A signal analyzer in which one or more analog inputs are sampled at regular intervals, converted to digital form, and fed to a memory. { 'dij·əd·əl 'sig·nəl ,an·ə ,liz·ər }

digital signal processing See signal processing. { ,dij·əd·əl ,sig·nəl 'prä·səs·iŋ }

digital signal processing chip [COMPUT SCI] A digital device for executing algorithms for the transformation or extraction of information from signals originally in analog form, such as audio or images. Abbreviated DSP chip. Also known as digital signal processor. { ,dij·əd·əl ,sig·nəl 'prä·səs·iŋ ,chip }

digital signal processor See digital signal processing chip. { ,dij·əd·əl 'sig·nəl ,prä,ses·ər }

digital signature [COMMUN] A set of alphabetic or numeric characters used to authenticate a cryptographic message by ensuring that the sender cannot later disavow the message, the receiver cannot forge the message or signature, and the receiver can prove to others that the contents of the message are genuine and originated with the sender. { 'dij·əd·əl 'sig·nə·chər }

digital simulation [COMPUT SCI] The representation of a system in a form acceptable to a digital

158

computer as opposed to an analog computer.
{ 'dij·əd·əl ˌsim·yə'lā·shən }
digital speech communications [COMMUN]
Transmission of voice in digitized or binary form
via landline or radio. { 'dij·əd·əl 'spēch kə
ˌmyün·əˌkā·shənz }
digital speech interpolation [COMMUN] In digital
speech communications, the use of periods of
inactivity or constant signal level to increase the
transmission efficiency by insertion of additional
signals. Abbreviated DSI. { ˌdij·əd·əl 'spēch ˌin·
tər·pəˌlā·shən }
digital subscriber line [COMMUN] A system that
provides subscribers with continuous, uninterrupted connections to the Internet over existing
telephone lines, offering a choice of speeds
ranging from 32 kilobits per second to more
than 50 megabits per second. Abbreviated DSL.
{ ¦dij·əd·əl səb'skrīb·ər ˌlīn }
digital synchronometer [ELECTR] A time comparator that provides a direct-reading digital
display of time with high precision by making accurate comparisons between its own digital clock
and high-accuracy time transmissions from radio
station WWV or a loran C station. { 'dij·əd·əl
ˌsiŋ·krə'näm·əd·ər }
digital system [COMPUT SCI] Any of the levels
of operation for a digital computer, including
the wires and mechanical parts, the logical
elements, and the functional units for reading,
writing, storing, and manipulating information.
{ 'dij·əd·əl 'sis·təm }
digital telemetering [COMPUT SCI] Conversion of a
continuous electrical analog signal into a digital
(number system) code prior to transmitting the
signal to a receiver. { 'dij·əd·əl ¦tel·əˌmēd·ər·iŋ }
digital television [COMMUN] Television in which
picture information is encoded into digital signals on the transmitter, and decoded at the
receiver. Abbreviated DTV. { 'dij·əd·əl 'tel·ə
ˌvizh·ən }
digital television converter [ELECTR] A converter
used to convert television programs from one
system to another, such as for converting 525-
line 60-field United States broadcasts to 625-line
50-field European PAL (phase-alternation line)
or SECAM (sequential couleur á memoire) standards; the video signal is digitized before conversion. { 'dij·əd·əl ¦tel·əˌvizh·ən kən'vərd·ər }
digital-to-analog converter [ELECTR] A converter
in which digital input signals are changed to
essentially proportional analog signals. Abbreviated DAC. { 'dij·əd·əl tü ¦an·əˌläg kən'vərd·ər }
digital-to-synchro converter [ELECTR] A converter that changes binary-coded decimal or
other digital input data to a three-wire synchro
output signal representing corresponding
angular data. { ¦dij·əd·əl tü ¦siŋ·krō kən'vərd·ər }
digital transducer [ELECTR] A transducer that
measures physical quantities and transmits the
information as coded digital signals rather than
as continuously varying currents or voltages.
{ 'dij·əd·əl tranz'dü·sər }
digital versatile disk See DVD. { ¦dij·əd·əl 'vər·
səd·əl ˌdisk }

digital video disk See DVD. { ¦dij·əd·əl 'vid·ē·ō
ˌdisk }
digital voltmeter [ELECTR] A voltmeter in which
the unknown voltage is compared with an internally generated analog voltage, the result being
indicated in digital form rather than by a pointer
moving over a meter scale. { 'dij·əd·əl 'vōlt
ˌmēd·ər }
digital watermark [COMPUT SCI] Invisible or inaudible data (a random pattern of bits or noise)
permanently embedded in a graphic, video, or audio file for protecting copyright or authenticating
data. { ˌdij·əd·əl 'wȯd·ərˌmärk }
digit-coded voice [COMPUT SCI] A limited, spoken
vocabulary, each word of which corresponds
to a code and which, upon keyed inquiry, can
be strung in meaningful sequence and can be
outputted as audio response to the inquiry.
{ 'dij·ət ˌkōd·əd 'vȯis }
digit compression [COMPUT SCI] Any process
which increases the number of digits stored at
a given location. { 'dij·ət kəm'presh·ən }
digit delay element [ELECTR] A logic element
that introduces a delay of one digit period in
a series of signals or pulses. { 'dij·ət di'lā
ˌel·ə·mənt }
digitize [COMPUT SCI] To convert an analog measurement of a quantity into a numerical value.
{ 'dij·əˌtīz }
digitizer [COMPUT SCI] A large drawing table connected to a computer video display and equipped
with a penlike or pucklike instrument whose
motions are reproduced on the screen. Also
known as digitizer tablet. { 'dij·əˌtīz·ər }
digitizer tablet See digitizer. { 'dij·əˌtīz·ər ˌtab·
lət }
digit period [ELECTR] The time interval between
successive pulses, usually representing binary
digits, in a computer or in pulse modulation,
determined by the pulse-repetition frequency.
Also known as digit time. { 'dij·ət ˌpir·ē·əd }
digit plane [COMPUT SCI] In a computer memory
consisting of magnetic cores arranged in a three-
dimensional array, a plane containing elements
for a particular digit position in various words.
{ 'dij·ət ˌplān }
digit pulse [ELECTR] An electrical pulse which
induces a magnetizing force in a number of
magnetic cores in a computer storage, all corresponding to a particular digit position in a
number of different words. { 'dij·ət ˌpəls }
digit rearrangement [COMPUT SCI] A method of
hashing which consists of selecting and shifting digits of the original key. { 'dij·ət ˌrē·
ə'rānj·mənt }
digit time See digit period. { 'dij·ət ˌtīm }
digram encoding [COMPUT SCI] A method of data
compression that relies on the fact that there
are unused characters in the alphabet and uses
these characters to represent common pairs of
characters. { 'dīˌgram inˌkōd·iŋ }
diheptal base [ELECTR] A tube base having 14
pins or 14 possible pin positions; used chiefly
on television cathode-ray tubes. { dī'hept·əl
'bās }

dimension

dimension |COMPUT SCI| A declarative statement that specifies the width and height of an array of data items. { də'men·chən }

dimension declaration statement |COMPUT SCI| A FORTRAN statement identifying arrays and specifying the number and bounds of the subscripts. { də'men·chən·əl dek·lə'rā·shən ˌstāt·mənt }

diminution |COMPUT SCI| Limiting the negative effect of an attack on a computer system. { ˌdim·ə'nü·shən }

DIMM |COMPUT SCI| A small circuit board that holds semiconductor memory chips with two independent rows of input/output contacts. Derived from dual in-line memory module.

dimmer |ELEC| An electrical or electronic control for varying the intensity of a lamp or other light source. { 'dim·ər }

dina |ELECTR| An airborne radar-jamming transmitter operating in the band from 92 to 210 megahertz with an output of 30 watts, radiating noise in one side band for spot or barrage jamming; the carrier and the other side band are suppressed. { 'dī·nə }

D-indicator See D-display. { 'dē ˌin·də,kād·ər }

diode |ELECTR| **1.** A two-electrode electron tube containing an anode and a cathode. **2.** See semiconductor diode. { 'dī,ōd }

diode alternating-current switch See trigger diode. { 'dī,ōd ˌȯl·tər,nād·iŋ ˌkər·ənt ˌswich }

diode amplifier |ELECTR| A microwave amplifier using an IMPATT, TRAPATT, or transferred-electron diode in a cavity, with a microwave circulator providing the input/output isolation required for amplification; center frequencies are in the gigahertz range, from about 1 to 100 gigahertz, and power outputs are up to 20 watts continuous-wave or more than 200 watts pulsed, depending on the diode used. { 'dī,ōd 'am·plə,fī·ər }

diode bridge |ELECTR| A series-parallel configuration of four diodes, whose output polarity remains unchanged whatever the input polarity. { 'dī,ōd ˌbrij }

diode-capacitor transistor logic |ELECTR| A circuit that uses diodes, capacitors, and transistors to provide logic functions. { ¦dī,ōd kə¦pas·əd·ər tran'zis·tər ˌläj·ik }

diode characteristic |ELECTR| The composite electrode characteristic of an electron tube when all electrodes except the cathode are connected together. { 'dī,ōd ˌkar·ik·tə·'ris·tik }

diode clamp See diode clamping circuit. { 'dī,ōd ˌklamp }

diode clamping circuit |ELECTR| A clamping circuit in which a diode provides a very low resistance whenever the potential at a certain point rises above a certain value in some circuits or falls below a certain value in others. Also known as diode clamp. { ¦dī,ōd 'klamp·iŋ ˌsər·kət }

diode clipping circuit |ELECTR| A clipping circuit in which a diode is used as a switch to perform the clipping action. { ¦dī,ōd 'klip·iŋ ˌsər·kət }

diode-connected transistor |ELECTR| A bipolar transistor in which two terminals are shorted to give diode action. { 'dī,ōd kə¦nek·təd tran'zis·tər }

diode demodulator |ELECTR| A demodulator using one or more diodes to provide a rectified output whose average value is proportional to the original modulation. Also known as diode detector. { 'dī,ōd dē'mäj·ə,lād·ər }

diode detector See diode demodulator. { 'dī,ōd di'tek·tər }

diode drop See diode forward voltage. { 'dī,ōd ˌdräp }

diode forward voltage |ELECTR| The voltage across a semiconductor diode that is carrying current in the forward direction; it is usually approximately constant over the range of currents commonly used. Also known as diode drop; diode voltage; forward voltage drop. { 'dī,ōd ¦fȯr·wərd 'vōl·tij }

diode function generator |ELECTR| A function generator that uses the transfer characteristics of resistive networks containing biased diodes; the desired function is approximated by linear segments. { 'dī,ōd 'feŋk·shən ˌjen·ə,rād·ər }

diode gate |ELECTR| An AND gate that uses diodes as switching elements. { 'dī,ōd ˌgāt }

diode laser See semiconductor laser. { 'dī,ōd ˌlāz·ər }

diode limiter |ELECTR| A peak-limiting circuit employing a diode that becomes conductive when signal peaks exceed a predetermined value. { ˌdī,ōd 'lim·əd·ər }

diode logic |ELECTR| An electronic circuit using current-steering diodes, such that the relations between input and output voltages correspond to AND or OR logic functions. { 'dī,ōd ˌläj·ik }

diode matrix |ELECTR| A two-dimensional array of diodes used for a variety of purposes such as decoding and read-only memory. { 'dī,ōd ˌmā·triks }

diode mixer |ELECTR| A mixer that uses a crystal or electron tube diode; it is generally small enough to fit directly into a radio-frequency transmission line. { 'dī,ōd ˌmik·sər }

diode modulator |ELECTR| A modulator using one or more diodes to combine a modulating signal with a carrier signal; used chiefly for low-level signaling because of inherently poor efficiency. { 'dī,ōd 'mäj·ə,lād·ər }

diode pack |ELECTR| Combination of two or more diodes integrated into one solid block. { 'dī,ōd ˌpak }

diode peak detector |ELECTR| Diode used in a circuit to indicate when peaks exceed a predetermined value. { 'dī,ōd 'pēk di,tek·tər }

diode-pentode |ELECTR| Vacuum tube having a diode and a pentode in the same envelope. { ¦dī,ōd ¦pen,tōd }

diode rectifier |ELECTR| A half-wave rectifier of two elements between which current flows in only one direction. { 'dī,ōd 'rek·tə,fī·ər }

diode rectifier-amplifier meter |ELECTR| The most widely used vacuum tube voltmeter for measurement of alternating-current voltage;

160

has separate tubes for rectification and direct-current amplification, permitting an optimum design for each. { ¦dī̩ōd ¦rek·tə̩fī·ər 'am·plə ̩fī·ər ̩mēd·ər }

diode switch |ELECTR| Diode which is made to act as a switch by the successive application of positive and negative biasing voltages to the anode (relative to the cathode), thereby allowing or preventing, respectively, the passage of other applied waveforms within certain limits of voltage. { 'dī̩ōd ̩swich }

diode theory |ELEC| The theory that in a semiconductor, when the barrier thickness is comparable to or smaller than the mean free path of the carriers, then the carriers cross the barrier without being scattered, much as in a vacuum tube diode. { 'dī̩ōd ̩thē·ə·rē }

diode transistor logic |ELECTR| A circuit that uses diodes, transistors, and resistors to provide logic functions. Abbreviated DTL. { ¦dī̩ōd tran'zis·tər ̩läj·ik }

diode-triode |ELECTR| Vacuum tube having a diode and a triode in the same envelope. { ¦dī ̩ōd 'trī̩ōd }

diode voltage See diode forward voltage. { 'dī ̩ōd ̩vōl·tij }

diode voltage regulator |ELECTR| A voltage regulator with a Zener diode, making use of its almost constant voltage over a range of currents. Also known as Zener diode voltage regulator. { ¦dī ̩ōd 'vōl·tij ̩reg·yə̩lād·ər }

DIP See dual in-line package. { dip }

diphase generator |ELEC| A generator that produces two alternating currents in quadrature. { 'dī̩fāz 'jen·ə̩rād·ər }

diplexer |ELECTR| A coupling system that allows two different transmitters to operate simultaneously or separately from the same antenna. { 'dī ̩plek·sər }

diplex operation |COMMUN| Simultaneous transmission or reception of two signals using a specified common element, such as a single antenna or a single carrier. { 'dī̩pleks ̩äp·ə̩rā·shən }

diplex radio transmission |COMMUN| The simultaneous transmission of two signals by using a common carrier wave. { ¦dī̩pleks 'rād·ē·ō tranz ̩mish·ən }

diplex reception |ELEC| Simultaneous reception of two signals which have some features in common, such as a single receiving antenna or a single carrier frequency. { 'dī̩pleks ri'sep·shən }

dipole antenna |ELECTROMAG| An antenna approximately one-half wavelength long, split at its electrical center for connection to a transmission line whose radiation pattern has a maximum at right angles to the antenna. Also known as doublet antenna; half-wave dipole. { 'dī̩pōl an'ten·ə }

dipole disk feed |ELECTROMAG| Antenna, consisting of a dipole near a disk, used to reflect energy to the disk. { 'dī̩pōl 'disk ̩fēd }

dipole moment See electric dipole moment. { 'dī ̩pōl ̩mō·mənt }

dipole polarization See orientation polarization. { 'dī̩pōl ̩pō·lə·rə'zā·shən }

dipole relaxation |ELEC| The process, occupying a certain period of time after a change in the applied electric field, in which the orientation polarization of a substance reaches equilibrium. { 'dī̩pōl ̩rē̩lak'sā·shən }

DIP switch |COMPUT SCI| A unit with several small rocker-type switches that plugs into a dual in-line package (DIP) on a printed circuit board. { 'dip ̩swich }

dipulse |COMMUN| Transmission of a binary code in which the presence of one cycle of a sine-wave tone represents a binary "1" and the absence of one cycle represents a binary "0." { 'dī̩pəls }

direct access See random access. { də'rekt 'ak·ses }

direct-access library |COMPUT SCI| A disk-stored set of programs, each of which is directly accessible without sequential search. { də¦rekt ¦ak·ses 'lī̩brer·ē }

direct-access memory See random-access memory. { də¦rekt ¦ak·ses 'mem·rē }

direct-access method |COMPUT SCI| A technique for directly determining the location of data on a disk (track and sector address) from an identifying key in the record. { də¦rekt 'ak̩ses ̩meth·əd }

direct-access storage See random-access memory. { də¦rekt ¦ak·ses 'stòr·ij }

direct-access storage device |COMPUT SCI| Any peripheral storage device, such as a disk or drum, that can be directly addressed by a computer. Abbreviated DASD. { də¦rekt ¦ak̩ses 'stòr·ij di ̩vīs }

direct-acting recorder |ENG| A recorder in which the marking device is mechanically connected to or directly operated by the primary detector. { də¦rekt ¦akt·iŋ ri'kòrd·ər }

direct address |COMPUT SCI| Any address specifying the location of an operand. { də¦rekt 'a ̩dres }

direct-address processing |COMPUT SCI| Any computer operation during which data are accessed by means of addresses rather than contents. { də¦rekt ̩a̩dres 'präs̩es·iŋ }

direct allocation |COMPUT SCI| A system in which the storage locations and peripheral units to be assigned to use by a computer program are specified when the program is written, in contrast to dynamic allocation. { də¦rekt ̩al·ə̩kā· shən }

direct-aperture antenna |ELECTROMAG| An antenna whose conductor or dielectric is a surface or solid, such as a horn, mirror, or lens. { də¦rekt ¦ap·ə·chər an'ten·ə }

direct audio radio service |COMMUN| Radio broadcasting from satellites directly to receivers on the ground. Abbreviated DARS. { də̩rekt ¦òd·ē·ō 'rād·ē·ō ̩sər·vəs }

direct broadcasting satellite system |COMMUN| A television broadcasting system in which program signals are transmitted from ground stations to satellite repeater stations in geostationary orbit, and from there directly to

direct broadcast radio satellite

home consumer terminals. Abbreviated DBS.
{ də¦rekt 'bröd,kast·iŋ 'sad·əl,īt ,sis·təm }

direct broadcast radio satellite | COMMUN | A satellite in geosynchronous orbit that broadcasts radio programming directly to inexpensive home, car-mounted, and portable radio receivers. { di¦rekt ¦bröd,kast 'rād·ē·ō ,sad·əl,īt }

direct code | COMPUT SCI | A code in which instructions are written in the basic machine language. { də¦rekt 'kōd }

direct connect modem | COMMUN | A device that transforms binary signals into electronic pulses (as opposed to sound modulations) that can be carried over a communications channel. { də'rekt kə¦nekt 'mō,dem }

direct control | COMPUT SCI | The control of one machine in a data-processing system by another, without human intervention. { də¦rekt kən'trōl }

direct control function See regulatory control function. { də¦rekt kən'trōl ,faŋk·shən }

direct-coupled amplifier | ELECTR | A direct-current amplifier in which a resistor or a direct connection provides the coupling between stages, so small changes in direct currents can be amplified. { də¦rekt ¦kəp·əld 'am·plə,fī·ər }

direct-coupled FET logic | ELECTR | A logic gate configuration used with gallium arsenide field-effect transistors operating in the enhancement mode, whose low power consumption and circuit simplicity lead to high packing density and potential use in very large-scale integrated circuits. Abbreviated DCFL. { də'rekt ¦kəp·əld ¦ef¦ē¦tē 'läj·ik }

direct-coupled transistor logic | ELECTR | Integrated-circuit logic using only resistors and transistors, with direct conductive coupling between the transistors; speed can be up to 1 megahertz. Abbreviated DCTL. { də¦rekt ¦kəp·əld tran¦zis·tər 'läj·ik }

direct coupling | ELEC | Coupling of two circuits by means of a non-frequency-sensitive device, such as a wire, resistor, or battery, so both direct and alternating current can flow through the coupling path. { də¦rekt 'kəp·liŋ }

direct current | ELEC | Electric current which flows in one direction only, as opposed to alternating current. Abbreviated dc. { də¦rekt 'kə·rənt }

direct-current amplifier | ELECTR | An amplifier that is capable of amplifying dc voltages and slowly varying voltages. { də¦rekt ¦kə·rənt 'am·plə,fī·ər }

direct-current circuit | ELEC | Any combination of dc voltage or current sources, such as generators and batteries, in conjunction with transmission lines, resistors, and power converters such as motors. { də¦rekt ¦kə·rənt 'sər·kət }

direct-current circuit theory | ELEC | An analysis of relationships within a dc circuit. { də¦rekt ¦kə·rənt 'sər·kət ,thē·ə·rē }

direct-current component | COMMUN | The average value of a signal; in television, it represents the average luminance of the picture being transmitted; in radar, the level from which the transmitted and received pulses rise. { də¦rekt ¦kə·rənt kəm'pō·nənt }

direct-current continuity | ELEC | Property of a circuit in which there is an established pathway for conduction of current from a direct-current source. { də¦rekt ¦kə·rənt ,känt·ən'ü·əd·ē }

direct-current coupling | ELECTR | That type of coupling in which the zero-frequency term of the Fourier series representing the input signal is transmitted. { də¦rekt ¦kə·rənt 'kəp·liŋ }

direct-current discharge | ELECTR | The passage of a direct current through a gas. { də¦rekt ¦kə·rənt 'dis,chärj }

direct-current dump | ELECTR | Removal of all direct-current power from a computer system or component intentionally, accidentally, or conditionally; in some types of storage, this results in loss of stored information. { də¦rekt ¦kə·rənt 'dəmp }

direct-current erase | ELECTR | Use of direct current to energize an erasing head of a tape recorder. { də¦rekt ¦kə·rənt ə'rās }

direct-current generator | ELEC | A rotating electric machine that converts mechanical power into dc power. { də¦rekt ¦kə·rənt 'jen·ə,rād·ər }

direct-current inserter | ELECTR | An analog television transmitter stage that adds to the video signal a dc component known as the pedestal level. { də¦rekt ¦kə·rənt in'sərd·ər }

direct-current motor | ELEC | An electric rotating machine energized by direct current and used to convert electric energy to mechanical energy. { də¦rekt ¦kə·rənt 'mōd·ər }

direct-current motor control See electronic motor control. { də¦rekt ¦kə·rənt 'mōd·ər kən,trōl }

direct-current offset | ELECTR | A direct-current level that may be added to the input signal of an amplifier or other circuit. { də¦rekt ¦kə·rənt 'öf,set }

direct-current picture transmission | COMMUN | Television transmission in which the signal contains a dc component that represents the average illumination of the entire scene. Also known as direct-current transmission. { də¦rekt ¦kə·rənt 'pik·chər tranz,mish·ən }

direct-current plate resistance | ELECTR | Value or characteristic used in vacuum-tube computations; it is equal to the direct-current plate voltage divided by the direct-current plate current. { də¦rekt ¦kə·rənt 'plāt ri,zis·təns }

direct-current power | ELEC | The power delivered by a dc power system, equal to the line voltage times the load current. { də¦rekt ¦kə·rənt 'paù·ər }

direct-current power supply | ELEC | A power supply that provides one or more dc output voltages, such as a dc generator, rectifier-type power supply, converter, or dynamotor. { də¦rekt ¦kə·rənt 'paù·ər sə,plī }

direct-current quadruplex system | COMMUN | Direct-current telegraph system which affords simultaneous transmission of two messages in each direction over the same line, achieved by superimposing neutral telegraph upon polar telegraph. { də¦rekt ¦kə·rənt 'kwä·drə,pleks ,sis·təm }

direct-current receiver |ELECTR| A radio receiver designed to operate directly from a 115-volt dc power line.   { də¦rekt ¦kə·rənt ri'sēv·ər }
direct-current reinsertion   See clamping.   { də¦rekt ¦kə·rənt ‚rē·in'sər·shən }
direct-current restoration   See clamping.   { də¦rekt ¦kə·rənt res·tə'rā·shən }
direct-current restorer |ELECTR| A clamp circuit used to establish a dc reference level in a signal without modifying to any important degree the waveform of the signal itself. Also known as clamper; reinserter.   { də¦rekt ¦kə·rənt ri'stȯr·ər }
direct-current signaling |ELEC| A transmission method that uses direct current.   { də¦rekt ¦kə·rənt 'sig·nəl·iŋ }
direct-current SQUID |ELECTR| A type of superconducting quantum interference device (SQUID) which contains two Josephson junctions in a superconducting loop; its state is determined from direct-current measurements.   { də¦rekt ¦kə·rənt 'skwid }
direct-current tachometer |ELEC| A dc generator operating with negligible load current and with constant field flux provided by a permanent magnet, so its dc output voltage is proportional to speed.   { də¦rekt ¦kə·rənt tə'käm·əd·ər }
direct-current telegraphy |COMMUN| Telegraphy in which direct current controlled by the transmitting apparatus is supplied to the line to form the transmitted signal.   { də¦rekt ¦kə·rənt tə'leg·rə·fē }
direct-current transducer |ELECTR| A transducer that requires dc excitation and provides a dc output that varies with the parameter being sensed.   { də¦rekt ¦kə·rənt tranz'düs·ər }
direct-current transmission   See direct-current picture transmission.   { də¦rekt ¦kə·rənt tranz'mish·ən }
direct-current vacuum-tube voltmeter |ELECTR| The amplifying and indicating portions of the diode rectifier-amplifier meter, which are usually designed so that the diode rectifier can be disconnected for dc measurements.   { də¦rekt ¦kə·rənt ¦vak·yəm ¦tüb 'vōlt‚mēd·ər }
direct-current voltage   See direct voltage.   { də¦rekt ¦kə·rənt 'vōl·tij }
direct-current working volts |ELEC| The maximum continuously applied dc voltage for which a capacitor is rated. Abbreviated dcwV.   { də¦rekt ¦kə·rənt 'wərk·iŋ ‚vōlts }
direct digital control |CONT SYS| The use of a digital computer generally on a time-sharing or multiplexing basis, for process control in petroleum, chemical, and other industries.   { də¦rekt ¦dij·əd·əl kən'trōl }
direct distance dialing |COMMUN| A telephone exchange service that allows a telephone user to dial subscribers outside the local area using a standard routing pattern from the local or end office.   { də¦rekt ¦dis·təns 'dīl·iŋ }
direct-drive arm |CONT SYS| A robot arm whose joints are directly coupled to high-torque motors.   { də'rekt ¦drīv ‚ärm }
direct electromotive force |ELEC| Unidirectional electromotive force in which the changes in

values are either zero or so small that they may be neglected.   { də'rekt i‚lek·trō'mōd·iv 'fȯrs }
direct-entry terminal |COMPUT SCI| A device from which data are received into a computer immediately, and which edits data at the time of receipt, allowing computer files to be accessed to validate the information entered, and allowing the terminal operator to be notified immediately of any errors.   { də¦rekt ¦en·trē 'term·ən·əl }
direct expert control system |CONT SYS| An expert control system that contains rules that directly associate controller output values with different values of the controller measurements and set points. Also known as rule-based control system.   { də¦rekt ‚eks·pərt kən'trōl ‚sis·təm }
direct-feedback system |CONT SYS| A system in which electrical feedback is used directly, as in a tachometer.   { də¦rekt 'fēd‚bak ‚sis·təm }
direct grid bias   See grid bias.   { də¦rekt ¦grid ‚bī·əs }
direct hierarchy control |COMPUT SCI| A method of manipulating data in a computer storage hierarchy in which data transfer is completely under the control of built-in algorithms and the user or programmer is not concerned with the various storage subsystems.   { də¦rekt 'hī·ər ‚är·kē kən‚trōl }
direct input/output |COMPUT SCI| The transfer of data to and from a computer's main storage by passing it through the central processing unit.   { də'rekt 'in‚pút 'aút‚pút }
direct-insert subroutine |COMPUT SCI| A body of coding or a group of instructions inserted directly into the logic of a program, often in multiple copies, whenever required.   { də¦rekt ¦in·sərt 'səb‚rü‚tēn }
direct instruction |COMPUT SCI| An instruction containing the address of the operand on which the operation specified in the instruction is to be performed.   { də'rekt in'strək·shən }
direct interelectrode capacitance   See interelectrode capacitance.   { də'rekt ‚in·tər·i'lek‚trōd kə'pas·əd·əns }
direct inward dialing |COMMUN| The capability for dialing individual telephone extensions in a large organization directly from outside, without going through a central switchboard.   { də'rekt ¦in·wərd 'dīl·iŋ }
directional antenna |ELECTROMAG| An antenna that radiates or receives radio waves more effectively in some directions than others.   { də'rek·shən·əl an'ten·ə }
directional beam |ELECTROMAG| A radio or radar wave that is concentrated in a given direction.   { də'rek·shən·əl 'bēm }
directional coupler |ELECTR| A device that couples a secondary system only to a wave traveling in a particular direction in a primary transmission system, while completely ignoring a wave traveling in the opposite direction. Also known as directive feed.   { də'rek·shən·əl 'kəp·lər }
directional filter |ELECTR| A low-pass, band-pass, or high-pass filter that separates the bands of frequencies used for transmission in opposite directions in a carrier system. Also known as

directional gain

directional separation filter. { də'rek·shən·əl 'fil·tər }

directional gain *See* directivity index. { də'rek·shən·əl 'gān }

directional microphone |ENG ACOUS| A microphone whose response varies significantly with the direction of sound incidence. { də'rek·shən·əl 'mī·krə‚fōn }

directional pattern *See* radiation pattern. { də'rek·shən·əl 'pad·ərn }

directional phase shifter |ELEC| Passive phase shifter in which the phase change for transmission in one direction differs from that for transmission in the opposite direction. { də'rek·shən·əl 'fāz ‚shif·tər }

directional relay |ELEC| Relay which functions in conformance with the direction of power, voltage, current, pulse, rotation, and so on. { də'rek·shən·əl 'rē‚lā }

directional response pattern *See* directivity pattern. { də'rek·shən·əl ri'späns ‚pad·ərn }

directional separation filter *See* directional filter. { də'rek·shən·əl sep·ə'rā·shən ‚fil·tər }

direction finder *See* radio direction finder. { də'rek·shən ‚fīnd·ər }

direction-independent radar |ENG| Doppler radar used in sentry applications. { də‚rek·shən ‚in·də¦pen·dənt 'rā‚där }

direction rectifier |ELECTR| A rectifier that supplies a direct-current voltage whose magnitude and polarity vary with the magnitude and relative polarity of an alternating-current synchro error voltage. { də'rek·shən 'rek·tə‚fī·ər }

directive |COMPUT SCI| An instruction in a source program that guides the compiler in making the translation to machine language, and is usually not translated into instructions in the object program. { də'rek·tiv }

directive feed *See* directional coupler. { də'rek·tiv ‚fēd }

directive gain |ELECTROMAG| Of an antenna in a given direction, 4π times the ratio of the radiation intensity in that direction to the total power radiated by the antenna. { də'rek·tiv ‚gān }

directivity |ELECTR| The ability of a logic circuit to ensure that the input signal is not affected by the output signal. |ELECTROMAG| **1.** The value of the directive gain of an antenna in the direction of its maximum value. **2.** The ratio of the power measured at the forward-wave sampling terminals of a directional coupler, with only a forward wave present in the transmission line, to the power measured at the same terminals when the direction of the forward wave in the line is reversed; the ratio is usually expressed in decibels. { də‚rek'tiv·əd·ē }

directivity factor |ENG ACOUS| **1.** The ratio of radiated sound intensity at a remote point on the principal axis of a loudspeaker or other transducer, to the average intensity of the sound transmitted through a sphere passing through the remote point and concentric with the transducer; the frequency must be stated. **2.** The ratio of the square of the voltage produced by sound waves arriving parallel to the principal

axis of a microphone or other receiving transducer, to the mean square of the voltage that would be produced if sound waves having the same frequency and mean-square pressure were arriving simultaneously from all directions with random phase; the frequency must be stated. { də‚rek'tiv·əd·ə ‚fak·tər }

directivity index |ENG ACOUS| The directivity factor expressed in decibels; it is 10 times the logarithm to the base 10 of the directivity factor. Also known as directional gain. { də‚rek'tiv·əd·ə ‚in‚deks }

directivity pattern |ENG ACOUS| A graphical or other description of the response of a transducer used for sound emission or reception as a function of the direction of the transmitted or incident sound waves in a specified plane and at a specified frequency. Also known as beam pattern; directional response pattern. { də‚rek'tiv·əd·ə ‚pad·ərn }

direct keying device |COMPUT SCI| A computer input device which enables direct entry of information by means of a keyboard. { də'rekt 'kē·iŋ di‚vīs }

directly heated cathode *See* filament. { də¦rect·lē ¦hēd·əd 'kā‚thōd }

direct-map cache |COMPUT SCI| A cache memory that is organized by linking it to locations in random-access memory. { də‚rekt ‚map 'kash }

direct memory access |COMPUT SCI| The use of special hardware for direct transfer of data to or from memory to minimize the interruptions caused by program-controlled data transfers. Abbreviated dma. { də¦rekt ¦mem·rē 'ak‚ses }

direct numerical control |COMPUT SCI| The use of a computer to program, service, and log a process such as a machine-tool cutting operation. { də ¦rekt nü¦mer·i·kəl kən'trōl }

director |ELECTR| Telephone switch which translates the digits dialed into the directing digits actually used to switch the call. |ELECTROMAG| A parasitic element placed a fraction of a wavelength ahead of a dipole receiving antenna to increase the gain of the array in the direction of the major lobe. { də'rek·tər }

direct organization |COMPUT SCI| A type of processing in which records within data sets stored on direct-access devices may be fetched directly if their physical locations are known. { də'rekt ór·gə·nə'zā·shən }

directory |COMPUT SCI| The listing and description of all the fields of the records making up a file. { də'rek·trē }

directory service |COMPUT SCI| **1.** A directory of the names and addresses of all the mail recipients on a particular network, which provides electronic mail addresses. **2.** A provider of online directories of Web sites and search engines. { də'rek·trē ‚sər·vəs }

directory tree |COMPUT SCI| A graphic representation of the hierarchical branching structure in which files are organized in a hard disk or other storage device. { də'rek·trē ‚trē }

direct outward dialing |COMMUN| A private automatic branch telephone exchange that permits all local stations to dial outside numbers. Abbreviated DOD. { də¦rekt ¦aut·wərd 'dīl·iŋ }

direct piezoelectricity [SOLID STATE] Name sometimes given to the piezoelectric effect in which an electric charge is developed on a crystal by the application of mechanical stress. { de'rekt pē ¦ā·zō,i,lek'tris·əd·ē }

direct point repeater [ELECTR] Telegraph repeater in which the receiving relay controlled by the signals received over a line repeats corresponding signals directly into another line or lines without the interposition of any other repeating or transmitting apparatus. { də¦rekt ¦point ri'pēd·ər }

direct-power generator [ENG] Any device which converts thermal or chemical energy into electric power by methods more direct than the conventional thermal cycle. { də¦rekt ¦pau·ər 'jen·ə ,rād·ər }

direct-radiator speaker [ENG ACOUS] A loudspeaker in which the radiating element acts directly on the air, without a horn. { də¦rekt ¦rād·ē,ād·ər ,spēk·ər }

direct read after write [COMPUT SCI] The reading of data immediately after the data have been written in order to check for errors in the recoding process. Abbreviated DRAW. { də¦rekt ¦rēd ,af·tər 'rīt }

direct realization [ELECTR] An active filter configuration that is derived by systematically replacing the elements of a passive RLC prototype filter (a filter that consists entirely of resistors, inductors, and capacitors) according to some rule. { di ¦rekt ,rē·ə·lə'zā·shən }

direct resistance-coupled amplifier [ELECTR] Amplifier in which the collector, drain, or plate of one stage is connected either directly or through a resistor to the base, gate, or control grid of the next stage; used to amplify small changes in direct current. { də¦rekt ri¦zis·təns ,kəp·əld 'am·plə,fī·ər }

direct route [ELEC] In wire communications, the trunks that connect a pair of switching centers, regardless of the geographical direction the actual trunk facilities may follow. { də¦rekt 'rüt }

direct sequence system [COMMUN] A system for generating spread spectrum transmissions by phase-modulating a sine wave pseudorandomly by an unending string of pseudonoise code symbols, each of duration much smaller than a bit. { də¦rekt 'sē·kwəns ,sis·təm }

direct stroke [ELEC] A lightning stroke that actually strikes some part of a power or communication system. { də¦rekt 'strōk }

direct symbol recognition [COMPUT SCI] Recognition by sensing the unique geometrical properties of symbols. { də¦rekt 'sim·bəl ,rek·ig ,nish·ən }

direct-view storage tube [ELECTR] A cathode-ray tube in which secondary emission of electrons from a storage grid is used to provide an intensely bright display for long and controllable periods of time. Also known as display storage tube; viewing storage tube. { də¦rekt ¦vyü 'stòr·ij ,tüb }

direct voltage [ELEC] A voltage that forces electrons to move through a circuit in the same direction continuously, thereby producing a direct current. Also known as direct-current voltage. { də¦rekt 'vōl·tij }

direct wave [COMMUN] A radio wave that is propagated directly through space from transmitter to receiver without being refracted by the ionosphere. { də¦rekt 'wāv }

direct-wire circuit [ELEC] Supervised protective signaling circuit usually consisting of one metallic conductor and a ground return and having signal-receiving equipment responsive to either an increase or a decrease in current. { də¦rekt ¦wīr 'sər,kət }

direct-writing galvanometer [ENG] A direct-writing recorder in which the stylus or pen is attached to a moving coil positioned in the field of the permanent magnet of a galvanometer. { də¦rekt ¦wrīd·iŋ ,gal·və'näm·əd·ər }

direct-writing recorder [ENG] A recorder in which the permanent record of varying electrical quantities or signals is made on paper, directly by a pen attached to the moving coil of a galvanometer or indirectly by a pen moved by some form of motor under control of the galvanometer. Also known as mechanical oscillograph. { də¦rekt ¦wrīd·iŋ ri'kòrd·ər }

disability glare See glare. { dis·ə'bil·əd·ē ,glär }

disable [COMPUT SCI] **1.** To prevent some action from being carried out. **2.** To turn off a computer system or a piece of equipment. { dis'ā·bəl }

disappearing filament pyrometer See optical-pyrometer. { 'dis·ə,pir·iŋ ,fil·ə·mənt pī'räm·əd·ər }

disassemble [COMPUT SCI] To translate a program from machine language to assembly language or to aid in its understanding. { ,dis·ə 'sem·bəl }

disassembler [COMPUT SCI] A program that translates machine language into assembly language. { ,dis·ə'sem·blər }

disaster dump [COMPUT SCI] A listing of the contents of a computer's central processing unit that is created when the computer detects an error that it cannot handle in the course of processing. { di'zas·tər ,dəmp }

disc See disk. { disk }

discharge [ELEC] To remove a charge from a battery, capacitor, or other electric-energy storage device. [ELECTR] The passage of electricity through a gas, usually accompanied by a glow, arc, spark, or corona. Also known as electric discharge. { 'dis,chärj }

discharge key [ELEC] Device for switching a capacitor suddenly from a charging circuit to a load through which it can discharge. { 'dis,chärj ,kē }

discharge lamp [ELECTR] A lamp in which light is produced by an electric discharge between electrodes in a gas (or vapor) at low or high pressure. Also known as electric-discharge lamp; gas-discharge lamp; vapor lamp. { 'dis,chärj ,lamp }

discharger [ELEC] A silver-impregnated cotton wick encased in a flexible plastic tube with an aluminum mounting lug, used on aircraft to reduce precipitation static. { 'dis‚chärj·ər }

discharge tube [ELECTR] An evacuated enclosure containing a gas at low pressure, through which current can flow when sufficient voltage is applied between metal electrodes in the tube. Also known as electric-discharge tube. { 'dis ‚chärj ‚tüb }

discomfort glare See glare. { dis'kəm·fərt ‚gler }

discone antenna [ELECTROMAG] A biconical antenna in which one of the cones is spread out to 180° to form a disk; the center conductor of the coaxial line terminates at the center of the disk, and the cable shield terminates at the vertex of the cone. { 'dis‚kōn an'ten·ə }

disconnect [ELEC] To open a circuit by removing wires or connections, as distinguished from opening a switch to stop current flow. [ENG] To sever a connection. { ‚dis·kə'nekt }

disconnect fitting [ELEC] An electrical connection that can be disconnected without tools. { ‚dis·kə'nekt ‚fid·iŋ }

disconnecting switch [ELEC] A switch that isolates a circuit or piece of electrical apparatus after interruption of the current. Also known as disconnector. { ‚dis·kə'nek·tiŋ ‚swich }

disconnector See disconnecting switch. { ‚dis·kə'nek·tər }

disconnector release [ELEC] Device which disengages the apparatus used in a telephone connection to restore it to its original condition when not in use. { ‚dis·kə'nek·tər ri'lēs }

discontinuous amplifier [ELECTR] Amplifier in which the input waveform is reproduced on some type of averaging basis. { ‚dis·kən'tin·yə·wəs 'am·plə‚fī·ər }

discrete address beacon system See Mode S. { di‚skrēt 'ad·res 'bē·kən ‚sis·təm }

discrete comparator See digital comparator. { di'skrēt kəm'par·əd·ər }

discrete cosine transform [COMMUN] A mathematical transform, used in bit rate reduction applications, in which the reconstructed bit stream is identical to the bit stream input to the system; in this regard, the transform is a mathematical process that can be perfectly undone. Abbreviated DCT. { di'skrēt 'kō‚sīn 'tranz‚fórm }

discrete sampling [ELECTR] Sampling in which the individual samples are of such long duration that the frequency response of the channel is not deteriorated by the sampling process. { di'skrēt 'sam·pliŋ }

discrete sound system [ENG ACOUS] A quadraphonic sound system in which the four input channels are preserved as four discrete channels during recording and playback processes; sometimes referred to as a 4-4-4 system. { di'skrēt 'saúnd ‚sis·təm }

discrete system [CONT SYS] A control system in which signals at one or more points may change only at discrete values of time. Also known as discrete-time system. { di'skrēt 'sis·təm }

discrete-time system See discrete system. { di'skrēt ‚tīm 'sis·təm }

discrete transfer function See pulsed transfer function. { di¦skrēt 'tranz·fər ‚faŋk·shən }

discrete-word intelligibility [COMMUN] The percent of intelligibility obtained when the speech units under consideration are words, usually presented so as to minimize the contextual relation between them. { di¦skrēt ‚wərd in ‚tel·ə·jə'bil·əd·ē }

discrimination [COMMUN] **1.** In frequency-modulated systems, the detection or demodulation of the imposed variations in the frequency of the carriers. **2.** In a tuned circuit, the degree of rejection of unwanted signals. **3.** Of any system or transducer, the difference between the losses at specified frequencies with the system or transducer terminated in specified impedances. [COMPUT SCI] See conditional jump. { di‚skrim·ə'nā·shən }

discriminator [ELECTR] A circuit in which magnitude and polarity of the output voltage depend on how an input signal differs from a standard or from another signal. { di'skrim·ə‚nād·ər }

discriminator transformer [ELECTR] A transformer designed to be used in a stage where frequency-modulated signals are converted directly to audio-frequency signals or in a stage where frequency changes are converted to corresponding voltage changes. { di'skrim·ə ‚nād·ər tranz'fór·mər }

disengage [ENG] To break the contact between two objects. { ‚dis·ən'gāj }

dish See parabolic reflector. { dish }

disintegration voltage [ELECTR] The lowest anode voltage at which destructive positive-ion bombardment of the cathode occurs in a hot-cathode gas tube. { dis‚in·tə'grā·shən ‚vōl·tij }

disjunctive search [COMPUT SCI] A search to find items that have at least one of a given set of characteristics. { dis'jəŋk·tiv 'sərch }

disk [COMPUT SCI] A rotating circular plate having a surface on which information may be stored as a pattern of magnetically polarized spots (on a magnetic disk) or holes (on an optical disk) on concentric recording tracks. Also known as magnetic disk. Also spelled disc. { disk }

disk armature [ELEC] The armature in a motor that has a disk winding or is made up of a metal disk. { 'disk ‚är·mə·chər }

disk cache [COMPUT SCI] A portion of random-access memory that contains the data most recently read from or written to the disk, allowing rapid access by the central-processing unit. { 'disk ‚kash }

disk capacitor [ELEC] A small, flat, circular capacitor that usually has a ceramic dielectric. { 'disk kə‚pas·əd·ər }

disk cartridge [COMPUT SCI] A removable module that contains a single magnetic disk platter which remains attached to the housing when placed into the disk drive. { 'disk ‚kär·trij }

disk crash See head crash. { 'disk ‚krash }

disk drive [COMPUT SCI] The physical unit that holds, spins, reads, and writes the magnetic disks. Also known as disk unit. { 'disk ‚drīv }

disk drive controller |COMPUT SCI| A device that enables a microcomputer to control the functioning of a disk drive. { 'disk ¦drīv kən'trō·lər }

diskette See floppy disk. { di'sket }

disk file |COMPUT SCI| An organized collection of records held on a magnetic disk. { 'disk ,fīl }

diskless work station |COMPUT SCI| A computer in a network that has no disk storage of its own. { ¦disk·ləs 'wərk ,stā·shən }

disk memory See disk storage. { 'disk ,mem·rē }

disk operating system |COMPUT SCI| An operating system which uses magnetic disks as its primary on-line storage. Abbreviated DOS. { ¦disk ¦äp·ə,rād·iŋ ,sis·təm }

disk pack |COMPUT SCI| A set of magnetic disks that can be removed from a disk drive as a unit. { 'disk ,pak }

disk recording |ENG ACOUS| 1. The process of inscribing suitably transformed acoustical or electrical signals on a phonograph record. 2. See phonograph record. { ¦disk ri'kórd·iŋ }

disk-seal tube |ELECTR| An electron tube having disk-shaped electrodes arranged in closely spaced parallel layers, to give low interelectrode capacitance along with high power output, up to 2500 megahertz. Also known as lighthouse tube; megatron. { 'disk ,sēl ,tüb }

disk storage |ELECTR| An external computer storage device consisting of one or more disks spaced on a common shaft, and magnetic heads mounted on arms that reach between the disks to read and record information on them. Also known as disk memory; magnetic disk storage. { ¦disk ¦stór·ij }

disk striping |COMPUT SCI| The distribution of a unit of data over two or more hard disks, enabling the data to be read more quickly. Also known as data striping. { 'disk ,strīp·iŋ }

disk thermistor |ELECTR| A thermistor which is produced by pressing and sintering an oxide binder mixture into a disk,0.2–0.6 inch (5–15 millimeters) in diameter and 0.04–0.5 inch (1.0–13 millimeters) thick, coating the major surfaces with conducting material, and attaching leads. { ¦disk thər'mis·tər }

disk unit See disk drive. { 'disk ,yü·nət }

dispatching |COMPUT SCI| The control of priorities in a queue of requests in a multiprogramming or multitasking environment. { dis'pach·iŋ }

dispatching priority |COMPUT SCI| In a multiprogramming or multitasking environment, the priority assigned to an active (non-real time, nonforeground) task. { dis'pach·iŋ prī,är·əd·ē }

dispenser cathode |ELECTR| An electron tube cathode having provisions for continuously replacing evaporated electron-emitting material. { də'spen·sər ,kath,ōd }

disperse |COMPUT SCI| A data-processing operation in which grouped input items are distributed among a larger number of groups in the output. { də'spərs }

dispersion |COMMUN| The entropy of the output of a communications channel when the input is known. |ELECTROMAG| Scattering of microwave radiation by an obstruction. { də'spər·zhən }

displacement |COMPUT SCI| The number of character positions or memory locations from some point of reference to a specified character or data item. Also known as offset. |ELEC| See electric displacement. { dis'plās·mənt }

displacement angle |ELEC| The change in the phase of an alternator's terminal voltage when a load is applied. { dis'plās·mənt ,aŋ·gəl }

display |ELECTR| 1. A visible representation of information, in words, numbers, or drawings, as on the cathode-ray tube screen of a radar set, navigation system, or computer console. 2. The device on which the information is projected. Also known as display device. 3. The image of the information. { di'splā }

display adapter See video display board. { di'splā ə,dap·tər }

display console |COMPUT SCI| A cathode-ray tube or other display unit on which data being processed or stored in a computer can be presented in graphical or character form; sometimes equipped with a light pen with which the user can alter the information displayed. { di'splā ,kän ,sōl }

display control |COMPUT SCI| A unit in a computer system consisting of channels and associated control circuitry that connect a number of visual display units with a central processor. { di'splā kən,trōl }

display cycle |COMPUT SCI| In computer graphics, the sequence of operations carried out to display an image. { di,splā ,sī·kəl }

display device See display. { di'splā di,vīs }

display element |COMPUT SCI| In computer graphics, a basic component of a display, such as a circle, line, or dot. { di'splā ,el·ə·mənt }

display entity |COMPUT SCI| In computer graphics, a group of display elements that can be manipulated as a unit. { di'splā ,en·təd·ē }

display formats See radar display formats. { di ,splā ,fór·matz }

display frame |COMPUT SCI| In computer graphics, one of a sequence of frames making up a computer-generated animation. { di'splā ,frām }

display information processor |COMPUT SCI| Computer used to generate situation displays in a combat operations center. { di'splā in·fər 'mā·shən ,präs,es·ər }

display list |COMPUT SCI| In computer graphics, a set of vectors that form an image stored in vectors graphics format. { di'splā ,list }

display packing |COMPUT SCI| An efficient means of transmitting the x and y coordinates of a point packed in a single word to halve the time required to freshen the spot on a cathode-ray tube display. { di'splā ,pak·iŋ }

display power management signaling |COMPUT SCI| Signaling whereby a video adapter can instruct a monitor to reduce its power level to conserve electricity. Abbreviated DPMS. { di ¦splā 'paů·ər ,man·ij·mənt ,sig·nəl·iŋ }

display primary |COMMUN| One of the primary colors produced in a video system that, when mixed in proper proportions, serve to produce the other desired colors. { di'splā 'prī,mer·ē }

display processor |COMPUT SCI| A section of a computer which handles the routines required to display an output on a cathode-ray tube. { di'splā ,präs,es·ər }

display screen *See* video monitor. { di'splā ,skrēn }

display storage tube *See* direct-view storage tube. { di'splā 'stȯr·ij ,tüb }

display system |COMPUT SCI| The total system, combining hardware and software, needed to achieve a visible representation of information in a data-processing system. { di'splā ,sis·təm }

display terminal |COMPUT SCI| A computer output device in which characters and sometimes graphic information appear on the screen of a cathode-ray tube; now largely replaced by monitors using bit-mapped displays. Also known as display unit; video display terminal (VDT). { di'splā ,tər·mən·əl }

display tube |ELECTR| A cathode-ray tube used to provide a visual display. Also known as visual display unit. { di'splā ,tüb }

display unit *See* display terminal. { di'splā ,yü·nət }

display window |COMMUN| Width of the portion of the frequency spectrum presented on panoramic presentation; expressed in frequency units, usually megahertz. { di'splā ,win,dō }

disposition |COMPUT SCI| The status of a file after it has been closed by a computer program, for example, retained or deleted. { ,dis·pə'zish·ən }

disruptive discharge |ELEC| A sudden and large increase in current through an insulating medium due to complete failure of the medium under electrostatic stress. { dis¦rəp·tiv 'dis,chärj }

dissector tube |ELECTR| Camera tube having a continuous photo cathode on which is formed a photoelectric emission pattern which is scanned by moving its electron-optical image over an aperture. { də'sek·tər ,tüb }

dissipation factor |ELEC| The inverse of Q, the storage factor. { ,dis·ə'pā·shən ,fak·tər }

dissipation line |ELECTROMAG| A length of stainless steel or Nichrome wire used as a noninductive terminating impedance for a rhombic transmitting antenna when several kilowatts of power must be dissipated. { ,dis·ə'pā·shən ,līn }

dissipation loss |ELEC| A measure of the power loss of a transducer in transmitting signals, expressed as the ratio of its input power to its output power. { ,dis·ə'pā·shən ,lȯs }

dissymmetrical network *See* dissymmetrical transducer. { ,dis·ə'me·trə·kəl 'net,wərk }

dissymmetrical transducer |ELECTR| A transducer whose input and output image impedances are not equal. Also known as dissymmetrical network. { ,dis·ə'me·trə·kəl tranz'dü·sər }

distance mark |ELECTR| A movable point produced on a radar display by a special signal generator, so that when the mark is moved to a target position on the screen the range to the target can be read on the calibrated dial of the signal generator; usually used for gun laying where highly accurate distance is important. { 'dis·təns ,märk }

distance marker |ENG| One of a series of concentric circles, painted or otherwise fixed on the screen of a plan position indicator, from which the distance of a target from the radar antenna can be read directly; used for surveillance and navigation where the relative distances between a number of targets are required simultaneously. Also known as radar range marker; range marker. { 'dis·təns ,märk·ər }

distance protection |ELEC| Effect of a device operative within a predetermined electrical distance on the protected circuit to cause and maintain an interruption of power in a faulty circuit. { 'dis·təns prə,tek·shən }

distance reception |COMMUN| Reception of messages from, or communication with, distant radio stations. Abbreviated DX. { 'dis·təns ri'sep·shən }

distance relay |ELEC| Protective relay, the operation of which is a function of the distance between the relay and the point of fault. { 'dis·təns ,rē,lā }

distance resolution |ENG| The minimum radial distance by which targets must be separated to be separately distinguishable by a particular radar. Also known as range discrimination; range resolution. { 'dis·təns ,rez·ə,lü·shən }

distance/velocity lag |CONT SYS| The delay caused by the amount of time required to transport material or propagate a signal or condition from one point to another. Also known as transportation lag; transport lag. { 'dis·təns və'läs·əd·ē ,lag }

distant field |ELECTROMAG| The electromagnetic field at a distance of five wavelengths or more from a transmitter, where the radial electric field becomes negligible. { ¦dis·tənt ¦fēld }

distortion |ELECTR| Any undesired change in the waveform of an electric signal passing through a circuit or other transmission medium. |ENG| In general, the extent to which a system fails to accurately reproduce the characteristics of an input signal at its output. |ENG ACOUS| Any undesired change in the waveform of a sound wave. { di'stȯr·shən }

distortion factor |COMMUN| Ratio of the effective value of the residue of a wave after elimination of the fundamental to the effective value of the original wave. { di'stȯr·shən ,fak·tər }

distortion meter |ENG| An instrument that provides a visual indication of the harmonic content of an audio-frequency wave. { di'stȯr·shən ,mēd·ər }

distress frequency |COMMUN| A frequency allotted to distress calls, generally by international agreement; for ships at sea and aircraft over the sea, it is 500 kilohertz. { də'stres ,frē·kwən·sē }

distributed amplifier |ELECTR| A wide-band amplifier in which tubes are distributed along artificial delay lines made up of coils acting with the input and output capacitances of the tubes. { di'strib·yəd·əd 'am·plə,fī·ər }

distributed bulletin board |COMPUT SCI| A collection of newsgroups on a wide-area network, whose postings are available to every user. { di,strib·yəd·əd 'bùl·ət·ən ,bórd }

distributed capacitance |ELEC| Capacitance that exists between the turns in a coil or choke, or between adjacent conductors or circuits, as distinguished from the capacitance concentrated in a capacitor. { di'strib·yəd·əd kə'pas·əd·əns }

distributed circuit |ELECTR| A film circuit whose effective components cannot be easily recognized as discrete. { di'strib·yəd·əd 'sər·kət }

distributed communications |COMMUN| Information transfer beyond the local level that may involve the originating source to transmit information to all communications centers on any one network, and may also cause an interchange of communications among several whole networks. { di'strib·yəd·əd kə'myü·nə'kā·shənz }

distributed computing |COMPUT SCI| The use of multiple network-connected computers for solving a problem or for information processing. { di,strib·yəd·əd kəm'pyüd·iŋ }

distributed control system |CONT SYS| A collection of modules, each with its own specific function, interconnected tightly to carry out an integrated data acquisition and control application. { di'strib·yəd·əd kən'trōl ,sis·təm }

distributed database |COMPUT SCI| A database maintained in physically separated locations and supported by a computer network so that it is possible to access all parts of the database from various points in the network. { di'strib·yəd·əd 'dad·ə ,bās }

distributed-emission photodiode |ELECTR| A broad-band photodiode proposed for detection of modulated laser beams at millimeter wavelengths; incident light falls on a photocathode strip that generates a traveling wave of photocurrent having the same wave velocity as the transmission line which the photodiode feeds. { di'strib·yəd·əd ə'mish·ən ,fōd·ō,dī,ōd }

distributed free space |COMPUT SCI| Empty spaces in a data layout to allow new data to be inserted at a future time. { di'strib·yəd·əd ¦frē ¦spās }

distributed intelligence |COMPUT SCI| The existence of processing capability in terminals and other peripheral devices of a computer system. Also known as distributed logic. { di'strib·yəd· əd in'tel·ə·jəns }

distributed logic See distributed intelligence. { di¦strib·yəd·əd 'läj·ik }

distributed logic cluster word processor |COMPUT SCI| A system of word processors each of which can operate independently, although printers are generally shared by a number of terminals. { di'strib·yəd·əd 'läj·ik,kləs·tər 'wərd ,präs,es·ər }

distributed network |COMMUN| A communications network in which there exist alternative routings between the various nodes. |COMPUT SCI| A computer network in which at least some of the processing is done at individual work stations and information is shared by and often

stored at the work stations. { di'strib·yəd·əd 'net,wərk }

distributed numerical control |CONT SYS| The use of central computers to distribute part-classification data to machine tools which themselves are controlled by computers or numerical control tapes. { di'strib·yəd·əd nü'mer·ə·kəl kən'trōl }

distributed-parameter system See distributed system. { di'strib·yəd·əd pə'ram·əd·ər ,sis·təm }

distributed paramp |ELECTR| Paramagnetic amplifier that consists essentially of a transmission line shunted by uniformly spaced, identical varactors; the applied pumping wave excites the varactors in sequence to give the desired traveling-wave effect. { di'strib·yəd·əd ¦par ¦amp }

distributed processing system |COMPUT SCI| An information processing system consisting of two or more programmable devices, connected so that information can be exchanged. { di'strib·yəd·əd 'präs,es·iŋ ,sis·təm }

distributed system |COMPUT SCI| A computer system consisting of a collection of autonomous computers linked by a network and equipped with software that enables the computers to coordinate their activities and to share the resources of system hardware, software, and data, so that users perceive a single, integrated computing facility. |CONT SYS| A collection of modules, each with its own specific function, interconnected to carry out integrated data acquisition and control in a critical environment. |SYS ENG| A system whose behavior is governed by partial differential equations, and not merely ordinary differential equations. Also known as distributed-parameter system. { di'strib·yəd·əd 'sis·təm }

distributing frame |ELECTR| Structure for terminating permanent wires of a central office, private branch exchange, or private exchange, and for permitting the easy change of connections between them by means of cross-connecting wires. { di'strib·yəd·iŋ ,frām }

distributing terminal assembly |ELECTR| Frame situated between each pair of selector bays to provide terminal facilities for the selector bank wiring and facilities for cross-connection to trunks running to succeeding switches. { di'strib·yəd·iŋ 'term·ən·əl ə,sem·blē }

distribution amplifier |ELECTR| A radio-frequency power amplifier used to feed television or radio signals to a number of receivers, as in an apartment house or a hotel. |ENG ACOUS| An audio-frequency power amplifier used to feed a speech or music distribution system and having sufficiently low output impedance so changes in load do not appreciably affect the output voltage. { ,dis·trə'byü·shən 'am·plə,fī·ər }

distribution cable |ELEC| Cable extending from a feeder cable into a specific area for the purpose of providing service to that area. { ,dis·trə'byü·shən ,kā·bəl }

distribution center |ELEC| In an alternating-current power system, the point at which control and routing equipment is installed. { ‚dis·trə'byü·shən ‚sen·tər }

distribution control See linearity control. { ‚dis·trə'byü·shən kən'trōl }

distribution frame [COMMUN] A place where a number of cables converge and signals are redistributed among them. { ‚dis·trə'byü·shən ‚frām }

distribution substation [ELEC] An electric power substation associated with the distribution system and the primary feeders for supply to residential, commercial, and industrial loads. { ‚dis·trə'byü·shən 'səb‚stā·shən }

distribution switchboard [ELEC] Power switchboard used for the distribution of electrical energy at the voltage common for each distribution within a building. { ‚dis·trə'byü·shən 'swich ‚bōrd }

distribution system [ELEC] Circuitry involving high-voltage switchgear, step-down transformers, voltage dividers, and related equipment used to receive high-voltage electricity from a primary source and redistribute it at lower voltages. Also known as electric distribution system. { ‚dis·trə'byü·shən ‚sis·təm }

distribution transformer [ELEC] An element of an electric distribution system located near consumers which changes primary distribution voltage to secondary distribution voltage. { ‚dis·trə'byü·shən tranz'fȯr·mər }

distributor [ELEC] **1.** Any device which allocates a telegraph line to each of a number of channels, or to each row of holes on a punched tape, in succession. **2.** A rotary switch that directs the high-voltage ignition current in the proper firing sequence to the various cylinders of an internal combustion engine. [ELECTR] The electronic circuitry which acts as an intermediate link between the accumulator and drum storage. { də'strib·yəd·ər }

distributor points [ELEC] Cam-operated contacts, the opening of which triggers the ignition pulse in an internal combustion engine. { də'strib·yəd·ər ‚pȯins }

disturbance [COMMUN] An undesired interference or noise signal affecting radio, television, or data reception. { də'stər·bəns }

disturbed-one output [ELECTR] One output of a magnetic cell to which partial-read pulses have been applied since that cell was last selected for writing. { də|stərbd |wən 'aùt‚pút }

dither [COMMUN] A technique for representing the entire gray scale of a picture by picture elements with only one of two levels ("white" and "black"), in which a multilevel input image signal is compared with a position-dependent set of thresholds, and picture elements are

set to "white" only where the image input signal exceeds the threshold. [CONT SYS] A force having a controlled amplitude and frequency, applied continuously to a device driven by a servomotor so that the device is constantly in small-amplitude motion and cannot stick at its null position. Also known as buzz. { 'dith·ər }

dither matrix [COMMUN] A square matrix of threshold values that is repeated as a regular array to provide a threshold pattern for an entire image in the dither method of image representation. { 'dith·ər ‚mā·triks }

divergence [ELECTR] The spreading of a cathode-ray stream due to repulsion of like charges (electrons). { də'vər·jəns }

diversity [COMMUN] Method of signal extraction by which an optimum resultant signal is derived from a combination of, or selection from, a plurality of transmission paths, channels, techniques, or physical arrangements; the system may employ space diversity, polarization diversity, frequency diversity, or any other arrangement by which a choice can be made between signals. { də'vər·səd·ē }

diversity factor [ELEC] Ratio of the sum of the individual maximum demands to total maximum demand, as applied to an electrical distribution system. { də'vər·səd·ē ‚fak·tər }

diversity gain [COMMUN] Gain in reception as a result of the use of two or more receiving antennas. { də'vər·səd·ē ‚gān }

diversity radar [ENG] A radar that uses two or more transmitters and receivers, each pair operating at a slightly different frequency but sharing a common antenna and video display, to obtain greater effective range and reduce susceptibility to jamming. { də'vər·səd·ē 'rā‚där }

diversity receiver [ELECTR] A radio receiver designed for space or frequency diversity reception. { də'vər·səd·ē ri'sē·vər }

diversity reception [COMMUN] Radio reception in which the effects of fading are minimized by combining two or more sources of signal energy carrying the same modulation. { də'vər·səd·ē ri'sep·shən }

diverter [ELEC] A low resistance which is connected in parallel with the series or commutating pole winding of a direct-current machine and diverts current from it, causing the magnetomotive force produced by the winding to vary. { də'vərd·ər }

diverter-pole generator [ELEC] Compound wound direct-current generator with the series winding of the diverter pole opposing the flux generated by the shunt wound main pole; provides a close voltage regulation. { də'vərd·ər‚pōl 'jen·ə ‚rād·ər }

divide check [COMPUT SCI] An error signal indicating that an illegal division (such as dividing by zero) was attempted. { də'vīd ‚chek }

divided slit scan |COMPUT SCI| In optical character recognition, a device consisting of a narrow column of photoelectric cells which scans as input character at given intervals for the purpose of obtaining its horizontal and vertical components. { də'vīd·əd 'slit ,skan }

dividing network See crossover network. { də'vīd·iŋ ,net,wərk }

division |COMPUT SCI| One of four required parts of a COBOL program, labeled identification, environment, data, and procedure, each with a set of rules governing the contents. { də'vizh·ən }

division subroutine |COMPUT SCI| A built-in program which achieves division by methods such as repetitive subtraction. { də'vizh·ən 'səb·rü,tēn }

dma See direct memory access.

DNS See domain name system.

Dobrowolsky generator |ELEC| Three-wire, direct-current generator with a balance coil connected across the armature; the coil's midpoint produces the midpoint voltage for the system. { ,dō·brə'väl·skē 'jen·ə,rād·ər }

docking station |COMPUT SCI| A device that connects a portable computer with peripherals such as an external monitor, keyboard, and so on, allowing a portable computer to function as a desktop computer. { 'däk·iŋ ,stā·shən }

document |COMPUT SCI| **1.** Any record, printed or otherwise, that can be read by a human or a machine. **2.** To prepare a written text and charts describing the purpose, nature, usage, and operation of a program or a system of programs. { 'däk·yə·mənt }

document alignment |COMPUT SCI| The phase of the reading process in which a transverse force is applied to a document to line up its reference edge with that of the reading station. { 'däk·yə·mənt ə,līn·mənt }

documentation |COMPUT SCI| The collection, organized and stored, of records that describe the purpose, use, structure, details, and operational requirements of a program, for the purpose of making this information easily accessible to the user. { ,däk·yə·mən'tā·shən }

document comparison utility |COMPUT SCI| A program that compares two documents created by word-processing programs and provides a display of the differences between them. { ,däk·yə·mənt kəm'par·ə·sən yü,til·əd·ē }

document flow |COMPUT SCI| The path taken by documents as they are processed through a record handling system. { 'däk·yə·mənt ,flō }

document handling |COMPUT SCI| In character recognition, the process of loading, feeding, transporting, and unloading a cut-form document that has been submitted for character recognition. { 'däk·yə·mənt ,hand·liŋ }

document image processing |COMPUT SCI| The scanning of paper documents followed by the storage, retrieval, display, and management of the resulting electronic images. Also known as document imaging. { ¦däk·yə·mənt 'im·ij ,prä ,ses·iŋ }

document imaging See document image processing. { 'däk·yə·mənt ,im·ij·iŋ }

document leading edge |COMPUT SCI| In character recognition, that edge which is the foremost one encountered during the reading process and whose relative position defines the document's direction of travel. { 'däk·yə·mənt ,lēd·iŋ 'ej }

document misregistration |COMPUT SCI| In character recognition, the improper state of appearance of a document, on site in a character reader, with respect to real or imaginary horizontal baselines. { 'däk·yə·mənt ,mis·rej·ə'strā·shən }

document number |COMPUT SCI| The number given to a document by its originators to be used as a means for retrieval; it will follow any one of various systems, such as chronological, subject area, or accession. { 'däk·yə·mənt ,nəm·bər }

document processing |COMPUT SCI| The creation, handling, labeling, and modification of text documents, such as in word processing and in the indexing of documents for retrieval based on their content. { ¦däk·yə·mənt 'präs·es·iŋ }

document reader |COMPUT SCI| An optical character reader which reads a limited amount of information (one to five lines) and generally operates from a predetermined format. { 'däk·yə·mənt ,rēd·ər }

document reference edge |COMPUT SCI| In character recognition, that edge of a source document which provides the basis of all subsequent reading processes, insofar as it indicates the relative position of registration marks, and the impending text. { 'däk·yə·mənt 'ref·rəns ,ej }

Document Type Definition |COMPUT SCI| In Standard Generalized Markup Language, a file that specifies the tags in a particular document and the relationships among the fields that they represent. Abbreviated DTD. { 'däk·yə·mənt ,tīp ,def·ə,nish·ən }

docuterm |COMPUT SCI| A word or phrase descriptive of the subject matter or concept of an item of information and considered important for later retrieval of information. { 'däk·yə,tərm }

DOD See direct outward dialing.

dog |COMPUT SCI| A name for the hexadecimal digit whose decimal equivalent is 13. { dòg }

doghouse |ELECTR| Small enclosure placed at the base of a transmitting antenna tower to house antenna tuning equipment. { 'dòg,haus }

Doherty amplifier |ELECTR| A linear radio-frequency power amplifier that is divided into two sections whose inputs and outputs are connected by quarter-wave networks; for all values of input signal voltage up to one-half maximum amplitude, section no. 1 delivers all the power to the load; above this level, section no. 2 comes into operation. { 'dō·ərd·ē ,am·plə ,fī·ər }

do loop |COMPUT SCI| A FORTRAN iterative technique which enables any number of instructions to be executed repeatedly. { 'dü ,lüp }

domain |COMPUT SCI| **1.** The set of all possible values contained in a particular field for every record of a file. **2.** The protected resources that are surrounded by the security perimeter of a distributed computer system. Also known as enclave; protected subnetwork. **3.** The final two or three letters of an Internet address, which specifies the highest subdivision; in the United States this is the type of organization, such as commercial, educational, or governmental, while outside the United States it is usually a country. { dō'mān }

domain name [COMPUT SCI] An alphanumeric string which identifies a particular computer or a network on the Internet. { dō'mān ,nām }

domain name system [COMPUT SCI] Abbreviated DNS. **1.** A system used on the Internet to map the easily remembered names of host computers (domain names) to their respective Internet Protocol (IP) numbers. **2.** A software database program that converts domain names to Internet Protocol addresses, and vice versa. { dō,mān 'nām ,sis·təm }

domain tip memory [COMPUT SCI] A computer memory in which the presence or absence of a magnetic domain in a localized region of a thin magnetic film designates a 1 or 0. Abbreviated DOT memory. Also known as magnetic domain memory. { dō'mān ,tip 'mem·rē }

domestic induction heater [ENG] A cooking utensil heated by current (usually of commercial power line frequency) induced in it by a primary inductor. { də'mes·tik in'dək·shən ,hēd·ər }

domestic public-frequency bands [COMMUN] Radio-frequency bands reserved for public service within the United States. { də'mes·tik ¦pəb·lik 'frē·kwən·sē ,banz }

domestic satellite [ENG] A satellite in stationary orbit 22,300 miles (35,680 kilometers) above the equator for handling 12 or more separate color television programs, thousands of private-line telephone calls, or an equivalent number of channels for other communication services within the United States. Abbreviated DOMSAT. { də'mes·tik 'sad·əl,īt }

dominant mode See fundamental mode. { 'däm·ə·nənt 'mōd }

DOMSAT See domestic satellite. { 'däm,sat }

dongle [COMPUT SCI] A hardware device that plugs into a computer or printer port and serves as a copy-protection device for certain software, which must verify its presence in order to run properly. Also known as hardware key. { 'daŋ·gəl }

donor [SOLID STATE] An impurity that is added to a pure semiconductor material to increase the number of free electrons.Also known as donor impurity; electron donor. { 'dō·nər }

donor impurity See donor. { 'dō·nər im,pyur·əde }

do-nothing instruction See NO OP. { 'dü ,nəth·iŋ in,strək·shən }

doorknob capacitor [ELEC] A high-voltage, plastic-encased capacitor resembling a doorknob in size and shape. { 'dȯr,näb kə,pas·əd·ər }

dopant See doping agent. { 'dō·pənt }

dope See doping agent. { dōp }

doped junction [ELECTR] A junction produced by adding an impurity to the melt during growing of a semiconductor crystal. { ¦dōpt 'jəŋk·shən }

doping [ELECTR] The addition of impurities to a semiconductor to achieve a desired charac-

teristic, as in producing an *n*-type or *p*-type material. Also known as semiconductor doping. { 'dōp·iŋ }

doping agent [ELECTR] An impurity element added to semiconductor materials used in crystal diodes and transistors. Also known as dopant; dope. { 'dōp·iŋ ,ā·jənt }

doping compensation [ELECTR] The addition of donor impurities to a *p*-type semiconductor or of acceptor impurities to an *n*-type semiconductor. { 'dōp·iŋ käm·pən'sā·shən }

Doppler filtering [ELECTR] A form of coherent signal processing in a Doppler radar involving, in a pulsed radar, multiple pulses in a coherent processing interval so that one Doppler shift, indicative of the target radial velocity, may be distinguished from another; similar Doppler-sensitive processing in a continuous-wave radar. { 'däp·lər ,fil·tər·iŋ }

Doppler radar [ENG] Coherent radar, either continuous wave or pulsed, capable of sensing the radial motion of targets by sensing the Doppler shift of the echoes. { 'däp·lər 'rā,där }

Doppler sonar [ENG] Sonar based on Doppler shift measurement technique. Abbreviated DS. { 'däp·lər 'sō,när }

Doppler tracking [ENG] Tracking of a target by using Doppler radar. { 'däp·lər ,trak·iŋ }

Doppler VOR [NAV] A ground-based navigational aid operating at very high frequency and using a wide-aperture radiation system to reduce azimuth errors caused by reflection from terrain and other obstacles; makes use of the Doppler principle to solve the problem of ambiguity that arises from the use of a radiation system with apertures that exceed one-half wavelength. { 'däp·lər ¦vē¦ō'är }

DOS See disk operating system. { däs }

dot See button. { dät }

dot-addressable [COMPUT SCI] The ability of an electronic display or a dot-matrix printer to specify the individual dots that form images of characters. { ¦dät ə'dres·ə·bəl }

dot character printer See dot matrix printer. { 'dät 'kar·ik·tər ,print·ər }

dot cycle [COMMUN] In teletypewriter systems, an on-off or mark-space cycle in which both mark and space have the same length as the unit pulse. { 'dät ,sī·kəl }

dot generator [ELECTR] A signal generator that produces a dot pattern on the screen of a color display device for use in convergence adjustments. { 'dät ,jen·ə,rād·ər }

dot matrix [COMPUT SCI] An array of dots that forms a character or graphic symbol. { ¦dät 'mā·triks }

dot matrix printer [COMPUT SCI] A type of printer that forms each character as a group of small dots, using a group of wires located in the printing element. Also known as dot character printer. { 'dät ¦mā·triks 'prin·tər }

dot-sequential color television [ELECTR] An analog color television system in which the red, blue, and green primary-color dots are formed in rapid succession along each scanning line. { ¦dät sə¦kwen·chəl 'kəl·ər 'tel·ə,vizh·ən }

dot system [ELECTR] Manufacturing technique for producing microelectronic circuitry. { 'dät ,sis·təm }

double-amplitude-modulation multiplier [ELECTR] A multiplier in which one variable is amplitude-modulated by a carrier, and the modulated signal is again amplitude-modulated by the other variable; the resulting double-modulated signal is applied to a balanced demodulator to obtain the product of the two variables. { ¦dəb·əl ¦am·plə,tüd ¦mäj·ə,lā·shən 'məl·tə,plī·ər }

double armature [ELEC] An armature with two separate windings on a single core. { 'dəb·əl 'är·mə·chər }

double-barrier resonant tunneling diode [ELECTR] A variant of the tunnel diode with thin layers of aluminum gallium arsenide and gallium arsenide that have sharp interfaces and have widths comparable to the Schrödinger wavelengths of the electrons, permitting resonant behavior. Abbreviated DBRT diode. { ¦dəb·əl ,bar·ē·ər ¦rez·ən·ənt ,tən·əl·iŋ 'dī,ōd }

double-base diode See unijunction transistor. { ¦dəb·əl ¦bās 'dī,ōd }

double-base junction diode See unijunction transistor. { ¦dəb·əl ¦bās 'jəŋk·shən 'dī,ōd }

double-base junction transistor [ELECTR] A tetrode transistor that is essentially a junction triode transistor having two base connections on opposite sides of the central region of the transistor. Also known as tetrode junction transistor. { ¦dəb·əl ¦bās 'jəŋk·shən tran'zis·tər }

double-beam cathode-ray tube [ELECTR] A cathode-ray tube having two beams and capable of producing two independent traces that may overlap; the beams may be produced by splitting the beam of one gun or by using two guns. { ¦dəb·əl ¦bēm ¦kāth,ōd 'rā ,tüb }

double-bounce calibration [ELECTR] Method of radar calibration which is used to determine the zero set error by using round-trip echoes; the correct range is the difference between the first and second echoes. { ¦dəb·əl ¦baúns kal·ə'brā·shən }

double-break switch [ELEC] Switch which opens the connected circuit at two points. { ¦dəb·əl ¦brāk 'swich }

double bridge See Kelvin bridge. { ¦dəb·əl 'brij }

double-buffered data transfer [COMPUT SCI] The transmission of data into the buffer register and from there into the device register proper. { ¦dəb·əl ¦bəf·ərd'dad·ə ,trans·fər }

double bus-double breaker [ELEC] A substation switching arrangement having two common buses and two breakers per connection. { ¦dəb·əl 'bəs ¦dəb·əl ,brāk·ər }

double bus-single breaker [ELEC] A substation switching arrangement that involves two common buses and only one breaker per connection. { ¦dəb·əl 'bəs ¦siŋ·gəl ,brāk·ər }

double-button microphone [ENG ACOUS] A carbon microphone having two carbon-filled buttonlike containers, one on each side of the diaphragm, to give twice the resistance change obtainable with a single button. Also known as differential microphone. { ¦dəb·əl ¦bət·ən 'mī·krə,fōn }

double-channel duplex [COMMUN] A method that provides for simultaneous communication between two stations through use of two radio-frequency channels, one in each direction. { ¦dəb·əl ¦chan·əl 'dü,pleks }

double-channel simplex [COMMUN] A method that provides for nonsimultaneous communication between two stations through use of two radio-frequency channels, one in each direction. { ¦dəb·əl ¦chan·əl 'sim,pleks }

double-click [COMPUT SCI] To depress and release a mouse button twice in quick succession; often used to initiate an action such as opening a file, and to extend actions that result from a single click. { ¦dəb·əl 'klik }

double-current cable code [COMMUN] A cable code in which characters are determined by bipolar characters of equal length. { ¦dəb·əl ¦kə·rənt 'kā·bəl ,kōd }

double-current generator [ELEC] Machine which supplies both direct and alternating current from the same armature winding. { ¦dəb·əl ¦kə·rənt 'jen·ə,rād·ər }

double-current signaling [COMMUN] A system of telegraph signaling that uses both positive and negative currents. { ¦dəb·əl ¦kə·rənt 'sig·nəl·iŋ }

double data rate [COMPUT SCI] A clocking technique that increases the transfer speeds of synchronous memories by using both the leading and trailing edges of the clock signal to transfer data, effectively doubling the transfer rate or bandwidth. { ¦dəb·əl 'dad·ə ,rāt }

double density [COMPUT SCI] Property of a computer storage medium that holds twice as much data per unit of storage space as the standard; applied particularly to floppy disks. { 'dəb·əl 'den·səd·ē }

double-diffused transistor [ELECTR] A transistor in which two pn junctions are formed in the semiconductor wafer by gaseous diffusion of both p-type and n-type impurities; an intrinsic region can also be formed. { ¦dəb·əl də¦fyüzd tran'zis·tər }

double-diode limiter [ELECTR] Type of limiter which is used to remove all positive signals from a combination of positive and negative pulses, or to remove all the negative signals from such a combination of positive and negative pulses. { ¦dəb·əl ¦dī¸ōd 'lim·əd·ər }

double-doped transistor [ELECTR] The original grown-junction transistor, formed by successively adding p-type and n-type impurities to the melt during growing of the crystal. { ¦dəb·əl ¸dōpt tran'zis·tər }

double-doublet antenna [ELECTROMAG] Two half-wave doublet antennas criss-crossed at their center, one being shorter than the other to give broader frequency coverage. { ¦dəb·əl ¦dəb·lət an'ten·ə }

double frequency shift keying [COMMUN] Multiplex system in which two telegraph signals are combined and transmitted simultaneously by a method of frequency shifting between four radio frequencies. { ¦dəb·əl 'frē·kwən·sē ¦shift 'kē·iŋ }

double image [ELECTR] A television picture consisting of two overlapping images due to reception of the analog signal over two paths of different length so that signals arrive at slightly different times. { ¦dəb·əl 'im·ij }

double-length number [COMPUT SCI] A number having twice as many digits as are ordinarily used in a given computer. Also known as double-precision number. { ¦dəb·əl ¦leŋkth 'nəm·bər }

double limiter See cascade limiter. { ¦dəb·əl 'lim·əd·ər }

double-list sorting [COMPUT SCI] A method of internal sorting in which the entire unsorted list is first placed in one portion of main memory and sorting action then takes place, creating a sorted list, generally in another area of memory. { ¦dəb·əl ¸list 'sórd·iŋ }

double moding [ELECTR] Undesirable shifting of a magnetron from one frequency to another at irregular intervals. { ¦dəb·əl 'mōd·iŋ }

double modulation [COMMUN] A method of modulation in which a subcarrier is first modulated with the desired intelligence, and the modulated subcarrier is then used to modulate a second carrier having a higher frequency. { ¦dəb·əl ¸mäj·ə'lā·shən }

double-polarity pulse-amplitude modulation [COMMUN] Pulse-amplitude modulation employing pulses of positive and negative polarity, the average value being equal to zero. Also known as bidirectional pulse-amplitude modulation. { ¦dəb·əl pə'lar·əd·ē 'pəls ¦am·plə ¸tüd ¸mäj·ə'lā·shən }

double-pole double-throw switch [ELEC] A six-terminal switch or relay contact arrangement that simultaneously connects one pair of terminals to either of two other pairs of terminals. Abbreviated dpdt switch. { ¦dəb·əl ¦pōl ¦dəb·əl ¦thrō 'swich }

double-pole single-throw switch [ELEC] A four-terminal switch or relay contact arrangement that simultaneously opens or closes two separate circuits or both sides of the same circuit. Abbreviated dpst switch. { ¦dəb·əl ¦pōl ¦siŋ·gəl ¦thrō 'swich }

double-pole switch [ELEC] A switch that operates simultaneously in two separate electric

circuits or in both lines of a single circuit. { ¦dəb·əl ¦pōl 'swich }

double précision [COMPUT SCI] The use of two computer words to represent a double-length number. { ¦dəb·əl prə'sizh·ən }

double-precision hardware [COMPUT SCI] Special arithmetic units in a computer designed to handle double-length numbers, employed in operations in which greater accuracy than normal is desired. { ¦dəb·əl prə¦sizh·ən 'härd¸wer }

double-precision number See double-length number. { ¦dəb·əl prə¦sizh·ən 'nəm·bər }

double-pulse recording [COMPUT SCI] A technique for recording binary digits in magnetic cells in which each cell consists of two regions that can be magnetized in opposite directions and the value of each bit (0 or 1) is determined by the order in which the regions occur. { ¦dəb·əl ¦pəls ri'kórd·iŋ }

doubler See frequency doubler. { 'dəb·lər }

double refraction See birefringence. { ¦dəb·əl ri'frak·shən }

double screen [ELECTR] Three-layer cathode-ray tube screen consisting of a two-layer screen with the addition of a second long-persistence coating having a different color and different persistence from the first. { ¦dəb·əl 'skrēn }

double-shield enclosure [ELEC] Type of shielded enclosure or room in which the inner wall is partially isolated electrically from the outer wall. { ¦dəb·əl ¦shēld in'klō·zhər }

double-sideband modulation [COMMUN] Amplitude modulation in which the modulated wave is composed of a carrier, an upper sideband whose frequency is the sum of the carrier and modulation frequencies, and a lower sideband whose frequency is the difference between the carrier and modulation frequencies. Abbreviated DSB. Also known as double-sideband transmitted-carrier modulation (DSB-TC modulation; DSTC modulation). { ¦dəb·əl ¦sīd¸band ¸mäj·ə'lā·shən }

double-sideband reduced-carrier modulation [COMMUN] A form of amplitude modulation in which both the upper and lower sidebands are transmitted but the power contained in the unmodulated carrier is reduced to a fixed level below that provided to the modulator. Abbreviated DSB-RC modulation. { ¸dəb·əl ¦sīd ¸band ri¸düst ¦kar·ē·ər ¸mä·jə¸lā·shən }

double-sideband suppressed-carrier modulation [COMMUN] A form of amplitude modulation in which both the upper and lower sidebands are transmitted but the power contained in the unmodulated carrier is reduced to a fixed level below that provided to the modulator. Abbreviated DSB-SC modulation. { ¸dəb·əl ¦sīd ¸band sə¸prest ¦kar·ē·ər ¸mäj·ə¸lā·shən }

double-sideband transmission [COMMUN] The transmission of a modulated carrier wave accompanied by both of the sidebands resulting from modulation; the upper sideband corresponds to the sum of the carrier and modulation frequencies, whereas the lower sideband corresponds to the difference between the carrier and modulation frequencies. { ¦dab·əl ¦sīd,band tranz'mish·ən }

double-sideband transmitted-carrier modulation See double-sideband modulation. { ¦dab·əl ¦sīd ,band tranz¦mid·əd ¦kar·ē·ər ,mäj·ə'lā·shən }

double-sided board [ELECTR] A printed wiring board that contains circuitry on both external layers. { ¦dab·əl ,sīd·əd 'bórd }

double-sided disk [COMPUT SCI] A diskette that can be written on both of its sides. { ¦dab·əl¦sīd· əd 'disk }

double-stream amplifier [ELECTR] Microwave traveling-wave amplifier in which amplification occurs through interaction of two electron beams having different average velocities. { ¦dab·əl ,strēm 'am·plə,fī·ər }

double-stub tuner [ELECTROMAG] Impedance-matching device, consisting of two stubs, usually fixed three-eighths of a wavelength apart, in parallel with the main transmission lines. { ¦dab·əl ,stəb 'tün·ər }

double-superheterodyne reception [COMMUN] Method of reception in which two frequency converters are employed before final detection. Also known as triple detection. { ¦dab·əl ,sü·pər¦het·rə,dīn ri'sep·shən }

doublet antenna See dipole antenna. { 'dab·lət an'ten·ə }

double-throw circuit breaker [ELEC] Circuit breaker by means of which a change in the circuit connections can be obtained by closing either of two sets of contacts. { ¦dab·əl ,thrō 'sər·kət ,brāk·ər }

double-throw switch [ELEC] A switch that connects one set of two or more terminals to either of two other similar sets of terminals. { ¦dab·əl ,thrō 'swich }

double-track tape recorder [ENG ACOUS] A tape recorder with a recording head that covers half the tape width, so two parallel tracks can be recorded on one tape. Also known as dual-track tape recorder; half-track tape recorder. { ¦dab·əl ,trak 'tāp ri,kórd·ər }

double triode [ELECTR] An electron tube having two triodes in the same envelope. Also known as duotriode. { ¦dab·əl 'trī,ōd }

doublet trigger [ELECTR] A trigger signal consisting of two pulses spaced a predetermined amount for coding purposes. { 'dab·lət ,trig·ər }

double-tuned amplifier [ELECTR] Amplifier of one or more stages in which each stage uses coupled circuits having two frequencies of resonance, to obtain wider bands than those obtainable with single tuning. { ¦dab·əl ,tünd 'am·plə,fī·ər }

double-tuned circuit [ELECTR] A circuit that is resonant to two adjacent frequencies, so that there are two approximately equal values of peak response, with a dip between. { ¦dab·əl ,tünd 'sər·kət }

double-tuned detector [ELECTR] A type of frequency-modulation discriminator in which the limiter output transformer has two secondaries, one tuned above the resting frequency and the other tuned an equal amount below. { ¦dab·əl ,tünd di'tek·tər }

double-winding synchronous generator [ELEC] Synchronous generator which has two similar windings, in phase with one another, mounted on the same magnetic structure but not connected electrically, designed to supply power to two independent external circuits. { ¦dab·əl ¦wīnd·iŋ ¦siŋ·krə·nəs 'jen·ə,rād·ər }

double word [COMPUT SCI] A unit containing twice as many bits as a word. { ¦dab·əl 'wərd }

double-word addressing [COMPUT SCI] An addressing mode in computers with short words (less than 16 bits) in which the second of two consecutive instruction words contains the address of a location. { ¦dab·əl ,wərd 'a ,dres·iŋ }

doubly linked ring [COMPUT SCI] A cycle arrangement of data elements in which searches are possible in both directions. { ¦dab·lē ¦liŋkt 'riŋ }

do-until structure [COMPUT SCI] A set of program statements that is executed once, and may then be executed repeatedly, depending on the results of a test specified in the first statement. { 'dü ən'til ,strək·chər }

do-while structure [COMPUT SCI] A set of program statements that is executed repeatedly, as long as some condition, specified in the first statement, remains in effect. { 'dü 'wīl ,strək·chər }

down-lead See lead-in. { 'daun ,lēd }

downlink [COMMUN] The radio or optical transmission path downward from a communications satellite to the earth or an aircraft, or from an aircraft to the earth. { 'daun,liŋk }

download [COMPUT SCI] To transfer a program or data file from a central computer to a remote computer or to the memory of an intelligent terminal. { 'daun,lōd }

downward compatibility [COMPUT SCI] The ability of an older or smaller computer to accept programs from a newer or larger one. Also known as backward compatibility. { 'daun·wərd kəm ,pad·ə'bil·əd·ē }

Dow oscillator See electron-coupled oscillator. { ¦daú 'äs·ə,lād·ər }

DPCM See differential pulse-code modulation.

dpdt switch See double-pole double-throw switch. { ¦dē¦pē¦dē'tē ,swich }

DPMS See display power management signaling.

dpst switch See double-pole single-throw switch. { ¦dē¦pē¦es'tē ,swich }

drag [COMPUT SCI] To move an object across a screen by moving a pointing device while holding down the control button. { drag }

drag and drop [COMPUT SCI] A feature whereby operations are performed on objects, such as icons or blocks of text, by dragging them across the screen to a particular spot. { ¦drag ən 'dräp }

drag-cup motor [ELEC] An induction motor having a cup-shaped rotor or conducting material, inside of which is a stationary magnetic core. { 'drag ,kəp 'mōd·ər }

drain [ELEC] See current drain. [ELECTR] The region into which majority carriers flow in a field-effect transistor; it is comparable to the collector of a bipolar transistor and the anode of an electron tube. { drān }

drain wire |ELEC| Metallic conductor frequently used in contact with foil-type signal-cable shielding to provide a low-resistance ground return at any point along the shield. { 'drān ,wīr }

DRAM *See* dynamic random-access memory. { 'dē,ram }

DRAW *See* direct read after write. { drȯ }

drawing program |COMPUT SCI| A graphics program that maintains images in vector graphics format, allowing the user to design and illustrate objects on the display screen. Also known as illustration program. { 'drȯ·iŋ ,prō·grəm }

dress |ELECTR| The arrangement of connecting wires in a circuit to prevent undesirable coupling and feedback.

drift |ENG| A gradual deviation from a set adjustment, such as frequency or balance current, or from a direction. { drift }

drift-corrected amplifier |ELECTR| A type of amplifier that includes circuits designed to reduce gradual changes in output, used in analog computers. { ¦drift kə¦rek·təd 'am·plə,fī·ər }

drift error |COMPUT SCI| An error arising in the use of an analog computer due to gradual changes in the output of circuits (such as amplifiers) in the computer. { 'drift ,er·ər }

drift space |ELECTR| A space in an electron tube which is substantially free of externally applied alternating fields and in which repositioning of electrons takes place. { 'drift ,spās }

drift speed |ELEC| Average speed at which electrons or ions progress through a medium. { 'drift ,spēd }

drift transistor |ELECTR| **1.** A transistor having two plane parallel junctions, with a resistivity gradient in the base region between the junctions to improve the high-frequency response. **2.** *See* diffused-alloy transistor. { 'drift tran,zis·tər }

drill circuit |COMMUN| A telegraph circuit used only to practice sending and receiving. { 'dril ,sər·kət }

drill down |COMPUT SCI| In data mining, viewing data at a greater level of detail; for example, viewing individual sales as opposed to viewing total sales. { ¦dril 'daȯn }

drill up |COMPUT SCI| In data mining, viewing data in less detail; for example, viewing total sales as opposed to individual sales. { ¦dril 'əp }

drive |ELECTR| *See* excitation. |ENG| The means by which a machine is given motion or power or by which power is transferred from one part of a machine to another. { drīv }

drive array |COMPUT SCI| A collection of hard disks organized to increase speed and improve reliability, often with the help of data stripping. { 'drīv ə,rā }

drive bay |COMPUT SCI| A space in the cabinet of a personal computer where disk drives, tape drives, and CD-ROM drives can be installed. Also known as bay. { 'drīv ,bā }

drive control *See* horizontal drive control. { 'drīv kən,trōl }

driveless work station |COMPUT SCI| A computer or terminal in a local area network that does not have its own disk drives and relies on a central mass storage facility for information storage. { 'drīv·ləs 'wərk ,stā·shən }

drive light |COMPUT SCI| A lamp on the front of a disk drive that lights to indicate when the unit is reading or writing data. { 'drīv ,līt }

driven array |ELECTROMAG| An antenna array consisting of a number of driven elements, usually half-wave dipoles, fed in phase or out of phase from a common source. { ¦drīv·ən ə'rā }

driven blocking oscillator *See* monostable blocking oscillator. { ¦drīv·ən 'bläk·iŋ 'äs·ə,lād·ər }

driven element |ELECTROMAG| An antenna element that is directly connected to the transmission line. { ¦drīv·ən 'el·ə·mənt }

drive pattern |COMMUN| In a facsimile system, undesired pattern of density variations caused by periodic errors in the position of the recording spot. { 'drīv ,pad·ərn }

drive pulse |ELECTR| An electrical pulse which induces a magnetizing force in an element of a magnetic core storage, reversing the polarity of the core. { 'drīv ,pəls }

driver |COMPUT SCI| A sequence of program instructions that controls an input/output device such as a tape drive or disk drive. |ELECTR| The amplifier stage preceding the output stage in a receiver or transmitter. |ENG ACOUS| The portion of a horn loudspeaker that converts electrical energy into acoustical energy and feeds the acoustical energy to the small end of the horn. { 'drī·vər }

driver element |ELECTROMAG| Antenna array element that receives power directly from the transmitter. { 'drī·vər ,el·ə·mənt }

driver sweep |ELECTR| Sweep triggered only by an incoming signal or trigger. { 'drī·vər ,swēp }

driver transformer |ELECTR| A transformer in the input circuit of an amplifier, especially in the transmitter. { 'drī·vər tranz'fȯr·mər }

drive winding |ELECTR| A coil of wire that is inductively coupled to an element of a magnetic memory. Also known as drive wire. { 'drīv ,wīn·diŋ }

drive wire *See* drive winding. { 'drīv ,wīr }

driving clock |ENG| A mechanism for driving an instrument at a required rate. { 'drīv·iŋ ,kläk }

driving-point function |CONT SYS| A special type of transfer function in which the input and output variables are voltages or currents measured between the same pair of terminals in an electrical network. { 'drīv·iŋ ,pȯint ,fəŋk·shən }

driving-point impedance |ELECTR| The complex ratio of applied alternating voltage to the resulting alternating current in an electron tube, network, or other transducer. { 'drīv·iŋ ,pȯint im'pēd·əns }

driving signal |ELECTR| Television signal that times the scanning at the pickup point. { 'drīv·iŋ ,sig·nəl }

drop bar |ELEC| Protective device used to ground a high-voltage capacitor when opening a door. { 'dräp ,bär }

drop bracket transposition |ELEC| Reversal of the relative positions of two parallel wire conductors while depressing one, so that the crossover is in a vertical plane. { 'dräp ,brak·ət tranz·pə'zish·ən }

drop-dead halt [COMPUT SCI] A machine halt from which there is no recovery; such a halt may occur through a logical error in programming; examples in which a drop-dead halt could occur are division by zero and transfer to a nonexistent instruction word. Also known as dead halt. { |dräp |ded 'hölt }

drop-in [COMPUT SCI] The accidental appearance of an unwanted bit, digit, or character on a magnetic recording surface or during reading from or writing to a magnetic storage device. { 'dräp ,in }

dropout [COMPUT SCI] The accidental disappearance of a valid bit, digit, or character from a storage medium or during reading from or writing to a storage device. [ELEC] Of a relay, the maximum current, voltage, power, or such, at which it will release from its energized position. [ELECTR] A reduction in output signal level during reproduction of recorded data, sufficient to cause a processing error. { 'dräp,aút }

dropout current [ELEC] The maximum current at which a relay or other magnetically operated device will release to its deenergized position. { 'dräp,aút ,kə·rənt }

dropout error [ELECTR] Loss of a recorded bit or any other error occurring in recorded magnetic tape due to foreign particles on or in the magnetic coating or to defects in the backing. { 'dräp,aút ,er·ər }

dropout fuse [ELEC] A fuse used on utility line poles which springs open when the fuse metal melts to provide rapid arc extinction, and which drops to an open-circuit position readily distinguishable from the ground. Also known as flip-open cutout fuse. { 'dräp,aút ,fyüz }

dropout voltage [ELEC] The maximum voltage at which a relay or other magnetically operated device will release to its deenergized position. { 'dräp,aút ,völ·tij }

dropping resistor [ELEC] A resistor used in series with a load to decrease the voltage applied to the load. { 'dräp·iŋ ri,zis·tər }

drop relay [ELEC] Relay activated by incoming ringing current to call an operator's attention to a subscriber's line. { |dräp 'rē,lā }

drop repeater [ELECTR] Microwave repeater that is provided with the necessary equipment for local termination of one or more circuits. { 'dräp ri,pēd·ər }

drop wire [ELEC] Wire suitable for extending an open wire or cable pair from a pole or cable terminal to a building. { 'dräp ,wīr }

drum [ELECTR] A computer storage device consisting of a rapidly rotating cylinder with a magnetizable external surface on which data can be read or written by many read/write heads floating a few millionths of an inch off the surface; once used as a primary storage device but now used as an auxiliary device. Also known as drum memory; drum storage; magnetic drum; magnetic drum storage. { drəm }

drum armature [ELEC] An armature that has a drum winding. { 'drəm ,ärm·ə·chər }

drum controller [ELEC] An electric device that has a drum switch for its main switching element; used to govern the way electric power is delivered to a motor. { 'drəm kən,trō·lər }

drum disk rectifier [ELEC] A mechanical rectifier using synchronous contacts and a copper oxide dry disk. { 'drəm ,disk 'rek·tə,fī·ər }

drum mark [COMPUT SCI] A character indicating the termination of a record on a magnetic drum. { 'drəm ,märk }

drum memory See drum. { |drəm 'mem·rē }

drum meter See liquid-sealed meter. { 'drəm ,mēd·ər }

drum parity error [COMPUT SCI] Parity error occurring during transfer of information onto or from drums. { |drəm 'par·əd·ē ,er·ər }

drum plotter [ENG] A graphics output device that draws lines with a continuously moving pen on a sheet of paper rolled around a rotating drum that moves the paper in a direction perpendicular to the motion of the pen. { 'drəm ,pläd·ər }

drum printer [COMPUT SCI] An impact printer in which a complete set of characters for each print position on a line is on a continuously rotating drum behind an inked ribbon, with paper in front of the ribbon; identical characters are printed simultaneously at all required positions on a line, on the fly, by signal-controlled hammers. { 'drəm ,print·ər }

drum recorder [ELECTR] A facsimile recorder in which the record sheet is mounted on a rotating drum or cylinder. { 'drəm ri,kòrd·ər }

drum storage See drum. { 'drəm ,stòr·ij }

drum switch [ELEC] A switch in which the electrical contacts are made on pins, segments, or surfaces on the periphery of a rotating cylinder or sector, or by the operation of a rotating cam. { 'drəm ,swich }

drum transmitter [ELECTR] A facsimile transmitter in which the subject copy is mounted on a rotating drum or cylinder. { |drəm tranz'mid·ər }

drum winding [ELEC] A type of winding in electric machines in which coils are housed in long, narrow gaps either in the outer surface of a cylindrical core or in the inner surface of a core with a cylindrical bore. { 'drəm ,wīnd·iŋ }

drunk mouse [COMPUT SCI] A mouse whose pointer jumps irrationally, usually as a result of dirt or grease on the rollers. { |drəŋk 'maús }

dry battery [ELEC] A battery made up of a series, parallel, or series-parallel arrangement of dry cells in a single housing to provide desired voltage and current values. { |drī 'bad·ə·rē }

dry cell [ELEC] A voltage-generating cell having an immobilized electrolyte. { 'drī ,sel }

dry-charged battery [ELEC] A storage battery in which the electrolyte is drained from the battery for storage, and which is filled with electrolyte and charged for a few minutes to prepare for use. { |drī ,chärjd 'bad·ə·rē }

dry circuit [ELEC] A relay circuit in which open-circuit voltages are very low and closed-circuit currents extremely small, so there is no arcing to roughen the contacts. { ¦drī ¦sər·kət }

dry contact [ELEC] A contact that does not break or make current. { ¦drī 'kän,takt }

dry-disk rectifier See metallic rectifier. { ¦drī ,disk 'rek·tə,fī·ər }

dry electrolytic capacitor [ELEC] An electrolytic capacitor in which the electrolyte is a paste rather than a liquid; the dielectric is a thin film of gas formed on one of the plates by chemical action. { ¦drī i¦lek·trə¦lid·ik kə'pas·əd·ər }

dry flashover voltage [ELECTR] Voltage at which the air surrounding a clean dry insulator or shell completely breaks down between electrodes. { ¦drī 'flash,ō·vər ,vōl·tij }

dry plasma etching See plasma etching. { ¦drī 'plaz·mə }

dry-plate rectifier See metallic rectifier. { 'drī ,plāt 'rek·tə,fī·ər }

dry reed relay [ELEC] Reed-type relay which does not use mercury at the relay contacts. { ¦drī ,rēd 'rē,lā }

dry reed switch [ELEC] A switch having contacts mounted on magnetic reeds in a vacuum enclosure, designed for reliable operation in dry circuits. { ¦drī ,rēd 'swich }

dry run [COMPUT SCI] A check of the logic and coding of a computer program in which the program's operations are followed from a flow chart and written instructions, and the results of each step are written down, before the program is run on a computer. Also known as desk check. [ENG] Any practice test or session. { ¦drī 'rən }

Drysdale ac polar potentiometer [ENG] A potentiometer for measuring alternating-current voltages in which the voltage is applied across a slide-wire supplied with current by a phase-shifting transformer; this current is measured by an ammeter and brought into phase with the unknown voltage by adjustment of the transformer rotor, and the unknown voltage is measured by observation of the slide-wire setting for a null indication of a vibration galvanometer. { 'drīz,dāl ¦ā¦sē ¦pō·lər pə,ten·chē'äm·əd·ər }

dry-tape fuel cell [ELEC] A fuel cell in which the fuel is in the form of a dry tape, coated with fuel, oxidant, and electrolyte, which is fed into the cell at a rate corresponding to the demand for electric energy. { 'drī ,tāp 'fyül ,sel }

DS See Doppler sonar.

DSB See double-sideband modulation.

DSB-RC modulation See double-sideband reduced-carrier modulation. { ¦dē¦es'bē ¦är'sē ,mäj·ə,lā·shən }

DSB-SC modulation See double-sideband suppressed-carrier modulation. { ¦dē,es,bē 'es'sē ,mäj·ə,lā·shən }

DSB-TC modulation See double-sideband modulation. { ¦de¦es¦bē ¦tē'sē ,mäj·ə,lā·shən }

D-scan See D-display. { 'dē ,skan }

D-scope See D-display. { 'dē ,skōp }

DSECT See dummy section. { ¦dē'sekt }

D-shell connector [COMPUT SCI] The connector at the end of the cable between a video adapter and a monitor that is plugged into the video adapter. { 'dē,shel kə,nek·tər }

DSI See digital speech interpolation.

DSL See digital subscriber line.

DSP chip See digital signal processing chip. { ¦de ¦es'pē ,chip }

DSS See decision support system.

DSTC modulation See double-sideband modulation. { ¦dē¦es¦tē'sē ,mäj·ə,lā·shən }

DTD See Document Type Definition.

DTL See diode transistor logic.

DTMF See dual-tone mulitfrequency.

DTMF dialing See push-button dialing. { ¦dē¦tē ¦em'ef ,dī·liŋ }

DTV See digital television.

D/U [COMMUN] Ratio of desired to undesired signals, usually expressed in decibels.

dual-actuator hard disk [COMPUT SCI] A hard disk that is equipped with two read/write heads. { ¦dül 'ak·chə,wād·ər ¦härd ,disk }

dual-channel amplifier [ENG ACOUS] An audio-frequency amplifier having two separate amplifiers for the two channels of a stereophonic sound system, usually operating from a common power supply mounted on the same chassis. { ¦dü·əl ¦chan·əl 'am·plə,fī·ər }

dual control [CONT SYS] An optimal control law for a stochastic adaptive control system that gives a balance between keeping the control errors and the estimation errors small. { ¦dü·əl kən'trōl }

dual diversity receiver [ELECTR] A diversity radio receiver in which the two antennas feed separate radio-frequency systems, with mixing occurring after the converter. { ¦dü·əl də'vər·səd·ē ri ,sē·vər }

dual-emitter transistor [ELECTR] A passivated pnp silicon planar epitaxial transistor having two emitters, for use in low-level choppers. { ¦dü·əl i'mid·ər tran,zis·tər }

dual-gun cathode-ray tube [ELECTR] A dual-trace oscilloscope in which beams from two electron guns are controlled by separate balanced vertical-deflection plates and also have separate brightness and focus controls. { ¦dü·əl ¦gən ,kath,ōd 'rā ,tüb }

dual in-line package [ELECTR] Microcircuit package with two rows of seven vertical leads that are easily inserted into an etched circuit board. Abbreviated DIP. { ¦dü·əl ¦in ,līn 'pak·ij }

duality principle Also known as principle of duality. [ELEC] The principle that for any theorem in electrical circuit analysis there is a dual theorem in which one replaces quantities with dual quantities; current and voltage, impedance and admittance, and meshes and nodes are examples of dual quantities. [ELECTR] The principle that analogies may be drawn between a transistor circuit and the corresponding vacuum tube circuit. {ELECTROMAG} The principle that one can obtain new solutions of Maxwell's equations from known solutions by replacing **E** with **H**, **H** with −**E**, ϵ with μ, and μ with ϵ. {MATH} A principle that if a theorem is true, it remains true if each object and operation is replaced by its dual; important in projective geometry and Boolean algebra. { dü'al·əd·ē ,prin·sə·pəl }

dual meter {ENG} Meter constructed so that two aspects of an electric circuit may be read simultaneously. { 'dü·əl ¦mēd·ər }

dual-mode control {CONT SYS} A type of control law which consists of two distinct types of operation; in linear systems, these modes usually consist of a linear feedback mode and a bang-bang-type mode. { 'dü·əl ,mōd kən'trōl }

dual modulation {COMMUN} The process of modulating a common carrier wave or subcarrier with two different types of modulation, each conveying separate information. { 'dü·əl ,mäj·ə'lā·shən }

dual network {ELEC} A network which has the same number of terminal pairs as a given network, and whose open-circuit impedance network is the same as the short-circuit admittance matrix of the given network, and vice versa. { 'dü·əl 'net ,wərk }

dual-scanned liquid-crystal display {ELECTR} A passive matrix liquid-crystal display that is improved by being refreshed twice as frequently as standard displays of this type. { ¦dül ,skand lik·wəd 'krist·əl di,splā }

dual-stripe magnetoresistive head {COMPUT SCI} A type of read/write head for hard disks that has separate areas for reading and writing, reduced vulnerability to outside interference, and the ability to pack data densely on disks { ¦dül ¦strīp mag,ned·ō·ri,zis·div 'hed }

dual-tone multifrequency {COMMUN} Signaling method employing set combinations of two specific frequencies used by subscribers and telephone private branch exchange attendants, if their switchboard positions are so equipped, to indicate telephone address digits, precedence ranks, and end of signaling. Abbreviated DTMF. { 'dü·əl ,tōn ,məl·tē'frē·kwən·sē }

dual-tone multifrequency dialing See pushbutton dialing. { 'dü·əl ,tōn ,məl·tē'frē·kwən·sē 'dī·liŋ }

dual-trace amplifier {ELECTR} An oscilloscope amplifier that switches electronically between two signals under observation in the interval between sweeps, so that waveforms of both signals are displayed on the screen. { 'dü·əl ,trās 'am·plə,fī·ər }

dual-trace oscilloscope {ELECTR} An oscilloscope which can compare two waveforms on the face of a single cathode-ray tube, using any one of several methods. { 'dü·əl ,trās ä'sil·ə ,skōp }

dual-track tape recorder See double-track tape recorder. { 'dü·əl ,trak 'tāp ri,kòrd·ər }

dual-use line {COMMUN} Communications link normally used for more than one mode of transmission, such as voice and data. { 'dü·əl ,yüs ,līn }

dual-use radar {ENG} Radar designed to perform both as surveillance radar and weather radar, of particular value in air traffic management where both the monitoring of aircraft and estimation of the weather environment are important. { 'dü·əl ,yüs 'rā,där }

dub {ENG ACOUS} **1.** To transfer recorded material from one recording to another, with or without the addition of new sounds, background music, or sound effects. **2.** To combine two or more sources of sound into one record. **3.** To add a new sound track or new sounds to a motion picture film, or to a recorded radio or television production. { dəb }

duct {COMMUN} An enclosed runway for cables. { dəkt }

dull emitter {ELECTR} An electron tube whose cathode is a filament that does not glow brightly. { ¦dəl ə'mid·ər }

dumb terminal {COMPUT SCI} A computer input/output device that lacks the capability to process or format data, and is thus entirely dependent on the main computer for these activities. { ¦dəm 'term·ən·əl }

dummy {COMMUN} Telegraphy network simulating a customer's loop for adjusting a telegraph repeater; the dummy side of the repeater is that toward the customer. {COMPUT SCI} An artificial address, instruction, or other unit of information inserted in a digital computer solely to fulfill prescribed conditions (such as word length or block length) without affecting operations. { 'dəm·ē }

dummy antenna {ELECTR} A device that has the impedance characteristic and power-handling capacity of an antenna but does not radiate or receive radio waves; used chiefly for testing a transmitter. Also known as artificial antenna. { ¦dəm·ē an'ten·ə }

dummy argument {COMPUT SCI} The variable appearing in the definition of a macro or function which will be replaced by an address at call time. { ¦dəm·ē 'är·gyə·mənt }

dummy file {COMPUT SCI} A nonexistent file which is treated by a computer program as if it were receiving its output data, when in fact the data are being ignored; used to suppress the creation of files that are needed only occasionally. { 'dəm·ē 'fīl }

dummy instruction {COMPUT SCI} An artificial instruction or address inserted in a list to serve a purpose other than the execution as an instruction. { ¦dəm·ē in'strək·shən }

dummy load {ELECTR} A dissipative device used at the end of a transmission line or waveguide to convert transmitted energy into heat, so that essentially no energy is radiated outward or reflected back to its source. { 'dəm·ē ,lōd }

dummy message {COMMUN} A message sent for some purpose other than its content, which may consist of dummy groups or may have a meaningless text. { ¦dəm·ē 'mes·ij }

dummy parameter |COMPUT SCI| A parameter whose value has no significance but which is included in an instruction or command to satisfy the requirements of the system. { 'dəm·ē pə'ram·əd·ər }

dummy record |COMPUT SCI| Meaningless information that is stored for some purpose such as fulfillment of a length requirement. { 'dəm·ē 'rek·ərd }

dummy section |COMPUT SCI| The part of an assembly language program in which the arrangement of the data in memory is specified. Abbreviated DSECT. { 'dəm·ē 'sek·shən }

dump |COMPUT SCI| To copy the contents of all or part of a storage, usually from an internal storage device into an external storage device. |ELECTR| To withdraw all power from a system or component accidentally or intentionally. { dəmp }

dump check |COMPUT SCI| A computer check that usually consists of adding all the digits during dumping, and verifying the sum when retransferring. { 'dəmp ‚chek }

dump power |ELEC| Electric power, generated by any source, which is in excess of the needs of the electric system and which cannot be stored or conserved. { 'dəmp ‚paů·ər }

dump routine |COMPUT SCI| A program within a computer's operating system that handles the processing of dumps. { 'dəmp rü‚tēn }

duodiode |ELECTR| An electron tube having two diodes in the same envelope, with either a common cathode or separate cathodes. Also known as double diode. { ‚dü·ō'dī‚ōd }

duodiode-pentode |ELECTR| An electron tube having two diodes and a pentode in the same envelope, generally with a common cathode. { ‚dü·ō'dī‚ōd 'pen‚tōd }

duodiode-triode |ELECTR| An electron tube having two diodes and a triode in the same envelope, generally with a common cathode. { ‚dü·ō'dī ‚ōd 'trī‚ōd }

duoplasmatron |ELECTR| An ion-beam source in which electrons from a hot filament are accelerated sufficiently to ionize a gas by impact; the resulting positive ions are drawn out by high-voltage electrons and focused into a beam by electrostatic lens action. { ‚dü·ō'plaz·mə‚trän }

duotriode See double triode. { ‚dü·ō'trī‚ōd }

duplex artificial line |ELEC| A balancing network, simulating the impedance of the real line and distant terminal apparatus, which is employed in a duplex circuit for the purpose of making the receiving device unresponsive to outgoing signal currents. { 'dü‚pleks ärd·ə‚fish·əl 'līn }

duplex cable |ELEC| Two insulated stranded conductors twisted together; they may have a common insulating covering. { ¦dü‚pleks 'kā·bəl }

duplex channel |COMMUN| A communication channel providing simultaneous transmission in both directions. { ¦dü‚pleks 'chan·əl }

duplex computer |COMPUT SCI| Two identical computers, either one of which can ensure

continuous operation of the system when the other is shut down. { ¦dü‚pleks kəm'pyüd·ər }

duplexed system |ENG| A system with two distinct and separate sets of facilities, each of which is capable of assuming the system function while the other assumes a standby status. Also known as redundant system. { 'dü‚plekst ‚sis·təm }

duplexer |ELECTR| A switching device used in radar to permit alternate use of the same antenna for both transmitting and receiving; other forms of duplexers serve for two-way radio communication using a single antenna at lower frequencies. Also known as duplexing assembly. { 'dü‚plek·sər }

duplexing |COMMUN| See duplex operation. |COMPUT SCI| The provision of redundant hardware or excess capacity which can pick up the work load in the event of failure of one part of a computer system. { 'dü‚pleks·iŋ }

duplexing assembly See duplexer. { 'dü‚pleks·iŋ ə‚sem·blē }

duplex operation |COMMUN| The operation of associated transmitting and receiving apparatus concurrently, as in ordinary telephones, without manual switching between talking and listening periods. Also known as duplexing; duplex transmission. |ENG| In radar, operation in which two identical and interchangeable equipments are provided, generally to enhance system reliability, one in an active state and the other immediately available for operation. { ¦dü‚pleks äp·ə'rā·shən }

duplex transmission See duplex operation. { ¦dü‚pleks tranz'mish·ən }

duplex tube |ELECTR| Combination of two vacuum tubes in one evelope. { ¦dü‚pleks 'tüb }

duplicate record |COMPUT SCI| An unwanted record that has the same key as another record in the same file. { 'düp·lə·kət 'rek·ərd }

duplication check |COMPUT SCI| A check based on the identity in results of two independent performances of the same task. { ‚düp·lə'kā·shən ‚chek }

duration control |ELECTR| Control for adjusting the time duration of reduced gain in a sensitivity-time control circuit. { də'rā·shən kən‚trōl }

Dushman equation See Richardson-Dushman equation. { 'dúsh·mən i‚kwā·zhən }

dust core See ferrite core. { 'dəst ‚kòr }

duty classification of a relay |ELEC| Expression of the frequency with which the relay may be required to operate without exceeding prescribed limitations. { 'düd·ē ‚klas·ə·fə‚kā·shən əv ə 'rē ‚lā }

duty cycle |ENG| **1.** The time intervals devoted to starting, running, stopping, and idling when a device is used for intermittent duty. **2.** The ratio of working time to total time for an intermittently operating device, usually expressed as a percent. Also known as duty factor. { 'düd·ē ‚sī·kəl }

duty factor |COMMUN| **1.** In a pulse radar or similar system, the ratio of average to pulse power; basically, the product of the pulse width (for square pulses) and the pulse repetition frequency. Also known as duty ratio.   **2.** *See* duty cycle.   { 'düd·ē ,fak·tər }

duty ratio   *See* duty factor.   { 'düd·ē ,rā·shō }

DUV   *See* data under voice.

DVD   [COMMUN] An optical disk that has formats for audio, video, and computer storage applications, and that uses the same basic structure as the compact disk (CD) to store data, but achieves a greater storage capability by using a track pitch less than half that of the CD, pits and lands as little as half as long as the shortest on a CD, and two substrates, bonded together. Derived from digital versatile disk; digital video disk.

DVD-audio   [COMMUN] A DVD format for digital storage of audio information. Also known as Book C.   { ¦dē¦vē¦dē 'ȯd·ē·ō }

DVD-RAM   *See* DVD-rewritable.   { ¦dē¦vē¦dē 'ram }

DVD-read-only   [COMMUN] A DVD format in which data written on the disk at the time of its manufacture are permanent, and the disk cannot be written or erased after that. Also known as Book A; DVD-ROM.   { ¦dē¦vē¦dē ,rēd 'ōn·lē }

DVD-rewritable   [COMMUN] A DVD format that allows audio or other digital data to be written, read, erased, and rewritten. Also known as Book E; DVD-RAM.   { ¦dē¦vē¦dē rē'rīd·ə·bəl }

DVD-ROM   *See* DVD-read-only.   { ¦dē¦vē¦dē 'räm }

DVD-video   [COMMUN] A DVD format for digital storage of video information. Also known as Book B.   { ¦dē¦vē¦dē 'vid·ē·ō }

DVD-write once   [COMMUN] A DVD format that allows users to record audio or other digital data in such a way that the recording is permanent and may be read indefinitely but cannot be erased. Also known as Book D.   { ¦dē¦vē¦dē ,rīt 'wəns }

dwell   [ELEC] The number of degrees through which the distributor cam rotates from the time that the contact points close to the time that they open again. Also known as dwell angle.   { dwel }

dwell angle   *See* dwell.   { 'dwel ,aŋ·gəl }

dwell time   [ELECTR] The length of time a radar examines a single target in making a single estimate about it; it is limited by the antenna rotation rate and beam width in simple radars, while in more flexible radars it is established by the computer-generated scheduling of operations. Also known as look time.   { 'dwel ,tīm }

DX   *See* distance reception.

dyadic processor   [COMPUT SCI] A type of multiprocessor that includes two processors which operate under control of the same copy of the operating system.   { dī'ad·ik 'präs,es·ər }

dye polymer recording   [COMPUT SCI] An optical recording technique in which dyed plastic layers are used as the recording medium.   { ¦dī 'päl·ə·mər ri'kȯrd·iŋ }

dynamic acceleration   *See* dynamic resolution.   { dī¦nam·ik ik,sel·ə'rā·shən }

dynamic address translator   [COMPUT SCI] A hardware device used in a virtual memory system to automatically identify a virtual address inquiry in terms of segment number, page number within the segment, and position of the record with reference to the beginning of the page.   { dī ¦nam·ik 'a,dres ,tranz,lād·ər }

dynamic algorithm   [COMPUT SCI] An algorithm whose operation is, to some extent, unpredictable in advance, generally because it contains logical decisions that are made on the basis of quantities computed during the course of the algorithm. Also known as heuristic algorithm.   { dī¦nam·ik 'al·gə,rith·əm }

dynamic beam forming   [ELECTR] A cathode-ray-tube design that ensures that the electron beam will impact a perfectly circular area of the display screen regardless of the location on the screen to which it is directed.   { dī¦nam·ik 'bēm ,form·iŋ }

dynamic behavior   [ENG] A description of how a system or an individual unit functions with respect to time.   { dī¦nam·ik bə'hāv·yər }

dynamic characteristic   *See* load characteristic.   { dī¦nam·ik kar·ik·tə'ris·tik }

dynamic check   [ENG] Check used to ascertain the correct performance of some or all components of equipment or a system under dynamic or operating conditions.   { dī¦nam·ik 'chek }

dynamic circuit   [ELECTR] A metal oxide semiconductor circuit designed to make use of its high input impedance to store charge temporarily at certain nodes of the circuit and thereby increase the speed of the circuit.   { dī¦nam·ik 'sər·kət }

dynamic condenser electrometer   [ELEC] A sensitive voltage-measuring instrument in which an object carrying charge resulting from the voltage is moved back and forth in an electrostatic field and the resulting alternating-current signal is observed.   { dī¦nam·ik kən¦den·sər i ,lek'träm·əd·ər }

dynamic convergence   [ELECTR] The process whereby the locus of the point of convergence of electron beams in a multibeam cathode-ray tube is made to fall on a specified surface during scanning.   { dī¦nam·ik kən'vər·jəns }

dynamic debugging routine   [COMPUT SCI] A debugging routine which operates in conjunction with the program being checked and interacts with it while the program is running.   { dī¦nam·ik dē'bəg·iŋ rü,tēn }

dynamic dump   [COMPUT SCI] A dump performed during the execution of a program.   { dī¦nam·ik 'dəmp }

dynamic error   [ELECTR] Error in a time-varying signal resulting from inadequate dynamic response of a transducer.   { dī¦nam·ik 'er·ər }

dynamic focusing   [ELECTR] The process of varying the focusing electrode voltage for a color picture tube automatically so the electron-beam spots remain in focus as they sweep over the flat surface of the screen.   { dī¦nam·ik 'fō·kəs·iŋ }

dynamic impedance   [ELEC] The impedance of a circuit having an inductance and a capacitance in parallel at the frequency at which this impedance has a maximum value. Also known as rejector impedance.   { dī¦nam·ik im'ped·əns }

dynamicizer   [COMPUT SCI] A device that converts a collection of data represented by a spatial arrangement of bits in a computer storage device into a series of signals occurring in time.   { dī'nam·ə,sīz·ər }

dynamic link [COMPUT SCI] A linking of data in two different programs, whereby modification in either program causes a similar change of the data in the other. { dī¦nam·ik 'liŋk }

dynamic loudspeaker [ENG ACOUS] A loudspeaker in which the moving diaphragm is attached to a current-carrying voice coil that interacts with a constant magnetic field to give the in-and-out motion required for the production of sound waves. Also known as dynamic speaker; moving-coil loudspeaker. { dī¦nam·ik 'laúd ˌspēk·ər }

dynamic memory See dynamic storage. { dī ¦nam·ik 'mem·rē }

dynamic memory allocation See dynamic storage allocation. { dī¦nam·ik 'mem·rē al·ə‚kā·shən }

dynamic microphone [ENG ACOUS] A moving-conductor microphone in which the flexible diaphragm is attached to a coil positioned in the fixed magnetic field of a permanent magnet. Also known as moving-coil microphone. { dī¦nam·ik 'mī·krə‚fōn }

dynamic noise suppressor [ENG ACOUS] An audio-frequency filter circuit that automatically adjusts its band-pass limits according to signal level, generally by means of reactance tubes; at low signal levels, when noise becomes more noticeable, the circuit reduces the low-frequency response and sometimes also reduces the high-frequency response. { dī¦nam·ik 'nóiz sə‚pres·ər }

dynamic pickup [ELECTR] A pickup in which the electric output is due to motion of a coil or conductor in a constant magnetic field. Also known as dynamic reproducer; moving-coil pickup. { dī¦nam·ik 'pik‚əp }

dynamic plate impedance [ELECTR] Internal resistance to the flow of alternating current between the cathode and plate of a tube. { dī ¦nam·ik 'plāt im‚pēd·əns }

dynamic plate resistance [ELECTR] Opposition that the plate circuit of a vacuum tube offers to a small increment of plate voltage; it is the ratio of a small change in plate voltage to the resulting change in the plate current, other tube voltages remaining constant. { dī¦nam·ik 'plāt ri‚zis·təns }

dynamic printout [COMPUT SCI] A printout of data which occurs during the machine run as one of the sequential operations. { dī¦nam·ik 'print ‚aút }

dynamic problem check [COMPUT SCI] Any dynamic check used to ascertain that the computer solution satisfies the given system of equations in an analog computer operation. { dī¦nam·ik 'präb·ləm ‚chek }

dynamic programming [MATH] A mathematical technique, more sophisticated than linear programming, for solving a multidimensional optimization problem, which transforms the problem into a sequence of single-stage problems having only one variable each. { dī¦nam·ik'prō·grə·miŋ }

dynamic program relocation [COMPUT SCI] The act of moving a partially executed program to another location in main memory, without hindering its ability to finish processing normally. { dī¦nam·ik 'prō·grəm ‚rē·lō‚kā·shən }

dynamic random-access memory [COMPUT SCI] A read-write random-access memory whose storage cells are based on transistor-capacitor combinations, in which the digital information is represented by charges that are stored on the capacitors and must be repeatedly replenished in order to retain the information. Abbreviated DRAM. { dī¦nam·ik 'ran·dəm 'ak·ses ‚mem·rē }

dynamic range [ELECTR] The ratio of the specified maximum signal level capability of a system or component to its noise level; usually expressed in decibels. { dī¦nam·ik 'rānj }

dynamic regulator [ELECTR] Transmission regulator in which the adjusting mechanism is in self-equilibrium at only one or a few settings and requires control power to maintain it at any other setting. { dī¦nam·ik 'reg·yə‚lād·ər }

dynamic relocation [COMPUT SCI] The ability to move computer programs or data from auxiliary memory into main memory at any convenient location. { dī¦nam·ik ‚rē·lō'kā·shən }

dynamic reproducer See dynamic pickup. { dī ¦nam·ik rē·prə'dü·sər }

dynamic resistance [ELEC] A device's electrical resistance when it is in operation. { dī¦nam·ik ri'zis·təns }

dynamic resolution [COMPUT SCI] A feature of some mice whereby the pointer moves a larger distance in proportion to the mouse's actual displacement when the mouse is moved quickly and a smaller distance when it is moved slowly. Also known as automatic acceleration; ballistic tracking; dynamic acceleration; variable acceleration. { dī¦nam·ik ‚rez·ə'lü·shən }

dynamic sequential control [COMPUT SCI] Method of operation of a digital computer through which it can alter instructions as the computation proceeds, or the sequence in which instructions are executed, or both. { dī¦nam·ik sə¦kwen·chəl kən'trōl }

dynamic shift register [COMPUT SCI] A shift register that stores information by using temporary charge storage techniques. { dī¦nam·ik 'shift ‚rej·‚ə·stər }

dynamic speaker See dynamic loudspeaker. { dī ¦nam·ik 'spēk·ər }

dynamic stop [COMPUT SCI] A loop in a computer program which is created by a branch instruction in the presence of an error condition, and which signifies the existence of this condition. { dī ¦nam·ik 'stäp }

dynamic storage [COMPUT SCI] **1.** Computer storage in which information at a certain position is not always available instantly because it is moving, as in an acoustic delay line or magnetic drum. Also known as dynamic memory. **2.** Computer storage consisting of capacitively charged circuit elements which must be continually refreshed or recharged at regular intervals. { dī¦nam·ik 'stòr·ij }

dynamic storage allocation [COMPUT SCI] A computer system in which memory capacity is made available to a program on the basis of actual, momentary need during program execution, and areas of storage may be reassigned at any time. Also known as dynamic allocation; dynamic memory allocation. { dī¦nam·ik ¦stór·ij ˌal·ə'kā·shən }

dynamic subroutine [COMPUT SCI] Subroutine that involves parameters, such as decimal point position or item size, from which a relatively coded subroutine is derived by the computer itself. { dī¦nam·ik 'səb·rü,tēn }

dynamic time warping [ENG ACOUS] In speech recognition, the operation of compressing or stretching the temporal pattern of speech signals to take speaker variations into account. { dī,nam·ik 'tīm ,wórp·iŋ }

dynamo *See* generator. { 'dī·nə,mō }

dynamoelectric amplifier generator [ELEC] A generator that serves as a power amplifier at low frequencies or direct current; the input signal is applied to the stationary field to change the excitation, and the amplified output is taken from the rotating armature. { ¦dī·nə,mō·i'lek·trik 'am·plə ˌfī·ər ,jen·ə,rād·ər }

dynamometer [ENG] **1.** An instrument in which current, voltage, or power is measured by the force between a fixed coil and a moving coil. **2.** A special type of electric rotating machine used to measure the output torque or driving torque of rotating machinery by the elastic deformation produced. { ,dī·nə'mäm·əd·ər }

dynamometer multiplier [ELEC] A multiplier in which a fixed and a moving coil are arranged

so that the deflection of the moving coil is proportional to the product of the currents flowing in the coils. { dī·nə'mäm·əd·ər 'məl·tə ,plī·ər }

dynamostatic [ELEC] Pertaining to a machine that uses direct or alternating current to produce static electricity. { ¦dī·nə,mō'stad·ik }

dynamotor [ELEC] A rotating electric machine having two or more windings on a single armature containing a commutator for direct-current operation and slip rings for alternating-current operation; when one type of power is fed in for motor operation, the other type is delivered by generator action. Also known as rotary converter; synchronous inverter. { 'dī·nə,mō·dər }

dynatron [ELECTR] A screen-grid tube in which secondary emission of electrons from the anode causes the anode current to decrease as anode voltage increases, resulting in a negative resistance characteristic. Also known as negatron. { 'dī·nə,trän }

dynatron oscillator [ELECTR] An oscillator in which secondary emission of electrons from the anode of a screen-grid tube causes the anode current to decrease as anode voltage is increased, giving the negative resistance characteristic required for oscillation. { 'dī·nə,trän ,äs·ə ,lād·ər }

dynode [ELECTR] An electrode whose primary function is secondary emission of electrons; used in multiplier phototubes and some types of television camera tubes. Also known as electron mirror. { 'dī,nōd }

E

E See electric-field vector.

EA See electronic attack.

EADI See electronic attitude directional indicator.

E and M lead signaling [COMMUN] Communications between a trunk circuit and a separate signaling unit over two leads: an M lead that transmits battery or ground signals to the signaling equipment, and an E lead which receives open or ground signals from the signaling unit. { ¦ē ən ¦em 'lēd ˌsig·nəl·iŋ }

early binding [COMPUT SCI] The assignment of data types (such as integer or string) to variables during the compilation of a computer program rather than at run time. { 'ər·lē ˈbīnd·iŋ }

early effect [ELECTR] A change in the base width of a bipolar transistor as a function of base-collector bias voltage. { 'ər·lē i,fekt }

Earnshaw's theorem [ELEC] The theorem that a charge cannot be held in stable equilibrium by an electrostatic field. { 'ərn,shȯz ,thir·əm }

EAROM See electrically alterable read-only memory. { 'ē,räm }

earphone [ENG ACOUS] **1.** An electroacoustical transducer, such as a telephone receiver or a headphone, actuated by an electrical system and supplying energy to an acoustical system of the ear, the waveform in the acoustical system being substantially the same as in the electrical system. **2.** A small, lightweight electroacoustic transducer that fits inside the ear, used chiefly with hearing aids. { 'ir,fōn }

earth See ground. { ərth }

earth current [ELEC] Return, fault, leakage, or stray current passing through the earth from electrical equipment. Also known as ground current. { 'ərth ,kə·rənt }

earth detector See leakage indicator. { 'ərth di'tek·tər }

earthed system See grounded system. { 'ərtht ,sis·təm }

earth electrode See ground electrode. { 'ərth i,lek ,trōd }

earthing reactor See grounding reactor. { 'ərth·iŋ rē,ak·tər }

earth station [COMMUN] A facility with a land-based antenna used to transmit and receive information to and from a communications satellite. { 'ərth ,stā·shən }

Easter-egging [ELECTR] An undirected procedure for checking electronic equipment, which derives its name from the children's activity of searching for hidden eggs at Eastertime. { 'ē·stər ,eg·iŋ }

easy [COMPUT SCI] A name for the hexadecimal digit whose decimal equivalent is 14. { 'ē·zē }

EBCDIC See extended binary-coded decimal interchange code. { 'eb·sə,dik }

E bend [ELECTROMAG] A smooth change in the direction of the axis of a waveguide, throughout which the axis remains in a plane parallel to the direction of polarization. Also known as E-plane bend. { 'ē ,bend }

EBIS See electron-beam ion source. { 'ē,bis }

EBIT See electron-beam ion trap. { 'ē,bit or ¦ē¦bē ¦ī'tē }

e-business See electronic commerce. { 'ē ,biz·nəs }

ECB See block encryption.

Eccles-Jordan circuit See bistable multivibrator. { ¦ek·əlz 'jȯrd·ən ,sər·kət }

Eccles-Jordan multivibrator See bistable multivibrator. { ¦ek·əlz 'jȯrd·ən ,məl·ti'vī,brād·ər }

ECDIS See electronic chart display and information system. { 'ek,dis or ¦ē¦sē¦dē¦ī'es }

E cell [ELEC] A timing device that converts the current-time integral of an electrical function into an equivalent mass integral (or the converse operation) up to a maximum of several thousand microampere-hours. { 'ē ,sel }

echo [ELECTR] **1.** The signal reflected, or backscattered, by a radar target, or that scattered in the receiver's direction in a bistatic radar; also, the indication of this signal on the radar display. Also known as echo pulse; radar echo; return. **2.** See ghost signal. { 'ek·ō }

echo amplitude [ELECTR] In radar, an empirical measure of the strength of a target signal as determined from the appearance of the echo; the amplitude of the echo waveform usually is measured by the deflection of the electron beam from the base line of an amplitude-modulated indicator. { '¦ek·ō 'am·plə,tüd }

echo area [ELECTROMAG] In radar, the area of a fictitious perfect reflector of electromagnetic waves that would reflect the same amount of energy back to the radar as the actual target. Also known as target cross section. { 'ek,ō ,er·ē·ə }

echo attenuation [ELECTR] The power transmitted at an output terminal of a transmission line, divided by the power reflected back to the same output terminal. { 'ek,ō ə,ten·yə'wā·shən }

echo box [ELECTR] A calibrated high-Q resonant cavity that stores part of the transmitted radar pulse power and gradually feeds this energy into the receiving system after completion of the pulse transmission; used to provide an artificial target signal for test and tuning purposes; being replace in design by other forms of built-in test equipment (BITE). { 'ek,ō ,bäks }

echo check [COMPUT SCI] A method of ascertaining the accuracy of transmission of data in which the transmitted data are returned to the sending end for comparison with original data. Also known as loopback check; loop check; readback check. { 'ek·ō ,chek }

echo contour [ELECTR] A trace of equal signal intensity of the radar echo displayed on a range height indicator or plan position indicator. { ¦ek·ō 'kän,túr }

echo frequency [ELECTR] The number of fluctuations, per unit time, in the power or amplitude of a radar target signal, often in reference to a moving target's echo going through cycles of constructive and destructive interference with coincident stationary clutter echo. { 'ek·ō ,frē·kwən·sē }

echo intensity [ELECTR] The brightness or brilliance of a radar echo as displayed on an intensity-modulated indicator; echo intensity is, within certain limits, proportional to the voltage of the target signal or to the square root of its power. { ¦ek·ō in'ten·səd·ē }

echo matching [ENG] Rotating an antenna to a position in which the pulse indications of an echo-splitting radar are equal. { 'ek·ō ,mach·iŋ }

echoplex technique [COMPUT SCI] A technique for detecting errors in a data communication system with full duplex lines, in which the signal generated when a character is typed on a keyboard is transmitted to a receiver and retransmitted to a display terminal, enabling the operator to check if the character displayed is the same as the character typed. { 'ek·ō,pleks tek,nēk }

echo power [ELECTR] The electrical strength, or power, of a radar target signal, normally measured in watts or dBm (decibels referred to 1 milliwatt). { 'ek·ō ,paú·ər }

echo pulse See echo. { 'ek·ō ,pəls }

echo recognition [ENG] Identification of a sonar reflection from a target, as distinct from energy returned by other reflectors. { 'ek·ō ,rek·ig ,nish·ən }

echo repeater [ENG ACOUS] In sonar calibration and training, an artificial target that returns a synthetic echo by receiving a signal and retransmitting it. { 'ek·ō ri,pēd·ər }

echo signal See target signal. { 'ek·ō ,sig·nəl }

echo-splitting radar [ENG] Radar in which the echo is split by special circuits associated with the antenna lobe-switching mechanism, to give two echo indications on the radarscope screen; when the two echo indications are equal in height, the target bearing is read from a calibrated scale. { ¦ek·ō ,splid·iŋ 'rā,där }

echo suppressor [ELECTR] **1.** A circuit that desensitizes radar navigation equipment for a fixed period after the reception of one pulse, for the purpose of rejecting delayed pulses arriving from longer, indirect reflection paths. **2.** A relay or other device used on a transmission line to prevent a reflected wave from returning to the sending end of the line. { 'ek·ō sə,pres·ər }

echo talker [COMPUT SCI] The interference created by the retransmission of a message back to its source while the source is still transmitting. { 'ek·ō ,tók·ər }

ECL See emitter-coupled logic.

ECM See embrittlement control message.

eco See electron-coupled oscillator.

e-commerce See electronic commerce. { 'ē ,käm ərs }

economy [COMPUT SCI] The ratio of the number of characters to be coded to the maximum number available with the code; for example, binary-coded decimal using 4 bits provides 16 possible characters but uses only 10 of them. { ē'kän·ə·mē }

ECRIS See electron cyclotron resonance source.

ECR source See electron cyclotron resonance source. { ¦ē¦sē¦är 'sórs }

ECSW See extended channel status word.

ED See electronic dummy.

eddy-current heating See induction heating. { 'ed·ē ,kə·rənt ,hēd·iŋ }

eddy-current sensor [ENG] A proximity sensor which uses an alternating magnetic field to create eddy currents in nearby objects, and then the currents are used to detect the presence of the objects. { 'ed·ē ,kə·rənt 'sen·sər }

eddy-current tachometer [ENG] A type of tachometer in which a rotating permanent magnet induces currents in a spring-mounted metal cylinder; the resulting torque rotates the cylinder and moves its attached pointer in proportion to the speed of the rotating shaft. Also known as drag-type tachometer. { 'ed·ē ,kə·rənt ta'käm·əd·ər }

EDEL room [ENG ACOUS] A control room in a sound-recording studio in which reflective or diffusive surfaces are placed near the loudspeaker and above the mixing console, while the rear wall behind the mixer is made absorptive. Derived from LEDE room (by reverse spelling). { 'ed·əl ,rüm or ¦ē¦dē¦ē'el ,rüm }

EDFA See erbium-doped fiber amplifier. { 'ed,fä or ¦ē¦dē¦ef'ä }

edgeboard connector See card-edge connector. { 'ej,bòrd kə,nek·tər }

edge connector [ELECTR] A row of etched lines on the edge of a printed circuit board that is inserted into a slot to establish a connection with another printed circuit board. { 'ej kə,nek·tər }

edge effect [ELECTR] An outward-curving distortion of lines of force near the edges of two parallel metal plates that form a capacitor. { 'ej i,fekt }

Edison battery [ELEC] A storage battery composed of cells having nickel and iron in an alkaline solution. Also known as nickel-iron battery. { ¦ed·ə·sən ¦bad·ə·rē }

Edison distribution system [ELEC] Three-wire direct-current distribution system, usually 120 to 240 volts, for combined light and power service from a single set of mains. { ¦ed·ə·sən ˌdis·trə'byü·shən ˌsis·təm }

Edison effect *See* thermionic emission. { 'ed·ə·sən i ˌfekt }

E-display [ELECTR] A radar display format in which the horizontal coordinate indicates range, the vertical indicates elevation, and the intensity of the target spot is proportional to signal strength. Also known as E-indicator; E-scan; E-scope. { 'ē di ˌsplā }

edit [COMPUT SCI] **1.** To modify the form or format of an output or input by inserting or deleting characters such as page numbers or decimal points. **2.** A computer instruction directing that this step be performed. { 'ed·ət }

edit capability [COMPUT SCI] The degree of sophistication available to the programmer to modify his or her statements while in the time-sharing mode. { 'ed·ət ˌkāp·ə ˌbil·əd·ē }

edit check [COMPUT SCI] A program instruction or subroutine that tests the validity of input in a data entry program. Also known as edit test. { 'ed·ət ˌchek }

edit mask [COMPUT SCI] The receiving word through which a source word is filtered, allowing for the suppression of leading zeroes, the insertion of floating dollar signs and decimal points, and other such formatting. { 'ed·ət ˌmask }

edit mode [COMPUT SCI] A software mode of operation in which previously entered text or data can be modified or replaced. { 'ed·ət ˌmōd }

editor program [COMPUT SCI] A special program by means of which a user can easily perform corrections, insertions, modifications, or deletions in an existing program or data file. { 'ed·ə·tər ˌprō·grəm }

edit test *See* edit check. { 'ed·ət ˌtest }

EDO RAM *See* extended data out random-access memory. { ˌā·dō 'ram *or* ¦ē¦dē¦ō }

EDP *See* electronic data processing.

EDP center *See* electronic data-processing center. { ¦ē¦dē'pē ˌsen·tər }

edulcorate [COMPUT SCI] To eliminate irrelevant data from a data file. { ē'dəl·kə ˌrāt }

EDVAC [COMPUT SCI] The first stored program computer, built in 1952. Derived from electron discrete variable automatic compiler. { 'ed ˌvak }

EEPROM *See* electrically erasable programmable read-only memory. { ¦ē¦ē ˌpräm }

EER *See* equal error rate.

effective address [COMPUT SCI] The address that is obtained by applying any specified indexing or indirect addressing rules to the specified address; the effective address is then used to identify the current operand. { ə¦fek·tiv 'a ˌdres }

effective ampere [ELEC] The amount of alternating current flowing through a resistance that produces heat at the same average rate as 1 ampere of direct current flowing in the same resistance. { ə¦fek·tiv 'am ˌpir }

effective bandwidth [ELECTR] The bandwidth of an assumed rectangular band-pass having the same transfer ratio at a reference frequency as a given actual band-pass filter, and passing the same mean-square value of a hypothetical current having even distribution of energy throughout that bandwidth. { ə¦fek·tiv 'band ˌwidth }

effective capacitance [ELEC] Total capacitance existing between any two given points of an electric circuit. { ə¦fek·tiv kə'pas·əd·əns }

effective center [ENG ACOUS] In a sonar projector, the point where lines coincident with the direction of propagation, as observed at different points some distance from the projector, apparently intersect. Also known as apparent source. { ə¦fek·tiv 'sen·tər }

effective confusion area [ENG] Amount of chaff whose radar cross-sectional area equals the radar cross-sectional area of the particular aircraft at a particular frequency. { ə¦fek·tiv kən'fyü·zhən ˌer·ē·ə }

effective current [ELEC] The value of alternating current that will give the same heating effect as the corresponding value of direct current. Also known as root-mean-square current. { ə¦fek·tiv 'kə·rənt }

effective earth radius [COMMUN] A radius value used in place of the geometric radius to correct for atmospheric refraction in estimating ranges of antennas when the index of refraction in the atmosphere changes linearly with height; under conditions of standard refraction it is ⁴⁄₃ the geometric radius. Also known as effective radius of the earth. { ə¦fek·tiv 'ərth ˌrād·ē·əs }

effective facsimile band [COMMUN] Frequency band of a facsimile signal wave equal in width to that between zero frequency and maximum keying frequency. { ə¦fek·tiv fak'sim·ə·lē ˌband }

effective horizon [COMMUN] A horizon whose distance at a given height above sea level is the distance to the horizon of a fictitious earth, having a radius ⁴⁄₃ times the earth's true radius; used to estimate ranges of antennas, taking atmospheric refraction into account. { ə¦fek·tiv hə'rīz·ən }

effective instruction [COMPUT SCI] The computer instruction that results from changing a basic instruction during program modification. Also known as actual instruction. { ə¦fek·tiv in'strək·shən }

effective isotropic radiated power [COMMUN] A measure of the strength of the signal leaving a satellite antenna in a particular direction, equal to the product of the power supplied to the satellite transmit antenna and its gain in that direction. Abbreviated eirp. { i ˌfek·tiv ˌī·sə ˌträp·ik ˌrād·ē ˌād·əd 'pau̇·ər }

effectively grounded [ELEC] Grounded through a connection of sufficiently low impedances (inherent or intentionally added) so that fault grounds which may occur cannot build up voltages dangerous to connected personnel or other equipment. { əˈfek·tiv·lē ˈgraund·əd }

effectiveness level [COMPUT SCI] A measure of the effectiveness of data-processing equipment, equal to the ratio of the operational use time to the total performance period, expressed as a percentage. Also known as average effectiveness level. { əˈfek·tiv·nəs ˌlev·əl }

effective percentage modulation [COMMUN] For a single sinusoidal input component, the ratio of the peak value of the fundamental component of the envelope to the average amplitude of the modulated wave expressed in percent. { əˈfek·tiv pərˈsent·ij ˌmäj·əˈlā·shən }

effective radiated power [ELECTROMAG] The product of antenna input power and antenna power gain, expressed in kilowatts. Abbreviated ERP. { əˈfek·tiv ˌrād·ē·ād·əd ˈpau̇·ər }

effective radius of the earth See effective earth radius. { əˈfek·tiv ˈrād·ē·əs əv the̱ ˈərth }

effective resistance See high-frequency resistance. { əˈfek·tiv riˈzis·təns }

effective speed [COMPUT SCI] The actual speed that a computer system can sustain over a period of time when the time devoted to various control, error-detection, and other overhead activities is taken into account. { əˈfek·tiv ˈspēd }

effective thermal resistance [ELECTR] Of a semiconductor device, the effective temperature rise per unit power dissipation of a designated junction above the temperature of a stated external reference point under conditions of thermal equilibrium. Also known as thermal resistance. { əˈfek·tiv ˈthər·məl riˈzis·təns }

effective time [COMPUT SCI] The time during which computer equipment is in actual use and produces useful results. { əˈfek·tiv ˈtīm }

effective value See root-mean-square value. { əˈfek·tiv ˈval·yü }

effector [CONT SYS] A motor, solenoid, or hydraulic piston that turns commands to a teleoperator into specific manipulatory actions. { əˈfek·tər }

EFL See error frequency limit.

e format [COMPUT SCI] A decimal, normalized form of a floating point number in FORTRAN in which a number such as 18.756 appears as .18756E + 02, which stands for .18756 × 10². { ˈē ˌfȯr‚mat }

EGNOS See European Geostationary Navigation Overlay System. { ˈegˌnōs }

E-HEMT See enhancement-mode high-electron-mobility transistor.

EHF See extremely high frequency.

EHSI See electronic horizontal-situation indicator.

E-H T junction [ELECTROMAG] In microwave waveguides, a combination of E- and H-plane T junctions forming a junction at a common point of intersection with the main waveguide. { ˈē ˌāch ˈtē ˌjəŋk·shən }

E-H tuner [ELECTROMAG] Tunable E-H T junction having two arms terminated in adjustable plungers used for impedance transformation. { ˈē ˌāch ˈtün·ər }

eight-level code [COMMUN] A teletypewriter code that uses eight impulses, in addition to the start and stop impulses, to define a character. { ˌāt ˌlev·əl ˈkōd }

E-indicator See E-display. { ˈē ˌin·dəˌkād·ər }

Einthoven galvanometer See string galvanometer. { ˈīntˌhō·vən ˌgal·vəˈnäm·əd·ər }

Einzel lens [ELECTR] An electrostatic lens that consists of three cylindrical tubes through which charged particles pass sequentially, the middle one of which is at a higher potential than the other two. { ˈīnt·səl ˌlenz }

eject [COMPUT SCI] To move the printing mechanism to the top of the following page, skipping the remainder of the current page. { ēˈjekt }

E-JFET See enhancement-mode junction field-effect transistor.

elaboration [COMPUT SCI] A technique, used chiefly in the Ada programming language, of setting up a hierarchy of calculated constants so that the values of one or more of them determine others further down in the hierarchy. { iˌlab·əˈrā·shən }

elastance [ELEC] The reciprocal of capacitance. { iˈlas·təns }

elastoresistance [ELEC] The change in a material's electrical resistance as it undergoes a stress within its elastic limit. { iˈlas·tō·riˈzis·təns }

elbow [ELECTROMAG] In a waveguide, a bend of comparatively short radius, normally 90°, and sometimes for acute angles down to 15°. { ˈel ˌbō }

electret [ELEC] A solid dielectric possessing persistent electric polarization, by virtue of a long time constant for decay of a charge instability. { iˈlekˌtret }

electret headphone [ENG ACOUS] A headphone consisting of an electret transducer, usually in the form of a push-pull transducer. { iˈlekˌtret ˈhedˌfōn }

electret microphone [ENG ACOUS] A microphone consisting of an electret transducer in which the foil electret diaphragm is placed next to a perforated, ridged, metal or metal-coated backplate, and output voltage, taken between diaphragm and backplate, is proportional to the displacement of the diaphragm. { iˈlekˌtret ˈmīˌkrəˌfōn }

electret transducer [ELECTR] An electroacoustic or electromechanical transducer in which a foil electret, stretched out to form a diaphragm, is placed next to a metal or metal-coated plate, and motion of the diaphragm is converted to voltage between diaphragm and plate, or vice versa. { iˈlekˌtret tranzˈdü·sər }

electric [ELEC] Containing, producing, arising from, or actuated by electricity; often used interchangeably with electrical. { iˈlek·trik }

electrical [ELEC] Related to or associated with electricity, but not containing it or having its properties or characteristics; often used interchangeably with electric. { əˈlek·trə·kəl }

electrical angle [ELEC] An angle that specifies a particular instant in an alternating-current cycle or expresses the phase difference between two alternating quantities; usually expressed in electrical degrees. { ə'lek·trə·kəl 'aŋ·gəl }

electrical breakdown See breakdown. { ə'lek·trə·kəl 'brākˌdaủn }

electrical center [ELEC] Point approximately midway between the ends of an inductor or resistor that divides the inductor or resistor into two equal electrical values. { ə'lek·trə·kəl 'sen·tər }

electrical circuit theory See circuit theory. { ə'lek·trə·kəl 'sər·kət ˌthē·ə·rē }

electrical code [ELEC] A systematic body of rules governing the practical application and installation of electrically operated equipment and devices and electric wiring systems. { ə'lek·trə·kəl 'kōd }

electrical conductance See conductance. { ə'lek·trə·kəl kən'dək·təns }

electrical conduction See conduction. { ə'lek·trə·kəl kən'dək·shən }

electrical conductivity See conductivity. { ə'lek·trə·kəl ˌkän,dək'tiv·əd·ē }

electrical conductivity analyzer [ELEC] Alternating-current, resistance-bridge device used to measure the electrical conductivity of solutions, slurries, or wet solids. { ə'lek·trə·kəl ˌkän,dək'tiv·əd·ē 'an·ə,līz·ər }

electrical degree [ELEC] A unit equal to ¹/₃₆₀ cycle of an alternating quantity. { i'lek·trə·kəl də'grē }

electrical drainage [ELEC] Diversion of electric currents from subterranean pipes to prevent electrolytic corrosion. { i'lek·trə·kəl 'drān·ij }

electrical engineer [ENG] An engineer whose training includes a degree in electrical engineering from an accredited college or university (or who has comparable knowledge and experience), to prepare him or her for dealing with the generation, transmission, and utilization of electric energy. { i'lek·trə·kəl ˌen·jə'nir }

electrical engineering [ENG] Engineering that deals with practical applications involving current flow through conductors, as in motors and generators. { i'lek·trə·kəl ˌen·jə'nir·iŋ }

electrical equipment [ELEC] Apparatus, appliances, devices, wiring, fixtures, fittings, and material used as a part of or in connection with an electrical installation. { i'lek·trə·kəl i'kwip·mənt }

electrical fault See fault. { i'lek·trə·kəl 'fȯlt }

electrical impedance Also known as impedance. [ELEC] **1.** The total opposition that a circuit presents to an alternating current, equal to the complex ratio of the voltage to the current in complex notation. Also known as complex impedance. **2.** The ratio of the maximum voltage in an alternating-current circuit to the maximum current; equal to the magnitude of the quantity in the first definition. { i'lek·trə·kəl im'pēd·əns }

electrical impedance meter [ELEC] An instrument which measures the complex ratio of voltage to current in a given circuit at a given

frequency. Also known as impedance meter. { i'lek·trə·kəl im'pēd·əns ˌmēd·ər }

electrical instability [ELEC] A persistent condition of unwanted self-oscillation in an amplifier or other electric circuit. { i'lek·trə·kəl ˌin·stə'bil·əd·ē }

electrical insulator See insulator. { i'lek·trə·kəl 'in·sə,lād·ər }

electrical interference See interference. { i'lek·trə·kəl ˌin·tər'fir·əns }

electrical length [ELECTROMAG] The length of a conductor expressed in wavelengths, radians, or degrees. { i'lek·trə·kəl 'leŋkth }

electrical loading See loading. { i'lek·trə·kəl 'lōd·iŋ }

electrically alterable read-only memory [COMPUT SCI] A read-only memory that can be reprogrammed electrically in the field a limited number of times, after the entire memory is erased by applying an appropriate electric field. Abbreviated EAROM. { i'lek·trə·klē 'ȯl·trə·bəl 'rēd ¦ōn·lē 'mem·rē }

electrically connected [ELEC] Connected by means of a conducting path, or through a capacitor, as distinguished from connection merely through electromagnetic induction. { i'lek·trə·klē kə'nek·təd }

electrically erasable programmable read-only memory [COMPUT SCI] An integrated-circuit memory chip that has an internal switch to permit a user to erase the contents of the chip and write new contents into it by means of electrical signals. Abbreviated EEPROM. { i'lek·trə·klē i'rās·ə·bəl prō'gram·ə·bəl 'rēd ¦ōn·lē 'mem·rē }

electrical measurement [ELEC] The measurement of any one of the many quantities by which electricity is characterized. { i'lek·trə·kəl 'mezh·ər·mənt }

electrical model [ELEC] A model in the form of a mathematical description or an electrical equivalent circuit that represents the behavior of an electrical device or system. { i'lek·trə·kəl 'mäd·əl }

electrical noise [ELEC] Noise generated by electrical devices, for example, motors, engine ignition, power lines, and so on, and propagated to the receiving antenna direct from the noise source. { i'lek·trə·kəl 'nȯiz }

electrical potential energy [ELEC] Energy possessed by electric charges by virtue of their position in an electrostatic field. { i'lek·trə·kəl pə'ten·chəl'en·ər·jē }

electrical pressure transducer See pressure transducer. { i'lek·trə·kəl 'presh·ər tranz,dü·sər }

electrical properties [ELEC] Properties of a substance which determine its response to an electric field, such as its dielectric constant or conductivity. { i'lek·trə·kəl 'präp·ərd·ēz }

electrical resistance See resistance. { i'lek·trə·kəl ri'zis·təns } •

electrical resistivity [ELEC] The electrical resistance offered by a material to the flow of current, times the cross-sectional area of current flow and per unit length of current path; the reciprocal of the conductivity. Also known as resistivity; specific resistance. { i'lek·trə·kəl ,rē·zis'tiv·əd·ē }

electrical resistor See resistor. { i'lek·trə·kəl ri 'zis·tər }

electrical resonator See tank circuit. { i'lek·trə· kəl 'rez·ən,ād·ər }

electrical symbol [ELEC] A simple geometrical symbol used to represent a component of a circuit in a schematic circuit diagram. { i'lek· trə·kəl 'sim·bəl }

electrical system [ELEC] System of wiring, switches, relays, and other equipment associated with receiving and distributing electricity. { i'lek·trə·kəl ,sis·təm }

electrical transcription See transcription. { i'lek· trə·kəl tranz'krip·shən }

electrical unit [ELEC] A standard in terms of which some electrical quantity is evaluated. { i'lek·trə·kəl 'yü·nət }

electrical zero [ELEC] A standard reference position from which rotor angles are measured in synchros and other rotating devices. { i'lek·trə·kəl 'zir·ō }

electric arc [ELEC] A discharge of electricity through a gas, normally characterized by a voltage drop approximately equal to the ionization potential of the gas. Also known as arc. { i¦lek·trik 'ärk }

electric-arc lamp See arc lamp. { i'lek·trik ,ärk 'lamp }

electric cell [ELEC] **1.** A single unit of a primary or secondary battery that converts chemical energy into electric energy. **2.** A single unit of a device that converts radiant energy into electric energy, such as a nuclear, solar, or photovoltaic cell. { i¦lek·trik 'sel }

electric charge See charge. { i¦lek·trik 'chärj }

electric circuit [ELEC] Also known as circuit. **1.** A path or group of interconnected paths capable of carrying electric currents. **2.** An arrangement of one or more complete, closed paths for electron flow. { i¦lek·trik 'sər·kət }

electric circuit theory See circuit theory. { i¦lek· trik 'sər·kət ,thē·ə·rē }

electric coil See coil. { i¦lek·trik 'kȯil }

electric comparator [ELEC] A comparator in which movement results in a change in some electrical quantity, which is then amplified by electrical means. { i¦lek·trik kəm'par·əd·ər }

electric condenser See capacitor. { i¦lek·trik kən'den·sər }

electric conductor See conductor. { i¦lek·trik kən'dək·tər }

electric connection [ELEC] A direct wire path for current between two points in a circuit. { i¦lek· trik kə'nek·shən }

electric connector [ELEC] A device that joins electric conductors mechanically and electrically to other conductors and to the terminals of apparatus and equipment. { i¦lek·trik kə'nek·tər }

electric constant [ELEC] The permittivity of empty space, equal to 1 in centimeter-gram-second electrostatic units and to $10^7/4\pi c^2$ farads per meter or, numerically, to 8.854×10^{-12} farad per meter in International System units, where

c is the speed of light in meters per second. Symbolized ϵ_0. { i¦lek·trik 'kän·stənt }

electric contact [ELEC] A physical contact that permits current flow between conducting parts. Also known as contact. { i¦lek·trik 'kän,takt }

electric contactor See contactor. { i¦lek·trik 'kän ,tak·tər }

electric control [ELEC] The control of a machine or device by switches, relays, or rheostats, as contrasted with electronic control by electron tubes or by devices that do the work of electron tubes. { i¦lek·trik kən'trōl }

electric controller [ELEC] A device that governs in some predetermined manner the electric power delivered to apparatus. { i¦lek·trik kən'trōl·ər }

electric converter See synchronous converter. { i¦lek·trik kən'vərd·ər }

electric corona See corona discharge. { i¦lek·trik kə'rō·nə }

electric current See current. { i¦lek·trik 'kə·rənt }

electric current density See current density. { i¦lek·trik ¦kə·rənt ,den·səd·ē }

electric current meter See ammeter. { i¦lek·trik ¦kə·rənt ,mēd·ər }

electric cutout See cutout. { i¦lek·trik 'kəd,aút }

electric delay line [ELECTR] A delay line using properties of lumped or distributed capacitive and inductive elements; can be used for signal storage by recirculating information-carrying wave patterns. { i¦lek·trik di'lā ,līn }

electric dipole [ELEC] A localized distribution of positive and negative electricity, without net charge, whose mean positions of positive and negative charges do not coincide. { i¦lek·trik 'dī ,pōl }

electric dipole moment [ELEC] A quantity characteristic of a charge distribution, equal to the vector sum over the electric charges of the product of the charge and the position vector of the charge. { i¦lek·trik 'dī,pōl ,mō·mənt }

electric discharge See discharge. { i¦lek·trik 'dis ,chärj }

electric-discharge lamp See discharge lamp. { i'lek·trik 'dis,chärj ,lamp }

electric-discharge tube See discharge tube. { i'lek·trik 'dis,chärj ,tüb }

electric displacement [ELEC] The electric field intensity multiplied by the permittivity. Symbolized D. Also known as dielectric displacement; dielectric flux density; displacement; electric displacement density; electric flux density; electric induction. { i'lek·trik dis'plās·mənt }

electric displacement density See electric displacement. { i'lek·trik dis'plās·mənt ,den·səd·ē }

electric distribution system See distribution system. { i'lek·trik ,dis·trə'byü·shən ,sis·təm }

electric energy measurement [ELEC] The measurement of the integral, with respect to time, of the power in an electric circuit. { i¦lek·trik 'en·ər·jē ,mezh·ər·mənt }

electric energy meter [ELEC] A device which measures the integral, with respect to time, of the power in an electric circuit. { i¦lek·trik ¦enər·jē ,mēd·ər }

electric eye   See photocell; phototube.   { i¦lek·trik 'ī }

electric field |ELEC| **1.** One of the fundamental fields in nature, causing a charged body to be attracted to or repelled by other charged bodies; associated with an electromagnetic wave or a changing magnetic field.   **2.** Specifically, the electric force per unit test charge.   { i¦lek·trik 'fēld }

electric-field intensity   See electric-field vector.   { i¦lek·trik ¦fēld in'ten·səd·ē }

electric-field strength   See electric-field vector.   { i¦lek·trik ¦fēld 'streŋkth }

electric-field vector |ELEC| The force on a stationary positive charge per unit charge at a point in an electric field. Designated **E**. Also known as electric-field intensity; electric-field strength; electric vector.   { i¦lek·trik ¦fēld 'vek·tər }

electric filter |ELECTR| **1.** A network that transmits alternating currents of desired frequencies while substantially attenuating all other frequencies. Also known as frequency-selective device.   **2.** See filter.   { i¦lek·trik 'fil·tər }

electric flowmeter |ELEC| Fluid-flow measurement device relying on an inductance or impedance bridge or on electrical-resistance rod elements to sense flow-rate variations.   { i¦lek·trik 'flō,mēd·ər }

electric flux |ELEC| **1.** The integral over a surface of the component of the electric displacement perpendicular to the surface; equal to the number of electric lines of force crossing the surface.   **2.** The electric lines of force in a region.   { i¦lek·trik 'fləks }

electric flux density   See electric displacement.   { i¦lek·trik 'fləks ,den·səd·ē }

electric flux line   See electric line of force.   { i¦lek·trik 'fləks ,līn }

electric forming |ELECTR| The process of applying electric energy to a semiconductor or other device to modify permanently its electrical characteristics.   { i¦lek·trik 'fȯr·miŋ }

electric fuse   See fuse.   { i¦lek·trik 'fyüz }

electric heating |ENG| Any method of converting electric energy to heat energy by resisting the free flow of electric current.   { i¦lek·trik 'hēd·iŋ }

electric hysteresis   See ferroelectric hysteresis.   { i¦lek·trik ,his·tə'rē·səs }

electrician |ENG| A skilled worker who installs, repairs, maintains, or operates electric equipment.   { i,lek'trish·ən }

electric image |ELEC| A fictitious charge used in finding the electric field set up by fixed electric charges in the neighborhood of a conductor; the conductor, with its distribution of induced surface charges, is replaced by one or more of these fictitious charges. Also known as image.   { i¦lek·trik 'im·ij }

electric induction   See electric displacement.   { i¦lek·trik in'dək·shən }

electric instrument |ENG| An electricity-measuring device that indicates, such as an ammeter or voltmeter, in contrast to an electric meter that totalizes or records.   { i¦lek·trik 'in·strə·mənt }

electric lamp |ELEC| A lamp in which light is produced by electricity, as the incandescent lamp, arc lamp, glow lamp, mercury-vapor lamp, and fluorescent lamp.   { i¦lek·trik 'lamp }

electric line of force |ELEC| An imaginary line drawn so that each segment of the line is parallel to the direction of the electric field or of the electric displacement at that point, and the density of the set of lines is proportional to the electric field or electrical displacement. Also known as electric flux line.   { i¦lek·trik ¦līn əv 'fȯrs }

electric main   See power transmission line.   { i¦lek·trik 'mān }

electric meter |ENG| An electricity-measuring device that totalizes with time, such as a watthour meter or ampere-hour meter, in contrast to an electric instrument.   { i¦lek·trik 'mēd·ər }

electric moment |ELEC| One of a series of quantities characterizing an electric charge distribution; an *l*-th moment is given by integrating the product of the charge density, the *l*-th power of the distance from the origin, and a spherical harmonic $Y*_{lm}$ over the charge distribution.   { i¦lek·trik 'mō·mənt }

electric monopole |ELEC| A distribution of electric charge which is concentrated at a point or is spherically symmetric.   { i¦lek·trik 'män·ə,pōl }

electric motor   See motor.   { i¦lek·trik 'mōd·ər }

electric network   See network.   { i¦lek·trik 'net ,wərk }

electric octupole moment |ELEC| A quantity characterizing an electric charge distribution; obtained by integrating the product of the charge density, the third power of the distance from the origin, and a spherical harmonic $Y*_{3m}$ over the charge distribution.   { i¦lek·trik 'äk·tə,pōl 'mō·mənt }

electric outlet   See outlet.   { i¦lek·trik 'aůt,let }

electric polarizability |ELEC| Induced dipole moment of an atom or molecule in a unit electric field.   { i¦lek·trik ,pō·lə,rī·zə'bil·əd·ē }

electric polarization   See polarization.   { i¦lek·trik ,pō·lə·rə'zā·shən }

electric potential |ELEC| The work which must be done against electric forces to bring a unit charge from a reference point to the point in question; the reference point is located at an infinite distance, or, for practical purposes, at the surface of the earth or some other large conductor. Also known as electrostatic potential; potential. Abbreviated V.   { i¦lek·trik pə'ten·chəl }

electric power |ELEC| The rate at which electric energy is converted to other forms of energy, equal to the product of the current and the voltage drop.   { i¦lek·trik 'paů·ər }

electric power line   See power line.   { i¦lek·trik 'paů·ər ,līn }

electric power meter |ENG| A device that measures electric power consumed, either at an instant, as in a wattmeter, or averaged over a time interval, as in a demand meter. Also known as power meter.   { i¦lek·trik 'paů·ər ,mēd·ər }

electric power station |ELEC| A generating station or an electric power substation.   { i¦lek·trik 'paů·ər ,stā·shən }

electric power substation

electric power substation [ELEC] An assembly of equipment in an electric power system through which electric energy is passed for transmission, transformation, distribution, or switching. Also known as substation. { i¦lek·trik ¦paů·ər 'səb ˌstā·shən }

electric power transmission [ELEC] Process of transferring electric energy from one point to another in an electric power system. { i¦lek·trik ¦paů·ər tranz₁mish·ən }

electric protective device [ELEC] A particular type of equipment used in electric power systems to detect abnormal conditions and to initiate appropriate corrective action. Also known as protective device. { i¦lek·trik prə'tek·tiv di₁vīs }

electric quadrupole [ELEC] A charge distribution that produces an electric field equivalent to that produced by two electric dipoles whose dipole moments have the same magnitude but point in opposite directions and which are separated from each other by a small distance. { i¦lek·trik 'kwä·drə₁pōl }

electric quadrupole lens [ELECTR] A device for focusing beams of charged particles which have four electrodes with alternately positive and negative polarity; used in electron microscopes and particle accelerators. { i¦lek·trik 'kwä·drə ₁pōl ₁lenz }

electric quadrupole moment [ELEC] A quantity characterizing an electric charge distribution, obtained by integrating the product of the charge density, the second power of the distance from the origin, and a spherical harmonic Y^*_{2m} over the charge distribution. { i¦lek·trik 'kwä·drə₁pōl ₁mō·mənt }

electric raceway See raceway. { i¦lek·trik 'rās₁wā }

electric reactor See reactor. { i¦lek·trik rē'ak·tər }

electric relay See relay. { i¦lek·trik 'rē₁lā }

electric rotating machinery [ELEC] Any form of apparatus which has a rotating member and generates, converts, transforms, or modifies electric power, such as a motor, generator, or synchronous converter. { i¦lek·trik ¦rō₁tād·iŋ mə ˌshēn·rē }

electric scanning [ELECTR] Scanning in which the required changes in radar beam direction are produced by variations in phase or amplitude of the currents fed to the various elements of the antenna array. { i¦lek·trik 'skan·iŋ }

electric shielding [ELECTROMAG] Any means of avoiding pickup of undesired signals or noise, suppressing radiation of undesired signals, or confining wanted signals to desired paths or regions, such as electrostatic shielding or electromagnetic shielding. Also known as screening; shielding. { i¦lek·trik 'shēld·iŋ }

electric shunt See shunt. { i¦lek·trik 'shənt }

electric solenoid See solenoid. { i¦lek·trik 'sō·lə ₁nóid }

electric spark See spark. { i¦lek·trik 'spärk }

electric strength See dielectric strength. { i¦lek· trik 'streŋkth }

electric susceptibility [ELEC] A dimensionless parameter measuring the ease of polarization of a dielectric, equal (in meter-kilogram-second units) to the ratio of the polarization to the product of the electric field strength and the vacuum permittivity. Also known as dielectric susceptibility. { i¦lek·trik sə₁sep·tə'bil·əd·ē }

electric switchboard See switchboard. { i¦lek·trik 'swich₁bórd }

electric telemetering [COMMUN] System to transmit electric impulses from the primary detector to a remote receiving station, with or without wire interconnections. { i¦lek·trik ₁tel· ə'mēd·ə·riŋ }

electric transducer [ELECTR] A transducer in which all of the waves are electric. { i¦lek·trik tranz'dü·sər }

electric transient [ELEC] A temporary component of current and voltage in an electric circuit which has been disturbed. { i¦lek·trik 'tran·zhənt }

electric tuning [ELECTR] Tuning a receiver to a desired station by switching a set of preadjusted trimmer capacitors or coils into the tuning circuits. { i¦lek·trik 'tün·iŋ }

electric vector See electric-field vector. { i¦lek·trik 'vek·tər }

electric-wave filter See filter. { i¦lek·trik ¦wāv 'fil· tər }

electric wind See convective discharge. { i¦lek·trik 'wind }

electric wire See wire. { i¦lek·trik 'wīr }

electric wiring See wiring. { i¦lek·trik 'wīr·iŋ }

electrification [ELEC] **1.** The process of establishing a charge in an object. **2.** The generation, distribution, and utilization of electricity. { i₁lek· trə·fə¦kā·shən }

electrization [ELEC] The electric polarization divided by the permittivity of empty space. { i₁lek· trə'zā·shən }

electroacoustic effect See acoustoelectric effect. { i¦lek·trō·ə¦kü·stik i'fekt }

electroacoustics [ENG ACOUS] The conversion of acoustic energy and waves into electric energy and waves, or vice versa. { i¦lek·trō·ə¦kü·stiks }

electroacoustic transducer [ENG ACOUS] A transducer that receives waves from an electric system and delivers waves to an acoustic system, or vice versa. Also known as sound transducer. { i¦lek·trō·ə¦kü·stik tranz'dü·sər }

electrochemical power generation [ENG] The direct conversion of chemical energy to electric energy, as in a battery or fuel cell. { i₁lek·trō 'kem·ə·kəl ¦paů·ər ₁jen·ə₁rā·shən }

electrochemical recording [ELECTR] Recording by means of a chemical reaction brought about by the passage of signal-controlled current through the sensitized portion of the record sheet. { i₁lek· trō¦kem·ə·kəl ri'kórd·iŋ }

electrochemical valve [ELEC] Electric valve consisting of a metal in contact with a solution or compound, across the boundary of which current flows more readily in one direction than in the other direction, and in which the valve action is accompanied by chemical changes. { i₁lek· trō'kem·ə·kəl 'valv }

electrochromic device [ENG] A self-contained, hermetically sealed, two-electrode electrolytic cell that includes one or more electrochromic materials and an electrolyte. { i‚lek·trə¦krōm·ik di'vīs }

electrochromic display [ELECTR] A solid-state passive display that uses organic or inorganic insulating solids which change color when injected with positive or negative charges. { i¦lek·trō¦krō·mik di'splā }

electrode [ELEC] An electric conductor through which an electric current enters or leaves a medium, whether it be an electrolytic solution, solid, molten mass, gas, or vacuum. { i'lek ‚trōd }

electrode admittance [ELECTR] Quotient of dividing the alternating component of the electrode current by the alternating component of the electrode voltage, all other electrode voltages being maintained constant. { i'lek‚trōd ad'mit·əns }

electrode capacitance [ELECTR] Capacitance between one electrode and all the other electrodes connected together. { i'lek‚trōd kə'pas·əd·əns }

electrode characteristic [ELECTR] Relation between the electrode voltage and the current to an electrode, all other electrode voltages being maintained constant. { i'lek‚trōd ‚kar·ik·tə'ris·tik }

electrode conductance [ELECTR] Quotient of the inphase component of the electrode alternating current by the electrode alternating voltage, all other electrode voltage being maintained constant; this is a variational and not a total conductance. Also known as grid conductance. { i'lek‚trōd kən'dək·təns }

electrode couple [ELEC] The pair of electrodes in an electric cell, between which there is a potential difference. { i'lek‚trōd ‚kə·pəl }

electrode current [ELECTR] Current passing to or from an electrode, through the interelectrode space within a vacuum tube. { i'lek‚trōd ‚kə·rənt }

electrode dark current [ELECTR] The electrode current that flows when there is no radiant flux incident on the photocathode in a phototube or camera tube. Also known as dark current. { i'lek ‚trōd ¦därk 'kə·rənt }

electrode dissipation [ELECTR] Power dissipated in the form of heat by an electrode as a result of electron or ion bombardment. { i'lek‚trōd ‚dis·ə'pā·shən }

electrode drop [ELECTR] Voltage drop in the electrode due to its resistance. { i'lek‚trōd ‚dräp }

electrode impedance [ELECTR] Reciprocal of the electrode admittance. { i'lek‚trōd im'pēd·əns }

electrode inverse current [ELECTR] Current flowing through an electrode in the direction opposite to that for which the tube is designed. { i'lek ‚trōd 'in·vərs ‚kə·rənt }

electrodeless discharge [ELECTR] An electric discharge generated by placing a discharge tube in a strong, high-frequency electromagnetic field. { i¦lek‚trōd·ləs 'dis‚chärj }

electrodeless lamp [ELECTR] A lamp based on an electrodeless discharge. { i¦lek‚trōd·ləs 'lamp }

electrode potential [ELECTR] The instantaneous voltage of an electrode with respect to the cathode of an electron tube. Also known as electrode voltage. { i'lek‚trōd pə'ten·chəl }

electrode resistance [ELECTR] Reciprocal of the electrode conductance; this is the effective parallel resistance and is not the real component of the electrode impedance. { i'lek‚trōd ri'zis·təns }

electrode voltage See electrode potential. { i'lek ‚trōd ‚vōl·tij }

electrodynamic ammeter [ENG] Instrument which measures the current passing through a fixed coil and a movable coil connected in series by balancing the torque on the movable coil (resulting from the magnetic field of the fixed coil) against that of a spiral spring. { i‚lek·trō·dī'nam·ik 'a‚mēd·ər }

electrodynamic instrument [ENG] An instrument that depends for its operation on the reaction between the current in one or more movable coils and the current in one or more fixed coils. Also known as electrodynamometer. { i‚lek·trō·dī'nam·ik 'in·strə·mənt }

electrodynamic loudspeaker [ENG ACOUS] Dynamic loudspeaker in which the magnetic field is produced by an electromagnet, called the field coil, to which a direct current must be furnished. { i‚lek·trō·dī'nam·ik 'laůd‚spēk·ər }

electrodynamic machine [ELEC] An electric generator or motor in which the output load current is produced by magnetomotive currents generated in a rotating armature. { i‚lek·trō·dī'nam·ik mə'shēn }

electrodynamic wattmeter [ENG] An electrodynamic instrument connected as a wattmeter, with the main current flowing through the fixed coil, and a small current proportional to the voltage flowing through the movable coil. Also known as moving-coil wattmeter. { i‚lek·trō·dī'nam·ik 'wät‚mēd·ər }

electrodynamometer See electrodynamic instrument. { i‚lek·trō‚dī·nə'mäm·əd·ər }

electroexplosive [ENG] An initiator or a system in which an electric impulse initiates detonation or deflagration of an explosive. { i‚lek·trō·ik 'splō·siv }

electrogram [ELECTR] A record of an image of an object made by sparking, usually on paper. { i'lek·trə‚gram }

electrograph [ENG] Any plot, graph, or tracing produced by the action of an electric current on prepared sensitized paper (or other chart material) or by means of an electrically controlled stylus or pen. { i'lek·trə‚graf }

electrographic pencil [ELECTR] A pencil used to make a conductive mark on paper, for detection by a conductive-mark sensing device. { i'lek·trə ‚graf·ik 'pen·səl }

electrokinetic transducer [ELEC] An instrument which converts dynamic physical forces, such as vibration and sound, into corresponding electric signals by measuring the streaming potential generated by passage of a polar fluid through a permeable refractory-ceramic or fritted-glass member between two chambers. { i¦lek·trō·kə'ned·ik tranz'dü·ser }

electroluminescence [ELECTR] The emission of light, not due to heating effects alone, resulting from application of an electric field to a material, usually solid. { i¦lek·trō,lü·mə'nes·əns }

electroluminescent cell See electroluminescent panel. { i¦lek·trō,lü·mə'nes·ənt 'sel }

electroluminescent display [ELECTR] A display in which various combinations of electroluminescent segments may be activated by applying voltages to produce any desired numeral or other character. { i¦lek·trō,lü·mə'nes·ənt di'splā }

electroluminescent lamp See electroluminescent panel. { i¦lek·trō,lü·mə'nes·ənt 'lamp }

electroluminescent panel [ELECTR] A surface-area light source employing the principle of electroluminescence; consists of a suitable phosphor placed between sheet-metal electrodes, one of which is essentially transparent, with an alternating current applied between the electrodes. Also known as electroluminescent cell; electroluminescent lamp; light panel; luminescent cell. { i¦lek·trō,lü·mə'nes·ənt 'pan·əl }

electrolyte-activated battery [ELEC] A reserve battery in which an aqueous electrolyte is stored in a separate chamber, and a mechanism, which may be operated from a remote location, drives the electrolyte out of the reservoir and into the cells of the battery for activation. { i¦lek·trə,līt ak·tə¦vād·əd 'bad·ə·rē }

electrolyte-MOSFET [ENG] A metal oxide semiconductor field-effect transistor (MOSFET) that is immersed in a solution to determine the concentrations of dissolved redox active species; the bulk part of the work function of the gate electrode of the transistor changes when the sensor membrane is oxidized or reduced. Abbreviated EMOSFET. { i¦lek·trə,līt 'mós,fet }

electrolytic arrester See aluminum-cell arrester. { i'lek·trə,lid·ik ə'res·tər }

electrolytic capacitor [ELEC] A capacitor consisting of two electrodes separated by an electrolyte; a dielectric film, usually a thin layer of gas, is formed on the surface of one electrode. Also known as electrolytic condenser. { i'lek·trə ,lid·ik kə'pas·əd·ər }

electrolytic condenser See electrolytic capacitor. { i'lek·trə,lid·ik kən'den·sər }

electrolytic interrupter [ELEC] An interrupter that consists of two electrodes in an electrolytic solution; bubbles formed in the solution continually interrupt the passage of current between the electrodes. { i'lek·trə,lid·ik ,int·ə'rəp·tər }

electrolytic recording [ELECTR] Electrochemical recording in which the chemical change is made possible by the presence of an electrolyte. { i'lek·trə,lid·ik ri'kórd·iŋ }

electrolytic rectifier [ELEC] A rectifier consisting of metal electrodes in an electrolyte, in which rectification of alternating current is accompanied by electrolytic action; polarizing film formed on one electrode permits current flow in one direction but not the other. { i'lek·trə,lid·ik 'rek·tə,fī·ər }

electrolytic rheostat [ELEC] A rheostat that consists of a tank of conducting liquid in which electrodes are placed, and resistance is varied by changing the distance between the electrodes, the depth of immersion of the electrodes, or the resistivity of the solution. Also known as water rheostat. { i'lek·trə,lid·ik 'rē·ə,stat }

electrolytic switch [ELEC] A switch having two electrodes projecting into a chamber partly filled with electrolyte, leaving an air bubble of predetermined width; the bubble shifts position and changes the amount of electrolyte in contact with the electrodes when the switch is tilted from true horizontal. { i'lek·trə,lid·ik 'swich }

electromagnetic cathode-ray tube [ELECTR] A cathode-ray tube in which electromagnetic deflection is used on the electron beam. { i¦lek·trō·mag'ned·ik 'ka,thōd 'rā ,tüb }

electromagnetic compatibility [ELECTR] The capability of electronic equipment or systems to be operated in the intended electromagnetic environment at design levels of efficiency. { i¦lek·trō·mag'ned·ik kəm,pat·ə'bil·əd·ē }

electromagnetic constant See speed of light. { i¦lek·trō·mag'ned·ik 'kän·stənt }

electromagnetic current [ELECTR] Motion of charged particles (for example, in the ionosphere) giving rise to electric and magnetic fields. { i¦lek·trō·mag'ned·ik 'kə·rənt }

electromagnetic damping [ELEC] Retardation of motion that results from the reaction between eddy currents in a moving conductor and the magnetic field in which it is moving. { i¦lek·trō·mag'ned·ik 'damp·iŋ }

electromagnetic deflection [ELECTR] Deflection of an electron stream by means of a magnetic field. { i¦lek·trō·mag'ned·ik di'flek·shən }

electromagnetic energy [ELECTROMAG] The energy associated with electric or magnetic fields. { i¦lek·trō·mag'ned·ik 'en·ər·jē }

electromagnetic environment [COMMUN] The radio-frequency fields existing in a given area. { i¦lek·trō·mag'ned·ik en'vi·rən·mənt }

electromagnetic field [ELECTROMAG] An electric or magnetic field, or a combination of the two, as in an electromagnetic wave. { i¦lek·trō·mag'ned·ik 'fēld }

electromagnetic field equations See Maxwell field equations. { i¦lek·trō·mag'ned·ik 'fēld i,kwā·zhənz }

electromagnetic focusing [ELECTR] Focusing the electron beam in a video display device by means of a magnetic field parallel to the beam; the field is produced by an adjustable value of direct current through a focusing coil mounted on the neck of the tube. { i¦lek·trō·mag'ned·ik 'fō·kəs·iŋ }

electromagnetic horn See horn antenna. { i¦lek·trō·mag'ned·ik 'hórn }

electromagnetic induction [ELECTROMAG] The production of an electromotive force either by motion of a conductor through a magnetic field so as to cut across the magnetic flux or by a change in the magnetic flux that threads a conductor. Also known as induction. { i¦lek·trō·mag'ned·ik in'dək·shən }

electromagnetic interference [ELEC] Interference, generally at radio frequencies, that is generated inside systems, as contrasted to radio-frequency interference coming from sources outside a system. Abbreviated emi. { i¦lek·trō·mag'ned·ik ˌin·tər'fir·əns }

electromagnetic interference [ELEC] Interference, generally at radio frequencies, that is generated inside systems, as contrasted to radio-frequency interference coming from sources outside a system. Abbreviated emi. { i¦lek·trō·mag'ned·ik ˌin·tər'fir·əns }

electromagnetic lens [ELECTR] An electron lens in which electron beams are focused by an electromagnetic field. { i¦lek·trō·mag'ned·ik 'lenz }

electromagnetic noise [ELEC] Noise in a communications system resulting from undesired electromagnetic radiation. Also known as radiation noise. { i¦lek·trō·mag'ned·ik 'nóiz }

electromagnetic pulse [ELECTROMAG] The pulse of electromagnetic radiation generated by a large thermonuclear explosion; althought not a direct threat to human health, it is a threat to electronic communications systems. { i¦lek·trō·mag'ned·ik 'pəls }

electromagnetic pump [ELEC] A pump in which a conductive liquid is made to move through a pipe by sending a large current transversely through the liquid; this current reacts with a magnetic field that is at right angles to the pipe and to current flow, to move the current-carrying liquid conductor. { i¦lek·trō·mag'ned·ik 'pəmp }

electromagnetic radiation [ELECTROMAG] Electromagnetic waves and, especially, the associated electromagnetic energy. { i¦lek·trō·mag'ned·ik ˌrād·ē'ā·shən }

electromagnetic reconnaissance [ELECTR] Reconnaissance for the purpose of locating and identifying potentially hostile transmitters of electromagnetic radiation, including radar, communication, missile-guidance, and navigation-aid equipment. { i¦lek·trō·mag'ned·ik ri'kän·ə·säns }

electromagnetic susceptibility [ELECTR] The tolerance of circuits and components to all sources of interfering electromagnetic energy. { i¦lek·trō·mag'ned·ik sə,sep·tə'bil·əd·ē }

electromagnetic transducer See electromechanical transducer. { i¦lek·trō·mag'ned·ik tranz'dü·sər }

electromagnetic wave [ELECTROMAG] A disturbance which propagates outward from any electric charge which oscillates or is accelerated; far from the charge it consists of vibrating electric and magnetic fields which move at the speed of light and are at right angles to each other and to the direction of motion. { i¦lek·trō·mag'ned·ik 'wāv }

electromechanical circuit [ELEC] A circuit containing both electrical and mechanical parameters of consequence in its analysis. { i¦lek·trō·mi'kan·ə·kəl 'sər·kə t }

electromechanical dialer [ELECTR] Telephone dialer which activates one of a set of desired numbers, precoded into it, when the user selects and presses a start button. { i¦lek·trō·mi'kan·ə·kəl 'dī·lər }

electromechanical plotter [COMPUT SCI] An automatic device used in conjunction with a digital computer to produce a graphic or pictorial representation of computer data on hard copy. { i¦lek·trō·mi'kan·ə·kəl 'pläd·ər }

electromechanical recording [ELECTR] Recording by means of a signal-actuated mechanical device, such as a pen arm or mirror attached to the moving coil of a galvanometer. { i¦lek·trō·mi'kan·ə·kəl ri'kòrd·iŋ }

electromechanical transducer [ELECTR] A transducer for receiving waves from an electric system and delivering waves to a mechanical system, or vice versa. Also known as electromagnetic transducer. { i¦lek·trō·mi'kan·ə·kəl tranz'dü·sər }

electrometer [ENG] An instrument for measuring voltage without drawing appreciable current. { i,lek'träm·əd·ər }

electrometer amplifier [ELECTR] A low-noise amplifier having sufficiently low current drift and other characteristics required for measuring currents smaller than 10^{-12} ampere. { i,lek'träm·əd·ər 'am·plə,fī·ər }

electrometer tube [ELECTR] A high-vacuum electron tube having a high input impedance (low control-electrode conductance) to facilitate measurement of extremely small direct currents or voltages. { i,lek'träm·əd·ər ,tüb }

electron acceptor See acceptor. { i'lek,trän ak 'sep·tər }

electron avalanche See avalanche. { i'lek,trän 'av·ə,lanch }

electron beam [ELECTR] A narrow stream of electrons moving in the same direction, all having about the same velocity. { i'lek,trän ,bēm }

electron-beam channeling [ELECTR] The technique of transporting high-energy, high-current electron beams from an accelerator to a target through a region of high-pressure gas by creating a path through the gas where the gas density may be temporarily reduced; the gas may be ionized; or a current may flow whose magnetic field focuses the electron beam on the target. { i'lek,trän ,bēm 'chan·əl·iŋ }

electron-beam drilling [ELECTR] Drilling of tiny holes in a ferrite, semiconductor, or other material by using a sharply focused electron beam to melt and evaporate or sublimate the material in a vacuum. { i'lek,trän ,bēm 'dril·iŋ }

electron-beam generator [ELECTR] Velocity-modulated generator, such as a klystron tube, used to generate extremely high frequencies. { i'lek,trän ,bēm 'jen·ə,rād·ər }

electron-beam ion source [ELECTR] A source of multiply charged heavy ions which uses an intense electron beam with energies of 5 to 10 kiloelectronvolts to successively ionize injected gas. Abbreviated EBIS. { i'lek,trän ,bēm 'ī,än ,sòrs }

electron-beam ion trap [ELECTR] A device for producing the highest possible charge states of heavy ions, in which impact ionization or excitation by successive electrons is efficiently achieved by causing the ions to be trapped in a compressed electron beam by the electron beam's space charge. Abbreviated EBIT { i¦lek ‚trän ‚bē 'i·ən ‚trap }

electron-beam lithography [ELECTR] Lithography in which the radiation-sensitive film or resist is placed in the vacuum chamber of a scanning-beam electron microscope and exposed by an electron beam under digital computer control. { i'lek‚trän ‚bēm li'thäg·rə·fē }

electron-beam magnetometer [ENG] A magnetometer that depends on the change in intensity or direction of an electron beam that passes through the magnetic field to be measured. { i'lek‚trän ‚bēm mag·nə'täm·əd·ər }

electron-beam parametric amplifier [ELECTR] A parametric amplifier in which energy is pumped from an electrostatic field into a beam of electrons traveling down the length of the tube, and electron couplers impress the input signal at one end of the tube and translate spiraling electron motion into electric output at the other. { i'lek‚trän ‚bēm ‚par·ə¦me·trik 'am·plə ‚fī·ər }

electron-beam pumping [ELECTR] The use of an electron beam to produce excitation for population inversion and lasing action in a semiconductor laser. { i'lek‚trän ‚bēm 'pəmp·iŋ }

electron-beam recorder [ELECTR] A recorder in which a moving electron beam is used to record signals or data on photographic or thermoplastic film in a vacuum chamber. { i'lek‚trän ‚bēm ri'kórd·ər }

electron-beam tube [ELECTR] An electron tube whose performance depends on the formation and control of one or more electron beams. { i'lek‚trän ‚bēm 'tüb }

electron-bombardment-induced conductivity [ELECTR] In a multimode display-storage tube, a process using an electron gun to erase the image on the cathode-ray tube interface. { i'lek ‚trän bäm¦bärd·mənt in‚düst kän·dək'tiv·əd·ē }

electron bunching See bunching. { i'lek‚trän 'bənch·iŋ }

electron collector See collector. { i'lek‚trän kə ‚lek·tər }

electron conduction [ELEC] Conduction of electricity resulting from motion of electrons, rather than from ions in a gas or solution, or holes in a solid. { i'lek‚trän kən‚dək·shən }

electron-coupled oscillator [ELECTR] An oscillator employing a multigrid tube in which the cathode and two grids operate as an oscillator; the anode-circuit load is coupled to the oscillator through the electron stream. Abbreviated eco. Also known as Dow oscillator. { i'lek‚trän ‚kəp·əld 'äs·ə‚lād·ər }

electron coupler [ELECTR] A microwave amplifier tube in which electron bunching is produced by an electron beam projected parallel to a magnetic field and, at the same time, subjected to a transverse electric field produced by a signal generator. Also known as Cuccia coupler. { i'lek ‚trän ‚kəp·lər }

electron coupling [ELECTR] A method of coupling two circuits inside an electron tube, used principally with multigrid tubes; the electron stream passing between electrodes in one circuit transfers energy to electrodes in the other circuit. Also known as electronic coupling. { i'lek‚trän ‚kəp·liŋ }

electron cyclotron resonance ion source See electron cyclotron resonance source. { i¦lek ‚trän ¦sī·klə‚trän 'rez·ə·nəns 'ī‚än ‚sórs }

electron cyclotron resonance source [ELECTR] A source of multiply charged heavy ions that uses microwave power to heat electrons to energies of tens of kilovolts in two magnetic mirror confinement chambers in series; ions formed in the first chamber drift into the second chamber, where they become highly charged. Abbreviated ECR source. Also known as electron cyclotron resonance ion source (ECRIS). { i'lek ‚trän 'sī·klə‚trän 'rez·ən·əns ‚sórs }

electron device [ELECTR] A device in which conduction is principally by electrons moving through a vacuum, gas, or semiconductor, as in a crystal diode, electron tube, transistor, or selenium rectifier. { i'lek‚trän di'vīs }

electron donor See donor. { i'lek‚trän ‚dō·nər }

electron efficiency [ELECTR] The power which an electron stream delivers to the circuit of an oscillator or amplifier at a given frequency, divided by the direct power supplied to the stream. Also known as electronic efficiency. { i'lek‚trän ə'fish·ən·sē }

electronegative [ELEC] **1.** Carrying a negative electric charge. **2.** Capable of acting as the negative electrode in an electric cell. { i¦lek·trō 'neg·əd·iv }

electron emission [ELECTR] The liberation of electrons from an electrode into the surrounding space, usually under the influence of heat, light, or a high electric field. { i'lek·trän i'mish·ən }

electron emitter [ELECTR] The electrode from which electrons are emitted. { i'lek‚trän i'mid·ər }

electron flow [ELEC] A current produced by the movement of free electrons toward a positive terminal; the direction of electron flow is opposite to that of current. { i'lek‚trän ‚flō }

electron gun [ELECTR] An electrode structure that produces and may control, focus, deflect, and converge one or more electron beams in an electron tube. { i'lek‚trän ‚gən }

electron-gun density multiplication [ELECTR] Ratio of the average current density at any specified aperture through which the electron stream passes to the average current density at the cathode surface. { i'lek‚trän ‚gən 'den·səd·ē ‚məl·tə·plə'kā·shən }

electron hole See hole. { i'lek,trän ¦hōl }
electron holography [ELECTR] An imaging technique using the wave nature of electrons and light, in which an interference pattern between an object wave and a reference wave is formed using a coherent field-emission electron beam from a sharp tungsten needle, and is recorded on film as a hologram, and the image of the original object is then reconstructed by illuminating a light beam equivalent to the reference wave onto the hologram. { i,lek,trän hō 'läg·rə·fē }
electronic [ELECTR] Pertaining to electron devices or to circuits or systems utilizing electron devices, including electron tubes, magnetic amplifiers, transistors, and other devices that do the work of electron tubes. { i,lek'trän·ik }
electronically agile radar [ENG] An airborne radar that uses a phased-array antenna which changes radar beam shapes and beam positions at electronic speeds. { i,lek'trän·ik·lē ,a·jəl 'rä ,där }
electronic alternating-current voltmeter [ELECTR] A voltmeter consisting of a direct-current milliammeter calibrated in volts and connected to an amplifier-rectifier circuit. { i,lek'trän·ik ¦al·tər,nād·iŋ ¦kə·rənt 'vōlt,mēd·ər }
electronic altimeter See radio altimeter. { i,lek 'trän·ik al'tim·əd·ər }
electronic attack [ELECTR] A term embracing all means in electronic warfare both to counter the enemy's electronic or electromagnetic sensing and communications and also to effect offense with high-power electromagnetic weaponry. Abbreviated EA. { i,lək'trän·ik ə'tak }
electronic attitude directional indicator [NAV] A multicolor cathode-ray-tube display of attitude information (roll and pitch) showing the aircraft's position in relation to the instrument landing system or a very high-frequency omnirange station. Abbreviated EADI. { i,lek'trän·ik 'ad·ə ,tüd də'rek·shən·əl 'in·də,kād·ər }
electronic azimuth marker [ELECTR] On an airborne radar plan position indicator (PPI) a bright rotatable radial line used for bearing determination. Also known as azimuth marker. { i,lek'trän·ik 'az·ə·məth ,märk·ər }
electronic bearing cursor [ELECTR] Of a marine radar set, the bright rotatable radial line on the plan position indicator used for bearing determination. Also known as electronic bearing marker. { i,lek'trän·ik 'ber·iŋ ,kər·sər }
electronic bearing marker See electronic bearing cursor. { i,lek'trän·ik 'ber·iŋ ,märk·ər }
electronic calculator [ELECTR] A calculator in which integrated circuits perform calculations and show results on a digital display; the displays usually use either seven-segment light-emitting diodes or liquid crystals. { i,lek'trän·ik 'kal·kyə ,lād·ər }
electronic camouflage [ELECTR] Use of electronic means, or exploitation of electronic characteristics to reduce, submerge, or eliminate the radar echoing properties of a target. { i,lek 'trän·ik 'kam·ə,fläzh }

electronic chart display and information system [NAV] A navigation information system with an electronic chart database, as well as navigational and piloting information (typically, vessel-route-monitoring, track-keeping, and track-planning information). Abbreviated ECDIS. { i·lek¦trän·ik 'chärt di¦splā ən ,in·fər'mā·shən ,sis·təm }
electronic chart reader [COMPUT SCI] A device which scans curves by a graphical recorder on a continuous paper form and converts them into digital form. { i,lek'trän·ik 'chärt ,rēd·ər }
electronic circuit [ELECTR] An electric circuit in which the equilibrium of electrons in some of the components (such as electron tubes, transistors, or magnetic amplifiers) is upset by means other than an applied voltage. { i,lek'trän·ik 'sər·kət }
electronic codebook mode See block encryption. { i,lek'trän·ik 'kōd,bùk ,mōd }
electronic commerce [COMPUT SCI] Business done on the Internet. Also known as e-business; e-commerce. { i·lek¦trän·ik 'kä·mərs }
electronic commutator [ELECTR] An electron-tube or transistor circuit that switches one circuit connection rapidly and successively to many other circuits, without the wear and noise of mechanical switches. { i,lek'trän·ik 'käm·yə ,tād·ər }
electronic component [ELECTR] A component which is able to amplify or control voltages or currents without mechanical or other nonelectrical command, or to switch currents or voltages without mechanical switches; examples include electron tubes, transistors, and other solid-state devices. { i,lek'trän·ik 'kämpō·nənt }
electronic computing units [ELECTR] The sensing sections of tabulating equipment which enable the machine to handle the contents of punched cards in a prescribed manner. { i,lek'trän·ik kəm'pyüd·iŋ ,yü·nəts }
electronic control [ELECTR] The control of a machine or process by circuits using electron tubes, transistors, magnetic amplifiers, or other devices having comparable functions. { i,lek'trän·ik kən'trōl }
electronic controller [ELECTR] Electronic device incorporating vacuum tubes or solid-state devices and used to control the action or position of equipment; for example, a valve operator. { i,lek'trän·ik kən'trōl·ər }
electronic counter [ELECTR] A circuit using electron tubes or equivalent devices for counting electric pulses. Also known as electronic tachometer. { i,lek'trän·ik 'kaünt·ər }
electronic countermeasure [ELECTR] An offensive or defensive tactic or device using electronic, electromagnetic, and reflecting apparatus to reduce the military effectiveness of enemy equipment involving electromagnetic radiation, such as radar, communication, guidance, or other radio-wave devices. Abbreviated ECM. { i,lek'trän·ik 'kaünt·ər,mezh·ər }
electronic coupling See electron coupling. { i,lek 'trän·ik 'kəp·liŋ }
electronic data processing [COMPUT SCI] Processing data by using equipment that is

predominantly electronic in nature, such as an electronic digital computer. Abbreviated EDP. { i,lek'trän·ik 'dad·ə ,prä·səs·iŋ }

electronic data-processing center [COMPUT SCI] The complex formed by the computer, its peripheral equipment, the personnel related to the operation of the center and control functions, and, usually, the office space housing hardware and personnel. Abbreviated EDP center. Also known as computer center. { i,lek'trän·ik 'dad·ə ,präs·əs·iŋ ,sen·tər }

electronic data-processing management science [COMPUT SCI] The field consisting of a class of management problems capable of being handled by computer programs. { i,lek'trän·ik 'dad·ə ,prä·səs·iŋ 'man·ij·mənt ,sī·əns }

electronic data-processing system [COMPUT SCI] A system for data processing by means of machines using electronic circuitry at electronic speed, as opposed to electromechanical equipment. { i,lek'trän·ik 'dad·ə ,prä·səs·iŋ ,sis·təm }

electronic defense evaluation [ELECTR] A mutual evaluation of radar and aircraft, with the aircraft trying to penetrate the radar's area of coverage in an electronic countermeasure environment. { i,lek'trän·ik di'fens i,val·yə'wā·shən }

electronic differential analyzer [COMPUT SCI] A form of analog computer using interconnected electronic integrators to solve differential equations. { i,lek'trän·ik ,dif·ə'ren·chəl 'an·ə,liz·ər }

electronic display [ELECTR] An electronic component used to convert electric signals into visual imagery in real time suitable for direct interpretation by a human operator. { i,lek'trän·ik di'splā }

electronic distance-measuring equipment [NAV] A navigation system consisting of airborne devices that transmit microsecond pulses to special ground beacons, which retransmit the signals to the aircraft; the length of expired time between transmission and reception is measured, converted to kilometers or miles, and presented to the pilot. { i,lek'trän·ik 'dis·təns ,mezh·ə·riŋ i,kwip·mənt }

electronic dummy [ENG ACOUS] A vocal simulator which is a replica of the head and torso of a person, covered with plastisol flesh that simulates the acoustical and mechanical properties of real flesh, and possessing an artificial voice and two artificial ears. Abbreviated ED. { i,lek'trän·ik 'dəm·ē }

electronic efficiency [ELECTR] Ratio of the power at the desired frequency, delivered by the electron stream to the circuit in an oscillator or amplifier, to the average power supplied to the stream. { i,lek'trän·ik i'fish·ən·sē }

electronic engineering [ENG] Engineering that deals with practical applications of electronics. { i,lek'trän·ik ,en·jə'nir·iŋ }

electronic fuse [ENG] A fuse, such as the radio proximity fuse, set off by an electronic device incorporated in it. { i,lek'trän·ik 'fyüz }

electronic heating [ENG] Heating by means of radio-frequency current produced by an electron-tube oscillator or an equivalent radio-frequency

power source. Also known as high-frequency heating; radio-frequency heating. { i,lek'trän·ik 'hēd·iŋ }

electronic horizontal-situation indicator [NAV] An integrated multicolor map display of an airplane's position combined with a color weather radar display, with a scale selected by the pilot, together with information on wind direction and velocity, horizontal situation, and deviation from the planned vertical path. Abbreviated EHSI. { i,lek'trän·ik ,här·ə'zänt·əl ,sich·ə¦wā·shən ,in·də ,kād·ər }

electronic interference [ELECTR] Any electrical or electromagnetic disturbance that causes undesirable response in electronic equipment. { i,lek'trän·ik ,int·ər·'fir·əns }

electronic jammer See jammer. { i,lek'trän·ik 'jam·ər }

electronic jamming See jamming. { i,lek'trän·ik 'jam·iŋ }

electronic line scanning [ELECTR] Method which provides motion of the scanning spot along the scanning line by electronic means. { i,lek'trän·ik 'līn ,skan·iŋ }

electronic listening device [ELECTR] A device used to capture the sound waves of conversation originating in an ostensibly private setting in a form, usually as a magnetic tape recording, which can be used against the target by adverse interests. { i,lek'trän·ik 'lis·niŋ di,vīs }

electronic locator See metal detector. { i,lek 'trän·ik 'lō,kād·ər }

electronic locking [ELECTR] A technique for preventing the operation of a switch until a specific electrical signal (the unlocking signal) is introduced into circuitry associated with the switch; usually, but not necessarily, the unlocking signal is a binary sequence. { i,lek'trän·ik 'läk·iŋ }

electronic logger See Geiger-Müller probe. { i,lek 'trän·ik 'läg·ər }

electronic mail [COMMUN] The electronic transmission of letters, messages, and memos through a communications network. Also known as e-mail. { i,lek'trän·ik 'māl }

electronic microradiography [ELECTR] Microradiography of very thin specimens in which the emission of electrons from an irradiated object, either the specimen or a lead screen behind it, is used to produce a photographic image of the specimen, which is then enlarged. Also known as e-mail. { i,lek'trän·ik 'mī·krō,rād·ē'äg·rə·fē }

electronic motor control [ELECTR] A control circuit used to vary the speed of a direct-current motor operated from an alternating-current power line. Also known as direct-current motor control; motor control. { i,lek'trän·ik 'mōd·ər kən,trōl }

electronic multimeter [ELECTR] A multimeter that uses semiconductor or electron-tube circuits to drive a conventional multiscale meter. { i,lek'trän·ik 'məl·tē,mēd·ər }

electronic music [ENG ACOUS] Music consisting of tones originating in electronic sound and noise generators used alone or in conjunction with electroacoustic shaping means and sound-recording equipment. { i,lek'trän·ik 'myü·zik }

electronic musical instrument [ENG ACOUS] A musical instrument in which an audio signal is produced by a pickup or audio oscillator and amplified electronically to feed a loudspeaker, as in an electric guitar, electronic carillon, electronic organ, or electronic piano. { i,lek'trän·ik ¦myü·zə·kəl 'in·strə·mənt }

electronic noise jammer [ELECTR] An electronic jammer which emits a radio-frequency carrier modulated with a white noise signal usually derived from a gas tube; used against enemy radar. { i,lek'trän·ik 'nȯiz ,jam·ər }

Electronic Numerical Integrator and Calculator See ENIAC. { i,lek'trän·ik nü'mer·ə·kəl 'int·ə, grād·ər ən 'kal·kyə,lād·ər }

electronic organ [ELECTR] A musical instrument which uses electronic circuits to produce music similar to that of a pipe organ. { i,lek'trän·ik 'ȯr·gən }

electronic packaging [ENG] The technology of packaging electronic equipment; in current usage it refers to inserting discrete components, integrated circuits, and MSI and LSI chips (usually attached to a lead frame by beam leads) into plates through holes on multilayer circuit boards (also called cards), where they are soldered in place. { i,lek'trän·ik 'pak·ij·iŋ }

electronic phase-angle meter [ELECTR] A phasemeter that makes use of electronic devices, such as amplifiers and limiters, that convert the alternating-current voltages being measured to square waves whose spacings are proportional to phase. { i,lek'trän·ik 'fāz ,aŋ·gəl ,mēd·ər }

electronic photometer See photoelectric photometer. { i,lek'trän·ik fō'täm·əd·ər }

electronic piano [ELECTR] A piano without a sounding board, in which vibrations of each string affect the capacitance of a capacitor microphone and thereby produce audio-frequency signals that are amplified and reproduced by a loudspeaker. { i,lek'trän·ik pē'an·ō }

electronic polarization [ELEC] Polarization arising from the displacement of electrons with respect to the nuclei with which they are associated, upon application of an external electric field. { i,lek'trän·ik ,pō·lə·rə'zā·shən }

electronic power supply See power supply. { i,lek'trän·ik 'paů·ər sə,plī }

electronic protection [ELECTR] Measures taken to counteract the effects of electronic attack. Abbreviated EP. { i,lək'trän·ik prə'tek·shən }

electronic publishing [COMMUN] The provision of information with high editorial and value-added content in electronic form, allowing the user some degree of control and interactivity. { i,lek¦trän·ik 'pəb·lish·iŋ }

electronic radiography [ELECTR] Radiography in which the image is detached by direct image converter tubes or by the use of television pickup or electronic scanning, and the resulant signals are amplified and presented for viewing on a kinescope. { i,lek'trän·ik rād·ē'äg·ra·fē }

electronic raster scanning See electronic scanning. { i,lek¦trän·ik 'ras·tər ,skan·iŋ }

electronic reconnaissance [ELECTR] The detection, identification, evaluation, and location of foreign, electromagnetic radiations emanating from other than nuclear detonations or radioactive sources. { i,lek'trän·ik ri'kän·ə·səns }

electronic recording [ELECTR] The process of making a graphical record of a varying quantity or signal (or the result of such a process) by electronic means, involving control of an electron beam by electric or magnetic fields, as in a cathode-ray oscillograph, in contrast to light-beam recording. { i,lek'trän·ik ri'kȯrd·iŋ }

electronic robot [CONT SYS] A robot whose motions are powered by a direct-current stepper motor. { i,lek'trän·ik 'rō,bät }

electronic scanning [ELECTR] Scanning in which the radar beam direction is determined by control of the relative phases of the signals fed to the elements of an otherwise stationary antenna array. { i,lek'trän·ik 'skan·iŋ }

electronic sculpturing [COMPUT SCI] Procedure for constructing a model of a system by using an analog computer, in which the model is devised at the console by interconnecting components on the basis of analogous configuration with real system elements; then, by adjusting circuit gains and reference voltages, dynamic behavior can be generated that corresponds to the desired response, or is recognizable in the real system. { i,lek'trän·ik 'skəlp·chə·riŋ }

electronic security [ELECTR] Protection resulting from all measures designed to deny to unauthorized persons information of value which might be derived from the possession and study of electromagnetic radiations. { i,lek'trän·ik sə'kyúr·əd·ē }

electronic sky screen equipment [ELECTR] Electronic device that indicates the departure of a missile from a predetermined trajectory. { i,lek'trän·ik 'skī ,skrēn i,kwip·mənt }

electronic spreadsheet [COMPUT SCI] A type of computer software for performing mathematical computations on numbers arranged in rows and columns, in which the numbers can depend on the values in other rows and columns, allowing large numbers of calculations to be carried out simultaneously. { i,lek'trän·ik 'spred,shēt }

electronic support measures See electronic warfare support measures. { i,lek'trän·ik sə'pȯrt ,mezh·ərz }

electronic surge arrester [ELECTR] Device used to switch to ground high-energy surges, thereby reducing transient energy to a level safe for secondary protectors, for example, Zener diodes, silicon rectifiers and so on. { i,lek'trän·ik 'sərj ə,res·tər }

electronic switch [ELECTR] **1.** Vacuum tube, crystal diodes, or transistors used as an on and off switching device. **2.** Test instrument used to present two wave shapes on a single gun cathode-ray tube. { i,lek'trän·ik 'swich }

electronic switching [COMMUN] Telephone switching using a computer with a storage containing program switching logic, whose output actuates switches that set up telephone

199

connections automatically. |ELECTR| The use of electronic circuits to perform the functions of a high-speed switch. { i,lek'trän·ik 'swich·iŋ }

electronic tablet [COMPUT SCI] A data-entry device consisting of stylus, writing surface, and circuitry that produces a pair of digital coordinate values corresponding continuously to the position of the stylus upon the surface. Also known as data tablet. { i,lek'trän·ik 'tab·lət }

electronic tachometer See electronic counter. { i,lek'trän·ik tə'käm·əd·ər }

electronic tuning |ELECTR| Tuning of a transmitter, receiver, or other tuned equipment by changing a control voltage rather than by adjusting or switching components by hand. { i,lek'trän·ik 'tün·iŋ }

electronic typewriter [COMPUT SCI] A typewriter whose operation is enhanced through the use of microprocessor technology to provide many of the functions of a word-processing system but which has at most a partial-line visual display. Also known as memory typewriter. { i'lek,trän·ik 'tīp,rīd·ər }

electronic video recording |ELECTR| The recording of black and white or color television visual signals on a reel of photographic film as coded black and white images. Abbreviated EVR. { i,lek'trän·ik 'vid·ē·ō ri,kórd·iŋ }

electronic voltage regulator |ELECTR| A device which maintains the direct-current power supply voltage for electronic equipment nearly constant in spite of input alternating-current line voltage variations and output load variations. { i,lek'trän·ik 'vōl·tij ,reg·yə,lād·ər }

electronic voltmeter [ENG] Voltmeter which uses the rectifying and amplifying properties of electron devices and their associated circuits to secure desired characteristics, such as high-input impedance, wide-frequency range, crest indications, and so on. { i,lek'trän·ik 'vōlt,mēd·ər }

electronic warfare |ELECTR| Military action involving the use of electromagnetic energy to determine, exploit, reduce, or prevent hostile use of the electromagnetic spectrum, and action which retains friendly use of electromagnetic spectrum. { i,lek'trän·ik 'wór,fer }

electronic warfare support measures |ELECTR| That division of electronic warfare involving actions taken to search for, intercept, locate, record, and analyze radiated electromagnetic energy for the purpose of exploiting such radiations in support of military operations. Also known as electronic support measures. { i,lek'trän·ik 'wór ,fer sə'pórt ,mezh·ərz }

electronic writing |ELECTR| The use of electronic circuits and electron devices to reproduce symbols, such as an alphabet, in a prescribed order on an electronic display device for the purpose of transferring information from a source to a viewer of the display device. { i,lek'trän·ik 'rīd·iŋ }

electron image tube See image tube. { i'lek,trän 'im·ij ,tüb }

electron injection |ELECTR| 1. The emission of electrons from one solid into another. 2. The process of injecting a beam of electrons with an electron gun into the vacuum chamber of a mass spectrometer, betatron, or other large electron accelerator. { i'lek,trän in'jek·shən }

electron lens |ELECTR| An electric or magnetic field, or a combination thereof, which acts upon an electron beam in a manner analogous to that in which an optical lens acts upon a light beam. Also known as lens. { i'lek,trän 'lenz }

electron microscope |ELECTR| A device for forming greatly magnified images of objects by means of electrons, usually focused by electron lenses. { i'lek,trän 'mī·krə,skōp }

electron mirror See dynode. { i'lek,trän mir·ər }

electron multiplier |ELECTR| An electron-tube structure which produces current amplification; an electron beam containing the desired signal is reflected in turn from the surfaces of each of a series of dynodes, and at each reflection an impinging electron releases two or more secondary electrons, so that the beam builds up in strength. Also known as multiplier. { i'lek ,trän 'məl·tə,plī·ər }

electron-multiplier phototube See multiplier phototube. { i'lek,trän 'məl·tə,plī·ər 'phōd·ō,tüb }

electronographic tube |ELECTR| An image tube used in astronomy in which the electron image formed by the tube is recorded directly upon film or plates. { i,lek¦trän·ə¦graf·ik 'tüb }

electronography |ELECTR| The use of image tubes to form intensified electron images of astronomical objects and record them directly on film or plates. { i,lek·trə'näg·rə·fē }

electronoluminescence See cathodoluminescence. { i,lek¦trän·ə,lü·mə'nes·əns }

electron optics |ELECTR| The study of the motion of free electrons under the influence of electric and magnetic fields. { i'lek,trän 'äp·tiks }

electron-ray indicator See cathode-ray tuning indicator. { i'lek,trän ,rā 'in·də,kād·ər }

electron-ray tube See cathode-ray tube. { i'lek ,trän ,rā ,tüb }

electron refraction |ELECTR| The bending of an electron beam passing from one region to another of different electric potential. { i¦lek ,trän ri'frak·shən }

electron-stream potential |ELECTR| At any point in an electron stream, the time average of the potential difference between that point and the electron-emitting surface. { i'lek,trän ,strēm pə'ten·chəl }

electron-stream transmission efficiency |ELECTR| At an electrode through which the electron stream (beam) passes, the ratio of the average stream current through the electrode to the stream current approaching the electrode. { i'lek,trän ,strēm tranz'mish·ən ə'fish·ən·sē }

electron telescope [ELECTR] A telescope in which an infrared image of a distant object is focused on the photosensitive cathode of an image converter tube; the resulting electron image is enlarged by electron lenses and made visible by a fluorescent screen. { i'lek,trän 'tel·ə,skōp }

electron tube [ELECTR] An electron device in which conduction of electricity is provided by electrons moving through a vacuum or gaseous medium within a gastight envelope. Also known as radio tube; tube; valve (British usage). { i'lek ,trän ,tüb }

electron-tube amplifier [ELECTR] An amplifier in which electron tubes provide the required increase in signal strength. { i'lek,trän ,tüb 'am·plə,fī·ər }

electron-tube generator [ELECTR] A generator in which direct-current energy is converted to radiofrequency energy by an electron tube in an oscillator circuit. { i'lek,trän ,tüb 'jen·ə,rād·ər }

electron-tube heater See heater. { i'lek,trän ,tüb 'hēd·ər }

electron-tube static characteristic [ELECTR] Relation between a pair of variables such as electrode voltage and electrode current with all other voltages maintained constant. { i'lek,trän ,tüb 'stad·ik kar·ik·tə'ris·tik }

electron voltaic effect [ELECTR] Sensitivity of photovoltaic cells to electron bombardment. { i¦lek,trän vōl'tā·ik i,fekt }

electrooptical birefringence See electrooptical Kerr effect. { i,lek·trō'äp·tə·kəl bī·ri'frin·jəns }

electrooptical character recognition See optical character recognition. { i,lek·trō'äp·tə·kəl 'kar·ik·tər ,rek·ig,nish·ən }

electrooptical Kerr effect [OPTICS] Birefringence induced by an electric field. Also known as electrooptical birefringence; Kerr effect. { i,lek·trō'äp·tə·kəl 'kər i,fekt }

electrooptical modulator [COMMUN] An optical modulator in which a Kerr cell, an electrooptical crystal, or other signal-controlled electrooptical device is used to modulate the amplitude, phase, frequency, or direction of a light beam. { i,lek·trō'äp·tə·kəl 'mäj·ə,lād·ər }

electrooptic material [OPTICS] A material in which the indices of refraction are changed by an applied electric field. { i,lek·trō'äp·tik mə'tir·ē·əl }

electrooptic radar [ENG] Radar system using electrooptic techniques and equipment instead of microwave to perform the acquisition and tracking operation. { i,lek·trō'äp·tik 'rā,där }

electrooptics [OPTICS] The study of the influence of an electric field on optical phenomena, as in the electrooptical Kerr effect and the Stark effect. Also known as optoelectronics. { i,lek·trō'äp·tiks }

electroosmotic driver [ELECTR] A type of solion for converting voltage into fluid pressure, which uses depolarizing electrodes sealed in an electrolyte and operates through the streaming potential effect. Also known as micropump. { i¦lek·trō·äz'mäd·ik 'drīv·ər }

electrophorus [ELEC] A device used to produce electric charges; it consists of a hard-rubber disk, which is negatively charged by rubbing with fur, and a metal plate, held by an insulating handle, which is placed on the disk; the plate is then touched with a grounded conductor, so that negative charge is removed and the plate has net positive charge. { i,lek'trä·fə·rəs }

electrophotoluminescence [ELECTR] Emission of light resulting from application of an electric field to a phosphor which is concurrently, or has been previously, excited by other means. { i,lek·trō¦fōd·ō,lü·mə'nes·ə ns }

electropositive [ELEC] 1. Carrying a positive electric charge. 2. Capable of acting as the positive electrode in an electric cell. { i,lek·trə 'päz·əd·iv }

electroresistive effect [ELECTR] The change in the resistivity of certain materials with changes in applied voltage. { i¦lek·tro·ri'zis·tiv i,fekt }

electroscope [ENG] An instrument for detecting an electric charge by means of the mechanical forces exerted between electrically charged bodies. { i'lek·trə,skōp }

electrosensitive recording [ELECTR] Recording in which the image is produced by passing electric current through the record sheet. { i¦lek·trō'sen·səd·iv ri'kórd·iŋ }

electrostatic [ELEC] Pertaining to electricity at rest, such as an electric charge on an object. { i,lek·trə'stad·ik }

electrostatic accelerator [ELECTR] Any instrument which uses an electrostatic field to accelerate charged particles to high velocities in a vacuum. { i,lek·trə'stad·ik ak'sel·ə,rād·ər }

electrostatic actuator See actuator. { i,lek·trə 'stad·ik 'ak·chə,wād·ər }

electrostatic analyzer [ELECTR] A device which filters an electron beam, permitting only electrons within a very narrow velocity range to pass through. { i,lek·trə'stad·ik 'an·ə,līz·ər }

electrostatic attraction See Coulomb attraction. { i,lek·trə'stad·ik ə'trak·shən }

electrostatic cathode-ray tube [ELECTR] A cathode-ray tube in which electrostatic deflection is used on the electron beam. { i,lek·trə'stad·ik ¦kath,ōd 'rā ,tüb }

electrostatic deflection [ELECTR] The deflection of an electron beam by means of an electrostatic field produced by electrodes on opposite sides of the beam; used chiefly in cathode-ray tubes for oscilloscopes. { i,lek·trə'stad·ik di'flek·shən }

electrostatic detection [ELECTR] The detection and location of any type of solid body, such as a mineral deposit or a mine, by measuring the associated electrostatic field which arises spontaneously or is induced by the detection equipment. { i,lek·trə'stad·ik di'tek·shən }

electrostatic energy [ELEC] The potential energy which a collection of electric charges possesses

electrostatic error

by virtue of their positions relative to each other. { i,lek·trə'stad·ik 'en·ər·jē }

electrostatic error See antenna effect. { i,lek·trə'stad·ik 'er·ər }

electrostatic field [ELEC] A time-independent electric field, such as that produced by stationary charges. { i,lek·trə'stad·ik 'fēld }

electrostatic focus [ELECTR] Production of a focused electron beam in a cathode-ray tube by the application of an electric field. { i,lek·trə'stad·ik 'fō·kəs }

electrostatic force [ELEC] Force on a charged particle due to an electrostatic field, equal to the electric field vector times the charge of the particle. { i,lek·trə'stad·ik ' fȯrs }

electrostatic force microscopy [ENG] The use of an atomic force microscope to measure electrostatic forces from electric charges on a surface. { i¦lek·trə,stad·ik ¦fȯrs mī'krä·skə·pē }

electrostatic generator [ELEC] Any machine which produces electric charges by friction or (more commonly) electrostatic induction. { i,lek·trə'stad·ik 'jen·ə,rād·ər }

electrostatic induction [ELEC] The process of charging an object electrically by bringing it near another charged object, then touching it to ground. Also known as induction. { i,lek·trə'stad·ik in'dək·shən }

electrostatic instrument [ELEC] A meter that depends for its operation on the forces of attraction and repulsion between electrically charged bodies. { i,lek·trə'stad·ik 'in·strə·mənt }

electrostatic interactions See Coulomb interactions. { i,lek·trə'stad·ik int·ə'rak·shənz }

electrostatic lens [ELECTR] An arrangement of electrostatic fields which acts upon beams of charged particles similar to the way a glass lens acts on light beams. { i,lek·trə'stad·ik 'lenz }

electrostatic loudspeaker [ENG ACOUS] A loudspeaker in which the mechanical forces are produced by the action of electrostatic fields; in one type the fields are produced between a thin metal diaphragm and a rigid metal plate. Also known as capacitor loudspeaker. { i,lek·trə'stad·ik 'laud ,spēk·ər }

electrostatic memory See electrostatic storage. { i'lek·trə,stad·ik 'mem·rē }

electrostatic microphone See capacitor microphone. { i'lek·trə,stad·ik 'mī·krə,fōn }

electrostatic octupole lens [ELECTR] A device for controlling beams of electrons or other charged particles, consisting of eight electrodes arranged in a circular pattern with alternating polarities; commonly used to correct aberrations of quadrupole lens systems. { i'lek·trə,stad·ik ¦äk·tə,pōl 'lenz }

electrostatic potential See electric potential. { i'lek·trə,stad·ik pə'ten·chəl }

electrostatic precipitator [ENG] A device which removes dust or other finely divided particles from a gas by charging the particles inductively with an electric field, then attracting them to highly charged collector plates. Also known as precipitator. { i'lek·trə,stad·ik prə'sip·ə ,tād·ər }

electrostatic quadrupole lens [ELECTR] A device for focusing beams of electrons or other charged particles, consisting of four electrodes arranged in a circular pattern with alternating polarities. { i'lek·trə,stad·ik ¦kwä·drə,pōl 'lenz }

electrostatic repulsion See Coulomb repulsion. { i'lek·trə,stad·ik ri'pəl·shən }

electrostatics [ELEC] The study of electric charges at rest, their electric fields, and potentials. { i,lek·trə'stad·iks }

electrostatic scanning [ELECTR] Scanning that involves electrostatic deflection of an electron beam. { i'lek·trə,stad·ik 'skan·iŋ }

electrostatic shielding [ELEC] The placing of a grounded metal screen, sheet, or enclosure around a device or between two devices to prevent electric fields from interacting. { i'lek·trə ,stad·ik 'shēld·iŋ }

electrostatic storage [ELECTR] A storage in which information is retained as the presence or absence of electrostatic charges at specific spot locations, generally on the screen of a special type of cathode-ray tube known as a storage tube. Also known as electrostatic memory. { i'lek·trə,stad·ik 'stȯr·ij }

electrostatic storage tube See storage tube. { i'lek·trə,stad·ik 'stȯr·ij ,tüb }

electrostatic stress [ELEC] An electrostatic field acting on an insulator, which produces polarization in the insulator and causes electrical breakdown if raised beyond a certain intensity. { i'lek·trə,stad·ik 'stres }

electrostatic transducer [ENG ACOUS] A transducer consisting of a fixed electrode and a movable electrode, charged electrostatically in opposite polarity; motion of the movable electrode changes the capacitance between the electrodes and thereby makes the applied voltage change in proportion to the amplitude of the electrode's motion. Also known as condenser transducer. { i'lek·trə,stad·ik tranz'dü·sər }

electrostatic tweeter [ENG ACOUS] A tweeter loudspeaker in which a flat metal diaphragm is driven directly by a varying high voltage applied between the diaphragm and a fixed metal electrode. { i'lek·trə,stad·ik 'twēd·ər }

electrostatic units [ELEC] A centimeter-gram-second system of electric and magnetic units in which the unit of charge is that charge which exerts a force of 1 dyne on another unit charge when separated from it by a distance of 1 centimeter in vacuum; other units are derived from this definition by assigning unit coefficients in equations relating electric and magnetic quantities. Abbreviated esu. { i'lek·trə,stad·ik 'yü·nəts }

electrostatic voltmeter [ENG] A voltmeter in which the voltage to be measured is applied between fixed and movable metal vanes; the resulting electrostatic force deflects the movable vane against the tension of a spring. { i'lek·trə ,stad·ik 'vōlt,mēd·ər }

electrostatic wattmeter [ENG] An adaptation of a quadrant electrometer for power measurements in which two quadrants are charged by

202

the voltage drop across a noninductive shunt resistance through which the load current passes, and the line voltage is applied between one of the quadrants and a moving vane. { i'lek·trə,stad·ik 'wät,mēd·ər }

electrostriction transducer [ENG ACOUS] A transducer which depends on the production of an elastic strain in certain symmetric crystals when an electric field is applied, or, conversely, which produces a voltage when the crystal is deformed. Also known as ceramic transducer. { i¦lek·trō'strik·shən tranz'dü·sər }

electrothermal ammeter See thermoammeter. { i¦lek·trō'thər·məl 'a,med·ər }

electrothermal energy conversion [ENG] The direct conversion of electric energy into heat energy, as in an electric heater. { i¦lek·trō'thər·məl 'en·ər·jē kən,vər·zhən }

electrothermal process [ENG] Any process which uses an electric current to generate heat, utilizing resistance, arcs, or induction; used to achieve temperatures higher than can be obtained by combustion methods. { i¦lek·trō'thər·məl 'präs·əs }

electrothermal recording [ELECTR] Type of electrochemical recording, used in facsimile equipment, wherein the chemical change is produced principally by signal-controlled thermal action. { i¦lek·trō'thər·məl ri'kȯrd·iŋ }

electrothermal voltmeter [ENG] An electrothermal ammeter employing a series resistor as a multiplier, thus measuring voltage instead of current. { i¦lek·trō'thər·məl 'vōlt,mēd·ər }

element [COMPUT SCI] A circuit or device performing some specific elementary data-processing function. [ELECTROMAG] Radiator, active or parasitic, that is a part of an antenna. { 'el·ə·mənt }

elemental area See picture element. { ,el·ə'ment·əl 'er·ē·ə }

elementary item [COMPUT SCI] An item considered to have no subordinate item in the COBOL language. { ,el·ə'men·trē ,īd·əm }

elementary stream [COMMUN] A generic term for one of the coded video, coded audio, or other coded bit streams in a digital television system. { ,el·ə'mən·trē 'strēm }

elevation angle [ELECTROMAG] The angle that a radio, radar, or other such beam makes with the horizontal. { ,el·ə'vā·shən ,aŋ·gəl }

elevation-angle error [ELECTROMAG] In radar, the error in the measurement of the elevation angle of a target resulting from the vertical bending or refraction of radio energy in traveling through the atmosphere. Also known as elevation error. { ,el·ə'vā·shən ,aŋ·gəl ,er·ər }

elevation error See elevation-angle error. { ,el·ə'vā·shən ,er·ər }

ELF See extremely low frequency.

elimination factor [COMPUT SCI] In information retrieval, the ratio obtained in dividing the number of documents that have not been retrieved by the total number of documents in the file. { ə,lim·ə'nā·shən ,fak·tər }

eliminator [ELECTR] Device that takes the place of batteries, generally consisting of a rectifier

operating from alternating current. { ə'lim·ə ,nād·ər }

E lines [ELEC] Contour lines of constant electrostatic field strength referred to some reference base. { 'ē ,līnz }

ellipsoidal floodlight [ELEC] A lighting unit used in theatrical lighting consisting of an ellipsoidal reflector with fixed spacing and a lamp; power requirements are 250–5000 watts and the reflector diameter is 10–24 inches (25–61 centimeters). Also known as scoop. { ə,lip'sȯid·əl 'flȯd·līt }

ellipsoidal spotlight [ELEC] A lighting unit consisting of a reflector, lamp, single or multiple lens system, and framing device; power requirements are 250-2000 watts. { ə,lip'sȯid·əl 'spät,līt }

elliptical system [ENG] A tracking or navigation system where ellipsoids of position are determined from time or phase summation relative to two or more fixed stations which are the focuses for the ellipsoids. { ə'lip·tə·kəl 'sis·təm }

elliptic-integral filter [ELECTR] An electronic filter whose gain characteristic has both an equal-ripple shape in the pass-band and equal minima of attenuation in the stop-band. Also known as Cauer filter. { ə,lip·tik ,int·ə·grəl 'fil·tər }

ellipticity See axial ratio. { ē,lip'tis·əd·ē }

elongation [COMMUN] The extension of the envelope of a signal due to delayed arrival of multipath components. { ē,lȯŋ'gä·shən }

ELSE instruction [COMPUT SCI] An instruction in a programming language which tells a program what actions to take if previously specified conditions are not met. { 'els in,strək·shən }

ELSE rule [COMPUT SCI] A convention in decision tables which spells out which action to take in the case specified conditions are not met. { 'els ,rül }

e-mail See electronic mail. { 'ē,māl }

emanation security [ELECTR] The protection resulting from all measures designed to deny unauthorized persons information of value which might be derived from unintentional emissions from other than telecommunications systems. { ,em·ə'nā·shən sə'kyùr·ə·dē }

embedded command [COMPUT SCI] In word processing, a code inserted in a text document that instructs the printer to change its print attributes. { em¦bed·əd kə'mand }

embedded pointer [COMPUT SCI] A pointer set in a data record instead of in a directory. { em'bed·əd 'pȯint·ər }

embedded system [COMPUT SCI] A computer system that cannot be programmed by the user because it is preprogrammed for a specific task and embedded within the equipment which it serves. { em'bed·əd 'sis·təm }

embossed plate printer [COMPUT SCI] In character recognition, a data preparation device which accomplishes printing by allowing a raised character behind the paper to push the paper against the printing ribbon in front of the paper. { em ¦bäst ¦plāt 'print·ər }

embossing stylus [ENG ACOUS] A recording stylus with a rounded tip that forms a groove by displacing material in the recording medium. { em'bäs·iŋ ,stī·ləs }

emergency alert system |COMMUN| A system of radio, television, and cable networks and wire services for communicating with the general public in emergency situations. { ə,mər·jən·sē ə'lərt ,sis·təm }

emergency broadcast system |COMMUN| A system of broadcast stations and interconnecting facilities authorized by the U.S. Federal Communications Commission to operate in a controlled manner during a war, threat of war, state of public peril or disaster, or other national emergency. { ə'mər·jən·sē 'bröd,kast ,sis·təm }

emergency power supply |ELEC| A source of power that becomes available, usually automatically, when normal power line service fails. { ə'mər·jən·sē 'pau·ər sə,plī }

emergency radio channel |COMMUN| Any radio frequency reserved for emergency use, particularly for distress signals. { ə'mər·jən·sē 'rād·ē·ō ,chan·əl }

emergency receiver |COMMUN| Receiver immediately available in a station for emergency communications and capable of being energized by self-contained or emergency power supply. { ə'mər·jən·sē ri'sē·vər }

emi See electromagnetic interference.

emission |ELECTROMAG| Any radiation of energy by means of electromagnetic waves, as from a radio transmitter. { i'mish·ən }

emission characteristics |ELECTR| Relation, usually shown by a graph, between the emission and a factor controlling the emission, such as temperature, voltage, or current of the filament or heater. { i'mish·ən ,kar·ik·tə'ris·tiks }

emission electron microscope |ELECTR| An electron microscope in which thermionic, photo, secondary, or field electrons emitted from a metal surface are projected on a fluorescent screen, with or without focusing. { i'mish·ən i ¦lek,trän 'mī·krə,skōp }

emission security |ELECTR| That component of communications security which results from all measures taken to protect any unintentional emissions of a telecommunications system from any form of exploitation other than cryptanalysis. { i'mish·ən sə'kyür·əd·ē }

emitter |ELECTR| A transistor region from which charge carriers that are minority carriers in the base are injected into the base, thus controlling the current flowing through the collector; corresponds to the cathode of an electron tube. Symbolized E. Also known as emitter region. { i'mid·ər }

emitter barrier |ELECTR| One of the regions in which rectification takes place in a transistor, lying between the emitter region and the base region. { i'mid·ər ,bar·ē·ər }

emitter bias |ELECTR| A bias voltage applied to the emitter electrode of a transistor. { i'mid·ər ,bī·əs }

emitter-coupled logic |ELECTR| A form of current-mode logic in which the emitters of two transistors are connected to a single current-carrying resistor in such a way that only one transistor conducts at a time. Abbreviated ECL. { i'mid·ər ¦kəp·əld 'läj·ik }

emitter follower |ELECTR| A grounded-collector transistor amplifier which provides less than unity voltage gain but high input resistance and low output resistance, and which is similar to a cathode follower in its operations. { i'mid·ər ,fäl·ə·wər }

emitter junction |ELECTR| A transistor junction normally biased in the low-resistance direction to inject minority carriers into a base. { i'mid·ər ,jəŋk·shən }

emitter region See emitter. { i'mid·ər ,rē·jən }

emitter resistance |ELECTR| The resistance in series with the emitter lead in an equivalent circuit representing a transistor. { i'mid·ər ri ,zis·təns }

EMM See entitlement management message.

E mode See transverse magnetic mode. { 'ē ,mōd }

EMOSFET See electrolyte-MOSFET.

emoticon |COMPUT SCI| A combination of keyboard characters that depicts a sideways face whose expression conveys an emotional response. Also known as smiley. { i'mōd·ə,kän }

emphasizer See preemphasis network. { 'em·fə ,sīz·ər }

empty-cell process |ENG| A wood treatment in which the preservative coats the cells without filling them. { 'em·tē ,sel 'präs·əs }

empty medium |COMPUT SCI| A material which has been prepared to have data recorded on it by the entry of some preliminary data, such as feed holes punched in a paper tape or header labels written on a magnetic tape; in contrast to a virgin medium. { 'em·tē 'mēd·ē·əm }

empty shell |COMPUT SCI| A room that has been fully prepared for the installation of computer and data-processing equipment. { 'em·tē 'shel }

emulation |COMPUT SCI| Imitation of one computer system by another so that the latter functions in exactly the same way and runs the same programs. { ,em·yə'lā·shən }

emulation mode |COMPUT SCI| A method of operation in which a computer actually executes the instructions of a different (simpler) computer, in contrast to normal mode. { ,em·yə'lā·shən ,mōd }

emulator |COMPUT SCI| The microprogram-assisted macroprogram which allows a computer to run programs written for another computer. { 'em·yə,lād·ər }

emulator circuit |COMPUT SCI| A circuit built into a computer's control section to enable it to process instructions that were written for another computer. { 'em·yə,lād·ər ,sər·kət }

enable |COMPUT SCI| **1.** To authorize an activity which would otherwise be suppressed, such as to write on a tape. **2.** To turn on a computer system or a piece of equipment. |ELECTR| To initiate the operation of a device or circuit by applying a trigger signal or pulse. { ə'nā·bəl }

enabled instruction |COMPUT SCI| An instruction in a program in data flow language, all of whose input values are present, so that the instruction may be carried out. { ə'nā·bəld in'strək·shən }

enabling pulse |ELECTR| A pulse that prepares a circuit for some subsequent action. { ə'nāb·liŋ ‚pəls }

encipher |COMMUN| To convert a plain-text message into unintelligible language by means of a cryptosystem. Also known as encrypt. { en'sī·fər }

enciphered facsimile communications |COMMUN| Communications in which security is accomplished by mixing pulses produced by a key generator with the output of the facsimile converter; plain text is recovered by subtracting the identical key at the receiving terminal; unauthorized listeners are unable to reconstruct the plain text unless they have an identical key generator and the daily key setting. { en'sī·fərd fak'sim·ə·lē kə ‚myün·ə'kā·shənz }

enclave *See* domain. { 'än‚klāv }

enclosed arc lamp |ELEC| An arc lamp in which the arc produced by carbon electrodes is protected from the atmosphere by a translucent enclosure. { in¦klōzd 'ärk ‚lamp }

encode |COMMUN| To express given information by means of a code. |COMPUT SCI| To prepare a routine in machine language for a specific computer. { en'kōd }

encoded abstract |COMPUT SCI| An abstract prepared to be scanned by automatic electronic machines. { en'kōd·əd 'ab‚strakt }

encoded question |COMPUT SCI| A question set up and encoded in the form appropriate for operating, programming, or conditioning a searching device. { en'kōd·əd 'kwes·chən }

encoder |COMMUN| An embodiment of an encoding process. |COMPUT SCI| In character recognition, that class of printer which is usually designed for the specific purpose of printing a particular type font in predetermined positions on certain size forms. |ELECTR| **1.** In an electronic computer, a network or system in which only one input is excited at a time and each input produces a combination of outputs. **2.** *See* matrix. { en'kōd·ər }

encoding strip |COMPUT SCI| In character recognition, the area reserved for the inscription of magnetic-ink characters, as in bank checks. { en'kōd·iŋ 'strip }

encrypt *See* encipher. { en'kript }

encryption |COMPUT SCI| The coding of a clear text message by a transmitting unit so as to prevent unauthorized eavesdropping along the transmission line; the receiving unit uses the same algorithm as the transmitting unit to decode the incoming message. { en'krip·shən }

end-around carry |COMPUT SCI| A carry from the most significant digit place to the least significant digit place. { ¦end ə¦raúnd 'kar·ē }

end-around shift *See* cyclic shift. { ¦end ə¦raúnd 'shift }

end cell |ELEC| One of a group of cells in series with a storage battery, which can be switched in to maintain the output voltage of the battery when it is not being charged. { 'end ‚sel }

end-cell rectifier |ELECTR| Small trickle charge rectifier used to maintain voltage of the storage battery end cells. { 'end ‚sel 'rek·tə‚fī·ər }

end distortion |COMMUN| The displacement of trailing edges of marking pulses transmitted over a teletypewriter circuit relative to the leading edge of the start pulse. { 'end di‚stòr·shən }

end effect |ELECTROMAG| The effect of capacitance at the ends of an antenna; it requires that the actual length of a half-wave antenna be about 5% less than a half wavelength. { 'end i‚fekt }

end effector |CONT SYS| The component of a robot that comes into contact with the workpiece and does the actual work on it. Also known as hand { 'end i‚fek·tər }

end-fire antenna *See* end-fire array. { 'end ‚fīr an'ten·ə }

end-fire array |ELECTROMAG| A linear array whose direction of maximum radiation is along the axis of the array; it may be either unidirectional or bidirectional; the elements of the array are parallel and in the same plane, as in a fishbone antenna. Also known as end-fire antenna. { 'end ‚fīr ə'rā }

end instrument |ELECTR| A pickup used in telemetering to convert a physical quantity to an inductance, resistance, voltage, or other electrical quantity that can be transmitted over wires or by radio. { 'end ‚in·strə·mənt }

endless loop |COMPUT SCI| A sequence of instructions in a computer program that is repeated over and over without end, due to a mistake in the programming. { 'end·ləs 'lüp }

end loss |ELECTROMAG| The difference between the actual and the effective lengths of a radiating antenna element. { 'end ‚lòs }

end mark |COMPUT SCI| A mark which signals the end of a unit of information. { 'end ‚märk }

end-of-arm speed |CONT SYS| The speed at which an end effector arrives at its desired position. { ¦end əv ¦ärm 'spēd }

end-of-block character |COMPUT SCI| A character that indicates the completion of a block of code. { ¦end əv ¦bläk 'kar·ik·tər }

end-of-data mark |COMPUT SCI| A character or word signaling the end of all data held in a particular storage unit. { ¦end əv 'dad·ə ‚märk }

end-of-field mark |COMPUT SCI| A data item signaling the end of a field of data, generally a variable-length field. { ¦end əv 'fēld ‚märk }

end of file |COMPUT SCI| **1.** Termination or point of completion of a quantity of data; end of file marks are used to indicate this point. **2.** Automatic procedures to handle tapes when the end of an input or output tape is reached; a reflective spot, called a record mark, is placed on the physical end of the tape to signal the end. { ¦end əv 'fīl }

end-of-file gap |COMPUT SCI| A gap of precise dimension to indicate the end of a file on tape. Abbreviated EOF gap. { ¦end əv 'fīl ‚gap }

end-of-file indicator *See* end-of-file mark. { ¦end əv 'fīl 'in·də‚kād·ər }

end-of-file mark |COMPUT SCI| A control character which signifies that the last record of a file has

been read. Also known as end-of-file indicator.
{ ¦end əv 'fīl ¸märk }

end-of-file routine |COMPUT SCI| A program which checks that the contents of a file read into the computer were correctly read; may also start the rewind procedure. { ¦end əv 'fīl rü¸tēn }

end-of-file spot |COMPUT SCI| A reflective piece of tape indicating the end of the tape. { ¦end əv 'fīl ¸spät }

end-of-message |COMMUN| A character or series of characters signifying the end of a message or record, such as a message sent by teletypewriter. { ¦end əv 'mes·ij }

end-of-record gap |COMPUT SCI| A gap of precise dimension (shorter than the end-of-file gap) which indicates the physical end of a record on a magnetic tape. Abbreviated EOR gap. { ¦end əv 're·kərd ¸gap }

end-of-record word |COMPUT SCI| The last word in a record, usually written in a special format that enables identification of the end of the record. { ¦end əv 're·kərd ¸wərd }

end-of-run routine |COMPUT SCI| A routine that carries out various housekeeping operations such as rewinding tapes and printing control totals before a run is completed. { ¦end əv 'rən rü¸tēn }

end-of-tape routine |COMPUT SCI| A program which is brought into play when the end of a tape is reached; may involve a series of validity checks and initiate the tape rewind. { ¦end əv 'tāp rü ¸tēn }

end-of-transmission card |COMMUN| Last card of each message; used to signal the end of a transmission and contains the same information as the header card, plus additional data for traffic analysis. { ¦end əv tranz'mish·ən ¸kärd }

end-of-transmission recognition |COMPUT SCI| The capability of a computer to recognize the end of transmission of a data string even if the buffer area is not filled. { ¦end əv tranz'mish·ən rek·ig ¸nish·ən }

endoradiosonde |ENG| A miniature battery-powered radio transmitter encapsulated like a pill, designed to be swallowed for measuring and transmitting physiological data from the gastrointestinal tract. { ¦en·dō'rād·ē·ō¸sänd }

endorser |COMPUT SCI| A special feature available on most magnetic-ink character-recognition readers that imprints a bank's endorsement on successful document reading. { en'dor·sər }

end point |COMPUT SCI| In vector graphics, one of the two ends of a line or vector. |CONT SYS| The point at which a robot stops along its path of motion. { 'end ¸point }

end-point rigidity |CONT SYS| The resistance of a robot to further movement after it has reached its end point. { 'en ¸point ri'jid·əd·ē }

end section |COMMUN| Additional portion of switchboard added to each end of a large multiple switchboard and used to extend some of the trunks or locals to these end positions to place all jacks within easy reach of the first and last operator. Also known as head section. { 'end ¸sek·shən }

end sentinel |COMPUT SCI| A character that indicates the end of a message or record. { 'end ¸sent·nəl }

end-to-end encryption |COMMUN| Encryption of a message at its point of origination so that it travels in encrypted form all the way to its destination. { ¦end·tü¦end in'krip·shən }

end user |COMPUT SCI| The person for whom the output of a computer is ultimately intended. { 'end ¸yüz·ər }

energized |ELEC| Electrically connected to a voltage source. Also known as alive; hot; live. { 'en·ər¸jīzd }

energy efficiency ratio |ELEC| A value that represents the relative electrical efficiency of air conditioners; it is the quotient obtained by dividing Btu-per-hour output by electrical-watts input during cooling. { 'en·ər·jē i'fish·ən·se ¸rā·shō }

energy of a charge |ELEC| Charge energy measured in ergs according to the equation $E = QV$, where Q is the charge and V is the potential in electrostatic units. { 'en·ər·jē əv ə 'chärj }

engineering channel circuit |COMMUN| Auxiliary circuit or channel (radio or wire) for use by operating or maintenance personnel for communications incident to the establishment, operation, maintenance, and control of communications facilities. { ¸en·jə'nir·iŋ 'chan·əl ¸sər·kət }

engineering time |COMPUT SCI| The nonproductive time of a computer, reserved for maintenance and servicing. { ¸en·jə'nir·iŋ ¸tīm }

engine starter |ELEC| The electric motor in the electrical system of an automobile that cranks the engine for starting. Also known as starter; starting motor. { 'en·jən ¸stärd·ər }

enhanceable language |COMPUT SCI| A computer language that has a modest degree of semantic extensibility. { en¦han·sə·bəl 'laŋ·gwij }

enhanced carrier demodulation |COMMUN| Amplitude demodulation system in which a synchronized local carrier of proper phase is fed into the demodulator to reduce demodulation distortion. { en¦hanst 'kar·ē·ər dē¸mäj·ə'lā·shən }

enhanced small device interface |COMPUT SCI| A standard method of connecting disk and tape drives to computers which allows for the transfer of 1–3 megabytes per second from disk drives holding up to 1 gigabyte of storage. Abbreviated ESDI. { en¦hanst ¦smól di¸vīs 'in·tər¸fās }

enhancement |COMPUT SCI| A substantial increase in the capabilities of hardware or software. |ELECTR| An increase in the density of charged carriers in a particular region of a semiconductor. { en'hans·mənt }

enhancement mode |ELECTR| Operation of a field-effect transistor in which no current flows when zero gate voltage is applied, and increasing the gate voltage increases the current. { en'hans·mənt ¸mōd }

enhancement-mode high-electron-mobility transistor |ELECTR| A high-electron-mobility transistor in which application of a positive bias to the gate electrode is required for current to flow between the source and drain electrodes.

Abbreviated E-HEMT. { en'hans·mənt ¦mōd 'hī i¦lek,trän mō¦bil·əd·ē tran'zis·tər }

enhancement-mode junction field-effect transistor [ELECTR] A type of gallium arsenide field-effect transistor in which the gate consists of the junction between the *n*-type gallium arsenide forming the conducting channel and *p*-type material implanted under a metal electrode. Abbrevate E-JFET. { en'hans·mənt ¦mōd 'jəŋk·shən 'fēld i,fekt tran'zis·tər }

ENIAC [COMPUT SCI] The first digital computer in the modern sense of the word, built 1942–1945. Derived from Electronic Numerical Integrator and Calculator. { 'ē·nē·ak }

E notation [COMPUT SCI] A type of scientific notation in which the phrase "times 10 to the power of" is replaced by the letter E; for example, 3.1×10^7 is written 3.1E+7 and 5.1×10^{-9} is written 5.1E−9. { 'ē nō,tā·shən }

enqueue [ENG] To place a data item in a queue. { en'kyü }

enquiry character [COMPUT SCI] A control character used to request a response from receiving equipment. { in'kwīr·ē ,kar·ik·tər }

enter key [COMPUT SCI] A key on a computer keyboard that corresponds to the return key on a typewriter and usually signals the computer to act on the information just entered on the keyboard. { 'en·tər ,kē }

entitlement control message [COMMUN] Private conditional access information which specifies control words and possibly other stream-specific, scrambling, or control parameters. { in'tī·təl·mənt kən'trōl ,məs·ij }

entitlement management message [COMMUN] Private conditional access information which specifies the authorization level or the services of specific decoders; addrressed to single decoders or groups of decoders. { in'tī·təl·mənt 'man·ij·mənt ,məs·ij }

entity See record. { 'ent·ə·tē }

entity type [COMPUT SCI] A particular kind of file in a database, such as an employee, customer, or product file. { 'ent·ə·tē ,tīp }

entrance [COMPUT SCI] The location of a program or subroutine at which execution is to start. Also known as entry point. { 'en·trəns }

entrance cable [ELEC] Cable that brings power from an outside power line into a building. { 'en·trəns ,kā·bəl }

entropy [COMMUN] A measure of the absence of information about a situation, or, equivalently, the uncertainty associated with the nature of a situation. { 'en·trə·pē }

entropy coding [COMMUN] Variable-length lossless coding of the digital representation of a signal to reduce redundancy. { 'en·trə·pē ,kōd·iŋ }

entry [COMPUT SCI] Input data fed during the execution of a program by means of a terminal. { 'en·trē }

entry block [COMPUT SCI] The area of main memory reserved for the data which will be introduced at execution time. { 'en·trē ,bläk }

entry condition [COMPUT SCI] A requirement that must be met before a program or routine can be entered by a computer program. Also known as initial condition. { 'en·trē kən,dish·ən }

entry instruction [COMPUT SCI] The first instruction to be executed in a subroutine. { 'en·trē in ,strək·shən }

entry point [COMMUN] A point in a coded bit stream after which a decoder can become properly initialized and commence syntactically correct decoding. The first transmitted picture after an entry point is either an I-picture or a P-picture. If the first transmitted picture is not an I-picture, the decoder may produce one or more pictures during acquisition. [COMPUT SCI] See entrance. { 'en·trē ,pôint }

entry portion [COMPUT SCI] The right-hand portion of a decision table, which comprises the condition entries and action entries, and whose columns are the decision rules. { 'en·trē ,pòr·shən }

entry sorting [COMPUT SCI] A method of internal sorting in which records or blocks of records are placed, one at a time, in a buffer area and then integrated into the sorted list before the next record is placed in the buffer. { 'en·trē ,sórd·iŋ }

envelope [COMMUN] A curve drawn to pass through the peaks of a graph, such as that of a moduated radio-frequency carrier signal. [ENG] The glass or metal housing of an electron tube or the glass housing of an incandescent lamp. { 'en·və,lōp }

envelope delay [COMMUN] The time required for the envelope of a modulated signal to travel between two points in a system. { 'en·və,lōp di,lā }

envelope delay distortion See delay distortion. { 'en·və,lōp di,lā di'stór·shən }

envelope detector See detector. { 'en·və,lōp di ,tek·tər }

environment [COMPUT SCI] The computer system in which an applications program is running, including the hardware and system software. { in'vī·ərn·mənt *or* in'vī·rən·ment }

environmental range [ENG] The range of environment throughout which a system or portion thereof is capable of operation at not less than the specified level of reliability. { in¦vī·ərn ¦mənt·əl 'rānj }

environmental test [ENG] A laboratory test conducted to determine the functional performance of a component or system under conditions that simulate the real environment in which the component or system is expected to operate. { in¦vī·ərn¦mənt·əl 'test }

environment division |COMPUT SCI| The section of a program written in COBOL which defines the hardware and files to be used by the program. { in¦vī·ərn¦mənt di'vizh·ən }

environment pointer |COMPUT SCI| **1.** A component of a task descriptor that designates where the instructions and data code for the task are located. **2.** A control component element belonging to the stack model of block structure execution that points to the current environment. { in¦vī·ərn¦mənt ‚póint·ər }

environment simulator |ENG| Any machine or artificial device that simulates all or some of the attributes of an environment. { in¦vī·ərn¦mənt 'sim·yə‚lād·ər }

EOF gap See end-of-file gap. { ¦ē¦ō'ef ‚gap }

EOR gap See end-of-record gap. { ¦ē¦ō'är ‚gap }

EP See electronic protection.

epitaxial diffused-junction transistor |ELECTR| A junction transistor produced by growing a thin, high-purity layer of semiconductor material on a heavily doped region of the same type. { ‚ep·ə'tak·sē·əl də¦fyüzd ¦jəŋk·shən tran'zis·tər }

epitaxial diffused-mesa transistor |ELECTR| A diffused-mesa transistor in which a thin, high-resistivity epitaxial layer is deposited on the substrate to serve as the collector. { ‚ep·ə'tak·sē·əl də¦fyüzd ¦mā·sə tran'zis·tər }

epitaxial layer |SOLID STATE| A semiconductor layer having the same crystalline orientation as the substrate on which it is grown. { ‚ep·ə'tak·sē·əl ‚lā·ər }

epitaxial transistor |ELECTR| Transistor with one or more epitaxial layers. { ‚ep·ə'tak·sē·əl tran'zis·tər }

E-plane antenna |ELECTROMAG| An antenna which lies in a plane parallel to the electric field vector of the radiation that it emits. { 'ē ‚plān an‚ten·ə }

E-plane bend See E bend. { 'ē ‚plān ‚bend }

E-plane T junction |ELECTROMAG| Waveguide T junction in which the change in structure occurs in the plane of the electric field. Also known as series T junction. { 'ē ‚plañ 'tē ‚jəŋk·shən }

EPROM See erasable programmable read-only memory. { 'ē‚präm }

equal error rate |COMMUN| The error rate of a verification system when the operating threshold for the accept/reject decision is adjusted such that the probability of false acceptance and that of false rejection become equal. Abbreviated EER. { ¦ē·kwəl 'er·ər ‚rāt }

equality gate See equivalence gate. { ē'kwal·əd·ē ‚gāt }

equalization |ELECTR| The effect of all frequency-discriminating means employed in transmitting, recording, amplifying, or other signal-handling systems to obtain a desired overall frequency response. Also known as frequency-response equalization. { ‚ē·kwə·lə'zā·shən }

equalizer |ELECTR| A network designed to compensate for an undesired amplitude-frequency or phase-frequency response of a system or component; usually a combination of coils, capacitors, and resistors. Also known as equalizing circuit. { 'ē·kwə‚līz·ər }

equalizer brake See equalizer. { 'ē·kwə‚līz·ər ‚brāk }

equalizing bar See equalizer. { 'ē·kwə‚līz·iŋ ‚bär }

equalizing circuit See equalizer. { 'ēkwə‚līz·iŋ ‚sər·kət }

equalizing current |ELEC| Current that circulates between two parallel-connected compound generators to equalize their output. { 'ē·kwə‚līz·iŋ ‚kər·ənt }

equalizing pulses |ELECTR| In analog television, pulses at twice the line frequency, occurring just before and after the vertical synchronizing pulses, which minimize the effect of line frequency pulses on the interlace. { 'ē·kwə‚līz·iŋ ‚pəl·səs }

equal ripple |ELECTR| Property of an amplitude or phase characteristic whose local maxima all have the same value, and whose local minima all have the same value, within a specified frequency range. { ‚ē·kwəl 'rip·əl }

equal-zero indicator |COMPUT SCI| A circuit component which is on when the result of an operation is zero. { ¦ē·kwəl ¦zir·ō 'in·də‚kād·ər }

equation solver |COMPUT SCI| A machine, usually analog, for solving systems of simultaneous equations, which may be linear, nonlinear, or differential, and for finding roots of polynomials. { i'kwā·zhən ‚sälv·ər }

equiangular spiral antenna |ELECTROMAG| A frequency-independent broad-band antenna, cut from sheet metal, that radiates a very broad, circularly polarized beam on both sides of its surface; this bidirectional radiation pattern is its chief limitation. { ¦ē·kwē¦aŋ·gyə·lər ‚spī·rəl an'ten·ə }

equilibrium brightness |ELECTR| Viewing screen brightness occurring when a display storage tube is in a fully written condition. { ‚ēkwə'lib·rē·əm 'brīt·nəs }

equipment |ENG| One or more assemblies capable of performing a complete function. { ə'kwip·mənt }

equipment augmentation |COMPUT SCI| **1.** Procuring additional automatic data-processing equipment capability to accommodate increased work load within an established data system. **2.** Obtaining additional sites or locations. { ə'kwip·mənt ‚ȯg·mən'tā·shən }

equipment chain |ENG| Group of equipments that are functionally in series; the failure of one or more of the equipments results in loss of the function. { ə'kwip·mənt ‚chān }

equipment compatibility |COMPUT SCI| The ability of a device to handle data prepared or handled by other equipment, without alteration of the code or of the form of the data. { ə'kwip·mənt kəm‚pad·ə'bil·əd·ē }

equipment failure |COMPUT SCI| A fault in equipment that results in its improper behavior or prevents the execution of a job as scheduled. { ə'kwip·mənt ,fāl·yər }

equipotential cathode *See* indirectly heated cathode. { ¦e·kwə·pə'ten·chəl 'kath,ōd }

equipotential surface [ELEC] A surface on which the electric potential is the same at every point. { ¦e·kwə·pə'ten·chəl 'sər·fəs }

equisignal [COMMUN] **1.** Pertaining to two signals of equal intensity, used particularly with reference to the signals of a radio range station. **2.** Referring to a radio system in which two identifiable separate radio signals are received with the same intensity. { ¦e·kwə¦sig·nəl }

equisignal surface [ELECTROMAG] Surface around an antenna formed by all points at which, for transmission, the field strength (usually measured in volts per meter) is constant. { ¦e·kwə ¦sig·nəl ,sər·fəs }

equivalence element *See* equivalence gate. { i'kwiv·ə·ləns ,el·ə·mənt }

equivalence gate [COMPUT SCI] A logic circuit that produces a binary output signal of 1 if its two binary input signals are the same, and an output signal of 0 if the input signals differ. Also known as biconditional gate; equality gate; equivalence element; exclusive-NOR gate; match gate. { i'kwiv·ə·ləns ,gāt }

equivalent binary digits [COMPUT SCI] The number of binary positions required to enumerate the elements of a given set. { i'kwiv·ə·lənt 'bī,ner·ē 'dij·əts }

equivalent circuit [ELEC] A circuit whose behavior is identical to that of a more complex circuit or device over a stated range of operating conditions. { i'kwiv·ə·lənt 'sər·kət }

equivalent four-wire system [COMMUN] A transmission system in which multiplex techniques are used to carry on duplex operation over a single pair of wires. { i'kwiv·ə·lənt ¦fōr ¦wīr 'sis·təm }

equivalent noise conductance [ELECTR] Spectral density of a noise current generator measured in conductance units at a specified frequency. { i'kwiv·ə·lənt 'nóiz kən,dək·təns }

equivalent noise pressure [ENG ACOUS] In an electroacoustic transducer or sound reception system, the root-mean-square sound pressure of a sinusoidal plane progressive wave, which when propagated parallel to the primary axis of the transducer, produces an open-circuit signal voltage equivalent to the root-mean-square of the inherent open-circuit noise voltage of the transducer in a transmission band with a bandwidth of 1 hertz and centered on the frequency of the plane sound wave. Also known as inherent noise pressure. { i'kwiv·ə·lənt 'nóiz ,presh·ər }

equivalent noise resistance [ELECTR] Spectral density of a noise voltage generator measured in ohms at a specified frequency. { i'kwiv·ə·lənt 'nóiz ri,zis·təns }

equivalent noise temperature [ELECTR] Absolute temperature at which a perfect resistor, of equal resistance to the component, would generate the same noise as does the component

at room temperature. { i'kwiv·ə·lənt 'nòiz ,tem·prə·chər }

equivalent periodic line [ELEC] Of a uniform line, a periodic line having the same electrical behavior, at a given frequency, as the uniform line when measured at its terminals or at corresponding section junctions. { i'kwiv·ə·lənt pir·ē¦äd·ik 'līn }

equivalent resistance [ELEC] Concentrated or lumped resistance that would cause the same power loss as the actual small resistance values distributed throughout a circuit. { i'kwiv·ə·lənt ri'zis·təns }

erasable programmable read-only memory [COMPUT SCI] A read-only memory in which stored data can be erased by ultraviolet light or other means and reprogrammed bit by bit with appropriate voltage pulses. Abbreviated EPROM. { i¦rās·ə·bəl prō¦gram·ə·bəl ¦rēd ,ōn·lē 'mem·rē }

erasable storage [COMPUT SCI] Any storage medium which permits new data to be written in place of the old, such as magnetic disk or tape. { i¦rās·ə·bəl 'stòr·ij }

erase [COMPUT SCI] To change all the binary digits in a digital computer storage device to binary zeros. [ELECTR] To remove recorded material from magnetic tape by passing the tape through a strong, constant magnetic field (dc erase) or through a high-frequency alternating magnetic field (ac erase). { i'rās }

erase character *See* ignore character. { i¦rās 'kar·ik·tər }

erase oscillator [ELECTR] The oscillator used in a magnetic recorder to provide the high-frequency signal needed to erase a recording on magnetic tape; the bias oscillator usually serves also as the erase oscillator. { i'rās ,äs·ə,lād·ər }

erasing head [ELECTR] A magnetic head used to obliterate material previously recorded on magnetic tape. { i'rās·iŋ ,hed }

erasing speed [ELECTR] In charge-storage tubes, the rate of erasing successive storage elements. { i'rās·iŋ ,spēd }

erbium-doped fiber amplifier [COMMUN] An optical-fiber amplifier whose fiber core is lightly doped with trivalent erbium ions which absorb light at pump wavelengths of 0.98 and 1.48 micrometers and emit it at a signal wavelength around 1.5 micrometers through stimulated emission. Abbreviated EDFA. { ¦ər·bē·əm ,dōpt ,fī·bər 'am·plə,fī·ər }

erlang [COMMUN] A unit of communication traffic load, equal to the traffic load whose calls, if placed end to end, will keep one path continuously occupied. { 'er,läŋ }

ERP *See* effective radiated power.

error [COMPUT SCI] An incorrect result arising from approximations used in numerical methods, rather than from a human mistake or computer malfunction. { 'er·ər }

error analysis [COMPUT SCI] In the solution of a problem on a digital computer, the estimation of the cumulative effect of rounding or truncation errors associated with basic arithmetic operations. { 'er·ər ə‚nal·ə·səs }

error burst |COMPUT SCI| The condition when more than one bit is in error in a given number of bits. { 'er·ər ‚bərst }

error character |COMPUT SCI| A character that indicates the existence of an error in the data being processed or transmitted, and usually specifies that a certain amount of preceding or following data is to be ignored. { 'er·ər ‚kar·ik·tər }

error checking and recovery |COMPUT SCI| An automatic procedure which checks for parity and will proceed with the execution after error correction. { 'er·ər ‚chek·iŋ ən ri'kəv·ə·rē }

error-checking code See self-checking code. { 'er·ər ‚chek·iŋ ‚kōd }

error coefficient |CONT SYS| The steady-state value of the output of a control system, or of some derivative of the output, divided by the steady-state actuating signal. Also known as error constant. { 'er·ər ‚kō·i'fish·ənt }

error constant See error coefficient. { 'er·ər ‚kän·stənt }

error-control procedures |COMMUN| Methods of detecting errors and correcting or recovering from those that occur in data transmission. { 'er·ər kən‚trōl prə‚sē·jərz }

error-correcting code |COMPUT SCI| Data representation that allows for error detection and error correction if the error is of a specific kind. Also known as error-correction code. Abbreviated ECC. { 'er·ər kə¦rek·tiŋ 'kōd }

error-correcting telegraph system |COMMUN| System employing an error-detecting code, and so conceived that any false signal initiates a repetition of the transmission of the character incorrectly received. { 'er·ər kə¦rek·tiŋ 'tel·ə‚graf ‚sis·təm }

error correction |COMMUN| Any system for reducing errors in an incoming message, such as sending redundant signals as a check. |COMPUT SCI| Computer device for automatically locating and correcting a machine error of dropping a bit or picking up an extraneous bit, without stopping the machine or having it go to a programmed recovery routine. { 'er·ər kə‚rek·shən }

error-correction code See error-correcting code. { 'er·ər kə‚rek·shən 'kōd }

error-correction routine |COMPUT SCI| A program which corrects specific error conditions in another program, routine, or subroutine. { 'er·ər kə‚rek·shən rü‚tēn }

error-detecting code See self-checking code. { 'er·ər di‚tek·tiŋ ‚kōd }

error-detecting system |COMPUT SCI| An automatic system which detects an error due to a lack of data, or erroneous data during transmission. { 'er·ər di‚tek·tiŋ‚sis·təm }

error detection and feedback system |COMPUT SCI| An automatic system which retransmits a piece of data detected by the computer as being in error. { 'er·ər di‚tek·shən ən 'fēd‚bak ‚sis·təm }

error detection routine See diagnostic routine. { 'er·ər di‚tek·shən rü‚tēn }

error diagnostic |COMPUT SCI| A computer printout of an instruction or data statement, pinpointing an error in the instruction or statement and spelling out the type of error involved. { 'er·ər ‚dī·əg'näs·tik }

error frequency limit |COMPUT SCI| The maximum number of single bit errors per unit of time that a computer will accept before a machine check interrupt is initiated. Abbreviated EFL. { 'er·ər ‚frē·kwən·sē ‚lim·ət }

error handling |COMPUT SCI| The ability of a computer program to deal with errors automatically. { 'er·ər ‚hand·liŋ }

error-indicating system |COMPUT SCI| Built-in circuits designed to indicate automatically that certain computational errors have occurred. { 'er·ər ‚in·də‚kād·iŋ ‚sis·təm }

error interrupt |COMPUT SCI| The halt in execution of a program because of errors which the computer is not capable of correcting. { 'er·ər 'int·ə‚rəpt }

error list |COMPUT SCI| A list generated by a compiler showing invalid or erroneous instructions in a source program. { 'er·ər ‚list }

error log |COMPUT SCI| A file that is created during data processing to hold data known to contain errors, and that is usually printed after completion of processing so that the errors can be corrected. { 'er·ər ‚läg }

error message |COMPUT SCI| A message indicating detection of an error. { 'er·ər ‚mes·ij }

error range |COMPUT SCI| A range of values such that an error condition will result if a specified data item falls within it. { 'er·ər ‚rānj }

error rate |COMMUN| The number of erroneous bits or characters received for some fixed number of bits transmitted. { 'er·ər ‚rāt }

error ratio |COMPUT SCI| The ratio of the number of erroneous items to the total number of bits or characters transmitted. { 'er·ər ‚rā·shō }

error recovery routine |COMPUT SCI| A part of a computer program that attempts to handle errors without terminating the program. { 'er·ər ri¦kəv·ə·rē rü‚tēn }

error report |COMPUT SCI| A list produced by a computer showing the error conditions, such as overflows and errors resulting from incorrect or unmatched data, that are generated during program execution. { 'er·ər ri‚pȯrt }

error routine |COMPUT SCI| A routine which takes control of a program and initiates corrective actions when an error is detected. { 'er·ər rü‚tēn }

error signal |CONT SYS| In an automatic control device, a signal whose magnitude and sign are used to correct the alignment between the controlling and the controlled elements. |ELECTR| A voltage that depends on the signal received from the target in a tracking system, having a polarity and magnitude dependent on the angle between the target and the center of the scanning beam. { 'er·ər ‚sig·nəl }

error tape |COMPUT SCI| The magnetic tape on which erroneous records are stored during processing. { 'er·ər ‚tāp }

error voltage [ELEC] A voltage, usually obtained from a selsyn, that is proportional to the difference between the angular positions of the input and output shafts of a servosystem; this voltage acts on the system to produce a motion that tends to reduce the error in position. Also known as error signal. { 'er·ər ˌvōl·tij }

ES See elementary stream.

Esaki tunnel diode See tunnel diode. { e'sä·kē ¦tən·əl 'dī,ōd }

E-scan See E-display. { 'ē ,skan }

escape [COMPUT SCI] To exit from a program, routine, or mode. { i'skāp }

escape character [COMPUT SCI] A character used to indicate that the succeeding character or characters are expressed in a code different from the code currently in use. { ə'skāp ,kar·ik·tər }

E-scope See E-display. { 'ē ,skōp }

ESD See external symbol dictionary.

ESDI See enhanced small device interface. { 'ez ,dē }

esoteric name [COMPUT SCI] A symbolic name that is chosen in a computer program to designate a collection of devices. { es·ə'ter·ik 'nām }

Essen coefficient [ELEC] The torque exerted on the moving part of an electric rotating machine divided by the volume enclosed by the air gap. { 'es·ən ,kō·i,fish·ənt }

esu See electrostatic units.

etched circuit [ENG] A printed circuit formed by chemical or electrolytic removal of unwanted portions of a layer of conductive material bonded to an insulating base. { ¦echt 'sər·kət }

Ethernet [COMPUT SCI] A protocol for interconnecting computers and peripheral devices in a local area network. { 'ē·thər,net }

EU See expected value.

European Geostationary Navigation Overlay System [NAV] A satellite-based augmentation system developed jointly by the European Union, European Space Agency, and EUCONTROL. Abbreviated EGNOS. { ,yùr·ə¦pē·ən ,jē·ō ¦stā·shə·ner·ē ,nav·ə¦gā·shən 'ō·vər,lay ,sis·təm }

EV See expected value.

even parity check [COMPUT SCI] A parity check in which the number of 0's or 1's in each word is expected to be even. { ¦ē·vən 'par·əd·ē ,chek }

event [COMMUN] A collection of elementary streams with a common time base, an associated start time, and an associated end time. [COMPUT SCI] The moment of time at which a specified change of state occurs; usually marks the completion of an asynchronous input/output operation. { i'vent }

event-driven monitor [COMPUT SCI] A computer program that measures the performance of a computer system by counting the tasks performed by the system. { i'vent ¦driv·ən 'män·əd·ər }

even-word boundary [COMPUT SCI] A storage address that is an integral multiple of the computer's word length. { 'ēv·ən ¦wərd 'baún·drē }

evolutionary computation See evolutionary programming. { ,ev·ə¦lü·sha,ner·ē ,kam·pyə'tā·shən }

evolutionary programming [COMPUT SCI] Computer programming with genetic algorithms. Also known as evolutionary computation; genetic programming. { ,ev·ə¦lü·sha,ner·ē 'prō,gram·iŋ }

evolutionary strategy See genetic algorithm. { ,ev·ə¦lü·sha,ner·ē 'strad·ə·jē }

EVR See electronic video recording.

E wave See transverse magnetic wave. { 'ē ,wāv }

exalted-carrier receiver [ELECTR] Receiver that counteracts selective fading by maintaining the carrier at a high level at all times; this minimizes the second harmonic distortion that would otherwise occur when the carrier drops out while leaving most of the sidebands at their normal amplitudes. { ig¦zòl·təd 'kar·ē·ər ri,sēv·ər }

except gate [ELECTR] A gate that produces an output pulse only for a pulse on one or more input lines and the absence of a pulse on one or more other lines. { ek'sept ,gāt }

exception handling [COMPUT SCI] Programming techniques for dealing with error conditions, generally without terminating execution of the program. [CONT SYS] The actions taken by a control system when unpredictable conditions or situations arise in which the controller must respond quickly. { ek'sep·shən ,hand·liŋ }

exception-item encoding [COMPUT SCI] A technique which allows the uninterrupted flow of a process by the automatic shunting of erroneous records to an error tape for later corrections. { ek'sep·shən ,īd·əm en'kōd·iŋ }

exception-principle system [COMPUT SCI] A technique which assumes no printouts except when an error is encountered. { ek'sep·shən ,prin·sə·pəl ,sis·təm }

exception reporting [COMPUT SCI] A form of programming in which only values that are outside predetermined limits, representing significant changes, are selected for printout at the output of a computer. { ek'sep·shən ri,pórd·iŋ }

excess-fifty code [COMPUT SCI] A number code in which the number n is represented by the binary equivalent of $n + 50$. { 'ek,ses 'fif·tē ,kōd }

excess-three code [COMPUT SCI] A number code in which the decimal digit n is represented by the four-bit binary equivalent of $n + 3$. Also known as XS-3 code. { ¦ek,ses 'thrē ,kōd }

exchange [COMMUN] **1.** A unit established by a telephone company for the administration of telephone service in a specified area, usually a town, a city, or a village and its environs, and consisting of one or more central offices together with the associated plant used in furnishing telephone service in that area. Also known as local exchange. **2.** Room or building equipped so telephone lines terminating there may be interconnected as required; equipment may include a switchboard or automatic switching apparatus. [COMPUT SCI] The

interchange of contents between two locations.
{ iks,chānj }

exchangeable disk storage [COMPUT SCI] A type
of disk storage, used as a backing storage,
in which the disks come in capsules, each
containing several disks; the capsules can be
replaced during operation of the computer and
can be stored until needed. { iks¦chānj·ə·bəl
'disk,stȯr·ij }

exchange buffering [COMPUT SCI] An input/
output buffering technique that avoids the
internal moving of data. { iks'chānj ,bəf·ə·riŋ }

exchange cable [ELEC] Lead covered, non-
quadded, paper-insulated cable used within a
given area to provide cable pairs between local
subscribers and a central office. { iks'chānj
,kā·bəl }

exchange current [ELEC] The magnitude of the
current which flows through a galvanic cell
when it is operating in a reversible manner.
{ iks'chānj ,kə·rənt }

exchange line [ELEC] Line joining a subscriber
or switchboard to a commercial exchange.
{ iks'chānj ,līn }

exchange message [COMPUT SCI] A device,
placed between a communication line and
a computer, in order to take care of certain
communication functions and thereby free the
computer for other work. { iks'chānj ,mes·ij }

exchange plant [COMMUN] Plant used to serve
subscriber's local needs as distinguished from
that used for long-distance communication.
{ iks'chānj ,plant }

exchange sort [COMPUT SCI] A method of arrang-
ing records or other types of data into a specified
order, in which adjacent pairs of records are
exchanged until the correct order is achieved.
{ iks'chānj ,sȯrt }

excitation [CONT SYS] The application of energy
to one portion of a system or apparatus in a
manner that enables another portion to carry out
a specialized function; a generalization of the
electricity and electronics definitions. [ELEC]
The application of voltage to field coils to
produce a magnetic field, as required for the
operation of an excited-field loudspeaker or a
generator. [ELECTR] **1.** The signal voltage that
is applied to the control electrode of an electron
tube. Also known as drive. **2.** Application of
signal power to a transmitting antenna. { ,ek
,sī'tā·shən }

excitation anode [ELECTR] An anode used to
maintain a cathode spot on a pool cathode of
a gas tube when output current is zero. { ,ek
,sī'tā·shən ,an,ōd }

excitation voltage [ELEC] Nominal voltage required for excitation of a circuit. { ,ek,sī'tā·shən
,vōl·tij }

exciter [ELEC] **1.** A small auxiliary generator that
provides field current for an alternating-current
generator. **2.** See exciter lamp. [ELECTR] A crystal oscillator or self-excited oscillator used to
generate the carrier frequency of a transmitter.
[ELECTROMAG] **1.** The portion of a directional
transmitting antenna system that is directly

connected to the transmitter. **2.** A loop or probe
extending into a resonant cavity or waveguide.
{ ek'sīd·ər }

exciter [ELECTR] A crystal oscillator or self-
excited oscillator used to generate the carrier
frequency of a transmitter. [ELECTROMAG] **1.**
The portion of a directional transmitting antenna
system that is directly connected to the transmitter. **2.** A loop or probe extending into a resonant
cavity or waveguide. { ek'sīd·ər }

exciter lamp [ELEC] A bright incandescent lamp
having a concentrated filament, used to excite a phototube or photocell in sound movie
and facsimile systems. Also known as exciter.
{ ek'sīd·ər ,lamp }

exciter response [ELEC] In electrical rotating
machinery, the rate of increase or decrease of the
main exciter voltage when resistance is suddenly
removed from or inserted in the main exciter field
circuit. { ek'sīd·ər ri'späns }

exciting current See magnetizing current.
{ ek'sīd·iŋ ,kə·rənt }

exciton-induced photoemission [ELECTR] A
two-stage process that takes place in an ionic
crystal in which color centers are present,
in which photon absorption leads to the
formation of an exciton, and the exciton then
transfers enough energy to color centers to eject
photoelectrons from the crystal. { ¦ek·si,tän n
,düst 'fōd·ō·i,mish·əm }

excitron [ELECTR] A single-anode mercury-pool
tube provided with means for maintaining a
continuous cathode spot. { 'ek·sə,trän }

exclusive-NOR gate See equivalence gate. { ik
¦sklü·siv 'nȯr ,gāt }

exclusive or [COMPUT SCI] An instruction which
performs the "exclusive or" operation on a bit-
by-bit basis for its two operand words, usually
storing the result in one of the operand locations.
Abbreviated XOR. { ik¦sklü·siv 'ȯr }

exclusive segments [COMPUT SCI] Parts of an
overlay program structure that cannot be resident
in main memory simultaneously. { ik'sklü·siv
'seg·mənts }

executable module [COMPUT SCI] A file holding a
computer program written in machine language
so that it is ready to run. { ,ek·sə'kyüd·ə·bəl
'mäj·yül }

executable program [COMPUT SCI] A program
that is ready to run on a computer. { ,ek·sə
¦kyüd·ə·bəl 'prō·grəm }

executable statement [COMPUT SCI] A program
statement that causes the computer to carry
out some operation, in contrast to a declarative
statement. { ,ek·sə'kyüd·ə·bəl 'stāt·mənt }

execute [COMPUT SCI] Usually, to run a compiled
or assembled program on the computer; by
extension, to compile or assemble and to run a
source program. { 'ek·sə,kyüt }

execute statement [COMPUT SCI] A program
statement that indicates the beginning of a job
statement in a job control language. { 'ek·sə
,kyüt ,stāt·mənt }

execution control program [COMPUT SCI] The
program delivered by the manufacturer which

permits the computer to handle the programs fed to it.   { ‚ek·sə¦kyü·shən kən'trōl ‚prō·grəm }

execution cycle |COMPUT SCI| The time during which an elementary operation takes place.   { ‚ek·sə'kyü·shən ‚sī·kəl }

execution error detection |COMPUT SCI| The detection of errors which become apparent only during execution time.   { ‚ek·sə¦kyü·shən 'er·ər di‚tek·shən }

execution time |COMPUT SCI| The time during which actual work, such as addition or multiplication, is carried out in the execution of a computer instruction.   { ‚ek·sə'kyü·shən ‚tīm }

executive communications |COMPUT SCI| The routine information transmitted to the operator on the status of programs being executed and of the requirements made by these programs of the various components of the system.   { ig¦zek·yəd·iv kə·‚myü·nə'kā·shənz }

executive control language |COMPUT SCI| The generic term for a finite set of instructions which enables the programmer to run a program more efficiently.   { ig¦zek·yəd·iv kən'trōl ‚laŋ·gwij }

executive file-control system |COMPUT SCI| The assignment of intermediate storage devices performed by the computer, and over which the programmer has no control.   { ig¦zek·yəd·iv 'fīl kən‚trōl ‚sis·təm }

executive guard mode |COMPUT SCI| A protective technique which prevents the programmer from accessing, or using, the executive instructions.   { ig¦zek·yəd·iv 'gärd ‚mōd }

executive instruction |COMPUT SCI| Instruction to determine how a specially written computer program is to operate.   { ig¦zek·yəd·iv in¦strək·shən }

executive logging |COMPUT SCI| The automatic bookkeeping of time utilization by programs of the various components of a computer system.   { ig¦zek·yəd·iv 'läg·iŋ }

executive routine |COMPUT SCI| A digital computer routine designed to process and control other routines. Also known as master routine; monitor routine.   { ig'zek·yəd·iv rü‚tēn }

executive schedule maintenance |COMPUT SCI| The scheduling of jobs to be run according to priorities as established and maintained by a computer's executive supervisor.   { ig¦zek·yəd·iv 'sked·jəl ‚mān·tə·nəns }

executive supervisor |COMPUT SCI| The component of the computer system which controls the sequencing, setup, and execution of the jobs presented to it.   { ig¦zek·yəd·iv 'sü·pər‚viz·ər }

executive system concurrency |COMPUT SCI| The capability of a computer system's executive supervisor to handle more than one job at the same time if these jobs do not require the same components at the same time.   { ig'zek·yəd·iv ‚sis·təm kən'kər·ən·sē }

executive system utilities |COMPUT SCI| The set of programs, such as diagnostic programs or file utility programs, which enables the executive supervisor to handle the jobs efficiently and completely.   { ig'zek·yəd·iv ‚sis·təm yü'til·əd·ēz }

exhaustion region |ELECTR| A layer in a semiconductor, adjacent to its contact with a metal,

in which there is almost complete ionization of atoms in the lattice and few charge carriers, resulting in a space-charge density.   { ig'zós·chən ‚rē·jən }

exit |COMPUT SCI| **1.** A way of terminating a repeated cycle of operations in a computer program. **2.** A place at which such a cycle can be stopped.   { 'eg·zət }

exogenous electrification |ELEC| The separation of electric charge in a conductor placed in a preexisting electric field, especially applied to the charge separation observed on metal-covered aircraft, resulting from induction effects, and by itself does not create any net total charge on the conductor.   { ‚ek'säj·ə·nəs i‚lek·trə·fə'kā·shən }

expanded batch |COMPUT SCI| A level of computer processing more complex than basic batch, in which computer programs perform complex computations and produce reports that analyze performance in addition to reporting it.   { ik'spand·əd 'bach }

expanded position indicator display |ELECTR| Display of an expanded sector from a plan position indicator presentation.   { ik'spand·əd pə'zish·ən ¦in·də‚kad·ər di‚splā }

expanded scope |ELECTR| Magnified portion of a given type of cathode-ray tube presentation.   { ik'spand·əd 'skōp }

expanded sweep |ELECTR| A cathode-ray sweep in which the movement of the electron beam across the screen is speeded up during a selected portion of the sweep time.   { ik'spand·əd 'swēp }

expander |ELECTR| A transducer that, for a given input amplitude range, produces a larger output range.   { ik'spand·ər }

expandor |ELECTR| The part of a compandor that is used at the receiving end of a circuit to return the compressed signal to its original form; attenuates weak signals and amplifies strong signals.   { ik'spand·dər }

expansion |ELECTR| A process in which the effective gain of an amplifier is varied as a function of signal magnitude, the effective gain being greater for large signals than for small signals; the result is greater volume range in an audio amplifier and greater contrast range in facsimile.   { ik'span·shən }

expansion board |COMPUT SCI| A printed circuit board that can be plugged into a computer to provide it with additional peripherals or enhancements, such as increased memory or communications facilities.   { ik'span·shən ‚bórd }

expansion bus |COMPUT SCI| The wiring and protocols that connect a computer's motherboard with the peripheral devices.   { ik'span·shən ‚bəs }

expansion slot |COMPUT SCI| A location in a computer system where additional facilities, especially circuit boards, can be plugged in to extend the computer's capability.   { ik'span·shən ‚slät }

expected utility See expected value.   { ek'spek·təd yü'til·əd·ē }

expected value |SYS ENG| In decision theory, a measure of the value or utility expected to result from a given strategy, equal to the sum over states of nature of the product of the probability of the state times the consequence or outcome of the strategy in terms of some value or utility parameter. Abbreviated EV. Also known as expected utility (EU). { ek'spek·təd 'val·yü }

expert control system [CONT SYS] A control system that uses expert systems to solve control problems. { ¦ek¦spərt kən'trōl ¸sis·təm }

expert system [COMPUT SCI] A computer system composed of algorithms that perform a specialized, usually difficult professional task at the level of (or sometimes beyond the level of) a human expert. { 'ek¸spərt ¸sis·təm }

explicit programming [CONT SYS] Robotic programming that employs detailed and exact descriptions of the tasks to be performed. { ik'splis·ət 'prō¸gram·iŋ }

exploded file [COMPUT SCI] A file in which more data have been added to each record in order to adapt it to a new application. { ik'splōd·əd 'fīl }

exponential amplifier [ELECTR] An amplifier capable of supplying an output signal proportional to the exponential of the input signal. { ¸ek·spə'nen·chəl 'am·plə¸fī·ər }

exponential horn [ENG ACOUS] A horn whose cross-sectional area increases exponentially with axial distance. { ¸ek·spə'nen·chəl 'hȯrn }

exponential transmission line [ELEC] A two-conductor transmission line whose characteristic impedance varies exponentially with electrical length along the line. { ¸ek·spə'nen·chəl tranz'mish·ən ¸līn }

exposure voltage [ELEC] The voltage at which the document-illuminating lamps are operated during exposure. { ik'spō·zhər ¸vōl·tij }

expression [COMPUT SCI] A mathematical or logical statement written in a source language, consisting of a collection of operands connected by operations in a logical manner. { ik'spresh·ən }

expulsion fuse See expulsion-fuse unit. { ik'spəl·shən ¸fyüz }

expulsion-fuse unit [ELEC] A vented fuse unit in which the arc is extinguished by the expulsion of gases generated by the arc and lining of the fuse holder, sometimes with the aid of a spring. Also known as expulsion fuse. { ik'spəl·shən ¸fyüz ¸yü·nət }

extended-area service [COMMUN] Telephone exchange service, without toll charges, that extends over an area where there is a community of interest, often in return for a somewhat higher exchange service rate. { ik¦stend·əd ¦er·ē·ə 'sər·vəs }

extended ASCII [COMMUN] An addition to the standard American Standard Code for Information Interchange, namely, characters 128 through 255; includes letters with diacritics, Greek letters, and special symbols. { ik¦sten·dəd 'as¸kē }

extended binary-coded decimal interchange code [COMPUT SCI] A computer code that uses eight binary positions to represent a single character, giving a possible maximum of 256 characters. Abbreviated EBCDIC. { ik¦stend·əd 'bī¸ner·ē ¦kōd·əd ¦des·məl 'int·ər¸chānj ¸kōd }

extended channel status word [COMPUT SCI] Stored information which follows an input/output interrupt. Abbreviated ECSW. { ik¦stend·əd 'chan·əl 'stad·əs ¸wərd }

extended data out random-access memory [COMPUT SCI] A type of dynamic random-access memory that was optimized for the 66-megahertz bus but largely has been replaced by faster systems. Abbreviated EDO RAM. { ik¦stend·əd ¸dad·ə ¦aút ¸ran·dəm 'ak¸ses ¸mem·rē }

extended-entry decision table [COMPUT SCI] A decision table in which the condition stub cites the identification of the condition but not the particular values, which are entered directly into the condition entries. { ik¦stend·əd 'en·trē di'sizh·ən ¸tā·bəl }

extended-hybrid FM IBOC [COMMUN] The second of three modes in the FM IBOC system approved by the Federal Communications Commission for use in the United States that increases data capacity by adding additional carriers closer to the analog host signal. The extended-hybrid IBOC mode adds two frequency partitions around the analog carrier, where digital audio date rate can range from 64 to 96 kbits/s, and the corresponding ancillary data rate will range from 83 kbits/s for 64-kbits/s audio to 51 kbits/s for 96-kbits/s audio. { ik'stend·əd 'hī·brəd 'ef¸em 'ī¸bäk }

extended-interaction tube [ELECTR] Microwave tube in which a moving electron stream interacts with a traveling electric field in a long resonator; bandwidth is between that of klystrons and traveling-wave tubes. { ik¦stend·əd int·ə'rak·shən ¸tüb }

extended-precision word [COMPUT SCI] A piece of data of 16 bytes in floating-point arithmetic when additional precision is required. { ik¦stend·əd prə'sizh·ən ¸wərd }

extended time scale See slow time scale. { ik¦stend·əd 'tīm ¸skāl }

extender [ELEC] A male or female receptacle connected by a short cable to make a test point more conveniently accessible to a test probe. { ik'sten·dər }

extend flip-flop [COMPUT SCI] A special flag set when there is a carry-out of the most significant bit in the register after an addition or a subtraction. { ik'stend 'flip¸fläp }

extensible language [COMPUT SCI] A programming language which can be modified by adding new features or changing existing ones. { ik'sten·sə·bəl 'laŋ·gwij }

Extensible Markup Language [COMMUN] A set of rules for writing markup languages which provides a robust, machine-readable information protocol that can handle complex objects. Abbreviated XML. { ik¦sten·sə·bəl 'märk¸əp ¸laŋ·gwij }

extensible system [COMPUT SCI] A computer system in which users may extend the basic system by implementing their own languages and subsystems and making them available for others to use. { ik'sten·sə·bəl 'sis·təm }

extension cord |ELEC| A line cord having a plug at one end and an outlet at the other end. { ik'sten·chən ‚kȯrd }

extension mechanism |COMPUT SCI| One of the components of an extensible language which allows the definition of new language features in terms of the primitive facilities of the base language. { ik'sten·chən ‚mek·ə‚niz·əm }

extension register |COMPUT SCI| A register that is combined with an accumulator register for calculations involving multiple precision arithmetic. { ik'sten·chən ‚rej·ə·stər }

extent |COMPUT SCI| The physical locations in a mass-storage device or volume allocated for use by a particular data set. { ik'stent }

extern |COMPUT SCI| A pseudoinstruction found in several assembly languages which explicitly tells an assembler that a symbol is external, that is, not defined in the program module. { ek'stərn }

external armature |ELEC| Armature for a machine of special design in which the armature is a ring which rotates around the magnetic poles. { ek'stərn·əl 'är·mə·chər }

external buffer |COMPUT SCI| A buffer storage located outside the computer's main storage, often within a control unit or other peripheral device. { ek'stərn·əl 'bəf·ər }

external declaration |COMPUT SCI| A declarative statement in a computer program that specifies that a symbolic name used in the program is defined in another program. { ek'stərn·əl ‚dek·lə'rā·shən }

external delay |COMPUT SCI| Time during which a computer cannot be operated due to circumstances beyond the reasonable control of the operators and maintenance engineers, such as a failure of the public power supply. { ek'stərn·əl di'lā }

external device |ENG| A piece of equipment that operates in conjunction with and under the control of a central system, such as a computer or control system, but is not part of the system itself. { ek'stərn·əl di'vīs }

external-device address |COMPUT SCI| The address of a component such as a tape drive. { ek‚stərn·əl di‚vīs 'a‚dres }

external-device control |COMPUT SCI| The capability of an external device to create an interrupt during the execution of a job. { ek‚stərn·əl di ‚vīs kən‚trōl }

external-device operands |COMPUT SCI| The part of an instruction referring to an external device such as a tape drive. { ek‚stərn·əl di‚vīs 'äp·ə ‚ranz }

external-device response |COMPUT SCI| The signal from an external device, such as a tape drive, that it is not busy. { ek‚stərn·əl di‚vīs ri‚späns }

external error |COMPUT SCI| An error sensed by the computer when this error occurs in a device such as a disk drive. { ek‚stərn·əl 'er·ər }

external interrupt |COMPUT SCI| Any interrupt caused by the operator or by some external device such as a tape drive. { ek‚stərn·əl 'int·ə‚rəpt }

external-interrupt status word |COMPUT SCI| The content of a special register which indicates, among other things, the source of the interrupt. { ek‚stərn·əl 'int·ə‚rəpt 'stad·əs ‚wərd }

external label |COMPUT SCI| A reference to a variable not defined in a program segment. { ek ‚stərn·əl 'lā·bəl }

externally stored program |COMPUT SCI| A program achieved by wiring plugboards, as in some tabulating equipment. { ek‚stərn·əl·ē ‚stȯrd 'prō·grəm }

external memory |COMPUT SCI| Any storage device not an integral part of a computer system, such as a magnetic tape or disk. { ek‚stərn·əl 'mem·rē }

external photoelectric effect See photoemission. { ek‚stərn·əl ‚fō·dō·i'lek·trik i‚fekt }

external Q |ELECTR| The inverse of the difference between the loaded and unloaded Q values of a microwave tube. { ek‚stərn·əl 'kyü }

external reference |COMPUT SCI| In a computer program, a branch or call to a separate independent program or routine. { ek‚stərn·əl 'ref·rəns }

external sensor |CONT SYS| A device that senses information about the environment of a control system but is not part of the system itself. { ek‚stərn·əl 'sen·sər }

external signal |COMPUT SCI| Any message to an operator for which no printout is required but which is self-explanatory, such as a light condition indicating whether the equipment is on or off. { ek‚stərn·əl 'sig·nəl }

external sorting |COMPUT SCI| The sorting of a list of items by a computer in which the list is too large to be brought into the memory at one time, and instead is brought into the memory a piece at a time so as to produce a collection of ordered sublists which are subsequently reordered by the computer to produce a single list. { ek‚stərn·əl 'sȯrd·iŋ }

external storage |COMPUT SCI| Large-capacity, slow-access data storage attached to a digital computer and used to store information that exceeds the capacity of main storage. { ek ‚stərn·əl 'stȯr·ij }

external symbol dictionary |COMPUT SCI| A list of external symbols and their relocatable addresses which allows the linkage editor to resolve interprogram references. Abbreviated ESD. { ek ‚stərn·əl 'sim·bəl ‚dik·shə‚ner·ē }

external table |COMPUT SCI| A table whose data are located outside a computer program, usually in a separate file. { ek'stərn·əl 'tā·bəl }

extinction voltage |ELECTR| The lowest anode voltage at which a discharge is sustained in a gas tube. { ek'stiŋk·shən ‚vōl·tij }

extract |COMPUT SCI| **1.** To form a new computer word by extracting and putting together selected segments of given words. **2.** To remove from a computer register or memory all items that meet a specified condition. { ik'strakt }

extract instruction |COMPUT SCI| An instruction that requests the formation of a new expression from selected parts of given expressions. { ik 'strakt in,strək·shən }

extra-high tension [ELECTR] British term for the high direct-current voltage applied to the second anode in a cathode-ray tube, ranging from about 4000 to 50,000 volts in various sizes of tubes. Abbreviated eht. { ¦ek·strə ¦hī 'ten·chən }

extra-high voltage [ELEC] A voltage above 345 kilovolts used for power transmission. Abbreviated ehv. { ¦ek·strə ¦hī 'vōl·tij }

extraneous emission [ELECTR] Any emission of a transmitter or transponder, other than the output carrier fundamental, plus only those sidebands intentionally employed for the transmission of intelligence. { ik'strän·ē·əs ə'mish·ən }

extraneous response [ELECTR] Any undersired response of a receiver, recorder, or other susceptible device, due to the desired signals, undersired signals, or any combination or interaction among them. { ik'strän·ē·əs ri'späns }

extranet [COMPUT SCI] A secure, Internet-based private network that allows organizations to share information with vendors, partners, customers, and so on; access requires either a password or digital encryption. { 'ek·strə,net }

extraterrestrial noise [ELECTROMAG] Cosmic and solar noise; radio disturbances from sources other than those related to the earth. { ¦ek·strə·tə'res·trē·əl 'nōiz }

extremely high frequency [COMMUN] The frequency band from 30,000 to 300,000 megahertz in the radio spectrum. Abbreviated EHF. { ek'strēm·lē 'hī 'frē·kwən·sē }

extremely low frequency [COMMUN] A frequency below 300 hertz in the radio spectrum. Apprevi-ated ELF. { ek'strēm·lē'lō 'frē·kwən·sē }

extrinsic detector [ENG] A semiconductor detector of electromagnetic radiation that is doped with an electrical impurity and utilizes transitions of charge carriers from impurity states in the band gap to nearby energy bands. { ek¦strinz·ik di'tek·tər }

extrinsic photoconductivity [ELECTR] Photoconductivity that occurs for photon energies smaller than the band gap and corresponds to optical excitation from an occupied imperfection level to the conduction band, or to an unoccupied imperfection level from the valence band, of a material. { ek¦strinz·ik ¸fō·dō·kän·dək'tiv·əd·ē }

extrinsic photoemission [ELECTR] Photoemission by an alkali halide crystal in which electrons are ejected directly from negative ion vacancies, forming color centers. Also known as direct ionization. { ek¦strin·sik ¸fōd·ō·i'mish·ən }

extrinsic properties [ELECTR] The properties of a semiconductor as modified by impurities or imperfections within the crystal. { ek¦strinz·ik 'präp·ərd·ēz }

extrinsic semiconductor [ELECTR] A semiconductor whose electrical properties are dependent on impurities added to the semiconductor crystal, in contrast to an intrinsic semiconductor, whose properties are characteristic of an ideal pure crystal. { ek¦strinz·ik 'sem·i·kən,dək·tər }

e-zine [COMPUT SCI] A Web-published magazine. { 'ē ,zēn }

F

F *See* farad.

fA *See* femtoampere.

fabrication |ENG| **1.** The manufacture of parts, usually structural or electromechanical parts. **2.** The assembly of parts into a structure. { ˌfab·ri'kā·shən }

face *See* faceplate. { fās }

face-bonding |ELECTR| Method of assembling hybrid microcircuits wherein semiconductor chips are provided with small mounting pads, turned facedown, and bonded directly to the ends of the thin-film conductors on the passive substrate. { 'fās ˌbänd·iŋ }

faceplate |ELECTR| The transparent or semitransparent glass front of a cathode-ray tube, through which the image is viewed or projected; the inner surface of the face is coated with fluorescent chemicals that emit light when hit by an electron beam. Also known as face. { 'fās,plāt }

facility assignment |COMPUT SCI| The allocation of core memory and external devices by the executive as required by the program being executed. { fə'sil·əd·ē ə,sīn·mənt }

facility dispersion |COMMUN| The distribution of circuits between two points over more than one physical or geographic route to reduce the likelihood of a trunk group being put completely out of service by facility damage or other circuit failure. { fə'sil·əd·ē di'spər·zhən }

facsimile |COMMUN| **1.** A system of communication in which a transmitter scans a photograph, map, or other fixed graphic material and converts the information into signal waves for transmission by wire or radio to a facsimile receiver at a remote point. Also known as fax; phototelegraphy; radiophoto; telephoto; telephotography; wirephoto. **2.** A photograph transmitted by radio to a facsimile receiver. Also known as radiophoto. { fak'sim·ə·lē }

facsimile modulation |COMMUN| Process in which the amplitude, frequency, or phase of a transmitted wave is varied with time in accordance with a facsimile transmission signal. { fak'sim·ə·lē ,mäj·ə·'lā·shən }

facsimile posting |COMPUT SCI| The process of transferring by a duplicating process a printed line of information from a report, such as a listing of transactions prepared on an accounting machine, to a ledger or other recorded sheet. { fak'sim·ə·lē 'pōst·iŋ }

facsimile receiver |ELECTR| The receiver used to translate the facsimile signal from a wire or radio communication channel into a facsimile record of the subject copy. { fak'sim·ə·lē ri'sē·vər }

facsimile recorder |ELECTR| The section of a facsimile receiver that performs the final conversion of electric signals to an image of the subject copy on the record medium. { fak'sim·ə·lē ri'kórd·ər }

facsimile signal |COMMUN| The picture signal produced by scanning the subject copy in a facsimile transmitter. { fak'sim·ə·lē ,sig·nəl }

facsimile signal level |ELECTR| Maximum facsimile signal power or voltage (root mean square or direct current) measured at any point in a facsimile system. { fak'sim·ə·lē 'sig·nəl ,lev·əl }

facsimile synchronizing |ELECTR| Maintenance of predetermined speed relations between the scanning spot and the recording spot within each scanning line. { fak'sim·ə·lē 'siŋ·krə,niz·iŋ }

facsimile telegraph |COMMUN| A telegraph system designed to transmit pictures. { fak'sim·ə·lē 'tel·ə,graf }

facsimile transmitter |ELECTR| The apparatus used to translate the subject copy into facsimile signals suitable for delivery over a communication system. { fak'sim·ə·lē tranz'mid·ər }

fade-out |COMMUN| A gradual and temporary loss of a received radio or television signal due to magnetic storms, atmospheric disturbances, or other conditions along the transmission path. { 'fād,aút }

fader |ELECTR| A multiple-unit level control used for gradual changeover from one audio video source { 'fād·ər }

fading |COMMUN| Variations in the field strength of a radio signal that are caused by changes in the transmission medium. { 'fād·iŋ }

fading margin |COMMUN| **1.** Number of decibels of attenuation which may be added to a specified radio-frequency propagation path before the signal-to-noise ratio of a specified channel falls below a specified minimum in order to avoid disruption of service. **2.** Allowance made in radio system planning to accommodate estimated fading. { 'fād·iŋ ,mär·jən }

Fahnestock clip |ELEC| A spring-type terminal to which a temporary connection can readily be made. { 'fan,stäk ,klip }

fail-safe system |ENG| A system designed so that failure of power, control circuits, structural members, or other components will not endanger people operating the system or other people in the vicinity.   { 'fāl ¦sāf ,sis·təm }

fail-safe tape   *See* incremental dump tape.   { 'fāl ¦sāf ,tāp }

fail soft   [ENG] A failure in the performance of a system component that neither results in immediate or major interruption of the system operation as a whole nor adversely affects the quality of its products.   { 'fāl ,sóft }

fail-soft system   [COMPUT SCI] A computer system with automatic controls that allow function to continue after a malfunction and, if necessary, permit the shutdown of the system without loss of data.   { 'fāl ¦sóft ,sis·təm }

failure logging   [COMPUT SCI] The automatic recording of the state of various components of a computer system following detection of a machine fault; used to initiate corrective procedures, such as repeating attempts to read or write a magnetic tape, and to aid customer engineers in diagnosing errors.   { 'fāl·yər ,läg·iŋ }

failure rate   [ENG] The probability of failure per unit of time of items in operation; sometimes estimated as a ratio of the number of failures to the accumulated operating time for the items.   { 'fāl·yər ,rāt }

fallback   [COMPUT SCI] The system, electronic or manual, which is substituted for the computer system in case of breakdown.   { 'fól,bak }

fallback switch   [COMMUN] A mechanical switch to transfer a communications path from a primary device to an identical standby device in the event of a primary device failure.   { 'fól,bak ,swich }

fallout   [ELECTR] Failure of electronic components during burn-in.   { 'fól,aüt }

fall time   [ELEC] Measure of time required for a circuit to change its output from a high level to a low level.   { 'fól ,tīm }

false alarm   [ELECTR] In radar, an indication of a detected target even though one does not exist, due to noise or interference levels exceeding the set threshold of detection.   { ¦fóls ə'lärm }

false drop   *See* false retrieval.   { ¦fóls 'dräp }

false retrieval   [COMPUT SCI] An item retrieved in an automatic library search which is unrelated or vaguely related to the subject of the search. Also known as false drop.   { ¦fóls ri'trē·vəl }

false sorts   [COMPUT SCI] Entries irrelevant to the subject sought which are retrieved in a search.   { ¦fóls 'sórts }

false target   [ELECTR] In radar, a contact (target) estimated to be where none exists, generally as the result of ambiguity in the data processing.   { ¦fóls 'tär·gət }

false-target generator   [ELECTR] An electronic countermeasure device that generates a delayed return signal on an enemy radar frequency to give erroneous position information.   { ¦fóls ¦tär·gət 'jen·ə,rād·ər }

FAMOS device   *See* floating-gate avalanche-injection metal-oxide semiconductor device.   { 'fā,mós di'vīs }

fan antenna   [ELECTROMAG] An array of folded dipoles of different length forming a wide-band

ultra-high-frequency or very-high-frequency antenna.   { 'fan an,ten·ə }

fan beam   [ELECTROMAG] **1.** A radio beam having an elliptically shaped cross section in which the ratio of the major to the minor axis usually exceeds 3 to 1; the beam is broad in the vertical plane and narrow in the horizontal plane.   **2.** A radar beam having the shape of a fan.   { 'fan ,bēm }

fanfold   [COMPUT SCI] Continuous paper that is perforated at page boundaries and can be folded back and forth at the perforations to form a stack.   { 'fan,fōld }

fan-in   [ELECTR] The number of inputs that can be connected to a logic circuit.   { 'fan,in }

fan marker   *See* fan-marker beacon.   { 'fan ,märk·ər }

fan-marker beacon   [NAV] A very-high frequency radio facility having a vertically directed fan beam interesecting an airway to provide a fix. Also known as fan marker; radio fan-marker beacon.   { 'fan ,märk·ər ,bē·kən }

fanned-beam antenna   [ELECTROMAG] Unidirectional antenna so designed that transverse cross sections of the major lobe are approximately elliptical.   { ¦fand ¦bēm an,ten·ə }

fanning beam   [ELECTROMAG] Narrow antenna beam which is repeatedly scanned over a limited arc.   { 'fan·iŋ ,bēm }

fanning strip   [ELEC] Insulated board, often of wood, which serves to spread out the wires of a cable for distribution to a terminal board.   { 'fan·iŋ ,strip }

fan-out   [ELECTR] The number of parallel loads that can be driven from one output mode of a logic circuit.   { 'fan,aüt }

FAQ   *See* Frequently Asked Questions.

farad   [ELEC] The unit of capacitance in the meter-kilogram-second system, equal to the capacitance of a capacitor which has a potential difference of 1 volt between its plates when the charge on one of its plates is 1 coulomb, there being an equal and opposite charge on the other plate. Symbolized F.   { 'fa,rad }

Faradaic current   *See* Faradic current.   { 'far·ə ,dā·ik ,kə·rənt }

Faraday birefringence   [OPTICS] Difference in the indices of refraction of left and right circularly polarized light passing through matter parallel to an applied magnetic field; it is responsible for the Faraday effect.   { 'far·ə,dā ,bī·ri'frin·jəns }

Faraday cage   *See* Faraday shield.   { 'far·ə,dā ,kāj }

Faraday cylinder   [ELEC] **1.** A closed, or nearly closed, hollow conductor, usually grounded, within which apparatus is placed to shield it from electrical fields.   **2.** A nearly closed, insulated, hollow conductor, usually shielded by a second grounded cylinder, used to collect and detect a beam of charged particles.   { 'far·ə,dā ,sil·ən·dər }

Faraday dark space   [ELECTR] The relatively nonluminous region that separates the negative glow from the positive column in a cold-cathode glow-discharge tube.   { 'far·ə,dā 'därk ,spās }

Faraday effect |OPTICS| Rotation of polarization of a beam of linearly polarized light when it passes through matter in the direction of an applied magnetic field; it is the result of Faraday birefringence. Also known as Faraday rotation; Kundt effect; magnetic rotation. { 'far·ə,dā i'fekt }

Faraday ice bucket experiment |ELEC| Experiment in which one lowers a charged metal body into a pail and observes the effect on an electroscope attached to the pail, with and without contact between body and pail; the experiment shows that charge resides on a conductor's outside surface. { 'far·ə,dā 'īs ,bək·ət ik ,spər·ə·mənt }

Faraday rotation See Faraday effect. { 'far·ə,dā rō'tā·shən }

Faraday rotation isolator See ferrite isolator. { 'far·ə,dā rō'tā·shən 'īs·əl,ād·ər }

Faraday screen See Faraday shield. { 'far·ə,dā ,skrēn }

Faraday shield |ELEC| Electrostatic shield composed of wire mesh or a series of parallel wires, usually connected at one end to another conductor which is grounded. Also known as Faraday cage; Faraday screen. { 'far·ə,dā ,shēld }

Faraday tube |ELEC| A tube of force for electric displacement which is of such size that the integral over any surface across the tube of the component of electric displacement perpendicular to that surface is unity. { 'far·ə,dā ,tüb }

Faradic current |ELEC| An intermittent and nonsymmetrical alternating current like that obtained from the secondary winding of an induction coil. Also spelled Faradaic current. { fə'rad·ik ,kə·rənt }

far-end crosstalk |COMMUN| Crosstalk that travels along the disturbed circuit in the same direction as desired signals in that circuit. { ¦fär ¦end 'krȯs,tȯk }

far field See Fraunhofer region. { ¦fär ¦fēld }

far-infrared maser |ENG| A gas maser that generates a beam having a wavelength well above 100 micrometers, and ranging up to the present lower wavelength limit of about 500 micrometers for microwave oscillators. { ¦fär in·frə'red 'mā·zər }

far-infrared radiation |ELECTROMAG| Infrared radiation the wavelengths of which are the longest of those in the infrared region, about 50–1000 micrometers; requires diffraction gratings for spectroscopic analysis. { ¦fär in·frə'red ,rād·ē'ā·shən }

Farnsworth image dissector tube See image dissector tube. { 'färnz,wərth 'im·ij di,sek·tər ,tüb }

far region See Fraunhofer region. { ¦fär ¦rē·jən }

far zone See Fraunhofer region. { ¦fär ¦zōn }

fast-access storage |COMPUT SCI| The section of a computer storage from which data can be obtained most rapidly. { ¦fast ¦ak·ses 'stȯr·ij }

fast automatic gain control |ELECTR| Radar automatic gain control method characterized by a response time that is long with respect to a pulse width, and short with respect to the time on target. { 'fast ,ȯd·ə,mad·ik 'gān kən,trōl }

fast time constant |ELECTR| Circuit with short time constant used to emphasize signals of short duration to produce discrimination against low-frequency components of clutter in radar. { 'fast 'tīm ,kän·stənt }

fast time scale |COMPUT SCI| In simulation by an analog computer, a scale in which the time duration of a simulated event is less than the actual time duration of the event in the physical system under study. { 'fast 'tīm ,skāl }

FAT See file allocation table. { fat or ¦ef¦ā'tē }

fatal error |COMPUT SCI| An error in a computer program which causes running of the program to be terminated. { ¦fād·əl 'er·ər }

father file |COMPUT SCI| A copy of the master file from the cycle or generation that precedes the one being updated. { 'fäth·ər ,fīl }

fatigue |ELECTR| The decrease of efficiency of a luminescent or light-sensitive material as a result of excitation. { fə'tēg }

fatware |COMPUT SCI| Software that is overly laden with features or is inefficiently designed, so that it occupies inordinate space in disk storage and random-access memory, and requires an inappropriate share of microprocessor power. Also known as bloatware. { 'fat,wer }

fault |ELEC| A defect, such as an open circuit, short circuit, or ground, in a circuit, component, or line. Also known as electrical fault; faulting. |ELECTR| Any physical condition that causes a component of a data-processing system to fail in performance. { fȯlt }

fault analysis |ENG| The detection and diagnosis of malfunctions in technical systems, in particular, by means of a scheme in which one or more computers monitor the technical equipment to signal any malfunction and designate the components responsible for it. { 'fȯlt ə,nal·ə·səs }

fault current See fault electrode current. { 'fȯlt ,kə·rənt }

fault electrode current |ELEC| The current to an electrode under fault conditions, such as during arc-backs and load short circuits. Also known as fault current; surge electrode current. { 'fȯlt i'lek,trōd ,kə·rənt }

fault finder |ENG| Test set for locating trouble conditions in communications circuits or systems. { 'fȯlt ,fīnd·ər }

faulting See fault. { 'fȯl·tiŋ }

fault masking |COMPUT SCI| Any type of hardware redundancy in which faults are corrected immediately and the operations of fault detection, location, and correction are indistinguishable. { 'fȯlt ,mask·iŋ }

fault monitoring |SYS ENG| A procedure for systematically checking for errors and malfunctions in the software and hardware of a computer or control system. { 'fȯlt ,män·ə·triŋ }

fault tolerance |SYS ENG| The capability of a system to perform in accordance with design specifications even when undesired changes in the internal structure or external environment occur. { 'fȯlt ,täl·ə·rəns }

Faure storage battery |ELEC| A storage battery in which the plates consist of lead-antimony supporting grids covered with a lead oxide paste, immersed in weak sulfuric acid. Also known as pasted-plate storage battery. { 'fȯr 'stȯr·ij ,bad·ə·rē }

fax See facsimile. { faks }

FB data set [COMPUT SCI] A data set which has F-format logical records and whose physical records are all some multiple of the size of the logical record, except possibly for a few truncated blocks. Also known as blocked F-format data set. { ¦ef¦bē ,dad·ə ,set }

FBM data set [COMPUT SCI] An FB data set which has a machine-control (M) character in its first byte of information. { ¦ef¦bē¦em ,dad·ə ,set }

FBSA data set [COMPUT SCI] An FBS data set which has an ASCII (American Standard Code for Information Interchange) control (A) character in its first byte of information. { ¦ef¦bē¦es¦ā ,dad·ə ,set }

FBS data set [COMPUT SCI] An FB data set which has at most one truncated block, which must be the last one in the data set. Also known as standard blocked F-format data set. { ¦ef¦bē¦es ,dad·ə ,set }

F connector [ELECTR] A plug and socket for interconnecting coaxial cables; commonly used to interconnect television receivers and cable or antenna sources. { 'ef kə,nek·tər }

FDDI See fiber-optic data distribution interface.

F-display [ELECTR] A radar display format in which the target appears as a spot in the center when the antenna of a tracking radar is aimed directly at it, with any displacement indicating pointing error. Also known as F-indicator; F-scan; F-scope. { 'ef di,splā }

FDM See frequency-division multiplexing.

FDMA See frequency-division multiple access.

feasibility study [SYS ENG] **1.** A study of applicability or desirability of any management or procedural system from the standpoint of advantages versus disadvantages in any given case. **2.** A study to determine the time at which it would be practicable or desirable to install such a system when determined to be advantageous. **3.** A study to determine whether a plan is capable of being accomplished successfully. { ,fēz·ə'bil·əd·ē ,stəd·ē }

feasibility test [SYS ENG] A test conducted to obtain data in support of a feasibility study or to demonstrate feasibility. { ,fēz·ə'bil·əd·ē ,test }

feasible solution [COMPUT SCI] In linear programming, any set of values for the variables x_j, $j = 1$, $2, \ldots, n$, that (1) satisfy the set of restrictions

$$\sum_{j=1}^{n} a_{ij} x_j \leq b_i, i = 1, 2, \ldots, m$$

$$\left(\text{alternatively, } \sum_{j=1}^{n} a_{ij} x_j \leq b_i, \text{ or } \sum_{j=1}^{n} a_{ij} x_j \leq b_i \right)$$

where the b_i are numerical constants known collectively as the right-hand side and the a_{ij}

are coefficients of the variables x_j, and (2) satisfy the restrictions $x_j \geq 0$. { 'fēz·ə·bəl sə'lü·shən }

feature [COMPUT SCI] In automatic pattern recognition, a property of an image that is useful for its interpretation. { 'fē·chər }

feature extraction-classification model [COMPUT SCI] A method of automatic pattern recognition in which recognition is achieved by making measurements on the patterns to be recognized, and then deriving features from these measurements. { 'fē·chər ik¦strak·shən ,klas·ə·fə¦kā·shən ,mäd·əl }

Federal Telecommunications System [COMMUN] System of commercial telephone lines, leased by the government, for use between major government installations for official telecommunications. { 'fed·rəl ,tel·ə·kə,myü·nə'kā·shənz ,sis·təm }

fedsim star [COMPUT SCI] The starlike shape that is characteristic of the Kiviat graph of a well-balanced computer system. { 'fed,sim ,stär }

feed [COMPUT SCI] **1.** To supply the material to be operated upon to a machine. **2.** A device capable of so feeding. [ELECTR] To supply a signal to the input of a circuit, transmission line, or antenna. [ELECTROMAG] The part of a radar antenna that is connected to or mounted on the end of the transmission line and serves to radiate radio-frequency electromagnetic energy to the reflector or receive energy therefrom; in multiple-element (array) antennas, the constrained network, radiation means or digital means for distributing the energy to the radiating elements and collecting the energy received by them. { fēd }

feedback [ELECTR] The return of a portion of the output of a circuit or device to its input. { 'fēd ,bak }

feedback admittance [ELECTR] Short-circuit transadmittance from the output electrode to the input electrode of an electron tube. { 'fēd ,bak əd'mit·əns }

feedback amplifier [ELECTR] An amplifier in which a passive network is used to return a portion of the output signal to its input so as to change the performance characteristics of the amplifier. { 'fēd,bak 'am·plə,fī·ər }

feedback branch [CONT SYS] A branch in a signal-flow graph that belongs to a feedback loop. { 'fēd,bak ,branch }

feedback circuit [ELECTR] A circuit that returns a portion of the output signal of an electronic circuit or control system to the input of the circuit or system. { 'fēd,bak ,sər·kət }

feedback compensation [CONT SYS] Improvement of the response of a feedback control system by placing a compensator in the feedback path, in contrast to cascade compensation. Also known as parallel compensation. { 'fēd,bak ,käm·pən,sā·shən }

feedback control loop See feedback loop. { 'fēd ,bak kən'trōl ,lüp }

feedback control signal |CONT SYS| The portion of an output signal which is retransmitted as an input signal. { 'fēd,bak kən'trōl ,sig·nəl }

feedback control system |CONT SYS| A system in which the value of some output quantity is controlled by feeding back the value of the controlled quantity and using it to manipulate an input quantity so as to bring the value of the controlled quantity closer to a desired value. Also known as closed-loop control system. { 'fēd ,bak kən'trōl ,sis·təm }

feedback factor |ELECTR| The fraction of the output voltage of an oscillator which is applied to the feedback network. { 'fēd,bak ,fak·tər }

feedback loop |CONT SYS| A closed transmission path or loop that includes an active transducer and consists of a forward path, a feedback path, and one or more mixing points arranged to maintain a prescribed relationship between the loop input signal and the loop output signal. Also known as feedback control loop. { 'fēd,bak ,lüp }

feedback oscillator |ELECTR| An oscillating circuit, including an amplifier, in which the output is fed back in phase with the input; oscillation is maintained at a frequency determined by the values of the components in the amplifier and the feedback circuits. { 'fēd,bak ,äs·ə,lād·ər }

feedback regulator |CONT SYS| A feedback control system that tends to maintain a prescribed relationship between certain system signals and other predetermined quantities. { 'fēd,bak ,reg·yə,lād·ər }

feedback transfer function |CONT SYS| In a feedback control loop, the transfer function of the feedback path. { 'fēd,bak 'tranz·fər ,faŋk·shən }

feedback winding |ELECTR| A winding to which feedback connections are made in a magnetic amplifier. { 'fēd,bak ,wīnd·iŋ }

feeder |ELEC| 1. A transmission line used between a transmitter and an antenna. 2. A conductor, or several conductors, connecting generating stations, substations, or feeding points in an electric power distribution system. 3. A group of conductors in an interior wiring system which link a main distribution center with secondary or branch-circuit distribution centers. { 'fēd·ər }

feeder cable |COMMUN| In communications practice, a cable extending from the central office along a primary route (main feeder cable) or from a main feeder cable along a secondary route (branch feeder cable) and providing connections to one or more distribution cables. { 'fēd·ər ,kā·bəl }

feeder distribution center |COMMUN| Distribution center at which feeders or subfeeders are connected. { 'fēd·ər dis·trə'byü·shən ,sen·tər }

feeder panel |ELEC| The part of a switchboard in an electric power distribution system where feeder connections are made. { 'fēd·ər ,pan·əl }

feeder reactor |ELEC| A small inductor connected in series with a feeder in order to limit and localize the disturbances due to faults on the feeder. { 'fēd·ər rē,ak·tər }

feedforward control |CONT SYS| Process control in which changes are detected at the process input and an anticipating correction signal is applied before process output is affected. { |fēd |fȯr·wərd kən,trōl }

feed holes |COMPUT SCI| Holes along the edges of continuous-feed computer paper that are engaged by sprockets to move the paper and maintain alignment during printing. { 'fēd ,hōlz }

feed horn |ELECTROMAG| A device located at the focus of a receiving paraboloidal antenna that acts as a receiver of radio waves which the antenna collects, focuses, and couples to transmission lines to the amplifier. { 'fēd ,hȯrn }

feed reel |ENG| The reel from which paper tape or magnetic tape is being fed. { 'fēd ,rēl }

feed shelf |COMPUT SCI| 1. A device for supporting documents for manual sensing. 2. The first few feet of a tape reel, used to prime the tape drive. { 'fēd ,shelf }

feed-tape |COMPUT SCI| A mechanism which will feed tape to be read or sensed. { 'fēd,tāp }

feedthrough |ELEC| A conductor that connects patterns on opposite sides of a printed circuit board. Also known as interface connection. { 'fēd ,thrü }

feedthrough capacitor |ELEC| A feedthrough terminal that provides a desired value of capacitance between the feedthrough conductor and the metal chassis or panel through which the conductor is passing; used chiefly for bypass purposes in ultra-high-frequency circuits. { 'fēd ,thrü kə'pas·əd·ər }

feedthrough insulator See feedthrough terminal. { 'fēd,thrü 'in·sə,lād·ər }

feedthrough terminal |ELEC| An insulator designed for mounting in a hole in a panel, wall, or bulkhead, with a conductor in the center on the insulator to permit feeding electricity through the partition. Also known as feedthrough insulator. { 'fēd,thrü 'tərm·ən·əl }

female connector |ELEC| A connector having one or more contacts set into recessed openings; jacks, sockets, and wall outlets are examples. { |fē,māl kə'nek·tər }

femitrons |ELECTR| Class of field-emission microwave devices. { 'fem·ə,tränz }

femtoampere |ELEC| A unit of current equal to 10^{-15} ampere. Abbreviated fA. { |fem·tō|am·pir }

femtovolt |ELEC| A unit of voltage equal to 10^{-15} volt. Abbreviated fV. { 'fem·tō,vōlt }

fence |ENG| 1. A line of data-acquisition or tracking stations used to monitor orbiting satellites. 2. A line of radar or radio stations for detection of satellites or other objects in orbit. 3. A line or network of early-warning radar stations. 4. A concentric steel fence erected around a ground radar transmitting antenna to serve as an artificial horizon and suppress ground clutter that would otherwise drown out weak signals returning at a low angle from a target. { fens }

fence cell |COMPUT SCI| A criterion for dividing a list into two equal or nearly equal parts in the course of a binary search. { 'fens ,sel }

Ferranti effect

Ferranti effect [ELEC] A rise in voltage occurring at the end of a long transmission line when its load is disconnected. { fə'ran·tē i,fekt }

ferreed switch [ELEC] A switch whose contacts are mounted on magnetic blades or reeds sealed into an evacuated tubular glass housing, the contacts being operated by external electromagnets or permanent magnets. { 'fe,rēd ,swich }

ferrimagnetic amplifier [ELECTR] A microwave amplifier using ferrites. { ,fe·ri·mag'ned·ik 'am·plə,fī·ər }

ferristor [ELECTR] A miniature, two-winding, saturable reactor that operates at a high carrier frequency and may be connected as a coincidence gate, current discriminator, free-running multivibrator, oscillator, or ring counter. { fə'ris·tər }

ferrite attenuator See ferrite limiter. { 'fe,rīt ə'ten·yə,wād·ər }

ferrite bead [ELECTR] Magnetic information storage device consisting of ferrite powder mixtures in the form of a bead fired on the current-carrying wires of a memory matrix. { 'fe,rīt 'bēd }

ferrite circulator [ELECTROMAG] A combination of two dual-mode transducers and a 45° ferrite rotator, used with rectangular waveguides to control and switch microwave energy. Also known as ferrite phase-differential circulator. { 'fe,rīt 'sər·kyə,lād·ər }

ferrite core [ELECTR] A magnetic core made of ferrite material. Also known as dust core; powdered-iron core. { 'fe,rīt 'kòr }

ferrite-core memory [ELECTR] A magnetic memory consisting of a matrix of tiny toroidal cores molded from a square-loop ferrite, through which are threaded the pulse-carrying wires and the sense wire. { 'fe,rīt ,kòr 'mem·rē }

ferrite device [ELEC] An electrical device whose principle of operation is based upon the use of ferrites in powdered, compressed, sintered form, making use of their ferrimagnetism and their high electrical resistivity, which makes eddy-current losses extremely low at high frequencies. { 'fe ,rīt di,vīs }

ferrite isolator [ELECTROMAG] A device consisting of a ferrite rod, centered on the axis of a short length of circular waveguide, located between rectangular-waveguide sections displaced 45° with respect to each other, which passes energy traveling through the waveguide in one direction while absorbing energy from the opposite direction. Also known as Faraday rotation isolator. { 'fe,rīt 'ī·sə,lād·ər }

ferrite limiter [ELECTROMAG] A passive, low-power microwave limiter having an insertion loss of less than 1 decibel when operating in its linear range, with minimum phase distortion; the input signal is coupled to a single-crystal sample of either yttrium iron garnet or lithium ferrite, which is biased to resonance by a magnetic field. Also known as ferrite attenuator. { 'fe,rīt 'lim·əd·ər }

ferrite phase-differential circulator See ferrite circulator. { 'fe,rīt ¦fāz dif·ə¦ren·chəl 'sər·kyə,lād·ər }

ferrite-rod antenna [ELECTROMAG] An antenna consisting of a coil wound on a rod of ferrite; used in place of a loop antenna in radio receivers. Also known as ferrod; loopstick antenna. { 'fe,rīt¦räd an'ten·ə }

ferrite rotator [ELECTROMAG] A gyrator consisting of a ferrite cylinder surrounded by a ring-type permanent magnet, inserted in a waveguide to rotate the plane of polarization of the electromagnetic wave passing through the waveguide. { 'fe,rīt 'rō,tād·ər }

ferrite switch [ELECTROMAG] A ferrite device that blocks the flow of energy through a waveguide by rotating the electric field vector 90°; the switch is energized by sending direct current through its magnetizing coil; the rotated electromagnetic wave is then reflected from a reactive mismatch or absorbed in a resistive card. { 'fe,rīt 'swich }

ferrite-tuned oscillator [ELECTR] An oscillator in which the resonant characteristic of a ferrite-loaded cavity is changed by varying the ambient magnetic field, to give electronic tuning. { 'fe ,rīt ¦tünd 'äs·ə,lād·ər }

ferroacoustic storage [ELECTR] A delay-line type of storage consisting of a thin tube of magnetostrictive material, a central conductor passing through the tube, and an ultrasonic driving transducer at one end of the tube. { ¦fe·rō·ə ¦küs·tik 'stòr·ij }

ferrod See ferrite-rod antenna. { 'fe,räd }

ferroelectric [SOLID STATE] A crystalline substance displaying ferroelectricity, such as barium titanate, potassium dihydrogen phosphate, and Rochelle salt; used in ceramic capacitors, acoustic tranducers, and dielectric amplifiers. Also known as Rochelle-electric. { ¦fe·rō·i'lek·trik }

ferroelectric converter [ELEC] A converter that transforms thermal energy into electric energy by utilizing the change in the dielectric constant of a ferroelectric material when heated beyond its Curie temperature. { ¦fe·rō·i'lek·trik kən 'vərd·ər }

ferroelectric hysteresis [ELEC] The dependence of the polarization of ferroelectric materials not only on the applied electric field but also on their previous history; analogous to magnetic hysteresis in ferromagnetic materials. Also known as dielectric hysteresis; electric hysteresis. { fe·rō·i'lek·trik ,his·tə'rē·səs }

ferroelectric hysteresis loop [ELEC] Graph of polarization or electric displacement versus applied electric field of a material displaying ferroelectric hysteresis. { ¦fe·rō·i'lek·trik ,his·tə'rē·səs ,lüp }

ferroelectricity [SOLID STATE] Spontaneous electric, polarization in a crystal; analogous to ferromagnetism { ¦fe·rō·i'lek·tris·əd·ē }

ferroelectric liquid-crystal display [ELECTR] An electronic display that employs a liquid crystal that is ferroelectric, such as smectic C*, which has two different stable molecular configurations; polarizers are positioned such that one state is optically transmissive while the other is dark. { ,fer·ō·i¦lek·trik ¦lik·wəd ¦kris·təl dis¦plä }

ferromagnetic amplifier [ELECTR] A parametric amplifier based on the nonlinear behavior of ferromagnetic resonance at high radio-frequency power levels; incorrectly known as garnet maser. { ¦fe·rō·mag¦ned·ik 'am·plə‚fī·ər }

ferromagnetic film See magnetic thin film.. { ¦fe·rō·mag¦ned·ik 'ilm }

ferromagnetic resonance [SOLID STATE] Magnetic resonance of a ferromagneti material { ¦fe·rō·mag¦ned·ik 'rez·ən·əns }

ferromagnetics [ELECTR] The science that deals with the storage of binary information and the logical control of pulse sequences through the utilization of the magnetic polarization properties of materials. { ¦fe·rō·mag¦ned·iks }

ferromagnetism [SOLID STATE] A property, exhibited by certain metals, alloys, and compounds of the transition (iron group rare-earth and actinide elements, in which the internal magnetic moments spontaneously organize in a common direction; gives rise to a permeability considerably greater than that of vaccum, and to magnetic hysteresis { ¦fe·rō·magnə‚tiz·əm }

ferroresonant circuit [ELECTR] A resonant circuit in which a saturable reactor provides nonlinear characteristics, with tuning being accomplished by varying circuit voltage or current. { ¦fe·rō'rez·ən·ənt 'sər·kət }

ferroresonant power supply [ELECTR] A transformer-based power supply, employed in high-current applications such as battery chargers, that uses nonlinear magnetic properties and a resonant circuit to regulate the output current. { ‚fe·rō‚rez·ən·ənt 'paú·ər sə‚plī }

ferroresonant static inverter [ELEC] A static inverter consisting of a simple square-wave inverter system and a tuned output transformer that performs filtering, voltage regulation, and current limiting. { ¦fe·rō'rez·ən·ənt ¦stad·ik in'vərd·ər }

FET See field-effect transistor.

fetch [COMPUT SCI] To locate and load into main memory a requested load module, relocating it as necessary and leaving it in a ready-to-execute condition. { fech }

fetch ahead See instruction lookahead. { ¦fech ə'hed }

fetch bit [COMPUT SCI] The fifth bit in a storage key; the value of the fetch bit can protect a stored block from destruction or from being accessed by unauthorized programs. { 'fech ‚bit }

fetch cycle [COMPUT SCI] The period during which a machine language instruction is read from memory into the control section of the central processing unit. { 'fech ‚sī·kəl }

F format [COMPUT SCI] **1.** In data management, a fixed-length logical record format. **2.** In FORTRAN, a real variable formatted as Fμ.d, where μ is the width of the field and d represents the number of digits to appear after the decimal point. { 'ef ‚fór·mat }

fiber [OPTICS] A transparent threadlike object made of glass or clear plastic, used to conduct light along selected paths. { 'fī·bər }

fiber bundle [OPTICS] A flexible bundle of glass or other transparent fibers, parallel to each other, used in fiber optics to transmit a complete image from one end of the bundle to the other. { 'fī·bər ‚bən·dəl }

fiber-optic circuit [COMMUN] A path for data transmission in which light acts as the information carrier and is transmitted by total internal reflection through a transparent optical waveguide. { ¦fī·bər ¦äp·tik 'sər·kət }

fiber-optic current sensor [ENG] An instrument for measuring currents on high-voltage lines, in which the magnetic field associated with the current changes the phase of light traveling through an optical fiber, and the phase change is measured in an interferometer. { 'fī·bər ¦äp·tik 'kə·rənt ‚sen·sər }

fiber-optic data distribution interface [COMMUN] A set of standards for high-speed fiber-optic local-area networks. Abbreviated FDDI. { ¦fī·bər ¦äp·tik ¦dad·ə ‚dis·trə¦byü·shən 'in·tər‚fās }

fiber optics [OPTICS] The technique of transmitting light through long, thin, flexible fibers of glass, plastic, or other transparent materials; bundles of parallel fibers can be used to transmit complete images. { 'fī·bər ‚äp·tiks }

fiber-optic sensor See optical-fiber sensor. { 'fī·bər ¦äp·tik 'sen·sər }

fiber waveguide See optical waveguide. { 'fī·bər 'wāv‚gīd }

fidelity [COMMUN] The degree to which a system accurately reproduces at its output the essential characteristics of the signal impressed on its input. { fə'del·əd·ē }

field [COMPUT SCI] **1.** A location in a record in a database that contains a specific piece of information **2.** A specified area on a geographical user interface for the input of a particular category of data. [ELECTR] One of the equal parts into which a frame is divided in interlaced scanning for television; includes one complete scanning operation from top to bottom of the picture and back again. { fēld }

fieldata code [COMMUN] A standardized military data transmission code, seven data bits plus one parity bit. { 'fēl‚dad·ə ‚kōd }

field delimiter [COMPUT SCI] Any symbol, such as a slash, colon, tab, or space, which enables an assembler to recognize the end of a field. { 'fēld də‚lim·əd·ər }

field designator [COMPUT SCI] A character generally placed at the beginning of a field to specify the nature of the data contained in it. { 'fēld ¦dez·ig‚nād·ər }

field-desorption microscope [ELECTR] A type of field-ion microscope in which the tip specimen is imaged by ions that are field-desorbed or field-evaporated directly from the surface rather than by ions obtained from an externally supplied gas. { 'fēld dē,sȯrp·shən ,mī·krə,skōp }

field discharge [ELECTR] A spark discharge due to high potential across a gap. { 'fēld ¦dis,chärj }

field-discharge switch [ELEC] A special type of switch that is connected in series with the field winding of an electrical machine, and that is operated to connect a resistor in parallel with the field winding before the main supply contacts are opened, in order to prevent the self-induced electromotive force in the field winding from reaching dangerous levels. { 'fēld ¦dis,chärj ,swich }

field effect [ELECTR] The local change from the normal value that an electric field produces in the charge-carrier concentration of a semiconductor. { 'fēld i,fekt }

field-effect capacitor [ELECTR] A capacitor in which the effective dielectric is a region of semiconductor material that has been depleted or inverted by the field effect. { 'fēld i,fekt kə'pas·əd·ər }

field-effect device [ELECTR] A semiconductor device whose properties are determined largely by the effect of an electric field on a region within the semiconductor. { 'fēld i,fekt di,vīs }

field-effect diode [ELECTR] A semiconductor diode in which the charge carriers are of only one polarity. { 'fēld i,fekt 'dī,ōd }

field-effect phototransistor [ELECTR] A field-effect transistor that responds to modulated light as the input signal. { 'fēld i,fekt ¦fōd·ō·tran'zis·tər }

field-effect tetrode [ELECTR] Four-terminal device consisting of two independently terminated semiconducting channels so displaced that the conductance of each is modulated along its length by the voltage conditions in the other. { 'fēld i,fekt 'te,trōd }

field-effect transistor [ELECTR] A transistor in which the resistance of the current path from source to drain is modulated by applying a transverse electric field between grid or gate electrodes; the electric field varies the thickness of the depletion layer between the gates, thereby reducing the conductance. Abbreviated FET. { 'fēld i,fekt tran'zis·tər }

field-effect-transistor resistor [ELECTR] A field-effect transistor in which the gate is generally tied to the drain; the resultant structure is used as a resistance load for another transistor. { 'fēld i,fekt tran¦zis·tər ri¦zis·tər }

field-effect varistor [ELECTR] A passive, two-terminal, nonlinear semiconductor device that maintains constant current over a wide voltage range. { 'fēld i,fekt və'ris·tər }

field emission [ELECTR] The emission of electrons from the surface of a metallic conductor into a vacuum (or into an insulator) under influence of a strong electric field; electrons penetrate through the surface potential barrier by virtue of the quantum-mechanical tunnel effect. Also known as cold emission. { 'fēld ə,mish·ən }

field-emission display [ELECTR] A flat-panel electronic display in which electrons are extracted from an array of cold-cathode emitters by applying a voltage between the cathode and a control electrode, and the electrons are then accelerated without deflection over a distance of less than 1 millimeter before colliding with a phosphor-coated flat faceplate. { 'fēld i ,mish·ən di,splā }

field-emission microscope [ELECTR] A device that uses field emission of electrons or of positive ions (field-ion microscope) to produce a magnified image of the emitter surface on a fluorescent screen. { 'fēld ə¦mish·ən 'mī·krə ,skōp }

field-emission tube [ELECTR] A vacuum tube within which field emission is obtained from a sharp metal point; must be more highly evacuated than an ordinary vacuum tube to prevent contamination of the point. { 'fēld ə¦mish·ən ,tüb }

field-emitter array [ELECTR] An array of pyramidal silicon structures, with spacing on the order of 10 micrometers, designed for field emission of electrons into a vacuum. { ¦fēld i¦mid·ər ə'rā }

field engineer [COMPUT SCI] A professional who installs computer hardware on customers' premises, performs routine preventive maintenance, and repairs equipment when it is out of order. Also known as field service representative. { 'fēld en·jə,nir }

field-enhanced emission [ELECTR] An increase in electron emission resulting from an electric field near the surface of the emitter. { ¦fēld in ¦hanst i'mish·ən }

field-free emission current [ELECTR] Electron current emitted by a cathode when the electric field at the surface of the cathode is zero. Also known as zero-field emission. { 'fēld ,frē i'mish·ən ,kə·rənt }

field frequency [ELECTR] The number of fields transmitted per second in a video system; equal to the frame frequency multiplied by the number of fields that make up one frame. Also known as field repetition rate. { 'fēld ,frē·kwən·sē }

field intensity [COMMUN] In Federal Communications Commission regulations, the electric field intensity in the horizontal direction. { 'fēld in ,ten·səd·ē }

field ionization [ELECTR] The ionization of gaseous atoms and molecules by an intense electric field, often at the surface of a solid. { 'fēld ,ī·ən·ə'zā·shən }

field-ion microscope [ELECTR] A microscope in which atoms are ionized by an electric field near a sharp tip; the field then forces the ions to a fluorescent screen, which shows an enlarged image of the tip, and individual atoms are made visible; this is the most powerful microscope yet produced. Also known as ion microscope. { ¦fēld ¦ī,än 'mī·krə,skōp }

field length [COMPUT SCI] The number of columns, characters, or bits in a specified field. { 'fēld ,leŋkth }

field of search |ELECTR| The space that a radar set or installation can cover effectively. { 'fēld əv 'sərch }

field of view |ELECTR| The space in which a radar can operate effectively. { 'fēld əv 'vyü }

field pattern *See* radiation pattern. { 'fēld ‚pad‚ərn }

field-programmable gate array |ELECTR| A gate-array device that can be configured and reconfigured by the system manufacturer and sometimes by the end user of the system. { ¦fēld prō‚gram‚ə‚bəl 'gāt ə‚rā }

field-programmable logic array |ELECTR| A programmed logic array in which the internal connections of the logic gates can be programmed once in the field by passing high current through fusible links, by using avalanche-induced migration to short base-emitter junctions at desired interconnections, or by other means. Abbreviated FPLA. Also known as programmable logic array. { 'fēld prō¦gram‚ə‚bəl 'läj‚ik ə'rā }

field repetition rate *See* field frequency. { 'fēld rep‚ə'tish‚ən ‚rāt }

field rheostat |ELEC| A rheostat used to adjust the current in the field winding of an electric machine. { 'fēld ‚rē‚ə‚stat }

field scan |ELECTR| Television term denoting the vertical excursion of an electron beam downward across a cathode-ray tube face, the excursion being made in order to scan alternate lines. { 'fēld ‚skan }

field section |COMPUT SCI| A portion of a field, such as the section formed by the second and third character of a 10-character field. { 'fēld ‚sek‚shən }

field separator |COMPUT SCI| A character that is used to mark the boundary between fields in a record. { 'fēld ‚sep‚ə‚rād‚ər }

field-sequential color television |COMMUN| An analog color television system in which the individual red, green, and blue primary colors are associated with successive fields. { ¦fēld sə ¦kwen‚chəl ¦kəl‚ər 'tel‚ə‚vizh‚ən }

field service representative *See* field engineer. { 'fēld ‚sər‚vəs ‚rep‚rə‚zent‚əd‚iv }

field squeeze |COMPUT SCI| In a mail merge operation, the elimination of extra blank spaces in a data field so that the data field is correctly printed within the text of the letter. { 'fēld ‚skwēz }

field-strength meter |ENG| A calibrated radio receiver used to measure the field strength of radiated electromagnetic energy from a radio transmitter. { 'fēld ‚streŋkth ‚mēd‚ər }

field telephone |COMMUN| A portable telephone designed for field or combat use. { 'fēld ‚tel‚ə ‚fōn }

field waveguide |ELECTROMAG| A single wire, threaded or coated with dielectric, which guides an electromagnetic field. Also known as G string. { 'fēld 'wāv‚gīd }

field wire |ELEC| An insulated flexible wire or cable used in field telephone and telegraph systems. { 'fēld ‚wīr }

fifth-generation computer |COMPUT SCI| A computer that would use artificial intelligence techniques to learn, reason, and converse in natural languages resembling human languages. { 'fifth ‚jen‚ə¦rā‚shən kəm'pyüd‚ər }

figurative constant |COMPUT SCI| A predefined constant in COBOL which does not require a description in data division, such as ZERO which stands for 0. { 'fig‚yə‚rəd‚iv 'kän‚stənt }

figure of merit |ELECTR| A performance rating that governs the choice of a device for a particular application; for example, the figure of merit of a magnetic amplifier is the ratio of usable power gain to the control time constant. { 'fig‚yər əv 'mer‚ət }

filament |ELEC| Metallic wire or ribbon which is heated in an incandescent lamp to produce light, by passing an electric current through the filament. |ELECTR| A cathode made of resistance wire or ribbon, through which an electric current is sent to produce the high temperature required for emission of electrons in a thermionic tube. Also known as directly heated cathode; filamentary cathode; filament-type cathode. { 'fil‚ə‚mənt }

filamentary cathode *See* filament. { ‚fil‚ə'ment‚ə‚rē }

filament current |ELECTR| The current supplied to the filament of an electron tube for heating purposes. { 'fil‚ə‚mənt ‚kə‚rənt }

filament emission |ELECTR| Liberation of electrons from a heated filament wire in an electron tube. { 'fil‚ə‚mənt i'mish‚ən }

filament lamp *See* incandescent lamp. { 'fil‚ə‚mənt ‚lamp }

filament saturation *See* temperature saturation. { 'fil‚ə‚mənt ‚sach‚ə'rā‚shən }

filament transformer |ELECTR| A small transformer used exclusively to supply filament or heater current for one or more electron tubes. { 'fil‚ə‚mənt tranz‚fór‚mər }

filament-type cathode *See* filament. { 'fil‚ə‚mənt ‚tīp 'kath‚ōd }

filament winding |ELECTR| The secondary winding of a power transformer that furnishes alternating-current heater or filament voltage for one or more electron tubes. { 'fil‚ə‚mənt ‚wīnd‚iŋ }

file |COMPUT SCI| A collection of related records treated as a unit. { fīl }

file allocation table |COMPUT SCI| A table stored on hard or removable disks used to locate files or sections of files if scattered about the disk. Abbreviated FAT. { ‚fīl ‚al‚ə'kā‚shən ‚tā‚bəl }

file compression program *See* file compression utility. { 'fīl kəm‚presh‚ən ‚prō‚grəm }

file compression utility |COMPUT SCI| A utility program that encodes files so that they take up less space in storage. Also known as file compression program. { 'fīl kəm‚presh‚ən yü ‚til‚əd‚ē }

file control system |COMPUT SCI| Software package which handles the transfer of data from any device into any device. { 'fīl kən‚trōl ‚sis‚təm }

file event |COMPUT SCI| A single access to any storage device for either input or output. { 'fīl i ‚vent }

file format [COMPUT SCI] The rules that determine the organization of data in a file. { 'fīl ,fȯr·mat }

file gap [COMPUT SCI] An area in a data storage medium which is used mainly to indicate the end of a file and sometimes the beginning of another. { 'fīl ,gap }

file-handling routine [COMPUT SCI] A part of a computer program that deals with reading and writing of data from and to a file. { 'fīl ,hand·liŋ rü,tēn }

file header [COMPUT SCI] A set of words comprising the file name and various characteristics of the file, found at the beginning of a file stored on magnetic tape or disk. { 'fīl ,hed·ər }

file identification [COMPUT SCI] A device, such as a label or tag, used to identify, describe, or name a physical medium, such as a disk or reel of magnetic tape, which contains data. { 'fīl ī,dent·ə·fə'kā·shən }

file layout [COMPUT SCI] A description of the arrangement of the data in a file. { 'fīl ,lā,aût }

file locking [COMPUT SCI] A technique that prevents processing of a file by more than one program or user at a time, ensuring that a file in use by one user is made unavailable to others. { 'fil ,läk·iŋ }

file maintenance [COMPUT SCI] Data-processing operation in which a master file is updated on the basis of one or more transaction files. { 'fīl ,mānt·ən·əns }

file management system [COMPUT SCI] Computer programs that control the space used for file storage and provide such services as input/output control and indexing. { 'fīl ¦man·ij·mənt ,sis·təm }

file manager [COMPUT SCI] Software for managing data that works only with single files and lacks relational capability. { 'fīl ,man·ə·jər }

file name [COMPUT SCI] The name given by the programmer to a specific set of data. { 'fīl ,nām }

file opening [COMPUT SCI] The process, carried out by computer software, of identifying a file and comparing the file header with specifications in the program being run to ensure that the file corresponds. { 'fīl ,ōp·ə·niŋ }

file organization [COMPUT SCI] The structure of a file meeting two requirements: to minimize the running time of the program, and to simplify the work involved in modifying the contents of the file. { ¦fīl ,ȯrg·ə·nə'zā·shən }

file organization routine [COMPUT SCI] A program which allocates data files into random-access storage devices. { ¦fīl ,ȯrg·ə·nə'zā·shən rü,tēn }

file-oriented system [COMPUT SCI] A computer configuration which considers a heavy, or exclusive, usage of data files. { 'fīl ,ȯr·ē,ent·əd ,sis·təm }

file printout [COMPUT SCI] Output from a computer printer consisting of a copy of the contents of a file held in some storage device, usually to assist in debugging a program. { 'fīl ,prin,taût }

file processing [COMPUT SCI] The job of updating, sorting, or validating a data file. { 'fīl ,präs ,es·iŋ }

file protection [COMPUT SCI] A mechanical device or a computer command which prevents erasing of or writing upon a magnetic tape but allows a program to read the data from the tape. { 'fīl prə,tek·shən }

file protection ring [COMPUT SCI] A ring that can be attached to, or detached from, the hub of a reel of magnetic tape, used to identify the reel's status and, in some computer systems, to prevent writing upon the tape when the ring is attached or detached. { 'fīl prə,tek·shən ,riŋ }

file reference [COMPUT SCI] An operation involving looking up and retrieving the information on file for a specified item or items. { 'fīl ,ref·rəns }

file reorganization [COMPUT SCI] An activity performed periodically on files and data bases, involving such operations as deletion of unneeded records, in order to minimize space requirements of files and improve efficiency of processing. { 'fīl rē,ōr·gə·nə'zā·shən }

file search [COMPUT SCI] An operation involving looking through the file for information on all items falling in a specified category, extracting the information for any item where the information recorded meets certain criteria, and determining whether or not there exists a specified pattern of information anywhere in the file. { 'fīl ,sərch }

file security See data security. { 'fīl sə,kyür·əd·ē }

file server [COMPUT SCI] A mass storage device that holds programs and data that can be accessed and shared by the workstations connected to a local-area network. Also known as network server. { 'fīl ,sər·vər }

file sharing [COMPUT SCI] The common use, by two or more users, of data and program files, usually located in a file server. { 'fīl ,sher·iŋ }

FileSize metric [COMPUT SCI] A measure of computer program size, equal to the total number of characters in the source file of the program. { ¦fīl'sīz ¦me·trik }

file specification [COMPUT SCI] A designation that enables a file to be located on a disk and includes the disk drive, name of the directory/subdirectory, and name of file. { 'fīl ,spes·ə·fə,kā·shən }

file storage unit [COMPUT SCI] The component of a computer system that stores information required for reference. { 'fīl ,stȯr·ij ,yü·nət }

file transfer [COMPUT SCI] The movement, under program control, of a file from one storage device to another. { 'fīl ,tranz·fər }

file transfer access and management [COMPUT SCI] A standard communications protocol for transferring files between systems of different vendors. Abbreviated FTAM. { ¦fīl ¦tranz·fər ¦ak ,ses ən 'man·ij·mənt }

file transfer protocol [COMPUT SCI] A set of standards that allows the user of any computer on the Internet to receive files from another computer, or to transmit files to another computer, after the user has specified a name and password for the other computer. Abbreviated FTP. { 'fīl ¦tranz·fər ,prōd·ə·kȯl }

file transfer utility |COMPUT SCI| A computer program specifically designed to handle file transfers. { 'fīl ¦tranz·fər yü,til·əd·ē }

file virus |COMPUT SCI| A computer virus that infects application files such as spreadsheets, computer games, or accounting software. { 'fīl ,vī·rəs }

fill characters |COMPUT SCI| Nondata characters or bits which are used to fill out a field on the left if data are right-justified or on the right if data are left-justified. { 'fil ,kar·ik·tərz }

filler |COMPUT SCI| Storage space that does not contain significant data but is needed to comply with length requirements or is reserved to fulfill some future need. { 'fil·ər }

film |ELEC| The layer adjacent to the valve metal in an electrochemical valve, in which is located the high voltage drop when current flows in the direction of high impedance. { film }

film integrated circuit |ELECTR| An integrated circuit whose elements are films formed in place on an insulating substrate. { ¦film int·ə¦grād·əd 'sər·kət }

film optical-sensing device |COMPUT SCI| A device capable of digitizing the information stored on a film. { 'film ¦äp·tə·kəl ¦sens·iŋ di,vīs }

film reader |ELECTR| A device for converting a pattern of transparent or opaque spots on a photographic film into a series of electric pulses. |OPTICS| A device for projecting or displaying microfilm so that an operator can read the data on the film; usually provided with equipment for moving or holding the film. { 'film ,rēd·ər }

film recorder |ELECTR| A device which places data, usually in the form of transparent and opaque spots or light and dark spots, on photographic film. { 'film ri,kòrd·ər }

film resistor |ELEC| A fixed resistor in which the resistance element is a thin layer of conductive material on an insulated form; the conductive material does not contain binders or insulating material. { 'film ri,zis·tər }

film scanning |ELECTR| The process of converting motion picture film into corresponding electric signals that can be transmitted by a video system. { 'film ,skan·iŋ }

filter |COMPUT SCI| A device or program that separates data or signals in accordance with specified criteria. |CONT SYS| See compensator. |ELECTR| Any transmission network used in electrical systems for the selective enhancement of a given class of input signals. Also known as electric filter; electric-wave filter. |ENG ACOUS| A device employed to reject sound in a particular range of frequencies while passing sound in another range of frequencies. Also known as acoustic filter. |OPTICS| An optical element that partially absorbs incident electromagnetic radiation in the visible, ultraviolet, or infrared spectra, consisting of a pane of glass or other partially transparent material, or of films separated by narrow layers; the absorption may be either selective or nonselective with respect to wavelength. Also known as optical filter. { 'fil·tər }

filter capacitor |ELEC| A capacitor used in a power-supply filter system to provide a low-reactance path for alternating currents and thereby suppress ripple currents, without affecting direct currents. { 'fil·tər kə,pas·əd·ər }

filter choke |ELEC| An iron-core coil used in a power-supply filter system to pass direct current while offering high impedance to pulsating or alternating current. { 'fil·tər ,chōk }

filter crystal |ELECTR| Quartz crystal which is used in an electrical circuit designed to pass energy of certain frequencies. { 'fil·tər ,krist·əl }

filter design |ELECTR| The design of electrical networks in which the principle of electrical resonance is used to make the network accept wanted frequencies while rejecting unwanted ones. { 'fil·tər di,zīn }

filter discrimination |ELECTR| Difference between the minimum insertion loss at any frequency in a filter attenuation band and the maximum insertion loss at any frequency in the operating range of a filter transmission band. { 'fil·tər di,skrim·ə'nā·shən }

filtered radar data |ELECTR| Radar data from which unwanted returns have been removed by mapping. { ¦fil·tərd 'rā,där ,dad·ə }

filter impedance compensator |ELECTR| Impedance compensator which is connected across the common terminals of electric wave filters when the latter are used in parallel to compensate for the effects of the filters on each other. { 'fil·tər im'ped·əns ,käm·pən'sād·ər }

filter pass band See filter transmission band. { 'fil·tər 'pas ,band }

filter reactor |ELEC| A reactor used for reducing the harmonic components of voltage in an alternating-current or direct-current circuit. { 'fil·tər rē,ak·tər }

filter section |ELEC| A simple RC, RL, or LC network used as a broad-band filter in a power supply, grid-bias feed, or similar device. { 'fil·tər ,sek·shən }

filter slot |ELECTROMAG| Choke in the form of a slot designed to suppress unwanted modes in a waveguide. { 'fil·tər ,slät }

filter transmission band |ELECTR| Frequency band of free transmission; that is, frequency band in which, if dissipation is neglected, the attenuation constant is zero. Also known as filter pass band. { ¦fil·tər tranz'mish·ən ,band }

final amplifier |ELECTR| The transmitter stage that feeds the antenna. { ¦fīn·əl 'am·plə,fī·ər }

financial planning system |COMPUT SCI| A decision-support system that allows the financial planner or manager to examine and evaluate many alternatives before making final decisions, and which employs the use of a model, usually a matrix of data elements which is constructed as a series of equations. { fī ¦nan·chəl 'plan·iŋ ,sis·təm }

finder |COMMUN| Switch or relay group in telephone switching systems that selects the path which the call is to take through the system; operates under the instruction of the calling station's dial. { 'fīnd·ər }

finder beam |COMPUT SCI| A beam of light projected by a light pen on the spot on the display screen where the light pen photodetector is focused, in order to aid the user in positioning the light pen. { 'fīnd·ər ,bēm }

F-indicator See F-display. { 'ef ,in·də,kād·ər }

finding circuit See lockout circuit. { 'fīnd·iŋ ,sər·kət }

fine index |COMPUT SCI| The more specific of two indices consulted to gain access to a record. { ¦fīn 'in,deks }

finger gripper |CONT SYS| A robot component that uses two or more joints for grasping objects. { 'fiŋ·gər ,grip·ər }

finite clipping |ELECTR| Clipping in which the threshold level is large but is below the peak input signal amplitude. { ¦fī,nīt 'klip·iŋ }

finite difference |MATH| The difference between the values of a function at two discrete points, used to approximate the derivative of the function. { ¦fī,nīt 'dif·rəns }

finite-difference equations |MATH| Equations arising from differential equations by substituting difference quotients for derivatives, and then using these equations to approximate a solution. { ¦fī,nīt ¦dif·rəns i,kwā·zhənz }

finite element method |ENG| A numerical analysis technique for obtaining approximate descriptions of continuous physical systems, used in structural mechanics, electrical field theory, and fluid mechanics; the system is broken into discrete elements interconnected at discrete node points, and the values of various physical quantities for the elements or node points are calculated numerically. { ¦fī,nīt 'el·ə·mənt ,meth·əd }

finite impulse response filter |ELECTR| An electric filter that will settle to a steady state within a finite amount of time after being exposed to a change in input. Abbreviated FIR filter. { ¦fī,nīt ,im,pəls ri'späns ,fil·tər }

finite precision number |COMPUT SCI| A number that can be represented by a finite set of symbols in a given numeration system. { ¦fī,nīt prə ¦sizh·ən 'nəm·bər }

finite-state machine |COMPUT SCI| An automaton that has a finite number of distinguishable internal configurations. { 'fī,nīt ¦stāt mə,shēn }

Finsen lamp |ELEC| A high-temperature carbon arc or mercury arc lamp that produces a mixture of blue, violet, and near-ultraviolet light; used to treat certain skin disorders and to test paints and other protective coatings. { 'fin·sən ,lamp }

fin waveguide |ELECTROMAG| Waveguide containing a thin longitudinal metal fin that serves to increase the wavelength range over which the waveguide will transmit signals efficiently; usually used with circular waveguides. { ¦fin 'wāv,gīd }

fire-control circuit |ELECTR| An electric circuit in a fire-control system. { 'fīr kən,trōl ,sər·kət }

fired state |ELECTR| The "on" state of a silicon controlled rectifier or other semiconductor switching device, occurring when a suitable triggering pulse is applied to the gate. { ¦fīrd 'stāt }

firewall |COMPUT SCI| Hardware and software programs that protect the resources of a private network from users in other networks, controlling all traffic according to a predefined access policy. { 'fī,wȯl }

firewire See IEEE 1394. { 'fīr,wīr }

FIR filter See finite impulse response filter. { 'fər or ¦ef¦ī'är ,fil·ter }

firing |ELECTR| **1.** The gas ionization that initiates current flow in a gas-discharge tube. **2.** Excitation of a magnetron or transmit-receive tube by a pulse. **3.** The transition from the unsaturated to the saturated state of a saturable reactor. { 'fīr·iŋ }

firing box |ELEC| A boxlike item in which are mounted switches, cables, fuses, plugs, indicator lights, batteries, and the like, specifically designed for firing a rocket or guided missile from a remote position. { 'fīr·iŋ ,bäks }

firing button |ELEC| A button or switch for firing guns or rockets. { 'fīr·iŋ ,bət·ən }

firing cable See shot-firing cable. { 'fīr·iŋ ,kā·bəl }

firing circuit |ELECTR| **1.** Circuit used with an ignitron to deliver a pulse of current of 5–50 amperes in the forward direction, from the igniter to the mercury, to start a cathode spot and to control the time of firing. **2.** By analogy, a similar control circuit of silicon-controlled rectifiers and like devices. { 'fīr·iŋ ,sər·kət }

firing point See critical grid voltage. { 'fīr·iŋ ,pȯint }

firing potential |ELECTR| Controlled potential at which conduction through a gas-filled tube begins. { 'fīr·iŋ pə,ten·chəl }

firmware |COMPUT SCI| A computer program or instruction, such as a microprogram, used so often that it is stored in a read-only memory instead of being included in software; often used in computers that monitor production processes. { 'fərm,wer }

first detector See mixer. { ¦fərst di'tek·tər }

first Fresnel zone |ELECTROMAG| Circular portion of a wavefront transverse to the line between an emitter and a more distant point, where the resultant disturbance is being observed, whose center is the intersection of the front with the direct ray, and whose radius is such that the shortest path from the emitter through the periphery to the receiving point is one-half wavelength longer than the direct ray. { ¦fərst frā'nel ,zōn }

first-generation |COMPUT SCI| Denoting electronic hardware, logical organization, and software characteristic of a first-generation computer. { ¦fərst jen·ə'rā·shən }

first-generation computer |COMPUT SCI| A computer from the earliest stage of computer development, ending in the early 1960s, characterized by the use of vacuum tubes, the performance of one operation at a time in strictly sequential fashion, and elementary software, usually including a program loader, simple utility routines, and an assembler to assist in program writing. { ¦fərst jen·ə¦rā·shən kəm'pyüd·ər }

first-item list [COMPUT SCI] A series of records that is printed with descriptive information from only the first record of each group. { ¦fərst 'ī·dəm ‚list }

first-level address [COMPUT SCI] The location of a referenced operand. { ¦fərst ¦lev·əl ə'dres }

first-level controller [CONT SYS] A controller that is associated with one of the subsystems into which a large-scale control system is partitioned by plant decomposition, and acts to satisfy local objectives and constraints. Also known as local controller. { ¦fərst ¦lev·əl kən'trōl·ər }

first-level interrupt handler [COMPUT SCI] A software or hardware routine that is activated by interrupt signals sent by peripheral devices and decides, based on the relative importance of the interrupts, how they should be handled. Abbreviated FLIH. { ¦fərst ¦lev·əl 'int·ə‚rəpt ‚hand·lər }

first-level packaging [ELECTR] Electronic packaging which provides interconnection directly to the integrated circuit chip. { ‚fərst ‚lev·əl 'pak·ij·iŋ }

first-order subroutine [COMPUT SCI] A subroutine which is entered directly from a main routine or program and which leads back to that program. Also known as first-remove subroutine. { ¦fərst ‚ȯrd·ər 'səb·rü‚tēn }

first-remove subroutine See first-order subroutine. { 'fərst 'rə‚müv 'səb·rü‚tēn }

first selector [ELECTR] Selector which immediately follows a line finder in a switch train and which responds to dial pulses of the first digit of the called telephone number. { 'fərst si'lek·tər }

Fischer-Hinnen method [ELEC] Method of analysis of a complex waveform which has like loops above and below the time axis, in which the amplitude and phase of the *n*-th harmonic is determined from the ordinates of the resultant wave at a series of times which divide the half wave into 2*n* equal time intervals. { ¦fish·ər ¦hin·ən ‚meth·əd }

fish-bone antenna [ELECTROMAG] **1.** Antenna consisting of a series of coplanar elements arranged in collinear pairs, loosely coupled to a balanced transmission line. **2.** Directional antenna in the form of a plane array of doublets arranged transversely along both sides of a transmission line. { 'fish ‚bōn an‚ten·ə }

fishpole antenna See whip antenna. { 'fish‚pōl an‚ten·ə }

five-level code [COMPUT SCI] A code which uses five bits to specify each character. { 'fīv ‚lev·əl 'kōd }

five-wire line [ELEC] A transmission line which has four conductors, all in phase, at the corners of a square and a fifth conductor at the center of the square which is out of phase with the others. { 'fīv ‚wīr 'līn }

fix [COMPUT SCI] A piece of coding that is inserted in a computer program to correct an error. { fiks }

fixed-active tooling [CONT SYS] Stationary equipment in a robotic system, such as numerical control equipment, sensors, cameras, conveying systems and parts feeders, that is activated and controlled by signals. { 'fikst ¦ak·tiv 'tül·iŋ }

fixed area [COMPUT SCI] That portion of the main storage occupied by the resident portion of the control program. { ¦fikst 'er·ē·ə }

fixed attenuator See pad. { ¦fikst ə'ten·yə‚wād·ər }

fixed bias [ELECTR] A constant value of bias voltage, independent of signal strength. { ¦fikst 'bī·əs }

fixed-bias transistor circuit [ELECTR] A transistor circuit in which a current flowing through a resistor is independent of the quiescent collector current. { ¦fikst ¦bī·əs tran'zis·tər ‚sər·kət }

fixed-block [COMPUT SCI] Pertaining to an arrangement of data in which all the blocks of data have the same number of words or characters, as determined by either the hardware requirements of the computer or the programmer. { ¦fikst 'bläk }

fixed capacitor [ELEC] A capacitor having a definite capacitance value that cannot be adjusted. { ¦fikst kə'pas·əd·ər }

fixed contact [ELEC] A relatively immovable contact that is engaged and disengaged by a moving contact to make and break a circuit, as in a switch or relay. { ¦fikst 'kän‚takt }

fixed-cycle operation [COMPUT SCI] An operation completed in a specified number of regularly timed execution cycles. { ¦fikst 'sī·kəl ‚äp·ə'rā·shən }

fixed disk [COMPUT SCI] A disk drive that permanently holds the disk platters. { ¦fikst 'disk }

fixed echo [ELECTR] A persistent echo indication that remains stationary on the radar display, indicating the presence of a fixed target. Also known as permanent echo. { ¦fikst 'ek·ō }

fixed-field method [COMPUT SCI] A method of data storage in which the same type of data is always placed in the same relative position. { 'fikst ‚fēld 'meth·əd }

fixed form coding [COMPUT SCI] Any method of coding a source language in which each part of the instruction appears in a fixed field. { 'fikst ‚fȯrm 'kōd·iŋ }

fixed-head disk [COMPUT SCI] A disk storage device in which the read-write heads are fixed in position, one to a track, and the arms to which they are attached are immovable. { 'fikst ‚hed 'disk }

fixed inductor [ELEC] An inductor whose coils are wound in such a manner that the turns remain fixed in position with respect to each other, and which either has no magnetic core or has a core whose air gap and position within the coil are fixed. { ¦fikst in'dək·tər }

fixed-length field [COMPUT SCI] A field that always has the same number of characters, regardless of its content. { ¦fikst ‚leŋkth 'fēld }

fixed-length operation [COMPUT SCI] A computer operation whose operands always have the same number of bits or characters. { ¦fikst ‚leŋkth ‚äp·ə'rā·shən }

fixed-length record [COMPUT SCI] One of a file of records, each of which must have the same specified number of data units, such as blocks, words, characters, or digits. { ¦fikst ‚leŋkth 'rek·ərd }

fixed logic [COMPUT SCI] Circuit logic of computers or peripheral devices that cannot be changed by external controls; connections must be physically broken to arrange the logic. { ¦fikst 'läj·ik }

fixed medium [COMPUT SCI] A data storage device in which the reading and writing of data do not involve mechanical motion. { 'fikst 'mē·dē·əm }

fixed memory [COMPUT SCI] Of a computer, a nondestructive readout memory that is only mechanically alterable. { ¦fikst 'mem·rē }

fixed-passive tooling [CONT SYS] Unpowered, accessory equipment in a robotic system, such as jigs, fixtures, and work-holding devices. { 'fikst ¦pas·iv 'tül·iŋ }

fixed-point arithmetic [COMPUT SCI] **1.** A method of calculation in which the computer does not consider the location of the decimal or radix point because the point is given a fixed position. **2.** A type of arithmetic in which the operands and results of all arithmetic operations must be properly scaled so as to have a magnitude between certain fixed values. { ¦fikst ˌpȯint ə'rith·mə·tik }

fixed-point calculation [COMPUT SCI] A calculation made with fixed-point arithmetic. { ¦fikst ˌpȯint ˌkal·kyə'lā·shən }

fixed-point computer [COMPUT SCI] A computer in which numbers in all registers and storage locations must have an arithmetic point which remains in the same fixed location. { ¦fikst ˌpȯint kəm'pyüd·ər }

fixed-point part See mantissa. { ¦fikst ˌpȯint 'pärt }

fixed-point representation [COMPUT SCI] Any method of representing a number in which a fixed-point convention is used. { ¦fikst ˌpȯint ˌrep·rə·zen'tā·shən }

fixed-point system [COMPUT SCI] A number system in which the location of the point is fixed with respect to one end of the numerals, according to some convention. { ¦fikst ˌpȯint 'sis·təm }

fixed-position addressing [COMPUT SCI] Direct access to an item in a data file on disk or drum, as opposed to a sequential search for this item starting with the first item in the file. { ¦fikst pə ˌzish·ən ə'dres·iŋ }

fixed-product area [COMPUT SCI] The area in core memory where multiplication takes place for certain types of computers. { ¦fikst ˌpräd·əkt 'er·ē·ə }

fixed-program computer [COMPUT SCI] A special-purpose computer having a program permanently wired in. { ¦fikst ˌprō·grəm kəm'pyüd·ər }

fixed resistor [ELEC] A resistor that has no provision for varying its resistance value. { ¦fikst ri'zis·tər }

fixed-satellite service [COMMUN] A radiocommunication service between earth stations at given positions that uses one or more satellites. Abbreviated FSS. { ¦fikst 'sad·əl‚īt ˌsər·vis }

fixed-sequence robot See fixed-stop robot. { 'fikst ¦sē·kwəns 'rō‚bät }

fixed service [COMMUN] Service providing radio communications between fixed points. { ¦fikst 'sər·vəs }

fixed-stop robot [CONT SYS] A robot in which the motion along each axis has a fixed limit, but the motion between these limits is not controlled and the robot cannot stop except at these limits. Also known as fixed-sequence robot; limited-sequence robot; nonservo robot. { 'fikst ¦stäp 'rō‚bät }

fixed storage [COMPUT SCI] A storage for data not alterable by computer instructions, such as magnetic-core storage with a lockout feature. { ¦fikst 'stȯr·ij }

fixed transmitter [ELECTR] Transmitter that is operated in a fixed or permanent location. { ¦fikst 'tranz'mid·ər }

fixed word length [COMPUT SCI] The length of a computer machine word that always contains the same number of characters or digits. { ¦fikst 'wərd ‚leŋkth }

flag [COMPUT SCI] Any of various types of indicators used for identification, such as a work mark, or a character that signals the occurrence of some condition, such as the end of a word. { flag }

flag flip-flop [COMPUT SCI] A one-bit register which indicates overflow, carry, or sign bit from past or current operations. { 'flag 'flip ‚fläp }

flag operand [COMPUT SCI] A part of the instruction of some assembly languages denoting which elements of the object instruction will be flagged. { 'flag 'äp·ə‚rand }

flame arc lamp [ELEC] An arc lamp in which carbon electrodes are impregnated with chemicals, such as calcium, barium, or titanium, which are more volatile than the carbon and radiate light when driven into the arc. { ¦flām 'ärk ‚lamp }

flame spraying [ENG] Deposition of a conductor on a board in molten form, generally through a metal mask or stencil, by means of a spray gun that feeds wire into a gas flame and drives the molten particles against the work. { 'flām ‚sprā·iŋ }

flange isolator See short waveguide isolator. { ¦flanj 'ī·sə‚lād·ər }

flap attenuator [ELECTROMAG] A waveguide attenuator in which a contoured sheet of dissipative material is moved into the guide through a nonradiating slot to provide a desired amount of power absorption. Also known as vane attenuator. { ¦flap ə'ten·yə‚wād·ər }

flare [ELECTR] A radar screen target indication having an enlarged and distorted shape due to excessive brightness. [ELECTROMAG] See horn antenna. { fler }

flare factor [ENG ACOUS] Number expressing the degree of outward curvature of the horn of a loudspeaker. { 'fler ‚fak·tər }

flash arc [ELECTR] A sudden increase in the emission of large thermionic vacuum tubes, probably due to irregularities in the cathode surface. { 'flash ‚ärk }

flashback voltage [ELECTR] Inverse peak voltage at which ionization takes place in a gas tube. { 'flash‚bak ‚vōl·tij }

flash barrier [ELEC] A fireproof structure between conductors of an electric machine, designed to minimize flashover or the damage caused by flashover. { 'flash ‚bar·ē·ər }

flasher [ELEC] A switch, generally either motor-driven or using a combination heater element and bimetallic strip, that turns lamps on and off rapidly. { 'flash·ər }

flashing over [ELEC] Accidental formation of an arc over the surface of a rotating commutator from brush-to-brush; usually caused by faulty insulation between commutator segments. { ¦flash·iŋ 'ō·vər }

flash lamp [ELECTR] A gaseous-discharge lamp used in a photoflash unit to produce flashes of light of short duration and high intensity for stroboscopic photography. Also known as stroboscopic lamp. { 'flash ‚lamp }

flash memory [COMPUT SCI] A type of electrically erasable programmable read-only memory (EE-PROM). While EPROM is reprogrammed bit-by-bit, flash memory is reprogrammed in blocks, making it faster. It is nonvolatile. { ¦flash 'mem·rē }

flashover [ELEC] An electric discharge around or over the surface of an insulator. { 'flash‚ō·vər }

flashover voltage [ELECTR] The voltage at which an electric discharge occurs between two electrodes that are separated by an insulator; the value depends on whether the insulator surface is dry or wet. Also known as sparkover voltage. { 'flash‚ō·vər ‚vōl·tij }

flash test [ELEC] A method of testing insulation by applying momentarily a voltage much higher than the rated working voltage. { 'flash ‚test }

flatbed plotter [ENG] A graphics output device that draws by moving a pen in both horizontal and vertical directions over a sheet of paper; the overall size of the drawing is limited by the height and width of this bed. { 'flat‚bed 'pläd·ər }

flat cable [ELEC] A cable made of round or rectangular, parallel copper wires arranged in a plane and laminated or molded into a ribbon of flexible insulating plastic. { ¦flat 'kā·bəl }

flat-conductor cable [ELEC] A cable made of wide, flat conductors arranged side by side in a plane and protected by ribbons of insulating plastic. { ¦flat kən‚dək·tər 'kā·bəl }

flat fading [COMMUN] Type of fading in which all components of the received radio signal fluctuate in the same proportion simultaneously. { ¦flat 'fād·iŋ }

flat file [COMPUT SCI] A two-dimensional array. { 'flat ‚fīl }

flat line [ELECTROMAG] A radio-frequency transmission line, or part thereof, having essentially 1-to-1 standing wave ratio. { 'flat ‚līn }

flatpack [ELECTR] Semiconductor network encapsulated in a thin, rectangular package, with the necessary connecting leads projecting from the edges of the unit. { 'flat‚pak }

flat-panel display See panel display. { 'flat ¦pan·əl di'splā }

flat-top antenna [ELECTROMAG] An antenna having two or more lengths of wire parallel to each other and in a plane parallel to the ground, each fed at or near its midpoint. { 'flat ‚täp an‚ten·ə }

flat-top response See band-pass response. { 'flat ¦täp ri'späns }

flat tuning [ELECTR] Tuning of a radio receiver in which a change in frequency of the received waves produces only a small change in the current in the tuning apparatus. { ¦flat 'tün·iŋ }

Fleming tube [ELECTR] The original diode, consisting of a heated filament and a cold metallic electrode in an evacuated glass envelope; negative current flows from the filament to the cold electrode, but not in the reverse direction. { 'flem·iŋ ‚tüb }

flexible circuit [ELECTR] A printed circuit made on a flexible plastic sheet that is usually die-cut to fit between large components. { ‚flek·sə·bəl 'sər‚kət }

flexible resistor [ELEC] A wire-wound resistor having the appearance of a flexible lead; made by winding the Nichrome resistance wire around a length of asbestos or other heat-resistant cord, then covering the winding with asbestos and braided insulating covering. { ‚flek·sə·bəl ri'zis·tər }

flexible waveguide [ELECTROMAG] A waveguide that can be bent or twisted without appreciably changing its electrical properties. { ‚flek·sə·bəl 'wāv‚gīd }

flexional symbols [COMPUT SCI] Symbols in which the meaning of each component digit is dependent on those which precede it. { ¦flek·shən·əl ¦sim·bəlz }

flexowriter [COMPUT SCI] A typewriterlike device to read in manually or to read out information of a computer to which it is connected; it can also be used to punch paper tape. { 'flek·sə‚wrīd·ər }

flicker effect [ELECTR] Random variations in the output current of an electron tube having an oxide-coated cathode, due to random changes in cathode emission. { 'flik·ər i‚fekt }

flight-path computer [COMPUT SCI] A computer that includes all of the functions of a course-line computer and also provides means for controlling the altitude of an aircraft in accordance with a desired plan of flight. { 'flīt ‚path kəm'pyüd·ər }

FLIH See first-level interrupt handler.

flip chip [ELECTR] A tiny semiconductor die having terminations all on one side in the form of solder pads or bump contacts; after the surface of the chip has been passivated or otherwise treated, it is flipped over for attaching to a matching substrate. Also known as solder-ball flip chip. { 'flip ‚chip }

flip-flop circuit See bistable multivibrator. { 'flip ‚fläp ‚sər·kət }

flip-open cutout fuse See dropout fuse. { ¦flip ¦ō·pən 'kəd‚aüt ‚fyüz }

floating [ELECTR] The condition wherein a device or circuit is not grounded and not tied to an established voltage supply. { 'flōd·iŋ }

floating address [COMPUT SCI] The symbolic address used prior to its conversion to a machine address. { ¦flōd·iŋ ə'dres }

floating battery [ELEC] A storage battery connected permanently in parallel with another power source; the battery normally handles only small charging or discharging currents, but takes over the entire load upon failure of the main supply. { ¦flōd·iŋ 'bad·ə·rē }

floating carrier modulation *See* controlled carrier modulation. { ¦flōd·iŋ ˌkar·ē·ər ˌmäj·ə'lā·shən }

floating charge [ELEC] Application of a constant voltage to a storage battery, sufficient to maintain an approximately constant state of charge while the battery is idle or on light duty. { ¦flōd·iŋ 'chärj }

floating control [ENG] Control device in which the speed of correction of the control element is proportional to the error signal. Also known as proportional-speed control. { ¦flōd·iŋ kən'trōl }

floating dollar sign [COMPUT SCI] A dollar sign used with an edit mask, allowing the sign to be inserted before the nonzero leading digit of a dollar amount. { ¦flōd·iŋ 'däl·ər ˌsīn }

floating-gate avalanche-injection metal-oxide semiconductor device [ELECTR] An erasable programmable read-only memory chip that holds its contents until they are erased by ultraviolet light. Abbreviated FAMOS device. { ¦flōd·iŋ ¦gāt ¦av·əˌlanch in¦jek·shən ¦med·əl ¦äk ˌsīd ˌsem·i·kən'dək·tər diˌvīs }

floating graphic [COMPUT SCI] A picture or graph that moves up or down on a page of a document as text is deleted or inserted above it. { ˌflōd·iŋ 'graf·ik }

floating grid [ELECTR] Vacuum-tube grid that is not connected to any circuit; it assumes a negative potential with respect to the cathode. Also known as free grid. { ¦flōd·iŋ 'grid }

floating input [ELEC] Isolated input circuit not connected to ground at any point. { ¦flōd·iŋ 'in ˌpút }

floating neutral [ELEC] Neutral conductor whose voltage to ground is free to vary when circuit conditions change. { ¦flōd·iŋ 'nü·trəl }

floating-point calculation [COMPUT SCI] A calculation made with floating-point arithmetic. { ¦flōd·iŋ ¦póint ˌkal·kyə'lā·shən }

floating-point coefficient *See* mantissa. { ¦flōd·iŋ ¦póint ˌkō·i'fish·ənt }

floating-point package [COMPUT SCI] A program which enables a computer to perform arithmetic operations when such capabilities are not wired into the computer. Also known as floating-point routine. { ¦flōd·iŋ ¦póint 'pak·ij }

floating-point processor [COMPUT SCI] A separate processor or a special section of a computer's main storage that is for the efficient handling of floating-point operations. { 'flōd·iŋ ¦póint 'präs ˌes·ər }

floating-point routine *See* floating-point package. { ¦flōd·iŋ ¦póint rü'tēn }

floating-point system [COMPUT SCI] A number system in which the location of the point does not remain fixed with respect to one end of the numerals. { ¦flōd·iŋ ¦póint 'sis·təm }

floating signal *See* differential signal. { 'flōd·iŋ 'sig·nəl }

flood [ELECTR] To direct a large-area flow of electrons toward a storage assembly in a charge storage tube. { fləd }

floodlight [ELEC] A light projector used for outdoor lighting of buildings, parking lots, sports fields, and the like, usually having a filament lamp or mercury-vapor lamp and a parabolic reflector. { 'fləd,līt }

floor outlet [ELEC] An electrical outlet whose face is level with or recessed into a floor. Also known as floor plug. { 'flōr ˌaút·lət }

flopover [ELECTR] A defect in television reception in which a series of frames move vertically up or down the screen, caused by lack of synchronization between the vertical and horizontal sweep frequencies. { 'fläp,ō·vər }

floppy disk [COMPUT SCI] A flexible plastic disk coated with magnetic oxide and used for data storage and data entry to a computer; a slot in its protective envelope or housing, which remains stationary while the disk rotates, exposes the track positions for the magnetic read/write head of the drive unit. Also known as diskette. { ¦fläp·ē 'disk }

flops [COMPUT SCI] A unit of computer speed, equal to one floating-point arithmetic operation per second. { fläps }

flow [COMPUT SCI] The sequence in which events take place or operations are carried out. { flō }

flow chart [ENG] A graphical representation of the progress of a system for the definition, analysis, or solution of a data-processing or manufacturing problem in which symbols are used to represent operations, data or material flow, and equipment, and lines and arrows represent interrelationships among the components. Also known as control diagram; flow diagram; flow sheet. { 'flō ˌchärt }

flow-chart symbol [ENG] Any of the existing symbols normally used to represent operations, data or materials flow, or equipment in a data-processing problem or manufacturing-process description. { 'flō ˌchärt ˌsim·bəl }

flow diagram *See* flow chart. { 'flō ˌdī·əˌgram }

flow direction [ENG] The antecedent-to-successor relation, indicated by arrows or other conventions, between operations on a flow chart. { 'flō dəˌrek·shən }

flow graph [COMPUT SCI] A directed graph that represents a computer program, wherein a node in the graph corresponds to a block of sequential code and branches correspond to decisions taken in the program. [SYS ENG] *See* signal-flow graph. { 'flō ˌgraf }

flow line [ENG] The connecting line or arrow between symbols on a flow chart or block diagram. { 'flō ˌlīn }

flow sheet *See* flow chart. { 'flō ˌshēt }

flow soldering [ENG] Soldering of printed circuit boards by moving them over a flowing wave of molten solder in a solder bath; the process permits precise control of the depth of immersion in the molten solder and minimizes heating of the board. Also known as wave soldering. { 'flō ˌsäd·ə·riŋ }

fluctuating current [ELEC] Direct current that changes in value but not at a steady rate. { ¦flək·chəˌwād·iŋ 'kə·rənt }

fluctuation [ELECTROMAG] The change in amplitude of a radar echo due to a target of some complexity changing its attitude or structural features. Fluctuations can be rapid (pulse-to-pulse) or somewhat slow (scan-to-scan). { ˌflək·chə'wā·shən }

fluid amplifier [ENG] An amplifier in which all amplification is achieved by interaction between jets of fluid, with no electronic circuit and usually no moving parts. { ¦flü·əd 'am·plə͵fī·ər }

fluid computer [COMPUT SCI] A digital computer constructed entirely from air-powered fluid logic elements; it contains no moving parts and no electronic circuits; all logic functions are carried out by interaction between jets of air. { ¦flü·əd kəm͵pyüd·ər }

fluorescent lamp [ELECTR] A tubular discharge lamp in which ionization of mercury vapor produces radiation that activates the fluorescent coating on the inner surface of the glass. { flü ¦res·ənt 'lamp }

fluorescent screen [ENG] A sheet of material coated with a fluorescent substance so as to emit visible light when struck by ionizing radiation such as x-rays or electron beams. { flü¦res·ənt 'skrēn }

fluoroscopic image intensifier [ELECTR] An electron-beam tube that converts a relatively feeble fluoroscopic image on the fluorescent input phosphor into a much brighter image on the output phosphor. { ˌflur·ə'skäp·ik 'im·ij in'ten·sə͵fī·ər }

flush left See left-justify. { ¦fləsh 'left }

flush right See right-justify. { ¦fləsh 'rīt }

flute storage [ELECTR] Ferrite storage consisting of a number of parallel lengths of fine prism-shaped tubing, each surrounding an insulated axial conductor that acts as a word line; the lengths of tubing are intersected at right angles by parallel sets of insulated wire bit lines that are displaced slightly from the word lines; each intersection stores one bit. { ¦flüt ¦stór·ij }

flutter [ELECTROMAG] A fast-changing variation in received signal strength, such as may be caused by antenna movements in a high wind or interaction with a signal or another frequency. { 'fləd·ər }

flutter echo [ELECTROMAG] A radar echo consisting of a rapid succession of reflected pulses resulting from a single transmitted pulse. { 'fləd·ər ͵ek·ō }

flux-compression generator [ELEC] A type of impulse generator in which megajoules of energy can be generated within microseconds by abruptly reducing the volume of a closed conducting cage that surrounds a region in which a magnetic field is established. Also known as magnetic cumulative generator. { ¦fləks kəm ͵presh·ən ¦jen·ə͵rād·ər }

flux gate [ENG] A detector that gives an electric signal whose magnitude and phase are proportional to the magnitude and direction of the external magnetic field acting along its axis; used to indicate the direction of the terrestrial magnetic field. { 'fləks ͵gāt }

flyback [ELECTR] The time interval in which the electron beam of a cathode-ray tube returns to its starting point after scanning one line or one field of a video. Also known as retrace; return trace. { 'flī͵bak }

flyback power supply [ELECTR] A high-voltage power supply used to produce the direct-current voltage of about 10,000–25,000 volts required for the second anode of a cathode-ray tube in a video display. { 'flī͵bak 'paür sə͵plī }

flyback transformer See horizontal output transformer. { 'flī͵bak tranz͵fór·mər }

flying-aperture scanner [ELECTR] An optical scanner, used in character recognition, in which a document is flooded with light, and light is collected sequentially spot by spot from the illuminated image. { ¦flī·iŋ ¦ap·ər·chər ͵skan·ər }

flying head [ELECTR] A read/write head used on magnetic disks and drums, so designed that it flies a microscopic distance off the moving magnetic surface and is supported by a film of air. { ¦flī·iŋ 'hed }

flying spot [ELECTR] A small point of light, controlled mechanically or electrically, which moves rapidly in a rectangular scanning pattern in a flying-spot scanner. { ¦flī·iŋ 'spät }

flying-spot scanner [ELECTR] A scanner used for video film and slide transmission, electronic writing, and character recognition, in which a moving spot of light, controlled mechanically or electrically, scans the image field, and the light reflected from or transmitted by the image field is picked up by a device that generates a corresponding electric signal output. Also known as optical scanner. { ¦flī·iŋ ͵spät 'skan·ər }

flywheel synchronization [ELECTR] Automatic frequency control of a scanning system by using the average timing of the incoming sync signals, rather than by making each pulse trigger the scanning circuit; used in analog television receivers designed for fringe-area reception, when noise pulses might otherwise trigger the sweep circuit prematurely. { 'flī͵wēl ͵siŋ·krə·nə'zā·shən }

FM See frequency modulation.

FM/AM multiplier [ELECTR] Multiplier in which the frequency deviation from the central frequency of a carrier is proportional to one variable, and its amplitude is proportional to the other variable; the frequency-amplitude-modulated carrier is then consecutively demodulated for frequency modulation (FM) and for amplitude modulation (AM); the final output is proportional to the product of the two variables. { 'ef͵em 'ā͵em 'məl·tə͵plī·ər }

focal-plane array [ELECTR] A photodetector that has up to a million photosensors on a single semiconductor silicon chip arranged in a rectangular grid matrix that is placed in the focal plane of an optical instrument. { ¦fō·kəl ͵plān ə'rā }

focus [ELECTR] To control convergence or divergence of the electron paths within one or more beams, usually by adjusting a voltage or current in a circuit that controls the electric or magnetic fields through which the beams pass, in order to obtain a desired image or a desired current density within the beam.   { 'fō·kəs }

focus control |ELECTR| A control that adjusts spot size at the screen of a cathode-ray tube to give the sharpest possible image; it may vary the current through a focusing coil or change the position of a permanent magnet.   { 'fō·kəs kən ,trōl }

focused-current log |ENG| A resistivity log that is obtained by means of a multiple-electrode arrangement.   { ¦fō·kəst ¦kə·rənt 'läg }

focusing anode |ELECTR| An anode used in a cathode-ray tube to change the size of the electron beam at the screen; varying the voltage on this anode alters the paths of electrons in the beam and thus changes the position at which they cross or focus.   { 'fō·kəs·iŋ ,an,ōd }

focusing coil |ELECTR| A coil that produces a magnetic field parallel to an electron beam for the purpose of focusing the beam.   { 'fō·kəs·iŋ ,kȯil }

focusing electrode |ELECTR| An electrode to which a potential is applied to control the cross-sectional area of the electron beam in a cathode-ray tube.   { 'fō·kəs·iŋ i,lek,trōd }

focusing magnet |ELECTR| A permanent magnet used to produce a magnetic field for focusing an electron beam.   { 'fō·kəs·iŋ ,mag·nət }

focus lamp |ELEC| **1.** A lamp whose filament has a spiral or zigzag form in order to reduce its size, so that it can be brought into the focus of a lens or mirror. **2.** An arc lamp whose feeding mechanism is designed to hold the arc in a constant position with respect to an optical system that is used to focus its rays.   { 'fō·kəs ,lamp }

focus projection and scanning |ELECTR| Method of magnetic focusing and electrostatic deflection of the electron beam of a hybrid vidicon; a transverse electrostatic field is used for beam deflection; this field is immersed with an axial magnetic field that focuses the electron beam.   { ¦fō·kəs prə,jek·shən ən 'skan·iŋ }

foil electret |ELEC| A thin film of strongly insulating material capable of trapping charge carriers, such as polyfluoroethylenepropylene, that is electrically charged to produce an external electric field; in the conventional design, charge carriers of one sign are injected into one surface, and a compensation charge of opposite sign forms on the opposite surface or an adjacent electrode.   { ¦fȯil i'lek·trət }

folded cavity |ELECTR| Arrangement used in a klystron repeater to make the incoming wave act on the electron stream from the cathode at several places and produce a cumulative effect.   { ¦fōld·əd 'kav·əd·ē }

folded dipole   *See* folded-dipole antenna.   { ¦fōld·əd 'dī,pōl }

folded-dipole antenna |ELECTROMAG| A dipole antenna whose outer ends are folded back and joined together at the center; the impedance is about 300 ohms, as compared to 70 ohms for a single-wire dipole; widely used with television and frequency-modulation receivers. Also known as folded dipole.   { ¦fōld·əd 'dī,pōl an'ten·ə }

folded horn |ENG ACOUS| An acoustic horn in which the path from throat to mouth is folded or curled to give the longest possible path in a given volume.   { ¦fōld·əd 'hȯrn }

folding |COMPUT SCI| A method of hashing which consists of splitting the original key into two or more parts and then adding the parts together.   { 'fōld·iŋ }

foldover |ELECTR| Picture distortion seen as a white line on the side, top, or bottom of a television picture; generally caused by nonlinear operation in either the horizontal or vertical deflection circuits of a receiver.   { 'fōl,dō·vər }

Foley pits |ENG ACOUS| Open boxes that are used in ADR studios and contain various materials (such as water, sand, gravel, rice, and nails) for generating sound effects that could not be recorded well during filming or video recording.   { 'fō·lē ,pits }

follow current |ELEC| The current at power frequency that passes through a surge diverter or other discharge path after a high-voltage surge has started the discharge.   { 'fäl·ō ,kə·rənt }

following error |CONT SYS| The difference between commanded and actual positions in contouring control.   { 'fäl·ə·wiŋ ,er·ər }

follow spot |ELEC| A high-intensity spotlight used to follow action in arenas and stadiums and on large stages; it is equipped with adjustable iris and shutter controls, and its light source is either a carbon arc or an incandescent bulb.   { 'fäl·ō ,spät }

font cartridge |COMPUT SCI| A removable module that can be plugged into a slot in a printer and has one or more fonts stored in a read-only memory chip.   { 'fänt ,kär·trij }

font compiler   *See* font generator.   { 'fänt kəm ,pīl·ər }

font generator |COMPUT SCI| A computer program that converts an outline font into the patterns of dots required for a particular size of font. Also known as font compiler.   { 'fänt ,jen·ə,rād·ər }

footprint |COMMUN| The area of the earth's surface that can be covered by a communications satellite at any given time. |COMPUT SCI| The amount and shape of the area occupied by equipment, such as a terminal or microcomputer, on desktop, floor, or other surface area.   { 'fůt ,print }

forbidden-character code |COMPUT SCI| A bit code which exists only when an error occurs in the binary coding of characters.   { fər¦bid·ən 'kar·ik·tər ,kōd }

forbidden-combination check |COMPUT SCI| A test for the occurrence of a nonpermissible code expression in a computer; used to detect computer errors.   { fər¦bid·ən ,käm·bə'na·shən ,chek }

force |COMPUT SCI| To intervene manually in a computer routine and cause the computer to execute a jump instruction.   { fȯrs }

force-controlled motion commands |CONT SYS| Robot control in which motion information is provided by computer software but sensing of forces or feedback is used by the robot to adapt this information to the environment. { 'fȯrs kən ¦trōld 'mō·shən kə͵manz }

forced programming See minimum-access programming. { ¦fȯrst 'prō͵gram·iŋ }

force feedback [CONT SYS] A method of error detection in which the force exerted on the effector is sensed and fed back to the control, usually by mechanical, hydraulic, or electric transducers. { ¦fȯrs ¦fēd͵bak }

forecasting [COMMUN] The prediction of conditions of radio propagation for a period extending anywhere from a few hours to a few months. { 'fȯr͵kast·iŋ }

foreground [COMPUT SCI] A program or process of high priority that utilizes machine facilities as needed, with less critical, background work performed in otherwise unused time. { 'fȯr ͵graünd }

fork-join model [COMPUT SCI] A method of programming on parallel machines in which one or more child processes branch out from the root task when it is time to do work in parallel, and end when the parallel work is done. { 'fȯrk ͵jȯin 'mäd·əl }

fork oscillator [ELECTR] An oscillator that uses a tuning fork as the frequency-determining element. { ¦fȯrk ͵äs·ə͵lād·ər }

formal language [COMPUT SCI] An abstract mathematical object used to model the syntax of a programming or natural language. { ¦fȯr·məl 'laŋ·gwij }

format [COMPUT SCI] **1.** The specific arrangement of data on a printed page, display screen, or such, or in a record, data file, or storage device. **2.** To prepare a disk to store information by using a special program that divides the disk into storage units such as tracks and sectors. { 'fȯr͵mat }

format effector See layout character. { 'fȯr͵mat i͵fek·tər }

formatted tape [COMPUT SCI] A magnetic tape which employs a prerecorded timing track by means of which blocks of data can be found after reference to a directory table. { fȯr¦mad·əd 'tāp }

formatting [COMPUT SCI] The preparation of a magnetic storage device to receive data structures; for example, the recording of track and sector information on a floppy disk. { 'fȯr ͵mad·iŋ }

form factor [ELEC] **1.** The ratio of the effective value of a periodic function, such as an alternating current, to its average absolute value. **2.** A factor that takes the shape of a coil into account when computing its inductance. Also known as shape factor. { 'fȯrm ͵fak·tər }

form feed character [COMPUT SCI] A control character that determines when a printer or display device moves to the next page, form, or equivalent unit of data. { 'fȯrm ¦fēd ͵kar·ik·tər }

form feeding [COMPUT SCI] The positioning of documents in order to move them past printing or sensing devices, either singly or in continuous rolls. { 'fȯrm ͵fēd·iŋ }

form feed printer [COMPUT SCI] A computer printer that accepts continuous forms or continuous sheets of paper. { 'fȯrm ¦fēd ͵print·ər }

forming [ELEC] Application of voltage to an electrolytic capacitor, electrolytic rectifier, or semiconductor device to produce a desired permanent change in electrical characteristics as a part of the manufacturing process. { 'fȯrm·iŋ }

forms [COMPUT SCI] Web pages that allow users to fill in and submit information, they are written in HTML and processed by CGI scripts. { fȯrmz }

forms control buffer [COMPUT SCI] A reserved storage containing coordinates for a page position on the printer; earlier printers utilized a carriage control tape, allowing the page to be set at a specific position. { 'fȯrmz kən͵trōl 'bəf·ər }

form stop [COMPUT SCI] A device which stops a machine when its supply of paper has run out. { 'fȯrm ͵stäp }

formula translation See FORTRAN. { ¦fȯr·myə·lə tranz¦lā·shən }

form-wound coil [ELEC] Armature coil that is formed or shaped over a fixture before being placed on the armature of a motor or generator. { 'fȯrm ¦waünd ͵kȯil }

for-next loop [COMPUT SCI] In computer programming, a high-level logic statement which defines a part of a computer program that will be repeated a certain number of times. { ¦fȯr¦nekst ͵lüp }

fors See G. { fȯrs }

FOR statement [COMPUT SCI] A statement in a computer program that is repeatedly executed a specified number of times, generally while a control variable takes on successive values over a specified range. { fȯr ͵stāt·mənt }

Forth [COMPUT SCI] A high-level programming language developed primarily for microcomputers and characterized by a number of features that make it highly adaptable and readily extensible, such as the ability to be used as an interpreter or an operating system. { fȯrth }

FORTRAN [COMPUT SCI] A family of procedure-oriented languages used mostly for scientific or algebraic applications; derived from formula translation. { 'fȯr͵tran }

forty-four-type repeater [ELECTR] Type of telephone repeater employing two amplifiers and no hybrid arrangements; used in a four-wire system. { ¦fȯrd·ē͵fȯr ¦tīp ri'pēd·ər }

forum See newsgroup. { 'fȯr·əm }

forward-acting regulator [ELECTR] Transmission regulator in which the adjustment made by the regulator does not affect the quantity which caused the adjustment. { 'fȯr·wərd ͵ak·tiŋ 'reg·yə͵lād·ər }

forward-backward counter [COMPUT SCI] A counter that has both an add and a subtract input so as to count in either an increasing or a decreasing direction. Also known as bidirectional counter. { ¦fȯr·wərd ¦bak·wərd 'kaünt·ər }

forward bias [ELECTR] A bias voltage that is applied to a pn-junction in the direction that causes a large current flow; used in some semiconductor diode circuits. { ¦fȯr·wərd 'bī·əs }

forward chaining |COMPUT SCI| In artificial intelligence, a method of reasoning which begins with a statement of all the relevant data and works toward the solution using the system's rules of inference. { ¦fȯr·wərd 'chān·iŋ }

forward compatibility *See* upward compatibility. { ¦fȯr·wərd kəm,pad·ə'bil·əd·ē }

forward coupler |ELECTR| Directional coupler used to sample incident power. { ¦fȯr·wərd 'kəp·lər }

forward current |ELECTR| Current which flows upon application of forward voltage. { ¦fȯr·wərd 'kə·rənt }

forward direction |ELECTR| Of a semiconductor diode, the direction of lower resistance to the flow of steady direct current. { ¦fȯr·wərd də'rek·shən }

forward drop |ELECTR| The voltage drop in the forward direction across a rectifier. { ¦fȯr·wərd 'dräp }

forward error analysis |COMPUT SCI| A method of error analysis based on the assumption that small changes in the input data lead to small changes in the results, so that bounds for the errors in the results caused by rounding or truncation errors in the input can be calculated. { ¦fȯr·wərd 'er·ər ə,nal·ə·səs }

forward error correction |COMMUN| The location and correction of errors occurring in data communications by the receiver without retransmission of data. { 'fȯr·wərd 'er·ər kə,rek·shən }

forward path |CONT SYS| The transmission path from the loop actuating signal to the loop output signal in a feedback control loop. { 'fȯr·wərd ,path }

forward propagation by ionospheric scatter |COMMUN| Radio communications technique using the scattering phenomenon exhibited by electromagnetic waves in the 30–100-megahertz region when passing through the ionosphere at an elevation of about 50 miles (85 kilometers). { ¦fȯr·wərd ,präp·ə,gā·shən bī ī'än·ə,sfir·ik ,skad·ər }

forward propagation by tropospheric scatter |COMMUN| Radio communications technique using high transmitting power levels, large antenna arrays, and the scattering phenomenon of the troposphere to permit communications far beyond line-of-sight distances. { ¦fȯr·wərd ,präp·ə ,gā·shən bī 'träp·ə,sfir·ik ,skad·ər }

forward recovery time |ELECTR| Of a semiconductor diode, the time required for the forward current or voltage to reach a specified value after instantaneous application of a forward bias in a given circuit. { ¦fȯr·wərd ri'kəv·ə·rē ,tīm }

forward reference |COMPUT SCI| Reference to a data element that has not yet been defined in the program being compiled. { 'fȯr·wərd 'ref·rəns }

forward resistance |ELECTR| The resistance of a semiconductor diode to current flow in the forward direction. { 'fȯr·wərd ri'zis·təns }

forward scatter |COMMUN| **1.** Propagation of electromagnetic waves at frequencies above the maximum usable high frequency through use of the scattering of a small portion of the transmitted energy when the signal passes from an unionized medium into a layer of the ionosphere. **2.** Collectively, the very-high-frequency forward propagation by ionospheric scatter and ultra-high-frequency forward propagation by tropospheric scatter communications techniques. { ¦fȯr·wərd 'skad·ər }

forward-scatter propagation *See* scatter propagation. { ¦fȯr·wərd ¦skad·ər präp·ə'gā·shən }

forward transfer function |CONT SYS| In a feedback control loop, the transfer function of the forward path. { ¦fȯr·wərd 'tranz·fər ,fəŋk·shən }

forward voltage drop *See* diode forward voltage. { ¦fȯr·wərd 'vōl·tij ,dräp }

forward wave |ELECTR| Wave whose group velocity is the same direction as the electron stream motion. { ¦fȯr·wərd ¦wāv }

Foster-Seely discriminator *See* phase-shift discriminator. { ¦fȯs·tər ¦sē·lē di'skrim·ə,nād·ər }

Foster's reactance theorem |CONT SYS| The theorem that the most general driving point impedance or admittance of a network, in which every mesh contains independent inductance and capacitance, is a meromorphic function whose poles and zeros are all simple and occur in conjugate pairs on the imaginary axis, and in which these poles and zeros alternate. { 'fȯs·tərz rē'ak·təns ,thir·əm }

four-address |COMPUT SCI| Pertaining to an instruction address which contains four address parts. { 'fȯr ə,dres }

four-channel sound system *See* quadraphonic sound system. { ¦fȯr ¦chan·əl 'saúnd ,sis·təm }

four-frequency diplex telegraphy |COMMUN| Frequency-shift telegraphy in which each of the four possible signal combinations corresponding to two telegraph channels is represented by a separate frequency. { ¦fȯr ¦frē·kwən·sē ¦dī,pleks tə'leg·rə·fē }

Fourier analyzer |ENG| A digital spectrum analyzer that provides push-button or other switch selection of averaging, coherence function, correlation, power spectrum, and other mathematical operations involved in calculating Fourier transforms of time-varying signal voltages for such applications as identification of underwater sounds, vibration analysis, oil prospecting, and brain-wave analysis. { 'fúr·ē,ā 'an·ə,līz·ər }

four-layer device |ELECTR| A *pnpn* semiconductor device, such as a silicon controlled rectifier, that has four layers of alternating *p*-and *n*-type material to give three *pn* junctions. { ¦fȯr ¦lā·ər di'vīs }

four-layer diode |ELECTR| A semiconductor diode having three junctions, terminal connections being made to the two outer layers that form the junctions; a Shockley diode is an example. { ¦fȯr ¦lā·ər 'dī,ōd }

four-layer transistor |ELECTR| A junction transistor having four conductivity regions but only three terminals; a thyristor is an example. { ¦fȯr ¦lā·ər tran'zis·tər }

four-phase modulation |COMMUN| Modulation in which data are encoded on a carrier frequency as a succession of phase shifts that will be 45,

135, 225, or 315°; each phase shift contains 2 bits of information called dibits, as follows: 225° represents 00, 315° is 01, 45° is 11, and 135° is 10.   { ¦fȯr ¦fāz ˌmäj·ə'lā·shən }

four-plus-one address   [COMPUT SCI] An instruction that contains four operand addresses and a control address.   { ¦fȯr ˌpləs ¦wən ə'dres }

four-pole double-throw   [ELEC] A 12-terminal switch or relay contact arrangement that simultaneously connects two pairs of terminals to either of two other pairs of terminals. Abbreviated 4PDT.   { ¦fȯr 'pōl ¦dəb·əl 'thrō }

four-quadrant multiplier   [COMPUT SCI] A multiplier in an analog computer in which both the reference signal and the number represented by the input may be bipolar, and the multiplication rules for algebraic sign are obeyed. Also known as quarter-square multiplier.   { ¦fȯr ¦kwäd·rənt 'məl·tə,plī·ər }

four-tape   [COMPUT SCI] To sort input data, supplied on two tapes, into incomplete sequences alternately on two output tapes; the output tapes are used for input on the succeeding pass, resulting in longer and longer sequences after each pass, until the data are all in one sequence on one output tape.   { ¦fȯr ¦tāp }

fourth-generation computer   [COMPUT SCI] A type of general-purpose digital computer used in the 1970s and 1980s that is characterized by increasingly advanced very large-scale integrated circuits and increasing use of a hierarchy of memory devices.   { 'fȯrth ˌjen·ə¦rā·shən kəm'pyüd·ər }

fourth-generation language   [COMPUT SCI] A higher-level programming language that automates many of the basic functions that must be spelled out in conventional languages, and can obtain results with an order-of-magnitude less coding because of its richer content of commands.   { 'fȯrth ˌjen·ə¦rā·shən 'laŋ·gwij }

four-track tape   [ENG ACOUS] Magnetic tape on which two tracks are recorded for each direction of travel, to provide stereo sound reproduction or to double the amount of source material that can be recorded on a given length of 1/4-inch (0.635-centimeter) tape.   { ¦fȯr ˌtrak 'tāp }

four-way switch   [ELEC] An electric switch employed in house wiring, that makes it possible to turn a light on or off at three or more places.   { 'fȯr ˌwā 'swich }

four-wire circuit   [COMMUN] A two-way circuit using two paths so arranged that communication currents are transmitted in one direction only on one path, and in the opposite direction on the other path; the transmission path may or may not employ four wires.   { 'fȯr ˌwīr 'sər·kət }

four-wire repeater   [ELECTR] Telephone repeater for use in a four-wire circuit and in which there are two amplifiers, one serving to amplify the telephone currents in one side of the fourwire circuit, and the other serving to amplify the telephone currents in the other side of the four-wire circuit.   { 'fȯr ˌwīr ri'pēd·ər }

four-wire subscriber line   [COMMUN] Four-wire circuit connecting a subscriber directly to a switching center.   { 'fȯr ˌwīr səb'skrīb·ər ˌlīn }

four-wire terminating set   [ELECTR] Hybrid arrangement by which four-wire circuits are terminated on a two-wire basis for interconnection with two-wire circuits.   { 'fȯr ˌwīr 'ter·mə,nād·iŋ ˌset }

fox   [COMPUT SCI] A name for the hexadecimal digit whose decimal equivalent is 15.   { fäks }

Fox broadcast   [COMMUN] Radio broadcast of messages for which receiving stations make no acknowledgment.   { 'fäks ˌbrȯd,kast }

FPLA   See field-programmable logic array.

Fr   See statcoulomb.

fractional horsepower motor   [ELEC] Any motor built into a frame smaller than that for a motor having an open construction and a continuous rating of 1 horsepower (745.7 watts) at 1800 revolutions per minute.   { ¦frak·shən·əl ¦hȯrs ˌpaủ·ər 'mōd·ər }

fractional quantum Hall effect   [ELECTR] The version of the quantum Hall effect in which the Hall resistance becomes precisely equal to $h/(p/q)e^2$, where h is Planck's constant, e is the electronic charge, q is an odd integer, and p is an integer not divisible by q.   { ¦frak·shən·əl 'kwän·təm 'hȯl i,fekt }

fragmentation   [COMPUT SCI] The tendency of files in disk storage to be divided up into many small areas scattered around the disk.   { ˌfrag·mən'tā·shən }

fragmenting   [COMPUT SCI] The breaking up of a document into its various components.   { 'frag ˌment·iŋ }

Frahm frequency meter   See vibrating-reed frequency metery.   { 'främ 'frē·kwən·sē ˌmēd·ər }

frame   [COMMUN] **1.** One cycle of a regularly recurring series of pulses. **2.** An elementary block of data for transmission over a network or communications system.   [COMPUT SCI] **1.** Subdivision of a browser window, with each section containing a separate Web page. **2.** See main frame.   [ELECTR] One complete representation of a video image.   { frām }

frame buffer   [COMPUT SCI] A device that stores a television picture or frame for processing.   { 'frām ˌbəf·ər }

frame frequency   [ELECTR] The number of times per second that the frame is completely scanned in a video system. Also known as picture frequency.   { 'frām ˌfrē·kwən·sē }

frame grabber   [COMPUT SCI] An external device that digitizes standard television video images for storage or processing in a computer.   { 'frām ˌgrab·ər }

frame period   [ELECTR] A time interval equal to the reciprocal of the frame frequency.   { 'frām ˌpir·ē·əd }

framer   [ELECTR] Device for adjusting facsimile equipment so that the start and end of a recorded line are the same as on the corresponding line of the subject copy.   { 'frām·ər }

framing   [ELECTR] Adjusting a facsimile picture to a desired position in the direction of line progression. Also known as phasing.   { 'frām·iŋ }

framing control   [ELECTR] **1.** A control that adjusts the centering, width, or height of the image on a video display device. **2.** A control that shifts a received facsimile picture horizontally. { 'frām·iŋ kən,trō }

Franck-Hertz experiment [ELECTR] Experiment for measuring the kinetic energy lost by electrons in inelastic collisions with atoms; it established the existence of discrete energy levels in atoms, and can be used to determine excitation and ionization potentials. { ¦fräŋk 'harts ik ,sper·ə·mənt }

franklin *See* statcoulomb. { 'fraŋk·lən }

franklin centimeter [ELEC] A unit of electric dipole moment, equal to the dipole moment of a charge distribution consisting of positive and negative charges of 1 statcoulomb separated by a distance of 1 centimeter. { 'fraŋk·lən 'sent·ə ,mēd·ər }

Franklin equation [ENG ACOUS] An equation for intensity of sound in a room as a function of time after shutting off the source, involving the volume and exposed surface area of the room, the speed of sound, and the mean sound-absorption coefficient. { 'fraŋk·lən i,kwā·zhən }

Fraunhofer region [ELECTROMAG] The region far from an antenna compared to the dimensions of the antenna and the wavelength of the radiation. Also known as far field; far region; far zone; radiation zone. { 'fraùn,hōf·ər ,rē·jən }

free admittance [ELEC] The reciprocal of the blocked impedance of a transducer. { ¦frē əd'mit·əns }

free charge [ELEC] Electric charge which is not bound to a definite site in a solid, in contrast to the polarization charge. { ¦frē 'chärj }

free field [COMPUT SCI] A property of information retrieval devices which permits recording of information in the search medium without regard to preassigned fixed fields. { 'frē ,fēld }

free-field storage [COMPUT SCI] Data storage that allows recording of the data without regard for fixed or preassigned fields. { 'frē ,fēld ,stór·ij }

freeform language [COMPUT SCI] A programming or command language that does not require rigid formatting. { ¦frē,fórm 'laŋ·gwij }

freeform text [COMPUT SCI] A record, or a variable-length portion of a record, that stores plain, unformatted English. { 'frē,fórm 'tekst }

free grid *See* floating grid. { 'frē ,grid }

free impedance [ELECTR] Impedance at the input of the transducer when the impedance of its load is made zero. Also known as normal impedance. { ¦frē im'pēd·əns }

free motional impedance [ELECTR] Of a transducer, the complex remainder after the blocked impedance has been subtracted from the free impedance. { ¦frē ¦mō·shən·əl im'pēd·əns }

freenet [COMPUT SCI] A bulletin board system, based in a public library or other community or government organization, that provides access to useful resources. { 'frē,net }

free-running frequency [ELECTR] Frequency at which a normally driven oscillator operates in the absence of a driving signal. { ¦frē ,rən·iŋ ,frē·kwən·sē }

free-running multivibrator *See* astable multivibrator. { 'frē ,rən·iŋ məl·tə'vī,brād·ər }

free-running sweep [ELECTR] Sweep triggered continuously by an internal trigger generator. { 'frē ,rən·iŋ 'swēp }

free-space field intensity [ELECTROMAG] Radio field intensity that would exist at a point in a uniform medium in the absence of waves reflected from the earth or other objects. { 'frē ,spās 'fēld in,ten·səd·ē }

free-space loss [ELECTROMAG] The theoretical radiation loss, depending only on frequency and distance, that would occur if all variable factors were disregarded when transmitting energy between two antennas. { 'frē ,spās ,lòs }

free-space propagation [ELECTROMAG] Propagation of electromagnetic radiation over a straight-line path in a vacuum or ideal atmosphere, sufficiently removed from all objects that affect the wave in any way. { 'frē ,spās ,präp·ə'gā·shən }

free-space radiation pattern [ELECTROMAG] Radiation pattern that an antenna would have if it were in free space where there is nothing to reflect, refract, or absorb the radiated waves. { 'frē ,spās rād·ē'ā·shən ,pad·ərn }

free symbol [COMPUT SCI] A contextual symbol preceded and followed by a space; it is always meaningful and always used to symbolize both grammatical and nongrammatical meaning; an example is the English "I." { ¦frē 'sim·bəl }

free symbol sequence [COMPUT SCI] A symbol sequence not preceded, or not followed, or neither preceded nor followed by space. { ¦frē 'sim·bəl 'sēk·wəns }

freeware [COMPUT SCI] Copyrighted software that is downloaded from the Internet for which there is no charge. { 'frē,wer }

frequency agility [ELECTR] A feature of modern radar permitting rapid changes of the carrier frequency within the band of operating frequencies for which the radar is designed; electronic rather than mechanical tuning permits pulse-to-pulse agility. { 'frē·kwən·sē ə,jil·əd·ē }

frequency allocation [COMMUN] Assignment of available frequencies in the radio spectrum to specific stations and for specific purposes, to give maximum utilization of frequencies with minimum interference between stations. { 'frē·kwən·sē ,al·ə'kā·shən }

frequency analysis [COMPUT SCI] A determination of the number of times certain parts of an algorithm are executed, indicating which parts of the algorithm consume large quantities of time and hence where efforts should be directed toward improving the algorithm. { 'frē·kwən·sē ə,nal·ə·səs }

frequency analyzer [ELECTR] A device which measures the intensity of many different frequency components in some oscillation, as in a radio band; used to identify transmitting sources. { 'frē·kwən·sē 'an·ə,līz·ər }

frequency-azimuth intensity [ELECTR] Type of radar display in which frequency, azimuth, and strobe intensity are correlated. { ¦frē·kwən·sē ¦az·ə·məth in,ten·səd·ē }

frequency bridge [ELECTR] A bridge in which the balance varies with frequency in a known manner, such as the Wien bridge; used to measure frequency. { 'frē·kwən·sē ,brij }

frequency carrier system [COMMUN] A form of frequency division multiplex in which intelligence is carried on subcarriers. { 'frē·kwən·sē ,kar·ē·ər ,sis·təm }

frequency changer *See* frequency converter. { 'frē·kwən·sē ,chānj·ər }

frequency-changer station [ELEC] An installation at which power is transmitted between two alternating-current electric power systems operating at different frequencies by a direct-current link. { 'frē·kwən·sē ,chānj·ər ,stā·shən }

frequency characteristic *See* frequency-response curve. { ¦frē·kwən·sē ,kar·ik·tə'ris·tik }

frequency compensation *See* compensation. { ¦frē·kwən·sē ,käm·pən'sā·shən }

frequency conversion [ELECTR] Converting the carrier frequency of a received signal from its original value to the intermediate frequency value in a superheterodyne receiver. { 'frē·kwən·sē kən ,vər·zhən }

frequency converter [ELEC] A circuit, device, or machine that changes an alternating current from one frequency to another, with or without a change in voltage or number of phases. Also known as frequency changer; frequency translator. { 'frē·kwən·sē kən,vərd·ər }

frequency counter [ELECTR] An electronic counter used to measure frequency by counting the number of cycles in an electric signal during a preselected time interval. { 'frē·kwən·sē ,kaunt·ər }

frequency cutoff [ELECTR] The frequency at which the current gain of a transistor drops 3 decibels below the low-frequency gain value. { ¦frē·kwən·sē 'kəd,óf }

frequency deviation [COMMUN] The peak difference between the instantaneous frequency of a frequency-modulated wave and the carrier frequency. { ¦frē·kwən·sē ,dē·vē'ā·shən }

frequency discriminator [ELECTR] A discriminator circuit that delivers an output voltage which is proportional to the deviations of a signal from a predetermined frequency value. { ¦frē·kwən·sē di'skrim·ə,nād·ər }

frequency distortion [ELECTR] Distortion in which the relative magnitudes of the different frequency components of a wave are changed during transmission or amplification. Also known as amplitude distortion; amplitude-frequency distortion; waveform-amplitude distortion. { ¦frē·kwən·sē di'stör·shən }

frequency diversity [COMMUN] Diversity reception involving the use of carrier frequencies separated 500 hertz or more and having the same modulation, to take advantage of the fact that fading does not occur simultaneously on different frequencies. { ¦frē·kwən·sē də'vər·səd·ē }

frequency divider [ELECTR] A harmonic conversion transducer in which the frequency of the output signal is an integral submultiple of the input frequency. Also known as counting-down circuit. { 'frē·kwən·sē di,vīd·ər }

frequency-division data link [COMMUN] Data link using frequency division techniques for channel spacing. { 'frē·kwən·sē di,vizh·ən 'dad·ə ,liŋk }

frequency-division multiple access [COMMUN] A technique by which multiple users who are geographically dispersed gain access to a communications channel to which they are assigned distinct and nonoverlapping sections of the electromagnetic spectrum. Abbreviated FDMA. { ¦frē·kwən·sē di¦vizh·ən ,məl·tə·pəl 'ak,ses }

frequency-division multiplexing [COMMUN] A multiplex system for transmitting two or more signals over a common path by using a different frequency band for each signal. Abbreviated fdm; FDM. Also known as frequency multiplexing. { 'frē·kwən·sē di,vizh·ən 'məl·tə,plek·siŋ }

frequency domain [COMMUN] A plane on which signal strength can be represented graphically as a function of frequency, instead of a function of time. [CONT SYS] Pertaining to a method of analysis, particularly useful for fixed linear systems in which one does not deal with functions of time explicitly, but with their Laplace or Fourier transforms, which are functions of frequency. { 'frē·kwən·sē də,mān }

frequency-domain optical storage [COMPUT SCI] A technique whereby up to 1000 bits of information would be stored at each spatial location in an optical storage medium by using persistent spectral holeburning. { 'frē·kwən·sē də¦mān 'äp·tə·kəl 'stör·ij }

frequency doubler [ELECTR] An amplifier stage whose resonant anode circuit is tuned to the second harmonic of the input frequency; the output frequency is then twice the input frequency. Also known as doubler. { 'frē·kwən·sē ,dəb·lər }

frequency drift [ELECTR] A gradual change in the frequency of an oscillator or transmitter due to temperature or other changes in the circuit components that determine frequency. { 'frē·kwən·sē ,drift }

frequency frogging [COMMUN] Interchanging of frequency allocations for carrier channels to prevent singing, reduce crosstalk, and reduce the need for equalization; modulators in each repeater translate a low-frequency group to a high-frequency group, and vice versa. { 'frē·kwən·sē ,fräg·iŋ }

frequency hopping [COMMUN] A spread-spectrum technique in which the frequency of the carrier changes pseudorandomly according to a pseudonoise code, with a consecutive group of code symbols defining a particular frequency. { 'frē·kwən·sē ,häp·iŋ }

frequency interlace [COMMUN] Carrier chrominance signal frequency chosen so I and Q sidebands are interwoven with luminance sidebands in the same bandwidth and in a manner that causes no mutual interference. { ¦frē·kwən·sē 'in·tər,lās }

frequency locus [CONT SYS] The path followed by the frequency transfer function or its inverse, either in the complex plane or on a graph of amplitude against phase angle; used in determining zeros of the describing function. { 'frē·kwən·sē ,lō·kəs }

frequency meter |ENG| **1.** An instrument for measuring the frequency of an alternating current; the scale is usually graduated in hertz, kilohertz, and megahertz. **2.** A device calibrated to indicate frequency of a radio wave. { 'frē·kwən·sē ,mēd·ər }

frequency-modulated carrier current telephony |COMMUN| Telephony involving the use of a frequency-modulated carrier signal transmitted over power-line wires or other wires. { 'frē·kwən·sē ,mäj·ə,lād·əd 'kar·ē·ər 'kə·rənt tə'lef·ə·nē }

frequency-modulated jamming |ELECTR| Jamming technique consisting of a constant amplitude radio-frequency signal that is varied in frequency about a center frequency to produce a signal over a band of frequencies. { 'frē·kwən·sē ,mäj·ə,lād·əd 'jam·iŋ }

frequency-modulated radar |ENG| Form of radar in which the radiated wave is frequency modulated, and the returning echo beats with the wave being radiated, thus enabling range to be measured. { 'frē·kwən·sē ,mäj·ə,lād·əd 'rā,där }

frequency modulation |COMMUN| Modulation in which the instantaneous frequency of the modulated wave differs from the carrier frequency by an amount proportional to the instantaneous value of the modulating wave. Abbreviated FM. { 'frē·kwən·sē ,mäj·ə,lā·shən }

frequency-modulation broadcast band |COMMUN| The band of frequencies extending from 88 to 108 megahertz; used for frequency-modulation radio broadcasting in the United States. { 'frē·kwən·sē ,mäj·ə,lā·shən 'bròd,kast ,band }

frequency-modulation detector |ELECTR| A device, such as a Foster-Seely discriminator, for the detection or demodulation of a frequency-modulated wave. { 'frē·kwən·sē ,mäj·ə,lā·shən di'tek·tər }

frequency-modulation Doppler |ENG| Type of radar involving frequency modulation of both carrier and modulation on radial sweep. { 'frē·kwən·sē ,mäj·ə,lā·shən 'däp·lər }

frequency modulation-frequency modulation |COMMUN| System in which frequency-modulated subcarriers are used to frequency-modulate a second carrier. { 'frē·kwən·sē ,mäj·ə,lā·shən 'frē·kwən·sē ,mäj·ə,lā·shən }

frequency-modulation noise level on carrier |COMMUN| Residual frequency modulation resulting from disturbance produced in an aural transmitter operating within the band of 50 to 15,000 hertz. { 'frē·kwən·sē ,mäj·ə,lā·shən 'nòiz ,lev·əl òn 'kar·ē·ər }

frequency modulation-phase modulation |COMMUN| System in which the several frequency-modulated subcarriers are used to phase modulate a second carrier. { 'frē·kwən·sē ,mäj·ə,lā·shən 'fāz ,mäj·ə,lā·shən }

frequency-modulation receiver |ELECTR| A radio receiver that receives frequency-modulated waves and delivers corresponding sound waves. { 'frē·kwən·sē ,mäj·ə,lā·shən ri'sē·vər }

frequency-modulation receiver deviation sensitivity |ELECTR| Least frequency deviation that produces a specified output power. { 'frē·kwən·sē ,mäj·ə ,lā·shən ri'sē·vər dē·vē|ā·shən sen·sə'tiv·əd·ē }

frequency-modulation synthesis |ENG ACOUS| A method of synthesizing musical tones which, in its simplest form, is carried out using two digital oscillators, with the output of one adding to the frequency (or phase) control of the other. { |frē·kwən·sē ,mä·jə'lā·shən ,sin·thə·səs }

frequency-modulation transmitter |ELECTR| A radio transmitter that transmits a frequency-modulated wave. { 'frē·kwən·sē ,mäj·ə,lā·shən tranz'mid·ər }

frequency-modulation tuner |ELECTR| A tuner containing a radio-frequency amplifier, converter, intermediate-frequency amplifier, and demodulator for frequency-modulated signals, used to feed a low-level audio-frequency signal to a separate audio-frequency amplifier and loudspeaker. { 'frē·kwən·sē ,mäj·ə,lā·shən 'tün·ər }

frequency modulator |ELECTR| A circuit or device for producing frequency modulation. { |frē·kwən·sē 'mäj·ə,lād·ər }

frequency monitor |ELECTR| An instrument for indicating the amount of deviation of the carrier frequency of a transmitter from its assigned value. { 'frē·kwən·sē ,män·əd·ər }

frequency multiplexing See frequency-division multiplexing. { |frē·kwən·sē 'məl·tə,plek·siŋ }

frequency multiplier |ELECTR| A harmonic conversion transducer in which the frequency of the output signal is an exact integral multiple of the input frequency. Also known as multiplier. { |frē·kwən·sē 'məl·tə,plī·ər }

frequency offset |COMMUN| A small difference in the carrier frequencies of television stations in adjacent cities operating on the same channel. { |frē·kwən·sē 'òf,set }

frequency-offset transponder |ELECTR| Transponder that changes the signal frequency by a fixed amount before retransmission. { |frē·kwən·sē |òf·set tran'spän·dər }

frequency optimum traffic See optimum working frequency. { 'frē·kwən·sē ,äp·tə·məm 'traf·ik }

frequency prediction chart |COMMUN| Graph showing curve for the maximum usable frequency, frequency optimum traffic, and lowest usable frequency between two specific points for various times throughout a 24-hour period. { 'frē·kwən·sē prə,dik·shən ,chärt }

frequency pulling |ELECTR| A change in the frequency of an oscillator due to a change in load impedance. { 'frē·kwən·sē ,púl·iŋ }

frequency recorder |ELEC| An instrument which uses a frequency bridge to sense the frequency of an alternating current, and which makes a graphical record of this frequency as a function of time. { 'frē·kwən·sē ri,kórd·ər }

frequency regulator |ELEC| A device that maintains the frequency of an alternating-current generator at a predetermined value. { 'frē·kwən·sē ,reg·yə,lā·dər }

frequency relay |ELECTR| Relay which functions at a predetermined value of frequency; may be an over-frequency relay, an under-frequency relay, or a combination of both. { 'frē·kwən·sē ,rē,lā }

frequency response |ENG| A measure of the effectiveness with which a circuit, device, or system transmits the different frequencies applied to it; it is a phasor whose magnitude is the ratio of the magnitude of the output signal to that of a sine-wave input, and whose phase is that of the output with respect to the input. Also known as amplitude-frequency response; sine-wave response. { 'frē·kwən·sē ri,späns }

frequency-response curve |ENG| A graph showing the magnitude or the phase of the freqency response of a device or system as a function of frequency. Also known as frequency characteristic. { 'frē·kwən·sē ri,späns ,kərv }

frequency-response equalization See equalization. { 'frē·kwən·sē ri,späns ,ē·kwə·lə'zā·shən }

frequency-response trajectory |CONT SYS| The path followed in the complex plane by the phasor that represents the frequency response as the frequency is varied. { 'frē·kwən·sē ri,späns trə'jek·trē }

frequency run |ELECTR| A series of tests made to determine the amplitude-frequency response characteristic of a transmission line, circuit, or device. { 'frē·kwən·sē ,rən }

frequency scan antenna |ELECTROMAG| A radar antenna similar to a phased array antenna in which one dimensional scanning is accomplished through frequency variation. { 'frē·kwən·sē ,skan an'ten·ə }

frequency scanning |ELECTR| Type of system in which output frequency is made to vary at a mechanical rate over a desired frequency band. { 'frē·kwən·sē ,skan·iŋ }

frequency-selective device See electric filter. { 'frē·kwən·sē si,lek·tiv di,vīs }

frequency separation multiplier |ELECTR| Multiplier in which each of the variables is split into a low-frequency part and a high-frequency part that are multiplied separately, and the results added to give the required product; this system makes it possible to get high accuracy and broad bandwidth. { 'frē·kwən·sē ,sep·ə,rā·shən 'məl·tə,plī·ər }

frequency separator |ELECTR| The circuit that separates the horizontal and vertical synchronizing pulses in an analog monochrome or color television receiver. { 'frē·kwən·sē ,sep·ə ,rād·ər }

frequency shift |ELECTR| A change in the frequency of a radio transmitter or oscillator. Also known as radio-frequency shift. { 'frē·kwən·sē ,shift }

frequency-shift converter |ELECTR| A device that converts a received frequency-shift signal to an amplitude-modulated signal or a direct-current signal. { 'frē·kwən·sē ,shift kən 'vərd·ər }

frequency-shift keyer |ELECTR| A lever to effect a frequency shift, that is, a change in the frequency of a radio transmitter, oscillator, or receiver. { 'frē·kwən·sē ,shift 'kē·ər }

frequency-shift keying |COMMUN| A form of frequency modulation used especially in telegraph, data, and facsimile transmission, in which the modulating wave shifts the output frequency between predetermined values corresponding to the frequencies of correlated sources. Abbreviated FSK. Also known as frequency-shift modulation; frequency-shift transmission. { 'frē·kwən·sē ,shift 'kē·iŋ }

frequency-shift modulation See frequency-shift keying. { 'frē·kwən·sē ,shift ,mäj·ə·shən }

frequency-shift transmission See frequency-shift keying. { 'frē·kwən·sē ,shift tranz'mish·ən }

frequency-slope modulation |COMMUN| Type of modulation in which the carrier signal is swept periodically over the entire width of the band, much as in chirp radar; modulation of the carrier with a voice or other communication signal changes the bandwidth of the system without affecting the uniform distribution of energy over the band. { 'frē·kwən·sē ,slōp ,mäj·ə'lā·shən }

frequency spectrum |SYS ENG| In the analysis of a random function of time, such as the amplitude of noise in a system, the limit as T approaches infinity of $1/(2\pi T)$ times the ensemble average of the squared magnitude of the amplitude of the Fourier transform of the function from −T to T. Also known as power-density spectrum; power spectrum; spectral density. { 'frē·kwən·sē ,spek·trəm }

frequency splitting |ELECTR| One condition of operation of a magnetron which causes rapid alternating from one mode of operation to another; this results in a similar rapid change in oscillatory frequency and consequent loss in power at the desired frequency. { 'frē·kwən·sē ,splid·iŋ }

frequency stability |ELECTR| The ability of an oscillator to maintain a desired frequency; usually expressed as percent deviation from the assigned frequency value. { 'frē·kwən·sē stə,bil·əd·ē }

frequency stabilization |COMMUN| Process of controlling the center or carrier frequency so that it differs from that of a reference source by not more than a prescribed amount. { 'frē·kwən·sē ,stā·bə·lə'zā·shən }

frequency standard |ELECTR| A stable oscillator, usually controlled by a crystal or tuning fork, that is used primarily for frequency calibration. { 'frē·kwən·sē ,stan·dərd }

frequency swing |COMMUN| **1.** Peak difference between the maximum and the minimum values of the instantaneous frequency. **2.** In frequency modulation, a term used to describe the change in frequency resulting from the modulation. { 'frē·kwən·sē ,swiŋ }

frequency synthesizer |ELECTR| A device that provides a choice of a large number of different frequencies by combining frequencies selected from groups of independent crystals, frequency dividers, and frequency multipliers. { ¦frē·kwən·sē ¦sin·thə,sīz·ər }

frequency telemetering |COMMUN| The transmittal of an alternating-current signal from a primary element by variations in the signal frequency, instead of amplitude. { ¦frē·kwən·sē ¦tel·ə¦mēd·ə·riŋ }

frequency-time-intensity |ELECTR| Type of radar display in which the frequency, time, and strobe intensity are correlated. { ¦frē·kwən·sē ¦tīm in'ten·səd·ē }

frequency tolerance |ELECTR| Of a radio transmitter, extent to which the carrier frequency of the transmitter may be permitted to depart from the frequency assigned. { 'frē·kwən·sē ,täl·ə·rəns }

frequency-to-voltage converter |ELECTR| A converter that provides an analog output voltage which is proportional to the frequency or repetition rate of the input signal derived from a flowmeter, tachometer, or other alternating-current generating device. Abbreviated F/V converter. { ¦frē·kwən·sē tə ¦vōl·tij kən'vərd·ər }

frequency transformation |CONT SYS| A transformation used in synthesizing a band-pass network from a low-pass prototype, in which the frequency variable of the transfer function is replaced by a function of the frequency. Also known as low-pass band-pass transformation. { ¦frē·kwən·sē ,tranz·fər'mā·shən }

frequency translation |COMMUN| Moving a modulated radio-frequency carrier signal to a new location in the frequency spectrum. { ¦frē·kwən·sē tranz'lā·shən }

frequency translator See frequency converter. { ¦frē·kwən·sē 'tranz,lād·ər }

frequency-type telemeter |ELECTR| Telemeter that employs frequency of an alternating current or voltage as the translating means. { 'frē·kwən·sē ,tīp 'tel·ə,mēd·ər }

frequency variation |ELECTR| The change over time of the deviation from assigned frequency of a radio-frequency carrier (or power supply system); usually tightly controlled because of national or industry standards. { ¦frē·kwən·sē ,ver·ē¦ā·shən }

Frequently Asked Questions |COMPUT SCI| Abbreviated FAQ. **1.** A document containing answers to common questions about the subjects of other documents to which it is linked. **2.** In particular, a document associated with a Web site that contains answers to common questions about the site. { ,frē·kwənt·lē ,askt 'kwes·chənz }

Fresnel region |ELECTROMAG| The region between the near field of an antenna (close to the antenna compared to a wavelength) and the Fraunhofer region. { frā'nel ,rē·jən }

Fresnel spotlight |ELEC| A lighting instrument that is composed of a lamp and a Fresnel (stepped planoconvex) lens; the unit can be made with or without reflectors and has a system to adjust the spacing between the lamp and the lens so as to control the light beam; models range from 100 to 5000 watts. { frā'nel 'spät,līt }

Fresnel zones |ELECTROMAG| Circular portions of a wavefront transverse to a line between an emitter and a point where the disturbance is being observed; the nth zone includes all paths whose lengths are between $n - 1$ and n half-

wavelengths longer than the line-of-sight path. Also known as half-period zones. { frā'nel ,zōnz }

frictional electricity |ELEC| The electric charges produced on two different objects, such as silk and glass or catskin and ebonite, by rubbing them together. Also known as triboelectricity. { 'frik·shən·əl i,lek'tri·səd·ē }

friction bonding |ENG| Soldering of a semiconductor chip to a substrate by vibrating the chip back and forth under pressure to create friction that breaks up oxide layers and helps alloy the mating terminals. { 'frik·shən ,bänd·iŋ }

friction-feed printer |COMPUT SCI| A computer printer in which a roller is used to hold and advance the paper, much as in an ordinary typewriter. { 'frik·shən ¦fēd ,print·ər }

fringe area |COMMUN| An area just beyond the limits of the reliable service area of a television or radio transmitter, in which signals are weak and the reception is erratic. { 'frinj ,er·ē·ə }

fringe howl |ENG ACOUS| Squeal or howl heard when some circuit in a receiver is on the verge of oscillation. { 'frinj ,haùl }

fringing fields |ELECTR| The electric fields produced by scattered electrons in an electron microscope. { 'frinj·iŋ ,fēlz }

frit seal |ENG| A seal made by fusing together metallic powders with a glass binder, for such applications as hermetically sealing ceramic packages for integrated circuits. { 'frit ,sēl }

frogging repeater |ELECTR| Carrier repeater having provisions for frequency frogging to permit use of a single multipair voice cable without having excessive crosstalk. { ¦fräg·iŋ ri¦pēd·ər }

from-to tester |ENG| Test equipment which checks continuity or impedance between points. { ¦frəm ,tü ,test·ər }

front-end |COMPUT SCI| Of a computer, under programmed instructions, performing data transfers and control operations to relieve a larger computer of these routines. { ¦frənt 'end }

front-end edit |COMPUT SCI| The process of checking and correcting data at the time it is entered into a computer system. { 'frənt ¦end 'ed·it }

front-end processor |COMPUT SCI| A computer which connects to the main computer at one end and communications channels at the other, and which directs the transmitting and receiving of messages, detects and corrects transmission errors, assembles and disassembles messages, and performs other processing functions so that the main computer receives pure information. { ¦frənt ¦end ,präs,es·ər }

front porch |COMMUN| Portion of a composite picture signal which lies between the leading edge of the horizontal blanking pulse and the leading edge of the corresponding synchronizing pulse. { ¦frənt 'pòrch }

front-to-back ratio |ELECTROMAG| Ratio of the effectiveness of a directional antenna, loudspeaker, or microphone toward the front and toward the rear. { ¦frənt tə ¦bak 'rā·shō }

fruit |ELECTR| Undesired signals received by a secondary radar from transponders responding to other radars. |NAV| Radar-beacon-system video display of a synchronous beacon return which results when several interrogator stations are located within the same general area; each interrogator receives its own interrogated reply as well as many synchronous replies resulting from interrogation of the airborne transponders by other ground stations. { 'früt }

frying noise |ELEC| Noise in telephone transmission even when no conversation is taking place; caused by signal current flowing across a resistance element having multiple intermittent paths. Also known as transmitter noise. { 'frī·iŋ 'nȯiz }

F-scan *See* F-display. { 'ef ˌskan }

F-scope *See* F-display. { 'ef ˌskōp }

FSK *See* frequency-shift keying.

FSS *See* fixed-satellite service.

FTAM *See* file transfer access and management. { 'ef₁tam }

FTP *See* file transfer protocol.

full adder |ELECTR| A logic element which operates on two binary digits and a carry digit from a preceding stage, producing as output a sum digit and a new carry digit. Also known as three-input adder. { ¦fül 'ad·ər }

full duplex |COMMUN| Data channel able to operate in both directions simultaneously. |COMPUT SCI| The complete duplication of any data-processing facility. { ¦fül ¦dü₁pleks }

full-duplex operation |COMMUN| Simultaneous communications in both directions between two points. { ¦fül ¦dü₁pleks ¸äp·ə'rā·shən }

full-featured software |COMPUT SCI| Software with the most advanced available functionality. { ¦fül ¦fē·chərd 'sȯft₁wer }

full load |ELEC| The greatest load that a circuit or piece of equipment is designed to carry under specified conditions. { ¦fül 'lōd }

full-load current |ELEC| The greatest current that a circuit or piece of equipment is designed to carry under specified conditions. { ¦fül ˌlōd 'kə·rənt }

full-motion video adapter |COMPUT SCI| A video adapter capable of displaying moving video images from a video cassette recorder, laser disk player, or camcorder on a computer screen. { ˌfül ¦mō·shən 'vid·ē·ō ə₁dap·tər }

full-pitch winding |ELEC| An armature winding in which the distance between two active conductors of a coil equals the pole pitch. { ¦fül ˌpich ˌwīnd·iŋ }

full-range fuse |ELEC| A high-voltage, current-limiting fuse that can safely interrupt any value of the fault current that causes the fuse elements (conductors) to melt. { ¦fül ¦ränj 'fyüz }

full-screen editor |COMPUT SCI| A computer program that allows the user to work with the computer in an interactive manner by using all or most of the area of a cathode-ray tube or similar electronic display. { ¦fül ¦skrēn 'ed·əd·ər }

full section filter |ELECTR| A filter network whose graphical representation has the shape of the Greek letter pi, connoting capacitance in the upright legs and inductance or reactance in the horizontal member. { 'fül ¸sek·shən 'fil·tər }

full subtracter |ELECTR| A logic element which operates on three binary input signals representing a minuend, subtrahend, and borrow digit, producing as output a difference digit and a new borrow digit. Also known as three-input subtracter. { ¦fül səb'trak·tər }

full-wave amplifier |ELECTR| An amplifier without any clipping. { 'fül ˌwāv 'am·plə₁fī·ər }

full-wave bridge |ELECTR| A circuit having a bridge with four diodes, which provides full-wave rectification and gives twice as much direct-current output voltage for a given alternating-current input voltage as a conventional full-wave rectifier. { 'fül ˌwāv 'brij }

full-wave control |ELECTR| Phase control that acts on both halves of each alternating-current cycle, for varying load power over the full range from 0 to the full-wave maximum value. { 'fül ˌwāv kən'trōl }

full-wave rectification |ELECTR| Rectification in which output current flows in the same direction during both half cycles of the alternating input voltage. { 'fül ˌwāv ˌrek·tə·fə'kā·shən }

full-wave rectifier |ELECTR| A double-element rectifier that provides full-wave rectification; one element functions during positive half cycles and the other during negative half cycles. { 'fül ˌwāv 'rek·tə₁fī·ər }

full-wave vibrator |ELEC| A vibrator having an armature that moves back and forth between two fixed contacts so as to change the direction of direct-current flow through a transformer at regular intervals and thereby permit voltage stepup by the transformer; used in battery-operated power supplies for mobile and marine radio equipment. { 'fül ˌwāv 'vī₁brād·ər }

full-word boundary |COMPUT SCI| In the IBM 360 system, any address which ends in 00, and is therefore a natural boundary for a four-byte machine word. { 'fül ˌwərd 'baün·drē }

fully populated board |COMPUT SCI| A printed circuit board on which no room remains to install additional chips or other electronic components that would provide additional capabilities. { 'fül·ē ¦päp·yə₁lād·əd 'bȯrd }

function |COMPUT SCI| In FORTRAN, a subroutine of a particular kind which returns a computational value whenever it is called. |MATH| A mathematical rule between two sets that assigns to each member of the first, exactly one member of the second. { 'fəŋk·shən }

functional |COMPUT SCI| In a linear programming problem involving a set of variables x_j, $j = 1, 2, \ldots, n$, a function of the form $c_1x_1 + c_2x_2 + \cdots + c_nx_n$ (where the c_j are constants) which one wishes to optimize (maximize or minimize, depending on the problem) subject to a set of restrictions. { 'faŋk·shən·əl }

functional analysis [SYS ENG] A part of the design process that addresses the activities that a system, software, or organization must perform to achieve its desired outputs, that is, the transformations necessary to turn available inputs into the desired outputs. { ¦'faŋk·shən·əl ə'nal·ə·səs }

functional analysis diagram [SYS ENG] A representation of functional analysis and, in particular, the transformations necessary to turn available inputs into the desired outputs, the flow of data or items between functions, the processing instructions that are available to guide the transformation, and the control logic that dictates the activation and termination of functions. { ¦faŋk·shən·əl ə'nal·ə·səs ¸dī·ə¸gram }

functional application [COMPUT SCI] A program or computer system, particularly a real-time system, that deals with the primary, ongoing operations of a business enterprise. { 'faŋk·shən·əl ¸ap·lə'kā·shən }

functional decomposition [CONT SYS] The partitioning of a large-scale control system into a nested set of generic control functions, namely the regulatory or direct control function, the optimizing control function, the adaptive control function, and the self-organizing function. { 'faŋk·shən·əl dē¸käm·pə'zish·ən }

functional design [COMPUT SCI] A level of the design process in which subtasks are specified and the relationships among them defined, so that the total collection of subsystems performs the entire task of the system. [SYS ENG] The aspect of system design concerned with the system's objectives and functions, rather than its specific components. { 'faŋk·shən·əl di'zīn }

functional diagram [COMPUT SCI] A diagram that indicates the functions of the principal parts of a total system and also shows the important relationships and interactions among these parts. { 'faŋk·shən·əl 'dī·ə¸gram }

functional error recovery [COMPUT SCI] A procedure whereby the operating system intervenes in certain common errors and attempts actions to allow execution of the computer program to continue. { 'faŋk·shən·əl 'er·ər ri¸kav·ə·rē }

functional failure [COMPUT SCI] Failure of a computer system to generate the correct results for a set of inputs. { ¸faŋk·shən·əl 'fāl·yər }

functional generator See function generator. { 'faŋk·shən·əl 'jen·ə¸rād·ər }

functional interleaving [COMPUT SCI] Alternating the parts of a number of sequences in a cyclic fashion, such as a number of accesses to memory followed by an access to a data channel. { ¦faŋk·shən·əl 'in·tər¸lēv·iŋ }

functional multiplier See function multiplier. { ¦faŋk·shən·əl ¦məl·tə¸plī·ər }

functional programming [COMPUT SCI] A type of computer programming in which functions are used to control the processing of logic. { 'faŋk·shən·əl 'prō¸gram·iŋ }

functional requirement [COMPUT SCI] The documentation which accompanies a program and states in detail what is to be performed by the system. { 'faŋk·shən·əl ri'kwīr·mənt }

functional specifications [COMPUT SCI] The documentation for the design of an information system, including the data base; the human and machine procedures; and the inputs, outputs, and processes for each data entry, query, update, and report program in the system. { 'faŋk·shən·əl ¸spes·ə·fə'kā·shənz }

functional switching circuit [ELECTR] One of a relatively small number of types of circuits which implements a Boolean function and constitutes a basic building block of a switching system; examples are the AND, OR, NOT, NAND, and NOR circuits. { ¦faŋk·shən·əl 'swich·iŋ ¸sər·kət }

functional unit [COMPUT SCI] The part of the computer required to perform an elementary process such as an addition or a pulse generation. { ¦faŋk·shən·əl 'yü·nət }

function code [COMPUT SCI] Special code which appears on a medium such as a paper tape and which controls machine functions such as a carriage return. { 'faŋk·shən ¸kōd }

function-evaluation routine [COMPUT SCI] A canned routine such as a log function or a sine function. { ¦faŋk·shən i¸val·yə'wā·shən rü¸tēn }

function generator Also known as functional generator. [ELECTR] **1.** An analog computer device that indicates the value of a given function as the independent variable is increased. **2.** A signal generator that delivers a choice of a number of different waveforms, with provisions for varying the frequency over a wide range. { 'faŋk·shən ¸jen·ə¸rād·ər }

function key [COMPUT SCI] A special key on a keyboard to control a mechanical function, initiate a specific computer operation, or transmit a signal that would otherwise require multiple key strokes. { 'faŋk·shən ¸kē }

function multiplier [ELECTR] An analog computer device that takes in the changing values of two functions and puts out the changing value of their product as the independent variable is changed. Also known as functional multiplier. { 'faŋk·shən ¦məl·tə¸plī·ər }

function switch [ELECTR] A network having a number of inputs and outputs so connected that input signals expressed in a certain code will produce output signals that are a function of the input information but in a different code. { 'faŋk·shən ¸swich }

function table [COMPUT SCI] **1.** Sets of computer information arranged so an entry in one set selects one or more entries in the other sets. **2.** A computer device that converts multiple inputs into a single output or encodes a single input into multiple outputs. { 'faŋk·shən ¸tā·bəl }

function unit [COMPUT SCI] In computer systems, a device which can store a functional relationship and release it continuously or in increments. { 'faŋk·shən ¸yü·nət }

functor See logic element. { 'faŋk·tər }

fundamental group [COMMUN] In wire communications, a group of trunks that connect each local or trunk switching center to a trunk switching center of higher rank on which it homes; the term also applies to groups that interconnect zone centers. { ¦fən·də¦ment·əl 'grüp }

fundamental mode [ELECTROMAG] The waveguide mode having the lowest critical frequency. Also known as dominant mode; principal mode. { ¦fən·də¦ment·əl 'mōd }

Funkel effect [ELECTR] Fluctuations in the current from an oxide cathode, or any cathode that does not consist of pure metal, due to fluctuations in the work function resulting from changes with time in the cathode surface. { 'fəŋ·kəl i ,fekt }

fuse [ELEC] An expendable device for opening an electric circuit when the current therein becomes excessive, containing a section of conductor which melts when the current through it exceeds a rated value for a definite period of time. Also known as electric fuse. { fyüz }

fuse alarm [ELEC] Circuit that produces a visual or audible signal to indicate a blown fuse. { 'fyüz ə,lärm }

fuse block [ELEC] An insulating base on which are mounted fuse clips or other contacts for fuses. Also known as fuseboard. { 'fyüz ,bläk }

fuseboard See fuse block. { 'fyüz,bórd }

fuse box See cutout box. { 'fyüz ,bäks }

fuse clip [ELEC] A spring contact used to hold and make connection to a cartridge-type fuse. { 'fyüz ,klip }

fuse cutout [ELEC] Assembly of a fuse support and a fuse holder which may or may not include the fuse link. { ¦fyüz 'kə,daút }

fused-electrolyte battery See thermal battery. { ¦fyüzd i'lek·trə,līt ,bad·ə·rē }

fuse diode [ELECTR] A diode that opens under specified current surge conditions. { 'fyüz ,dī ,ōd }

fuse disconnecting switch [ELEC] Disconnecting switch in which a fuse unit forms a part of the blade. { ¦fyüz dis·kə'nek·tiŋ ,swich }

fused junction See alloy junction. { ¦füzd 'jəŋk·shən }

fused-junction diode See alloy-junction diode. { ¦fyüzd ¦jəŋk·shən 'dī,ōd }

fused-junction transistor See alloy-junction transistor. { ¦fyüzd ¦jəŋk·shən tran'zis·tər }

fused semiconductor [ELECTR] Junction formed by recrystallization on a base crystal from a liquid phase of one or more components and the semiconductor. { ¦fyüzd 'sem·i·kən ,dək·tər }

fuse link [ELEC] Part of a fuse that carries the current of the circuit and all or part of which melts when the current exceeds a predetermined value. { 'fyüz ,liŋk }

fuse PROM [COMPUT SCI] A programmable read-only memory in which the programming is carried out either by blowing open microscopic fuse links to define a logic one or zero for each cell in the memory array, or by causing metal to short

out base-emitter transistor junctions to program the ones or zeros into the memory. { 'fyüz ,präm }

fuse wire [ELEC] Wire made from an alloy that melts at a relatively low temperature and overheats to this temperature when carrying a particular value of overload current. { 'fyüz ,wīr }

fusible resistor [ELEC] A resistor designed to protect a circuit against overload; its resistance limits current flow and thereby protects against surges when power is first applied to a circuit; its fuse characteristic opens the circuit when current drain exceeds design limits. { ¦fyü·zə·bəl ri'zis·tər }

future address patch [COMPUT SCI] A computer output containing the address of a symbol and the address of the last reference to that symbol. { ¦fyü·chər a'dres ,pach }

future label [COMPUT SCI] An address referenced in the operand field of an instruction, but which has not been previously defined. { 'fyü·chər ,lā·bəl }

fuze See fuse. { fyüz }

fuzzy [MATH] Property of objects or processes that are not amenable to precise definition or precise measurement. { ¦fəz·ē }

fuzzy algorithm [COMPUT SCI] An ordered set of instructions, comprising fuzzy assignment statements, fuzzy conditional statements, and fuzzy unconditional action statements, that, upon execution, yield an approximate solution to a specified problem. { ¦fəz·ē 'al·gə,rith·əm }

fuzzy assignment statement [COMPUT SCI] An instruction in a fuzzy algorithm that assigns a possibly fuzzy value to a variable. { ¦fəz·ē ə'sīn·mənt ,stāt·mənt }

fuzzy conditional statement [COMPUT SCI] An instruction in a fuzzy algorithm that assigns a possibly fuzzy value to a variable or causes an action to be executed, provided that a fuzzy condition holds. { ¦fəz·ē kən'dish·ən·əl ,stāt·mənt }

fuzzy controller [CONT SYS] An automatic controller in which the relation between the state variables of the process under control and the action variables, whose values are computed from observations of the state variables, is given as a set of fuzzy implications or as a fuzzy relation. { ¦fəz·ē kən'trōl·ər }

fuzzy logic [MATH] The logic of approximate reasoning, bearing the same relation to approximate reasoning that two-valued logic does to precise reasoning. { ¦fəz·ē 'läj·ik }

fuzzy mathematics [MATH] A methodology for systematically handling concepts that embody imprecision and vagueness. { ¦fəz·ē ,math·ə'mad·iks }

fuzzy model [MATH] A finite set of fuzzy relations that form an algorithm for determining the outputs of a process from some finite number of past inputs and outputs. { ¦fəz·ē 'mäd·əl }

fuzzy relation [MATH] A fuzzy subset of the cartesian product X × Y, denoted as a relation from a set X to a set Y. { ¦fəz·ē ri'lā·shən }

fuzzy relational equation   [MATH] An equation of the form A · R = B, where A and B are fuzzy sets, R is a fuzzy relation, and A · R stands for the composition of A with R.   { ¦fəz·ē ri¦lā·shən·əl i'kwā·zhən }

fuzzy set   [MATH] An extension of the concept of a set, in which the characteristic function which determines membership of an object in the set is not limited to the two values 1 (for membership in the set) and 0 (for nonmembership), but can take on any value between 0 and 1 as well.   { 'fəz·ē 'set }

fuzzy system   [SYS ENG] A process that is too complex to be modeled by using conventional mathematical methods, and that gives rise to data that are, in general, soft, with no precise boundaries; examples are large-scale engineering complex systems, social systems, economic systems, management systems, medical diagnostic processes, and human perception.   { ¦fəz·ē 'sis·təm }

fuzzy unconditional action statement   [COMPUT SCI] An instruction in a fuzzy algorithm that specifies a possibly fuzzy mathematical operation or an action to be executed.   { ¦fəz·ē ən·kən ¦dish·ən·əl 'ak·shən ‚stāt·mənt }

fV   See femtovolt.

F/V converter   See frequency-to-voltage converter.   { ¦ef¦vē kən'vərd·ər }

G

G *See* conductance.

GaAs FET *See* gallium arsenide field-effect transistor. { 'gas‚fet }

gain The increase in signal power that is produced by an amplifier; usually given as the ratio of output to input voltage, current, or power, expressed in decibels. Also known as transmission gain. |ELECTROMAG| *See* antenna gain. { gān }

gain asymptotes |CONT SYS| Asymptotes to a logarithmic graph of gain as a function of frequency. { 'gān 'as‚əm‚tōts }

gain-bandwidth product |ELECTR| The midband gain of an amplifier stage multiplied by the bandwidth in megacycles. { ¦gān ¦band‚width ‚präd‚əkt }

gain control |ELECTR| A device for adjusting the gain of a system or component. { ¦gān kən‚trōl }

gain-crossover frequency |CONT SYS| The frequency at which the magnitude of the loop ratio is unity. { ¦gān ¦kròs‚ō‚vər ‚frē‚kwən‚sē }

gain margin |CONT SYS| The reciprocal of the magnitude of the loop ratio at the phase crossover frequency, frequently expressed in decibels. { 'gān ¦mär‚jən }

gain reduction |ELECTR| Diminution of the output of an amplifier, usually achieved by reducing the drive from feed lines by use of equalizer pads or reducing amplification by a volume control. { 'gān ri‚dək‚shən }

gain scheduling |CONT SYS| A method of eliminating influences of variations in the process dynamics of a control system by changing the parameters of the regulator as functions of auxiliary variables which correlate well with those dynamics. { 'gān ‚skej‚ə‚liŋ }

gain sensitivity control *See* differential gain control. { ¦gān ‚sen‚sə'tiv‚əd‚ē kən‚trōl }

gain turndown |ELEC| A receiver gain control incorporated in a transponder to protect the transmitter from overload. { ¦gān 'tərn‚daún }

gallium arsenide field-effect transistor |ELECTR| A field-effect transistor in which current between the ohmic source and drain contacts is carried by free electrons in a channel consisting of *n*-type gallium arsenide, and this current is modulated by a Schottky-barrier rectifying contact called the gate that varies the cross-sectional area of the channel. Abbreviated GaAs FET. { 'gal‚ē‚əm 'ärs‚ən‚īd 'fēld i¦fekt tran'zis‚tər }

galvanic |ELEC| Pertaining to electricity flowing as a result of chemical action. { gal'van‚ik }

galvanic battery |ELEC| A galvanic cell, or two or more such cells electrically connected to produce energy. { gal'van‚ik 'bad‚ə‚rē }

galvanic cell |ELEC| An electrolytic cell that is capable of producing electric energy by electrochemical action. { gal'van‚ik 'sel }

galvanic couple |ELEC| A pair of unlike substances, such as metals, which generate a voltage when brought in contact with an electrolyte. { gal'van‚ik 'kəp‚əl }

galvanic current |ELEC| A steady direct current. { gal'van‚ik 'kə‚rənt }

galvanometer |ENG| An instrument for indicating or measuring a small electric current by means of a mechanical motion derived from electromagnetic or electrodynamic forces produced by the current. { ‚gal‚və'näm‚əd‚ər }

galvanometer constant |ELEC| Number by which a certain function of the reading of a galvanometer must be multiplied to obtain the current value in ordinary units. { ‚gal‚və'näm‚əd‚ər 'kän‚stənt }

galvanometer recorder |ENG ACOUS| A sound recorder in which the audio signal voltage is applied to a coil suspended in a magnetic field; the resulting movements of the coil cause a tiny attached mirror to move a reflected light beam back and forth across a slit in front of a moving photographic film. { ‚gal‚və'näm‚əd‚ər ri'kòrd‚ər }

galvanometer shunt |ELEC| Resistor connected in parallel with a galvanometer to increase its range under certain conditions; it allows only a known fraction of the current to pass through the galvanometer. { ‚gal‚və'näm‚əd‚ər ‚shənt }

galvanostat |ELEC| A device to deliver constant current from a high-voltage battery. { gal'van‚ə ‚stat }

game theory |MATH| The mathematical study of games or abstract models of conflict situations from the viewpoint of determining an optimal policy or strategy. Also known as theory of games. { 'gām ‚thē‚ə‚rē }

game tree |MATH| A tree graph used in the analysis of strategies for a game, in which the vertices of the graph represent positions in the game, and a given vertex has as its successors all vertices that can be reached in one move from the given position. Also known as lookahead tree. { 'gām ,trē }

gang |ELEC| A mechanical connection of two or more circuit devices so that they can be varied at the same time. { gaŋ }

gang capacitor |ELEC| A combination of two or more variable capacitors mounted on a common shaft to permit adjustment by a single control. { ¦gaŋ kə'pas·əd·ər }

ganged control |ELECTR| Controls of two or more circuits mounted on a common shaft to permit simultaneous control of the circuits by turning a single knob. { ¦gaŋd kən'trōl }

gang switch |ELEC| A combination of two or more switches mounted on a common shaft to permit operation by a single control. Also known as deck switch. { 'gaŋ ,swich }

gantry-type robot |CONT SYS| A continuous-path, Cartesian-coordinate robot constructed in a bridge shape that uses rails to move along a single horizontal axis or along either of two perpendicular horizontal axes. { 'gan·trē ¦tīp 'rō ,bät }

gap |COMMUN| A region not adequately covered by the main lobes of a radar antenna, or in a larger area, not well covered by the fields of view of the radars of a network. |COMPUT SCI| A uniformly magnetized area in a magnetic storage device (tape, disk), used to indicate the end of an area containing information. |ELEC| The spacing between two electric contacts. { gap }

gap coding |COMMUN| A process for conveying information by inserting gaps or periods of non-transmission in a system that normally transmits continuously. { 'gap ,kōd·iŋ }

gap digit |COMPUT SCI| A digit in a machine word that does not represent data or instructions, such as a parity bit or a digit included for engineering purposes. { 'gap ,dij·it }

gap factor |ELECTR| Ratio of the maximum energy gained in volts to the maximum gap voltage in a tube employing electron accelerating gaps, that is, a traveling-wave tube. { 'gap ,fak·tər }

gap-filler radar |ENG| Radar used to fill gaps in radar coverage of other radar. { 'gap ,fil·ər 'rā ,där }

gapless tape |COMPUT SCI| A magnetic tape upon which raw data is recorded in a continuous manner; the data are streamed onto the tape without the word gaps; the data still may contain signs and end-of-record marks in the gapless form. { 'gap·ləs 'tāp }

gapped tape |COMPUT SCI| A magnetic tape upon which blocked data has been recorded; it contains all of the flag bits and format to be read directly into a computer for immediate use. { ¦gapt 'tāp }

gap scatter |COMPUT SCI| The deviation from the exact distance required between read/write heads and the magnetized surface. { 'gap ,skad·ər }

garbage See hash. { 'gär·bij }

garbage collection |COMPUT SCI| In a computer program with dynamic storage allocation, the automatic process of identifying those memory cells whose contents are no longer useful for the computation in progress and then making them available for some other use. { 'gär·bij kə ,lek·shən }

garbage in, garbage out |COMPUT SCI| A phrase often stressed during introductory courses in computer utilization as a reminder that, regardless of the correctness of the logic built into the program, no answer can be valid if the input is erroneous. Abbreviated GIGO. { 'gär·bij 'in ¦gär·bij 'aút }

garble |COMMUN| To alter a message intentionally or unintentionally so that it is difficult to understand. { 'gär·bəl }

garbling |ELECTR| Confusion resulting from a secondary radar receiving overlapping coded responses from transponders in a dense target environment. { 'gär·bliŋ }

garnet maser |ELECTR| A name incorrectly applied to a ferromagnetic amplifier. { 'gär·nət 'mā·zər }

gas-activated battery |ELEC| A reserve battery which is activated by introducing a gas which reacts with a material between the electrodes of the battery to form an electrolyte. { ¦gas ak·tə ¦vād·əd 'bad·ə·rē }

gas-bubble protective device See Buchholz protective device. { 'gas ,bəb·əl prə'tek·tiv di,vīs }

gas capacitor |ELEC| A capacitor consisting of two or more electrodes separated by a gas, other than air, that serves as a dielectric. { ¦gas kə'pas·əd·ər }

gas cell |ELEC| Cell in which the action depends on the absorption of gases by the electrodes. { 'gas ,sel }

gas current |ELECTR| A positive-ion current produced by collisions between electrons and residual gas molecules in an electron tube. Also known as ionization current. { 'gas ,kə·rənt }

gas discharge |ELECTR| Conduction of electricity in a gas, due to movements of ions produced by collisions between electrons and gas molecules. { ¦gas 'dis,chärj }

gas-discharge display |ELECTR| A display in which seven or more cathode elements form the segments of numerical or alphameric characters when energized by about 160 volts direct current; the segments are vacuum-sealed in a neon-mercury gas mixture. { ¦gas 'dis,chärj di,splā }

gas-discharge lamp See discharge lamp. { ¦gas 'dis,chärj ,lamp }

gas doping |ELECTR| The introduction of impurity atoms into a semiconductor material by epitaxial growth, by using streams of gas that are mixed before being fed into the reactor vessel. { 'gas ,dōp·iŋ }

GasFET |ENG| A gas sensor based on changes, upon exposure to hydrogen, in the surface part of the work function of a palladium component that serves as the gate contact of a metal oxide semiconductor field-effect transistor (MOSFET). { 'gas,fet }

gas-filled cable |ELEC| A coaxial or other cable containing gas under pressure to serve as insulation and keep out moisture. { 'gas ,fild 'kā·bəl }

gas-filled diode [ELECTR] A gas tube which is a diode, such as a cold-cathode rectifier or phanotron. { 'gas ,fild 'dī,ōd }

gas-filled rectifier *See* cold-cathode rectifier. { 'gas ,fild 'rek·tə,fī·ər }

gas-filled triode [ELECTR] A gas tube which has a grid or other control element, such as a thyratron or ignitron. { 'gas ,fild 'trī,ōd }

gas focusing [ELECTR] A method of concentrating an electron beam by utilizing the residual gas in a tube; beam electrons ionize the gas molecules, forming a core of positive ions along the path of the beam which attracts beam electrons and thereby makes the beam more compact. Also known as ionic focusing. { 'gas ,fō·kəs·iŋ }

gas-insulated substation [ELEC] An electric power substation in which all live equipment and busbars are housed in grounded metal enclosures sealed and filled with sulfur hexafluoride gas. { 'gas ,in·sə,lād·əd 'səb,stā·shən }

gas ionization [ELECTR] Removal of the planetary electrons from the atoms of gas filling an electron tube, so that the resulting ions participate in current flow through the tube. { ¦gas ,ī·ə·nə'zā·shən }

gas magnification [ELECTR] Increase in current through a phototube due to ionization of the gas in the tube. { 'gas ,mag·nə·fə'kā·shən }

gas phototube [ELECTR] A phototube into which a quantity of gas has been introduced after evacuation, usually to increase its sensitivity. { ¦gas 'fōd·ō,tüb }

gas scattering [ELECTR] The scattering of electrons or other particles in a beam by residual gas in the vacuum system. { 'gas ,skad·ə·riŋ }

gas-sensitive field-effect transistor [ELECTR] A field-effect transistor whose gate electrode is composed of a material, such as palladium, that is sensitive to a particular gas, such as hydrogen, so that the gain of the transistor depends on the concentration of this gas. { 'gas ¦sen·səd·iv 'fēld i,fekt tran,zis·tər }

gassiness [ELECTR] Presence of unwanted gas in a vacuum tube, usually in relatively small amounts, caused by the leakage from outside or evolution from the inside walls or elements of the tube. { 'gas·ē·nəs }

gassing [ELEC] The evolution of gas in the form of small bubbles in a storage battery when charging continues after the battery has been completely charged. { 'gas·iŋ }

gassy tube [ELECTR] A vacuum tube that has not been fully evacuated or has lost part of its vacuum due to release of gas by the electrode structure during use, so that enough gas is present to impair operating characteristics appreciably. Also known as soft tube. { ¦gas·ē 'tüb }

gas tetrode *See* tetrode thyratron. { ¦gas 'te,trōd }

gas thermostatic switch [ELEC] A thermostatic switch in which heat causes the pressure of gas in a sealed metal bellows to increase, thereby moving the bellows and closing the contacts of a switch. { ¦gas 'thər,mə,stad·ik 'swich }

gas tube [ELECTR] An electron tube into which a small amount of gas or vapor is admitted after the tube has been evacuated; ionization of gas molecules during operation greatly increases current flow. { 'gas ,tüb }

gas vacuum breakdown [ELECTR] Ionization of residual gas in a vacuum, causing reverse conduction in an electron tube. { ¦gas ¦vak·yəm 'brāk ,daün }

gate [ELECTR] **1.** A circuit having an output and a multiplicity of inputs and so designed that the output is energized only when a certain combination of pulses is present at the inputs. **2.** A circuit in which one signal, generally a square wave, serves to switch another signal on and off. **3.** One of the electrodes in a field-effect transistor. **4.** To control the passage of a pulse or signal. **5.** In radar, an electric waveform which is applied to the control point of a circuit to alter the mode of operation of the circuit at the time when the waveform is applied. Also known as gating waveform. **6.** In radar, an electronic waveform applied to a circuit or a timing cue applied to logic to alter the operation of the circuit or logic at the appropriate time; generally used in anticipation of an input of particular interest. { gāt }

gate-array device [ELECTR] An integrated logic circuit that is manufactured by first fabricating a two-dimensional array of logic cells, each of which is equivalent to one or a few logic gates, and then adding final layers of metallization that determine the exact function of each cell and interconnect the cells to form a specific network when the customer orders the device. { 'gāt ə ,rā di,vīs }

gate-controlled rectifier [ELECTR] A three-terminal semiconductor device, such as a silicon controlled rectifier, in which the unidirectional current flow between the rectifier terminals is controlled by a signal applied to a third terminal called the gate. { 'gāt kən,trōld 'rek·tə,fī·ər }

gate-controlled switch [ELECTR] A semiconductor device that can be switched from its nonconducting or "off" state to its conducting or "on" state by applying a negative pulse to its gate terminal and that can be turned off at any time by applying reverse drive to the gate. Abbreviated GCS. { 'gāt kən,trōld 'swich }

gated-beam tube [ELECTR] A pentode electron tube having special electrodes that form a sheet-shaped beam of electrons; this beam may be deflected away from the anode by a relatively small voltage applied to a control electrode, thus giving extremely sharp cutoff of anode current. { ¦gād·əd ¦bēm 'tüb }

gated sweep [ELECTR] Sweep in which the duration as well as the starting time is controlled to exclude undesired echoes from the indicator screen. { ¦gād·əd 'swēp }

gate equivalent circuit [ELECTR] A unit of measure for specifying relative complexity of digital circuits, equal to the number of individual logic gates that would have to be interconnected to perform the same function as the digital circuit under evaluation. { 'gāt i¦kwiv·ə·lənt ¦sər·kət }

gate generator |ELECTR| A circuit used to generate gate pulses; in one form it consists of a multivibrator having one stable and one unstable position. { 'gāt ¦jen·ə·rād·ər }

gate multivibrator |ELECTR| Rectangular-wave generator designed to produce a single positive or negative gate voltage when triggered and then to become inactive until the next trigger pulse. { 'gāt ¦məl·ti'vī¦brād·ər }

gate pulse |ELECTR| A pulse that triggers a gate circuit so it will pass a signal. { 'gāt ¦pəls }

gate turnoff |ELECTR| A *pnpn* switching device comparable to a silicon-controlled rectifier, but having a more complex gate structure that permits easy and fast turnoff as well as turn-on from its gate input terminal, at frequencies up to 100 kilohertz. { 'gāt ¦tər¦nȯf }

gate-turnoff silicon controlled rectifier |ELECTR| A silicon controlled rectifier that can be turned off by applying a current to its gate; used largely for direct-current switching, because turnoff can be achieved in a fraction of a microsecond. { ¦gāt ¦tər¦nȯf 'sil·ə·kən kən¦trōld 'rek·tə¦fī·ər }

gateway |COMMUN| A point of entry and exit to another system, such as the connection point between a local-area network and an external-communications network. { 'gāt¦wā }

gate winding |ELECTR| A winding used in a magnetic amplifier to produce on-off action of load current. { 'gāt ¦wīnd·iŋ }

gather write |COMPUT SCI| An operation that creates a single output record from data items gathered from nonconsecutive locations in main memory. { ¦gath·ər 'wrīt }

gating |ELECTR| The process of selecting those portions of a wave that exist during one or more selected time intervals or that have magnitudes between selected limits. { 'gād·iŋ }

gating waveform *See* gate. { ¦gād·iŋ 'wāv¦fȯrm }

Gaussian beam |ELECTROMAG| A beam of electromagnetic radiation whose wave front is approximately spherical at any point along the beam and whose transverse field intensity over any wave front is a Gaussian function of the distance from the axis of the beam. { 'gaús·ē·ən 'bēm }

Gaussian noise |COMMUN| Random electromagnetic signals inherent in nature, both in the surroundings of a receiver and produced in the receiver itself; typically produced by the thermal agitation of molecular structures, and having Gaussian statistics in its components. Also known as thermal noise. { ¦gaú·sē·ən 'nȯiz }

Gaussian noise generator |ELECTR| A signal generator that produces a random noise signal whose frequency components have a Gaussian distribution centered on a predetermined frequency value. { ¦gaú·sē·ən 'nȯiz ¦jen·ə¦rād·ər }

Gauss' law of flux |ELEC| The law that the total electric flux which passes out from a closed surface equals (in rationalized units) the total charge within the surface. { 'gaús ¦lȯ əv 'flȯks }

gc *See* gigahertz.

GCS *See* gate-controlled switch.

G-display |ELECTR| A radar display format in which the target of a tracking radar appears as a spot, as in an F-display, with "wings" (horizontal extensions of the plot) that increase in length as the range decreases. Also known as G-indicator; G-scan; G-scope. { 'jē di¦splā }

gearbox *See* transmission. { 'gir¦bäks }

Geissler tube |ELECTR| An experimental discharge tube with two electrodes at opposite ends, used to demonstrate and study the luminous effects of electric discharges through various gases at low pressures. { 'gīs·lər ¦tüb }

gelled cell |ELEC| A lead-acid cell with a nonspillable gelled electrolyte for portable use. { ¦jeld ¦sel }

gemmho |ELEC| A unit of conductance, equal to 10^{-6} mho, being the conductance of a substance which has a resistance of 10^6 ohms. { 'je¦mō }

gen |COMPUT SCI| To install an operating system or a systems software package for a particular configuration of computer equipment. Abbreviation for generate. { jen }

gender |ELEC| The classification of a connector as female or male. { 'jen·dər }

gender changer |ELEC| A small passive device that is placed between two connectors of the same gender to enable them to be joined. Also known as cable matcher. { 'jen·dər ¦chān·jər }

general address |COMMUN| Group of characters included in the heading of a message that causes the message to be routed to all addresses included in the general address category. { ¦jen·rəl ə'dres }

generalized routine |COMPUT SCI| A routine which can process a wide variety of jobs; for example, a generalized sort routine which will sort in ascending or descending order on any number of fields whether alphabetic or numeric, or both, and whether binary coded decimals or pure binaries. { 'jen·rə¦līzd rü'tēn }

generalized system |COMPUT SCI| A computer system developed for a broad range of users. { ¦jen·rə¦līzd 'sis·təm }

general program |COMPUT SCI| A computer program designed to solve a specific type of problem when values of appropriate parameters are supplied. { ¦jen·rəl 'prō·grəm }

general-purpose automatic test system |ELECTR| Modular, computer-type, automatic electronic checkout system capable of finding faults in electronic equipment at the system, subsystem, line replaceable unit, module, and piece part levels. { ¦jen·rəl ¦pər·pəs ¦ȯd·ə¦mad·ik 'test ¦sis·təm }

general-purpose computer |COMPUT SCI| A device that manipulates data without detailed, step-by step control by human hand and is designed to be used for many different types of problems.   { ¦jen·rəl ¦pər·pəs kəm'pyüd·ər }

general-purpose function generator |COMPUT SCI| A function generator which can be adjusted to generate many different functions, rather than being designed for a particular function. Also known as arbitrary function generator.   { ¦jen·rəl ¦pər·pəs 'faŋk·shən ‚jen·ə‚rād·ər }

general-purpose language |COMPUT SCI| A computer programming language whose use is not restricted to a particular type of computer or a specialized application.   { 'jen·rəl ¦pər·pəs 'laŋ· gwij }

general-purpose systems simulation *See* GPSS.   { 'jen·rəl ¦pər·pəs 'sis·təmz ‚sim·yə‚lā·shən }

general register *See* local register.   { ¦jen·rəl 'rej· ə·stər }

general routine |COMPUT SCI| In computers, a routine, or program, applicable to a class of problems; it provides instructions for solving a specific problem when appropriate parameters are supplied.   { ¦jen·rəl rü'tēn }

generate |COMPUT SCI| **1.** To create a particular program by selecting parts of a general-program skeleton (or outline) and specializing these parts into a cohesive entity.   **2.** *See* gen.   { 'jen·ə‚rāt }

generate and test |COMPUT SCI| A computer problem-solving method in which a sequence of candidate solutions is generated, and each is tested to determine if it is an appropriate solution.   { 'jen·ə‚rāt ən 'test }

generated address |COMPUT SCI| An address calculated or determined by instructions contained in a computer program for subsequent use by that program. Also known as calculated address; synthetic address.   { ¦jen·ə‚rād·əd ə'dres }

generating area *See* fetch.   { 'jen·ə‚rād·iŋ ‚er·ē·ə }

generating magnetometer |ENG| A magnetometer in which a coil is rotated in the magnetic field to be measured with the resulting generated voltage being proportional to the strength of the magnetic field.   { 'jen·ə‚rād·iŋ mag·nə'täm·əd·ər }

generating routine *See* generator.   { 'jen·ə‚rād·iŋ rü‚tēn }

generation |COMPUT SCI| **1.** Any one of three groups used to historically classify computers according to their electronic hardware components, logical organization and software, or programming techniques; computers are thus known as first-, second-, or third-generation; a particular computer may possess characteristics of all generations simultaneously.   **2.** One of a family of data sets, related to one another in that each is a modification of the next most recent data set.   { ‚jen·ə'rā·shən }

generation data group |COMPUT SCI| A collection of files, each a modification of the previous one, with the newest numbered 0, the next −1, and so forth, and organized so that each time a new file is added the oldest is deleted. Abbreviated GDG.   { ‚jen·ə'rā·shən 'dad·ə ‚grüp }

generation number |COMPUT SCI| A number contained in the file label of a reel of magnetic tape

that indicates the generation of the data set of the tape.   { ‚jen·ə'rā·shən ‚nəm·bər }

generation rate |ELECTR| In a semiconductor, the time rate of creation of electron-hole pairs.   { ‚jen·ə'rā·shən ‚rāt }

generative grammar |COMPUT SCI| A set of rules that describes the valid expressions in a formal language on the basis of a set of the parts of speech (formally called the set of metavariables or phrase names) and the alphabet or character set of the language.   { 'jen·rəd·iv 'gram·ər }

generator |COMPUT SCI| A program that produces specific programs as directed by input parameters. Also known as generating routine. |ELEC| A machine that converts mechanical energy into electrical energy; in its commonest form, a large number of conductors are mounted on an armature that is rotated in a magnetic field produced by field coils. Also known as dynamo; electric generator. |ELECTR| **1.** A vacuum-tube oscillator or any other nonrotating device that generates an alternating voltage at a desired frequency when energized with direct-current power or low-frequency alternating-current power.   **2.** A circuit that generates a desired repetitive or nonrepetitive waveform, such as a pulse generator.   { 'jen·ə‚rād·ər }

generator field control |ELEC| Method of regulating the output voltage of a generator by controlling the voltage which excites the field of the generator.   { 'jen·ə‚rād·ər 'fēld kən‚trōl }

generator lock |ELECTR| Circuitry that synchronizes two video signals so that they can be mixed. Abbreviated genlock.   { 'jen·ə‚rād·ər ‚läk }

generator reactor |ELEC| A small inductor connected between power-plant generators and the rest of an electric power system in order to limit and localize the effects of voltage transients.   { 'jen·ə‚rād·ər rē‚ak·tər }

generator resistance |ELEC| The resistance of the current source in a network; usually much smaller than the load but taken into account in some network calculations.   { 'jen·ə‚rād·ər ri ‚zis·təns }

genetic algorithm |COMPUT SCI| A search procedure based on the mechanics of natural selection and genetics. Also known as evolutionary strategy.   { jə‚ned·ik 'al·gə‚rith·əm }

genetic programming *See* evolutionary programming.   { jə‚ned·ik 'prō‚gram·iŋ }

genlock *See* generator lock.   { 'gen‚läk }

geomagnetic noise |COMMUN| Interference in radio communications arising from terrestrial magnetism.   { ¦jē·ō·mag¦ned·ik 'nóiz }

geometrical distortion |COMPUT SCI| A discrepancy between the horizontal and vertical dimensions of the picture elements on an electronic display, causing, for example, circles to appear as ovals unless corrected for in software.   { ¦jē·ə ¦me·trə·kəl di'stòr·shən }

geometric programming |SYS ENG| A nonlinear programming technique in which the relative contribution of each of the component costs is first determined; only then are the variables in the component costs determined.   { ¦¦ē·ə¦me·trik 'prō ¸gram·iŋ }

geophone |ELECTR| A transducer, used in seismic work, that responds to motion of the ground at a location on or below the surface of the earth.   { 'jē·ə¸fōn }

germanium diode |ELECTR| A semiconductor diode that uses a germanium crystal pellet as the rectifying element. Also known as germanium rectifier.   { jər'mān·ē·əm 'dī¸ōd }

germanium rectifier   See germanium diode.   { jər'mān·ē·əm 'rek·tə¸fī·ər }

germanium transistor |ELECTR| A transistor in which the semiconductor material is germanium, to which electric contacts are made.   { jər'mān·ē·əm tran'zis·tər }

get |COMPUT SCI| An instruction in a computer program to read data from a file.   { get }

getmain |COMPUT SCI| An instruction used in some programming languages to request dynamic allocation of additional storage space to the program.   { 'get¸mān }

getter sputtering |ELECTR| The deposition of high-purity thin films at ordinary vacuum levels by using a getter to remove contaminants remaining in the vacuum.   { 'gəd·ər ¸spəd·ə·riŋ }

ghost |COMPUT SCI| To display a menu option in a dimmed, fuzzy typeface to indicate that this option is no longer available. |ELECTR| In radar, a contact generated where in fact no target exists, resulting from measurement ambiguity or attempts to resolve ambiguities with multiple observations in a multiple-target situation.   { gōst }

ghost algebraic manipulation language |COMPUT SCI| An algebraic manipulation language which externally gives the appearance of manipulating quite general mathematical expressions, although internally it is functioning with canonically represented data, much like the simpler seminumerical languages.   { ¦gōst al·jə'brā·ik mə¸nip·yə'lā·shən ¸laŋ·gwij }

ghost image |ELECTR| An undesired duplicate image offset from the desired image on a video display device.   { 'gōst ¸im·ij }

ghost mode |ELECTROMAG| Waveguide mode having a trapped field associated with an imperfection in the wall of the waveguide; a ghost mode can cause trouble in a waveguide operating close to the cutoff frequency of a propagation mode.   { 'gōst ¸mōd }

ghost pulse |ELECTR| An unwanted signal appearing on the screen of a radar indicator and caused by echoes which have a basic repetition frequency differing from that of the desired signals. Also known as ghost image; ghost signal.   { 'gōst ¸pəls }

ghost signal |ELECTR| The reflection-path signal that produces a ghost image on an analog television receiver. Also known as echo.   { 'gōst ¸sig·nəl }

GHz   See gigahertz.

gibberish   See hash.   { 'jib·rish }

GIF   See graphics interchange format.   { gif }

gigabit |COMMUN | One billion bits, or 1,000,000,000 bits.   { 'gig·ə¸bit }

gigacycle   See gigahertz.   { 'gig·ə¸sī·kəl }

gigaflops |COMPUT SCI| A unit of computer speed, equal to 10^9 flops.   { 'gig·ə¸fläps }

gigahertz |COMMUN| Unit of frequency equal to 10^9 hertz. Abbreviated GHz. Also known as gigacycle (gc); kilomegacycle; kilomegahertz.   { 'gig·ə¸hərts }

gigawatt |ELEC| One billion watts, or 10^9 watts. Abbreviated GW.   { 'gig·ə¸wät }

GIGO   See garbage in, garbage out.   { 'gī¸gō }

gigohm |ELEC| One thousand megohms, or 10^9 ohms.   { 'gig¸ōm }

Gilbert circuit |ELECTR| A circuit that compensates for nonlinearities and instabilities in a monolithic variable-transconductance circuit by using the logarithmic properties of diodes and transistors.   { 'gil·bərt ¸sər·kət }

Gilbert-Varshamov bound |COMPUT SCI| In the theory of quantum computation, a sufficient condition for an algorithm that encodes N logical qubits into N' carrier qubits (with N' larger than N) to correct any error on any M carrier qubits; namely, that N/N' be smaller than

$$1 - 2| - \chi \log_{2\chi} - (1 - \chi) \log_2(1 - X)|,$$

where $\chi = 2M/N'$.   { ¦gil·bərt ¸vär¦sha·məv ¦baund }

gimmick |ELEC| Length of twisted two-conductor cable, used as a variable capacitive load, in which the capacitance is varied by untwisting and separating the individual conductors.   { 'gim·ik }

G-indicator   See G-display.   { 'jē ¸in·də¸kād·ər }

GKS   See graphical kernel system.

glare |COMMUN | The interference that arises when an attempt is made to place a telephone call just as an incoming call is arriving; in the case of data transmission under the control of a computer, this can render the line or even the computer temporarily inoperative.   { gler }

glare filter |ENG| A screen that is placed over the face of a cathode-ray tube to reduce glare from ambient and overhead light.   { 'gler ¸fil·tər }

glass capacitor |ELEC| A capacitor whose dielectric material is glass.   { ¦glas kə'pas·əd·ər }

glassivation |ELECTR| Method of transistor passivation by a pyrolytic glass-deposition technique, whereby silicon semiconductor devices, complete with metal contact systems, are fully encapsulated in glass.   { ¸glas·ə'vā·shən }

glass-plate capacitor |ELEC| High-voltage capacitor in which the metal plates are separated by sheets of glass serving as the dielectric, with the complete assembly generally immersed in oil.   { 'glas ¸plāt kə'pas·əd·ər }

glass resistor |ELEC| A glass tube with a helical carbon resistance element painted on it.   { ¦glas ri'zis·tər }

glass switch |ELECTR| An amorphous solid-state device used to control the flow of electric current. Also known as ovonic device.   { ¦glas 'swich }

glass-to-metal seal |ELECTR| An airtight seal between glass and metal parts of an electron tube, made by fusing together a special glass and special metal alloy having nearly the same temperature coefficients of expansion. { ¦glas tə ¦med·əl 'sēl }

glint |ELECTR| **1.** Pulse-to-pulse variation in the apparent angular center of a target, due to target scattering complexity and dynamics; causes angle errors in tracking radars using either conical scan or monopulse techniques. **2.** The use of this effect to degrade tracking or seeking functions of an enemy weapons system. { glint }

glitch |ELECTR| **1.** An undesired transient voltage spike occurring on a signal being processed. **2.** A minor technical problem arising in electronic equipment. { glich }

global format |COMPUT SCI| A choice of label alignment or numeric format in a spreadsheet program that applies to all the cells of the spreadsheet. { ¦glō-bəl 'fȯr,mat }

global memory |COMPUT SCI| Computer storage that can be used by a number of processors connected together in a multiprocessor system. { 'glō-bəl 'mem·rē }

global orbiting navigation satellite system See GLONASS. { ¦glō-bəl ¦ȯrb·əd·iŋ ,nav·ə¦gā-shən 'sad·ə,līt ,sis·təm }

Global Positioning System |NAV| A positioning or navigation system designed to use 24 satellites, each carrying atomic clocks, to provide a receiver anywhere on earth with extremely accurate measurements of its three-dimensional position, velocity, and time. Abbreviated GPS. { 'glō-bəl pə'zish·niŋ ,sis·təm }

global resource sharing |COMPUT SCI| The ability of all of the users of a local-area network to share any of the resources (storage devices, input/output devices, and so forth) connected to the network. { 'glō-bəl ri'sȯrs ,sher·iŋ }

global search and replace |COMPUT SCI| A text-editing function of a word-processing system in which text is scanned for a given combination of characters, and each such combination is replaced by another set of characters. { ¦glō-bəl ¦sərch ən ri'plās }

global system for mobile communications See GSM. { ¦glō-bəl ¦sis·təm fər ,mō-bəl kə,myü-nə'kā-shənz }

global variable |COMPUT SCI| A variable which can be accessed (used or changed) throughout a computer program and is not confined to a single block. { ¦glō-bəl 'ver·ē·ə·bəl }

Globar lamp |ELEC| A lamp whose illuminating element is a silicon carbide rod which gives off blackbody radiation when heated. { 'glō,bär ,lamp }

GLONASS |NAV| A worldwide Russian navigation system designed to use 24 satellites in three uniformly spaced orbital planes to provide three-dimensional position and velocity data to equipped users on or above the earth's surface. Acronym for global orbiting navigation satellite system. { 'glō,nas }

glossary |COMPUT SCI| A file of commonly used phrases that can be retrieved in a word-processing program, usually through use of a command and a keyword. { 'gläs·ə·rē }

glow discharge |ELECTR| A discharge of electricity through gas at relatively low pressure in an electron tube, characterized by several regions of diffuse, luminous glow and a voltage drop in the vicinity of the cathode that is much higher than the ionization voltage of the gas. Also known as cold-cathode discharge. { ¦glō ¦dis,chärj }

glow-discharge cold-cathode tube See glow-discharge tube. { ¦glō ¦dis,chärj ¦kōld 'kath,ōd ,tüb }

glow-discharge microphone |ENG ACOUS| Microphone in which the action of sound waves on the current forming a glow discharge between two electrodes causes corresponding variations in the current. { ¦glō ¦dis,chärj 'mī·krə,fōn }

glow-discharge tube |ELECTR| A gas tube that depends for its operation on the properties of a glow discharge. Also known as glow-discharge cold-cathode tube; glow tube. { ¦glō ¦dis,chärj ,tüb }

glow-discharge voltage regulator |ELECTR| Gas tube that varies in resistance, depending on the value of the applied voltage; used for voltage regulation. { ¦glō ¦dis,chärj 'vōl·tij ,reg·yə,lād·ər }

glow lamp |ELECTR| A two-electrode electron tube containing a small quantity of an inert gas, in which light is produced by a negative glow close to the negative electrode when voltage is applied between the electrodes. { 'glō ,lamp }

glow potential |ELECTR| The potential across a glow discharge, which is greater than the ionization potential and less than the sparking potential, and is relatively constant as the current is varied across an appreciable range. { 'glō pə ,ten·chəl }

glow tube See glow-discharge tube. { 'glō ,tüb }

glow-tube oscillator |ELECTR| A circuit using a glow-discharge tube which functions as a simple relaxation oscillator, generating a fixed-amplitude periodic sawtooth waveform. { 'glō ¦tüb 'äs·ə,lād·ər }

GNU |COMPUT SCI| Freely distributed software for producing and distributing nonproprietary software that is compatible with Unix, but is not Unix. { gə'nü }

gobo |ENG| A panel used to shield a television camera lens from direct light. |ENG ACOUS| A sound-absorbing shield used with a microphone to block unwanted sounds. { 'gō,bō }

Golay code |COMMUN| A linear, block-based error-correcting code that is particularly suited to applications where short code word length and low latency are important. { gə'lā ,kōd }

gold doping |ELECTR| A technique for controlling the lifetime of minority carriers in a transistor; gold is diffused into the base and collector regions to reduce storage time in transistor circuits. { 'gōl ,dōp·iŋ }

golden-section search |COMPUT SCI| A dichotomizing search in which, in each step, the remaining items are divided as closely as possible according to the golden section. { 'gōl·dən 'sek·shən ,sərch }

gold-leaf electroscope |ELEC| An electroscope in which two narrow strips of gold foil or leaf suspended in a glass jar spread apart when charged; the angle between the strips is related to the charge. { 'gōld ¦lēf i'lek·trə‚skōp }

goovoo |COMPUT SCI| A file within a generation data group, so called because of the notation used in some systems in which, for example, G003 V001 is volume 1, generation −3 of a generation data group. { 'gü‚vü }

GOP See group of pictures.

Gopher |COMPUT SCI| A menu-based program for browsing the Internet and finding and gaining access to files, programs, definitions, and other Internet resources. { 'gō·fər }

GOTO-less programming |COMPUT SCI| The writing of computer programs without the use of GOTO statements. { ¦gō·tü‚les 'prō‚gram·iŋ }

Goto pair |ELECTR| Two tunnel diodes connected in series in such a way that when one is in the forward conduction region, the other is in the reverse tunneling region; used in high-speed gate circuits. { 'gō·dō ‚per }

GOTO statement |COMPUT SCI| A statement in a computer program that provides for the direct transfer of control to another statement with the identifier that is the argument of the GOTO statement. { 'gō‚tü ‚stāt·mənt }

government frequency bands |COMMUN| Radio-frequency bands which are allotted to various departments and services of the federal government. { 'gəv·ər·mənt ‚fre·kwən·se ‚banz }

g parameter |ELECTR| One of a set of four transistor equivalent-circuit parameters; they are the inverse of the h parameters. { 'jē pə‚ram·əd·ər }

GPS See Global Positioning System.

GPSS |COMPUT SCI| A problem-oriented programming language designed to assist the user in developing models. Acronym for general-purpose systems simulation.

graceful degradation |COMPUT SCI| A programming technique to prevent catastrophic system failure by allowing the machine to operate, though in a degraded mode, despite failure or malfunction of several integral units or subsystems. { 'grās·fúl ‚deg·rə'dā·shən }

graceful exit |COMPUT SCI| The ability to escape from a problem situation in a computer program without having to reboot the computer. { ¦grās·fəl 'eg·zət }

grade |COMMUN| One of two types of television service, designated grade A and grade B, each having a specified signal strength, that of grade A being several timeshigher than B. { grād }

graded-junction transistor See rate-grown transistor. { ¦grād·əd ¦jəŋk·shən tran'zis·tər }

graded periodicity technique |ELECTR| A technique for modifying the response of a surface acoustic wave filter by varying the spacing between successive electrodes of the interdigital transducer. { ¦grād·əd ‚pir·ē·ə'dis·əd·ē tek ‚nēk }

grain direction |COMPUT SCI| In character recognition, the arrangement of paper fibers in relation to a document's travel through a character reader. { 'grān də‚rek·shən }

grandfather |COMPUT SCI| A data set that is two generations earlier than the data set under consideration. { 'gran‚fath·ər }

grandfather cycle |COMPUT SCI| The period during which records are kept but not used except to reconstruct other records which are accidentally lost. { 'gran‚fath·ər ‚sī·kəl }

granularity |SYS ENG| The degree to which a system can be broken down into separate components, making it customizable and flexible. { ‚gran·yə'lar·əd·ē }

graphechon |ELECTR| A storage tube having two electron guns, one for writing and the other for reading and simultaneous erasing, on opposite sides of the storage medium, which consists of an insulator or semiconductor deposited on a thin substratum of metal supported by a fine mesh. { 'graf·ə‚kän }

graphical design |ELECTR| Methods of obtaining operating data for an electron tube or semiconductor circuit by using graphs which plot the relationship between two variables, such as plate voltage and grid voltage, while another variable, such as plate current, is held constant. { ¦graf·ə·kəl di'zīn }

graphical kernel system |COMPUT SCI| A standard system and language for creating two- and three-dimensional master graphics images on many types of display devices. Abbreviated GKS. { ¦graf·i·kəl 'kər·nəl ‚sis·təm }

graphical symbol |ELEC| A true symbol, rather than a coarse picture, representing an element in an electrical diagram. { ¦graf·ə·kəl 'sim·bəl }

graphical user interface |COMPUT SCI| A user interface in which program features are represented by icons that the user can access and manipulate with a pointing device. Abbreviated GUI. { ¦graf·ə·kəl ‚yü·zər 'in·tər‚fās }

graphical visual display device |COMPUT SCI| A computer input-output device which enables the user to manipulate graphic material in a visible two-way, real-time communication with the computer, and which consists of a light pen, keyboard, or other data entry devices, and a visual display unit monitored by a controller. Also known as graphoscope. { ¦graf·ə·kəl ¦vizh·ə·wəl di'spla·ə‚vīs }

graphic display |ELECTR| The display of data in graphical form on the screen of a cathode-ray tube. { ¦graf·ik di'splā }

graphic equalizer |ENG ACOUS| A device that allows the response of audio equipment to be modified independently in several frequency bands through the use of a bank of slide controls whose positions form a graph of the frequency response. { ¦graf·ik ē·kwə‚lī·zər }

graphic panel |CONT SYS| A master control panel which indicates the status of equipment and operations in a system, and their relationships. { ¦graf·ik 'pan·əl }

graphics driver |COMPUT SCI| A series of instructions that activates a graphics device, such as a display screen or plotter. { 'graf·iks ‚drīv·ər }

graphics engine [COMPUT SCI] A specialized processor that carries out graphics processing independently of the main central processing unit. Also known as graphics processor. { 'graf·iks ¦en·jən }

graphics interchange format [COMPUT SCI] Common file format for compressed graphic images on the World Wide Web that is limited to 256 colors. Abbreviated GIF. { ¦graf·iks 'in·tər,chānj ,fȯr,mat }

graphics interface [COMPUT SCI] A user interface that displays icons to represent objects. { 'graf·iks ¦in·tər,fās }

graphics primitive [COMPUT SCI] A basic building block for graphic images, such as a dot, line, or curve. { 'graf·iks ¦prim·əd·iv }

graphics processor See graphics engine. { 'graf·iks ¦prä·ses·ər }

graphics program [COMPUT SCI] A program for the generation of images, ranging in complexity from simple line drawings to realistically shaded pictures that resemble photographs. { 'graf·iks ,prō·grəm }

graphics tablet [COMPUT SCI] A padlike peripheral device which is designed so that shapes appear on the monitor's screen when the tablet is drawn upon with a pointed device. { 'graf·iks ,tab·lət }

graphics terminal [COMPUT SCI] An input/output device that can accept and display picture images. { 'graf·iks ¦tər·mən·əl }

graphite anode [ELECTR] **1.** The rod of graphite which is inserted into the mercury-pool cathode of an ignitron to start current flow. **2.** The collector of electrons in a beam power tube or other high-current tube. { 'gra,fīt 'an,ōd }

graphite resistor [ELEC] A resistor made of carbon for resistance heating. { 'gra,fīt ri'zis·tər }

graphoscope See graphical visual display device. { 'graf·ə,skōp }

graph theory [MATH] **1.** The mathematical study of the structure of graphs and networks. **2.** The body of techniques used in graphing functions in the plane. { 'graf ,thē·ə·rē }

grass [ELECTR] Clutter due to circuit noise in a radar receiver, seen on an A scope as a pattern resembling a cross section of turf. Also known as hash. { gras }

grasshopper fuse [ELEC] Small fuse incorporating a spring which, upon release by the fusing wire, connects an auxiliary circuit to operate an alarm. { 'gras,häp·ər ,fyüz }

Grassot fluxmeter [ENG] A type of fluxmeter in which a light coil of wire is suspended in a magnetic field in such a way that it can rotate; the ends of the suspended coil are connected to a search coil of known area penetrated by the magnetic flux to be measured; the flux is determined from the rotation of the suspended coil when the search coil is moved. { ,grä,sō 'fləks,mēd·ər }

Gratz rectifier [ELECTR] Three-phase, full-wave rectifying circuit using six rectifiers connected in a bridge circuit. { 'grats ,rek·tə,fī·ər }

grav See G. { grav }

Gray code [COMMUN] A modified binary code in which sequential numbers are represented by expressions that differ only in one bit, to minimize errors. Also known as reflective binary code. { 'grā ,kōd }

greeking [COMPUT SCI] The display of the format of a document without displaying the characters. { 'grēk·iŋ }

grid [COMPUT SCI] In optical character recognition, a system of two groups of parallel lines, perpendicular to each other, used to measure or specify character images. [ELEC] A metal plate with holes or ridges, used in a storage cell or battery as a conductor and a support for the active material. [ELECTR] An electrode located between the cathode and anode of an electron tube, which has one or more openings through which electrons or ions can pass, and serves to control the flow of electrons from cathode to anode. { grid }

grid-anode transconductance See transconductance. { 'grid ¦an,ōd ,tranz·kən'dək·təns }

grid battery See C battery. { 'grid ,bad·ə·rē }

grid bias [ELECTR] The direct-current voltage applied between the control grid and cathode of an electron tube to establish the desired operating point. Also known as bias; C bias; direct grid bias. { 'grid ,bī·əs }

grid-bias cell See bias cell. { 'grid ,bī·əs ,sel }

grid blocking [ELECTR] **1.** Method of keying a circuit by applying negative grid bias several times cutoff value to the grid of a tube during key-up conditions; when the key is down, the blocking bias is removed and normal current flows through the keyed circuit. **2.** Blocking of capacitance-coupled stages in an amplifier caused by the accumulation of charge on the coupling capacitors due to grid current passed during the reception of excessive signals. { 'grid ,bläk·iŋ }

grid blocking capacitor See grid capacitor. { 'grid ¦bläk·iŋ kə,pas·əd·ər }

grid cap [ELECTR] A top-cap terminal for the control grid of an electron tube. { 'grid ,kap }

grid capacitor [ELECTR] A small capacitor used in the grid circuit of an electron tube to pass signal current while blocking the direct-current anode voltage of the preceding stage. Also known as grid blocking capacitor; grid condenser. { 'grid kə,pas·əd·ər }

grid-cathode capacitance [ELECTR] Capacitance between the grid and the cathode in a vacuum tube. { 'grid ¦kath,ōd kə,pas·əd·əns }

grid characteristic [ELECTR] Relationship of grid current to grid voltage of a vacuum tube. { 'grid ¦kar·ik·tə,ris·tik }

grid circuit [ELECTR] The circuit connected between the grid and cathode of an electron tube. { 'grid ,sər·kət }

grid condenser See grid capacitor. { 'grid kən,den·sər }

grid conductance See electrode conductance. { 'grid kən'dək·təns }

grid control [ELECTR] Control of anode current of an electron tube by variation (control) of the control grid potential with respect to the cathode of the tube. { 'grid kən‚trōl }

grid-controlled mercury-arc rectifier [ELECTR] A mercury-arc rectifier in which one or more electrodes are employed exclusively to control the starting of the discharge. Also known as grid-controlled rectifier. { 'grid kən‚trōld 'mər‚kyə‚re ¦ärk 'rek‚tə‚fī‚ər }

grid-controlled rectifier See grid-controlled mercury-arc rectifier. { 'grid kən‚trōld 'rek‚tə ‚fī‚ər }

grid control tube [ELECTR] Mercury-vapor-filled thermionic vacuum tube with an external grid control. { 'grid kən‚trōl ‚tüb }

grid current [ELECTR] Electron flow to a positive grid in an electron tube. { 'grid ‚kə‚rənt }

grid-dip meter [ELECTR] A multiple-range electron-tube oscillator incorporating a meter in the grid circuit to indicate grid current; the meter reading dips (reads lower grid current) when an external resonant circuit is tuned to the oscillator frequency. Also known as grid-dip oscillator. { 'grid ‚dip ‚mēd‚ər }

grid-dip oscillator See grid-dip meter. { 'grid ‚dip ¦äs‚ə‚lād‚ər }

grid drive [ELECTR] A signal applied to the grid of a transmitting tube. { 'grid ‚drīv }

grid driving power [ELECTR] Average product of the instantaneous value of the grid current and of the alternating component of the grid voltage over a complete cycle; this comprises the power supplied to the biasing device and to the grid. { 'grid ¦drīv‚iŋ ‚paú‚ər }

grid element [ELEC] A sinuous resistor used to heat a furnace, made of heavy wire, strap, or casting and suspended from refractory or stainless supports built into the furnace walls, floor, and roof. { 'grid ‚el‚ə‚mənt }

grid-glow tube [ELECTR] A glow-discharge tube in which one or more control electrodes initiate but do not limit the anode current except under certain operating conditions. { 'grid ¦glō ‚tüb }

gridistor [ELECTR] Field-effect transistor which uses the principle of centripetal striction and has a multichannel structure, combining advantages of both field effect transistors and minority carrier injection transistors. { gri'dis‚tər }

grid leak [ELECTR] A resistor used in the grid circuit of an electron tube to provide a discharge path for the grid capacitor and for charges built up on the control grid. { 'grid ‚lēk }

grid-leak detector [ELECTR] A detector in which the desired audio-frequency voltage is developed across a grid leak and grid capacitor by the flow of modulated radio-frequency current; the circuit provides square-law detection on weak signals and linear detection on strong signals, along with amplification of the audio-frequency signal. { 'grid ‚lēk di‚tek‚tər }

grid limiter [ELECTR] Limiter circuit which operates by limiting positive grid voltages by means of a large ohmic value resistor; as the exciting signal moves in a positive direction with respect to the

cathode, current through the resistor causes an IR drop which holds the grid voltage essentially at cathode potential; during negative excursions no current flows in the grid circuit, so no voltage drop occurs across the resistor. { 'grid ‚lim‚əd‚ər }

grid locking [ELECTR] Defect of tube operation in which the grid potential becomes continuously positive due to excessive grid emission. { 'grid ‚läk‚iŋ }

grid modulation [ELECTR] Modulation produced by feeding the modulating signal to the control-grid circuit of any electron tube in which the carrier is present. { 'grid ‚mäj‚ə'lā‚shən }

grid neutralization [ELECTR] Method of amplifier neutralization in which a portion of the grid-cathode alternating-current voltage is shifted 180° and applied to the plate-cathode circuit through a neutralizing capacitor. { 'grid ‚nü‚trə‚lə'zā‚shən }

grid-plate capacitance [ELECTR] Direct capacitance between the grid and the plate in a vacuum tube. { 'grid ¦plāt kə'pas‚əd‚əns }

grid-plate transconductance See transconductance. { 'grid ¦plāt tranz‚kən'dək‚təns }

grid-pool tube [ELECTR] An electron tube having a mercury-pool cathode, one or more anodes, and a control electrode or grid that controls the start of current flow in each cycle; the excitron and ignitron are examples. { 'grid ¦pül ‚tüb }

grid pulse modulation [ELECTR] Modulation produced in an amplifier or oscillator by applying one or more pulses to a grid circuit. { 'grid ¦pəls ‚mäj‚ə'lā‚shən }

grid pulsing [ELECTR] Circuit arrangement of a radio-frequency oscillator in which the grid of the oscillator is biased so negatively that no oscillation takes place even when full plate voltage is applied; pulsing is accomplished by removing this negative bias through the application of a positive pulse on the grid. { 'grid ‚pəls‚iŋ }

grid resistor [ELECTR] A general term used to denote any resistor in the grid circuit. { 'grid ri‚zis‚tər }

grid return [ELECTR] External conducting path for the return grid current to the cathode. { 'grid ri ‚tərn }

grid suppressor [ELECTR] Resistor of low ohmic value inserted in the grid circuit of a radio-frequency amplifier to prevent low-frequency parasitic oscillations. { 'grid sə‚pres‚ər }

grid swing [ELECTR] Total variation in grid-cathode voltage from the positive peak to the negative peak of the applied signal voltage. { 'grid ‚swiŋ }

grid transformer [ELECTR] Transformer to supply an alternating voltage to a grid circuit or circuits. { 'grid tranz‚fōr‚mər }

grid-type level detector [ELECTR] A detector using a vacuum tube with input applied to a grid. { 'grid ‚tip 'lev‚əl di‚tek‚tər }

grid voltage [ELECTR] The voltage between a grid and the cathode of an electron tube. { 'grid ‚vōl‚tij }

Griebhard's rings [ELEC] A method of producing lines of constant color on a copper sheet, coinciding with the equipotential lines of an electric field. { 'grēb·härts ‚riŋz }

grinding |ELECTR| **1.** A mechanical operation performed on silicon substrates of semiconductors to provide a smooth surface for epitaxial deposition or diffusion of impurities. **2.** A mechanical operation performed on quartz crystals to alter their physical size and hence their resonant frequencies. { 'grīn·diŋ }

grip vector |CONT SYS| A vector from a point on the wrist socket of a robot to the point where the end effector grasps an object; describes the orientation of the object in space. { 'grip ‚vek·tər }

groover |ENG| A tool for forming grooves in a slab of concrete not yet hardened. { 'grüv·ər }

Grosh's law |COMPUT SCI| The law that the processing power of a computer is proportional to the square of its cost. { 'grōsh·əz ‚lò }

gross index |COMPUT SCI| The first of two indexes consulted to gain access to a record. { ‚grōs 'in ‚deks }

gross information content |COMMUN| Measure of the total information, redundant or otherwise, contained in a message; expressed as the number of bits, nits, or Hartleys required to transmit the message with specified accuracy over a noiseless medium without coding. { ¦grōs ‚in·fər'mā·shən ‚kän·tent }

ground |ELEC| **1.** A conducting path, intentional or accidental, between an electric circuit or equipment and the earth, or some conducting body serving in place of the earth. Abbreviated gnd. Also known as earth (British usage); earth connection. **2.** To connect electrical equipment to the earth or to some conducting body which serves in place of the earth. { graùnd }

ground absorption |ELECTROMAG| Loss of energy in transmission of radio waves, due to dissipation in the ground. { 'graùnd əb‚sórp·shən }

ground cable |ELEC| A heavy cable connected to earth for the purpose of grounding electric equipment. { 'graùnd ‚kā·bəl }

ground circuit |ELEC| A telephone or telegraph circuit part of which passes through the ground. { 'graùnd ‚sər·kət }

ground clutter |ELECTROMAG| Clutter on a ground or airborne radar due to reflection of signals from the ground or objects on the ground. Also known as ground flutter; ground return; land return; terrain echoes. { 'graùnd ‚kləd·ər }

ground conductivity |ELEC| The effective conductivity of the ground, used in calculating the attenuation of radio waves. { 'graùnd ‚kän·dək ¦tiv·əd·ē }

ground current See earth current. { 'graùnd ‚kə·rənt }

ground detector |ELEC| An instrument or equipment used for indicating the presence of a ground on an ungrounded system. Also known as ground indicator. { 'graùnd di‚tek·tər }

ground dielectric constant |ELEC| Dielectric constant of the earth at a given location. { 'graùnd di·ə¦lek·trik 'kän·stənt }

grounded-anode amplifier See cathode follower. { ¦graùnd·əd 'an‚ōd ‚am·plə‚fī·ər }

grounded-base amplifier |ELECTR| An amplifier that uses a transistor in a grounded-base connection. { ¦graùnd·əd 'bās ‚am·plə‚fī·ər }

grounded-base connection |ELECTR| A transistor circuit in which the base electrode is common to both the input and output circuits; the base need not be directly connected to circuit ground. Also known as common-base connection. { ¦graùnd·əd 'bās kə‚nek·shən }

grounded-cathode amplifier |ELECTR| Electron-tube amplifier with a cathode at ground potential at the operating frequency, with input applied between control grid and ground, and with the output load connected between plate and ground. { ¦graùnd·əd 'kath‚ōd ‚am·plə‚fī·ər }

grounded-collector connection |ELECTR| A transistor circuit in which the collector electrode is common to both the input and output circuits; the collector need not be directly connected to circuit ground. Also known as common-collector connection. { ¦graùnd·əd kə'lek·tər kə ‚nek·shən }

grounded-emitter amplifier |ELECTR| An amplifier that uses a transistor in a grounded-emitter connection. { ¦graùnd·əd i'mid·ər ‚am·plə ‚fī·ər }

grounded-emitter connection |ELECTR| A transistor circuit in which the emitter electrode is common to both the input and output circuits; the emitter need not be directly connected to circuit ground. Also known as common-emitter connection. { ¦graùnd·əd i'mid·ər kə‚nek·shən }

grounded-gate amplifier |ELECTR| Amplifier that uses thin-film transistors in which the gate electrode is connected to ground; the input signal is fed to the source electrode and the output is obtained from the drain electrode. { ¦graùnd·əd 'gāt ‚am·plə‚fī·ər }

grounded-grid amplifier |ELECTR| An electron-tube amplifier circuit in which the control grid is at ground potential at the operating frequency; the input signal is applied between cathode and ground, and the output load is connected between anode and ground. { ¦graùnd·əd 'grid ‚am·plə‚fī·ər }

grounded-grid-triode circuit |ELECTR| Circuit in which the input signal is applied to the cathode and the output is taken from the plate; the grid is at radio-frequency ground and serves as a screen between the input and output circuits. { ¦graùnd·əd ¦grid ¦trī‚ōd 'sər·kət }

grounded-grid-triode mixer |ELECTR| Triode in which the grid forms part of a grounded electrostatic screen between the anode and cathode, and is used as a mixer for centimeter wavelengths. { ¦graùnd·əd ¦grid ¦trī‚ōd 'mik·sər }

grounded-plate amplifier See cathode follower. { ¦graùnd·əd 'plāt ‚am·plə‚fī·ər }

grounded system |ELEC| Any conducting apparatus connected to ground. Also known as earthed system. { ¦graùnd·əd 'sis·təm }

ground effect |COMMUN| The effect of ground conditions on radio communications. { 'graùnd i‚fekt }

ground electrode |ELEC| A conductor buried in the ground, used to maintain conductors connected to it at ground potential and dissipate current conducted to it into the earth, or to provide a return path for electric current in a direct-current power transmission system. Also known as earth electrode; grounding electrode. { 'graund i¦lek,trōd }

ground equalizer inductors [ELECTROMAG] Coils, having relatively low inductance, inserted in the circuit to one or more of the grounding points of an antenna to distribute the current to the various points in any desired manner. { ¦graund 'ē·kwə,līz·ər in,dək·tərz }

ground fault [ELEC] Accidental grounding of a conductor. { 'graund ,fólt }

ground fault interrupter [ELEC] A fast-acting circuit breaker that also senses very small ground fault currents such as might flow through the body of a person standing on damp ground while touching a hot alternating-current line wire. { 'graund ,fólt ,int·ə,rəp·tər }

ground flutter See ground clutter. { 'graund ,fləd·ər }

ground indicator See ground detector. { 'graund ,in·də,kād·ər }

grounding [ELEC] Intentional electrical connection to a reference conducting plane, which may be earth, but which more generally consists of a specific array of interconnected electrical conductors referred to as the grounding conductor. { 'graund·iŋ }

grounding conductor [ELEC] An array of interconnected electric conductors at a uniform potential, to which electrical connections are made for the purpose of grounding. { 'graund·iŋ kən ,dək·tər }

grounding electrode See ground electrode. { 'graund·iŋ i,lek,trōd }

grounding plate [ELEC] An electrically grounded metal plate on which a person stands to discharge static electricity picked up by his body, or a similar plate buried in the ground to act as a ground rod. { 'graund·iŋ ,plāt }

grounding reactor [ELEC] A reactor sometimes used in a grounded alternating-current system which joins a conductor or neutral point to ground and serves to limit ground current in case of a fault. Also known as earthing reactor (British usage). { 'graund·iŋ rē,ak·tər }

grounding receptacle [ELEC] A receptacle which has an extra contact that accepts the third round or U-shaped prong of a grounding attachment plug and is connected internally to a supporting strap, providing a ground both through the outlet box and the grounding conductor, armor, or raceway of the wiring system. { 'graund·iŋ ri ,sep·tə·kəl }

grounding transformer [ELEC] Transformer intended primarily for the purpose of providing a neutral point for grounding purposes. { 'graund·iŋ tranz,fór·mər }

ground junction See grown junction. { 'graund ,jəŋk·shən }

ground loop [COMMUN] Return currents or magnetic fields from relatively high-powered circuits or components which generate unwanted noisy

signals in the common return of relatively low-level signal circuits. { 'graund ,lüp }

ground lug [ELEC] A lug that connects a grounding conductor to a grounding electrode. { 'graund ,ləg }

ground noise [ENG ACOUS] The residual system noise in the absence of the signal in recording and reproducing; usually caused by inhomogeneity in the recording and reproducing media, but may also include tube noise and noise generated in resistive elements in the amplifier system. { 'graund ,nóiz }

ground outlet [ELEC] Outlet equipped with a receptacle of the polarity type having, in addition to the current-carrying contacts, one grounded contact which can be used for the connection of an equipment-grounding conductor. { 'graund ,aút·let }

ground-penetrating radar See ground-probing radar. { ¦graund ,pen·ə¦trād·iŋ 'rā,där }

ground plane [ELEC] A grounding plate, above-ground counterpoise, or arrangement of buried radial wires required with a ground-mounted antenna that depends on the earth as the return path for radiated radio-frequency energy. { 'graund ,plān }

ground-plane antenna [ELECTROMAG] Vertical antenna combined with a grounded horizontal disk, turnstile element, or similar ground-plane simulation; such antennas may be mounted several wavelengths above the ground, and provide a low radiation angle. { 'graund ,plān an'ten·ə }

ground plate [ELEC] A plate of conducting material embedded in the ground to act as a ground electrode. { 'graund ,plāt }

ground potential [ELEC] Zero potential with respect to the ground or earth. { 'graund pə,ten·chəl }

ground-probing radar [ENG] A nondestructive technique using electromagnetic waves to locate objects or interfaces buried beneath the earth's surface or located within a visually opaque structure. Also known as ground-penetrating radar; subsurface radar; surface-penetrating radar. { ¦graund ¦prōb·iŋ 'rā,där }

ground protection [ELEC] Protection provided a circuit by a device which opens the circuit when a fault to ground occurs. { 'graund prə,tek·shən }

ground recharge [ELEC] The flow of electrons from the ground, in reference to lightning effects. { 'graund ¦rē,chärj }

ground-reflected wave [ELECTROMAG] Component of the ground wave that is reflected from the ground. { 'graund ri¦flek·təd 'wāv }

ground resistance [ELEC] Opposition of the earth to the flow of current through it; its value depends on the nature and moisture content of the soil, on the material, composition, and nature of connections to the earth, and on the electrolytic action present. { 'graund ri ,zis·təns }

ground return [ELEC] Use of the earth as the return path for a transmission line. [ELECTRO-MAG] **1.** An echo received from the ground by

an airborne radar set. **2.** *See* ground clutter.
{ 'graund ri,tərn }

ground-return circuit |ELEC| Circuit which has a conductor (or two or more in parallel) between two points and which is completed through the ground or earth. { 'graund ri,tərn ,sər·kət }

ground rod |ELEC| A rod that is driven into the earth to provide good grounding. { 'graund ,räd }

groundscatter propagation |COMMUN| Multihop ionospheric radio propagation along other than the great-circle path between transmitting and receiving stations; radiation from the transmitter is first reflected back to earth from the ionosphere, then scattered in many directions from the earth's surface. { 'graund ,skad·ər ,präp·ə'gā·shən }

ground system |ELECTROMAG| The portion of an antenna that is closely associated with an extensive conducting surface, which may be the earth itself. { 'graund ,sis·təm }

ground-up read-only memory |COMPUT SCI| A read-only memory which is designed from the bottom up, and for which all fabrication masks used in the multiple mask process are custom-generated. { 'graund ¦əp 'rēd ¦ōn·lē 'mem·rē }

ground wave |COMMUN| A radio wave that is propagated along the earth and is ordinarily affected by the presence of the ground and the troposphere; includes all components of a radio wave over the earth except ionospheric and tropospheric waves. Also known as surface wave. { 'graund ,wāv }

ground wire |ELEC| A conductor used to connect electric equipment to a ground rod or other grounded object. { 'graund ,wīr }

group |COMMUN| A communication transmission subdivision containing a number of voice channels, either within a supergroup or separately, normally comprised of up to 12 voice channels occupying the frequency band 60–108 kilohertz; each voice channel may be multiplexed for teletypewriter operation, if required. { grüp }

group A kits |ELECTR| Normally those items of electronic equipment which may be permanently or semipermanently installed in an aircraft for supporting, securing, or interconnecting the components and controls of the equipment, and which will not in any manner compromise the security classification of the equipment. { ¦grüp 'ā ,kits }

group B kits |ELECTR| Normally, the operating or operable component of the electronic equipment in an aircraft which, when installed on or in connection with group A parts, constitute the complete operable equipment. { ¦grüp 'bē ,kits }

group bus |ELEC| A scheme of electrical connections for a generating station in which more than two feeder lines are supplied by two bus-selector circuit breakers which lead to a main bus and an auxiliary bus. { 'grüp ¦bəs }

group busy tone |COMMUN| High tone connected to the jack sleeves of an outgoing trunk group as an indication that all trunks in the group are busy. { 'grüp ¦biz·ē ,tōn }

group code *See* systematic error-checking code. { 'grüp ,kōd }

group-coded record |COMPUT SCI| A method of recording data on magnetic tape with eight tracks of data and one parity track, in which every eighth byte in conjunction with the parity track is used for detection and correction of all single-bit errors. { 'grüp ,kōd·əd 'rek·ərd }

group communications software *See* groupware. { ,grüp kə,myü·nə,kā·shənz 'sóf,wer }

grouped-frequency operation |COMMUN| Use of different frequency bands for channels in opposite directions in a two-wire carrier system. { 'grüpt ¦frē·kwən·sē ,äp·ə,rā·shən }

grouped records |COMPUT SCI| Two or more records placed together and identified by a single key, to save storage space or reduce access time. { 'grüpt ¦rek·ərdz }

group-indicate |COMPUT SCI| To print indicative information from only the first record of a group. { 'grüp ¦in·də,kāt }

grouping |COMMUN| Periodic error in the spacing of recorded lines in a facsimile system. { 'grüp·iŋ }

grouping circuits |COMMUN| Circuits used to interconnect two or more switchboard positions together, so that one operator may handle the several switchboard positions from one operator's set. { 'grüp·iŋ ,sər·kəts }

grouping of records |COMPUT SCI| Placing records together in a group to either conserve storage space or reduce access time. { 'grüp·iŋ əv 'rek·ərdz }

group mark |COMPUT SCI| A character signaling the beginning or end of a group of data. { 'grüp ¦märk }

group modulation |COMMUN| Process by which a number of channels, already separately modulated to a specific frequency range, are again modulated to shift the group to another range. { 'grüp ,mäj·ə'lā·shən }

group of pictures |COMMUN| In MPEG-2, a group of pictures consists of one or more pictures in sequence. Abbreviated GOP. { 'grüp əv 'pik·chərz }

group printing |COMPUT SCI| The printing of information summarizing the data on a group of cards or other records when a key change occurs. { 'grüp ¦print·iŋ }

groupware |COMPUT SCI| Multiuser software that supports information sharing through digital media, such as electronic mail and messaging, electronic meeting systems and audio conferencing, group calendaring and scheduling, workflow process diagramming and analysis tools, and group document handling including group editing. { 'grüp,wer }

grove cell |ELEC| Primary cell, having a platinum electrode in an electrolyte of nitric acid within a porous cup, outside of which is a zinc electrode in an electrolyte of sulfuric acid; it normally operates on a closed circuit. { 'grōv ,sel }

Grover's algorithm |COMPUT SCI| An algorithm for finding an item in a database of 2N items, using a quantum computer, in a time of order $2^{N/2}$

steps instead of order 2^N steps. { ¦grō·vərz'al·gə ,rith·əm }

growler |ELEC| An electromagnetic device consisting essentially of two field poles arranged as in a motor, used for locating short-circuited coils in the armature of a generator or motor and for magnetizing or demagnetizing objects; a growling noise indicates a short-circuited coil. { 'graúl·ər }

grown-diffused transistor [ELECTR] A junction transistor in which the final junctions are formed by diffusion of impurities near a grown junction. { ¦grōn di¦fyüzd tran'zis·tər }

grown junction |ELECTR| A junction produced by changing the types and amounts of donor and acceptor impurities that are added during the growth of a semiconductor crystal from a melt. Also known as ground junction. { ¦grōn ¦jəŋk·shən }

grown-junction photocell [ELECTR] A photodiode consisting of a bar of semiconductor material having a *pn* junction at right angles to its length and an ohmic contact at each end of the bar. { ¦grōn ¦jəŋk·shən 'fōd·ō,sel }

grown-junction transistor |ELECTR| A junction transistor in which different impurities are placed in the melt in sequence as the silicon or germanium seed crystal is slowly withdrawn, to produce the alternate *pn* and *np* junctions. { ¦grōn ¦jəŋk·shən tran'zis·tər }

G-scan *See* G-display. { 'jē ,skan }

G-scope *See* G-display. { 'jē ,skōp }

GSM |COMMUN| A digital cellular telephone technology that is based on time-division multiple access; it operates on the 900-megahertz and 1.8-gigahertz bands in Europe, where it is the predominant cellular system, and on the 1.9-gigahertz band in the United States. Derived from global system for mobile communications.

G string *See* field waveguide. { 'jē ,striŋ }

guard arm |ELEC| **1.** Crossarm placed across and in line with a cable to prevent damage to the cable. **2.** Crossarm located over wires to prevent foreign wires from falling into them. { 'gärd ,ärm }

guard band |ELECTR| A narrow frequency band provided between adjacent channels in certain portions of the radio spectrum to prevent interference between stations. { 'gärd ,band }

guarded command |COMPUT SCI| A program statement within a group of such statements that determines whether the other statements will be executed by the computer. { 'gärd·əd kə'mand }

guarding |ELEC| A method of eliminating surface-leakage effects from measurements of electrical resistance which employs a low-resistance conductor in the vicinity of one of the terminals or a portion of the measuring circuit. { 'gärd·iŋ }

guard relay |ELEC| Used in the linefinder circuit to make sure that only one linefinder can be connected to any line circuit when two or more line relays are operated simultaneously. { 'gärd ¦rē,lā }

guard ring |ELEC| A ring-shaped auxiliary electrode surrounding one of the plates of a parallel-plate capacitor to reduce edge effects. |ELECTR| A ring-shaped auxiliary electrode used in an electron tube or other device to modify the electric field or reduce insulator leakage; in a counter tube or ionization chamber a guard ring may also serve to define the sensitive volume. |PHYS| A device used in heat flow experiments to ensure an even distribution of heat, consisting of a ring that surrounds the specimen and is made of a similar material. { 'gärd ,riŋ }

guard shield |ELECTR| Internal floating shield that surrounds the entire input section of an amplifier; effective shielding is achieved only when the absolute potential of the guard is stabilized with respect to the incoming signal. { 'gärd ,shēld }

guard signal |COMPUT SCI| A signal used in digital-to-analog converters, analog-to-digital converters, or other converters which permits values to be read or converted only when the values are not changing, usually to avoid ambiguity error. { 'gärd ,sig·nəl }

guard wire |ELEC| A grounded conductor placed beneath an overhead transmission line in order to ground the line, in case it breaks, before reaching the ground. { 'gärd ,wīr }

Gudden-Pohl effect |ELECTR| The momentary illumination produced when an electric field is applied to a phosphor previously excited by ultraviolet radiation. { 'gúd·ən 'pōl i,fekt }

guest computer |COMPUT SCI| A computer that operates under the control of another computer (the host). { 'gest kəm,pyüd·ər }

GUI *See* graphical user interface. { 'gü,ē *or* ¦jē¦yü'ī }

guided propagation |COMMUN| Type of radio-wave propagation in which radiated rays are bent excessively by refraction in the lower layers of the atmosphere; this bending creates an effect much as if a duct or waveguide has been formed in the atmosphere to guide part of the radiated energy over distances far beyond the normal range. Also known as trapping. { 'gīd·əd präp·ə'gā·shən }

guided wave |ELECTROMAG| A wave whose energy is concentrated near a boundary or between substantially parallel boundaries separating materials of different properties and whose direction of propagation is effectively parallel to these boundaries; waveguides transmit guided waves. { 'gīd·əd 'wāv }

guide wavelength [ELECTROMAG] Wavelength of electromagnetic energy conducted in a waveguide; guide wavelength for all air-filled guides is always longer than the corresponding free-space wavelength. { 'gīd ¦wāv,leŋkth }

guidewire [ENG] A wire embedded in the surface of the path traveled by an electromagnetically guided automated guided vehicle. { 'gīd,wīr }

Guillemin line [ELECTR] A network or artificial line used in high-level pulse modulation to generate a nearly square pulse, with steep rise and fall; used in radar sets to control pulse width. { gē·yə'ma ,līn }

gulp [COMPUT SCI] A series of bytes considered as a unit. { gəlp }

Gunn amplifier [ELECTR] A microwave amplifier in which a Gunn oscillator functions as a negative-resistance amplifier when placed across the terminals of a microwave source. { 'gən ¦am·plə,fī·ər }

Gunn diode See Gunn oscillator. { 'gən ¦dī,ōd }

Gunn effect [ELECTR] Development of a rapidly fluctuating current in a small block of a semiconductor (perhaps n-type gallium arsenide) when a constant voltage above a critical value is applied to contacts on opposite faces. { 'gən i,fekt }

Gunn oscillator [ELECTR] A microwave oscillator utilizing the Gunn effect. Also known as Gunn diode. { 'gən ¦äs·ə,läd·ər }

GW See gigawatt.

gyrator [ELECTROMAG] A waveguide component that uses a ferrite section to give zero phase shift for one direction of propagation and 180° phase shift for the other direction; in other words, it causes a reversal of signal polarity for one direction of propagation but not for the other direction. Also known as microwave gyrator. { 'jī,rād·ər }

gyrator filter [ELECTR] A highly selective active filter that uses a gyrator which is terminated in a capacitor so as to have an inductive input impedance. { 'jī,rād·ər ,fil·tər }

gyroklystron [ELECTR] A microwave tube which, like the gyrotron, is based on cyclotron resonance coupling between microwave fields and an electron beam in vacuum, but which employs two or more cavities, and in which electrons give up their energy to an alternating electric field in a circuit separate from the one that supports the field that bunches the electrons. { ,jī·rə'klī,strän }

gyromagnetic coupler [ELECTR] A coupler in which a single-crystal yig (yttrium iron garnet) resonator provides coupling at the required low signal levels between two crossed stripline resonant circuits. { ¦jī·rō·mag'ned·ik 'kəp·lər }

gyrotron [ELECTR] **1.** A device that detects motion of a system by measuring the phase distortion that occurs when a vibrating tuning fork is moved. **2.** A type of microwave tube in which microwave amplification or generation results from cyclotron-resonance coupling between microwave fields and an electron beam in vacuum. Also known as cyclotron-resonance maser. { 'jī·rə,trän }

H

hacker |COMPUT SCI| A person who uses a computer system without a specific, constructive purpose or without proper authorization. { 'hak·ər }

hacking |COMPUT SCI| Use of a computer system without a specific, constructive purpose, or without proper authorization. { 'hak·iŋ }

halation |ELECTR| An area of glow surrounding a bright spot on a fluorescent screen, due to scattering by the phosphor or to multiple reflections at front and back surfaces of the glass faceplate. { hā'lā·shən }

half-adder |ELECTR| A logic element which operates on two binary digits (but no carry digits) from a preceding stage, producing as output a sum digit and a carry digit. { ¦haf ¦ad·ər }

half-adjust |COMPUT SCI| A rounding process in which the least significant digit is dropped and, if the least significant digit is one-half or more of the number base, one is added to the next more significant digit and all carries are propagated. { ¦haf ə¦jəst }

half block |COMPUT SCI| The unit of transfer between main storage and the buffer control unit; it consists of a column of 128 elements, each element 16 bytes long. { 'haf ¦bläk }

half-bridge |ELEC| A bridge having two power supplies, located in two of the bridge arms, to replace the single power supply of a conventional bridge. { 'haf ¦brij }

half carry |COMPUT SCI| A flag used in the central processing unit of some computers to indicate that a carry has occurred from the low-order N bits of a 2N-bit number to the high-order N bits. { 'haf ¦kar·ē }

half cycle |ENG| The time interval corresponding to half a cycle, or 180°, at the operating frequency of a circuit or device. { 'haf ¦sī·kəl }

half-cycle transmission |COMMUN| Data transmission and control system that uses synchronized sources of 60-hertz power at the transmitting and receiving ends; either of two receiver relays can be actuated by choosing the appropriate half-cycle polarity of the 60-hertz transmitter power supply. { 'haf ¦sī·kəl tranz'mish·ən }

half-duplex circuit |COMMUN| A circuit designed for half-duplex operation. Abbreviated HDX. { 'haf ¦dü,pleks ,sər·kət }

half-duplex operation |COMMUN| Operation of a telegraph system in either direction over a single channel, but not in both directions simultaneously. { 'haf ¦dü,pleks äp·ə'rā·shən }

half-duplex repeater |ELECTR| Duplex telegraph repeater provided with interlocking arrangements which restrict the transmission of signals to one direction at a time. { 'haf ¦dü,pleks ri'pēd·ər }

half-height drive |COMPUT SCI| A personal-computer disk drive whose height is half that of earlier disk drives. { ¦haf ,hīt 'drīv }

half-period zones See Fresnel zones. { 'haf ,pir·ē·əd ,zōnz }

half-power beamwidth |ELECTROMAG| The angle across the main lobe of an antenna pattern between the two directions at which the antenna's sensitivity is half its maximum value at the center of the lobe. Abbreviated HPBW. { 'haf ¦paů·ər 'bēm,width }

half-power frequency |ELECTR| One of the two values of frequency, on the sides of an amplifier response curve, at which the voltage is $1/\sqrt{2}$ (70.7%) of a midband or other reference value. Also known as half-power point. { 'haf ¦paů·ər 'frē·kwən·sē }

half-power point |ELECTR| **1.** A point on the graph of some quantity in an antenna, network, or control system, versus frequency, distance, or some other variable at which the power is half that of a nearby point at which power is a maximum. **2.** See half-power frequency. { 'haf ¦paů·ər ,pȯint }

half-pulse-repetition-rate delay |ELECTR| In the loran navigation system, an interval of time equal to half the pulse repetition rate of a pair of loran transmitting stations, introduced as a delay between transmission of the master and slave signals, to place the slave station signal on the B trace when the master station signal is mounted on the A trace pedestal. { 'haf ,pəls ,rep·ə'tish·ən ,rāt di,lā }

half-shift register |ELECTR| Logic circuit consisting of a gated input storage element, with or without an inverter. { 'haf ,shift ,rej·ə·stər }

half-subtracter |ELECTR| A logic element which operates on two digits from a preceding stage, producing as output a difference digit and a borrow digit. Also known as one-digit subtracter; two-input subtracter. { 'haf səb'trak·tər }

half tap |ELEC| Bridge placed across conductors without disturbing their continuity. { 'haf ,tap }

half-track tape recorder See double-track tape recorder. { 'haf ,trak 'tāp ri,kȯrd·ər }

half-wave [ELEC] Pertaining to half of one cycle of a wave. [ELECTROMAG] Having an electrical length of a half wavelength. { 'haf ¦wāv }

half-wave amplifier [ELECTR] A magnetic amplifier whose total induced voltage has a frequency equal to the power supply frequency. { 'haf ¦wāv 'am·plə,fī·ər }

half-wave antenna [ELECTROMAG] An antenna whose electrical length is half the wavelength being transmitted or received. { 'haf ¦wāv an'ten·ə }

half-wave dipole See dipole antenna. { 'haf ¦wāv 'dī,pōl }

half-wavelength [ELECTROMAG] The distance corresponding to an electrical length of half a wavelength at the operating frequency of a transmission line, antenna element, or other device. { 'haf ¦wāv,leŋkth }

half-wave rectification [ELECTR] Rectification in which current flows only during alternate half cycles. { 'haf ¦wāv ,rek·tə·fə'kā·shən }

half-wave rectifier [ELECTR] A rectifier that provides half-wave rectification. { 'haf ¦wāv 'rek·tə ,fī·ər }

half-wave transmission line [ELECTROMAG] Transmission line which has an electrical length equal to one-half the wavelength of the signal being transmitted or received. { 'haf ¦wāv tranz'mish·ən ,līn }

half-wave vibrator [ELEC] A vibrator having only one pair of contacts; interrupts the flow of direct current through the primary of a power transformer, but does not reverse the current. { 'haf ¦wāv 'vī,brād·ər }

half-word I/O buffer [COMPUT SCI] A buffer, the upper half being used to store the upper half of a word for both input and output characters, the lower half of the buffer being used for purposes such as the storage of constants. { 'haf ,wərd ¦ī ¦ō ,bəf·ər }

Hall-effect modulator [ELECTR] A Hall-effect multiplier used as a modulator to give an output voltage that is proportional to the product of two input voltages or currents. { 'hȯl i,fekt 'mäj·ə,lād·ər }

Hall-effect multiplier [ELECTR] A multiplier based on the Hall effect, used in analog computers to solve such problems as finding the square root of the sum of the squares of three independent variables. { 'hȯl i,fekt 'məl·tə,plī·ər }

Hall-effect switch [ELECTR] A magnetically activated switch that uses a Hall generator, trigger circuit, and transistor amplifier on a silicon chip. { 'hȯl i,fekt ,swich }

Hall resistance [ELECTR] The ratio of the transverse voltage developed across a current-carrying conductor, due to the Hall effect, to the current itself. { 'hȯl ri,zis·təns }

Hall voltage [ELECTR] The no-load voltage developed across a semiconductor plate due to the Hall effect, when a specified value of control current flows in the presence of a specified magnetic field. { 'hȯl ,vōl·tij }

Hallwachs effect [ELECTR] The discharge of a negatively charged metal plate caused by photoemission when the plate is exposed to ultraviolet light. { 'häl,väks i,fekt }

halo [ELECTR] An undesirable bright or dark ring surrounding an image on the fluorescent screen of a television cathode-ray tube; generally due to overloading or maladjustment of the camera tube. { 'hā·lō }

halt [COMPUT SCI] The cessation of the execution of the sequence of operations in a computer program resulting from a halt instruction, hang-up, or interrupt. { 'hȯlt }

Hamming code [COMMUN] An error-correcting code used in data transmission. { 'ham·iŋ ,kōd }

Hamming distance See signal distance. { 'ham·iŋ ,dis·təns }

ham radio See amateur radio. { ¦ham 'rād·ē·ō }

hand See end effector. { hand }

hand generator [ELEC] A manually cranked dynamo or alternator, usually used as the prime mover for emergency radio transmitters. { 'hand ¦jen·ə,rād·ər }

hand-held computer [COMPUT SCI] A small, battery-powered mobile computer for personal or business use. Also known as palmtop, personal digital assistant (PDA). { ¦hand,held kəm'pyüd·ər }

hand-held scanner [ENG] An image-reading device that is held and operated by a person. { ¦hand ,held 'skan·ər }

handie-talkie [ELECTR] Two-way radio communications unit small enough to be carried in the hand. { 'hand·dē 'tȯk·ē }

handle [COMPUT SCI] **1.** One of several small squares that appear around a selected object in an object-oriented computer-graphics program, and can be dragged with a mouse to move, enlarge, reduce, or change the shape of the object. **2.** In particular, one of the two interior points on a Bézier curve that can be dragged to alter its shape. Also known as control handle. { 'han·dəl }

handler [COMPUT SCI] A computer program developed to perform one particular function, such as control of input from, and output to, a specific peripheral device. { ,hand·lər }

hand-reset [ELEC] Pertaining to a relay in which the contacts must be reset manually to their original positions when normal conditions are resumed. { 'hand 'rē,set }

handset [ENG] A combination of a telephone-type receiver and transmitter, and sometimes also other components, designed for holding in one hand. { 'hand,set }

handshaking [COMMUN] The establishment of synchronization between sending and receiving equipment by means of exchange of specific character configurations. { 'hand,shak·iŋ }

hang-up [COMPUT SCI] A nonprogrammed stop in a computer routine caused by a human mistake or a computer malfunction. { 'haŋ,əp }

haptic interface [COMPUT SCI] A device that allows a user to interact with a computer by receiving tactile feedback; for example, glove or pen devices that allow users to touch and manipulate three-dimensional virtual objects. { ‚hap·tik 'in·tər‚fās }

haptics |COMPUT SCI| The study of the use of touch in order to produce computer interfaces that will allow users to interact with digital objects by means of force feedback and tactile feedback. { 'hap·tiks }

hard code |COMPUT SCI| Program statements that are written into the computer program itself, in contrast to external tables and files to hold values and parameters used by the program. { 'härd 'kōd }

hard-coded program |COMPUT SCI| A software program or program subroutine that is designed to perform a specific task and is not easily modified. { ¦härd ‚kōd·əd 'prō·grəm }

hard copy |COMPUT SCI| Human-readable type-written or printed characters produced on paper at the same time that information is being keyboarded in a coded machine language, as when punching cards or paper tape. { 'härd ¦käp·ē }

hard crash |COMPUT SCI| An abrupt halting of operations by a computer due to a malfunction, allowing the users or operators of the computer little or no time to minimize its effects. { 'härd 'krash }

hard disk |COMPUT SCI| A magnetic disk made of rigid material, providing high-capacity random-access storage. { 'härd ¦disk }

hard disk drive |COMPUT SCI| A high-capacity magnetic storage device that holds one or more hard disks and controls their positioning, reading, and writing; used to store programs and data, and to transfer instructions or information to the computer's working memory for use or further processing. Also known as hard drive. { 'härd ¦disk ¦drīv }

hard drive See hard disk drive. { 'härd ‚drīv }

hard edit |COMPUT SCI| The process of checking and correction that causes data containing errors to be rejected by a computer system. { 'härd 'ed·it }

hardened circuit |ELECTR| A circuit that uses components whose tolerance to radiation released by a nuclear explosion has been increased by various radiation-hardening procedures. { 'härd·ənd ¦sər·kət }

hardened links |COMMUN| Transmission links that require special construction or installation to assure a high probability of survival under nuclear attack. { 'härd·ənd ¦liŋks }

hard error |COMPUT SCI| Any error that results from malfunctioning of hardware, including storage devices and data transmission equipment. { 'härd 'er·ər }

hard failure |COMPUT SCI| Equipment failure that requires repair by a person with specialized knowledge before the equipment can be put back into operation. { 'härd 'fāl·yər }

hard-limiting |COMMUN| Limiting condition for which there is little variation in the output signal over the input signal range where the input is subject to limiting. { 'härd ‚lim·əd·iŋ }

hard page |COMPUT SCI| A page break that is inserted in a document by the user, and whose location is not changed by the addition, deletion, or reformatting of text. { 'härd ‚pāj }

hard patch |COMPUT SCI| A modification of a computer program, generally to repair a software error, which is applied to a stored copy of the program in machine language, so that recompilation of the source program is unnecessary and the change is permanent. { 'härd 'pach }

hard return |COMPUT SCI| A control code that is entered into a document by pressing the enter key. { 'härd ri‚tərn }

hard-sectored disk |COMPUT SCI| A disk whose sectors are set up during manufacture. { 'härd ¦sek·tərd 'disk }

hard tube See high-vacuum tube. { 'härd ‚tüb }

hardware |COMPUT SCI| The physical, tangible, and permanent components of a computer or a data-processing system. { 'härd‚wer }

hardware check See machine check. { 'härd‚wer ‚chek }

hardware compatibility |COMPUT SCI| Property of two computers such that the object code from one machine can be loaded and executed on the other to produce exactly the same results. { 'härd‚wer kəm‚pad·ə'bil·əd·ē }

hardware control |COMPUT SCI| The control of, and communications between, the various parts of a computer system. { 'härd‚wer kən‚trōl }

hardware description language |COMPUT SCI| A computer language that facilitates the documentation, design, and manufacturing of digital systems, particularly very large-scale integrated circuits, and combines program verification techniques with expert system design methodologies. { 'här‚dwer di'skrip·shən ‚laŋ·gwij }

hardware diagnostic |COMPUT SCI| A computer program designed to determine whether the components of a computer are operating properly. { 'härd‚wer dī·əg'näs·tik }

hardware division |COMPUT SCI| Mathematical division performed by electronic circuitry on a large computer as a result of a single machine instruction. { 'här‚dwer di‚vizh·ən }

hardware floating point |COMPUT SCI| Complex circuitry within a central processing unit that carries out floating-point arithmetic. { 'här ‚dwer 'flōd·iŋ 'pȯint }

hardware key See dongle. { 'här‚dwer ‚kē }

hardware monitor |COMPUT SCI| A system used to evaluate the performance of computer hardware; it collects information such as central processing unit usage from voltage level sensors that are attached to the circuitry and measure the length of time or the number of times various signals occur, and displays this information or stores it on a medium that is then fed into a special data-reduction program. { 'härd‚wer ‚män·əd·ər }

hardware multiplexing |COMPUT SCI| A procedure in which a servicing unit interleaves its attention among a family of serviced units in such a way that the serviced units appear to be receiving constant attention. { 'härd‚wer 'məl·tə‚plek·siŋ }

hardware multiplication |COMPUT SCI| Multiplication performed by electronic circuitry on a large computer as a result of a single machine instruction. { 'här‚dwer ‚məl·tə·plə‚kā·shən }

hard-wire |ELEC| To connect electric components with solid, metallic wires as opposed to radio links and the like. { 'härd ¦wīr }

hard-wired |COMPUT SCI| Having a fixed wired program or control system built in by the manufacturer and not subject to change by programming. { 'härd ¦wīrd }

hard-wire telemetry See wire-link telemetry. { 'härd ‚wīr tə'lem·ə·trē }

hard x-ray |ELECTR| An x-ray having high penetrating power. { 'härd ¦eks‚rā }

harmful interference |COMMUN| Radiation, emission, or induction which endangers the functioning of a radionavigation broadcasting service or of a safety broadcasting service, or obstructs or repeatedly interrupts a radio service operating in accordance with the appropriate regulations. { 'härm·fúl ‚int·ə'fir·əns }

harmonica bug |ELECTR| A surreptitious interception technique applied to telephone lines; the target instrument is modified so that a tuned relay bypasses the switch hook and ringing circuit when a 500-hertz tone is received; this tone was originally generated by use of a harmonica. { här'män·ə·kə ‚bəg }

harmonic analyzer |ELECTR| An instrument that measures the strength of each harmonic in a complex wave. Also known as harmonic wave analyzer. { här'man·ik 'an·ə‚liz·ər }

harmonic antenna |ELECTROMAG| An antenna whose electrical length is an integral multiple of a half-wavelength at the operating frequency of the transmitter or receiver. { här'män·ik an'ten·ə }

harmonic attenuation |ELECTR| Attenuation of an undesired harmonic component in the output of a transmitter. { här'män·ik ə‚ten·yə'wā·shən }

harmonic conversion transducer |ELECTR| A conversion transducer of which the useful output frequency is a multiple or a submultiple of the input frequency. { här'män·ik kən¦vər·zhən tranz‚dü·sər }

harmonic detector |ELECTR| Voltmeter circuit so arranged as to measure only a particular harmonic of the fundamental frequency. { här'män·ik di'tek·tər }

harmonic distortion |ELECTR| Nonlinear distortion in which undesired harmonics of a sinusoidal input signal are generated because of circuit nonlinearity. { här'män·ik di'stòr·shən }

harmonic filter |ELECTR| A filter that is tuned to suppress an undesired harmonic in a circuit. { här'män·ik 'fil·tər }

harmonic generator |ELECTR| A generator operated under conditions such that it generates strong harmonics along with the fundamental frequency. { här'män·ik 'jen·ə‚rād·ər }

harmonic interference |COMMUN| Interference due to the presence of harmonics in the output of a radio transmission. { här'män·ik ‚in·tər'fir·əns }

harmonic oscillator See sinusoidal oscillator. { här'män·ik 'äs·ə‚lād·ər }

harmonic producer |ELECTR| Tuning-fork controlled oscillator device capable of producing odd and even harmonics of the fundamental tuning-fork frequency; used to provide carrier frequencies for broad-band carrier systems. { här'män·ik prə‚dü·sər }

harmonic selective ringing |COMMUN| Selective ringing which employs currents of several frequencies and ringers, each tuned mechanically or electrically to the frequency of one of the ringing currents, so that only the desired ringer responds. { här'män·ik si¦lek·tiv 'riŋ·iŋ }

harmonic telephone ringer |ELECTR| Telephone ringer which responds only to alternating current within a very narrow frequency band. { här'män·ik 'tel·ə‚fōn ‚riŋ·ər }

harmonic wave analyzer See harmonic analyzer. { här'män·ik 'wāv ‚an·ə‚liz·ər }

harness |ELEC| Wire and cables so arranged and tied together that they may be inserted and connected, or may be removed after disconnection, as a unit. { 'här·nəs }

Harris flow |ELECTR| Electron flow in a cylindrical beam in which a radial electric field is used to overcome space charge divergence. { 'har·əs ‚flō }

hartley |COMMUN| A unit of information content, equal to the designation of 1 of 10 possible and equally likely values or states of anything used to store or convey information. { 'härt·lē }

Hartley oscillator |ELECTR| A vacuum-tube oscillator in which the parallel-tuned tank circuit is connected between grid and anode; the tank coil has an intermediate tap at cathode potential, so the grid-cathode portion of the coil provides the necessary feedback voltage. { 'härt·lē ¦äs·ə ‚läd·ər }

Hartley principle |COMMUN| The principle that the total number of bits of information that can be transmitted over a channel in a given time is proportional to the product of channel bandwidth and transmission time. { 'härt·lē ‚prin·sə·pəl }

Hartree equation |ELECTR| An equation which gives the lowest anode voltage at which it is theoretically possible to maintain oscillation in the different modes of a magnetron. { 'här·trē i ‚kwā·zhən }

hash |COMPUT SCI| Data which are obviously meaningless, caused by human mistakes or computer malfunction. Also known as garbage; gibberish. |ELECTR| See grass. { hash }

hash coding See hashing. { 'hash ‚kōd·iŋ }

hashing |COMPUT SCI| **1.** A method for converting representations of values within fields, usually keys, to a more compact form. **2.** An addressing technique that uses keys to store and retrieve data in a file. { 'hash·iŋ }

hash total |COMPUT SCI| A sum obtained by adding together numbers having different meanings; the sole purpose is to ensure that the correct number of data have been read by the computer. { 'hash ¦tōd·əl }

HASP |COMPUT SCI| A technique used on some types of larger computers to control input and output between a computer and its peripheral devices by utilizing mass-storage devices to temporarily store data. Acronym for Houston Automatic Spooling Processor. { hasp }

hatted code [COMMUN] Randomized code consisting of an encoding section; the plain text groups are arranged in alphabetical or other significant order, accompanied by their code groups arranged in a nonalphabetical or random order. { 'had·əd ¦kōd }

H attenuator *See* H network. { 'āch ə'ten·yə ‚wād·ər }

Hay bridge [ELEC] A four-arm alternating-current bridge used to measure inductance in terms of capacitance, resistance, and frequency; bridge balance depends on frequency. { 'hā ‚brij }

H bend *See* H-plane bend. { 'āch ‚bend }

HBT *See* heterojunction bipolar transistor.

HDA *See* head/disk assembly.

H-display [NAV] A radar display format in which a short cursor is added to the target spot in a B-display format, the slope of which is proportional to the sine of the elevation angle. Also known as H-indicator; H-scan; H-scope. { 'āch di‚splā }

HDLC *See* high-level data-link control.

HDTV *See* high-definition television.

HDX *See* half-duplex circuit.

head [COMPUT SCI] A device that reads, records, or erases data on a storage medium such as a drum or tape; examples are a small electromagnet or a sensing or punching device. [ELECTR] The photoelectric unit that converts the sound track on motion picture film into corresponding audio signals in a motion picture projector. [ENG ACOUS] *See* cutter. { hed }

head crash [COMPUT SCI] The collision of the read-write head and the magnetic recording surface of a hard disk. Also known as disk crash. { 'hed ‚krash }

head/disk assembly [COMPUT SCI] An airtight assembly including a disk pack and read/write heads. Abbreviated HDA. { 'hed 'disk ə'sem·blē }

header [COMMUN] The first section of a message, which contains information such as the addressee, routing, data, and origination time. [COMPUT SCI] *See* header label. [ELEC] A mounting plate through which the insulated terminals or leads are brought out from a hermetically sealed relay, transformer, transistor, tube, or other device. { 'hed·ər }

header label [COMPUT SCI] A block of data at the beginning of a magnetic tape file containing descriptive information to identify the file. Also known as header. { 'hed·ər ‚lā·bəl }

header record [COMPUT SCI] Computer input record containing common, constant, or identifying information for records that follow. { 'hed·ər ‚rek·ərd }

head gap [COMPUT SCI] The space between the read/write head and the recording medium, such as a disk in a computer. { 'hed ‚gap }

heading-upward plan position indicator [ELECTR] A plan position indicator in which the

heading of the craft appears at the top of the indicator at all times. { ¦hed·iŋ 'əp·wərd ¦plan pə¦zish·ən 'ind·ə‚kād·ər }

heading-upward plan position indicator [ELECTR] A plan position indicator in which the heading of the craft appears at the top of the indicator at all times. { ¦hed·iŋ 'əp·wərd ¦plan pə¦zish·ən 'ind·ə ‚kād·ər }

headlight [ELEC] A lamp, usually fitted with a reflector and a special lens, that is mounted on the front of a locomotive or automotive vehicle to illuminate the road ahead. { 'hed‚līt }

head-mounted display [COMPUT SCI] A tracking device incorporating liquid-crystal displays or miniature cathode-ray tubes worn on a user's head to simulate a virtual environment (a three-dimensional sensation of depth) and to provide information on head movements for updating visual images. { ¦hed ‚maúnt·əd di'splā }

head parking [COMPUT SCI] The positioning of the read/write head of a hard disk over the landing zone to ensure against head crashes. { 'hed ‚pärk·iŋ }

head-per-track [COMPUT SCI] An arrangement having one read/write head for each magnetized track on a disk or drum to eliminate the need to move a single head from track to track. { 'hed pər 'trak }

headphone [ENG ACOUS] An electroacoustic transducer designed to be held against an ear by a clamp passing over the head, for private listening to the audio output of a communications, radio, or television receiver or other source of audio-frequency signals. Also known as phone. { 'hed‚fōn }

head section *See* end section. { 'hed ‚sek·shən }

headset [ENG ACOUS] A single headphone or a pair of headphones, with a clamping strap or wires holding them in position. { 'hed‚set }

head stepping rate [COMPUT SCI] The rate at which the read/write head of a disk drive moves from one track to another on the disk surface. { 'hed ¦step·iŋ ‚rāt }

hearing aid [ENG ACOUS] A miniature, portable sound amplifier for persons with impaired hearing, consisting of a microphone, audio amplifier, earphone, and battery. { 'hir·iŋ ‚ād }

heat-activated battery *See* thermal battery. { 'hēt ¦ak·tə‚vād·əd 'bad·ə·rē }

heat coil [ELEC] Protective device which uses a mechanical element which is allowed to move when the fusible substance that holds it in place is heated above a predetermined temperature by current in the circuit. { 'hēt ‚kóil }

heater [ELECTR] An electric heating element for supplying heat to an indirectly heated cathode in an electron tube. Also known as electron-tube heater. { 'hēd·ər }

heater-type cathode *See* indirectly heated cathode. { 'hēd·ər ‚tīp 'kath‚ōd }

heating element [ELEC] The part of a heating appliance in which electrical energy is transformed into heat. { 'hēd·iŋ ‚el·ə·mənt }

heat lamp [ELEC] An infrared lamp used for brooders in farming, for drying paint or ink, for keeping food warm, and for therapeutic and other applications requiring heat with or without some visible light. { 'hēt ,lamp }

heat of emission [ELECTR] Additional heat energy that must be supplied to an electron-emitting surface to maintain it at a constant temperature. { 'hēt əv i'mish·ən }

heat run [ELEC] A series of temperature measurements made on an electric device during operating tests under various conditions. { 'hēt ,rən }

heatsink [ELEC] A mass of metal that is added to a device for the purpose of absorbing and dissipating heat; used with power transistors and many types of metallic rectifiers. Also known as dissipator. { 'hēt,siŋk }

heavy-ion source [ELECTR] Any source of ionized molecules or atoms of elements heavier than helium. { 'hev·ē ¦ī,än 'sȯrs }

hectometric wave [COMMUN] A radio wave between the wavelength limits of 100 and 1000 meters, corresponding to the frequency range of 3000 to 300 kilohertz. { ,hek·tə'me·trik 'wāv }

Heidelberg capsule [ELECTR] A radio pill for telemetering pH values of gastric acidity. { 'hīd·əl·bərg ,kap·səl }

height control [ELECTR] The video display control that adjusts picture height. { 'hīt kən,trōl }

height gain [ELECTR] A radio-wave interference phenomenon which results in a more or less periodic signal strength variation with height; this specifically refers to interference between direct and surface-reflected waves; maxima or minima in these height-gain curves occur at those elevations at which the direct and reflected waves are exactly in phase or out of phase respectively. { 'hīt ,gān }

height input [ELECTR] Radar height information on target received by a computer from height finders and relayed via ground-to-ground data link or telephone. { 'hīt 'in,pút }

height overlap coverage [ELECTR] Height-finder coverage within which there is an area of overlapping coverage from adjacent height finders or other radar stations. { 'hīt 'ō·vər,lap ,kəv·rij }

height-position indicator [ELECTR] Radar display which shows simultaneously angular elevation, slant range, and height of objects detected in the vertical sight plane. { 'hīt pə¦zish·ən 'in·də,kād·ər }

height-range indicator [ELECTR] **1.** Radar display which shows an echo as a bright spot on a rectangular field, slant range being indicated along the X axis, height above the horizontal plane being indicated (on a magnified scale) along the Y axis, and height above the earth being shown by a cursor. **2.** Cathode-ray tube from which altitude and range measurements of flightborne objects may be viewed. { 'hīt ¦rānj 'in·də,kād·ər }

height-range indicator display See range-height indicator display. { 'hīt ¦rānj 'in·də,kād·ər }

Heising modulation See constant-current modulation. { 'hī·ziŋ ,mäj·ə,lā·shən }

helical antenna [ELECTROMAG] An antenna having the form of a helix. Also known as helix antenna. { 'hel·ə·kəl an'ten·ə }

helical line [ELECTROMAG] A transmission line with a helical inner conductor. { 'hel·ə·kəl 'līn }

helical potentiometer [ELEC] A multiturn precision potentiometer in which a number of complete turns of the control knob are required to move the contact arm from one end of the helically wound resistance element to the other end. { 'hel·ə·kəl pə,ten·chē'äm·əd·ər }

helical resonator [ELECTROMAG] A cavity resonator with a helical inner conductor. { 'hel·ə·kəl 'rez·ən,ād·ər }

helical scanning [COMMUN] A method of facsimile scanning in which a single-turn helix rotates against a stationary bar to give horizontal movement of an elemental area. [ELECTR] A method of recording on videotape and digital audio tape in which the tracks are recorded diagonally from top to bottom by wrapping the tape around the rotating-head drum in a helical path. { 'hel·ə·kəl 'skan·iŋ }

helical traveling-wave tube See helix tube. { 'hel·ə·kəl ¦trav·ə·liŋ 'wāv ,tüb }

helitron [ELECTR] An electrostatically focused, low-noise backward-wave oscillator; the microwave output signal frequency can be swept rapidly over a wide range by varying the voltage applied between the cathode and the associated radio-frequency circuit. { 'hel·ə,trän }

helix [ELEC] A spread-out, single-layer coil of wire, either wound around a supporting cylinder or made of stiff enough wire to be self-supporting. { 'hē,liks }

helix antenna See helical antenna. { 'hē,liks an'ten·ə }

helix tube [ELECTR] A traveling-wave tube in which the electromagnetic wave travels along a wire wound in a spiral about the path of the beam, so that the wave travels along the tube at a velocity approximately equal to the beam velocity. Also known as helical traveling-wave tube. { 'hē,liks ,tüb }

helmet-mounted display [ELECTR] An electronic display that presents, on a combining glass within the visor of the helmet of a helicopter gunner, primary information for directing firepower; the angular direction of the helmet is sensed and used to control weapons to point in the same direction as the gunner is looking. Also known as visually coupled display. { 'hel·mət ¦maúnt·əd di'splā }

helmholtz [ELEC] A unit of dipole moment per unit area, equal to 1 Debye unit per square angstrom, or approximately 3.335×10^{-10} coulomb per meter. { 'helm,hōlts }

Helmholtz resonator [ENG ACOUS] An enclosure having a small opening consisting of a straight tube of such dimensions that the enclosure resonates at a single frequency determined by the geometry of the resonator. { 'helm,hōlts ¦rez·ən ,äd·ər }

Helmholtz's theorem See Thévenin's theorem. { 'helm,hōlt·səz ,thir·əm }

help screen |COMPUT SCI| Instructions that explain how to use the software of a computer system and that can be presented on the screen of a video display terminal at any time. { 'help ˌskrēn }

HEMT *See* high-electron-mobility transistor.

heptode |ELECTR| A seven-electrode electron tube containing an anode, a cathode, a control electrode, and four additional electrodes that are ordinarily grids. Also known as pentagrid. { 'hep,tōd }

hermaphroditic connector [ELEC| A connector in which both mating parts are exactly alike at their mating surfaces. Also known as sexless connector. { hər,ma·frə'did·ik kə'nek·tər }

herringbone pattern |ELECTR| An interference pattern sometimes seen on television receiver screens, consisting of a horizontal band of closely spaced V- or S-shaped lines. { 'her·iŋ,bōn ˌpad·ərn }

Hertz antenna |ELECTROMAG| An ungrounded half-wave antenna. { 'hərts anˌten·ə }

Hertz effect |ELECTR| Increase in the length of a spark induced across a spark gap when the gap is irradiated with ultraviolet light. |ELECTROMAG| A dependence of the attenuation of a linearly polarized electromagnetic wave passing through a grating of metal rods on the angle between the electric vector and the rod direction, with the attenuation being a minimum when the two are perpendicular. { 'hərts i,fekt }

hesitation |COMPUT SCI| A brief automatic suspension of the operations of a main program in order to perform all or part of another operation, such as rapid transmission of data to or from a peripheral unit. { ˌhez·ə'tā·shən }

Hesser's variation |COMPUT SCI| A variation of a Kiviat graph in which all variables are arranged so that their plots approach the circumference of the graph as the system being evaluated approaches saturation, and the scales on the various axes may not cover the full 0–100% range, or may be in units other than percent. { 'hes·ərz ˌver·ē ˌā·shən }

heterodyne |ELECTR| To mix two alternating-current signals of different frequencies in a nonlinear device for the purpose of producing two new frequencies, the sum of and difference between the two original frequencies. { 'hed·ə·rə ˌdīn }

heterodyne analyzer |ENG ACOUS| A type of constant-bandwidth analyzer in which the electric signal from a microphone beats with the signal from an oscillator, and one of the side bands produced by this modulation is then passed through a fixed filter and detected. { 'hed·ə·rə,dīn 'an·ə,līz·ər }

heterodyne conversion transducer *See* converter. { 'hed·ə·rə,dīn kənˌvər·zhən tranz,dü·sər }

heterodyne detector |ELECTR| A detector in which an unmodulated carrier frequency is combined with the signal of a local oscillator having a slightly different frequency, to provide an audio-frequency beat signal that can be heard with a loudspeaker or headphones; used chiefly for code reception. { 'hed·ə·rə,dīn di'tek·tər }

heterodyne frequency |COMMUN| Either of the two new frequencies resulting from heterodyne action between the two input frequencies of a heterodyne detector. { 'hed·ə·rə,dīn 'frē·kwən·sē }

heterodyne frequency meter |ELECTR| A frequency meter in which a known frequency, which may be adjustable or fixed, is heterodyned with an unknown frequency to produce a zero beat or an audio-frequency signal whose value is measured by other means. Also known as heterodyne wavemeter. { 'hed·ə·rə,dīn 'frē·kwən·sē ˌmēd·ər }

heterodyne interference *See* heterodyne whistle. { 'hed·ə·rə,dīn ˌin·tər'fir·əns }

heterodyne measurement |ELECTR| A measurement carried out by a type of harmonic analyzer which employs a highly selective filter, at a frequency well above the highest frequency to be measured, and a heterodyning oscillator. { 'hed·ə·rə,dīn 'mezh·ər·mənt }

heterodyne modulator *See* mixer. { 'hed·ə·rə,dīn 'mäj·ə,lād·ər }

heterodyne oscillator |ELECTR| **1.** A separate variable-frequency oscillator used to produce the second frequency required in a heterodyne detector for code reception. **2.** *See* beat-frequency oscillator. { 'hed·ə·rə,dīn 'äs·ə,lād·ər }

heterodyne reception |ELECTR| **1.** Radio reception in which the incoming radio-frequency signal is combined with a locally generated RF signal of different frequency, followed by detection. Also known as beat reception. **2.** In radar, use of a receiver that is tuned by adjusting a local oscillator signal within the receiver to a frequency differing from the frequency desired to be received by a fixed amount. When received signals are mixed with the reference signal, the difference frequency, called the intermediate frequency, is produced, permitting further signal processing at that convenient fixed frequency. { 'hed·ə·rə,dīn ri'sep·shən }

heterodyne repeater |ELECTR| A radio repeater in which the received radio signals are converted to an intermediate frequency, amplified, and reconverted to a new frequency band for transmission over the next repeater section. { 'hed·ə·rə,dīn ri'pēd·ər }

heterodyne wavemeter *See* heterodyne frequency meter. { 'hed·ə·rə,dīn 'wāv,mēd·ər }

heterodyne whistle |COMMUN| A steady, high-pitched audio tone heard in an ordinary amplitude-modulation radio receiver under certain conditions when two signals that differ slightly in carrier frequency enter the receiver and heterodyne to produce an audio beat. Also known as heterodyne interference. { 'hed·ə·rə ˌdīn 'wis·əl }

heterojunction |ELECTR| The boundary between two different semiconductor materials, usually with a negligible discontinuity in the crystal structure. { ˌhed·ə·rō'jəŋk·shən }

heterojunction bipolar transistor |ELECTR| A bipolar transistor that has two or more materials

heterojunction field-effect transistor

making up the emitter, base, and collector regions, giving it a much higher maximum frequency than a silicon bipolar transistor. Abbreviated HBT. { ¦hed·ə·rə,jəŋk·shən 'bī,pōl· ər tran,zis·tər }

heterojunction field-effect transistor See high-electron-mobility transistor. { ¦hed·ə·rə,jəŋk·shən 'fēld i,fekt tran,zis·tər }

heterostatic [ELEC] Pertaining to the measurement of one electrostatic potential by means of a different potential. { ¦hed·ə·rō'stad·ik }

heterostatic connection [ELEC] An arrangement of a quadrant electrometer in which the vane is maintained at a high potential with respect to one of the quadrant pairs and the deflection of the vane is linearly proportional to the unknown voltage applied across the quadrant pairs. { ¦hed·ə·rə,stad·ik kə'nek·shən }

heuristic algorithm See dynamic algorithm. { hyü'ris·tik 'al·gə,rith·əm }

heuristic program [COMPUT SCI] A program in which a computer tries each of several methods of solving a problem and judges whether the program is closer to solution after each attempt. Also known as heuristic routine. { hyü'ris·tik 'prō·grəm }

heuristic routine See heuristic program. { hyü 'ris·tik rü'tēn }

hexadecimal notation [COMPUT SCI] A notation in the scale of 16, using decimal digits 0 to 9 and six more digits that are sometimes represented by A, B, C, D, E, and F. { ,hek·sə'des·məl nō'tā·shən }

hexode [ELECTR] A six-electrode electron tube containing an anode, a cathode, a control electrode, and three additional electrodes that are ordinarily grids. { 'hek,sōd }

HF See high frequency.

HFET See high-electron-mobility transistor. { 'āch,fet }

HH beacon [NAV] Nondirectional radio homing beacon which has a power output of 2000 watts or greater. { ,āch'āch ,bē·kən }

hickey [ELEC] A threaded coupling for attaching an electrical fixture to an outlet box, used when wires from the fixture come out of the end of a stem on the fixture, rather than through an opening in the side of the stem. { 'hik·ē }

hidden file [COMPUT SCI] A disk file that does not appear in a directory listing and cannot be displayed, changed, or deleted. { ¦hid·ən ¦fīl }

hierarchical control [CONT SYS] The organization of controllers in a large-scale system into two or more levels so that controllers in each level send control signals to controllers in the level below and feedback or sensing signals to controllers in the level above. Also known as control hierarchy. { ¦hī·ər¦är·kə·kəl kən'trōl }

hierarchical distributed processing system [COMPUT SCI] A type of distributed processing system in which processing functions are distributed outward from a central computer to intelligent terminal controllers or satellite information processors. Also known as host-centered system; host/satellite system. { ¦hī·ər¦är·kə·kəl di ¦strib·yəd·əd 'prä,ses·iŋ ,sis·təm }

hierarchical file [COMPUT SCI] A file with a grandfather-father-son structure. { ¦hī·ər¦är·kə· kəl 'fīl }

hierarchical level I [ELEC] The level of reliability evaluation of an electric power system that is concerned only with the generation facilities. { ,hī·ər'är·ə·kəl ,lev·əl 'wən }

hierarchical level II [ELEC] The level of reliability evaluation of an electric power system that is concerned only with the generation and transmission facilities. { ,hī·ər'ärk·ə·kəl ,lev·əl 'tü }

hierarchical level III [ELEC] The level of reliability evaluation of an electric power system that is concerned with all three functional zones of the system, that is, generation, transmission, and distribution facilities. { ,hī·ər'ärk·ə·kəl ,lev·əl 'thrē }

hierarchical storage management [COMPUT SCI] A method of managing large amounts of data in which files are assigned to various storage media based on how soon or how frequently they will be needed. { ¦hī·ər¦är·kə·kəl 'stór·ij ,man·ij·mənt }

hi-fi See high fidelity. { 'hī'fī }

high-altitude radio altimeter See radar altimeter. { 'hī ¦al·tə,tüd 'rād·ē·ō al'tim·əd·ər }

high boost See high-frequency compensation. { 'hī ¦büst }

high core [COMPUT SCI] The locations with higher addresses in a computer's main storage, usually occupied by the operating system. { 'hī 'kòr }

high-current rectifier [ELECTR] A solid-state device, gas tube, or vacuum tube used to convert alternating to direct current for powering low-impedance loads. { 'hī ,kə·rənt 'rek·tə,fī·ər }

high-current switch [ELEC] A switch used to redirect heavy current flow; usually has a make-before-break feature to prevent excessive arcing. { 'hī ,kə·rənt 'swich }

high definition [COMMUN] Television or facsimile equivalent of high fidelity, in which the reproduced image contains such a large number of accurately reproduced elements that picture details approximate those of the original scene. { 'hī ,def·ə'nish·ən }

high-definition television [COMMUN] A television system with a resolution of more than 1000 scan lines, as compared to 525–625 scan lines in conventional systems. Abbreviated HDTV. { ¦hī def·ə¦nish·ən 'tel·ə,vizh·ən }

high-density disk [COMPUT SCI] A diskette that holds two or more times as much data per unit area as a double-density disk of the same size. { ¦hī ¦den·səd·ē 'disk }

high-density drive [COMPUT SCI] A disk drive that accepts both high-density and double-density disks. { ¦hī ¦den·səd·ē 'drīv }

270

high-electron-mobility transistor |ELECTR| A type of field-effect transistor consisting of gallium arsenide and gallium aluminum arsenide, with a Schottky metal contact on the gallium aluminum arsenide layer and two ohmic contacts penetrating into the gallium arsenide layer, serving as the gate, source, and drain respectively. Abbreviated HEMT. Also known as heterojunction field-effect transistor (HFET); modulation-doped field-effect transistor (MODFET); selectively doped heterojunction transistor (SDHT); two-dimensional electron gas field-effect transistor (TEGFET). { 'hī i'lek,trän mō¦bil·əd·ē tran,zis·tər }

higher-level language See high-level language. { 'hī·ər ,lev·əl ,laŋ·gwij }

higher-order language See high-level language. { 'hī·ər ,ȯr·dər ,laŋ·gwij }

higher-order software |COMPUT SCI| Software for designing and documenting an information system by decomposing the system into elementary components that are mathematically correct and error-free. Abbreviated HOS. { ¦hī·ər ,ȯr·dər 'sȯft,wer }

higher than high-level language |COMPUT SCI| A programming language, such as an application development language, report program, or financial planning language, that is oriented toward a particular application and is much easier to use for that application than a conventional programming language. { 'hī·ər than 'hī ,lev·əl ,laŋ·gwij }

high fidelity |ENG ACOUS| Audio reproduction that closely approximates the sound of the original performance. Also known as hi-fi. { ¦hī fi 'del·əd·ē }

high frequency |COMMUN| Federal Communications Commission designation for the band from 3 to 30 megahertz in the radio spectrum. Abbreviated HF. { 'hī ¦frē·kwən·sē }

high-frequency carrier telegraphy |COMMUN| Form of carrier telegraphy in which the carrier currents have their frequencies above the range transmitted over a voice-frequency telephone channel. { 'hī ¦frē·kwən·sē 'kar·ē·ər tə'leg·rə·fē }

high-frequency compensation |ELECTR| Increasing the amplification at high frequencies with respect to that at low and middle frequencies in a given band, such as in a video band or an audio band. Also known as high boost. { 'hī ¦frē·kwən·sē ,käm·pən'sā·shən }

high-frequency propagation |COMMUN| Propagation of radio waves in the high-frequency band, which depends entirely on reflection from the ionosphere. { 'hī ¦frē·kwən·sē ,präp·ə'gā·shən }

high-frequency resistance |ELEC| The total resistance offered by a device in an alternating-current circuit, including the direct-current resistance and the resistance due to eddy current, hysteresis, dielectric, and corona losses. Also known as alternating-current resistance;

effective resistance; radio-frequency resistance. { 'hī ¦frē·kwən·sē ri'zis·təns }

high-frequency transformer |ELECTR| A transformer which matches impedances and transmits a frequency band in the carrier (or higher) frequency ranges. { 'hī ¦frē·kwən·sē tranz'fȯr·mər }

high-frequency triode |ELECTR| A triode designed for operation at high frequency, having small spacings between the grid and the cathode and anode, large emission and power densities, and low active and inactive capacitances. { 'hī ¦frē·kwən·sē 'trī,ōd }

high-frequency voltmeter |ELECTR| A voltmeter designed to measure currents alternating at high frequencies. { 'hī ¦frē·kwən·sē 'vōlt,mēd·ər }

high-impedance voltmeter |ELEC| A voltage-measuring device with a high-impedance input to reduce load on the unit under test; a vacuum-tube voltmeter is one type. { 'hī im ¦pēd·əns 'vōlt,mēd·ər }

high-information-content display |ELECTR| An electronic display that has a sufficient number of pixels (75,000 to 2,000,000) to show standard or high-definition television images or comparable computer images. { ¦hī ,in·fər'mā·shən ,kän ,tent di,splā }

high-K capacitor |ELEC| A capacitor whose dielectric material is a ferroelectric having a high dielectric constant, up to about 6000. { 'hī ,kā kə'pas·əd·ər }

high level |COMMUN| A range of allowed picture parameters defined by the MPEG-2 video coding specifications which corresponds to high-definition television. |ELECTR| The more positive of the two logic levels or states in a binary digital logic system. { 'hī ¦lev·əl }

high-level data-link control Abbreviated HDLC. |COMMUN| A bit-oriented protocol for managing information flow in a data communications channel that supports both full-duplex and half-duplex transmission, and both point-to-point and multipoint communications using synchronous data transmission. |COMPUT SCI| A communications protocol that allows devices from different manufacturers to interface with each other and standardizes the transmission of packets of information between them. { 'hī ,lev·əl 'dad·ə ,liŋk kən,trōl }

high-level index |COMPUT SCI| The first part of a file name, which frequently specifies the category of data to which it belongs. { 'hī ,lev·əl 'in,deks }

high-level language |COMPUT SCI| A computer language whose instructions or statements each correspond to several machine language instructions, designed to make coding easier. Also known as higher-level language; higher-order language. { 'hī ,lev·əl 'laŋ·gwij }

high-level modulation |COMMUN| Modulation produced at a point in a system where the power

highlights

level approximates that at the output of the system. { 'hī ,lev·əl ,mäj·ə'lā·shən }

highlights |ELECTR| Bright areas occurring in a video image. { 'hī,līts }

high-low bias test |ELECTR| A routine maintenance procedure that tests equipment over and under normal operating conditions in order to detect defective units. { ¦hī ¦lō 'bī·əs ,test }

high-mu tube |ELECTR| A tube having a very high amplification factor. { 'hī ,myü 'tüb }

high-order |COMPUT SCI| Pertaining to a digit location in a numeral, the leftmost digit being the highest-order digit. { 'hī ¦ór·dər }

high-pass filter |ELECTR| A filter that transmits all frequencies above a given cutoff frequency and substantially attenuates all others. { 'hī ,pas 'fil·tər }

high-positive indicator |COMPUT SCI| A component in some computers whose status is "on" if the number tested is positive and nonzero. { 'hī ¦päz·əd·iv ¦in·də,kād·ər }

high-potting |ELEC| Testing with a high voltage, generally on a production line. { 'hī ¦päd·iŋ }

high-pressure mercury-vapor lamp |ELECTR| A discharge tube containing an inert gas and a small quantity of liquid mercury; the initial glow discharge through the gas heats and vaporizes the mercury, after which the discharge through mercury vapor produces an intensely brilliant light. { 'hī ¦presh·ər 'mər·kyə·rē ¦vā·pər 'lamp }

high Q |ELECTR| A characteristic wherein a component has a high ratio of reactance to effective resistance, so that its Q factor is high. { ¦hī 'kyü }

high-Q cavity |ELECTROMAG| A cavity resonator which has a large Q factor, and thus has a small energy loss. Also known as high-Q resonator. { 'hī ,kyü 'kav·əd·ē }

high-Q resonator See high-Q cavity. { 'hī ,kyü 'rez·ən,ād·ər }

high-resistance voltmeter |ELEC| A voltmeter having a resistance considerably higher than 1000 ohms per volt, so that it draws little current from the circuit in which a measurement is made. { 'hī ri,zis·təns 'vōlt,mēd·ər }

high-resolution electron microscope |ELECTR| An electron microscope in which lens aberrations are minimized and lens currents and the accelerating voltage are maintained with a high degree of stability, in order to achieve extremely high resolution. { ¦hī ,rez·ə¦lü·shən i¦lek,trän 'mī·krə ,skōp }

high side |COMPUT SCI| The part of a remote device that communicates with a computer. { 'hī ,sīd }

high-side capacitance coupling |ELECTR| Taking the output of an oscillator or amplifier from a point of high potential, using a capacitor to block direct current flow. { 'hī ,sīd kə'pas·əd·əns ,kəp·liŋ }

high-speed carry |COMPUT SCI| A technique in parallel addition to speed up the propagation of carries. { 'hī ,spēd'kar·ē }

high-speed data acquisition system |COMPUT SCI| A system which collects and transmits data

rapidly to a monitoring and controlling center. { 'hī ,spēd 'dad·ə ¦ak·wə,zish·ən ,sis·təm }

high-speed excitation system |ELEC| Excitation system capable of changing its voltage rapidly in response to a change in the excited generator field circuit. { 'hī ,spēd ,ek·sə'tā·shən ,sis·təm }

high-speed oscilloscope |ELECTR| An oscilloscope with a very fast sweep, capable of observing signals with rise times or periods on the order of nanoseconds. { 'hī ,spēd ä'sil·ə,skōp }

high-speed printer |COMPUT SCI| A printer which can function at a high rate, relative to the state of the art; 600 lines per minute is considered high speed. Abbreviated HSP. { 'hī ,spēd 'print·ər }

high-speed reader |COMPUT SCI| The fastest input device existing at a particular time in the state of the technology. { 'hī ,spēd 'rēd·ər }

high-speed relay |ELECTR| A relay specifically designed for short operate time, short release time, or both. { 'hī ,spēd 'rē,lā }

high-speed storage See rapid storage. { 'hī ,spēd 'stór·ij }

high-technology robot |CONT SYS| A robot equipped with feedback, vision, real-time data acquisition, and powerful controllers. { 'hī tek 'näl·ə·jē 'rō,bät }

high-temperature fuel cell |ELEC| A fuel cell which operates at temperatures above about 550°C, can use inexpensive hydrocarbon fuels, and usually uses a molten salt as an electrolyte. { 'hī ,tem·prə·chər 'fyül ,sel }

high tension See high voltage. { 'hī ¦ten·chən }

high-tension separation See electrostatic separation. { 'hī ,ten·chən ,sep·ə'rā·shən }

high-tier system |COMMUN| A wireless telephone system that supports base stations with large coverage areas and low traffic densities, but provides low-quality voice service and has limited data-service capabilities with high delays. { ,hī 'tir ,sis·təm }

high-vacuum rectifier |ELECTR| Vacuum-tube rectifier in which conduction is entirely by electrons emitted from the cathode. { 'hī ¦vak·yüm 'rek·tə,fī·ər }

high-vacuum switching tube |ELECTR| A microwave transmit-receive (TR) tube of the high-vacuum variety, as contrasted with gas-tube or semiconductor devices. { 'hī ¦vak·yüm 'swich·iŋ ,tüb }

high-vacuum tube |ELECTR| Electron tube evacuated to such a degree that its electrical characteristics are essentially unaffected by gaseous ionization. Also known as hard tube. { 'hī ¦vak·yüm ,tüb }

high voltage |ELEC| A voltage on the order of thousands of volts. Also known as high tension. { 'hī ¦vōl·tij }

high-voltage direct current |ELEC| A long-distance direct-current power transmission system that uses direct-current voltages up to about 1 megavolt to keep transmission losses down. Abbreviated HVDC. { 'hī ¦vōl·tij di¦rekt 'kə·rənt }

high-voltage electron microscope |ELECTR| An electron microscope whose accelerating voltage is on the order of 10^6 volts, as compared with 40–100 kilovolts for an ordinary electron microscope; it has the advantages of increased specimen penetration, reduced specimen damage, better theoretical resolution, and more efficient dark-field operation. { 'hī ¦vōl·tij i¦lek,trän 'mī·krə ,skōp }

high-voltage insulation [ELEC] Electrical insulation designed to prevent breakdown in a circuit operating at high voltages. { 'hī ¦vōl·tij ,in·sə'lā·shən }

high-water mark [COMPUT SCI] The maximum number of jobs that are in a queue awaiting execution by a large computer system during a specified period of observation. { ¦hī 'wȯd·ər ,märk }

Hilbert transformer [ELECTR] An electric filter whose gain is $-j$ for positive frequencies and j for negative frequencies, where j is the square root of -1. { 'hil·bərt tranz,fȯr·mər }

hill-and-dale recording See vertical recording. { ¦hil ən ¦dāl ri'kȯrd·iŋ }

hill bandwidth [ELECTR] The difference between the upper and lower frequencies at which the gain of an amplifier is 3 decibels less than its maximum value. { ¦hil 'band,width }

H-indicator See H-display. { 'āch ¦in·də,kād·ər }

hi pot [ELEC] High potential voltage applied across a conductor to test the insulation or applied to an etched circuit to burn out tenuous conducting paths that might later fail in service. { 'hī ,pät }

hiss [COMMUN] Random noise in the audio-frequency range, similar to prolonged sibilant sounds. { his }

historical data [COMPUT SCI] Any data that is not actively maintained by a computer system and cannot be readily revised or updated. { hi'stär·ə·kəl 'dad·ə }

hit [COMPUT SCI] **1.** The obtaining of a correct answer in an information-retrieval system. **2.** An attempt to access a specified piece of information on a website; a count of the number of such attempts is an indicator or the usage or popularity of the site. { hit }

hit-on-the-fly system [COMPUT SCI] A printer in a computer system where either the print roller or the paper is in continuous motion. { ¦hid ȯn the 'flī ,sis·təm }

hit rate [COMPUT SCI] The ratio of the number of records found and processed during a particular processing run, to the total number of records available. { 'hit ,rāt }

Hittorf dark space See cathode dark space. { 'hi·dȯrf 'därk ,spās }

Hittorf principle [ELECTR] The principle that a discharge between electrodes in a gas at a given pressure does not necessarily occur between the closest points of the electrodes if the distance between these points lies to the left of the minimum on a graph of spark potential versus distance. Also known as short-path principle. { 'hi·dȯrf ,prin·sə·pəl }

H mode See transverse electric mode. { 'āch ,mōd }

H network [ELECTR] An attenuation network composed of five branches and having the form of the letter H. Also known as H attenuator; H pad. { 'āch ¦net,wərk }

hog [COMPUT SCI] A computer program that uses excessive computer resources, such as memory or processing power, or requires excessive time to execute. { häg }

hoghorn antenna See horn antenna. { 'häg,hȯrn an'ten·ə }

hold [COMPUT SCI] To retain information in a computer storage device for further use after it has been initially utilized. [ELECTR] To maintain storage elements at equilibrium voltages in a charge storage tube by electron bombardment. { hōld }

hold circuit [ELECTR] A circuit in a sampled-data control system that converts the series of impulses, generated by the sampler, into a rectangular function, in order to smooth the signal to the motor or plant. { 'hōld ,sər·kət }

hold control [ELECTR] A manual control that changes the frequency of the horizontal or vertical sweep oscillator in an analog television receiver, so that the frequency more nearly corresponds to that of the incoming synchronizing pulses. { 'hōld kən,trōl }

holder [ELEC] A device that mechanically and electrically accommodates one or more crystals, fuses, or other components in such a way that they can readily be inserted or removed. { 'hōl·dər }

hold facility [COMPUT SCI] The ability of a computer to operate in a hold mode. { 'hōld fə ,sil·əd·ē }

holding anode [ELECTR] A small auxiliary anode used in a mercury-pool rectifier to keep a cathode spot energized during the intervals when the main-anode current is zero. { 'hōl·diŋ ,an,ōd }

holding beam [ELECTR] A diffused beam of electrons used to regenerate the charges stored on the dielectric surface of a cathode-ray storage tube. { 'hōl·diŋ ,bēm }

holding coil [ELECTR] A separate relay coil that is energized by contacts which close when a relay pulls in, to hold the relay in its energized position after the orginal operating circuit is opened. { 'hōl·diŋ ,kȯil }

holding current [ELECTR] The minimum current required to maintain a switching device in a closed or conducting state after it is energized or triggered. { 'hōl·diŋ ,kə·rənt }

holding time [COMMUN] Period of time a trunk or circuit is in use on a call, including operator's time in connecting and subscriber's or user's conversation time. { 'hōl·diŋ ,tīm }

hold lamp [ELEC] Indicating lamp which remains lighted while a telephone connection is being held. { 'hōld ,lamp }

hold mode [COMPUT SCI] The state of an analog computer in which its operation is interrupted without altering the values of the variables it is handling, so that computation can continue when the interruption is over. Also known as interrupt mode. { 'hōld ,mōd }

hold queue |COMPUT SCI| A queue consisting of jobs that have been submitted for execution by a large computer system and are waiting to be run. { 'hōld ‚kyü }

hole |SOLID STATE| A vacant electron energy state near the top of an energy band in a solid; behaves as though it were a positively charged particle. Also known as electron hole. { hōl }

hole conduction |ELECTR| Conduction occurring in a semiconductor when electrons move into holes under the influence of an applied voltage and thereby create new holes. { 'hōl kən¦dək·shən }

hole injection |ELECTR| The production of holes in an n-type semiconductor when voltage is applied to a sharp metal point in contact with the surface of the material. { 'hōl in‚jek·shən }

hole mobility |ELECTR| A measure of the ability of a hole to travel readily through a semiconductor, equal to the average drift velocity of holes divided by the electric field. { 'hōl mō‚bil·əd·ē }

hole trap |ELECTR| A semiconductor impurity capable of releasing electrons to the conduction or valence bands, equivalent to trapping a hole. { 'hōl ‚trap }

holistic masks |COMPUT SCI| In character recognition, that set of characters which resides within a character reader and theoretically represents the exact replicas of all possible input characters. { hō'lis·tik 'masks }

Hollerith string |COMPUT SCI| A sequence of characters preceded by an H and a character count in FORTRAN, as 4HSTOP. { 'häl·ə·rəth ‚striŋ }

hollow cathode |ELECTR| A cathode which is hollow and closed at one end in a discharge tube filled with inert gas, designed so that radiation is emitted from the cathode glow inside the cathode. { 'häl·ō 'kath‚ōd }

hollow-pipe waveguide |ELECTROMAG| A waveguide consisting of a hollow metal pipe; electromagnetic waves are transmitted through the interior and electric currents flow on the inner surfaces. { 'häl·ō ‚pīp 'wāv‚gīd }

holographic memory |COMPUT SCI| A memory in which information is stored in the form of holographic images on thermoplastic or other recording films. { ‚häl·ə'graf·ik 'mem·rē }

holographic storage |COMMUN| A form of data storage in which bits of information are distributed throughout the storage volume and recorded interferometrically, rather than being stored at discrete locations in the medium. { ¦häl·ə‚graf·ik 'stör·ij }

Holtz machine See Toepler-Holtz machine. { 'hōlts mə‚shēn }

home |COMPUT SCI| The location at the upper lefthand corner of an electronic display. |NAV| To navigate toward a point by maintaining constant some navigational parameter other than altitude. { hōm }

home address |COMPUT SCI| A technique used to identify each disk track uniquely by means of a 9-byte record immediately following the index marker; the record contains a flag (good or defective track), cylinder number, head number,

cyclic check, and bit count appendage. { ¦hōm 'ad‚res }

home-on-jam |ELECTR| A feature that permits radar or a passive seeker to track a jamming source in angle, to guide a weapon to the jammer. { ¦hōm ‚ön ¦jam }

home page |COMPUT SCI| A document in a hypertext system that serves as the point of entry to a web of related documents, and generally contains introductory information and hyperlinks to other documents in the web. Also known as welcome page. { ‚hōm 'pāj }

home record |COMPUT SCI| The first record in the chaining method of file organization. { 'hōm ¦rek·ərd }

hometaxial-base transistor |ELECTR| Transistor manufactured by a single-diffusion process to form both emitter and collector junctions in a uniformly doped silicon slice; the resulting homogeneously doped base region is free from accelerating fields in the axial (collector-to-emitter) direction, which could cause undesirable high current flow and destroy the transistor. { 'häm·ə‚tak·sē·əl ‚bās tran'zis·tər }

homing antenna |ELECTROMAG| A directional antenna array used in flying directly to a target that is emitting or reflecting radio or radar waves. { 'hōm·iŋ an‚ten·ə }

homing beacon |NAV| A radio beacon, either airborne or on the ground, toward which an aircraft can fly if equipped with a radio compass or homing adapter. Also known as radio homing beacon. { 'hōm·iŋ ‚bē·kən }

homing device |ELECTR| A control device that automatically starts in the correct direction of motion or rotation to achieve a desired change, as in a remote-control tuning motor for a television receiver. |NAV| A transmitter, receiver, or adapter used for homing aircraft or used by aircraft for homing purposes. { 'hōm·iŋ di‚vīs }

homing relay |ELEC| A stepping relay that returns to a specified starting position before each operating cycle. { 'hōm·iŋ ‚rē‚lā }

homodyne reception |ELECTR| **1.** A system of radio reception for suppressed-carrier systems of radiotelephony, in which the receiver generates a voltage having the original carrier frequency and combines it with the incoming signal. Also known as zero-beat reception. **2.** Referring to a radio or radar receiver in which received signals are mixed with a reference signal at the same frequency as the signal intended to be received; the mixing produces a voltage output dependent only on the phase difference of the two inputs, hence a voltage at the "beat" frequency if there is a slight difference. { 'hä·mə‚dīn ri'sep·shən }

homogeneous network |COMPUT SCI| A computer network consisting of fairly similar computers from a single manufacturer. { ¦hō·mə ¦jē·nē·əs 'net‚wərk }

homojunction bipolar transistor |ELECTR| Any bipolar transistor that is composed entirely of one type of semiconductor. { ¦hō·mō‚jəŋk·shən bī‚pō·lər tran'zis·tər }

homopolar |ELEC| **1.** Electrically symmetrical. **2.** Having equal distribution of charge. { ¦hä·mə'pō·lər }

homopolar generator |ELECTR| A direct-current generator in which the poles presented to the armature are all of the same polarity, so that the voltage generated in active conductors has the same polarity at all times; a pure direct current is thus produced, without commutation. Also known as acyclic machine; homopolar machine; unipolar machine. { ¦hä·mə'pō·lər 'jen·ə ‚rād·ər }

homopolar machine *See* homopolar generator. { ¦hä·mə'pō·lər mə'shēn }

hook |COMPUT SCI| A modification of a computer program to add instructions to an existing part of the program. |ELECTR| A circuit phenomenon occurring in four-zone transistors, wherein hole or electron conduction can occur in opposite directions to produce voltage drops that encourage other types of conduction. { hùk }

hook collector transistor |ELECTR| A transistor in which there are four layers of alternating *n*- and *p*-type semiconductor material and the two interior layers are thin compared to the diffusion length. Also known as hook transistor; *pn* hook transistor. { ¦hùk kə'lek·tər tran‚zis·tər }

hook transistor *See* hook collector transistor. { 'hùk tran‚zis·tər }

hookup |ELEC| An arrangement of circuits and apparatus for a particular purpose. { 'hùk‚əp }

hookup wire |ELEC| Tinned and insulated solid or stranded soft-drawn copper wire used in making circuit connections. { 'hùk‚əp ‚wïr }

hoot stop |COMPUT SCI| A closed loop that generates an audible signal; usually employed to signal an error or for operating convenience. { 'hüt ‚stäp }

hop |COMMUN| A single reflection of a radio wave from the ionosphere back to the earth in traveling from one point to another. { häp }

horizontal blanking |ELECTR| Blanking of a video picture tube during the horizontal retrace. { ‚här·ə'zänt·əl 'blaŋk·iŋ }

horizontal blanking pulse |ELECTR| The rectangular pulse that forms the pedestal of the composite video signal between active horizontal lines and causes the display device to be cut off during retrace. Also known as line-frequency blanking pulse. { ‚här·ə'zänt·əl 'blaŋk·iŋ ‚pəls }

horizontal centering control |ELECTR| The centering control provided in a video display to shift the position of the entire image horizontally in either direction on the screen. { ‚här·ə'zänt·əl 'sen·tə·riŋ kən‚trōl }

horizontal convergence control |ELECTR| The control that adjusts the amplitude of the horizontal dynamic convergence voltage in a video display device. { ‚här·ə'zänt·əl kən'vər·jəns kən ‚trōl }

horizontal definition *See* horizontal resolution. { ‚här·ə'zänt·əl ‚def·ə'nish·ən }

horizontal deflection electrode |ELECTR| One of a pair of electrodes that move the electron beam horizontally from side to side on the fluorescent screen of a cathode-ray tube employing electrostatic deflection. { ‚här·ə'zänt·əl di'flek·shən i'lek‚trōd }

horizontal deflection oscillator |ELECTR| The oscillator that produces, under control of the horizontal synchronizing signals, the sawtooth voltage waveform that is amplified to feed the horizontal deflection coils on the picture tube of a video display. Also known as horizontal oscillator. { ‚här·ə'zänt·əl di'flek·shən 'äs·ə‚lād·ər }

horizontal distributed processing system |COMPUT SCI| A type of distributed system in which two or more computers which are logically equivalent are connected together, with no hierarchy or master/slave relationship. { ‚här·ə'zänt·əl di ¦strib·yəd·əd 'prä‚ses·iŋ ‚sis·təm }

horizontal drive control |ELECTR| The control in a television receiver, usually at the rear, that adjusts the output of the horizontal oscillator. Also known as drive control. { ‚här·ə'zänt·əl 'drïv kən‚trōl }

horizontal flyback |ELECTR| Flyback in which the electron beam of a picture tube returns from the end of one scanning line to the beginning of the next line. Also known as horizontal retrace. { ‚här·ə'zänt·əl 'flï‚bak }

horizontal frequency *See* line frequency. { ‚här· ə'zänt·əl 'frē·kwən‚sē }

horizontal hold control |ELECTR| The hold control that changes the free-running period of the horizontal deflection oscillator in an analog television receiver, so that the picture remains steady in the horizontal direction. { ‚här·ə'zänt·əl 'hōld kən‚trōl }

horizontal instruction |COMPUT SCI| An instruction in machine language to carry out independent operations on various operands in parallel or in a well-defined time sequence. { ‚här·ə'zänt·əl in'strək·shən }

horizontal linearity control |ELECTR| A linearity control that permits narrowing or expanding of the width of the left half of a television receiver image, to give linearity in the horizontal direction so that circular objects appear as true circles. { ‚här·ə'zänt·əl ‚lin·ē'ar·əd·ē kən‚trōl }

horizontal line frequency *See* line frequency. { ‚här·ə'zänt·əl 'lïn ‚frē·kwən·sē }

horizontal oscillator *See* horizontal deflection oscillator. { ‚här·ə'zänt·əl 'äs·ə‚lād·ər }

horizontal output stage |ELECTR| The television receiver stage that feeds the horizontal deflection coils of the picture tube through the horizontal output transformer; may also include a part of the second-anode power supply for the picture tube. { ‚här·ə'zänt·əl 'aùt‚pùt ‚stāj }

horizontal output transformer |ELECTR| A transformer used in a television receiver to provide the horizontal deflection voltage, the high voltage for the second-anode power supply of the picture tube, and the filament voltage for the high-voltage rectifier tube. Also known as flyback transformer; horizontal sweep transformer. { ‚här·ə'zänt·əl 'aùt‚pùt tranz‚fór·mər }

horizontal parity check *See* longitudinal parity check. { ‚här·ə'zänt·əl 'par·əd·ē ‚chek }

horizontal polarization |COMMUN| Transmission of linear polarized radio waves whose electric field vector is parallel to the earth's surface. { ,här·ə'zänt·əl ,pō·lə·rə'zā·shən }

horizontal resolution |ELECTR| The number of individual picture elements or dots that can be distinguished in a horizontal scanning line of a video display. Also known as horizontal definition. { ,här·ə'zänt·əl ,rez·ə'lü·shən }

horizontal retrace See horizontal flyback. { ,här·ə'zänt·əl 'rē,trās }

horizontal scanning frequency |ELECTR| The number of horizontal lines scanned by the electron beam in a video system in 1 second. { ,här·ə'zänt·əl 'skan·iŋ ,frē·kwən·sē }

horizontal sweep |ELECTR| The sweep of the electron beam from left to right across the screen of a cathode-ray tube. { ,här·ə'zänt·əl 'swēp }

horizontal sweep transformer See horizontal output transformer. { ,här·ə'zänt·əl 'swēp tranz,fȯr·mər }

horizontal synchronizing pulse |ELECTR| The rectangular pulse transmitted at the end of each line in an analog television system, to keep the receiver in line-by-line synchronism with the transmitter. Also known as line synchronizing pulse. { ,här·ə'zänt·əl 'siŋ·krə,niz·iŋ ,pəls }

horizontal system |COMPUT SCI| A programming system in which instructions are written horizontally, that is, across the page. { ,här·ə'zänt·əl 'sis·təm }

horizontal vee |ELECTROMAG| An antenna consisting of two linear radiators in the form of the letter V, lying in a horizontal plane. { ,här·ə'zänt·əl 'vē }

horn |ELECTROMAG| See horn antenna. |ENG ACOUS| A tube whose cross-sectional area increases from one end to the other, used to radiate or receive sound waves and to intensify and direct them. Also known as acoustic horn. { hȯrn }

horn antenna |ELECTROMAG| A microwave antenna produced by flaring out the end of a circular or rectangular waveguide into the shape of a horn, for radiating radio waves directly into space. Also known as electromagnetic horn; flare (British usage); hoghorn antenna (British usage); horn; horn radiator. { 'hȯrn an'ten·ə }

horn arrester |ELEC| A lightning arrester in which the spark gap has thick wire horns that spread outward and upward; the arc forms at the narrowest bottom part of the gap, travels upward, and extinguishes itself when it reaches the widest part of the gap. Also known as horn lightning arrester. { 'hȯrn ə'res·tər }

horn gap |ELEC| Type of spark gap which is provided with divergent electrodes. { 'hȯrn ,gap }

horn-gap switch |ELEC| Form of air switch provided with arcing horns. { 'hȯrn ,gap ,swich }

horn lightning arrester See horn arrester. { 'hȯrn 'līt·niŋ ə,res·tər }

horn-loaded speaker |ENG ACOUS| A loudspeaker that has an acoustic horn between the diaphragm and the air load. { ¦hȯrn ,lōd·əd 'spēk·ər }

horn loudspeaker |ENG ACOUS| A loudspeaker in which the radiating element is coupled to the air

or another medium by means of a horn. { 'hȯrn 'laůd,spēk·ər }

horn radiator See horn antenna. { 'hȯrn 'rād·ē ,ād·ər }

HOS See higher-order software.

hospital information system |COMPUT SCI| The collection, evaluation or verification, storage, and retrieval of information about a patient. { 'häs,pid·əl ,in·fər'mā·shən ,sis·təm }

host-based system |COMMUN| A communications system that is controlled by a central computer system. { 'hōst ,bāst ,sis·təm }

host-centered system See hierarchical distributed processing system. { 'hōst ,sen·tərd ,sis·təm }

host computer |COMPUT SCI| **1.** The central or controlling computer in a time-sharing or distributed-processing system. **2.** The computer upon which depends a specialized computer handling the input/output functions in a real-time system. **3.** A computer that can function as the source or recipient of data transfers on a network. { 'hōst kəm¦pyüd·ər }

host language database management system |COMPUT SCI| A database management system that, from a programmer's point of view, represents an extension of an existing programming language. { 'hōst ,laŋ·gwij 'dad·ə,bās 'man·ij·mənt ,sis·təm }

host processor |COMPUT SCI| The central computer in a hierarchical distributed processing system, which is typically located at some central site where it serves as a focal point for the collection of data, and often for the provision of services which cannot economically be distributed. { 'hōst ¦prä,ses·ər }

host/satellite system See hierarchical distributed processing system. { 'hōst 'sad·əl,īt ,sis·təm }

hot See energized. { hät }

hot carrier |ELECTR| A charge carrier, which may be either an electron or a hole, that has relatively high energy with respect to the carriers normally found in majority-carrier devices such as thin-film transistors. { 'hät ¦kar·ē·ər }

hot-carrier diode See Schottky barrier diode. { 'hät ,kar·ē·ər 'dī,ōd }

hot cathode |ELECTR| A cathode in which electron or ion emission is produced by heat. Also known as thermionic cathode. { 'hät 'kath,ōd }

hot-cathode gas-filled tube See thyratron. { 'hät ¦kath,ōd ¦gas 'fild 'tüb }

hot-cathode tube See thermionic tube. { 'hät ¦kath,ōd ,tüb }

hot editing |CONT SYS| A method for detecting errors in the programming of a robot in which as many errors as possible are identified and resolved during testing, without setting the robotic program to its starting condition. { 'hät 'ed·əd·iŋ }

hot electron |ELECTR| An electron that is in excess of the thermal equilibrium number and, for metals, has an energy greater than the Fermi level; for semiconductors, the energy must be a definite amount above that of the edge of the conduction band. { 'hät i'lek,trän }

hot-electron transistor |ELECTR| A transistor in which electrons tunnel through a thin emitter-

hum

base barrier ballistically (that is, without scattering), traverse a very narrow base region, and cross a barrier at the base-collector interface whose height, controlled by the collector voltage, determines the fraction of electrons coming to the collector.   { ¦hät i'lek,trän ,tran'zis·tər }

hot-electron triode |ELECTR| Solid-state, evaporated thin-film structure directly equivalent to a vacuum triode.   { 'hät i,lek,trän 'trī,ōd }

hot-filament ionization gage |ELECTR| An ionization gage in which electrons emitted by an incandescent filament, and attracted toward a positively charged grid electrode, collide with gas molecules to produce ions which are then attracted to a negatively charged electrode; the ion current is a measure of the number of gas molecules.   { 'hät ¦fil·ə·mənt ,ī·ə·nə'zā·shən ,gāj }

hot hole |ELECTR| A hole that can move at much greater velocity than normal holes in a semiconductor.   { 'hät ,hōl }

hot junction |ELECTR| The heated junction of a thermocouple.   { 'hät 'jəŋk·shən }

hot key |COMPUT SCI| A computer key or key combination that causes a specified action to occur, regardless of what else the computer is currently doing.   { 'hät ,kē }

hot line |COMMUN| Direct circuit between two points, available for immediate use without patching or switching.   { 'hät 'līn }

hot link |COMPUT SCI| A linking of information in two documents such that modification of the information in the source document results in the same change in the destination document.   { ¦hät 'liŋk }

hot spot |COMPUT SCI| A word in a multiprocessor memory that several processors attempt to access simultaneously, creating a conflict or bottleneck.   { 'hät ,spät }

hot wire |ELEC| **1.** A resistive wire in an electric relay that expands when heated and contracts when cooled. **2.** An electrical lead that has an electric potential with respect to the ground.   { 'hät 'wīr }

hot-wire ammeter |ENG| An ammeter which measures alternating or direct current by sending it through a fine wire, causing the wire to heat and to expand or sag, deflecting a pointer. Also known as thermal ammeter.   { 'hät ¦wīr 'a ,med·ər }

hot-wire microphone |ENG ACOUS| A velocity microphone that depends for its operation on the change in resistance of a hot wire as the wire is cooled by varying particle velocities in a sound wave.   { 'hät ¦wīr 'mī·krə,fōn }

housekeeping |COMPUT SCI| Those operations or routines which do not contribute directly to the solution of a computer program, but rather to the organization of the program.   { 'haus,kēp·iŋ }

housekeeping run |COMPUT SCI| The performance of a program or routine to maintain the structure of files, such as sorting, merging, addition of new records, or deletion or modification of existing records.   { 'haus,kēp·iŋ ,rən }

Houston Automatic Spooling Processor   *See* HASP.   { hyüs·tən ¦öd·ə¦mad·ik 'spül·iŋ ,prä ,ses·ər }

howl |ENG ACOUS| Undesirable prolonged sound produced by a radio receiver or audio-frequency amplifier system because of either electric or acoustic feedback.   { haul }

howler |COMMUN| In telephone practice, an associated unit by which the test desk operator may connect a high tone of varying loudness to a subscriber's line to call the subscriber's attention to the fact that the phone receiver is off the hook.   { 'haul·ər }

howl repeater |COMMUN| Condition in telephone repeater operation where more energy is returned than sent, resulting in an oscillation being set up on the circuit.   { 'haul ri'pēd·ər }

H pad   *See* H network.   { ¦ach ,pad }

h parameter |ELECTR| One of a set of four transistor equivalent-circuit parameters that conveniently specify transistor performance for small voltages and currents in a particular circuit. Also known as hybrid parameter.   { ¦ach pə ,ram·əd·ər }

HPBW   *See* half-power beamwidth.

H plane |ELECTROMAG| The plane of an antenna in which lies the magnetic field vector of linearly polarized radiation.   { ¦ach ,plān }

H-plane bend |ELECTROMAG| A rectangular waveguide bend in which the longitudinal axis of the waveguide remains in a plane parallel to the plane of the magnetic field vector throughout the bend. Also known as H bend.   { ¦ach ,plān ,bend }

H-plane T junction |ELECTROMAG| Waveguide T junction in which the change in structure occurs in the plane of the magnetic field. Also known as shunt T junction.   { ¦ach ,plān 'tē ,jəŋk·shən }

H-scan   *See* H-display.   { 'ach ,skan }

H-scope   *See* H-display.   { 'ach ,skōp }

HSP   *See* high-speed printer.

HTML   *See* Hypertext Markup Language.

HTTP   *See* Hypertext Transfer Protocol.

hub |COMPUT SCI| An electric socket in a plugboard into which one may insert or connect leads or may plug wires.   { həb }

hub ring |COMPUT SCI| A thin plastic ring placed around the center hole of a floppy disk to prevent the disk from warping and damaging its contents if it is improperly inserted in a disk drive.   { 'həb ,riŋ }

hue control |ELECTR| A control that varies the phase of the chrominance signals with respect to that of the burst signal in an analog color television receiver, in order to change the hues in the image. Also known as phase control.   { 'hyü kən,trōl }

Huffman method |COMPUT SCI| A data compression technique in which a bit representation for each character is determined that is as close as possible to the character's predicted information content, based on its frequency of occurrence.   { 'həf·mən ,meth·əd }

hum |ELECTR| An electrical disturbance occurring at the power supply frequency or its harmonics, usually 60 or 120 hertz in the United States. { həm }

human-computer interaction |COMPUT SCI| The processes through which human users work with interactive computer systems. { ¦yü·mən kəm ¦pyüd·ər ‚in·tər'ak·shən }

hum bar |ELECTR| A dark horizontal band extending across a television picture due to excessive hum in the video signal applied to the input of the picture tube. { 'həm ‚bär }

hum-bucking coil |ENG ACOUS| A coil wound on the field coil of an excited-field loudspeaker and connected in series opposition with the voice coil, so that hum voltage induced in the voice coil is canceled by that induced in the hum-bucking coil. { 'həm ‚bək·iŋ ‚kóil }

humidity capacitor |ELECTR| A device for measuring ambient relative humidity by sensing a change in capacitance. { hyü'mid·əd·ē kə'pas·əd·ər }

hum modulation |ELECTR| Modulation of a radio-frequency signal or detected audio-frequency signal by hum; heard in a radio receiver only when a station is tuned in. { 'həm ‚mäj·ə‚lā·shən }

hunting |CONT SYS| Undesirable oscillation of an automatic control system, wherein the controlled variable swings on both sides of the desired value. |ELECTR| Operation of a selector in moving from terminal to terminal until one is found which is idle. { 'hənt·iŋ }

hunting circuit See lockout circuit. { 'hənt·iŋ ‚sər·kət }

HVDC See high-voltage direct current.

H wave See transverse electric wave. { 'āch ‚wāv }

hybrid algebraic manipulation language |COMPUT SCI| The most ambitious type of algebraic manipulation language, which accepts the broadest spectrum of mathematical expressions but possesses, in addition, special representations and special algorithms for particular special classes of expressions. { 'hī·brəd ‚al·jə'brā·ik mə‚nip·yə'lā·shən ‚laŋ·gwij }

hybrid AM IBOC |COMMUN| The initial mode of the AM IBOC system approved by the Federal Communications Commission for use in the United States that adds digital audio capacity to an AM signal by inserting digital sidebands in the spectrum above, below, and within the analog AM signal. The digital audio data rate can range from 36 to 56 kbits/s, and the corresponding ancillary data rate is 0.4 kbits/s in both cases. { 'hī·brəd 'ā‚em 'ī‚bäk }

hybrid balance |ELEC| Loss between two conjugate sides of a hybrid set less the same loss when one of the other sides is open or shorted. { 'hī·brəd 'bal·əns }

hybrid circuit |ELEC| A circuit in which two or more basically different types of components, such as tubes and transistors, performing similar functions are used together. { 'hī·brəd 'sər·kət }

hybrid coil See hybrid transformer. { 'hī·brəd 'kóil }

hybrid computer |COMPUT SCI| A computer designed to handle both analog and digital data. Also known as analog-digital computer; hybrid system. { 'hī·brəd kəm'pyüd·ər }

hybrid distributed processing system |COMPUT SCI| A distributed processing system that includes both horizontal and hierarchical distribution. { 'hī·brəd di¦strib·yəd·əd 'prä‚ses·iŋ ‚sis·təm }

hybrid electromagnetic wave |ELECTROMAG| Wave which has both transverse and longitudinal components of displacement. { 'hī·brəd i¦lek·trō·mag¦ned·ik 'wāv }

hybrid FM IBOC |COMMUN| The first of three modes in the FM IBOC system approved by the Federal Communications Commission for use in the United States that increases data capacity by a adding additional carriers closer to the analog host signal. The hybrid IBOC mode adds one frequency partition around the analog carrier and is characterized by the highest possible digital and analog audio quality with a limited amount of ancillary data available to the broadcaster. Digital audio data rates can range from 64 to 96 kbits/s, and the corresponding ancillary data rate can range from 33 kbits/s for 64-kbits/s audio and 1 kbit/s for 96-kbits/s audio. { 'hī·brəd 'ef‚em 'ī ‚bäk }

hybrid hardware control |COMPUT SCI| The control of and communication between the various parts of a hybrid computer. { 'hī·brəd 'härd‚wer kən‚tról }

hybrid input/output |COMPUT SCI| The routines required to handle inputs to and outputs from a computer system comprising digital and analog computers. { 'hī·brəd ¦in‚pút ¦aút‚pút }

hybrid integrated circuit |ELECTR| A circuit in which one or more discrete components are used in combination with integrated-circuit construction. { 'hī·brəd ¦int·ə‚grād·əd 'sər·kət }

hybrid interface |COMPUT SCI| A device that joins a digital to an analog computer, converting digital signals transmitted serially by the digital computer into analog signals that are transmitted simultaneously to the various units of the analog computer, and vice versa. { 'hī·brəd 'in·tər‚fās }

hybrid junction |ELECTR| A transformer, resistor, or waveguide circuit or device that has four pairs of terminals so arranged that a signal entering at one terminal pair divides and emerges from the two adjacent terminal pairs, but is unable to reach the opposite terminal pair. Also known as bridge hybrid. { 'hī·brəd 'jəŋk·shən }

hybrid microcircuit |ELECTR| Microcircuit in which thin-film, thick-film, or diffusion techniques are combined with separately attached semiconductor chips to form the circuit. { 'hī·brəd 'mī·krō‚sər·kət }

hybrid network |COMMUN| Nonhomogeneous communications network required to operate with signals of dissimilar characteristics (such as analog and digital modes). |ELECTR| A four-port circuit, useful in radar and other microwave applications as a power switch or signal comparator, in which two inputs add constructively at one output and destructively at the other, with good isolation between the two iinputs and between the two outputs; the waveguide magic tee and the rat race are types of hybrid networks. { 'hī·brəd 'net,wərk }

hybrid parameter *See* h parameter. { 'hī·brəd pə'ram·əd·ər }

hybrid problem analysis |COMPUT SCI| The determination of the parts of a problem best suited for the digital computer. { 'hī·brəd ¦präb·ləm ə'nal·ə·səs }

hybrid programming |COMPUT SCI| Hybrid system routines that handle timing, function generation, and simulation. { 'hī·brəd 'prō,gram·iŋ }

hybrid redundancy |COMPUT SCI| A synthesis of triple modular redundancy and standby replacement redundancy, consisting of a triple modular redundancy system (or, in general, an N-modular redundancy system) with a bank of spares so that when one of the units in the triple modular redundancy system fails it is replaced by a spare unit. { 'hī·brəd ri'dən·dən·sē }

hybrid relay |ELEC| A relay in which solid-state elements are combined with moving contacts. { 'hī·brəd 'rē,lā }

hybrid repeater *See* hybrid transformer. { 'hī·brəd ri'pēd·ər }

hybrid set |ELEC| Two or more transformers interconnected to form a hybrid junction. Also known as transformer hybrid. { 'hī·brəd 'set }

hybrid simulation |COMPUT SCI| The use of a hybrid computer for purposes of simulation. { 'hī·brəd ,sim·yə'lā·shən }

hybrid system |COMPUT SCI| **1.** A computer system that performs two or more functions, such as data processing and word processing. **2.** *See* hybrid computer. { 'hī,brid ,sis·təm }

hybrid system checkout |COMPUT SCI| The static check of a hybrid system and of the digital program and analog wiring required to solve a problem. { 'hī·brəd ¦sis·təm 'chek,aüt }

hybrid tee |ELECTROMAG| A microwave hybrid junction composed of an E-H tee with internal matching elements; it is reflectionless for a wave propagating into the junction from any arm when the other three arms are match-terminated. Also known as magic tee. { 'hī·brəd 'tē }

hybrid thin-film circuit |ELECTR| Microcircuit formed by attaching discrete components and semiconductor devices to networks of passive components and conductors that have been vacuum-deposited on glazed ceramic, sapphire, or glass substrates. { 'hī·brəd ¦thin ,film 'sər·kət }

hybrid transformer |ELEC| A single transformer that performs the essential functions of a hybrid set. Also known as bridge transformer; hybrid coil; hybrid repeater. { 'hī·brəd tranz'fór·mər }

hydroelectricity |ELEC| Electric power produced by hydroelectric generators. Also known as hydropower. { ¦hī·drō·i,lek'tris·əd·ē }

hydrogen discharge lamp |ELECTR| A discharge lamp containing hydrogen and used as a source of ultraviolet radiation. { 'hī·drə·jən ¦dis,chärj ,lamp }

hydrogen thyratron |ELECTR| A thyratron containing hydrogen instead of mercury vapor to give freedom from effects of changes in ambient temperature; used in radar pulse circuits and stroboscopic photography. { 'hī·drə·jən 'thī·rə ,trän }

hydrophone |ENG| A device which receives underwater sound waves and converts them to electric waves. { 'hī·drə,fōn }

hydrophone array |COMMUN| A group of two or more hydrophones which feed into a common receiver. { 'hī·drə,fōn ə,rā }

hydrophone noise |ELEC| Any unwanted disturbance in the electric waves delivered by a hydrophone. { 'hī·drə,fōn ,nòiz }

hydrophone response |ELEC| The electric waves delivered by a hydrophone in response to waterborne sound waves. { 'hī·drə,fōn ri¦späns }

hydropower *See* hydroelectricity. { 'hī·drə,paü· ər }

hygristor |ELECTR| A resistor whose resistance varies with humidity; used in some types of recording hygrometers. { hī'gris·tər }

hyperbolic amplitude |COMMUN| Excursion of a signal measured along hyperbolic rather than Cartesian coordinates. { ¦hī·pər¦bäl·ik 'am·plə ,tüd }

hyperbolic antenna |ELECTROMAG| A radiator whose reflector in cross section describes a half hyperbola. { ¦hī·pər¦bäl·ik an'ten·ə }

hyperbolic horn |ENG| Horn whose equivalent cross-sectional radius increases according to a hyperbolic law. { ¦hī·pər¦bäl·ik 'hòrn }

hyperbolic sweep generator |ELECTR| A sweep generator that generates a waveform resembling a hyperbola. { ¦hī·pər¦bäl·ik 'swēp ,jen·ə,rād· ər }

hyperbolic waveform |ELECTR| A waveform which is an approximate hyperbola. { ¦hī·pər ¦bäl·ik 'wāv,fórm }

hypercube |COMPUT SCI| A configuration of parallel processors in which the locations of the processors correspond to the vertices of a mathematical hypercube and the links between them correspond to its edges. { 'hī·pər ,kyüb }

hyperdisk |COMPUT SCI| A mass-storage technique which uses a large-capacity storage and a disk for overflow. { 'hī·pər,disk }

hyperlink |COMPUT SCI| A highlighted word, phrase, or image in the display of a computer document which, when chosen, connects the user to another part of the same document or to different document (text, image, audio, video, or animation). In electronic documents, these cross references can be followed by a mouse click, and the target of the hyperlink may be on a physically distant computer connected by a network or the Internet. { 'hī·pər,liŋk }

hypermedia |COMPUT SCI| Hypertext-based systems that combine data, text, graphics, video, and sound. { 'hī·pər,mē·dē·ə }

hyperpure germanium detector |ELECTR| A variant of the lithium-drifted germanium crystal which uses high-purity germanium, making it possible to store the detector at room temperature rather than liquid nitrogen temperature. { ¦hī·pər¦pyür jər¦mā·nē·əm di'tek·tər }

hypersensor |ELECTR| Single-component, resettable circuit breaker which operates as a majority-carrier tunneling device, and is used for overcurrent or overvoltage protection of integrated circuits. { ¦hī·pər¦sen·sər }

hypertape control unit *See* tape control unit. { 'hī·pər,tāp kən'trōl ,yü·nət }

hypertape drive *See* cartridge tape drive. { 'hī·pər ,tāp ,drīv }

hypertext |COMPUT SCI| A data structure in which there are links between words, phrases, graphics, or other elements and associated information so that selection of a key object can activate a linkage and reveal the information. { 'hī·pər ,tekst }

Hypertext Markup Language |COMPUT SCI| The language used to specifically encode the content and format of a document and to link documents on the World Wide Web. Abbreviated HTML. { ¦hī·pər,tekst 'märk,əp ,laŋ·gwij }

Hypertext Transfer Protocol |COMPUT SCI| The communication protocol for transmitting linked documents between computers; it is the basis for the World Wide Web and follows the TCP/IP protocol for the client-server model of computing. Abbreviated HTTP. { 'hī·pər,tekst 'tranz·fər ,prōd·ə,kól }

hypervisor |COMPUT SCI| A control program enabling two operating systems to share a common computing system. { 'hī·pər,vīz·ər }

hyphenation zone |COMPUT SCI| In word processing, the area adjacent to the right margin consisting of those positions at which words may be hyphenated. { 'hī·fə,nā·shən ,zōn }

hysteresimeter |ENG| A device for measuring hysteresis. { his,ter·ə'sim·əd·ər }

hysteresis |ELECTR| An oscillator effect wherein a given value of an operating parameter may result in multiple values of output power or frequency. { ,his·tə'rē·səs }

hysteresis motor |ELEC| A synchronous motor without salient poles and without direct-current excitation which utilizes the hysteresis and eddy-current losses induced in its hardened-steel rotor to produce rotor torque. { ,his·tə'rē·səs ¦mōd·ər }

IBOC *See* in-band/on-channel. { ¦ī¦bē¦ō'sē *or* 'ī,bäk }

IC *See* integrated circuit.

icon |COMPUT SCI| A symbolic representation of a computer function that appears on an electronic display and makes it possible to command this function by selecting the symbol. { 'ī,kän }

iconoscope |ELECTR| A television camera tube in which a beam of high-velocity electrons scans a photoemissive mosaic that is capable of storing an electric charge pattern corresponding to an optical image focused on the mosaic. Also known as storage camera; storage-type camera tube. { ī'kän·ə,skōp }

ICS system *See* intercarrier sound system. { ¦ī¦sē'es ,sis·təm }

ICW *See* interrupted continuous wave.

ideal bunching |ELECTR| Theoretical condition in which the bunching of electrons in a velocity-modulated tube would give a single infinitely large current peak during each cycle. { ī'dēl 'bən·chiŋ }

ideal dielectric |ELEC| Dielectric in which all the energy required to establish an electric field in the dielectric is returned to the source when the field is removed. Also known as perfect dielectric. { ī'dēl ,dī·i'lek·trik }

ideal network |ELECTR| An interconnection of lumped, constant electrical quantities analyzed without consideration of noise and distributed parameters that would exist in actual settings. { ī'dēl 'net,wərk }

ideal transducer |ELEC| Hypothetical passive transducer which transfers the maximum possible power from the source to the load. { ī'dēl tranz'dü·sər }

ideal transformer |ELEC| A hypothetical transformer that neither stores nor dissipates energy, has unity coefficient of coupling, and has pure inductances of infinitely great value. { ī'dēl tranz'fȯr·mər }

I demodulator |ELECTR| Stage of an analog color television receiver which combines the chrominance signal with the color oscillator output to restore the I signal. { 'ī dē'mäj·ə,lād·ər }

identification |CONT SYS| The procedures for deducing a system's transfer function from its response to a step-function input or to an impulse. { ī,dent·ə·fə'kā·shən }

identification and authentication |COMPUT SCI| The process of determining with high assurance the identity of a person who is seeking access to a computing system. { ī,den·tə·fə¦kā·shən ən ə,then·tə'kā·shən }

identification division |COMPUT SCI| The section of a program, written in the COBOL language, which contains the name of the program and the name of the programmer. { ī,dent·ə·fə'kā·shən di¦vizh·ən }

identifier |COMPUT SCI| A symbol whose purpose is to specify a body of data. { ī'dent·ə,fī·ər }

identifier word |COMPUT SCI| A full-length computer word associated with a search function. { ī'dent·ə,fī·ər ,wərd }

identity gate *See* identity unit. { ī'den·ə,dē ,gāt }

identity unit |COMPUT SCI| A logic element with several binary input signals and a single binary output signal whose value is 1 if all the input signals have the same value and 0 if they do not. Also known as identity gate. { ī'den·ə,dē ,yü·nət }

idiostatic connection |ELEC| An arrangement of a quadrant electrometer in which the vane is electrically connected to one of the quadrant pairs and the deflection of the vane is proportional to the square of the unknown voltage applied across the quadrant pairs. { ,id·ē·ə ,stad·ik kə'nek·shən }

I-display |ELECTR| A radar display format in which the target appears as a circle, of radius proportional to range, when a tracking radar antenna is pointed at it exactly and as a segment of the circle when there is a pointing error. Also known as I-indicator; I-scan; I-scope. { 'ī di,splā }

idle component *See* reactive component. { 'īd·əl kəm'pō·nənt }

idle current *See* reactive current. { 'īd·əl ,kə·rənt }

idler frequency |ELECTR| Of a parametric device, a sum or difference frequency generated within the parametric device other than the input, output, or pump frequencies which require specific circuit consideration to achieve the desired device performance; it is called an idler frequency since, in conventional parametric amplifiers, it is more or less a useless byproduct of the parametric process. { 'īd·lər ,frē·kwən·sē }

idle time |COMPUT SCI| The time during which a piece of hardware in good operating condition is unused. { 'īd·əl ,tīm }

idle trunk lamp |ELEC| Signal lamp associated with an outgoing trunk to indicate that the trunk is not busy. { 'īd·əl 'trəŋk ,lamp }

IDP *See* integrated data processing.

IEEE 1394 |COMPUT SCI| The standard for connecting storage, digital audio and video, and other peripheral devices to personal computers at data transfer rates up to 400 million bits per second. Also known as firewire.

IF *See* intermediate frequency.

i-f amplifier *See* intermediate-frequency amplifier. { ¦ī'ef 'am·plə,fī·ər }

IF canceler |ELECTR| In radar, a moving-target indicator canceler operating at the intermediate frequency using an internal phase reference as in a coherent radar, as opposed to a video canceler. { ¦ī'ef 'kans·lər }

I-frame *See* intra-coded picture. { 'ī ,frām }

IF statement *See* conditional jump. { 'if ,stāt·mənt }

if then else |COMPUT SCI| A logic statement in a high-level programming language that defines the data to be compared and the actions to be taken as the result of a comparison. { ¦if then 'els }

i-f transformer *See* intermediate-frequency transformer. { ¦ī ¦ef tranz'fȯr·mər }

IGBT *See* insulated-gate bipolar transistor.

IGES *See* initial graphics exchange specification.

IGFET *See* metal oxide semiconductor field-effect transistor. { 'ig,fet }

ignition interference |COMMUN| Radio interference due to the spark discharges in an automotive or other ignition system. { ig'nish·ən ,int·ə'fir·əns }

ignition reserve |ELEC| In an ignition system for an internal combustion engine, the difference between the minimum voltage available and the maximum voltage required by the system. { ig'nish·ən ri,zərv }

ignition timing |ELEC| The time of delivery of the spark from the coil to the spark plug in relation to the time the piston reaches the correct position for the power stroke in an internal combustion engine. { ig'nish·ən ,tīm·iŋ }

ignitor |ELECTR| **1.** An electrode used to initiate and sustain the discharge in a switching tube. Also known as keep-alive electrode (deprecated). **2.** A pencil-shaped electrode, made of carborundum or some other conducting material that is not wetted by mercury, partly immersed in the mercury-pool cathode of an ignitron and used to initiate conduction at the desired point in each alternating-current cycle. { ig'nīd·ər }

ignitron |ELECTR| A single-anode pool tube in which an ignitor electrode is employed to initiate the cathode spot on the surface of the mercury pool before each conducting period. { 'ig·nə ,trän }

ignitron contactor |ELECTR| A circuit containing an ignitron and control contacts that serves as a heavy-duty switch in the primary of a resistance-welding transformer. { 'ig·nə,trän 'kän,tak·tər }

ignore character |COMPUT SCI| Also known as erase character. **1.** A character indicating that no action whatever is to be taken, that is, a character to be ignored; often used to obliterate an erroneous character. **2.** A character indicating that the preceding or following character is to be ignored, as specified. **3.** A character indicating that some specified action is not to be taken. { ig'nȯr ,kar·ik·tər }

I-indicator *See* I-display. { 'ī ¦in·də,kād·ər }

IIR filter *See* infinite impulse response filter. { ¦ī ¦ī'är ,fil·tər }

I²L *See* integrated injection logic.

ILF *See* infralow frequency.

ill-conditioned problem |COMPUT SCI| A problem in which a small error in the data or in subsequent calculation results in much larger errors in the answers. { ¦il kən¦dish·ənd 'präb·ləm }

illegal character |COMPUT SCI| A character or combination of bits that is not accepted as a valid representation by a computer or by a specific routine; commonly detected and used as an indication of a machine malfunction. { i'lē·gəl 'kar·ik·tər }

illegal operation |COMPUT SCI| An operation specified by a program instruction that cannot be carried out by the computer. { i'lē·gəl äp·ə'rā·shən }

illumination control |ELECTR| A photoelectric control that turns on lights when outdoor illumination decreases below a predetermined level. { ə,lü·mə,nā·shən kən,trōl }

illustration program *See* drawing program. { ,il·ə'strā·shən ,prō·grəm }

image |COMMUN| **1.** One of two groups of side bands generated in the process of modulation; the unused group is referred to as the unwanted image. **2.** The scene reproduced by a video display. |COMPUT SCI| A copy of the information contained in one medium recorded on a different data medium. |ELECT| *See* electric image. |ELECTROMAG| The input reflection coefficient corresponding to the reflection coefficient of a specified load when the load is placed on one side of a waveguide junction and a slotted line is placed on the other. { 'im·ij }

image admittance |ELECTR| The reciprocal of image impedance. { 'im·ij ad,mit·əns }

image antenna |ELECTROMAG| A fictitious electrical counterpart of an actual antenna, acting mathematically as if it existed in the ground directly under the real antenna and served as the direct source of the wave that is reflected from the ground by the actual antenna. { 'im·ij an ¦ten·ə }

image attenuation constant |ELECTR| The real part of the image transfer constant. { 'im·ij ə ,ten·yə'wā·shən ,kän·stənt }

image converter |ELECTR| *See* image tube. |OPTICS| A converter that uses a fiber optic bundle to change the form of an image, for more convenient recording and display or for the coding of secret messages. { 'im·ij kən,vərd·ər }

image converter camera |ELECTR| A camera consisting of an image tube and an optical system which focuses the image produced on the phosphorescent screen of the tube onto photographic film. { 'im·ij ¦kən¦vərd·ər ¦kam·rə }

image dissection photography |ELECTR| A method of high-speed photography in which an image is split in any one of various ways into interlaced space and time elements which can be unscrambled or played back through the system either to be viewed or to give a master negative. { 'im·ij di¦sek·shən fə¦täg·rə·fē }

image dissector |COMPUT SCI| In optical character recognition, a device that optically examines an input character for the purpose of breaking it down into its prescribed elements. { 'im·ij di ¦sek·tər }

image dissector tube |ELECTR| A television camera tube in which an electron image produced by a photoemitting surface is focused in the plane of the defining aperture and is scanned past that aperture. Also known as Farnsworth image dissector tube. { 'im·ij di¦sek·tər ¦tüb }

image effect |ELECTROMAG| Effect produced on the field of an antenna due to the presence of the earth; electromagnetic waves are reflected from the earth's surface, and these reflections often are accounted for by an image antenna at an equal distance below the earth's surface. { 'im·ij i¦fekt }

image enhancement |COMPUT SCI| Improvement of the quality of a picture, with the aid of a computer, by giving it higher contrast or making it less blurred or less noisy. { 'im·ij in'hans·mənt }

image force |ELEC| The electrostatic force on a charge in the neighborhood of a conductor, which may be thought of as the attraction to the charge's electric image. { 'im·ij ¦fórs }

image frequency |ELECTR| An undesired carrier frequency that differs from the frequency to which a superheterodyne receiver is tuned by twice the intermediate frequency. { 'im·ij ¦frē·kwən·sē }

image iconoscope |ELECTR| A camera tube in which an optical image is projected on a semitransparent photocathode, and the resulting electron image emitted from the other side of the photocathode is focused on a separate storage target; the target is scanned on the same side by a high-velocity electron beam, neutralizing the elemental charges in sequence to produce the camera output signal at the target. Also known as superemitron camera (British usage). { 'im·ij ī'kän·ə¦skōp }

image impedance |ELECTR| One of the impedances that, when connected to the input and output of a transducer, will make the impedances in both directions equal at the input terminals and at the output terminals. { 'im·ij im¦pēd·əns }

image intensifier See light amplifier. { 'im·ij in'ten·sə¦fī·ər }

image interference |COMMUN| Interference occurring in a superheterodyne receiver when a station broadcasting on the image frequency is received along with the desired station. { 'im·ij ¦in·tər'fir·əns }

image isocon |ELECTR| A television camera tube which is similar to the image orthicon but whose return beam consists of scanning beam electrons that are scattered by positive stored charges on the target. { 'im·ij 'ī·sə¦kän }

image load |ELECTR| Load parameters reflected back to the source by line discontinuities. { 'im·ij ¦lōd }

image orthicon |ELECTR| A television camera tube in which an electron image is produced by a photoemitting surface and focused on one side of a separate storage tube that is scanned on its opposite side by a beam of low-velocity electrons; electrons that are reflected from the storage tube, after positive stored charges are neutralized by the scanning beam, form a return beam which is amplified by an electron multiplier. { 'im·ij 'ór·thə¦kän }

image parameter design |ELECTR| A method of filter design using image impedance and image transfer functions as the fundamental network functions. { 'im·ij pə¦ram·əd·ər di¦zīn }

image parameter filter |ELECTR| A filter constructed by image parameter design. { 'im·ij pə ¦ram·əd·ər ¦fil·tər }

image phase constant |ELECTR| The imaginary part of the image transfer constant. { 'im·ij 'fāz ¦kän·stənt }

image potential |ELEC| The potential set up by an electric image. { 'im·ij pə¦ten·chəl }

image processing |COMPUT SCI| A technique in which the data from an image are digitized and various mathematical operations are applied to the data, generally with a digital computer, in order to create an enhanced image that is more useful or pleasing to a human observer, or to perform some of the interpretation and recognition tasks usually performed by humans. Also known as picture processing. { 'im·ij ¦prä·ses·iŋ }

image ratio |ELECTR| In a heterodyne receiver, the ratio of the image frequency signal input at the antenna to the desired signal input for identical outputs. { 'im·ij ¦rā·shō }

image reject mixer |ELECTR| Combination of two balanced mixers and associated hybrid circuits designed to separate the image channel from the signal channels normally present in a conventional mixer; the arrangement gives image rejection up to 30 decibels without the use of filters. { 'im·ij 'rē¦jekt ¦mik·sər }

image response |ELECTR| The response of a superheterodyne receiver to an undesired signal at its image frequency. { 'im·ij ri¦späns }

image restoration |COMPUT SCI| Operation on a picture with a digital computer to make it more closely resemble the original object. { 'im·ij ¦res·tə'rā·shən }

image-storage array |ELECTR| A solid-state panel or chip in which the image-sensing elements may be a metal oxide semiconductor or a charge-coupled or other light-sensitive device that can be manufactured in a high-density configuration. { 'im·ij ¦stór·ij ə¦rā }

image table [CONT SYS] A data table that contains the status of all inputs, registers, and coils in a programmable controller. { 'im·ij ,tā·bəl }

image transfer constant [ELECTR] One-half the natural logarithm of the complex ratio of the steady-state apparent power entering and leaving a network terminated in its image impedance. { 'im·ij 'tranz·fər ,kän·stənt }

image tube [ELECTR] An electron tube that reproduces on its fluorescent screen an image of the optical image or other irradiation pattern incident on its photosensitive surface. Also known as electron image tube; image converter. { 'im·ij ,tüb }

IMAP See Internet Mail Access Protocol. { 'ī,map }

imitative deception [ELECTR] Introduction of electromagnetic radiations into enemy channels which imitate their own emissions, in order to mislead them. { 'im·ə,tād·iv di'sep·shən }

immediate-access [COMPUT SCI] 1. Pertaining to an access time which is relatively brief, or to a relatively fast transfer of information. 2. Pertaining to a device which is directly connected with another device. { i'mē·dē·ət 'ak·ses }

immediate address [COMPUT SCI] The value of an operand contained in the address part of an instruction and used as data by this instruction. { i'mē·dē·ət 'a,dres }

immediate data [COMPUT SCI] Data that appears in an instruction exactly as it is to be processed. { i'mēd·ē·ət 'dad·ə }

immediate instruction [COMPUT SCI] A computer program instruction, part of which contains the actual data to be operated upon, rather than the address of that data. { i'mēd·ē·ət in'strək·shən }

immediate operand [COMPUT SCI] An operand contained in the instruction which specifies the operation. { i'mē·dē·ət 'äp·ə,rand }

immediate processing See demand processing. { i'mē·dē·ət 'präs,es·iŋ }

immersion electron lens [ELECTR] An electron lens in which the object, usually the cathode, lies deep within the electric field so that the index of refraction varies rapidly in its vicinity. { i¦mər·zhən i'lk,trän ,lenz }

immersion electron microscope [ELECTR] An emission electron microscope in which the specimen is a flat conducting surface which may be heated, illuminated, or bombarded by high-velocity electrons or ions so as to emit low-velocity thermionic, photo-, or secondary electrons; these are accelerated to a high velocity in an immersion objective or cathode lens and imaged as in a transmission electron microscope. { ə'mər·zhən i¦lek,trän 'mī·krə,skōp }

immersion electrostatic lens See bipotential electrostatic lens. { ə'mər·zhən i¦lek·trə,stad·ik 'lenz }

immersion heater [ELEC] An electric device for heating a liquid by direct immersion in the liquid. { ə'mər·zhən ,hēd·ər }

immersive simulation See virtual reality. { i¦mər·siv sim·yə'lā·shən }

immittance [ELEC] A term used to denote both impedance and admittance, as commonly applied to transmission lines, networks, and certain types of measuring instruments. { i'mit·əns }

impact avalanche and transit time diode See IMPATT diode. { 'im,pakt ¦av·ə,lanch ən 'tran·zit ,tīm 'dī,ōd }

impact excitation [ELEC] Starting of damped oscillations in a radio circuit by a sudden surge, such as that produced by a spark discharge. { 'im,pakt ,ek·sə'tā·shən }

impact ionization [ELECTR] Ionization produced by the impact of a high-energy charge carrier on an atom of semiconductor material; the effect is an increase in the number of charge carriers. { 'im,pakt ,ī·ə·nə'zā·shən }

impact microphone [ENG ACOUS] An instrument that picks up the vibration of an object impinging upon another, used especially on space probes to record the impact of small meteoroids. { 'im ,pakt 'mī·krə,fōn }

impact-noise analyzer [ENG] An analyzer used with a sound-level meter to evaluate the characteristics of impact-type sounds and electric noise impulses that cannot be measured accurately with a noise meter alone. { 'im,pakt ,nòiz 'an·ə ,līz·ər }

IMPATT amplifier [ELECTR] A diode amplifier that uses an IMPATT diode; operating frequency range is from about 5 to 100 gigahertz, primarily in the C and X bands, with power output up to about 20 watts continuous-wave or 100 watts pulsed. { 'im,pat ,am·plə,fī·ər }

IMPATT diode [ELECTR] A *pn* junction diode that has a depletion region adjacent to the junction, through which electrons and holes can drift, and is biased beyond the avalanche breakdown voltage. Derived from impact avalanche and transit time diode. { 'im,pat ,dī,ōd }

impedance See electrical impedance. { im'pēd·əns }

impedance-admittance matrix [ELECTR] A four-element matrix used to describe analytically a transistor in terms of impedances or admittances. { im'pēd·əns ad'mit·əns 'mā·triks }

impedance bridge [ELEC] A device similar to a Wheatstone bridge, used to compare impedances which may contain inductance, capacitance, and resistance. { im'pēd·əns ,brij }

impedance coil [ELEC] A coil of wire designed to provide impedance in an electric circuit. { im'pēd·əns ,kòil }

impedance compensator [ELEC] Electric network designed to be associated with another network or a line with the purpose of giving the impedance of the combination a desired characteristic with frequency over a desired frequency range. { im'pēd·əns 'käm·pən,sād·ər }

impedance component [ELEC] 1. Resistance or reactance. 2. A device such as a resistor, inductor, or capacitor designed to provide impedance in an electric circuit. { im'pēd·əns kəm,pō·nənt }

impedance coupling |ELEC| Coupling of two signal circuits with an impedance. { im'pēd·əns ,kəp·liŋ }

impedance drop |ELEC| The total voltage drop across a component or conductor of an alternating-current circuit, equal to the phasor sum of the resistance drop and the reactance drop. { im'pēd·əns ,dräp }

impedance irregularity |ELEC| A discontinuity or abrupt change which results from a junction between unlike sections of a transmission line or an irregularity on a line. { im'pēd·əns i,reg·yə'lar·əd·ē }

impedance match |ELEC| The condition in which the external impedance of a connected load is equal to the internal impedance of the source or to the surge impedance of a transmission line, thereby giving maximum transfer of energy from source to load, minimum reflection, and minimum distortion. { im'pēd·əns ,mach }

impedance-matching network |ELEC| A network of two or more resistors, coils, or capacitors used to couple two circuits in such a manner that the impedance of each circuit will be equal to the impedance into which it locks. Also known as line-building-out network. { im'pēd·əns ¦mach·iŋ ,net,wərk }

impedance matrix |ELEC| A matrix Z whose elements are the mutual impedances between the various meshes of an electrical network; satisfies the matrix equation $V = ZI$, where V and I are column vectors whose elements are the voltages and currents in the meshes. { im'pēd·əns ,mā·triks }

impedance meter See electrical impedance meter. { im'pēd·əns ,mēd·ər }

imperative language |COMPUT SCI| A programming language in which programs largely consist of a series of commands to assign values to objects. { im'per·əd·iv ,laŋ·gwij }

imperative statement |COMPUT SCI| A statement in a symbolic program which is translated into actual machine-language instructions by the assembly routine. { im'per·əd·iv ,stāt·mənt }

implanted atom |ELECTR| An atom introduced into semiconductor material by ion implantation. { im'plant·əd 'ad·əm }

implanted device |ELECTR| A resistor or other device that is fabricated within a silicon or other semiconducting substrate by ion implantation. { im'plant·əd di'vīs }

implementation |COMPUT SCI| **1.** The installation of a computer system or an information system. **2.** The use of software on a particular computer system. { ,im·plə,men'tā·shən }

implicit programming |CONT SYS| Robotic programming that uses descriptions of the tasks at hand which are less exact than in explicit programming. { im'plis·ət 'prō,gram·iŋ }

impressed current |ELEC| Direct current supplied by an external power source in a cathodic protection installation. { im¦prest 'kər·ənt }

impressed voltage |ELEC| Voltage applied to a circuit or device. { im'prest 'vōl·tij }

improvement factor |COMMUN| See noise improvement factor. |ELECTR| In radar, a measure of the effectiveness of Doppler-sensitive processes, given by the ratio of the signal-to-clutter power ratios with and without the use of the processing, averaged over all target velocities. { im 'prüv·mənt ,fak·tər }

improvement threshold |COMMUN| The condition of unity for the ratio of peak carrier voltage to peak noise voltage after selection and before any nonlinear process such as amplitude limiting. { im'prüv·mənt ,thresh,hōld }

impulse excitation See shock excitation. { 'im ,pəls ,ek·sə'tā·shən }

impulse generator |ELEC| An apparatus which produces very short surges of high-voltage or high-current power by discharging capacitors in parallel or in series. Also known as pulse generator. { 'im,pəls ,jen·ə,rād·ər }

impulse modulation |CONT SYS| Modulation of a signal in which it is replaced by a series of impulses, equally spaced in time, whose strengths (integrals over time) are proportional to the amplitude of the signal at the time of the impulse. { 'im,pəls ,mäj·ə,lā·shən }

impulse noise |ELEC| Noise characterized by transient short-duration disturbances distributed essentially uniformly over the useful passband of a transmission system. { 'im,pəls ,nóiz }

impulse period See pulse period. { 'im,pəls ,pir·ē·əd }

impulse relay |ELEC| A relay that stores the energy of a short pulse, to operate the relay after the pulse ends. { 'im,pəls ¦rē,lā }

impulse response |CONT SYS| The response of a system to an impulse which differs from zero for an infinitesimal time, but whose integral over time is unity; this impulse may be represented mathematically by a Dirac delta function. { 'im ,pəls ri,späns }

impulse separator |ELECTR| In an analog television receiver, the circuit that separates the horizontal synchronizing impulses in the received signal from the vertical synchronizing impulses. { 'im,pəls ,sep·ə,rād·ər }

impulse signaling |COMMUN| Conveying information by means of on-off conditions transmitted down a line or over free space. { 'im,pəls ¦sig·nə·liŋ }

impulse strength |ELEC| Voltage breakdown of insulation under voltage surges on the order of microseconds in duration. { 'im,pəls ,streŋkth }

impulse train |CONT SYS| An input consisting of an infinite series of unit impulses, equally separated in time. { 'im,pəls ,trān }

impulse transmission |COMMUN| Form of signaling which employs impulses of either or both polarities for transmission to indicate the occurrence of transitions in the signals; used principally to reduce the effects of low-frequency interference; the impulses are generally formed by suppressing the low-frequency components, including direct current, of the signals. { 'im ,pəls tranz'mish·ən }

impulse-type telemeter [COMMUN] A telemeter that employs electric impulses as the translating means. { 'im‚pəls ‚tīp tə'lem·əd·ər }

impulse voltage [ELEC] A unidirectional voltage that rapidly rises to a peak value and then drops to zero more or less rapidly. Also known as pulse voltage. { 'im‚pəls ‚vōl·tij }

in-band/on-channel [COMMUN] A system of digital radio where the digital signals are placed within the current AM and FM bands and within the FCC-assigned channel of a radio station. Abbreviated IBOC. { ¦in ¦band ¦ón 'chan·əl }

incandescent lamp [ELEC] An electric lamp that produces light when a metallic filament is heated white-hot in a vacuum by passing an electric current through the filament. Also known as filament lamp; light bulb. { ‚in·kən'des·ənt 'lamp }

incandescent readout [ELECTR] A readout in which each character is formed by energizing an appropriate combination of seven bar-shaped incandescent lamps. { ‚in·kən'des·ənt 'rēd‚aút }

inching See jogging. { 'inch·iŋ }

incident power [ELEC] Product of the outgoing current and voltage, from a transmitter, traveling down a transmission line to the antenna. { 'in·sə·dənt ¦paú·ər }

incident wave [ELECTR] A current or voltage wave that is traveling through a transmission line in the direction from source to load. { 'in·sə·dənt ¦wāv }

incoming first selector [ELEC] Connects incoming calls from outlying dial offices to local second selectors. { 'in‚kəm·iŋ ¦fərst si'lek·tər }

incoming selector [ELEC] Selector associated with trunk circuits from another central office. { 'in‚kəm·iŋ si'lek·tər }

incorporate [COMPUT SCI] To place in storage. { in'kòr·pə‚rāt }

incremental compiler [COMPUT SCI] A compiler that generates code for a statement, or group of statements, which is independent of the code generated for other statements. { ‚iŋ·krə'ment·əl kəm'pīl·ər }

incremental computer [COMPUT SCI] A special-purpose computer designed to process changes in variables as well as absolute values; for instance, a digital differential analyzer. { ‚iŋ·krə'ment·əl kəm'pyüd·ər }

incremental digital recorder [COMPUT SCI] Magnetic tape recorder in which the tape advances across the recording head step by step, as in a punched-paper-tape recorder; used for recording an irregular flow of data economically and reliably. { ‚iŋ·krə'ment·əl ¦dij·əd·əl ri'kórd·ər }

incremental dump tape [COMPUT SCI] A safety technique used in time-sharing which consists in copying all files (created or modified by a user during a day) on a magnetic tape; in case of system failure, the file storage can then be reconstructed. Also known as failsafe tape. { ‚iŋ·krə'ment·əl 'dəmp ‚tāp }

incremental frequency shift [COMMUN] Method of superimposing incremental intelligence on another intelligence by shifting the center frequency of an oscillator a predetermined amount. { ‚iŋ·krə'ment·əl 'frē·kwən‚sē ‚shift }

incremental mode [COMPUT SCI] The plotting of a curve on a cathode-ray tube by illuminating a fixed number of points at a time. { ‚iŋ·krə'ment·əl ¦mōd }

incremental representation [COMPUT SCI] A way of representing variables used in incremental computers, in which changes in the variables are represented instead of the values of the variables themselves. { ‚iŋ·krə'ment·əl ‚rep·rə·sən'tā·shən }

independent-sideband modulation [COMMUN] Modulation in which the radio-frequency carrier is reduced or eliminated and two channels of information are transmitted, one on an upper and one on a lower sideband. Abbreviated ISB modulation. { ‚in·də'pen·dənt ¦sīd‚band ‚mäj·ə'lā·shən }

independent-sideband receiver [ELECTR] A radio receiver designed for the reception of independent-sideband modulation, having provisions for restoring the carrier. { ‚in·də'pen·dənt ¦sīd‚band ri'sē·vər }

independent-sideband transmitter [ELECTR] A transmitter which produces independent-sideband modulated signals. { ‚in·də'pen·dənt ¦sīd‚band tranz'mid·ər }

index [COMPUT SCI] **1.** A list of record surrogates arranged in order of some attribute expressible in machine-orderable form. **2.** To produce a machine-orderable set of record surrogates, as in indexing a book. **3.** To compute a machine location by indirection, as is done by index registers. **4.** The portion of a computer instruction which indicates what index register (if any) is to be used to modify the address of an instruction. { 'in‚deks }

index arithmetic unit [COMPUT SCI] A section of some computers that performs addition or subtraction operations on address parts of instructions for the purpose of indexing, boundary tests for memory protection, and so forth. { 'in‚deks ə'rith·mə·tik ‚yü·nət }

indexed address [COMPUT SCI] An address which is modified, generally by means of index registers, before or during execution of a computer instruction. { 'in‚dekst ə'dres }

indexed array [COMPUT SCI] An array of data items in which the individual items can be accessed by specifying their position through use of a subscript. { 'in‚dekst ə'rā }

indexed sequential data set [COMPUT SCI] A collection of related data items that are stored sequentially on a key, but are also accessible through index tables maintained by the system. { 'in‚dekst si¦kwen·chəl 'dad·ə ‚set }

indexed sequential organization [COMPUT SCI] A sequence of records arranged in collating sequence used with direct-access devices. { 'in‚dekst si¦kwen·chəl ‚ór·gə·nə'zā·shən }

index marker [COMPUT SCI] The beginning (and end) of each track in a disk, which is recognized by a special sensing device within the disk mechanism. { 'in‚deks ‚märk·ər }

index of cooperation |COMMUN| In rectilinear scanning or recording, the product of the total length of a scanning or recording line by the number of scanning or recording lines per unit length divided by pi. { 'in,deks əv kō,äp·ə'rā·shən }

index of modulation *See* modulation factor. { 'in ,deks əv ,mäj·ə'lā·shən }

index point |COMPUT SCI| A hardware reference mark on a disk or drum for use in timing. { 'in ,deks ,póint }

index register |COMPUT SCI| A hardware element which holds a number that can be added to (or, in some cases, subtracted from) the address portion of a computer instruction to form an effective address. Also known as base register; B box; B line; B register; B store; modifier register. { 'in ,deks ,rej·ə·stər }

index word *See* modifier. { 'in,deks ,wərd }

indicative data |COMPUT SCI| Data which describe a specific item. { in'dik·əd·iv 'dad·ə }

indicator |COMPUT SCI| A device announcing an error or failure. |ELECTR| A cathode-ray tube or other device that presents information transmitted or relayed from some other source, as from a radar receiver. { 'in·də,kād·ər }

indicator element |ELECTR| A component whose variability under conditions of manufacture or use is likely to cause the greatest variation in some measurable parameter. { 'in·də,kād·ər 'el·ə·mənt }

indicator gate |ELECTR| Rectangular voltage waveform which is applied to the grid or cathode circuit of an indicator cathode-ray tube to sensitize or desensitize it during a desired portion of the operating cycle. { 'in·də,kād·ər ,gāt }

indicator lamp |ELEC| A neon lamp whose on-off condition is used to convey information. { 'in·də,kād·ər ,lamp }

indicator tube |ELECTR| An electron-beam tube in which useful information is conveyed by the variation in cross section of the beam at a luminescent target. { 'in·də,kād·ər ,tüb }

indirect address |COMPUT SCI| An address in a computer instruction that indicates a location where the address of the referenced operand is to be found. Also known as multilevel address. { ,in·də'rekt ə'dres }

indirect addressing |COMPUT SCI| A programming device whereby the address part of an instruction is not the address of the operand but rather the location in storage where the address of the operand may be found. { ,in·də'rekt ə'dres·iŋ }

indirect control |COMPUT SCI| The control of one peripheral unit by another through some sequence of events that involves human intervention. { ,in·də'rekt kən'trōl }

indirectly heated cathode |ELECTR| A cathode to which heat is supplied by an independent heater element in a thermionic tube; this cathode has the same potential on its entire surface, whereas the potential along a directly heated filament varies from one end to the other. Also known as equipotential cathode; heater-type cathode;

unipotential cathode. { ,in·də'rek·lē ¦hēd·əd 'kath,ōd }

indirect-path echo |ELECTROMAG| An echo resulting from radar transmission and reception, not via the direct path to the target but rather via reflections, for example, from a large building very near the radar, resulting in incorrect bearing estimation. { ,in·də'rekt ¦path 'ek·ō }

indirect stroke |ELEC| A lightning stroke that induces a voltage in a power or communications system without actually striking it. { ,in·də'rekt 'strōk }

individual distributed numerical control |CONT SYS| A form of distributed numerical control involving only a few machines, each of which operates independently of the others and is unaffected by their failures. { ,in·də'vij·ə·wəl di 'strib·yəd·əd nü'mer·ə·kəl kən'trōl }

individual line |COMMUN| Subscriber line arranged to serve only one main station, although additional stations may be connected to the line as extensions; an individual line is not arranged for discriminatory ringing with respect to the stations on that line. { ,in·də'vij·ə·wəl 'līn }

induced dipole |ELEC| An electric dipole produced by application of an electric field. { in 'düst 'dī,pōl }

induced moment |ELEC| The average electric dipole moment per molecule which is produced by the action of an electric field on a dielectric substance. { in'düst 'mō·mənt }

inductance |ELECTROMAG| **1.** That property of an electric circuit or of two neighboring circuits whereby an electromotive force is generated (by the process of electromagnetic induction) in one circuit by a change of current in itself or in the other. **2.** Quantitatively, the ratio of the emf (electromotive force) to the rate of change of the current. { in'dək·təns }

inductance coil *See* coil. { in'dək·təns ,kóil }

induction *See* electrostatic induction; electromaagnetic induction. { in'dək·shən }

induction charging |ELEC| Production of electric charge on a body by means of electrostatic induction. { in'dək·shən ,chär·jiŋ }

induction field |ELECTROMAG| A component of an electromagnetic field associated with an alternating current in a loop, coil, or antenna which carries energy alternately away from and back into the source, with no net loss, and which is responsible for self-inductance in a coil or mutual inductance with neighboring coils. { in'dək·shən ,fēld }

induction frequency converter |ELEC| Slip-ring induction machine which is driven by an external source of mechanical power and whose primary circuits are connected to a source of electric energy having a fixed frequency; the secondary circuits deliver energy at a frequency proportional to the relative speed of the primary magnetic field and the secondary member. { in'dək·shən 'frē·kwən,sē kən,vərd·ər }

induction generator |ELEC| A nonsynchronous alternating-current generator whose construction is identical to that of an ac motor, and which

induction instrument

is driven above synchronous speed by external sources of mechanical power. { in'dək·shən ¦jen·ə·rād·ər }

induction instrument [ENG] Meter that depends for its operation on the reaction between magnetic flux set up by current in fixed windings, and other currents set up by electromagnetic induction in conducting parts of the moving system. { in'dək·shən ‚in·strə·mənt }

induction loudspeaker [ENG ACOUS] Loudspeaker in which the current which reacts with the steady magnetic field is induced in the moving member. { in'dək·shən ¦laüd‚spēk·ər }

induction machine [ELEC] An asynchronous alternating-current machine, such as an induction motor or induction generator, in which the windings of two electric circuits rotate with respect to each other and power is transferred from one circuit to the other by electromagnetic induction. { in'dək·shən mə‚shēn }

induction motor [ELEC] An alternating-current motor in which a primary winding on one member (usually the stator) is connected to the power source, and a secondary winding on the other member (usually the rotor) carries only current induced by the magnetic field of the primary. { in'dək·shən ‚mōd·ər }

induction problem [ELECTROMAG] An effect of potentials and currents induced in conductors of a telephone system by paralleling power facilities or power lines. { in'dək·shən ‚präb·ləm }

induction regulator [ELEC] A transformer in which the voltage produced in a secondary winding is varied by changing the position of the primary winding. { in'dək·shən 'reg·yə‚lād·ər }

induction watthour meter [ELEC] A watthour meter used with alternating current; the energy taken by a circuit over a period of time is proportional to the rotation in that period of a light aluminum disk, in which a driving torque is developed by the joint action of the alternating magnetic flux produced by the potential circuit and by the load current. { in'dək·shən 'wät‚aúr ‚mēd·ər }

inductive charge [ELEC] The charge that exists on an object as a result of its being near another charged object. { in'dək·tiv 'chärj }

inductive circuit [ELEC] A circuit containing a higher value of inductive reactance than capacitive reactance. { in'dək·tiv 'sər·kət }

inductive coordination [ELECTROMAG] Measures to reduce induction problems. { in'dək·tiv kō ‚órd·ən'ā·shən }

inductive coupler [ELEC] A mutual inductance that provides electrical coupling between two circuits; used in radio equipment. { in'dək·tiv 'kəp·lər }

inductive coupling [ELEC] Coupling of two circuits by means of the mutual inductance provided by a transformer. Also known as transformer coupling. { in'dək·tiv 'kəp·liŋ }

inductive fault analysis [ELECTR] A method of analyzing the effects of defects on an integrated circuit, in which a computer simulates an electron

that scatters at random faults in the form of additional or missing areas of material on the set of drawings of the masks from which the circuits are fabricated. { in‚dək·tiv 'fólt ə‚nal·ə·səs }

inductive feedback [ELECTR] 1. Transfer of energy from the plate circuit to the grid circuit of a vacuum tube by means of induction. 2. Transfer of energy from the output circuit to the input circuit of an amplifying device through an inductor, or by means of inductive coupling. { in'dək·tiv 'fēd‚bak }

inductive filter [ELECTR] A low-pass filter used for smoothing the direct-current output voltage of a rectifier; consists of one or more sections in series, each section consisting of an inductor on one of the pair of conductors in series with a capacitor between the conductors. Also known as LC filter. { in'dək·tiv 'fil·tər }

inductive grounding [ELEC] Use of grounding connections containing an inductance in order to reduce the magnitude of short-circuit currents created by line-to-ground faults. { in'dək·tiv 'graúnd·iŋ }

inductive interference [COMMUN] Effect arising from the characteristics and inductive relations of electric supply and communications systems of such character and magnitude as would prevent the communications circuits from rendering service satisfactorily and economically if methods of inductive coordination were not applied. { in'dək·tiv ‚in·tər'fir·əns }

inductive line pair [COMMUN] A telephone line displaying induction whose effects are of consequence, as in crosstalk; opposed to twisted pair. { in'dək·tiv 'līn ‚per }

inductive load [ELEC] A load that is predominantly inductive, so that the alternating load current lags behind the alternating voltage of the load. Also known as lagging load. { in'dək·tiv 'lōd }

inductive neutralization [ELECTR] Neutralizing an amplifier whereby the feedback susceptance due to an interelement capacitance is canceled by the equal and opposite susceptance of an inductor. Also known as coil neutralization; shunt neutralization. { in'dək·tiv ‚nü·trə·lə'zā·shən }

inductive-output tube [ELECTR] A tube in which output energy is obtained from the electron stream by electric induction between a cylindrical output electrode and the electron stream that flows through but does not touch the electrode. { in¦dək·tiv 'aút‚pút ‚tüb }

inductive reactance [ELEC] Reactance due to the inductance of a coil or circuit. { in'dək·tiv rē'ak·təns }

inductive superconducting fault-current limiter See shielded-core superconducting fault-current limiter. { in¦dək·tiv ‚sü·pər·kən¦dək·tiŋ 'fólt ‚cər·ənt ‚lim·əd·ər }

inductive susceptance [ELEC] In a circuit containing almost no resistance, the part of the susceptance due to inductance. { in'dək·tiv sə'sep·təns }

288

inductive tuning [ELECTR] Tuning involving the use of a variable inductance. { in'dək·tiv 'tün·iŋ }

inductive voltage divider [ELEC] An autotransformer that has its winding subdivided into 10 equal turn sections so that when an alternating voltage V is applied to the whole winding the voltage across each section is nominally V/10; used as a ratio standard for electrical measurements. { in'dək·tiv 'vōl·tij di,vīd·ər }

inductive waveform [ELEC] A graph or trace of the effect of current buildup across an inductive network; proportional to the exponential of the product of a negative constant and the time. { in'dək·tiv 'wāv,fȯrm }

inductor See coil. { in'dək·tər }

inductor alternator [ELEC] A synchronous generator in which the field winding is fixed in magnetic position relative to the armature conductors. { in'dək·tər 'ȯl·tər,nād·ər }

inductor generator [ELEC] An alternating-current generator in which all the windings are fixed, and the flux linkages are varied by rotating an appropriately toothed ferromagnetic rotor; sometimes used for generating high power at frequencies up to several thousand hertz for induction heating. { in'dək·tər 'jen·ə,rād·ər }

inductor microphone [ENG ACOUS] Moving-conductor microphone in which the moving element is in the form of a straight-line conductor. { in'dək·tər 'mī·krə,fōn }

industrial frequency bands [COMMUN] The radio-frequency bands allocated in the United States for land mobile communications of private industries other than transportation. { in'dəs·trē·əl 'frē·kwən·sē ,banz }

industrial television [COMMUN] Closed-circuit video system used for remote viewing of industrial processes and operations; may also be used for training purposes. Abbreviated ITV. { in'dəs·trē·əl 'tel·ə,vizh·ən }

ineffective time [COMPUT SCI] Time during which a computer can operate normally but which is not used effectively because of mistakes or inefficiency in operating the installation or for other reasons. { ,in·i'fek·tiv 'tīm }

inertia switch [ELEC] A switch that is actuated by an abrupt change in the velocity of the item on which it is mounted. { i'nər·shə ,swich }

inference control [COMPUT SCI] A method of preventing data about specific individuals from being inferred from statistical information in a data base about groups of people. { 'in·frəns kən,trōl }

inference program [COMPUT SCI] A computer program that uses certain facts provided as input to reach conclusions. { 'in·frəns ,prō·grəm }

infinite baffle [ENG ACOUS] A loudspeaker baffle which prevents interaction between the front and back radiation of the loudspeaker. { 'in·fə·nət 'baf·əl }

infinite-capacity loading [CONT SYS] The deliberate overloading of a robotic work center with excessive force or weight in order to determine the overload protection necessary to maintain proper load conditions. { 'in·fə·nət kə'pas·əd·ē ,lōd·iŋ }

infinite impulse response filter [ELECTR] An electronic filter that will continue oscillating in a decaying manner forever after being exposed to a change in input. Abbreviated IIR filter. { ,in·fə·nət ¦im,pəls ri¦späns ,fil·tər }

infinite sequence See sequence. { 'in·fə·nət 'sē·kwəns }

infinity [COMPUT SCI] Any number larger than the maximum number that a computer is able to store in any register. { in'fin·əd·ē }

infinity transmitter [ELECTR] A device used to tap a telephone; the telephone instrument is so modified that an interception device can be actuated from a distant source without the caller's becoming aware. { in'fin·əd·ē tranz'mid·ər }

infix operation [COMPUT SCI] An operation carried out within an operation, as the addition of a and b prior to the multiplication by c or division by d in the operation $(a+b)c/d$. { 'in,fiks ,äp·ə,rā·shən }

influence diagram [SYS ENG] A graph-theoretic representation of a decision, which may include four types of nodes (decision, chance, value, and deterministic), directed arcs between the nodes (which identify dependencies between them), a marginal or conditional probability distribution defined at each chance node, and a mathematical function associated with each of the other types of node. { 'in,flü·əns ,dī·ə,gram }

influence factor See telephone influence factor. { 'in,flü·əns ,fak·tər }

information [COMMUN] Data which has been recorded, classified, organized, related, or interpreted within a framework so that meaning emerges. { ,in·fər'mā·shən }

information architecture [COMPUT SCI] The organization of large bodies of content, as well as the organization and labeling (tagging) of content at the document level to make information easy to search, navigate, and manage. { ,in·fər ,mā·shən 'är·kə,tek·chər }

information bit [COMMUN] Bit that is generated by the data source but is not used by the data-transmission system. { ,in·fər'mā·shən ,bit }

information center [COMMUN] Center designed specifically for storing, processing, and retrieving information for dissemination at regular intervals, on demand or selectively, according to express needs of users. { ,in·fər'mā·shən ,sen·tər }

information channel [COMMUN] A facility used to transmit information between data-processing terminals separated by large distances. { ,in·fər'mā·shən ¦chan·əl }

information content [COMMUN] A numerical measure of the information generated in selecting a specific symbol (or message), equal to the negative of the logarithm of the probability of the symbol (or message) selected. Also known as negentropy. { ,in·fər'mā·shən ¦kän,tent }

information engineering [COMPUT SCI] The process of networking, collecting, analyzing, and reporting information, as well as controlling business, manufacturing, or service operations. { ,in·fər'mā·shən ,en·jə,nir·iŋ }

information feedback system [COMMUN] An information transmission system in which a return transmission is used to verify the accuracy of the sent transmission. { ,in·fər'mā·shən 'fēd ,bak ,sis·təm }

information float [COMPUT SCI] Information that is not located in a file or data base but is traveling between systems or is not assigned to a particular computer system. { ,in·fər'mā·shən ,flōt }

information flow [COMPUT SCI] The graphic representation of data collection, data processing, and report distribution throughout an organization. { ,in·fər'mā·shən ,flō }

information flow control [COMPUT SCI] A restriction on the use of information generated by a computer system that is consistent with the access controls on the resources of the system itself. { ,in·fər'mā·shən 'flō kən,trōl }

information interchange [COMMUN] The exchange of information between machines. { ,in·fər'mā·shən 'in·tər,chānj }

information link See data link. { ,in·fər'mā·shən ,liŋk }

information management [COMMUN] The science that deals with definitions, uses, value and distribution of information that is processed by an organization, whether or not it is handled by a computer. { ,in·fər'mā·shən 'man·ij·mənt }

information network [COMPUT SCI] A service that provides a variety of information services to subscribers on a dial-up basis. Also known as subscription database. { ,in·fər'mā·shən 'net ,wərk }

information precedence relation [COMPUT SCI] A statement that some specified piece of data is required for the production of another piece of data. { ,in·fər'mā·shən 'pres·ə·dəns ri,lā·shən }

information processing [COMPUT SCI] **1.** The manipulation of data so that new data (implicit in the original) appear in a useful form. **2.** See data processing. { ,in·fər'mā·shən 'prä·ses·iŋ }

Information Processing Language See IPL. { ,in·fər'mā·shən 'prä·ses·iŋ 'laŋ·gwij }

information rate [COMMUN] The information content generated per symbol or per second by an information source. { ,in·fər'mā·shən ,rāt }

information redundancy [COMPUT SCI] The use of more information than is absolutely necessary, such as the application of error-detection and error-correction codes, in order to increase the reliability of a computer system. { ,in·fər'mā·shən rə'dən·dən·sē }

information requirements [COMPUT SCI] Actual or anticipated questions which may be posed to an information retrieval system. { ,in·fər'mā·shən rə'kwīr·məns }

information resources management [COMPUT SCI] A concept for processing information that focuses on the information and places data-processing technology (software and hardware) in a secondary role. { ,in·fər'mā·shən ri'sòr·səz ,man·ij·mənt }

information retrieval [COMPUT SCI] The technique and process of searching, recovering, and interpreting information from large amounts of stored data. { ,in·fər'mā·shən ri,trē·vəl }

information selection systems [COMPUT SCI] A class of information processing systems which carry out a sequence of operations necessary to locate in storage one or more items assumed to have certain specified characteristics, and to retrieve such items directly or indirectly, in whole or in part. { ,in·fər'mā·shən si'lek·shən ,sis·təmz }

information separator [COMPUT SCI] A character that separates items or fields of information in a record, especially a variable-length record. { ,in·fər'mā·shən 'sep·ə,rād·ər }

information source [COMMUN] A system which produces messages by making successive selections from a group of symbols. { ,in·fər'mā·shən ,sòrs }

information system [COMMUN] Any means for communicating knowledge from one person to another, ranging from simple verbal communication to completely computerized methods of storing, searching, and retrieving of information. { ,in·fər'mā·shən ,sis·təm }

information system architecture [COMPUT SCI] The study of the structure of both computer systems and the organizations that use them, in order to develop computer systems that support the objectives of the organizations more effectively. { ,in·fər'mā·shən 'sis·təm 'är·kə,tek·chər }

information technology [COMPUT SCI] The collection of technologies that deal specifically with processing, storing, and communicating information, including all types of computer and communications systems as well as reprographics methodologies. { ,in·fər'mā·shən tek 'näl·ə·jē }

information theory [COMMUN] A branch of theory which is devoted to problems in communications, and which provides criteria for comparing different communications systems on the basis of signaling rate, using a numerical measure of the amount of information gained when the content of a message is learned. { ,in·fər'mā·shən ,thē·ə·rē }

information unit [COMMUN] A unit of information content, equal to a bit, nit, or hartley, according to whether logarithms are taken to base 2, e, or 10. { ,in·fər'mā·shən ,yü·nət }

information utility [COMPUT SCI] An information network that specializes in supplying information to businesses and other organizations. { ,in·fər'mā·shən yü,til·əd·ē }

information word See data word. { ,in·fər'mā·shən ,wərd }

infradyne receiver [ELECTR] A superheterodyne receiver in which the intermediate frequency is higher than the signal frequency, so as to obtain high selectivity. { 'in·frə,dīn ri'sē·vər }

infralow frequency [COMMUN] A designation for the band from 0.3 to 3 kilohertz in the radio spectrum. Abbreviated ILF. { 'in·frə,lō 'frē·kwən·sē }

infrared bolometer [ELECTR] A bolometer adapted to detecting infrared radiation, as opposed to microwave radiation. { ¦in·fra¦red bə'läm·əd·ər }

infrared communications set [ELECTR] Components required to operate a two-way electronic system using infrared radiation to carry intelligence. { ¦in·fra¦red kə,myü·nə'kā·shənz ,set }

infrared detector [ELECTR] A device responding to infrared radiation, used in detecting fires, or overheating in machinery, planes, vehicles, and people, and in controlling temperature-sensitive industrial processes. { ¦in·fra¦red di'tek·tər }

infrared-emitting diode [ELECTR] A light-emitting diode that has maximum emission in the near-infrared region, typically at 0.9 micrometer for *pn* gallium arsenide. { ¦in·fra¦red i¦mid·iŋ 'dī,ōd }

infrared heterodyne detector [ELECTR] A heterodyne detector in which both the incoming signal and the local oscillator signal frequencies are in the infrared range and are combined in a photodetector to give an intermediate frequency in the kilohertz or megahertz range for conventional amplification. { ¦in·fra¦red ¦hed·ə·rə,dīn di'tek·tər }

infrared image converter [ELECTR] A device for converting an invisible infrared image into a visible image, consisting of an infrared-sensitive, semitransparent photocathode on one end of an evacuated envelope and a phosphor screen on the other, with an electrostatic lens system between the two. Also known as infrared image tube. { ¦in·fra¦red 'im·ij kən,vərd·ər }

infrared image tube *See* infrared image converter. { ¦in·fra¦red 'im·ij ,tüb }

infrared jamming [ELECTR] An attempt to confuse heat-seeking missiles by emissions which overload their inputs or misdirect them. { ¦in·fra¦red 'jam·iŋ }

infrared lamp [ELEC] An incandescent lamp which operates at reduced voltage with a filament temperature of 4000°F (2200°C) so that it radiates electromagnetic energy primarily in the infrared region. { ¦in·fra¦red 'lamp }

infrared photoconductor [ELECTR] A conductor whose conductivity increases when it is exposed to infrared radiation. { ¦in·fra¦red ¦fōd·ō·kən ¦dək·tər }

infrared radiation [ELECTROMAG] Electromagnetic radiation whose wavelengths lie in the range from 0.75 or 0.8 micrometer (the long-wavelength limit of visible red light) to 1000 micrometers (the shortest microwaves). { ¦in·fra¦red ,rād·ē'ā·shən }

infrared receiver [ELECTR] A device that intercepts or demodulates infrared radiation that may carry intelligence. Also known as nancy receiver. { ¦in·fra¦red ri'sē·vər }

infrared scanner [ELECTR] An infrared detector mounted on a motor-driven platform which causes it to scan a field of view line by line, much as in television. { ¦in·fra¦red 'skan·ər }

infrared thermistor [ELECTR] A thermistor used to measure the power of infrared radiation. { ¦in·fra¦red thər'mis·tər }

infrared transmitter [ELECTR] A transmitter that emits energy in the infrared spectrum; may be modulated with intelligence signals. { ¦in·fra ¦red tranz'mid·ər }

infrared vidicon [ELECTR] A vidicon whose photoconductor surface is sensitive to infrared radiation. { ¦in·fra¦red 'vid·ə,kän }

inherent noise pressure *See* equivalent noise pressure. { in'hir·ənt 'nóiz ,presh·ər }

inherent storage [COMPUT SCI] Any type of storage in which the storage medium is part of the hardware of the computer medium. { in'hir·ənt 'stòr·ij }

inheritance [COMPUT SCI] A feature of object-oriented programming that allows a new class to be defined simply by stating how it differs from an existing class. { in'her·əd·əns }

inherited error [COMPUT SCI] The error existing in the data supplied at the beginning of a step in a step-by-step calculation as executed by a program. { in'her·əd·əd 'er·ər }

inhibit-gate [ELECTR] Gate circuit whose output is energized only when certain signals are present and other signals are not present at the inputs. { in'hib·ət,gāt }

inhibiting input [ELECTR] A gate input which, if in its prescribed state, prevents any output which might otherwise occur. { in'hib·əd·iŋ 'in ,pùt }

inhibiting signal [ELECTR] A signal, which when entered into a specific circuit will prevent the circuit from exercising its normal function; for example, an inhibit signal fed into an AND gate will prevent the gate from yielding an output when all normal input signals are present. { in'hib· əd·iŋ ,sig·nəl }

inhibit pulse [ELECTR] A drive pulse that tends to prevent flux reversal of a magnetic cell by certain specified drive pulses. { in'hib·ət ,pəls }

initial condition *See* entry condition. { i'nish·əl kən'dish·ən }

initial condition mode *See* reset mode. { i'nish·əl kən'dish·ən ,mōd }

initial graphics exchange specification [COMPUT SCI] A standard graphics file format for three-dimensional wire-frame models. Abbreviated IGES. { i'nish·əl 'graf·iks iks,chānj ,spes·ə·fə'kā·shən }

initial instructions [COMPUT SCI] A routine stored in a computer to aid in placing a program in memory. Also known as initial orders. { i'nish· əl in'strək·shənz }

initial inverse voltage [ELECTR] Of a rectifier tube, the peak inverse anode voltage immediately following the conducting period. { i'nish·əl ¦in ,vərs 'vōl·tij }

initialize [COMPUT SCI] **1.** To set counters, switches, and addresses to zero or other starting values at the beginning of, or at prescribed points in, a computer routine. **2.** To begin an operation, and more specifically, to adjust the environment to the required starting configuration. { i'nish·ə,līz }

initial orders *See* initial instructions. { i'nish·əl 'òr·dərz }

initial program load |COMPUT SCI| A routine, used in starting up a computer, that loads the operating system from a direct-access storage device, usually a disk or diskette, into the computer's main storage. Abbreviated IPL. { i'nish·əl 'prō·grəm ‚lōd }

initial program load button See bootstrap button. { i'nish·əl 'prō·grəm ‚lōd ¦bət·ən }

initial surge voltage |ELEC| A spike of voltage experienced when a noncompensated load is first connected to a generator. { i'nish·əl 'sərj ‚vōl·tij }

initiate See trigger. { i'nish·ē‚āt }

initiator |COMPUT SCI| A part of an operating system of a large computer that runs several jobs at the same time, setting up the job, monitoring its progress, and performing any necessary cleanup after the job's completion. { i'nish·ē‚ād·ər }

injection |ELECTR| **1.** The method of applying a signal to an electronic circuit or device. **2.** The process of introducing electrons or holes into a semiconductor so that their total number exceeds the number present at thermal equilibrium. { in'jek·shən }

injection efficiency |ELECTR| A measure of the efficiency of a semiconductor junction when a forward bias is applied, equal to the current of injected minority carriers divided by the total current across the junction. { in'jek·shən ə‚fish·ən·sē }

injection electroluminescence |ELECTR| Radiation resulting from recombination of minority charge carriers injected in a *pn* or *pin* junction that is biased in the forward direction. Also known as Lossev effect; recombination electroluminescence. { in'jek·shən i¦lek·trō‚lü·mə 'nes·əns }

injection grid |ELECTR| Grid introduced into a vacuum tube in such a way that it exercises control over the electron stream without causing interaction between the screen grid and control grid. { in'jek·shən ‚grid }

injection locking |ELECTR| The capture or synchronization of a free-running oscillator by a weak injected signal at a frequency close to the natural oscillator frequency or to one of its subharmonics; used for frequency stabilization in IMPATT or magnetron microwave oscillators, gas-laser oscillators, and many other types of oscillators. { in'jek·shən ¦läk·iŋ }

injection luminescent diode |ELECTR| Gallium arsenide diode, operating in either the laser or the noncoherent mode, that can be used as a visible or near-infrared light source for triggering such devices as light-activated switches. { in 'jek·shən ‚lü·mə¦nes·ənt 'dī‚ōd }

injection signal |ENG ACOUS| The sawtooth frequency-modulated signal which is added to the first detector circuit for mixing with the incoming target signal. { in'jek·shən ‚sig·nəl }

injector |ELECTR| An electrode through which charge carriers (holes or electrons) are forced to enter the high-field region in a spacistor. { in'jek·tər }

ink bleed |COMPUT SCI| In character recognition, the capillary extension of ink beyond the original

edges of a printed or handwritten character. { 'iŋk ‚blēd }

ink smudge |COMPUT SCI| In character recognition, the overflow of ink beyond the original edges of a printed or handwritten character. { 'iŋk ‚sməj }

ink squeezeout |COMPUT SCI| In character recognition, the overflow of ink from the stroke centerline to the edges of a printed or handwritten character. { 'iŋk 'skwē‚zaút }

in-line coding |COMPUT SCI| Any group of instructions within the main body of a program. { 'in ¦līn 'kōd·iŋ }

in-line guns |ELECTR| An arrangement of three electron guns in a horizontal line; used in color picture tubes that have a slot mask in front of vertical color phosphor stripes. { 'in ¦līn 'gənz }

in-line procedure |COMPUT SCI| A short body of coding or instruction which accomplishes some purpose. { 'in ¦līn prə'sē·jər }

in-line processing |COMPUT SCI| The processing of data in random order, not subject to preliminary editing or sorting. { 'in ¦līn 'prä·ses·iŋ }

in-line subroutine |COMPUT SCI| A subroutine which is an integral part of a program. { 'in ¦līn 'səb·rü‚tēn }

in-line tuning |ELECTR| Method of tuning the intermediate-frequency strip of a superheterodyne receiver in which all the intermediate-frequency amplifier stages are made resonant to the same frequency. { 'in ¦līn 'tün·iŋ }

in-phase and quadrature video |ELECTR| The pair of video signals produced in a radar receiver using two homodyne reception channels in which the reference signal iin one has shifted by ninety degrees of phase from the reference signal in the other; the process overcomes certain limiting conditions in the use of just one homodyne channel. { 'in ‚fāz ənd 'kwä·drə·chər 'vid·ē·ō }

in-phase component |ELEC| The component of the phasor representing an alternating current which is parallel to the phasor representing voltage. { 'in ‚fāz kəm'pō·nənt }

in-phase rejection See common-mode rejection. { 'in ‚fāz ri'jek·shən }

in-phase signal See common-mode signal. { 'in ‚fāz 'sig·nəl }

input |COMPUT SCI| The information that is delivered to a data-processing device from the external world, the process of delivering this data, or the equipment that performs this process. |ELECTR| **1.** The power or signal fed into an electrical or electronic device. **2.** The terminals to which the power or signal is applied. { 'in‚pút }

input admittance |ELEC| The admittance measured across the input terminals of a four-terminal network with the output terminals short-circuited. { 'in‚pút əd‚mit·əns }

input area |COMPUT SCI| A section of internal storage reserved for storage of data or instructions received from an input unit such as cards or tape. Also known as input block; input storage. { 'in ‚pút ‚er·ē·ə }

input capacitance |ELECTR| The short-circuited transfer capacitance that exists between the input terminals and all other terminals of an electron tube (except the output terminal) connected together. { 'in‚pût kə'pas‚ad‚əns }

input data [COMPUT SCI] Data employed as input. { 'in‚pût ‚dad‚ə }

input equipment [COMPUT SCI] **1.** The equipment used for transferring data and instructions into an automatic data-processing system. **2.** The equipment by which an operator transcribes original data and instructions to a medium that may be used in an automatic data-processing system. { 'in‚pût i‚kwip‚mənt }

input gap [ELECTR] An interaction gap used to initiate a variation in an electron stream; in a velocity-modulated tube it is in the buncher resonator. { 'in‚pût ‚gap }

input impedance [ELEC] The impedance across the input terminals of a four-terminal network when the output terminals are short-circuited. { 'in‚pût im‚pēd‚əns }

input-limited [COMPUT SCI] Pertaining to a system or operation whose speed or efficiency depends mainly on the speed of input into the machine rather than the speed of the machine itself. { 'in‚pût ‚lim‚əd‚əd }

input magazine [COMPUT SCI] A part of a card-handling device which supplies the cards to the processing portion of the machine. Also known as magazine. { 'in‚pût ‚mag‚ə‚zēn }

input/output [COMPUT SCI] Pertaining to all equipment and activity that transfers information into or out of a computer. Abbreviated I/O. { 'in‚pût 'aut‚pût }

input/output adapter [COMPUT SCI] A circuitry which allows input/output devices to be attached directly to the central processing unit. { 'in‚pût 'aut‚pût ə‚dap‚tər }

input/output area [COMPUT SCI] A portion of computer memory that is reserved for accepting data from input devices and holding data for transfer to output devices. { 'in‚pût 'aut‚pût ‚er‚ē‚ə }

input/output bound [COMPUT SCI] Pertaining to a system or condition in which the time for input and output operation exceeds other operations. Also known as input/output limited. { 'in‚pût 'aut‚pût ‚baûnd }

input/output buffer [COMPUT SCI] An area of a computer memory used to temporarily store data and instructions transferred into and out of a computer, permitting several such transfers to take place simultaneously with processing of data. { 'in‚pût 'aut‚pût ‚bəf‚ər }

input/output channel [COMPUT SCI] The physical link connecting the computer to an input device or to an output device. { 'in‚pût 'aut‚pût ‚chan‚əl }

input/output controller [COMPUT SCI] An independent processor which provides the data paths between input and output devices and main memory. { 'in‚pût 'aut‚pût kən‚trōl‚ər }

input/output control system [COMPUT SCI] A set of flexible routines that supervise the input and output operations of a computer at the detailed machine-language level. Abbreviated IOCS. { 'in‚pût 'aut‚pût kən‚trōl ‚sis‚təm }

input/output control unit [COMPUT SCI] The piece of hardware which controls the operation of one or more of a type of devices such as tape drives or disk drives; this unit is frequently an integral part of the input/output device itself. { 'in‚pût 'aut‚pût kən'trōl ‚yü‚nət }

input/output device See peripheral device. { 'in‚pût 'aut‚pût di‚vīs }

input/output generation [COMPUT SCI] A procedure involved in installing an operating system on a large computer in which addresses and attributes of peripheral equipment under the computer's control are described in a language that can be read by the operating system. Abbreviated IOGEN. { 'in‚pût 'aut‚pût ‚jen‚ə‚rā‚shən }

input/output instruction [COMPUT SCI] An instruction in a computer program that causes transfer of data between peripheral devices and main memory, and enables the central processing unit to control the peripheral devices connected to it. { 'in‚pût 'aut‚pût in‚strək‚shən }

input/output interrupt [COMPUT SCI] A technique by which the central processor needs only initiate an input/output operation and then handle other matters, while other units within the system carry out the rest of the operation. { 'in‚pût 'aut‚pût 'int‚ə‚rəpt }

input/output interrupt identification [COMPUT SCI] The ascertainment of the device and channel taking part in the transfer of information into or out of a computer that causes a particular input/output interrupt, and of the status of the device and channel. { 'in‚pût 'aut‚pût 'int‚ə‚rəpt ī‚dent‚ə‚fə‚kā‚shən }

input/output interrupt indicator [COMPUT SCI] A device which registers an input/output interrupt associated with a particular input/output channel; it can be used in input/output interrupt identification. { 'in‚pût 'aut‚pût 'int‚ə‚rəpt ‚in‚də‚kād‚ər }

input/output library [COMPUT SCI] A set of programs which take over the job from the programmer of creating the required instructions to access the various peripheral devices. Also known as input/output routines. { 'in‚pût 'aut‚pût ‚lī‚brer‚ē }

input/output limited See input/output bound. { 'in‚pût 'aut‚pût ‚lim‚əd‚əd }

input/output order [COMPUT SCI] A procedure of transferring data between main memory and peripheral devices which is assigned to and performed by an input/output controller. { 'in‚pût 'aut‚pût ‚ór‚dər }

input/output processor [COMPUT SCI] A hardware device or software processor whose sole function is to handle input and output operations. { 'in‚pût 'aut‚pût ‚prä‚ses‚ər }

input/output referencing [COMPUT SCI] The use of symbolic names in a computer program to indicate data on input/output devices, the actual devices allocated to the program being determined when the program is executed. { 'in‚pût 'aut‚pût ‚ref‚rən‚siŋ }

input/output register [COMPUT SCI] Computer register that provides the transfer of information from inputs to the central computer, or from it to output equipment. { 'in,pút 'aút,pút ,rėj·ə·stər }

input/output relation [SYS ENG] The relation between two vectors whose components are the inputs (excitations, stimuli) of a system and the outputs (responses) respectively. { 'in,pút 'aút ,pút ri,lā·shən }

input/output routines See input/output library. { 'in,pút 'aút,pút rü,tēnz }

input/output statement [COMPUT SCI] A statement in a computer language that summons data or stores data in a peripheral device. { 'in,pút 'aút,pút ,stāt·mənt }

input/output switching [COMPUT SCI] A technique in which a number of channels can connect input and output devices to a central processing unit; each device may be assigned to any available channel, so that several different channels may service a particular device during the execution of a program. { 'in,pút 'aút,pút ,swich·iŋ }

input/output traffic control [COMPUT SCI] The coordination, by both hardware and software facilities, of the actions of a central processing unit and the input, output, and storage devices under its control, in order to permit several input/output devices to operate simultaneously while the central processing unit is processing data. { 'in,pút 'aút,pút 'traf·ik kən,trōl }

input/output wedge [COMPUT SCI] The characteristic shape of a Kiviat graph of a system which is approaching complete input/output boundedness. { 'in,pút 'aút,pút ,wej }

input program See data entry program. { 'in,pút ,prō·grəm }

input record [COMPUT SCI] **1.** A record that is read from an input device into a computer memory during the performance of a program or routine. **2.** A record that has been stored in an input area and is ready to be processed. { 'in,pút ,rek·ərd }

input register [COMPUT SCI] A register that accepts input information from a computer at one speed and supplies the information to the central processing unit at another speed, usually much greater. { 'in,pút ,rej·ə·stər }

input resistance See transistor input resistance. { 'in,pút ri,zis·təns }

input resonator See buncher resonator. { 'in,pút 'rēz·ən,ād·ər }

input routine [COMPUT SCI] A routine which controls the loading and reading of programs, data, and other routines into a computer for storage or immediate use. Also known as loading routine. { 'in,pút rü,tēn }

input section [COMPUT SCI] The part of a program which controls the reading of data into a computer memory from external devices. { 'in,pút ,sek·shən }

input station [COMPUT SCI] A terminal in an in-plant communications system at which data can be entered into the system directly as events take place, enabling files to be immediately updated. { 'in,pút ,stā·shən }

input storage See input area. { 'in,pút ,stor·ij }

inquiry [COMPUT SCI] A request for the retrieval of a particular item or set of items from storage. { in'kwī·ə·rē }

inquiry and communications system [COMPUT SCI] A computer system in which centralized records are maintained with data transmitted to and from terminals at remote locations or in an in-plant system, and which immediately responds to inquiries from remote terminals. { in'kwī·ə·rē ən kə,myü·nə'kā·shənz ,sis·təm }

inquiry and subscriber display [COMPUT SCI] An inquiry display unit that is distant from its computer and communicates with it over wire lines. { in'kwī·ə·rē ən səb'skrīb·ər di,splā }

inquiry display terminal [COMPUT SCI] A cathode-ray-tube terminal which allows the user to query the computer through a keyboard, the answer appearing on the screen. { in'kwī·ə·rē di'splā ,tər·mən·əl }

inquiry station [COMPUT SCI] A remote terminal from which an inquiry may be sent to a computer over wire lines. { in'kwī·ə·rē ,stā·shən }

inquiry unit [COMPUT SCI] Any terminal which enables a user to query a computer and get a hard-copy answer. { in'kwī·ə·rē ,yü·nət }

inscribe [COMPUT SCI] To rewrite data on a document in a form which can be read by an optical or magnetic ink character recognition machine. { in'skrīb }

insensitive time See dead time. { in'sen·sə·tiv ,tīm }

insertion gain [ELECTR] The ratio of the power delivered to a part of the system following insertion of an amplifier, to the power delivered to that same part before insertion of the amplifier; usually expressed in decibels. { in'sər·shən ,gān }

insertion loss [ELECTR] The loss in load power due to the insertion of a component or device at some point in a transmission system; generally expressed as the ratio in decibels of the power received at the load before insertion of the apparatus, to the power received at the load after insertion. { in'sər·shən ,lós }

insertion switch [COMPUT SCI] Process by which information is inserted into the computer by an operator who manually operates switches. { in'sər·shən ,swich }

instability [CONT SYS] A condition of a control system in which excessive positive feedback causes persistent, unwanted oscillations in the output of the system. { ,in·stə'bil·əd·ē }

installation processing control [COMPUT SCI] A system that automatically schedules the processing of jobs by a computer installation, in order to minimize waiting time and time taken to prepare equipment for operation. { ,in·stə'lā·shən 'präs,es·iŋ kən,trōl }

installation specification [COMPUT SCI] The criteria defined by a computer manufacturer for specifying correct physical installation. { ,in·stə'lā·shən ,spes·ə·fə,kā·shən }

installation tape number [COMPUT SCI] A number that is permanently assigned to a reel of magnetic tape to identify it.   { ‚in·stə'lā·shən ¦tāp ‚nəm·bər }

installed capacity |ELEC| The maximum runoff of a hydroelectric facility that can be constantly maintained and utilized by equipment.   { in'stöld kə'pas·əd·ē }

install program |COMPUT SCI| A computer program that adapts a software package for use on a particular computer system.   { in'stöl ‚prō·grəm }

instance variable |COMPUT SCI| The data in an object of an object-oriented program.   { 'in·stəns ‚ver·ē·ə·bəl }

instantaneous automatic gain control |ELECTR| Portion of a radar system that automatically adjusts the gain of an amplifier for each pulse to obtain a substantially constant output-pulse peak amplitude with different input-pulse peak amplitudes; the circuit is fast enough to act during the time a pulse is passing through the amplifier.   { ¦in·stən¦tā·nē·əs ‚öd·ə¦mad·ik 'gān kən‚trōl }

instantaneous bandwidth |COMMUN| A term in radar for the modulation bandwidth of the signal being used, so as not to be confused with the radar's agility bandwidth, or its approved operating bandwidth.   { ¦in·stən¦tā·nē·əs 'band ‚width }

instantaneous carrying current |ELEC| The maximum value of current which a switch, circuit breaker, or similar apparatus can carry instantaneously.   { ¦in·stən¦tā·nē·əs 'kar·ē·iŋ ‚kə·rənt }

instantaneous companding |ELECTR| Companding in which the effective gain variations are made in response to instantaneous values of the signal wave.   { ¦in·stən¦tā·nē·əs kəm'pan·diŋ }

instantaneous description |COMPUT SCI| For a Turing machine, the set of machine conditions at a given point in the computation, including the contents of the tape, the position of the read-write head on the tape, and the internal state of the machine.   { ¦in·stən¦tā·nē·əs di'skrip·shən }

instantaneous effects |COMMUN| Impairment of telephone or telegraph transmission caused by instantaneous changes in phase or amplitude of the wave in a transmission line.   { ¦in·stən ¦tā·nē·əs i'feks }

instantaneous frequency |COMMUN| The time rate of change of the angle of an angle-modulated wave.   { ¦in·stən¦tā·nē·əs 'frē·kwən·sē }

instantaneous frequency-indicating receiver |ELECTR| A radio receiver with a digital, cathode-ray, or other display that shows the frequency of a signal at the instant it is picked up anywhere in the band covered by the receiver.   { ¦in·stən ¦tā·nē·əs ¦frē·kwən·sē ‚in·də‚kād·iŋ ri'sē·vər }

instantaneous power |ELEC| The product of the instantaneous voltage and the instantaneous current for a circuit or component.   { ¦in·stən ¦tā·nē·əs 'pau̇·ər }

instantaneous recording |ENG ACOUS| A recording intended for direct reproduction without further processing.   { ¦in·stən¦tā·nē·əs ri'körd·iŋ }

instantaneous sample |COMMUN| One of a sequence of instantaneous values of a wave taken at regular intervals.   { ¦in·stən¦tā·nē·əs 'sam·pəl }

instantiation |COMPUT SCI| **1.** An external declaration or a reference to another program or subprogram in the Ada programming language.   **2.** The deduction of omitted values in a set of data from the known values.   **3.** The creation of an object of a specific class in an object-oriented program.   { in‚stan·chē'ā·shən }

instant-on switch |ELECTR| A switch that applies a reduced filament voltage to all tubes in a television receiver continuously, so the picture appears almost instantaneously after the set is turned on.   { 'in·stənt 'ön ‚swich }

instant replay   *See* video replay.   { 'in·stənt 'rē ‚plā }

instruction |COMPUT SCI| A pattern of digits which signifies to a computer that a particular operation is to be performed and which may also indicate the operands (or the locations of operands) to be operated on.   { in'strək·shən }

instruction address |COMPUT SCI| The address of the storage location in which a given instruction is stored.   { in'strək·shən ə'dres }

instruction address register |COMPUT SCI| A special storage location, forming part of the program controller, in which addresses of instructions are stored in order to control their sequential retrieval from memory during the execution of a program.   { in'strək·shən ə¦dres 'rej·ə·stər }

instruction area |COMPUT SCI| A section of storage used for storing program instructions.   { in'strək·shən ‚er·ē·ə }

instruction code |COMPUT SCI| That part of an instruction which distinguishes it from all other instructions and specifies the action to be performed.   { in'strək·shən ‚köd }

instruction constant |COMPUT SCI| A dummy instruction of the type K = I, where K is irrelevant to the program.   { in'strək·shən ‚kän·stənt }

instruction counter |COMPUT SCI| A counter that indicates the location of the next computer instruction to be interpreted. Also known as location counter; program counter; sequence counter.   { in'strək·shən ‚kau̇nt·ər }

instruction cycle |COMPUT SCI| The steps involved in carrying out an instruction.   { in'strək·shən ‚sī·kəl }

instruction format |COMPUT SCI| Any rule which assigns various functions to the various digits of an instruction.   { in'strək·shən ‚för‚mat }

instruction length |COMPUT SCI| The number of bits or bytes (eight bits per byte) which defines an instruction.   { in'strək·shən ‚leŋkth }

instruction lookahead |COMPUT SCI| A technique for speeding up the process of fetching and decoding instructions in a computer program, and of computing addresses of required operands and fetching them, in which the control unit fetches any unexecuted instructions on hand, to the extent this is feasible. Also known as fetch ahead.   { in'strək·shən 'lu̇k·ə‚hed }

instruction mix |COMPUT SCI| The proportion of various types of instructions that appear in a particular computer program, or in a benchmark representing a class of programs. { in'strək·shən ˌmiks }

instruction modification [COMPUT SCI] A change, carried out by the program, in an instruction so that, upon being repeated, this instruction will perform a different operation. { in'strək·shən ˌmäd·ə·fə'kā·shən }

instruction pointer [COMPUT SCI] **1.** A component of a task descriptor that designates the next instruction to be executed by the task. **2.** An element of the control component of the stack model of block structure execution, which points to the current instruction. { in'strək·shən ˌpȯint·ər }

instruction register [COMPUT SCI] A hardware element that receives and holds an instruction as it is extracted from memory; the register either contains or is connected to circuits that interpret the instruction (or discover its meaning). Also known as current-instruction register. { in'strək·shən ˌrej·ə·stər }

instruction repertory See instruction set. { in'strək·shən ˌrep·ə₁tȯr·ē }

instruction set [COMPUT SCI] **1.** The set of instructions which a computing or data-processing system is capable of performing. **2.** The set of instructions which an automatic coding system assembles. **3.** Also known as instruction repertory. { in'strək·shən ˌset }

instruction time [COMPUT SCI] The time required to carry out an instruction having a specified number of addresses in a particular computer. { in'strək·shən ˌtīm }

instruction transfer [COMPUT SCI] An instruction which transfers control to one or another subprogram, depending upon the value of some operation. { in'strək·shən ˌtranz·fər }

instruction word [COMPUT SCI] A computer word containing an instruction rather than data. Also known as coding line. { in'strək·shən ˌwərd }

instrumentation amplifier [ELECTR] An amplifier that accepts a voltage signal as an input and produces a linearly scaled version of this signal at the output; it is a closed-loop fixed-gain amplifier, usually differential, and has high input impedance, low drift, and high common-mode rejection over a wide range of frequencies. { ˌin·strə·men'tā·shən 'am·plə₁fī·ər }

instrumented range [ELECTR] In radar, the distance from which an echo might be received just before the subsequent pulse is transmitted; sometimes called unambiguous range. Echoes from targets at greater ranges are liable to be associated with the subsequent pulse and thought to be coming from a target at a much shorter range. { 'in·strə｜mən·təd ˌrānj }

instrument multiplier [ELEC] A highly accurate resistor used in series with a voltmeter to extend its voltage range. Also known as voltage multiplier; voltage-range multiplier. { 'in·strə·mənt 'məl·tə₁plī·ər }

instrument resistor [ELEC] A high-accuracy, four-terminal resistor used to bypass the major portion of currents around the low-current elements of an instrument, such as a direct-current ammeter. { 'in·strə·mənt ri₁zis·tər }

instrument shunt [ELEC] A resistor designed to be connected in parallel with an ammeter to extend its current range. { 'in·strə·mənt ˌshənt }

instrument transformer [ELEC] A transformer that transfers primary current, voltage, or phase values to the secondary circuit with sufficient accuracy to permit connecting an instrument to the secondary rather than the primary; used so only low currents or low voltages are brought to the instrument. { 'in·strə·mənt tranz₁fȯr·mər }

instrument-type relay [ELEC] A relay constructed like a meter, with one adjustable contact mounted on the scale and the other contact mounted on the pointer. Also known as contact-making meter. { 'in·strə·mənt ˌtīp 'rē₁lā }

insulated [ELEC] Separated from other conducting surfaces by a nonconducting material. { 'in·sə₁lād·əd }

insulated conductor [ELEC] A conductor surrounded by insulation to prevent current leakage or short circuits. Also known as insulated wire. { 'in·sə₁lād·əd kən'dək·tər }

insulated-gate bipolar transistor [ELECTR] A power semiconductor device that combines low forward voltage drop, gate-controlled turnoff, and high switching speed. It structurally resembles a vertically diffused MOSFET, featuring a double diffusion of a *p*-type region and an *n*-type region, but differs from the MOSFET in the use of a *p*+ substrate layer (in the case of an *n*-channel device) for the drain. The effect is to change the transistor into a bipolar device, as this *p*-type region injects holes into the *n*-type drift region. Abbreviated IGBT. { ｜in·sə₁lād·əd₁gāt bī₁pō·lər tran'zis·tər }

insulated-gate field-effect transistor See metal oxide semiconductor field-effect transistor. { 'in·sə₁lād·əd ｜gāt ｜fēld i₁fekt tran'zis·tər }

insulated-return power system [ELEC] A system for distributing electric power to trains or other vehicles, in which both the outgoing and return conductors are insulated, in contrast to a track-return system. { 'in·sə₁lād·əd ri｜tərn 'pau̇·ər ˌsis·təm }

insulated-substrate monolithic circuit [ELECTR] Integrated circuit which may be either an all-diffused device or a compatible structure so constructed that the components within the silicon substrate are insulated from one another by a layer of silicon dioxide, instead of reverse-biased *pn* junctions used for isolation in other techniques. { 'in·sə₁lād·əd ｜səb₁strāt ｜män·ə ｜lith·ik 'sər·kət }

insulated wire *See* insulated conductor. { 'in·sə ˌlād·əd 'wīr }

insulating strength [ELEC] Measure of the ability of an insulating material to withstand electric stress without breakdown; it is defined as the voltage per unit thickness necessary to initiate a disruptive discharge; usually measured in volts per centimeter. { 'in·sə,lād·iŋ ,streŋkth }

insulation [ELEC] A material having high electrical resistivity and therefore suitable for separating adjacent conductors in an electric circuit or preventing possible future contact between conductors. Also known as electrical insulation. { ,in·sə'lā·shən }

insulation coordination [ELEC] Steps taken to ensure that electric equipment is not damaged by overvoltages and that flashovers are localized in regions where no damage results from them. { ,in·sə'lā·shən kō'ȯrd·ən,ā·shən }

insulation protection [ELEC] Use of devices to protect insulators of power transmission lines from damage by heavy arcs. { ,in·sə'lā·shən prə ¦tek·shən }

insulation resistance [ELEC] The electrical resistance between two conductors separated by an insulating material. { ,in·sə'lā·shən ri¦zis·təns }

insulator [ELEC] A device having high electrical resistance and used for supporting or separating conductors to prevent undesired flow of current from them to other objects. Also known as electrical insulator. { 'in·sə,lād·ər }

insulator arc-over [ELEC] Discharge of power current in the form of an arc, following a surface discharge over an insulator. { 'in·sə,lād·ər 'ärk ,ō·vər }

integer constant [COMPUT SCI] A constant that uses the values 0, 1, . . . , 9 with no decimal point in FORTRAN. { 'int·ə·jər ¦kän·stənt }

integer data type [COMPUT SCI] A scalar date type which is used to represent whole numbers, that is, values without fractional parts. { 'int·ə·jər 'dad·ə ,tāp }

integer programming [SYS ENG] A series of procedures used in operations research to find maxima or minima of a function subject to one or more constraints, including one which requires that the values of some or all of the variables be whole numbers. { 'int·ə·jər 'prō,gram·iŋ }

integer variable [COMPUT SCI] A variable in FORTRAN whose first character is normally I, J, K, L, M, or N. { 'int·ə·jər 'ver·ē·ə·bəl }

integral action [CONT SYS] A control action in which the rate of change of the correcting force is proportional to the deviation. { 'int·ə·grəl ,ak·shən }

integral compensation [CONT SYS] Use of a compensator whose output changes at a rate proportional to its input. { 'int·ə·grəl ,käm·pən'sā·shən }

integral control [CONT SYS] Use of a control system in which the control signal changes at a rate proportional to the error signal. { 'int·ə·grəl kən,trōl }

integral discriminator [ELECTR] A circuit which accepts only pulses greater than a certain minimum height. { 'int·ə·grəl di'skrim·ə,nād·ər }

integral-mode controller [CONT SYS] A controller which produces a control signal proportional to the integral of the error signal. { 'int·ə·grəl ¦mōd kən,trōl·ər }

integral modem [COMMUN] A modem built directly into a machine to enable it to communicate over a telephone line. Also known as internal modem. { 'int·ə·grəl ¦mō,dem }

integral network [CONT SYS] A compensating network which produces high gain at low input frequencies and low gain at high frequencies, and is therefore useful in achieving low steady-state errors. Also known as lagging network; lag network. { 'int·ə·grəl 'net,wərk }

integral quantum Hall effect [ELECTR] The version of the quantum Hall effect in which the Hall resistance becomes precisely equal to $(h/e^2)/n$, where h is Planck's constant, e is the electronic charge, and n is an integer. { 'int·ə·grəl ¦kwän·təm 'hȯl i,fekt }

integral square error [CONT SYS] A measure of system performance formed by integrating the square of the system error over a fixed interval of time; this performance measure and its generalizations are frequently used in linear optimal control and estimation theory. { 'int·ə·grəl ¦skwer ,er·ər }

integrated circuit [ELECTR] An interconnected array of active and passive elements integrated with a single semiconductor substrate or deposited on the substrate by a continuous series of compatible processes, and capable of performing at least one complete electronic circuit function. Abbreviated IC. Also known as integrated semiconductor. { 'int·ə,grād·əd 'sər·kət }

integrated-circuit capacitor [ELECTR] A capacitor that can be produced in a silicon substrate by conventional semiconductor production processes. { 'int·ə,grād·əd ¦sərkət kə'pas·əd·ər }

integrated-circuit filter [ELECTR] An electronic filter implemented as an integrated circuit, rather than by interconnecting discrete electrical components. { ¦int·i,grād·əd 'sər·kət ,fil·tər }

integrated-circuit memory *See* semiconductor memory. { 'int·ə,grād·əd ¦sər·kət 'mem·rē }

integrated-circuit resistor [ELECTR] A resistor that can be produced in or on an integrated-circuit substrate as part of the manufacturing process. { 'int·ə,grād·əd ¦sərkət ri'zis·tər }

integrated communications system [COMMUN] Communications system on either a unilateral or joint basis in which a message can be filed at any communications center in that system and be delivered to the addressee by any other appropriate communications center in that system without reprocessing enroute. { 'int·ə,grād·əd kə,myü·nə'kā·shənz ,sis·təm }

integrated console [COMMUN] Computer control console that is capable of controlling the operation of the switching center equipment of an integrated communications system. { 'int·ə ˌgrād·əd 'kän·ˌsōl }

integrated data dictionary [COMPUT SCI] An index or catalog of information about a data base that is physically and logically integrated into the data base. { 'in·tə·ˌgrād·əd 'dad·ə 'dik·shə·ˌner·ē }

integrated data processing [COMPUT SCI] Data processing that has been organized and carried out as a whole, so that intermediate outputs may serve as inputs for subsequent processing with no human copying required. Abbreviated IDP. { 'in·tə·ˌgrād·əd ˈdad·ə 'prä·ses·iŋ }

integrated data retrieval system [COMPUT SCI] A section of a data-processing system that provides facilities for simultaneous operation of several video-data interrogations in a single line and performs required communications with the rest of the system; it provides storage and retrieval of both data subsystems and files and standard formats for data representation. { 'in·tə·ˌgrād·əd ˈdad·ə ri'trē·vəl ˌsis·təm }

integrated electronics [ELECTR] A generic term for that portion of electronic art and technology in which the interdependence of material, device, circuit, and system-design consideration is especially significant; more specifically, that portion of the art dealing with integrated circuits. { 'in·tə,grād·əd i,lek'trän·iks }

integrated information processing [COMPUT SCI] System of computers and peripheral systems arranged and coordinated to work concurrently or independently on different problems at the same time. { 'in·tə,grād·əd ,in·fər'mā·shən ,prä·ses·iŋ }

integrated information system [COMMUN] An expansion of a basic information system achieved through system design of an improved or broader capability by functionally or technically relating two or more information systems, or by incorporating a portion of the functional or technical elements of one information system into another. { 'in·tə,grād·əd ,in·fər'mā·shən ,sis·təm }

integrated injection logic [ELECTR] Integrated-circuit logic that uses a simple and compact bipolar transistor gate structure which makes possible large-scale integration on silicon for logic arrays, memories, watch circuits, and various other analog and digital applications. Abbreviated I²L. Also known as merged-transistor logic. { 'in·tə ˌgrād·əd in'jek·shən 'läj·ik }

integrated optics [OPTICS] A thin-film device containing tiny lenses, prisms, and switches to transmit very thin laser beams, and serving the same purposes as the manipulation of electrons in thin-film devices of integrated electronics. { 'in·tə,grād·əd 'äp·tiks }

integrated semiconductor See integrated circuit. { 'in·tə,grād·əd ˈsem·i·kən‚dək·tər }

integrated services digital network [COMMUN] A public end-to-end digital communications network which has capabilities of signaling, switching, and transport over facilities such as wire pairs, coaxial cables, optical fibers, microwave radio, and satellites, and which supports a wide range of services, such as voice, data, video, facsimile, and music, over standard interfaces. Abbreviated ISDN. { 'in·tə,grād·əd ˈsər·vəs·əz 'dij·əd·əl 'net,wərk }

integrated software [COMPUT SCI] **1.** A collection of computer programs designed to work together to handle an application, either by passing data from one to another or as components of a single system. **2.** A collection of computer programs that work as a unit with a unified command structure to handle several applications, such as word processing, spread sheets, data-base management, graphics, and data communications. { 'in·tə,grād·əd 'sóft,wer }

integrated thermionic circuit [ELECTR] A circuit fabricated from subminiature thin-film metal patterns on two planar substrates separated by an evacuated space about 1 millimeter in thickness to form miniature planar, thermionic, vacuum-tube devices, with densities approaching those of conventional integrated circuits. { 'in·tə,grād·əd ˈthər·mē‚än·ik 'sər·kət }

integrating amplifier [ELECTR] An operational amplifier with a shunt capacitor such that mathematically the waveform at the output is the integral (usually over time) of the input. { 'int·ə ˌgrād·iŋ 'am·plə,fī·ər }

integrating detector [ELECTR] A frequency-modulation detector in which a frequency-modulated wave is converted to an intermediate-frequency pulse-rate modulated wave, from which the original modulating signal can be recovered by use of an integrator. { 'int·ə ˌgrād·iŋ di'tek·tər }

integrating filter [ELECTR] A filter in which successive pulses of applied voltage cause cumulative buildup of charge and voltage on an output capacitor. { 'int·ə,grād·iŋ 'fil·tər }

integrating frequency meter [ENG] An instrument that measures the total number of cycles through which the alternating voltage of an electric power system has passed in a given period of time, enabling this total to be compared with the number of cycles that would have elapsed if the prescribed frequency had been maintained. Also known as master frequency meter. { 'int·ə,grād·iŋ 'frē·kwən·sē ,mēd·ər }

integrating galvanometer [ENG] A modification of the d'Arsonval galvanometer which measures the integral of current over time; it is designed to be able to measure changes of flux in an exploring coil which last over periods of several minutes. { 'int·ə,grād·iŋ ,gal·və'näm·əd·ər }

integrating meter [ENG] An instrument that totalizes electric energy or some other quantity consumed over a period of time. { 'int·ə,grād·iŋ 'mēd·ər }

integrating network [ELECTR] A circuit or network whose output waveform is the time integral of its input waveform. Also known as integrator. { 'int·ə,grād·iŋ 'net,wərk }

integration [SYS ENG] The arrangement of components in a system so that they function together in an efficient and logical way. { ,int·ə'grā·shən }

integration test [COMPUT SCI] A stage in testing a computer system in which a collection of modules in the system is tested as a group. { ˌint·ə'grā·shən ˌtest }

integrator [ELECTR] **1.** A computer device that approximates the mathematical process of integration. **2.** See integrating network. { 'int·ə‚grād·ər }

integrity [COMPUT SCI] Property of data which can be recovered in the event of its destruction through failure of the recording medium, user carelessness, program malfunction, or other mishap. { in'teg·rəd·ē }

intelligence [COMMUN] Data, information, or messages that are to be transmitted. { in'tel·ə·jəns }

intelligent agent See knowbot. { in¦tel·ə·jənt 'ā·jənt }

intelligent cable [COMMUN] A multiline communications cable that is equipped with a microprocessor to analyze or convert signals. { in'tel·ə·jənt ¦kā·bəl }

intelligent controller [COMPUT SCI] A peripheral control unit whose operation is controlled by a built-in microprocessor. { in'tel·ə·jənt kən'trō·lər }

intelligent database [COMPUT SCI] **1.** A database that can respond to queries in a high-level, interactive language. **2.** A database that can store validation criteria with each item of data, so that all programs entering or updating the data must conform to these criteria. { in'tel·ə·jənt 'dad·ə‚bās }

intelligent machine [COMPUT SCI] A machine that uses sensors to monitor the environment and thereby adjust its actions to accomplish specific tasks in the face of uncertainty and variability. { in'tel·ə·jənt mə'shēn }

intelligent robot [CONT SYS] A robot that functions as an intelligent machine, that is, it can be programmed to take actions or make choices based on input from sensors. { in'tel·ə·jənt 'rō‚bät }

intelligent sensor See smart sensor. { in¦tel·ə·jənt 'sen·sər }

intelligent terminal [COMPUT SCI] A computer input/output device with its own memory and logic circuits which can perform certain operations normally carried out by the computer. Also known as smart terminal. { in'tel·ə·jənt 'ter·mən·əl }

intelligent work station [COMPUT SCI] A work station that has an intelligent terminal to carry out a variety of functions independently. { in'tel·ə·jənt 'wərk ‚stā·shən }

intelligibility [COMMUN] The percentage of speech units understood correctly by a listener in a communications system; customarily used for regular messages where the context aids the listener, in distinction to articulation. Also known as speech intelligibility. { in‚tel·ə·jə'bil·əd·ē }

intelligible crosstalk [COMMUN] Crosstalk which is sufficiently understandable under pertinent circuit and room noise conditions that meaningful information can be obtained by more sensitive listeners. { in'tel·ə·jə·bəl 'kròs‚tók }

Intelsat [COMMUN] A satellite network, formerly under international control, used for global communication by more than 100 countries; the system uses geostationary satellites over the Atlantic, Pacific, and Indian oceans and highly directional antennas at earth stations. Derived from International Telecommunications Satellite. { in'tel‚sat }

intensifier electrode [ELECTR] An electrode used to increase the velocity of electrons in a beam near the end of their trajectory, after deflection of the beam. Also known as post-accelerating electrode; post-deflection accelerating electrode. { in'ten·sə‚fī·ər i¦lek‚trōd }

intensifier image orthicon [ELECTR] An image orthicon combined with an image intensifier that amplifies the electron stream originating at the photocathode before it strikes the target. { in'ten·sə‚fī·ər ¦im·ij 'ór·thə‚kän }

intensity control See brightness control. { in'ten·səd·ē kən‚trōl }

intensity modulation [ELECTR] Modulation of electron beam intensity in a cathode-ray tube in accordance with the magnitude of the received signal. { in'ten·səd·ē ‚mäj·ə'lā·shən }

interaction space [ELECTR] A region of an electron tube in which electrons interact with an alternating electromagnetic field. { ¦in·tə¦rak·shən ‚spās }

interactive graphical input [COMPUT SCI] Information which is delivered to a computer by using hand-held devices, such as writing styli used with electronic tablets and light-pens used with cathode-ray tube displays, to sketch a problem description in an on-line interactive mode in which the computer acts as a drafting assistant with unusual powers, such as converting rough freehand motions of a pen or stylus to accurate picture elements. { ¦in·tə¦rak·tiv ¦graf·ə·kəl 'in ‚pùt }

interactive information system [COMPUT SCI] An information system in which the user communicates with the computing facility through a terminal and receives rapid responses which can be used to prepare the next input. { ¦in·tə¦rak·tiv ‚in·fər'mā·shən ‚sis·təm }

interactive language [COMPUT SCI] A programming language designed to operate in an environment in which the user and computer communicate as transactions are being processed. { ¦in·tə¦rak·tiv 'laŋ·gwij }

interactive processing [COMPUT SCI] Computer processing in which the user can modify the operation appropriately while observing results at critical steps. { ¦in·tə¦rak·tiv 'prä·ses·iŋ }

interactive television [COMMUN] A form of television in which the content is personalized and the viewer can control its various parameters. { ¦in·tə¦rak·tiv 'tel·ə‚vizh·ən }

interactive terminal [COMPUT SCI] A computer terminal designed for two-way communication between operator and computer. { ¦in·tə¦rak·tiv 'ter·mən·əl }

interbase current [ELECTR] The current that flows from one base connection of a junction tetrode transistor to the other, through the base region. { 'in·tər,bäs 'kə·rənt }

interblock [COMPUT SCI] A device or system that prevents one part of a computing system from interfering with another. { 'in·tər,bläk }

interblock gap [COMPUT SCI] A space separating two blocks of data on a magnetic tape. { 'in·tər ,bläk ,gap }

intercarrier channel [COMMUN] A carrier telegraph channel in the available frequency spectrum between carrier telephone channels. { ¦in·tər¦kar·ē·ər ,chan·əl }

intercarrier noise suppression [ELECTR] Means of suppressing the noise resulting from increased gain when a high-gain receiver with automatic volume control is tuned between stations; the suppression circuit automatically blocks the audio-frequency input of the receiver when no signal exists at the second detector. Also known as interstation noise suppression. { ¦in·tər¦kar·ē·ər 'nȯiz sə,presh·ən }

intercarrier sound system [ELECTR] An analog television receiver arrangement in which the television picture carrier and the associated sound carrier are amplified together by the video intermediate-frequency amplifier and passed through the second detector, to give the conventional video signal plus a frequency-modulated sound signal whose center frequency is the 4.5 megahertz difference between the two carrier frequencies. Abbreviated ICS system. { ¦in·tər¦kar·ē·ər 'saȯnd ,sis·təm }

intercept call [COMMUN] In telephone practice, routing of a call placed to a disconnected or nonexisting telephone number, to an operator, or to a machine answering device, or to a tone. { 'in·tər·sept ,ȯl }

interception [COMMUN] Tapping or tuning in to a telephone or radio message not intended for the listener. { ,in·tər'sep·shən }

intercept station [COMMUN] Provides service for subscribers whereby calls to disconnected stations or dead lines are either routed to an intercept operator for explanation, or the calling party receives a distinctive tone that informs the party that he has made such a call has been made. { 'in·tər,sept ,stā·shən }

intercept tape [COMMUN] A tape used for temporary storage of messages for trunk channels and tributary stations that are having equipment or circuit trouble. { 'in·tər,sept ,tāp }

intercept trunk [COMMUN] Trunk to which a call for a vacant number, a changed number, or a line out of order is connected for action by an operator. { 'in·tər,sept ,trəŋk }

interchange [ELEC] The current flowing into or out of a power system which is interconnected with one or more other power systems. { 'in·tər ,chānj }

interchangeability [ENG] The ability to replace the components, parts, or equipment of one manufacturer with those of another, without losing function or suitability. { ,in·tər ,chānj·ə'bil·əd·ē }

interchannel crosstalk [COMMUN] Crosstalk between channels in a multiplex system. { ¦in·tər ¦chan·əl 'krȯs,tȯk }

intercom See intercommunicating system. { 'in·tər,käm }

intercommunicating system [COMMUN] **1.** A telephone system providing direct communication between telephones on the same premises. **2.** A two-way communication system having a microphone and loudspeaker at each station and providing communication within a limited area. **3.** Also known as intercom. { ¦in·tər·kə'myü·nə,kād·iŋ ,sis·təm }

interconnected multiple processor [COMPUT SCI] A collection of computers that are physically separated but linked by communication channels to handle distributed data processing. { ,in·tər·kə'nek·təd 'məl·tə·pəl 'präs,es·ər }

interconnection [ELEC] A link between power systems enabling them to draw on one another's reserves in time of need and to take advantage of energy cost differentials resulting from such factors as load diversity, seasonal conditions, time-zone differences, and shared investment in larger generating units. { ¦in·tər·kə'nek·shən }

interconversion [COMMUN] Changing the representation of information from one code to another, as from six-bit to ASCII. { ¦in·tər·kən'vər· zhən }

interdigital magnetron [ELECTR] Magnetron having axial anode segments around the cathode, alternate segments being connected together at one end, remaining segments connected together at the opposite end. { ,in·tər'dij·əd·əl 'mag·nə,trän }

interdigital structure [ELECTR] A structure in which the length of the region between two electrodes is increased by an interlocking-finger design for metallization of the electrodes. Also known as interdigitated structure. { ,in·tər'dij·əd·əl ,strək·chər }

interdigital transducer [ELECTR] Two interlocking comb-shaped metallic patterns applied to a piezoelectric substrate such as quartz or lithium niobate, used for converting microwave voltages to surface acoustic waves, or vice versa. { ,in·tər'dij·əd·əl tranz'dü·sər }

interdigitated structure See interdigital structure. { ,in·tər'dij·ə,tād·əd ,strək·chər }

interelectrode capacitance [ELECTR] The capacitance between one electrode of an electron tube and the next electrode on the anode side. Also known as direct interelectrode capacitance. { ,in·tər·i'lek,trōd kə'pas·əd·əns }

interelectrode transit time [ELECTR] Time required for an electron to traverse the distance between the two electrodes. { ,in·tər·i'lek,trōd 'tran·zət ,tīm }

interface [COMPUT SCI] **1.** Some form of electronic device that enables one piece of gear to communicate with or control another. **2.** A device linking two otherwise incompatible devices, such as an editing terminal of one manufacturer to typesetter of another. { 'in·tər,fās }

interface adapter |COMMUN| A device that connects a terminal or computer to a network. { 'in·tər,fās ə,dap·tər }

interface card |COMPUT SCI| A card containing circuits that allow a device to interface with other devices. { 'in·tər,fās ,kärd }

interface connection See feedthrough. { 'in·tər ,fās kə,nek·shən }

interface control module |COMPUT SCI| Relocatable modularized compiler allowing for efficient operation and easy maintenance. { 'in·tər,fās kən'trōl ,mäj·ül }

interfacial polarization See space-charge polarization. { 'in·tər,fā·shəl ,pō·lə·rə'zā·shən }

interfacility transfer trunk |COMMUN| Trunk interconnecting switching centers of two different facilities. { ,in·tər·fə'sil·əd·ē 'tranz·fər ,trəŋk }

interference |COMMUN| Any undesired energy that tends to interfere with the reception of desired signals. Also known as electrical interference; radio interference. { ,in·tər'fir·əns }

interference analyzer |ELECTR| An instrument that discloses the frequency and amplitude of unwanted input. { ,in·tər'fir·əns ,an·ə,līz·ər }

interference blanker |ELECTR| Device that permits simultaneous operation of two or more pieces of radio or radar equipment without confusion of intelligence, or that suppresses undesired signals when used with a single receiver. { ,in·tər'fir·əns ,blaŋ·kər }

interference fading |COMMUN| Fading of the signal produced by different wave components traveling slightly different paths in arriving at the receiver (often termed multipath). { ,in·tər'fir·əns ,fād·iŋ }

interference filter |ELECTR| **1.** A filter used to attenuate artificial interference signals entering a receiver through its power line. **2.** A filter used to attenuate unwanted carrier-frequency signals in the tuned circuits of a receiver. |OPTICS| An optical filter in which the wavelengths that are not transmitted are removed by interference phenomena rather then by absorption or scattering. { ,in·tər'fir·əns ,fil·tər }

interference pattern |ELECTR| Pattern produced on a radar display by undesired signals. Also, the vertical coverage pattern of a radar antenna resulting from the interference of the direct-path and earth-reflected path signals. { ,in·tər'fir·əns ,pad·ərn }

interference prediction |ELECTR| Process of estimating the interference level of a particular equipment as a function of its future electromagnetic environment. { ,in·tər'fir·əns prə'dik·shən }

interference reduction |ELECTR| Reduction of interference from such causes as power lines and equipment, radio transmitters, and lightning, usually through the use of electric filters. Also known as interference suppression. { ,in·tər'fir·əns ri'dək·shən }

interference region |COMMUN| That region in space in which interference between wave trains occurs; in microwave propagation, it refers to the region bounded by the ray path and the surface of the earth which is above the radio horizon. { ,in·tər'fir·əns ,rē·jən }

interference rejection |ELECTR| Use of a filter to reject (to bypass to ground) unwanted input. { ,in·tər'fir·əns ri'jek·shən }

interference source suppression |ELECTR| Techniques applied at or near the source to reduce its emission of undesired signals. { ,in·tər'fir·əns 'sórs sə,presh·ən }

interference spectrum |ELECTR| Frequency distribution of the jamming interference in the propagation medium external to the receiver. { ,in·tər'fir·əns ,spek·trəm }

interference suppression See interference reduction. { ,in·tər'fir·əns sə'presh·ən }

interference wave |COMMUN| A radio wave reflected by the lower atmosphere which produces an interference pattern when combined with the direct wave. { ,in·ter'fir·əns ,wāv }

interferometer systems |ELECTR| Method of determining the position of a target in azimuth by using an interferometer to compare the phases of signals at the output terminals of a pair of antennas receiving a common signal from a distant source. { ,in·tə·fə'räm·əd·ər ,sis·təmz }

interfix |COMPUT SCI| A technique for describing relationships of key words in an item or document in a way which prevents crosstalk from causing false retrievals when very specific entries are made. { 'in·tər,fiks }

interior distribution |ELEC| Distribution of electric power within a building or plant. { in'tir·ē·ər ,di·strə'byü·shən }

interior label |COMPUT SCI| A label attached to the data that it identifies. { in'tir·ē·ər 'lā·bəl }

interlace |COMPUT SCI| To assign successive memory location numbers to physically separated locations on a storage tape or magnetic drum of a computer, usually to reduce access time. { in·tər,läs }

interlaced scanning |ELECTR| A scanning process in which the distance from center to center of successively scanned lines is two or more times the nominal line width, so that adjacent lines belong to different fields. Also known as line interlace. { ,in·tər,läst 'skan·iŋ }

interlace operation |COMPUT SCI| System of computer operation where data can be read out or copied into memory without interfering with the other activities of the computer. { ,in·tər,läs ,äp·ə'rā·shən }

interleave |COMPUT SCI| **1.** To alternate parts of one sequence with parts of one or more other sequences in a cyclic fashion such that each sequence retains its identity. **2.** To arrange the members of a sequence of memory addresses in different memory modules of a computer system, in order to reduce the time taken to access the sequence. { ,in·tər'lēv }

interleaved windings |ELEC| An arrangement of winding coils around a transformer core in which the coils are wound in the form of a disk, with a group of disks for the low-voltage windings stacked alternately with a group of disks for the high-voltage windings. { |in·tər¦lēvd 'wīn·diŋz }

interlock |COMPUT SCI| **1.** A mechanism, implemented in hardware or software, to coordinate the activity of two or more processes within a computing system, and to ensure that one process has reached a suitable state such that the other may proceed. **2.** See deadlock. { 'in·tər ‚läk }

interlock relay |ELEC| A relay composed of two or more coils, each with its own armature and associated contacts, so arranged that movement of one armature or the energizing of its coil is dependent on the position of the other armature. { 'in·tər‚läk ‚rē‚lā }

interlock switch |ELEC| A switch designed for mounting on a door, drawer, or cover so that it opens automatically when the door or other part is opened. { 'in·tər‚läk ‚swich }

interlude |COMPUT SCI| A small routine or program which is designed to carry out minor preliminary calculations or housekeeping operations before the main routine begins to operate, and which can usually be overwritten after it has performed its function. { 'in·tər‚lüd }

intermediate control data |COMPUT SCI| Control data at a level which is neither the most nor the least significant, or which is used to sort records into groups that are neither the largest nor the smallest used; for example, if control data are used to specify state, town, and street, then the data specifying town would be intermediate control data. { ‚in·tər'mēd·ē·ət kən'trōl ‚dad·ə }

intermediate distributing frame |ELEC| Frame in a local telephone central office, the primary purpose of which is to cross-connect the subscriber line multiple to the subscriber line circuit; in a private exchange, the intermediate distributing frame is for similar purposes. { ‚in·tər'mēd·ē·ət di'strib·yəd·iŋ ‚frām }

intermediate frequency |ELECTR| In radio or radar heterodyne receivers, that frequency produced by mixing the received signal, presumably at the intended carrier frequency, with a local-oscillator signal offset from the carrier frequency, the intermediate frequency being the difference of the two. Abbreviated IF. { ‚in·tər'mēd·ē·ət 'frē·kwən·sē }

intermediate-frequency amplifier |ELECTR| The section of a superheterodyne receiver that amplifies signals after they have been converted to the fixed intermediate-frequency value by the frequency converter. Abbreviated i-f amplifier. { ‚in·tər'mēd·ē·ət ¦frē·kwən·sē 'am·plə‚fī·ər }

intermediate-frequency jamming |ELECTR| Form of continuous wave jamming that is accomplished by transmitting two continuous wave signals separated by a frequency equal to the center frequency of the radar receiver intermediate-frequency amplifier, expecting the radar's own mixer to produce the obscuring intermediate-frequency signal. { ‚in·tər'mēd·ē·ət ¦frē·kwən·sē 'jam·iŋ }

intermediate-frequency response ratio |ELECTR| In a superheterodyne receiver, the ratio of the intermediate-frequency signal input at the antenna to the desired signal input for identical outputs. Also known as intermediate-interference ratio. { ‚in·tər'mēd·ē·ət ¦frē·kwən·sē ri'späns ‚rä·shō }

intermediate-frequency signal |ELECTR| A modulated or continuous-wave signal whose frequency is the intermediate-frequency value of a superheterodyne receiver and is produced by frequency conversion before demodulation. { ‚in·tər'mēd·ē·ət ¦frē·kwən·sē ‚sig·nəl }

intermediate-frequency stage |ELECTR| One of the stages in the intermediate-frequency amplifier of a superheterodyne receiver. { ‚in·tər'mēd·ē·ət ¦frē·kwən·sē ‚stāj }

intermediate-frequency strip |ELECTR| A receiver subassembly consisting of the intermediate-frequency amplifier stages, installed or replaced as a unit. { ‚in·tər'mēd·ē·ət ¦frē·kwən·sē ‚strip }

intermediate-frequency transformer |ELECTR| The transformer used at the input and output of each intermediate-frequency amplifier stage in a superheterodyne receiver for coupling purposes and to provide selectivity. Abbreviated i-f transformer. { ‚in·tər'mēd·ē·ət ¦frē·kwən·sē tranz'fȯr·mər }

intermediate-infrared radiation |ELECTROMAG| Infrared radiation having a wavelength between about 2.5 micrometers and about 50 micrometers; this range includes most molecular vibrations. Also known as mid-infrared radiation. { ‚in·tər'mēd·ē·ət ¦in·frə‚red ‚rād·ē'ā·shən }

intermediate-interference ratio See intermediate-frequency response ratio. { ‚in·tər 'mēd·ē·ət ‚in·tər'fir·əns ‚rä·shō }

intermediate language level |COMPUT SCI| A computer program that has been converted by a compiler into a form that does not resemble the original program but that still requires further processing by an interpreter at run time before it can be executed. { ‚in·tər'mēd·ē·ət 'laŋ·gwij ‚lev·əl }

intermediate memory storage |COMPUT SCI| An electronic device for holding working figures temporarily until needed and for releasing final figures to the output. { ‚in·tər'mēd·ē·ət 'mem·rē ‚stȯr·ij }

intermediate repeater |ELECTR| Repeater for use in a trunk or line at a point other than an end. { ‚in·tər'mēd·ē·ət ri'pēd·ər }

intermediate result |COMPUT SCI| A quantity or value derived from an operation performed in the course of a program or subroutine which is itself used as an operand in further operations. { ‚in·tər'mēd·ē·ət ri'zəlt }

intermediate storage |COMPUT SCI| The portion of the computer storage facilities that usually

stores information in the processing stage.
{ ,in·tər'mēd·ē·ət 'stór·ij }

intermediate total [COMPUT SCI] A sum that is
produced when there is a change in the value of
control data at a level that is neither the most nor
the least significant. { ,in·tər'mēd·ē·ət ¦tōd·əl }

intermediate trunk distributing frame [ELEC]
A frame which mounts terminal blocks for
connecting linefinders and first selectors.
{ ,in·tər'mēd·ē·ət 'traŋk di'strib·yəd·iŋ ,frām }

intermittent current [ELEC] A unidirectional current that flows and ceases to flow at irregular or
regular intervals. { ¦in·tər¦mit·ənt 'kə·rənt }

intermittent scanning [ELECTR] Scans of an antenna beam at irregular intervals to increase
difficulty of detection by intercept receivers.
{ ¦in·tər¦mit·ənt 'skan·iŋ }

intermodulation [ELECTR] Modulation of the
components of a complex wave by each other,
producing new waves whose frequencies are
equal to the sums and differences of integral
multiples of the component frequencies of the
original complex wave. { ,in·tər,mäj·ə'lā·shən }

intermodulation distortion [ELECTR] Nonlinear
distortion characterized by the appearance of
output frequencies equal to the sums and differences of integral multiples of the input
frequency components; harmonic components
also present in the output are usually not
included as part of the intermodulation
distortion. { ,in·tər,mäj·ə'lā·shən di,stór·shən }

intermodulation interference [ELECTR] Interference that occurs when the signals from two undesired stations differ by exactly the intermediate-
frequency value of a superheterodyne receiver,
and both signals are able to pass through the
preselector due to poor selectivity. { ,in·tər
,mäj·ə'lā·shən ,in·tər'fir·əns }

internal arithmetic [COMPUT SCI] Arithmetic operations carried out in a computer's arithmetic unit
within the central processing unit. { in'tərn·əl
ə'rith·mə,tik }

internal buffer [COMPUT SCI] A portion of a computer's main storage used to temporarily hold
data that is being transferred into and out of main
storage. { in'tərn·əl 'bəf·ər }

internal cache *See* primary cache. { in¦tərn·əl
'kash }

internal clocking [COMPUT SCI] Synchronization
of the electronic circuitry of a device by a
timing clock within the device itself. { in'tərn·əl
'kläk·iŋ }

internal cycle time [COMPUT SCI] The time required to change the information in a single
register of a computer, usually a fraction of the
cycle time of the main memory. Also known as
clock time. { in'tərn·əl 'sī·kəl ,tīm }

internal data transfer [COMPUT SCI] The movement of data between registers in a computer's
central processing unit or between a register and
main storage. { in'tərn·əl 'dad·ə ,tranz·fər }

internal dielectric field *See* dielectric field.
{ in'tərn·əl ,dī·ə'lek·trik 'fēld }

internal hemorrhage [COMPUT SCI] A condition
in which a computer program continues to run

following an error but produces dubious results
and may adversely affect other programs or the
performance of the entire system. { in'tərn·əl
'hem·rij }

internal interrupt [COMPUT SCI] A signal for attention sent to a computer's central processing
unit by another component of the computer.
{ in'tərn·əl 'int·ə,rəpt }

internal label [COMPUT SCI] An identifier providing a name for data that is recorded with the data
in a storage medium. { in'tərn·əl 'lā·bəl }

internal loss *See* loss. { in'tərn·əl 'lòs }

internally stored program [COMPUT SCI] A sequence of instructions stored inside the computer in the same storage facilities as the
computer data, as opposed to external storage
on tape, disk, or drum. { in'tərn·əl·ē ¦stórd
'prō·grəm }

internal memory *See* internal storage. { in'tərn·əl
'mem·rē }

internal modem *See* integral modem. { in'tərn·əl
'mō,dem }

internal resistance [ELEC] The resistance within
a voltage source, such as an electric cell or
generator. { in'tərn·əl ri'zis·təns }

internal schema [COMPUT SCI] The physical configuration of data in a data base. { in'tərn·əl
'skē·mə }

internal sorting [COMPUT SCI] The sorting of a list
of items by a computer in which the entire list
can be brought into the main computer memory
and sorted in memory. { in'tərn·əl 'sórd·iŋ }

internal storage [COMPUT SCI] The total memory
or storage that is accessible automatically to
a computer without human intervention. Also
known as internal memory. { in'tərn·əl 'stór·ij }

internal storage capacity [COMPUT SCI] The quantity of data that can be retained simultaneously in internal storage. { in'tərn·əl 'stór·ij kə
,pas·əd·ē }

internal table [COMPUT SCI] A table or array that
is coded directly into a computer program and
is compiled along with the rest of the program.
{ in'tərn·əl 'tā·bəl }

international ampere [ELEC] The current that,
when flowing through a solution of silver nitrate
in water, deposits silver at a rate of 0.001118
gram per second; it has been superseded by the
ampere as a unit of current, and is equal to
approximately 0.999850 ampere. { ¦in·tər¦nash·
ən·əl 'am,pir }

international broadcasting [COMMUN] Radio
broadcasting for public entertainment between
different countries, on frequency bands
between 5950 and 21,750 kilohertz, assigned by
international agreement. { ¦in·tər¦nash·ən·əl
'bròd,kast·iŋ }

international cable code *See* Morse cable code.
{ ¦in·tər¦nash·ən·əl 'kā·bəl ,kōd }

international call sign [COMMUN] Call sign assigned according to the provisions of the International Telecommunication Union to identify a
radio station; the nationality of the radio station
is identified by the first or the first two characters.
{ ¦in·tər¦nash·ən·əl 'kòl ,sīn }

international code signal

international code signal [COMMUN] Code adopted by many nations for international communications; it uses combinations of letters in lieu of words, phrases, and sentences; the letters are transmitted by the hoisting of international alphabet flags or by transmitting their dot and dash equivalents in the international Morse code. Also known as international signal code. { ¦in·tər¦nash·ən·əl 'kōd ˌsig·nəl }

international control frequency bands [COMMUN] Radio-frequency bands assigned in the United States to links between stations used for international communication and their associated control centers. { ¦in·tər¦nash·ən·əl kən¦trōl 'frē·kwən·sē ˌbanz }

international control station [COMMUN] Fixed station in the fixed public control service associated directly with the international fixed public radio communications service. { ¦in·tər¦nash·ən·əl kən'trōl ˌstā·shən }

international fixed public radio communications service [COMMUN] Fixed service, the stations of which are open to public correspondence; this service is intended to provide radio communications between the United States and its territories and foreign or overseas points. { ¦in·tər¦nash·ən·əl ¦fixt ¦pəb·lik ¦rād·ē·ō kə ˌmyü·nə'kā·shən ˌsər·vəs }

international Morse code See continental code. { ¦in·tər¦nash·ən·əl ¦mors 'kōd }

international ohm [ELEC] A unit of resistance, equal to that of a column of mercury of uniform cross section that has a length of 160.3 centimeters and a mass of 14.4521 grams at the temperature of melting ice; it has been superseded by the ohm, and is equal to 1.00049 ohms. { ¦in·tər¦nash·ən·əl 'ōm }

international radio silence [COMMUN] Three-minute periods of radio silence, on the frequency of 500 kilohertz only, commencing 15 and 45 minutes after each hour, during which all marine radio stations must listen on that frequency for distress signals of ships and aircraft. { ¦in·tər ¦nash·ən·əl ¦rād·ē·ō ¦sī·ləns }

international signal code See international code signal. { ¦in·tər¦nash·ən·əl 'sig·nəl ˌkōd }

international system of electrical units [ELEC] System of electrical units based on agreed fundamental units for the ohm, ampere, centimeter, and second, in use between 1893 and 1947, inclusive; in 1948, the Giorgi, or meter-kilogram-second-absolute system, was adopted for international use. { ¦in·tər¦nash·ən·əl ¦sistəm əv i ¦lek·trə·kəl 'yü·nəts }

international telegraph alphabet See CCIT 2 code. { ¦in·tər¦nash·ən·əl 'tel·ə,graf ,al·fə,bet }

International Telegraphic Consultative Committee code 2 See CCIT 2 code. { ¦in·tər ¦nash·ən·əl ¦tel·ə¦graf·ik kən¦səl·tə·div kə¦mid·ē ¦kōd 'tü }

international volt [ELEC] A unit of potential difference or electromotive force, equal to 1/1.01858 of the electromotive force of a Weston cell at 20°C; it has been superseded by the volt, and is equal to 1.00034 volts. { ¦in·tər¦nash·ən·əl 'vōlt }

internet [COMMUN] A system of local area networks that are joined together by a common communications protocol. { 'in·tər,net }

Internet [COMPUT SCI] A worldwide system of interconnected computer networks, communicating by means of TCP/IP and associated protocols. { 'in·tər,net }

Internet Mail Access Protocol [COMPUT SCI] An Internet standard for directly reading and manipulating e-mail messages stored on remote servers. { ¦in·tər,net ¦māl 'ak,ses ,prōd·ə,kól }

Internet protocol [COMMUN] The set of standards responsible for ensuring that data packets transmitted over the Internet are routed to their intended destinations. Abbreviated IP. { ¦in·tər ,net 'prōd·ə,kól }

Internet telephony [COMMUN] Phone calls routed over the Internet by analog-to-digital conversion of speech signals. { ,in·tər,net tə'lef·ə·nē }

internetting [COMPUT SCI] Connections and communications paths between separate data communications networks that allow transfer of messages. { ¦in·tər¦ned·iŋ }

interoffice trunk [COMMUN] A direct trunk between local central offices in the same exchange. { ¦in·tər'óf·əs 'trəŋk }

interphase reactor [ELEC] A type of current-equalizing reactor that is connected between two parallel silicon controlled rectifier converters and provides balanced system operation when both converters are conducting by acting as an inductive voltage divider. { 'in·tər,fāz rē,ak·tər }

interphase transformer [ELECTR] Autotransformer or a set of mutually coupled reactors used in conjunction with three-phase rectifier transformers to modify current relations in the rectifier system to increase the number of anodes of different phase relations which carry current at any instant. { 'in·tər,fāz tranz,fór· mər }

interphone [COMMUN] An intercommunication system using headphones and microphones for communication between adjoining or nearby studios or offices, or between crew locations on an aircraft, vessel, or tank or other vehicle. Also known as talk-back circuit. { 'in·tər,fōn }

interpolation [MATH] A process used to estimate an intermediate value of one (dependent) variable which is a function of a second (independent) variable when values of the dependent variable corresponding to several discrete values of the independent variable are known. { in ,tər·pə'lā·shən }

interposition trunk [COMMUN] Trunk which connects two positions of a large switchboard so that a line on one position can be connected to a line on another position. { ,in·tər·pə'zish·ən ,trəŋk }

interpreter [COMPUT SCI] **1.** A program that translates and executes each source program statement before proceeding to the next one. Also known as interpretive routine. **2.** See conversational compiler. { in'tər·prəd·ər }

interpretive code See interpretive language. { in'tər·prəd·iv ,kōd }

304

interpretive language |COMPUT SCI| A computer programming language in which each instruction is immediately translated and acted upon by the computer, as opposed to a compiler which decodes a whole program before a single instruction can be executed. Also known as interpretive code. { in'tər·prəd·iv 'laŋ·gwij }

interpretive programming |COMPUT SCI| The writing of computer programs in an interpretive language, which generally uses mnemonic symbols to represent operations and operands and must be translated into machine language by the computer at the time the instructions are to be executed. { in'tər·prəd·iv 'prō,gram·iŋ }

interpretive trace program |COMPUT SCI| An interpretive routine that provides a record of the machine code into which the source program is translated and of the result of each step, or of selected steps, of the program. { in'tər·prəd·iv 'trās ,prō·grəm }

interprocedure metric |COMPUT SCI| A software metric that estimates the complexity of a module or computer program based on the way that the data are used, organized, and allocated in relationship with some other modules. { ,in·tər·prə¦sē·jər 'me·trik }

interprocess communication |COMPUT SCI| The communication between computer programs running concurrently under the control of the same operating system. { ¦in·tər,prä·səs kə,myü·ə'kā·shən }

interrecord gap See record gap. { ¦in·tər'rek·ərd ,gap }

interrogation |COMMUN| The transmission of a radio-frequency pulse, or combination of pulses, intended to trigger a transponder or group of transponders, a racon system, or an IFF system, in order to elicit an electromagnetic reply. Also known as challenging signal. { in,ter·ə'gā·shən }

interrogation suppressed time delay |COMMUN| Overall fixed time delay between transmission of an interrogation and reception of the reply to this interrogation at zero distance. { in,ter·ə'gā·shən sə¦prest 'tīm di,lā }

interrogator |ELECTR| 1. A radar transmitter which sends out a pulse that triggers a transponder; usually combined in a single unit with a responsor, which receives the reply from a transponder and produces an output suitable for actuating a display of some navigational parameter. Also known as challenger; interrogator-transmitter. 2. See interrogator-responsor. { in'ter·ə,gād·ər }

interrogator-responsor |ELECTR| A transmitter and receiver combined, used for sending out pulses to interrogate a radar beacon and for receiving and displaying the resulting replies. Also known as interrogator. { in'ter·ə,gād·ər ri'spän·sər }

interrogator-transmitter See interrogator. { in 'ter·ə,gād·ər tranz'mid·ər }

interrupt |COMPUT SCI| 1. To stop a running program in such a way that it can be resumed at a later time, and in the meanwhile permit some other action to be performed. 2. The action of such a stoppage. { 'int·ə,rəpt }

interrupt-driven system |COMPUT SCI| An operating system in which the interrupt system is the mechanism for reporting all changes in the states of hardware and software resources, and such changes are the events that induce new assignments of these resorces to meet work-load demands. { 'int·ə,rəpt ,driv·ən ,sis·təm }

interrupted continuous wave |COMMUN| A continuous wave that is interrupted at a constant audio-frequency rate high enough to give several interruptions for each keyed code dot. Abbreviated ICW. { 'int·ə,rəp·təd kən¦tin·yə·wəs 'wāv }

interrupted current |ELEC| A current produced by opening and closing at regular intervals a circuit that would otherwise carry a steady current or one that varied continuously with time. { 'int·ə ,rəp·təd 'kə·rənt }

interrupter |ELEC| An electric, electronic, or mechanical device that periodically interrupts the flow of a direct current so as to produce pulses. { 'int·ə,rəp·tər }

interrupter vibrator |ELEC| A mechanical device used to change direct current to alternating current. { 'int·ə,rəp·tər 'vī,brād·ər }

interrupt handler |COMPUT SCI| A section of a computer program or of the operating system that takes control when an interrupt is received and performs the operations required to service the interrupt. { 'int·ə,rəpt ,hand·lər }

interrupting capacity |ELEC| Maximum power in the arc that a circuit breaker or fuse can successfully interrupt without restrike or violent failure; rated in volt-amperes for alternating-current circuits and watts for direct-current circuits. { ,int·ə'rəp·tiŋ kə,pas·əd·ē }

interrupt mask |COMPUT SCI| A technique of suppressing certain interrupts and allowing the control program to handle these masked interrupts at a later time. { 'int·ə,rəpt ,mask }

interrupt mode See hold mode. { 'int·ə,rəpt ,mōd }

interrupt priorities |COMPUT SCI| The sequence of importance assigned to attending to the various interrupts that can occur in a computer system. { 'in·tə,rəpt prī,är·ə·dēz }

interrupt routine |COMPUT SCI| A program that responds to an interrupt by carrying out prescribed actions. { 'int·ə,rəpt rü,tēn }

interrupt signal |COMPUT SCI| A control signal which requests the immediate attention of the central processing unit. { 'int·ə,rəpt ,sig·nəl }

interrupt system |COMPUT SCI| The means of interrupting a program and proceeding with it at a later time; this is accomplished by saving the contents of the program counter and other specific registers, storing them in reserved areas, starting the new instruction sequence, and upon completion, reloading the program counter and registers to return to the original program, and reenabling the interrupt. { 'int·ə,rəpt ,sis·təm }

interrupt trap |COMPUT SCI| A program-controlled technique which either recognizes or ignores an interrupt, depending upon a switch setting. { 'int·ə‚rəpt ‚trap }

interrupt vector |COMPUT SCI| A list comprising the locations of various interrupt handlers. { 'int·ə‚rəpt ‚vek·tər }

intersection data |COMPUT SCI| Data which are meaningful only when associated with the concatenation of two segments. { ‚in·tər'sek·shən ‚dad·ə }

interstage transformer |ELECTR| A transformer used to provide coupling between two stages. { 'in·tər‚stāj tranz‚fór·mər }

interstation noise suppression See intercarrier noise suppression. { ¦in·tər'stā·shən 'nóiz sə ‚presh·ən }

intersymbol interference |COMMUN| In a transmission system, extraneous energy from the signal in one or more keying intervals which tends to interfere with the reception of the signal in another keying interval, or the disturbance which results. { ¦in·tər'sim·bəl ‚in·tər'fir·əns }

intersystem communications |COMPUT SCI| The ability of two or more computer systems to share input, output, and storage devices, and to send messages to each other by means of shared input and output channels or by channels that directly connect central processors. { ¦in·tər¦sis·təm kə ‚myü·nə'kā·shənz }

intertoll trunk |COMMUN| A trunk between toll offices in different telephone exchanges. { 'in·tər ‚tōl 'trəŋk }

interval arithmetic |COMPUT SCI| A method of numeric computation in which each variable is specified as lying within some closed interval, and each arithmetic operation computes an interval containing all values that can result from operating on any numbers selected from the intervals associated with the operands. Also known as range arithmetic. { 'in·tər·vəl ə'rith·mə· tik }

intra-coded picture |COMMUN| A MPEG-2 picture that is coded using information present only in the picture itself and not depending on information from other pictures; provides a mechanism for random access into the compressed video data; employs transform coding of the pel blocks and provides only moderate compression. Also known as I-frame; I-picture. { ¦in·trə 'kōd·əd 'pik·chər }

intranet |COMPUT SCI| A private network, based on Internet protocols, that is accessible only within an organization. Intranets are set up for many purposes, including e-mail, access to corporate databases and documents, and videoconferencing, as well as buying and selling goods and services. { 'in·trə‚net }

intraprocedure metric |COMPUT SCI| A software metric that determines the complexity of a computer program as a function of the relationships of the different modules constituting the program, generally by constructing a flow graph and deriving the complexity from this graph. { ‚in·trə·prə¦sē·jər 'me·trik }

intrinsic-barrier diode |ELECTR| A *pin* diode, in which a thin region of intrinsic material sepa-

rates the *p*-type region and the *n*-type region. { in'trin·sik ¦bar·ē·ər 'dī‚ōd }

intrinsic-barrier transistor |ELECTR| A *pnip* or *npin* transistor, in which a thin region of intrinsic material separates the base and collector. { in'trin·sik ¦bar·ē·ər tran'zis·tər }

intrinsic conductivity |SOLID STATE| The conductivity of a semiconductor or metal in which impurities and structural defects are absent or have a very low concentration. { in'trin·sik ‚kän ‚dək'tiv·əd·ē }

intrinsic contact potential difference |ELEC| True potential difference between two perfectly clean metals in contact. { in'trin·sik ‚kän‚takt pə¦ten·chəl 'dif·ərns }

intrinsic detector |ENG| A semiconductor detector of electromagnetic radiation that utilizes the generation of electron-hole pairs across the semiconductor band gap. { in'trin·sik di'tek· tər }

intrinsic electric strength |ELEC| The extremely high dielectric strength displayed by a substance at low temperatures. { in¦trin·sik i¦lek· trik ‚streŋkth }

intrinsic layer |ELECTR| A layer of semiconductor material whose properties are essentially those of the pure undoped material. { in'trin·sik 'lā·ər }

intrinsic procedure See built-in function. { in 'trin·sik prə'sē·jər }

inverse current |ELECTR| The current resulting from an inverse voltage in a contact rectifier. { 'in‚vərs 'kə·rənt }

inverse direction |ELECTR| The direction in which the electron flow encounters greater resistance in a rectifier, going from the positive to the negative electrode; the opposite of the conducting direction. Also known as reverse direction. { 'in‚vərs də'rek·shən }

inverse electrode current |ELECTR| Current flowing through an electrode in the direction opposite to that for which the tube is designed. { 'in ‚vərs i'lek‚trōd ‚kə·rənt }

inverse feedback See negative feedback. { 'in ‚vərs 'fēd‚bak }

inverse limiter |ELECTR| A transducer, the output of which is constant for input of instantaneous values within a specified range and a linear or other prescribed function of the input for inputs above and below that range. { 'in‚vərs 'lim·əd·ər }

inverse network |ELEC| Two two-terminal networks are said to be inverse when the product of their impedances is independent of frequency within the range of interest. { 'in‚vərs 'net ‚wərk }

inverse neutral telegraph transmission |COMMUN| Form of transmission in which marking signals are zero current intervals and spacing signals are current pulses of either polarity. { 'in ‚vərs ‚nü·trəl 'tel·ə‚graf tranz‚mish·ən }

inverse peak voltage |ELECTR| **1.** The peak value of the voltage that exists across a rectifier tube or x-ray tube during the half cycle in which current does not flow. **2.** The maximum instantaneous voltage value that a rectifier tube or x-ray tube can withstand in the inverse direction (with anode negative) without breaking down and becoming conductive. { 'in,vərs ¦pēk 'vōl·tij }

inverse problem |CONT SYS| The problem of determining, for a given feedback control law, the performance criteria for which it is optimal. { 'in ,vərs 'präb·ləm }

inverse video *See* reverse video. { 'in,vərs ¦vid·ē·ō }

inverse voltage |ELECTR| The voltage that exists across a rectifier tube or x-ray tube during the half cycle in which the anode is negative and current does not normally flow. { 'in,vərs 'vōl·tij }

inversion |COMMUN| The process of scrambling speech for secrecy by beating the voice signal with a fixed, higher audio frequency and using only the difference frequencies. |ELEC| The solution of certain problems in electrostatics through the use of the transformation in Kelvin's inversion theorem. |OPTICS| The formation of an inverted image by an optical system. { in'vər·zhən }

inversion temperature |ENG| The temperature to which one junction of a thermocouple must be raised in order to make the thermoelectric electromotive force in the circuit equal to zero, when the other junction of the thermocouple is held at a constant low temperature. { in'vər·zhən ,tem·prə·chər }

inverted amplifier |ELECTR| A two-tube amplifier in which the control grids are grounded and the input signal is applied between the cathodes; the grid then serves as a shield between the input and output circuits. { in'vərd·əd 'am·plə,fī·ər }

inverted file |COMPUT SCI| **1.** A file, or method of file organization, in which labels indicating the locations of all documents of a given type are placed in a single record. **2.** A file whose usual order has been inverted. { in'vərd·əd 'fīl }

inverted L antenna |ELECTROMAG| An antenna consisting of one or more horizontal wires to which a connection is made by means of a vertical wire at one end. { in'vərd·əd ¦el an,ten·ə }

inverted vee |ELECTROMAG| **1.** A directional antenna consisting of a conductor which has the form of an inverted V, and which is fed at one end and connected to ground through an appropriate termination at the other. **2.** A center-fed horizontal dipole antenna whose arms have ends bent downward 45°. { in'vərd·əd 'vē }

inverter |ELEC| A device for converting direct current into alternating current; it may be electromechanical, as in a vibrator or synchronous inverter, or electronic, as in a thyratron inverter circuit. Also known as dc-to-ac converter; dc-to-ac inverter. |ELECTR| *See* phase inverter. { in'vərd·ər }

inverter circuit *See* NOT circuit. { in'vərd·ər ,sər·kət }

inverting amplifier |ELECTR| Amplifier whose output polarity is reversed as compared to

its input; such an amplifier obtains its negative feedback by a connection from output to input, and with high gain is widely used as an operational amplifier. { in'vərd·iŋ 'am·plə,fī·ər }

inverting function |ELECTR| A logic device that inverts the input signal, so that the output is out of phase with the input. { in'vərd·iŋ ,fəŋk·shən }

inverting parametric device |ELECTR| Parametric device whose operation depends essentially upon three frequencies, a harmonic of the pump frequency and two signal frequencies, of which the higher signal frequency is the difference between the pump harmonic and the lower signal frequency. { in'vərd·iŋ ¦par·ə¦me·trik di'vīs }

inverting terminal |ELECTR| The negative input terminal of an operational amplifier; a positive-going voltage at the inverting terminal gives a negative-going output voltage. { in'vərd·iŋ 'tər·mən·əl }

inward-outward dialing system |COMMUN| Dialing system whereby calls within the local exchange area may be dialed directly to and from base private branch exchange telephone stations without the assistance of the base private branch exchange operator; CENTREX, a service offered by some telephone companies, is a form of inward-outward dialing. { ¦in·wərd ¦aut,wərd 'dī·liŋ ,sis·təm }

I/O *See* input/output.

IOCS *See* input/output control system.

IOGEN *See* input/output generation. { 'ī¦ō,jen }

ion-beam scanning |ELECTR| The process of analyzing the mass spectrum of an ion beam in a mass spectrometer either by changing the electric or magnetic fields of the mass spectrometer or by moving a probe. { 'ī,än ,bēm ,skan·iŋ }

ion burn *See* ion spot. { 'ī,än ,bərn }

ion-exchange electrolyte cell |ELEC| Fuel cell which operates on hydrogen and oxygen in the air, similar to the standard hydrogen-oxygen fuel cell with the exception that the liquid electrolyte is replaced by an ion-exchange membrane; operation is at atmospheric pressure and room temperature. { 'ī,än iks,chānj i'lek·trə,līt ,sel }

ion gage *See* ionization gage. { 'ī,än ,gāj }

ion gun *See* ion source. { 'ī,än ,gən }

ionic focusing *See* gas focusing. { ī'än·ik 'fō·kəs·iŋ }

ionic heated cathode |ELECTR| Hot cathode heated primarily by ionic bombardment of the emitting surface. { ī'än·ik ,hēd·əd 'kath,ōd }

ionization arc-over |ELEC| **1.** Arcing across terminals or contacts due to ionization of the adjacent air or gas. **2.** Arcing across satellite antenna terminals as the satellite passes through the ionized regions of the ionosphere. { ,ī·ə·nə'zā·shən 'är ,kō·vər }

ionization current *See* gas current. { ,ī·ə·nə'zā·shən ¦kə·rənt }

ionization density [ELECTR] The density of ions in a gas. { ,ī·ə·nə'zā·shən |den·səd·ē }

ionization gage [ELECTR] An instrument for measuring low gas densities by ionizing the gas and measuring the ion current. Also known as ion gage; ionization vacuum gage. { ,ī·ə·nə'zā·shən ,gāj }

ionization source See ion source. { ,ī·ə·nə'zā·shən ,sòrs }

ionization time [ELECTR] Of a gas tube, the time interval between the initiation of conditions for and the establishment of conduction at some stated value of tube voltage drop. { ,ī·ə·nə'zā·shən ,tīm }

ionization vacuum gage See ionization gage. { ,ī·ə·nə'zā·shən 'vak·yəm ,gāj }

ion microscope See field-ion microscope. { 'ī,än 'mī·krə,skōp }

ion migration [ELEC] Movement of ions produced in an electrolyte, semiconductor, and so on, by the application of an electric potential between electrodes. { 'ī,än mī'grā·shən }

ionophone [ENG ACOUS] A high-frequency loudspeaker in which the audio-frequency signal modulates the radio-frequency supply to an arc maintained in a quartz tube, and the resulting modulated wave acts directly on ionized air to create sound waves. { ī'än·ə,fōn }

ionosphere [GEOPHYS] That part of the earth's upper atmosphere which is sufficiently ionized by solar ultraviolet radiation so that the concentration of free electrons affects the propagation of radio waves; its base is at about 40 or 50 miles (70 or 80 kilometers) and it extends to an indefinite height. { ī'an·ə,sfir }

ionospheric error [COMMUN] Variation in the character of the ionospheric transmission path or paths used by the radio waves of electronic navigation systems which, if not compensated, will produce an error in the information generated by the system. { ,ī,än·ə'sfir·ik 'er·ər }

ionospheric propagation [COMMUN] Propagation of radio waves over long distances by reflection from the ionosphere, useful at frequencies up to about 25 megahertz. { ,ī,än·ə'sfir·ik ,präp·ə'gā·shən }

ionospheric recorder [ELECTR] A radio device for determining the distribution of virtual height with frequency, and the critical frequencies of the various layers of the ionosphere. { ,ī,än·ə'sfir·ik ri'kòrd·ər }

ionospheric scatter [COMMUN] A form of scatter propagation in which radio waves are scattered by the lower E layer of the ionosphere to permit communication over distances from 600 to 1400 miles (1000 to 2250 kilometers) when using the frequency range of about 25 to 100 megahertz. { ,ī,än·ə'sfir·ik 'skad·ər }

ionospheric wave See sky wave. { ,ī,än·ə'sfir·ik 'wāv }

ion pump [ELECTR] A vacuum pump in which gas molecules are first ionized by electrons that have been generated by a high voltage and are spiraling in a high-intensity magnetic field, and the molecules are then attracted to a cathode, or propelled by electrodes into an auxiliary pump or an ion trap. { 'ī,än ,pəmp }

ion-selective field-effect transistor [ELECTR] A field-effect transistor whose gate electrode is sensitive to certain ions in an electrolyte, so that the gain of the transistor depends on the concentration of these ions. Abbreviated ISFET. { 'ī,än si|lek·tiv 'fēld i,fekt tran'zis·tər }

ion source [ELECTR] A device in which gas ions are produced, focused, accelerated, and emitted as a narrow beam. Also known as ion gun; ionization source. { 'ī,än ,sòrs }

ion spot [ELECTR] Of a cathode-ray tube screen, an area of localized deterioration of luminescence caused by bombardment with negative ions. Also known as ion burn. { 'i,än ,spät }

ion trap [ELECTR] **1.** An arrangement whereby ions in the electron beam of a cathode-ray tube are prevented from bombarding the screen and producing an ion spot, usually employing a magnet to bend the electron beam so that it passes through the tiny aperture of the electron gun, while the heavier ions are less affected by the magnetic field and are trapped inside the gun. **2.** A metal electrode, usually of titanium, into which ions in an ion pump are absorbed. { 'ī,än ,trap }

IP address [COMPUT SCI] A computer's numeric address, such as 128.201.86.290, by which it can be located within a network. { |ī'pē ə,dres }

I-picture See intra-coded picture. { 'ī |pik·chər }

IPL [COMPUT SCI] **1.** Collective term for a series of list-processing languages developed principally by A. Newell, H. A. Simon, and J. C. Shaw. Derived from Information Processing Language. **2.** See initial program load.

IPL button See bootstrap button. { |ī|pē'el 'bət· ən }

IR drop See resistance drop. { |ī|är 'dräp }

IRG See record gap.

iris [ELECTROMAG] A conducting plate mounted across a waveguide to introduce impedance; when only a single mode can be supported, an iris acts substantially as a shunt admittance and may be used for matching the waveguide impedance to that of a load. Also known as diaphragm; waveguide window. { 'ī·rəs }

I²R loss See copper loss. { 'ī,skwerd'är ,lòs }

ISB modulation See independent-sideband modulation. { |ī|es|bē ,mäj·ə'lā·shən }

I-scan See I-display. { 'ī ,skan }

I-scope See I-display. { 'ī ,skōp }

ISDN *See* integrated services digital network.

ISDN modem |ELECTR| A device that converts signals used in a computer to signals that can be transmitted over the integrated services digital network, and vice versa. { ,ī,es,dē,en 'mō,dem }

ISFET *See* ion-selective field-effect transistor. { 'is,fet }

I signal |ELECTR| The in-phase component of the chrominance signal in color television, having a bandwidth of 0 to 1.5 megahertz, and consisting of +0.74(R − Y) and −0.27(B − Y), where Y is the luminance signal, R is the red camera signal, and B is the blue camera signal. { 'ī ,sig·nəl }

isobits |COMPUT SCI| Binary digits having the same value. { 'ī·sə,bits }

isochronous circuits |ELEC| Circuits having the same resonant frequency. { ī'sä·krə·nəs 'sər·kəts }

isochronous communications |COMMUN| Synchronization of a data communications network from timing signals provided by the network itself. { ī'sä·krə·nəs kə,myü·nə'kā·shənz }

isocirculator |ELECTROMAG| A circulator that has an absorber in one of its terminals and thereby acts as an isolator. { ,ī·sō'sər·kyə,lād·ər }

isoelectric |ELEC| Pertaining to a constant electric potential. { ¦ī·sō·i'lek·trik }

isograph |ELECTR| An electronic calculator that ascertains both real and imaginary roots for algebraic equations. { 'ī·sə,graph }

isolate |ELEC| To disconnect a circuit or piece of equipment from an electric supply system. { 'ī·sə,lāt }

isolated camera |ELECTR| **1.** A television camera that views a particular portion of a scene of action and produces a tape which can then be used either immediately for instant replay or for video replay at a later time. **2.** The technique of video replay involving such a camera. { 'ī·sə,lād·əd 'kam·rə }

isolated location |COMPUT SCI| A location in a computer memory which is protected by some hardware device so that it cannot be addressed by a computer program and its contents cannot be accidentally altered. { 'ī·sə,lād·əd lō'kā·shən }

isolating switch |ELEC| A switch intended for isolating an electric circuit from the source of power; it has no interrupting rating and is intended to be operated only after the circuit has been opened by some other means. { 'ī·sə ,lād·iŋ ,swich }

isolation |COMPUT SCI| The ability of a logic circuit having more than one input to ensure that each input signal is not affected by any of the others. { ,ī·sə'lā·shən }

isolation amplifier |ELECTR| An amplifier used to minimize the effects of a following circuit on the preceding circuit. { ,ī·sə'lā·shən 'am·plə,fī·ər }

isolation diode |ELECTR| A diode used in a circuit to allow signals to pass in only one direction. { ,ī·sə'lā·shən 'dī,ōd }

isolation network |ELEC| A network inserted in a circuit or transmission line to prevent interaction

between circuits on each side of the insertion point. { ,ī·sə'lā·shən 'net,wərk }

isolation transformer |ELEC| A transformer inserted in a system to separate one section of the system from undesired influences of other sections. { ,ī·sə'lā·shən tranz'fór·mər }

isolator |ELECTR| A passive attenuator in which the loss in one direction is much greater than that in the opposite direction; a ferrite isolator for waveguides is an example. { 'ī·sə,lād·ər }

isolith |ELECTR| Integrated circuit of components formed on a single silicon slice, but with the various components interconnected by beam leads and with circuit parts isolated by removal of the silicon between them. { 'ī·sə,lith }

isopulse system |COMMUN| In adaptive communications, a pulse coding system wherein the number of information pulses transmitted is indicated by special inserted pulses. { 'ī·sə,pəls ,sis·təm }

isotope lamp |ELECTR| A discharge lamp containing gas of a single isotope and thus producing highly monochromatic light. { 'ī·sə,tōp ,lamp }

isotropic antenna *See* unipole. { ¦ī·sə¦trä·pik an'ten·ə }

isotropic dielectric |ELEC| A dielectric whose polarization always has a direction that is parallel to the applied electric field, and a magnitude which does not depend on the direction of the electric field. { ¦ī·sə¦trä·pik ,dī·ə'lek·trik }

isotropic gain of an antenna *See* absolute gain of an antenna. { ¦ī·sə¦trä·pik 'gān əv ən an'ten·ə }

item |COMPUT SCI| A set of adjacent digits, bits, or characters which is treated as a unit and conveys a single unit of information. { 'īd·əm }

item advance |COMPUT SCI| A technique of efficiently grouping records to optimize the overlap of read, write, and compute times. { 'īd·əm əd ,vans }

item design |COMPUT SCI| The specification of what fields make up an item, the order in which the fields are to be recorded, and the number of characters to be allocated to each field. { 'īd·əm di,zīn }

item size |COMPUT SCI| The length of an item expressed in characters, words, or blocks. { 'īd·əm ,sīz }

iteration process |COMPUT SCI| The process of repeating a sequence of instructions with minor modifications between successive repetitions. { ,īd·ə'rā·shən ,prä·səs }

iterations per second |COMPUT SCI| In computers, the number of approximations per second in iterative division; the number of times an operational cycle can be repeated in 1 second. { ,īd·ə'rā·shənz pər 'sek·ənd }

iterative array |COMPUT SCI| In a computer, an array of a large number of interconnected identical processing modules, used with appropriate driver and control circuits to permit a large number of simultaneous parallel operations. { 'īd·ə,rād·iv ə'rā }

iterative division |COMPUT SCI| In computers, a method of dividing by use of the operations of addition, subtraction, and multiplication; a quotient of specified precision is obtained by a series of successively better approximations. { 'īd·ə,rād·iv di'vizh·ən }

iterative filter [ELECTR] Four-terminal filter that provides iterative impedance. { 'īd·ə,rād·iv 'fil·tər }

iterative impedance [ELECTR] Impedance that, when connected to one pair of terminals of a four-terminal transducer, will cause the same impedance to appear between the other two terminals. { 'īd·ə,rād·iv im'pēd·əns }

iterative routine [COMPUT SCI] A computer program that obtains a result by carrying out a series of operations repetitiously until some specified condition is met. { 'īd·ə,rād·iv rü'tēn }

ITV *See* industrial television.

J

Jablochkoff candle |ELECTR| An early type of arc lamp in which carbons were placed side by side and separated by plaster of paris.   { yə'bläch‚kȯf ‚kand·əl }

jack |ELEC| A connecting device into which a plug can be inserted to make circuit connections; may also have contacts that open or close to perform switching functions when the plug is inserted or removed.   { jak }

jammer |ELECTR| One who, or equipment which, transmits electromagnetic signals to obscure or deceive radio or radar receivers, preventing them from receiving the intended signals clearly.   { 'jam·ər }

jammer finder |ELECTR| Radar which attempts to obtain the range of the target by training a highly directional pencil beam on a jamming source. Also known as burnthrough.   { 'jam·ər ‚fīn·dər }

jamming |ELECTR| Radiation, reradiation, or reflection of electromagnetic waves so as to impair the usefulness of the radio spectrum for military purposes including communication and radar. Also known as active jamming; electronic jamming.   { 'jam·iŋ }

J antenna |ELECTROMAG| Antenna having a configuration resembling a J, consisting of a half-wave antenna end-fed by a parallel-wire quarter-wave section.   { 'jā ant‚en·ə }

jar |ELEC| A unit of capacitance equal to 1000 statfarads, or approximately 1.11265×10^{-9} farad; it is approximately equal to the capacitance of a Leyden jar; this unit is now obsolete.   { jär }

Java |COMPUT SCI| An object-oriented programming language based on C++ that was designed to run in a network such as the Internet; mostly used to write programs, called applets, that can be run on Web pages.   { 'jäv·ə }

JavaScript |COMPUT SCI| A scripting language that is added to standard HTML to create interactive documents.   { 'jäv·ə‚skript }

Java virtual machine |COMPUT SCI| An interpreter that translates Java bytecode into actual machine instructions in real time. Abbreviated JVM.   { 'jäv·ə ‚vər·chə·wəl mə'shēn }

J box See junction box.   { 'jā ‚bäks }

J-display |ELECTR| A radar display format in which range is presented as a reference circle with radial projections from it indicating echo strength; a circular A-display. Also known as J-indicator; J-scan; J-scope.   { 'jā di‚splā }

JFET See junction field-effect transistor.   { 'jā ‚fet }

J-indicator See J-display.   { 'jā ‚in·də‚kād·ər }

jitter |COMMUN| In facsimile, distortion in the received copy caused by momentary errors in synchronism between the scanner and recorder mechanisms; does not include slow errors in synchronism due to instability of the frequency standards used in the facsimile transmitter and recorder. |ELECTR| Small, rapid variations in a waveform due to mechanical vibrations, fluctuations in supply voltages, control-system instability, and other causes.   { 'jid·ər }

jittered pulse recurrence frequency |COMMUN| Random variation of the pulse repetition period; provides a discrimination capability against repeater-type jammers.   { 'jid·ərd ¦pəls ri¦kə·rəns 'frē·kwən‚sē }

J-K flip-flop |ELECTR| A storage stage consisting only of transistors and resistors connected as flip-flops between input and output gates, and working with charge-storage transistors; gives a definite output even when both inputs are 1.   { ¦jā¦kā 'flip‚fläp }

job |COMPUT SCI| A unit of work to be done by the computer; it is a single entity from the standpoint of computer installation management, but may consist of one or more job steps.   { jäb }

job class |COMPUT SCI| The set of jobs on a computer system whose resource requirements (for the central processing unit, memory, and peripheral devices) fall within specified ranges.   { 'jäb ‚klas }

job control block |COMPUT SCI| A group of data containing the execution-control data and the job identification when the job is initiated as a unit of work to the operating system.   { 'jäb kən‚trōl ‚bläk }

job control language See command language.   { 'jäb kən‚trōl ‚laŋ·gwij }

job control statement |COMPUT SCI| Any of the statements used to direct an operating system in its functioning, as contrasted to data, programs, or other information needed to process a job but not intended directly for the operating system itself. Also known as control statement. { 'jäb kən¦trōl ¦stāt·mənt }

job entry system |COMPUT SCI| A part of the operating system of a large computer system that accepts and schedules jobs for execution and controls the printing of output. { 'jäb ¦en·trē ¦sis·təm }

job family See job class. { 'jäb ¦fam·lē }

job flow control |COMPUT SCI| Control over the order in which jobs are handled by a computer in order to use the central processing units and the units under the computer's control as efficiently as possible. { 'jäb 'flō kən¦trōl }

job grade See job class. { 'jäb ¦grād }

job library |COMPUT SCI| A partitioned data set, or a concatenation of partitioned data sets, used as the primary source of object programs (load modules) for a particular job, and more generally, as a source of runnable programs from which all or most of the programs for a given job will be selected. { 'jäb ¦li¦brer·ē }

job management program |COMPUT SCI| A control program in a computer's operating system that initials and schedules jobs. { 'jäb ¦man·ij·mənt ¦prō·grəm }

job mix |COMPUT SCI| The distribution of the jobs handled by a computer system among the various job classes. { 'jäb ¦miks }

job-oriented terminal |COMPUT SCI| A terminal, such as a point-of-sale terminal, at which data taken directly from a source can enter a communication network directly. { ¦jäb ¦ȯr·ē¦ent·əd 'tər·mən·əl }

job processing control |COMPUT SCI| The section of the control program responsible for initiating operations, assigning facilities, and proceeding from one job to the next. { 'jäb ¦prä·ses·iŋ kən¦trōl }

job queue |COMPUT SCI| A set of computer programs that are ready to be executed in a prescribed order. { 'jäb ¦kyü }

job schedule |CONT SYS| A control program that selects from a job queue the next job to be processed. { 'jäb ¦sked·yül }

job stacking |COMPUT SCI| The presentation of jobs to a computer system, each job followed by another. { 'jäb ¦stak·iŋ }

job step |COMPUT SCI| A unit of work in a job stream. { 'jäb ¦step }

job stream |CONT SYS| A collection of jobs in a job queue. { 'jäb ¦strēm }

job swapping |COMPUT SCI| Temporary suspension of job processing by a computer so that higher-priority jobs can be handled. { 'jäb ¦swäp·iŋ }

jogging |ELEC| Quickly repeated opening and closing of a circuit to produce small movements

of the driven machine. Also known as inching. { 'jäg·iŋ }

Johnson noise See thermal noise. { 'jän·sən ¦nȯiz }

join |COMPUT SCI| A portion of a robotic control program that directs an activity to resume after it has been interrupted. { ¦jȯin }

joint |ELEC| A juncture of two wires or other conductive paths for current. |ENG| The surface at which two or more mechanical or structural components are united. { ¦jȯint }

Joint Photographic Experts Group |COMPUT SCI| An international group that sets standards for continuous-tone image (still and video) coding. { ¦jȯint ¦fōd·ə¦graf·iks 'ek¦spərts ¦grüp }

joint pole |ELEC| Pole used in common by two or more utility companies. { ¦jȯint 'pōl }

joint space |CONT SYS| The space defined by a vector whose components are the translational and angular displacements of each joint of a robotic link. { ¦jȯint ¦spās }

Joshi effect |ELECTR| The change in the current passing through a gas or vapor when the gas or vapor is irradiated with visible light. { 'jō·shē i ¦fekt }

Joule heat |ELEC| The heat which is evolved when current flows through a medium having electrical resistance, as given by Joule's law. { 'jül ¦hēt }

Joule's law |ELEC| The law that when electricity flows through a substance, the rate of evolution of heat in watts equals the resistance of the substance in ohms times the square of the current in amperes. { 'jülz ¦lȯ }

journaling |COMPUT SCI| Recording processes or transactions for backup or accounting purposes. { 'jər·nəl·iŋ }

joystick |ENG| A two-axis displacement control operated by a lever or ball, for XY positioning of a device or an electron beam. { 'jȯi¦stik }

JPEG |ENG| Graphics file format for compressed still images, particularly photographic images found on the World Wide Web; developed by the Joint Photographic Experts Group. { 'jā¦peg }

J-scan See J-display. { 'jā ¦skan }

J-scope See J-display. { 'jā ¦skōp }

jump |COMPUT SCI| A transfer of control which terminates one sequence of instructions and begins another sequence at a different location. Also known as branch; transfer. { ¦jəmp }

jumper |ELEC| A short length of conductor used to close a circuit between two electircal terminals. { 'jəm·pər }

jumping trace routine |COMPUT SCI| A trace routine which is primarily concerned with providing a record of jump instructions in order to show the sequence of program steps that the computer followed. { 'jəm·piŋ ¦trās rü¦tēn }

jump phenomenon [CONT SYS] A phenomenon occurring in a nonlinear system subjected to a sinusoidal input at constant frequency, in which the value of the amplitude of the forced oscillation can jump upward or downward as the input amplitude is varied through either of two fixed values, and the graph of the forced amplitude versus the input amplitude follows a hysteresis loop. { 'jəmp fə,näm·ə·nən }

jump resonance [CONT SYS] A jump discontinuity occurring in the frequency response of a nonlinear closed-loop control system with saturation in the loop. { 'jəmp ,rez·ən·əns }

jump vector [COMPUT SCI] A list of entry-point addresses for various sections of a computer program; used by the program to branch to a section that performs a desired function. Also known as vector; vector table. { 'jəmp ,vek·tər }

junction [ELEC] *See* major node. [ELECTR] A region of transition between two different semiconducting regions in a semiconductor device, such as a *pn* junction, or between a metal and a semiconductor. [ELECTROMAG] A fitting used to join a branch waveguide at an angle to a main waveguide, as in a tee junction. Also known as waveguide junction. { 'jəŋk·shən }

junction box [ENG] A protective enclosure into which wires or cables are led and connected to form joints. Also known as J box. { 'jəŋk·shən ,bäks }

junction capacitance *See* barrier capacitance. { 'jəŋk·shən kə'pas·əd·əns }

junction capacitor [ELECTR] An integrated-circuit capacitor that uses the capacitance of a reverse-biased *pn* junction. { 'jəŋk·shən kə'pas·əd·ər }

junction diode [ELECTR] A semiconductor diode in which the rectifying characteristics occur at an alloy, diffused, electrochemical, or grown junction between *n*-type and *p*-type semiconductor materials. Also known as junction rectifier. { 'jəŋk·shən ¦dī,ōd }

junction field-effect transistor [ELECTR] A field-effect transistor in which there is normally a channel of relatively low-conductivity semiconductor joining the source and drain, and this channel is reduced and eventually cut off by junction depletion regions, reducing the conductivity, when a voltage is applied between the gate electrodes. Abbreviated JFET. { 'jəŋk·shən 'fēld i,fekt tran¦zis·tər }

junction filter [ELECTR] A combination of a high-pass and a low-pass filter that is used to separate frequency bands for transmission over separate paths. { 'jəŋk·shən ,fil·tər }

junction isolation [ELECTR] Electrical isolation of a component on an integrated circuit by surrounding it with a region of a conductivity type that forms a junction, and reverse-biasing the junction so it has extremely high resistance. { 'jəŋk·shən ,ī·sə'lā·shən }

junction loss [COMMUN] In telephone circuits, that part of the repetition equivalent assignable to interaction effects arising at trunk terminals. { 'jəŋk·shən ,lós }

junction phenomena [ELECTR] Phenomena which occur at the boundary between two semiconductor materials, or a semiconductor and a metal, such as the existence of an electrostatic potential in the absence of current flow, and large injection currents which may arise when external voltages are applied across the junction in one direction. { 'jəŋk·shən fə,näm·ə·nə }

junction point *See* branch point. { 'jəŋk·shən ,póint }

junction pole [ELEC] Pole at the end of a transposition section of an open-wire line or the pole common to two adjacent transposition sections. { 'jəŋk·shən ,pōl }

junction rectifier *See* junction diode. { 'jəŋk·shən ¦rek·tə,fī·ər }

junction station [ELECTR] Microwave relay station that joins a microwave radio leg or legs to the main or through route. { 'jəŋk·shən ,stā·shən }

junction transistor [ELECTR] A transistor in which emitter and collector barriers are formed between semiconductor regions of opposite conductivity type. { 'jəŋk·shən tran¦zis·tər }

junction transposition [ELEC] Transposition located at the junction pole between two transposition sections of an open-wire line. { 'jəŋk·shən ,tranz·pə'zish·ən }

junctor [ELEC] In crossbar systems, a circuit extending between frames of a switching unit and terminating in a switching device on each frame. { 'jəŋk·tər }

justify [COMPUT SCI] To shift data so that they assume a particular position relative to one or more reference points, lines, or marks in a storage medium. { 'jəs·tə,fī }

JVM *See* Java virtual machine.

K

k   *See* kilobit.

K   *See* cathode; kilobyte.

kA   *See* kiloampere.

Ka band   [COMMUN] A band of frequencies extending from 33 to 36 gigahertz, corresponding to wavelengths of 9.09 to 8.34 millimeters.   { kā'ā ,band }

Kalman filter   [CONT SYS] A linear system in which the mean squared error between the desired output and the actual output is minimized when the input is a random signal generated by white noise.   { 'kal·mən ,fil·tər }

Kanji   [COMPUT SCI] A set of Chinese characters that are employed by users of the Chinese language to code information in computer programs and on visual displays.   { 'kän·jē }

Karnaugh map   [ELECTR] A truth table that has been rearranged to show a geometrical pattern of functional relationships for gating configurations; with this map, essential gating requirements can be recognized in their simplest form.   { 'kär·nò ,map }

Karp circuit   [ELECTR] A slow-wave circuit used at millimeter wavelengths for backward-wave oscillators.   { 'kärp ,sər·kət }

K band   [COMMUN] A band of radio frequencies extending from 10,900 to 36,000 megahertz, corresponding to wavelengths of 2.75 to 0.834 centimeters.   { 'kā ,band }

K-band single-access service   [COMMUN] A service provided by the Tracking and Data Relay Satellite System, with return-link data rates up to 300 and 800 megabits per second for the Ku and Ka bands, respectively, and forward-link data at 25 megabits per second in both bands. Abbreviated KSA.   { ¦kā,band ¦siŋ·gəl 'ak ,ses ,sər·vəs }

kbit   *See* kilobit.   { 'kā,bit }

Kbit   *See* kilobit.   { 'kā,bit }

kbyte   *See* kilobyte.   { 'kā,bīt }

Kbyte   *See* kilobyte.   { 'kā,bīt }

KDD   *See* knowledge discovery in databases.

K-display   [ELECTR] A radar display format in which an echo signal appears on side-by-side A-displays, as from a two-beam tracking radar antenna, with equal amplitudes indicating no pointing error. Also known as K-indicator; K-scan; K-scope.   { 'kā di,splā }

keep-alive circuit   [ELECTR] A circuit used with a transmit-receive (TR) tube or anti-TR tube to produce residual ionization for the purpose of reducing the initiation time of the main discharge.   { ¦kēp ə'līv ,sər·kət }

keep-alive electrode   *See* ignitor.   { ¦kēp ə'līv i'lek ,trōd }

kelvin   [ELEC] A name formerly given to the kilowatt-hour. Also known as thermal volt.   { 'kel·vən }

Kelvin bridge   [ELEC] A specialized version of the Wheatstone bridge network designed to eliminate, or greatly reduce, the effect of lead and contact resistance, and thus permit accurate measurement of low resistance. Also known as double bridge; Kelvin network; Thomson bridge.   { 'kel·vən ,brij }

Kelvin guard-ring capacitor   [ELEC] A capacitor with parallel circular plates, one of which has a guard ring separated from the plate by a narrow gap; it is used as a standard, whose capacitance can be accurately calculated from its dimensions.   { 'kel·vən 'gärd ,riŋ kə,pas·əd·ər }

Kelvin network   *See* Kelvin bridge.   { 'kel·vən ,net ,wərk }

Kelvin replenisher   [ELEC] A simple electrostatic generator in which curved metal plates attached to an insulating arm rotate between larger curved plates, and the contacts of the smaller plates with wipers connecting them to the larger plates and to each other result in the accumulation of charge on the smaller plates, energy being supplied by the rotation of the arm.   { ¦kel·vən ri'plen·əsh·ər }

Kelvin skin effect   *See* skin effect.   { 'kel·vən 'skin i,fekt }

Kendall effect   [COMMUN] A spurious pattern or other distortion in a facsimile record caused by unwanted modulation products arising from the transmission of a carrier signal; occurs principally when the width of one side band is greater than half the facsimile carrier frequency.   { 'kend·əl i ,fekt }

kenotron |ELEC| A high-vacuum diode designed to serve as a rectifier in appliances requiring high voltage and low current.   { 'ken·ə,trän }

kernel |COMPUT SCI| **1.** A computer program that must be modified before it can be used on a particular computer.   **2.** The programs that form the most essential part of a computer's operating system.   { 'kərn·əl }

Kerr cell |OPTICS| A glass cell containing a dielectric liquid that exhibits the Kerr effect, such as nitrobenzene, in which is inserted the two plates of a capacitor, used to observe the Kerr effect on light passing through the cell.   { 'kər ,sel }

Kerr effect   *See* electrooptical Kerr effect.   { 'kər i,fekt }

key |COMMUN| A telephone term for an on-off switch in the subscriber loop, either at a manual switchboard or in the telephone set. |COMPUT SCI| A data item that serves to uniquely identify a data record. |ELEC| **1.** A hand-operated switch used for transmitting code signals. Also known as signaling key.   **2.** A special lever-type switch used for opening or closing a circuit only as long as the handle is depressed. Also known as switching key.   { kē }

key access |COMPUT SCI| Locating data in a file by using the value of a key.   { 'kē ,ak,ses }

key auto-key cipher |COMMUN| A stream cipher in which the cryptographic bit stream generated at a given time is determined by the cryptographic bit stream generated at earlier times.   { 'kē 'όd·ō ,kē ,sī·fər }

keyboard |ENG| A set of keys or control levers having a systematic arrangement and used to operate a machine or other piece of equipment such as a typewriter, typesetter, processing unit of a computer, or piano.   { 'kē,bόrd }

keyboard enhancer |COMPUT SCI| Software that expands the functions of a computer keyboard by allowing the user to implement functions or enter predefined segments of text with a single keystroke. Also known as keyboard processor.   { 'kē,bόrd in,han·sər }

keyboard entry |COMPUT SCI| A piece of information fed manually into a computing system by means of a set of keys, such as a typewriter.   { 'kē ,bόrd 'en·trē }

keyboard inquiry |COMPUT SCI| A question asked a computer concerning the status of a program being run, or concerning the value achieved by a specific variable, by means of a console typewriter.   { 'kē,bόrd ¦in·kwə·rē }

keyboard lockout |COMPUT SCI| An arrangement for preventing transmission from a particular keyboard while other transmissions are taking place on the same circuit.   { 'kē,bόrd 'läk, aút }

keyboard lockup |COMPUT SCI| A condition in which entries typed on a keyboard are ignored by a terminal.   { 'kē,bόrd 'läk,əp }

keyboard mapping |COMPUT SCI| The process of assigning the meaning of keys on a computer keyboard.   { 'kē,bόrd ,map·iŋ }

keyboard printer |COMPUT SCI| A computer input device that includes a keyboard and a printer that prints the keyed-in data and often also prints computer output information.   { 'kē,bόrd 'print·ər }

keyboard processor |COMPUT SCI| **1.** The circuitry in a computer keyboard that converts keystrokes into the appropriate character codes.   **2.** *See* keyboard enhancer.   { 'kē,bόrd ,prä,ses·ər }

keyboard send/receive |ELECTR| A manual teleprinter that can transmit or receive. Abbreviated KSR. Also known as keyboard teleprinter.   { 'kē,bόrd ¦send·ri'sēv }

keyboard teleprinter   *See* keyboard send/receive.   { 'kē,bόrd 'tel·ə,print·ər }

keyboard template |COMPUT SCI| A card that is placed adjacent to the function keys of a computer keyboard and identifies their use for a particular software environment.   { 'kē,bόrd ,tem· plət }

key cabinet |ELECTR| A case, installed on a customer's premises, to permit different lines to the control office to be connected to various telephone stations; it has signals to indicate originating calls and busy lines.   { 'kē ,kab· ə·nət }

key change |COMPUT SCI| The occurrence, in a file of records which have been sorted according to their keys and are being read into a computer, of a record whose key differs from that of its immediate predecessor.   { 'kē ,chānj }

key compression |COMPUT SCI| A technique used to reduce the number of bits contained in a key.   { 'kē kəm,presh·ən }

key-disk machine |COMPUT SCI| A keyboard machine used to record data directly on a magnetic disk.   { 'kē ,disk mə,shēn }

keyed clamp |ELECTR| Clamping circuit in which the time of clamping is determined by a control signal.   { 'kēd 'klamp }

keyed clamp circuit |ELECTR| A clamp circuit in which the time of clamping is controlled by separate voltage or current sources, rather than by the signal itself. Also known as synchronous clamp circuit.   { 'kēd 'klamp ,sər·kət }

keyed sequential access method |COMPUT SCI| A method for locating data in a file either directly, by using the value of a key within a particular record, or sequentially, according to the values of the keys in all the records of the file. Abbreviated KSAM.   { 'kēd si'kwen·chəl 'ak,ses ,meth·əd }

key entry |COMPUT SCI| The entering of data into a computer by means of a keyboard.   { 'kē ,en·trē }

keyer |ELECTR| Device which changes the output of a transmitter from one condition to another according to the intelligence to be transmitted.   { 'kē·ər }

keyer adapter |ELECTR| Device which detects a modulated signal and produces the modulating frequency as a direct-current signal of varying amplitude.   { 'kē·ər ə,dap·tər }

key field |COMPUT SCI| A field in a segment or record that holds the value of a key to that record.   { 'kē ,fēld }

keying |COMMUN| The forming of signals by modulating a direct currect or other carrier between discrete values of some characteristic. { 'kē·iŋ }

keying error rate |COMMUN| The ratio of the number of characters incorrectly transmitted to the total number of characters in a message. { 'kē·iŋ 'er·ər ,rāt }

keying frequency |COMMUN| In facsimile, the maximum number of times a second that a black-line signal occurs when scanning the subject copy. { 'kē·iŋ ,frē·kwən·sē }

keying interval |COMMUN| In a periodically keyed transmission system, one of the set of intervals starting from a change in state and equal in length to the shortest time between changes of state. { 'kē·iŋ ,int·ər·vəl }

keying sequence |COMMUN| A sequence of letters or numbers that enciphers or deciphers a polyalphabetic substitution cipher character by character. { 'kē·iŋ ,sē·kwəns }

keying wave *See* marking wave. { 'kē·iŋ ,wāv }

keyless ringing |COMMUN| Form of machine ringing on a manual telephone switchboard which is started automatically by the insertion of the calling plug into the jack of the called line. { 'kē·ləs 'riŋ·iŋ }

keylock switch |ELEC| A switch that can be operated only by inserting and turning a key such as that used in ordinary locks. { 'kē,läk ,swich }

keypad |COMPUT SCI| A cluster of special-purpose keys to one side of the regular typing keys on a terminal keyboard. { 'kē,pad }

key pulse |COMMUN| System of signaling where numbered keys are depressed instead of using a dial. { 'kē ,pəls }

key punch |COMPUT SCI| A keyboard-actuated device that punches holes in a card; it may be a hand-feed punch or an automatic feed punch. { 'kē ,pənch }

keystone distortion |COMMUN| Distortion produced by scanning in a rectilinear manner, with constant-amplitude sawtooth waves, a plane target area which is not normal to the average direction of the beam. { 'kē,stōn di'stȯr·shən }

keystoning |ELECTR| Producing a keystone-shaped (wider at the top than at the bottom, or vice versa) scanning pattern resulting from an off-axis condition between an image-projection device and the display surface. { 'kē,stōn·iŋ }

keyswitch |COMPUT SCI| A switch that is operated by depressing a key on the keyboard of a data entry terminal. { 'kē,swich }

key telephone system |COMMUN| A telephone system consisting of phones with several keys, connecting cables, and relay switching apparatus, which does not need a special operator to handle incoming or outgoing calls and which generally permits users to select one of several possible lines and to hold calls. { 'kē 'tel·ə,fōn ,sis·təm }

key telephone unit |COMMUN| A small mounting plate with relays which performs pickup and hold switching functions in a key telephone system. { 'kē 'tel·ə,fōn ,yü·nət }

key-to-disk system |COMPUT SCI| A data-entry system in which information entered on several keyboards is collected on different sections of a magnetic disk, and the data are extracted from the disk when complete, and are copied onto a magnetic tape or another disk for further processing on the main computer. { 'kē tə 'disk ,sis·təm }

key-to-tape system |COMPUT SCI| A data-entry system, predecessor to the modern key-to-disk system, consisting of several keyboards connected to a central controlling unit, typically a minicomputer, which collected information from each keyboard and then directed it to a magnetic tape. { 'kē tə 'tāp ,sis·təm }

key transformation |COMPUT SCI| A function that assigns integer values to keys. { 'kē ,tranz·fər'mā·shən }

key value |COMPUT SCI| The actual characters contained in a key. { 'kē 'val·yü }

keyword |COMPUT SCI| A group of letters and numbers in a specific order that has special significance in a computer system. { 'kē,wərd }

keyword-in-context index |COMPUT SCI| A computer-generated listing of titles of documents, produced on a line printer, with the keywords lined up vertically in a fixed position within the title and arranged in alphabetical order. Abbreviated KWIC index. { 'kē,wərd in 'kän ,tekst ,in,deks }

keyword-out-of-context index |COMPUT SCI| A computer-generated listing of document titles with their keywords listed separately, arranged in the alphabetical order of the keywords. Abbreviated KWOC index. { 'kē,wərd aút əv 'kän,tekst ,in,deks }

keyword parameter |COMPUT SCI| A parameter whose significance is indicated by a keyword, usually with an equal sign linking the two. { 'kē ,wərd pə'ram·əd·ər }

keyword search |COMPUT SCI| A method of filing and locating information through the use of keywords that describe the content of records. { 'kē,wərd ,sərch }

keyword spotting |ENG ACOUS| An approach to task-oriented speech understanding through detecting a limited number of keywords that would most likely express the intent of a speaker, rather than attempting to recognize every word in an utterance. { 'kē,wərd ,spät·iŋ }

KHN filter *See* state-variable filter. { ¦kā¦äch'en ,fil·tər }

kidney joint |ELECTROMAG| Flexible joint, or air-gap coupling, used in the waveguide of certain radars and located near the transmitting-receiving position. { 'kid·nē ,jȯint }

killer circuit |ELECTR| Vacuum tube or tubes and associated circuits in which are generated the blanking pulses used to temporarily disable a radar set. { 'kil·ər ,sər·kət }

killer pulse |ELECTR| Blanking pulse generated by a killer circuit. { 'kil·ər ,pəls }

killer stage *See* color killer circuit. { 'kil·ər ,stāj }

kiloampere |ELEC| A metric unit of current flow equal to 1000 amperes. Abbreviated kA. { 'ki·lō'am,pir }

kilobit [COMPUT SCI] A unit of information content equal to 1024 bits. Abbreviated kbit; Kbit. Symbolized k. { 'kil·ə,bit }

kilobyte [COMPUT SCI] A unit of information content equal to 1024 bytes. Abbreviated kbyte; Kbyte. Symbolized K. { 'kil·ə,bīt }

kilohm [ELEC] A unit of electrical resistance equal to 1000 ohms. Abbreviated KΩ; kohm. { 'kil,ōm }

kilomegacycle See gigahertz. { ˌkil·ə'meg·ə,sī·kəl }

kilomegahertz See gigahertz. { ˌkil·ə'meg·ə,hərtz }

kilovar [ELEC] A unit equal to 1000 volt-amperes reactive. Abbreviated kvar. { 'kil·ə,vär }

kilovolt [ELEC] A unit of potential difference equal to 1000 volts. Abbreviated kV. { 'kil·ə ,vōlt }

kilovolt-ampere [ELEC] A unit of apparent power in an alternating-current circuit, equal to 1000 volt-amperes. Abbreviated kVA. { 'kil·ə,vōlt 'am ,pir }

kilovoltmeter [ELEC] A voltmeter which measures potential differences on the order of several kilovolts. { ˌkil·ə,vōlt,mēd·ər }

kilovolts peak [ELECTR] The peak voltage applied to an x-ray tube, expressed in kilovolts. Abbreviated kVp. { 'kil·ə,vōlts ˌpēk }

kilowatt-hour [ELEC] A unit of energy or work equal to 1000 watt-hours. Abbreviated kWh; kW-hr. Also known as Board of Trade Unit. { 'kil·ə ,wät ,aúr }

K-indicator See K-display. { 'kā ,in·də,kād·ər }

kinescope See picture tube. { 'kin·ə,skōp }

Kingdon trap [ELEC] A thin charged wire for confining charged particles; ions are attracted toward the wire, but their angular momentum causes them to spiral around the wire in trajectories that have a low probability of hitting the wire. { 'kin·dən trap }

Kirchhoff's current law [ELEC] The law that at any given instant the sum of the instantaneous values of all the currents flowing toward a point is equal to the sum of instantaneous values of all the currents flowing away from the point. Also known as Kirchhoff's first law. { 'kərk,hōfs 'kə·rənt ,lȯ }

Kirchhoff's first law See Kirchhoff's current law. { 'kərk,hōfs 'fərst ,lȯ }

Kirchhoff's law [ELEC] Either of the two fundamental laws dealing with the relation of currents at a junction and voltages around closed loops in an electric network; they are known as Kirchhoff's current law and Kirchhoff's voltage law. { 'kərk ,hōfs ,lȯ }

Kirchhoff's second law See Kirchhoff's voltage law. { 'kərk,hōfs 'sek·ənd ,lȯ }

Kirchhoff's voltage law [ELEC] The law that at each instant of time the algebraic sum of the voltage rises around a closed loop in a network is equal to the algebraic sum of the voltage drops, both being taken in the same direction around the loop. Also known as Kirchhoff's second law. { 'kərk,hōfs 'vōl·tij ,lȯ }

Kiviat graph [COMPUT SCI] A circular diagram used in computer performance evaluation, in which variables are plotted on axes of the circle with 0% at the center of the circle and 100% at the circumference, and variables which are "good" and "bad" as they approach 100% are plotted on alternate axes. { 'kiv·ē·ət ,graf }

klaxon [ENG ACOUS] A diaphragm horn sometimes operated by hand. { 'klak·sən }

kludge [COMPUT SCI] A poorly designed data-processing system composed of ill-fitting mismatched components. { klüj }

klystron [ELECTR] A type of beam power tube, used often in radar and other microwave applications, in which the beam of electrons passes through radio-frequency resonant cavities, or variations, to effect the interaction between the electrons and the signal being amplified or produced. { 'klī,strän }

klystron generator [ELECTR] Klystron tube used as a generator, with its second cavity or catcher directly feeding waves into a waveguide. { 'klī ,strän ˌjen·ə,rād·ər }

klystron oscillator See velocity-modulated oscillator. { 'klī,strän ˌäs·ə,lād·ər }

klystron repeater [ELECTR] Klystron tube operated as an amplifier and inserted directly in a waveguide in such a way that incoming waves velocity-modulate the electron stream emitted from a heated cathode; a second cavity converts the energy of the electron clusters into waves of the original type but of greatly increased amplitude and feeds them into the outgoing guide. { 'klī,strän riˌpēd·ər }

knee frequency See break frequency. { 'nē ,frē·kwən·sē }

knife switch [ELEC] An electric switch consisting of a metal blade hinged at one end to a stationary jaw, so that the blade can be pushed over to make contact between spring clips. { 'nīf ,swich }

Knill-Laflamme bound |COMPUT SCI| In the theory of quantum computation, a necessary condition for an algorithm that encodes N logical qubits into N' carrier qubits (with N' larger than N) to correct any error on any M carrier qubits; namely, that N' be equal to or larger than 4M + N. { kə ¦nil lə¦fläm ¦baünd }

knob-and-tube wiring |ELEC| An electric wiring method used for light and power circuits that uses open insulated wiring on solid insulators; now obsolete and illegal in most countries. { ¦näb ən 'tüb ,wir·iŋ }

knowbot |COMPUT SCI| A program which, when given a request, searches and retrieves information on the Internet. Also known as intelligent agent; knowledge robot. { 'nō,bät }

knowledge base |COMPUT SCI| A collection of facts, assumptions, beliefs, and heuristics that are used in combination with a database to achieve desired results, such as a diagnosis, an interpretation, or a solution to a problem. { 'näl·ij ,bās }

knowledge-based system |COMPUT SCI| A computer system whose usefulness derives primarily from a data base containing human knowledge in a computerized format. { 'näl·ij ,bāst ,sis·təm }

knowledge discovery in databases |COMPUT SCI| The process of identifying valid, novel, potentially useful, and ultimately under-standable structure in data. Abbreviated KDD. { ¦näl·ij di¦skəv·ə·rē in 'dad·ə,bās·əs }

knowledge engineer |COMPUT SCI| An individual who constructs the knowledge base of an expert system. { 'näl·ij ,en·jə,nir }

knowledge robot *See* knowbot. { 'näl·ij ¦rō,bät }

known-good die |ELECTR| An unpackaged, fully tested integrated circuit chip. { ,nōn ¦güd 'dī }

kohm *See* kilohm. { ¦kā¦ōm }

krypton lamp |ELEC| An arc lamp filled with krypton; one type pierces fog for 1000 feet (300 meters) or more and is used to light airplane runways at night. { 'krip·tän ,lamp }

KSAM *See* keyed sequential access method. { 'kā ,sam }

KSA service *See* K-band single-access service. { ¦kā¦es'ā ,sər·vəs }

K-scan *See* K-display. { 'kā ,skan }

K-scope *See* K-display. { 'kā ,skōp }

KSR *See* keyboard send/receive.

Ku band |COMMUN| A band of frequencies extending from 15.35 to 17.25 gigahertz, corresponding to wavelengths of 1.95 to 1.74 centimeters. { 'kyü ,band *or* ¦kā¦yü ,band }

Ku-band fixed satellite service |COMMUN| Satellite communication at and near the Ku band, with the uplink frequency in bands from 12.75 to 13.25 gigahertz and 14.0 to 14.5 gigahertz and the downlink frequency in a band from 10.7 to 11.7 gigahertz. { 'kyü band ,fikst 'sad·əl,īt sər·vis }

Kundt effect |OPTICS| **1.** The occurrence of a very large magnetic rotation when polarized light passes through very thin films of pure ferromagnetic materials. **2.** *See* Faraday effect. { 'künt i,fekt }

kV *See* kilovolt.

kVA *See* kilovolt-ampere.

kvar *See* kilovar. { 'kā,vär }

kVp *See* kilovolts peak.

kWh *See* kilowatt-hour.

kW-hr *See* kilowatt-hour.

L

label |COMPUT SCI| A data item that serves to identify a data record (much in the same way as a key is used), or a symbolic name used in a program to mark the location of a particular instruction or routine. { 'lā·bəl }

label alignment |COMPUT SCI| The manner in which text is aligned in the cells of a particular spreadsheet. { 'lā·bəl ə,līn·mənt }

label constant See location constant. { 'lā·bəl ,kän·stənt }

label data type |COMPUT SCI| A scalar data type that refers to locations in the computer program. { 'lā·bəl 'dad·ə ,tīp }

label record |COMPUT SCI| A tape record containing information concerning the file on that tape, such as format, record length, and block size. { 'lā·bəl ,re·kərd }

labile oscillator |ELECTR| An oscillator whose frequency is controlled from a remote location by wire or radio. { 'lā,bīl 'äs·ə,lād·ər }

labor grade See job class. { 'lā·bər ,grād }

lacing |ELEC| Tying insulated wires together to support each other and form a single neat cable, with separately laced branches. { 'lās·iŋ }

ladder attenuator |ELECTR| A type of ladder network designed to introduce a desired, adjustable loss when working between two resistive impedances, one of which has a shunt arm that may be connected to any of various switch points along the ladder. { 'lad·ər ə'ten·yə,wād·ər }

ladder diagram |CONT SYS| A diagram used to program a programmable controller, in which power flows through a network of relay contacts arranged in horizontal rows between two vertical rails on the side of the diagram containing the symbolic power. { 'lad·ər ,dī·ə ,gram }

ladder network |ELECTR| A network composed of a sequence of H, L, T, or pi networks connected in tandem; chiefly used as an electric filter. Also known as series-shunt network. { 'lad·ər 'net ,wərk }

laddic |ELECTR| Multiaperture magnetic structure resembling a ladder, used to perform logic functions; operation is based on a flux change in the shortest available path when adjacent rungs of the ladder are initially magnetized with opposite polarity. { 'lad·ik }

lag |ELECTR| A persistence of the electric charge image in a camera tube for a small number of frames. { lag }

lagging coil |ELEC| A small coil used to compensate for the lagging current in the voltage coil of an alternating-current watthour meter. { 'lag·iŋ ,kȯil }

lagging current |ELEC| An alternating current that reaches its maximum value up to 90° behind the voltage that produces it. { 'lag·iŋ ,kə·rənt }

lagging load See inductive load. { 'lag·iŋ ,lōd }

lag-lead network See lead-lag network. { 'lag 'lēd ,net,wərk }

lag network See integral network. { 'lag ,net,wərk }

lag time |ELEC| The time between the application of current and rupture of the circuit within the detonator. { 'lag ,tīm }

Lalande cell |ELEC| A type of wet cell that uses a zinc anode and cupric oxide cathode cast as flat plates or hollow cylinders, and an electrolyte of sodium hydroxide in aqueous solution (caustic soda). { lə'länd ,sel }

laminated contact |ELEC| Switch contact made up of a number of laminations, each making individual contact with the opposite conducting surface. { 'lam·ə,nād·əd 'kän,takt }

laminography See sectional radiography. { ,lam·ə 'näg·rə·fē }

lamp |ENG| A device that produces light, such as an electric lamp. { lamp }

lamp bank |ELEC| A number of incandescent lamps connected in parallel or series to serve as a resistance load for full-load tests of electric equipment. { 'lamp ,baŋk }

lamp cord |ELEC| Two twisted or parallel insulated wires, usually no. 18 or no. 20, used chiefly for connecting electric equipment to wall outlets. { 'lamp ,kȯrd }

lamp depreciation |ELEC| The decrease in amount of light emitted by a lamp during its operating life. { 'lamp di,prē·shē,ā·shən }

lampholder |ELEC| A device designed to connect an electric lamp to a circuit and to support it mechanically. { 'lamp,hōld·ər }

lamp inrush current |ELEC| The surge of current that occurs when an incandescent lamp is turned on. { ¦lamp 'in,rəsh ,kər·ənt }

LAN See local-area network. { lan }

land |ELECTR| **1.** One of the regions between pits on a track on an optical disk. **2.** See terminal area. { land }

land-earth station |COMMUN| A facility that routes calls from mobile stations via satellite to and from terrestrial telephone networks. Abbreviated LES. { ¦land ¦ərth ¸stā·shən }

land effect *See* coastal refraction. { 'land i¸fekt }

landing zone [COMPUT SCI] The data-free area on the surface of a hard disk over which the read-write head comes to rest when the computer is shut off and the disk stops rotating. { 'land·iŋ ¸zōn }

landline [ELEC] A communications cable on or under the earth's surface, in contrast to a submarine cable. { 'lan¸līn }

land mobile-satellite service [COMMUN] A mobile-satellite service in which the mobile earth stations are located on land. Abbreviated LMSS. { ¦land ¸mō·bəl 'sad·əl¸īt ¸sər·vəs }

land mobile service [COMMUN] Mobile service between base stations and mobile stations, or between land mobile stations. { 'land ¦mō·bəl ¦sər·vəs }

land mobile station [COMMUN] Mobile station in the land mobile service, capable of surface movement within the geographical limits of a country or continent. { 'land ¦mō·bəl ¦stā·shən }

land return *See* ground clutter. { 'land ri¸tərn }

land station [COMMUN] Station in the mobile service not intended for operation while in motion. { 'land ¸stā·shən }

land transportation frequency bands [COMMUN] A group of radio-frequency bands between 25 megahertz and 30,000 megahertz allocated for use by taxicabs, railroads, buses, and trucks. { 'land ¸tranz·pər¦tā·shən 'frē·kwən¸sē ¸banz }

land transportation radio services [COMMUN] Any service of radio communications operated by and for the sole use of certain land transportation carriers, the radio transmitting facilities of which are defined as fixed, land, or mobile stations. { 'land ¸tranz·pər¦tā·shən 'rād·ē·ō ¸sər·vəs·əz }

Langevin ion-mobility theories [ELECTR] Two theories developed to calculate the mobility of ions in gases; the first assumes that atoms and ions interact through a hard-sphere collision and have a constant mean free path, while the second assumes that there is an attraction between atoms and ions arising from the polarization of the atom in the ion's field, in addition to hard-sphere repulsion for close distances of approach. { länzh·van ¦ī¸än mō'bil·əd·ē ¸thē·ə·rēz }

Langevin ion-recombination theory [ELECTR] A theory predicting the rate of recombination of negative with positive ions in an ionized gas on the assumption that ions of opposite sign approach one another under the influence of mutual attraction, and that their relative velocities are determined by ion mobilities; applicable at high pressures, above 1 or 2 atmospheres. { länzh·van ¦ī¸än rē¸käm·bə'nā·shən ¸thē·ə·rē }

Langmuir-Child equation *See* Child's law. { 'laŋ ¸myùr 'chīld i¸kwā·zhən }

Langmuir dark space [ELECTR] A nonluminous region surrounding a negatively charged probe inserted in the positive column of a glow discharge. { 'laŋ¸myùr 'därk ¸spās }

Langmuir diffusion pump [ENG] A type of diffusion pump in which the mercury vapor emerges

from a nozzle, giving it motion in a direction away from the high-vacuum side of the pump. { ¦laŋ·myùr di'fyü·zhən ¸pəmp }

language [COMPUT SCI] The set of words and rules used to construct sentences with which to express and process information for handling by computers and associated equipment. { 'laŋ·gwij }

language converter [COMPUT SCI] A device which translates a form of data (such as that on microfilm) into another form of data (such as that on magnetic tape). { 'laŋ·gwij kən¸vərd·ər }

language subset [COMPUT SCI] A portion of a programming language that can be used alone; usually applied to small computers that do not have the capability of handling the complete language. { 'laŋ·gwij 'səb¸set }

language translator [COMPUT SCI] **1.** Any assembler or compiler that accepts human-readable statements and produces equivalent outputs in a form closer to machine language. **2.** A program designed to convert one computer language to equivalent statements in another computer language, perhaps to be executed on a different computer. **3.** A routine that performs or assists in the performance of natural language translations, such as Russian to English, or Chinese to Russian. { 'laŋ·gwij ¸tranz¸lād·ər }

L antenna [ELECTROMAG] An antenna that consists of an elevated horizontal wire having a vertical down-lead connected at one end. { 'el an¸ten·ə }

lap dissolve [ELECTR] Changeover from one video scene to another so that the new picture appears gradually as the previous picture simultaneously disappears. { 'lap di¸zälv }

lapel microphone [ENG ACOUS] A small microphone that can be attached to a lapel or pocket on the clothing of the user, to permit free movement while speaking. { lə'pel ¦mī·krə¸fōn }

lapping [ELECTR] Moving a quartz, semiconductor, or other crystal slab over a flat plate on which a liquid abrasive has been poured, to obtain a flat polished surface or to reduce the thickness a carefully controlled amount. { 'lap·iŋ }

laptop computer *See* notebook computer. { 'lap ¸täp kəm¸pyüd·ər }

lap winding [ELEC] A two-layer winding in which each coil is connected in series to the adjacent coil. { 'lap ¸wīn·diŋ }

large-scale integrated circuit [ELECTR] A very complex integrated circuit, which contains well over 100 interconnected individual devices, such as basic logic gates and transistors, placed on a single semiconductor chip. Abbreviated LSI circuit. Also known as chip circuit; multiple-function chip. { 'lärj¦skāl ¸int·ə¸grād·əd 'sər·kət }

large-scale integrated memory *See* semiconductor memory. { 'lärj¦skāl ¸int·ə¸grād·əd 'mem·rē }

large-systems control theory [CONT SYS] A branch of the theory of control systems concerned with the special problems that arise in the design of control algorithms (that is, control policies and strategies) for complex systems. { 'lärj ¸sis·təmz kən'trōl ¸thē·ə·rē }

laryngophone [ENG ACOUS] A microphone designed to be placed against the throat of a speaker, to pick up voice vibrations directly without responding to background noise. { lə'riŋ·gə ‚fōn }

LASCR See light-activated silicon controlled rectifier.

LASCS See light-activated silicon controlled switch.

laser [OPTICS] An active electron device that converts input power into a very narrow, intense beam of coherent visible or infrared light; the input power excites the atoms of an optical resonator to a higher energy level, and the resonator forces the excited atoms to radiate in phase. Derived from light amplification by stimulated emission of radiation. { 'lā·zər }

laser amplifier [ELECTR] A laser which is used to increase the output of another laser. Also known as light amplifier. { 'lā·zər ‚am·plə‚fī·ər }

laser communication [COMMUN] Optical communication in which the light source is a laser whose beam is modulated for voice, video, or data communication over wide information bandwidths, typically 1 gigahertz or more. { 'lā·zər kə‚myü·nə'kā·shən }

laser diode See semiconductor laser. { 'lā·zər ‚dī ‚ōd }

laser disk storage See optical disk storage. { 'lā·zər ‚disk ‚stȯr·ij }

laser flash tube [ELECTR] A high-power, air-cooled or water-cooled xenon flash tube designed to produce high-intensity flashes for pumping applications. { 'lā·zər 'flash ‚tüb }

laser-holography storage [COMPUT SCI] A computer storage technology in which information is stored in microscopic spots burned in a holographic substrate by a laser beam, and is read by sensing a lower-energy laser beam that is transmitted through these spots. { 'lā·zər hō ‚läg·rə·fē ‚stȯr·ij }

laser memory [COMPUT SCI] A computer memory in which a controlled laser beam acts on individual and extremely small areas of a photosensitive or other type of surface, for storage and subsequent readout of digital data or other types of information. { 'lā·zər ‚mem·rē }

laser radiation detector [ELECTR] A photodetector that responds primarily to the coherent visible, infrared, or ultraviolet light of a laser beam. { 'lā·zər ‚rād·ē'ā·shən di‚tek·tər }

laser recorder [COMMUN] An image reproducer that resembles a facsimile system, in which a laser beam is initially modulated by the video signal and swept over photographic film or paper to reproduce an image received over wire or radio communication systems. { 'lā·zər ri'kȯrd·ər }

laser scriber [ENG] A laser-cutting setup used in place of a diamond scriber for dicing thin slabs of silicon, gallium arsenide, and other semiconductor materials used in the production of semiconductor diodes, transistors, and integrated circuits; also used for scribing sapphire and ceramic substrates. { 'lā·zər ‚skrīb·ər }

laser threshold [ELECTR] The minimum pumping energy required to initiate lasing action in a laser. { 'lā·zər ‚thresh·hōld }

laser-triggered switch [ELEC] A high-voltage high-power switch that consists of a spark gap triggered into conduction by a laser beam. { 'lā·zər ‚trig·ərd 'swich }

last-mask read-only memory [COMPUT SCI] A read-only memory in which the final mask used in the fabrication process determines the connections to the internal transistors, and these connections in turn determine the data pattern that will be read out when the cell is accessed. Also known as contact-mask read-only memory. { 'last ‚mask ‚rēd ‚ōn·lē 'mem·rē }

latch [ELECTR] An electronic circuit that reverses and maintains its state each time that power is applied. { lach }

latch-in relay [ELEC] A relay that maintains its contacts in the last position assumed, even without coil energization. { 'lach‚in 'rē‚lā }

latch-up phenomenon [ELECTR] In a bipolar or MOS integrated circuit, the generation of photocurrents by ionizing radiation which can provide a trigger signal for a parasitic pnpn circuit and possibly result in permanent damage or operational failure if the circuit remains in this state. { 'lach ‚əp fə‚näm·ə‚nän }

late binding [COMPUT SCI] The assignment of data types (such as integer or string) to variables at the time of execution of a computer program, rather than during the compilation phase. { ‚lāt 'bīnd·iŋ }

latency [COMPUT SCI] The waiting time between the order to read/write some information from/to a specified place and the beginning of the data read/write operation. { 'lat·ən·sē }

lateral parity check [COMPUT SCI] The number of one bits counted across the width of the magnetic tape; this number plus a one or a zero must always be odd (or even), depending upon the manufacturer. { 'lad·ə·rəl 'par·əd·ē ‚chek }

lateral recording [ENG ACOUS] A type of disk recording in which the groove modulation is parallel to the surface of the recording medium so that the cutting stylus moves from side to side during recording. { 'lad·ə·rəl ri'kȯrd·iŋ }

lattice filter [ELECTR] An electric filter consisting of a lattice network whose branches have L-C parallel-resonant circuits shunted by quartz crystals. { 'lad·əs ‚fil·tər }

lattice network [ELEC] A network that is composed of four branches connected in series to form a mesh; two nonadjacent junction points serve as input terminals, and the remaining two junction points serve as output terminals. { 'lad·əs 'net‚wərk }

lattice winding [ELEC] A winding made of lattice coils and used for electric machines. { 'lad·əs ‚wīn·diŋ }

launching [ELECTROMAG] The process of transferring energy from a coaxial cable or transmission line to a waveguide. { 'lȯn·chiŋ }

Lauritsen electroscope [ELEC] A rugged and sensitive electroscope in which a metallized

quartz fiber is the sensitive element. { 'laù·
rət·sən i'lek·trə,skōp }

lawnmower |ELECTR| Type of radio-frequency
preamplifier used with radar receivers. { 'lón
,mō·ər }

law of electric charges |ELEC| The law that like
charges repel, and unlike charges attract. { 'lò
əv i̧lek·trik 'chärj·əz }

law of electrostatic attraction See Coulomb's law.
{ 'lò əv i̧lek·trə̧stad·ik ə'trak·shən }

Lawrence tube See chromatron. { 'lär·əns ,tüb }

layer |COMPUT SCI| One of the divisions within
which components or functions are isolated in
a computer system with layered architecture or a
communications system with layered protocols.
{ 'lā·ər }

layer capacitance See cathode interface capacitance. { 'lā·ər kə'pas·əd·əns }

layered architecture |COMPUT SCI| A technique
used in designing computer software, hardware,
and communications in which system or network components are isolated in layers so that
changes can be made in one layer without
affecting the others. { 'lā·ərd 'är·kə,tek·chər }

layered-protocols technique |COMMUN| A tech-
nique for isolating the functions required in
a data communications network so that these
functions can be set up in a modular fashion and
changes can be made in one area without affecting the others. { 'lā·ərd 'prōd·ə,kòlz tek'nēk }

layer impedance See cathode interface impedance.
{ 'lā·ər im'pēd·əns }

layer winding |ELEC| Coil-winding method in
which adjacent turns are laid evenly and side by
side along the length of the coil form; any number
of additional layers may be wound over the
first, usually with sheets of insulating material
between the layers. { 'lā·ər ,wīn·diŋ }

layout character |COMPUT SCI| A control charac-
ter that determines the form in which the output
data generated by a printer or display device are
arranged. Also known as format effector. { 'lā
,aùt ,kar·ik·tər }

lay ratio |ELEC| The ratio of the axial length of one
complete turn of the helix formed by the core of a
cable or the wire of a stranded conductor, to the
mean diameter of the cable. { 'lā ,rā·shō }

lazy evaluation See demand-driven execution.
{ 'lā·zē i,val·yə'wā·shən }

lazy H antenna |ELECTROMAG| An antenna array
in which two or more dipoles are stacked one
above the other to obtain greater directivity.
{ ¦lā·zē 'āch an,ten·ə }

L band |COMMUN| A band of radio frequencies
between 1 and 2 gigahertz. { 'el ,band }

LCD See liquid crystal display.

LC filter See inductive filter. { ¦el¦sē ,fil·tər }

LC ratio |ELEC| The inductance of a circuit in
henrys divided by capacitance in farads. { ¦el
¦sē ,rā·shō }

L-display |ELECTR| A radar display format in
which an echo signal appears as two horizontal
deflections, left and right, from a vertical reference line, as associated with a two-beam tracking
radar antenna, equal amplitudes indicating no
pointing error and position on the vertical axis
indicating range. Also known as L-indicator; L-
scan; L-scope. { 'el di,splā }

LDM See limited-distance modem.

lead |ELEC| A wire used to connect two points in
a circuit. { led }

lead-acid battery |ELEC| A storage battery in
which the electrodes are grids of lead containing
lead oxides that change in composition during
charging and discharging, and the electrolyte is
dilute sulfuric acid. { 'led ,as·əd 'bad·ə·rē }

lead angle |ENG| The angle that the tangent to a
helix makes with the plane normal to the axis of
the helix. { 'lēd ,aŋ·gəl }

lead compensation |CONT SYS| A type of feed-
back compensation primarily employed for
stabilization or for improving a system's transient
response; it is generally characterized by a series
compensation transfer function of the type

$$G_c(s) = K \frac{(s-z)}{(s-p)}$$

where $z < p$ and K is a constant. { 'lēd ,käm·
pən'sā·shən }

lead-covered cable |ELEC| A cable whose con-
ductors are protected from moisture and mechanical damage by a lead sheath. { 'led
¦kəv·ərd 'kā·bəl }

leader |COMPUT SCI| A record which precedes a
group of detail records, giving information about
the group not present in the detail records; for
example, "beginning of batch 17." |ENG| The
unrecorded length of magnetic tape that enables
the operator to thread the tape through the drive
and onto the take-up reel without losing data or
recorded music, speech, or such. { 'lēd·ər }

leader label |COMPUT SCI| A record appearing at
the beginning of a magnetic tape to uniquely
identify the tape as one required by the system.
{ 'lēd·ər ,lā·bəl }

lead-in |ELEC| A single wire used to connect a
single-terminal outdoor antenna to a receiver or
transmitter. Also known as down-lead. { 'lēd
,in }

leading character elimination |COMPUT SCI| A
method of data compression used for dictionaries that are stored in alphabetical order, in
which the coding for each word has two parts:
the number of characters in common with the
previous word, and the unique suffix. { ¦lēd·iŋ
'kar·ik·tər i,lim·ə'nā·shən }

leading current |ELEC| An alternating current
that reaches its maximum value up to 90°

ahead of the voltage that produces it.   { 'lēd·iŋ ¦kə·rənt }

leading load |ELEC| Load that is predominately capacitive, so that its current leads the voltage applied to the load.   { 'lēd·iŋ ¦lōd }

leading pad |COMPUT SCI| Characters that fill unused space at the left end of a data field.   { 'lēd·iŋ 'pad }

leading phase |ELEC| In three-phase power measurement, the phase whose voltage is leading upon that of one of the other phases by 120°.   { 'lēd·iŋ ¦fāz }

lead-in insulator |ELEC| A tubular insulator inserted in a hole drilled through a wall, through which the lead-in wire can be brought into a building.   { 'lēd,in 'in·sə,lād·ər }

lead-lag ballast |ELEC| A ballast for a pair of fluorescent lamps, one operating on leading current and the other on lagging current, to diminish the stroboscopic effect.   { ¦lēd 'lag ,bal·əst }

lead-lag network |CONT SYS| Compensating network which combines the characteristics of the lag and lead networks, and in which the phase of a sinusoidal response lags a sinusoidal input at low frequencies and leads it at high frequencies. Also known as lag-lead network.   { 'lēd 'lag 'net ,wərk }

lead network   See derivative network.   { 'lēd ,net ,wərk }

lead sulfide cell |ELECTR| A cell used to detect infrared radiation; either its generated voltage or its change of resistance may be used as a measure of the intensity of the radiation.   { 'led ¦səl,fīd 'sel }

leaf   See terminal vertex.   { lēf }

leakage conductance |ELEC| The conductance of the path over which leakage current flows; it is normally a low value.   { 'lēk·ij kən¦dək·təns }

leakage current |ELEC| **1.** Undesirable flow of current through or over the surface of an insulating material or insulator. **2.** The flow of direct current through a poor dielectric in a capacitor. |ELECTR| The alternating current that passes through a rectifier without being rectified.   { 'lēk·ij ,kə·rənt }

leakage current |ELEC| **1.** Undesirable flow of current through or over the surface of an insulating material or insulator. **2.** The flow of direct current through a poor dielectric in a capacitor. |ELECTR| The alternating current that passes through a rectifier without being rectified.   { 'lēk·ij ,kə·rənt }

leakage indicator |ELEC| An instrument used to measure or detect current leakage from an electric system to earth. Also known as earth detector.   { 'lēk·ij ,in·də,kād·ər }

leakage radiation |ELECTROMAG| In a radio transmitting system, radiation from anything other than the intended radiating system.   { 'lēk·ij ,rād·ē'ā·shən }

leakage resistance |ELEC| The resistance of the path over which leakage current flows; it is normally high.   { 'lēk·ij ri¦zis·təns }

leaky |ELEC| Pertaining to a condition in which the leakage resistance has dropped so much

below its normal value that excessive leakage current flows; usually applied to a capacitor.   { 'lēk·ē }

leaky-wave antenna |ELECTROMAG| A wide-band microwave antenna that radiates a narrow beam whose direction varies with frequency; it is fundamentally a perforated waveguide, thin enough to permit flush mounting for aircraft and missile radar applications.   { 'lēk·ē ¦wāv an 'ten·ə }

leapfrog test |COMPUT SCI| A computer test using a special program that performs a series of arithmetical or logical operations on one group of storage locations, transfers itself to another group, checks the correctness of the transfer, then begins the series of operations again; eventually, all storage positions will have been tested.   { 'lēp,fräg ,test }

learning control |CONT SYS| A type of automatic control in which the nature of control parameters and algorithms is modified by the actual experience of the system.   { 'lərn·iŋ kən,trōl }

learning machine |COMPUT SCI| A machine that is capable of improving its future actions as a result of analysis and appraisal of past actions.   { 'lər·niŋ mə,shēn }

leased facility |COMMUN| A collection of communication lines dedicated to a particular service; sometimes the lines have a predetermined path through system switching equipment.   { 'lēst fə'sil·əd·ē }

least frequently used |COMPUT SCI| A technique for using main storage efficiently, in which new data replace data in storage locations that have been used least often, as determined by an algorithm.   { 'lēst ¦frē·kwənt·lē 'yüzd }

least recently used |COMPUT SCI| A technique for using main storage efficiently, in which new data replace data in storage locations that have not been accessed for the longest period, as determined by an algorithm.   { 'lēst ¦rē·sənt·lē ,yüzed }

least significant bit |COMPUT SCI| The bit that carries the lowest value or weight in binary notation for a numeral; for example, when 13 is represented by binary 1101, the 1 at the right is the least significant bit. Abbreviated LSB.   { ¦lēst sig¦nif·i·kənt 'bit }

least significant character |COMPUT SCI| The character in the rightmost position in a number or word.   { ¦lēst sig¦nif·i·kənt 'kar·ik·tər }

Leclanché cell |ELEC| The common dry cell, which is a primary cell having a carbon positive electrode and a zinc negative electrode in an electrolyte of sal ammoniac and a depolarizer.   { lə¦klan¦shā ,sel }

LED   See light-emitting diode.

Leduc current |ELEC| An asymmetrical alternating current obtained from, or similar to that obtained from, the secondary winding of an induction coil; used in electrobiology.   { lə'dük ,kə·rənt }

left-hand taper |ELEC| A taper in which there is greater resistance in the counterclockwise half of the operating range of a rheostat or potentiometer (looking from the shaft end) than in the clockwise half.   { 'left ¦hand 'tā·pər }

left-justify |COMPUT SCI| To shift the contents of a register so that the left, or most significant, digit is at some specified position.   { 'left 'jəs·tə·fī }

left value |COMPUT SCI| The memory address of a symbolic variable in a computer program. Abbreviated lvalue.   { 'left ‚val·yü }

leg |COMPUT SCI| The sequence of instructions that is followed in a computer routine from one branch point to the next.   { leg }

legacy system |COMPUT SCI| A computer system that has been in operation for a long time, and whose functions are too essential to be disrupted by upgrading or integration with another system.   { 'leg·ə·sē ‚sis·təm }

LEIT   See light emission via inelastic tunneling.

LEIT device |ELECTR| A light source consisting of two crossed, thin metal-film strips separated by a very thin insulating layer and attached to a battery to produce light emission via inelastic tunneling (LEIT).   { ‚el¦ē¦ī'tē di‚vīs }

Lenard rays |ELECTR| Cathode rays produced in air by a Lenard tube.   { 'lā‚närt ‚rāz }

Lenard tube |ELECTR| An early experimental electron-beam tube that had a thin glass or metallic foil window at the end opposite the cathode, through which the electron beam could pass into the atmosphere.   { 'lā‚närt ‚tüb }

length block |COMPUT SCI| The total number of records, words, or characters contained in one block.   { 'leŋkth ‚bläk }

lengthened dipole |ELECTROMAG| An antenna element with lumped inductance to compensate an end loss.   { 'leŋk·thənd 'dī‚pōl }

lens |COMMUN| A dielectric or metallic structure that is highly transparent to radio waves and can bend them to produce a desired radiation pattern; used with antennas for radar and microwave relay systems.   { lenz }

lens antenna |ELECTROMAG| A microwave antenna in which a dielectric lens is placed in front of the dipole or horn radiator to concentrate the radiated energy into a narrow beam or to focus received energy on the receiving dipole or horn.   { 'lenz an‚ten·ə }

LES   See land-earth station.

Leslie effect |ENG ACOUS| A dynamic timbre-changing effect created by rotating one or more directional speakers inside a cabinet such that a mixture of Doppler-shifted reflections is generated in the output of an electronic instrument.   { 'lez·lē i‚fekt }

letter |COMMUN| A character used in an alphabet generally representing one or more sounds of a spoken language.   { 'led·ər }

letter code |COMPUT SCI| A Baudot code function which cancels errors by causing the receiving terminal to print nothing.   { 'led·ər ‚kōd }

letter-perfect printer   See letter-quality printer.   { 'led·ər ¦pər·fekt 'print·ər }

letter-quality printer |COMPUT SCI| A printer that produces high-quality output. Also known as correspondence printer; letter-perfect printer.   { 'led·ər ¦kwäl·əd·ē 'print·ər }

letters shift |COMMUN| **1.** A movement of a teletypewriter carriage which permits printing of alphabetic characters in an appropriate, generally linear sequence.   **2.** The control which actuates this movement. Abbreviated LTRS.   { 'led·ərz 'shift }

level |COMMUN| **1.** A specified position on an amplitude scale (for example, magnitude) applied to a signal waveform, such as reference white level and reference black level in a video signal. **2.** A range of allowed picture parameters and combinations of picture parameters in the digital television system. |COMPUT SCI| **1.** The status of a data item in COBOL language indicating whether this item includes additional items. **2.** See channel. |ELEC| A single bank of contacts, as on a stepping relay. |ELECTR| **1.** The difference between a quantity and an arbitrarily specified reference quantity, usually expressed as the logarithm of the ratio of the quantities.   **2.** A charge value that can be stored in a given storage element of a charge storage tube and distinguished in the output from other charge values.   { 'lev·əl }

level 1 cache   See primary cache.   { ¦lev·əl 'wən ‚kash }

level 2 cache   See secondary cache.   { ‚lev·əl ‚tü 'kash }

level compensator |ELECTR| **1.** Automatic transmission-regulating feature or device used to minimize the effects of variations in amplitude of the received signal.   **2.** Automatic gain control device used in the receiving equipment of a telegraph circuit.   { 'lev·əl 'käm·pən‚sād·ər }

level converter |ELECTR| An amplifier that converts nonstandard positive or negative logic input voltages to standard DTL or other logic levels.   { 'lev·əl kən‚vərd·ər }

level indicator |ENG ACOUS| An indicator that shows the audio voltage level at which a recording is being made; may be a volume-unit meter, neon lamp, or cathode-ray tuning indicator.   { 'lev·əl 'in·də‚kād·ər }

level set |COMPUT SCI| A revision of a software package in which most or all of the executable programs are replaced with improved versions.   { 'lev·əl ‚set }

level shifting |ELECTR| Changing the logic level at the interface between two different semiconductor logic systems.   { 'lev·əl ‚shif·tiŋ }

lever switch |ELEC| A switch having a lever-shaped operating handle.   { 'lev·ər ‚swich }

Lewis-Rayleigh afterglow |ELECTR| A golden yellow light emitted by nitrogen gas following the passage of an electric discharge, associated with recombination of nitrogen atoms.   { ¦lü·əs ¦rā·lē 'af·tər‚glō }

Leyden jar |ELEC| An early type of capacitor, consisting simply of metal foil sheets on the inner and outer surfaces of a glass jar.   { 'līd·ən ‚jär }

LF   See low-frequency.

LF loran   See low-frequency loran.   { ¦el¦ef 'lôr‚an }

Liapunov function   See Lyapunov function.   { 'lyä·pú·nòf ‚fəŋk·shən }

librarian |COMPUT SCI| The program which maintains and makes available all programs and routines composing the operating system. { 'lī'brer·ē·ən }

library |COMPUT SCI| **1.** A computerized facility containing a collection of organized information used for reference. **2.** An organized collection of computer programs together with the associated program listings, documentation, users' directions, decks, and tapes. { 'lī,brer·ē }

library routine |COMPUT SCI| A computer program that is part of some program library. { 'lī,brer·ē rü,tēn }

library software |COMPUT SCI| The collection of programs and routines in the library of a computer system. { 'lī,brer·ē 'sȯft,wer }

library tape |COMPUT SCI| A magnetic tape that is kept in a stored, indexed collection for ready use and is made generally available. { 'lī,brer·ē ,tāp }

Lichenberger figures See Lichtenberg figures. { 'lī·kən,bər·gər ,fig·yərz }

Lichtenberg figures |ELEC| Patterns produced on a photographic emulsion, or in fine powder spread over the surface of a solid dielectric, by an electric discharge produced by a high transient voltage. Also known as Lichenberger figures. { 'lik·tən·bərg ,fig·yərz }

light-activated silicon controlled rectifier |ELECTR| A silicon controlled rectifier having a glass window for incident light that takes the place of, or adds to the action of, an electric gate current in providing switching action. Abbreviated LASCR. Also known as photo-SCR; photothyristor. { 'līt |ak·tə,vād·əd |sil·ə·kən kən |trōld 'rek·tə,fī·ər }

light-activated silicon controlled switch |ELECTR| A semiconductor device that has four layers of silicon alternately doped with acceptor and donor impurities, but with all four of the *p* and *n* layers made accessible by terminals; when a light beam hits the active light-sensitive surface, the photons generate electron-hole pairs that make the device turn on; removal of light does not reverse the phenomenon; the switch can be turned off only by removing or reversing its positive bias. Abbreviated LASCS. { 'līt |ak·tə,vād·əd |sil·ə·kən kən|trōld 'swich }

light amplification by stimulated emission of radiation See laser. { 'līt ,am·plə·fə'kā·shən bī |stim·yə,lād·əd i|mish·ən əv ,rād·ē'ā·shən }

light amplifier |ELECTR| **1.** Any electronic device which, when actuated by a light image, reproduces a similar image of enhanced brightness, and which is capable of operating at very low light levels without introducing spurious brightness variations (noise) into the reproduced image. Also known as image intensifier. **2.** See laser amplifier. { 'līt ,am·plə,fī·ər }

light-beam galvanometer See d'Arsonval galvanometer. { 'līt ,bēm ,gal·və'näm·əd·ər }

light bulb See incandescent lamp. { 'līt ,bəlb }

light carrier injection |ELECTR| A method of introducing the carrier in a facsimile system by periodic variation of the scanner light beam, the average amplitude of which is varied by the density changes of the subject copy. Also known as light modulation. { 'līt 'kar·ē·ər in,jek·shən }

light chopper |ELECTR| A rotating fan or other mechanical device used to interrupt a light beam that is aimed at a phototube, to permit alternating-current amplification of the phototube output and to make its output independent of strong, steady ambient illumination. { 'līt 'chäp·ər }

light emission via inelastic tunneling |ELECTR| A process in which electrons tunneling through a thin insulating layer separating two metals excite surface plasmons which then scatter from surface and structural discontinuities, radiating visible light. Abbreviated LEIT. { |līt i,mish·ən ,vē·ə |in·ə,las·tik 'tən·əl·iŋ }

light-emitting diode |ELECTR| A rectifying semiconductor device which converts electrical energy into electromagnetic radiation. The wavelength of the emitted radiation ranges from the near-ultraviolet to the near-infrared, that is, from about 400 to over 1500 nanometers. Abbreviated LED. { 'līt i,mid·iŋ 'dī,ōd }

light-gating cathode-ray tube |ELECTR| A cathode-ray tube in which the electron beam varies the transmission or reflection properties of a screen that is positioned in the beam of an external light source. { 'līt ,gād·iŋ |kath,ōd 'rā ,tüb }

light guide See optical fiber. { 'līt ,gīd }

light gun |ELECTR| A light pen mounted in a gun-type housing. { 'līt ,gən }

lighthouse tube See disk-seal tube. { 'līt,hau̇s ,tüb }

lighting branch circuit |ELEC| A circuit that supplies power to outlets for lighting fixtures only. { 'līd·iŋ 'branch ,sər·kət }

light meter |ENG| A small, portable device for measuring illumination; an exposure meter is a specific application, being calibrated to give photographic exposures. { 'līt ,mēd·ər }

light modulation See light carrier injection. { 'līt ,mäj·ə,lā·shən }

light modulator |ELECTR| The combination of a source of light, an appropriate optical system, and a means for varying the resulting light beam to produce an optical sound track on motion picture film. { 'līt ,mäj·ə,lād·ər }

light-negative |ELECTR| Having negative photoconductivity, hence decreasing in conductivity (increasing in resistance) under the action of light. { 'līt ¦neg·ə·tiv }

lightning arrester |ELEC| A protective device designed primarily for connection between a conductor of an electrical system and ground to limit the magnitude of transient overvoltages on equipment. Also known as arrester; surge arrester. { 'līt·niŋ ə,res·tər }

lightning conductor |ELEC| A conductor designed to carry the current of a lightning discharge from a lightning rod to ground. { 'līt·niŋ kən,dək·tər }

lightning generator |ELEC| A high-voltage power supply used to generate surge voltages resembling lightning, for testing insulators and other high-voltage components. { 'līt·niŋ ,jen·ə ,rād·ər }

lightning protection |ELEC| Means, such as lightning rods and lightning arresters, of protecting electrical systems, buildings, and other property from lightning. { 'līt·niŋ prə,tek·shən }

lightning recorder See sferics receiver. { 'līt·niŋ ri,kórd·ər }

lightning rod |ELEC| A metallic rod set up on an exposed elevation of a structure and connected to a low-resistance ground to intercept lightning discharges and to provide a direct conducting path to ground for them. { 'līt·niŋ ,räd }

lightning surge |ELEC| A transient disturbance in an electric circuit due to lightning. { 'līt·niŋ ,sərj }

lightning switch |ELEC| A manually operated switch used to connect a radio antenna to ground during electrical storms, rather than to the radio receiver. { 'līt·niŋ ,swich }

light-operated switch |ELECTR| A switch that is operated by a beam or pulse of light, such as a light-activated silicon controlled rectifier. { 'līt ¦äp·ə,rād·əd 'swich }

light panel See electroluminescent panel. { 'līt ,pan·əl }

light pen |ELECTR| A tiny photocell or photomultiplier, mounted with or without fiber or plastic light pipe in a pen-shaped housing; it is held against a cathode-ray screen to make measurements from the screen or to change the nature of the display. { 'līt ,pen }

light-positive |ELECTR| Having positive photoconductivity; selenium ordinarily has this property. { 'līt ¦päz·ə·tiv }

light relay See photoelectric relay. { 'līt ,rē,lā }

light-sensitive |ELECTR| Having photoconductive, photoemissive, or photovoltaic characteristics. Also known as photosensitive. { 'līt 'sen·səd·iv }

light-sensitive cell See photodetector. { 'līt ¦sen· səd·iv 'sel }

light-sensitive detector See photodetector. { 'līt ¦sen·səd·iv di'tek·tər }

light-sensitive tube See phototube. { 'līt ¦sen· səd·iv 'tüb }

light sensor photodevice See photodetector. { 'līt ,sen·sər 'fōd·ō·di,vīs }

light stability |COMPUT SCI| In optical character recognition, the ability of an image to retain its spectral appearance when exposed to radiant energy. { 'līt stə,bil·əd·ē }

light valve |ELECTR| **1.** A device whose light transmission can be made to vary in accordance with an externally applied electrical quantity, such as volatage, current, electric field, or magnetic field, or an electron beam. **2.** Any direct-view electronic display optimized for reflecting or transmitting an image with an independent collimated light source for projection purposes. { 'līt,valv }

limit check |COMPUT SCI| A check to determine if a value entered into a computer system is within acceptable minimum and maximum values. { 'lim·ət ,chek }

limited-access data |COMPUT SCI| Data to which only authorized users have access. { 'lim·əd·əd ¦ak·ses 'dad·ə }

limited-degree-of-freedom robot |CONT SYS| Robot whose end effector can be positioned and oriented in fewer than six degrees of freedom. { 'lim·əd·əd di'grē əv 'frē·dəm 'rō,bät }

limited-distance modem |COMMUN| A modem used only for communications within a building in order to improve the signal quality where a long distance exists between the terminal and the computer. |COMPUT SCI| A device designed to transmit and receive signals over relatively short distances, typically less than 5 miles (8 kilometers). Abbreviated LDM. Also known as line driver. { 'lim·əd·əd ¦dis·təns 'mō,dem }

limited-entry decision table |COMPUT SCI| A decision table in which the condition stub specifies exactly the condition or the value of the variable. { 'lim·əd·əd ¦en·trē di'sizh·ən ,tā·bəl }

limited integrator |ELECTR| A device used in analog computers that has two input signals and one output signal whose value is proportional to the integral of one of the input signals with respect to the other as long as this output signal does not exceed specified limits. { 'lim·əd·əd 'int·ə ,grād·ər }

limited-sequence robot See fixed-stop robot. { 'lim·əd·əd ¦sē·kwəns 'rō,bät }

limited signal |ELECTR| Radar signal that is intentionally limited in amplitude by the dynamic range of the circuits involved; useful in some radio and radar processing. { 'lim·əd·əd 'sig·nəl }

limited space-charge accumulation mode |ELECTR| A mode of operation of a Gunn diode in which the frequency of operation is set by a resonant circuit to be much higher than the transit-time frequency so that domains have insufficient time to form while the field is above threshold and, as a result, the sample is maintained in the negative conductance state during a large fraction of the voltage cycle. Abbreviated LSA mode. { ¦lim·əd·əd ,spās ,chärj ə,kyü·myə'lā·shən,mōd }

limiter |ELECTR| An electronic circuit used to prevent the amplitude of an electronic waveform from exceeding a specified level while preserving the shape of the waveform at amplitudes less than the specified level. Also known as amplitude limiter; amplitude-limiting circuit; automatic peak limiter; clipper; clipping circuit; limiter circuit; peak limiter. { 'lim·əd·ər }

limiter circuit *See* limiter. { 'lim·əd·ər ,sər·kət }

limiting |ELECTR| A desired or undesired amplitude-limiting action performed on a signal by a limiter. Also known as clipping; peak clipping. { 'lim·əd·iŋ }

limit priority |COMPUT SCI| An upper bound to the dispatching priority that a task can assign to itself or any of its subtasks. { 'lim·ət prī'är·əd·ē }

limit ratio |ELECTR| Ratio of peak value to limited value, or comparison of such ratios. { 'lim·ət ,rā·shō }

limit switch |ELEC| A switch designed to cut off power automatically at or near the limit of travel of a moving object controlled by electrical means. { 'lim·ət ,swich }

Lindeck potentiometer |ELEC| A potentiometer in which an unknown potential difference is balanced against a known potential difference derived from a fixed resistance carrying a variable current; the converse of most potentiometers. { 'lin,dek pə,ten·tē·äm·əd·ər }

Lindemann electrometer |ELEC| A variant of the quadrant electrometer, designed for portability and insensitivity to changes in position, in which the quadrants are two sets of plates about 6 millimeters apart, mounted on insulating quartz pillars; a needle rotates about a taut silvered quartz suspension toward the oppositely charged plates when voltage is applied to it, and its movement is observed through a microscope. { 'lin·də·mən ,ē,lek'träm·əd·ər }

L-indicator *See* L-display. { 'el ,in·də,kād·ər }

line |ELECTR| **1.** The path covered by the electron beam of a picture tube in one sweep from left to right across the screen. **2.** One horizontal scanning element in a facsimile system. **3.** *See* trace. { līn }

line and trunk group |COMMUN| A group consisting of four-wire circuits, incoming private automatic branch exchange trunks, and intertoll trunk groups. { ¦līn ən ¦trəŋk ,grüp }

linear |CONT SYS| Having an output that varies in direct proportion to the input. { 'lin·ē·ər }

linear amplifier |ELECTR| An amplifier in which changes in output current are directly proportional to changes in applied input voltage. { 'lin·ē·ər 'am·plə,fī·ər }

linear array |ELECTROMAG| An antenna array in which the dipole or other half-wave elements are arranged end to end on the same straight line. Also known as collinear array. { 'lin·ē·ər ə'rā }

linear-array camera |ELECTR| A solid-state television camera that has only a single row of light-sensitive elements or pixels. { 'lin·ē·ər ə¦rā 'kam·rə }

linear bounded automaton |COMPUT SCI| A nondeterministic, one-tape Turing machine whose read/write head is confined to move only on a restricted section of tape initially containing the input. { 'lin·ē·ər ¦baúnd·əd ó'täm·ə,tän }

linear circuit *See* linear network. { 'lin·ē·ər 'sər·kət }

linear comparator |ELECTR| A comparator circuit which operates on continuous, or nondiscrete, waveforms. Also known as continuous comparator. { 'lin·ē·ər kəm'par·əd·ər }

linear computing element |ELEC| A linear circuit in an analog computer. { 'lin·ē·ər kəm'pyüd·iŋ ,el·ə·mənt }

linear conductor antenna |ELECTROMAG| An antenna consisting of one or more wires which all lie along a straight line. { 'lin·ē·ər kən'dək·tər an,ten·ə }

linear control |ELEC| Rheostat or potentiometer having uniform distribution of graduated resistance along the entire length of its resistance element. { 'lin·ē·ər kən'trōl }

linear control system |CONT SYS| A linear system whose inputs are forced to change in a desired manner as time progresses. { 'lin·ē·ər kən'trōl ,sis·təm }

linear detection |ELECTR| Detection in which the output voltage is substantially proportional, over the useful range of the detecting device, to the voltage of the input wave. { 'lin·ē·ər di'tek·shən }

linear distortion |ELECTR| Amplitude distortion in which the output signal envelope is not proportional to the input signal envelope and no alien frequencies are involved. { 'lin·ē·ər di'stór·shən }

linear electrical constants of a uniform line |ELEC| Series resistance, series inductance, shunt conductance, and shunt capacitance per unit length of line. { 'lin·ē·ər i¦lek·tra·kəl 'kän·stəns əv ə ¦yü·nə,fórm 'līn }

linear electrical parameters *See* transmission-line parameters. { 'lin·ē·ər i¦lek·tra·kəl pə'ram·əd·ərz }

linear feedback control |CONT SYS| Feedback control in a linear system. { 'lin·ē·ər 'fēd,bak kən,trōl }

linear integrated circuit |ELECTR| An integrated circuit that provides linear amplification of signals. { 'lin·ē·ər ¦int·ə,grād·əd 'sər·kət }

linearity control |ELECTR| A cathode-ray-tube control which varies the distribution of scanning speed throughout the trace interval. Also known as distribution control. { ,lin·ē'ar·əd·ē kən ,trōl }

linearization |CONT SYS| **1.** The modification of a system so that its outputs are approximately linear functions of its inputs, in order to facilitate analysis of the system. **2.** The mathematical approximation of a nonlinear system, whose departures from linearity are small, by a linear system corresponding to small changes in the variables about their average values. { ‚lin·ē·ər·ə'zā·shən }

linear-logarithmic intermediate-frequency amplifier [ELECTR] Amplifier used to avoid overload or saturation as a protection against jamming in a radar receiver. { 'lin·ē·ər ‚läg·ə 'rith·mik ‚in·tər¦mē·dē·ət ¦frē·kwən·sē 'am·plə ‚fī·ər }

linearly graded junction [ELECTR] A *pn* junction in which the impurity concentration does not change abruptly from donors to acceptors, but varies smoothly across the junction, and is a linear function of position. { 'lin·ē·ər·lē ¦grād·əd 'jəŋk·shən }

linear magnetic amplifier [ELECTR] A magnetic amplifier employing negative feedback to make its output load voltage a linear function of signal current. { 'lin·ē·ər mag¦ned·ik 'am·plə‚fī·ər }

linear modulation [COMMUN] Modulation in which the amplitude of the modulation envelope (or the deviation from the resting frequency) is directly proportional to the amplitude of the intelligence signal at all modulation frequencies. { 'lin·ē·ər ‚mäj·ə'lā·shən }

linear motor [ELEC] An electric motor that has in effect been split and unrolled into two flat sheets, so that the motion between rotor and stator is linear rather than rotary. { 'lin·ē·ər 'mōd·ər }

linear network [ELEC] A network in which the parameters of resistance, inductance, and capacitance are constant with respect to current or voltage, and in which the voltage or current of sources is independent of or directly proportional to other voltages and currents, or their derivatives, in the network. Also known as linear circuit. { 'lin·ē·ər 'net‚wərk }

linear oscillator *See* harmonic oscillator. { 'lin·ē· ər 'äs·ə‚lād·ər }

linear-phase [ELECTR] Pertaining to a filter or other network whose image phase constant is a linear function of frequency. { 'lin·ē·ər ‚fāz }

linear polarization [OPTICS] Polarization of an electromagnetic wave in which the electric vector at a fixed point in space remains pointing in a fixed direction, although varying in magnitude. Also known as plane polarization. { 'lin·ē·ər ‚pō·lə·rə'zā·shən }

linear power amplifier [ELECTR] A power amplifier in which the signal output voltage is directly proportional to the signal input voltage. { 'lin·ē·ər 'paù·ər ‚am·plə‚fī·ər }

linear programming [MATH] The study of maximizing or minimizing a linear function $f(x_1, \ldots, x_n)$ subject to given constraints which are linear inequalities involving the variables x_i. { 'lin·ē·ər 'prō‚gram·iŋ }

linear-quadratic-Gaussian problem [CONT SYS] An optimal-state regulator problem, containing Gaussian noise in both the state and measurement equations, in which the expected value of the quadratic performance index is to be minimized. Abbreviated LQG problem. { 'lin·ē·ər kwə'drad·ik 'gaús·ē·ən ‚präb·ləm }

linear rectifier [ELECTR] A rectifier, the output current of voltage of which contains a wave having a form identical with that of the envelope of an impressed signal wave. { 'lin·ē·ər 'rek·tə‚fī·ər }

linear regulator problem [CONT SYS] A type of optimal control problem in which the system to be controlled is described by linear differential equations and the performance index to be minimized is the integral of a quadratic function of the system state and control functions. Also known as optimal regulator problem; regulator problem. { 'lin·ē·ər 'reg·yə‚lād·ər ‚präb·ləm }

linear repeater [ELECTR] A repeater used in communication satellites to amplify input signals a fixed amount, generally with traveling-wave tubes or solid-state devices operating in their linear region. { 'lin·ē·ər ri'pēd·ər }

linear sweep [ELECTR] A cathode-ray sweep in which the beam moves at constant velocity from one side of the screen to the other, then suddenly snaps back to the starting side. { 'lin·ē·ər ¦swēp }

linear-sweep delay circuit [ELECTR] A widely used form of linear time-delay circuit in which the input signal initiates action by a linear sawtooth generator, such as the bootstrap or Miller integrator, whose output is then compared with a calibrated direct-current reference voltage level. { 'lin·ē·ər ¦swēp di‚lā ‚sər·kət }

linear-sweep generator [ELECTR] An electronic circuit that provides a voltage or current that is a linear function of time; the waveform is usually recurrent at uniform periods of time. { 'lin·ē·ər ¦swēp ‚jen·ə‚rād·ər }

linear system [CONT SYS] A system in which the outputs are components of a vector which is equal to the value of a linear operator applied to a vector whose components are the inputs. { 'lin·ē·ər 'sis·təm }

linear system analysis [CONT SYS] The study of a system by means of a model consisting of a linear mapping between the system inputs (causes or excitations), applied at the input terminals, and the system outputs (effects or responses), measured or observed at the output terminals. { 'lin·ē·ər ¦sis·təm ə'nal·ə·səs }

linear taper [ELEC] A taper that gives the same change in resistance per degree of rotation over the entire range of a potentiometer. { 'lin·ē·ər 'tā·pər }

linear time base [ELECTR] A time base that makes the electron beam of a cathode-ray tube move at a constant speed along the horizontal time scale. { 'lin·ē·ər 'tīm ‚bās }

linear transducer |ELECTR| A transducer for which the pertinent measures of all the waves concerned are linearly related. { 'lin·ē·ər tranz'dü·sər }

linear unit |ELECTR| An electronic device used in analog computers in which the change in output, due to any change in one of two or more input signals, is proportional to the change in that input and does not depend upon the values of the other inputs. { 'lin·ē·ər ¦yü·nət }

linear variable-differential transformer |ELECTR| A transformer in which a diaphragm or other transducer sensing element moves an armature linearly inside the coils of a differential transformer, to change the output voltage by changing the inductances of the coils in equal but opposite amounts. Abbreviated LVDT. { 'lin·ē·ər ¦ver·ē·ə·bəl ,dif·ə¦ren·chəl tranz¦fōr·mər }

line balance |ELEC| 1. Degree of electrical similarity of the two conductors of a transmission line. 2. Matching impedance, equaling the impedance of the line at all frequencies, that is used to terminate a two-wire line. { 'līn ,bal·əns }

line-balance converter See balun. { 'līn ,bal·əns kən,vərd·ər }

line-building-out network See impedance-matching network. { 'līn 'bild·iŋ ¦aút ,net,wərk }

line circuit |ELEC| 1. Equipment associated with each station connected to a dial or manual switchboard. 2. A circuit to interconnect an individual telephone and a channel terminal. { 'līn ,sər·kət }

line code |COMPUT SCI| The single instruction required to solve a specific type of problem on a special-purpose computer. { 'līn ,kōd }

line conditioning |COMMUN| The addition of compensating reactances to a data transmission line to reduce amplitude and phase delays over certain frequency bands. { 'līn kən,dish·ə·tyniŋ }

line conductor |ELEC| A metal used as a conductor in a power line; the most frequently used conductors are copper and aluminum. { 'līn kən ,dək·tər }

line-controlled blocking oscillator |ELECTR| A circuit formed by combining a monostable blocking oscillator with an open-circuit transmission line in the regenerative circuit; it is capable of generating pulses with large amounts of power. { 'līn kən,trōld ¦bläk·iŋ 'äs·ə,lād·ər }

line cord |ELEC| A two-wire cord terminating in a two-prong plug at one end and connected permanently to a radio receiver or other appliance at the other end; used to make connections to a source of power. Also known as power cord. { 'līn ,kórd }

line-cord resistor |ELEC| An asbestos-enclosed wire-wound resistor incorporated in a line cord along with the two regular wires. { 'līn ,kórd ri ,zis·tər }

line discipline |COMPUT SCI| The rules that govern exactly how data are transferred between locations in a communications network. { 'līn ,dis·ə·plən }

line dot matrix |COMPUT SCI| A line printer that uses the dot matrix printing technique. Also known as parallel dot character printer. { 'līn ¦dät 'mā,triks }

line driver |COMMUN| See limited-distance modem. |ELECTR| An integrated circuit that acts as the interface between logic circuits and a two-wire transmission line. { 'līn ,drīv·ər }

line drop |ELEC| The voltage drop existing between two points on a power line or transmission line, due to the impedance of the line. { 'līn ,dräp }

line-drop compensator |ELEC| A device that restores the voltage lost when electricity is transmitted along a wire. { 'līn ¦dräp 'käm·pən ,sād·ər }

line-drop signal |COMMUN| Signal associated with a subscriber line on a manual switchboard. { 'līn ¦dräp ,sig·nəl }

line editor |COMPUT SCI| A text-editing system that stores a file of discrete lines of text to be printed out on the console (or displayed) and manipulated on a line-by-line basis, so that editing operations are limited and are specified for lines identified by a specific number. { 'līn ,ed·əd·ər }

line equalizer |ELEC| An equalizer containing inductance or capacitance, inserted in a transmission line to modify the frequency response of the line. { 'līn ¦ē·kwə,līz·ər }

line facility |COMMUN| A transmission line in a communication system, together with amplifiers spaced at regular intervals to offset attenuation in the line. { 'līn fə,sil·əd·ē }

line fault |ELEC| A defect, such as an open circuit, short circuit, or ground, in an electric line for transmission or distribution of power or of speech, music, or other content. { 'līn ,fólt }

line feed |COMPUT SCI| 1. Signal that causes a printer to feed the paper up a discrete number of lines. 2. Rate at which paper is fed through a printer. { 'līn ,fēd }

line fill |COMMUN| Ratio of the number of connected main telephone stations on a line to the nominal main station capacity of that line. { 'līn ,fil }

line filter |ELEC| 1. A filter inserted between a power line and a receiver, transmitter, or other unit of electric equipment to prevent passage of noise signals through the power line in either direction. Also known as power-line filter. 2. A filter inserted in a transmission line or high-voltage power line for carrier communication purposes. { 'līn ,fil·tər }

line filter balance |COMMUN| Network designed to maintain phantom group balance when one side of the group is equipped with a carrier system. { 'līn ¦fil·tər ,bal·əns }

line finder |COMMUN| A switching device that automatically locates an idle telephone or telegraph circuit going to the desired destination. |COMPUT SCI| A device that automatically advances the platen of a line printer or typewriter. { 'līn ,fīn·dər }

line-finder switch |COMMUN| In telephony, an automatic switch for seizing selector apparatus which provides dial tone to the calling party. { 'līn ‚fīn·dər ‚swich }

line frequency |ELECTR| The number of times per second that the scanning spot sweeps across the screen in a horizontal direction in a video system. Also known as horizontal frequency; horizontal line frequency. { 'līn ‚frē·kwən·sē }

line-frequency blanking pulse See horizontal blanking pulse. { 'līn ‚frē·kwən·sē 'blaŋk·iŋ ‚pəls }

line interlace See interlaced scanning. { 'līn 'in·tər‚lās }

line item |COMPUT SCI| Any data that is considered to be of equal importance to other data in the same file. { 'līn ‚īd·əm }

line level |COMMUN| Signal level in decibels at a particular position on a transmission line. { 'līn ‚lev·əl }

line location |ELEC| The location of power and communications lines when two or more such lines run along the same route; they should either be used jointly, or located with respect to each other so as to avoid unnecessary crossings, conflicts, and inductive exposures. { 'līn lō ‚kā·shən }

line loop |COMMUN| Portion of a telephone circuit that includes a user's telephone set and the pair of wires that connect it with the distributing frame of a central office. { 'līn ‚lüp }

line-loop resistance |ELEC| Metallic resistance of the line wires that extend from an individual telephone set to the dial central office. { 'līn ‚lüp ri‚zis·təns }

line loss |ELEC| Total of the various energy losses occurring in a transmission line. { 'līn ‚lòs }

line microphone |ENG ACOUS| A highly directional microphone consisting of a single straight-line element or an array of small parallel tubes of different lengths, with one end of each abutting a microphone element. Also known as machine-gun microphone. { 'līn ‚mī·krə‚fōn }

line misregistration |COMPUT SCI| In character recognition, the improper appearance of a line of characters, on site in a character reader, with respect to a real or imaginary horizontal line. { 'līn ‚mis‚rej·ə'strā·shən }

line noise |COMMUN| Noise originating in a transmission line from such causes as poor joints and inductive interference from power lines. { 'līn ‚nòiz }

line number |COMPUT SCI| A number at the beginning or end of each line of a computer program that specifies its position in a sequence. { 'līn ‚nəm·bər }

line of code |COMPUT SCI| A single statement in a programming language. { 'līn əv 'kōd }

line of electrostatic induction |ELEC| A unit of electric flux equal to the electric flux associated with a charge of 1 statcoulomb. { 'līn əv i‚lek·trə ‚stad·ik in'dək·shən }

line of sight |ELECTROMAG| The straight line for a transmitting radar antenna in the direction of the beam. { 'līn əv 'sīt }

line pad |ELECTR| Pad inserted between a program amplifier and a transmission line, to isolate the amplifier from impedance variations of the line. { 'līn ‚pad }

line parameters See transmission-line parameters. { 'līn pə‚ram·əd·ərz }

line printer |COMPUT SCI| A device that prints an entire line in a single operation, without necessarily printing one character at a time. { 'līn ‚print·ər }

line printing |COMPUT SCI| The printing of an entire line of characters as a unit. { 'līn ‚print·iŋ }

line pulsing |ELECTR| Method of pulsing a transmitter in which an artificial line is charged over a relatively long period of time and then discharged through the transmitter tubes in a short interval determined by the line characteristic. { 'līn ‚pəls·iŋ }

line radiation |ELECTROMAG| Electromagnetic radiation from a power line caused mainly by corona pulses; gives rise to radio interference. { 'līn ‚rād·ē‚ā·shən }

line reflection |COMMUN| Reflection of a signal at the end of a transmission line, at the junction of two or more lines, or at a substation. { 'līn ri ‚flek·shən }

line regulation |ELEC| The maximum change in the output voltage or current of a regulated power supply for a specified change in alternating-current line voltage, such as from 105 to 125 volts. { 'līn ‚reg·yə'lā·shən }

line relay |ELEC| Relay which is controlled over a subscriber line or trunkline. { 'līn rē‚lā }

line-sequential color television |COMMUN| An analog color television system in which an entire line is one color, with colors changing from line to line in a red, blue, and green sequence. { 'līn si‚kwən·chəl 'kəl·ər 'tel·ə‚vizh·ən }

line side |ELEC| Terminal connections to an external or outstation source, such as data terminal connections to a communications circuit connecting to another data terminal. { 'līn ‚sīd }

line skew |COMPUT SCI| In character recognition, a form of line misregistration, when the string of characters to be recognized appears in a uniformly slanted condition with respect to a real or imaginary baseline. { 'līn ‚skyü }

line speed |COMMUN| Maximum rate at which signals may be transmitted over a given channel, usually in bauds or bits per second. { 'līn ‚spēd }

lines per minute |COMPUT SCI| A measure of the speed of the printer. Abbreviated LPM. { 'līnz pər 'min·ət }

line-stabilized oscillator |ELECTR| Oscillator in which a section of line is used as a sharply selective circuit element for the purpose of controlling the frequency. { 'līn ‚stā·bə‚līzd 'äs·ə‚lād·ər }

line stretcher |ELECTROMAG| Section of waveguide or rigid coaxial line whose physical length is variable to provide impedance matching. { 'līn ‚strech·ər }

line switching |COMMUN| A telephone switching system in which a switch attached to a subscriber line connects an originating call to an idle part of the switching apparatus. |ELECTR| Connecting or disconnecting the line voltage from a piece of electronic equipment.   { 'līn ,swich·iŋ }

line switching concentrator |COMMUN| Switching center used between a group of users and the switching center to reduce the number of trunks and increase efficiency of switching equipment usage (sometimes referred to as statistical multiplexing).   { līn ¦swich·iŋ ¦kän·sən,trād·ər }

line synchronizing pulse   See horizontal synchronizing pulse.   { 'līn ,siŋ·krə,nīz·iŋ ,pəls }

line-to-ground fault |ELEC| A defect in a power or communications line in which faulty insulation allows the conductor to make contact with the earth.   { 'līn tə 'graùnd ,fólt }

line transducer |ELECTR| A special type of electret transducer consisting essentially of a coaxial cable with polarized dielectric, and with the center conductor and shield serving as electrodes; mechanical excitation resulting in a deformation of the shield at any point along the length of the cable produces an electrical output signal.   { 'līn tranz,dü·sər }

line transformer |ELEC| Transformer connecting a transmission line to terminal equipment; used for isolation, line balance, impedance matching, or additional circuit connections.   { 'līn tranz ,fór·mər }

line trap |ELEC| A filter consisting of a series inductance shunted by a tuning capacitor, inserted in series with the power or telephone line for a carrier-current system to minimize the effects of variations in line attenuation and reduce carrier energy loss.   { 'līn ,trap }

line tuning |ELEC| Adjustment of the frequency of carrier current of a communication system to tune out the reactance of a capacitor with suitable inductance.   { 'līn ,tün·iŋ }

line turnaround |COMMUN| The time required for a half-duplex circuit to reverse the direction of transmission.   { 'līn 'tərn·ə,raùnd }

line unit |ELECTR| Electric control device used to send, receive, and control the impulses of a teletypewriter.   { 'līn ,yü·nət }

line-use ratio |COMMUN| As applied to facsimile broadcasting, the ratio of the available line to the total length of scanning line.   { 'līn ,yüs ,rā·shō }

line voltage |ELEC| The voltage provided by a power line at the point of use.   { 'līn ,vōl·tij }

line-voltage regulator |ELEC| A regulator that counteracts variations in power-line voltage, so as to provide an essentially constant voltage for the connected load.   { 'līn ,vōl·tij 'reg·yə,lād·ər }

linguistic model |COMPUT SCI| A method of automatic pattern recognition in which a class of patterns is defined as those patterns satisfying a certain set of relations among suitably defined primitive elements. Also known as syntactic model.   { liŋ'gwis·tik 'mäd·əl }

link |COMMUN| General term used to indicate the existence of communications facilities between two points.   { liŋk }

linkage |COMPUT SCI| In programming, coding that connects two separately coded routines.   { 'liŋ·kij }

linkage editor |COMPUT SCI| A service routine that converts the output of assemblers and compilers into a form that can be loaded and executed.   { 'liŋ·kij ,ed·əd·ər }

link control message |COMMUN| **1.** Message sent over a link of a network to condition the link to handle transmissions in a prearranged manner. **2.** Message used only between a pair of terminals for the conditioning of the link for digital system control.   { 'liŋk kən'trōl ,mes·ij }

linked list   See chained list.   { 'liŋkt 'list }

link encryption |COMMUN| The application of on-line crypto-operation to the individual links of relay systems so that all messages passing over the link are encrypted in their entirety.   { 'liŋk en'krip·shən }

link field |COMPUT SCI| The first word of a message buffer, used to point to the next buffer on the message queue.   { 'liŋk ,fēld }

link group |COMMUN| A collection of links that employ the same multiplex terminal equipment.   { 'liŋk ,grüp }

linking loader |COMPUT SCI| A loader which combines the functions of a relocating loader with the ability to combine a number of program segments that have been independently compiled into an executable program.   { 'liŋk·iŋ 'lōd·ər }

lin-log amplifier   See linear-logarithmic intermediate-frequency amplifier.   { ¦lin ¦läg 'am·plə ,fī·ər }

Linux |COMPUT SCI| A freely available, open-source operating system kernel capable of running on many different types of computer hardware; first released in 1991.   { 'lin·əks }

LIOCS |COMPUT SCI| Set of routines handling buffering, blocking, label checking, and overlap of input/output with processing. Derived from logical input/output control system.   { 'lī,äks }

lip-sync |COMMUN| Synchronization of sound and motion picture so that facial movements of speech coincide with the sounds.   { 'lip ,siŋk }

liquid crystal display |ELECTR| A digital display that consists of two sheets of glass separated by a sealed-in, normally transparent, liquid crystal material; the outer surface of each glass sheet has a transparent conductive coating such as tin oxide or indium oxide, with the viewing-side coating etched into character-forming segments that have leads going to the edges of the display; a voltage applied between front and back electrode coatings disrupts the orderly arrangement of the molecules, darkening the liquid enough to form visible characters even though no light is generated. Abbreviated LCD.   { 'lik·wəd 'krist·əl di'splā }

liquid-dielectric capacitor |ELEC| A capacitor in which the plate assemblies are mounted in a tank filled with a suitable oil or liquid dielectric.   { 'lik·wəd ,dī·ə¦lek·trik kə'pas·əd·ər }

liquid fuse unit |ELEC| Fuse unit in which the fuse link is immersed in a liquid, or provision is made for drawing the arc into the liquid when the fuse link melts.   { 'lik·wəd 'fyüz ,yü·nət }

liquid-metal fuel cell |ELEC| A fuel cell that uses molten potassium and bismuth as reactants and a molten salt electrolyte; has very high power output, but a relatively short life. { 'lik·wəd ¦med·əl 'fyül ˌsel }

liquid-metal MHD generator [ELEC] A system for generating electric power in which the kinetic energy of a flowing, molten metal is converted to electric energy by magnetohydrodynamic (MHD) interaction. { 'lik·wəd ¦med·əl ¦em¦āch¦dē 'jen·ə ˌrād·ər }

liquid rheostat [ELECTR] A variable-resistance type of voltage regulator in which the variable-resistance element is liquid, usually water; carbon electrodes are raised or lowered in the liquid to change resistance ratings and control voltage flow. { 'lik·wəd 'rē·ə,stat }

liquid semiconductor [ELECTR] An amorphous material in solid or liquid state that possesses the properties of varying resistance induced by charge carrier injection. { 'lik·wəd 'sem·i·kən ˌdək·tər }

LISP [COMPUT SCI] An interpretive language developed for the manipulation of symbolic strings of recursive data; can also be used to manipulate mathematical and arithmetic logic. Derived from list processing language. { lisp }

list [COMPUT SCI] **1.** A last-in, first-out storage organization, usually implemented by software, but sometimes implemented by hardware. **2.** In FORTRAN, a set of data items to be read or written. { list }

list processing [COMPUT SCI] A programming technique in which list structures are used to organize memory. { 'list ¦prä,ses·iŋ }

list processing language See LISP. { 'list ¦prä ˌses·iŋ ˌlaŋ·gwij }

listserv [COMPUT SCI] The software (server) used to maintain an electronic mailing list. Also known as list server. { 'list,sərv }

list server See listserv. { 'list ˌsər·vər }

list structure [COMPUT SCI] A set of data items, connected together because each element contains the address of a successor element (and sometimes of a predecessor element). { 'list ˌstrək·chər }

literal operand [COMPUT SCI] An operand, usually occurring in a source language instruction, whose value is specified by a constant which appears in the instruction rather than by an address where a constant is stored. { 'lid·ə·rəl ¦äp·ə¦rand }

lithium battery [ELEC] A solid-state battery with a lithium anode, an iodine-polyvinyl pyridine cathode, and an electrolyte consisting of a layer of lithium iodide; used in cardiac pacemakers. { 'lith·ē·əm 'bad·ə·rē }

lithium cell [ELEC] A primary cell for producing electrical energy by using lithium metal for one electrode immersed in usually an organic electrolyte. { 'lith·ē·əm ˌsel }

lithium-drifted germanium crystal [ELECTR] A high-resolution junction detector, used especially for more penetrating gamma-radiation and higher-energy electrons, produced by drifting lithium ions through a germanium crystal to produce an intrinsic region where impurity-based carrier generation centers are deactivated,

sandwiched between a p layer and an n layer. { 'lith·ē·əm ¦drif·təd jər'mā·nē·əm 'krist·əl }

lithium-sulfur battery [ELEC] A storage battery in which the cells use a molten lithium cathode and a molten sulfur anode separated by a molten salt electrolyte that consists of lithium iodide, potassium iodide, and lithium chloride. { 'lith·ē·əm ¦səl·fər 'bad·ə·rē }

lithography [ELECTR] A technique used for integrated circuit fabrication in which a silicon slice is coated uniformly with a radiation-sensitive film, the resist, and an exposing source (such as light, x-rays, or an electron beam) illuminates selected areas of the surface through an intervening master template for a particular pattern. { ¦ə'thäg·rə·fē }

little LEO system [COMMUN] A system of small satellites in low earth orbit (LEO) that provides messaging, data, and location services but does not have the capability of voice transmission. { ¦lit·əl ¦lē·o ˌsis·təm }

litzendraht wire See litz wire. { 'lits·ən,drät ˌwīr }

litz wire [ELEC] Wire consisting of a number of separately insulated strands woven together so each strand successively takes up all possible positions in the cross section of the entire conductor, to reduce skin effect and thereby reduce radio-frequency resistance. Derived from litzendraht wire. { 'lits ˌwīr }

live [COMMUN] Being broadcast directly at the time of production, instead of from recorded or filmed program material. { līv }

live chassis [ELECTR] A radio, television, or other chassis that has a direct chassis connection to one side of the alternating-current line. { 'līv 'cha·sē }

live data [COMPUT SCI] Actual data that are employed during the final testing of a computer system, as opposed to test data. { 'līv 'dad·ə }

live system [COMPUT SCI] A computer system on which all testing has been completed so that it is fully operational and ready for production work. { 'līv 'sis·təm }

liveware [COMPUT SCI] The people involved in the operation of a computer system, thought of as a component of the system along with hardware and software. { 'līv,wer }

LLL circuit See low-level logic circuit. { ¦el¦el'el ˌsər·kət }

L network [ELECTR] A network composed of two branches in series, with the free ends connected to one pair of terminals; the junction point and one free end are connected to another pair of terminals. { 'el ˌnet,wərk }

load [COMPUT SCI] **1.** To place data into an internal register under program control. **2.** To place a program from external storage into central memory under operator (or program) control, particularly when loading the first program into an otherwise empty computer. **3.** An instruction, or operator control button, which causes the computer to initiate the load action. **4.** The amount of work scheduled on a computer system, usually expressed in hours of work. [ELEC] **1.** A device that consumes electric power. **2.** The amount of electric power that is drawn from a power line, generator, or other power source. **3.** The material to be heated by an induction heater or dielectric heater. Also known as work. |ELECTR| The device that receives the useful signal output of an amplifier, oscillator, or other signal source. { lōd }

load-and-go |COMPUT SCI| An operating technique with no stops between the loading and execution phases of a program; may include assembling or compiling. { ¦lōd ən 'gō }

load-break switch |ELEC| An electric switch in a circuit with several hundred thousand volts, designed to carry a large amount of current without overheating the open position, having enough insulation to isolate the circuit in closed position, and equipped with arc interrupters to interrupt the load current. { 'lōd ¦brāk ,swich }

load cell |ELEC| A device which measures large pressures by applying the pressure to a piezoelectric crystal and measuring the voltage across the crystal; the cell plus a recording mechanism constitutes a strain gage. { 'lōd ,sel }

load characteristic |ELECTR| Relation between the instantaneous values of a pair of variables such as an electrode voltage and an electrode current, when all direct electrode supply voltages are maintained constant. Also known as dynamic characteristic. { 'lōd ,kar·ik·tə'ris·tik }

load circuit |ELECTR| Complete circuit required to transform power from a source such as an electron tube to a load. { 'lōd ,sər·kət }

load circuit efficiency |ELECTR| Ratio between useful power delivered by the load circuit to the load and the load circuit power input. { 'lōd ¦sər·kət i'fish·ən·sē }

load compensation |CONT SYS| Compensation in which the compensator acts on the output signal after it has generated feedback signals. Also known as load stabilization. { 'lōd käm·pən'sā·shən }

load curve |ELEC| A graph that plots the power supplied by an electric power system versus time. { 'lōd ,kərv }

load divider |ELEC| Unit for distributing power to various units. { 'lōd di,vīd·ər }

loaded line |ELEC| Wire line in which loading coils have been inserted at regular intervals to reduce attenuation and phase lag at the frequencies within the band used. { 'lōd·əd līn }

loaded motional impedance See motional impedance. { 'lōd·əd ¦mō·shən·əl im'pēd·əns }

loaded Q |ELECTROMAG| The Q factor of a specific mode of resonance of a microwave tube or resonant cavity when there is external coupling to that mode. { 'lōd·əd kyü }

loader |COMPUT SCI| A computer program that takes some other program from an input or storage device and places it in memory at some predetermined address. { 'lōd·ər }

load factor |ELEC| The ratio of average electric load to peak load, usually calculated over a 1-hour period. { 'lōd ,fak·tər }

load impedance |ELECTR| The complex impedance presented to a transducer by its load. { 'lōd im,pēd·əns }

loading |ELEC| The addition of inductance to a transmission line to improve its transmission characteristics throughout a given frequency band. Also known as electrical loading. |ENG ACOUS| Placing material at the front or rear of a loudspeaker to change its acoustic impedance and thereby alter its radiation. { 'lōd·iŋ }

loading coil |ELECTROMAG| **1.** An iron-core coil connected into a telephone line or cable at regular intervals to lessen the effect of line capacitance and reduce distortion. Also known as Pupin coil; telephone loading coil. **2.** A coil inserted in series with a radio antenna to increase its electrical length and thereby lower the resonant frequency. { 'lōd·iŋ ,köil }

loading device |COMPUT SCI| Equipment from which programs or other data can be transferred or copied into a computer. { 'lōd·iŋ di,vīs }

loading disk |ELECTROMAG| Circular metal piece mounted at the top of a vertical antenna to increase its natural wavelength. { 'lōd·iŋ ,disk }

loading program |COMPUT SCI| Program used to load other programs into computer memory. Also known as bootstrap program. { 'lōd·iŋ ,prō·grəm }

loading routine See input routine. { 'lōd·iŋ rü ,tēn }

load isolator |ELECTROMAG| Waveguide or coaxial device that provides a good energy path from a signal source to a load, but provides a poor energy path for reflections from a mismatched load back to the signal source. { 'lōd ,ī·sə ,lād·ər }

load leveling |ELEC| A method for reducing the large fluctuations that occur in electricity demand, for example by storing excess electricity during periods of low demand for use during periods of high demand. { 'lōd ,lev·ə·liŋ }

load line |ELECTR| A straight line drawn across a series of tube or transistor characteristic curves to show how output signal current will change with input signal voltage when a specified load resistance is used. { 'lōd ,līn }

load loss |ELEC| The sum of the copper loss of a transformer, due to resistance in the windings, plus the eddy current loss in the winding, plus the stray loss. { 'lōd ,lòs }

load module |COMPUT SCI| A program in a form suitable for loading into memory and executing. { 'lōd ,mä·jül }

load point |COMPUT SCI| Preset point on a magnetic tape from which reading or writing will start. { 'lōd ,pòint }

load power |ELEC| Of an energy load, the average rate of flow of energy through the terminals of that load when connected to a specified source. { 'lōd ,paù·ər }

load regulation |ELEC| The maximum change in the output voltage or current of a regulated power supply for a specified change in load conditions. { 'lōd ,reg·yə,lā·shən }

load shedding |ELEC| A procedure in which parts of an electric power system are disconnected in an attempt to prevent failure of the entire system due to overloading. { 'lōd ,shed·iŋ }

load shifting |ELEC| In an electric power system, the transfer of loads from times of peak demand to off-peak time periods. { 'lōd ,shift·iŋ }

load stabilization *See* load compensation. { 'lōd ,stā·bə·lə,zā·shən }

lobe |ELECTROMAG| A part of the radiation pattern of a directional antenna representing an area of stronger radio-signal transmission. Also known as radiation lobe. |ENG ACOUS| A portion of the directivity pattern of a transducer representing an area of increased emission or response. { lōb }

lobe-half-power width |ELECTROMAG| In a plane containing the direction of the maximum energy of a lobe, the angle between the two directions in that plane about the maximum in which the radiation intensity is one-half the maximum value of the lobe. { ¦lōb ¦haf ¦pau̇·ər ,width }

lobe switching *See* beam switching. { 'lōb ,swich·iŋ }

lobing |ELECTROMAG| Formation of maxima and minima at various angles of the vertical plane antenna pattern by the reflection of energy from the surface surrounding the radar antenna; these reflections reinforce the main beam at some angles and detract from it at other angles, producing fingers of energy. { 'lōb·iŋ }

local action |ELEC| **1.** Internal losses of a battery caused by chemical reactions producing local currents between different parts of a plate. **2.** Quantitatively, the percentage loss per month in the capacity of a battery on open circuit, or the amount of current needed to keep the battery fully charged. { 'lō·kəl 'ak·shən }

local-area network |COMPUT SCI| A communications network connecting various hardware devices together within a building by means of a continuous cable, an in-house voice-data telephone system, or a radio-based system. Abbreviated LAN. { 'lō·kəl¦er·ē·ə 'net,wərk }

local battery |ELEC| Battery that actuates the telegraphic station recording instruments, as distinguished from the battery furnishing current to the line. { 'lō·kəl 'bad·ə·rē }

local-battery telephone set |ELECTR| Telephone set for which the transmitter current is supplied from a battery, or other current supply circuit, individual to the telephone set; the signaling current may be supplied from a local hand generator or from a centralized power source. { 'lō·kəl 'bad·ə·rē 'tel·ə,fōn ,set }

local cell |ELEC| A galvanic cell resulting from differences in potential between adjacent areas on the surface of a metal immersed in an electrolyte. { 'lō·kəl 'sel }

local central office |COMMUN| A telephone central office, which terminates subscriber lines and makes connections with other central offices, usually equipped to serve 10,000 main telephones of its immediate community. { 'lō·kəl ¦sen·trəl 'ȯf·əs }

local circuit |COMMUN| Circuit to a main or auxiliary circuit which can be made available at any station or patched from point to point through one or more stations. { 'lō·kəl 'sər·kət }

local control |COMMUN| System or method of radio-transmitter control whereby the control functions are performed directly at the transmitter. { 'lō·kəl kən'trōl }

local controller *See* first-level controller. { 'lō·kəl kən'trōl·ər }

local device |COMPUT SCI| Peripheral equipment that is linked directly to a computer or other supporting equipment, without an intervening communications channel. { 'lō·kəl di'vīs }

local exchange *See* exchange. { 'lō·kəl iks'chānj }

localization |COMPUT SCI| Imposing some physical order upon a set of objects, so that a given object has a greater probability of being in some particular regions of space than in others. { ,lō·kə·lə'zā·shən }

local line *See* local loop. { 'lō·kəl 'līn }

local loop |COMMUN| A telephone line from the user's location that terminates at the local central office. Also known as local line. { 'lō·kəl 'lüp }

local networking |CONT SYS| The system of communication linking together the components of a single robot. { 'lō·kəl 'net,wərk·iŋ }

local oscillator |ELECTR| The oscillator in a superheterodyne receiver, whose output is mixed with the incoming modulated radio-frequency carrier signal in the mixer to give the frequency conversions needed to produce the intermediate-frequency signal. { 'lō·kəl 'äs·ə,lād·ər }

local-oscillator injection |ELECTR| Adjustment used to vary the magnitude of the local oscillator signal that is coupled into the mixer. { 'lō·kəl 'äs·ə,lād·ər in¦jek·shən }

local-oscillator radiation |ELECTR| Radiation of the fundamental or harmonics of the local oscillator of a superheterodyne receiver. { 'lō·kəl 'äs·ə,lād·ər ,rād·ē'ā·shən }

local register |COMPUT SCI| One of a relatively small number (usually less than 32) of high-speed storage elements in a computer system which may be directly referred to by the instructions in the programs. Also known as general register. { 'lō·kəl 'rej·ə·stər }

local side |COMMUN| Terminal connections to an internal or in-station source such as data terminal connections to input or output devices. { 'lō·kəl ,sīd }

local storage |COMPUT SCI| The collection of local registers in a computer system. { 'lō·kəl 'stȯr·ij }

local trunk |COMMUN| Trunk between local and long-distance switchboards, or between local and private branch exchange switchboards. { 'lō·kəl 'trəŋk }

local variable |COMPUT SCI| A variable which can be accessed (used or changed) only in one block of a computer program. { 'lō·kəl 'ver·ē·ə·bəl }

locate mode |COMPUT SCI| A method of communicating with an input/output control system (IOCS), in which the address of the data involved, but not the data themselves, is transferred between the IOCS routine and the program. { 'lō,kāt ,mōd }

location |COMPUT SCI| Any place in which data may be stored; usually expressed as a number. { lō'kā·shən }

location constant |COMPUT SCI| A number that identifies an instruction in a computer program, written in a higher-level programming language, and used to refer to this instruction at other points in the program. Also known as label constant. { lō'kā·shən ˌkän·stənt }

location counter *See* instruction counter. { lō'kā·shən ˌkaunt·ər }

lock |ELECTR| **1.** To fasten onto and automatically follow a target by means of a radar beam; similarly to acquire and maintain attention to the signal from a single source by anticipation of some feature of it. **2.** To control the frequency of an oscillator by means of an applied signal of constant frequency. { läk }

locked-in line |COMMUN| A telephone line that remains established after the caller has hung up. { 'läktˌin ˌlīn }

locked oscillator |ELECTR| A sine-wave oscillator whose frequency can be locked by an external signal to the control frequency divided by an integer. { 'läkt 'äs·əˌlād·ər }

locked-oscillator detector |ELECTR| A frequency-modulation detector in which a local oscillator follows, or is locked to, the input frequency; the phase difference between local oscillator and input signal is proportional to the frequency deviation, and an output voltage is generated proportional to the phase difference. { 'läkt ¦äs·əˌlād·ər di,tek·tər }

locked-rotor current |ELEC| The current drawn by a stalled electric motor. { 'läkt ¦rōd·ər ˌkə·rənt }

lock-in |ELECTR| Shifting and automatic holding of one or both of the frequencies of two oscillating systems which are coupled together, so that the two frequencies have the ratio of two integral numbers. { 'läk ˌin }

lock-in amplifier |ELECTR| An amplifier that uses some form of automatic synchronization with an external reference signal to detect and measure very weak electromagnetic radiation at radio or optical wavelengths in the presence of very high noise levels. { 'läkˌin 'am·pləˌfī·ər }

locking |ELECTR| Controlling the frequency of an oscillator by means of an applied signal of constant frequency. |ENG| Automatic following of a target by a radar antenna. { 'läk·iŋ }

lock-on |ELECTR| **1.** The procedure wherein a target-seeking system (such as some types of radars) is continuously and automatically following a target in one or more coordinates (for example, range, bearing, elevation), or wherein a signal intercept system isolates signals from a single source. **2.** The instant at which radar begins to track a target automatically. { 'läk ¦ón }

lockout |COMMUN| **1.** In a telephone circuit controlled by two voice-operated devices, the inability of one or both subscribers to get through, because of either excessive local circuit noise or continuous speech from one or both subscribers. Also known as receiver lockout system. **2.** In

mobile communications, an arrangement of control circuits whereby only one receiver can feed the system at one time to avoid distortion. Also known as receiver lockout system. |COMPUT SCI| **1.** In computer communications, the inability of a remote terminal to achieve entry to a computer system until project programmer number, processing authority code, and password have been validated against computer-stored lists. **2.** The precautions taken to ensure that two or more programs executing simultaneously in a computer system do not access the same data at the same time, make unauthorized changes in shared data, or otherwise interfere with each other. **3.** Preventing the central processing unit of a computer from accessing storage because input/output operations are taking place. **4.** Preventing input and output operations from taking place simultaneously. { 'läkˌaut }

lockout circuit |ELECTR| A switching circuit which responds to concurrent inputs from a number of external circuits by responding to one, and only one, of these circuits at any time. Also known as finding circuit; hunting circuit. { 'läk ˌaut ˌsər·kət }

lock-up relay |ELEC| A relay that locks in its energized position either by permanent magnetic biasing which can be released only by applying a reverse magnetic pulse or by auxiliary contacts that keep its coil energized until the circuit is interrupted. { 'läkˌəp rē,lā }

lodar |NAV| A direction finder used to determine the direction of arrival of loran signals, free of night effect, by observing the separately distinguishable ground and sky-wave loran signals on a cathode-ray oscilloscope and positioning a loop antenna to obtain a null indication of the component selected to be most suitable. Also known as lorad. { 'lō,där }

log |COMMUN| A record of radio and television station operating data. |COMPUT SCI| A record of computer operating runs, including tapes used, control settings, halts, and other pertinent data. { läg }

logarithmic amplifier |ELECTR| An amplifier whose output signal is a logarithmic function of the input signal. { 'läg·əˌrith·mik 'am·pləˌfī·ər }

logarithmic diode |ELECTR| A diode that has an accurate semilogarithmic relationship between current and voltage over wide and forward dynamic ranges. { 'läg·əˌrith·mik 'dī,ōd }

logarithmic fast time constant |ELECTR| Constant false alarm rate scheme which has a logarithmic intermediate-frequency amplifier followed by a fast time constant circuit. { 'läg·ə ˌrith·mik 'fast ¦tīm ˌkän·stənt }

logarithmic multiplier |ELECTR| A multiplier in which each variable is applied to a logarithmic function generator, and the outputs are added together and applied to an exponential function generator, to obtain an output proportional to the product of two inputs. { 'läg·ə‚ri<u>th</u>·mik 'məl·tə‚plī·ər }

logbook |COMPUT SCI| A bound volume in which operating data of a computer is noted. { 'läg ‚bůk }

logic |ELECTR| **1.** The basic principles and applications of truth tables, interconnections of on/off circuit elements, and other factors involved in mathematical computation in a computer. **2.** General term for the various types of gates, flip-flops, and other on/off circuits used to perform problem-solving functions in a digital computer. { 'läj·ik }

logical comparison |COMPUT SCI| The operation of comparing two items in a computer and producing a one output if they are equal or alike, and a zero output if not alike. { 'läj·ə·kəl kəm'par·ə·sən }

logical construction |COMPUT SCI| A simple logical property that determines the type of characters which a particular code represents; for example, the first two bits can tell whether a character is numeric or alphabetic. { 'läj·ə·kəl kən'strək·shən }

logical data independence |COMPUT SCI| A data base structured so that changing the logical structure will not affect its accessibility by the program reading it. { 'läj·ə·kəl ¦dad·ə ‚in·də'pen·dəns }

logical data type |COMPUT SCI| A scalar data type in which a data item can have only one of two values: true or false. Also known as Boolean data type. { 'läj·ə·kəl 'dad·ə ‚tīp }

logical decision |COMPUT SCI| The ability to select one of many paths, depending upon intermediate programming data. { 'läj·ə·kəl di'sizh·ən }

logical device table |COMPUT SCI| A table that is used to keep track of information pertaining to an input/output operation on a logical unit, and that contains such information as the symbolic name of the logical unit, the logical device type and the name of the file currently attached to it, the logical input/output request currently pending on the device, and a pointer to the buffers currently associated with the device. { 'läj·ə·kəl di¦vīs ‚tā·bəl }

logical drive |COMPUT SCI| A data storage unit, such as a subpartition of a hard drive or an array of storage units, recognized and handled according to the logic of the operating system like a single physical drive. { 'läj·i·kəl ¦drīv }

logical expression |COMPUT SCI| Two arithmetic expressions connected by a relational operator indicating whether an expression is greater than, equal to, or less than the other, or connected by a logical variable, logical constant (true or false), or logical operator. { 'läj·ə·kəl ik'spresh·ən }

logical field |COMPUT SCI| A data field whose variables can take on only two values, which are designated yes and no, true and false, or 0 and 1. { 'läj·ə·kəl 'fēld }

logical file |COMPUT SCI| A file as seen by the program accessing it. { 'läj·ə·kəl 'fīl }

logical flow chart |COMPUT SCI| A detailed graphic solution in terms of the logical operations required to solve a problem. { 'läj·ə·kəl 'flō ‚chärt }

logical gate See switching gate. { 'läj·ə·kəl 'gāt }

logical instruction |COMPUT SCI| A digital computer instruction which forms a logical combination (on a bit-by-bit basis) of its operands and leaves the result in a known location. { 'läj·ə·kəl in'strək·shən }

logical network |COMPUT SCI| **1.** A collection of computers that is presented as a single network to the user, although it may encompass more than one physical network. **2.** A part of a network of computers that is set up to function as a separate network. { 'läj·i·kəl 'net‚wərk }

logical page |COMPUT SCI| A unit of computer storage consisting of a specified number of bytes. { 'läj·ə·kəl 'pāj }

logical record |COMPUT SCI| A group of adjacent, logically related data items. { 'läj·ə·kəl 'rek·ərd }

logical security |COMPUT SCI| Mechanisms internal to a computing system that are used to protect against internal misuse of computing time and unauthorized access to data. { 'läj·ə·kəl sə'kyür·əd·ē }

logical shift |COMPUT SCI| A shift operation that treats the operand as a set of bits, not as a signed numeric value or character representation. { 'läj·ə·kəl 'shift }

logical sum |COMPUT SCI| A computer addition in which the result is 1 when either one or both input variables is 1, and the result is 0 when the input variables are both 0. { 'läj·ə·kəl 'səm }

logical symbol |COMPUT SCI| A graphical symbol used to represent a logic element. { 'läj·ə·kəl 'sim·bəl }

logical unit |COMPUT SCI| An abstraction of an input/output device in the form of an additional name given to the device in a computer program. { 'läj·ə·kəl 'yü·nət }

logic-arithmetic unit See arithmetical unit. { 'läj·ik·ə'rith·mə·tik ‚yü·nət }

logic bomb |COMPUT SCI| A computer program that destroys data, generally immediately after it has been loaded. { 'läj·ik ‚bäm }

logic card |ELECTR| A small fiber chassis on which resistors, capacitors, transistors, magnetic cores, and diodes are mounted and interconnected in such a way as to perform some computer function; computers employing this type of construction may be repaired by removing the faulty card and replacing it with a new card. { 'läj·ik ‚kärd }

logic chip |COMPUT SCI| An integrated circuit that performs logic functions. { 'läj·ik ‚chip }

logic circuit |COMPUT SCI| A computer circuit that provides the action of a logic function or logic operation. Also known as logic gate. { 'läj·ik ‚sär·kət }

logic design |COMPUT SCI| The design of a computer at the level which considers the operation of each functional block and the relationships between the functional blocks. { 'läj·ik di‚zīn }

logic diagram |COMPUT SCI| A graphical representation of the logic design or a portion thereof; displays the existence of functional elements and the paths by which they interact with one another. { 'läj·ik ,dī·ə,gram }

logic element [COMPUT SCI] A hardware circuit that performs a simple, predefined transformation on its input and presents the resulting signal as its output. Occasionally known as functor. { 'läj·ik ,el·ə·mənt }

logic error [COMPUT SCI] An error in programming that is caused by faulty reasoning, resulting in the program's functioning incorrectly if the instructions containing the error are encountered. { 'läj·ik ,er·ər }

logic gate See logic circuit. { 'läj·ik ,gāt }

logic high [ELECTR] The electronic representation of the binary digit 1 in a digital circuit or device. { 'läj·ik 'hī }

logic level [ELECTR] One of the two voltages whose values have been arbitrarily chosen to represent the binary numbers 1 and 0 in a particular data-processing system. { 'läj·ik ,lev·əl }

logic low [ELECTR] The electronic representation of the binary digit 0 in a digital circuit or device. { 'läj·ik ,lō }

logic operation [COMPUT SCI] A nonarithmetical operation in a computer, such as comparing, selecting, making references, matching, sorting, and merging, where logical yes-or-no quantities are involved. { 'läj·ik ,äp·ə'rā·shən }

logic operator [COMPUT SCI] A rule which assigns, to every combination of the values "true" and "false" among one or more independent variables, the value "true" or "false" to a dependent variable. { 'läj·ik 'äp·ə,rād·ər }

logic section See arithmetical unit. { 'läj·ik ,sek·shən }

logic-seeking printer [COMPUT SCI] A line printer that examines each line to be printed so that it can save time by skipping over blank spaces. { 'läj·ik ¦sēk·iŋ 'print·ər }

logic swing [ELECTR] The voltage difference between the logic levels used for 1 and 0; magnitude is chosen arbitrarily for a particular system and is usually well under 10 volts. { 'läj·ik ,swiŋ }

logic switch [ELECTR] A diode matrix or other switching arrangement that is capable of directing an input signal to one of several outputs. { 'läj·ik ,swich }

logic unit [COMPUT SCI] A separate unit which exists in some computer systems to carry out logic (as opposed to arithmetic) operations. { 'läj·ik ,yü·nət }

logic word [COMPUT SCI] A machine word which represents an arbitrary set of digitally encoded symbols. { 'läj·ik ,wərd }

log-in See log-on. { 'läg,in }

logo [COMPUT SCI] A high-level, interactive programming language that features a triangular shape called a turtle which can be moved about an electronic display through the use of familiar English-word commands. { 'lō·gō }

log-off [COMPUT SCI] The procedure for a user to disconnect from a computer system, including the release of resources that were assigned to the user. { 'läg ,óf }

log-on [COMPUT SCI] The procedure for users to identify themselves to a computer system for authorized access to their programs and information. { 'läg ,ón }

log-out See log-off. { 'läg ,aút }

log-periodic antenna [ELECTROMAG] A broadband antenna which consists of a sheet of metal with two wedge-shaped cutouts, each with teeth cut into its radii along circular arcs; characteristics are repeated at a number of frequencies that are equally spaced on a logarithmic scale. { 'läg ,pir·ē¦ad·ik an'ten·ə }

loktal base [ELECTR] A special base for small vacuum tubes, so designed that it locks the tube firmly in a corresponding special eight-pin loktal socket; the tube pins are sealed directly into the glass envelope. { 'läk·təl ,bās }

long-base-line system [COMMUN] System in which the distance separating ground stations approximates the distance to the target being tracked. { 'lóŋ 'bās ,līn ,sis·təm }

long card [COMPUT SCI] A full-size printed circuit board that is plugged into an expansion slot in a microcomputer. { 'lóŋ ¦kärd }

long-conductor antenna See long-wire antenna. { 'lóŋ kən¦dək·tər an,ten·ə }

long discharge [ELEC] **1.** A capacitor or other electrical charge accumulator which takes a long time to leak off. **2.** A gaseous electrical discharge in which the length of the discharge channel is very long compared with its diameter; lightning discharges are natural examples of long discharges. Also known as long spark. { 'lóŋ 'dis,chärj }

long-distance loop [COMMUN] Line from a subscriber's station directly to a long-distance switchboard. { 'lóŋ ,dis·təns 'lüp }

long-distance xerography [COMMUN] A facsimile system that uses a cathode-ray scanner at the microwave transmitting terminal; at the receiving terminal, a lens projects the received cathode-ray image onto the selenium-coated drum of a xerographic copying machine. { 'lóŋ ,dis·təns zi'räg·rə,fē }

long-haul carrier system [COMMUN] An intercity telephone communication system; it may use a frequency-division multiplexed signal modulating subcarrier or it may use digital technology. { 'lóŋ ,hól 'kar·ē·ər ,sis·təm }

long-haul radio [COMMUN] A microwave radio system capable of transmitting telephone, video, data, and telegraph signals over distances on the order of 4000 miles (6500 kilometers) or more on line-of-sight paths between a series of repeaters that demodulate the signal to an intermediate frequency and then remodulate it. { 'lóŋ ,hól 'räd·ē·ō }

longitudinal circuit [ELEC] Circuit formed by one telephone wire (or by two or more telephone wires in parallel) with return through the earth or through any other conductors except those which are taken with the original wire or wires to form a metallic telephone circuit. { ,län·jə'tüd·ən·əl 'sər·kət }

longitudinal current [ELEC] Current which flows in the same direction in the two wires of a parallel pair using the earth or other conductors for a return path. { ˌlän·jə'tüd·ən·əl 'kə·rənt }

longitudinal-mode delay line [ELECTR] A magnetostrictive delay in which signals are propagated by means of longitudinal vibrations in the magnetostrictive material. { ˌlän·jə'tüd·ən·əl ¦mōd di'lā ˌlīn }

longitudinal parity [COMMUN] Parity associated with bits recorded on one track in a data block, to indicate whether the number of recorded bits in the block is even or odd. { ˌlän·jə'tüd·ən·əl 'par·əd·ē }

longitudinal parity check [COMMUN] The count for even or odd parity of all the bits in a message as a precaution against transmission error. Also known as horizontal parity check. { ˌlän·jə'tüd·ən·əl 'par·əd·ē ˌchek }

longitudinal redundancy check [COMMUN] A method of checking for errors, in which data are arranged in blocks according to some rule, and the correctness of each character in the block is determined according to the rule. Abbreviated LRC. { ˌlän·jə'tüd·ən·əl ri'dən·dən·sē ˌchek }

Longley-Rice [COMMUN] A model used to predict the long-term median transmission loss over irregular terrain that is applied to predicting signal strength at one or more locations. Longley-Rice computations are employed both by the FCC allocations rules for FM stations to predict signal strength contours and by propagation modeling software to predict signal strengths in a two-dimensional grid on a map. The FCC implementation of Longley-Rice computations employs average terrain computations and an assumed 30-ft receive antenna height. { 'lȯŋ·lē 'rīs }

long-line current [ELEC] A current that flows through the earth from an anodic to a cathodic area and returns along an underground pipe or other metal structure, often over a considerable distance and as the result of concentration cell action. { 'lȯŋ ˌlīn ˌkə·rənt }

long-line effect [ELECTR] An effect occurring when an oscillator is coupled to a transmission line with a bad mismatch; two or more frequencies may then be equally suitable for oscillation, and the oscillator jumps from one of these frequencies to another as its load changes. { 'lȯŋ ˌlīn i‚fekt }

long-lines engineering [COMMUN] Engineering performed to develop, modernize, or expand long-haul, point-to-point communications facilities using radio, microwave, or wire circuits. { 'lȯŋ ˌlīnz 'en·jə'nir·iŋ }

long-persistence screen [ELECTR] A fluorescent screen containing phosphorescent compounds that increase the decay time, so a pattern may be seen for several seconds after it is produced by the electron beam. { 'lȯŋ pər‚sis·təns 'skrēn }

long spark See long discharge. { 'lȯŋ ¦spärk }

long-tail pair [ELECTR] A two-tube or transistor circuit that has a common resistor (tail resistor) which gives strong negative feedback. { 'lȯŋ ˌtāl ˌper }

long-term predictor [COMMUN] An electric filter that removes redundancies in a signal associated

with long-term correlations so that information can be transmitted more efficiently. { ¦lȯŋ ˌtərm prə'dik·tər }

long-term repeatability [CONT SYS] The close agreement of positional movements of a robotic system repeated under identical conditions over long periods of time. { 'lȯŋ ˌtərm ri‚pēd·ə'bil·əd·ē }

long wave [COMMUN] An electromagnetic wave having a wavelength longer than the longest broadcast-band wavelength of about 545 meters, corresponding to frequencies below about 550 kilohertz. { 'lȯŋ ¦wāv }

long-wave radio [COMMUN] A radio which can receive frequencies below the lowest broadcast frequency of 550 kilohertz. { 'lȯŋ ˌwāv 'rād·ē·ō }

long-wire antenna [ELECTROMAG] An antenna whose length is a number of times greater than its operating wavelength, so as to give a directional radiation pattern. Also known as long-conductor antenna. { 'lȯŋ ˌwīr an'ten·ə }

lookahead [COMPUT SCI] A procedure in which a processor is preparing one instruction in a computer program while executing its predecessor. { 'lúk·ə‚hed }

lookahead tree See game tree. { 'lúk·ə‚hed ‚trē }

look-through [ELECTR] **1.** When jamming, a technique whereby the jamming emission is interrupted irregularly for extremely short periods to allow monitoring of the victim signal during jamming operations. **2.** When being jammed, the technique of observing or monitoring a desired signal during interruptions in the jamming signal. { 'lúk ‚thrü }

look time See dwell time. { 'lük ‚tīm }

look-up [COMPUT SCI] An operation or process in which a table of stored values is scanned (or searched) until a value equal to (or sometimes, greater than) a specified value is found. { 'lúk‚əp }

look-up table [COMPUT SCI] A stored matrix of data for reference purpose. { 'lúk ‚əp ˌtā·bəl }

loop [COMPUT SCI] A sequence of computer instructions which are executed repeatedly, but usually with address modifications changing the operands of each iteration, until a terminating condition is satisfied. [ELEC] **1.** A closed path or circuit over which a signal can circulate, as in a feedback control system. **2.** Commercially, the portion of a connection from central office to subscriber in a telephone system. [ELECTROMAG] See coupling loop; loop antenna. { lüp }

loop antenna [ELECTROMAG] A directional-type antenna consisting of one or more complete turns of a conductor, usually tuned to resonance by a variable capacitor connected to the terminals of the loop. Also known as loop. { 'lüp ‚ten·ə }

loopback check See echo check. { 'lüp‚bak ‚chek }

loopback switch [ELECTR] A switch at the end of a telephone line that is used to test the line and, when closed, reflects received signals to the sender. { 'lüp‚bak ‚swich }

loop body [COMPUT SCI] The set of statements to be performed iteratively with the range of a loop. { 'lüp ‚bäd·ē }

loop check *See* echo check. { 'lüp ‚chek }

loop checking |COMMUN| Sending signals from the central office to test the integrity of local loops. { 'lüp ‚chek·iŋ }

loop circuit |COMMUN| Common communications circuit shared by more than two parties; when applied to a teletypewriter operation, all machines print all data entered on the loop. { 'lüp ‚sər·kət }

loop coupling |ELECTROMAG| A method of transferring energy between a waveguide and an external circuit, by inserting a conducting loop into the waveguide, oriented so that electric lines of flux pass through it. { 'lüp ‚käp·liŋ }

loop dialing |COMMUN| Return-path method of dialing in which the dial pulses are sent out over one side of the interconnecting line or trunk and are returned over the other side; limited to short-haul traffic. { 'lüp ‚dī·liŋ }

loop filter |ELECTR| A low-pass filter, which may be a simple RC filter or may include an amplifier, and which passes the original modulating frequencies but removes the carrier-frequency components and harmonics from a frequency-modulated signal in a locked-oscillator detector. { 'lüp ‚fil·tər }

loop flow *See* parallel flow. { 'lüp ‚flō }

loop gain |CONT SYS| The ratio of the magnitude of the primary feedback signal in a feedback control system to the magnitude of the actuating signal. |ELECTR| Total usable power gain of a carrier terminal or two-wire repeater; maximum usable gain is determined by, and may not exceed, the losses in the closed loop. { 'lüp ‚gān }

loop head |COMPUT SCI| The first instruction of a loop, which contains the mode of execution, induction variable, and indexing parameters. { 'lüp ‚hed }

loop-mile |ELEC| Length of wire in a mile of two-wire line. { 'lüp ¦mīl }

loop network *See* ring network. { 'lüp 'net‚wərk }

loop pulsing |COMMUN| Regular, momentary interruptions of the direct-current path at the sending end of a transmission line. Also known as dial pulsing. { 'lüp ‚pəls·iŋ }

loop ratio *See* loop transfer function. { 'lüp ‚rā·shō }

loopstick antenna *See* ferrite-rod antenna. { 'lüp ‚stik an‚ten·ə }

loop stop |COMPUT SCI| A small closed loop that is entered to stop the progress of a computer program, usually when some condition occurs that requires intervention by the operator or that should be brought to the operator's attention. Also known as stop loop. { 'lüp ‚stäp }

loop test |ELEC| A telephone or telegraph line test that is made by connecting a faulty line to good lines in such a way as to form a loop in which measurements can be made to determine the position of the fault. { 'lüp ‚test }

loop transfer function |CONT SYS| For a feedback control system, the ratio of the Laplace transform of the primary feedback signal to the Laplace transform of the actuating signal. Also known as loop ratio. { 'lüp 'tranz·fər ‚faŋk·shən }

loop transmittance |CONT SYS| **1.** The transmittance between the source and sink created by the splitting of a specified node in a signal flow graph. **2.** The transmittance between the source and sink created by the splitting of a node which has been inserted in a specified branch of a signal flow graph in such a way that the transmittance of the branch is unchanged. { 'lüp tranz‚mit·əns }

loose coupling |ELEC| Coupling of a degree less than the critical coupling. { 'lüs 'kəp·liŋ }

loose list |COMPUT SCI| A list, some of whose cells are empty and thus do not contain records of the file. Also known as thin list. { 'lüs 'list }

loosely coupled computer |COMPUT SCI| A computer that can function by itself and can also be connected to other computers to exchange data when necessary. { 'lüs·lē ¦kəp·əld kəm'pyüd·ər }

lorad *See* lodar. { 'lór‚ad }

loran |NAV| The designation of a family of radio navigation systems by which hyperbolic lines of position are determined by measuring the difference in the times of reception of synchronized pulse signals from two or more fixed transmitters. Derived from long-range navigation. { 'lór‚an }

Lorentz gas |ELECTR| A model of completely ionized gas in which ions are assumed to be stationary and interactions between electrons are neglected. { 'lór‚ens ‚gas }

Lorentz local field |ELEC| In a theory of electric polarization, the average electric field due to the polarization at a molecular site that is calculated under the assumption that the field due to polarization by molecules inside a small sphere centered at the site may be neglected. Also known as Mossotti field. { 'lór‚ens ¦lō·kəl 'fēld }

loss |COMMUN| *See* transmission loss. |ENG| Power that is dissipated in a device or system without doing useful work. Also known as internal loss. { lós }

loss current |ELEC| The current which passes through a capacitor as a result of the conductivity of the dielectric and results in power loss in the capacitor. |ELECTROMAG| The component of the current across an inductor which is in phase with the voltage (in phasor notation) and is associated with power losses in the inductor. { lós ‚kə·rənt }

losser circuit |ELEC| Resonant circuit having sufficient high-frequency resistance to prevent sustained oscillation at the resonant frequency. { lós·ər ‚sər·kət }

loss evaluation |ELEC| A method of achieving an economic balance between buyer and seller in adding material to a transformer design to get lower losses, in which one calculates a value in dollars per kilowatt for load loss and for no-load loss. { lós i‚val·yə'wā·shən }

Lossev effect *See* injection electroluminescence. { ‚lò‚sef i‚fekt }

loss factor |ELEC| The power factor of a material multiplied by its dielectric constant; determines the amount of heat generated in a material. { lós ‚fak·tər }

lossless data compression |COMMUN| Data compression in which the recovered data are

assured to be identical to the source. { ¦lȯs¦les
'dad·ə kəm,presh·ən }

lossless junction |ELECTROMAG| A waveguide
junction in which all the power incident on
the junction is reflected from it. { 'lȯs·ləs
,jəŋk·shən }

loss modulation See absorption modulation.
{ 'lȯs ,mäj·ə,lā·shən }

loss of information See walk down. { 'lȯs əv ,in·fər
,mā·shən }

lossy attenuator |ELECTROMAG| In waveguide
technique, a length of waveguide deliberately
introducing a transmission loss by the use of some
dissipative material. { 'lȯs·ē ə'ten·yə,wād·ər }

lossy data compression |COMMUN| Data com-
pression in which controlled degradation of the
data is allowed. { ¦lȯs·ē 'dad·ə kəm,presh·ən }

lossy line |ELEC| **1.** Cable used in test measure-
ments which has a large attenuation per unit
length. **2.** Transmission line designed to have
a high degree of attenuation. { 'lȯs·ē 'līn }

lost cluster |COMPUT SCI| Disk records that are
not associated with a file name in a disk directory.
{ 'lȯst 'kləs·tər }

loudness analyzer |ELECTR| An instrument that
produces a cathode-ray display which shows the
loudness of airborne sounds at a number of
subdivisions of part or all of the audio spectrum.
{ 'laud·nəs ,an·ə,līz·ər }

loudness control |ENG ACOUS| A combination
volume and tone control that boosts bass
frequencies when the control is set for low
volume, to compensate automatically for the
reduced response of the ear to low frequencies at
low volume levels. Also known as compensated
volume control. { 'laud·nəs kən,trōl }

loudspeaker |ENG ACOUS| A device that converts
electrical signal energy into acoustical energy,
which it radiates into a bounded space, such as
a room, or into outdoor space. Also known as
speaker. { 'laud,spēk·ər }

loudspeaker dividing network See crossover
network. { 'laud,spēk·ər di'vīd·iŋ ,net,wərk }

loudspeaker voice coil See voice coil. { 'laud
,spēk·ər 'vȯis ,kȯil }

low core |COMPUT SCI| The locations with the
lower addresses in a computer's main storage,
usually used to store control values needed to
run the system and other critical information and
instructions. { 'lō ,kȯr }

low-definition television |COMMUN| Television
that involves less than about 200 scanning lines
per complete image. { 'lō ,def·ə,nish·ən 'tel·ə
,vizh·ən }

lower half-power frequency |ELECTR| The fre-
quency on an amplifier response curve which
is smaller than the frequency for peak response
and at which the output voltage is $1/\sqrt{2}$ of its
midband or other reference value. { 'lō·ər 'haf
,pau̇·ər 'frē·kwən,sē }

lower sideband |COMMUN| The sideband contain-
ing all frequencies below the carrier-frequency value
that are produced by an amplitude-modulation
process. { 'lō·ər 'sīd,band }

lower-sideband upconverter |ELECTR| Paramet-
ric amplifier in which the frequency, power,
impedance, and gain considerations are the
same as for the nondegenerate amplifier; here,
however, the output is taken at the difference
frequency, or the lower sideband, rather than the
signal-input frequency. { 'lō·ər ¦sīd,band 'əp·
kən,vərd·ər }

lowest required radiating power |COMMUN| The
smallest power output of an antenna which will
suffice to maintain a specified grade of broadcast
service. Abbreviated LRRP. { 'lō·əst ri¦kwīrd
'rād·ē,ād·iŋ ,pau̇·ər }

lowest useful high frequency |COMMUN| The
lowest high frequency that is effective at a specified time for ionospheric propagation of radio
waves between two specified points. Abbreviated
LUHF. { 'lō·əst ¦yüs·fəl 'hī ,frē·kwən,sē }

low-frequency |COMMUN| A Federal Communi-
cations Commission designation for the band
from 30 to 300 kilohertz in the radio spectrum.
Abbreviated LF. { 'lō ,frē·kwən,sē }

low-frequency antenna |ELECTROMAG| An an-
tenna designed to transmit or receive radiation
at frequencies of less than about 300 kilohertz.
{ 'lō ,frē·kwən,sē an'ten·ə }

low-frequency compensation |ELECTR| Compen-
sation that serves to extend the frequency range
of a broad-band amplifier to lower frequencies.
{ 'lō ,frē·kwən,sē ,käm·pə'sā·shən }

low-frequency current |ELEC| An alternating
current having a frequency of less than about
300 kilohertz. { 'lō ,frē·kwən·sē 'kə·rənt }

low-frequency cutoff |ELECTR| A frequency be-
low which the gain of a system or device
decreases rapidly. { 'lō ,frē·kwən·sē 'kə,dȯf }

low-frequency gain |ELECTR| The gain of the
voltage amplifier at frequencies less than those
frequencies at which this gain is close to its
maximum value. { 'lō ,frē·kwən,sē 'gān }

low-frequency impedance corrector |ELEC|
Electric network designed to be connected to
a basic network, or to a basic network and a
building-out network, so that the combination
will simulate, at low frequencies, the sending-
end impedance, including dissipation, of a line.
{ 'lō ,frē·kwən,sē im'pēd·əns kə,rek·tər }

low-frequency loran |NAV| A modification of
standard loran, which operates in the low-
frequency range of approximately 100 to 200
kilohertz to increase range over land and during
daytime, and which matches cycles rather than
envelopes of pulses to obtain a more accurate
fix. Abbreviated LF loran. Also known as cycle-
matching loran. { 'lō ,frē·kwən·sē 'lȯr,an }

low-frequency padder |ELECTR| In a super-
heterodyne receiver, a small adjustable
capacitor connected in series with the oscillator
tuning coil and adjusted during alignment to
obtain correct calibration of the circuit at the

low-frequency end of the tuning range. { 'lō ˌfrē·
kwən·sē 'pad·ər }

low-frequency propagation [ELECTROMAG] Propagation of radio waves at frequencies between 30 and 300 kilohertz. { 'lō ˌfrē·kwən·sē
ˌpräp·ə'gā·shən }

low-frequency transconductance [ELECTR] The
change in the plate current of a vacuum tube
divided by the change in the control-grid voltage
that produces it, at frequencies small enough for
these two quantities to be considered in phase.
{ 'lō ˌfrē·kwən·sē ˌtranz·kən'dək·təns }

low-frequency tube [ELECTR] An electron tube
operated at frequencies small enough so that the
transit time of an electron between electrodes is
much smaller than the period of oscillation of the
voltage. { 'lō ˌfrē·kwən·sē 'tüb }

low-impedance measurement [ELECTR] The
measurement of an impedance which is small
enough to necessitate use of indirect methods.
{ 'lō im,pēd·əns 'mezh·ər·mənt }

low-impedance switching tube [ELECTR] A gas
tube which has a static impedance on the order
of 10,000 ohms, but zero or negative dynamic
impedance, and therefore can be used as a relay
and transmits information with negligible loss as
well. { 'lō im,pēd·əns 'swich·iŋ ˌtüb }

low level [ELECTR] The less positive of the two
logic levels or states in a digital logic system.
{ 'lō ˌlev·əl }

low-level language [COMPUT SCI] A computer language consisting of mnemonics that directly
correspond to machine language instructions;
for example, an assembler that converts the
interpreted code of a higher-level language to
machine language. { 'lō ˌlev·əl 'laŋ·gwij }

low-level logic circuit [ELECTR] A modification of
a diode-transistor logic circuit in which a resistor
and capacitor in parallel are replaced by a diode,
with the result that a relatively small voltage
swing is required at the base of the transistor to
switch it on or off. Abbreviated LLL circuit. { 'lō
ˌlev·əl 'läj·ik ˌsər·kət }

low-level modulation [ELECTR] Modulation produced at a point in a system where the power level
is low compared with the power level at the output of the system. { 'lō ˌlev·əl ˌmäj·ə'lā·shən }

low-loss [ELEC] Having a small dissipation of
electric or electromagnetic power. { 'lō ¦lòs }

low-noise amplifier [ELECTR] An amplifier having
very low background noise when the desired
signal is weak or absent; field-effect transistors
are used in audio preamplifiers for this purpose.
{ 'lō ˌnȯiz 'am·plə,fī·ər }

low-noise preamplifier [ELECTR] A low-noise amplifier placed in a system prior to the main
amplifier, sometimes close to the source; used
to establish a satisfactory noise figure at an
early point in the system. { 'lō ˌnȯiz prē'am·
plə,fī·ər }

low-order [COMPUT SCI] Pertaining to the digit
which contributes the smallest amount to the
value of a numeral, or to its position, or to the
rightmost position of a word. { 'lō ˌȯr·dər }

low-pass band-pass transformation See frequency transformation. { 'lō ˌpas 'band ˌpas
ˌtranz·fər,mā·shən }

low-pass filter [ELEC] A filter that transmits alternating currents below a given cutoff frequency
and substantially attenuates all other currents.
{ 'lō ˌpas 'fil·tər }

low-power television station [COMMUN] A television broadcasting facility limited in transmitter
output so as to provide reception in only a local
area, with a typical service area radius of 3–
16 miles (5–26 kilometers). Abbreviated LPTV
station. { 'lō ¦paủ·ər 'tel·ə,vish·ən ˌstā·shən }

low-Q filter [ELECTR] A filter in which the energy
dissipated in each cycle is a fairly large fraction of
the energy stored in the filter. { 'lō ¦kyü 'fil·tər }

low-reactance grounding [ELEC] Use of grounding connections with a moderate amount of
inductance to effect a moderate reduction in the
short-circuit current created by a line-to-ground
fault. { 'lō rē¦ak·təns 'graủnd·iŋ }

low side [COMPUT SCI] The part of a controller
or other remote device that communicates with
terminals or other remote devices, rather than
with the host computer. { 'lō ˌsīd }

low-technology robot [CONT SYS] The simplest
type of robot, with only two or three degrees
of freedom, and only the end points of motion
specified, using fixed and adjustable stops.
{ 'lō tek¦näl·ə·jē 'rō,bät }

low-tier system [COMMUN] A wireless telephone
system that provides high quality and low-delay
voice and data capabilities but has small cells.
{ ˌlō 'tēr ˌsis·təm }

low voltage [ELEC] **1.** Voltage which is small
enough to be regarded as safe for indoor use,
usually 120 volts in the United States. **2.** Voltage which is less than that needed for normal
operation; a result of low voltage may be burnout
of electric motors due to loss of electromotive
force. { 'lō 'vōl·tij }

low-voltage relay [COMMUN] A relay that responds to the drop in voltage (increase in
current) when a telephone line becomes active;
used to activate interception and eavesdropping
equipment. { 'lō ¦vōl·tij 'rē,lā }

L pad [ENG ACOUS] A volume control having
essentially the same impedance at all settings.
{ 'el ˌpad }

LPM See lines per minute.

LPTV station See low-power television station.
{ ¦el¦pē¦tē'vē ˌstā·shən }

LQG problem See linear-quadratic-Gaussian
problem. { ¦el¦kyü¦jē ˌpräb·ləm }

LRC See longitudinal redundancy check.

LRRP See lowest required radiating power.

LSA diode [ELECTR] A microwave diode in which
a space charge is developed in the semiconductor
by the applied electric field and is dissipated
during each cycle before it builds up appreciably,
thereby limiting transit time and increasing the
maximum frequency of oscillation. Derived from
limited space-charge accumulation diode. { ¦el
¦es¦ā 'dī,ōd }

LSA mode See limited space-charge accumulation mode. { ¦el¦es'ā ‚mōd }

LSB See least significant bit.

L-scan See L-display. { 'el ‚skan }

L-scope See L-display. { 'el ‚skōp }

LSI circuit See large-scale integrated circuit. { ¦el ¦es'ī ‚sər·kət }

Luenberger observer [CONT SYS] A compensator driven by both the inputs and measurable outputs of a control system. { 'lün‚bərg·ər əb'zər·vər }

LUHF See lowest useful high frequency.

Lukasiewicz notation See Polish notation. { lü ‚kä·shē'ā‚vits nō‚tā·shən }

luminaire [ELEC] An electric lighting fixture, wall bracket, portable lamp, or other complete lighting unit designed to contain one or more electric lighting sources and associated reflectors, refractors, housing, and such support for those items as necessary. { ¦lü·mə¦ner }

luminance carrier See picture carrier. { 'lü·mə·nəns ‚kar·ē·ər }

luminance channel [COMMUN] A path intended primarily for the luminance signal in an analog color television system. { 'lü·mə·nəns ‚chan·əl }

luminance primary [COMMUN] One of the three transmission primaries whose amount determines the luminance of a color in a color video system. { 'lü·mə·nəns 'prī‚mer·ē }

luminance signal [COMMUN] The color video signal that is intended to have exclusive control of the luminance of the picture. Also known as Y signal. { 'lü·mə·nəns ‚sig·nəl }

luminescent cell See electroluminescent panel. { ‚lü·mə'nes·ənt 'sel }

luminescent screen [ELECTR] The screen in a cathode-ray tube, which becomes luminous when bombarded by an electron beam and maintains its luminosity for an appreciable time. { ‚lü·mə'nes·ənt 'skrēn }

luminous sensitivity [ELECTR] For a phototube, the quotient of the anode current by the incident luminous flux. { 'lü·mə·nəs ‚sen·sə'tiv·əd·ē }

lumped constant [ELEC] A single constant that is electrically equivalent to the total of that type of distributed constant existing in a coil or circuit. Also known as lumped parameter. { 'ləmpt 'kän·stənt }

lumped-constant network [ELEC] An analytical tool in which distributed constants (inductance, capacitance, and resistance) are represented as hypothetical components. { 'ləmpt ¦kän·stənt 'net‚wərk }

lumped discontinuity [ELECTROMAG] An analytical tool in the study of microwave circuits in which the effective values of inductance, capacitance, and resistance representing a discontinuity in a waveguide are shown as discrete components of equivalent value. { 'ləmpt ‚dis‚känt·ən'ü·əd·ē }

lumped element [ELECTROMAG] A section of a transmission line designed so that electric or magnetic energy is concentrated in it at specified frequencies, and inductance or capacitance may therefore be regarded as concentrated in it, rather than distributed over the length of the line. { 'ləmpt 'el·ə·mənt }

lumped impedance [ELECTROMAG] An impedance concentrated in a single component rather than distributed throughout the length of a transmission line. { 'ləmpt im'pēd·əns }

lumped parameter See lumped constant. { 'ləmpt pə'ram·əd·ər }

Luneberg lens [ELECTROMAG] A type of antenna consisting of a dielectric sphere whose index of refraction varies with distance from the center of the sphere so that a beam of parallel rays falling on the lens is focused at a point on the lens surface diametrically opposite from the direction of incidence, and, conversely, energy emanating from a point on the surface is focused into a plane wave. Accurately spelled Luneburg lens. { 'lü·nə‚bərg ‚lenz }

Luneburg lens See Luneberg lens. { 'lü·nə‚bərg ‚lenz }

Luxemburg effect [COMMUN] Cross modulation between two radio signals during their passage through the ionosphere, due to the nonlinearity of the propagation characteristics of free charges in space. { 'lùk·səm‚bərg i‚fekt }

l value See left value. { 'el ‚val·yü }

LVDT See linear variable-differential transformer.

Lyapunov function [MATH] A function of a vector and of time which is positive-definite and has a negative-definite derivative with respect to time for nonzero vectors, is identically zero for the zero vector, and approaches infinity as the norm of the vector approaches infinity; used in determining the stability of control systems. Also spelled Liapunov function. { lē'äp·ə‚nóf ‚faŋk·shən }

Lyapunov stability criterion [CONT SYS] A method of determining the stability of systems (usually nonlinear) by examining the sign-definitive properties of an associated Lyapunov function. { lē'äp·ə‚nóf stə'bil·əd·ē krī‚tir·ē·ən }

M

M *See* megabyte.

ma *See* milliampere.

MAC *See* message authentication code.

machine |COMPUT SCI| **1.** A mechanical, electric, or electronic device, such as a computer, tabulator, sorter, or collator. **2.** A simplified, abstract model of an internally programmed computer, such as a Turing machine. { mə'shēn }

machine address |COMPUT SCI| The actual and unique internal designation of the location at which an instruction or datum is to be stored or from which it is to be retrieved. { mə'shēn ə'dres }

machine available time |COMPUT SCI| The time during which a computer has its power turned on, is not undergoing maintenance, and is thought to be operating properly. { mə'shēn ə'vāl·ə·bəl ˌtīm }

machine check |COMPUT SCI| A check that tests whether the parts of equipment are functioning properly. Also known as hardware check. { mə'shēn ˌchek }

machine-check indicator |COMPUT SCI| A protective device which turns on when certain conditions arise within the computer; the computer can be programmed to stop or to run a separate correction routine or to ignore the condition. { mə'shēn ˌchek ˌin·də̇ˌkād·ər }

machine code |COMPUT SCI| **1.** A computer representation of a character, digit, or action command in internal form. **2.** A computer instruction in internal format, or that part of the instruction which identifies the action to be performed. **3.** The set of all instruction types that a particular computer can execute. { mə'shēn ˌkōd }

machine conditions |COMPUT SCI| A component of a task descriptor that specifies the contents of all programmable registers in the processor, such as arithmetic and index registers. { mə'shēn kən'dish·ənz }

machine cycle |COMPUT SCI| **1.** The shortest period of time at the end of which a series of events in the operation of a computer is repeated. **2.** The series of events itself. { mə'shēn ˌsī·kəl }

machine-dependent |COMPUT SCI| Referring to programming languages, programs, systems, and procedures that can be used only on a particular computer or on a line of computers manufactured by a single company. { mə'shēn di,pen·dənt }

machine error |COMPUT SCI| A deviation from correctness in computer-processed data, caused by equipment failure. { mə'shēn ˌer·ər }

machine-gun microphone *See* line microphone. { mə'shēn ˌgən 'mī·krə,fōn }

machine-independent |COMPUT SCI| Referring to programs and procedures which function in essentially the same manner regardless of the machine on which they are carried out. { mə'shēn ˌin·də'pen·dənt }

machine instruction |COMPUT SCI| A set of digits, binary bits, or characters that a computer can recognize and act upon, and that, when interpreted or decoded, indicates the action to be performed and which operand is to be involved in the action. { mə'shēn in,strək·shən }

machine instruction statement |COMPUT SCI| A statement consisting usually of a tag, an operating code, and one or more addresses. { mə'shēn inˈstrək·shən ˌstāt·mənt }

machine interruption |COMPUT SCI| A halt in computer operations followed by the beginning of a diagnosis procedure, as a result of an error detection. { mə'shēn ˌint·ə'rəp·shən }

machine language |COMPUT SCI| The set of instructions available to a particular digital computer, and by extension the format of a computer program in its final form, capable of being executed by a computer. { mə'shēn ˌlaŋ·gwij }

machine language code |COMPUT SCI| A set of instructions appearing as combinations of binary digits. { mə'shēn ˈlaŋ·gwij 'kōd }

machine learning |COMPUT SCI| The process or technique by which a device modifies its own behavior as the result of its past experience and performance. { mə'shēn ˌlərn·iŋ }

machine logic |COMPUT SCI| The structure of a computer, the operation it performs, and the type and form of data used internally. { mə'shēn ˌläj·ik }

machine operator |COMPUT SCI| The person who manipulates the computer controls, brings up and closes down the computer, and can override a number of computer decisions. { mə'shēn ˌäp·ə,rād·ər }

machine-oriented language *See* computer-oriented language. { mə'shēn ˌȯr·ē̇en·təd 'laŋ·gwij }

machine-oriented programming system |COMPUT SCI| A system written in assembly language (or macro code) directly oriented toward the computer's internal language. { mə'shēn ¦ȯr·ē,ent·əd 'prō,gram·iŋ ,sis·təm }

machine processible form [COMPUT SCI] Any input medium such as a punch card, paper tape, or magnetic tape. { mə'shēn ¦präs,es·ə·bəl 'fȯrm }

machine-readable See machine-sensible. { mə'shēn 'rēd·ə·bəl }

machine-recognizable See machine-sensible. { mə'shēn ,rek·ig'nīz·ə·bəl }

machine ringing [COMMUN] In a telephone system, ringing which is started either mechanically or by an operator, after which it continues automatically until the call is answered or abandoned. { mə'shēn ,riŋ·iŋ }

machine run See run. { mə'shēn ¦rən }

machine script [COMPUT SCI] Any data written in a form that can immediately be used by a computer. { mə'shēn ,skript }

machine-sensible [COMPUT SCI] Capable of being read or sensed by a device, usually by one designed and built specifically for this task. Also known as machinable; machine-readable; machine-recognizable; mechanized. { mə'shēn ¦sen·sə·bəl }

machine-sensible information [COMPUT SCI] Information in a form which can be read by a specified machine. { mə'shēn ¦sen·sə·bəl ,in·fər'mā·shən }

machine-spoiled time [COMPUT SCI] Computer time wasted on production runs that cannot be completed or whose results are made worthless by a computer malfunction, plus extensions of running time on runs that are hampered by a malfunction. { mə'shēn ¦spȯild ,tīm }

machine switching system See automatic exchange. { mə'shēn 'swich·iŋ ,sis·təm }

machine-tool control [COMPUT SCI] The computer control of a machine tool for a specific job by means of a special programming language. { mə'shēn ,tül kən,trōl }

machine translation See mechanical translation. { mə'shēn tranz'lā·shən }

machine vision See computer vision. { mə'shēn ,vizh·ən }

machine word [COMPUT SCI] The fundamental unit of information in a word-organized digital computer, consisting of a fixed number of binary bits, decimal digits, characters, or bytes. { mə'shēn ,wərd }

macro See macroinstruction. { 'mak·rō }

macroassembler [COMPUT SCI] A program made up of one or more sequences of assembly language statements, each sequence represented by a symbolic name. { ¦mak·rō·ə'sem·blər }

macrocode [COMPUT SCI] A coding and programming language that assembles groups of computer instructions into single instructions. { 'mak·rə,kōd }

macrodefinition [COMPUT SCI] A statement that defines a macroinstruction and the set of ordinary instructions which it replaces. { ¦mak·rō ,def·ə'nish·ən }

macroexpansion [COMPUT SCI] Instructions generated by a macroinstruction and inserted into an assembly language program. { ¦mak·rō·ik'span·chən }

macro flow chart [COMPUT SCI] A graphical representation of the overall logic of a computer program in which entire segments or subroutines of the program are represented by single blocks and no attempt is made to specify the detailed operation of the program. { 'mak·rō 'flō ,chärt }

macrogeneration [COMPUT SCI] The creation of many machine instructions from one macroword. { ¦mak·rō,jen·ə¦rā·shən }

macrogenerator See macroprocessor. { ¦mak·rō'jen·ə,rād·ər }

macroinstruction [COMPUT SCI] An instruction in a higher-level language which is equivalent to a specific set of one or more ordinary instructions in the same language. Also known as macro. { ¦mak·rō·in'strək·shən }

macrolanguage [COMPUT SCI] A computer language that manipulates stored strings in which particular sites of the string are marked so that other strings can be inserted in these sites when the stored string is brought forth. { 'mak·rō ,laŋ·gwij }

macrolibrary [COMPUT SCI] A collection of prewritten specialized but unparticularized routines (or sets of statements) which reside in mass storage. { 'mak·rō,lī,brer·ē }

macroparameter [COMPUT SCI] The character in a macro operand which will complete an open subroutine created by the macroinstruction. { ,mak·rō·pə'ram·əd·ər }

macroprocessor [COMPUT SCI] A piece of software which replaces each macroinstruction in a computer program by the set of ordinary instructions which it stands for. Also known as macrogenerator. { ¦mak·rə'präs,es·ər }

macroprogram [COMPUT SCI] A computer program that consists of macroinstructions. { ma'krō'prō,grəm }

macroprogramming [COMPUT SCI] The process of writing machine procedure statements in terms of macroinstructions. { ¦mak·rō'prō,gram·iŋ }

macroskeleton [COMPUT SCI] A definition of a macroinstruction in a precise but content-free way, which can be particularized by a processor as directed by macroinstruction parameters. Also known as model. { ¦mak·rō'skel·ə·tən }

macrosystem [COMPUT SCI] A language in which words represent a number of machine instructions. { 'mak·rō,sis·təm }

macro virus [COMPUT SCI] A virus that hides inside document and spreadsheet files used by popular word processing and spreadsheet applications. { ¦mak·rō 'vī·rəs }

madistor [ELECTR] A cryogenic semiconductor device in which injection plasma can be steered or controlled by transverse magnetic fields, to give the action of a switch. { ma'dis·tər }

MADT See microalloy diffused transistor.

MAG See maximum available gain.

magamp See magnetic amplifier. { 'mag,amp }

magazine [COMPUT SCI] A holder of microfilm or magnetic recording media strips. { ¦mag·ə¦zēn }

magic eye *See* cathode-ray tuning indicator. { 'maj·ik 'ī }

magic tee *See* hybrid tee. { 'maj·ik 'tē }

magnesium anode |ELEC| Bar of magnesium buried in the earth, connected to an underground cable to prevent cable corrosion due to electrolysis. { mag'nē·zē·əm 'an,ōd }

magnesium cell |ELEC| A primary cell in which the negative electrode is made of magnesium or one of its alloys. { mag'nē·zē·əm ¦sel }

magnesium-copper sulfide rectifier |ELECTR| Dry-disk rectifier consisting of magnesium in contact with copper sulfide. { mag'nē·zē·əm 'käp·ər ¦səl,fīd 'rek·tə,fī·ər }

magnesium-manganese dioxide cell |ELEC| Type of electrochemical (dry) cell battery in which the active elements are magnesium and manganese dioxide. { mag'nē·zē·əm 'maŋ·gə ,nēs dī'äk,sīd ,sel }

magnesium-silver chloride cell |ELEC| A reserve primary cell that is activated by adding water; active elements are magnesium and silver chloride. { mag'nē·zē·əm 'sil·vər 'klȯr,īd ,sel }

magnesyn |ELEC| A portion of a repeater unit; a two-pole permanently magnetized rotor within a three-phase two-pole delta-connected stator which carries the indicating pointer and is free to rotate in any direction. { 'mag·nə,sin }

magnetically focused tube |ELECTR| An image tube in which electrons from the photocathode are accelerated by electric fields and forced into tight spiral paths as they are further accelerated by a uniform magnetic field down the center of the tube. { mag'ned·ə·klē ¦fō·kəst 'tüb }

magnetic amplifier |ELECTR| A device that employs saturable reactors to modulate the flow of alternating-current electric power to a load in response to a lower-energy-level direct-current input signal. Abbreviated magamp. Also known as transductor. { mag'ned·ik 'am·plə,fī·ər }

magnetic bubble memory *See* bubble memory. { mag'ned·ik ¦bəb·əl ,mem·rē }

magnetic card |COMPUT SCI| A card with a magnetic surface on which data can be stored by selective magnetization. { mag'ned·ik 'kärd }

magnetic card file |COMPUT SCI| A direct-access storage device in which units of data are stored on magnetic cards contained in one or more magazines from which they are withdrawn, when addressed, to be carried at high speed past a read/write head. { mag'ned·ik 'kärd ,fīl }

magnetic cell |ELECTR| One unit of a magnetic memory, capable of storing one bit of information as a zero state or a one state. { mag'ned·ik 'sel }

magnetic character |COMPUT SCI| A character printed with magnetic ink, as on bank checks, for reading by machines as well as by humans. { mag'ned·ik 'kar·ik·tər }

magnetic character reader |COMPUT SCI| A character reader that reads special type fonts printed in magnetic ink, such as those used on bank checks, and feeds the character data directly to a computer for processing. { mag'ned·ik ¦kar·ik·tər ,rēd·ər }

magnetic character sorter |COMPUT SCI| A device that reads documents printed with magnetic ink; all data read are stored, and records are sorted on any required field. Also known as magnetic document sorter-reader. { mag'ned·ik ¦kar·ik·tər ,sȯrd·ər }

magnetic core Also known as core. |ELECTR| A configuration of magnetic material, usually a mixture of iron oxide or ferrite particles mixed with a binding agent and formed into a tiny doughnutlike shape, that is placed in a spatial relationship to current-carrying conductors, and is used to maintain a magnetic polarization for the purpose of storing data, or for its nonlinear properties as a logic element. Also known as memory core. |ELECTROMAG| A quantity of ferrous material placed in a coil or transformer to provide a better path than air for magnetic flux, thereby increasing the inductance of the coil and increasing the coupling between the windings of a transformer. { mag'ned·ik 'kȯr }

magnetic core multiplexer |COMPUT SCI| A device which channels many bit inputs into a single output. { mag'ned·ik ¦kȯr 'məl·tə,plek·sər }

magnetic core storage |COMPUT SCI| A computer storage system in which each of thousands of magnetic cores stores one bit of information; current pulses are sent through wires threading through the cores to record or read out data; used extensively in the 1950s and 1960s, and still used in specialized military applications and in space vehicles. Also known as core memory; core storage. { mag'ned·ik ¦kȯr 'stȯr·ij }

magnetic cumulative generator *See* flux-compression generator. { mag¦ned·ik ¦kyü·myə ,lād·iv 'jen·ə,rād·ər }

magnetic deflection |ELECTR| Deflection of an electron beam by the action of a magnetic field, as in a television picture tube. { mag'ned·ik di'flek·shən }

magnetic delay line |ELECTR| Delay line, used for the storage of data in a computer, consisting essentially of a metallic medium along which the velocity of the propagation of magnetic energy is small compared to the speed of light; storage is accomplished by the recirculation of wave patterns containing information, usually in binary form. { mag'ned·ik di'lā ,līn }

magnetic dipole antenna |ELECTROMAG| Simple loop antenna capable of radiating an electromagnetic wave in response to a circulation of electric current in the loop. { mag'ned·ik 'dī,pōl an,ten·ə }

magnetic disk |COMPUT SCI| A rotating circular plate having a magnetizable surface on which information may be stored as a pattern of polarized spots on concentric recording tracks. { mag'ned·ik 'disk }

magnetic document sorter-reader *See* magnetic character sorter. { mag'ned·ik ¦däk·yə·mənt 'sȯrd·ər ,rēd·ər }

magnetic domain memory *See* domain-tip memory. { mag'ned·ik də'mān ,mem·rē }

magnetic drum *See* drum. { mag'ned·ik 'drəm }

magnetic drum receiving equipment |ELECTR| Radar developed for detection of targets beyond line of sight using ionospheric reflection and very low power. { mag'ned·ik 'drəm ri'sēv·iŋ i ,kwip·mənt }

magnetic drum storage See drum. { mag'ned·ik ¦drəm 'stȯr·ij }

magnetic earphone |ENG ACOUS| An earphone in which variations in electric current produce variations in a magnetic field, causing motion of a diaphragm. { mag'ned·ik 'ir,fōn }

magnetic element |ENG| That part of an instrument producing or influenced by magnetism. { mag'ned·ik 'el·ə·mənt }

magnetic film See magnetic thin film. { mag'ned·ik 'film }

magnetic firing circuit |ELECTR| A type of firing circuit in which the capacitor is discharged through the igniter by saturating a reactor, which is connected in series with the capacitor; often used in ignitron rectifiers to obtain longer life and greater reliability than is possible with thyratron firing tubes. { mag'ned·ik 'fīr·iŋ ,sər·kət }

magnetic flux quantum |ELEC| A fundamental unit of magnetic flux, the total magnetic flux in a fluxoid in a type II superconductor, equal to $h/(2e)$, where h is Planck's constant and e is the magnitude of the electron charge, or approximately 2.07×10^{-15} weber. { mag,ned·ik 'fləks ,kwän·təm }

magnetic head |ELECTR| The electromagnet used for reading, recording, or erasing signals on a magnetic disk, drum, or tape. Also known as magnetic read/write head. { mag'ned·ik 'hed }

magnetic-ink character recognition |COMPUT SCI| That branch of character recognition which involves the sensing of magnetic-ink characters for the purpose of determining the character's most probable identity. Abbreviated MICR. { mag'ned·ik ¦iŋk 'kar·ik·tər ,rek·ig,nish·ən }

magnetic loudspeaker |ENG ACOUS| Loudspeaker in which acoustic waves are produced by mechanical forces resulting from magnetic reactions. Also known as magnetic speaker. { mag'ned·ik 'laùd,spēk·ər }

magnetic memory See magnetic storage. { mag 'ned·ik 'mem·rē }

magnetic memory plate |ELECTR| Magnetic memory consisting of a ferrite plate having a grid of small holes through which the read-in and read-out wires are threaded; printed wiring may be applied directly to the plate in place of conventionally threaded wires, permitting mass production of plates having a high storage capacity. { mag'ned·ik ¦mem·rē ,plāt }

magnetic microphone |ENG ACOUS| A microphone consisting of a diaphragm acted upon by sound waves and connected to an armature which varies the reluctance in a magnetic field surrounded by a coil. Also known as reluctance microphone; variable-reluctance microphone. { mag'ned·ik 'mī·krə,fōn }

magnetic modulator |ELECTR| A modulator in which a magnetic amplifier serves as the mod-

ulating element for impressing an intelligence signal on a carrier. { mag'ned·ik 'mäj·ə,lād·ər }

magnetic pinch See pinch effect. { mag'ned·ik 'pinch }

magnetic printing |ELECTR| The permanent and usually undesired transfer of a recorded signal from one section of a magnetic recording medium to another when these sections are brought together, as on a reel of tape. Also known as crosstalk; magnetic transfer. { mag'ned·ik 'print·iŋ }

magnetic random access memory |COMPUT SCI| A nonvolatile memory in which submicrometer-sized magnetic structures store digital information in their magnetic orientation. Abbreviated MRAM. { mag¦ned·ik ,ran·dəm 'ak,ses ,mem·rē }

magnetic read/write head See magnetic head. { mag'ned·ik ¦rēd ¦rīt ,hed }

magnetic recorder |ELECTR| An instrument that records information, generally in the form of audio-frequency or digital signals, on magnetic tape or magnetic wire as magnetic variations in the medium. { mag'ned·ik ri'kȯrd·ər }

magnetic recording |ELECTR| Recording by means of a signal-controlled magnetic field. { mag'ned·ik ri'kȯrd·iŋ }

magnetic reproducer |ELECTR| An instrument which moves a magnetic recording medium, such as a tape, wire, or disk, past an electromagnetic transducer that converts magnetic signals on the medium into electric signals. { mag'ned·ik ,rē·prə'dü·sər }

magnetic reproducing |ELECTR| The conversion of information on magnetic tape or magnetic wire, which was originally produced by electric signals, back into electric signals. { mag'ned·ik ,rē·prə'dü·siŋ }

magnetic rotation |OPTICS| **1.** In a weak magnetic field, the rotation of the plane of polarization of fluorescent light emitted perpendicular to the field and perpendicular to the propagation direction of the incident light. **2.** See Faraday effect. { mag'ned·ik rō'tā·shən }

magnetic shift register |COMPUT SCI| A shift register in which the pattern of settings of a row of magnetic cores is shifted one step along the row by each new input pulse; diodes in the coupling loops between cores prevent backward flow of information. { mag'ned·ik 'shift ,rej·ə·stər }

magnetic sound track |ENG ACOUS| A magnetic tape, attached to a motion picture film, on which a sound recording is made. { mag'ned·ik 'saùn ,trak }

magnetic speaker See magnetic loudspeaker. { mag'ned·ik 'spēk·ər }

magnetic spin transistor See magnetic switch. { mag,ned·ik ,spin tran'zis·tər }

magnetic stepping motor See stepper motor. { mag'ned·ik 'step·iŋ ,mōd·ər }

magnetic storage |COMPUT SCI| A device utilizing magnetic properties of materials to store data; may be roughly divided into two categories, moving (drum, disk, tape) and static (core,

thin film). Also known as magnetic memory.
{ mag'ned·ik 'stȯr·ij }

magnetic stripe |COMPUT SCI| A small length of magnetic tape on a card or badge, containing data that is machine-readable. { mag¦ned·ik 'strīp }

magnetic striped ledger |COMPUT SCI| A ledger sheet used on a special typing device which stores the coded data on a magnetic strip on the sheet while typing out the data on the sheet; the magnetic strip can be read directly by a special reader linked to a computer. { mag'ned·ik ¦strīpt 'lej·ər }

magnetic switch |ELECTR| A switching device consisting of three metallic layers (a paramagnetic layer between two ferromagnetic layers), whose action is based on electron spin and is controlled by a small magnetic field. Also known as bipolar spin device; bipolar spin switch; magnetic spin transistor; spin transistor; spin valve. { mag¦ned·ik 'swich }

magnetic tape |ELECTR| A plastic, paper, or metal tape that is coated or impregnated with magnetizable iron oxide particles; used in magnetic recording and in computer storage chiefly for archiving and backup. { mag'ned·ik 'tāp }

magnetic tape core |ELECTR| Toroidal core formed by winding a strip of thin magnetic core material around a form. { mag'ned·ik ¦tāp 'kȯr }

magnetic tape file operation |COMPUT SCI| All the jobs related to creating, sorting, inputting, and maintenance of magnetic tapes in a magnetic tape environment. { mag'ned·ik ¦tāp 'fīl ¸äp·ə ¸rā·shən }

magnetic tape group |COMPUT SCI| A cabinet containing two or more magnetic tape units, each of which can operate independently, but which sometimes share one or more channels with which they communicate with a central processor. Also known as tape cluster; tape group. { mag'ned·ik ¦tāp ¸grüp }

magnetic tape librarian |COMPUT SCI| Routine which provides a computer the means to automatically run a sequence of programs. { mag'ned·ik ¦tāp lī¸brer·ē·ən }

magnetic tape master file |COMPUT SCI| A magnetic tape consisting of a set of related elements such as is found in a payroll, an inventory, or an accounts receivable; a master file is, as a rule, periodically updated. { mag'ned·ik ¦tāp ¦mas·tər 'fīl }

magnetic tape parity |COMPUT SCI| A check performed on the data bits on a tape; usually an odd (or even) condition is expected and the occurrence of the wrong parity indicates the presence of an error. { mag'ned·ik ¦tāp 'par·əd·ē }

magnetic tape reader |ELECTR| A computer device that is capable of reading information recorded on magnetic tape by transforming this information into electric pulses. { mag'ned·ik ¦tāp ¸rēd·ər }

magnetic tape station |COMPUT SCI| On-line device that provides write, read, and erase data on

magnetic tape to permit high-speed storage of data. { mag'ned·ik ¦tāp ¸stā·shən }

magnetic tape storage |COMPUT SCI| Storage of binary information on magnetic tape, generally on 5 to 10 tracks, with up to several thousand bits per inch (more than a thousand bits per centimeter) on each track. { mag'ned·ik ¦tāp ¸stȯr·ij }

magnetic tape switching unit |COMPUT SCI| A device which permits the computer operator to bring into play any number of tape drives as required by the system. { mag'ned·ik ¦tāp 'swich·iŋ ¸yü·nət }

magnetic tape terminal |COMPUT SCI| Device which converts pulses in series to pulses in parallel while checking for bit parity prior to the entry in buffer storage. { mag'ned·ik ¦tāp 'tər·mən·əl }

magnetic tape unit |COMPUT SCI| A computer unit that usually consists of a tape transport, reading and recording heads, and associated electric and electronic equipment. { mag'ned·ik ¦tāp ¸yü·nət }

magnetic thin film |SOLID STATE| A sheet or cylinder of magnetic material less than 5 micrometers thick, usually possessing uniaxial magnetic anisotropy; used mainly in computer storage and logic elements. Also known as ferromagnetic film; magnetic film. { mag'ned·ik 'thin ¦film }

magnetic transfer See magnetic printing. { mag'ned·ik 'tranz·fər }

magnetic tunnel junction |ELECTR| A magnetic storage and switching device in which two magnetic layers are separated by an insulating barrier, typically aluminum oxide, that is only 1–2 nanometers thick, allowing an electronic current whose magnitude depends on the orientation of both magnetic layers to tunnel through the barrier when it is subject to a small electric bias. { mag¦ned·ik 'tən·əl ¸jəŋk·shən }

magnetizing current |ELEC| The current that flows through the primary winding of a power transformer when no loads are connected to the secondary winding; this current establishes the magnetic field in the core and furnishes energy for the no-load power losses in the core. Also known as exciting current. { 'mag·nə¸tiz·iŋ ¸kə·rənt }

magneto |ELEC| An alternating-current generator that uses one or more permanent magnets to produce its magnetic field; frequently used as a source of ignition energy on tractor, marine, industrial, and aviation engines. Also known as magnetoelectric generator. { mag'nēd·ō }

magnetoelectric generator See magneto. { mag ¦nēd·ō·i'lek·trik 'jen·ə¸rād·ər }

magnetoelectronics |ELECTR| The use of electron spin (as opposed to charge) in electronic devices. Also known as spin electronics; spintronics. { mag¸ned·ō·i·lek'trän·iks }

magnetohydrodynamic generator |ELEC| A system for generating electric power in which the kinetic energy of a flowing conducting fluid is converted to electric energy by a magnetohydrodynamic interaction. Abbreviated MHD generator. { mag¦nēd·ō,hī·drə·dī'nām·ik 'jen·ə ,rād·ər }

magnetooptical switch |COMPUT SCI| A thin-film modulator which acts on a laser beam by polarization, causing the beam to emerge from the output prism at a different angle. { mag ¦nēd·ō¦äp·tə·kəl 'swich }

magnetooptic disk |COMMUN| A data storage device in which information is stored in small magnetic marks along tracks on a rotating disk; the information is read by sensing the change in polarization of reflected focused light and can be altered by using a higher-power focused light spot to locally heat the medium and, with the application of an external magnetic field, switch the magnetic domains of the material. { mag ,ned·ō,äp·tik 'disk }

magnetooptic Kerr effect |OPTICS| Changes produced in the optical properties of a reflecting surface of a ferromagnetic substance when the substance is magnetized; this applies especially to the elliptical polarization of reflected light, when the ordinary rules of metallic reflection would give only plane polarized light. Also known as Kerr magnetooptical effect. { mag¦nēd·ō ¦äp·tik 'kər i,fekt }

magnetooptic material |OPTICS| A material whose optical properties are changed by an applied magnetic field. { mag¦nēd·ō¦äp·tik mə'tir·ē·əl }

magnetooptic recording |ENG| An erasable data storage technology in which data are stored on a rotating disk in a thin magnetic layer that may be switched between two magnetization states by the combination of a magnetic field and a pulse of light from a diode laser. { mag,ned·ō,äp·tik ri'kòrd·iŋ }

magnetooptics |OPTICS| The study of the effect of a magnetic field on light passing through a substance in the field. { mag¦nēd·ō¦äp·tiks }

magnetoresistance |ELECTR| The change in the electrical resistance of a material when it is subjected to an applied magnetic field, this property has widespread application in sensors and magnetic read heads. |ELECTROMAG| The change in electrical resistance produced in a current-carrying conductor or semiconductor on application of a magnetic field. { mag¦nēd·ō· ri'zis·təns }

magnetoresistive memory |ELECTR| A random-access memory that uses the magnetic state of small ferromagnetic regions to store data, plus magnetoresistive devices to read the data, all integrated with silicon integrated-circuit electronics. { mag,ned·ō·ri,zis·tiv 'mem·rē }

magnetoresistor |ELECTR| Magnetic field-controlled variable resistor. { mag¦nēd·ō·ri'zis·tər }

magnetostrictive filter |ELECTR| Filter network which uses the magnetostrictive phenomena to form high-pass, low-pass, band-pass, or band-elimination filters; the impedance characteristic is the inverse of that of a crystal. { mag¦nēd·ō ¦strik·tiv 'fil·tər }

magnetostrictive loudspeaker |ENG ACOUS| Loudspeaker in which the mechanical forces result from the deformation of a material having

magnetostrictive properties. { mag¦nēd·ō¦strik· tiv 'laủd,spēk·ər }

magnetostrictive microphone |ENG ACOUS| Microphone which depends for its operation on the generation of an electromotive force by the deformation of a material having magneto-strictive properties. { mag¦nēd·ō¦strik·tiv 'mī·krə,fōn }

magnetostrictive oscillator |ELECTR| An oscillator whose frequency is controlled by a magnetostrictive element. { mag¦nēd·ō¦strik·tiv 'äs·ə ,lād·ər }

magneto telephone set |ELEC| Local battery telephone set in which current for signaling by the telephone station is supplied from a local hand generator, usually a magneto. { mag'nēd·ō 'tel·ə,fōn ,set }

magnetovision |ENG| A method of measuring and displaying magnetic field distributions in which scanning results from a thin-film Permalloy magnetoresistive sensor are processed numerically and presented in the form of a color map on a video display unit. { mag'ned·ə,vizh·ən }

magnetron |ELECTR| One of a family of crossed-field microwave tubes, wherein electrons, generated from a heated cathode, move under the combined force of a radial electric field and an axial magnetic field in such a way as to produce a bunching of electrons and hence microwave radiation. Useful in the frequency range 1–40 gigahertz; a pulsed microwave radiation source for radar and continuous source for microwave cooking. { 'mag·nə,trän }

magnetron oscillator |ELECTR| Oscillator circuit employing a magnetron tube. { 'mag·nə,trän 'äs·ə,lād·ər }

magnetron pulling |ELECTR| Frequency shift of a magnetron caused by factors which vary the standing waves or the standing-wave ratio on the radio-frequency lines. { 'mag·nə,trän 'pùl·iŋ }

magnetron pushing |ELECTR| Frequency shift of a magnetron caused by faulty operation of the modulator. { 'mag·nə,trän 'pùsh·iŋ }

magnetron vacuum gage |ELECTR| A vacuum gage that is essentially a magnetron operated beyond cutoff in the vacuum being measured. { 'mag·nə,trän 'vak·yəm ,gāj }

magnet wire |ELEC| The insulated copper or aluminum wire used in the coils of all types of electromagnetic machines and devices. { 'mag·nət ,wīr }

magnistor |ELECTR| A device that utilizes the effects of magnetic fields on injection plasmas in semiconductors such as indium antimonide. { mag'nis·tər }

mag-slip See synchro. { 'mag,slip }

mail box |COMPUT SCI| **1.** A portion of a computer's main storage that can be used to hold information about other devices. **2.** Computer storage facilities designed to hold electronic mail. { 'māl ,bäks }

mailbox name |COMPUT SCI| The first part of an electronic mail address, which identifies the storage space that has been set aside in a computer to receive a user's electronic mail messages. Also known as username. { 'māl ,bäks ,nām }

mailing list |COMMUN | A list of users of the Internet or another computer network who all receive copies of electronic mail messages.   { 'māl·iŋ ,list }

mail merge |COMPUT SCI| The process of combining a form letter with a list of names and addresses to produce individualized letters.   { 'māl ,mərj }

main |ELEC| **1.** One of the conductors extending from the service switch, generator bus, or converter bus to the main distribution center in interior wiring.   **2.** *See* power transmission line.   { mān }

main-and-transfer bus |ELEC| A substation switching arrangement similar to a single bus but with an additional transfer bus provided.   { 'mān ən 'tranz·fər ,bəs }

main bang |ELECTR| In colloquial usage, a transmitted pulse within a radar system.   { 'mān 'baŋ }

main clock *See* master clock.   { 'mān 'kläk }

main controller |COMPUT SCI| A control unit assigned to direct the other control units in a computer system.   { 'mān kən'trō·lər }

main distributing frame |ELEC| Frame which terminates the permanent outside lines entering the central office building on one side and the subscriber-line multiple cabling, trunk multiple cabling, and so on, used for associating an outside line with any desired terminal on the other side; it usually carries the control-office protective devices, and functions as a test point between line and office. Also known as main frame.   { 'mān di'strib·yəd·iŋ ,frām }

main exciter |ELEC| Exciter which supplies energy for the field excitation of a principal electric machine.   { 'mān ik'sīd·ər }

main frame |COMPUT SCI| **1.** A large computer.   **2.** The part of a computer that contains the central processing unit, main storage, and associated control circuitry. Also known as frame.   { 'mān ,frām }

main instruction buffer |COMPUT SCI| A section of storage in the instruction unit, 16 bytes in length, used to hold prefetched instructions.   { 'mān in'strək·shən ,bəf·ər }

main level |COMMUN | A range of allowed picture parameters defined by the MPEG-2 video coding specification.   { 'mān 'lev·əl }

main lobe *See* major lobe.   { 'mān 'lōb }

main loop |COMPUT SCI| A set of instructions that constitute the primary structure of a repetitive computer program.   { 'mān ,lüp }

main memory *See* main storage.   { 'mān 'mem·rē }

main path |COMPUT SCI| The principal branch of a routine followed by a computer in the process of carrying out the routine.   { 'mān 'path }

main profile |COMMUN | A subset of the syntax of the MPEG-2 video coding specification that is supported over a large range of applications.   { 'mān 'prō,fīl }

main program |COMPUT SCI| **1.** The central part of a computer program, from which control may be transferred to various subroutines and to which

control is eventually returned. Also known as main routine.   **2.** *See* executive routine.   { 'mān 'prō·grəm }

main routine *See* executive routine; main program.   { 'mān rü'tēn }

main station |COMMUN | Telephone station with a distinct call number designation, directly connected to a central office.   { 'mān 'stā·shən }

main storage |COMPUT SCI| A digital computer's principal working storage, from which instructions can be executed or operands fetched for data manipulation. Also known as main memory.   { 'mān 'stor·ij }

main sweep |ELECTR| On certain fire-control radar, the longest range scale available.   { 'mān 'swēp }

maintenance pack |COMPUT SCI| A disk drive that is used to store copies of computer programs for the purpose of applying and testing changes made in the course of software maintenance.   { 'mānt·ən·əns ,pak }

maintenance routine |COMPUT SCI| A computer program designed to detect conditions which may give rise to a computer malfunction in order to assist a service engineer in performing routine preventive maintenance.   { 'mānt·ən·əns rü ,tēn }

maintenance time |COMPUT SCI| The time required for both corrective and preventive maintenance of a computer or other components of a computer system.   { 'mānt·ən·əns ,tīm }

main vector |COMMUN | A pair of numbers that represent the vertical and horizontal displacement of a region of a reference picture for MPEG-2 prediction.   { 'mān 'vek·tər }

major cycle |COMPUT SCI| The time interval between successive appearances of a given storage position in a serial-access computer storage.   { 'mā·jər 'sī·kəl }

majority carrier |ELECTR| The type of charge carrier, that is, electron or hole, that constitutes more than half the carriers in a semiconductor.   { mə'jär·əd·ē 'kar·ē·ər }

majority element *See* majority gate.   { mə'jär·əd·ē 'el·ə·mənt }

majority emitter |ELECTR| Of a transistor, an electrode from which a flow of minority carriers enters the interelectrode region.   { mə'jär·əd·ē i'mid·ər }

majority gate |COMPUT SCI| A logic circuit which has one output and several inputs, and whose output is energized only if a majority of its inputs are energized. Also known as majority element; majority logic.   { mə'jär·əd·ē 'gāt }

majority logic *See* majority gate.   { mə'jär·əd·ē 'läj·ik }

major key |COMPUT SCI| The primary key for identifying a record.   { 'mā·jər ,kē }

major lobe |ELECTROMAG| Antenna lobe indicating the direction of maximum radiation or reception. Also known as main lobe.   { 'mā·jər 'lōb }

major node |ELEC| A point in an electrical network at which three or more elements are connected together. Also known as junction.   { 'mā· jər 'nōd }

major relay station |ELECTR| Tape relay station which has two or more trunk circuits connected thereto to provide an alternate route or to meet command requirements. { 'mā·jər 'rē,lā ,stā·shən }

major wave See long wave. { 'mā·jər 'wāv }

make |ELEC| Closing of relay, key, or other contact. { 'māk }

make-and-break circuit |ELEC| A circuit that is alternately opened and closed. { 'māk ən 'brāk ,sər·kət }

make-break operation |COMMUN| A circuit operation in which there is a cessation of current flow as a pulse transmission occurs. { 'māk 'brāk ,äp·ə,rā·shən }

make-busy |COMMUN| A switch whose activation makes a dial telephone line or group of telephone lines appear to be busy and thereby prevents completion of incoming calls. { 'māk 'biz·ē }

make contact |ELEC| Contact of a device which closes a circuit upon the operation of the device (normally open). { 'māk ,kän,takt }

makeup time |COMPUT SCI| The time required to rerun programs on a computer because of operator errors and other problems. { 'māk,əp ,tīm }

making current |ELEC| The peak value attained by the current during the first cycle after a switch, circuit breaker, or similar apparatus is closed. { 'māk·iŋ ,kə·rənt }

male connector |ELEC| An electrical connector with protruding contacts for joining with a female connector. { 'māl kə'nek·tər }

malfunction routine |COMPUT SCI| A program used in troubleshooting. { mal'fəŋk·shən rü,tēn }

malicious code |COMPUT SCI| Programming code that is capable of causing harm to availability, integrity of code or data, or confidentiality in a computing system; encompasses Trojan horses, viruses, worms, and trapdoors. { mə|lish·əs 'kōd }

management information system |COMMUN| A communication system in which data are recorded and processed to form the basis for decisions by top management of an organization. Abbreviated MIS. { 'man·ij·mənt ,in·fər'mā·shən ,sis·təm }

Manchester coding See phase encoding. { 'man·chə·stər ,kōd·iŋ }

Manchester plate |ELEC| A storage battery consisting of a heavy alloy grid with circular openings into which are pressed pure lead buttons that are made from lead tape by crimping and rolling to develop a large surface area and are coated with lead peroxide, PbO_2. { 'man·chə·stər ,plāt }

manifest constant |COMPUT SCI| A value that is assigned to a symbolic name at the beginning of a computer program and is not subject to change during execution. { 'man·ə,fest 'kän·stənt }

manipulated variable |COMPUT SCI| Variable whose value is being altered to bring a change in some condition. { mə'nip·yə,lād·əd 'ver·ē·ə·bəl }

manipulator |CONT SYS| An armlike mechanism on a robotic system that consists of a series of segments, usually sliding or jointed which grasp and move objects with a number of degrees of freedom, under automatic control. { mə'nip·yə ,lād·ərz }

man-machine system See human-machine system. { 'man mə|shēn 'sis·təm }

mantissa |COMPUT SCI| A fixed point number composed of the most significant digits of a given floating-point number. Also known as fixed-point part; floating-point coefficient. { man'tis·ə }

manual central office |COMMUN| Central office of a manual telephone system. { 'man·yə·wəl ¦sen·trəl 'óf·əs }

manual control unit |CONT SYS| A portable, hand-held device that allows an operator to program and store instructions related to robot motions and positions. Also known as programming unit. { 'man·yə·wəl kən'trōl ,yü·nət }

manual exchange |COMMUN| Any exchange where calls are completed by an operator. { 'man·yə·wəl iks'chānj }

manual input |COMPUT SCI| The entry of data by hand into a device at the time of processing. { 'man·yə·wəl 'in,pút }

manual number generator See manual word generator. { 'man·yə·wəl 'nəm·bər ,jen·ə,rād·ər }

manual operation |COMPUT SCI| Any processing operation performed by hand. { 'man·yə·wəl ,äp·ə'rā·shən }

manual rate-aided tracking |ELECTR| Radar circuit which tracks individual targets by computing the velocity from position fixes inserted manually into the circuitry. { 'man·yə·wəl 'rāt ¦ād·əd 'trak·iŋ }

manual ringing |COMMUN| Ringing which is started by the manual operation of a key and continues only while the key is held in operation. { 'man·yə·wəl 'riŋ·iŋ }

manual switchboard |ELEC| Telephone switchboard in which the connections are made manually, by plugs and jacks, or by keys. { 'man·yə·wəl 'swich,bórd }

manual switching |ELECTR| Method by which manual connection is made between two or more teletypewriter circuits. { 'man·yə·wəl 'swich·iŋ }

manual telephone set |ELECTR| Telephone set not equipped with a dial. { 'man·yə·wəl 'tel·ə ,fōn ,set }

manual telephone system |COMMUN| A telephone system in which connections between customers are ordinarily established manually by telephone operators in accordance with orders given verbally by calling parties. { 'man·yə·wəl 'tel·ə,fōn ,sis·təm }

manual word generator |COMPUT SCI| A device into which an operator can enter a computer word by hand, either for direct insertion into memory or to be held until it is read during the execution of a program. Also known as manual number generator. { 'man·yə·wəl 'wərd ,jen·ə,rād·ər }

many-to-many correspondence |COMPUT SCI| A structure that establishes relationships between items in a data base, such that one unit of data can relate to many units, and many units can relate back to one unit and to other units as well. { 'men·ē tə 'men·ē ,kär·ə'spän·dəns }

map |COMPUT SCI| **1.** An output produced by an assembler, compiler, linkage editor, or relocatable loader which indicates the (absolute or relocatable) locations of such elements as programs, subroutines, variables, or arrays. **2.** By extension, an index of the storage allocation on a magnetic disk or drum. { map }

Marconi antenna |ELECTROMAG| Antenna system of which the ground is an essential part, as distinguished from a Hertz antenna. { mär′kō·nē an′ten·ə }

marginal checking |ELECTR| A preventive-maintenance procedure in which certain operating conditions, such as supply voltage or frequency, are varied about their normal values in order to detect and locate incipient defective units. { ′mär·jən·əl ′chek·iŋ }

marginal relay |ELEC| Relay with a small margin between its nonoperative current value (maximum current applicable without operation) and its operative value (minimum current that operates the relay). { ′mär·jən·əl ′rē,lā }

marginal test |ELECTR| A test of electronic equipment in which conditions are varied until failures occur or faults can be detected, allowing measurement of permissible operating margins. { ′mär·jən·əl ′test }

maritime frequency bands |COMMUN| In the United States, a collection of radio frequencies allocated for communication between coast stations and ships or between ships. { ′mar·ə,tīm ′frē·kwən·sē ,banz }

maritime mobile satellite service |COMMUN| A mobile satellite service in which the mobile earth stations are located on board ships. Abbreviated MMSS. { ,mar·ə,tīm ′mō·bəl ′sad·əl,t ,sər·vəs }

maritime mobile service |COMMUN| A mobile service between coast stations and ship stations, or between ship stations, in which survival craft stations may also participate. { ′mar·ə,tīm ′mō·bəl ′sər·vəs }

mark |COMMUN| The closed-circuit condition in telegraphic communication, during which the signal actuates the printer; the opposite of space. |COMPUT SCI| A distinguishing feature used to signal some particular location or condition. { märk }

mark detection |COMPUT SCI| That class of character recognition systems which employs coded documents, in the form of boxes or windows, in order to convey intended information by means of pencil or ink marks made in specific boxes. { ′märk di,tek·shən }

mark-hold |COMMUN| The transmission of a steady mark to indicate that there is no traffic over a telegraph channel; the upper marking frequency of a duplex channel (2225 hertz) is used to disable echo suppressors which may interfere with data communications. { ¦märk ¦hōld }

marking and spacing intervals |COMMUN| Intervals of closed and open conditions in transmission circuits. { ¦märk·iŋ ən ¦spās·iŋ ′in·tər·vəlz }

marking bias |COMMUN| Bias distortion that lengthens the marking impulse. { ′märk·iŋ ,bī·əs }

marking current |ELEC| Magnitude and polarity of current in the line when the receiving mechanism is in the operating position. { ′märk·iŋ ,kə·rənt }

marking-end distortion |COMMUN| End distortion that lengthens the marking impulse. { ′märk·iŋ ¦end di,stór·shən }

marking pulse |ELEC| In a teletypewriter, the signal interval during which time the teletypewriter selector unit is operated. { ′märk·iŋ ,pəls }

marking wave |ELEC| In telegraphic communications, that portion of the emission during which the active portions of the code character are being transmitted. Also known as keying wave. { ′märk·iŋ ,wāv }

Markov-based model |COMPUT SCI| A model that represents a computer system by a Markov chain, which represents the set of all possible states of the system, with the possible transitions between these states. { ′mär,kóf ,bāst ,mäd·əl }

mark reading |COMPUT SCI| In character recognition, that form of mark detection which employs a photoelectric device to locate and convey intended information; the information appears as special marks on sites (windows) within the document coding area. { ′märk ,rēd·iŋ }

mark sensing |COMPUT SCI| In character recognition, that form of mark detection which depends on the conductivity of graphite pencil marks to locate and convey intended information; the information appears as special marks on sites (windows) within the document coding area. { ′märk ,sens·iŋ }

mark-space multiplier |ELECTR| A multiplier used in analog computers in which one input controls the mark-to-space ratio of a square wave while the other input controls the amplitude of the wave, and the output, obtained by a smoothing operation, is proportional to the average value of the signal. Also known as time-division multiplier. { ¦märk ¦spās ′məl·tə ,plī·ər }

mark-space ratio See mark-to-space ratio. { ¦märk ¦spās ′rā·shō }

mark-to-space ratio |ELECTR| The ratio of the duration of the positive-amplitude part of a square wave to that of the negative-amplitude part. Also known as mark-space ratio. { ¦märk ¦tə ¦spās ′rā·shō }

mark-to-space transition |COMMUN| The process of switching from a mark to a space. { ¦märk ¦tə ¦spās tran′zish·ən }

markup |COMPUT SCI| The process of adding information (tags) to an electronic document that are not part of the content but describe its structure or elements. { ′märk,əp }

markup language |COMPUT SCI| A set of rules and procedures for markup. { ′märk,əp ,laŋ·gwij }

Marx circuit |ELEC| An electric circuit used in an impulse generator in which capacitors are charged in parallel through charging resistors, and then connected in series and discharged through the test piece by the simultaneous sparkover of spark gaps. { ′märks ,sər·kət }

maser amplifier |ELECTR| A maser which is used to increase the power produced by another maser. { 'mā·zər 'am·plə,fīər }

MA service *See* multiple-access service. { ¦em'ā ,sər·vəs }

mask |ELECTR| A thin sheet of metal or other material containing an open pattern, used to shield selected portions of a semiconductor or other surface during a deposition process. { mask }

maskable interrupt |COMPUT SCI| An interrupt that can be allowed to occur or prevented from occurring by software. { ¦mas·kə·bəl 'int·ə,rəpt }

masking |COMPUT SCI| **1.** Replacing specific characters in one register by corresponding characters in another register. **2.** Extracting certain characters from a string of characters. |ELECTR| **1.** Using a covering or coating on a semiconductor surface to provide a masked area for selective deposition or etching. **2.** A programmed procedure for eliminating radar coverage in areas where such transmissions may be of use to the enemy for navigation purposes, by weakening the beam in appropriate directions or by use of additional transmitters on the same frequency at suitable sites to interfere with homing; also used to suppress the beam in areas where it would interfere with television reception. { 'mask·iŋ }

mask matching |COMPUT SCI| In character recognition, a method employed in character property detection in which a correlation or match is attempted between a specimen character and each of a set of masks representing the characters to be recognized. { 'mask ,mach·iŋ }

mask register |COMPUT SCI| Filter which determines the parts of a word which are to be tested. { 'mask ,rej·ə·stər }

mask word |COMPUT SCI| A word modifier used in a logical AND operation. { 'mask ,wərd }

Mason's theorem |CONT SYS| A formula for the overall transmittance of a signal flow graph in terms of transmittances of various paths in the graph. { 'mās·ənz ,thir·əm }

massage |COMPUT SCI| To process data, primarily to convert it into a more useful form or into a form that will simplify processing. { mə'säzh }

mass communication |COMMUN| Communication which is directed to or reaches an appreciable fraction of the population. { 'mas kə ,myü·nə'kā·shən }

mass conversion |COMPUT SCI| The transfer of data from one computer system to another, in which all the data is converted in a single operation, rather than in gradual increments. { 'mas kən,vər·zhən }

mass data multiprocessing |COMPUT SCI| The basic concept of time sharing, with many inquiry stations to a central location capable of on-line data retrieval. { 'mas ¦dad·ə ,məl·ti'prä,ses·iŋ }

mass-memory unit |COMPUT SCI| Drum or disk memory that provides rapid access bulk storage for messages that are awaiting availability of outgoing channels. { 'mas 'mem·rē ,yü·nət }

mass resistivity |ELEC| The product of the electrical resistance of a conductor and its mass, divided by the square of its length; the product of the electrical resistivity and the density. { 'mas ,rē,zis'tiv·əd·ē }

mass storage |COMPUT SCI| A computer storage with large capacity, especially one whose contents are directly accessible to a computer's central processing unit. { 'mas 'stȯr·ij }

mass-storage system |COMPUT SCI| A computer system containing a large number of storage devices, with one of these devices containing the master file of the operating system, routines, and library routines. { 'mas ¦stȯr·ij ,sis·təm }

master antenna television system |COMMUN| A network that distributes television signals from a common antenna to apartments or dwellings under collective ownership. Abbreviated MATV system. { 'mas·tər an¦ten·ə 'tel·ə,vizh·ən ,sis·təm }

master arm |ENG| A component of a remote manipulator whose motions are automatically duplicated by a slave arm, sometimes with changes of scale in displacement or force. { 'mas·tər 'ärm }

master clock |COMPUT SCI| The electronic or electric source of standard timing signals, often called clock pulses, required for sequencing the operation of a computer. Also known as main clock; master synchronizer; master timer. { 'mas·tər 'kläk }

master console *See* console. { 'mas·tər 'kän,sōl }

master control |COMMUN| The control console that contains the main program controls for a radio or television transmission system or network. |COMPUT SCI| A computer program, oriented toward applications, which carries out the highest level of control in a hierarchy of programs, routines, and subroutines. { 'mas·tər kən¦trōl }

master control interrupt |COMPUT SCI| A signal which causes the master control program to take over control of a computer system. { 'mas·tər kən¦trōl 'in·tə,rəpt }

master data |COMPUT SCI| A set of data which are rarely changed, or changed in a known and constant manner. { 'mas·tər 'dad·ə }

master file |COMPUT SCI| **1.** A computer file containing relatively permanent information, usually updated periodically, such as subscriber records or payroll data other than time worked. **2.** A computer file that is used as an authoritative source of data in carrying out a particular job on the computer. { 'mas·tər 'fīl }

master frequency meter *See* integrating frequency meter. { 'mas·tər 'frē·kwən·sē ,mēd·ər }

master gain |ELECTR| Control of overall gain of an amplifying system as opposed to varying the gain of several individual inputs. { 'mas·tər 'gān }

master group |COMMUN| In carrier telephony, ten supergroups (600 voice channels) multiplexed together and treated as a unit. { 'mas·tər ,grüp }

master instruction tape |COMPUT SCI| A computer magnetic tape on which all programs for a system of runs are recorded. { 'mas·tər in'strək·shən ,tāp }

master mode |COMPUT SCI| The mode of operation of a computer system exercised by the operating system or executive system, in which a privileged class of instructions, which user programs cannot execute, is permitted. Also known as monitor mode; privileged mode. { 'mas·tər ˌmōd }

master multivibrator |ELECTR| Master oscillator using a multivibrator unit. { 'mas·tər ˌməl·ti'vī ˌbrād·ər }

master oscillator |ELECTR| An oscillator that establishes the carrier frequency of the output of an amplifier or transmitter. { 'mas·tər 'äs·ə,lād·ər }

master-oscillator power amplifier |ELECTR| Transmitter using an oscillator followed by one or more stages of radio-frequency amplification. { 'mas·tər 'äs·ə,lād·ər 'pau·ər ,am·plə,fī·ər }

master plan position indicator |ELECTR| In a radar system, a plan position indicator which controls remote indicators or repeaters. { 'mas·tər 'plan pə¦zish·ən 'in·də,kād·ər }

master program file |COMPUT SCI| The tape record of all programs for a system of runs. { 'mas·tər 'prō·grəm ,fīl }

master record |COMPUT SCI| The basic updated record which will be used for the next run. { 'mas·tər 'rek·ərd }

master routine See executive routine. { 'mas·tər rü'tēn }

master scheduler |COMPUT SCI| A program in a job entry system that assigns priorities to jobs submitted for execution. { 'mas·tər 'sked·yə·lər }

master/slave manipulator |ENG| A mechanical, electromechanical, or hydromechanical device which reproduces the hand or arm motions of an operator, enabling the operator to perform manual motions while separated from the site of the work. { 'mas·tər 'slāv mə'nip·yə,lād·ər }

master/slave mode |COMPUT SCI| The feature ensuring the protection of each program when more than one program resides in memory. { 'mas·tər 'slāv ,mōd }

master/slave system |COMPUT SCI| A system of interlinked computers under the control of one computer (master computer). { 'mas·tər 'slāv ,sis·təm }

master switch |ELEC| **1.** Switch that dominates the operation contactors, relays, or other magnetically operated devices. **2.** Switch electrically ahead of a number of individual switches. { 'mas·tər ,swich }

master synchronization pulse |COMMUN| In telemetry, a pulse distinguished from other telemetering pulses by amplitude and duration, used to indicate the end of a sequence of pulses. { 'mas·tər ,siŋ·krə·nə'zā·shən ,pəls }

master synchronizer See master clock. { 'mas·tər 'siŋ·krə,nīz·ər }

master system tape |COMPUT SCI| A monitor program centralizing the control of program operation by loading and executing any program on a system tape. { 'mas·tər 'sis·təm ,tāp }

master tape |COMPUT SCI| A magnetic tape that contains data which must not be overwritten,

such as an executive routine or master file; updating a master tape means generating a new master tape onto which supplementary data have been added. { 'mas·tər 'tāp }

master terminal |COMPUT SCI| A computer terminal that is used to monitor and control a computer system. { 'mas·tər 'tər·mən·əl }

master timer See master clock. { 'mas·tər'tīm·ər }

match |COMPUT SCI| A data-processing operation similar to a merge, except that instead of producing a sequence of items made up from the input sequences, the sequences are matched against each other on the basis of some key. { mach }

matched filter |COMPUT SCI| In character recognition, a method employed in character property detection in which a vertical projection of the input character produces an analog waveform which is then compared to a set of stored waveforms for the purpose of determining the character's identity. |ELECTR| A filter with the property that, when the input consists of noise in addition to a specified desired signal, the signal-to-noise ratio is the maximum which can be obtained in any linear filter. { 'macht 'fil·tər }

matched impedance |ELEC| An impedance of a load which is equal to the impedance of a generator, so that maximum power is delivered to the load. { 'macht im'pēd·əns }

matched load |ELECTR| A load having the impedance value that results in maximum absorption of energy from the signal source. { 'macht 'lōd }

matched transmission line |ELEC| Transmission line terminated with a load equivalent to its characteristic impedance. { 'macht tranz'mish·ən ,līn }

match gate See equivalence gate. { 'mach ,gāt }

matching |COMPUT SCI| A computer problem-solving method in which the current situation is represented as a schema to be mapped into the desired situation by putting the two in correspondence. { 'mach·iŋ }

matching diaphragm |ELECTROMAG| Diaphragm consisting of a slit in a thin sheet of metal, placed transversely across a waveguide for matching purposes; the orientation of the slit with respect to the long dimension of the waveguide determines whether the diaphragm acts as a capacitive or inductive reactance. { 'mach·iŋ 'dī·ə,fram }

matching impedance |ELEC| Impedance value that must be connected to the terminals of a signal-voltage source for proper matching. { 'mach·iŋ im¦pēd·əns }

matching section |ELECTROMAG| A section of transmission line, a quarter or half wavelength long, inserted between a transmission line and a load to obtain impedance matching. { 'mach·iŋ ¦sek·shən }

matching stub |ELECTROMAG| Device placed on a radio-frequency transmission line which varies the impedance of the line; the impedance of the line can be adjusted in this manner. { 'mach·iŋ ,stəb }

match processing |COMPUT SCI| The checking of two or more units of data for common characteristics. { 'mach 'prä,ses·iŋ }

math coprocessor *See* numeric processor extension. { ¦math 'kō,prä,ses·ər }

mathematical check |COMPUT SCI| A programmed computer check of a sequence of operations, using the mathematical properties of that sequence. { ¦math·ə¦mad·ə·kəl 'chek }

mathematical function program |COMPUT SCI| A set of routinely used mathematical functions, such as square root, which are efficiently coded and called for by special symbols. { ¦math·ə ¦mad·ə·kəl 'faŋk·shən ,prō·grəm }

mathematical software |COMPUT SCI| The set of algorithms used in a computer system to solve general mathematical problems. { ¦math·ə ¦mad·ə·kəl 'sȯft,wer }

mathematical subroutine |COMPUT SCI| A computer subroutine in which a well-defined mathematical function, such as exponential, logarithm, or sine, relates the output to the input. { ¦math·ə¦mad·ə·kəl 'səb·rü,tēn }

matrix |COMPUT SCI| A latticework of input and output leads with logic elements connected at some of their intersections. |ELECTR| **1.** The section of an analog video system that transforms the red, green, and blue source signals into color-difference signals and combines them with the chrominance subcarrier. Also known as color coder; color encoder; encoder. **2.** The section of an analog color television receiver that transforms the color-difference signals into the red, green, and blue signals needed to drive the display device. Also known as color decoder; decoder. { 'mā·triks }

matrix algebra tableau |COMPUT SCI| The current matrix at the end of an iteration while running a linear program. { 'mā·triks ¦al·jə·brə ta'blō }

matrix-array camera |ELECTR| A solid-state video camera that has a rectangular array of light-sensitive elements or pixels. { 'mā·triks ə'rā ,kam·rə }

matrix printing |COMPUT SCI| High-speed printing in which characterlike configurations of dots are printed through the proper selection of wire ends from a matrix of wire ends. Also known as stylus printing; wire printing. { 'mā·triks 'print·iŋ }

matrix sound system |ENG ACOUS| A quadraphonic sound system in which the four input channels are combined into two channels by a coding process for recording or for stereo frequency-modulation broadcasting and decoded back into four channels for playback of recordings or for quadraphonic stereo reception. { 'mā·triks 'saund ,sis·təm }

matrix storage |COMPUT SCI| A computer storage in which coordinates are used to address the locations or circuit elements. Also known as coordinate storage. { 'mā·triks ,stȯr·ij }

mattress array *See* billboard array. { 'ma·trəs ə'rā }

MATV system *See* master antenna television system. { ¦em¦ā¦tē¦vē 'sis·təm }

mavar *See* parametric amplifier. { 'mā,vär }

maximum available gain |ELECTR| The theoretical maximum power gain available in a transistor stage; it is seldom achieved in practical circuits because it can be approached only when feedback is negligible. Abbreviated MAG. { 'mak·sə·məm ə¦vāl·ə·bəl 'gān }

maximum average power output |ELECTR| In television, the maximum radio-frequency output power that can occur under any combination of signals transmitted, averaged over the longest repetitive modulation cycle. { 'mak·sə·məm ¦av·rəj ¦paü·ər 'aut,püt }

maximum demand |ELEC| The greatest average value of the power, apparent power, or current consumed by a customer of an electric power system, the averages being taken over successive time periods, usually 15 or 30 minutes in length. { 'mak·sə·məm di'mand }

maximum keying frequency |ELECTR| In facsimile, the frequency in hertz that is numerically equal to the spot speed divided by twice the horizontal dimension of the spot. { 'mak·sə·məm 'kē·iŋ ,frē·kwən·sē }

maximum modulating frequency |ELECTR| Highest picture frequency required for a facsimile transmission system; the maximum modulating frequency and the maximum keying frequency are not necessarily equal. { 'mak·sə·məm 'mäj·ə,lād·iŋ ,frē·kwən·sē }

maximum operating frequency |COMPUT SCI| The highest rate at which the modules perform iteratively and reliably. { 'mak·sə·məm 'äp·ə ,rād·iŋ ,frē·kwən·sē }

maximum retention time |ELECTR| Maximum time between writing into and reading an acceptable output from a storage element of a charge storage tube. { 'mak·sə·məm ri'ten· chən ,tīm }

maximum signal level |ELECTR| In an amplitude-modulated facsimile system, the level corresponding to copy black or copy white, whichever has the highest amplitude. { 'mak·sə·məm 'sig·nəl ,lev·əl }

maximum unambiguous range |ELECTROMAG| The range beyond which the echo from a pulsed radar signal returns after generation of the next pulse, and can thus be mistaken as a short-range echo of the next cycle. { 'mak·sə·məm ,ən·am ¦big·yə·wəs 'rānj }

maximum undistorted power output |ELECTR| Of a transducer, the maximum power delivered under specified conditions with a total harmonic output not exceeding a specified percentage. { 'mak·sə·məm ,ən·di¦stȯrd·əd 'paü·ər ,aut,püt }

maximum usable frequency |COMMUN| The upper limit of the frequencies that can be used at a specified time for point-to-point radio transmission involving propagation by reflection from the regular ionized layers of the ionosphere. Abbreviated MUF. { 'mak·sə·məm ¦yü·zə·bəl 'frē·kwən·sē }

Maxwell bridge |ELEC| A four-arm alternating-current bridge used to measure inductance (or capacitance) in terms of resistance and capacitance (or inductance); bridge balance is independent of frequency. Also known as Maxwell-Wien bridge; Wien-Maxwell bridge. { 'mak,swel ,brij }

Maxwell equations *See* Maxwell field equations. { 'mak,swel i,kwā·zhənz }

Maxwell field equations [ELECTROMAG] Four differential equations which relate the electric and magnetic fields to electric charges and currents, and form the basis of the theory of electromagnetic waves. Also known as electromagnetic field equations; Maxwell equations. { 'mak ,swel 'fēld i,kwā·zhənz }

Maxwell's cyclic currents *See* mesh currents. { 'mak,swelz 'sī·klik 'kə·rəns }

Maxwell's electromagnetic theory [ELECTROMAG] A mathematical theory of electric and magnetic fields which predicts the propagation of electromagnetic radiation, and is valid for electromagnetic phenomena where effects on an atomic scale can be neglected. { 'mak,swelz i,lek·trō·mag'ned·ik 'thē·ə·rē }

Maxwell-Wagner mechanism [ELEC] A capacitor consisting of two parallel metal plates with two layers of material between them, one with vanishing conductivity, the other with finite conductivity and vanishing electric susceptibility. { 'mak,swel 'wag·nər 'mek·ə,niz·əm }

Maxwell-Wien bridge *See* Maxwell bridge. { 'mak ,swel 'wēn 'brij }

Mbit *See* megabit. { 'em,bit }

Mbyte *See* megabyte. { 'em,bīt }

McCabe's cyclomatic number [COMPUT SCI] The total number of decision statements in a computer program plus one; a measure of the complexity of the program. { mə,kābz ,sī·klə ,mad·ik 'nəm·bər }

McNally tube [ELECTR] Reflex klystron tube, the frequency of which may be electrically controlled over a wide range; used as a local oscillator. { mik'nal·ē ,tüb }

M contour [CONT SYS] A line on a Nyquist diagram connecting points having the same magnitude of the primary feedback ratio. { 'em ,kän·túr }

MCT *See* MOS-controlled thyristor.

M-derived filter [ELECTR] A filter consisting of a series of T or pi sections whose impedances are matched at all frequencies, even though the sections may have different resonant frequencies. { 'em di,rīvd 'fil·tər }

M-display [ELECTR] A radar display format in which a trace-deflecting pulse can be moved along the range axis of an A-display to assist the operator in determining and reporting the range of a target. Also known as M-indicator; M-scan; M-scope. { 'em di,splā }

MDS *See* minimum discernible signal.

meaconing [ELECTROMAG] A system for receiving electromagnetic signals and rebroadcasting them with the same frequency so as, for instance, to confuse navigation; a confusion reflector, such as chaff, is an example. { 'mē·kə·niŋ }

Mealy machine [COMPUT SCI] A sequential machine in which the output depends on both the current state of the machine and the input. { 'mē·lē mə,shēn }

mean carrier frequency [ELECTR] Average carrier frequency of a transmitter corresponding to the resting frequency in a frequency-modulated system. { 'mēn 'kar·ē·ər ,frē·kwən·sē }

mean power [ELECTR] For a radio transmitter, the power supplied to the antenna transmission line by a transmitter during normal operation, averaged over a time sufficiently long compared with the period of the lowest frequency encountered in the modulation; a time of 1/10 second during which the mean power is greatest will be selected normally. { 'mēn 'paú·ər }

means-ends analysis [COMPUT SCI] A method of problem solving in which the difference between the form of the data in the present and desired situations is determined, and an operator is then found to transform from one into the other, or, if this is not possible, objects between the present and desired objects are created, and the same procedure is then repeated on each of the gaps between them. { 'mēnz 'enz ə,nal·ə·səs }

mean-square-error criterion [CONT SYS] Evaluation of the performance of a control system by calculating the square root of the average over time of the square of the difference between the actual output and the output that is desired. { 'mēn 'skwer 'er·ər krī,tir·ē·ən }

mean time between failures [COMPUT SCI] A measure of the reliability of a computer system, equal to average operating time of equipment between failures, as calculated on a statistical basis from the known failure rates of various components of the system. Abbreviated MTBF. { 'mēn 'tīm bi ,twēn 'fāl·yərz }

measurand transmitter [COMMUN] A telemetry transmitter that transmits a signal modulated according to the values of the quantity being measured. { 'mezh·ə,rand tranz,mid·ər }

measured service [COMMUN] Telephone service for which charge is made according to the measured amount of usage. { 'mezh·ərd 'sər·vəs }

mechanical bearing cursor *See* bearing cursor. { mi'kan·ə·kəl 'ber·iŋ ,kər·sər }

mechanical damping [ENG ACOUS] Mechanical resistance which is generally associated with the moving parts of an electromechanically transducer such as a cutter or a reproducer. { mi'kan·ə·kəl 'damp·iŋ }

mechanical dialer *See* automatic dialer. { mi'kan·ə·kəl 'dī·lər }

mechanical filter [ELECTR] Filter, used in intermediate-frequency amplifiers of highly selective superheterodyne receivers, consisting of shaped metal bars, rods, or disks that act as coupled mechanical resonators when used with piezoelectric or magnetostrictive input and output transducers and coupled by small-diameter wires. Also known as mechanical wave filter. { mi'kan·ə·kəl 'fil·tər }

mechanical jamming See passive jamming. { mi'kan·ə·kəl 'jam·iŋ }

mechanical modulator [ELEC] A device that varies a carrier wave by moving some part of a circuit element. { mi'kan·ə·kəl 'mäj·ə,lād·ər }

mechanical oscillograph See direct-writing recorder. { mi'kan·ə·kəl ä'sil·ə,graf }

mechanical rectifier [ELEC] A rectifier in which rectification is accomplished by mechanical action, as in a synchronous vibrator. { mi'kan·ə·kəl 'rek·tə,fī·ər }

mechanical replacement [COMPUT SCI] The replacement of one piece of hardware by another piece of hardware at the instigation of the manufacturer. { mi'kan·ə·kəl ri'plās·mənt }

mechanical resistance See resistance. { mi'kan ·ə·kəl ri'zis·təns }

mechanical scanner [COMPUT SCI] In optical character recognition, a device that projects an input character into a rotating disk, on the periphery of which is a series of small, uniformly spaced apertures; as the disk rotates, a photocell collects the light passing through the apertures. { mi'kan·ə·kəl 'skan·ər }

mechanical stepping motor [ELEC] A device in which a voltage pulse through a solenoid coil causes reciprocating motion by a solenoid plunger, and this is transformed into rotary motion through a definite angle by ratchet-and-pawl mechanisms or other mechanical linkages. { mi'kan·ə·kəl 'step·iŋ ,mōd·ər }

mechanical tilt [ELECTR] **1.** Vertical tilt of the mechanical axis of a radar antenna. **2.** The angle indicated by the tilt indicator dial. { mi'kan·ə·kəl 'tilt }

mechanical translation [COMPUT SCI] Automatic translation of one language into another by means of a computer or other machine that contains a dictionary look-up in its memory, along with the programs needed to make logical choices from synonyms, supply missing words, and rearrange word order as required for the new language. Also known as machine translation. { mi'kan·ə·kəl tranz'lā·shən }

mechanical wave filter See mechanical filter. { mi'kan·ə·kəl 'wāv ,fil·tər }

mechanized See machine-sensible. { 'mek·ə ,nīzd }

mechatronics [ENG] A branch of engineering that incorporates the ideas of mechanical and electronic engineering into a whole, and, in particular, covers those areas of engineering concerned with the increasing integration of mechanical, electronic, and software engineering into a production process. { ,mek·ə'trän·iks }

media conversion [COMPUT SCI] The transfer of data from one storage type (such as magnetic tape) to another storage type (such as magnetic or optical disk). { 'mē·dē·ə kən,vər·zhən }

media conversion buffer [COMPUT SCI] Large storage area, such as a drum, on which data may be stored at low speed during nonexecution time, to be later transferred at high speed into core memory during execution time. { 'mē·dē·ə kən ,vər·zhən ,bəf·ər }

medical electronics [ELECTR] A branch of electronics in which electronic instruments and equipment are used for such medical applications as diagnosis, therapy, research, anesthesia control, cardiac control, and surgery. { 'med·ə·kəl i,lek'trän·iks }

medical frequency bands [COMMUN] A collection of radio frequency bands allocated to medical equipment in the United States. { 'med·ə·kəl 'frē·kwən·sē ,banz }

medium [COMPUT SCI] The material, or configuration thereof, on which data are recorded; usually applied to storable, removable media, such as disks and magnetic tape. { 'mē·dē·əm }

medium frequency [COMMUN] A Federal Communications Commission designation for the band from 300 to 3000 kilohertz in the radio spectrum. Abbreviated MF. { 'mē·dē·əm 'frē·kwən·sē }

medium-frequency propagation [COMMUN] Radio propagation at broadcast frequencies where skip is not an important factor. { 'mē·dē·əm ¦frē·kwən·sē ,präp·ə'gā·shən }

medium-frequency tube [ELECTR] An electron tube operated at frequencies between 300 and 3000 kilohertz, at which the transit time of an electron between electrodes is much smaller than the period of oscillation of the voltage. { 'mē·dē·əm ¦frē·kwən·sē ,tüb }

medium-scale integration [ELECTR] Fabrication of solid-state integrated circuits having more than about 12 gate-equivalent circuits. Abbreviated MSI. { 'mē·dē·əm ¦skäl ,int·ə'grā·shən }

medium-technology robot [CONT SYS] An automatically controlled machine that employs servomechanisms and microprocessor control units. { 'mē·dē·əm tek,näl·ə·jē 'rō,bät }

megabit [COMPUT SCI] A unit of information content equal to 1,048,576 (1024 × 1024) bits. Abbreviated Mbit. { 'meg·ə,bit }

megabyte [COMPUT SCI] A unit of information content equal to 1,048,576 (1024 × 1024) bytes. Abbreviated Mbyte. Symbolized M. { 'meg·ə ,bīt }

megaflops [COMPUT SCI] A unit of computer speed, equal to 10^6 flops. { 'meg·ə,fläps }

megapel display [COMPUT SCI] A computer graphics display that handles 10^6 or more pixels (pels). { 'meg·ə,pel di,splā }

megatron See disk-seal tube. { 'meg·ə,trän }

megavolt [ELEC] A unit of potential difference or emf (electromotive force), equal to 1,000,000 volts. Abbreviated MV. { 'meg·ə,vōlt }

megawatt year of electricity [ELEC] A unit of electric energy, equal to the energy from a power of 1,000,000 watts over a period of 1 tropical year, or to 3.1557 × 10^{13} joules. Abbreviated MWYE. { 'meg·ə,wät ¦yir əv i,lek'tris·əd·ē }

megohm [ELEC] A unit of resistance, equal to 1,000,000 ohms. { 'me,gōm }

megohmmeter [ELEC] An instrument which is used for measuring the high resistance of electrical materials of the order of 20,000 megohms at 1000 volts; one direct-reading type employs a permanent magnet and a moving coil. { 'me ,gōmē,mēd·ər }

Meissner oscillator |ELECTR| An electron-tube oscillator in which the grid and plate circuits are inductively coupled through an independent tank circuit which determines the frequency. { 'mīs·nər ,äs·ə,lād·ər }

melodeon |ELECTR| Broadband panoramic receiver used for countermeasures reception; all types of received electromagnetic radiation are presented as vertical pips on a frequency-calibrated cathode-ray indicator screen. { mə'lōd·ē·ən }

meltback transistor |ELECTR| A junction transistor in which the junction is made by melting a properly doped semiconductor and allowing it to solidify again. { 'melt'back tran'zist·ər }

membrane keyboard |COMPUT SCI| A flat keyboard, used with microcomputers and hand-held calculators, that consists of two closely spaced membranes separated by a flat sheet called a spacer with holes corresponding to the keys. { 'mem,brān 'kē,bȯrd }

memex |COMPUT SCI| A hypothetical machine described by Vannevar Bush, which would store written records so that they would be available almost instantly by merely pushing the right button for the information desired. { 'me,meks }

memistor |ELEC| Nonmagnetic memory device consisting of a resistive substrate in an electrolyte; when used in an adaptive system, a direct-current signal removes copper from an anode and deposits it on the substrate, thus lowering the resistance of the substrate; reversal of the current reverses the process, raising the resistance of the substrate. { me'mis·tər }

memory |COMPUT SCI| Any apparatus in which data may be stored and from which the same data may be retrieved; especially, the internal, high-speed, large-capacity working storage of a computer, as opposed to external devices. Also known as computer memory. { 'mem·rē }

memory address register |COMPUT SCI| A special register containing the address of a word currently required. { 'mem·rē 'ad,res ,rej·ə·stər }

memory bank |COMPUT SCI| A physical section of a computer memory, which may be designed to handle information transfers independently of other such transfers in other such sections. { 'mem·rē ,baŋk }

memory buffer register |COMPUT SCI| A special register in which a word is stored as it is read from memory or just prior to being written into memory. { 'mem·rē 'bəf·ər ,rej·ə·stər }

memory capacity See storage capacity. { 'mem·rē kə'pas·əd·ē }

memory card |COMPUT SCI| A small card, typically with dimensions of about 2 × 3 inches (5 × 8 centimeters), that can store information, usually in integrated circuits or magnetic strips. { 'mem·rē ,kärd }

memory cell |COMPUT SCI| A single storage element of a memory, together with associated circuits for storing and reading out one bit of information. { 'mem·rē ,sel }

memory chip See semiconductor memory. { 'mem·rē ,chip }

memory contention |COMPUT SCI| A situation in which two different programs, or two parts of a program, try to read items in the same block of memory at the same time. { 'mem·rē kən'ten·chən }

memory core See magnetic core. { 'mem·rē ,kȯr }

memory cycle See cycle time. { 'mem·rē ,sī·kəl }

memory dump See storage dump. { 'mem·rē ,dəmp }

memory dump routine |COMPUT SCI| A debugging routine which produces a listing of a consecutive section of memory, either numbers or instructions, at selected points in a program. { 'mem·rē ¦dəmp rü,tēn }

memory element |COMPUT SCI| Any component part of core memory. { 'mem·rē ,el·ə·mənt }

memory expansion card |COMPUT SCI| A printed circuit board that contains additional storage and can be plugged into a computer to increase its storage capacity. { 'mem·rē ik'span·chən ,kärd }

memory fill See storage fill. { 'mem·rē ,fil }

memory gap |COMPUT SCI| A gulf in access time, capacity, and cost of computer storage technologies between fast, expensive, main-storage devices and slow, high-capacity, inexpensive secondary-storage devices. Also known as access gap. { 'mem·rē ,gap }

memory guard |COMPUT SCI| Built-in safety devices which prevent a program or a programmer from accessing certain memory areas reserved for the central processor. Also known as memory protect. { 'mem·rē ,gärd }

memory hierarchy |COMPUT SCI| A ranking of computer memory devices, with devices having the fastest access time at the top of the hierarchy, and devices with slower access times but larger capacity and lower cost at lower levels. { 'mem·rē 'hī·ər,är·kē }

memory lockout register |COMPUT SCI| A special register containing the limiting addresses of an area in memory which may not be accessed by the program. { 'mem·rē 'läk,aut ,rej·ə·stər }

memory management |COMPUT SCI| **1.** The allocation of computer storage in a multiprogramming system so as to maximize processing efficiency. **2.** The collection of routines for placing, fetching, and removing pages or segments into or out of the main memory of a computer system. { 'mem·rē ,man·ij·mənt }

memory map |COMPUT SCI| The list of variables, constants, identifiers, and their memory locations when a FORTRAN program is being run. Also known as memory map list. { 'mem·rē ,map }

memory map list See memory map. { 'mem·rē ,map ,list }

memory mapping |COMPUT SCI| The method by which a computer translates between its logical address space and its physical address space. { 'mem·rē ,map·iŋ }

memory overlay |COMPUT SCI| The efficient use of memory space by allowing for repeated use of the same areas of internal storage during the different stages of a program; for instance, when a subroutine is no longer required, another routine can replace all or part of it. { 'mem·rē 'ō·vər,lā }

memory port |COMPUT SCI| A logical connection through which data are transferred in or out of main memory under control of the central processing unit. { 'mem·rē ,pȯrt }

memory power |COMPUT SCI| A relative characteristic pertaining to differences in access time speeds in different parts of memory; for instance, access time from the buffer may be a tenth of the access time from core. { 'mem·rē ,paủ·ər }

memory print See storage dump. { 'mem·rē ,print }

memory printout |COMPUT SCI| A listing of the contents of memory. { 'mem·rē 'print,aủt }

memory protect See memory guard. { 'mem·rē prə,tekt }

memory protection See storage protection. { 'mem·rē prə'tek·shən }

memory-reference instruction |COMPUT SCI| A type of instruction usually requiring two machine cycles, one to fetch the instruction, the other to fetch the data at an address (part of the instruction itself) and to execute the instruction. { 'mem·rē ¦ref·rəns in,strək·shən }

memory register See storage register. { 'mem·rē ,rej·ə·stər }

memory search routine |COMPUT SCI| A debugging routine which has as an essential feature the scanning of memory in order to locate specified instructions. { 'mem·rē 'sərch rü,tēn }

memory-segmentation control |COMPUT SCI| Address-computing logic to address words in memory with dynamic allocation and protection of memory segments assigned to different users. { 'mem·rē ,seg·mən'tā·shən kən,trōl }

memory sniffer |COMPUT SCI| A diagnostic routine that continually tests the computer memory while the machine is in operation. { 'mem·rē ,snif·ər }

memory storage |COMPUT SCI| The sum total of the computer's storage facilities, that is, core, drum, disk, cards, and paper tape. { 'mem·rē ,stȯr·ij }

memory switch See ovonic memory switch. { 'mem·rē ,swich }

memory tube See storage tube. { 'mem·rē ,tüb }

memory typewriter See electronic typewriter. { 'mem·rē ,tīp·rīd·ər }

memotron |ELECTR| An electrical-visual storage tube which is capable of bistable visual-signal display, controllable in duration from a few milliseconds to infinity, and which is suited to specialized oscillography. { 'mem·ə,trän }

MEMS |COMPUT SCI| Micro-electro-mechanical system. { memz or ¦em¦ē¦em'es }

MEMS microphone |ENG ACOUS| A very small microphone, generally less than 1 millimeter, that can be used directly onto an electronic chip and commonly uses a small thin membrane fabricated on the chip to detect sound. { ¦memz or ¦em¦ē¦em¦es 'mī·krə,fōn }

menu |COMPUT SCI| A list of computer functions appearing on a video display terminal which indicates the possible operations that a computer can perform next, only one of which can be selected by the operator. { 'men·yü }

menu bar |COMPUT SCI| **1.** In a graphical user interface, a horizontal strip near the top of the screen or a window, containing the titles of available pull-down menus. **2.** A horizontal or vertical strip containing the names of currently available commands. { 'men·yü ,bär }

menu-driven system |COMPUT SCI| An interactive computer system in which the operator requests the processing to be performed by making selections from a series of menus. { 'men·yü ¦driv·ən 'sis·təm }

mercury arc |ELECTR| An electric discharge through ionized mercury vapor, giving off a brilliant bluish-green light containing strong ultraviolet radiation. { 'mər·kyə·rē 'ärk }

mercury-arc rectifier |ELECTR| A gas-filled rectifier tube in which the gas is mercury vapor; small sizes use a heated cathode, while larger sizes rated up to 8000 kilowatts and higher use a mercury-pool cathode. Also known as mercury rectifier; mercury-vapor rectifier. { 'mər·kyə·rē ¦ärk 'rek·tə,fī·ər }

mercury cell |ELEC| A primary dry cell that delivers an essentially constant output voltage throughout its useful life by means of a chemical reaction between zinc and mercury oxide; widely used in hearing aids. Also known as mercury oxide cell. { 'mər·kyə·rē ,sel }

mercury delay line |ELECTR| An acoustic delay line in which mercury is the medium for sound transmission. Also known as mercury memory; mercury storage. { 'mər·kyə·rē di'lā ,līn }

mercury lamp See mercury-vapor lamp. { 'mər·kyə·rē ,lamp }

mercury memory See mercury delay line. { 'mər·kyə·rē 'mem·rē }

mercury oxide cell See mercury cell. { 'mər·kyə·rē ¦äk,sīd ,sel }

mercury-pool cathode |ELECTR| A cathode of a gas tube consisting of a pool of mercury; an arc spot on the pool emits electrons. { 'mər·kyə·rē ,pül 'kath,ōd }

mercury-pool rectifier See pool-cathode mercury-arc rectifier. { 'mər·kyə·rē ,pül 'rek·tə,fī·ər }

mercury storage See mercury delay line. { 'mər·kyə·rē 'stȯr·ij }

mercury switch |ELEC| A switch that is closed by making a large globule of mercury move up to the contacts and bridge them; the mercury is usually moved by tilting the entire switch. { 'mər·kyə·rē ,swich }

mercury tank |ELECTR| A container of mercury, with pairs of transducers at opposite ends, used in a mercury delay line. { 'mər·kyə·rē ,taŋk }

mercury tube See mercury-vapor tube; pool tube. { 'mər·kyə·rē ,tüb }

mercury-vapor lamp |ELECTR| A lamp in which light is produced by an electric arc between two electrodes in an ionized mercury-vapor atmosphere; it gives off a bluish-green light rich in ultraviolet radiation. Also known as mercury lamp. { 'mər·kyə·rē ¦vā·pər ,lamp }

mercury-vapor rectifier See mercury-arc rectifier. { 'mər·kyə·rē ¦vā·pər 'rek·tə,fī·ər }

mercury-vapor tube |ELECTR| A gas tube in which the active gas is mercury vapor. Also known as mercury tube. { 'mər·kyə·rē ¦vā·pər ¦tüb }

mercury-wetted reed switch |ELEC| A reed switch containing a pool of mercury at one end and normally operated vertically; the contacts on the reeds are covered with a mercury film by capillary action; each operation of the switch renews this mercury film contact, thereby increasing the operating life of the switch many times. { 'mər·kyə·rē ¦wed·əd 'rēd ‚swich }

merge |COMPUT SCI| To create an ordered set of data by combining properly the contents of two or more sets of data, each originally ordered in the same manner as the output data set. Also known as mesh. { mərj }

merged-transistor logic See integrated injection logic. { ¦mərjd tran¦zis·tər 'läj·ik }

merge search |COMPUT SCI| A procedure for searching a table in which both the table and file records must first be ordered in the same sequence on the key involved, and the table is searched sequentially until a table-record key equal to or greater than the file-record key is found, upon which the file record is processed if its key is equal, and the process is repeated with the next file record, starting at the table position where the previous search terminated. { 'mərj ‚sərch }

merge sort |COMPUT SCI| To produce a single sequence of items ordered according to some rule, from two or more previously ordered or unordered sequences, without changing the items in size, structure, or total number; although more than one pass may be required for a complete sort, items are selected during each pass on the basis of the entire key. { 'mərj ‚sȯrt }

merging routine |COMPUT SCI| A program that creates a single sequence of items, ordered according to some rule, out of two or more sequences of items, each sequence ordered according to the same rule. { 'mərj·iŋ rü‚tēn }

merit |ELECTR| A performance rating that governs the choice of a device for a particular application; it must be qualified to indicate type of rating, as in gain-bandwidth merit or signal-to-noise merit. { 'mer·ət }

mesa device |ELECTR| Any device produced by diffusing the surface of a germanium or silicon wafer and then etching down all but selected areas, which then appear as physical plateaus or mesas. { 'mā·sə di‚vīs }

mesa diode |ELECTR| A diode produced by diffusing the entire surface of a large germanium or silicon wafer and then delineating the individual diode areas by a photoresist-controlled etch that removes the entire diffused area except the island or mesa at each junction site. { 'mā·sə ‚dī‚ōd }

mesa transistor |ELECTR| A transistor in which a germanium or silicon wafer is etched down in steps so the base and emitter regions appear as physical plateaus above the collector region. { 'mā·sə tran'zis·tər }

MESFET See metal semiconductor field-effect transistor. { 'mes‚fet }

mesh |COMPUT SCI| See merge. |ELEC| A set of branches forming a closed path in a network so that if any one branch is omitted from the set, the remaining branches of the set do not form a closed path. Also known as loop. { mesh }

mesh analysis |ELEC| A method of electrical circuit analysis in which the mesh currents are taken as independent variables and the potential differences around a mesh are equated to 0. { 'mesh ə'nal·ə·səs }

mesh connection See delta connection. { 'mesh kə‚nek·shən }

mesh currents |ELEC| The currents which are considered to circulate around the meshes of an electric network, so that the current in any branch of the network is the algebraic sum of the mesh currents of the meshes to which that branch belongs. Also known as cyclic currents; Maxwell's cyclic currents. { 'mesh kə·rəns }

mesh impedance |ELEC| The ratio of the voltage to the current in a mesh when all other meshes are open. Also known as self-impedance. { 'mesh im'pēd·əns }

mesh network |COMMUN| A communications network in which each node has at least two links to other nodes. { 'mesh ‚net‚wərk }

mesomerism See resonance. { mə'säm·ə‚riz·əm }

message |COMMUN| A series of words or symbols, transmitted with the intention of conveying information. |COMPUT SCI| An arbitrary amount of information with beginning and end defined or implied: usually, it originates in one place and is intended to be transmitted to another place. { 'mes·ij }

message accounting |COMMUN| Use of equipment to make records of telephone calls for billing purposes. { 'mes·ij ə‚kaúnt·iŋ }

message authentication |COMMUN| Security measure designed to establish the authenticity of a message by means of an authenticator within the transmission derived from certain predetermined elements of the message itself. { 'mes·ij ȯ‚then·tə'kā·shən }

message authentication code |COMPUT SCI| The encrypted personal identification code appended to the message transmitted to a computer; the message is accepted only if the decrypted code is recognized as valid by the computer. Abbreviated MAC. { 'mes·ij ȯ‚then·tə'kā·shən ‚kōd }

message blocking |COMMUN| The division of messages into blocks having a fixed number of bytes in order to provide consistent work units and thereby simplify the design of data communications networks. { 'mes·ij ‚bläk·iŋ }

message buffer |COMPUT SCI| One of a number of sections of computer memory, which contains a message that can be transmitted between tasks in the computer system to request service and receive replies from tasks, and which is stored in a system buffer area, outside the address spaces of tasks. { 'mes·ij ‚bəf·ər }

message center |COMMUN| A communications facility charged with the responsibility for acceptance, preparation for transmission,

transmission, receipt and delivery of messages. { 'mes·ij ‚sen·tər }

message display console |COMPUT SCI| A cathode-ray tube on which is displayed information requested by the user. { 'mes·ij di'splā ‚kän‚sōl }

message exchange |COMPUT SCI| A device which acts as a buffer between a communication line and a computer and carries out communication functions. { 'mes·ij iks‚chānj }

message indicator |COMMUN| Element placed within a message to serve as a guide to the selection or derivation and application of the correct key to facilitate the prompt decryption of the message. { 'mes·ij ‚in·də'kād·ər }

message interpolation |COMMUN| Data message insertion during intersyllable periods or speech pauses on a busy voice channel without breaking down the voice connection or noticeably affecting the voice transmission. { 'mes·ij in ‚tər·pə'lā·shən }

message keying element |COMMUN| That part of the key which changes with every message. { 'mes·ij 'kē·iŋ ‚el·ə·mənt }

message-oriented applications |COMMUN| Applications of data communications that involve medium-size data transfers in the range of hundreds to a few thousand bytes or characters, and are usually unidirectional information flows from source to destination. { 'mes·ij ‚òr·ē‚ent·əd ‚ap·lə'kā·shənz }

message queuing |COMPUT SCI| The stacking of messages according to some priority rule as the messages await processing. { 'mes·ij ‚kyü·iŋ }

message reference block |COMMUN| A set of signals denoting the beginning or end of a message. { 'mes·ij 'ref·rəns ‚bläk }

message registration |COMMUN| A method for counting the number of completed charged calls which originate from a particular telephone line, making one scoring for each local call and more than one scoring for calls between zones. { 'mes·ij ‚rej·ə‚strā·shən }

message routing |COMMUN| Selection of the communication path over which a message is sent. { 'mes·ij ‚rüd·iŋ }

message switching |COMMUN| A system in which data transmitted between stations on different circuits within a network are routed through central points. { 'mes·ij ‚swich·iŋ }

message trailer |COMMUN| The last part of a data communications message that signals the end of the message and may also contain control information such as a check character. { 'mes·ij ‚trā·lər }

messaging |COMMUN| Electronic communication in which a message is sent directly to its destination without being stored en route. { 'mes·ij·iŋ }

meta character |COMPUT SCI| A character in a computer programming language system that has some controlling role with respect to other characters with which it may be associated. { 'med·ə ‚kar·ik·tər }

metacompiler |COMPUT SCI| A compiler that is used chiefly to construct compilers for other programming languages. { 'med·ə·kəm ‚pī·lər }

metadata |COMPUT SCI| A description of the data in a source, distinct from the actual data; for example, the currency by which prices are measured in a data source for purchasing goods. { 'med·ə‚dad·ə }

metadyne |ELECTR| A type of rotating magnetic amplifier having more than one brush per pole, used for voltage regulation or transformation. { 'med·ə‚dīn }

metal-air battery See air-depolarized battery. { 'med·əl ¦er 'bad·ə·rē }

metalanguage |COMPUT SCI| A programming language that uses symbols to represent the syntax of other programming languages, and is used chiefly to write compilers for those languages. { 'med·ə‚laŋ·gwij }

metal antenna |ELECTROMAG| An antenna which has a relatively small metal surface, in contrast to a slot antenna. { 'med·əl an'ten·ə }

metal-clad substation |ELEC| An electric power substation housed in a metal cabinet, either indoors or outdoors. { 'med·əl ‚klad 'səb ‚stā·shən }

metal detector |ELECTR| An electronic device for detecting concealed metal objects, such as guns, knives, or buried pipelines, generally by radiating a high-frequency electromagnetic field and detecting the change produced in that field by the ferrous or nonferrous metal object being sought. Also known as electronic locator; metal locator; radio metal locator. { 'med·əl di‚tek·tər }

metal-film resistor |ELEC| A resistor in which the resistive element is a thin film of metal or alloy, deposited on an insulating substrate of an integrated circuit. { 'med·əl ¦film ri'zis·tər }

metal halide lamp |ELECTR| A discharge lamp in which metal halide salts are added to the contents of a discharge tube in which there is a high-pressure arc in mercury vapor; the added metals generate different wavelengths, to give substantially white light at an efficiency approximating that of high-pressure sodium lamps. { 'med·əl 'ha‚līd ‚lamp }

metal-in-gap head |ELECTR| A ring head in which the gap in the ring is lined with a metallic material having a higher saturation magnetization in order to extend the maximum field of the head. Abbreviated MIG head. { ‚med·əl in ‚gap 'hed }

metal-insulator semiconductor |SOLID STATE| Semiconductor construction in which an insulating layer, generally a fraction of a micrometer thick, is deposited on the semiconducting substrate before the pattern of metal contacts is applied. Abbreviated MIS. { 'med·əl ¦in·sə ‚lād·ər 'sem·i·kən‚dək·tər }

metallic circuit |ELEC| Wire circuit of which the ground or earth forms no part. { mə'tal·ik 'sər·kət }

metallic disk rectifier See metallic rectifier. { mə'tal·ik ¦disk 'rek·tə‚fī·ər }

metallic electrode arc lamp |ELEC| A type of arc lamp in which light is produced by luminescent vapor introduced into the arc by evaporation from the cathode; the anode is solid copper, and the cathode is formed of magnetic iron oxide with titanium as the light-producing element and other chemicals to control steadiness and vaporization.   { mə'tal·ik i₁lek₁trōd ärk ₁lamp }

metallic insulator [ELECTROMAG] Section of transmission line used as a mechanical support device; the section is an odd number of quarter-wavelengths long at the frequency of interest, and the input impedance becomes high enough so that the section effectively acts as an insulator.   { mə'tal·ik 'in·sə₁lād·ər }

metallic rectifier [ELECTR] A rectifier consisting of one or more disks of metal under pressure-contact with semiconductor coatings or layers, such as a copper oxide, selenium, or silicon rectifier. Also known as contact rectifier; dry-disk rectifier; dry-plate rectifier; metallic-disk rectifier; semiconductor rectifier.   { mə'tal·ik 'rek·tə₁fī·ər }

metallized capacitor [ELEC] A capacitor in which a film of metal is deposited directly on the dielectric to serve in place of a separate foil strip; has self-healing characteristics.   { 'med·əl₁īzd kə'pas·əd·ər }

metallized-paper capacitor [ELEC] A modification of a paper capacitor in which metal foils are replaced by extremely thin films of metal deposited on the paper; if a breakdown occurs, these films burn away in the area of the breakdown.   { 'med·əl₁īzd ₁pāp·ər kə'pas·əd·ər }

metallized resistor [ELEC] A resistor made by depositing a thin film of high-resistance metal on the surface of a glass or ceramic rod or tube.   { 'med·əl₁īzd ri'zis·tər }

metal locator *See* metal detector.   { 'med·əl ₁lō ₁kād·ər }

metal-nitride-oxide semiconductor [SOLID STATE] A semiconductor structure that has a double insulating layer; typically, a layer of silicon dioxide (SiO_2) is nearest the silicon substrate, with a layer of silicon nitride (Si_3N_4) over it. Abbreviated MNOS.   { 'med·əl ₁nī₁trīd ₁äk₁sīd 'sem·i·kən₁dək·tər }

metal oxide resistor [ELEC] A metal-film resistor in which an oxide of a metal such as tin is deposited as a film onto an insulating substrate.   { 'med·əl ₁äk₁sīd ri'zis·tər }

metal oxide semiconductor [SOLID STATE] A metal insulator semiconductor structure in which the insulating layer is an oxide of the substrate material; for a silicon substrate, the insulating layer is silicon dioxide (SiO_2). Abreviated MOS.   { 'med·əl ₁äk₁sīd 'sem·i·kən₁dək·tər }

metal oxide semiconductor field-effect transistor [ELECTR] A field-effect transistor having a gate that is insulated from the semiconductor substrate by a thin layer of silicon dioxide. Abbreviated MOSFET; MOST; MOS transistor. Formerly known as insulated-gate field-effect transistor (IGFET).   { 'med·əl ₁äk₁sīd 'sem·i·kən₁dək·tər 'fēld i₁fekt tran'zis·tər }

metal oxide semiconductor integrated circuit [ELECTR] An integrated circuit using metal oxide semiconductor transistors; it can have a higher density of equivalent parts than a bipolar integrated circuit.   { 'med·əl ₁äk₁sīd 'sem·i·kən ₁dək·tər 'int·ə₁grād·əd 'sər·kət }

metal semiconductor field-effect transistor [ELECTR] A field-effect transistor that uses a thin film of gallium arsenide, with a Schottky barrier gate formed by depositing a layer of metal directly onto the surface of the film. Abbreviated MESFET.   { 'med·əl 'sem·i·kən₁dək·tər 'fēld i₁fekt tran'zis·tər }

metascope [ELECTR] An infrared receiver used for converting pulsed invisible infrared rays into visible signals for communication purposes; also used with an infrared source for reading maps in darkness.   { 'med·ə₁skōp }

metavariable [COMPUT SCI] One of the elements of a formal language, corresponding to the parts of speech of a natural language. Also known as component name; phrase name.   { 'med·ə'ver·ē·ə·bəl }

meteoric scatter [COMMUN] A form of scatter propagation in which meteor trails serve to scatter radio waves back to earth.   { ₁mēd·ē'ȯr·ik 'skad·ər }

meteorological frequency bands [COMMUN] A collection of radio and microwave frequency bands allocated for use by radiosondes and ground-based radars used in weather forecasting in the United States.   { ₁med·ē·ə·rə'läj·ə·kəl 'frē·kwən₁sē ₁banz }

meter [ENG] A device for measuring the value of a quantity under observation; the term is usually applied to an indicating instrument alone.   { 'mēd·ər }

meter bridge [ELEC] A uniform resistance wire 1 meter in length, mounted above a scale marked in millimeters, with terminals added to make the device usable as either part of a Wheatstone bridge or of a potentiometer.   { 'mēd·ər ₁brij }

meter sensitivity [ENG] The accuracy with which a meter can measure a voltage, current, resistance, or other quantity.   { 'mēd·ər ₁sen·sə'tiv·əd·ē }

meter-type relay [ELEC] A relay that uses a meter movement having a contact-bearing pointer which moves toward or away from a fixed contact mounted on the meter scale.   { 'mēd·ər ₁tīp ₁rē·lā }

method of images |ELEC| In electrostatics, a method of determining the electric fields and potentials set up by charges in the vicinity of a conductor, in which the conductor and its induced surface charges are replaced by one or more fictitious charges. { 'meth·əd əv 'im·ij·əz }

metric waves |ELECTROMAG| Radio waves having wavelengths between 1 and 10 meters, corresponding to frequencies between 30 and 300 megahertz (the very-high-frequency band). { 'me·trik 'wāvz }

mF See millifarad.

MF See medium frequency.

Mflop See million floating-point operations per second. { 'em,fläp }

MFSK See multiple-frequency-shift keying.

MHD generator See magnetohydrodynamic generator. { |em|āch'dē 'jen·ə,rād·ər }

mho See siemens. { mō }

MIC See microwave integrated circuit.

mica capacitor |ELEC| A capacitor whose dielectric consists of thin rectangular sheets of mica and whose electrodes are either thin sheets of metal foil stacked alternately with mica sheets, or thin deposits of silver applied to one surface of each mica sheet. { 'mī·kə kə'pas·əd·ər }

mickey-mouse |COMPUT SCI| To play with something new, such as hardware, software, or a system, until a feel is gotten for it and the proper operating procedure is discovered, understood, and mastered. { |mik·ē 'maús }

MICR See magnetic-ink character recognition.

microactuator |ENG| A very small actuator, with physical dimensions in the submicrometer to millimeter range, generally batch-fabricated from silicon wafers. { ,mī·krō'ak·chə,wād·ər }

microalloy diffused transistor |ELECTR| A microalloy transistor in which the semiconductor wafer is first subjected to gaseous diffusion to produce a nonuniform base region. Abbreviated MADT. { |mī·krō'al,ói də'fyüzd tran'zis·tər }

microalloy transistor |ELECTR| A transistor in which the emitter and collector electrodes are formed by etching depressions, then electroplating and alloying a thin film of the impurity metal to the semiconductor wafer, somewhat as in a surface-barrier transistor. { |mī·krō'al,ói tran 'zis·tər }

microammeter |ELEC| An ammeter whose scale is calibrated to indicate current values in microamperes. { |mī·krō'a,mēd·ər }

microampere |ELEC| A unit of current equal to one-millionth of an ampere. Abbreviated μA. { |mī·krō'am,pir }

microcapacitor |ELECTR| Any very small capacitor used in microelectronics, usually consisting of a thin film of dielectric material sandwiched between electrodes. { |mī·krō·kə'pas·əd·ər }

microchannel plate |ELECTR| A plate that consists of extremely small cylinder-shaped electron multipliers mounted side by side, to provide image intensification factors as high as 100,000. Also known as channel plate multiplier. { |mī·krō|chan·əl 'plāt }

microchip See chip. { 'mī·krō,chip }

microcircuitry |ELECTR| Electronic circuit structures that are orders of magnitude smaller and lighter than circuit structures produced by the most compact combinations of discrete components. Also known as microelectronic circuitry; microminiature circuitry. { |mī·krō'sər·kə·trē }

microcode |COMPUT SCI| A code that employs microinstructions; not ordinarily used in programming. { 'mī·krō,kōd }

microcomputer |COMPUT SCI| **1.** A digital computer whose central processing unit resides on a single semiconductor integrated circuit chip, a microprocessor. **2.** An electronic device, typically consisting of a microprocessor central processing unit, semiconductor memory (RAM), graphics display, and keyboard. Typical configurations also include a hard disk for persistent memory, a compact disk drive, a disk drive which allows removable disks to be used to move data in and out of the machine, and a pointing device. { |mī·krō·kəm'pyüd·ər }

microcomputer development system |COMPUT SCI| A complete microcomputer system that is used to test both the software and hardware of other microcomputer-based systems. { |mī·krō·kəm'pyüd·ər di'vel·əp·mənt ,sis·təm }

microcontroller |ELECTR| A microcomputer, microprocessor, or other equipment used for precise process control in data handling, communication, and manufacturing. { |mī·krō·kən'trōl·ər }

microcoulomb |ELEC| A unit of electric charge equal to one-millionth of a coulomb. Abbreviated μC. { |mī·krō'kü,läm }

microdiagnostic program |COMPUT SCI| A microprogram that tests a specific hardware component, such as a bus or store location, for faults. { |mī·krō,dī·əg'näs·tik 'prō·grəm }

microdisk |COMPUT SCI| A small floppy disk with a diameter between 3 and 4 inches (7 and 10 centimeters). Also known as microfloppy disk. { 'mī·krō,disk }

micro-electro-mechanical system |ENG| A system in which micromechanisms are coupled with microelectronics, most commonly fabricated as microsensors or microactuators. Abbreviated MEMS. Also known as microsystem. { |mī·krōi ,lek·trə·mə'kan·ə·kəl ,sis·təm }

microelectronic circuitry See microcircuitry. { |mī·krō·i,lek'trän·ik 'sər·kə·trē }

microelectronics |ELECTR| The technology of constructing circuits and devices in extremely small packages by various techniques. Also known as microminiaturization; microsystem electronics. { |mī·krō·i,lek'trän·iks }

microelement |ELECTR| Resistor, capacitor, transistor, diode, inductor, transformer, or other electronic element or combination of elements mounted on a ceramic wafer 0.025 centimeter thick and about 0.75 centimeter square; individual microelements are stacked, interconnected, and potted to form micromodules. |ENG| An element of a work cycle whose time span is too short to be observed by the unaided eye. { ¦mī·krō'el·ə·mənt }

microfabrication |ENG| The technology of fabricating microsystems from silicon wafers, using standard semiconductor process technologies in combination with specially developed processes. { ¦mī·krō¦fab·rə'kā·shən }

microfarad |ELEC| A unit of capacitance equal to one-millionth of a farad. Abbreviated μF. { ¦mī·krō'far·əd }

microfloppy disk See microdisk. { ¦mī·krō¦fläp·ē ¸disk }

microhm |ELEC| A unit of resistance, reactance, and impedance, equal to 10^{-6} ohm. { 'mī·krōm }

microimage |COMPUT SCI| A single image stored on a microform medium. { 'mī·krō¸im·ij }

microinstruction |COMPUT SCI| The portion of a microprogram that specifies the operation of individual computing elements and such related subunits as the main memory and the input/output interfaces; usually includes a next-address field that eliminates the need for a program counter. { ¦mī·krō·in'strək·shən }

microlock |ELECTR| **1.** Satellite telemetry system that uses phase-lock techniques in the ground receiving equipment to achieve extreme sensitivity. **2.** A lock by a tracking station upon a minitrack radio transmitter. **3.** The system by which this lock is effected. { 'mī·krō¸läk }

micromainframe |COMPUT SCI| A main frame of a computer placed on one or more integrated circuit chips. { ¦mī·krō'mān¸frām }

micromechanical display |ENG| A video display based on an array of mirrors on a silicon chip that can be deflected by electrostatic forces. Abbreviated MMD. { ¸mī·krō·mə¸kan·i·kəl di'splā }

micromechanics |ENG| **1.** The design and fabrication of micromechanisms. **2.** See composite micromechanics. { ¦mī·krō·mə'kan·iks }

micromechatronics |ENG| The branch of engineering concerned with micro-electromechanical systems. { ¦mī·krō·mek·ə'trän·iks }

microminiature circuitry See microcircuitry. { ¦mī·krō¦min·ə·char 'sər·kə·trē }

microminiaturization See microelectronics. { ¦mī·krō¸min·ə·chə·rə'zā·shən }

micromodule |ELECTR| Cube-shaped, plug-in, miniature circuit composed of potted microelements; each microelement can consist of a resistor, capacitor, transistor, or other element, or a combination of elements. { 'mī·krō¸mäj·ül }

microoperation |COMPUT SCI| Any clock-timed step of an operation. { ¦mī·krō¸äp·ə'rā·shən }

micro-opto-electro-mechanical system |ENG| A microsystem that combines the functions of optical, mechanical, and electronic components in a single, very small package or assembly. Abbreviated MOEMS. { ¦mī·krō¦äp·tō i¦lek·trō mə¦kan·ə·kəl 'sis·təm }

micro-opto-mechanical system |ENG| A microsystem that combines optical and mechanical functions without the use of electronic de-

vices or signals. Abbreviated MOMS. { ¦mī·krō ¦op·to·mə'kan·ə·kəl ¸sis·təm }

microperf |COMPUT SCI| A type of continuous-feed computer paper having extremely small perforations along the separations and edges which give separated pages the appearance of standard typewriter paper. { 'mī·krə¸pərf }

microphone |ENG ACOUS| An electroacoustic device containing a transducer which is actuated by sound waves and delivers essentially equivalent electric waves. { 'mī·krə¸fōn }

microphone transducer |ENG ACOUS| A device which converts variation in the position or velocity of some body into corresponding variations of some electrical quantity, in a microphone. { 'mī·krə¸fōn tranz'dü·sər }

microphonics |ELECTR| Noise caused by mechanical vibration of the elements of an electron tube, component, or system. Also known as microphonism. { ¸mī·krə'fän·iks }

microphonism See microphonics. { mī'krä·fə ¸niz·əm }

microprocessing unit |ELECTR| A microprocessor with its external memory, input/output interface devices, and buffer, clock, and driver circuits. Abbreviated MPU. { ¦mī·krō'prä¸ses·iŋ ¸yü·nət }

microprocessor |ELECTR| A single silicon chip on which the arithmetic and logic functions of a computer are placed. { ¦mī·krō'prä¸ses·ər }

microprocessor intertie and communication system |COMMUN| A data communications system which provides the communication network with its own dedicated processing resources and reduces in-terminal response time, compensating for the capacity used up by communications terminals. Abbreviated MICS. { mī·krō'prä¸ses·ər 'in·tər¸tī ən kə¸myü·nə'kā·shən ¸sis·təm }

microprogram |COMPUT SCI| A computer program that consists only of basic elemental commands which directly control the operation of each functional element in a microprocessor. { ¦mī·krō'prō·grəm }

microprogrammable instruction |COMPUT SCI| An instruction that does not refer to a core memory address and that can be microprogrammed, thus specifying various commands within one instruction. { ¦mī·krō·prə'gram·ə·bəl in'strək·shən }

microprogramming |COMPUT SCI| Transformation of a computer instruction into a sequence of elementary steps (microinstructions) by which the computer hardware carries out the instruction. { ¦mī·krō'prō¸gram·iŋ }

micropump See electroosmotic driver. { 'mī·krə ¸pəmp }

microradiometer |ELECTR| A radiometer used for measuring weak radiant power, in which a thermopile is supported on and connected directly to the moving coil of a galvanometer. Also known as radiomicrometer. { ¦mīkrō¸rād·ē'äm·əd·ər }

microspec function |COMPUT SCI| The set of microinstructions which performs a specific operation in one or more machine cycles. { 'mī·krə ¸spek ¸fəŋk·shən }

microstrip [ELECTROMAG] A strip transmission line that consists basically of a thin-film strip in intimate contact with one side of a flat dielectric substrate, with a similar thin-film ground-plane conductor on the other side of the substrate. { 'mī·krə‚strip }

microsystem See micro-electro-mechanical system. { 'mī·krō‚sis·təm }

microsystem electronics See microelectronics. { 'mī·krə‚sis·təm i‚lek'trän·iks }

microvolt [ELEC] A unit of potential difference equal to one-millionth of a volt. Abbreviated μV. { 'mī·krə‚vōlt }

microvoltmeter [ELECTR] A voltmeter whose scale is calibrated to indicate voltage values in microvolts. { ¦mī·krō'vōlt‚mēd·ər }

microwave [ELECTROMAG] An electromagnetic wave which has a wavelength between about 0.3 and 30 centimeters, corresponding to frequencies of 1–100 gigahertz; however, there are no sharp boundaries distinguishing microwaves from infrared and radio waves. { 'mī·krə‚wāv }

microwave amplifier [ELECTR] A device which increases the power of microwave radiation. { 'mī·krə‚wāv 'am·plə‚fī·ər }

microwave antenna [ELECTROMAG] A combination of an open-end waveguide and a parabolic reflector or horn, used for receiving and transmitting microwave signal beams at microwave repeater stations. { 'mī·krə‚wāv an'ten·ə }

microwave attenuator [ELECTROMAG] A device that causes the field intensity of microwaves in a waveguide to decrease by absorbing part of the incident power; usually consists of a piece of lossy material in the waveguide along the direction of the electric field vector. { 'mī·krə ‚wāv ə'ten·yə‚wād·ər }

microwave cavity See cavity resonator. { 'mī·krə ‚wāv ‚kav·ə·dē }

microwave circuit [ELECTROMAG] Any particular grouping of physical elements, including waveguides, attenuators, phase changers, detectors, wavemeters, and various types of junctions, which are arranged or connected together to produce certain desired effects on the behavior of microwaves. { 'mī·krə‚wāv ‚sər·kət }

microwave circulator See circulator. { 'mī·krə ‚wāv 'sər·kyə‚lād·ər }

microwave communication [COMMUN] Transmission of messages using highly directional microwave beams, which are generally relayed by a series of microwave repeaters spaced up to 50 miles (80 kilometers) apart. { 'mī·krə‚wāv kə ‚myü·nə'kā·shən }

microwave detector [ELECTR] A device that can demonstrate the presence of a microwave by a specific effect that the wave produces, such as a bolometer, or a semiconductor crystal making a pinpoint contact with a tungsten wire. { 'mī·krə ‚wāv di‚tek·tər }

microwave device [ELECTR] Any device capable of generating, amplifying, modifying, detecting, or measuring microwaves, or voltages having microwave frequencies. { 'mī·krə‚wāv di‚vīs }

microwave filter [ELECTROMAG] A device which passes microwaves of certain frequencies in a transmission line or waveguide while rejecting or absorbing other frequencies; consists of resonant cavity sections or other elements. { 'mī·krə‚wāv ‚fil·tər }

microwave generator See microwave oscillator. { 'mī·krə‚wāv 'jen·ə‚rād·ər }

microwave gyrator See gyrator. { 'mī·krə‚wāv 'jī ‚rād·ər }

microwave hop [COMMUN] A microwave communications channel between two stations with directive antennas that are aimed at each other. { 'mī·krə‚wāv 'häp }

microwave integrated circuit [ELECTR] A microwave circuit that uses integrated-circuit production techniques involving such features as thin or thick films, substrates, dielectrics, conductors, resistors, and microstrip lines, to build passive assemblies on a dielectric. Abbreviated MIC. { 'mī·krə‚wāv 'int·ə‚grād·əd 'sər·kət }

microwave link See microwave repeater. { 'mī· krə‚wāv ‚liŋk }

microwave network [COMMUN] A series of microwave repeaters, spaced up to 50 miles (80 kilometers) apart, which relay messages over long distances using highly directional microwave beams. { 'mī·krə‚wāv 'net‚wərk }

microwave noise standard [ENG] An electrical noise generator of calculable intensity that is used to calibrate other noise sources by using comparison methods. { 'mī·krə‚wāv 'noiz ‚stan· dərd }

microwave oscillator [ELECTR] A type of electron tube or semiconductor device used for generating microwave radiation or voltage waveforms with microwave frequencies. Also known as microwave generator. { 'mī·krə‚wāv 'äs·ə‚lād·ər }

microwave radiometer See radiometer. { 'mī·krə ‚wāv ‚rād·ē'äm·əd·ər }

microwave receiver [ELECTR] Complete equipment that is needed to convert modulated microwaves into useful information. { 'mī·krə ‚wāv ri'sē·vər }

microwave relay See microwave repeater. { 'mī· krə‚wāv 'rē‚lā }

microwave repeater [COMMUN] A tower that is equipped with a receiver and transmitter for picking up, amplifying, and passing on in either direction the signals sent over a microwave network by highly directional microwave beams. Also known as microwave link; microwave relay. { 'mī·krə‚wāv ri'pēd·ər }

microwave resonance cavity See cavity resonator. { 'mī·krə‚wāv 'rez·ən·əns ‚kav·əd·ē }

microwave solid-state device [ELECTR] A semiconductor device for the generation or amplification of electromagnetic energy at microwave frequencies. { 'mī·krə‚wāv ¦säl·əd ¦stāt di'vīs }

microwave transmission line [ELECTROMAG] A material structure forming a continuous path from one place to another and capable of directing the transmission of electromagnetic energy along this path. { 'mī·krə‚wāv tranz'mish·ən ‚līn }

microwave tube |ELECTR| A high-vacuum tube designed for operation in the frequency region from approximately 3000 to 300,000 megahertz. { 'mī·krə,wāv ,tüb }

microwave waveguide *See* waveguide. { 'mī·krə ,wāv 'wāv,gīd }

MICS *See* microprocessor intertie and communication system.

middle core |COMPUT SCI| The locations with medium addresses in a computer's main storage; usually assigned to workspace for application programs. { 'mid·əl 'kȯr }

middle-ultraviolet lamp |ELECTR| A mercury-vapor lamp designed to produce radiation in the wavelength band from 2800 to 3200 angstrom units (280 to 320 nanometers) such as sunlamps and photochemical lamps. { 'mid·əl ¦əl·trə ¦vī·lət 'lamp }

mid-frequency gain |ELECTR| The maximum gain of an amplifier, when this gain depends on the frequency; for an RC-coupled voltage amplifier the gain is essentially equal to this value over a large range of frequencies. { 'mid¦frē·kwən·sē ,gān }

MIDI *See* musical instrument digital interface. { 'mid·ē }

midicomputer |COMPUT SCI| A computer having greater performance and capacity than a minicomputer and less than that of a mainframe. { 'mid·ē·kəm,pyüd·ər }

midrange |ENG ACOUS| A loudspeaker designed to reproduce medium audio frequencies, generally used in conjunction with a crossover network, a tweeter, and a woofer. Also known as squawker. { 'mid,rānj }

mid-square generator |COMPUT SCI| A procedure for generating a sequence of random numbers, in which a member of a sequence is squared and the middle digits of the resulting number form the next member of the sequence. { 'mid¦skwər 'jen·ə,rād·ər }

Mie's double plate |ELEC| A device consisting of two small metal disks with insulating handles; they are held in contact in an electric field and then separated, and the charge on one of the disks is then measured to determine the electric displacement. { 'mēz ¦dəb·əl 'plāt }

MIG head *See* metal-in-gap head. { ¦em¦ī'jē ,hed or 'mig ,hed }

migration |COMPUT SCI| Movement of frequently used data items to more accessible storage locations, and of infrequently used data items to less accessible locations. { mī'grā·shən }

Miller bridge |ELECTR| Type of bridge circuit for measuring amplification factors of vacuum tubes. { 'mil·ər ,brij }

Miller code |COMPUT SCI| A code used internally in some computers, in which a binary 1 is represented by a transition in the middle of a bit (either up or down), and a binary 0 is represented by no transition following a binary 1; a transition between bits represents successive 0's; in this code, the longest period possible without a transition is two bit times. { 'mil·ər ,kōd }

Miller effect |ELECTR| The increase in the effective grid-cathode capacitance of a vacuum tube due to the charge induced electrostatically on the grid by the anode through the grid-anode capacitance. { 'mil·ər i,fekt }

Miller generator *See* bootstrap integrator. { 'mil·ər 'jen·ə ,rād·ər }

Miller integrator |ELECTR| A resistor-capacitor charging network having a high-gain amplifier paralleling the capacitor; used to produce a linear time-base voltage. Also known as Miller time-base. { 'mil·ər 'int·ə,grād·ər }

Miller time-base *See* Miller integrator. { 'mil·ər 'tīm ,bās }

milliammeter |ELEC| An ammeter whose scale is calibrated to indicate current values in milliamperes. { ,mil·ē'am,ēd·ər }

milliampere |ELEC| A unit of current equal to one-thousandth of an ampere. Abbreviated mA. { ¦mil·ē'am,pir }

millifarad |ELEC| A unit of capacitance equal to one-thousandth of a farad. Abbreviated mF. { ¦mil·ē'far·əd }

Millikan meter |ELECTR| An integrating ionization chamber in which a gold-leaf electroscope is charged a known amount and ionizing events reduce this charge, so that the resulting angle through which the gold leaf is repelled at any given time indicates the number of ionizing events that have occurred. { 'mil·ə·kən ,mēd·ər }

millimeter wave |ELECTROMAG| An electromagnetic wave having a wavelength between 1 millimeter and 1 centimeter, corresponding to frequencies between 30 and 300 gigahertz. Also known as millimetric wave. { 'mil·ə,mēd·ər 'wāv }

millimetric wave *See* millimeter wave. { ¦mil·ə ¦me·trik 'wāv }

million floating-point operations per second |COMPUT SCI| A unit used to measure the processing speed or throughput of supercomputers or array processors. Abbreviated Mflop. { 'mil·yən ¦flōd·iŋ ¦pȯint ,äp·ə'rā·shənz pər 'sek·ənd }

million instructions per second |COMPUT SCI| A unit used to measure the speed at which a computer's central processing unit can process instructions. Abbreviated MIPS. { 'mil·yən in'strək·shənz pər 'sek·ənd }

millivolt |ELEC| A unit of potential difference or emf equal to one-thousandth of a volt. Abbreviated mV. { 'mil·ə,vōlt }

millivoltmeter |ELEC| A voltmeter whose scale is calibrated to indicate voltage values in millivolts. { ,mil·ə'vōlt,mēdər }

Mills cross |ELECTROMAG| An antenna array that consists of two antennas oriented perpendicular to each other and that produces a narrow pencil beam. { milz 'krȯs }

MIMD |COMPUT SCI| A type of multiprocessor architecture in which several instruction cycles may be active at any given time, each independently fetching instructions and operands into multiple processing units and operating on them in a concurrent fashion. Acronym for multiple-instruction-stream, multiple-datastream.

MIME |COMPUT SCI| The Multimedia Internet Mail Enhancements standard, describing a way of encoding binary files, such as pictures, videos, sounds, and executable files, within a normal text message in an operating-system-independent manner. { mīm }

M-indicator See M-display. { 'em ,in·də,kād·ər }

miniature electron tube |ELECTR| A small electron tube having no base, with tube electrode leads projecting through the glass bottom in positions corresponding to those of pins for either a seven-pin or nine-pin tube base. { 'min·ə·chər i'lek,trän ,tüb }

miniaturization |ELECTR| Reduction in the size and weight of a system, package, or component by using small parts arranged for maximum utilization of space. { ,min·ə·chə·rə'zā·shən }

minicartridge |COMPUT SCI| A self-contained package of reel-to-reel magnetic tape that resembles a cassette or cartridge but is slightly different in design and dimensions. { 'min·ē,kär·trij }

minicomputer |COMPUT SCI| A relatively small general-purpose digital computer, intermediate in size between a microcomputer and a main frame. { 'min·ē·kəm,pyüd·ər }

minidisk |COMPUT SCI| A floppy disk that has a diameter of 5.25 inches (approximately 13 centimeters). Also known as minifloppy disk. { 'min·ē,disk }

minifloppy disk See minidisk. { 'min·ē,fläp·ē ,disk }

minimal-latency coding See minimum-access coding. { 'min·ə·məl |lāt·ən·sē ,kōd·iŋ }

minimal realization |CONT SYS| In linear system theory, a set of differential equations, of the smallest possible dimension, which have an input/output transfer function matrix equal to a given matrix function G(s). { 'min·ə·məl ,rē·ə·lə'zā·shən }

minimize |COMMUN| Condition when normal message and telephone traffic is drastically reduced so messages connected with an actual or simulated emergency will not be delayed. |COMPUT SCI| In a graphical user interface environment, to reduce a window to an icon that represents the application running in the window. { 'min·ə,mīz }

minimum-access coding |COMPUT SCI| Coding in such a way that a minimum time is required to transfer words to and from storage, for a computer in which this time depends on the location in storage. Also known as minimal-latency coding; minimum-delay coding; minimum-latency coding. { 'min·ə·məm |ak,ses ,kōd·iŋ }

minimum-access programming |COMPUT SCI| The programming of a digital computer in such a way that minimum waiting time is required to obtain information out of the memory. Also known as forced programming; minimum-latency programming. { 'min·ə·məm |ak,ses 'prō,gram·iŋ }

minimum-access routine See minimum-latency routine. { 'min·ə·məm |ak,ses rü,tēn }

minimum configuration |COMPUT SCI| **1.** A computer system that has only essential hardware

components. **2.** The smallest assortment of hardware and software components required to carry out a particular data-processing function. { 'min·ə·məm kən,fig·yə'rā·shən }

minimum-delay coding See minimum-access coding. { 'min·ə·məm di,lā 'kōd·iŋ }

minimum detectable signal See threshold signal. { 'min·ə·məm di'tek·tə·bəl 'sig·nəl }

minimum discernible signal |ELECTR| **1.** Receiver input power level that is just sufficient to produce a discernible signal in the receiver output; a receiver sensitivity test. **2.** In radar, the minimum echo power level at the input to the receiver that results in an output that can be confidently seem relative to the noise, usually determined with test equipment in the field. Abbreviated MDS. { 'min·ə·məm di'sər·nə·bəl 'sig·nəl }

minimum-distance code |COMMUN| A binary code in which the signal distance does not fall below a specified minimum value. { 'min·ə·məm |dis·təns 'kōd }

minimum firing current |ELEC| The limit below which firing will not occur in electric blasting caps. { 'min·ə·məm 'fīr·iŋ ,kə·rənt }

minimum-latency coding See minimum-access coding. { 'min·ə·məm |lat·ən·sē ,kōd·iŋ }

minimum-latency programming See minimum-access programming. { 'min·ə·məm |lat·ən·sē 'prō,gram·iŋ }

minimum-latency routine |COMPUT SCI| A computer routine that is constructed so that the latency in serial-access storage is less than the random latency that would be expected if storage locations were chosen without regard for latency. Also known as minimum-access routine. { 'min·ə·məm |lat·ən·sē rü,tēn }

minimum-loss attenuator |ELECTR| A section linking two unequal resistive impedances which is designed to introduce the smallest attenuation possible. Also known as minimum-loss pad. { 'min·ə·məm |lòs ə'ten·yə,wād·ər }

minimum-loss matching |ELECTR| Design of a network linking two resistive impedances so that it introduces a loss which is as small as possible. { 'min·ə·məm |lòs ,mach·iŋ }

minimum-loss pad See minimum-loss attenuator. { 'min·ə·məm |lòs ,pad }

minimum-phase system |CONT SYS| A linear system for which the poles and zeros of the transfer function all have negative or zero real parts. { 'min·ə·məm 'fāz ,sis·təm }

minimum signal level |ELECTR| In facsimile, level corresponding to the copy white or copy black signal, whichever is the lower. { 'min·ə·məm 'sig·nəl ,lev·əl }

mini-supercomputer |COMPUT SCI| A supercomputer that is about a quarter to a half as fast in vector processing as the most powerful supercomputers. { ,min·ē'sü·pər·kəm,püd·ər }

minitrack |ELECTR| A subminiature radio transmitter capable of sending data over 4000 miles (6500 kilometers) on extremely low power. { 'min·ē,trak }

minor bend |ELECTROMAG| Rectangular waveguide bent so that throughout the length of a bend a longitudinal axis of the guide lies in one plane which is parallel to the narrow side of the waveguide. { 'mīn·ər 'bend }

minor control data |COMPUT SCI| Control data which are at the least significant level used, or which are used to sort records into the smallest groups used; for example, if control data are used to specify state, town, and street, then the data specifying street would be minor control data. { 'mīn·ər kən'trōl ,dad·ə }

minor cycle |COMPUT SCI| The time required for the transmission or transfer of one machine word, including the space between words, in a digital computer using serial transmission. Also known as word time. { 'mīn·ər ,sī·kəl }

minority carrier |SOLID STATE| The type of carrier, electron, or hole that constitutes less than half the total number of carriers in a semiconductor. { mə'när·əd·ē 'kar·ē·ər }

minority emitter |ELECTR| Of a transistor, an electrode from which a flow of minority carriers enters the interelectrode region. { mə'när·əd·ē i'mid·ər }

minor key |COMPUT SCI| A secondary key for identifying a record. { 'mīn·ər ,kē }

minor lobe |ELECTROMAG| Any lobe except the major lobe of an antenna radiation pattern. Also known as secondary lobe; side lobe. { 'mīn·ər ¦lōb }

minor loop |CONT SYS| A portion of a feedback control system that consists of a continuous network containing both forward elements and feedback elements. { 'mīn·ər ¦lüp }

minor relay station |ELECTR| A tape relay station which has tape relay responsibility but does not provide an alternate route. { 'mīn·ər 'rē,lā ,stā·shən }

minor switch |ELEC| Single-motion stepping switch mounted atop the telephone connectors and most commonly used for the party-line selection. { 'mīn·ər ¦swich }

minus zone |COMPUT SCI| The bit positions in a computer code that represent the algebraic minus sign. { 'mī·nəs ,zōn }

MIPS See million instructions per second.

mirage effect |COMMUN| Reception of radio waves at distances far beyond the normally expected range due to abnormal refraction caused by meteorological conditions such as abnormal vertical water-vapor and temperature gradients. { mə'räzh i,fekt }

mirror galvanometer |ELEC| A galvanometer having a small mirror attached to the moving element, to permit use of a beam of light as an indicating pointer. Also known as reflecting galvanometer. { 'mir·ər ,gal·və'näm·əd·ər }

mirroring |COMPUT SCI| The recording of the same data in two or more locations to ensure continuous operation of a computer system in the event of a failure. { 'mir·ər·iŋ }

MIS See management information system; metal-insulator semiconductor.

misfire |ELECTR| Failure to establish an arc between the main anode and the cathode of an ignitron or other mercury-arc rectifier during a scheduled conducting period. { 'mis,fīr }

mismatch |ELEC| The condition in which the impedance of a source does not match or equal the impedance of the connected load or transmission line. { 'mis,mach }

mismatch factor See reflection factor. { 'mis ,mach ,fak·tər }

mismatch loss |ELECTR| Loss of power delivered to a load as a result of failure to make an impedance match of a transmission line with its load or with its source. { 'mis,mach ,lòs }

mismatch slotted line |ELECTROMAG| A slotted line linking two waveguides which is not properly designed to minimize the power reflected or transmitted by it. { 'mis,mach 'släd·əd 'līn }

misregistration |COMPUT SCI| In character recognition, the improper state of appearance of a character, line, or document, on site in a character reader, with respect to a real or imaginary horizontal baseline. { ,mis,rej·ə'strā·shən }

missing error |COMPUT SCI| The result of calling for a subroutine not available in the library. { ,mis·iŋ 'er·ər }

mistake |COMPUT SCI| A human action producing an unintended result, in contrast to an error in a computer operation. { mə'stāk }

misuse detection |COMPUT SCI| The technology that seeks to identify an attack on a computer system by its attempted effect on sensitive resources. { mis'yüs di,tek·shən }

mixed congruential generator |COMPUT SCI| A congruential generator in which the constant *b* in the generating formula is not equal to zero. { 'mikst ,kän·grü'en·chəl 'jen·ə,rād·ər }

mixed-entry decision table |COMPUT SCI| A decision table in which the action entries may be either sequenced or unsequenced. { 'mikst ¦en·trē di'sizh·ən ,tā·bəl }

mixed highs |COMMUN| In analog color television, a method of reproducing very fine picture detail by transmitting high-frequency components as part of luminance signals for achromatic reproduction in color pictures. { 'mikst 'hīz }

mixed-mode expression |COMPUT SCI| An expression involving operands of more than one data type. { 'mikst ¦mōd ik'spresh·ən }

mixer |ELECTR| **1.** A device having two or more inputs, usually adjustable, and a common output; used to combine separate audio or video signals linearly in desired proportions to produce an output signal. **2.** The stage in a superheterodyne receiver in which the incoming modulated radio-frequency signal is combined with the signal of a local r-f oscillator to produce a modulated intermediate-frequency signal. Also known as first detector; heterodyne modulator; mixer-first detector. { 'mik·sər }

mixer-first detector See mixer. { 'mik·sər ¦fərst di'tek·tər }

mixer tube |ELECTR| A multigrid electron tube, used in a superheterodyne receiver, in which control voltages of different frequencies are impressed upon different control grids, and the nonlinear properties of the tube cause the generation of new frequencies equal to the sum and difference of the impressed frequencies. { 'mik·sər ,tüb }

mixing |ELECTR| Combining two or more signals, such as the outputs of several microphones. { 'mik·siŋ }

MMSS See maritime mobile satellite service.

mnemonic code |COMPUT SCI| A programming code that is easy to remember because the codes resemble the original words, such as MPY for multiply and ACC for accumulator. { nə'män·ik 'kōd }

MNOS See metal-nitride-oxide semiconductor.

mobile code |COMPUT SCI| Code that can be transmitted across, and executed at the other end of, a network, and is capable of running on multiple platforms, for example, Java. { ,mō·bəl 'kōd }

mobile digital computer |COMPUT SCI| Large, mobile, fixed-point operation, one-address, parallel-mode type digital computer. { 'mō·bəl ¦dij·əd·əl kəm'pyüd·ər }

mobile earth station |COMMUN| An earth station intended to be used while in motion or at halts at unspecified points. { ,mō·bəl 'ərth ,stā·shən }

mobile earth terminal |COMMUN| An antenna small enough to fit in a briefcase or suitcase, used for satellite communications, especially by news-service reporters at locations that cannot be accessed by conventional transportable satellite news-gathering terminals. Abbreviated MET. { ,mō·bəl 'ərth ,tər·mən·əl }

mobile radio |COMMUN| Radio communication in which the transmitter is installed in a vessel, vehicle, or airplane and can be operated while in motion. { 'mō·bəl 'rād·ē·ō }

mobile-relay station |COMMUN| Base station in which the base receiver automatically tunes on the base station transmitter and which retransmits all signals received by the base station receiver; used to extend the range of mobile units, and requires two frequencies for operation. { 'mō·bəl 'rē,lā ,stā·shən }

mobile robot |CONT SYS| A robot mounted on a movable platform that transports it to the area where it carries out tasks. { 'mō·bəl 'rō,bät }

mobile satellite service |COMMUN| A radiocommunication service between mobile earth stations by means of one or more space stations. Abbreviated MSS. { ¦mō·bəl 'sad·əl,īt ,sər·vəs }

mobile service |COMMUN| A radiocommunication service between mobile and land stations or between mobile stations. { ¦mō·bəl 'sər·vəs }

mobile station |COMMUN| **1.** Station in the mobile service intended to be used while in motion or during halts at unspecified points. **2.** One or more transmitters that are capable of transmission while in motion. { 'mō·bəl 'stā·shən }

mobile systems equipment |COMPUT SCI| Computers located on planes, ships, or vans. { 'mō·bəl 'sis·təmz i,kwip·mənt }

Möbius resistor |ELEC| A nonreactive resistor made by placing strips of aluminum or other metallic tape on opposite sides of a length of dielectric ribbon, twisting the strip assembly half a turn, joining the ends of the metallic tape, then soldering leads to opposite surfaces of the resulting loop. { 'mər·bē·əs ri,zis·tər }

modal distortion See modal noise. { ¦mōd·əl di'stór·shən }

modal noise |COMMUN| Interference of a multimode optical communications fiber with a laser light source when a speckle pattern in the light intensity in the fiber alters because of motion of the fiber or changes in the laser spectrum. Also known as modal distortion. { ¦mōd·əl 'nói·z }

mode |COMMUN| Form of the information in a communication such as literal language, digital data, and video. |COMPUT SCI| One of several alternative conditions or methods of operation of a device. |ELECTROMAG| A form of propagation of guided waves that is characterized by a particular field pattern in a plane transverse to the direction of propagation. Also known as transmission mode. { mōd }

mode converter See mode transducer. { 'mōd kən,vərd·ər }

mode filter |ELECTROMAG| A waveguide filter designed to separate waves of the same frequency but of different transmission modes. { 'mōd ,fil·tər }

mode jump |ELECTR| Change in mode of magnetron operation from one pulse to the next; each mode represents a different frequency and power level. { 'mōd ,jəmp }

model See macroskeleton. { 'mäd·əl }

model-based expert system |COMPUT SCI| An expert system that is based on knowledge of the structure and function of the object for which the system is designed. { ¦mäd·əl ,bāst 'ek·spərt ,sis·təm }

model-following problem |CONT SYS| The problem of determining a control that causes the response of a given system to be as close as possible to the response of a model system, given the same input. { 'mäd·əl ¦fäl·ə·wiŋ ,präb·ləm }

model reduction |CONT SYS| The process of discarding certain modes of motion while retaining others in the model used by an active control system, in order that the control system can compute control commands with sufficient rapidity. { 'mäd·əl ri'dək·shən }

model reference system |CONT SYS| An ideal system whose response is agreed to be optimum; computer simulation in which both the model system and the actual system are subjected to the same stimulus is carried out, and parameters of the actual system are adjusted to minimize the difference in the outputs of the model and the actual system. { 'mäd·əl 'ref·rəns ,sis·təm }

model symbol |COMPUT SCI| The standard usage of geometrical figures, such as squares, circles, or triangles, to help illustrate the various working parts of a model; each symbol must, nevertheless, be footnoted for complete clarification. { 'mäd·əl ,sim·bəl }

modem |ELECTR| A combination modulator and demodulator at each end of a telephone line to convert binary digital information to audio tone signals suitable for transmission over the

modulating electrode

line, and vice versa. Derived from modulator-demodulator. { 'mō,dem }

modem eliminator |COMPUT SCI| A device that is used to connect two computers in proximity and that mimics the action of two modems and a telephone line. { 'mō,dem ə'lim·ə,nād·ər }

mode number |ELECTR| **1.** The number of complete cycles during which an electron of average speed is in the drift space of a reflex klystron. **2.** The number of radians of phase in the microwave field of a magnetron divided by 2π as one goes once around the anode. { 'mōd ,nəm·bər }

moder See coder. { 'mōd·ər }

modern control |CONT SYS| A control system that takes account of the dynamics of the processes involved and the limitations on measuring them, with the aim of approaching the condition of optimal control. { 'mäd·ərn kən'trōl }

mode shift |ELECTR| Change in mode of magnetron operation during a pulse. { 'mōd ,shift }

mode skip |ELECTR| Failure of a magnetron to fire on each successive pulse. { 'mōd ,skip }

mode switch |COMPUT SCI| A preset control which affects the normal response of various components of a mechanical desk calculator. |ELECTR| A microwave control device, often consisting of a waveguide section of special cross section, which is used to change the mode of microwave power transmission in the waveguide. { 'mōd ,swich }

mode transducer |ELECTR| Device for transforming an electromagnetic wave from one mode of propagation to another. Also known as mode converter; mode transformer. { 'mōd tranz ,dü·sər }

mode transformer See mode transducer. { 'mōd tranz,fór·mər }

MODFET See high-electron-mobility transistor. { 'mäd,fet }

modified constant-voltage charge |ELEC| Charging of a storage battery in which the voltage of the charging circuit is held substantially constant, but a fixed resistance is inserted in the battery circuit producing a rising voltage characteristic at the battery terminals as the charge progresses. { 'mäd·ə,fīd |kän·stənt ¦vōl·tij 'chärj }

modifier |COMPUT SCI| A quantity used to alter the address of an operand in a computer, such as the cycle index. Also known as index word. { 'mäd·ə ,fī·ər }

modifier register See index register. { 'mäd·ə ,fī·ər ,rej·ə·stər }

modify |COMPUT SCI| **1.** To alter a portion of an instruction so its interpretation and execution will be other than normal; the modification may permanently change the instruction or leave it unchanged and affect only the current execution; the most frequent modification is that of the effective address through the use of index registers. **2.** To alter a subroutine according to a defined parameter. { 'mäd·ə,fī }

modify structure |COMPUT SCI| A statement in a database language that allows changes to be made in the structure of the records in a file. { 'mäd·ə,fī ,strək·chər }

moding |ELECTR| Defect of magnetron oscillation in which it oscillates in one or more undesired modes. { 'mōd·iŋ }

modula-2 |COMPUT SCI| A general-purpose programming language that allows a computer program to be written as separate modules which can be compiled separately but can share a common code. { 'mäj·ə·lə 'tü }

modular circuit |ELECTR| Any type of circuit assembled to form rectangular or cubical blocks that perform one or more complete circuit functions. { 'mäj·ə·lər 'sər·kət }

modular compilation |COMPUT SCI| The separate translation into machine language of the individual parts of a computer program, which are then combined into a single program by a linkage editor. { 'mäj·ə·lər ,käm·pə'lā·shən }

modularity |COMPUT SCI| The property of functional flexibility built into a computer system by assembling discrete units which can be easily joined to or arranged with other parts or units. { ,mäj·ə'lar·əd·ē }

modular programming |COMPUT SCI| The construction of a computer program from a collection of modules, each of workable size, whose interactions are rigidly restricted. { 'mäj·ə·lər 'prō,gram·iŋ }

modular structure |ELECTR| **1.** An assembly involving the use of integral multiples of a given length for the dimensions of electronic components and electronic equipment, as well as for spacings of holes in a chassis or printed wiring board. **2.** An assembly made from modules. { 'mäj·ə·lər 'strək·chər }

modulate |ELECTR| To vary the amplitude, frequency, or phase of a wave, or vary the velocity of the electrons in an electron beam in some characteristic manner. { 'mäj·ə,lāt }

modulated amplifier |ELECTR| Amplifier stage in a transmitter in which the modulating signal is introduced and modulates the carrier. { 'mäj·ə ,lād·əd 'am·plə,fī·ər }

modulated carrier |COMMUN| Radio-frequency carrier wave whose amplitude, phase, or frequency has been varied according to the intelligence to be conveyed. { 'mäj·ə,lād·əd 'kar·ē·ər }

modulated continuous wave |COMMUN| Wave in which the carrier is modulated by a constant audio-frequency tone. { 'mäj·ə,lād·əd kən ¦tin·yə·wəs 'wāv }

modulated stage |ELECTR| Radio-frequency stage to which the modulator is coupled and in which the continuous wave (carrier wave) is modulated according to the system of modulation and the characteristics of the modulating wave. { 'mäj·ə ,lād·əd 'stāj }

modulating electrode |ELECTR| Electrode to which a potential is applied to control the magnitude of the beam current. { 'mäj·ə,lād·iŋ i'lek,tröd }

modulating signal [COMMUN] Signal which causes a variation of some characteristics of a carrier. { 'mäj·ə,lād·iŋ 'sig·nəl }

modulation [COMMUN] The process or the result of the process by which some parameter of one wave is varied in accordance with some parameter of another wave. { ,mäj·ə'lā·shən }

modulation capability [ELECTR] Of an aural transmitter, the maximum percentage modulation that can be obtained without exceeding a given distortion figure. { ,mäj·ə'lā·shən ,kā·pə·'bil·ə·dē }

modulation code [COMMUN] A code used to cause variations in a signal in accordance with a predetermined scheme; normally used to alter or modulate a carrier wave to transmit data. { ,mäj·ə'lā·shən ,kōd }

modulation crest [COMMUN] The peak amplitude of an amplitude-modulated wave. { ,mäj·ə'lā·shən 'krest }

modulation-doped field-effect transistor See high-electron-mobility transistor. { ,mäj·ə'lā·shən ¦dōpt 'fēld i¦fekt tran'zis·tər }

modulation envelope [COMMUN] The peaks of the waveform of a modulated signal. { ,mäj·ə'lā·shən 'en·və,lōp }

modulation factor [COMMUN] **1.** In general, the ratio of the peak variation in the modulation actually used in a transmitter to the maximum variation for which the transmitter was designed. **2.** In an amplitude-modulated wave, the ratio (usually expressed in percent) of the peak variation of the envelope from its reference value, to the reference value. Also known as index of modulation. **3.** In a frequency-modulated wave, the ratio of the actual frequency swing to the frequency swing required for 100% modulation. { ,mäj·ə'lā·shən ,fak·tər }

modulation index [COMMUN] The ratio of the frequency deviation to the frequency of the modulating wave in a frequency-modulation system when using a sinusoidal modulating wave. Also known as ratio deviation. { ,mäj·ə'lā·shən ,in ,deks }

modulation meter [ENG] Instrument for measuring the degree of modulation (modulation factor) of a modulated wave train, usually expressed in percent. { ,mäj·ə'lā·shən ,mēd·ər }

modulation rise [ELECTR] Increase of the modulation percentage caused by nonlinearity of any tuned amplifier, usually the last intermediate-frequency stage of a receiver. { ,mäj·ə'lā·shən ,rīz }

modulation transformer [ENG ACOUS] An audio-frequency transformer which matches impedances and transmits audio frequencies between one or more plates of an audio output stage and the grid or plate of a modulated amplifier. { ,mäj·ə'lā·shən tranz,fór·mər }

modulator [ELECTR] **1.** The transmitter component that supplies the modulating signal to the amplifier stage or that triggers the modulated amplifier stage to produce pulses at desired instants as in radar. **2.** A device that produces modulation by any means, such as by virtue of a nonlinear characteristic or by controlling some circuit quantity in accordance with the waveform of a modulating signal. { 'mäj·ə,lād·ər }

modulator-demodulator See modem. { 'mäj·ə ,lād·ər dē'mäj·ə,lād·ər }

modulator glow tube [ELECTR] Cold cathode recorder tube that is used for facsimile and sound-on-film recording; provides a modulated high-intensity point source of light. { 'mäj·ə ,lād·ər 'glō ,tüb }

module [COMPUT SCI] **1.** A distinct and identifiable unit of computer program for such purposes as compiling, loading, and linkage editing. **2.** One memory bank and associated electronics in a computer. [ELECTR] A packaged assembly of wired components, built in a standardized size and having standardized plug-in or solderable terminations. { 'mäj·ül }

modulo N check [COMPUT SCI] A procedure for verification of the accuracy of a computation by repeating the steps in modulo N arithmetic and comparing the result with the original result (modulo N). Also known as residue check. { 'mäj·ə,lō ¦en 'chek }

modulo-two adder [COMPUT SCI] A logical circuit for adding one-digit binary numbers. { 'mäj·ə ,lō ¦tü 'ad·ər }

MOEMS See micro-opto-electro-mechanical system. { 'mō,emz }

moiré [COMMUN] In a video system, the spurious pattern in the reproduced picture resulting from interference beats between two sets of periodic structures in the image. { mò'rā }

molded capacitor [ELEC] Capacitor, usually mica, that has been encased in a molded plastic insulating material. { 'mōl·dəd kə'pas·əd·ər }

molecular circuit [ELECTR] A circuit in which the individual components are physically indistinguishable from each other. { mə'lek·yə·lər 'sər·kət }

molecular electronics [ELECTR] The use of biological or organic molecules for fabricating electronic materials with novel electronic, optical, or magnetic properties; applications include polymer light-emitting diodes, conductive-polymer sensors, pyroelectric plastics, and, potentially, molecular computational devices. { mə'lek·yə·lər i,lek'trän·iks }

molecular engineering [ELECTR] The use of solid-state techniques to build, in extremely small volumes, the components necessary to provide the functional requirements of overall equipments, which when handled in more conventional ways are vastly bulkier. { mə'lek·yə·lər ,en·jə'nir·iŋ }

moment sensor [ENG] A device that measures the force applied at a remote point in a robotic system. { 'mō·mənt ,sen·sər }

MOMS See micro-opto-mechanical system. { mämz or ¦em¦ō¦em'es }

monadic operation [COMPUT SCI] An operation on one operand, such as a negation. { mō'nad·ik ,äp·ə·'rā·shən }

monaural sound [ENG ACOUS] Sound produced by a system in which one or more microphones are connected to a single transducing channel which is coupled to one or two earphones worn by the listener. { män'ör·əl 'saůnd }

monitor [COMPUT SCI] 1. To supervise a program, and check that it is operating correctly during its execution, usually by means of a diagnostic routine. 2. *See* video monitor. { 'män·əd·ər }

monitor board [COMMUN] A console at which a supervising telephone operator sits and from which she or he can intercept calls being handled by other operators. { 'män·əd·ər ,bôrd }

monitor control dump [COMPUT SCI] A memory dump routinely carried out by the system once a program has been run. { 'män·əd·ər kən'trōl ,dəmp }

monitor display [COMPUT SCI] The facility of stopping the central processing unit and displaying information of main storage and internal registers; after manual intervention, normal instruction execution can be initiated. { 'män·əd·ər di ,splā }

monitoring amplifier [ELECTR] A power amplifier used primarily for evaluation and supervision of a program. { 'män·ə·triŋ ,am·plə,fī·ər }

monitoring key [ELECTR] Key which, when operated, makes it possible for an attendant or operator to listen on a telephone circuit without appreciably impairing transmission on the circuit. { 'män·ə·triŋ ,kē }

monitor mode *See* master mode. { 'män·əd·ər ,mōd }

monitor operating system [COMPUT SCI] The control of the routines which achieves efficient use of all the hardware components. { 'män·əd·ər 'äp·ə,rād·iŋ ,sis·təm }

monitor printer [COMMUN] A teleprinter used in a technical control facility or communications center for checking incoming teletypewriter signals. [COMPUT SCI] Input-output device, capable of receiving coded signals from the computer, which automatically operates the keyboard to print a hard copy and, when desired, to punch paper tape. { 'män·əd·ər ,print·ər }

monitor routine *See* executive routine. { 'män·əd·ər rü,tēn }

monobrid circuit [ELECTR] Integrated circuit using a combination of monolithic and multichip techniques by means of which a number of monolithic circuits, or a monolithic device in combination with separate diffused or thin-film components, are interconnected in a single package. { 'män·ə,brid ,sər·kət }

monocharge electret [ELECTR] A type of foil electret that carries electrical charge of the same sign on both surfaces. { 'män·ō,chärj i'lek·trət }

monochromatic radiation [ELECTROMAG] Electromagnetic radiation having wavelengths confined to an extremely narrow range. { män·ə·krə'mad·ik ,rād·ē'ā·shən }

monochrome channel [ELECTR] In a color television system, any path which is intended to carry the monochrome signal; the monochrome channel may also carry other signals. { 'män·ə ,krōm ¦chan·əl }

monochrome signal [ELECTR] 1. A signal waveform used for controlling luminance values in monochrome television. 2. The portion of a signal wave that has major control of the luminance values in a color television system, regardless of whether the picture is displayed in color or in monochrome. Also known as M signal. { 'män·ə,krōm ,sig·nəl }

monochrome television [COMMUN] Television in which the final reproduced picture is monochrome, having only shades of gray between black and white. Also known as black-and-white television. { 'män·ə,krōm 'tel·ə,vizh·ən }

monocord switchboard [ELEC] Local battery switchboard in which each telephone line terminates in a single jack and plug. { 'män·ə ,kôrd 'swich,bôrd }

monofier [ELECTR] Complete master oscillator and power amplifier system in a single evacuated tube envelope; electrically, it is equivalent to a stable low-noise oscillator, an isolator, and a two- or three-cavity klystron amplifier. { 'män·ə ,fī·ər }

monolithic ceramic capacitor [ELECTR] A capacitor that consists of thin dielectric layers interleaved with staggered metal-film electrodes; after leads are connected to alternate projecting ends of the electrodes, the assembly is compressed and sintered to form a solid monolithic block. { ,män·ə'lith·ik sə'ram·ik kə'pas·əd·ər }

monolithic filter [COMMUN] A device used to separate telephone communications sent simultaneously over the transmission line, consisting of a series of electrodes vacuum-deposited on a crystal plate so that the plated sections are resonant with ultrasonic sound waves, and the effect of the device is similar to that of an electric filter. { ,män·ə'lith·ik 'fil·tər }

monolithic integrated circuit [ELECTR] An integrated circuit having elements formed in place on or within a semiconductor substrate, with at least one element being formed within the substrate. { ,män·ə'lith·ik 'int·ə,grād·əd 'sər·kət }

monophonic sound [ENG ACOUS] Sound produced by a system in which one or more microphones feed a single transducing channel which is coupled to one or more loudspeakers. { ¦män·ə¦fän·ik ,saůnd }

monopinch [ELECTR] Antijam application of the monopulse technique where the error signal is used to provide discrimination against jamming signals. { 'män·ō,pinch }

monopole antenna [ELECTROMAG] An antenna, usually in the form of a vertical tube or helical whip, on which the current distribution forms a standing wave, and which acts as one part of a dipole whose other part is formed by its electrical image in the ground or in an effective ground plane. Also known as spike antenna. { 'män·ə ,pōl an'ten·ə }

monopulse [ELECTR] A radar technique for accurate estimation of target position in angle and range on each pulse transmitted, without requiring sequential transmission as in, for example, conical scanning radars; reduces errors due to echo fluctuations. { 'män·ə,pəls }

monoscope [ELECTR] A signal-generating electron-beam tube in which a picture signal is produced by scanning an electrode that has a predetermined pattern of secondary-emission response over its surface. Also known as monotron; phasmajector. { 'män·ə,skōp }

monostable [ELECTR] Having only one stable state. { ¦män·ō¦stā·bəl }

monostable blocking oscillator [ELECTR] A blocking oscillator in which the electron tube or other active device carries no current unless positive voltage is applied to the grid. Also known as driven blocking oscillator. { ¦män·ō ¦stā·bəl 'bläk·iŋ ¦äs·ə,lād·ər }

monostable circuit [ELECTR] A circuit having only one stable condition, to which it returns in a predetermined time interval after being triggered. { ¦män·ō¦stā·bəl 'sər·kət }

monostable multivibrator [ELECTR] A multivibrator with one stable state and one unstable state; a trigger signal is required to drive the unit into the unstable state, where it remains for a predetermined time before returning to the stable state. Also known as one-shot multivibrator; single-shot multivibrator; start-stop multivibrator; univibrator. { ¦män·ō¦stā·bəl ,məl·tə'vī ,brād·ər }

monostatic radar [ENG] A radar with transmitter and receiver in the same place, whether or not it uses a duplexed antenna. { ¦män·ə¦stad·ik 'rā ,där }

monotonicity [ELECTR] In an analog-to-digital converter, the condition wherein there is an increasing output for every increasing value of input voltage over the full operating range. { ,män·ə·tə'nis·əd·ē }

monotron See monoscope. { 'män·ə,trän }

Moore code [COMMUN] A binary teleprinter code with seven binary digits for each letter. { 'mur ,kōd }

Moore machine [COMPUT SCI] A sequential machine in which the output depends uniquely on the current state of the machine, and not on the input. { 'mur mə,shēn }

Moore's law [COMPUT SCI] The prediction by Gordon Moore (cofounder of the Intel Corporation) that the number of transistors on a microprocessor would double periodically (approximately every 18 months). { 'murz ,lo }

Moore-Smith sequence See net. { ¦mur 'smith ,sē·kwəns }

Morse cable code [COMMUN] A code used chiefly in submarine cable telegraphy, in which positive and negative current impulses of equal length represent dots and dashes, and a space is represented by the absence of current. Also known as cable code; international cable code. { 'mors 'kā·bəl ,kōd }

Morse code [COMMUN] **1.** A telegraph code for manual operating, consisting of short (dot) and long (dash) signals and various-length spaces. Also known as American Morse code. **2.** Collective term for Morse code (American Morse code)

and continental code (International Morse code). { 'mors 'kōd }

MOS See metal oxide semiconductor.

mosaic [ELECTR] A light-sensitive surface used in video camera tubes, consisting of a thin mica sheet coated on one side with a large number of tiny photosensitive silver-cesium globules, insulated from each other. { mō'zā·ik }

MOS controlled thyristor [ELECTR] A type of thyristor in which there is a very thin metal oxide semiconductor (MOS) integrated circuit in the top surface of the high-power thyristor components, so that only a small gate current is needed to turn the entire device off or on. Abbreviated MCT. { ¦em¦ō¦es kən,trōld thī'ris·tər }

MOSFET See metal oxide semiconductor field-effect transistor. { 'mos,fet }

MOSFET-C filter [ELECTR] An active integrated-circuit filter in which the resistors of an active-RC filter are replaced with metal oxide semiconductor field-effect transistors (MOSFETs). { ¦mos ,fet 'sē ,fil·tər }

Mosotti field See Lorentz local field. { mō'säd·ē ,fēld }

MOST See metal oxide semiconductor field-effect transistor.

MOS transistor See metal oxide semiconductor field-effect transistor. { ¦em¦ō¦es tran'zis·tər }

most significant bit [COMPUT SCI] The left-most bit in a word. Abbreviated msb. { 'mōst sig 'nif·i·gənt 'bit }

most significant character [COMPUT SCI] The character in the leftmost position in a number or word. { 'mōst sig'nif·i·gənt 'kar·ik·tər }

motherboard [COMPUT SCI] A common pathway over which information is transmitted between the hardware devices (the central processing unit, memory, and each of the peripheral control units) in a microcomputer. { 'məth·ər,bord }

motional impedance [ELECTR] Of a transducer, the complex remainder after the blocked impedance has been subtracted from the loaded impedance. Also known as loaded motional impedance. { 'mō·shən·əl im'pēd·əns }

motion-compensated coding scheme [COMMUN] A form of differential pulse-code modulation in which the motions of objects are estimated and comparisons of intensities are carried out between picture elements in successive frames spatially displaced by an amount equal to the motion of an object. { 'mō·shən ¦käm·pən ,sād·əd 'kōd·iŋ ,skēm }

motion picture pickup [ELECTR] Use of a television camera to pick up scenes directly from motion picture film. { 'mō·shən ¦pik·chər 'pik ,əp }

motion register [COMPUT SCI] The register which controls the go/stop, forward/reverse motion of a tape drive. { 'mō·shən ,rej·ə·stər }

motion vector [COMMUN] A pair of numbers which represent the vertical and horizontal displacement of a region of a reference picture for MPEG-2 prediction. { 'mō·shən 'vek·tər }

motor [ELEC] A machine that converts electric energy into mechanical energy by utilizing forces produced by magnetic fields on current-carrying conductors. Also known as electric motor. { 'mōd·ər }

motorboating [ELECTR] Undesired oscillation in an amplifying system or transducer, usually of a pulse type, occurring at a subaudio or low-audio frequency. { 'mōd·ər,bōd·iŋ }

motor branch circuit [ELEC] A branch circuit that terminates at a motor; it must have conductors with current-carrying capacity at least 125% of the motor full-load current rating, and overcurrent protection capable of carrying the starting current of the motor. { 'mōd·ər 'branch ,sər·kət }

motor control See electronic motor control. { 'mōd·ər kən,trōl }

motor-converter [ELEC] Induction motor and a synchronous converter with their rotors mounted on the same shaft and with their rotor windings connected in series; such converter operates synchronously at a speed corresponding to the sum of the numbers of poles of the two machines. { 'mōd·ər kən'vərd·ər }

motor element [ENG ACOUS] That portion of an electroacoustic receiver which receives energy from the electric system and converts it into mechanical energy. { 'mōd·ər ,el·ə·mənt }

motor-generator set [ELEC] A motor and one or more generators that are coupled mechanically for use in changing one power-source voltage to other desired voltages or frequencies. { 'mōd·ər 'jen·ə,rād·ər ,set }

mountain effect [ELECTROMAG] The effect of rough terrain on radio-wave propagation, causing reflections that produce errors in radio direction-finder indications. { 'maùnt·ən i,fekt }

mousable interface [COMPUT SCI] A user interface that responds to input from a mouse for various functions. { ¦maùs·ə·bəl 'in·tər,fās }

mouse [COMPUT SCI] A small device with a rubber-coated ball that is moved about by hand over a flat surface and generates signals to control the position of a cursor or pointer on a computer display. { maùs }

movable contact [ELEC] The relay contact that is mechanically displaced to engage or disengage one or more stationary contacts. Also known as armature contact. { 'mü·və·bəl 'kän,takt }

movable-head disk drive [COMPUT SCI] A type of disk drive in which read/write heads are moved over the surface of the disk, toward and away from the center, so that they are correctly positioned to read or write the desired information. { 'mü·və·bəl ¦hed 'disk ,drīv }

move mode [COMPUT SCI] A method of communicating between an operating program and an input/output control system in which the data records to be read or written are actually moved into and out of program-designated memory areas. { 'müv ,mōd }

move operation [COMPUT SCI] An operation in which data is moved from one storage location to another. { 'müv ,äp·ə,rā·shən }

moving-coil galvanometer [ENG] Any galvanometer, such as the d'Arsonval galvanometer, in which the current to be measured is sent through

a coil suspended or pivoted in a fixed magnetic field, and the current is determined by measuring the resulting motion of the coil. { 'müv·iŋ ¦kòil ,gal·və'näm·əd·ər }

moving-coil instrument [ELEC] Any instrument in which current is sent through one or more coils suspended or pivoted in a magnetic field, and the motion of the coils is used to measure either the current in the coils or the strength of the field. { 'müv·iŋ ¦kòil 'in·strə·mənt }

moving-coil loudspeaker See dynamic loudspeaker. { 'müv·iŋ ¦kòil 'laùd,spēk·ər }

moving-coil meter [ELEC] A meter in which a pivoted coil is the moving element. { 'müv·iŋ ¦kòil 'mēd·ər }

moving-coil microphone See dynamic microphone. { 'müv·iŋ ¦kòil 'mī·krə,fōn }

moving-coil pickup See dynamic pickup. { 'müv·iŋ ¦kòil 'pik,əp }

moving-coil voltmeter [ENG] A voltmeter in which the current, produced when the voltage to be measured is applied across a known resistance, is sent through coils pivoted in the magnetic field of permanent magnets, and the resulting torque on the coils is balanced by control springs so that the deflection of a pointer attached to the coils is proportional to the current. { 'müv·iŋ ¦kòil 'vōlt,med·ər }

moving-conductor loudspeaker [ENG ACOUS] A loudspeaker in which the mechanical forces result from reactions between a steady magnetic field and the magnetic field produced by current flow through a moving conductor. { 'müv·iŋ kən ¦dək·tər 'laùd,spēk·ər }

moving-head disk [COMPUT SCI] A disk-storage device in which one or more read-write heads are attached to a movable arm which allows each head to cover many tracks of information. { 'müv·iŋ ¦hed 'disk }

moving-iron voltmeter [ENG] A voltmeter in which a field coil is connected to the voltage to be measured through a series resistor; current in the coil causes two vanes, one fixed and one attached to the shaft carrying the pointer, to be similarly magnetized; the resulting torque on the shaft is balanced by control springs. { 'müv·iŋ ¦ī·ərn 'vōlt,med·ər }

moving-magnet voltmeter [ENG] A voltmeter in which a permanent magnet aligns itself with the resultant magnetic field produced by the current in a field coil and another permanent control magnet. { 'müv·iŋ ¦mag·nət 'vōlt,med·ər }

Moving Picture Experts Group See MPEG. { 'müv·iŋ 'pik·chər 'ek,spərts 'grüp }

moving-target indicator [ELECTR] A feature that limits the display of radar information primarily to moving targets; signals due to reflections from stationary objects are canceled by a memory circuit. Abbreviated MTI. { 'müv·iŋ ¦tär·gət 'in·də ,kād·ər }

MPEG [COMMUN] Standards for compression of digitized audio and video signals developed by the ISO/IEC JTC1/SC29 WG11; may also refer to the group itself (Moving Picture Experts Group). { 'em,peg }

MPEG-1 [COMMUN] Standards for compression of digitized audio and video signals that were developed by the Moving Picture Experts Group with relatively low-bit-rate systems in mind, such as video from a CD-ROM, and continue to be used, although most video applications currently use MPEG-2; they comprise ISO/IEC standards 11172-1 (systems), 11172-2 (video), 11172-3 (audio), 11172-4 (compliance testing), and 11172-5 (technical report). { 'em,peg 'wən }

MPEG-2 [COMMUN] Standards for compression of digitized audio and video signals that were developed by the Moving Picture Experts Group primarily for professional video applications, such as video production, and distribution and storage applications on media ranging from tape to DVD, and which form the basis of digital television systems; they comprise ISO/IEC standards 13818-1 (systems), 13818-2 (video), 13818-3 (audio), and 13818-4 (compliance). { 'em,peg 'tü }

MPEG-2 AAC [COMMUN] An advanced audio coder; a high-quality, low-bit-rate perceptual audio coding system. { 'em,peg 'tü ¦ā ¦ā ¦sē }

MPU See microprocessing unit.

MQ register [COMPUT SCI] Temporary-storage register whose contents can be transferred to or from, or swapped with, the accumulator. { 'em ¦kyü 'rej·ə·stər }

MRAM See magnetic random access memory. { 'em,ram }

msb See most significant bit.

M-scan See M-display. { 'em ,skan }

M-scope See M-display. { 'em ,skōp }

MSI See magnetic source imaging medium-scale integration.

M signal See monochrome signal. { 'em ,sig·nəl }

MSS See mobile satellite service.

MTBF See mean time between failures.

MTI See moving-target indicator.

M-type backward-wave oscillator [ELECTR] A backward-wave oscillator in which focusing and interaction are through magnetic fields, as in a magnetron. Also known as M-type carcinotron; type-M carcinotron. { 'em ,tīp 'bak·wərd ¦wāv 'äs·ə,lād·ər }

M-type carcinotron See M-type backward-wave oscillator. { 'em ,tīp kär'sin·ə,trän }

MUF See maximum usable frequency.

mu factor [ELECTR] Ratio of the change in one electrode voltage to the change in another electrode voltage under the conditions that a specified current remains unchanged and that all other electrode voltages are maintained constant; a measure of the relative effect of the voltages on two electrodes upon the current in the circuit of any specified electrode. { 'myü ,fak·tər }

multiaccess computer [COMPUT SCI] A computer system in which computational and data resources are made available simultaneously to a number of users who access the system through terminal devices, normally on an interactive or conversational basis. { ¦məl·tē'ak,ses kəm,pyüd·ər }

multiaccess network See multiple-access network. { ¦məl·tē'ak,ses 'net,wərk }

multiaddress [COMPUT SCI] Referring to an instruction that has more than one address part. { ¦məl·tē'a,dres }

multianode tube [ELECTR] Electron tube having two or more main anodes and a single cathode. { ¦məl·tē'an,ōd ,tüb }

multiaperture reluctance switch [ELECTR] Two-aperture ferrite storage core which may be used to provide a nondestructive readout computer memory. { ¦məl·tē'ap·ə·chər ri'lək·təns ,swich }

multiaspect [COMPUT SCI] Pertaining to searches or systems which permit more than one aspect, or facet, of information to be used in combination, one with the other to effect identifying or selecting operations. { ¦məl·tē'as,pekt }

multicavity klystron [ELECTR] A klystron in which there is at least one cavity between the input and output cavities, each of which remodulates the beam so that electrons are more closely bunched. { ¦məl·tē'kav·əd·ē 'klī,strän }

multicavity magnetron [ELECTR] A magnetron in which the circuit includes a plurality of cavities, generally cut into the solid cylindrical anode so that the mouths of the cavities face the central cathode. { ¦məl·tē'kav·əd·ē 'mag·nə,trän }

multicellular horn [ELECTROMAG] A cluster of horn antennas having mouths that lie in a common surface and that are fed from openings spaced one wavelength apart in one face of a common waveguide. [ENG ACOUS] A combination of individual horn loudspeakers having individual driver units or joined in groups to a common driver unit. Also known as cellular horn. { ¦məl·tē'sel·yə·lər 'hòrn }

multichannel communication [COMMUN] Communication in which there are two or more communication channels over the same path, such as a communication cable, or a radio transmitter which can broadcast on two different frequencies, either individually or simultaneously. { ¦məl·tē'chan·əl kə,myü·nə'kā·shən }

multichannel field-effect transistor [ELECTR] A field-effect transistor in which appropriate voltages are applied to the gate to control the space within the current flow channels. { ¦məl·tē'chan·əl 'fēld i¦fekt tran'zis·tər }

multichannel loading [COMMUN] Behavior of a multichannel communications system with all channels active. { ¦məl·tē'chan·əl 'lōd·iŋ }

multichannel telephone system [COMMUN] A telephone system in which two or more communications channels are carried over a single telephone cable or radio link. { ¦məl·tē'chan·əl 'tel·ə,fōn ,sis·təm }

multichip microcircuit [ELECTR] Microcircuit in which discrete, miniature, active electronic elements (transistor or diode chips) and thin-film or diffused passive components or component clusters are interconnected by thermocompression bonds, alloying, soldering, welding, chemical deposition, or metallization. { 'məl·tē,chip 'mī·krō,sər·kət }

multicollector electron tube [ELECTR] An electron tube in which electrons travel to more than one electrode. { ¦məl·tē·kə¦lek·tər i'lek ,trän ,tüb }

multicomputer system [COMPUT SCI] A system consisting of more than one computer, usually under the supervision of a master computer, in which smaller computers handle input/output and routine jobs while the large computer carries out the more complex computations. { ¦məl·tē·kəm¦pyüd·ər ,sis·təm }

multicoupler [ELECTR] A device for connecting several receivers to one antenna and properly matching the impedances of the receivers to the antenna. { 'məl·tə,kəp·lər }

multidimensional Turing machine [COMPUT SCI] A variation of a Turing machine in which tapes are replaced by multidimensional structures. { ¦məl·tə·di'men·shən·əl 'tur·iŋ mə,shēn }

multidrop line [COMMUN] A telephone pair which terminates at several locations. { 'məl·tē,dräp ,līn }

multielectrode tube [ELECTR] Electron tube containing more than three electrodes associated with a single electron stream. { ¦məl·tē·i'lek ,trōd ,tüb }

multielement array [ELECTROMAG] An antenna array having a large number of antennas. { ¦məl·tē'el·ə·mənt ə'rā }

multielement parasitic array [ELECTROMAG] Antennas consisting of an array of driven dipoles and parasitic elements, arranged to produce a beam of high directivity. { ¦məl·tē'el·ə·mənt ,par·ə'sid·ik ə,rā }

multielement vacuum tube [ELECTR] A vacuum tube which has one or more grids in addition to the cathode and plate electrodes. { ¦məl·tē'el·ə·mənt 'vak·yəm ,tüb }

multigrid tube [ELECTR] An electron tube having two or more grids between cathode and anode, as a tetrode or pentode. { 'məl·tə,grid ,tüb }

multigun tube [ELECTR] A cathode-ray tube having more than one electron gun. { 'məl·tə,gən ,tüb }

multihead Turing machine [COMPUT SCI] A variation of a Turing machine in which more than one head is allowed per tape. { 'məl·tē,hed 'tur·iŋ mə,shēn }

multijob operation [COMPUT SCI] The concurrent or interleaved execution of job steps from more than one job. { 'məl·tē,jäb ,äp·ə'rā·shən }

multijunction solar cell [ELECTR] A solar cell made of two or more materials, each optimally efficient over a limited spectral range. Also known as multiple-junction solar cell. { ,məl·tē ¦jəŋk·shən ¦sō·lər 'sel }

multilayer board [ELECTR] A printed wiring board that contains circuitry on internal layers throughout the cross section of the board as well as on the external layers. { ,məl·tē,lā·ər 'bȯrd }

multilayer optical storage [COMPUT SCI] An extension of optical disk storage technology to the third dimension by stacking data layers one above another, with each layer separated by a spacer region. { ,məl·tə¦lā·ər ¦äp·ti·kəl 'stȯr·ij }

multilevel address See indirect address. { ¦məl·tə 'lev·əl a,dres }

multilevel indirect addressing [COMPUT SCI] A programming device whereby the address retrieved in the memory word may itself be an indirect address that points to another memory location, which in turn may be another indirect address, and so forth. { ¦məl·tə'lev·əl ,in·də,rekt ə'dres·iŋ }

multilevel transmission [COMMUN] Transmission of digital information in which three or more levels of voltage are recognized as meaningful, as 0,1,2 instead of simply 0,1. { ¦məl·tə'lev·əl tranz'mish·ən }

multiline appearances [COMMUN] **1.** The ability of a telephone to receive or originate additional voice or data calls at the terminal while it is still engaged in the primary voice call. **2.** The ability to bring additional parties to a primary telephone call. { ¦məl·tē'līn ə'pir·ən·səz }

multilist organization [COMPUT SCI] A chained file organization in which each segment is indexed. { 'məl·tē,list ,ȯr·gə·nə'zā·shən }

multimedia technology [COMPUT SCI] The synergistic union of digital video, audio, computer, information, and telecommunication technologies. { ¦məl·tə,mēd·ē·ə tek'näl·ə·jē }

multimeter See volt-ohm-milliammeter. { 'məl·tə ,mēd·ər or məl'tim·əd·ər }

multipactor [ELECTR] A high-power, high-speed microwave switching device in which a thin electron cloud is driven back and forth between two parallel plane surfaces in a vacuum by a radio-frequency electric field. { 'məl·tə,pak·tər }

multipass sort [COMPUT SCI] Computer program designed to sort more data than can be contained within the internal storage of a computer; intermediate storage, such as disk, tape, or drum, is required. { 'məl·tē,pas 'sȯrt }

multipath [COMMUN] A radio frequency reception condition in which a radio signal reaching a receiving antenna arrives by multiple paths due to reflections of the signal off of various surfaces in the environment. By traveling different distances to the receiver, the reflections arrive with different time delays and signal strengths. When multiplath conditions are great enough, analog reception of radio broadcasts is affected in a variety of ways, including stop-light fades, picket fencing, and distortion received audio. [ELECTROMAG] **1.** In radar, a propagation situation wherein the direct-path signal from radar to a target is interfered with by the reflected-path signal; usually refers to reflections from earth's surface. Both surveillance sensitivity and tracking accuracy, particularly at low elevation angles, are affected. **2.** See multipath transmission. { 'məl·tē,path }

multipath cancellation [COMMUN] Occurrence of essentially complete cancellation of radio signals because of the relative amplitude and phase differences of the components arriving over separate paths. { 'məl·tə,path ,kan·sə'lā·shən }

multipath transmission [ELECTROMAG] The propagation phenomenon that results in signals reaching a radio receiving antenna by two or more paths, causing distortion in radio and ghost images in television. Also known as multipath. { 'məl·tə,path tranz'mish·ən }

multiple See parallel. { 'məl·tə·pəl }

multiple access [COMMUN] Multiplexing schemes by which multiple users who are geographically dispersed gain access to a shared telecommunications facility or channel. { ,məl·tə·pəl 'ak ,ses }

multiple-access computer [COMPUT SCI] A computer system whose facilities can be made available to a number of users at essentially the same time, normally through terminals, which are often physically far removed from the central computer and which typically communicate with it over telephone lines. { 'məl·tə·pəl ¦ak,ses kəm ,pyüd·ər }

multiple-access network [COMPUT SCI] A computer network that permits every computer on it to communicate with the network at any time during operation. Also known as multiaccess network. { 'məl·tə·pəl ¦ak,ses 'net,wərk }

multiple-access service [COMMUN] One of the services of the Tracking and Data Relay Satellite System, which provides simultaneous return-link service from as many as 20 low-earth-orbiting user spacecraft, with data rates up to 3 megabits per second for each user, and a time-shared forward-link service to the user spacecraft with a maximum data rate of 300 kilobits per second, one user at a time. Abbreviated MA service. { ,məl·tə·pəl 'ak,ses ,sər·vəs }

multiple accumulating registers [COMPUT SCI] Special registers capable of handling factors larger than one computer word in length. { 'məl·tə·pəl ə'kyü·mya,lād·iŋ 'rej·ə,stərz }

multiple-address code [COMPUT SCI] A computer instruction code in which more than one address or storage location is specified; the instruction may give the locations of the operands, the destination of the result, and the location of the next instruction. { 'məl·tə·pəl ¦a,dres ,kōd }

multiple-address computer [COMPUT SCI] A computer whose instruction contains more than one address, for example, an operation code and three addresses A, B, C, such that the content of A is multiplied by the content of B and the product stored in location C. { 'məl·tə·pəl ¦a,dres kəm ,pyüd·ər }

multiple-address instruction [COMPUT SCI] An instruction which has more than one address in a computer; the addresses give locations of other instructions, or of data or instructions that are to be operated upon. { 'məl·tə·pəl ¦a,dres in ,strak·shən }

multiple appearance [ELEC] Jack arrangement in telephone switchboards whereby a single-line circuit appears before two or more operators. { 'məl·tə·pəl ə'pir·əns }

multiple-beam antenna [ELECTROMAG] An antenna or antenna array which radiates several beams in different directions. { 'məl·tə·pəl ¦bēm an'ten·ə }

multiple computer operation [COMPUT SCI] The utilization of any one computer of a group of computers by means of linkages provided by multiplexor channels, all computers being linked through their channels or files. { 'məl·tə·pəl kəm'pyüd·ər ,äp·ə,rā·shən }

multiple-contact switch See selector switch. { 'məl·tə·pəl ¦kän,takt ,swich }

multiple decay See branching. { 'məl·tə·pəl di'kā }

multiple disintegration See branching. { 'məl·tə·pəl di,sin·tə'grā·shən }

multiple-frequency-shift keying [COMMUN] A modulation scheme in which a number of carrier frequencies (2, 4, 8, and so forth) are transmitted according to a group of consecutive data bits (n bits producing 2^n frequencies). Abbreviated MFSK. { 'məl·tə·pəl ¦frē·kwən·sē ¦shift ,kē·iŋ }

multiple-function chip See large-scale integrated circuit. { 'məl·tə·pəl ¦fəŋk·shən ,chip }

multiple-instruction-stream, multiple-data-stream See MIMD. { ¦məl·tə·pəl in'strak·shən ,strēm ¦məl·tə·pəl 'dad·ə ,strēm }

multiple jacks [ELEC] Series of jacks with tip, ring, and sleeve, respectively connected in parallel, and appearing in different panels of the face equipment of a telephone exchange. { 'məl·tə·pəl 'jaks }

multiple-junction solar cell See multijunction solar cell. { ¦məl·tə·pəl ¦jəŋk·shən ¦sō·lər ,sel }

multiple-key access [COMPUT SCI] A technique for locating stored data in a computer system by using the values contained in two or more separate key fields. { 'məl·tə·pəl ¦kē 'ak,ses }

multiple lamp holder [ELEC] A device that can be inserted in a lamp holder to act as two or more lamp holders. Also known as current tap. { 'məl·tə·pəl 'lamp ,hōl·dər }

multiple-length arithmetic [COMPUT SCI] Arithmetic performed by a computer in which two or more machine words are used to represent each number in the calculations, usually to achieve higher precision in the result. { 'məl·tə·pəl ¦leŋkth ə'rith·mə,tik }

multiple-length number [COMPUT SCI] A number having two or more times as many digits as are ordinarily used in a given computer. { 'məl·tə·pəl ¦leŋkth 'nəm·bər }

multiple-length working [COMPUT SCI] Any processing of data by a computer in which two or more machine words are used to represent each data item. { 'məl·tə·pəl ¦leŋkth 'wərk·iŋ }

multiple modulation [COMMUN] A succession of modulating processes in which the modulated wave from one process becomes the modulating wave for the next. Also known as compound modulation. { 'məl·tə·pəl ,mäj·ə'lā·shən }

multiple module access [COMPUT SCI] Device which establishes priorities in storage access in a multiple computer environment. { 'məl·tə·pəl ¦mäj·yül ,ak,ses }

multiple precision arithmetic |COMPUT SCI| Method of increasing the precision of a result by increasing the length of the number to encompass two or more computer words in length. { 'məl·tə·pəl prə¦sizh·ən ə'rith·mə,tik }

multiple programming |COMPUT SCI| The execution of two or more operations simultaneously. { 'məl·tə·pəl 'prō,gram·iŋ }

multiple-purpose tester See volt-ohm-milliammeter. { 'məl·tə·pəl ¦pər·pəs 'tes·tər }

multiple resonance |ELEC| Two or more resonances at different frequencies in a circuit consisting of two or more coupled circuits which are resonant at slightly different frequencies. { 'məl·tə·pəl 'rez·ən·əns }

multiple switchboard |ELEC| Manual telephone switchboard in which each subscriber line is attached to two or more jacks, to be within reach of several operators. { 'məl·tə·pəl 'swich,bórd }

multiple target generator |ELECTR| An electronic countermeasures device that produces several false responses in a hostile radar set. { 'məl·tə·pəl 'tär·gət ,jen·ə,rād·ər }

multiple-tuned antenna |ELECTROMAG| Low-frequency antenna having a horizontal section with a multiplicity of tuned vertical sections. { 'məl·tə·pəl ¦tünd an'ten·ə }

multiple twin quad |ELEC| Quad cable in which the four conductors are arranged in two twisted pairs, and the two pairs twisted together. { 'məl·tə·pəl 'twin 'kwäd }

multiple-unit semiconductor device |ELECTR| Semiconductor device having two or more seats of electrodes associated with independent carrier streams. { 'məl·tə·pəl ¦yü·nət 'sem·i·kən,dək·tər di,vīs }

multiple-unit steerable antenna See musa. { 'məl·tə·pəl ¦yü·nət 'stir·ə·bəl an'ten·ə }

multiple-unit tube See multiunit tube. { 'məl·tə·pəl ¦yü·nət 'tüb }

multiple winding |ELEC| A winding composed of several circuits connected in parallel. { 'məl·tə·pəl 'wīnd·iŋ }

multiplexer |ELECTR| A device for combining two or more signals, as for multiplex, or for creating the composite color video signal from its components in color television. Also spelled multiplexor. { 'məl·tə,plek·sər }

multiplexing |COMMUN| **1.** A set of techniques that enable the sharing of the usable electromagnetic spectrum of a telecommunications channel (the channel pass-band) among multiple users for the transfer of individual information streams. **2.** In particular, the case in which the user information streams join at a common access point to the channel. { 'məl·tə,pleks·iŋ }

multiplex mode |COMPUT SCI| The utilization of differences in operating speeds between a computer and transmission lines; the multiplexor channel scans each line in sequence, and any transmitted pulse on a line is assembled in an area reserved for this line; consequently, a number of users can be handled by the computer simultaneously. Also known as multiplexor channel operation. { 'məl·tə,pleks ,mōd }

multiplex operation |COMMUN| Simultaneous transmission of two or more messages in either or both directions over a carrier channel. { 'məl·tə,pleks ,äp·ə'rā·shən }

multiplexor See multiplexer. { 'məl·tə,plek·sər }

multiplexor channel operation See multiplex mode. { 'məl·tə,plek·sər ¦chan·əl ,äp·ə,rā·shən }

multiplexor terminal unit |COMPUT SCI| Device which permits a large number of data transmission lines to access a single computer. { 'məl·tə ,plek·sər 'ter·mən·əl ,yü·nət }

multiplex transmission |COMMUN| The simultaneous transmission of two or more programs or signals over a single radio-frequency channel, such as by time division, frequency division, code division, or phase division. { 'məl·tə,pleks tranz'mish·ən }

multiplication |ELECTR| An increase in current flow through a semiconductor because of increased carrier activity. { ,məl·tə·pli'kā·shən }

multiplication table |COMPUT SCI| In certain computers, a part of memory holding a table of numbers in which the computer looks up values in order to perform the multiplication operation. { ,məl·tə·pli'kā·shən ,tā·bəl }

multiplication time |COMPUT SCI| The time required for a computer to perform a multiplication; for a binary number it will be equal to the total of all the addition times and all the shift times involved in the multiplication. { ,məl·tə·pli'kā·shən ,tīm }

multiplicative congruential generator |COMPUT SCI| A congruential generator in which the constant *b* in the generating formula is equal to zero. { ,məl·tə¦plik·əd·iv ,kän,grü'en·chəl 'jen·ə,rād·ər }

multiplier |ELEC| A resistor used in series with a voltmeter to increase the voltage range. Also known as multiplier resistor. |ELECTR| **1.** A device that has two or more inputs and an output that is a representation of the product of the quantities represented by the input signals; voltages are the quantities commonly multiplied. **2.** See electron multiplier; frequency multiplier. { 'məl·tə,plī·ər }

multiplier field |COMPUT SCI| The area reserved for a multiplication, equal to the length of multiplier plus multiplicand plus one character. { 'məl·tə ,plī·ər ,fēld }

multiplier phototube |ELECTR| A phototube with one or more dynodes between its photocathode and the output electrode; the electron stream from the photocathode is reflected off each dynode in turn, with secondary emission adding electrons to the stream at each reflection. Also known as electron-multiplier phototube; photoelectric electron-multiplier tube; photomultiplier; photomultiplier tube. { 'məl·tə,plī·ər 'fōd·ō,tüb }

multiplier-quotient register |COMPUT SCI| A register equal to two words in length in which the quotient is developed and in which the multiplier is entered for multiplication. { 'məl·tə,plī·ər 'kwō·shənt ,rej·ə·stər }

multiplier resistor See multiplier. { 'məl·tə,plī·ər ri,zis·tər }

multiplier traveling-wave photodiode |ELECTR| Photodiode in which the construction of a traveling-wave tube is combined with that of a multiplier phototube to give increased sensitivity. { 'məl·tə,plī·ər |trav·ə·liŋ 'wāv ,fōd·ō'dī,ōd }

multiplier tube |ELECTR| Vacuum tube using secondary emission from a number of electrodes in sequence to obtain increased output current; the electron stream is reflected, in turn, from one electrode of the multiplier to the next. { 'məl·tə ,plī·ər ,tüb }

multipling |COMMUN| Use of multidrop lines to provide for changes in telephone service patterns or requirements; unused terminals afford convenient access to wiretappers. { 'məl·tə·pliŋ }

multiply defined symbol |COMPUT SCI| Common assembler or compiler error printout indicating that a label has been used more than once. { 'məl·tə·plē di¦fīnd 'sim·bəl }

multipoint line |COMMUN| A line which is shared by two or more different tributary stations. { 'məl·tə,pȯint ,līn }

multiport memory |COMPUT SCI| A memory shared by many processors to communicate among themselves. { 'məl·tə,pȯrt 'mem·rē }

multiprecision arithmetic |COMPUT SCI| A form of arithmetic similar to double precision arithmetic except that two or more words may be used to represent each number. { |məl·tə·prə'sizh·ən ə'rith·mə,tik }

multiprocessing |COMPUT SCI| Carrying out of two or more sequences of instructions at the same time in a computer. { ,məl·tə'prä,ses·iŋ }

multiprocessing system See multiprocessor. { ,məl·tə'prä,ses·iŋ ,sis·təm }

multiprocessor |COMPUT SCI| A data-processing system that can carry out more than one program, or more than one arithmetic operation, at the same time. Also known as multiprocessing system. { ,məl·tə'prä,ses·ər }

multiprocessor interleaving |COMPUT SCI| Technique used to speed up processing time; by splitting banks of memory each with x microseconds access time and accessing each one in sequence 1/n-th of a cycle later, a reference to memory can be had every x/n microseconds; this speed is achieved at the cost of hardware complexity. { ,məl·tə'prä,ses·ər ,in·tər'lēv·iŋ }

multiprogramming |COMPUT SCI| The interleaved execution of two or more programs by a computer, in which the central processing unit executes a few instructions from each program in succession. { ,məl·tə'prō,gram·iŋ }

multiprogramming executive control |COMPUT SCI| Control program structure required to handle multiprogramming with either a fixed or a variable number of tasks. { ,məl·tə'prō,gram·iŋ ig¦zek·yəd·iv kən'trōl }

multirole programmable device |CONT SYS| A device that contains a programmable memory to store data on positioning robots and sequencing their motion. { 'məl·tə,rōl prō¦gram·ə·bəl di'vīs }

multisegment magnetron |ELECTR| Magnetron with an anode divided into more than two segments, usually by slots parallel to its axis. { 'məl·tə,seg·mənt 'mag·nə,trän }

multispeed motor |ELEC| An induction motor that can rotate at any one of two or more speeds, independent of the load. { 'məl·tə,spēd 'mōd·ər }

multistable circuit |ELECTR| A circuit having two or more stable operating conditions. { 'məl·tə ,stā·bəl 'sər·kət }

multistage amplifier See cascade amplifier. { 'məl·tē,stāj 'am·plə,fī·ər }

multistatic radar |ENG| Radar in which successive antenna lobes are sequentially engaged to provide a tracking capability without physical movement of the antenna. { 'məl·tē,stad·ik 'rā ,där }

multistation |COMMUN| Pertaining to a network in which each station can communicate with each of the other stations. { 'məl·tē,stā·shən }

multistator watt-hour meter |ELEC| An induction type of watt-hour meter in which several stators exert a torque on the rotor. { 'məl·tē,stād·ər 'wat ,au̇r ,mēd·ər }

multistrip coupler |ELECTR| A series of parallel metallic strips placed on a surface acoustic wave filter between identical apodized interdigital transducers; it converts the spatially nonuniform surface acoustic wave generated by one transducer into a spatially uniform wave received at the other transducer, and helps to reject spurious bulk acoustic modes. { 'məl·tə,strip 'kəp·lər }

multisync monitor |COMPUT SCI| A video display monitor that automatically adjusts to the synchronization frequency of the video source from which it is receiving signals. { |məl·ti,siŋk 'män·ə·tər }

multisystem coupling |COMPUT SCI| The electronic connection of two or more computers in proximity to make them act as a single logical machine. { 'məl·tə,sis·təm 'kəp·liŋ }

multisystem network |COMPUT SCI| A data communications network that has two or more host computers with which the various terminals in the system can communicate. { 'məl·tə,sis·təm 'net,wərk }

multitape Turing machine |COMPUT SCI| A variation of a Turing machine in which more than one tape is permitted, each tape having its own read-write head. { 'məl·tē,tāp 'tu̇r·iŋ mə,shēn }

multitasking |COMPUT SCI| The simultaneous execution of two or more programs by a single central processing unit. { |məl·tē'task·iŋ }

multitask operation |COMPUT SCI| A sophisticated form of multijob operation in a computer which allows a single copy of a program module to be used for more than one task. { |məl·tē'task ,äp·ə'rā·shən }

multithreading |COMPUT SCI| A processing technique that allows two or more of the same type of transaction to be carried out simultaneously. { |məl·tē'thred·iŋ }

multitrack operation |COMPUT SCI| The selection of the next read/write head in a cylinder, usually indicated by bit zero of the operation code in the channel command word. { |məl·tē'trak ,äp·ə'rā·shən }

multitrack recording system |ENG| Recording system which provides two or more recording paths on a medium, which may carry either related or unrelated recordings in common time relationship. { ¦məl·tē'trak ri'kórd·iŋ ¸sis·təm }

multiturn potentiometer |ELEC| A precision wire-wound potentiometer in which the resistance element is formed into a helix, generally having from 2 to 10 turns. { 'məl·tē¸tərn pə¸ten·chē'äm·əd·ər }

multiunit tube |ELECTR| Electron tube containing within one glass or metal envelope, two or more groups of electrodes, each associated with separate electron streams. Also known as multiple-unit tube. { ¦məl·tē¦yü·nət 'tüb }

multiuser system |COMPUT SCI| A computer system with multiple terminals, enabling several users, each at their own terminal, to use the computer. { ¦məl·tē¦yü·zər 'sis·təm }

multivariable system |CONT SYS| A dynamical system in which the number of either inputs or outputs is greater than 1. { ¦məl·tē'ver·ē·ə·bəl ¸sis·təm }

multivator |ELEC| Type of dc-to-dc up-converter. |ELECTR| An automatic device for analyzing a number of dust samples that might be collected by spacecraft on the moon, Mars, and other planets, to detect the presence of microscopic organisms with a multiplier phototube that measures the fluorescence given off. { 'məl·tə ¸vād·ər }

multivibrator |ELECTR| A relaxation oscillator using two tubes, transistors, or other electron devices, with the output of each coupled to the input of the other through resistance-capacitance elements or other elements to obtain in-phase feedback voltage. { ¸məl·tə'vī¸brād·ər }

multivolume file |COMPUT SCI| A file that consists of more than one physical unit of storage medium. { ¦məl·tē¦väl·yəm 'fīl }

multiway merge |COMPUT SCI| A computer operation in which three or more lists are merged into a single list. { 'məl·tē¸wā 'mərj }

Murray code |COMMUN| A binary code with five binary digits per letter which was developed to be used with a typewriterlike device which would punch holes in paper tape, and is now the basis of the widely used CCIT 2 code. { 'mər·ē ¸kōd }

Murray loop test |ELEC| A method of localizing a fault in a cable by replacing two arms of a Wheatstone bridge with a loop formed by the cable under test and a good cable connected to the far end of the defective cable. { 'mər·ē 'lüp ¸test }

musa |ELECTROMAG| An electrically steerable receiving antenna whose directional pattern can be rotated by varying the phases of the contributions of the individual units. Derived from multiple-unit steerable antenna. { 'myü·sə }

musical instrument digital interface |COMPUT SCI| **1.** The digital standard for connecting computers, musical instruments, and synthesizers. **2.** A compression format for encoding music. Abbreviated MIDI. { ¸myü·zi·kəl ¦in·strə·mənt ¸dij·ə·dəl 'in·tər¸fās }

muting circuit |ELECTR| **1.** Circuit which cuts off the output of a receiver when no radio-frequency carrier greater than a predetermined intensity is reaching the first detector. **2.** Circuit for making a receiver insensitive during operation of its associated transmitter. { 'myüd·iŋ ¸sər·kət }

muting switch |ELEC| **1.** A switch used in connection with automatic tuning systems to silence the receiver while tuning from one station to another. **2.** A switch used to ground the output of a phonograph pickup automatically while a record changer is in its change cycle. { 'myüd·iŋ ¸swich }

mutual admittance |ELEC| For two meshes of a network carrying alternating current, the ratio of the complex current in one mesh to the complex voltage in the other, when the voltage in all meshes besides these two is 0. { 'myü·chə·wəl ad'mit·əns }

mutual branch See common branch. { 'myü·chə ·wəl 'branch }

mutual capacitance |ELEC| The accumulation of charge on the surfaces of conductors of each of two circuits per unit of potential difference between the circuits. { 'myü·chə·wəl kə'pas·əd· əns }

mutual conductance See transconductance. { 'myü·chə·wəl kən'dək·təns }

mutual deadlock |COMPUT SCI| A condition in which deadlocked tasks are awaiting resource assignments, and each task on a list awaits release of a resource held by the following task, with the last task awaiting release of a resource held by the first task. Also known as circular wait. { 'myü·chə·wəl 'ded¸läk }

mutual impedance |ELEC| For two meshes of a network carrying alternating current, the ratio of the complex voltage in one mesh to the complex current in the other, when all meshes besides the latter one carry no current. { 'myü·chə·wəl im'pēd·əns }

mutual interference |COMMUN| Interference from two or more electrical or electronic systems which affects these systems on a reciprocal basis. { 'myü·chə·wəl ¸in·tər'fir·əns }

mV See millivolt.

MV See megavolt.

MWYE See megawatt year of electricity.

myriametric waves |ELECTROMAG| Electromagnetic waves having wavelengths between 10 and 100 kilometers, corresponding to the very low frequency band. { ¦mir·ē·ə¦me·trik 'wāvz }

N

nacelle |ELEC| An enclosure containing the electric generating equipment in a wind-energy conversion system. { nə'sel }

NAK See negative acknowledgement. { nak or ,en ,ā'kā }

nancy receiver See infrared receiver. { 'nan·sē ri ,sēv·ər }

NAND circuit |ELECTR| A logic circuit whose output signal is a logical 1 if any of its inputs is a logical 0, and whose output signal is a logical 0 if all of its inputs are logical 1. { 'nand ,sər·kət }

nanoelectronics |ELECTR| The technology of electronic devices whose dimensions range from atoms up to 100 nanometers. { ,nan·ō·i,lek 'trän·iks }

nanotechnology |ENG| **1.** Systems for transforming matter, energy, and information that are based on nanometer-scale components with precisely defined molecular features. **2.** Techniques that produce or measure features less than 100 nanometers in size. { ,nan·ō·tek'näl· ə·jē }

NAPLPS See North American presentation-level protocol syntax. { 'nap,lips }

narrow-band amplifier |ELECTR| An amplifier which increases the magnitude of signals over a band of frequencies whose bandwidth is small compared to the average frequency of the band. { 'nar·ō ¦band 'am·plə,fī·ər }

narrow-band frequency modulation |COMMUN| Frequency-modulated broadcasting system used primarily for two-way voice communication, typically having a maximum deviation of 15 kilohertz or less. { 'nar·ō ¦band 'frē·kwən·sē ,mäj·ə'lā· shən }

narrow-band-pass filter |ELECTR| A band-pass filter in which the band of frequencies transmitted by the filter has a bandwidth which is small compared to the average frequency of the band. { 'nar·ō ¦ban ,pas ,fil·tər }

narrow-band path |COMMUN| A communications path having a bandwith typically of less than 20 kilohertz. { 'nar·ō ¦band 'path }

narrow-beam antenna |ELECTROMAG| An antenna which radiates most of its power in a cone having a radius of only a few degrees. { 'nar·ō ¦bēm an'ten·ə }

narrow-sector recorder |ELECTR| A radio direction finder with which atmospherics are received from a limited sector related to the position

of the antenna; this antenna is usually rotated continuously and the bearings of the atmospherics recorded automatically. { 'nar·ō ¦sek·tər ri'kȯrd·ər }

N-ary code |COMMUN| Code employing N distinguishable types of code elements. { 'en·ə·rē ,kōd }

N-ary pulse-code modulation |COMMUN| Pulse-code modulation in which the code for each element consists of any one of N distinguishable types of elements. { 'en·ə·rē 'pəls ,kōd ,mäj·ə'lā·shən }

National Radio Systems Committee |COMMUN| Abbreviated NRSC. A technical standards setting body of the radio broadcasting industry, co-sponsored by the Consumer Electronics Association (CEA) and the National Association of Broadcasters (NAB). { ¦nash·ən·əl ¦rād·ē,ō ¦sis·təmz kə,mid·ē }

National Television Systems Committee |COMMUN| Abbreviated NTSC. The organization that developed the transmission standard for color television broadcasting in the United States, and the black-and-white system that preceded it. The NTSC color system was adopted by the Federal Communications Commission in 1953 and remains in the FCC Rules today. NTSC was also adopted in a number of other countries around the world for distribution of color video programming. The NTSC standard provides for a screen density of 525 scan lines per picture. For U.S. television service, NTSC has a field repetition rate of just under 60 fields/second, and a frame rate of just under 30 frames/second; one frame is composed of two fields for interlace scanning systems. { ¦nash·ən·əl 'tel·ə,vizh·ən ¦sis·təmz kə ,mid·ē }

native language |COMPUT SCI| Machine language that is executed by the computer for which it is specifically designed, in contrast to a computer using an emulator. { 'nād·iv 'laŋ·gwij }

native mode |COMPUT SCI| **1.** The mode of operation of a software product that is being used on a computer for which it was specifically designed, without use of an emulator. **2.** The mode of operation of a device that is carrying out the function for which it was designed and is not emulating another device. { 'nād·iv 'mōd }

natural antenna frequency |ELECTROMAG| Lowest resonant frequency of an antenna without added inductance or capacitance. { 'nach·rəl an'ten·ə ,frē·kwən·sē }

natural binary coded decimal system |COMPUT SCI| A particular binary coded decimal system that uses the first ten binary numbers in sequence to represent the digits 0 through 9. { 'nach·rəl 'bī,ner·ē ¦kōd·əd ¦des·məl ,sis·təm }

natural frequency |ELECTR| The lowest resonant frequency of an antenna, circuit, or component. { 'nach·rəl 'frē·kwən·sē }

natural function generator See analytical function generator. { 'nach·rəl 'fəŋk·shən ,jen·ə,rād·ər }

natural interference |COMMUN| Electromagnetic interference arising from natural terrestrial phenomena (called atmospheric interference), or electromagnetic interference caused by natural disturbances originating outside the atmosphere of the earth (called galactic and solar noise). { 'nach·rəl ,in·tər'fir·əns }

natural language |COMPUT SCI| A computer language whose rules reflect and describe current rather than prescribed usage; it is often loose and ambiguous in interpretation, meaning different things to different hearers. { 'nach·rəl 'laŋ·gwij }

natural language interaction |COMPUT SCI| The interaction of users with computer systems through the medium of natural languages. { 'nach·rəl ¦laŋ·gwij ,in·tər'ak·shən }

natural language processing |COMPUT SCI| Computer analysis and generation of natural language text; encompasses natural language interaction and natural language text processing. { 'nach·rəl ¦laŋ·gwij 'prä,ses·iŋ }

natural language text processing |COMPUT SCI| Computer processing of natural language text into a more useful form, as in automatic text translation or text summarization. { 'nach·rəl ¦laŋ·gwij 'tekst ,prä,ses·iŋ }

natural wavelength |ELECTROMAG| Wavelength corresponding to the natural frequency of an antenna or circuit. { 'nach·rəl 'wāv,leŋkth }

navigation |COMPUT SCI| In a database management system, the techniques provided for locating information within the system. { ,nav·ə'gā·shən }

navigation receiver |ELECTR| An electronic device that determines a ship's position by receiving and comparing radio signals from transmitters at known locations. { ,nav·ə'gā·shən ri ,sē·vər }

n-channel |ELECTR| A conduction channel formed by electrons in an *n*-type semiconductor, as in an *n*-type field-effect transistor. { 'en ,chan·əl }

n-channel metal-oxide semiconductor See NMOS. { ¦en ,chan·əl ,med·əl ¦äk,sīd 'sem·i·kən ,dək·tər }

N curve |ELECTR| A plot of voltage against current for a negative-resistance device; its slope is negative for some values of current or voltage. { 'en ,kərv }

N-display |ELECTR| A radar display format in which a trace-deflecting pulse can be moved along the vertical range axis of an L-display to assist an operator in determining and reporting the range of a target. Also known as N-indicator; N-scan; N-scope. { 'en di,splā }

NDRO See nondestructive readout.

NEA material See negative-electron affinity material. { ¦en¦ē'ā mə,tir·ē·əl }

near-end crosstalk |COMMUN| A type of interference that may occur at carrier telephone repeater stations when output signals of one repeater leak into the same end of the other repeater. { 'nir ,end 'krós,tók }

near field |ELECTROMAG| The electromagnetic field that exists within one wavelength of a source of electromagnetic radiation, such as a transmitting antenna. { 'nir ,fēld }

near-infrared radiation |ELECTROMAG| Infrared radiation having a relatively short wavelength, between 0.75 and about 2.5 micrometers (some scientists place the upper limit from 1.5 to 3 micrometers), at which radiation can be detected by photoelectric cells, and which corresponds in frequency range to the lower electronic energy levels of molecules and semiconductors. Also known as photoelectric infrared radiation. { 'nir ,in·frə'red ,rād·ē'ā·shən }

necessary bandwidth |COMMUN| For a given class of emission, the minimum value of the occupied bandwidth sufficient to ensure the transmission of information at the rate and with the quality required for the system employed, under specified conditions. { 'nes·ə,ser·ē 'band ,width }

needle gap |ELECTR| Spark gap in which the electrodes are needle points. { 'nēd·əl ,gap }

needle scratch See surface noise. { 'nēd·əl ,skrach }

needle test point |ELEC| A sharp steel probe connected to a test cord for making contact with a conductor. { 'nēd·əl 'test ,póint }

negative |ELEC| Having a negative charge. { 'neg·əd·iv }

negative acknowledgement |COMPUT SCI| In a data communications network, a control character returned from a receiving machine to a sending machine to indicate the presence of errors in the preceding block of data. Abbreviated NAK. { 'neg·əd·iv ik'näl·ij·mənt }

negative booster |ELEC| Booster used in connection with a ground-return system to reduce the difference of potential between two points on the grounded return. { 'neg·əd·iv 'bü·stər }

negative charge |ELEC| The type of charge which is possessed by electrons in ordinary matter, and which may be produced in a resin object by rubbing with wool. Also known as negative electricity. { 'neg·əd·iv 'chärj }

negative conductor |ELEC| The conductor that is connected to the negative terminal of a voltage source. { 'neg·əd·iv kən'dək·tər }

negative effective mass amplifiers and generators [ELECTR] Class of solid-state devices for broad-band amplification and generation of electrical waves in the microwave region; these devices use the property of the effective masses of charge carriers in semiconductors becoming negative with sufficiently high kinetic energies. { 'neg·əd·iv i¦fek·tiv ¦mas 'am·plə ‚fī·ərz ən 'jen·ə‚rād·ərz }

negative electricity See negative charge. { 'neg·əd·iv ‚i‚lek'tris·əd·ē }

negative electrode See cathode. { 'neg·əd·iv i'lek ‚trōd }

negative electron-affinity material [ELECTR] A material, such as gallium phosphide, whose surface has been treated with a substance, such as cesium, so that the surface barrier is reduced, band-bending occurs so that the top of the conduction band lies above the vacuum level, and the electron affinity of the substance in negative. Abbreviated NEA material. { 'neg·əd·iv i'lek‚trän ə'fin·əd·ē mə‚tir·ē·əl }

negative feedback [CONT SYS] Feedback in which a portion of the output of a circuit, device, or machine is fed back 180° out of phase with the input signal, resulting in a decrease of amplification so as to stabilize the amplification with respect to time or frequency, and a reduction in distortion and noise. Also known as inverse feedback; reverse feedback; stabilized feedback. { 'neg·əd·iv 'fēd‚bak }

negative glow [ELECTR] The luminous flow in a glow-discharge cold-cathode tube occurring between the cathode dark space and the Faraday dark space. { 'neg·əd·iv 'glō }

negative-grid generator [ELECTR] Conventional oscillator circuit in which oscillation is produced by feedback from the plate circuit to a grid which is normally negative with respect to the cathode, and which is designed to operate without drawing grid current at any time. { 'neg·əd·iv ¦grid 'jen·ə ‚rād·ər }

negative-grid thyratron [ELECTR] A thyratron with only one grid, which serves to prevent the flow of current until its potential relative to the cathode is made less negative than a certain critical value. { 'neg·əd·iv ¦grid 'thī·rə‚trän }

negative impedance [ELECTR] An impedance such that when the current through it increases, the voltage drop across the impedance decreases. { 'neg·əd·iv im'pēd·əns }

negative-impedance repeater [ELECTR] A telephone repeater that provides an effective gain for voice-frequency signals by insertion into the line of a negative impedance that cancels out line impedances responsible for transmission losses. { 'neg·əd·iv im¦pēd·əns ri'pēd·ər }

negative logic [ELECTR] Logic circuitry in which the more positive voltage (or current level) represents the 0 state; the less positive level represents the 1 state. { 'neg·əd·iv ‚läj·ik }

negative modulation [ELECTR] **1.** Television modulation system in which an increase in scene brightness corresponds to a decrease in amplitude-modulated transmitter power; used in United States analog television transmitters. **2.** Modulation in which an increase in brightness corresponds to a decrease in the frequency of a frequency-modulated facsimile transmitter. Also known as negative transmission. { 'neg·əd·iv ‚mäj·ə'lā·shən }

negative phase sequence [ELEC] The phase sequence that corresponds to the reverse of the normal order of phases in a polyphase system. { 'neg·əd·iv ¦fāz 'sē·kwəns }

negative-phase-sequence relay [ELEC] Relay which functions in conformance with the negative-phase-sequence component of the current, voltage, or power of the circuit. { 'neg· əd·iv ¦fāz ¦sē·kwəns 'rē‚lā }

negative picture phase [ELECTR] The video signal phase in which the signal voltage swings in a negative direction for an increase in brilliance. { 'neg·əd·iv 'pik·chər ‚fāz }

negative plate [ELEC] The internal plate structure that is connected to the negative terminal of a storage battery. Also known as negative electrode. { 'neg·əd·iv 'plāt }

negative potential [ELEC] An electrostatic potential which is lower than that of the ground, or of some conductor or point in space that is arbitrarily assigned to have zero potential. { 'neg·əd· iv pə'ten·chəl }

negative resistance [ELECTR] The resistance of a negative-resistance device. { 'neg·əd·iv ri'zis· təns }

negative-resistance device [ELECTR] A device having a range of applied voltages within which an increase in this voltage produces a decrease in the current. { 'neg·əd·iv ri¦zis·təns di'vīs }

negative-resistance oscillator [ELECTR] An oscillator in which a parallel-tuned resonant circuit is connected to a vacuum tube so that the combination acts as the negative resistance needed for continuous oscillation. { 'neg·əd·iv ri¦zis·təns 'äs·ə‚lād·ər }

negative-resistance repeater [ELECTR] Repeater in which gain is provided by a series negative resistance or a shunt negative resistance, or both. { 'neg·əd·iv ri¦zis·təns ri'pēd·ər }

negative terminal [ELEC] The terminal of a battery or other voltage source that has more electrons than normal; electrons flow from the negative terminal through the external circuit to the positive terminal. { 'neg·əd·iv 'tər·mən·əl }

negative thermion See thermoelectron. { 'neg· əd·iv 'thər‚mē‚än }

negative-transconductance oscillator [ELECTR] Electron-tube oscillator in which the output of the tube is coupled back to the input without phase shift, the phase condition for oscillation being satisfied by the negative transconductance of the tube. { 'neg·əd·iv ‚tranz·kən'dək·təns 'äs· tə‚lād·ər }

negative transmission See negative modulation. { 'neg·əd·iv tranz'mish·ən }

negatron See dynatron. { 'neg·ə‚trän }

negentropy See information content. { nə'gen ·trə‚pē }

neon glow lamp | ELECTR | A glow lamp containing neon gas, usually rated between ½₅ and 3 watts, and producing a characteristic red glow; used as an indicator light and electronic circuit component. { 'nē,än 'glō ,lamp }

neon oscillator | ELECTR | Relaxation oscillator in which a neon tube or lamp serves as the switching element. { 'nē,än 'äs·ə,lād·ər }

neon tube | ELECTR | An electron tube in which neon gas is ionized by the flow of electric current through long lengths of gas tubing, to produce a luminous red glow discharge; used chiefly in outdoor advertising signs. { 'nē,än ,tüb }

Nernst bridge | ELEC | A four-arm bridge containing capacitors instead of resistors, used for measuring capacitance values at high frequencies. { 'nernst ,brij }

Nernst glower See Nernst lamp. { 'nernst ,glō·ər }

Nernst lamp | ELEC | An electric lamp consisting of a short, slender rod of zirconium oxide in open air, heated to brilliant white incandescence by current. Also known as Nernst glower. { 'nernst ,lamp }

nesistor | ELECTR | A negative-resistance semiconductor device that is basically a bipolar field-effect transistor. { ne'zis·tər }

nest | COMPUT SCI | To include data or subroutines in other items of a similar nature with a higher hierarchical level so that it is possible to access or execute various levels of data or routines recursively. { nest }

nesting | COMPUT SCI | **1.** Inclusion of a routine wholly within another routine. **2.** Inclusion of a DO statement within a DO statement in FORTRAN. { 'nest·iŋ }

nesting storage See push-down storage. { 'nest·iŋ 'stȯr·ij }

net | COMMUN | A number of communication stations equipped for communicating with each other, often on a definite time schedule and in a definite sequence. { net }

net call sign | COMMUN | A call sign that represents all stations within a net. { 'net 'kȯl ,sīn }

net control station | COMMUN | Communications station having the responsibility of clearing traffic and exercising circuit discipline within a net. { 'net kən'trōl ,stā·shən }

net loss | COMMUN | The ratio of the power at the input of a transmission system to the power at the output; expressed in nepers, it is one-half the natural logarithm of this ratio, and in decibels it is 10 times the common logarithm of the ratio. { 'net 'lȯs }

network | COMMUN | A number of radio or television broadcast stations connected by fiber-optic cable, coaxial cable, radio, or wire lines, so all stations can broadcast the same program simultaneously. | COMPUT SCI | See computer network. | ELEC | A collection of electric elements, such as resistors, coils, capacitors, and sources of energy, connected together to form several interrelated circuits. Also known as electric network. { 'net,wərk }

network admittance | ELEC | The admittance between two terminals of a network under specified conditions. { 'net,wərk ad'mit·əns }

network analysis | ELEC | Derivation of the electrical properties of a network, from its configuration, element values, and driving forces. | ENG | An analytic technique used during project planning to determine the sequence of activities and their interrelationship within the network of activities that will be required by the project. Also known as network planning. { 'net,wərk ə'nal·ə·səs }

network analyzer | COMPUT SCI | An analog computer in which networks are used to simulate power line systems or physical systems and obtain solutions to various problems before the systems are actually built. { 'net,wərk 'an·ə ,līz·ər }

network architecture | COMMUN | The high-level design of a communications system, including the choice of hardware, software, and protocols. { 'net,wərk 'är·kə,tek·chər }

network constant | ELEC | One of the resistance, inductance, mutual inductance, or capacitance values involved in a circuit or network; if these values are constant, the network is said to be linear. { 'net,wərk 'kän·stənt }

network control program | COMPUT SCI | A computer program that controls communications between multiple terminals and a mainframe. { |net,wərk kən'trōl ,prō·grəm }

network data structure | COMPUT SCI | The arrangement of data in a computer system into interconnected groupings of information according to relationships between groupings. { 'net,wərk 'dad·ə ,strək·chər }

network filter | ELEC | A combination of electrical elements (for example, interconnected resistors, coils, and capacitors) that represents relatively small attenuation to signals of a certain frequency, and great attenuation to all other frequencies. { 'net,wərk 'fil·tər }

network flow | ELEC | Flow of current in a network. { 'net,wərk 'flō }

networking | COMPUT SCI | The use of transmission lines to join geographically separated computers. { 'net,wərk·iŋ }

network input impedance | ELEC | The impedance between the input terminals of a network under specified conditions. { 'net ,wərk 'in·pùt im,pēd·əns }

network master relay | ELEC | Relay that performs the chief functions of closing and tripping an alternating-current low-voltage network protector. { 'net,wərk 'mas·tər 'rē,lā }

network operating system | COMPUT SCI | The system software of a local-area network, which manages the network's resources, handling multiple inputs concurrently and providing necessary security. Abbreviated NOS. { ,net,wərk 'äp·ə ,rād·iŋ ,sis·təm }

network phasing relay | ELEC | Relay which functions in conjunction with a master relay to limit closure of the network protector to a predetermined relationship between the voltage and the network voltage. { 'net,wərk 'fāz·iŋ 'rē,lā }

network planning See network analysis. { 'net ,wərk ,plan·iŋ }

network relay |ELEC| Form of voltage, power, or other type of relay used in the protection and control of alternating-current low-voltage networks. { 'net,wərk 'rē,lā }

network server See file server. { 'net,wərk ,sər·vər }

network synthesis |ELEC| Derivation of the configuration and element values of a network with given electrical properties. { 'net,wərk ,sin·thə·səs }

network system |COMPUT SCI| A type of data-base management system in which data records can be related in more general structures than in a hierarchical file, permitting a given record to have more than one parent. { 'net,wərk ,sis·təm }

network terminal protocol |COMMUN| A set of standards that allows the user of a computer connected to a network to log in on any other computer on the network. Also known as TELNET. { ¦net,wərk ¦tərm·ən·əl 'prōd·ə,kȯl }

network theory |ELEC| The systematizing and generalizing of the relations between the currents, voltages, and impedances associated with the elements of an electrical network. { 'net ,wərk ,thē·ə·rē }

network transfer admittance |ELEC| The current that would flow through a short circuit between one pair of terminals in a network if a unit voltage were applied across the other pair. { 'net,wərk 'trans·fər ad,mit·əns }

network vulnerability scan |COMPUT SCI| The process of determining the connectivity of the protected subnetwork within a security perimeter of a distributed computing system, and then testing the strength of protection at all access points to the subnetwork. { ¦net,wərk ,vəl·nər·ə'bil·əd·ē ,skan }

Neugebauer effect |ELEC| A small change in the polarization of an optically isotropic medium in an external electric field, related to the electrooptical Kerr effect. { 'nȯi·gə,baů·ər i,fekt }

neural network |COMPUT SCI| An information-processing device that utilizes a very large number of simple modules, and in which information is stored by components that at the same time effect connections between these modules. { 'nůr·əl 'net,wərk }

neuristor |ELECTR| A device that behaves like a nerve fiber in having attenuationless propagation of signals; one goal of research is development of a complete artificial nerve cell, containing many neuristors, that could duplicate the function of the human eye and brain in recognizing characters and other visual images. { nů'ris·tər }

neuromorphic engineering |ENG| Use of the functional principles of biological nervous systems to inspire the design and fabrication of artificial nervous systems, such as vision chips and roving robots. { ¦nů·rō,mȯr·fik ,en·jə'nir·iŋ }

neuronal interface |ENG| An artificial synapse capable of reversible chemical-to-electrical transduction processes between neural tissue and conventional solid-state electronic devices for applications such as aural, visual, and

mechanical prostheses, as well as expanding human memory and intelligence. { nů¦rōn·əl 'in·tər,fās }

neurotechnology |ENG| The application of microfabricated devices to achieve direct contact with the electrically active cells of the nervous system (neurons). { ,nů·rō·tek'näl·ə·jē }

neutral |ELEC| Referring to the absence of a net electric charge. { 'nü·trəl }

neutral conductor |ELEC| A conductor of a polyphase circuit or of a single-phase, three-wire circuit which is intended to have a potential such that the potential differences between it and each of the other conductors are approximately equal in magnitude and are also equally spaced in phase. { 'nü·trəl kən'dək·tər }

neutral ground |ELEC| Ground connected to the neutral point or points of an electric circuit, transformer, rotating machine, or system. { 'nü·trəl 'graůnd }

neutralize |ELECTR| To nullify oscillation-producing voltage feedback from the output to the input of an amplifier through tube interelectrode capacitances; an external feedback path is used to produce at the input a voltage that is equal in magnitude but opposite in phase to that fed back through the interelectrode capacitance. { 'nü·trə,līz }

neutralized radio-frequency stage |ELECTR| Stage having an additional circuit connected to feed back, in the opposite phase, an amount of energy equivalent to what is causing the oscillation, thus neutralizing any tendency to oscillate and making the circuit function strictly as an amplifier. { 'nü·trə,līzd ¦rād·ē·ō 'frē·kwən·sē ,stāj }

neutralizing capacitor |ELECTR| Capacitor, usually variable, employed in a radio receiving or transmitting circuit to feed a portion of the signal voltage from the plate circuit of a stage back to the grid circuit. { 'nü·trə,līz·iŋ kə,pas·əd·ər }

neutralizing circuit |ELECTR| Portion of an amplifier circuit which provides an intentional feedback path from plate to grid to prevent regeneration. { 'nü·trə,līz·iŋ ,sər·kət }

neutralizing voltage |ELECTR| Voltage developed in the plate circuit (Hazeltine neutralization) or in the grid circuit (Rice neutralization), used to nullify or cancel the feedback through the tube. { 'nü·trə,līz·iŋ ¦vōl·tij }

neutral point |ELEC| Point which has the same potential as the point of junction of a group of equal nonreactive resistances connected at their free ends to the appropriate main terminals or lines of the system. { 'nü·trəl ,pȯint }

neutral relay |ELEC| Relay in which the movement of the armature does not depend upon the direction of the current in the circuit controlling the armature. Also known as nonpolarized relay. { 'nü·trəl 'rē,lā }

neutral return path |ELEC| A route from the load back to the power source, completing a circuit in an electric power distribution system, which is grounded, usually by connections to water pipes. { 'nü·trəl ri'tərn ,path }

neutral safety switch [ELEC] An electric switch that is connected to the ignition switch of an internal combustion engine and prevents starting the engine unless the transmission shift lever is in the neutral or park position, or the clutch pedal is depressed. { ¦nü·trəl ′sāf·tē ‚swich }

neutral stability [CONT SYS] Condition in which the natural motion of a system neither grows nor decays, but remains at its initial amplitude. { ′nü·trəl stə′bil·əd·ē }

neutral temperature [ELECTR] The temperature of the hot junction of a thermocouple at which the electromotive force of the thermocouple attains its maximum value, when the cold junction is maintained at a constant temperature of 0°C. { ′nü·trəl ′tem·prə·chər }

neutral zone *See* dead band. { ′nü·trəl ‚zōn }

new-band service [COMMUN] A broadcasting service that is allocated a portion of the radio frequency spectrum that was not previously used. { ′nü ‚band ‚sər·vis }

newsgroup [COMPUT SCI] A collection of computers on a wide-area network that form a discussion group on a particular topic, such that a message generated by any computer in the group is automatically distributed over the network to all the others. Also known as forum. { ′nüz‚grüp }

next-event file [COMPUT SCI] A portion of a computer simulation program which maintains a list of all events to be processed and updates the simulated time. { ′nekst i′vent ‚fīl }

nexus [COMMUN] A connection or interconnection of a communications system, such as a data link or a network of branches and nodes. { ′nek·səs }

nibble [COMPUT SCI] A unit of computer storage or information equal to one-half a byte. { ′nib·əl }

Nichol's chart [CONT SYS] A plot of curves along which the magnitude M or argument α of the frequency control ratio is constant on a graph whose ordinate is the logarithm of the magnitude of the open-loop transfer function, and whose abscissa is the open-loop phase angle. { ′nik·əlz ‚chärt }

nickel-cadmium battery [ELEC] A sealed storage battery having a nickel anode, a cadmium cathode, and an alkaline electrolyte; widely used in cordless appliances; without recharging, it can serve as a primary battery. Also known as cadmium-nickel storage cell. { ′nik·əl ¦kad·mē·əm ′bad·ə·rē }

nickel delay line [ELECTR] An acoustic delay line in which nickel is used to transmit sound signals. { ′nik·əl di′lā ‚līn }

nickel-iron battery *See* Edison battery. { ′nik·əl ¦ī·ərn ′bad·ə·rē }

NIF *See* noise improvement factor.

N-indicator *See* N-display. { ′en ‚in·də‚kād·ər }

nine's complement [COMPUT SCI] The radix-minus-1 complement of a numeral whose radix is 10. { ′nīnz ′käm·plə·mənt }

Nipkow disk [COMPUT SCI] In optical character recognition, a disk having one or more spirals of holes around the outer edge, with successive openings positioned so that rotation of the disk provides mechanical scanning, as of a document. { ′nip·kō ‚disk }

nit [COMMUN] A unit of information content such that the information content of a symbol or message in nits is the negative of the natural logarithm of the probability of selecting that symbol or message from all the symbols or messages which could have been chosen. Also known as nepit. { nit }

n-key rollover [COMPUT SCI] The ability of a computer-terminal keyboard to remember the order in which keys were operated and pass this information to the computer even when several keys are depressed before other keys have been released. { ′en ‚kē ′rōl‚ō·vər }

N-level address [COMPUT SCI] A multilevel address specifying N levels of addressing. { ′en ‚lev·əl ′ad‚res }

N-level logic [ELECTR] An arrangement of gates in a digital computer in which not more than N gates are connected in series. { ′en ‚lev·əl ′läj·ik }

N-modular redundancy [COMPUT SCI] A generalization of triple modular redundancy in which there are N identical units, where N is any odd number. { ′en ¦mäj·ə·lər ri′dən·dən·sē }

NMOS [ELECTR] Metal-oxide semiconductors that are made on *p*-type substrates, and whose active carriers are electrons that migrate between *n*-type source and drain contacts. Derived from *n*-channel metal-oxide semiconductor. { ′en ‚mòs }

NMRR *See* normal-mode rejection ratio.

nn junction [ELECTR] In a semiconductor, a region of transition between two regions having different properties in *n*-type semiconducting material. { ¦en¦en ‚jənk·shən }

no-address instruction [COMPUT SCI] An instruction which a computer can carry out without using an operand from storage. { ′nō ′ad‚res in ‚strək·shən }

no-break power [ELEC] Power system designed to fulfill load requirements during the interval between the failure of the primary power and the time the auxiliary power can be made available. { ′nō ′brāk ′pau̇·ər }

nodal analysis [ELEC] A method of electrical circuit analysis in which potential differences are taken as independent variables and the sum of the currents flowing into a node is equated to 0. { ′nōd·əl ə′nal·ə·səs }

nodal points [ELEC] Junction points in a transmission system; the automatic switches and switching centers are the nodal points in automated systems. { ′nōd·əl ‚pȯins }

node [ELEC] *See* branch point. [ELECTR] A junction point within a network. { nōd }

node voltage [ELEC] The voltage at a given point in an electric network with respect to that at a node. { ′nōd ‚vōl·tij }

noise [COMMUN] Unwanted electrical signal disturbances. [ELEC] Interfering and unwanted currents or voltages in an electrical device or system. { nȯiz }

noise analyzer |ELECTR| A device used for noise analysis. { 'nȯiz ,an·ə,līz·ər }

noise-canceling microphone *See* close-talking microphone. { 'nȯiz ¦kans·liŋ 'mī·krə,fōn }

noise digit |COMPUT SCI| A digit, usually 0, inserted into the rightmost position of the mantissa of a floating point number during a left-shift operation associated with normalization. Also known as noisy digit. { 'nȯiz ,dij·ət }

noise distortion |COMMUN| Noise on a communications facility which exceeds standards governing acceptable levels and which negatively affects the signal. { 'nȯiz di,stȯr·shən }

noise factor |ELECTR| The ratio of the total noise power per unit bandwidth at the output of a system to the portion of the noise power that is due to the input termination, at the standard noise temperature of 290 K. Also known as noise figure. { 'nȯiz ,fak·tər }

noise figure *See* noise factor. { 'nȯiz ,fig·yər }

noise filter |ELECTR| **1.** A filter that is inserted in an alternating-current power line to block noise interference that would otherwise travel through the line in either direction and affect the operation of receivers. **2.** A filter used in a radio receiver to reduce noise, usually an auxiliary low-pass filter which can be switched in or out of the audio system. { 'nȯiz ,fil·tər }

noise generator |ELECTR| A device which produces (usually random) electrical noise, for use in tests of the response of electrical systems to noise, and in measurements of noise intensity. Also known as noise source. { 'nȯiz ,jen·ə,rād·ər }

noise grade |COMMUN| Number which defines the relative noise at a particular location with respect to other locations throughout the world. { 'nȯiz ,grād }

noise improvement factor |COMMUN| In pulse modulation, the receiver output signal-to-noise ratio divided by the receiver input signal-to-noise ratio. Abbreviated NIF. Also known as improvement factor; signal-to-noise improvement factor. { 'nȯiz im'prüv·mənt ,fak·tər }

noise jammer |ELECTR| An electronic jammer that emits a carrier modulated with recordings or synthetic reproductions of natural atmospheric noise; the radio-frequency carrier may be suppressed; used to discourage the enemy by simulating naturally adverse communications conditions. { 'nȯiz ,jam·ər }

noise jamming |ELECTR| The emission of a radio-frequency carrier modulated with a white noise signal, derived from a gas-discharge tube or other broadband noise source, appearing in an enemy radar as background noise, tending to mask the radar echo or, in communications, the radio signal of interest. { 'nȯiz ,jam·iŋ }

noise killer |ELEC| **1.** Device installed in a circuit to reduce its interference to other circuits. **2.** *See* noise suicide circuit. { 'nȯiz ,kil·ər }

noiseless channel |COMMUN| In information theory, a communications channel in which the effects of random influences are negligible, and there is essentially no random error. { 'nȯiz·ləs 'chan·əl }

noise limiter |ELECTR| A limiter circuit that cuts off all noise peaks that are stronger than the highest peak in the desired signal being received, thereby reducing the effects of atmospheric or human-produced interference. Also known as noise silencer; noise suppressor. { 'nȯiz ,lim·əd·ər }

noise measurement |ELECTR| Any of a wide range of measurements of random and nonrandom electrical noise, but usually noise-power measurement. { 'nȯiz ,mezh·ər·mənt }

noise-metallic |ELECTR| In telephone communications, weighted noise current in a metallic circuit at a given point when the circuit is terminated at that point in the nominal characteristic impedance of the circuit. { 'nȯiz mə'tal·ik }

noise-modulated jamming |ELECTR| Random electronic noise that appears at the radar receiver as background noise and tends to mask the desired radar echo or radio signal. { 'nȯiz ¦mäj·ə¦lād·əd 'jam·iŋ }

noise-power measurement |ELECTR| Measurement of the power carried by electrical noise averaged over some brief interval of time, usually by amplifying noise from the source in a linear amplifier and then using a quadratic detector followed by a low-pass filter and an indicating device. { 'nȯiz ¦paù·ər ,mezh·ər·mənt }

noise-reducing antenna system |ELECTROMAG| Receiving antenna system so designed that only the antenna proper can pick up signals; it is placed high enough to be out of the noise-interference zone, and is connected to the receiver with a shielded cable or twisted transmission line that is incapable of picking up signals. { 'nȯiz ri¦düs·iŋ an'ten·ə ,sis·təm }

noise silencer *See* noise limiter. { 'nȯiz ,sī·lən·sər }

noise source *See* noise generator. { 'nȯiz ,sȯrs }

noise suicide circuit |ELECTR| A circuit which reduces the gain of an amplifier for a short period whenever a sufficiently large noise pulse is received. Also known as noise killer. { 'nȯiz 'sü·ə,sīd ,sər·kət }

noise suppression |ELECTR| Any method of reducing or eliminating the effects of undesirable electrical disturbances, as in frequency modulation whenever the signal carrier level is greater than the noise level. { 'nȯiz sə,presh·ən }

noise suppressor **1.** A circuit that blocks the audio-frequency amplifier of a radio receiver automatically when no carrier is being received, to eliminate background noise. Also known as squelch circuit. **2.** A circuit that reduces record surface noise when playing phonograph records, generally by means of a filter that blocks out the higher frequencies where such noise predominates. **3.** *See* noise limiter. { 'nȯiz sə,pres·ər }

noise temperature |ELEC| The temperature at which the thermal noise power of a passive system per unit bandwidth would be equal to the actual noise at the actual terminals; the standard reference temperature for noise measurements is 290 K. { 'nȯiz ,tem·prə·chər }

noise testing [ELECTR] The measurement of the power dissipated in a resistance termination of given value joined to one end of a telephone or telegraph circuit when no test power is applied to the circuit. { 'nȯiz ˌtest·iŋ }

noise tube [ELECTR] A gas tube used as a source of white noise. { 'nȯiz ˌtüb }

noise weighting [ELECTR] Use of an electrical network to obtain a weighted average over frequency of the noise power, which is representative of the relative disturbing effects of noise in a communications system at various frequencies. { 'nȯiz ˌwād·iŋ }

noisy channel [COMMUN] In information theory, a communications channel in which the effects of random influences cannot be dismissed. { 'nȯiz·ē 'chan·əl }

noisy digit See noise digit. { 'nȯiz·ē 'dij·ət }

noisy mode [COMPUT SCI] A floating-point arithmetic procedure associated with normalization in which "1" bits, rather than "0" bits, are introduced in the low-order bit position during the left shift. { 'nȯiz·ē ˌmōd }

no-load current [ELEC] The current which flows in a network when the output is open-circuited. { 'nō ˈlōd 'kə·rənt }

no-load loss [ELEC] The power loss of a device that is operated at rated voltage and frequency but is not supplying power to a load. { 'nō ˈlōd 'lȯs }

no-load voltage See open-circuit voltage. { 'nō ˈlōd 'vōl·tij }

nominal band [COMMUN] Frequency band of a facsimile-signal wave equal in width to that between zero frequency and maximum modulating frequency; the frequency band occupied in the transmitting medium will, in general, be greater than the nominal band. { 'näm·ə·nəl ˌband }

nominal bandwidth [COMMUN] The interval between the assigned frequency limits of a channel. [ENG] The difference between the nominal upper and lower cutoff frequencies of an acoustic or electric filter. { 'näm·ə·nəl 'band,width }

nominal impedance [ELEC] Impedance of a circuit under conditions at which it was designed to operate; normally specified at center of operating frequency range. { 'näm·ə·nəl im'pēd·əns }

nominal value [ELEC] The value of some property (such as resistance, capacitance, or impedance) of a device at which it is supposed to operate, under normal conditions, as opposed to actual value. { 'näm·ə·nəl 'val·yü }

nonacoustic coupler [ELECTR] A type of modem that is built into a microcomputer or terminal and connects it directly to a telephone line. { ˌnän·ə'kü·stik 'kəp·lər }

nonambiguity [COMMUN] The property of a code in which any character can be recognized uniquely without reference to preceding characters or the spatial position of a character. { ˌnän ˌam·bə'gyü·əd·ē }

nonanticipatory system See causal system. { ˌnän·an'tis·ə·pə,tȯr·ē ˌsis·təm }

nonarithmetic shift See cyclic shift. { ˌnän,a ˌrith'med·ik 'shift }

nonblocking access [COMMUN] Connection of the incoming line or trunk made within the switching center at all times, provided that the required outgoing line or trunk is not busy. { 'nän,bläk·iŋ 'ak,ses }

noncoherent integration [ELECTR] A radar signal processing technique in which the amplitudes of successive pulses from a single scene or target location are added for increased sensitivity; such integration in the excited phosphor of a cathode-ray-tube radar display representing one point in space is an elementary example of this process. { ¦nän·kō'hir·ənt 'in·tə¦grā·shən }

noncoincident demand [ELEC] The sum of the peak demands of all the utilities in a specified region, regardless of the times at which they occurred. { ¦nän·kō'in·sə·dənt di'mand }

noncomposite color picture signal [COMMUN] The signal in analog color television transmission that represents complete color picture information but excludes the line- and field-synchronizing signals. { ¦nän·kəm'päz·ət ¦kəl·ər 'pik·chər ,sig·nəl }

nondegenerate amplifier [ELECTR] Parametric amplifier that is characterized by a pumping frequency considerably higher than twice the signal frequency; the output is taken at the signal input frequency; the amplifier exhibits negative impedance characteristics, indicative of infinite gain, and is therefore capable of oscillation. { ¦nän·di'jen·ə·rət 'am·plə,fī·ər }

nondegenerative basic feasible solution [COMPUT SCI] In linear programming, a basic feasible solution with exactly m positive variables x_i, where m is the number of constraint equations. { ¦nän·di'jen·rəd·iv 'bā·sik 'fēz·ə·bəl sə'lü·shən }

nondeletable message [COMPUT SCI] A message that appears on a computer display which can be removed only by entering a specific command. { ˌnän·di'lēd·ə·bəl 'mes·ij }

nondestructive breakdown [ELECTR] Breakdown of the barrier between the gate and channel of a field-effect transistor without causing failure of the device; in a junction field-effect transistor, avalanche breakdown occurs at the pn junction. { ¦nän·di'strək·div 'brāk,daún }

nondestructive read [COMPUT SCI] A reading process that does not erase the data in memory; the term sometimes includes a destructive read immediately followed by a restorative write-back. Also known as nondestructive readout (NDRO). { ¦nän·di'strək·div 'rēd }

nondirectional See omnidirectional. { ¦nän·di 'rek·shən·əl }

nondirectional antenna See omnidirectional antenna. { ¦nän·di'rek·shən·əl an'ten·ə }

nonerasable storage See read-only memory. { ¦nän·i'rās·ə·bəl 'stȯr·ij }

nonexecutable statement [COMPUT SCI] A statement in a higher-level programming language which cannot be related to the instructions in the machine language program ultimately produced, but which provides the compiler with essential information from which it may determine the allocation of storage and other organizational characteristics of the final program. (¦nän ‚ek·sə'kyüd·ə·bəl 'stāt·mənt }

nonfatal error [COMPUT SCI] An error in a computer program which does not result in termination of execution, but which causes the processor to invent an interpretation, issue a warning, and continue processing. ('nän‚fād·əl 'er·ər }

nonfunctional packages software [COMPUT SCI] General-purpose software which permits the user to handle her or his particular applications requirements with little or no additional program or systems design work, or to perform certain specialized computational functions. (nän'fəŋk·shən·əl 'pak·ij·əz 'sóft‚wer }

nongraphic character [COMPUT SCI] A set of signals that, when sent to a printer, results in a control action, such as carriage return, line feed, or tab, rather than the generation of a printed character. (¦nän'graf·ik 'kar·ik·tər }

nonhoming tuning system [ELECTR] Motor-driven automatic tuning system in which the motor starts up in the direction of previous rotation; if this direction is incorrect for the new station, the motor reverses, after turning to the end of the dial, then proceeds to the desired station. (¦nän'hōm·iŋ 'tün·iŋ ‚sis·təm }

noninductive [ELEC] Having negligible or zero inductance. (‚nän·in'dək·tiv }

noninductive capacitor [ELEC] A capacitor constructed so it has practically no inductance; foil layers are staggered during winding, so an entire layer of foil projects at either end for contact-making purposes; all currents then flow laterally rather than spirally around the capacitor. (‚nän·in'dək·tiv kə'pas·əd·ər }

noninductive resistor [ELEC] A wire-wound resistor constructed to have practically no inductance, either by using a hairpin winding or be reversing connections to adjacent sections of the winding. (‚nän·in'dək·tiv ri'zis·tər }

noninductive winding [ELEC] A winding constructed so that the magnetic field of one turn or section cancels the field of the next adjacent turn or section. (‚nän·in'dək·tiv 'wīn·diŋ }

nonintelligible crosstalk [COMMUN] Crosstalk which cannot be understood regardless of its received volume, but which because of its syllabic nature is more annoying subjectively than thermal-type noise. (‚nän·in'tel·ə·jə·bəl 'krós‚tók }

noninteracting control [CONT SYS] A feedback control in a system with more than one input and more than one output, in which feedback transfer functions are selected so that each input influences only one output. (¦nän‚in·tər'ak·tiŋ kən'trōl }

noninverting amplifier [ELECTR] An operational amplifier in which the input signal is applied to the ungrounded positive input terminal to give a gain greater than unity and make the output voltage change in phase with the input voltage. (¦nän·in'vərd·iŋ 'am·plə‚fī·ər }

noninverting parametric device [ELECTR] Parametric device whose operation depends essentially upon three frequencies, a harmonic of the pump frequency and two signal frequencies, of which one is the sum of the other plus the pump harmonic. (¦nän·in'vərd·iŋ ‚par·ə¦me·trik di'vīs }

nonlinear amplifier [ELECTR] An amplifier in which a change in input does not produce a proportional change in output. ('nän‚lin·ē·ər 'am·plə‚fī·ər }

nonlinear capacitor [ELEC] Capacitor having a mean charge characteristic or a peak charge characteristic that is not linear, or a reversible capacitance that varies with bias voltage. ('nän ‚lin·ē·ər kə'pas·əd·ər }

nonlinear circuit [ELEC] A circuit in which the current and voltage in any element that results from two sources of energy acting together is not equal to the sum of the currents or voltages that result from each of the sources acting alone. ('nän‚lin·ē·ər 'sər·kət }

nonlinear circuit component [ELECTR] An electrical device for which a change in applied voltage does not produce a proportional change in current. Also known as nonlinear device; nonlinear element. ('nän‚lin·ē·ər¦sər·kət kəm ('pō·nənt }

nonlinear control system [CONT SYS] A control system that does not have the property of superposition, that is, one in which some or all of the outputs are not linear functions of the inputs. ('nän‚lin·ē·ər kən'trōl ‚sis·təmz }

nonlinear coupler [ELECTR] A type of frequency multiplier which uses the nonlinear capacitance of a junction diode to couple energy from the input circuit, which is tuned to the fundamental, to the output circuit, which is tuned to the desired harmonic. ('nän‚lin·ē·ər 'kəp·lər }

nonlinear crosstalk [COMMUN] Interaction between channels occupying different wavelengths in a wavelength-division-multiplexed system because of optical nonlinearities in the transmission medium. (‚nän‚lin·ē·ər 'krós‚tók }

nonlinear detection [ELECTR] Detection based on the curvature of a tube characteristic, such as square-law detection. ('nän‚lin·ē·ər di'tek·shən }

nonlinear device See nonlinear circuit component. ('nän‚lin·ē·ər di'vīs }

nonlinear dielectric [ELEC] A dielectric whose polarization is not proportional to the applied electric field. ('nän‚lin·ē·ər ‚dī·ə'lek·trik }

nonlinear distortion [ELECTR] Distortion in which the output of a system or component does not have the desired linear relation to the input. [ENG ACOUS] The ratio of the total root-mean-square (rms) harmonic distortion output of a microphone to the rms value of the fundamental component of the output. ('nän‚lin·ē·ər di'stór·shən }

nonlinear element See nonlinear circuit component. ('nän‚lin·ē·ər 'el·ə·mənt }

nonlinear feedback control system [CONT SYS] Feedback control system in which the relationships between the pertinent measures of the system input and output signals cannot be adequately described by linear means. ('nän ‚lin·ē·ər 'fēd‚bak kən'trōl ‚sis·təm }

nonlinear fiber amplifier [COMMUN] An optical amplifier in which nonlinear interactions (stimulated Raman and Brillouin scattering and four-wave mixing) between pump light and the signal cause transfer of power to the signal, resulting in fiber gain. { ˌnänˈlin·ē·ər ˌfī·bər ˈam·plə͵fī·ər }

nonlinear inductance [ELEC] The behavior of an inductor for which the voltage drop across the inductor is not proportional to the rate of change of current, such as when the inductor has a core of magnetic material in which magnetic induction is not proportional to magnetic field strength. { ˈnänˌlin·ē·ər inˈdək·təns }

nonlinear network [ELEC] A network in which the current or voltage in any element that results from two sources of energy acting together is not equal to the sum of the currents or voltages that result from each of the sources acting alone. { ˈnänˌlin·ē·ər ˈnet͵wərk }

nonlinear oscillator [ELECTR] A radio-frequency oscillator that changes frequency in response to an audio signal; it is the basic circuit used in eavesdropping devices. { ˈnänˌlin·ē·ər ˈäs·ə͵lād·ər }

nonlinear programming [MATH] A branch of applied mathematics concerned with finding the maximum or minimum of a function of several variables, when the variables are constrained to yield values of other functions lying in a certain range, and either the function to be maximized or minimized, or at least one of the functions whose value is constrained, is nonlinear. { ˈnän ˌlin·ē·ər ˈprō͵gram·iŋ }

nonlinear reactance [ELECTR] The behavior of a coil or capacitor whose voltage drop is not proportional to the rate of change of current through the coil, or the charge on the capacitor. { ˈnänˌlin·ē·ər rēˈak·təns }

nonlinear resistance [ELECTR] The behavior of a substance (usually a semiconductor) which does not obey Ohm's law but has a voltage drop across it that is proportional to some power of the current. { ˈnänˌlin·ē·ər riˈzis·təns }

nonlinear taper [ELEC] Nonuniform distribution of resistance throughout the element of a potentiometer or rheostat. { ˈnänˌlin·ē·ər ˈtā·pər }

nonloaded Q [ELEC] Of an electric impedance, the Q value of the impedance without external coupling or connection. Also known as basic Q. { ˈnänˌlōd·əd ˈkyü }

nonmaintenance time [COMPUT SCI] The elapsed time during scheduled working hours between the determination of a machine failure and placement of the equipment back into operation. { ¦nänˈmānt·ən·əns ͵tīm }

nonmetallic sheathed cable [ELEC] Assembly of two or more rubber-covered conductors in an outer sheath of nonconducting fibrous material that has been treated to make it flame-resistant and moisture-repellent. { ¦nän·məˈtal·ik ˈshēt͟hd ˈkā·bəl }

non-minimum-phase system [CONT SYS] A linear system whose transfer function has one or more poles or zeros with positive, nonzero real parts. { ¦nän¦min·ə·məm ˈfāz ͵sis·təm }

nonmultiple switchboard [ELEC] Manual telephone switchboard in which each subscriber line is attached to only one jack. { ¦nänˈməl·tə·pəl ˈswich͵bȯrd }

nonnumeric character [COMPUT SCI] Any character except a digit. { ¦nän·nüˈmer·ik ˈkar·ik·tər }

nonnumeric programming [COMPUT SCI] Computer programming that deals with objects other than numbers. { ¦nän·nü͵mer·ik ˈprō͵gram·iŋ }

non-ohmic [ELEC] Pertaining to a substance or circuit component that does not obey Ohm's law. { ¦nänˈō·mik }

nonpolarized relay See neutral relay. { ¦nänˈpō·lə ͵rīzd ˈrē͵lā }

nonpreemptive multitasking See cooperative multitasking. { ͵nän·prē¦em·tiv ˈməl·tē͵task·iŋ }

nonprint code [COMPUT SCI] A bit combination which is interpreted as no printing, no spacing. { ¦nän¦print ˈkōd }

nonpriority interrupt [COMPUT SCI] Any one of a group of interrupts which may be disregarded by the central processing unit. { ¦nän·priˈär·əd·ē ˈint·ə͵rəpt }

nonprocedural language [COMPUT SCI] A programming language in which the program does not follow the actual steps a computer follows in executing a program. { ͵nän·prəˈsē·jə·rəl ˈlaŋ ͵gwij }

nonreactive [ELEC] Pertaining to a circuit, component, or load that has no capacitance or impedance, so that an alternating current is in phase with the corresponding voltage. { ͵nän·rēˈak·tiv }

nonreactive load See resistive load. { ͵nän·rēˈak·tiv ˈlōd }

nonrecoverable error [COMPUT SCI] An error detected during computer processing that cannot be handled by the computer system and therefore causes processing to be interrupted. { ¦nän·riˈkəv·rə·bəl ˈer·ər }

nonrecursive filter [ELECTR] A digital filter that lacks feedback; that is, its output depends on present and past input values only and not on previous output values. { ͵nän·ri͵kər·siv ˈfil·tər }

nonredundant system [COMPUT SCI] A computer system designed in such a way that only the absolute minimum amount of hardware is utilized to implement its function. { ¦nän·riˈdən·dənt ˈsis·təm }

nonrenewable fuse unit [ELEC] Fuse unit that cannot be readily restored for service after operation. { ¦nän·riˈnü·ə·bəl ˈfyüz ͵yü·nət }

nonreproducing code [COMPUT SCI] A code which normally does not appear as such in a generated output but will result in a function such as paging or spacing. { ¦nän͵rē·prəˈdü·siŋ ˈkōd }

nonresident routine [COMPUT SCI] Any computer routine which is not stored permanently in the memory but must be read into memory from a data carrier or external storage device. { ¦nänˈrez·ə·dənt rü͵tēn }

nonresonant antenna [ELECTROMAG] A long-wire or traveling-wave antenna which does not have natural frequencies of oscillation, and responds equally well to radiation over a broad range of frequencies. { ¦nän'rez·ən·ənt an'ten·ə }

non-return-to-zero |COMPUT SCI| A mode of recording and readout in which it is not necessary for the signal to return to zero after each item of recorded data. Abbreviated NRZ. { 'nän ri,tərn tə 'zir·ō }

nonrotating disk See semiconductor disk. { ¦nän'rō,tād·iŋ 'disk }

nonscrollable message |COMPUT SCI| A message on a computer display that does not scroll off the top of the display as new information is written at the bottom. { ¦nän'skrō·lə·bəl 'mes·ij }

nonservo robot See fixed-stop robot. { ¦nän'sər·vō 'rō,bät }

nonshared control unit |COMPUT SCI| A control unit relating to only one device. Also known as unipath. { ¦nän¦sherd kən'trōl ,yü·nət }

nonshorting contact switch |ELEC| Selector switch in which the width of the movable contact is less than the distance between contact clips, so that the old circuit is broken before the new circuit is completed. { 'nän,shörd·iŋ 'kän,takt ,swich }

nonsinusoidal waveform |ELEC| The representation of a wave which does not vary in a sinusoidal manner, and which therefore contains harmonics. { ¦nän,sī·nə'sȯid·əl 'wāv,förm }

nonstop computer |COMPUT SCI| A computer system that is equipped with duplicate components or excess capacity so that a hardware or software failure will not interrupt processing. { ¦nän'stäp kəm'pyüd·ər }

nonstop switch |ELEC| A manual switch in an elevator car that can prevent the car from stopping at a specified floor. { ¦nän,stäp 'swich }

nonstorage camera tube |ELECTR| Television camera tube in which the picture signal is, at each instant, proportional to the intensity of the illumination on the corresponding area of the scene. { ¦nän'stȯr·ij 'kam·rə ,tüb }

nonswappable program |COMPUT SCI| A program that is given priority status so that its execution cannot be suspended to allow execution of other programs. { ¦nän'swäp·ə·bəl 'prō·gram }

nonsynchronous |ELEC| Not related in phase, frequency, or speed to other quantities in a device or circuit. { ¦nän'siŋ·krə·nəs }

nonsynchronous timer |ELECTR| A circuit at the receiving end of a communications link which restores the time relationship between pulses when no timing pulses are transmitted. { ¦nän'siŋ·krə·nəs 'tīm·ər }

nonsynchronous transmission |ELECTR| A data transmission process in which a clock is not used to control the unit intervals within a block or a group of data signals. { ¦nän'siŋ·krə·nəs tranz'mish·ən }

nonsynchronous vibrator |ELECTR| Vibrator that interrupts a direct-current circuit at a frequency unrelated to the other circuit constants and does not rectify the resulting stepped-up alternating voltage. { ¦nän'siŋ·krə·nəs 'vī,brād·ər }

nonuniform memory access machine |COMPUT SCI| A multiprocessor in which the memory is spread out over memory modules, which are attached to the processors, so that each processor has its own memory module. Abbreviated NUMA machine. { ,nän¦yün·i,förm ¦mem·rē 'ak·ses mə,shēn }

nonvolatile memory See nonvolatile storage. { ¦nän'väl·ə·təl 'mem·rē }

nonvolatile random-access memory |COMPUT SCI| A semiconductor storage device which has two memory cells for each bit, one of which is volatile, as in a static RAM (random-access memory), and provides unlimited read and write operations, while the other is nonvolatile, and provides the ability to retain information when power is removed. Abbreviated NV RAM. { ¦nän'väl·ə·təl ¦ran·dəm ¦ak,ses 'mem·rē }

nonvolatile storage |COMPUT SCI| A computer storage medium that retains information in the absence of power, such as a magnetic tape, drum, or core. Also known as nonvolatile memory. { ¦nän'väl·ə·təl 'stȯr·ij }

NO OP |COMPUT SCI| An instruction telling the computer to do nothing, except to proceed to the next instruction in sequence. Also known as do-nothing instruction; no-operation instruction. { 'nō ,äp }

no-operation instruction See NO OP. { ¦nō ,äp·ə'rā·shən in'strək·shən }

NOR circuit |ELECTR| A circuit in which output voltage appears only when signal is absent from all of its input terminals. { 'nȯr ,sər·kət }

normal direction flow |COMPUT SCI| The direction from left to right or top to bottom in flow charting. { 'nȯr·məl di'rek·shən ,flō }

normal electrode |ELEC| Standard electrode used for measuring electrode potentials. { 'nȯr·məl i'lek,trōd }

normal form |COMPUT SCI| The form of a floating-point number whose mantissa lies between 0.1 and 1.0. { 'nȯr·məl 'förm }

normal impedance See free impedance. { 'nȯr·məl im'pēd·əns }

normalization |COMPUT SCI| Breaking down of complex data structures into flat files. { ,nȯr·mə·lə'zā·shən }

normalize |COMPUT SCI| **1.** To adjust the representation of a quantity so that this representation lies within a prescribed range. **2.** In particular, to adjust the exponent and mantissa of a floating point number so that the mantissa falls within a prescribed range. { 'nȯr·mə,līz }

normalized Q |ELEC| The ratio of the reactive component of the impedance of a filter section to the resistive component. { 'nȯr·mə,līzd 'kyü }

normal mode |COMPUT SCI| Operation of a computer in which it executes its own instructions rather than those of a different computer. { 'nȯr·məl ,mōd }

normal-mode helix |ELECTROMAG| A type of helical antenna whose diameter and electrical length are considerably less than a wavelength, and which has a radiation pattern with greatest intensity normal to the helix axis. { 'nȯr·məl ,mōd 'hē,liks }

normal-mode rejection ratio |ELECTR| The ability of an amplifier to reject spurious signals at the power-line frequency or at harmonics of the line frequency. Abbreviated NMRR. { 'nȯr·məl ¦mōd ri'jek·shən ˌrā·shō }

normal orientation [COMPUT SCI] In optical character recognition, that determinate position which indicates that the line elements of an inputted source document appear parallel with the document's leading edge. { 'nȯr·məl ˌȯr·ē·ən'tā·shən }

normal range [COMPUT SCI] An interval within which results are expected to fall during normal operations. { 'nȯr·məl 'rānj }

North American presentation-level protocol syntax [COMMUN] A format for transmitting text and graphics that allows the transmission of large amounts of information over narrow-bandwidth transmission lines. Abbreviated NAPLPS. { ¦nȯrth ə¦mer·ə·kən ˌprē·zən ¦tā·shən ˌlev·əl ˌprōd·ə·kȯl 'sin,taks }

Norton equivalent circuit [ELEC] An equivalent circuit that consists of a parallel connection of a current source and a two-terminal circuit, where the current source is usually dependent on the electric signals applied to the input terminals. { ¦nȯrt·ən i'kwiv·ə·lənt ˌsər·kət }

Norton's theorem [ELEC] The theorem that the voltage across an element that is connected to two terminals of a linear network is equal to the short-circuit current between these terminals in the absence of the element, divided by the sum of the admittances between the terminals associated with the element and the network respectively. { 'nȯrt·ənz ˌthir·əm }

NOS *See* network operating system. { ¦en¦ō'es *or* 'näs }

notation *See* positional notation. { nō'tā·shən }

notch [ELECTR] Rectangular depression extending below the sweep line of the radar indicator in some types of equipment. { näch }

notch antenna [ELECTROMAG] Microwave antenna in which the radiation pattern is determined by the size and shape of a notch or slot in a radiating surface. { 'näch an,ten·ə }

notch filter [ELECTR] A band-rejection filter that produces a sharp notch in the frequency response curve of a system; used in television transmitters to provide attenuation at the low-frequency end of the channel, to prevent possible interference with the sound carrier of the next lower channel. { 'näch ,fil·tər }

notching [ELEC] Term indicating that a predetermined number of separate impulses are required to complete operation of a relay. { 'näch·iŋ }

NOT circuit [ELECTR] A logic circuit with one input and one output that inverts the input signal at the output; that is, the output signal is a logical 1 if the input signal is a logical 0, and vice versa. Also known as inverter circuit. { 'nät ,sər·kət }

notebook computer [COMPUT SCI] A portable computer typically weighing less than 6 pounds (3 kilograms) that has a flat-panel display and miniature hard disk drives, and is powered by rechargeable batteries. Also known as laptop computer. { 'nōt,buk kəm,pyüd·ər }

nought state *See* zero condition. { 'nȯt ,stāt }

novar [ELECTR] Beam-power tube having a nine-pin base. { 'nō,vär }

NP [COMPUT SCI] The class of decision problems for which solutions can be checked in polynomial time.

NP-complete problem [COMPUT SCI] One of the hardest problems in class NP, such that, if there are any problems in class NP but not in class P, this is one of them. { ¦en¦pē kəm'plēt ˌpräb·ləm }

NP-hard [COMPUT SCI] Referring to problems at least as hard as or harder than any problem in NP. Given a method for solving an NP-hard problem, any problem in NP can be solved with only polynomially more work. { ¦en¦pē 'härd }

npin transistor [ELECTR] An *npn* transistor which has a layer of high-purity germanium between the base and collector to extend the frequency range. { 'en,pin tran'zis·tər }

N-plus-one address instruction [COMPUT SCI] An instruction with N + 1 address parts, one of which gives the location of the next instruction to be carried out. { 'en pləs 'wən 'ad,res in ,strək·shən }

npnp diode *See* pnpn diode. { ¦en,pē¦en,pē 'dī ,ōd }

npnp transistor [ELECTR] An *npn*-junction transistor having a transition or floating layer between *p* and *n* regions, to which no ohmic connection is made. Also known as *pnpn* transistor. { ¦en,pē¦en,pē tran'zis·tər }

npn semiconductor [ELECTR] Double junction formed by sandwiching a thin layer of *p*-type material between two layers of *n*-type material of a semiconductor. { 'en,pē'en 'sem·i·kən,dək·tər }

npn transistor [ELECTR] A junction transistor having a *p*-type base between an *n*-type emitter and an *n*-type collector; the emitter should then be negative with respect to the base, and the collector should be positive with respect to the base. { 'en,pē'en tran'zis·tər }

NPO-body [ELEC] Referring to a series of temperature-compensating capacitors that have an invariant dielectric constant over a specified temperature range. { ¦en¦pē¦ō 'bäd·ē }

np semiconductor [ELECTR] Region of transition between *n*- and *p*-type material. { ¦en¦pē 'sem·i·kən,dək·tər }

NPX *See* numeric processor extension.

NRSC *See* National Radio Systems Committee.

NRZ *See* non-return-to-zero.

N-scan *See* N-display. { 'en ,skan }

N-scope *See* N-display. { 'en ,skōp }

nt *See* nit.

NTSC *See* National Television System Committee.

n-type conduction [ELECTR] The electrical conduction associated with electrons, as opposed to holes, in a semiconductor. { 'en ,tīp kən ,dək·shən }

N-type crystal rectifier [ELECTR] Crystal rectifier in which forward current flows when the semiconductor is negative with respect to the metal.   { 'en ˌtīp ¦krist·əl 'rek·tə₊fī·ər }

n-type germanium |ELECTR| Germanium to which more impurity atoms of donor type (with valence 5, such as antimony) than of acceptor type (with valence 3, such as indium) have been added, with the result that the conduction electron density exceeds the hole density.   { 'en ˌtīp jər'mā·nē·əm }

n-type semiconductor |ELECTR| An extrinsic semiconductor in which the conduction electron density exceeds the hole density.   { 'en ˌtīp 'sem·i·kən₊dək·tər }

nuclear electric power generation |ELEC| Large-scale generation of electric power in which the source of energy is nuclear fission, generally in a nuclear reactor, or nuclear fusion.   { 'nü·klē·ər i ¦lek·trik 'paú·ər ˌjen·ə₊rā·shən }

nuclear triode detector |ELECTR| A type of junction detector that has two outputs which together determine the precise location on the detector where the ionizing radiation was incident, as well as the energy of the ionizing particle.   { 'nü·klē·ər ¦trī₊ōd di'tek·tər }

nucleus |COMPUT SCI| **1.** That portion of the control program that must always be present in main storage. **2.** The main storage area used in the nucleus (first definition) and other transient control program routines.   { 'nü·klē·əs }

null character |COMPUT SCI| A control character used as a filler in data processing; may be inserted or removed from a sequence of characters without affecting the meaning of the sequence, but may affect format or control equipment.   { 'nəl ˌkar·ik·tər }

null-current circuit |ELECTR| A circuit used to measure current, in which the unknown current is opposed by a current resulting from applying a voltage controlled by a slide wire across a series resistor, and the slide wire is continuously adjusted so that the resulting current, as measured by a direct-current detector amplifier, is equal to zero.   { 'nəl ¦kə·rənt 'sər·kət }

null-current measurement |ELECTR| Measurement of current using a null-current circuit.   { 'nəl ¦kə·rənt 'mezh·ər·mənt }

null detection |ELEC| Altering of adjustable bridge circuit components, to obtain zero current.   { 'nəl di₊tek·shən }

null detector See null indicator.   { 'nəl di₊tek·tər }

null indicator |ENG| A galvanometer or other device that indicates when voltage or current is zero; used chiefly to determine when a bridge circuit is in balance. Also known as null detector.   { 'nəl ˌin·də₊kād·ər }

null modem cable |COMPUT SCI| A cable that connects two local computers via serial ports without the use of a modem.   { ¦nəl 'mō₊dem ˌkā·bəl }

NUMA machine See nonuniform memory access machine.   { 'nü·mə mə'shēn }

number cruncher |COMPUT SCI| A computer with great power to carry out computations, designed to maximize this ability rather than to process large amounts of data.   { 'nəm·bər ˌkrən·chər }

number record printer |COMMUN| A printer in a relay station that provides a complete automatic written record of channel numbers and the fixed routing line associated with each message that is relayed through that particular station.   { 'nəm·bər ¦rek·ərd ˌprint·ər }

numeric |COMPUT SCI| In computers, pertaining to data composed wholly or partly of digits, as distinct from alphabetic.   { nü'mer·ik }

numerical decrement See decrement.   { nü'mer·i·kəl 'dek·rə·mənt }

numerical display device |ELECTR| Any device for visually displaying numerical figures, such as a numerical indicator tube, a device utilizing electroluminescence, or a device in which any one of a stack of transparent plastic strips engraved with digits can be illuminated by a small light at the edge of the strip.   { nü'mer·i·kəl di'splā di₊vīs }

numerical indicator tube |ELECTR| An electron tube capable of visually displaying numerical figures; some varieties also display alphabetical characters and commonly used symbols.   { nü'mer·i·kəl 'in·də₊kād·ər ˌtüb }

numerical tape |COMPUT SCI| The tape required by a computer operating a machine tool.   { nü'mer·i·kəl 'tāp }

numeric character See digit.   { nü'mer·ik 'kar·ik·tər }

numeric character set |COMPUT SCI| A character set that includes only digits and certain special characters, such as plus and minus signs and control characters.   { nü'mer·ik 'kar·ik·tər ˌset }

numeric coding |COMPUT SCI| Code in which only digits are used, usually binary or octal.   { nü'mer·ik 'kōd·iŋ }

numeric control |COMPUT SCI| The action of programs written for specialized computers which operate machine tools.   { nü'mer·ik kən'trōl }

numeric coprocessor See numeric processor extension.   { nü¦mer·ik kō'prä₊ses·ər }

numeric data |COMPUT SCI| Data consisting of digits and not letters of the alphabet or special characters.   { nü'mer·ik 'dad·ə }

numeric format |COMPUT SCI| The manner in which numbers are displayed in the cells of a particular spreadsheet.   { nü¦mer·ik 'fȯr₊mat }

numeric keypad |COMPUT SCI| A section of a computer keyboard that contains a group of keys, usually about 12, arranged in compact fashion for entering numeric characters efficiently. Also known as numeric pad.   { nü'mer·ik 'kē₊pad }

numeric pad See numeric keypad.   { nü'mer·ik 'pad }

numeric pager |COMMUN| A receiver in a radio paging system that contains a display device that can show numeric messages, most commonly a telephone number.   { nü'mer·ik 'pā·jər }

numeric printer |COMPUT SCI| Old type of printer which positioned its keys to print a field in one operation, rather than one digit at a time.   { nü'mer·ik 'print·ər }

numeric processor extension |COMPUT SCI| A specialized integrated circuit that is added to a computer to perform high-speed floating-point mathematical calculations. Abbreviated NPX. Also known as arithmetic processor; math coprocessor; numeric coprocessor. { nü¦mer·ik 'prä‚ses·ər ik ‚sten·chən }

numeric variable |COMPUT SCI| The symbolic name of a data element whose value changes during the carrying out of a computer program. { nü'mer·ik 'ver·ē·ə·bəl }

nutating antenna |ENG| An antenna system used in conical scan radar, in which a dipole or feed horn moves in a small circular orbit about the axis of a paraboloidal reflector without changing its polarization. { 'nü‚tād·iŋ an'ten·ə }

nutator |ENG| A mechanical or electrical device used to move a radar beam in a circular, conical, spiral, or other manner periodically to obtain greater air surveillance than could be obtained with a stationary beam. { 'nü‚tād·ər }

nuvistor |ELECTR| Electron tube in which all electrodes are cylindrical, placed one inside the other with close spacing, in a ceramic envelope. { nü'vis·tər }

NV RAM See nonvolatile random-access memory. { ¦en¦vē 'ram }

nybble |COMPUT SCI| A string of bits, smaller than a byte, operated on as a unit. { 'nib·əl }

Nyquist contour |CONT SYS| A directed closed path in the complex frequency plane used in constructing a Nyquist diagram, which runs upward, parallel to the whole length of the imaginary axis at an infinitesimal distance to the right of it, and returns from $+j_\infty$ to $-j_\infty$ along a semicircle of infinite radius in the right half-plane. { 'nī‚kwist ‚kän‚túr }

Nyquist diagram |CONT SYS| A plot in the complex plane of the open-loop transfer function

as the complex frequency is varied along the Nyquist contour; used to determine stability of a control system. { 'nī‚kwist ‚dī·ə‚gram }

Nyquist interval |COMMUN| Maximum separation in time which can be given to regularly spaced instantaneous samples of a wave of specified bandwidth for complete determination of the waveform of the signal. { 'nī‚kwist ‚in·tər·vəl }

Nyquist rate |COMMUN| The maximum rate at which code elements can be unambiguously resolved in a communications channel with a limited range of frequencies; equal to twice the frequency range. { 'nī‚kwist ‚rāt }

Nyquist sampling |COMMUN| The periodic sampling of audio or video signals, in order to preserve their information content, at a rate equal to twice the highest frequency to be preserved. { 'nī‚kwist ‚sam·pliŋ }

Nyquist stability criterion See Nyquist stability theorem. { 'nī‚kwist stə'bil·əd·ē krī‚tir·ē·ən }

Nyquist stability theorem |CONT SYS| The theorem that the net number of counterclockwise rotations about the origin of the complex plane carried out by the value of an analytic function of a complex variable, as its argument is varied around the Nyquist contour, is equal to the number of poles of the variable in the right half-plane minus the number of zeros in the right half-plane. Also known as Nyquist stability criterion. { 'nī‚kwist stə'bil·əd·ē ‚thir·əm }

Nyquist's theorem |ELECTR| The mean square noise voltage across a resistance in thermal equilibrium is four times the product of the resistance, Boltzmann's constant, the absolute temperature, and the frequency range within which the voltage is measured. { 'nī‚kwists ‚thir·əm }

O

OASIS *See* Open-Access Same-Time Information System. { ,ō'ā·səs }

O attenuator |ELECTR| A dissipative attenuator in which the circuit has the form of a ladder with two rungs, and the resistances across the rungs are unequal, so that the impedances across the two pairs of terminals are unequal. { 'ō ə'ten·yə ,wād·ər }

OBA *See* octave-band analyzer.

object |COMPUT SCI| **1.** Any collection of related items. **2.** The name of a single element in an object-oriented programming language. { 'äb·jekt }

object code |COMPUT SCI| The statements generated from source code by a compiler, constituting an intermediate step in the translation of source code into executable machine language. { 'äb ,jekt ,kōd }

object computer |COMPUT SCI| The computer processing an object program; the same computer compiling the source program could, therefore, be called the source computer; such terminology is seldom used in practice. { 'äb·jekt kəm ,pyüd·ər }

object deck |COMPUT SCI| The set of machine-readable computer instructions produced by a compiler, either in absolute format (that is, containing only fixed addresses) or, more frequently, in relocatable format. { 'äb·jekt ,dek }

object language |COMPUT SCI| The intended and desired output language in the translation or conversion of information from one language to another. { 'äb·jekt ,laŋ·gwij }

object library *See* object program library. { 'äb·jekt'lī,brer·ē }

Object Management Group object model |COMPUT SCI| A model that defines common object semantics in an object-oriented computer system. Abbreviated OMG object model. { 'äb·jikt ¦man·ij·mənt ¦grüp 'äb·jikt ,mäd·əl }

object module |COMPUT SCI| The computer language program prepared by an assembler or a compiler after acting on a programmer-written source program. { 'äb·jekt ,mäj·ül }

object-oriented graphics *See* vector graphics. { 'äb,jekt ,ȯr·ē,en·təd 'graf·iks }

object-oriented interface |COMPUT SCI| A user interface that employs icons and a mouse. { 'äb ,jekt ,ȯr·ē,en·təd 'in·tər,fās }

object-oriented language |COMPUT SCI| A programming language consisting of a sequence of commands directed at objects. { 'äb,jekt ,ȯr·ē ,en·təd 'laŋ·gwij }

object-oriented programming |COMPUT SCI| A computer programming methodology that focuses on data rather than processes, with programs composed of self-sufficient modules (objects) containing all the information needed to manipulate a data structure. Abbreviated OOP. { 'äb,jekt ,ȯr·ē,en·təd 'prō,gram·iŋ }

object program |COMPUT SCI| The computer language program prepared by an assembler or a compiler after acting on a programmer-written source program. Also known as object routine; target program; target routine. { 'äb·jekt ,prō ,gram }

object program library |COMPUT SCI| A collection of computer programs in the form of relocatable instructions, which reside on, and may be read from, a mass storage device. Also known as object library. { 'äb·jekt ,prō,gram ,lī,brer·ē }

object request broker |COMPUT SCI| The central component of CORBA, which passes requests from clients to the objects on which they are invoked. Abbreviated ORB. { 'äb·jikt ri¦kwest ,brō·kər }

object routine *See* object program. { 'äb·jekt rü ,tēn }

object time |COMPUT SCI| The time during which execution of an object program is carried out. { 'äb·jekt ,tīm }

oblique-incidence transmission |COMMUN| Transmission of a radio wave obliquely up to the ionosphere and down again. { ə'blēk ¦in·sə·dəns tranz'mish·ən }

observability |CONT SYS| Property of a system for which observation of the output variables at all times is sufficient to determine the initial values of all the state variables. { əb,zər·və'bil·əd·ē }

observation spillover |CONT SYS| The part of the sensor output of an active control system caused by modes that have been omited from the control algorithm in the process of model reduction. { ,äb·zər'vā·shən 'spil,ō·vər }

observer |CONT SYS| A linear system B driven by the inputs and outputs of another linear system A which produces an output that converges to some linear function of the state of system A.

Also known as state estimator; state observer. { əb'zər·vər }

obsolescence [ENG] Decreasing value of functional and physical assets or value of a product or facility from technological changes rather than deterioration. { ,äb·sə'les·əns }

occlusion [COMPUT SCI] In computer vision, the obstruction of a view. { ə'klü·zhən }

occupied bandwidth [COMMUN] Frequency bandwidth such that, below its lower and above its upper frequency limits, the mean powers radiated are each equal to 0.5% of the total mean power radiated by a given emission. { 'äk·yə ,pīd 'band,width }

OCR *See* optical character recognition.

octal base [ELECTR] Tube base having a central aligning key and positioned for eight equally spaced pins. { 'äkt·əl ,bās }

octal debugger [COMPUT SCI] A simple debugging program which permits only octal (instead of symbolic) address references. { 'äkt·əl dē'bag·ər }

octave-band analyzer [ENG ACOUS] A portable sound analyzer which amplifies a microphone signal, feeds it into one of several band-pass filters selected by a switch, and indicates the magnitude of sound in the corresponding frequency band on a logarithmic scale; all the bands except the highest and lowest span an octave in frequency. Abbreviated OBA. { 'äk·tiv ¦band 'an·ə,līz·ər }

octave-band filter [ENG ACOUS] A band-pass filter in which the upper cutoff frequency is twice the lower cutoff frequency. { 'äk·tiv ¦band 'fil·tər }

octave-band oscillator [ELECTR] An oscillator that can be tuned over a frequency range of 2 to 1, so that its highest frequency is twice its lowest frequency. { 'äk·tiv ¦band 'äs·ə,lād·ər }

octode [ELECTR] An eight-electrode electron tube containing an anode, a cathode, a control electrode, and five additional electrodes that are ordinarily grids. { 'äk,tōd }

octonary signaling [COMMUN] A communications mode in which information is passed by the presence and absence of plus and minus variation of eight discrete levels of one parameter of the signaling medium. { 'äk·tə,ner·ē 'sig·nə·liŋ }

odd-even check [COMPUT SCI] A means of detecting certain kinds of errors in which an extra bit, carried along with each word, is set to zero or one so that the total number of zeros or ones in each word is always made even or always made odd. Also known as parity check. { 'äd 'ē·vən ,chek }

odd parity [COMPUT SCI] Property of an expression in binary code which has an odd number of ones. { 'äd 'par·əd·ē }

odd parity check [COMPUT SCI] A parity check in which the number of 0's or 1's in each word is expected to be odd; if the number is even, the check bit is 1, and if the number is odd, the check bit is 0. { 'äd 'par·əd·ē ,chek }

O-display [ELECTR] A radar display format in which an adjustable notch, absenting any trace, is moved in an A-display to assist the operator in determining and reporting the range of a target. Also known as O-indicator; O-scan; O-scope. { 'ō di,splā }

odoriferous homing [ELECTR] Homing on the ionized air produced by the exhaust gases of a snorkeling submarine. { ,ō·də'rif·ə·rəs 'hōm·iŋ }

OEM [ELECTR] Abbreviation for original equipment manufacturer. Generally describes original, factory-installed equipment.

off-center plan position indicator [ELECTR] A plan position indicator in which the center of the display that represents the location of the radar can be moved from the center of the screen to any position on the face of the PPI. { 'óf ,sent·ər 'plan pə,zish·ən 'in·də,kād·ər }

off-hook [COMMUN] The active state (closed loop) of a subscriber or PBX user loop. { 'óf ,húk }

off-hook service [COMMUN] Priority telephone service for key personnel that affords a connection from caller to receiver by the simple expedient of removing the phone from its cradle or hook. { 'óf ,húk ,sər·vəs }

off-line [COMPUT SCI] Describing equipment not connected to a computer, or temporarily disconnected from one. { 'óf ¦līn }

off-line cipher [COMMUN] Method of encrypting which is not associated with a particular transmission system and in which the resulting encrypted message can be transmitted by any means. { 'óf ¦līn 'sī·fər }

off-line equipment [COMPUT SCI] Peripheral equipment or devices not in direct communication with the central processing unit of a computer. Also known as auxiliary equipment. { 'óf ¦līn i'kwip·mənt }

off-line mode [COMPUT SCI] Any operation, such as printing, which does not involve the main computer. { 'óf ¦līn 'mōd }

off-line operation [COMPUT SCI] Operation of peripheral equipment in conjunction with, but not under the control of, the central processing unit. { 'óf ¦līn ,äp·ə'rā·shən }

off-line processing [COMPUT SCI] Any processing which takes place independently of the central processing unit. { 'óf ¦līn 'prä,ses·iŋ }

off-line storage [COMPUT SCI] A storage device not under control of the central processing unit. { 'óf ¦līn 'stór·ij }

off-line unit [COMPUT SCI] Any operation device which is not attached to the main computer. { 'óf ¦līn 'yü·nət }

offload [COMPUT SCI] To transfer operations from one computer to another, usually from a large computer to a smaller one. { 'óf,lōd }

offset [COMPUT SCI] *See* displacement. [CONT SYS] The steady-state difference between the desired control point and that actually obtained in a process control system. { 'óf,set }

offset-center plan position indicator *See* off-center plan position indicator. { 'óf,set ¦sen·tər 'plan pə,zish·ən 'in·də,kād·ər }

offset plan position indicator *See* off-center plan position indicator. { 'óf,set 'plan pə,zish·ən 'in·də,kād·ər }

offset voltage |ELECTR| The differential input voltage that must be applied to an operational amplifier to return the zero-frequency output voltage to zero volts, due to device mismatching at the input stage. { 'óf,set ,vōl·tij }

ohm |ELEC| The unit of electrical resistance in the rationalized meter-kilogram-second system of units, equal to the resistance through which a current of 1 ampere will flow when there is a potential difference of 1 volt across it. Symbolized Ω. { ōm }

ohmic |ELEC| Pertaining to a substance or circuit component that obeys Ohm's law. { 'ō·mik }

ohmic contact |ELEC| A region where two materials are in contact, which has the property that the current flowing through it is proportional to the potential difference across it. { 'ō·mik 'kän ,takt }

ohmic dissipation |ELECTR| Loss of electric energy when a current flows through a resistance due to conversion into heat. Also known as ohmic loss. { 'ō·mik ,dis·ə'pā·shən }

ohmic loss See ohmic dissipation. { 'ō·mik 'lòs }

ohmic resistance |ELEC| Property of a substance, circuit, or device for which the current flowing through it is proportional to the potential difference across it. { 'ō·mik ri'zis·təns }

ohmmeter |ENG| An instrument for measuring electric resistance; scale may be graduated in ohms or megohms. { 'ō,mēd·ər }

Ohm's law |ELEC| The law that the direct current flowing in an electric circuit is directly proportional to the voltage applied to the circuit; it is valid for metallic circuits and many circuits containing an electrolytic resistance. { 'ōmz ,lò }

Ohm's law |ELEC| The law that the direct current flowing in an electric circuit is directly proportional to the voltage applied to the circuit; it is valid for metallic circuits and many circuits containing an electrolytic resistance. { 'ōmz ,lò }

ohms per volt |ENG| Sensitivity rating for measuring instruments, obtained by dividing the resistance of the instrument in ohms at a particular range by the full-scale voltage value at that range. { 'ōmz pər 'vōlt }

oil-break |ELEC| Property of an electrical switch, circuit breaker, or similar apparatus whose contacts separate in oil. { 'òil ,brāk }

oil circuit breaker |ELECTR| A high-voltage circuit breaker in which the arc is drawn in oil to dissipate the heat and extinguish the arc; the intense heat of the arc decomposes the oil, generating a gas whose high pressure produces a flow of fresh fluid through the arc that furnishes the necessary insulation to prevent a restrike of the arc. { 'òil 'sər·kət ,brāk·ər }

oil-filled cable |ELEC| Cable having insulation impregnated with an oil which is fluid at all operating temperatures and provided with facilities such as longitudinal ducts or channels and with reservoirs; by this means positive oil pressure can be maintained within the cable at all times, incipient voids are promptly filled during periods of expansion, and all surplus oil is adequately taken care of during periods of contraction. { 'òil ¦fild ,kā·bəl }

oil-immersed |ELEC| Property of a transformer, reactor, regulator, or similar apparatus whose coils are immersed in an insulating liquid that is usually, but not necessarily, oil. { 'òil i,mərst }

oil switch |ELEC| A switch whose contacts are immersed in oil in order to suppress the arc and prevent the contacts from being damaged. { 'òil ,swich }

O-indicator See O-display. { 'ō ,in·də,kād·ər }

OL See only loadable.

olivette |ELEC| Standing floodlight used in the wings for lighting stage entrances and acting areas at fairly close range; bulb wattage ranges from 500 to 1500 watts. { ¦äl·ə¦vet }

OLRT system See on-line real-time system. { ,ō ,el,är'tē ,sis·təm }

omegatron |ELECTR| A miniature mass spectrograph, about the size of a receiving tube, that can be sealed to another tube and used to identify the residual gases left after evacuation. { ō'meg·ə ,trän }

OMG object model See Object Management Group object model. { ¦ō¦em¦jē 'äb·jikt ,mäd·əl }

omission factor |COMPUT SCI| In information retrieval, the ratio obtained in dividing the number of nonretrieved relevant documents by the total number of relevant documents in the file. { ō'mish·ən ,fak·tər }

omnidirectional |ELECTR| Radiating or receiving equally well in all directions. Also known as nondirectional. { ¦äm·nə·di'rek·shən·əl }

omnidirectional antenna |ELECTROMAG| An antenna that has an essentially circular radiation pattern in azimuth and a directional pattern in elevation. Also known as nondirectional antenna. { ¦äm·nə·di'rek·shən·əl an'ten·ə }

OMR See optical mark reading.

OMS See ovonic memory switch.

onboard |COMPUT SCI| Referring to a computer hardware component that is built directly into the computer. { 'òn'bórd }

on-call circuit |COMMUN| A permanently designated circuit that is activated only upon request of the user; this type of circuit is usually provided when a full-period circuit cannot be justified and the duration of use cannot be anticipated; during unactivated periods, the communications facilities required for the circuit are available for other requirements. { òn 'kól ,sər·kət }

ondograph |ELECTR| An instrument that draws the waveform of an alternating-current voltage step by step; a capacitor is charged momentarily to the amplitude of a point on the voltage wave, then discharged into a recording galvanometer, with the action being repeated a little further along on the waveform at intervals of about 0.01 second. { 'än·də,graf }

ondoscope |ELECTR| A glow-discharge tube used to detect high-frequency radiation, as in the vicinity of a radar transmitter; the radiation ionizes the gas in the tube and produces a visible glow. { 'än·də,skōp }

one-address code [COMPUT SCI] In computers, a code using one-address instructions. { 'wən ə ,dres 'kōd }

one-address instruction [COMPUT SCI] A digital computer programming instruction that explicitly describes one operation and one storage location. Also known as single-address instruction. { 'wən ə,dres in'strək·shən }

one condition [COMPUT SCI] The state of a magnetic core or other computer memory element in which it represents the value 1. Also known as one state. { 'wən kən,dish·ən }

one-digit subtracter See half-subtracter. { 'wən ,dij·ət səb'trak·tər }

one-dimensional array [COMPUT SCI] A group of related data elements arranged in a single row or column. { ¦wən də¦men·chən·əl ə'rā }

one-ended tape Turing machine [COMPUT SCI] A variation of a Turing machine in which the tape can be extended to the right, but not to the left. { 'wən ¦end·əd ¦tāp 'tùr·iŋ mə,shēn }

one-level address [COMPUT SCI] In digital computers, an address that directly indicates the location of an instruction or some data. { 'wən ,lev·əl ə'dres }

one-level code [COMPUT SCI] Any code using absolute addresses and absolute operation codes. { 'wən ,lev·əl 'kōd }

one-level subroutine [COMPUT SCI] A subroutine that does not use other subroutines during its execution. { 'wən ,lev·əl 'səb·rü,tēn }

one-line adapter [COMPUT SCI] A unit connecting central processes and permitting high-speed transfer of data under program control. { 'wən ,līn ə'dap·tər }

one-part code [COMMUN] Code in which the plain text elements are arranged in alphabetical or numerical order, accompanied by their code groups also arranged in alphabetical, numerical, or other systematic order. { 'wən ,pärt 'kōd }

one-pass operation [COMPUT SCI] An operating method, now standard, which produces an object program from a source program in one pass. { 'wən ,pas ,äp·ə'rā·shən }

one-plus-one address instruction [COMPUT SCI] A digital computer instruction whose format contains two address parts; one address designates the operand to be involved in the operation; the other indicates the location of the next instruction to be executed. { 'wən ,pləs 'wən ə¦dres in,strək·shən }

one-quadrant multiplier [ELECTR] Of an analog computer, a multiplier in which operation is restricted to a single sign of both input variables. { 'wən ,kwä·drənt 'məl·tə,plī·ər }

one's complement [COMPUT SCI] A numeral in binary notation, derived from another binary number by simply changing the sense of every digit. { 'wən 'käm·plə·mənt }

ones-complement code [COMPUT SCI] A number coding system used in some computers, where, for any number x, $x = (1 - 2^{n-1}) \cdot a_0 + 2^{n-2}a_1 + \cdots + a_{n-1}$, where $a_i = 1$ or 0. { 'wənz 'käm·plə·mənt ,kōd }

one-shot multivibrator See monostable multivibrator. { 'wən ,shät ,məl·tə'vī,brād·ər }

one-shot operation See single-step operation. { 'wən ,shät ,äp·ə'rā·shən }

one-sided abrupt junction [ELECTR] An abrupt junction that is realized by giving one side of the junction a high doping level compared with the other; that is, an n^+p or p^+n junction. { 'wən ,sīd·əd ə'brəpt 'jəŋk·shən }

one state See one condition. { 'wən ,stāt }

one-step operation See single-step operation. { 'wən ,step ,äp·ə'rā·shən }

one-time pad [COMMUN] A keying sequence based on random numbers that is used to code a single message and is then destroyed. { ¦wən ¦tīm 'pad }

one-to-many correspondence [COMPUT SCI] A structure that establishes relationships between two types of items in a data base such that one item of the first type can relate to several items of the second type, but items of the second type can relate back to only one item of the first type. { ¦wən tə ¦men·ē ,kär·ə'spän·dəns }

one-to-one assembler [COMPUT SCI] An assembly program which produces a single instruction in machine language for each statement in the source language. Also known as one-to-one translator. { ¦wən tə ¦wən ə'sem·blər }

one-to-one translator See one-to-one assembler. { ¦wən tə ¦wən 'tranz,lād·ər }

O network [ELEC] Network composed of four impedance branches connected in series to form a closed circuit, two adjacent junction points serving as input terminals, the remaining two junction points serving as output terminals. { 'ō ,net,wərk }

one-way trunk [ELEC] Trunk between two central offices, used for calls that originate at one of those offices, but not for calls that originate at the other. Also known as outgoing trunk. { 'wən ,wā 'trəŋk }

on-hook [COMMUN] The idle state (open loop) of a subscriber or PBX user loop. { 'ón ,húk }

onion diagram [SYS ENG] A schematic diagram of a system that is composed of concentric circles, with the innermost circle representing the core, and all the outer layers dependent on the core. { 'ən·yən ,dī·ə,gram }

on-line [COMPUT SCI] Pertaining to equipment capable of interacting with a computer. [ELECTR] The state in which a piece of equipment or a subsystem is connected and powered to deliver its proper output to the system. { 'ón ,līn }

on-line central file [COMPUT SCI] An organized collection of data, such as an on-line disk file, in a storage device under direct control of a central processing unit, that serves as a continually available source of data in applications where real-time or direct-access capabilites are required. { 'ón ,līn 'sen·trəl 'fīl }

on-line cipher [COMMUN] A method of encryption directly associated with a particular transmission system, whereby messages may be encrypted and simultaneously transmitted from one station to one or more stations where reciprocal equipment is automatically operated. { 'ȯn ˌlīn 'sī·fər }

on-line computer system |COMPUT SCI| A computer system which is adapted to on-line operation. { 'ȯn ˌlīn kəm'pyüd·ər ˌsis·təm }

on-line cryptographic operation See on-line operation. { 'ȯn ˌlīn ˌkrip·tə'graf·ik ˌäp·ə'rā·shən }

on-line data reduction |COMPUT SCI| The processing of information as rapidly as it is received by the computing system. { 'ȯn ˌlīn ˌdad·ə ri ˌdək·shən }

on-line disk file |COMPUT SCI| A magnetic disk directly connected to the central processing unit, thereby increasing the memory capacity of the computer. { 'ȯn ˌlīn 'disk ˌfīl }

on-line equipment |COMPUT SCI| The equipment or devices in a system whose operation is under control of the central processing unit, and in which information reflecting current activity is introduced into the data-processing system as soon as it occurs. { 'ȯn ˌlīn i'kwip·mənt }

on-line inquiry |COMPUT SCI| A level of computer processing that results from adding to an expanded batch system the capability to immediately access, from any terminal, any record that is stored in the disk files attached to the computer. { 'ȯn ˌlīn 'in·kwə·rē }

on-line mode |COMPUT SCI| Mode of operation in which all devices are responsive to the central processor. { 'ȯn ˌlīn ˌmōd }

on-line operation |COMMUN| A method of operation whereby messages are encrypted and simultaneously transmitted from one station to one or more other stations where reciprocal equipment is automatically operated to permit reception and simultaneous decryptment of the message. Also known as on-line cryptographic operation. |COMPUT SCI| Computer operation in which input data are fed into the computer directly from observing instruments or other input equipment, and computer results are obtained during the progress of the event. { 'ȯn ˌlīn ˌäp·ə'rā·shən }

on-line real-time system |COMPUT SCI| A computer system that communicates interactively with users, and immediately returns to them the results of data processing during an interaction. Abbreviated OLRT system. { 'ȯn ˌlīn 'rēl ˌtīm 'sis·təm }

on-line secured communications system |COMMUN| Any combination of interconnected communications centers partially or wholly equipped for on-line cryptographic operation and capable of relaying or switching message traffic using on-line cryptographic procedures. { 'ȯn ˌlīn si'kyürd kə,myü·nə'kā·shənz ˌsis·təm }

on-line storage |COMPUT SCI| Storage controlled by the central processing unit of a computer. { 'ȯn ˌlīn 'stȯr·ij }

on-line tab setting |COMPUT SCI| A feature in some computer printers which allows the computer that controls the printer to issue commands to set and change the tab stops. { 'ȯn ˌlīn 'tab ˌsed·iŋ }

on-line typewriter |COMPUT SCI| A typewriter which transmits information into and out of a computer, and which is controlled by the central processing unit and thus by whatever program the computer is carrying out. { 'ȯn ˌlīn 'tīp ˌrīd·ər }

only loadable |COMPUT SCI| Attribute of a load module which can be brought into main memory only by a LOAD macroinstruction given from another module. Abbreviated OL. { 'ōn·lē 'lōd·ə·bəl }

on-off control |CONT SYS| A simple control system in which the device being controlled is either full on or full off, with no intermediate operating positions. Also known as on-off system. { 'ȯn 'ȯf kən,trōl }

on-off keying |COMMUN| Binary form of amplitude modulation in which one of the states of the modulated wave is the absence of energy in the keying interval. { 'ȯn 'ȯf ˌkē·iŋ }

on-off switch |ELEC| A switch used to turn a receiver or other equipment on or off; often combined with a volume control in radio and television receivers. { 'ȯn 'ȯf ˌswich }

on-off system See on-off control. { 'ȯn 'ȯf ˌsis·təm }

on-off tests |ELEC| Tests conducted to determine the source of interference by switching various suspected sources on and off while observing the victim receiver. { 'ȯn 'ȯf ˌtests }

Onsager theory of dielectrics |ELEC| A theory for calculating the dielectric constant of a material with polar molecules in which the local field at a molecule is calculated for an actual spherical cavity of molecular size in the dielectric using Laplace's equation, and the polarization catastrophe of the Lorentz field theory is thereby avoided. { 'ȯn,säg·ər ,thē·ə·rē əv ,dī·ə'lek·triks }

on the beam |ELECTR| Centered on a beam of, or on an equisignal zone of, radiant energy, as a radio range. { 'ȯn thə 'bēm }

OOP See object-oriented programming.

open |ELEC| **1.** Condition in which conductors are separated so that current cannot pass. **2.** Break or discontinuity in a circuit which can normally pass a current. { 'ō·pən }

Open-Access Same-Time Information System |ELEC| An electronic system that uses Internet Web nodes to communicate to everyone in a fair and equitable manner information on available transmission capability and the cost of purchasing transmission services on the electric power transmission system, and allows for purchasing and reselling of transmission rights. Abbreviated OASIS. { ˌō·pən¦ak,ses ¦sām¦tīm ,in·fər'mā·shən ˌsis·təm }

open architecture |COMPUT SCI| A computer architecture whose specifications are made widely available to allow third parties to develop add-on peripherals for it. { 'ō·pən 'ar·kə,tek·chər }

open-bus system |COMPUT SCI| A computer with an expansion bus that is designed to easily accept expansion boards. { ¦ō·pən 'bəs ˌsis·təm }

open-center plan position indicator |ENG| A plan position indicator on which no signal is displayed within a set distance from the center. { 'ō·pən ˌsen·tər 'plan pə,zish·ən 'in·də,kād·ər }

open circuit |ELEC| An electric circuit that has been broken, so that there is no complete path for current flow. { 'ō·pən 'sər·kət }

open-circuit impedance |ELEC| Of a line or four-terminal network, the driving-point impedance when the far end is open. { 'ō·pən ¦sər·kət im'pēd·əns }

open-circuit jack |ELEC| Jack that normally leaves its circuit open; the circuit can be closed only by a circuit connected to the plug that is inserted in the jack. { 'ō¦·pən ¦sər·kət 'jak }

open-circuit signaling |COMMUN| Type of signaling in which no current flows while the circuit is in the idle condition. { 'ō·pən ¦sər·kət 'sig·nə·liŋ }

open-circuit voltage |ELEC| The voltage at the terminals of a source when no appreciable current is flowing. Also known as no-load voltage. { 'ō·pən ¦sər·kət 'vōl·tij }

open-delta connection |ELEC| An unsymmetrical transformer connection which is employed when one transformer of a bank of three single-phase delta-connected units must be cut out, because of failure. Also known as V connection. { 'ō·pən ¦del·tə kə,nek·shən }

open-ended |COMPUT SCI| Of techniques, designed to facilitate or permit expansion, extension, or increase in capability; the opposite of closed-in and artificially constrained. { ¦ō·pən ¦en·dəd }

open-ended system |COMPUT SCI| In character recognition, a system in which the input data to be read are derived from sources other than the computer with which the character reader is associated. { 'ō·pən ¦en·dəd 'sis·təm }

open file |COMPUT SCI| A file that can be accessed for reading, writing, or both. { 'ō·pən 'fīl }

open-flame arc |ELECTR| An electric arc which causes the anode to evaporate and be ejected as a flame. { 'ō·pən ¦flām 'ärk }

open-fuse cutout |ELEC| Enclosed fuse cutout in which the fuse support and fuse holder are exposed. { 'ō·pən ˌfyüz 'kə,daút }

open-link fuse |ELEC| A simple type of fuse that consists of a strip of fuse material bolted to open terminal blocks. { 'ō·pən ¦liŋk 'fyüz }

open-loop control system |CONT SYS| A control system in which the system outputs are controlled by system inputs only, and no account is taken of actual system output. { 'ō·pən ¦lüp kən'trōl ,sis·təm }

open-phase protection |ELEC| Effect of a device operating on the loss of current in one phase of a polyphase circuit to cause and maintain the interruption of power in the circuit. { ¦ō·pən ¦fāz prə'tek·shən }

open-phase relay |ELEC| Relay which functions by reason of the opening of one or more phases of a polyphase circuit, when sufficient current is flowing in the remaining phase or phases. { ¦ō·pən ¦fāz 'rē,lā }

open plug |ELEC| Plug designed to hold jack springs in their open position. { 'ō·pən ,pləg }

open routine |COMPUT SCI| **1.** A routine which can be inserted directly into a larger routine without

a linkage or calling sequence. **2.** A computer program that changes the state of a file from closed to open. { 'ō·pən rü,tēn }

open shop |COMPUT SCI| A data-processing-center organization in which individuals from outside the data-processing community are permitted to implement their own solutions to problems. { 'ō·pən ¦shäp }

open source software |COMPUT SCI| Software that is written in such a way that others are encouraged to freely redistribute it, and all changes to the code must be made freely available. { ¦ō·pən ,sörs 'sóf,twer }

open standard |COMPUT SCI| Freely distributed. { 'ō·pən 'stan·dərd }

open subroutine |COMPUT SCI| A set of computer instructions that collectively perform some particular function and are inserted directly into the program each and every time that particular function is required. { 'ō·pən 'səb·rü,tēn }

open system |COMPUT SCI| A computer system whose key software interfaces are specified, documented, and made publicly available. { 'ō·pən 'sis·təm }

open-system architecture |COMPUT SCI| The structure of a computer network that allows different types of computers and peripheral devices from different manufacturers to be connected together. { 'ō·pən ¦sis·təm 'är·kə ,tek·chər }

open wire |ELEC| A conductor supported above the ground, separate from other conductors. { 'ō·pən 'wīr }

open-wire carrier system |COMMUN| A system for carrier telephony using an open-wire line. { 'ō·pən ¦wir 'kar·ē·ər ,sis·təm }

open-wire feeder See open-wire transmission line. { 'ō·pən ¦wir 'fēd·ər }

open-wire loop |ELEC| Branch line on a main open-wire line. { 'ō·pən ¦wir 'lüp }

open-wire transmission line |ELEC| A transmission line consisting of two spaced parallel wires supported by insulators, at the proper distance to give a desired value of surge impedance. Also known as open-wire feeder. { 'ō·pən ¦wir tranz'mish·ən ,līn }

operand |COMPUT SCI| Any one of the quantities entering into or arising from an operation. { 'äp·ə,rand }

operate time |COMPUT SCI| The phase of computer operation when an instruction is being carried out. { 'äp·ə,rāt ,tīm }

operating angle |ELECTR| Electrical angle of the input signal (for example, portion of a cycle) during which plate current flows in a vacuum tube amplifier. { 'äp·ə,rād·iŋ ,aŋ·gəl }

operating delay |COMPUT SCI| Computer time lost because of mistakes or inefficiency of operating personnel or users of the system, excluding time lost because of defects in programs or data. { 'äp·ə,rād·iŋ di,lā }

operating instructions |COMPUT SCI| A detailed description of the actions that must be carried out by a computer operator in running a program or group of interrelated programs, usually included in the documentation of a program supplied by a programmer or systems analyst, along with the source program and flow charts. { 'äp·ə,rād·iŋ in,strək·shənz }

operating point |ELECTR| Point on a family of characteristic curves of a vacuum tube or transistor where the coordinates of the point represent the instantaneous values of the electrode voltages and currents for the operating conditions under study or consideration. { 'äp·ə,rād·iŋ ˌpȯint }

operating position |COMMUN| Terminal of a communications channel which is attended by an operator; usually the term refers to a single operator, such as a radio operator's position or a telephone operator's position; however, certain terminals may require more than one operating position. { 'äp·ə,rād·iŋ pə,zish·ən }

operating power |ELECTROMAG| Power that is actually supplied to a radio transmitter antenna. { 'äp·ə,rād·iŋ ˌpaů·ər }

operating range |ELECTR| The frequency range over which a reversible transducer is operable. { 'äp·ə,rād·iŋ ˌrānj }

operating ratio |COMPUT SCI| The time during which computer hardware operates and gives reliable results divided by the total time scheduled for computer operation. { 'äp·ə,rād·iŋ ˌrā·shō }

operating system |COMPUT SCI| A set of programs and routines which guides a computer or network in the performance of its tasks, assists the programs (and programmers) with certain supporting functions, and increases the usefulness of the computer or network hardware. { 'äp·ə ˌrād·iŋ ˌsis·təm }

operating system supervisor |COMPUT SCI| The control program of a set of programs which guide a computer in the performance of its tasks and which assist the program with certain supporting functions. { 'äp·ə,rād·iŋ ˌsis·təm 'sü·pər,vīz·ər }

operational |ENG| Of equipment such as aircraft or vehicles, being in such a state of repair as to be immediately usable. { ˌäp·ə'rā·shən·əl }

operational amplifier |ELECTR| An amplifier having high direct-current stability and high immunity to oscillation, generally achieved by using a large amount of negative feedback; used to perform analog-computer functions such as summing and integrating. { ˌäp·ə'rā·shən·əl 'am·plə,fī·ər }

operational label |COMPUT SCI| A combination of letters and digits at the beginning of the tape which uniquely identify the tape required by the system. { ˌäp·ə'rā·shən·əl 'lā·bəl }

operational standby program |COMPUT SCI| The program operating in the standby computer when in the duplex mode of operation. { ˌäp·ə'rā·shən·əl 'stand,bī ˌprō,gram }

operation code |COMPUT SCI| A field or portion of a digital computer instruction that indicates which action is to be performed by the computer. Also known as command code. { ˌäp·ə'rā·shən ˌkōd }

operation cycle |COMPUT SCI| The portion of a memory cycle required to perform an operation; division and multiplication usually require more than one memory cycle to be completed. { ˌäp·ə'rā·shən ˌsī·kəl }

operation decoder |COMPUT SCI| A device that examines the operation contained in an instruction of a computer program and sends signals to the circuits required to carry out the operation. { ˌäp·ə'rā·shən dē'kōd·ər }

operation number |COMPUT SCI| **1.** Number designating the position of an operation, or its equivalent subroutine, in the sequence of operations composing a routine. **2.** Number identifying each step in a program stated in symbolic code. { ˌäp·ə'rā·shən ˌnəm·bər }

operation part |COMPUT SCI| That portion of a digital computer instruction which is reserved for the operation code. { ˌäp·ə'rā·shən ˌpärt }

operation register |COMPUT SCI| A register used to store and decode the operation code for the next instruction to be carried out by a computer. { ˌäp·ə'rā·shən ˌrej·ə·stər }

operations research |MATH| The mathematical study of systems with input and output from the viewpoint of optimization subject to given constraints. { ˌäp·ə'rā·shənz ri,sərch }

operations sequence |CONT SYS| The logical series of procedures that constitute the task for a robot. { ˌäp·ə'rā·shənz ˌsē·kwəns }

operation time |COMPUT SCI| The time elapsed during the interpretation and execution of an arithmetic or logic operation by a computer. { ˌäp·ə'rā·shən ˌtīm }

operator |COMPUT SCI| Anything that designates an action to be performed, especially the operation code of a computer instruction. { 'äp·ə ˌrād·ər }

operator hierarchy |COMPUT SCI| A sequence of mathematical operators which designates the order in which these operators are to be applied to any mathematical expression in a given programming language. { 'äp·ə,rād·ər 'hī·ər,är·kē }

operator interrupt |COMPUT SCI| A step whereby control is passed to the monitor, and a message, usually requiring a typed answer, is printed on the console typewriter. { 'äp·ə,rād·ər 'in·tə,rəpt }

operator's console |COMPUT SCI| Equipment which provides for manual intervention and monitoring computer operation. { 'äp·ə,rād·ərz 'kän,sōl }

operator subgoaling |COMPUT SCI| A computer problem-solving method in which the inability of the computer to take the desired next step at any point in the problem-solving process leads to a subgoal of making that step feasible. { 'äp·ə ˌrād·ər ˌsəb'gōl·iŋ }

optical amplifier |ENG| An optoelectronic amplifier in which the electric input signal is converted to light, amplified as light, then converted back to an electric signal for the output. { 'äp·tə·kəl 'am·plə,fī·ər }

optical bar-code reader |COMPUT SCI| A device which uses any of various photoelectric methods to read information which has been coded by placing marks in prescribed boxes on documents with ink, pencil, or other means. { 'äp·tə·kəl 'bär ˌkōd ˌrēd·ər }

optical character recognition |COMPUT SCI| That branch of character recognition concerned with the automatic identification of handwritten or printed characters by any of various photoelectric methods. Abbreviated OCR. Also known as electrooptical character recognition. { 'äp·tə·kəl 'kar·ik·tər ˌrek·ig͵nish·ən }

optical communication |COMMUN| The use of electromagnetic waves in the region of the spectrum near visible light for the transmission of signals representing speech, pictures, data pulses, or other information, usually in the form of a laser beam modulated by the information signal. { 'äp·tə·kəl kə͵myü·nə'kā·shən }

optical computer |COMPUT SCI| A computer that uses various combinations of holography, lasers, and mass-storage memories for such applications as ultra-high-speed signal processing, image deblurring, and character recognition. { 'äp·tə·kəl kəm'pyüd·ər }

optical coupler See optoisolator. { 'äp·tə·kəl 'kəp·lər }

optical coupling |ELECTR| Coupling between two circuits by means of a light beam or light pipe having transducers at opposite ends, to isolate the circuits electrically. { 'äp·tə·kəl 'kəp·liŋ }

optical data storage |COMPUT SCI| The technology of placing information in a medium so that, when a light beam scans the medium, the reflected light can be used to recover the information. { ¦äp·tə·kəl 'dad·ə ͵stȯr·ij }

optical disk |COMPUT SCI| A type of video disk storage device consisting of a pressed disk with a spiral groove at the bottom of which are submicrometer-sized depressions that are sensed by a laser beam. { 'äp·tə·kəl 'disk }

optical disk storage |COMPUT SCI| A computer storage technology in which information is stored in submicrometer-sized holes on a rotating disk, and is recorded and read by laser beams focused on the disk. Also known as laser disk storage; video disk storage. { 'äp·tə·kəl ¦disk 'stȯr·ij }

optical electronic reproducer See optical sound head. { 'äp·tə·kəl i͵lek'trän·ik ͵rē·prə'dü·sər }

optical encoder |ELECTR| An encoder that converts positional information into corresponding digital data by interrupting light beams directed on photoelectric devices. { 'äp·tə·kəl in'kōd·ər }

optical fiber |OPTICS| A long, thin thread of fused silica, or other transparent substance, used to transmit light. Also known as light guide. { 'äp·tə·kəl 'fī·bər }

optical-fiber amplifier |COMMUN| A device for amplifying signals transmitted over optical fibers, consisting of a low-loss single-mode fiber made of basic silica glass, along whose length gain is generated by coupling pump light at either or both fiber ends, or at periodic locations in between. { ͵äp·tə·kəl ͵fī·bər 'am·plə͵fī·ər }

optical-fiber cable See optical waveguide. { 'äp·tə·kəl ¦fī·bər 'kā·bəl }

optical-fiber sensor |ENG| An instrument in which the physical quantity to be measured is made to modulate the intensity, spectrum, phase, or polarization of light from a light-emitting diode or laser diode traveling through an optical fiber; the modulated light is detected by a photodiode. Also known as fiber-optic sensor. { 'äp·tə·kəl ¦fī·bər 'sen·sər }

optical filter See filter. { 'äp·tə·kəl 'fil·tər }

optical information processor See optical information system. { 'äp·tə·kəl ͵in·fər'mā·shən ͵prä ͵ses·ər }

optical information system |COMPUT SCI| A device that uses light to process information; consists of one or several light sources, a one- or two-dimensional plane of data such as a film transparency, lens, or other optical component, and a detector. Also known as optical information processor. { 'äp·tə·kəl ͵in·fər'mā·shən ͵sis·təm }

optical isolator See optoisolator. { 'äp·tə·kəl 'ī·sə ͵lād·ər }

optical lithography |ELECTR| Lithography in which an integrated circuit pattern is first created on a glass plate or mask and is then transferred to the resist by one of a number of optical techniques by using visible or ultraviolet light. { 'äp·tə·kəl li'thäg·rə·fē }

optically coupled isolator See optoisolator. { 'äp·tə·klē ¦kup·əld 'ī·sə͵lād·ər }

optical mark reading |COMPUT SCI| Optically sensing information encoded as a series of marks, such as lines or filled-in boxes on a test answer sheet, or some special pattern, such as the Universal Product Code. Abbreviated OMR. { 'äp·tə·kəl ¦märk ͵rēd·iŋ }

optical mask |ELECTR| A thin sheet of metal or other substance containing an open pattern, used to suitably expose to light a photoresistive substance overlaid on a semiconductor or other surface to form an integrated circuit. { 'äp·tə·kəl 'mask }

optical memory |COMPUT SCI| A computer memory that uses optical techniques which generally involve an addressable laser beam, a storage medium which responds to the beam for writing and sometimes for erasing, and a detector which reacts to the altered character of the medium when it uses the beam to read out stored data. { 'äp·tə·kəl 'mem·rē }

optical microphone |ENG ACOUS| A microphone in which the motion of a membrane is detected using a light beam reflected from it, either with the aid of an interferometer or by detecting the deflection of the beam. { ¦äp·tə·kəl 'mī·krə͵fōn }

optical modulator |COMMUN| A device used for impressing information on a light beam. { 'äp·tə·kəl 'mäj·ə͵lād·ər }

optical mouse |COMPUT SCI| A mouse that emits a light signal and uses its reflection from a reflective grid to determine position and movement. { 'äp·tə·kəl 'maüs }

optical processing |COMPUT SCI| The use of light, including visible and infrared, to handle data-processing information. { 'äp·tə·kəl 'prä͵ses·iŋ }

optical proximity sensor |ENG| A device that uses the principle of triangulation of reflected infrared or visible light to measure small distances in a robotic system. { 'äp·tə·kəl präk 'sim·əd·ē ͵sen·sər }

optical reader |COMPUT SCI| A computer data-entry machine that converts printed characters, bar or line codes, and pencil-shaded areas into a computer-input code format. { 'äp·tə·kəl 'rēd·ər }

optical relay |ELECTR| An optoisolator in which the output device is a light-sensitive switch that provides the same on and off operations as the contacts of a relay. { 'äp·tə·kəl 'rē,lā }

optical scanner *See* flying-spot scanner. { 'äp·tə·kəl 'skan·ər }

optical sound head |ELECTR| The assembly in motion picture projection which reproduces photographically recorded sound; light from an incandescent lamp is focused on a slit, light from the slit is in turn focused on the optical sound track of a film, and the light passing through the film is detected by a photoelectric cell. Also known as optical electronic reproducer. { 'äp·tə·kəl 'saund ,hed }

optical sound recorder *See* photographic sound recorder. { 'äp·tə·kəl 'saund ri,kòrd·ər }

optical sound reproducer *See* photographic sound reproducer. { 'äp·tə·kəl 'saund ,rē·prə 'dü·sər }

optical storage |COMPUT SCI| Storage of large amounts of data in permanent form on photographic film or its equivalent, for nondestructive readout by means of a light source and photodetector. { 'äp·tə·kəl 'stòr·ij }

optical tape storage |COMMUN| A data storage technology in which information is stored on a tape that is wound on a spool and has a large number of parallel channels, and information is retrieved by sensing the reflected light when a light beam scans the medium. { ¦äp·tə·kəl 'tāp ,stòr·ij }

optical type font |COMPUT SCI| A special type font whose characters are designed to be easily read by both people and optical character recognition machines. { 'äp·tə·kəl ¦tīp ,fänt }

optical waveguide |ELECTROMAG| A waveguide in which a light-transmitting material such as a glass or plastic fiber is used for transmitting information from point to point at wavelengths somewhere in the ultraviolet, visible-light, or infrared portions of the spectrum. Also known as fiber waveguide; optical-fiber cable. { 'äp·tə·kəl 'wāv,gīd }

optimal control theory |CONT SYS| An extension of the calculus of variations for dynamic systems with one independent variable, usually time, in which control (input) variables are determined to maximize (or minimize) some measure of the performance (output) of a system while satisfying specified constraints. { 'äp·tə·məl kən'trōl ,thē·ə·rē }

optimal feedback control |CONT SYS| A subfield of optimal control theory in which the control variables are determined as functions of the current state of the system. { 'äp·tə·məl 'fēd ,bak kən,trōl }

optimal programming |CONT SYS| A subfield of optimal control theory in which the control variables are determined as functions of time for a specified initial state of the system. { 'äp·tə·məl 'prō,gram·iŋ }

optimal regulator problem *See* linear regulator problem. { 'äp·tə·məl 'reg·yə,lād·ər ,präb·ləm }

optimal smoother |CONT SYS| An optimal filer algorithm which generates the best estimate of a dynamical variable at a certain time based on all available data, both past and future. { 'äp·tə·məl 'smüth·ər }

optimization |SYS ENG| **1.** Broadly, the efforts and processes of making a decision, a design, or a system as perfect, effective, or functional as possible. **2.** Narrowly, the specific methodology, techniques, and procedures used to decide on the one specific solution in a defined set of possible alternatives that will best satisfy a selected criterion. Also known as system optimization. { ,äp·tə·mə'zā·shən }

optimize |COMPUT SCI| To rearrange the instructions or data in storage so that a minimum number of time-consuming jumps or transfers are required in the running of a program. { 'äp·tə,mīz }

optimized code |COMPUT SCI| A machine-language program that has been revised to remove inefficiencies and unused or unnecessary instructions so that the program is executed more quickly and occupies less storage space. { 'äp·tə,mīzd 'kōd }

optimizer |COMPUT SCI| A utility program that processes machine-language programs and generates optimized code. { 'äp·tə,mīz·ər }

optimum array current |ELECTROMAG| The current distribution in a broadside antenna array which is such that for a specified side-lobe level the beam width is as narrow as possible, and for a specified first null the side-lobe level is as small as possible. { 'äp·tə·məm ə'rā ,kə·rənt }

optimum bunching |ELECTR| Bunching condition required for maximum output in a velocity modulation tube. { 'äp·tə·məm 'bənch·iŋ }

optimum code |COMPUT SCI| A computer code which is particularly efficient with regard to a particular aspect; for example, minimum time of execution, minimum or efficient use of storage space, and minimum coding time. { 'äp·tə·məm 'kōd }

optimum coupling *See* critical coupling. { 'äp·tə·məm 'kəp·liŋ }

optimum filter |ELECTR| An electric filter in which the mean square value of the error between a desired output and the actual output is at a minimum. { 'äp·tə·məm 'fil·tər }

optimum programming |COMPUT SCI| Production of computer programs that maximize efficiency with respect to some criteria such as least cost, least use of storage, least time, or least use of time-sharing peripheral equipment. { 'äp·tə·məm 'prō,gram·iŋ }

optimum traffic frequency *See* optimum working frequency. { 'äp·tə·məm 'traf·ik ,frē·kwən·sē }

optimum working frequency |COMMUN| The most effective frequency at a specified time for ionospheric propagation of radio waves between two specified points. Also known as frequency optimum traffic; optimum traffic frequency. { 'äp·tə·məm 'wərk·iŋ ,frē·kwən·sē }

optional halt instruction |COMPUT SCI| A halt instruction that can cause a computer program to stop either before or after the instruction is obeyed if certain criteria are met. Also known as optional stop instruction. { 'äp·shən·əl 'hölt in ,strək·shən }

optional product |COMPUT SCI| Any of various forms of documentation that may be made available with a software product, such as source code, manuals, and instructions. { 'äp·shən·əl 'präd·əkt }

optional stop instruction See optional halt instruction. { 'äp·shən·əl 'stäp in,strək·shən }

option switch |COMPUT SCI| **1.** A DIP switch or jumper that activates an optional feature. **2.** A software parameter that overrides a default value and thereby activates an optional feature. Also known as option toggle. { 'äp·shən ,swich }

option toggle See option switch. { 'äp·shən ,täg·əl }

optoacoustic modulator See acoustooptic modulator. { ¦äp·tō·ə¦küs·tik 'mäj·ə,lād·ər }

optocoupler See optoisolator. { ¦äp·tō'kəp·lər }

optoelectronic amplifier |ENG| An amplifier in which the input and output signals and the method of amplification may be either electronic or optical. { ¦äp·tō·i,lek'trän·ik 'am·plə,fī·ər }

optoelectronic integration |ELECTR| A technology that combines optical components with electronic components such as transistors on a single wafer to obtain highly functional circuits. { ,äp·tō,i·lek¦trän·ik ,in·tə'grā·shən }

optoelectronic isolator See optoisolator. { ¦äp·tō·i,lek'trän·ik 'ī·sə,lād·ər }

optoelectronics |ELECTR| **1.** The branch of electronics that deals with solid-state and other electronic devices for generating, modulating, transmitting, and sensing electromagnetic radiation in the ultraviolet, visible-light, and infrared portions of the spectrum. **2.** See photonics. { ¦äp·tō·i,lek'trän·iks }

optoelectronic scanner |ELECTR| A scanner in which lenses, mirrors, or other optical devices are used between a light source or image and a photodiode or other photoelectric device. { ¦äp·tō·i,lek'trän·ik 'skan·ər }

optoisolator |ELECTR| A coupling device in which a light-emitting diode, energized by the input signal, is optically coupled to a photodetector such as a light-sensitive output diode, transistor, or silicon controlled rectifier. Also known as optical coupler; optical isolator; optically coupled isolator; optocoupler; optoelectronic isolator; photocoupler; photoisolator. { ¦äp·tō'ī·sə,lād·ər }

optophone |ENG ACOUS| A device with a photoelectric cell to convert ordinary printed letters into a series of sounds; used by the blind. { 'äp·tə,fōn }

or |COMPUT SCI| An instruction which performs the logical operation "or" on a bit-by-bit basis for its two or more operand words, usually storing the result in one of the operand locations. Also known as OR function. { ör }

ORB See object request broker. { örb or ¦ō¦är'bē }

ORB core |COMPUT SCI| The part of an object request broker that is responsible for the communication of requests. { 'örb ,kör or ¦ō¦är'bē ,kör }

orbitron |ELECTR| A maser that uses synthetic atoms composed of free electrons orbiting long, thin, positively charged, metal wires. { 'ör·bə,trän }

OR circuit See OR gate. { 'ör ,sər·kət }

ordered array |COMPUT SCI| A set of data elements that has been arranged in rows and columns in a specified order so that each element can be individually accessed. { 'örd·ərd ə'rā }

ordered list |COMPUT SCI| A set of data items that has been arranged in a specified sequence to aid in processing its contents. { 'örd·ərd 'list }

orderly shutdown |COMPUT SCI| The procedures for shutting off a computer system in an organized manner, normally after all work in progress has been completed, permitting restarting of the systems without loss of transactions or data. { 'örd·ər·lē 'shət,daůn }

order tone |COMMUN| Tone sent over a trunk to indicate that the trunk is ready to receive an order or, to the receiving operator, that an order is about to arrive. { 'örd·ər ,tōn }

ordinal type |COMPUT SCI| A data type whose possible values are sequential in the manner of the integers 1, 2, 3, and so forth; for example, the months January, February, and so forth. { 'örd·nəl 'tīp }

OR function See or. { 'ör ,fəŋk·shən }

organic electrolyte cell |ELEC| A type of wet cell that is based on the use of particularly reactive metals such as lithium, calcium, or magnesium in conjunction with organic electrolytes; the best-known type is the lithium-cupric fluoride cell. { ör'gan·ik i'lek·trə,līt ,sel }

OR gate |ELECTR| A multiple-input gate circuit whose output is energized when any one or more of the inputs is in a prescribed state; performs the function of the logical inclusive-or; used in digital computers. Also known as OR circuit. { 'ör,gāt }

orient |COMPUT SCI| To change relative and symbolic addresses to absolute form. { 'ör·ē·ənt }

orientation |ELECTROMAG| The physical positioning of a directional antenna or other device having directional characteristics. { ,ör·ē·ən'tā·shən }

orientation effect |ELEC| Those bulk properties of a material which result from orientation polarization. { ,ör·ē·ən'tā·shən i,fekt }

orientation polarization |ELEC| Polarization arising from the orientation of molecules which have permanent dipole moments arising from an asymmetric charge distribution. Also known as dipole polarization. { ,ör·ē·ən'tā·shən ,pō·lə·rə ,zā·shən }

orifice |ELECTROMAG| Opening or window in a side or end wall of a waveguide or cavity resonator through which energy is transmitted. { 'ör·ə·fəs }

origin |COMPUT SCI| Absolute storage address in relative coding to which addresses in a region are referenced. { 'är·ə·jən }

original document *See* source document. { ə'rij·ən·əl 'däk·yə·mənt }

original equipment manufacturer *See* OEM. { ə·'rij·ə·nəl i'kwip·mənt man·yə·'fak·chər·ər }

orthicon |ELECTR| A camera tube in which a beam of low-velocity electrons scans a photoemissive mosaic that is capable of storing a pattern of electric charges; has higher sensitivity than the iconoscope. { 'ȯr·thə,kän }

orthogonal |COMPUT SCI| **1.** An area of a computer display in which units of distance are the same horizontally and vertically so that there is no distortion. **2.** A viewing area in which positions are determined by using a cartesian coordinate system with horizontal and vertical axes. { ȯr'thäg·ən·əl }

orthogonal antennas |ELECTROMAG| In radar, a pair of transmitting and receiving antennas, or a single transmitting-receiving antenna, designed for the detection of a difference in polarization between the transmitted energy and the energy returned from the target. { ȯr'thäg·ən·əl an'ten·əz }

orthogonal parity check |COMPUT SCI| A parity checking system involving both a lateral and a longitudinal parity check. { ȯr'thäg·ən·əl 'par·əd·ē ,chek }

orthotronic error control |COMPUT SCI| An error check carried out to ensure correct transmission, which uses lateral and longitudinal parity checks. { |ȯr·thə|trän·ik 'er·ər kən,trōl }

O-scan *See* O-display. { 'ō ,skan }

osciducer |ELECTR| Transducer in which information pertaining to the stimulus is provided in the form of deviation from the center frequency of an oscillator. { |äs·ə|dü·sər }

oscillation *See* cycling. { ,äs·ə'lā·shən }

oscillator |ELECTR| **1.** An electronic circuit that converts energy from a direct-current source to a periodically varying electric output. **2.** The stage of a superheterodyne receiver that generates a radio-frequency signal of the correct frequency to mix with the incoming signal and produce the intermediate-frequency value of the receiver. **3.** The stage of a transmitter that generates the carrier frequency of the station or some fraction of the carrier frequency. { 'äs·ə,lād·ər }

oscillator harmonic interference |ELECTR| Interference occurring in a superheterodyne receiver due to the interaction of incoming signals with harmonics (usually the second harmonic) of the local oscillator. { 'äs·ə,lād·ər här'män·ik ,in·tər'fir·əns }

oscillator-mixer-first detector *See* converter. { 'äs·ə,lād·ər 'mik·sər ,fərst di'tek·tər }

oscillatory circuit |ELEC| Circuit containing inductance or capacitance, or both, and resistance, connected so that a voltage impulse will produce an output current which periodically reverses or oscillates. { 'äs·ə·lə,tȯr·ē 'sər·kət }

oscillatory discharge |ELEC| Alternating current of gradually decreasing amplitude which, under certain conditions, flows through a circuit containing inductance, capacitance, and resistance when a voltage is applied. { 'äs·ə·lə,tȯr·ē 'dis ,chärj }

oscillatory surge |ELEC| Surge which includes both positive and negative polarity values. { 'äs·ə·lə,tȯr·ē 'sərj }

oscillistor |ELECTR| A bar of semiconductor material, such as germanium, that will oscillate much like a quartz crystal when it is placed in a magnetic field and is carrying direct current that flows parallel to the magnetic field. { |äs·ə'lis·tər }

oscillograph tube |ELECTR| Cathode-ray tube used to produce a visible pattern, which is the graphical representation of electric signals, by variations of the position of the focused spot or spots according to these signals. { ə'sil·ə,graf ,tüb }

oscilloscope *See* cathode-ray oscilloscope. { ə'sil·ə,skōp }

O-scope *See* O-display. { 'ō ,skōp }

OTH radar *See* over-the-horizon radar. { ō|tē|āch 'rä,där }

OTS *See* ovonic threshold switch.

O-type backward-wave oscillator |ELECTR| A backward-wave tube in which an electron gun produces an electron beam focused longitudinally throughout the length of the tube, a slow-wave circuit interacts with the beam, and at the end of the tube a collector terminates the beam. Also known as O-type carcinotron; type-O carcinotron. { 'ō ,tīp 'bak·wərd |wāv 'äs·ə ,lad·ər }

O-type carcinotron *See* O-type backward-wave oscillator. { 'ō |tīp kär'sin·ə,trän }

outage |ELEC| A failure in an electric power system. { 'aúd·ij }

outgoing trunk *See* one-way trunk. { 'aút,gō·iŋ 'traŋk }

outlet |ELEC| A power line termination from which electric power can be obtained by inserting the plug of a line cord. Also known as convenience receptacle; electric outlet; receptacle. { 'aút ,let }

outlet box |ELEC| A box at which lines in an electric wiring system terminate, so that electric appliances or fixtures may be connected. { 'aút ,let ,bäks }

outline processor |COMPUT SCI| A software system that organizes notes in ordinary English into an outline that serves as the basis for a document. { 'aút,līn ,prä,ses·ər }

out-of-line coding |COMPUT SCI| Instructions in a routine that are stored in a different part of computer storage from the rest of the instructions. { 'aút əv |līn 'kōd·iŋ }

out-of-service jack |ELEC| Jack associated with a test jack which removes the circuit from service when a shorted plug is inserted. { 'aút əv |sər·vəs 'jak }

out-plant system |COMPUT SCI| A data-processing system that has one or more remote terminals from which information is transmitted to a central computer. { 'aut ,plant ,sis·təm }

output |COMPUT SCI| **1.** The data produced by a data-processing operation, or the information that is the objective or goal in data processing. **2.** The data actively transmitted from within the computer to an external device, or onto a permanent recording medium (paper, microfilm). **3.** The activity of transmitting the generated information. **4.** The readable storage medium upon which generated data are written, as in hardcopy output. |ELECTR| **1.** The current, voltage, power, driving force, or information which a circuit or device delivers. **2.** Terminals or other places where a circuit or device can deliver current, voltage, power, driving force, or information. { 'aut,put }

output area |COMPUT SCI| A part of storage that has been reserved for output data. Also known as output block. { 'aut,put ,er·ē·ə }

output block |COMPUT SCI| **1.** A portion of the internal storage of a computer that is reserved for receiving, processing, and transmitting data to be transferred out. **2.** See output area. { 'aut,put ,bläk }

output-bound computer |COMPUT SCI| A computer that is slowed down by its output functions. { 'aut,put ,baund kəm,pyüd·ər }

output bus driver |ELECTR| A device that power-amplifies output signals from a computer to allow them to drive heavy circuit loads. { 'aut ,put 'bəs ,drīv·ər }

output capacitance |ELECTR| Of an *n*-terminal electron tube, the short-circuit transfer capacitance between the output terminal and all other terminals, except the input terminal, connected together. { 'aut,put kə,pas·əd·əns }

output class |COMPUT SCI| An indicator of the priority of output from a computer that determines the order in which it is printed from a spool file. { 'aut,put ,klas }

output device See output unit. { 'aut,put di,vīs }

output gap |ELECTR| An interaction gap by means of which usable power can be abstracted from an electron stream in a microwave tube. { 'aut,put ,gap }

output impedance |ELECTR| The impedance presented by a source to a load. { 'aut,put im ,pēd·əns }

output indicator |ENG| A meter or other device that is connected to a radio receiver to indicate variations in output signal strength for alignment and other purposes, without indicating the exact value of output. { 'aut,put ,in·də,kād·ər }

output link |COMMUN| The last link in a communications chain. { 'aut,put ,liŋk }

output meter |ENG| An alternating-current voltmeter connected to the output of a receiver or amplifier to measure output signal strength in volume units or decibels. { 'aut,put ,mēd·ər }

output-meter adapter |ENG| Device that can be slipped over the plate prong of the output tube of a radio receiver to provide a conventional terminal to which an output meter can be connected during alignment. { 'aut,put ,mēd·ər ə,dap·tər }

output monitor interrupt |COMPUT SCI| A data-processing step in which control is passed to the monitor to determine the precedence order for two requests having the same priority level. { 'aut,put ,man·əd·ər 'int·ə,rəpt }

output power |ELEC| Power delivered by a system or transducer to its load. { 'aut,put ,pau·ər }

output program See output routine. { 'aut,put ,prō,gram }

output rating See carrier power output rating. { 'aut,put ,rād·iŋ }

output record |COMPUT SCI| **1.** A unit of data that has been transcribed from a computer to an external medium or device. **2.** The unit of data that is currently held in the output area of a computer before being transcribed to an external medium or device. { 'aut,put ,rek·ərd }

output resistance |ELECTR| The resistance across the output terminals of a circuit or device. { 'aut,put ri,zis·təns }

output routine |COMPUT SCI| A series of computer instructions which organizes and directs all operations associated with the transcription of data from a computer to various media and external devices by various types of output equipment. Also known as output program. { 'aut,put rü ,tēn }

output stage |ELECTR| The final stage in any electronic equipment. { 'aut,put ,stāj }

output transformer |ELECTR| The iron-core audio-frequency transformer used to match the output stage of a radio receiver or an amplifier to its loudspeaker or other load. { 'aut,put tranz,fór·mər }

output tube |ELECTR| Power-amplifier tube designed for use in an output stage. { 'aut,put ,tüb }

output unit |COMPUT SCI| In computers, a unit which delivers information from the computer to an external device or from internal storage to external storage. { 'aut,put ,yü·nət }

output word |COMPUT SCI| Any running word into which an input word is to be translated. { 'aut ,put ,wərd }

outside extension |COMMUN| Telephone extension on premises separated from the main station. { 'aut,sīd ik'sten·chən }

overall response |ELECTR| The ratio between system input and output. { |ō·vər|ól ri'späs }

overbunching |ELECTR| In velocity-modulated streams of electrons, the bunching condition produced by the continuation of the bunching process beyond the optimum condition. { |ō·vər|bənch·iŋ }

overcompound |ELEC| To use sufficiently many series turns in a compound-wound generator so that the terminal voltage at rated load is greater than at no load, usually to compensate for increased line drop. { |ō·vər|käm,paund }

overcoupled circuits |ELECTR| Two resonant circuits which are tuned to the same frequency but coupled so closely that two response peaks are obtained; used to attain broad-band response with substantially uniform impedance. { 'ō·vər ,kəp·əld 'sər·kəts }

overcurrent |ELECTR| An abnormally high current, usually resulting from a short circuit. { ¦ō·vər¦kə·rənt }

overcurrent protection See overload protection. { ¦ō·vər¦kə·rənt prə'tek·shən }

overdriven amplifier |ELECTR| Amplifier stage which is designed to distort the input-signal waveform by permitting the grid signal to drive the stage beyond cutoff or plate-current saturation. { ¦ō·vər¦driv·ən 'am·plə‚fī·ər }

overflow |COMPUT SCI| **1.** The condition that arises when the result of an arithmetic operation exeeds the storage capacity of the indicated result-holding storage. **2.** That part of the result which exceeds the storage capacity. { 'ō·vər‚flō }

overflow bucket |COMPUT SCI| A unit of storage in a direct-access storage device used to hold an overflow record. { 'ō·vər‚flō ‚bək·ət }

overflow check indicator See overflow indicator. { 'ō·vər‚flō ‚chek ‚in·də‚kād·ər }

overflow error |COMPUT SCI| The condition in which the numerical result of an operation exceeds the capacity of the register. { 'ō·vər‚flō 'er·ər }

overflow indicator |COMPUT SCI| A bistable device which changes state when an overflow occurs in the register associated with it, and which is designed so that its condition can be determined, and its original condition restored. Also known as overflow check indicator. { 'ō·vər‚flō ‚in·də ‚kād·ər }

overflow record |COMPUT SCI| A unit of data whose length is too great for it to be stored in an assigned section of a direct-access storage, and which must be stored in another area from which it may be retrieved by means of a reference stored in the original assigned area in place of the record. { 'ō·vər‚flō ‚rek·ərd }

overflow storage |COMMUN| Additional storage provided in a store-and-forward-switching center to prevent the loss of messages (or parts of messages) offered to the switching center when it is fulfilled. |COMPUT SCI| Extra storage capacity in a computer or calculator that allows a small amount of overflow. { 'ō·vər‚flō ‚stȯr·ij }

overhead |COMPUT SCI| The time a computer system spends doing computations that do not contribute directly to the progress of any user tasks in the system, such as allocation of resources, responding to exceptional conditions, providing protection and reliability, and accounting. { 'ō·vər‚hed }

overlap |COMMUN| **1.** In teletypewriter practice, the selecting of another code group while the printing of a previously selected code group is taking place. **2.** Amount by which the effective height of the scanning facsimile spot exceeds the nominal width of the scanning line. |COMPUT SCI| To perform some or all of an operation concurrently with one or more other operations. { 'ō·vər‚lap }

overlapped memories |COMPUT SCI| An arrangement of computer memory banks in which, to cut down access time, successive words are taken from different memory banks, rewriting in one bank being overlapped by logic operations in

another bank, with memory access in still another bank. { 'ō·vər‚lapt 'mem·rēz }

overlapping |COMPUT SCI| An operation whereby, if the processor determines that the current instruction and the next instruction lie in different storage modules, the two words may be retrieved in parallel. { ¦ō·vər¦lap·iŋ }

overlapping input/output |COMPUT SCI| A procedure in which a computer system works on several programs, suspending work on a program and moving to another when it encounters an instruction for input/output operation, which is then executed when input/output operations from other programs have been carried out. { ¦ō·vər¦lap·iŋ 'in‚pút 'aút‚pút }

overlap radar |ENG| Radar located in one sector whose area of useful radar coverage includes a portion of another sector. { 'ō·vər‚lap 'rā‚där }

overlay |COMPUT SCI| A technique for bringing routines into high-speed storage from some other form of storage during processing, so that several routines will occupy the same storage locations at different times; overlay is used when the total storage requirements for instructions exceed the available main storage. { 'ō·vər‚lā }

overlay transistor |ELECTR| Transistor containing a large number of emitters connected in parallel to provide maximum power amplification at extremely high frequencies. { 'ō·vər‚lā tran'zis·tər }

overload |ELECTR| A load greater than that which a device is designed to handle; may cause overheating of power-handling components and distortion in signal circuits. { 'ō·vər‚lōd }

overload capacity |ELEC| Current, voltage, or power level beyond which permanent damage occurs to the device considered. { 'ō·vər‚lōd kə ‚pas·əd·ē }

overload current |ELECTR| A current greater than that which a circuit is designed to carry; may melt wires or damage elements of the circuit. { 'ō·vər ‚lōd ‚kə·rənt }

overloading |COMPUT SCI| The use, in some advanced programming languages, of two or more variables or subroutines with the same name; the compiler determines by inference which entity is referred to each time the name occurs. { ¦ō·vər ¦lōd·iŋ }

overload level |ELEC| Level above which operation ceases to be satisfactory as a result of signal distortion, overheating, damage, and so forth. { 'ō·vər‚lōd ‚lev·əl }

overload protection |ELEC| Effect of a device operative on excessive current, but not necessarily on short circuit, to cause and maintain the interruption of current flow to the device governed. Also known as overcurrent protection. { 'ō·vər‚lōd prə‚tek·shən }

overload relay |ELEC| A relay that opens a circuit when the load in the circuit exceeds a preset value, in order to provide overload protection; usually responds to excessive current, but may respond to excessive values of power, temperature, or other quantities. Also known as overload release. { 'ō·vər‚lōd ‚rē‚lā }

overload release See overload relay. { 'ō·vər‚lōd ri‚lēs }

overmodulation |COMMUN| Amplitude modulation greater than 100%, causing distortion because the carrier voltage is reduced to zero during portions of each cycle. { ¦ō·vər‚mäj·ə'lā·shən }

overpotential See overvoltage. { ¦ō·vər·pə'ten·chəl }

override |CONT SYS| To cancel the influence of an automatic control by means of a manual control. { 'ō·və‚rīd }

overriding process control |CONT SYS| Process control in which any one of several controllers associated with one control valve can be made to override another in accordance with a priority requirement of the process. { 'ō·və‚rīd·iŋ 'prä·səs kən‚trōl }

overrun |COMPUT SCI| The arrival of an amount of data greater than the space allocated to it. { 'ō·və‚rən }

overshoot |ELECTROMAG| The reception of microwave signals where they were not intended, due to an unusual atmospheric condition that sets up variations in the index of refraction. { 'ō·vər‚shüt }

over-the-horizon propagation See scatter propagation. { 'ō·vər thə hə'rīz·ən ‚präp·ə'gā·shən }

over-the-horizon radar |ELECTROMAG| Radar operating in such a way that targets otherwise shielded from view by earth's curvature are detected; the use of carrier frequencies at which the ionosphere is particularly reflective, so that radar signals are reflected back to the surface at great ranges, or use of signal characteristics exploiting surface-coupled propagation as example techniques. Abbreviated OTH radar. { 'ō·vər thə hə'rīz·ən 'rā‚där }

overthrow distortion |COMMUN| Distortion caused when the maximum amplitude of the signal wavefront exceeds the steady state of amplitude of the signal wave. { 'ō·vər‚thrō di ‚stór·shən }

overtone crystal |ELECTR| Quartz crystal cut in such a manner that it will operate at a higher order than its fundamental frequency, or operate at two frequencies simultaneously as in a synthesizer. { 'ō·vər‚tōn ‚krist·əl }

overvoltage |ELEC| A voltage greater than that at which a device or circuit is designed to operate. Also known as overpotential. |ELECTR| The amount by which the applied voltage exceeds the Geiger threshold in a radiation counter tube. { ¦ō·vər¦vōl·tij }

overvoltage crowbar |ELEC| A circuit that monitors the output of a power supply and prevents the output voltage from exceeding a preset voltage, under any failure condition, by having a low resistance (crowbar) placed across the output terminals when an overvoltage occurs. { ¦ō·vər ¦vōl·tij 'krō‚bär }

overwrite |COMPUT SCI| To enter information into a storage location and destroy the information previously held there. { ¦ō·vər¦rīt }

ovonic device See glass switch. { ō'vän·ik di‚vīs }

ovonic memory switch |ELECTR| A glass switch which, after being brought from the highly resistive state to the conducting state, remains in the conducting state until a current pulse returns it to its highly resistive state. Abbreviated OMS. Also known as memory switch. { ō'vän·ik 'mem·rē ‚swich }

ovonic threshold switch |ELECTR| A glass switch which, after being brought from the highly resistive state to the conducting state, returns to the highly resistive state when the current falls below a holding current value. Abbreviated OTS. { ō'vän·ik 'thresh‚hōld ‚swich }

Ovshinsky effect |ELECTR| The characteristic of a special thin-film solid-state switch that responds identically to both positive and negative polarities so that current can be made to flow in both directions equally. { ōv'shin·skē i‚fekt }

Owen bridge |ELECTR| A four-arm alternating-current bridge used to measure self-inductance in terms of capacitance and resistance; bridge balance is independent of frequency. { 'ō·wən ‚brij }

own coding |COMPUT SCI| A series of instructions added to a standard software routine to change or extend the routine so that it can carry out special tasks. { 'ōn 'kōd·iŋ }

owned program See proprietary program. { 'ōnd 'prō‚gram }

oxide-coated cathode |ELECTR| A cathode that has been coated with oxides of alkaline-earth metals to improve electron emission at moderate temperatures. Also known as Wehnelt cathode. { 'äk‚sīd ‚kōd·əd 'kath‚ōd }

oxide isolation |ELECTR| Isolation of the elements of an integrated circuit by forming a layer of silicon oxide around each element. { 'äk‚sīd ‚ī·sə'lā·shən }

oxide passivation |ELECTR| Passivation of a semiconductor surface by producing a layer of an insulating oxide on the surface. { 'äk‚sīd ‚pas·ə·vā·shən }

P

pA See picoampere.

PABX See private automatic branch exchange.

PAC See perceptual audio coding.

pack |COMPUT SCI| To reduce the amount of storage required to hold information by changing the method of encoding the data. { pak }

package |COMPUT SCI| A program that is written for a general and widely used application in such a way that its usefulness is not impaired by the problems of data or organization of a particular user. { 'pak·ij }

packaged circuit See rescap. { 'pak·ijd ¦sər·kət }

packaged magnetron |ELECTR| Integral structure comprising a magnetron, its magnetic circuit, and its output matching device. { 'pak·ijd 'mag·nə ‚trän }

packaging |ELEC| The process of physically locating, connecting, and protecting devices or components. { 'pak·ə·jiŋ }

packaging density |ELECTR| The number of components per unit volume in a working system or subsystem. { 'pak·ə·jiŋ ‚den·səd·ē }

packed decimal |COMPUT SCI| A means of representing two digits per character, to reduce space and increase transmission speed. { 'pakt 'des· məl }

packed file |COMPUT SCI| A file that has been encoded so that it takes up less space in storage. Also known as compressed file. { ¦akt 'fīl }

packet |COMMUN| A short section of data of fixed length that is transmitted as a unit; consists of a header followed by a number of contiguous bytes from an elementary data stream. { 'pak·ət }

packetized elementary stream |COMMUN| A generic term for a coded bit stream in a digital transport system. In a digital television system, one coded video, coded audio, or other coded elementary stream is carried in a sequence of PES packets with one stream identification code. { 'pak·ə¦tīzd ‚el·ə'men·trē 'strēm }

packet switching See packet transmission. { 'pak· ət ‚swich·iŋ }

packet transmission |COMMUN| Transmission of standardized packets of data over transmission lines rapidly by networks of high-speed switching computers that have the message packets stored in fast-access core memory. Also known as packet switching. { 'pak·ət tranz‚mish·ən }

packing density |COMPUT SCI| The amount of information per unit of storage medium, as characters per inch on tape, bits per inch or drum, or bits per square inch in photographic storage. |ELECTR| The number of devices or gates per unit area of an integrated circuit. { 'pak·iŋ ‚den·səd·ē }

packing routine |COMPUT SCI| A subprogram which compresses data so as to eliminate blanks and reduce the storage needed for a file. { 'pak·iŋ rü‚tēn }

pad |ELECTR| **1.** An arrangement of fixed resistors used to reduce the strength of a radio-frequency or audio-frequency signal by a desired fixed amount without introducing appreciable distortion. Also known as fixed attenuator. **2.** See terminal area. { pad }

padder |ELECTR| A trimmer capacitor inserted in series with the oscillator tuning circuit of a superheterodyne receiver to control calibration at the low-frequency end of a tuning range. { 'pad·ər }

padding |COMPUT SCI| The adding of meaningless data (usually blanks) to a unit of data to bring it up to some fixed size. { 'pad·iŋ }

page |COMPUT SCI| **1.** A standard quantity of main-memory capacity, usually 512 to 4096 bytes or words, used for memory allocation and for partitioning programs into control sections. **2.** A standard quantity of source program coding, usually 8 to 64 lines, used for displaying the coding on a cathode-ray tube. { pāj }

pageable memory |COMPUT SCI| The part of a computer's main storage that is subject to paging in a virtual storage system. { 'pāj·ə·bəl 'mem·rē }

page boundary |COMPUT SCI| The address of the first (lowest) word or byte within a page of memory. { 'pāj ‚baún·drē }

page data set |COMPUT SCI| A file for storing images of pages in a virtual storage system, so that they can be returned to main storage for further processing when needed. { 'pāj 'dad·ə ‚set }

page description language |COMPUT SCI| A high-level language that specifies the format of a page generated by a printer; it is translated into specific codes by any printer that supports the language. Abbreviated PDL. { 'pāj di‚skrip·shən ‚laŋ·gwij }

page fault |COMPUT SCI| An interruption that occurs while a page which is referred to by the program is being read into memory. { 'pāj ‚fólt }

page printer |COMPUT SCI| A computer output device which composes a full page of characters before printing the page. { 'pāj ,print·ər }

pager |COMMUN| A receiver in a radio paging system. { 'pāj·ər }

page reader |COMPUT SCI| In character recognition, a character reader capable of processing cut-form documents of varying sizes; sometimes capable of reading information in reel forms. { 'pāj ,rēd·ər }

page skip |COMPUT SCI| A control character that causes a printer to skip over the remainder of the current page and move to the beginning of the following page. { 'pāj ,skip }

page table |COMPUT SCI| A key element in the virtual memory technique; a table of addresses where entries are adjusted for easy relocation of pages. { 'pāj ,tā·bəl }

page turning |COMPUT SCI| **1.** The process of moving entire pages of information between main memory and auxiliary storage, usually to allow several concurrently executing programs to share a main memory of inadequate capacity. **2.** In conversational time-sharing systems, the moving of programs in and out of memory on a round-robin, cyclic schedule so that each program may use its allotted share of computer time. { 'pāj ,tərn·iŋ }

paging |COMPUT SCI| The scheme used to locate pages, to move them between main storage and auxiliary storage, or to exchange them with pages of the same or other computer programs; used in computers with virtual memories. { 'pāj·iŋ }

paging rate |COMPUT SCI| The number of pages per second moved by virtual storage between main storage and the page data set. { 'pāj·iŋ ,rāt }

paging system |COMMUN| A system which gives an indication to a particular individual that he or she is wanted at the telephone, such as by sounding a number, calling by name over a loudspeaker, or producing an audible signal in a radio receiver carried in the individual's pocket. { 'pāj·iŋ ,sis·təm }

paint |COMPUT SCI| To fill an area of a display screen or printed output with a color, shade of gray, or image. |ELECTR| In radar, a colloqual term for an echo signal or its display; sometimes called the "skin paint," as of an aircraft. { pānt }

paint program |COMPUT SCI| A graphics program that maintains images in raster format, allowing the user to simulate painting with the aid of a mouse or a graphics tablet. { 'pānt ,prō·grəm }

pair |ELEC| Two like conductors employed to form an electric circuit. { per }

paired cable |ELEC| Cable in which the single conductors are twisted together in groups of two, none of which is arranged with others to form quads. { 'perd 'kā·bəl }

paired synchronous detection |ELECTR| The arrangement of two homodyne channels in a radar receiver such that both the phase and the amplitude of a received signal is preserved in the two video signals produced. { ¦perd 'siŋ·krə·nəs di,tek·shən }

pairing |ELECTR| In television, imperfect interlace of lines composing the two fields of one frame of the picture; instead of having the proper equal spacing, the lines appear in groups of two. { 'per·iŋ }

palette |COMPUT SCI| In computer graphics, the set of colors that can be shown on a display monitor. { 'pal·ət }

Palmer scan |ELECTR| Combination of circular or raster and conical radar scans; the beam is swung around the horizon, and at the same time a conical scan is performed. { 'päm·ər ,skan }

palmtop See hand-held computer. { 'päm,täp }

PAL system See phase-alternation line system. { 'pal ,sis·təm }

PAM See pulse-amplitude modulation.

panadapter See panoramic adapter. { 'pan·ə ,dap·tər }

pancake coil |ELEC| A coil having the shape of a pancake, usually with the turns arranged in the form of a flat spiral. { 'pan,kāk ¦kȯil }

panel |COMPUT SCI| The face of the console, which is normally equipped with lights, switches, and buttons to control the machine, correct errors, determine the status of the various CPU (central processing unit) parts, and determine and revise the contents of various locations. Also known as control panel; patch panel. { 'pan·əl }

panel board See control board. { 'pan·əl ,bȯrd }

panel display |ELECTR| An electronic display in which a large orthogonal array of display devices, such as electroluminescent devices or light-emitting diodes, form a flat screen. Also known as flat-panel display. { 'pan·əl di,splā }

panoramic adapter |ELECTR| A device designed to operate with a search receiver to provide a visual presentation on an oscilloscope screen of a band of frequencies extending above and below the center frequency to which the search receiver is tuned. Also known as panadapter. { ¦pan·ə ¦ram·ik ə¦dap·tər }

panoramic display |ELECTR| A display that simultaneously shows the relative amplitudes of all signals received at different frequencies. { ¦pan·ə¦ram·ik di'splā }

panoramic radar |ENG| Nonscanning radar which transmits signals over a wide beam in the direction of interest. { ¦pan·ə¦ram·ik 'rā,där }

panoramic receiver |ELECTR| Radio receiver that permits continuous observation on a cathode-ray-tube screen of the presence and relative strength of all signals within a wide frequency range. { ¦pan·ə¦ram·ik ri'sē·vər }

pan-range |ELECTR| Intensity-modulated, A-type radar indication with a slow vertical sweep applied to video; stationary targets give solid vertical deflection, and moving targets give broken vertical deflection. { 'pan ,rānj }

pantograph |ENG| A device that sits on the top of an electric locomotive or cars in an electric train and picks up electricity from overhead wires to run the train. { 'pan·tə,graf }

pantography |ENG| System for transmitting and automatically recording radar data from an indicator to a remote point. { pan'täg·rə·fē }

paper capacitor |ELEC| A capacitor whose dielectric material consists of oiled paper sandwiched between two layers of metallic foil. { 'pā·pər kə'pas·əd·ər }

paper-tape Turing machine |COMPUT SCI| A variation of a Turing machine in which a blank square can have a nonblank symbol written on it, but this symbol cannot be changed thereafter. { 'pā·pər ¦tāp 'tür·iŋ mə,shēn }

paper throw |COMPUT SCI| The movement of paper through a computer printer for a purpose other than printing, in which the distance traveled, and usually the speed, is greater than that of a single line spacing. { 'pā·pər ,thrō }

paraballoon |ELECTROMAG| Air-inflated radar antenna. { ¦par·ə·bə'lün }

parabolic antenna |ELECTROMAG| Antenna with a radiating element and a parabolic reflector that concentrates the radiated power into a beam. { ¦par·ə¦bäl·ik an'ten·ə }

parabolic microphone |ENG ACOUS| A microphone used at the focal point of a parabolic sound reflector to give improved sensitivity and directivity, as required for picking up a band marching down a football field. { ¦par·ə¦bäl·ik 'mī·krə,fōn }

parabolic reflector |ELECTROMAG| **1.** An antenna having a concave surface which is generated either by translating a parabola perpendicular to the plane in which it lies (in a cylindrical parabolic reflector), or rotating it about its axis of symmetry (in a paraboloidal reflector). Also known as dish. **2.** See paraboloidal reflector. { ¦par·ə¦bäl·ik ri'flek·tər }

paraboloidal antenna See paraboloidal reflector. { pə¦rab·ə¦lȯid·əl an'ten·ə }

paraboloidal reflector |ELECTROMAG| An antenna having a concave surface which is a paraboloid of revolution; it concentrates radiation from a source at its focal point into a beam. Also known as paraboloidal antenna. Also known as parabolic reflector. { pə¦rab·ə¦lȯid·əl ri'flek·tər }

paragraph |COMPUT SCI| A complete, logical sequence of instructions in the COBOL programming language, required to carry out a definable program or task. { 'par·ə,graf }

parallel |COMPUT SCI| Simultaneous transmission of, storage of, or logical operations on the parts of a word, character, or other subdivision of a word in a computer, using separate facilities for the various parts. |ELEC| Connected to the same pair of terminals. Also known as multiple; shunt. { 'par·ə,lel }

parallel access |COMPUT SCI| Transferral of information to or from a storage device in which all elements in a unit of information are transferred simultaneously. Also known as simultaneous access. { 'par·ə,lel 'ak,ses }

parallel addition |COMPUT SCI| A method of addition by a computer in which all the corresponding pairs of digits of the addends are processed at the same time during one cycle, and one or more subsequent cycles are used for propagation and adjustment of any carries that may have been generated. { 'par·ə,lel ə'dish·ən }

parallel algorithm |COMPUT SCI| An algorithm in which several computations are carried on simultaneously. { 'par·ə,lel 'al·gə,rith·əm }

parallel buffer |ELECTR| Electronic device (magnetic core or flip-flop) used to temporarily store digital data in parallel, as opposed to series storage. { 'par·ə,lel 'bəf·ər }

parallel by character |COMPUT SCI| The handling of all the characters of a machine word simultaneously in separate lines, channels, or storage cells. { 'par·ə,lel bī 'kar·ik·tər }

parallel circuit |ELEC| An electric circuit in which the elements, branches (having elements in series), or components are connected between two points, with one of the two ends of each component connected to each point. { 'par·ə ,lel 'sər·kət }

parallel communications |COMMUN| The simultaneous transmission of data over two or more communications channels. { 'par·ə,lel kə ,myü·nə'kā·shənz }

parallel compensation See feedback compensation. { 'par·ə,lel ,käm·pən'sā·shən }

parallel computation |COMPUT SCI| The simultaneous computation of several parts of a problem. { 'par·ə,lel ,käm·pyü'tā·shən }

parallel computer |COMPUT SCI| **1.** A computer that can carry out more than one logic or arithmetic operation at one time. **2.** See parallel digital computer. { 'par·ə,lel kəm'pyüd·ər }

parallel conversion |COMPUT SCI| The process of transferring operations from one computer system to another, during which both systems are run together for a period of time to ensure that they are producing identical results. { 'par·ə,lel kən'vər·zhən }

parallel digital computer |COMPUT SCI| Computer in which the digits are handled in parallel; mixed serial and parallel machines are frequently called serial or parallel, according to the way arithmetic processes are performed; an example of a parallel digital computer is one which handles decimal digits in parallel, although it might handle the bits constituting a digit either serially or in parallel. { 'par·ə,lel 'dij·əd·əl kəm'pyüd·ər }

parallel dot character printer See line dot matrix. { 'par·ə,lel ¦dät 'kar·ik·tər ,print·ər }

parallel element-processing ensemble |COMPUT SCI| A powerful electronic computer used by the U.S. Army to simulate tracking and discrimination of reentry vehicles as part of the ballistic missile defense research program. Abbreviated PEPE. { 'par·ə,lel 'el·ə·mənt ¦prä,ses·iŋ än ,säm·bəl }

parallel feed |ELECTR| Application of a direct-current voltage to the plate or grid of a tube in parallel with an alternating-current circuit, so that the direct-current and the alternating-current components flow in separate paths. Also known as shunt feed. { 'par·ə,lel 'fēd }

parallel flow |ELEC| Also known as loop flow. **1.** The flow of electric current from one point to another in an electric network over multiple paths, in accordance with Kirchhoff's laws. **2.** In particular, the flow of electric current through electric power systems over paths other than the contractual path. { 'par·ə,lel 'flō }

parallel gripper [CONT SYS] A robot end effector made up of two jawlike components that grasp objects. { 'par·ə,lel 'grip·ər }

parallel impedance [ELEC] One of two or more impedances that are connected to the same pair of terminals. { 'par·ə,lel im'pēd·əns }

parallel input/output [COMPUT SCI] Data that are transmitted into and out of a computer over several conductors simultaneously. { 'par·ə,lel 'in,pút 'aút,pút }

parallel interface [ELECTR] A link between two devices in which all the information transferred between them is transmitted simultaneously over separate conductors. Also known as parallel port. { 'par·ə,lel 'in·tər,fās }

parallel operation [COMPUT SCI] Performance of several actions, usually of a similar nature, by a computer system simultaneously through provision of individual similar or identical devices. { 'par·ə,lel ,äp·ə'rā·shən }

parallel padding [ELEC] Method of parallel operation for two or more power supplies in which their current limiting or automatic crossover output characteristic is employed so that each supply regulates a portion of the total current, each parallel supply adding to the total and padding the output only when the load current demand exceeds the capability, or limit setting, of the first supply. { 'par·ə,lel 'pad·iŋ }

parallel-plate capacitor [ELEC] A capacitor consisting of two parallel metal plates, with a dielectric filling the space between them. { 'par·ə,lel ¦plāt kə'pas·əd·ər }

parallel-plate waveguide [ELECTROMAG] Pair of parallel conducting planes used for propagating uniform circularly cylindrical waves having their axes normal to the plane. { 'par·ə,lel ¦plāt 'wāv ,gīd }

parallel port See parallel interface. { 'par·ə,lel ,pórt }

parallel processor See multiprocessor. { 'par·ə ,lel 'prä,ses·ər }

parallel programming [COMPUT SCI] A method for performing simultaneously the normally sequential steps of a computer program, using two or more processors. { 'par·ə,lel 'prō,gram·iŋ }

parallel radio tap [COMMUN] A telephone tapping procedure in which a battery-powered miniature radio transmitter is bridged across the target pair. { 'par·ə,lel 'rād·ē·ō ,tap }

parallel rectifier [ELECTR] One of two or more rectifiers that are connected to the same pair of terminals, generally in series with small resistors or inductors, when greater current is desired than can be obtained with a single rectifier. { 'par·ə ,lel 'rek·tə,fī·ər }

parallel reliability [SYS ENG] Property of a system composed of functionally parallel elements in such a way that if one of the elements fails, the parallel units will continue to carry out the system function. { 'par·ə,lel ri'lī·ə,bil·əd·ē }

parallel representation [COMPUT SCI] The simultaneous appearance of the different bits of a digital variable on parallel bus lines. { 'par·ə ,lel ,rep·ri,zen'tā·shən }

parallel resonance [ELEC] Also known as antiresonance. **1.** The frequency at which the inductive and capacitive reactances of a parallel resonant circuit are equal. **2.** The frequency at which the parallel impedance of a parallel resonant circuit is a maximum. **3.** The frequency at which the parallel impedance of a parallel resonant circuit has a power factor of unity. { 'par·ə,lel 'rez·ən·əns }

parallel resonant circuit [ELEC] A circuit in which an alternating-current voltage is applied across a capacitor and a coil in parallel. Also known as antiresonant circuit. { 'par·ə,lel 'rez·ən·ənt ,sər·kət }

parallel resonant interstage [ELECTR] A coupling between two amplifier stages achieved by means of a parallel-tuned LC circuit. { 'par·ə,lel 'rez·ən·ənt 'in·tər,stāj }

parallel-rod oscillator [ELECTR] Ultra-high-frequency oscillator circuit in which parallel rods or wires of required length and dimensions form the tank circuits. { 'par·ə,lel ¦räd 'äs·ə,lād·ər }

parallel running [COMPUT SCI] **1.** The running of a newly developed system in a data-processing area in conjunction with the continued operation of the current system. **2.** The final step in the debugging of a system; this step follows a system test. { 'par·ə,lel 'rən·iŋ }

parallel search storage [COMPUT SCI] A device for very rapid search of a volume of stored data to permit finding a specific item. { 'par·ə,lel ¦sərch ,stór·ij }

parallel series [ELEC] Circuit in which two or more parts are connected together in parallel to form parallel circuits, and in which these circuits are then connected together in series so that both methods of connection appear. { 'par·ə,lel 'sir·ēz }

parallel storage [COMPUT SCI] A storage device in which words (or characters or digits) can be read in or out simultaneously. { 'par·ə,lel 'stór·ij }

parallel-T network [ELEC] A network used in capacitance measurements at radio frequencies, having two sets of three impedances, each in the form of the letter T, with the arms of the two T's joined to common terminals, and the source and detector each connected between two of these terminals. Also known as twin-T network. { 'par·ə,lel ¦tē 'net,wərk }

parallel transfer [COMPUT SCI] Simultaneous transfer of all bits in a storage location constituting a character or word. { 'par·ə,lel 'tranz·fər }

parallel transmission [COMPUT SCI] The transmission of characters of a word over different lines, usually simultaneously; opposed to serial transmission. { 'par·ə,lel tranz'mish·ən }

parallel-tuned circuit [ELEC] A circuit with two parallel branches, one having an inductance and a resistance in series, the other a capacitance and a resistance in series. { 'par·ə,lel ,tünd 'sər·kət }

parallel wires [ELEC] Two conductors which are parallel to each other; often used in transmission lines. { 'par·ə,lel 'wīrz }

parameter |ELEC| **1.** The resistance, capacitance, inductance, or impedance of a circuit element. **2.** The value of a transistor or tube characteristic. |MATH| An arbitrary constant or variable so appearing in a mathematical expression that changing it gives various cases of the phenomenon represented. { pə'ram·əd·ər }

parameter-driven system |COMPUT SCI| A software system whose functions and operations are controlled mainly by parameters. { pə'ram·əd·ər ¦driv·ən 'sis·təm }

parameter identification |SYS ENG| The problem of estimating the values of the parameters that govern a dynamical system from data on the observed behavior of the system. { 'pə'ram·əd·ər ī, dent·ə·fə'kā·shən }

parameter tags |COMPUT SCI| Constants that are used by several computer programs. { pə'ram·əd·ər ,tagz }

parameter word |COMPUT SCI| A word in a computer storage containing one or more parameters that specify the action of a routine or subroutine. { pə'ram·əd·ər ,wərd }

parametric amplifier |ELECTR| A highly sensitive ultra-high-frequency or microwave amplifier having as its basic element an electron tube or solid-state device whose reactance can be varied periodically by an alternating-current voltage at a pumping frequency. Also known as mavar; paramp; reactance amplifier. { ¦par·ə¦me·trik 'am·plə,fī·ər }

parametric converter |ELECTR| Inverting or non-inverting parametric device used to convert an input signal at one frequency into an output signal at a different frequency. { ¦par·ə¦me·trik kən'vərd·ər }

parametric device |ELECTR| Electronic device whose operation depends essentially upon the time variation of a characteristic parameter usually understood to be a reactance. { ¦par·ə ¦me·trik di'vīs }

parametric down-converter |ELECTR| Parametric converter in which the output signal is at a lower frequency than the input signal. { ¦par·ə¦me·trik 'daún kən,vərd·ər }

parametric equalizer |ENG ACOUS| A device that allows control over the center frequencies, bandwidths, and amplitudes (parameters) of band-pass filters that determine the frequency response of audio equipment. { ¦par·ə¦me·trik ,ē·kwə'līz·ər }

parametric excitation |ENG| The method of exciting and maintaining oscillations in either an electrical or mechanical dynamic system, in which excitation results from a periodic variation in an energy storage element in a system such as a capacitor, inductor, or spring constant. { ¦par·ə ¦me·trik ,ek·si'tā·shən }

parametric oscillator |ELECTR| An oscillator in which the reactance parameter of an energy-storage device is varied to obtain oscillation. |OPTICS| A device consisting of an optically nonlinear crystal surrounded by a pair of mirrors to which is applied a relatively high-frequency laser beam and a relatively low-frequency signal, resulting in a low-

frequency output whose frequency can be varied, usually by varying the indices of refraction. { ¦par·ə ¦me·trik 'äs·ə,lād·ər }

parametric phase-locked oscillator See parametron. { ¦par·ə¦me·trik 'fāz ,läkt 'äs·ə,lād·ər }

parametric programming |COMPUT SCI| A programming approach in which data are stored in external tables or files, rather than within the program itself, and accessed by the program when needed, so that the values of these data can be changed with relative ease. { ¦par·ə¦me·trik 'prō,gram·iŋ }

parametric up-converter |ELECTR| Parametric converter in which the output signal is at a higher frequency than the input signal. { ¦par·ə ¦me·trik 'əp kən,vərd·ər }

parametrized voice response system |ENG ACOUS| A voice response system which first extracts informative parameters from human speech, such as natural resonant frequencies (formants) of the speaker's vocal tract and the fundamental frequency (pitch) of the voice, and which later reconstructs speech from such stored parameters. { pə'ram·ə,trīzd 'vóis ri ,späns ,sis·təm }

parametron |ELECTR| A resonant circuit in which either the inductance or capacitance is made to vary periodically at one-half the driving frequency; used as a digital computer element, in which the oscillation represents a binary digit. Also known as parametric phase-locked oscillator; phase-locked oscillator; phase-locked subharmonic oscillator. { pə'ram·ə,trän }

paramp See parametric amplifier. { 'par,amp }

paraphase amplifier |ELECTR| An amplifier that provides two equal output signals 180° out of phase. { 'par·ə,fāz 'am·plə,fī·ər }

parasite |ELEC| Current in a circuit, due to some unintentional cause, such as inequalities of temperature or of composition; particularly troublesome in electrical measurements. { 'par·ə ,sīt }

parasitic |ELECTR| An undesired and energy-wasting signal current, capacitance, or other parameter of an electronic circuit. { ¦par·ə ¦sid·ik }

parasitic antenna See parasitic element. { ¦par·ə ¦sid·ik an'ten·ə }

parasitic current |ELEC| An eddy current in a piece of electrical machinery; gives rise to energy losses. { ¦par·ə¦sid·ik 'kə·rənt }

parasitic element |ELECTROMAG| An antenna element that serves as part of a directional antenna array but has no direct connection to the receiver or transmitter and reflects or reradiates the energy that reaches it, in a phase relationship such as to give the desired radiation pattern. Also known as parasitic antenna; parasitic reflector; passive element. { ¦par·ə¦sid·ik 'el·ə·mənt }

parasitic oscillation |ELECTR| An undesired self-sustaining oscillation or a self-generated transient impulse in an oscillator or amplifier circuit, generally at a frequency above or below the correct operating frequency. { ¦par·ə¦sid·ik ,äs·ə'lā·shən }

parasitic reflector See parasitic element. { ¦par·ə ¦sid·ik ri'flek·tər }

parasitic suppressor |ELECTR| A suppressor, usually in the form of a coil and resistor in parallel, inserted in a circuit to suppress parasitic high-frequency oscillations. { ¦par·ə¦sid·ik sə' pres·ər }

paraxial trajectory |ELEC| A trajectory of a charged particle in an axially symmetric electric or magnetic field in which both the distance of the particle from the axis of symmetry and the angle between this axis and the tangent to the trajectory are small for all points on the trajectory. { par'ak·sē·əl trə'jek·trē }

parent |COMPUT SCI| An element that precedes a given element in a data structure. { 'per·ənt }

parenthesis-free notation See Polish notation. { pə'ren·thə·səs ¦frē nō'tā·shən }

parity |COMPUT SCI| The use of a self-checking code in a computer employing binary digits in which the total number of 1's or 0's in each permissible code expression is always even or always odd. { 'par·əd·ē }

parity bit |COMMUN| An additional nondata bit that is attached to a set of data bits to check their validity; it is set so that the sum of one-bits in the augmented set is always odd or always even. { 'par·əd·ē ‚bit }

parity check See odd-even check. { 'par·əd·ē ‚chek }

parity error |COMPUT SCI| A machine error in which an odd number of bits are accidentally changed, so that the error can be detected by a parity check. { 'par·əd·ē ‚er·ər }

parity transformation |COMMUN| A change in value of a transmitted character denoting the number of one-bits. { 'par·əd·ē ‚tranz·fər'mä·shən }

parser |COMPUT SCI| The portion of a computer program that carries out parsing operations. { 'pär·sər }

parsing |COMPUT SCI| A process whereby phrases in a string of characters in a computer language are associated with the component names of the grammar that generated the string. { 'pärs·iŋ }

partial carry |COMPUT SCI| A word composed of the carries generated at each position when adding many digits in parallel. { 'pär·shəl 'kar·ē }

partial common battery |COMMUN| Type of telephone system in which the talking battery is supplied by each individual telephone, and the signaling and supervisory battery is supplied by the switchboard. { 'pär·shəl 'käm·ən 'bad·ə·rē }

partial function |COMPUT SCI| A partial function from a set A to a set B is a correspondence between some subset of A and B which associates with each element of the subset of A a unique element of B. { 'pär·shəl 'fəŋk·shən }

partially populated board |COMPUT SCI| A printed circuit board on which some but not all of the possible electronic components are mounted, leaving room for additional components. { 'pär·shə·lē ¦päp·yə‚lād·əd 'bórd }

partial-read pulse |ELECTR| Current pulse that is applied to a magnetic memory to select a specific magnetic cell for reading. { 'pär·shəl 'rēd 'pəls }

partial-response maximum-likelihood technique |COMMUN| A method of constructing a digital data stream from an analog signal by using information acquired by sampling the analog waveform at selected instants of time rather than using the entire waveform, and then applying the Viterbi algorithm to find the most likely sequence of bits. Abbreviated PRML technique. { ¦pär·shəl ri‚späns 'mak·sə·məm ¦līk·lē‚húd tek ‚nēk }

partial-select output |ELECTR| The voltage response produced by applying partial-read or partial-write pulses to an unselected magnetic cell. { 'pär·shəl sə‚lekt 'aút‚pút }

partition |COMPUT SCI| **1.** A reserved portion of a computer memory, sometimes used for the execution of a single computer program. **2.** One of a number of fixed portions into which a computer memory is divided in certain multiprogramming systems. { pär'tish·ən }

partitioned data set |COMPUT SCI| A single data set, divided internally into a directory and one or more sequentially organized subsections called members, residing on a direct access for each device, and commonly used for storage or program libraries. { pär'tish·ənd 'dad·ə ‚set }

partitioned display |COMPUT SCI| An electronic display that can be divided into two or more viewing areas under user or program control. Also known as split screen. { pär'tish·ənd di'splä }

partitioned file |COMPUT SCI| A file on disk storage that is divided into subdivisions, each of which constitutes a complete file. { pär'tish·ənd 'fīl }

partition noise |ELECTR| Noise that arises in an electron tube when the electron beam is divided between two or more electrodes, as between screen grid and anode in a pentode. { pär'tish·ən ‚nóiz }

part operation |COMPUT SCI| The part in an instruction that specifies the kind of arithmetical or logical operation to be performed, but not the address of the operands. { 'pärt ‚äp·ə‚rā·shən }

part programming |CONT SYS| The planning and specification of the sequence of steps or events in the operation of a numerically controlled machine tool. { 'pärt ‚prō‚gram·iŋ }

party line |COMMUN| A subscriber line arranged to serve more than one station, with discriminatory ringing for each station. { 'pärd·ē 'līn }

party-line bus |COMPUT SCI| Parallel input/output bus lines to which are wired all external devices, connected to a processor register by suitable logic. { 'pärd·ē ¦līn 'bəs }

party-line carrier system |COMMUN| A single-frequency carrier telephone system in which the carrier energy is transmitted directly to all other carrier terminals of the same channel. { 'pärd·ē ¦līn 'kar·ē·ər ‚sis·təm }

parylene capacitor |ELEC| A highly stable fixed capacitor using parylene film as the dielectric; it can be operated at temperatures up to 170°C, as well as at cryogenic temperatures. { 'par·ə‚lēn kə'pas·əd·ər }

Pascal |COMPUT SCI| A procedure-oriented programming language whose highly structured design facilitates the rapid location and correction of coding errors. { pa'skal }

Paschen's law [ELECTR] The law that the sparking potential between two parallel plate electrodes in a gas is a function of the product of the gas density and the distance between the electrodes. Also known as Paschen's rule. { 'päsh·ənz ,lȯ }

Paschen's rule See Paschen's law. { 'päsh·ənz ,rül }

pass [COMPUT SCI] A complete cycle of reading, processing, and writing in a computer. { pas }

passband [ELECTR] A frequency band in which the attenuation of a filter is essentially zero. { 'pas,band }

pass element [ELECTR] Controlled variable resistance device, either a vacuum tube or power transistor, in series with the source of direct-current power; the pass element is driven by the amplified error signal to increase its resistance when the output needs to be lowered or to decrease its resistance when the output must be raised. { 'pas ,el·ə·mənt }

passivation [ELECTR] Growth of an oxide layer on the surface of a semiconductor to provide electrical stability by isolating the transistor surface from electrical and chemical conditions in the environment; this reduces reverse-current leakage, increases breakdown voltage, and raises power dissipation rating. { ,pas·ə'vā·shən }

passive AND gate See AND gate. { 'pas·iv 'and ,gāt }

passive antenna [ELECTROMAG] An antenna which influences the directivity of an antenna system but is not directly connected to a transmitter or receiver. { 'pas·iv an'ten·ə }

passive component See passive element. { 'pas·iv kəm'pō·nənt }

passive corner reflector [ELECTROMAG] A corner reflector that is energized by a distant transmitting antenna; used chiefly to improve the reflection of radar signals from objects that would not otherwise be good radar targets. { 'pas·iv 'kȯr·nər ri,flek·tər }

passive device [COMPUT SCI] A unit of a computer which cannot itself initiate a request for communication with another device, but which honors such a request from another device. { 'pas·iv di'vīs }

passive double reflector [ELECTROMAG] A combination of two passive reflectors positioned to bend a microwave beam over the top of a mountain or ridge, generally without appreciably changing the general direction of the beam. { 'pas·iv 'dəb·əl ri'flek·tər }

passive electronic countermeasures [ELECTR] Electronic countermeasures that do not radiate energy, including reconnaissance or surveillance equipment that detects and analyzes electromagnetic radiation from radar and communications transmitters, and devices such as chaff which return confusing or obscuring echoes to enemy radar; passive electronic attack. { 'pas·iv i,lek'trän·ik 'kaȯnt·ər,mezh·ərz }

passive element [ELEC] An element of an electric circuit that is not a source of energy, such as a resistor, inductor, or capacitor. Also known as passive component. [ELECTROMAG] See parasitic element. { 'pas·iv 'el·ə·mənt }

passive filter [ELEC] An electric filter composed of passive elements, such as resistors, inductors, or capacitors, without any active elements, such as vacuum tubes or transistors. { 'pas·iv 'fil·tər }

passive jamming [ELECTR] Use of confusion reflectors to return spurious and confusing signals to enemy radars. Also known as mechanical jamming. { 'pas·iv 'jam·iŋ }

passive-matrix liquid-crystal display See supertwisted nematic liquid-crystal display. { ¦pas·iv ¦mā·triks ¦lik·wəd ¦krist·əl di'splā }

passive network [ELEC] A network that has no source of energy. { 'pas·iv 'net,wərk }

passive radar [ENG] A technique for detecting objects at a distance by picking up the microwave electromagnetic energy that is both radiated and reflected by all bodies. { 'pas·iv 'rā,där }

passive radiator [ENG ACOUS] A loudspeaker driver with no voice-coil or magnet assemblies that is mounted in a box with a woofer and exhibits a resonance that can be used to improve the low-frequency response of the system. { ¦pas·iv 'rād·ē,ād·ər }

passive-radiator system [ELECTR] A loudspeaker system in which the woofer is mounted in a box that also has a second speaker with no voice-coil or magnet assemblies. { ¦pas·iv 'rād·ē,ād·ər ,sis·təm }

passive reflector [ELECTROMAG] A flat reflector used to change the direction of a microwave or radar beam; often used on microwave relay towers to permit placement of the transmitter, repeater, and receiver equipment on the ground, rather than at the tops of towers. Also known as plane reflector. { 'pas·iv ri'flek·tər }

passive system [ELECTR] Electronic system which emits no energy, and does not give away its position or existence. { 'pas·iv ,sis·təm }

passive termination [COMPUT SCI] The simplest means of ending a chain of peripheral devices connected to a small computer system interface (SCSI) port, suitable for chains with no more than four devices. { ,pas·iv ,tər·mə'nā·shən }

passive transducer [ELECTR] A transducer containing no internal source of power. { 'pas·iv tranz'dü·sər }

passthrough [COMPUT SCI] A procedure that allows a user to communicate with a computer through the use of the operating system of a second computer. { 'pas,thrü }

password [COMPUT SCI] A unique word or string of characters that must be supplied to meet security requirements before a program, computer operator, or user can gain access to data. { 'pas ,wərd }

password guessing [COMPUT SCI] A method of gaining unauthorized access to a computing system by using computers and dictionaries or large word lists to try likely passwords. { 'pas ,wərd ,ges·iŋ }

paste [ELEC] In batteries, the medium in the form of a paste or jelly, containing an electrolyte; it is positioned adjacent to the negative electrode of a dry cell; in an electrolytic cell, the paste serves as one of the conducting plates. { pāst }

pasted-plate storage battery *See* Faure storage battery. { 'pās·təd ¦plāt 'stȯr·ij ‚bad·ə·rē }

patch |COMPUT SCI| **1.** To modify a program or routine by inserting a machine language correction in an object deck, or by inserting it directly into the computer through the console. **2.** The section of coding inserted in this way. |ELEC| A temporary connection between jacks or other terminations on a patch board. { pach }

patch board |ELEC| A board or panel having a number of jacks at which circuits are terminated; patch cords are plugged into the jacks to connect various circuits temporarily as required in broadcast, communication, and computer work. { 'pach ‚bȯrd }

patch cord |ELEC| A cord equipped with plugs at each end, used to connect two jacks on a patch board. { 'pach ‚kȯrd }

patch panel *See* control panel; panel. { 'pach ‚pan·əl }

path |COMPUT SCI| **1.** The logical sequence of instructions followed by a computer in carrying out a routine. **2.** A series of physical or logical connections between records or segments in a database management system, generally involving the use of pointers. { path }

path attenuation |COMMUN| Power loss between transmitter and receiver, due to any cause. { 'path ə‚ten·yə'wā·shən }

path computation |CONT SYS| The calculations involved in specifying the trajectory followed by a robot. { 'path ‚käm·pyə‚tā·shən }

path length *See* physical path length; software path length. { 'path ‚leŋkth }

path plotting |ELECTROMAG| In laying out a microwave system, the plotting of the path followed by the microwave beam on a profile chart which indicates the earth's curvature. { 'path ‚pläd·iŋ }

pattern analysis |COMPUT SCI| The phase of pattern recognition that consists of using whatever is known about the problem at hand to guide the gathering of data about the patterns and pattern classes, and then applying techniques of data analysis to help uncover the structure present in the data. { 'pad·ərn ə‚nal·ə·səs }

pattern generator |ELECTR| A signal generator used to produce a test waveform for service work on a display device. { 'pad·ərn ‚jen·ə‚rād·ər }

pattern recognition |COMPUT SCI| The automatic identification of figures, characters, shapes, forms, and patterns without active human participation in the decision process. { 'pad·ərn ‚rek·ig'nish·ən }

pattern-sensitive fault |COMPUT SCI| A fault that appears only in response to one pattern or sequence of data, or certain patterns or sequences. { 'pad·ərn ¦sen·səd·iv 'fȯlt }

PAX *See* private automatic exchange. { paks }

payload |COMMUN| Referring to the bytes which follow the header byte in a packet; the transport stream packet header and adaptation fields are not payload. { 'pā‚lȯd }

pay television *See* subscription television. { 'pā 'tel·ə‚vizh·ən }

P band |COMMUN| A band of radio frequencies extending from 225 to 390 megahertz, corresponding to wavelengths of 133.3 to 76.9 centimeters. { 'pē ‚band }

PBX *See* private branch exchange.

p-channel metal-oxide semiconductor *See* PMOS. { ¦pē ‚chan·əl ‚med·əl ¦äk‚sīd 'sem·i·kən‚dək·tər }

PCI *See* peripheral component interconnect.

P class |COMPUT SCI| The class of decision problems that can be solved in polynomial time. { ¦pē klas }

PCM *See* pulse-code modulation.

PCN *See* personal communications network.

PCP *See* primary control program.

PCR *See* program clock reference.

PCS *See* personal communications service.

PCSB *See* pulse-coded scanning beam.

PD *See* potential difference.

PDA *See* postacceleration.

PDF *See* portable document format.

P display *See* plan position indicator. { 'pē di ‚splā }

PDL *See* page description language.

PDM *See* pulse-duration modulation.

4PDT *See* four-pole double-throw.

PDU *See* power distribution unit.

peak attenuation |COMMUN| The diminution of response to a modulated wave experienced on modulation crests. { 'pēk ə‚ten·yə'wā·shən }

peak cathode current |ELECTR| **1.** Maximum instantaneous value of a periodically recurring cathode current. **2.** Highest instantaneous value of a randomly recurring pulse of cathode current. **3.** Highest instantaneous value of a nonrecurrent pulse of cathode current occurring under fault conditions. { 'pēk 'kath‚ōd ‚kə·rənt }

peak clipper *See* limiter. { 'pēk ‚klip·ər }

peak clipping |ELEC| Reduction of the maximum demand for electric power from an electrical utility, often achieved by direct control of customer loads by signals directed to customer appliances. |ELECTR| *See* limiting. { 'pēk ‚klip·iŋ }

peak detector |ELECTR| A detector whose output voltage approximates the true peak value of an applied signal; the detector tracks the signal in its sample mode and preserves the highest input signal in its hold mode. { 'pēk di‚tek·tər }

peak distortion |COMMUN| Largest total distortion of telegraph signals noted during a period of observation. { 'pēk di'stȯr·shən }

peak envelope power |ELECTR| Of a radio transmitter, the average power supplied to the antenna transmission line by a transmitter during one radio-frequency cycle at the highest crest of the modulation envelope, taken under conditions of normal operation. { 'pēk 'en·və‚lōp ‚paů·ər }

peaker |ELECTR| A small fixed or adjustable inductance used to resonate with stray and distributed capacitances in a broad-band amplifier to increase the gain at the higher frequencies. { 'pēk·ər }

peak forward voltage |ELECTR| The maximum instantaneous voltage applied to an electronic device in the direction of lesser resistance to current flow. { 'pēk 'fȯr·wərd 'vōl·tij }

peaking circuit |ELECTR| A circuit used to improve the high-frequency response of a broadband amplifier; in shunt peaking, a small coil is placed in series with the anode load; in series peaking, the coil is placed in series with the grid of the following stage. { 'pēk·iŋ ,sər·kət }

peaking network |ELECTR| Type of interstage coupling network in which an inductance is effectively in series (series-peaking network), or in shunt (shunt-peaking network), with the parasitic capacitance to increase the amplification at the upper end of the frequency range. { 'pēk·iŋ ,net ,wərk }

peaking transformer |ELEC| A transformer in which the number of ampere-turns in the primary is high enough to produce many times the normal flux density values in the core; the flux changes rapidly from one direction of saturation to the other twice per cycle, inducing a highly peaked voltage pulse in a secondary winding. { 'pēk·iŋ tranz,fȯr·mər }

peak inverse anode voltage |ELECTR| Maximum instantaneous anode voltage in the direction opposite to that in which the tube or other device is designed to pass current. { 'pēk 'in,vərs 'an ,ōd ,vōl·tij }

peak inverse voltage |ELECTR| Maximum instantaneous anode-to-cathode voltage in the reverse direction which is actually applied to the diode in an operating circuit. { 'pēk 'in,vərs ,vōl· tij }

peak limiter See limiter. { 'pēk ,lim·əd·ər }

peak load |ELEC| The maximum instantaneous load or the maximum average load over a designated interval of time. Also known as peak power. { 'pēk ,lōd }

peak power See peak load. { 'pēk 'pau̇·ər }

peak second algorithm |COMMUN| A set of mathematical procedures for attempting to predict the number of transmissions that will be carried out in a communications system during the busiest 1-second interval during some study period. { 'pēk 'sek·ənd 'al·gə,rith·əm }

peak signal level |ELECTR| Expression of the maximum instantaneous signal power or voltage as measured at any point in a facsimile transmission system; this includes auxiliary signals. { 'pēk 'sig·nəl ,lev·əl }

peak-to-valley ratio |COMMUN| The ratio of the largest amplitude of a modulated wave to its smallest value. { ¦pēk tə 'val·ē ,rā·shō }

peak value |ELEC| The maximum instantaneous value of a varying current, voltage, or power during the time interval under consideration. Also known as crest value. { 'pēk 'val·yü }

pedestal See blanking level. { 'ped·əst·əl }

pedestal level See blanking level. { 'ped·əst·əl ,lev·əl }

peek |COMPUT SCI| An instruction that causes the contents of a specific storage location in a computer to be displayed. { pēk }

peephole masks |COMPUT SCI| In character recognition, a set of characters (each character residing in the character reader in the form of strategically placed points) which theoretically render all input characters as being unique regardless of their style. { 'pēp,hōl ,masks }

peer |COMMUN| A functional unit in a communications system that is in the same protocol layer as another such unit. { pir }

peer-to-peer network |COMMUN| A local-area network in which there is no central controller and all the nodes have equal access to the resources of the network. { ¦pir tə ¦pir 'net ,wərk }

pel See pixel. { pel }

pencil beam |ELECTROMAG| A beam of radiant energy concentrated in an approximately conical or cylindrical portion of space of relatively small diameter; this type of beam is used for many revolving navigational lights and radar beams. { 'pen·səl ,bēm }

pencil beam antenna |ELECTROMAG| Unidirectional antenna designed so that cross sections of the major lobe formed by planes perpendicular to the direction of maximum radiation are approximately circular. { 'pen·səl ,bēm an,ten·ə }

pencil follower |COMPUT SCI| A device for converting graphic images to digital form; the information to be analyzed appears on a reading table where a reading pencil is made to follow the trace, and a mechanism beneath the table surface transmits position signals from the pencil to an electronic console for conversion to digital form. { 'pen·səl ,fäl·ə·wər }

pencil tube |ELECTR| A small tube designed especially for operation in the ultra-high-frequency band; used as an oscillator or radio-frequency amplifier. { 'pen·səl ,tüb }

pending input/output |COMPUT SCI| An input/output operation that has been initiated but not yet carried out, so that the central processing unit either is temporarily idle or services other programs and tasks until the operation is completed. { 'pend·iŋ 'in,pu̇t 'au̇t,pu̇t }

penetration depth |ELEC| In induction heating, the thickness of a layer, extending inward from a conductor's surface, whose resistance to direct current equals the resistance of the whole conductor to alternating current of a given frequency. { ,pen·ə'trā·shən ,depth }

penetration frequency See critical frequency. { ,pen·ə'trā·shən ,frē·kwən·sē }

penetration phosphors |ELECTR| Phosphors of two different colors that are placed in separate layers on the screen of a cathode-ray tube to form a system for creating color displays in which a high-energy beam penetrates the first layer and excites the second, while a low-energy beam is stopped by the first layer and excites it. { ,pen·ə'trā·shən ,fäs·fərz }

penetration testing |COMPUT SCI| An activity that is intended to determine if there is a way to cause a computer program to fail to perform in the expected manner; it involves hypothesizing flaws that would prevent the program from enforcing security, and conducting experiments to confirm or refute the hypothesized flaws. { ,pen·ə'trā·shən ,test·iŋ }

Penning gage See Philips ionization gage. { 'pen·iŋ ,gāj }

pentagrid See heptode. { 'pen·tə,grid }

pentode [ELECTR] A five-electrode electron tube containing an anode, a cathode, a control electrode, and two additional electrodes that are ordinarily grids. { 'pen,tōd }

pentode transistor [ELECTR] Point-contact transistor with four-point-contact electrodes; the body serves as a base with three emitters and one collector. { 'pen,tōd tran'zis·tər }

PEPE See parallel element-processing ensemble. { 'pe,pē }

percentage differential relay [ELECTR] Differential relay which functions when the difference between two quantities of the same nature exceeds a fixed percentage of the smaller quantity. Also known as biased relay; ratio-balance relay; ratio-differential relay. { pər'sen·tij ,dif·ə'ren·chəl 'rē ,lā }

percentage modulation See percent modulation. { pər'sen·tij ,maj·ə'lā·shən }

percentage ripple [ELECTR] Ratio of the effective value of the ripple voltage to the average value of the total voltage, expressed as a percentage. { pər'sen·tij 'rip·əl }

percent distortion [COMMUN] The ratio of the amplitude of a harmonic component to the fundamental component multiplied by 100. { pər'sent di'stòr·shən }

percent make [ELECTR] **1.** In pulse testing, the length of time a circuit stands closed compared to the length of the test signal. **2.** Percentage of time during a pulse period that telephone dial pulse springs are making contact. { pər'sent 'māk }

percent modulation [COMMUN] The modulation factor expressed as a percentage. Also known as percentage modulation. { pər'sent ,māj·ə'lā·shən }

perceptron [COMPUT SCI] A pattern recognition machine, based on an analogy to the human nervous system, capable of learning by means of a feedback system which reinforces correct answers and discourages wrong ones. { pər'sep ,trän }

perceptual audio coding [COMMUN] The process of representing an audio signal with fewer bits while still preserving audio quality. The coding schemes are based on the perceptual characteristics of the human ear; some examples of these coders are PAC, AAC, MPEG-2, and AC-3. Also known as audio bit rate reduction; audio compression. Abbreviated PAC. { pər ¦səp·chə·wəl 'òd·ē·ō ,kōd·iŋ }

percolation [COMPUT SCI] The transfer of needed data back from secondary storage devices to main storage. { pər·kə'lā·shən }

perfect dielectric See ideal dielectric. { 'pər·fikt ,dī·ə'lek·trik }

perforator [COMMUN] In telegraph practice, a device for punching code signals in paper tape

for application to a tape transmitter. { 'pər·fə ,rād·ər }

perform [COMPUT SCI] A subroutine in the COBOL programming language that allows a portion of a program to be executed on command by other portions of the same program. { pər'fórm }

performance failure [COMPUT SCI] Failure of a computer system in which the system operates correctly but fails to deliver the results in a timely fashion. { pər'fòr·məns ,fāl·yər }

perfory [COMPUT SCI] The removable edges of computer paper containing holes engaged by the pin-feed mechanism. { 'pər·fə·rē }

periodic antenna [ELECTROMAG] An antenna in which the input impedance varies as the frequency is altered. { ¦pir·ē¦äd·ik an'ten·ə }

periodic duty [ELEC] Intermittent duty in which the load conditions are regularly recurrent. { ¦pir·ē¦äd·ik 'düd·ē }

periodic field focusing [ELECTR] Focusing of an electron beam where the electrons follow a trochoidal path and the focusing field interacts with them at selected points. { ¦pir·ē¦äd·ik 'fēld ,fō·kə·siŋ }

periodic line [ELEC] Line consisting of successive and identical sections, similarly oriented, the electrical properties of each section not being uniform throughout; the periodicity is in space and not in time; an example of a periodic line is the loaded line with loading coils uniformly spaced. { ¦pir·ē¦äd·ik ¦līn }

peripheral See peripheral device. { pə'rif·ə·rəl }

peripheral buffer [COMPUT SCI] A device acting as a temporary storage when transmission occurs between two devices operating at different transmission speeds. { pə'rif·ə·rəl 'bəf·ər }

peripheral component interconnect [COMPUT SCI] A bus standard for connecting additional input/output devices (such as graphics or modem cards) to a personal computer. Abbreviated PCI. { pə,rif·ə·rəl kəm,pō·nənt 'in·tər,kə·nek }

peripheral control unit [COMPUT SCI] A device which connects a unit of peripheral equipment with the central processing unit of a computer and which interprets and responds to instructions from the central processing unit. { pə'rif·ə·rəl kən'trōl ,yü·nət }

peripheral device [COMPUT SCI] Any device connected internally or externally to a computer and used to enter or display data, such as the keyboard, mouse, monitor, scanner, and printer. { pə'rif·ərəl di,vīs }

peripheral equipment [COMPUT SCI] Equipment that works in conjunction with a computer but is not part of the computer itself. { pə'rif·ə·rəl i'kwip·mənt }

peripheral interface channel [COMPUT SCI] A path along which information can flow between a unit of peripheral equipment and the central processing unit of a computer. { pə'rif·ə·rəl 'in·tər,fās ,chan·əl }

peripheral-limited [COMPUT SCI] Property of a computer system whose processing time is determined by the speed of its peripheral equipment rather than by the speed of its central processing unit. { pə'rif·ə·rəl ¦lim·əd·əd }

peripheral operation [COMPUT SCI] An operation in which an input or output device is used, and which is not directly controlled by a computer while the operation is being carried out. { pə'rif·ə·rəl ‚äp·ə'rā·shən }

peripheral processing [COMPUT SCI] Processing that is carried out by peripheral equipment or by an auxiliary computer. { pə'rif·ə·rəl 'prä‚ses·iŋ }

peripheral processor [COMPUT SCI] Auxiliary computer performing specific operations under control of the master computer. { pə'rif·ə·rəl 'prä‚ses·ər }

peripheral transfer [COMPUT SCI] The transmission of data between two units of peripheral equipment or between a peripheral unit and the central processing unit of a computer. { pə'rif·ə·rəl 'tranz·fər }

peripheral units See peripheral equipment. { pə'rif·ə·rəl ‚yü·nəts }

peristaltic charge-coupled device [ELECTR] A high-speed charge-transfer integrated circuit in which the movement of the charges is similar to the peristaltic contractions and dilations of the digestive system. { ¦per·ə¦stäl·tik 'chärj ¦kəp·əld di'vīs }

Perl See Practical Extraction and Reporting Language. { pərl }

permanent echo [ELECTR] See fixed echo. [ELECTROMAG] A signal reflected from an object that is fixed with respect to a radar site. { 'pər·mə·nənt 'ek·ō }

permanent error [COMPUT SCI] An error that occurs when a sector mark on disk pack or floppy disk is incorrectly modified by writing data over it, and that can be corrected only by clearing the entire disk and rewriting the track and sector marks. { 'pər·mə·nənt 'er·ər }

permanent fault [COMPUT SCI] A hardware malfunction that always occurs when a particular set of conditions exists, and that can be made to occur deliberately, in contrast to a sporadic fault. { 'pər·mə·nənt 'fòlt }

permanent-magnet dynamic loudspeaker See permanent-magnet loudspeaker. { 'pər·mə·nənt ¦mag·nət dī¦nam·ik 'laùd‚spēk·ər }

permanent-magnet focusing [ELECTR] Focusing of the electron beam in a cathode-ray tube by means of the magnetic field produced by one or more permanent magnets mounted around the neck of the device. { 'pər·mə·nənt ¦mag·nət 'fō·kəs·iŋ }

permanent-magnet loudspeaker [ENG ACOUS] A moving-conductor loudspeaker in which the steady magnetic field is produced by a permanent magnet. Also known as permanent-magnet dynamic loudspeaker. { 'pər·mə·nənt ¦mag·nət 'laùd‚spēk·ər }

permanent-magnet stepper motor [ELEC] A stepper motor in which the rotor is a powerful permanent magnet and each stator coil is energized independently in sequence; the rotor aligns itself with the stator coil that is energized. { 'pər·mə·nənt ¦mag·nət 'step·ər ‚mōd·ər }

permanent-split capacitor motor [ELEC] A capacitor motor in which the starting capacitor and the auxiliary winding remain in the circuit for both starting and running. Abbreviated PSC motor. Also known as capacitor start-run motor. { 'pər·mə·nənt ¦split kə'pas·əd·ər ‚mōd·ər }

permanent storage [COMPUT SCI] A means of storing data for rapid retrieval by a computer; does not permit changing the stored data. { 'pər·mə·nənt 'stòr·ij }

Permasyn motor [ELEC] A synchronous motor which has permanent magnets embedded in the squirrel-cage rotor to provide an equivalent direct-current field. { 'pər·mə·sən 'mōd·ər }

permatron [ELECTR] Thermionic gas-discharge diode in which the start of conduction is controlled by an external magnetic field. { 'pər·mə‚trän }

permeability tuning [ELEC] Process of tuning a resonant circuit by varying the permeability of an inductor; it is usually accomplished by varying the amount of magnetic core material of the inductor by slug movement. { ‚pər·mē·ə'bil·əd·ē ‚tün·iŋ }

permittivity [ELEC] The dielectric constant multiplied by the permittivity of empty space, where the permittivity of empty space (ϵ_0) is a constant appearing in Coulomb's law, having the value of 1 in centimeter-gram-second electrostatic units, and of 8.854 × 10^{-12} farad/meter in rationalized meter-kilogram-second units. Symbolized ϵ. { ‚pər·mə'tiv·əd·ē }

permutation modulation [COMMUN] Proposed method of transmitting digital information by means of band-limited signals in the presence of additive white gaussian noise; pulse-code modulation and pulse-position modulation are considered simple special cases of permutation modulation. { ‚pər·mya'tā·shən ‚mäj·ə‚lā·shən }

permutation table [COMMUN] In computers, a table designed for the systematic construction of code groups; it may also be used to correct garbles in groups of code text. { ‚pər·mya'tā·shən ‚tā·bəl }

perpendicular recording See vertical recording. { ‚pər·pən¦dik·yə·lər ri'kòrd·iŋ }

persistence [ELECTR] **1.** A measure of the length of time that the screen of a cathode-ray tube remains luminescent after excitation is removed; ranges from 1 for short persistence to 7 for long persistence. **2.** A faint luminosity displayed by certain gases for some time after the passage of an electric discharge. { pər'sis·təns }

persistent-image device [ELECTR] An optoelectronic amplifier capable of retaining an image for a definite length of time. { pər'sis·tənt ¦im·ij di ‚vīs }

persistron [ELECTR] A device in which electroluminescence and photoconductivity are used in a single panel capable of producing a steady or persistent display with pulsed signal input. { pər'sis‚trän }

personal communications network [COMMUN] The series of small low-power antennas that support a personal communications service, and are linked to a master telephone switch that is connected to the main telephone network. Abbreviated PCN. { ˌpərs·ən·əl kəˌmyü·nə ¦kā·shənz ˌnet¸wərk }

personal communications service [COMMUN] A mobile telephone service in which pocket-sized telephones carried by the users communicate via small low-power transmitter-receiver antennas that are installed throughout a city or community. Abbreviated PCS. { ˌpərs·ən·əl kə ˌmyü·nə¦kā·shənz ˌsər·vəs }

personal computer [COMPUT SCI] A computer for home or personal use. { ˈpər·sən·əl kəmˈpyüd·ər }

personal digital assistant See hand-held computer. { ˌpərs·ən·əl ˌdij·əd·əl əˈsis·tənt }

personal identification code [COMPUT SCI] A special number up to six characters in length on a strip of magnetic tape embedded in a plastic card which identifies a user accessing a special-purpose computer. Abbreviated PIC. { ˈpər·sən·əl ī¸den·tə·fəˈkā·shən ˌkōd }

personal information manager [COMPUT SCI] Software that combines the functions of word-processing, database, and desktop accessory programs, making it possible to organize information that is relatively loosely structured. Abbreviated PIM. { ¦pər·sən·əl ˌin·fər¦mā·shən ˌman·ij·ər }

persuader [ELECTR] Element of storage tube which directs secondary emission to electron multiplier dynodes. { pərˈswād·ər }

pertinency factor [COMPUT SCI] In information retrieval, the ratio obtained in dividing the total number of relevant documents retrieved by the total number of documents retrieved. { ˈpər·tə·nən·sē ˌfak·tər }

perveance [ELECTR] The space-charge-limited cathode current of a diode divided by the $\frac{3}{2}$ power of the anode voltage. { ˈpər·vē·əns }

PES See packetized elementary stream. { ¦pē¦ē ¦es }

PES packet [COMMUN] The data structure used to carry elementary stream data; consists of a packet header followed by PES packet payload. { ¦pē¦ē¦es ¦pak·ət }

PES stream [COMMUN] Referring to a stream consisting of PES packets, all of whose payloads consists of data from a single elementary stream, and all of which have the same stream ID number. { ¦pē¦ē¦es ¦strēm }

Petersen coil See arc-suppression coil. { ˈpēd·ər·sən ˌkȯil }

Petri net [COMMUN] An abstract, formal model of information flow, which is used as a graphical language for modeling systems with interacting concurrent components; in mathematical terms, a structure with four parts or components: a finite set of places, a finite set of transitions, an input function, and an output function. { ˈpē·trē ˌnet }

petticoat insulator [ELEC] Insulator having an outward-flaring lower part that is hollow inside to increase the length of the surface leakage path and keep part of the path dry at all times. { ˈped·i¸kōt ˈin·sə¸lād·ər }

pf See power factor.

pF See picofarad.

PF key See programmed function key. { ¸pē'ef ˌkē }

PFM See pulse-frequency modulation.

P-frame See predicted picture. { ˈpē ˌfrām }

phanotron [ELECTR] A hot-filament diode rectifier tube utilizing an arc discharge in mercury vapor or an inert gas, usually xenon. { ˈfan·ə ˌträn }

phantastran [ELECTR] A solid-state phantastron. { fanˈtasˌträn }

phantastron [ELECTR] A monostable pentode circuit used to generate sharp pulses at an adjustable and accurately timed interval after receipt of a triggering signal. { fanˈtasˌträn }

phantom circuit [COMMUN] A communication circuit derived from two other communication circuits or from one other circuit and ground, with no additional wire lines. { ˈfan·təm ˈsər·kət }

phantom-circuit loading coil [ELEC] Loading coil for introducing a desired amount of inductance into a phantom circuit, and a minimum amount of inductance into its constituent circuits. { ˈfan·təm ¦sər·kət ˈlōd·iŋ ˌkȯil }

phantom-circuit repeating coil [ELEC] Repeating coil used at a terminal of a phantom circuit, in the terminal circuit extending from the midpoints of the associated side-circuit repeating coils. { ˈfan·təm ¦sər·kət riˈpēd·iŋ ˌkȯil }

phantom group [ELEC] **1.** Group of four open-wire conductors suitable for the derivation of a phantom circuit. **2.** Three circuits which are derived from simplexing two physical circuits to form a phantom circuit. { ˈfan·təm ˈgrüp }

phantom repeating coil [ELEC] A side-circuit repeating coil or a phantom-circuit repeating coil when discrimination between these two types is not necessary. { ˈfan·təm riˈpēd·iŋ ˌkȯil }

phantom signals [ELECTR] Signals appearing on a radar display, the cause of which cannot readily be determined and which may be caused by circuit fault, interference, propagation anomalies, measurement ambiguities, jamming, and so on. { ˈfan·təm ˈsig·nəlz }

phantom target See echo box. { ˈfan·təm ˈtär·gət }

phase advancer [ELEC] Phase modifier which supplies leading reactive volt-amperes to the system to which it is connected; may be either synchronous or asynchronous. { ˈfāz id ˌvan·sər }

phase-alternation line system [COMMUN] A color television system used in Europe and other parts of the world, in which the phase of the color subcarrier is changed from scanning line to scanning line, requiring transmission of a line switching signal as well as a color burst. Abbreviated PAL system. { ˈfāz ˌȯl·tər¦nā·shən ˌlīn ˌsis·təm }

phase-angle meter See phase meter. { ˈfāz ¦aŋ·gəl ˌmēd·ər }

phase-balance relay [ELEC] Relay which functions by reason of a difference between two quantities associated with different phases of a polyphase circuit. { ˈfāz ¦bal·əns ˈrē¸lā }

phase change *See* phase shift. { 'fāz ,chānj }

phase-change coefficient *See* phase constant. { ¦fāz ,chānj ,kō·i,fish·ənt }

phase-change recording [COMPUT SCI] An optical recording technique that uses a laser to alter the crystalline structure of a metallic surface to create bits that reflect or absorb light when they are illuminated during the read operation. { ¦fāz ,chānj ri'kórd·iŋ }

phase comparator [COMPUT SCI] A comparator that accepts two radio-frequency input signals of the same frequency and provides two video outputs which are proportional, respectively, to the sine and cosine of the phase difference between the two inputs. { 'fāz kəm,par·əd·ər }

phase-comparison relaying [ELEC] A method of detecting faults in an electric power system in which signals are transmitted from each of two terminals every half cycle so that a continuous signal is received at an intermediate point if there is no fault between the terminals, while a periodic signal is received if there is a fault. { 'fāz kəm ,par·ə·sən 'rē,lā·iŋ }

phase conductor [ELEC] In a polyphase circuit, any conductor other than the neutral conductor. { 'fāz kən,dək·tər }

phase constant [ELECTROMAG] A rating for a line or medium through which a plane wave of a given frequency is being transmitted; it is the imaginary part of the propagation constant, and is the space rate of decrease of phase of a field component (or of the voltage or current) in the direction of propagation, in radians per unit length. Also known as phase-change coefficient; wavelength constant. { 'fāz ,kän·stənt }

phase control *See* hue control. { 'fāz kən,trōl }

phase converter [ELEC] A converter that changes the number of phases in an alternating-current power source without changing the frequency. { 'fāz kən,vərd·ər }

phase-correcting network *See* phase equalizer. { 'fāz kə¦rek·tiŋ 'net,wərk }

phase correction [COMMUN] Process of keeping synchronous telegraph mechanisms in substantially correct phase relationship. { 'fāz kə ,rek·shən }

phase crossover [CONT SYS] A point on the plot of the loop ratio at which it has a phase angle of 180°. { 'fāz 'kròs,ō·vər }

phased array [ELECTROMAG] An array of dipoles on a radar antenna in which the signal feeding each dipole is varied so that antenna beams can be formed in space and scanned very rapidly in azimuth and elevation. { 'fāzd ə'rā }

phased-array radar [ENG] Radar using an antenna of the multiple-element array type in which the relative phasing of the elements, electronically controlled, positions the main beam in angle without need of moving the antenna. { 'fāzd ə'rā 'rā,där }

phase delay [COMMUN] Ratio of the total phase shift (radians) of a sinusoidal signal in transmission through a system or transducer, to the frequency (radians/second) of the signal. { 'fāz di,lā }

phase detector [ELECTR] **1.** A circuit that provides a direct-current output voltage which is related to the phase difference between an oscillator signal and a reference signal, for use in controlling the oscillator to keep it in synchronism with the reference signal. Also known as phase discriminator. **2.** A circuit or device in a radar receiver giving a voltage output dependent upon the phase difference of two inputs; used in Doppler sensing in a coherent radar. { 'fāz di,tek·tər }

phase deviation [COMMUN] The peak difference between the instantaneous angle of a modulated wave and the angle of the sine-wave carrier. { 'fāz ,dē·vē'ā·shən }

phase discriminator *See* phase detector. { 'fāz di,skrim·ə,nād·ər }

phase distortion [COMMUN] **1.** The distortion which occurs in an instrument when the relative phases of the input signal differ from those of the output signal. **2.** *See* phase-frequency distortion. { 'fāz di,stór·shən }

phase encoding [COMPUT SCI] A method of recording data on magnetic tape in which a logical 1 is defined as the transition from one magnetic polarity to another positioned at the center of the bit cell, and 0 is defined as the transition in the opposite direction, also at the center of the cell. Also known as Manchester coding. { 'fāz in'kōd·iŋ }

phase equalizer [ELECTR] A network designed to compensate for phase-frequency distortion within a specified frequency band. Also known as phase-correcting network. { 'fāz 'ē·kwə,liz·ər }

phase excursion [COMMUN] In angle modulation, the difference between the instantaneous angle of the modulated wave and the angle of the carrier. { 'fāz ik,skər·zhən }

phase factor *See* power factor. { 'fāz ,fak·tər }

phase-frequency distortion [COMMUN] Distortion occurring because phase shift is not proportional to frequency over the frequency range required for transmission. Also known as phase distortion. { 'fāz ¦fre·kwən·sē di,stór·shən }

phase generator [ELECTR] An instrument that accepts single-phase input signals over a given frequency range, or generates its own signal, and provides continuous shifting of the phase of this signal by one or more calibrated dials. { 'fāz ,jen·ə,rād·ər }

phase inversion [ELECTR] Production of a phase difference of 180° between two similar wave shapes of the same frequency. { 'fāz in,vər·zhən }

phase inverter [ELECTR] A circuit or device that changes the phase of a signal by 180°, as required for feeding a push-pull amplifier stage without using a coupling transformer, or for changing the polarity of a pulse; a triode is commonly used as a phase inverter. Also known as inverter. { 'fāz in,vərd·ər }

phase jitter [ELECTR] Jitter that undesirably shortens or lengthens pulses intermittently during data processing or transmission. { 'fāz ,jid·ər }

phase lock [ELECTR] Technique of making the phase of an oscillator signal follow exactly the phase of a reference signal by comparing the phases between the two signals and using the resultant difference signal to adjust the frequency of the reference oscillator. { 'fāz ˌläk }

phase-locked communication |COMMUN| Systems in which oscillators at the receiver and transmitter are locked in phase. { 'fāz ˌlläkt kə ˌmyü·nə'kā·shən }

phase-locked loop [ELECTR] A circuit that consists essentially of a phase detector which compares the frequency of a voltage-controlled oscillator with that of an incoming carrier signal or reference-frequency generator; the output of the phase detector, after passing through a loop filter, is fed back to the voltage-controlled oscillator to keep it exactly in phase with the incoming or reference frequency. Abbreviated PLL. { 'fāz ˌläkt 'lüp }

phase-locked oscillator See parametron. { 'fāz ˌläkt 'äs·ə,läd·ər }

phase-locked subharmonic oscillator See parametron. { 'fāz ˌläkt ˌsəb·här'män·ik 'äs·ə ˌläd·ər }

phase-locked system [ENG] A radar system, having a stable local oscillator, in which information regarding the target is gained by measuring the phase shift of the echo. { 'fāz ˌläkt ˌsis·təm }

phase magnet |COMMUN| Magnetically operated latch used to phase a facsimile transmitter or recorder. Also known as trip magnet. { 'fāz ˌmag·nət }

phase margin [CONT SYS] The difference between 180° and the phase of the loop ratio of a stable system at the gain-crossover frequency. { 'fāz ˌmär·jən }

phase meter |ENG| An instrument for the measurement of electrical phase angles. Also known as phase-angle meter. { 'fāz ˌmēd·ər }

phase modifier |ELEC| Machine whose chief purpose is to supply leading or lagging reactive voltamperes to the system to which it is connected; may be either synchronous or asynchronous. { 'fāz ˌmäd·ə,fī·ər }

phase modulation |COMMUN| Modulation in which the linearly increasing angle of a sine wave has added to it a phase angle that is proportional to the instantaneous value of the modulating signal (message to be communicated). Abbreviated PM. { 'fāz ˌmäj·ə,lā·shən }

phase-modulation detector [ELECTR] A device which recovers or detects the modulating signal from a phase-modulated carrier. { 'fāz ˌmäj·ə ˌlā·shən di,tek·tər }

phase-modulation transmitter [ELECTR] A radio transmitter used to broadcast a phase-modulated signal. { 'fāz ˌmäj·ə,lā·shən tranz ˌmid·ər }

phase modulator |ELECTR| An electronic circuit that causes the phase angle of a modulated wave to vary (with respect to an unmodulated carrier) in accordance with a modulating signal. { 'fāz ˌmäj·ə,läd·ər }

phase plane analysis |CONT SYS| A method of analyzing systems in which one plots the time derivative of the system's position (or some other quantity characterizing the system) as a function of position for various values of initial conditions. { 'fāz ˌplān ə'nal·ə·səs }

phase portrait |CONT SYS| A graph showing the time derivative of a system's position (or some other quantity characterizing the system) as a function of position for various values of initial conditions. { 'fāz ˌpȯr·trət }

phaser |COMMUN| Facsimile device for adjusting equipment so the recorded elemental area bears the same relation to the record sheet as the corresponding transmitted elemental area bears to the subject copy in the direction of the scanning line. |ELECTROMAG| Microwave ferrite phase shifter employing a longitudinal magnetic field along one or more rods of ferrite in a waveguide. { 'fāz·ər }

phase response |ELECTR| A graph of the phase shift of a network as a function of frequency. { 'fāz ri,späns }

phase reversal modulation |COMMUN| Form of pulse modulation in which reversal of signal phase serves to distinguish between the two binary states used in data transmission. { 'fāz ri,vər·səl ˌmäj·ə'lā·shən }

phase-rotation relay See phase-sequence relay. { 'fāz rō¦tā·shən 'rē,lā }

phase-sensitive detector |ELECTR| An electronic circuit that consists essentially of a multiplier and a low-pass circuit and that produces a direct-current output signal that is proportional to the product of the amplitudes of two alternating-current input signals of the same frequency and to the cosine of the phase between them. { 'fāz ˌsen·səd·iv di,tek·tər }

phase-sequence relay |ELEC| Relay which functions according to the order in which the phase voltages successively reach their maximum positive values. Also known as phase-rotation relay. { 'fāz ¦sē·kwəns 'rē,lā }

phase shift |ELECTR| The phase angle between the input and output signals of a network or system. { 'fāz ˌshift }

phase-shift circuit |ELECTR| A network that provides a voltage component which is shifted in phase with respect to a reference voltage. { 'fāz ¦shift ˌsər·kət }

phase-shift control See phase control. { 'fāz ¦shift kən,trōl }

phase-shift discriminator |ELECTR| A discriminator that uses two similarly connected diodes, fed by a transformer that is tuned to the center frequency; when the frequency-modulated or phase-modulated input signal swings away from this center frequency, one diode receives a stronger signal than the other; the net output of the diodes is then proportional to the frequency displacement. Also known as Foster-Seely discriminator. { 'fāz ¦shift di,skrim·ə,nād·ər }

phase shifter [ELEC] A device used to change the phase relation between two alternating-current values. { 'fāz ‚shif·tǝr }

phase-shifting transformer [ELEC] A transformer which produces a difference in phase angle between two circuits. { 'fāz ¦shif·tiŋ tranz ‚fór·mǝr }

phase-shift keying [COMMUN] A form of phase modulation in which the modulating function shifts the instantaneous phase of the modulated wave between predetermined discrete values. Abbreviated PSK. { 'fāz ¦shift ‚kē·iŋ }

phase-shift oscillator [ELECTR] An oscillator in which a network having a phase shift of 180° per stage is connected between the output and the input of an amplifier. { 'fāz ¦shift 'äs·ǝ‚lād·ǝr }

phase splitter [ELEC] A circuit that takes a single input alternating voltage and produces two or more output alternating voltages that differ in phase from one another. { 'fāz ‚splid·ǝr }

phase transformation [ELEC] A change of polyphase power from three-phase to six-phase, from three-phase to twelve-phase, and so forth, by use of transformers. { 'fāz ‚tranz·fǝr‚mā·shǝn }

phase transformer [ELEC] A transformer for changing a two-phase current to a three-phase current, or vice versa. { 'fāz tranz‚fór·mǝr }

phase undervoltage relay [ELEC] Relay which functions by reason of the reduction of one phase voltage in a polyphase circuit. { 'fāz 'ǝn·dǝr ‚vōl·tij 'rē‚lā }

phase winding [ELEC] One of the individual windings on the armature of a polyphase motor or generator. { 'fāz ‚wīnd·iŋ }

phasing *See* framing. { 'fāz·iŋ }

phasing line [ELECTR] That portion of the length of scanning line set aside for the phasing signal in a video system. { 'fāz·iŋ ‚līn }

phasing signal [ELECTR] A signal used to adjust the picture position along the scanning line in a facsimile system. { 'fāz·iŋ ‚sig·nǝl }

phasitron [ELECTR] An electron tube used to frequency-modulate a radio-frequency carrier; internal electrodes are designed to produce a rotating disk-shaped corrugated sheet of electrons; audio input is applied to a coil surrounding the glass envelope of the tube, to produce a varying axial magnetic field that gives the desired phase or frequency modulation of the RF carrier input to the tube. { 'fāz·ǝ‚trän }

phasmajector *See* monoscope. { 'faz·mǝ‚jek·tǝr }

Philips ionization gage [ELECTR] An ionization gage in which a high voltage is applied between two electrodes, and a strong magnetic field deflects the resulting electron stream, increasing the length of the electron path and thus increasing the chance for ionizing collisions of electrons with gas molecules. Abbreviated pig. Also known as cold-cathode ionization gage; Penning gage. { 'fil·ǝps ‚ī·ǝ·nǝ'zā·shǝn ‚gāj }

phonation [ENG ACOUS] Production of speech sounds. { fō'nā·shǝn }

phone *See* headphone; telephone set. { fōn }

phonemic synthesizer [ENG ACOUS] A voice response system in which each word is abstractly represented as a sequence of expected vowels and consonants, and speech is composed by juxtaposing the expected phonemic sequence for each word with the sequences for the preceding and following words. { fǝ'nē·mik 'sin·thǝ‚sīz·ǝr }

phone patch [ELECTR] A device connecting an amateur or citizens'-band transceiver temporarily to a telephone system. { 'fōn ‚pach }

phone plug [ELEC] A standard plug having a ¾-inch-diameter (19-millimeter) shank, used with headphones, microphones, and other audio equipment; usually designed for use with either two or three conductors. Also known as telephone plug. { 'fōn ‚plǝg }

phonetic alphabet [COMMUN] A list of standard words used for positive identification of letters in a voice message transmitted by radio or telephone. { fǝ'ned·ik 'al·fǝ‚bet }

phonetic search [COMPUT SCI] A method of locating information in a file in which an algorithm is used to locate combinations of characters that sound similar to a specified combination. { fǝ'ned·ik 'sǝrch }

phonic motor [ELEC] A small synchronous motor which is driven by the current of an accurate oscillator, such as a crystal oscillator, and whose frequency is thus constant to a high degree of accuracy; used in astronomical instruments where a driving speed of great accuracy is required. { 'fän·ik 'mōd·ǝr }

phonograph [ENG ACOUS] An instrument for recording or reproducing acoustical signals, such as voice or music, by transmission of vibrations from or to a stylus that is in contact with a groove in a rotating disk. { 'fō·nǝ‚graf }

phono jack [ELECTR] A jack designed to accept a phono plug and provide a ground connection for the shield of the conductor connected to the plug. { 'phō·nō ‚jak }

phono plug [ELECTR] A plug designed for attaching to the end of a shielded conductor, for feeding audio-frequency signals from a phonograph or other audio-frequency source to a mating phono jack on a preamplifier or amplifier. { 'fō·nō ‚plǝg }

phosphor dot [ELECTR] One of the tiny dots of phosphor material that are used in groups of three, one group for each primary color, on the screen of a color video picture tube. { 'fäs·fǝr ‚dät }

photocapacitative effect [ELEC] A change in the capacitance of a bulk semiconductor or semiconductor surface film upon exposure to light. { ‚fōd·ō·ō·kǝ'pas·ǝ‚tā·tiv i‚fekt }

photocathode [ELECTR] A photosensitive surface that emits electrons when exposed to light or other suitable radiation; used in phototubes, video camera tubes, and other light-sensitive devices. { 'fōd·ō'kath‚ōd }

photocell [ELECTR] A solid-state photosensitive electron device whose current-voltage characteristic is a function of incident radiation. Also known as electric eye; photoelectric cell. { 'fōd·ǝ‚sel }

photocell relay [ELECTR] A relay actuated by a signal received when light falls on, or is prevented from falling on, a photocell. { 'fōd·ǝ‚sel 'rē‚lā }

photocomposition |COMPUT SCI| Composition of type using electrophotographic techniques such as phototypesetters and laser printers. { ¦fōd·ō ¦käm·pə'zish·ən }

photoconduction |SOLID STATE| An increase in conduction of electricity resulting from absorption of electromagnetic radiation { ¦fōd·ō·kən'dək·shən }

photoconductive cell |ELECTR| A device for detecting or measuring electromagnetic radiation by variation of the conductivity of a substance (called a photoconductor) upon absorption of the radiation by this substance. Also known as photoresistive cell; photoresistor. { ¦fōd·ō·kən'dək·tiv 'sel }

photoconductive device |ELECTR| A photoelectric device which utilizes the photoinduced change in electrical conductivity to provide an electrical signal. { ¦fōd·ō·kən'dək·tiv di'vīs }

photoconductive film |ELECTR| A film of material whose current-carrying ability is enhanced when illuminated. { ¦fōd·ō·kən'dək·tiv 'film }

photoconductive gain factor |ELECTR| The ratio of the number of electrons per second flowing through a circuit containing a cube of semiconducting material, whose sides are of unit length, to the number of photons per second absorbed in this volume. { ¦fōd·ō·kən'dək·tiv 'gān ¦fak·tər }

photoconductive meter |ELECTR| An exposure meter in which a battery supplies power through a photoconductive cell to a milliammeter. { ¦fōd·ō·kən'dək·tiv 'mēd·ər }

photoconductivity |SOLID STATE| The increase in electrical conductivity displayed by many nonmetallic solids when they absorb electromagnetic radiation. { ¦fōd·ō·¦kän¦dək'tiv·əd·ē }

photoconductivity gain |ELECTR| The number of charge carriers that circulate through a circuit involving a photoconductor for each charge carrier generated by light. { fōd·ō¦kän¦dək'tiv·əd·ē ¦gān }

photoconductor |SOLID STATE| A nonmetallic solid whose conductivity increases when it is exposed to electromagnetic radiation. { ¦fōd·ō· kən'dək·tər }

photoconductor diode See photodiode. { fōd· ō·kən'dək·tər 'dī¦ōd }

photocoupler See optoisolator. { ¦fōd·ō'kəp·lər }

photodarlington |ELECTR| A Darlington amplifier in which the input transistor is a phototransistor. { ¦fōd·ō'där·liŋ·tən }

photodetector |ELECTR| A detector that responds to radiant energy; examples include photoconductive cells, photodiodes, photoresistors, photoswitches, phototransistors, phototubes, and photovoltaic cells. Also known as light-sensitive cell; light-sensitive detector; light sensor photodevice; photodevice; photoelectric detector; photosensor. { ¦fōd·ō·di'tek· tər }

photodevice See photodetector. { ¦fōd·ō·di¦vīs }

photodiffusion effect See Dember effect. { ¦fōd· ō·di'fyü·zhən i¦fekt }

photodiode |ELECTR| A semiconductor diode in which the reverse current varies with illumina-

tion; examples include the alloy-junction photocell and the grown-junction photocell. Also known as photoconductor diode. { ¦fōd·ō'dī ¦ōd }

photoelectric |ELECTR| Pertaining to the electrical effects of light, such as the emission of electrons, generation of voltage, or a change in resistance when exposed to light. { ¦fōd·ō·i 'lek·trik }

photoelectric absorption |ELECTR| Absorption of photons in one of the several photoelectric effects. { ¦fōd·ō·i'lek·trik əb'sórp·shən }

photoelectric cell See photocell. { ¦fōd·ō·i'lek· trik 'sel }

photoelectric constant |ELECTR| The ratio of the frequency of radiation causing emission of photoelectrons to the voltage corresponding to the energy absorbed by a photoelectron; equal to Planck's constant divided by the electron charge. { ¦fōd·ō·i'lek·trik 'kän·stənt }

photoelectric control |ELECTR| Control of a circuit or piece of equipment by changes in incident light. { ¦fōd·ō·i'lek·trik kən'trōl }

photoelectric counter |ELECTR| A photoelectrically actuated device used to record the number of times a given light path is intercepted by an object. { ¦fōd·ō·i'lek·trik 'kaúnt·ər }

photoelectric cutoff register control |ELECTR| Use of a photoelectric control system as a longitudinal position regulator to maintain the position of the point of cutoff with respect to a repetitive pattern of moving material. { ¦fōd·ō·i'lek·trik ¦kət¸óf ¦rej·ə·stər kən¸trōl }

photoelectric detector See photodetector. { ¦fōd·ō·i'lek·trik di'tek·tər }

photoelectric device |ELECTR| A device which gives an electrical signal in response to visible, infrared, or ultraviolet radiation. { ¦fōd·ō· i'lek·trik di¦vīs }

photoelectric effect See photoelectricity. { ¦fōd·ō·i'lek·trik i¸fekt }

photoelectric electron-multiplier tube See multiplier phototube. { ¦fōd·ō·i'lek·trik i¦lek¸trän 'məl·tə¸plī·ər ¸tüb }

photoelectric infrared radiation See near-infrared radiation. { ¦fōd·ō·i'lek·trik ¦in·frə¦red ¸rā·dē'ā·shən }

photoelectric intrusion detector |ELECTR| A burglar-alarm system in which interruption of a light beam by an intruder reduces the illumination on a phototube and thereby closes an alarm circuit. { ¦fōd·ō·i'lek·trik in'trü·zhən di¸tek·tər }

photoelectricity |ELECTR| The liberation of an electric charge by electromagnetic radiation incident on a substance; includes photoemission, photoionization, photoconduction, the photovoltaic effect, and the Auger effect (an internal photoelectric process). Also known as photoelectric effect; photoelectric process. { ¦fōd·ō¸i ¸lek'tris·əd·ē }

photoelectric lighting control |ELECTR| Use of a photoelectric relay actuated by a change in illumination in a given area or at a given point. { ¦fōd·ō·i'lek·trik 'līd·iŋ kən¸trōl }

photoelectric loop control [CONT SYS] A photoelectric control system used as a position regulator for a loop of material passing from one strip-processing line to another that may travel at a different speed. Also known as loop control. { ¦fōd·ō·i'lek·trik 'lüp kən,trōl }

photoelectric process *See* photoelectricity. { ¦fōd·ō·i'lek·trik 'prä·səs }

photoelectric register control [CONT SYS] A register control using a light source, one or more phototubes, a suitable optical system, an amplifier, and a relay to actuate control equipment when a change occurs in the amount of light reflected from a moving surface due to register marks, dark areas of a design, or surface defects. Also known as photoelectric scanner. { ¦fōd·ō·i'lek·trik 'rej·ə·stər kən,trōl }

photoelectric relay [ELECTR] A relay combined with a phototube and amplifier, arranged so changes in incident light on the phototube make the relay contacts open or close. Also known as light relay. { ¦fōd·ō·i'lek·trik 'rē,lā }

photoelectric scanner *See* photoelectric register control. { ¦fōd·ō·i'lek·trik 'skan·ər }

photoelectric sorter [CONT SYS] A photoelectric control system used to sort objects according to color, size, shape, or other light-changing characteristics. { ¦fōd·ō·i'lek·trik 'sórd·ər }

photoelectric tube *See* phototube. { ¦fōd·ō·i'lek·trik 'tüb }

photoelectromagnetic effect [ELECTR] The effect whereby, when light falls on a flat surface of an intermetallic semiconductor located in a magnetic field that is parallel to the surface, excess hole-electron pairs are created, and these carriers diffuse in the direction of the light but are deflected by the magnetic field to give a current flow through the semiconductor that is at right angles to both the light rays and the magnetic field. { ¦fōd·ō·i¦lek·trō·mag'nedik i'fekt }

photoelectromotive force [ELECTR] Electromotive force caused by photovoltaic action. { ¦fōd·ō·i¦lek·trō'mōd·iv 'fórs }

photoelectron [ELECTR] An electron emitted by the photoelectric effect. { ¦fōd·ō·i'lek,trän }

photoemission [ELECTR] The ejection of electrons from a solid (or less commonly, a liquid) by incident electromagnetic radiation. Also known as external photoelectric effect. { ¦fōd·ō·i'mish·ən }

photoemission threshold [ELECTR] The energy of a photon which is just sufficient to eject an electron from a solid or liquid in photoemission. { ¦fōd·ō·i'mish·ən 'thresh,hōld }

photoemissive cell [ELECTR] A device which detects or measures radiant energy by measurement of the resulting emission of electrons from the surface of a photocathode. { ¦fōd·ō·i'mis·iv 'sel }

photoemissivity [ELECTR] The property of a substance that emits electrons when struck by light. { ¦fōd·ō,ē·mə'siv·əd·ē }

photofabrication [ELECTR] In manufacturing circuit boards and integrated circuits, a process in which the etching pattern is placed over the circuit board or semiconductor material, the board or chip is placed in a special solution, and the assembly is exposed to light. { ,fōd·ō ,fab·rə'kā·shən }

photoflash lamp [ELEC] A lamp consisting of a glass bulb filled with finely shredded aluminum foil in an atmosphere of oxygen; when the foil is ignited by a low-voltage dry cell, it burns with a burst of high-intensity light of short time duration and with definitely regulated time characteristics. { 'fōd·ə,flash ,lamp }

photoflash unit [ELECTR] A portable electronic light source for photographic use, consisting of a capacitor-discharge power source, a flash tube, a battery for charging the capacitor, and sometimes also a high-voltage pulse generator to trigger the flash. { 'fōd·ə,flash ,yü·nət }

photoflood lamp [ELEC] An incandescent lamp used in photography which has a high-temperature filament, so that it gives high illumination and high color temperature for a short lifetime. { 'fōd·ə,fləd ,lamp }

photoglow tube [ELECTR] Gas-filled phototube used as a relay by making the operating voltage sufficiently high so that ionization and a flow discharge occur, with considerable current flow, when a certain illumination is reached. { 'fōd·ō ,glō ,tüb }

photographic recording [COMMUN] Facsimile recording in which a photosensitive surface is exposed to a signal-controlled light beam or spot. { ¦fōd·ə¦graf·ik ri'kórd·iŋ }

photographic sound recorder [ELECTR] A sound recorder having means for producing a modulated light beam and means for moving a light-sensitive medium relative to the beam to give a photographic recording of sound signals. Also known as optical sound recorder. { ¦fōd·ə ¦graf·ik 'saúnd ri,kórd·ər }

photographic sound reproducer [ELECTR] A sound reproducer in which an optical sound record on film is moved through a light beam directed at a light-sensitive device, to convert the recorded optical variations back into audio signals. Also known as optical sound reproducer. { ¦fōd·ə¦graf·ik 'saúnd ,rē·prə,düs·ər }

photoisland grid [ELECTR] Photosensitive surface in the storage-type, Farnsworth dissector tube for television cameras. { 'fōd·ō,ī·lənd ,grid }

photoisolator *See* optoisolator. { ¦fōd·ō'ī·sə ,lād·ər }

photomask [ELECTR] A film or glass negative that has many high-resolution images, used in the production of semiconductor devices and integrated circuits. { 'fōd·ō,mask }

photometer [ENG] An instrument used for making measurements of light or electromagnetic radiation, in the visible range. { fō'täm·əd·ər }

photomultiplier *See* multiplier phototube. { ¦fōd·ō'məl·tə,plī·ər }

photomultiplier cell [ELECTR] A transistor whose *pn*-junction is exposed so that it conducts more readily when illuminated. { ¦fōd·ō'məl·tə,plī·ər ¦sel }

photomultiplier counter [ELECTR] A scintillation counter that has a built-in multiplier phototube. { ¦fōd·ō'məl·tə,plī·ər ¦kaůnt·ər }

photomultiplier tube See multiplier phototube. { ¦fōd·ō'məl·tə,plī·ər ¦tüb }

photon coupled isolator [ELECTR] Circuit coupling device, consisting of an infrared emitter diode coupled to a photon detector over a short shielded light path, which provides extremely high circuit isolation. { 'fō,tän ¦kəp·əld 'ī·sə,lād·ər }

photon coupling [ELECTR] Coupling of two circuits by means of photons passing through a light pipe. { 'fō,tän ¦kəp·liŋ }

photonegative [ELECTR] Having negative photoconductivity, hence decreasing in conductivity (increasing in resistance) under the action of light; selenium sometimes exhibits photonegativity. { ¦fōd·ō'neg·ə·tiv }

photonics [ELECTR] The electronic technology involved with the practical generation, manipulation, analysis, transmission, and reception of electromagnetic energy in the visible, infrared, and ultraviolet portions of the light spectrum. It contributes to many fields, including astronomy, biomedicine, data communications and storage, fiber optics, imaging, optical computing, optoelectronics, sensing, and telecommunications. Also known as optoelectronics. { fō'tän·iks }

photopositive [ELECTR] Having positive photoconductivity, hence increasing in conductivity (decreasing in resistance) under the action of light; selenium ordinarily has photopositivity. { ¦fōd·ō'päz·əd·iv }

photoresistive cell See photoconductive cell. { ¦fōd·ō·ri'zis·tiv 'sel }

photoresistor See photoconductive cell. { ¦fōd·ō·ri'zis·tər }

photo-SCR See light-activated silicon controlled rectifier. { ¦fōd·ō ¦es¦sē'är }

photosensitive See light-sensitive. { ¦fōd·ō'sen·səd·iv }

photosensor See photodetector. { ¦fōd·ō'sen·sər }

phototelegraphy See facsimile. { ¦fōd·ō·tə'leg·rə·fē }

photothyristor See light-activated silicon controlled rectifier. { ¦fōd·ō·thī'ris·tər }

phototransistor [ELECTR] A junction transistor that may have only collector and emitter leads or also a base lead, with the base exposed to light through a tiny lens in the housing; collector current increases with light intensity, as a result of amplification of base current by the transistor structure. { ¦fōd·ō·tran'zis·tər }

phototronic photocell See photovoltaic cell. { ¦fōd·ə¦trän·ik 'fōd·ə,sel }

phototube [ELECTR] An electron tube containing a photocathode from which electrons are emitted when it is exposed to light or other electromagnetic radiation. Also known as electric eye; light-sensitive tube; photoelectric tube. { 'fōd·ō ,tüb }

phototube cathode [ELECTR] The photoemissive surface which is the most negative element of a phototube. { 'fōd·ō,tüb 'kath,ōd }

phototube relay [ELECTR] A photoelectric relay in which a phototube serves as the light-sensitive device. { 'fōd·ō,tüb 'rē,lā }

photovaristor [ELECTR] Varistor in which the current-voltage relation may be modified by illumination, for example, one in which the semiconductor is cadmium sulfide or lead telluride. { ¦fōd·ō·və'ris·tər }

photovoltaic [ELECTR] Capable of generating a voltage as a result of exposure to visible or other radiation. { ¦fōd·ō·vōl'tā·ik }

photovoltaic cell [ELECTR] A device that detects or measures electromagnetic radiation by generating a potential at a junction (barrier layer) between two types of material, upon absorption of radiant energy. Also known as barrier-layer cell; barrier-layer photocell; boundary-layer photocell; photronic photocell. { ¦fōd·ō·vōl'tā·ik ,sel }

photovoltaic effect [ELECTR] The production of a voltage in a nonhomogeneous semiconductor, such as silicon, or at a junction between two types of material, by the absorption of light or other electromagnetic radiation. { ¦fōd·ō·vōl'tā·ik i ,fekt }

photovoltaic meter [ELECTR] An exposure cell in which a photovoltaic cell produces a current proportional to the light falling on the cell, and this current is measured by a sensitive microammeter. { ¦fōd·ō·vōl'tā·ik ,mēd·ər }

photox cell [ELECTR] Type of photovoltaic cell in which a voltage is generated between a copper base and a film of cuprous oxide during exposure to visible or other radiation. { 'fō,täks ,sel }

photronic cell [ELECTR] Type of photovoltaic cell in which a voltage is generated in a layer of selenium during exposure to visible or other radiation. { fō'trän·ik ,sel }

photronic photocell See photovoltaic cell. { fō'trän·ik 'fōd·ə,sel }

phrase name See metavariable. { 'frāz ,nām }

physical data independence [COMPUT SCI] A file structure such that the physical structure of the data can be modified without changing the logical structure of the file. { 'fiz·ə·kəl ¦dad·ə ,in·di'pen·dəns }

physical data structure [COMPUT SCI] The manner in which data are physically arranged on a storage medium, including various indices and pointers. { 'fiz·ə·kəl 'dad·ə ,strək·chər }

physical device table [COMPUT SCI] A table associated with a physical input/output unit containing such information as the device type, an indication of data paths that may be used to transfer information to and from the device, status information on whether the device is busy, the input/output operation currently pending on the device, and the availability of any storage contained in the device. { 'fiz·ə·kəl di¦vīs ,tā·bəl }

physical drive [COMPUT SCI] An operational hard disk, which may be formatted to include more than one logical drive. { 'fiz·i·kəl ¦drīv }

physical electronics [ELECTR] The study of physical phenomena basic to electronics, such as discharges, thermionic and field emission,

428

and conduction in semiconductors and metals. { 'fiz·ə·kəl ˌiˌlek'trän·iks }

physical input/output control system See PIOCS. { 'fiz·ə·kəl ˌin‚pu̇t ˌau̇t‚pu̇t kən'trōl ˌsis·təm }

physical network |COMPUT SCI| A system of computers that communicate via cabling, modems, or other hardware, and may include more than one logical network or form part of a logical network. { 'fiz·i·kəl ˈnet‚wərk }

physical path length |COMPUT SCI| The physical distance that an electronic signal must travel between two points. Also known as path length. { 'fiz·ə·kəl ˈpath ˌleŋkth }

physical realizability |CONT SYS| For a transfer function, the possibility of constructing a network with this transfer function. { 'fiz·ə·kəl ˌrē·ə ˌlīz·ə'bil·əd·ē }

physical record |COMPUT SCI| A set of adjacent data characters recorded on some storage medium, physically separated from other physical records that may be on the same medium by means of some indication that can be recognized by a simple hardware test. Also known as record block. { 'fiz·ə·kəl 'rek·ərd }

physical system See causal system. { 'fiz·ə·kəl 'sis·təm }

pi attenuator |ELEC| An attenuator consisting of a pi network whose impedances are all resistances. { 'pī ə'ten·yəˌwād·ər }

PIC See personal identification code. { ˌpēˈīˈsē or pik }

pick-and-place robot |CONT SYS| A simple robot, often with only two or three degrees of freedon and little or no trajectory control, whose sole function is to transfer items from one place to another. { ˈpik ən ˈplās 'rō‚bät }

pick device See pointing device. { 'pik di‚vīs }

picking |COMPUT SCI| Identification of information displayed on a screen for subsequent computer processing, by pointing to it with a lightpen. { 'pik·iŋ }

pickoff |ELECTR| A device used to convert mechanical motion into a proportional electric signal. { 'pik‚ȯf }

pickup |ELEC| 1. A device that converts a sound, scene, measurable quantity, or other form of intelligence into corresponding electric signals, as in a microphone, phonograph pickup, or television camera. 2. The minimum current, voltage, power, or other value at which a relay will complete its intended function. 3. Interference from a nearby circuit or system. { 'pik‚əp }

pickup tube See camera tube. { 'pik‚əp ‚tüb }

pickup voltage |ELEC| Of a magnetically operated device, the voltage at which the device starts to operate. { 'pik‚əp ‚vōl·tij }

picoammeter |ENG| An ammeter whose scale is calibrated to indicate current values in picoamperes. { ˌpē·kō'am‚ēd·ər }

picoampere |ELEC| A unit of current equal to 10^{-12} ampere, or one-millionth of a microampere. Abbreviated pA. { ˌpē·kō'am‚pir }

picofarad |ELEC| A unit of capacitance equal to 10^{-12} farad, or one-millionth of a microfarad. Also known as micromicrofarad (deprecated

usage); puff (British usage). Abbreviated pF. { ˌpē·kō'far·əd }

picture |COMMUN| 1. The image on the screen of a video display. 2. Source, coded, or reconstructed image data; a source or reconstructed picture consists of three rectangular matrices representing the luminance and two chrominance signals. |COMPUT SCI| In COBOL, a symbolic description of each data element or item according to specified rules concerning numerals, alphanumerics, location of decimal points, and length. { 'pik·chər }

picture black See black signal. { 'pik·chər ˈblak }

picture carrier |COMMUN| A carrier frequency located 1.25 megahertz above the lower frequency limit of a standard National Television Systems Committee television signal; in color television, it is used for transmitting color information. Also known as luminance carrier. { 'pik·chər ˌkar·ē·ər }

picture compression |COMPUT SCI| The elimination of redundant information from a digital picture through the use of efficient encoding techniques in which frequently occurring gray levels or blocks of gray levels are represented by short codes and infrequently occurring ones by longer codes. { 'pik·chər kəmˌpresh·ən }

picture element |ELECTR| 1. That portion, in facsimile, of the subject copy which is seen by the scanner at any instant; it can be considered a square area having dimensions equal to the width of the scanning line. 2. In video, any segment of a scanning line, the dimension of which along the line is exactly equal to the nominal line width; the area which is being explored at any instant in the scanning process. Also known as critical area; elemental area; pixel; recording spot; scanning spot. { 'pik·chər ˌel·ə·mənt }

picture frequency |COMMUN| A frequency that results solely from scanning of subject copy in a facsimile system. |ELECTR| See frame frequency. { 'pik·chər ˌfrē·kwən·sē }

picture grammar |COMPUT SCI| A formalism for carrying out computations on pictures and describing picture structure. { 'pik·chər ˌgram·ər }

picture processing See image processing. { 'pik·chər ˌprä‚ses·iŋ }

picture segmentation |COMPUT SCI| The division of a complex picture into parts corresponding to regions or objects, so that the picture can then be described in terms of the parts, their properties, and their spatial relationships. Also known as scene analysis; segmentation. { 'pik·chər ˌseg·mən'tā·shən }

picture signal |COMMUN| The signal resulting from the scanning process in a video system. { 'pik·chər ˌsig·nəl }

picture synchronizing pulse See vertical synchronizing pulse. { 'pik·chər 'siŋ·krəˌnīz·iŋ ˌpəls }

picture transmission |COMMUN| Electric transmission of a picture having a gradation of shade values. { 'pik·chər tranz'mish·ən }

picture transmitter See visual transmitter. { 'pik·chər tranzˌmid·ər }

picture tube [ELECTR] A cathode-ray tube used in video displays to produce an image by varying the electron-beam intensity as the beam is deflected from side to side and up and down to scan a raster on the fluorescent screen at the large end of the tube. Also known as kinescope; television picture tube. { 'pik·chər ‚tüb }

picture-tube brightener [ELECTR] A small step-up transformer that can be inserted between the socket and base of a picture tube to increase the heater voltage and thereby increase picture brightness to compensate for normal aging of the tubes. { 'pik·chər ‚tüb ‚brīt·ən·ər }

picture white See white signal. { 'pik·chər ¦wīt }

Pierce oscillator [ELECTR] Oscillator in which a piezoelectric crystal unit is connected between the grid and the plate of an electron tube, in what is basically a Colpitts oscillator, with voltage division provided by the grid-cathode and plate-cathode capacitances of the circuit. { 'pirs 'äs·ə ‚lād·ər }

piezoelectric [SOLID STATE] Having the ability to generate a voltage when mechanical force is applied, or to produce a mechanical force when a voltage is applied, as in a piezoelectric crystal. { pē¦ā·zō·ə'lek·trik }

piezoelectric crystal [SOLID STATE] A crystal which exhibits the piezoelectric effect; used in crystal loudspeakers, crystal microphones, and crystal cartridges. { pē¦ā·zō·ə'lek·trik 'krist·əl }

piezoelectric effect [SOLID STATE] **1.** The generation of electric polarization in certain dielectric crystal as a result of the application of mechanical stress. **2.** The reverse effect, in which application of a voltage between certain faces of the crystal produces a mechanical distortion of the meterial. { pē¦ā·zō·ə'lek·trik i'fekt }

piezoelectric element [ELECTR] A piezoelectric crystal used in an electric circuit, for example, as a transducer to convert mechanical or acoustical signals to electric signals, or to control the frequency of a crystal oscillator. { pē¦ā·zō·ə'lek·trik 'el·ə·mənt }

piezoelectricity [SOLID STATE] Electricity or electric polarization resulting from the piezoelectric effect. { pē¦ā·zō·ə‚lek'tris·əd·ē }

piezoelectric loudspeaker See crystal loudspeaker. { pē¦ā·zō·ə'lek·trik 'laúd‚spēk·ər }

piezoelectric microphone See crystal microphone. { pē¦ā·zō·ə'lek·trik 'mī·krə‚fōn }

piezoelectric oscillator See crystal oscillator. { pē¦ā·zō·ə'lek·trik 'äs·ə‚lād·ər }

piezoelectric resonator See crystal resonator. { pē¦ā·zō·ə'lek·trik 'rez·ən‚ād·ər }

piezoelectric semiconductor [SOLID STATE] A semiconductor exhibiting the piezoelectric effect, such as quartz, Rochelle salt, and barium titanate. { pē¦ā·zō·ə'lek·trik 'sem·i·kən‚dək·tər }

piezoelectric transducer [ELECTR] A piezoelectric crystal used as a transducer, either to convert mechanical or acoustical signals to electric signals, as in a microphone, or vice versa, as in ultrasonic metal inspection. { pē¦ā·zō·ə'lek·trik tranz'dü·sər }

piezojunction effect [ELECTR] A change in the current-voltage characteristic of a *pn* junction that is produced by a mechanical stress. { pē ‚ā·zō'jəŋk·shən i‚fekt }

piezoresistive microphone [ENG ACOUS] A microphone in which a piezoresistive material is deposited on the edges of a membrane, and variations in the resistance of this material resulting from motion of the membrane are sensed, typically in a Wheatstone bridge. { pē ‚ā·zō·ri¦zis·tiv 'mī·krə‚fōn }

pi filter [ELECTR] A filter that has a series element and two parallel elements connected in the shape of the Greek letter pi (π). { 'pī ‚fil·tər }

pig [ELECTR] **1.** An ion source based on the same principle as the Philips ionization gage. **2.** See Philips ionization gage. { pig }

piggyback board [ELECTR] A small printed circuit board that is mounted on a larger board to provide additional circuitry. { 'pig·ē‚bak ‚bôrd }

piggyback twistor [ELECTR] Electrically alterable nondestructive-readout storage device that uses a thin narrow tape of magnetic material wound spirally around a fine copper conductor to store information; another similar tape is wrapped on top of the first, piggyback fashion, to sense the stored information; a binary digit or bit is stored at the intersection of a copper strap and a pair of these twistor wires. { 'pig·ē‚bak ¦twis·tər }

pigtail [ELEC] A short, flexible wire, usually stranded or braided, used between a stationary terminal and a terminal having a limited range of motion, as in relay armatures. { 'pig‚tāl }

pigtail splice [ELEC] A splice made by twisting together the bared ends of parallel conductors. { 'pig‚tāl ‚splīs }

pileup [ELECTR] A set of moving and fixed contacts, insulated from each other, formed as a unit for incorporation in a relay or switch. Also known as stack. { 'pīl‚əp }

pill [ELECTROMAG] A microwave stripline termination. { pil }

pillbox antenna [ELECTROMAG] Cylindrical parabolic reflector enclosed by two plates perpendicular to the cylinder, spaced to permit the propagation of only one mode in the desired direction of polarization. { 'pil‚bäks an'ten·ə }

pilot [COMMUN] **1.** In a transmission system, a signal wave, usually single frequency, transmitted over the system to indicate or control its characteristics. **2.** Instructions, in tape relay, appearing in routing line, relative to the transmission or handling of that message. [COMPUT SCI] A model of a computer system designed to test its design, logic, and data flow under operating conditions. { 'pī·lət }

PILOT [COMPUT SCI] A programming language designed for applications to computer-aided instruction and the question-and-answer type of interaction that occurs in that environment. { 'pī·lät }

pilot cell [ELEC] Selected cell of a storage battery whose temperature, voltage, and specific gravity are assumed to indicate the condition of the entire battery. { 'pī·lət ‚sel }

pilot lamp [ELEC] A small lamp used to indicate that a circuit is energized. Also known as pilot light. { 'pī·lət ,lamp }

pilot light *See* pilot lamp. { 'pī·lət ,līt }

pilot motor [ELEC] A small motor used in the automatic control of an electric current. { 'pī·lət ,mōd·ər }

pilot relaying [ELEC] A system for protecting transmission consisting of protective relays at line terminals and a communication channel between relays which is used by the relays to determine if a fault is within the protected line section, in which case all terminals are tripped simultaneously at high speed, or outside it, in which case tripping is blocked. { 'pī·lət rē,lā·iŋ }

pilot system [COMPUT SCI] A system for evaluating new procedures for handling data in which a sample that is representative of the data to be handled is processed. { 'pī·lət ,sis·təm }

pilot test [COMPUT SCI] A test of a computer system under operating conditions and in the environment for which the system was designed. { 'pī·lət ,test }

pilot tone [COMMUN] Single frequency transmitted over a channel to operate an alarm or automatic control. { 'pī·lət ,tōn }

pilot wire regulator [CONT SYS] Automatic device for controlling adjustable gains or losses associated with transmission circuits to compensate for transmission changes caused by temperature variations, the control usually depending upon the resistance of a conductor or pilot wire having substantially the same temperature conditions as the conductors of the circuits being regulated. { 'pī·lət ,wīr 'reg·yə,lād·ər }

PIM *See* personal information manager. { ¦pē ¦ī'em *or* pim }

pi mode [ELECTR] Of a magnetron, the mode of operation for which the phases of the fields of successive anode openings facing the interaction space differ by pi radians. { 'pī ,mōd }

pin [ELECTR] A terminal on an electron tube, semiconductor, integrated circuit, plug, or connector. Also known as base pin; prong. { pin }

pinch effect [ELEC] Manifestation of the magnetic self-attraction of parallel electric currents, such as constriction of ionized gas in a discharge tube, or constriction of molten metal through which a large current is flowing. Also known as cylindrical pinch; magnetic pinch; rheostriction. { 'pinch i,fekt }

pinch-off voltage [ELECTR] Of a field-effect transistor, the voltage at which the current flow between source and drain is blocked because the channel between these electrodes is completely depleted. { 'pinch,òf ,vōl·tij }

pinch resistor [ELECTR] A silicon integrated-circuit resistor produced by diffusing an *n*-type layer over a *p*-type resistor; this narrows or pinches the resistive channel, thereby increasing the resistance value. { 'pinch ri'zis·tər }

pinch roller [ELECTR] A small, freely turning wheel that presses the magnetic tape against the capstan in order to move the tape. { 'pinch ,rō·lər }

pincushion distortion [ELECTR] Distortion in which all four sides of a video image are concave (curving inward). { 'pin,kúsh·ən di,stòr·shən }

pin diode [ELECTR] A diode consisting of a silicon wafer containing nearly equal *p*-type and *n*-type impurities, with additional *p*-type impurities diffused from one side and additional *n*-type impurities from the other side; this leaves a lightly doped intrinsic layer in the middle, to act as a dielectric barrier between the *n*-type and *p*-type regions. Also known as power diode. { 'pin 'dī,ōd }

pine-tree array [ELECTROMAG] Array of dipole antennas aligned in a vertical plane known as the radiating curtain, behind which is a parallel array of dipole antennas forming a reflecting curtain. { 'pīn ,trē ə,rā }

pi network [ELEC] An electrical network which has three impedance branches connected in series to form a closed circuit, with the three junction points forming an output terminal, an input terminal, and a common output and input terminal. { 'pī ,net,wərk }

pin-feed printer [COMPUT SCI] A computer printer in which the paper is aligned and advanced by protrusions on two wheels which engage evenly spaced holes along the edges of the paper. Also known as tractor-feed printer. { 'pin ¦fēd 'print·ər }

ping [ELECTR] A sonic or ultrasonic pulse sent out by an echo-ranging sonar. { piŋ }

pinger [ENG ACOUS] A battery-powered, low-energy source for an echo sounder. { 'piŋ·ər }

ping-pong [COMMUN] To switch a transmission so that it travels in the opposite direction. [COMPUT SCI] The programming technique of using two magnetic tape units for multiple reel files and switching automatically between the two units until the complete file is processed. { 'piŋ,päŋ }

pin jack [ELEC] Single conductor jack having an opening for the insertion of a plug of very small diameter. { 'pin ,jak }

pin junction [ELECTR] A semiconductor device having three regions: *p*-type impurity, intrinsic (electrically pure), and *n*-type impurity. { 'pin ,jəŋk·shən }

pinout [ELECTR] A graphic or text description of the function of electronic signals transmitted through each pin and receptacle in a connector. { 'pin,aút }

PIOCS [COMPUT SCI] An extension of the hardware, constituting an interface between programs and data channels; opposed to LIOCS, logical input/output control system. Derived from physical input/output control system. { 'pī,äks }

pip *See* blip. { pip }

pipe [COMPUT SCI] Any software-controlled technique for transfering data from one program or task to another during processing. { pīp }

pipelining [COMPUT SCI] A procedure for processing instructions in a computer program more rapidly, in which each instruction is divided into numerous small stages, and a population of instructions are in various stages at any given time. { 'pīp,līn·iŋ }

pipe-to-soil potential |ELEC| The voltage potential (emf) generated between a buried pipe and its surrounding soil, the result of electrolytic action and a cause of electrolytic corrosion of the pipe. { 'pīp tə ¦sȯil pə,ten·chəl }

pi point |ELEC| Frequency at which the insertion phase shift of an electric structure is 180° or an integral multiple of 180°. { 'pī ,pȯint }

pi section filter |ELEC| An electric filter made of several pi networks connected in series. { 'pī ,sek·shən ,fil·tər }

piston |ELECTROMAG| A sliding metal cylinder used in waveguides and cavities for tuning purposes or for reflecting essentially all of the incident energy. Also known as plunger; waveguide plunger. { 'pis·tən }

piston attenuator |ELECTROMAG| A microwave attenuator inserted in a waveguide to introduce an amount of attenuation that can be varied by moving an output coupling device along its longitudinal axis. { 'pis·tən ə'ten·yə,wād·ər }

pitch |COMPUT SCI| The distance between the centerlines of adjacent rows of hole positions in punched paper tape. { pich }

pitch-row |COMPUT SCI| The distance between two adjacent holes in a paper tape. { 'pich ,rō }

pi-T transformation See Y-delta transformation. { ¦pī 'tē ,tranz·fər,mā·shən }

pixel |COMPUT SCI| The smallest part of an electronically coded picture image. |ELECTR| The smallest addressable element in an electronic display; a short form for picture element. Also known as pel. { pik'sel }

PL/1 |COMPUT SCI| A multipurpose programming language, developed by IBM for the Model 360 systems, which can be used for both commercial and scientific applications. { ¦pē¦el'wən }

PLA See programmed logic array.

placeholder |COMPUT SCI| A section of computer storage reserved for information that will be provided later. { 'plās,hōl·dər }

plaintext |COMMUN| The form of a message in which it can be generally understood, before it has been transformed by a code or cipher into a form in which it can be read only by those privy to the secrets of the cipher. |COMPUT SCI| Data that are to be encrypted. { 'plān,tekst }

plain vanilla See vanilla. { 'plān və'nil·ə }

planar area |COMPUT SCI| In computer graphics, an object with boundaries, such as a circle or polygon. { 'plān·ər ,er·ē·ə }

planar array |ELECTR| An array of ultrasonic transducers that can be mounted in a single plane or sheet, to permit closer conformation with the hull design of a sonar-carrying ship. { 'plā·nər ə¦rā }

planar-array antenna |ELECTROMAG| An array antenna in which the centers of the radiating elements are all in the same plane. { 'plā·nər ə¦rā an'ten·ə }

planar ceramic tube |ELECTR| Electron tube having parallel planar electrodes and a ceramic envelope. { 'plā·nər sə¦ram·ik 'tüb }

planar device |ELECTR| A semiconductor device having planar electrodes in parallel planes, made

by alternate diffusion of p- and n-type impurities into a substrate. { 'plā·nər di,vīs }

planar diode |ELECTR| A diode having planar electrodes in parallel planes. { 'plā·nər 'dī,ōd }

planar photodiode |ELECTR| A vacuum photodiode consisting simply of a photocathode and an anode; light enters through a window sealed into the base, behind the photocathode. { 'plā·nər ¦fōd·ō'dī,ōd }

planar process |ENG| A silicon-transistor manufacturing process in which a fractional-micrometer-thick oxide layer is grown on a silicon substrate; a series of etching and diffusion steps is then used to produce the transistor inside the silicon substrate. { 'plā·nər ,prä·səs }

planar transistor |ELECTR| A transistor constructed by an etching and diffusion technique in which the junction is never exposed during processing, and the junctions reach the surface in one plane; characterized by very low leakage current and relatively high gain. { 'plā·nər tran'zis·tər }

plane |ELECTR| Screen of magnetic cores; planes are combined to form stacks. { plān }

plane earth |ELECTROMAG| Earth that is considered to be a plane surface as used in ground-wave calculations. { 'plān ,ərth }

plane-earth attenuation |ELECTROMAG| Attenuation of an electromagnetic wave over an imperfectly conducting plane earth in excess of that over a perfectly conducting plane. { 'plān ,ərth ə,ten·yə'wā·shən }

plane of polarization |ELECTROMAG| Plane containing the electric vector and the direction of propagation of electromagnetic wave. { 'plān əv ,pō·lə·rə'zā·shən }

plane polarization See linear polarization. { 'plān ,pō·lə·rə'zā·shən }

plane-polarized wave |ELECTROMAG| An electromagnetic wave whose electric field vector at all times lies in a fixed plane that contains the direction of propagation through a homogeneous isotropic medium. { 'plān ¦pō·lə,rīzd ,wāv }

plane reflector See passive reflector. { 'plān ri ¦flek·tər }

planetary wave See long wave. { 'plan·ə,ter·ē 'wāv }

planigraphy See sectional radiography. { plə 'nig·rə·fē }

planoconvex spotlight |ELEC| A light that can be used as a sharply defined spotlight or for soft-edged lighting; ranges in power from 100 to 2000 watts. { ¦plā·nō'kän,veks 'spät,līt }

plan position indicator |ELECTR| A radar display in which echoes from various targets appear as bright spots at the same locations as they would on a circular map of the area being scanned, the radar antenna being at the center of the map. Variations of the plan position indicator format include limited-sector display with the radar location offset from the center appropriately, the orientation to true or magnetic north or the radar-vehicle heading at the top, and so on. Abbreviated PPI. { 'plan pə'zish·ən 'in·də,kād·ər }

plan position indicator repeater |ELECTR| Unit which repeats a plan position indicator (PPI) at a location remote from the radar console. Also known as remote plan position indicator. { 'plan pə'zish·ən 'in·də,kād·ər ri,pēd·ər }

plant |COMPUT SCI| To place a number or instruction that has been generated in the course of a computer program in a storage location where it will be used or obeyed at a later stage of the program. { plant }

Plante cell |ELEC| A type of lead-acid cell in which the active material is formed on the plates by electrochemical means during repeated charging and discharging, instead of being applied as a prepared paste. { plän'tā ,sel }

plant factor |ELEC| The ratio of the average power load of an electric power plant to its rated capacity. Also known as capacity factor. { 'plant ,fak·tər }

plasma cathode |ELECTR| A cathode in which the source of electrons is a gas plasma rather than a solid. { 'plaz·mə 'kath,ōd }

plasma diode |ELECTR| A diode used for converting heat directly into electricity; it consists of two closely spaced electrodes serving as cathode and anode, mounted in an envelope in which a low-pressure cesium vapor fills the interelectrode space; heat is applied to the cathode, causing emission of electrons. { 'plaz·mə 'dī,ōd }

plasma display |ELECTR| A display in which sets of parallel conductors at right angles to each other are deposited on glass plates, with the very small space between the plates filled with a gas; each intersection of two conductors defines a single cell that can be energized to produce a gas discharge forming one element of a dot-matrix display. { 'plaz·mə di'splā }

plasma etching |ELECTR| A method of forming integrated-circuit patterns on a surface, in which charged species in a plasma formed above a masked surface are directed to impact the nonmasked regions of the surface and knock out substrate atoms. Also known as dry plasma etching. { 'plaz·mə 'ech·iŋ }

plasma generator |ELECTR| Any device that produces a high-velocity plasma jet, such as a plasma accelerator, engine, oscillator, or torch. { 'plaz·mə 'jen·ə,rād·ər }

plasma gun |ELECTR| A machine, such as an electric-arc chamber, that will generate very high heat fluxes to convert neutral gases into plasma. |ELECTROMAG| An electromagnetic device which creates and accelerates bursts of plasma. { 'plaz·mə ,gən }

plasma sheath |ELECTR| An envelope of ionized gas that surrounds a spacecraft or other body moving through an atmosphere at hypersonic velocities; affects transmission, reception, and diffraction of radio waves. { 'plaz·mə ,shēth }

plasmatron |ELECTR| A gas-discharge tube in which independently generated plasma serves as a conductor between a hot cathode and an anode; the anode current is modulated by varying either the conductivity or the effective cross section of the plasma. { 'plaz·mə,trän }

plastic film capacitor |ELEC| A capacitor constructed by stacking, or forming into a roll, alternate layers of foil and a dielectric which consists of a plastic, such as polystyrene or Mylar, either alone or as a laminate with paper. { 'plas·tik ¦film kə'pas·əd·ər }

plastic plate |ELECTR| A plate of plastic dielectric material used as a base for a semiconductor device. { 'plas·tik 'plāt }

plate |ELEC| **1.** One of the conducting surfaces in a capacitor. **2.** One of the electrodes in a storage battery. |ELECTR| See anode. { plāt }

plateau |ELECTR| The portion of the plateau characteristic of a counter tube in which the counting rate is substantially independent of the applied voltage. { pla'tō }

plateau characteristic |ELECTR| The relation between counting rate and voltage for a counter tube when radiation is constant, showing a plateau after the rise from the starting voltage to the Geiger threshold. Also known as counting rate-voltage characteristic. { pla'tō ,kar·ik·tə'ris·tik }

plate circuit See anode circuit. { plāt ¦sər·kət }

plate-circuit detector See anode-circuit detector. { 'plāt ¦sər·kət di,tek·tər }

plate current See anode current. { 'plāt ,kə·rənt }

plated circuit |ELECTR| A printed circuit produced by electrodeposition of a conductive pattern on an insulating base. Also known as plated printed circuit. { 'plād·əd 'sər·kət }

plate detector See anode detector. { 'plāt di,tek·tər }

plate dissipation See anode dissipation. { 'plāt ,dis·ə,pā·shən }

plated printed circuit See plated circuit. { 'plād·əd 'print·əd 'sər·kət }

plated wire memory |COMPUT SCI| A nonvolatile magnetic memory utilizing small zones of thin films plated on wires; such memories are characterized by very fast access and nondestructive readout. { 'plād·əd ¦wīr 'mem·rē }

plate efficiency See anode efficiency. { 'plāt i,fish·ən·sē }

plate impedance See anode impedance. { 'plāt im,pēd·əns }

plate input power See anode input power. { 'plāt 'in,pút ,pau·ər }

plate-load impedance See anode impedance. { 'plāt ,lōd im,pēd·əns }

plate modulation See anode modulation. { 'plāt ,mäj·ə,lā·shən }

plate neutralization See anode neutralization. { 'plāt ,nü·trə·lə,zā·shən }

plate pulse modulation See anode pulse modulation. { 'plāt 'pəls ,mäj·ə,lā·shən }

plate resistance See anode resistance. { 'plāt ri,zis·təns }

plate saturation See anode saturation. { 'plāt ,sach·ə,rā·shən }

platform |COMPUT SCI| The hardware system and the system software used by a computer program. { 'plat,fôrm }

platinotron |ELECTR| A microwave tube that may be used as a high-power saturated amplifier or oscillator in pulsed radar applications; requires permanent magnet just as does a magnetron. { plə'tin·ə,trän }

platter   [COMPUT SCI] One of the disks in a hard-disk drive or disk pack.   { 'plad·ər }

playback   [ENG ACOUS] Reproduction of a sound recording.   { 'plā,bak }

playback head   [ELECTR] A head that converts a changing magnetic field on a moving magnetic tape into corresponding electric signals. Also known as reproduce head.   { 'plā,bak ,hed }

playback robot   [CONT SYS] A robot that repeats the same sequence of motions in all its operations, and is first instructed by an operator who puts it through this sequence.   { 'plā,bak 'rō ,bät }

pliotron   [ELECTR] Any hot-cathode vacuum tube having one or more grids.   { 'plī·ə,trän }

PLL   See phase-locked loop.

plug   [ELEC] The half of a connector that is normally movable and is generally attached to a cable or removable subassembly; inserted in a jack, outlet, receptacle, or socket.   { pləg }

plug adapter lamp holder   [ELEC] A device that can be inserted in a lamp holder to act as a lamp holder and one or more receptacles. Also known as current tap.   { 'pləg ə,dap·tər 'lamp ,hōld·ər }

plugboard   See control panel.   { 'pləg,bórd }

plugboard chart   See plugging chart.   { 'pləg,bórd ,chärt }

plug-compatible hardware   [COMPUT SCI] A piece of equipment which can be immediately connected to a computer manufactured by another company.   { 'pləg kəm,pad·ə·bəl 'härd·wer }

plug fuse   [ELEC] A fuse designed for use in a standard screw-base lamp socket.   { 'pləg ,fyüz }

plugging   [ELEC] Braking an electric motor by reversing its connections, so it tends to turn in the opposite direction; the circuit is opened automatically when the motor stops, so the motor does not actually reverse.   { 'pləg·iŋ }

plugging chart   [COMPUT SCI] A printed chart of the sockets in a plugboard on which may be shown the jacks or wires connecting these sockets. Also known as plugboard chart.   { 'pləg·iŋ ,chärt }

plug-in   [COMPUT SCI] A small software application that extends the capabilities (such as multimedia, audio, or video) of a browser.   { 'pləg ,in }

plug-in unit   [ELEC] A component or subassembly having plug-in terminals so all connections can be made simultaneously by pushing the unit into a suitable socket.   { 'pləg·in ,yü·nət }

plug program patching   [COMPUT SCI] A relatively small auxiliary plugboard patched with a specific variation of a portion of a program and designed to be plugged into a relatively larger plugboard patched with the main program.   { 'pləg 'prō ,gram ,pach·iŋ }

plug-to-plug compatibility   [COMPUT SCI] Property of a peripheral device that can be made to operate with a computer merely by attachment of a plug or a relatively small number of cables.   { 'pləg tə 'pləg kəm,pad·ə'bil· əd·ē }

plunger   See piston.   { 'plən·jər }

plus-90 orientation   [COMPUT SCI] In optical character recognition, that determinate position which indicates that the line elements of an inputted source document appear perpendicular with the leading edge of the optical reader.   { 'pləs 'nīn·tē ,ór·ē·ən,tā·shən }

plus zone   [COMPUT SCI] The bit positions in a computer code which represent the algebraic plus sign.   { 'pləs ,zōn }

PM   See phase modulation.

PMLCD   See supertwisted nematic liquid-crystal display.

PMOS   [ELECTR] Metal-oxide semiconductors that are made on n-type substrates, and whose active carriers are holes that migrate between p-type source and drain contacts. Derived from p-channel metal-oxide semiconductor.   { 'pē ,mòs }

PMS notation   [COMPUT SCI] A notation that provides a clear, concise description of the physical structure of computer systems, and that contains only a few primitive components, namely symbols for memory, link, switch, data operation, control unit, and transducer. Acronym for processor-memory-switch notation.   { 'pē ¦em¦es nō'tā·shən }

PN code   See pseudorandom noise code.   { ,pē'en 'kōd }

pneumatic transmission lag   [ELEC] The time delay in a pneumatic transmission line between the generation of an impulse at one end and the resultant reaction at the other end.   { nü'mad·ik ¦tranz,mish·ən ,lag }

pn hook transistor   See hook collector transistor.   { ¦pē¦en 'hùk tran,zis·tər }

pnip transistor   [ELECTR] An intrinsic junction transistor in which the intrinsic region is sandwiched between the n-type base and the p-type collector.   { ¦pē,en,ī¦pē tran,zis·tər }

pn junction   [ELECTR] The interface between two regions in a semiconductor crystal which have been treated so that one is a p-type semiconductor and the other is an n-type semiconductor; contains a permanent dipole charge layer.   { ¦pē ¦en ¦jəŋk·shən }

pnpn diode   [ELECTR] A semiconductor device consisting of four layers of p-type and n-type semiconductor material, with terminal connections to the two outer layers. Also known as $npnp$ diode.   { ¦pē¦en¦pē¦en ,dī,ōd }

pnpn transistor   See $npnp$ transistor.   { ¦pē¦en¦pē ¦en tran,zis·tər }

pnp transistor   [ELECTR] A junction transistor having an n-type base between a p-type emitter and a p-type collector.   { ¦pē¦en¦pē tran,zis·tər }

Pockels readout optical modulator   [ELECTR] A device for storing data in the form of images; it consists of bismuth silicon oxide crystal coated with an insulating layer of parylene and transparent electrodes evaporated on the surfaces; a blue laser is used for writing and a red laser is used for nondestructive readout or processing. Abbreviated PROM. { 'päk·əlz ¦rēd ‚aút ‚äp·tə·kəl 'mäj·ə‚lād·ər }

pocket |COMPUT SCI| One of the several receptacles into which punched cards are fed by a card sorter. { 'päk·ət }

Poggendorff's first method *See* constant-current dc potentiometer. { 'päg·ən‚dórfs 'first ‚meth·əd }

Poggendorff's second method *See* constant-resistance dc potentiometer. { 'päg·ən‚dórfs 'sek·ənd ‚meth·əd }

point contact |ELECTR| A contact between a specially prepared semiconductor surface and a metal point, usually maintained by mechanical pressure but sometimes welded or bonded. { 'póint 'kän‚takt }

point-contact diode |ELECTR| A semiconductor rectifier that uses the barrier formed between a specially prepared semiconductor surface and a metal point to produce the rectifying action. { 'póint ¦kän‚takt ‚dī‚ōd }

point-contact silicon cell |ELECTR| A type of solar cell whose efficiency is enhanced by a combination of tiny doped-silicon dots scattered across the lower surface of the silicon crystal and fine aluminum threads that penetrate the silicon layer to collect current from each point. { 'póint ¦kän ‚takt 'sil·ə·kən ‚sel }

point-contact transistor |ELECTR| A transistor having a base electrode and two or more point contacts located near each other on the surface of an *n*-type semiconductor. { 'póint ¦kän‚takt tran‚zis·tər }

pointer |COMPUT SCI| The part of an instruction which contains the address of the next record to be accessed. { 'póint·ər }

pointing device |COMPUT SCI| A handheld device, such as a mouse, puck, or stylus, that controls a position indicator on a display screen. Also known as pick device. { 'póint·iŋ di‚vīs }

pointing stick |COMPUT SCI| A small rubberized device located in the center of a computer keyboard, which is moved with a finger tip to position a pointer. { 'póint·iŋ ‚stik }

point jammer |ELECTR| Any electronic jammer directed against a specific enemy installation operating on a specific frequency. { 'póint ‚jam·ər }

point-junction transistor |ELECTR| Transistor having a base electrode and both point-contact and junction electrodes. { 'póint ‚jəŋk·shən tran‚zis·tər }

point-mode display |COMPUT SCI| A method of representing information in the form of dots on the face of a cathode-ray tube. { 'póint ‚mōd di ‚splā }

point-of-origin system |COMPUT SCI| A computer system in which data collection occurs at the point where the data are actually created, as in a point-of-sale terminal. { ¦póint əv 'är·ə·jən ‚sis·təm }

point-of-sale terminal |COMPUT SCI| A computer-connected terminal used in place of a cash register in a store, for customer checkout and such added functions as recording inventory data, transferring funds from the customer's bank account to the merchant's bank account, and checking credit on charged or charge-card purchases; the terminals can be modified for many nonmerchandising applications, such as checkout of books in libraries. Abbreviated POS terminal. { 'póint əv ¦sāl 'term·ən·əl }

point projection electron microscope |ELECTR| An electron microscope in which a real or virtual point source of electrons produces a highly magnified shadow. { 'póint prə¦jek·shən i¦lek ‚trän 'mī·krə‚skōp }

point-source light |ELEC| A special lamp in which the radiating element is concentrated in a small physical area. { 'póint ‚sórs ‚līt }

point target |ELECTROMAG| In radar, an object which returns a target signal by reflection from a relatively simple discrete surface; such targets are ships, aircraft, projectiles, missiles, and buildings. { 'póint ‚tär·gət }

point-to-point communication |COMMUN| Radio communication between two fixed stations. { 'póint tə 'póint kə‚myü·nə'kā·shən }

point-to-point programming |CONT SYS| A method of programming a robot in which each major change in the robot's path of motion is recorded and stored for later use. { ¦póint tə ¦póint 'prō‚gram·iŋ }

Point-to-Point Protocol |COMMUN| A standard governing dial-up connections of computers to the Internet via a telephone modem. Abbreviated PPP. { ‚póin·tü ‚póint 'prōd·ə‚kól }

point transposition |ELEC| Transposition, usually in an open-wire line, which is executed within a distance comparable to the wire separation, without material distortion of the normal wire configuration outside this distance. { 'póint ‚tranz·pə‚zish·ən }

poison |ELECTR| A material which reduces the emission of electrons from the surface of a cathode. { 'póiz·ən }

poke |COMPUT SCI| An instruction that causes a value in a storage location in a microcomputer's main storage to be replaced. { pōk }

polar-coordinate navigation system |NAV| A system in which one or more signals are emitted from a facility (or co-located facilities) to produce simultaneous indication of bearing and distance. { 'pō·lər kō¦órd·ən·ət ‚nav·ə'gā·shən ‚sis·təm }

polarity |COMMUN| **1.** The direction in which a direct current flows, in a teletypewriter system. **2.** The sense of the potential of a portion of a video signal representing a dark area of a scene relative to the potential of a portion of the signal representing a light area. { pə'lar·əd·ē }

polarity effect |ELECTR| An effect for which the breakdown voltage across a vacuum separating two electrodes, one of which is pointed, is much higher when the pointed electrode is the anode. { pə'lar·əd·ē i‚fekt }

polarizability [ELEC] The electric dipole moment induced in a system, such as an atom or molecule, by an electric field of unit strength. { ‚pō·lə‚rīz·ə'bil·əd·ē }

polarizability catastrophe [ELEC] According to a theory using the Lorentz field concept, the phenomenon where, at a certain temperature, the dielectric constant of a material becomes infinite. { ‚pō·lə‚rīz·ə'bil·əd·ē kə'tas·trə·fē }

polarization [ELEC] **1.** The process of producing a relative displacement of positive and negative bound charges in a body by applying an electric field. **2.** A vector quantity equal to the electric dipole moment per unit volume of a material. Also known as dielectric polarization; electric polarization. **3.** A chemical change occurring in dry cells during use, increasing the internal resistance of the cell and shortening its useful life. { ‚pō·lə·rə'zā·shən }

polarization charge See bound charge. { ‚pō·lə·rə'zā·shən ‚chärj }

polarization diversity [COMMUN] A method of transmission and reception used to minimize the effects of selective fading of the horizontal and vertical components of a radio signal; it is usually accomplished through the use of separate vertically and horizontally polarized receiving antennas. { ‚pō·lə·rə'zā·shən də'ver·səd·ē }

polarization division multiple access [COMMUN] A technique for allowing multiple users at geographically dispersed locations to gain access to a shared communications channel by assigning them electric fields of different polarization. { ‚pō·lə·rə‚zā·shən də‚vizh·ən ¦məl·tə·pəl 'ak‚ses }

polarization division multiplexing [COMMUN] The sharing of a communications channel among multiple users by assigning them electric fields of different polarization. { ‚pō·lə·rə'zā·shən di ‚vizh·ən 'məl·tə‚pleks·iŋ }

polarization fading [COMMUN] Fading as the result of changes in the direction of polarization in one or more of the propagation paths of waves arriving at a receiving point. { ‚pō·lə·rə'zā·shən ‚fād·iŋ }

polarized electrolytic capacitor [ELEC] An electrolytic capacitor in which the dielectric film is formed adjacent to only one metal electrode; the impedance to the flow of current is then greater in one direction than in the other. { 'pō·lə‚rīzd i¦lek·trə¦lid·ik kə'pas·əd·ər }

polarized electromagnetic radiation [ELECTROMAG] Electromagnetic radiation in which the direction of the electric field vector is not random. { 'pō·lə‚rīzd i¦lek·trō·mag¦ned·ik ‚rād·ē'ā·shən }

polarized ion source [ELECTR] A device that generates ion beams in such a manner that the spins of the ions are aligned in some direction. { 'pō·lə‚rīzd 'ī‚än ‚sȯrs }

polarized meter [ENG] A meter having a zero-center scale, with the direction of deflection of the pointer depending on the polarity of the voltage or the direction of the current being measured. { 'pō·lə‚rīzd 'mēd·ər }

polarized plug [ELEC] A plug that can be inserted in its receptacle only when in a predetermined position. { 'pō·lə‚rīzd 'pləg }

polarized receptacle [ELEC] A receptacle designed for use with a polarized plug, to ensure that the grounded side of an alternating-current line or the positive side of a direct-current line is always connected to the same terminal on a piece of equipment. { 'pō·lə‚rīzd ri'sep·tə·kəl }

polarized relay [ELEC] Relay in which the movement of the armature depends upon the direction of the current in the circuit controlling the armature. Also known as polar relay. { 'pō·lə ‚rīzd 'rē‚lā }

polar keying [COMMUN] Telegraph signal in which circuit current flows in one direction for spacing. { 'pō·lər 'kē·iŋ }

polar modulation [COMMUN] Amplitude modulation in which the positive excursions of the carrier are modulated by one signal and the negative excursions by another. { 'pō·lər ‚mäj·ə'lā·shən }

polar radiation pattern [ELECTROMAG] Diagram showing the relative strength of the radiation from an antenna in all directions in a given plane. [ENG ACOUS] Diagram showing the strength of sound waves radiated from a loudspeaker in various directions in a given plane, or a similar response pattern for a microphone. { 'pō·lər ‚rād·ē'ā·shən ‚pad·ərn }

polar relay See polarized relay. { 'pō·lər 'rē‚lā }

polar resolution [COMPUT SCI] Given the x and y components of a vector, the process of finding the magnitude of the vector and the angle it makes with the x axis. { 'pō·lər ‚rez·ə'lü·shən }

polar transmission [COMMUN] **1.** A method of signaling in teletypewriter transmission in which direct currents flowing in opposite directions represent a mark and a space respectively, and absence of current indicates a no-signal condition. **2.** By extension, any system of signaling that uses three conditions, representing a mark, a space, or a no-signal condition. { 'pō·lər tranz'mish·ən }

pole [ELEC] **1.** One of the electrodes in an electric cell. **2.** An output terminal on a switch; a double-pole switch has two output terminals. { pōl }

pole-positioning [CONT SYS] A design technique used in linear control theory in which many or all of a system's closed-loop poles are positioned as required, by proper choice of a linear state feedback law; if the system is controllable, all of the closed-loop poles can be arbitrarily positioned by this technique. { 'pōl pə‚zish·ən·iŋ }

pole-zero configuration [CONT SYS] A plot of the poles and zeros of a transfer function in the complex plane; used to study the stability of a system, its natural motion, its frequency response, and its transient response. { 'pōl ¦zir·ō kən‚fig·yə'rā·shən }

poling [ELEC] Adjustment of polarity; specifically, in wire-line practice, the use of transpositions between transposition sections of open wire or between lengths of cable, to cause the residual cross-talk couplings in individual sections or lengths to oppose one another. { 'pōl·iŋ }

Polish notation [COMPUT SCI] **1.** A notation system for digital-computer or calculator logic in which there are no parenthetical expressions and each operator is a binary or unary operator in the sense that it operates on not more than two operands. Also known as Lukasiewicz notation; parenthesis-free notation. **2.** The version of this notation in which operators precede the operands with which they are associated. Also known as prefix notation. { 'pō·lish nō'tā·shən }

polling |COMMUN| A process that involves interrogating in succession every terminal on a shared communications line to determine which of the terminals require service. { 'pōl·iŋ }

polling list |COMMUN| A roster of transmitting devices sequentially scanned in a time-sharing system. { 'pōl·iŋ ,list }

polyalphabetic substitution cipher |COMMUN| A cipher that uses several substitution alphabets in turn. { ¦päl·ē,al·fə'bed·ik ¸səb·stə'tü·shən ¸sī·fər }

polychromatic radiation |ELECTROMAG| Electromagnetic radiation that is spread over a range of frequencies. { ¦päl·i·krō'mad·ik ¸rād·ē'ā·shən }

polyline |COMPUT SCI| In computer graphics, a series of connected line segments and arcs that are treated as a single entity. { 'päl·ē,līn }

polymer-dispersed liquid-crystal display |ELECTR| An electronic display in which the display elements have micrometer sized-diameter, have nearly spherical liquid-crystal droplets surrounded by a solid polymer, and the display is switched from a white opaque appearance to a clear transparent appearance by applying an electric field. { ¦päl·ə·mərdi,spərst ¸lik·wəd ¸krist·əl di'splā }

polymorphic system |COMPUT SCI| A computer system that is organized around a central pool of shared software modules which are selected as they are needed for processing. { ¦päl·i¦mór·fik 'sis·təm }

polymorphism |COMPUT SCI| A property of object-oriented programming that allows many different types of objects to be treated in a uniform manner by invoking the same operation on each object. { ¸päl·i'mór,fiz·əm }

polynomial time |COMPUT SCI| The property of the time required to solve a problem on a computer for which there exist constants c and k such that, if the input to the problem can be specified in N bits, the problem can be solved in $c \times N^k$ elementary operations. { ¦päl·ə¦nō·mē·əl 'tīm }

polyphase |ELEC| Having or utilizing two or more phases of an alternating-current power line. { 'päl·i,fāz }

polyphase circuit |ELEC| Group of alternating-current circuits (usually interconnected) which enter (or leave) a delimited region at more than two points of entry; they are intended to be so energized that, in the steady state, the alternating currents through the points of entry, and the alternating potential differences between them, all have exactly equal periods, but have differences in phase, and may have differences in waveform. { 'päl·i,fāz 'sər·kət }

polyphase meter |ENG| An instrument which measures some electrical quantity, such as power factor or power, in a polyphase circuit. { 'päl·i ,fāz 'mēd·ər }

polyphase rectifier |ELECTR| A rectifier which utilizes two or more diodes (usually three), each of which operates during an equal fraction of an alternating-current cycle to achieve an output current which varies less than that in an ordinary half-wave or full-wave rectifier. { 'päl·i ,fāz 'rek·tə,fī·ər }

polyphase synchronous generator |ELEC| Generator whose alternating-current circuits are so arranged that two or more symmetrical alternating electromotive forces with definite phase relationships are produced at its terminals. { 'päl·i,fāz 'sin·krə·nəs 'jen·ə,rād·ər }

polyphase transformer |ELEC| A transformer with multiple sets of primary and secondary windings on a single core; used in a polyphase circuit. { 'päl·i,fāz tranz'fór·mər }

polyphase wattmeter |ENG| An instrument that measures electric power in a polyphase circuit. { 'päl·i,fāz 'wät,mēd·ər }

polyrod antenna |ELECTROMAG| End-fire directional dielectric antenna consisting of a polystyrene rod energized by a section of waveguide. { 'päl·i,räd an'ten·ə }

polystyrene capacitor |ELEC| A capacitor that uses film polystyrene as a dielectric between rolled strips of metal foil. { 'päl·i'stī,rēn kə'pas·əd·ər }

polystyrene dielectric |ELEC| Polystyrene used in applications where its very high resistivity, good dielectric strength, and other electrical properties are important, such as for electrical insulation or in dielectrics. { 'päl·i'stī,rēn ¦dī·ə'lek·trik }

polyvalent number |COMPUT SCI| A number, consisting of several figures, used for description, wherein each figure represents one of the characteristics being described. { 'päl·i'vā·lənt 'nəm·bər }

pool cathode |ELECTR| A cathode at which the principal source of electron emission is a cathode spot on a liquid-metal electrode, usually mercury. { 'pül ,kath,ōd }

pool-cathode mercury-arc rectifier |ELECTR| A pool tube connected in an electric circuit; its rectifying properties result from the fact that only the mercury-pool cathode, and not the anode, can emit electrons. Also known as mercury-pool rectifier. { 'pül ,kath,ōd 'mər·kyə·rē ¦ärk 'rek·tə ,fī·ər }

pool-cathode tube See pool tube. { 'pül ,kath ,ōd ,tüb }

Poole-Frenkel effect |ELEC| An increase in the electrical conductivity of insulators and semiconductors in strong electric fields. { ¦pül 'freŋ·kəl i,fekt }

pool tube |ELECTR| A gas-discharge tube having a mercury-pool cathode. Also known as mercury tube; pool-cathode tube. { 'pül ,tüb }

pop |COMPUT SCI| To obtain information from the top of a stack and then reset a pointer to the next item in the stack. { päp }

POP See Post Office Protocol. { päp or ¦pē¦ō'pē }

popcorn noise |ELECTR| Noise that is produced by erratic jumps of bias current between two levels at random intervals in operational amplifiers and other semiconductor devices. { 'päp ‚körn }

pop hole *See* pop. { 'päp ‚hōl }

Popov's stability criterion |CONT SYS| A frequency domain stability test for systems consisting of a linear component described by a transfer function preceded by a nonlinear component characterized by an input-output function, with a unity gain feedback loop surrounding the series connection. { pä'pófs stə'bil· əd·ē krī ‚tir·ē·ən }

popping |COMPUT SCI| The deletion of the top element of a stack. { 'päp·iŋ }

pop shot *See* pop. { 'päp ‚shät }

populate |COMPUT SCI| To add electronic components, such as memory chips, to a circuit board. { 'päp·yə‚lāt }

population |COMPUT SCI| A collection of records in a data base that share one or more characteristics in common. |ELECTR| The set of electronic components on a printed circuit board. { ‚päp·yə'lā·shən }

porcelain capacitor |ELEC| A fixed capacitor in which the dielectric is a high grade of porcelain, molecularly fused to alternate layers of fine silver electrodes to form a monolithic unit that requires no case or hermetic seal. { 'pòrs·lən kə'pas·əd·ər }

port |COMPUT SCI| **1.** An interface between a communications channel and a unit of computer hardware. **2.** To modify an application program, developed to run with a particular operating system, so that it can run with another operating system. **3.** A designation which a program on a client computer uses to specify a server program on a computer in a network. |ELEC| An entrance or exit for a network. |ELECTROMAG| An opening in a waveguide component, through which energy may be fed or withdrawn, or measurements made. { pòrt }

portability |COMPUT SCI| Property of a computer program that is sufficiently flexible to be easily transferred to run on a computer of a type different from the one for which it was designed. { ‚pòrd·ə'bil·əd·ē }

portable audio terminal |COMPUT SCI| A lightweight, self-contained computer terminal with a typewriter keyboard, which can be attached to a telephone line by placing the telephone handset in a receptacle in the terminal. { 'pòrd·ə·bəl 'òd·ē·ō ‚tərm·ən·əl }

portable data terminal |COMPUT SCI| A computer terminal that can be carried about by hand to collect data from remote locations and to transfer this data to a computer system. { 'pòrd·ə·bəl 'dad·ə ‚tər·mən·əl }

portable document format |COMPUT SCI| A computer file format for publishing and distributing electronic documents (text, image, or multimedia) with the same layout, formatting, and font attributes as in the original. The files can be opened and viewed on any computer or operating system; however, special software is required. Abbreviated PDF. { ¦pòrd·ə·bəl ‚däk·yə·mənt 'fòr‚mat }

ported system *See* vented-box system. { 'pòrt·əd ‚sis·təm }

port expander |COMPUT SCI| Equipment that connects links to several other devices to one port in a computer. { 'pòrt ik‚span·dər }

porting |COMPUT SCI| The process of converting software to run on a computer other than the one for which it was originally written. { 'pòrd·iŋ }

posistor |ELECTR| A thermistor having a large positive resistance-temperature characteristic. { pä'zis·tər }

positional-error constant |CONT SYS| For a stable unity feedback system, the limit of the transfer function as its argument approaches zero. { pə'zish·ən·əl ¦er·ər ‚kän·stənt }

positional notation |MATH| Any of several numeration systems in which a number is represented by a sequence of digits in such a way that the significance of each digit depends on its position in the sequence as well as its numeric value. Also known as notation. { pə'zish·ən·əl nō'tā·shən }

positional parameter |COMPUT SCI| One of a number of parameters in a group, whose significance is determined by its position within the group. { pə'zish·ən·əl pə'ram·əd·ər }

positional servomechanism |CONT SYS| A feedback control system in which the mechanical position (as opposed to velocity) of some object is automatically maintained. { pə'zish·ən·əl ¦sər·vō'mek·ə‚niz·əm }

position control |CONT SYS| A type of automatic control in which the input commands are the desired position of a body. { pə'zish·ən kən ‚tròl }

position indicator |ENG| An electromechanical dead-reckoning computer, either an air-position indicator or a ground-position indicator. { pə'zish·ən ‚in·də‚kād·ər }

positioning action |CONT SYS| Automatic control action in which there is a predetermined relation between the value of a controlled variable and the position of a final control element. { pə'zish·ən·iŋ ‚ak·shən }

positioning time |COMPUT SCI| The time required for a storage medium such as a disk to be positioned and for read/write heads to be properly located so that the desired data can be read or written. { pə'zish·ən·iŋ ‚tīm }

position pulse *See* commutator pulse. { pə 'zish·ən ‚pəls }

position sensor |ENG| A device for measuring a position and converting this measurement into a form convenient for transmission. Also known as position transducer. { pə'zish·ən ‚sen·sər }

position telemetering |ENG| A variation of voltage telemetering in which the system transmits the measurand by positioning a variable resistor or other component in a bridge circuit so as to produce relative magnitudes of electrical quantities or phase relationships. { pə'zish·ən ¦tel·ə'mēd·ə·riŋ }

position transducer *See* position sensor. { pə'zish·ən tranz‚dü·sər }

438

positive |ELEC| Having fewer electrons than normal, and hence having ability to attract electrons. { 'päz·əd·iv }

positive bias |ELECTR| A bias such that the control grid of an electron tube is positive with respect to the cathode. { 'päz·əd·iv 'bī·əs }

positive charge |ELEC| The type of charge which is possessed by protons in ordinary matter, and which may be produced in a glass object by rubbing with silk. { 'päz·əd·iv 'chärj }

positive column |ELECTR| The luminous glow, often striated, that occurs between the Faraday dark space and the anode in a glow-discharge tube. Also known as positive glow. { 'päz·əd·iv 'käl·əm }

positive electrode See anode. { 'päz·əd·iv i'lek ,trōd }

positive feedback |CONT SYS| Feedback in which a portion of the output of a circuit or device is fed back in phase with the input so as to increase the total amplification. Also known as reaction (British usage); regeneration; regenerative feedback; retroaction (British usage). { 'päz·əd·iv 'fēd,bak }

positive glow See positive column. { 'päz·əd·iv 'glō}

positive-grid oscillator See retarding-field oscillator. { 'päz·əd·iv 'grid 'äs·ə,lād·ər }

positive-ion sheath |ELECTR| Collection of positive ions on the control grid of a gas-filled triode tube. { 'päz·əd·iv ‖ī,än ,shēth }

positive logic |ELECTR| Logic circuitry in which the more positive voltage (or current level) represents the 1 state; the less positive level represents the 0 state. { 'päz·əd·iv 'läj·ik }

positive modulation |ELECTR| In an amplitude-modulated analog television system, that form of television modulation in which an increase in brightness corresponds to an increase in transmitted power. { 'päz·əd·iv ,mäj·ə'lā·shən }

positive phase sequence |ELEC| The phase sequence that corresponds to the normal order of phases in a polyphase system. { 'päz·əd·iv ‖fāz ,sē·kwəns }

positive-phase-sequence relay |ELEC| Relay which functions in conformance with the positive-phase-sequence component of the current, voltage, or power of the circuit. { 'päz·əd·iv ,fāz ,sē·kwəns ,rē·lā }

positive ray |ELECTR| A stream of positively charged atoms or molecules, produced by a suitable combination of ionizing agents, accelerating fields, and limiting apertures. { 'päz·əd·iv 'rā }

positive terminal |ELEC| The terminal of a battery or other voltage source toward which electrons flow through the external circuit. { 'päz·əd·iv 'tərm·ən·əl }

positive transmission |COMMUN| Transmission of analog television signals in such a way that an increase in initial light intensity causes an increase in the transmitted power. { 'päz·əd·iv tranz'mish·ən }

positive zero |COMPUT SCI| The zero value reached by counting down from a positive number in the binary system. { 'päz·əd·iv 'zir·ō }

post |COMPUT SCI| To add or update records in a file. { pōst }

postaccelerating electrode See intensifier electrode. { ,pōst·ak'sel·ə,rād·iŋ i'lek,trōd }

postacceleration |ELECTR| Acceleration of beam electrons after deflection in an electron-beam tube. Also known as postdeflection acceleration (PDA). { ,pōst·ak,sel·ə'rā·shən }

postdecrementing See autodecrement addressing. { ‖pōst'dek·rə,ment·iŋ }

postdeflection accelerating electrode See intensifier electrode. { ,pōst·di'flek·shən ak'sel·ə ,rād·iŋ i'lek,trōd }

postdeflection acceleration See postacceleration. { ,pōst·di'flek·shən ak'sel·ə,rā·shən }

postedit |COMPUT SCI| To edit the output data of a computer. { 'pōst,ed·ət }

postemphasis See deemphasis. { ‖pōst'em·fə· səs }

postequalization See deemphasis. { ‖pōst,ē·kwə· lə'zā·shən }

POS terminal See point-of-sale terminal. { ‖pē ‖ō'es ,term·ən·əl }

postfix notation See reverse Polish notation. { 'pōst,fiks nō'tā·shən }

postincrementing See autoincrement addressing. { ‖pōst'in·krə,ment·iŋ }

postindexing |COMPUT SCI| Operation in which the contents of a register indicated by the index bits of an indirect address are added to the indirect address to form the effective address. { pōst'in,dek·siŋ }

posting See update. { 'pōst·iŋ }

posting interpreter See transfer interpreter. { 'pōst·iŋ in'tər·prəd·ər }

postmortem |COMPUT SCI| Any action taken after an operation is completed to help analyze that operation. { pōst'mórd·əm }

postmortem dump |COMPUT SCI| **1.** The printout showing the state of all registers and the contents of main memory, taken after a computer run terminates normally or terminates owing to fault. **2.** The program which generates this printout. { pōst'mórd·əm 'dəmp }

postmortem program See postmortem routine. { pōst'mórd·əm 'prō·grəm }

postmortem routine |COMPUT SCI| A computer routine designed to provide information about the operation of a program after the program is completed. Also known as postmortem program. { pōst'mórd·əm rü,tēn }

post office |COMPUT SCI| The software and files in an electronic mail system that receive messages and deliver them to recipients. { 'pōst ,óf·əs }

Post Office Protocol |COMPUT SCI| An Internet standard for delivering e-mail from a server to an e-mail client on a personal computer. Abbreviated POP. { ‖pōst ,óf·əs 'prōd·ə,kúl }

postprocessor |COMPUT SCI| A program that converts graphical output data to a form that can be used by computing equipment. { ‖pōst'prä ,ses·ər }

posttuning drift |ELECTR| In a frequency-agile source such as the fast-tuning oscillators used in set-on jammers for electronic warfare equipment, the increase in frequency brought about by the drop in temperature of the varactor after warm-up time, settling time, and the time when the oscillator has reached a new frequency. Abbreviated PTD. { 'pōs,tün·iŋ 'drift }

pot See potentiometer. { pät }

potential See electric potential. { pə'ten·chəl }

potential difference |ELEC| Between any two points, the work which must be done against electric forces to move a unit charge from one point to the other. Abbreviated PD. { pə'ten·chəl ¦dif·rəns }

potential divider See voltage divider. { pə'ten·chəl di'vīd·ər }

potential drop |ELEC| The potential difference between two points in an electric circuit. { pə'ten·chəl ¦dräp }

potential gradient |ELEC| Difference in the values of the voltage per unit length along a conductor or through a dielectric. { pə'ten·chəl 'grād·ē·ənt }

potential sputtering |ELECTR| The ejection of mainly neutral atoms from the surface of a solid insulator due to the impact of slow, multiply charged ions whose kinetic energy alone is incapable of initiating sputtering. { pə'ten·chəl ,spəd·ə·riŋ }

potential transformer See voltage transformer. { pə'ten·chəl tranz'fór·mər }

potential transformer phase angle |ELEC| Angle between the primary voltage vector and the secondary voltage vector reversed; this angle is conveniently considered as positive when the reversed, secondary voltage vector leads the primary voltage vector. { pə'ten·chəl tranz'fór·mər 'fāz ,aŋ·gəl }

potentiometer |ELEC| A resistor having a continuously adjusted sliding contact that is generally mounted on a rotating shaft; used chiefly as a voltage divider. Also known as pot (slang). |ENG| A device for the measurement of an electromotive force by comparison with a known potential difference. { pə,ten·chē'äm·əd·ər }

potentiometric controller |CONT SYS| A controller that operates on the null balance principle, in which an error signal is produced by balancing the sensor signal against a set-point voltage in the input circuit; the error signal is amplified for use in keeping the load at a desired temperature or other parameter. { pə¦ten·chē·ə¦me·trik kən'trōl·ər }

potentiometric electrode |ELEC| An electrode that produces a voltage logarithmically dependent on the concentration of a selected ionic substance. { pə,ten·chē·ə¦me·trik i'lek,trōd }

potentiometry |ELEC| Use of a potentiometer to measure electromotive forces, and the applications of such measurements. { pə,ten·chē 'äm·ə·trē }

Potier diagram |ELEC| Vector diagram showing the voltage and current relations in an alternating-current generator. { pō'tyä ,dī·ə,gram }

potted circuit |ELEC| A pulse-forming network immersed in oil and enclosed in a metal container. { 'päd·əd 'sər·kət }

potted line |ELEC| Pulse-forming network immersed in oil and enclosed in a metal container. { 'päd·əd 'līn }

potting |ELECTR| Process of filling a complete electronic assembly with a thermosetting compound for resistance to shock and vibration, and for exclusion of moisture and corrosive agents. { 'päd·iŋ }

powdered-iron core See ferrite core. { ¦paùd·ərd 'ī·ərn 'kòr }

power amplification See power gain. { 'paù·ər ,am·plə·fə'kā·shən }

power amplifier |ELECTR| The final stage in multistage amplifiers, such as audio amplifiers and radio transmitters, designed to deliver maximum power to the load, rather than maximum voltage gain, for a given percent of distortion. { 'paù·ər ¦am·plə,fī·ər }

power amplifier tube See power tube. { 'paù·ər ¦am·plə,fī·ər ,tüb }

power attenuation See power loss. { 'paù·ər ə ,ten·yə'wā·shən }

power bandwidth |COMMUN| The frequency range for which half the rated power of an audio amplifier is available at rated distortion. { 'paù·ər ¦band,width }

power check |COMPUT SCI| An automatic suspension of computer operations resulting from a significant fluctuation in internal electric power. { 'paù·ər ,chek }

power circuit |ELEC| The wires that carry current to electric motors and other devices that use electric power. { 'paù·ər ,sər·kət }

power component See active component. { 'paù·ər kəm,pō·nənt }

power cord See line cord. { 'paù·ər ,kòrd }

power-density spectrum See frequency spectrum. { 'paù·ər ¦den·səd·ē ,spek·trəm }

power detection |ELECTR| Form of detection in which the power output of the detecting device is used to supply a substantial amount of power directly to a device such as a loudspeaker or recorder. { 'paù·ər di,tek·shən }

power detector |ELECTR| Detector capable of handling strong input signals without appreciable distortion. { 'paù·ər di,tek·tər }

power diode See pin diode. { 'paù·ər ,dī,ōd }

power distribution unit |COMPUT SCI| Equipment located in or near a computer room which breaks down electric power from a high-voltage source to appropriate levels for distribution to the central processing unit and peripheral devices. Abbreviated PDU. { 'paù·ər ,di·strə'byü·shən ,yü·nət }

power down |COMPUT SCI| To exit from any running programs and remove floppy- and hard-disk cartridges before switching the computer off. { ¦paù·ər ¦daùn }

power factor |ELEC| The ratio of the average (or active) power to the apparent power (root-mean-square voltage times rms current) of an alternating-current circuit. Abbreviated pf. Also known as phase factor. { 'paù·ər ,fak·tər }

power-factor controller |ELECTR| A solid-state electronic device that reduces excessive energy waste in alternating-current induction motors by holding constant the phase angle between current and voltage. { 'paů·ər ,fak·tər kən ,trōl·ər }

power-factor meter |ENG| A direct-reading instrument for measuring power factor. { 'paů·ər ,fak·tər ,mēd·ər }

power-factor regulator |ELEC| Regulator which functions to maintain the power factor of a line or an apparatus at a predetermined value, or to vary it according to a predetermined plan. { 'paů·ər ,fak·tər ,reg·yə,lād·ər }

power frequency |ELEC| The frequency at which electric power is generated and distributed; in most of the United States it is 60 hertz. { 'paů·ər ,frē·kwən·sē }

power gain |ELECTR| The ratio of the power delivered by a transducer to the power absorbed by the input circuit of the transducer. Also known as power amplification. |ELECTROMAG| An antenna ratio equal to 4π (12.57) times the ratio of the radiation intensity in a given direction to the total power delivered to the antenna. { 'paů·ər ,gān }

power generator |ELEC| A device for producing electric energy, such as an ordinary electric generator or a magnetohydrodynamic, thermionic, or thermoelectric power generator. { 'paů·ər,jen·ə ,rād·ər }

power level |ELEC| The ratio of the amount of power being transmitted past any point in an electric system to a reference power value; usually expressed in decibels. { 'paů·ər ,lev·əl }

power line |ELEC| Two or more wires conducting electric power from one location to another. Also known as electric power line. { 'paů·ər ,līn }

power-line carrier |ELEC| The use of transmission lines to transmit speech, metering indications, control impulses, and other signals from one station to another, without interfering with the lines' normal function of transmitting power. { 'paů·ər ,līn ,kar·ē·ər }

power-line filter See line filter. { 'paů·ər ,līn ,fil·tər }

power-line interference |COMMUN| Interference caused by radiation from high-voltage power lines. { 'paů·ər ,līn ,in·tər,fir·əns }

power-line monitor |ELECTR| A device that continuously observes and records levels of electric power on a power line. { 'paů·ər¦līn ¦män·əd·ər}

power loss |ELECTR| The ratio of the power absorbed by the input circuit of a transducer to the power delivered to a specified load; usually expressed in decibels. Also known as power attenuation. { 'paů·ər ,lòs }

power meter See electric power meter. { 'paů·ər ,mēd·ər }

power output |ELECTR| The alternating-current power in watts delivered by an amplifier to a load. { 'paů·ər ¦aůt,půt }

power output tube See power tube. { 'paů·ər¦aůt ,půt ,tüb }

power pack |ELECTR| Unit for converting power from an alternating- or direct-current supply into an alternating- or direct-current power at

voltages suitable for supplying an electronic device. { 'paů·ər ,pak }

power rating |ELEC| The power available at the output terminals of a component or piece of equipment that is operated according to the manufacturer's specifications. { 'paů·ər ,rād·iŋ }

power rectifier |ELEC| A device which converts alternating current to direct current and operates at high power loads. { 'paů·ər 'rek·tə,fī·ər }

power relay |ELEC| Relay that functions at a predetermined value of power; may be an overpower relay, an underpower relay, or a combination of both. { 'paů·ər 'rē,lā }

power resistor |ELEC| A resistor used in electric power systems, ranging in size from 5 watts to many kilowatts, and cooled by air convection, air blast, or water. { 'paů·ər ri,zis·tər }

power semiconductor |ELECTR| A semiconductor device capable of dissipating appreciable power (generally over 1 watt) in normal operation; may handle currents of thousands of amperes or voltages up into thousands of volts, at frequencies up to 10 kilohertz. { 'paů·ər 'sem·i·kən,dək·tər }

power spectrum See frequency spectrum. { 'paů·ər ,spek·trəm }

power supply |ELECTR| A source of electrical energy, such as a battery or power line, employed to furnish the tubes and semiconductor devices of an electronic circuit with the proper electric voltages and currents for their operation. Also known as electronic power supply. { 'paů·ər sə ,plī }

power supply circuit |ELEC| An electrical network used to convert alternating current to direct current. { 'paů·ər sə,plī ,sər·kət }

power-supply rejection ratio |ELECTR| The ratio between the gain of an amplifier for difference signals between the input terminals, and the gain for variations of the power-supply voltages. Abbreviated PSRR. { ¦paů·ər sə,plī ri'jek·shən ,rā·shō }

power switch |ELEC| An electric switch which energizes or deenergizes an electric load; ranges from ordinary wall switches to load-break switches and disconnecting switches in power systems operating at voltages of hundreds of thousands of volts. { 'paů·ər ,swich }

power switchboard |ELEC| Part of a switch gear which consists of one or more panels upon which are mounted the switching control, measuring, protective, and regulatory equipment; the panel or panel supports may also carry the main switching and interrupting devices together with their connection. { 'paů·ər 'swich,bórd }

power switching |ELEC| Switching between supplies of electrical energy at high levels of current and voltage. { 'paů·ər ,swich·iŋ }

power transfer equation |ELEC| An equation for the power flow across a transmission line in terms of the relative magnitudes and phases of the terminal voltages, and the inductive reactance component and resistive component of the line. { ¦päů·ər 'tranz·fər i,kwā·zhən }

power transfer theorem |ELEC| The theorem that, in an electrical network which carries direct or sinusoidal alternating current, the greatest possible power is transferred from one section to another when the impedance of the section that acts as a load is the complex conjugate of the impedance of the section that acts as a source, where both impedances are measured across the pair of terminals at which the power is transferred, with the other part of the network disconnected. { ¦paü·ər 'tranz·fər ‚thir·əm }

power transformer |ELEC| An iron-core transformer having a primary winding that is connected to an alternating-current power line and one or more secondary windings that provide different alternating voltage values. { 'paü·ər tranz‚fôr·mər }

power transistor |ELECTR| A junction transistor designed to handle high current and power; used chiefly in audio and switching circuits. { 'paü·ər tran‚zis·tər }

power transmission line |ELEC| The facility in an electric power system used to transfer large amounts of power from one location to a distant location; distinguished from a subtransmission or distribution line by higher voltage, greater power capability, and greater length. Also known as electric main; main (both British usages). { 'paü·ər tranz'mish·ən ‚līn }

power transmission tower |ELEC| A rigid steel tower supporting a high-voltage electric power transmission line, having a large enough spacing between conductors, and between conductors and ground, to prevent corona discharge. { 'paü·ər tranz'mish·ən ‚taü·ər }

power tube |ELECTR| An electron tube capable of handling more current and power than an ordinary voltage-amplifier tube; used in the last stage of an audio-frequency amplifier or in high-power stages of a radio-frequency amplifier. Also known as power amplifier tube; power output tube. { 'paü·ər ‚tüb }

power typing |COMPUT SCI| A word-processing technique that allows the automatic typing of repetitious text, such as appears in a form letter. { 'paü·ər ‚tīp·iŋ }

power up |COMPUT SCI| To check that the computer memory, peripherals, and input/output channels are working properly before the operating system is loaded. { ¦paü·ər ¦əp }

power winding |ELEC| In a saturable reactor, a winding to which is supplied the power to be controlled; commonly the functions of the output and power windings are accomplished by the same winding, which is then termed the output winding. { 'paü·ər ‚wīnd·iŋ }

PPI See plan position indicator.

P-picture See predicted picture. { 'pē ‚pik·chər }

pp junction |ELECTR| A region of transition between two regions having different properties in p-type semiconducting material. { ¦pē¦pē ‚jəŋk·shən }

PPM See pulse-position modulation.

PPP See Point-to-Point Protocol.

P pulse See commutator pulse. { 'pē ‚pəls }

Practical Extraction and Reporting Language |COMPUT SCI| A scripting language often used for creating CGI programs. Abbreviated Perl. { ¦prak·ti·kəl ik‚strak·shən and ri'pòrt·iŋ ‚laŋ·gwij }

pragma |COMPUT SCI| A directive inserted into a computer program to prevent the automatic execution of certain error checking and reporting routines which are no longer necessary when the program has been perfected. { 'prag·mə }

pragmatics |COMMUN| The branch of semiotics that treats the relation of symbols to behavior and the meaning received by the listener or reader of a statement. |COMPUT SCI| The fourth and final phase of natural language processing, following contextual analysis, that takes into account the speaker's goal in uttering a particular thought in a particular way in determining what constitutes an appropriate response. { prag'mad·iks }

preamble |COMMUN| The portion of a commercial radiodata message that is sent first, containing the message number, office of origin, date, and other numerical data not part of the following message text. { 'prē‚am·bəl }

preamplifier |ELECTR| An amplifier whose primary function is to boost the output of a low-level audio-frequency, radio-frequency, or microwave source to an intermediate level so that the signal may be further processed without appreciable degradation of the signal-to-noise ratio of the system. Also known as preliminary amplifier. { prē'am·plə‚fī·ər }

precedence |COMPUT SCI| The order in which operators are processed in a programming language. { 'pres·əd·əns }

precedence relation |COMPUT SCI| A rule stating that, in a given programming language, one of two operators is to be applied before the other in any mathematical expression. { 'pres·əd·əns ri ‚lā·shən }

precipitation attenuation |ELECTROMAG| Loss of radio energy due to the passage through a volume of the atmosphere containing precipitation; part of the energy is lost by scattering, and part by absorption. { prə‚sip·ə'tā·shən ə‚ten·yə'wā·shən }

precipitation clutter suppression |ELECTR| Technique of reducing, by one of the various devices integral to the radar system, clutter caused by rain in the radar range. { prə ‚sip·ə'tā·shən ¦kləd·ər sə‚presh·ən }

precipitation noise |ELECTR| Noise generated in an antenna circuit, generally in the form of a relaxation oscillation, caused by the periodic discharge of the antenna or conductors in the vicinity of the antenna into the atmosphere. { prə‚sip·ə'tā·shən ‚nòiz }

precipitation static |COMMUN| Static interference due to the discharge of large charges built up on an aircraft or other object by rain, sleet, snow, or electrically charged clouds. { prə ‚sip·ə'tā·shən ‚stad·ik }

precipitator See electrostatic precipitator. { prə 'sip·ə‚tād·ər }

precision attribute |COMPUT SCI| A set of one or more integers that denotes the number of symbols used to represent a given number and positional information for determining the base point of the number. { prə'sizh·ən 'a·trə,byüt }

precision-balanced hybrid circuit [ELEC] Circuit used to interconnect a four-wire telephone circuit to a particular two-wire circuit, in which the impedance of the balancing network is adjusted to give a relatively high degree of balance. { prə'sizh·ən ¦bal·ənst 'hī·brəd 'sər·kət }

precision net [ELEC] In a four-wire terminating set or similar device employing a hybrid coil, an artificial line designed and adjusted to provide an accurate balance for the loop and subscribers set or line impedance. { prə'sizh·ən ,net }

precision sweep [ELECTR] Delayed and expanded sweep as in an analog radar display, or similar selection and timing of a digital display, permitting closer examination of received signals of high resolution. { prə'sizh·ən ,swēp }

precompiled module [COMPUT SCI] A standardized subroutine that is separately developed and compiled for use in many different computer programs. { ¦prē·kəm'pīld 'mäj·yül }

precompiler [COMPUT SCI] A computer program that indentifies syntax errors and other problems in a program before it is converted to machine language by a compiler. { ¦prē·kəm'pīl·ər }

preconduction current [ELECTR] Low value of plate current flowing in a thyratron or other grid-controlled gas tube prior to the start of conduction. { ¦prē·kən'dək·shən ,kə·rənt }

predecessor job [COMPUT SCI] A job whose output is used as input to another job, and which must therefore be completed before the second job is started. { ¦pred·ə,ses·ər ,jäb }

predefined function [COMPUT SCI] A sequence of instructions that is identified by name in a computer program but is built into the high-level programming language from which the program is complied or is retrieved from somewhere outside the program, such as a subroutine library. { ¦prē·di'fīnd 'fəŋk·shən }

predetection combining [ELECTR] Method used to produce an optimum signal from multiple receivers involved in diversity reception of signals. { ¦prē·di'tek·shən kəm'bīn·iŋ }

predicate [COMPUT SCI] A statement in a computer program that evaluates an expression in order to arrive at a true or false answer. { 'pred·ə ,kāt }

predicted picture [COMMUN] A MPEG-2 picture that is coded with respect to the nearest previous intra-coded picture. This technique is termed forward prediction. Predicted pictures provide more compression than intra-coded pictures and serve as a reference for future predicted pictures or bidirectional pictures. Predicted pictures can propagate coding errors when they (or bidirectional pictures) are predicted from prior predicted pictures where the prediction is flawed. Also known as P-frame; P-picture. { pri'dikt·əd 'pik·chər }

predicted-wave signaling [COMMUN] Communications system in which detection is optimized in the presence of severe noise by using mechanical

resonator filters and other circuits in the detector to take advantage of known information on the arrival and completion times of each pulse, as well as on pulse shape, pulse frequency and spectrum, and possible data content. { prə'dik·təd ,wāv 'sig·nəl·iŋ }

predictive coder [COMMUN] Any technique for compressing audio or video signals in which a synthesizer at the receiver is controlled by signal parameters extracted at the transmitter to remake the signal. Also known as predictive encoder. { prə,dik·tiv 'kō·dər }

predictive coding [COMMUN] In data compression, a method of coding information in which a sample value is presented as the error term formed by the difference between the sample and its prediction. { prə¦dik·tiv 'kōd·iŋ }

predictive encoder See predictive coder. { prə ,dik·tiv in'kō·dər }

preece [ELEC] A unit of electrical resistivity equal to 10^{13} times the product of 1 ohm and 1 meter. { prēs }

preedit [COMPUT SCI] To edit data before feeding it to a computer. { prē'ed·ət }

preemphasis [ELECTR] A process which increases the magnitude of some frequency components with respect to the magnitude of others to reduce the effects of noise introduced in subsequent parts of the system. { prē'em·fə·səs }

preemphasis network [ELECTR] An RC (resistance-capacitance) filter inserted in a system to emphasize one range of frequencies with respect to another. Also known as emphasizer. { prē'em·fə·səs ,net,wərk }

preemptive multitasking [COMPUT SCI] A method of running more than one program on a computer at a time, in which control of the processor is decided by the operating system, which allocates each program a recurring time segment. { prē ¦emp·tiv 'məl·tē,task·iŋ }

preferred numbers [ELECTR] A series of numbers adopted by the Electronic Industries Association and the military services for use as nominal values of resistors and capacitors, to reduce the number of different sizes that must be kept in stock for replacements. Also known as preferred values. { pri'fərd 'nəm·bərz }

preferred values See preferred numbers. { pri 'fərd 'val·yüz }

prefix notation See Polish notation. { 'prē,fiks nō,tā·shən }

prefocus lamp [ELEC] A light bulb whose filaments are precisely positioned with respect to the lamp socket. { prē'fō·kəs ,lamp }

preheat fluorescent lamp [ELECTR] A fluorescent lamp in which a manual switch or thermal starter is used to preheat the cathode for a few seconds before high voltage is applied to strike the mercury arc. { 'prē,hēt flü¦res·ənt 'lamp }

preindexing [COMPUT SCI] Operation in which the address bits of a word are added to the contents of a specified register to determine the pointer address. { prē'in,deks·iŋ }

preliminary amplifier See preamplifier. { pri' lim·ə,ner·ē 'am·plə,fī·ər }

preprocessor |COMPUT SCI| A program that converts data into a format suitable for computer processing. { ¦prē'prä,ses·ər }

preprogrammed robot |CONT SYS| A robot that cannot adapt itself to the task it is carrying out, and must follow a built-in program. Also known as sequence robot. { ¦prē'prō₀gramd 'rō,bät }

preprogramming |COMPUT SCI| The prerecording of instructions or commands for a machine, such as an automated tool in a factory. { prē'prō ,gram·iŋ }

preread head |COMPUT SCI| A read head that is placed near another read head in such a way that it can read data stored on a moving medium such as a tape or disk before these data reach the second head. { 'prē,rēd ,hed }

prescaler |ELECTR| A scaler that extends the upper frequency limit of a counter by dividing the input frequency by a precise amount, generally 10 or 100. { 'prē,skāl·ər }

preselection |COMPUT SCI| A technique for saving computation time in buffered computers in which a block of data is read into computer storage from the next input tape to be called upon before the data are required in the computer; the selection of the next input tape is determined by instructions to the computer. { ¦prē·si'lek·shən }

preselector |ELEC| Device in automatic switching which performs its selecting operation before seizing an idle trunk. |ELECTR| A tuned radio-frequency amplifier stage used ahead of the frequency converter in a superheterodyne receiver to increase the selectivity and sensitivity of the receiver. { ¦prē·si'lek·tər }

presentation See radar display. { ,prez·ən'tā·shən }

presentation graphics program |COMPUT SCI| An application program for creating and enhancing the visual appeal and understandability of charts and graphs, with the aid of a library or predrawn images that can be combined with other artwork. { ,prez·ən¦tā·shən 'graf·iks ,prō·grəm }

preset |COMPUT SCI| **1.** Of a variable, having a value established before the first time it is used. **2.** To initialize a value of a variable before the value of the variable is used or tested. { 'prē ,set }

preset parameter |COMPUT SCI| In computers, a parameter which is fixed for each problem at a value set by the programmer. { 'prē,set pə'ram·əd·ər }

presort |COMPUT SCI| **1.** The first part of a sort program in which data items are arranged into strings that are equal to or greater than some prescribed length. **2.** The sorting of data on off-line equipment before it is processed by a computer. { prē'sórt }

press teletype network |COMMUN| A large teletypewriter network employed by a press association or other news distributing organization, usually employing modern carrier telegraph circuits operating over both wire and radio facilities, and transmitting to as many as 2000 stations simultaneously. { 'pres 'tel·ə,tīp ,net,wərk }

press-to-talk switch |ELECTR| A switch mounted directly on a microphone to provide a convenient means for switching two-way radiotelephone equipment or electronic dictating equipment to the talk position. { 'pres tə 'tók ,swich }

pressure cable |ELEC| A cable in which a fluid such as oil or gas, at greater than atmospheric pressure, surrounds the conductors and insulation and keeps their temperature down. { 'presh·ər ,kā·bəl }

pressure microphone |ENG ACOUS| A microphone whose output varies with the instantaneous pressure produced by a sound wave acting on a diaphragm; examples are capacitor, carbon, crystal, and dynamic microphones. { 'presh·ər 'mī·krə·fōn }

pressure pad |ENG ACOUS| A felt pad mounted on a spring arm, used to hold magnetic tape in close contact with the head on some tape recorders. { 'presh·ər ,pad }

pressure pickup |ELECTR| A device that converts changes in the pressure of a gas or liquid into corresponding changes in some more readily measurable quantity such as inductance or resistance. { 'presh·ər 'pik,əp }

pressure switch |ELEC| A switch that is actuated by a change in pressure of a gas or liquid. { 'presh·ər ,swich }

presumptive address See address constant. { pri'zəm·tiv ə'dres }

presumptive instruction See basic instruction. { pri'zəm·tiv in'strək·shən }

pretersonics See acoustoelectronics. { ¦prēd·ər ¦sän·iks }

pre-transmit-receive tube See pre-TR tube. { ¦prē 'tranz,mit ri'sēv ,tüb }

pretravel |CONT SYS| The distance or angle through which the actuator of a switch moves from the free position to the operating position. { 'prē,trav·əl }

pretrigger |ELECTR| Trigger used to initiate sweep ahead of transmitted pulse. { prē'trig·ər }

pre-TR tube |ELECTR| Gas-filled radio-frequency switching tube used in some radar systems to protect the transmit-receive tube from excessively high power and the receiver from frequencies other than the fundamental. Derived from pre-transmit-receive tube. { prē¦tē'är ,tüb }

previewing |COMPUT SCI| In character recognition, a process of attempting to gain prior information about the characters that appear on an incoming source document; this information, which may include the range of ink density, relative positions, and so forth, is used as an aid in the normalization phase of character recognition. { 'prē,vyü·iŋ }

previous element coding |COMMUN| System of signal coding, used for digital television transmission, whereby each transmitted picture element is dependent upon the similarity of the preceding picture element. { 'prē·vē·əs 'el·ə·mənt ,kō·diŋ }

prewhitening filter See whitening filter. { prē'wīt·ən·iŋ ,fil·tər }

PRF See pulse repetition rate.

pri See primary winding. { prī }

primary |ELEC| One of the high-voltage conductors of a power distribution system. *See* primary winding. { 'prī ,mer·ē }

primary battery |ELEC| A battery consisting of one or more primary cells. { 'prī,mer·ē 'bad·ə·rē }

primary cache |COMPUT SCI| A cache memory located within a microprocessor chip itself. Also known as internal cache; level I cache. { ¦prī ,mer·ē 'kash }

primary cell |ELEC| A cell that delivers electric current as a result of an electrochemical reaction that is not efficiently reversible, so that the cell cannot be recharged efficiently. { 'prī,mer·ē 'sel }

primary center |COMMUN| A telephone office having lower rank than a sectional center and higher rank than a toll center; connects toll centers and may also serve as a toll center for nearby end offices. { 'prī,mer·ē 'sen·tər }

primary circuit |ELEC| One of a collection of coupled coils or circuits that receives electric power from a source and transfers it to the secondary circuit by electromagnetic induction. { 'prī,mer·ē 'sər·kət }

primary coil |ELEC| The input coil in an induction coil or transformer. { 'prī,mer·ē 'kȯil }

primary control program |COMPUT SCI| The program which provides the sequential scheduling of jobs and basic operating systems functions. Abbreviated PCP. { 'prī,mer·ē kən'trōl ,prō·grəm }

primary detector *See* sensor. { 'prī,mer·ē di'tek·tər }

primary electron |ELECTR| An electron which bombards a solid surface, causing secondary emission. { 'prī,mer·ē i'lek,trän }

primary emission |ELECTR| Emission of electrons due to primary causes, such as heating of a cathode, and not to secondary effects, such as electron bombardment. { 'prī,mer·ē i'mish·ən }

primary fault |ELEC| In an electric circuit, the initial breakdown of the insulation of a conductor, usually followed by a flow of power current. { 'prī,mer·ē ,fȯlt }

primary flow |ELECTR| The current flow that is responsible for the major properties of a semiconductor device. { 'prī,mer·ē 'flō }

primary frequency |COMMUN| Frequency assigned for normal use on a particular circuit or communications channel. { 'prī,mer·ē 'frē·kwən·sē }

primary-frequency standard |COMMUN| One of the standards of frequency maintained by various governments; the operating frequency of a radio station is determined by comparison with multiples of this standard frequency. { 'prī,mer·ē 'frē·kwən·sē ,stan·dərd }

primary fuel cell |ELEC| A fuel cell in which the fuel and oxidant are continuously consumed. { 'prī,mer·ē 'fyül ,sel }

primary index |COMPUT SCI| An index that holds the values of primary keys, in sequence. { 'prī ,mer·ē 'in,deks }

primary key |COMPUT SCI| A key that identifies a record or portion of a record and determines the sequence of records in a file or other data structure. { 'prī,mer·ē 'kē }

primary photocurrent |ELECTR| A photocurrent resulting from nonohmic contacts unable to replenish charge carriers which pass out of the opposite contact, and whose maximum gain is unity. { 'prī,mer·ē 'fōd·ō,kə·rənt }

primary power cable |ELEC| Power service cables connecting the outside power source to the main-office switch and metering equipment. { 'prī ,mer·ē 'paů·ər ,kā·bəl }

primary radar |ENG| A radar that receives and interprets the reflected signal from scattering objects (targets and clutter) in its view. { 'prī ,mer·ē 'rā,där }

primary register |COMPUT SCI| A general-purpose register in a central processing unit that is available for direct utilization by computer programs. { 'prī,mer·ē 'rej·ə,stər }

primary relay |ELEC| Relay that produces the initial action in a sequence of operations. { 'prī ,mer·ē 'rē,lā }

primary service area |COMMUN| The area in which the ground wave of a broadcast station is not subject to objectionable interference or fading. { 'prī,mer·ē 'sər·vəs ,er·ē·ə }

primary skip zone |ELECTROMAG| Area around a transmitter beyond the ground wave but within the skip distance. { 'prī,mer·ē 'skip ,zōn }

primary storage |COMPUT SCI| Main internal storage of a computer. { 'prī,mer·ē 'stȯr·ij }

primary surveillance radar *See* primary radar. { 'prī,mer·ē sər'vā·ləns ,rā,där }

primary voltage |ELEC| The voltage applied to the terminals of the primary winding of a transformer. { 'prī,mer·ē 'vōl·tij }

primary wave |COMMUN| A radio wave traveling by a direct path, as contrasted with skips. { 'prī ,mer·ē 'wāv }

primary winding |ELEC| The transformer winding that receives signal energy or alternating-current power from a source. Also known as primary. Abbreviated pri. Symbolized P. { 'prī,mer·ē 'wīnd·iŋ }

prime register |COMPUT SCI| One of the registers that is inactive at any given time in a central processing unit with duplicate general-purpose registers. { 'prīm 'rej·ə,stər }

primitive |COMPUT SCI| A sketchy specification, omitting details, of some action in a computer program. |CONT SYS| A basic operation of a robot, initialized by a single command statement in the program that controls the robot. { 'prim·əd·iv }

primitive abstract data type |COMPUT SCI| A simple abstract data type that is typically implemented directly in a high-level programming language; examples include integers and real numbers (with appropriate arithmetic operators), booleans (with appropriate logical operators), text strings, and pointers. { 'prim·əd·iv 'ab,strakt 'dad·ə ,tīp }

principal axis |ENG ACOUS| A reference direction for angular coordinates used in describing the directional characteristics of a transducer; it is usually an axis of structural symmetry or the direction of maximum response. { 'prin·sə·pəl 'ak·səs }

principal E plane |ELECTROMAG| Plane containing the direction of radiation of electromagnetic waves and arranged so that the electric vector everywhere lies in the plane. { 'prin·sə·pəl 'ē ‚plān }

principal H plane |ELECTROMAG| Plane that contains the direction of radiation and the magnetic vector, and is everywhere perpendicular to the E plane. { 'prin·sə·pəl 'āch ‚plān }

principle of duality See duality principle. { 'prin·sə·pəl əv dü'al·əd·ē }

principle of optimality |CONT SYS| A principle which states that for optimal systems, any portion of the optimal state trajectory is optimal between the states it joins. { 'prin·sə·pəl əv ‚äp·tə'mal·əd·ē }

principle of reciprocity See reciprocity theorem. { 'prin·sə·pəl əv ‚res·ə'präs·əd·ē }

principle of superposition |ELEC| **1.** The principle that the total electric field at a point due to the combined influence of a distribution of point charges is the vector sum of the electric field intensities which the individual point charges would produce at that point if each acted alone. **2.** The principle that, in a linear electrical network, the voltage or current in any element resulting from several sources acting together is the sum of the voltages or currents resulting from each source acting alone. Also known as superposition theorem. { 'prin·sə·pəl əv ‚sü·pər·pə 'zish·ən }

print driver |COMPUT SCI| The portion of a computer program that directs output to a printer and usually also controls printer functions such as pagination and the setting of the margins and page headers. { 'print ‚drī·vər }

printed circuit |ELECTR| A conductive pattern that may or may not include printed components, formed in a predetermined design on the surface of an insulating base in an accurately repeatable manner. { 'print·əd 'sər·kət }

printed circuit board |ELECTR| A flat board whose front contains slots for integrated circuit chips and connections for a variety of electronic components, and whose back is printed with electrically conductive pathways between the components. Also known as circuit board. { 'print·əd 'sər·kət ‚bórd }

printed-wiring armature |ELEC| An armature in which the conductors consist of printed-wiring strips on both sides of a thin insulating disk, to give a low-inertia armature for servomotors and other variable high-speed applications. { 'print·əd ‚wīr·iŋ 'ärm·ə‚chür }

printed wiring board |ELECTR| A copper-clad dielectric material with conductors etched on the external or internal layers. { ¦print·əd 'wīr·iŋ ‚bórd }

printer |COMPUT SCI| A computer output mechanism that prints characters one at a time or one line at a time. { 'print·ər }

printer file |COMPUT SCI| **1.** A file that contains the information that the printer driver needs in order

to generate the codes required by the printer. **2.** A document in print image format. { 'prin·tər ‚fīl }

print head |COMPUT SCI| The mechanism that generates the characters to be reproduced by a character printer. { 'print ‚hed }

print image format |COMPUT SCI| The format of a document that has been prepared for output on the printer. { 'print ‚im·ij ‚fór‚mat }

printing calculator |COMPUT SCI| A desk-model electronic calculator that provides a printed record on paper tape with or without a digital display. { 'print·iŋ 'kal·kyə‚lād·ər }

printing element |COMPUT SCI| The part of the print head mechanism that comes into contact with the paper to print characters or other images. { 'print·iŋ ‚el·ə·mənt }

printing-telegraph code |COMMUN| A five- or seven-unit code used for operation of a teleprinter, teletypewriter, and similar telegraph printing devices. { 'print·iŋ ¦tel·ə‚graf ‚kōd }

printing telegraphy |COMMUN| Method of telegraph operation in which the received signals are automatically recorded in printed characters. { 'print·iŋ tə'leg·rə·fē }

print member |COMPUT SCI| The part of a computer printer that determines the form of a printed character, such as a print wheel or type bar. { 'print ‚mem·bər }

printout |COMPUT SCI| A printed output of a data-processing machine or system. { 'print‚aút }

print position |COMPUT SCI| One of the positions on a printer at which a character can be printed. { 'print pə‚zish·ən }

print queue |COMPUT SCI| A prioritized list, maintained by the operating system, of the output from a computer system waiting on a spool file to be printed. { 'print ‚kyü }

print server |COMPUT SCI| A computer controlling a series of printers. { 'print ‚sər·vər }

printthrough |ELECTR| Transfer of signals from one recorded layer of magnetic tape to the next on a reel. { 'print‚thrü }

print train |COMPUT SCI| **1.** The chain in a chain printer or the drum in a drum printer that holds the type slugs to make impressions on paper. **2.** The electronic character set that serves a similar function in a laser printer. { 'print ‚trān }

print wheel |COMPUT SCI| A disk which has around its rim the letters, numerals, and other characters that are used in printing in a wheel printer. { 'print ‚wēl }

priority-arbitration circuit |COMPUT SCI| A logic circuit which combines all interrupts but allows only the highest-priority request to enable its active flipflop. { prī'är·əd·ē ‚är·bə'trā·shən ‚sər·kət }

priority indicator |COMMUN| Data attached to a message to indicate its relative priority and hence the order in which it will be transmitted. |COMPUT SCI| Data attached to a computer program or job which are used to determine the order in which it will be processed by the computer. { prī'är·əd·ē 'in·də‚kād·ər }

priority interrupt |COMPUT SCI| An interrupt procedure in which control is passed to the monitor, the required operation is initiated, and then control returns to the running program, which never knows that it has been interrupted.   { prī'är·əd·ē 'int·ə,rəpt }

priority phase |COMPUT SCI| Phase consisting of execution of operations in response to instruments or process interrupts other than clock interrupts.   { prī'är·əd·ē 'fāz }

priority polling |COMMUN| In a data communications network, a system in which nodes with high activity are interrogated more frequently than those with only occasional traffic.   { prī'är·əd·ē 'pōl·iŋ }

priority processing |COMPUT SCI| A method of computer time-sharing in which the order in which programs are processed is determined by a system of priorities, involving such factors as the length, nature, and source of the programs.   { prī'är·əd·ē 'prä,ses·iŋ }

priority queueing |COMPUT SCI| The arrangement of jobs to be carried out in a list according to their relative importance, with the most important first.   { prī'är·əd·ē 'kyü·iŋ }

privacy system |COMMUN| A device or method for scrambling overseas telephone conversations handled by radio links in order to make them unintelligible to outside listeners. Also known as privacy transformation; secrecy system.   { 'prī·və·sē ,sis·təm }

privacy transformation See privacy system.   { 'prī·və·sē ,tranz·fər'mā·shən }

private automatic branch exchange |COMMUN| A private branch exchange in which connections are made by remote-controlled switches. Abbreviated PABX.   { 'prī·vət ¦ȯd·ə,mad·ik 'branch iks ,chānj }

private automatic exchange |COMMUN| A private telephone exchange in which connections are made by remote-controlled switches. Abbreviated PAX.   { 'prī·vət ¦ȯd·ə,mad·ik iks,chānj }

private branch exchange |COMMUN| A telephone exchange serving a single organization, having a switchboard and associated equipment, usually located on the customer's premises; provides for switching calls between any two extensions served by the exchange or between any extension and the national telephone system via a trunk to a central office. Abbreviated PBX.   { 'prī·vət 'branch iks,chānj }

private branch exchange access line |ELEC| Circuit that connects a main private branch exchange (PBX) to a switching center.   { 'prī·vət 'branch iks,chānj 'ak,ses ,līn }

private data |COMPUT SCI| Data that are open to a single user only.   { 'prī·vət 'dad·ə }

private exchange |COMMUN| Telephone exchange serving a single organization and having no means for connecting to a public telephone system.   { 'prī·vət iks'chānj }

private library |COMPUT SCI| An organized collection of programs and other software that is the property of a single user of a computer system and is not generally available to other users.   { 'prī·vət 'lī,brer·ē }

private line |COMMUN| A line, channel, or service reserved solely for one user.   { 'prī·vət 'līn }

private line arrangement |COMPUT SCI| The structure of a computer system in which each input/output device has a set of lines leading to the central processing unit for the device's own private use. Also known as radial selector.   { 'prī·vət ¦līn ə,rānj·mənt }

private line service |COMMUN| Service provided by United States common carriers engaged in domestic or international wire, radio, and cable communications for the intercity communications purposes of a customer; this service is provided over integrated communications pathways, including facilities or local channels, which are integrated components of intercity private line services, and station equipment between specified locations for a continuous period or for regularly recurring periods at stated hours.   { 'prī·vət ¦līn ,sər·vəs }

private pack |COMPUT SCI| A disk pack assigned exclusively to one application or one user so that the operating system does not try to allocate space on the device to others.   { 'prī·vət 'pak }

privileged instruction |COMPUT SCI| A class of instructions, usually including storage protection setting, interrupt handling, timer control, input/output, and special processor status-setting instructions, that can be executed only when the computer is in a special privileged mode that is generally available to an operating or executive system, but not to user programs.   { 'priv·ə·lijd in'strək·shən }

privileged mode See master mode.   { 'priv·ə·lijd ,mōd }

PRML technique See partial-response maximum-likelihood technique.   { ¦pē¦är¦em'el tek,nēk }

probabilistic automaton |COMPUT SCI| A device, with a finite number of internal states, which is capable of scanning input words over a finite alphabet and responding by successively changing its internal state in a probabilistic way. Also known as stochastic automaton.   { ‚präb·ə·bə'lis·tik ȯ'täm·ə,tän }

probabilistic sequential machine |COMPUT SCI| A probabilistic automaton that has the capability of printing output words probabilistically, over a finite output alphabet. Also known as stochastic sequential machine.   { ‚präb·ə·bə'lis·tik si'kwen·chəl mə'shēn }

probe |COMMUN| To determine a radio interference by obtaining the relative interference level in the immediate area of a source by the use of a small, insensitive antenna in conjunction with a receiving device. |ELECTROMAG| A metal rod that projects into but is insulated from a waveguide or resonant cavity; used to provide coupling to an external circuit for injection or extraction of energy or to measure the standing-wave ratio. Also known as waveguide probe.   { prōb }

problem check |COMPUT SCI| One or more tests used to assist in obtaining the correct machine solution to a problem.   { 'präb·ləm ,chek }

problem-defining language |COMPUT SCI| A programming language that literally defines a problem and may specifically define the input and output, but does not define the method of transforming one to the other. Also known as problem-specification language. { 'präb·ləm di ¦fīn·iŋ ˌlaŋ·gwij }

problem definition |COMPUT SCI| The art of compiling logic in the form of general flow charts and logic diagrams which clearly explain and present the problem to the programmer in such a way that all requirements involved in the run are presented. { 'präb·ləm ˌdef·ə¸nish·ən }

problem-describing language |COMPUT SCI| A programming language that describes, in the most general way, the problem to be solved, but gives no indication of the problem's detailed characteristics or its solution. { 'präb·ləm di ¦skrīb·iŋ ˌlaŋ·gwij }

problem file *See* run book. { 'präb·ləm ˌfīl }

problem folder *See* run book. { 'präb·ləm ˌfōld·ər }

problem mode |COMPUT SCI| A condition of computer operation in which, in contrast to supervisor mode, the privileged instructions cannot be executed, preventing the program from upsetting the supervisor program or any other program. { 'präb·ləm ˌmōd }

problem-oriented language |COMPUT SCI| A language designed to facilitate the accurate expression of problems belonging to specific sets of problem types. { 'präb·ləm ˌór·ē¦ent·əd ˌlaŋ·gwij }

problem-solving language |COMPUT SCI| A programming language that can be used to specify a complete solution to a problem. { 'präb·ləm ¦sälv·iŋ ˌlaŋ·gwij }

problem-specification language *See* problem-defining language. { 'präb·ləm ˌspes·ə·fə¦kā·shən ˌlaŋ·gwij }

procedural programming |COMPUT SCI| A list of instructions telling a computer, step-by-step, what to do, usually having a linear order of execution from the first statement to the second and so forth with occasional loops and branches. Procedural programming languages include C, C++, Fortran, Pascal, and Basic. { prə¸sē·jə·rəl 'prō¸gram·iŋ }

procedural representation |COMPUT SCI| The representation of certain concepts in a computer by procedures or programs in some appropriate language, rather than by static data items such as numbers or lists. { prə'sē·jə·rəl ˌrep·rə·zen'tā·shən }

procedure |COMPUT SCI| **1.** A sequence of actions (or computer instructions) which collectively accomplish some desired task. **2.** In particular, a subroutine that causes an effect external to itself. { prə'sē·jər }

procedure declaration |COMPUT SCI| A statement that causes a procedure to be given a name and written as a segment of a computer program. { prə'sē·jər ˌdek·lə¸rā·shən }

procedure division |COMPUT SCI| The section of a program (written in the COBOL language) in which a programmer specifies the operations to be performed with the data names appearing in the program. { prə'sē·jər di¸vizh·ən }

procedure library |COMPUT SCI| A collection of job control language routines that are stored on a disk file and can be executed by entering a command naming the routine. Abbreviated PROCLIB. { prə'sē·jər ˌlī¸brer·ē }

procedure-oriented language |COMPUT SCI| A language designed to facilitate the accurate description of procedures, algorithms, or routines belonging to a certain set of procedures. { prə'sē·jər ˌór·ē¦ent·əd ˌlaŋ·gwij }

proceed-to-select signal |COMMUN| Signal returned from distant automatic equipment over the backward signaling path, in response to a calling signal, to indicate that selecting information can be transmitted; in certain signaling systems, both signals can be the same. { prə'sēd tə si'lekt ˌsig·nəl }

proceed-to-transmit signal |COMMUN| Signal returned from a distant manual switchboard over the backward signaling path, in response to a calling signal, to indicate that the teleprinter of the distant operator is connected to the circuit. { prə'sēd tə tranz'mit ˌsig·nəl }

process |COMPUT SCI| **1.** To assemble, compile, generate, interpret, compute, and otherwise act on information in a computer. **2.** A program that is running on a computer. { 'prä¸ses }

process-bound program *See* CPU-bound program. { 'prä¸ses ¦baúnd 'prō·grəm }

process control system |CONT SYS| The automatic control of a continuous operation. { 'prä ¸səs kən¸trōl ¸sis·təm }

processing |COMMUN| Further handling, manipulation, consolidation, compositing, and so on, of information to convert it from one format to another or to reduce it to manageable or intelligible information. { 'prä¸ses·iŋ }

processing interrupt |COMPUT SCI| The interruption of the batch processing mode in a real-time system when live data are entered in the system. { 'prä¸ses·iŋ 'int·ə¸rəpt }

processing program |COMPUT SCI| Any computer program that is not a control program, such as an application program, or a noncontrolling part of the operating system, such as a sort-merge program or language translator. { 'prä¸ses·iŋ ¸prō·gram }

processing section |COMPUT SCI| The computer unit that does the actual changing of input into output; includes the arithmetic unit and intermediate storage. { 'prä¸ses·iŋ ¸sek·shən }

process-limited *See* processor-limited. { 'prä¸səs ¦lim·əd·əd }

processor |COMPUT SCI| **1.** A device that performs one or many functions, usually a central processing unit. **2.** A program that transforms some input into some output, such as an assembler, compiler, or linkage editor. { 'prä¸ses·ər }

processor complex |COMPUT SCI| The central portion of a very large computer consisting of several central processing units working in concert. { 'prä¸ses·ər ˌkäm¸pleks }

processor error interrupt |COMPUT SCI| The interruption of a computer program because a parity check indicates an error in a word that has been transferred to or within the central processing unit.   { 'prä,ses·ər ¦er·ər ,int·ə,rəpt }

processor-limited |COMPUT SCI| Property of a computer system whose processing time is determined by the speed of its central processing unit rather than by the speed of its peripheral equipment. Also known as process-limited.   { 'prä,ses·ər ,lim·əd·əd }

processor-memory-switch notation   See PMS notation.   { 'prä,ses·ər 'mem·rē ,swich nō,tā·shən }

processor stack pointer |COMPUT SCI| A programmable register used to access all temporary-storage words related to an interrupt-service routine which was halted when a new service routine was called in.   { 'prä,ses·ər 'stak ,pȯint·ər }

processor status word |COMPUT SCI| A word comprising a set of flag bits and the interrupt-mask status.   { 'prä,ses·ər 'stad·əs ,wərd }

process simulation |COMPUT SCI| The use of computer programming, computer vision, and feedback to simulate manufacturing techniques.   { 'prä,ses ,sim·yə,lā·shən }

PROCLIB   See procedure library.   { 'präk,līb }

prod   See test prod.   { präd }

product demodulator |ELECTR| A receiver demodulator whose output is the product of the input signal voltage and a local oscillator signal voltage at the input frequency. Also known as product detector.   { 'präd·əkt di,mäj·ə,lād·ər }

product detector   See product demodulator.   { 'präd·əkt di,tek·tər }

production |COMPUT SCI| 1. The processing of useful work by a computer system, excluding the development and testing of new programs. 2. A rule in a grammar of a formal language that describes how parts of a string (or word, phrase, or construct) can be replaced by other strings. Also known as rule of inference.   { prə'dək·shən }

production program |COMPUT SCI| A proprietary program used primarily for internal processing in a business and not generally made available to third parties for profit.   { prə'dək·shən ,prō ,gram }

production test |COMPUT SCI| A test of a computer system with actual data in the environment where it will be used.   { prə'dək·shən ,test }

production time |COMPUT SCI| Good computing time, including occasional duplication of one case for a check or rerunning of the test run; also including duplication requested by the sponsor, any reruns caused by misinformation or bad data supplied by sponsor, and error studies using different intervals, convergence criteria, and so on.   { prə'dək·shən ,tīm }

product modulator |ELECTR| Modulator whose modulated output is substantially equal to the carrier and the modulating wave; the term implies a device in which intermodulation between components of the modulating wave does not occur.   { 'präd·əkt ,mäj·ə,lād·ər }

profile |COMMUN| A defined subset of the syntax specified in the MPEG-2 video coding specification.   { 'prō,fīl }

program |COMMUN| 1. A sequence of audio signals alone, or audio and video signals, transmit-

ted for entertainment or information. 2. A collection of program elements. Program elements may be elementary streams, and need not have any defined time base. Those that do have a common time base are intended for synchronized presentation. |COMPUT SCI| A detailed and explicit set of directions for accomplishing some purpose, the set being expressed in some language suitable for input to a computer, or in machine language.   { 'prō·grəm or 'prō,gram }

program analysis |COMPUT SCI| The process of determining the functions to be carried out by a computer program.   { 'prō·grəm ə,nal·ə·səs }

program block |COMPUT SCI| A division or section of a computer program that functions to a large extent as if it were a separate program.   { 'prō·grəm ,bläk }

program check |COMPUT SCI| A built-in check system in a program to determine that the program is running correctly.   { 'prō·grəm ,chek }

program clock reference |COMMUN| A time stamp in the transport stream from which decoder timing is derived. Abbreviated PCR.   { 'prō·grəm ¦kläk 'ref·rəns }

program compatibility |COMPUT SCI| The type of compatibility shared by two computers that can process the identical program or programs written in the same source language or machine language.   { 'prō·grəm kəm,pad·ə'bil·əd·ē }

program control |CONT SYS| A control system whose set point is automatically varied during definite time intervals in order to make the process variable vary in some prescribed manner.   { 'prō·grəm kən,trōl }

program conversion |COMPUT SCI| The changing of the source language of a computer program from one dialect to another, or the modification of the program to operate with a different operating system or data-base management system.   { 'prō·grəm kən,vər·zhən }

program counter   See instruction counter.   { 'prō· grəm ,kau̇nt·ər }

program design |COMPUT SCI| The phase of computer program development in which the hardware and software resources needed by the program are identified and the logic to be used by the program is determined.   { 'prō·grəm di ,zīn }

program development time |COMPUT SCI| The total time taken on a computer to produce operating programs, including the time taken to compile, test, and debug programs, plus the time taken to develop and test new procedures and techniques.   { 'prō·grəm di'vel·əp·mənt ,tīm }

program editor |COMPUT SCI| A computer routine used in time-sharing systems for on-line modification of computer programs.   { 'prō·grəm ,ed·ə·tər }

program element |COMMUN| A generic term for one of the elementary streams or other data streams that may be included in the program of a digital video system. |COMPUT SCI| Part of a central computer system that carries out the instruction sequence scheduled by the programmer.   { 'prō·grəm ,el·ə·mənt }

program failure alarm |COMMUN| Signal-operated radio or television relay that gives a visual and/or aural alarm when the program fails on the line being monitored; a time delay is provided to prevent the relay from operating and giving a false alarm during station identification periods or other short periods of silence in program continuity. { 'prō·grəm 'fāl·yər ə,lärm }

program generator |COMPUT SCI| A program that permits a computer to write other programs automatically. { 'prō·grəm ,jen·ə,rād·ər }

program library |COMPUT SCI| An organized set of computer routines and programs. { 'prō·grəm ,lī·brer·ē }

program listing |COMPUT SCI| A list of the statements in a computer program, usually produced as a by-product of the compilation of the program. { 'prō·grəm ,list·iŋ }

program logic |COMPUT SCI| A particular sequence of instructions in a computer program. { 'prō·grəm ,läj·ik }

programmable calculator |COMPUT SCI| An electronic calculator that has some provision for changing its internal program, usually by inserting a new magnetic card on which the desired calculating program has been stored. { prō'gram·ə·bəl 'kal·kyə,lād·ər }

programmable controller |CONT SYS| A control device, normally used in industrial control applications, that employs the hardware architecture of a computer and a relay ladder diagram language. Also known as programmable logic controller. { prō'gram·ə·bəl kən'trōl·ər }

programmable counter |ELECTR| A counter that divides an input frequency by a number which can be programmed into decades of synchronous down counters; these decades, with additional decoding and control logic, give the equivalent of a divide-by-N counter system, where N can be made equal to any number. { prō'gram·ə·bəl 'kaúnt·ər }

programmable decade resistor |ELECTR| A decade box designed so that the value of its resistance can be remotely controlled by programming logic as required for the control of load, time constant, gain, and other parameters of circuits used in automatic test equipment and automatic controls. { prō'gram·ə·bəl 'de,kād ri ,zis·tər }

programmable device |COMPUT SCI| Any device whose operation is controlled by a stored program that can be changed or replaced. { prō'gram·ə·bəl di'vīs }

programmable electronic system |SYS ENG| A system based on a computer and connected to sensors or actuators for the purpose of control, protection, or monitoring. { prō'gram·ə·bəl i'lek,trän·ik ,sis·təm }

programmable logic array See field-programmable logic array. { prō'gram·ə·bəl ¦läj·ik ə,rā }

programmable logic controller See programmable controller. { prō'gram·ə·bəl ¦läj·ik kən,trōl·ər }

programmable power supply |ELEC| A power supply whose output voltage can be changed by digital control signals. { prō'gram·ə·bəl 'paú·ər sə,plī }

programmable read-only memory |COMPUT SCI| An integrated-circuit memory chip which can be programmed only once by the user after which the information stored in the chip cannot be altered. Abbreviated PROM. { prō'gram·ə·bəl ¦rēd¦ōn·lē 'mem·rē }

program maintenance |COMPUT SCI| The updating of computer programs both by error correction and by alteration of programs to meet changing needs. { 'prō·grəm 'mānt·ən·əns }

programmatic interface See application program interface. { ,prō·grə¦mad·ik 'in·tər,fās }

programmed check |COMPUT SCI| **1.** An error-detecting operation programmed by instructions rather than built into the hardware. **2.** A computer check in which a sample problem with known answer, selected for having a program similar to that of the next problem to be run, is put through the computer. { 'prō,gramd 'chek }

programmed dump |COMPUT SCI| A storage dump which results from an instruction in a computer program at a particular point in the program. { 'prō,gramd 'dəmp }

programmed function key |COMPUT SCI| A key on the keyboard of a computer terminal that lacks a predefined function but can be assigned a function by a computer program. Abbreviated PF key. { 'prō,gramd 'fəŋk·shən ,kē }

programmed halt |COMPUT SCI| A halt that occurs deliberately as the result of an instruction in the program. Also known as programmed stop. { 'prō,gramd 'hòlt }

programmed logic array |ELECTR| An array of AND/OR logic gates that provides logic functions for a given set of inputs programmed during manufacture and serves as a read-only memory. Abbreviated PLA. { 'prō,gramd ¦läj·ik ə,rā }

programmed marginal check |COMPUT SCI| Computer program that varies its own voltage to check some piece of electronic computer equipment during a preventive maintenance check. { 'prō,gramd 'mär·jən·əl 'chek }

programmed operators |COMPUT SCI| Computer instructions which enable subroutines to be accessed with a single programmed instruction. { 'prō,gramd 'äp·ə,rād·ərz }

programmed stop See programmed halt. { 'prō ,gramd 'stäp }

programmer |COMPUT SCI| A person who prepares sequences of instructions for a computer, without necessarily converting them into the detailed codes. { 'prō,gram·ər }

programmer analyst |COMPUT SCI| A person who both writes computer programs and analyzes and designs information systems. { 'prō,gram·ər 'an·əl,ist }

programmer-defined macroinstruction |COMPUT SCI| A macroinstruction which is equivalent to a set of ordinary instructions as specified by the programmer for use in a particular computer program. { 'prō ,gram·ər di¦fīnd 'ma·krō·in'strək·shən }

programming |COMPUT SCI| Preparing a detailed sequence of operating instructions for a particular problem to be run on a digital computer. Also known as computer programming. { 'prō‚gram·iŋ }

programming language |COMPUT SCI| The language used by a programmer to write a program for a computer. { 'prō‚gram·iŋ ‚laŋ·gwij }

programming panel |CONT SYS| A device used to edit a program or insert and monitor it in a programmable controller. { 'prō‚gram·iŋ ‚pan·əl }

programming unit *See* manual control unit. { 'prō‚gram·iŋ ‚yü·nət }

program module |COMPUT SCI| A logically self-contained and discrete part of a larger computer program, for example, a subroutine or a coroutine. { 'prō·grəm ‚mäj·yül }

program monitor |COMMUN| A monitor used to observe the quality of a radio or television broadcast. { 'prō·grəm 'män·əd·ər }

program parameter |COMPUT SCI| In computers, an adjustable parameter in a subroutine which can be given a different value each time the subroutine is used. { 'prō·grəm pə'ram·əd·ər }

program register |COMPUT SCI| The register in the control unit of a digital computer that stores the current instruction of the program and controls the operation of the computer during the execution of that instruction. Also known as computer control register. { 'prō·grəm ‚rej·ə·stər }

program scan |CONT SYS| The span of time during which a programmable controller processor executes all the instructions of a given program. { 'prō·grəm ‚skan }

program-sensitive fault |COMPUT SCI| A hardware malfunction that appears only in response to a particular sequence (or kind of sequence) of program instructions. { 'prō·grəm ‚sen·səd·iv 'fôlt }

program specification |COMPUT SCI| A statement of the precise functions which are to be carried out by a computer program, including descriptions of the input to be processed by the program, the processing needed, and the output from the program. { 'prō·grəm ‚spes·ə·fə'kā·shən }

program specific information |COMMUN| Normative data that is necessary for the demultiplexing of transport streams and the successful regeneration of programs. Abbreviated PSI. { ¦prō·grəm spə¦sif·ik 'in·fər'mā·shən }

program state |COMPUT SCI| The mode of operation of a computer during the execution of instructions in an application program. { 'prō·grəm ‚stāt }

program status word |COMPUT SCI| An internal register to the central processing unit denoting the state of the computer at a moment in time. { 'prō·grəm ‚stad·əs ‚wərd }

program step |COMPUT SCI| In computers, some part of a program, usually one instruction. { 'prō·grəm ‚step }

program stop |COMPUT SCI| An instruction built into a computer program that will automatically stop the machine under certain conditions, or upon reaching the end of processing or completing the solution of a program. Also known as halt instruction; stop instruction. { 'prō·grəm ‚stäp }

program storage |COMPUT SCI| Portion of the internal storage reserved for the storage of programs, routines, and subroutines; in many systems, protection devices are used to prevent inadvertent alteration of the contents of the program storage; contrasted with temporary storage. { 'prō·grəm ‚stòr·ij }

program tape |COMPUT SCI| Tape containing the sequence of computer instructions for a given problem. { 'prō·grəm ‚tāp }

program test |COMPUT SCI| A system of checking before running any problem in which a sample problem of the same type with a known answer is run. { 'prō·grəm ‚test }

program testing time |COMPUT SCI| The machine time expended for program testing, debugging, and volume and compatibility testing. { 'prō·grəm 'test·iŋ ‚tīm }

program time |COMPUT SCI| The phase of computer operation when an instruction is being interpreted so that it can be carried out. { 'prō·grəm ‚tīm }

progressive overflow |COMPUT SCI| Retrieval of a randomly stored overflow record by a forward serial search from the home address. { prə'gres·iv 'ō·vər‚flō }

progressive scanning |COMMUN| Scanning all lines in sequence, without interlace, so all picture elements are included during one vertical sweep of the scanning beam. Also known as sequential scanning. { prə'gres·iv 'skan·iŋ }

progressive-wave antenna *See* traveling-wave antenna. { prə'gres·iv ‚wāv an'ten·ə }

projection cathode-ray tube |ELECTR| A cathode-ray tube designed to produce an intensely bright but relatively small image that can be projected onto a large viewing screen by an optical system. { prə'jek·shən ¦kath‚ōd 'rā ‚tüb }

projection display |ELECTR| An electronic system in which an image is generated on a high-brightness cathode-ray tube or similar electronic image generator and then optically projected onto a larger screen. { prə'jek·shən di'splā }

projection net *See* net. { prə'jek·shən ‚net }

projection plan position indicator |ELECTR| Unit in which the image of a 4-inch (10-centimeter) dark-trace cathode-ray tube is projected on a 24-inch (61-centimeter) horizontal plotting surface; the echoes appear as magenta-colored arcs on white background. { prə'jek·shən 'plan pə'zish·ən 'in·də‚kād·ər }

projector |ENG ACOUS| **1.** A horn designed to project sound chiefly in one direction from a loudspeaker. **2.** An underwater acoustic transmitter. { prə'jek·tər }

PROLOG |COMPUT SCI| A programming language that is for artificial intelligence applications, and uses problem descriptions to reach solutions, based on precise rules. { 'prō‚läg }

PROM *See* programmable read-only memory. { präm }

PROM burner |COMPUT SCI| A special device used to write on a programmable read-only memory (PROM). { 'präm ‚bər·nər }

PROM programmer |ELECTR| A device that holds several programmable read-only memory

(PROM) chips and writes instructions and data into them by melting connections in their circuitry. { 'präm 'prō‚gram·ər }

prompt |COMPUT SCI| A message or format displayed on the screen of a computer terminal that requires the user to respond in some way before processing can continue. { prämpt }

pronate |CONT SYS| To orient a robot toward a position in which the back or protected side of a manipulator faces up and is exposed. { 'prō‚nāt }

prong *See* pin. { prän }

proof plane |ELEC| A small metal plane supported by an insulating handle and used to transfer a small fraction of the electric charge on a body to an electrometer to investigate the charge distribution on the body. { 'prüf ‚plān }

proof total |COMPUT SCI| One of a group of totals which are compared with each other to check their consistency. { 'prüf ‚tōd·əl }

propagated error |COMPUT SCI| An error which takes place in one operation and spreads through succeeding operations. { 'präp·ə‚gād·əd 'er·ər }

propagation constant |ELECTROMAG| A rating for a line or medium along or through which a wave of a given frequency is being transmitted; it is a complex quantity; the real part is the attenuation constant in nepers per unit length, and the imaginary part is the phase constant in radians per unit length. { ‚präp·ə'gā·shən ‚kän·stənt }

propagation delay |ELECTR| The time required for a signal to pass through a given complete operating circuit; it is generally of the order of nanoseconds, and is of extreme importance in computer circuits. { ‚präp·ə'gā·shən di‚lā }

propagation loss |COMMUN| The attenuation of signals passing between two points of a transmission path. { ‚präp·ə'gā·shən ‚lós }

propagation mode |ELECTROMAG| A form of propagation of electromagnetic radiation in a periodic beamguide in which the field distributions over cross sections of the beam are identical at positions separated by one period of the guide. { 'präp·ə'gā·shən ‚mōd }

propagation notice |COMMUN| A forecast of propagation conditions for long-distance radio communications, broadcast at regular intervals over radio stations operated by the National Institute of Standards and Technology. { ‚präp·ə'gā·shən ‚nōd·əs }

propagation path |COMMUN| A path between receiver and transmitter including direct tropospheric scatter, ionospheric scatter, E-layer skip, and F_1-layer and F_2-layer skip and echo. { ‚präp·ə'gā·shən ‚path }

propagation time delay |COMMUN| The time required for a wave to travel between two points of a transmission path. { ‚präp·ə'gā·shən 'tīm di‚lā }

propagation velocity |ELECTROMAG| Velocity of electromagnetic wave propagation in the medium under consideration. { ‚präp·ə'gā·shən və‚läs·əd·ē }

property detector |COMPUT SCI| In character recognition, that electronic component of a character reader which processes the normalized

signal for the purpose of extracting from it a set of characteristic properties on the basis of which the character can be subsequently identified. { 'präp·ərd·ē di‚tek·tər }

property list |COMPUT SCI| A list for describing some object or concept, in which odd-numbered items name a property or attribute of a relevant class of objects, and the item following the property name is the property's value for the described objects. { 'präp·ərd·ē ‚list }

proportional control |CONT SYS| Control in which the amount of corrective action is proportional to the amount of error. { prə'pór·shən·əl kən'trōl }

proportional ionization chamber |ELECTR| An ionization chamber in which the initial ionization current is amplified by electron multiplication in a region of high electric-field strength, as in a proportional counter; used for measuring ionization currents or charges over a period of time, rather than for counting. { prə'pór·shən·əl ‚ī·ə·nə'zā·shən ‚chām·bər }

proportional-plus-derivative control |CONT SYS| Control in which the control signal is a linear combination of the error signal and its derivative. { prə'pór·shən·əl ‚pləs də'riv·əd·iv kən‚trōl }

proportional-plus-integral control |CONT SYS| Control in which the control signal is a linear combination of the error signal and its integral. { prə'pór·shən·əl ‚pləs 'int·ə·grəl kən‚trōl }

proportional-plus-integral-plus-derivative control |CONT SYS| Control in which the control signal is a linear combination of the error signal, its integral, and its derivative. { prə'pór·shən·əl ‚pləs 'int·ə·grəl ‚pləs də'riv·əd·iv kən‚trōl }

proportional-speed control *See* floating control. { prə'pór·shən·əl 'spēd kən‚trōl }

proprietary program |COMPUT SCI| **1.** A computer program that is owned by someone, and whose use may thus be restricted in some manner or entail payment of a fee. Also known as owned program. **2.** More narrowly, a program that is exploited commercially as a separate product. { prə'prī·ə‚ter·ē 'prō·grəm }

proprioceptor |CONT SYS| A device that senses the position of an arm or other computer-controlled articulated mechanism of a robot and provides feedback signals. { ‚prō·prē·ə'sep·tər }

protected contour |COMMUN| A representation of the theoretical signal strength of a radio station that appears on a map as a closed polygon surrounding the station's transmitter site. The FCC defines a particular signal strength contour such as 60 dBuV/m, for certain classes of station, as the protected contour. In allocating the facilities of other radio stations, the protected contour of an existing station may not be overlapped by certain interferring contours of other stations. The protected contour coarsely represents the primary coverage area of a station, within which there is little likelihood that the signals of another station will cause interference with its reception. { prə'tek·təd 'kän·túr }

protected format |COMPUT SCI| Parts of a computer display that cannot be altered by typing from the keyboard. { prə'tek·təd 'fór‚mat }

protected location |COMPUT SCI| A storage cell arranged so that access to its contents is denied under certain circumstances, in order to prevent programming accidents from destroying essential programs and data. { prə'tek·təd lō'kā·shən }

protected-logic module |COMPUT SCI| A module that stores selected computer programs that must remain unaltered. { prə'tek·təd ¦läj·ik 'māj·yül }

protected subnetwork See domain. { prə¦tek·təd səb'net,wərk }

protection code |COMPUT SCI| A component of a task descriptor that specifies the protection domain of the task, that is, the authorizations it has to perform certain actions. { prə'tek·shən ,kōd }

protection key |COMPUT SCI| An indicator, usually 1 to 6 bits in length, associated with a program and intended to grant the program access to those sections of memory which the program can use but to deny the program access to all other parts of memory. { prə'tek·shən ,kē }

protection profile |COMPUT SCI| A structure for defining the security and functionality requirements of a computing system. { prə'tek·shən ,prō·fīl }

protective device See electric protective device. { prə'tek·tiv di'vīs }

protective grounding |ELEC| Grounding of the neutral conductor of a secondary power-distribution system, and of all metal enclosures for conductors, to protect persons from dangerous currents. { prə'tek·tiv 'graünd·iŋ }

protective relay |ELEC| A relay whose principal function is to protect service from interruption or to prevent or limit damage to apparatus. { prə'tek·tiv 'rē,lā }

protective resistance |ELECTR| Resistance used in series with a gas tube or other device to limit current flow to a safe value. { prə'tek·tiv ri'zis·təns }

protector |ELEC| Device to protect equipment or personnel from high voltages or currents. { prə'tek·tər }

protector block |ELEC| Rectangular piece of carbon with an insulated metal insert, or porcelain with a carbon insert, constituting an element of a protector; it forms a gap which will break down and provide a path to ground for voltages over 350 volts. { prə'tek·tər ,bläk }

protector gap |ELEC| A device designed to limit voltage, usually from lightning strkes, in order to protect telephone and telegraph equipment; consists of two carbon blocks with an air gap between them. { prə'tek·tər ,gap }

protector tube |ELECTR| A glow-discharge cold-cathode tube that becomes conductive at a predetermined voltage, to protect a circuit against overvoltage. { prə'tek·tər ,tüb }

Proteus See advanced signal-processing system. { 'prōd·ē·əs }

protocol |COMPUT SCI| **1.** A set of hardware and software interfaces in a terminal or computer which allows it to transmit over a communications network, and which collectively forms a

communications language. **2.** See communication protocol. { 'prōd·ə,kȯl }

protocol-level timer |COMMUN| A time-measuring unit within a communicating device that issues high-priority interrupts which synchronize and set deadlines for protocol-related activities. { 'prōd·ə,kȯl ¦lev·əl 'tīm·ər }

proton microscope |ELECTR| A microscope that is similar to the electron microscope but uses protons instead of electrons as the charged particles. { 'prō,tän 'mī·krə,skōp }

prototype |ENG| A model suitable for use in complete evaluation of form, design, and performance. { 'prōd·ə,tīp }

proving |COMPUT SCI| Testing whether a computer is free of faults and capable of functioning normally, usually by having it carry out a check routine or diagnostic routine. { 'prüv·iŋ }

proximity detector |ENG| A sensing device that produces an electrical signal when approached by an object or when approaching an object. { präk'sim·əd·ē di,tek·tər }

proximity effect |ELEC| Redistribution of current in a conductor brought about by the presence of another conductor. { präk'sim·əd·ē i,fekt }

proximity-focused tube |ELECTR| A type of image tube in which electrons are rapidly accelerated across a narrow gap, 1.5 to 3.5 millimeters wide, between the photocathode and the phosphor screen, both deposited on plane-parallel optical windows. { präk'sim·əd·ē ¦fō·kəst 'tüb }

proximity sensor |CONT SYS| Any device that measures short distances within a robotic system. Also known as noncontact sensor. { präk 'sim·əd·ē 'sen·sər }

proxy server |COMPUT SCI| Software for caching and filtering Web content to reduce network traffic on intranets, and for increasing security by filtering content and restricting access. { 'präk·sē ,sər·vər }

PRR See pulse repetition rate.

PSC motor See permanent-split capacitor motor. { ¦pē¦es¦sē 'mōd·ər }

pseudoanalog display |ELECTR| An electronic display consisting of a dedicated arrangement of discrete pixels used to present analog or quantitative information. { ¦süd·ō¦an·ə,läg di'splā }

pseudocode |COMPUT SCI| In software engineering, an outline of a program written in English or the user's natural language; it is used to plan the program, and also serves as a source for test engineers doing software maintenance; it cannot be compiled. { 'süd·ō,kōd }

pseudocoloring |COMPUT SCI| A method of assigning arbitrary colors to the gray levels of a black-and-white image. It is popular in thermography (the imaging of heat), where hotter objects (with high pixel values) are assigned one color (for example, red), and cool objects (with low pixel values) are assigned another color (for example, blue), with other colors assigned to intermediate values. { ,süd·ō'kəl·ər·iŋ }

pseudoinstruction |COMPUT SCI| **1.** A symbolic representation in a compiler or interpreter. **2.** See quasi-instruction. { 'sü·dō·in,strək·shən }

pseudonoise code See pseudorandom noise code.
{ ¦süd·ō'nóiz 'kōd }

pseudo-operation [COMPUT SCI] An operation which is not part of the computer's operation repertoire as realized by hardware; hence, an extension of the set of machine operations.
{ ¦sü·dō ,äp·ə'rā·shən }

pseudorandom noise code [COMMUN] A method of transmitting messages in the presence of interference or noise, in which each binary digit in the original message is encoded by a long series of binary digits with desirable autocorrelation properties. Also known as pseudonoise code. Abbreviated PN code.
{ ¦süd·ō¦ran·dəm 'nóiz ,kōd }

pseudorandom numbers [COMPUT SCI] Numbers produced by a definite arithmetic process, but satisfying one or more of the standard tests for randomness. { ,sü·dō'ran·dəm 'nəm·bərz }

PSI See program specific information.

PSK See phase-shift keying.

psophometer [ENG] An instrument for measuring noise in electric circuits; when connected across a 600-ohm resistance in the circuit under study, the instrument gives a reading that by definition is equal to half of the psophometric electromotive force actually existing in the circuit. { sō'fäm·əd·ər }

psophometric electromotive force [ELECTR] The true noise voltage that exists in a circuit. { ¦säf·ə¦me·trik i¦lek·trə¦mōd·iv 'fórs }

psophometric voltage [ELECTR] The noise voltage as actually measured in a circuit under specified conditions. { ¦säf·ə¦me·trik 'vōl·tij }

PSR See primary radar.

PSRR See power-supply rejection ratio.

PSTN See public switched telephone network.

PTD See posttuning drift.

PTM See pulse-time modulation.

p-type conductivity [ELECTR] The conductivity associated with holes in a semiconductor, which are equivalent to positive charges. { 'pē ¦tīp ,kän,dək'tiv·əd·ē }

p-type crystal rectifier [ELECTR] Crystal rectifier in which forward current flows when the semiconductor is positive with respect to the metal. { 'pē ¦tīp 'krist·əl 'rek·tə,fī·ər }

p-type semiconductor [ELECTR] An extrinsic semiconductor in which the hole density exeeds the conduction electron density. { 'pē ¦tīp 'sem·i·kən,dək·tər }

p⁺-type semiconductor [ELECTR] A p-type semiconductor in which the excess mobile hole concentration is very large. { 'pē¦pləs ,tīp 'sem·i·kən,dək·tər }

p-type silicon [ELECTR] Silicon to which more impurity atoms of acceptor type (with valence of 3, such as boron) than of donor type (with valence of 5, such as phosphorus) have been added, with the result that the hole density exceeds the conduction electron density. { 'pē ¦tīp 'sil·ə ,kän }

public address system See sound-reinforcement system. { 'pəb·lik ə'dres ,sis·təm }

public communications service [COMMUN] Telephone or telegraph service provided for the transmission of unofficial communications for the public. { 'pəb·lik kə,myü·nə'kā·shənz ,sər·vəs }

public correspondence [COMMUN] Any telecommunications which offices and stations at the disposal of the public must accept for transmission. { 'pəb·lik ,kär·ə'spän·dəns }

public data [COMPUT SCI] Data that are open to all users, with no security measures necessary as far as reading is concerned. { 'pəb·lik 'dad·ə }

public-key algorithm [COMMUN] A cryptographic algorithm in which one key (usually the enciphering key) is made public and a different key (usually the deciphering key) is kept secret; it must not be possible to deduce the private key from the public key. { 'pəb·lik¦kē 'al·gə,rith·əm }

public network [COMMUN] A communications network that can be used by anyone, usually on a fee basis. { 'pəb·lik 'net,wərk }

public pack [COMPUT SCI] A disk pack that can be used by any program and any application in a computer system. { 'pəb·lik 'pak }

public radio communications services [COMMUN] Land, mobile, and fixed services, the stations of which are open to public correspondence. { 'pəb·lik 'rād·ē·ō kə,myü·nə'kā·shənz ,sər·və·səz }

public-safety frequency bands [COMMUN] Radio-frequency bands allocated in the United States for communication on land between base stations and mobile stations or between mobile stations by police, fire, highway, forestry, and emergency services. { 'pəb·lik ¦sāf·tē 'frē·kwən·sē ,banz }

public-safety radio service [COMMUN] Any service of radio communication essential to either the discharge of non-Federal governmental functions relating to public safety responsibilities or the alleviation of an emergency endangering life or property, the radio transmitting facilities of which are defined as fixed, land, or mobile stations. { 'pəb·lik ¦sāf·tē 'rād·ē·ō ,sər·vəs }

public switched telephone network [COMMUN] The worldwide voice telephone network. Abbreviated PSTN. { ¦pəb·lik ¦swicht 'tel·ə,fōn ,net ,wərk }

puff See picofarad. { pəf }

pull-down menu [COMPUT SCI] A list of options for action that appears near the top of a display screen, usually overlaying the current contents of the screen without disrupting them, and usually in response to an indicator being pointed at an icon. { 'púl ¦daún 'men·yü }

pulling [ELECTR] An effect that forces the frequency of an oscillator to change from a desired value; causes include undesired coupling to another frequency source or the influence of changes in the oscillator load impedance. { 'púl·iŋ }

pulling figure [ELECTR] The total frequency change of an oscillator when the phase angle of the reflection coefficient of the load impedance varies through 360°, the absolute value of this reflection coefficient being constant at 0.20. { 'púl·iŋ ,fig·yər }

pulsating current |ELEC| Periodic direct current. { 'pəl,sād·iŋ 'kə·rənt }

pulsating electromotive force |ELEC| Sum of a direct electromotive force and an alternating electromotive force. Also known as pulsating voltage. { 'pəl,sād·iŋ i¦lek·trə¦mōd·iv 'fȯrs }

pulsating voltage *See* pulsating electromotive force. { 'pəl,sād·iŋ 'vōl·tij }

pulse amplifier |ELEC| An amplifier designed specifically to amplify electric pulses without appreciably changing their waveforms. { 'pəls ¦am·plə,fī·ər }

pulse-amplitude modulation |COMMUN| Amplitude modulation of a pulse carrier. Abbreviated PAM. { 'pəls ¦am·plə,tüd ,mäj·ə,lā·shən }

pulse-amplitude modulation-frequency modulation |COMMUN| System in which pulse-amplitude-modulated subcarriers are used to frequency-modulate a second carrier; binary digits are formed by the absence or presence of a pulse in an assigned position. { 'pəls ¦am·plə,tüd ,mäj·ə ,lā·shən 'frē·kwən·sē ,mäj·ə,lā·shən }

pulse analyzer |ELECTR| An instrument used to measure pulse widths and repetition rates, and to display on a cathode-ray screen the waveform of a pulse. { 'pəls ,an·ə,līz·ər }

pulse bandwidth |COMMUN| The bandwidth outside of which the amplitude of a pulse-frequency spectrum is below a prescribed fraction of the peak amplitude. { 'pəls 'band,width }

pulse cable |COMMUN| A communications cable, capable of transmitting pulses without unacceptable distortion. { 'pəls ,kā·bəl }

pulse carrier |COMMUN| A pulse train used as a carrier. { 'pəls ,kar·ē·ər }

pulse circuit |ELECTR| An active electrical network designed to respond to discrete pulses of current or voltage. { 'pəls ,sər·kət }

pulse code |COMMUN| A code consisting of various combinations of pulses, such as the Morse code, Baudot code, and the binary code used in computers. { 'pəls ,kōd }

pulse-coded scanning beam |NAV| **1.** A radio or radar beam which is swept over a sector of space and is accompanied by a repeated pattern of pulses that is varied to indicate the position of the beam in space. **2.** A system of ground equipment that generates such beams at microwave frequencies to furnish guidance to aircraft making microwave landings. Abbreviated PCSB. { 'pəls ¦kōd·əd 'skan·iŋ ,bēm }

pulse-code modulation |COMMUN| Modulation in which the peak-to-peak amplitude range of the signal to be transmitted is divided into a number of standard values, each having its own code; each sample of the signal is then transmitted as the code for the nearest standard amplitude. Abbreviated PCM. { 'pəls ¦kōd ,mäj·ə'lā·shən }

pulse coder *See* coder. { 'pəls ,kōd·ər }

pulse coding and correlation |COMMUN| A general technique concerning a variety of methods used to change the transmitted waveform and then decode upon its reception; pulse compression is a special form of pulse coding and correlation. { 'pəls ¦kōd·iŋ ən ,kär·ə'lā·shən }

pulse communication |COMMUN| Radio communication using pulse modulation. { 'pəls kə ,myü·nə,kā·shən }

pulse compression |ELECTR| **1.** A matched filter technique used to discriminate against signals which do not correspond to the transmitted signal. **2.** In radar, a process in which a relatively long pulse is frequency- or phase-modulated so that a properly designed receiver produces an output with a very narrow peak response much as though a very narrow pulse had been transmitted; valuable in achieving high range resolution in long transmitted pulses. { 'pəls kəm,presh·ən }

pulse-compression radar |ENG| A radar system in which the transmitted signal is linearly frequency-modulated or otherwise spread out in time to reduce the peak power that must be handled by the transmitter; signal amplitude is kept constant; the receiver uses a linear filter to compress the signal and thereby reconstitute a short pulse for the radar display. { 'pəls kəm ,presh·ən 'rā,där }

pulse counter |ELECTR| A device that indicates or records the total number of pulses received during a time interval. { 'pəls ,kaúnt·ər }

pulse decay time |COMMUN| The interval of time required for the trailing edge of a pulse to decay from 90% to 10% of the peak pulse amplitude. { 'pəls di'kā ,tīm }

pulse-delay network |ELECTR| A network consisting of two or more components such as resistors, coils, and capacitors, used to delay the passage of a pulse. { 'pəls di'lā 'net,wərk }

pulse demodulator |ELECTR| A device that recovers the modulating signal from a pulse-modulated wave. { ¦pəls dē¦mäj·ə,lād·ər }

pulse-density modulation *See* pulse-frequency modulation. { ¦pəls ,den·sət·ē ,mäj·əlā·shən }

pulse discriminator |ELECTR| A discriminator circuit that responds only to a pulse having a particular duration or amplitude. { 'pəls di ,skrim·ə,nād·ər }

pulsed oscillator |ELECTR| An oscillator that generates a carrier-frequency pulse or a train of carrier-frequency pulses as the result of self-generated or externally applied pulses. { 'pəlst 'äs·ə,lād·ər }

pulse droop |ELECTR| A distortion of an otherwise essentially flat-topped rectangular pulse, characterized by a decline of the pulse top. { 'pəls ,drüp }

pulsed transfer function |CONT SYS| The ratio of the z-transform of the output of a system to the z-transform of the input, when both input and output are trains of pulses. Also known as discrete transfer function; z-transfer function. { 'pəlst 'tranz·fər ,faŋk·shən }

pulse duration |COMMUN| The time interval between the first and last instants at which the instantaneous amplitude reaches a stated fraction of the peak pulse amplitude. Also known as pulse length; pulse width (both deprecated usages). { 'pəls dü'rā·shən }

pulse-duration coder *See* coder. { 'pəls dü¦rā·shən 'kōd·ər }

pulse-duration discriminator [ELECTR] A circuit in which the sense and magnitude of the output are a function of the deviation of the pulse duration from a reference. { 'pəls du̇ǃrā·shən di'skrim·ə,nād·ər }

pulse-duration modulation [COMMUN] Modulation of a pulse carrier wherein the value of each instantaneous sample of a modulating wave produces a pulse of proportional duration by varying the leading, trailing, or both edges of a pulse. Abbreviated PDM. Also known as pulse-length modulation; pulse-width modulation. { 'pəls du̇ǃrā·shən ,mäj·ə,lā·shən }

pulse-duration modulation-frequency modulation [COMMUN] System in which pulse-duration-modulated subcarriers are used to frequency-modulate a second carrier. Also known as pulse-width modulation-frequency modulation. { 'pəls du̇ǃrā·shən ,mäj·ə,lā·shən 'frē·kwən·sē ,mäj·ə,lā·shən }

pulse-forming network [ELECTR] A network used to shape the leading or trailing edge of a pulse. { 'pəls ǃfȯrm·iŋ 'net,wərk }

pulse-frequency modulation [COMMUN] A form of pulse-time modulation in which the pulse repetition rate is the characteristic that is varied. Abbreviated PFM. { 'pəls ǃfrē·kwən·sē ,mäj·ə,lā·shən }

pulse generator [ELEC] See impulse generator. [ELECTR] A generator that produces repetitive pulses or signal-initiated pulses. { 'pəls ,jen·ə,rād·ər }

pulse height [ELECTR] The strength or amplitude of a pulse, measured in volts. { 'pəls ,hīt }

pulse-height discriminator [ELECTR] A circuit that produces a specified output pulse when and only when it receives an input pulse whose amplitude exceeds an assigned value. Also known as amplitude discriminator. { 'pəls ,hīt di'skrim·ə ,nād·ər }

pulse-height selector [ELECTR] A circuit that produces a specified output pulse only when it receives an input pulse whose amplitude lies between two assigned values. Also known as amplitude selector; diffractional pulse-height discriminator. { 'pəls ,hīt si'lek·tər }

pulse improvement threshold [COMMUN] In a constant-amplitude pulse-modulation system, the condition in which the peak pulse voltage is greater than twice the peak noise voltage, after selection and before nonlinear processes such as amplitude clipping and limiting. { 'pəls im ǃprüv·mənt 'thresh,hōld }

pulse integrator [ELECTR] An RC (resistance-capacitance) circuit which stretches in time duration a pulse applied to it. { 'pəls ,int·ə ,grād·ər }

pulse interference eliminator [ELECTR] Device which removes pulsed signals which are not precisely on the radar operating frequency. { 'pəls ,in·tərǃfir·əns i,lim·ə,nād·ər }

pulse interference separator and blanker [ELECTR] Automatic interference blanker that will blank all video signals not synchronous with the radar pulse-repetition frequency. { 'pəls ,in·tər ǃfir·əns 'sep·ə,rād·ər ən 'blaŋk·ər }

pulse interference suppression [ELECTR] Means employed in radar, such as noting asynchronous returns or pulses clearly of unlikely widths or pulses at frequencies other than the operating frequency, to reduce confusion from pulses of other radars or pulsed deceptive countermeasures. { 'pəls ,in·tərǃfir·əns sə'presh·ən }

pulse interleaving [COMMUN] A process in which pulses from two or more sources are combined in time-division multiplex for transmission over a common path. { 'pəls ,in·tər'lēv·iŋ }

pulse-interval modulation See pulse-spacing modulation. { 'pəls ǃin·tər·vəl ,mäj·ə,lā·shən }

pulse jitter [COMMUN] A relatively small variation of the pulse spacing in a pulse train; the jitter may be random or systematic, depending on its origin, and is generally not coherent with any pulse modulation imposed. { 'pəls ,jid·ər }

pulse length See pulse duration. { 'pəls ,leŋkth }

pulse-length modulation See pulse-duration modulation. { 'pəls ǃleŋkth ,mäj·ə,lā·shən }

pulse-link repeater [ELECTR] Arrangement of apparatus used in telephone signaling systems for receiving pulses from one E and M signaling circuit, and retransmitting corresponding pulses into another E and M signaling circuit. { 'pəls ǃliŋk ri'pēd·ər }

pulse-mode multiplexing [COMMUN] A type of time-division multiplexing employing pulse-amplitude modulation in which a sequence of pulses is repeatedly transmitted, and the amplitude of each pulse in the sequence is modulated by a different communication channel. { 'pəls ǃmōd 'məl·tə,pleks·iŋ }

pulse-modulated jamming [COMMUN] Use of jamming pulses of various widths and repetition rates. { 'pəls ,mäj·ə¦lād·əd 'jam·iŋ }

pulse-modulated radar [ENG] Form of radar in which the radiation consists of a series of discrete pulses. { 'pəls ,mäj·ə¦lād·əd 'rā,där }

pulse modulation [COMMUN] A system of modulation in which the amplitude, duration, position, or mere presence of discrete pulses may be so controlled as to represent the message to be communicated. { 'pəls ,mäj·ə,lā·shən }

pulse modulator [ELECTR] A device for carrying out the pulse modulation of a radio-frequency carrier signal. { 'pəls ,mäj·ə,lād·ər }

pulse-numbers modulation [COMMUN] Modulation in which a pulse carrier's pulse density per unit time varies in accordance with a modulating wave, by making systematic omissions without changing the phase or amplitude of the transmitted pulses; as an example, the omission of every other pulse could correspond to zero modulation; the reinsertion of some or all pulses then corresponds to positive modulation, and the omission of more than every other pulse corresponds to negative modulation. { 'pəls ǃnəm·bərz ,mäj·ə,lā·shən }

pulse operation [ELECTR] For microwave tubes, a method of operation in which the energy is delivered in pulses. { 'pəls ,äp·ə,rā·shən }

pulse period [COMMUN] In telephony, time required for one opening and closing of the loop of a calling telephone; for example, the time required to open and close the dial pulse springs once. Also known as impulse period. { 'pəls ,pir·ē·əd }

pulse-phase modulation See pulse-position modulation. { 'pəls ¦fāz ,mäj·ə,lā·shən }

pulse-position modulation [COMMUN] Modulation of a pulse carrier wherein the value of each instantaneous sample of a modulating wave varies the position in time of a pulse relative to its unmodulated time of occurrence. Abbreviated PPM. Also known as pulse-phase modulation. { 'pəls pə¦zish·ən ,mäj·ə,lā·shən }

pulse power [ELECTR] In radar, the average power transmitted during a pulse. While often called the radar's peak power, it is not to be confused with the instantaneous peak power in each cycle of the carrier frequency. { 'pəls ,pau̇·ər }

pulser [ELECTR] A modulator of the energy-storage type, using a pulse-forming network, to produce the pulsed voltage and current required by a microwave oscillator, such as a magnetron, in radar transmitters. { 'pəl·sər }

pulse radar [ENG] Radar in which the transmitter sends out high-power pulses that are spaced far apart in comparison with the duration of each pulse; the receiver is active for reception of echoes in the interval following each pulse. { 'pəls 'rā,där }

pulse-rate telemetering [ELECTR] Telemetering in which the number of pulses per unit time is proportional to the magnitude of the measured quantity. { 'pəls ¦rāt ,tel·ə,mēd·ə·riŋ }

pulse recurrence rate See pulse repetition rate. { 'pəls ri'kə·rəns ,rāt }

pulse recurrence time [COMMUN] Time elapsing between the start of one transmitted pulse and the next pulse; the reciprocal of the pulse repetition rate. { 'pəls ri'kə·rəns ,tīm }

pulse regeneration [ELECTR] The process of restoring pulses to their original relative timings, forms, and magnitudes. { 'pəls ri,jen·ə,rā·shən }

pulse repeater [ELECTR] Device used for receiving pulses from one circuit and transmitting corresponding pulses into another circuit; it may also change the frequencies and waveforms of the pulses and perform other functions. { 'pəls ri ,pēd·ər }

pulse repetition frequency See pulse repetition rate. { 'pəls ,rep·ə¦tish·ən ,frē·kwən·sē }

pulse repetition rate [ELECTR] The number of times per second that a pulse is transmitted. Abbreviated PRR. Also known as pulse recurrence rate; pulse repetition frequency (PRF). { 'pəls ,rep·ə¦tish·ən ,rāt }

pulse rise time [COMMUN] The interval of time required for the leading edge of a pulse to rise from 10% to 90% of the peak pulse amplitude. { 'pəls 'rīz ,tīm }

pulse scaler [ELECTR] A scaler that produces an output signal when a prescribed number of input pulses has been received. { 'pəls ,skāl·ər }

pulse selector [ELECTR] A circuit or device for selecting the proper pulse from a sequence of telemetering pulses. { 'pəls si,lek·tər }

pulse shaper [ELECTR] A transducer used for changing one or more characteristics of a pulse, such as a pulse regenerator or pulse stretcher. { 'pəls ,shāp·ər }

pulse-spacing modulation [COMMUN] A form of pulse-time modulation in which the pulse spacing is varied. Also known as pulse-interval modulation. { 'pəls ¦spās·iŋ ,mäj·ə,lā·shən }

pulse stretcher [ELECTR] A pulse shaper that produces an output pulse whose duration is greater than that of the input pulse and whose amplitude is proportional to the peak amplitude of the input pulse. { 'pəls ,strech·ər }

pulse subcarrier [COMMUN] One of a number of frequency-modulation carriers modulating a radio-frequency carrier, each of which is in turn pulse-modulated. { 'pəls 'səb,kar·ē·ər }

pulse synthesizer [ELECTR] A circuit used to supply pulses that are missing from a sequence due to interference or other causes. { 'pəls ,sin·thə ,sīz·ər }

pulse-time modulation [COMMUN] Modulation in which the time of occurrence of some characteristic of a pulse carrier is varied from the unmodulated value; examples include pulse-duration, pulse-interval, and pulse-position modulation. Abbreviated PTM. { 'pəls ¦tīm ,mäjə,lā·shən }

pulse-train analysis [COMMUN] A Fourier analysis of a pulse train. { 'pəls ¦trān ə,nal·ə·səs }

pulse transformer [ELECTR] A transformer capable of operating over a wide range of frequencies, used to transfer nonsinusoidal pulses without materially changing their waveforms. { 'pəls tranz,fȯr·mər }

pulse transmitter [ELECTR] A pulse-modulated transmitter whose peak-power-output capabilities are usually large with respect to the average-power-output rating. { 'pəls tranz,mid·ər }

pulse-type telemetering [COMMUN] Signal transmission system with pulses as a function of time, but independent of electrical magnitude; in a pulse-counting system the number of pulses per unit time corresponds to the measured variable; in pulse-width or pulse-duration types, the length of the pulse is controlled by the measured variable. { 'pəls ¦tīp ,tel·ə,mēd·ə· riŋ }

pulse voltage See impulse voltage. { 'pəls ,vōl·tij }

pulse width See pulse duration. { 'pəls ,width }

pulse-width discriminator [ELECTR] Device that measures the pulse length of video signals and passes only those whose time duration falls into some predetermined design tolerance. { 'pəls ¦width di'skrim·ə,nād·ər }

pulse-width modulated static inverter [ELEC] A variation of the quasi-square-wave static inverter, operating at high frequency, in which the pulse width, and not the amplitude, of the square wave is adjusted to approximate the sine wave. { 'pəls ¦width ¦mäj·ə,lād·əd 'stad·ik in,vərd·ər }

pulse-width modulation See pulse-duration modulation. { 'pəls ¦width ˌmäj·ə¦lā·shən }

pulse-width modulation-frequency modulation See pulse-duration modulation-frequency modulation. { 'pəls ¦width ˌmäj·ə¦lā·shən 'frē·kwən·sē ˌmäj·ə¦lā·shən }

pulsing key |COMMUN| **1.** Method of passing voice frequency pulses over the line under control of a key at the original office; used with E and M supervision on intertoll dialing. **2.** System of signaling where numbered keys are depressed instead of using a dial. { 'pəls·iŋ ˌkē }

pulsing transformer |ELEC| Transformer that is designed to supply pulses of voltage or current. { 'pəls·iŋ tranzˌfȯr·mər }

pump |ELECTR| Of a parametric device, the source of alternating-current power which causes the nonlinear reactor to behave as a time-varying reactance. { pəmp }

pumped hydroelectric storage |ELEC| A method of energy storage in which excess electrical energy produced at times of low demand is used to pump water into a reservoir, and this water is released at times of high demand to operate hydroelectric generators. { 'pəmpt ¦hī·drō·i'lek·trik 'stȯr·ij }

pumped tube |ELECTR| An electron tube that is continuously connected to evacuating equipment during operation; large pool-cathode tubes are often operated in this manner. { 'pəmpt ˌtüb }

pumping frequency |ELECTR| Frequency at which pumping is provided in a maser, quadrupole amplifier, or other amplifier requiring high-frequency excitation. { 'pəmp·iŋ ˌfrē·kwən·sē }

pump oscillator |ELECTR| Alternating-current generator that supplies pumping energy for maser and parametric amplifiers; operates at twice or some higher multiple of the signal frequency. { 'pəmp ˌäs·ə¦lād·ər }

punch |COMPUT SCI| **1.** A device for making holes representing information in a medium such as cards or paper tape, in response to signals sent to it. **2.** A hole in a medium such as a card or paper tape, generally made in an array with other holes (or lack of holes) to represent information. { pənch }

punch card |COMPUT SCI| A medium by means of which data are fed into a computer in the form of rectangular holes punched in the card; once the primary data-output medium, it is now largely obsolete. Also known as card; punched card. { 'pənch ˌkärd }

punched card See punch card. { 'pəncht ˌkärd }

punch-through |ELECTR| An emitter-to-collector breakdown which can occur in a junction transistor with very narrow base region at sufficiently high collector voltage when the space-charge layer extends completely across the base region. { 'pənchˌthrü }

punctuation bit |COMPUT SCI| A binary digit used to indicate the beginning or end of a variable-length record. { ˌpəŋk·chə'wā·shən ˌbit }

puncture |ELEC| Disruptive discharge through insulation involving a sudden and large increase in current through the insulation due to complete failure under electrostatic stress. { 'pəŋk·chər }

puncture voltage |ELEC| The voltage at which a test specimen is electrically punctured. { 'pəŋk·chər ˌvōl·tij }

Pupin coil See loading coil. { pyü'pēn ˌkȯil }

pup jack See tip jack. { 'pəp ˌjak }

pure procedure |COMPUT SCI| A procedure that never modifies any part of itself during execution. { 'pyür prə¦sē·jər }

pure vanilla See vanilla. { 'pyür və'nil·ə }

purge |COMPUT SCI| To remove data from computer storage so that space occupied by the data can be reused. { pərj }

purge date |COMPUT SCI| The date after which data are released and the storage area can be used for storing other data. { 'pərj ˌdāt }

purify |COMPUT SCI| To remove errors from data. { 'pyür·əˌfī }

purity coil |ELECTR| A coil mounted on the neck of a color picture tube, used to produce the magnetic field needed for adjusting color purity; the direct current through the coil is adjusted to a value that makes the magnetic field orient the three individual electron beams so each strikes only its assigned color of phosphor dots. { 'pyür·əd·ē ˌkȯil }

purity control |ELECTR| A potentiometer or rheostat used to adjust the direct current through the purity coil. { 'pyür·əd·ē kənˌtrōl }

purity magnet |ELECTR| An adjustable arrangement of one or more permanent magnets used in place of a purity coil in a color cathode ray. { 'pyür·əd·ē ˌmag·nət }

purple plague |ELECTR| A compound formed by intimate contact of gold and aluminum, which appears on silicon planar devices and integrated circuits using gold leads bonded to aluminum thin-film contacts and interconnections, and which seriously degrades the reliability of semiconductor devices. { 'pər·pəl 'plāg }

push |COMPUT SCI| To add an item to a stack. { push }

push button |COMPUT SCI| A small area delineated on a graphical user interface whose selection by the user instructs the computer to perform a specific task. { 'push ˌbət·ən }

push-button dialing |ELECTR| Dialing a number by pushing buttons on the telephone rather than turning a circular wheel; each depressed button causes an oscillator to oscillate simultaneously at two different frequencies, generating a pair of audio tones which are recognized by central-office (or PBX) switching equipment as digits of a telephone number. Also known as dual-tone multifrequency dialing; tone dialing; touch call. { 'push ¦bət·ən 'dī·liŋ }

push-button switch |ELEC| A master switch that is operated by finger pressure on the end of an operating button. { 'push ¦bət·ən 'swich }

push-button tuner |ELECTR| A device that automatically tunes a radio receiver or other piece of equipment to a desired frequency when the button assigned to that frequency is pressed. { 'push ¦bət·ən 'tün·ər }

push-down automaton [COMPUT SCI] A nondeterministic, finite automaton with an auxiliary tape having the form of a push-down storage. { 'púsh ¦daún ȯ'täm·ə¸tän }

push-down list [COMPUT SCI] An ordered set of data items so constructed that the next item to be retrieved is the item most recently stored; in other words, last-in, first-out (LIFO). { 'púsh ¸daún ¸list }

push-down storage [COMPUT SCI] A computer storage in which each new item is placed in the first location in the storage and all the other items are moved back one location; it thus follows the principle of a push-down list. Also known as cellar; nesting storage; running accumulator. { 'púsh¸daún ¸stȯr·ij }

pushing [COMPUT SCI] The placing of a data element at the top of a stack. { 'púsh·iŋ }

push-pull amplifier [ELECTR] A balanced amplifier employing two similar electron tubes or equivalent amplifying devices working in phase opposition. { 'púsh ¦púl 'am·plə¸fī·ər }

push-pull currents See balanced currents. { 'púsh ¦púl 'kə·rəns }

push-pull electret transducer [ELECTR] A type of transducer in which a foil electret is sandwiched between two electrodes and is specially treated or arranged so that the electrodes exert forces in opposite directions on the diaphragm, and the net force is a linear function of the applied voltage. { 'púsh ¦púl i'lek·trət tranz'dü·sər }

push-pull magnetic amplifier [ELECTR] A realization of a push-pull amplifier using magnetic amplifiers. { 'púsh ¦púl mag'ned·ik 'am·plə¸fī·ər }

push-pull oscillator [ELECTR] A balanced oscillator employing two similar electron tubes or equivalent amplifying devices in phase opposition. { 'púsh ¦púl 'äs·ə¸lād·ər }

push-pull transformer [ELECTR] An audio-frequency transformer having a center-tapped winding and designed for use in a push-pull amplifier. { 'púsh ¦púl tranz'fór·mər }

push-pull transistor [ELECTR] **1.** A realization of a push-pull amplifier using transistors. **2.** A Darlington circuit in which the two transistors required for a push-pull amplifier exist in a single substrate. { 'púsh ¦púl tran'zis·tər }

push-pull voltages See balanced voltages. { 'púsh ¦púl 'vōl·tij·əz }

push-push amplifier [ELECTR] An amplifier employing two similar electron tubes with grids connected in phase opposition and with anodes connected in parallel to a common load; usually used as a frequency multiplier to emphasize even-order harmonics; transistors may be used in place of tubes. { 'púsh ¦púsh 'am·plə¸fī·ər }

push-to-talk circuit [ELEC] Simplex circuit in which changeover from the receive to transmit state is accomplished by depressing a single spring-return switch, and releasing the switch returns the circuit to the receive state; the push-to-talk switch is located on microphones and telephone handsets; it is most often applied to radio circuits. { 'púsh tə 'tȯlk ¸sər·kət }

push-up list [COMPUT SCI] An ordered set of data items so constructed that the next item to be retrieved will be the item that was inserted earliest in the list, resulting in a first-in, first-out (FIFO) structure. { 'púsh¸əp ¸list }

put [COMPUT SCI] A programming instruction that causes data to be written from computer storage into a file. { pút }

pyrometer [ENG] Any of a broad class of temperature-measuring devices; they were originally designed to measure high temperatures, but some are now used in any temperature range; includes radiation pyrometers, thermocouples, resistance pyrometers, and thermistors. { pī 'räm·əd·ər }

pyrone detector [ELECTR] Crystal detector in which rectification occurs between iron pyrites and copper or other metallic points. { 'pī¸rōn di¸tek·tər }

Q

Q |PHYS| A measure of the ability of a system with periodic behavior to store energy equal to 2π times the average energy stored in the system divided by the energy dissipated per cycle. Also known as Q factor; quality factor; storage factor.

QAM See quadrature amplitude modulation.

QBE See query by example.

Q factor See Q. { 'kyü ,fak·tər }

Q meter |ENG| A direct-reading instrument which measures the Q of an electric circuit at radio frequencies by determining the ratio of inductance to resistance, and which has also been developed to measure many other quantities. Also known as quality-factor meter. { 'kyü ,mēd·ər }

Q multiplier |ELECTR| A filter that gives a sharp response peak or a deep rejection notch at a particular frequency, equivalent to boosting the Q of a tuned circuit at that frequency. { 'kyü 'məl·tə,plī·ər }

Q point See quiescent operating point. { 'kyü ,póint }

QPSK See quadrature phase-shift keying.

Q signal |COMMUN| A three-letter abbreviation starting with Q, used in the International List of Abbreviations for radiotelegraphy to represent complete sentences. |ELECTR| The quadrature component of the chrominance signal in analog color television, having a bandwidth of 0 to 0.5 megahertz; it consists of +0.48(R-Y) and +0.41(B-Y), where Y is the luminance signal, R is the red camera signal, and B is the blue camera signal. { 'kyü ,sig·nəl }

quad |ELEC| A series of four separately insulated conductors, generally twisted together in pairs. |ELECTR| A series-parallel combination of transistors; used to obtain increased reliability through double redundancy, because the failure of one transistor will not disable the entire circuit. { kwäd }

quadded cable |ELEC| Cable in which at least some of the conductors are arranged in the form of quads. { 'kwäd·əd 'kā·bəl }

quadded redundancy |COMPUT SCI| A form of redundancy in which each logic gate is quadruplicated, and the outputs of one stage are interconnected to the inputs of the succeeding stage by a connection pattern so that errors made in earlier stages are overridden in later stages, where the original correct signals are restored. { 'kwäd·əd ri'dən·dən·sē }

quad density |COMPUT SCI| A format for floppy-disk storage that holds four times as much data as would normally be contained. { 'kwäd 'den·səd·ē }

quad in-line |ELECTR| An integrated-circuit package that has two rows of staggered pins on each side, spaced closely enough together to permit 48 or more pins per package. Abbreviated QUIL. { 'kwäd ,in'līn }

quadraphonic sound system |ENG ACOUS| A system for reproducing sound by means of four loudspeakers properly situated in the listening room, usually at the four corners of a square, with each loudspeaker being fed its own identifiable segment of the program signal. Also known as four-channel sound system. { ¦kwä·drə¦fän·ik 'saúnd }

quadrature amplifier |ELECTR| An amplifier that shifts the phase of a signal 90°; used in an analog color television receiver to amplify the 3.58-megahertz chrominance subcarrier and shift its phase 90° for use in the Q demodulator. { 'kwä·drə·chər ,am·plə,fī·ər }

quadrature amplitude modulation |COMMUN| 1. Quadrature modulation in which the two carrier components are amplitude-modulated. 2. A digital modulation technique in which digital information is encoded in bit sequences of specified length and these bit sequences are represented by discrete amplitude levels of an analog carrier, by a phase shift of the analog carrier from the phase that represented the previous bit sequence by a multiple of 90°, or by both. 3. Abbreviated QAM. { ¦kwäd·rə·chər ,am·plə,tüd ,mäj·ə'lā·shən }

quadrature component |ELEC| A vector representing an alternating quantity which is in quadrature (at 90°) with some reference vector. See reactive component. { 'kwä·drə·chər kəm ,pō·nənt }

quadrature current See reactive current. { 'kwä·drə·chər ,kə·rənt }

quadrature modulation |COMMUN| Modulation of two carrier components 90° apart in phase by separate modulating functions. { 'kwä·drə·chər ,mäj·ə'lā·shən }

quadrature partial-response keying |COMMUN| A modulation technique in which two orthogonally phased carriers are combined; each carrier is modulated by one of the digital bit streams to one of three levels. Abbreviated QPRK. { 'kwä·drə·chər ¦pär·shəl ri'späns ‚kē·iŋ }

quadrature phase-shift keying [COMMUN] Phase-shift keying in which four different phase angles are used, usually spaced 90° apart. Abbreviated QPSK. Also known as quadriphase; quaternary phase-shift keying. { ¦kwäd·rə·chər 'fāz‚shift ‚kē·iŋ }

quadriphase See quadrature phase-shift keying. { 'kwäd·rə‚fāz }

quadruplex circuit [ELEC] Telegraph circuit designed to carry two messages in each direction at the same time. { 'kwä·drə‚pleks ‚sər·kət }

quadrupole amplifier [ELECTR] A low-noise parametric amplifier consisting of an electron-beam tube in which quadrupole fields act on the fast cyclotron wave of the electron beam to produce high amplification at frequencies in the range of 400-800 megahertz. { 'kwä·drə‚pōl 'am·plə ‚fī·ər }

quad word [COMPUT SCI] A word 16 bytes long. { 'kwäd ‚wərd }

qualified name [COMPUT SCI] A name that is further identified by associating it with additional names, usually the names of things that contain the thing being named. { 'kwäl·ə‚fīd ¦nām }

qualifier [COMPUT SCI] A name that is associated with another name to give additional information about the latter and distinguish it from other things having the same name. { 'kwäl·ə‚fī·ər }

quality factor See Q. { 'kwäl·əd·ē ‚fak·tər }

quality-factor meter See Q meter. { 'kwäl·əd·ē ‚fak·tər ‚mēd·ər }

quality program [COMPUT SCI] A computer program that is correct, reliable, efficient, maintainable, flexible, testable, portable, and reusable. { ¦kwäl·əd·ē 'prō·grəm }

quantity [COMPUT SCI] In computers, a positive or negative real number in the mathematical sense; the term quantity is preferred to the term number in referring to numerical data; the term number is used in the sense of natural number and reserved for "the number of digits," the "number of operations," and so forth. { 'kwän·əd·ē }

quantity of electricity See charge. { 'kwän·əd·ē əv ‚i‚lek'tris·əd·ē }

quantization [COMMUN] Division of the range of values of a wave into a finite number of subranges, each of which is represented by an assigned or quantized value within the subrange. { ‚kwän·tə'zā·shən }

quantization distortion [COMMUN] Inherent distortion introduced in the process of quantization of a waveform. Also known as quantization noise; quantumization distortion; quantumization noise. { ‚kwän·tə'zā·shən di‚stòr·shən }

quantization level [COMMUN] Discrete value of the output designating a particular subrange of the input. { ‚kwän·tə'zā·shən ‚lev·əl }

quantization noise See quantization distortion. { ‚kwän·tə'zā·shən ‚nòiz }

quantized electronic structure [ELECTR] A material that confines electrons in such a small space that their wave-like behavior becomes important

and their properties are strongly modified by quantum-mechanical effects. { ¦kwän‚tīzd i·lek· ¦trän·ik 'strək·chər }

quantized frequency modulation [COMMUN] Frequency modulation that involves quantization; it uses time and frequency redundancy within a voice frequency channel during each transmitted symbol; used to combat distortion due to multipath, selection fading, and noise spikes. { 'kwän‚tīzd 'frē·kwən·sē ‚mäj·ə‚lā·shən }

quantized pulse modulation [COMMUN] Pulse modulation that involves quantization, such as pulse-numbers modulation and pulse-code modulation. { 'kwän‚tīzd 'pəls ‚mäj·ə‚lā·shən }

quantizer [COMMUN] A processing step that intentionally reduces the precision of discrete cosine transform coefficients. [ELECTR] A device that measures the magnitude of a time-varying quantity in multiples of some fixed unit, at a specified instant or specified repetition rate, and delivers a proportional response that is usually in pulse code or digital form. { 'kwän'tīz·ər }

quantum [COMMUN] One of the subranges of possible values of a wave which is specified by quantization and represented by a particular value within the subrange. { 'kwän·təm }

quantum computer [COMPUT SCI] A computer in which the time evolution of the state of the individual switching elements of the computer is governed by the laws of quantum mechanics. { 'kwän·təm kəm¦pyüd·ər }

quantum dot [ELECTR] A quantized electronic structure in which electrons are confined with respect to motion in all three dimensions. { ‚kwänt·əm 'dät }

quantum efficiency [ELECTR] The average number of electrons photoelectrically emitted from a photocathode per incident photon of a given wavelength in a phototube. { 'kwän·təm i‚fish·ən·sē }

quantum electronics [ELECTR] The branch of electronics associated with the various energy states of matter, motions within atoms or groups of atoms, and various phenomena in crystals; examples of practical applications include the atomic hydrogen maser and the cesium atomic-beam resonator. { 'kwän·təm ‚i‚lek'trän·iks }

quantum Hall effect [ELECTR] A phenomenon exhibited by certain semiconductor devices at low temperatures and high magnetic fields, whereby the Hall resistance becomes precisely equal to $(h/e^2)/n$, where h is Planck's constant, e is the electronic charge, and n is either an integer or a rational fraction. Also known as von Klitzing effect. { 'kwän·təm 'hòl i‚fekt }

quantumization distortion See quantization distortion. { ‚kwän·tə·mə'zā·shən di‚stòr·shən }

quantumization noise See quantization distortion. { ‚kwän·tə·mə'zā·shən ‚nòiz }

quantum well [ELECTR] A thin layer of material (typically between 1 and 10 nanometers thick) within which the potential energy of an electron is less than outside the layer, so that the motion of the electron perpendicular to the layer is quantized. { ¦kwän·təm 'wel }

quantum well infrared photodetector [ELECTR] A detector of infrared radiation composed of numerous alternating layers of controlled thickness of gallium arsenide and aluminum gallium arsenide; the spectral response of the device can be tailored within broad limits by adjusting the aluminum-to-gallium ratio and the thicknesses of the layers during growth. Abbreviated QWIP. { ¦kwänt·əm ¦wel ¦in·frə'red ¦fōd·ō·di'tek·tər }

quantum well injection transit-time diode [ELECTR] An active microwave diode that employs resonant tunneling through a gallium arsenide quantum well located between two aluminum gallium arsenide barriers to inject electrons into an undoped gallium arsenide drift region. Abbreviated QWITT diode. { ¦kwän·təm ¦wel in ¦jek·shən ¦tranz·it ¦tīm 'dī̇,ōd }

quantum well injection transit-time diode [ELECTR] An active microwave diode that employs resonant tunneling through a gallium arsenide quantum well located between two aluminum gallium arsenide barriers to inject electrons into an undoped gallium arsenide drift region. Abbreviated QWITT diode. { ¦kwän·təm ¦wel in,jek·shən ¦tranz·it ¦tīm 'dī̇,ōd }

quantum wire [ELECTR] A strip of conducting material about 10 nanometers or less in width and thickness that displays quantum-mechanical effects such as the Aharanov-Bohm effect and universal conductance fluctuations. { 'kwän·təm 'wīr }

quarternary phase-shift keying [ELECTR] Modulation of a microwave carrier with two parallel streams of nonreturn-to-zero data in such a way that the data is transmitted as 90° phase shifts of the carrier; this gives twice the message channel capacity of binary phase-shift keying in the same bandwidth. Abbreviated QPSK. { 'kwät·ə,ner·ē 'fāz ,shift ,kē·iŋ }

quarter-square multiplier [COMPUT SCI] A device used to carry out function multiplication in an analog computer by implementing the algebraic identity $xy = \frac{1}{4}[(x + y)^2 - (x - y)^2]$. { 'kwȯrd·ər ,skwer 'məl·tə,plī·ər }

quarter-wave [ELECTROMAG] Having an electrical length of one quarter-wavelength. { 'kwȯrd·ər ,wāv }

quarter-wave antenna [ELECTROMAG] An antenna whose electrical length is equal to one quarter-wavelength of the signal to be transmitted or received. { 'kwȯrd·ər ,wāv an'ten·ə }

quarter-wave attenuator [ELECTROMAG] Arrangement of two wire gratings, spaced an odd number of quarter-wavelengths apart in a waveguide, used to attenuate waves traveling through in one direction. { 'kwȯrd·ər ,wāv ə'ten·yə,wād·ər }

quarter-wave line See quarter-wave stub. { 'kwȯrd·ər ,wāv ,līn }

quarter-wave matching section See quarter-wave transformer. { 'kwȯrd·ər ,wāv 'mach·iŋ ,sek·shən }

quarter-wave stub [ELECTROMAG] A section of transmission line that is one quarter-wavelength long at the fundamental frequency being transmitted; when shorted at the far end, it has a high

impedance at the fundamental frequency and all odd harmonics, and a low impedance for all even harmonics. Also known as quarter-wave line; quarter-wave transmission line. { 'kwȯrd·ər ,wāv ¦stəb }

quarter-wave termination [ELECTROMAG] Metal plate and a wire grating spaced about one-fourth of a wavelength apart in a waveguide, with the plate serving as the termination of the guide; waves reflected from the metal plate are canceled by waves reflected from the grating so that all energy is absorbed (none is reflected) by the quarter-wave termination. { 'kwȯrd·ər ,wāv tər·mə'nā·shən }

quarter-wave transformer [ELECTROMAG] A section of transmission line approximately one quarter-wavelength long, used for matching a transmission line to an antenna or load. Also known as quarter-wave matching section. { 'kwȯrd·ər ,wāv tranz'fȯr·mər }

quarter-wave transmission line See quarter-wave stub. { 'kwȯrd·ər ,wāv tranz'mish·ən ,līn }

quartz crystal [ELECTR] A natural or artificially grown piezoelectric crystal composed of silicon dioxide, from which thin slabs or plates are carefully cut and ground to serve as a crystal plate. { 'kwȯrts ¦krist·əl }

quartz-crystal filter [ELECTR] A filter which utilizes a quartz crystal; it has a small bandwidth, a high rate of cutoff, and a higher unloaded Q than can be obtained in an ordinary resonator. { 'kwȯrts ,krist·əl 'fil·tər }

quartz-crystal resonator [ELECTR] A quartz plate whose natural frequency of vibration is used to control the frequency of an oscillator. Also known as quartz resonator. { 'kwȯrts ,krist·əl 'rez·ən ,äd·ər }

quartz delay line [ELECTR] An acoustic delay line in which quartz is used as the medium of sound transmission. { 'kwȯrts di'lā ,līn }

quartz-fiber electroscope [ELECTR] Electroscope in which a gold-plated quartz fiber serves the same function as the gold leaf of a conventional electroscope. { ¦kwȯrts ¦fī·bər i'lek·trə,skōp }

quartz-iodine lamp [ELECTR] An electric lamp having a tungsten filament and a quartz envelope filled with iodine vapor. { 'kwȯrts 'ī·ə,dīn ,lamp }

quartz lamp [ELECTR] A mercury-vapor lamp having a transparent envelope made from quartz instead of glass; quartz resists heat, permitting higher currents, and passes ultraviolet rays that are absorbed by ordinary glass. { 'kwȯrts ¦lamp }

quartz oscillator [ELECTR] An oscillator in which the frequency of the output is determined by the natural frequency of vibration of a quartz crystal. { 'kwȯrts 'äs·ə,lād·ər }

quartz plate See crystal plate. { 'kwȯrts ¦plāt }

quartz resonator See quartz-crystal resonator. { 'kwȯrts 'rez·ən,äd·ər }

quartz strain gage [ELECTR] A device used to measure small deformations of a substance by determining the resulting voltage that develops in a quartz attached to it. { 'kwȯrts 'strān ,gāj }

quasi-instruction [COMPUT SCI] An expression in a source program which resembles an instruction in form, but which does not have a corresponding machine instruction in the object program, and is directed to the assembler or compiler. Also known as pseudoinstruction. { ¦kwä·zē in'strak·shən }

quasi-linear feedback control system [CONT SYS] Feedback control system in which the relationships between the pertinent measures of the system input and output signals are substantially linear despite the existence of nonlinear elements. { ¦kwä·zē 'lin·ē·ər 'fēd,bak kən'trol ,sis·təm }

quasi-linear system [CONT SYS] A control system in which the relationships between the input and output signals are substantially linear despite the existence of nonlinear elements. { ¦kwä·zē 'lin·ē·ər 'sis·təm }

quasi-parallel execution [COMPUT SCI] The execution of a collection of coroutines by a single processor that can work on only one coroutine at a time; the order of execution is arbitrary and each coroutine is executed independently of the rest. { ¦kwä·zē 'par·ə,lel ,ek·sə'kyü·shən }

quasi-random code generator [COMMUN] High-speed coded information source used in the design and evaluation of wide-band communications links by providing a means of closed-loop testing. { ¦kwä·zē 'ran·dəm 'kōd ,jen·ə,rād·ər }

quasi-square-wave static inverter [ELEC] A static inverter that generates two square waves superimposed on one another to approximate an ac sine wave, using a silicon-controlled rectifier bridge and control circuit to control the pulse width and amplitude of the resulting wave, thereby achieving regulation. { ¦kwä·zē 'skwer ,wāv 'stad·ik in'vərd·ər }

quaternary phase-shift keying See quadrature phase-shift keying. { ¦kwät·ər,ner·ē 'fāz ,shift ,kē·iŋ }

quaternary signaling [COMMUN] An electrical communications mode in which information is passed by the presence and absence, or plus and minus variations, of four discrete levels of one parameter of the signaling medium. { 'kwät·ən ,er·ē 'sig·nə·liŋ }

qubit [COMPUT SCI] In quantum computation, a superposition of the ground state and the excited state of an elementary two-level quantum system (such as a two-level atom or a nuclear spin), corresponding to a classical bit that is either 0 (corresponding to the ground state) or 1 (corresponding to the excited state). { 'kyü·bit }

quenched spark gap [ELEC] A spark gap having provisions for rapid deionization; one form consists of many small gaps between electrodes that have relatively large mass and are good radiators of heat; the electrodes serve to cool the gaps rapidly and thereby stop conduction. { 'kwencht 'spärk ,gap }

quench frequency [ELECTR] Number of times per second that a circuit is caused to go in and out of oscillation. { 'kwench ,frē·kwən·sē }

quenching [ELECTR] **1.** The process of terminating a discharge in a gas-filled radiation-counter tube by inhibiting reignition. **2.** Reduction of

the intensity of resonance radiation resulting from deexcitation of atoms, which would otherwise have emitted this radiation, in collisions with electrons or other atoms in a gas. { 'kwench·iŋ }

quenching frequency [ELECTR] The frequency of an alternating voltage that is applied to a super-regenerative detector stage to prevent sustained oscillation. { 'kwench·iŋ ,frē·kwən·sē }

quench oscillator [ELECTR] Circuit in a superre-generative receiver which produces the frequency signal. { 'kwench ,äs·ə,lād·ər }

query [COMPUT SCI] A computer instruction to interrogate a database. { 'kwir·ē }

query by example [COMPUT SCI] A software product used to search a database for information having formats or ranges of values specified by English-like statements that indicate the desired results. Abbreviated QBE. { 'kwir·ē bī ig'zam·pəl }

query language [COMPUT SCI] A generalized computer language that is used to interrogate a database. { 'kwir·ē ,laŋ·gwij }

query layer [COMPUT SCI] A program that mediates between data sources on the World Wide Web and a user's query by breaking the query into subqueries against each information source and then gathering together the results for presentation to the user. { 'kwir·ē ,lā·ər }

query program [COMPUT SCI] A computer program that allows a user to retrieve information from a database and have it displayed on a terminal or printed out. { 'kwir·ē ,prō·grəm }

QUEST See quantized electronic structure. { kwest }

question-answering system [COMPUT SCI] An information retrieval system in which a direct answer is expected in response to a submitted query, rather than a set of references that may contain the answers. { 'kwes·chən 'an·sə·riŋ ,sis·təm }

queue [COMPUT SCI] **1.** A list of items waiting for attention in a computer system, generally ordered according to some criteria. **2.** A linear list whose elements are inserted and deleted in a first-in-first-out order. { kyü }

queued access method [COMPUT SCI] A set of predecures controlled by queues for efficient transfer of data between a computer and input-output devices. { 'kyüd 'ak,ses ,meth·əd }

queue-driven system [COMPUT SCI] A software system that uses many queues for tasks in various phases of processing. { 'kyü ¦driv·ən ,sis·təm }

queuing network model [COMPUT SCI] A model that represents a computer system by a network of devices through which customers (such as transactions, processes, or server requests) flow, and queues may form at each device due to its finite service rate. { 'kyü·iŋ ,net,wərk ,mäd·əl }

quibinary [COMPUT SCI] A numeration system, used in data processing, in which each decimal digit is represented by seven binary digits, a group of five which are coefficients of 8, 6, 4, 2, and 0, and a group of two which are coefficients of 1 and 0. { 'kwib·ə,ner·ē }

quick-break fuse [ELEC] A fuse designed to draw out the arc and break the circuit rapidly when the fuse wire melts, generally by separating the broken ends with a spring. { 'kwik ¦brāk 'fyüz }

quick-break switch [ELEC] A switch that breaks a circuit rapidly, independently of the rate at which the switch handle is moved, to minimize arcing. { 'kwik ¦brāk 'swich }

quick-make switch [ELEC] Switch or circuit breaker which has a high contact-closing speed, independent of the operator. { 'kwik ¦māk 'swich }

quiesce [COMPUT SCI] To prevent a computer system from starting new jobs so that the system gradually winds down as current jobs are completed, usually in preparation for a planned outage. { kwē'es }

quiescent [ELECTR] Pertaining to a circuit element which has no input signal, so that it does not perform its active function. { kwē'es·ənt }

quiescent-carrier telephony [COMMUN] A radiotelephony system in which the carrier is suppressed whenever there are no voice signals to be transmitted. { kwē'es·ənt ¦kar·ē·ər tə'lef·ə·nē }

quiescent operating point [ELECTR] The currents and voltages in an electronic circuit when the input signal is replaced by its average value, so that all currents and voltages can be approximated by series expansions around this point. Also known as Q point. { kwē,es·ənt 'äp·ə,rād·iŋ ,póint }

quiescent period [COMMUN] Resting period, or the period between pulse transmissions. { kwē'es·ənt ¦pir·ē·əd }

quiescent point [ELECTR] The point on the characteristic curve of an amplifier representing the conditions that exist when the input signal equals zero. { kwē'es·ənt ¦póint }

quiescent push-pull [ELECTR] Push-pull output stage so arranged in a radio receiver that practically no current flows when an input signal is not present.. { kwē'es·ənt ¦push ¦púl }

quiet automatic volume control See delayed automatic gain control. { 'kwī·ət ¦ód·ə¦mad·ik 'väl·yəm kən,trōl }

quiet battery [ELECTR] Source of energy of special design or with added filters which is sufficiently quiet and free from interference that it may be used for speech transmission. Also known as talking battery. { 'kwī·ət 'bad·ə·rē }

quieting sensitivity [ELECTR] Minimum signal input to a frequency-modulated receiver which is required to give a specified output signal-to-noise ratio under specified conditions. { 'kwī·əd·iŋ ,sen·sə,tiv·əd·ē }

quiet tuning [ELECTR] Circuit arrangement for silencing the output of a radio receiver, except when it is accurately tuned to an incoming carrier wave. { 'kwī·ət 'tün·iŋ }

QWITT diode See quantum well injection transit-time diode. { ¦kyü¦dəb·əl,yü¦ī¦tē¦tē 'dī,ōd }

465

R

race condition [ELEC] An ambiguous condition occurring in control counters when one flip-flop changes to its next state before a second one has had sufficient time to latch. { 'rās kən‚dish·ən }

raceway [ELEC] A channel used to hold and protect wires, cables, or busbars. Also known as electric raceway. { 'rās‚wā }

rack panel [ELECTR] A panel designed for mounting on a relay rack; its width is 19 inches (48.26 centimeters), height is a multiple of 1 ¾ inches (4.445 centimeters), and the mounting notches are standardized as to size and position. { 'rak ‚pan·əl }

racon See radar beacon. { 'rā‚kän }

radar [ENG] A system using beamed and reflected radio-frequency energy for detecting, locating, and examining objects, measuring distance or altitude, assisting in navigation, military operations, air traffic management, and weather appraisal, and many other military and civil purposes. Timing of the return of reflected energy and examination of its nature are fundamental to all radar applications. Derived from radio detection and ranging. { 'rā‚där }

radar altimeter [NAV] A radio altimeter, useful at altitudes much greater than the 5000-foot (1500-meter) limit of frequency-modulated radio altimeters, in which simple pulse-type radar equipment is used to send a pulse straight down from an aircraft and to measure its total time of travel to the surface and back to the aircraft. Also known as high-altitude radio altimeter; pulse-type altimeter. { 'rā‚där al'tim·əd·ər }

radar antenna [ELECTROMAG] A device which radiates radio-frequency energy in a radar system, concentrating the transmitted power in the direction of the target, and which provides a large area to collect the echo power of the returning wave. { 'rā‚där an'ten·ə }

radar antijamming [ELECTR] Measures taken to counteract radar jamming (electronic attack). { 'rā‚där ‚ant·i'jam·iŋ }

radar attenuation [ELECTROMAG] Ratio of the power delivered by the transmitter to the transmission line connecting it with the transmitting antenna, to the power reflected from the target which is delivered to the receiver by the transmission line connecting it with the receiving antenna. { 'rā‚där ə‚ten·yə'wā·shən }

radar beacon [NAV] A radar receiver-transmitter that transmits a strong coded radar signal whenever its radar receiver is triggered by an interrogating radar on an aircraft or ship; the coded beacon reply can be used by the navigator to determine his own position in terms of bearing and range from the beacon. Also known as racon; radar transponder. { 'rā‚där ‚bē·kən }

radar beam [ELECTROMAG] The movable beam of radio-frequency energy produced by a radar transmitting antenna; its shape is commonly defined as the loci of all points at which the power has decreased to one-half of that at the center of the beam. { 'rā‚där ‚bēm }

radar cell [ELECTROMAG] Volume whose dimensions are one radar pulse length by one radar beam width. { 'rā‚där ‚sel }

radar clutter See clutter. { 'rā‚där ‚kləd·ər }

radar command guidance [ENG] A missile guidance system in which radar equipment at the launching site determines the positions of both target and missile continuously, computes the missile course corrections required, and transmits these by radio to the missile as commands. { 'rā‚där kə'mand ‚gīd·əns }

radar constant [ELECTR] The product of the factors of radar performance equation that describe characteristics of the particular radar to which the equations are applied; these include peak power, antenna gain or aperture, beam width, pulse length, pulse repetition frequency, wavelength, polarization, and noise level of the receiver. { 'rā‚där ‚kän·stənt }

radar contact [ENG] Recognition and identification of an echo on a radar screen; an aircraft is said to be on radar contact when its radar echo can be seen and identified on a PPI (plan-position indicator) display. { 'rā‚där ‚kän‚takt }

radar control [ELECTR] Guidance, direction, or employment exercised over an aircraft, guided missile, gun battery, or the like, by means of, or with the aid of, radar. { 'rā‚där kən‚trōl }

radar control and interface apparatus [ELECTR] That subsystem of a radar that acts on the output of the receiver to provide significant reports to the system using that radar and also to control the radar in ways appropriate to the situation; constituted of a human operator and visual display in elementary radar, and of computer operations and data displays for human management in more modern radar. { 'rā,där kən,trōl ənd in·tər ,fās ,ap·ə,rad·əs }

radar countermeasure [ELECTR] Electronic and electromagnetic actions used against enemy radar, such as jamming and confusion reflectors. Abbreviated RCM. { 'rā,där 'kaunt·ər,mezh·ər }

radar cross section [ELECTROMAG] In representing a radar target, a convenient expression of the incident-signal intercept area that, if the intercepted signal were reradiated isotropically, would return to the radar the same signal strength as the target actually does. { 'rā,där 'krós ,sek·shən }

radar data filtering [ELECTR] Quality analysis process that causes the computer to reject certain radar data and to alert personnel of mapping and surveillance consoles to the rejection. { 'rā ,där 'dad·ə ,fil·triŋ }

radar display [ELECTR] Visual presentation of the output of a radar receiver produced either on the screen of a cathode-ray tube or in computer-generated displays of symbols and notations based on that output in more automated systems. Also known as radar presentation. { 'rā ,där di,splā }

radar display formats [ELECTR] Any of a variety of visual representations of radar receiver output to assist the operator in interpreting the data, managing the radar, and making reasonable reports to the user system. Many of the formats have been given letter names, such as the A-display (or A-scope), and so on; the PPI (plan position indicator), RHI (range-height indicator), A-scope, and B-scope are among the most frequently used. Also known as display formats. { 'rā,där di,splā ,fòr,matz }

radar distribution switchboard [ELECTR] Switching panel for connecting video, trigger, and bearing from any one of five systems, to any or all of 20 repeaters; also contains order lights, bearing cutouts, alarms, test equipment, and so forth. { 'rā,där ,dis·trə'byü·shən ,swich,bòrd }

radar echo See echo. { 'rā,där ,ek·ō }

radar equation [ELECTROMAG] An equation that relates the transmitted and received powers and antenna gains of a primary radar system to the echo area and distance of the radar target. { 'rā ,där i,kwā·zhən }

radar frequency band [ELECTROMAG] A frequency band of microwave radiation in which radar operates. { 'rā,där 'frē·kwən·sē ,band }

radar image [ELECTR] The image of an object, a vehicle or an entire scene, which is produced on a radar display or in an appropriate medium. { 'rā ,där ,im·ij }

radar indicator [ELECTR] A cathode-ray tube and associated equipment to provide a visual indication of the echo signals picked up by a radar set. { 'rā,där ,in·də,kād·ər }

radar intelligence item [ELECTR] A feature which is radar significant but which cannot be identified exactly at the moment of its appearance as homogeneous. { 'rā,där in¦tel·ə·jəns ,īd·əm }

radar jamming [ELECTR] Radiation, reradiation, or reflection of electromagnetic waves so as to impair the usefulness of radar used by the enemy. { 'rā,där ,jam·iŋ }

radar netting unit [ELECTR] Optional electronic equipment that converts the operations central of certain air defense fire distribution systems to a radar netting station. { 'rā,där¦ned·iŋ ,yü·nət }

radar presentation See radar display. { 'rā,där ,prē,zen'tā·shən }

radar range [ELECTROMAG] The maximum distance at which a radar set is ordinarily effective in detecting objects. { 'rā,där ,rānj }

radar range equation [ELECTROMAG] An equation which expresses radar range in terms of transmitted power, minimum detectable signal, antenna gain, and the target's radar cross section. { 'rā,där ,rānj i,kwā·zhən }

radar receiver [ELECTR] That subsystem of a radar that is designed to amplify, enhance as appropriate with signal processing, and demodulate radar echo signals and feed them to a radar display or similar data processer. { 'rā,där ri ,sēv·ər }

radar receiver-transmitter [ELECTR] A single component having the dual functions of generating electromagnetic energy for transmission, and of receiving, demodulating, and sometimes presenting intelligence from the reflected electromagnetic energy. { 'rā,där ri ,sēv·ər tranz'mid·ər }

radar reflection [ELECTROMAG] The return of electromagnetic waves, generated by a radar installation, from an object on which the waves are incident. { 'rā,där ri,flek·shən }

radar reflection interval [ELECTROMAG] The time required for a radar pulse to travel from the source to the target and return to the source, taking the velocity of radio propagation to be equal to the velocity of light. { 'rā,där ri,flek·shən ,in·tər·vəl }

radar reflectivity [ELECTROMAG] The fraction of electromagnetic energy generated by a radar installation which is reflected by an object. { 'rā ,där ,rē,flek'tiv·əd·ē }

radar relay [ENG] **1.** Equipment for relaying the radar video and appropriate synchronizing signal to a remote location. **2.** Process or system by which radar echoes and synchronization data are transmitted from a search radar installation to a receiver at a remote point. { 'rā,där 'rē,lā }

radar repeater [ELECTR] A radar indicator used to reproduce the radar's own display at a remote position; with proper selection, the display of any one of several radar systems can be reproduced. { 'rā,där ri,pēd·ər }

radar return [NAV] The signal indication of an object which has reflected energy that was transmitted by a primary radar. Also known as radio echo. { 'rā,där ri,tərn }

radar scanning |ENG| The process or action of directing a radar beam through a space search pattern for the purpose of locating a target. { 'rā ,där ,skan·iŋ }

radarscope |ELECTR| An older term for a radar display, connoting usually the use of a cathode-ray tube serving as an oscilloscope, the face of which is the radar viewing screen. Also known as scope. { 'rā,där,skōp }

radar selector switch |ELECTR| Manual or motor-driven switch which transfers a plan-position indicator repeater from one system to another, switching video, trigger, and bearing data. { 'rā ,där si'lek·tər ,swich }

radar set |ENG| A complete assembly of radar equipment, consisting of a transmitter, antenna, receiver, and signal processor, and appropriate control and interface apparatus. The term radar alone is often used. { 'rā,där ,set }

radar signal spectrograph |ELECTR| An electronic device in the form of a scanning filter which provides a frequency analysis of the amplitude-modulated back-scattered signal. { 'rā,där ¦sig·nəl 'spek·trə,graf }

radar transmitter |ELECTR| That subsystem of a radar that converts electrical power to the radio-frequency electromagnetic signals desired, then sends them to the antenna. { 'rā,där tranz,mid·ər }

radar transponder See radar beacon. { 'rā,där tranz'pän·dər }

radar volume |ELECTROMAG| The volume in space that is irradiated by a given radar; for a continuous-wave radar it is equivalent to the antenna radiation pattern; for a pulse radar it is a function of the cross-section area of the beam of the antenna and the pulse length of the transmitted pulse. { 'rā,där ,väl·yəm }

radechon |ELECTR| A storage tube having a single electron gun and a dielectric storage medium consisting of a sheet of mica sandwiched between a continuous metal backing plate and a fine-mesh screen; used in simple delay schemes, signal-to-noise improvement, signal comparison, and conversion of signal-time bases. Also known as barrier-grid storage tube. { 'rad·ə,kän }

radial-beam tube |ELECTR| A vacuum tube in which a radial beam of electrons is rotated past circumferentially arranged anodes by an external rotating magnetic field; used chiefly as a high-speed switching tube or commutator. { ¦rād·ē·əl ¦bēm ,tüb }

radial grating |ELECTROMAG| Conformal wire grating consisting of wires arranged radially in a circular frame, like the spokes of a wagon wheel, and placed inside a circular waveguide to obstruct E waves of zero order while passing the corresponding H waves. { 'rād·ē·əl 'grād·iŋ }

radial lead |ELEC| A wire lead coming from the side of a component rather than axially from the end. { 'rād·ē·əl 'lēd }

radial selector See private line arrangement. { 'rād·ē·əl si'lek·tər }

radiant reflectance |ELECTROMAG| Ratio of reflected radiant power to incident radiant power. { 'rād·ē·ənt ri'flek·təns }

radiant transmittance |ELECTROMAG| Ratio of transmitted radiant power to incident radiant power. { 'rād·ē·ənt tranz'mit·əns }

radiated interference |COMMUN| Interference which is transmitted through the atmosphere according to the laws of electromagnetic wave propagation; the term is generally considered to include the transfer of interfering energy in inductive or capacitive coupling. { 'rād·ē,ād·əd ,in·tər'fir·əns }

radiated power |ELECTROMAG| The total power emitted by a transmitting antenna. { 'rād·ē ,ād·əd 'pau·ər }

radiating curtain |ELECTROMAG| Array of dipoles in a vertical plane, positioned to reinforce each other; it is usually placed one-fourth wavelength ahead of a reflecting curtain of corresponding half-wave reflecting antennas. { 'rād·ē,ād·iŋ 'kərt·ən }

radiating element |ELECTROMAG| Basic subdivision of an antenna which in itself is capable of radiating or receiving radio-frequency energy. { 'rād·ē,ād·iŋ 'el·ə·mənt }

radiating guide |ELECTROMAG| Waveguide designed to radiate energy into free space; the waves may emerge through slots or gaps in the guide, or through horns inserted in the wall of the guide. { 'rād·ē,ād·iŋ 'gīd }

radiation angle |ELECTROMAG| The vertical angle between the line of radiation emitted by a directional antenna and the horizon. { ,rād·ē'ā·shən ,aŋ·gəl }

radiation characteristic |COMMUN| One of the identifying features of a radiating signal, such as frequency and pulse width. { ,rād·ē'ā·shən ,kar·ik·tə'ris·tik }

radiation cooling |ELECTR| Cooling of an electrode resulting from its emission of heat radiation. { ,rād·ē'ā·shən ,kül·iŋ }

radiation counter tube See counter tube. { ,rād·ē'ā·shən ¦kaunt·ər ,tüb }

radiation efficiency |ELECTROMAG| Of an antenna, the ratio of the power radiated to the total power supplied to the antenna at a given frequency. { ,rād·ē'ā·shən i,fish·ən·sē }

radiation-enhanced diffusion |ELEC| A mechanism for ion-beam mixing of a film and a substrate in which lattice defects that are formed by the atomic displacements produced by ion bombardment result in an increase in interdiffusion coefficients. { ,rād·ē¦ā·shən in,hanst də'fyü·zhən }

radiation field |ELECTROMAG| The electromagnetic field that breaks away from a transmitting antenna and radiates outward into space as electromagnetic waves; the other type of electromagnetic field associated with an energized antenna is the induction field. { ,rād·ē'ā·shən ,fēld }

radiation intensity |ELECTROMAG| The power radiated from an antenna per unit solid angle in a given direction. { ,rād·ē'ā·shən in,ten·səd·ē }

radiation lobe See lobe. { ,rād·ē'ā·shən ,lōb }

radiation noise See electromagnetic noise. { ,rād·ē'ā·shən ,nȯiz }

radiation pattern |ELECTROMAG| Directional dependence of the radiation of an antenna. Also known as antenna pattern; directional pattern; field pattern. { ˌrād·ē′ā·shən ˌpad·ərn }

radiation thermocouple |ELEC| An infrared detector consisting of several thermocouples connected in series, arranged so that the radiation falls on half of the junctions, causing their temperature to increase so that a voltage is generated. { ˌrād·ē′ā·shən ′thər·mə‚kəp·əl }

radiation zone See Fraunhofer region. { ˌrād·ē′ā·shən ‚zōn }

radiator |ELECTROMAG| **1.** The part of an antenna or transmission line that radiates electromagnetic waves either directly into space or against a reflector for focusing or directing. **2.** A body that emits radiant energy. { ′rād·ē‚ād·ər }

radio- |ELECTROMAG| A prefix denoting the use of radiant energy, particularly radio waves. { ′rād·ē·ō }

radio |COMMUN| The transmission of signals through space by means of electromagnetic waves. |ELECTR| See radio receiver. { ′rād·ē·ō }

radioacoustics |COMMUN| Study of the production, transmission, and reproduction of sounds carried from one place to another by radiotelephony. { ¦rād·ē·ō·ə′küs·tiks }

radioactive fallout See fallout. { ¦rād·ē·ō′ak·tiv ′fȯl‚aut }

radio aid to navigation |ELECTR| An aid to navigation which utilizes the propagation characteristics of radio waves to furnish navigation information. { ′rād·ē·ō ′ād tə ‚nav·ə′gā·shən }

radio altimeter |ENG| An absolute altimeter that depends on the reflection of radio waves from the earth for the determination of altitude, as in a frequency-modulated radio altimeter and a radar altimeter. Also known as electronic altimeter; reflection altimeter. { ′rād·ē·ō al′tim·əd·ər }

radio altitude See radar altitude. { ′rād·ē·ō ′al·tə ‚tüd }

radio and wire integration |COMMUN| The combining of wire circuits with radio facilities. { ′rād·ē·ō ən ′wīr ‚int·ə′grā·shən }

radio antenna See antenna. { ′rād·ē·ō an′ten·ə }

radio attenuation |ELECTROMAG| For one-way propagation, the ratio of the power delivered by the transmitter to the transmission line connecting it with the transmitting antenna to the power delivered to the receiver by the transmission line connecting it with the receiving antenna. { ′rād·ē·ō ə‚ten·yə′wā·shən }

radio aurora See artificial radio aurora. { ′rād·ē·ō ə′rȯr·ə }

radio autopilot coupler |ENG| Equipment providing means by which an electrical navigational signal operates an automatic pilot. { ′rād·ē·ō ′ȯd·ō‚pī·lət ′kəp·lər }

radio B battery |ELEC| A B-type battery used in a radio set, usually consisting of 15 to 30 permanently connected cells. { ′rād·ē·ō ′bē ‚bad·ə·rē }

radio beacon |NAV| A nondirectional radio transmitting station in a fixed geographic location, emitting a characteristic signal from which bearing information can be obtained by a radio direction finder on a ship or aircraft. Also known as aerophare; radiophare. { ′rād·ē·ō ′bē· kən }

radio bearing |NAV| The bearing of a radio transmitter from a receiver as determined by a radio direction finder. { ′rād·ē·ō ‚ber·iŋ }

radio blackout |COMMUN| A fadeout that may last several hours or more at a particular frequency. Also known as blackout. { ′rād·ē·ō ′blak ‚aut }

radio broadcasting |COMMUN| Radio transmission intended for general reception. { ′rād·ē·ō ′brȯd‚kast·iŋ }

radio button |COMPUT SCI| In a graphical user interface, one of a group of small circles that represent a set of choices (indicated by text next to the circles) from which only one can be selected; the selected choice is indicated by a partly filled circle. { ′rād·ē·ō ‚bət·ən }

radio command |ELECTR| A radio control signal to which a guided missile or other remote-controlled vehicle or device responds. { ′rād·ē·ō kə‚mand }

radio communication |COMMUN| Communication by means of radio waves. { ′rād·ē·ō kə‚myü· nə′kā·shən }

radiocommunication service |COMMUN| A service involving the emission, transmission, or reception of radio waves for specific telecommunications purposes. { ˌrād·ē·ō·kə‚myü·nə′kā·shən ‚sər·vəs }

radio compass See automatic direction finder. { ′rād·ē·ō ′käm·pəs }

radio control |ELECTR| The control of stationary or moving objects by means of signals transmitted through space by radio. { ′rād·ē·ō kən′trōl }

radio countermeasures |ELECTR| Electrical or other techniques depriving the enemy of the benefits which would ordinarily accrue to him through the use of any technique employing the radiation of radio waves; it includes benefits derived from radar and intercept services. { ′rād·ē·ō ′kaunt·ər‚mezh·ərz }

radio data system |COMMUN| The radio data system (RDS) signal is a low-bit-rate data stream transmitted on the 57-kHz subcarrier of an FM radio signal. Radio listeners know that radio data system through its ability to permit RDS radios to display call letters and search for stations based on their programming format. Special traffic announcements can be transmitted to RDS radios, as well as emergency alerts. { ′rād·ē·ō ′dad·ə ‚sis·təm }

radio detection and ranging See radar. { ′rād·ē·ō di′tek·shən ən ′rānj·iŋ }

radiodetermination satellite service |COMMUN| A system that employs at least two geosynchronous satellites, a central ground station, and hand-held or vehicle-mounted transceivers to enable users to determine and transmit their precise position. Abbreviated RDSS. { ′rād·ē·ōdi ‚tər·mə′nā·shən ′sad·əl‚īt ‚sər·vəs }

radio direction finder |NAV| A radio aid to navigation that uses a rotatable loop or other highly

470

directional antenna arrangement to determine the direction of arrival of a radio signal. Abbreviated RDF. Also known as direction finder. { 'rād·ē·ō di'rek·shən ˌfīn·dər }

radio echo *See* radar return. { 'rād·ē·ō ˌek·ō }

radio facsimile system |COMMUN| A facsimile system in which signals are transmitted by radio rather than by wire. { 'rād·ē·ō fak'sim·ə·lē ˌsis·təm }

radio fadeout |COMMUN| Increased absorption of radio waves passing through the lower layers of the ionosphere due to a sudden and abnormal increase in ionization in these regions; signals at receivers then fade out or disappear. { 'rād·ē·ō 'fād,aüt }

radio fan-marker beacon *See* fan-marker beacon. { 'rād·ē·ō 'fan ˌmär·kər ˌbē·kən }

radio fix |COMMUN| Determination of the position of the source of radio signals by obtaining cross bearings on the transmitter with two or more radio direction finders in different locations, then computing the position by triangulation. |NAV| **1.** Determination of the position of a vessel or aircraft equipped with direction-finding equipment by ascertaining the direction of radio signals received from two or more transmitting stations of known location and then computing the position by triangulation. **2.** Determination of position of an aircraft in flight by identification of a radio beacon or by locating the intersection of two radio beams. { 'rād·ē·ō ˌfiks }

radio-frequency alternator |ELEC| A rotating-type alternator designed to produce high power at frequencies above power-line values but generally lower than 100,000 hertz; used chiefly for high-frequency heating. { 'rād·ē·ō ¦frē·kwən·sē 'ȯl·tə,nād·ər }

radio-frequency amplifier |ELECTR| An amplifier that amplifies the high-frequency signals commonly used in radio communications. { 'rād·ē·ō ¦frē·kwən·sē 'am·plə,fī·ər }

radio-frequency bandwidth |COMMUN| Band of frequencies comprising 99% of the total radiated power of the signal transmission extended to include any discrete frequency on which the power is at least 0.25% of the total radiated power. { 'rād·ē·ō ¦frē·kwən·sē 'band,width }

radio-frequency cable |ELECTROMAG| A cable having electric conductors separated from each other by a continuous homogeneous dielectric or by touching or interlocking spacer beads; designed primarily to conduct radio-frequency energy with low losses. Also known as RG line. { 'rād·ē·ō ¦frē·kwən·sē ˌkā·bəl }

radio-frequency choke |ELEC| A coil designed and used specifically to block the flow of radio-frequency current while passing lower frequencies or direct current. { 'rād·ē·ō ¦frē·kwən·sē ˌchōk }

radio-frequency component |COMMUN| Portion of a signal or wave which consists only of the radio-frequency alternations, and not including its audio rate of change in amplitude frequency. { 'rād·ē·ō ¦frē·kwən·sē kəm,pō·nənt }

radio-frequency current |ELEC| Alternating current having a frequency higher than 10,000 hertz. { 'rād·ē·ō ¦frē·kwən·sē ˌkə·rənt }

radio-frequency filter |ELECTR| An electric filter which enhances signals at certain radio frequencies or attenuates signals at undesired radio frequencies. { 'rād·ē·ō ¦frē·kwən·sē ˌfil·tər }

radio-frequency generator |ELECTR| A generator capable of supplying sufficient radio-frequency energy at the required frequency for induction or dielectric heating. { 'rād·ē·ō ¦frē·kwən·sē 'jen·ə ˌrād·ər }

radio-frequency head |ENG| Unit consisting of a radar transmitter and part of a radar receiver, the two contained in a package for ready removal and installation. { 'rād·ē·ō ¦frē·kwən·sē 'hed }

radio-frequency heating *See* electronic heating. { 'rād·ē·ō ¦frē·kwən·sē 'hēd·iŋ }

radio-frequency interference |COMMUN| Interference from sources of energy outside a system or systems, as contrasted to electromagnetic interference generated inside systems. Abbreviated RFI. { 'rād·ē·ō ¦frē·kwən·sē ˌin·tər'fir·əns }

radio-frequency measurement |ELECTR| The precise measurement of frequencies above the audible range by any of various techniques, such as a calibrated oscillator with some means of comparison with the unknown frequency, a digital counting or scaling device which measures the total number of events occurring during a given time interval, or an electronic circuit for producing a direct current proportional to the frequency of its input signal. { 'rād·ē·ō ¦frē·kwən·sē 'mezh·ər·mənt }

radio-frequency oscillator |ELECTR| An oscillator that generates alternating current at radio frequencies. { 'rād·ē·ō ¦frē·kwən·sē 'äs·ə,lād·ər }

radio-frequency power supply |ELECTR| A high-voltage power supply in which the output of a radio-frequency oscillator is stepped up by an air-core transformer to the high voltage required for the second anode of a cathode-ray tube, then rectified to provide the required high direct-current voltage; used in some television receivers. { 'rād·ē·ō ¦frē·kwən·sē 'paů·ər sə,plī }

radio-frequency pulse |COMMUN| A radio-frequency carrier that is amplitude-modulated by a pulse; the amplitude of the modulated carrier is zero before and after the pulse. Also known as radio pulse. { 'rād·ē·ō ¦frē·kwən·sē ˌpəls }

radio-frequency reactor |ELECTR| A reactor used in electronic circuits to pass direct current and offer high impedance at high frequencies. { 'rād·ē·ō ¦frē·kwən·sē rē'ak·tər }

radio-frequency resistance *See* high-frequency resistance. { 'rād·ē·ō ¦frē·kwən·sē ri'zis·təns }

radio-frequency sensor |ENG| A device that uses radio signals to determine the position of objects to be manipulated by a robotic system. { 'rād·ē·ō ¦frē·kwən·sē ˌsen·sər }

radio-frequency shift *See* frequency shift. { 'rād·ē·ō ¦frē·kwən·sē ˌshift }

radio-frequency signal generator |ELECTR| A test instrument that generates the various radio

radio-frequency spectrum

frequencies required for alignment and servicing of electronics equipment. Also known as service oscillator. { 'rād·ē·ō ¦frē·kwən·sē 'sig·nəl ‚jen·ə ‚rād·ər }

radio-frequency spectrum *See* radio spectrum. { 'rād·ē·ō ¦frē·kwən·sē 'spek·trəm }

radio-frequency SQUID [ELECTR] A type of SQUID which has only one Josephson junction in a superconducting loop; its state is determined from radio-frequency measurements of the impedance of the ring. { 'rād·ē·ō ¦frē·kwən·sē 'skwid }

radiogoniometer [ELECTR] A goniometer used as part of a radio direction finder. { ¦rād·ē·ō ‚gō·nē'äm·əd·ər }

radiogoniometry [ENG] Science of locating a radio transmitter by means of taking bearings on the radio waves emitted by such a transmitter. { ¦rād·ē·ō‚gō·nē'äm·ə·trē }

radio guidance [ELECTR] Guidance of a flight-borne missile or other vehicle from a ground station by means of radio signals. { 'rād·ē·ō 'gīd·əns }

radio homing beacon *See* homing beacon. { 'rād·ē·ō 'hōm·iŋ ‚bē·kən }

radio horizon [COMMUN] The locus of points at which direct rays from a transmitter become tangential to the surface of the earth; the distance to the radio horizon is affected by atmospheric refraction. { 'rād·ē·ō hə'rīz·ən }

radio interference *See* interference. { 'rād·ē·ō ‚in·tər'fir·əns }

radio metal locator *See* metal detector. { 'rād·ē·ō 'med·əl 'lō‚kād·ər }

radiometer [ELECTR] A receiver for detecting microwave thermal radiation and similar weak wide-band signals that resemble noise and are obscured by receiver noise; examples include the Dicke radiometer, subtraction-type radiometer, and two-receiver radiometer. Also known as microwave radiometer; radiometer-type receiver. [ENG] An instrument for measuring radiant energy; examples include the bolometer, microradiometer, and thermopile. { ‚rād·ē'äm·əd·ər }

radiometer-type receiver *See* radiometer. { ‚rād·ē'äm·əd·ər ¦tīp ri'sē·vər }

radiomicrometer *See* microradiometer. { ¦rād·ē·ō·mī'kräm·əd·ər }

radio net [COMMUN] System of radio stations operating with each other; a military net usually consists of a radio station of a superior unit and stations of all subordinate or supporting units. { 'rād·ē·ō ‚net }

radio-paging system [COMMUN] A system consisting of personal paging receivers, radio transmitters, and an encoding device, designed to alert an individual, or group of individuals, and deliver a short message. { 'rād·ē·ō ¦pāj·iŋ ‚sis·təm }

radiophare *See* radio beacon. { 'rād·ē·ō‚fer }

radiophone *See* radiotelephone. { 'rād·ē·ō‚fōn }

radiophoto *See* facsimile. { ¦rād·ē·ō'fōd·ō }

radio pill [ELECTR] A device used in biotelemetry for monitoring the physiologic activity of an animal, such as pH values of stomach acid; an

example is the Heidelberg capsule. { 'rād·ē·ō ‚pil }

radio receiver [ELECTR] A device that converts radio waves into intelligible sounds or other perceptible signals. Also known as radio; radio set; receiving set. { 'rād·ē·ō ri‚sēv·ər }

radio relay satellite *See* communications satellite. { 'rād·ē·ō 'rē‚lā ‚sad·əl‚īt }

radio relay system [COMMUN] A radio transmission system in which intermediate radio stations or radio repeaters receive and retransmit radio signals. Also known as relay system. { 'rād·ē·ō 'rē‚lā ‚sis·təm }

radio repeater [COMMUN] A repeater that acts as an intermediate station in transmitting radio communications signals or radio programs from one fixed station to another; serves to extend the reliable range of the originating station; a microwave repeater is an example. { 'rād·ē·ō ri‚pēd·ər }

radio scanner *See* scanning radio. { 'rād·ē·ō 'skan·ər }

radio scattering *See* scattering. { 'rād·ē·ō 'skad·ə·riŋ }

radio set *See* radio transmitter. { 'rād·ē·ō ‚set }

radio shielding [ELEC] Metallic covering over all electric wiring and ignition apparatus, which is grounded at frequent intervals for the purpose of eliminating electric interference with radio communications. { 'rād·ē·ō ‚shēld·iŋ }

radio signal [COMMUN] A signal transmitted by radio. { 'rād·ē·ō ‚sig·nəl }

radio silence [COMMUN] Period during which all or certain radio equipment capable of radiation is kept inoperative. { 'rād·ē·ō 'sī·ləns }

radiosonde commutator [ELECTR] A component of a radiosonde consisting of a series of alternate electrically conducting and insulating strips; as these are scanned by a contact, the radiosonde transmits temperature and humidity signals alternately. { 'rād·ē·ō‚sänd 'käm·yə‚tād·ər }

radio spectrum [COMMUN] The entire range of frequencies in which useful radio waves can be produced, extending from the audio range to about 300,000 megahertz. Also known as radio-frequency spectrum. { 'rād·ē·ō 'spek·trəm }

radio spectrum allocation [COMMUN] The specification of the frequencies of the radio spectrum which are available for use by the various radio services. { 'rād·ē·ō ¦spek·trəm ‚al·ə'kā·shən }

radio station [COMMUN] A station equipped to engage in radio communication or radio broadcasting. { 'rād·ē·ō ‚stā·shən }

radiotelemetry [COMMUN] The reception of data at a location remote from the source of the data, using radio-frequency electromagnetic radiation as the means of transmission. { 'rād·ē·ō tə'lem·ə·trē }

radiotelephone [COMMUN] **1.** Pertaining to telephony over radio channels. **2.** A radio transmitter and radio receiver used together for two-way telephone communication by radio. Also known as radiophone. { ¦rād·ē·ō'tel·ə‚fōn }

radiotelephony [COMMUN] Two-way transmission of sounds by means of modulated radio waves, without interconnecting wires. { ¦rād·ē·ō·tə'lef·ə·nē }

radio time signal |COMMUN| A time signal sent by radio broadcast. { 'rād·ē·ō 'tīm ,sig·nəl }

radio tower |COMMUN| A tower, usually several hundred meters tall, either guyed or freestanding, on which a transmitting antenna is mounted to increase the range of radio transmission; in some cases, the tower itself may be the antenna. { 'rād·ē·ō ,tau̇·ər }

radio tracking [ENG] The process of keeping a radio or radar beam set on a target and determining the range of the target continuously. { 'rād·ē·ō 'trak·iŋ }

radio transmission |COMMUN| The transmission of signals through space at radio frequencies by means of radiated electromagnetic waves. { 'rād·ē·ō tranz'mish·ən }

radio transmitter [ELECTR] The equipment used for generating and amplifying a radio-frequency carrier signal, modulating the carrier signal with intelligence, and feeding the modulated carrier to an antenna for radiation into space as electromagnetic waves. Also known as radio set; transmitter. { 'rād·ē·ō 'tranz,mid·ər }

radio transponder [ELECTR] A transponder which receives and transmits radio waves. { 'rād·ē·ō tran'spän·dər }

radio tube See electron tube. { 'rād·ē·ō ,tüb }

radio wave [ELECTROMAG] An electromagnetic wave produced by reversal of current in a conductor at a frequency in the range from about 10 kilohertz to about 300,000 megahertz. { 'rād·ē·ō ,wāv }

radix See root. { 'rād·iks }

radix transformation |COMPUT SCI| A method of transformation that involves changing the radix or base of the original key and either discarding excess high-order digits (that is, digits in excess of the number desired in the key) or extracting some part of the transformed number. { 'rād·iks ,tranz·fər'mā·shən }

radome [ELECTROMAG] A strong, thin shell, made from a dielectric material that is transparent to radio-frequency radiation, and used to house a radar antenna, or a space communications antenna of similar structure. { 'rā,dōm }

RAID [COMPUT SCI] A group of hard disks that operate together to improve performance or provide fault tolerance and error recovery through data striping, mirroring, and other techniques. Derived from redundant array of inexpensive disks. { rād }

rail-fence jammer See continuous-wave jammer. { 'rāl ¦fens ,jam·ər }

railing |ELECTR| Simple radar pulse jamming at high recurrence rates (50 to 150 kilohertz); it results in an image on a radar indicator resembling fence railing. { 'rāl·iŋ }

rain attenuation |COMMUN| Attenuation of radio waves when passing through moisture-bearing cloud formations or areas in which rain is falling; increases with the density of the moisture in the transmission path. { 'rān ə,ten·yə,wā·shən }

rainbow |ELECTR| Technique which applies pulse-to-pulse frequency changing to identifying and discriminating against decoys and chaff. { 'rān,bō }

RAM See random-access memory. { ram }

Rambus dynamic random-access memory |COMPUT SCI| High-performance memory that can transfer data at rates of 800 megahertz and higher. Abbreviated RDRAM. { ¦ram,bəs dī ¦nam·ik ,ran·dəm 'ak,ses ,mem·rē }

RAM disk See RAM drive. { 'ram ,disk }

RAM drive |COMPUT SCI| A portion of a computer's random-access memory (RAM) that is made to simulate a disk drive. Also known as RAM disk. { 'ram ,drīv }

rampage through core |COMPUT SCI| Action of a computer program that writes data in incorrect locations or otherwise alters storage locations improperly, because of a program error. { 'ram ,pāj thrü 'kȯr }

ramp generator |ELECTR| A circuit that generates a sweep voltage which increases linearly in value during one cycle of sweep, then returns to zero suddenly to start the next cycle. { 'ramp ,jen·ə ,rād·ər }

RAM resident |COMPUT SCI| A program that remains stored in a computer's random-access memory (RAM) at all times. Also known as terminate and stay resident (TSR). { ¦ram 'rez·ə·dənt }

random access |COMMUN| The process of beginning to read and decode the coded bit stream at an arbitrary point. |COMPUT SCI| **1.** The ability to read or write information anywhere within a storage device in an amount of time that is constant regardless of the location of the information accessed and of the location of the information previously accessed. Also known as direct access. **2.** A process in which data are accessed in nonsequential order and possibly at irregular intervals of time. Also known as single reference. { 'ran·dəm 'ak,ses }

random-access discrete address |COMMUN| Communications technique in which radio users share one wide band instead of each user getting an individual narrow band. { 'ran·dəm ¦ak,ses di'skrēt ə'dres }

random-access disk file |COMPUT SCI| A file which is contained on a disk having one head per track and in which consecutive records are not necessarily in consecutive locations. { 'ran·dəm ¦ak,ses 'disk ,fīl }

random-access input/output |COMPUT SCI| A technique which minimizes seek time and overlaps with processing. { 'ran·dəm ¦ak,ses 'in,pu̇t 'au̇t,pu̇t }

random-access memory |COMPUT SCI| A data storage device having the property that the time required to access a randomly selected datum does not depend on the time of the last access or the location of the most recently accessed datum. Abbreviated RAM. Also known as direct-access memory; direct-access storage; random-access storage; random storage; uniformly accessible storage.   { 'ran·dəm ¦ak,ses 'mem·rē }

random-access programming   [COMPUT SCI] Programming without regard for the time required for access to the storage positions called for in the program, in contrast to minimum-access programming.   { 'ran·dəm ¦ak,ses 'prō,gram·iŋ }

random-access storage   See random-access memory.   { 'ran·dəm ¦ak,ses 'stȯr·ij }

randomized jitter   [ELECTR] Jitter by means of noise modulation.   { 'ran·də,mīzd 'jid·ər }

randomizing scheme   [COMPUT SCI] A technique of distributing records among storage modules to ensure even distribution and seek time.   { 'ran·də,mīz·iŋ ,skēm }

random number generator   [COMPUT SCI] **1.** A mathematical program which generates a set of numbers which pass a randomness test. **2.** An analog device that generates a randomly fluctuating variable, and usually operates from an electrical noise source.   { 'ran·dəm 'nəm·bər ,jen·ə,rād·ər }

random pulsing   [COMMUN] Continuous, varying, pulse-repetition rate, accomplished by noise modulation or continuous frequency change.   { 'ran·dəm 'pəls·iŋ }

random-sampling voltmeter   [ENG] A sampling voltmeter which takes samples of an input signal at random times instead of at a constant rate; the synchronizing portions of the instrument can then be simplified or eliminated.   { 'ran·dəm ¦sam·pliŋ 'vōlt,mēd·ər }

random storage   See random-access memory.   { 'ran·dəm 'stȯr·ij }

random superimposed coding   [COMPUT SCI] A system of coding in which a set of random numbers is assigned to each concept to be encoded; with punched cards, each number corresponds to some one hole to be punched in a given field.   { 'ran·dəm ¦sü·pər·im'pōzd 'kōd·iŋ }

random winding   [ELEC] A coil winding in which the turns are positioned haphazardly rather than in layers.   { 'ran·dəm 'wīnd·iŋ }

range   [COMMUN] **1.** In printing telegraphy, that fraction of a perfect signal element through which the time of selection may be varied to occur earlier or later than the normal time of selection without causing errors while signals are being received. **2.** Upper and lower limits through which the index arm of the rangefinder mechanism of a teletypewriter may be moved and still receive correct copy.   [CONT SYS] **1.** The maximum distance a robot's arm or wrist can travel. Also known as reach. **2.** The volume comprising the locations to which a robot's arm or wrist can travel.   [ENG] **1.** The distance capability of a radio or radar system. **2.** In radar measurement, the distance to a target measured usually by the time elapsed between the transmission of a pulse and the receipt of the target's echo.   { rānj }

range-amplitude display   [ELECTR] Radar display in which a base provides the range scale from which echoes appear as deflections normal to the base.   { 'rānj 'am·plə,tüd di,splā }

range arithmetic   See interval arithmetic.   { 'rānj ə¦rith·mə·tik }

range attenuation   [ELECTROMAG] In radar terminology, the decrease in power density (flux density) caused by the divergence of the flux lines with distance, this decrease being in accordance with the inverse-square law.   { 'rānj ə,ten·yə¦wā·shən }

range-bearing display   See B display.   { 'rānj 'ber·iŋ di,splā }

range calibrator   [ELECTR] **1.** A device with which the operator of a transmitter calculates the distance over which the signal will extend intelligibly. **2.** A device for adjusting radar range indications by use of known range targets or delayed signals; particularly useful in radars using analog echo timing.   { 'rānj¦kal·ə,brād·ər }

range check   [COMPUT SCI] A method of checking the validity of input data by determining whether the values fall within an expected range.   { 'rānj ,chek }

range comprehension   [ELECTR] In a frequency-modulation sonor system, valves between the maximum and the minimum ranges.   { 'rānj ,käm·pri¦hen·shən }

range delay   [ELECTROMAG] A control used in radars which permits the operator to present on the radarscope only those echoes from targets which lie beyond a certain distance from the radar; by using range delay, undesired echoes from nearby targets may be eliminated while the indicator range is increased.   { 'rānj di,lā }

rangefinder   [COMMUN] A movable, calibrated unit of the receiving mechanism of a teletypewriter by means of which the selecting interval may be moved with respect to the start signal.   [ELECTR] A device which determines the distance to an object by measuring the time it takes for a radio wave to travel to the object and return.   { 'rānj,fīnd·ər }

range gate   [ELECTR] A gate voltage that is used to select radar echoes from a very narrow interval of ranges.   { 'rānj ¦gāt }

range gate capture   [ELECTR] Electronic countermeasure technique using a spoofer radar transmitter to produce a false target echo that can make a fire-control tracking radar move off the real target and follow the false one.   { 'rānj ¦gāt ,kap·chər }

range gating   [ELECTR] The process of selecting, for further use, only those radar echoes that lie within a small interval of ranges.   { 'rānj ,gād·iŋ }

range-height indicator display   [ELECTR] A radar display showing the distance between a reference point, usually the radar, and a target, along with the vertical distance between a horizontal reference plane, usually containing the radar, and the target. Abbreviated RHI.   { 'rānj 'hīt 'in·də ,kād·ər di,splā }

range-imaging sensor   [ENG] A robotic device that makes precise measurements, by using the principles of algebra, trigonometry, and geometry, of the distance from a robot's end effector to various parts of an object, in order to form an image of the object.   { 'rānj ¦im·ij·iŋ ,sen·sər }

range mark offset |ELECTR| Displacement of range mark on a type B indicator. { 'rānj ¦märk 'óf,set }

range of a loop |COMPUT SCI| The set of instructions contained between the opening and closing statements of a do loop. { 'rānj əv ə 'lüp }

range rate |ELECTR| The rate at which the distance from the measuring equipment to the target or signal source that is being tracked is changing with respect to time. { 'rānj ,rāt }

range ring |ELECTR| Accurate, adjustable ranging mark on a plan position indicator; such marks at set range intervals are displayed as concentric rings as the display is generated. { 'rānj ,riŋ }

range selection |ELECTR| Control on a radar indicator for selection of range scale. { 'rānj si ,lek·shən }

range sensing |ENG| The precise measurement of the distance of a device from a robot's end effector. { 'rānj ,sens·iŋ }

range step |ELECTR| Vertical displacement on M-indicator sweep to measure range. { 'rānj ,step }

range strobe |ELECTROMAG| An index mark which may be displayed on various types of radar indicators to assist in the determination of the exact range of a target. { 'rānj ,strōb }

range sweep |ELECTR| A sweep intended primarily for measurement of range. { 'rānj ,swēp }

range-tracking element |ELECTR| An element in a radar set that measures range and its time derivative, by means of which a range gate is actuated slightly before the predicted instant of signal reception. { 'rānj ¦trak·iŋ ,el·ə·mənt }

range unit |ELECTR| Radar system component used for control and indication (usually counters) of range measurements. { 'rānj ,yü·nət }

range zero |ELECTR| Alignment of start sweep trace with zero range. { 'rānj ,zir·ō }

ranging oscillator |ELECTR| Oscillator circuit containing an LC (inductor-capacitor) resonant combination in the cathode circuit, usually used in radar equipment to provide range marks. { 'rānj·iŋ 'äs·ə,lād·ər }

rapid access loop |COMPUT SCI| A small section of storage, particularly in drum, tape, or disk storage units, which has much faster access than the remainder of the storage. { 'rap·əd ¦ak,ses ,lüp }

rapid memory See rapid storage. { 'rap·əd 'mem· rē }

rapid selector |COMPUT SCI| A device which scans codes recorded on microfilm; microimages of the documents associated with the codes may also be recorded on the film. { 'rap·əd si'lek·tər }

rapid storage |COMPUT SCI| In computers, storage with a very short access time; rapid access is generally gained by limiting storage capacity. Also known as high-speed storage; rapid memory. { 'rap·əd 'stór·ij }

rare-earth-doped fiber amplifier |COMMUN| An optical fiber amplifier whose fiber core is lightly doped with trivalent rare-earth ions, which absorb light at certain pump wavelengths and emit it at some signal wavelength through stimulated emission. { ¦rär ,ərth ,dōpt ,fī·bər 'am·plə,fī·ər }

raster |ELECTR| A predetermined pattern of scanning lines that provides substantially uniform coverage of an area; in video the raster is seen as closely spaced parallel lines, most evident when there is no picture. { 'ras·tər }

raster graphics |COMPUT SCI| A computer graphics coding technique which codes each picture element of the picture area in digital form. Also known as bit-mapped graphics. { 'ras·tər ¦graf·iks }

rasterization |COMPUT SCI| The conversion of graphics objects composed of vectors or line segments into dots for transmission to raster graphics displays and to dot matrix and laser printers. { ,ras·tə·rə'zā·shən }

raster scanning |ELECTR| Radar scan very similar to electron-beam scanning in an ordinary television set; horizontal sector scan that changes in elevation. { 'ras·tər ¦skan·iŋ }

rate action See derivative action. { 'rāt ,ak·shən }

rated speed |COMPUT SCI| The maximum operating speed that can be sustained by a data-processing device or communications line, not allowing for periodic pauses for various reasons such as carriage return on a printer. { 'rād·əd 'spēd }

rate effect |ELECTR| The phenomenon of a *pnpn* device switching to a high-conduction mode when anode voltage is applied suddenly or when high-frequency transients exist. { 'rāt i,fekt }

rate feedback |ELECTR| The return of a signal, proportional to the rate of change of the output of a device, from the output to the input. { 'rāt 'fēd,bak }

rate-grown transistor |ELECTR| A junction transistor in which both impurities (such as gallium and antimony) are placed in the melt at the same time and the temperature is suddenly raised and lowered to produce the alternate *p*-type and *n*-type layers of rate-grown junctions. Also known as graded-junction transistor. { 'rāt ¦grōn tran 'zis·tər }

rate multiplier |COMPUT SCI| An integrator in which the quantity to be integrated is held in a register and is added to the number standing in an accumulator in response to pulses which arrive at a constant rate. { 'rāt ,məl·tə,plī·ər }

rate servomechanism See velocity servomechanism. { 'rāt ¦sər·vō'mek·ə,niz·əm }

rate test |COMPUT SCI| A test that verifies that the time constants of the integrators are correct; used in analog computers. { 'rāt ,test }

rate transmitter |ELECTR| A transmitter in a missile being launched, used with a ground receiver to indicate the rate of speed increase. { 'rāt tranz,mid·ər }

ratio arm circuit |ELEC| Two adjacent arms of a Wheatstone bridge, designed so they can be set to provide a variety of indicated resistance ratios. { 'rā·shō ¦ärm ,sər·kət }

ratio-balance relay See percentage differential relay. { 'rā·shō ¦bal·əns ,rē,lā }

ratio control system |CONT SYS| Control system in which two process variables are kept at a fixed ratio, regardless of the variation of either of the variables, as when flow rates in two separate fluid conduits are held at a fixed ratio. { 'rā·shō kən'trōl ,sis·təm }

ratio detector |ELECTR| A frequency-modulation detector circuit that uses two diodes and requires no limiter at its input; the audio output is determined by the ratio of two developed intermediate-frequency voltages whose relative amplitudes are a function of frequency. { 'rā·shō di,tek·tər }

ratio deviation *See* modulation index. { 'rā·shō ,dē·vē'ā·shən }

ratio-differential relay *See* percentage differential relay. { 'rā·shō ,dif·ə¦ren·chəl 'rē,lā }

ratio meter |ENG| A meter that measures the quotient of two electrical quantities; the deflection of the meter pointer is proportional to the ratio of the currents flowing through two coils. { 'rā·shō ,mēd·ər }

rationalized units |ELEC| A system of electrical units, such as occurs in the International System, in which the factor of 4π is removed from the field equations and appears instead in the explicit expressions for the fields of a point charge and current element. { 'rash·ən·əl,īzd 'yü·nəts }

ratio of transformation |ELEC| Ratio of the secondary voltage of a transformer to the primary voltage under no-load conditions, or the corresponding ratio of currents in a current transformer. { 'rā·shō əv ,tranz·fər'mā·shən }

ratio of transformer |ELEC| Ratio of the number of turns in one winding of a transformer to the number of turns in the other, unless otherwise specified. { 'rā·shō əv tranz'fȯr·mər }

ratio resistor |ELEC| One of the resistors in a Wheatstone or Kelvin bridge whose resistances appear in a pair of ratios which are equal in a balanced bridge. { 'rā·shō ri,zis·tər }

rat race |ELECTR| A hybrid network in the form of a ring in microwave circuitry. { 'rat ,rās }

Rayleigh video |ELECTR| Referring to the video and its particular probability density produced by an amplitude detector (demodulato) when a Gaussian radio noise is incident to it. { 'rā·lē ¦vid·ē·ō }

ray path |COMMUN| Geometric path between signal transmitting and receiving locations. { 'rā ,path }

ray tracing |COMPUT SCI| The creation of reflections, refractions, and shadows in a graphics image by following a series of rays from a light source and determining the effect of light on each pixel in the image. { 'rā ,trās·iŋ }

R-C amplifier *See* resistance-capacitance coupled amplifier. { ¦är¦sē 'am·plə,fī·ər }

R-C circuit *See* resistance-capacitance circuit. { ¦är¦sē 'sər·kət }

R-C constant *See* resistance-capacitance constant. { ¦är¦sē 'kän·stənt }

R-C coupled amplifier *See* resistance-capacitance coupled amplifier. { ¦är¦sē ¦kəp·əld 'am·plə,fī·ər }

R-C coupling *See* resistance coupling. { ¦är¦sē 'kəp·liŋ }

RCM *See* radar countermeasure.

R-C network *See* resistance-capacitance network. { ¦är¦sē 'net,wərk }

R-C oscillator *See* resistance-capacitance oscillator. { ¦är¦sē 'äs·ə,lād·ər }

R-DAT system *See* rotary digital audio tape system. { 'är ,dat ,sis·təm *or* ¦är ¦dē¦ā'tē ,sis·təm }

RDF *See* radio direction finder.

R-display |ELECTR| A radar display format in which only the display around a target of interest is expanded in range in an A-display format, to improve the accuracy of range estimation and to permit closer examination of the target signal. Also known as R-indicator; R-scan; R-scope. { 'är di,splā }

RDRAM *See* Rambus dynamic random-access memory. { ¦är¦dē'ram }

RDS *See* radio data system.

RDSS *See* radiodetermination satellite service.

reach *See* range. { rēch }

reactance |ELEC| The imaginary part of the impedance of an alternating-current circuit. { rē'ak·təns }

reactance amplifier *See* parametric amplifier. { rē'ak·təns 'am·plə,fī·ər }

reactance drop |ELEC| The component of the phasor representing the voltage drop across a component or conductor of an alternating-current circuit which is perpendicular to the current. { rē'ak·təns ,dräp }

reactance frequency multiplier |ELECTR| Frequency multiplier whose essential element is a nonlinear reactor. { rē'ak·təns 'frē·kwən·sē 'məl·tə,plī·ər }

reactance grounded |ELEC| Grounded through a reactance. { rē'ak·təns ,graùn·dəd }

reactance relay |ELEC| Form of impedance relay, the operation of which is a function of the reactance of a circuit. { rē'ak·təns ,rē,lā }

reactance tube |ELECTR| Vacuum tube operated in a way that it presents almost a pure reactance to the circuit. { rē'ak·təns ,tüb }

reactance-tube modulator |ELECTR| An electron-tube circuit, used to produce phase or frequency modulation, in which the reactance is varied in accordance with the instantaneous amplitude of the modulating voltage. { rē'ak·təns ,tüb 'mäj·ə ,lād·ər }

reaction *See* positive feedback. { rē'ak·shən }

reaction motor |ELEC| A synchronous motor whose rotor contains salient poles but which has no windings and no permanent magnets. { rē'ak·shən ,mōd·ər }

reactive |ELEC| Pertaining to either inductive or capacitance reactance; a reactive circuit has a high value of reactance in comparison with resistance. { rē'ak·tiv }

reactive component |ELEC| In the phasor representation of quantities in an alternating-current circuit, the component of current, voltage, or apparent power which does not contribute power, and which resultsfrom inductive or capacitive reactance in the circuit, namely, the reactive current, reactive voltage, or reactive power. Also known as idle component; quadrature component; wattless component.   { rē'ak·tiv kəm'pō·nənt }

reactive current |ELEC| In the phasor representation of alternating current, the component of the current perpendicular to the voltage, which contributes no power but increases the power losses of the system. Also known as idle current; quadrature current; wattless current.   { rē'ak·tiv 'kə·rənt }

reactive factor |ELEC| The ratio of reactive power to apparent power.   { rē'ak·tiv ˌfak·tər }

reactive ion etching |ELECTR| A directed chemical etching process used in integrated circuit fabrication in which chemically active ions are accelerated along electric field lines to meet a substrate perpendicular to its surface.   { rē'ak·tiv 'ī ˌän ˌech·iŋ }

reactive load |ELEC| A load having inductive or capacitive reactance.   { rē'ak·tiv 'lōd }

reactive power |ELEC| The power value obtained by multiplying together the effective value of current in amperes, the effective value of voltage in volts, and the sine of the angular phase difference between current and voltage. Also known as wattless power.   { rē'ak·tiv 'pau̇·ər }

reactive voltage |ELEC| In the phasor representation of alternating current, the voltage component that is perpendicular to the current.   { rē'ak·tiv 'vōl·tij }

reactive volt-ampere   See volt-ampere reactive.   { rē'ak·tiv 'vōlt 'am,pir }

reactive volt-ampere hour   See var hour.   { rē'ak·tiv 'vōlt 'am,pir 'au̇·ər }

reactive volt-ampere meter   See varmeter.   { rē'ak·tiv 'vōlt 'am,pir ,mēd·ər }

reactor |ELEC| A device that introduces either inductive or capacitive reactance into a circuit, such as a coil or capacitor. Also known as electric reactor.   { rē'ak·tər }

read |COMPUT SCI| **1.** To acquire information, usually from some form of storage in a computer. **2.** To convert magnetic spots, characters, or punched holes into electrical impulses. |ELECTR| To generate an output corresponding to the pattern stored in a charge storage tube.   { rēd }

read-around number   See read-around ratio.   { 'rēd əˌrau̇nd ˌnəm·bər }

read-around ratio |COMPUT SCI| The number of times that a particular bit in electrostatic storage may be read without seriously affecting nearby bits. Also known as read-around number.   { 'rēd əˌrau̇nd ˌrā·shō }

read-back check   See echo check.   { 'rēd ˌbak ˌchek }

Read diode |ELECTR| A high-frequency semiconductor diode consisting of an avalanching *pn* junction, biased to fields of several hundred thousand volts per centimeter, at one end of a high-resistance carrier serving as a drift space for the charge carriers.   { 'rēd ˌdī,ōd }

reader |COMPUT SCI| A device that converts information from one form to another, as

from punched paper tape to magnetic tape.   { 'rēd·ər }

reader-interpreter |COMPUT SCI| A service routine that reads an input string, stores programs and data on random-access storage for later processing, identifies the control information contained in the input string, and stores this control information separately in the appropriate control lists.   { 'rēd·ər in'tər·prəd·ər }

read error |COMPUT SCI| A condition in which the content of a storage device cannot be electronically identified.   { 'rēd ,er·ər }

read head |COMPUT SCI| A device that converts digital information stored on a magnetic tape, drum, or disk into electrical signals usable by the computer arithmetic unit.   { 'rēd ,hed }

read-in |COMPUT SCI| To sense information contained in some source and transmit this information to an internal storage.   { 'rēd ,in }

readiness review |COMPUT SCI| An on-site examination of the adequacy of preparations for effective utilization upon installation of a computer, and to identify any necessary corrective actions.   { 'red·i·nəs ri,vyü }

reading rate |COMPUT SCI| Number of characters, words, or fields sensed by an input sensing device per unit of time.   { 'rēd·iŋ ,rāt }

read-in program |COMPUT SCI| Computer program that can be put into a computer in a simple binary form and allows other programs to be read into the computer in more complex forms.   { 'rēd ,in ,prō·grəm }

read-only memory |COMPUT SCI| A device for storing data in permanent, or nonerasable, form; usually an optical, static electronic, or magnetic device allowing extremely rapid access to data. Abbreviated ROM. Also known as nonerasable storage; read-only storage.   { 'rēd ¦ōn·lē 'mem·rē }

read-only storage   See read-only memory.   { 'rēd ¦ōn·lē 'stòr·ij }

read-only terminal |COMPUT SCI| A peripheral device, such as a printer, that can only receive signals.   { 'rēd ¦ōn·lē 'tər·mən·əl }

readout |COMPUT SCI| **1.** The presentation of output information by means of lights a display, printout, or other methods. **2.** To sense information contained in some computer internal storage and transmit this information to a storage external to the computer.   { 'rēd,au̇t }

readout station |COMMUN| A recording or receiving radio station at which data are received.   { 'rēd,au̇t ,stā·shən }

read screen |COMPUT SCI| In optical character recognition (OCR), the transparent component part of most character readers through which appears the input document to be recognized.   { 'rēd ,skrēn }

read time |COMPUT SCI| The time interval between the instant at which information is called for from storage and the instant at which delivery is completed in a computer.   { 'rēd ,tīm }

read-while-writing |COMPUT SCI| The reading of a record or group of records into storage from tape at the same time another record or group of records is written from storage to tape. { 'rēd ‚wīl 'rīd·iŋ }

read/write channel [COMPUT SCI] A path along which information is transmitted between the central processing unit of a computer and an input, output, or storage unit under the control of the computer. { 'rēd 'rīt ‚chan·əl }

read/write check indicator [COMPUT SCI] A device incorporated in certain computers to indicate upon interrogation whether or not an error was made in reading or writing; the machine can be made to stop, retry the operation, or follow a special subroutine, depending upon the result of the interrogation. { 'rēd 'rīt 'chek ‚in·də‚kād·ər }

read/write comb [COMPUT SCI] The set of arms mounted with magnetic heads that reach between the disks of a disk storage device to read and record information. { 'rēd 'rīt ‚kōm }

read/write head [COMPUT SCI] A magnetic head that both senses and records data. Also known as combined head. { 'rēd 'rīt ‚hed }

read/write memory [COMPUT SCI] A computer storage in which data may be stored or retrieved at comparable intervals. { 'rēd 'rīt ‚mem·rē }

read/write random-access memory [COMPUT SCI] A random access memory in which data can be written into memory as well as read out of memory. { 'rēd 'rīt 'ran·dəm 'ak‚ses ‚mem·rē }

ready-to-receive signal [COMMUN] Signal sent back to a facsimile transmitter to indicate that a facsimile receiver is ready to accept the transmission. { 'red·ē tə ri'sēv ‚sig·nəl }

real data type [COMPUT SCI] A scalar data type which contains a normalized fraction (mantissa) and an exponent (characteristic) and is used to represent floating-point data, usually decimal. { 'rēl 'dad·ə ‚tīp }

realizability [CONT SYS] Property of a transfer function that can be realized by a network that has only resistances, capacitances, inductances, and ideal transformers. { ‚rē·ə‚līz·ə'bil·əd·ē }

real power [ELEC] The component of apparent power that represents true work; expressed in watts, it is equal to volt-amperes multiplied by the power factor. { 'rēl ‚pau·ər }

real-space-transfer transistor [ELECTR] A transistor that utilizes the effect of the increase in electron energy and temperature in high electric fields. { ‚rēl ‚spās 'tranz·fər tran‚zis·tər }

real storage [COMPUT SCI] Actual physical storage of data and instructions. { 'rēl 'stór·ij }

real-time [COMPUT SCI] Pertaining to a data-processing system that controls an ongoing process and delivers its outputs (or controls its inputs) not later than the time when these are needed for effective control; for instance, airline reservations booking and chemical processes control. { 'rēl ‚tīm }

real-time clock [COMPUT SCI] A pulse generator which operates at precise time intervals to determine time intervals between events and initiate specific elements of processing. { 'rēl ‚tīm 'kläk }

real-time control system [COMPUT SCI] A computer system which controls an operation in real

time, such as a rocket flight. { 'rēl ‚tīm kən'trōl ‚sis·təm }

real-time operation [COMPUT SCI] **1.** Of a computer or system, an operation or other response in which programmed responses to an event are essentially simultaneous with the event itself. **2.** An operation in which information obtained from a physical process is processed to influence or control the physical process. { 'rēl ‚tīm ‚äp·ə'rā·shən }

real-time processing [COMPUT SCI] The handling of input data at a rate sufficient to ensure that the instructions generated by the computer will influence the operation under control at the required time. { 'rēl ‚tīm 'prä‚ses·iŋ }

real-time programming [COMPUT SCI] Programming for a situation in which results of computations will be used immediately to influence the course of ongoing physical events. { 'rēl ‚tīm 'prō‚gram·iŋ }

real-time system [COMPUT SCI] A system in which the computer is required to perform its tasks within the time restraints of some process or simultaneously with the system it is assisting. { 'rēl ‚tīm 'sis·təm }

rear-projection [ELECTR] Pertaining to video system in which the picture is projected on a ground-glass screen for viewing from the opposite side of the screen. { 'rir prə'jek·shən }

reasonableness [COMPUT SCI] A measure of the extent to which data processed by a computer falls within an acceptable allowance for errors, as determined by quantitative tests. { 'rēz·nə·bəl·nəs }

reboot [COMPUT SCI] To reload systems software into a computer so that it makes a new start. { rē'büt }

rebroadcast [COMMUN] Repetition of a radio or television program at a later time. { rē'bródˌkast }

recall factor [COMPUT SCI] A measure of the efficiency of an information retrieval system, equal to the number of retrieved relevant documents divided by the total number of relevant documents in the file. { 'rē‚kól ‚fak·tər }

received power [ELECTROMAG] **1.** The total power received at an antenna from a signal, such as a radar target signal. **2.** In a mobile communications system, the root-mean-square value of power delivered to a load which properly terminates an isotropic reference antenna. { ri'sēvd 'pau·ər }

receive-only [COMMUN] A teleprinter which has no keyboard, and thus can receive but not transmit. Abbreviated RO. { ri'sēv 'ōn·lē }

receiver [ELECTR] The complete equipment required for receiving modulated radio waves and converting them into the original intelligence, such as into sounds or pictures, or converting to desired useful information as in a radar receiver. { ri'sē·vər }

receiver bandwidth [ELECTR] Spread, in frequency, between the halfpower points on the receiver response curve. { ri'sē·vər 'band‚width }

receiver gating [ELECTR] Application of operating voltages to one or more stages of a receiver only during that part of a cycle of operation when reception is desired. { ri'sē·vər ,gād·iŋ }

receiver incremental tuning [ELECTR] Control feature to permit receiver tuning (of a transceiver) up to 3 kilohertz to either side of the transmitter frequency. { ri'sē·vər ,in·krə,ment·əl 'tün·iŋ }

receiver lockout system See lockout. { ri'sē·vər 'läk,aút ,sis·təm }

receiver noise threshold [ELECTR] External noise appearing at the front end of a receiver, plus the noise added by the receiver itself, whichdetermines a noise threshold that has to be exceeded by the minimum discernible signal. { ri'sē·vər 'nóiz ,thresh,hōld }

receiver radiation [ELECTROMAG] Radiation of interfering electromagnetic fields by the oscillator of a receiver. { ri'sē·vər ,rād·ē'ā·shən }

receiver synchro See synchro receiver. { ri'sē·vər 'siŋ·krō }

receiving antenna [ELECTROMAG] An antenna used to convert electromagnetic waves to modulated radio-frequency currents. { ri'sēv·iŋ an,ten·ə }

receiving area [ELECTROMAG] The factor by which the power density must be multiplied to obtain the received power of an antenna, equal to the gain of the antenna times the square of the wavelength divided by 4π. { ri'sēv·iŋ ,er·ē·ə }

receiving loop loss [COMMUN] In telephones, that part of the repetition equivalent assignable to the station set, subscriber line, and battery supply circuit that are on the receiving end. { ri'sēv·iŋ ,lüp ,lós }

receiving set See radio receiver. { ri'sēv·iŋ ,set }

receiving tube [ELECTR] A low-voltage and low-power vacuum tube used in radio receivers, computers, and sensitive control and measuring equipment. { ri'sēv·iŋ ,tüb }

receptacle See outlet. { ri'sep·tə·kəl }

reception [COMMUN] The conversion of modulated electromagnetic waves or electric signals, transmitted through the air or over wires or cables, into the original intelligence, or into desired useful information (as in radar), by means of antennas and electronic equipment. { ri'sep·shən }

recharge [ELEC] To restore a cell or battery to a charged condition by sending a current through it in a direction opposite to that of the discharging current. { rē'chärj }

rechargeable battery See storage battery. { rē'chär·jə·bəl'bad·ə·rē }

reciprocal ferrite switch [ELECTROMAG] A ferrite switch that can be inserted in a waveguide to switch an input signal to either of two output waveguides; switching is done by a Faraday rotator when acted on by an external magnetic field. { ri'sip·rə·kəl 'fe,rīt ,swich }

reciprocal impedance [ELEC] Two impedances Z_1 and Z_2 are said to be reciprocal impedances with respect to an impedance Z (invariably a resistance) if they are so related as to satisfy the equation $Z_1Z_2 = Z^2$. { ri'sip·rə·kəl im'pēd·əns }

reciprocal ohm See siemens. { ri'sip·rə·kəl 'ōm }

reciprocal ohm centimeter See roc. { ri'sip·rə·kəl 'ōm 'sent·i,mēd·ər }

reciprocal ohm meter See rom. { ri'sip·rə·kəl 'ōm ,mēd·ər }

reciprocal transducer [ELECTR] Transducer which satisfies the principle of reciprocity. { ri'sip·rə·kəl tranz'dü·sər }

reciprocation [ELECTR] In electronics, a process of deriving a reciprocal impedance from a given impedance, or finding a reciprocal network for a given network. { ri,sip·rə'kā·shən }

reciprocity calibration [ENG ACOUS] A measurement of the projector loss and hydrophone loss of a reversible transducer by means of the reciprocity theorem and comparisons with the known transmission loss of an electric network, without knowing the actual value of either the electric power or the acoustic power. { ,res·ə'präs·əd·ē ,kal·ə,brā·shən }

reciprocity theorem Also known as principle of reciprocity. [ELEC] **1.** The electric potentials V_1 and V_2 produced at some arbitrary point, due to charge distributions having total charges of q_1 and q_2 respectively, are such that $q_1V_2 = q_2V_1$. **2.** In an electric network consisting of linear passive impedances, the ratio of the electromotive force introduced in any branch to the current in any other branch is equal in magnitude and phase to the ratio that results if the positions of electromotive force and current are exchanged. [ELECTROMAG] Given two loop antennas, a and b, then $I_{ab}/V_a = I_{ba}/V_b$, where I_{ab} denotes the current received in b when a is used as transmitter, and V_a denotes the voltage applied in a; I_{ba} and V_b are the corresponding quantities when b is the transmitter, a the receiver; it is assumed that the frequency and impedances remain unchanged. [ENG ACOUS] The sensitivity of a reversible electroacoustic transducer when used as a microphone divided by the sensitivity when used as a source of sound is independent of the type and construction of the transducer. { ,res·ə'präs·əd·ē ,thir·əm }

reclaimer [COMPUT SCI] A device that performs dynamic storage allocation, periodically searching memory to locate cells whose contents are no longer useful for computation, and making them available for other uses. { rē'klām·ər }

reclosing relay [ELEC] Form of voltage, current, power, or other type of relay which functions to reclose a circuit. { 'rē,klōz·iŋ 'rē,lā }

recognition [COMPUT SCI] The act or process of identifying (or associating) an input with one of a set of possible known alternatives, as in character recognition and pattern recognition. { ,rek·ig'nish·ən }

recognition gate [COMPUT SCI] A logic circuit used to select devices identified by a binary address code. Also known as decoding gate. { ,rek·ig'nish·ən ,gāt }

recoil implantation [ELECTR] A mechanism for ion-beam mixing of a film and a substrate in which atoms are driven from the film into the substrate as a result of direct collisions with incident ions. { ¦rē,kóil ,im·plan'tā·shən }

recombination coefficient [ELECTR] The rate of recombination of positive ions with electrons or negative ions in a gas, per unit volume, divided by the product of the number of positive ions per unit volume and the number of electrons or negative ions per unit volume.   { ‚rē‚käm‚bə'nā·shən ‚kō·i‚fish·ənt }

recombination electroluminescence   See injection electroluminescence.   { ‚rē‚käm‚bə'nā·shən i¦lek·trō‚lü·mə'nes·əns }

recombination velocity |ELECTR| On a semiconductor surface, the ratio of the normal component of the electron (or hole) current density at the surface to the excess electron (or hole) charge density at the surface.   { ‚rē‚käm·bə'nā·shən və ‚läs·əd·ē }

reconditioned carrier reception |ELECTR| Method of reception in which the carrier is separated from the sidebands to eliminate amplitude variations and noise, and is then added at an increased level to the sideband, to obtain a relatively undistorted output.   { ‚rē·kən'dish·ənd 'kar·ē·ər ri‚sep·shən }

reconditioned carrier reception |ELECTR| Method of reception in which the carrier is separated from the sidebands to eliminate amplitude variations and noise, and is then added at an increased level to the sideband, to obtain a relatively undistorted output.   { ‚rē· kən'dish·ənd 'kar·ē·ər ri‚sep·shən }

reconstitution |COMPUT SCI| The conversion of tokens back to the keywords they represent in a programming language, before generation of the output of an interpreted program.   { rē‚kän· stə'tü·shən }

recontrol time   See deionization time.   { ‚rē·kən 'trōl ‚tīm }

record |COMPUT SCI| A group of adjacent data items in a computer system, manipulated as a unit. Also known as entity.   { 'rek·ərd }

record block   See physical record.   { 'rek·ərd ‚bläk }

record density   See bit density; character density.   { 'rek·ərd ‚den·səd·ē }

recorder   See recording instrument.   { ri'kórd·ər }

record gap |COMPUT SCI| An area in a storage medium, such as magnetic tape or disk, which is devoid of information; it delimits records, and, on tape, allows the tape to stop and start between records without loss of data. Also known as interrecord gap (IRG).   { 'rek·ərd ‚gap }

record head   See recording head.   { ri'kórd ‚hed }

recording-completing trunk |ELEC| Trunk for extending a connection from a local line to a toll operator, used for recording the call and for completing the toll connection.   { ri'kórd·iŋ kəm'plēd·iŋ ‚trəŋk }

recording density |COMPUT SCI| The amount of data that can be stored in a unit length of magnetic tape, usually expressed in bits per inch or characters per inch.   { ri'kórd·iŋ ‚den·səd·ē }

recording head |ELECTR| A magnetic head used only for recording. Also known as record head.   { ri'kórd·iŋ ‚hed }

recording instrument |ENG| An instrument that makes a graphic or acoustic record of one or more variable quantities. Also known as recorder.   { ri'kórd·iŋ ‚in·strə·mənt }

recording lamp |ELECTR| A lamp whose intensity can be varied at an audio-frequency rate, for exposing variable-density sound tracks on motion picture film and for exposing paper or film in photographic facsimile recording.   { ri'kórd·iŋ ‚lamp }

recording level |ELECTR| Amplifier output level required to secure a satisfactory recording.   { ri'kórd·iŋ ‚lev·əl }

recording noise |ELECTR| Noise that is introduced during a recording process.   { ri'kórd·iŋ ‚nóiz }

recording spot   See picture element.   { ri'kórd·iŋ ‚spät }

recording storage tube |ELECTR| Type of cathode-ray tube in which the electric equivalent of an image can be stored as an electrostatic charge pattern on a storage surface; there is no visual display, but the stored information can be read out at any later time as an electric output signal.   { ri'kórd·iŋ 'stór·ij ‚tüb }

recording trunk |ELEC| Trunk extending from a local central office or private branch exchange to a toll office, which is used only for communications with toll operators and not for completing toll connections.   { ri'kórd·iŋ ‚trəŋk }

record layout |COMPUT SCI| A form showing how fields are positioned within a record, usually with information about each field.   { 'rek·ərd ‚lā‚aút }

record length |COMPUT SCI| The number of characters required for all the information in a record.   { 'rek·ərd ‚leŋkth }

record locking |COMPUT SCI| Action of a computer system that makes a record that is being processed by one user unavailable to other users, to prevent more than one user from attempting to update the same information simultaneously.   { 'rek·ərd ‚läk·iŋ }

record mark |COMPUT SCI| A symbol that signals a record's beginning or end.   { 'rek·ərd ‚märk }

record variable |COMPUT SCI| A group of related but dissimilar data items that can be worked on as a single unit. Also known as structured variable.   { 'rek·ərd ‚ver·ē·ə·bəl }

recovery interrupt |COMPUT SCI| A type of interruption of program execution which provides the computer with access to subroutines to handle an error and, if successful, to continue with the program execution.   { ri'kəv·ə·rē 'int·ə‚rəpt }

recovery routine |COMPUT SCI| A computer routine that attempts to resolve automatically conditions created by errors, without causing the computer system to shut down or otherwise do serious damage.   { ri'kəv·ə·rē rü‚tēn }

recovery system |COMPUT SCI| A system for recognizing a malfunction in a database management system, reporting it, reconstructing the damaged part of the database, and resuming processing.   { ri'kəv·ə·rē ‚sis·təm }

recovery time |ELECTR| **1.** The time required for the control electrode of a gas tube to regain control after anode-current interruption. **2.** The time required for a fired TR (transmit-receive) or pre-TR tube to deionize to such a level that the attenuation of a low-level radio-frequency signal transmitted through the tube is decreased to a specified value. **3.** The time required for a fired ATR (anti-transmit-receive) tube to deionize to such a level that the normalized conductance and susceptance of the tube in its mount are within specified ranges. **4.** The interval required, after a sudden decrease in input signal amplitude to a system or component, to attain a specified percentage (usually 63%) of the ultimate change in amplification or attenuation due to this decrease. **5.** The time required for a radar receiver to recover to half sensitivity after the end of the transmitted pulse, so it can effectively receive a return echo; a consequence of duplexed operation. { ri'kəv·ə·rē ˌtīm }

rectangular pulse [ELECTR] A pulse in which the wave amplitude suddenly changes from zero to another value at which it remains constant for a short period of time, and then suddenly changes back to zero. { rek'taŋ·gyə·lər 'pəls }

rectangular scanning [ELECTR] Two-dimensional sector scanning in which a slow sector scanning in one direction is superimposed on a rapid sector scanning in a perpendicular direction. { rek'taŋ·gyə·lər 'skan·iŋ }

rectangular scanning [ELECTR] Two-dimensional sector scanning in which a slow sector scanning in one direction is superimposed on a rapid sector scanning in a perpendicular direction. { rek'taŋ·gyə·lər 'skan·iŋ }

rectangular wave [ELECTR] A periodic wave that alternately and suddenly changes from one to the other of two fixed values. Also known as rectangular wave train. { rek'taŋ·gyə·lər 'wāv }

rectangular waveguide [ELECTROMAG] A waveguide having a rectangular cross section. { rek'taŋ·gyə·lər 'wāvˌgīd }

rectangular wave train See rectangular wave. { rek'taŋ·gyə·lər 'wāv ˌtrān }

Rectenna [ELECTR] A device that converts microwave energy in direct-current power; consists of a number of small dipoles, each having its own diode rectifier network, which are connected to direct-current buses. { rek'ten·ə }

rectification [ELEC] The process of converting an alternating current to a unidirectional current. { ˌrek·tə·fə'kā·shən }

rectification factor [ELECTR] Quotient of the change in average current of an electrode by the change in amplitude of the alternating sinusoidal voltage applied to the same electrode, the direct voltages of this and other electrodes being maintained constant. { ˌrek·tə·fə'kā·shən ˌfak·tər }

rectified value [ELEC] For an alternating quantity, the average of all the positive (or negative) values of the quantity during an integral number of periods. { 'rek·tə·fīd 'val·yü }

rectifier [ELEC] A nonlinear circuit component that allows more current to flow in one direction than the other; ideally, it allows current to flow in one direction unimpeded but allows no current to flow in the other direction. { 'rek·tə·fī·ər }

rectifier filter [ELECTR] An electric filter used in smoothing out the voltage fluctuation of an electron tube rectifier, and generally placed between the rectifier's output and the load resistance. { 'rek·tə·fī·ər ˌfil·tər }

rectifier instrument [ENG] Combination of an instrument sensitive to direct current and a rectifying device whereby alternating current (or voltages) may be rectified for measurement. { 'rek·tə·fī·ər ˌin·strə·mənt }

rectifier rating [ELECTR] A performance rating for a semiconductor rectifier, usually on the basis of the root-mean-square value of sinusoidal voltage that it can withstand in the reverse direction and the average current density that it will pass in the forward direction. { 'rek·tə·fī·ər ˌrād·iŋ }

rectifier stack [ELECTR] A dry-disk rectifier made up of layers or stacks of disks of individual rectifiers, as in a selenium rectifier or copper-oxide rectifier. { 'rek·tə·fī·ər ˌstak }

rectifier transformer [ELECTR] Transformer whose secondary supplies energy to the main anodes of a rectifier. { 'rek·tə·fī·ər tranz'fór·mər }

rectilinear scanning [ELECTR] Process of scanning an area in a predetermined sequence of narrow parallel strips. { ˌrek·tə'lin·ē·ər 'skan·iŋ }

recuperability [COMMUN] Ability to continue to operate after a partial or complete loss of the primary communications facility resulting from sabotage, enemy attack, or other disaster. { rē ˌküp·rə'bil·əd·ē }

recurrence rate See repetition rate. { ri'kər·əns ˌrāt }

recursion [COMPUT SCI] A technique in which an apparently circular process is used to perform an iterative process. { ri'kər·zhən }

recursive filter [ELECTR] A digital filter that has feedback; that is, its output depends not only on present and past input values but on past output values as well. { riˌkər·siv 'fil·tər }

recursive macro call [COMPUT SCI] A call to a macroinstruction already called when used in conjunction with conditional assembly. { ri'kər·siv ˌmak·rō ˌkól }

recursive procedure [COMPUT SCI] A method of calculating a function by deriving values of it which become more accurate at each step; recursive procedures are explicitly outlawed in most systems with the exception of a few which use languages such as ALGOL and LISP. { ri'kər·siv prə'sē·jər }

recursive subroutine [COMPUT SCI] A reentrant subroutine whose partial results are stacked, with a processor stack pointer advancing and retracting as the subroutine is called and completed. { ri'kər·siv 'səb·rüˌtēn }

recycling [ELECTR] Returning to an original condition, as to 0 or 1 in a counting circuit. { rē'sīk·liŋ }

redefine [COMPUT SCI] A procedure used in certain programming languages to specify different utilizations of the same storage area at different times. { ˌrē·di'fīn }

redistribution |ELECTR| The alteration of charges on an area of a storage surface by secondary electrons from any other area of the surface in a charge storage tube or television camera tube. { rē,dis·trəˈbyü·shən }

redox cell |ELEC| Cell designed to convert the energy of reactants to electrical energy; an intermediate reductant, in the form of liquid electrolyte, reacts at the anode in a conventional manner; it is then regenerated by reaction with a primary fuel. { ˈrē,däks ,sel }

red-tape operation See bookkeeping operation. { ˈred ¦tāp ,äp·ə,rā·shən }

reduced instruction set computer |COMPUT SCI| A computer in which the compiler and hardware are interlocked, and the compiler takes over some of the hardware functions of conventional computers and translates high-level-language programs directly into low-level machine code. Abbreviated RISC. { ri¦düst inˈstrək·shən ,set kəmˈpyüd·ər }

reduced-order controller |CONT SYS| A control algorithm in which certain modes of the structure to be controlled are ignored, to enable control commands to be computed with sufficient rapidity. { riˈdüst ¦ȯr·dər kənˈtrōl·ər }

reduced telemetry |COMMUN| Raw telemetry data transformed into a usable form. { riˈdüst təˈlem·ə·trē }

reduction |COMPUT SCI| Any process by which data are condensed, such as changing the encoding to eliminate redundancy, extracting significant details from the data and eliminating the rest, or choosing every second or third out of the totality of available points. { riˈdək·shən }

reduction rule |COMPUT SCI| The principal computation rule in the lambda calculus; it states that an operator-operand combination of the form $(\lambda x MA)$ may be transformed into the expression $S^x_A MA$, obtained by substituting the lambda expression A for all instances of x in M, provided there are no conflicts of variable names. Also known as beta rule. { riˈdək·shən ,rül }

reductive grammar |COMPUT SCI| A set of syntactic rules for the analysis of strings to determine whether the strings exist in a language. { riˈdək·tiv ˈgram·ər }

redundancy |COMMUN| In the transmission of information, the fraction of the gross information content of a message which can be eliminated without loss of essential information. |COMPUT SCI| Any deliberate duplication or partial duplication of circuitry or information to decrease the probability of a system or communication failure. { riˈdən·dən·sē }

redundancy bit |COMPUT SCI| A bit which carries no information but which is added to the information-carrying bits of a character or stream of characters to determine their accuracy. { riˈdən·dən·sē ,bit }

redundancy check |COMPUT SCI| A forbidden-combination check that uses redundant digits called check digits to detect errors made by a computer. { riˈdən·dən·sē ,chek }

redundant array of inexpensive disks See RAID. { ri,dən·dənt ə¦rā əv ,in·ik,spen·siv ˈdisks }

redundant character |COMPUT SCI| A character specifically added to a group of characters to ensure conformity with certain rules which can be used to detect computer malfunction. { riˈdən·dənt ˈkar·ik·tər }

redundant code |COMMUN| A code which uses more signal elements than are needed to represent the information it transmits. { riˈdən·dənt ˈkōd }

redundant digit |COMPUT SCI| Digit that is not necessary for an actual computation but serves to reveal a malfunction in a digital computer. { riˈdən·dənt ˈdij·it }

redundant system See duplexed system. { riˈdən·dənt ,sis·təm }

reed frequency meter See vibrating-reed frequency meter. { ˈrēd ˈfrē·kwən·sē ,mēd·ər }

Reed-Solomon code |COMMUN| A linear, block-based error-correcting code with wide-ranging applications, which is based on the mathematics of finite fields. { ¦rēd ¦säl·ə·mən ,kōd }

reel number |COMPUT SCI| A number identifying a reel of magnetic tape in a file containing more than one reel and indicating the order in which the reel is to be used. Also known as reel sequence number. { ˈrēl ,nəm·bər }

reel sequence number See reel number. { ˈrēl ˈsē·kwəns ,nəm·bər }

reenterable |COMPUT SCI| The attribute that describes a program or routine which can be shared by several tasks concurrently. { rēˈen·trə·bəl }

reentrant code See reentrant program. { rēˈen·trənt ,kōd }

reentrant program |COMPUT SCI| A subprogram in a time-sharing or multiprogramming system that can be shared by a number of users, and can therefore be applied to a given user program, interrupted and applied to some other user program, and then reentered at the point of interruption of the original user program. Also known as reentrant code. { rēˈen·trənt ,prō,gram }

reentrant winding |ELEC| Armature winding that returns to its starting point, thus forming a closed circuit. { rēˈen·trənt ,wīnd·iŋ }

reentry point |COMPUT SCI| The instruction in a computer program at which execution is resumed after the program has jumped to another place. { rēˈen·trē ,pȯint }

reentry system See turnaround system. { rēˈen·trē ,sis·təm }

reference address See address constant. { ˈref·rəns ˈad,res }

reference block |COMPUT SCI| A block within a computer program governing a numerically controlled machine which has enough data to allow resumption of the program following an interruption. { ˈref·rəns ,bläk }

reference burst See color burst. { ˈref·rəns ,bərst }

reference frequency |COMMUN| Frequency having a fixed and specified position with respect to the assigned frequency. { ˈref·rəns ,frē·kwən·sē }

reference level |ENG ACOUS| The level used as a basis of comparison when designating the level of an audio-frequency signal in decibels or volume units. Also known as reference signal level. { 'ref·rəns ,lev·əl }

reference listing |COMPUT SCI| A list printed by a compiler showing the instructions in the machine language program which it generates. { 'ref·rəns ,list·iŋ }

reference mark |ELECTR| One of the marks used in a design of a printed circuit, giving scale dimensions and indicating the edges of the circuit board. { 'ref·rəns ,märk }

reference monitor |COMPUT SCI| A means of checking that a particular user is allowed access to a specified object in a computing system. Also known as access-control mechanism; reference validation mechanism. { 'ref·rəns ,män·əd·ər }

reference noise |ELECTR| The power level used as a basis of comparison when designating noise power expressed in decibels above reference noise (dBrn); the reference usually used is 10^{-12} watt (-90 decibels above 1 milliwatt; dBm) at 1000 hertz. { 'ref·rəns ,nȯiz }

reference record |COMPUT SCI| Output of a compiler that lists the operations and their positions in the final specific routine and contains information describing the segmentation and storage allocation of the routine. { 'ref·rəns ,rek·ərd }

reference signal level See reference level. { 'ref·rəns 'sig·nəl ,lev·əl }

reference supply |ELECTR| A source of stable and constant voltage, such as a Zener diode, used in analog computers, regulated power supplies, and a variety of other circuits for comparison with a varying voltage. { 'ref·rəns sə,plī }

reference tone |ENG| Stable tone of known frequency continuously recorded on one track of multitrack signal recordings and intermittently recorded on signal track recordings by the collection equipment operators for subsequent use by the data analysts as a frequency reference. { 'ref·rəns ,tōn }

reference validation mechanism See reference monitor. { ¦ref·rəns ,val·ə'dā·shən ¦mek·ə,niz·əm }

reference voltage |ELEC| An alternating-current voltage used for comparison, usually to identify an in-phase or out-of-phase condition in an ac circuit. { 'ref·rəns ,vōl·tij }

reference white |COMMUN| 1. In a scene viewed by video camera, the color of light from a nonselective diffuse reflector that is lighted by the normal illumination of the scene. 2. The color by which this color is simulated on a video screen or other display device. { 'ref·rəns ,wīt }

reference white level |ELECTR| In television, the level at the point of observation corresponding to the specified maximum excursion of the picture signal in the white direction. { 'ref·rəns 'wīt ,lev·əl }

reflectance |COMPUT SCI| In optical character recognition, the relative brightness of the inked area that forms the printed or handwritten char-

acter; distinguished from background reflectance and brightness. { ri'flek·təns }

reflected binary ! |COMPUT SCI| A particular form ofGray code which is constructed according to the following rule: Let the first 2^N code patterns be given, for any N greater than 1; the next 2^N code patterns are derived by changing the (N + 1)-th bit from the right from 0 to 1 and repeating the original 2^N patterns in reverse order in the N rightmost positions. Also known as reflected code. { ri'flek·təd 'bī,ner·ē }

reflected code See reflected binary. { ri'flek·təd 'kōd }

reflected impedance |ELEC| 1. Impedance value that appears to exist across the primary of a transformer due to current flowing in the secondary. 2. Impedance which appears at the input terminals as a result of the characteristics of the impedance at the output terminals. { ri'flek·təd im'pēd·əns }

reflected resistance |ELEC| Resistance value that appears to exist across the primary of a transformer when a resistive load is across the secondary. { ri'flek·təd ri'zis·təns }

reflecting antenna |ELECTROMAG| An antenna used to achieve greater directivity or desired radiation patterns, in which a dipole, slot, or horn radiates toward a larger reflector which shapes the radiated wave to produce the desired pattern; the reflector may consist of one or two plane sheets, a parabolic or paraboloidal sheet, or a paraboloidal horn. { ri'flek·tiŋ an'ten·ə }

reflecting curtain |ELECTROMAG| A vertical array of half-wave reflecting antennas, generally used one quarter-wavelength behind a radiating curtain of dipoles to form a high-gain antenna. { ri'flek·tiŋ 'kərt·ən }

reflecting electrode |ELECTR| Tabular outer electrode or the repeller plate in a microwave oscillator tube, corresponding in construction but not in function to the plate of an ordinary triode; used for generating extremely high frequencies. { ri'flek·tiŋ i'lek,trōd }

reflecting galvanometer See mirror galvanometer. { ri'flek·tiŋ ,gal·və'näm·əd·ər }

reflecting grating |ELECTROMAG| Arrangement of wires placed in a waveguide to reflect one desired wave while allowing one or more other waves to pass freely. { ri'flek·tiŋ 'grād·iŋ }

reflection altimeter See radio altimeter. { ri'flek·shən al'tim·əd·ər }

reflection factor |ELEC| Ratio of the load current that is delivered to a particular load when the impedances are mismatched to that delivered under conditions of matched impedances. Also known as mismatch factor; reflectance; transition factor. { ri'flek·shən ,fak·tər }

reflection lobes |ELECTROMAG| Three-dimensional sections of the radiation pattern of a directional antenna, such as a radar antenna, which results from reflection of radiation from the earth's surface. { ri'flek·shən ,lōbz }

reflection loss |ELEC| 1. Reciprocal of the ratio, expressed in decibels, of the scalar values of the volt-amperes delivered to the load to the volt-amperes that would be delivered to a load of the same impedance as the source. **2.** Apparent transmission loss of a line which results from a portion of the energy being reflected toward the source due to a discontinuity in the transmission line. { ri'flek·shən ,lòs }

reflective binary code *See* reflected binary. { ri'flek·tiv 'bī,ner·ē 'kōd }

reflective code *See* Gray code. { ri'flek·tiv 'kōd }

reflective spot [COMPUT SCI] A piece of metallic foil that is embedded in a magnetic tape to indicate the end of a reel. { ri'flek·tiv ,spät }

reflector [ELECTROMAG] **1.** A single rod, system of rods, metal screen, or metal sheet used behind an antenna to increase its directivity. **2.** A metal sheet or screen used as a mirror to change the direction of a microwave radio beam. { ri'flek·tər }

reflector characteristic [ELECTR] A chart of power output and frequency deviation of a reflex klystron as a function of reflector voltage. { ri'flek·tər ,kar·ik·tə'ris·tik }

reflector microphone [ENG ACOUS] A highly directional microphone which has a surface that reflects the rays of impinging sound from a given direction to a common point at which a microphone is located, and the sound waves in the speech-frequency range are in phase at the microphone. { ri'flek·tər ,mī·krə,fōn }

reflector voltage [ELECTR] Voltage between the reflector electrode and the cathode in a reflex klystron. { ri'flek·tər ,vōl·tij }

reflex baffle [ENG ACOUS] A loudspeaker baffle in which a portion of the radiation from the rear of the diaphragm is propagated forward after controlled shift of phase or other modification, to increase the overall radiation in some portion of the audio-frequency spectrum. Also known as vented baffle. { 'rē,fleks ,baf·əl }

reflex bunching [ELECTR] The bunching that occurs in an electron stream which has been made to reverse its direction in the drift space. { 'rē ,fleks ,bənch·iŋ }

reflex circuit [ELECTR] A circuit in which the signal is amplified twice by the same amplifier tube or tubes, once as an intermediate-frequency signal before detection and once as an audio-frequency signal after detection. { 'rē,fleks ,sər·kət }

reflexive processing [COMPUT SCI] Information processing in which two or more computers connected by communications channels run identical programs and take the same actions at the same time, so that users in different locations can work on the same programs at the same time. { ri'flek·siv 'prä,ses·iŋ }

reflex klystron [ELECTR] A single-cavity klystron in which the electron beam is reflected back through the cavity resonator by a repelling electrode having a negative voltage; used as a microwave oscillator. Also known as reflex oscillator. { 'rē,fleks 'klī,strän }

reflex oscillator *See* reflex klystron. { 'rē,fleks 'äs·ə,lād·ər }

reformat [COMPUT SCI] To change the arrangement of data in a storage device. { rē'fòr·mat }

refraction [COMMUN] That property of earth's atmosphere that, due to its density profile, causes radio waves to propagate generally with a downward curve, sometimes rivaling the curvature of the earth; in radar height estimation, corrections for estimated refraction must be made. [ELECTROMAG] The change in direction of lines of force of an electric or magnetic field at a boundary between media with different permittivities or permeabilities. { ri'frak·shən }

refraction loss [ELECTROMAG] Portion of the transmission loss that is due to refraction resulting from nonuniformity of the medium. { ri'frak·shən ,lòs }

refractive constant *See* index of refraction. { ri 'frak·tiv 'kän·stənt }

refractive index *See* index of refraction. { ri'frak· tiv ,in,deks }

refresh [COMPUT SCI] A process of periodically replacing data to prevent the data from decaying, as on a cathode-ray-tube display or in a dynamic random-access memory. { ri'fresh }

regenerate [ELECTR] **1.** To restore pulses to their original shape. **2.** To restore stored information to its original form in a storage tube in order to counteract fading and disturbances. { rē'jen·ə ,rāt }

regeneration [CONT SYS] *See* positive feedback. [ELECTR] Replacement or restoration of charges in a charge storage tube to overcome decay effects, including loss of charge by reading. { rē ,jen·ə'rā·shən }

regenerative amplifier [ELECTR] An amplifier that uses positive feedback to give increased gain and selectivity. { rē'jen·rəd·iv 'am·plə ,fī·ər }

regenerative braking [ELEC] A system of dynamic braking in which the electric drive motors are used as generators and return the kinetic energy of the motor armature and load to the electric supply system. { rē'jen·rəd·iv 'brāk·iŋ }

regenerative clipper [ELECTR] A type of monostable multivibrator which is a modification of a Schmitt trigger; used for pulse generation. { rē'jen·rəd·iv 'klip·ər }

regenerative detector [ELECTR] A vacuum-tube detector circuit in which radio-frequency energy is fed back from the anode circuit to the grid circuit to give positive feedback at the carrier frequency, thereby increasing the amplification and sensitivity of the circuit. { rē'jen·rəd·iv di'tek·tər }

regenerative divider [ELECTR] Frequency divider which employs modulation, amplification, and selective feedback to produce the output wave. { rē'jen·rəd·iv di'vīd·ər }

regenerative feedback *See* positive feedback. { rē'jen·rəd·iv 'fēd,bak }

regenerative fuel cell [ELEC] A fuel cell in which the reaction product is processed to regenerate the reactants. { rē'jen·rəd·iv 'fyül ,sel }

regenerative read [COMPUT SCI] A read operation in which the data are automatically written back into the locations from which they are taken. { rē'jen·rəd·iv 'rēd }

regenerative receiver |ELECTR| A radio receiver that uses a regenerative detector. { rē'jen·rəd·iv ri'sē·vər }

regenerative repeater |COMMUN| A repeater that performs pulse regeneration to restore the original shape of a pulse signal used in teletypewriter and other code circuits. { rē'jen·rəd·iv ri'pēd·ər }

regenerator |ELECTR| **1.** A circuit that repeatedly supplies current to a display or memory device to prevent data from decaying. **2.** See repeater. { rē'jen·ə,rād·ər }

region |COMPUT SCI| A group of machine addresses which refer to a base address. { 'rē·jən }

regional address |COMPUT SCI| An address of a machine instruction within a series of consecutive addresses; for example, R18 and R19 are specific addresses in an R region of N consecutive addresses, where all addresses must be named. { 'rēj·ən·əl ə'dres }

regional center |COMMUN| A long-distance telephone office which has the highest rank in routing of telephone calls. { 'rēj·ən·əl 'sen·tər }

register |COMMUN| **1.** The accurate matching or superimposition of two or more images, such as the three color images on the screen of a color display. Also known as registration. **2.** The alignment of positions relative to a specified reference or coordinate, such as hole alignments in punched cards, or positioning of images in an optical character recognition device. **3.** Part of an automatic switching telephone system that receives and stores the dialing pulses that control the further operations necessary in establishing a telephone connection. |COMPUT SCI| The computer hardware for storing one machine word. { 'rej·ə·stər }

register capacity |COMPUT SCI| The upper and lower limits of the numbers which may be processed in a register. { 'rej·ə·stər kə'pas·əd·ē }

register circuit |ELECTR| A switching circuit with memory elements that can store from a few to millions of bits of coded information; when needed, the information can be taken from the circuit in the same code as the input, or in a different code. { 'rej·ə·stər ,sər·kət }

register control |CONT SYS| Automatic control of the position of a printed design with respect to reference marks or some other part of the design, as in photoelectric register control. { 'rej·ə·stər kən,trōl }

register length |COMPUT SCI| The number of digits, characters, or bits, which a register can store. { 'rej·ə·stər ¦leŋkth }

register-level compatibility |COMPUT SCI| Property of hardware components that are totally compatible, having registers with the same type, size, and names. { ¦rej·ə·stər ,lev·əl kəm,pad·ə'bil·əd·ē }

register-sender |COMMUN| A unit that generates and recognizes the supervisory signals to make connection to a circuit switching unit. { 'rej·ə·stər 'sen·dər }

register variable |COMPUT SCI| A variable in a computer program that is assigned to a register in the central processing unit instead of to a location in main storage. { 'rej·ə·stər ,ver·ē·ə·bəl }

registration See register. { ,rej·ə'strā·shən }

registration mark |COMPUT SCI| In character recognition, a preprinted indication of the relative position and direction of various elements of the source document to be recognized. { ,rej·ə'strā·shən ,märk }

regular |ELECTROMAG| In a definite direction; not diffused or scattered, when applied to reflection, refraction, or transmission. { 'reg·yə·lər }

regular expression |COMPUT SCI| A formal description of a language acceptable by a finite automaton or for the behavior of a sequential switching circuit. { 'reg·yə·lər ik'spresh·ən }

regulated power supply |ELEC| A power supply containing means for maintaining essentially constant output voltage or output current under changing load conditions. { 'reg·yə,lād·əd 'paú·ər sə,plī }

regulating system See automatic control system. { 'reg·yə,lād·iŋ ,sis·təm }

regulating transformer |ELEC| Transformer having one or more windings excited from the system circuit or a separate source and one or more windings connected in series with the system circuit for adjusting the voltage or the phase relation or both in steps, usually without interrupting the load. { 'reg·yə,lād·iŋ tranz,fȯr·mər }

regulating winding |ELEC| Of a transformer, a supplementary winding connected in series with one of the main windings to change the ratio of transformation or the phase relation, or both, between circuits. { 'reg·yə,lād·iŋ ,wīnd·iŋ }

regulation |CONT SYS| The process of holding constant a quantity such as speed, temperature, voltage, or position by means of an electronic or other system that automatically corrects errors by feeding back into the system the condition being regulated; regulation thus is based on feedback, whereas control is not. |ELEC| The change in output voltage that occurs between no load and full load in a transformer, generator, or other source. |ELECTR| The difference between the maximum and minimum tube voltage drops within a specified range of anode current in a gas tube. { ,reg·yə'lā·shən }

regulation of constant-current transformer |ELEC| Maximum departure of the secondary current from its rated value expressed in percent of the rated secondary current, with rated primary voltage and frequency applied. { ,reg·yə'lā·shən əv ¦kän·stənt ¦kə·rənt tranz'fȯr·mər }

regulator |CONT SYS| A device that maintains a desired quantity at a predetermined value or varies it according to a predetermined plan. { 'reg·yə,lād·ər }

regulator problem See linear regulator problem. { 'reg·yə,lād·ər ,präb·ləm }

reimbursed time |COMPUT SCI| The machine time which is loaned or rented to another office, agency, or organization, either on a reimbursable or reciprocal basis. { rē·əm,bərst 'tīm }

Reinartz crystal oscillator [ELECTR] Crystal-controlled vacuum-tube oscillator in which the crystal current is kept low by placing a resonant circuit in the cathode lead tuned to half the crystal frequency; the resulting regeneration at the crystal frequency improves efficiency without the danger of uncontrollable oscillation at other frequencies. { 'rīn₁ärts 'krist·əl 'äs·ə₁läd·ər }

reinitialize [COMPUT SCI] To return a computer program to the condition it was in at the start of processing, so that nothing remains from previous executions of the program. { ¦rē·i'nish·əl ₁īz }

reinserter See direct-current restorer. { ¦rē·ən 'sərd·ər }

reinsertion of carrier [ELECTR] Combining a locally generated carrier signal in a receiver with an incoming signal of the suppressed carrier type. { ¦rē·ən'sər·shən əv 'kar·ē·ər }

rejection band [ELECTROMAG] The band of frequencies below the cutoff frequency in a uniconductor waveguide. Also known as stop band. { ri'jek·shən ₁band }

rejector See trap. { ri'jek·tər }

rejector circuit See band-stop filter. { ri'jek·tər ₁sər·kət }

rejector impedance See dynamic impedance. { ri'jek·tər im₁pēd·əns }

relation [COMPUT SCI] A two-dimensional table in which data are arranged in a relational data structure. { ri'lā·shən }

relational algebraic language [COMPUT SCI] A low-level procedural language for carrying out fundamental algebraic operations on a database of relations. { ri'lā·shən·əl 'al·jə₁brā·ik ₁laŋ·gwij }

relational calculus language [COMPUT SCI] A higher-level nonprocedural language for operating on a database of relations, containing statements that can be mapped to the fundamental algebraic operations on the database. { ri'lā·shən·əl 'kal·kyə·ləs ₁laŋ·gwij }

relational capability [COMPUT SCI] Property of two or more data files that can be joined together for viewing, editing, or creation of reports. { ri ¦lā·shən·əl ₁kāp·ə'bil·əd·ē }

relational database See relational system. { ri'lā·shən·əl 'dad·ə₁bās }

relational data structure [COMPUT SCI] A type of data structure in which data are represented as tables in which no entry contains more than one value. { ri'lā·shən·əl 'dad·ə ₁strək·chər }

relationally complete [COMPUT SCI] Property of a programming language that provides for the construction of all relations derivable from some set of base relations by the application of the primitive algebraic operations. { ri'lā·shən·əl·ē kəm'plēt }

relational operator [COMPUT SCI] An operator that indicates whether one quantity is equal to, greater than, or less than another. { ri'lā·shən·əl 'äp·ə₁rād·ər }

relational spreadsheet [COMPUT SCI] A spreadsheet whose data are stored in a central database and are copied from the database into the spreadsheet when the spreadsheet is called up. { ri¦lā·shən·əl 'spred₁shēt }

relational system [COMPUT SCI] A database management system in which a relational data structure is used. Also known as relational database. { ri'lā·shən·əl ₁sis·təm }

relative address [COMPUT SCI] The numerical difference between a desired address and a known reference address. { 'rel·əd·iv ə'dres }

relative attenuation [ELECTR] The ratio of the peak output voltage of an electric filter to the voltage at the frequency being considered. { 'rel·əd·iv ə₁ten·yə'wā·shən }

relative bandwidth [ELECTR] For an electric filter, the ratio of the bandwidth being considered to a specified reference bandwidth, such as the bandwidth between frequencies at which there is an attenuation of 3 decibels. { 'rel·əd·iv 'band ₁width }

relative byte address [COMPUT SCI] A relative address expressed as the number of bytes from a point of reference to the desired address. { 'rel·ə·tiv 'bīt ₁ad₁res }

relative coding [COMPUT SCI] A form of computer programming in which the address part of an instruction indicates not the desired address but the difference between the location of the instruction and the desired address. { 'rel·əd·iv 'kōd·iŋ }

relative dielectric constant See dielectric constant. { 'rel·əd·iv ¦dī·i'lek·trik 'kän·stənt }

relative gain [ELECTROMAG] The gain of an antenna in a given direction when the reference antenna is a half-wave, loss-free dipole isolated in space whose equatorial plane contains the given direction. { 'rel·əd·iv ¦gān }

relative interference effect [ENG ACOUS] Of a single-frequency electric wave in an electroacoustic system, the ratio, usually expressed in decibels, of the amplitude of a wave of specified reference frequency to that of the wave in question when the two waves are equal in interference effects. { 'rel·əd·iv ₁in·tər'fir·əns i₁fekt }

relative permittivity See dielectric constant. { 'rel·əd·iv ₁pər·mə'tiv·əd·ē }

relative power gain [ELECTROMAG] Of one transmitting or receiving antenna over another, the measured ratio of the signal power one produces at the receiver input terminals to that produced by the other, the transmitting power level remaining fixed. { 'rel·əd·iv 'paů·ər ₁gān }

relative resistance [ELEC] The ratio of the resistance of a piece of a material to the resistance of a piece of specified material, such as annealed copper, having the same dimensions and temperature. { 'rel·əd·iv ri'zis·təns }

relative response [ELECTR] In a transducer, the amount (in decibels) by which the response under some particular condition exceeds the response under a reference condition. { 'rel·əd·iv ri'späns }

relative triple precision [COMPUT SCI] The retention of three times as many digits of a quantity as the computer normally handles; for example, a computer whose basic word consists of 10 decimal digits is called upon to handle 30 decimal digit quantities. { 'rel·əd·iv 'trip·əl prə'sizh·ən }

relative vector [COMPUT SCI] In computer graphics, a vector whose end points are given in relative coordinates. { 'rel·əd·iv 'vek·tər }

relaxation circuit [ELECTR] Circuit arrangement, usually of vacuum tubes, reactances, and resistances, which has two states or conditions, one, both, or neither of which may be stable; the transient voltage produced by passing from one to the other, or the voltage in a state of rest, can be used in other circuits. { ,rē,lak'sā·shən ,sər·kət }

relaxation inverter [ELECTR] An inverter that uses a relaxation oscillator circuit to convert direct-current power to alternating-current. { ,rē ,lak'sā·shən in,vərd·ər }

relaxation oscillator [ELECTR] An oscillator whose fundamental frequency is determined by the time of charging or discharging a capacitor or coil through a resistor, producing waveforms that may be rectangular or sawtooth. { ,rē ,lak'sā·shən ,äs·ə,lād·ər }

relay [COMMUN] A microwave or other radio system used for passing a signal from one radio communication link to another. [ELEC] A device that is operated by a variation in the conditions in one electric circuit and serves to make or break one or more connections in the same or another electric circuit. Also known as electric relay. { 'rē,lā }

relay center [COMMUN] A switching center in which messages are automatically routed according to data contained in the messages or message headers. { 'rē,lā ,sen·tər }

relay contact [ELEC] One of the pair of contacts that are closed or opened by the movement of the armature of a relay. { 'rē,lā ,kän,takt }

relay control system [CONT SYS] A control system in which the error signal must reach a certain value before the controller reacts to it, so that the control action is discontinuous in amplitude. { 'rē,lā kən'trōl ,sis·təm }

relay satellite See communications satellite. { 'rē ,lā ,sad·əl,īt }

relay selector [ELEC] Relay circuit associated with a selector, consisting of a magnetic impulse counter, for registering digits and holding a circuit. { 'rē,lā si,lek·tər }

relay station See repeater station. { 'rē,lā ,stā·shən }

relay system [COMMUN] See radio relay system. [ELEC] Dial-switching equipment that does not use mechanical switches, but is made up principally of relays. { 'rē,lā ,sis·təm }

reliability [ENG] The probability that a component part, equipment, or system, including computer hardware and software, will satisfactorily perform its intended function under given circumstances, such as environmental conditions, limitations as to operating time, and frequency and thoroughness of maintenance for a specified period of time. { ri,lī·ə'bil·əd·ē }

relieving anode [ELECTR] Of a pool-cathode tube, an auxiliary anode which provides an alternative conducting path for reducing the current to another electrode. { ri'lēv·iŋ ,an,ōd }

relocatable code [COMPUT SCI] A code generated by an assembler or compiler, and in which all memory references needing relocation are either specially marked or relative to the current program-counter reading. { ¦rē·lō¦kād·ə·bəl 'kōd }

relocatable emulator [COMPUT SCI] An emulator which does not require a stand-alone machine but executes in a multiprogramming environment. { ¦rē·lō¦kād·ə·bəl 'em·yə,lād·ər }

relocatable program [COMPUT SCI] A program coded in such a way that it may be located and executed in any part of memory. { ¦rē·lō¦kād·ə·bəl 'prō,gram }

relocate [COMPUT SCI] To establish or change the location of a program routine while adjusting or modifying the address references within the instructions to correctly indicate the new locations. { rē'lō,kāt }

relocating loader [COMPUT SCI] A loader in which some of the addresses in the program to be loaded are expressed relative to the start of the program rather than in absolute form. { ¦rē·lō ¦kād·iŋ 'lōd·ər }

relocation hardware [COMPUT SCI] Equipment in a multiprogramming system which allows a computer program to be run in any available space in memory. { ,rē·lō'kā·shən ,härd,wer }

relocation register [COMPUT SCI] A hardware element that holds a constant to be added to the address of each memory location in a computer program running in a multiprogramming system, as determined by the location of the area in memory assigned to the program. { ,rē·lō'kā·shən ,rej·ə·stər }

reluctance microphone See magnetic microphone. { ri'lək·təns ,mī·krə,fōn }

reluctance motor [ELEC] A synchronous motor, similar in construction to an induction motor, in which the member carrying the secondary circuit has salient poles but no direct-current excitation; it starts as an induction motor but operates normally at synchronous speed. { ri'lək·təns ,mōd·ər }

reluctance pressure transducer [ENG] Pressure-measurement transducer in which pressure changes activate equivalent magnetic-property changes. { ri'lək·təns 'presh·ər tranz,dü·sər }

remedial maintenance See corrective maintenance. { ri'mēd·ē·əl 'mānt·ən·əns }

remember condition [ELECTR] Condition of a flip-flop circuit in which no change takes place between a given internal state and the next state. { ri'mem·bər kən,dish·ən }

remodulator [ELECTR] A circuit that converts amplitude modulation to audio frequency-shift modulation for transmission of data signals over a radio channel. Also known as converter. { rē'mäj·ə,lād·ər }

remote access [COMPUT SCI] Ability to gain entry to a computer system from a location some distance away. { ri'mōt 'ak,ses }

remote batch computing [COMPUT SCI] The running of programs, usually during nonprime hours, or whenever the demands of real-time or time-sharing computing slacken sufficiently to allow less pressing programs to be run. { ri'mōt 'bach kəm,pyüd·iŋ }

remote batch processing [COMPUT SCI] Batch processing in which an input device is located at a distance from the main installation and has access to a computer through a communication link. { ri'mōt 'bach ,präˌses·iŋ }

remote calculator [COMPUT SCI] A keyboard device that can be connected to the central processing unit of a distant computer over an ordinary telephone channel, enabling the user to present programs to the computer. { ri'mōt 'kal·kyə ,lād·ər }

remote communications software [COMPUT SCI] Software that allows a microcomputer to control or duplicate the operation of another microcomputer at a distant location, using the standard telephone system. { riˌmōt kə,myü·nəˈkā·shənz 'söf,wer }

remote computing system [COMPUT SCI] A data-processing system that has terminals distant from the central processing unit, from which users can communicate with the central processing unit and compile, debug, test, and execute programs. { ri'mōt kəm'pyüd·iŋ ,sis·təm }

remote computing system exchange [COMPUT SCI] A device that handles communications between the central processing unit and remote consoles of a remote computing system, and enables several remote consoles to operate at the same time without interfering with each other. { ri'mōt kəm'pyüd·iŋ ,sis·təm iks,chānj }

remote computing system language [COMPUT SCI] A computer language used for communications between the central processing unit and remote consoles of a remote computer system, generally incorporating a procedure-oriented language such as FORTRAN, but also containing operating statements, such as instructions to debug or execute programs. { ri'mōt kəm'pyüd·iŋ ,sis·təm ,laŋ·gwij }

remote computing system log [COMPUT SCI] A record of the volumes of data transmitted and of the frequency of various types of events during the operation of remote consoles in a remote computing system. { ri'mōt kəm'pyüd·iŋ ,sis·təm ,läg }

remote console [COMPUT SCI] A terminal in a remote computing system that has facilities for communicating with, and exerting control over, the central processing unit, and which may have any of various types of display units, printers, and data entry devices for direct communication with the central processing unit. { ri'mōt 'kän,sōl }

remote control [CONT SYS] Control of a quantity which is separated by an appreciable distance from the controlling quantity; examples include telemetering, telephone, and television. { ri'mōt kən'trōl }

remote-cutoff tube See variable-mu tube. { ri'mōt ¦kəd,óf ,tüb }

remote debugging [COMPUT SCI] 1. The testing and correction of computer programs at a remote console of a remote computing system. 2. See remote testing. { ri'mōt dē'bəg·iŋ }

remote indicator [ELECTR] 1. An indicator located at a distance from the data-gathering sensing element, with data being transmitted to the indicator mechanically, electrically over wires, or by means of light, radio, or sound waves. 2. See repeater. { ri'mōt 'in·də,kād·ər }

remote inquiry [COMPUT SCI] Interrogation of the content of an automatic data-processing equipment storage unit from a device remotely displaced from the storage unit site. { ri'mōt 'iŋ ,kwə·rē }

remote manipulator [ENG] A mechanical, electromechanical, or hydromechanical device that enables a person, directly controlling the device through handles or switches, to perform manual operations while separated from the site of the work. Also known as manipulator; teleoperator. { ri'mōt mə'nip·yə,lād·ər }

remote metering See telemetering. { ri'mōt 'mēd·ə·riŋ }

remote pickup [COMMUN] Picking up a radio or television program at a remote location and relaying it to the studio or transmitter over wire lines or a radio link. { ri'mōt 'pik,əp }

remote plan position indicator See plan position indicator repeater. { ri'mōt ¦plan pə¦zish·ən 'in·də,kād·ər }

remote sensing [ELEC] Sensing, by a power supply, of voltage directly at the load, so that variations in the load lead drop do not affect load regulation. { ri'mōt 'sens·iŋ }

remote subscriber [COMMUN] Subscriber to a network that does not have direct access to the switching center, but has access to the circuit through a facility such as a base message center. { ri'mōt səb'skrīb·ər }

remote terminal [COMPUT SCI] A computer terminal which is located away from the central processing unit of a data-processing system, at a location convenient to a user of the system. { ri'mōt 'tər·mən·əl }

remote testing [COMPUT SCI] A method of testing and correcting computer programs; programmers do not go to the computer center but provide detailed instructions to be carried out by computer operators along with the programs and associated test data. Also known as remote debugging. { ri'mōt 'test·iŋ }

removable medium [COMPUT SCI] A data storage medium, such as magnetic tape or floppy disk, that can be physically removed from the unit that reads and writes on it. { ri'müv·ə·bəl 'mē·dē·əm }

removable plugboard See detachable plugboard. { ri'müv·ə·bəl 'pləg,bórd }

REM statement [COMPUT SCI] A statement in a computer program that consists of remarks or comments that document the program, and contains no executable code. { 'rem ,stāt·mənt }

repeatability [CONT SYS] The ability of a robot to reposition itself at a location to which it is directed or at which it is commanded to stop. { ri,pēd·ə'bil·əd·ē }

repeat accuracy |CONT SYS| The variations in the actual position of a robot manipulator from one cycle to the next when the manipulator is commanded to repeatedly return to the same point or position. { ri'pēt 'ak·yə·rə·sē }

repeater |ELEC| *See* repeating coil. |ELECTR| **1.** An amplifier or other device that receives weak signals and delivers corresponding stronger signals with or without reshaping of waveforms; may be either a one-way or two-way repeater. Also known as regenerator. **2.** An indicator that shows the same information as is shown on a master indicator. Also known as remote indicator. { ri'pēd·ər }

repeater jammer |ELECTR| A jammer that intercepts an enemy radar signal and reradiates the signal after modifying it to incorporate erroneous data on azimuth, range, or number of targets. { ri'pēd·ər ˌjam·ər }

repeater station |COMMUN| A station containing one or more repeaters. Also known as relay station. { ri'pēd·ər ˌstā·shən }

repeating coil |ELEC| A transformer used to provide inductive coupling between two sections of a telephone line when a direct connection is undesirable. Also known as repeater. { ri 'pēd·iŋ ˌkȯil }

repeating-coil bridge cord |ELEC| In telephony, a method of connecting the common office battery to the cord circuits by connecting the battery to the midpoints of a repeating coil, bridged across the cord circuit. { ri'pēd·iŋ ˌkȯil 'brij ˌkȯrd }

repeat key |COMPUT SCI| A key on a typewriter or computer keyboard that, when depressed at the same time as a character key, causes repeated printing or generation of the character until one of the keys is released. { ri'pēt ˌkē }

repeat operator |COMPUT SCI| A pseudo instruction using two arguments, a count *p* and an increment *n*: the word immediately following the instruction is repeated *p* times, with the values 0, *n*, 2*n*, . . . , (*p* − 1)*n* added to the successive words. { ri'pēt ˌäp·əˌrād·ər }

repeller |ELECTR| An electrode whose primary function is to reverse the direction of an electron stream in an electron tube. Also known as reflector. { ri'pel·ər }

repetition equivalent |COMMUN| In a complete telephone connection, a measure of the grade of transmission experienced by the subscribers using the connection; it includes the combined effects of volume, distortion, noise, and all other subscriber reactions and usages. { ˌrep·ə'tish·ən i'kwiv·ə·lənt }

repetition frequency *See* repetition rate. { ˌrep·ə'tish·ən ˌfrē·kwən·sē }

repetition instruction |COMPUT SCI| An instruction that causes one or more other instructions to be repeated a specified number of times, usually with systematic address modification occurring between repetitions. { ˌrep·ə'tish·ən in,strək·shən }

repetition rate |COMMUN| The rate at which recurrent signals are produced or transmitted. Also known as recurrence rate; repetition frequency. { ˌrep·ə'tish·ən ˌrāt }

repetitive addressing |COMPUT SCI| A system used on some computers in which, under certain conditions, an instruction is written without giving the address of the operand, and the operand address is automatically that of the location addressed by the last previous instruction. { rə'ped·əd·iv ə'dres·iŋ }

repetitive analog computer |COMPUT SCI| An analog computer which repeatedly carries out the solution of a problem at a rapid rate (10 to 60 times a second) while an operator may vary parameters in the problem. { rə'ped·əd·iv 'an·ə,läg kəm'pyüd·ər }

repetitive statement |COMPUT SCI| A statement in a computer program that is repeatedly executed for a specified number of times or for as long as a specified condition holds true. { ri'ped·əd·iv 'stāt·mənt }

repetitive unit |COMPUT SCI| A type of circuit which appears more than once in a computer. { rə'ped·əd·iv yü·nət }

reply |COMMUN| A radio-frequency signal or combination of signals transmitted by a transponder in response to an interrogation. Also known as response. { ri'plī }

report |COMPUT SCI| An output document prepared by a data-processing system. { ri'pȯrt }

report generator |COMPUT SCI| A routine which produces a complete data-processing report, given only a description of the desired content and format, plus certain information concerning the input file. Also known as report writer. { ri'pȯrt ˌjen·ə,rād·ər }

reporting time interval |COMMUN| The time for transmission of data or a report from the originating terminal to the end receiver. { ri'pȯrd·iŋ 'tīm ˌin·tər·vəl }

report program |COMPUT SCI| A program that prints out an analysis of a file of records, usually arranged by keys, each analysis or total being produced when a key change takes place. { ri'pȯrt ˌprō,gram }

report program generator |COMPUT SCI| A nonprocedural programming language that provides a convenient method of producing a wide variety of reports. Abbreviated RPG. { ri'pȯrt ¦prō,gram ˌjen·ə,rād·ər }

report writer *See* report generator. { ri'pȯrt ˌrīd·ər }

representation condition |COMPUT SCI| The condition that, if one software entity is less than another entity in terms of a selected attribute, then any software metric for that attribute must associate a smaller number to the first entity than it does to the second entity. { ˌrep·rə·zen'tā·shən kən,dish·ən }

representative calculating time |COMPUT SCI| The time required to perform a specified operation or series of operations. { ¦rep·ri¦zen·təd·iv 'kal·kyə,lād·iŋ ˌtīm }

reproduce head *See* playback head. { ¦rē·prə¦düs ˌhed }

reproducing system *See* sound-reproducing system. { ¦rē·prə¦düs·iŋ ˌsis·təm }

reproduction speed |COMMUN| Area of copy recorded per unit time in facsimile transmission. { ¦rē·prə¦dək·shən ˌspēd }

repulsion-induction motor |ELEC| A repulsion motor that has a squirrel-cage winding in the rotor in addition to the repulsion-motor winding. { ri'pəl·shən in'dək·shən ˌmōd·ər }

repulsion motor |ELEC| An alternating-current motor having stator windings connected directly to the source of ac power and rotor windings connected to a commutator; brushes on the commutator are short-circuited and are positioned to produce the rotating magnetic field used for starting and running. { ri'pəl·shən ˌmōd·ər }

repulsion-start induction motor |ELEC| An alternating-current motor that starts as a repulsion motor; at a predetermined speed the commutator bars are short-circuited to give the equivalent of a squirrel-cage winding for operation as an induction motor with constant-speed characteristics. { ri'pəl·shən ¦stärt in'dək·shən ˌmōd·ər }

request/grant logic |COMPUT SCI| Logic circuitry which, in effect, selects the interrupt line with highest priority. { ri'kwest 'grant ˌläj·ik }

request repeat system |COMMUN| System using an error-detecting code, and so arranged that a signal detected as being in error automatically initiates a request for retransmission. { ri'kwest ri¦pēt ˌsis·təm }

reradiation |COMMUN| Undesirable radiation of signals generated locally in a radio receiver, causing interference or revealing the location of the receiver. { rē,rā·dē'ā·shən }

rerun |COMPUT SCI| To run a program or a portion of it again on a computer. Also known as rollback. { 'rē,rən }

rerun point |COMPUT SCI| A location in a program from which the program may be started anew after an interruption of the computer run. { 'rē ˌrən ˌpóint }

rerun routine |COMPUT SCI| A routine designed to be used in the wake of a computer malfunction or a coding or operating mistake to reconstitute a routine from the last previous rerun point. { 'rē ˌrən ˌrü,tēn }

rescap |ELEC| A capacitor and resistor assembly manufactured as a packaged encapsulated circuit. Also known as capacitor-resistor unit; capristor; packaged circuit; resistor-capacitor unit. { 'res,kap }

rescue dump |COMPUT SCI| The copying of the entire contents of a computer memory into auxiliary storage devices, carried out periodically during the course of a computer program so that in case of a machine failure the program can be reconstituted at the last point at which this operation was executed. { 'res·kyü ˌdəmp }

reserve |COMPUT SCI| To assign portions of a computer memory and of input/output and storage devices to a specific computer program in a multiprogramming system. { ri'zərv }

reserve battery |ELEC| A battery which is inert until an operation is performed which brings all the cell components into the proper state and location to become active. { ri'zərv 'bad·ə·rē }

reserved word |COMPUT SCI| A word which cannot be used in a programming language to represent an item of data because it has some particular significance to the compiler, or which can be used only in a particular context. { ri'zərvd 'wərd }

reset See clear. { 'rē,set }

reset action |CONT SYS| Floating action in which the final control element is moved at a speed proportional to the extent of proportional-position action. { 'rē,set ˌak·shən }

reset condition |ELECTR| Condition of a flip-flop circuit in which the internal state of the flip-flop is reset to zero. { 'rē,set kən,dish·ən }

reset cycle |COMPUT SCI| The return of a cycle index counter to its initial value. { 'rē,set ˌsī·kəl }

reset input |COMPUT SCI| The act of resetting the original conditions of a problem after a program is run on an analog computer. { 'rē,set 'in,pút }

reset mode |COMPUT SCI| The phase of operation of an analog computer during which the required initial conditions are entered into the system and the computing units are inoperative. Also known as initial condition mode. { 'rē,set ,mōd }

reset pulse |ELECTR| **1.** A drive pulse that tends to reset a magnetic cell in the storage section of a digital computer. **2.** A pulse used to reset an electronic counter to zero or to some predetermined position. { 'rē,set ,pəls }

resettability |ELECTR| The ability of the tuning element of an oscillator to retune the oscillator to the same operating frequency for the same set of input conditions. { ri,sed·ə'bil·əd·ē }

resident executive |COMPUT SCI| The portion of the executive routine that is permanently stored in a computer's main memory. Also known as resident monitor. { 'rez·ə·dənt ig'zek·yəd·iv }

resident module See resident routine. { 'rez·ə·dənt 'mä·jəl }

resident monitor See resident executive. { 'rez·ə·dənt 'män·əd·ər }

resident routine |COMPUT SCI| Any computer routine which is stored permanently in the memory, such as the resident executive. Also known as resident module. { 'rez·ə·dənt rü'tēn }

residual charge |ELEC| The charge remaining on the plates of a capacitor after initial discharge. { rə'zij·ə·wəl 'chärj }

residual current |ELECTR| Current flowing through a thermionic diode when there is no anode voltage, due to the velocity of the electrons emitted by the heated cathode. { rə'zij·ə·wəl ¦kə·rənt }

residual error rate See undetected error rate. { rə'zij·ə·wəl 'er·ər ,rāt }

residual modulation See carrier noise. { rə'zij·ə·wəl ˌmäj·ə'lā·shən }

residual voltage |ELEC| Vector sum of the voltages to ground of the several phase wires of an electric supply circuit. { rə'zij·ə·wəl 'vōl·tij }

residue check See modulo N check. { 'rez·ə,dü ˌchek }

residue system |COMPUT SCI| A number system in which each digit position corresponds to a different radix, all pairs of radices are relatively prime, and the value of a digit with radix *r* for an integer A is equal to the remainder when A is divided by *r*. { 'rez·ə‚dü ‚sis·təm }

resilience |COMPUT SCI| The ability of computer software to be used for long periods of time. { rə'zil·yəns }

resistance |ELEC| **1.** The opposition that a device or material offers to the flow of direct current, equal to the voltage drop across the element divided by the current through the element. Also known as electrical resistance. **2.** In an alternating-current circuit, the real part of the complex impedance. { ri'zis·təns }

resistance box |ELEC| A box containing a number of precision resistors connected to panel terminals or contacts so that a desired resistance value can be obtained by withdrawing plugs (as in a post-office bridge) or by setting multicontact switches. { ri'zis·təns ‚bäks }

resistance bridge *See* Wheatstone bridge. { ri'zis·təns ‚brij }

resistance-capacitance circuit |ELEC| A circuit which has a resistance and a capacitance in series, and in which inductance is negligible. Abbreviated R-C circuit. { ri'zis·təns kə'pas·əd·əns ‚sər·kət }

resistance-capacitance constant |ELEC| Time constant of a resistive-capacitive circuit, equal in seconds to the resistance value in ohms multiplied by the capacitance value in farads. Abbreviated R-C constant. { ri'zis·təns kə'pas·əd·əns ‚kän·stənt }

resistance-capacitance coupled amplifier |ELECTR| An amplifier in which a capacitor provides a path for signal currents from one stage to the next, with resistors connected from each side of the capacitor to the power supply or to ground; it can amplify alternating-current signals but cannot handle small changes in direct currents. Also known as R-C amplifier; R-C coupled amplifier; resistance-coupled amplifier. { ri'zis·təns kə'pas·əd·əns ‚kəp·əld 'am·plə‚fī·ər }

resistance-capacitance network |ELEC| Circuit containing resistances and capacitances arranged in a particular manner to perform a specific function. Abbreviated R-C network. { ri'zis·təns kə'pas·əd·əns 'net‚wərk }

resistance-capacitance oscillator |ELECTR| Oscillator in which the frequency is determined by resistance and capacitance elements. Abbreviated R-C oscillator. { ri'zis·təns kə'pas·əd·əns 'äs·ə‚lād·ər }

resistance commutation |ELEC| Commutation of an electric rotating machine in which brushes with relatively high resistance span at least one commutator segment, in order to achieve a linear variation of current with time, and thereby minimize self-inductive voltage in the coils. { ri'zis·təns ‚kam·yə'tā·shən }

resistance-coupled amplifier *See* resistance-capacitance coupled amplifier. { ri'zis·təns ‚kəp·əld 'am·plə‚fī·ər }

resistance coupling |ELECTR| Coupling in which resistors are used as the input and output impedances of the circuits being coupled; a coupling capacitor is generally used between the resistors to transfer the signal from one stage to the next. Also known as R-C coupling; resistance-capacitance coupling; resistive coupling. { ri'zis·təns ‚kəp·liŋ }

resistance drop |ELEC| The voltage drop occurring between two points on a conductor due to the flow of current through the resistance of the conductor; multiplying the resistance in ohms by the current in amperes gives the voltage drop in volts. Also known as IR drop. { ri'zis·təns ‚dräp }

resistance element |ELEC| An element of resistive material in the form of a grid, ribbon, or wire, used singly or built into groups to form a resistor for heating purposes, as in an electric soldering iron. { ri'zis·təns ‚el·ə·mənt }

resistance grounding |ELEC| Electrical grounding in which lines are connected to ground by a resistive (totally dissipative) impedance. { ri'zis·təns ‚graúnd·iŋ }

resistance heating |ELEC| The generation of heat by electric conductors carrying current; degree of heating is proportional to the electrical resistance of the conductor; used in electrical home appliances, home or space heating, and heating ovens and furnaces. { ri'zis·təns ‚hēd·iŋ }

resistance lamp |ELEC| Electric lamp used to prevent the current in a circuit from exceeding a desired limit. { ri'zis·təns ‚lamp }

resistance loss |ELEC| Power loss due to current flowing through resistance; its value in watts is equal to the resistance in ohms multiplied by the square of the current in amperes. { ri'zis·təns ‚lòs }

resistance material |ELEC| Material having sufficiently high resistance per unit length or volume to permit its use in the construction of resistors. { ri'zis·təns mə'tir·ē·əl }

resistance measurement |ELEC| The quantitative determination of that property of an electrically conductive material, component, or circuit called electrical resistance. { ri'zis·təns ‚mezh·ər·mənt }

resistance meter |ENG| Any instrument which measures electrical resistance. Also known as electrical resistance meter. { ri'zis·təns ‚mēd·ər }

resistance noise *See* thermal noise. { ri'zis·təns ‚nòiz }

resistance-start motor |ELEC| A split-phase motor having a resistance connected in series with the auxiliary winding; the auxiliary circuit is opened when the motor attains a predetermined speed. { ri'zis·təns ¦stärt ‚mōd·ər }

resistance strain gage [ELECTR] A strain gage consisting of a strip of material that is cemented to the part under test and that changes in resistance with elongation or compression. { ri'zis·təns 'strān ‚gāj }

resistive coupling See resistance coupling. { ri 'zis·tiv 'kəp·liŋ }

resistive load [ELEC] A load whose total reactance is zero, so that the alternating current is in phase with the terminal voltage. Also known as nonreactive load. { ri'zis·tiv 'lōd }

resistive superconducting fault-current limiter [ELEC] A fault-current limiter in which a superconductor is directly connected in series to the line to be protected and is immersed in a coolant which is chilled by a refrigerant, and the connection from the line at room temperature to the superconductor is provided by special current leads, which are designed to minimize the heat transfer to the coolant. { ri¦zis·tiv ¦sü·pər·kən ‚dək·tiŋ 'fȯlt‚ kər·ənt ‚lim·əd·ər }

resistive unbalance [ELEC] Unequal resistance in the two wires of a transmission line. { ri'zis·tiv ən'bal·əns }

resistivity See electrical resistivity. { ‚rē‚zis'tiv· əd·ē }

resistor [ELEC] A device designed to have a definite amount of resistance; used in circuits to limit current flow or to provide a voltage drop. Also known as electrical resistor. { ri'zis·tər }

resistor-capacitor-transistor logic [ELECTR] A resistor-transistor logic with the addition of capacitors that are used to enhance switching speed. { ri'zis·tər kə'pas·əd·ər tran'zis·tər ‚läj·ik }

resistor-capacitor unit See rescap. { ri'zis·tər kə'pas·əd·ər ‚yü·nət }

resistor color code [ELEC] Code adopted by the Electronic Industries Association to mark the values of resistance on resistors in a readily recognizable manner; the first color represents the first significant figure of the resistor value, the second color the second significant figure, and the third color represents the number of zeros following the first two figures; a fourth color is sometimes added to indicate the tolerance of the resistor. { ri'zis·tər 'kəl·ər ‚kōd }

resistor core [ELEC] Insulating support on which a resistor element is wound or otherwise placed. { ri'zis·tər ‚kȯr }

resistor element [ELEC] That portion of a resistor which possesses the property of electric resistance. { ri'zis·tər ‚el·ə·mənt }

resistor network [ELEC] An electrical network consisting entirely of resistances. { ri'zis·tər 'net‚wərk }

resistor termination [ELECTR] A thick-film conductor pad overlapping and contacting a thick-film resistor area. { ri'zis·tər ‚tər·mə'nā·shən }

resistor-transistor logic [ELECTR] One of the simplest logic circuits, having several resistors, a transistor, and a diode. Abbreviated RTL. { ri'zis·tər tran'zis·tər ‚läj·ik }

resnatron [ELECTR] A microwave-beam tetrode containing cavity resonators, used chiefly for generating large amounts of continuous power at high frequencies. { 'rez·nə‚trän }

resolution [CONT SYS] The smallest increment in distance that can be distinguished and acted upon by an automatic control system. [ELECTR] In television, the maximum number of lines that can be discerned on the screen at a distance equal to screen height. [ELECTROMAG] In radar, the minimum separation between two targets or features thereof, in angle, range, cross range, or range rate, at which they can be distinguished on a radar display or in the data processing. Also known as resolving power. { ‚rez·ə'lü·shən }

resolution chart See test pattern. { ‚rez·ə'lü·shən ‚chärt }

resolution error [COMPUT SCI] An error of an analog computing unit that results from its inability to respond to changes of less than a given magnitude. { ‚rez·ə'lü·shən ‚er·ər }

resolution factor [COMPUT SCI] In information retrieval, the ratio obtained in dividing the total number of documents retrieved (whether relevant or not to the user's needs) by the total number of documents available in the file. { ‚rez·ə'lü·shən ‚fak·tər }

resolution wedge [COMMUN] On a video test pattern, a group of gradually converging lines used to measure resolution. { ‚rez·ə'lü·shən ‚wej }

resolve motion-rate control [CONT SYS] A form of robotic control in which the controlled variables are the velocity vectors of the end points of a manipulator, and the angular velocities of the joints are determined to obtain the desired results. { ri'zolv 'mō·shən ¦rāt kən‚trōl }

resolver [ELEC] A synchro or other device whose rotor is mechanically driven to translate rotor angle into electrical information corresponding to the sine and cosine of rotor angle; used for interchanging rectangular and polar coordinates. Also known as sine-cosine generator; synchro resolver. [ELECTR] **1.** A synchro or other device whose input is the angular position of an object, such as the rotor of an electric machine, and whose output is electric signals, usually proportional to the sine and cosine of an angle, and often in digital form; used to interchange rectangular and polar coordinates, and in servomechanisms to report the orientation of controlled objects. Also known as angular resolver. **2.** A device that accepts a single vector-valued analog input and produces for output either analog or digital signals proportional to two or three orthogonal components of the vector. Also known as vector resolver. { ri'zäl·vər }

resolving cell [ELECTROMAG] In radar, volume in space whose diameter is the product of slant range and beam width, and whose length is the pulse length. { ri'zälv·iŋ ‚sel }

resolving power [ELECTROMAG] See resolution. { ri'zälv·iŋ ‚paů·ər }

resolving time [COMPUT SCI] In computers, the shortest permissible period between trigger pulses for reliable operation of a binary cell. [ENG] Minimum time interval, between events, that can be detected; resolving time may refer to an electronic circuit, to a mechanical recording device, or to a counter tube. { ri'zälv·iŋ ,tīm }

resonance [ELEC] A phenomenon exhibited by an alternating-current circuit in which there are relatively large currents near certain frequencies, and a relatively unimpeded oscillation of energy from a potential to a kinetic form; a special case of the physics definition. { 'rez·ən·əns }

resonance bridge [ELEC] A four-arm alternating-current bridge used to measure inductance, capacitance, or frequency; the inductor and the capacitor, which may be either in series or in parallel, are tuned to resonance at the frequency of the source before the bridge is balanced. { 'rez·ən·əns ,brij }

resonance curve [ELEC] Graphical representation illustrating the manner in which a tuned circuit responds to the various frequencies in and near the resonant frequency. { 'rez·ən·əns ,kərv }

resonance method [ELEC] A method of determining the impedance of a circuit element, in which resonance frequency of a resonant circuit containing the element is measured. { 'rez·ən·əns ,meth·əd }

resonance transformer [ELEC] A high-voltage transformer in which the secondary circuit is tuned to the frequency of the power supply. [ELECTR] An electrostatic particle accelerator, used principally for acceleration of electrons, in which the high-voltage terminal oscillates between voltages which are equal in magnitude and opposite in sign. { 'rez·ən·əns tranz,för·mər }

resonant antenna [ELECTROMAG] An antenna for which there is a sharp peak in the power radiated or intercepted by the antenna at a certain frequency, at which electric currents in the antenna form a standing-wave pattern. { 'res·ən·ənt an 'ten·ə }

resonant capacitor [ELEC] A tubular capacitor that is wound to have inductance in series with its capacitance. { 'res·ən·ənt kə'pas·əd·ər }

resonant cavity See cavity resonator. { 'res·ən·ənt 'kav·əd·ē }

resonant chamber See cavity resonator. { 'res·ən·ənt 'chām·bər }

resonant circuit [ELEC] A circuit that contains inductance, capacitance, and resistance of such values as to give resonance at an operating frequency. { 'res·ən·ənt 'sər·kət }

resonant coupling [ELEC] Coupling between two circuits that reaches a sharp peak at a certain frequency. { 'res·ən·ənt 'kəp·liŋ }

resonant diaphragm [ELECTROMAG] A diaphragm, in waveguide technique, so proportioned as to introduce no reactive impedance at the design frequency. { 'res·ən·ənt 'dī·ə,fram }

resonant element See cavity resonator. { 'res·ən·ənt 'el·ə·mənt }

resonant gate transistor [ELECTR] Surface field-effect transistor incorporating a cantilevered beam which resonates at a specific frequency

to provide high-Q-frequency discrimination. { 'res·ən·ənt 'gāt tran,zis·tər }

resonant helix [ELECTROMAG] An inner helical conductor in certain types of transmission lines and resonant cavities, which carries currents with the same frequency as the rest of the line or cavity. { 'res·ən·ənt 'hē·liks }

resonant iris [ELECTROMAG] A resonant window in a circular waveguide; it resembles an optical iris. { 'res·ən·ənt 'ī·rəs }

resonant line [ELECTROMAG] A transmission line having values of distributed inductance and distributed capacitance so as to make the line resonant at the frequency it is handling. { 'res·ən·ənt 'līn }

resonant-line oscillator [ELECTR] Oscillator in which one or more sections of transmission lines are employed as resonant elements. { 'res·ən·ənt ¦līn 'äs·ə,lād·ər }

resonant-line tuner [ELECTR] A device in which resonant lines are used to tune the antenna, radio-frequency amplifier, or radio-frequency oscillator circuits; tuning is achieved by moving shorting contacts that change the electrical lengths of the lines. { 'res·ən·ənt ¦līn 'tün·ər }

resonant-mode power supply [ELECTR] An electronic power supply in which the current and voltage waveforms are shaped to sinusoids by a small inductor and capacitor inserted in the current path. { ¦rez·ən·ənt ¦mōd 'pau·ər sə,plī }

resonant-reed relay [ELEC] A reed relay in which the reed switch closes only when the required frequency is applied to the operating coil, to make one of the reeds vibrate until its amplitude is sufficient to make contact with the other reed; used in selective paging systems. { 'res·ən·ənt ¦rēd 'rē,lā }

resonant resistance [ELEC] Resistance value to which a resonant circuit is equivalent. { 'res·ən·ənt ri'zis·təns }

resonant voltage step-up [ELEC] Ability of an inductor and a capacitor in a series resonant circuit to deliver a voltage several times greater than the input voltage of the circuit. { 'res·ən·ənt ¦vōl·tij 'step,əp }

resonant wavelength [ELECTROMAG] The wavelength in free space of electromagnetic radiation having a frequency equal to a natural resonance frequency of a cavity resonator. { 'res·ən·ənt 'wāv,leŋkth }

resonant window [ELECTROMAG] A parallel combination of inductive and capacitive diaphragms, used in a waveguide structure to provide transmission at the resonant frequency and reflection at other frequencies. { 'res·ən·ənt 'win·dō }

resonate [ELEC] To bring to resonance, as by tuning. { 'rez·ən,āt }

resonating cavity [ELECTROMAG] Short piece of waveguide of adjustable length, terminated at either or both ends by a metal piston, an iris diaphragm, or some other wave-reflecting device; it is used as a filter, as a means of coupling between guides of different diameters, and as impedance networks corresponding to those used in radio circuits. { 'rez·ən,ād·iŋ 'kav·əd·ē }

resonator grid |ELECTR| Grid that is attached to a cavity resonator in velocity-modulated tubes to provide coupling between the resonator and the electron beam. { 'rez·ən‚äd·ər ‚grid }

responder |ELECTR| The transmitter section, including the appropriate encoder, of a radar transponder. { ri'spän·dər }

responder beacon |ELECTR| The radar beacon that serves to emit the signals of the responder in a transponder. { ri'spän·dər ‚bē·kən }

response |COMMUN| See reply. |CONT SYS| A quantitative expression of the output of a device or system as a function of the input. Also known as system response. { ri'späns }

response characteristic |CONT SYS| The response as a function of an independent variable, such as direction or frequency, often presented in graphical form. { ri'späns ‚kar·ik·tə‚ris·tik }

response time |COMPUT SCI| The delay experienced in time sharing between request and answer, a delay which increases when the number of users on the system increases. |CONT SYS| The time required for the output of a control system or element to reach a specified fraction of its new value after application of a step input or disturbance. |ELEC| The time it takes for the pointer of an electrical or electronic instrument to come to rest at a new value, after the quantity it measures has been abruptly changed. { ri'späns ‚tīm }

responsor |ELECTR| The receiving section of an interrogator-responsor. { ri'spän·sər }

restart |COMPUT SCI| To go back to a specific planned point in a routine, usually in the case of machine malfunction, for the purpose of rerunning the portion of the routine in which the error occurred; the length of time between restart points in a given routine should be a function of the mean free error time of the machine itself. { 'rē‚stärt }

resting frequency See carrier frequency. { 'rest·iŋ ‚frē·kwən·sē }

restore |COMPUT SCI| In computers, to regenerate, to return a cycle index or variable address to its initial value, or to store again. |ELECTR| Periodic charge regeneration of volatile computer storage systems. { ri'stȯr }

restorer See direct-current restorer. { ri'stȯr·ər }

restorer pulses |ELECTR| In computers, pairs of complement pulses, applied to restore the coupling-capacitor charge in an alternating-current flip-flop. { ri'stȯr·ər ‚pəls·əz }

restoring logic |ELECTR| Circuitry designed so that even with an imperfect input pulse a standard output occurs at the exit of each successive logic gate. { ri'stȯr·iŋ ‚läj·ik }

rest potential |ELEC| Residual potential difference remaining between an electrode and an electrolyte after the electrode has become polarized. { 'rest pə‚ten·chəl }

restricted function |COMPUT SCI| A function of the operating system that cannot be used by application programs. { ri'strik·təd ‚fəŋk·shən }

retarding-field oscillator |ELECTR| An oscillator employing an electron tube in which the elec-

trons oscillate back and forth through a grid that is maintained positive with respect to both the cathode and anode; the field in the region of the grid exerts a retarding effect through the grid in either direction. Also known as positive-grid oscillator. { ri'tärd·iŋ ¦fēld 'äs·ə ‚lād·ər }

retard transmitter |ELECTR| Transmitter in which a delay period is introduced between the time of actuation and the time of transmission. { ‚ri'tärd tranz‚mid·ər }

retention period |COMPUT SCI| The length of time that data must be kept on a reel of magnetic tape before it can be destroyed. { ri'ten·chən ‚pir·ē·əd }

retention time |ELECTR| The maximum time between writing into a storage tube and obtaining an acceptable output by reading. Also known as storage time. { ri'ten·chən ‚tīm }

retina |COMPUT SCI| In optical character recognition, a scanning device. { 'ret·ən·ə }

retina character reader |COMPUT SCI| A character reader that operates in the manner of the human retina in recognizing identical letters in different type fonts. { 'ret·ən·ə 'kar·ik·tər ‚rēd·ər }

retrace See flyback. { 'rē‚trās }

retrace blanking |ELECTR| Blanking a video display during vertical retrace intervals to prevent retrace lines from showing on the screen. { 'rē ‚trās ‚blaŋk·iŋ }

retrace line |ELECTR| The line traced by the electron beam in a cathode-ray tube in going from the end of one line or field to the start of the next line or field. Also known as return line. { 'rē‚trās ‚līn }

retransmission unit |ELECTR| Control unit used at an intermediate station for feeding one radio receiver-transmitter unit for two-way communication. { ¦rē·tranz'mish·ən ‚yü·nət }

retrieve |COMPUT SCI| To find and select specific information. { ri'trēv }

retroaction See positive feedback. { ¦re·trō'ak·shən }

retrofit |ENG| A modification of equipment to incorporate changes made in later production of similar equipment; it may be done in the factory or field. Derived from retroactive refit. { 're·trō ‚fit }

retry |COMPUT SCI| When a central processing unit error is detected during execution of an instruction, the computer will execute this instruction unless a register was altered by the operation. { 'rē‚trī }

return |COMPUT SCI| **1.** To return control from a subroutine to the calling program. **2.** To go back to a planned point in a computer program and rerun a portion of the program, usually when an error is detected; rerun points are usually not more than 5 minutes apart. |ELECTR| See echo. { ri'tərn }

return address |COMPUT SCI| The address in storage to which a computer program is directed upon completion of a subroutine. { ri'tərn 'ad ‚res }

return busy tone |COMMUN| A signal returned to the register-sender that, in turn, returns a busy indication to the calling station. { ri'tərn 'biz·ē ,tōn }

return code |COMPUT SCI| An indicator that is issued by a computer upon completion of a subroutine or function, or of the entire program, that indicates the result of the processing and, in particular, whether the processing was successful or ended abnormally because of an error. { ri'tərn ,kōd }

return interval |ELECTR| Interval corresponding to the direction of sweep not used for delineation. { ri'tərn ,in·tər·vəl }

return jump |COMPUT SCI| A jump instruction in a subroutine which passes control to the first statement in the program which follows the instruction called the subroutine. { ri'tərn ,jəmp }

return key |COMPUT SCI| A key on a typewriter or a computer keyboard that, when depressed, causes a print mechanism or cursor to move to the beginning of the next line. { ri'tərn ,kē }

return line See retrace line. { ri'tərn ,līn }

return loss |COMMUN| **1.** The difference between the power incident upon a discontinuity in a transmission system and the power reflected from the discontinuity. **2.** The ratio in decibels of the power incident upon a discontinuity to the power reflected from the discontinuity. { ri'tərn ,lòs }

return to zero mode |COMPUT SCI| Computer readout mode in which the signal returns to zero between each bit indication. { ri'tərn tə 'zir·ō ,mōd }

return trace See flyback. { ri'tərn ,trās }

return wire |ELEC| The ground wire, common wire, or negative wire of a direct-current power circuit. { ri'tərn ,wīr }

reusable |COMPUT SCI| Of a program, capable of being used by several tasks without having to be reloaded; it is a generic term, including reenterable and serially reusable. { rē'yü·zə·bəl }

reverse bias |ELECTR| A bias voltage applied to a diode or a semiconductor junction with polarity such that little or no current flows; the opposite of forward bias. { ri'vərs 'bī·əs }

reverse-blocking tetrode thyristor See silicon controlled switch. { ri'vərs ¦bläk·iŋ 'te,trōd thī'ris·tər }

reverse-blocking triode thyristor See silicon controlled rectifier. { ri'vərs ¦bläk·iŋ 'trī,ōd thī'ris·tər }

reverse code dictionary |COMPUT SCI| Alphabetic or alphanumeric arrangement of codes associated with their corresponding English words or terms. { ri'vərs ¦kōd 'dik·shə,ner·ē }

reverse current |ELECTR| Small value of direct current that flows when a semiconductor diode has reverse bias. { ri'vərs 'kə·rənt }

reverse-current protection |ELEC| A device which senses when there is a reversal in the normal direction of current in an electric power system, indicating an abnormal condition of the system, and which initiates appropriate action to prevent damage to the system. { ri'vərs ¦kə·rənt prə,tek·shən }

reverse-current relay |ELEC| Relay that operates whenever current flows in the reverse direction. { ri'vərs ¦kə·rənt 'rē,lā }

reverse direction See inverse direction. { ri'vərs di'rek·shən }

reverse-direction flow |COMPUT SCI| A logical path that runs upward or to the left on a flowchart. { ri'vərs di¦rek·shən ,flō }

reverse feedback See negative feedback. { ri'vərs 'fēd,bak }

reverse key |ELEC| Key used in a circuit to reverse the polarity of that circuit. { ri'vərs 'kē }

reverse Polish notation |COMPUT SCI| The version of Polish notation, used in some calculators, in which operators follow the operators with which they are associated. Abbreviated RPN. Also known as postfix notation; suffix notation. { ri'vərs 'pō·lish nō'tā·shən }

reverse power |ELEC| Transmission of electric energy through a circuit in a direction opposite to the usual direction. { ri'vərs 'paů·ər }

reverse video |COMPUT SCI| An electronic display mode in which the normal properties of the display are reversed; for example, normally white characters on a black background will appear as black characters on a white background. Also known as inverse video. { ri'vərs 'vid·ē·ō }

reverse voltage |ELEC| In the case of two opposing voltages, voltage of that polarity which produces the smaller current. { ri'vərs 'vōl·tij }

reversible booster |ELEC| Booster capable of adding to and subtracting from the voltage of a circuit. { ri'vər·sə·bəl 'büs·tər }

reversible capacitance |ELECTR| Limit, as the amplitude of an applied sinusoidal capacitor voltage approaches zero, of the ratio of the amplitude of the resulting in-phase fundamental-frequency component of transferred charge to the amplitude of the applied voltage, for a given constant bias voltage superimposed on the sinusoidal voltage. { ri'vər·sə·bəl kə'pas·əd·əns }

reversible counter |COMPUT SCI| A counter which stores a number whose value can be decreased or increased in response to the appropriate control signal. { ri'vər·sə·bəl 'kaůnt·ər }

reversible motor |ELEC| A motor in which the direction of rotation can be reversed by means of a switch that changes motor connections when the motor is stopped. { ri'vər·sə·bəl 'mōd·ər }

reversible transducer |ELECTR| Transducer whose loss is independent of transmission direction. { ri'vər·sə·bəl tranz'düs·ər }

reversing motor |ELEC| A motor for which the direction of rotation can be reversed by changing electric connections or by other means while the motor is running at full speed; the motor will then come to a stop, reverse, and attain full speed in the opposite direction. { ri'vərs·iŋ ,mōd·ər }

reversing switch |ELEC| A switch intended to reverse the connections of one part of a circuit. { ri'vərs·iŋ ,swich }

revolute-coordinate robot See jointed-arm robot. { 'rev·ə¡lüt kō¡ȯrd·ən·ət 'rō,bät }

rewind |ELECTR| **1.** The components on a magnetic tape recorder that serve to return the tape to the supply reel at high speed. **2.** To return a magnetic tape to its starting position. { 'rē,wīnd }

rewrite |COMPUT SCI| The process of restoring a storage device to its state prior to reading; used when the information-storing state may be destroyed by reading. { 'rē,rīt }

RFI See radio-frequency interference.

RGB monitor |COMPUT SCI| A video display screen that requires separate red, green, and blue signals from a computer or other source. { 'är ¡jē¡bē 'män·əd·ər }

RG line See radio-frequency cable. { 'är'jē ,līn }

rheostat |ELEC| A resistor constructed so that its resistance value may be changed without interrupting the circuit to which it is connected. Also known as variable resistor. { 'rē·ə,stat }

rheostatic control |ELEC| A method of controlling the speed of electric motors that involves varying the resistance or reactance in the armature or field circuit; used in motors that drive elevators. { ¡rē·ə¡stad·ik kən'trōl }

rheostriction See pinch effect. { 'rē·ə,strik·shən }

rheotaxial growth |ENG| A chemical vapor deposition technique for producing silicon diodes and transistors on a fluid layer having high surface mobility. { ¡rē·ə¡tak·sē·əl 'grōth }

RHI display See range-height indicator display. { ¡är¡āch'ī di,splā }

rhombic antenna |ELECTROMAG| A horizontal antenna having four conductors forming a diamond or rhombus; usually fed at one apex and terminated with a resistance or impedance at the opposite apex. Also known as diamond antenna. { 'räm·bik an'ten·ə }

rhumbatron See cavity resonator. { 'rəm·bə,trän }

ribbon cable |ELEC| A cable made of normal, round, insulated wires arranged side by side and fastened together by a cohesion process to form a flexible ribbon. { 'rib·ən ,kā·bəl }

ribbon conductor |ELEC| A thin, flat piece of metal suitable for carrying electric current. { 'rib·ən kən,dək·tər }

ribbon microphone |ENG ACOUS| A microphone whose electric output results from the motion of a thin metal ribbon mounted between the poles of a permanent magnet and driven directly by sound waves; it is velocity-actuated if open to sound waves on both sides, and pressure-actuated if open to sound waves on only one side. { 'rib·ən 'mī·krə,fōn }

Rice neutralization |ELECTR| Development of voltage in the grid circuit of a vacuum tube in order to nullify or cancel feedback through the tube. { 'rīs ,nü·trə·lə'zā·shən }

Rice neutralizing circuit |ELECTR| Radio-frequency amplifier circuit that neutralizes the grid-to-plate capacitance of an amplifier tube. { 'rīs 'nü·trə,līz·iŋ ,sər·kət }

Rice video |ELECTR| Referring to the video and its particular probability density produced by an amplitude detector (demodulator) when the Gaussian radio noise and a signal of a known and constant amplitude are together incident to it. { 'rīs ¡vid·ē·ō }

Richardson-Dushman equation |ELECTR| An equation for the current density of electrons that leave a heated conductor in thermionic emission. Also known as Dushman equation. { 'rich·ərd·sən 'dəsh·mən i,kwā·zhən }

Richardson effect See thermionic emission. { 'rich·ərd·sən i,fekt }

Richardson plot |ELECTR| A graph of log (J/T^2) against $1/T$, where J is the current density of electrons leaving a heated conductor in thermionic emission, and T is the temperature of the conductor; according to the Richardson-Dushman equation, this is a straight line. { 'rich·ərd·sən ,plät }

ridge waveguide |ELECTROMAG| A circular or rectangular waveguide having one or more longitudinal internal ridges that serve primarily to increase transmission bandwidth by lowering the cutoff frequency. { 'rij 'wāv,gīd }

Rieke diagram |ELECTR| A chart showing contours of constant power output and constant frequency for a microwave oscillator, drawn on a Smith chart or other polar diagram whose coordinates represent the components of the complex reflection coefficient at the oscillator load. { 'rē·kə ,dī·ə,gram }

right-hand taper |ELEC| Taper in which there is greater resistance in the clockwise half of the operating range of a rheostat or potentiometer (looking from the shaft end) than in the counterclockwise half. { 'rīt 'hand 'tā·pər }

right-justify |COMPUT SCI| To shift the contents of a register so that the right or least significant digit is at some specified position. { 'rīt 'jəs·tə,fī }

right value |COMPUT SCI| The actual data content of a symbolic variable in a computer program; it is one of two components of the symbolic variable, the other being the memory address. Abbreviated rvalue. { 'rīt 'val·yü }

rigid copper coaxial line |ELECTROMAG| A coaxial cable in which the central conductor and outer conductor are formed by joining rigid pieces of copper. { 'rij·id 'käp·ər kō'ak·sē·əl ,līn }

rigid insulation |ELEC| Electrical insulation that is part of a rigid structure, and must provide mechanical strength and stability of form as well as a dielectric barrier; mica, glass, porcelain, and thermosetting resins are the principal materials used. { 'rij·id ,in·sə'lā·shən }

R-indicator See R-display. { 'är ,in·də,kād·ər }

ring |COMPUT SCI| A cyclic arrangement of data elements, usually including a specified entry pointer. { riŋ }

ring-around |COMMUN| **1.** Improper routing of a call back through a switching center already trying to complete the same call, thus tying up the trunks by repeating the cycle. **2.** Oscillation of a repeater caused by leakage of the transmitter signal into the receiver. { 'riŋ ə,raůnd }

ring bus |ELEC| A substation switching arrangement that may consist of four, six, or more breakers connected in a closed loop, with the same number of connection points. { 'riŋ ‚bəs }

ring circuit [ELECTROMAG] In waveguide practice, a hybrid T junction having the physical configuration of a ring with radial branches. { 'riŋ ‚sər·kət }

ring counter [ELECTR] A loop of binary scalers or other bistable units so connected that only one scaler is in a specified state at any given time; as input signals are counted, the position of the one specified state moves in an ordered sequence around the loop. { 'riŋ ‚kaunt·ər }

ring data structure [COMPUT SCI] Stored data that is organized by a chain of pointers so that the last pointer is directed back to the beginning of the chain. { 'riŋ 'dad·ə ‚strək·chər }

ring discharge [ELECTR] A ring-shaped discharge generated by a high-frequency oscillating electromagnetic field produced by an external coil. Also known as toroidal discharge. { ¦riŋ 'dis ‚chärj }

ring head [ELECTR] A recording and playback head in a magnetic recording system which has the form of a ring with a gap at one point, and on which the coils are wound. { 'riŋ ‚hed }

ringing [COMMUN] The production of an audible or visible signal at a station or switchboard by means of an alternating or pulsating current. [CONT SYS] An oscillatory transient occurring in the output of a system as a result of a sudden change in input. { 'riŋ·iŋ }

ringing circuit [ELECTR] A circuit which has a capacitance in parallel with a resistance and inductance, with the whole in parallel with a second resistance; it is highly underdamped and is supplied with a step or pulse input. { 'riŋ·iŋ ‚sər·kət }

ring modulator [ELECTR] A modulator in which four diode elements are connected in series to form a ring around which current flows readily in one direction; input and output connections are made to the four nodal points of the ring; used as a balanced modulator, demodulator, or phase detector. { 'riŋ 'mäj·ə‚läd·ər }

ring network [COMMUN] A communications network in which the nodes can be considered to be on a circle, about which messages must be routed. Also known as loop network. { 'riŋ 'net ‚wərk }

ring power transmission line [ELEC] A power transmission line that is closed upon itself to form a ring; provides two paths between the power station and any customer, and enables a faulty section of the line to be disconnected without interrupting service to customers. { 'riŋ 'paù·ər tranz'mish·ən ‚līn }

ring shift See cyclic shift. { 'riŋ ¦shift }

ring structure [COMPUT SCI] A chained file organization such that the end of the chain points to its beginning. { 'riŋ ‚strək·chər }

ring time [ELECTR] The length of time in microseconds required for a pulse of energy transmitted into an echo box to die out; a measurement of the performance of radar. { 'riŋ ‚tīm }

ripple [ELEC] The alternating-current component in the output of a direct-current power supply, arising within the power supply from incomplete filtering or from commutator action in a dc generator. { 'rip·əl }

ripple-carry adder [COMPUT SCI] A device for addition of two n-bit binary numbers, formed by connecting n full adders in cascade, with the carry output of each full adder feeding the carry input of the following full adder. { 'rip·əl ¦kar·ē ‚ad·ər }

ripple filter [ELECTR] A low-pass filter designed to reduce ripple while freely passing the direct current obtained from a rectifier or direct-current generator. Also known as smoothing circuit; smoothing filter. { 'rip·əl ‚fil·tər }

ripple voltage [ELEC] The alternating component of the unidirectional voltage from a rectifier or generator used as a source of direct-current power. { 'rip·əl ‚vōl·tij }

RISC See reduced instruction set computer. { risk }

rise time [CONT SYS] The time it takes for the output of a system to change from a specified small percentage (usually 5 or 10) of its steady-state increment to a specified large percentage (usually 90 or 95). [ELEC] The time for the pointer of an electrical instrument to make 90% of the change to its final value when electric power suddenly is applied from a source whose impedance is high enough that it does not affect damping. { 'rīz ‚tīm }

rising-sun magnetron [ELECTR] A multicavity magnetron in which resonators having two different resonant frequencies are arranged alternately for the purpose of mode separation; the cavities appear as alternating long and short radial slots around the perimeter of the anode structure, resembling the rays of the sun. { 'rīz·iŋ ¦sən 'mag·nə‚trän }

Rivest-Shamir-Adleman algorithm [COMMUN] A public-key algorithm whose strength is based on the fact that factoring large composite prime numbers into their prime factors involves an overwhelming amount of computation. Abbreviated RSA algorithm. { ri'vest shə'mir 'ad·əl·mən ‚al·gə‚rith·əm }

RLL code See run-length-limited code. { ¦är¦el'el ‚kōd }

rms value See root-mean-square value. { ¦är¦em 'es ‚val·ü }

RO See receive-only.

robot [CONT SYS] A mechanical device that can be programmed to perform a variety of tasks of manipulation and locomotion under automatic control. { 'rō‚bät }

robust program [COMPUT SCI] **1.** A computer program using an iterative process that converges rapidly to the solution being sought. **2.** A computer program that performs well even under unusual conditions. { ¦rō·bəst 'prō·grəm }

roc [ELEC] A unit of electrical conductivity equal to the conductivity of a material in which an electric field of 1 volt per centimeter gives rise to a current density of 1 ampere per square centimeter. Derived from reciprocal ohm centimeter. { räk }

Rochelle-electric See ferroelectric. { rō'shel·i‚lek·trik }

rocket antenna |ELECTROMAG| An antenna carried on a rocket, to receive signals controlling the rocket or to transmit measurements made by instruments aboard the rocket. { 'räk·ət an ‚ten·ə }

rocky point effect |ELECTR| Transient but violent discharges between electrodes in high-voltage transmitting tubes. { 'räk·ē ¦póint i‚fekt }

rod gap |ELEC| **1.** A device that is usually formed of two $1/2$-square-inch (3-square-centimeter) rods, one grounded and the other connected to the line conductor, but may also have the shape of rings or horns, used to limit the magnitude of transient overvoltages on an electrical system as a result of lightning strikes. **2.** Spark gap in which the electrodes are two coaxial rods, with ends between which the discharge takes place, cut perpendicularly to the axis. { 'räd ‚gap }

rod thermistor |ELECTR| A type of thermistor that has high resistance, long time constant, and moderate power dissipation; it is extruded as a long vertical rod 0.250–2.0 inches (0.63–5.1 centimeters) long and 0.050–0.110 inch (0.13–0.28 centimeter) in diameter, of oxide-binder mix and sintered; ends are coated with conducting paste and leads are wrapped on the coated area. { 'räd thər'mis·tər }

roentgen current |ELEC| An electric current arising from the motion of polarization charges, as in the rotation of a dielectric in a charged capacitor. { 'rent·gən ‚kər·ənt }

Roget's spiral |ELEC| A spiral wire, suspended vertically with the lower end in mercury, that is made to go through a cycle in which an electric current passing through the wire produces mutual attraction between the coils, causing the wire to lift out of the mercury and breaking the current; the spiral then expands under its own weight, so that the lower end drops back into the mercury and the current is reestablished. { rō ¦zhäz 'spī·rəl }

role indicator |COMPUT SCI| In information retrieval, a code assigned to a key word to indicate its part of speech, nature, or function. { 'rōl ‚in·də‚kād·ər }

rollback See rerun. { 'rōl‚bak }

roll in |COMPUT SCI| To restore to main memory a section of program or data that had previously been rolled out. { 'rōl ‚in }

rolling transposition |ELEC| Transposition in which the conductors of an open wire circuit are physically rotated in a substantially helical manner; with two wires, a complete transposition is usually executed in two consecutive spans. { 'rōl·iŋ ‚tranz·pə'zish·ən }

roll-off |ELECTR| Gradually increasing loss or attenuation with increase or decrease of frequency beyond the substantially flat portion of the amplitude-frequency response characteristic of a system or transducer. { 'rōl ‚óf }

roll out |COMPUT SCI| **1.** To make available additional main memory for one task by copying another task onto auxiliary storage. **2.** To read a computer register or counter by adding a one to each digit column simultaneously until all have

returned to zero, with a signal being generated at the instant a column returns to zero. { 'rōl ‚aút }

rollover |COMPUT SCI| A keyboard feature that allows more than one key to be depressed simultaneously, enabling the keys to be depressed more rapidly in sequence. { 'rōl‚ō·vər }

roll your own See user program. { 'rōl yər 'ōn }

rom |ELEC| A unit of electrical conductivity, equal to the conductivity of a material in which an electric field of 1 volt per meter gives rise to a current density of 1 ampere per square meter. Derived from reciprocal ohm meter. { räm }

ROMable code |COMPUT SCI| A computer program developed to be stored permanently in a read-only memory (ROM). { 'räm·ə·bəl 'kōd }

roof filter |ELECTR| Low-pass filter used in carrier telephone systems to limit the frequency response of the equipment to frequencies needed for normal transmission, thereby blocking unwanted higher frequencies induced in the circuit by external sources; improves runaround crosstalk suppression and minimizes high-frequency singing. { 'rüf ‚fil·tər }

room noise |COMMUN| Ambient noise in a telephone station. { 'rüm ‚nóiz }

room power |ELECTR| The electric power that is fed to the machinery in a computer room after passing through a power distribution unit, motor-generator set, or other conditioning and isolating device. { 'rüm ‚paú·ər }

root |COMPUT SCI| The origin or most fundamental point of a tree diagram. Also known as base. { rüt }

root component See root symbol. { 'rüt kəm 'pō·nənt }

root directory |COMPUT SCI| The starting point in a hierarchical file system, where the system operates when it is first started. { 'rüt di‚rek·trē }

root locus plot |CONT SYS| A plot in the complex plane of values at which the loop transfer function of a feedback control system is a negative number. { 'rüt ‚lō·kəs ‚plät }

root-mean-square current See effective current. { 'rüt ‚mēn 'skwer ‚kə·rənt }

root-mean-square value |PHYS| The square root of the time average of the square of a quantity; for a periodic quantity, such as a sine wave used for audio measurements, the average is taken over one complete cycle. Abbreviated rms value. Also known as effective value. { 'rüt ‚mēn 'skwer 'val·yü }

root segment |COMPUT SCI| The master or controlling segment of an overlay structure which always resides in the main memory of a computer. { 'rüt ‚seg·mənt }

root sum square |COMMUN| A method of combining the power of multiple signals by taking the square root of the sum of the squares of all the signals. Abbreviated RSS. { 'rüt ‚səm 'skwər }

root-sum-square value |PHYS| The square root of the sum of the squares of a series of related values; commonly used to express total harmonic distortion. { 'rüt ‚səm 'skwər 'val·yü }

root symbol |COMPUT SCI| An element of a formal language, generally unique, that is not derivable from other language elements. Also known as root component. { 'rüt ,sim·bəl }

root task |COMPUT SCI| The initial program on a parallel machine from which one or more child processes branch out in the fork-join model. { 'rüt ,task }

rope-lay conductor |ELEC| Cable composed of a central core surrounded by one or more layers of helically laid groups of wires. { 'rōp ¦lā kən ,dək·tər }

Rosenberg crossed-field generator |ELEC| A type of dynamoelectric amplifier which is self-regulating and can operate while the rotor varies in speed, the current never rising above a certain value. { 'rōz·ən,bərg ¦kròst ¦fēld 'jen·ə,rād·ər }

rosin joint |ELEC| A soldered joint in which one of the wires is surrounded by an almost invisible film of insulating rosin, making the joint intermittently or continuously open even though it looks good. { 'räz·ən ,jóint }

rotary amplifier See rotating magnetic amplifier. { 'rōd·ə·rē 'am·plə,fī·ər }

rotary beam |ELECTROMAG| Short-wave antenna system highly directional in azimuth and altitude, mounted in such a manner that it can be rotated to any desired position, either manually or by an electric motor drive. { 'rōd·ə·rē 'bēm }

rotary converter See dynamotor. { 'rōd·ə·rē kən'vərd·ər }

rotary coupler See rotating joint. { 'rōd·ə·rē 'kəp·lər }

rotary digital audio tape system |ELECTR| A digital audio tape system that uses the helical-scan technology developed for video systems, with a rotating drum containing two metal-in-gap heads. Abbreviated R-DAT system. { ¦rōd·ə·rē ,dij·əd·əl ,ód·ē·ō 'tāp ,sis·təm }

rotary gap See rotary spark gap. { 'rōd·ə·rē 'gap }

rotary joint See rotating joint. { 'rōd·ə·rē 'jóint }

rotary phase converter |ELEC| Machine which converts power from an alternating-current system of one or more phases to an alternating-current system of a different number of phases, but of the same frequency. { 'rōd·ə·rē 'fāz kən,vərd·ər }

rotary power source |ELEC| An uninterruptible power system in which a battery driven dc motor mechanically drives an ac generator in the event of a power outage. { 'rōd·ə·rē 'paù·ər ,sórs }

rotary spark gap |ELEC| A spark gap in which sparks occur between one or more fixed electrodes and a number of electrodes projecting outward from the circumference of a motor-driven metal disk. Also known as rotary gap. { 'rōd·ə·rē 'spärk ,gap }

rotary stepping relay See stepping relay. { 'rōd·ə·rē 'step·iŋ ,rē,lā }

rotary stepping switch See stepping relay. { 'rōd·ə·rē 'step·iŋ ,swich }

rotary switch |ELEC| A switch that is operated by rotating its shaft. { 'rōd·ə·rē 'swich }

rotary system |COMMUN| A telephone switching system that uses unidirectional, rotary switches that carry ten sets of brushes (wipers), only one of which is tripped as part of the control and selection process. { 'rōd·ə·rē 'sis·təm }

rotary transformer |ELEC| A rotating machine used to transform direct-current power from one voltage to another. { 'rōd·ə·rē tranz'fór·mər }

rotary-vane attenuator |ELECTROMAG| Device designed to introduce attenuation into a waveguide circuit by varying the angular position of a resistive material in the guide. { 'rōd·ə·rē ¦vān ə'ten·yə,wād·ər }

rotary voltmeter |ENG| Type of electrostatic voltmeter used for measuring high voltages. { 'rōd·ə·rē 'vōlt,mēd·ər }

rotating amplifier See rotating magnetic amplifier. { 'rō,tād·iŋ 'am·plə,fī·ər }

rotating-anode tube |ELECTR| An x-ray tube in which the anode rotates continuously to bring a fresh area of its surface into the beam of electrons, allowing greater output without melting the target. { 'rō,tād·iŋ ¦an,ōd ,tüb }

rotating-coil gaussmeter |ENG| An instrument for measuring low magnetic field strengths and flux densities by measuring the voltage induced in a search coil that is rotated in the field at constant speed. { ,rō,tād·iŋ ,kóil 'gaùs,mēd·ər }

rotating joint |ELECTROMAG| A joint that permits one section of a transmission line or waveguide to rotate continuously with respect to another while passing radio-frequency energy. Also known as rotary coupler; rotary joint. { 'rō ,tād·iŋ 'jóint }

rotating magnetic amplifier |ELEC| A prime-mover-driven direct-current generator whose power output can be controlled by small field input powers, to give power gain as high as 10,000. Also known as rotary amplifier; rotating amplifier. { 'rō,tād·iŋ mag'ned·ik 'am·plə ,fī·ər }

rotation |COMPUT SCI| An operation performed on data in a register of the central processing unit, in which all the bits in the register are shifted one position to the right or left, and the endmost bit, which is shifted out of the register, is carried around to the position at the opposite end of the register. { rō'tā·shən }

rotational delay See rotational latency. { rō'tā·shən·əl di'lā }

rotational latency |COMPUT SCI| The time required, following an order to read or write information in disk storage, for the location of the information to revolve beneath the appropriate read/write head. Also known as rotational delay. { rō'tā·shən·əl 'lāt·ən·sē }

rotational position sensing |COMPUT SCI| A fast disk search method whereby the control unit looks for a specified sector, and then receives the sector number required to access the record. { rō'tā·shən·əl pə'zish·ən ,sens·iŋ }

rotator |ELECTROMAG| A device that rotates the plane of polarization of a plane-polarized electromagnetic wave, such as a twist in a waveguide. { 'rō,tād·ər }

rotoflector |ELECTROMAG| In radar, elliptically shaped, rotating reflector used to reflect a vertically directed radar beam at right angles so that it radiates in a horizontal direction. { 'rōd·ə,flek·tər }

rotor |COMMUN| **1.** Disk with a set of input contacts and a set of output contacts, connected by any prearranged scheme designed to rotate within an electrical cipher machine. **2.** Disk whose rotation produces a variation of some cryptographic element in a cipher machine usually by means of lugs (or pins) in or on its periphery. |ELEC| The rotating member of an electrical machine or device, such as the rotating armature of a motor or generator, or the rotating plates of a variable capacitor. { 'rōd·ər }

rotor plate |ELEC| One of the rotating plates of a variable capacitor, usually directly connected to the metal frame. { 'rōd·ər ,plāt }

round-robin scheduling |COMPUT SCI| A scheduling algorithm which repeatedly runs through a list of users, giving each user the opportunity to use the central processing unit in succession. { 'raúnd ¦räb·ən 'skej·ə·liŋ }

round-the-world echo |COMMUN| A signal occurring every ¹/₇ second when a radio wave repeatedly encircles the earth at its speed of 186,000 miles (300,000 kilometers) per second. { 'raúnd thə 'wərld 'ek·ō }

round-trip echoes |ELECTROMAG| Multiple reflection echoes produced when a radar pulse is reflected from a target strongly enough so that the echo is reflected back to the target where it produces a second echo. { 'raúnd ¦trip 'ek·ōz }

router |COMMUN| A device that selects an appropriate pathway for a message and routes the message accordingly. { 'raúd·ər }

routine |COMPUT SCI| A set of digital computer instructions designed and constructed so as to accomplish a specified function. { rü'tēn }

routine library |COMPUT SCI| Ordered set of standard and proven computer routines by which problems or parts of problems may be solved. { rü'tēn ,lī,brer·ē }

routing |COMMUN| The assignment of a path by which a message will travel to its destination. { 'rüd·iŋ }

routing indicator |COMMUN| **1.** A group of letters, engineered and assigned, to identify a station within a digital communications network. **2.** A group of letters assigned to indicate the geographic location of a station; a fixed headquarters of a command, activity, or unit at a geographic location; or the general location of a tape relay or tributary station to facilitate the routing of traffic over tape relay networks. { 'rüd·iŋ ,in·də ,kād·ər }

routing message |COMMUN| The function performed at a central message processor of selecting the route, or alternate route required, by which a message will proceed to the next point in reaching its destination. { 'rüd·iŋ ,mes·ij }

row |COMPUT SCI| **1.** The characters, or corresponding bits of binary-coded characters, in a computer word. **2.** Equipment which simultaneously processes the bits of a character, the characters of a word, or corresponding bits of binary-coded characters in a word. **3.** Corresponding positions in a group of columns. { rō }

row address |COMPUT SCI| An index array entry field which contains the main storage address of a data block. { 'rō 'ad,res }

Rowland current |ELEC| A convection current that arises when a charged capacitor plate is rotated. { 'rō·lənd ,kər·ənt }

row order |COMPUT SCI| The storage of a matrix $a(m,n)$ as $a(1,1),a(1,2),\ldots,a(1,n),a(2,1), a(2,2),\ldots$ { 'rō ,ór·dər }

RPG See report program generator.

RPN See reverse Polish notation.

RS-232 |COMMUN| A standard developed by the Electronic Industries Association that governs the interface between data processing and data communications equipment, and is widely used to connect microcomputers to peripheral devices.

RSA algorithm See Rivest-Shamir-Adleman algorithm. { ¦är¦es¦ā 'al·gə,rith·əm }

R-scan See R-display. { 'är ,skan }

R-scope See R-display. { 'är ,skōp }

RSS See root sum square.

RTL See resistor-transistor logic.

rubber banding |COMPUT SCI| In computer graphics, the moving of a line or object, with one end held fixed in position. { ¦rəb·ər 'band·iŋ }

ruggedization |ELECTR| Making electronic equipment and components resistant to severe shock, temperature changes, high humidity, or other detrimental environmental influences. { ,rəg·ə·də'zā·shən }

rule-based control system See direct expert control system. { ¦rül ,bäst kən'trōl ,sis·təm }

rule-based expert system |COMPUT SCI| An expert system based on a collection of rules that a human expert would follow in dealing with a problem. { ¦rül ¦bäst 'ek,spərt ,sis·təm }

rule of inference See production. { 'rül əv 'in·frəns }

run |COMPUT SCI| A single, complete execution of a computer program, or one continuous segment of computer processing, used to complete one or more tasks for a single customer or application. Also known as machine run. { rən }

runaround crosstalk |COMMUN| Crosstalk resulting from coupling between the high-level end of one repeater and the low-level end of another repeater, as at a carrier telephone repeater station. { 'rən·ə,raúnd 'krós,tók }

runaway effect |ELECTR| The phenomenon whereby an increase in temperature causes an increase in a collector-terminal current in a transistor, which in turn results in a higher temperature and, ultimately, failure of the transistor; the effect limits the power output of the transistor. { 'rən·ə,wā i,fekt }

runaway electron |ELECTR| An electron, in an ionized gas to which an electric field is applied, that gains energy from the field faster than it loses energy by colliding with other particles in the gas. { 'rən·ə,wā i'lek,trän }

runaway tape |COMPUT SCI| A tape reel that spins rapidly and out of control as the result of a hardware malfunction. { ¦rən·ə¦wā 'tāp }

run book |COMPUT SCI| The collection of materials necessary to document a program run on a computer. Also known as problem file; problem folder. { 'rən ,bùk }

run chart |COMPUT SCI| A flow chart for one or more computer runs which shows input, output, and the use of peripheral units, but no details of the execution of the run. Also known as run diagram. { 'rən ,chärt }

run diagram *See* run chart. { 'rən ,dī·ə,gram }

run documentation |COMPUT SCI| Detailed instructions to the operator on how to run a particular computer program. { 'rən ,däk·yə·men'tā·shən }

run-length encoding |COMPUT SCI| A method of data compression that encodes strings of the same character as a single number. { 'rən ¦leŋkth in'kōd·iŋ }

run-length-limited code |COMMUN| A binary code in which a 1 is inserted after a certain number of 0's, in order to avoid long strings of 0's, which would require very accurate clocking in order to ensure that a bit was not lost. Abbreviated RLL code. { ¦rən ,leŋkth ,lim·əd·əd 'kōd }

run motor |ELEC| In facsimile equipment, a motor which supplies the power to drive the scanning or recording mechanisms; a synchronous motor is used to limit the speed. { 'rən ,mōd·ər }

running accumulator *See* push-down storage. { 'rən·iŋ ə'kyü·mə,lād·ər }

run-time error |COMPUT SCI| An error in a computer program that is not detected until the program is executed, and then causes a processing error to occur. { 'rən ¦tīm 'er·ər }

run-time error handler |COMPUT SCI| A system control program that detects and diagnoses run-time errors and issues messages concerning them. { 'rən ¦tīm 'er·ər ,hand·lər }

run-time library |COMPUT SCI| A collection of general-purpose routines that form part of a language translator and allow computer programs to be run with a particular operating system. { 'rən ¦tīm 'lī,brer·ē }

rvalue *See* right value. { 'är,val·yü }

S

S *See* siemens.

sacrificial compliant substrate *See* compliant substrate. { ˌsak·rəˌfishˈəl kəmˈplī·ənt 'səb ˌstrāt }

safety factor |ELEC| The amount of load, above the normal operating rating, that a device can handle without failure. |MECH| *See* factor of safety. { 'sāf·tē ˌfak·tər }

sag |ELEC| Slack introduced in an aerial cable or open-wire line to compensate for contraction during cold weather. { sag }

Saint Elmo's fire |ELEC| A visible electric discharge, sometimes seen on the mast of a ship, on metal towers, and on projecting parts of aircraft, due to concentration of the atmospheric electric field at such projecting parts. { 'sānt 'el·mōz 'fīr }

salammoniac cell |ELEC| Cell in which the electrolyte consists primarily of a solution of ammonium chloride. { ˌsal·əˈmō·nē·ak ˌsel }

salient-pole field winding |ELEC| A type of field winding in electric machinery where the winding turns are concentrated around the pole core. { 'sāl·yənt ˌpōl 'fēld ˌwīnd·iŋ }

Salisbury dark box |ELECTR| Isolating chamber used for test work in connection with radar equipment; the walls of the chamber are specially constructed to absorb all impinging microwave energy at a certain frequency. { 'sȯlz,ber·ē 'därk 'bäks }

Sallen-Key filter |ELECTR| An electric filter that uses a single amplifier of positive low gain, realized by an operational amplifier and two feedback resistors. { ˌsal·ən 'kē ˌfil·tər }

sample-and-hold circuit |ELECTR| A circuit that measures an input signal at a series of definite times, and whose output remains constant at a value corresponding to the most recent measurement until the next measurement is made. { ˌsam·pəl ən 'hōld ˌsər·kət }

sampled-data control system |CONT SYS| A form of control system in which the signal appears at one or more points in the system as a sequence of pulses or numbers usually equally spaced in time. { 'sam·pəld ˌdad·ə kən'trōl ˌsis·təm }

sampler |CONT SYS| A device, used in sampled-data control systems, whose output is a series of impulses at regular intervals in time; the height of each impulse equals the value of the

continuous input signal at the instant of the impulse. { 'sam·plər }

sampling |ENG| Process of obtaining a sequence of instantaneous values of a wave. { 'sam·pliŋ }

sampling gate |ELECTR| A gate circuit that extracts information from the input waveform only when activated by a selector pulse. { 'sam·pliŋ ˌgāt }

sampling interval |CONT SYS| The time between successive sampling pulses in a sampled-data control system. { 'sam·pliŋ ˌin·tər·vəl }

sampling process |ENG| The process of obtaining a sequence of instantaneous values of some quantity that varies continuously with time. { 'sam·pliŋ ˌprä·səs }

sampling switch *See* commutator switch. { 'sam·pliŋ ˌswich }

sampling synthesis |ENG ACOUS| Any method of synthesizing musical tones that is based on playing back digitally recorded sounds. { 'sam·pliŋ ˌsin·thə·səs }

sampling theorem |COMMUN| The theorem that a signal that varies continuously with time is completely determined by its values at an infinite sequence of equally spaced times if the frequency of these sampling times is greater than twice the highest frequency component of the signal. Also known as Shannon's sampling theorem. { 'sam·pliŋ ˌthir·əm }

sampling time |ENG| The time between successive measurements of a physical quantity. { 'sam·pliŋ ˌtīm }

sampling voltmeter |ENG| A special type of voltmeter that detects the instantaneous value of an input signal at prescribed times by means of an electronic switch connecting the signal to a memory capacitor; it is particularly effective in detecting high-frequency signals (up to 12 gigahertz) or signals mixed with noise. { 'sam·pliŋ 'vōlt,mēd·ər }

sanatron circuit |ELECTR| A variable time-delay circuit having two pentodes and two diodes, used to produce very short gate waveforms having time durations that vary linearly with a reference voltage. { 'san·ə,trän ˌsər·kət }

sand boil *See* blowout. { 'san ˌbȯil }

sand load |ELECTROMAG| An attenuator used as a power-dissipating terminating section for a coaxial line or waveguide; the dielectric space in the line is filled with a mixture of sand and graphite that acts as a matched-impedance load, preventing standing waves. { 'san ,lōd }

SANTA See systematic analog network testing approach. { 'san·tə }

SAR See synthetic-aperture radar.

SASAR See segmented aperture-synthetic aperture radar. { 'sā,sär }

satellite communication |COMMUN| Communication that involves the use of an active or passive satellite to extend the range of a communications, radio, television, or other transmitter by returning signals to earth from an orbiting satellite. { 'sad·əl,īt kə,myü·nə,kā·shən }

satellite computer |COMPUT SCI| A computer which, under control of the main computer, handles the input and output routines, thereby allowing the main computer to be fully dedicated to computations. { 'sad·əl,īt kəm,pyüd·ər }

Satellite Digital Audio Radio Service |COMMUN| Referring to satellite-delivered digital audio systems. The digital audio data rate in these systems is specified as being 64 kbits/s. Abbreviated SDARS. { 'sad·əl,īt 'dij·əd·əl 'ȯd·ē·ō'rād·ē·ō ,sər·vəs }

satellite master antenna television system |COMMUN| A master antenna television system equipped with a television receive-only antenna and associated electronics to receive broadcasts relayed by geostationary satellites. Abbreviated SMATV system. { 'sad·əl,īt 'mas·tər an¦ten·ə 'tel·ə,vizh·ən ,sis·təm }

satellite processor |COMPUT SCI| One of the outlying processors in a hierarchical distributed processing system, typically placed at or near point-of-transaction locations, and designed to serve the users at those locations. { 'sad·əl,īt ,prä,ses·ər }

saturated diode |ELECTR| A diode that is passing the maximum possible current, so further increases in applied voltage have no effect on current. { 'sach·ə,rād·əd 'dī,ōd }

saturating signal |ELECTR| In radar, a signal of an amplitude greater than the dynamic range of the receiving system. { 'sach·ə,rād·iŋ 'sig·nəl }

saturation |ELECTR| **1.** The condition that occurs when a transistor is driven so that it becomes biased in the forward direction (the collector becomes positive with respect to the base, for example, in a *pnp* type of transistor). **2.** See anode saturation; temperature saturation. { ,sach·ə'rā·shən }

saturation current |ELECTR| **1.** In general, the maximum current which can be obtained under certain conditions. **2.** In a vacuum tube, the space-charge-limited current, such that further increase in filament temperature produces no specific increase in anode current. **3.** In a vacuum tube, the temperature-limited current, such that a further increase in anode-cathode potential difference produces only a relatively small increase in current. **4.** In a gaseous-discharge device, the maximum current which can be obtained for a given mode of discharge. **5.** In a semiconductor, the maximum current which just precedes a change in conduction mode. { ,sach·ə'rā·shən ¦kə·rənt }

saturation limiting |ELECTR| Limiting the minimum output voltage of a vacuum-tube circuit by operating the tube in the region of plate-current saturation (not to be confused with emission saturation). { ,sach·ə'rā·shən 'lim·əd·iŋ }

saturation signal |ELECTROMAG| A radio signal (or radar echo) which exceeds a certain power level fixed by the design of the receiver equipment; when a receiver or indicator is "saturated," the limit of its power output has been reached. { ,sach·ə'rā·shən ¦sig·nəl }

sawtooth generator |ELECTR| A generator whose output voltage has a sawtooth waveform; used to produce sweep voltages for cathode-ray tubes. { 'sȯ,tüth 'jen·ə,rād·ər }

sawtooth modulated jamming |ELECTR| Electronic countermeasure technique when a high-level jamming signal is transmitted, thus causing large automatic gain control voltages to be developed at the radar receiver that, in turn, cause target pip and receiver noise to completely disappear. { 'sȯ,tüth ¦mäj·ə,lād·əd 'jam·iŋ }

sawtooth pulse |ELECTR| An electric pulse having a linear rise and a virtually instantaneous fall, or conversely, a virtually instantaneous rise and a linear fall. { 'sȯ,tüth ¦pəls }

sawtooth waveform |ELECTR| A waveform characterized by a slow rise time and a sharp fall, resembling a tooth of a saw. { 'sȯ,tüth ¦wāv,fȯrm }

saxophone |ELECTROMAG| Vertex-fed linear array antenna giving a cosecant-squared radiation pattern. { 'sak·sə,fōn }

S band |COMMUN| A band of radio frequencies extending from 1550 to 5200 megahertz, corresponding to wavelengths of 19.37 to 5.77 centimeters. { 'es ,band }

S-band hiran See shiran. { 'es ¦band 'hī,ran }

S-band single-access service |COMMUN| One of the services provided by the Tracking and Data Relay Satellite System, which provides return-link data rates up to 6 megabits per second for each user spacecraft and forward-link data at 300 kilobits per second. Abbreviated SSA { ¦es ,band ,siŋ·gəl 'ak,ses ,sər·vəs }

S-100 bus |ELECTR| A bus assembly with 100 conductors; widely used in microcomputer-based systems. { ,es ¦wən'hən·drəd 'bəs }

SC See sectional center.

SCADA See supervisory control and data acquisition. { 'skad·ə or ¦es¦skē¦ā¦dē'ā }

scalar |COMPUT SCI| A single value or item. { 'skā·lər }

scalar data type |COMPUT SCI| The manner in which a sequence of bits represents a single data item in a computer program. Also known as aggregate data type. { 'skā·lər 'dad·ə ,tīp }

scalar processor |COMPUT SCI| A computer that carries out computations on one number at a time. { 'skā·lər 'prä,ses·ər }

scalar quantization |COMPUT SCI| A data compression technique in which a value is presented (in approximation) by the closest, in some mathematical sense, of a predefined set of allowable values. { 'skā·lər ,kwän·tə'zā·shən }

scale-of-ten circuit *See* decade scaler. { ¦skāl əv ¦ten 'sər·kət }

scale-of-two circuit *See* binary scaler. { ¦skāl əv ¦tü 'sər·kət }

scaler [ELECTR] A circuit that produces an output pulse when a prescribed number of input pulses is received. Also known as counter; scaling circuit. { skāl·ər }

scaling [ELECTR] Counting pulses with a scaler when the pulses occur too fast for direct counting by conventional means. { 'skāl·iŋ }

scaling circuit *See* scaler. { 'skāl·iŋ ‚sər·kət }

scaling factor [ELECTR] The number of input pulses per output pulse of a scaling circuit. Also known as scaling ratio. { 'skāl·iŋ ‚fak·tər }

scaling ratio *See* scaling factor. { 'skāl·iŋ ‚rā·shō }

scan [COMPUT SCI] To examine information, following a systematic, predetermined sequence, for some particular purpose. [ELECTR] The motion, usually periodic, given to the major lobe of an antenna; the process of directing the radio-frequency beam successively over all points in a given region of space. [ENG] **1.** To examine an area, a region in space, or a portion of the radio spectrum point by point in an ordered sequence; for example, conversion of a scene or image to an electric signal or use of radar to monitor an airspace for detection, navigation, or traffic control purposes. **2.** One complete circular, up-and-down, or left-to-right sweep of the radar, light, or other beam or device used in making a scan. { skan }

scan converter [ELECTR] **1.** Equipment that converts radar date images to data at a sampling rate suitable for transmission over telephone lines or narrow-band radio circuits for use at remote locations. Scan converters may work digitally with quantized data; analog ones often use a "memory" scope, a cathode-ray tube of long persistence, permitting nondestructive readout of radar, television, and data displays. **2.** A cathode-ray tube that is capable of storing radar, television, and data displays for nondestructive readout over prolonged periods of time. { 'skan kən‚vərd·ər }

scan head [ELECTR] A sensing device that is moved across the image being scanned. { 'skan ‚hed }

scanistor [ELECTR] Integrated semiconductor optical-scanning device that converts images into electrical signals; the output analog signal represents both amount and position of light shining on its surface. { skə'nis·tər }

scan line [ELECTR] A horizontal row of pixels on a video screen that are examined or refreshed in succession in one sweep across the screen during the scanning process. { 'skan ‚līn }

scanner [COMMUN] That part of a facsimile transmitter which systematically translates the densities of the elemental areas of the subject copy into corresponding electric signals. [COMPUT SCI] A device that converts an image of something outside a computer, such as text, a drawing, or a photograph, into a digital image that it sends into the computer for display or further processing. { 'skan·ər }

scanner selector [COMPUT SCI] An electronic device interfacing computer and multiplexers when more than one multiplexer is used. { 'skan·ər si ‚lek·tər }

scanning circuit *See* sweep circuit. { 'skan·iŋ ‚sər·kət }

scanning electron microscope [ELECTR] A type of electron microscope in which a beam of electrons, a few hundred angstroms in diameter, systematically sweeps over the specimen; the intensity of secondary electrons generated at the point of impact of the beam on the specimen is measured, and the resulting signal is fed into a cathode-ray-tube display which is scanned in synchronism with the scanning of the specimen. Abbreviated SEM. { 'skan·iŋ i'lek‚trän 'mī·krə‚skōp }

scanning frequency *See* stroke speed. { 'skan·iŋ ‚frē·kwən‚sē }

scanning head [ELECTR] Light source and phototube combined as a single unit for scanning a moving strip of paper, cloth, or metal in photoelectric side-register control systems. { 'skan·iŋ ‚hed }

scanning line [COMMUN] **1.** In a video system, a single, continuous, narrow strip which is determined by the process of scanning. **2.** Path traced by the scanning or recording spot in one sweep across the subject copy or record sheet. { 'skan·iŋ ‚līn }

scanning linearity [ELECTR] In a video system, the uniformity of scanning speed during the trace interval. { 'skan·iŋ ‚lin·ē'ar·əd·ē }

scanning line frequency *See* stroke speed. { 'skan·iŋ ‚līn ‚frē·kwən‚sē }

scanning loss [ELECTROMAG] In a radar system employing a scanning antenna, the reduction in sensitivity (usually expressed in decibels) due to scanning across the target, compared with that obtained when the beam is directed constantly at the target. { 'skan·iŋ ‚lòs }

scanning radio [ELECTR] A radio receiver that automatically scans across public service, emergency service, or other radio bands and stops at the first preselected station which is on the air. Also known as radio scanner. { 'skan·iŋ 'rād·ē·ō }

scanning sequence [ENG] The order in which the points in a region are scanned; for example, in television the picture is scanned horizontally from left to right and vertically from top to bottom. { 'skan·iŋ ‚sēk·wəns }

scanning speed *See* spot speed. { 'skan·iŋ ‚spēd }

scanning spot *See* picture element. { 'skan·iŋ ‚spät }

scanning switch *See* commutator switch. { 'skan·iŋ ‚swich }

scanning transmission electron microscope [ELECTR] A type of electron microscope which scans with an extremely narrow beam that is transmitted through the sample; the detection apparatus produces an image whose brightness depends on atomic number of the sample. Abbreviated STEM. { 'skan·iŋ tranz'mish·ən i'lek,trän 'mī·krə,skōp }

scanning tunneling microscope [ELECTR] An instrument for producing surface images with atomic-scale lateral resolution, in which a fine probe tip is raster-scanned over the surface at a distance of 0.5–1 nanometer, and the resulting tunneling current, or the position of the tip required to maintain a constant tunneling current, is monitored. Also known as tunneling microscope. { 'skan·iŋ ¦tən·əl·iŋ 'mī·krə,skōp }

scanning yoke See deflection yoke. { 'skan·iŋ ‚yōk }

scatter band [COMMUN] In pulse interrogation systems, the total bandwidth occupied by the frequency spread by numerous interrogations operating on the same nominal radio frequency. { 'skad·ər ‚band }

scatterer [ELECTROMAG] Object in an otherwise relatively homogeneous propagation medium that intercepts electromagnetic waves such as radar signals and reflects them in directions associated with the shape and composition of the object. Examples include individual raindrops, earth surface features, sea-wave crests, buildings, and vehicles. { 'skad·ər·ər }

scattering [ELECTROMAG] Diffusion of electromagnetic waves in a random manner by air masses in the upper atmosphere, permitting long-range reception, as in scatter propagation. Also known as radio scattering. { 'skad·ə·riŋ }

scattering coefficient [ELECTROMAG] One of the elements of the scattering matrix of a waveguide junction; that is, a transmission or reflection coefficient of the junction. { 'skad·ə·riŋ ‚kō·i ‚fish·ənt }

scattering cross section [ELECTROMAG] The power of electromagnetic radiation scattered by an antenna divided by the incident power. { 'skad·ə·riŋ 'kròs ‚sek·shən }

scattering matrix [ELECTROMAG] A square array of complex numbers consisting of the transmission and reflection coefficients of a waveguide junction. { 'skad·ə·riŋ ‚mā·triks }

scatter loading [COMPUT SCI] The process of loading a program into main memory such that each section or segment of the program occupies a single, connected memory area but the several sections of the program need not be adjacent to each other. { 'skad·ər ‚lōd·iŋ }

scatter propagation [ELECTROMAG] Transmission of radio waves far beyond line-of-sight distances by using high power and a large transmitting antenna to beam the signal upward into the atmosphere and by using a similar large receiving antenna to pick up the small portion of the signal that is scattered by the atmosphere. Also known as beyond-the-horizon communication; forward-scatter propagation; over-the-horizon propagation. { 'skad·ər ‚präp· ə,gā·shən }

scatter read [COMPUT SCI] An input operation that places various segments of an input record into noncontiguous areas in central memory. { 'skad·ər ‚rēd }

scatter reflections [ELECTROMAG] Reflections from portions of the ionosphere having different virtual heights, which mutually interfere and cause rapid fading. { 'skad·ər ri,flek·shənz }

scene analysis See picture segmentation. { 'sēn ə,nal·ə·səs }

scheduled down time [COMPUT SCI] A period of time designated for closing down a computer system for preventive maintenance. { 'skej·əld 'daùn,tīm }

scheduler [COMPUT SCI] A system control program that determines the sequence in which programs will be processed by a computer and automatically submits them for execution at predetermined times. { 'skej·ə·lər }

scheduling algorithm [COMPUT SCI] A systematic method of determining the order in which tasks will be performed by a computer system, generally incorporated into the operating system. { 'skej·ə·liŋ ‚al·gə,rith·əm }

schema [COMPUT SCI] A logical description of the data in a data base, including definitions and relationships of data. { 'skē·mə }

schematic circuit diagram See circuit diagram. { ski'mad·ik 'sər·kət ‚dī·ə,gram }

Schering bridge [ELEC] A four-arm alternating-current bridge used to measure capacitance and dissipation factor; bridge balance is independent of frequency. { 'sher·iŋ ‚brij }

Schmitt circuit [ELECTR] A bistable pulse generator in which an output pulse of constant amplitude exists only as long as the input voltage exceeds a certain value. Also known as Schmitt limiter; Schmitt trigger. { 'shmit ‚sər·kət }

Schmitt limiter See Schmitt circuit. { 'shmit 'lim·əd·ər }

Schmitt trigger See Schmitt circuit. { 'shmit 'trig·ər }

Schottky barrier [ELECTR] A transition region formed within a semiconductor surface to serve as a rectifying barrier at a junction with a layer of metal. { 'shät·kē ‚bar·ē·ər }

Schottky barrier diode [ELECTR] A semiconductor diode formed by contact between a semiconductor layer and a metal coating; it has a nonlinear rectifying characteristic; hot carriers (electrons for *n*-type material or holes for *p*-type material) are emitted from the Schottky barrier of the semiconductor and move to the metal coating that is the diode base; since majority carriers predominate, there is essentially no injection or storage of minority carriers to limit switching speeds. Also known as hot-carrier diode; Schottky diode. { 'shät·kē ¦bar·ē·ər 'dī ‚ōd }

Schottky diode See Schottky barrier diode. { 'shät·kē 'dī,ōd }

Schottky-diode FET logic [ELECTR] A logic gate configuration used with gallium-arsenide field-effect transistors operating in the depletion mode, in which very small Schottky diodes at the gate input provide the logical OR function and the level shifting required to make the input and output voltage levels compatible. Abbreviated SDFL. { 'shät·kē ¦dī,ōd ¦ef¦ē¦tē 'läj·ik }

Schottky effect [SOLID STATE] The enhancement of the termionic emission of a conductor resulting from an electric field at the conductor surface. { 'shät·kē i,fekt }

Schottky noise See shot noise. { 'shät·kē ,nóiz }

Schottky theory [SOLID STATE] A theory describing the rectification properties of junction between a semiconductor and a metal that result from formation of a depletion layer at the surface of fontact. { 'shät·kē ,thē·ə·rē }

Schottky transistor-transistor logic [ELECTR] A transistor-transistor logic circuit in which a Schottky diode with forward diode voltage is placed across the base-collector junction of the output transistor in order to improve the speed of the circuit. { 'shät·kē tran¦zis·tər tran¦zis·tər 'läj·ik }

Schrage motor [ELEC] A type of alternating-current commutator motor whose speed is controlled by varying the position of sets of brushes on the commutator. { 'shräg·ə ,mōd·ər }

Schwinger critical field [ELEC] That electric field at which an electron is accelerated from rest to a velocity at which its kinetic energy equals its rest energy over a distance of one Compton wavelength. { ¦shviŋ·ər ¦krid·ə·kəl 'fēld }

scientific calculator [COMPUT SCI] An electronic calculator that has provisions for handling exponential, trigonometric, and sometimes other special functions in addition to performing arithmetic operations. { ,sī·ən'tif·ik 'kal·kyə,lād·ər }

scientific computer [COMPUT SCI] A computer which has a very large memory and is capable of handling extremely high-speed arithmetic and a very large variety of floating-point arithmetic commands. { ,sī·ən'tif·ik kəm'pyüd·ər }

scientific notation [COMPUT SCI] The display of numbers in which a base number, representing the significant digits, is followed by a number representing the power of 10 to which the base number is raised. { ,sī·ən¦tif·ik nō'tā·shən }

scientific system [COMPUT SCI] A system devoted principally to computations as opposed to business and data-processing systems, the main emphasis of which is on the updating of data records and files rather than the performance of calculations. { ,sī·ən'tif·ik 'sis·təm }

scintillation [ELECTROMAG] **1.** Fluctuation in radar echo amplitude, usually that associated with atmospheric irregularities in the propagation path. **2.** Random fluctuation, in radio propagation, of the received field about its mean value, the deviations usually being relatively small. { ,sint·əl'ā·shən }

scissoring [COMPUT SCI] In computer graphics, the deletion of those parts of an image that fall outside a window that has been placed over the original image. Also known as clipping. { 'siz·ər·iŋ }

scoop See ellipsoidal floodlight. { sküp }

scope [COMPUT SCI] For a variable in a computer program, the portion of the computer program within which the variable can be accessed (used or changed). [ELECTR] See radarscope. { skōp }

scotoscope [ELECTR] A telescope which employs an image intensifier to see in the dark. { 'skäd·ə ,skōp }

Scott connection [ELECTR] A type of transformer which transmits power from two-phase to three-phase systems, or vice versa. { 'skät kə,nek·shən }

Scott top [ELEC] Transformers arranged in the Scott connection for converting electrical power from two-phase to three-phase, or vice versa. { 'skät ,täp }

SCR See system clock reference.

scramble [COMMUN] To mix, in cryptography, in random or quasi-random fashion. { 'skram·bəl }

scrambler [ELECTR] A circuit that divides speech frequencies into several ranges by means of filters, then inverts and displaces the frequencies in each range so that the resulting reproduced sounds are unintelligible; the process is reversed at the receiving apparatus to restore intelligible speech. Also known as speech inverter; speech scrambler. { 'skram·blər }

scrambling [COMMUN] The alteration of the characteristics of a video, audio, or coded data stream in order to prevent unauthorized reception of the information in a clear form. { 'skram·bliŋ }

scratch [COMPUT SCI] To remove data or to set up its identifying labels so that new data can be written over it. { skrach }

scratch file [COMPUT SCI] A temporary file for future use, created by copying all or part of a data set to an auxiliary memory device. { 'skrach ,fīl }

scratch-pad memory [COMPUT SCI] A very fast intermediate storage (in the form of flip-flop register or semiconductor memory) which often supplements main core memory. { 'skrach ,pad ,mem·rē }

scratch tape [COMPUT SCI] A reel of magnetic tape containing data that may now be destroyed. { 'skrach ,tāp }

screed wire See ground wire. { 'skrēd ,wīr }

screen [COMPUT SCI] To make a preliminary selection from a set of entities, selection criteria being based on a given set of rules or conditions. [ELECTR] **1.** The surface on which an image is made visible for viewing; it may be a fluorescent screen with a phosphor layer that converts the energy of an electron beam to visible light, or a translucent or opaque screen on which the optical image is projected, or a display surface of the types commonly used in computers. **2.** See screen grid. [ELECTROMAG] Metal partition or shield which isolates a device from external magnetic or electric fields. { skrēn }

screen angle [ELECTROMAG] Vertical angle bounded by a straight line from the radar antenna to the horizon and the horizontal at the antenna assuming a ⁴⁄₃ earth's radius. { 'skrēn ,aŋ·gəl }

screen capture See screen shot. { 'skrēn ,kap·chər }

screen dissipation [ELECTR] Power dissipated in the form of heat on the screen grid as the result of bombardment by the electron stream. { 'skrēn ,dis·ə,pā·shən }

screen dump |COMPUT SCI| **1.** The printing of everything that appears on a computer screen. **2.** The printed copy that results from this action. { 'skrēn ,dəmp }

screened trailing cable |ELEC| A flexible cable provided with a protective screen of conducting material, so applied as to enclose each power core separately or to enclose together all the cores of the cable. { 'skrēnd 'trāl·iŋ 'kā·bəl }

screen format |COMPUT SCI| The manner in which information is arranged and presented on a cathode-ray tube or other electronic display. { 'skrēn ,fȯr,mat }

screen formatter |COMPUT SCI| A computer program that enables the user to design and set up screen formats. Also known as screen generator; screen painter. { 'skrēn ,fȯr,mad·ər }

screen generator *See* screen formatter. { 'skrēn ,jen·ə,rād·ər }

screen grid |ELECTR| A grid placed between a control grid and an anode of an electron tube, and usually maintained at a fixed positive potential, to reduce the electrostatic influence of the anode in the space between the screen grid and the cathode. Also known as screen. { 'skrēn ,grid }

screen image buffer |COMPUT SCI| A section of computer storage that contains a representation of the information that appears on an electronic display. Abbreviated SIB. { 'skrēn ,im·ij ,bəf·ər }

screening *See* electric shielding. { 'skrēn·iŋ }

screen memory |COMPUT SCI| The portion of a microcomputer storage that is reserved for setting up screen formats. { 'skrēn ¦mem·rē }

screen overlay |COMPUT SCI| **1.** An array of cells on a video display screen that allow a user to command a computer by touching buttons displayed on the screen at the locations of the cells. **2.** A window of data that is temporarily displayed on a screen, leaving the original display intact when the window is removed. { ¦skrēn 'ō·vər,lā }

screen painter *See* screen formatter. { 'skrēn ,pān·tər }

screen saver |COMPUT SCI| A program that launches when a computer is not in use for a predetermined period, displaying various transient or moving images on a computer screen. Originally used to prevent computer screen damage from prolonged display of a static image, screen savers are now more of an amusement or security feature as modern monitors are less susceptible to screen burning. { 'skrēn ,sāv·ər }

screen shot |COMPUT SCI| A digital image or file containing all or part of what is seen on a computer display. Also known as screen capture. { 'skrēn ,shät }

scribing |ELECTR| Cutting a grid pattern of deep grooves with a diamond-tipped tool in a slice of semiconductor material containing a number of devices, so that the slice can be easily broken into individual chips. { 'skrīb·iŋ }

script |COMPUT SCI| An executable list of commands written in a programming language. { skript }

scripting language |COMPUT SCI| An interpreted language (for example, JavaScript and Perl)

used to write simple programs, called scripts. { 'skrip·tiŋ ,laŋ·gwij }

scroll |COMPUT SCI| To move information in an electronic display up, down, left, or right, so that new information appears and some of the existing information is moved away. { skrōl }

scroll arrow |COMPUT SCI| An arrow on a video display screen that is clicked in order to scroll the screen in the corresponding direction. { 'skrōl ,a·rō }

scroll bar |COMPUT SCI| A horizontal or vertical bar that contains a box that is clicked and dragged up, down, left, or right in order to scroll the screen. { 'skrōl ,bär }

scrolling |COMPUT SCI| The continuous movement of information either vertically or horizontally on a video screen. { 'skrōl·iŋ }

scrub |COMPUT SCI| To examine a large amount of data and eliminate duplicate or unneeded items. { skrəb }

SCS *See* silicon controlled switch.

SCSI *See* small computer system interface. { 'skəz·ē }

scuzzy *See* small computer system interface. { 'skəz·ē }

SDARS *See* Satellite Digital Audio Radio Service.

SDFL *See* Schottky-diode FET logic.

SDHT *See* high-electron-mobility transistor.

SDMA *See* space-division multiple access.

SDRAM *See* synchronous dynamic random access memory. { ¦es¦dē'ram }

SDTV *See* standard definition television.

sea clutter |ELECTROMAG| A clutter on an airborne radar due to reflection of signals from the sea. Also known as sea return; wave clutter. { 'sē ,kləd·ər }

sealed-beam headlight |ELEC| A headlight in which the filament, reflector, and lens are contained in a single sealed unit. { ¦sēld ,bēm 'head,līt }

sealed tube |ELECTR| Electron tube which is hermetically sealed. { 'sēld 'tüb }

sealing compound |ELEC| A compound used in dry batteries, capacitor blocks, transformers, and other components to keep out air and moisture. { 'sēl·iŋ ,käm,paúnd }

seamless integration |COMPUT SCI| The addition of a routine or program that works smoothly with an existing system and can be activated and used as if it had been built into the system when the system was put together. { ¦sēm·ləs ,int·ə'grā·shən }

search |COMPUT SCI| To seek a desired item or condition in a set of related or similar items or conditions, especially a sequentially organized or nonorganized set, rather than a multidimensional set. |ENG| To explore a region in space with radar. { sərch }

search antenna |ELECTROMAG| A radar antenna or antenna system designed for search. { 'sərch an,ten·ə }

search argument |COMPUT SCI| The item or condition that is desired in a search procedure. { 'sərch ,är·gyə·mənt }

search engine |COMPUT SCI| **1.** Any software that locates and retrieves information in a database.

2. A server with a stored index of Web pages that is capable of returning lists of pages that match keyword queries. { 'sərch ,en·jən }

search field |COMPUT SCI| A field in a record or segment whose value is examined in a search. { 'sərch ,fēld }

search gate |ELECTR| A gate pulse used to search back and forth over a certain range. { 'sərch ,gāt }

searching lighting *See* horizontal scanning. { 'sərch·iŋ ,līd·iŋ }

search key |COMPUT SCI| A data item, or the value of a data item, that is used in carrying out a search. { 'sərch ,kē }

search time |COMPUT SCI| Time required to locate a particular field of data in a computer storage device; requires a comparison of each field with a predetermined standard until an identity is obtained. { 'sərch ,tīm }

sea return *See* sea clutter. { 'sē ri,tərn }

seasonal factors |COMMUN| Factors that are used to adjust skywave absorption data for seasonal variations; these variations are due primarily to seasonal fluctuations in the heights of the ionospheric layers. { 'sēz·ən·əl 'fak·tərz }

seasoning |ELECTR| Overcoming a temporary unsteadiness of a component that may appear when it is first installed. { 'sēz·ən·iŋ }

SEC *See* secondary-electron conduction.

secondary |ELEC| Low-voltage conductors of a power distributing system. { 'sek·ən,der·ē }

secondary allocation |COMPUT SCI| An area of disk storage that is assigned to a file which has become too large for the area originally assigned to it. { 'sek·ən,der·ē ,al·ə'kā·shən }

secondary battery *See* storage battery. { 'sek·ən ,der·ē 'bad·ə·rē }

secondary cache |COMPUT SCI| High-speed memory between the primary cache and main memory that supplies the processor with the most frequently requested data and instructions. Also known as level 2 cache. { ,sek·ən,der·ē 'kash }

secondary cell *See* storage cell. { 'sek·ən,der·ē 'sel }

secondary circuit |ELEC| The wiring connected to the secondary winding of a transformer, induction coil, or similar device. { 'sek·ən,der·ē 'sər·kət }

secondary electron |ELECTR| **1.** An electron emitted as a result of bombardment of a material by an incident electron. **2.** An electron whose motion is due to a transfer of momentum from primary radiation. { 'sek·ən,der·ē i'lek,trän }

secondary-electron conduction |ELECTR| Transport of charge by secondary electrons moving through the interstices of a porous material under the influence of an externally applied electric field. Abbreviated SEC. { 'sek·ən,der·ē i'lek,trän kən,dək·shən }

secondary emission |ELECTR| The emission of electrons from the surface of a solid or liquid into a vacuum as a result of bombardment by electrons or other charged particles. { 'sek·ən ,der·ē i'mish·ən }

secondary grid emission |ELECTR| Electron emission from a grid resulting directly from bombardment of its surface by electrons or other charged particles. { 'sek·ən,der·ē 'grid i,mish·ən }

secondary index |COMPUT SCI| An index that provides an alternate method of accessing records or portions of records in a data base or file. Also known as alternate index. { 'sek·ən,der·ē 'in,deks }

secondary key |COMPUT SCI| A key that holds the physical location of a record or a portion of a record in a file or database, and provides an alternative means of accessing data. Also known as alternate key. { 'sek·ən,der·ē 'kē }

secondary lobe *See* minor lobe. { 'sek·ən,der·ē 'lōb }

secondary photocurrent |ELECTR| A photocurrent resulting from ohmic contacts that are able to replenish charge carriers which pass out of the opposite contact in order to maintain charge neutrality, and whose maximum gain is much greater than unity. { 'sek·ən,der·ē 'fōd·ō,kə·rənt }

secondary radar |ELECTR| A radar system in which the transmitted signal from its interrogator causes a transponder borne by a cooperative aircraft to transmit a response on a separate frequency that is received and interpreted by the interrogating radar. { 'sek·ən,der·ē 'rā,där }

secondary station |COMMUN| Any station in a radio network other than the net control station. { 'sek·ən,der·ē 'stā·shən }

secondary storage |COMPUT SCI| Any means of storing and retrieving data external to the main computer itself but accessible to the program. { 'sek·ən,der·ē 'stòr·ij }

second breakdown |ELECTR| Destructive breakdown in a transistor, wherein structural imperfections cause localized current concentrations and uncontrollable generation and multiplication of current carriers; reaction occurs so suddenly that the thermal time constant of the collector regions is exceeded, and the transistor is irreversibly damaged. { 'sek·ənd 'brāk,daùn }

second-channel interference *See* alternate-channel interference. { 'sek·ənd ¦chan·əl ,in·tər'fir·əns }

second detector |ELECTR| The detector that separates the intelligence signal from the intermediate-frequency signal in a superheterodyne receiver. { 'sek·ənd di'tek·tər }

second-generation computer |COMPUT SCI| A computer characterized by the use of transistors rather than vacuum tubes, the execution of input/output operations simultaneously with calculations, and the use of operating systems. { 'sek·ənd ,jen·ə¦rā·shən kəm'pyüd·ər }

second-order subroutine |COMPUT SCI| A subroutine that is entered from another subroutine, in contrast to a first-order subroutine; it constitutes the second level of a two-level or higher-level routine. Also known as second-remove subroutine. { 'sek·ənd ¦òr·dər 'səb·rü,tēn }

second-remove subroutine *See* second-order subroutine. { 'sek·ənd ri¦müv 'səb·rü,tēn }

second-time-around echo [ELECTR] A radar echo received from one pulse after the transmission of a subsequent pulse and liable to be associated with the latter, giving an erroneous indication of range. { 'sek·ənd ¦tīm ə'raund ‚ek·ō }

second-trip echo See second-time-around echo. { 'sek·ənd ¦trip ‚ek·ō }

secrecy system See privacy system. { 'sē·krə·sē ‚sis·təm }

secret-key algorithm [COMPUT SCI] A cryptographic algorithm which uses the same cryptographic key for encryption and decryption, requiring that the key first be transmitted from the sender to the recipient via a secure channel. { ‚sē·krət ‚kē 'al·gə‚rith·əm }

section [COMMUN] Each individual transmission span in a radio relay system; a system has one more section than it has repeaters. { 'sek·shən }

sectional center [COMMUN] A long-distance telephone office which connects several primary centers and which is in class number 2; only a regional center has greater importance in routing telephone calls. Abbreviated SC. { 'sek·shən·əl 'sen·tər }

sectionalized vertical antenna [ELECTROMAG] Vertical antenna that is insulated at one or more points along its length; the insertion of suitable reactances or applications of a driving voltage across the insulated points results in a modified current distribution giving a more desired radiation pattern in the vertical plane. { 'sek·shən·əl‚īzd 'vərd·ə·kəl an‚ten·ə }

sectional radiography [ELECTR] The technique of making radiographs of plane sections of a body or an object; its purpose is to show detail in a predetermined plane of the body, while blurring the images of structures in other planes. Also known as laminography; planigraphy; tomography. { 'sek·shən·əl ‚rād·ē'äg·rə·fē }

sector [COMPUT SCI] **1.** A portion of a track on a magnetic disk or a band on a magnetic drum. **2.** A unit of data stored in such a portion. [ELECTROMAG] Coverage of a radar as measured in azimuth. { 'sek·tər }

sectoral horn [ELECTROMAG] Horn with two opposite sides parallel and the two remaining sides which diverge. { 'sek·tə·rəl 'hȯrn }

sector display [ELECTR] A display in which only a sector of the total service area of a radar system is shown; usually the sector is selectable. { 'sek·tər di‚splā }

sector interleave [COMPUT SCI] A sequence indicating the order in which sectors are arranged on a hard disk, generally so as to minimize access times. Also known as sector map. { 'sek·tər 'in·tər‚lēv }

sector map See sector interleave. { 'sek·tər ‚map }

sector mark [COMPUT SCI] A location on each sector of each track of a disk pack or floppy disk that gives the sector's address, tells whether the sector is in use, and gives other control information. { 'sek·tər ‚märk }

sector scan [ELECTR] A radar scan through a limited angle, as distinguished from complete rotation. { 'sek·tər ‚skan }

secure visual communications [COMMUN] The transmission of an encrypted digital signal consisting of animated visual and audio information; the distance may vary from a few hundred feet to thousands of miles. { si'kyu̇r 'vizh·ə·wəl kə‚myü·nə'kā·shənz }

secure voice [COMMUN] Voice message that is scrambled or coded, therefore not transmitted in the clear. { si'kyu̇r 'vȯis }

security [COMPUT SCI] The existence and enforcement of techniques which restrict access to data, and the conditions under which data may be obtained. { si'kyu̇r·əd·ē }

security kernel [COMPUT SCI] A portion of an operating system into which all security-related functions have been concentrated, forming a small, certifiably secure nucleus which is separate from the rest of the system. { si'kyu̇r·əd·ē ‚kər·nəl }

security perimeter [COMPUT SCI] A logical boundary of a distributed computer system, surrounding all the resources that are controlled and protected by the system. { sə'kyu̇r·əd·ē pə‚rim·əd·ər }

security target [COMPUT SCI] A description of a product meeting the security and functionality requirements of a computing system. { sə'kyu̇r·əd·ē ‚tär·gət }

Seebeck coefficient [ELECTR] The ratio of the open-circuit voltage to the temperature difference between the hot and cold junctions of a circuit exhibiting the Seebeck effect. { 'zā‚bek ‚kō·ē'fish·ənt }

Seebeck effect [ELECTR] The development of a voltage due to differences in temperature between two junctions of dissimilar metals in the same circuit. { 'zā‚bek i‚fekt }

seed [COMPUT SCI] An initial number used by an algorithm such as a random number generator. { sēd }

seeding [ELECTR] The introduction of atoms with a low ionization potential into a hot gas to increase electrical conductivity. { 'sēd·iŋ }

seek [COMPUT SCI] **1.** To position the access mechanism of a random-access storage device at a designated location or position. **2.** The command that directs the positioning to take place. { sēk }

seek area [COMPUT SCI] An area of a direct-access storage device, such as a magnetic disk file, assigned to hold records to which rapid access is needed, and located so that the physical characteristics of the device permit such access. Also known as cylinder. { 'sēk ‚er·ē·ə }

seek time [COMPUT SCI] The time required for the access mechanism of a random-access storage device to be properly positioned. { 'sēk ‚tīm }

segment [COMPUT SCI] **1.** A single section of an overlay program structure, which can be loaded into the main memory when and as needed. **2.** In some direct-access storage devices, a hardware-defined portion of a track having fixed data capacity. { 'seg·mənt }

segmentation [COMMUN] The division of a long communications message into smaller messages that can be transmitted intermittently. { COMPUT SCI } **1.** The division of virtual storage into identifiable functional regions, each having enough addresses so that programs or data stored in them will not assign the same addresses more than once. **2.** The division of a large computer program into smaller units, called segments. *See* picture segmentation. { ˌseg·mənˈtā·shən }

segmented aperture-synthetic aperture radar [ENG] An enhancement of synthetic aperture radar that overcomes restrictions on the effective length of the receiving antenna by using a receiving antenna array composed of a set of contiguous subarrays and employing signal processing to provide the proper phase corrections for each subarray. Abbreviated SASAR. { ˈseg ˌment·əd ˈap·ə·chər sinˈthed·ik ˈap·ə·chər ˈrāˌdär }

segment mark [COMPUT SCI] A special character written on tape to separate one section of a tape file from another. { ˈseg·mənt ˌmärk }

select [COMPUT SCI] **1.** To choose a needed subroutine from a file of subroutines. **2.** To take one alternative if the report on a condition is of one state, and another alternative if the report on the condition is of another state. **3.** To pull from a mass of data certain items that require special attention. { siˈlekt }

select bit [COMPUT SCI] The bit (or bits) in an input/output instruction word which selects the function of a specified device. Also known as subdevice bit. { siˈlekt ˌbit }

selecting circuit [ELEC] A simple switching circuit that receives the identity (the address) of a particular item and selects that item from among a number of similar ones. { siˈlek·tiŋ ˌsər·kət }

selection [COMMUN] The process of addressing a call to a specific station in a selective calling system. { siˈlek·shən }

selection check [COMPUT SCI] Electronic computer check, usually automatic, to verify that the correct register, or other device, is selected in the performance of an instruction. { siˈlek·shən ˌchek }

selection sort [COMPUT SCI] A sorting routine that scans a list of items repeatedly and, on each pass, selects the item with the lowest value and places it in its final position. { siˈlek·shən ˌsórt }

selective absorption [ELECTROMAG] A greater absorption of electromagnetic radiation at some wavelengths (or frequencies) than at others. { siˈlek·tiv abˈsórp·shən }

selective calling system [COMMUN] A radio communications system in which the central station transmits a coded call that activates only the receiver to which that code is assigned. { siˈlek·tiv ˌkȯl·iŋ ˌsis·təm }

selective circuit [ELEC] A circuit that transmits certain types of signals and fails to transmit or attenuates others. { siˈlek·tiv ˈsər·kət }

selective dump [COMPUT SCI] An edited or nonedited listing of the contents of selected areas of memory or auxiliary storage. { siˈlek·tiv ˈdəmp }

selective fading [COMMUN] Fading that is different at different frequencies in a frequency band occupied by a modulated wave, causing distortion that varies in nature from instant to instant. { siˈlek·tiv ˈfād·iŋ }

selective identification feature [ELECTR] Airborne pulse-type transponder which provides automatic selective identification of aircraft in which it is installed to ground, shipboard, or airborne recognition installations. { siˈlek·tiv ī ˌden·tə·fəˈkā·shən ˌfē·chər }

selective interference [COMMUN] Interference whose energy is concentrated in a narrow band of frequencies. { siˈlek·tiv ˌin·tərˈfir·əns }

selective jamming [ELECTR] Jamming in which only a single radio channel is jammed. { siˈlek·tiv ˈjam·iŋ }

selectively doped heterojunction transistor *See* high-electron-mobility transistor. { siˈlek·tiv·lē ˈdōpt ˌhed·ə·rōˈjəŋk·shən tranˈzis·tər }

selective photoelectric effect [ELECTR] A resonance in the dependence of photoemission on the incident photon energy that is displayed when light is incident on a thin-metal film and the light vector has a component perpendicular to a crystal plane. Also known as spectral selective photoelectric effect; vector effect. { siˈlek·tiv ˌfōd·ō·iˈlek·trik iˌfekt }

selective reflection [ELECTROMAG] Reflection of electromagnetic radiation more strongly at some wavelengths (or frequencies) than at others. { siˈlek·tiv riˈflek·shən }

selective ringing [COMMUN] Telephone arrangement on party lines, in which only the bell of the called subscriber rings, with other bells on the party line remaining silent. { siˈlek·tiv ˈriŋ·iŋ }

selective scattering [ELECTROMAG] Scattering of electromagnetic radiation more strongly at some wavelengths than at others. { siˈlek·tiv ˈskad·ə·riŋ }

selective trace [COMPUT SCI] A tracing routine wherein only instructions satisfying certain specified criteria are subject to tracing. { siˈlek·tiv ˈtrās }

selectivity [ELECTR] **1.** The ability of a radio receiver to separate a desired signal frequency from other signal frequencies, some of which may differ only slightly from the desired value. **2.** The inverse of the shape factor of a bandpass filter. { səˌlekˈtiv·əd·ē }

selector [COMPUT SCI] Computer device which interrogates a condition and initiates a particular operation dependent upon the report. [ELECTR] An automatic or other device for making connections to any one of a number of circuits, such as a selector relay or selector switch. { siˈlek·tər }

selector channel [COMPUT SCI] A unit which connects high-speed input/output devices, such as magnetic tapes, disks, and drums, to a computer memory. { siˈlek·tər ˈchan·əl }

selector switch [ELEC] A manually operated multiposition switch. Also called multiple-contact switch. { siˈlek·tər ˌswich }

selenium cell [ELECTR] A photoconductive cell in which a thin film of selenium is used between suitable electrodes; the resistance of the cell decreases when the illumination is increased. { səˈlē·nē·əm ˌsel }

selenium diode |ELECTR| A small area selenium rectifier which has characteristics similar to those of selenium rectifiers used in power systems. { sə'lē·nē·əm 'dī,ōd }

selenium rectifier |ELECTR| A metallic rectifier in which a thin layer of selenium is deposited on one side of an aluminum plate and a conductive metal coating is deposited on the selenium. { sə'lē·nē·əm 'rek·tə,fī·ər }

self-adapting system |SYS ENG| A system which has the ability to modify itself in response to changes in its environment. { ¦self ə¦dap·tiŋ 'sis·təm }

self-adjusting communications See adaptive communications. { ¦self ə¦jəst·iŋ kə,myü·nə'kā·shənz }

self-bias |ELECTR| A grid bias provided automatically by the resistor in the cathode or grid circuit of an electron tube; the resulting voltage drop across the resistor serves as the grid bias. Also known as automatic C bias; automatic grid bias. { ¦self ¦bī·əs }

self-bias transistor circuit |ELECTR| A transistor with a resistance in the emitter lead that gives rise to a voltage drop which is in the direction to reverse-bias the emitter junction; the circuit can be used even if there is zero direct-current resistance in series with the collector terminal. { ¦self ¦bī·əs tran'zis·tər ,sər·kət }

self-checking code |COMPUT SCI| An encoding of data so designed and constructed that an invalid code can be properly detected; this permits the detection, but not the correction, of almost all errors. Also known as error-checking code; error-detecting code. { ¦self ¦chek·iŋ 'kōd }

self-checking number |COMPUT SCI| A number with a suffix figure related to the figure of the number, used to check the number after it has been transferred from one medium or device to another. { ¦self ¦chek·iŋ 'nəm·bər }

self-cleaning contact See wiping contact. { ¦self 'klēn·iŋ 'kän,takt }

self-complementing code |COMPUT SCI| A binary-coded-decimal code in which the combination for the complement of a digit is the complement of the combination for that digit. { ¦self ¦käm·plə,ment·iŋ 'kōd }

self-contained database management system |COMPUT SCI| A database management system that is in no way an extension of any programming language, and is usually quite independent of any language. { ¦self kən¦tānd ¦dad·ə¦bās 'man·ij·mənt ,sis·təm }

self-diagnostic routine |COMPUT SCI| A test of an electronic device that is performed automatically, usually when the device is turned on. Also known as self-test. { ¦self ¦dī·əg¦näs·tik rü'tēn }

self-documenting code |COMPUT SCI| A sequence of programming statements that are simple and straightforward and can be readily implemented by another programmer. { ¦self ¦däk·yə,ment·iŋ 'kōd }

self-excited |ELEC| Operating without an external source of alternating-current power. { ¦self ik'sīd·əd }

self-excited oscillator |ELECTR| An oscillator that depends on its own resonant circuits for initiation of oscillation and frequency determination. { ¦self ik'sīd·əd 'äs·ə,lād·ər }

self-extracting file |COMPUT SCI| A compressed (zipped) file that unzips itself when it is executed. { ¦self ik,strak·tiŋ 'fīl }

self-healing dielectric breakdown |ELECTR| A dielectric breakdown in which the breakdown process itself causes the material to become insulating again. { ¦self ¦hēl·iŋ ,dī·ə¦lek·trik 'brāk,daủn }

self-impedance See mesh impedance. { ¦self im ¦pēd·əns }

self-optimizing communications See adaptive communications. { ¦self ¦äp·tə,mīz·iŋ kə,myü·nə'kā·shəns }

self-pulsing |ELECTR| Special type of grid pulsing which automatically stops and starts the oscillations at the pulsing rate by a special circuit. { ¦self ¦pəls·iŋ }

self-quenched detector |ELECTR| Superregenerative detector in which the time constant of the grid leak and grid capacitor is sufficiently large to cause intermittent oscillation above audio frequencies, serving to stop normal regeneration each time just before it spills over into a squealing condition. { ¦self ¦kwencht di'tek·tər }

self-quenching oscillator |ELECTR| Oscillator producing a series of short trains of radio-frequency oscillations separated by intervals of quietness. { 'self ¦kwench·iŋ 'äs·ə,lād·ər }

self-repair |COMPUT SCI| Any type of hardware redundancy in which faults are selectively masked and are detected, located, and subsequently corrected by the replacement of the failed unit by an unfailed replica. { ¦self ri¦per }

self-reset |ELEC| Automatically returning to the original position when normal conditions are resumed; applied chiefly to relays and circuit breakers. { ¦self 'rē,set }

self-resetting loop |COMPUT SCI| A loop whose termination causes the numbers stored in all locations affected by the loop to be returned to the original values which they had upon entry into the loop. { ¦self ri¦sed·iŋ 'lüp }

self-saturation |ELECTR| The connection of half-wave rectifiers in series with the output windings of the saturable reactors of a magnetic amplifier, to give higher gain and faster response. { ¦self ,sach·ə¦rā·shən }

self-scanned image sensor |ELECTR| A solid-state device, still in the early stages of development, which converts an optical image into a television signal without the use of an electron beam; it consists of an array of photoconductor diodes, each located at the intersection of mutually perpendicular address strips respectively connected to horizontal and vertical scan generators and video coupling circuits. { 'self ¦skand 'im·ij ,sen·sər }

self-starting synchronous motor |ELEC| A synchronous motor provided with the equivalent of a squirrel-cage winding, to permit starting as an induction motor. { 'self ¦stärd·iŋ 'siŋ·krə·nəs 'mōd·ər }

self-steering microwave array [ELECTROMAG] An antenna array used with electronic circuitry that senses the phase of incoming pilot signals and positions the antenna beam in their direction of arrival. { 'self ¦stir·iŋ 'mī·krō,wāv ə'rā }

self-synchronous device See synchro. { ¦self ¦siŋ·krə·nəs di'vīs }

self-synchronous repeater See synchro. { ¦self ¦siŋ·krə·nəs ri'pēd·ər }

self-test See self-diagnostic routine. { 'self ¦test }

self-triggering program [COMPUT SCI] A computer program which automatically commences execution as soon as it is fed into the central processing unit. { self ¦trig·ə·riŋ 'prō·grəm }

self-tuning regulator [CONT SYS] A type of adaptive control system composed of two loops, an inner loop which consists of the process and an ordinary linear feedback regulator, and an outer loop which is composed of a recursive parameter estimator and a design calculation, and which adjusts the parameters of the regulator. Abbreviated STR. { self ¦tün·iŋ 'reg·yə,lād·ər }

selsyn See synchro. { 'sel·sin }

selsyn generator See synchro transmitter. { 'sel·sin ¦jen·ə,rād·ər }

selsyn motor See synchro receiver. { 'sel·sin ,mōd·ər }

selsyn receiver See synchro receiver. { 'sel·sin ri,sē·vər }

selsyn system See synchro system. { 'sel·sin ,sis·təm }

selsyn transmitter See synchro transmitter. { 'sel·sin tranz,mid·ər }

SEM See scanning electron microscope.

semantic analysis [COMPUT SCI] A phase of natural language processing, following parsing, that involves extraction of context-independent aspects of a sentence's meaning, including the semantic roles of entities mentioned in the sentence, and quantification information, such as cardinality, iteration, and dependency. { si'man·tik ə'nal·ə·səs }

semantic error [COMPUT SCI] The use of an incorrect symbolic name in a computer program. { si'man·tik 'er·ər }

semantic extension [COMPUT SCI] An extension mechanism which introduces new kinds of objects into an extensible language, such as additional data types or operations. { si'man·tik ik'sten·shən }

semantic gap [COMPUT SCI] The difference between a data or language structure and the objects that it models. { si'man·tik 'gap }

semaphore [COMPUT SCI] A memory cell that is shared by two parallel processes which rely on each other for their continued operation, and that provides an elementary form of communication between them by indicating when significant events have taken place. { 'sem·ə,fōr }

semialgorithm [COMPUT SCI] A procedure for solving a problem that will continue endlessly if the problem has no solution. { ,sem·ē'al·gə ,rith·əm }

semiautomatic telephone system [COMMUN] Telephone system that limits automatic dialing to only those subscribers who are served by the same exchange as the calling subscriber. { ¦sem·ē,öd·ə'mad·ik 'tel·ə,fōn ,sis·təm }

semiconducting compound [SOLID STATE] A compound which is a semiconductor, such as copper oxide, mercury indium telluride, zinc sulfide, cadmium selenide, and magnesium iodide. { ¦sem·i·kən¦dək·tiŋ 'käm,paúnd }

semiconducting crystal [SOLID STATE] A crystal of a semiconductor, such as silicon, germanium, or gray tin. { ¦sem·i·kən¦dək·tiŋ ¦krist·əl }

semiconductor [ELECTR] A solid crystalline material whose conductivity is intermediate between that of a metal and an insulator and may depend on temperature or voltage; by making suitable contacts to the material or by making the material suitably inhomogenous, electrical rectification and amplification may be obtained. { ¦sem·i·kən¦dək·tər }

semiconductor device [ELECTR] Electronic device in which the characteristic distinguishing electronic conduction takes place within a semiconductor. { ¦sem·i·kən¦dək·tər di,vīs }

semiconductor diode [ELECTR] **1.** A two-electrode semiconductor device that utilizes the rectifying properties of a *pn* junction or a point contact. **2.** More generally, any two-terminal electronic device that utilizes the properties of the semiconductor from which it is constructed. Also known as crystal diode; crystal rectifier; diode. { ¦sem·i·kən¦dək·tər 'dī,ōd }

semiconductor-diode parametric amplifier [ELECTR] Parametric amplifier using one or more varactors. { ¦sem·i·kən¦dək·tər ¦dī,ōd ¦par·ə¦me·trik 'am·plə ,fī·ər }

semiconductor disk [COMPUT SCI] A large semiconductor memory that imitates a disk drive in that the operating system can read and write to it as though it were an ordinary disk, but at a much faster rate. Also known as nonrotating disk. { 'sem·i·kən,dək·tər ,disk }

semiconductor doping See doping. { ¦sem·i·kən ¦dək·tər 'dōp·iŋ }

semiconductor heterostructure [ELECTR] A structure of two different semiconductors in junction contact having useful electrical or electrooptical characteristics not achievable in either conductor separately; used in certain types of lasers and solar cells. { ¦sem·i·kən¦dək· tər 'hed·ə·rō,strək·chər }

semiconductor junction [ELECTR] Region of transition between semiconducting regions of different electrical properties, usually between *p*-type and *n*-type material. { ¦sem·i·kən¦dək·tər ,jəŋk·shən }

semiconductor laser [OPTICS] A laser in which stimulated emission of coherent light occurs at a *pn* junction when electrons and holes are driven into the junction by carrier injection, electron-beam excitation, impact ionization, optical excitation, or other means; used as light transmitters and modulators in optical communications and integrated optics. Also known as diode laser; laser diode. { ¦sem·i·kən¦dək·tər 'lā·zər }

semiconductor memory |COMPUT SCI| A device for storing digital information that is fabricated by using integrated circuit technology. Also known as integrated-circuit memory; large-scale integrated memory; memory chip; semiconductor storage; transistor memory. { ¦sem·i·kən ¦dək·tər ‚mem·rē }

semiconductor rectifier See metallic rectifier. { ¦sem·i·kən¦dək·tər 'rek·tə‚fī·ər }

semiconductor storage See semiconductor memory. { 'sem·i·kən‚dək·tər ‚stór·ij }

semiconductor thermocouple |ELECTR| A thermocouple made of a semiconductor, which offers the prospect of operation with high-temperature gradients, because semiconductors are good electrical conductors but poor heat conductors. { ¦sem·i·kən¦dək·tər 'thər·mə‚kəp·əl }

semidense list |COMPUT SCI| A list that can be divided into two contiguous portions, with all the cells in the larger portion filled and all the other cells empty. { ¦sem·i'dens 'list }

semimagnetic controller |ELEC| Electrical controller having only part of its basic functions performed by devices that are operated by electromagnets. { ¦sem·i·mag'ned·ik kən'trōl·ər }

seminumerical algebraic manipulation language |COMPUT SCI| The most elementary type of algebraic manipulation language, constructed to manipulate data from rigid classes of mathematical objects possessing strictly canonical forms. { ¦sem·i·nü'mer·ə·kəl ‚al·jə'brā·ik mə ‚nip·yə'lā·shən ‚laŋ·gwij }

semiselective ringing |COMMUN| In telephone service, party line ringing wherein the bells of two stations are rung simultaneously; the differentiation is made by the number of rings. { ¦sem·i·si'lek·tiv 'riŋ·iŋ }

semitransparent photocathode |ELECTR| Photocathode in which radiant flux incident on one side produces photoelectric emission from the opposite side. { ¦sem·i·tranz'par·ənt ¦fōd·ō'kath ‚ōd }

sender |COMMUN| Part of an automatic-switching telephone system that receives pulses from a dial or other source and, in accordance with them, controls the further operations necessary in establishing a telephone connection. { 'sen·dər }

sending-end impedance |ELEC| Ratio of an applied potential difference to the resultant current at the point where the potential difference is applied; the sending-end impedance of a line is synonymous with the driving-point impedance of the line. { 'send·iŋ ¦end im'pēd·əns }

sense amplifier |ELECTR| Circuit used to determine either a phase or voltage change in communications-electronics equipment and to provide automatic control function. { 'sens ‚am·plə‚fī·ər }

sense antenna |ELECTROMAG| An auxiliary antenna used with a directional receiving antenna to resolve a 180° ambiguity in the directional indication. Also known as sensing antenna. { 'sens an‚ten·ə }

sense light |COMPUT SCI| A light which can be turned on or off, its status being the determinant as to which path a program will select. { 'sens ‚līt }

sensing antenna See sense antenna. { 'sens·iŋ an‚ten·ə }

sensing element See sensor. { 'sens·iŋ ‚el·ə·mənt }

sensing signal |COMMUN| A special signal that is transmitted to alert the receiving station at the beginning of a message. { 'sens·iŋ ‚sig·nəl }

sensistor |ELECTR| Silicon resistor whose resistance varies with temperature, power, and time. { sen'zis·tər }

sensitive data |COMPUT SCI| Data that can be read or processed in specified transactions by a specified program, device, or user. { 'sen·səd·iv 'dad·ə }

sensitive switch See snap-action switch. { 'sen·səd·iv 'swich }

sensitivity |ELECTR| **1.** The minimum input signal required to produce a specified output signal, for a radio receiver or similar device. **2.** Of a camera tube, the signal current developed per unit incident radiation, that is, per watt per unit area. { ‚sen·sə'tiv·əd·ē }

sensitivity function |CONT SYS| The ratio of the fractional change in the system response of a feedback-compensated feedback control system to the fractional change in an open-loop parameter, for some specified parameter variation. { ‚sen·sə'tiv·əd·ē ‚fəŋk·shən }

sensitivity time control |ELECTR| A controlled reduction in sensitivity of a radar receiver immediately after the transmission of a pulse, with a programmed restoration of full sensitivity as returns come from greater ranges; done to prevent the reception of a multitude of tiny targets close to the radar, such as birds and insects, and to prevent receiver saturation by large targets at very short range. { ‚sen·sə'tiv·əd·ē 'tīm kən‚trōl }

sensitization See activation. { ‚sen·səd·ə'zā·shən }

sensor |ENG| The generic name for a device that senses either the absolute value or a change in a physical quantity such as temperature, pressure, flow rate, or pH, or the intensity of light, sound, or radio waves and converts that change into a useful input signal for an information-gathering system; a television camera is therefore a sensor, and a transducer is a special type of sensor. Also known as primary detector; sensing element. { 'sen·sər }

sensory control |CONT SYS| Control of a robot's actions on the basis of its sensor readings. { 'sen·sə·rē kən'trōl }

sensory controlled robot |CONT SYS| A robot whose programmed sequence of instructions can be modified by information about the environment received by the robot's sensors. { 'sen·sə·rē kən'trōld 'rō‚bät }

sentence |COMPUT SCI| An entire instruction in the COBOL programming language. { 'sent·əns }

sentinel |COMPUT SCI| Symbol marking the beginning or end of an element of computer information such as an item or a tape. { 'sent·ən·əl }

separately excited |ELEC| Obtaining excitation from a source other than the machine or device itself. { 'sep·rət·lē ik'sīd·əd }

separation |ENG ACOUS| The degree, expressed in decibels, to which left and right stereo channels are isolated from each other. { ˌsep·ə'rā·shən }

separation filter |ELECTR| Combination of filters used to separate one band of frequencies from another. { ˌsep·ə'rā·shən ˌfil·tər }

separation theorem |CONT SYS| A theorem in optimal control theory which states that the solution to the linear quadratic Gaussian problem separates into the optimal deterministic controller (that is, the optimal controller for the corresponding problem without noise) in which the state used is obtained as the output of an optimal state estimator. { ˌsep·ə'rā·shən ˌthir·əm }

separator |COMPUT SCI| A datum or character that denotes the beginning or ending of a unit of data. |ELEC| A porous insulating sheet used between the plates of a storage battery. |ELECTR| A circuit that separates one type of signal from another by clipping, differentiating, or integrating action. { 'sep·ə,rād·ər }

separator page |COMPUT SCI| A page preceding or following a report in a computer printout giving all information needed to identify the report. { 'sep·ə,rād·ər ,pāj }

separatrix |CONT SYS| A curve in the phase plane of a control system representing the solution to the equations of motion of the system which would cause the system to move to an unstable point. { 'sep·ə,triks }

septate coaxial cavity |ELECTROMAG| Coaxial cavity having a vane or septum, added between the inner and outer conductors, so that it acts as a cavity of a rectangular cross section bent transversely. { 'sep,tāt kō'ak·sē·əl 'kav·əd·ē }

septate waveguide |ELECTROMAG| Waveguide with one or more septa placed across it to control microwave power transmission. { 'sep ,tāt 'wāv,gīd }

septum |ELECTROMAG| A metal plate placed across a waveguide and attached to the walls by highly conducting joints; the plate usually has one or more windows, or irises, designed to give inductive, capacitive, or resistive characteristics. { 'sep·təm }

sequence |COMPUT SCI| To put a set of symbols into an arbitrarily defined order; that is, to select A if A is greater than or equal to B, or to select B if A is less than B. { 'sē·kwəns }

sequence calling |COMPUT SCI| The instructions used for linking a closed subroutine with a main routine; that is, standard linkage and a list of the parameters. { 'sē·kwəns ,kȯl·iŋ }

sequence check |COMPUT SCI| To verify that correct precedence relationships are obeyed, usually by checking for ascending sequence numbers. { 'sē·kwəns ,chek }

sequence checking routine |COMPUT SCI| In computers, a checking routine which records specified data regarding the operations resulting from each instruction. { 'sē·kwəns ¦chek·iŋ rü,tēn }

sequence counter See instruction counter. { 'sē·kwəns ,kaȯnt·ər }

sequence error |COMPUT SCI| An error that arises when the arrangement of items in a set does not follow some specified order. { 'sē·kwəns ,er·ər }

sequence monitor |COMPUT SCI| The automatic step-by-step check by a computer of the manual actions required for the starting and shutdown of a computer. { 'sē·kwəns ,män·əd·ər }

sequence number |COMPUT SCI| A number assigned to an item to indicate its relative position in a series of related items. { 'sē·kwəns ,nəm·bər }

sequence pointer |COMPUT SCI| For a list that is stored in computer memory, the portion of a list item that gives the storage location of the subsequent item on the list (or the locations of the subsequent and previous items of a symmetric list). Also known as sequencing pointer. { 'sē·kwəns 'pȯint·ər }

sequencer |COMPUT SCI| A machine which puts items of information into a particular order, for example, it will determine whether A is greater than, equal to, or less than B, and sort or order accordingly. Also known as sorter. |ENG| A mechanical or electronic device that may be set to initiate a series of events and to make the events follow in a given sequence. { 'sē·kwən·sər }

sequence register |COMPUT SCI| A counter which contains the address of the next instruction to be carried out. { 'sē·kwəns ,rej·ə·stər }

sequence robot See preprogrammed robot. { 'sē·kwəns ,rō,bät }

sequencing equipment |COMMUN| Special selecting device that permits messages received from several teletypewriter circuits to be subsequently selected and retransmitted over a reduced number of trunks or circuits. { 'sē·kwəns·iŋ i,kwip·mənt }

sequencing pointer See sequence pointer. { 'sē·kwəns·iŋ 'pȯint·ər }

sequential access |COMPUT SCI| A process that involves reading or writing data serially and, by extension, a data-recording medium that must be read serially, as a magnetic tape. { si'kwen·chəl 'ak,ses }

sequential batch operating system |COMPUT SCI| Software equipment that automatically begins running a new job on a computer system as soon as the current job is completed. { si'kwen·chəl 'bach 'äp·ə,rād·iŋ ,sis·təm }

sequential circuit |ELEC| A switching circuit whose output depends not only upon the present state of its input, but also on what its input conditions have been in the past. { si'kwen·chəl 'sər·kət }

sequential color television |COMMUN| A color television system in which the primary color components of a picture are transmitted one after the other; the three basic types are the line-sequential, dot-sequential, and field-sequential color television systems. Also known as sequential system. { si'kwen·chəl ¦kəl·ər 'tel·ə,vizh·ən }

sequential control |COMPUT SCI| Manner of operating a computer by feeding orders into the computer in a given order during the solution of a problem. { si'kwen·chəl kən'trōl }

sequential logic element [ELECTR] A circuit element having at least one input channel, at least one output channel, and at least one internal state variable, so designed and constructed that the output signals depend on the past and present states of the inputs. { si'kwen·chəl ¦läj·ik ¦el·ə·mənt }

sequential machine [COMPUT SCI] A mathematical model of a certain type of sequential circuit, which has inputs and outputs that can each take on any value from a finite set and are of interest only at certain instants of time, and in which the output depends on previous inputs as well as the concurrent input. { si'kwen·chəl mə'shēn }

sequential network [COMPUT SCI] An idealized model of a sequential circuit that reflects its logical but not its electronic properties. { si'kwen·chəl 'net,wərk }

sequential operation [COMPUT SCI] The consecutive or serial execution of operations, without any simultaneity or overlap. { si'kwen·chəl ,äp·ə'rā·shən }

sequential organization [COMPUT SCI] The write and read of records in a physical rather than a logical sequence. { si'kwen·chəl ,ór·gə·nə'zā·shən }

sequential processing [COMPUT SCI] Processing items in a collection of data according to some specified sequence of keys, in contrast to serial processing. { si'kwen·chəl 'prä,ses·iŋ }

sequential scanning See progressive scanning. { si'kwen·chəl 'skan·iŋ }

sequential scheduling system [COMPUT SCI] A first-come, first-served method of selecting jobs to be run. { si'kwen·chəl 'skej·ə·liŋ ,sis·təm }

sequential search [COMPUT SCI] A procedure for searching a table that consists of starting at some table position (usually the beginning) and comparing the file-record key in hand with each table-record key, one at a time, until either a match is found or all sequential positions have been searched. { si'kwen·chəl 'sərch }

sequential selection [COMMUN] The selection of the elements of a message (such as letters) from a set of possible elements (such as the alphabet), one after another. { si'kwen·chəl si'lek·shən }

sequential system See sequential color television. { si'kwen·chəl 'sis·təm }

serial [COMPUT SCI] Pertaining to the internal handling of data in sequential fashion. { 'sir·ē·əl }

serial-access [COMPUT SCI] **1.** Pertaining to memory devices having structures such that data storage sites become accessible for read/write in time-sequential order; circulating memories and magnetic tapes are examples of serial-access memories. **2.** Pertaining to a particular process or program that accesses data items sequentially, without regard to the capability of the memory hardware. **3.** Pertaining to character-by-character transmission from an on-line real-time keyboard. { 'sir·ē·əl 'ak,ses }

serial addition [COMPUT SCI] An arithmetic operation in which two numbers are added one digit at a time. { 'sir·ē·əl ə'dish·ən }

serial bit [COMPUT SCI] Digital computer storage in which the individual bits that make up a computer word appear in time sequence. { 'sir·ē·əl ,bit }

serial communications [COMMUN] The transmission of digital data over a single channel. { 'sir·ē·əl kə,myü·nə'kā·shənz }

serial digital computer [COMPUT SCI] A digital computer in which the digits are handled serially, although the bits that make up a digit may be handled either serially or in parallel. { 'sir·ē·əl 'dij·əd·əl kəm'pyüd·ər }

serial dot character printer [COMPUT SCI] A computer printer in which the dot matrix technique is used to print characters, one at a time, with a movable print head that is driven back and forth across the page. { 'sir·ē·əl ¦dät 'kar·ik·tər ,print·ər }

serial file [COMPUT SCI] The simplest type of file organization, in which no subsets are defined, no directories are provided, no particular file order is specified, and a search is performed by sequential comparison of the query with identifiers of all stored items. { 'sir·ē·əl 'fīl }

serial input/output [COMPUT SCI] Data that are transmitted into and out of a computer over a single conductor, one bit at a time. { 'sir·ē·əl 'in,pút 'aút,pút }

serial interface [COMPUT SCI] A link between a microcomputer and a peripheral device in which data is transmitted over a single conductor, one bit at a time. Also known as serial port. { 'sir·ē·əl 'in·tər,fās }

serialize [COMPUT SCI] To convert a signal suitable for parallel transmission into a signal suitable for serial transmission, consisting of a sequence of bits. { 'sir·ē·əl,līz }

serially reusable [COMPUT SCI] An attribute possessed by a program that can be used for several tasks in sequence without having to be reloaded into main memory for each additional use. { 'sir·ē·ə·lē rē'yü·zə·bəl }

serial memory [COMPUT SCI] A computer memory in which data are available only in the same sequence as originally stored. { 'sir·ē·əl 'mem·rē }

serial operation [COMPUT SCI] The flow of information through a computer in time sequence, using only one digit, word, line, or channel at a time. { 'sir·ē·əl ,äp·ə'rā·shən }

serial-parallel [COMPUT SCI] **1.** A combination of serial and parallel; for example, serial by character, parallel by bits comprising the character. **2.** Descriptive of a device which converts a serial input into a parallel output. { 'sir·ē·əl 'par·ə,lel }

serial-parallel conversion [COMPUT SCI] The transformation of a serial data representation as found on a disk or drum into the parallel data representation as exists in core. { 'sir·ē·əl 'par·ə,lel kən'vər·zhən }

serial port See serial interface. { 'sir·ē·əl ,pórt }

serial processing [COMPUT SCI] Processing items in a collection of data in the order that they appear in a storage device, in contrast to sequential processing. { 'sir·ē·əl 'prä,ses·iŋ }

serial processor [COMPUT SCI] A computer in which data are handled sequentially by separate units of the system. { 'sir·ē·əl 'prä,ses·ər }

serial programming |COMPUT SCI| In computers, programming in which only one operation is executed at one time. { 'sir·ē·əl 'prō₁gram·iŋ }

serial storage |COMPUT SCI| Computer storage in which time is one of the coordinates used to locate any given bit, character, or word; access time, therefore, includes a variable waiting time, ranging from zero to many word times. { 'sir·ē·əl 'stȯr·ij }

serial transfer |COMPUT SCI| Transfer of the characters of an element of information in sequence over a single path in a digital computer. { 'sir·ē·əl 'tranz·fər }

serial transmission |COMMUN| Transmission of groups of elements of a signal in time intervals that follow each other without overlapping. { 'sir·ē·əl tranz'mish·ən }

series |ELEC| An arrangement of circuit components end to end to form a single path for current. { 'sir·ēz }

series circuit |ELEC| A circuit in which all parts are connected end to end to provide a single path for current. { 'sir·ēz ₁sər·kət }

series compensation |CONT SYS| See cascade compensation. |ELEC| The insertion of variable, controlled, high-voltage series capacitors into transmission lines in order to modify the impedance structure of a transmission network so as to adjust the power-flow distribution on individual lines and thus increase the power flow across such compensated lines. { 'sir·ēz ₁käm·pən'sā·shən }

series connection |ELEC| A connection that forms a series circuit. { 'sir·ēz kə₁nek·shən }

series excitation |ELEC| The obtaining of field excitation in a motor or generator by allowing the armature current to flow through the field winding. { 'sir·ēz ₁ek·sə'tā·shən }

series-fed vertical antenna |ELECTROMAG| Vertical antenna which is insulated from the ground and energized at the base. { 'sir·ēz ₁fed 'vərd·i·kəl an'ten·ə }

series feed |ELECTR| Application of the direct-current voltage to the plate or grid of a vacuum tube through the same impedance in which the alternating-current flows. { 'sir·ēz 'fēd }

series generator |ELEC| A generator whose armature winding and field winding are connected in series. Also known as series-wound generator. { 'sir·ēz ₁jen·ə₁rād·ər }

series loading |ELECTR| Loading in which reactances are inserted in series with the conductors of a transmission circuit. { 'sir·ēz ₁lōd·iŋ }

series modulation |ELECTR| Modulation in which the plate circuits of a modulating tube and a modulated amplifier tube are in series with the same plate voltage supply. { 'sir·ēz ₁mäj·ə'lā·shən }

series motor |ELEC| A commutator-type motor having armature and field windings in series; characteristics are high starting torque, variation of speed with load, and dangerously high speed on no-load. Also known as series-wound motor. { 'sir·ēz ₁mōd·ər }

series multiple |ELEC| Type of switchboard jack arrangement in which a single line circuit appears before two or more operators, all appearances being connected in series. { 'sir·ēz ₁məl·tə·pəl }

series-parallel circuit |ELEC| A circuit in which some of the components or elements are connected in parallel, and one or more of these parallel combinations are in series with other components of the circuit. { 'sir·ēz ₁par·ə₁lel ₁sər·kət }

series-parallel control |ELEC| A method of controlling the speed of electric motors in which the motors, or groups of motors, are connected in series at some times and in parallel at other times. { 'sir·ēz ₁par·ə₁lel kən₁trōl }

series-parallel switch |ELEC| A switch used to change the connections of lamps or other devices from series to parallel, or vice versa. { 'sir·ēz ₁par·ə₁lel 'swich }

series peaking |ELECTR| Use of a peaking coil and resistor in series as the load for a video amplifier to produce peaking at some desired frequency in the passband, such as to compensate for previous loss of gain at the high-frequency end of the passband. { 'sir·ēz 'pēk·iŋ }

series radio tap |COMMUN| A telephone tapping procedure in which a miniature radio transmitter is inserted in series with one wire of the target pair so that the transmitter derives its power from the telephone central battery. { 'sir·ēz 'rād·ē·ō ₁tap }

series reactor |ELEC| A reactor used in alternating-current power systems for protection against excessively large currents under short-circuit or transient conditions; it consists of coils of heavy insulated cable either cast in concrete columns or supported in rigid frames and mounted on insulators. Also known as current-limiting reactor. { 'sir·ēz rē₁ak·tər }

series regulator |ELEC| A regulator that controls output voltage or current by automatically varying a resistance in series with the voltage source. { 'sir·ēz 'reg·yə₁lād·ər }

series repeater |ELEC| A type of negative impedance telephone repeater which is stable when terminated in an open circuit and oscillates when it is connected to a low impedance, in contrast to a shunt repeater. { 'sir·ēz ri'pēd·ər }

series resonance |ELEC| Resonance in a series resonant circuit, wherein the inductive and capacitive reactances are equal at the frequency of the applied voltage; the reactances then cancel each other, reducing the impedance of the circuit to a minimum, purely resistive value. { 'sir·ēz 'rez·ən·əns }

series resonant circuit |ELEC| A resonant circuit in which the capacitor and coil are in series with the applied alternating-current voltage. { 'sir·ēz ₁rez·ən·ənt ₁sər·kət }

series-shunt network See ladder network. { 'sir·ēz ₁shənt 'net₁wərk }

series T junction See E-plane T junction. { 'sir·ēz 'tē ₁jəŋk·shən }

series transistor regulator |ELECTR| A voltage regulator whose circuit has a transistor in series with the output voltage, a Zener diode, and a resistor chosen so that the Zener diode is

517

approximately in the middle of its operating range.   { 'sir·ēz tran'zis·tər 'reg·yə,lād·ər }

series-tuned circuit |ELEC| A simple resonant circuit consisting of an inductance and a capacitance connected in series.   { 'sir·ēz |tünd ,sər·kət }

series winding |ELEC| A winding in which the armature circuit and the field circuit are connected in series with the external circuit.   { 'sir·ēz ,wīnd·iŋ }

series-wound generator   See series generator. { 'sir·ēz |waúnd 'jen·ə,rād·ər }

series-wound motor   See series motor.   { 'sir·ēz |waúnd 'mōd·ər }

serrated pulse |ELECTR| Vertical and horizontal synchronizing pulse divided into a number of small pulses, each of which acts for the duration of half a line in an analog television system. { se,rād·əd 'pəls }

serrodyne |ELECTR| Phase modulator using transit time modulation of a traveling-wave tube or klystron.   { 'ser·ə,dīn }

server |COMPUT SCI| A computer or software package that sends requested information to a client or clients in a network.   { 'sər·vər }

service area |COMMUN| The area that is effectively served by a given radio or television transmitter, navigation aid, or other type of transmitter. Also known as coverage.   { 'sər·vəs ,er·ē·ə }

service band |COMMUN| Band of frequencies allocated to a given class of radio service.   { 'sər·vəs ,band }

service bit |COMMUN| A bit used in data transmission to monitor the transmission rather than to convey information, such as a request that part of a message be repeated.   { 'sər·vəs ,bit }

service bureau |COMPUT SCI| An organization that offers time sharing and software services to its users who communicate with a computer in the bureau from terminals on their premises. { 'sər·vəs ,byúr·ō }

service oscillator   See radio-frequency signal generator.   { 'sər·vəs 'äs·ə,lād·ər }

service program |COMPUT SCI| A computer program that is used in a computer system to support the functioning of the system, such as a librarian or a utility program.   { 'sər·vəs ,prō ,gram }

service provider |COMPUT SCI| An organization that provides access to a wide-area network, such as the Internet.   { 'sər·vəs prə,vīd·ər }

service routine |COMPUT SCI| A section of a computer code that is used in so many different jobs that it cannot belong to any one job.   { 'sər·vəs rü,tēn }

service wires |ELEC| The conductors that bring the electric power into a building.   { 'sər·vəs ,wīrz }

servicing time |COMPUT SCI| Machine down-time necessary for routine testing, for machine servicing due to breakdown, or for preventive servicing measures; includes all test time (good or bad) following breakdown and subsequent repair or preventive servicing.   { 'sər·vəs·iŋ ,tīm }

serving |ELEC| A covering, such as thread or tape, that protects a winding from mechanical damage. Also known as coil serving.   { 'sərv·iŋ }

servo   See servomotor.   { 'sər·vō }

servo amplifier |ELECTR| An amplifier used in a servomechanism.   { 'sər·vō 'am·plə,fī·ər }

servolink |CONT SYS| A power amplifier, usually mechanical, by which signals at a low power level are made to operate control surfaces requiring relatively large power inputs, for example, a relay and motor-driven actuator.   { 'sər·vō,liŋk }

servo loop   See single-loop servomechanism. { 'sər·vō ,lüp }

servomechanism |CONT SYS| An automatic feedback control system for mechanical motion; it applies only to those systems in which the controlled quantity or output is mechanical position or one of its derivatives (velocity, acceleration, and so on). Also known as servo system.   { |sər·vō'mek·ə,niz·əm }

servomotor |CONT SYS| The electric, hydraulic, or other type of motor that serves as the final control element in a servomechanism; it receives power from the amplifier element and drives the load with a linear or rotary motion. Also known as servo.   { 'sər·vō,mōd·ər }

servomultiplier |ELECTR| An electromechanical multiplier in which one variable is used to position one or more ganged potentiometers across which the other variable voltages are applied.   { |sər·vō'məl·tə,plī·ər }

servo system   See servomechanism.   { 'sər·vō ,sis·təm }

sesquisideband transmission |COMMUN| Transmission of a carrier modulated by one full sideband and half of the other sideband. { |ses·kwē'sīd,band tranz'mish·ən }

set |COMPUT SCI| A collection of record types. |ELECTR| The placement of a storage device in a prescribed state, for example, a binary storage cell in the high or 1 state. |ENG| A combination of units, assemblies, and parts connected or otherwise used together to perform an operational function, such as a radar set.   { set }

set analyzer   See analyzer.   { 'set ,an·ə,līz·ər }

set-associative cache |COMPUT SCI| A cache memory in which incoming data are distributed in sequence to each of two to eight areas or sets, and is generally read out in the same manner, allowing each set to prepare for the next input/output operation.   { 'set ə,sōs·ē,ād·iv ,kash }

set class |COMPUT SCI| The collection of set occurrences that have been or may be created in accordance with a particular set description. { 'set ,klas }

set composite |ELEC| Signaling circuit in which two signaling or telegraph legs may be superimposed on a two-wire, interoffice trunk by means of one of a balanced pair of high-impedance coils connected to each side of the line with an associated capacitor network.   { 'set kəm,päz·ət }

set condition |ELECTR| Condition of a flip-flop circuit in which the internal state of the flip-flop is set to 1.   { 'set kən,dish·ən }

set description |COMPUT SCI| For a specified data set, a definition of the set class name, set-owner selection criteria, set-member eligibility rules, and set-member ordering rules. { 'set di ‚skrip·shən }

set occurrence |COMPUT SCI| An instance of a set created in accordance with a set description. { 'set ə‚kə·rəns }

set point |CONT SYS| The value selected to be maintained by an automatic controller. { 'set ‚pȯint }

set pulse |ELECTR| An electronic pulse designed to place a memory cell in a specified state. { 'set ‚pəls }

settling time See correction time. { 'set·liŋ ‚tīm }

setup |ELECTR| The ratio between the reference black level and the reference white level in analog television, both measured from the blanking level; usually expressed as a percentage. { 'sed ‚əp }

sexless connector See hermaphroditic connector. { 'seks·ləs kə'nek·tər }

sferics receiver |ELECTR| An instrument which measures, electronically, the direction of arrival, intensity, and rate of occurrence of atmospherics; in its simplest form, the instrument consists of two orthogonally crossed antennas, whose output signals are connected to an oscillograph so that one loop measures the north-south component while the other measures the east-west component; the signals are combined vertically to give the azimuth. Also known as lightning recorder. { 'sfir·iks ri‚sē·vər }

SGML See Standard Generalized Markup Language.

shaded-pole motor |ELEC| A single-phase induction motor having one or more auxiliary short-circuited windings acting on only a portion of the magnetic circuit; generally, the winding is a closed copper ring embedded in the face of a pole; the shaded pole provides the required rotating field for starting purposes. { 'shād·əd ¦pōl 'mōd·ər }

shading |ELECTR| Television process of compensating for the spurious signal generated in a camera tube during trace intervals. { 'shād·iŋ }

shading ring |ENG ACOUS| A heavy copper ring sometimes placed around the central pole of an electrodynamic loudspeaker to serve as a shorted turn that suppresses the hum voltage produced by the field coil. { 'shād·iŋ ‚riŋ }

shading signal |ELECTR| Television camera signal that serves to increase the gain of the amplifier in the camera during those intervals of time when the electron beam is on an area corresponding to a dark portion of the scene being televised. { 'shād·iŋ ‚sig·nəl }

shadow attenuation |ELECTROMAG| Attenuation of radio waves over a sphere in excess of that over a plane when the distance over the surface and other factors are the same. { 'shad·ō ə‚ten·yə'wā·shən }

shadow batch system |COMPUT SCI| An online data collection system that initially only stores transactions in the computer system for refer-

ence, and updates the master files only at the end of the day or processing period. { 'shad·ō ‚bach ‚sis·təm }

shadow effect |COMMUN| Reduction in the strength of an ultra-high-frequency signal caused by some object (such as a mountain or a tall building) between the points of transmission and reception. { 'shad·ō i‚fekt }

shadow factor |ELECTROMAG| The ratio of the electric-field strength that would result from propagation of waves over a sphere to that which would result from propagation over a plane under comparable conditions. { 'shad·ō ‚fak·tər }

shadow mask |ELECTR| A thin, perforated metal mask mounted just back of the phosphor-dot faceplate in a three-gun color picture tube; the holes in the mask are positioned to ensure that each of the three electron beams strikes only its intended color phosphor dot. Also known as aperture mask. { 'shad·ō ‚mask }

shadow region |ELECTROMAG| Region in which, under normal propagation conditions, the field strength from a given transmitter is reduced by some obstruction which renders effective radio reception of signals or radar detection of objects in this region improbable. { 'shad·ō ‚rē·jən }

shaft coupling See coupling. { 'shaft ‚kəp·liŋ }

shaft-position encoder |ELECTR| An analog-to-digital converter in which the exact angular position of a shaft is sensed and converted to digital form. { 'shaft pə¦zish·ən in'kōd·ər }

shannon |COMMUN| A unit of information content, equal to the designation of one of two possible and equally likely values or states of anything used to store or convey information. { 'shan·ən }

Shannon formula |COMMUN| A theorem in information theory which states that the highest number of binary digits per second which can be transmitted with arbitrarily small frequency of error is equal to the product of the bandwidth and $\log_2 (1 + R)$, where R is the signal-to-noise ratio. { 'shan·ən ‚fȯr·myə·lə }

Shannon limit |COMMUN| Maximum signal-to-noise ratio improvement which can be achieved by the best modulation technique as implied by Shannon's theorem relating channel capacity to signal-to-noise ratio. { 'shan·ən ‚lim·ət }

Shannon's sampling theorem See sampling theorem. { 'shan·ənz 'sam·pliŋ ‚thir·əm }

shaped-beam antenna |ELECTROMAG| Antenna with a directional pattern which, over a certain angular range, is of special shape for some particular use. { 'shāpt ¦bēm an'ten·ə }

shape factor |ELEC| See form factor. |ELECTR| The ratio of the 60-decibel bandwidth of a band-pass filter to the 3-decibel bandwidth. { 'shāp ‚fak·tər }

shape-fill |COMPUT SCI| The filled-in areas on a graphic electronic display. { 'shāp ‚fil }

shaping circuit See corrective network. { 'shāp·iŋ ‚sər·kət }

shaping network See corrective network.
{ 'shāp·iŋ ,net,wərk }

shared control unit [COMPUT SCI] A control unit which controls several devices with similar characteristics, such as tape devices. { 'sherd kən ¦trōl ,yü·nət }

shared file [COMPUT SCI] A direct-access storage device that is used by more than one computer or data-processing system. { 'sherd 'fīl }

shared load [COMPUT SCI] A workload that can be shared by more than one computer, particularly during peak periods. { 'sherd 'lōd }

shared logic [COMPUT SCI] **1.** The simultaneous use of a single computer by multiple users. **2.** An arrangement of computers or computerized equipment in which the processing capabilities of one computer, including the ability to use peripheral devices, can be distributed to the other computers. { 'sherd 'läj·ik }

shared-logic cluster word processor [COMPUT SCI] A system of terminals lacking word-processing capability and printers joined to a single computer designed to carry out word-processing functions. { 'sherd ¦läj·ik ¦kləs·tər 'wərd ,prä,ses·ər }

shared resource [COMPUT SCI] Peripheral equipment that is simultaneously shared by several users. { 'sherd 'rē,sòrs }

shareware [COMPUT SCI] Copyrighted software that can be tried before buying. { 'sher,wer }

sharing device [COMPUT SCI] A small, inexpensive multiplexer that combines two independent data signals, which are then transmitted over the same communications line. { 'sher·iŋ di,vīs }

sharp-cutoff tube [ELECTR] An electron tube in which the control-grid openings are uniformly spaced; the anode current then decreases linearly as the grid voltage is made more negative, and cuts off sharply at a particular grid voltage. { 'shärp ¦kəd,óf ,tüb }

sharpness of resonance [ELEC] The narrowness of the frequency band around the resonance at which the response of an electric circuit exceeds an arbitrary fraction of its maximum response, often 70.7%. { 'shärp·nəs əv 'rez·ən·əns }

sharp tuning [ELEC] Having high selectivity; responding only to a desired narrow range of frequencies. { 'shärp 'tün·iŋ }

sheath [ELEC] A protective outside covering on a cable. [ELECTR] A space charge formed by ions near an electrode in a gas tube. [ELECTROMAG] The metal wall of a waveguide. { shēth }

sheath-reshaping converter [ELECTROMAG] In a waveguide, a mode converter in which the change of wave pattern is achieved by gradual reshaping of the sheath of the waveguide and of conducting metal sheets mounted longitudinally in the guide. { 'shēth rē¦shāp·iŋ kən'vərd·ər }

sheet feeder [COMPUT SCI] A device that feeds noncontinuous forms or sheets of paper into a printer. { 'shēt ,fēd·ər }

sheet grating [ELECTROMAG] Three-dimensional grating consisting of thin, longitudinal, metal sheets extending along the inside of a waveguide for a distance of about a wavelength, and used to stop all waves except one predetermined wave that passes unimpeded. { 'shēt ,grād·iŋ }

shell [COMPUT SCI] A program that provides an interface between a user and the computer's operating system by reading commands and sending them to the operating system for execution. { shel }

shell account [COMPUT SCI] A type of limited access to the Internet in which the user is connected to the Internet indirectly through a second computer on which the user has established an account. { 'shel ə,kaúnt }

Shenstone effect [ELECTR] An increase in photoelectric emission of certain metals following passage of an electric current. { 'shen,stōn i,fekt }

SHF See superhigh frequency.

shielded-conductor cable [ELEC] Cable in which the insulated conductor or conductors are enclosed in a conducting envelope or envelopes, constructed so that substantially every point on the surface of the insulation is at ground potential or at some predetermined potential with respect to ground. { 'shēl·dəd kən¦dək·tər 'kā·bəl }

shielded-core superconducting fault-current limiter [ELEC] A limiter which is essentially a transformer, with its primary normal conducting coil connected in series to the line to be protected, while the secondary side is a superconducting tube (that is, a one-turn coil). Also known as inductive superconducting fault-current limiter; shorted-transformer superconducting fault-current limiter. { ¦shēl·dəd ,cór ¦sü·pər·kən,dək·tiŋ 'fólt ,kər·ant ,lim·əd·ər }

shielded joint [ELEC] Cable joint having its insulation so enveloped by a conducting shield that substantially every point on the surface of the insulation is at ground potential, or at some predetermined potential with respect to ground. { 'shēl·dəd 'jóint }

shielded line [ELECTROMAG] Transmission line, the elements of which confine the propagated waves to an essentially finite space; the external conducting surface is called the sheath. { 'shēl·dəd 'līn }

shielded pair [ELEC] A pair of wires within a cable that is individually covered by a conducting shield. { 'shēld·əd 'per }

shielded wire [ELEC] Insulated wire covered with a metal shield, usually of tinned braided copper wire. { 'shēl·dəd 'wīr }

shield factor [COMMUN] Ratio of noise (or induced current or voltage) in a telephone circuit when a source of shielding is present to the corresponding quantity when the shielding is absent. { 'shēld ,fak·tər }

shield grid [ELECTR] A grid that shields the control grid of a gas tube from electrostatic fields, thermal radiation, and deposition of thermionic emissive material; it may also be used as an additional control electrode. { 'shēld ‚grid }

shield-grid thyratron |ELECTR| A thyratron having a shield grid, usually operated at cathode potential. { 'shēld ¦grid 'thī·rə‚trän }

shielding See electric shielding. { 'shēld·iŋ }

shielding ratio |ELECTROMAG| The ratio of a field in a specified region when electrical shielding is in place to the field in that region when the shielding is removed. { 'shēld·iŋ ‚rā·shō }

shift |COMPUT SCI| A movement of data to the right or left, in a digital-computer location, usually with the loss of characters shifted beyond a boundary. { shift }

shift register |COMPUT SCI| A computer hardware element constructed to perform shifting of its contained data. { 'shift ‚rej·ə·stər }

shift-register generator |COMPUT SCI| A random-number generator which consists of a sequence of shift operations and other operations, such as no-carry addition. { 'shift ¦rej·ə·stər 'jen·ə ‚rād·ər }

shiran |ELECTR| Specially designed frequency-modulation continuous-wave distance-measuring equipment used for performing distance measurements of an accuracy comparable to first-order triangulation. Derived from S-band hiran. { 'shī‚ran }

shock excitation |ELEC| Excitation produced by a voltage or current variation of relatively short duration; used to initiate oscillation in the resonant circuit of an oscillator. Also known as impulse excitation. { 'shäk ‚ek‚sī'tā·shən }

Shockley diode |ELECTR| A *pnpn* silicon controlled switch having characteristics that permit operation as a unidirectional diode switch. { 'shäk·lē 'dī·ōd }

shore effect |ELECTROMAG| Bending of radio waves toward the shoreline when traveling over water near a shoreline, due to the slightly greater velocity of radio waves over water than over land; this effect causes errors in radio-direction-finder indications. { 'shȯr i‚fekt }

Shor's algorithm |COMPUT SCI| An algorithm for factoring a large number within a reasonable amount of time, using a quantum computer. { ¦shȯrz 'al·gə‚rith·əm }

short See short circuit. { shȯrt }

short antenna |ELECTROMAG| An antenna shorter than about one-tenth of a wavelength, so that the current may be assumed to have constant magnitude along its length, and the antenna may be treated as an elementary dipole. { 'shȯrt an‚ten·ə }

short card |COMPUT SCI| A printed circuit board that is plugged into an expansion slot in a microcomputer and is only half the length of a full-size card. { ¦shȯrt ¦kärd }

short circuit |ELEC| A low-resistance connection across a voltage source or between both sides of a circuit or line, usually accidental and usually resulting in excessive current flow that may cause damage. Also known as short. { 'shȯrt 'sər·kət }

short-circuit impedance |ELEC| Of a line or four-terminal network, the driving point impedance when the far-end is short-circuited. { 'shȯrt ¦sər·kət im'pēd·əns }

short-circuiting transfer |ENG| Transfer of melted material from a consumable electrode during short circuits. { 'shȯrt ¦sər·kəd·iŋ 'tranz·fər }

short-circuit transition See shunt transition. { 'shȯrt ¦sər·kət tranz'zish·ən }

short-contact switch |ELEC| Selector switch in which the width of the movable contact is greater than the distance between contact clips, so that the new circuit is contacted before the old one is broken; this avoids noise during switching. { 'shȯrt ¦kän‚tak ‚swich }

shorted-transformer superconducting fault-current limiter See shielded-core superconducting fault-current limiter. { ¦shȯrd·əd tranz¦fȯrm·ər ‚sü·pər·kən‚dək·tiŋ 'fȯlt ‚kər·ənt ‚lim·əd·ər }

short-gate gain |ELECTR| Video gain on short-range gate. { 'shȯrt ¦gāt 'gān }

short-haul |COMMUN| Pertaining to devices capable of transmitting and receiving signals over distances up to about 1 mile (1.6 kilometers). { 'shȯrt ‚hȯl }

short-line seeking |COMPUT SCI| A method of accelerating the operation of a computer printer, in which the printer is sent directly to the beginning of the next line to be printed without going to the left margin of the paper. { 'shȯrt ¦līn 'sēk·iŋ }

short-path principle See Hittorf principle. { 'shȯrt ¦path 'prin·sə·pəl }

short-precision number See single-precision number. { 'shȯrt pri¦sizh·ən 'nəm·bər }

short-range radar |ENG| Radar whose maximum line-of-sight range, for a reflecting target having 1 square meter of area perpendicular to the beam, is between 50 and 150 miles (80 and 240 kilometers). { 'shȯrt ¦rānj 'rā‚där }

short shot See short. { 'shȯrt 'shät }

short-term predictor |COMMUN| An electric filter that removes redundancies in a signal associated with short-term correlations so that information can be transmitted more efficiently. { ‚shȯrt ‚tərm prə'dik·tər }

short-term repeatability |CONT SYS| The close agreement of positional movements of a robotic system repeated under identical conditions over a short period of time and at the same location. { 'shȯrt ‚tərm ri‚pēd·ə'bil·əd·ē }

short-time rating |ELEC| A rating defining the load that a machine, apparatus, or device can carry for a specified short time. { 'shȯrt ¦tīm 'rād·iŋ }

shortwave broadcasting |COMMUN| Radio broadcasting at frequencies in the range from about 1600 to 30,000 kilohertz, above the standard broadcast band. { 'shȯrt¦wāv 'brȯd‚kast·iŋ }

shortwave converter |ELECTR| Electronic unit designed to be connected between a receiver and its antenna system to permit reception of frequencies higher than those the receiver ordinarily handles. { 'shȯrt¦wāv kən'vərd·ər }

short waveguide isolator |ELECTR| A device that functions as an isocirculator in a miniature microwave circuit and consists of a waveguide T junction with a magnetized cylinder of ferrite at the center and an absorber on the side arm of the T. Also known as flange isolator. { 'shórt ¦wāv ‚gīd 'ī·sə‚lād·ər }

shortwave propagation [COMMUN] Propagation of radio waves at frequencies in the range from about 1600 to 30,000 kilohertz. { 'shórt'wāv ‚präp·ə'gā·shən }

short word [COMPUT SCI] The fixed word of lesser length in computers capable of handling words of two different lengths; in many computers this is referred to as a half-word because the length is exactly the half-length of the full word. { 'shórt 'wərd }

shot effect See shot noise. { 'shät i‚fekt }

shot-firing cable [ELEC] A two-conductor cable which leads from the exploder to the detonator wires. Also known as firing cable. { 'shät ¦fīr·iŋ ‚kā·bəl }

shot-firing circuit [ELEC] The path taken by the electric current from the exploder along the shot-firing cable, the detonator wires, and finally the detonator when a shot is detonated. { 'shät ¦fīr·iŋ ‚sər·kət }

shot noise [ELECTR] Noise voltage developed in a thermionic tube because of the random variations in the number and the velocity of electrons emitted by the heated cathode; the effect causes sputtering or popping sounds in radio receivers and snow effects in analog television pictures. Also known as Schottky noise; shot effect. { 'shät ‚nóiz }

shunt [ELEC] **1.** A precision low-value resistor placed across the terminals of an ammeter to increase its range by allowing a known fraction of the circuit current to go around the meter. Also known as electric shunt. **2.** To place one part in parallel with another. **3.** See parallel. { shənt }

shunt-excited antenna [ELECTROMAG] A tower antenna, not insulated from the ground at the base, whose feeder is connected at a point about one-fifth of the way up the antenna and usually slopes up to this point from a point some distance from the antenna's base. { 'shənt ik ¦sīd·əd an'ten·ə }

shunt-fed vertical antenna [ELECTROMAG] Vertical antenna connected to the ground at the base and energized at a point suitably positioned above the grounding point. { 'shənt ¦fed ¦vərd·ə·kəl an'ten·ə }

shunt generator [ELEC] A generator whose field winding and armature winding are connected in parallel, and in which the armature supplies both the load current and the field current. { 'shənt ¦jen·ə‚rād·ər }

shunting [ELEC] The act of connecting one device to the terminals of another so that the current is divided between the two devices in proportion to their respective admittances. { 'shənt·iŋ }

shunt loading [ELEC] Loading in which reactances are applied in shunt across the conductors. { 'shənt ¦lōd·iŋ }

shunt motor [ELEC] A direct-current motor whose field circuit and armature circuit are connected in parallel. { 'shənt ¦mōd·ər }

shunt neutralization See inductive neutralization. { 'shənt‚nü·trə·lə'zā·shən }

shunt peaking [ELECTR] The use of a peaking coil in a parallel circuit branch connecting the output load of one stage to the input load of the following stage, to compensate for high-frequency loss due to the distributed capacitances of the two stages. { 'shənt ¦pēk·iŋ }

shunt reactor [ELEC] A reactor that has a relatively high inductance and is wound on a magnetic core containing an air gap; used to neutralize the charging current of the line to which it is connected. { 'shənt rē¦ak·tər }

shunt regulator [ELEC] A regulator that maintains a constant output voltage by controlling the current through a dropping resistance in series with the load. { 'shənt ¦reg·yə‚lād·ər }

shunt repeater [ELEC] A type of negative impedance telephone repeater which is stable when it is short-circuited, but oscillates when terminated by a high impedance, in contrast to a series repeater; it can be thought of as a negative admittance. { 'shənt ri¦pēd·ər }

shunt T junction See H-plane T junction. { 'shənt 'tē ‚jəŋk·shən }

shunt transition [ELEC] A method of changing the connection of motors from series to parallel in which one motor, or group of motors, is first short-circuited, then disconnected, and finally connected in parallel with the other motor or motors. Also known as short-circuit transition. { 'shənt tran¦zish·ən }

shunt-wound [ELEC] Having armature and field windings in parallel, as in a direct-current generator or motor. { 'shənt ¦waúnd }

shut-down circuit [ENG] An electronic, electric, or pneumatic system designed to shut off and close down process systems or equipment; can be used for routine or emergency situations. { 'shət‚daún ‚sər·kət }

shuttered image converter [ELECTR] An image tube whose photoelectrons can be rapidly switched off to allow a camera to record the image on its screen. { ¦shəd·ərd 'im·ij kən‚vərd·ər }

SIB See screen image buffer.

SIC See dielectric constant.

sideband [ELECTROMAG] **1.** The frequency band located either above or below the carrier frequency, within which fall the frequency components of the wave produced by the process of modulation. **2.** The wave components lying within such bands. { 'sīd‚band }

side circuit [COMMUN] One of the circuits arranged to derive a phantom circuit. { 'sīd ‚sər·kət }

side echo [ELECTROMAG] Echo due to a side lobe of an antenna. { 'sīd ‚ek·ō }

side effect [COMPUT SCI] A consistent result of a procedure that is in addition to or peripheral to the basic result. { 'sīd i‚fekt }

side lobe See minor lobe. { 'sīd ‚lōb }

side-lobe blanking [ELECTR] Radar technique that compares the signal strength in the main antenna with the echo received in an auxiliary antenna of gain between the side-lobe level and the main-beam gain of the main antenna; done to determine if the echo is coming from the main-beam direction, and blanking the echoes whenever they are stronger in the auxiliary channel. { 'sīd ¦lōb 'blaŋk·iŋ }

side-lobe suppression [ELECTR] Design or techniques in radar intended to reduce the effect of side lobes in the antenna's pattern. { 'sīd ¦lōbe sə‚presh·ən }

sidetone [COMMUN] The sound of the speaker's own voice as heard in his or her telephone receiver; the effect is undesirable if excessive and is usually reduced by special circuits. { 'sīd‚tōn }

sidetone level [COMMUN] The ratio of the volume of the sidetone to the volume of the speaker's voice, usually expressed in decibels. { 'sīd‚tōn ‚lev·əl }

sidetone ranging [COMMUN] A method of measuring time delay, and thereby range, by sending a radio signal to a satellite, in which several audio tones of different frequencies are broadcast, and the phases of the tones transmitted from the satellite are compared with the sent tone phases. { 'sīd‚tōn 'rānj·iŋ }

siemens [ELEC] A unit of conductance, admittance, and susceptance, equal to the conductance between two points of a conductor such that a potential difference of 1 volt between these points produces a current of 1 ampere; the conductance of a conductor in siemens is the reciprocal of its resistance in ohms. Formerly known as mho (℧); reciprocal ohm. Symbolized S. { 'sē·mənz }

sift [COMPUT SCI] To extract certain desired information items from a large quantity of data. { sift }

sigma-delta analog-to-digital converter [ELECTR] A converter that uses an analog circuit to generate a single-valued pulse stream in which the frequency of pulses is determined by the analog source, and then uses a digital circuit to repeatedly sum the number of these pulses over a fixed time interval, converting the pulses to numeric values. { ¦sig·mə ¦del·tə ‚an·ə‚läg tü ‚dij·əd·əl kən‚vərd·ər }

sigma-delta converter [ELECTR] A class of electronic systems containing both analog and digital subsystems whose most common application is the conversion of analog signals to digital form, and vice versa, using pulse density modulation to create a high-rate stream of single-amplitude pulses in either case. Also known as delta-sigma converter. { ‚sig·mə ‚del·tə kən'vərd·ər }

sigma-delta digital-to-analog converter [ELECTR] A converter that uses a digital circuit to convert numeric values from a digital processor to a pulse stream and then uses an analog low-pass filter to

produce an analog waveform. { ¦sig·mə ¦del·tə ‚dij·əd·əl tü ‚an·ə‚läg kən'vərd·ər }

sigma-delta modulator [ELECTR] The circuit used to generate a pulse stream in a sigma-delta converter. Also known as delta-sigma modulator. { ‚sig·mə ‚del·tə 'mäj·ə‚lād·ər }

signal [COMMUN] **1.** A visual, aural, or other indication used to convey information. **2.** The intelligence, message, or effect to be conveyed over a communication system. **3.** *See* signal wave. { 'sig·nəl }

signal bias [COMMUN] Form of teletypewriter signal distortion brought about by the lengthening or shortening of pulses during transmission; when marking pulses are all lengthened, a marking signal bias results; when marking pulses are all shortened, a spacing signal bias results. { 'sig·nəl ‚bī·əs }

signal carrier *See* carrier. { 'sig·nəl ‚kar·ē·ər }

signal channel [COMMUN] A signal path for transmitting electric signals; such paths may be separated by frequency division or time division. { 'sig·nəl ‚chan·əl }

signal conditioning [COMMUN] Processing the form or mode of a signal so as to make it intelligible to or compatible with a given device, such as a data transmission line, including such manipulation as pulse shaping, pulse clipping, digitizing, and linearizing. { 'sig·nəl kən‚dish·ən·iŋ }

signal distance [COMPUT SCI] The number of bits that are not the same in two binary words of equal length. Also known as hamming distance. { 'sig·nəl ‚dis·təns }

signal distortion generator [ELECTR] Instrument designed to apply known amounts of distortion on a signal for the purpose of testing and adjusting communications equipment such as teletypewriters. { 'sig·nəl di'stòr·shən ‚jen·ə‚rād·ər }

signal-flow graph [SYS ENG] An abbreviated block diagram in which small circles, called nodes, represent variables of the system, and the nodes are connected by lines, called branches, which represent one-way signal multipliers; an arrow on the line indicates direction of signal flow, and a letter near the arrow indicates the multiplication factor. Also known as flow graph. { 'sig·nəl ¦flō 'graf }

signal generator [ENG] An electronic test instrument that delivers a sinusoidal output at an accurately calibrated frequency that may be anywhere from the audio to the microwave range; the frequency and amplitude are adjustable over a wide range, and the output usually may be amplitude- or frequency-modulated. Also known as test oscillator. { 'sig·nəl ‚jen·ə‚rād·ər }

signal in band [COMMUN] To send control signals at frequencies within the frequency range of the data signal. { ¦sig·nəl in ¦band }

signaling key *See* key. { 'sig·nə·liŋ ‚kē }

signaling rate |COMMUN| The rate at which signals are transmitted. { 'sig·nə·liŋ ‚rāt }

signal intensity |COMMUN| The electric-field strength of the electromagnetic wave transmitting a signal. { 'sig·nəl in‚ten·səd·ē }

signal level |COMMUN| The difference between the level of a signal at a point in a transmission system and the level of an arbitrarily specified reference signal. { 'sig·nəl ‚lev·əl }

signal light |COMMUN| A light specifically designed for the transmission of code messages by means of visible light rays that are interrupted or deflected by electric or mechanical means. |ENG| A signal, illumination, or any pyrotechnic light used as a sign. { 'sig·nəl ‚līt }

signal normalization *See* signal standardization. { 'sig·nəl ‚nȯr·mə·lə'zā·shən }

signal out of band |COMMUN| To send control signals at frequencies outside the frequency range of the data signal. { ¦sig·nəl aút əv ¦band }

signal processing |COMMUN| The extraction of information from complex signals in the presence of noise, generally by conversion of the signals into digital form followed by analysis using various algorithms. Also known as digital signal processing (DSP). { 'sig·nəl ‚prä‚ses·iŋ }

signal regeneration |COMMUN| The restoration of a waveform representing a signal to approximate its original amplitude and shape. Also known as signal reshaping. { 'sig·nəl rē‚jen·ə'rā·shən }

signal reshaping *See* signal regeneration. { 'sig·nəl rē‚shāp·iŋ }

signal-shaping network |ELECTR| Network inserted in a telegraph circuit, usually at the receiving end, to improve the waveform of the code signals. { 'sig·nəl ¦shāp·iŋ ‚net‚wərk }

signal speed |COMMUN| The rate at which code elements are transmitted by a communications system. { 'sig·nəl ‚spēd }

signal standardization |COMMUN| The use of one signal to generate another which meets specified requirements for shape, amplitude, and timing. Also known as signal normalization. { 'sig·nəl ‚stan·dər·də'zā·shən }

signal strength |ELECTROMAG| The strength of the signal produced by a radio transmitter at a particular location, usually expressed as microvolts or millivolts per meter of effective receiving antenna height. { 'sig·nəl ‚straŋkth }

signal-strength meter |ELECTR| A meter that is connected to the automatic volume-control circuit of a communication receiver and calibrated in decibels or arbitrary S units to read the strength of a received signal. Also known as S meter; S-unit meter. { 'sig·nəl ¦straŋkth ‚mēd·ər }

signal-to-interference ratio |ELECTR| The relative magnitude of signal waves and waves which interfere with signal-wave reception. { 'sig·nəl tü ‚in·tər'fir·əns ‚rā·shō }

signal-to-noise improvement factor *See* noise improvement factor. { 'sig·nəl tə 'nȯiz im'prüv·mənt ‚fak·tər }

signal-to-noise ratio |ELECTR| The ratio of the amplitude of a desired signal at any point to the

amplitude of noise signals at that same point; often expressed in decibels; the peak value is usually used for pulse noise, while the root-mean-square (rms) value is used for random noise. Abbreviated S/N; SNR. { 'sig·nəl tə 'nȯiz ‚rā·shō }

signal tracer |ELECTR| An instrument used for tracing the progress of a signal through a radio receiver or an audio amplifier to locate a faulty stage. { 'sig·nəl ‚trā·sər }

signal voltage |ELEC| Effective (root-mean-square) voltage value of a signal. { 'sig·nəl ‚vōl·tij }

signal wave |COMMUN| A wave whose characteristics permit some intelligence, message, or effect to be conveyed. Also known as signal. { 'sig·nəl ‚wāv }

signal-wave envelope |COMMUN| Contour of a signal wave which is composed of a series of wave cycles. { 'sig·nəl ¦wāv 'en·və‚lōp }

signal winding |ELEC| Control winding, of a saturable reactor, to which the independent variable (signal wave) is applied. { 'sig·nəl ‚wīnd·iŋ }

sign-and-magnitude code |COMPUT SCI| The representation of an integer X by $(-1)^{a_0} (2^{n-2} a_1 + 2^{n-3} a_2 + \cdots + a_{n-1})$, where a_0 is 0 for X positive, and a_0 is 1 for X negative, and any a_j is either 0 or 1. { 'sīn ən 'mag·nə‚tüd ‚kōd }

signature |ELECTR| The characteristic pattern of a target as displayed by detection and classification equipment. { 'sig·nə·chər }

sign bit |COMPUT SCI| A sign digit consisting of one bit. { 'sīn ‚bit }

sign check indicator |COMPUT SCI| An error checking device, indicating no sign or improper signing of a field used for arithmetic processes; the machine can, upon interrogation, be made to stop or enter into a correction routine. { 'sīn ¦chek 'in·də‚kād·ər }

sign digit |COMPUT SCI| A digit containing one to four binary bits, associated with a data item and used to denote an algebraic sign. { 'sīn ‚dij·ət }

signed decimal |COMPUT SCI| A form of packed decimal representation in which the low-order nibble of the last byte has a sign bit that specifies whether the number is positive or negative. { 'sīnd 'des·məl }

signed field |COMPUT SCI| A field of data that contains a number which includes a sign digit indicating the number's sign. { 'sīnd 'fēld }

signed integer |COMPUT SCI| A whole number whose value lies anywhere in a domain that extends from a negative to a positive integer, and which therefore carries a sign. { 'sīnd 'int·ə·jər }

sign flag |COMPUT SCI| A bit in a status byte in a computer's central processing unit that indicates whether the result of an arithmetic operation is positive or negative. { 'sīn ‚flag }

significance arithmetic |COMPUT SCI| A rough technique for estimating the numbers and positions of the significant digits of the radix approximation that results when an arithmetic operation is applied to operands in radix approximation form. { sig'nif·i·kəns ə‚rith·mə·tik }

sign position |COMPUT SCI| That position, always at or near the left or right end of a numeral, in which the algebraic sign of the number is represented. { 'sīn pə,zish·ən }

silent discharge [ELECTR] An inaudible electric discharge in air that occurs at high voltage and consumes a relatively large amount of energy. { ¦sī·lənt 'dis,chärj }

silent period [COMMUN] Period during each hour in which ship and shore radio stations must remain silent and listen for distress calls. { 'sī·lənt 'pir·ē·əd }

silicide resistor [ELECTR] A thin-film resistor that uses a silicide of molybdenum or chromium, deposited by direct-current sputtering in an integrated circuit when radiation hardness or high resistance values are required. { 'sil·ə,sīd ri'zis·tər }

silicon capacitor [ELECTR] A capacitor in which a pure silicon-crystal slab serves as the dielectric; when the crystal is grown to have a *p* zone, a depletion zone, and an *n* zone, the capacitance varies with the externally applied bias voltage, as in a varactor. { 'sil·ə·kən kə'pas·əd·ər }

silicon controlled rectifier [ELECTR] A semiconductor rectifier that can be controlled; it is a *pnpn* four-layer semiconductor device that normally acts as an open circuit, but switches rapidly to a conducting state when an appropriate gate signal is applied to the gate terminal. Abbreviated SCR. Also known as reverse-blocking triode thyristor. { 'sil·ə·kən kən'trōld 'rek·tə,fī·ər }

silicon controlled switch [ELECTR] A four-terminal switching device having four semiconductor layers, all of which are accessible; it can be used as a silicon controlled rectifier, gate-turnoff switch, complementary silicon controlled rectifier, or conventional silicon transistor. Abbreviated SCS. Also known as reverse-blocking tetrode thyristor. { 'sil·ə·kən kən'trōld 'swich }

silicon detector See silicon diode. { 'sil·ə·kən di'tek·tər }

silicon diode [ELECTR] A crystal diode that uses silicon as a semiconductor; used as a detector in ultra-high- and super-high-frequency circuits. Also known as silicon detector. { 'sil·ə·kən 'dī,ōd }

silicon homojunction See bipolar junction transistor. { ¦sil·ə·kən 'hä·mə,jəŋk·shən }

silicon image sensor [ELECTR] A video camera in which the image is focused on an array of individual light-sensitive elements formed from a charge-coupled-device semiconductor chip. Also known as silicon imaging device. { 'sil·ə·kən 'im·ij ,sen·sər }

silicon imaging device See silicon image sensor. { 'sil·ə·kən 'im·ij·iŋ di,vīs }

silicon-on-insulator [ELECTR] A semiconductor manufacturing technology in which thin films of single-crystalline silicon are grown over an electrically insulating substrate. { 'sil·ə·kən ȯn 'in·sə,lād·ər }

silicon-on-sapphire [ELECTR] A semiconductor manufacturing technology in which metal oxide semiconductor devices are constructed in a thin single-crystal silicon film grown on an electrically insulating synthetic sapphire substrate. Abbreviated SOS. { 'sil·ə·kən ȯn 'sa,fīr }

silicon rectifier [ELECTR] A metallic rectifier in which rectifying action is provided by an alloy junction formed in a high-purity silicon slab. { 'sil·ə·kən 'rek·tə,fī·ər }

silicon resistor [ELECTR] A resistor using silicon semiconductor material as a resistance element, to obtain a positive temperature coefficient of resistance that does not appreciably change with temperature; used as a temperature-sensing element. { 'sil·ə·kən ri'zis·tər }

silicon retina [ELECTR] An analog very large scale integrated circuit chip that performs operations which resemble some of the functions performed by the retina of the human eye. { ,sil·ə,kän 'ret·ən·ə }

silicon solar cell [ELECTR] A solar cell consisting of *p* and *n* silicon layers placed one above the other to form a *pn* junction at which radiant energy is converted into electricity. { 'sil·ə·kən 'sō·lər 'sel }

silicon-symmetrical switch [ELECTR] Thyristor modified by adding a semiconductor layer so that the device becomes a bidirectional switch; used as an alternating-current phase control, for synchronous switching and motor speed control. { 'sil·ə·kən si'me·trə·kəl 'swich }

silicon transistor [ELECTR] A transistor in which silicon is used as the semiconducting-material. { 'sil·ə·kən tran'zis·tər }

silver battery [ELEC] A solid-state battery based on an Ag_4RbI_5 electrolyte that conducts positive silver ions. { 'sil·vər 'bad·ə·rē }

silver-cadmium storage battery [ELEC] A storage battery that combines the excellent space and weight characteristics of silver-zinc batteries with long shelf life and other desirable properties of nickel-cadmium batteries. { 'sil·vər 'kad·mē·əm 'stȯr·ij ,bad·ə·rē }

silvered mica capacitor [ELECTR] A mica capacitor in which a coating of silver is deposited directly on the mica sheets to serve in place of conducting metal foil. { 'sil·vərd ¦mī·kə kə'pas·əd·ər }

silver migration [ELEC] A process, causing reduction in insulation resistance and dielectric failure; silver, in contact with an insulator, at high humidity, and subjected to an electrical potential, is transported ionically from one location to another. { 'sil·vər mī'grā·shən }

silver oxide cell [ELEC] A primary cell in which depolarization is accomplished by an oxide of silver. { 'sil·vər ¦äk,sīd ,sēl }

silverstat regulator [ELEC] Multitapped resistor, the taps of which are connected to single-leaf silver contacts; variation of voltage causes a solenoid to open or close these contacts, shorting out more or less of the resistance in the exciter circuit as a means of regulating the output voltage to the desired value. { 'sil·vər ,stat 'reg·yə,lād·ər }

silver-zinc storage battery [ELEC] A storage battery that gives higher current output and greater watt-hour capacity per unit of weight and volume than most other types, even at high discharge rates; used in missiles and torpedoes, where its high cost can be tolerated. { 'sil·vər ¦ziŋk 'stȯr·ij ¦bad·ə·rē }

SIMD [COMPUT SCI] A type of multiprocessor architecture in which there is a single instruction cycle, but multiple sets of operands may be fetched to multiple processing units and may be operated upon simultaneously within a single instruction cycle. Acronym for single-instruction-stream, multiple-data-stream. { ¦es¦ī¦em¦dē }

SIMM [COMPUT SCI] A printed circuit board that holds several semiconductor memory chips and is used to add memory to a computer. Acronym for single in-line memory module. { sim }

simple buffering [COMPUT SCI] A technique for obtaining simultaneous performance of input/output operations and computing; it involves associating a buffer with only one input or output file (or data set) for the entire duration of the activity on that file (or data set). { 'sim·pəl 'bəf·ə·riŋ }

simple data structure [COMPUT SCI] An arrangement of data in a database or file in which each grouping of data, such as a record, is of equal importance or significance. { 'sim·pəl 'dad·ə ¦strək·chər }

simple electrostatic lens [ELECTR] An electrostatic lens that consists of a circular hole in a conducting plate with different electrostatic fields on the two sides. { ¦sim·pəl i¦lek·trə ¦stad·ik 'lenz }

simple harmonic current [ELEC] Alternating current, the instantaneous value of which is equal to the product of a constant, and the cosine of an angle varying linearly with time. Also known as sinusoidal current. { 'sim·pəl här'män·ik 'kə·rənt }

simple harmonic electromotive force [ELEC] An alternating electromotive force which is equal to the product of a constant and the cosine or sine of an angle which varies linearly with time. { 'sim·pəl här'män·ik i¦lek·trə¦mōd·iv 'fȯrs }

Simple Mail Transfer Protocol [COMPUT SCI] An Internet standard for sending e-mail messages. Abbreviated SMTP. { ¦sim·pəl 'māl ¦tranz·fər ¦prōd·ə,kȯl }

simple oscillator See harmonic oscillator. { 'sim·pəl 'äs·ə,lād·ər }

simplex channel [COMMUN] A channel which permits transmission in one direction only. { 'sim,pleks ¦chan·əl }

simplex structure [COMPUT SCI] The structure of an information processing system designed in such a way that only the minimum amount of hardware is utilized to implement its function. { 'sim,pleks ¦strək·chər }

simplex transmission [COMMUN] A mode of radio transmission in which communication takes place between two stations in only one direction at a time. { 'sim,pleks tranz¦mish·ən }

SIMSCRIPT [COMPUT SCI] A high-level programming language used in simulation, in which systems are described in terms of sets, entities, which are groups of sets, and attributes, which are properties associated with entities. { 'sim ¦skript }

simulation [COMPUT SCI] The development and use of computer models for the study of actual or postulated dynamic systems. { ¦sim·yə'lā·shən }

simulation language [COMPUT SCI] A computer language used to write programs for the simulation of the behavior through time of such things as transportation and manufacturing systems; SIMSCRIPT is an example. { ¦sim·yə'lā·shən ¦laŋ·gwij }

simulator [COMPUT SCI] A routine which is executed by one computer but which imitates the operations of another computer. [ENG] A computer or other piece of equipment that simulates a desired system or condition and shows the effects of various applied changes, such as a flight simulator. { 'sim·yə,lād·ər }

simultaneous access See parallel access. { ¦sī·məl'tā·nē·əs 'ak,ses }

simultaneous color television [ELECTR] A color television system in which the phosphors for the three primary colors are excited at the same time, not one after another; the shadow-mask color picture tube gives a simultaneous display. { ¦sī·məl'tā·nē·əs 'kəl·ər 'tel·ə,vizh·ən }

simultaneous computer [COMPUT SCI] **1.** A computer, usually of the analog or hybrid type, in which separate units of hardware are used to carry out the various parts of a computation, the execution of different parts usually overlap in time, and the various hardware units are interconnected in a manner determined by the computation. **2.** A computer that serves to back up another computer and can replace it when it is not operating effectively. { ¦sī·məl'tā·nē·əs kəm'pyüd·ər }

simultaneous lobing [ELECTR] A radar direction-finding technique in which the signals received by two partly overlapping antenna beams are compared in phase or power to obtain a measure of the angular displacement of a target from the equisignal direction; arrangement of (usually) four such beams to effect measurement in both angle directions. { ¦sī·məl'tā·nē·əs 'lōb·iŋ }

simultaneous peripheral operations on line See spooling. { ¦sī·məl'tā·nē·əs pə'rif·ə·rəl ¦äp·ə'rā·shənz ȯn 'līn }

sine-cosine encoder [ELECTR] A shaft-position encoder having a special type of angle-reading code disk that gives an output which is a binary representation of the sine of the shaft angle. { 'sīn 'kō,sīn in'kōd·ər }

sine-cosine generator See resolver. { 'sīn 'kō,sīn 'jen·ə,rād·ər }

sine potentiometer [ELECTR] A potentiometer whose direct-current output voltage is proportional to the sine of the shaft angle; used as a resolver in computer and radar systems. { 'sīn pə,ten·chē'äm·əd·ər }

sine-wave modulated jamming [ELECTR] Jamming signal produced by modulating a continuous wave signal with one or more sine waves. { 'sīn ¦wāv ,mäj·ə,lād·əd 'jam·iŋ }

sine-wave oscillator *See* sinusoidal oscillator. { 'sīn ¦wāv 'as·ə,lād·ər }

sine-wave response *See* frequency response. { 'sīn ¦wāv ri'späns }

singing [CONT SYS] An undesired, self-sustained oscillation in a system or component, at a frequency in or above the passband of the system or component; generally due to excessive positive feedback. { 'siŋ·iŋ }

singing margin [CONT SYS] The difference in level, usually expressed in decibels, between the singing point and the operating gain of a system or component. { 'siŋ·iŋ ,mär·jən }

singing point [CONT SYS] The minimum value of gain of a system or component that will result in singing. { 'siŋ·iŋ ,pöint }

singing-stovepipe effect [ELEC] Reception and reproduction of radio signals by ordinary pieces of metal in contact with each other, such as sections of stovepipe; it occurs when rusty bolts, faulty welds, or mechanically loose connections within strong radiated fields near transmitters produce intermodulation interference; the mechanically poor connections serve as nonlinear diodes. { 'siŋ·iŋ ¦stōv,pīp i,fekt }

single-address instruction *See* one-address instruction. { 'siŋ·gəl ¦ad,res in'strək·shən }

single-board computer [COMPUT SCI] A computer consisting of a processor and memory on a single printed circuit board. { 'siŋ·gəl ¦bȯrd kəm'pyüd·ər }

single bus [ELEC] A substation switching arrangement that involves one common bus for all connections and one breaker per connection. { 'siŋ·gəl 'bəs }

single-button carbon microphone [ENG ACOUS] Microphone having a carbon-filled buttonlike container on only one side of its flexible diaphragm. { 'siŋ·gəl ¦bət·ən ¦kär·bən 'mī·krə,fōn }

single-channel multiplier [ELECTR] A type of photomultiplier tube in which electrons travel down a cylindrical channel coated on the inside with a resistive secondary-emitting layer, and gain is achieved by multiple electron impacts on the inner surface as the electrons are directed down the channel by an applied voltage over the length of the channel. { 'siŋ·gəl ¦chan·əl 'məl·tə,plī·ər }

single-channel simplex [COMMUN] Simplex operation that provides nonsimultaneous radio communications between stations using the same frequency channel. { 'siŋ·gəl ¦chan·əl 'sim,pleks }

single-chip computer [COMPUT SCI] A computer whose processor consists of a single integrated circuit. { 'siŋ·gəl ¦chip kəm'pyüd·ər }

single-current transmission [COMMUN] Telegraph transmission in which a current flows, in only one direction, during marking intervals, and no current flows during spacing intervals. { 'siŋ·gəl ¦kə·rənt tranz'mish·ən }

single density [COMPUT SCI] Property of computer storage which holds the standard amount of data per unit of storage space. { 'siŋ·gəl 'den·səd·ē }

single-edged push-pull amplifier circuit [ELECTR] Amplifier circuit having two transmission paths designed to operate in a complementary manner and connected to provide a single unbalanced output without the use of an output transformer. { 'siŋ·gəl ¦ejd ¦push ¦pül 'am·plə,fī·ər }

single-electron transistor [ELECTR] A transistor whose dimensions are extremely small, in the nanometer range, causing it to exhibit characteristics that are sensitive to the transport and storage of single electrons. { ,siŋ·gəl i,lek·trän tran'zis·tər }

single-end amplifier [ELECTR] Amplifier stage which normally employs only one tube or semiconductor or, if more than one tube or semiconductor is used, they are connected in parallel so that operation is asymmetric with respect to ground. Also known as single-sided amplifier. { 'siŋ·gəl ¦end 'am·plə,fī·ər }

single-ended [ELEC] Unbalanced, as when one side of a transmission line or circuit is grounded. { 'siŋ·gəl 'end·əd }

single-ended signal [ELECTR] A circuit signal that is the voltage difference between two nodes, one of which can be defined as being at ground or reference voltage. { ¦siŋ·gəl ¦en·dəd 'sig·nəl }

single-event upset [ELECTR] A change in the state of a logic device from 0 to 1 or vice versa, as the result of the passage of a single cosmic ray. { ¦siŋ·gəl i¦vent 'əp,set }

single-frequency duplex [COMMUN] Duplex carrier communications that provide communications in opposite directions, but not simultaneously, over a single-frequency carrier channel, the transfer between transmitting and receiving conditions being automatically controlled by the voices or other signals of the communicating parties. { 'siŋ·gəl 'frē·kwən·sē 'dü,pleks }

single-frequency simplex [COMMUN] Single-frequency carrier communications in which manual rather than automatic switching is used to change over from transmission to reception. { 'siŋ·gəl 'frē·kwən·sē 'sim,pleks }

single-gun color tube [ELECTR] A color picture tube having only one electron gun and one electron beam; the beam is sequentially deflected across phosphors for the three primary colors to form each color picture element, as in the chromatron. { 'siŋ·gəl ¦gən 'kəl·ər ,tüb }

single-hop transmission [COMMUN] Radio transmission in which radio waves are reflected from the ionosphere only once along their path from the transmitter to the receiver. { 'siŋ·gəl ¦häp trans'mish·ən }

single in-line memory module *See* SIMM. { ¦siŋ·gəl ¦in ,līn 'mem·rē ,mä·jəl }

single in-line package [ELECTR] A packaged resistor network or other assembly that has a single row of terminals or lead wires along one edge of the package. Abbreviated SIP. { 'siŋ·gəl 'in,līn 'pak·ij }

single-instruction-stream, multiple-data-stream
See SIMD. { ¦siŋ·gəl in¦strək·shən ¦strēm ¦məl·tə·pəl 'dad·ə ¸strēm }

single-instruction-stream, single-data-stream
See SISD. { ¦siŋ·gəl in¦strək·shən ¸strēm ¦siŋ·gəl 'dad·ə ¸strēm }

single-keyboard point-of-sale system [COMPUT SCI] A point-of-sale system based upon electronic cash registers as stand-alone units, each equipped with a few internal registers and some programming capability. { 'siŋ·gəl ¦kē ¸bȯrd ¦pȯint əv 'sāl ¸sis·təm }

single-length [COMPUT SCI] Pertaining to the expression of numbers in binary form in such a way that they can be included in a single computer word. { 'siŋ·gəl 'leŋkth }

single-loop feedback [CONT SYS] A system in which feedback may occur through only one electrical path. { 'siŋ·gəl ¦lüp 'fēd¸bak }

single-loop servomechanism [CONT SYS] A servomechanism which has only one feedback loop. Also known as servo loop. { 'siŋ·gəl ¦lüp 'sər·vō ¸mek·ə¸niz·əm }

single-phase [ELEC] Energized by a single-alternating voltage. { 'siŋ·gəl 'fāz }

single-phase circuit [ELEC] Either an alternating-current circuit which has only two points of entry, or one which, having more than two points of entry, is intended to be so energized that the potential differences between all pairs of points of entry are either in phase or differ in phase by 180°. { 'siŋ·gəl ¦fāz 'sər·kət }

single-phase circuit [ELEC] Either an alternating-current circuit which has only two points of entry, or one which, having more than two points of entry, is intended to be so energized that the potential differences between all pairs of points of entry are either in phase or differ in phase by 180°. { 'siŋ·gəl ¦fāz 'sər·kət }

single-phase meter [ENG] A type of power-factor meter that contains a fixed coil that carries the load current, and crossed coils that are connected to the load voltage; there is no spring to restrain the moving system, which takes a position to indicate the angle between the current and voltage. { 'siŋ·gəl ¦fāz 'mēd·ər }

single-phase motor [ELEC] A motor energized by a single alternating voltage. { 'siŋ·gəl ¦fāz 'mōd·ər }

single-phase rectifier [ELECTR] A rectifier whose input voltage is a single sinusoidal voltage, in contrast to a polyphase rectifier. { 'siŋ·gəl ¦fāz 'rek·tə¸fī·ər }

single-point grounding [ELEC] Grounding system that attempts to confine all return currents to a network that serves as the circuit reference; to be effective, no appreciable current is allowed to flow in the circuit reference, that is, the sum of the return currents is zero. { 'siŋ·gəl 'pȯint 'graund·iŋ }

single-polarity pulse [ELEC] Pulse in which the sense of the departure from normal is in one direction only. { 'siŋ·gəl pə¦lar·əd·ē 'pəls }

single-polarity pulse-amplitude modulation
See unidirectional pulse-amplitude modulation. { 'siŋ·gəl pə¦lar·əd·ē 'pəls 'am·plə¸tüd ¸mäj·ə'lā·shən }

single-pole double-throw [ELEC] A three-terminal switch or relay contact arrangement that connects one terminal to either of two other terminals. Abbreviated SPDT. { 'siŋ·gəl 'pōl 'dəb·əl 'thrō }

single-pole single-throw [ELEC] A two-terminal switch or relay contact arrangement that opens or closes one circuit. Abbreviated SPST. { 'siŋ·gəl 'pōl 'siŋ·gəl 'thrō }

single-precision number [COMPUT SCI] A number having as many digits as are ordinarily used in a given computer, in contrast to a double-precision number. Also known as short-precision number. { 'siŋ·gəl prə¦sizh·ən 'nəm·bər }

single-program, multiple-data *See* SPMD. { ¦siŋ·gəl ¦prō·gram ¦məl·tə·pəl 'dad·ə }

single reference *See* random access. { 'siŋ·gəl 'ref·rəns }

singlesheet feed [COMPUT SCI] Equipment for feeding one sheet of paper to a computer printer at a time. { 'siŋ·gəl¸shēt 'fēd }

single-shot blocking oscillator [ELECTR] Blocking oscillator modified to operate as a single-shot trigger circuit. { 'siŋ·gəl ¦shät 'bläk·iŋ 'äs·ə¸lād·ər }

single-shot multivibrator *See* monostable multivibrator. { 'siŋ·gəl ¦shät ¦məl·ti'vī¸brād·ər }

single-shot operation *See* single-step operation. { 'siŋ·gəl ¦shät ¸äp·ə'rā·shən }

single-shot trigger circuit [ELECTR] Trigger circuit in which one triggering pulse initiates one complete cycle of conditions ending with a stable condition. Also known as single-trip trigger circuit. { 'siŋ·gəl ¦shät 'trig·ər ¸sər·kət }

single-sideband [COMMUN] Pertaining to single-sideband communication. Abbreviated SSB. { 'siŋ·gəl 'sīd¸band }

single-sideband communication [COMMUN] A communication system in which one of the two sidebands used in amplitude-modulation is suppressed; the carrier wave may be either transmitted, suppressed, or partially suppressed. { 'siŋ·gəl ¦sīd¸band kə¸myü·nə'kā·shən }

single-sideband modulation [COMMUN] Modulation resulting from elimination of all components of one sideband from an amplitude-modulated wave. { 'siŋ·gəl ¦sīd¸band ¸mäj·ə'lā·shən }

single-sideband transmission [COMMUN] Transmission of a carrier and substantially only one sideband of modulation frequencies, as in television where only the upper sideband is transmitted completely for the picture signal; the carrier wave may be either transmitted or suppressed, partially or totally. { 'siŋ·gəl ¦sīd ¸band tranz'mish·ən }

single-sided [COMPUT SCI] Pertaining to storage media that use only one of two sides for recording data. { 'siŋ·gəl 'sīd·əd }

single-sided amplifier *See* single-end amplifier. { 'siŋ·gəl ¦sīd·əd 'am·plə¸fī·ər }

single-sided board |ELECTR| A printed wiring board that contains all of the interconnect material on one of the external layers. { ˌsiŋ·gəl ˌsīd·əd 'bȯrd }

single-signal receiver |ELECTR| A highly selective superheterodyne receiver for code reception, having a crystal filter in the intermediate-frequency amplifier. { 'siŋ·gəl ¦sig·nəl ri'sē·vər }

single-step operation |COMPUT SCI| A method of computer operation, used in debugging or detecting computer malfunctions, in which a program is carried out one instruction at a time, each instruction being performed in response to a manual control device such as a switch or button. Also known as one-shot operation; one-step operation; single-shot operation; step-by-step operation. { 'siŋ·gəl ¦step ˌäp·ə'rā·shən }

single-stub transformer |ELECTROMAG| Shorted section of a coaxial line that is connected to a main coaxial line near a discontinuity to provide impedance matching at the discontinuity. { 'siŋ·gəl ¦stəb tranz'fȯr·mər }

single-stub tuner |ELECTROMAG| Section of transmission line terminated by a movable short-circuiting plunger or bar, attached to a main transmission line for impedance-matching purposes. { 'siŋ·gəl ¦stəb 'tün·ər }

single threading |COMPUT SCI| Transaction processing in which one transaction is completed before another is begun. { 'siŋ·gəl 'thred·iŋ }

single-throw switch |ELEC| A switch in which the same pair of contacts is always opened or closed. { 'siŋ·gəl ¦thrō 'swich }

single-tone keying |COMMUN| Form of keying in which the modulating function causes the carrier to be modulated with a single tone for one condition, which may be either marking or spacing, and the carrier is unmodulated for the other condition. { 'siŋ·gəl ¦tōn 'kē·iŋ }

single-trip trigger circuit See single-shot trigger circuit. { 'siŋ·gəl ¦trip 'trig·ər ˌsər·kət }

single-tuned amplifier |ELECTR| An amplifier characterized by resonance at a single frequency. { 'siŋ·gəl ¦tünd 'am·plə‚fī·ər }

single-tuned circuit |ELEC| A circuit whose behavior is the same as that of a circuit with a single inductance and a single capacitance, together with associated resistances. { 'siŋ·gəl ¦tünd 'sər·kət }

single-tuned interstage |ELECTR| An interstage circuit which is resonant at a single frequency. { 'siŋ·gəl ¦tünd 'in·tər‚stāj }

single-unit semiconductor device |ELECTR| Semiconductor device having one set of electrodes associated with a single carrier stream. { 'siŋ·gəl ¦yü·nət 'sem·i·kən‚dək·tər di‚vīs }

single-wire line |ELEC| **1.** Transmission line that uses the ground as one side of the circuit. **2.** A surface-wave transmission line that consists of a single conductor which has a dielectric coating or other treatment that confines the propagated energy close to the wire. { 'siŋ·gəl ¦wīr 'līn }

singly linked ring |COMPUT SCI| A cyclic arrangement of data elements in which searches may be performed in either a clockwise or a coun-terclockwise direction, but not both. { 'siŋ·glē ¦liŋkt 'riŋ }

sink |COMMUN| Equipment at the end of a communications channel that receives signals and may perform other functions such as error detection. |ELECTROMAG| The region of a Rieke diagram where the rate of change of frequency with respect to phase of the reflection coefficient is maximum for an oscillator; operation in this region may lead to unsatisfactory performance by reason of cessation or instability of oscillations. { siŋk }

sinusoidal angular modulation See angle modulation. { ˌsī·nə'sȯid·əl 'aŋ·gyə·lər ˌmäj·ə'lā·shən }

sinusoidal current See simple harmonic current. { ˌsī·nə'sȯid·əl 'kə·rənt }

sinusoidal oscillator |ELECTR| An oscillator circuit whose output voltage is a sine-wave function of time. Also known as harmonic oscillator; sine-wave oscillator. { ˌsī·nə'sȯid·əl 'as·ə‚lād·ər }

SIP See single in-line package. { sip }

SISD |COMPUT SCI| A type of computer architecture in which there is a single instruction cycle, and operands are fetched in serial fashion into a single processing unit before execution. Acronym for single-instruction-stream, single-data-stream. { ¦es‚ī¦es'dē }

SIT See static induction transistor.

site |COMPUT SCI| A position available for the symbols of an inscription, for example, a digital place. { sīt }

situation-display tube |ELECTR| Large cathode-ray tube used to display tabular and vector messages pertinent to the various functions of an air defense mission. { ‚sich·ə'wā·shən di'splā ‚tüb }

six-phase circuit |ELEC| Combination of circuits energized by alternating electromotive forces which differ in phase by one-sixth of a cycle (60°). { 'siks ¦fāz 'sər·kət }

six-phase rectifier |ELECTR| A rectifier in which transformers are used to produce six alternating electromotive forces which differ in phase by one-sixth of a cycle, and which feed six diodes. { 'siks ¦fāz 'rek·tə‚fī·ər }

size control |ELECTR| A control provided on a video display device for changing the size of a picture either horizontally or vertically. { 'sīz kən‚trōl }

skeletal coding |COMPUT SCI| A set of incomplete instructions in symbolic form, intended to be completed and specialized by a processing program written for that purpose. { 'skel·əd·əl 'kōd·iŋ }

skew |COMPUT SCI| In character recognition, a condition arising at the read station whereby a character or a line of characters appears in a "twisted" manner in relation to a real or imaginary horizontal baseline. |ELECTR| **1.** The deviation of a received facsimile frame from rectangularity due to lack of synchronism between scanner and recorder; expressed numerically as the tangent of the angle of this deviation. **2.** The degree of nonsynchronism of supposedly parallel bits when bit-coded characters are read from magnetic tape. { skyü }

skew failure |COMPUT SCI| In character recognition, the condition that exists during document alignment whereby the document reference edge is not parallel to that of the read station. { 'skyü ,fāl·yər }

skiatron See dark-trace tube. { 'skī·ə,trän }

skin antenna |ELECTROMAG| Flush-mounted aircraft antenna made by using insulating material to isolate a portion of the metal skin of the aircraft. { 'skin an,ten·ə }

skin depth |ELECTROMAG| The depth beneath the surface of a conductor, which is carrying current at a given frequency due to electromagnetic waves incident on its surface, at which the current density drops to one neper below the current density at the surface. { 'skin ,depth }

skin effect |ELEC| The tendency of alternating currents to flow near the surface of a conductor thus being restricted to a small part of the total sectional area and producing the effect of increasing the resistance. Also known as conductor skin effect; Kelvin skin effect. { 'skin i,fekt }

skin resistance |ELEC| For alternating current of a given frequency, the direct-current resistance of a layer at the surface of a conductor whose thickness equals the skin depth. { 'skin ri ,zis·təns }

skin tracking |ELECTROMAG| Tracking of an object by means of radar without using a beacon or other signal device on board the object being tracked. { 'skin ,trak·iŋ }

skiograph |ELECTR| An instrument used to measure the intensity of x-rays. { 'skī·ə,graf }

skip |COMPUT SCI| **1.** In fixed-instruction-length digital computers, to bypass or ignore one or more instructions in an otherwise sequential process. **2.** Action of a computer printer that moves rapidly over a line so that a blank line appears in the printout. { skip }

skip chain |COMPUT SCI| A programming technique which matches a word against a set of test words; if there is a match, control is transferred (skipped) to a routine, otherwise the word is matched with the next test word in sequence. { 'skip ,chān }

skip distance |ELECTROMAG| The minimum distance that radio waves can be transmitted between two points on the earth by reflection from the ionosphere, at a specified time and frequency. { 'skip ,dis·təns }

skip effect |COMMUN| The existence of a circular-shaped area around a radio transmitter within which no radio signals are received, because ground signals are received only inside the oval and sky-wave signals are received only outside the oval. { 'skip i,fekt }

skip fading |ELECTROMAG| Fading due to fluctuations of ionization density at the place in the ionosphere where the wave is reflected which causes the skip distance to increase or decrease. { 'skip ,fād·iŋ }

skip flag |COMPUT SCI| The thirty-fifth bit of a channel command word which suppresses the transfer of data to main storage. { 'skip ,flag }

skip keying |ELECTR| Reduction of radar pulse repetition frequency to submultiple of that normally used, to reduce mutual interference between radar or to increase the length of radar time base. { 'skip ,kē·iŋ }

skip-searched chain |COMPUT SCI| A chain which has pointers and can therefore be searched without examining each link. { 'skip ,sərcht ,chān }

skip zone |COMMUN| The area between the outer limit of reception of radio high-frequency ground waves and the inner limit of reception of sky waves, where no signal is received. { 'skip ,zōn }

sky wave |ELECTROMAG| A radio wave that travels upward into space and may or may not be returned to earth by reflection from the ionosphere. Also known as ionospheric wave. { 'skī ,wāv }

sky-wave correction |ELECTR| The correction to be applied to the time difference readings of received sky waves to convert them to an equivalent ground-wave reading. { 'skī ¦wāv kə'rek·shən }

sky-wave transmission delay |ELECTROMAG| Amount by which the time of transit from transmitter to receiver of a pulse carried by sky waves reflected once from the E layer exceeds the time of transit of the same pulse carried by ground waves. { 'skī ¦wāv tranz'mish·ən di,lā }

slab |ELECTR| A relatively thick-cut crystal from which blanks are obtained by subsequent transverse cutting. { slab }

Slater's rule |ELECTR| The ratio of the cathode radius to the anode radius of a magnetron is approximately equal to $(N - 4)/(N + 4)$, where N is the number of resonators. { 'slād·ərz ,rül }

slave |COMPUT SCI| A terminal or computer that is controlled by another computer. |CONT SYS| A device whose motions are governed by instructions from another machine. { slāv }

slave antenna |ELECTROMAG| A directional antenna positioned in azimuth and elevation by a servo system; the information controlling the servo system is supplied by a tracking or positioning system. { 'slāv an,ten·ə }

slave mode See user mode. { 'slāv ,mōd }

slave tube |ELECTR| A display monitor that is connected to another monitor and provides an identical display. { 'slāv ,tüb }

sleep |COMPUT SCI| State of a computer system that halts, or a program that appears to be doing nothing because the program is caught in an endless loop. { 'slēp }

sleeve |ELEC| **1.** The cylindrical contact that is farthest from the tip of a phone plug. **2.** Insulating tubing used over wires or components. Also known as bushing; sleeving. |ENG| A cylindrical part designed to fit over another part. { slēv }

sleeve antenna |ELECTROMAG| A single vertical half-wave radiator, the lower half of which is a metallic sleeve through which the concentric feed line runs; the upper radiating portion, one quarter-wavelength long, connects to the center of the line. { 'slēv an,ten·ə }

sleeve dipole antenna |ELECTROMAG| Dipole antenna surrounded in its central portion by a coaxial cable. { 'slēv 'dī,pōl an'ten·ə }

sleeving See sleeve. { 'slēv·iŋ }

slewing motor |ELEC| A motor used to drive a radar antenna at high speed for slewing to pick up or track a target. { 'slü·iŋ ‚mōd·ər }

slew rate |COMPUT SCI| The speed at which a logic-seeking print head advances to the succeeding line and finds the position where it is to start printing. |CONT SYS| The maximum rate at which a system can follow a command. |ELECTR| The maximum rate at which the output voltage of an operational amplifier changes for a square-wave or step-signal input; usually specified in volts per microsecond. { 'slü ‚rāt }

slicer See amplitude gate. { 'slīs·ər }

slicer amplifier See amplitude gate. { 'slīs·ər ‚am·plə‚fī·ər }

slicing |ELECTR| Transmission of only those portions of a waveform lying between two amplitude values. { 'slīs·iŋ }

slide-back voltmeter |ELECTR| An electronic voltmeter in which an unknown voltage is measured indirectly by adjusting a calibrated voltage source until its voltage equals the unknown voltage. { 'slīd ‚bak 'vōlt‚mēd·ər }

slider |ELEC| Sliding type of movable contact. { 'slīd·ər }

slide-wire bridge |ELEC| A bridge circuit in which the resistance in one or more branches is controlled by the position of a sliding contact on a length of resistance wire stretched along a linear scale. { 'slīd ¦wīr ‚brij }

slide-wire potentiometer |ELEC| A potentiometer (variable resistor) which employs a movable sliding connection on a length of resistance wire. { 'slīd ¦wīr pə‚ten·chē'äm·əd·ər }

sliding contact See wiping contact. { 'slīd·iŋ 'kän ‚takt }

slip |ELEC| **1.** The difference between synchronous and operating speeds of an induction machine. Also known as slip speed. **2.** Method of interconnecting multiple wiring between switching units by which trunk number 1 becomes the first choice for the first switch, trunk number 2 first choice for the second switch, trunk number 3 first choice for the third switch, and so on. |ELECTR| Distortion produced in the recorded facsimile image which is similar to that produced by skew but is caused by slippage in the mechanical drive system. { slip }

slip ring |ELEC| A conductive rotating ring which, in combination with a stationary brush, provides a continuous electrical connection between rotating and stationary conductors; used in electric rotating machinery, synchros, gyroscopes, and scanning radar antennas. { 'slip ‚riŋ }

slit scan |COMPUT SCI| In character recognition, a magnetic or photoelectric device that obtains the horizontal structure of an inputted character by vertically projecting its component elements at given intervals. { 'slit ‚skan }

slot |COMPUT SCI| A connection to a computer bus into which printed ciruit boards or integrated circuit boards can be inserted. |ELEC| One of the conductor-holding grooves in the face of the rotor or stator of an electric rotating machine. { slät }

slot antenna |ELECTROMAG| An antenna formed by cutting one or more narrow slots in a large metal surface fed by a coaxial line or waveguide. { 'slät an‚ten·ə }

slot-bound |COMPUT SCI| Condition of a computer when all the slots in the machine's bus are filled with printed circuit boards, so that it is not possible to expand the machine's capacity by plugging in additional boards. { 'slät ‚baund }

slot coupling |ELECTROMAG| Coupling between a coaxial cable and a waveguide by means of two coincident narrow slots, one in a waveguide wall and the other in the sheath of the coaxial cable. { 'slät ‚käp·liŋ }

slot-mask picture tube |ELECTR| An in-line gun-type color picture tube in which the shadow mask is perforated by short, vertical slots, and the screen is painted with vertical phosphor stripes. { 'slät ‚mask 'pik·chər ‚tüb }

slot radiator |ELECTROMAG| Primary radiating element in the form of a slot cut in the walls of a metal waveguide or cavity resonator or in a metal plate. { 'slät ‚rād·ē‚ād·ər }

slotted line See slotted section. { 'släd·əd 'līn }

slotted section |ELECTROMAG| A section of waveguide or shielded transmission line in which the shield is slotted to permit the use of a movable probe for examination of standing waves. Also known as slotted line; slotted waveguide. { 'släd·əd ‚sek·shən }

slotted waveguide See slotted section. { 'släd·əd 'wāv‚gīd }

slot wedge |ELEC| The wedge that holds the windings in a slot in the rotor or stator core of an electrical machine. { 'slät ‚wej }

slow-blow fuse |ELEC| A fuse that can withstand up to 10 times its normal operating current for a brief period, as required for circuits and devices which draw a very heavy starting current. { 'slō ¦blō 'fyüz }

slow death |ELECTR| The gradual change of transistor characteristics with time; this change is attributed to ions which collect on the surface of the transistor. { 'slō 'deth }

slowed-down video |ELECTR| Technique or method of transmitting radar data over narrow-bandwidth circuits; the procedure involves storing the radar video over the time required for the antenna to move through the beam width, and the subsequent sampling of this stored video at some periodic rate at which all of the range intervals of interest are sampled at least once each beam width or per azimuth quantum; the radar returns are quantized at the gap-filler radar site. { 'slōd ¦daun 'vid·ē·ō }

slow memory See slow storage. { 'slō 'mem·rē }

slow-motion video disk recorder |ELECTR| A magnetic disk recorder that stores one field of video information per revolution, for instant replay at normal speed or any degree of slow motion down to complete stopping of action. { 'slō ¦mō·shən 'vid·ē·o ¦disk ri'kórd·ər }

slow-scan television |COMMUN| Television system that uses a slow rate of horizontal scanning, requiring typically 8 seconds for each complete scan of the scene; suitable for transmitting printed matter, photographs, and illustrations. Abbreviated SSTV. { 'slō ¦skan 'tel·ə,vizh·ən }

slow storage |COMPUT SCI| In computers, storage with a relatively long access time. Also known as slow memory. { 'slō 'stór·ij }

slow time scale |COMPUT SCI| In simulation by an analog computer, a time scale in which the time duration of a simulated event is greater than the actual time duration of the event in the physical system under study. Also known as extended time scale. { 'slō 'tīm ,skāl }

slow wave |ELECTROMAG| A wave having a phase velocity less than the velocity of light, as in a ridge wave guide. { 'slō 'wāv }

SLSI circuit See super-large-scale integrated circuit. { ,es,el,es'ī ,sər·kət }

slug tuner |ELECTROMAG| Waveguide tuner containing one or more longitudinally adjustable pieces of metal or dielectric. { 'sləg ¦tün·ər }

slug tuning |ELECTROMAG| Means of varying the frequency of a resonant circuit by introducing a slug of material into either the electric field or magnetic field, or both. { 'sləg ¦tün·iŋ }

small computer system interface |COMPUT SCI| An interface standard or format for personal computers that allows the connection of up to seven peripheral devices. Abbreviated SCSI (scuzzy). { ¦smól kəm¦pyüd·ər ,sis·təm 'in·tər ,fās }

small-scale integration |ELECTR| Integration in which a complete major subsystem or system is fabricated on a single integrated-circuit chip that contains integrated circuits which have appreciably less complexity than for medium-scale integration. Abbreviated SSI. { 'smól ¦skāl ,int·ə'grā·shən }

small-signal parameter |ELECTR| One of the parameters characterizing the behavior of an electronic device at small values of input, for which the device can be represented by an equivalent linear circuit. { 'smól ¦sig·nəl pə'ram·əd·ər }

small talk |COMPUT SCI| A high-level, user-friendly programming language that incorporates the functions of an operating system. { 'smól ,tók }

smart card |COMPUT SCI| A plastic card in which is embedded a microprocessor that is usually programmed to hold information about the card holder or user. Also known as chip card. { 'smärt ,kärd }

smart sensor |ENG| A microsensor integrated with signal-conditioning electronics such as analog-to-digital converters on a single silicon chip to form an integrated microelectromechanical component that can process information itself or communicate with an embedded microprocessor. Also known as intelligent sensor. { ,smärt 'sen·sər }

smart structures |ENG| Structures that are capable of sensing and reacting to their environment in a predictable and desired manner, through the integration of various elements, such as sensors, actuators, power sources, signal processors, and communications network. In addition to carrying mechanical loads, smart structures may alleviate vibration, reduce acoustic noise, monitor their own condition and environment, automatically perform precision alignments, or change their shape or mechanical properties on command. { ,smärt 'strək·chərz }

smart terminal See intelligent terminal. { 'smärt 'tər·mən·əl }

smart tool |CONT SYS| A robot end effector or fixed tool that uses sensors to measure the tool's position relative to reference markers or a workpiece or jig, and an actuator to adjust the tool's position with respect to the workpiece. { 'smärt ,tül }

SMATV system See satellite master antenna television system. { ¦es,em,ā,tē'vē ,sis·təm }

smear |ELECTR| A video picture defect in which objects appear to be extended horizontally beyond their normal boundaries in a blurred or smeared manner; one cause is excessive attenuation of high video frequencies in an analog television receiver. { smir }

S meter See signal-strength meter. { 'es ,mēd·ər }

smiley See emoticon. { 'smīl·ē }

Smithell's burner |ENG| Two concentric tubes that can be added to a bunsen burner to separate the inner and outer flame cones. { 'smith·əl ,bər·nər }

smoke |ENG| Dispersions of finely divided (0.01–5.0 micrometers) solids or liquids in a gaseous medium. { smōk }

smoke chamber |ENG| That area in a fireplace directly above the smoke shelf. { 'smōk ,chām·bər }

smoke detector |ENG| A photoelectric system for an alarm when smoke in a chimney or other location exceeds a predetermined density. { 'smōk di,tek·tər }

smoke point |ENG| The maximum flame height in millimeters at which kerosine will burn without smoking, tested under standard conditions; used as a measure of the burning cleanliness of jet fuel and kerosine. { 'smōk ,póint }

smoke shelf |ENG| A horizontal surface directly behind the throat of a fireplace to prevent downdrafts. { 'smōk ,shelf }

smokestack |ENG| A chimney for the discharge of flue gases from a furnace operation such as in a steam boiler, powerhouse, heating plant, ship, locomotive, or foundry. { 'smōk,stak }

smoke test |ENG| A test used on kerosine to determine the highest point to which the flame can be turned before smoking occurs. { 'smōk ,test }

smoke washer |ENG| A device for removing particles from smoke by forcing it through a spray of water. { 'smōk ,wäsh·ər }

smooth blasting [ENG] Blasting to ensure even faces without cracks in the rock. { 'smü<u>th</u> 'blast·iŋ }

smooth drilling [ENG] Drilling in a rock formation in which a fast rotation of the drill stem, a fast rate of penetration, and a high recovery of core can be achieved with vibration-free rotation of the drill stem. { 'smü<u>th</u> 'dril·iŋ }

smoothing [ENG] Making a level, or continuously even, surface. { 'smü<u>th</u>·iŋ }

smoothing choke [ELECTR] Iron-core choke coil employed as a filter to remove fluctuations in the output current of a vacuum-tube rectifier or direct-current generator. { 'smü<u>th</u>·iŋ ‚chōk }

smoothing circuit See ripple filter. { 'smü<u>th</u>·iŋ ‚sər·kət }

smoothing filter See ripple filter. { 'smü<u>th</u>·iŋ ‚fil·tər }

smoothing plane [ENG] A finely set hand tool, usually 5.5–10 inches (14–25.4 centimeters) long, for finishing small areas on wood. { 'smü<u>th</u>·iŋ ‚plān }

smother kiln [ENG] A kiln into which smoke can be introduced for blackening pottery. { 'smə<u>th</u>·ər ‚kil }

SMPT See Simple Mail Transfer Protocol.

smudging [ENG] A frost-preventive measure used in orchards; properly, it means the production of heavy smoke, supposed to prevent radiational cooling, but it is generally applied to both heating and smoke production. { 'sməj·iŋ }

S/N See signal-to-noise ratio.

snake hole [ENG] **1.** A blasting hole bored directly under a boulder. **2.** A drill hole used in quarrying or bench blasting. { 'snāk ‚hōl }

snaking [ENG] Towing a load with a long cable. { 'snāk·iŋ }

snap-action switch [ELEC] A switch that responds to very small movements of its actuating button or lever and changes rapidly and positively from one contact position to the other; the trademark of one version is Micro Switch. Also known as sensitive switch. { 'snap ¦ak·shən ‚swich }

snap-back forming [ENG] A plastic-sheet-forming technique in which an extended, heated, plastic sheet is allowed to contract over a form shaped to the desired final contour. { 'snap ‚bak ‚fȯrm·iŋ }

snap fastener [ENG] A fastener consisting of a ball on one edge of an article that fits in a socket on an opposed edge, and used to hold edges together, such as those of a garment. { 'snap ‚fas·ən·ər }

snap gage [ENG] A device with two flat, parallel surfaces spaced to control one limit of tolerance of an outside diameter or a length. { 'snap ‚gāj }

snap hook See spring hook. { 'snap ‚hu̇k }

snap-off diode [ELECTR] Planar epitaxial passivated silicon diode that is processed so a charge is stored close to the junction when the diode is conducting; when reverse voltage is applied, the stored charge then forces the diode to snap off or switch rapidly to its blocking state. { 'snap ‚ȯf 'dī‚ōd }

snap-on ammeter [ELEC] An ac ammeter having a magnetic core in the form of hinged jaws that can be snapped around the current-carrying wire. Also known as clamp-on ammeter. { 'snap¦ón 'am‚ēd·ər }

snapper [ENG] A device for collecting samples from the ocean bottom, and which closes to prevent the sample from dropping out as it is raised to the surface. { 'snap·ər }

snap ring [ENG] A form of spring used as a fastener; the ring is elastically deformed, put in place, and allowed to snap back toward its unstressed position into a groove or recess. { 'snap ‚riŋ }

snapshot [COMPUT SCI] The storing of the entire contents of the memory, including status indicators and hardware registers. { 'snap‚shät }

snapshot dump [COMPUT SCI] An edited printout of selected parts of the contents of main memory, performed at one or more times during the execution of a program without materially affecting the operation of the program. { 'snap‚shät ‚dəmp }

snapshot program [COMPUT SCI] A program that provides dumps of certain portions of memory when certain instructions are executed or when certain conditions are fulfilled. { 'snap‚shät ‚prō·grəm }

snatch block [ENG] A pulley frame or sheave with an eye through which lashing can be passed to fasten it to a scaffold or pole. { 'snach ‚bläk }

snatch plate [ENG] A thick steel plate through which a hole about one-sixteenth of an inch larger than the outside diameter of the drill rod on which it is to be used is drilled; the plate is slipped over the drill rod and one edge is fastened to a securely anchored chain, and if rods must be pulled because high-pressure water is encountered, the eccentric pull of the chain causes the outside of the rods to be gripped and held against the pressure of water; the rod is moved a short distance out of the hole each time the plate is tapped. { 'snach ‚plāt }

S-N diagram [ENG] In fatigue testing, a graphic representation of the relationship of stress S and the number of cycles N before failure of the material. { ¦es¦en 'dī·ə‚gram }

sneak path [COMPUT SCI] In computers, an undesired circuit through a series-parallel configuration. { 'snēk ‚path }

snifter valve [ENG] A valve on a pump that allows air to enter or escape, and accumulated water to be released. { 'snif·tər ‚valv }

snivet [ELECTR] Straight, jagged, or broken vertical black line appearing near the right-hand edge of a television receiver screen. { 'sniv·ət }

SNOBOL [COMPUT SCI] A computer programming language that has significant applications in program compilation and generation of symbolic equations. Derived from String-Oriented-Symbolic Language. { 'snō‚bȯl }

snooperscope [ELECTR] An infrared source, an infrared image converter, and a battery-operated high-voltage direct-current source constructed in portable form to permit a foot soldier or other user to see objects in total darkness; infrared radiation sent out by the infrared source is reflected

snorkel

back to the snooperscope and converted into a visible image on the fluorescent screen of the image tube. { 'snüp·ər,skōp }

snorkel [ENG] Any tube which supplies air for an underwater operation, whether it be for material or personnel. { 'snȯr·kəl }

snow [ELECTR] Small, random, white spots produced on an analog television or radar screen by inherent noise signals originating in the receiver. { snō }

snow bin [ENG] A box for measuring the amount of snowfall; a type of snow gage. { 'snō ,bin }

snow mat [ENG] A device used to mark the surface between old and new snow, consisting of a piece of white duck 28 inches (71 centimeters) square, having in each corner triangular pockets in which are inserted slats placed diagonally to keep the mat taut and flat. { 'snō ,mat }

snow pillow [ENG] A device used to record the changing weight of the snow cover at a point, consisting of a fluid-filled bladder lying on the ground with a pressure transducer or a vertical pipe and float connected to it. { 'snō ,pil·ō }

snow resistograph [ENG] An instrument for recording a hardness profile of a snow cover by recording the force required to move a blade up through the snow. { 'snō ri'zis·tə,graf }

snow sampler [ENG] A hollow tube for collecting a sample of snow in place. Also known as snow tube. { 'snō ,sam·plər }

snow scale See snow stake. { 'snō ,skāl }

snow stake [ENG] A wood scale, calibrated in inches, used in regions of deep snow to measure its depth; it is bolted to a wood post or angle iron set in the ground. Also known as snow scale. { 'snō ,stāk }

snow static [ELECTROMAG] Precipitation static caused by falling snow. { 'snō ,stad·ik }

snow tube See snow sampler. { 'snō ,tüb }

SNR See signal-to-noise ratio.

Snyder sampler [ENG] A mechanical device for obtaining small representative quantities from a moving stream of pulverized or granulated solids; it consists of a cast-iron plate revolving in a vertical plane on a horizontal axis with an inclined sample spout; the material to be sampled comes to the sampler by way of an inclined chute whenever the sample spout comes in line with the moving stream. { 'snī·dər 'sam·plər }

soap bubble test [ENG] A leak test in which a soap solution is applied to the surface of the vessel under internal pressure test; soap bubbles form if the tracer gas leaks from the vessel. { 'sōp ,bəb·əl ,test }

socket [ELEC] A device designed to provide electric connections and mechanical support for an electronic or electric component requiring convenient replacement. [ENG] A device designed to receive and grip the end of a tubular object, such as a tool or pipe. { 'säk·ət }

socket-head screw [ENG] A screw fastener with a geometric recess in the head into which an appropriate wrench is inserted for driving and turning, with consequent improved nontamperability. { 'säk·ət ¦hed ,skrü }

socket wrench [ENG] A wrench with a socket to fit the head of a bolt or a nut. { 'säk·ət ,rench }

soda-acid extinguisher [ENG] A fire-extinguisher from which water is expelled at a high rate by the generation of carbon dioxide, the result of mixing (when the extinguisher is tilted) of sulfuric acid and sodium bicarbonate. { 'sōd·ə 'as·əd ik'stiŋ·gwə·shər }

sodar [ENG] Sound-wave transmitting and receiving equipment that is used to remotely measure the vertical turbulence structure and wind profile of the lower layer of the atmosphere by analyzing sound reflected in scattering by atmospheric turbulence. Derived from sonic detection and ranging. { 'sō,där }

sodium amalgam-oxygen cell [ELEC] Fuel cell system in which materials functioning in the dual capacity of fuel and anode are consumed continuously; low operating temperatures and high power-to-weight ratios are significant characteristics of the system. { 'sōd·ē·əm ə'mal·gəm 'äk·sə·jən ,sel }

sodium/sulfur battery [ELEC] A storage battery that operates at temperatures of 300–350°C (570–660°F) and has a liquid sodium anode and liquid sulfur cathode separated by a solid ceramic electrolyte that conducts sodium ions. { 'sōd·ē·əm 'səl·fər 'bad·ə·rē }

sodium-vapor lamp [ELECTR] A discharge lamp containing sodium vapor, used chiefly for outdoor illumination. { 'sōd·ē·əm ¦vā·pər 'lamp }

soft automation [ENG] Automatic control, chiefly through the use of computer processing, with relatively little reliance on computer hardware. { 'sȯft ,ȯd·ə'mā·shən }

soft computing [COMPUT SCI] A family of methods that imitate human intelligence with the goal of creating tools provided with some humanlike capabilities (such as learning, reasoning, and decision making), and are based on fuzzy logic, neural networks, and probabilistic reasoning techniques such as genetic algorithms. { ,sȯft kəm'pyüd·iŋ }

soft copy [COMPUT SCI] Information that is displayed on a screen, given by voice, or stored in a form that cannot be read directly by a person, as on magnetic tape, disk, or microfilm. { 'sȯft 'käp·ē }

soft-copy terminal [COMPUT SCI] A computer terminal that presents its output through an electronic display, rather than printing it on paper. { 'sȯft ¦käp·ē 'tər·mən·əl }

soft crash [COMPUT SCI] A halt in computer operations in which the computer operator has enough warning time to take action to minimize the effects of the stoppage. { 'sȯft 'krash }

soft edit [COMPUT SCI] A checking and correction process that allows data in which problems have been identified to be accepted by a computer system. { 'sȯft 'ed·it }

soft error [COMPUT SCI] An error that occurs in automatic operations but does not recur when the operation is attempted a second time. { 'sȯft 'er·ər }

soft failure [COMPUT SCI] A failure that can be overcome without the assistance of a person with specialized knowledge to repair the device. { 'sȯft 'fāl·yər }

soft flow [ENG] The free-flowing characteristics of a plastic material under conventional molding conditions. { 'sȯft 'flō }

soft font [COMPUT SCI] A typeface or set of typefaces that is contained in the software of a computer system and is transmitted to the printer before printing. Also known as downloadable font. { 'sȯft 'fänt }

soft limiting [ELECTR] Limiting in which there is still an appreciable increase in output for increases in input signal strength up into the range at which limiting action occurs. { 'sȯft 'lim·əd·iŋ }

soft page break [COMPUT SCI] A page break that is inserted in a document by a word-processing program, and can move if text is added, deleted, or reformatted above it. { ¦sȯft 'pāj ¦brāk }

soft patch [COMPUT SCI] A temporary change in a computer program's machine language that is carried out while the program is in memory, and thus prevails only for the duration of a single run of the program. { 'sȯft 'pach }

soft return [COMPUT SCI] A control code that is automatically entered into a text document by the word-processing program to mark the end of a line, based on the current right margin. { 'sȯft ri'tərn }

soft sector [COMPUT SCI] A disk or drum format in which the locations of sectors are determined by control information written on the storage medium rather than by some physical means. { 'sȯft 'sek·tər }

soft tube [ELECTR] **1.** An x-ray tube having a vacuum of about 0.000002 atmosphere (0.2 pascal), the remaining gas being left in intentionally to give less-penetrating rays than those of a more completely evacuated tube. **2.** See gassy tube. { 'sȯft ,tüb }

software [COMPUT SCI] The totality of programs usable on a particular kind of computer, together with the documentation associated with a computer or program, such as manuals, diagrams, and operating instructions. { 'sȯf,wer }

software compatibility [COMPUT SCI] Property of two computers, with respect to a particular programming language, in which a source program from one machine in that language will compile and execute to produce acceptably similar results in the other. { 'sȯf,wer kəm,pad·ə'bil·əd·ē }

software driver [COMPUT SCI] Software that is designed to handle the interaction between a computer and its peripheral equipment, changing the format of data as necessary. { 'sȯf,wer 'drīv·ər }

software engineering [COMPUT SCI] The systematic application of scientific and technological knowledge, through the medium of sound engineering principles, to the production of computer programs, and to the requirements definition, functional specification, design description, program implementation, and test methods that lead up to this code. { 'sȯf,wer ,en·jə 'nir·iŋ }

software flexibility [COMPUT SCI] The ability of software to change easily in response to different user and system requirements. { 'sȯf,wer ,flek·sə'bil·əd·ē }

software floating point [COMPUT SCI] Special routines that allow high-level programming languages to perform floating-point arithmetic on computer hardware designed for integer arithmetic. { 'sȯf,wer 'flōd·iŋ 'pȯint }

software interface [COMPUT SCI] A computer language whereby computer programs can communicate with each other, and one language can call upon another for assistance. { 'sȯf,wer 'in·tər·fās }

software maintenance [COMPUT SCI] The correction of errors in software systems and the remedying of inadequacies in running the software. { 'sȯf,wer ,mānt·ən,əns }

software metric [COMPUT SCI] **1.** A rule for quantifying some characteristic or attribute of a computer software entity. **2.** One of a set of techniques whose aim is to measure the quality of a computer program. { ¦sȯf,wer 'me·trik }

software monitor [COMPUT SCI] A system, used to evaluate the performance of computer software, that is similar to accounting packages, but can collect more data concerning usage of various components of a computer system and is usually part of the control program. { 'sȯf,wer ,män·ə·dər }

software multiplexing [COMPUT SCI] A procedure used in a time-sharing or multiprogrammed system in which the central processing unit, acting under control of a software algorithm, interleaves its attention between a family of programs waiting for service, in such a way that the programs appear to be processed in parallel. { 'sȯf,wer 'məl·ti,pleks·iŋ }

software package [COMPUT SCI] A program for performing some specific function or calculation which is useful to more than one computer user and is sufficiently well documented to be used without modification on a defined configuration of some computer system. { 'sȯf,wer ,pak·ij }

software path length [COMPUT SCI] The number of machine-language instructions required to carry out some specified task. Also known as path length. { 'sȯf,wer 'path ,leŋkth }

software protection [COMPUT SCI] The use of various techniques to prevent the unauthorized duplication of software. Also known as copy protection. { 'sȯf,wer prə,tek·shən }

soft-wired numerical control See computer numerical control. { 'sȯf ,wīrd nü'mer·ə·kəl kən'trōl }

solar battery [ELECTR] An array of solar cells, usually connected in parallel and series. { 'sō·lər 'bad·ə·rē }

solar cell [ELECTR] A pn-junction device which converts the radiant energy of sunlight directly and efficiently into electrical energy. { 'sō·lər 'sel }

solar generator [ELEC] An electric generator powered by radiation from the sun and used in some satellites. { 'sō·lər 'jen·ə,rād·ər }

solar noise See solar radio noise. { 'sō·lər 'nȯiz }

solar radio noise [ELECTROMAG] Radio noise originating at the sun, and increasing greatly in intensity during sunspots and flares; it is heard as a hissing noise on shortwave radio receivers. Also known as solar noise. { 'sō·lər 'rād·ē·ō ,nȯiz }

solar sensor [ELECTR] A light-sensitive diode that sends a signal to the attitude-control system of a spacecraft when it senses the sun. Also known as sun sensor. { 'sō·lər 'sen·sər }

solder-ball flip chip *See* flip chip. { ¦säd·ər ,bȯl 'flip ,chip }

soldering lug [ELEC] A stamped metal strip used as a terminal to which wires can be soldered. { 'säd·ə·riŋ ,ləg }

solderless contact *See* crimp contact. { 'säd·ər·ləs 'kän,takt }

solderless wrapped connection *See* wire-wrap connection. { 'säd·ər·ləs 'rapt kə'nek·shən }

solder track [ELECTR] A conducting path on a printed circuit board that is formed by applying molten solder to the board. { 'säd·ər ,trak }

sole [ELECTR] Electrode used in magnetrons and backward-wave oscillators to carry a current that generates a magnetic field in the direction wanted. { sōl }

solenoid [ELECTROMAG] **1.** Also known as electric solenoid. **2.** An electrically energized coil of insulated wire which produces a magnetic field within the coil. **3.** In particular, a coil that surrounds a movable iron core which is pulled to a central position with respect to the coil when the coil is energized by sending current through it. { 'säl·ə,nȯid }

solid-dielectric capacitor [ELEC] A capacitor whose dielectric is one of several solid materials such as ceramic, mica, glass, plastic film, or paper. { 'säl·əd ¦dī·ə¦lek·trik kə'pas·əd·ər }

solid-electrolyte battery [ELEC] A primary battery whose electrolyte is either a solid crystalline salt, such as silver iodide or lead chloride, or an ion-exchange membrane; in either case, conductivity is almost entirely ionic. { 'säl·əd i'lek·trə,līt 'bad·ə·rē }

solid-electrolyte fuel cell [ELEC] Self-contained fuel cell in which oxygen is the oxidant and hydrogen is the fuel; the oxidant and fuel are kept separated by a solid electrolyte which has a crystalline structure and a low conductivity. { 'säl·əd i'lek·trə,līt 'fyül ,sel }

solid electrolytic capacitor [ELEC] An electrolytic capacitor in which the dielectric is an anodized coating on one electrode, with a solid semiconductor material filling the rest of the space between the electrodes. { 'säl·əd i¦lek·trə¦lid·ik kə'pas·əd·ər }

solid insulator [ELEC] An electric insulator made of a solid substance, such as sulfur, polystyrene, rubber, or porcelain. { 'säl·əd 'in·sə,lād·ər }

solid logic technology [ELECTR] A method of computer construction that makes use of miniaturized modules, resulting in faster circuitry because of the reduced distances that current must travel. { 'säl·əd ¦läj·ik tek'näl·ə·jē }

solid state [ENG] Pertaining to a circuit, device, or system that depends on some combination of electrical, magnetic, and optical phenomena within a solid that is usually a crystalline semiconductor material. { 'säl·əd 'stāt }

solid-state battery [ELEC] A battery in which both the electrodes and the electrolyte are solid-state materials. { 'säl·əd ¦stāt 'bad·ə·rē }

solid-state circuit [ELECTR] Complete circuit formed from a single block of semiconductor material. { 'säl·əd ¦stāt 'sər·kət }

solid-state circuit breaker [ELECTR] A circuit breaker in which a Zener diode, silicon controlled rectifier, or solid-state device is connected to sense when load terminal voltage exceeds a safe value. { 'säl·əd ¦stāt 'sər·kət ,brāk·ər }

solid-state component [ELECTR] A component whose operation depends on the control of electrical or magnetic phenomena in solids, such as a transistor, crystal diode, or ferrite device. { 'säl·əd ¦stāt kəm'pō·nənt }

solid-state device [ELECTR] A device, other than a conductor, which uses magnetic, electrical, and other properties of solid materials, as opposed to vacuum or gaseous devices. { 'säl·əd ¦stāt di'vīs }

solid-state image sensor *See* charge-coupled image sensor. { 'säl·əd ¦stāt 'im·ij ,sen·sər }

solid-state lamp *See* light-emitting diode. { 'säl·əd ¦stāt 'lamp }

solid-state laser [OPTICS] A laser in which a semiconductor material produces the coherent output beam. { 'säl·əd ¦stāt 'lā·zər }

solid-state memory [COMPUT SCI] A computer memory whose elements consist of integrated-circuit bistable multivibrators in which bits of information are stored as one of two states. { 'säl·əd ¦stāt 'mem·rē }

solid-state power amplifier [ELECTR] An amplifier that uses field-effect transistors to provide useful amplification at gigahertz frequencies. { ,säl·əd ¦stāt 'pau̇·ər ,am·plə,fī·ər }

solid-state relay [ELECTR] A relay that uses only solid-state components, with no moving parts. Abbreviated SSR. { 'säl·əd ¦stāt 'rē,lā }

solid-state switch [ELECTR] A microwave switch in which a semiconductor material serves as the switching element; a zero or negative potential applied to the control electrode will reverse-bias the switch and turn it off, and a slight positive voltage will turn it on. { 'säl·əd ¦stāt 'swich }

solid-state thyratron [ELECTR] A semiconductor device, such as a silicon controlled rectifier, that approximates the extremely fast switching speed and power-handling capability of a gaseous thyratron tube. { 'säl·əd ¦stāt 'thī·rə,trän }

solid-state uninterruptible power system [ELEC] An uninterruptible power system in which the load operates continuously from the output of a dc-to-ac static inverter powered by a battery. { 'säl·əd ¦stāt ,ən,int·ə'rəp·tə·bəl 'pau̇·ər ,sis·təm }

solid tantalum capacitor [ELEC] An electrolytic capacitor in which the anode is a porous pellet of tantalum; the dielectric is an extremely thin layer of tantalum pentoxide formed by anodization of the exterior and interior surfaces of the pellet; the cathode is a layer of semiconducting manganese dioxide that fills the pores of the anode over the dielectric. { 'säl·əd ¦tant·əl·əm kə'pas·əd·ər }

solion |ELEC| An electrochemical device in which amplification is obtained by controlling and monitoring a reversible electrochemical reaction. { ¦säl'ī,än }

solution ceramic |ELEC| A nonbrittle, inorganic ceramic insulating coating that can be applied to wires at a low temperature; examples include ceria, chromia, titania, and zirconia. { sə'lü·shən sə'ram·ik }

solvent welding |ENG| A technique for joining plastic pipework in which a mixture of solvent and cement is applied to the pipe end and to the socket, with the parts then being joined and allowed to set. { 'säl·vənt ,weld·iŋ }

Sommerfeld equation See Sommerfeld formula. { 'zòm·ər,felt i,kwä·zhən }

Sommerfeld formula |ELECTROMAG| An approximate formula for the field strength of electromagnetic radiation generated by an antenna at distances small enough so that the curvature of the earth may be neglected, in terms of radiated power, distance from the antenna, and various constants and parameters. Also known as Sommerfeld equation. { 'zòm·ər,felt ,fòr·myə·lə }

sonar |ENG| **1.** A system that uses underwater sound, at sonic or ultrasonic frequencies, to detect and locate objects in the sea, or for communication; the commonest type is echo-ranging sonar; other versions are passive sonar, scanning sonar, and searchlight sonar. Derived from sound navigation and ranging. **2.** See sonar set. { 'sō,när }

sonar array |ELECTR| An arrangement of several sonar transducers or sonar projectors, appropriately spaced and energized to give proper directional characteristics. { 'sō,när ə,rā }

sonar detector See sonar receiver. { 'sō,när di ,tek·tər }

sonar projector |ENG ACOUS| An electromechanical device used under water to convert electrical energy to sound energy; a crystal or magnetostriction transducer is usually used for this purpose. { 'sō,när prə,jek·tər }

sonar receiver |ELECTR| A receiver designed to intercept and amplify the sound signals reflected by an underwater target and display the accompanying intelligence in useful form; it may also pick up other underwater sounds. Also known as sonar detector. { 'sō,när ri'sē·vər }

sonar resolver |ELECTR| A resolver used with echo-ranging and depth-determining sonar to calculate and record the horizontal range of a sonar target, as required for depth-bombing. { 'sō,när ri,zäl·vər }

sonar self-noise |ELECTR| Unwanted sonar signals generated in the sonar equipment itself. { 'sō,när 'self'nòiz }

sonar set |ENG| A complete assembly of sonar equipment for detecting and ranging or for communication. Also known as sonar. { 'sō,när ,set }

sonar transducer |ENG ACOUS| A transducer used under water to convert electrical energy to sound energy and sound energy to electrical energy. { 'sō,när tranz,dü·sər }

sonar transmitter |ELECTR| A transmitter that generates electrical signals of the proper frequency and form for application to a sonar transducer or sonar projector, to produce sound waves of the same frequency in water; the sound waves may carry intelligence. { 'sō,när tranz ,mid·ər }

son file |COMPUT SCI| The master file that is currently being updated. { 'sən ,fīl }

sonic delay line See acoustic delay line. { 'sän·ik di'lā ,līn }

sophisticated robot |CONT SYS| A robot that can be programmed and is controlled by a microprocessor. { sə'fis·tə,kād·əd 'rō,bät }

sophisticated vocabulary |COMPUT SCI| An advanced and elaborate set of instructions; a computer with a sophisticated vocabulary can go beyond the more common mathematical calculations such as addition, multiplication, and subtraction, and perform operations such as linearize, extract square root, and select highest number. { sə'fis·tə,kād·əd və'kab·yə,ler·ē }

sort |COMPUT SCI| **1.** To rearrange a set of data items into a new sequence, governed by specific rules of precedence. **2.** The program designed to perform this activity. { sòrt }

sort algorithm |COMPUT SCI| The methods followed in arranging a set of data items into a sequence according to precise rules. { 'sòrt ¦al·gə,rith·əm }

sorter See sequencer. { 'sòrd·ər }

sort field |COMPUT SCI| A field in a record that is used in determining the final sorted sequence of the records. { 'sòrt ,fēld }

sort generator |COMPUT SCI| A computer program that produces other programs which arrange collections of items into sequences as specified by parameters in the original program. { 'sòrt ,jen·ə,rād·ər }

sort key |COMPUT SCI| A key used as a basis for determining the sequence of items in a set. { 'sòrt ,kē }

sort/merge |COMPUT SCI| To combine two or more similar files, with the records arranged in the appropriate order, according to precise rules. { 'sòrt 'mərj }

sort/merge package |COMPUT SCI| A set of programs capable of sorting and merging data files. { 'sòrt 'mərj ,pak·ij }

sort order |COMPUT SCI| The sequence into which a collection of records are arranged after they have been sorted. { 'sòrt ,òr·dər }

sort pass |COMPUT SCI| Any one of a collection of similar procedures carried out during a sort operation in which a part of the sort is completed. { 'sòrt ,pas }

sortworker |COMPUT SCI| A file created temporarily by a computer program to hold intermediate results when the amount of data to be sorted exceeds the available storage space. { 'sòrt ,wər·kər }

SOS |COMMUN| The distress signal in radio-telegraphy, consisting of the letters S, O, and S of the international Morse code.

sound analyzer |ENG| An instrument which measures the amount of sound energy in various frequency bands; it generally consists of a set of fixed electrical filters or a tunable electrical filter, along with associated amplifiers and a meter which indicates the filter output. { 'saund ,an·ə ,līz·ər }

sound board |COMPUT SCI| An adapter which provides a computer with the capability of reproducing and recording digitally encoded sound. Also known as audio adapter; sound card. { 'saún ,bȯrd }

sound card *See* sound board. { 'saún kärd }

sound carrier |COMMUN| The analog television carrier that is frequency-modulated by the sound portion of a television program; the unmodulated center frequency of the sound carrier is 4.5 megahertz higher than the video carrier frequency for the same television channel. { 'saund ,kar·ē·ər }

sound channel |ELECTR| The series of stages that handles only the sound signal in a television receiver. { 'saund ,chan·əl }

sound filmstrip |ENG ACOUS| A filmstrip that has accompanying sound on a separate disk or tape, which is manually or automatically synchronized with projection of the pictures in the strip. { 'saund 'film,strip }

sound gate |ENG ACOUS| The gate through which film passes in a sound-film projector for conversion of the sound track into audio-frequency signals that can be amplified and reproduced. { 'saund ,gāt }

sound head |ENG ACOUS| **1.** The section of a sound motion picture projector that converts the photographic or magnetic sound track to audible sound signals. **2.** In a sonar system, the cylindrical container for the transmitting projector and the receiving hydrophone. { 'saund ,hed }

sound-level meter |ENG| An instrument used to measure noise and sound levels in a specified manner; the meter may be calibrated in decibels or volume units and includes a microphone, an amplifier, an output meter, and frequency-weighting networks. { 'saund ¦lev·əl ,mēd·ər }

sound navigation and ranging *See* sonar. { 'saund ,nav·ə'gā·shən ən 'rānj·iŋ }

sound-powered telephone |ENG ACOUS| A telephone operating entirely on current generated by the speaker's voice, with no external power supply; sound waves cause a diaphragm to move a coil back and forth between the poles of a powerful but small permanent magnet, generating the required audio-frequency voltage in the coil. { 'saund ¦paú·ərd 'tel·ə,fōn }

sound production |ENG ACOUS| Conversion of energy from mechanical or electrical into acoustical form, as in a siren or loudspeaker. { 'saund prə,dək·shən }

sound reception |ENG ACOUS| Conversion of acoustical energy into another form, usually electrical, as in a microphone. { 'saund ri,sep·shən }

sound recording |ENG ACOUS| The process of recording sound signals so they may be reproduced at any subsequent time, as on a disk, sound track, or magnetic tape. { 'saund ri,kȯrd·iŋ }

sound-reinforcement system |ENG ACOUS| An electronic means for augmenting the sound output of a speaker, singer, or musical instrument in cases where it is either too weak to be heard above the general noise or too reverberant; basic elements of such a system are microphones, amplifiers, volume controls, and loudspeakers. Also known as public address system. { 'saund ,rē·in'fȯrs·mənt ,sis·təm }

sound-reproducing system |ENG ACOUS| A combination of transducing devices and associated equipment for picking up sound at one location and time and reproducing it at the same or some other location and at the same or some later time. Also known as audio system; reproducing system; sound system. { 'saund ,rē·prə'düs·iŋ ,sis·təm }

sound spectrograph |ENG ACOUS| An instrument that records and analyzes the spectral composition of audible sound. { 'saund 'spek·trə ,graf }

soundstripe |ENG ACOUS| A longitudinal stripe of magnetic material placed on some motion picture films for recording a magnetic sound track. { 'saund,strīp }

sound system *See* sound-reproducing system. { 'saund ,sis·təm }

sound track |ENG ACOUS| A narrow band, usually along the margin of a sound film, that carries the sound record; it may be a variable-width or variable-density optical track or a magnetic track. { 'saund ,trak }

sound transducer *See* electroacoustic transducer. { 'saund tranz,düs·ər }

sound trap |ELECTR| A wave trap in an analog television receiver circuit that prevents sound signals from entering the picture channels. { 'saund ,trap }

source |ELEC| The circuit or device that supplies signal power or electric energy or charge to a transducer or load circuit. |ELECTR| The terminal in a field-effect transistor from which majority carriers flow into the conducting channel in the semiconductor material. { sȯrs }

source address |COMPUT SCI| The first address of a two-address instruction (the sound address is known as the destination address). { 'sȯrs 'ad,res }

source code |COMPUT SCI| The statements in which a computer program is initially written before translation into machine language. { 'sȯrs ,kōd }

source data automation equipment |COMPUT SCI| Equipment (except paper tape and magnetic tape cartridge typewriters acquired separately and not operated in support of a computer) which, as a by-product of its operation, produces a record in a medium which is acceptable by automatic data-processing equipment. { 'sȯrs ¦dad·ə ,ȯd·ə'mā·shən i,kwip·mənt }

source data capture |COMPUT SCI| The procedures for entering source data into a computer system. { 'sȯrs 'dad·ə ,kap·chər }

source data entry |COMPUT SCI| Entry of data into a computer system directly from its source, without transcription. { 'sȯrs 'dad·ə ‚en·trē }

source degeneration |ELECTR| The addition of a circuit element between a transistor source and ground, with several effects, including a reduction in gain. { 'sȯrs di‚jen·ə'rā·shən }

source document |COMPUT SCI| The original medium containing the basic data to be used by a data-processing system, from which the data are converted into a form which can be read into a computer. Also known as original document. { 'sȯrs ‚däk·yə·mənt }

source-follower amplifier See common-drain amplifier. { 'sȯrs 'fäl·ə·wər 'am·plə‚fī·ər }

source impedance |ELEC| Impedance presented by a source of energy to the input terminals of a device. { 'sȯrs im‚pēd·əns }

source language |COMPUT SCI| The language in which a program (or other text) is originally expressed. { 'sȯrs ‚laŋ·gwij }

source library |COMPUT SCI| A collection of computer programs in compiler language or assembler language. { 'sȯrs ‚lī‚brer·ē }

source listing |COMPUT SCI| A printout of a source program. { 'sȯrs ‚list·iŋ }

source module |COMPUT SCI| An organized set of statements in any source language recorded in machine-readable form and suitable for input to an assembler or compiler. { 'sȯrs ‚mäj·ül }

source program |COMPUT SCI| The form of a program just as the programmer has written it, often on coding forms or machine-readable media; a program expressed in a source-language form. { 'sȯrs ‚prō‚gram }

source program optimizer |COMPUT SCI| A routine for examining the source code of a program under development and providing information about use of the various portions of the code, enabling the programmer to modify those sections of the target program that are most heavily used in order to improve performance of the final, operational program. { 'sȯrs ‚prō‚gram ‚äp·tə‚mīz·ər }

source stream |COMMUN| A single, nonmultiplexed stream of samples before compression coding. { 'sȯrs ‚strēm }

source time |COMPUT SCI| The time involved in fetching the contents of the register specified by the first address of a two-address instruction. { 'sȯrs ‚tīm }

source transition loss |ELECTR| The transmission loss at the junction between an energy source and a transducer connecting that source to an energy load; measured by the ratio of the source power to the input power. { 'sȯrs tran'zish·ən ‚lȯs }

sourcing |ELECTR| Redesign or the modification of existing equipment to eliminate a source of radio-frequency interference. { 'sȯrs·iŋ }

space |COMMUN| The open-circuit condition or the signal causing the open-circuit condition in telegraphic communication; the closed-circuit condition is called the mark. { spās }

space character See blank character. { 'spās ‚kar·ik·tər }

space charge |ELEC| The net electric charge within a given volume. { 'spās ‚chärj }

space-charge balanced flow |ELECTR| A method of focusing an electron beam in the interaction region of a traveling-wave tube; there is an axial magnetic field in the interaction region which is stronger than that in the gun region; at the transition between the two values of magnetic field strength, the beam is given a rotation in such a direction as to produce an inward force that counterbalances the outward forces from space charge and from the centrifugal forces set up by rotation. { 'spās ‚chärj 'bal·ənst 'flō }

space-charge debunching |ELECTR| A process in which the mutual interactions between electrons in a stream spread out the electrons of a bunch. { 'spās ‚chärj di'bənch·iŋ }

space-charge effect |ELECTR| Repulsion of electrons emitted from the cathode of a thermionic vacuum tube by electrons accumulated in the space charge near the cathode. { 'spās ‚chärj i‚fekt }

space-charge grid |ELECTR| Grid operated at a low positive potential and placed between the cathode and control grid of a vacuum tube to reduce the limiting effect of space charge on the current through the tube. { 'spās ‚chärj ‚grid }

space-charge layer See depletion layer. { 'spās ‚charj ‚lā·ər }

space-charge limitation |ELECTR| The current flowing through a vacuum between a cathode and an anode cannot exceed a certain maximum value, as a result of modification of the electric field near the cathode due to space charge in this region. { 'spās ‚charj ‚lim·ə'tā·shən }

space-charge polarization |ELEC| Polarization of a dielectric which occurs when charge carriers are present which can migrate an appreciable distance through the dielectric but which become trapped or cannot discharge at an electrode. Also known as interfacial polarization. { 'spās ‚chärj ‚pō·lə·rə'zā·shən }

space-charge region |ELECTR| Of a semiconductor device, a region in which the net charge density is significantly different from zero. { 'spās ‚chärj ‚rē·jən }

space communication |COMMUN| Communication between a vehicle in outer space and the earth, using high-frequency electromagnetic radiation. { 'spās kə‚myü·nə'kā·shən }

spacecraft ground instrumentation |ENG| Instrumentation located on the earth for monitoring, tracking, and communicating with manned spacecraft, satellites, and space probes. Also known as ground instrumentation. { 'spās‚kraft 'graůnd ‚in·strə‚mən'tā·shən }

space current |ELECTR| Total current flowing between the cathode and all other electrodes in a tube; this includes the plate current, grid current, screen grid current, and any other electrode current which may be present. { 'spās ‚kə·rənt }

spaced antenna |ELECTROMAG| Antenna system consisting of a number of separate antennas spaced a considerable distance apart, used to minimize local effects of fading at short-wave receiving stations. { 'spāst an'ten·ə }

space diversity reception [ELECTROMAG] Radio reception involving the use of two or more antennas located several wavelengths apart, feeding individual receivers whose outputs are combined; the system gives an essentially constant output signal despite fading due to variable propagation characteristics, because fading affects the spaced-out antennas at different instants of time. { 'spās di'vər·səd·ē ri'sep·shən }

space-division multiple access [COMMUN] The use of the same portion of the electromagnetic spectrum over two or more transmission paths; in most applications, the paths are formed by multibeam antennas, and each beam is directed toward a different geographic area. Abbreviated SDMA. { ¦spās də‚vizh·ən ‚məl·tə·pəl 'ak‚ses }

space reflection symmetry See parity. { 'spās ri‚flek·shən 'sim·ə·trē }

space request [COMPUT SCI] A parameter that specifies the amount of storage space required by a new file at the time the file is created. { 'spās ri‚kwest }

space suppression [COMPUT SCI] Prevention of the normal movement of paper in a computer printer after the printing of a line of characters. { 'spās sə‚presh·ən }

space-time adaptive processing [ELECTR] Radar techniques in which the antenna is subject to automatic pattern shaping to counter angularly displace noise sources (such as jammers), and the coherent signal processing is subject to automatic processes in which Doppler filters are optimally shaped to counter nonuniform distribution of background signals (such as surface clutter in airborne radar) in Doppler. { 'spās 'tīm ə'dap·tiv 'präs‚es·iŋ }

space-to-mark transition [COMMUN] The transition from the space condition to the mark condition in telegraphic communication. { ¦spās tə ¦märk tran'zish·ən }

space wave [ELECTROMAG] The component of a ground wave that travels more or less directly through space from the transmitting antenna to the receiving antenna; one part of the space wave goes directly from one antenna to the other; another part is reflected off the earth between the antennas. { 'spās ‚wāv }

spacing pulse [COMMUN] In teletypewriter operation, the signal interval during which the selector unit is not operated. { 'spās·iŋ ‚pəls }

spacistor [ELECTR] A multiple-terminal solid-state device, similar to a transistor, that generates frequencies up to about 10,000 megahertz by injecting electrons or holes into a space-charge layer which rapidly forces these carriers to a collecting electrode. { spā'sis·tər }

spaghetti [ELEC] Insulating tubing used over bare wires or as a sleeve for holding two or more insulated wires together; the tubing is usually made of a varnished cloth or a plastic. { spə'ged·ē }

spaghetti code [COMPUT SCI] Computer program code that lacks a coherent structure, and in which the sequence of program execution frequently jumps to a distant instruction in the program listing, making the program very difficult to follow. { spə'ged·ē ‚kōd }

spanned record [COMPUT SCI] A logical record which covers more than one block, used when the size of a data buffer is fixed or limited. { 'spand 'rek·ərd }

spark [ELEC] A short-duration electric discharge due to a sudden breakdown of air or some other dielectric material separating two terminals, accompanied by a momentary flash of light. Also known as electric spark; spark discharge; sparkover. { spärk }

spark arrester [ELEC] A device that reduces or eliminates electric sparks at a point where a circuit is opened and closed. { 'spärk ə‚res·tər }

spark capacitor [ELEC] Capacitor connected across a pair of contact points, or across the inductance which causes the spark, for the purpose of diminishing sparking at these points. { 'spärk kə‚pas·əd·ər }

spark discharge See spark. { 'spärk 'dis‚chärj }

spark gap [ELEC] An arrangement of two electrodes between which a spark may occur; the insulation (usually air) between the electrodes is self-restoring after passage of the spark; used as a switching device, for example, to protect equipment against lightning or to switch a radar antenna from receiver to transmitter and vice versa. { 'spärk ‚gap }

spark-gap generator [ELEC] A high-frequency generator in which a capacitor is repeatedly charged to a high voltage and allowed to discharge through a spark gap into an oscillatory circuit, generating successive trains of damped high-frequency oscillations. { 'spärk ¦gap ‚jen·ə‚rād·ər }

sparking potential See breakdown voltage. { 'spärk·iŋ pə‚ten·chəl }

sparking voltage See breakdown voltage. { 'spärk·iŋ ‚vōl·tij }

spark killer See spark suppressor. { 'spärk ‚kil·ər }

sparkover See spark. { 'spärk‚ō·vər }

sparkover voltage See flashover voltage. { 'spärk ‚ō·vər ‚vōl·tij }

spark plate [ELEC] A metal plate insulated from the chassis of an auto radio by a thin sheet of mica, and connected to the battery lead to bypass noise signals picked up by battery wiring in the engine compartment. { 'spärk ‚plāt }

spark plug [ELEC] A device that screws into the cylinder of an internal combustion engine to provide a pair of electrodes between which an electrical discharge is passed to ignite the explosive mixture. { 'spärk ‚pləg }

spark suppressor [ELEC] A device used to prevent sparking between a pair of contacts when the contacts open, such as a resistor and capacitor in series between the contacts, or, in the case of an inductive circuit, a rectifier in parallel with the inductor. Also known as spark killer. { 'spärk sə ‚pres·ər }

spark transmitter [ELECTR] A radio transmitter that utilizes the oscillatory discharge of a capacitor through an inductor and a spark gap as the source of radio-frequency power. { 'spärk tranz'mid·ər }

spark voltage |ELEC| The voltage required to create an arc across the gap of a spark plug. { 'spärk ,vōl·tij }

spatial data management |COMPUT SCI| A technique whereby users retrieve information in databases, document files, or other sources by making contact with picture symbols displayed on the screen of a video terminal through the use of such devices as light pens, joy sticks, and heat-sensitive screens for finger-touch activation. { 'spā·shəl 'dad·ə ,man·ij·mənt }

SPC See stored-program control.

SPDT See single-pole double-throw.

speaker See loudspeaker. { 'spēk·ər }

speaker identification |ENG ACOUS| The use of automated equipment to find the identity of a talker, in a known population of talkers, using the speech input. { ,spēk·ər ī,dent·ə·tə'kā·shən }

speaker verification |ENG ACOUS| The use of automated equipment to authenticate a claimed speaker identity from a voice signal based on speaker-specific characteristics reflected in spoken words or sentences. Abbreviated SV. { ,spēk·ər ,ver·i·fə'kā·shən }

special character |COMPUT SCI| A computer-representable character that is not alphabetic, numeric, or blank. { 'spesh·əl 'kar·ik·tər }

special-purpose computer |COMPUT SCI| A digital or analog computer designed to be especially efficient in a certain class of applications. { 'spesh·əl ¦pər·pəs kəm'pyüd·ər }

special-purpose language |COMPUT SCI| A programming language designed to solve a particular type of problem. { 'spesh·əl ¦pər·pəs 'laŋ·gwij }

specific charge |ELEC| The ratio of a particle's charge to its mass. { spə'sif·ik 'chärj }

specific conductance See conductivity. { spə'sif·ik kən'dək·təns }

specific cryptosystem |COMMUN| A general cryptosystem and a cryptographic key or set of keys for controlling the cryptographic process. { spə'sif·ik 'krip·tō,sis·təm }

specific inductive capacity See dielectric constant. { spə'sif·ik in'dək·tiv kə'pas·əd·ē }

specific insulation resistance See volume resistivity. { spə'sif·ik ,in·sə'lā·shən ri,zis·təns }

specific repetition rate |ELECTR| The pulse repetition rate of a pair of transmitting stations of an electronic navigation system using various rates differing slightly from each other, as in loran. { spə'sif·ik ,rep·ə'tish·ən ,rāt }

specific resistance See electrical resistivity. { spə'sif·ik ri'zis·təns }

specific routine |COMPUT SCI| Computer routine to solve a particular data-handling problem in which each address refers to explicitly stated registers and locations. { spə'sif·ik rü'tēn }

spectral pyrometer See narrow-band pyrometer. { 'spek·trəl pī'räm·əd·ər }

spectral response See spectral sensitivity. { 'spek·trəl ri'späns }

spectral selective photoelectric effect See selective photoelectric effect. { ¦spek·trəl si ¦lek·tiv ,fōd·ō·i'lek·trik i,fekt }

spectral sensitivity |ELECTR| Radiant sensitivity, considered as a function of wavelength. { 'spek·trəl ,sen·sə'tiv·əd·ē }

spectrum level |COMMUN| The level of the part of a specified signal at a specified frequency that is contained within a specified frequency bandwidth, centered at the particular frequency. { 'spek·trəm ,lev·əl }

spectrum-selectivity characteristic |ELECTR| Measure of the increase in the minimum input signal power over the minimum detectable signal required to produce an indication on a radar indicator, if the received signal has a spectrum different from that of the normally received signal. { 'spek·trəm ,si,lek'tiv·əd·ē ,kar·ik·tə ,ris·tik }

spectrum signature |ELECTR| The spectral characteristics of the transmitter, receiver, and antenna of an electronic system, including emission spectra, antenna patterns, and other characteristics. { 'spek·trəm ,sig·nə·chər }

spectrum signature analysis |ELECTR| The evaluation of electromagnetic interference from transmitting and receiving equipment to determine operational and environment compatibility. { 'spek·trəm ,sig·nə·chər ə,nal·ə·səs }

speech amplifier |ENG ACOUS| An audio-frequency amplifier designed specifically for amplification of speech frequencies, as for public-address equipment and radiotelephone systems. { 'spēch ,am·plə,fī·ər }

speech bandwidth |COMMUN| The range of speech frequencies that can be transmitted by a carrier telephone system. { 'spēch 'band ,width }

speech clipper |ENG ACOUS| A clipper used to limit the peaks of speech-frequency signals, as required for increasing the average modulation percentage of a radiotelephone or amateur radio transmitter. { 'spēch ,klip·ər }

speech coder |COMMUN| A device that uses data-compression techniques to convert a high-bit-rate signal resulting from digital pulse-code modulation of speech to a low-rate digital signal that can be transmitted or stored. { 'spēch ,kōd·ər }

speech coil See voice coil. { 'spēch ,kȯil }

speech compression |COMMUN| Modulation technique that takes advantage of certain properties of the speech signal to permit adequate information quality, characteristics, and the sequential pattern of a speaker's voice to be transmitted over a narrower frequency band than would otherwise be necessary. { 'spēch kəm,presh·ən }

speech frequency See voice frequency. { 'spēch ,frē·kwən·sē }

speech intelligibility See intelligibility. { 'spēch in,tel·ə·jə'bil·əd·ē }

speech interpolation |COMMUN| Method of obtaining more than one voice channel per voice circuit by giving each subscriber a speech path in the proper direction only at times when the subscriber's speech requires it. { 'spēch ,in·tər·pəl¦ā·shən }

speech inverter See scrambler. { 'spēch in,vərd·ər }

speech recognition [ENG ACOUS] The process of analyzing an acoustic speech signal to identify the linguistic message that was intended, so that a machine can correctly respond to spoken commands. { 'spēch ,rek·ig'nish·ən }

speech scrambler See scrambler. { 'spēch ,skram·blər }

speech synthesis See voice response. { 'spēch 'sin·thə·səs }

speed control [ELEC] A control that changes the speed of a motor or other drive mechanism, as for a phonograph or magnetic tape recorder. { 'spēd kən,trōl }

speed-matching buffer [COMPUT SCI] A small computer storage unit that connects two devices operating at different data transfer rates; each device writes into and reads from the buffer at its own rate. { 'spēd ¦mach·iŋ 'bəf·ər }

speed of light [ELECTROMAG] The speed of propagation of electromagnetic waves in a vacuum, which is a physical constant equal to exactly 299,792.458 kilometers per second. Also known as electromagnetic constant; velocity of light. { 'spēd əv 'līt }

speed-power product [ELECTR] The product of the gate speed or propagation delay of an electronic circuit and its power dissipation. { 'spēd¦pau·ər ,präd·əkt }

speed regulator [ELEC] A device that maintains the speed of a motor or other device at a predetermined value or varies it in accordance with a predetermined plan. { 'spēd ,reg·yə ,lād·ər }

spelling checker [COMPUT SCI] A program, used in conjunction with word-processing software, which automatically checks words in a text against a dictionary of commonly used words and identifies words that appear to be misspelled. { 'spel·iŋ ,chek·ər }

sphere gap [ELEC] A spark gap between two equal-diameter spherical electrodes. { 'sfir ,gap }

spherical capacitor [ELEC] A capacitor made of two concentric metal spheres with a dielectric filling the space between the spheres. { 'sfir·ə·kəl kə'pas·əd·ər }

spherical-coordinate robot [CONT SYS] A robot in which the degrees of freedom of the manipulator arm are defined primarily by spherical coordinates. { 'sfir·ə·kəl kō¦órd·ən·ət 'rō,bät }

spherical-earth attenuation [ELECTROMAG] Attenuation over an imperfectly conducting spherical earth in excess of that over a perfectly conducting plane. { 'sfir·ə·kəl ¦ərth ə,ten·yə ,wā·shən }

spherical-earth factor [ELECTROMAG] The ratio of the electric field strength that would result from propagation over an imperfectly conducting spherical earth to that which would result from propagation over a perfectly conducting plane. { 'sfir·ə·kəl ¦ərth ,fak·tər }

spider [COMPUT SCI] A program that searches the Internet for new, publicly accessible resources and transmits its findings to a database that is accessible to search engines. [ELEC] A structure on the shaft of an electric rotating machine that supports the core or poles of the rotor, consisting of a hub, spokes, and rim, or some similar arrangement. [ENG ACOUS] A highly flexible perforated or corrugated disk used to center the voice coil of a dynamic loudspeaker with respect to the pole piece without appreciably hindering in-and-out motion of the voice coil and its attached diaphragm. { 'spīd·ər }

spiderweb antenna [ELECTROMAG] All-wave receiving antenna having several different lengths of doublets connected somewhat like the web of a spider to give favorable pickup characteristics over a wide range of frequencies. { 'spīd·ər,web an,ten·ə }

spike antenna See monopole antenna. { 'spīk an ,ten·ə }

spike microphone [ENG ACOUS] A device for clandestine aural surveillance in which the sensor is a spike driven into the wall of the target area and mechanically coupled to the diaphragm of a microphone on the other side of the wall. { 'spīk ,mī·krə,fōn }

spillover [COMMUN] The receiving of a radio signal of a different frequency from that to which the receiver is tuned, due to broad tuning characteristics. { 'spil,ō·vər }

spillover positions [COMMUN] When a transmitting channel is unusually busy or inoperative, the resulting backlogged traffic can be switched to spillover (storage) positions where it is held for immediate transmission when a channel becomes available. { 'spil,ō·vər pə,zish·ənz }

spin electronics See magnetoelectronics. { 'spin ,i·lek,trän·iks }

spin filter [ELECTR] A device used in a Lamb-shift polarized ion source to cause those atoms having an undesired nuclear spin orientation to decay from their metastable state to the ground state, while those with the desired spin orientation are allowed to pass through without decay. { 'spin ,fil·tər }

spinthariscope [ELECTR] An instrument for viewing the scintillations of alpha particles on a luminescent screen, usually with the aid of a microscope. { spin'thar·ə,skōp }

spin transistor See magnetic switch. { 'spin tran ,zis·tər }

spintronics See magnetoelectronics. { spin 'trän·iks }

spin valve See magnetic switch. { 'spin ,valv }

spiral delay line [ELECTROMAG] A transmission line which has a helical inner conductor. { 'spī·rəl di'lā ,līn }

spiral four cable [ELEC] A quad cable in which the four conductors are twisted about a common axis, the two sets of opposite conductors being used as pairs. { 'spī·rəl 'fór ,kā·bəl }

spiral scanning [ENG] Scanning in which the direction of maximum radiation describes a portion of a spiral; the rotation is always in one direction; used with some types of radar antennas. { 'spī·rəl 'skan·iŋ }

splatter |COMMUN| Distortion due to overmodulation of a transmitter by peak signals of short duration, particularly sounds containing high-frequency harmonics; it is a form of adjacent-channel interference. { 'splad·ər }

splice |ELEC| A joint used to connect two lengths of conductor with good mechanical strength and good conductivity. { splīs }

splicing |COMMUN| The concatenation, performed on the system level, of two different elementary streams. { 'splīs·iŋ }

split |COMPUT SCI| To divide a database, file, or other data set into two or more separate parts. { split }

split-anode magnetron |ELECTR| A magnetron in which the cylindrical anode is divided longitudinally into halves, between which extremely high-frequency oscillations are produced. { 'split ¦an ‚ōd 'mag·nə‚trän }

split-phase motor |ELEC| A single-phase induction motor having an auxiliary winding connected in parallel with the main winding, but displaced in magnetic position from the main winding so as to produce the required rotating magnetic field for starting; the auxiliary circuit is generally opened when the motor has attained a predetermined speed. { 'split ¦fāz 'mōd·ər }

split screen *See* partitioned display. { 'split 'skrēn }

split-stator variable capacitor |ELECTR| Variable capacitor having a rotor section that is common to two separate stator sections; used in grid and plate tank circuits of transmitters for balancing purposes. { 'split ¦stād·ər 'ver·ē·ə·bəl kə'pas·əd·ər }

splitting |ELECTR| In the scope presentation of the standard loran (2000 kilohertz), signals the slow diminution of the leading or lagging edge of the pulse so that it resembles two pulses and eventually a single pulse, which appears to be normal but which may be displaced in time by as much as 10,000 microseconds; this phenomenon is caused by shifting of the E_1 reflections from the ionosphere, and if the deformation is that of the leading edge and is not detected, it will cause serious errors in the reading of the navigational parameter. { 'splid·iŋ }

split transducer |ENG| A directional transducer with electroacoustic transducing elements which are divided and arranged so that there is an electrical separation of each division. { 'split tranz'dü·sər }

split-word operation |COMPUT SCI| A computer operation performed with portions of computer words rather than whole words as is normally done. { 'split ¦wərd ‚äp·ə'rā·shən }

SPMD |COMPUT SCI| A type of programming on a multiprocessor in which parallel programs all run the same subroutine but operate on different data. Acronym for single-program, multiple-data.

spoiler |ELECTROMAG| Rod grating mounted on a parabolic reflector to change the pencil-beam pattern of the reflector to a cosecant-squared pattern; rotating the reflector and grating 90° with respect to the feed antenna changes one pattern to the other. { 'spȯi·lər }

spontaneous polarization |ELEC| Electric polarization that a substance possesses in the absence of an external electric field. { spän'tā·nē·əs ‚pō·lə·rə'zā·shən }

spoofing |COMPUT SCI| A method of gaining unauthorized access to computers or networkds by sending messages with someone else's IP address, so that the message appears, to the targeted system, to be coming from a trusted host. |ELECTR| Deceiving or misleading an enemy in electronic operations, as by continuing transmission on a frequency after it has been effectively jammed by the enemy, using decoy radar transmitters to lead the enemy into a useless jamming effort, or transmitting radio messages containing false information for intentional interception by the enemy. { 'spüf·iŋ }

spooling |COMPUT SCI| The temporary storage of input and output on high-speed input-output devices, typically magnetic disks and drums, in order to increase throughput. Acronym for simultaneous peripheral operations on line. { 'spül·iŋ }

sporadic fault |COMPUT SCI| A hardware malfunction that occurs intermittently and at unpredictable times. { spə'rad·ik 'fȯlt }

sporadic reflections |ELECTROMAG| Sharply defined reflections of substantial intensity from the sporadic E layer at frequencies greater than the critical frequency of the layer; they are variable with respect to time of occurrence, geographic location, and range of frequencies at which they are observed. { spə'rad·ik ri'flek·shənz }

spot |ELECTR| In a cathode-ray tube, the area instantaneously affected by the impact of an electron beam. { spät }

spot beam |COMMUN| A beam generated by a communications satellite antenna of sufficient size that the angular spread of energy in the beam is small, always smaller than the earth's angular beam width as seen from the satellite. { 'spät ‚bēm }

spot jammer |ELECTR| A jammer that interferes with reception of a specific channel or frequency. { 'spät ‚jam·ər }

spot jamming |ELECTR| An electronic attack technique in which a continuous narrow-band signal is transmitted, giving a stronger jamming signal to a particular victim radar than had a wide-band transmission been used. { 'spät ‚jam·iŋ }

spotlight |ELEC| **1.** A strong beam of light that illuminates only a small area about an object. **2.** A lamp that has a strongly focused beam. { 'spät‚līt }

spot noise figure |ELECTR| Of a transducer at a selected frequency, the ratio of the output noise power per unit bandwidth to a portion thereof attributable to the thermal noise in the input termination per unit bandwidth, the noise temperature of the input termination being standard (290 K). { 'spät 'nȯiz ‚fig·yər }

spot-size error |ELECTR| The distortion of the radar returns on the radarscope presentation caused by the diameter of the electron beam which displays the returns of the scope and the lateral radiation across the scope of part of the glow produced when the electron beam strikes the phosphorescent coating of the cathode-ray tube.   { 'spät ¦sīz ,er·ər }

spot speed |COMMUN| **1.** In a video system, the product of the length (in units of elemental area, that is, in spots) of scanning line by the number of scanning lines per second.   **2.** In facsimile transmission, the speed of the scanning or recording spot within the available line. Also known as scanning speed.   { 'spät ,spēd }

spottiness |ELECTR| Bright spots scattered irregularly over the reproduced image in a television receiver, due to man-made or static interference entering the television system at some point.   { 'späd·ē·nəs }

spray point |ELEC| One of the sharp points arranged in a row and charged to a high direct-current potential, used to charge and discharge the conveyor belt in a Van de Graaff generator.   { 'sprā ,point }

spread   See sensitivity.   { spred }

spreader |ELEC| An insulating crossarm used to hold apart the wires of a transmission line or multiple-wire antenna.   { 'spred·ər }

spreading method |ELEC| A method of calculating the potential due to a set of point charges by replacing them with a continuous distribution of charge or a distribution of charge and polarization.   { 'spred·iŋ ,meth·əd }

spreadsheet program |COMPUT SCI| A computer program that simulates an accountant's worksheet on screen as an array of rows (usually numbered) and columns (usually assigned alphabetical letters) whose intersections are called cells; the program allows the user to enter data in the cells and to embed formulas which relate the values in different cells.   { 'spred ,shēt 'prō·grəm }

spread spectrum transmission |ELECTR| Communications technique in which many different signal waveforms are transmitted in a wide band; power is spread thinly over the band so narrow-band radios can operate within the wide-band without interference; used to achieve security and privacy, prevent jamming, and utilize signals buried in noise.   { 'spred 'spek·trəm tranz,mish·ən }

spring contact |ELEC| A relay or switch contact mounted on a flat spring, usually of phosphor bronze.   { 'spriŋ ¦kän,takt }

sprocket pulse |COMPUT SCI| **1.** A pulse generated by a magnetized spot which accompanies every character recorded on magnetic tape; this pulse is used during read operations to regulate the timing of the read circuits, and also to provide a count on the number of characters read from the tape.   **2.** A pulse generated by the sprocket or driving hole in paper tape which serves as the timing pulse for reading or punching the paper tape.   { 'spräk·ət ,pəls }

SPST   See single-pole single-throw.

spurious emission   See spurious radiation.   { 'spyúr·ē·əs i'mish·ən }

spurious modulation |ELECTR| Undesired modulation occurring in an oscillator, such as

frequency modulation caused by mechanical vibration.   { 'spyúr·ē·əs ,mäj·ə'lā·shən }

spurious radiation |ELECTROMAG| Any emission from a radio transmitter at frequencies outside its frequency band. Also known as spurious emission.   { 'spyúr·ē·əs ,rād·ē'ā·shən }

spurious response |ELECTR| Response of a radio receiver to a frequency different from that to which the receiver is tuned.   { 'spyúr·ē·əs ri'späns }

spurt tone |COMMUN| Short audio-frequency tone used for signaling or dialing selection.   { 'spərt ,tōn }

sputtering |ELECTR| Also known as cathode sputtering. **1.** The ejection of atoms or groups of atoms from the surface of the cathode of a vacuum tube as the result of heavy-ion impact.   **2.** The use of this process to deposit a thin layer of metal on a glass, plastic, metal, or other surface in vacuum.   { 'spəd·ə·riŋ }

SQL   See Structured Query Language.

square-law demodulator   See square-law detector.   { 'skwer ¦lò dē'mäj·ə,lād·ər }

square-law detector |ELECTR| A demodulator whose output voltage is proportional to the square of the amplitude-modulated input voltage. Also known as square-law demodulator.   { 'skwer ¦lò di,tek·tər }

square wave |ELEC| An oscillation the amplitude of which shows periodic discontinuities between two values, remaining constant between jumps.   { 'skwer 'wāv }

square-wave amplifier |ELECTR| Resistance-coupled amplifier, the circuit constants of which are to amplify a square wave with the minimum amount of distortion.   { 'skwer ¦wāv 'am·plə,fī·ər }

square-wave generator |ELECTR| A signal generator that generates a square-wave output voltage.   { 'skwer ¦wāv 'jen·ə,rād·ər }

square-wave response |ELECTR| The response of a circuit or device when a square wave is applied to the input.   { 'skwer ¦wāv ri,späns }

squaring circuit |ELECTR| **1.** A circuit that reshapes a sine or other wave into a square wave.   **2.** A circuit that contains nonlinear elements proportional to the square of the input voltage.   { 'skwer·iŋ ,sər·kət }

squawker   See midrange.   { 'skwók·ər }

squealing |ELECTR| A condition in which a radio receiver produces a high-pitched note or squeal along with the desired radio program, due to interference between stations or to oscillation in some receiver circuit.   { 'skwēl·iŋ }

squeezable waveguide |ELECTROMAG| A waveguide whose dimensions can be altered periodically; used in rapid scanning.   { 'skwēz·ə·bəl 'wāv ,gīd }

squeeze section |ELECTROMAG| Length of waveguide constructed so that alteration of the critical dimension is possible with a corresponding alteration in the electrical length.   { 'skwēz ,sek·shən }

squegger   See blocking oscillator.   { 'skweg·ər }

squegging |ELECTR| Condition of self-blocking in an electron-tube-oscillator circuit.   { 'skweg·iŋ }

squegging oscillator   See blocking oscillator. { 'skweg·iŋ ,äs·ə,lād·ər }

squelch |ELECTR| To automatically quiet a receiver by reducing its gain in response to a specified characteristic of the input.   { skwelch }

squelch circuit   See noise suppressor.   { 'skwelch ,sər·kət }

SQUID   See superconducting quantum interference device.   { skwid }

squint |ELECTROMAG| **1.** The angle between the two major lobe axes in a radar lobe-switching antenna.   **2.** The angular difference between the axis of radar antenna radiation and a selected geometric axis, such as the axis of the reflector.   **3.** The angle between the full-right and full-left positions of the beam of a conical-scan radar antenna.   { skwint }

squirrel-cage motor |ELEC| An induction motor in which the secondary circuit consists of a squirrel-cage winding arranged in slots in the iron core.   { 'skwərl ¦kāj ,mōd·ər }

squirrel-cage rotor   See squirrel-cage winding.   { 'skwərl ¦kāj ,rōd·ər }

squirrel-cage winding |ELEC| A permanently short-circuited winding, usually uninsulated, around the periphery of the rotor and joined by continuous end rings. Also known as squirrel-cage rotor.   { 'skwərl ¦kāj ,wīnd·iŋ }

squishing   See compaction.   { 'skwish·iŋ }

squitter |ELECTR| Random firing, intentional or otherwise, of the transponder transmitter in the absence of interrogation.   { 'skwid·ər }

SRAM   See static random-access memory.   { 'es ,ram }

SRC   See stored response chain.

SSA Service   See S-band single-access service.   { ¦es¦es'ā ,sər·vəs }

SSB   See single-sideband.

SSI   See small-scale integration.

SSR   See solid-state relay.

SSTV   See slow-scan television.

stability |CONT SYS| The property of a system for which any bounded input signal results in a bounded output signal.   { stə'bil·əd·ē }

stability criterion |CONT SYS| A condition which is necessary and sufficient for a system to be stable, such as the Nyquist criterion, or the condition that poles of the system's overall transmittance lie in the left half of the complex-frequency plane.   { stə'bil·əd·ē krī,tir·ē·ən }

stability exchange principle |CONT SYS| In a linear system, which is either dynamically stable or unstable depending on the value of a parameter, the complex frequency varies with the parameter in such a way that its real and imaginary parts pass through zero simultaneously; the principle is often violated.   { stə'bil·əd·ē iks'chānj ,prin·sə·pəl }

stability factor |ELECTR| A measure of a transistor amplifier's bias stability, equal to the rate of change of collector current with respect to reverse saturation current.   { stə'bil·əd·ē ,fak·tər }

stabilivolt |ELECTR| Gas tube that maintains a constant voltage drop across its terminals, essentially independent of current, over a relatively wide range.   { stə'bil·ə,vōlt }

stabilization |CONT SYS| See compensation. |ELECTR| Feedback introduced into transistor amplifier stages to reduce distortion by making the amplification substantially independent of electrode voltages.   { ,stā·bə·lə'zā·shən }

stabilized feedback   See negative feedback.   { 'stā·bə,līzd 'fēd,bak }

stabilized winding |ELEC| Auxiliary winding used particularly in star-connected transformers to stabilize the neutral point of the fundamental frequency voltages, to protect the transformer and the system from excessive third-harmonic voltages; and to prevent telephone interference caused by third-harmonic currents and voltages in the lines and earth. Also known as tertiary winding.   { 'stā·bə,līzd 'wīnd·iŋ }

stabistor |ELECTR| A diode component having closely controlled conductance, controlled storage charge, and low leakage, as required for clippers, clamping circuits, bias regulators, and other logic circuits that require tight voltage-level tolerances.   { stā'bis·tər }

stable local oscillator   See stalo.   { 'stā·bəl 'lō·kəl 'äs·ə,lād·ər }

stable strobe |ELECTR| Series of strobes which behaves as if caused by a single jammer.   { 'stā·bəl 'strōb }

stack |COMPUT SCI| A portion of a computer memory used to temporarily hold information, organized as a linear list for which all insertions and deletions, and usually all accesses, are made at one end of the list.   { stak }

stack automaton |COMPUT SCI| A variation of a pushdown automaton in which the read-only head of the input tape is allowed to move both ways, and the read-write head on the pushdown storage is allowed to scan the entire pushdown list in a read-only mode.   { 'stak ȯ'täm·ə,tän }

stacked array |ELECTROMAG| An array in which the antenna elements are stacked one above the other and connected in phase to increase the gain.   { 'stakt ə'rā }

stacked-dipole antenna |ELECTROMAG| Antenna in which directivity is increased by providing a number of identical dipole elements, excited either directly or parasitically; the resultant radiation pattern depends on the number of dipole elements used, the spacing and phase difference between the elements, and the relative magnitudes of the currents.   { 'stakt 'dī,pōl an ,ten·ə }

stacked-job processing |COMPUT SCI| A technique of automatic job-to-job transition, with little or no operator intervention.   { 'stakt 'jäb ,prä,ses·iŋ }

stacked loops |ELECTROMAG| Two or more loop antennas arranged above each other on a vertical supporting structure and connected in phase to increase the gain. Also known as vertically stacked loops.   { 'stakt 'lüps }

stacking |ELECTROMAG| The placing of antennas one above the other, connecting them in phase to increase the gain.   { 'stak·iŋ }

stack model |COMPUT SCI| A model for describing the run-time execution of programs written in block-structured languages, consisting of a program component, which remains unchanged throughout the execution of the program; a control component, consisting of an instruction pointer and an environment pointer; and a stack of records containing all the data the program operates on. { 'stak ˌmäd·əl }

stack operation | COMPUT SCI | A computer system in which flags, return address, and all temporary addresses are saved in the core in sequential order for any interrupted routine so that a new routine (including the interrupted routine) may be called in. { 'stak ˌäp·ə‚rā·shən }

stack pointer | COMPUT SCI | A register which contains the last address of a stack of addresses. { 'stak ˌpȯint·ər }

stadiometry | COMPUT SCI | In computer vision, the determination of the distance to an object based on the size of its image. { ˌstād·ē'äm·ə·trē }

stage gain | ELECTR | The ratio of the output power of an amplifier stage to the input power, usually expressed in decibels. { 'stāj ˌgān }

stagger | COMMUN | Periodic error in the position of the recorded spot along a recorded facsimile line. { 'stag·ər }

staggered tuning | ELECTR | Alignment of successive tuned circuits to slightly different frequencies in order to widen the overall amplitude-frequency response curve. { 'stag·ərd 'tün·iŋ }

staggering | COMMUN | Offsetting of two channels of different carrier systems from exact sideband frequency coincidence to avoid mutual interference. { 'stag·ə·riŋ }

staggering advantage | COMMUN | Effective reduction of interference between carrier channels, due to staggering. { 'stag·ə·riŋ ad‚van·tij }

stagger-tuned amplifier | ELECTR | An amplifier that uses staggered tuning to give a wide bandwidth. { 'stag·ər ‚tünd 'am·plə‚fī·ər }

stagger-tuned filter | ELECTR | A filter consisting of a cascade of amplifier stages with tuned coupling networks whose resonant frequencies and bandwidths may be easily adjusted to achieve an overall transmission function of desired shape (maximally flat or equal ripple). { 'stag·ər ‚tünd 'fil·tər }

staging | COMPUT SCI | Moving blocks of data from one storage device to another. { 'stāj·iŋ }

staircase signal | COMMUN | In analog television transmissions, a waveform that consists of a series of discrete steps resembling a staircase. { 'ster‚kās ‚sig·nəl }

stake | ELEC | An iron peg used as a power electrode to transfer current into the ground in electrical prospecting. { stāk }

stale link | COMPUT SCI | A hyperlink to a document that has been erased or removed from the World Wide Web. Also known as black hole. { ‚stāl 'liŋk }

stalo | ELECTR | A highly stable local radio-frequency oscillator used in coherent radar both for up-converting the transmit signal to the carrier frequency and down-converting the received signals to the intermediate frequency. { 'stā‚lō }

stamping | ELECTR | A transformer lamination that has been cut out of a strip or sheet of metal by a punch press. { 'stam·piŋ }

stand-alone machine | COMPUT SCI | A machine capable of functioning independently of a master computer, either part of the time or all of the time. { 'stand ə‚lōn mə'shēn }

standard antenna | ELECTROMAG | An open single-wire antenna (including the lead-in wire) having an effective height of 4 meters. { 'stan·dərd an'ten·ə }

standard blocked F-format data set See FBS data set. { 'stan·dərd 'bläkt ‚ef'fȯr‚mat 'dad·ə ‚set }

standard broadcast band See broadcast band. { 'stan·dərd 'brȯd‚kast ‚band }

standard broadcast channel | COMMUN | Band of frequencies occupied by the carrier and two side bands of a radio broadcast signal, with the carrier frequency at the center. { 'stan·dərd 'brȯd‚kast ‚chan·əl }

standard broadcasting | COMMUN | Radio broadcasting using amplitude modulation in the band of frequencies from 535 to 1605 kilohertz; carrier frequencies are placed 10 kilohertz apart. { 'stan·dərd 'brȯd‚kast·iŋ }

standard capacitor | ELEC | A capacitor constructed in such a manner that its capacitance value is not likely to vary with temperature and is known to a high degree of accuracy. Also known as capacitance standard. { 'stan·dərd kə'pas·əd·ər }

standard cell | ELEC | A primary cell whose voltage is accurately known and remains sufficiently constant for instrument calibration purposes; the Weston standard cell has a voltage of 1.018636 volts at 20°C. { 'stan·dərd 'sel }

standard definition television | COMMUN | Term used to signify a digital television system in which the quality is approximately equivalent to that of NTSC. Also called standard digital television. Abbreviated SDTV. { 'stan·dərd def·ə¦nish·ən 'tel·ə‚vizh·ən }

standard digital television See standard definition television. { 'stan·dərd 'dij·əd·əl 'tel·ə‚vizh·ən }

standard form | COMPUT SCI | The form of a floating point number whose mantissa lies within a standard specified range of values. { 'stan·dərd 'fȯrm }

standard-frequency signal | COMMUN | One of the highly accurate signals broadcast by government radio stations and used for testing and calibrating radio equipment all over the world; in the United States signals are broadcast by the National Bureau of Standards' radio stations WWV, WWVH, WWVB, and WWVL. { 'stan·dərd ¦frē·kwən‚sē ‚sig·nəl }

standard function See built-in function. { 'stan·dərd 'fəŋk·shən }

Standard Generalized Markup Language | COMPUT SCI | A system that encodes the logical structure and content of a document rather than its display formatting, or even the medium in which the document will be displayed; widely used in the publishing business and for producing technical documentation. Abbreviated SGML. { ¦stan·dərd ‚jen·rə‚līzd 'märk‚əp ‚laŋ·gwij }

standard interface |COMPUT SCI| **1.** A joining place of two systems or subsystems that has a previously agreed-upon form, so that two systems may be readily connected together. **2.** In particular, a system of uniform circuits and input/output channels connecting the central processing unit of a computer with various units of peripheral equipment. { 'stan·dərd 'in·tər,fās }

standardize |COMPUT SCI| To replace any given floating point representation of a number with its representation in standard form; that is, to adjust the exponent and fixed-point part so that the new fixed-point part lies within a prescribed standard range. { 'stan·dər,dīz }

standard noise temperature |ELECTR| The standard reference temperature for noise measurements, equal to 290 K. { 'stan·dərd 'nȯiz ,tem·prə·chər }

standard parallel port |COMPUT SCI| A parallel port that can transfer data in only one direction. { ¦stan·dərd ,par·ə,lel 'pȯrt }

standard preemphasis |COMMUN| Preemphasis in frequency-modulation and analog television aural broadcasting whose level lies between upper and lower limits specified by the Federal Communications Commission. { 'stan·dərd prē'em·fə·səs }

standard propagation |ELECTROMAG| Propagation of radio waves over a smooth spherical earth of specified dielectric constant and conductivity, under conditions of standard refraction in the atmosphere. { 'stan·dərd ,präp·ə'gā·shən }

standard refraction |ELECTROMAG| Refraction which would occur in an idealized atmosphere in which the index of refraction decreases uniformly with height at a rate of 39×10^{-6} per kilometer; standard refraction may be included in ground wave calculations by use of an effective earth radius of 8.5×10^6 meters, or $\frac{4}{3}$ the geometrical radius of the earth. { 'stan·dərd ri'frak·shən }

standard subroutine |COMPUT SCI| In computers, a subroutine which is applicable to a class of problems. { 'stan·dərd 'səb·rü,tēn }

standard test-tone power |ELECTR| One milliwatt (0 decibels above one milliwatt) at 1000 hertz. { 'stan·dərd 'test ¦tōn ,paü·ər }

standby battery |ELEC| A storage battery held in reserve as an emergency power source in event of failure of regular power facilities at a radio station or other location. { 'stand¦bī ,bad·ə·rē }

standby computer |COMPUT SCI| A computer in a duplex system that takes over when the need arises. { 'stand¦bī kəm,pyüd·ər }

standby mode |ELEC| The operation of a circuit or device with unused portions of the circuit disconnected to reduce power consumption. { 'stan,bī mōd }

standby power source |ELEC| An uninterruptible power system in which the load normally operated from the commercial power line is switched to the output of a dc-to-ac static inverter powered by a battery in the event of a power failure. { 'stand¦bī 'paü·ər ,sȯrs }

standby register |COMPUT SCI| In computers, a register into which information can be copied to be

available in case the original information is lost or mutilated in processing. { 'stand¦bī ,rej·ə·stər }

standby replacement redundancy |COMPUT SCI| A form of redundancy in which there is a single active unit and a reserve of spare units, one of which replaces the active unit if it fails. { 'stand ¦bī ri'plās·mənt ri,dən·dən·sē }

standby time |COMPUT SCI| **1.** The time during which two or more computers are tied together and available to answer inquiries or process intermittent actions on stored data. **2.** The elapsed time between inquiries when the equipment is operating on an inquiry application. { 'stand¦bī ,tīm }

standing-on-nines carry |COMPUT SCI| In high-speed parallel addition of decimal numbers, an arrangement that causes carry digits to pass through one or more nine digits, while signaling that the skipped nines are to be reset to zero. { ¦stand·iŋ ȯn ¦nīnz 'kar·ē }

standing wave |PHYS| A wave in which the ratio of an instantaneous value at one point to that at any other point does not vary with time. Also known as stationary wave. { 'stand·iŋ 'wāv }

standing-wave loss factor |ELECTROMAG| The ratio of the transmission loss in an unmatched waveguide to that in the same waveguide when matched. { 'stand·iŋ ¦wāv 'lȯs ,fak·tər }

standoff insulator |ELEC| An insulator used to support a conductor at a distance from the surface on which the insulator is mounted. { 'stan,dȯf 'in·sə,lād·ər }

standoff jammer |ELECTR| An aircraft that patrols the target air space and engages in high-power jamming of both the acquisition or tracking devices and the closing vehicles, by using powerful transmitters excited by travelling-wave tubes. { 'stan,dȯf 'jam·ər }

standstill feature |CONT SYS| A device which insures that false signals such as fluctuations in the power supply do not cause a controller to be altered. { 'stan,stil ,fē·chər }

star-connected circuit |ELEC| Polyphase circuit in which all the current paths within the region that delimits the circuit extend from each of the points of entry of the phase conductors to a common conductor (which may be the neutral conductor). { 'stär kə¦nek·təd 'sər·kət }

star-delta switching starter |ELEC| A type of motor starter, used with three-phase induction motors, that switches the stator windings from a star connection to a delta connection. { 'stär 'del·tə ¦swich·iŋ ,stärd·ər }

star-free expression |COMPUT SCI| An expression containing only Boolean operations and concatenation, used to define the language corresponding to a counter-free machine. { 'stär ¦frē ik'spresh·ən }

star lamp |ELEC| A high-pressure xenon arc, used in a planetarium, which produces a tiny, intense point of light focused through thousands of individual lenses and pinholes, and projected to the planetarium's dome. { 'stär ,lamp }

star network |COMMUN| A communications network in which all communications between any two points must pass through a central node. Also known as centralized configuration. { 'stär ¦net,wȯrk }

start bit |COMPUT SCI| The first bit transmitted in asynchronous data transmission to unequivocally indicate the start of the word. { 'stärt ,bit }

start codes |COMMUN| 32-bit codes embedded in the coded bit stream that are unique; used for several purposes including identifying some of the layers in the coding syntax. { 'stärt ,kōdz }

start dialing signal |COMMUN| Signal transmitted from the incoming end of a circuit, following the receipt of a seizing signal, to indicate that the necessary circuit conditions have been established for receiving the numerical routine information. { 'stärt 'dīl·iŋ ,sig·nəl }

started task |COMPUT SCI| A computer program that is kept permanently in main storage and, though not a part of the operating system, is treated as though it were. { 'stärd·əd 'task }

start element |COMMUN| The first element of a character in certain serial transmissions, used to permit synchronization. { 'stärt ,el·ə·mənt }

starter |ELEC| 1. A device used to start an electric motor and to accelerate the motor to normal speed. 2. *See* engine starter. |ELECTR| An auxiliary control electrode used in a gas tube to establish sufficient ionization to reduce the anode breakdown voltage. Also known as trigger electrode. { 'stär·dər }

starting box |ELEC| A device for providing extra resistance in the armature of a motor while it is being started. { 'stärd·iŋ ,bäks }

starting motor *See* engine starter. { 'stärd·iŋ ,mōd·ər }

starting reactor |ELEC| A reactor that is used to limit the starting current of electric motors, and usually consists of an iron-core inductor connected in series with the machine stator winding. { 'stärd·iŋ rē,ak·tər }

startover |COMPUT SCI| Program function that causes a computer that is not active to become active. { 'stär,dō·vər }

startover data transfer and processing program |COMPUT SCI| Program which controls the transfer of startover data from the active to the standby machine and their subsequent processing by the standby machine. { 'stär,dō·vər 'dad·ə ,tranz·fər ən 'prä,ses·iŋ ,prō,gram }

start-stop multivibrator *See* monostable multivibrator. { 'stärt 'stäp ,məl·ti'vī,brād·ər }

start-stop printing telegraph |COMMUN| Form of printing telegraph in which the signal-receiving mechanisms, normally at rest, are started in operation at the beginning and stopped at the end of each character transmitted over the channel. { 'stärt 'stäp 'print·iŋ 'tel·ə·graf }

start-stop system |COMMUN| A telegraph system in which each group of code elements corresponding to a character is preceded by a start signal that prepares the receiving mechanism to receive and register a character, and is followed by a stop signal that brings the receiving mechanism to rest in preparation for the reception of the next character. { 'stärt 'stäp ,sis·təm }

stat- |ELEC| A prefix indicating an electrical unit in the electrostatic centimeter-gram-second system of units; it is attached to the corresponding SI unit. { stat }

stat℧ *See* statmho.

statΩ *See* statohm.

statA *See* statampere. { 'stat¦ā }

statampere |ELEC| The unit of electric current in the electrostatic centimeter-gram-second system of units, equal to a flow of charge of 1 statcoulomb per second; equal to approximately 3.3356×10^{-10} ampere. Abbreviated statA. { stad'am,pir }

statC *See* statcoulomb. { 'stat,sē }

statcoulomb |ELEC| The unit of charge in the electrostatic centimeter-gram-second system of units, equal to the charge which exerts a force of 1 dyne on an equal charge at a distance of 1 centimeter in a vacuum; equal to approximately 3.3356×10^{-10} coulomb. Abbreviated statC. Also known as franklin (Fr); unit charge. { 'stat¦kü ,läm }

state |CONT SYS| A minimum set of numbers which contain enough information about a system's history to enable its future behavior to be computed. { stāt }

state equations |CONT SYS| Equations which express the state of a system and the output of a system at any time as a single valued function of the system's input at the same time and the state of the system at some fixed initial time. { 'stāt i,kwā·zhənz }

state estimator *See* observer. { 'stāt ,es·tə,mād·ər }

state feedback |CONT SYS| A class of feedback control laws in which the control inputs are explicit memoryless functions of the dynamical system state, that is, the control inputs at a given time t_a are determined by the values of the state variables at t_a and do not depend on the values of these variables at earlier times $t \geq t_a$. { 'stāt 'fēd,bak }

state graph |COMPUT SCI| A directed graph whose nodes correspond to internal states of a sequential machine and whose edges correspond to transitions among these states. { 'stāt ,graf }

statement |COMPUT SCI| An elementary specification of a computer action or process, complete and not divisible into smaller meaningful units; it is analogous to the simple sentence of a natural language. { 'stāt·mənt }

statement editor |COMPUT SCI| A text editor in which the text is divided into superlines, that is, units greater than ordinary lines, resulting in easier editing and freedom from truncation problems. { 'stāt·mənt ,ed·əd·ər }

state observer *See* observer. { 'stāt əb,zər·vər }

state space |CONT SYS| The set of all possible values of the state vector of a system. { 'stāt ,spās }

state table |COMPUT SCI| A table that represents a sequential machine, in which the rows correspond to the internal states, the columns to the input combinations, and the entries to the next state. { 'stāt ,tā·bəl }

state transition equation [CONT SYS] The equation satisfied by the $n \times n$ state transition matrix $\Phi(t,t_0)$: $\partial\Phi(t,t_0)/\partial t = A(t)\,\Phi(t,t_0)$, $\Phi(t_0,t_0) = I$; here I is the unit $n \times n$ matrix, and $A(t)$ is the $n \times n$ matrix which appears in the vector differential equation $dx(t)/dt = A(t)x(t)$ for the n-component state vector $x(t)$. { 'stāt tran'zish·ən i‚kwā·zhən }

state transition matrix [CONT SYS] A matrix $\Phi(t,t_0)$ whose product with the state vector x at an initial time t_0 gives the state vector at a later time t; that is, $x(t) = \Phi(t,t_0)x(t_0)$. { 'stāt tran'zish·ən ‚mā·triks }

state variable [CONT SYS] One of a minimum set of numbers which contain enough information about a system's history to enable computation of its future behavior. { 'stāt ‚ver·ē·ə·bəl }

state-variable filter [ELECTR] A multiple-amplifier active filter that has three outputs for highpass, band-pass, and low-pass transfer functions respectively. Also known as KHN filter. { 'stāt ‚ver·ē·ə·bəl ‚fil·tər }

state vector [COMPUT SCI] See task descriptor. [CONT SYS] A column vector whose components are the state variables of a system. { 'stāt ‚vek·tər }

statF See statfarad. { 'stad‚ef }

statfarad [ELEC] Unit of capacitance in the electrostatic centimeter-gram-second system of units, equal to the capacitance of a capacitor having a charge of 1 statcoulomb, across the plates of which the charge is 1 statvolt; equal to approximately 1.1126×10^{-12} farad. Abbreviated statF. { 'stat¦fa‚rad }

static [COMMUN] A hissing, crackling, or other sudden sharp sound that tends to interfere with the reception, utilization, or enjoyment of desired signals or sounds. { 'stad·ik }

static algorithm [COMPUT SCI] An algorithm whose operation is known in advance. Also known as deterministic algorithm. { 'stad·ik 'al‚gə‚rith·əm }

static breeze See convective discharge. { 'stad·ik 'brēz }

static characteristic [ELECTR] A relation between a pair of variables, such as electrode voltage and electrode current, with all other operating voltages for an electron tube, transistor, or other amplifying device maintained constant. { 'stad·ik ‚kar·ik·tə'ris·tik }

static charge [ELEC] An electric charge accumulated on an object. { 'stad·ik 'chärj }

static check [COMPUT SCI] Of a computer, one or more tests of computing elements, their interconnections, or both, performed under static conditions. { 'stad·ik 'chek }

static debugging routine [COMPUT SCI] A debugging routine which is used after the program being checked has been run and has stopped. { 'stad·ik dē'bəg·iŋ rü‚tēn }

static discharger [ELEC] A rubber-covered cloth wick about 6 inches (15 centimeters) long, sometimes attached to the trailing edges of the surfaces of an aircraft to discharge static electricity in flight. { 'stad·ik 'dis‚chär·jər }

static dump [COMPUT SCI] An edited printout of the contents of main memory or of the auxiliary

storage, performed in a fixed way; it is usually taken at the end of a program run either automatically or by operator intervention. { 'stad·ik 'dəmp }

static electricity [ELEC] **1.** The study of the effects of macroscopic charges, including the transfer of a static charge from one object to another by actual contact or by means of a spark that bridges an air gap between the objects. **2.** See electrostatics. { 'stad·ik ‚i‚lek'tris·əd·ē }

static eliminator [ELECTR] Device intended to reduce the effect of atmospheric static interference in a radio receiver. { 'stad·ik i‚lim·ə‚nād·ər }

static induction transistor [ELECTR] A type of transistor capable of operating at high current and voltage, whose current-voltage characteristics do not saturate, and are similar in form to those of a vacuum triode. Abbreviated SIT. { 'stad·ik in'dək·shən tran‚zis·tər }

static inverter [ELEC] A device that converts a dc voltage to a stable ac voltage for use in an uninterruptible power system. { 'stad·ik in'vərd·ər }

staticize [COMPUT SCI] **1.** To capture transient data in stable form, thus converting fleeting events into examinable information. **2.** To extract an instruction from the main computer memory and store the various component parts of it in the appropriate registers, preparatory to interpreting and executing it. { 'stad·ə‚sīz }

static machine [ELEC] A machine for generating electric charges, usually by electric induction, sometimes used to build up high voltages for research purposes. { 'stad·ik mə‚shēn }

static random-access memory [COMPUT SCI] A read-write random-access memory that uses either four transistors and two resistors to form a passive-load flip-flop, or six transistors to form a flip-flop with dynamic loads, for each cell in an array. Once data are loaded into the flip-flop storage elements, the flip-flop will indefinitely remain in that state until the information is intentionally changed or the power to the memory circuit is shut off. Abbreviated SRAM. { 'stad·ik 'rand·əm ¦ak‚ses 'mem·rē }

static reactive compensator [ELEC] A thyristor-controlled generator of reactive power that is used to compensate for reactive power in an electric power system in order to limit voltage variations. Also known as static var compensator. { 'stad·ik rē¦ak·tiv ‚käm·pən'sād·ər }

static regulator [ELECTR] Transmission regulator in which the adjusting mechanism is in self-equilibrium at any setting and requires control power to change the setting. { 'stad·ik ‚reg·yə‚lād·ər }

static sensitivity [ELECTR] In phototubes, quotient of the direct anode current divided by the incident radiant flux of constant value. { 'stad·ik ‚sen·sə'tiv·əd·ē }

static storage [COMPUT SCI] Computer storage such that information is fixed in space and available at any time, as in flip-flop circuits, electrostatic memories, and coincident-current magnetic-core storage. { 'stad·ik 'stór·ij }

static subroutine [COMPUT SCI] In computers, a subroutine which involves no parameters other than the addresses of the operands. { 'stad·ik 'səb·rü,tēn }

static switching [ELEC] Switching of circuits by means of magnetic amplifiers, semiconductors, and other devices that have no moving parts. { 'stad·ik 'swich·iŋ }

static var compensator See static reactive compensator. { 'stad·ik 'vär ,käm·pən,sād·ər }

static variable [COMPUT SCI] A local variable that does not cease to exist upon termination of the block in which it can be accessed, but instead retains its most recent value until the next execution of this block. { 'stad·ik 'ver·ē·ə·bəl }

station [COMMUN] See broadcast station. [COMPUT SCI] One of a series of essentially similar positions or facilities occurring in a data-processing system. [ELEC] An assembly line or assembly machine location at which a wiring board or chassis is stopped for insertion of one or more parts. [ELECTR] A location at which radio, television, radar, or other electric equipment is installed. { stā·shən }

stationary ergodic noise [ELECTR] A stationary noise for which the probability that the noise voltage lies within any given interval at any time is nearly equal to the fraction of time that the noise voltage lies within this interval if a sufficiently long observation interval is recorded. { 'stā·sha ,ner·ē ər'gäd·ik 'nȯiz }

stationary noise [ELECTR] A random noise for which the probability that the noise voltage lies within any given interval does not change with time. { 'stā·sha,ner·ē 'nȯiz }

stationary wave See standing wave. { 'stā·sha ,ner·ē 'wāv }

station authentication [COMMUN] Security measure designed to establish the authenticity of a transmitting or receiving station. { 'stā·shən ȯ,then·tə'kā·shən }

statistical monitor [COMPUT SCI] A software monitor that collects information by periodically sampling activity in the system. { stə'tis·tə·kəl 'män·əd·ər }

statistical multiplexer [ELECTR] A device which combines several low-speed communications channels into a single high-speed channel, and which can manage more communications traffic than a standard multiplexer by analyzing traffic and choosing different transmission patterns. { stə'tis·tə·kəl 'məl·tə,plek·sər }

statistical multiplexing [COMMUN] Time-division multiplexing in which time on a communications channel is assigned to multiple users on a demand basis, rather than periodically to each user. { stə,tis·ti·kəl 'məl·tə,pleks·iŋ }

statmho [ELEC] The unit of conductance, admittance, and susceptance in the electrostatic centimeter-gram-second system of units, equal to the conductance between two points of a conductor when a constant potential difference of 1 statvolt applied between the points produces in this conductor a current of 1 statampere, the conductor not being the source of any electro-motive force; equal to approximately 1.1126×10^{-12} mho. Abbreviated stat℧. Also known as statsiemens (statS). { 'stat,mō }

statohm [ELEC] The unit of resistance, reactance, and impedance in the electrostatic centimeter-gram-second system of units, equal to the resistance between two points of a conductor when a constant potential difference of 1 statvolt between these points produces a current of 1 statampere; it is equal to approximately 8.9876×10^{11} ohms. Abbreviated statΩ. { 'stad,ōm }

stator [ELEC] The portion of a rotating machine that contains the stationary parts of the magnetic circuit and their associated windings. { 'stād·ər }

stator armature [ELEC] A stator which includes the main current-carrying winding in which electromotive force produced by magnetic flux rotation is induced; it is found in most alternating-current machines. { 'stād·ər 'är·mə·chər }

stator plate [ELEC] One of the fixed plates in a variable capacitor; stator plates are generally insulated from the frame of the capacitor. { 'stād·ər ,plāt }

statS See statmho. { ¦stat'es }

statsiemens See statmho. { ¦stat'sē·mənz }

status byte [COMPUT SCI] A byte of storage whose contents indicate the activities currently taking place in some part of the computer or various conditions governing the execution of a computer program; often, each bit is assigned a particular meaning. { 'stad·əs ,bīt }

status check [COMPUT SCI] The detection of software failures and verification of programs through the use of redundant computers. { 'stad·əs ,chek }

status line [COMPUT SCI] A conductor on the bus of a computer over which an addressed storage location or component transmits its status to the central processing unit. { 'stad·əs ,līn }

status register [COMPUT SCI] A register maintained by the central processing unit that contains a status byte with information about activities currently taking place there. { 'stad·əs ,rej·ə·stər }

status word [COMPUT SCI] A word indicating the state of the system or the diagnosis of a state into which the system has entered. { 'stad·əs ,wərd }

statV See statvolt.

statvolt [ELEC] The unit of electric potential and electromotive force in the electrostatic centimeter-gram-second system of units, equal to the potential difference between two points such that the work required to transport 1 statcoulomb of electric charge from one to the other is equal to 1 erg; equal to approximately 299.79 volts. Abbreviated statV. { 'stat,vōlt }

STD See system target decoder.

STD input buffer [COMMUN] A first-in, first-out buffer at the input of a system target decoder for storage of compressed data from elementary streams before decoding. { ¦es¦tē¦dē 'in,pút ,bəf·ər }

STDM See synchronous time-division multiplexing.

steady-state current [ELEC] An electric current that does not change with time. { 'sted·ē ¦stāt 'kə·rənt }

steady-state error |CONT SYS| The error that remains after transient conditions have disappeared in a control system. { 'sted·ē ¦stāt 'er·ər }

steam-electric generator |ELEC| An electric generator driven by a steam turbine. { ¦stēm i¦lek·trik 'jen·ə,rād·ər }

steerable antenna |ELECTROMAG| A directional antenna whose major lobe can be readily shifted in direction. { 'stir·ə·bəl an'ten·ə }

steganography |COMPUT SCI| The art and science of hiding a message in a medium, such as a digital picture or audio file, so as to defy detection. { ,steg·ə'näg·rə·fē }

STEM See scanning transmission electron microscope. { stem }

stenode circuit |ELECTR| Superheterodyne receiving circuit in which a piezoelectric unit is used in the intermediate-frequency amplifier to balance out all frequencies except signals at the crystal frequency, thereby giving very high selectivity. { 'ste,nōd ,sər·kət }

step |COMPUT SCI| A single computer instruction or operation. { step }

step angle |ELEC| The angle between two successive positions of a stepping motor. { 'step ,aŋ·gəl }

step attenuator |ELECTR| An attenuator in which the attenuation can be varied in precisely known steps by means of switches. { 'step ə,ten·yə ,wād·ər }

step-by-step operation See single-step operation. { ¦step bī ¦step ,äp·ə'rā·shən }

step-by-step switch |ELEC| A bank-and-wiper switch in which the wipers are moved by electromagnet ratchet mechanisms individual to each switch. { ¦step bī ¦step 'swich }

step-by-step system |COMMUN| See Strowger system. |CONT SYS| A control system in which the drive motor moves in discrete steps when the input element is moved continuously. { ¦step bī ¦step 'sis·təm }

step change |ELECTR| The change of a variable from one value to another in a single process, taking a negligible amount of time. { 'step ,chānj }

step counter |COMPUT SCI| In computers, a counter in the arithmetic unit used to count the steps in multiplication, division, and shift operations. { 'step ,kaunt·ər }

step-down transformer |ELEC| A transformer in which the alternating-current voltages of the secondary windings are lower than those applied to the primary winding. { 'step ¦daun tranz'fȯr·mər }

step-function generator |ELECTR| A function generator whose output waveform increases and decreases suddenly in steps that may or may not be equal in amplitude. { 'step ¦fəŋk·shən 'jen·ə,rād·ər }

stepped-wave static inverter |ELEC| A static inverter that generates several pulses in each half cycle and combines them to achieve an output voltage which needs very little filtering. { 'stept ¦wāv 'stad·ik in'vərd·ər }

stepper motor |ELEC| A motor that rotates in short and essentially uniform angular movements rather than continuously; typical steps are

30, 45, and 90°; the angular steps are obtained electromagnetically rather than by the ratchet and pawl mechanisms of stepping relays. Also known as magnetic stepping motor; stepping motor; step-servo motor. { 'step·ər ,mōd·ər }

stepping See zoning. { 'step·iŋ }

stepping motor See stepper motor. { 'step·iŋ ,mōd·ər }

stepping relay |ELEC| A relay whose contact arm may rotate through 360° but not in one operation. Also known as rotary stepping relay; rotary stepping switch; stepping switch. { 'step·iŋ ,rē ,lā }

stepping switch See stepping relay. { 'step·iŋ ,swich }

step-recovery diode |ELECTR| A varactor in which forward voltage injects carriers across the junction, but before the carriers can combine, voltage reverses and carriers return to their origin in a group; the result is abrupt cessation of reverse current and a harmonic-rich waveform. { 'step ri¦kəv·rē 'dī,ōd }

step response |CONT SYS| The behavior of a system when its input signal is zero before a certain time and is equal to a constant nonzero value after this time. { 'step ri,späns }

step-servo motor See stepper motor. { 'step 'sər·vō ,mōd·ər }

step strobe marker |ELECTR| Form of strobe marker in which the discontinuity is in the form of a step in the time base. { 'step 'strōb ,mär·kər }

step-up transformer |ELEC| Transformer in which the energy transfer is from a low-voltage winding to a high-voltage winding or windings. { 'step¦əp tranz,fȯr·mər }

step voltage regulator |ELEC| A type of voltage regulator used on distribution feeder lines; it provides increments or steps of voltage change. { 'step 'vōl·tij ,reg·yə,lād·ər }

sterba curtain |ELECTROMAG| Type of stacked dipole antenna array consisting of one or more phased half-wave sections with a quarter-wave section at each end; the array can be oriented for either vertical or horizontal radiation, and can be either center or end fed. { 'stər·bə ,kərt·ən }

stereo See stereophonic; stereo sound system. { 'ste·rē·ō }

stereo amplifier |ENG ACOUS| An audio-frequency amplifier having two or more channels, as required for use in a stereo sound system. { 'ste·rē·ō 'am·plə ,fī·ər }

stereo broadcasting |COMMUN| Broadcasting two sound channels for reproduction by a stereo sound system having a stereo tuner at its input, to afford a listener a sense of the spatial distribution of the sound sources. { 'ster·ē·ō 'brȯd,kast·iŋ }

stereofluoroscopy |ELECTR| A fluoroscopic technique that gives three-dimensional images. { ¦ster·ē·ə·flü'räs·kə·pē }

stereo multiplex |COMMUN| Stereo broadcasting by a frequency-modulation station, in which the outputs of two channels are transmitted on the same carrier by frequency-division multiplexing. { 'ster·ē·ō 'məl·tə,pleks }

stereophonic [ENG ACOUS] Pertaining to three-dimensional pickup or reproduction of sound, as achieved by using two or more separate audio channels. Also known as stereo. { ¦ster·ē·ə 'fän·ik }

stereophonics [ENG ACOUS] The study of reproducing or reinforcing sound in such a way as to produce the sensation that the sound is coming from sources whose spatial distribution is similar to that of the original sound sources. { ¦ster·ē·ə'fän·iks }

stereophonic sound system See stereo sound system. { ¦ster·ē·ə'fan·ik 'saúnd ˌsis·təm }

stereo preamplifier [ENG ACOUS] An audio-frequency preamplifier having two channels, used in a stereo sound system. { 'ster·ē·ō ¦prē'am·plə,fī·ər }

stereo recorded tape [ENG ACOUS] Recorded magnetic tape having two separate recordings, one for each channel of a stereo sound system. { 'ster·ē·ō ri¦kȯrd·əd 'tāp }

stereo sound system [ENG ACOUS] A sound reproducing system in which a stereo pickup, stereo tape recorder, stereo tuner, or stereo microphone system feeds two independent audio channels, each of which terminates in one or more loudspeakers arranged to give listeners the same audio perspective that they would get at the original sound source. Also known as stereo; stereophonic sound system. { 'ster·ē·ō 'saúnd ˌsis·təm }

stereo subcarrier [COMMUN] A subcarrier whose frequency is the second harmonic of the pilot subcarrier frequency used in frequency-modulation stereo broadcasting. { 'ster·ē·ō ¦səb'kar·ē·ə·r }

stereo tape recorder [ENG ACOUS] A magnetic-tape recorder having two stacked playback heads, used for reproduction of stereo recorded tape. { 'ster·ē·ō 'tāp ri,kȯrd·ər }

stereo tuner [ENG ACOUS] A tuner having provisions for receiving both channels of a stereo broadcast. { 'ster·ē·ō 'tün·ər }

sticking [COMPUT SCI] In computers, the tendency of a flip-flop to remain in, or to spontaneously switch to, one of its two stable states. { 'stik·iŋ }

stigmator [ELECTR] A device that corrects asymmetries in an electron lens by superposing on the field of the lens a second adjustable field. { stig'mäd·ər }

stiletto [ELECTR] An advanced electronic subsystem contained in United States strike aircraft type F-4D for detection, identification, and location of ground-based radars; the location of radar targets is determined by direction finding and passive ranging techniques; it is used for the delivery of guided and unguided weapons against the target radars under all weather conditions. { stə'led·ō }

stimulated-emission device [ELECTR] A device that uses the principle of amplification of electromagnetic waves by stimulated emission, namely, a maser or a laser. { 'stim·yə,lād·əd i'mish·ən di ,vīs }

stimulus [CONT SYS] A signal that affects the controlled variable in a control system. { 'stim·yə·ləs }

STN LCD See supertwisted nematic liquid-crystal display.

stochastic automaton See probabilistic automaton. { stō'kas·tik ȯ'täm·ə,tän }

stochastic control theory [CONT SYS] A branch of control theory that aims at predicting and minimizing the magnitudes and limits of the random deviations of a control system through optimizing the design of the controller. { stō'kas·tik kən'trōl ,thē·ə·rē }

stochastic sequential machine See probabilistic sequential machine. { stō'kas·tik si'kwen·chəl mə'shēn }

stop [CONT SYS] A bound or final position of a robot's movement. { stäp }

stop band See rejection band. { 'stäp ,band }

stop bits [COMPUT SCI] The last two bits transmitted in asynchronous data transmission to unequivocally indicate the end of a word. { 'stäp ,bits }

stop code [COMPUT SCI] A character that is placed in a storage medium and, when encountered, causes the computer system to cease processing until it is directed to continue. { 'stäp ,kȯd }

stop element [COMMUN] The last element of a character in certain serial transmissions, used to ensure the recognition of the next start element. { 'stäp ,el·ə·mənt }

stop instruction [COMPUT SCI] An instruction in a computer program that causes execution of the program to stop. { 'stäp in,strək·shən }

stoplight [ELEC] One of the lights that are installed at the rear of an automotive vehicle and are automatically turned on when the driver applies the brakes. { 'stäp,līt }

stop loop See loop stop. { 'stäp ,lüp }

stopping capacitor See coupling capacitor. { 'stäp·iŋ kə,pas·əd·ər }

stopping potential [ELECTR] Voltage required to stop the outward movement of electrons emitted by photoelectric or thermionic action. { 'stäp·iŋ pə,ten·chəl }

stop signal [COMMUN] Signal that initiates the transfer of facsimile equipment from active to standby conditions. { 'stäp ,sig·nəl }

storage [COMPUT SCI] Any device that can accept, retain, and read back one or more times; the means of storing data may be chemical, electrical, magnetic, mechanical, or sonic. { 'stȯr·ij }

storage address register [COMPUT SCI] A register used to hold the address of a location in storage containing data that is being processed. { 'stȯr·ij ¦ad,res ,rej·ə·stər }

storage allocation [COMPUT SCI] The process of assigning storage locations to data or instructions in a digital computer. { 'stȯr·ij ,al·ə'kā·shən }

storage and retrieval system [COMPUT SCI] An organized method of putting items away in a manner which permits their recall or retrieval from storage. Also known as storetrieval system. { 'stȯr·ij ən ri'trē·vəl ,sis·təm }

storage area [COMPUT SCI] A specified set of locations in a storage unit. Also known as zone. { 'stȯr·ij ,er·ē·ə }

storage battery [ELEC] A connected group of two or more storage cells or a single storage cell. Also known as accumulator; accumulator battery; rechargeable battery; secondary battery. { 'stȯr·ij ˌbad·ə·rē }

storage block [COMPUT SCI] A contiguous area of storage whose contents can be handled in a single operation. { 'stȯr·ij ˌbläk }

storage buffer register [COMPUT SCI] A register used in some microcomputers during input or output operations to temporarily hold a copy of the contents of a storage location. { 'stȯr·ij ¦bəf·ər ˌrej·ə·stər }

storage calorifier See cylinder. { 'stȯr·ij kə'lȯr·ə ˌfī·ər }

storage camera See iconoscope. { 'stȯr·ij ˌkam·rə }

storage capacity [COMPUT SCI] The quantity of data that can be retained simultaneously in a storage device; usually measured in bits, digits, characters, bytes, or words. Also known as capacity; memory capacity. { 'stȯr·ij kə,pas·əd·ē }

storage cell [COMPUT SCI] An elementary (logically indivisible) unit of storage; the storage cell can contain one bit, character, byte, digit (or sometimes word) of data. [ELEC] An electrolytic cell for generating electric energy, in which the cell after being discharged may be restored to a charged condition by sending a current through it in a direction opposite to that of the discharging current. Also known as secondary cell. { 'stȯr·ij ˌsel }

storage compacting [COMPUT SCI] The practice, followed on multiprogramming computers which use dynamic allocation, of assigning and reassigning programs so that the largest possible area of adjacent locations remains available for new programs. { 'stȯr·ij kəm,pakt·iŋ }

storage cycle [COMPUT SCI] **1.** Periodic sequence of events occurring when information is transferred to or from the storage device of a computer. **2.** Storing, sensing, and regeneration from parts of the storage sequence. { 'stȯr·ij ˌsī·kəl }

storage cycle time [COMPUT SCI] The time required to read and restore one word from a computer storage, or to write one word in computer storage. { 'stȯr·ij ˌsī·kəl ˌtīm }

storage density [COMPUT SCI] The number of characters stored per unit-length of area of storage medium (for example, number of characters per inch of magnetic tape). { 'stȯr·ij ˌden·səd·ē }

storage device [COMPUT SCI] A mechanism for performing the function of data storage: accepting, retaining, and emitting (unchanged) data items. Also known as computer storage device. { 'stȯr·ij di,vīs }

storage dump [COMPUT SCI] A printout of the contents of all or part of a computer storage. Also known as memory dump; memory print. { 'stȯr·ij ˌdəmp }

storage element [COMPUT SCI] Smallest part of a digital computer storage used for storing a single bit. { 'stȯr·ij ˌel·ə·mənt }

storage factor See Q. { 'stȯr·ij ˌfak·tər }

storage fill [COMPUT SCI] Storing a pattern of characters in areas of a computer storage that are not intended for use in a particular machine run; these characters cause the machine to stop if one of these areas is erroneously referred to. Also known as memory fill. { 'stȯr·ij ˌfil }

storage hierarchy [COMPUT SCI] The sequence of storage devices, characterized by speed, type of access, and size for the various functions of a computer; for example, core storage for programs and data, disks or drums for temporary storage of massive amounts of data, magnetic tapes and disks for backup storage. { 'stȯr·ij 'hī·ər,är·kē }

storage integrator [COMPUT SCI] In an analog computer, an integrator used to store a voltage in the hold condition for future use while the rest of the computer assumes another computer control state. { 'stȯr·ij ˌint·ə,grād·ər }

storage key [COMPUT SCI] A special set of bits associated with every word or character in some block of storage, which allows tasks having a matching set of protection key bits to use that block of storage. { 'stȯr·ij ˌkē }

storage location [COMPUT SCI] A digital-computer storage position holding one machine word and usually having a specific address. { 'stȯr·ij lō,kā·shən }

storage mark [COMPUT SCI] The name given to a point location which defines the character space immediately to the left of the most significant character in accumulator storage. { 'stȯr·ij ˌmärk }

storage medium [COMPUT SCI] Any device or recording medium into which data can be copied and held until some later time, and from which the entire original data can be obtained. { 'stȯr·ij ˌmēd·ē·əm }

storage oscilloscope [ELECTR] An oscilloscope that can retain an image for a period of time ranging from minutes to days, or until deliberately erased to make room for a new image. { 'stȯr·ij ə¦sil·ə,skōp }

storage pool [COMPUT SCI] A collection of similar data storage devices. { 'stȯr·ij ˌpül }

storage print [COMPUT SCI] In computers, a utility program that records the requested core image, core memory, or drum locations in absolute or symbolic form either on the line-printer or on the delayed-printer tape. { 'stȯr·ij ˌprint }

storage protection [COMPUT SCI] Any restriction on access to storage blocks, with respect to reading, writing, or both. Also known as memory protection. { 'stȯr·ij prə,tek·shən }

storage register [COMPUT SCI] A register in the main internal memory of a digital computer storing one computer word. Also known as memory register. { 'stȯr·ij rej·ə·stər }

storage-retrieval machine [CONT SYS] A computer-controlled machine for an automated storage and retrieval system that operates on rails and moves material either vertically or horizontally between a storage compartment and a transfer station. { ¦stȯr·ij ri'trēv·əl mə ˌshēn }

storage ripple [COMPUT SCI] A hardware function, used during maintenance periods, which reads or writes zeros or ones through available storage locations to detect a malfunctioning storage unit. { 'stȯr·ij ˌrip·əl }

storage surface [COMPUT SCI] In computers, the surface (screen), in an electrostatic storage tube, on which information is stored. { 'stȯr·ij ˌsər·fəs }

storage tank See tank. { 'stȯr·ij ˌtaŋk }

storage time [ELECTR] **1.** The time required for excess minority carriers stored in a forward-biased *pn* junction to be removed after the junction is switched to reverse bias, and hence the time interval between the application of reverse bias and the cessation of forward current. **2.** The time required for excess charge carriers in the collector region of a saturated transistor to be removed when the base signal is changed to cut-off level, and hence for the collector current to cease. { 'stȯr·ij ˌtīm }

storage-to-register instruction [COMPUT SCI] A machine-language instruction to move a word of data from a location in main storage to a register. { 'stȯr·ij tə 'rej·ə·stər in,strək·shən }

storage-to-storage instruction [COMPUT SCI] A machine-language instruction to move a word of data from one location in main storage to another. { 'stȯr·ij tə 'stȯr·ij in,strək·shən }

storage tube [ELECTR] An electron tube employing cathode-ray beam scanning and charge storage for the introduction, storage, and removal of information. Also known as electrostatic storage tube; memory tube (deprecated usage). { 'stȯr·ij ˌtüb }

storage-type camera tube See iconoscope. { 'stȯr·ij ˌtīp 'kam·rə ˌtüb }

store [COMPUT SCI] **1.** To record data into a (static) data storage device. **2.** To preserve data in a storage device. { stȯr }

store and forward [COMMUN] A procedure in data communications in which data are stored at some point between the sender and the receiver and are later forwarded to the receiver. { 'stȯr ən 'fȯr·wərd }

stored program [COMPUT SCI] A computer program that is held in a computer's main storage and carried out by a central processing unit that reads and acts on its instructions. { 'stȯrd 'prō ˌgram }

stored-program computer [COMPUT SCI] A digital computer which executes instructions that are stored in main memory as patterns of data. { 'stȯrd ˌprō,gram kəm'pyüd·ər }

stored-program control [COMMUN] Electronic control of a telecommunications switching system by means of a program of instructions stored in bulk electronic memory. Abbreviated SPC. { 'stȯrd 'prō,gram kən'trōl }

stored-program logic [COMPUT SCI] Program that is stored in a memory unit containing logical commands in order to perform the same processes on all problems. { 'stȯrd 'prō,gram ,läj·ik }

stored-program numerical control See computer numerical control. { 'stȯrd ˌprō,gram nü'mer·ə·kəl kən,trōl }

stored response chain [COMPUT SCI] A fixed sequence of instructions that are stored in a file and acted on by an interactive computer program at a point where it would normally request instructions from the user, in order to save the user the trouble of repeatedly keying the same commands for a frequently used function. Abbreviated SRC. { 'stȯrd ri'späns ˌchän }

stored routine [COMPUT SCI] In computers, a series of instructions in storage to direct the step-by-step operation of the machine. { 'stȯrd rü'tēn }

stored word [COMPUT SCI] The actual linear combination of letters (or their machine equivalents) to be placed in the machine memory; this may be physically quite different from a dictionary word. { 'stȯrd 'wərd }

storethrough [COMPUT SCI] The process of updating data in main memory each time the central processing unit writes into a cache. { 'stȯr,thrü }

store transmission bridge [ELEC] Transmission bridge, which consists of four identical impedance coils (the two windings of the back-bridge relay and live relay of a connector, respectively) separated by two capacitors, which couples the calling and called telephones together electrostatically for the transmission of voice-frequency (alternating) currents, but separates the two lines for the transmission of direct current for talking purposes (talking current). { 'stȯr tranz'mish·ən ˌbrij }

storetrieval system See storage and retrieval system. { 'stȯ·ri,trē·vəl ˌsis·təm }

STR See self-tuning regulator.

straightforward circuit [COMMUN] Circuit in which signaling is automatic and in one direction. { 'strāt¦fȯr·wərd 'sər·kət }

straight-line coding [COMPUT SCI] A digital computer program or routine (section of program) in which instructions are executed sequentially, without branching, looping, or testing. { 'strāt ¦līn 'kōd·iŋ }

strained-layer superlattice [ELECTR] A structure consisting of alternating layers of two different semiconducting materials, each several nanometers thick, in which a mismatch between the lattice spacings of the two materials of up to several percent is accommodated by elastic strains in the thin layers without the generation of mismatch defects. { 'strānd ¦lā·ər ¦sü·pər'lad·əs }

strain insulator [ELEC] An insulator used between sections of a stretched wire or antenna to break up the wire into insulated sections while withstanding the total pull of the wire. { 'strān ˌin·sə,lād·ər }

stranded conductor See stranded wire. { 'stran·dəd kən'dək·tər }

stranded wire [ELEC] A conductor composed of a group of wires or a combination of groups of wires, usually twisted together. Also known as stranded conductor. { 'stran·dəd 'wīr }

strapped magnetron [ELECTR] A multicavity magnetron in which resonator segments having the same polarity are connected together by small conducting strips to suppress undesired modes of oscillation. { 'strapt 'mag·nə,trän }

strapping [ELEC] Connecting two or more points in a circuit or device with a short piece of wire or metal. |ELECTR| Connecting together resonator segments having the same polarity in a multicavity magnetron to suppress undesired modes of oscillation. { 'strap·iŋ }

strapping option |COMPUT SCI| The rearrangement of jumpers on a printed circuit board to render a hardware feature operative or inoperative. { 'strap·iŋ ˌäp·shən }

stray capacitance |ELECTR| Undesirable capacitance between circuit wires, between wires and the chassis, or between components and the chassis of electronic equipment. { 'strā kə'pas·əd·əns }

stray current |ELEC| **1.** A portion of a current that flows over a path other than the intended path, and may cause electrochemical corrosion of metals in contact with electrolytes. **2.** An undesirable current generated by discharge of static electricity; it commonly arises in loading and unloading petroleum fuels and some chemicals, and can initiate explosions. { 'strā ˌkə·rənt }

stream |COMPUT SCI| A collection of binary digits that are transmitted in a continuous sequence, and from which extraneous data such as control information or parity bits are excluded. { strēm }

stream cipher |COMMUN| A cipher that makes use of an algorithmic procedure to produce an unending sequence of binary digits which is then combined either with plaintext to produce ciphertext or with ciphertext to recover plaintext. { 'strēm ˌsī·fər }

stream editor |COMPUT SCI| A modification of a statement editor to allow superlines that expand and contract as necessary; the most powerful type of text editor. Also known as string editor. { 'strēm ˌed·əd·ər }

streaming |COMPUT SCI| A malfunction in which a communicating device constantly transmits worthless data and thereby locks out all other devices on the line. { 'strēm·iŋ }

streaming current |ELEC| The electric current which is produced when a liquid is forced to flow through a diaphragm, capillary, or porous solid. { 'strēm·iŋ ˌkə·rənt }

streaming media |COMPUT SCI| Audio or video files that can begin playing as they are being downloaded to a computer. { ¦strēm·iŋ 'mēd·ē·ə }

streaming potential |ELEC| The difference in electric potential between a diaphragm, capillary, or porous solid and a liquid that is forced to flow through it. { 'strēm·iŋ pə,ten·chəl }

streaming tape |COMPUT SCI| A type of high-speed magnetic tape that is used as a backup storage for disks, particularly hard disks in microcomputer systems. { 'strēm·iŋ 'tāp }

STRESS |COMPUT SCI| A problem-oriented programming language used to solve structural engineering problems. Derived from structural engineering system solver. { stres }

stress sensor |CONT SYS| A contact sensor that responds to the forces produced by mechanical contact. { 'stres ˌsen·sər }

stress test |COMPUT SCI| A test of new software or hardware under unusually heavy work loads. { 'stres ˌtest }

striation |ELECTR| A succession of alternately luminous and dark regions sometimes observed in the positive column of a glow-discharge tube near the anode. { strī'ā·shən }

striking potential |ELECTR| **1.** Voltage required to start an electric arc. **2.** Smallest grid-cathode potential value at which plate current begins flowing in a gas-filled triode. { 'strīk·iŋ pə ˌten·chəl }

string |COMPUT SCI| A set of consecutive, adjacent items of similar type; normally a bit string or a character string. { striŋ }

string break |COMPUT SCI| In the sorting of records, the situation that arises when there are no records having keys with values greater than the highest key already written in the sequence of records currently being processed. { 'striŋ ˌbrāk }

string constant |COMPUT SCI| An arbitrary combination of letters, digits, and other symbols that is treated in a manner completely analogous to numeric constants. { 'striŋ ˌkän·stənt }

string editor See stream editor. { 'striŋ ˌed·əd·ər }

string electrometer |ENG| An electrometer in which a conducting fiber is stretched midway between two oppositely charged metal plates; the electrostatic field between the plates displaces the fiber laterally in proportion to the voltage between the plates. { 'striŋ ˌi,lek'träm·əd·ər }

string galvanometer |ENG| A galvanometer consisting of a silver-plated quartz fiber under tension in a magnetic field, used to measure oscillating currents. Also known as Einthoven galvanometer. { 'striŋ ˌgal·və'näm·əd·ər }

string manipulation |COMPUT SCI| The handling of strings of characters in a computer storage as though they were single units of data. { 'striŋ mə,nip·yə,lā·shən }

string manipulation language See string processing language. { 'striŋ mə,nip·yə,lā·shən ˌlaŋ·gwij }

String-Oriented-Symbolic Language See SNOBOL. { 'striŋ ¦ór·ē,ent·əd sim'bäl·ik 'laŋ·gwij }

string processing language |COMPUT SCI| A higher-level programming language equipped with facilities to synthesize and decompose character strings, search them in response to arbitrarily complex criteria, and perform a variety of other manipulations. Also known as string manipulation language. { 'striŋ 'prä ˌses·iŋ ˌlaŋ·gwij }

stringy floppy |COMPUT SCI| A peripheral storage device for microcomputers that uses a removable magnetic tape cartridge with a $\frac{1}{16}$-inchwide (1.5875-millimeter) loop of magnetic tape. { 'striŋ·ē 'fläp·ē }

strip-line circuit |ELECTROMAG| A circuit in which one or more strip transmission lines serve as filters or other circuit components. { 'strip ¦līn ˌsər·kət }

strip transmission line |ELECTROMAG| A microwave transmission line consisting of a thin, narrow, rectangular metal strip that is supported above a ground-plane conductor or between two wide ground-plane conductors and is usually separated from them by a dielectric material. { 'strip tranz'mish·ən ˌlīn }

strobe |ELECTR| **1.** Intensified spot in the sweep of a deflection-type indicator, used as a reference mark for ranging or expanding the presentation. **2.** Intensified sweep on a radar's plan-position indicator or B-scope; such a strobe may result from certain types of interference, or it may be purposely applied as a bearing or heading marker, or to show the estimated azimuth of a jamming source, as a "jam strobe." **3.** A signaling pulse of very short duration. { strōb }

strobe circuit |ELECTR| A circuit that produces an output pulse only at certain times or under certain conditions, such as a gating circuit or a coincidence circuit. { 'strōb ˌsər·kət }

strobe marker |ELECTR| A small bright spot, or a short gap, or other discontinuity produced on the trace of a radar display to indicate that part of the time base which is receiving attention. { 'strōb 'mär·kər }

strobe pulse |ELECTR| Pulse of duration less than the time period of a recurrent phenomenon used for making a close investigation of that phenomenon; the frequency of the strobe pulse bears a simple relation to that of the phenomenon, and the relative timing is usually adjustable. { 'strōb ˌpəls }

strobing |COMPUT SCI| The technique required to time-synchronize data appearing as pulses at the output of a computer memory. { 'strōb·iŋ }

stroboscopic lamp See flash lamp. { ˌsträb·ə ¦skäp·ik 'lamp }

stroboscopic tube See strobotron. { ˌsträb·ə ¦skäp·ik 'tüb }

strobotron |ELECTR| A cold-cathode gas-filled arc-discharge tube having one or more internal or external grids to initiate current flow and produce intensely bright flashes of light for a stroboscope. Also known as stroboscopic tube. { 'strō·bə ˌträn }

stroke |COMPUT SCI| **1.** In optical character recognition, straight or curved portion of a letter, such as is commonly made with one smooth motion of a pen. Also known as character stroke. **2.** That segment of a printed or handwritten character which has been temporarily isolated from other segments for the purpose of analyzing it, particularly with regard to its dimensions and relative reflectance. Also known as character stroke. |ELECTR| The penlike motion of a focused electron beam in cathode-ray-tube diplays. { strōk }

stroke analysis |COMPUT SCI| In character recognition, a method employed in character property detection in which an input specimen is dissected into certain prescribed elements; the sequence, relative positions, and number of detected elements are then used to identify the characters. { 'strōk ə ˌnal·ə·səs }

stroke center line |COMPUT SCI| In character recognition, a line midway between the two average-edge lines; the center line describes the stroke's direction of travel. Also known as center line. { 'strōk 'sen·tər ˌlīn }

stroke edge |COMPUT SCI| In character recognition, a continuous line, straight or otherwise, which traces the outermost part of intersection of the stroke along the two sides of its greatest dimension. { 'strōk ˌej }

stroke speed |COMMUN| Number of times per minute that a fixed line, perpendicular to the direction of scanning, is crossed in one direction by a scanning or recording spot in a facsimile system. Also known as scanning frequency; scanning line frequency. { 'strōk ˌspēd }

stroke width |COMPUT SCI| In character recognition, the distance that obtains, at a given location, between the points of intersection of the stroke edges and a line drawn perpendicular to the stroke center line. { 'strōk ˌwidth }

strong algorithm |COMMUN| A cryptographic algorithm for which the cost or time required to obtain the message or key is prohibitively great in practice even though the message may be obtainable in theory. { 'strȯŋ 'al·gə ˌrith·əm }

strongly typed language |CONT SYS| A high-level programming language in which the type of each variable must be declared at the beginning of the program, and the language itself then enforces rules concerning the manipulation of variables according to their types. { 'strȯŋ·lē ¦tīpt 'laŋ·gwij }

Strowger system |COMMUN| An automatic telephone switching system that uses successive step-by-step selector switches actuated by current pulses produced by rotation of a telephone dial. Also known as step-by-step system. { 'strō·gər ˌsis·təm }

structural engineering system solver See STRESS. { 'strək·chə·rəl ˌen·jə'nir·iŋ 'sis·təm ˌsäl·vər }

structural information |COMPUT SCI| Information specifying the number of independently variable features or degrees of freedom of a pattern. { 'strək·chə·rəl ˌin·fər'mā·shən }

structure |COMPUT SCI| For a data-processing system, the nature of the chain of command, the origin and type of data collected, the form and destination of results, and the procedures used to control operations. { 'strək·chər }

structured analysis |SYS ENG| A method of breaking a large problem or process into smaller components to aid in understanding, and then identifying the components and their interrelationships and reassembling them. { 'strək·chərd ə'nal·ə·səs }

structured data type |COMPUT SCI| The manner in which a collection of data items, which may have the same or different scalar data types, is represented in a computer program. { 'strək·chərd 'dad·ə ˌtīp }

structured programming |COMPUT SCI| The use of program design and documentation techniques that impose a uniform structure on all computer programs. { 'strək·chərd 'prō ˌgram· iŋ }

Structured Query Language |COMPUT SCI| The standard language for accessing relational databases. Abbreviated SQL. { ˌstrək·chərd 'kwir·ē ˌlaŋ·gwij }

structured variable See record variable. { 'strək·chərd 'ver·ē·ə·bəl }

structured walkthrough [COMPUT SCI] A formal method of debugging a computer system or program, involving a systematic review to search for errors and inefficiencies. { 'strək·chərd 'wòk ,thrü }

stub [COMPUT SCI] **1.** The left-hand portion of a decision table, consisting of a single column, and comprising the condition stub and the action stub. **2.** A program module that is only partly completed, to the extent needed to fulfill the requirements of other modules in the computer system. [ELECTROMAG] **1.** A short section of transmission line, open or shorted at the far end, connected in parallel with a transmission line to match the impedance of the line to that of an antenna or transmitter. **2.** A solid projection one-quarter-wavelength long, used as an insulating support in a waveguide or cavity. { stəb }

stub angle [ELECTROMAG] Right-angle elbow for a coaxial radio-frequency transmission line which has the inner conductor supported by a quarter-wave stub. { 'stəb ,aŋ·gəl }

stub cable [ELEC] Short branch off a principal cable; the end is often sealed until it is used at a later date; pairs in the stub are referred to as stubbed-out pairs. { 'stəb ¦kā·bəl }

stub matching [ELECTROMAG] Use of a stub to match a transmission line to an antenna or load; matching depends on the spacing between the two wires of the stub, the position of the shorting bar, and the point at which the transmission line is connected to the stub. { 'stəb ,mach·iŋ }

stub-supported line [ELECTROMAG] A transmission line that is supported by short-circuited quarter-wave sections of coaxial line; a stub exactly a quarter-wavelength long acts as an insulator because it has infinite reactance. { 'stəb səl¦pòrd·əd 'līn }

stub tuner [ELECTROMAG] Stub which is terminated by movable short-circuiting means and used for matching impedance in the line to which it is joined as a branch. { 'stəb ,tün·ər }

studio [COMMUN] A facility in which video or audio programs are produced. { 'stüd·ē·ō }

stunt box [ELEC] A device to control the nonprinting functions of a teletypewriter terminal. { 'stənt ,bäks }

stutter [COMMUN] Series of undesired black and white lines sometimes produced when a facsimile signal undergoes a sharp amplitude change. { 'stəd·ər }

stylus [COMPUT SCI] The pointed device used to draw images on a graphics tablet. { 'stī·ləs }

stylus printing See matrix printing. { 'stī·ləs ,print·iŋ }

subalphabet [COMPUT SCI] A subset of an alphabet. { səb'al·fə,bet }

subaperture [ENG] Any subset of an array of transmitters of acoustic or electromagnetic radiation. { səb'ap·ə·chər }

subassembly [ELECTR] Two or more components combined into a unit for convenience in assembling or servicing equipment; an intermediate-frequency strip for a receiver is an example. { ¦səb·ə'sem·blē }

subcarrier [ELECTR] **1.** A carrier that is applied as a modulating wave to modulate another carrier. **2.** See chrominance subcarrier. { ¦səb'kar·ē·ər }

subcarrier oscillator [ELECTR] **1.** The crystal oscillator that operates at the chrominance subcarrier or burst frequency of 3.579545 megahertz in an analog color television receiver; this oscillator, synchronized in frequency and phase with the transmitter master oscillator, furnishes the continuous subcarrier frequency required for demodulators in the receiver. **2.** An oscillator used in a telemetering system to translate variations in an electrical quantity into variations of a frequency-modulated signal at a subcarrier frequency. { ¦səb'kar·ē·ər 'äs·ə,lād·ər }

subchannel [COMPUT SCI] The portion of an input/output channel associated with a specific input/output operation. { ¦səb'chan·əl }

subclutter visibility [ELECTR] A measure of the effectiveness of moving-target indicator radar, equal to the ratio of the signal from a fixed target that can be canceled to the signal from a just visible moving target; often calculated for a target moving at an optimum velocity (unlike improvement factor). { 'səb¦kləd·ər ,viz·ə'bil·əd·ē }

subcommutation [COMMUN] In telemetry, commutation of additional channels with output applied to individual channels of the primary commutator. { ¦səb,käm·yə'tā·shən }

subcycle generator [ELECTR] Frequency-reducing device used in telephone equipment which furnishes ringing power at a submultiple of the power supply frequency. { 'səb,sī·kəl 'jen·ə,rād·ər }

subdivided capacitor [ELEC] Capacitor in which several capacitors known as sections are mounted so that they may be used individually or in combination. { ¦səb·di'vīd·əd kə'pas·əd·ər }

subframe [COMMUN] In telemetry, a complete sequence of frames during which all subchannels of a specific channel are sampled once. { 'səb ,frām }

subharmonic triggering [ELECTR] A method of frequency division which makes use of a triggered multivibrator having a period of one cycle which allows triggering only by a pulse that is an exact integral number of input pulses from the last effective trigger. { ¦səb·här'män·ik 'trig·ə·riŋ }

submarine cable [ELEC] A cable designed for service under water; usually a lead-covered cable with steel armor applied between layers of jute. { ¦səb·mə'rēn 'kā·bəl }

submillimeter wave [ELECTROMAG] An electromagnetic wave whose wavelength is less than 1 millimeter, corresponding to frequencies above 300 gigahertz. { ,səb¦mil·ə,mēd·ər ,wāv }

subminiature tube [ELECTR] An extremely small electron tube designed for use in hearing aids and other miniaturized equipment; a typical subminiature tube is about 1½ inches (4 centimeters) long and 0.4 inch (1 centimeter) in diameter, with the pins emerging through the glass base. { ¦səb'min·yə·chər 'tüb }

sub-Nyquist sampling [COMMUN] **1.** Any technique of sampling an analog signal at a rate lower than the Nyquist rate in such a way as to preserve signal content without aliasing distortion. **2.** In particular, the sampling of video signals at a rate lower than the Nyquist rate and at an odd multiple of the frame rate, so that the aliasing components are placed into periodically spaced voids in the video spectrum where they can be removed by a comb filter at the receiver. { ¦səb 'nī,kwist ,sam·pliŋ }

suboptimization [SYS ENG] The process of fulfilling or optimizing some chosen objective which is an integral part of a broader objective; usually the broad objective and lower-level objective are different. { ¦səb,äp·tə·mə'zā·shən }

subprogram [COMPUT SCI] A part of a larger program which can be converted independently into machine language. { ¦səb'prō,gram }

subrefraction [ELECTROMAG] Atmospheric refraction which is less than standard refraction. { ¦səb·ri'frak·shən }

subroutine [COMPUT SCI] **1.** A body of computer instruction (and the associated constants and working-storage areas, if any) designed to be used by other routines to accomplish some particular purpose. **2.** A statement in FORTRAN used to define the beginning of a closed subroutine (first definition). { 'səb·rü,tēn }

subroutine library [COMPUT SCI] A collection of subroutines that is stored on a disk or other direct-access storage device and can be used by a programmer through facilities of the computer's operating system. { 'səb·rü,tēn lī,brēr·ē }

subschema [COMPUT SCI] An individual user's partial view of a database. { 'səb,skē·mə }

subscriber line [ELEC] A telephone line between a central office and a telephone station, private branch exchange, or other end equipment. Also known as central office line; subscriber loop. { səb'skrīb·ər ,līn }

subscriber loop See subscriber line. { səb 'skrīb·ər ,lüp }

subscriber multiple [ELEC] Bank of jacks in a manual switchboard providing outgoing access to subscriber lines, and usually having more than one appearance across the face of the switchboard. { səb'skrīb·ər 'məl·tə·pəl }

subscriber set See subset. { səb'skrīb·ər ,set }

subscriber station [COMMUN] The connection between a central office and an outside location, including the circuit, some circuit termination equipment, and possibly some associated input/output equipment. { səb'skrīb·ər ,stā·shən }

subscription database See information network. { səb'skrip·shən 'dad·ə,bās }

subscription television [COMMUN] A television service in which programs are broadcast in coded or scrambled form, for reception only by subscribers who make payments for use of the decoding or unscrambling devices required to obtain a clear program. Also known as pay television. { səb'skrip·shən 'tel·ə,vizh·ən }

subset [COMMUN] A telephone or other subscriber equipment connected to a communication system, such as a modem. Derived from subscriber set. { 'səb,set }

substandard propagation [ELECTROMAG] The propagation of radio energy under conditions of substandard refraction in the atmosphere; that is, refraction by an atmosphere or section of the atmosphere in which the index of refraction decreases with height at a rate of less than 12 N units (unit of index of refraction) per 1000 feet (304.8 meters). { ¦səb¦stan·dərd ,präp·ə'gā·shən }

substation See electric power substation. { 'səb ,stā·shən }

substitute mode [COMPUT SCI] One method of exchange buffering, in which segments of storage function alternately as buffer and as program work area. { 'səb·stə,tüt ,mōd }

substitution alphabet [COMMUN] An alphabet used in a coded message in which each letter in the original message is replaced by another letter in the coded message, according to a set of rules. { ,səb·stə¦tü·shən 'al·fə,bet }

substitution cipher [COMMUN] A cipher in which the characters of the original message are replaced by other characters according to a key. { ,səb·stə'tü·shən ,sī·fər }

substrate [ELECTR] The physical material on which a microcircuit is fabricated; used primarily for mechanical support and insulating purposes, as with ceramic, plastic, and glass substrates; however, semiconductor and ferrite substrates may also provide useful electrical functions. [ENG] Basic surface on which a material adheres, for example, paint or laminate. { 'səb ,strāt }

substring [COMPUT SCI] A sequence of successive characters within a string. { 'səb,striŋ }

subsurface radar See ground-probing radar. { ,səb,sər·fəs 'rā·dar }

subsurface wave [ELECTROMAG] Electromagnetic wave propagated through water or land; operating frequencies for communications may be limited to approximately 35 kilohertz due to attenuation of high frequencies. { ¦səb¦sər·fəs 'wāv }

subsynchronous [ELEC] Operating at a frequency or speed that is related to a submultiple of the source frequency. { ¦səb¦siŋ·krə·nəs }

subsynchronous resonance [ELEC] An electrical resonant frequency on an alternating-current transmission line that is less than the line frequency, and results from the insertion of series capacitors to cancel out part of the line and system reactance. { səb¦siŋ·krə·nəs 'rez·ən·əns }

subsystem [ENG] A major part of a system which itself has the characteristics of a system, usually consisting of several components. { 'səb ,sis·təm }

subtracter [COMPUT SCI] A computer device that can form the difference of two numbers or quantities. { səb'trak·tər }

subtractive synthesis |ENG ACOUS| A method of synthesizing musical tones, in which an electronic circuit produces a standard waveform (such as a sawtooth wave), which contains a very large number of harmonics at known relative amplitudes, and this circuit is followed by a variety of electric or electronic filters to convert the basic tone signals into the desired musical waveforms. { səb,trak·tiv 'sin·thə·səs }

subtractor |ELECTR| A circuit whose output is determined by the differences in analog or digital input signals. { səb'trak·tər }

subvoice-grade channel |COMMUN| A channel whose bandwidth is smaller than the bandwidth of a voice-grade channel; it is usually a sub-channel of a voice-grade line. { ¦səb'vȯis ,grād ,chan·əl }

subway-type transformer |ELEC| Transformer of submersible construction. { 'səb,wā ¦tīp tranz'fȯr·mər }

subwoofer |ENG ACOUS| A loudspeaker designed to reproduce extremely low audio frequencies, extending into the infrasonic range, generally used in conjunction with a crossover network, a woofer, and a tweeter. { 'səb,wüf·ər }

successive approximation converter |COMPUT SCI| An analog-to-digital converter which operates by successively considering each bit position in the digital output and setting that bit equal to 0 or 1 on the basis of the output of a comparator. { sək'ses·iv ə,präk·sə'mā·shən kən,vərd·ər }

successor job |COMPUT SCI| A job that uses the output of another job (predecessor) as its input, so that it cannot start until the other job has been successfully completed. { sək'ses·ər ,jäb }

suffix notation See reverse Polish notation. { 'səf ,iks nō,tā·shən }

Suhl effect |ELECTR| When a strong transverse magnetic field is applied to an *n*-type semiconducting filament, holes injected into the filament are deflected to the surface, where they may recombine rapidly with electrons or be withdrawn by a probe. { 'sül i,fekt }

suite |COMPUT SCI| A collection of related computer programs run one after another. { swēt }

sulfating |ELEC| The formation of lead sulfate on the plates of lead-acid storage batteries reducing the energy-storing ability of the battery and eventually causing failure. { 'səl,fād·iŋ }

summary recorder |COMPUT SCI| In computers, output equipment which records a summary of the information handled. { 'səm·ə·rē ri'kȯrd·ər }

summation check |COMPUT SCI| An error-detecting procedure involving adding together all the digits of some number and comparing this sum to a previously computed value of the same sum. { sə'mā·shən ,chek }

summation network See summing network. { sə'mā·shən ,net,wərk }

summing amplifier |ELECTR| An amplifier that delivers an output voltage which is proportional to the sum of two or more input voltages or currents. { 'səm·iŋ 'am·plə,fī·ər }

summing network |ELEC| A passive electric network whose output voltage is proportional to the sum of two or more input voltages. Also known as summation network. { 'səm·iŋ 'net,wərk }

sun follower |ELECTR| A photoelectric pickup and an associated servomechanism used to maintain a sun-facing orientation, as for a space vehicle. Also known as sun seeker. { 'sən ,fäl·ə·wər }

S-unit meter See signal-strength meter. { 'es ,yü·nət ,mēd·ər }

sunlamp |ELEC| A mercury-vapor gas-discharge tube used to produce ultraviolet radiation for therapeutic or cosmetic purposes. { 'sən,lamp }

sun seeker See sun follower. { 'sən ,sēk·ər }

sun sensor See solar sensor. { 'sən ,sen·sər }

sun strobe |ELECTR| The signal display seen on a radar plan-position-indicator screen when the radar antenna is aimed at the sun; the pattern resembles that produced by continuous-wave interference, and is due to radio-frequency energy radiated by the sun. { 'sən ,strōb }

supercardioid microphone |ENG ACOUS| A microphone whose response pattern resembles a cardioid but is exaggerated along the axis of maximum response, so that it is highly sensitive in one direction and insensitive in all others. Also known as superdirectional microphone. { ¦sü·pər,kärd·ē,ȯid 'mī·krə,fōn }

superchip See super-large-scale integrated circuit. { 'sü·pər,chip }

supercomputer |COMPUT SCI| A computer which is among those with the highest speed, largest functional size, biggest physical dimensions, or greatest monetary cost in any given period of time. { 'sü·pər,kəm,pyüd·ər }

superconducting computer |COMPUT SCI| A high-performance computer whose circuits employ superconductivity and the Josephson effect to reduce computer cycle time. { ¦sü·pər·kən'dəkt·iŋ kəm'pyüd·ər }

superconducting fault-current limiter |ELEC| A device which uses the transition of superconductors from zero to finite resistance to limit the fault current that results from a short circuit in an electric power system to a value that is not much higher than the nominal current. { ¦sü·pər·kən ,duk·tiŋ ,fȯlt ,kər·ənt ,lim·əd·ər }

superconducting magnetic energy storage |ELEC| The storing of electrical energy, generally for use by an electrical utility during peak load period, as a circulating current in a large superconducting coil or magnet. { ¦sü·pər·kən ¦dəkt·iŋ mag¦ned·ik ¦en·ər·jē 'stȯr·ij }

superconducting material See superconductor. { ¦sü·pər·kən'dəkt·iŋ mə'tir·ē·əl }

superconducting quantum interference device |ELECTR| A superconducting ring that couples with one or two Josephson junctions; applications include high-sensitivity magnetometers, near-magnetic-field antennas, and measurement of very small currents or voltages. Abbreviated SQUID. { ¦sü·pər·kən'dəkt·iŋ 'kwän·təm ,in·tər ¦fir·əns di,vīs }

superconductivity |SOLID STATE| A property of many metals, alloys, and chemical compounds at temperatures near absolute zero by virtue of which their electrical resistivity vanishes and they become strongly diammagnetic. { |sü·pər·kän'dək'tiv·əd· }

superconductor |SOLID STATE| Any material capable of exhibiting superconductivity; examples include iridium, lead, mercury, niobium, tin, tantalum, vanadium, and many alloys. Also known as cryogenic conductor; superconducting material { |sü·pər·kən'dək·r }

superdirectional microphone See supercardioid microphone. { ,sü·pər·di,rek·shən·əl 'mī·krə ,fōn }

superemitron camera See image iconoscope. { |sü·pər'em·ə,trän ,kam·rə }

supergroup |COMMUN| In carrier telephony, five groups (60 voice channels) multiplexed together and treated as a unit; a basic supergroup occupies the band between 312 and 552 kilohertz. { 'sü·pər,grüp }

superhet See superheterodyne receiver. { 'sü·pər ,het }

superheterodyne receiver |ELECTR| A receiver in which all incoming modulated radio-frequency carrier signals are converted to a common intermediate-frequency carrier value for additional amplification and selectivity prior to demodulation, using heterodyne action; the output of the intermediate-frequency amplifier is then demodulated in the second detector to give the desired audio-frequency signal. Also known as superhet. { |sü·pər'he·trə,dīn ri'sē·vər }

superhigh frequency |COMMUN| A frequency band from 3000 to 30,000 megahertz, corresponding to wavelengths from 1 to 10 centimeters. Abbreviated SHF. { |sü·pər'hī |frē·kwən·sē }

super-large-scale integrated circuit |ELECTR| A very complex integrated circuit that has a high density of transistors and other components, for a total of 10^6 or more components. Also known as superchip. Abbreviated SLSI circuit. { |sü·pər |lärj |skāl ,in·tə'grād·əd 'sər·kət }

superlattice |ELECTR| A structure consisting of alternating layers of two different semiconductor materials, each several nanometers thick. { |sü·pər'lad·əs }

superline |COMPUT SCI| A unit of text longer than an ordinary line, used in some of the more powerful text editors. { 'sü·pər,līn }

supermicro |COMPUT SCI| A computer resembling a supermini in design but scaled down to the size of a microcomputer, usually capable of working with a small number of users at once. { |sü·pər'mī·krō }

superposed circuit |COMMUN| Additional channel obtained from one or more circuits, normally provided for other channels, in a way that all channels can be used simultaneously without mutual interference. { |sü·pər'pōzd 'sər·kət }

superposition integral |CONT SYS| An integral which expresses the response of a linear system to some input in terms of the impulse response or step response of the system; it may be thought of as the summation of the responses to impulses or step functions occurring at various times. { ,sü·pər·pə'zish·ən 'int·ə·grəl }

superposition theorem See principle of superposition. { ,sü·pər·pə'zish·ən 'thir·əm }

superregeneration |ELECTR| Regeneration in which the oscillation is broken up or quenched at a frequency slightly above the upper audibility limit of the human ear by a separate oscillator circuit connected between the grid and anode of the amplifier tube, to prevent regeneration from exceeding the maximum useful amount. { |sü·pər·ri,jen·ə'rā·shən }

superscalar architecture |COMPUT SCI| A design that enables a central processing unit to send several instructions to different execution units simultaneously, allowing it to execute several instructions in each clock cycle. { |sü·pər,skä·lər 'är·kə,tek·chər }

supersensitive relay |ELEC| A relay that operates on extremely small currents, generally below 250 microamperes. { |sü·pər'sen·səd·iv 'rē,lā }

superset |COMPUT SCI| A programming language that contains all the features of a given language and has been expanded or enhanced to include other features as well. { 'sü·pər,set }

superstandard propagation |ELECTROMAG| The propagation of radio waves under conditions of superstandard refraction in the atmosphere, that is, refraction by an atmosphere or section of the atmosphere in which the index of refraction decreases with height at a rate of greater than 12 N units (unit of index of refraction) per 1000 feet (304.8 meters). { |sü·pər'stan·dərd ,präp·ə'gā·shən }

supertweeter |ENG ACOUS| A loudspeaker designed to reproduce extremely high audio frequencies, extending into the ultrasonic range, generally used in conjunction with a crossover network, a tweeter, and a woofer. { 'süp·ər ,twēd·ər }

supertwisted nematic liquid-crystal display |ELECTR| A display in which nematic liquid-crystal molecules are twisted more than 90°, and the picture elements respond to the average (root-mean-square) voltage applied by transistors connected to each row and column to switch the liquid. Abbreviated STN LCD. Also known as passive-matrix liquid-crystal display (PM LCD). { |sü·pər,twis·təd nə|mad·ik ,lik·wəd 'krist·əl di ,splā }

supervisor |COMPUT SCI| A collection of programs, forming part of the operating system, that provides services for and controls the running of user programs. { 'sü·pər,vī·zər }

supervisor call |COMPUT SCI| A mechanism whereby a computer program can interrupt the normal flow of processing and ask the supervisor to perform a function for the program that the program cannot or is not permitted to perform for itself. Also known as system call. { 'sü·pər ,vī·zər ,kòl }

supervisor interrupt |COMPUT SCI| An interruption caused by the program being executed which issues an instruction to the master control program. { 'sü·pər,vī·zər 'int·ə,rəpt }

supervisor mode |COMPUT SCI| A method of computer operation in which the computer can execute all its own instructions, including the privileged instruction not normally allowed to the programmer, in contrast to problem mode. { 'sü·pər,vī·zər ,mōd }

supervisory computer |COMPUT SCI| A computer which accepts test results from satellite computers, transmits new programs to the satellite computers, and may further communicate with a larger computer. { 'sü·pər¦vīz·ə·rē kəm'pyüd·ər }

supervisory control and data acquisition |ENG| A version of telemetry commonly used in wide-area industrial applications, such as electrical power generation and distribution and water distribution, which includes supervisory control of remote stations as well as data acquisition from those stations over a bidirectional communications link. Abbreviated SCADA. { ,sü·pər ¦vīz·ə·rē kən,trōl ən 'dad·ə ,ak·wə,zish·ən }

supervisory controlled manipulation |ENG| A form of remote manipulation in which a computer enables the operator to teach the manipulator motion patterns to be remembered and repeated later. { 'sü·pər¦vīz·ə·rē kən'trōld mə ,nip·yə'lā·shən }

supervisory expert control system |CONT SYS| A control system in which an expert system is used to supervise a set of control, identification, and monitoring algorithms. { ,sü·pər¦vīz·ə·rē ,ek,spərt kən'trōl ,sis·təm }

supervisory program |COMPUT SCI| A program that organizes and regulates the flow of work in a computer system, for example, it may automatically change over from one run to another and record the time of the run. { ¦sü·pər ¦vīz·ə·rē 'prō,gram }

supervisory routine |COMPUT SCI| A program or routine that initiates and guides the execution of several (or all) other routines and programs; it usually forms part of (or is) the operating system. { ¦sü·pər¦vīz·ə·rē rü'tēn }

supervisory signal |ELEC| A signal which indicates the operating condition of a circuit or a combination of circuits in a switching apparatus or other electrical equipment to an attendant. { ¦sü·pər¦vīz·ə·rē 'sig·nəl }

supervisory system |ELEC| A system of control, indicating, and telemetry devices which operates between the stations of an electric power distribution system, using a single common channel to transmit signals. { ¦sü·pər¦vīz·ə·rē 'sis·təm }

supervoltage |ELEC| A voltage in the range of 500 to 2000 kilovolts, used for some x-ray tubes. { ¦sü·pər'vōl·tij }

supplementary group |ELEC| In wire communications, a group of trunks that directly connects local or trunk switching centers over other than a fundamental (or backbone) route. { ¦sap·lə ¦men·trē 'grüp }

supply voltage |ELEC| The voltage obtained from a power source for operation of a circuit or device. { sə'plī ,vōl·tij }

suppressed carrier |COMMUN| A carrier in a modulated signal that is suppressed at the transmitter; the chrominance subcarrier in an analog color television transmitter is an example. { sə'prest 'kar·ē·ər }

suppressed-carrier modulation |COMMUN| Modulation resulting from elimination or partial suppression of the carrier component from an amplitude modulated wave. { sə'prest 'kar·ē·ər ,mäj·ə'lā·shən }

suppressed-carrier transmission |COMMUN| Transmission in which the carrier component of the modulated wave is eliminated or partially suppressed, leaving only the side bands to be transmitted. { sə'prest 'kar·ē·ər tranz'mish·ən }

suppression |COMPUT SCI| **1.** Removal or deletion usually of insignificant digits in a number, especially zero suppression. **2.** Optional function in either on-line or off-line printing devices that permits them to ignore certain characters or groups of characters which may be transmitted through them. |ELECTR| Elimination of any component of an emission, as a particular frequency or group of frequencies in a radio-frequency signal. { sə'presh·ən }

suppressor |ELEC| **1.** In general, a device used to reduce or eliminate noise or other signals that interfere with the operation of a communication system, usually at the noise source. **2.** Specifically, a resistor used in series with a spark plug or distributor of an automobile engine or other internal combustion engine to suppress spark noise that might otherwise interfere with radio reception. |ELECTR| See suppressor grid. { sə'pres·ər }

suppressor grid |ELECTR| A grid placed between two positive electrodes in an electron tube primarily to reduce the flow of secondary electrons from one electrode to the other; it is usually used between the screen grid and the anode. Also known as suppressor. { sə'pres·ər ,grid }

suppressor pulse |ELECTR| Pulse used to disable an ionized flow field or beacon transponder during intervals when interference would be encountered. { sə'pres·ər ,pəls }

surface-acoustic-wave device |ELECTR| Any device, such as a filter, resonator, or oscillator, which employs surface acoustic waves with frequencies in the range $10^7 - 10^9$ hertz, traveling on the optically polished surface of a piezoelectric substrate, to process electronic signals. { 'sər·fəs ə'kü·stik 'wāv di,vīs }

surface-acoustic-wave filter |ELECTR| An electric filter consisting of a piezoelectric bar with a polished surface along which surface acoustic waves can propagate, and on which are deposited metallic transducers, one of which is connected, via thermocompression-bonded leads, to the electric source, while the other drives the load. { 'sər·fəs ə'kü·stik 'wāv ˌfiltər }

surface analysis |COMPUT SCI| A procedure in which a computer program writes a series of test characters onto a magnetic data storage medium and then reads them back to determine the location of any flaws in the medium. { 'sər·fəs ə,nal·ə·səs }

surface barrier |ELECTR| A potential barrier formed at a surface of a semiconductor by the trapping of carriers at the surface. { 'sər·fəs ,bar·ē·ər }

surface-barrier diode |ELECTR| A diode utilizing thin-surface layers, formed either by deposition of metal films or by surface diffusion, to serve as a rectifying junction. { 'sər·fəs ¦bar·ē·ər 'dī,ōd }

surface-barrier transistor |ELECTR| A transistor in which the emitter and collector are formed on opposite sides of a semiconductor wafer, usually made of *n*-type germanium, by training two jets of electrolyte against its opposite surfaces to etch and then electroplate the surfaces. { 'sər·fəs ¦bar·ē·ər tran'zis·tər }

surface-charge transistor |ELECTR| An integrated-circuit transistor element based on controlling the transfer of stored electric charges along the surface of a semiconductor. { 'sər·fəs ¦chärj tran'zis·tər }

surface-controlled avalanche transistor |ELECTR| Transistor in which avalanche breakdown voltage is controlled by an external field applied through surface-insulating layers, and which permits operation at frequencies up to the 10-gigahertz range. { 'sər·fəs kən¦trōld 'av·ə,lanch tran,zis·tər }

surface leakage |ELEC| The passage of current over the surface of an insulator. { 'sər·fəs ,lē·kij }

surface micromachining |ENG| A set of processes based upon deposition, patterning, and selective etching of thin films to form a free-standing microsensor on the surface of a silicon wafer. { ¦sər·fəs ,mī·krə·mə'shēn·iŋ }

surface-mount technology |ELECTR| The technique of mounting electronic circuit components and their electrical connections on the surface of a printed board, rather than through holes. { 'sər·fəs ¦maúnt tek'näl·ə·jē }

surface noise |ELECTR| The noise component in the electric output of a phonograph pickup due to irregularities in the contact surface of the groove. Also known as needle scratch. { 'sər·fəs ,nóiz }

surface passivation |ELECTR| A method of coating the surface of a *p*-type wafer for a diffused junction transistor with an oxide compound, such as silicon oxide, to prevent penetration of the impurity in undesired regions. { 'sər·fəs ,pas·ə'vā·shən }

surface-penetrating radar *See* ground-probing radar. { ,sər·fəs ,pen·ə,trād·iŋ 'rā,där }

surface resistivity |ELEC| The electric resistance of the surface of an insulator, measured between the opposite sides of a square on the surface; the value in ohms is independent of the size of the square and the thickness of the surface film. { 'sər·fəs ,rē,zis'tiv·əd·ē }

surface wave |COMMUN| *See* ground wave. |ELECTROMAG| A wave that can travel along an interface between two different mediums without radiation; the interface must be essentially straight in the direction of propagation; the commonest interface used is that between air and the surface of a circular wire. { 'sər·fəs ,wāv }

surface-wave transmission line |ELECTROMAG| A single conductor transmission line energized in such a way that a surface wave is propagated along the line with satisfactorily low attenuation. { 'sər·fəs ¦wāv tranz'mish·ən ,līn }

surge |ELEC| A momentary large increase in the current or voltage in an electric circuit. |ENG| **1.** An upheaval of fluid in a processing system, frequently causing a carryover (puking) of liquid through the vapor lines. **2.** The peak system pressure. **3.** An unstable pressure buildup in a plastic extruder leading to variable throughput and waviness of the hollow plastic tube. { sərj }

surge admittance |ELEC| Reciprocal of surge impedance. { 'sərj ad,mit·əns }

surge arrester |ELEC| A protective device designed primarily for connection between a conductor of an electrical system and ground to limit the magnitude of transient overvoltages on equipment. Also known as arrester; lightning arrester. { 'sərj ə,res·tər }

surge current |ELEC| A short-duration, high-amperage electric current wave that may sweep through an electrical network, as a power transmission network, when some portion of it is strongly influenced by the electrical activity of a thunderstorm. { 'sərj ,kə·rənt }

surge electrode current *See* fault electrode current. { 'sərj i'lek,trōd ,kə·rənt }

surge generator |ELEC| A device for producing high-voltage pulses, usually by charging capacitors in parallel and discharging them in series. { 'sərj ,jen·ə,rād·ər }

surge impedance *See* characteristic impedance. { 'sərj im,pēd·əns }

surge protector |ELEC| A device placed in an electrical circuit to prevent the passage of surges and spikes that could damage electronic equipment. { 'sərj prə,tek·tər }

surge suppressor |ELECTR| A circuit that responds to the rate of change of a current or voltage to prevent a rise above a predetermined value; it may include resistors, capacitors, coils, gas tubes, and semiconducting disks. Also known as transient suppressor. { 'sərj sə,pres·ər }

surveillance radar |ENG| A search radar that includes significant means of associating

detections of targets of interest (contacts) into
tracks with additional sorting and labeling of
data as the user system may require; normally
more highly automated and equipped with
data-processing computers than the simpler
search radar. { sər'vā·ləns ˌrā̇ˌdär }

survivable route |COMMUN| A communication
cable system begun in 1960 in which the cable,
main stations, amplifiers, and power feed stations are placed underground; it incorporates the
latest techniques of protection against natural
disasters and nuclear blasts, and avoids possible
target areas. { sər'vī·və·bəl 'rüt }

susceptance |ELEC| The imaginary component
of admittance. { sə'sep·təns }

susceptance standard |ELEC| Standard that in-
troduces calibrated small values of shunt capacitance into 50-ohm coaxial transmission arrays.
{ sə'sep·təns ˌstan·dərd }

susceptibility See electric susceptibility. { sə
ˌsep·tə'bil·əd·ē }

susceptometer |ENG| An instrument that mea-
sures paramagnetic, diamagnetic, or ferromagnetic susceptibility. { ˌsə'sep'täm·əd·ər }

suspension insulator |ELEC| A type of insulator
used to support a conductor of an overhead
transmission line, consisting of one or a string
of insulating units suspended from a pole or
tower, with the conductor attached to the end.
{ sə'spen·shən |in·sə,lād·ər }

sustained oscillation |CONT SYS| Continued os-
cillation due to insufficient attenuation in the
feedback path. { sə'stānd ˌäs·ə'lā·shən }

SV See speaker verification.

swamping resistor |ELECTR| Resistor placed in
the emitter lead of a transistor circuit to minimize
the effects of temperature on the emitter-base
junction resistance. { 'swämp·iŋ ri,zis·tər }

swap out |COMPUT SCI| The action of an operating
system on a process wherein it blocks the process
and writes the contents of its memory onto a disk
in order to make available more memory for other
current processes. { 'swäp ˌaüt }

swapping |COMPUT SCI| A procedure in which a
running program is temporarily suspended and
moved onto secondary storage, and primary
storage is reassigned to a more pressing job, in
order to maximize the efficient use of primary
storage. { 'swäp·iŋ }

sweep |ELECTR| 1. The steady movement of the
electron beam across the screen of a cathode-
ray tube, producing a steady bright line when no
signal is present; the line is straight for a linear
sweep and circular for a circular sweep. 2. The
steady change in the output frequency of a signal
generator from one limit of its range to the other.
{ swēp }

sweep amplifier |ELECTR| An amplifier used with
a cathode-ray tube, such as in a television receiver or cathode-ray oscilloscope, to amplify the
sawtooth output voltage of the sweep oscillator,
to shape the waveform for the deflection circuits
of a television picture tube, or to provide balanced signals to the deflection plates. { 'swēp
ˌam·plə,fī·ər }

sweep circuit |ELECTR| The sweep oscillator,
sweep amplifier, and any other stage used to
produce the deflection voltage or current for a
cathode-ray tube. Also known as scanning circuit.
{ 'swēp ˌsər·kət }

sweep generator |ELECTR| 1. An electronic circuit
that generates a voltage or current, usually
recurrent, as a prescribed function of time; the
resulting waveform is used as a time base to be
applied to the deflection system of an electron-
beam device, such as a cathode-ray tube. Also
known as time-base generator; timing-axis oscillator. 2. A test instrument that generates a
radio-frequency voltage whose frequency varies
back and forth through a given frequency range
at a rapid constant rate; used to produce an input
signal for circuits or devices whose frequency
response is to be observed on an oscilloscope.
Also known as sweep oscillator. { 'swēp ˌjen·ə
ˌrād·ər }

sweeping receivers |ELECTR| Automatically and
continuously tuned receivers designed to stop
and lock on when a signal is found, or to
continually plot band occupancy. { 'swēp·iŋ ri
ˌsē·vərz }

sweep jamming |ELECTR| Jamming with a rel-
atively narrow-band continuous signal being
varied in frequency (swept) so that pulselike
signals are produced in a radar as the jamming
passes through its passband. { 'swēp ˌjam·iŋ }

sweep oscillator See sweep generator. { 'swēp
ˌäs·ə,lad·ər }

sweep rate |ELECTR| The number of times a radar
radiation pattern rotates during 1 minute; sometimes expressed as the duration of one complete
rotation in seconds. { 'swēp ˌrāt }

sweep test |ELECTR| Test given coaxial cable with
an oscilloscope to check attenuation. { 'swēp
ˌtest }

sweep-through jammer |ELECTR| A jamming
transmitter which is swept through a radio-
frequency band in short steps to jam each
frequency briefly. { 'swēp|thrü 'jam·ər }

sweep voltage |ELECTR| Periodically varying volt-
age applied to the deflection plates of a cathode-
ray tube to give a beam displacement that is a
function of time, frequency, or other data base.
{ 'swēp ˌvōl·tij }

swept-frequency analyzer |ELECTR| A spectrum
analyzer in which a ramp generator simultaneously moves a spot horizontally across an
electronic display and increases the frequency
of a local oscillator; and any signal at the
input, at a frequency such that the difference
between its frequency and the local oscillator
is within the bandwidth of an intermediate-
frequency filter, vertically deflects the spot on
the display by an amount proportional to the
amplitude of the input signal being analyzed.
{ |swept ˌfrē·kwən·sē 'an·ə,līz·ər }

swing |ELEC| Variation in frequency or amplitude
of an electrical quantity. { swiŋ }

swinging choke |ELEC| An iron-core choke hav-
ing a core that can be operated almost at
magnetic saturation; the inductance is then a

maximum for small currents, and swings to a lower value as current increases. Also known as swinging reactor. { 'swiŋ·iŋ 'chōk }

swinging reactor See swinging choke. { 'swiŋ·iŋ rē'ak·tər }

switch |COMPUT SCI| **1.** A hardware or programmed device for indicating that one of several alternative states or conditions have been chosen, or to interchange or exchange two data items. **2.** A symbol used to indicate a branch point, or a set of instructions to condition a branch. |ELEC| A manual or mechanically actuated device for making, breaking, or changing the connections in an electric circuit. Also known as electric switch. Symbolized SW. { swich }

switchboard |COMMUN| A manually or automatically operated apparatus at a telephone exchange, on which the various circuits from subscribers and other exchanges are terminated to enable communication either between two subscribers on the same exchange, or between subscribers on different exchanges. Also known as telephone switchboard. |ELEC| A single large panel or assembly of panels on which are mounted switches, circuit breakers, meters, fuses, and terminals essential to the operation of electric equipment. Also known as electric switchboard. { 'swich,bòrd }

switched capacitor |ELECTR| An integrated circuit element, consisting of a capacitor with two metal oxide semiconductor (MOS) switches, whose function is approximately equivalent to that of a resistor. { 'swicht kə'pas·əd·ər }

switched-capacitor filter |ELECTR| An integrated-circuit filter in which a resistor is simulated by a combination of a capacitor and metal oxide semiconductor switches that are turned on and off periodically at a high frequency. Also known as switched-C filter. { ,swicht kə'pas·əd·ər ,fil·tər }

switched-C filter See switched-capacitor filter. { ,swicht 'sē ,fil·tər }

switched circuit |COMMUN| A communications circuit or channel that can be turned on and off and made to serve various users. { 'swicht 'sər·kət }

switched line |COMMUN| A communications line, such as a dial telephone line, whose path can vary each time the line is used. { 'swicht 'līn }

switched-message network |COMPUT SCI| A data transmission system in which a user can communicate with any other user of the network. { 'swicht ¦mes·ij 'net,wərk }

switched network |COMMUN| A communications network, such as the dial telephone network, in which any station may be connected with any other through the use of switching and control devices. { 'swicht 'net,wərk }

switch function |ELECTR| A circuit having a fixed number of inputs and outputs designed such that the output information is a function of the input information, each expressed in a certain code or signal configuration or pattern. { 'swich ,fəŋk·shən }

switchgear |ELEC| The aggregate of switching devices for a power or transforming station, or for electric motor control. { 'swich,gir }

switch hook |ELECTR| A switch on a telephone set that closes the circuit when the receiver is removed from the hook or cradle. { 'swich ,hùk }

switching |ELEC| Making, breaking, or changing the connections in an electrical circuit. { 'swich·iŋ }

switching center |COMMUN| The equipment in a relay station for automatically or semiautomatically relaying communications traffic. { 'swich·iŋ ,sen·tər }

switching circuit |ELEC| A constituent electric circuit of a switching or digital processing system which receives, stores, or manipulates information in coded form to accomplish the specified objectives of the system. { 'swich·iŋ ,sər·kət }

switching device |ENG| An electrical or mechanical device or mechanism, which can bring another device or circuit into an operating or nonoperating state. Also known as switching mechanism. { 'swich·iŋ di,vīs }

switching diode |ELECTR| A crystal diode that provides essentially the same function as a switch; below a specified applied voltage it has high resistance corresponding to an open switch, while above that voltage it suddenly changes to the low resistance of a closed switch. { 'swich·iŋ ,dī,ōd }

switching gate |ELECTR| An electronic circuit in which an output having constant amplitude is registered if a particular combination of input signals exists; examples are the OR, AND, NOT, and INHIBIT circuits. Also known as logical gate. { 'swich·iŋ ,gāt }

switching key See key. { 'swich·iŋ ,kē }

switching mechanism See switching device. { 'swich·iŋ ,mek·ə,niz·əm }

switching node |COMMUN| A location in a communications network where messages or lines are routed. { 'swich·iŋ ,nōd }

switching pad |ELECTR| Transmission-loss pad automatically cut in and out of a toll circuit for different desired operating conditions. { 'swich·iŋ ,pad }

switching substation |ELEC| An electric power substation whose equipment is mainly for connections and interconnections, and does not include transformers. { 'swich·iŋ 'səb ,stā·shən }

switching surface |CONT SYS| In feedback control systems employing bang-bang control laws, the surface in state space which separates a region of maximum control effort from one of minimum control effort. { 'swich·iŋ ,sər·fəs }

switching system |COMMUN| An assembly of switching and control devices provided so that any station in a communications system may be connected as desired with any other station. { 'swich·iŋ ,sis·təm }

switching theory |ELECTR| The theory of circuits made up of ideal digital devices; included are the theory of circuits and networks for telephone switching, digital computing, digital control, and data processing. { 'swich·iŋ ¦thē·ə·rē }

switching-through relay |ELEC| Control relay of a line-finder selector, connector, or other stepping switch, which extends the loop of a calling telephone through to the succeeding switch in a switch train. { 'swich·iŋ ¦thrü 'rē̱,lā }

switching time |ELECTR| **1.** The time interval between the reference time and the last instant at which the instantaneous voltage response of a magnetic cell reaches a stated fraction of its peak value. **2.** The time interval between the reference time and the first instant at which the instantaneous integrated voltage response of a magnetic cell reaches a stated fraction of its peak value. { 'swich·iŋ ,tīm }

switching transistor |ELECTR| A transistor designed for on/off switching operation. { 'swich·iŋ tran'zis·tər }

switching trunk |ELEC| Trunk from a long-distance office to a local exchange office used for completing a long-distance call. { 'swich·iŋ ,trəŋk }

switching tube |ELECTR| A gas tube used for switching high-power radio-frequency energy in the antenna circuits of radar and other pulsed radio-frequency systems; examples are those used in some radar modulators (pulsers) and those used for receiver protection in radar duplexers. { 'swich·iŋ ,tüb }

switch jack |ELEC| Any of the devices that provide terminals for the control circuits of the switch. { 'swich ,jak }

switch-over travel |ELEC| That movement of a switch-operating lever which takes place after the switch has been actuated either to close or open its contacts. { 'swich ,ō·vər 'trav·əl }

switch pretravel |ELEC| That movement of a switch-operating level that takes place before the switch is actuated either to close or to open its contacts. { 'swich 'prē,trav·əl }

switch register |COMPUT SCI| A manual switch on the control panel by means of which a bit may be entered in a processor register. { 'swich ,rej·ə·stər }

switch room |COMMUN| Part of a central office building that houses switching mechanisms and associated apparatus. { 'swich ,rüm }

switch selectable addressing |COMPUT SCI| The setting of DIP switches in a peripheral or terminal device to determine the address that identifies the device to the computer system. { 'swich si ¦lek·tə·bəl 'ad,res·iŋ }

switch train |ELEC| A series of switches in tandem. { 'swich ,trān }

syllabic compandor |ELECTR| A compandor in which the effective gain variations are made at speeds allowing response to the syllables of speech but not to individual cycles of the signal wave. { si'lab·ik kəm'pan·dər }

symbolic address |COMPUT SCI| In coding, a programmer-defined symbol that represents the

location of a particular datum item, instruction, or routine. Also known as symbolic number. { sim'bäl·ik 'ad,res }

symbolic algebraic manipulation language |COMPUT SCI| An algebraic manipulation language which admits the most general species of mathematical expressions, usually representing them as general tree structures, but which lacks certain special algorithms. { sim'bäl·ik ,al·jə'brā·ik mə ,nip·yə'lā·shən ,laŋ·gwij }

symbolic assembly language listing |COMPUT SCI| A list that may be produced by a computer during the compilation of a program showing the source language statements together with the corresponding machine language instructions generated by them. { sim'bäl·ik ə'sem·blē ,laŋ·gwij ,list·iŋ }

symbolic assembly system |COMPUT SCI| A system for forming programs that can be run on a computer, consisting of an assembly language and an assembler. { sim'bäl·ik ə'sem·blē ,sis·təm }

symbolic coding |COMPUT SCI| Instruction written in an assembly language, using symbols for operations and addresses. Also known as symbolic programming. { sim'bäl·ik 'kōd·iŋ }

symbolic computation system See symbolic system. { sim¦bäl·ik kəm'pyü'tā·shən ,sis·təm }

symbolic computing |COMPUT SCI| The development and use of symbolic systems. { sim¦bäl·ik kəm'pyüd·iŋ }

symbolic debugging |COMPUT SCI| A method of correcting known errors in a computer program written in a source language, in which certain statements are compiled together with the program. { sim'bäl·ik dē'bəg·iŋ }

symbolic language |COMPUT SCI| A language which expresses addresses and operation codes of instructions in symbols convenient to humans rather than in machine language. { sim'bäl·ik 'laŋ·gwij }

symbolic mathematical computation |COMPUT SCI| The manipulation of symbols, representing variables, functions, and other mathematical objects, and combinations of these symbols, representing formulas, equations, and expressions, according to mathematical rules, for example, the rules of algebra or calculus. { sim'bäl·ik¦math·ə ¦mad·ə·kəl ,käm·pyə'tā·shən }

symbolic name |COMPUT SCI| A name given to some entity that is actually something else; for example, the name of a table in a computer program actually represents the physical storage locations used to hold the data stored in that table, as well as the values stored in those locations. { sim'bäl·ik 'nām }

symbolic number See symbolic address. { sim 'bäl·ik 'nəm·bər }

symbolic programming See symbolic coding. { sim'bäl·ik 'prō,gram·iŋ }

symbolic system |COMPUT SCI| A computer program that performs computations with constants and variables according to the rules of algebra, calculus, and other branches of mathematics. Also known as algebraic computation system;

computer algebra system; symbolic computation system. { sim¦bäl·ik 'sis·təm }

symbol input [COMPUT SCI] Includes all contextual symbols that may appear in a source text. { 'sim·bəl 'in,pút }

symbol sequence [COMPUT SCI] A sequence of contextual symbols not interrupted by space. { 'sim·bəl 'sē·kwəns }

symbol table [COMPUT SCI] A mapping for a set of symbols to another set of symbols or numbers. { 'sim·bəl 'tā·bəl }

symmetrical architecture [COMPUT SCI] A type of computer design that allows any type of data to be used with any type of instruction. { si'me·trə·kəl 'ärk·ə,tek·chər }

symmetrical avalanche rectifier [ELECTR] Avalanche rectifier that can be triggered in either direction, after which it has a low impedance in the triggered direction. { sə'me·trə·kəl 'av·ə ,lanch ,rek·tə,fī·ər }

symmetrical band-pass filter [ELECTR] A band-pass filter whose attenuation as a function of frequency is symmetrical about a frequency at the center of the pass band. { sə'me·trə·kəl 'band ,pas ,fil·tər }

symmetrical band-reject filter [ELECTR] A band-rejection filter whose attenuation as a function of frequency is symmetrical about a frequency at the center of the rejection band. { sə'me·trə·kəl 'band ri,jekt ,fil·tər }

symmetrical clipper [ELECTR] A clipper in which the upper and lower limits on the amplitude of the output signal are positive and negative values of equal magnitude. { sə'me·trə·kəl 'klip·ər }

symmetrical deflection [ELECTR] A type of electrostatic deflection in which voltages that are equal in magnitude and opposite in sign are applied to the two deflector plates. { sə'me·trə·kəl di'flek·shən }

symmetrical H attenuator [ELECTR] An H attenuator in which the impedance near the input terminals equals the corresponding impedance near the output terminals. { sə'me·trə·kəl 'āch ə,ten·yə'wād·ər }

symmetrical inductive diaphragm [ELECTROMAG] A waveguide diaphragm which consists of two plates that leave a space at the center of the waveguide, and which introduces an inductance in the waveguide. { sə'me·trə·kəl in'dək·tiv 'dī·ə,fram }

symmetrical O attenuator [ELECTR] An O attenuator in which the impedance near the input terminals equals the corresponding impedance near the output terminals. { sə'me·trə·kəl 'ō ə ,ten·yə,wād·ər }

symmetrical pi attenuator [ELECTR] A pi attenuator in which the impedance near the input terminals equals the corresponding impedance near the output terminals. { sə'me·trə·kəl 'pī ə ,ten·yə,wād·ər }

symmetrical T attenuator [ELECTR] A T attenuator in which the impedance near the input terminals equals the corresponding impedance near the output terminals. { sə'me·trə·kəl 'tē ə ,ten·yə,wād·ər }

symmetrical transducer [ELECTR] A transducer is symmetrical with respect to a specified pair of terminations when the interchange of that pair of terminations will not affect the transmission. { sə'me·trə·kəl tranz'dü·sər }

symmetric list [COMPUT SCI] A list with sequencing pointers to previous as well as subsequent items. { sə'me·trik 'list }

sync See synchronization. { siŋk }

sync generator See synchronizing generator. { 'siŋk ,jen·ə,rād·ər }

synchro [ELEC] Any of several devices which are used for transmitting and receiving angular position or angular motion over wires, such as a synchro transmitter or synchro receiver. Also known as mag-slip (British usage); self-synchronous device; self-synchronous repeater; selsyn. { 'siŋ·krō }

synchro control transformer [ELEC] A transformer having its secondary winding on a rotor; when its three input leads are excited by angle-defining voltages, the two output leads deliver an alternating-current voltage that is proportional to the sine of the difference between the electrical input angle and the mechanical rotor angle. { 'siŋ·krō kən¦trōl tranz,fȯr·mər }

synchro control transmitter [ELEC] A high-accuracy synchro transmitter, having high-impedance windings. { 'siŋ·krō kən¦trōl tranz,mid·ər }

synchro differential motor [ELEC] Motor which is electrically similar to the synchro differential generator except that a damping device is added to prevent oscillations; both its rotor and stator are connected to synchro generators, and its function is to indicate the sum or difference between the two signals transmitted by the generators. { 'siŋ·krō ,dif·ə'ren·chəl ,mōd·ər }

synchro differential receiver [ELEC] A synchro receiver that subtracts one electrical angle from another and delivers the difference as a mechanical angle. Also known as differential synchro. { 'siŋ·krō ,dif·ə'ren·chəl ri'sē·vər }

synchro differential transmitter [ELEC] A synchro transmitter that adds a mechanical angle to an electrical angle and delivers the sum as an electrical angle. Also known as differential synchro. { 'siŋ·krō ,dif·ə'ren·chəl tranz'mid·ər }

synchro generator See synchro transmitter. { 'siŋ·krō 'jen·ə,rād·ər }

synchro motor See synchro receiver. { 'siŋ·krō ,mōd·ər }

synchronism [ELEC] Of a synchronous motor, the condition under which the motor runs at a speed which is directly related to the frequency of the power applied to the motor and is not dependent upon variables. { 'siŋ·krə,niz·əm }

synchronization [ENG] The maintenance of one operation in step with another, as in keeping the electron beam of a television picture tube in step with the electron beam of the television camera tube at the transmitter. Also known as sync. { ,siŋ·krə·nə'zā·shən }

synchronized blocking oscillator [ELECTR] A blocking oscillator which is synchronized with pulses occurring at a rate slightly faster than its own natural frequency. { 'siŋ·krə,nīzd 'bläk·iŋ 'äs·ə,lād·ər }

synchronizer [COMPUT SCI] A computer storage device used to compensate for a difference in rate of flow of information or time of occurrence of events when transmitting information from one device to another. [ELECTR] The component of a radar set which generates the timing voltage for the complete set. { 'siŋ·krə,nīz·ər }

synchronizing generator [ELECTR] An electronic generator that supplies synchronizing pulses to television studio and transmitter equipment. Also known as sync generator; sync-signal generator. { 'sin·krə,nīz·iŋ 'jen·ə,rād·ər }

synchronizing pulse [COMMUN] In pulse modulation, a pulse which is transmitted to synchronize the transmitter and the receiver; it is usually distinguished from signal-carrying pulses by some special characteristic. { 'sin·krə,nīz·iŋ ,pəls }

synchronizing reactor [ELEC] Current-limiting reactor for connecting momentarily across the open contacts of a circuit-interrupting device for synchronizing purposes. { 'sin·krə,nīz·iŋ rē'ak·tər }

synchronizing relay [ELEC] Relay which functions when two alternating-current sources are in agreement within predetermined limits of phase angle and frequency. { 'sin·krə,nīz·iŋ 'rē,lā }

synchronizing signal See sync signal. { 'sin·krə ,nīz·iŋ ,sig·nəl }

synchronous [ENG] In step or in phase, as applied to two or more circuits, devices, or machines. { 'siŋ·krə·nəs }

synchronous booster converter [ELEC] Synchronous converter having an alternating-current generator mounted on the same shaft and connected in series with it to adjust the voltage at the commutator of the converter. { 'siŋ·krə·nəs 'büs·tər kən'vərd·ər }

synchronous capacitor [ELEC] A synchronous motor running without mechanical load and drawing a large leading current, like a capacitor; used to improve the power factor and voltage regulation of an alternating-current power system. { 'siŋ·krə·nəs kə'pas·əd·ər }

synchronous clamp circuit See keyed clamp circuit. { 'siŋ·krə·nəs 'klamp ,sər·kət }

synchronous communications [COMPUT SCI] The high-speed transmission and reception of long groups of characters at a time, requiring synchronization of the sending and receiving devices. { 'siŋ·krə·nəs kə,myü·nə'kā·shənz }

synchronous computer [COMPUT SCI] A digital computer designed to operate in sequential elementary steps, each step requiring a constant amount of time to complete, and being initiated by a timing pulse from a uniformly running clock. { 'siŋ·krə·nəs kəm'pyüd·ər }

synchronous converter [ELEC] A converter in which motor and generator windings are combined on one armature and excited by one magnetic field; normally used to change alternating to direct current. Also known as converter; electric converter. { 'siŋ·krə·nəs kən'vərd·ər }

synchronous data-link control [COMMUN] A bit-oriented protocol for managing the flow of information in a data-communications system, in full, half-duplex, or multipoint modes, that uses an error-check algorithm. { 'siŋ·krə·nəs 'dad·ə ,liŋk kən,trōl }

synchronous data transmission [COMMUN] Data transmission in which a clock defines transmission times for data; since start and stop bits for each character are not needed, more of the transmission bandwidth is available for message bits. { 'siŋ·krə·nəs 'dad·ə tranz ,mish·ən }

synchronous demodulator See synchronous detector. { 'siŋ·krə·nəs dē'mäj·ə,lād·ər }

synchronous detection [ELECTR] The act of mixing two nearly identical frequencies, such as the oscillator reference signal and the signal received in a coherent radar, producing a voltage output sinusoidally related to the phase difference of the two. { 'siŋ·krə·nəs di,tek·shən }

synchronous detector [ELECTR] **1.** A detector that inserts a missing carrier signal in exact synchronism with the original carrier at the transmitter; when the input to the detector consists of two suppressed-carrier signals in phase quadrature, as in the chrominance signal of an analog color television receiver, the phase of the reinserted carrier can be adjusted to recover either one of the signals. Also known as synchronous demodulator. **2.** See cross-correlator. { 'siŋ·krə·nəs di'tek·tər }

synchronous dynamic random access memory [COMPUT SCI] High-speed memory that is controlled by the system clock and can run at bus speeds up to 100 megahertz. Abbreviated SDRAM. { ¦siŋ·krə·nəs dī,nam·ik ,ran·dəm 'ak ,ses ,mem·rē }

synchronous gate [ELECTR] A time gate in which the output intervals are synchronized with an incoming signal. { 'siŋ·krə·nəs 'gāt }

synchronous generator [ELEC] A machine that generates an alternating voltage when its armature or field is rotated by a motor, an engine, or other means. The output frequency is exactly proportional to the speed at which the generator is driven. { 'siŋ·krə·nəs 'jen·ə,rād·ər }

synchronous inverter See dynamotor. { 'siŋ· krə·nəs in'vərd·ər }

synchronous machine [ELEC] An alternating-current machine whose average speed is proportional to the frequency of the applied or generated voltage. { 'siŋ·krə·nəs mə'shēn }

synchronous motor [ELEC] A synchronous machine that transforms alternating-current electric power into mechanical power, using field magnets excited with direct current. { 'siŋ·krə·nəs 'mōd·ər }

synchronous operation [ELECTR] **1.** An operation that takes place regularly or predictably with respect to the occurrence of a particular event in another process. **2.** In particular, an operation whose timing is controlled by pulses generated by an electronic clock. { 'siŋ·krə·nəs ,äp·ə'rā·shən }

synchronous phase modifier |ELEC| A synchronous motor that runs without mechanical load, and is provided with means for varying its power factor to simulate a capacitive or inductive reactor; used in voltage regulation of alternating-current power systems. { 'siŋ·krə·nəs 'fāz ,shif·tər }

synchronous rectifier |ELECTR| A rectifier in which contacts are opened and closed at correct instants of time for rectification by a synchronous vibrator or by a commutator driven by a synchronous motor. { 'siŋ·krə·nəs 'rek·tə,fī·ər }

synchronous switch |ELECTR| A thyratron circuit used to control the operation of ignitrons in such applications as resistance welding. { 'siŋ·krə·nəs 'swich }

synchronous system |COMMUN| A telecommunication system in which transmitting and receiving apparatus operate continuously at substantially the same rate, and correction devices are used, if necessary, to maintain them in a fixed time relationship. { 'siŋ·krə·nəs 'sis·təm }

synchronous time-division multiplexing |COMMUN| A data transmission technique in which several users make use of a single channel by means of a system in which time slots are allotted on a fixed basis, usually in round-robin fashion. Abbreviated STDM. { 'siŋ·krə·nəs 'tīm də,vizh·ən 'məl·tə,pleks·iŋ }

synchronous working |COMPUT SCI| The mode of operation of a synchronous computer, in which the starting of each operation is clock-controlled. { 'siŋ·krə·nəs 'wərk·iŋ }

synchro receiver |ELEC| A synchro that provides an angular position related to the applied angle-defining voltages; when two of its input leads are excited by an alternating-current voltage and the other three input leads are excited by the angle-defining voltages, the rotor rotates to the corresponding angular position; the torque of rotation is proportional to the sine of the difference between the mechanical and electrical angles. Also known as receiver synchro; selsyn motor; selsyn receiver; synchro motor. { 'siŋ·krō ri'sē·vər }

synchro resolver See resolver. { 'siŋ·krō ri'zäl·vər }

synchroscope |ELECTR| A cathode-ray oscilloscope designed to show a short-duration pulse by using a fast sweep that is synchronized with the pulse signal to be observed. { 'siŋ·krə,skōp }

synchro system |ELEC| An electric system for transmitting angular position or motion; in the simplest form it consists of a synchro transmitter connected by wires to a synchro receiver; more complex systems include synchro control transformers and synchro differential transmitters and receivers. Also known as selsyn system. { 'siŋ·krō ,sis·təm }

synchro transmitter |ELEC| A synchro that provides voltages related to the angular position of its rotor; when its two input leads are excited by an alternating-current voltage, the magnitudes and polarities of the voltages at the three output leads define the rotor position. Also known as selsyn generator; selsyn transmitter; synchro generator; transmitter; transmitter synchro. { 'siŋ·krō tranz'mid·ər }

sync separator |ELECTR| A circuit that separates synchronizing pulses from the video signal in an analog television receiver. { 'siŋk ,sep·ə,rād·ər }

sync signal |COMMUN| A signal transmitted after each line and field to synchronize the scanning process in a video system. Also known as synchronizing signal. { 'siŋk ,sig·nəl }

sync-signal generator See synchronizing generator. { 'siŋk ,sig·nəl 'jen·ə,rād·ər }

syntactic analysis |COMPUT SCI| The problem of associating a given string of symbols through a grammar to a programming language, so that the question of whether the string belongs to the language may be answered. { sin'tak·tik ə'nal·ə·səs }

syntactic error See syntax error. { sin'tak·tik 'er·ər }

syntactic extension |COMPUT SCI| An extension mechanism which creates new notations for existing or user-defined mechanisms in an extensible language. { sin'tak·tik ik'sten·shən }

syntactic model See linguistic model. { sin 'tak·tik 'mäd·əl }

syntactic semigroup |SYS ENG| For a sequential machine, the set of all transformations performed by all input sequences. { sin'tak·tik 'sem·i,grüp }

syntax |COMPUT SCI| The set of rules needed to construct valid expressions or sentences in a language. { 'sin,taks }

syntax checker See syntax scanner. { 'sin,taks ,chek·ər }

syntax diagram |COMPUT SCI| A pictorial diagram showing the rules for forming an instruction in a computer programming language, and how the components of the statement are related. { 'sin ,taks 'dī·ə,gram }

syntax-directed compiler |COMPUT SCI| A general-purpose compiler that can service a family of languages by providing the syntactic rules for language analysis in the form of data, typically in tabular form, rather than using a specific parsing algorithm for a particular language. Also known as syntax-oriented compiler. { 'sin,taks di¦rek·təd kəm'pīl·ər }

syntax error |COMPUT SCI| An error in the format of a statement in a computer program that violates the rules of the programming language employed. Also known as syntactic error. { 'sin ,taks ,er·ər }

syntax-oriented compiler See syntax-directed compiler. { 'sin,taks ¦ör·ē,ent·əd kəm'pīl·ər }

syntax scanner |COMPUT SCI| A subprogram of a compiler or interpreter that checks the source program for syntax errors, and reports any such errors by printing the erroneous statement together with a diagnostic message. Also known as syntax checker. { 'sin,taks ,skan·ər }

synthesis See system design. { 'sin·thə·səs }

synthesizer |ELECTR| **1.** An electronic instrument which combines simple elements to generate more complex entities; examples are frequency synthesizer and sound synthesizer. **2.** Circuitry generating multiple frequencies at very low power that are used in radar transmissions, particularly in frequency-agile radars. { 'sin·thə,sīz·ər }

synthetic address *See* generated address. { sin'thed·ik 'ad,res }

synthetic aperture [ENG] A method of increasing the ability of an imaging system, such as radar or acoustical holography, to resolve small details of an object, in which a receiver of large size (or aperture) is in effect synthesized by the motion of a smaller receiver and the proper correlation of the detected signals. { sin'thed·ik 'ap·ə·chər }

synthetic-aperture radar [ENG] A radar system in which an aircraft moving along a very straight path emits microwave pulses continuously at a frequency constant enough to be coherent for a period during which the aircraft may have traveled about 1 kilometer; all echoes returned during this period can then be processed as if a single antenna as long as the flight path had been used. { sin'thed·ik ¦ap·ə·chər 'rā,där }

synthetic language [COMPUT SCI] A pseudocode or symbolic language; fabricated language. { sin'thed·ik 'laŋ·gwij }

syntony [ELEC] Condition in which two oscillating circuits have the same resonant frequency. { 'sin·tə·nē }

sysgen *See* system generation. { 'sis,jen }

SYSIN [COMPUT SCI] The principal input stream of an operating system. Derived from system input. { 'sis,in }

system [ELECTR] A combination of two or more sets generally physically separated when in operation, and such other assemblies, subassemblies, and parts necessary to perform an operational function or functions. [ENG] An assemblage of interrelated components designed to perform prescribed functions. { 'sis·təm }

system analysis [CONT SYS] The use of mathematics to determine how a set of interconnected components whose individual characteristics are known will behave in response to a given input or set of inputs. { 'sis·təm ə,nal·ə·səs }

systematic analog network testing approach [ELECTR] An on-line minicomputer-based system with an integrated data-based and optimal human intervention, which provides computer printouts used in automatic testing of electronic systems; aimed at maximizing cost effectivity. Abbreviated SANTA. { ,sis·tə'mad·ik 'an·ə,läg 'net,wərk ,test·iŋ ə,prōch }

systematic distortion [ELEC] Periodic or constant distortion, such as bias or characteristic distortion; the direct opposite of fortuitous distortion. { ,sis·tə'mad·ik di'stòr·shən }

systematic error-checking code [COMPUT SCI] A type of self-checking code in which a valid character consists of the minimum number of digits needed to identify the character and distinguish it from any other valid character, and a set of check digits which maintain a minimum specified signal distance between any two valid characters.

Also known as group code. { ,sis·tə'mad·ik 'er·ər ¦chek·iŋ ,kōd }

system bandwidth [CONT SYS] The difference between the frequencies at which the gain of a system is $\sqrt{2}/2$ (that is, 0.707) times its peak value. { 'sis·təm 'band,width }

system calendar [COMPUT SCI] A register in a computer system that holds the date and year and provides them in response to supervisor calls to the operating system. { 'sis·təm 'kal·ən·dər }

system call *See* supervisor call. { 'sis·təm ,kòl }

system catalog [COMPUT SCI] An index of all files controlled by the operating system of a large computer. { 'sis·təm 'kad·əl,äg }

system chart [COMPUT SCI] A flowchart that emphasizes the component operations which make up a system. { 'sis·təm ,chärt }

system check [COMPUT SCI] A check on the overall performance of the system, usually not made by built-in computer check circuits; for example, control total, hash totals, and record counts. { 'sis·təm ,chek }

system clock [COMPUT SCI] A circuit that emits regularly timed pulses that are used to synchronize the operations of all the circuits of a computer. { 'sis·təm 'kläk }

system clock reference [COMMUN] A time stamp in the program stream from which decoder timing is derived. Abbreviated SCR. { 'sis·təm ¦kläk 'ref·rəns }

system command [COMPUT SCI] A special instruction to a computer system to carry out a particular processing function, such as allowing a user to gain access to the system, running a program, activating a translator, or issuing a status report. { 'sis·təm kə,mand }

system design [COMPUT SCI] Determination in detail of the exact operational requirements of a system, resolution of these into file structures and input/output formats, and relation of each to management tasks and information requirements. [CONT SYS] A technique of constructing a system that performs in a specified manner, making use of available components. Also known as synthesis. { 'sis·təm di,zīn }

system designer [COMPUT SCI] A person who prepares final system documentation, analyzes findings, and synthesizes new system design. { 'sis·təm di,zīn·ər }

system documentation [COMPUT SCI] Detailed information, in either written or computerized form, about a computer system, including its architecture, design, data flow, and programming logic. { 'sis·təm ,däk·yə·mən'tā·shən }

system evaluation [COMPUT SCI] A periodic evaluation of the system to assess its status in terms of original or current expectations and to chart its future direction. { 'sis·təm i,val·yə'wā·shən }

system flowchart *See* data flow diagram. { 'sis·təm 'flō,chärt }

system generation [COMPUT SCI] A process that creates a particular and uniquely specified operating system; it combines user-specified options and parameters with manufacturer-supplied general-purpose or nonspecialized program subsections to produce an operating system (or other complex software) of the desired form and capacity. Abbreviated sysgen. { 'sis·təm ˌjen·ə'rā·shən }

system header |COMMUN| A data structure that carries information summarizing the system characteristics of the digital television multiplexed bit stream. { 'sis·təm ˌhed·ər }

system improvement time |COMPUT SCI| The machine downtime needed for the installation and testing of new components, large or small, and machine downtime necessary for modification of existing components; this includes all programming tests following the above actions to prove the machine is operating properly. { 'sis·təm im'prüv·mənt ˌtīm }

system input See SYSIN. { 'sis·təm 'in,pút }

system integration |COMPUT SCI| The procedures involved in combining separately developed modules of components so that they work together as a complete computer system. { 'sis·təm ˌin·tə'grā·shən }

system-level timer |COMPUT SCI| A hardware device that is set by the operating system to interrupt it after a specified time interval, either to set deadlines for events or to remind the operating system to take some action. { 'sis·təm ¦lev·əl 'tīm·ər }

system library |COMPUT SCI| An organized collection of computer programs that is maintained on-line with a computer system by being held on a secondary storage device and is managed by the operating system. { 'sis·təm 'lī,brer·ē }

system loader |COMPUT SCI| A computer program that loads all the other programs, including the operating system, into a computer's main storage. { 'sis·təm ˌlōd·ər }

system master tapes |COMPUT SCI| Magnetic tapes containing programmed instructions necessary for preparing a computer prior to running programs. { 'sis·təm 'mas·tər 'tāps }

system operation |COMPUT SCI| The administration and operation of an automatic data-processing equipment-oriented system, including staffing, scheduling, equipment and service contract administration, equipment utilization practices, and time-sharing. { 'sis·təm ˌäp·ə'rā·shən }

system optimization See optimization. { 'sis·təm ˌäp·tə·mə'zā·shən }

system response See response. { 'sis·təm ri'späns }

systems management |SYS ENG| The manage-ment of information technology systems in an organization or commercial enterprise, including all activities involved in configuring, installing, maintaining, and updating these systems. { 'sis·təmz 'man·ij·mənt }

system software |COMPUT SCI| Computer software involved with data and program management, including operating systems, control programs, and database management systems. { 'sis·təm 'sȯft,wer }

systems programming |COMPUT SCI| The development and production of programs that have to do with translation, loading, supervision, maintenance, control, and running of computers and computer programs. { 'sis·təmz ˌprō,gram·iŋ }

systems specification See systems definition. { 'sis·təmz ˌspes·ə·fə,kā·shən }

systems test |COMPUT SCI| The running of whole computer system against test data; a complete simulation of the actual running system for purposes of testing the adequacy of the system. { 'sis·təmz ˌtest }

system study |COMPUT SCI| A detailed study to determine whether, to what extent, and how automatic data-processing equipment should be used; it usually includes an analysis of the existing system and the design of the new system, including the development of system specifications which provide a basis for the selection of equipment. { 'sis·təm ˌstəd·ē }

system supervisor |COMPUT SCI| A control program which ensures an efficient transition in running program after program and accomplishing setups and control functions. { 'sis·təm 'sü·pər,vīz·ər }

system target decoder |COMMUN| A hypothetical reference model of a decoding process used to describe the semantics of the digital television multiplexed bit stream. Abbreviated STD. { 'sis·təm ¦tär·gət dē'kōd·ər }

system unit |COMPUT SCI| **1.** An individual card, section of tape, or the like, which is manipulated during operation of the system; class 1 systems have one unit per document; class 2 systems have one unit per vocabulary term or concept. **2.** See case. { 'sis·təm ˌyü·nət }

systolic array |COMPUT SCI| An array of processing elements of cells connected to a memory which pulses data through the array in such a way that each data item can be used effectively at each cell it passes while being pumped from cell to cell along the array. { si'stäl·ik ə'rā }

T

TΩ *See* teraohm.

table |COMPUT SCI| A set of contiguous, related items, each uniquely identified either by its relative position in the set or by some label. { 'tā-bəl }

table-driven compiler |COMPUT SCI| A compiler in which the source language is described by a set of syntax rules. { 'tā·bəl ¦driv·ən kəm'pī·lər }

table-driven program |COMPUT SCI| A computer program that relies on tables stored outside of the program in the computer's memory to furnish data. { 'tā·bəl ¦driv·ən 'prō₁gram }

table look-up |COMPUT SCI| A procedure for calculating the location of an item in a table by means of an algorithm, rather than by conducting a search for the item. { 'tā·bəl 'lùk ₁əp }

table look-up device |ELECTR| A logic circuit in which the input signals are grouped as address digits to a memory device, and, in response to any particular combination of inputs, the memory device location that is addressed becomes the output. { 'tā·bəl ¦lùk ₁əp di₁vīs }

table management program |COMPUT SCI| A computer program that handles the creation and maintenance of tables and access to data stored in them. { 'tā·bəl ¦man·ij·mənt ₁prō₁gram }

tabular language |COMPUT SCI| A part of a program which represents the composition of a decision table required by the problem considered. { 'tab·yə·lər 'laŋ·gwij }

tabulate |COMPUT SCI| To order a set of data into a table form, or to print a set of data as a table, usually indicating differences and totals, or just totals. { 'tab·yə₁lāt }

tabulation character |COMPUT SCI| A character that controls the action of a computer printer and is not itself printed, although it forms part of the data to be printed. { ₁tab·yə'lā·shən ₁kar·ik·tər }

tactical electronic warfare |ELECTR| The application of electronic warfare to tactical air operations; tactical electronic warfare encompasses the three major subdivisions of electronic warfare: electronic warfare support measures, electronic countermeasures, and electronic countercountermeasures. { 'tak·tə·kəl ₁i₁lek'trän·ik 'wòr ₁fer }

tactical frequency |COMMUN| Radio frequency assigned to a military unit to be used in the accomplishment of a tactical mission. { 'tak·tə·kəl 'frē·kwən·sē }

tactile feedback |COMPUT SCI| In haptics, devices that provide a user with the sensations of heat, pressure, and texture. { ₁tak·təl 'fēd₁bak }

tactile sensor |CONT SYS| A transducer, usually associated with a robot end effector, that is sensitive to touch; comprises stress and touch sensors. { 'tak·təl 'sen·sər }

Tafel slope |ELEC| The slope of a curve of overpotential or electrolytic polarization in volts versus the logarithm of current density. { 'tä·fəl ₁slōp }

tag |COMPUT SCI| **1.** A unit of information used as a label or marker. **2.** The symbol written in the location field of an assembly-language coding form, and used to define the symbolic address of the data or instruction written on that line. { 'tag }

tag converting unit |COMPUT SCI| A device capable of reading the perforations of a price tag as input data. { 'tag kən'vərd·iŋ ₁yü·nət }

tag field |COMPUT SCI| A data item within a variant record that identifies the format to be used in the record. { 'tag ₁fēld }

tag format |COMPUT SCI| The arrangement of data in a short record inserted in a direct-access storage to indicate the location of an overflow record. { 'tag ₁fòr₁mat }

tag image file format |COMPUT SCI| File format used for storing bitmap images at any resolution. Abbreviated TIFF. { ¦tag ₁im·ij 'fīl ₁fòr₁mat }

tag sort |COMPUT SCI| A method of sorting data in which the addresses of records rather than the records themselves are used to determine the sequence. { 'tag ₁sòrt }

tail |ELECTR| **1.** A small pulse that follows the main pulse of a radar set and rises in the same direction. **2.** The trailing edge of a pulse. { tāl }

tail clipping |ELECTR| Method of sharpening the trailing edge of a pulse. { 'tāl ₁klip·iŋ }

takedown |COMPUT SCI| The actions performed at the end of an equipment operating cycle to prepare the equipment for the next setup; for example, to remove the tapes from the tape handlers at the end of a computer run is a takedown operation. { 'tāk₁daùn }

takedown time |COMPUT SCI| The time required to take down a piece of equipment. { 'tāk₁daùn ₁tīm }

talk-back circuit *See* interphone. { 'tòk ₁bak ₁sər·kət }

talking battery *See* quiet battery. { 'tòk·iŋ ₁bad·ə·rē }

talk-listen switch [ENG ACOUS] A switch provided on intercommunication units to permit using the loudspeaker as a microphone when-desired. { 'tók 'lis·ən ,swich }

tandem [ELEC] Two-terminal pair networks are in tandem when the output terminals of one network are directly connected to the input terminals of the other network. { 'tan·dəm }

tandem central office [COMMUN] A telephone office that makes connections between local offices in an area where there is such a high density of local offices that it would be uneconomical to make direct connections between each of them. Also known as tandem office. { 'tan·dəm 'sen·trəl 'óf·əs }

tandem compensation See cascade compensation. { 'tan·dəm ,käm·pən'sā·shən }

tandem connection See cascade connection. { 'tan·dəm kə'nek·shən }

tandem distributed numerical control [CONT SYS] A form of distributed numerical control involving a series of machines connected by a conveyor and automatic loading and unloading devices that are under control of the central computers. { 'tan·dəm di¦strib·yəd·əd nú¦mer·ə·kəl kən'trōl }

tandem office See tandem central office. { 'tan·dəm 'óf·əs }

tandem switching [COMMUN] System of routing telephone calls in which calls do not travel directly between local offices, but rather through a tandem central office. { 'tan·dəm 'swich·iŋ }

tandem system [COMPUT SCI] A computing system in which there are two central processing units, usually with one controlling the other, and with data proceeding from one processing unit into the other. { 'tan·dəm ,sis·təm }

tank [ELECTR] **1.** A unit of acoustic delay-line storage containing a set of channels, each forming a separate recirculation path. **2.** The heavy metal envelope of a large mercury-arc rectifier or other gas tube having a mercury-pool cathode. **3.** See tank circuit. { taŋk }

tank circuit [ELECTR] A circuit which exhibits resonance at one or more frequencies, and which is capable of storing electric energy over a band of frequencies continuously distributed about the resonant frequency, such as a coil and capacitor in parallel. Also known as electrical resonator; tank. { 'taŋk ,sər·kət }

tantalum capacitor [ELEC] An electrolytic capacitor in which the anode is some form of tantalum; examples include solid tantalum, tantalum-foil electrolytic, and tantalum-slug electrolytic capacitors. { 'tant·əl·əm kə'pas·əd·ər }

tantalum-foil electrolytic capacitor [ELEC] An electrolytic capacitor that uses plain or etched tantalum foil for both electrodes, with a weak acid electrolyte. { 'tant·əl·əm ¦fóil i¦lek·trə¦lid·ik kə'pas·əd·ər }

tantalum nitride resistor [ELECTR] A thin-film resistor consisting of tantalum nitride deposited on a substrate, such as industrial sapphire. { 'tant·əl·əm 'nī,trīd ri'zis·tər }

tantalum-slug electrolytic capacitor [ELEC] An electrolytic capacitor that uses a sintered slug of tantalum as the anode, in a highly conductive acid electrolyte. { 'tant·əl·əm ¦sləg i¦lek·trə ¦lid·ik kə'pas·əd·ər }

T antenna [ELECTROMAG] An antenna consisting of one or more horizontal wires, with a lead-in connection being made at the approximate center of each wire. { 'tē an,ten·ə }

tap [ELEC] A connection made at some point other than the ends of a resistor or coil. { tap }

tap changer [ELEC] A device which is used to change the ratio of the input and output voltages of a transformer over any one of a definite number of steps. { 'tap ,chān·jər }

tap crystal [ELECTR] Compound semiconductor that stores current when stimulated by light and then gives up energy as flashes of light when it is physically tapped. { 'tap ,krist·əl }

tape [COMPUT SCI] A ribbonlike material used to store data in lengthwise sequential position. { tāp }

tape alternation [COMPUT SCI] The switching of a computer program back and forth between two tape units in order to avoid interruption of the program during mounting and removal of tape reels. { 'tāp ,ól·tər,nā·shən }

tape-automated bonding [ELECTR] A semiconductor chip (die) assembly method, where the chips are connected to polyimide (tape) carriers, complete with circuitry for attachment to a printed circuit board. The chip-bonded tape carriers typically are supplied on a reel (like a roll of film) for automated circuit assembly processes. { ¦tāp ,ód·ə,mād·əd 'bän·diŋ }

tape bootstrap routine [COMPUT SCI] A computer routine stored in the first block of a magnetic tape that instructs the computer to read certain programs from the tape. { 'tāp 'büt,strap rü ,tēn }

tape cluster See magnetic tape group. { 'tāp ,kləs·tər }

tape control unit [COMPUT SCI] A device which senses which tape unit is to be accessed for read or write purpose and opens up the necessary electronic paths. Formerly known as hypertape control unit. { 'tāp kən,trōl ,yü·nət }

tape crease [COMPUT SCI] A fold or wrinkle in a magnetic tape that results in an error in the reading or writing of data at that point. { 'tāp ,krēs }

tape deck [ENG ACOUS] A tape-recording mechanism that is mounted on a motor board, including the tape transport, electronics, and controls, but no power amplifier or loudspeaker. { 'tāp ,dek }

tape drive [COMPUT SCI] A tape reading or writing device consisting of a tape transport, electronics, and controls; it usually refers to magnetic tape exclusively. { 'tāp ,drīv }

tape editor [COMPUT SCI] A routine designed to help edit, revise, and correct a routine contained on a tape. { 'tāp ,ed·əd·ər }

tape group See magnetic tape group. { 'tāp ,grüp }

tape label [COMPUT SCI] A record appearing at the beginning or at the end of a magnetic tape to uniquely identify the tape as the one required by the system. { 'tāp ,lā·bəl }

tape library [COMPUT SCI] A special area, most often a room within a computer installation, used to store magnetic tapes. { 'tāp ,lī,brer·ē }

tape-limited [COMPUT SCI] Pertaining to a computer operation in which the time required to read and write tapes exceeds the time required for computation. { 'tāp ¦lim·əd·əd }

tape mark [COMPUT SCI] **1.** A special character or coding, an attached piece of reflective material, or other device that indicates the physical end of recording on a magnetic tape. Also known as destination warning mark; end-of-tape mark. **2.** A special character that divides a file of magnetic tape into sections, usually followed by a record with data describing the particular section of the file. Also known as control mark. { 'tāp ,märk }

tape operating system [COMPUT SCI] A computer operating system in which source programs and sometimes incoming data are stored on magnetic tape, rather than in the computer memory. Abbreviated TOS. { 'tāp 'äp·ə,rād·iŋ ,sis·təm }

tape player [ENG ACOUS] A machine designed only for playback of recorded magnetic tapes. { 'tāp ,plā·ər }

tape plotting system [COMPUT SCI] A digital incremental plotter in which the digital data are supplied from a magnetic or paper tape. { 'tāp 'pläd·iŋ ,sis·təm }

tape pool [COMPUT SCI] A collection of tape drives. { 'tāp ,pül }

tape-processing simultaneity [COMPUT SCI] A feature of some computer systems whereby reading or writing of data can be carried out on all the tape units at the same time, while the central processing unit continues to process data. { 'tāp ¦prä,ses·iŋ ,sī·məl·tə'nē·əd·ē }

taper [ELEC] Continuous or gradual change in electrical properties with mechanical position such as rotation or length; for example, continuous change of cross section of a waveguide, or distribution of resistance in a potentiometer. { 'tā·pər }

tape recorder [ENG ACOUS] A device that records audio signals and other information on magnetic tape by selective magnetization of iron oxide particles that form a thin film on the tape; a recorder usually also includes provisions for playing back the recorded material. { 'tāp ri ,kȯrd·ər }

tape recording [ENG ACOUS] The record made on a magnetic tape by a tape recorder. { 'tāp ri,kȯrd·iŋ }

tapered transmission line See tapered waveguide. { 'tā·pərd tranz'mish·ən ,līn }

tapered waveguide [ELECTROMAG] A waveguide in which a physical or electrical characteristic changes continuously with distance along the axis of the waveguide. Also known as tapered transmission line. { 'tā·pərd 'wāv,gīd }

tape search unit [COMPUT SCI] Small, fully transistorized, special-purpose, digital data-processing system using a stored program to perform logical functions necessary to search a magnetic tape in off-line mode, in response to a specific request. { 'tā·pər 'sərch ,yü·nət }

tape serial number [COMPUT SCI] A number identifying a magnetic tape which remains unchanged throughout the time the tape is used, even though all other information about the tape may change. { 'tā·pər 'sir·ē·əl ,nəm·bər }

tape skip [COMPUT SCI] A machine instruction to space forward and erase a portion of tape when a defect on the tape surface causes a write error to persist. { 'tāp ,skip }

tape station [COMPUT SCI] A tape reading or writing device consisting of a tape transport, electronics, and controls; it may use either magnetic tape or paper tape. { 'tāp ,stā·shən }

tape-to-tape conversion [COMPUT SCI] A routine which directs a computer to copy information from one tape to another tape of a different kind; for example, from a seven-track onto a nine-track tape. { 'tāp tə ¦tāp kən'vər·zhən }

tape transport [COMPUT SCI] The mechanism that physically moves a tape past a stationary head. Also known as transport. { 'tāp ,tranz,pȯrt }

tape unit [COMPUT SCI] A tape reading or writing device consisting of a tape transport, electronics, controls, and possibly a cabinet; the cabinet may contain one or more magnetic tape stations. { 'tāp ,yü·nət }

tapped control [ELECTR] A rheostat or potentiometer having one or more fixed taps along the resistance element, usually to provide a fixed grid bias or for automatic bass compensation. { 'tapt kən'trōl }

tapped-potentiometer function generator [ELECTR] A device used in analog computers for representing a function of one variable, consisting of a potentiometer with a number of taps held at voltages determined by a table of values of the variable; the input variable sets the angular position of a shaft that moves a slide contact, and the output voltage is taken from the slide contact. { 'tapt pə,ten·chē¦äm·əd·ər 'fəŋk·shən ,jen·ə,rād·ər }

tapped resistor [ELEC] A wire-wound fixed resistor having one or more additional terminals along its length, generally for voltage-divider applications. { 'tapt ri'zis·tər }

tap switch [ELEC] Multicontact switch used chiefly for connecting a load to any one of a number of taps on a resistor or coil. { 'tap ,swich }

target [ELECTR] **1.** In a television camera tube, the storage surface that is scanned by an electron beam to generate an output signal current corresponding to the charge-density pattern stored there. **2.** In radar and sonar, any object capable of reflecting the transmitted beam; depending on context, often connotes an object of interest as opposed to clutter. [ENG] In radar and sonar, any object capable of reflecting the transmitted beam. { 'tär·gət }

target acquisition [ELECTR] **1.** The first appearance of a recognizable and useful echo signal from a new target in radar and sonar. **2.** See acquisition. { 'tär·gət ,ak·wə'zish·ən }

target central processing unit [COMPUT SCI] The type of central processing unit for which a language processor (assembler, compiler, or interpreter) generates machine language output. { 'tär·gət ¦sen·trəl 'prä,ses·iŋ ,yü·nət }

573

target configuration |COMPUT SCI| The combination of input, output, and storage units and the amount of computer memory required to carry out an object program. { 'tär·gət kən,fig·yə ,rā·shən }

target cross section See echo area. { 'tär·gət 'krós ,sek·shən }

target-designating system |ELECTR| A system for designating to one instrument a target which has already been located by a second instrument; it employs electrical data transmitters and receivers which indicate on one instrument the pointing of another. { 'tär·gət ¦dez·ig,nād·iŋ ,sis·təm }

target discrimination |ELECTR| The ability of a detection or guidance system to distinguish a target from its background or to discriminate between two or more targets that are close acquisition. { 'tär·gət di,skrim·ə,nā·shən }

target language |COMPUT SCI| The language into which a program (or text) is to be converted. { 'tär·gət ¦laŋ·gwij }

target pack |COMPUT SCI| A disk pack that is used to maintain systems software and, in particular, to hold a copy of a system control program on which modifications are made and tested. { 'tär·gət ,pak }

target phase |COMPUT SCI| The stage of handling a computer program at which the object program is first carried out after it has been compiled. { 'tär·gət ,fāz }

target program See object program. { 'tär·gət ¦prō ,gram }

target routine See object program. { 'tär·gət rü ,tēn }

target signal |ELECTROMAG| The radio energy returned to a radar by a target. Also known as echo signal; video signal. { 'tär·gət ,sig·nəl }

target signature |ELECTR| Characteristic pattern of the target displayed by detection and classification equipment. { 'tär·gət ,sig·nə·chər }

task |COMPUT SCI| A set of instructions, data, and control information capable of being executed by the central processing unit of a digital computer in order to accomplish some purpose; in a multiprogramming environment, tasks compete with one another for control of the central processing unit, but in a nonmultiprogramming environment a task is simply the current work to be done. { task }

task descriptor |COMPUT SCI| The vital information about a task in a multitask system which must be saved when the task is interrupted. Also known as state vector. { 'task di,skrip·tər }

task management |COMPUT SCI| The functions, assumed by the operating system, of switching the processor among tasks, scheduling, sending messages or timing signals between tasks, and creating or removing tasks. { 'task ,man·ij·mənt }

task programmer |COMPUT SCI| A person who writes applications programs for controlling a robotic system. { 'task ,prō,gram·ər }

task switching |COMPUT SCI| Switching back and forth between two or more active programs

without having to close or open any of them. Also known as context switching. { 'task ,swich·iŋ }

T attenuator |ELEC| **1.** A resistive attenuator with three resistors forming a T network. **2.** A power-tap type of attenuator which removes part of the power from a main line through a T connection and dissipates the power, without reflection into the main line. { 'tē ə,ten·yə,wād·ər }

Taylor connection |ELEC| A transformer connection for converting three-phase power to two-phase power, or vice versa. { 'tā·lər kə ,nek·shən }

T circulator |ELECTROMAG| A circulator in which three identical rectangular waveguides are joined asymmetrically to form a T-shaped structure, with a ferrite post or wedge at its center; power entering any waveguide emerges from only one adjacent waveguide. { 'tē ,sər·kyə,lād·ər }

T connector |ELEC| A type of electric connector that joins a through conductor to another conductor at right angles to it. { 'tē kə,nek·tər }

TCP See Transmission Control Protocol.

TCP/IP See Transmission Control Protocol/Internet Protocol.

TD See transmitter-distributor.

TDD See display device.

TDM See time-division multiplexing.

TDMA See time-division multiple access.

TDR See time-domain reflectometer.

TDRSS See Tracking and Data Relay Satellite System.

TEA See transferred-electron amplifier.

teach |CONT SYS| To program a robot by guiding it through its motions, which are then recorded and stored in its computer. { tēch }

teach box See teach pendant. { 'tēch ,bäks }

teach-by-doing |CONT SYS| A method of programming a robot in which the operator guides the robot through its intended motions by holding it and performing the work. { ¦tēch ·bī 'dü· iŋ }

teach-by-driving |CONT SYS| Programming a robot by using a teach pendant. { ¦tēch ·bī 'drīv·iŋ }

teach gun See teach pendant. { 'tēch ,gən }

teaching interface |CONT SYS| The devices and hardware that are used to instruct robots and other machinery how to operate, and to specify their motions. { 'tēch·iŋ 'in·tər,fās }

teach mode |CONT SYS| The mode of operation in which a robot is instructed in its motions, usually by guiding it through these motions using a teach pendant. { 'tēch ,mōd }

teach pendant |CONT SYS| A hand-held device used to instruct a robot, specifying the character and types of motions it is to undertake. Also known as teach box; teach gun. { 'tēch ,pen·dənt }

tears |COMMUN| In an analog television picture, a horizontal disturbance caused by noise, in which the picture appears to be torn apart. { tirz }

teaser transformer |ELEC| Transformer, of two T-connected, single-phase units for three-phase to two-phase or two-phase to three-phase operation, which is connected between the midpoint of the main transformer and the third wire of the three-phase system. { 'tēz·ər tranz,fór·mər }

technetron [ELECTR] High-power multichannel field-effect transistor. { 'tek·nə,trän }

technical control board [ELEC] Testing position in a switch center or relay station with provisions for testing switches and associated access lines and trunks. { 'tek·nə·kəl kən'trōl ,bórd }

technical load [ELEC] Portion of a communications-electronics facility operational power load required for primary and ancillary equipment, including necessary lighting and air conditioning or ventilation required for full continuity of operation. { 'tek·nə·kəl 'lōd }

TEGFET See high-electron-mobility transistor. { 'teg,fet }

telautograph [COMMUN] A writing telegraph instrument, the forerunner of the facsimile machine, in which manual movement of a pen at the transmitting position varies the current in two circuits in such a way as to cause corresponding movements of a pen at the remote receiving instrument; ordinary handwriting can thus be transmitted over wires. { te'lód·ə,graf }

telecast [COMMUN] A television broadcast intended for reception by the general public, involving the transmission of the picture and sound portions of the program. { 'tel·ə,kast }

telechir [CONT SYS] A handlike remote manipulator. { 'tel·ə,kir }

telechirics [CONT SYS] The use of teleoperators or remote manipulators. { ¦tel·ə¦kir·iks }

telecine camera [ELECTR] A video camera used in conjunction with film or slide projectors to televise motion pictures and still images. { ¦tel·ə¦sin·ē 'kam·rə }

telecommunicating device for the deaf See telecommunications display device. { ¦tel·ə·kə'myü·nə·kād·iŋ di'vīs ,fór thə 'def }

telecommunications [COMMUN] Communication over long distances. { ¦tel·ə·kə,myü·nə'kā·shənz }

Telecommunications Coordinating Committee [COMMUN] Committee organized by the U.S. State Department and composed of major government departments, agencies, and industrial organizations; makes recommendations on telecommunications matters affecting international telecommunications. { ¦tel·ə·kə,myü·nə'kā·shənz kō'órd·ən,ād·iŋ kə,mid·ē }

telecommunications display device [COMMUN] A telephone equipped with a keyboard and display for users who have hearing or speech impairments. Also known as telecommunications device for the deaf; text telephone. Abbreviated TDD. { ¦tel·ə·kə,myü·nə'kā·shənz di'splā di,vīs }

teleconference [COMMUN] **1.** A two-way interactive meeting between relatively small groups of people remote from one another but linked by telecommunication facilities involving audio communication, and possibly also video, graphics, or facsimile. **2.** More broadly, any of various facilities allowing people to communicate among each other over some distance,

encompassing teleseminars and telemeetings. { ¦tel·ə'kän·frəns }

telegram [COMMUN] A message sent by telegraphy. { 'te·lə,gram }

telegraph alphabet See telegraph code. { 'tel·ə,graf 'al·fə,bet }

telegraph bandwidth [COMMUN] The difference between the limiting frequencies of a channel used to transmit telegraph signals. { 'tel·ə,graf 'band,width }

telegraph cable [ELEC] A uniform conductive circuit consisting of twisted pairs of insulated wires or coaxially shielded wires or combinations of each, used to carry telegraph signals. { 'tel·ə,graf ,kā·bəl }

telegraph carrier [COMMUN] The single-frequency wave which is modulated by transmitting apparatus in carrier telegraphy. { 'tel·ə,graf ,kar·ē·ər }

telegraph circuit [COMMUN] The complete wire or radio circuit over which signal currents flow between transmitting and receiving apparatus in a telegraph system. { 'tel·ə,graf ,sər·kət }

telegraph code [COMMUN] A system of symbols for transmitting telegraph messages in which each letter or other character is represented by a set of long and short electrical pulses, or by pulses of opposite polarity, or by time intervals of equal length in which a signal is present or absent. Also known as telegraph alphabet. { 'tel·ə,graf ,kōd }

telegraph concentrator [ELEC] Switching arrangement by means of which a number of branch or subscriber lines or station sets may be connected to a lesser number of trunklines, operating positions, or instruments through the medium of manual or automatic switching devices to obtain more efficient use of facilities. { 'tel·ə,graf ,kän·sən,trād·ər }

telegraph distributor [ELEC] Device which effectively associates one direct-current or carrier-telegraph channel in rapid succession with the elements of one or more sending or receiving devices. { 'tel·ə,graf di,strib·yəd·ər }

telegraph emission [COMMUN] The signal transmitted by a telegraph system, classified by type of transmission, type of modulation, bandwidth, and supplementary characteristics. { 'tel·ə,graf i,mish·ən }

telegraph grade [COMMUN] The class of communication circuits that can transmit only telegraphic signals, comprising the lowest types of circuits in regard to speed, accuracy, and cost. { 'tel·ə,graf ,grād }

telegraph interference [COMMUN] Any undesired electrical energy that tends to interfere with the reception of telegraph signals. { 'tel·ə,graf ,in·tər'fir·əns }

telegraph receiver [ELEC] A tape reperforator, teletypewriter, or other equipment which converts telegraph signals into a pattern of holes on a tape, printed letters, or other forms of information. { 'tel·ə,graf ri,sē·vər }

telegraph repeater [ELEC] A repeater inserted at intervals in long telegraph lines to amplify weak code signals, with or without reshaping of pulses, and to retransmit them automatically over the next section of the line.   { 'tel·ə‚graf ri‚pēd·ər }

telegraph signal distortion   [COMMUN] Time displacement of transitions between conditions, such as marking and spacing, with respect to their proper relative positions in perfectly timed signals; the total distortion is the algebraic sum of the bias and the characteristic and fortuitous distortions.   { 'tel·ə‚graf ¦sig·nəl di‚stòr·shən }

telegraph transmitter   [ELEC] A device that controls an electric power source in order to form telegraph signals.   { 'tel·ə‚graf tranz‚mid·ər }

telegraphy   [COMMUN] Communication at a distance by means of code signals consisting of current pulses sent over wires or by radio; it is the oldest form of electrical digital communication.   { tə'leg·rə·fē }

telemeeting   [COMMUN] A meeting between people remote from one another, but linked by audio and video telecommunications facilities that provide primarily one-way communication from a few people at one location to large numbers of people at other locations, and use temporary equipment or circuits.   { 'tel·ə‚mēd·iŋ }

telemetering   [ENG] Transmitting the readings of instruments to a remote location by means of wires, radio waves, or other means. Also known as remote metering; telemetry.   { ‚tel·ə'mēd·ə·riŋ }

telemetering antenna   [ELECTROMAG] A highly directional antenna, generally mounted on a servo-controlled mount for tracking purposes, used at ground stations to receive telemetering signals from a guided missile or spacecraft.   { ‚tel·ə'mēd·ə·riŋ an'ten·ə }

telemetering receiver   [ELECTR] A device in a telemetering system which converts electrical signals into an indication or recording of the value of the quantity being measured at a distance.   { ‚tel·ə'mēd·ə·riŋ ri'sē·vər }

telemetering transmitter   [ELECTR] A device which converts the readings of instruments into electrical signals for transmission to a remote location by means of wires, radio waves, or other means.   { ‚tel·ə'mēd·ə·riŋ tranz'mid·ər }

telemetry   See telemetering.   { tə'lem·ə·trē }

teleoperation   [ENG] **1.** The real-time control of remotely located machines that act as the eyes and hands of a person located elsewhere, it has been used in undersea and lunar exploration, mining, and microsurgery. **2.** Operation from a remote location. Also known as remote manipulation.   { ‚tel·ē‚äp·ə'rā·shən }

teleoperator   See remote manipulator.   { ‚tel·ē ‚äp·ə‚rād·ər }

telephone   [COMMUN] A system of converting sound waves into variations in electric current or other electrical quantities that can be transmitted and reconverted into sound waves at a distant point, used primarily for voice communication; it consists essentially of a telephone transmitter and receiver at each station, interconnecting wires, cables, optical fibers, or terrestrial or satellite radio transmission systems, signaling devices, a central power supply, and switching

facilities. Also known as telephone system.   [ENG ACOUS]   See telephone set.   { 'tel·ə‚fōn }

telephone-answering system   [COMMUN] A special type of private branch exchange system used by a telephone-answering service bureau to provide secretarial service for its customers.   { 'tel·ə‚fōn ¦an·sə·riŋ ‚sis·təm }

telephone carrier current   [ELEC] A carrier current used for telephone communication over power lines or to obtain more than one channel on a single pair of wires.   { 'tel·ə‚fōn 'kar·ē·ər ‚kə·rənt }

telephone central office   See central office.   { 'tel·ə‚fōn 'sen·trəl 'óf·əs }

telephone channel   [COMMUN] A one-way or two-way path suitable for the transmission of audio signals between two stations.   { 'tel·ə‚fōn ‚chan·əl }

telephone circuit   [ELEC] The complete circuit over which audio and signaling currents travel in a telephone system between the two telephone subscribers in communication with each other; the circuit usually consists of insulated conductors, a radio link, or a fiber-optic cable.   { 'tel·ə ‚fōn ‚sər·kət }

telephone data set   [COMPUT SCI] Equipment interfacing a data terminal with a telephone circuit.   { 'tel·ə‚fōn 'dad·ə ‚set }

telephone dial   [ENG] **1.** A switch operated by a finger wheel, used to make and break a pair of contacts the required number of times for setting up a telephone circuit to the party being called. **2.** By extension, the push-button apparatus used to generate dual-tone multifrequency (DTMF) signals.   { 'tel·ə‚fōn ‚dīl }

telephone emission   See telephone signal.   { 'tel·ə ‚fōn i‚mish·ən }

telephone induction coil   [ELEC] A coil used in a telephone circuit to match the impedance of the line to that of a telephone transmitter or receiver.   { 'tel·ə‚fōn in'dək·shən ‚kòil }

telephone influence factor   [COMMUN] A measure of the interference of power-line harmonics with telephone lines, which is derived by weighting the terms in the mathematical expression for the total harmonic distortion of the power-line voltage.   { 'tel·ə‚fōn 'in·flü·əns ‚fak·tər }

telephone line   [ELEC] The conductors extending between telephone subscriber stations and central offices.   { 'tel·ə‚fōn ‚līn }

telephone loading coil   See loading coil.   { 'tel·ə ‚fōn 'lōd·iŋ ‚kòil }

telephone modem   [ELECTR] A piece of equipment that modulates and demodulates one or more separate telephone circuits, each containing one or more telephone channels; it may include multiplexing and demultiplexing circuits, individual amplifiers, and carrier-frequency sources.   { 'tel·ə‚fōn 'mō‚dem }

telephone pickup   [ELEC] A large flat coil placed under a telephone set to pick up both voices during a telephone conversation for recording purposes.   { 'tel·ə‚fōn ‚pik·əp }

telephone plug   See phone plug.   { 'tel·ə‚fōn ‚pləg }

telephone receiver   [ENG ACOUS] The portion of a telephone set that converts the audio-frequency current variations of a telephone line into sound waves, by the motion of a diaphragm activated by a magnet whose field is varied by the electrical impulses that come over the telephone wire. { 'tel·ə,fōn ri,sē·vər }

telephone relay [ELEC] A relay having a multiplicity of contacts on long spring strips mounted parallel to the coil, actuated by a lever arm or other projection of the hinged armature; used chiefly for switching in telephone circuits. { 'tel·ə,fōn ,rē,lā }

telephone repeater [ELECTR] A repeater inserted at one or more intermediate points in a long telephone line to amplify telephone signals so as to maintain the required current strength. { 'tel·ə,fōn ri,pēd·ər }

telephone repeating coil [ELEC] A coil used in a telephone circuit for inductively coupling two sections of a line when a direct connection is undesirable. { 'tel·ə,fōn ri'pēd·iŋ ,kȯil }

telephone ringer [ELECTROMAG] **1.** An electromagnetic device that actuates a clapper which strikes one or more gongs to produce a ringing sound; used with a telephone set to signal a called party. **2.** By extension, the electronic device that performs the same function. { 'tel·ə ,fōn ,riŋ·ər }

telephone set [ENG ACOUS] An assembly including a telephone transmitter, a telephone receiver, and associated switching and signaling devices. Also known as phone; telephone. { 'tel·ə ,fōn ,set }

telephone signal [COMMUN] The electrical signal transmitted by a telephone system, classified by type of transmission, type of modulation, bandwidth, and supplementary characteristics. Also known as telephone emission. { 'tel·ə,fōn ,sig·nəl }

telephone switchboard *See* switchboard. { 'tel·ə ,fōn 'swich,bȯrd }

telephone system *See* telephone. { 'tel·ə,fōn ,sis·təm }

telephone transmitter [ENG ACOUS] The microphone used in a telephone set to convert speech into audio-frequency electric signals. { 'tel·ə ,fōn tranz,mid·ər }

telephony [COMMUN] The transmission of speech to a distant point by means of electric signals. { tə'lef·ə·nē }

telephoto *See* facsimile. { ¦tel·ə'fōd·ō }

telephotography *See* facsimile. { ¦tel·ə·fə'täg·rə·fē }

teleport [COMMUN] A planned business development area that features direct and economic access to a large number of domestic and international satellites for users in the surrounding region, with the aid of a regional distribution network. { 'tel·ə,pȯrt }

telepresence [CONT SYS] The quality of sensory feedback from a teleoperator or telerobot to a human operator such that the operator feels present at the remote site. { ¦tel·ə'prez·əns }

teleprinter [COMPUT SCI] Any typewriter-type device capable of being connected to a computer and of printing out a set of messages under computer control. { 'tel·ə,print·ər }

teleprinting [COMMUN] Telegraphy in which the transmitter and receiver are teletypewriters. { 'tel·ə,print·iŋ }

teleprocessing [COMPUT SCI] **1.** The use of telecommunications equipment and systems by a computer. **2.** A computer service involving input/output at locations remote from the computer itself. { 'tel·ə,prä,ses·iŋ }

teleprocessing monitor [COMPUT SCI] A computer program that manages the transfer of information between local and remote terminals. Abbreviated TP monitor. { 'tel·ə,prä,ses·iŋ 'män·əd·ər }

telering [ELECTR] In telephony, a frequency-selector device for the production of ringing power. { 'tel·ə,riŋ }

telerobot [CONT SYS] A type of teleoperator that embodies features of a robot and is programmed for communication with a human operator in a high-level language but can revert to direct control in the event of unplanned contingencies. { ,tel·ə'rō,bät }

teleseminar [COMMUN] A form of long-distance, electronic communication, primarily one-way, to many destinations from one source, for educational purposes, involving audio communication, and possibly also video and some form of graphics. { ¦tel·ə'sem·ə,när }

telesynd [ELECTR] Telemeter or remote-control equipment which is synchronous in both speed and position. { 'tel·ə,sind }

teleterminal [COMPUT SCI] An instrument that integrates the functions of a telephone set and a computer terminal with keyboard and video screen. { ¦tel·ə'tər·mən·əl }

telethesis [ENG] A robotic manipulation aid for the physically disabled that may be located remote from the body. There are two forms, operated by voice command, or operated through a body-powered prosthesis or a joystick. { tə'le·th·ə·səs }

teletypewriter [COMMUN] A special electric typewriter that produces coded electric signals corresponding to manually typed characters, and automatically types messages when fed with similarly coded signals produced by another machine; it allows access to telephone services for people who are deaf, or who have a hearing, speech, or communication impairment. Also known as TWX machine. Abbreviated TTY. { ¦tel·ə'tīp,rīd·ər }

teletypewriter code [COMMUN] Special code in which each code group is made up of five units, or elements, of equal length which are known as marking or spacing impulses; the five-unit start-stop code consists of five signal impulses preceded by a start impulse and followed by a stop impulse. { ¦tel·ə'tīp,rīd·ər ,kōd }

teletypewriter exchange service [COMMUN] A service furnished by telephone companies to subscribers in the United States, whereby any of the subscribers can communicate directly with any other subscriber via teletypewriter. Also known as TWX service. { ¦tel·ə'tīp,rīd·ər iks ¦chānj ,sər·vəs }

teletypewriter signal distortion | COMMUN | Of a start-stop teletypewriter signal, the shifting of the transition points of the signal pulses from their proper positions relative to the beginning of the start pulse; the magnitude of the distortion is expressed in percent of a perfect unit pulse length. { ¦tel·ə'tīp ‚rīd·ər ¦sig·nəl di‚stȯr·shən }

televise | COMMUN | To pick up a scene with a video camera and convert it into corresponding electric signals for transmission by a television station. { 'tel·ə‚vīz }

television | COMMUN | A system for converting a succession of visual images into corresponding electric signals and transmitting these signals by radio or over wires to distant receivers at which the signals can be used to reproduce the original images. Abbreviated TV. { 'tel·ə‚vizh·ən }

television antenna | ELECTROMAG | An antenna suitable for transmitting or receiving television broadcasts; since television transmissions in the United States are horizontally polarized, the most basic type of receiving antenna is a horizontally mounted half-wave dipole. { 'tel·ə‚vizh·ən an ‚ten·ə }

television bandwidth | COMMUN | The difference between the limiting frequencies of a television channel; in the United States, this is 6 megahertz. { 'tel·ə‚vizh·ən 'band‚width }

television broadcast band | COMMUN | Several groups of channels, each containing a number of 6-megahertz channels, that are available for assignment to television broadcast stations. { 'tel·ə‚vizh·ən 'brȯd‚kast ‚band }

television broadcasting | COMMUN | Transmission of television programs by means of radio waves for reception by the public. { 'tel·ə‚vizh·ən 'brȯd‚kast·iŋ }

television camera | ELECTR | The pickup unit used to convert a scene into corresponding electric signals; optical lenses focus the scene to be televised on the photosensitive surface of a camera tube, and the tube breaks down the visual image into small picture elements and converts the light intensity of each element in turn into a corresponding electric signal. Also known as camera. { 'tel·ə‚vizh·ən ‚kam·rə }

television camera tube See camera tube. { 'tel·ə ‚vizh·ən 'kam·rə ‚tüb }

television channel | COMMUN | A band of frequencies 6 megahertz wide in the television broadcast band, available for assignment to a television broadcast station. { 'tel·ə‚vizh·ən ‚chan·əl }

television emission See television signal. { 'tel·ə ‚vizh·ən i‚mish·ən }

television interference | COMMUN | Interference produced in television receivers by other transmitting devices. Abbreviated TVI. { 'tel·ə ‚vizh·ən ‚in·tər'fir·əns }

television monitor | ELECTR | A display device used to continuously check the image picked up by a television camera and the sound picked up by video camera or other source to provide continuous observation of image content and/or quality. { 'tel·ə‚vizh·ən ‚man·əd·ər }

television network | COMMUN | An arrangement of communication channels, suitable for transmission of video and accompanying audio signals, which link together groups of television broadcasting stations or closed-circuit television users in different cities so that programs originating at one point can be fed simultaneously to all others. { 'tel·ə‚vizh·ən ‚net‚wərk }

television pickup station | COMMUN | A land mobile station used for the transmission of television program material and related communications from the scene of an event occurring at a point remote to a television broadcast station. { 'tel·ə‚vizh·ən 'pik·əp ‚stā·shən }

television picture tube See picture tube. { 'tel·ə ‚vizh·ən 'pik·chər ‚tüb }

television receive only antenna | COMMUN | A parabolic reflector or dish with sufficient gain to receive signals from geostationary satellites, together with a feed horn that collects the signals reflected by the dish, a low-noise amplifier for preamplification, and a tunable satellite receiver. Abbreviated TVRO. { 'tel·ə‚vizh·ən ri'sēv ¦ōn·lē an'ten·ə }

television receiver | ELECTR | A receiver that converts incoming television signals into the original scenes along with the associated sounds. Also known as television set. { 'tel·ə‚vizh·ən ri ‚sē·vər }

television relay system See television repeater. { 'tel·ə‚vizh·ən 'rē‚lā ‚sis·təm }

television repeater | ELECTR | A repeater that transmits television signals from point to point by using radio waves in free space as a medium, such transmission not being intended for direct reception by the public. Also known as television relay system. { 'tel·ə‚vizh·ən ri‚pēd·ər }

television screen | ELECTR | The fluorescent screen of the picture tube in a television receiver. { 'tel·ə‚vizh·ən ‚skrēn }

television set See television receiver. { 'tel·ə ‚vizh·ən ‚set }

television signal | COMMUN | A general term for the aural and visual signals that are broadcast together to provide the sound and picture portions of an analog television program. Also known as television emission. { 'tel·ə‚vizh·ən ‚sig·nəl }

television station | COMMUN | The installation, assemblage of equipment, and location where radio transmissions are sent or received. { 'tel·ə ‚vizh·ən ‚stā·shən }

television studio | COMMUN | A complex of rooms specifically designed for the origination of live or taped television programs. { 'tel·ə‚vizh·ən ‚stüd·ē·ō }

television transmitter | ELECTR | An electronic device that converts the audio and video signals of a television program into modulated radio-frequency energy that can be radiated from an antenna and received on a television receiver. { 'tel·ə‚vizh·ən tranz‚mid·ər }

television tuner |ELECTR| A component in a television receiver that selects the desired channel and converts the frequencies received to lower frequencies within the passband of the intermediate-frequency chain. { 'tel·ə,vizh·ən ,tü·nər }

telewriter |COMMUN| System in which writing movement at the transmitting end causes corresponding movement of a writing instrument at the receiving end. { 'tel·ə,rīd·ər }

Telex |COMMUN| A worldwide teleprinter exchange service providing direct send and receive teleprinter connections between subscribers. Abbreviated TEX. { 'te,leks }

telluric current *See* earth current. { tə'lūr·ik ,kə·rənt }

TELNET *See* network terminal protocol. { 'tel ,net }

TEM mode *See* transverse electromagnetic mode. { ,tē,ē'em ,mōd }

TE mode *See* transverse electric mode. { ¦tē'ē ,mōd }

temperature-compensated Zener diode |ELECTR| Positive-temperature-coefficient reversed-bias Zener diode (*pn* junction) connected in series with one or more negative-temperature forward-biased diodes within a single package. { 'tem·prə·chər ¦kām·pən,sād·əd 'zē·nər 'dī,ōd }

temperature-compensating capacitor |ELEC| Capacitor whose capacitance varies with temperature in a known and predictable manner; used extensively in oscillator circuits to compensate for changes in the values of other parts with temperatures. { 'tem·prə·chər ¦käm·pən,sād·iŋ kə'pas·əd·ər }

temperature compensation |ELECTR| The process of making some characteristic of a circuit or device independent of changes in ambient temperature. { 'tem·prə·chər ,käm·pən,sā·shən }

temperature resistance coefficient |ELEC| The ratio of the change of electrical resistance in a wire caused by a change in its temperature of 1°C as related to its resistance at 0°C. { 'tem·prə·chər ri'zis·təns ,kō·i,fish·ənt }

temperature saturation |ELECTR| The condition in which the anode current of a thermionic vacuum tube cannot be further increased by increasing the cathode temperature at a given value of anode voltage; the effect is due to the space charge formed near the cathode. Also known as filament saturation; saturation. { 'tem·prə·chər ,sach·ə,rā·shən }

temperature sensor |ENG| A device designed to respond to temperature stimulation. { 'tem·prə·chər ,sen·sər }

temperature transducer |ENG| A device in an automatic temperature-control system that converts the temperature into some other quantity such as mechanical movement, pressure, or electric voltage; this signal is processed in a controller, and is applied to an actuator which controls the heat of the system. { 'tem·prə·chər tranz ,dü·sər }

template |COMPUT SCI| **1.** A prototype pattern against which observed patterns are matched in a pattern recognition system. **2.** A computer program that is used in conjunction with an electronic spreadsheet to solve a particular type of problem. { 'tem·plət }

template matching |COMPUT SCI| The comparison of a picture or other data with a stored program or template, for purposes of identification or inspection. { 'tem·plət ,mach·iŋ }

temporary file |COMPUT SCI| A file that is created during the execution of a computer program to hold interim results and is erased before the program is completed. { 'tem·pə,rer·ē 'fīl }

temporary storage |COMPUT SCI| The storage capacity reserved or used for retention of temporary or transient data. { 'tem·pə,rer·ē 'stōr·ij }

TEM wave *See* transverse electromagnetic wave. { ,tē,ē'em ,wāv }

terahertz technology |ENG| The generation, detection, and application (such as in communications and imaging) of electromagnetic radiation roughly in the frequency range from 0.05 to 20 terahertz, corresponding to wavelengths from 6 millimeters down to 15 micrometers. { ,ter·ə ,harts tek'näl·ə·jē }

teraohm |ELEC| A unit of electrical resistance, equal to 10^{12} ohms, or 1,000,000 megohms. Abbreviated TΩ. { 'ter·ə,ōm }

teraohmmeter |ENG| An ohmmeter having a teraohm range for measuring extremely high insulation resistance values. { ¦ter·ə'ōm,mēd·ər }

terminal |COMPUT SCI| A site or location at which data can leave or enter a system. |ELEC| **1.** A screw, soldering lug, or other point to which electric connections can be made. Also known as electric terminal. **2.** The equipment at the end of a microwave relay system or other communication channel. **3.** One of the electric input or output points of a circuit or component. { 'tər·mən·əl }

terminal area |ELECTR| The enlarged portion of conductor material surrounding a hole for a lead on a printed circuit. Also known as land; pad. { 'tər·mən·əl ¦er·ē·ə }

terminal block |COMMUN| **1.** A cluster of five captive screw terminals at which a telephone pair terminates; the center terminal is for the ground wire, and two other terminals are used for the tip and ring wires. **2.** By extension, a similar cluster of any number of screw terminals. { 'tər·mən·əl ,bläk }

terminal board |ELEC| An insulating mounting for terminal connections. Also known as terminal strip. { 'tər·mən·əl ,bōrd }

terminal box |ELEC| An enclosure which includes, mounts, and protects one or more terminals or terminal boards; it may include a cover and such accessories as mounting hardware, brackets, locks, and conduit fittings. { 'tər·mən·əl ,bäks }

terminal cutout pairs |ELEC| Numbered, designated pairs brought out of a cable at a terminal. { 'tər·mən·əl ,kəd,aút ,perz }

terminal equipment |COMMUN| **1.** Assemblage of communications-type equipment required to transmit or receive a signal on a channel or circuit, whether it be for delivery or relay. **2.** In radio relay systems, equipment used at points where intelligence is inserted or derived, as distinct from equipment used to relay a reconstituted signal. **3.** Telephone and teletypewriter switchboards and other centrally located equipment at which wire circuits are terminated. { 'tər·mən·əl i,kwip·mənt }

terminal leg See terminal stub. { 'tər·mən·əl ,leg }

terminal network |COMPUT SCI| A system that links intelligent terminals through a communications channel. { 'tər·mən·əl 'net,wərk }

terminal pair |ELEC| An associated pair of accessible terminals, such as the input or output terminals of a device or network. { 'tər·mən·əl 'per }

terminal repeater |COMMUN| **1.** Assemblage of equipment designed specifically for use at the end of a communications circuit, as contrasted with the repeater designed for an intermediate point. **2.** Two microwave terminals arranged to provide for the interconnection of separate systems, or separate sections of a system. { 'tər·mən·əl ri'pēd·ər }

terminal room |COMMUN| In telephone practice, a room associated with a central office, private branch exchange, or private exchange, which contains distributing frames, relays, and similar apparatus, except that mounted in the switchboard section. { 'tər·mən·əl ,rüm }

terminal station |COMMUN| Receiving equipment and associated multiplex equipment used at the ends of a radio-relay system. { 'tər·mən·əl ,stā·shən }

terminal strip See terminal board. { 'tər·mən·əl ,strip }

terminal stub |ELEC| Piece of cable that comes with a cable terminal for splicing into the main cable. Also known as terminal leg. { 'tər·mən·əl ,stəb }

terminal vertex |MATH| A vertex in a rooted tree that has no successor. Also known as leaf. { 'tər·mən·əl 'vər,teks }

terminal voltage |ELEC| The voltage at the terminals connected to the source of electricity for an electric machine. { 'tər·mən·əl ,vōl·tij }

terminate and stay resident See RAM resident. { |tər·mə,nāt ən ,stā 'rez·ə·dənt }

terminated line |ELEC| Transmission line terminated in a resistance equal to the characteristic impedance of the line, so there is no reflection and no standing waves. { 'tər·mə,nād·əd 'līn }

terminating |ELEC| Closing of the circuit at either end of a line or transducer by connecting some device thereto; terminating does not imply any special condition such as the elimination of reflection. { 'tər·mə,nād·iŋ }

ternary code |COMMUN| Code in which each code element may be any one of three distinct kinds or values. { 'tər·nə·rē 'kōd }

ternary incremental representation |COMPUT SCI| A type of incremental representation in which the value of the change in a variable is defined as $+1$, -1, or 0. { 'tər·nə·rē ,iŋ·krə'ment·əl ,rep·ri·zən'tā·shən }

ternary pulse code modulation |COMMUN| Pulse code modulation in which each code element may be any one of three distinct kinds or values. { 'tər·nə·rē 'pəls ¦kōd ,mäj·ə'lā·shən }

terrain echoes See ground clutter. { tə'rān 'ek· ōz }

tertiary storage |COMPUT SCI| Any of several types of computer storage devices, usually consisting of magnetic tape transports and mass storage tape systems, which have slower access times, larger capacity, and lower cost than main storage or secondary storage. { 'tər·shē,er·ē 'stōr·ij }

tertiary winding See stabilized winding. { 'tər·shē ,er·ē 'wīnd·iŋ }

testboard |ELEC| Switchboard equipped with testing apparatus, arranged so that connections can be made from it to telephone lines or central-office equipment for testing purposes. { 'test,bórd }

test clip |ELEC| A spring clip used at the end of an insulated wire lead to make a temporary connection quickly for test purposes. { 'test ,klip }

test data |COMPUT SCI| A set of data developed specifically to test the adequacy of a computer run or system; the data may be actual data that has been taken from previous operations, or artificial data created for this purpose. { 'test ,dad·ə }

test file |COMPUT SCI| A file consisting of test data. { 'test ,fīl }

testing level |ELEC| Value of power used for reference represented by 0.001 watt working in 600 ohms. { 'test·iŋ ,lev·əl }

test jack |ELEC| **1.** Appearance of a circuit or circuit element in jacks for testing purposes. **2.** In recent practice, a jack multipled with the switchboard operating jack. { 'test ,jak }

test lead |ELEC| A flexible insulated lead, usually with a test prod at one end, used for making tests, connecting instruments to a circuit temporarily, or making other temporary connections. { 'test ,lēd }

test oscillator See signal generator. { 'test ,äs·ə ,lād·ər }

test pattern |COMMUN| A chart having various combinations of lines, squares, circles, and graduated shading used to check definition, linearity, and contrast of a video system. Also known as resolution chart. { 'test ,päd·ərn }

test point |ELEC| A terminal or plug-in connector provided in a circuit to facilitate monitoring, calibration, or trouble-shooting. { 'test ,póint }

test prod |ELEC| A metal point attached to an insulating handle and connected to a test lead for convenience in making a temporary connection to a terminal while tests are being made. Also known as prod. { 'test ,präd }

test program See check routine. { 'test ,prō ,gram }

test record |COMPUT SCI| A record within a test file. { 'test ,rek·ərd }

test routine See check routine. { 'test rü,tēn }

test run |COMPUT SCI| The performance of a computer program to check that it is operating correctly, by using test data to generate results that can be compared with expected answers. { 'test ¦rən }

test set |ELECTR| A combination of instruments needed for servicing a particular type of electronic equipment. { 'test ,set }

test system |COMPUT SCI| **1.** A computer system that is being tested before being used for production work. **2.** A version of a computer system that is retained, even after a live system is in use, chiefly to diagnose problems without interfering with the work of the live system. { 'test ,sis·təm }

test under mask |COMPUT SCI| A procedure for checking the status of selected bits in a byte by comparing the byte with another byte in which these selected bits are set to one and the other bits are set to zero. { 'test ,ən·dər 'mask }

tetrode |ELECTR| A four-electrode electron tube containing an anode, a cathode, a control electrode, and one additional electrode that is ordinarily a grid. { 'te,trōd }

tetrode junction transistor See double-base junction transistor. { 'te,trōd 'jəŋk·shən tran,zis·tər }

tetrode thyratron |ELECTR| A thyratron with two control electrodes. Also known as gas tetrode. { 'te,trōd 'thī·rə,trän }

tetrode transistor |ELECTR| A four-electrode transistor, such as a tetrode point-contact transistor or double-base junction transistor. { 'te ,trōd tran'zis·tər }

TE wave See transverse electric wave. { ¦tē'ē ,wāv }

text |COMMUN| The part of a message that conveys information, excluding bits or characters needed to facilitate transmission of the message. { tekst }

text-editing system |COMPUT SCI| A computer program, together with associated hardware, for the on-line creation and modification of computer programs and ordinary text. { 'tekst ¦ed·əd·iŋ ,sis·təm }

text-to-speech synthesizer |ENG ACOUS| A voice response system that provides an automatic means to take a specification of any English text at the input and generate a natural and intelligible acoustic speech signal at the output by using complex sets of rules for predicting the needed phonemic states directly from the input message and dictionary pronunciations. { ¦tekst tə ¦spēch 'sin·thə,sīz·ər }

TFT See thin-film transistor.

thallofide cell |ELECTR| A photoconductive cell in which the active light-sensitive material is thallium oxysulfide in a vacuum; it has maximum response at the red end of the visible spectrum and in the near infrared. { 'thal·ə,fīd ,sel }

theater television |ELECTR| A large projection-type television receiver used in theaters, generally for closed-circuit showing of important sport events. { 'thē·ə·dər 'tel·ə,vizh·ən }

theoretical cutoff frequency |ELEC| Of an electric structure, a frequency at which, disregarding the effects of dissipation, the attenuation constant changes from zero to a positive value or vice versa. { ,thē·ə'red·ə·kəl 'kəd,óf ,frē·kwən·sē }

theory of games See game theory. { 'thē·ə·rē əv 'gāmz }

thermal agitation |SOLID STATE| Random movements of the free electrons in a conductor, producing noise signals that may become noticeable when they occur at the input of a high-gain amplifier. Also known as thermal effect. { 'thər·məl ,aj·ə'tād·shən }

thermal ammeter See hot-wire ammeter. { 'thər·məl 'am,ēd·ər }

thermal battery |ELEC| **1.** A combination of thermal cells. Also known as fused-electrolyte battery; heat-activated battery. **2.** A voltage source consisting of a number of bimetallic junctions connected to produce a voltage when heated by a flame. { 'thər·məl ¦bad·ə·rē }

thermal cell |ELEC| A reserve cell that is activated by applying heat to melt a solidified electrolyte. { 'thər·məl ¦sel }

thermal converter |ELECTR| A device that converts heat energy directly into electric energy by using the Seebeck effect; it is composed of at least two dissimilar materials, one junction of which is in contact with a heat source and the other junction of which is in contact with a heat sink. Also known as thermocouple converter; thermoelectric generator; thermoelectric power generator; thermoelement. { ENG| An instrument used with external resistors for ac current and voltage measurements over wide ranges, consisting of a conductor heated by an electric current, with one or more hot junctions of a thermocouple attached to it, so that the output emf responds to the temperature rise, and hence the current. { 'thər·məl kən'vərd·ər }

thermal cutout |ELEC| A heat-sensitive switch that automatically opens the circuit of an electric motor or other device when the operating temperature exceeds a safe value. { 'thər·məl 'kəd ,aút }

thermal drift |ELECTR| Drift caused by internal heating of equipment during normal operation or by changes in external ambient temperature. { 'thər·məl 'drift }

thermal effect See thermal agitation. { 'thər·məl i'fekt }

thermal flasher |ELEC| An electric device that opens and closes a circuit automatically at regular intervals because of alternate heating and cooling of a bimetallic strip that is heated by a resistance element in series with the circuit being controlled. { 'thər·məl 'flash·ər }

thermal horsepower |ELEC| Electrical motor horsepower as determined by current readings from a thermal-type ammeter; will be higher than load horsepower determined from kilowatt-input methods. Also known as true motor load. { 'thər·məl 'hórs,paú·ər }

thermal imagery |ELECTR| Imagery produced by measuring and recording electronically the thermal radiation of objects. { 'thər·məl 'im·ij·rē }

thermal instrument [ENG] An instrument that depends on the heating effect of an electric current, such as a thermocouple or hot-wire instrument. { 'thər·məl 'in·strə·mənt }

thermal limit [ELEC] A limit on the power carried by an electric power system that results from the heating effects of the power carried by the devices. { 'thər·məl 'lim·ət }

thermal microphone [ENG ACOUS] Microphone depending for its action on the variation in the resistance of an electrically heated conductor that is being alternately increased and decreased in temperature by sound waves. { 'thər·məl 'mī·krə,fōn }

thermal noise [COMMUN] See Gaussian noise. [ELECTR] Electric noise produced by thermal agitation of electrons in conductors and semiconductors. Also known as Johnson noise; resistance noise. { 'thər·məl 'nȯiz }

thermal noise generator [ELECTR] A generator that uses the inherent thermal agitation of an electron tube to provide a calibrated noise source. { 'thər·məl ,nȯiz ,jen·ə,rād·ər }

thermal power plant [ENG] A facility to produce electric energy from thermal energy released by combustion of a fuel or consumption of a fissionable material. { 'thər·məl 'pau̇·ər ,plant }

thermal regenerative cell [ELEC] Fuel-cell system in which the reactants are regenerated continuously from the products formed during the cell reaction. { 'thər·məl rē'jen·rəd·iv 'sel }

thermal relay [ELEC] A relay operated by the heat produced by current flow. { 'thər·məl 'rē,lā }

thermal resistance See effective thermal resistance. { 'thər·məl ri'zis·təns }

thermal resistor [ELEC] A resistor designed so its resistance varies in a known manner with changes in ambient temperature. { 'thər·məl ri'zis·tər }

thermal runaway [ELECTR] A condition that may occur in a power transistor when collector current increases collector junction temperature, reducing collector resistance and allowing a greater current to flow, which, in turn, increases the heating effect. { 'thər·məl 'rən·ə,wā }

thermal switch [ELEC] A temperature-controlled switch. Also known as thermoswitch. { 'thər·məl 'swich }

thermal tuning [ELEC] The process of changing the operating frequency of a system by using controlled thermal expansion to alter the geometry of the system. { 'thər·məl 'tün·iŋ }

thermal volt See kelvin. { 'thər·məl 'vōlt }

thermal wattmeter [ENG] A wattmeter in which thermocouples are used to measure the heating produced when a current is passed through a resistance. { 'thər·məl 'wät,mēd·ər }

thermion [ELECTR] A charged particle, either negative or positive, emitted by a heated body, as by the hot cathode of a thermionic tube. { |thərm'ī,än }

thermionic [ELECTR] Pertaining to the emission of electrons as a result of heat. { ,thər·mē'än·ik }

thermionic cathode See hot cathode. { ,thər·mē'än·ik 'ka,thōd }

thermionic converter [ELECTR] A device in which heat energy is directly converted to electric energy; it has two electrodes, one of which is raised to a sufficiently high temperature to become a thermionic electron emitter, while the other, serving as an electron collector, is operated at a significantly lower temperature. Also known as thermionic generator; thermionic power generator; thermoelectric engine. { ,thər·mē'än·ik kən'vərd·ər }

thermionic current [ELECTR] Current due to directed movements of thermions, such as the flow of emitted electrons from the cathode to the plate in a thermionic vacuum tube. { ,thər·mē'än·ik 'kə·rənt }

thermionic detector [ELECTR] A detector using a hot-cathode tube. { ,thər·mē'än·ik di'tek·tər }

thermionic diode [ELECTR] A diode electron tube having a heated cathode. { ,thər·mē'än·ik 'dī,ōd }

thermionic emission [ELECTR] **1.** The outflow of electrons into vacuum from a heated electric conductor. Also known as Edison effect; Richardson effect. **2.** More broadly, the liberation of electrons or ions from a substance as a result of heat. { ,thər·mē'än·ik i'mish·ən }

thermionic fuel cell [ELECTR] A thermionic converter in which the space between the electrodes is filled with cesium or other gas, which lowers the work functions of the electrodes, and creates an ionized atmosphere, controlling the electron space charge. { ,thər·mē'än·ik 'fyül ,sel }

thermionic generator See thermionic converter. { ,thər·mē'än·ik 'jen·ə,rād·ər }

thermionic power generator See thermionic converter. { ,thər·mē'än·ik 'pau̇·ər 'jen·ə,rād·ər }

thermionics [ELECTR] The study and applications of thermionic emission. { ,thər·mē'än·iks }

thermionic triode [ELECTR] A three-electrode thermionic tube, containing an anode, a cathode, and a control electrode. { ,thər·mē'än·ik 'trī,ōd }

thermionic tube [ELECTR] An electron tube that relies upon thermally emitted electrons from a heated cathode for tube current. Also known as hot-cathode tube. { ,thər·mē'än·ik 'tüb }

thermionic work function [ELECTR] Energy required to transfer an electron from the fermi energy in a given metal through the surface to the vacuum just outside the metal. { ,thər·mē'än·ik 'wərk ,fəŋk·shən }

thermistor [ELECTR] A resistive circuit component, having a high negative temperature coefficient of resistance, so that its resistance decreases as the temperature increases; it is a stable, compact, and rugged two-terminal ceramiclike semiconductor bead, rod, or disk. Derived from thermal resistor. { thər'mis·tər }

thermoammeter [ENG] An ammeter that is actuated by the voltage generated in a thermocouple through which is sent the current to be measured; used chiefly for measuring radio-frequency currents. Also known as electrothermal ammeter; thermocouple ammeter. { |thər·mō'am ,ēd·ər }

thermocompression bonding [ENG] Use of a combination of heat and pressure to make connections, as when attaching beads to integrated-circuit chips; examples include wedge bonding and ball bonding. { ¦thər·mō·kəm'presh·ən 'bänd·iŋ }

thermocouple [ENG] A device consisting basically of two dissimilar conductors joined together at their ends; the thermoelectric voltage developed between the two junctions is proportional to the temperature difference between the junctions, so the device can be used to measure the temperature of one of the junctions when the other is held at a fixed, known temperature, or to convert radiant energy into electric energy. { 'thər·mə,kəp·əl }

thermocouple ammeter See thermoammeter. { 'thər·mə,kəp·əl 'am,ēd·ər }

thermocouple converter See thermal converter. { 'thər·mə,kəp·əl kən'vərd·ər }

thermoelectric converter [ELECTR] A converter that changes solar or other heat energy to electric energy; used as a power source on spacecraft. { ¦thər·mō·i'lek·trik kən'vərd·ər }

thermoelectric engine See thermionic converter. { ¦thər·mō·i'lek·trik 'en·jən }

thermoelectric generator See thermal converter. { ¦thər·mō·i'lek·trik 'jen·ə,rād·ər }

thermoelectric junction See thermojunction. { ¦thər·mō·i'lek·trik 'jəŋk·shən }

thermoelectric material [ELECTR] A material that can be used to convert thermal energy into electric energy or provide refrigeration directly from electric energy; good thermoelectric materials include lead telluride, germanium telluride, bismuth telluride, and cesium sulfide. { ¦thər·mō·i'lek·trik mə'tir·ē·əl }

thermoelectric power generator See thermal converter. { ¦thər·mō·i'lek·trik 'paú·ər ,jen·ə,rād·ər }

thermoelectric solar cell [ELECTR] A solar cell in which the sun's energy is first converted into heat by a sheet of metal, and the heat is converted into electricity by a semiconductor material sandwiched between the first metal sheet and a metal collector sheet. { ¦thər·mō·i'lek·trik 'sō·lər 'sel }

thermoelectromotive force [ELEC] Voltage developed due to differences in temperature between parts of a circuit containing two or more different metals. { ¦thər·mō·i¦lek·trə¦mōd·iv 'fórs }

thermoelectron [ELECTR] An electron liberated by heat, as from a heated filament. Also known as negative thermion. { ¦thər·mō·i'lek,trän }

thermoelement See thermal converter. { ¦thər·mō'el·ə·mənt }

thermogalvanometer [ENG] Instrument for measuring small high-frequency currents by their heating effect, generally consisting of a direct-current galvanometer connected to a thermocouple that is heated by a filament carrying the current to be measured. { ¦thər·mō·gal·və'näm·əd·ər }

thermojunction [ELECTR] One of the surfaces of contact between the two conductors of a thermocouple. Also known as thermoelectric junction. { ¦thər·mō'jəŋk·shən }

thermojunction battery [ELEC] Nuclear-type battery which converts heat into electrical energy directly by the thermoelectric or Seebeck effect. { ¦thər·mō'jəŋk·shən 'bad·ə·rē }

thermomigration [ELECTR] A technique for doping semiconductors in which exact amounts of known impurities are made to migrate from the cool side of a wafer of pure semiconductor material to the hotter side when the wafer is heated in an oven. { ¦thər·mō·mī'grā·shən }

thermopile [ENG] An array of thermocouples connected either in series to give higher voltage output or in parallel to give higher current output, used for measuring temperature or radiant energy or for converting radiant energy into electric power. { 'thər·mə,pīl }

thermopile generator [ELEC] An electricity source powered by the heating of an electrical resistor that can be connected to a thermopile to generate small amounts of electric current. { 'thər·mə,pīl 'jen·ə,rād·ər }

thermoplastic recording [ELECTR] A recording process in which a modulated electron beam deposits charges on a thermoplastic film, and application of heat by radio-frequency heating electrodes softens the film enough to produce deformation that is proportional to the density of the stored electrostatic charges; an optical system is used for playback. { ¦thər·mə¦plas·tik ri'kórd·iŋ }

thermopower [ELEC] A measure of the temperature-induced voltage in a conductor. { 'thər·mə,paú·ər }

thermoregulator [ENG] A high-accuracy or high-sensitivity thermostat; one type consists of a mercury-in-glass thermometer with sealed-in electrodes, in which the rising and falling column of mercury makes and breaks an electric circuit. { ¦thər·mō'reg·yə,lād·ər }

thermorelay See thermostat. { ¦thər·mō'rē,lā }

thermostat [ENG] An instrument which measures changes in temperature and directly or indirectly controls sources of heating and cooling to maintain a desired temperature. Also known as thermorelay. { 'thər·mə,stat }

thermostatic switch [ELEC] A temperature-operated switch that receives its operating energy by thermal conduction or convection from the device being controlled or operated. { ¦thər·mə¦stad·ik 'swich }

thermoswitch See thermal switch. { 'thər·mə ,swich }

thermovoltmeter [ENG] A voltmeter in which a current from the voltage source is passed through a resistor and a fine vacuum-enclosed platinum heater wire; a thermocouple, attached to the midpoint of the heater, generates a voltage of a few millivolts, and this voltage is measured by a direct-current millivoltmeter. { ¦thər·mō 'vōlt ,mēd·ər }

Thévenin equivalent circuit [ELEC] An equivalent circuit that consists of a series connection of a voltage source and a two-terminal circuit, where the voltage source is usually dependent on the electric signals applied to the input terminals. { tā·vō¦na i¸kwiv·ə·lənt 'sər·kət }

Thévenin generator |ELEC| The voltage generator in the equivalent circuit of Thévenin's theorem. { tā·vō'na ¸jen·ə¸rād·ər }

Thévenin's theorem |ELEC| A theorem in network problems which allows calculation of the performance of a device from its terminal properties only: the theorem states that at any given frequency the current flowing in any impedance, connected to two terminals of a linear bilateral network containing generators of the same frequency, is equal to the current flowing in the same impedance when it is connected to a voltage generator whose generated voltage is the voltage at the terminals in question with the impedance removed, and whose series impedance is the impedance of the network looking back from the terminals into the network with all generators replaced by their internal impedances. Also known as Helmholtz's theorem. { tā·vō'naz ¸thir·əm }

thick-film capacitor |ELEC| A capacitor in a thick-film circuit, made by successive screen-printing and firing processes. { 'thik ¦film kə 'pas·əd·ər }

thick-film circuit |ELECTR| A microcircuit in which passive components, of a ceramic-metal composition, are formed on a ceramic substrate by successive screen-printing and firing processes, and discrete active elements are attached separately. { 'thik ¦film 'sər¦kət }

thick-film hybrid |ELECTR| An assembly consisting of a thick-film circuit pattern with mounting positions for the insertion of conventional silicon devices. { ¸thik ¸film 'hī·brəd }

thick-film resistor |ELEC| Fixed resistor whose resistance element is a film well over 0.001 inch (25 micrometers) thick. { 'thik ¦film ri'zis·tər }

thimble |COMPUT SCI| A cone-shaped, rotating printing element on an impact printer having character slugs around the perimeter and a hammer that drives the appropriate slug forward to print the impression on paper. { 'thim·bəl }

thin film |ELECTR| A film a few molecules thick deposited on a glass, ceramic, or semiconductor substrate to form a capacitor, resistor, coil, cryotron, or other circuit component. { 'thin 'film }

thin-film capacitor |ELEC| A capacitor that can be constructed by evaporation of conductor and dielectric films in sequence on a substrate; silicon monoxide is generally used as the dielectric. { 'thin ¦film kə'pas·əd·ər }

thin-film circuit |ELECTR| A circuit in which the passive components and conductors are produced as films on a substrate by evaporation or sputtering; active components may be similarly produced or mounted separately. { 'thin ¦film 'sər·kət }

thin-film cryotron |ELECTR| A cryotron in which the transition from superconducting to normal resistivity of a thin film of tin or indium, serving as a gate, is controlled by current in a film of lead that crosses and is insulated from the gate. { 'thin ¦film 'krī·ə¸trän }

thin-film field-emitter cathode |ELECTR| A sharply pointed microminiature electron field emitter with an integral low-voltage extraction gate. { ¦thin ¸film ¸fēld i¸mid·ər 'kath¸ōd }

thin-film integrated circuit |ELECTR| An integrated circuit consisting entirely of thin films deposited in a patterned relationship on a substrate. { 'thin ¦film 'int·ə¸grād·əd 'sər·kət }

thin-film material |ELECTR| A material that can be deposited as a thin film in a desired pattern by a variety of chemical, mechanical, or high-vacuum evaporation techniques. { 'thin ¦film mə'tir·ē·əl }

thin-film memory See thin-film storage. { 'thin ¦film 'mem·rē }

thin-film resistor |ELEC| A fixed resistor whose resistance element is a metal, alloy, carbon, or other film having a thickness of about 0.000001 inch (25 nanometers). { 'thin ¦film ri'zis·tər }

thin-film semiconductor |ELECTR| Semiconductor produced by the deposition of an appropriate single-crystal layer on a suitable insulator. { 'thin ¦film 'sem·i·kən¸dək·tər }

thin-film solar cell |ELECTR| A solar cell in which a thin film of gallium arsenide, cadmium sulfide, or other semiconductor material is evaporated on a thin, flexible metal or plastic substrate; the rather low efficiency (about 2%) is compensated by the flexibility and light weight, making these cells attractive as power sources for spacecraft. { 'thin ¦film 'sō·lər 'sel }

thin-film storage |COMPUT SCI| A high-speed storage device that is fabricated by depositing layers, one molecule thick, of various materials which, after etching, provide microscopic circuits which can move and store data in small amounts of time. Also known as thin-film memory. { 'thin ¦film 'stòr·ij }

thin-film transistor |ELECTR| A field-effect transistor constructed entirely by thin-film techniques, for use in thin-film circuits. Abbreviated TFT. { 'thin ¦film tran'zis·tər }

think time |COMPUT SCI| Idle time between time intervals in which transmission takes place in a real-time system. { 'thiŋk ¸tīm }

thin list See loose list. { 'thin 'list }

third-generation computer |COMPUT SCI| One of the general purpose digital computers introduced in the late 1960s; it is characterized by integrated circuits and has logical organization and software which permit the computer to handle many programs at the same time, allow one to add or remove units from the computer, permit some or all input/output operations to occur at sites remote from the main processor, and allow conversational programming techniques. { 'thərd ¸jen·ə¦rā·shən kəm'pyüd·ər }

Thomson bridge See Kelvin bridge. { 'täm·sən ¸brij }

thoriated emitter See thoriated tungsten filament. { 'thòr·ē¸ād·əd i'mid·ər }

thoriated tungsten filament [ELECTR] A vacuum-tube filament consisting of tungsten mixed with a small quantity of thorium oxide to give improved electron emission. Also known as thoriated emitter. { 'thȯr·ē,ād·əd ¦təŋ·stən 'fil·ə·mənt }

thrashing [COMPUT SCI] An undesirable condition in a multiprogramming system, due to overcommitment of main memory, in which the various tasks compete for pages and none can operate efficiently. { 'thrash·iŋ }

thread [COMPUT SCI] A sequence of beads that are strung together. { thred }

threat [COMPUT SCI] An event that can cause harm to computers, to their data or programs, or to computations. { thret }

three-address code [COMPUT SCI] In computers, a multiple-address code which includes three addresses, usually two addresses from which data are taken and one address where the result is entered; location of the next instruction is not specified, and instructions are taken from storage in preassigned order. { 'thrē 'ad,res ,kōd }

three-address instruction [COMPUT SCI] In computers, an instruction which includes an operation and specifies the location of three registers. { 'thrē 'ad,res in'strək·shən }

three-dimensional display system [ELECTR] A radar display showing range, azimuth, and elevation simultaneously. { 'thrē di¦men·chən·əl di'splä ,sis·təm }

three-dimensional sound See virtual acoustics. { ¦thrē də,men·shən·əl 'saúnd }

three-input adder See full adder. { 'thrē ¦in,pút 'ad·ər }

three-input subtracter See full subtracter. { 'thrē ¦in,pút səb'trak·tər }

three-junction transistor [ELECTR] A pnpn transistor having three junctions and four regions of alternating conductivity; the emitter connection may be made to the p region at the left, the base connection to the adjacent n region, and the collector connection to the n region at the right, while the remaining p region is allowed to float. { 'thrē ¦jəŋk·shən tran'zis·tər }

three-layer diode [ELECTR] A junction diode with three conductivity regions. { 'thrē ¦lā·ər 'dī,ōd }

three-level subroutine [COMPUT SCI] A subroutine in which a second subroutine is called, and a third subroutine is called by the second subroutine. { 'thrē ¦lev·əl 'səb·rü,tēn }

three-phase circuit [ELEC] A circuit energized by alternating-current voltages that differ in phase by one-third of a cycle or 120°. { 'thrē ¦fāz 'sər·kət }

three-phase current [ELEC] Current delivered through three wires, with each wire serving as the return for the other two and with the three current components differing in phase successively by one-third cycle, or 120 electrical degrees. { 'thrē ¦fāz 'kə·rənt }

three-phase four-wire system [ELEC] System of alternating-current supply comprising four conductors, three of which are connected as in a three-phase, three-wire system, the fourth being connected to the neutral point of the supply, which may be grounded. { 'thrē ¦fāz 'fȯr ¦wīr 'sis·təm }

three-phase magnetic amplifier [ELECTR] A magnetic amplifier whose input is the sum of three alternating-current voltages that differ in phase by 120°. { 'thrē ¦fāz mag'ned·ik 'am·plə,fī·ər }

three-phase motor [ELEC] An alternating-current motor operated from a three-phase circuit. { 'thrē ,fāz 'mōd·ər }

three-phase rectifier [ELEC] A rectifier supplied by three alternating-current voltages that differ in phase by one-third of a cycle or 120°. { 'thrē ¦fāz 'rek·tə,fī·ər }

three-phase seven-wire system [ELEC] System of alternating-current supply from groups of three single-phase transformers connected in Y to obtain a three-phase, four-wire grounded neutral system of higher voltage for power, the neutral wire being common to both systems. { 'thrē¦fāz 'sev·ən ¦wīr 'sis·təm }

three-phase three-wire system [ELEC] System of alternating-current supply comprising three conductors between successive pairs of which are maintained alternating differences of potential successively displaced in phase by one-third cycle. { 'thrē ¦fāz 'thrē ¦wīr 'sis·təm }

three-phase transformer [ELEC] A transformer used in a three-phase circuit, with three sets of primary and secondary windings on a single core. { 'thrē ,fāz tranz'fȯr·mər }

three-plus-one address [COMPUT SCI] An instruction format containing an operation code, three operand address parts, and a control address. { 'thrē ,pləs ¦wən 'ad,res }

three-pulse canceler [ELECTR] A moving-target indicator technique in which two "two-pulse cancelers" are cascaded together, improving the velocity response by widening the rejection around zero Doppler and, unavoidably, around each associated ambiguity. { 'thrē ¦pəls 'kan·slər }

three-pulse cascaded canceler [ELECTR] A moving-target indicator technique in which two "two-pulse cancelers" are cascaded together; this improves the velocity response. { 'thrē ¦pəls kas'kād·əd 'kan·slər }

three-way switch [ELEC] An electric switch with three terminals used to control a circuit from two different points. { 'thrē ¦wā 'swich }

three-wire generator [ELEC] Electric generator with a balance coil connected across the armature, the midpoint of the coil providing the potential of the neutral wire in a three-wire system. { 'thrē ¦wir ¦jen·ə,rād·ər }

three-wire system [ELEC] System of electric supply comprising three conductors, one of which (known as the neutral wire) is maintained at a potential midway between the potential of the other two (referred to as the outer conductors); part of the load may be connected directly between the outer conductors, the remainder being divided as evenly as possible into two parts, each of which is connected between the neutral and one outer conductor; there are thus two distinct supply voltages, one being twice the other. { 'thrē ¦wir 'sis·təm }

threshold [ELECTR] In a modulation system, the smallest value of carrier-to-noise ratio at the input to the demodulator for all values above which a small percentage change in the input carrier-to-noise ratio produces a substantially equal or smaller percentage change in the output signal-to-noise-ratio. [ENG] The least value of a current, voltage, or other quantity that produces the minimum detectable response in an instrument or system. { 'thresh,hōld }

threshold element [COMPUT SCI] A logic circuit which has one output and several weighted inputs, and whose output is energized if and only if the sum of the weights of the energized inputs exceeds a prescribed threshold value. { 'thresh ,hōld ,el·ə·mənt }

threshold frequency [ELECTR] The frequency of incident radiant energy below which there is no photoemissive effect. { 'thresh,hōld ,frē·kwən·sē }

thresholding [COMPUT SCI] In machine vision, the comparison of an element's brightness or other characteristic with a set value or threshold. { 'thresh,hōld·iŋ }

threshold signal [ELECTROMAG] A received radio signal (or radar echo) whose power is just above the noise level of the receiver. Also known as minimum detectable signal. { 'thresh,hōld ,sig·nəl }

threshold switch [ELECTR] A voltage-sensitive alternating-current switch made from a semiconductor material deposited on a metal substrate; when the alternating-current voltage acting on the switch is increased above the threshold value, the number of free carriers present in the semiconductor material increases suddenly, and the switch changes from a high resistance of about 1 megohms to a low resistance of less than 1 ohm; in other versions of this switch, the threshold voltage is controlled by heat, pressure, light, or moisture. { 'thresh,hōld ,swich }

threshold value [COMPUT SCI] A point beyond which there is a change in the manner a program executes; in particular, an error rate above which the operating system shuts down the computer system on the assumption that a hardware failure has occurred. [CONT SYS] The minimum input that produces a corrective action in an automatic control system. { 'thresh,hōld ,val·yü }

threshold voltage [ELECTR] **1.** In general, the voltage at which a particular characteristic of an electronic device first appears. **2.** The voltage at which conduction of current begins in a *pn* junction. **3.** The voltage at which channel formation occurs in a metal oxide semiconductor field-effect transistor. **4.** The voltage at which a solid-state lamp begins to emit light. { 'thresh,hōld ,vōl·tij }

throttling [CONT SYS] Control by means of intermediate steps between full on and full off. { 'thräd·əl·iŋ }

throughput [COMMUN] A measure of the effective rate of transmission of data by a communications system. [COMPUT SCI] The productivity of a data-processing system, as expressed in computing work per minute or hour. { 'thrü,pùt }

through repeater [ELECTR] Microwave repeater that is not equipped to provide for connections to any local facilities other than the service channel. { 'thrü ri,pēd·ər }

throw-away device [ELECTR] An electronic component that is not serviced and is discarded and replaced upon failure. { 'thrō ə,wā di,vīs }

thump [ENG ACOUS] Low-frequency transient disturbance in a system or transducer characterized audibly by the vocal imitation of the word. { thəmp }

thunk [COMPUT SCI] An additional subprogram created by the compiler to represent the evaluation of the argument of an expression in the call-by-name procedure. { thəŋk }

thyratron [ELECTR] A hot-cathode gas tube in which one or more control electrodes initiate but do not limit the anode current except under certain operating conditions. Also known as hot-cathode gas-filled tube. { 'thī·rə,trän }

thyratron gate [ELECTR] In computers, an AND gate consisting of a multielement gas-filled tube in which conduction is initiated by the coincident application of two or more signals; conduction may continue after one or more of the initiating signals are removed. { 'thī·rə,trän ,gāt }

thyratron inverter [ELECTR] An inverter circuit that uses thyratrons to convert direct-current power to alternating-current power. { 'thī·rə ,trän in,vərd·ər }

thyrector [ELECTR] Silicon diode that acts as an insulator up to its rated voltage, and as a conductor above rated voltage; used for alternating-current surge voltage protection. { 'thī'rek·tər }

thyristor [ELECTR] A transistor having a thyratronlike characteristic; as collector current is increased to a critical value, the alpha of the unit rises above unity to give high-speed triggering action. { 'thī'ris·tər }

tick [COMMUN] A pulse broadcast at 1-second intervals by standard frequency- and time-broadcasting stations to indicate the exact time. [COMPUT SCI] A time interval equal to $\frac{1}{60}$ second, used primarily in discussing computer operations. { tik }

tickler coil [ELECTR] Small coil connected in series with the plate circuit of an electron tube and inductively coupled to a grid-circuit coil to establish feedback or regeneration in a radio circuit; used chiefly in regenerative detector circuits. { 'tik·lər ,kòil }

tie [ELEC] **1.** Electrical connection or strap. **2.** *See* tie wire. { tī }

tie cable [ELEC] **1.** Cable between two distributing frames or distributing points. **2.** Cable between two private branch exchanges. **3.** Cable between a private branch exchange switchboard and main office. **4.** Cable connecting two other cables. { 'tī ,kā·bəl }

tie line [COMMUN] **1.** A leased communication channel or circuit. **2.** *See* data link. { 'tī ,līn }

tie point [ELEC] Insulated terminal to which two or more wires may be connected. { 'tī ,pòint }

tie trunk [ELEC] Telephone line or channel directly connecting two private branch exchanges. { 'tī ,trəŋk }

tie wire [ELEC] A short piece of wire used to tie an open-line wire to an insulator. Also known as tie. { 'tī ,wīr }

TIF See telephone influence factor.

TIFF See tag image file format. { tif }

tight coupling See close coupling. { 'tīt 'kəp·liŋ }

tightly coupled computer [COMPUT SCI] A computer linked to another computer in a manner that requires both computers to function as a single unit. { 'tīt·lē ¦kəp·əld kəm'pyüd·ər }

tile painting [COMPUT SCI] **1.** The use of patterns to create shadings that fill shapes and areas on a monochrome display. **2.** The use of very small dots of two or more colors to make blends or shades that fill shapes and areas on a color display. { 'tīl ,pānt·iŋ }

tiling [COMPUT SCI] Dividing an electronic display into two or more nonoverlapping areas that display the outputs of different programs being run concurrently on a computer. { 'tīl·iŋ }

time assignment speech interpolation [COMMUN] Modulation technique based on the fact that speech is never a continuous stream of information, but consists of a large number of short signals; therefore, the period between the speech signals is used for transmitting other data including additional speech signals. { 'tīm ə¦sīn·mənt 'spēch ,in·tər·pə,lā·shən }

time base [ELECTR] A device which moves the fluorescent spot rhythmically across the screen of the cathode-ray tube. { 'tīm ,bās }

time-base generator See sweep generator. { 'tīm ¦bās ,jen·ə,rād·ər }

time-code generator [ELECTR] A crystal-controlled pulse generator that produces a train of pulses with various predetermined widths and spacings, from which the time of day and sometimes also day of year can be determined; used in telemetry and other data-acquisition systems to provide the precise time of each event. { 'tīm ¦kōd ,jen·ə ,rā·dər }

time-controlled system See clock control system. { 'tīm kən¦trōld ,sis·təm }

time-current characteristics [ELEC] Of a fuse, the relation between the root-mean-square alternating current or direct current and the time for the fuse to perform the whole or some specified part of its interrupting function. { 'tīm 'kə·rənt ,kar·ik·tə,ris·tiks }

time-delay circuit [ELECTR] A circuit in which the output signal is delayed by a specified time interval with respect to the input signal. Also known as delay circuit. { 'tīm di¦lā ,sər·kət }

time-delay fuse [ELEC] A fuse in which the burnout action depends on the time it takes for the overcurrent heat to build up in the fuse and melt the fuse element. { 'tīm di¦lā ,fyüz }

time-delay relay [ELEC] A relay in which there is an appreciable interval of time between energizing or deenergizing of the coil and movement of

the armature, such as a slow-acting relay and a slow-release relay. { 'tīm di¦lā ,rē,lā }

time-derived channel [COMMUN] Any of the channels which result from time-division multiplexing of a channel. { 'tīm di¦rīvd ,chan·əl }

time-division data links [COMMUN] Radio communications which use time-division techniques for channel separation. { 'tīm di,vizh·ən 'dad·ə ,liŋks }

time-division multiple access [COMMUN] A technique that allows multiple users who are geographically dispersed to gain access to a communications channel, by permitting each user access to the full pass-band of the channel for a limited time, after which the access right is assigned to another user. Abbreviated TDMA. { ¦tīm də ,vizh·ən ,məl·tə·pəl 'ak,ses }

time-division multiplexing [COMMUN] A process for transmitting two or more signals over a common path by using successive time intervals for different signals. Also known as time multiplexing. Abbreviated TDM. [COMPUT SCI] The interleaving of bits or characters in time to compensate for the slowness of input devices as compared to data transmission lines. { 'tīm di ¦vizh·ən ,məl·tə,pleks·iŋ }

time-division multiplier See mark-space multiplier. { 'tīm di¦vizh·ən ,məl·tə,plī·ər }

time-division switching system [ELECTR] A type of electronic switching system in which input signals on lines and trunks are sampled periodically, and each active input is associated with the desired output for a specific phase of the period. { 'tīm di¦vizh·ən 'swich·iŋ ,sis·təm }

time-domain reflectometer [ELECTR] An instrument that measures the electrical characteristics of wideband transmission systems, subassemblies, components, and lines by feeding in a voltage step and displaying the superimposed reflected signals on an oscilloscope equipped with a suitable time-base generator. Abbreviated TDR. { 'tīm də¦mān ,rē,flek'täm· əd·ər }

time factor See time scale. { 'tīm ,fak·tər }

time gate [ELECTR] A circuit that gives an output only during chosen time intervals. { 'tīm ,gāt }

time-height section [ELECTR] A facsimile trace of a vertically directed radar; specifically, a cloud-detection radar. { 'tīm ·hīt ,sek·shən }

time hopping [COMMUN] A spread spectrum technique, usually used in combination with other methods, in which the transmitted pulse occurs in a manner determined by a pseudorandom code which places the pulse in one of several possible positions per frame. { 'tīm ,häp·iŋ }

time-invariant system [CONT SYS] A system in which all quantities governing the system's behavior remain constant with time, so that the system's response to a given input does not depend on the time it is applied. { 'tīm in,ver·ē· ənt ,sis·təm }

time-mark generator | ELECTR | A signal generator that produces highly accurate clock pulses which can be superimposed as pips on a cathode-ray screen for timing the events shown on the display. { 'tīm ¦märk ˌjen·ə·rād·ər }

time modulation | COMMUN | Modulation in which the time of occurrence of a definite portion of a waveform is varied in accordance with a modulating signal. { 'tīm ˌmäj·ə͵lā·shən }

time multiplexing See multiprogramming time-division multiplexing. { 'tīm ˌməl·tə͵pleks·iŋ }

time-of-day clock | COMPUT SCI | An electronic device that registers the actual time, generally accurate to 0.1 second, through a 24-hour cycle, and transmits its reading to the central processing unit of a computer upon demand. { ¦tīm əv ¦dā ˌkläk }

time of delivery | COMMUN | The time at which the addressee or responsible relay agency provides a receipt for a message. { 'tīm əv di'liv·ə·rē }

time of origin | COMMUN | The time at which a message is released for transmission. { 'tīm əv 'är·ə·jən }

time of receipt | COMMUN | The time at which a receiving station completes reception of a message. { 'tīm əv ri'sēt }

time-pulse distributor | ELECTR | A device or circuit for allocating timing pulses or clock pulses to one or more conducting paths or control lines in specified sequence. { 'tīm ˌpəls di͵strib·yəd·ər }

time quantum See time slice. { 'tīm ˌkwän·təm }

timer | COMPUT SCI | A hardware device that can interrupt a computer program after a time interval specified by the program, generally to remind the program to take some action. | ELECTR | A circuit used in radar and in electronic navigation systems to start pulse transmission and synchronize it with other actions, such as the start of a cathode-ray sweep. { 'tīm·ər }

timer clock | COMPUT SCI | An electronic device in the central processing unit of a computer which times events that occur during the operation of the system in order to carry out such functions as changing computer time, detecting looping and similar error conditions, and keeping a log of operations. { 'tī·mər ˌkläk }

time redundancy | COMPUT SCI | Performing a computation more than once and checking the results in order to increase reliability. { 'tīm ri ˌdən·dən·sē }

time scale | COMPUT SCI | The ratio of the time duration of an event as simulated by an analog computer to the actual time duration of the event in the physical system under study. Also known as time factor. { 'tīm ˌskāl }

time-share | COMPUT SCI | To perform several independent processes almost simultaneously by interleaving the operations of the processes on a single high-speed processor. { 'tīm ˌsher }

time-shared amplifier | ELECTR | An amplifier used with a synchronous switch to amplify signals from different sources one after another. { 'tīm ¦sherd ˌam·plə͵fī·ər }

time-sharing | COMPUT SCI | The simultaneous utilization of a computer system from multiple terminals. { 'tīm ˌsher·iŋ }

time signal | COMMUN | An accurate signal which is broadcast by radio and marks a specified time or time interval, used for setting timepieces and for determining their errors; in particular, a radio signal broadcast at accurately known times each day on a number of different frequencies by WWV and other stations. { 'tīm ˌsig·nəl }

time signal service | COMMUN | Radio communications service for the transmission of time signals of stated high precision, intended for general reception. { 'tīmd 'sig·nəl ˌsər·vəs }

time slice | COMPUT SCI | A time interval during which a time-sharing system is processing one particular computer program. Also known as time quantum. { 'tīm ˌslīs }

time-stamp | COMMUN | A term that indicates the time of a specific action such as the arrival of a byte or the presentation of a presentation unit. { 'tīm ˌstamp }

time switch | ENG | A clock-controlled switch used to open or close a circuit at one or more predetermined times. { 'tīm ˌswich }

time-varying system | CONT SYS | A system in which certain quantities governing the system's behavior change with time, so that the system will respond differently to the same input at different times. { 'tīm ¦ver·ē·iŋ ˌsis·təm }

timing-axis oscillator See sweep generator. { 'tīm·iŋ ˌak·səs ˌäs·ə͵lād·ər }

timing circuit See clock. { 'tīm·iŋ ˌsər·kət }

timing error | COMPUT SCI | An error made in planning or writing a computer program, usually in underestimating the time that will be taken by input/output or other operations, which causes unnecessary delays in the execution of the program. { 'tīm·iŋ ˌer·ər }

timing loop | COMPUT SCI | A set of instructions in a computer program whose execution time is known and whose only function is to cause a delay in processing by causing the loop to be executed an appropriate number of times. { 'tīm·iŋ ˌlüp }

timing motor | ELEC | A motor which operates from an alternating-current power system synchronously with the alternating-current frequency, used in timing and clock mechanisms. Also known as clock motor. { 'tīm·iŋ ˌmōd·ər }

timing relay | ELEC | Form of auxiliary relay used to introduce a definite time delay in the performance of a function. { 'tīm·iŋ ˌrē͵lā }

timing signal | COMPUT SCI | A pulse generated by the clock of a digital computer to provide synchronization of its activities. | ELECTR | Any signal recorded simultaneously with data on magnetic tape for use in identifying the exact time of each recorded event. { 'tīm·iŋ ˌsig·nəl }

tinsel cord | ELEC | A highly flexible cord used for headphone leads and test leads, in which the conductors are strips of thin metal foil or tinsel wound around a strong but flexible central cord. { 'tin·səl ˌkȯrd }

tip | ELEC | The contacting part at the end of a phone plug. | ELECTR | A small protuberance on the envelope of an electron tube, resulting from the closing of the envelope after evacuation. { tip }

tip jack |ELEC| A small single-hole jack for a single-pin contact plug. Also known as pup jack. { 'tip ,jak }

tip side |ELEC| Conductor of a circuit which is associated with the tip of a plug or the top spring of a jack; by extension, it is common practice to designate by these terms the conductors having similar functions or arrangements in circuits where plugs or jacks may not be involved. { 'tip ,sīd }

Tirrill regulator |ELEC| A device for regulating the voltage of a generator, in which the field resistance of the exciter is short-circuited temporarily when the voltage drops. { 'tir·əl ,reg·yə,lād·ər }

title bar |COMPUT SCI| An area at the top of a window that contains the name of the file or application in the window. { 'tīd·əl ,bär }

T junction |ELECTR| A network of waveguides with three waveguide terminals arranged in the form of a letter T; in a rectangular waveguide a symmetrical T junction is arranged by having either all three broadsides in one plane or two broadsides in one plane and the third in a perpendicular plane. { 'tē ,jəŋk·shən }

T²L See transistor-transistor logic.

T1 line |COMMUN| High-speed digital connection that transmits data at 1.5 million bits per second through the telephone-switching network. { ,tē'wən ,līn }

T3 line |COMMUN| High-speed digital connection that transmits data at 45 million bits per second through the telephone-switching network. { ,tē'thrē ,līn }

TM mode See transverse magnetic mode. { ¦tē'em ,mōd }

TM wave See transverse magnetic wave. { ¦tē'em ,wāv }

T network |ELEC| A network composed of three branches, with one end of each branch connected to a common junction point, and with the three remaining ends connected to an input terminal, an output terminal, and a common input and output terminal, respectively. { 'tē ,net,wərk }

Toepler-Holtz machine |ELEC| An early type of machine for continuously producing electrical charges at high voltage by electrostatic induction, superseded by the Wimhurst machine. Also known as Holtz machine. { 'tep·lər 'hōlts mə ,shēn }

toggle |COMPUT SCI| **1.** To switch back and forth between two stable states or modes of operation. **2.** A hardware or software device that carries out this switching action. |ELECTR| To switch over to an alternate state, as in a flip-flop. { 'täg·əl }

toggle condition |ELECTR| Condition of a flip-flop circuit in which the internal state of the flip-flop changes from 0 to 1 or from 1 to 0. { 'täg·əl kən ,dish·ən }

toggle switch |ELEC| A small switch that is operated by manipulation of a projecting lever that is combined with a spring to provide a snap action for opening or closing a circuit quickly. |ELECTR| An electronically operated circuit that holds either of two states until changed. { 'täg·əl ,swich }

token |COMMUN| A unique grouping of bits that is transmitted as a unit in a communications network and used as a signal to notify stations in the network when they have control and are free to send information or take other specified actions. |COMPUT SCI| **1.** A distinguishable unit in a sequence of characters. **2.** A single byte that is used to represent a keyword in a programming language in order to conserve storage space. **3.** A physical object, such as a badge or identity card, issued to authorized users of a computing system, building, or area. { 'tō·kən }

tokenization |COMPUT SCI| The conversion of keywords of a programming language to tokens in order to conserve storage space. { ,tō·kən· ə'zā·shən }

token-passing protocol |COMMUN| The assignment of data communications channels to units which communicate according to a fixed priority sequence. { 'tō·kən ¦pas·iŋ 'prōd·ə,kòl }

token-sharing network |COMMUN| A communications network in which all the stations are linked to a common bus and control is determined by a group of bits (token) that is passed along the bus from station to station. { 'tō·kən ¦sher·iŋ 'net,wərk }

toll |COMMUN| **1.** Charge made for a connection beyond an exchange boundary. **2.** Any part of telephone plant, circuits, or services for which toll charges are made. { tōl }

toll call |COMMUN| Telephone call to points beyond the area within which telephone calls are covered by a flat monthly rate or are charged for on a message unit basis. { 'tōl ,kòl }

toll center |COMMUN| A telephone central office where trunks from end offices are joined to the long-distance system, and operators are present; it is a class-4 office. { 'tōl ,sen·tər }

toll line |COMMUN| A telephone line or channel that connects different telephone exchanges. { 'tōl ,līn }

toll office |COMMUN| A telephone central office which serves mainly to terminate and interconnect toll lines and various types of trunks. { 'tōl ,òf·əs }

toll terminal loss |COMMUN| The part of the overall transmission loss on a toll connection that is attributable to the facilities from the toll center through the tributary office, to and including the subscriber's equipment. { 'tōl 'tər·mən·əl ,lòs }

Tolman and Stewart effect |ELEC| The development of negative charge at the forward end of a metal rod which is suddenly stopped after rapid longitudinal motion. { ¦täl·mən ən 'stü·ərt i,fekt }

tomography See sectional radiography. { tə'mäg· rə·fē }

tone control |ELECTR| A control used in an audio-frequency amplifier to change the frequency response so as to secure the most pleasing proportion of bass to treble; individual bass and treble controls are provided in some amplifiers. { 'tōn kən,trōl }

tone dialing See push-button dialing. { 'tōn ,dīl· iŋ }

tone generator |ELECTR| A signal generator used to generate an audio-frequency signal suitable for signaling purposes or for testing audio-frequency equipment. { 'tōn ,jen·ə,rād·ər }

tone-modulated waves |COMMUN| Waves obtained from continuous waves by amplitude-modulating them at audio frequency in a substantially periodic manner. { 'tōn ¦mäj·ə,lād·əd ,wāvz }

tone modulation |COMMUN| Type of code-signal transmission obtained by causing the radio-frequency carrier amplitude to vary at a fixed audio frequency. { 'tōn ,mäj·ə,lā·shən }

tone-only pager |COMMUN| A receiver in a radio paging system that alerts the user to call a specific telephone number. { 'tōn ¦on·lē 'pāj·ər }

tone-operated net-loss adjuster |COMMUN| System for stabilizing the net loss of a telephone circuit by a tone transmitted between conversations. { 'tōn ¦äp·ə,rād·əd 'net ¦lós·ə,jəs·tər }

tone reversal |COMMUN| Distortion of the recorder copy in facsimile which causes the various shades of black and white not to be in the proper order. { 'tōn ri,vər·səl }

toolbar |COMPUT SCI| A row or column of on-screen push buttons containing icons that represent frequently accessed commands. { 'tül,bär }

top-down analysis |COMPUT SCI| A predictive method of syntactic analysis which, starting from the root symbol, attempts to predict the means by which a string was generated. { ¦täp ¦daún ə'nal·ə·səs }

top-loaded vertical antenna |ELECTROMAG| Vertical antenna constructed so that, because of its greater size at the top, there results modified current distribution, giving a more desirable radiation pattern in the vertical plane. { ¦täp ¦lōd·əd 'vərd·ə·kəl an'ten·ə }

topological shielding |ELEC| An optimal lightning protection system in which a series of shields (such as a building's sheet metal or a metal cabinet), each one surrounding the next, are connected so that deleterious voltage and power levels are reduced at each successive inner shield. { ¦täp·ə¦läj·ə·kəl 'shēld·iŋ }

topology |COMPUT SCI| The physical or logical arrangement of the stations (nodes) in a communications network. { tə'päl·ə·jē }

topology of circuits |ELEC| The study of electric networks in terms of the geometry of their connections only; used in finding such properties of circuits as equivalence and duality, and in analyzing and synthesizing complex circuits. { tə¦päl·ə·jē əv 'sər·kəts }

tornadotron |ELECTR| Millimeter-wave device which generates radio-frequency power from an enclosed, orbiting electron cloud, excited by a radio-frequency field, when subjected to a strong, pulsed magnetic field. { tór'nād·ə ,trän }

toroidal discharge See ring discharge. { tə¦ròid·əl 'dis,chärj }

torque amplifier |COMPUT SCI| An analog computer device having input and output shafts and supplying work to rotate the output shaft in positional correspondence with the input shaft without imposing any significant torque on the input shaft. { 'tórk ,am·plə,fī·ər }

torque constant |ELEC| The ratio of the torque delivered by a motor to the current supplied to it. { 'tórk ,kän·stənt }

torque-speed characteristic |ELEC| For electric motors, the relationship of developed torque to armature speed. { 'tórk 'spēd ,kar·ik·tə,ris·tik }

torsional mode delay line |COMPUT SCI| A device in which torsional vibrations are propagated through a solid material to make use of the propagation time of the vibrations to obtain a time delay for the signals. { 'tór·shən·əl ¦mōd di'lā,līn }

torsion galvanometer |ENG| A galvanometer in which the force between the fixed and moving systems is measured by the angle through which the supporting head of the moving system must be rotated to bring the moving system back to its zero position. { 'tór·shən ,gal·və'näm·əd·ər }

torsion-string galvanometer |ENG| A sensitive galvanometer in which the moving system is suspended by two parallel fibers that tend to twist around each other. { 'tór·shən ¦striŋ ,gal·və'näm·əd·ər }

TOS See tape operating system.

total deadlock |COMPUT SCI| A deadlock that involves all the tasks in a multiprogramming system. { 'tōd·əl 'ded,läk }

total harmonic distortion |ELECTR| Ratio of the power at the fundamental frequency, measured at the output of the transmission system considered, to the power of all harmonics observed at the output of the system because of its nonlinearity, when a single frequency signal of specified power is applied to the input of the system; it is expressed in decibels. { 'tōd·əl här'män·ik di'stór·shən }

touch call See push-button dialing. { 'təch ,kòl }

touch control |ELEC| A circuit that closes a relay when two metal areas are bridged by a finger or hand. { 'təch kən,trōl }

touchpad |COMPUT SCI| A small, touch-sensitive pad that enables the user to move the pointer on the display screen of a personal computer by moving a finger or other object along the pad, and to click by tapping the pad. { 'təch·,pad }

touch screen |COMPUT SCI| An electronic display that allows a user to send signals to a computer by touching an area on the display with a finger, pencil, or other object. { 'təch ,skrēn }

touch sensor |CONT SYS| A device such as a small, force-sensitive switch that uses contact to generate feedback in robotic systems. { 'təch ,sen·sər }

tower |ELECTROMAG| A tall metal structure used as a transmitting antenna, or used with another such structure to support a transmitting antenna wire. { taú·ər }

tower case |COMPUT SCI| A system unit that stands in a vertical position. { 'taú·ər ,kās }

tower loading |ELEC| Load placed on a tower by its own weight, the weight of the wires with or without ice covering, the insulators, the wind pressure normal to the line acting both on the tower and the wires, and the pull from the wires. { 'taú·ər ‚lōd·iŋ }

tower radiator [ELECTROMAG] Metal structure used as a transmitting antenna. { 'taú·ər ¦rād·ē ‚ād·ər }

Townsend avalanche *See* avalanche. { 'taún‧zənd ‚av·ə‚lanch }

Townsend characteristic [ELECTR] Current-voltage characteristic curve for a phototube at constant illumination and at voltages below that at which a glow discharge occurs. { 'taún‧zənd ‚kar·ik·tə‚ris·tik }

Townsend coefficient [ELECTR] The number of ionizing collisions by an electron per centimeter of path length in the direction of the applied electric field in a radiation counter. { 'taún‧zənd ‚kō·i‚fish·ənt }

Townsend discharge [ELECTR] A discharge which occurs at voltages too low for it to be maintained by the electric field alone, and which must be initiated and sustained by ionization produced by other agents; it occurs at moderate pressures, above about 0.1 torr, and is free of space charges. { 'taún‧zənd ‚dis‚chärj }

Townsend ionization *See* avalanche. { 'taún‧zənd ‚ī·ə·nə‚zā·shən }

Tow-Thomas filter [ELECTR] A multiple-amplifier active filter that has the advantage of ease of design but the disadvantage of lacking a high-pass output in its basic configuration. { ¦tō 'täm·əs ‚fil·tər }

T pad [ELEC] A pad made up of resistance elements arranged in a T network (two resistors inserted in one line, with a third between their junction and the other line). { 'tē ‚pad }

TP monitor *See* teleprocessing monitor. { ¦tē'pē ‚män·əd·ər }

trace [COMPUT SCI] To provide a record of every step, or selected steps, executed by a computer program, and by extension, the record produced by this operation. [ELECTR] The visible path of a moving spot on the screen of a cathode-ray tube. Also known as line. { trās }

trace interval [ELECTR] Interval corresponding to the direction of sweep used for delineation. { 'trās ‚in·tər·vəl }

trace routine [COMPUT SCI] A routine which tracks the execution of a program, step by step, to locate a program malfunction. Also known as tracing routine. { 'trās rü‚tēn }

trace sensitivity [ELECTR] The ability of an oscilloscope to produce a visible trace on the scope face for a specified input voltage. { 'trās ‚sen·sə‚tīv·əd·ē }

trace statement [COMPUT SCI] A statement, included in certain programming languages, that causes certain error-checking procedures to be carried out on specified segments of a source program. { 'trās ‚stāt·mənt }

tracing routine *See* trace routine. { 'trās·iŋ rü‚tēn }

track [ELECTR] **1.** A path for recording one channel of information on a magnetic tape, drum, or other magnetic recording medium; the location of the

track is determined by the recording equipment rather than by the medium. **2.** The trace on a plan-position indicator or similar display resulting from the association of successive detections presumed to be from the same moving target; or the same information from an appropriate radar data processor. { trak }

trackball [COMPUT SCI] A ball inset in the console of a video display terminal, the keyboard of a personal computer, or a small box-shaped holder, which can be rotated by the operator, and whose motion is followed by a cursor on the display screen. { 'trak‚bȯl }

tracker [COMPUT SCI] An input device used in a virtual environment, which is capable of reporting its location in space and its orientation. { 'trak·ər }

track filtering [ELECTR] In radar data processing, the treatment of each subsequent measurement of a target's position, generally by weighting factors, to reduce the effects of measurement error, resulting in a "smoothing" of the track. { 'trak ‚fil·tər·iŋ }

tracking [ELEC] A leakage or fault path created across the surface of an insulating material when a high-voltage current slowly but steadily forms a carbonized path. [ELECTR] The condition in which all tuned circuits in a receiver accurately follow the frequency indicated by the tuning dial over the entire tuning range. [ENG] **1.** A motion given to the major lobe of a radar or radio antenna such that some preassigned moving target in space is always within the major lobe. **2.** The process of following the movements of an object; may be accomplished by keeping the reticle of an optical system or a radar beam on the object, by plotting its bearing and distance at frequent intervals, or by a combination of techniques. { 'trak·iŋ }

Tracking and Data Relay Satellite System [COMMUN] A system providing telecommunication services between low-earth-orbiting user spacecraft and user control centers; it consists of a series of geostationary spacecraft and an earth terminal located at White Sands, New Mexico. Abbreviated TDRSS. { ¦trak·iŋ an ¦dad·ə ¦rē‚lā 'sad·ə‚līt ‚sis·təm }

tracking cross [COMPUT SCI] A cross displayed on the screen of a video terminal which automatically follows a light pen. Also known as tracking cursor. { 'trak·iŋ ‚krȯs }

tracking cursor *See* tracking cross. { 'trak·iŋ ‚kər·sər }

tracking filter [ELECTR] Electronic device for attenuating unwanted signals while passing desired signals, by phase-lock techniques that reduce the effective bandwidth of the circuit and eliminate amplitude variations. { 'trak·iŋ ‚fil·tər }

tracking problem [CONT SYS] The problem of determining a control law which when applied to a dynamical system causes its output to track a given function; the performance index is in many cases taken to be of the integral square error variety. { 'trak·iŋ ‚präb·ləm }

track in range [ELECTR] To adjust the gate of a radar set so that it opens at the correct instant to accept the signal from a target of changing range from the radar. { 'trak in 'rānj }

track pitch [ELECTR] The physical distance between track centers. { 'trak ‚pich }

track-return power system [ELEC] A system for distributing electric power to trains or other vehicles, in which the track rails are used as an uninsulated return conductor. { 'trak ri¦tərn 'paü·ər ‚sis·təm }

track-to-track access time [COMPUT SCI] The time required for a read-write head to move between the adjacent cylinders of a disk. { ¦trak tə ¦trak 'ak‚ses ‚tīm }

track-while-scan [ELECTR] Radar operation used to detect a radar target, compute its velocity, and predict its future position without interfering with continuous radar scanning. { ¦trak ‚wīl 'skan }

tractor-feed printer See pin-feed printer. { 'trak·tər ¦fēd 'print·ər }

traffic [COMMUN] The messages transmitted and received over a communication channel. { 'traf·ik }

traffic diagram [COMMUN] Chart or illustration used to show the movement and control of traffic over a communications system. { 'traf·ik ‚dī·ə ‚gram }

traffic distribution [COMMUN] Routing of communications traffic through a terminal to a switchboard or dialing center. { 'traf·ik ‚di·strə ‚byü·shən }

traffic flow security [COMMUN] Transmission of an uninterrupted flow of random text on a wire or radio link between two stations with no indication to an interceptor of what portions of this steady stream constitute encrypted message text and what portions are merely random filler. { 'traf·ik ¦flō si‚kyùr·əd·ē }

traffic forecast [COMMUN] Traffic level prediction on which communications system management decisions and engineering effort are based. { 'traf·ik ‚fòr‚kast }

trailer [ELECTR] A bright streak at the right of a dark area or dark line in an analog television picture, or a dark area or streak at the right of a bright part; usually due to insufficient gain at low video frequencies. { 'trā·lər }

trailer label [COMPUT SCI] A record appearing at the end of a magnetic tape that uniquely identifies the tape as one required by the system. { 'trā·lər ‚lā·bəl }

trailer record [COMPUT SCI] A record which contains data pertaining to an associated group of records immediately preceding it. { 'trā·lər ‚rek·ərd }

trailing antenna [ELECTROMAG] An aircraft radio antenna having one end weighted and trailing free from the aircraft when in flight. { 'trāl·iŋ an ¦ten·ə }

trailing edge [ELECTR] The major portion of the decay of a pulse. { 'trāl·iŋ 'ej }

trailing pad [COMPUT SCI] Characters placed to the right of information in a field of data to fulfill length requirements or for cosmetic purposes. { 'trāl·iŋ ‚pad }

trainer [ELECTR] A piece of equipment used for training operators of radar, sonar, and other electronic equipment by simulating signals received under operating conditions in the field. { 'trā·nər }

training data [CONT SYS] Data entered into a robot's computer at the beginning of an operation. { 'trān·iŋ ‚dad·ə }

training time [COMPUT SCI] The machine time expended in training employees in the use of the equipment, including such activities as mounting, console operation, converter operation, and printing operation, and time spent in conducting required demonstrations. { 'trān·iŋ ¦tīm }

train printer [COMPUT SCI] A computer printer in which the characters are carried in a track and a hammer strikes the proper character against the paper as it passes the print position. { 'trān ‚print·ər }

trajectory control [CONT SYS] A type of continuous-path control in which a robot's path is calculated based on mathematical models of joint acceleration, arm loads, and actuating signals. { trə'jek·trē kən‚trōl }

transacter [COMPUT SCI] A system in which data from sources in a number of different locations, as in a factory, are transmitted to a data-processing center and immediately processed by a computer. { tran'sak·tər }

transaction [COMPUT SCI] General description of updating data relevant to any item. { tran'sak·shən }

transaction data [COMPUT SCI] A set of data in a data-processing area in which the incidence of the data is essentially random and unpredictable; hours worked, quantities shipped, and amounts invoiced are examples from, respectively, the areas of payroll, accounts receivable, and accounts payable. { tran'sak·shən ‚dad·ə }

transaction file See detail file. { tran'sak·shən ‚fīl }

transaction processing system [COMPUT SCI] A system which processes predefined transactions, one at a time, with direct, on-site entry of the transactions into a terminal, and which produces predefined outputs and maintains the necessary data base. { tran'sak·shən ‚prä‚ses·iŋ ‚sis·təm }

transaction record See change record. { tran'sak·shən ‚rek·ərd }

transaction tape See change tape. { tran'sak·shən ‚tāp }

transadmittance [ELECTR] A specific measure of transfer admittance under a given set of conditions, as in forward transadmittance, interelectrode transadmittance, short-circuit transadmittance, small-signal forward transadmittance, and transadmittance compression ratio. { ¦tranz·ad'mit·əns }

transceiver [COMPUT SCI] A computer terminal that can transmit and receive information to and from an input/output channel. [ELECTR] A radio transmitter and receiver combined in one unit and having switching arrangements such as to permit both transmitting and receiving. Also known as transmitter-receiver. { tran'sē·vər }

transconductance [ELECTR] **1.** An electron-tube rating, equal to the change in plate current divided by the change in control-grid voltage that causes it, when the plate voltage and all other voltages are maintained constant. Also known as grid-anode transconductance; grid-plate transconductance; mutual conductance. Symbolized G_m; g_m. **2.** A field-effect-transistor rating, equal to the change in drain current divided by the change in gate-to-source voltage that causes it, when the drain voltage and all other voltages are maintained constant. Symbolized g_{fs}. **3.** An amplifier parameter, equal to the change in output current divided by the change in input voltage that causes it. Symbolized g_m. { ¦tranz·kən'dək·təns }

transconductance amplifier [ELECTR] An amplifier whose output current (rather than output voltage) is proportional to its input voltage. { ¸tranz·kən¸duk·təns 'am·plə¸fī·ər }

transconductance-C filter [ELECTR] An integrated-circuit filter that combines the functions of an amplifier and a simulated resistor into a transconductance amplifier. { 'tranz·kən ¸duk·təns 'sē ¸fil·tər }

transconductor See transconductance amplifier. { ¸tranz·kən'dək·tər }

transcribe [COMPUT SCI] To copy, with or without translating, from one external computer storage medium to another. [ELECTR] To record, as to record a radio program by means of electric transcriptions or magnetic tape for future re-broadcasting. { tranz'krīb }

transcriber [COMPUT SCI] The equipment used to convert information from one form to another, as for converting computer input data to the medium and language used by the computer. { tranz'krī·bər }

transducer [ENG] Any device or element which converts an input signal into an output signal of a different form; examples include the microphone, loudspeaker, barometer, photoelectric cell, automobile horn, doorbell, and underwater sound transducer. { tranz'dü·sər }

transducer loss [ELECTR] The ratio of the power available to a transducer from a specified source to the power that the transducer delivers to a specified load; usually expressed in decibels. { tranz'dü·sər ¸lòs }

transductor See magnetic amplifier. { tranz' dək·tər }

transfer See jump. { 'tranz·fər }

transfer admittance [ELECTR] An admittance rating for electron tubes and other transducers or networks; it is equal to the complex alternating component of current flowing to one terminal from its external termination, divided by the complex alternating component of the voltage applied to the adjacent terminal on the cathode or reference side; all other terminals have arbitrary external terminations. { 'tranz·fər ad ¸mit·əns }

transfer characteristic [ELECTR] **1.** Relation, usually shown by a graph, between the voltage of one electrode and the current to another electrode, with all other electrode voltages being maintained constant. **2.** Function which, multiplied by an input magnitude, will give a resulting output magnitude. **3.** Relation between the illumination on a camera tube and the corresponding output-signal current, under specified conditions of illumination. { 'tranz·fər ¸kar·ik· tə¸ris·tik }

transfer check [COMPUT SCI] Check (usually automatic) on the accuracy of the transfer of a word in a computer operation. { 'tranz·fər ¸chek }

transfer conditionally [COMPUT SCI] To copy, exchange, read, record, store, transmit, or write data or to change control or jump to another location according to a certain specified rule or in accordance with a certain criterion. { tranz'fər kən'dish·ən·ə·lē }

transfer constant [ENG] A transducer rating, equal to one-half the natural logarithm of the complex ratio of the product of the voltage and current entering a transducer to that leaving the transducer when the latter is terminated in its image impedance; alternatively, the product may be that of force and velocity or pressure and volume velocity; the real part of the transfer constant is the image attenuation constant, and the imaginary part is the image phase constant. Also known as transfer factor. { 'tranz·fər ¸kän·stənt }

Transfer Control Protocol See Transmission Control Protocol. { ¸tranz·fər kən'trōl ¸prōd·ə¸kòl }

transfer factor See transfer constant. { 'tranz·fər ¸fak·tər }

transfer function [CONT SYS] The mathematical relationship between the output of a control system and its input: for a linear system, it is the Laplace transform of the output divided by the Laplace transform of the input under conditions of zero initial-energy storage. { 'tranz·fər ¸faŋk·shən }

transfer impedance [ELEC] The ratio of the voltage applied at one pair of terminals of a network to the resultant current at another pair of terminals, all terminals being terminated in a specified manner. { 'tranz·fər im¸pēd·əns }

transfer-in-channel command [COMPUT SCI] A command used to direct channel control to a specified location in main storage when the next channel command word is not stored in the next location in sequence. { ¦tranz·fər in 'chan·əl kə ¸mand }

transfer instruction [COMPUT SCI] Step in computer operation specifying the next operation to be performed, which is not necessarily the next instruction in sequence. { 'tranz·fər in ¸strək·shən }

transfer interpreter [COMPUT SCI] A variation of a punched-card interpreter that senses a punched card and prints the punched information on the

following card. Also known as posting interpreter.
{ 'tranz·fər in,tər·prəd·ər }

transfer matrix [CONT SYS] The generalization of the concept of a transfer function to a multivariable system; it is the matrix whose product with the vector representing the input variables yields the vector representing the output variables. { 'tranz·fər ,mā·triks }

transfer operation [COMPUT SCI] An operation which moves information from one storage location or one storage medium to another (for example, read, record, copy, transmit, exchange). { 'tranz·fər ,äp·ə,rā·shən }

transfer rate [COMPUT SCI] The speed at which data are moved from a direct-access device to a central processing unit. { 'tranz·fər ,rāt }

transfer ratio [ENG] From one point to another in a transducer at a specified frequency, the complex ratio of the generalized force or velocity at the second point to the generalized force or velocity applied at the first point; the generalized force or velocity includes not only mechanical quantities, but also other analogous quantities such as acoustical and electrical; the electrical quantities are usually electromotive force and current. { 'tranz·fər ,rā·shō }

transferred-electron amplifier [ELECTR] A diode amplifier, which generally uses a transferred-electron diode made from doped *n*-type gallium arsenide, that provides amplification in the gigahertz range to well over 50 gigahertz at power outputs typically below 1 watt continuous-wave. Abbreviated TEA. { 'tranz'fərd i¦lek,trän 'am·plə ,fī·ər }

transferred-electron device [ELECTR] A semiconductor device, usually a diode, that depends on internal negative resistance caused by transferred electrons in gallium arsenide or indium phosphide at high electric fields; transit time is minimized, permitting oscillation at frequencies up to several hundred megahertz. { 'tranz'fərd i ¦lek,trän di'vīs }

transfer robot [CONT SYS] A fixed-sequence robot that moves parts from one location to another. { 'tranz·fər 'rō,bät }

transfer switch [ELEC] A switch for transferring one or more conductor connections from one circuit to another. { 'tranz·fər ,swich }

transfer test [COMMUN] Verification of transmitted information by temporary storing, retransmitting, and comparing. { 'tranz·fər ,test }

transform [COMPUT SCI] To change the form of digital-computer information without significantly altering its meaning. { tranz'förm }

transformation [ELEC] For two networks which are equivalent as far as conditions at the terminals are concerned, a set of equations giving the admittances or impedances of the branches of one circuit in terms of the admittances or impedances of the other. { ,tranz·fər'mā·shən }

transformation matrix [ELECTROMAG] A two-by-two matrix which relates the amplitudes of the traveling waves on one side of a waveguide junction to those on the other. { ,tranz·fər'mā·shən ,mā·triks }

transformer [ELECTROMAG] An electrical component consisting of two or more multiturn coils of wire placed in close proximity to cause the magnetic field of one to link the other; used to transfer electric energy from one or more alternating-current circuits to one or more other circuits by magnetic induction. { tranz'för·mər }

transformer bridge [ELEC] A network consisting of a transformer and two impedances, in which the input signal is applied to the transformer primary and the output is taken between the secondary center-tap and the junction of the impedances that connect to the outer leads of the secondary. { tranz'för·mər ,brij }

transformer-coupled amplifier [ELECTR] Audio-frequency amplifier that uses untuned iron-core transformers to provide coupling between stages. { tranz'för·mər ¦kəp·əld 'am·plə,fī·ər }

transformer coupling [ELEC] See inductive coupling. [ELECTR] Interconnection between stages of an amplifier which employs a transformer for connecting the plate circuit of one stage to the grid circuit of the following stage; a special case of inductive coupling. { tranz'för·mər ,kəp·liŋ }

transformer hybrid See hybrid set. { tranz'för·mər 'hī·brəd }

transformer load loss [ELEC] Losses in a transformer which are incident to the carrying of the load; load losses include resistance loss in the windings due to load current, stray loss due to stray fluxes in the windings, core clamps, and so on, and to circulating current, if any, in parallel windings. { tranz'för·mər 'lōd ,lòs }

transformer loss [ELEC] Ratio of the signal power that an ideal transformer of the same impedance ratio would deliver to the load impedance, to the signal power that the actual transformer delivers to the load impedance; this ratio is usually expressed in decibels. { tranz' för·mər ,lòs }

transformer read-only store [COMPUT SCI] In computers, read-only store in which the presence or absence of mutual inductance between two circuits determines whether a binary 1 or 0 is stored. { tranz'för·mər 'rēd ¦ōn·lē 'stòr }

transformer rectifier [ELEC] A combination of a transformer and a rectifier that allows input alternating current to be varied and then rectified into direct current. { 'tranz,för·mər 'rek·tə,fī·ər }

transformer substation [ELEC] An electric power substation whose equipment includes transformers. { tranz'för·mər 'səb,stā·shən }

transformer voltage ratio [ELEC] Ratio of the root-mean-square primary terminal voltage to the root-mean-square secondary terminal voltage under specified conditions of load. { tranz'för·mər 'vōl·tij ,rā·shō }

transforming section [ELECTROMAG] Length of waveguide or transmission line of modified cross section, or with a metallic or dielectric insert, used for impedance transformation. { tranz'förm·iŋ ,sek·shən }

transhybrid loss [ELEC] In a carrier telephone system, the transmission loss at a given frequency measured across a hybrid circuit joined to a given two-wire termination and balancing network.   { ˌtranzˈhī-brəd 'lȯs }

transient   [PHYS] A pulse, damped oscillation, or other temporary phenomenon occurring in a system prior to reaching a steady-state condition.   { 'tranch·ənt }

transient analyzer   [ELECTR] An analyzer that generates transients in the form of a succession of equal electric surges of small amplitude and adjustable waveform, applies these transients to a circuit or device under test, and shows the resulting output waveforms on the screen of an oscilloscope.   { 'tranch·ənt ˌan·ə‚līz·ər }

transient distortion   [ELECTR] Distortion due to inability to amplify transients linearly.   { 'tranch·ənt di‚stȯr·shən }

transient phenomena   [ELEC] Rapidly changing actions occurring in a circuit during the interval between closing of a switch and settling to a steady-state condition, or any other temporary actions occurring after some change in a circuit.   { 'tranch·ənt fə‚näm·ə·nä }

transient program   [COMPUT SCI] A computer program that is stored in a computer's main memory only while it is being executed.   { 'tranch·ənt 'prō·grəm }

transient suppressor   See surge suppressor.   { 'tranch·ənt sə'pres·ər }

transistance   [ELECTR] The characteristic that makes possible the control of voltages or currents so as to accomplish gain or switching action in a circuit; examples of transistance occur in transistors, diodes, and saturable reactors.   { tran'zis·təns }

transistor   [ELECTR] An active component of an electronic circuit consisting of a small block of semiconducting material to which at least three electrical contacts are made, usually two closely spaced rectifying contacts and one ohmic (nonrectifying) contact; it may be used as an amplifier, detector, or switch.   { tran'zis·tər }

transistor amplifier   [ELECTR] An amplifier in which one or more transistors provide amplification comparable to that of electron tubes.   { tran'zis·tər ‚am·plə‚fī·ər }

transistor biasing   [ELECTR] Maintaining a direct-current voltage between the base and some other element of a transistor.   { tran'zis·tər ‚bī·əs·iŋ }

transistor characteristics   [ELECTR] The values of the impedances and gains of a transistor.   { tran'zis·tər ‚kar·ik·tə‚ris·tiks }

transistor chip   [ELECTR] An unencapsulated transistor of very small size used in microcircuits.   { tran'zis·tər ‚chip }

transistor circuit   [ELECTR] An electric circuit in which a transistor is connected.   { tran'zis·tər ‚sər·kət }

transistor clipping circuit   [ELECTR] A circuit in which a transistor is used to achieve clipping action; the bias at the input is set at such a level that output current cannot flow during a portion of the amplitude excursion of the input voltage or current waveform.   { tran'zis·tər 'klip·iŋ ‚sər·kət }

transistor gain   [ELECTR] The increase in signal power produced by a transistor.   { tran'zis·tər ‚gān }

transistor input resistance   [ELECTR] The resistance across the input terminals of a transistor stage. Also known as input resistance.   { tran'zis·tər 'in‚pu̇t ri‚zis·təns }

transistor magnetic amplifier   [ELECTR] A magnetic amplifier together with a transistor preamplifier, the latter used to make the signal strong enough to change the flux in the core of the magnetic amplifier completely during a half-cycle of the power supply voltage.   { tran'zis·tər mag'ned·ik 'am·plə‚fī·ər }

transistor memory   See semiconductor memory.   { tran'zis·tər ‚mem·rē }

transistor radio   [ELECTR] A radio receiver in which transistors are used in place of electron tubes.   { tran'zis·tər ‚rād·ē·ō }

transistor-transistor logic   [ELECTR] A logic circuit containing two transistors, for driving large output capacitances at high speed. Abbreviated T²L; TTL.   { tran'zis·tər tran'zis·tər 'läj·ik }

transition   [COMMUN] Change from one circuit condition to the other; for example, the change from mark to space or from space to mark.   { tran'zish·ən }

transition element   [ELECTROMAG] An element used to couple one type of transmission system to another, as for coupling a coaxial line to a waveguide.   { tran'zish·ən ‚el·ə·mənt }

transition factor   See reflection factor.   { tran'zish·ən ‚fak·tər }

transition function   [COMPUT SCI] A function which determines the next state of a sequential machine from the present state and the present input.   { tran'zish·ən ‚fəŋk·shən }

transition loss   [ELEC] At a junction between a source and a load, the ratio of the available power to the power delivered to the load.   { tran'zish·ən ‚lȯs }

transition point   [ELECTROMAG] A point at which the constants of a circuit change in such a way as to cause reflection of a wave being propagated along the circuit.   { tran'zish·ən ‚pȯint }

transitron   [ELECTR] Thermionic-tube circuit whose action depends on the negative transconductance of the suppressor grid of a pentode with respect to the screen grid.   { 'tran·sə‚trän }

transitron oscillator   [ELECTR] A negative-resistance oscillator in which the screen grid is more positive than the anode, and a capacitor is connected between the screen grid and the suppressor grid; the suppressor grid periodically divides the current between the screen grid and the anode, thereby producing oscillation.   { 'tran·sə‚trän 'äs·ə‚lād·ər }

transit time   [ELECTR] The time required for an electron or other charge carrier to travel between two electrodes in an electron tube or transistor.   { 'trans·ət ‚tīm }

transit-time microwave diode   [ELECTR] A solid-state microwave diode in which the transit time of charge carriers is short enough to permit operation in microwave bands.   { 'trans·ət ‚tīm 'mī·krə‚wāv 'dī‚ōd }

transit-time mode   [ELECTR] A mode of operation of a Gunn diode in which a charge dipole, consisting of an electron accumulation and a depletion layer, travels through the semiconductor at a frequency dependent on the length of the semiconductor layer and the drift velocity. { 'trans·ət ,tīm ,mōd }

translate |COMPUT SCI| To convert computer information from one language to another, or to convert characters from one representation set to another, and by extension, the computer instruction which directs the latter conversion to be carried out. { tran'slāt }

translating circuit See translator. { tran'slād·iŋ ,sər·kət }

translation algorithm |COMPUT SCI| A specific, effective, essentially computational method for obtaining a translation from one language to another. { tran'slā·shən 'al·gə,rith·əm }

translator |COMPUT SCI| A computer network or system having a number of inputs and outputs, so connected that when signals representing information expressed in a certain code are applied to the inputs, the output signals will represent the same information in a different code. Also known as translating circuit. |ELECTR| A combination television receiver and low-power television transmitter, used to pick up television signals on one frequency and retransmit them on another frequency to provide reception in areas not served directly by television stations. { tran'slād·ər }

translator routine |COMPUT SCI| A program which accepts statements in one language and outputs them as statements in another language. { tran'slād·ər rü,tēn }

transliterate |COMPUT SCI| To represent the characters or words of one language by corresponding characters or words of another language. { tran'slid·ə,rāt }

transmission |ELECTR| **1.** The process of transferring a signal, message, picture, or other form of intelligence from one location to another location by means of wire lines, radio waves, light beams, infrared beams, or other communication systems. **2.** A message, signal, or other form of intelligence that is being transmitted. **3.** See transmittance. { tranz'mish·ən }

transmission access |ELEC| The use of electric power lines and other power transmitting facilities by parties other than the owners of the lines. Also known as common carriage. { tranz'mish·ən 'ak,ses }

transmission band |ELECTROMAG| Frequency range above the cutoff frequency in a waveguide, or the comparable useful frequency range for any other transmission line, system, or device. { tranz'mish·ən ,band }

transmission control character |COMMUN| A character included in a message to control its routing to the intended destination. { tranz'mish·ən kən'trōl ,kar·ik·tər }

Transmission Control Protocol |COMMUN| The set of standards that is responsible for breaking down and reassembling the data packets transmitted on the Internet, for ensuring complete delivery of the packets and for controlling data

flow. Abbreviated TCP. { tranz,mish·ən kən'trōl ,prōd·ə,kól }

Transmission Control Protocol/Internet Protocol |COMPUT SCI| The Internet's principal communication standard, dictating how packets of information are sent and received across multiple networks. TCP breaks down and reassembles packets, and IP ensures that the packets are sent to the correct destination. Abbreviated TCP/IP. { tranz,mish·ən kən'trōl ,prōd·ə,kól 'in·tər,net ,prōd·ə,kól }

transmission electron microscope |ELECTR| A type of electron microscope in which the specimen transmits an electron beam focused on it, image contrasts are formed by the scattering of electrons out of the beam, and various magnetic lenses perform functions analogous to those of ordinary lenses in a light microscope. { tranz'mish·ən i'lek,trän 'mī·krə,skōp }

transmission electron radiography |ELECTR| A technique used in microradiography to obtain radiographic images of very thin specimens; the photographic plate is in close contact with the specimen, over which is placed a lead foil and then a light-tight covering; hardened x-rays shoot through the light-tight covering. { tranz'mish·ən i,lek,tran ,rād·ē'äg·rə·fē }

transmission facilities |COMMUN| All equipment and the medium required to transmit a message. { tranz'mish·ən fə,sil·əd·ēz }

transmission gain See gain. { tranz'mish·ən ,gān }

transmission gate |ELECTR| A gate circuit that delivers an output waveform that is a replica of a selected input during a specific time interval which is determined by a control signal. { tranz'mish·ən ,gāt }

transmission interface converter |COMPUT SCI| A device that converts data to or from a form suitable for transfer over a channel connecting two computer systems or connecting a computer with its associated data terminals. { tranz'mish·ən 'in·tər,fās kən,vərd·ər }

transmission level |COMMUN| The ratio of the signal power at any point in a transmission system to the signal power at some point in the system chosen as a reference point; usually expressed in decibels. { tranz'mish·ən ,lev·əl }

transmission line |ELEC| A system of conductors, such as wires, waveguides, or coaxial cables, suitable for conducting electric power or signals efficiently between two or more terminals. { tranz'mish·ən ,līn }

transmission-line admittance |ELEC| The complex ratio of the current flowing in a transmission line to the voltage across the line, where the current and voltage are expressed in phasor notation. { tranz'mish·ən ¦līn ad,mit·əns }

transmission-line attenuation |ELEC| The decrease in power of a transmission-line signal from one point to another, expressed as a ratio or in decibels. { tranz'mish·ən ¦līn ə,ten·yə,wā·shən }

transmission-line cable |ELEC| The coaxial cable, waveguide, or microstrip which forms a transmission line; a number of standard types have been designated, specified by size and materials. { tranz'mish·ən ¦līn ,kā·bəl }

transmission-line constants See transmission-line parameters. { tranz'mish·ən ¦līn ,kän·stəns }

transmission-line current |ELEC| The amount of electrical charge which passes a given point in a transmission line per unit time. { tranz 'mish·ən ¦līn ,kə·rənt }

transmission-line efficiency |ELEC| The ratio of the power of a transmission-line signal at one end of the line to that at the other end where the signal is generated. { tranz'mish·ən ¦līn i,fish·ən·sē }

transmission-line impedance |ELEC| The complex ratio of the voltage across a transmission line to the current flowing in the line, where voltage and current are expressed in phasor notation. { tranz'mish·ən ¦līn im,pēd·əns }

transmission-line parameters |ELEC| The quantities which are necessary to specify the impedance per unit length of a transmission line, and the admittance per unit length between various conductors of the line. Also known as linear electrical parameters; line parameters; transmission line constants. { tranz'mish·ən ¦līn pə,ram·əd·ərz }

transmission-line power |ELEC| The amount of energy carried past a point in a transmission line per unit time. { tranz'mish·ən ¦līn ,paú·ər }

transmission-line reflection coefficient |ELEC| The ratio of the voltage reflected from the load at the end of a transmission line to the direct voltage. { tranz'mish·ən ¦līn ri'flek·shən ,kō·i ,fish·ənt }

transmission-line theory |ELEC| The application of electrical and electromagnetic theory to the behavior of transmission lines. { tranz'mish·ən ¦līn ,thē·ə·rē }

transmission-line transducer loss |ELEC| The ratio of the power delivered by a transmission line to a load to that produced at the generator, expressed in decibels; equal to the sum of the attenuation of the line and the mismatch loss. { tranz'mish·ən ¦līn trans'dü·sər,lòs }

transmission-line voltage |ELEC| The work that would be required to transport a unit electrical charge between two specified conductors of a transmission line at a given instant. { tranz 'mish·ən ¦līn ,vōl·tij }

transmission loss |COMMUN| **1.** The ratio of the power at one point in a transmission system to the power at a point farther along the line; usually expressed in decibels. **2.** The actual power that is lost in transmitting a signal from one point to another through a medium or along a line. Also known as loss. { tranz'mish·ən ,lòs }

transmission mode See mode. { tranz'mish·ən ,mōd }

transmission modulation |ELECTR| Amplitude modulation of the reading-beam current in a charge storage tube as the beam passes through apertures in the storage surface; the degree of modulation is controlled by the stored charge pattern. { tranz'mish·ən ,mäj·ə'lā·shən }

transmission primaries |COMMUN| The set of three color primaries that correspond to the three independent signals contained in the color signal. { tranz'mish·ən 'prī,mer·ēz }

transmission regulator |ELECTR| In electrical communications, a device that maintains substantially constant transmission levels over a system. { tranz'mish·ən ,reg·yə,lād·ər }

transmission security |COMMUN| Component of communications security which results from all measures designed to protect transmissions from unauthorized interception, traffic analysis, and imitative deception. { tranz'mish·ən si ,kyúr·əd·ē }

transmission speed |COMMUN| The number of information elements sent per unit time; usually expressed as bits, characters, bands, word groups, or records per second or per minute. { tranz'mish·ən ,spēd }

transmission substation |ELEC| An electric power substation associated with high voltage levels. { tranz'mish·ən 'səb,stā·shən }

transmission time |COMMUN| Absolute time interval from transmission to reception of a signal. { tranz'mish·ən ,tīm }

transmissivity |ELECTROMAG| The ratio of the transmitted radiation to the radiation arriving perpendicular to the boundary between two mediums. { ,tranz·mə'siv·əd·ē }

transmit |COMMUN| To send a message, program, or other information to a person or place by wire, radio, or other means. |COMPUT SCI| To move data from one location to another. { tranz'mit }

transmit-receive module |ELECTR| Microwave circuitry providing signal amplification on transmit, elemental duplexing, receiver functions, and phase control, usually featuring solid-state devices and compact packaging, for use at every element of a phased array radar antenna, the entire assembly constituting an "active" phase array. { tranz'mit ri'sēv¦mäj·ül }

transmittability |COMMUN| The ability of standard electronic and mechanical elements and automatic communications equipment to handle a code under various signal-to-noise ratios; for example, a code with a variable number of elements such as Morse presents technical problems in automatic interpretation not encountered in a fixed-length code. { tranz,mid·ə'bil·əd·ē }

transmittance |ELECTROMAG| The radiant power transmitted by a body divided by the total radiant power incident upon the body. Also known as transmission. { tranz'mid·əns }

transmitted-carrier operation |COMMUN| Form of amplitude-modulated carrier transmission in which the carrier wave is transmitted. { tranz 'mid·əd ¦kar·ē·ər ,äp·ə,rā·shən }

transmitter |COMMUN| **1.** In telephony, the microphone that converts sound waves into audio-frequency signals. **2.** See radio transmitter. { tranz'mid·ər }

transmitter-distributor |ELEC| In teletypewriter operations, a motor-driven device which translates teletypewriter code combinations form perforated tape into electrical impulses, and transmits these impulses to one or more receiving stations. Abbreviated TD. { tranz'mid·ər di'strib·yəd·ər }

transmitter noise See frying noise. { tranz'mid·ər ‚nȯiz }

transmitter off | COMMUN | A signal sent by a receiving device to a transmitter, directing it to stop sending information if it is doing so, or not to send information if it is preparing to do so. Abbreviated XOFF. { tranz'mid·ər 'ȯf }

transmitter on | COMMUN | A signal sent by a receiving device to a transmitter, directing it to transmit any information it has to send. Abbreviated XON. { tranz'mid·ər 'ȯn }

transmitter-receiver See transceiver. { tranz' mid·ər ri'sē·vər }

transmitter synchro See synchro transmitter. { tranz'mid·ər ‚siŋ·krō }

transmitting loop loss | COMMUN | That part of the repetition equivalent assignable to the station set, subscriber line, and battery supply circuit which is on the transmitting end. { tranz'mid·iŋ ¦lüp ‚lȯs }

transmitting mode | COMPUT SCI | Condition of an input/output device, such as a magnetic tape when it is actually reading or writing. { tranz'mid·iŋ ‚mōd }

transolver | ELEC | A synchro having a two-phase cylindrical rotor within a three-phase stator, for use as a transmitter or a control transformer with no degradation of accuracy or nulls. { tran'säl·vər }

transparent | COMPUT SCI | Pertaining to a device or system that processes data without the user being aware of or needing to understand its operation. { tranz'par·ənt }

transpolarizer | ELEC | An electrostatically controlled circuit impedance that can have about 30 discrete and reproducible impedance values: two capacitors, each having a crystalline ferroelectric dielectric with a nearly rectangular hysteresis loop, are connected in series and act as a single low impedance to an alternating-current sensing signal when both capacitors are polarized in the same direction; application of 1-microsecond pulses of appropriate polarity increases the impedance in steps. { tranz'pō·lə‚rīz·ər }

transponder | COMMUN | **1.** A transmitter-receiver capable of accepting the challenge of an interrogator and automatically transmitting an appropriate reply. **2.** A receiver-transmitter, such as on satellites, which receives a transmission and retransmits it at another radio frequency. { tranz'pän·dər }

transponder beacon See responder beacon. { tranz'pän·dər ‚bē·kən }

transponder dead time | ELECTR | Time interval between the start of a pulse and the earliest instant at which a new pulse can be received or produced by a transponder. { tranz'pän·dər 'ded ‚tīm }

transponder set | ELECTR | A complete electronic set which is designed to receive an interrogation signal, and which retransmits coded signals that

can be interpreted by the interrogating station; it may also utilize the received signal for actuation of additional equipment such as local indicators or servo amplifiers. { tranz'pän·dər ‚set }

transponder suppressed time delay | ELECTR | Overall fixed time delay between reception of an interrogation and transmission of a reply to this interrogation. { tranz'pän·dər sə'prest 'tīm di‚lā }

transport | COMPUT SCI | **1.** To convey as a whole from one storage device to another in a digital computer. **2.** See tape transport. { trans'pȯrt (verb), 'tranz‚pȯrt (noun) }

transportable computer | COMPUT SCI | A microcomputer that can be carried about conveniently but, in contrast to a portable computer, requires an external power source. { tranz'pȯrd·ə·bəl kəm'pyüd·ər }

transportation lag See distance/velocity lag. { ‚tranz·pər'tā·shən ‚lag }

transport delay unit | COMPUT SCI | A device used in analog computers which produces an output signal as a delayed form of an input signal. Also known as delay unit; transport unit. { 'tranz ‚pȯrt di'lā ‚yü·nət }

transport lag See distance/velocity lag. { 'tranz ‚pȯrt ‚lag }

transport unit See transport delay unit. { 'tranz ‚pȯrt ‚yü·nət }

transposition | COMMUN | Interchanging the relative positions of conductors at regular intervals along a transmission line to reduce cross talk. { ‚tranz·pə'zish·ən }

transposition cipher | COMMUN | A cipher in which the order of the characters in the original message is changed. { ‚tranz·pə'zish·ən ‚sī·fər }

transradar | COMMUN | Bandwidth compression system developed for long-range narrow-band transmission of radio signals from a radar receiver to a remote location. { 'tränz‚rā‚där }

transrectification | ELEC | Rectification that occurs in one circuit when an alternating voltage is applied to another circuit. { tranz‚rek·tə·fə'kā·shən }

transrectification characteristic | ELECTR | Graph obtained by plotting the direct-voltage values for one electrode of a vacuum tube as abscissas against the average current values in the circuit of that electrode as ordinates, for various values of alternating voltage applied to another electrode as a parameter; the alternating voltage is held constant for each curve, and the voltages on other electrodes are maintained constant. { tranz ‚rek·tə·fə'kā·shən ‚kar·ik·tə‚ris·tik }

transrectifier | ELECTR | Device, ordinarily a vacuum tube, in which rectification occurs in one electrode circuit when an alternating voltage is applied to another electrode. { tranz'rek·tə ‚fī·ər }

transresistance | ELEC | The ratio of the voltage between any two connections of a four-terminal junction to the current passing between the other two connections. { ‚tranz·ri'zis·təns }

transresistance amplifier | ELECTR | An amplifier whose output voltage is proportional to its input current. { ‚tranz·ri‚zis·təns 'am·plə‚fī·ər }

transverse electric mode |ELECTROMAG| A mode in which a particular transverse electric wave is propagated in a waveguide or cavity. Abbreviated TE mode. Also known as H mode (British usage). { trans¦vərs i¦lek·trik ‚mōd }

transverse electric wave |ELECTROMAG| An electromagnetic wave in which the electric field vector is everywhere perpendicular to the direction of propagation. Abbreviated TE wave. Also known as H wave (British usage). { trans¦vərs i'lek·trik 'wāv }

transverse electromagnetic mode |ELECTROMAG| A mode in which a particular transverse electromagnetic wave is propagated in a waveguide or cavity. Abbreviated TEM mode. { trans¦vərs i ¦lek·trō·mag'ned·ik 'mōd }

transverse electromagnetic wave |ELECTROMAG| An electromagnetic wave in which both the electric and magnetic field vectors are everywhere perpendicular to the direction of propagation. Abbreviated TEM wave. { trans¦vərs i¦lek·trō·mag'ned·ik 'wāv }

transverse interference |ELEC| Interference occurring across terminals or between signal leads. { trans¦vərs ‚in·tər'fir·əns }

transverse magnetic mode |ELECTROMAG| A mode in which a particular transverse magnetic wave is propagated in a waveguide or cavity. Abbreviated TM mode. Also known as E mode (British usage). { trans¦vərs mag'ned·ik 'mōd }

transverse magnetic wave |ELECTROMAG| An electromagnetic wave in which the magnetic field vector is everywhere perpendicular to the direction of propagation. Abbreviated TM wave. Also known as E wave (British usage). { trans ¦vərs mag'ned·ik 'mōd 'wāv }

transverse recording |ELECTR| Technique for recording video signals on magnetic tape using a four-transducer rotating head. { trans¦vərs ri'kórd·iŋ }

trap |COMPUT SCI| An automatic transfer of control of a computer to a known location, this transfer occurring when a specified condition is detected by hardware. |ELECTR| **1.** A tuned circuit used in the radio-frequency or intermediate-frequency section of a receiver to reject undesired frequencies; traps in analog television receiver video circuits keep the sound signal out of the picture channel. Also known as rejector. **2.** *See* wave trap. { trap }

trap address |COMPUT SCI| The location at which control is transferred in case of an interrupt as soon as the current instruction is completed. { 'trap 'ad‚res }

TRAPATT diode |ELECTR| A *pn* junction diode, similar to the IMPATT diode, but characterized by the formation of a trapped space-charge plasma within the junction region; used in the generation and amplification of microwave power. Derived from trapped plasma avalanche transit time diode. { 'tra‚pat ‚dī‚ōd }

trapezium distortion |ELECTR| A defect in a cathode-ray tube in which the trace is confined within a trapezium rather than a rectangle, usually as a result of interaction between the

two pairs of deflection plates. { trə'pē·zē·əm di ‚stór·shən }

trapezoidal generator |ELECTR| Electronic stage designed to produce a trapezoidal voltage wave. { ¦trap·ə¦zóid·əl 'jen·ə‚rād·ər }

trapezoidal pulse |ELECTR| An electrical pulse in which the voltage rises linearly to some value, remains constant at this value for some time, and then drops linearly to the original value. { ¦trap·ə¦zóid·əl 'pəls }

trapezoidal wave |ELECTR| A wave consisting of a series of trapezoidal pulses. { ¦trap·ə¦zóid·əl 'wāv }

trapped plasma avalanche transit time diode *See* TRAPATT diode. { 'trapt 'plaz·mə 'av·ə‚lanch 'trans·ət ‚tīm 'dī‚ōd }

trapping *See* guided propagation. { 'trap·iŋ }

trapping mode |COMPUT SCI| A procedure by means of which the computer, upon encountering a predetermined set of conditions, saves the program in its present status, executes a diagnostic procedure, and then resumes the processing of the program as of the moment of interruption. { 'trap·iŋ ‚mōd }

trash heap |COMPUT SCI| An area in a computer's memory that has been assigned to a program but contains data which are no longer useful and are therefore wasteful of storage space. { 'trash ‚hēp }

traveling cable |ELEC| A cable that provides electrical contact between a fixed electrical outlet and an elevator or dumbwaiter car in the hoistway. { 'trav·əl·iŋ 'kā·bəl }

traveling-wave amplifier |ELECTR| An amplifier that uses one or more traveling-wave tubes to provide useful amplification of signals at frequencies of the order of thousands of megahertz. Also known as traveling-wave-tube amplifier (TWTA). { 'trav·əl·iŋ ¦wāv 'am·plə‚fī·ər }

traveling-wave antenna |ELECTROMAG| An antenna in which the current distributions are produced by waves of charges propagated in only one direction in the conductors. Also known as progressive-wave antenna. { 'trav·əl·iŋ ¦wāv an'ten·ə }

traveling-wave magnetron |ELECTR| A traveling-wave tube in which the electrons move in crossed static electric and magnetic fields that are substantially normal to the direction of wave propagation, as in practically all modern magnetrons. { 'trav·əl·iŋ ¦wāv 'mag·nə‚trän }

traveling-wave magnetron oscillations |ELECTR| Oscillations sustained by the interaction between the space-charge cloud of a magnetron and a traveling electromagnetic field whose phase velocity is approximately the same as the mean velocity of the cloud. { 'trav·əl·iŋ ¦wāv 'mag·nə ‚trän ‚äs·ə‚lā·shənz }

traveling-wave parametric amplifier |ELECTR| Parametric amplifier which has a continuous or iterated structure incorporating nonlinear reactors and in which the signal, pump, and difference-frequency waves are propagated along the structure. { 'trav·əl·iŋ ¦wāv ¦par·ə¦me·trik 'am·plə,fī·ər }

traveling-wave phototube [ELECTR] A traveling-wave tube having a photocathode and an appropriate window to admit a modulated laser beam; the modulated laser beam causes emission of a current-modulated photoelectron beam, which in turn is accelerated by an electron gun and directed into the helical slow-wave structure of the tube. { 'trav·əl·iŋ ¦wāv 'fōd·ə,tüb }

traveling-wave tube [ELECTR] An electron tube in which a stream of electrons interacts continuously or repeatedly with a guided electromagnetic wave moving substantially in synchronism with it, in such a way that there is a net transfer of energy from the stream to the wave; the tube is used as an amplifier or oscillator at frequencies in the microwave region. { 'trav·əl·iŋ ¦wāv ,tüb }

traveling-wave-tube amplifier See traveling-wave amplifier. { ¦trav·əl·iŋ ,wāv ,tüb 'am·plə,fī·ər }

tree [COMPUT SCI] A data structure in which each element may be logically followed by two or more other elements, there is one element with no predecessor, every other element has a unique predecessor, and there are no circular lists. [ELECTR] A set of connected circuit branches that includes no meshes; responds uniquely to each of the possible combinations of a number of simultaneous inputs. Also known as decoder. { trē }

tree automaton [COMPUT SCI] An automaton that processes inputs in the form of trees, usually trees associated with parsing expressions in context-free languages. { 'trē ,ȯd·ə,mä·shən }

tree diagram [COMPUT SCI] A flow diagram which has no closed paths. { 'trē ,dī·ə,gram }

tree pruning [COMPUT SCI] In computer programming, a strategy for eliminating branches of the complete game tree associated with a given position in a game such as chess or checkers, creating subtrees that explore a limited number of continuations for a limited number of moves. { 'trē ,prün·iŋ }

TRF receiver See tuned-radio-frequency receiver. { ,tē,är'ef ri,sē·vər }

triad [COMPUT SCI] A group of three bits, pulses, or characters forming a unit of data. [ELECTR] A triangular group of three small phosphor dots, each emitting one of the three primary colors on the screen of a three-gun color picture tube. { 'trī,ad }

triangular pulse [ELECTR] An electrical pulse in which the voltage rises linearly to some value, and immediately falls linearly to the original value. { trī'aŋ·gyə·lər 'pəls }

triangular wave [ELECTR] A wave consisting of a series of triangular pulses. { trī'aŋ·gyə·lər 'wāv }

triboelectricity See frictional electricity. { ¦trī·bō ,i,lek'tris·əd·ē }

triboelectric series [ELEC] A list of materials that produce an electrostatic charge when rubbed together, arranged in such an order that a material has a positive charge when rubbed with a material below it in the list, and has a negative

charge when rubbed with a material above it in the list. { ¦trī·bō·i¦lek·trik 'sir·ēz }

triboelectrification [ELEC] The production of electrostatic charges by friction. { ¦trī·bō·i ,lek·trə·fə'kā·shən }

tributary station [COMMUN] Communications terminal consisting of equipment compatible for the introduction of messages into or reception from its associated relay station. { 'trib·yə ,ter·ē 'stā·shən }

trickle charge [ELEC] A continuous charge of a storage battery at a low rate to maintain the battery in a fully charged condition. { 'trik·əl ,chärj }

trickling [COMPUT SCI] The temporary transfer of momentarily unneeded data from main storage to secondary storage devices. { 'trik·liŋ }

tricolor picture tube See color picture tube. { 'trī ,kəl·ər 'pik·chər ,tüb }

triductor [ELEC] Arrangement of iron-core transformers and capacitors used to triple a power-line frequency. { trī'dək·tər }

trigatron [ELECTR] Gas-filled, spark-gap switch used in line pulse modulators. { 'trig·ə,trän }

trigger [COMPUT SCI] To execute a jump to the first instruction of a program after the program has been loaded into the computer. Also known as initiate. [ELECTR] **1.** To initiate an action, which then continues for a period of time, as by applying a pulse to a trigger circuit. **2.** The pulse used to initiate the action of a trigger circuit. **3.** See trigger circuit. { 'trig·ər }

trigger action [ELECTR] Use of a weak input pulse to initiate main current flow suddenly in a circuit or device. { 'trig·ər ,ak·shən }

trigger circuit [ELECTR] **1.** A circuit or network in which the output changes abruptly with an infinitesimal change in input at a predetermined operating point. Also known as trigger. **2.** A circuit in which an action is initiated by an input pulse, as in some radar modulators. **3.** See bistable multivibrator. { 'trig·ər ,sər·kət }

trigger control [ELECTR] Control of thyratrons, ignitrons, and other gas tubes in such a way that current flow may be started or stopped, but not regulated as to rate. { 'trig·ər kən,trōl }

trigger diode [ELECTR] A symmetrical three-layer avalanche diode used in activating silicon-controlled rectifiers; it has a symmetrical switching mode, and hence fires whenever the breakover voltage is exceeded in either polarity. Also known as diode ac switch (diac). { 'trig·ər 'dī,ōd }

triggered spark gap [ELEC] A fixed spark gap in which the discharge passes between two electrodes but is initiated by an auxiliary trigger electrode to which low-power pulses are applied at regular intervals by a pulse amplifier. { 'trig·ərd 'spärk ,gap }

trigger electrode See starter. { 'trig·ər i,lek·trōd }

triggering [ELECTR] Phenomenon observed in some high-performance magnetic amplifiers with very low leakage rectifiers; as the input current is decreased in magnitude, the amplifier remains at cutoff for some time, and the output then suddenly shoots upward. { 'trig·ə·riŋ }

trigger level |ELECTR| In a transponder, the minimum input to the receiver which is capable of causing a transmitter to emit a reply. { 'trig·ər ,lev·əl }

trigger pulse |ELECTR| A pulse that starts a cycle of operation. Also known as tripping pulse. { 'trig·ər ,pəls }

trigger switch |ELEC| A switch that is actuated by pulling a trigger, and is usually mounted in a pistol-grip handle. { 'trig·ər ,swich }

trigger tube |ELECTR| A cold-cathode gas-filled tube in which one or more auxiliary electrodes initiate the anode current but do not control it. { 'trig·ər ,tüb }

trigistor |ELECTR| A *pnpn* device with a gating control acting as a fast-acting switch similar in nature to a thyratron. { tri'gis·tər }

trim |ELECTR| Fine adjustment of capacitance, inductance, or resistance of a component during manufacture or after installation in a circuit. { trim }

trimmer capacitor |ELEC| A relatively small variable capacitor used in parallel with a larger variable or fixed capacitor to permit exact adjustment of the capacitance of the parallel combination. { 'tim·ər kə,pas·əd·ər }

trimmer potentiometer |ELEC| A potentiometer which is used to provide a small-percentage adjustment and is often used with a coarse control. { 'tim·ər pə,ten·chē'äm·əd·ər }

triode |ELECTR| A three-electrode electron tube containing an anode, a cathode, and a control electrode. { 'trī,ōd }

triode clamp |ELECTR| A keyed clamp circuit utilizing triodes, such as a circuit which contains a complementary pair of bipolar transistors. { 'trī ,ōd 'klamp }

triode clipping circuit |ELECTR| A clipping circuit that utilizes a transistor or vacuum triode. { 'trī ,ōd 'klip·iŋ ,sər·kət }

triode laser |ELECTR| Gas laser whose light output may be modulated by signal voltages applied to an integral grid. { 'trī,ōd 'lā·zər }

triode transistor |ELECTR| A transistor that has three terminals. { 'trī,ōd tran'zis·tər }

triple-conversion receiver |ELECTR| Communications receiver having three different intermediate frequencies to give higher adjacent-channel selectivity and greater image-frequency suppression. { 'trip·əl kən,vər·zhən ri,sē·vər }

triple detection *See* double-superheterodyne reception. { 'trip·əl di,tek·shən }

triple-length working |COMPUT SCI| Processing of data by a computer in which three machine words are used to represent each data item, in order to achieve the desired precision in the results. { 'trip·əl ¦leŋkth 'wərk·iŋ }

triple modular redundancy |COMPUT SCI| A form of redundancy in which the original computer unit is triplicated and each of the three independent units feeds into a majority voter, which outputs the majority signal. { 'trip·əl 'mäj·ə·lər ri'dən·dən·sē }

triple-stub transformer |ELECTROMAG| Microwave transformer in which three stubs are placed a quarter-wavelength apart on a coaxial line and adjusted in length to compensate for impedance mismatch. { 'trip·əl ¦stab tranz'for·mər }

triplex cable |ELEC| An electrical cable consisting of three individually insulated wires that are twisted together and covered by an outer layer of protective material. { 'trip,leks ,kā·bəl }

triplexer |ELECTR| Dual duplexer that permits the use of two receivers simultaneously and independently in a radar system. { 'tri,plek·sər }

triplex system |COMMUN| Telegraph system in which two messages in one direction and one message in the other direction can be sent simultaneously over a single circuit. { 'tri,pleks ,sis·təm }

trip magnet *See* phase magnet. { 'trip ,mag·nət }

tripping device |ELEC| Mechanical or electromagnetic device used to bring a circuit breaker or starter to its off or open position, either when certain abnormal electrical conditions occur or when a catch is actuated manually. { 'trip·iŋ di ,vīs }

tripping pulse *See* trigger pulse. { 'trip·iŋ ,pəls }

trisistor |ELECTR| Fast-switching semiconductor consisting of an alloyed junction *pnp* device in which the collector is capable of electron injection into the base; characteristics resemble those of a thyratron electron tube, and switching time is in the nanosecond range. { tri'zis·tər }

tristate logic |ELECTR| A form of transistor-transistor logic in which the output stages or input and output stages can assume three states; two are the normal low-impedance 1 and 0 states, and the third is a high-impedance state that allows many tristate devices to time-share bus lines. { 'trī,stāt 'läj·ik }

tri-tet oscillator |ELECTR| Crystal-controlled, electron-coupled, vacuum-tube oscillator circuit which is isolated from the output circuit through use of the screen grid electrode as the oscillator anode; used for multiband operation because it generates strong harmonics of the crystal frequency. { 'trī,tet 'äs·ə,lād·ər }

troffer |ELEC| A long, recessed lighting unit having its opening flush with the surface of the ceiling and serving as a support and reflector for lamps. { 'träf·ər }

Trojan horse |COMPUT SCI| A computer program that has an unannounced (usually undesirable) function in addition to a desirable apparent function. { ,trō·jən 'hors }

trolley pole |ELEC| The pole which conducts electricity from the trolley wire to the trolley. { 'träl·ē ,pōl }

trolley wire |ELEC| The means by which power is conveyed to an electric trolley locomotive; it is an overhead wire which conducts power to the locomotive by the trolley pole. { 'träl·ē ,wīr }

trombone |ELECTROMAG| U-shaped, adjustable, coaxial-line matching assembly. { 'träm'bōn }

troposcatter *See* tropospheric scatter. { 'trōp·ō ,skad·ər }

tropospheric scatter |COMMUN| Scatter propagation of radio waves caused by irregularities in the refractive index of air in the troposphere; used for long-distance communications, with the aid of relay facilities, 180–300 miles (300–500 kilometers) apart. Also known as troposcatter. { ¦träp·ə¦sfir·ik 'skad·ər }

tropospheric wave [COMMUN] A radio wave that is propagated by reflection from a region of abrupt change in dielectric constant or its gradient in the troposphere. { ¦träp·ə¦sfir·ik 'wāv }

trouble-location problem [COMPUT SCI] In computers, a test problem used in a diagnostic routine. { 'trəb·əl lō¦kā·shən ‚präb·ləm }

troubleshoot [COMPUT SCI] To find and correct errors and faults in a computer, usually in the hardware. { 'trəb·əl‚shüt }

true-motion radar [ELECTR] A radar set which provides a true-motion radar presentation on the plan-position indicator, as opposed to the relative-motion, true-or-relative-bearing, presentation most commonly used. { 'trü ¦mō·shən 'rā‚där }

true-motion radar presentation [ELECTR] A radar plan-position indicator presentation in which the center of the scope represents the same geographic position, until reset, with all moving objects, including the user's own craft, moving on the scope. { 'trü ¦mō·shən 'rā‚där ‚pres·ən ‚tā·shən }

true motor load See thermal horsepower. { 'trü 'mōd·ər ‚lōd }

truncate [CONT SYS] To stop a robotic process before it has been completed. { 'trəŋ‚kāt }

truncated paraboloid [ELECTROMAG] Paraboloid antenna in which a portion of the top and bottom have been cut away to broaden the main lobe in the vertical plane. { 'trəŋ‚kād·əd pə'rab·ə ‚lóid }

truncation error [ENG] The error resulting from the analysis of a partial set of data in place of a complete or infinite set. { trəŋ'kā·shən ‚er·ər }

trunk [COMMUN] A telephone line connecting two central offices. Also known as trunk circuit. [COMPUT SCI] A path over which information is transferred in a computer. { trəŋk }

trunk circuit See trunk. { 'trəŋk ‚sər·kət }

trunk exchange [COMMUN] A telephone exchange whose main function is to interconnect trunks. { 'trəŋk iks'chānj }

trunk feeder [ELEC] An electric power transmission line that connects two generating stations, or a generating station and an important substation, or two electrical distribution networks. { 'trəŋk ‚fēd·ər }

trunk group [COMMUN] The collection of trunks of a given type or characteristic that connect two switching points. { 'trəŋk ‚grüp }

T-section filter [ELEC] T network used as an electric filter. { 'tē ‚sek·shən ‚fil·tər }

TSR See RAM resident.

T switch [ELECTR] An electrical switch that joins a machine to either of two other devices. { 'tē ‚swich }

TTL See transistor-transistor logic.

TTY See teletypewriter.

tube See electron tube. { 'tüb }

tube coefficient [ELECTR] Any of the constants that describe the characteristics of a thermionic vacuum tube, such as amplification factor, mutual conductance, or alternating-current plate resistance. { 'tüb ‚kō·i‚fish·ənt }

tube heating time [ELECTR] Time required for a tube to attain operating temperature. { 'tüb ¦hēd·iŋ ‚tīm }

tube noise [ELECTR] Noise originating in a vacuum tube, such as that due to shot effect and thermal agitation. { 'tüb ‚nóiz }

tube of flux See tube of force. { 'tüb əv 'fləks }

tube of force [ELEC] A region of space bounded by a tubular surface consisting of the lines of force which pass through a given closed curve. Also known as tube of flux. { 'tüb əv 'fórs }

tube tester [ELECTR] A test instrument designed to measure and indicate the condition of electron tubes used in electronic equipment. { 'tüb ‚tes·tər }

tube voltage drop [ELECTR] In a gas tube, the anode voltage during the conducting period. { 'tüb 'vōl·tij ‚dräp }

tube voltmeter See vacuum-tube voltmeter. { 'tüb 'vōlt‚mēd·ər }

tubular capacitor [ELEC] A paper or electrolytic capacitor having the form of a cylinder, with leads usually projecting axially from the ends; the capacitor plates are long strips of metal foil separated by insulating strips, rolled into a compact tubular shape. { 'tü·byə·lər kə'pas·əd·ər }

tunable echo box [ELECTROMAG] Echo box consisting of an adjustable cavity operating in a single mode; if calibrated, the setting of the plunger at resonance will indicate the wavelength. { 'tü·nə·bəl 'ek·ō ‚bäks }

tunable filter [ELECTR] An electric filter in which the frequency of the passband or rejection band can be varied by adjusting its components. { 'tü·nə·bəl 'fil·tər }

tunable magnetron [ELECTR] Magnetron which can be tuned mechanically or electronically by varying its capacitance or inductance. { 'tü·nə·bəl 'mag·nə‚trän }

tune [ELECTR] To adjust for resonance at a desired frequency. { tün }

tuned amplifier [ELECTR] An amplifier in which the load is a tuned circuit; load impedance and amplifier gain then vary with frequency. { ¦tünd 'am·plə‚fī·ər }

tuned-anode oscillator [ELECTR] A vacuum-tube oscillator whose frequency is determined by a tank circuit in the anode circuit, coupled to the grid to provide the required feedback. Also known as tuned-plate oscillator. { ¦tünd 'an‚ōd ‚äs·ə ‚lād·ər }

tuned-anode tuned-grid oscillator See tuned-grid tuned-anode oscillator. { ¦tünd 'an‚ōd ¦tünd 'grid ‚äs·ə‚lād·ər }

tuned-base oscillator [ELECTR] Transistor oscillator in which the frequency-determining resonant circuit is located in the base circuit; comparable to a tuned-grid oscillator. { ¦tünd 'bäs ‚äs·ə‚lād·ər }

tuned cavity See cavity resonator. { ¦tünd 'kav· əd·ē }

tuned circuit |ELECTR| A circuit whose components can be adjusted to make the circuit responsive to a particular frequency in a tuning range. Also known as tuning circuit. { ¦tünd 'sər·kət }

tuned-collector oscillator |ELECTR| A transistor oscillator in which the frequency-determining resonant circuit is located in the collector circuit; this is comparable to a tuned-anode electron-tube oscillator. { ¦tünd kə'lek·tər ‚äs·ə‚lād·ər }

tuned filter |ELECTR| Filter that uses one or more tuned circuits to attenuate or pass signals at the resonant frequency. { ¦tünd 'fil·tər }

tuned-grid oscillator |ELECTR| Oscillator whose frequency is determined by a parallel-resonant circuit in the grid coupled to the plate to provide the required feedback. { ¦tünd 'grid ‚äs·ə‚lād·ər }

tuned-grid tuned-anode oscillator |ELECTR| A vacuum-tube oscillator whose frequency is determined by a tank circuit in the grid circuit, coupled to the anode to provide the required feedback. Also known as tuned-anode tuned-grid oscillator. { ¦tünd 'grid ¦tünd 'an‚ōd ‚äs·ə‚lād·ər }

tuned-plate oscillator See tuned-anode oscillator. { ¦tünd 'plāte ‚äs·ə‚lād·ər }

tuned-radio-frequency receiver |ELECTR| A radio receiver consisting of a number of amplifier stages that are tuned to resonance at the carrier frequency of the desired signal by a gang capacitor; the amplified signals at the original carrier frequency are fed directly into the detector for demodulation, and the resulting audio-frequency signals are amplified by an audio-frequency amplifier and reproduced by a loudspeaker. Abbreviated TRF receiver. { ¦tünd 'räd·ēo ¦frē·kwən·sē ri‚sē·vər }

tuned-radio-frequency transformer |ELECTR| A transformer used for selective coupling in radio-frequency stages. { ¦tünd 'räd·ēo ¦frē·kwən·sē tranz‚fór·mər }

tuned-reed frequency meter See vibrating-reed-frequency meter. { ¦tünd 'rēd 'frē·kwən·sē ‚mēd·ər }

tuned relay |ELEC| A relay having mechanical or other resonating arrangements that limit response to currents at one particular frequency. { ¦tünd 'rē‚lā }

tuned resonating cavity |ELECTROMAG| Resonating cavity half a wavelength long or some multiple of a half wavelength, used in connection with a waveguide to produce a resultant wave with the amplitude in the cavity greatly exceeding that of the wave in the waveguide. { ¦tünd 'rez·ən‚äd·iŋ ‚kav·əd·ē }

tuned transformer |ELEC| Transformer whose associated circuit elements are adjusted as a whole to be resonant at the frequency of the alternating current supplied to the primary, thereby causing the secondary voltage to build up to higher values than would otherwise be obtained. { ¦tünd tranz'fór·mər }

tuner |ELECTR| The portion of a receiver that contains circuits which can be tuned to accept the carrier frequency of a desired transmitter while rejecting the carrier frequencies of all other stations on the air at that time. { 'tü·nər }

tungar tube |ELECTR| A gas tube having a heated thoriated tungsten filament serving as cathode and a graphite disk serving as anode in an argon-filled bulb at a low pressure; used chiefly as a rectifier in battery chargers. { 'təŋ‚är ‚tüb }

tungsten filament |ELEC| A filament used in incandescent lamps, and as an incandescent cathode in many types of electron tubes, such as thermionic vacuum tubes. { 'təŋ·stən 'fil·ə·mənt }

tungsten-halogen lamp |ELECTR| A lamp containing a halogen, usually iodine or bromine, which combines with tungsten evaporated from the filament. { ¦təŋ·stən 'hal·ə·jən ‚lamp }

tuning |COMPUT SCI| The use of various techniques involving adjustments to both hardware and software to improve the operating efficiency of a computer system. |ELECTR| The process of adjusting the inductance or the capacitance (or both) in a tuned circuit, for example, in a radio, television, or radar receiver or transmitter, so as to obtain optimum performance at a selected frequency. { 'tün·iŋ }

tuning capacitor |ELEC| A variable capacitor used for tuning purposes. { 'tün·iŋ kə‚pas·əd·ər }

tuning circuit See tuned circuit. { 'tün·iŋ ‚sər·kət }

tuning coil |ELEC| A variable inductance coil for adjusting the frequency of an oscillator or tuned circuit. { 'tün·iŋ ‚kóil }

tuning core |ELECTROMAG| A ferrite core that is designed to be moved in and out of a coil or transformer to vary the inductance. { 'tün·iŋ ‚kór }

tuning indicator |ELECTR| A device that indicates when a radio receiver is tuned accurately to a station; it is connected to a circuit having a direct-current voltage that varies with the strength of the incoming carrier signal. { 'tün·iŋ ‚in·də‚kād·ər }

tuning range |ELECTR| The frequency range over which a receiver or other piece of equipment can be adjusted by means of a tuning control. { 'tün·iŋ ‚rānj }

tuning screw |ELECTROMAG| A screw that is inserted into the top or bottom wall of a waveguide and adjusted as to depth of penetration inside for tuning or impedance-matching purposes. { 'tün·iŋ ‚skrü }

tuning stub |ELECTROMAG| Short length of transmission line, usually shorted at its free end, connected to a transmission line for impedance-matching purposes. { 'tün·iŋ ‚stəb }

tuning susceptance |ELECTR| Normalized susceptance of an anti-transmit-receive tube in its mount due to the deviation of its resonant frequency from the desired resonant frequency. { 'tün·iŋ sə‚sep·təns }

tuning wand |ELEC| Rod of insulating material having a brass plug at one end and a powered iron core at the other end; used for checking receiver alignment. { 'tün·iŋ ‚wänd }

tunnel diode |ELECTR| A heavily doped junction diode that has a negative resistance at very low voltage in the forward bias direction, due to quantum-mechanical tunneling, and a short circuit in the negative bias direction. Also known as Esaki tunnel diode. { 'tən·əl ,dī,ōd }

tunneling cryotron |ELECTR| A low-temperature current-controlled switching device that has two electrodes of superconducting material separated by an insulating film, forming a Josephson junction, and a control line whose currents generate magnetic fields that switch the device between two states characterized by the presence or absence of electrical resistance. { 'tən·əl·iŋ 'krī·ə,trän }

tunneling microscope See scanning tunneling microscope. { 'tən·əl·iŋ 'mī·krə,skōp }

tunnel junction |ELECTR| A two-terminal electronic device having an extremely thin potential barrier to electron flow, so that the transport characteristic (the current-voltage curve) is primarily governed by the quantum-mechanical tunneling process which permits electrons to penetrate the barrier. { 'tən·əl ,jəŋk·shən }

tunnel rectifier |ELECTR| Tunnel diode having a relatively low peak-current rating as compared with other tunnel diodes used in memory-circuit applications. { 'tən·əl ,rek·tə,fī·ər }

tunnel resistor |ELECTR| Resistor in which a thin layer of metal is plated across a tunneling junction, to give the combined characteristics of a tunnel diode and an ordinary resistor. { 'tən·əl ri,zis·tər }

tunnel triode |ELECTR| Transistorlike device in which the emitter-base junction is a tunnel diode and the collector-base junction is a conventional diode. { 'tən·əl ,trī,ōd }

tuple |COMPUT SCI| A horizontal row of data items in a relational data structure; corresponds to a record or segment in other types of data structures. { 'tü·pəl }

turbine generator |ELEC| An electric generator driven by a steam, hydraulic, or gas turbine. { 'tər·bən ,jen·ə,rād·ər }

turboalternator |ELEC| An alternator, such as a synchronous generator, which is driven by a steam turbine. { ¦tər·bō'ȯl·tər,nād·ər }

Turing machine |COMPUT SCI| A mathematical idealization of a computing automaton similar in some ways to real computing machines; used by mathematicians to define the concept of computability. { 'tür·iŋ mə,shēn }

turn |ELEC| One complete loop of wire. { 'tərn }

turnaround system |COMPUT SCI| In character recognition, a system in which the input data to be read have previously been printed by the computer with which the reader is associated; an application is invoice billing and the subsequent recording of payments. Also known as reentry system. { 'tərn·ə,raund ,sis·təm }

turnaround time |COMPUT SCI| The delay between submission of a job for a data-processing system and its completion. { 'tərn·ə,raund ,tīm }

turnkey |COMPUT SCI| A complete computer system delivered to a customer in running condition, with all necessary premises, hardware and software equipment, supplies, and operating personnel. { 'tərn,kē }

turn-off time |ELECTR| The time that is takes a gate circuit to shut off a current. { 'tərn,ȯf ,tīm }

turn-on time |ELECTR| The time that it takes a gate circuit to allow a current to reach its full value. { 'tərn,ȯn ,tīm }

turns ratio |ELEC| The ratio of the number of turns in a secondary winding of a transformer to the number of turns in the primary winding. { 'tərnz ,rā·shō }

turnstile antenna |ELECTROMAG| An antenna consisting of one or more layers of crossed horizontal dipoles on a mast, usually energized so the currents in the two dipoles of a pair are equal and in quadrature; used with television, frequency modulation, and other very-high-frequency or ultra-high-frequency transmitters to obtain an essentially omnidirectional radiation pattern. { 'tərn,stīl an,ten·ə }

turntable |ENG ACOUS| The rotating platform on which a disk record is placed for recording or playback. { 'tərn,tā·bəl }

turret tuner |ELECTR| A television tuner having one set of pretuned circuits for each channel, mounted on a drum that is rotated by the channel selector; rotation of the drum connects each set of tuned circuits in turn to the receiver antenna circuit, radio-frequency amplifier, and r-f oscillator. { 'tə·rət ,tü·nər }

turtle |COMPUT SCI| A cursor with the attributes of both position and direction; usually, an arrow that points in the direction it is about to move and generates a line along its path. { 'tərd·əl }

tutorial |COMPUT SCI| A method of computer-assisted instruction that involves a collection of screen formats, generally arranged in sequences that can be selected from a menu, and presented in response to the terminal operator's request. { tü'tȯr·ē·əl }

TV See television.

TV camera scanner |COMPUT SCI| In optical character recognition, a device that images an input character onto a sensitive photoconductive target of a camera tube, thereby developing an electric charge pattern on the inner surface of the target; this pattern is then explored by a scanning beam which traces out a rectangular pattern with the result that a waveform is produced which represents the character's most probable identity. { tē'vē ¦kam·rə ,skan·ər }

TVI See television interference.

TVRO See television receive only antenna.

tweeter |ENG ACOUS| A loudspeaker designed to handle only the higher audio frequencies, usually those well above 3000 hertz; generally used in conjunction with a crossover network and a woofer. { 'twēd·ər }

twin arithmetic units |COMPUT SCI| A feature of some computers where the essential portions of the arithmetic section are virtually duplicated. { 'twin ə'rith·mə,tik ,yü·nəts }

twin axial cable |COMMUN| A transmission line consisting of two coaxial cables enclosed within a single sheath, each used to transmit signals in one direction. { 'twin 'ak·sē·əl 'kā·bəl }

twin check [COMPUT SCI] Continuous check of computer operation, achieved by the duplication of equipment and automatic comparison of results. { 'twin 'chek }

twin-T filter [ELEC] An electric filter consisting of a parallel-T network with values of network elements chosen in such a way that the outputs due to each of the paths precisely cancel at a specified frequency. { 'twin 'tē ,fil·tər }

twin-T network See parallel-T network. { 'twin 'tē ,net,wərk }

twist [ELECTROMAG] A waveguide section in which there is a progressive rotation of the cross section about the longitudinal axis of the waveguide. { twist }

twisted pair [ELEC] A cable composed of two small insulated conductors twisted together without a common covering. Also known as copper pair. { 'twis·təd 'per }

twist-lock connector [ELEC] A power plug and receptacle in which the plug must be twisted after insertion to lock it in place, to guard against the plug accidentally being knocked loose. { 'twist ,läk kə'nek·tər }

two-address code [COMPUT SCI] In computers, a code using two-address instructions. { 'tü 'ad ,res ,kōd }

two-address instruction [COMPUT SCI] In computers, an instruction which includes an operation and specifies the location of two registers. { 'tü 'ad,res in,strək·shən }

two-dimensional electron gas field-effect transistor See high-electron-mobility transistor. { ,tü di'men·shən·əl i¦lek,trän 'gas ¦fēld i¦fekt tran 'zis·tər }

two-dimensional storage [COMPUT SCI] A direct-access storage device in which the storage locations assigned to a particular file do not have to be physically adjacent, but instead may be taken from one or more seek areas. { 'tü ¦di ¦men·shən·əl 'stȯr·ij }

two-gap head [COMPUT SCI] One of two separate magnetic tape heads, one for reading and the other for recording data. { 'tü ¦gap 'hed }

two-hop transmission [COMMUN] Propagation of radio waves in which the waves are reflected from the ionosphere, then reflected from the ground, and then reflected from the ionosphere again before reaching the receiver. { 'tü ¦häp tranz'mish·ən }

two-input subtracter See half-subtracter. { 'tü ¦in,pút səb'trak·tər }

two-level subroutine [COMPUT SCI] A subroutine in which entry is made to a second, lower-level subroutine. { 'tü ¦lev·əl 'səb·rü,tēn }

two-out-of-five code [COMPUT SCI] An encoding of the decimal digits using five binary bits and having the property that every code element contains two 1's and three 0's. { 'tü aúd əv ¦fīv 'kōd }

two-part code [COMMUN] Randomized code consisting of an encoding section in which the plain text groups are arranged in alphabetical or other significant order accompanied by their code groups in nonalphabetical or random order, and a decoding section in which the code groups are arranged in alphabetical or numerical order and are accompanied by their meanings given in the encoding section. { 'tü ¦pärt 'kōd }

two-pass compiler [COMPUT SCI] A language processor that goes through the program to be translated twice; on the first pass it checks the syntax of statements and constructs a table of symbols, while on the second pass it actually translates program statements into machine language. { 'tü ¦pas kəm'pīl·ər }

two-phase alternating-current circuit [ELEC] A circuit in which there are two alternating currents on separate wires, the two currents being 90° out of phase. { 'tü ¦fāz 'ȯl·tər,nād·iŋ kə·rənt ,sər·kət }

two-phase current [ELEC] Current delivered through two pairs of wires or at a phase difference of one-quarter cycle (90°) between the current in the two pairs. { 'tü ¦fāz 'kə·rənt }

two-phase five-wire system [ELEC] System of alternating-current supply comprising five conductors, four of which are connected as in a two-phase four-wire system, the fifth being connected to the neutral points of each phase. { 'tü ¦fāz 'fīv ¦wīr 'sis·təm }

two-phase four-wire system [ELEC] System of alternating-current supply comprising two pairs of conductors, between one pair of which is maintained an alternating difference of potential displaced in phase by one-quarter of a period from an alternating difference of potential of the same frequency maintained between the other pair. { 'tü ¦fāz 'fȯr ¦wīr 'sis·təm }

two-phase three-wire system [ELEC] System of alternating-current supply comprising three conductors, between one of which (known as the common return) and each of the other two are maintained alternating difference of potential displaced in phase by one-quarter of a period with relation to each other. { 'tü ¦fāz 'thrē ¦wīr 'sis·təm }

two-plus-one address instruction [COMPUT SCI] An instruction in a computer program which has two addresses specifying the locations of operands and one address specifying the location in which the result is to be entered. { 'tü ,pləs ¦wən 'ad,res in,strək·shən }

two-port junction [ELECTROMAG] A waveguide junction with two openings; it can consist either of a discontinuity or obstacle in a waveguide, or of two essentially different waveguides connected together. { 'tü ¦pȯrt 'jəŋk·shən }

two-port system [CONT SYS] A system which has only one input or excitation and only one response or output. { 'tü ¦pȯrt 'sis·təm }

two-pulse canceler [ELECTR] A moving-target indicator canceler which compares the phase variation of two successive pulses received from a target; discriminates against signals with radial velocities which produce a Doppler frequency equal to a multiple of the pulse repetition frequency. { 'tü ¦pəls 'kan·slər }

two-quadrant multiplier [COMPUT SCI] Of an analog computer, a multiplier in which operation is restricted to a single sign of one input variable only. { 'tü ¦kwäd·rənt 'məl·tə,plī·ər }

two-source frequency keying [COMMUN] Keying in which the modulating wave shifts the output frequency between predetermined values derived from independent sources. { 'tü ¦sórs 'frē·kwən·sē ,kē·iŋ }

two-state Turing machine [COMPUT SCI] A variation of a Turing machine in which only two states are allowed, although the number of symbols may be large. { 'tü ¦stāt 'túr·iŋ mə,shēn }

two-symbol Turing machine [COMPUT SCI] A variation of a Turing machine in which only two symbols are permitted, although the number of states may be large. { 'tü ¦sim·bəl 'túr·iŋ mə ,shēn }

two-tone keying [COMMUN] Keying in which the modulating wave causes the carrier to be modulated with one frequency for the marking condition and modulated with a different frequency for the spacing condition. { 'tü ¦tōn 'kē·iŋ }

two-tone modulation [COMMUN] In teletype-writer operation, a method of modulation in which two different carrier frequencies are employed for the two signaling conditions; the transition from one frequency to the other is abrupt, with resultant phase discontinuities. { 'tü ¦tōn ,mäj·ə'lā·shən }

two-wire circuit [ELEC] A metallic circuit formed by two conductors insulated from each other; in contrast with a four-wire circuit, it uses only one line or channel for transmission of electric waves in both directions. { 'tü ¦wīr 'sər·kət }

two-wire repeater [ELECTR] Repeater that provides for transmission in both directions over a two-wire circuit; in carrier transmission, it usually operates on the principle of frequency separation for the two directions of transmission. { 'tü ¦wīr ri'pēd·ər }

TWTA See traveling-wave amplifier.

TWX machine See teletypewriter. { ,tē ,dəb·əl·yü 'eks mə,shēn }

TWX service See teletypewriter exchange service. { ,tē ,dəb·əl·yü 'eks ,sər·vəs }

Twystron [ELECTR] Very-high-power, hybrid microwave tube, combining the input section of a high-power klystron with the output section of a traveling wave tube, characterized by high operating efficiency and wide bandwidths. { 'twī ,strän }

typeahead buffer [COMPUT SCI] A temporary storage device in a keyboard or microcomputer that holds information typed on the keyboard before the central processing unit is ready to accept it. { 'tīp·ə,hed 'bəf·ər }

type A wave See continuous wave. { 'tīp ¦ā ,wāv }

type A1 wave [COMMUN] An unmodulated, keyed, continuous wave. { 'tīp ¦ā¦wən ,wāv }

type A2 wave [COMMUN] A modulated, keyed, continuous wave. { 'tīp ¦ā¦tü ,wāv }

type A3 wave [COMMUN] A continuous wave modulated by music, speech, or other sounds. { 'tīp ¦ā¦thrē ,wāv }

type A4 wave [COMMUN] A superaudio frequency-modulated continuous wave, as used in facsimile systems. { 'tīp ¦ā¦fór ,wāv }

type A9 wave [COMMUN] A composite transmission and continuous wave that is not type A1, A2, A3, A4, or A5 wave. { 'tīp ¦ā¦nīn ,wāv }

type B wave [COMMUN] A keyed, damped wave. { 'tīp ¦bē ,wāv }

type drum [COMPUT SCI] A steel cylinder containing 128 to 144 lateral bands, each band containing the alphabet, the digits 0–9, and the standard set of punctuation marks such as commas and periods, and revolving at high speed; printing is achieved by a hammer facing each band and activated at the right time to cause a character to be printed on the paper flowing between hammers and drum. { 'tīp ,drəm }

type I assembly [ELECTR] An assembly consisting entirely of surface-mounted electronic components, on either one or both sides of a printed board. { 'tīp ¦wən ə'sem·blē }

type II assembly [ELECTR] An assembly of both surface-mounted and leaded electronic components, in which the surface-mounted components are on both sides of the printed board. { 'tīp ¦tü ə'sem·blē }

type III assembly [ELECTR] An assembly of both surface-mounted and leaded electronic components, in which the surface-mounted components are only on the bottom side of the printed board. { 'tīp ¦thrē ə'sem·blē }

type-M carcinotron See M-type backward-wave oscillator. { 'tīp ¦em 'kärs·ən·ə,trän }

type-O carcinotron See O-type backward-wave oscillator. { 'tīp ¦ō 'kärs·ən·ə,trän }

typewriter terminal [COMPUT SCI] An electric typewriter combined with an ASCII or other code generator that provides code output for feeding a computer, calculator, or other digital equipment; the terminal also produces hard copy when driven by incoming code signals. { 'tīp,rīd·ər ,tər·mən·əl }

U

UART See universal asynchronous receiver transmitter. { 'yü,ärt }

Uda antenna See Yagi-Uda antenna. { 'ü·də an ,ten·ə }

UDP See user datagram protocol.

U format [COMPUT SCI] A record format which the input/output control system treats as completely unknown and unpredictable. { 'yü ,fȯr,mat }

UHF See ultrahigh frequency.

UJT See unijunction transistor.

ULSA See ultra-low-side-lobe antenna.

ULSI circuit See ultra-large-scale integrated circuit. { 'yü¦el¦es¦ī 'sər·kət }

ultra-audion circuit [ELECTR] Regenerative detector circuit in which a parallel resonant circuit is connected between the grid and the plate of a vacuum tube, and a variable capacitor is connected between the plate and cathode to control the amount of regeneration. { 'əl·trə 'ȯd·ē·,än ,sər·kət }

ultra-audion oscillator [ELECTR] Variation of the Colpitts oscillator circuit; the resonant circuit employs a section of transmission line. { 'əl·trə 'ȯd·ē·,än ,äs·ə,lād·ər }

ultrahigh frequency [COMMUN] The band of frequencies between 300 and 3000 megahertz in the radio spectrum, corresponding to wavelengths of 10 centimeters to 1 meter. Abbreviated UHF. { ¦əl·trə'hī 'frē·kwən·sē }

ultrahigh-frequency tuner [ELECTR] A tuner in a television receiver for reception of stations transmitting in the ultrahigh-frequency band. { ¦əl·trə'hī 'frē·kwən·sē 'tü·nər }

ultra-large-scale integrated circuit [ELECTR] A complex integrated circuit that contains more than 1,000,000 elements. Abbreviated ULSI circuit. { ¦əl·trə'lärj ¦skāl 'int·ə,gräd·əd 'sər·kət }

ultra-low-side-lobe antenna [ELECTROMAG] A radar antenna so carefully designed and constructed that side lobes are all lower than about 45 decibels below the main beam; particularly valuable in modern airborne radar. Abbreviated ULSA. { ¦əl·trə ,lō 'sīd ¦lōb an,ten·ə }

ultrashort waves [COMMUN] Radio waves shorter than 10 meters in wavelength; corresponding to frequencies above 30 megahertz. { ¦əl·trə'shȯrt 'wāvz }

ultra-small-aperture terminal [COMMUN] An antenna less than 20 inches (0.5 meter) in diameter

that is used for reception of direct broadcasts from geosynchronous satellites. Abbreviated USAT. { ¦əl·trə ¦smȯl ¦ap·ə·chər 'tərm·ə·nəl }

ultrasonic camera [ELECTR] A device which produces a picture display of ultrasonic waves sent through a sample to be inspected or through live tissue; a piezoelectric crystal is used to convert the ultrasonic waves to voltage differences, and the voltage pattern on the crystal modulates the intensity of an electronic beam scanning the crystal; this beam in turn controls the intensity of a beam in a display device. { ¦əl·trə'sän·ik 'kam·rə }

ultrasonic communication [COMMUN] Communication accomplished through water by keying the sound output of echo-ranging sonar on ships or submarines or by using other such devices. { ¦əl·trə'sän·ik kə,myü·nə'kā·shən }

ultrasonic delay line [ENG ACOUS] A delay line in which use is made of the propagation time of sound through a medium such as fused quartz, barium titanate, or mercury to obtain a time delay of a signal. Also known as ultrasonic storage cell. { ¦əl·trə'sän·ik di'lā ,līn }

ultrasonic generator [ENG ACOUS] A generator consisting of an oscillator driving an electroacoustic transducer, used to produce acoustic waves above about 20 kilohertz. { ¦əl·trə'sän·ik 'jen·ə,rād·ər }

ultrasonic storage cell See ultrasonic delay line. { ¦əl·trə'sän·ik 'stȯr·ij ,sel }

ultrasonic transducer [ENG ACOUS] A transducer that converts alternating-current energy above 20 kilohertz to mechanical vibrations of the same frequency; it is generally either magnetostrictive or piezoelectric. { ¦əl·trə'sän·ik tranz 'dü·sər }

ultrasonic transmitter [ENG ACOUS] A device used to track seals, fish, and other aquatic animals; the device is fastened to the outside of the animal or fed to it, and has a loudspeaker which is made to vibrate at an ultrasonic frequency, propagating ultrasonic waves through the water to a special microphone or hydrophone. { ¦əl·trə'sän·ik tranz'mid·ər }

ultraviolet-erasable programmable read-only memory [COMPUT SCI] An integrated-circuit memory chip in which the stored information can be erased only by ultraviolet light and the circuit can be reprogrammed with new information

ultraviolet-erasable programmable read-only memory eraser

that can be stored indefinitely. Abbreviated UV EPROM; UVPROM. { ¦əl·trə'vī·lət i'rās·ə·bəl 'prō‚gram·ə·bəl ¦rēd ‚ōn·lē 'mem·rē }

ultraviolet-erasable programmable read-only memory eraser [COMPUT SCI] A device that removes the contents of ultraviolet-erasable programmable read-only memory chips by exposing them to ultraviolet light. { ¦əl·trə'vī·lət i'rās·ə·bəl 'prō‚gram·ə·bəl ¦rēd ‚ōn·lē 'mem·rē i'rā·sər }

ultraviolet lamp [ELECTR] A lamp providing a high proportion of ultraviolet radiation, such as various forms of mercury-vapor lamps. { ¦əl·trə'vī·lət 'lamp }

umbrella antenna [ELECTROMAG] Antenna in which the wires are guyed downward in all directions from a central pole or tower to the ground, somewhat like the ribs of an open umbrella. { əm'brel·ə an‚ten·ə }

unambiguous name [COMPUT SCI] The name of a file or other data item that completely specifies the item to a computer system. { ¦ən·am'big·yə·wəs 'nām }

unamplified back bias [ELECTR] Degenerative voltage developed across a fast time constant circuit within an amplifier stage itself. { ¦ən'am·plə‚fīd 'bak ‚bī·əs }

unattended operation [COMPUT SCI] An operation in which components in the hardware of a communications terminal or data-processing system operate automatically, allowing handling of signals or data without human intervention. { ¦ən·ə'ten·dəd ‚äp·ə'rā·shən }

unattended time [COMPUT SCI] Time during which a computer is turned off but is not undergoing maintenance. Also known as unused time. { ¦ən·ə'ten·dəd ‚tīm }

unbalanced line [ELEC] A transmission line in which the voltages on the two conductors are not equal with respect to ground; a coaxial line is an example. { ¦ən'bal·ənst 'līn }

unbalanced output [ELEC] An output in which one of the two input terminals is substantially at ground potential. { ¦ən'bal·ənst 'aút‚pút }

unbalanced wire circuit [ELEC] A wire circuit whose two sides are inherently electrically unlike. { ¦ən'bal·ənst 'wīr ‚sər·kət }

unblanking pulse [ELECTR] Voltage applied to a cathode-ray tube to overcome bias and cause trace to be visible. { ¦ən'blaŋk·iŋ ‚pəls }

unbreakable cipher [COMMUN] A cipher for which the message or key cannot be obtained through cryptanalysis, even with an unlimited amount of computational power, data storage, and calendar time. { ¦ən'brāk·ə·bəl 'sī·fər }

unbundling [COMPUT SCI] The separate provision of software products and services versus computer hardware (equipment). { ¦ən'bənd·liŋ }

uncatalog [COMPUT SCI] To remove an entry from the system catalog so that the file named in the entry can no longer be accessed by the operating system. { ¦ən'kad·əl‚äg }

uncharged [ELEC] Having no electric charge. { ¦ən'chärjd }

unconditional [COMPUT SCI] Not subject to conditions external to the specific instruction. { ¦ən·kən'dish·ən·əl }

unconditional jump [COMPUT SCI] A digital-computer instruction that interrupts the normal process of obtaining instructions in an ordered sequence, and specifies the address from which the next instruction must be taken. Also known as unconditional transfer. { ¦ən·kən'dish·ən·əl 'jəmp }

underbunching [ELECTR] In velocity-modulated electron streams, a condition representing less than the optimum bunching. { ¦ən·dər ¦bənch·iŋ }

undercurrent relay [ELEC] A relay designed to operate when its coil current falls below a predetermined value. { 'ən·dər‚kə·rənt 'rē‚lā }

undercut [ELECTR] Undesirable lateral etching by chemicals in the fabrication of semiconductor devices. { 'ən·dər‚kət }

underflow [COMPUT SCI] The generation of a result whose value is smaller than the smallest quantity that can be represented or stored by a computer. { 'ən·dər‚flō }

underlap [COMMUN] 1. In facsimile transmission, the space between the recorded elemental area in one recording line and the adjacent elemental area in the next recording line, when these areas are smaller than normal; or the space between the elemental areas in the direction of the recording line. 2. The amount by which the effective height of the scanning spot falls short of the nominal width of the scanning line. { 'ən·dər‚lap }

undershoot [CONT SYS] The amount by which a system's response to an abrupt change in input falls short of that desired. { 'ən·dər‚shüt }

underthrow distortion [COMMUN] Distortion occurring in facsimile when the maximum signal amplitude is too low. { ¦ən·dər¦thrō di'stòr·shən }

undervoltage protection [ELEC] An undervoltage relay which removes a motor from service when a low-voltage condition develops, so that the motor will not draw excessive current, or which prevents a large induction or synchronous motor from starting under low-voltage conditions. { ¦ən·dər‚vōl·tij prə'tek·shən }

undervoltage relay [ELEC] A relay designed to operate when its coil voltage falls below a predetermined value. { ¦ən·dər‚vōl·tij 'rē‚lā }

undetected error rate [COMMUN] The number of bits (or other units of information) which are received but are not detected or corrected by error-control equipment, divided by the total number of bits (or other units of information) transmitted. Also known as residual error rate. { ¦ən·di'tek·təd 'er·ər ‚rāt }

undistorted wave [COMMUN] Periodic wave in which both the attenuation and velocity of propagation are the same for all sinusoidal components, and in which no sinusoidal component is present at one point that is not present at all points. { ¦ən·di'stòrd·əd 'wāv }

undisturbed-one output |ELECTR| "One" output of a magnetic cell to which no partial-read pulses have been applied since that cell was last selected for writing. { ¦ən·di'stərbd 'wən 'aút,pút }

undisturbed-zero output |ELECTR| "Zero" output of a magnetic cell to which no partial-write pulses have been applied since that cell was last selected for reading. { ¦ən·di'stərbd 'zir·ō 'aút ,pút }

unfired tube |ELECTR| Condition of a TR (transmit-receive), ATR, or pre-TR tube in which there is no radio-frequency glow discharge at either the resonant gap or resonant window. { ¦ən'fīrd 'tüb }

unformatted file |COMPUT SCI| Any data file, such as a text file, that does not have various properties such as a consistent structure with regard to record length and order of data elements. { ¦ən'fòr,mad·əd 'fīl }

unidirectional antenna |ELECTROMAG| An antenna that has a single well-defined direction of maximum gain. { ¦yü·nə·də'rek·shən·əl an'ten·ə }

unidirectional coupler |ELECTR| Directional coupler that samples only one direction of transmission. { ¦yü·nə·də'rek·shən·əl 'kəp·lər }

unidirectional log-periodic antenna |ELECTROMAG| A broad-band antenna in which the cut-out portions of a log-periodic antenna are mounted at an angle to each other, to give a unidirectional radiation pattern in which the major radiation is in the backward direction, off the apex of the antenna; impedance is essentially constant for all frequencies, as is the radiation pattern. { ¦yü·nə·də'rek·shən·əl 'läg ,pir·ē'äd·ik an'ten·ə }

unidirectional microphone |ENG ACOUS| A microphone that is responsive predominantly to sound incident from one hemisphere, without picking up sounds from the sides or rear. { ¦yü·nə·də'rek·shən·əl 'mī·krə,fōn }

unidirectional pulse-amplitude modulation |COMMUN| Modulation of pulse-amplitude type in which all pulses rise in the same direction. Also known as single-polarity pulse-amplitude modulation. { ¦yü·nə·də'rek·shən·əl ¦pəls ¦am·plə,tüd ,mäj·ə'lā·shən }

unidirectional pulses |ELECTR| Single polarity pulses which all rise in the same direction. { ¦yü·nə·də'rek·shən·əl 'pəl·səz }

unidirectional transducer |ELECTR| Transducer that measures stimuli in only one direction from a reference zero or rest position. Also known as unilateral transducer. { ¦yü·nə·də'rek·shən·əl tranz'dü·sər }

uniform line |ELEC| Line which has substantially identical electrical properties through its length. { 'yü·nə,fórm 'līn }

uniformly accessible storage See random-access memory. { ,yü·nə'fórm·lē ak'ses·ə·bəl 'stór·ij }

uniform plane wave |ELECTROMAG| Plane wave in which the electric and magnetic intensities have constant amplitude over the equiphase surfaces; such a wave can only be found in free space at an infinite distance from the source. { 'yü·nə,fórm 'plän 'wāv }

uniform resource locator |COMPUT SCI| The unique Internet address assigned to a Web document or resource by which it can be accessed by all Web browsers. The first part of the address specifies the applicable Internet protocol (IP), for example, http or ftp; the second part provides the IP address or domain name of the location. Abbreviated URL. { ,yü·nə,fórm ri,sórs 'lō·kād·ər }

unijunction transistor |ELECTR| An n-type bar of semiconductor with a p-type alloy region on one side; connections are made to base contacts at either end of the bar and to the p-region. Abbreviated UJT. Formely known as double-base diode; double-base junction diode. { 'yü·nə ,jəŋk·shən tran'zis·tər }

unilateral conductivity |ELECTR| Conductivity in only one direction, as in a perfect rectifier. { ¦yü·nə'lad·ə·rəl ,kän·dək'tiv·əd·ē }

unilateralization |ELECTR| Use of an external feedback circuit in a high-frequency transistor amplifier to prevent undesired oscillation by canceling both the resistive and reactive changes produced in the input circuit by internal voltage feedback; with neutralization, only the reactive changes are canceled. { ,yü·nə ,lad·ə·rə·lə'zā·shən }

unilateral transducer See unidirectional transducer. { ¦yü·nə'lad·ə·rəl tranz'dü·sər }

unimorph |ELECTR| A piezoelectric microphone that consists of a single piezoelectric disk cemented to a thin metal plate. { 'yü·nə,mórf }

uninterruptible power system |ELEC| A system that provides protection against primary alternating-current power failure and variations in power-line frequency and voltage. Abbreviated UPS. { ¦ən¦in·tə'rəp·tə·bəl 'paú·ər ,sis·təm }

union |COMPUT SCI| A data structure that can store items of different types, but can store only one item at a time. { 'yün·yən }

union catalog |COMPUT SCI| A merged listing of the contents of two or more catalogs (of libraries, for example). { 'yün·yən 'kad·əl,äg }

unipath See nonshared control unit. { 'yü·nə ,path }

unipolar |ELEC| Having but one pole, polarity, or direction; when applied to amplifiers or power supplies, it means that the output can vary in only one polarity from zero and, therefore, must always contain a direct-current component. { 'yü·nə,pō·lər }

unipolar machine See homopolar generator. { ¦yü·nə'pō·lər mə'shēn }

unipolar transistor |ELECTR| A transistor that utilizes charge carriers of only one polarity, such as a field-effect transistor. { ¦yü·nə'pō·lər tran'zis·tər }

unipole |ELECTROMAG| A hypothetical antenna that radiates or receives signals equally well in all directions. Also known as isotropic antenna. { 'yü·nə,pōl }

unipotential cathode See indirectly heated cathode. { ¦yü·nə·pə'ten·chəl 'kath,ōd }

unipotential electrostatic lens |ELECTR| An electrostatic lens in which the focusing is produced by application of a single potential difference; in its simplest form it consists of three apertures of which the outer two are at a common potential, and the central aperture is at a different, generally lower, potential. { ¦yü·nə·pə'ten·chəl i¦lek·trə ¦stad·ik 'lenz }

uniprocessor |COMPUT SCI| A computer that has a single central processing unit and works sequentially on only one program at a time. { 'yü·nə,prä,ses·ər }

unit charge See statcoulomb. { 'yü·nət 'chärj }

unit delay |ELECTR| A network whose output is equal to the input delayed by one unit of time. { 'yü·nət di'lā }

uniterm |COMPUT SCI| A word, symbol, or number used as a description for retrieval of information from a collection; especially, such a description used in a coordinate indexing system. { 'yü·nə ,tərm }

unit length |COMMUN| Basic element of time used in determining signaling speeds in message transmission. { 'yü·nət 'leŋkth }

unitor |COMPUT SCI| In computers, a device or circuit which performs a function corresponding to the Boolean operation of union. { 'yü·nə·tər }

unit record |COMPUT SCI| Any of a collection of records, all of which have the same form and the same data elements. { 'yü·nət 'rek·ərd }

unit string |COMPUT SCI| A string that has only one element. { 'yü·nət 'striŋ }

unit test |COMPUT SCI| The testing of a module within a computer system. { 'yü·nət ,test }

unity gain bandwidth |ELECTR| Measure of the gain-frequency product of an amplifier; unity gain bandwidth is the frequency at which the open-loop gain becomes unity, based on 6 decibels per octave crossing. { 'yü·nəd·ē ¦gān 'band ,width }

unity power factor |ELEC| Power factor of 1.0, obtained when current and voltage are in phase, as in a circuit containing only resistance or in a reactive circuit at resonance. { 'yü·nəd·ē 'paủ·ər ,fak·tər }

universal asynchronous receiver transmitter |COMPUT SCI| An electronic circuit that converts bytes of data between the parallel format in which bits are stored side by side within a device and the serial format whereby bits are transmitted sequentially over a communications line. Abbreviated UART. { ¦yü·nə¦vər·səl ā'siŋ·krə·nəs ri'sē·vər tranz'mid·ər }

universal conductance fluctuations |ELECTR| Fluctuations in the conductance of a quantum wire, as a function of applied voltage, whose root-mean-square value depends only on the geometry of the device, and which are reproducible in any one sample but vary in detail from one sample to another. { ¦yü·nə¦vər·səl kən'dək·təns ,flək·chə,wā·shənz }

universal language |COMPUT SCI| A programming language that is widely employed to write programs that can be run on a wide variety of computers. { ¦yü·nə¦vər·səl 'laŋ·gwij }

universal motor |ELEC| A motor that may be operated at approximately the same speed and output on either direct current or single-phase

alternating current. Also known as ac/dc motor. { ¦yü·nə¦vər·səl 'mōd·ər }

universal output transformer |ENG ACOUS| An output transformer having a number of taps on its winding, to permit its use between the audio-frequency output stage and the loudspeaker of practically any radio receiver by proper choice of connections. { ¦yü·nə¦vər·səl 'aủt,pủt tranz ,fòr·mər }

universal product code |COMPUT SCI| **1.** A 10-digit bar code on the outside of a package for electronic scanning at supermarket checkout counters; each digit is represented by the ratio of the widths of adjacent stripes and white areas. **2.** The corresponding combinations of binary digits into which the scanned bars are converted for computer processing that provides continuously updated inventory data and printout of the register tape at the checkout counter. { ¦yü·nə ¦vər·səl 'präd·əkt ,kōd }

universal receiver See ac/dc receiver. { ¦yü·nə ¦vər·səl ri'sē·vər }

universal resonance curve |ELEC| A plot of Y/Y₀ against $Q_0\delta$ for a series-resonant circuit, or of Z/Z_0 against $Q_0\delta$ for a parallel-resonant circuit, where Y and Z are the admittance and impedance of a circuit, Y_0 and Z_0 are the values of these quantities at resonance, Q_0 is the Q value of the circuit at resonance, and δ is the deviation of the frequency from resonance divided by the resonant frequency; it can be applied to all resonant circuits. { ¦yü·nə¦vər·səl 'rez·ən·əns ,kərv }

universal robot |CONT SYS| A robot whose end effector would be flexible enough to perform any desired task. { ¦yü·nə¦vər·səl 'rō,bät }

universal serial bus |COMPUT SCI| A serial interface that can transfer data at up to 480 million bits per second and connect up to 127 daisy-chained peripheral devices. Abbreviated USB. { ,yü·nə ,vər·səl ,sir·ē·əl 'bəs }

universal shunt See Ayrton shunt. { ¦yü·nə¦vər·səl 'shənt }

universal Turing machine |COMPUT SCI| A Turing machine that can simulate any Turing machine. { ¦yü·nə¦vər·səl 'tủr·iŋ mə,shēn }

univibrator See monostable multivibrator. { ¦yü· nə'vī,brād·ər }

Unix |COMPUT SCI| An operating system that was designed for use with microprocessors and with the C programming language, and that has been adopted for use with several 16-bit-microprocessor microcomputers. { 'yü·niks }

unload |COMPUT SCI| To remove or copy data from a computer system. { ən'lōd }

unloaded Q |ELECTR| The Q of a system when there is no external coupling to it. { ¦ən'lōd·əd 'kyü }

unloading amplifier |ELECTR| Amplifier that is capable of reproducing or amplifying a given voltage signal while drawing negligible current from the voltage source. { ¦ən'lōd·iŋ 'am·plə,fī·ər }

unloading circuit |COMPUT SCI| In an analog computer, a computing element or combination of computing elements capable of reproducing or amplifying a given voltage signal while drawing negligible current from the voltage source. { |ən'lōd·iŋ ,sər·kət }

unloading device |COMPUT SCI| Equipment that holds programs and other data that have been copied or removed from a computer system. { ən'lōd·iŋ di,vīs }

unmodified instruction *See* basic instruction. { |ən'mäd·ə,fīd in'strək·shən }

unpack |COMPUT SCI| **1.** To recover the individual data items contained in packed data. **2.** More specifically, to convert a packed decimal number into individual digits (and sometimes a sign). { |ən'pak }

unprotect |COMPUT SCI| To remove restrictions on access to a file so that any computer program can read and alter the data contained in it. { |ən·prə'tekt }

unsaturated standard cell |ELEC| One of two types of Weston standard cells (batteries); used for voltage calibration work not requiring an accuracy greater than 0.01%. { |ən'sach·ə,rād·əd 'stan·dərd 'sel }

unsolicited message |COMPUT SCI| A warning or error message that is automatically issued by a computer program when it detects a problem, and that does not depend on the operator making a query. { ,ən·sə'lis·əd·əd 'mes·ij }

untuned |ELEC| Not resonant at any of the frequencies being handled. { |ən'tünd }

unused time *See* unattended time. { |ən'yüzd 'tīm }

unwind |COMPUT SCI| In computers, to rearrange and code a sequence of instructions to eliminate red-tape operations. { |ən'wīnd }

up-converter |ELECTR| Type of parametric amplifier which is characterized by the frequency of the output signal being greater than the frequency of the input signal. { 'əp kən,vərd·ər }

update |COMPUT SCI| **1.** In computers, to modify an instruction so that the address numbers it contains are increased by a stated amount each time the instruction is performed. **2.** To change a record by entering current information; for example, to enter a new address or account number in the record pertaining to an employee or customer. Also known as posting. { |əp|dāt }

update service |COMPUT SCI| A service that guarantees installation of updates to products within a certain period of time after they become available. { 'əp,dāt ,sər·vəs }

uplink |COMMUN| The radio transmission path upward from the earth to a communications satellite, or from the earth to aircraft. { 'əp,liŋk }

upload |COMPUT SCI| To transfer or copy data from a smaller computer, such as a microcomputer, to a larger computer. { 'əp,lōd }

upper half-power frequency |ELECTR| The frequency on an amplifier response curve which is greater than the frequency for peak response and at which the output voltage is $1/\sqrt{2}$ (that is, 0.707) of its midband or other reference value. { 'əp·ər |haf |pau̇·ər 'frē·kwən·sē }

upper sideband |COMMUN| The higher of two frequencies or groups of frequencies produced by a modulation process. { 'əp·ər 'sīd,band }

UPS *See* uninterruptible power system.

up time |COMPUT SCI| The time during which equipment is either producing work or is available for productive work. Also known as available time. { 'əp ,tīm }

upward compatibility |COMPUT SCI| The ability of a newer or larger computer to accept programs from an older or smaller one. { 'əp·wərd kəm,pad·ə'bil·əd·ē }

URL *See* uniform resource locator.

usable |COMPUT SCI| Pertaining to a computer system that is easy for all users to work with. { 'yüz·ə·bəl }

USAT *See* ultra-small-aperture terminal. { 'yü ,sat or |yü|es|ā'tē }

USB *See* universal serial bus.

UseNet |COMPUT SCI| A global network of newsgroups that is linked by the Internet and other wide-area networks. { 'yüz,net }

user |COMMUN| An individual, installation, or activity having access to a switching center through a local private branch exchange, or by dialing an access code. |COMPUT SCI| Anyone who requires the use of services of a computing system or its products. { 'yü·zər }

user datagram protocol |COMMUN| A communications protocol providing a direct way to send and receive datagrams over an IP network but with few error recovery resources, used mainly for broadcasting over a network, for example, with streaming media. Abbreviated UDP. { ,yüz·ər 'dad·ə,gram ,prōd·ə,kȯl }

user-defined function |COMPUT SCI| A subroutine written by the programmer to calculate and return the value of a mathematical function. { |yü·zər di'fīnd 'fəŋk·shən }

user-defined type |COMPUT SCI| A data type that is not provided by a strongly typed language but is instead created by the programmer for a particular computer program. { |yü·zər di'fīnd 'tīp }

user exit |COMPUT SCI| A point in a computer program at which a user can cause control to be transferred outside the program. { 'yü·zər 'eg·zət }

user friendly |COMPUT SCI| Property of a system that is easy for an untrained person to use and sets up an easily understood dialog between the user and the computer. { 'yü·zər 'frend·lē }

user group |COMPUT SCI| An organization of users of the computers of a particular vendor, which shares information and ideas, and may develop system software and influence vendors to change their products. { 'yü·zər ,grüp }

userID |COMPUT SCI| The name used to log in to a network, remote server, and so on. { ,yüz·ər |ī'dē }

user interface |COMPUT SCI| **1.** The point at which a user or a user department or organization interacts with a computer system. **2.** The part of an interactive computer program that sends messages to and receives instructions from a terminal user. { 'yü·zər 'in·tər,fās }

user mode |COMPUT SCI| The mode of operation exercised by the user programs of a computer system in which there is a class of privileged instructions that is not permitted, since these can be executed only by the operating system or executive system. Also known as slave mode. { 'yü·zər ‚mōd }

username *See* mailbox name. { 'yüz·ər‚nām }

user program |COMPUT SCI| A computer program written by the person who uses it or by personnel of the organization that will use it. Also known as roll your own; user-written code. { 'yü·zər 'prō ‚gram }

user-programmable memory |COMPUT SCI| That portion of the internal storage of a microcomputer that is available for programs entered or loaded in by the user. { 'yü·zər prō'gram·ə·bəl 'mem·rē }

user-to-user service |COMMUN| Method of switching that enables direct user-to-user connection which does not include message store-and-forward service. { ¦yü·zər tə ¦yü·zər 'sər·vəs }

user-written code *See* user program. { 'yü·zər ¦writ·ən 'kōd }

utility routine |COMPUT SCI| A program or routine of general usefulness, usually not very complicated, and applicable to many jobs or purposes. { yü'til·əd·ē rü‚tēn }

utilization factor |ELEC| In electric power distribution, the maximum demand of a system or part of a system divided by its rated capacity. { ‚yüd·əl·ə'zā·shən ‚fak·tər }

utilization ratio |COMPUT SCI| The ratio of the effective time on a computer to the total up time. { ‚yüd·əl·ə'zā·shən ‚rā·shō }

uuencode |COMPUT SCI| A protocol for sending binary files in ASCII text format over the Internet, particularly e-mail attachments. { ‚yü'yü·in‚kōd }

UV EPROM *See* ultraviolet-erasable programmable read-only memory. { ‚yü‚vē 'ē‚präm }

UVPROM *See* ultraviolet-erasable programmable read-only memory. { ‚yü‚vē'präm }

V

V *See* electric potential; volt.

VA *See* volt-ampere.

vacuum capacitor |ELEC| A capacitor with separated metal plates or cylinders mounted in an evacuated glass envelope to obtain a high breakdown voltage rating. { 'vak·yəm kə'pas·əd·ər }

vacuum circuit breaker |ELEC| A circuit breaker in which a pair of contacts is hermetically sealed in a vacuum envelope; the contacts are separated by using a bellows to move one of them; an arc is produced by metallic vapor boiled from the electrodes, and is extinguished when the vapor particles condense on solid surfaces. { 'vak·yəm 'sər·kət ,brā·kər }

vacuum diffusion |ELECTR| Diffusion of impurities into a semiconductor material in a continuously pumped hard vacuum. { 'vak·yəm di'fyü·zhən }

vacuum fluorescent lamp |ELECTR| An evacuated display tube in which the anodes are coated with a phosphor that glows when electrons from the cathode strike it, to create a display. { 'vak·yəm flü'res·ənt 'lamp }

vacuum microelectronics |ELECTR| The technology of the vacuum transistor and similar microminiature devices based on electron field emission into a vacuum. { ¦vak·yüm ,mī·krō·i ,lek'trän·iks }

vacuum phototube |ELECTR| A phototube that is evacuated to such a degree that its electrical characteristics are essentially unaffected by gaseous ionization; in a gas phototube, some gas is intentionally introduced. { 'vak·yəm 'fōd·ō ,tüb }

vacuum relay |ELEC| A sensitive relay having its contacts mounted in a highly evacuated glass housing, to permit handling radio-frequency voltages as high as 20,000 volts without flashover between contacts even though contact spacing is but a few hundredths of an inch when open. { 'vak·yəm 'rē,lā }

vacuum switch |ELEC| A switch having its contacts in an evacuated envelope to minimize sparking. { 'vak·yəm ,swich }

vacuum transistor |ELECTR| A microminiature electronic device based on the control of electron emission from a field-emitter array into a vacuum. { 'vak·yəm tran'zis·tər }

vacuum tube |ELECTR| An electron tube evacuated to such a degree that its electrical characteristics are essentially unaffected by the presence of residual gas or vapor. { 'vak·yəm ,tüb }

vacuum-tube amplifier |ELECTR| An amplifier employing one or more vacuum tubes to control the power obtained from a local source. { 'vak·yəm ¦tüb 'am·plə¦fī·ər }

vacuum-tube circuit |ELECTR| An electric circuit in which a vacuum tube is connected. { 'vak·yəm ¦tüb 'sər·kət }

vacuum-tube clipping circuit |ELECTR| A circuit in which a vacuum tube is used to achieve clipping action; the bias at the input is set at such a level that output current cannot flow during a portion of the amplitude excursion of the input voltage or current waveform. { 'vak·yəm ¦tüb 'klip·iŋ 'sər·kət }

vacuum-tube electrometer |ELECTR| An electrometer in which the ionization current in an ionization chamber is amplified by a special vacuum triode having an input resistance above 10,000 megohms. { 'vak·yəm ¦tüb ,i,lek 'träm·əd·ər }

vacuum-tube keying |ELECTR| Code-transmitter keying system in which a vacuum tube is connected in series with the plate supply lead of a frequency-controlling stage of the transmitter; when the key is open, the tube blocks, interrupting the plate supply to the output stage; closing the key allows the plate current to flow through the keying tube and the output tubes. { 'vak·yəm ¦tüb 'kē·iŋ }

vacuum-tube modulator |ELECTR| A modulator employing a vacuum tube as a modulating element for impressing an intelligence signal on a carrier. { 'vak·yəm ¦tüb 'mäj·ə,lād·ər }

vacuum-tube oscillator |ELECTR| A circuit utilizing a vacuum tube to convert direct-current power into alternating-current power at a desired frequency. { 'vak·yəm ¦tüb 'äs·ə,lād·ər }

vacuum-tube rectifier |ELECTR| A rectifier in which rectification is accomplished by the unidirectional passage of electrons from a heated electrode to one or more other electrodes within an evacuated space. { 'vak·yəm ¦tüb 'rek·tə ,fī·ər }

vacuum-tube voltmeter |ENG| Any of several types of instrument in which vacuum tubes, acting as amplifiers or rectifiers, are used in circuits for the measurement of alternating-current or direct-current voltage. Abbreviated VTVM. Also known as tube voltmeter. { 'vak·yəm ˈtüb 'vōlt ˌmēd·ər }

validation [COMPUT SCI] The act of testing for compliance with a standard. { ˌval·ə'dā·shən }

validity check [COMPUT SCI] Computer check of input data, based on known limits for variables in given fields. { və'lid·əd·ē ˌchek }

valid program [COMPUT SCI] A computer program whose statements, individually and together, follow the syntactical rules of the programming language in which it is written, so that they are capable of being translated into a machine language program. { 'val·əd 'prō,gram }

valley attenuation [ELECTR] For an electric filter with an equal ripple characteristic, the maximum attenuation occurring at a frequency between two frequencies where the attenuation reaches a minimum value. { 'val·ē ə,ten·yə,wā·shən }

valley filling [ELEC] The addition of loads to an electric power system in off-peak periods. { 'val·ē ,fil·iŋ }

value-added network [COMMUN] A communications network that provides not only communications channels but also other services such as automatic error detection and correction, protocol conversions, and store-and-forward message services. { 'val·yü ˈad·əd 'net,wərk }

value parameter [COMPUT SCI] A parameter whose value is copied by a subprogram which can then alter its copy without affecting the original. { 'val·yü pə'ram·əd·ər }

valve See electron tube. { valv }

valve arrester [ELEC] A type of lightning arrester which consists of a single gap or multiple gaps in series with current-limiting elements; gaps between spaced electrodes prevent flow of current through the arrester except when the voltage across them exceeds the critical gap flashover. { valv ə,res·tər }

Van Atta array [ELECTROMAG] Antenna array in which pairs of corner reflectors or other elements equidistant from the center of the array are connected together by a low-loss transmission line in such a way that the received signal is reflected back to its source in a narrow beam to give signal enhancement without amplification. { va'nad·ə ə,rā }

Van de Graaff accelerator [ELECTR] A Van de Graaff generator equipped with an evacuated tube through which charged particles may be accelerated. { 'van də ,graf ak,sel·ə,rād·ər }

Van de Graaff generator [ELECTR] A high-voltage electrostatic generator in which electrical charge is carried from ground to a high-voltage terminal by means of an insulating belt and is discharged onto a large, hollow metal electrode. { 'van də ,graf ,jen·ə,rād·ər }

Van der Pol oscillator [ELECTR] A type of relaxation oscillator which has a single pentode tube and an external circuit with a capacitance that causes the device to switch between two values of the screen voltage. { 'van dər ,pōl ,äs·ə,lād·ər }

vane-anode magnetron [ELECTR] Cavity magnetron in which the walls between adjacent cavities have parallel plane surfaces. { 'vān 'an ,ōd 'mag·nə,trän }

vanilla [COMPUT SCI] Referring to a generalized system, usually software, that has not been subjected to special modifications, enhancements, or customization. Also known as plain vanilla; pure vanilla. { və'nil·ə }

V antenna [ELECTROMAG] An antenna having a V-shaped arrangement of conductors fed by a balanced line at the apex; the included angle, length, and elevation of the conductors are proportioned to give the desired directivity. Also spelled vee antenna. { 'vē an,ten·ə }

vapor lamp See discharge lamp. { 'vā·pər ,lamp }

var See volt-ampere reactive.

varactor [ELECTR] A semiconductor device characterized by a voltage-sensitive capacitance that resides in the space-charge region at the surface of a semiconductor bounded by an insulating layer. Also known as varactor diode; variable-capacitance diode; varicap; voltage-variable capacitor. { va'rak·tər }

varactor diode See varactor. { va'rak·tər 'dī,ōd }

varactor tuning [ELECTR] A method of tuning in which varactor diodes are used to vary the capacitance of a tuned circuit. { va'rak·tər 'tün·iŋ }

var hour [ELEC] A unit of the integral of reactive power over time, equal to a reactive power of 1 var integrated over 1 hour; equal in magnitude to 3600 joules. Also known as reactive volt-ampere hour; volt-ampere-hour reactive. { 'vär ,aúr }

var hour meter [ENG] An instrument that measures and registers the integral of reactive power over time in the circuit to which it is connected. { 'var ˈaúr ,mēd·ər }

variable [COMPUT SCI] A data item, or specific area in main memory, that can assume any of a set of values. { 'ver·ē·ə·bəl }

variable attenuator [ELECTR] An attenuator for reducing the strength of an alternating-current signal either continuously or in steps, without causing appreciable signal distortion, by maintaining a substantially constant impedance match. { 'ver·ē·ə·bəl ə'ten·yə,wād·ər }

variable-bandwidth filter [ELECTR] An electric filter whose upper and lower cutoff frequencies may be independently selected, so that almost any bandwidth may be obtained; it usually consists of several stages of RC filters, each separated by buffer amplifiers; tuning is accomplished by varying the resistance and capacitance values. { 'ver·ē·ə·bəl ˈband,width ,fil·tər }

variable bit rate [COMMUN] Operation in a digital system where the bit rate varies with time during the decoding of a compressed bit stream. { 'vər·ē·ə·bəl 'bit ,rāt }

variable-block [COMPUT SCI] Pertaining to an arrangement of data in which the number of words or characters in a block can vary, as determined by the programmer. { 'ver·ē·ə·bəl 'bläk }

variable-capacitance diode See varactor. { 'ver·ē·ə·bəl kəˈpas·əd·əns 'dī,ōd }

variable capacitor [ELEC] A capacitor whose capacitance can be varied continuously by moving one set of metal plates with respect to another. { 'ver·ē·ə·bəl kə'pas·əd·ər }

variable carrier modulation *See* controlled carrier modulation. { 'ver·ē·ə·bəl ¦kar·ē·ər ˌmäj·ə'lā·shən }

variable connector [COMPUT SCI] A flow chart symbol representing a sequence connection which is not fixed, but which can be varied by the flow-charted procedure itself; it corresponds to an assigned GO TO in a programming language such as FORTRAN. { 'ver·ē·ə·bəl kə,nek·tər }

variable coupling [ELEC] Inductive coupling that can be varied by moving one coil with respect to another. { 'ver·ē·ə·bəl 'kəp·liŋ }

variable-cycle operation [COMPUT SCI] An operation that requires a variable number of regularly timed execution cycles for its completion. { 'ver·ē·ə·bəl ¦sī·kəl ˌäp·ə'rā·shən }

variable diode function generator [ELECTR] An improvement of a diode function generator in which fully adjustable potentiometers are used for breakpoint and slope resistances, permitting the programming of analytic, arbitrary, and empirical functions, including inflections. Abbreviated VDFG. { 'ver·ē·ə·bəl ¦dī,ōd 'fəŋk·shən ,jen·ə,rād·ər }

variable field [COMPUT SCI] A field of data whose length is allowed to vary within certain specified limits. { 'ver·ē·ə·bəl 'fēld }

variable-length field [COMPUT SCI] A data field in which the number of characters varies, the length of the field being stored within the field itself. { 'ver·ē·ə·bəl ¦leŋkth 'fēld }

variable-length operation [COMPUT SCI] A computer operation whose operands are allowed to have a variable number of bits or characters. { 'ver·ē·ə·bəl ¦leŋkth ,äp·ə'rā·shən }

variable-length record [COMPUT SCI] A data or file format that allows each record to be exactly as long as needed. { 'ver·ē·ə·bəl ¦leŋkth 'rek·ərd }

variable-length word [COMPUT SCI] A computer word whose length is determined by the programmer. { 'ver·ē·ə·bəl ¦leŋkth 'wərd }

variable-mu tube [ELECTR] An electron tube in which the amplification factor varies in a predetermined manner with control-grid voltage; this characteristic is achieved by making the spacing of the grid wires vary regularly along the length of the grid, so that a very large negative grid bias is required to block anode current completely. Also known as remote-cutoff tube. { 'ver·ē·ə·bəl ¦myü 'tüb }

variable parameter [COMPUT SCI] A parameter whose storage address is passed to a subprogram so that the subprogram can alter its value. { 'ver·ē·ə·bəl pə'ram·əd·ər }

variable point [COMPUT SCI] A system of numeration in which the location of the decimal point is indicated by a special character at that position. { 'ver·ē·ə·bəl 'point }

variable-reluctance microphone *See* magnetic microphone. { 'ver·ē·ə·bəl ri¦lək·təns 'mī·krə ,fōn }

variable-reluctance stepper motor [ELEC] A stepper motor having a soft iron rotor with

teeth or poles so positioned that they cannot simultaneously align with all the stator poles. { 'ver·ē·ə·bəl ri¦lək·təns 'step·ər ,mōd·ər }

variable resistor *See* rheostat. { 'ver·ē·ə·bəl ri'zis·tər }

variable-sequence robot [CONT SYS] A robot controlled by instructions that can be modified. { 'ver·ē·ə·bəl ¦sē·kwəns 'rō,bät }

variable speech control [ELECTR] A method of removing small portions of speech from a tape recording at regular intervals and stretching the remaining sounds to fill the gaps, so that recorded speech can be played back at twice or even 2½ times the original speed without changing pitch and without significant loss of intelligibility. Abbreviated VSC. { 'ver·ē·ə·bəl 'spēch kən,trōl }

variable-speed generator [ELEC] A generator whose speed can be adjusted within certain limits, with a method of regulation that causes it to deliver a constant voltage. { 'ver·ē·ə·bəl ¦spēd 'jen·ə,rād·ər }

variable-speed motor [ELEC] A motor whose speed depends upon the load that it carries. { 'ver·ē·ə·bəl ¦spēd 'mōd·ər }

variable-speed scanning [ELECTR] Scanning method whereby the speed of deflection of the scanning beam in the cathode-ray tube of a television camera is governed by the optical density of the film being scanned. { 'ver·ē·ə·bəl ¦spēd 'skan·iŋ }

variable-transconductance circuit [ELECTR] A circuit used in four-quadrant multipliers that employs a simple differential transistor pair in which one variable input to the base of one transistor controls the device's gain; and one transistor amplifies the other's variable input, applied to the common emitter point, in proportion to the control input. { 'ver·ē·ə·bəl ¦tranz·kən¦dək·təns 'sər·kət }

variable transformer [ELEC] An iron-core transformer having provisions for varying its output voltage over a limited range or continuously from zero to maximum output voltage, generally by means of a contact arm moving along exposed turns of the secondary winding. Also known as adjustable transformer; continuously adjustable transformer. { 'ver·ē·ə·bəl tranz'fór·mər }

variable waveguide attenuator [ELECTROMAG] Device designed to introduce a variable amount of attenuation into a waveguide circuit by moving a lossy vane (a component that absorbs electromagnetic energy) either sideways across the waveguide or into the waveguide through a longitudinal slot. { 'ver·ē·ə·bəl 'wāv,gīd ə,ten·yə,wād·ər }

variable-word-length [COMPUT SCI] A phrase referring to a computer in which the number of characters addressed is not a fixed number but is varied by the data or instruction. { 'ver·ē·ə·bəl 'wərd ,leŋkth }

variant [COMMUN] **1.** One of two or more cipher or code symbols which have the same plain text equivalent. **2.** One of several plain text meanings that are represented by a single code group. { 'ver·ē·ənt }

variant record [COMPUT SCI] A record variable whose format is made to depend on some circumstance; for example, a record dealing with rates of pay might contain information on hourly rates for some employees and weekly or monthly salaries for others. { 'ver·ē·ənt 'rek·ərd }

varicap *See* varactor. { 'var·ə,kap }

variety [SYS ENG] The logarithm (usually to base 2) of the number of discriminations that an observer or a sensing system can make relative to a system. { və'rī·əd·ē }

variocoupler [ELECTROMAG] In radio practice, a transformer in which the self-impedance of windings remains essentially constant while the mutual impedance between the windings is adjustable. { ¦ver·ē·ō 'kəp·lər }

variolosser [ELEC] Device in which loss can be controlled by a voltage or current. { 'ver·ē·ə ,lòs·ər }

varioplex [ELEC] Telegraph switching system that establishes connections on a circuit-sharing basis between a multiplicity of telegraph transmitters in one locality and respective corresponding telegraph receivers in another locality over one or more intervening telegraph channels; maximum use of channel capacity is secured by momentarily storing the signals and allocating circuit time in rotation among the transmitters having information in storage. { 'ver·ē·ə,pleks }

varistor [ELECTR] A two-electrode semiconductor device having a voltage-dependent nonlinear resistance; its resistance drops as the applied voltage is increased. Also known as voltage-dependent resistor. { və'ris·tər }

Varley loop test [ELEC] A method of using a Wheatstone bridge to determine the distance from the test point to a fault in a telephone or telegraph line or cable. { 'vär·lē 'lüp ,test }

var measurement [ELEC] The measurement of reactive power in a circuit. { 'vär ,mezh·ər·mənt }

varmeter [ENG] An instrument for measuring reactive power in vars. Also known as reactive voltampere meter. { 'vär,mēd·ər }

V-beam radar [ELECTROMAG] A volumetric radar system that uses two fan beams to determine the distance, bearing, and height of a target: one beam is vertical and the other inclined; the beams intersect at ground level and rotate continuously about a vertical axis; the time difference between the arrivals of the echoes of the two beams is a measure of target elevation. { 'vē ¦bēm 'rā,där }

VBV *See* video buffering verifier.

V-chip [COMMUN] An electronic device and protocol intended to exert control over program reception based on program content rating. { 'vē ,chip }

VCO *See* voltage-controlled oscillator.

V connection *See* open-delta connection. { 'vē kə ,nek·shən }

VCR *See* videocassette recorder.

V cut [ENG] In mining and tunneling, a cut where the material blasted out in plan is like the letter V;

usually consists of six or eight holes drilled into the face, half of which form an acute angle with the other half. { 'vē ,kət }

VDFG *See* variable diode function generator.

VDT *See* display terminal.

vector *See* jump vector. { 'vek·tər }

vector canceler [ELECTR] A canceler used in moving-target-indication radar, often using digital methods, operating on both the in-phase and quadrature video signals produced by the phase detectors. { 'vek·tər 'kans·lər }

vectored interrupt [COMPUT SCI] A signal that instructs a computer program to temporarily halt the processing it is doing and transfer control to a routine whose address is given by an entry in a jump vector specified by a value included in the signal. { 'vek·tərd 'in·tə·rəpt }

vector effect *See* selective photoelectric effect. { 'vek·tər i,fekt }

vector graphics [COMPUT SCI] A computer graphics image-coding technique which codes only the image itself as a series of lines, according to the cartesian coordinates of the lines' origins and terminations. Also known as object-oriented graphics. { 'vek·tər ¦graf·iks }

vector impedance meter [ENG] An instrument that not only determines the ratio between voltage and current, to give the magnitude of impedance, but also determines the phase difference between these quantities, to give the phase angle of impedance. { 'vek·tər im'pēd·əns ,mēd·ər }

vector power [ELEC] Vector quantity equal in magnitude to the square root of the sum of the squares of the active power and the reactive power. { 'vek·tər ,paú·ər }

vector-power factor [ELEC] Ratio of the active power to the vector power; it is the same as power factor in the case of simple sinusoidal quantities. { 'vek·tər ¦paú·ər ,fak·tər }

vector processing [COMPUT SCI] A procedure for speeding the processing of information by a computer, in which pipelined units perform arithmetic operations on uniform, linear arrays of data values, and a single instruction involves the execution of the same operation on every element of the array. { 'vek·tər 'prä,ses·iŋ }

vector quantization [COMPUT SCI] A data compression technique in which a finite sequence of values is presented as resembling the template (from among the choices available to a given code book) that minimizes a distortion measure. { ¦vek·tər ,kwän·tə'zā·shən }

vector resolver *See* resolver. { 'vek·tər ri,zäl·vər }

vector table *See* jump vector. { 'vek·tər ,tā·bəl }

vector voltmeter [ENG] A two-channel high-frequency sampling voltmeter that measures phase as well as voltage of two input signals of the same frequency. { 'vek·tər 'vōlt,mēd·ər }

vee antenna *See* V antenna. { 'vē an,ten·ə }

velocity constant [CONT SYS] The ratio of the rate of change of the input command signal to the steady-state error, in a control system where these two quantities are proportional. { və'läs·əd·ē ,kän·stənt }

velocity control *See* rate control. { və'läs·əd·ē kən,trōl }

velocity error |CONT SYS| The difference between the rate of change of the actual position of a control system component and the rate of change of the desired position. { və'läs·əd·ē ,er·ər }

velocity filter |ELECTR| Storage tube device which blanks all targets that do not move more than one resolution cell in less than a predetermined number of antenna scans. { və'läs·əd·ē ,fil·tər }

velocity microphone |ENG ACOUS| A microphone whose electric output depends on the velocity of the air particles that form a sound wave; examples are a hot-wire microphone and a ribbon microphone. { və'läs·əd·ē 'mī·krə,fōn }

velocity-modulated oscillator |ELECTR| Oscillator which employs velocity modulation to produce radio-frequency power. Also known as klystron oscillator. { və'läs·əd·ē ,mäj·ə¦läd·əd 'äs·ə,lād·ər }

velocity modulation |ELECTR| **1.** Modulation in which a time variation in velocity is impressed on the electrons of a stream. **2.** A television system in which the intensity of the electron beam remains constant throughout a scan, and the velocity of the spot at the screen is varied to produce changes in picture brightness (not in general use). { və'läs·əd·ē ,mäj·ə'lā·shən }

velocity of light *See* speed of light. { və'läs·əd·ē əv 'līt }

velocity pickup |ELEC| A device that generates a voltage proportional to the relative velocity between two principal elements of the pickup, the two elements usually being a coil of wire and a source of magnetic field. { və'läs·əd·ē 'pik,əp }

velocity servomechanism |CONT SYS| A servomechanism in which the feedback-measuring device generates a signal representing a measured value of the velocity of the output shaft. Also known as rate servomechanism. { və'läs·əd·ē 'sər·vō,mek·ə,niz·əm }

velocity shaped canceler *See* cascaded feedback canceler. { və'läs·əd·ē ¦shāpt 'kan·slər }

vented baffle *See* reflex baffle. { 'ven·təd 'baf·əl }

vented battery |ELEC| A nickel-cadmium or other battery which lacks provisions for recombination of gases produced during normal operation, so that these gases must be vented to the atmosphere to avoid rupture of the cell case. { 'ven·təd 'bad·ə,rē }

vented-box system |ELECTR| A loudspeaker system in which the woofer is mounted in a box with a vent connecting the air inside the box to the outside. Also known as ported system. { ¦ven·təd 'bäks ,sis·təm }

ventilation |ENG| Provision for the movement, circulation, and quality control of air in an enclosed space. { ,vent·əl'ā·shən }

verb |COMPUT SCI| In COBOL, the action indicating part of an unconditional statement. { vərb }

verbal information verification |ENG ACOUS| A method of talker authentication that involves checking the content of a spoken password or pass-phrase, such as a personal identification number, a social security number, or a mother's

maiden name. Abbreviated VIV. { ,vər·bəl ,in·fər ¦mā·shən ,ver·i·fə'kā·shən }

verification |COMPUT SCI| The process of checking the results of one data transcription against the results of another data transcription; both transcriptions usually involve manual operations. { ,ver·ə·fə'kā·shən }

verify |COMMUN| To ensure that the meaning and phraseology of the transmitted message convey the exact intention of the originator. |COMPUT SCI| To determine whether an operation has been completed correctly. { 'ver·ə,fī }

vernier capacitor |ELEC| Variable capacitor placed in parallel with a larger tuning capacitor to provide a finer adjustment after the larger unit has been set approximately to the desired position. { 'vər·nē·ər kə'pas·əd·ər }

vernitel |ELECTR| Precision device which makes possible the transmission of data with high accuracy over standard frequency modulated-frequency modulated telemetering systems. { 'vər·nə,tel }

versatile automatic test equipment |ELECTR| Computer-controlled tester, for missile electronic systems, that trouble-shoots faults by deductive logic and isolates them to the plug-in module or component level. { 'vər·səd·əl ¦öd·ə ¦mad·ik 'test i,kwip·mənt }

vertical antenna |ELECTROMAG| A vertical metal tower, rod, or suspended wire used as an antenna. { 'vərd·ə·kəl an'ten·ə }

vertical ballistic transistor |ELECTR| A transistor in which, ideally, electrons traverse ballistically (that is without scattering) a very short region that separates a cathode from an anode, and whose effective cross section is modulated by the potential of metal gate contacts. { ¦vərd·i·kəl bə ,lis·tik tran'zis·tər }

vertical blanking |ELECTR| Blanking of a video display device during the vertical retrace. { 'vərd·ə·kəl 'blaŋk·iŋ }

vertical centering control |ELECTR| The centering control provided in a video display device to shift the position of the entire image vertically in either direction on the screen. { 'vərd·ə·kəl 'sen·tər·iŋ kən,trōl }

vertical component effect *See* antenna effect. { 'vərd·ə·kəl kəm'pō·nənt i,fekt }

vertical definition *See* vertical resolution. { 'vərd·ə·kəl ,def·ə'nish·ən }

vertical deflection oscillator |ELECTR| The oscillator that produces, under control of the vertical synchronizing signals, the sawtooth voltage waveform that is amplified to feed the vertical deflection coils on the picture tube of an analog television receiver. Also known as vertical oscillator. { 'vərd·ə·kəl di'flek·shən 'äs·ə,lād·ər }

vertical hold control |ELECTR| The hold control that changes the free-running period of the vertical deflection oscillator in an analog television receiver, so the picture remains steady in the vertical direction. { 'vərd·ə·kəl ¦hōld kən,trōl }

vertical instruction |COMPUT SCI| An instruction in machine language to carry out a single operation or a time-ordered series of a fixed number and type of operation on a single set of operands. { 'vərd·ə·kəl in'strək·shən }

vertical interval reference |ELECTR| A reference signal inserted into an analog television program signal every $\frac{1}{60}$ second, in line 19 of the vertical blanking period between NTSC television frames, to provide references for luminance amplitude, black-level amplitude, sync amplitude, chrominance amplitude, and color-burst amplitude and phase. Abbreviated VIR. { 'vərd·ə·kəl ¦in·tər·vəl 'ref·rəns }

vertical linearity control |ELECTR| A linearity control that permits narrowing or expanding the height of the image on the upper half of the screen of a video display, to give linearity in the vertical direction so that circular objects appear as true circles. { 'vərd·ə·kəl ˌlin·ē¦ar·əd·ē kən,trōl }

vertically stacked loops See stacked loops. { 'vərd·ə·klē ¦stakt 'lüps }

vertical metal oxide semiconductor technology |ELECTR| For semiconductor devices, a technology that involves essentially the formation of four diffused layers in silicon and etching of a V-shaped groove to a precisely controlled depth in the layers, followed by deposition of metal over silicon dioxide in the groove to form the gate electrode. Abbreviated VMOS technology. { 'vərd·ə·kəl ¦med,əl ¦äk,sīd ¦sem·i·kən,dək·tər tek'näl·ə·jē }

vertical oscillator See vertical deflection oscillator. { 'vərd·ə·kəl 'äs·ə,lād·ər }

vertical parity check See lateral parity check. { 'vərd·ə·kəl 'par·əd·ē ,chek }

vertical polarization |COMMUN| Transmission of linear polarized radio waves whose electric field vector is perpendicular to the earth's surface. { 'vərd·ə·kəl ˌpō·lə·rə'zā·shən }

vertical recording |ELECTR| Magnetic recording in which bits are magnetized in directions perpendicular to the surface of the recording medium, allowing the bits to be smaller. Also known as perpendicular recording. { 'vərd·ə·kəl ri'kórd·iŋ }

vertical redundancy check See lateral parity check. { 'vərd·ə·kəl ri'dən·dən·sē ,chek }

vertical resolution |ELECTR| The number of distinct horizontal lines, alternately black and white, that can be seen in the reproduced image of a video image test pattern; it is primarily fixed by the number of horizontal lines used in scanning. Also known as vertical definition. { 'vərd·ə·kəl ˌrez·ə'lü·shən }

vertical retrace |ELECTR| The return of the electron beam to the top of the screen at the end of each video field. { 'vərd·ə·kəl 'rē,trās }

vertical sweep |ELECTR| The downward movement of the scanning beam from top to bottom of the picture being televised. { 'vərd·ə·kəl 'swēp }

vertical synchronizing pulse |ELECTR| One of the six pulses that are transmitted at the end of each field in an analog television system to keep the receiver in field-by-field synchronism with the transmitter. Also known as picture synchronizing pulse. { 'vərd·ə·kəl 'siŋ·krə,nīz·iŋ ,pəls }

vertical tab |COMPUT SCI| A control character that causes a computer printer to jump from its current line to another preset line further down the page. { 'vərd·ə·kəl 'tab }

very high frequency |COMMUN| The band of frequencies from 30 to 300 megahertz in the radio spectrum, corresponding to wavelengths of 1 to 10 meters. Abbreviated VHF. { ¦ver·ē ¦hī 'frē·kwən·sē }

very high frequency oscillator |ELECTR| An oscillator whose frequency lies in the range from a few to several hundred megahertz; it uses distributed, rather than lumped, impedances, such as parallel wire transmission lines or coaxial cables. { ¦ver·ē ¦hī 'frē·kwən·sē 'äs·ə,lad·ər }

very high frequency tuner |ELECTR| A tuner in a television receiver for reception of stations transmitting in the very high frequency band; it generally has 12 discrete positions corresponding to channels 2--13. { ¦ver·ē ¦hī 'frē·kwən·sē 'tün·ər }

very large scale integrated circuit |ELECTR| A complex integrated circuit that contains between 20,000 and 1,000,000 transistors. Abbreviated VLSI circuit. { ¦ver·ē ¦lärj ¦skāl 'int·ə,grād·əd 'sər·kət }

very long baseline interferometry |ELECTR| A method of improving angular resolution in the observation of radio sources; these are simultaneously observed by two radio telescopes which are very far apart, and the signals are recorded on magnetic tapes which are combined electronically or on a computer. Abbreviated VLBI. { ¦ver·ē ¦loŋ ¦bās·līn ,in·tər·fə'räm·ə·trē }

very long range radar |ELECTR| Equipment whose maximum range on a reflecting target of 10.76 square feet (1 square meter) normal to the signal path exceeds 800 miles (1300 kilometers), provided line of sight exists between the target and the radar. { ¦ver·ē ¦lōŋ ¦rānj 'rä,där }

very low frequency |COMMUN| The band of frequencies from 3 to 30 kilohertz in the radio spectrum, corresponding to wavelengths of 10 to 100 kilometers. Abbreviated VLF. { ¦ver·ē ¦lō 'frē·kwən·sē }

very short range radar |ELECTR| Equipment whose range on a reflecting target of 10.76 square feet (1 square meter) normal to the signal path is less than 50 miles (80 kilometers), provided line of sight exists between the target and the radar. { ¦ver·ē ¦shórt ¦rānj 'rä,där }

very small aperture terminal |COMMUN| An antenna approximately 6 feet (1.8 meters) or less in diameter, that is used for both broadcast reception and interactive communications via geosynrchronous satellites. Abbreviated VSAT. { ¦ver·ē ¦smól ¦ap·ə·chər 'tərm·ə·nəl }

vestigial sideband [COMMUN] The transmitted portion of an amplitude-modulated sideband that has been largely suppressed by a filter having a gradual cutoff in the neighborhood of the carrier frequency; the other sideband is transmitted without much suppression. Abbreviated VSB. { vəˈstij·ē·əl ˈsīd‚band }

vestigial-sideband filter [ELECTR] A filter that is inserted between a transmitter and its antenna to suppress part of one of the sidebands. { vəˈstij·ē·əl ¦sīd‚band ‚fil·tər }

vestigial-sideband transmission [COMMUN] A type of radio signal transmission for amplitude modulation in which the normal complete sideband on one side of the carrier is transmitted, but only a part of the other sideband is transmitted. Also known as asymmetrical-sideband transmission. { vəˈstij·ē·əl ¦sīd‚band tranz‚mish·ən }

VF *See* voice frequency.

V format [COMPUT SCI] A data record format in which the logical records are of variable length and each record begins with a record length indication. { ¦vē ‚fȯr‚mat }

VHF *See* very high frequency.

via [ELECTR] A pathway that is etched to allow electrical contact between different layers of a semiconductor device. { ˈvē·ə *or* ˈvī·ə }

vibrating capacitor [ELEC] A capacitor whose capacitance is varied in a cyclic manner to produce an alternating voltage proportional to the charge on the capacitor; used in a vibrating-reed electrometer. { ˈvī‚brād·iŋ kəˈpas·əd·ər }

vibrating-reed electrometer [ENG] An instrument using a vibrating capacitor to measure a small charge, often in combination with an ionization chamber. { ˈvī‚brād·iŋ ¦rēd ‚i‚lekˈträm·əd·ər }

vibrating-reed frequency meter [ENG] A frequency meter consisting of steel reeds having different and known natural frequencies, all excited by an electromagnet carrying the alternating current whose frequency is to be measured. Also known as Frahm frequency meter; reed frequency meter; tuned-reed frequency meter. { ˈvī‚brād·iŋ ¦rēd ˈfrē·kwən‚sē ‚mēd·ər }

vibration galvanometer [ENG] An alternating-current galvanometer in which the natural oscillation frequency of the moving element is equal to the frequency of the current being measured. { vīˈbrā·shən ‚gal·vəˈnäm·əd·ər }

vibration pickup [ELEC] An electromechanical transducer capable of converting mechanical vibrations into electrical voltages. { vīˈbrā·shən ‚pik‚əp }

vibrator [ELEC] An electromechanical device used primarily to convert direct current to alternating current but also used as a synchronous rectifier; it contains a vibrating reed which has a set of contacts that alternately hit stationary contacts attached to the frame, reversing the direction of current flow; the reed is activated when a soft-iron slug at its tip is attracted to the pole piece of a driving coil. { ˈvī‚brād·ər }

vibrator power supply [ELEC] A power supply using a vibrator to produce the varying current necessary to actuate a transformer, the output of which is then rectified and filtered. { ˈvī‚brād·ər ˈpau̇·ər sə‚plī }

vibrator-type inverter [ELEC] A device that uses a vibrator and an associated transformer or other inductive device to change direct-current input power to alternating-current output power. { ˈvī‚brād·ər ¦tīp inˈvərd·ər }

vibrotron [ELECTR] A triode electron tube having an anode that can be moved or vibrated by an externally applied force. { ˈvī·brə‚trän }

video [ELECTR] **1.** Pertaining to picture signals or to the sections of a television system that carry these signals in either unmodulated or modulated form. **2.** Pertaining to the demodulated radar receiver output that is applied to a radar indicator or otherwise treated in the radar's own data processing. { ˈvid·ē·ō }

video adapter [COMPUT SCI] A printed circuit board that is plugged into a computer and generates the text and graphics images on a monitor. Also known as video board; video display board. { ˈvid·ē·ō ə‚dap·tər }

video amplifier [ELECTR] A low-pass amplifier having a band width on the order of 2–10 megahertz, used in television and radar transmission and reception; it is a modification of an RC-coupled amplifier, such that the high-frequency half-power limit is determined essentially by the load resistance, the internal transistor capacitances, and the shunt capacitance in the circuit. { ˈvid·ē·ō ˈam‚plə‚fī·ər }

video board *See* video adapter. { ˈvid·ē·ō ‚bȯrd }

video buffering verifier [COMMUN] A hypothetical decoder that is conceptually connected to the output of an encoder; provides a constraint on the variability of the data rate that an encoder can produce. Abbreviated VBV. { ˈvid·ē·ō ˈbəf·ə·riŋ ‚ver·ə‚fī·ər }

video canceler [ELECTR] A canceler used in moving-target-indication radar operating on the video signals produced by the phase detector of a coherent radar, or on the video signal in a simpler radar, to detect fluctuation of the video signal caused by a target moving over stationary clutter. { ˈvid·ē·ō ˈkans·lər }

videocassette [ELECTR] A compact plastic case containing a magnetic tape for video recording and playing. { ¦vid·ē·ō‚kə'set }

videocassette recorder [ELECTR] A device for video recording and playing of magnetic tapes that are contained in plastic cases. Abbreviated VCR. { ¦vid·ē·ō‚kə'set ri‚kȯrd·ər }

videoconference [COMMUN] A teleconference that employs some type of video camera or other audio or video equipment to convey pictures and sound from one location to another, either in a one-way or two-way fashion. { ‚vid·ē·ō'kän·frəns }

videoconferencing [COMPUT SCI] Two-way interactive, digital communication through video streaming on the Internet, or by communications satellite, video telephone, and so forth. { ‚vid·ē·ō'kän·frəns·iŋ }

video correlator |ELECTR| Radar circuit that enhances automatic target detection capability, provides data for digital target plotting, and gives improved immunity to noise, interference, and jamming. { 'vid·ē·ȯ 'kär·ə,lād·ər }

video dial-tone |COMMUN| A service that enables a subscriber to select a video information provider from among many such providers offering service in a neighborhood, and to access on demand, from this provider, a movie or multimedia content, or similar services such as games and home shopping. { ,vid·ē·ȯ 'dīl ,tōn }

video discrimination |ELECTR| Radar circuit used to reduce the frequency band of the video amplifier stage in which it is used. { 'vid·ē·ȯ di,skrim·ə'nā·shən }

video disk recorder |ELECTR| A video recorder that records television visual signals and sometimes aural signals on a magnetic, optical, or other type of disk which is usually about the size of a long-playing phonograph record. { 'vid·ē·ȯ ¦disk ri,kȯrd·ər }

video disk storage *See* optical disk storage. { 'vid·ē·ȯ ¦disk ,stȯr·ij }

video display board *See* video adapter. { ¦vid·ē·ō di'splā ,bȯrd }

video display terminal *See* display terminal. { 'vid·ē·ȯ di'splā ,tər·mən·əl }

video frequency |COMMUN| One of the frequencies output by a video camera when an image is scanned; for standard-definition systems, it may be any value from almost zero to well over 4 megahertz. { 'vid·ē·ȯ 'frē·kwən,sē }

video game |ELECTR| A form of interactive entertainment in which the player responds to electronically generated images that appear on a video display screen. { 'vid·ē·ō ,gām }

video integrator |ELECTR| **1.** Electric counter-countermeasures device that is used to reduce the response to nonsynchronous signals such as noise, and is useful against random pulse signals and noise. **2.** Device which uses the redundancy of repetitive signals to improve the output signal-to-noise ratio, by summing the successive video signals. { 'vid·ē·ȯ 'int·ə,grād·ər }

video masking |ELECTR| Method of removing chaff echoes and other extended clutter from radar displays. { 'vid·ē·ȯ 'mask·iŋ }

video monitor |COMPUT SCI| The cathode-ray-tube screen of a video terminal. Also known as display screen; monitor. { 'vid·ē·ō 'män·əd·ər }

videophone *See* video telephone. { 'vid·ē·ə,fōn }

video player |ELECTR| A playback device that converts a video disk, videotape, or other type of recorded video or audio program into signals suitable for driving a home display. { 'vid·ē·ō ,plā·ər }

video RAM |COMPUT SCI| Dynamic random-access memory optimized for use with video displays. { ,vid·ē·ō 'ram }

video recorder |ELECTR| A magnetic tape recorder capable of storing video and audio signals for playback at a later time. { 'vid·ē·ō ri ,kȯrd·ər }

video replay |ELECTR| **1.** Also known as video-tape replay. **2.** A procedure in which the audio and video signals of a television program are recorded on magnetic tape and then the tape is run through equipment later to rebroadcast the live scene. **3.** A similar procedure in which the scene is rebroadcast almost immediately after it occurs. Also known as instant replay. { 'vid·ē·ō 'rē,plā }

video sensing |COMPUT SCI| In optical character recognition, a scanning technique in which the document is flooded with light from an ordinary light source, and the image of the character is reflected onto the face of a cathode-ray tube, where it is scanned by an electron beam. { 'vid·ē·ō 'sens·iŋ }

video signal |COMMUN| In analog television, the signal containing all of the visual information together with blanking and synchronizing pulses. |ELECTROMAG| *See* target signal. { 'vid·ē·ō ,sig·nəl }

videotape |ELECTR| A magnetic tape designed primarily for recording of television programs. { 'vid·ē·ō ,tāp }

videotape recorder |ELECTR| A device for video recording and playing of a magnetic tape either in a video cassette or on an open reel. { ¦vid·ē·ō 'tāp ri,kȯrd·ər }

videotape recording |ELECTR| A method of recording video programs on magnetic tape for later rebroadcasting or replay; may also refer to the device that performs this function. Abbreviated VTR. { 'vid·ē·ō ¦tāp ri,kȯrd·iŋ }

videotape replay *See* video replay. { 'vid·ē·ō ¦tap 'rē,plā }

video telephone |COMMUN| A communication instrument which transmits visual images along with the attendant speech. Also known as video-phone. { 'vid·ē·ō 'tel·ə,fōn }

videotex |COMMUN| An electronic home information delivery system, either teletext or videotext. { 'vid·ē·ō,teks }

videotext |COMMUN| A computer communication service which uses information from a database, and which allows the user, equipped with a limited computer terminal, to interact with the service in selecting information to be displayed, so as to provide electronic mail, teleshopping, financial services, calculation services, and such. { ¦vid·ē·ō'tekst }

video transformer |ELECTR| A transformer designed to transfer, from one circuit to another, the signals containing picture information in a video system. { 'vid·ē·ō tranz'fȯr·mər }

video transmitter *See* visual transmitter. { 'vid·ē·ō tranz'mid·ər }

vidicon |ELECTR| A camera tube in which a charge-density pattern is formed by photoconduction and stored on a photoconductor surface that is scanned by an electron beam, usually of low-velocity electrons. { 'vid·ə,kän }

Vienna definition language |COMPUT SCI| A language for defining the syntax and semantics of programming languages; consists of a syntactic metalanguage for defining the syntax of programming and data structures, and a semantic metalanguage which specifies programming language semantics operationally in terms of the computations to which programs give rise during execution.   { vē'en·ə ,def·ə'nish·ən ,laŋ·gwij }

viewfinder |ELECTR| An auxiliary optical or electronic device attached to a video camera so the operator can see the scene as the camera sees it.   { 'vyü,fīn·dər }

viewing storage tube   *See* direct-view storage tube.   { 'vyü·iŋ 'stór·ij ,tüb }

viewing time |ELECTR| Time during which a storage tube is presenting a visible output corresponding to the stored information.   { 'vyü·iŋ ,tīm }

viewport   *See* window.   { 'vyü,pórt }

viologen display |ELECTR| An electrochromic display based on an electrolyte consisting of an aqueous solution of a dipositively charged organic salt, containing a colorless cation that undergoes a one-electron reduction process to produce a purple radical cation, upon application of a negative potential to the electrode.   { vē'äl·ə·jən di,splā }

VIR   *See* vertical interval reference.

virgin medium |COMPUT SCI| A material designed to have data recorded on it which is as yet completely lacking any information, such as a paper tape without any punched holes, not even feed holes; in contrast to an empty medium.   { 'vər·jən 'mēd·ē·əm }

virtual acoustics |ENG ACOUS| Digitally processing sounds so that they appear to come from particular locations in three-dimensional space, with the goal of simulating the complex acoustic field experienced by a listener within a natural environment. Also known as auralization; three-dimensional sound.   { ,vər·chə·wəl ə'küs· tiks }

virtual address |COMPUT SCI| A symbol that can be used as a valid address part but does not necessarily designate an actual location.   { 'vər·chə·wəl 'ad,res }

virtual cathode |ELECTR| The locus of a space-charge-potential minimum such that only some of the electrons approaching it are transmitted, the remainder being reflected back to the electron-emitting cathode.   { 'vər·chə·wəl 'kath ,ōd }

virtual decimal point   *See* assumed decimal point.   { 'vər·chə·wəl 'des·məl ,póint }

virtual direct-access storage |COMPUT SCI| A device used with mass-storage systems, whereby data are retrieved prior to usage by a batch-processing program and automatically transcribed onto disk storage.   { 'vər·chə·wəl də¦rekt ¦ak,ses 'stór·ij }

virtual environment   *See* virtual reality.   { ¦vər·chə·wəl in'vī·rən·mənt }

virtual machine |COMPUT SCI| A portion of a computer system or of a computer's time that is controlled by an operating system and functions as though it were a complete system, although in reality the computer is shared with other independent operating systems.   { 'vər·chə·wəl mə'shēn }

virtual memory |COMPUT SCI| A combination of primary and secondary memories that can be treated as a single memory by programmers because the computer itself translates a program or virtual address to the actual hardware address.   { 'vər·chə·wəl 'mem·rē }

virtual private network |COMMUN| A wide-area network whose links are provided by a common carrier although they appear to the users to behave like dedicated lines, and whose computers use a common cryptographic key to send messages from one computer in the network to another. Abbreviated VPN.   { ¦vər·chə·wəl ¦prī· vət 'net,wərk }

virtual reality |COMPUT SCI| A simulation of an environment that is experienced by a human operator provided with a combination of visual (computer-graphic), auditory, and tactile presentations generated by a computer program. Also known as artificial reality; immersive simulation; virtual environment; virtual world.   { ¦vər·chə·wəl rē'al·əd·ē }

Virtual Reality Modeling Language |COMPUT SCI| The markup specification for three-dimentional (virtual reality) objects and environments on the Web. Abbreviated VRML.   { ,vər·chə·wəl rē ,al·əd·ē 'mäd·əl·iŋ ,laŋ·gwij }

virtual world |COMPUT SCI| A navigable visual digital environment.   { ¦vər·chə·wəl 'wərld }

virus |COMPUT SCI| A computer program that replicates itself and transfers itself to another computing system.   { 'vī·rəs }

visibility factor |ELECTR| The ratio of the minimum signal input detectable by ideal instruments connected to the output of a receiver, to the minimum signal power detectable by a human operator through a display connected to the same receiver. Also known as display loss.   { ,viz·ə'bil·əd·ē ,fak·tər }

visual display unit   *See* display tube.   { 'vizh·ə·wəl di'splā ,yü·nət }

visualization |COMPUT SCI| The process of converting data into a geometric or graphic representation.   { ,vizh·ə·lə'zā·shən }

visually coupled display   *See* helmet-mounted display.   { 'vizh·ə·lē ¦kəp·əld di'splā }

visual radio range |NAV| Any radio range through which aircraft are flown by visual instrumentation not associated with aural reception.   { 'vizh·ə·wəl 'rād·ē·ō ,rānj }

visual scanner |ELECTR| Device that optically scans printed or written data and generates an analog or digital signal.   { 'vizh·ə·wəl 'skan·ər }

visual servoing |CONT SYS| The use of a solid-state camera on the end effector of a robot to provide feedback.   { 'vizh·ə·wəl 'sər·vō·iŋ }

visual storage tube |ELECTR| Any electrostatic storage tube that also provides a visual readout.   { 'vizh·ə·wəl 'stór·ij ,tüb }

visual transmitter |COMMUN| The system of devices used to transmit the visual portion of the television signal in an analog television service.   { 'vizh·ə·wəl tranz'mid·ər }

Viterbi algorithm |COMMUN| A decoding procedure for convolutional codes that uses the maximum-likelihood method. { vi¦ter·bē 'al·gə ₁rith·əm }

VIV See verbal information verification.

VLBI See very long baseline interferometry.

VLF See very low frequency.

VLSI circuit See very large scale integrated circuit. { ¦vē¦el¦es¦ī 'sər·kət }

VMOS technology See vertical metal oxide semiconductor technology. { 'vē₁mós tek₁näl·ə·jē }

vocoder |ELECTR| A system of electronic apparatus for synthesizing speech according to dynamic specifications derived from an analysis of that speech. { vō'kōd·ər }

vodas |ELECTR| A voice-operated switching device used in transoceanic radiotelephone circuits to suppress echoes and singing sounds automatically; it connects a subscriber's line automatically to the transmitting station as soon as he starts speaking and simultaneously disconnects it from the receiving station, thereby permitting the use of one radio channel for both transmitting and receiving without appreciable switching delay as the parties alternately talk. Derived from voice-operated device anti-singing. { 'vō₁das }

voder |ELECTR| An electronic system that uses electron tubes and filters, controlled through a keyboard, to produce voice sounds artificially. Derived from voice operation demonstrator. { 'vō·dər }

vogad |ELECTR| An automatic gain control circuit used to maintain a constant speech output level in long-distance radiotelephony. Derived from voice-operated gain-adjusted device. { 'vō ₁gad }

voice call sign |COMMUN| A call sign provided primarily for voice communications. { 'vóis 'kól ₁sīn }

voice channel |COMMUN| A communication channel having sufficient bandwidth to carry voice frequencies intelligibly; the minimum bandwidth for an analog voice channel is about 3000 hertz for good intelligibility. { 'vóis ₁chan·əl }

voice coder |ELECTR| Device that converts speech input into digital form prior to encipherment for secure transmission and converts the digital signals back to speech at the receiver. { 'vóis ₁kō·dər }

voice coil |ENG ACOUS| The coil that is attached to the diaphragm of a moving-coil loudspeaker and moves through the air gap between the pole pieces due to interaction of the fixed magnetic field with that associated with the audio-frequency current flowing through the voice coil. Also known as loudspeaker voice coil; speech coil (British usage). { 'vóis ₁kóil }

voice/data system |COMMUN| Integrated communications system for transmitting both voice and digital data. { 'vóis 'dad·ə ₁sis·təm }

voice digitization |ELECTR| The conversion of analog voice signals to digital signals. { 'vóis ₁dij·əd·ə'zā·shən }

voice frequency |COMMUN| An audio frequency in the range essential for transmission of speech of commercial quality, from about 300 to 3400 hertz. Abbreviated VF. Also known as speech frequency. { 'vóis ₁frē·kwən·sē }

voice-frequency carrier telegraphy |COMMUN| Carrier telegraphy in which the carrier currents have frequencies such that the modulated currents may be transmitted over a voice-frequency telephone channel. { 'vóis ¦frē·kwən·sē ¦kar·ē·ər tə'leg·rə·fē }

voice-frequency dialing |ELECTR| Method of dialing by which the direct-current pulses from the dial are transformed into voice-frequency alternating-current pulses. { 'vóis ¦frē·kwən·sē ₁dī·liŋ }

voice-frequency telegraph system |COMMUN| Telegraph system permitting the use of many channels on a single circuit; a different audio frequency is used for each channel, being keyed in the conventional manner; the various audio frequencies at the receiving end are separated by suitable filter circuits and fed to their respective receiving circuits. { 'vóis ¦frē·kwən·sē 'tel·ə₁graf ₁sis·təm }

voice-grade channel |COMMUN| A channel whose bandwidth is large enough to transmit voice-frequency signals. { 'vóis ¦grād ₁chan·əl }

voice mail |COMMUN| A method of storing voice-recorded messages and delivering them electronically to an intended receiver. { 'vóis ₁māl }

voice-operated device |ELECTR| Any of several devices that are brought into operation by a sound signal, or some characteristic of such a signal. { 'vóis ¦äp·ə₁rād·əd di₁vīs }

voice-operated device anti-singing See vodas. { 'vóis ¦äp·ə₁rād·əd di₁vīs ¦an·tē'siŋ·iŋ }

voice-operated gain-adjusted device See vogad. { 'vóis ¦äp·ə₁rād·əd 'gān ə¦jəs·təd di₁vīs }

voice-operated loss control and suppressor |ELECTR| Voice-operated device which switches loss out of the transmitting branch and inserts loss in the receiving branch under control of the subscriber's speech. { 'vóis ¦äp·ə ₁rād·əd 'lós kən₁tról ən sə'pres·ər }

voice operation demonstrator See voder. { 'vóis ₁äp·ə¦rā·shən 'dem·ən₁strād·ər }

voice print |ENG ACOUS| A voice spectrograph that has individually distinctive patterns of voice characteristics that can be used to identify one person's voice from other voice patterns. { 'vóis ₁print }

voice recognition unit |COMMUN| A computer peripheral device that recognizes a limited number of spoken words and converts them into equivalent digital signals which can serve as computer input or initiate other desired actions. { 'vóis ₁rek·ig'nish·ən ₁yü·nət }

voice response |ENG ACOUS| A computer-controlled recording system in which basic sounds, numerals, words, or phrases are individually stored for playback under computer control as the reply to a keyboarded query. { 'vóis ri₁späns }

voice store and forward |COMMUN| A computer-supported system that converts spoken messages to digital format, stores them temporarily, and then transmits them to a receiver where they are converted back to sound. { 'vȯis ¦stȯr ən 'fȯr·wərd }

voice synthesizer [ELECTR] A synthesizer that simulates speech in any language by assembling a language's elements or phonemes under digital control, each with the correct inflection, duration, pause, and other speech characteristics. { 'vȯis ¦sin·thə₁sīz·ər }

void [COMPUT SCI] In optical character recognition, an island of insufficiently inked paper within the area of the intended character stroke. { vȯid }

volatile file [COMPUT SCI] Any file in which data are rapidly added or deleted. { 'väl·əd·əl 'fīl }

volatile memory See volatile storage. { 'väl·əd·əl 'mem·rē }

volatile storage [COMPUT SCI] A storage device that must be continuously supplied with energy, or it will lose its retained data. Also known as volatile memory. { 'väl·əd·əl 'stȯr·ij }

volt [ELEC] The unit of potential difference or electromotive force in the meter-kilogram-second system, equal to the potential difference between two points for which 1 coulomb of electricity will do 1 joule of work in going from one point to the other. Symbolized V. { vōlt }

Volta effect See contact potential difference. { 'vōl·tə i₁fekt }

voltage [ELEC] Potential difference or electromotive force measured in volts. { 'vōl·tij }

voltage amplification [ELECTR] The ratio of the magnitude of the voltage across a specified load impedance to the magnitude of the input voltage of the amplifier or other transducer feeding that load; often expressed in decibels by multiplying the common logarithm of the ratio by 20. { 'vōl·tij ₁am·plə·fə'kā·shən }

voltage amplifier [ELECTR] An amplifier designed primarily to build up the voltage of a signal, without supplying appreciable power. { 'vōl·tij 'am·plə₁fī·ər }

voltage-amplitude-controlled clamp [ELECTR] A single diode clamp in which the diode functions as a clamp whenever the potential at point A rises above V_R; the diode is then in its forward-biased condition and acts as a very low resistance. { 'vōl·tij ¦am·plə₁tüd kən¦trōld 'klamp }

voltage coefficient [ELEC] For a resistor whose resistance varies with voltage, the ratio of the fractional change in resistance to the change in voltage. { 'vōl·tij ₁kō·i₁fish·ənt }

voltage-controlled oscillator [ELECTR] An oscillator whose frequency of oscillation can be varied by changing an applied voltage. Abbreviated VCO. { 'vōl·tij kən¦trōld 'äs·ə₁lād·ər }

voltage corrector [ELECTR] Active source of regulated power placed in series with an unregulated supply to sense changes in the output voltage (or current), and to correct for these changes by automatically varying its own output in the opposite direction, thereby maintaining the total output voltage (or current) constant. { 'vōl·tij kə₁rek·tər }

voltage-current dual [ELEC] A pair of circuits in which the elements of one circuit are replaced by their dual elements in the other circuit according to the duality principle; for example, currents are replaced by voltages, capacitances by resistances. { 'vōl·tij 'kə·rənt 'dül }

voltage-dependent resistor See varistor. { 'vōl·tij di¦pen·dənt ri'zis·tər }

voltage derating [ELEC] The reduction of a voltage rating to extend the lifetime of an electric device or to permit operation at a high ambient temperature. { 'vōl·tij ¦dē'rād·iŋ }

voltage divider [ELEC] A tapped resistor, adjustable resistor, potentiometer, or a series arrangement of two or more fixed resistors connected across a voltage source; a desired fraction of the total voltage is obtained from the intermediate tap, movable contact, or resistor junction. Also known as potential divider. { 'vōl·tij di₁vīd·ər }

voltage doubler [ELECTR] A transformerless rectifier circuit that gives approximately double the output voltage of a conventional half-wave vacuum-tube rectifier by charging a capacitor during the normally wasted half-cycle and discharging it in series with the output voltage during the next half-cycle. Also known as doubler. { 'vōl·tij ₁dəb·lər }

voltage drop [ELEC] The voltage developed across a component or conductor by the flow of current through the resistance or impedance of that component or conductor. { 'vōl·tij ₁dräp }

voltage feed [ELECTROMAG] Excitation of a transmitting antenna by applying voltage at a point of maximum potential (at a voltage loop or antinode). { 'vōl·tij ₁fēd }

voltage flare [ELEC] A higher than normal voltage purposely supplied to exposure lamps for a short period to produce full brilliance. { 'vōl·tij ₁fler }

voltage gain [ELECTR] The difference between the output signal voltage level in decibels and the input signal voltage level in decibels; this value is equal to 20 times the common logarithm of the ratio of the output voltage to the input voltage. { 'vōl·tij ₁gān }

voltage generator [ELECTR] A two-terminal circuit element in which the terminal voltage is independent of the current through the element. { 'vōl·tij ₁jen·ə₁rād·ər }

voltage gradient [ELEC] The voltage per unit length along a resistor or other conductive path. { 'vōl·tij ₁grād·ē·ənt }

voltage level [ELEC] At any point in a transmission system, the ratio of the voltage existing at that point to an arbitrary value of voltage used as a reference. { 'vōl·tij ₁lev·əl }

voltage measurement [ELEC] Determination of the difference in electrostatic potential between two points. { 'vōl·tij ₁mezh·ər·mənt }

voltage multiplier [ELEC] See instrument multiplier. [ELECTR] A rectifier circuit capable of supplying a direct-current output voltage that is two or more times the peak value of the alternating-current voltage. { 'vōl·tij ₁məl·tə₁plī·ər }

voltage-multiplier circuit [ELEC] A rectifier circuit capable of supplying a direct-current output

623

voltage node

voltage that is two or more times the peak value of the alternating-current input voltage; useful for high-voltage, low-current supplies. { 'vōl·tij ¦mәl·tә‚plī·ər ‚sәr·kәt }

voltage node [ELECTROMAG] Point having zero voltage in a stationary wave system, as in an antenna or transmission line; for example, a voltage node exists at the center of a half-wave antenna. { 'vōl·tij ‚nōd }

voltage phasor [ELEC] A line whose length represents the magnitude of a sinusoidally varying voltage and whose angle with the positive x-axis represents its phase. { 'vōl·tij ‚fā·zər }

voltage quadrupler [ELECTR] A rectifier circuit, containing four diodes, which supplies a direct-current output voltage which is four times the peak value of the alternating-current input voltage. { 'vōl·tij kwә‚drüp·lər }

voltage-range multiplier See instrument multiplier. { 'vōl·tij ¦rānj ‚mәl·tә‚plī·ər }

voltage rating [ELEC] The maximum sustained voltage that can safely be applied to an electric device without risking the possibility of electric breakdown. Also known as working voltage. { 'vōl·tij ‚rād·iŋ }

voltage ratio [ELEC] The root-mean-square primary terminal voltage of a transformer divided by the root-mean-square secondary terminal voltage under a specified load. { 'vōl·tij ‚rā·shō }

voltage reflection coefficient [ELECTROMAG] The ratio of the phasor representing the magnitude and phase of the electric field of the backward-traveling wave at a specified cross section of a waveguide to the phasor representing the forward-traveling wave at the same cross section. { 'vōl·tij ri¦flek·shən ‚kō·i‚fish·ənt }

voltage regulation [ELEC] The ratio of the difference between no-load and full-load output voltage of a device to the full-load output voltage, expressed as a percentage. { 'vōl·tij ‚reg·yә ‚lā·shən }

voltage regulator [ELECTR] A device that maintains the terminal voltage of a generator or other voltage source within required limits despite variations in input voltage or load. Also known as automatic voltage regulator; voltage stabilizer. { 'vōl·tij ‚reg·yә‚lād·ər }

voltage-regulator diode [ELECTR] A diode that maintains an essentially constant direct voltage in a circuit despite changes in line voltage or load. { 'vōl·tij ‚reg·yә‚lād·ər ‚dī‚ōd }

voltage-regulator tube [ELECTR] A glow-discharge tube in which the tube voltage drop is approximately constant over the operating range of current; used to maintain an essentially constant direct voltage in a circuit despite changes in line voltage or load. Also known as VR tube. { 'vōl·tij ‚reg·yә‚lād·ər ‚tüb }

voltage saturation See anode saturation. { 'vōl·tij ‚sach·ә‚rā·shən }

voltage stabilizer See voltage regulator. { 'vōl·tij ‚stā·bә‚līz·ər }

voltage step electronics [ELECTR] A sudden change in a voltage from one specified value to another at a particular instant in time. { 'vōl·tij ¦step i‚lek'trän·iks }

voltage transformer [ELEC] An instrument transformer whose primary winding is connected in parallel with a circuit in which the voltage is to be measured or controlled. Also known as potential transformer. { 'vōl·tij tranz‚fór·mər }

voltage-tunable tube [ELECTR] Oscillator tube whose operating frequency can be varied by changing one or more of the electrode voltages, as in a backward-wave magnetron. { 'vōl·tij ¦tün·ә·bәl 'tüb }

voltage-variable capacitor See varactor. { 'vōl·tij ¦ver·ē·ә·bәl kә'pas·әd·ər }

voltaic cell [ELEC] A primary cell consisting of two dissimilar metal electrodes in a solution that acts chemically on one or both of them to produce a voltage. { vōl'tā·ik 'sel }

voltaic pile [ELEC] An early form of primary battery, consisting of a pile of alternate pairs of dissimilar metal disks, with moistened pads between pairs. { vōl'tā·ik 'pīl }

voltammeter [ELEC] An instrument that may be used either as a voltmeter or ammeter. { väl 'tam·әd·ər }

volt-ampere [ELEC] The unit of apparent power in the International System; it is equal to the apparent power in a circuit when the product of the root-mean-square value of the voltage, expressed in volts, and the root-mean-square value of the current, expressed in amperes, equals 1. Abbreviated VA. { 'vōlt 'am‚pir }

volt-ampere hour [ELEC] A unit for expressing the integral of apparent power over time, equal to the product of 1 volt-ampere and 1 hour, or to 3600 joules. { 'vōlt 'am‚pir 'aůr }

volt-ampere-hour reactive See var hour. { 'vōlt 'am‚pir 'rē'ak·tiv }

volt-ampere reactive [ELEC] The unit of reactive power in the International System; it is equal to the reactive power in a circuit carrying a sinusoidal current when the product of the root-mean-square value of the voltage, expressed in volts, by the root-mean-square value of the current, expressed in amperes, and by the sine of the phase angle between the voltage and the current, equals 1. Abbreviated var. Also known as reactive volt-ampere. { 'vōlt 'am‚pir rē'ak·tiv }

volt box [ELEC] A series of resistors arranged so that a desired fraction of a voltage can be measured, and the voltage thereby computed. { 'vōlt ‚bäks }

voltmeter [ENG] An instrument for the measurement of potential difference between two points, in volts or in related smaller or larger units. { 'vōlt‚mēd·ər }

voltmeter-ammeter [ENG] A voltmeter and an ammeter combined in a single case but having separate terminals. { 'vōlt‚mēd·ər 'am‚ēd·ər }

voltmeter-ammeter method [ELEC] A method of measuring resistance in which simultaneous readings of the voltmeter and ammeter are taken, and the unknown resistance is calculated from Ohm's law. { 'vōlt‚mēd·ər 'am‚ēd·ər ‚meth·әd }

624

voltmeter sensitivity [ELEC] Ratio of the total resistance of the voltmeter to its full scale reading in volts, expressed in ohms per volt. { 'vōlt ,mēd·ər ,sen·sə'tiv·əd·ē }

volt-ohm-milliammeter [ENG] A test instrument having a number of different ranges for measuring voltage, current, and resistance. Also known as circuit analyzer; multimeter; multiple-purpose tester. { 'vōlt 'ōm ¦mil·ē'am ,ēd·ər }

volume [COMPUT SCI] A single unit of external storage, all of which can be read or written by a single access mechanism or input/output device. [ENG ACOUS] The magnitude of a complex audio-frequency current as measured in volume units on a standard volume indicator. { 'väl·yəm }

volume compressor [ENG ACOUS] An audio-frequency circuit that limits the volume range of a radio program at the transmitter, to permit using a higher average percent modulation without risk of overmodulation; also used when making disk recordings, to permit a closer groove spacing without overcutting. Also known as automatic volume compressor. { 'väl·yəm kəm ,pres·ər }

volume control [ENG ACOUS] A potentiometer used to vary the loudness of a reproduced sound by varying the audio-frequency signal voltage at the input of the audio amplifier. { 'väl·yəm kən ,trōl }

volume control system [ENG ACOUS] An electronic system that regulates the signal amplification or limits the output of a circuit, such as a volume compressor or a volume expander. { 'väl·yəm kən,trōl ,sis·təm }

volume expander [ENG ACOUS] An audio-frequency control circuit sometimes used to increase the volume range of a radio program or recording by making weak sounds weaker and loud sounds louder; the expander counteracts volume compression at the transmitter or recording studio. Also known as automatic volume expander. { 'väl·yəm ik,span·dər }

volume indicator [ENG ACOUS] A standardized instrument for indicating the volume of a complex electric wave such as that corresponding to speech or music; the reading in volume units is equal to the number of decibels above a reference level which is realized when the instrument is connected across a 600-ohm resistor that is dissipating a power of 1 milliwatt at 100 hertz. Also known as volume unit meter. { 'väl·yəm ,in·də,kād·ər }

volume label [COMPUT SCI] A record that contains information about the contents of a particular storage device, usually a disk or magnetic tape, and is written somewhere on that device. { 'väl·yəm ,lā·bəl }

volume-limiting amplifier [ELECTR] Amplifier containing an automatic device that functions only when the input signal exceeds a predetermined level, and then reduces the gain so the output volume stays substantially constant despite further increases in input volume; the normal gain of the amplifier is restored when the input volume returns below the predetermined limiting level. { 'väl·yəm ¦lim·əd·iŋ 'am·plə,fī·ər }

volume range [ELEC] In a transmission system, the difference, expressed in decibels, between the maximum and minimum volumes that can be satisfactorily handled by the system. [ENG ACOUS] The difference, expressed in decibels, between the maximum and minimum volumes of a complex audio-frequency signal occurring over a specified period of time. { 'väl·yəm ,rānj }

volume resistivity [ELEC] Electrical resistance between opposite faces of a 1-centimeter cube of insulating material, commonly expressed in ohm-centimeters. Also known as specific insulation resistance. { 'väl·yəm ,rē,zis'tiv·əd·ē }

volume table of contents [COMPUT SCI] A list of all the files in a volume, usually with descriptions of their contents and locations. Abbreviated VTOC. { 'väl·yəm ¦tā·bəl əv 'kän,tens }

volume target [ELECTROMAG] A radar target composed of a large number of objects too close together to be resolved. { 'väl·yəm 'tär·gət }

volume test [COMPUT SCI] The processing of a volume of actual data to check for program malfunction. { 'väl·yəm ,test }

volumetric storage [COMMUN] Any data storage technology in which information is stored throughout a three-dimensional volume rather than merely on a surface. { ¦väl·yə¦me·trik 'stòr·ij }

volume unit [ENG ACOUS] A unit for expressing the audio-frequency power level of a complex electric wave, such as that corresponding to speech or music; the power level in volume units is equal to the number of decibels above a reference level of 1 milliwatt as measured with a standard volume indicator. Abbreviated VU. { 'väl·yəm ,yü·nət }

volume unit meter See volume indicator. { 'väl·yəm ,yü·nət ,mēd·ər }

von Klitzing effect See quantum Hall effect. { fòn 'klit·siŋ i,fekt }

von Neumann bottleneck [COMPUT SCI] An inefficiency inherent in the design of any von Neumann machine that arises from the fact that most computer time is spent in moving information between storage and the central processing unit rather than operating on it. { fòn 'nòi,män 'bäd·əl,nek }

von Neumann machine [COMPUT SCI] A stored-program computer equipped with a program counter. { fòn 'nòi,män mə,shēn }

VOR See very high frequency omnidirectional radio range.

voxel [COMPUT SCI] The smallest box-shaped part of a three-dimensional image or scan. Derived from volume pixel. { 'väks·əl }

VPN See virtual private network.

VRC See lateral parity check.

VRML See Virtual Reality Modeling Language.

VR tube See voltage-regulator tube. { ¦vē'är ,tüb }

VSAT See very small aperture terminal. { 'vē,sat }

VSB *See* vestigial sideband.
8 VSB |COMMUN| Vestigial sideband modulation with 8 discrete amplitude levels.
16 VSB |COMMUN| Vestigial sideband modulation with 16 discrete amplitude levels.
VSC *See* variable speech control.
VTOC *See* volume table of contents.
VTR *See* videotape recording.

VTVM *See* vacuum-tube voltmeter.
vulnerability |COMPUT SCI| A weakness in a computing system that can result in harm to the system or its operations, especially when this weakness is exploited by a hostile person or organization or when it is present in conjunction with particular events or circumstances. { ˌvəl·nə·rə'bil·əd·ē }

W

wafer |ELECTR| A thin semiconductor slice on which matrices of microcircuits can be fabricated, or which can be cut into individual dice for fabricating single transistors and diodes. |ENG| A flat element for a process unit, as in a series of stacked filter elements. { 'wā·fər }

wafer lever switch |ELECTR| A lever switch in which a number of contacts are arranged on one or both sides of one or more wafers, for engaging one or more contacts on a movable wafer segment actuated by the operating lever. { 'wā·fər 'lev·ər ,swich }

wafer socket |ELECTR| An electron-tube socket consisting of one or two wafers of insulating material having holes in which are spring metal clips that grip the terminal pins of a tube. { 'wā·fər ,säk·ət }

Wagner earth connection *See* Wagner ground. { 'wag·nər 'ərth kə,nek·shən }

Wagner ground |ELEC| A ground connection used with an alternating-current bridge to minimize stray capacitance errors when measuring high impedances; a potentiometer is connected across the bridge supply oscillator, with its movable tap grounded. Also known as Wagner earth connection. { 'wag·nər ,graůnd }

wait |CONT SYS| Cessation of motion of a robot manipulator, under computer control, until further notice. { wāt }

waiting time *See* idle time. { 'wād·iŋ tīm }

wait state |COMPUT SCI| The state of a computer program in which it cannot use the central processing unit normally because the unit is waiting to complete an input/output operation. { 'wāt ,stāt }

walk down |ELECTR| A malfunction in a magnetic core of a computer storage in which successive drive pulses or digit pulses cause charges in the magnetic flux in the core that persist after the magnetic fields associated with pulses have been removed. Also known as loss of information. { 'wȯk ,daůn }

walkthrough |COMPUT SCI| A step-by-step review of a computer program or system during its design to search for errors and problems. { 'wȯk ,thrü }

walkthrough method |CONT SYS| The instruction of a robot by taking it through its sequences of motions, so that these actions are stored in its memory and recalled when necessary. { 'wȯk ¦thrü ,meth·əd }

wall box |ELEC| A metal box set into a wall to hold switches, receptacles, or similar electrical wiring components. { 'wȯl ,bäks }

wall effect |ELECTR| The contribution to the ionization in an ionization chamber by electrons liberated from the walls. { 'wȯl i,fekt }

wall frame *See* wall box. { 'wȯl ,frām }

wall outlet |ELEC| An outlet mounted on a wall, from which electric power can be obtained by inserting the plug of a line cord. { 'wȯl ¦aůt·lət }

wallpaper |COMPUT SCI| The design or image used as a computer monitor background. { 'wȯl,pā·pər }

WAN *See* wide-area network.

wand |COMPUT SCI| A hand-held device that contains an optical scanner to sense bar codes and other patterns and transmits the data to a computer. { wänd }

Ward-Leonard speed-control system |CONT SYS| A system for controlling the speed of a direct-current motor in which the armature voltage of a separately excited direct-current motor is controlled by a motor-generator set. { 'wȯrd 'len·ərd 'spēd kən¦trōl ,sis·təm }

warm boot |COMPUT SCI| To boot a computer system after it has been running. { ¦wȯrm 'büt }

warm start |COMPUT SCI| A resumption of computer operation, following a problem-generated shutdown, in which programs running on the system can resume at the point they were at when the shutdown occurred and data is not lost. { 'wȯrm 'stärt }

warm-up time |ENG| A span of time between the first application of power to a system and the moment when the system can function fully. { 'wȯrm,əp ,tīm }

warning device |COMPUT SCI| A visible or audible alarm to inform the operator of a machine condition. { 'wȯrn·iŋ di,vīs }

warning message |COMPUT SCI| A diagnostic message that is issued when a computer program detects an error or potential problem but continues processing. { 'wȯrn·iŋ ,mes·ij }

warning-receiver system |ELECTR| An electronic countermeasure system, carried on a tactical or transport aircraft, which is programmed to alert a pilot when his aircraft is being illuminated by a specific radar signal above predetermined power thresholds. { 'wȯrn·iŋ ri¦sē·vər ,sis·təm }

washer thermistor [ELECTR] A thermistor in the shape of a washer, which may be as large as 0.75 inch (1.9 centimeters) in diameter and 0.50 inch (1.3 centimeters) thick; it is formed by pressing and sintering an oxide-binder mixture. { 'wäsh·ər thər,mis·tər }

watchdog timer [CONT SYS] In a flexible manufacturing system, a safety device in the form of a control interface on an automated guided vehicle that shuts down part or all of the system under certain conditions. { 'wäch,dóg ,tīm·ər }

water-activated battery [ELEC] A primary battery that contains the electrolyte but requires the addition of or immersion in water before it is usable. { 'wód·ər ¦ak·tə,vād·əd 'bad·ə·rē }

water-cooled tube [ELECTR] An electron tube that is cooled by circulating water through or around the anode structure. { 'wód·ər ¦küld 'tüb }

water cooling [ELECTR] Cooling the electrodes of an electron tube by circulating water through or around them. [ENG] Cooling in which the primary coolant is water. { 'wód·ər ,kül·iŋ }

water dropper [ELEC] A simple electrostatic generator in which each of two series of water drops falls through cylindrical metal cans into lower cans with funnels, and the cans are electrically connected in such a way that charge accumulates on them, energy being supplied by the gravitational force on the water drops. { 'wód·ər ,dräp·ər }

Waterloo Fortran IV See WATFIV. { 'wód·ər,lü 'fór ,tran 'fór }

water rheostat See electrolytic rheostat. { 'wód·ər 'rē·ə,stat }

WATFIV [COMPUT SCI] A programming language based on FORTRAN that is used in learning environments and is characterized by fast compilation and excellent diagnostic messages and debugging aids. Acronym for Waterloo Fortran IV. { 'wät,fīv }

WATS See Wide Area Telephone Service. { wäts }

wattage rating [ELEC] A rating expressing the maximum power that a device can safely handle continuously. { 'wäd·ij ,rād·iŋ }

watt current See active current. { 'wät ,kə·rənt }

watt-hour [ELEC] A unit of energy used in electrical measurements, equal to the energy converted or consumed at a rate of 1 watt during a period of 1 hour, or to 3600 joules. Abbreviated Wh. { 'wät ¦aúr }

watt-hour capacity [ELEC] Number of watt-hours which can be delivered from a storage battery under specified conditions as to temperature, rate of discharge, and final voltage. { 'wät ¦aúr kə,pas·əd·ē }

watt-hour meter [ENG] A meter that measures and registers the integral, with respect to time, of the active power of the circuit in which it is connected; the unit of measurement is usually the kilowatt-hour. { 'wät ¦aúr ,mēd·ər }

wattless component See reactive component. { 'wät,ləs kəm'pō·nənt }

wattless current See reactive current. { 'wät,ləs 'kə·rənt }

wattless power See reactive power. { 'wät,ləs 'paú·ər }

wattmeter [ENG] An instrument that measures electric power in watts ordinarily. { 'wät,mēd·ər }

wave analyzer See harmonic analyzer. { 'wāv ,an·ə,līz·ər }

wave angle [ELECTROMAG] The angle, either in bearing or elevation, at which a radio wave leaves a transmitting antenna or arrives at a receiving antenna. { 'wāv ,aŋ·gəl }

wave antenna [ELECTROMAG] Directional antenna composed of a system of parallel, horizontal conductors, varying from a half to several wavelengths long, terminated to ground at the far end in its characteristic impedance. { 'wāv an,ten·ə }

wave clutter See sea clutter. { 'wāv ,kləd·ər }

wave converter [ELECTROMAG] Device for changing a wave of a given pattern into a wave of another pattern, for example, baffle-plate converters, grating converters, and sheath-reshaping converters for waveguides. { 'wāv kən,vərd·ər }

wave duct [ELECTROMAG] **1.** Waveguide, with tubular boundaries, capable of concentrating the propagation of waves within its boundaries. **2.** Natural duct, formed in air by atmospheric conditions, through which waves of certain frequencies travel with more than average efficiency. { 'wāv ,dəkt }

wave filter [ELEC] A transducer for separating waves on the basis of their frequency; it introduces relatively small insertion loss to waves in one or more frequency bands and relatively large insertion loss to waves of other frequencies. { 'wāv ,fil·tər }

waveform-amplitude distortion See frequency distortion. { 'wāv,fórm ¦am·plə,tüd di,stòr·shən }

waveform coder See waveform compression. { 'wāv,fórm ,kōd·ər }

waveform compression [COMMUN] Any technique for compressing audio or video signals in which a facsimile of the source-signal waveform is replicated at the receiver with a level of distortion that is judged acceptable. Also known as waveform coder. { 'wāv,fórm kəm,presh·ən }

wavefront reversal See optical phase conjugation. { 'wāv,frənt ri'vər·səl }

waveguide [ELECTROMAG] **1.** Broadly, a device which constrains or guides the propagation of electromagnetic waves along a path defined by the physical construction of the waveguide; includes ducts, a pair of parallel wires, and a coaxial cable. Also known as microwave waveguide. **2.** More specifically, a metallic tube which can confine and guide the propagation of electromagnetic waves in the lengthwise direction of the tube. { 'wāv,gīd }

waveguide bend [ELECTROMAG] A section of waveguide in which the direction of the longitudinal axis is changed; an **E**-plane bend in a rectangular waveguide is bent along the narrow dimension, while an **H**-plane bend is bent along the wide dimension. Also known as waveguide elbow. { 'wāv,gīd ¦bend }

waveguide cavity [ELECTROMAG] A cavity resonator formed by enclosing a section of

628

waveguide between a pair of waveguide windows. { 'wāv‚gīd ¦kav·əd·ē }

waveguide connector [ELECTROMAG] A mechanical device for electrically joining and locking together separable mating parts of a waveguide system. Also known as waveguide coupler. { 'wāv‚gīd kə¦nek·tər }

waveguide coupler See waveguide connector. { 'wāv‚gīd ¦kəp·lər }

waveguide critical dimension [ELECTROMAG] Dimension of waveguide cross section which determines the cutoff frequency. { 'wāv‚gīd 'krid·ə·kəl də'men·shən }

waveguide cutoff frequency [ELECTROMAG] Frequency limit of propagation along a waveguide for waves of a given field configuration. { 'wāv‚gīd 'kəd‚óf ‚frē·kwən·sē }

waveguide elbow See waveguide bend. { 'wāv‚gīd ¦el·bō }

waveguide filter [ELECTROMAG] A filter made up of waveguide components, used to change the amplitude-frequency response characteristic of a waveguide system. { 'wāv‚gīd ¦fil·tər }

waveguide hybrid [ELECTROMAG] A waveguide circuit that has four arms so arranged that a signal entering through one arm will divide and emerge from the two adjacent arms, but will be unable to reach the opposite arm. { 'wāv‚gīd 'hī·brəd }

waveguide junction See junction. { 'wāv‚gīd ¦jəŋk·shən }

waveguide plunger See piston. { 'wāv‚gīd ¦plən·jər }

waveguide probe See probe. { 'wāv‚gīd ¦prōb }

waveguide propagation [COMMUN] Long-range communications in the 10- to 35-kilohertz frequency range that are made possible by the waveguide characteristics of the atmospheric wave duct formed by the ionospheric D layer and the surface of the earth. { 'wāv‚gīd ‚präp·ə'gā·shən }

waveguide resonator See cavity resonator. { 'wāv‚gīd ¦rez·ən‚ād·ər }

waveguide shim [ELECTROMAG] Thin resilient metal sheet inserted between waveguide components to ensure electrical contact. { 'wāv‚gīd ‚shim }

waveguide slot [ELECTROMAG] A slot in a waveguide wall, either for coupling with a coaxial cable or another waveguide, or to permit the insertion of a traveling probe for examination of standing waves. { 'wāv‚gīd ‚slät }

waveguide switch [ELECTROMAG] A switch designed for mechanically positioning a waveguide section so as to couple it to one of several other sections in a waveguide system. { 'wāv‚gīd ‚swich }

waveguide synthesis [ENG ACOUS] A method of synthesizing the sounds of a string or wind instrument that simulates traveling waves on a string or inside a bore or horn using digital delay lines. { ‚wāv‚gīd 'sin·thə·səs }

waveguide window See iris. { 'wāv‚gīd ¦win·dō }

wave impedance [ELECTROMAG] The ratio, at every point in a specified plane of a waveguide, of the transverse component of the electric field to the transverse component of the magnetic field. { 'wāv im‚pēd·əns }

wavelength constant See phase constant. { 'wāv ‚leŋkth ‚kän·stənt }

wavelength-division multiplexing [COMMUN] The sharing of the total available pass-band of a transmission medium in the optical portion of the electromagnetic spectrum by assigning individual information streams to signals of different wavelengths. Abbreviated WDM. { ¦wāv‚leŋth də¦vizh·ən 'məl·tə‚pleks·iŋ }

wavelength shifter [ELECTR] A photofluorescent compound used with a scintillator material to increase the wavelengths of the optical photons emitted by the scintillator, thereby permitting more efficient use of the photons by the phototube or photocell. { 'wāv‚leŋkth ‚shif·tər }

wave microphone [ENG ACOUS] Any microphone whose directivity depends upon some type of wave interference, such as a line microphone or a reflector microphone. { 'wāv 'mī·krə‚fōn }

wave noise [ELECTR] Noise in the electric current of a detector that results from fluctuations in the intensity of electromagnetic radiation falling on the detector. { 'wāv ‚nóiz }

wave polarization See polarization. { 'wāv ‚pō·lə·rə‚zā·shən }

wave-shaping circuit [ELECTR] An electronic circuit used to create or modify a specified time-varying electrical quantity, usually voltage or current, using combinations of electronic devices, such as vacuum tubes or transistors, and circuit elements, including resistors, capacitors, and inductors. { 'wāv ¦shāp·iŋ ‚sər·kət }

wave soldering See flow soldering. { 'wāv ‚säd·ə·riŋ }

wave tail [ELECTR] Part of a signal-wave envelope (in time or distance) between the steady-state value (or crest) and the end of the envelope. { 'wāv ‚tāl }

wave tilt [ELECTROMAG] Forward inclination of a radio wave due to its proximity to ground. { 'wāv ‚tilt }

wave trap [ELECTR] A resonant circuit connected to the antenna system of a receiver to suppress signals at a particular frequency, such as that of a powerful local station that is interfering with reception of other stations. Also known as trap. { 'wāv ‚trap }

waxy-electrolyte battery [ELEC] A primary battery in which the electrolyte is a waxy material, such as polyethylene glycol, in which is dissolved a small amount of a salt, such as zinc chloride; the electrodes are frequently made of zinc and manganese dioxide, and the electrolyte is melted and painted on a paper sheet to form the separator. { 'wak·sē i¦lek·trə¦līt 'bad·ə·rē }

WDM See wavelength-division multiplexing.

weather observation radar See weather radar. { 'weth·ər ‚äb·zər‚vā·shən 'rā‚där }

weather radar [ENG] Radar specifically designed to enhance the echo from rain, other precipitates, and weather conditions in general, and with processing techniques by which reflectivity of the scatterers, wind speeds, and direction all can be accurately estimated. Also known as weather observation radar. { 'we<u>th</u>·ər 'rā,där }

Web See World Wide Web. { web }

Web browser See browser. { 'web ,braúz·ər }

Web page [COMPUT SCI] A document written in HTML and available for viewing on the World Wide Web; it may contain images, sound, video, formatted text, and hyperlinks. { 'web ,pāj }

Web server [COMPUT SCI] A program that processes document requests; it also has a database, which is a repository of data and content. { 'web ,sər·vər }

Web services [COMPUT SCI] A collection of SML-based standards that enable electronic communication and interaction independently of the computer platforms or specific technologies used by the communication parties. { 'web ,serv·ə·səs }

Web site [COMPUT SCI] A collection of thematically related, hyperlinked World Wide Web services, mainly HTML documents, usually located on a specific Web server and reachable through a URL assigned to the site. { 'web ,sīt }

wedge [COMMUN] A convergent pattern of equally spaced black and white lines, used in a television test pattern to indicate resolution. [ELECTRO-MAG] A waveguide termination consisting of a tapered length of dissipative material introduced into the guide, such as carbon. { wej }

wedge-base lamp [ELEC] A small indicator lamp that has wire leads folded back on opposite sides of a flat glass base. { 'wej ¦bās 'lamp }

Wehnelt cathode See oxide-coated cathode. { 'vān,elt ,kath,ōd }

weighted area masks [COMPUT SCI] In character recognition, a set of characters (each character residing in the character reader in the form of weighted points) which theoretically render all input specimens unique, regardless of the size or style. { 'wād·əd 'er·ē·ə ,masks }

weighted code [COMPUT SCI] A method of representing a decimal digit by a combination of bits, in which each bit is assigned a weight, and the value of the decimal digit is found by multiplying each bit by its weight and then summing the results. { 'wād·əd 'kōd }

weighted quasi peak [ELECTR] Referring to a fast attack, slow-decay detector circuit that approximately responds to signal peaks, and that has varying attenuation as a function of frequency so as to produce a measurement that approximates the human hearing system. Abbreviated WQP. { 'wād·əd 'kwä·zē 'pēk }

weighting [ENG] The artificial adjustment of measurements to account for factors that, in the normal use of the device, would otherwise be different from conditions during the measurements. { 'wād·iŋ }

weighting network [ENG ACOUS] One of three or more circuits in a sound-level meter designed to adjust its response; the A and B weighting networks provide responses approximating the 40- and 70-phon equal loudness contours, respectively, and the C weighting network provides a flat response up to 8000 hertz. { 'wād·iŋ ,net ,wərk }

weightlessness switch See zero-gravity switch. { 'wāt·ləs·nəs ,swich }

welcome page See home page. { 'wel·kəm ,pāj }

welding current [ELEC] The current that flows through a circuit while a weld is being made. { 'weld·iŋ ,kə·rənt }

welding generator [ELEC] A generator used for supplying the welding current. { 'weld·iŋ ,jen·ə ,rād·ər }

welding transformer [ELEC] A high-current, low-voltage power transformer used to supply current for welding. { 'weld·iŋ tranz,fór·mər }

well-regulated system [CONT SYS] A system with a regulator whose action, together with that of the environment, prevents any disturbance from permanently driving the system from a state in which it is stable, that is, a state in which it retains its structure and survives. { wel ¦reg·yə,lād·əd ,sis·təm }

Weston standard cell [ELEC] A standard cell used as a highly accurate voltage source for calibrating purposes; the positive electrode is mercury, the negative electrode is cadmium, and the electrolyte is a saturated cadmium sulfate solution; the Weston standard cell has a voltage of 1.018636 volts at 20°C. { 'wes·tən 'stan·dərd 'sel }

wet cell [ELEC] A primary cell in which there is a substantial amount of free electrolyte in liquid form. { 'wet ,sel }

wet contact [ELEC] Contact through which direct current flows. { 'wet 'kän,takt }

wet electrolytic capacitor [ELEC] An electrolytic capacitor employing a liquid electrolyte. { 'wet i¦lek·trə¦lid·ik kə¦pas·əd·ər }

wet flashover voltage [ELECTR] The voltage at which an electric discharge occurs between two electrodes that are separated by an insulator whose surface has been sprayed with water to simulate rain. { 'wet 'flash,ō·vər ,vōl·tij }

wet-reed relay [ELEC] Reed-type relay containing mercury at the relay contacts to reduce arcing and contact bounce. { 'wet ¦rēd 'rē,lā }

wetting [ELECTR] The coating of a contact surface with an adherent film of mercury. { 'wed·iŋ }

Wheatstone bridge [ELEC] A four-arm bridge circuit, all arms of which are predominately resistive; used to measure the electrical resistance of an unknown resistor by comparing it with a known standard resistance. Also known as resistance bridge; Wheatstone network. { 'wēt ,stōn 'brij }

Wheatstone network See Wheatstone bridge. { 'wēt,stōn 'net,wərk }

wheeling [ELEC] The transfer of one utility's energy over another utility's lines for delivery to a third utility. { wēl·iŋ }

wheel printer [COMPUT SCI] A line printer that prints its characters from the rim of a wheel around which is the type for the alphabet, numerals, and other characters. { 'wēl 'print·ər }

wheel static [ELECTR] Interference encountered in automobile-radio installations due to static electricity developed by friction between the tires and the street. { 'wēl ,stad·ik }

whiffletree switch [ELECTR] In computers, a multiposition electronic switch composed of gate tubes and flip-flops, so named because its circuit diagram resembles a whiffletree. { 'wif·əl,trē ,swich }

WHILE statement [COMPUT SCI] A statement in a computer program that is executed repeatedly, as long as a specified condition holds true. { 'wīl ,stāt·mənt }

whip antenna [ELECTROMAG] A flexible vertical rod antenna, used chiefly on vehicles. Also known as fishpole antenna. { 'wip an,ten·ə }

white compression [COMMUN] In analog television the reduction in picture-signal gain at levels corresponding to light areas, with respect to the gain at the level for midrange light values; the overall effect of white compression is to reduce contrast in the highlights of the picture. { 'wīt kəm,presh·ən }

white level [COMMUN] The carrier signal level corresponding to maximum picture brightness in analog television. { 'wīt ,lev·əl }

whitening filter [ELECTR] An electrical filter which converts a given signal to white noise. Also known as prewhitening filter. { 'wīt·niŋ ,fil·tər }

white noise [COMMUN] See additive white Gaussian noise. [PHYS] Random noise that has a constant energy per unit bandwidth at every frequency in the range of interest. { 'wīt ,nȯiz }

white signal [COMMUN] Signal at any point in a facsimile system produced by the scanning of a minimum density area of the subject copy. { 'wīt ,sig·nəl }

white transmission [COMMUN] **1.** In an amplitude-modulated system, that form of transmission in which the maximum transmitted power corresponds to the minimum density of the subject copy. **2.** In a frequency-modulation system, that form of transmission in which the lowest transmitted frequency corresponds to the minimum density of the subject copy. { 'wīt tranz'mish·ən }

whr See watt-hour.

wide-area network [COMMUN] A computer or telecommunication system consisting of a set of nodes that are interconnected by a set of links, and generally covers a large geographic area, usually on the order of hundreds of miles. Abbreviated WAN. { 'wīd ¦er·ē·ə 'net,wərk }

Wide Area Telephone Service [COMMUN] A special telephone service that allows a customer to call anyone in one or more of six regions into which the continental United States has been divided, on a direct dialing basis, for a flat monthly charge related to the number of regions to be called. Abbreviated WATS. { 'wīd ¦er·ē·ə 'tel·ə,fōn ,sər·vəs }

wideband [ELECTR] Property of a tuner, amplifier, or other device that can pass a broad range of frequencies. { 'wīd¦band }

wideband amplifier [ELECTR] An amplifier that will pass a wide range of frequencies with substantially uniform amplification. { 'wīd¦band 'am·plə,fī·ər }

wideband communications system [COMMUN] Communications system which provides numerous channels of communications on a highly reliable and secure basis which are relatively invulnerable to interruption by natural phenomena or countermeasures; included are multichannel telephone cable, tropospheric scatter, multichannel line-of-sight radio system such as microwave, and satellites. { 'wīd ¦band kə ,myü·nə'kā·shənz ,sis·təm }

wideband ratio [COMMUN] Ratio in a system of the occupied frequency bandwidth to the intelligence bandwidth. { 'wīd ¦band 'rā·shō }

wideband repeater [ELECTR] Airborne system that receives a radio-frequency signal and relays it to another facility via a separate RF channel. { 'wīd¦band ri'pēd·ər }

wideband switching [ELECTR] Basically, four-wire circuits using correed matrices with electronic controls capable of switching wideband facilities up to 50 kilohertz in bandwidth. { 'wīd¦band 'swich·iŋ }

wideband transformer [ELEC] A transformer that can transfer electric energy from one circuit to another at any of a broad range of frequencies. { 'wīd¦band tranz'fȯr·mər }

wide-open [ELECTR] Refers to the untuned characteristic or lack of frequency selectivity. { 'wīd 'ō·pən }

Widrow-Hoff least-mean-squares algorithm [COMMUN] An algorithm that is widely used in adaptive signal processing; for time-discrete analysis with a finite-response filter, it is represented by the first-order difference equation $\mathbf{W}_{k+1} = \mathbf{W}_k + 2\mu e_k \mathbf{X}_k$, where k is a time index that takes on integral values; $\mathbf{W}$ is a vector whose components are the coefficients of the filter; μ is the convergence coefficient; e_k is the residual signal or error, equal to $y_k - \hat{y}_k$, where y_k and $\hat{y}_k$ are the outputs of the plant (where the unprocessed signal is generated) and the filter, respectively; and $\mathbf{X}$ is a vector whose components are the present value of the input and $L - 1$ past values of the input, where L is the number of filter coefficients. { ¦wi·drō ¦häf 'lēst ,mēn 'skwerz 'al·gə,rith·əm }

width [COMMUN] **1.** The horizontal dimension of a video picture. **2.** The time duration of a pulse. { width }

width control [ELECTR] Control that adjusts the width of the image on the screen of a video display. { 'width kən,trōl }

Wiegand effect [ELEC] The generation of an electrical pulse in a coil wrapped around or located near a Wiegand wire subjected to a changing magnetic field. { 'vē·gänt i,fekt }

Wiegand module [ELEC] The apparatus for generating an electrical pulse by means of the Wiegand effect, consisting of a Wiegand wire, two small magnets, and a pickup coil.   { 've·gänt ˌmäj·əl }

Wiegand wire |ELEC| A work-hardened wire whose magnetic permeability is much greater near its surface than at its center.   { 've·gänt ˌwīr }

Wien bridge oscillator |ELECTR| A phase-shift feedback oscillator that uses a Wien bridge as the frequency-determining element.   { 'vēn ˌbrij 'äs·ə‚lād·ər }

Wien capacitance bridge |ELEC| A four-arm alternating-current bridge used to measure capacitance in terms of resistance and frequency; two adjacent arms contain capacitors respectively in parallel and in series with resistors, while the other two arms are nonreactive resistors; bridge balance depends on frequency.   { 'vēn kə'pas·əd·əns ˌbrij }

Wien-DeSauty bridge   See DeSauty's bridge.   { 'vēn də·sō'tē ˌbrij }

Wien frequency bridge |ELEC| A modification of the Wien capacitance bridge, used to measure frequencies.   { 'vēn 'frē·kwən·sē ˌbrij }

Wien inductance bridge |ELEC| A four-arm alternating-current bridge used to measure inductance in terms of resistance and frequency; two adjacent arms contain inductors respectively in parallel and in series with resistors, while the other two arms are nonreactive resistors; bridge balance depends on frequency.   { 'vēn in'dək·təns ˌbrij }

Wien-Maxwell bridge   See Maxwell bridge.   { 'vēn 'maks‚wel ˌbrij }

Wierl equation |ELECTR| A formula for the intensity of an electron beam scattered through a specified angle by diffraction from the molecules in a gas.   { 'virl i'kwā·zhən }

WiFi™ |COMMUN| An acronym for Wireless Fidelity that denotes equipment conforming to the Institute of Electrical and Electronic Engineers technical specifications for wireless Local Area Networks (IEEE 802.11 standards). Wi-Fi networks utilize unlicensed radio frequencies but are compatible with and may be connected to a wired Ethernet Local Area Network. A typical application is the wireless, high speed connection of a portable computer to the Internet.   { 'wī·'fī }

wild card |COMPUT SCI| A symbolic character in a search argument such that any character will satisfy it.   { 'wīld ˌkärd }

Williams tube |ELECTR| A cathode-ray storage tube in which information is stored as a pattern of electric charges produced, maintained, read, and erased by suitably controlled scanning of the screen by the electron beam.   { 'wil·yəmz ˌtüb }

Wilson electroscope |ELEC| An electroscope that has a single gold leaf which, when charged, is attracted to a grounded metal plate inclined at an angle that maximizes the instrument's sensitivity.   { 'wil·sən i'lek·trə‚skōp }

Wimshurst machine |ELEC| An electrostatic generator consisting of two glass disks rotating in opposite directions, having sectors of tinfoil and collecting combs so arranged that static electricity is produced for charging Leyden jars or discharging across a gap.   { 'wimz‚hərst mə‚shēn }

Winchester disk |COMPUT SCI| A type of disk storage device characterized by nonremovable or sealed disk packs; extremely narrow tracks; a lubricated surface that allows the head to rest on the surface during start and stop operations; and servomechanisms which utilize a magnetic pattern, recorded on the medium itself, to position the head.   { 'win·ches·tər ˌdisk }

Winchester technology |COMPUT SCI| Innovations designed to achieve disks with up to 6 × 10^8 bytes per disk drive; the technology includes nonremovable or sealed disk packs, a read/write head that weighs only 0.25 gram and floats above the surface, magnetic orientation of iron oxide particles on the disk surface, and lubrication of the disk surface.   { 'win·ches·tər tek'näl·ə·jē }

wind |ELECTR| The manner in which magnetic tape is wound onto a reel; in an A wind, the coated surface faces the hub; in a B wind, the coated surface faces away from the hub.   { wind }

wind charger |ELEC| A wind-driven direct-current generator used for charging storage batteries.   { 'win 'chär·jər }

winding |ELEC| **1.** One or more turns of wire forming a continuous coil for a transformer, relay, rotating machine, or other electric device. **2.** A conductive path, usually of wire, that is inductively coupled to a magnetic storage core or cell.   { 'wīnd·iŋ }

window |COMPUT SCI| A separate viewing area on a display screen that is established by the computer software. Also known as viewport. |ELECTR| A material having minimum absorption and minimum reflection of radiant energy, sealed into the vacuum envelope of a microwave or other electron tube to permit passage of the desired radiation through the envelope to the output device. |ELECTROMAG| A hole in a partition between two cavities or waveguides, used for coupling.   { 'win·dō }

window editor |COMPUT SCI| An interactive program that allows the user to view and alter stored information by using the video display.   { 'win·dō ˌed·əd·ər }

windowing |COMPUT SCI| **1.** The procedure of selecting a portion of a large drawing to be displayed on the screen of a computer graphics system, usually by placing a rectangular window over a compressed version of the entire drawing displayed on the screen. **2.** Dividing an electronic display into areas that display the outputs of different programs and can overlap in the same manner as pieces of paper on a desk, partially concealing the contents of pages underneath.   { 'win‚dō·iŋ }

wind turbine |ELEC| An advanced type of windmill designed to convert wind energy into electrical energy.   { 'win 'tər‚bīn }

wing spot generator |ELECTR| Electronic circuit that grows wings on the video target signal of a type G indicator; these wings are inversely proportional in size to the range.   { 'wiŋ 'spät ‚jen·ə‚rād·ər }

wiper |ELEC| That portion of the moving member of a selector, or other similar device, in communications practice, which makes contact with the terminals of a bank.   { 'wī·pər }

wiping contact |ELEC| A switch or relay contact designed to move laterally with a wiping motion after it touches a mating contact. Also known as self-cleaning contact; sliding contact.   { 'wīp·iŋ ‚kän‚takt }

wire |ELEC| A single bare or insulated metallic conductor having solid, stranded, or tinsel construction, designed to carry current in an electric circuit. Also known as electric wire.   { wīr }

wire bonding |ELEC| Lead-covered tie used to connect two cable sheaths until a splice is permanently closed and covered. |ELECTR| **1.** A method of connecting integrated-circuit chips to their substrate, using ultrasonic energy to weld very fine wires mechanically from metallized terminal pads along the periphery of the chip to corresponding bonding pads on the substrate. **2.** The attachment of very fine aluminum or gold wire (by thermal compression or ultrasonic welding) from metallized terminal pads along the periphery of an integrated circuit chip to corresponding bonding pads on the surface of the package leads.   { 'wīr‚bänd·iŋ }

wired-program computer |COMPUT SCI| A computer in which the sequence of instructions that form the operating program is created by interconnection of wires on a removable control panel.   { 'wīrd ¦prō‚gram kəm'pyüd·ər }

wire facsimile system |COMMUN| A facsimile system in which messages are sent over wires or cables, rather than by radio.   { 'wīr fak'sim·ə·lē ‚sis·təm }

wireframe |COMPUT SCI| **1.** In computer-aided design, a line-drawn model. **2.** In computer graphics, an image-rendering technique in which only edges and vertices are shown.   { ‚wīr'frām }

wireframe model |COMPUT SCI| In computer-aided design, the representation of all surfaces of a three-dimensional object in outline form.   { 'wīr‚frām ‚mäd·əl }

wire fusing current |ELEC| The electric current which will cause a wire to melt.   { 'wīr ¦fyüz·iŋ ‚kə·rənt }

wiregrating |ELECTROMAG| A series of wires placed in a waveguide that allow one or more types of waves to pass and block all others.   { 'wīr‚grād·iŋ }

wire holder |ELEC| A special type of electrical insulator fitted with a mounting screw or mounting bolt and having a hole for securing an electrical wire or cable.   { 'wīr ‚hōl·dər }

wireless cable |COMMUN| A television broadcasting system in which signals are collected and transmitted to towers for net transmission to homes outfitted with appropriate antennas and receiving equipment.   { ‚wīr·ləs 'kā·bəl }

wireless LAN |COMMUN| A local-area network whose devices or telephones communicate by radio transmissions.s Abbreviated WLAN.   { 'wīr·ləs 'lan }

wire line |ELECTR| One or more current-conducting wires or cables, used for communication, control, or telemetry.   { 'wīr ‚līn }

wire-link telemetry |COMMUN| Telemetry in which electric signals are sent over transmission lines, rather than by radio. Also known as hard-wire telemetry.   { 'wīr ¦liŋk tə'lem·ə·trē }

wire mile |ELEC| Unit of measure of the length of two-conductor wire between two points; the length of the route multiplied by the number of circuits gives the number of wire miles.   { 'wīr 'mīl }

wirephoto |COMMUN| **1.** A photograph transmitted over wires to a facsimile receiver. **2.** See facsimile.   { 'wīr‚fōd·ō }

wire printing   See matrix printing.   { 'wīr ‚print·iŋ }

wire recorder |ENG ACOUS| A magnetic recorder that utilizes a round stainless steel wire about 0.004 inch (0.01 centimeter) in diameter instead of magnetic tape.   { 'wīr ri‚kòrd·ər }

wire recording |ENG ACOUS| Magnetic recording by use of a magnetized wire.   { 'wīr ri ‚kòrd·iŋ }

wiretap |COMMUN| A concealed connection to a telephone line, office intercommunication line, or other wiring system, for the purpose of monitoring conversations and activities in a room from a remote location without knowledge of the participants, legally or illegally.   { 'wīr ‚tap }

wire telegraphy |COMMUN| Telegraphy in which messages are sent over wires or cables, rather than by radio.   { 'wīr tə'leg·rə·fē }

wire-wound potentiometer |ELEC| A potentiometer which is similar to a slide-wire potentiometer, except that the resistance wire is wound on a form and contact is made by a slider which moves along an edge from turn to turn.   { 'wīr ¦waùnd pə‚ten·chē'äm·əd·ər }

wire-wound resistor |ELEC| A resistor employing as the resistance element a length of high-resistance wire or ribbon, usually Nichrome, wound on an insulating form.   { 'wīr ¦waùnd ri'zis·tər }

wire-wound rheostat |ELEC| A rheostat in which a sliding or rolling contact moves over resistance wire that has been wound on an insulating core.   { 'wīr ¦waùnd 'rē·ə‚stat }

wire-wrap connection |ELEC| A solderless connection made by wrapping several turns of bare wire around a sharp-corner rectangular terminal under tension, using either a power tool or hand tool. Also known as solderless wrapped connection; wrapped connection.   { 'wīr ¦wrap kə‚nek·shən }

wiring |ELEC| The installation and utilization of a system of wire for conduction of electricity. Also known as electric wiring.   { 'wīr·iŋ }

wiring board   See control panel.   { 'wīr·iŋ ‚bòrd }

wiring diagram   See circuit diagram.   { 'wīr·iŋ ‚dī·ə ‚gram }

wiring harness |ELEC| An array of insulated conductors bound together by lacing cord, metal bands, or other binding, in an arrangement suitable for use only in specific equipment for which the harness was designed; it may include terminations.   { 'wīr·iŋ ‚här·nəs }

WLAN *See* wireless LAN.

wobbulator |ELECTR| A signal generator in which a motor-driven variable capacitor is used to vary the output frequency periodically between two known limits, as required for displaying a frequency-response curve on the screen of a cathode-ray oscilloscope.   { 'wäb·yə,läd·ər }

woofer |ENG ACOUS| A large loudspeaker designed to reproduce low audio frequencies at relatively high power levels; usually used in combination with a crossover network and a high-frequency loudspeaker called a tweeter.   { 'wüf·ər }

word |COMPUT SCI| The fundamental unit of storage capacity for a digital computer, almost always considered to be more than eight bits in length. Also known as computer word.   { wȯrd }

word-addressable computer *See* word-oriented computer.   { 'wȯrd ə¦dres·ə·bəl kəm'pyüd·ər }

word boundary |COMPUT SCI| A storage address that is a multiple of the word length of a computer.   { 'wȯrd ,baún·drē }

word concatenation system |ENG ACOUS| The simplest form of voice response system, which retrieves previously spoken versions of words or phrases and carefully forms them into a sequence without pauses, to approximate normally spoken word sequences.   { 'wȯrd kən,kat·ən'ā·shən ,sis·təm }

word format |COMPUT SCI| Arrangement of characters in a word, with each position or group of positions in the word containing certain specified data.   { 'wȯrd 'fȯr,mat }

word length |COMPUT SCI| The number of bits, digits, characters, or bytes in one word.   { 'wȯrd ,leŋkth }

word mark |COMPUT SCI| A nondata punctuation bit used to delimit a word in a variable-word-length computer.   { 'wȯrd ,märk }

word-oriented computer |COMPUT SCI| A computer in which the locations of words are addressed, and the bits and characters within the words can be addressed only through use of special instructions. Also known as word-addressable computer.   { 'wȯrd ¦ȯr·ē·təd kəm'pyüd·ər }

word processing |COMPUT SCI| The use of computers or computerlike equipment to write, edit, and format text.   { 'wȯrd ¦prä,ses·iŋ }

word processor |COMPUT SCI| **1.** A computer that is either dedicated to word processing or is used with a software package that supports word processing, together with a printer. **2.** A person who operates such a device.   { 'wȯrd ,prä,ses·ər }

word rate |COMPUT SCI| In computer operations, the frequency derived from the elapsed period between the beginning of the transmission of one word and the beginning of the transmission of the next word.   { 'wȯrd ,rāt }

word time *See* minor cycle.   { 'wȯrd ,tīm }

word wrap |COMPUT SCI| A procedure whereby a word processor automatically ends each line when it is full and starts the next line with the next word, never breaking a word. Also known as wrap mode.   { 'wȯrd ,rap }

work *See* load.   { wȯrk }

work assembly |COMPUT SCI| The clerical activities related to organizing collections of data records and computer programs or series of related programs.   { 'wərk ə,sem·blē }

work file |COMPUT SCI| A file created to hold data temporarily during processing.   { 'wərk ,fīl }

working program |COMPUT SCI| A valid program which, when translated into machine language, can be executed on a computer.   { 'wərk·iŋ 'prō ,gram }

working Q *See* loaded Q.   { 'wərk·iŋ 'kyü }

working set |COMPUT SCI| The smallest collection of instruction and data words of a given computer program which should be loaded into the main storage of a computer system so that efficient processing is possible.   { 'wərk·iŋ 'set }

working-set window |COMPUT SCI| A fixed time interval during which the working set is referenced.   { 'wərk·iŋ ¦set 'win·dō }

working space *See* working storage.   { 'wərk·iŋ ,spās }

working storage |COMPUT SCI| **1.** An area of main memory that is reserved by the programmer for storing temporary or intermediate values. Also known as working space. **2.** In COBOL (computer language), a section in the data division used for describing the name, structure, usage, and initial value of program variables that are neither constants nor records of input/output files.   { 'wərk·iŋ 'stȯr·ij }

working voltage *See* voltage rating.   { 'wərk·iŋ ,vōl·tij }

workspace |COMPUT SCI| In a string processing language, the portion of computer memory that contains the string currently being processed.   { 'wərk,spās }

work station |COMPUT SCI| A workplace where a person can interact with a computer on a conversational basis, either a microcomputer and printer or a terminal connected to a remote computer.   { 'wərk ,stā·shən }

work tape |COMPUT SCI| A magnetic tape that is available for general use during data processing.   { 'wərk ,tāp }

world coordinates |CONT SYS| A robotic coordinate system that is fixed with respect to the Earth.   { 'wərld kō'ȯrd·ən·əts }

world modeling |CONT SYS| Robot programming that allows the system to perform complex tasks, based on stored data.   { 'wərld 'mäd·əl·iŋ }

World Wide Web |COMPUT SCI| A part of the Internet that contains linked text, image, sound, and video documents. Abbreviated WWW. Also known as Web.   { ¦wərld ¦wīd 'web }

worm |COMPUT SCI| A computer program that seeks to replicate itself and to spread, with the goal of consuming and exhausting computer resources, thereby causing computing systems to fail.   { wərm }

WORM |COMPUT SCI| Pertaining to a storage device, such as an optical disk, that allows the user to record data only once and to read back the data an unlimited number of times. Abbreviation for write-once, read-many.   { wərm }

worst case evaluation |COMPUT SCI| A testing situation in which the most unfavorable possible combination of circumstances is evaluated. { 'wərst ¦kās i,val·yə'wā·shən }

woven-screen storage |COMPUT SCI| Digital storage plane made by weaving wires coated with thin magnetic films; when currents are sent through a selected pair of wires that are at right angles in the screen, storage and readout occur at the intersection of the two wires. { 'wō·vən ¦skrēn 'stȯr·ij }

wow |ENG ACOUS| A low-frequency flutter; when caused by an off-center hole in a disk record, occurs once per revolution of the turntable. { wau̇ }

WQP See weighted quasi peak.

wrap mode See word wrap. { 'rap ,mōd }

wrapped connection See wire-wrap connection. { 'rapt kə·'nek·shən }

writable control storage |COMPUT SCI| A section of the control storage holding microprograms which can be loaded from a console file or under microprogramming control. { 'rīd·ə·bəl kən'trōl ,stȯr·ij }

write |COMPUT SCI| 1. To transmit data from any source onto an internal storage medium. 2. A command directing that an output operation be performed. { rīt }

write enable ring |COMPUT SCI| A file protection ring that must be attached to the hub of a reel of magnetic tape in order to physically allow data to be transcribed onto the reel. Also known as write ring. { 'rīt i¦nā·bəl ,riŋ }

write error |COMPUT SCI| 1. A condition in which information cannot be written onto or into a storage device, due to dust, dirt, damage to the recording surface, or damaged electronic components. 2. A condition in which there is an inconsistency between the pattern of bits transmitted to the write head of a magnetic tape drive and the pattern sensed immediately afterward by the read head. { 'rīt ,er·ər }

write head |ELECTR| Device that stores digital information as coded electrical pulses on a storage medium, such as a disk or tape. { 'rīt ,hed }

write inhibit ring |COMPUT SCI| A file protection ring that physically prevents data from being written on a reel of magnetic tape when it is attached to the hub of the reel. { 'rīt in¦hib·ət ,riŋ }

write-once, read-many See WORM. { ¦rīt 'wəns ¦rēd 'men·ē }

write protection |COMPUT SCI| 1. Any procedure used to prevent writing on storage media. 2. Any software technique that allows a computer program to read from any area in storage but not to write outside its own area. { 'rīt prə,tek·shən }

writer |COMPUT SCI| The part of a job entry system

that controls output, in particular, the printer and the spool file. { 'rīd·ər }

write ring See write enable ring. { 'rīt ,riŋ }

write time |COMPUT SCI| The time required to transcribe a data item into a computer storage device. { 'rīt ,tīm }

write to operator |COMPUT SCI| A message issued by a computer program and displayed on the system console that provides information or indicates the status of the program and requires no action by the operator. Abbreviated WTO. { 'rīt tü 'äp·ə,rād·ər }

write to operator with reply |COMPUT SCI| A message issued by a computer program and displayed on the system console that requires action by the operator in order for execution of the program to continue. Abbreviated WTOR. { 'rīt tü 'äp·ə,rād·ər with ri'plī }

writing speed |ELECTR| Lineal scanning rate of the electron beam across the storage surface in writing information on a cathode-ray storage tube. { 'rīd·iŋ ,spēd }

WTO See write to operator.

WTOR See write to operator with reply.

Wulf electrometer |ENG| 1. A variant of the string electrometer in which charged metal plates are replaced by charged knife-edges. 2. An electrometer in which two conducting fibers are placed side by side, and their separation upon charging is measured. { ¦wu̇lf i,lek'träm·əd·ər }

Wullenweber antenna |ELECTROMAG| An antenna array consisting of two concentric circles of masts, connected to be electronically steerable; used for ground-to-air communication at Strategic Air Command bases. { 'wu̇l·ən,web·ər an ,ten·ə }

WWV |COMMUN| The call letters of a radio station maintained by the National Institute of Standards and Technology to provide standard radio and audio frequencies and other technical services, such as precision time signals and radio propagation disturbance warnings; the station broadcasts on 2.5, 5, 10, 15, and 20 megahertz 24 hours per day, 7 days per week at various times.

WWVH |COMMUN| Maintained by the National Institute of Standards and Technology, the radio station at Maui, Hawaii, broadcasting services similar to those of WWV on 5, 10, and 15 megahertz.

WWW See World Wide Web.

wye |ELEC| Polyphase circuit whose phase differences are 120° and which when drawn resembles the letter Y. |ENG| A pipe branching off a straight main run at an angle of 45°. Also known as Y; yoke. { 'wī }

wye branch See Y branch. { 'wī ,branch }

wye connection See Y network. { 'wī kə,nek· shən }

wye level See Y level. { 'wī ,lev·əl }

X

X band |COMMUN| A radio-frequency band extending from 8 to 12 gigahertz. { 'eks ,band }

xenon arc lamp |ELEC| An arc lamp filled with xenon giving a light intensity approaching that of the carbon arc; particularly valuable in projecting motion pictures. { 'zē,nän 'ärk ,lamp }

xenon flash lamp |ELEC| A flash tube containing xenon gas, which produces an intense peak of radiant energy at a wavelength of 566 nanometers when a high direct-current pulsed voltage is applied between electrodes at opposite ends of the tube. { 'zē,nän 'flash ,lamp }

XML *See* Extensible Markup Language.

XOFF *See* transmitter off. { eks'óf }

XON *See* transmitter on. { eks'ön }

XOR *See* exclusive or. { eks'ör }

x-ray generator |ELECTR| A metal from whose surface large amounts of x-rays are emitted when it is bombarded with high-velocity electrons; metals with high atomic weight are the most efficient generators. { 'eks ,rā ,jen·ə,rād·ər }

x-ray lithography |ELECTR| Lithography in which the resist is exposed to a well-collimated, high-intensity x-ray beam projected through a special mask in close proximity to the silicon slice. { 'eks ,rā li'thäg·rə·fē }

x-ray target |ELECTR| The metal body with which high-velocity electrons collide, in a vacuum tube designed to produce x-rays. { 'eks ,rā ,tär·gət }

x-ray tube |ELECTR| A vacuum tube designed to produce x-rays by accelerating electrons to a high velocity by means of an electrostatic field, then suddenly stopping them by collision with a target. { 'eks ,rā ,tüb }

XS-3 code *See* excess-three code. { 'ek,ses 'thrē ,kōd }

X server |COMPUT SCI| Software that draws the screen image and handles standard input in an X Windows System; in contrast to typical usage of the term "server"; an X server is located on the user's computer; the client is the application that is displayed, which may be located on a remote node of the network. { 'eks ,ser·vər }

X Windows System |COMPUT SCI| A graphical environment providing window management for computer applications; originally developed to provide a graphical user interface for Unix systems, it has been ported to other platforms. { ¦eks 'win,dōz ,sis·təm }

XY switching system |ELECTR| A telephone switching system consisting of a series of flat bank and wiper switches in which the wipers move in a horizontal plane, first in one direction and then in another under the control of pulses from a subscriber's dial; the switches are stacked on frames, and are operated one after another. { ¦eks¦wī 'swich·iŋ ,sis·təm }

Y

Y *See* wye.

Yagi antenna *See* Yagi-Uda antenna.   { 'yäg·ē an ˌten·ə }

Yagi-Uda antenna [ELECTROMAG] An end-fire antenna array having maximum radiation in the direction of the array line; it has one dipole connected to the transmission line and a number of equally spaced unconnected dipoles mounted parallel to the first in the same horizontal plane to serve as directors and reflectors. Also known as Uda antenna; Yagi antenna.   { 'yäg·ē 'üd·ə an ˌten·ə }

Y circulator [ELECTROMAG] Circulator in which three identical rectangular waveguides are joined to form a symmetrical Y-shaped configuration, with a ferrite post or wedge at its center; power entering any waveguide will emerge from only one adjacent waveguide.   { 'wī 'sər·kyəˌlād·ər }

Y connection *See* Y network.   { 'wī kəˌnek·shən }

Y-delta transformation [ELEC] One of two electrically equivalent networks with three terminals, one being connected internally by a Y configuration and the other being connected internally by a delta transformation. Also known as delta-Y transformation; pi-T transformation.   { 'wī 'del·tə ˌtranz·fərˌmā·shən }

yig device [ELECTR] A filter, oscillator, parametric amplifier, or other device that uses an yttrium-iron-garnet crystal in combination with a variable magnetic field to achieve wide-band tuning in microwave circuits. Derived from yttrium-iron-garnet device.   { 'yig diˌvīs }

yig filter [ELECTR] A filter consisting of an yttrium-iron-garnet crystal positioned in a magnetic field provided by a permanent magnet and a solenoid; tuning is achieved by varying the amount of direct current through the solenoid; the bias magnet serves to tune the filter to the center of the band, thus minimizing the solenoid power required to tune over wide bandwidths.   { 'yig ˌfil·tər }

yig-tuned parametric amplifier [ELECTR] A parametric amplifier in which tuning is achieved by varying the amount of direct current flowing through the solenoid of a yig filter.   { 'yig ˈtünd ˌpar·əˈme·trik 'am·pləˌfī·ər }

yig-tuned tunnel-diode oscillator [ELECTR] Microwave oscillator in which precisely controlled wide-band tuning is achieved by varying the current through a tuning solenoid that acts on a yig filter in the tunnel-diode oscillator circuit.   { 'yig ˈtünd ˈtən·əl ˈdīˌōd 'äs·əˌlād·ər }

YIQ [COMMUN] The method of representing colors in color video systems, where Y represents the luminosity, which is also the black-and-white signal, I represents red minus the luminosity, and Q represents blue minus the luminosity.

Y junction [ELECTROMAG] A waveguide in which the longitudinal axes of the waveguide form a Y.   { 'wī ˌjəŋk·shən }

Y network [ELEC] A star network having three branches. Also known as wye connection; Y connection.   { 'wī ˌnetˌwərk }

yoke [COMPUT SCI] Two or more read/write heads that are physically joined together and move as a unit over a disk, so that it is possible to read from or write to adjacent tracks without moving the head. [ELECTR] *See* deflection yoke.   { yōk }

y parameter [ELECTR] One of a set of four transistor equivalent-circuit parameters, used especially with field-effect transistors, that conveniently specify performance for small voltage and current in an equivalent circuit; the equivalent circuit is a current source with shunt impedance at both input and output.   { 'wī pə ˌram·əd·ər }

Y signal *See* luminance signal.   { 'wī ˌsig·nəl }

yttrium-iron-garnet device *See* yig device.   { ˈi·trē·əm ˈī·ərn ˈgär·nət diˌvīs }

Z

Zener breakdown [ELECTR] Nondestructive breakdown in a semiconductor, occurring when the electric field across the barrier region becomes high enough to produce a form of field emission that suddenly increases the number of carriers in this region. Also known as Zener effect.   { 'zē·nər 'brāk‚dau̇n }

Zener diode [ELECTR] A semiconductor breakdown diode, usually constructed of silicon, in which reverse-voltage breakdown is based on the Zener effect.   { 'zē·nər 'dī‚ōd }

Zener diode voltage regulator *See* diode voltage regulator.   { 'zē·nər 'dī‚ōd 'vōl·tij ‚reg·yə‚lād·ər }

Zener effect *See* Zener breakdown.

Zener voltage *See* breakdown voltage.   { 'zē·nər ‚vōl·tij }

Zepp antenna [ELECTROMAG] Horizontal antenna which is a multiple of a half-wavelength long and is fed at one end by one lead of a two-wire transmission line that is some multiple of a quarter-wavelength long.   { 'zep an‚ten·ə }

zero-access instruction [COMPUT SCI] An instruction consisting of an operation which does not require the designation of an address in the usual sense; for example, the instruction, "shift left 0003," has in its normal address position the amount of the shift desired.   { 'zir·ō ¦ak‚ses in'strək·shən }

zero-access storage [COMPUT SCI] Computer storage for which waiting time is negligible.   { 'zir·ō ¦ak‚ses 'stȯr·ij }

zero-address instruction format [COMPUT SCI] An instruction format in which the instruction contains no address, used when an address is not needed to specify the location of the operand, as in repetitive addressing. Also known as addressless instruction format.   { 'zir·ō ¦ad ‚res in¦strək·shən ‚fȯr‚mat }

zero beat [ELEC] The condition in which a circuit is oscillating at the exact frequency of an input signal, so no beat tone is produced or heard.   { 'zir·ō ‚bēt }

zero-beat reception *See* homodyne reception.   { 'zir·ō ¦bēt ri'sep·shən }

zero bias [ELECTR] The condition in which the control grid and cathode of an electron tube are at the same direct-current voltage.   { 'zir·ō 'bī·əs }

zero-bias tube [ELECTR] Vacuum tube which is designed so that it may be operated as a class B amplifier without applying a negative bias to its control grid.   { 'zir·ō ¦bī·əs 'tüb }

zero compression [COMPUT SCI] Any of a number of techniques used to eliminate the storage of nonsignificant leading zeros during data processing in a computer.   { 'zir·ō kəm'presh·ən }

zero condition [COMPUT SCI] The state of a magnetic core or other computer memory element in which it represents the value 0. Also known as nought state; zero state.   { 'zir·ō kən‚dish·ən }

zero error [ELECTR] Delay time occurring within the transmitter and receiver circuits of a radar system; for accurate range data, this delay time must be compensated for in the calibration of the radar display or other time-based determination of range.   { 'zir·ō 'er·ər }

zero-field emission *See* field-free emission current.   { 'zir·ō ¦fēld i'mish·ən }

zero fill [COMPUT SCI] To place leading zeros in the portion of a field to the left of a numeric value.   { 'zir·ō ‚fil }

zero flag [COMPUT SCI] A bit in a status register that is set to 1 to indicate that another register in the central processing unit contains all zeros or that two compared values are equal, and is set to 0 to indicate the contrary.   { 'zir·ō ‚flag }

zero-gravity switch [ELEC] A switch that closes as weightlessness or zero gravity is approached; in one version, a conductive sphere of mercury encompasses two contacts at zero gravity but flattens away from the upper contact under the influence of gravity. Also known as weightlessness switch.   { 'zir·ō ¦grav·əd·ē ‚swich }

zero level [ENG ACOUS] Reference level used for comparing sound or signal intensities; in audio-frequency work, a power of 0.006 watt is generally used as zero level; in sound, the threshold of hearing is generally assumed as the zero level.   { 'zir·ō ‚lev·əl }

zero-level address [COMPUT SCI] The operand contained in an instruction so structured as to make immediate use of the operand.   { 'zir·ō ¦lev·əl 'ad‚res }

zero method *See* null method.   { 'zir·ō ‚meth·əd }

zero-order hold [CONT SYS] A device which converts a sampled output into an output which is held constant between samples at the last sampled value.   { 'zir·ō ¦ȯrd·ər 'hōld }

zero output [ELECTR] **1.** Voltage response obtained from a magnetic cell in a zero state by a reading or resetting process. **2.** Integrated voltage response obtained from a magnetic cell in a zero state by a reading or resetting process; a ratio of a one output to a zero output is a one-to-zero ratio. { 'zir·ō 'aùt,pùt }

zero phase-sequence relay |ELEC| Relay which functions in conformance with the zero phase-sequence component of the current, voltage, or power of the circuit. { 'zir·ō ¦fāz ¦sē·kwəns 'rē ,lā }

zero potential |ELEC| Expression usually applied to the potential of the earth, as a convenient reference for comparison. { 'zir·ō pə'ten·chəl }

zero state *See* zero condition. { 'zir·ō ,stāt }

zero subcarrier chromaticity |COMMUN| Chromaticity, in color television, which is intended to be displayed when the subcarrier amplitude is zero. { 'zir·ō səb¦kar·ē·ər ,krō·mə'tis·əd·ē }

zero suppression |COMPUT SCI| A process of replacing leading (nonsignificant) zeros in a numeral by blanks; it is an editing operation designed to make computable numerals easily readable to the human eye. { 'zir·ō sə'presh·ən }

zero time reference |ELECTR| Reference point in time from which the operations of various radar circuits are measured. { 'zir·ō 'tīm ,ref·rəns }

zigzag reflections |ELECTROMAG| From a layer of the ionosphere, high-order multiple reflections which may be of abnormal intensity; they occur in waves which travel by multihop ionosphere reflections and finally turn back toward their starting point by repeated reflections from a slightly curved or sloping portion of an ionized layer. { 'zig,zag ri,flek·shənz }

zinc/air cell |ELEC| A primary battery with very high energy density that uses atmospheric oxygen as the active cathode material and zinc as the anode; cans containing the anode and cathode are separated by an annular insulator. { ¦ziŋk 'er ,sel }

zinc/alkaline/manganese dioxide battery |ELEC| The most widely used primary battery, with an anode consisting of zinc powder, an electrolyte consisting of an aqueous solution of potassium hydroxide, and a cathode consisting of a blend of electrolytic manganese dioxide and graphite. { ¦ziŋk ¦al·kə,līn ¦maŋ·gə,nēs dī'äk,sīd ,bad·ə·rē }

zinc/bromine battery |ELEC| A storage battery that is based on the reaction of zinc (Zn) and bromine (Br₂) to form zinc bromide (ZnBr₂), and that uses liquid electrolytes which circulate through stacks of bipolar cells. { ¦ziŋk ¦brō,mēn 'bad·ə·rē }

zinc-silver chloride primary cell |ELEC| A reserve primary cell that is activated by adding water; it can have a high capacity, up to 40 watt-hours per pound, and long life after activation. { 'ziŋk 'sil·vər ¦klór,īd 'prī,mer·ē 'sel }

zip |COMPUT SCI| Open standard for file compression and decompression used with personal computers. { zip }

zirconium lamp |ELECTR| A high-intensity point-source lamp having a zirconium oxide cathode in an argon-filled bulb, used because of its low emanation of long-wavelength light and its concentrated source. { ,zər'kō·nē·əm 'lamp }

Ziv-Lempel compression |COMPUT SCI| A data compression technique in which data is represented by a sequence of numbers standing for the positions of character strings in a dictionary; this dictionary initially contains every character in the alphabet and is continually enlarged by forming new strings from the string just compressed and the upcoming character in the text. { 'ziv 'lem·pəl kəm'presh·ən }

zone |COMPUT SCI| **1.** One of the top three rows of a punched card, namely, the 11, 12, and zero rows. **2.** *See* storage area. { zōn }

zone bit |COMPUT SCI| One of a set of bits used to indicate some grouping of characters. { 'zōn ,bit }

zone blanking |ELECTR| Method of turning off the cathode-ray tube during part of the sweep of an antenna. { 'zōn 'blaŋk·iŋ }

zoned decimal |COMPUT SCI| A format for use with EBCDIC input and output in which the first four bits of each character are called the zone portion, the second four bits, called the data portion, contain a hexadecimal digit, and the zone portion of the lowest-order character may indicate the sign of an integer; thus, + 1234 (in hexadecimal notation) would be represented as 1111/0001/1111/0010/1111/0011/1100/0100. { 'zōnd 'des·məl }

zoning |ELECTROMAG| The displacement of various portions of the lens or surface of a microwave reflector so the resulting phase front in the near field remains unchanged. Also known as stepping. { 'zōn·iŋ }

Z parameter |ELECTR| One of a set of four transistor equivalent-circuit parameters; they are the inverse of the Y parameters. { 'zē pə,ram·əd·ər }

z-transfer function *See* pulsed transfer function. { 'zē 'tranz·fər ,faŋk·shən }

z-transform |MATH| The z-transform of a sequence whose general term is f_n is the sum of a series whose general term is $f_n z^{-n}$, where z is a complex variable; n runs over the positive integers for a one-sided transform, over all the integers for a two-sided transform. { 'zē 'tranz ,fòrm }

Z variometer *See* vertical intensity variometer. { 'zē ,ver·ē'äm·əd·ər }

Appendix

Equivalents of commonly used units for the U.S. Customary System and the metric system

1 inch = 2.5 centimeters (25 millimeters)	1 centimeter = 0.4 inch	1 inch = 0.083 foot
1 foot = 0.3 meter (30 centimeters)	1 meter = 3.3 feet	1 foot = 0.33 yard (12 inches)
1 yard = 0.9 meter	1 meter = 1.1 yards	1 yard = 3 feet (36 inches)
1 mile = 1.6 kilometers	1 kilometer = 0.62 mile	1 mile = 5280 feet (1760 yards)
1 acre = 0.4 hectare	1 hectare = 2.47 acres	
1 acre = 4047 square meters	1 square meter = 0.00025 acre	
1 gallon = 3.8 liters	1 liter = 1.06 quarts = 0.26 gallon	1 quart = 0.25 gallon (32 ounces; 2 pints)
1 fluid ounce = 29.6 milliliters	1 milliliter = 0.034 fluid ounce	1 pint = 0.125 gallon (16 ounces)
32 fluid ounces = 946.4 milliliters		1 gallon = 4 quarts (8 pints)
1 quart = 0.95 liter	1 gram = 0.035 ounce	1 ounce = 0.0625 pound
1 ounce = 28.35 grams	1 kilogram = 2.2 pounds	1 pound = 16 ounces
1 pound = 0.45 kilogram	1 kilogram = 1.1×10^{-3} ton	1 ton = 2000 pounds
1 ton = 907.18 kilograms		
$°F = (1.8 \times °C) + 32$	$°C = (°F - 32) \div 1.8$	

Appendix

Conversion factors for the U.S. Customary System, metric system, and International System

A. Units of length

Units	cm	m	in.	ft	yd	mi
1 cm = 1	0.01	0.3937008	0.03280840	0.01093613	6.213712×10^{-6}
1 m = 100.	1	39.37008	3.280840	1.093613	6.213712×10^{-4}
1 in. = 2.54	0.0254	1	0.08333333....	0.02777777....	1.578283×10^{-5}
1 ft = 30.48	0.3048	12.	1	0.3333333....	$1.893939... \times 10^{-4}$
1 yd = 91.44	0.9144	36.	3.	1	$5.68181... \times 10^{-4}$
1 mi = 1.609344×10^5	1.609344×10^3	6.336×10^4	5280.	1760.	1

B. Units of area

Units	cm²	m²	in.²	ft²	yd²	mi²
1 cm² = 1	10^{-4}	0.1550003	1.076391×10^{-3}	1.195990×10^{-4}	3.861022×10^{-11}
1 m² = 10^4	1	1550.003	10.76391	1.195990	3.861022×10^{-7}
1 in.² = 6.4516	6.4516×10^{-4}	1	$6.944444... \times 10^{-3}$	7.716049×10^{-4}	2.490977×10^{-10}
1 ft² = 929.0304	0.09290304	144.	1	0.1111111...	3.587007×10^{-8}
1 yd² = 8361.273	0.8361273	1296.	9.	1	3.228306×10^{-7}
1 mi² = 2.589988×10^{10}	2.589988×10^6	4.014490×10^9	2.78784×10^7	3.0976×10^6	1

Appendix

C. Units of volume

Units	m^3	cm^3	liter	$in.^3$	ft^3	qt	gal
1 m^3	$=1$	10^6	10^3	6.102374×10^4	35.31467×10^{-3}	1.056688	264.1721
1 cm^3	$=10^{-6}$	1	10^{-3}	0.06102374	3.531467×10^{-5}	1.056688×10^{-3}	2.641721×10^{-4}
1 liter	$=10^{-3}$	1000.	1	61.02374	0.03531467	1.056688	0.2641721
1 $in.^3$	$=1.638706 \times 10^{-5}$	16.38706	0.01638706	1	5.787037×10^{-4}	0.01731602	4.329004×10^{-3}
1 ft^3	$=2.831685 \times 10^{-2}$	28316.85	28.31685	1728.	1	2.992208	7.480520
1 qt	$=9.463529 \times 10^{-4}$	946.3529	0.9463529	57.75	0.03342014	1	0.25
1 gal (U.S.)	$=3.785412 \times 10^{-3}$	3785.412	3.785412	231.	0.1336806	4.	1

D. Units of mass

Units	g	kg	oz	lb	metric ton	ton
1 g	$=1$	10^{-3}	0.03527396	2.204623×10^{-3}	10^{-6}	1.102311×10^{-6}
1 kg	$=1000.$	1	35.27396	2.204623	10^{-3}	1.102311×10^{-3}
1 oz (avdp)	$=28.34952$	0.02834952	1	0.0625	2.834952×10^{-5}	3.125×10^{-5}
1 lb (avdp)	$=453.5924$	0.4535924	16.	1	4.535924×10^{-4}	$5. \times 10^{-4}$
1 metric ton	$=10^6$	1000.	35273.96	2204.623	1	1.102311
1 ton	$=907184.7$	907.1847	32000.	2000.	0.9071847	1

Conversion factors for the U.S. Customary System, metric system, and International System (cont.)

E. Units of density

Units	$g \cdot cm^{-3}$	$g \cdot L^{-1}, kg \cdot m^{-3}$	$oz \cdot in.^{-3}$	$lb \cdot in.^{-3}$	$lb \cdot ft^{-3}$	$lb \cdot gal^{-1}$
1 g·cm⁻³ = 1	1000.	0.5780365	0.03612728	62.42795	8.345403
1 g·L⁻¹, kg·m⁻³ = 10^{-3}	1	5.780365×10^{-4}	3.612728×10^{-5}	0.06242795	8.345403×10^{-3}
1 oz·in.⁻³ = 1.729994	1729.994	1	0.0625	108.	14.4375
1 lb·in.⁻³ = 27.67991	27679.91	16.	1	1728.	231.
1 lb·ft⁻³ = 0.01601847	16.01847	9.259259×10^{-3}	5.787037×10^{-4}	1	0.1336806
1 lb·gal⁻¹ = 0.1198264	119.8264	4.749536×10^{-3}	4.329004×10^{-3}	7.480519	1

F. Units of force

Units	N	dyn	kgf	$poundal$	lbf	$ton\text{-}force$
1 N = 1	10^5	0.1019716	7.233014	0.2248089	1.124045×10^{-4}
1 dyn = 10^{-5}	1	1.019716×10^{-6}	7.233014×10^{-5}	2.248089×10^{-6}	1.124045×10^{-9}
1 kgf = 9.80665	980665.	1	70.93164	2.204623	1.102311×10^{-3}
1 poundal = 0.1382550	13825.50	1.409808×10^{-2}	1	0.03108095	1.554048×10^{-5}
1 lbf = 4.448222	444822.2	0.4535924	32.17405	1	$5. \times 10^{-4}$
1 ton-force = 8896.443	8.896443×10^8	907.1847	64348.10	2000.	1

Appendix

G. Units of pressure

Units	Pa, 1 N·m⁻²	dyn·cm⁻²	bar	atm	kgf·cm⁻²	mmHg (torr)	in. Hg	lbf·in.⁻²
1 Pa, 1 $N \cdot m^{-2}$ =	1	10	10^{-5}	9.869233×10^{-6}	1.019716×10^{-5}	7.500617×10^{-3}	2.952999×10^{-4}	1.450377×10^{-4}
1 $dyn \cdot cm^{-2}$ =	0.1	1	10^{-6}	9.869233×10^{-7}	1.019716×10^{-6}	7.500617×10^{-4}	2.952999×10^{-5}	1.450377×10^{-5}
1 bar =	10^5	10^6	1.	0.9869233	1.019716	750.0617	29.52999	14.50377
1 atm =	101325	1013250	1.01325	1	1.033227	760.	29.92126	14.69595
1 $kgf \cdot cm^{-2}$ =	98066.5	980665	0.980665	0.9678411	1	735.5592	28.95903	14.22334
1 mmHg (torr) =	133.3224	1333.224	1.333224×10^{3}	1.315789×10^{-3}	1.359510×10^{-3}	1	0.03937008	0.01933678
1 in. Hg =	3386.388	33863.88	0.03386388	0.03342105	0.03453155	25.4	1	0.4911541
1 $lbf \cdot in.^{-2}$ =	6894.757	68947.57	0.06894757	0.06804596	0.07030696	51.71493	2.036021	1

H. Units of energy

Units	g mass (energy equiv)	J	eV	cal	cal_IT	Btu_IT	kWh	hp·h	ft·lbf	ft³·lbf·in.⁻²	liter·atm
1 g mass (energy equiv) =	1	8.987552×10^{13}	5.609589×10^{32}	2.148076×10^{13}	2.146640×10^{13}	8.518555×10^{10}	2.496542×10^{7}	3.347918×10^{7}	6.628878×10^{13}	4.603388×10^{11}	8.870024×10^{11}
1 J =	1.112650×10^{-14}	1	6.241509×10^{18}	0.2390057	0.2388459	9.478172×10^{-4}	$2.777777\ldots \times 10^{-7}$	3.725062×10^{-7}	0.7375622	5.121960×10^{-3}	9.869233×10^{-3}
1 eV =	1.782662×10^{-33}	1.602177×10^{-19}	1	3.829294×10^{-20}	3.826733×10^{-20}	1.518570×10^{-22}	4.450490×10^{-26}	5.968206×10^{-26}	1.181705×10^{-19}	8.206283×10^{-22}	1.581225×10^{-21}
1 cal =	4.655328×10^{-14}	4.184	2.611448×10^{19}	1	0.9993312	3.965667×10^{-3}	$1.1622222\ldots \times 10^{-6}$	1.558562×10^{-6}	3.085960	2.143028×10^{-2}	0.04129287

Conversion factors for the U.S. Customary System, metric system, and International System (cont.)

H. Units of energy (cont.)

Units	g mass (energy equiv)	J	eV	cal	cal_IT	Btu_IT	kWh	hp-h	ft-lbf	ft³·lbf·in.⁻²	liter-atm
1 cal_IT	$= 4.658443 \times 10^{-14}$	4.1868	2.613195×10^{19}	1.000669	1	3.968321×10^{-3}	1.163×10^{-6}	1.559609×10^{-6}	3.088025	2.144462×10^{-2}	0.04132050
1 Btu_IT	$= 1.173908 \times 10^{-11}$	1055.056	6.585141×10^{21}	252.1644	251.9958	1	2.930711×10^{-4}	3.930148×10^{-4}	778.1693	5.403953	10.41259
1 kWh	$= 4.005540 \times 10^{-8}$	3600000.*	2.246943×10^{25}	860420.7	859845.2	3412.142	1	1.341022	2655224.	18349.06	35529.24
1 hp-h	$= 2.986931 \times 10^{-8}$	2384519.	1.675545×10^{25}	641615.6	641186.5	2544.33	0.7456998	1	1980000.	13750.	26494.15
1 ft-lbf	$= 1.508551 \times 10^{-14}$	1.355818	8.462351×10^{18}	0.3240483	0.3238315	1.285067×10^{-3}	3.766161×10^{-7}	$5.050505\ldots \times 10^{-7}$	1	$6.944444\ldots \times 10^{-3}$	0.01338088
1 ft³·lbf·in.⁻²	$= 2.172313 \times 10^{-12}$	195.2378	1.218578×10^{21}	46.66295.	46.63174	0.1850497	5.423272×10^{-5}	$7.272727\ldots \times 10^{-5}$	144.*	1	1.926847
1 liter-atm	$= 1.127393 \times 10^{-12}$	101.325	6.324209×10^{20}	24.21726	24.20106	0.09603757	2.814583×10^{-5}	3.774419×10^{-5}	74.73349	0.5189825	1

I. Units of power

Units	W	kW	Btu_IT/h	ft-lbf/s	hp	metric hp
1 W	$= 1$	10^{-3}	3.412142	0.7375621	1.341022×10^{-3}	1.359622×10^{-3}
1 kW	$= 10^3$	1	3412.142	737.5621	1.341022	1.359622
1 Btu_IT/h	$= 0.2930711$	2.930711×10^{-4}	1	0.2161581	3.930148×10^{-4}	3.984658×10^{-4}
1 ft-lbf/s	$= 1.355818$	1.355818×10^{-3}	4.626243	1	$1.818181\ldots \times 10^{-3}$	1.843399×10^{-3}
1 hp	$= 745.6999$	0.7456999	2544.434	550.	1	1.013870
1 metric hp	$= 735.4988$	0.7354988	2509.626	542.4760	0.9863201	1

Standard equations

Coulomb's law
$$F = \frac{1}{4\pi\varepsilon_0}\frac{q_1 q_2}{r^2}$$

Maxwell's equations:
Gauss's law for electrostatics $\nabla \cdot \mathbf{D} = \rho$
Gauss's law for magnetostatics $\nabla \cdot \mathbf{B} = 0$

Faraday's law of induction $\nabla \times \mathbf{E} = -\dfrac{\partial \mathbf{B}}{\partial t}$

Ampère's law $\nabla \times \mathbf{H} = \mathbf{J} + \dfrac{\partial \mathbf{D}}{\partial t}$

Lorentz force equation $\mathbf{F} = q\mathbf{v} \times \mathbf{B} + q\mathbf{E}$
Polarization $\mathbf{P} = \mathbf{D} - \varepsilon_0 \mathbf{E}$

Magnetization $\mathbf{M} = \dfrac{\mathbf{B}}{\mu_0} - \mathbf{H}$

Curie's law $M = K\dfrac{B}{T}$

Poynting vector $\mathbf{S} = \dfrac{1}{\mu_0}\mathbf{E} \times \mathbf{B}$

Velocity-wavelength relationship $v = \lambda \nu$
Energy of light $e = h\nu$

Planck's radiation formula $p_{\text{radiated}} = \displaystyle\int_0^\bullet \frac{8\pi h\nu^3}{c^3 (e^{h\nu/kT} - 1)}\,d\nu$

Stefan-Boltzmann equation $p_{\text{radiated}} = \sigma T^4$
Ohm's law $V = IR \;\; or \;\; \mathbf{J} = \sigma \mathbf{E}$

Kirchhoff's current law $\displaystyle\sum_{n=1}^{N} I_n = 0$ at a node

Kirchhoff's voltage law $\displaystyle\sum_{m=1}^{M} V_m = 0$ around a loop

Instantaneous power equation $p(t) = V(t)I(t)$

Tellegen's theorem for an isolated circuit $p(t) = \displaystyle\sum_{n=1}^{M} V_n(t)I_n(t) = 0$

Thévenin-Norton theorem $V_{\text{Thevenin}} = I_{\text{Norton}}R_{\text{equivalent}}$ or $V_{OC} = I_{SC}R_{eq}$

Current-charge relationship $I(t) = \dfrac{dq(t)}{dt}$

Power-energy relationship $p(t) = \dfrac{de(t)}{dt}$

Current-voltage relationship for a capacitor $I(t) = C\dfrac{dV(t)}{dt}$

Energy stored in a capacitor $e(t) = \dfrac{1}{2}CV^2(t)$

Current-voltage relationship for an inductor $V(t) = L\dfrac{dI(t)}{dt}$

Energy stored in an inductor $e(t) = \dfrac{1}{2}LI^2(t)$

Complex AC power $S = VI^* = |S| \angle \phi = P_{\text{avg}} + jQ = I_{\text{RMS}}^2 R + jI_{\text{RMS}}^2 X$

Joule's law for a resistor $P_{\text{avg}} = I_{\text{RMS}}^2 R$ = heat rate

Flux-field relationship $B = \dfrac{\phi}{A}$

Geometric resistance relationship $R = \dfrac{\rho L}{A}$

Geometric capacitance relationship $C = \dfrac{\varepsilon A}{d}$

Appendix

Standard equations (cont.)

Geometric inductance relationship	$L = \dfrac{N\phi}{I}$
Resonant frequency	$\omega_r = \dfrac{1}{\sqrt{LC}}$
Transformer voltage gain equation	$V_{2_{RMS}} = \dfrac{n_2}{n_1} V_{1_{RMS}}$
RMS voltage	$V_{RMS} = \sqrt{\dfrac{1}{V} \int_0^T V^2(t)\, dt}$
Diode equation	$I(t) = I_S \left(e^{qV(t)/kT} - 1\right) \cong I_S\left(e^{40V(t)} - 1\right)$
BJT current gain equation	$I_C(t) = \beta I_B(t)$
Einstein equation for Brownian motion of charged particles	$\dfrac{\mu}{D} = \dfrac{e}{kT}$
Faraday's law of electrochemistry	$m(t) = \dfrac{w}{nF}\int_0^t I(\tau)\, d\tau$
Butler-Vollmer electrode equation	$I(t) = nFAk^0 \left(c_o e^{-\alpha nF(V(t)-V^o)/RT} - c_R e^{(1-\alpha)nF(V(t)-V^o)/RT}\right)$
Newton's law of cooling	$Q(t) = -K\dfrac{dT(t)}{dt}$
Nyquist-Shannon sampling theorem	$\omega_{sampling} \geq 2\omega_{system}$
Parseval's theorem	$\int_{-\infty}^{+\infty} |f(t)|^2\, dt = \dfrac{1}{2\pi}\int_{-\infty}^{+\infty} |F(j\omega)|^2\, d\omega$

Standard equations symbology

A = area
B = magnitude of magnetic flux density
B = magnetic flux density
c = speed of light = 299,792,458 m/s
c_R, c_o = surface concentrations of the redox partners at the electrodes
C = capacitance
d = separation of capacitor plates
d/dt = differentiation with respect to time
D = diffusion coefficient
D = electric displacement
e = energy [in energy of light]
e = 2.71828... [in Planck's radiation formula; Diode equation; Butler-Vollmer electrode equation]
e = electron charge $\cong 1.602 \times 10^{-19}$ C [in Einstein equation for Brownian motion of charged particles]
$e(t)$ = instantaneous value of energy at time t
E = electric field strength
$f(t)$ = value of function f of a real variable t
F = magnitude of force between electric charges q_1 and q_2 [in Coulomb's law]
F = Faraday constant $\cong 96,485$ C/mol [in Faraday's law of electrochemistry; Butler-Vollmer electrode equation]
F = force acting on charge q
$F(j\omega)$ = Fourier transform of $f(t)$
h = Planck's constant $\cong 6.626 \times 10^{-34}$ J · s
H = magnetic field strength or field intensity
I = electric current
$I(t), I(\tau)$ = instantaneous value of electric current at time t or τ

Standard equations symbology (*cont.*)

I_R = electric current flowing towards (or negative of electric current flowing away from) a particular node in an electric network through branch n, adjacent to that node

$I_n(t)$ = instantaneous value of electric current flowing through branch n of an electric network at time t

$I_{\text{Norton}} = I_{SC}$ = the short-circuit current, obtained by shorting the terminals looking into a one-port circuit

I^* = complex conjugate of electric current phasor

I_{RMS} = root-mean-square current (square root of the mean value of the square of the electric current, averaged over one cycle in an alternating-current circuit)

I_S = reverse bias saturation current of a diode

$I_B(t)$ = instantaneous value of the base current of a bipolar junction transistor at time t

$I_C(t)$ = instantaneous value of the collector current of a bipolar junction transistor at time t

j = square root of -1

$\mathbf{J}$ = electric current density

k = Boltzmann's constant $\cong 1.38 \times 10^{-23}$ J/K

k^0 = standard rate constant

K = Curie constant of a material |in Curie's law|

K = constant of proportionality |in Newton's law of cooling|

L = inductance |in current-voltage relationship for an inductor; energy stored in an inductor; geometric inductance relationship; resonant frequency|

L = length of a resistor |in geometric resistance relationship|

$m(t)$ = quantity of substance produced during electrolysis up to time t

M = magnitude of magnetization

$\mathbf{M}$ = magnetization

n = number of electrons acquired or released per molecule during electrolysis

n_1, n_2 = number of turns in the primary and secondary windings of a transformer

$p(t)$ = instantaneous value of power at time t

P_{radiated} = total power radiated, per unit area, from surface of a blackbody

$\mathbf{P}$ = dielectric polarization

P_{avg} = power averaged over one cycle in an alternating-current circuit

q = electric charge |in Lorentz force equation|

q = electron charge $\cong 1.602 \times 10^{-19}$ C |in diode equation|

$q(t)$ = instantaneous value of electric charge at time t

q_1, q_2 = electric charges 1 and 2

Q = reactive power in an alternating-current circuit (units are VAR)

$Q(t)$ = instantaneous value of total heat flow out of a body at time t

r = distance between electric charges q_1 and q_2

R = electrical resistance |in Ohm's law; complex AC power; geometric resistance relationship|

R = molar gas constant $\cong 8.31$ J/(mol·K) |in Butler-Vollmer electrode equation|

$R_{\text{equivalent}} = R_{\text{eq}}$ = the equivalent resistance looking into the terminals of a one-port circuit obtained by zeroing all the sources

$\mathbf{S}$ = Poynting vector (density of energy flow associated with an electromagnetic field)

S = complex alternating-current power

$|S|$ = magnitude of complex alternating-current power, usually referred to as the apparent power (units are VA)

t = time

T = absolute (thermodynamic) temperature |in Curie's law; Planck's radiation formula; Stefan-Boltzmann equation; diode equation; Einstein equation for Brownian motion of charged particles; Faraday's law of electrochemistry; Butler-Vollmer electrode equation; Newton's law of cooling|

T = period of an alternating-current circuit |in RMS voltage|

$T(t)$ = instantaneous value of temperature at time t

$\partial/\partial t$ = partial differentiation with respect to time

v = speed of wave motion

$\mathbf{v}$ = velocity of charge q

V = voltage or potential difference |in Ohm's law|

V = voltage phasor |in complex AC power|

$V(t)$ = instantaneous value of voltage (potential difference) at time t

V_m = voltage (potential difference) across one of the branches, m, forming a closed loop in an electric network, the voltages all being taken in the same direction around the loop

$V_n(t)$ = instantaneous value of voltage (potential difference) across branch n in an electric network at time t

Appendix

$V_{\text{Thevenin}} = V_{OC}$ = the open-circuit voltage looking into the terminals of a one-port circuit

V_{RMS} = root-mean-square voltage

$V_{1_{\text{RMS}}}, V_{2_{\text{RMS}}}$ = root-mean-square voltages across the primary and secondary windings of a transformer

V^0 = standard potential

w = atomic weight of accumulated material

X = reactance

α = transfer coefficient ($0 < \alpha < 1$; usually $\alpha \approx 0.5$)

β = base-to-collector current gain of a bipolar junction transistor

ε = permittivity of the dielectric in a capacitor

ε_0 = electric constant (permittivity of free space)

λ = wavelength

μ = electron mobility

μ_0 = magnetic constant (permeability of free space)

ν = frequency

$\pi = 3.14159...$

ρ = electric charge density

σ = Stefan-Boltzmann constant $\cong 5.67 \times 10^{-8}$, when p_{radiated} is expressed in W/m² and T is expressed in kelvins (K) |in Stefan-Boltzmann equation|

σ = electric conductivity |in Ohm's law|

τ = time of electrolysis

ϕ = magnetic flux |in flux-field relationship; geometric inductance relationship|

$\angle\phi = e^{j\phi}$, where $e = 2.71828...$, j = square root of -1, and ϕ = angle of the complex alternating-current power; it turns the phasor S by the angle ϕ in the positive (counterclockwise) direction

ω_r = resonant frequency

ω_{sampling} = sampling frequency

ω_{system} = highest frequency of the sampled system or signal

$\nabla\cdot$ = divergence operator

$\nabla\times$ = curl operator

$\times$ = cross product of vectors

$\int_0^\infty d\nu$ = integration over the variable ν (frequency) from 0 to ∞

$\int_0^T dt$ = integration over the variable t (time) from 0 to T (the period of an alternating-current circuit)

$\int_0^t d\tau$ = integration over the variable τ (time of electrolysis) from 0 to t (the duration of electrolysis)

$\int_{-\infty}^{+\infty} dt$ = integration over the variable t from $-\infty$ to $+\infty$

$\int_{-\infty}^{+\infty} d\omega$ = integration over the variable ω from $-\infty$ to $+\infty$

$\sum_{n=1}^{N}$ = summation of a quantity indexed by a discrete variable n (here labeling the branches of an electric network adjacent to a particular node |in Kirchhoff's current law| or all the branches in a network |in Tellegen's theorem for an isolated circuit|) from $n = 1$ to $n = N$

$\sum_{m=1}^{M}$ = summation of a quantity indexed by a discrete variable m (here labeling branches that form a closed loop in an electric network) from $m = 1$ to $m = M$

Special constants

$\pi = 3.14159\ 26535\ 89793\ 23846\ 2643\ ...$

$e = 2.71828\ 18284\ 59045\ 23536\ 0287\ ... = \lim_{n\to\infty}\left(1 + \frac{1}{n}\right)^n$

= natural base of logarithms

$\sqrt{2} = 1.41421\ 35623\ 73095\ 0488\ ...$

$\sqrt{3} = 1.73205\ 08075\ 68877\ 2935\ ...$

$\sqrt{5} = 2.23606\ 79774\ 99789\ 6964\ ...$

$\sqrt[3]{2} = 1.25992\ 1050\ ...$

$\sqrt[3]{3} = 1.44224\ 9570\ ...$

$\sqrt[5]{2} = 1.14869\ 8355\ ...$

$\sqrt[5]{3} = 1.24573\ 0940\ ...$

$e^\pi = 23.14069\ 26327\ 79269\ 006\ ...$

$\pi^e = 22.45915\ 77183\ 61045\ 47342\ 715\ ...$

$e^e = 15.15426\ 22414\ 79264\ 190\ ...$

$\log_{10} 2 = 0.30102\ 99956\ 63981\ 19521\ 37389\ ...$

$\log_{10} 3 = 0.47712\ 12547\ 19662\ 43729\ 50279\ ...$

$\log_{10} e = 0.43429\ 44819\ 03251\ 82765\ ...$

$\log_{10} \pi = 0.49714\ 98726\ 94133\ 85435\ 12683\ ...$

$\log_e 10 = \ln 10 = 2.30258\ 50929\ 94045\ 68401\ 7991\ ...$

$\log_e 2 = \ln 2 = 0.69314\ 71805\ 59945\ 30941\ 7232\ ...$

$\log_e 3 = \ln 3 = 1.09861\ 22886\ 68109\ 69139\ 5245\ ...$

$\gamma = 0.57721\ 56649\ 01532\ 86060\ 6512\ ... = $ Euler's constant

$= \lim_{n\to\infty}\left(1 + \frac{1}{2} + \frac{1}{3} + \cdots + \frac{1}{n} - \ln n\right)$

$e^\lambda = 1.78107\ 24179\ 90197\ 9852\ ...$

$\sqrt{e} = 1.64872\ 12707\ 00128\ 1468\ ...$

$\sqrt{\pi} = \Gamma(\frac{1}{2}) = 1.77245\ 38509\ 05516\ 02729\ 8167\ ...$

where Γ is the gamma function

$\Gamma(\frac{1}{3}) = 2.67893\ 85347\ 07748\ ...$

$\Gamma(\frac{1}{4}) = 3.62560\ 99082\ 21908\ ...$

1 radian $= 180°/\pi = 57.29577\ 95130\ 8232\ ...°$

$1° = \pi/180$ radians $= 0.01745\ 32925\ 19943\ 29576\ 92\ ...$ radians

SOURCE: Murray R. Spiegel and John Liu. *Mathematical Handbook of Formulas and Tables*, 2d ed., Schaum's Outline Series, McGraw-Hill, 1999.

Appendix

Recommended values (2002) of selected fundamental physical constants

Quantity	Symbol*	Numerical value†	Units	Relative uncertainty (standard deviation)
Speed of light in vacuum	c	299792458	m/s	(defined)
Permeability of vacuum	μ_0	$4\pi \times 10^{-7}$	N/A²	(defined)
Permittivity of vacuum	ε_0	8.854187817...	10^{-12} F/m	(defined)
Constant of gravitation	G	6.6742 (10)	10^{-11} m³/(kg · s²)	1.5×10^{-4}
Planck constant	$\hbar$	6.6260693 (11)	10^{-34} J · s	1.7×10^{-7}
Elementary charge	e	1.60217653 (14)	10^{-19} C	8.5×10^{-8}
Magnetic flux quantum, $h/(2e)$	Φ_0	2.06783372 (18)	10^{-15} Wb	8.5×10^{-8}
Fine-structure constant, $\mu_0 c e^2/(2h)$	α	7.297352568 (24)	10^{-3}	3.3×10^{-9}
	α^{-1}	137.03599911 (46)		3.3×10^{-9}
Electron mass	m_e	9.1093826 (16)	10^{-31} kg	1.7×10^{-7}
Proton mass	m_p	1.67262171 (29)	10^{-27} kg	1.7×10^{-7}
Neutron mass	m_n	1.67492728 (29)	10^{-27} kg	1.7×10^{-7}
Proton-electron mass ratio	m_p/m_e	1836.15267261 (85)		4.6×10^{-10}
Rydberg constant, $m_e c \alpha^2/(2h)$	R_∞	10973731.568525 (73)	m⁻¹	6.6×10^{-12}
Bohr radius, $\alpha/(4\pi R_\infty)$	a_0	5.291772108 (18)	10^{-11} m	3.3×10^{-9}
Compton wavelength of the electron, $h/(m_e c) = \alpha^2/(2R_\infty)$	λ_c	2.426310238 (16)	10^{-12} m	6.7×10^{-9}
Classical electron radius, $\mu_0 e^2/(4\pi m_e) = \alpha^3/(4\pi R_\infty)$	r_e	2.817940325 (28)	10^{-15} m	1.0×10^{-8}
Bohr magneton, $e\hbar/(4\pi m_e)$	μ_B	9.27400949 (80)	10^{-24} J/T	8.6×10^{-8}
Electron magnetic moment	μ_e	-9.28476412 (80)	10^{-24} J/T	8.6×10^{-8}
Electron magnetic moment/Bohr magneton ratio	μ_e/μ_B	-1.0011596521859 (38)		3.8×10^{-12}
Nuclear magneton, $e\hbar/(4\pi m_p)$	μ_N	5.05078343 (43)	10^{-27} J/T	8.6×10^{-8}
Proton magnetic moment/nuclear magneton ratio	μ_p/μ_N	2.792847351 (28)		1.0×10^{-8}
Avogadro constant	N_A	6.0221415 (10)	10^{23}	1.7×10^{-7}
Faraday constant, $N_A e$	F	96485.3383 (83)	C/mol	8.6×10^{-8}
Molar gas constant	R	8.314472 (15)	J/(mol · K)	1.7×10^{-6}
Boltzmann constant, R/N_A	k	1.3806505 (24)	10^{-23} J/K	1.7×10^{-6}

*A = ampere, C = coulomb, F = farad, J = joule, K = kelvin, kg = kilogram, m = meter, mol = mole, N = newton, s = second, T = tesla, Wb = weber.
†Recommended by CODATA Task Group on Fundamental Constants. Digits in parentheses represent one-standard-deviation uncertainties in final two digits of quoted value.

Electrical and magnetic units

Quantity	Unit and symbol	Derivation	Common conversions
SI base units			
Mass	kilogram, kg		
Time	second, s		
Length	meter, m		
Electric current	ampere, A		
Thermodynamic temperature	kelvin, K		
Luminous intensity	candela, cd		
Amount of substance	mole, mol		
Derived units			
Energy	joule, J	$N \cdot m = m^2 \cdot kg \cdot s^{-2}$	
Power	watt, W	$J \cdot s^{-1} = m^2 \cdot kg \cdot s^{-3}$	$1\ W \cdot h = 3600\ J$
Potential difference, emf	volt, V	$W \cdot A^{-1} = m^2 \cdot kg \cdot s^{-3} \cdot A^{-1}$	
Resistance	ohm, Ω	$V \cdot A^{-1} = m^2 \cdot kg \cdot s^{-3} \cdot A^{-2}$	
Electric charge	coulomb, C	$s \cdot A$	$1\ A \cdot h = 3600\ C$
Capacitance	farad, F	$C \cdot V^{-1} = m^{-2} \cdot kg^{-1} \cdot s^4 \cdot A^2$	
Conductance	siemens, S	$A \cdot V^{-1} = m^{-2} \cdot kg^{-1} \cdot s^3 \cdot A^2$	
Magnetic flux	weber, Wb	$V \cdot s = m^2 \cdot kg \cdot s^{-2} \cdot A^{-1}$	
Inductance	henry, H	$Wb \cdot A^{-1} = m^2 \cdot kg \cdot s^{-2} \cdot A^{-1}$	
Magnetic flux density	tesla, T	$Wb \cdot m^{-2} = kg \cdot s^{-2} \cdot A^{-1}$	$1\ T = 10^4\ gauss$
Magnetic field strength	ampere per meter	$m^{-1} \cdot A$	
Current density	ampere per square meter	$m^{-2} \cdot A$	
Electric field strength	volt per meter	$V \cdot m^{-1} = m \cdot kg \cdot s^{-3} \cdot A^{-1}$	
Permittivity	farad per meter	$F \cdot m^{-1} = m^{-3} \cdot kg^{-1} \cdot s^4 \cdot A^2$	
Permeability	henry per meter	$H \cdot m^{-1} = m \cdot kg \cdot s^{-2} \cdot A^{-2}$	

Appendix

Dimensional formulas of common quantities

Quantity	Definition	Dimensional formula
Mass	Fundamental	M
Length	Fundamental	L
Time	Fundamental	T
Velocity	Distance/time	LT^{-1}
Acceleration	Velocity/time	LT^{-2}
Force	Mass × acceleration	MLT^{-2}
Momentum	Mass × velocity	MLT^{-1}
Energy	Force × distance	ML^2T^{-2}
Power	Force × velocity	ML^2T^{-3}
Angle	Arc/radius	I
Angular velocity	Angle/time	T^{-1}
Angular acceleration	Angular velocity/time	T^{-2}
Torque	Force × lever arm	ML^2T^{-2}
Angular momentum	Momentum × lever arm	ML^2T^{-1}
Moment of inertia	Mass × radius squared	ML^2
Area	Length squared	L^2
Volume	Length cubed	L^3
Density	Mass/volume	ML^{-3}
Pressure	Force/area	$ML^{-1}T^{-2}$
Action	Energy × time	ML^2T^{-1}
Viscosity	Force per unit area per unit velocity gradient	$ML^{-1}T^{-1}$

Internal energy and generalized work

Type of energy	Intensive factor	Extension factor	Element of work
Mechanical			
Expansion	Pressure (P)	Volume (V)	$-P\,dV$
Stretching	Surface tension (γ)	Area (A)	$\gamma\,dA$
Extension	Tensile stretch (F)	Length (l)	$F\,dl$
Thermal	Temperature (T)	Entropy (S)	$T\,dS$
Chemical	Chemical potential (gm)	Moles (n)	$\mu\,dn$
Electrical	Electric potential (E)	Charge (Q)	$E\,dQ$
Gravitational	Gravitational field strength (mg)	Height (h)	$mg\,dh$
Polarization			
Electrostatic	Electric field strength (E)	Total electric polarization (P)	$E\,dP$
Magnetic	Magnetic field strength (H)	Total magnetic polarization (M)	$H\,dM$

Formulas for trigonometric (circular) functions*

Definitions

$$\sin z = \frac{e^{iz} - e^{-iz}}{2i} \quad (z = x + iy)$$

$$\cos z = \frac{e^{iz} + e^{-iz}}{2}$$

$$\tan z = \frac{\sin z}{\cos z}$$

$$\csc z = \frac{1}{\sin z}$$

$$\sec z = \frac{1}{\cos z}$$

$$\cot z = \frac{1}{\tan z}$$

Periodic properties

$$\sin(z + 2k\pi) = \sin z \quad (k \text{ any integer})$$

$$\cos(z - 2k\pi) = \cos z$$

$$\tan(z - k\pi) = \tan z$$

Relations between circular functions

$$\sin^2 z + \cos^2 z = 1$$

$$\sec^2 z + \tan^2 z = 1$$

$$\csc^2 z + \cot^2 z = 1$$

Negative angle formulas

$$\sin(-z) = -\sin z$$

$$\cos(-z) = \cos z$$

$$\tan(-z) = -\tan z$$

Addition formulas

$$\sin(z_1 + z_2) = \sin z_1 \cos z_2 + \cos z_1 \sin z_2$$

$$\cos(z_1 + z_2) = \cos z_1 \cos z_2 - \sin z_1 \sin z_2$$

$$\tan(z_1 + z_2) = \frac{\tan z_1 + \tan z_2}{1 - \tan z_1 \tan z_2}$$

$$\cot(z_1 + z_2) = \frac{\cot z_1 \cot z_2 - 1}{\cot z_2 + \cot z_1}$$

Half-angle formulas

$$\sin \frac{z}{2} = \pm \left(\frac{1 - \cos z}{2} \right)^{\frac{1}{2}}$$

$$\cos \frac{z}{2} = \pm \left(\frac{1 + \cos z}{2} \right)^{\frac{1}{2}}$$

Half-angle formulas (*cont.*)

$$\tan \frac{z}{2} = \pm \left(\frac{1 - \cos z}{1 + \cos z} \right)^{\frac{1}{2}}$$

$$= \frac{1 - \cos z}{\sin z} = \frac{\sin z}{1 + \cos z}$$

The ambiguity in sign may be resolved with the aid of a diagram.

Transformation of trigonometric integrals

If $\tan \frac{u}{2} = z$ then

$$\sin u = \frac{2z}{1 + z^2}, \quad \cos u = \frac{1 - z^2}{1 + z^2},$$

$$du = \frac{2}{1 + z^2} \, dz$$

Multiple-angle formulas

$$\sin 2z = 2 \sin z \cos z = \frac{2 \tan z}{1 + \tan^2 z}$$

$$\cos 2z = 2 \cos^2 z - 1 = 1 - 2 \sin^2 z$$

$$= \cos^2 z - \sin^2 z = \frac{1 - \tan^2 z}{1 + \tan^2 z}$$

$$\tan 2z = \frac{2 \tan z}{1 - \tan^2 z} = \frac{2 \cot z}{\cot^2 z - 1}$$

$$= \frac{2}{\cot z - \tan z}$$

$$\sin 3z = 3 \sin z - 4 \sin^3 z$$

$$\cos 3z = -3 \cos z + 4 \cos^3 z$$

$$\sin 4z = 8 \cos^3 z \sin z - 4 \cos z \sin z$$

$$\cos 4z = 8 \cos^4 z - 8 \cos^2 z + 1$$

Products of sines and cosines

$$2 \sin z_1 \sin z_2 = \cos(z_1 - z_2) - \cos(z_1 + z_2)$$

$$2 \cos z_1 \cos z_2 = \cos(z_1 - z_2) + \cos(z_1 + z_2)$$

$$2 \sin z_1 \cos z_2 = \sin(z_1 - z_2) + \sin(z_1 + z_2)$$

Addition and subtraction of two functions

$$\sin z_1 + \sin z_2$$

$$= 2 \sin \left(\frac{z_1 + z_2}{2} \right) \cos \left(\frac{z_1 - z_2}{2} \right)$$

$$\sin z_1 - \sin z_2$$

$$= 2 \sin \left(\frac{z_1 + z_2}{2} \right) \sin \left(\frac{z_1 - z_2}{2} \right)$$

Appendix

Formulas for trigonometric (circular) functions* (*cont.*)

Addition and subtraction of two functions (*cont.*)

$$\cos z_1 + \cos z_2$$

$$= 2 \cos\left(\frac{z_1 + z_2}{2}\right) \cos\left(\frac{z_1 - z_2}{2}\right)$$

$$\cos z_1 - \cos z_2$$

$$= -2 \sin\left(\frac{z_1 + z_2}{2}\right) \sin\left(\frac{z_1 - z_2}{2}\right)$$

$$\tan z_1 \pm \tan z_2 = \frac{\sin (z_1 \pm z_2)}{\cos z_1 \cos z_2}$$

$$\cot z_1 \pm \cot z_2 = \frac{\sin (z_2 \pm z_1)}{\sin z_1 \sin z_2}$$

Relations between squares of sines and cosines

$$\sin^2 z_1 - \sin^2 z_2 = \sin (z_1 + z_2) \sin (z_1 - z_2)$$

$$\cos^2 z_1 - \cos^2 z_2 = -\sin (z_1 + z_2) \sin (z_1 - z_2)$$

$$\cos^2 z_1 - \sin^2 z_2 = \cos (z_1 + z_2) \cos (z_1 - z_2)$$

Formulas for solution of plane triangles

In a triangle with angles A, B, and C and sides opposite a, b, and c respectively,

$$\frac{a}{\sin A} = \frac{b}{\sin B} = \frac{c}{\sin C}$$

Formulas for solution of plane triangles

$$\cos A = \frac{c^2 + b^2 - a^2}{2bc}$$

$$a = b \cos C + c \cos B$$

$$\frac{a + b}{a - b} = \frac{\tan \frac{1}{2} (A + B)}{\tan \frac{1}{2} (A - B)}$$

$$\text{area} = \frac{bc \sin A}{2} = [s(s - a)(s - b)(s - c)]^{1/2}$$

$$s = \frac{1}{2} (a + b + c)$$

Formulas for solution of spherical triangles

If A, B, and C are the three angles and a, b, and c the opposite sides,

$$\frac{\sin A}{\sin a} = \frac{\sin B}{\sin b} = \frac{\sin C}{\sin c}$$

$$\cos a = \cos b \cos c + \sin b \sin c \cos A$$

$$= \frac{\cos b \cos (b \pm \theta)}{\cos \theta}$$

where $\tan \theta = \tan b \cos A$

$$\cos A = -\cos B \cos C + \sin B \sin C \cos a$$

*From M. Abramowitz and I. A. Stegun (eds.), *Handbook of Mathematical Functions (with Formulas, Graphs, and Mathematical Tables)*, 10th printing, National Bureau of Standards, 1972.

General rules of differentiation and integration*

Differentiation

$$\frac{d}{dx}(c) = 0$$

$$\frac{d}{dx}(cx) = c$$

$$\frac{d}{dx}(cx^n) = ncx^{n-1}$$

$$\frac{d}{dx}(u \pm v \pm w \pm \cdots) = \frac{du}{dx} \pm \frac{dv}{dx} \pm \frac{dw}{dx} \pm \cdots$$

$$\frac{d}{dx}(cu) = c\frac{du}{dx}$$

$$\frac{d}{dx}(uv) = u\frac{dv}{dx} + v\frac{du}{dx}$$

$$\frac{d}{dx}(uvw) = uv\frac{dw}{dx} + uw\frac{dv}{dx} + vw\frac{du}{dx}$$

$$\frac{d}{dx}\left(\frac{u}{v}\right) = \frac{v(du/dx) - u(dv/dx)}{v^2}$$

$$\frac{d}{dx}(u^n) = nu^{n-1}\frac{du}{dx}$$

$$\frac{dy}{dx} = \frac{dy}{du}\frac{du}{dx} \quad \text{(Chain rule)}$$

$$\frac{du}{dx} = \frac{1}{dx/du}$$

$$\frac{dy}{dx} = \frac{dy/du}{dx/du}$$

Integration

$$\int a\, dx = ax$$

$$\int af(x)\, dx = a\int f(x)\, dx$$

$$\int (u \pm v \pm w \pm \cdots)\, dx = \int u\, dx \pm \int v\, dx \pm \int w\, dx \pm \cdots$$

$$\int u\, dv = uv - \int v\, du \quad \text{[integration by parts]}$$

$$\int f(ax)\, dx = \frac{1}{a}\int f(u)\, du$$

$$\int F\{f(x)\}\, dx = \int F(u)\frac{dx}{du}\, du = \int \frac{F(u)}{f'(x)}du \quad \text{where } u = f(x)$$

$$\int u^n\, du = \frac{u^{n+1}}{n+1}, \quad n \neq -1 \quad \text{[for } n = -1\text{]}$$

$$\int \frac{du}{u} = \ln u \quad \text{if } u > 0 \text{ or } \ln(-u) \text{ if } u < 0$$
$$= \ln |u|$$

$$\int e^u\, du = e^u$$

$$\int a^u\, du = \int e^{u\ln a}du = \frac{e^{u\ln a}}{\ln a} = \frac{a^u}{\ln a}, \quad a > 0, a \neq 1$$

$$\int \sin u\, du = -\cos u$$

$$\int \cos u\, du = \sin u$$

Appendix

General rules of differentiation and integration* (cont.)

Integration (cont.)

$$\int \tan u \, du = \ln \sec u = -\ln \cos u$$

$$\int \cot u \, du = \ln \sin u$$

$$\int \sec u \, du = \ln (\sec u + \tan u) = \ln \tan \left(\frac{u}{2} + \frac{\pi}{4} \right)$$

$$\int \csc u \, du = \ln (\csc u - \cot u) = \ln \tan \frac{u}{2}$$

$$\int \sec^2 u \, du = \tan u$$

$$\int \csc^2 u \, du = -\cot u$$

$$\int \tan^2 u \, du = \tan u - u \int \cot^2 u \, du = -\cot u - u$$

$$\int \sin^2 u \, du = \frac{u}{2} - \frac{\sin 2u}{4} = \frac{1}{2}(u - \sin u \cos u)$$

$$\int \cos^2 u \, du = \frac{u}{2} + \frac{\sin 2u}{4} = \frac{1}{2}(u + \sin u \cos u)$$

$$\int \sec u \tan u \, du = \sec u$$

$$\int \csc u \cot u \, du = -\csc u$$

$$\int \sinh u \, du = \cosh u$$

$$\int \cosh u \, du = \sinh u$$

$$\int \tanh u \, du = \ln \cosh u$$

$$\int \coth u \, du = \ln \sinh u$$

$$\int \text{sech} \, u \, du = \sin^{-1}(\tanh u) \quad \text{or} \quad 2 \tan^{-1} e^u$$

$$\int \text{csch} \, u \, du = \ln \tanh \frac{u}{2} \quad \text{or} \quad -\coth^{-1} e^u$$

$$\int \text{sech}^2 u \, du = \tanh u$$

$$\int \text{csch}^2 u \, du = -\coth u$$

$$\int \tanh^2 u \, du = u - \tanh u$$

$$\int \coth^2 u \, du = u - \coth u$$

$$\int \sinh^2 u \, du = \frac{\sinh 2u}{4} - \frac{u}{2} = \frac{1}{2}(\sinh u \cosh u - u)$$

$$\int \cosh^2 u \, du = \frac{\sinh 2u}{4} + \frac{u}{2} = \frac{1}{2}(\sinh u \cosh u + u)$$

$$\int \text{sech} \, u \tanh u \, du = -\text{sech} \, u$$

$$\int \text{csch} \, u \coth u \, du = -\text{csch} \, u$$

$$\int \frac{du}{u^2 + a^2} = \frac{1}{a} \tan^{-1} \frac{u}{a}$$

General rules of differentiation and integration* (cont.)

Integration (cont.)

$$\int \frac{du}{u^2 - a^2} = \frac{1}{2a} \ln\left(\frac{u-a}{u+a}\right) = -\frac{1}{a}\coth^{-1}\frac{u}{a} \quad u^2 > a^2$$

$$\int \frac{du}{a^2 - u^2} = \frac{1}{2a} \ln\left(\frac{a+u}{a-u}\right) = \frac{1}{a}\tanh^{-1}\frac{u}{a} \quad u^2 < a^2$$

$$\int \frac{du}{\sqrt{a^2 - u^2}} = \sin^{-1}\frac{u}{a}$$

$$\int \frac{du}{\sqrt{u^2 + a^2}} = \ln(u + \sqrt{u^2 + a^2}) \quad \text{or} \quad \sinh^{-1}\frac{u}{a}$$

$$\int \frac{du}{\sqrt{u^2 - a^2}} = \ln(u + \sqrt{u^2 - a^2})$$

$$\int \frac{du}{u\sqrt{u^2 - a^2}} = \frac{1}{a}\sec^{-1}\left|\frac{u}{a}\right|$$

$$\int \frac{du}{u\sqrt{u^2 + a^2}} = -\frac{1}{a}\ln\left(\frac{a + \sqrt{u^2 + a^2}}{u}\right)$$

$$\int \frac{du}{u\sqrt{a^2 - u^2}} = -\frac{1}{a}\ln\left(\frac{a + \sqrt{a^2 - u^2}}{u}\right)$$

$$\int f^{(n)}g \, dx = f^{(n-1)}g - f^{(n-2)}g' + f^{(n-3)}g'' - \cdots (-1)^n \int fg^{(n)} \, dx$$

This is called generalized integration by parts.

---

*Here, u, v, w are functions of x; a, b, c, p, q, n are constants, restricted if indicated; $e = 2.71828\ldots$ is the natural base of logarithms; ln u denotes the natural logarithm of u (that is, the logarithm to the base e) where it is assumed that $u > 0$ [in general, to extend formulas to cases where $u < 0$ as well, replace ln u by ln $|u|$]; all angles are in radians; all constants of integration are omitted but implied.

SOURCE: Murray R. Spiegel and John Liu, *Mathematical Handbook of Formulas and Tables*, 2d ed., Schaum's Outline Series, McGraw-Hill, 1999.

Appendix

Basic integral transforms

Type of transform	Definition of transform	Definition of inverse transform
Fourier transform	$F(s) = \int_{-\infty}^{\infty} f(x)e^{-i2\pi sx}\,dx$	$f(x) = \int_{-\infty}^{\infty} F(s)e^{i2\pi sx}\,ds$
Fourier cosine transform	$F_c(s) = 2\int_{0}^{\infty} f(x)\cos 2\pi sx\,dx$	$f(x) = 2\int_{0}^{\infty} F_c(s)\cos 2\pi sx\,ds$
Fourier sine transform	$F_s(s) = 2\int_{0}^{\infty} f(x)\sin 2\pi sx\,dx$	$f(x) = 2\int_{0}^{\infty} F_s(s)\sin 2\pi sx\,dx$
Alternative definitions:		
Fourier transform	$F(k) = \dfrac{1}{\sqrt{2\pi}} \int_{-\infty}^{\infty} e^{-ikx}f(x)\,dx$	$f(x) = \dfrac{1}{\sqrt{2\pi}} \int_{-\infty}^{\infty} e^{ikx}F(k)\,dk$
Fourier cosine transform	$F_c(k) = \sqrt{\dfrac{2}{\pi}} \int_{0}^{\infty} \cos kx\, f(x)\,dx$	$f(x) = \sqrt{\dfrac{2}{\pi}} \int_{0}^{\infty} \cos kx\, F_c(k)\,dk$
Fourier sine transform	$F_s(k) = \sqrt{\dfrac{2}{\pi}} \int_{0}^{\infty} \sin kx\, f(x)\,dx$	$f(x) = \sqrt{\dfrac{2}{\pi}} \int_{0}^{\infty} \sin kx\, F_s(k)\,dk$
Laplace transform[a]	$f(s) = \int_{0}^{\infty} e^{-st}\phi(t)\,dt$	$\phi(t) = \dfrac{1}{2\pi i} \int_{c-i\infty}^{c+i\infty} f(s)e^{st}\,ds \quad 0 < t < \infty$
z transform[b]	$X(z) = x_0 + x_1 z^{-1} + x_2 z^{-2} + x_3 z^{-3} + \cdots$ $= \displaystyle\sum_{n=0}^{\infty} x_n z^{-n}$	$x_n = \dfrac{1}{2\pi i f} \oint_C X(z)z^{n-1}\,dz$

[a] In the definition of the inverse Laplace transform, the integration is along any line Re $s = c$ of the complex s plane on which the integral in the definition of the Laplace transform converges absolutely.

[b] In the definition of the inverse z transform, the integration is counterclockwise about a closed path in the complex z plane which encloses all singularities of $X(z)$.

SOURCE: After R. N. Bracewell, *The Fourier Transform and Its Applications*, 3d ed., McGraw-Hill, 1999; L. Debnath, *Integral Transforms and Their Applications*, CRC Press, 1995; articles on "Fourier series and transforms," "Laplace transform," and "Z transform" in *McGraw-Hill Encyclopedia of Science & Technology*, 9th ed., 2002.

Mathematical notation, with definitions

Signs and symbols

$+$	Plus (sign of addition)
$+$	Positive
$-$	Minus (sign of subtraction)
$-$	Negative
$\pm\,(\mp)$	Plus or minus (minus or plus)
$\times$	Times, by (multiplication sign)
$\cdot$	Multiplied by
$\div$	Sign of division
$/$	Divided by
$:$	Ratio sign, divided by, is to
$::$	Equals, as (proportion)
$<$	Less than
$>$	Greater than
$\ll$	Much less than
$\gg$	Much greater than
$=$	Equals
$\equiv$	Identical with
$\sim$	Similar to
$\approx$	Approximately equals
$\cong$	Approximately equals, congruent
$\leq$	Equal to or less than
$\geq$	Equal to or greater than
$\neq$	Not equal to
$\rightarrow \doteq$	Approaches
μ	Varies as
$\bullet$	Infinity
$\sqrt{}$	Square root of
$\sqrt[3]{}$	Cube root of
$\therefore$	Therefore
$\parallel$	Parallel to
$(\,)\,[\,]\,\{\,\}$	Parentheses, brackets, and braces; quantities enclosed by them to be taken together in multiplying, dividing, etc.
$\overline{AB}$	Length of line from A to B
π	pi $= 3.14159\ldots$
$\circ$	Degrees
$'$	Minutes
$''$	Seconds
$\angle$	Angle

Appendix

Mathematical notation, with definitions (*cont.*)	
	Signs and symbols (cont.)
dx	Differential of x
Δ	(delta) difference
Δx	Increment of x
$\partial u/\partial x$	Partial derivative of u with respect to x
$\int$	Integral of
$\int_b^a$	Integral of, between limits a and b
$\oint$	Line integral around a closed path
Σ	(sigma) summation of
$f(x), F(x)$	Functions of x
∇	Del or nabla, vector differential operator
∇^2	Laplacian operator
£	Laplace operational symbol
$4!$	Factorial $4 = 1 \times 2 \times 3 \times 4$
$|x|$	Absolute value of x
$\dot{x}$	First derivative of x with respect to time
$\ddot{x}$	Second derivative of x with respect to time
$\mathbf{A} \times \mathbf{B}$	Vector-product; magnitude of $\mathbf{A}$ times magnitude of $\mathbf{B}$ times sine of the angle from $\mathbf{A}$ to $\mathbf{B}$; $AB \sin \overline{AB}$
$\mathbf{A} \cdot \mathbf{B}$	Scalar product of $\mathbf{A}$ and $\mathbf{B}$; magnitude of $\mathbf{A}$ times magnitude of $\mathbf{B}$ times cosine of the angle from $\mathbf{A}$ to $\mathbf{B}$; $AB \cos \overline{AB}$
	Mathematical logic
$p, q, P(x)$	Sentences, propositional functions, propositions
$-p, \sim p,$ non p, Np	Negation red "not p" ($\neq$: read "not equal")
$p \vee q, p + q, Apq$	Disjunction, read "p or q," "p, q," or both
$p \wedge q, p \cdot q, p\&q, Kpq$	Conjunction, read "p and q"
$p \rightarrow q, p \supset q, p \Rightarrow q, Cpq$	Implication, read "p implies q" or "if p then q"
$p \leftrightarrow q, p \equiv q, p \Leftrightarrow q, Epq,$ p iff$q$	Equivalence, read "p is equivalent to q" or "p if and only if q"
n.a.s.c.	Read "necessary and sufficient condition"
$( ), [ ], \{ \}, ..,\cdot\cdot$	Parentheses
$\forall, \Sigma$	Universal quantifier, read "for all" or "for every"
$\exists, \Pi$	Existential quantifier, read "there is a" or "there exists"
$\vdash$	Assertion sign ($p \vdash q$: read "q follows from p"; $\vdash p$; read "p is or follows from an axiom," or "p is a tautology")
$0, 1$	Truth, falsity (values)
$=$	Identity
$\overset{Df}{=}, \overset{df}{=}, \overline{\frac{}{df}}, \equiv$	Definitional identity
∎	"End of proof"; "QED"

Mathematical notation, with definitions (*cont.*)

Set theory, relations, functions

X, Y	Sets
$x \in X$	x is a member of the set X
$x \notin X$	x is not a member of X
$A \subset X, A \subseteq X$	Set A is contained in set X
$A \not\subset X, A \not\subseteq X$	A is not contained in X
$X \cup Y, X + Y$	Union of sets X and Y
$X \cap Y, X \cdot Y$	Intersection of sets X and Y
$+, \dotplus, \circ$	Symmetric difference of sets
$\cup X_i, \Sigma X_i$	Union of all the sets X_i
$\cap X_i, \Pi X_i$	Intersection of all the sets X_i
$\varnothing, 0, \Lambda$	Null set, empty set
X', CX, CX	Complement of the set X
$X - Y, X \backslash Y$	Difference of sets X and Y
$x(P(x)); \{x|P(x)\}, \{x:P(x)\}$	The set of all x with the property P
$(x,y,z), \langle x,y,z \rangle$	Ordered set of elements x, y, and z; to be distinguished from (x,z,y) for example
$\{x,y,z\}$	Unordered set, the set whose elements are x, y, z and no others
$\{a_1, a_2, \ldots, a_n\}$, $\{a_1\}_{i=1,2,\ldots,n}, \{a_1\}_{i=1}^n$	The set whose numbers are a_i, where i is any number whole from 1 to n
$\{a_1, a_2, \ldots\}$, $\{a_1\}_{i=1,2,\ldots}, \{a_1\}_{i=1}^{\infty}$	The set whose members are a_i, where i is any whole positive number
$X \times Y$	Cartesian product, set of all (x,y) such that $x \in X$, $y \in Y$
$\{a_i\}_{i \in I}$	The set whose elements are a_i, where $i \in I$
$xRy, R(x,y)$	Relation
$\equiv, \cong, \sim, \approx$	Equivalence relations, for example, congruence
$\geqq, \geq, \&, \geqslant, \leqq, \leq, <$	Transitive relations, for example, numerical order
$f \cdot X \to Y, X \xrightarrow{f} Y,$ $X \to Y, f \in Y^x$	Function, mapping, transformation
$f^{-1}, \overset{-1}{f}, X \xleftarrow{f^{-1}} Y$	Inverse mapping
$g \circ f$	Composite functions: $(g \circ f)(x) = g(f(x))$
$f(X)$	Image of X by f
$f^{-1}(X)$	Inverse-image set, counter image
1-1, one-one	Read "one-to-one correspondence"
$X \xrightarrow{\phi} Y$ $\phi \downarrow \quad \downarrow \psi$ $W \xrightarrow{g} Z$	Diagram: the diagram is commutative in case $\psi \circ f = g \circ \phi$
f/A	Partial mapping, restriction of function f to set A
$\overline{\overline{X}}, \operatorname{card} X, /X/$	Cardinal of the set A
$\aleph_0, \mathrm{d}$	Denumerable infinity

Appendix

Mathematical notation, with definitions (*cont.*)

	Set theory, relations, functions (cont.)
$c, \mathfrak{c}, 2^{\aleph_0}$	Power of continuum
ω	Order type of the set of positive integers
σ^-	Read "countably"
	Number, numerical functions
$1.4; 1,4; 1 \cdot 4$	Read "one and four-tenths"
$1(1)20(10)100$	Read "from 1 to 20 in intervals of 1, and from 20 to 100 in intervals of 10"
const	Constant
$A \geqq 0$	The number A is nonnegative, or, the matrix A is positive definite, or, the matrix A has nonnegative entries
$x \vert y$	Read "x divides y"
$x \equiv y \bmod p$	Read "x congruent to y modulo p"
$a_0 + \dfrac{1}{a_1} + \dfrac{1}{a_2} + \cdots,$ $\quad a_0 + \dfrac{1\vert}{\vert a_1\vert} + \cdots$	Continued fractions
$[a,b]$	Closed interval
$[a,b), [a,b\vert$	Half-open interval (open at the right)
$(a,b), \vert a,b\vert$	Open interval
$[a,\infty), [a,\rightarrow\vert$	Interval closed at the left, infinite to the right
$(-\infty,\infty), \vert\leftarrow, \rightarrow\vert$	Set of all real numbers
$\max_{x \in X} f(x),$ $\max\{f(x)\vert x \in X\}$	Maximum of $f(x)$ when x is in the set X
min	Minimum
sup, l.u.b.	Supremum, least upper bound
inf, g.l.b.	Infimum, greatest lower bound
$\lim_{x \to a} f(x) = b,$ $\lim_{x=a} f(x) = b,$ $f(x) \to b \text{ as } x \to a$	b is the limit of $f(x)$ as x approaches a
$\lim_{x \to a^-} f(x),$ $\lim_{x=a-0} f(x), f(a-)$	Limit of $f(x)$ as x approaches a from the left
$\limsup, \overline{\lim}$	Limit superior
$\liminf, \underline{\lim}$	Limit inferior
l.i.m.	Limit in the mean
$z = x + iy = re^{i\theta},$ $\zeta = \xi + i\eta,$ $w = u + iv = \rho e^{i\phi}$	Complex variables
$\bar{z}, z^*$	Complex conjugate
Re, $\neg$	Real part
Im, $\mathcal{I}$	Imaginary part
arg	Argument

Mathematical notation, with definitions (*cont.*)

	Number, numerical functions (cont.)
$\frac{\partial(u,v)}{\partial(x,y)}, \frac{D(u,v)}{D(x,y)};$	Jacobian, functional determinant
$\int_E f(x)d\mu(x)$	Integral (for example, Lebesgue integral) of function f over set E with respect to measure μ
$f(n) \sim \log n$ as $n \to \infty$	$f(n)/\log n$ approaches 1 as $n \to \infty$
$f(n) = O(\log n)$ as $n \to \infty$	$f(n)/\log n$ is bounded as $n \to \infty$
$f(n) = o(\log n)$	$f(n)/\log n$ approaches zero
$f(n) \nearrow b, f(x) \uparrow b$	$f(x)$ increases, approaching the limit b
$f(n) \downarrow b, f(x) \searrow b$	$f(x)$ decreases, approaching the limit b
a.e., p.p.	Almost everywhere
ess sup	Essential supremum
$C^0, C^0(X), C(X)$	Space of continuous functions
$C^k, C^k[a, b]$	The class of functions having continuous kth derivative (on $[a,b]$)
C'	Same as C^1
$\text{Lip}_\alpha, \text{Lip } \alpha$	Lipschitz class of functions
$L^p, L_p, L^p[a, b]$	Space of functions having integrable absolute pth power (on $[a,b]$)
L'	Same as L^1
$(C, \mu), (C, p)$	Cesàro summability

	Special functions
$[x]$	The integral part of x
$\binom{n}{k}, {}^nC_k, {}_nC_k$	Binomial coefficient $n!/k!(n-k)!$
$\left(\frac{n}{p}\right)$	Legendre symbol
$e^x, \exp x$	Exponential function
$\sinh x, \cosh x, \tanh x$	Hyperbolic functions
sn x, cn x, dn x	Jacobi elliptic functions
$\wp(x)$	Weierstrass elliptic function
$\Gamma(x)$	Gamma function
$I_v(x)$	Bessel function
$X_x(x)$	Characteristic function of the set X; $\chi_x(x) = 1$ in case $x \in X$, otherwise $\chi_x(x) = 0$
sgn x	Signum: sgn $=0$, while sgn $x = x/|x|$ for $x \neq 0$
$\delta(x)$	Dirac delta function

	Algebra, tensors, operators
$+, \cdot, \times, \circ, \top, \tau$	Laws of composition in algebraic systems
$e, 0$	Identity, unit, neutral element (of an additive system)
$e, 1, I$	Identity, unit, neutral element (of a general algebraic system)

Appendix

Mathematical notation, with definitions (*cont.*)

	Algebra, tensors, operators (*cont.*)
$e, \mathfrak{e}, E, P$	Idempotent
a^{-1}	Inverse of a
$\text{Hom}(M,N)$	Group of all homomorphisms of M into N
G/H	Factor group, group of cosets
$[K{:}k]$	Dimension of K over k
$\oplus, +$	Direct sum
f	Tensor product, Kronecker product
$\wedge$	Exterior product, Grassmann product
$\vec{x}, \mathbf{x}, \mathfrak{x}, x$	Vector
$\vec{x} \cdot \vec{r}, \mathbf{x} \cdot \mathbf{y}, (\mathfrak{x},\mathfrak{h})$	Inner product, scalar product, dot product
$\mathbf{x} \times \nabla, [\mathfrak{x},\mathfrak{h}], \mathbf{x} \wedge \mathbf{y}$	Outer product, vector product, cross product
$\|x\|, |x|, \|x\|, \|x\|_p$	Norm of the vector x
Ax, xA	The image of x under the transformation A
δ_{ij}	Kronecker delta: $\delta_{ij} = 1$, while $\delta_{ij} = 0$ for $i \neq j$
$A', tA, A^t, {}^tA$	Transpose of the matrix A
$A^*, \bar{A}$	Adjoint, Hermitian conjugate of A
$\text{tr}\, A, \text{Sp}\, A$	Trace of the matrix A
$\det A, |A|$	Determinant of the matrix A
$\Delta^n f(x), \Delta_h^n f, \underset{h}{\Delta^n} f(x)$	Finite differences
$[x_0,x_1], [x_0,x_1x_2],$ $\underset{x_1 x_0}{\Delta u}, [x_0 x_1] f$	Divided differences
$\nabla f, \text{grad}\, f$	Read "gradient of f"
$\nabla \cdot \mathbf{v}, \text{div}\, \mathbf{v}$	Read "divergence of $\mathbf{v}$"
$\nabla \times \mathbf{v}, \text{curl}\, \mathbf{v}, \text{root}\, \mathbf{v}$	Read "curl of $\mathbf{v}$"
$\nabla^2, \Delta, \text{div grad}$	Laplacian
$[X,Y]$	Poisson bracket, or commutator, or Lie product
$GL(n,R)$	Full linear group of degree n over field R
$O(n,R)$	Full orthogonal group
$SO(n,R)\, O^+(n,R)$	Special orthogonal group
	Topology
E^n	Euclidean n space
S^n	n sphere
$\rho(p,q), d(p,q)$	Metric, distance (between points p and q)
$\bar{X}, X^-, \text{cl}\, X, X^c$	Closure of the set X
$\text{Fr}X, \text{fr}X, \partial X, \text{bdry}\, X$	Frontier, boundary of X
$\text{int}\, X, \mathring{X}$	Interior of X
T_2 space	Hausdoff space

Mathematical notation, with definitions (*cont.*)

Topology (cont.)

F_σ	Union of countably many closed sets
G_δ	Intersection of countably many open sets
dim X	Dimensionality, dimension of X
$\pi_1(X)$	Fundamental group of the space X
$\pi_n(X)$, $\pi_n(X, A)$	Homotopy groups
$H_n(X)$, $H_n(X, A; G)$, $H_*(X)$	Homology groups
$H^n(X)$, $H^n(X, A; G)$, $H^*(X)$	Cohomology groups

Probability and statistics

X, Y	Random variables
$P(X \leq 2)$, $\Pr(X \leq 2)$	Probability that $X \leq 2$
$P(X \leq 2 \mid Y \geq 1)$	Conditional probability
$E(X)$, $\mathcal{E}(X)$	Expectation of X
$E(X \mid Y \geq 1)$	Conditional expectation
c.d.f.	Cumulative distribution function
p.d.f.	Probability density function
c.f.	Characteristic function
$\bar{x}$	Mean (especially, sample mean)
σ, s.d.	Standard deviation
σ^2 Var, var	Variance
μ_1, μ_2, μ_3, μ_i, μ_{ij}	Moments of a distribution
ρ	Coefficient of correlation
$\rho_{12.34}$	Partial correlation coefficient

Appendix

Schematic electronic symbols*

Ammeter		Coaxial cable	
Amplifier, general		Crystal, piezoelectric	
Amplifier, inverting		Delay line	
Amplifier, operational			
AND gate		Diac	
Antenna, balanced		Diode, field-effect	
Antenna, general		Diode, general	
Antenna, loop		Diode, Gunn	
Antenna, loop, multiturn		Diode, light-emitting	
Battery		Diode, photosensitive	
Capacitor, feedthrough			
Capacitor, fixed		Diode, PIN	
Capacitor, variable		Diode, Schottky	
Capacitor, variable, split-rotor		Diode, tunnel	
Capacitor, variable, split-stator		Diode, varactor	
Cathode, electron-tube, cold		Diode, Zener	
Cathode, electron-tube, directly heated		Directional coupler	
Cathode, electron-tube, indirectly heated		Directional wattmeter	
Cavity resonator		Exclusive-OR gate	
Cell, electrochemical		Female contact, general	
Circuit breaker		Ferrite bead	

*From S. Gibilisco, *The Illustrated Dictionary of Electronics*, 8th ed., McGraw-Hill, 2001.

Filament, electron-tube		Inductor, powdered-iron core	
		Inductor, powdered-iron core, bifilar	
Fuse		Inductor, powdered-iron core, tapped	
Galvanometer		Inductor, powdered-iron core, variable	
			or
Grid, electron-tube			
Ground, chassis		Integrated circuit, general	
		Jack, coaxial or photo	
Ground, earth			
		Jack, phone, two-conductor	
Headset		Jack, phone, three-conductor	
Handset, double		Key, telegraph	
Headset, single		Lamp, incandescent	
Headset, stereo		Lamp, neon	
Inductor, air core		Male contact, general	
Inductor, air core, bifilar		Meter, general	
Inductor, air core, tapped		Microammeter	
Inductor, air core, variable		Microphone	
Inductor, iron core		Microphone, directional	
Inductor, iron core, bifilar		Milliammeter	
Inductor, iron core, tapped		NAND gate	
Inductor, iron core, variable		Negative voltage connection	
		NOR gate	

673

Appendix

NOT gate		Rectifier, gas-filled	
Optoisolator			
		Rectifier, high-vacuum	
OR gate			
Outlet, two-wire, nonpolarized		Rectifier, semiconductor	
Outlet, two-wire, polarized		Rectifier, silicon-controlled	
Outlet, three-wire		Relay, double-pole, double-throw	
Outlet, 234-V			
Plate, electron-tube		Relay, double-pole, single-throw	
Plug, two-wire, nonpolarized			
Plug, two-wire, polarized		Relay, single-pole, double-throw	
Plug, three-wire		Relay, single-pole, single-throw	
Plug, 234-V		Resistor, fixed	
Plug, coaxial or phono		Resistor, preset	
		Resistor, tapped	
Plug, phone, two-conductor		Resonator	
Plug, phone, three-conductor		Rheostat	
Positive voltage connection		Saturable reactor	
Potentiometer		Signal generator	
Probe, radio-frequency		Solar battery	

674

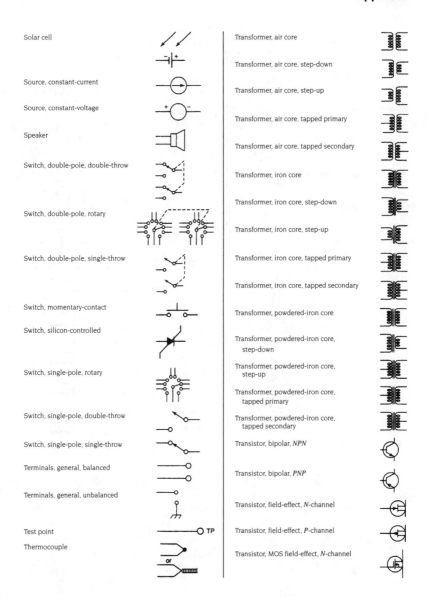

Solar cell

Source, constant-current

Source, constant-voltage

Speaker

Switch, double-pole, double-throw

Switch, double-pole, rotary

Switch, double-pole, single-throw

Switch, momentary-contact

Switch, silicon-controlled

Switch, single-pole, rotary

Switch, single-pole, double-throw

Switch, single-pole, single-throw

Terminals, general, balanced

Terminals, general, unbalanced

Test point

Thermocouple

or

Transformer, air core

Transformer, air core, step-down

Transformer, air core, step-up

Transformer, air core, tapped primary

Transformer, air core, tapped secondary

Transformer, iron core

Transformer, iron core, step-down

Transformer, iron core, step-up

Transformer, iron core, tapped primary

Transformer, iron core, tapped secondary

Transformer, powdered-iron core

Transformer, powdered-iron core, step-down

Transformer, powdered-iron core, step-up

Transformer, powdered-iron core, tapped primary

Transformer, powdered-iron core, tapped secondary

Transistor, bipolar, *NPN*

Transistor, bipolar, *PNP*

Transistor, field-effect, *N*-channel

Transistor, field-effect, *P*-channel

Transistor, MOS field-effect, *N*-channel

675

Appendix

Transistor, MOS field-effect,
  P-channel

Transistor, photosensitive, *NPN*

Transistor, photosensitive, *PNP*

Transistor, photosensitive, field-effect,
  N-channel

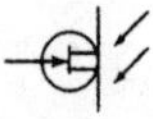

Transistor, photosensitive, field-effect,
  P-channel

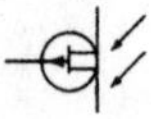

Transistor, unijunction

Triac

Tube, diode

Tube, heptode

Tube, hexode

Tube, pentode

Tube, photosensitive

Tube, tetrode

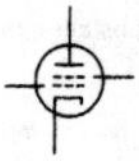

Tube, triode

Voltmeter

Wattmeter

Waveguide, circular

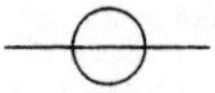

Waveguide, flexible

Waveguide, rectangular

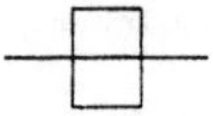

Waveguide, twisted

Wires, crossing, connected

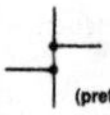

(preferred)

or

(alternative)

Wires, crossing, not connected

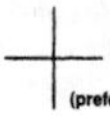

(preferred)

or

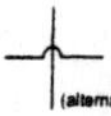

(alternative)

Partial family tree of programming languages

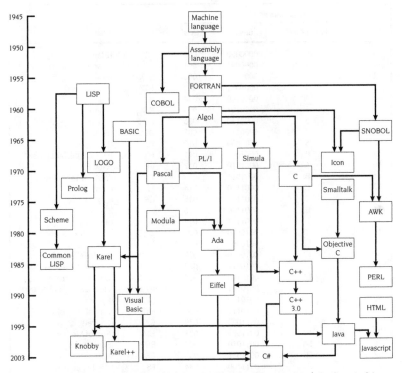

From Glenn D. Blank, Robert F. Barnes, and Edwin J. Kay, *The Universal Computer: Introducing Computer Science with Multimedia*, McGraw-Hill, New York, 2003.

Appendix

ASCII code*

Order number				Character code	
Decimal	Hexadecimal	Binary	Octal		
0	0	0000000	000	NUL	(Blank)
1	1	0000001	001	SOH	(Start of Header)
2	2	0000010	002	STX	(Start of Text)
3	3	0000011	003	ETX	(End of Text)
4	4	0000100	004	EOT	(End of Transmission)
5	5	0000101	005	ENQ	(Enquiry)
6	6	0000110	006	ACK	(Acknowledge (Positive))
7	7	0000111	007	BEL	(Bell)
8	8	0001000	010	BS	(Backspace)
9	9	0001001	011	HT	(Horizontal Tabulation)
10	A	0001010	012	LF	(Line Feed)
11	B	0001011	013	VT	(Vertical Tabulation)
12	C	0001100	014	FF	(Form Feed)
13	D	0001101	015	CR	(Carriage Return)
14	E	0001110	016	SO	(Shift Out)
15	F	0001111	017	SI	(Shift In)
16	10	0010000	020	DLE	(Data Link Escape)
17	11	0010001	021	DC1	(Device Control 1)
18	12	0010010	022	DC2	(Device Control 2)
19	13	0010011	023	DC3	(Device Control 3)
20	14	0010100	024	DC4	(Device Control 4-Stop)
21	15	0010101	025	NAK	(Negative Acknowledge)
22	16	0010110	026	SYN	(Synchronization)
23	17	0010111	027	ETB	(End of Text Block)
24	18	0011000	030	CAN	(Cancel)
25	19	0011001	031	EM	(End of Medium)
26	1A	0011010	032	SUB	(Substitute)
27	1B	0011011	033	ESC	(Escape)
28	1C	0011100	034	FS	(File Separator)
29	1D	0011101	035	GS	(Group Separator)
30	1E	0011110	036	RS	(Record Separator)
31	1F	0011111	037	US	(Unit Separator)
32	20	0100000	040	SP	(Space)
33	21	0100001	041	!	
34	22	0100010	042	''	
35	23	0100011	043	#	
36	24	0100100	044	$	
37	25	0100101	045	%	
38	26	0100110	046	&	
39	27	0100111	047	'	(Closing Single Quote)
40	28	0101000	050	(	
41	29	0101001	051	)	
42	2A	0101010	052	*	
43	2B	0101011	053	+	
44	2C	0101100	054	,	(Comma)
45	2D	0101101	055	-	(Hyphen)
46	2E	0101110	056	.	(Period)
47	2F	0101111	057	/	
48	30	0110000	060	0	
49	31	0110001	061	1	
50	32	0110010	062	2	
51	33	0110011	063	3	

ASCII code* (cont.)

	Order number				Character code
Decimal	Hexadecimal	Binary	Octal		
52	34	0110100	064	4	
53	35	0110101	065	5	
54	36	0110110	066	6	
55	37	0110111	067	7	
56	38	0111000	070	8	
57	39	0111001	071	9	
58	3A	0111010	072	:	
59	3B	0111011	073	;	
60	3C	0111100	074	<	(Less Than)
61	3D	0111101	075	=	
62	3E	0111110	076	>	(Greater Than)
63	3F	0111111	077	?	
64	40	1000000	100	@	
65	41	1000001	101	A	
66	42	1000010	102	B	
67	43	1000011	103	C	
68	44	1000100	104	D	
69	45	1000101	105	E	
70	46	1000110	106	F	
71	47	1000111	107	G	
72	48	1001000	110	H	
73	49	1001001	111	I	
74	4A	1001010	112	J	
75	4B	1001011	113	K	
76	4C	1001100	114	L	
77	4D	1001101	115	M	
78	4E	1001110	116	N	
79	4F	1001111	117	O	
80	50	1010000	120	P	
81	51	1010001	121	Q	
82	52	1010010	122	R	
83	53	1010011	123	S	
84	54	1010100	124	T	
85	55	1010101	125	U	
86	56	1010110	126	V	
87	57	1010111	127	W	
88	58	1011000	130	X	
89	59	1011001	131	Y	
90	5A	1011010	132	Z	
91	5B	1011011	133	[	(Opening Bracket)
92	5C	1011100	134	\	(Reverse Slant)
93	5D	1011101	135	]	(Closing Bracket)
94	5E	1011110	136	^	(Circumflex)
95	5F	1011111	137	—	(Underline)
96	60	1100000	140	`	(Opening Single Quote)
97	61	1100001	141	a	
98	62	1100010	142	b	
99	63	1100011	143	c	
100	64	1100100	144	d	
101	65	1100101	145	e	
102	66	1100110	146	f	
103	67	1100111	147	g	
104	68	1101000	150	h	

ASCII code* (cont.)

Decimal	Order number Hexadecimal	Binary	Octal	Character code	
105	69	1101001	151	i	
106	6A	1101010	152	j	
107	6B	1101011	153	k	
108	6C	1101100	154	l	
109	6D	1101101	155	m	
110	6E	1101110	156	n	
111	6F	1101111	157	o	
112	70	1110000	160	p	
113	71	1110001	161	q	
114	72	1110010	162	r	
115	73	1110011	163	s	
116	74	1110100	164	t	
117	75	1110101	165	u	
118	76	1110110	166	v	
119	77	1110111	167	w	
120	78	1111000	170	x	
121	79	1111001	171	y	
122	7A	1111010	172	z	
123	7B	1111011	173	{	(Opening Brace)
124	7C	1111100	174	l	(Vertical Line)
125	7D	1111101	175	}	(Closing Brace)
126	7E	1111110	176	-	(Overline (Tilde))
127	7F	1111111	177	DEL	(Delete/Rubout)

* ASCII (American Standard Code for Information Interchange) is a code for representing English characters as numbers, with each character assigned an order number from 0 through 127. In this table the order numbers are given in decimal, hexadecimal, binary, and octal notations. The first 32 character codes are control codes (non-printable) and the other 96 are representable characters. Since ASCII was developed in the 1960s for use with teletypewriters, the control codes are now rarely used for their original purpose, and their descriptions given here are somewhat obscure.

Electromagnetic spectrum

Frequency, Hz	Wavelength, m	Nomenclature	Typical source
10^{23}	3×10^{-15}	Cosmic photons	Astronomical
10^{22}	3×10^{-14}	X-rays	Radioactive nuclei
10^{21}	3×10^{-13}	X-rays	
10^{20}	3×10^{-12}	X-rays	Atomic inner shell, positron-electron annihilation
10^{19}	3×10^{-11}	Soft x-rays	Electron impact on a solid
10^{18}	3×10^{-10}	Ultraviolet, x-rays	Atoms in sparks
10^{17}	3×10^{-9}	Ultraviolet	Atoms in sparks and arcs
10^{16}	3×10^{-8}	Ultraviolet	Atoms in sparks and arcs
10^{15}	3×10^{-7}	Visible spectrum	Atoms, hot bodies, molecules
10^{14}	3×10^{-6}	Infrared	Hot bodies, molecules
10^{13}	3×10^{-5}	Infrared	Hot bodies, molecules
10^{12}	3×10^{-4}	Far-infrared	Hot bodies, molecules
10^{11}	3×10^{-3}	Microwaves	Electronic devices
10^{10}	3×10^{-2}	Microwaves, radar	Electronic devices
10^{9}	3×10^{-1}	Radar	Electronic devices, interstellar hydrogen
10^{8}	3	Television, FM radio	Electronic devices
10^{7}	30	Short-wave radio	Electronic devices
10^{6}	300	AM radio	Electronic devices
10^{5}	3000	Long-wave radio	Electronic devices
10^{4}	3×10^{4}	Induction heating	Electronic devices
10^{3}	3×10^{5}		Electronic devices
100	3×10^{6}	Power	Rotating machinery
10	3×10^{7}	Power	Rotating machinery
1	3×10^{8}		Communicated direct current
0	Infinity	Direct current	Batteries

Microwave frequency bands

Microwave band	Frequency range, GHz	Approximate wavelength range, mm
L	1–2	150–300
S	2–4	75–150
C	4–8	37–75
X	8–12	25–37
K_u	12–18	17–25
K	18–27	11–17
K_a	27–40	7.5–11
V	40–75	4–7.5
W	75–110	2.7–4

Radio spectrum

Band number[a]	Frequency range[b]	Approximate wavelength range	Band designation	Some applications[e] Application	Some applications[e] Frequency range[b]
4	3–30 kHz[d]	10–100 km	Very low frequency (VLF)	VLF radio navigation	9–14 kHz[e]
5	30–300 kHz	1–10 km	Low frequency (LF)	Loran C Longwave broadcasting[f]	90–110 kHz 150–290 kHz
6	300 kHz to 3 MHz	100–1000 m	Medium frequency (MF)	AM broadcasting	535–1705 kHz
7	3–30 MHz	10–100 m	High frequency (HF)	Shortwave broadcasting Citizens band (CB)	5.95–26.1 MHz (8 frequency bands) 26.96–27.41 MHz
8	30–300 MHz	1–10 m	Very high frequency (VHF)	Cordless telephones Television channels 2–4 Television channels 5–6 FM broadcasting Instrument landing system (ILS) localizer VOR (VHF omnidirectional range) Television channels 7–13 Terrestrial digital audio broadcasting (DAB)[g]	43.71–44.49 and 46.60–46.98 MHz (base transmitters), 48.75–49.51 and 49.66–50.0 MHz (handset transmitters) 54–72 MHz 76–88 MHz 88–108 MHz 108–112 MHz 108–118 MHz 174–216 MHz 174–240 MHz

Radio spectrum (cont.)

Band number[a]	Frequency range[b]	Approximate wavelength range	Band designation[b]	Some applications[c]	
				Application	Frequency range[b]
9	300 MHz to 3 GHz[h]	100–1000 mm	Ultrahigh frequency (UHF)	ILS glide slope	329–335 MHz
				Television channels 14–69	470–806 MHz
				Cellular telephones	824–849 MHz (mobile transmitters). 869–894 MHz (base transmitters)
				"900-MHz" cordless telephones	902–928 Mhz
				Distance measuring equipment (DME), Tacan	960–1215 MHz
				Secondary surveillance radar (SSR)	1030 MHz (beacons), 1090 MHz (transponders)
				Global Positioning System (GPS)	1176.45 MHz (L3), 1227.6 MHz (L2), 1575.42 MHz (L1)
				Radio astronomy[i]	1400–1427 and 1660–1670 MHz
				Terrestrial DAB[g]	1452–1492 MHz
				Inmarsat system	1535–1543.5 MHz (downlink), 1636.5–1645 MHz (uplink)
				Personal communications systems (PCS)	1850–1910 and 1930–1990 MHz (licensed), 1910–1930 (unlicensed)
				Deep space communications	2110–2120 MHz (uplink), 2290–2300 MHz (downlink)
				Satellite DAB	2310–2360 MHz
				Microwave ovens	2400–2500 MHz
10	3–30 GHz[h]	10–100 mm	Superhigh frequency (SHF)	Satellite communications, C band	3.7–4.2 GHz (downlink), 5.9–6.4 GHz (uplink)
				Satellite communications, Ku band	10.7–12.57 GHz (downlink), 12.7–14.8 GHz (uplink)

Radio spectrum (cont.)

Band number[a]	Frequency range[b]	Approximate wavelength range	Band designation	Application	Some applications[e] Frequency range[b]
11	30–300 GHz[h,i]	1–10 mm	Extremely high frequency (EHF), or millimeter-wave	Direct broadcasting satellite (DBS) systems	12.2–12.7 GHz
				Satellite communications, K_a band	18.55–18.8 and 19.2–20.2 GHz (downlink), 28.35–28.6 and 29.0–30.0 GHz (uplink)
12	300 GHz to 3 THz[j]	0.1–1 mm	Submillimeter		

[a]"Band number N" extends from $0.3 \times 10N$ to $3 \times 10N$ hertz. The upper limit is included in each band; the lower limit is excluded.

[b]kHz = kilohertz (10^3 Hz), MHz = megahertz (10^6 Hz), GHz = gigahertz (10^9 Hz), THz = terahertz (10^{12} Hz).

[c]Radio spectrum allocations can vary in different parts of the world. The allocations given here pertain to the United States unless otherwise noted.

[d]Frequencies below 9 kHz have not been allocated by the International Telecommunications Union (ITU) and the U.S. Federal Communications Commission (FCC).

[e]An important user of this frequency band was the Omega navigation system (4 frequencies, 10.2–13.6 kHz), which was terminated in 1997.

[f]Longwave broadcasting is permitted in Europe.

[g]Terrestrial digital audio broadcasting (DAB) is allocated the 174–240-MHz and 1452–1492-MHz bands in Europe, Canada, and other countries. In the United States, these bands are unavailable for this service, and DAB in the AM and FM broadcasting bands is under development.

[h]Microwave frequencies (1–110 GHz) are also classified according to an alternative set of designations given in the table titled "Microwave frequency bands."

[i]Many frequency bands between 13 36 MHz and 275 GHz are allocated to radio astronomy, most of them shared with other services. Two of the most important, listed here, are used to study the hydrogen line at 1420.4 MHz and hydroxyl lines near 1665 and 1667 MHz.

[j]Frequencies above 300 GHz in the United States and above 275 GHz elsewhere have not been allocated by the FCC and the ITU, but allocation of frequencies up to 1000 GHz is anticipated. In the meantime, the ITU and the FCC have urged the protection from harmful interference of observations for radio astronomy, earth exploration satellites, and space research in various frequency bands above 275 GHz around spectral lines that are of interest in these fields.